Lecture Notes on Data Engineering and Communications Technologies

Volume 35

The aim of the book series is to present cutting edge engineering approaches to data technologies and communications. It will publish latest advances on the engineering task of building and deploying distributed, scalable and reliable data infrastructures and communication systems.

The series will have a prominent applied focus on data technologies and communications with aim to promote the bridging from fundamental research on data science and networking to data engineering and communications that lead to industry products, business knowledge and standardisation.

**** Indexing: The books of this series are submitted to ISI Proceedings, MetaPress, Springerlink and DBLP ****

More information about this series at http://www.springer.com/series/15362

D. Jude Hemanth · V. D. Ambeth Kumar ·
S. Malathi · Oscar Castillo · Bogdan Patrut
Editors

Emerging Trends in Computing and Expert Technology

Set 1

Springer

Editors
D. Jude Hemanth
Department of ECE
Karunya University
Coimbatore, India

V. D. Ambeth Kumar
Department of Computer
Science and Engineering
Panimalar Engineering College
Chennai, Tamil Nadu, India

S. Malathi
Department of Computer
Science and Engineering
Panimalar Engineering College
Chennai, Tamil Nadu, India

Oscar Castillo
Division of Graduate
Studies and Research
Tijuana Institute of Technology
Tijuana, Baja California, Mexico

Bogdan Patrut
Faculty of Computer Science
"Alexandru Ioan Cuza"
University of Iasi
Iasi, Romania

ISSN 2367-4512 ISSN 2367-4520 (electronic)
Lecture Notes on Data Engineering and Communications Technologies
ISBN 978-3-030-32149-9 ISBN 978-3-030-32150-5 (eBook)
https://doi.org/10.1007/978-3-030-32150-5

This Springer imprint is published by the registered company Springer Nature Switzerland AG
The registered company address is: Gewerbestrasse 11, 6330 Cham, Switzerland

Preface

Dear distinguished delegates and guests,

The organizing committee is extremely privileged to welcome our distinguished delegates and guests to the International Conference for "**P**hoenixes on **E**merging **C**urrent **T**rends in **E**ngineering **A**nd **M**anagement (**PECTEAM 2K19**) organizing Conference on **COMET** (Emerging Current Trends in **COM**puting and **Expert T**echnology) which will be held from 22 to 23 March 2019, in Chennai, Tamil Nadu, India.

COMET is well supported by IEEE Madras section, CSI, ICTACT, ISTE and IETE. The objective of this conference is to bring together numerous problems encountered in the current as well as future high technologies under a common platform. It has been specifically organized to congregate international researchers and technocrats for presenting their innovative work, share and gain knowledge to have a proper insight into the significant challenges currently being addressed in various fields of Computing and Expert Technology.

The conference proceedings has a complete track record of the papers that are reviewed and presented at the conference. The principle objective is to provide an international scientific forum wherein the participants can mutually exchange their innovative ideas in relevant fields and interact in depth through discussion with peer groups. Both inward research and core areas of Computing and Expert Technology and its applications will be covered during these events.

The conference has received 540 papers all over India and abroad from various institutions in different domains, out of which 163 papers are selected for presentation. The submitted papers were peer-reviewed by external reviewers and advisory board based on the subject of interest and selected on the basis of originality, clarity and significance in line with the theme of the conference.

It is guaranteed that a high-quality program will be conducted as the conference will be graced by an unparalleled number of professional speakers from multidisciplinary fields, from both India and abroad.

We would like to thank the chief patron, patrons, conveners, organizing committee members, advisory members for their immense help, motivation, constant cooperation and continuous encouragement for the conference. We deeply appreciate the editorial board for their wonderful editorial service for the successful outcome of this proceedings. We are very much grateful to all those who have contributed in both the front desk and the backyard to make COMET an astounding success.

COMET, featuring high-impact presentations, would be enriching and undoubtedly an unique, rewarding and memorable experience being hosted at Panimalar Engineering College, Chennai, India.

Organization

Steering Committee Members

Chief Patron

P. Chinnadurai (Secretary and Correspondent)

Patrons

C. Vijayarajeswari (Director)
C. Sakthikumar (Director)
Saranya Sree Sakthikumar (Director)

Co-patrons

K. Mani (Principal)

Convener

S. Malathi

Co-convener

V. D. Ambeth Kumar

Organizing Committees

D. Karunkuzhali
S. Maheswari

R. Manikandan
D. Silas Stephen

Internal Advisory Board

P. Kannan
S. Murugavalli
M. Helda Mercy
S. Selvi
C. Esakkiappan
R. Manmohan

International Advisory Board

Vincenzo Piuri	University of Milan, Italy
K. C. Santosh	University of South Dakota, USA
Rajkumar Buyya	University of Melbourne, Victoria
Prasanta K. Ghosh	Syracuse University, Syracuse
James Geller	New Jersey Institute of Technology, USA
Claire Monteleoni	George Washington University, USA
Brian Borowski	Stevens Institute of Technology, USA
Thomas Augustine	University of Colorado Denver, USA
Ellen Gethner	University of Colorado Denver, USA
Mona Diab	George Washington University, USA
Stefano Basagni	Northeastern University, USA
Yasir Ibrahim	F.Engg, Jerash University, Jordan
Young-Bae Ko	Ajou University, South Korea
Riichiro Mizoguchi	Osaka University, Japan
Sattar J. Aboud	University of Bedfordshire, UK
Ab. Halim Bin Abu Bakar	University of Malaya, Malaysia
Akhtar Kalam	Victoria University, Australia
Saad Mekhilef	University of Malaya, Malaysia
Varatharaju M. V.	Ibra College of Technology, Oman
B. Rampriya	Debre Markos University, Debre, Ethiopia
Ahmed Faheem Zobaa	Brunel University London, Uxbridge, UK
Easwaramoorthy Rangaswamy	Amity University, Singapore
Anbalagan Krishnan	Xiamen University, Malaysia
Shanmugam Joghee	Skyline University College, Sharjah, UAE
J. Paulo Davim	University of Aveiro, Campus Santiago, Portugal
Talal Yusaf	University of Southern Queensland, Australia
Fehim Findik	Sakarya University, Turkey
A. S. Shaja	Senior Data Scientist, Yodlee, California, USA
Junzo Watada	Waseda University, Wakamatsu, Japan

National Advisory Board

L. Ganesan	IIT Madras, Chennai
A. Kannan	Anna University, Chennai
S. Nirmala Devi	Anna University, Chennai
K. Sridharan	Anna University, Chennai
Prakash	Karpagam University, Coimbatore
S. Pandurangan	IIT Madras, Chennai
K. Selvamani	Anna University, Chennai
A. Suresh	NIET, Coimbatore
L. Arun Raj	Crescent Institute of Science and Technology, Chennai
K. Ramalakshmi	Karunya Institute of Technology and Sciences, Coimbatore
R. Venkatesan	Karunya Institute of Technology and Sciences, Coimbatore
Arun Raj Kumar	NIT, Puducherry
C. Swarnalatha	Anna University, Madurai
K. Panneerselvam	IGNOU, Kozhikode, Kerala
G. Kumaresan	CEG, Anna University, Chennai
V. S. Senthil Kumar	CEG, Anna University, Chennai
Indra Rajasingh	VIT University, Chennai
V. Santhi	VIT University, Chennai
K. Chandrasekaran	NITK, Mangalore
Sipra Das Bit	LIEST, West Bengal
S. Bhaskar	Anna University, Chennai
Uma Bhattacharya	Besu, Howrah, West Bengal
S. Amalraj	Anna University, Chennai
G. Ravikumar	Anna University, Chennai
K. Malathi	CEG-AU, Chennai
M. Ganesh Madhan	MIT-AU, Chennai
E. Logashanmugam	Sathyabama University, Chennai
B. Sheela Rani	Sathyabama University, Chennai

Web Design Committee

R. Priya
Rahul Chiranjeevi Veluri (PG Scholar)
P. Naveen Teja (UG Scholar)
P. Yaagesh Prasad (PG Scholar)

Publication Committee

S. Vimala
S. Sathya

B. Akshaya (PG Scholar)
R. Jayashree (PG Scholar)
R. Christina Rini (PG Scholar)
N. Savitha (PG Scholar)

Proceeding Committee

K. Valarmathi
M. Anitha (PG Scholar)
K. P. Ashvitha (PG Scholar)
J. Jaya Sruthi (PG Scholar)
R. Vidhya (PG Scholar)

Publicity Committee

S. Sathya Venkateshwaren (PG Scholar)
M. Gobinath (PG Scholar)
R. B. Aarthinivashini (PG Scholar)
S. Divya (PG Scholar)
R. Vidhya (PG Scholar)

Hospitality Committee

M. Rajendiran
D. Elangovan
M. Dhivya (PG Scholar)
K. Priyanka (PG Scholar)
M. Priyanga (PG Scholar)

Registration Committee

J. Josepha Menandas
G. Kumutha Rajeswari (PG Scholar)
M. R. Tamizhkkanal (PG Scholar)
J. Freeda (PG Scholar)
M. Shilpa Aarthi (PG Scholar)
R. Monisha (PG Scholar)

Finance Committee

Sofia Vincent
Lakshmi

Contents

Advances in Control and Soft Computing

Advances in Circuits and Systems in Computing

Synthesis and Characterization of Sol-Gel Spin Coated ZnO Thin Films

G. Divya(✉), S. Sindhu, and K. Shreekrishna Kumar

Department of Electronics, School of Technology and Applied Sciences, Mahatma Gandhi University, Pullarikkunnu, Kottayam, Kerala, India
divyajishnu45@gmail.com

Abstract. In this paper, spin coating process is adopted for the preparation of thin films of ZnO over the glass substrate. The key components used for the fabrication of ZnO solution are zinc acetate dehydrate, 2-Methoxy ethanol and mono ethanol amine. Structural morphology of the samples are estimated using the technique of XRD (X-ray Diffraction) and SEM (Scanning Electron microscopy) respectively. The results indicates that precise control of reactance, spinning speed and heat treatment of the films have great impact on the crystal growth and orientation. The XRD results demonstrates c-axis orientation of ZnO film, having the preferred peak along (002) plane and the grain size is 15.8 nm. the optical transmittance spectrum reviles that the average value of transmittance is about 95% and bandgap is assessed to be 3.34 eV.

Keywords: Spin coating · Structural characterization · Optical characterization

1 Introduction

The hexagonal wurtzite and zinc blend are the two different standard forms of Zinc Oxide. Out of which, the wurtzite structure is more stable and common. Zinc Oxide being a versatile semiconducting material with typical properties like hexagonal wurtzite structure [1], wide band-gap of 3.37 eV, large exitonic binding energy of 60 mV, n-type conductivity and resistivity control over 10^{-3} to 10^{-5} Ω cm range [2]. Due to the large and direct band gap, the break down voltage is high, electronic noise is low, operate in high temperature and high power and able to withstand high electric field. Moreover, it shows non-toxicity, electrochemical stability and it is very abundant in nature.

In ZnO, a coordination between Zn atom and four O atoms exist in tetrahedral direction, where the d electrons of Zn atoms are hybridized with p electrons of the O atoms. Even though the stoichiometry of ZnO is an insulator, it consist of large number of voids due to the presence of excess Zn atoms. This may have great influence on the properties like electrical property, defect structure etc. The crystal structure of ZnO makes it suitable for the fabricating of high quality oriented thin films. There are a number of useful properties that help in the fabrication of thin film devices like wide band gap, high electron mobility, good transparency and strong room-temperature luminescence. These properties are applicable in liquid crystal displays as transparent electrodes, energy saving in heat protecting windows, light emitting diodes and thin

D. J. Hemanth et al. (Eds.): COMET 2019, LNDECT 35, pp. 3–9, 2020.
https://doi.org/10.1007/978-3-030-32150-5_1

film transistors also it is very well suitable for photocatalyst, gas sensors, light-emitting diodes, nano-lasers etc. [3–6].

Commercialization of the transparent conductive oxides (TCOs) in the present era leads to higher industrial values regarding new design of TCOs. Newfangled electronic structures have been widely developed by incorporating the TCOs [7]. Recent years, design and production of new transparent conductive oxides (TCOs) resulted in improved production of various optoelectronic devices like transparent Thin Film Transistors, Light Emitting Diodes, gas sensors, solar cells, liquid crystal displays etc. For the production of TCOs, several metal oxides like In_2O_3, SnO_3, ZnO, and TiO_2 have been used extensively. ZnO have some special interest in transparent electronics as they are the one among the materials which possess transparent and conductive properties and is a best example of TCOs [8]. ZnO also shows some particularly attractive character due to its low cost, greater abundance and chemical stability.

Deposition techniques such as pulsed laser deposition [9], magnetron sputtering [10], electron beam evaporation [11], spray pyrolysis and sol-gel methods [12] are available for ZnO thin film deposition. Here the thin films are prepared by the simple and economical method of spin coating technique. This technique have some advantages like ease of compositional modification, possibility of large area deposition and also high vacuum conditions are not required. One of the key factor among this deposition technique is controlling the size and shape of coated materials which are subjected to the environmental conditions. The microstructure, surface morphology and optical properties of ZnO transparent conducting films are investigated in this paper.

2 Experimental Techniques

Spin coating is the fabrication technique used to deposit ZnO thin film on the substrate. At first, a glass substrate coated with Indium Tin Oxide (ITO) is cleaned ultrasonically. The glass substrate is sonicated using acetone and methanol for 5 min each. Which is then systematically rinsed with distilled water and finally dried at 660 °C for 2 h. To prepare ZnO solution, Zinc acetate dehydrate $Zn(CH_3COO)_2.2H_2O$ is added to 2-Methoxyethanol ($(CH)_3CHOH$) containing monoethanolamine (MEA) (H_2NCH_2 CH_2OH). Molar ratio of Zinc acetate dehydrate, the precursor and MEA, the stabilizer is taken as 1:1. The precursor concentration is maintained at 0.5 mol/L. the solvent, 2-Methoxy ethanol provides the temperature control by absorbing the heat that is generated during the exothermic reaction, while MEA is used to prevent the formation of colloidal from aggregating the solution. The solution is then stirred continuously for 3 h. at 60 °C using a magnetic stirrer, which helps to increase the reaction process between all the materials in the solution. The solution is again stirred at room temperature for 6 h. to produce a clear homogeneous and transparent solution.

The solution is poured on to the glass substrate a rotational speed of 3000 rpm for 30 s and preheated at 250 °C for 5 min. This helps to evaporate the solvents and organic residuals. The process is repeated for ten times and then the films are post-heated for 3 h at 400 °C.

X-ray diffraction (XRD) is used for structural characterization of ZnO thin films. The XRD pattern is obtained with XPERT-PRO diffractometer having Cu Kα

(λ = 1.54060Å) radiation and scanning range 2θ is set between 20° and 90°. At the time of measurement, the current and the voltage of XRD are maintained at 30 mA, 45 kV respectively. Surface morphology is obtained from Scanning Electron Microscopy (SEM). Optical characteristics are analyzed using UV-Vis Spectrometer with a wavelength range 300 nm–900 nm and the optical band gap is measured from the transmission spectra. The thickness of ZnO thin film is measured using a surface profilometer.

3 Results and Discussions

The crystal quality and orientation of synthesized ZnO thin film can be studied using XRD analysis. The XRD pattern of spin coated ZnO thin film with 0.50 μm thickness is shown in Fig. 1. (100), (002) and (101) directions are the diffraction peaks of the sample, obtained from XRD pattern. A strong preferential growth is found along the c axis in (002) plane conforms that ZnO shows hexagonal wurtzite structure and the peak appeared at 2θ = 34.52° [13].

The unit cells "a", "b" and "c" of the polycrystalline ZnO thin film having (002) orientation are computed by the equations [14, 15]:

$$a = b = \sqrt{\frac{1}{3}}\lambda \Big/ sin\theta \quad (1)$$

$$c = \frac{\lambda}{sin\theta} \quad (2)$$

The lattice parameters for the unit cell are tabulated in the Table 1, which are in good agreement with the values reported in JCPDS Card No. 36–1451.

Table 1. Lattice parameters of unit cell

	a(Å)	c(Å)
Standard	3.253	5.215
Calculated	3.01164	5.21633

To calculate the particle size (D) of ZnO thin film Scherrer formulae is used [16]

$$D = \frac{k\lambda}{\beta_{2\theta} cos\theta} \quad (3)$$

Where k is a constant and its value is near to unity which is taken as 0.94, λ is the wavelength of X-Ray used and the value is 1.54 Å, $\beta_{2\theta}$ is the full width at half maximum of (002) peak of XRD pattern and 2θ is the Bragg angle [17]. From the calculation, the average value of grain size is obtained to be 15.84 nm and is listed in Table 2.

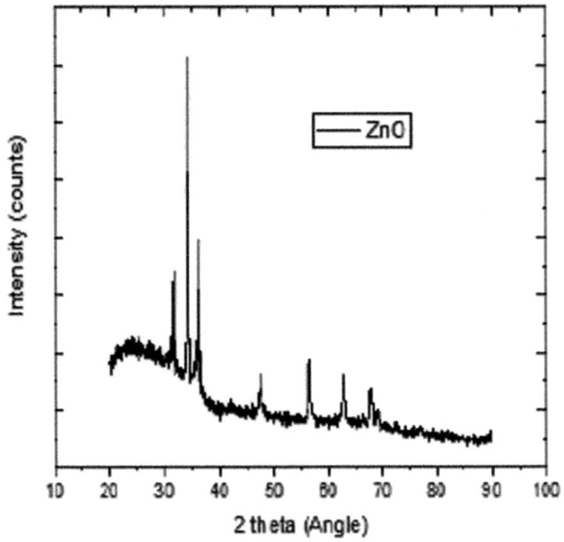

Fig. 1. XRD pattern of spin coated ZnO thin film

The extent of crystal defects is called dislocation density (δ) and the strain (ε) of thin film can be estimated by the following equations [18, 19]:

$$\delta = \frac{1}{D^2} \tag{4}$$

$$\varepsilon = \beta Cos\theta/4 \tag{5}$$

Table 2. Estimated structural parameters

Plane	β	2θ°	D (nm)	δ (nm)	$\varepsilon \times 10^{-2}$
(002)	0.5	34.3561	15.84838217	0.003981	2.0851

To investigate surface topology of the microstructure Scanning Electron Micrograph (SEM) can be used. SEM micrograph of ZnO thin film obtained by spin coating method is shown in Fig. 2. From the SEM image it is obvious that the morphology of ZnO grains are continuous and are tightly packed. The approximate crystallite size of the thin film crystals that are regularly distributed on the glass surface is 16 nm. It is also observed that the smaller grains makes the surface smooth and transparent.

Figure 3 shows the optical transmission spectrum of the ZnO thin film. From the transmission spectra it is obvious that a very good transmittance in the UV-Vis range from 300 to 800 nms. The average value of transmittance from 400 to 800 nm region is found to be 91–95%. Since ZnO is a direct band gap semiconducting material, a sharp absorption edge of ZnO is observed at 370 nm region in the spectrum. Band gap of

Fig. 2. SEM image of ZnO thin film

coated ZnO thin film is obtained by prognosticating the direct relationship between $(\alpha h\upsilon)^2$ and $h\upsilon$ according to the equation [20]:

$$\alpha h\upsilon = A\left(h\upsilon - E_g\right)^{1/2} \tag{6}$$

Where α is the absorption coefficient, υ is the frequency of the incident radiation and h is Planck's constant, $h\upsilon$ is the photon energy, Eg is the optical band gap and A is a constant.

Figure 4 represents the plot of $(\alpha h\upsilon)^2$ versus $h\upsilon$. A line tangential to the curve is plotted in $(\alpha h\upsilon)^2$ vs $h\upsilon$ graph to zero absorption coefficient in order to obtain the optical band gap (Eg) values. Only a single slope exist in the graph which also recommends that the ZnO thin film has direct and allowed transition. The band gap value of the ZnO thin film is obtained to be 3.34 eV which is nearly equal to band gap of bulk ZnO.

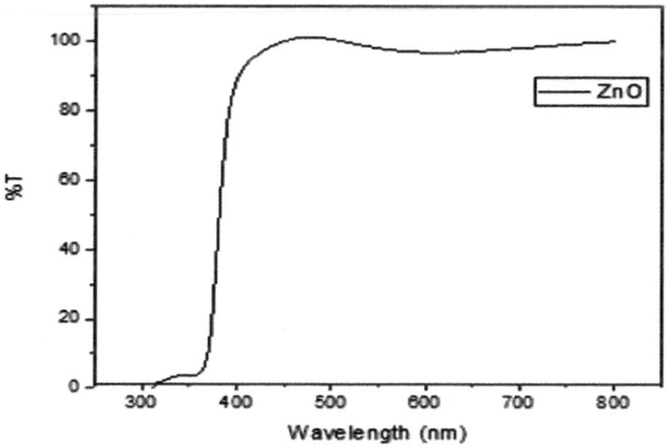

Fig. 3. Transfer spectra of ZnO thin film

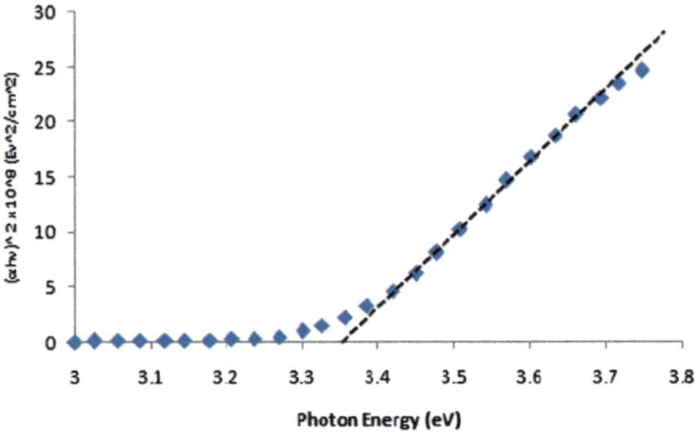

Fig. 4. Tauc Plot of Spin coated ZnO thin film

4 Conclusion

It has been found that thin film of ZnO deposited on glass substrate coated with ITO by sol-gel process is polycrystalline in nature. In this process, many factors influences the quality of the film and we have thoroughly optimized parameters such as precursor concentration, rotation speed, annealing temperature, etc. to obtain a better crystalline structure of the ZnO thin film. The XRD results provides a film with peak oriented along (002) direction. SEM micrograph revels that the small grains forms a flat and apparent surface for ZnO thin film. The optical transmittance of ZnO thin film is about 95% in the 400 nm–800 nm range and the energy band gap is obtained as 3.34 eV. The work presents an ongoing research attempt for improving the efficiency and a cost effective technique for developing transparent conducting ZnO thin films. The high crystallinity and transmission gives us a positive sign in the development of ZnO films for emerging thin-film sensors, transistors and solar cells.

References

1. Özgür, Ü., Alivov, Y.I., Liu, C., Teke, A., Reshchikov, M.A., Özgür, Ü., Alivov, Y.I., Liu, C., Teke, A., Reshchikov, M.A., Do, S., Avrutin, V.: A comprehensive review of ZnO materials and devices. J. Appl. Phys. **98**, 105 (2005). https://doi.org/10.1063/1.1992666
2. Ismail, B., Abaab, M., Rezig, B.: Structural and electrical properties of ZnO films prepared by screen printing technique. Thin Solid Films **383**, 92–94 (2001). https://doi.org/10.1016/S0040-6090(00)01787-9
3. Chakrabarti, S., Dutta, B.K.: Photocatalytic degradation of model textile dyes in wastewater using ZnO as semiconductor catalyst. J. Hazard. Mater. **112**, 269–278 (2004). https://doi.org/10.1016/j.jhazmat.2004.05.013
4. Hyung, J., Yun, J., Cho, K., Hwang, I., Lee, J., Kim, S.: Necked ZnO nanoparticle-based NO_2 sensors with high and fast response. Sens. Actuators B Chem. J. **140**, 412–417 (2009). https://doi.org/10.1016/j.snb.2009.05.019
5. Saito, N., Haneda, H., Sekiguchi, T., Ohashi, N., Sakaguchi, I., Koumoto, K.: Low-temperature fabrication of light-emitting zinc oxide micropatterns using self-assembled

monolayers. Adv. Mater. **14**, 418–421 (2002). https://doi.org/10.1002/1521-4095(20020318)14:6%3c418:AID-ADMA418%3e3.0.CO;2-K
6. Huang, M.H., Mao, S., Feick, H., Yan, H., Wu, Y., Kind, H., Weber, E., Russo, R., Yang, P.: Room-temperature ultraviolet nanowire nanolasers (2001). https://doi.org/10.1126/science.1060367
7. Kumomi, H., Nomura, K., Kamiya, T., Hosono, H.: Amorphous oxide channel TFTs. Thin Solid Films **516**, 1516–1522 (2008). https://doi.org/10.1016/j.tsf.2007.03.161
8. Hosono, H.: Recent progress in transparent oxide semiconductors: materials and device application. Thin Solid Films **515**, 6000–6014 (2007). https://doi.org/10.1016/j.tsf.2006.12.125
9. Ryu, Y.R., Zhu, S., Budai, J.D., Chandrasekhar, H.R., Miceli, P.F., White, H.W.: Optical and structural properties of ZnO films deposited on GaAs by pulsed laser deposition. J. Appl. Phys. **88**, 201–204 (2000). https://doi.org/10.1063/1.373643
10. Ismail, A., Abdullah, M.J.: The structural and optical properties of ZnO thin films prepared at different RF sputtering power. J. King Saud Univ. Sci. **25**, 209–215 (2013). https://doi.org/10.1016/j.jksus.2012.12.004
11. Agarwal, D.C., Chauhan, R.S., Kumar, A., Kabiraj, D., Singh, F., Khan, S.A., Avasthi, D.K., Pivin, J.C., Kumar, M., Ghatak, J., Satyam, P.V.: Synthesis and characterization of ZnO thin film grown by electron beam evaporation. J. Alloys Compd. **123105**, 7 (2006). https://doi.org/10.1063/1.2204333
12. Liu, A., Zhang, J., Wang, Q.: Structural and optical properties of ZnO thin films prepared by different sol-gel processes. Chem. Eng. Comm. **198**, 494–503 (2011). https://doi.org/10.1080/00986445.2010.500168
13. Anand, V.K., Sood, S.C., Sharma, A.: Characterization of ZnO thin film deposited by sol-gel process. In: AIP Conference Proceedings, vol. 1324, pp. 399–401 (2010). https://doi.org/10.1063/1.3526243
14. Saleem, M., Fang, L., Wakeel, A., Rashad, M., Kong, C.Y.: Simple preparation and characterization of nano-crystalline zinc oxide thin films by sol-gel method on glass substrate. World J. Condens. Matter Phys. **2012**, 10–15 (2012)
15. Amutha, C., Dhanalakshmi, A., Lawrence, B., Kulathuraan, K., Ramadas, V., Natarajan, B.: Influence of concentration on structural and optical characteristics of nanocrystalline ZnO thin films synthesized by sol-gel dip coating method. Prog. Nanotechnol. Nanomater. **3**, 13–18 (2014)
16. Khan, Z.R., Zulfequar, M., Khan, M.S.: Optical and structural properties of thermally evaporated cadmium sulphide thin films on silicon (1 0 0) wafers. Mater. Sci. Eng. B Solid-State Mater. Adv. Technol. **174**, 145–149 (2010). https://doi.org/10.1016/j.mseb.2010.03.006
17. Kamaruddin, S.A., Chan, K.Y., Yow, H.K., Zainizan Sahdan, M., Saim, H., Knipp, D.: Zinc oxide films prepared by sol-gel spin coating technique. Appl. Phys. A Mater. Sci. Process. **104**, 263–268 (2011). https://doi.org/10.1007/s00339-010-6121-2
18. Wang, X.S., Wu, Z.C., Webb, J.F., Liu, Z.G.: Ferroelectric and dielectric properties of Li-doped ZnO thin films prepared by pulsed laser deposition. Appl. Phys. A Mater. Sci. Process. **77**, 561–565 (2003). https://doi.org/10.1007/s00339-002-1497-2
19. Nagayasamy, N., Gandhimathination, S., Veerasamy, V.: The effect of ZnO thin film and its structural and optical properties prepared by sol-gel spin coating method. Open J. Metal **3**, 8–11 (2013)
20. Caglar, M., Ilican, S., Caglar, Y.: Influence of dopant concentration on the optical properties of ZnO: in films by sol-gel method. Thin Solid Films **517**, 5023–5028 (2009). https://doi.org/10.1016/j.tsf.2009.03.037

Investigation and Experimental Evaluation of Vapor Compression Refrigeration System by Means of Alternative Refrigerants

Nilam P. Jadhav[1(✉)], V. K. Bupesh Raja[1], Suhas P. Deshmukh[3], and Mandar M. Lele[2]

[1] Department of Mechanical Engineering, Sathyabama University, Chennai, India
jadhavnilamp@gmail.com
[2] Department of Mechanical Engineering, MIT COE, SPPU, Pune, India
[3] Department of Mechanical Engineering, Govt. COE, Karad, Karad, India

Abstract. A theoretical investigation is considered to discover possible alternatives on behalf of R134a, R12 and R22 for traditional vapor compression refrigeration system. For the same two azeotrope blends, R404A and R407C are nominated. The theoretical analysis is conceded for the refrigerating system at a continual condensation temperature of 50 °C and −10 °C to 10 °C evaporating temperature. Results to investigate alternative refrigerant have slightly lower performance coefficient than HFC134a, CFC22 and CFC12 respectively. Further, it is noticed the average performance coefficient of alternative refrigerants is 7.26% for R404A and 1.41% for R407C enhanced for the degree of superheating and sub cooling. The experimental setup is developed for R404A and R407C to observed different parameters such as pull down time for evaporator temperature, discharge pressure, power consumption, performance coefficient and energy efficiency ratio and compared with R134a. Smaller pull down time, lesser power consumption and isentropic work done improves the performance coefficient of the system, similar or maybe higher behavior like R134a proved that the nominated refrigerants used as an alternative to R12 and R22.

Keywords: Refrigeration · Alternative refrigerants · Performance · Degrees of superheat · Degree of sub cooling

1 Introduction

Recently refrigeration and air conditioning find major role in different application areas like in domestic, commercial, industrial, transport and pharmaceutical and food preservation industries. Due to increasing demand of this system leads towards more energy consumption. To run these systems effectively, the working medium utilized known as refrigerants, have a major role towards the performance of the system. Due to global awareness associated to (GWP) global warming potential and (ODP) ozone depletion potential, the vital role focuses further attention towards selection of refrigerant. Most of the industrial sectors realized that chlorofluorocarbons (CFCs) such as R12 and R22 were major role concerning the destruction of the ozone layer. This leads

D. J. Hemanth et al. (Eds.): COMET 2019, LNDECT 35, pp. 10–26, 2020.
https://doi.org/10.1007/978-3-030-32150-5_2

the industrial and domestic sectors to replace CFCs by hydro chlorofluorocarbons (HCFCs) and hydrofluorocarbons (HFCs). HFC refrigerants have similar vapor pressure, stable and non-flammable compared to HCFCs and CFCs refrigerants. Apart from all these combinations the hydrocarbon refrigerants such as isobutene, n-butane, propane or hydrocarbon mixtures are also utilized for most of the refrigeration and air conditioning applications. As per ASHRAE standard guideline hydrocarbon refrigerants are highly flammable indexed (A3), however they are non-toxic, low GWP and higher performance compared to other category refrigerants.

The present work emphases towards the investigation of vapor compression refrigeration system performance by utilizing zeotropic refrigerant blends, which can be suitable alternate for R134a, R12 and R22. HFCs and its blends such as Zoetrope, finding its application in most of residential and commercial refrigeration sectors. For the present revision R404A, zeotropic blend of R134a/R125/R143a (4%, 44%, 52% by weight.) and R407C, zeotropic blend of R134a/R32/R125 (52%, 23%, 25% by weight.) are nominated to conduct theoretic investigation. Likewise R134a, R404A may be utilized in commercial refrigeration system for low and medium temperature range and also used in new equipment to replace R502. For old systems based on R22 may also retrofitted by R404A with suitable POE.

Whereas in case of air conditioning systems R407C have similar properties like R22. Similar to R404A, it is suitable for new equipment and used to retrofit old R22 schemes. R407C is also meet in most of the medium temperature applications, direct expansion and light and residential air conditioning applications. Both selected candidates are non-flammable and non-toxic. They have zero ODP and having slightly more GWP value compared to R134a. R404A have adjacent performance comparing to R502 with low evaporating temperature. Whereas, R407C have similar evaporation and condensation pressure parallel to R22 without any modification towards system components.

Domanski et al. [1] developed evolutionary algorithm to explore the performance of R404a and R410a and observed that R404a had increased performance for evaporator and condenser COP. Alike Ferreira et al. [2] carried out the experiment to observe the magnetic field effect on the thermos-physical properties of four refrigerants, including R404a and R410a. He got the similar results for the evaporator and condenser COP as Domanski. Patil [3] worked on U tube micro fin and U tube smooth condensers at various operating condition to monitor the performance of the system and Chinnaraj et al. [4] focused attention on the window air conditioner for smaller capacity based on R22 with electronic expansion valve retrofitted with R407C and R290. Both the authors noted the ratio of energy efficiency and performance of the system was improved for R404A. Jerald et al. [5] investigate the performance of R12 vapor compression systems retrofitted with zoetrope blend R404a. He utilized five different diameters for capillaries to observed performance of R404a. His experimental study concluded that R404a have better cooling capacity, faster pull-down time and miscibility of oil with R404a was better than R134a which results better efficiency in the system. The amount of refrigerant charged required for R404a (600 g) is also very little compared to R134a (1 kg) to attain same cooling capacity. Li and Zhao [12] has compared operating characteristics and performance amongst R404A and R502 and reported that R502 have more discharge temperature compared to R404A which leads to improve the reliability and life of the compressor unit. Though the pressure ratio of R404A is slightly 7% to 26% more compared to R502 and may be improved later by redesign and optimization of the refrigeration system. Shilliday et al. [6] studied the refrigeration cycles operating at 40 °C and −10 °C condensing and evaporating temperatures for exergy and energy analysis of R290, R744 and R404A respectively. He reported that both R404A and R290 show higher COPs related with R744. Chen et al. [7] utilized rotary compressor specially designed for R407C to examine the theoretical and experimental performance of R407C and R161/R32/R125. Several parameters are evaluated by the authors and reported that R161/R32/R125 have slightly more discharge temperature than that of R407C. Dalkilic et al. [9] carried out theoretical analysis that include R290/R600a, R290/R1270 refrigerant blends and compared with the traditional refrigerants R12, R22 and R134a. Entire theoretical analysis is performed to monitor various parameters and effect of degree of super heating and sub cooling such as refrigerant type, coefficient of performance and volumetric refrigeration capacity for −30 °C to 10 °C and 50 °C evaporating and condensing temperatures. Lastly, he concluded that HC1270/HC290 (80:20 wt%) and HC260a/HC290 (60:40 wt %) blends have improved performance compared to R22 and R12 respectively. Thangavel et al. [10] monitor performance of the system by applying different loads on evaporator by utilizing hydrocarbon refrigerants. The author assumed evaporator temperature as −10 °C and the entire analysis is performed for 30 °C to 65 °C range of condenser temperatures to conduct computational and experimental analysis to determine that the hydrocarbon mixture of propane and iso-butane (each 50% by wt.) is one of the possible alternatives for R12 and R134a. Tiwari et al. [11] proposed experimental study of R404A and R134a in domestic refrigerator. The author concluded that the pull-down time of R404A was earlier and miscibility of oil with R404A increased the life of compressor compared with R134a. Li et al. [12] proposed experimental setup for analysis of operating characteristics of low evaporation temperature R404A and R502 refrigeration system. The author develops refrigeration system considering R502 as a base model to analyze the behavior of R04A. He concludes that the discharge temperature of R404A is lower than R502 which leads towards improvement in reliability and life of the compressor unit. Also, the author notified that the ration of pressure for R404A is somewhat higher compare to R502 and hence may be neglected due to very small difference.

This paper deals with the use of zeotropic refrigerant blends to investigate vapor compression system performance. The zeotropic refrigerants R404A (R134a 4%/R125 44%/R143a 52% by wt.) and R407C (R134a 52%/R32 23%/R125 25% by wt.) are used to notice the performance against the predictable refrigerants R22, R134a and R12. For the same −10 °C to 10 °C evaporator temperature range is carefully chosen to investigate the effect of sub cooling and superheating on the pressure ratio (PR), isentropic compression work (W), cooling effect (RE) and performance coefficient (COP). Apart from this the other parameters like suction vapor flow rate (SVFR), volumetric refrigeration capacity (VRC) and power per ton of refrigeration (PTR) are also observed in the same evaporating temperature range. Correspondingly test rig was developed for R134a, R404A and R407C to monitor the performance of refrigeration system.

2 Theoretical Cycle Performance

Theoretical cycles without sub-cooling and superheating and with sub-cooling and superheating for theoretical analysis as shown in Fig. 1(a) and (b) respectively.

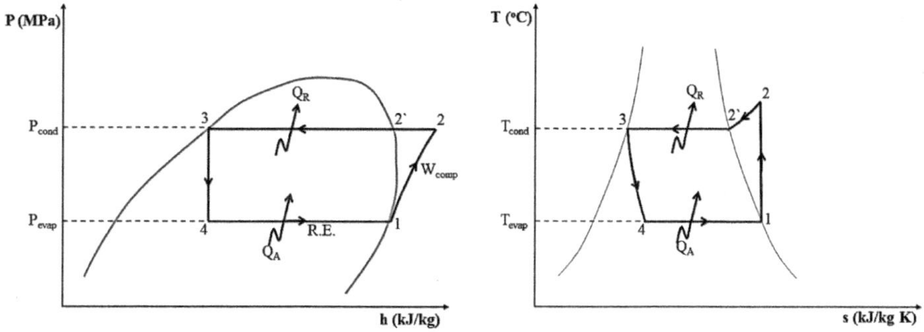

(a) without sub-cooling and superheating

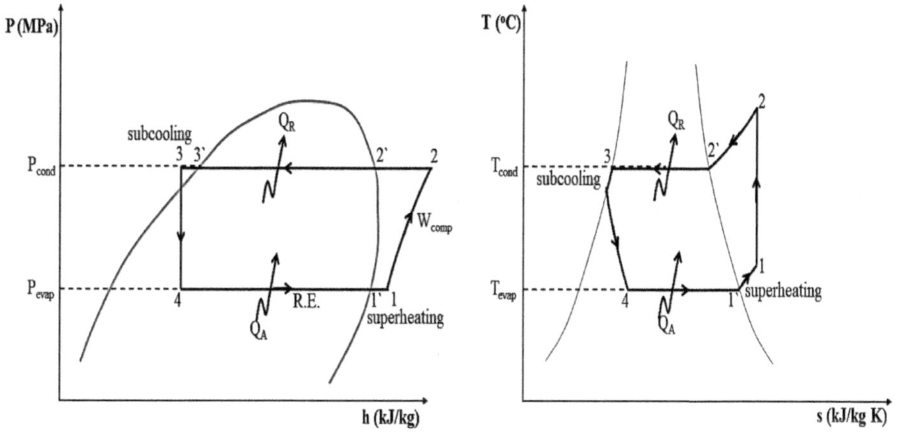

(b) with sub-cooling and superheating

Fig. 1. Theoretical vapor compression refrigeration cycle in case of (a) without sub-cooling and superheating and (b) with sub-cooling and superheating

Standard reference database, Refprop 8.0 is utilized for thermos physical properties and cycle performance parameters of R134a, R404A and R407C. It is evident that there are definite deviations between ideal and actual refrigeration cycle associated with pressure drop due to heat exchange and fluid flow of equipment that of surrounding. Concerning to this certain assumption are considered for theoretical analysis, such as: pressure drop is negligible, heat loss at the liquid line heat exchanger, isentropic compression work, expansion is isenthalpic, steady state for individual component and ambient temperature selected is suited for Indian climate conditions which is 35 °C. In perfect energy assessment of refrigeration system, heat balance and energy balance of respective component are considered. The performance characteristics of the system are based on refrigerating capacity and the COP which is given as follows:

$$Q_{evap} = \dot{m}_r(h_1 - h_4)\left(\mathrm{kJ\,kg^{-1}}\right) \tag{1}$$

where, $\dot{m}_r$ = mass flow rate (kg/s); h_1, h_4 = enthalpies of refrigerant at evaporator outlet and inlet (kJ/kg) respectively.

$$W = \dot{m}_r(h_2 - h_1)\left(\mathrm{kJ\,kg^{-1}}\right) \tag{2}$$

where, W = isentropic compressor work; h2, h4 = enthalpies of refrigerant at compressor outlet and inlet (kJ/kg) respectively.

Assuming the steady state analysis the COP is calculated as:

$$COP = \frac{Q_{evap}}{W} \tag{3}$$

The refrigerant condenser pressure is reduced extremely when it is admitted in the capillary tube and achieved the evaporator pressure. Henceforth the pressure ratio based on the condenser and evaporator pressure is given as:

$$P_r = P_c/P_e \tag{4}$$

where, P_c, P_e = condenser and evaporator pressure (MPa) respectively

Power per ton of refrigeration is give as:

$$PTR = 3.5\,\mathrm{W}/RE\left(\mathrm{kW\,TR^{-1}}\right) \tag{5}$$

For non-superheating/sub cooling and superheating/sub cooling vapor compression refrigeration cycle, the volumetric refrigerating capacity is given as:

$$VRC = \rho_1 \cdot RE\left(\mathrm{kJ\,m^{-3}}\right) \tag{6}$$

where, ρ_1 = density of refrigerant at compressor inlet

Suction flow of vapor per kW of refrigeration is given as:

$$SVFR = 1/(\rho_1 \cdot RE)\left(\mathrm{L\,s^{-1}}\right) \tag{7}$$

3 Experimental Setup

The experimental test rig is complete vapor compression refrigeration system developed for R134a. The schematic diagram is shown in Fig. 2. The refrigerator has a capacity of 1TR equipped with air cooled, four row, staggered tube arrangement, single pass, tube-fin condenser unit, hermetically sealed compressor having capacity of 1TR and displacement 12.58 cc/rev., capillary tube and the evaporator which is submerged in the calorimeter. Mono ethylene glycol mixed with water (each 50% by wt.) in the calorimeter and maintained at a temperature of 45 °C with the help of electrical heater. Seven K type thermocouples are used to measure various temperatures at different locations. Two pressure gauges are instrumented at outlet and inlet of compressor unit for measuring suction and discharge pressure respectively. Mass flow meter is connected among the capillary and dryer unit at the suction line to note the flow rate of refrigerant. One energy meter is linked to measure energy consumption rate for compressor while other energy meter is linked to measure the amount of heat input given to the heater. Previously, filling refrigerant charge, soap bubble test is carried out to evacuate the system. Initially the system is charged with 640 g of R134a to monitor the performance so that the available data can be used for comparison purpose. The suction line heat exchanger is utilized for sub cooling and superheating of refrigerant. Separate water-glycol re-circulation pump is used for to maintain desired temperature of the calorimeter and proper mixing of water-glycol mixture to attain the uniform cooling effect. As the evaporator is submerged is the water-glycol mixture in the calorimeter, the initial and final temperature of water indicates the refrigerating effect produced by the refrigeration system.

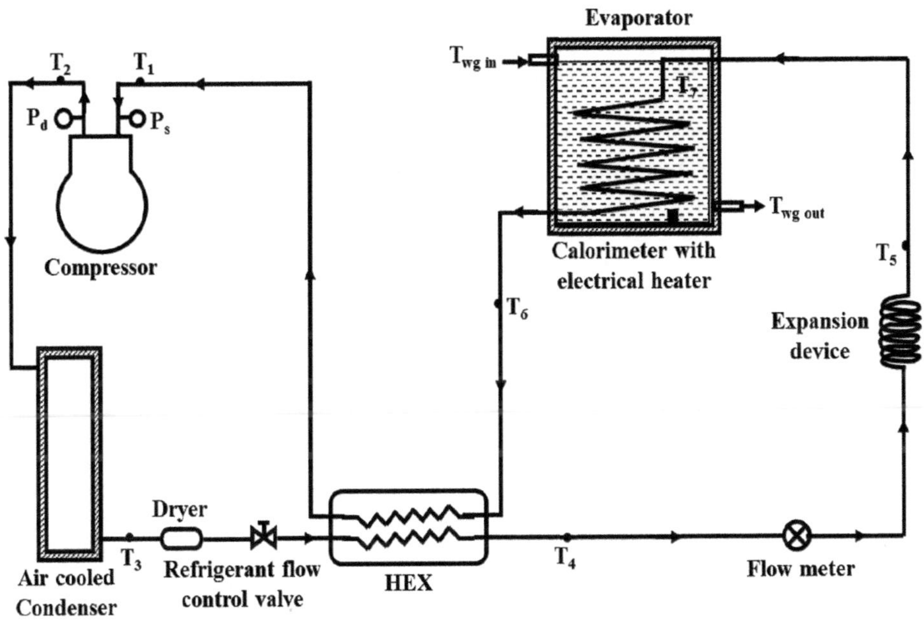

Fig. 2. Schematic diagram of experimental refrigeration system. P_d, P_s = discharge and suction pressure gauge; T_1–T_7 = K type thermocouples; $T_{wg\ in}$, $T_{wg\ out}$ = water-mono ethylene glycol solution in and out; HEX = suction line heat exchanger

4 Theoretical Results

In most of the cooling systems R12 and R22 refrigerants are widely used. Both of these candidates have good performance along with higher values of ODP and GWP. Due to prohibition by the Montreal Protocol against R12 and R22 and common uses in cooling system these two candidates are chosen as reference fluids. Whereas owing to very good thermo physical properties of R134a, is also considered for comparative analysis. The performance of R404A and R407C refrigerant blends are investigated and are compared with R12, R22 and R134a.

Various functioning properties of pure and blend refrigerants such as pressure ratio, isentropic compression work, power per ton of refrigeration, evaporation pressure, refrigerating effect, suction vapor flow rate, volumetric refrigeration capacity and performance coefficient are investigated theoretically and measure the performance for various evaporating temperature ranges. The plots are divided into two groups as no superheating/sub cooling for all selected candidates and R404A, R407C compared with R12, R22 and R134a. In the second group 5 °C superheat/sub cooling is considered for all selected candidates and R404A, R407C compared only with R134a due to wide uses in refrigeration systems and good thermo physical properties.

The variation of functional properties of pure and blend refrigerants such as pressure ratio (P_r), isentropic compression work (W), power per ton of refrigeration (PTR), evaporation pressure (P_{evp}), refrigerating effect (RE), coefficient of performance (COP), suction vapor flow rate (SVFR) and volumetric refrigeration capacity (VRC) are investigated theoretically and plotted against the evaporating temperature (T_{evap}) as shown in Figs. 3, 4 and 5 considering no superheating and sub cooling of refrigerant for constant condensing temperature of 50 °C and evaporation temperature range of −10 °C to 10 °C respectively. Results shown in Table 1 is example for case study to compare traditional pure refrigerants R12 and R22 with alternative refrigerant blends. Tables 2, 3 and 4 shows the deviation values of alternative refrigerant blends with respect to R12 and R22 at 50 °C condensing temperature and −10 °C evaporation temperature with no superheating and sub cooling of refrigerants. It has been seen from Fig. 3(a) and (b) that the saturation vapor pressures for R404A (R134a 4%/R125 44%/R143a 52% by wt.) and R407C (R134a 52%/R32 23%/R125 25% by wt.) is much more compared to R12 and R22. It has been observed that at −10 °C evaporating temperature, R404A and R407C have nearly 49.11% and 45.49% more evaporation pressure compared to R12 and 17.48% and 11.61% more evaporation pressure compared to R22 respectively. Further the evaporation pressure increases with reducing evaporation pressure. The same is also noticed in the deviation tables of refrigerants. Figure 3(c) shows the effect of evaporation temperature on pressure ratio. The pressure ratio curves for both the blends R404A and R407C lies below R12, R22 and R134a, which indicates that the pressure ratio for both the two blends is very less. The pressure ratio for R404A is 4.14% and 4% less compared to R12 and R22 and that of R407C is 2.55% and 2.41% less compared to R12 and R22 respectively at the evaporation temperature of 10 °C. The pressure ratio goes on increasing for reduction in evaporation temperature for all candidates, but seems very less pressure ratio for R404A and

R407C for the entire range of evaporation temperature. Also, the pressure ratio of R404A and R407C have 23% and 22% less value when compared with R134a.

The improved efficiency of the refrigeration system is due to a reduction in pressure ratio of the compressor and the same is reflected in the Fig. 4 (c, d). The reduction in pressure ratio and increase in evaporation pressure at higher value of evaporation temperature (10 °C) for R404A and R407C results reduction in isentropic compressor work. Also, for decreased evaporation temperature (−10 °C) the isentropic compressor work increases for all candidates. It has been observed that the isotropic compressor work for R407C have nearer values with R12 and R22 whereas the compressor work for R404A have 4.13%, 7.48% and 33.59% lowest value compared to R12, R22 and R134a respectively. For reduction in evaporator temperature the same trend is observed for R404A compared to other candidates.

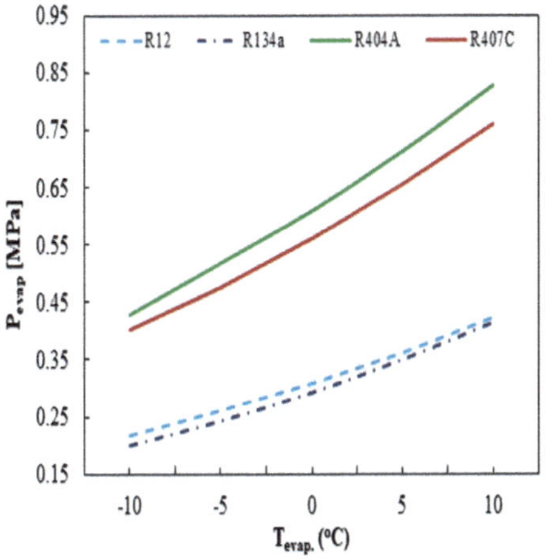

a) R12 vs alternative refrigerants

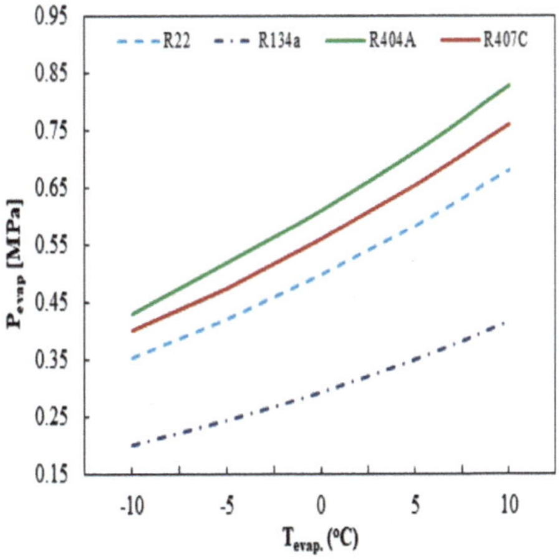

b) R22 vs alternative refrigerants

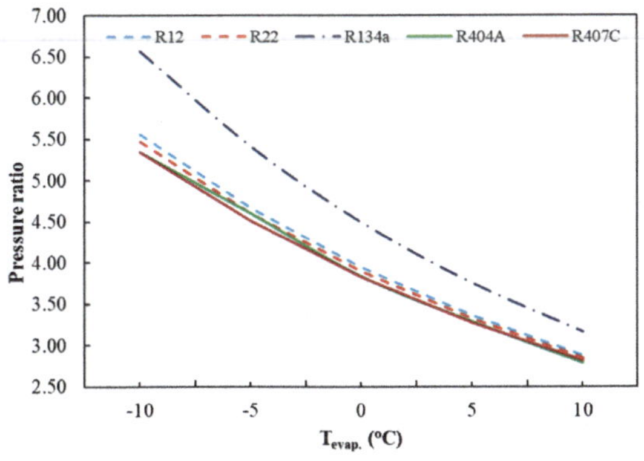

c) Pressure ratio vs alternative refrigerants

Fig. 3. Evaporating pressure (a, b) and pressure ratio (c) vs evaporating temperature

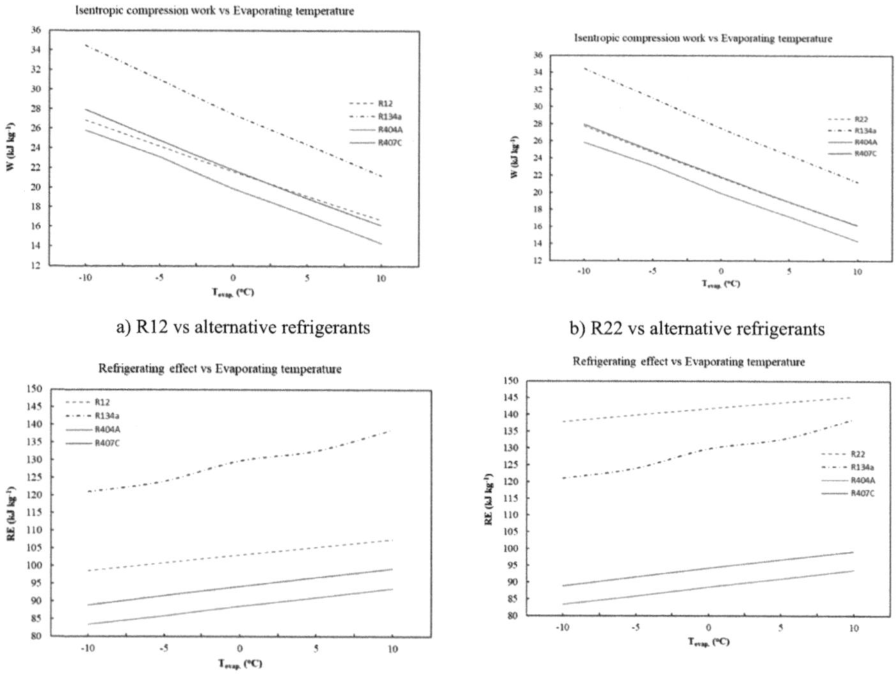

a) R12 vs alternative refrigerants

b) R22 vs alternative refrigerants

c) R12 vs alternative refrigerants

d) R22 vs alternative refrigerants

Fig. 4. Refrigerating effect (a, b) and isentropic compression work (c, d) vs evaporating temperature

Figure 5 shows the effect of evaporating temperature on the performance of coefficient and observed that the refrigerating effect of R404A is 18.17%, 65% and 45% less than R12, R22 and R134a respectively for 50 °C condensing and −10 °C evaporating temperature. Also, the same result observed for R407C that the refrigerating effect is 11.05%, 55% and 36.35% less than R12, R22 and R134a for the same condensing and evaporating temperature. Whereas the refrigerating effect for R404A and R407C is improved by 16.74% and 16.01% compared with R134a for 5 °C degree of superheat and sub cooling. The effect of superheating and sub cooling results the closer values of R404A compared with R12 and R134a.

Figure 6(a, b, c) indicates different parameters that are considered to investigate power per ton of refrigeration, volumetric refrigeration capacity and suction vapor flow rate. PTR should be minimized as it is concerned with required power for per ton of refrigeration to run a system. Figure 6(a) indicates that both the candidates R404A and R407C have more value of PTR compared with R12, R22 and R134a. The relation of volumetric refrigeration capacity related to evaporating temperature is as shown in Fig. 6(b), which concern to cooling capacity per unit of vapor volume at the exit of the evaporator. It is known that for given swept volume in the compressor, high capacity of cooing can be obtained by high volumetric refrigerant capacity. However, it is observed that the VRC value of nominated refrigerants is much below compared to R12, R22 and R134a. For increased evaporator temperature the decrease in a specific volume of the

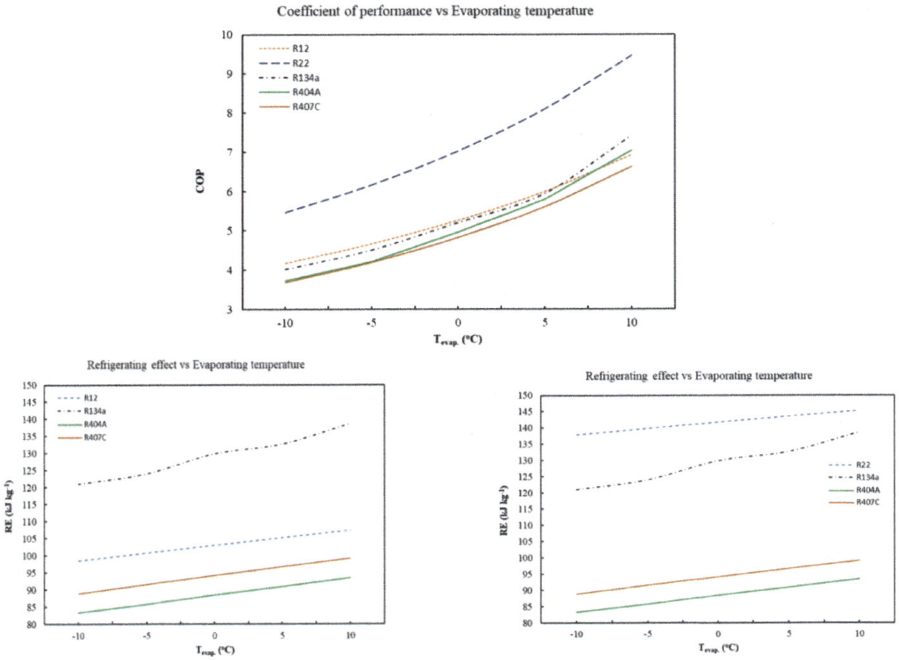

Fig. 5. Evaporating temperature vs performance coefficient (COP)

Table 1. Action on standard vapor compression cycle using various refrigerants at T_{cond} = 50 °C and T_{evap} = −10 °C (no superheating/sub cooling)

Refrigerant	P_{evap}	P_{cond}	P_r	W_{comp}	RE	PTR	VRC	SVFR	COP
(wt%)	(MPa)	(MPa)		(kJ kg^{-1})	(kJ kg^{-1})	(kW TR^{-1})	(kJ m^{-3})	(L s^{-1})	
R 12	0.218	1.216	5.560	26.873	98.58	0.9541	143.733	0.7105	3.668
R 22	0.354	1.942	5.475	27.739	137.95	0.7037	181.363	0.5513	4.973
R 134a	0.200	1.317	6.569	34.476	121.04	0.9969	160.620	0.6225	3.510
R 404A (4%R134a+44%R125 +52%R143a)	0.430	2.300	5.338	25.807	83.42	1.082	98.994	1.0101	3.232
R 407C (52%R134a+23%R32 +25%R125)	0.401	2.146	5.346	27.933	88.77	1.101	108.308	0.9232	3.177

suction vapor causes an increase in mass flow rate of refrigerant. Hence, to reduce the mass flow rate SVFR value should be lesser. Yet again in Fig. 6(c) shows that the SVFR value for R404A and R407C is much more compared to R12, R22 and R134a. The opposite behavior of R404A and R407C associated to PTR, VRC and SVFR is avoided by considering the effect of degree of superheating and sub cooling. The effect of 5 °C superheating and sub cooling on R404A and R407C and evaluation with R134a is demonstrated in Fig. 7(a, b, c, d) respectively. Also, the effect of superheating and sub

Table 2. Nonconformity values of alternative refrigerants from R12 at T_{cond} = 50 °C and T_{evap} = −10 °C (no superheating/sub cooling)

Refrigerant	P_r	W_{comp}	RE	PTR	VRC	SVFR	COP
(wt%)		(kJ kg^{-1})	(kJ kg^{-1})	(kW TR^{-1})	(kJ m^{-3})	(L s^{-1})	
	%	%	%	%	%	%	%
R 134a	15.37	22.05	18.56	4.29	12.38	−14.13	−4.49
R 404A (4%R134a+44% R125+52%R143a)	−4.148	−4.131	−18.173	11.882	−42.162	29.658	−13.485
R 407C (52%R134a+23% R32+25%R125)	−4.006	3.795	−11.051	13.369	−29.938	23.039	−15.432

Table 3. Nonconformity values of alternative refrigerants from R22 at T_{cond} = 50 °C and T_{evap} = −10 °C (no superheating/sub cooling)

Refrigerant	P_r	W_{comp}	RE	PTR	VRC	SVFR	COP
(wt%)		(kJ kg^{-1})	(kJ kg^{-1})	(kW TR^{-1})	(kJ m^{-3})	(L s^{-1})	
	%	%	%	%	%	%	%
R 134a	16.66	19.54	−13.97	29.40	−12.91	11.44	−41.65
R 404A (4%R134a+44% R125+52%R143a)	−2.559	−7.486	−65.368	35.001	−83.205	45.416	−53.85
R 407C (52%R134a+23% R32+25%R125)	−2.419	0.695	−55.401	36.098	−67.451	40.280	−56.489

Table 4. Nonconformity values of alternative refrigerants from R134a at T_{cond} = 50 °C and T_{evap} = −10 °C (no superheating/sub cooling)

Refrigerant	P_r	W_{comp}	RE	PTR	VRC	SVFR	COP
(wt%)		(kJ kg^{-1})	(kJ kg^{-1})	(kW TR^{-1})	(kJ m^{-3})	(L s^{-1})	
	%	%	%	%	%	%	%
R 404A (4%R134a+44% R125+52%R143a)	−23.057	−33.593	−45.097	7.928	−62.251	38.367	−8.611
R 407C (52%R134a+23% R32+25%R125)	−22.888	−23.424	−36.352	9.482	−48.299	32.568	−10.475

Table 5. Nonconformity values of alternative refrigerants from R134a at T_{cond} = 50 °C and T_{evap} = −10 °C (with superheating/sub cooling by 5 °C)

Refrigerant	P_r	W_{comp}	RE	PTR	VRC	SVFR	COP
(by wt%)		(kJ kg^{-1})	(kJ kg^{-1})	(kW TR^{-1})	(kJ m^{-3})	(L s^{-1})	
	%	%	%	%	%	%	%
R 404A (4%R134a+44% R125+52%R143a)	−24.859	−37.85	−37.545	−0.221	−54.279	35.182	0.2206
R 407C (52%R134a+23% R32+25%R125)	−22.888	−25.900	−30.538	3.553	−42.353	29.751	−3.6848

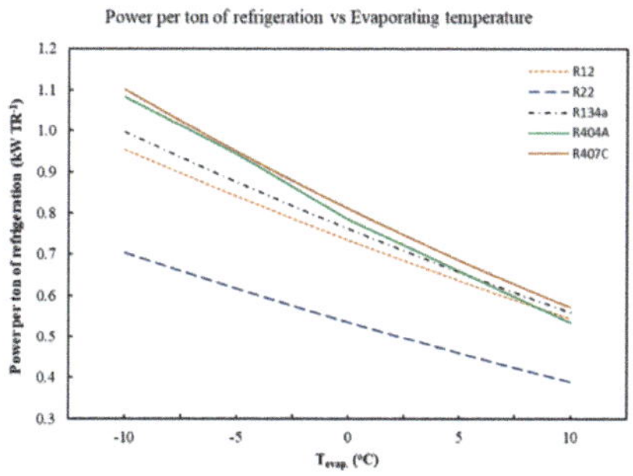

a) Effect of T_{evap} on PTR

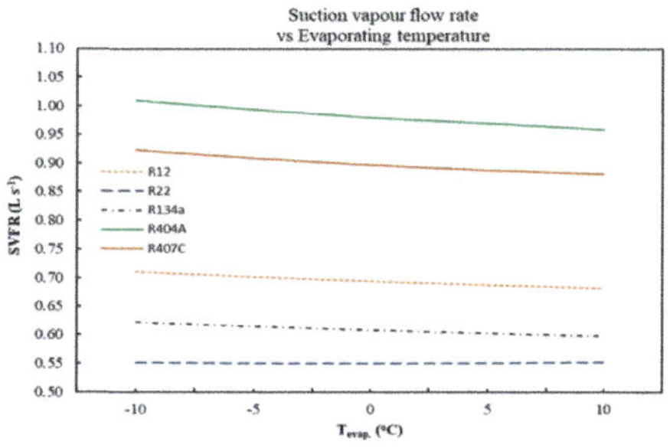

b) Effect of T_{evap} on VRC

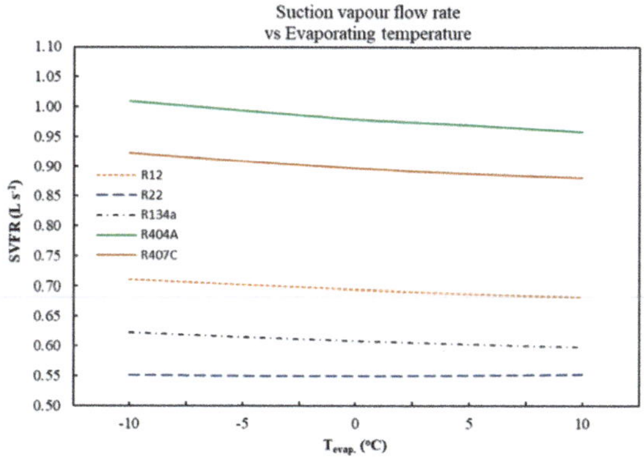

c) Effect of T_{evap} on SVFR

Fig. 6. Effect of evaporating temperature on PTR (a), VRC (b) and SVFR (c)

cooling at a constant condensation temperature for the performance of coefficient is illustrated in Fig. 7(a). The effect of superheat and sub cooling on all three candidates R404A, R407C and R134a indicates higher value for performance. At −10 °C evaporating temperature performance of R404A is nearer to R134a and having higher value for increased evaporating temperature. However, at the same evaporating temperature R407C have 3.6% lower performance value compared with R134a and showing equal performance to R134a at 0 °C. Hence, the effect of superheating and sub cooling positively increased the performance of R404A and R407C and proved very good results compared to R134a. Also, from Fig. 7(b, c, d) and from Table 5 it has been observed that the power required per ton to run the system using R404A have 0.22% low value compared with R134a and it decreased as the evaporating temperature increases. Whereas for R407C have 3.5% more power required per ton that of R134a and for increased evaporating temperature the power required is very less that of R134a. Although the required power is low for R404A and R407C, the volumetric refrigeration capacity is 54.17% and 42.35% less and suction vapor flow needed for refrigeration have 35.18% and 29.75% higher values respectively compared with R134a.

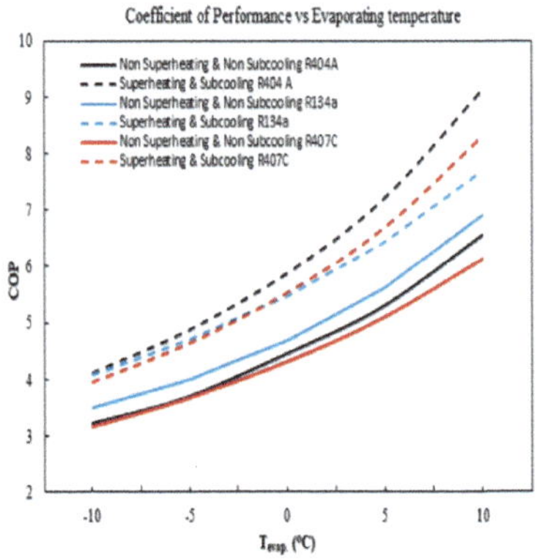

a) Effect of T_{super} and T_{sub} on performance

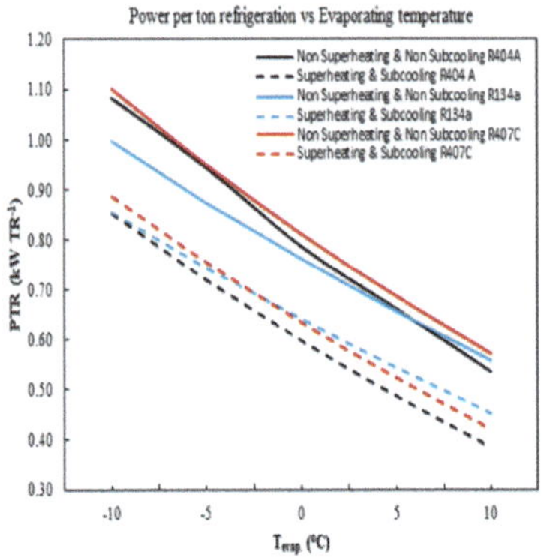

b) Effect of T_{super} and T_{sub} on PTR

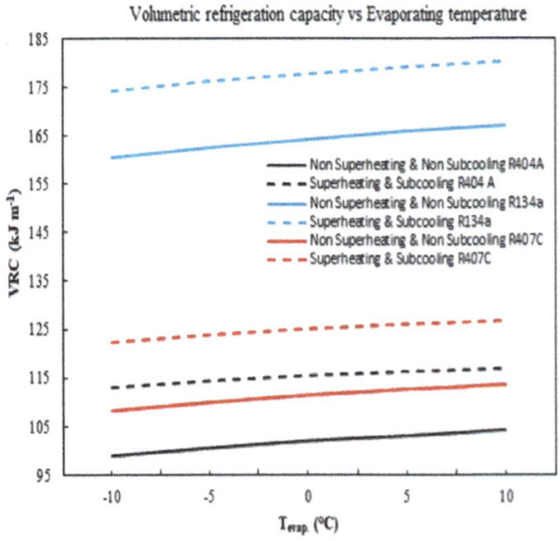

c) Effect of T_{super} and T_{sub} on VRC

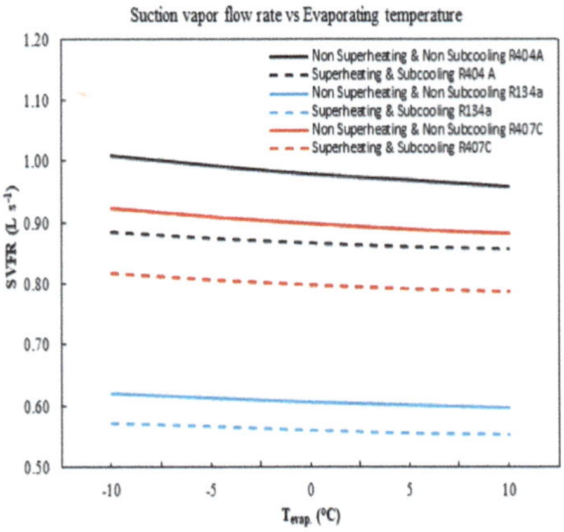

d) Effect of T_{super} and T_{sub} on SVFR

Fig. 7. Effect of superheating and sub cooling on performance (a), PTR (b), VRC (c) and SVFR (d)

5 Experimental Results

The experimental setup as shown in Fig. 2 is utilized to monitor the performance of R134a, R404A and R407C. The parameters such as pull-down time required to achieve the desired evaporator temperature, discharge pressure, compressor discharge temperature, power consumption for compressor, isentropic compression work, cooling effect, performance coefficient and energy efficiency ratio are discussed for experimental analysis. The experimental results for pull down time and discharge pressure are plotted in Figs. 8 and 9 respectively. According to International Standard Organization [8], the pull-down time is the time required for the evaporator to achieve the desired temperature. Aimed at better performance the pull time should be minimized which is further depends on the thermo physical properties of refrigerant used. From Fig. 8 it is observed that the pull-down time required for both R404A and R407C is faster compared to R134a. The behavior of discharge pressure is observed after the system undergone steady state condition. Discharge pressure is one of the important parameters that affects the performance of a refrigeration system. It should be minimized to influence the stability of the lubricants and allow the lightweight construction of the compressor and condenser units. From Fig. 9 it is observed that the average discharge pressure for both R404A is 41.95% and R407Cis 37.50% more compared to R134a. Hence modifications in the compressor and condenser unit and lubrication system is required for R404A and R407C.

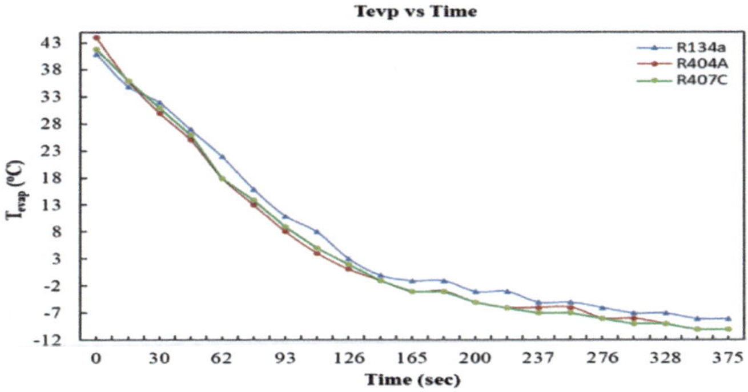

Fig. 8. Pull down time of R404A, R407C and R134a

The excessive discharge for R404A and R407C leads to less power consumption as shown in Fig. 10. The average power consumption rate for R404A is 19.35% and that of R407C is 3.11% less compared to R134a. Further the power consumption rate is 16.78% less for R404A that of R407C. The effect of performance coefficient against the water temperature in the evaporator is shown in Fig. 11. The average performance coefficient of R407C is 18.32% and R404A is 7.14% more that of R134a. The average decrease in refrigerating effect of R404A is 29.92% and of R407C is 2.74%, however, the isentropic work of R407C is 33.90% and of R404A is 38.27% less that of R134a. This results improvement in the performance of the system.

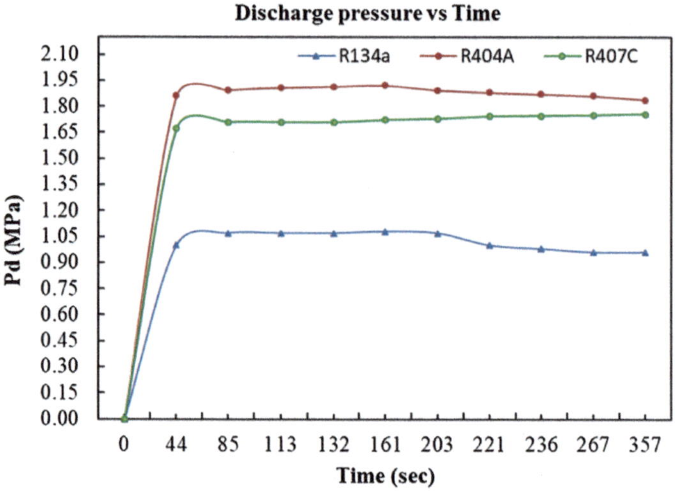

Fig. 9. Discharge pressure of R404A, R407C and R134a

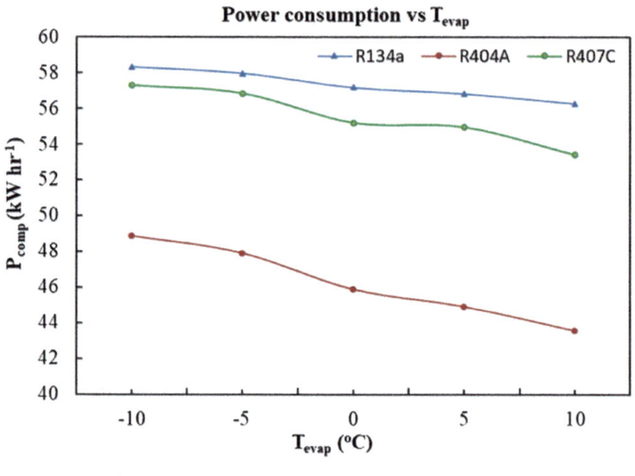

Fig. 10. Power consumption vs evaporation temperature for R404A, R407C and R134a

The effect of evaporating temperature on (EER) energy efficiency ratio is as shown in Fig. 12. It is observed that the average energy efficiency ratio of R404A and R407C is 18.98% and 2.97% more that of R134a. This is due to the less power consumption and isentropic work of the compressor unit neglecting the power required for fans coil unit.

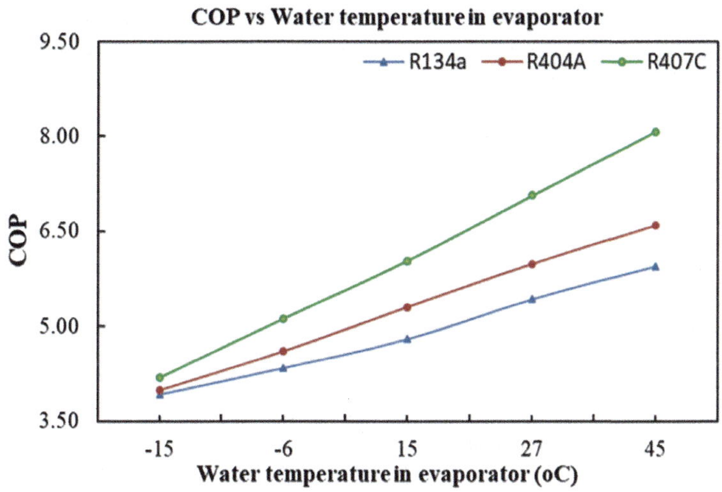

Fig. 11. Coefficient of performance vs water temperature in evaporator for R404A, R407C and R134a

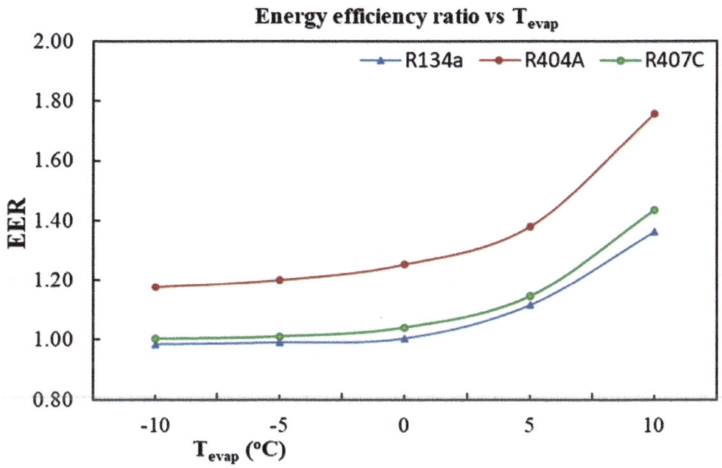

Fig. 12. Effect of EER vs evaporation temperature for R404A, R407C and R134a

6 Conclusions

The performance of alternative refrigerants, R404A and R407C intended for ideal vapor compression refrigeration system is investigated for replacement of CFC12, HFC134a and CFC22. Further, experimental analysis is done to observe the performance of alternative refrigerants. The performance of the two candidates R404A and R407C have lesser however, is improved by considering the degree of superheating and sub cooling. The pulldown time is similar to R134a and power consumption to run the system is much below that of R134a. As the isentropic work done is lower, the performance coefficient is more and hence the efficiency ratio is more. It is proved that R40A and R407C are the promising refrigerants as an alternative to R12 and R22.

References

1. Domanski, P.A., Brown, J.S., Heo, J., Wojtusiak, J., McLinden, M.O.: A thermodynamic analysis of refrigerants: performance limits of the vapor compression cycle. Int. J. Refrig **38**, 71–79 (2013)
2. Ferreira, C.A.I., Newell, T.A., Chato, J.C., Nan, X.: R404A condensing under forced flow conditions inside smooth, micro fin and cross-hatched horizontal tubes. Int. J. Refrig **26**, 433–441 (2003)
3. Patil, P.A.: Performance analysis of HFC-404A vapor compression refrigeration system using shell and U-tube smooth and micro fin tube condensers. J. Thermal Energy Gener. **25**, 77–91 (2012)
4. Chinnaraj, C., Vijayan, R., Govindarajan, P.: Analysis of eco-friendly refrigerants usage in air conditioner. Am. J. Environ Sci. **7**, 510–514 (2011)
5. Jerald, A.L., Senthilkumaran, D.: Investigations on the performance of vapor compression system retrofitted with zeotropic refrigerant R404A. Am. J. Environ Sci. **10**(1), 35–43 (2014)
6. Shilliday, J.A., Tassou, S.A., Shilliday, N.: Comparative energy and exergy analysis of R744, R404A and R290 refrigeration cycles. Int. J. Low-Carbon Technol. **4**, 1–8 (2009)
7. Chen, G.M., Han, X.H., Wang, Q., Zhu, Z.W.: Cycle performance study on R32/R125/R161 as an alternative refrigerant to R407C. Appl. Therm. Eng. **17**, 2559–2565 (2007)
8. ISO, International Standard Organization, International Standard-8187, household refrigerating applications (refrigerators/freezers) characteristics and test methods (1991)
9. Dalkilic, A.S., Wongwises, S.: A performance comparison of vapor-compression refrigeration system using various alternative refrigerants. Int. Commun. Heat Mass Transf. **37**, 1340–1349 (2010)
10. Thangavel, P., Somasundaram, P.: Part load performance analysis of vapor compression refrigeration system with hydrocarbon refrigerants. J. Sci. Ind. Res. **72**, 454–460 (2013)
11. Tiwari, A., Gupta, R.C.: Experimental study of R404A and R134a in domestic refrigerator. Int. J. Eng. Sci. Technol. **3**, 6390–6393 (2001)
12. Li, H., Zhao, Z.: Analysis of the operating characteristics of a low evaporation temperature R404A refrigeration system. In: International Refrigeration and Air conditioning conference, Purdue University (2008). (2221-1-6)

A Novel and Customizable Framework for IoT Based Smart Home Nursing for Elderly Care

J. Boobalan(✉) and M. Malleswaran

Department of Electronics and Communication Engineering, University College of Engineering Kancheepuram, Kanchipuram 631552, Tamilnadu, India
research.boo@gmail.com

Abstract. Massive growth in the Internet of Things (IoT) Technology tends it possible to recognize and analyze the health conditions of the patients in a ubiquitous manner for every time instances. The IoT technology enables the seamless platform to connect people and things to one another for elevating and making our lives easier. The integration of IoT into healthcare applications leads researchers to develop an intelligent healthcare system. Inspired by these aspects, a novel architecture is proposed in this paper for an integrated system to monitor the isolated older patients and surveillance of the smart home. This approach grants a venue for adjustable experimentations which is capable of responding to quick changes in the environment. This concentrate on the security mechanism of the E-healthcare and the environment to detect the hazardous situations. The results of the proposed approach, absorbing the specific levels of the patient, suggests what sort of medications prescribed by the physician should be followed and the behavior of the smart home with Wireless Sensor Networks (WSN) with necessary actuations. Based on the datasets of UCI Repository and the real health logs obtained from various hospitals, the testing has been performed. The main outcome of this research is the localizing the network information is less than a minute and mobility absorption is 90%. The experimental results of the proposed work prove that it outperforms the existing systems for Smart Home healthcare services.

Keywords: Smart healthcare monitoring · Surveillance of smart home · Sensor network

1 Introduction

The Internet of Things (IoT) is the network of physical devices, vehicles, home appliances and other things integrated with electronics, software, sensors, actuators, and connectivity which facilitate these objects to connect and exchange data [1]. Though it is embedded computing system, each thing is uniquely identifiable but capable of interoperate within the internet infrastructure. The inclusion of Information and Communication Technologies in healthcare has resulted in flawless healthcare service delivery anytime and anywhere. Wireless sensor network has broad application prospect in a variety of fields such as medical and health, military, defense and home automation. When it comes to healthcare, security and reliability are crucial issues to

D. J. Hemanth et al. (Eds.): COMET 2019, LNDECT 35, pp. 27–38, 2020.
https://doi.org/10.1007/978-3-030-32150-5_3

the general public. One of the most challenging goals of modern society is to improve the efficiency of healthcare infrastructures and biomedical systems. In fact, a core issue is need to provide patients with all of quality care while lowering medical costs and, at the same time, dealing with the problem of nursing staff rationing. As per a global report on aging and health in 2015 [2], the world's elderly population is growing rapidly, and most people expect to live for the first time more than 60 years.

Although human functionality at an older age is more likely to weaken, this can eventually lead to various diseases such as heart issues, stroke or heart attack, bloodstream infections and Alzheimer. As highlighted [3] in fact, current procedures for patient monitoring, care, management, and supervision are often manually executed by nursing staff. Automatic identification and monitoring of people and biomedical devices in hospitals, correct associations of drug patients, real time monitoring of patients and early discovery of clinical deterioration are just a few possible examples. As per the survey conducted by several organizations smart home market revenue and the reason for the adoption of technology are described in Fig. 1.

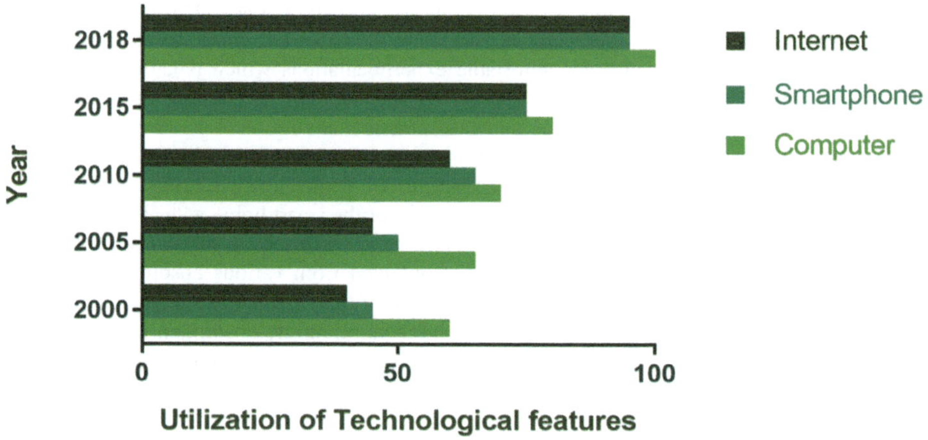

Fig. 1. Adaptation of technology

2 Related Works

Various researchers suggested variety of healthcare and home security systems. Following are the contributions made in this field by various researches. The digital home services provided to the public with multimedia entertainment, communication and health services makes their lives more comfortable [5]. Several researchers foreseen that the market for modern digital home service would be massive and that many would benefit from this predicament. The intention of this research was to explore the existing vital information and literature to clearly identify the future demand for digital home services. The paper also examined the attributes of customers and their relationship with digital home service acceptance. Another system [6] implemented IoT based smart home security system with Smart-phone alert and door access control. Passive motion Infra Red sensor (PIR) and camera interfaces were used respectively to track movement and capture images. Features such as viewing video streams via mobile phones have been added to this system. Furthermore, when a burglar is detected, voice alert or siren is activated to alert neighbors. The Liquid Crystal Display (LCD) could be used to setup the web server. An IoT based Home Security [7] presented that the system will send an alert to the user via internet if any infringement occurs. This alert system also included internet voice calls. If the person entered in the house is not an intruder but an unexpected guest, the owner arranges to greet his guest. In order to provide better service for the householder, the system must be integrated with the cloud.

The Sensor networks based healthcare monitoring systems with Wireless Sensor Network technology which monitors the various healthcare parameters such as body temperature and Electrocardiogram (ECG). The proposed prototype reduces the burden on patients to visit the doctor every time for monitoring of these health parameters. The constant, ease and flexible monitoring of the patient are made possible when the reliable wireless sensor network is deployed. The system was mainly focused on remote monitoring of the patient, inside and outside the hospital room and in ICU. A cloud-based framework [8] that manages the health-related large data effectively and benefits from the internet and social media ubiquity. The framework provides the mobile and desktop users with (a) risk assessment service to disease and (b) a health expert consultation service. The energy-efficient mechanisms for health applications for recognition of human contexts are proposed and the solutions for energy consumption, accuracy of recognition and latency are qualitatively compared [9].

The classification of movement in the past has been widely considered. Most of the existing systems are based on low cost and small axial accelerometers, both due to gravity and body motion accelerations. This makes them suitable for postural orientation and body movement monitoring, to evaluate metabolic energy expenditure indirectly, several solutions are available. Trust is one of the major influence for acceptance of new particularly automated technologies [10, 11] in technology acceptance literature. Lack of trust in the system is a crucial obstacle to the acceptance and use of IT services in consumer health. [12], and timely reactions to customer's requirements have been significant in establishing reliance. Infection is a major problem encountered in health care delivery services worldwide.

System failures may occur due to hardware failure, software bugs, power shortages, or natural effects. Most IoT systems studies have been done on the assumption that there are few faults disrupting the operation of an IoT system. These systems are increasingly vulnerable to failures such as power deficiencies or environmental hazards as other systems due to their geographical distribution and scarcely maintained sensors and devices. Furthermore, with the increasing number of nodes in large systems, the probability of failure rises and the system works inaccurately. Besides, many of the data contained in this proposed system is too valuable to lost in e-health circumstances due to system failure. However many studies in IoT environments have until now focussed on fault tolerant data devices. Access control, privacy, permission, integrity, availability and the reliability of the application layer expected to fulfill high security requirements are the main safety concerns. The information sharing capabilities offered by the application layer raises security concern on data privacy, access control and disclosure of information. Security concern of this system requires information transmission and quick response. Customers can unreservedly place gadgets in important areas to obtain beneficial information. This system must be compact and easy to use at the end of the day. It seemed to be a slow system that was direct and straightforward, with the goal that customers could make the quick move. The system mustn't be compromised regardless of its different routes including information source control, data transfer substance, and data acquisition the primary processor of the security sensor gadget is removed. Similarly the system should have attributes e.g.: safe, strong and high temperature, so that data transfer process and data acceptance is not shortened. The use of sensor devices is limited by security systems. These problems will cause the security systems to be constrained. In any event, it is essential for a security system to require a wide use of sensors for efficient operation and for the identification of protest in every area of the home. In security systems, the use of gadgets is also essential, sensors need to be in the correct range, not very close or not too far so that the development cannot be identified and should be based on human instinct. [14].

In recent years, the use of portable computers and smartphones for executing health information searches through the internet has been improved extraordinarily. There may be hundreds of IoT servers in this world, and every year there will be a considerable increase in the number of IoT servers. And if every IoT server supports its own proprietary communication protocol, it wouldn't be feasible to use IoT systems. Thus typical communing protocols have been enacted in IoT environments for interoperability and many of the IoT servers have been built to support the standard communicating protocols [15].

A questionnaire of internet projects undertaken in 2013 disclosed that approximately 72% of the web subscribers contacted the web to discover health material during the year 2012. This survey inspires to develop a system which enables the technique to monitor patients remotely and prescribe relevant medicines or exercises by the concerned physician. This proposed research is mainly focused on an integrated solution to the home automation with primary health care issues. For this reason, the system chose Raspberry pi as the main component to perform multiprocessing and response time of this is very fast compared to other traditional microcontrollers. This proposed article describes A Novel and Customizable Framework for IoT Based Smart Home Nursing for Elderly Care (SHNEC). The paper is organized as follows: Sect. 3 provides the architectural views, component description, Algorithms and implementation of the system. Section 4 describes the experimental results and discussion. Finally, in Sect. 5, conclusion and future scope are discussed followed by the references.

3 Proposed System Architecture and Implementation

This paper presents a 4-tiered architecture and the four stages of the system implementations are;

(a) Physical sensing layer
(b) Data processing layer

(c) Network layer
(d) Application layer.

Three key elements of healthcare-IoT systems are (i) body area sensor network (i.e. heartbeat, Blood Pressure (BP), temperature, movement of the patient (captured by PIR)) (ii) Internet-connected smart gateways, or a local access network, and (iii) cloud support. Innumerable technologies supply facilities to special stockholders in the framework through this ideology. The mooted mechanism affords a general service of patient's tutelage which comprises of two central functionalities:

1. Patient tracking and Monitoring: The precious resource of this approach is the precise knowledge of the location of patient as it allows for prompt reaction when urgent assistance is needed. The current status of critically ill patients must be available at any time to medical staff if patients can wander around the premises. But when the patient is immovable, the acquired data from the sensor network like abnormal variations in the heartbeat, temperature, Blood Pressure (BP), oxygen level will be intimated through SMS and E-Mail to the authenticated person to take necessary action against the abnormalities.
2. Monitoring of smart home: since the proposed system is an integrated application the surveillance system deployed needs to synchronize with the E-healthcare services. Depending on the hazardous conditions such as intruder detection, LPG detection security breach, the status of the environment may need to be responding in such a way that perhaps implementing automatic detection of aberrant variations in such parameters.

3.1 Component Description

In addition to the conventional devices for the design of smart home this research facilitate some more advantageous functions such as Gas leakage detection, Fire detection, Security of the Home as illustrated in Fig. 5. The Raspberry Pi 3 with an integrated IEEE 802.11n is employed in the designed system, to perform fetching, processing and communication processes. It fetches the signal from webcams and Passive Infrared Sensor (PIR) sensors and captured the images of the environment. The major feature of the PIR sensor is it can also work in darkness, so for security, the developed system utilizes the PIR and web camera for detection of abnormalities.

The MQ-2 Gas sensor used to detect or measure gasses like LPG, Alcohol, Propane, Hydrogen, CO and even methane. When the sensor measured the gas in 'part per million (ppm)' the analog pin to be used is TTL steered and operates on 5 V and is therefore suitable for most common microcontrollers. If the leakage of Gas is beyond the certain limited prescribed the petroleum corporation the servo motor which is connected at the Gas regulator knob moves 60° clockwise to turn off the Gas. In case of heavy leakage of gas causes fire, the LM35 series sensors are used to protect the environment from fire. When it comes to healthcare the LM35 sensor can also be used for measuring the temperature level of the patient where the sensor capable of measure ranges from −55 °C to 150 °C. The main importance of the LM35 sensor is the temperature can be directly calibrated in Celsius and is ideal for remote applications. The efficiency of this sensor is significantly better than thermistor.

The blood pressure sensor is intended for blood pressure measurement. This also records the systolic and diastolic pressure and pulse rate. The instrument attached to the inflatable air bladder cuff and used with a stethoscope to measure blood pressure in an artery is more accurate and reliable than the sphygmomanometer. Simply stated, blood pressure is measured using BP sensors against the blood vessel walls or arteries. The heartbeat sensor is used to detect the heartbeat of the patient which results in the digital output of the heart rate if the finger is placed on the sensor. The operating range of heartbeat sensor is +5 V DC and measures within the range of 60–100 bpm, operation is based on the principle of light modulation by blood flow through finger at each pulse [16].

3.2 Algorithm

On the basis of the discussion above, the practical application of the system proposed is portrayed as an algorithm [14]. Initially, the signals from the sensors (PIR, gas, fire, Heartbeat and Blood pressure) values are taken from the corresponding GPIO pins are connected with the Raspberry Pi.

If the Earlier state (ES) and the Present state (PS) are remains equal therefore there will be no interrupt so as the control function leave from the algorithm. The existence of an intruder, fire, and gas, heartbeat and blood pressure abnormalities are identified by when the detected signal ES and PS are not similar. Then the images are captured by the camera connected with the Raspberry pi are saved in the cache memory. At the end, the system constructs the E-mail which includes the whole information about the environment to the end-user. The workflow of the smart home is illustrated in the Fig. 2.

Algorithm

```
IN: INT ; P; S; F; Hb; Bp
Output: Em
CS= GPIO input (SN)
if PS = CS then
exit
else
Capture image (USB)
Connect = PN(USB)
email (M; To; Txt)
Warning!!
user action
end if
```

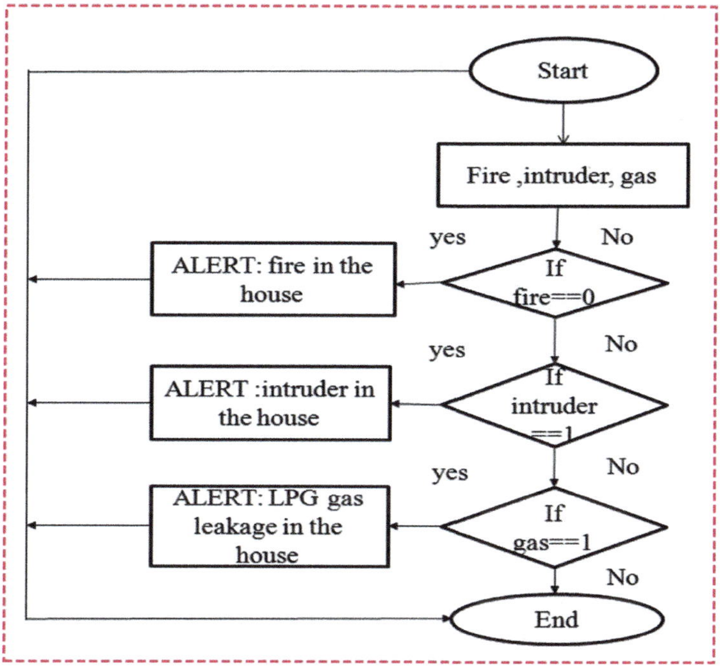

Fig. 2. Flowchart of the smart home monitoring

3.3 System Design

The distributed and integrated system of the Smart Home nursing for elderly Care has been designed and implemented in order to achieve the task for different paradigms. The functional architecture of the developed system is depicted in Fig. 3 comprises three sections such as; a Base station where the physical components have been placed on specific topology and the received data were preprocessed with respect to the prior knowledge of the information suggested by the clinician. These corresponding logs were stored in the Web server through the WiFi for Global access and the final section of the system design is remote access where the physician or the authenticated person can view the activities and actions suggested against the behavioral changes of the patient.

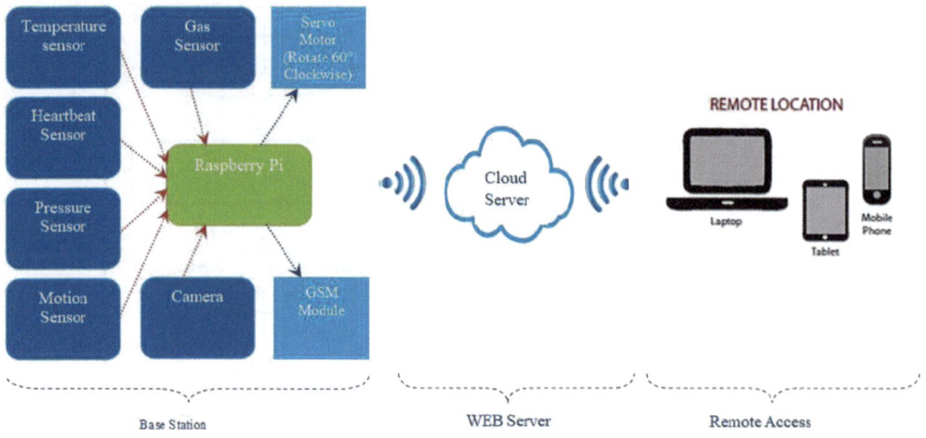

Fig. 3. Functional architecture of smart home nursing for elderly care

3.4 Implementation

The end-user application of the proposed system for Smart home is illustrated in Fig. 4. The sensors involved in home monitoring are an MQ-2 gas sensor for LPG leakage detection if the leakage is exceeded certain level as described by the safety management studies, the servo motor connected to the system is for turning LPG cylinder OFF by turning 60° to clockwise and counterclockwise to turn OFF. PIR sensor for motion detection for intruders, if any intrusion detection or any malfunction spotted is captured by the camera associated with the system. The LM35 sensor measures the temperature level and detects fire, the DC motor in the system behaves like the fire extinguisher to protect the environment from the fire.

When it comes to the healthcare, the sensors associated with the system are as depicted in Fig. 5. The prime health factors of a human being such as Heartbeat, Blood pressure, temperature, the oxygen level are measured by the respective sensors and compared with the preset values recommended by the physician and the system makes the necessary arrangements for the good care of the patient and the logs are updated to the web server. If any discrepancy identified in the sensor values, the system alerts the concerned clinician or the guardian through SMS and E-mail. A dedicated GSM module is installed with the system for the SMS feature. The developed system has the improved security scheme to protect the user information from the security threats and loss by providing separate login credentials. The main advantage of the proposed system is the response to the behavioral changes in the system function is prioritized, hence whichever sensor becomes wild and causes critical damage to the system will be served first.

Fig. 4. Implementation of smart home

1. Pressure sensor
2. Heartbeat sensor
3. Gas sensor
4. ADC

Fig. 5. Implementation of smart healthcare

4 Results and Discussion

The system can be accessed and monitored through online using the specified IP address to yield the better and reliable performance. The system can be logged in using the given login credentials as illustrated in Fig. 6(a). After the successful login, the information can be viewed as in Fig. 6(b) and (c). And, if any hazardous behavior is sensed, the camera capture the image as depicted in the Fig. 6(d).

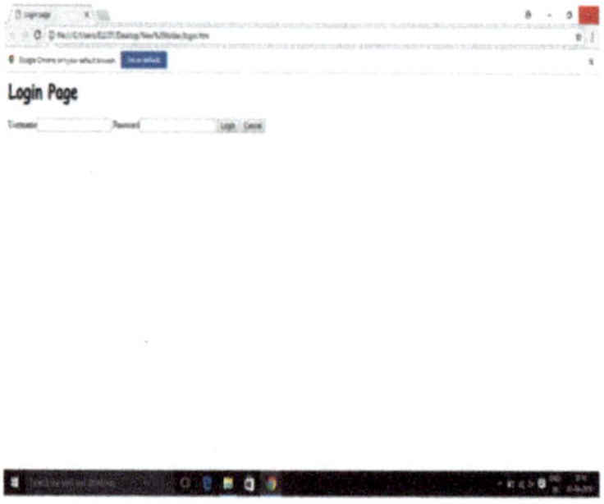

(a)

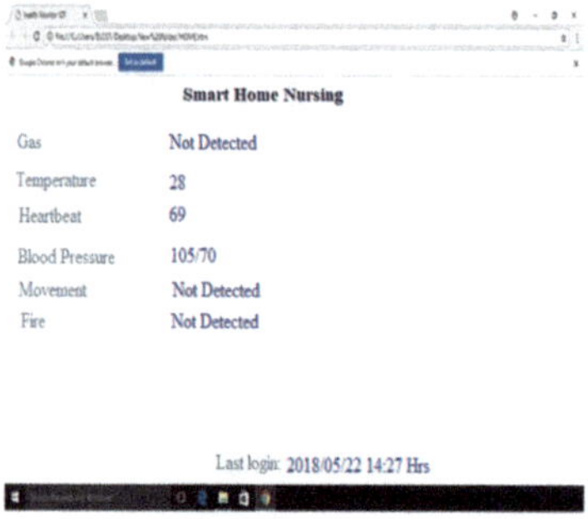

(b)

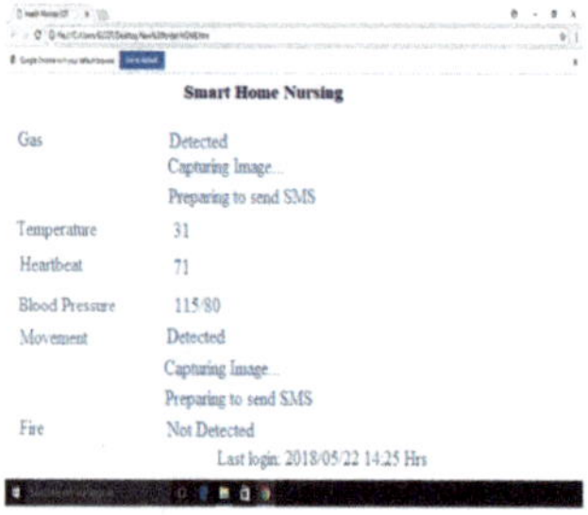

(c)

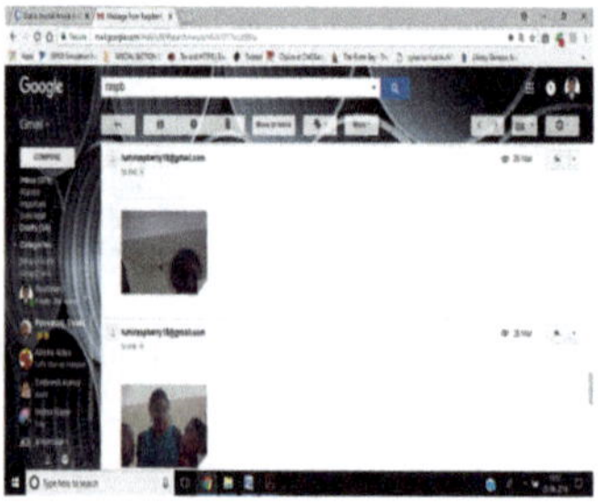

(d)

Fig. 6. (a) Login page. (b) The system information when everything is normal. (c) The system information when (d) Captured images of hazards detected received in the E-mail.

5 Conclusion and Future Work

In this paper, A Novel and Customizable Framework for IoT Based Smart Home Nursing for Elderly Care (SHNEC) were described. This system can be widely installed in any indoor environment due to the self-customizable feature; with the improvement of sensor technology, the system will become more efficient and useful. The proposed system was designed using Raspberry Pi and had been experimentally proven to yield confidentiality, integrity, scalability, and reliability of the system. The key advantages of the developed system are quick responses to the issues occurred and the reliable communication over the internet and the intimation to the guardian or the physician is done by SMS as well as the E-mail hence either one fails the trustworthy of the system is achieved. In future, the research work will include different sensor network for the various environment to enhance and to make a better Smart Environments.

Ethical Approval. "All procedures performed in studies involving human participants were in accordance with the ethical standards of the institutional and/or national research committee and with the 1964 Helsinki declaration and its later amendments or comparable ethical standards."

References

1. http://www.ijarcs.info
2. World Report on Ageing and Health, World Health Organization (2015)
3. Redondi, A., Chirico, M., Borsani, L., Cesana, M., Tagliasacchi, M.: An integrated system based on wireless sensor networks for patient monitoring, localization and tracking. Ad Hoc Netw. **11**(1), 39–53 (2013)
4. Tanwar, S., Tyagi, S., Kumar, S.: The role of the internet of things and smart grid for the development of a smart city. In: International Conference on Internet of Things for Technological Development (IoT4TD), pp. 1–10 (2017)
5. Noh, M.L., Kim, J.S.: Factors influencing the user acceptance of digital home services. Telecommun. Policy. **34**(11), 672–682 (2010)
6. Anwar, S., Kishore, D.: IOT based smart home security system with alert and door access control using smart phone. Int. J. Eng. Res. Technol. (IJERT) **5**(12), 1–5 (2016)
7. Kodali, R.K., Jain, V., Bose, S., Boppana, L.: IoT based smart security and home automation system. In: International Conference on Computing, Communication and Automation (ICCCA), pp. 126–132 (2016)
8. Abbas, A., Ali, M., Shahid Khan, M., Khan, S.: Personalized healthcare cloud services for disease risk assessment and wellness management using social media. Pervasive Mobile Comput. **28**, 81–99 (2016)
9. Rault, T., Bouabdallah, A., Challal, Y., Marin, F.: A survey of energy-efficient context recognition systems using wearable sensors for healthcare applications. Pervasive Mobile Comput. **37**, 23–44 (2017)
10. Dzindolet, M., Peterson, S., Pomranky, R., Pierce, L., Beck, H.: The role of trust in automation reliance. Int. J. Hum Comput Stud. **58**(6), 697–718 (2003)
11. Lippert, S., Davis, M.: A conceptual model integrating trust into planned change activities to enhance technology adoption behavior. J. Inf. Sci. **32**(5), 434–448 (2006)

12. Jimison, H., Gorman, P., Woods, S., Nygren, P., Walker, M., Norris, S., Hersh, W.: Barriers and drivers of health information technology use for the elderly, chronically Ill, and underserved Rockville. Agency for Healthcare Research and Quality, Maryland (Evidence Report/Technology Assessment No. 175175) (2008)
13. Adegboye, M., Zakaria, S., Ahmed, B., Olufemi, G.: Knowledge, awareness and practice of infection control by healthcare workers in the intensive care units of a tertiary hospital in Nigeria. Afr. Health Sci. **18**(1), 72 (2018)
14. Tanwar, S., Patel, P., Patel, K., Tyagi, S., Kumar, N., Obaidat, M.S.: An advanced internet of thing based security alert system for smart home. In: 2017 International Conference on Computer, Information and Telecommunication Systems (CITS) (2017)
15. Woo, M.W., Lee, J., Park, K.: A reliable IoT system for personal healthcare devices. Future Gener. Comput. Syst. **78**, 626–640 (2018)
16. Pardeshi, V., Sagar, S., Murmurwar, S., Hage, P.: Health monitoring systems using IoT and Raspberry Pi—a review. In: 2017 International Conference on Innovative Mechanisms for Industry Applications (ICIMIA) (2017)

Design and Implementation of Greenhouse Monitoring System Using Zigbee Module

S. Mani Rathinam(✉) and V. Chamundeeswari

Department of Electrical and Electronics Engineering,
St. Joseph's College of Engineering, Chennai, Tamil Nadu, India
manirathinam786@gmail.com

Abstract. The checking and control of nursery condition assume a vital job in nursery creation and the board. To screen the nursery condition parameters adequately, it is important to plan an estimation and control framework. This paper introduces a control structure of remote sensor organize framework dependent on Zigbee handset for nursery, which comprises of some sensor hubs set in the nursery and an ace hub associated with upper PC in the checking focus. The sensor hubs gather signs of nursery temperature, moistness, and light and soil dampness, control the actuators, and transmit the information through the remote Zigbee handset; the ace hub gets the information through the Zigbee handset and sends the information to the upper PC for continuous observing. To make an ideal situation the fundamental climatic and ecological parameters, for example, temperature, moistness, light force and soil dampness should be controlled. On the off chance that any of the Greenhouse parameters surpasses the edge esteem set by the client, essential control move will make put consequently and furthermore ready will be given to the client through Zigbee. The controlling move will make put with the assistance of fan, water sprayer and so forth. If the Greenhouse parameter falls beneath the edge esteem, the controllers will be killed consequently. Result demonstrates that the framework is reasonable and dependable, and has wide application later on.

Keywords: Green house · Atmega MCU · Zigbee · Temperature · Humidity · Soil moisture · Light

1 Introduction

Nursery is a sort of cutting-edge agricultural office controlling and recreating characteristic atmosphere in plant development, changing the plant development condition, and making the reasonable conditions for plant development, staying away from the outside season change and the antagonistic impacts caused by terrible climate. With the advancement of establishment farming, the cutting-edge vast scale nursery has been broadly utilized in the exactness horticulture [1–3]. The need of its condition quality has turned out to be ever more elevated. Nursery is assuming a critical job in the creation of out-of-season vegetables and blooms, and additionally high esteem and fragile plants. The reason for nursery ecological control is to get the best conditions for harvest development, increment trim yields, enhance nature of products, and direct the

D. J. Hemanth et al. (Eds.): COMET 2019, LNDECT 35, pp. 39–47, 2020.
https://doi.org/10.1007/978-3-030-32150-5_4

development cycle of products, through changing ecological factors, for example, temperature, stickiness, light and the dimension of carbon dioxide. With the quick advancement of the sensor, PC and correspondence advances, conventional nursery condition control technique has been supplanted by canny nursery checking and control innovation. Correspondence innovation of the insightful checking and control frameworks for nursery has turned into a concentration in the field of organized farming. Correspondence framework in canny nursery observing and control framework can be actualized in two essential ways, which are wired correspondence and remote correspondence or Zigbee correspondence. Physical topology and attributes of link and obstruction created by current may cause issues for wired correspondence, which convey high power utilization contrasted with remote radios. Moreover, wired media, for example, coaxial link and curved match have poor adaptability particularly in nursery trim redesign [4]. Contrasted and wired frameworks, remote frameworks empower less demanding gadget establishment, organize expansion and reconfigurations. Hence, remote arrangements give adaptability and cost not exactly wired frameworks. Likewise, propels in gadgets and remote correspondence bring littler chips with diminished power utilization and diminished costs which empower the improvement of low-power and ease remote control applications [5]. So, the checking framework dependent on Zigbee module, microcontroller unit (MCU) and sensor innovation can give another approach to ongoing gathering nursery condition parameters in long-go [6]. In this paper, we present a plan strategy for a nursery control framework utilizing remote sensor arrange dependent on Zigbee module and give its execution.

This paper comprises of Sect. 2 explains with General Scheme of Greenhouse Wireless Sensor, Sect. 3 explains with Existing technique, Sect. 4 explains with Proposed strategy, Sect. 5 explains with Hardware Description, Sect. 6 portrays the Hardware Results, Sect. 7 portrays the Simulation Results, Sect. 8 deals with Conclusion, Sect. 9 deals with future work.

2 General Scheme of Greenhouse Wireless Sensor Network Monitoring System

The scheme of this paper is intended for vegetable nursery checking and control framework. Structure of remote sensor systems framework is shown in Fig. 1. The nursery observing framework embracing the ace slave structure comprises primarily of two sections: an upper PC and a few remote sensor hubs for nursery [7]. The upper PC laid at the control focus is an ace hub controlled by microcontroller with a remote Zigbee module associated with a PC, which speaks with the sensor hubs through remote channel. The sensor hubs are set at each zone of the nursery, which are principally made out of the hub microcontroller, different sensors and remote Zigbee interface module. The upper PC in control focus is in charge of sending the control outline, accepting and preparing information from the slave hub sensors, and showing and putting away the handling results [8]. Each sensor hub is doled out an alternate deliver to separate itself from others. All sensor hubs will get the control outline from the ace hub of the control focus, and identify the location in the control outline. On the off chance that the location of one hub is predictable with the location of got control outline, the sensor hub will start to gather the signs of temperature, mugginess, light and carbon dioxide focus, and transmit them to the observing focus [9]. The sensor hubs that are not picked won't gather and transmit the information to the host PC.

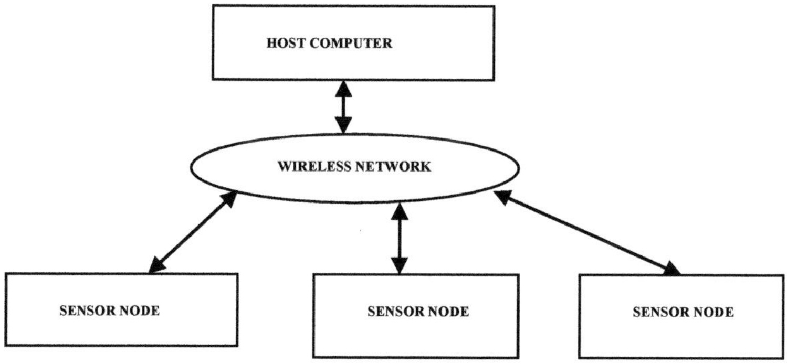

Fig. 1. Wireless module data transferring

3 Existing Method

The present endeavor is watching using RF headways of remote correspondence in low range multi-sensor blend advancement to structure nursery checking sensor center points, the temperature control and normally alert isn't in the current and data checking control.

4 Proposed Method

The limit natural parameters, for example, temperature, stickiness, light power and soil dampness should be controlled showcase the information in LED Display. The solenoid valve and sprayer are to control the temperature in the dampness air and stepper engine to control the rooftop top in the nursery observing (Figs. 2 and 3).

5 Hardware Description

The proto model has two segments receiver side and transmitter side. The Transmitter side gathers the information from the sensor and makes the restorative activity as needs be the point at which it crosses the limit level it is possible that it changes the rooftop top dimension or alters fake lighting. In the receiver side, the observing of the sensor status and Alert framework is done by Atmega microcontroller (Table 1).

The transmitter side uses Atmega16 microcontroller where it is feed information from Temperature sensor (LM35), Humidity sensor (DHT11), Light Dependent Resistor (LDR), Soil Moisture sensor.

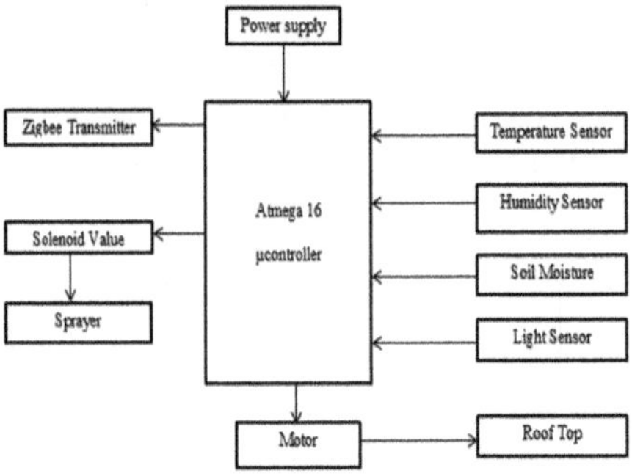

Fig. 2. Block diagram - Transmitter

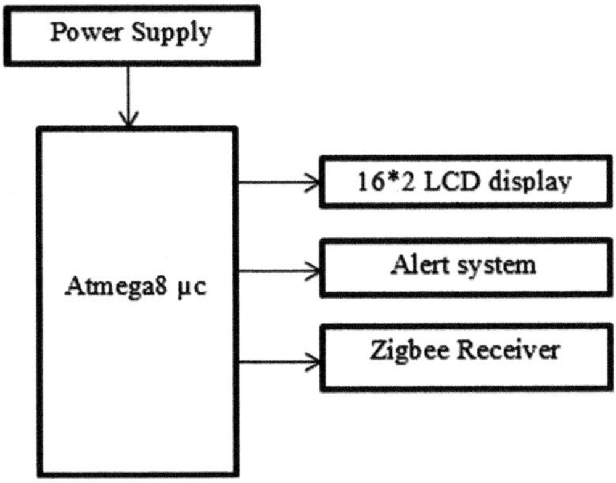

Fig. 3. Block diagram - Receiver

Table 1. Hardware and software tools

Hardware tools	Software tools
Atmega 16 Microcontroller board	AVR studio 4.0
Zigbee Module – Transmitter and Receiver module	WinAVR
16*2 graphic liquid crystal display	Extreme burner
Temperature sensor – LM35	
Humidity sensor - DHT11	

The LM35 arrangement are accuracy coordinated circuit temperature gadgets with a yield voltage directly corresponding to the Centigrade temperature [10].

The DHT11 is a fundamental, ultra minimal effort computerized temperature and mugginess sensor. It utilizes a capacitive moistness sensor and a thermistor to quantify the encompassing air, and releases an advanced flag on the information pin. This sensor incorporates a resistive-type stickiness estimation part and a NTC temperature estimation segment [16].

The Soil Moisture Sensor utilizes capacitance to gauge the volumetric water substance of soil by estimating the dielectric permittivity of the soil, which is a component of the water content.

A light-dependent resistor is a light-controlled variable resistor. In the dim, a photo resistor can have an opposition as high as a few Megohms (MΩ), while in the light, a photo resistor can have an obstruction as low as a couple of hundred ohms.

Relay module is connected to input for the relay module is +5 V digital signal from the microcontroller and the other side of the relay is connected to +12 V buzzer and +12 V solenoid value. The Power supply is provided to the proto kind model by +12 V, 2A DC connector.

LM35 is used to obtain the temperature of the plant. DHT11 is used to obtain the ambient humidity condition. LDR produces dependent resistance value according to the light ambient conditions. Soil moisture sensor is used to find the level of wet and dry soil conditions. Solenoid value is used to spray the water in the plant at dry conditions. Motor is used for roof top adjustment conditions for sunlight to fall on the plant.

The programming has set to a limit level when it crosses the dimension a corrective action has to encounter. For Example, let us say it is 30 °C. When it crosses the 30 °C, the 4 channel Relay module will be activated and the solenoid esteem opens the water siphon.

The Receiver is to screen the information in the 16*2 LCD show and when the esteem is expanded a past more prominent than the edge level, the ready framework is activated to alarm the harm.

Atmega16 is a 8-bit microcontroller dependent on the AVR RISC design. [10–15] (Table 2).

Table 2. Comparison of the microcontrollers

	8051	PIC	AVR
Speed	Slow	Moderate	Fast
Memory	Small	Large	Large
Architecture	Cisc	Risc	Risc
Adc	Not Present	Inbuilt	Inbuilt
Timers	Inbuilt	Inbuilt	Inbuilt

6 Hardware Results

See Figs. 4 and 5.

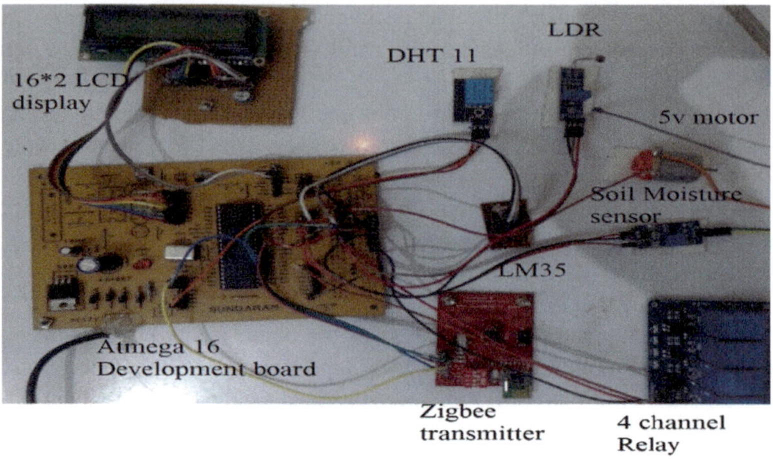

Fig. 4. A prototype of Transmitter

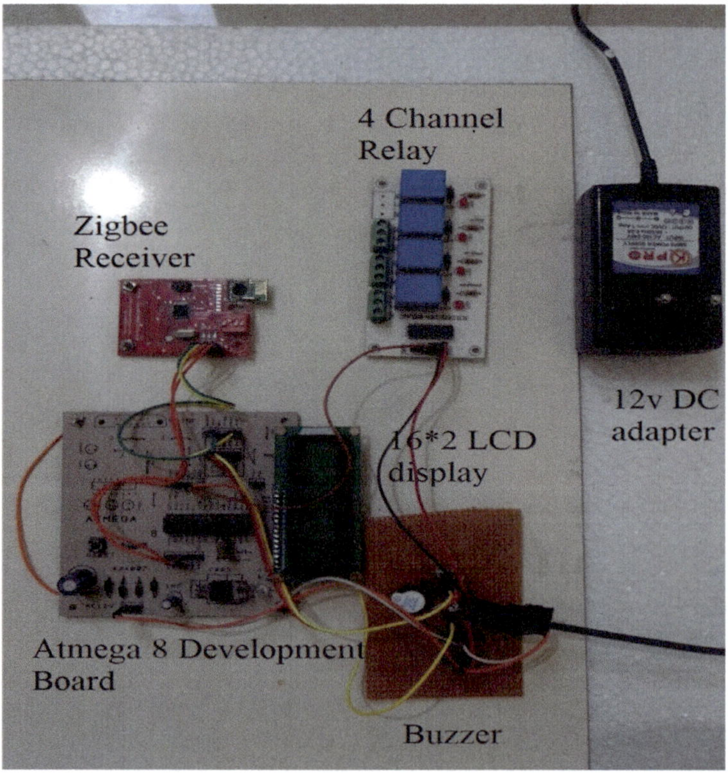

Fig. 5. A prototype of Receiver

7 Proteus - Simulation Results

Atmega 16 microcontroller is the master of the activity. It drives control signals from the LM 35 temperature sensor and LDR light dependent resistor (Fig. 6).

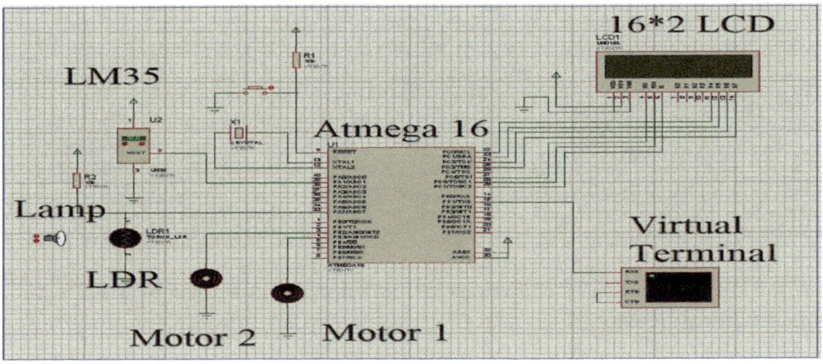

Fig. 6. Initial conditions of the setup

Motor 1 is the corrective action for the LM35, when the temperature is increased beyond 34 °C the motor 1 rotates.

When the lamp is brought near LDR, the light conditions is increased beyond 70 units, motor 2 rotates (Fig. 7).

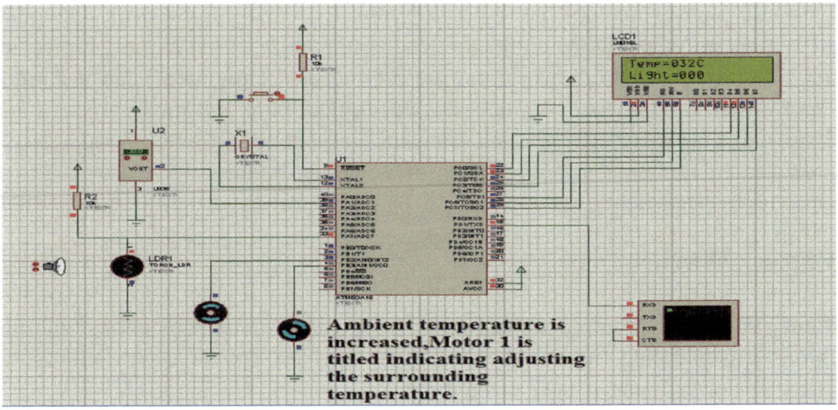

Fig. 7. When the ambient temperature is increased it is sensed by LM35 and feedback action is encountered by Motor1.

The temperature of the module is 32 °C is detected by LM35 and showed in the module. The light beginning conditions are given as 000 (Fig. 8).

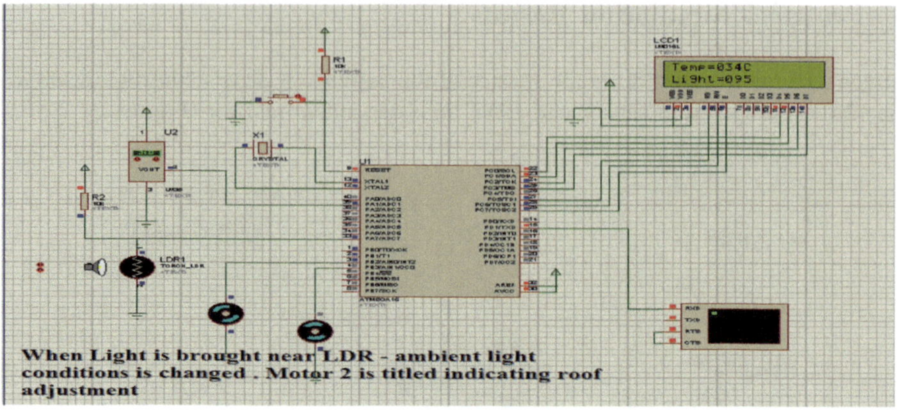

Fig. 8. When the ambient light conditions are increased it is sensed by LM35 and feedback action is encountered by Motor1

The temperature of the module is 34 °C is detected by LM35 and the engine has changed the position. The light introductory conditions are given as 000 and it is expanded to 095 by bringing the light closer.

8 Conclusion

As per the attributes of the nursery ecological checking, the article advances a sort of configuration plan of nursery natural data remote observing framework dependent on ZIGBEE innovation and GSM correspondence innovation. What's more, it presents the general structure of the framework and the product and equipment plan technique for each part in detail. It gives a practical answer for the little and medium-sized nursery checking and corrective actions and precautionary actions to be encountered. This simulation is encountered to overall idea of the green-house.

9 Future Work

The future work which can be enhanced is grow a plant under greenhouse monitoring conditions and planning to have detailed case study on a plant which will grow in a particular season to yield it year throughout.

References

1. Erazo, M., Rivas, D., Pérez, M., Galarza, O., Bautista, V.: Design and implementation of a wireless sensor network for rose greenhouses monitoring. (978-1-4799-6466-6/15)

2. Kampianakis, E., Kimionis, J., Tountas, K., Konstantopoulos, C., Koutroulis, E., Bletsas, A.: Wireless environmental sensor networking with analog scatter radio and timer principles. https://doi.org/10.1109/jsen.2014.2331704
3. Yu, C., Cui, Y., Zhang, L., Yang, S.: ZigBee wireless sensor network in environmental monitoring applications. (978-1-4244-3693-4/09)
4. Aher, M.P., Nikam, S.M., Parbat, R.S., Chandre, V.S.: A hybrid wired/wireless infrastructure networking for green house management (978-1-5090-2080-5/16)
5. Baviskar, J., Mulla, A., Baviskar, A., Ashtekar, S., Chintawar, A.: Real time monitoring and control system for green house based on 802.15.4 wireless sensor network. (978-1-4799-3070-8/14)
6. Rangan, K., Vigneswaran, T.: An embedded systems approach to monitor green house. (978-1-4244-9182-7/10)
7. Krishna, K.L., Madhuri, J., Anuradha, K.: A ZigBee based energy efficient environmental monitoring alerting and controlling system. (978-1-5090-2552-7/16)
8. Liu, Y., Hassan, K.A., Karlsson, M., Weister, O., Gong, S.: Active plant wall for green indoor climate based on cloud and internet of things. https://doi.org/10.1109/access.2018.2847440
9. Vatari, S., Bakshi, A., Thakur, T.: Green house by using IOT and cloud computing. (978-1-5090-0774-5/16)
10. http://www.avr-tutorials.com/projects/atmega16-microcontroller-digital-lcd-thermometer
11. http://extremeelectronics.co.in/avr-tutorials/interfacing-temperature-sensor-lm35/
12. http://www.electronicwings.com/avr-atmega/xbee-interfacing-with-atmega32
13. https://www.gme.cz/data/attachments/dsh.958-112.1.pdf
14. https://www.engineersgarage.com/proteus/tags/proteus
15. http://kannupandiyan.blogspot.com/p/avr.html
16. https://www.instructables.com/id/Measuring-Humidity-Using-Sensor-DHT11/

Power Efficient Pulse Triggered Flip-Flop Design Using Pass Transistor Logic

C. S. Manju(✉), N. Poovizhi, and R. Rajkumar

Department of ECE, Dr. N.G.P. Institute of Technology, Coimbatore, Tamil Nadu, India
{manju, poovizhi, rajkumar.r}@drngpit.ac.in

Abstract. In today's scenario modern electronic systems are in need of compact digital circuits. So, miniaturization of circuits is rapidly growing area. The speed and power consumption of a digital circuit is determined by Flip-flops. The total number of transistors in the clock generation circuit indirectly reduced by the introduction of the pass transistor logic in the existing flip flop design, which also reduces the power consumption of the circuit due to transistor reduction. Reduction of number of transistors in flip-flop design leads to miniaturization of digital circuits. In this paper, the pulse triggered Flip-Flop (FF) for power efficiency is designed and discussed. The AND function in the clock generation circuitry is replaced with the PTL based AND gate circuit. The n-mos transistors are arranged in parallel using the PTL based AND gate and due to faster discharge of the pulse less power is consumed.

Keywords: Clock generation circuitry · Digital circuits · Flip-flops · PTL · n-mos transistors

1 Introduction

For all the VLSI designers the major considerable parameters were the performance, area, cost and reliability of the designed circuits. The low attraction parameter in the past days is the design of the circuits with low power consideration. But, power is consumption is considered as equal as area and speed, in the recent research articles. The excessive power consumption is the most important factor which is becoming the limiting factor in fabrication of single chip or on a multiple chip module (Lin 2014) by incorporating more number of transistors. Reducing number of transistors in flip-flop design leads to miniaturization of the digital circuits. The feature size of CMOS technology process shrinks according to Moore's Law, (the number of transistors per square inch on IC had doubled every year since their creation), and designers are able to integrate more number of transistors onto the same die. If the number of transistors increases then both switching activity and the amount of power dissipation will increase in the form of heat, where heat is one of the most important packaging challenges in this era; it is one of the main point that leads to design of low power methodologies. Another important aspect of research in low power area is the reliability of the integrated circuit. Reliability issue occurs if higher average current is flowing in the circuit due to more switching. In particular, digital designs now a-days often adopt

D. J. Hemanth et al. (Eds.): COMET 2019, LNDECT 35, pp. 48–61, 2020.
https://doi.org/10.1007/978-3-030-32150-5_5

intensive pipelining techniques and employ many Flip Flops. It is also estimated that the power consumption of the clock system is as high as 50% of the total system power. Flip Flop thus contribute a substantial segment of the chip area and power consumption to the overall system design.

The organization of paper is as follows. In Sect. 2, literature survey of different existing flip-flop design techniques are presented. In Sect. 3, the proposed low power flip-flop architecture is given. In Sect. 4, the results of the proposed technique, comparison of various flip-flop designs and the conclusion are presented.

2 Literature Review

In (Saranya and Arumugam 2013), the intricacy of the locking mechanism is reduced by using the Hybrid Latch Flip-Flop which decreases the area of the chip and the delay time is also reduced. The hybrid latch flip-flop which consumes low power was designed by placing the input node and the output node at a shorter distance. This reduction in the distance between the input and output minimizes the delay time (Karimi et al. 2018). In (Gupta and Mehra 2012), A dual edge triggered flip flop is designed. A comparison is made among the existing designs of dual edge triggered flip-flop such as EP_CDFF, EP_CPFF and DET-SAFF with the proposed design of the dual edge triggered flip-flop (DET-FF). Pulse generator and conditional discharge is incorporated in EP_CDFF type of design. The function of pulse generator is to generate the dual pulse which is active at both rising and falling edge of the clock. To eliminate unwanted transitions of the flip-flop there by reducing the power dissipation, conditional precharge technique is included in EP_CDFF type of design. In another type of design called DET-SAFF sense amplifier is used in the design of flip-flop. By incorporating sense amplifier in, this design drastically reduces the power dissipation.

In (Sadrossadat et al. 2011), a statistical design of the flip-flop is proposed to attain better performance by decreasing the power leakage, switching and area. The proposed design showed that for the flip-flops which are designed using statistical tools (Teh et al. 2006) have designed the D-flip-flop (Zhao et al. 2004), called adaptive-coupling flip-flop (ACFF), which uses less number of transistors, when compared to other flip-flop designs with low power consumptions. This ACFF uses 2 lesser transistors than the transmission-gate flip-flop (TGFF). In another technique called Modified sense amplifier flip-flop a pre-charge sense amplifier, a set and reset latch is included to hold the data. The latency of SAFF is little bit higher than other flip-flop designs due to delay of one output from other output in the output stage. This problem is overcome in the design, where it supports completely symmetric output transitions (Mahmoodi et al. 2009).

An adapted version of ip-DCO design is the SCCER (Phyu and Goh 2005). Conditional discharged technique is used in the above design in which, if the input stays HIGH the switching activity is controlled by the reduction in the discharge paths. Here, the pull up resistors is replaced by the inverters which are connected back to back. A weak pull up transistor and inverter is used in place of pull down resistor to reduce the load capacitance at a node point. But this pull down resistor needs to be powerful to make sure that the node can be discharged properly. The pull down circuit requires more area and consumes more power, which is the major drawback of the above design. This in turn increases delay because of the increase in area the discharge path takes longer time. Plus, to carry out the discharge operation, wider pulse width is required.

The hard edge property is used in master–slave design (Teh et al. 2006). The skew tolerant concept and cycle stealing is allowed in the pulsed flip-flops designing. A pulse generator is used outside the latching part in the explicit DEFF in which duplication is not required the data latch part. The XOR using a floating inverter using pMOS, nMOS pair that does not have a direct connection with Vdd or ground is designed using pass transistors the transmission gate (TG), PASS, TSPC-SPLIT, etc. can be used as the latching part of the flip flops. This explicitly generated pulse achieves a transparency window in the design process.

Based on the transmission gate based XOR logic the pulse generator designed. The design has low capacitive load on the critical path by placing a small simple structure on the critical path. But this produces a noise when exposed to the diffusion input. Also the problem of ratio in size was produced in the theep-DSFF. In order to improve the driving ability and robustness of the transmission gates an inverter is added at the input terminal of the design circuits.

By studying the different existing types of flip-flops it is observed that there are some drawbacks regarding these flip-flops as below:

- Due to large switching and in the internal nodes high power consumed.
- Noise occurs due to the appearance of glitches at the output.
- Discharging occurs on every raising edge of the clock pulse.
- Delay is caused by a discharging of stacked transistors.
- Longer input to output delay during 0 to 1 transitions.
- The internal nodes become floating when the input and output is equal to 1.

The pulse triggered flip-flop is proposed in such a way that it avoids switching at an internal node thereby lowering the 0 to 1 delay and reducing the power consumption. Pulse-triggered flip-flops are classified based on the pulse generators, as implicit-pulsed & explicit-pulsed, static or semi static or dynamic or semi dynamic and single-edge triggered & double-edge triggered flip-flops. It is called implicit-pulse triggered flip-flops (ip-FF) due to the internal generation of pulse in the flip-flop. Some of the examples are hybrid latch flip-flip (HLFF), semi-dynamic flip-flop (SDFF), and implicit-pulsed data-close-to-output flip-flop (ip- DCO). Whereas in explicit-pulse triggered flip-flops (ep-FF), the pulse is generated outside the flipflop. Example-explicit-pulse data-close-to-output flip-flop (ep-DCO) (Alioto et al. 2010). Comparatively, power consumption is less in Implicit types P-FFs, the main drawback is the poor timing characteristics due to the large discharging path. In Explicit pulse generation, the power consumption is high but the logic partition from the latch design speeds up the circuit. The drawback in the explicit type can be overcome if single pulse generator shares a group of FFs. The explicit type P-FF is focused for the design in this paper.

2.1 Explicit Pulse Triggered Flip-Flop

Due to the semi dynamic structure the Ep-DCO is mentioned as one of the fastest flip-flops (Sadrossadat et al. 2011). Since the is a lower delay this particular type of flip-flop is used in high-performance applications where low delay is attained. With the help of pulse triggering mechanism, this allows more independence in cycle budget with its negative setup time feature. The schematic diagram for the ep-DCO flip-flop is shown in Fig. 1.

It is a semi-dynamic structure that has dynamic (first stage) and a static stage (second). After the rising edge of the clock, for a small period of time MN2 and MN3 transistors are turned on. The flip-flop act as a transparent latch during this interval, so input data is passed to the output. Once this transparent period is over, the above mentioned transistors turn off the pull-down paths in two stages. Hence whatever changes happen at the input side will not reflected at the output side. When the circuit is in hold mode, the output and internal node states are maintained in their respective states by keepers.

Upon the detailed analysis of the ep-DCO circuit, due to charging and discharging at the internal node X, notify able power is disbursed. This charging and discharging occurs at every clock cycle when the data at the input side stable. These operations are not significant; hence the functioning of the circuit does not contribute for the power dissipation of the same. The internal node X during the charging and discharging produces glitches, when the output is HIGH. It creates a discharge path for internal node X is recharged HIGH the output node that stays on for a short period of time after the start of the evaluation period (Chandrakasan et al. 1992); this path causes glitches. The drawback is, these glitches increase the switching power consumption and noise appears at the output which makes the system to be faulty.

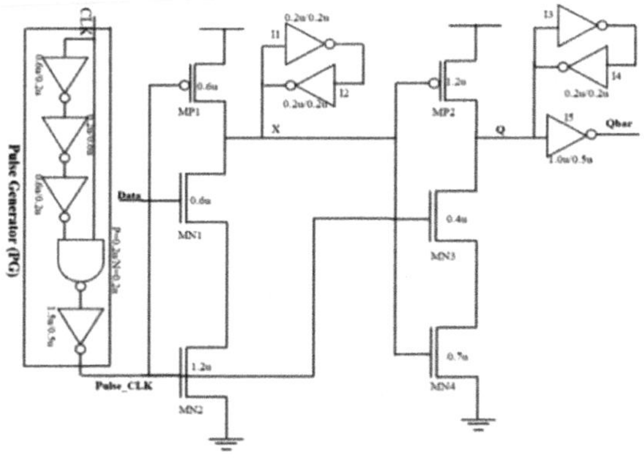

Fig. 1. Explicit-pulse data-close-to-output (Sadrossadat et al. 2011).

2.2 Conditional Discharge Flip-Flop

For the double-edge sampling, a suitable circuit the pulse generator (Sadrossadat et al. 2011). The flip-flop consists of two stages. The transition from LOW-to-HIGH is captured in stage 1. Assuming that the value of Q and Q_fdbk were LOW and HIGH respectively when input is HIGH, internal node X is discharged. Because of this, output node gets charged to HIGH via transistor P2. The stage 2 is for capturing the transition from HIGH-to-LOW. Stage 1 gets disabled when there is low input and node X maintains the same state. During the same sampling period, HIGH value will be present at node Y and the second stage's path of discharge gets enabled, which makes the output node to accurately capture the input data.

The purpose of using transistor N5 at the discharge path of stage 1 is to reduce the switching power. When Q_fdbk = HIGH, whereas Q = LOW and X = HIGH, MN3 transistor will be turned on, and this enables discharge path. When the value of input goes from LOW-to-HIGH and clock is HIGH, MN1, MN2, and MN3 transistors will be in on state, node X will be LOW, Q and Q_fdbk will have the value of HIGH and LOW respectively. This turns off the nMOS transistor in stage 1. Node X gets discharge once when data goes from LOW to HIGH. The purpose of dual path is to make sure that the flip-flop samples the HIGH-to-LOW transition. With the help of dual path, the LOW-to-HIGH delay is reduced (Fig. 2).

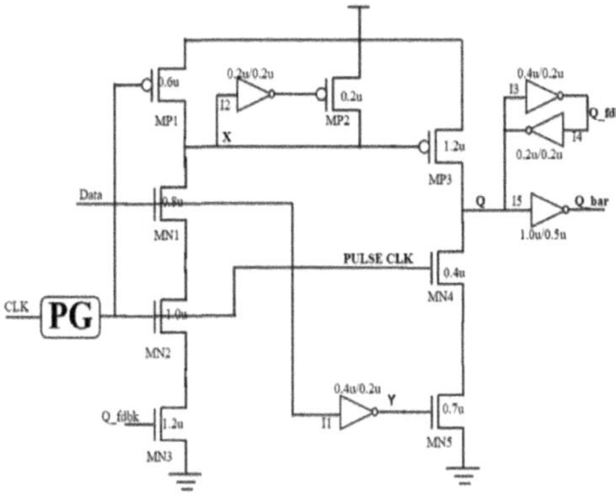

Fig. 2. Conditional discharge flip-flop (Chandrakasan et al. 1992).

2.3 Static-Conditional Discharge Flip-Flop

Figure 3, shows an analogous P-FF design (SCDFF) which employs a technique called static conditional discharge (Tschanz et al. 2001). It utilizes static latch which is the main difference from CDFF design. So node X is free from periodical precharges.

But it has more D to Q delay when compared with CDFF design. Both CDFF and SCDFF designs suffer from delay due to discharging path which consists of transistors

MN1–MN3. To avoid this delay for improved speed, an efficient pull-down circuit is required, which leads to more area and power.

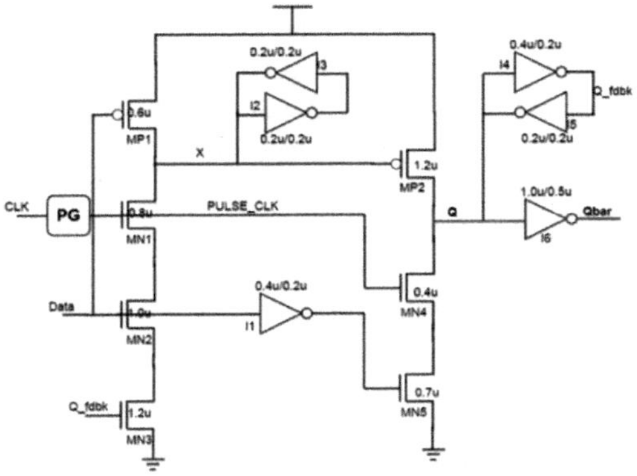

Fig. 3. Static conditional discharge flip-flop (Tschanz et al. 2001).

2.4 Modified Hybrid Latch Flip-Flop

MHLLF (Weste and Harris 2011) is an adapted and enhanced version of P-FF design Fig. 4, Static latch structure is utilized. Delay is reduced by precharging the node but this action increases the power consumption. Hence precharging of node is not employed in this design.

The weak pull-up transistor P1 makes the node to be maintained at high value when Q = LOW. This also eliminates the unneeded discharging problem at node. But the input changes from 0 to 1, the design takes long time to transmit the data from input to output. This occurs primarily due to the fact that the node is not pre-discharged. More transistors are required in this design to improve the discharging capability. More power is consumed when input is one due to the floating nodes.

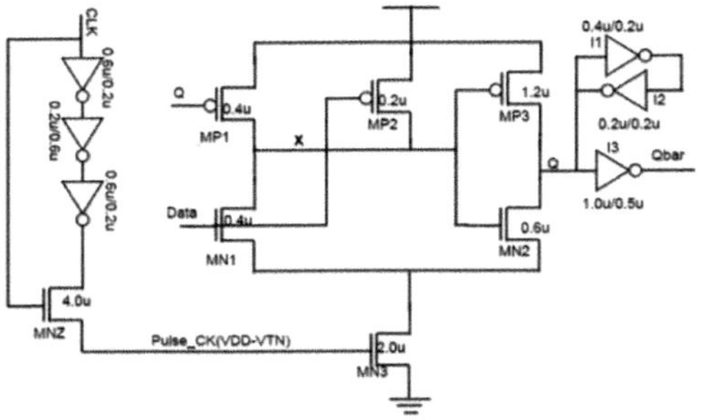

Fig. 4. Modified hybrid latch flip-flop (Weste and Harris 2011).

2.5 True Single Phase Clocked Latch Flip-Flop

The true single phase clocked latch flip-flop (Phyu and Goh 2005) design employs a signal feed-through technique to have improved delay. Like the SCDFF design, the TPSC design too adopts a static latch structure and a conditional discharge scheme to overcome redundant state change at an internal node. But TSPC latch structure has three major differences and this leads to unique design. The stage one consists of weak pull-up pMOS transistor in which gate is grounded. This kind of connection gives rise to pseudo-nMOS design and saves the charge keeper circuit. This makes the circuit design simple and also has low load capacitance of node X (Phyu and Goh 2005; Rasouli et al. 2005). The second stage uses a pass transistor MNx is employed to make input data drive directly node Q. Pull up transistor MP2 and inverter in second stage creates an extra path for the auxiliary signal to reach node Q from input.

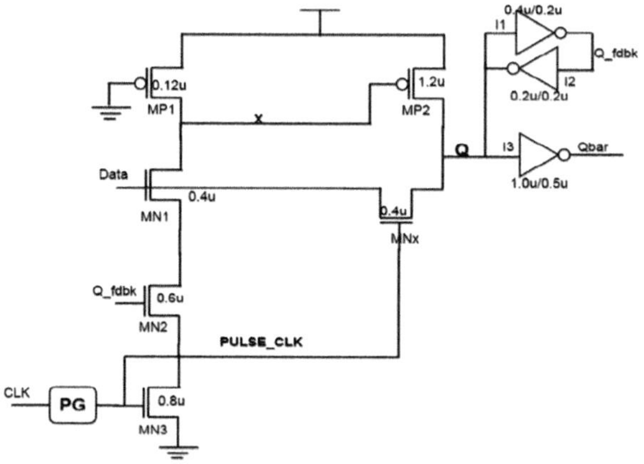

Fig. 5. True single phase clocked latch flip-flop (Rasouli et al. 2005).

Delay of data transition can be reduced by pulling up the node level. Third difference is that stage two inverter's pull down network is removed. The role of MNx transistor is to provide a path for discharging to drive node Q during LOW to HIGH transitions and to discharge node Q during HIGH to LOW transitions. The TFSC design has a charge keeper (two inverters), a pull-down network (two nMOS transistors), and a control inverter. To support feed through an extra pass transistor-nMOS is added to support signal feed through. The advantage of this design is that delay is reduced. The operation principle is discussed in brief. When data does not changes upon the arrival of clock pulse, ON current passes through MNx, which keeps the input stage from any driving effort. At the same instant, data at input and feedback output will have complement signal levels and node X is turned off. As a result, switching of signal does not occur in any internal nodes. In addition to that, if input changes from LOW to HIGH, transistor MP2 is turned ON due to discharge of node X. thus action also makes the node Q high. Referring to Fig. 5, this gives rise to the worst case timing of the Flip-flop operations as the conduction takes place in discharging path only for short period of time. On the other hand, input source provides the boost and is passed through transistor MNx which greatly reduces the delay with the signal feed through scheme. This action does not burden the input source in this design which is the case of pass transistor logic because conduction takes place for very short period. When data changes from 1 to 0, clock pulse turns on the transistor MNx and node Q discharges through this route. The input source is responsible for discharging. But the transistor remains on for a short period of time. So, this increases the loading of input source. The delay in the critical path does not depend on the discharging and there is no need to change the transistor size to improve the speed. When the value of the keeper logic is complemented, the discharging duty of input source is lifted.

3 Proposed Pass Transistor Logic Flip-Flop

The proposed design is an implicit type pulse triggered flip-flop with a conditional pulse enhancement scheme. Two methods are used in this design to overcome the disadvantages in the existing designs.

In existing designs, transistors are used in more number in discharging path which leads to high delay and more consumption of power while powering up the transistors. The solution is that number of transistors must be reduced in discharging path. So if value 1 is given as input, the strength of pull down transistor must be improved.

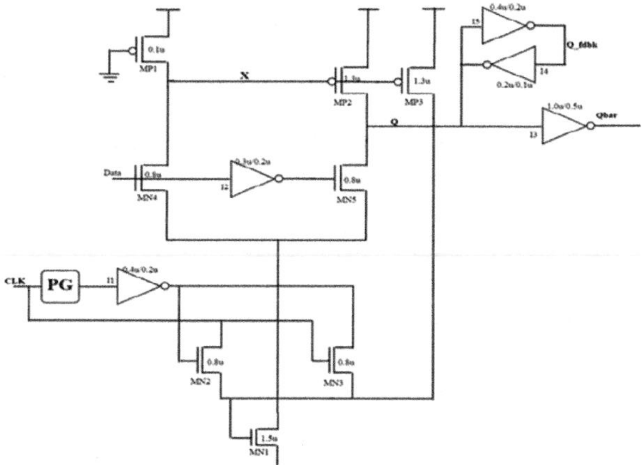

Fig. 6. Proposed pass transistor logic flip-flop

This design utilizes the upper part of the SCCER design. PTL based AND logic is formed by connecting two transistors MN2 and MN3 in parallel. The discharging operation in transistor MN1 is controlled by this logic. Complementary inputs are applied to the AND logic. As an outcome, zero value is maintained at the output node. When "0" is applied to both inputs, floating node occurs. But this floating node does not do any damage to the performance of the circuit. For every rising edge of the clock pulse, critical condition occurs. Weak logic high is passed to a node by turning on transistors N2 and N3. To enhance the strength of this weak pulse the transistor N1 is turned on for a time interval which is equal to the inverter I1 delay. Because of the presence of minimized voltage swing, the node's switching power is reduced. In MHLFF design, a single transistor drives the discharge control signal but in this design tow nMOS transistors connected in parallel enhance the speed of pulse generation.

This design reduces the count of transistors in the discharging path. Since less number of transistors is used, speed is enhanced, delay is reduced and area occupied is less. The flip-flop using the conditional enhancement scheme is illustrated in the Fig. 6. Pulses that generate discharging are activated only when it is needed, so glitches which occur due to unwanted circuitry are not present which minimizes the power consumption. PMOS transistors replace the delay inverters because they consume more power. This PMOS transistor increases the strength of the pull down for a loner discharge path. To reduce and power size of the transistor is reduced.

4 Results and Discussion

Proposed and existing designs' performance are compared. To demonstrate the proposed design's performance, a TSMC 90-nm CMOS is used to analyze power, area and delay. Since the design of pulse width is important for the correct capture of data and for the power consumption (Zhao et al. 2009), the size of the transistors in the pulse generator logic are designed for a 120 ps in pulse width in the TT case. The sizing also make sure that pulse generator function properly. By generating the input signals through buffers, the rise and fall time delays of signal are minimized. Since the proposed design requires pass transistors in order to reduce the power consumption. Five test pattern each with various data switching probability are given as input. Four of them are deterministic patterns, with 0% (all-0 or all-1), 25%, 50%, and 100% data transition probabilities, respectively.

To simulate the results TANNER EDA tool version 13.0 is used.

4.1 Comparative Analysis of Power Consumed by Various Flip-Flop Designs

Table 1 gives the features and results of simulation. It can be seen that less number of transistors are used in proposed design This is mainly attributed to the pass transistor logic, which mainly reduces the size of transistor on the discharging path. The Average power consumed is reduced by 65% when compared to other existing techniques. In addition to this, number of transistors is also reduced.

Table 1. Feature comparison of various FF designs

FF Design	ep-DCO	CDFF	SCDFF	MHLFF	TPSCFF
No. of transistor	28	30	31	19	24
No. of nodes	17	19	19	13	17
Average power (in mw)	9.50	14.79	13.06	13.64	13.91
Delay (in μs)	86.09	78.41	130.39	237.73	375.80

Table 2 summarizes the average comparison of power of various FF designs for various switching activity. It can be seen that proposed design consumes least amount of power in four out of the five test patterns. The amount of power consumed by the design will change for different test patterns. The proposed design consumes least power for a 25% data switching pattern when compared with all the existing FF design. More power is consumed by ep-DCO design due to unnecessary discharging of power.

Table 2. Average power comparison of various ff designs for various switching activity

Average power (in mw)	ep-DCO	CDFF	SCDFF	MHLFF	TPSCFF	PTLFF
100% activity	9.70	7.05	19.31	19.26	18.09	6.17
50% activity	8.19	19.50	14.12	13.87	16.25	3.11
25% activity	7.12	16.31	11.99	13.70	14.28	2.93
0% all-0	1.81	8.85	5.92	11	7.66	1.94
0% all-1	5.0	7.93	7.21	12.24	11.66	6.58

4.2 Simulation Results

The flip-flop designs are designed and simulated and the corresponding waveforms are obtained. By simulating the design the average power and delay can be measured. The static power is measured from the waveform. Power is calculated at various nodes of the flip-flop design. The pulse width of the flip-flop is calculated from the delay of three inverters used in a pulse generator. Simulation results for the proposed and existing flip-flop designs are shown in Figs. 7, 8, 9, 10, 11 and 12. From these simulation results delay is calculated.

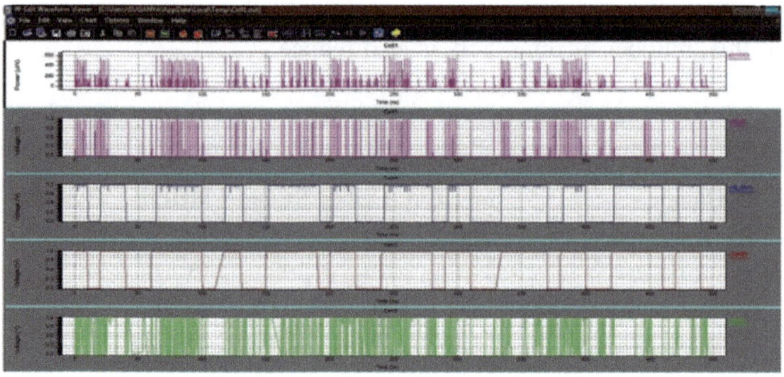

Fig. 7. Explicit-pulse Data-close-to-output waveform

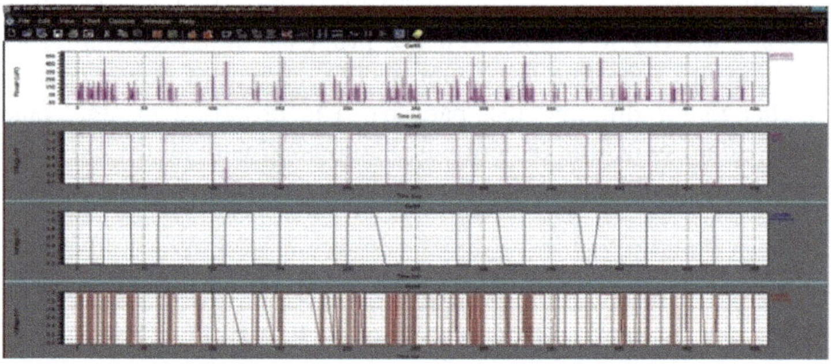

Fig. 8. Conditional Discharge Flip-flop waveform

From the Figs. 7 and 8 it is inferred that the trans-measured delay between data to output Q for the Explicit-pulse type flip-flop is 86.09 μs and for the Conditional Discharge flip-flop is 78.41 μs respectively.

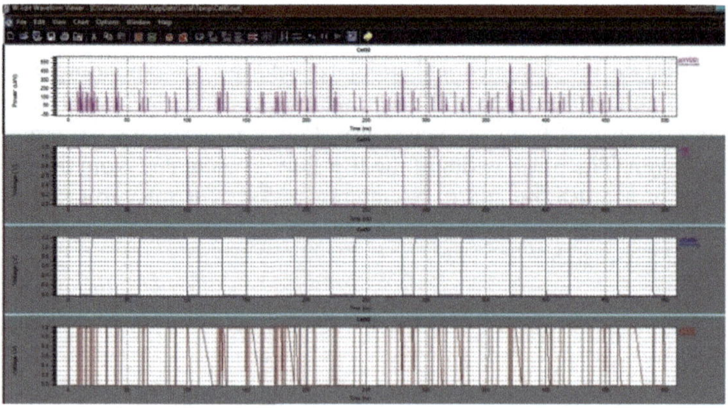

Fig. 9. Static Conditional Discharge Flip-flop waveform

From the Fig. 10 it is inferred that the trans-measured delay between data to output Q for this Static Conditional Discharge flip-flop is 130.39 μs.

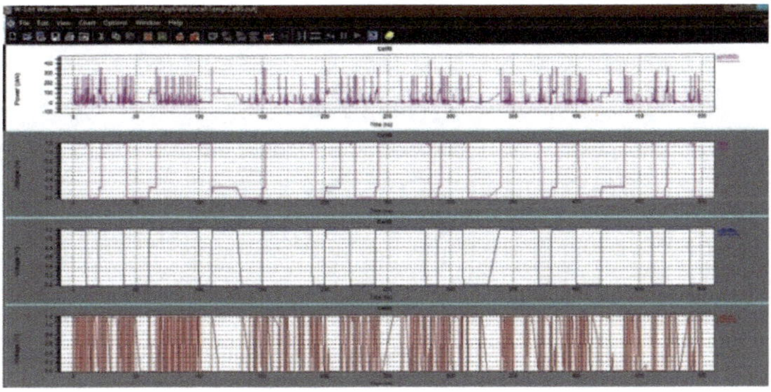

Fig. 10. Modified Hybrid Latch Flip-flop waveform

From the Fig. 11 it is inferred that the trans-measured delay between data to output Q for this Modified Hybrid Latch Flip-flop is 237.73 μs.

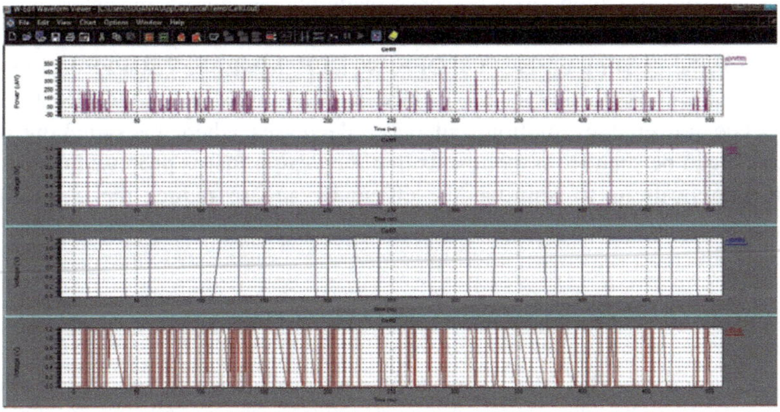

Fig. 11. True Single Phase Clocked Latch Flip-Flop waveform

From the Figs. 11 and 12, it is inferred that the trans-measured delay between data to output Q for the True Single Phase Clocked Latch Flip-Flop is 375.80 μs and for the Proposed Pass-Transistor Logic Flip-Flop is 124 μs respectively.

Above analysis of various types of flip-flops convey that the proposed Pass Transistor Logic flip-flop is found to be highly power efficient. Area complexity also reduces along with the power in PTLFF as the number of transistors required is less compared to other existing flip-flop techniques.

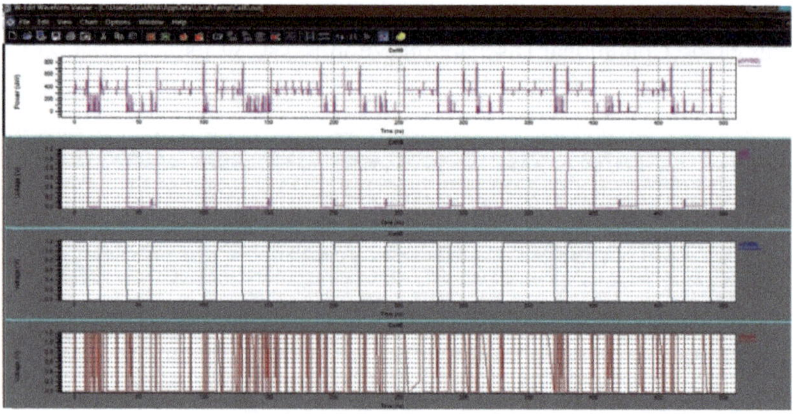

Fig. 12. Simulation result of proposed pass-transistor logic flip-flop

5 Conclusion

In this paper, a power efficient pulse-triggered Flip-Flop (FF) design using pass transistor logic is presented. Here, Pass Transistor logic based AND gate replaces an AND function in the clock generation circuitry. Since in the PTL-style based AND gate, the n-mos transistors are arranged in parallel and due to faster discharge of the pulse less power is consumed. Comparison has been made for number of transistors and average power consumed for 100% activity, 50% activity, 25% activity, 0% activity and delay are done for the existing ep-DCO, CDFF, SCDFF, MHLFF, TSCPFF techniques and the proposed PTLFF technique by TANNER EDA tools using MOSIS 90 nm technology. The power consumption and delay decreases when the switching activities decrease. It can be seen from the results that the design that is proposed can be used in real time applications to improve the efficiency and to reduce the consumption of power.

References

Karimi, A., Rezai, A., Hajhashemkhani, M.M.: A novel design for ultra-low power pulse-triggered D-Flip-Flop with optimized leakage power. Integration **60**, 160–166 (2018)

Lin, J.-F.: Low-power pulse-triggered flip-flop design based on a signal feed-through. IEEE Trans. Very Large Scale Integr. (VLSI) Syst. **22**(1), 181–185 (2014)

Saranya, L., Arumugam, S.: Optimization of power for sequential elements in pulse triggered flip-flop using low power topologies. Int. J. Sci. Technol. Res. **2**(3), 140–145 (2013)

Gupta, T., Mehra, R.: Efficient explicit pulsed double edge triggered flip-flop by using dependency on data. IOSR J. Electron. Commun. Eng. (IOSRJECE) **2**(1), 01–07 (2012)

Sadrossadat, S., Mostafa, H., Anis, M.: Statistical design framework of sub-micron flip-flop circuits considering die-to-die and within-die variations. IEEE Trans. Semicond. Manuf. **24** (2), 69–79 (2011)

Zhao, P., Darwish, T., Bayoumi, M.: High-performance and low power conditional discharge flip-flop. IEEE Trans. Very Large Scale Integr. (VLSI) Syst. **12**(5), 477–484 (2004)

Zhao, P., McNeely, J.B., Golconda, P.K., Venigalla, S., Wang, N., Downey, L.: Clocked-pseudo-NMOS flip-flops for level conversion in dual supply systems. IEEE Trans. Very Large Scale Integr. (VLSI) Syst. **17**(9), 1196–1202 (2009)

Mahmoodi, H., Tirumalashetty, V., Cooke, M., Roy, K.: Ultra low power clocking scheme using energy recovery and clock gating. IEEE Trans. Very Large Scale Integr. (VLSI) Syst. **17**(1), 33–44 (2009)

Rasouli, S.H., Khademzadeh, A., Afzali-Kusha, A., Nourani, M.: Low-power single-and double-edge-triggered flip-flops for high-speed applications. IEEE Proc. Circuits Devices Syst. **152** (2), 118–122 (2005)

Phyu, M.W., Goh, W.L., Yeo, K.S.: A low-power static dual edge-triggered flip-flop using an output-controlled discharge configuration. In: 2005 IEEE International Symposium on Circuits and Systems, pp. 2429–2432. IEEE (2005)

Tschanz, T., Narendra, S., Chen, Z., Borkar, S., Sachdev, M., De, V.: Comparative delay and energy of single edge-triggered and dual edge triggered pulsed flip-flops for high-performance microprocessors. In: Proceedings International Symposium on Low Power Electronics and Design, pp. 207–212. IEEE (2001)

Weste, N., Harris, D.: CMOS VLSI Design: A Circuits and Systems Perspective, 3rd edn. Pearson, New York (2011)

Teh, C.K., Hamada, M., Fujita, T., Hara, H., Ikumi, N., Oowaki, Y.: Conditional data mapping flip-flops for low-power and high-performance systems. IEEE Trans. Very Large Scale Integr. (VLSI) Syst. **14**(12), 1379–1383 (2006)

Chandrakasan, P., Sheng, S., Brodersen, R.W.: Low-power CMOS digital design. IEEE J. Solid-State Circuits **27**, 473–484 (1992)

Alioto, M., Consoli, E., Palumbo, G.: General strategies to design nanometer flip-flops in the energy-delay space. IEEE Trans. Circuits Syst. I Regul. Pap. **57**(7), 1583–1596 (2010)

International Water Border Detection System for Maritime Navigation

T. Lavanya(✉), Kavitha Subramani, M. S. Vinmathi, and S. Murugavalli

Panimalar Engineering College, Chennai, India
lavanyatr99@gmail.com, kavitha_bhskr@yahoo.com, vinmathis@gmail.com, murugavalli26@rediffmail.com

Abstract. Maritime navigation has been challenging for a lot of centuries. International borders have not been clear in case of Sea routes. This affects chiefly fishing community who has to go to deep sea on regular basis. In maritime conditions, finding borders can be tricky given a lot of factors. We design GPS-based location for alerting users from crossing international waters. The users can access their location using GPS and the alert system is activated for border crossing across international waters. Though many solutions have been proposed for specific location based issues, they provide solutions for a specific group of users such as Tamil fishermen at Rameshwaram. The proposed system is not confined to one specific location but to every border on this world. It can be configured to a specific set of borders depending on the state or country etc. The border data can be feed from Internet sources. since these data is subject to change due to various political, geographical and various other reasons, it has to be updated regularly by experts in the field. We use an API called GEOCoder which is on-line/off-line Geo-location API which accurately determines the location.

Keywords: Border detection · International waters · GPS · GEOCoder

1 Introduction

Maritime borders are not recognizable like the land borders. There are so many practical difficulties in maritime border marking due to the ocean's geographical features. Many illicit activities happen via sea and oceans which include organized crimes, smuggling of drugs and illicit materials, human trafficking etc. So, every nation ensures a rigid organized border security system for maintaining the peace inside the nation as well as between the neighbouring countries. But in some extreme cases, poor innocent people like fishermen who are unaware of the maritime border line gets imprisoned or even executed while crossing the ocean. As a matter of lives and our country's economic status, there should a proper system to alert these fisher-men about their location and the maritime border. Many border alert systems for marine navigation have been proposed which utilizes Global Positioning System.

D. J. Hemanth et al. (Eds.): COMET 2019, LNDECT 35, pp. 62–69, 2020.
https://doi.org/10.1007/978-3-030-32150-5_6

2 Existing System

GPS based border alert system [1] to put an end for the slaughtering of the fishermen. The purpose behind this the fishermen crossing the region or limit. This incident because of absence of mindfulness where their limit is found and the expense of this slip-up is their life. The GPS antenna present in the GPS module gets the data from the GPS satellite and it uncovers the position data. This data got from the GPS antenna is sent to the controlling station where it is decoded. In this manner, the total information identified with the vehicle is accessible at the controlling unit. This data is sent to the proprietor or to the concerned individual constantly utilizing a GPS modem. On the off chance that the individual crosses the outskirt, he checks the information originating from GPS and he is alarmed.

The solution [2] is given to save fisherman from the danger which they face in their day to day life. This consists of a boat detection and monitoring gadget where it uses the GPS technique to identify the boundary and intimates the fisherman from trespassing. The RFID (Radio Frequency Identification) detects the maritime boundary and intimates the PIC (Peripheral Interface Controller) to take the necessary action according to the problem information the controller receives.

The GPS is employed to search out the situation of the boat. If the boat is identified near the boundary, GPS notifies to coastal office and gives warning signal to fisherman via GSM communication [3]. The signal is sent to the Engine Control Unit to control the pace of the moving boat when it further closes to the maritime boundary. Prohibited activities such as smuggling, prowler are monitored and notified as alerts to fisherman.

Location Based Services using Android [4] realizes three types of LBS services where mobile is configured as a server and SQLite database is used to store information. Fisherman receives alerts about ice berg, Tsunami, Cyclones and many others to enable safe routing. A controlling device is designed to overcome the danger with Global positioning system (GPS) which gives current region of fishing vessel in water. The microcontroller checks whether the boat has crossed the coastal border using GPS. The nearest coast defends ship and the mariner is informed about the coastal border crossing via RF alerts at VHF (30–300 MHz) which includes extensive region. The coastguard informed with boat statistics taken from the GPS receiver to safeguard the boat from the peril.

The border protection forces will first receive a notification which will be sent to all other devices who are sailing in the ship. The software [5] will help to find the facts about gadgets used by the opponents and will intimate about the danger. To handle these danger this software can be used to save the situation. This is mainly applicable for Tamil Fishermen who work in the borders.

There are facilities like Location Manager, Location Provider, Geocoding, GoogleMap available in android operating system for implementing LBS services [6] (geo-services). This application can also be used by normal people to know the location and to reach their destination correctly. The border security forces will receive the message and will intimate the people regarding their border crossing and their opponent forces. The application [7] may be extensively used by human beings in the border to locate the correct path to reach the destination. The notification will be sent to the border security forces which act as the server to all different devices that are operated by people in ships. The software will notify the statistics of wherein the devices are being positioned and intimate them about the problems that arise because of opponent forces in ships to server. This can act as an incident control utility to keep away from conflicts at various situations.

The framework comprises of three noteworthy modules all together evade all the inadequacies of the border alert system. It incorporates Vessel tracking module, RADAR Identification module, up degradation in GPS72H to altogether functioning the Automatic Identification System [8]. In the event that a fisherman navigates border, a caution is produced demonstrating that the border has been crossed. Moreover, a message will be sent to the station with respect to the shore stating that the border has been crossed via a GSM transmitter interface. In this way guards can help and offer extra help to that fisherman if necessary. Keeping in contemplations about existences of Indian fisherman, the device [9] has been made to help them not to transport beyond. It likewise cautions and keeps the fishermen crossing the national ocean outskirt. The methodology depicts the intruder situating is being executed with the assistance of GPS by coordinating the code given by it and the code produced in arm microcontroller through timer. It is a valuable gadget [10] for more secure navigation, particularly for fishermen.

3 Proposed System

The Proposed system uses GPS embedded in border alert which protects the fishermen from being killed or prosecuted for crossing the border unaware of it. It alerts them using an alarm system. The current location is found using GPS receiver in a Smart Phone (Fig. 1).

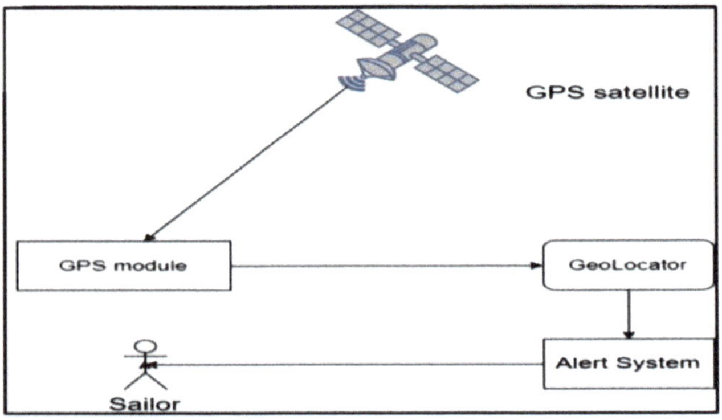

Fig. 1. Architecture diagram

The instantaneous values of latitude and longitude are the current latitude and longitude values are estimated. Those values are compared to a predefined value and calculate the current location. With the help of this comparison, the distance of boat from the maritime border is calculated and the fishermen are alerted about their location before they reach 5 km of the boundary and are notified for every 30 s.

GPS Satellite provides location information to the Android device and the mobile device can retrieve the country code for the location. Comparing with the nearby locations' data, it can detect if the border is approaching or not (Fig. 2).

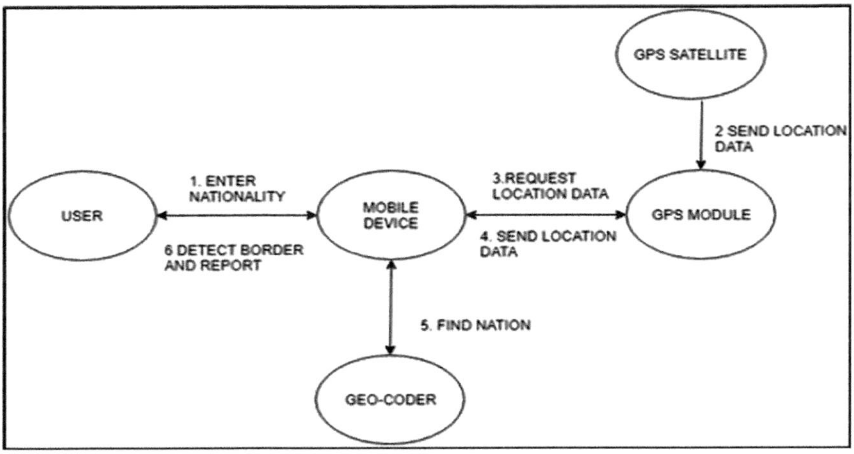

Fig. 2. Functional block diagram

The user gives the input as country name and GPS satellite gives information about the location to GPS module. The mobile device requests the fisherman current location and GPS module sends location data and identifies the country and notifies whether the border is approaching.

The retrieval of GPS location from the GPS satellite and send it to Activity class to geolocate. This data is processed and finally displayed to user via Android UI. This processing is done by Geocoder API. This system includes the modules

(i) GPS Module: current location is read from GPS satellite by GPS module and is printed on the Android UI. This requires permission to use GPS access in Android Manifest file. This information is collected from satellite and displayed on Android screen.
(ii) GEO Coder: The Border information of the current location is send to Geocoder API in which the actual location to country mapping is performed. It also finds country mapping data from nearby GPS locations
(iii) Check for User input: It can find border for user-entered locations. This follows similar procedure as in previous process except that it takes custom locations. This can be used a metric to keep track of end points in the current path.

Steps

```
Start
  check GPS enabled
  if (GPS not enabled),
      alert user (" GPS not enabled ")
  if (GPS enabled), loc = getLocation ()
  lat=loc.getLatitude() ,
  lon=loc.getLongitude()
  country[0] =gc.getAddress.getCountry(lat,lon)
  country[1-4]=gc.getAddress.getCountry(lat+/-
                                0.1,lon+/-0.1)
  for(i=0;i<=4;i++)
    if country[i]!=other values ,
         alert("Borderapproaching")
    ifcountry[i]==null,
         alert("Complete International waters")
 end
```

4 Experimental Results

GET LOCATION

A user can check his current location in GET LOCATION Module with GPS enabled mobile. Latitude and Longitude values for the current location are displayed and alert is given if it is near the border (Figs. 3, 4 and 5).

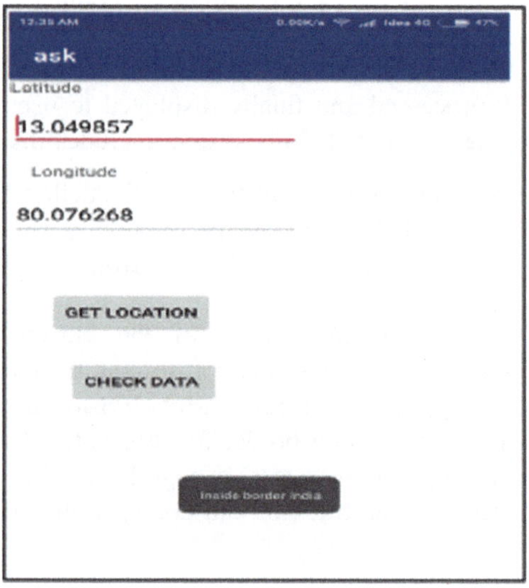

Fig. 3. Get Location

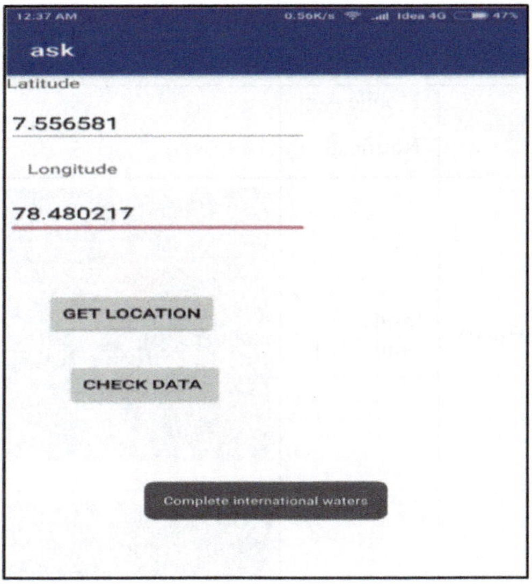

Fig. 4. International waters

CHECK FOR USER INPUT

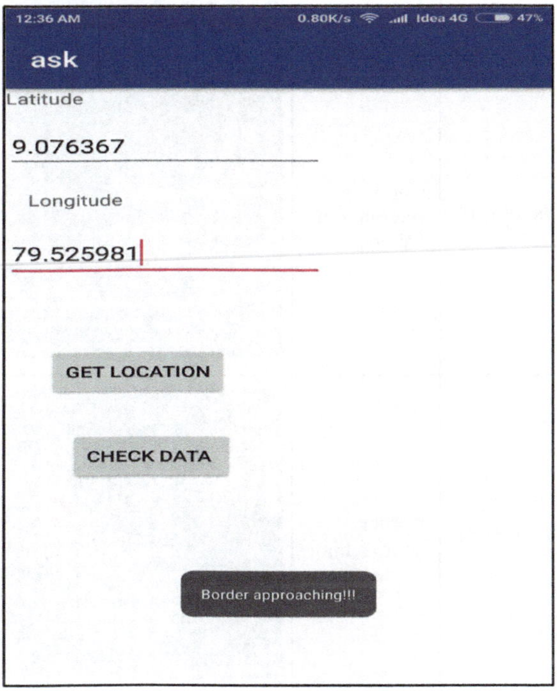

Fig. 5. Border warning

Table 1. Maritime boundary

Sl. no	Latitude	Longitude	Mobile App Notification	Map Locations
1.	13.0498 57	80.076268	Inside Border India	
2.	12.9829 40	81.032567	Coastal Area - India	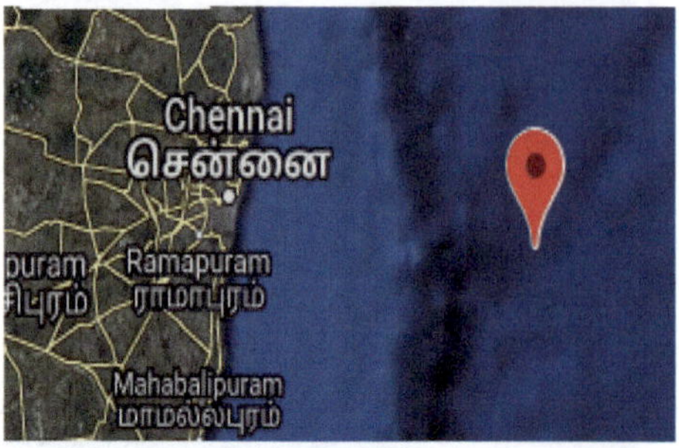
3.	7.55658 1	78.480217	Complete Internationa l Waters	
4.	9.07636 7	79.525981	Border approaching	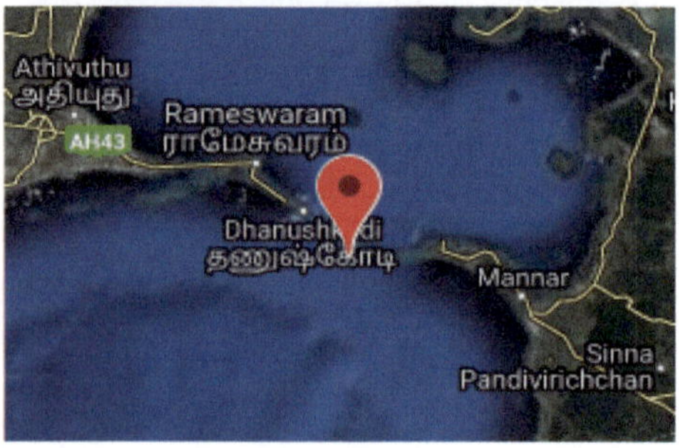

Enter latitude and longitude values for the location to be checked. Click ‘Check Data’. The given input is read and whether the given location is near border is detected as shown in Table 1.

5 Conclusion

The border detection algorithm is realized with the proposed state machine that investigates the features in a sequential manner. The border detection system can be used by fishermen and sailors during maritime travel in deep waters. It uses GPS module in the device which can be accurate. Future enhancements can include improved speed in retrieval and provide more accurate since Geocoder is accurate for about 3–4 km offline. SDK’s like geoFabrik, OSMgeocode can also be used for better performance. The Mobile app can be made available in other languages and Help line numbers is displayed in the app during emergency situation.

References

1. Jaganath, K., Sunilkumar, A.: GPS based border alert system for fisherman. Int. J. Technol. Res. Eng. **4**(5), 778–780 (2009)
2. Bhavani, R.G., Samuel, F.: GPS based system for detection and control of maritime boundary. In: 59th International Midwest Symposium on Circuits and Systems (MWSCAS), pp. 601–604. IEEE, Khalifa University (2016)
3. Isaac, J.: Advanced border alert system using GPS and with intelligent engine control unit. Int. J. Electr. Comput. Eng. (IJECE) **1**(4), 11–14 (2015)
4. Kiruthika, S., Rajasekaran, N.: A wireless mode of protected defence mechanism to mariners using GSM technology. Int. J. Emerg. Technol. Innovative Eng. **1**(5), 33–37 (2015)
5. Naveen Kumar, M., Ranjith, R.: Border alert and smart tracking system with alarm using DGPS and GSM. Int. J. Emerg. Technol. Comput. Sci. Electron. (IJETCSE) **8**(1), 45–51 (2014)
6. Kumar, S., Qadeer, M.A., Gupta, A.: Location based services using android (LBSOID). In: 2009 IEEE International Conference on Internet Multimedia Services Architecture and Applications (IMSAA). IEEE (2009)
7. Kumar, R.D., Aldo, M.S., Joseph, J.C.F.: Alert system for fishermen crossing border using Android. In: 2016 International Conference on Electrical, Electronics and Optimization Techniques (ICEEOT), pp. 4791–4795. IEEE (2016)
8. Karthikeyan, R., Dhandapani, A., Mahalingham, U.: Protecting of fishermen on Indian maritime boundaries. J. Comput. Appl. **5**(3), 0974–1925 (2012)
9. Suresh Kumar, K.: Design of low cost maritime boundary identification device using GPS system. Int. J. Eng. Sci. Technol. **2**(9), 4665–4672 (2010)
10. Sivagnanam, G., Midhun, A.J., Krishna, N., Anguraj, G.M.S.R.A.: Coast guard alert and rescue system for international maritime line crossing of fisherman. Int. J. Innovative Res. Adv. Eng. (IJIRAE) **2**(2), 82–86 (2015)

A Comprehensive Survey on Hybrid Electric Vehicle Technology with Multiport Converters

Damarla Indira(✉) and M. Venmathi

Department of EEE, St. Joseph's College of Engineering, Chennai, India
indira.damarla@gmail.com

Abstract. This paper presents a Comprehensive survey of multiport converters for Hybrid Electric Vehicles (HEVs). HEV technology provides an effective solution for achieving higher fuel economy and better performances with reduced greenhouse gas emissions. HEVs are more popular due to their low carbon footprint and ease of integration with renewable energy with energy storage systems. Increasing the driving miles and flexible battery charging are the biggest challenges facing the HEV Technology. A literature review has been conducted to identify the different techniques for improving the driving miles and to achieve flexible charging capability. The most promising concept to give a proper solution for the challenges in HEV is the development of multiport converters. Many researchers are proposed different topologies of multiport converters and control strategies to improve the performance of the drive system. The high efficiency of conversion makes a multiport converter very attractive in HEVs. The survey emphasis energy storage devices, different motor configurations, multiport converter topologies and control strategies for the HEV propulsion system to improve the performance of the drive system.

Keywords: Multiport converters · Electric Vehicles · Hybrid Electric Vehicles · Renewable energy sources · DC-DC converter

1 Introduction

The automotive industry is the major contributor to Global warming. In Internal Combustion Engine (ICE) Vehicles only 25% of the energy produced from the reaction is converted into mechanical power and remaining 75% of the energy is lost which is shown in Fig. 1. This wasted energy is converted into harmful gasses called greenhouse gasses like carbon dioxide (CO_2), Nitrous oxide (N_2O) etc. To protect the environment from such harmful gasses, fossil fuels and to enable an environmental friendly transportation, the ICE vehicles are replaced by Electric Vehicles.

The automotive industry is the major contributor to greenhouse effect. So it is necessary to develop a new technology that would enable an environmental friendly transportation sector. Electric Vehicles (EVs) were introduced in the year 1834 [1]. The operation of EV involves two stages of energy conversion. The first stage of conversion is to get the electrical energy from fossil fuel and the second stage is to obtain the kinetic energy from electrical energy. This double conversion makes economically expensive. In order to reach the cost effective and energy efficient transportation

D. J. Hemanth et al. (Eds.): COMET 2019, LNDECT 35, pp. 70–85, 2020.
https://doi.org/10.1007/978-3-030-32150-5_7

gasoline powered vehicles were invented. Now the biggest problems facing are shortages in the gasoline and air contamination. These issues are mainly due to emissions of several toxic elements grouped under the generic term greenhouse gasses and heavy usage of non-renewable resources. To reduce the air pollution some efforts are being carried out to improve the carbon footprint. In order to protect the environment from these issues many researchers are focusing to develop the use of renewable energy sources and EVs [2, 3]. The combination of these two is strongly connected. Recently renewable energy sources are most attractive due to generating more power with low cost. For complete utilization of these sources it requires storage technologies. At this point of storage requirement EVs are playing best role.

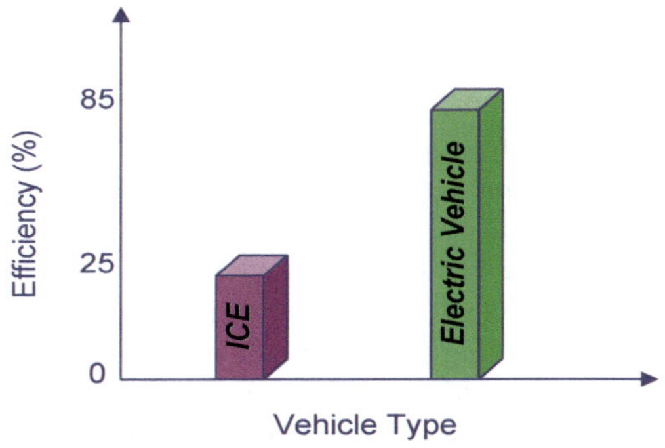

Fig. 1. Performance curve

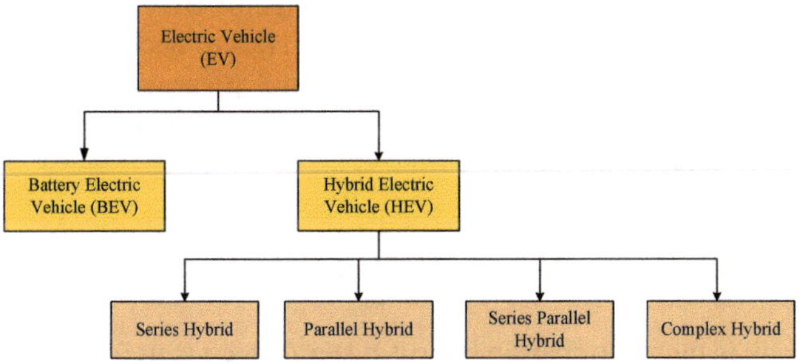

Fig. 2. Classifications of electric vehicles

According to energy converter type used to propel the vehicle and vehicle power and function, EVs are classified in to 3 types which are shown in Fig. 2.

- Battery based Electric Vehicle (BEV)
- Hybrid Electric Vehicle (HEV)
- Plug-in Hybrid Electric Vehicles (PHEV)

Battery based Electric Vehicle (BEV) are also called as pure electric vehicle. It uses rechargeable battery to drive the vehicle, ICE is not present. After the drive battery needs to charge and stores energy for next drive.

Hybrid Electric Vehicle (HEV) uses both ICE and battery. The battery used in HEV does not need separate charging as it gets charged from the vehicle stopping and it is also known as regenerative braking. The different ways of hybridization density can give different architectures are series HEV, parallel HEV, series parallel HEV, and complex HEV.

The Plug-in Hybrid Electric Vehicles (PHEV) is similar to that of HEV but the batteries used in PEV are charged from either electricity power outlet in a house or any commercial place.

Nowadays the researchers mainly focused on HEVs and energy storage vehicles which have the capability to run on both electricity and conventional fuels even though grid connected electric vehicles are present [4].

Various technologies are implemented based on the type of electric vehicle. These technologies are mainly focused on driving the motor [5]. To improves the overall performance of an electric vehicle researchers are working on various techniques and control structures. The main area of technology in electric vehicles is the development of novel converter with centralized control structures to improve the overall efficiency and performance. Another area of research is to implementation of battery charging circuits. Different battery charging circuits are developed [6, 7]. The issues which are primarily in grid connected electric vehicles are electrical load management and grid stability [8]. Implementation of new integrated multiport converter for HEV makes to improve the driving miles and enhances the self-charging capability.

The main aim of this paper is to present a survey of various multiport converters researchers have developed to improve the performance of HEV. To achieve the self-charging capability and to improve the driving range of HEV, it is necessary to know the structure of HEV and different batteries used. Hence, Sect. 2 discusses the energy storage devices or battery banks in HEV. Section 3 presents various electric motors used in HEV. In Sect. 4 various multiport converters used to improve the driving range. Finally Sect. 5 concludes the paper.

2 Energy Storage Devices in HEVS

The battery bank is the major part in hybrid system which is used to stores the electrical energy during regenerative braking [9]. In electric vehicles Lithium-Ion (Li-Ion) batteries are dominating in the market [10] where as in some aspects the best choice is Nickel-Metal Hydride batteries [11]. Commercial batteries used in electric vehicles are Li-Ion battery bank (388 V, 360 Ah), Lead-Acid batteries (12 V, 170 Ah), Li-Ion batteries (30 kWh) and Sodium-Sulphate batteries [12]. The capacitance of carbon ultra-capacitors is as high as 4000 F with voltage rating up to 3 V per cell [13]. These are high energy storage devices that can be fully charged within a few seconds and most suitable for regenerative braking. Unfortunately these are not preferred because of their low energy density.

The size of the battery bank is based on manufacturer specifications depending upon what extent the vehicle will be operated in electric mode. The vehicle size is as light as possible by using smaller battery banks. An electric vehicle utilizes environmental friendly renewable energy sources as the power source. This clean energy reduces the emissions and it consumes higher energy due to added weight of vehicle. Toyota Prius3 uses smaller battery pack compared to other vehicles. It uses Li-Ion battery bank capacity of 4.4 kWh and weighs 80 kg [14]. The Chevrolet volt uses much bigger battery bank that has a capacity of 14 kWh, which add 183 kg to the vehicle [15].

The battery bank consists of number of battery cells that are connected in series and parallel to meet the desired capacity. The other measures from battery bank are voltage, current, temperature measurement, a cell balancing circuit, and a cooling system [16]. These measures ensure that it can operate in optimum efficiency and prevent the battery from damaging of cells.

State of Charge (SOC) decides the operating range of a battery and it is from 25% to 95% [17]. This makes the usable capacity is 70%. The battery banks should be oversized to have a sufficient usable capacity.

In HEVs, most commonly used batteries are Nickel-Metal Hydride and the Li-Ion. These are mostly preferred in automotive industry due its merits of high power to weight ratio, high capacity, fast charging, and long lifecycle. Nickel-Metal Hydride does not contain any toxic materials and both are rechargeable [18]. The Li-Ion battery requires protection circuits to prevent from overcharge and over discharge conditions whereas Nickel-Metal Hydride contains simple circuitry [19, 20]. The Nickel-Metal Hydride battery is mostly preferred compared to Li-Ion because of its merits but Li-Ion battery is the choice for HEVs and purely electric vehicles if the price of production reduces.

3 Electric Motors for HEVs

Generally Electric motors are used in EVs for electric propulsion because of their significant performance. Varies researchers are investigating in the several areas of motor propulsion, methods to eliminate position sensors, motor control techniques, inverter current sensors. The technological challenges for the electric motors will be wide speed range, light weight, maximum torque, high efficiency, and long life. EVs use three types of motors. They are AC Induction Motors (ACIMs), Permanent Magnet Synchronous Motors (PMSMs), Brushless DC Motors (BLDCs) and Switched Reluctance Motors (SRMs).

ACIMs are mostly dominating in cars for various reasons [21]. These motors are best choice for driving EVs because of their low production cost, ease of manufacturing, less maintenance due to lack of brushes, good efficiency at all load conditions and at different speed ranges, high robustness and good dynamic performance. However, to achieve the good dynamic performance it is necessary to apply highly complex vector control technique which increases the price of the vehicle [13].

The PMSMs are widely used in both HEV and EV applications. This type of machine has high power to weight ratio, high torque, and has high peak efficiency. The speed of PMSM is controlled by handling field oriented control. At low speeds the maximum amount of torque is generated. As the speed increases, the power increase to a maximum and the torque decreases, as shown in Fig. 3.

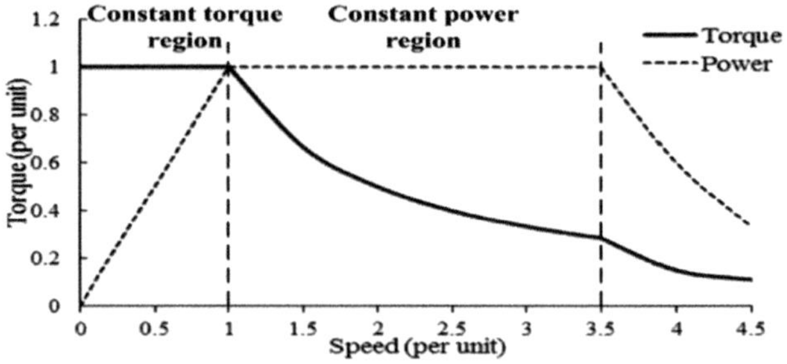

Fig. 3. Torque speed characteristics of PMSM

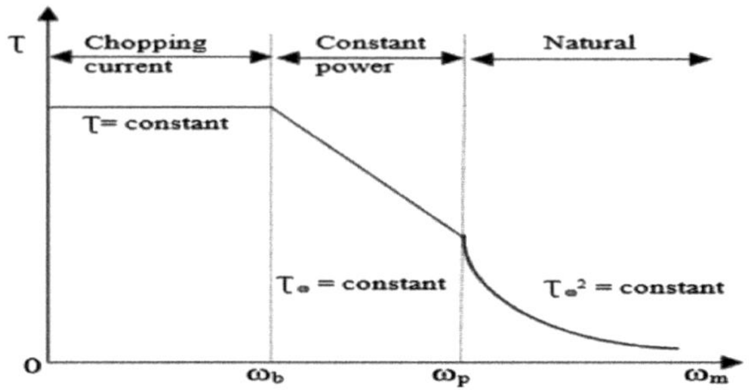

Fig. 4. Torque speed characteristics of SRM

To achieve safe vehicle propulsion these properties are well suitable. From these properties highest torque is maintained during acceleration and steady torque at low speeds. There is no need for a transmission system in traction motor application due to these beneficial properties. Motor can operate efficiently during crucial speeds by choosing a suitable gear ratio even though sufficient torque available. They are mainly used in medium weight or traction applications.

The primary choice of motor in EVs and HEVs are PMSM and BLDC due to its high power density. But these motors have problems with high cost, demagnetization, and fault tolerance.

Compared to PMSM and BLDC, SRM has good efficiency improved power density. The SRM have low losses, high efficiency, no permanent magnets on the rotor, high reliability, low acoustic noise, excellent fault-tolerance ability and higher torque to power ratio [22]. Due to these advantages SRMs are mostly preferred in EVs and HEVs. The main important features in SRM for the selection of EVs and HEVs are less weight due to no winding in rotor, less cost and high efficiency [23]. The weight of the overall system increases due to heavy motors which results lower acceleration and decreases the performance of overall system. The speed torque characteristics of SRM are shown in Fig. 4. The current in the stator winding are switched ON and OFF based on its rotor position. The speed at which peak current is applied to the motor at rated voltage with constant switching angle is called base speed (ω_b).

4 Multiport Converters for HEVs

The power electronic converter plays a major role in EVs and HEVs. The conventional converters suffer from the following drawbacks.

- High cost due to the more number of switches.
- More Switching Losses.
- Complex structure.

The efficiency is lower because of more conversion steps. If the voltage levels for different renewable energy sources are quite different, and the DC bus voltage is much higher than that of the renewable energy sources, the DC-DC converters are operated in extreme case, where duty ratio is closed either to 1 or 0. The control is separated. The controller only controls the DC-DC converter without considering the overall system performance.

To overcome the above issues faced by conventional converters many researchers are continued their research towards novel converters called multiport converters. Multiport converters are the best promising choices for EVs and HEVs [24]. They are much beneficial for the following reasons.

Single power conversion stage reduces component count on semiconductor switches, drive circuits.

Reduced size due to reduced component count when compared to dc link based conventional converter.

- The system efficiency can be improved.
- Low cost due to fewer power components.
- Due to single stage power conversion, the converter has a centralized control for regulating the output voltage and determining the power sharing ratio.
- The converter naturally yields to bi-directional power flow in all ports.
- High productivity because of minimized transformation steps.
- Compact packaging.
- Less switching losses because of less no of switches.

In recent years multiport converters are proposed which consists of more than one input source like PV, wind, fuel cell etc. and one regulated output voltage. The block diagram of multiport converter is shown in Fig. 5.

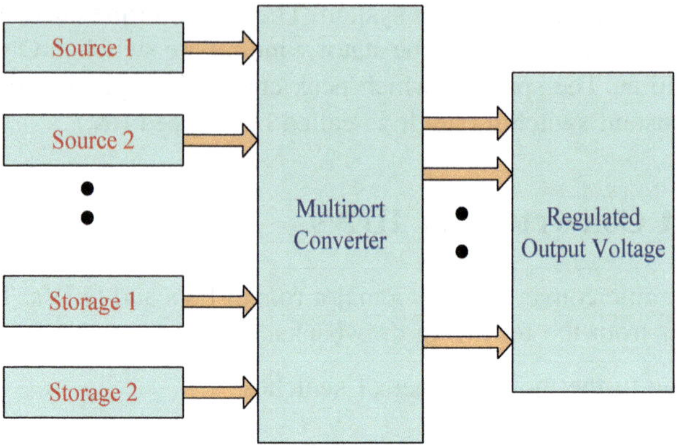

Fig. 5. Multiport converters

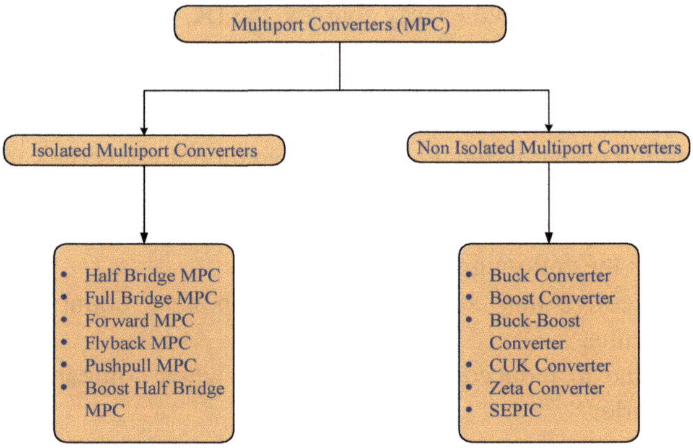

Fig. 6. Classification of multiport converters

Multiport converter has many advantages against conventional structure in terms of the number of power devices and conversion steps used because the system resources are shared which in turn improves the system efficiency. The comparison between conventional and multiport converters is given in Table 1.

Table 1. Comparison of conventional and multiport converters.

Category	Conventional structure	Multiport structure
Need a common DC bus	Yes	No
Conversion steps	More than once	Minimized
Control Scheme	Separate control	Centralized control
Power flow management	Complicated, slow	Simple, fast
Transformer	Multiple	Single, Multi winding
Implementation effort	High	Low
Need a common DC bus	More than once	Minimized

Multiport converters are broadly classified into two types given in Fig. 6.

a. Isolated Multiport converters.
b. Non Isolated Multiport converters.

4.1 Literature Review on Multiport Converters for EVs and HEVs

The transformer coupled triple half bridge three port bi directional converters controlled by phase shift and PWM is proposed shown in Fig. 7. This multiport converter is mainly used in Fuel cell Vehicle (FCV) applications. It uses battery/supercapacitor as a storage port. The phase shifting is extended by zero voltage switching. Compared to other topologies it reduces the conduction losses and current stress of the power switches [25].

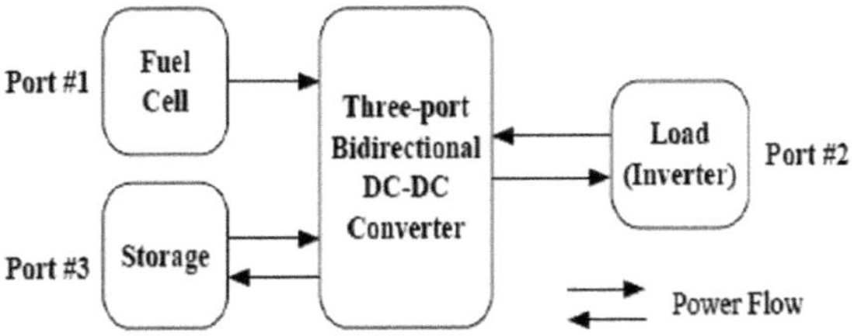

Fig. 7. Three port bidirectional converter

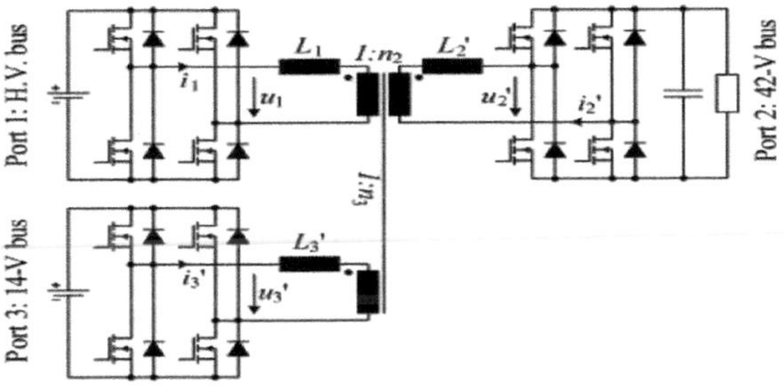

Fig. 8. Isolated bidirectional multiport converter

In [26] isolated three port bidirectional multiport converter for HEVs and FCVs shown in Fig. 8. It implements the multiple voltage levels power distribution system. Generally in EVs during start-up and acceleration storage element enables the regenerative brake and energy releasing functions. With this multiport converter bidirectional power flow and different voltage levels is possible.

The characteristics of HEVs, FCVs and more electric vehicles are presented in [27]. It also focused the future challenges in automobile industry. In [28] a review of current and future scenario of EVs and the importance of power electronic converters, electric motors in EVs and HEVs are presented. The comparison of various control techniques and suitable power electronic converter configurations are discussed in [29]. It also represents that which topology is better suitable for PHEVs.

A four level flying capacitor dc to dc converter interfacing with inverter and battery is shown in Fig. 9 has been proposed for HEVs in [30]. The limitations of boost converter used in HEVs are minimized by using this novel converter.

The inverters used in HEVs are more expensive and gives low efficiency due to heavy size of inductor shown in Fig. 10. To avoid these limitations a new multilevel boost inverter without an inductor has been designed for HEV technology [31]. In this paper a cascaded H-bridge multi-level boost inverter is proposed for EVs and HEVs which will produces high efficiency.

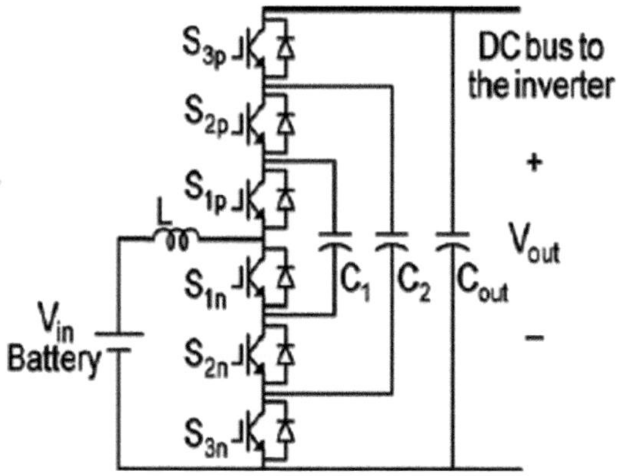

Fig. 9. Four level flying capacitor dc-dc converter

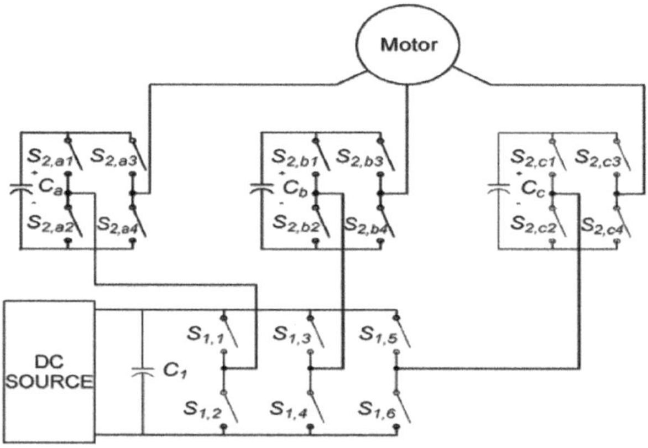

Fig. 10. Multi-level boost inverter without an inductor

In [32] an integrated converter is proposed for HEVs and PHEVs shown in Fig. 11. This converter is able to charge the battery and transfer the electrical energy from battery bank to bus system. It also achieves the fault tolerance capability by using reduced number of inductors and transducers.

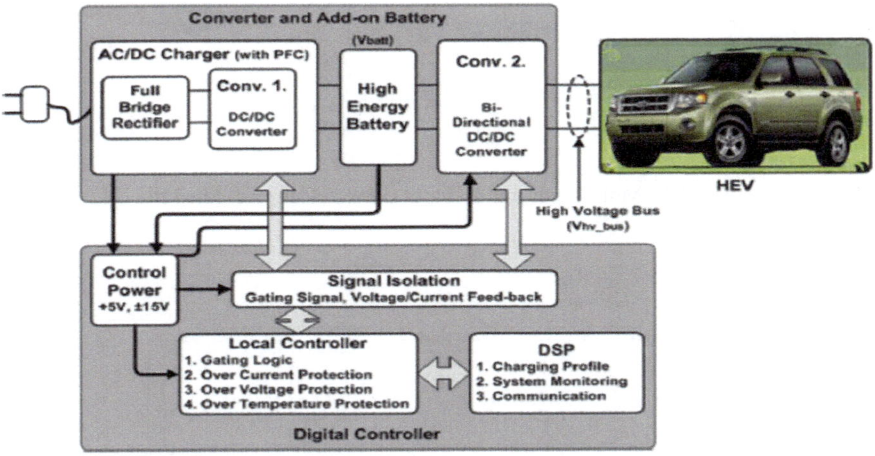

Fig. 11. Integrated converter

A novel half bridge integrated zero voltage switched full bridge converter is implemented for battery charging in EVs and HEVs has been proposed in [33]. This resonant converter gives the merits of reduced filter size, less switching and conduction losses, improves the power factor and overall system performance.

To reduce the greenhouse gas emissions and to improve the efficiency, EVs and HEVs are the best choice. Nowadays renewable energy sources are encouraged. Multiport power electronic interface for renewable energy sources and storages is proposed in this paper. It is multi input multi output power electronic converter which is capable of interfacing with different sources, storages, and loads shown in Fig. 12. It exhibits excellent steady state and dynamic performance, optimal energy and power management [34] (Fig. 13).

Generally EVs and HEVs use batteries for storing electrical energy. These batteries are capable of recharging and discharging. A novel dc to dc multiport bidirectional converter is implemented for parking lot integrating EVs as either energy source or electric load shown in Fig. 14. The main aim of this work is to design compact multiport bidirectional converter which can be able to respond various power transactions in parking lot [36].

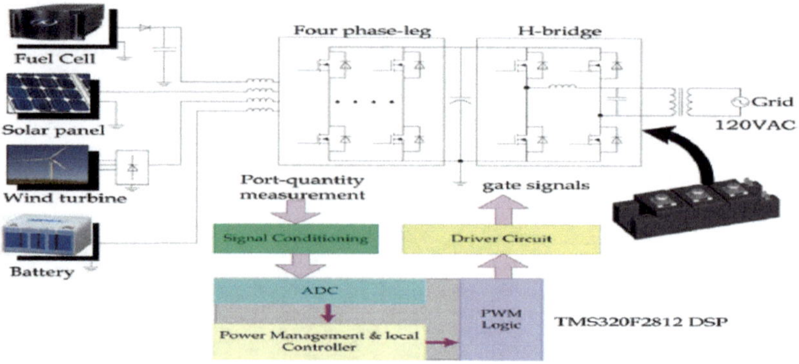

Fig. 12. Multi input multi output power electronic converter

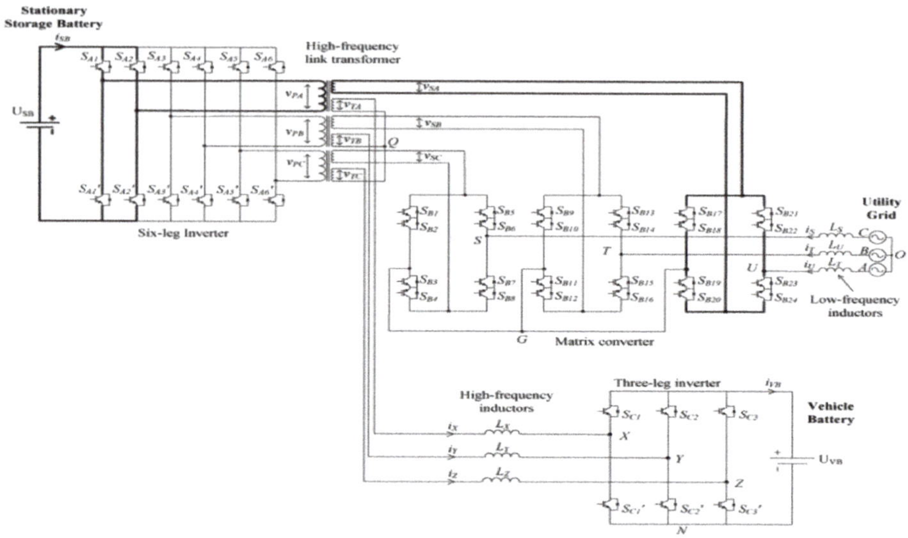

Fig. 13. Schematic diagram of Multiport converter with storage port

The lack of properties in hybrid energy storage systems are integration of inductively coupled power transfer, flexible structure and a unified controller. This leads to resulting in battery currents which are not completely decoupled from high frequency and high magnitude current and state of charge of ultra-capacitor not controlled properly. In this paper, a multiport power electronic interface is implemented which acts as an energy router for on board electric and plug in hybrid electric vehicles with inductively coupled power transfer and hybrid energy storage systems is shown in Fig. 15. A central controller is designed which can completely resolves aforementioned drawbacks [37].

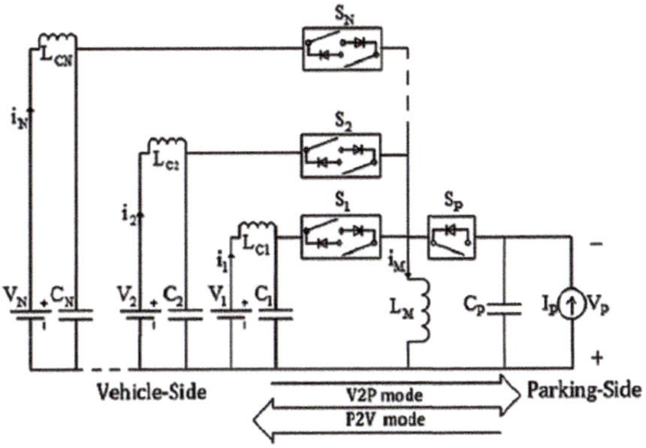

Fig. 14. Multiport bidirectional converter for parking lot

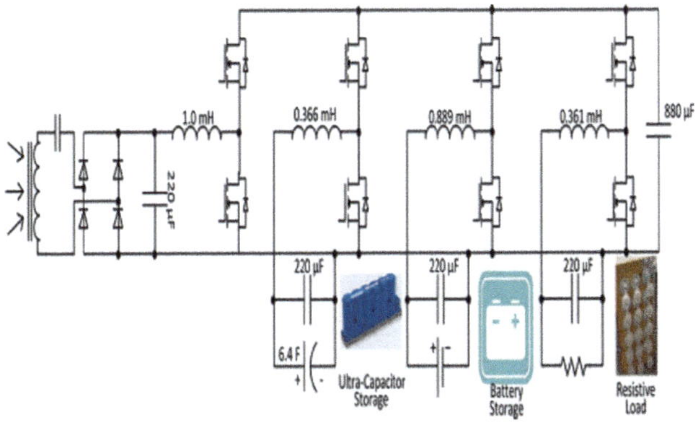

Fig. 15. Multiport power electronic interface

Nowadays SRM drives are playing a vital role in EVs and HEVs because of special features. The mechanical volume of SRM is low due to no rotor windings and no permanent magnets required. SRM has several advantages as compared to other competing machines high reliability, low cost, robust structure, wide speed range, and good fault tolerance ability, which give these motors the ability to work in high temperature, high speed, and safety critical applications. The challenging issue in SRM drive is the large power ripple due current commutation which reduces the overall efficiency and shortens the battery life. To overcome these limitations an integrated multiport power converter (IMPC) with small ripple and bidirectional power flow shown in Fig. 16 is proposed in this paper. To restrain battery current ripple a novel multi objective power flow control method with repetitive controller is also proposed in [38].

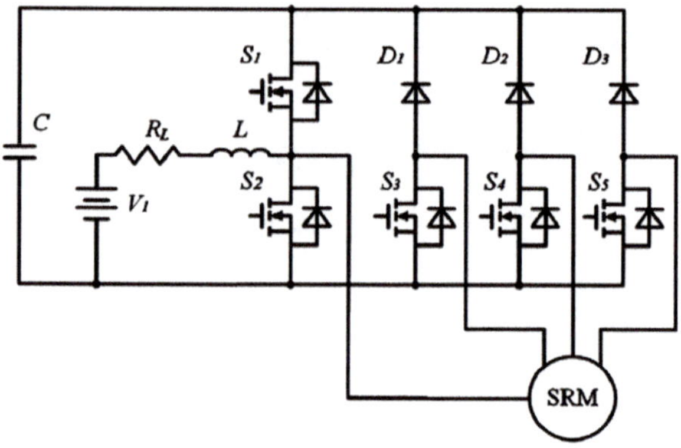

Fig. 16. Integrated Multiport power converters

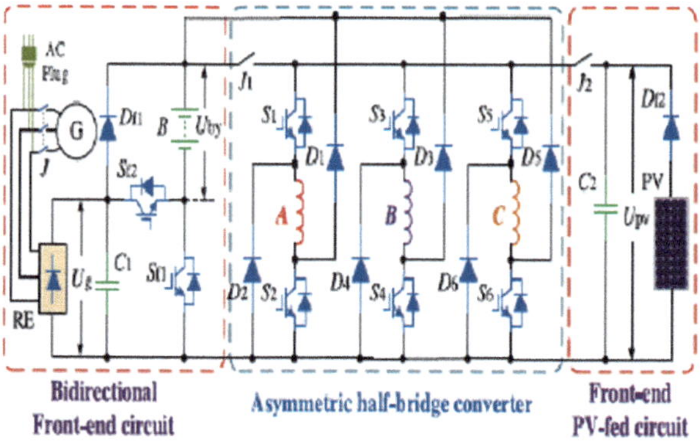

Fig. 17. Schematic diagram of the multiport bidirectional SRM drive for solar assisted HEVs

In construction point of view internal combustion engine (ICE) based vehicles are costly. With this reason all the ICE vehicles are replaced by electrical based. Due to limitations in current battery technologies the driving distance is short in pure battery operated vehicles. To improve the motoring performance and to achieve self-charging capability, a multiport bidirectional SRM drive for solar assisted hybrid electric vehicle power train has been proposed. The schematic diagram of the multiport bidirectional SRM drive for solar assisted HEVs is shown in Fig. 17. The photovoltaic (PV) panels are installed on the top of vehicle to achieve self-charging there by it reduces the usage of charging stations [39].

4.2 Comparison of Different Multiport Converters

A comparison is made between various converter topologies mentioned in literature review and the best suitable multiport converter for EVs and HEVs is the Integrated Multiport bidirectional converter with switched multiplexing technique which can

Table 2 Comparison of various multiport converter topologies

	CT1 [25]	CT2 [26]	CT3 [31]	CT4 [32]	CT5 [34]	CT6 [35]	CT7 [39]	CT8 [38]
Isolation	Yes	Yes	No	No	Yes	Yes	No	No
Renewable energy sources	No	No	No	No	Yes	No	No	No
Dependence on energy storage devices	High	High	High	High	High	Medium	Low	Low
Number of devices	Less	Less	More	More	More	More	More	Less
Fault tolerance capability	Not Achieved	Not Achieved	Not Achieved	Achieved	Achieved	Not Achieved	Achieved	Achieved
Inverter size & Cost	Less	Less	Less	More	More	More	Less	Less
Dependence on charging stations	More	More	More	More	Low	Low	Low	Low
Self-charging capability	Not Provided	Not Provided	Not Provided	Not Provided	Not Provided	Not Provided	Provided	Not Provided
Conduction loss & Stress on power switches	Min	Min	Max	Max	Max	Max	Max	Min
Efficiency & overall performance	High	Moderate	High	Moderate	High	Less	High	High

CT: Converter Topology

achieve better response in all aspects except self-charging capability. Self-charging capability can be achieved by inserting PV module on the top of the vehicle. The comparison of various multiport converter topologies is given in Table 2.

5 Conclusion

In this research paper, a comprehensive survey on multiport converters for EVs and HEVs are presented. The first section contains an introduction about EVs, classification of different types of EV architectures and the importance of those vehicles towards global warming problem. Second section is focused on the review of various energy storage devices used in EVs, HEVs. In third section the different electric motors used in EVs for electric propulsion to achieve their significant performance is mentioned. Various multiport converters for EVs proposed by various research papers with different techniques are analysed in Sect. 4. Comparisons are made between various topologies which are addressed in literature and identified the best suitable converter for EVs and HEVs with good performance. Finally concludes the paper in Sect. 5.

References

1. Chan, C.C.: The state of the art of electric, hybrid, and fuel cell vehicles. Proc. IEEE **95**(4), 704–718 (2007)
2. Rajashekara, K.: History of electric vehicles in general motors. In: Annual Meeting, pp. 447–454. Industry Applications Society (1993)
3. Eberle, U., Helmolt, R.V.: Sustainable transportation based on electric vehicle concepts. Energy Environ. Sci. **3**, 689–699 (2010)
4. Lulhe, A.M., Oate, T.N.: A technology review paper for drives used in electrical vehicle (EV) & hybrid electrical vehicles (HEV). In: International Conference on Control, Instrumentation, Communication and Computational Technologies (2015)
5. Wang, S., Zhou, D., Cheng, H.: The optimized design of power conversion circuit and drive circuit of switched reluctance drive. In: IEEE International Conference on Control & Automation (ICCA) (2016)
6. Hua, C.C., Fang, Y.H., Lin, C.W.: LLC resonant converter for electric vehicle battery chargers. IET Power Electron. **9**, 2369–2376 (2016)
7. Lee, I.O.: Hybrid PWM-resonant converter for electric vehicle on-board battery chargers. IEEE Trans. Power Electron. **31**, 3639–3649 (2016)
8. Abdulaal, A., Cintuglu, M.H., Asfour, S., Mohammed, O.: Solving the multivariant EV routing problem incorporating V2G and G2V options. IEEE Trans. Transp. Electrif. **3**(1), 238–248 (2016)
9. http://autocaat.org/Technologies/Hybrid_and_Battery_Electric_Vehicles/HEV_Levels/
10. Burke, A.F.: Batteries and ultracapacitors for electric, hybrid, and fuel cell vehicles. Proc. IEEE **95**(4), 806–820 (2007)
11. Wu, X., Cao, B., Li, X., Xu, J., Ren, X.: Component sizing optimization of plug-in hybrid electric vehicles. Appl. Energy **88**, 799–804 (2011)
12. Zhang, X., Wang, J., Yang, J., Cai, Z., He, Q., Hou, Y.: Prospects of new energy vehicles for China market. In: Proceedings of Hybrid and Eco-Friendly Vehicle Conference, pp. 1–8 (2008)
13. Gulhane, V., Tarambale, M.R., Nerkar, Y.P.: A scope for the research and development activities on electric vehicle technology in Pune City. In: 2006 Proceedings of IEEE Conference on Electric and Hybrid Vehicles, pp. 1–8 (2006)
14. http://www.greencarcongress.com/2011/09/toyota-introduces-2012-prius-plug-in-hybrid.html
15. https://media.gm.com/content/dam/Media/microsites/product/Volt_2016/doc/VOLT_BATTERY.pdf
16. Rajasekhar, M.V., Gorre, P.: High voltage battery pack design for hybrid electric vehicles. In: 2015 IEEE International Transportation Electrification Conference (ITEC), pp. 1–17 (2015)
17. Marano, V., Onori, S., Guezennec, Y., Rizzoni, G., Madella, N.: Lithium-ion batteries life estimation for plug-in hybrid electric vehicles. In: 2009 IEEE Vehicle Power and Propulsion Conference, pp. 536–543 (2009)
18. http://batteryuniversity.com/learn/archive/whats_the_best_battery
19. Aditya, J.P., Ferdowsi, M.: Comparison of NiMH and Li-ion batteries in automotive applications. In: 2008 IEEE Vehicle Power and Propulsion Conference, pp. 1–6 (2008)
20. Layte, H.L., Zerbel, D.W.: Battery cell control and protection circuits. In: 1972 IEEE Power Processing and Electronics Specialists Conference, pp. 106–110 (1972)
21. Rippel, W.: Induction Versus DC Brushless Motors (2007)

22. Lin, J., Schofield, N., Emadi, A.: External-rotor 6–10 switched reluctance motor for an electric bicycle. IEEE Trans. Transp. Electrif. **1**(4), 348–356 (2015)
23. Bostanci, E., Moallem, M., Parsapour, A., Fahimi, B.: Opportunities and challenges of switched reluctance motor drives for electric propulsion: a comparative study. IEEE Trans. Transp. Electrif. **3**(1), 58–75 (2017)
24. AL-Chlaihawi, S.J.M.: Multiport converter in electrical vehicles-a review. Int. J. Sci. Res. Publ. **6**, 378–382 (2016)
25. Tao, H., Kotsopoulos, A., Duarte, J.L., Hendrix, M.A.M.: Triple-half-bridge bidirectional converter controlled by phase shift and PWM. In: Proceedings of IEEE Applied Power Electronics Conference, pp. 1256–1262, March 2006
26. Zhao, C., Round, S.D., Kolar, J.W.: An isolated three-port bidirectional DC-DC converter with decoupled power flow management. IEEE Trans. Power Electron. **23**(5), 2443–2453 (2008)
27. Lukic, S.M., Emadi, A., Rajashekara, K., Williamson, S.: Topological overview of hybrid electric and fuel cell vehicular power system architectures and configurations. IEEE Trans. Veh. Technol. **54**(3), 763–770 (2005)
28. Emadi, A., Rajashekara, K.: Power electronics and motor drives in electric, hybrid electric, and plug-in hybrid electric vehicles. IEEE Trans. Ind. Electron. **55**(6), 2237–2245 (2008)
29. Amjadi, Z., Williamson, S.S.: Power-electronics-based solutions for plug-in hybrid electric vehicle energy storage and management systems. IEEE Trans. Ind. Electron. **57**(2), 608–616 (2010)
30. Qian, W., Cha, H., Peng, F.Z., Tolbert, L.M.: 55-kW variable 3X DC-DC converter for plug-in hybrid electric vehicles. IEEE Trans. Power Electron. **27**(4), 1668–1678 (2012)
31. Du, Z., Ozpineci, B., Tolbert, L.M., Chiasson, J.N.: DC–AC cascaded H-Bridge multilevel boost inverter with no inductors for electric/hybrid electric vehicle applications. IEEE Trans. Ind. Appl. **45**(3), 963–970 (2009)
32. Lee, Y.J., Khaligh, A., Emadi, A.: Advanced integrated bidirectional AC/DC and DC/DC converter for plug-in hybrid electric vehicles. IEEE Trans. Veh. Technol. **58**(8), 3970–3980 (2009)
33. Lee, I.O., Moon, G.W.: Half-bridge integrated ZVS full-bridge converter with reduced conduction loss for electric vehicle battery chargers. IEEE Trans. Ind. Electron. **61**(8), 3978–3988 (2014)
34. Jiang, W., Fahimi, B.: Multiport power electronic interface—concept, modeling, and design. IEEE Trans. Power Electron. **26**(7), 1890–1900 (2010)
35. Waltrich, G., Duarte, J.L., Hendrix, M.A.: Multiport converter for fast charging of electrical vehicle battery. IEEE Trans. Ind. Appl. **48**(6), 2129–2139 (2012)
36. Rezaee, S., Farjah, E.: A DC–DC multiport module for integrating plug-in electric vehicles in a parking lot: topology and operation. IEEE Trans. Power Electron. **29**(11), 5688–5695 (2014)
37. McDonough, M.: Integration of inductively coupled power transfer and hybrid energy storage system: A multiport power electronics interface for battery-powered electric vehicles. IEEE Trans. Power Electron. **30**(11), 6423–6433 (2015)
38. Yi, F., Cai, W.: Modeling, control, and seamless transition of the bidirectional battery-driven switched reluctance motor/generator drive based on integrated multiport power converter for electric vehicle applications. IEEE Trans. Power Electron. **31**(10), 7099–7111 (2015)
39. Gan, C., Jin, N., Sun, Q., Kong, W., Hu, Y., Tolbert, L.M.: Multiport bidirectional SRM drives for solar-assisted hybrid electric bus powertrain with flexible driving and self-charging functions. IEEE Trans. Power Electron. **33**(10), 8231–8245 (2018)

Study of Various Algorithms on PAPR Reduction in OFDM System

R. Raja Kumar[1(✉)], R. Pandian[1], and P. Indumathi[2]

[1] Sathyabama Institute of Science and Technology (SIST), Chennai, India
rrkmird@gmail.com, rpandianme@rediffmail.com
[2] MIT Campus, Anna University, Chennai, India
indu@mitindia.edu

Abstract. Artificial Bee Colony algorithm (ABC) invented in 2005 by Karaboga, mimics the honey bees. In this paper, we use ABC algorithm to reduce one of the deadly problems of OFDM, namely, peak to average power ratio (PAPR). Orthogonal frequency division multiplexing (OFDM) is a type of Multicarrier modulation that reduces fading due to frequency-selective fading channels. However, OFDM has few setbacks. A major demerit of OFDM system is that it has higher PAPR. The high PAPR causes inter-modulation and out-of-band radiation due to power amplifier nonlinearity. Therefore, it is important to decrease PAPR of an OFDM signal. Many methods such as Partial Transmit Sequence (PTS) and Selective Level Mapping (SLM) have been proposed. A new method called artificial bee colony combined with Partial Transmit Sequence (ABC-PTS) algorithm that is suboptimal is proposed in our work to search for right mix of phase factors.

Keywords: OFDM block · PAPR · SLM · PTS · PTS-ABC

1 Introduction

Orthogonal frequency division multiplexing (OFDM) is the foundation of numerous telecommunication fields. OFDM is a "Multi-Carrier Modulation Scheme" which provides an excellent solution for high speed digital communications. In this technology the data which is required to be transferred can run across a huge quantity of orthogonal carriers and the data is being regulated at a low rate. By selecting appropriate frequency space among them, the carriers can be brought into orthogonal position. Hence, OFDM is considered as an advanced & excellent modulation technique. OFDM is excellent in handling high speed data transmission, multipath propagation issues, and bandwidth efficiency. These positive attributes of OFDM makes it suitable for high rate data transmission compared to other data transmission methodologies. Flexibility, resistance, robustness, scalability, efficient spectrum usage, easy recovery of lost symbols are the main advantages of OFDM. The main setbacks of OFDM are large Peak to Average Power Ratio (PAPR) and high sensitivity to carrier synchronization. In order to decrease PAPR, various methodologies such as Selective Level Mapping, Partial Transmit Sequence, PTS-ABC [1–13] are used.

D. J. Hemanth et al. (Eds.): COMET 2019, LNDECT 35, pp. 86–93, 2020.
https://doi.org/10.1007/978-3-030-32150-5_8

2 Spectral Efficiency of OFDM

OFDM is a notable variant of Frequency Division Multiplexing (FDM). The transmission rate is shared among the subcarriers, on account of FDM and no affiliation exists among the subcarriers. For instance, as appeared in Fig. 1, think about that the assigned data rate must be shared among 5 bearers (state a, b, c, d, e). There is no relationship among the bearers; a, b, c, d and e and we can exchange anything inside the given transmission rate. In the event that the subcarriers are harmonics, say (b = 2a, c = 3a, d = 4a, e = 5a, multiples of a) at that point they are unrelated. This is a special case of FDM, namely OFDM. Figure 2 shows a typical OFDM transmitter.

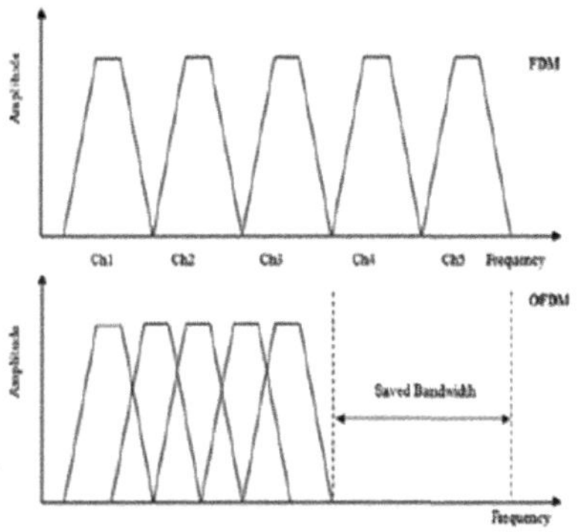

Fig. 1. Comparing OFDM with FDM

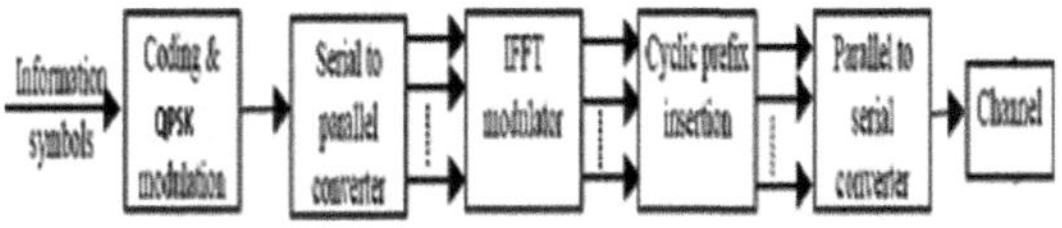

Fig. 2. OFDM Transmitter block diagram

3 PAPR Problem

The large Peak to Average Power Ratio (PAPR) is a significant issue and is the main setback of OFDM systems. The IFFT has uniform power spectrum as the input symbol and a non-uniform power spectrum as the output stream. Instead of assigning higher quantity of transmission energy to major subcarriers, large quantity of energy is assigned to the minority ones. This issue can be quantified as the PAPR measure, which in turn leads to other issues in the OFDM system.

The PAPR is the ratio of the highest power of a sample in the OFDM transmit symbol to the mean power of that OFDM symbol. In a multicarrier system, PAPR will happen when the diverse sub-carriers have stage variations or out of phase among themselves. At each moment the sub-carriers are differ from one another orthogonally, since the phase values are distinct. While each and every point of the constituent sub-carriers can obtain the topmost value at the same time, the immediate rise up makes the 'peak' output envelope. In an OFDM system the existence of huge quantity of the modulated subcarriers leads to the peak value of the system in comparison with the average value of an entire system. This ratio is widely known as Peak-to-Average Power Ratio. An OFDM signal has a number of self-sustained modulated sub-carriers, which results in a high PAPR. While N number of signals are being adjoined with the same phase, the average or mean power of the signal is N times when the peak power is generated. Hence OFDM signal has a very high PAPR, which is very sensitive to nonlinearity of the high-power amplifier.

4 Effect of PAPR

The main setback of the OFDM signal is the very large Peak to Average Power Ratio (PAPR). Hence, a huge linear-region is required for the functioning of RF power amplifiers. Else it results in the signal distortion if the highest points of the OFDM signal acquired the non-linear region. The inter-modulation and out of band radiation between the number of subcarriers occurs due to the signal distortion and it needs a large power back-offs can be functioning with power amplifiers. Also, this results in the exorbitant transmitters and an incompetent amplification on the another end which is highly favorable to decrease the PAPR. It is highly recommended to reduce the PAPR, since the tremendous peak leads to the saturation point in the amplifiers and generates the inter-modulation effect between the subcarriers.

5 Signal Scrambling Techniques

There are various signal scrambling methodologies available to scramble the OFDM signal, the basic concept of these techniques is to choose the one which produces the least PAPR values for transmission. These methodologies cannot reduce the PAPR values below the particular threshold, but it can decrease the PAPR value to the maximum extreme. The various types of approaches in scrambling techniques are Selective Mapping (SLM) and Partial Transmit Sequences (PTS).

5.1 Selective Level Mapping (SLM)

In order to obtain the substitute sequences for the input symbol, the input data sequences in the SLM are being multiplied by every phase. In the IFFT operation, there are various input data sequences available and the sequence with least PAPR value can be selected for data transmission.

The SLM technique first generates the statistically independent input data sequences, and the resultant M statistically independent data blocks for m = 1,2,…M are being progressive into the IFFT operation at the same time. The discrete time-domain sequences are obtained using the SLM technique and then the PAPR values of these M vectors are computed individually. Finally, the data sequences Xd with the minimum PAPR value is chosen for data transmission (Fig. 3).

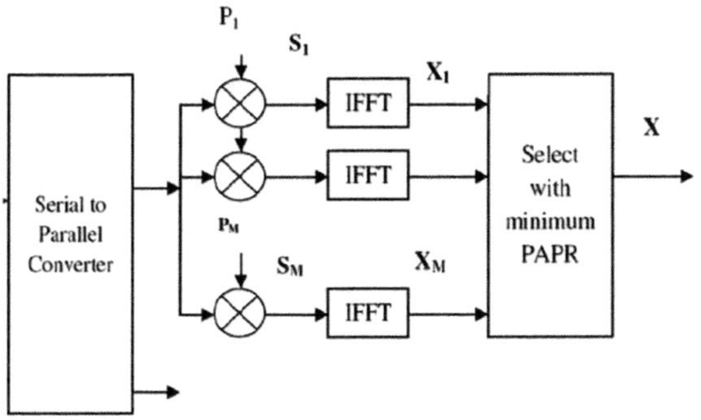

Fig. 3. Block diagram of SLM

5.2 Partial Transmit Sequence

The main methodology to reduce the PAPR value is the Partial Transmit Sequence. This technique partitions the data block into non-overlapping sub data blocks. In partial transmit sequence, each sub-blocks have the independent rotation factor. The main concept of this methodology is to partition the actual OFDM input data symbol into sub-data which can be transferred via the sub-blocks and then multiplied by the weighing value with distinct phase rotation factor till the best value with least PAPR will be selected (Fig. 4).

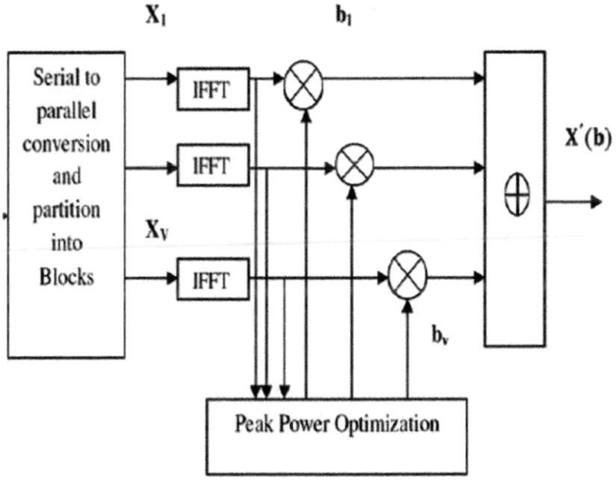

Fig. 4. Block diagram of PTS

In the frequency domain the data sequence X is partitioned into V sub-sequences which were being transferred through the sub-blocks without any overlapping and also of equal size N which have N/V non-zero values in every sub-block. The signal which having the best and least PAPR value is nominated and then transmitted. The main setback of this method is the big complexity in computation. Since the number of subcarriers increases, the complexity in computation also increases. Owing to this issue, the complexity to obtain the optimal phase factor increases, and the PAPR value decreases. The input data sequence with the least PAPR value is selected for the data transmission and the likelihood of incidence of large PAPR value is also decreased.

5.3 Partial Transmit Sequence - Artificial Bee Colony

The ABC is studied here and this algorithm mimics honey bees. In the ABC algorithm, the location and quality (nectar) of a food source relate to optimization solution. The ABC algorithm involves employed bees, onlooker bees and scout bees. To start with, a beginning populace of honey bees got going. An employed bee changes the position (solution) in its memory, by visually inspecting, and checking the nectar (fitness value) content of the new source. On the off chance the new nectar sum is higher than that of the past source, the honey bee retains the new position and overlooks the past position. Else, she keeps the situation of the past source in her memory. After every utilized honey bee completes the pursuit procedure, they share the nectar data of the nourishment sources and their position data with the spectator honey bees.

An onlooker bee determines the nectar information obtained from the employed bees and selects the food source with a probability related to its nectar amount. On account of the utilized honey bee, the passerby honey bee creates a modification of the situation in her memory and checks the nectar measure of the potential source. Given that its nectar is higher than that of the past source, the honey bee retains the current position and overlooks the past one. After the passerby honey bees complete their ventures, scouts are resolved. The utilized honey bee of a drained source turns into a scout and begins to search arbitrarily for another sustenance source. These means are rehashed through a fixed number of cycles, called greatest number of cycles, or until an end rule is fulfilled.

6 PTS Based ABC

The algorithm involves the following procedure:

- Initialize phase vector b_i and arrive at the maximum cycle number.
- Compute the fitness function for each phase vector.
- Repeat.
- Let Employed bees look out for new food sources, and calculate the fitness function of each new source.
- Let Onlooker bees choose food sources.
- Onlooker bees look searches for new food sources and evaluate the fitness of each b'_i
- If a Limit Value is reached, go to step 6. Otherwise, continue
- Assign scout bees to randomly find new phase vector
- Memorize the best food source
- Until ¼ of the maximum cycle number is reached.

7 Simulation Results and Discussions

The complementary cumulative distribution function (CCDF) of the PAPR is the preferred performance measure for PAPR reduction techniques. CCDF is the probability how much the instantaneous PAPR exceeds the Average PAPR. Plots for PTS-ABC-1 and PTS-ABC-2 correspond to the number of sub-blocks 2 and 4 respectively. We understand as number of sub-blocks increases, PAPR reduction improves (Figs. 5, 6, 7 and 8).

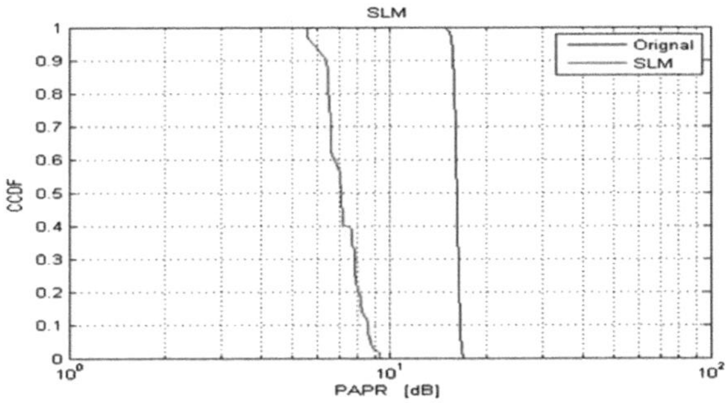

Fig. 5. SLM with 64 sub-carriers

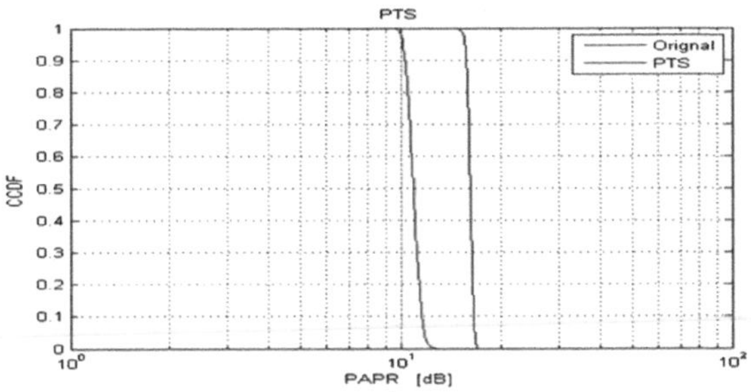

Fig. 6. PTS with 64 sub-carriers

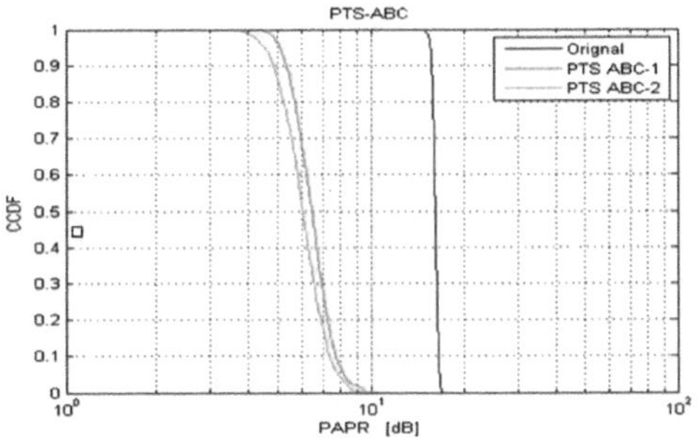

Fig. 7. PTS-ABC with 64 sub-carriers

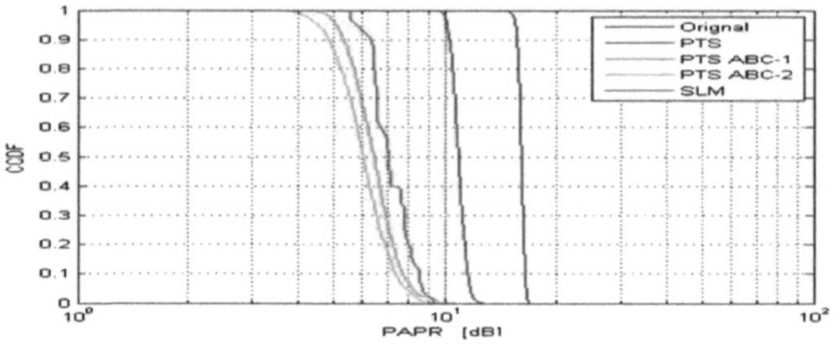

Fig. 8. Comparison plot of different algorithms for PAPR reduction

Also, as the number of sub-carriers increase, there is an increase in the PAPR. SLM, PTS and PTS-ABC algorithms show appreciable reduction in PAPR which can be understood from the Table 1 given below.

Table 1. PAPR for different algorithms

Number of subcarriers	Original PAPR in dB	PAPR of PTS in dB	PAPR of SLM in dB	PAPR of PTS-ABC in dB
8	7.1162	3.5468	4.4788	3.6571
16	10.1235	5.8058	5.4013	4.4735
32	13.1229	8.3082	6.3704	5.2959
64	16.1327	10.986	7.2599	4.929
128	19.1521	13.728	8.0569	6.7485
256	22.1654	16.558	8.6381	6.5143
512	25.1783	19.443	9.224	7.812
1024	28.1845	22.369	9.7697	8.3497

8 Conclusion

Thus, here PAPR reduction techniques (SLM, PTS, PTS-ABC) in OFDM system are analyzed and found that the PTS-ABC has shown better performance with less computational complexity as compared to other techniques. SLM and PTS are critical probabilistic plans for PAPR decrease, SLM can create autonomous different recurrence space OFDM signals, while the option OFDM signals produced by PTS are free. PTS isolates the recurrence vector into some sub-obstructs before applying the stage change. In this way, a portion of the unpredictability of a few full IFFT tasks can be kept away from in PTS, so it is more profitable than SLM if measure of computational intricacy is restricted. PTS strategy is uncommon instance of SLM technique. For PTS strategy, the quantity of turn variables might be restricted in certain range. A suboptimal phase optimization scheme based on artificial bee colony (ABC-PTS) algorithm, has shown efficient PAPR reduction in OFDM system with less complexity when compared with others.

References

1. Baxley, R.J., Zhou, G.T.: Comparing selected mapping and partial transmit sequence for PAPR reduction. IEEE Trans. Broadcast. **53**(4), 797–803 (2007)
2. Han, S.H., Lee, J.H.: An overview of peak-to-average power ratio reduction techniques for multicarrier transmission. IEEE Wirel. Commun. **12**(2), 56–65 (2005)
3. Heo, S.J., Noh, H.S., No, J.S. Shin, D.J.: A modified SLM scheme with low complexity for PAPR reduction of OFDM systems. In: IEEE 18th International Symposium on Personal, Indoor and Mobile Radio Communications, pp. 1–5 (2007)
4. Jiang, T., Wu, Y.: An overview: peak-to-average power ratio reduction techniques for OFDM signals. IEEE Trans. Broadcast. **54**(2), 257–268 (2008)
5. Le Goff, S.Y., Khoo, B.K., Tsimenidis, C.C., Sharif, B.S.: A novel selected mapping technique for PAPR reduction in OFDM systems. IEEE Trans. Commun. **56**(11), 1775–1779 (2008)
6. Li, X., Cimini, L.J.: Effects of clipping and filtering on the performance of OFDM. IEEE Commun. Lett. **2**(5), 1634–1638 (1998)
7. Lim, D.W., Heo, S.J., No, J.S., Chung, H.A.: New PTS OFDM scheme with low complexity for PAPR reduction. IEEE Trans. Broadcast. **52**(1), 77–82 (2006)
8. Muller, S.H., Huber, J.B.: OFDM with reduced peak-to-average power ratio by optimum combination of partial transmit sequences. Electron. Lett. **33**(5), 368–369 (1997)
9. Ochiai, H., Imai, H.: On the distribution of the peak-to-average power ratio in OFDM signals. IEEE Trans. Commun. **49**(2), 282–289 (2001)
10. Wang, Y., Chen, W., Tellambura, C.: A PAPR reduction method based on artificial bee colony algorithm for OFDM signals. IEEE Trans. Wirel. Commun. **9**(10), 2994–2999 (2010)
11. Yang, L., Soo, K.K., Siu, Y.M., Li, S.Q.: A low complexity selected mapping scheme by use of time domain sequence superposition technique for PAPR reduction in OFDM system. IEEE Trans. Broadcast. **54**(4), 821–824 (2008)
12. Yang, L., Soo, K.K., Li, S.Q., Siu, Y.M.: PAPR reduction using low complexity PTS to construct of OFDM signals without side information. IEEE Trans. Broadcast. **57**(2), 284–290 (2011)
13. Zhou, G.T., Peng, L.: Optimality condition for selected mapping in OFDM. IEEE Trans. Signal Process. **54**(8), 3159–3165 (2006)

Corrosion Studies on Induction Furnace Steel Slag Reinforced Aluminium A356 Composite

K. S. Sridhar Raja and V. K. Bupesh Raja(✉)

School of Mechanical Engineering,
Sathyabama Institute of Science and Technology, Chennai, India
sridhar2raja@gmail.com, bupeshvk@gmail.com

Abstract. In this study Aluminium A356 alloy is reinforced with induction furnace steel slag for improving the mechanical and corrosion properties inorder to meet the functional requirements in automotive brake shoe. The Metal Matrix Composite (MMC) was prepared by using stir casting technique. The metal matrix composite was subjected to salt spray test to study the corrosion behavior in marine environment. The minimal weight loss due to corrosion indicated that the addition of steel slag reinforcement improved the corrosion rate of the metal matrix composite. The samples subjected to corrosion were characterized using optical microscope and SEM. It was observed that the addition of the steel slag not only improved the corrosion resistance, it also improved the hardness of the material, which is desirable for brake shoe application.

Keywords: A356 · Steel slag · Salt spray test · SEM · Corrosion · Brake shoe · Automobile

1 Introduction

Aluminium matrixes reinforced with hard ceramic particles are most widely used in marine, aerospace, and automotive application. Due to its weight to high strength ratio, low density, and good wear behavior the metal matrix composite is replaced in conventional alloys [1]. The ceramic particle has greater significance on the mechanical properties of composites like tensile strength, corrosion resistance, and plastic deformation. Cast Aluminium alloy matrix like A356, have been widely prepared with many ceramic particles like SiC, TiB_2, Basalt, fly ash [2–5] as the reinforcing material. Various factors selected such as reinforcement percentage, microstructure of the matrix aluminium, particle size and distribution etc. are entrenched by the researchers. The corrosion behavior of the composite material gets changed even when there is a small change in any one of the above factors [6–10]. The metal matrix composite by virtue of the reinforcement particles used has direct influence on the nature of the protective oxide film layer formed and the corrosion resistance of the material. Similarly, the reinforcing particles have an influence on the formation of discontinuities and defects like porosity, cracks, etc. in the passive oxide layer and thus triggering the corrosion attack [10, 11].

D. J. Hemanth et al. (Eds.): COMET 2019, LNDECT 35, pp. 94–100, 2020.
https://doi.org/10.1007/978-3-030-32150-5_9

It was observed from the above literature that many researchers had thoroughly studied the corrosion resistance imparted by reinforcing different ceramic particles. The main objective of this work was to analyze the effect of reinforcing steel slag particle with different weight percentages (3%, 6%, 9% and 12%). The corrosion behavior study of the aluminium composite was carried out by exposing the specimens in freely aerated 5% NaCl fogin order to observe the behavior of the material in marine condition. The characterization of the corroded specimens was done with optical microstructure, SEM examinations.

2 Materials and Methods

The material used in this study was A356 and its chemical composition is shown in Table 1. The steel slag that was obtained during casting of steel in induction furnace was crushed to small size and it was pulverized in the ball mill to reach the particle size of 1–10 μm. The chemical composition of the steel slag is shown in Table 2. The steel slag was added along with the aluminum matrix with different weight proportion (viz. 3%, 6%, 9% and 12%) before casting.

Table 1. Chemical composition of A356 alloy [12]

Si	Fe	Cu	Mn	Mg	Ni	Zn	Ti	Al
6.58	0.16	0.06	0.06	0.57	0.01	0.01	0.14	Bal

Table 2. Chemical composition of steel slag.

Fe	Al	Mg	Si	Mn	Ca	Cr	Ti	Others
30.0	8.37	0.32	47.44	9.66	2.45	0.58	0.63	Bal

3 Experimental Methodology

The composite was prepared by liquid stir casting technique. The A356 alloy was melted in the electrical muffle furnace at 650 °C. The pulverized steel slag particle along with potassium Titanate (K_2TiF_6) was preheated at about 350 °C to remove moisture from the particles. After the aluminium alloy gets molten, the preheated steel slag along with K_2TiF_6 [13] was added to the molten metal. With the help of a mechanical stirrer the molten metal was stirred during the addition of particles and then it was poured into a permanent metal mold.

3.1 Salt Spray Technique

According to ASTM B117 M standards, the composite material was cut into test coupons of size 20 × 20 × 6 mm, rinsed with distilled water and cleaned with Acetone to remove grease and oil, and polished by using 1200 grit size emery sheets. The salt spray test was carried out by mixing 5% of NaCl solution with 95% of distilled water and the test was done in the room temperature. The specimens were tied in a nylon wire and immersed in the chamber [12]. The NaCl solution was then sprayed on the specimen in the form of fog inside the test chamber continuously for 48 h.

By using De-Wintor Inverted Trinocular metallurgical microscope the microstructure of the corroded specimens was studied. The Carl Zeiss-Supra 55 Field-Emission Scanning Electron Microscope (FESEM) was used to investigate the corroded surface morphology.

The polished specimens were subjected to Vickers Microhardness test, which was carried out by making three indentations under a load of 300 gm conforming to ASTM E10 standards.

4 Results and Discussions

4.1 MicroHardness

The Microhardness test was carried out on both cast and MMC materials, to study the effect of steel slag particle in the aluminum matrix. According to ASTM E10 standards a minor load and major load of 10 kg and 60 kg were applied with 10 s delay. The average hardness values were measured at four different regions on the composite material and shown in Fig. 1.

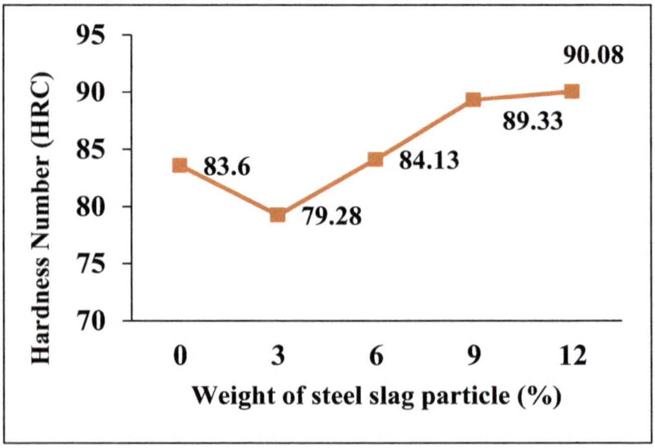

Fig. 1. Hardness of steel slag reinforced composite

From Fig. 1 it has been observed that the hardness of the composite initially decreased at 3% reinforcement. After 3% reinforcement on further increase the hardness increased rapidly.

4.2 Salt Spray Corrosion Test

The salt spray corrosion test was conducted by using 5% NaCl solution in a fog medium. The NaCl reacts with iron and with aluminum metal matrix and forms corrosion products on the surface. The weight loss method was adopted to calculate the corrosion rate, over an interval of 120 h. The iron content in the steel slag react with NaCl solution and forms a white corrosion layer. The Fig. 2 shows the relation between the weight loss and the corrosion rate. It was observed that the corrosion rate increases by the increase in weight loss.

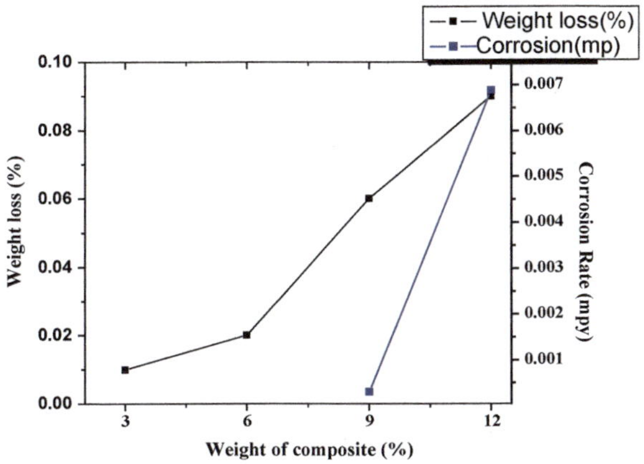

Fig. 2. Corrosion rate and weight loss in aluminium metal matrix composite

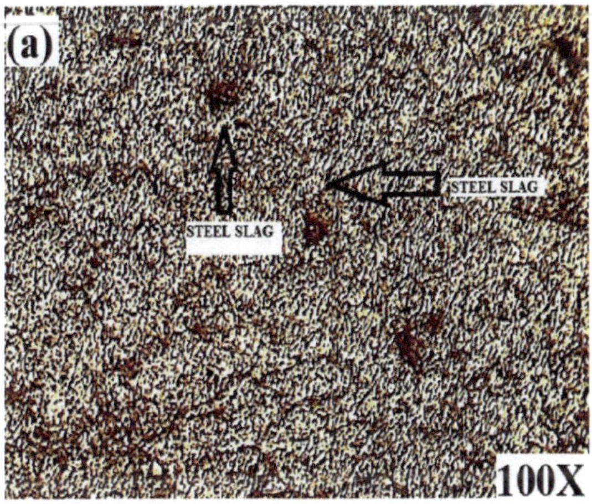

Fig. 3. Microstructure of corroded steel slag composite

The Fig. 3 shows the microstructure of the corroded composite specimen. The microstructure of the corroded surface shows steel slag in dark brown spots indicating the presence of steel slag particles embedded in the aluminium matrix. The grains are acicular in nature having iron in the grain boundaries indicating the distribution of the steel slag with the increase of the percentage of steel slag particles, the risk of agglomeration increases.

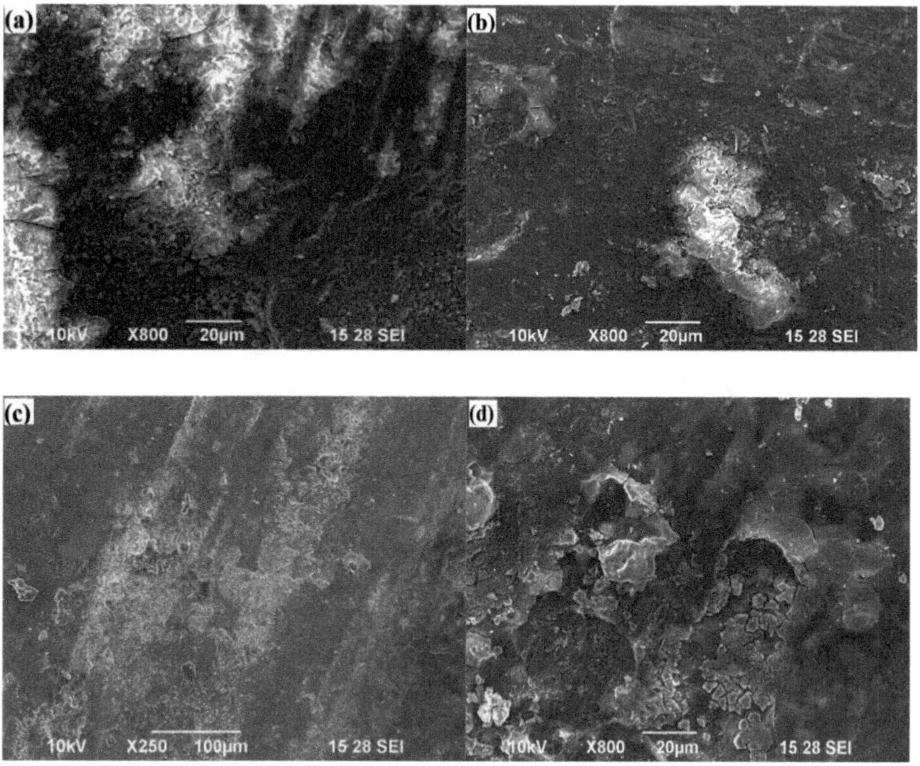

Fig. 4. SEM morphology of the corroded specimens with (a) 3% (b) 6% (c) 9% (d) 12%

The Fig. 4(a) shows the salt spray exposed specimens were subjected to SEM analysis to study the morphology changes. The aluminium A356 metal matrix composite having 3% weight of steel slag showed a uniform layer of oxides, indicating a minimum susceptibility to corrosion or in other words this material shows fairly good immunity to corrosion.

The Fig. 4(b) shows the specimen having 6% of steel slag showed morphology indicating localized pitting associated with oxides formation. The Fig. 4(c) shows the specimen with 9% steel slag matrix showed a pronounced corrosion on the surface with flaking of a corrosion product, and indicating the formation of pores which shall leads to further pitting.

The Fig. 4(d) shows the specimen with 12% steel slag addition shows severe corrosion in terms of pitting, formation of flakey layers of cracked regions of spread out across the surface of the material. Hence the SEM analysis indicates that with the increase of slag content the risk of susceptibility to corrosion is pronounced.

5 Conclusion

The corrosion behavior of Al-Si-Mg alloy (A356) was studied by using salt spray test. The results that were observed shows that the corrosion rate increases as the reinforcement particles increases, due to the presence of iron in the steel slag. The SEM structure also shows the white corrosion surrounding the particle. The surface morphology of the 3% steel slag MMC shows the least corrosion whereas the 12% steel slag MMC shows more corrosion which may be attributed to the agglomeration of steel slag particles. Hence this study shows that the addition of 3% slag with A356 gives a corrosion resistant metal matrix composite for marine applications with no additional expenditure for reinforcing particles.

References

1. Ravikumar, K., Kiran, K., Sreebalaji, V.S.: Characterization of mechanical properties of aluminium/tungsten carbide composites. Measurement **102**, 142–149 (2017)
2. Dwivedi, S.P., Sharma, S., Mishra, R.K.: Microstructure and mechanical properties of A356/SiC composites fabricated by electromagnetic stir casting. Procedia Mater. Sci. **6**, 1524–1532 (2014)
3. Mazahery, A., Shabani, M.O.: Mechanical properties of A356 matrix composites reinforced with nano-SiC particles. Strength Mater. **44**(6), 686–692 (2012)
4. Venkatachalam, G., Kumaravel, A.: Fabrication and characterization of A356-basalt ash-fly ash composites processed by stir casting method. Polym. Polym. Compos. **25**(3), 209–214 (2017)
5. Bhiftime, E.I., Gueterres, N.F.D.S.: Investigation on the mechanical properties of A356 alloy reinforced AlTiB/SiCp composite by semi-solid stir casting method. In: IOP Conference Series: Materials Science and Engineering, vol. 202, p. 012081 (2017). https://doi.org/10.1088/1757-899x/202/1/012081
6. Seah, K.H.W., Sharma, S.C., Girish, B.M.: Corrosion characteristics of ZA-27-graphite particulate composites. Corros. Sci. **39**, 1–7 (1997)
7. Pinto, G.M., Nayak, J., Shetty, A.N.: Corrosion behaviour of 6061 Al-15vol. Pct. SiC composite and its base alloy in a mixture of 1: 1 hydrochloric and sulphuric acid medium. Int. J. Electrochem. Sci. **4**(10), 1452–1468 (2009)
8. Pohlman, S.L.: Corrosion and electrochemical behavior of boron/aluminum composites. Corrosion **34**(5), 156–159 (1978)
9. Aylor, D.M., Moran, P.J.: Effect of reinforcement on the pitting behavior of aluminum-base metal matrix composites. J. Electrochem. Soc. **132**(6), 1277–1281 (1985)
10. Sherif, E.S.M., Almajid, A.A., Latif, F.H., Junaedi, H.: Effects of graphite on the corrosion behavior of aluminum-graphite composite in sodium chloride solutions. Int. J. Electrochem. Sci. **6**, 1085–1099 (2011)

11. Kumari, P.R., Nayak, J., Shetty, A.N.: Corrosion behavior of 6061/Al-15 vol. pct. SiC composite and the base alloy in sodium hydroxide solution. Arab. J. Chem. **9**, 1144–1154 (2016)
12. Raja, K.S., Raja, V.K., Vignesh, K.R., Rao, S.N.: Effect of steel slag on the impact strength of aluminium metal matrix composite. Appl. Mech. Mater. **766–767**, 240–245 (2015)
13. Sridhar Raja, K.S., Bupesh Raja, V.K.: Corrosion behaviour of boron carbide reinforced aluminium metal matrix composite. ARPN J. Eng. Appl. Sci. **10**(2), 10392–10394 (2015)

Performance Appraisal System and Its Effectiveness on Employee's Efficiency in Dairy Product Company

Manjula Pattnaik[1(✉)] and Balachandra Pattanaik[2]

[1] Department of Accounting, Princess Nourah bint Abdulrahman University, Riyadh, Saudi Arabia
drmanjula23@gmail.com
[2] Department of ECE, Mettu University, Mettu, Ethiopia
balapk1971@gmail.com

Abstract. Performance appraisal is a systematic description. Performance appraisal is the most essential need for the company's development, so that's why it is considered as one of the important and essential topic in recent trend of research. It is the basis of employee's development. It is essential to plan, design and implement a useful and suitable performance appraisal system in every company. This study is conducted to analyze the effectiveness of performance appraisal system in Dairy Product Company in Chennai, India. The sampling technique used in this research is convenient sampling. Both primary and secondary data are used in this study. The collected data were analyzed and tabulated in a sequential manner in proper tables and charts. Employee's effective performance is estimated by using various Statistical tools used for analyzing the data are Karl Pearson's correlation, Chi-square test, Spearman's rank correlation coefficient, Mann Whitney test, Cohen's kappa.

Keywords: Performance appraisal · Effectiveness · Mann Whitney test · Cohen's Kappa test

1 Introduction

Performance Appraisal is a continuous process to evaluate the employee's progress, successes and failures. Performance Appraisal to analyze the strengths and weaknesses, and eligibility for their promotion or requirement of further training. Performance Appraisals had four major objectives including the training and development of employees, salary reviews, planning job rotation and assistance promotions [1]. There are traditional as well as modern methods of performance appraisals such as Straight ranking method, Paired comparison analysis, Field Review, Critical incident method, Forced ranking, Management By Objectives, 360° performance appraisal, Behaviorally anchored rating scales, Human resource accounting etc. It was unknown exactly how the performance of the turnover of the firm which was influenced and effected by contextual factors as per context–emergent turnover theory (CETT) [2]. The HRM concept comprise all people under employment emphasized on employee attitude, knowledge, ability, experience [4]. The previous researcher have used regression

D. J. Hemanth et al. (Eds.): COMET 2019, LNDECT 35, pp. 101–108, 2020.
https://doi.org/10.1007/978-3-030-32150-5_10

analysis to relate farms productivity with farms age, experience and efficiency measures [5]. Producer spend maximum time on employees training, supervision providing all new facilities spent less time on frame work, but more productive [3]. The most profitable expansions were highly correlated with modernizing facilities. But adding too many to increase in size of herd bring decline in return on asset and by expansion, the dairy firms were increase milk production and decrease in management and labor cost [6]. Employee skills are developed through acquisition and development of a firms human capital by adopting good Human Resource Management Practices [7]. This research focus on the employees performance of the production site of the dairy product company in Chennai. In Tamil Nadu especially in Chennai milk product industry is dominated by Tamil Nadu Co Operative Milk producers Federation Ltd. Arokya, Tirumala, GRB, Hatsun and Aavin are some of the popular milk product companies in Chennai are considered for this study. Here the analysis of performance appraisal of these employees has been done to give necessary suggestion to improve the productivity of the firms. In India the packaged milk segment is dominated by Gujarat Cooperative Milk Marketing Federation (GCMMF), which is the largest player. All other local dairy cooperatives have their local brands (For e.g. Gokul, Warana in Maharashtra, Saras in Rajasthan, Verka in Punjab, Vijaya in Andhra Pradesh, Aavin in Tamil Nadu, etc.

2 Conceptual Frame Work

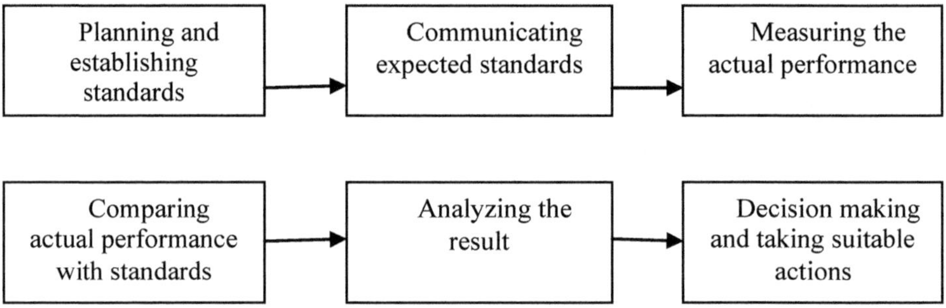

3 Objective of the Study

Primary Objective:
To study and analyze the effectiveness of performance appraisal system on employee's efficiency in Dairy product company in Chennai.

Secondary Objectives:

- To determine the awareness level among the employees about the performance appraisal system adapted in the company.

- To verify the employees opinion towards the existing performance appraisal system.
- To evaluate whether the performance appraisal system helps to improve the employee's performance.
- To give suggestions to improve the performance appraisal system in Dairy product company.

Calculation of Sample size:
Sample size is determined by Slovin's Formula: Sample Size = N/(1 + N*e2)
Where N = population size
e = margin of error
Here, the population is the number of employees working in production site = 250, e = margin of error = 5%
Sample size = 250/[1 + 250*(0.05)2] = 153, but as per the convenience 150 samples has been taken for the research.

4 Research Methodology

In a well-written standard operating procedures (SOPs) provide direction, improve communication, reduce training time, and improve work consistency. The SOP development process is an excellent way for managers, workers, and technical advisers to cooperate for everyone's benefit [10].

Descriptive research design is adopted in this study. Survey method used for primary data collection by distributing questionnaire to employees. Convenient sampling technique is used for this study. Secondary data was collected from magazines, journals and reference materials from internet. There are 150 Samples collected from the total population of 250 employees working on the production site of the dairy products. Analysis of the collected data was done with the help of tables, graphs and statistical analysis tools like Karl Pearson's correlation, Chi-square test, Spearman's rank correlation coefficient, Mann Whitney, Cohen's kappa [9].

5 Literature Review

In 1960s the trend of processing milk into various products initiated in Assam. Apart from Assam Tripura and Manipur were the highest milk producing states and with highest cross bred animals [12]. Performance appraisal is a positive approach towards motivation of employees as well as the management. In a firm Performance Standards should be defined and communicated to the employees for monitoring and comparing actual performance with expected standards throughout the year [11]. In 2015 there is a study has been done to intended and understand the issues, challenges faced by dairy stakeholders in Indian Dairy Industry by Professor Sanjay and Bhagyasree which too focus on performance appraisal. However level of education, position in firm, gender and year of experience making a significant difference in turnover intention [8].

6 Statistical Analysis

6.1 Analysis Using Karl Pearson's Correlation

To determine the significant difference between the benefits and efficiency in performance appraisal system.

Null hypothesis (Ho): There is no significant difference between benefits and efficiency in performance appraisal system.

Alternate hypothesis (H1): There is a significant difference between benefits and efficiency in performance appraisal system.

Observed data (Table 1):

Table 1. Observed data: benefits and efficiency in performance appraisal system

Particulars	Benefits (X)	Efficiency (Y)
Strongly agree	10	10
Agree	95	95
Neutral	45	40
Disagree	0	5
Strongly disagree	0	0

Calculation (Table 2):

Table 2. Calculation: benefits and efficiency in performance appraisal system

X	X^2	Y	Y^2	XY
10	100	10	100	100
95	9025	95	9025	9025
45	2025	40	1600	1800
0	0	5	25	0
0	0	0	0	0

$$r = \frac{N\sum XY - \sum X \sum Y}{\sqrt{N\sum X^2 - (\sum X)^2}\sqrt{N\sum Y^2 - (\sum Y)^2}}$$

$$= 32125/32650$$

$$r = 0.983$$

Since r is positive, there is no significant difference between benefits and efficiency of employees through Performance Appraisal system.

6.2 Analysis Using Mann Whitney

To determine the relationship between improvement in employees through performance appraisal system and meeting out the business goals and objectives.

Null Hypothesis (Ho): There is a positive relationship between improvement in employees through performance appraisal system and meeting out the business goals and objectives.

Alternative Hypothesis (H_1): There is no positive relationship between improvement in employees through performance appraisal system and meeting out the business goals and objectives (Table 3).

Table 3. Improvement in employees through performance appraisal system.

Particulars	Improvement through PAS	Meeting out the business goal & objective
Strongly agree	25	10
Agree	95	100
Neutral	25	30
Disagree	5	10
Strongly disagree	0	0
Total	150	150

Calculation (Table 4):

Table 4. Calculation of ranks

Order	Rank
25	6.5
95	9
25	6.5
5	3
0	1.5
10	4.5
100	10
30	8
10	4.5
0	1.5

Formulae:

$$\mathrm{U} = n1n2 + \frac{n1(n1+1)}{2} - R1$$

$$\text{Mean of U} = \frac{n1n2}{2}$$

$$\text{Variance of U} = \frac{(n1n2)(n1 + n2 + 1)}{12}$$

$$Z = \frac{U - E(U)}{\sqrt{V(U)}}$$
$$= 2.92$$
$$Z\alpha(\text{tab}) = 7.815$$
$$Z\alpha(\text{tab}) > Z\ \text{cal}$$

Null hypothesis (H_0) is accepted as the table value is more than the result obtained. This indicates that business goals and objectives are aligning with the continuous improvement of employee's performance.

6.3 Analysis Using Cohen's Kappa

To determine the relationship between interaction among appraiser - appraise and satisfaction on current performance of the system.

Null hypothesis (Ho): There is positive relationship between interaction among appraiser - appraise and satisfaction on current performance of the system.

Alternate hypothesis (H1): There is negative relationship between interaction among appraiser - appraise and satisfaction on current performance of the system.

$$k = \frac{\Pr(a) - \Pr(e)}{1 - \Pr(e)}$$

Calculation of k value (Tables 5, 6 and 7):

Table 5. Calculation of k value

Particular	Yes	No	Total
Yes	60	40	100
No	50	0	50
Total	110	40	150

Table 6. Calculation of k value in percentage

Particular	Yes	No	Total
Yes	.40	.26	.66
No	.33	0	.33
Total	.77	.26	.99

Table 7. Summary of Calculation of k value

Particular	Yes	No
Yes	.4818	–
No	–	.0858

Expected Frequency = 0.5672

$$k = \frac{Pr(a) - Pr(e)}{1 - Pr(e)}$$

$$k = \frac{0.60 - 0.5672}{1 - 0.5672}$$
$$= 0.0757$$

Null hypotheses is accepted, as the result of k is positive, which means its good for the firm to keep continuous interaction between the appraiser & appraise in order to satisfy and meet expected performance standards.

7 Findings

Employees agreed that the performance appraisal is followed periodically in their organization. Employees felt qualification, experience, job knowledge and service are the basis of their grading system and they understood its importance. But only few employees aware about the performance appraisal system in the beginning of their service. Most of the employees believe that this appraisal helps to improve their potential and agree that appraisal system provides opportunity for each employee to express their developmental needs. Overall employees are recognised and satisfied with their performance appraisal system. Collection of feedback from performance appraisal plays very important role in further improvement in their performance.

8 Suggestions

The firm's management should be more interactive and discuss the employee's performance in order to aware about their challenges and motivate them for best performance in future. The managements in the organization should be aware of every individual's inner emotions so that they can have good relationship with their employees [8]. In order to make it an ongoing process the time period for conducting the appraisal should be revised. Very selective persons should be appointed as performance appraisal panel that can do it in neutral and avoid subjectivity, as there is a positive relationship between the appraise – appraiser. Seniority has to be considered for promotional activities so that the employees would not have any dispute among them as employees benefited by performance appraisal. The employee should be well informed about his duties, obligations, role in job expected by the employer.

9 Conclusions

From all above analysis, it is clear that there is a very good appraisal system was followed by the firms. There is strong link between interaction among appraiser - appraise and satisfaction on current performance of the system. It was observed that performance appraisal help the firm to decide about employee's promotion or transfer as well as their salary determination. So importance and satisfaction are two different aspects of Performance Appraisal system. The performance appraisal of the employees in Dairy milk products has been conducted up to the satisfactory level of the employees and as well as aware of the system. The employees are made to perform fair and equitable compensation based on their performance in their work.

References

1. Stup, R.E., Hyde, J., Holden, L.A.: Relationships between selected human resource management practices and dairy farm performance. J. Dairy Sci. **89**(3), 1116–1120 (2006)
2. Brymer, R.A., Sirmon, D.G.: Pre-exit bundling, turnover of professionals, and firm performance. J. Manag. Stud. **55**(1), 146–173 (2018)
3. Bewley, J., Palmer, R.W., Jackson-Smith, D.B.: An overview of experiences of Wisconsin dairy farmers who modernized their operations. J. Dairy Sci. **84**, 717–729 (2001)
4. Hazlauskatte, R., Buciuniene, I.: The role of human resource and their management in the establishment of sustainable competitive advantage. Eng. Econ. **5**(60), 78–84 (2008)
5. Ford, S.A., Shonkwiler, J.S.: The effect of managerial ability on farm financial success. Agric. Resource Econ. Rev. **23**, 150–157 (1994)
6. Hadley, G.L., Harsh, S.B., Wolf, C.A.: Managerial and financial implications of major dairy farm expansions in Michigan and Wisconsin. J. Dairy Sci. **85**, 2053–2064 (2002)
7. Huselid, M.A.: The impact of human resource management practices on turnover, productivity, and corporate financial performance. Academy of management journal **38**, 635–672 (1995)
8. Xu, Y., Jiang, J.: Empirical research on relationship of caddies' reward satisfaction, organizational commitment and turnover intention. Chin. Stud. **4**(02), 56 (2015)
9. SAS institute, SAS/STAT version 9.1, SAS Inc. Cary, NC (2005)
10. Stup, R.E.: Standard operating procedure: a writing guide. Penn State University Cooperative Extension, University Park
11. Banerjee, S.: Performance appraisal practice and its effect on employees motivation, A study on agro based organization. IJMS **V**(3), 4 (2018)
12. Paula, D., Chanel, B.S.: Improving milk yield performance of crossbred cattle in North-eastern state of India. Agric. Econ. Res. Rev. **23**, 69–75 (2010)

N-2 Contingency Screening and Ranking of an IEEE Test System

Poonam Upadhyay(✉) and B. Vamshi Ram

Department of EEE, VNRVJIET, Hyderabad 500090, Telangana, India
upadhyay_p@vnrvjiet.in

Abstract. With the increasing load demand and the increase of interconnections to meet the load demand, the system has to be operated not only economically but also securely. Nowadays in deregulated market the transmission lines are stressed heavily because of load demand and to operate economically. N-2 contingencies are important enough to study for on line security assessment. In this paper, we had focused on contingency selection of N-2 transmission line contingencies by using contingency screening and ranking methods, so we can differentiate between critical and non critical contingencies and also measure the severity of the contingency.

Keywords: Deregulated market · N-2 contingency · Critical contingencies

1 Introduction

This instruction file for Word users (there is a separate instruction file for LaTeX users) may be used as a template. Kindly send the final and checked Word and PDF files of your paper to the Contact Volume Editor. This is usually one of the organizers of the conference. You should make sure that the Word and the PDF files are identical and correct and that only one version of your paper is sent. It is not possible to update files at a later stage. Please note that we do not need the printed paper. The ability of the system to operate in normal state during an event i.e. contingency is called as power system security. Power system security is important both during the planning and operational phase of the system. The power system network is classified into five operating states namely Normal, Alert, Emergency, Extremis and Restorative states based on, whether the Equality 'E' and Inequality constraints 'I' and also the security constraints of the system are satisfied which is represented in the below table [8, 9] (Table 1).

Power system security can be classified into static steady state system security, transient stability system security and dynamic stability system security. In this paper static steady state system security is considered and we will consider line flows and voltage profile at buses and in Transient and dynamic stability security we focus on angle and frequency stability besides line flows and voltages [2, 7].

The power system equipment are designed to operate at certain limits and are protected by using automatic devices [6]. In case of any disturbance, which results in violation of the limits, it will make the protective device to operate and if this

D. J. Hemanth et al. (Eds.): COMET 2019, LNDECT 35, pp. 109–116, 2020.
https://doi.org/10.1007/978-3-030-32150-5_11

Table 1. Operating states of PSS

State	Condition	Comment
Normal	'E', 'I' is satisfied	Security constraints are satisfied, Secure state
Alert	'E' is satisfied and 'I' is in danger of violation	Decrease in Reserve margins, Preventive control action and Unsecure state
Emergency	'E' is satisfied and 'I' is violated	Severe disturbance, Heroic actions should be initiated
Extremis	'E', 'I' is violated	Emergency control action
Restorative	'I' is satisfied. 'E' is not satisfied	System transits to Normal or Alert state

disturbance causes any further switches to operate, other equipment will be out of service. If this process of cascading events continues, the complete system or parts of it may collapse, which is referred as system Blackout.

Contingency is defined as removal or outage of power system equipment when it is failed to operate. We have power outage which means removal of both real power i.e. generators and reactive power i.e. shunt compensators and we also have branch outage i.e. the outage of transmission line and transformers. In this work transmission line outages are considered.

Multiple contingencies are given importance in a deregulated environment; these cascaded events are havoc to the system [3], and the number of possible multiple contingencies is greater for the system when the size of the system increases which is given by the formula

$$\left\langle \begin{matrix} N \\ k \end{matrix} \right\rangle = \frac{N!}{(N-k)!k!} \tag{1}$$

where

'N' No. of Power system components
'k' No. of outages

In this work we have considered N-2 Contingency analysis of transmission lines which is given by the formula,

$$\binom{L}{2} = \frac{L(L-1)}{2} \tag{2}$$

Where 'L' is the no. of transmission lines.

Contingencies are caused by faults, improper function of protective equipment, insufficient reserve margins etc. Multiple transmission line outages are due to bad weather conditions (lightening), accidents (falling of towers) and faults [2]. Number of N-2 contingencies increase with size of the system and based on contingency having limited geographical effect, two screening algorithms are developed and they are presented in next section. Further the screened contingencies are ranked using performance index [1, 3, 5, 10].

Fast decoupled load flow method has been used during ranking as it is able to perform for three simultaneous outages i.e. (N-3) and we can use base case for calculating different outages instead of flat voltage start and it has also got more speed and less memory when compared with other load flow methods [4, 11].

The paper is structured as follows: In Sect. 2 we will see Contingency selection. In Sect. 3 we will see Algorithm and Flowchart. In Sect. 4 we will see Test case and Results. In Sect. 5, we will see Conclusion and Future scope respectively.

2 Contingency Selection

Based on the literature survey done this methods are best for N-2 contingency Selection. Contingency screening and ranking falls under this category, screening algorithms and real power performance index are explained below. In order to avoid calculation of large number of possible N-2 contingencies in which most of them will be Non-critical and to reduce the computational effort involved we are using these screening algorithms to screen critical contingencies from the list of possible contingencies and to know the severity of contingencies we are ranking the contingencies. As these algorithms are applicable only for change in MW flows in line we have considered real performance index to rank the contingencies in later part. LODF algorithm uses sensitivity data, whereas Line overload algorithm uses line flows and line limits also. The advantage of LODF algorithm is it detects the pairs which result in violation only after the second outage. We are doing this without solving the full set.

2.1 LODF Screening Algorithm

1. Calculate L^2 LODF values for N-1 contingency.
2. Choose a value of LODF, d^* and all the values above d^* are recorded and make a list.
3. From the above list all the possible N-2 Contingencies are combined.
4. Remove the non unique elements from the above list and the remaining combinations are critical contingencies.

By varying the value of d^*, size of the structure/list can be varied. For a well chosen value of d^* the list will be sparse.

2.2 LINE OVERLOAD Screening Algorithm

1. Calculate L^2 LODF values for N-1 contingency and line flow and limit information.
2. Choose a value for line overload, O^* and all the values above O^* are recorded and make a list.
3. From the above list, N-2 Contingencies are made by combining the entries in the tracking list with every other line in the system.

4. Remove the non unique elements from the above list and the remaining combinations are critical contingencies.

2.3 Contingency Ranking

1. Real Performance Index (PI_{MW})
   To measure the change in MW line flows of the system during a contingency is given by Real performance index.

$$PI_{MW} = \sum_{i=1}^{L} \frac{W}{2n} \left(\frac{P_i}{P_{max}}\right)^{2n} \quad (3)$$

Where,

L is the number of transmission lines of the system
n is an exponent and value is between 1 to 5
W is a real non-negative weighting factor
P_i is the line flow through the line i
P_{max} is the maximum flow through the respective line.

3 Algorithm and Flowchart for N-2 Contingency Analysis

3.1 Algorithm of the N-2 Contingency Screening and Ranking of an IEEE Test System

1. Read the given load flow data of the system.
2. Run the load flow for the base case without any outages on the system. and Calculate the LODF values and line flows for all possible N-1 contingencies.
3. Apply the LODF screening algorithm and Line overload screening algorithm for different values of d* and O* respectively.
4. Choose a best value for d* and O* and select all the pairs that are obtained by the above process.
5. Model a N-2 contingency from the above list and run the load flow for the case.
6. Line flows for the above case are calculated and real performance index (PIMW) is calculated.
7. Repeat the steps 5 and 6 for the remaining cases that are screened from above process.
8. Rank the contingencies based on their values.
9. We obtain the critical contingencies based on their severity from the above list.

3.2 Flow Chart of the N-2 Contingency Screening and Ranking of an IEEE Test System

See Fig. 1.

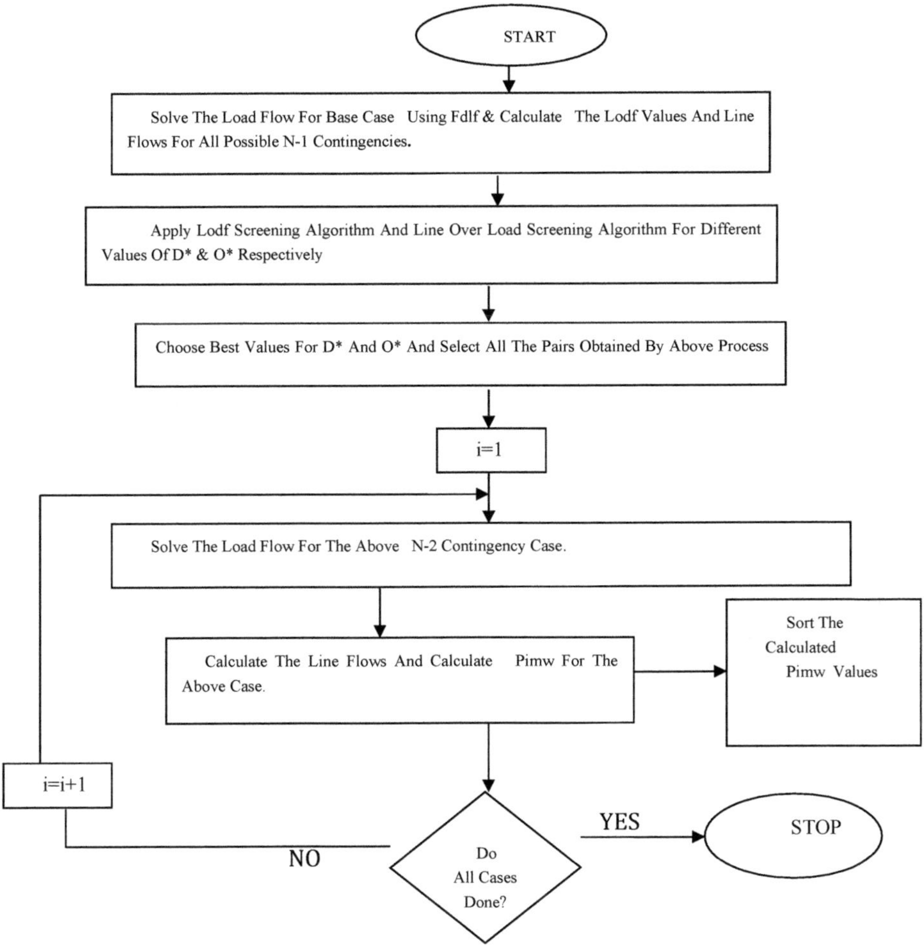

Fig. 1. Flowchart of N-2 contingency screening and ranking algorithm

4 Test Case and Results

The above screening algorithms and ranking is applied for an IEEE 6 bus system and number of possible N-2 contingencies are 55 as 11 transmission lines are available on this system (Fig. 2).

The results of LODF screening and line over load screening algorithm are presented below, in IEEE 6 bus system 52 contingencies are credible contingencies. Screening algorithms are implemented in Power simulator package and ranking methods are implemented in Mipower software package.

4.1 Contingency Screening

See Tables 2 and 3.

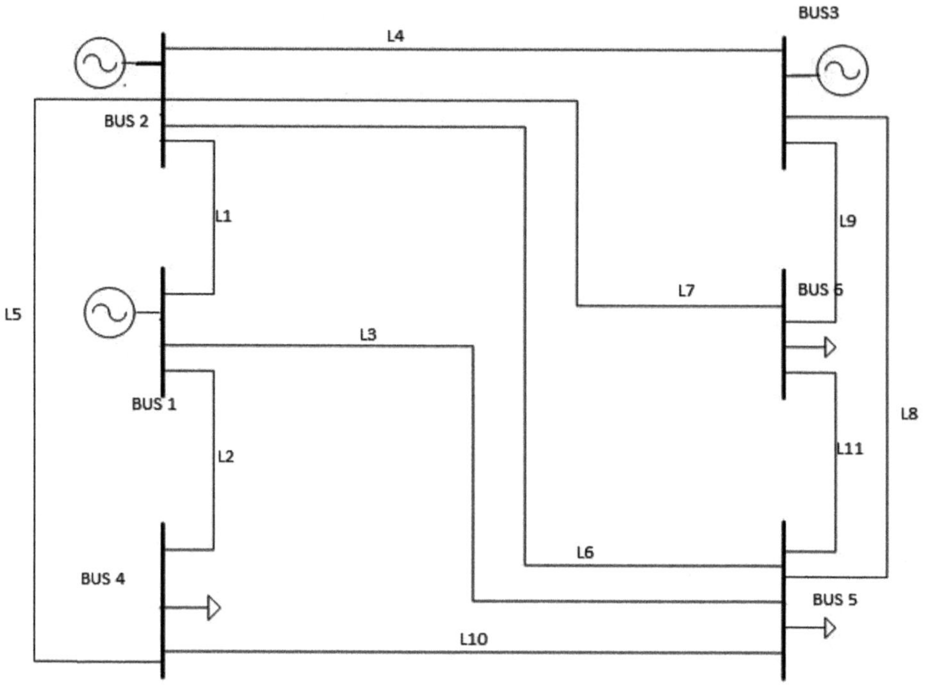

Fig. 2. Single line diagram of an IEEE 6 bus system.

Table 2. LODF screening algorithm for different d*

S. No.	d* Value	Captured	Missed
1	15	52	0
2	17.5	50	2
3	20	44	8
4	22.5	39	13
5	25	30	22

Table 3. Line overload screening algorithm for different O*

S. No.	Over load O*	Captured	Missed
1	95	49	3
2	98	45	7
3	100	40	12
4	105	34	18
5	110	27	25

4.2 Contingency Ranking

See Table 4.

Table 4. PI_{MW} values of N-2 contingency

LINES	PIMW	RANK
L4L9	6.53E+01	1
L8L9	6.16E+01	2
L2L5	6.11E+01	3
L7L9	2.92E+01	4
L5L9	1.55E+01	5
L2L9	1.43E+01	6
L5L8	1.40E+01	7
L3L9	1.24E+01	8
L1L5	1.05E+01	9
L3L8	1.05E+01	10
L9L11	1.03E+01	11
L1L9	1.01E+01	12
L9L10	1.00E+01	13
L6L9	9.94E+00	14
L5L11	9.78E+00	15
L5L7	9.70E+00	16
L5L10	9.30E+00	17
L3L6	5.72E+00	35
L3L11	5.62E+00	36
L2L6	5.32E+00	37
L3L10	4.96E+00	38
L7L11	4.91E+00	39
L1L3	4.89E+00	40
L3L4	4.79E+00	41
L4L7	4.75E+00	42
L6L7	4.70E+00	43

LINES	PIMW	RANK
L5L6	9.24E+00	18
L4L5	8.75E+00	19
L3L5	8.47E+00	20
L2L8	8.29E+00	21
L8L11	8.17E+00	22
L6L8	7.92E+00	23
L2L3	7.74E+00	24
L7L8	6.97E+00	25
L4L8	6.68E+00	26
L1L2	6.62E+00	27
L8L10	6.60E+00	28
L1L8	6.25E+00	29
L2L10	5.96E+00	30
L2L4	5.95E+00	31
L2L7	5.89E+00	32
L2L11	5.89E+00	33
L3L7	5.75E+00	34
L1L7	4.69E+00	44
L7L10	4.68E+00	45
L10L11	4.56E+00	46
L6L11	4.55E+00	47
L4L11	4.46E+00	48
L1L11	4.39E+00	49
L4L6	4.31E+00	50
L4L10	4.25E+00	51
L1L6	4.24E+00	52

5 Conclusion and Future Scope

Contingency analysis is important part of the PSS. It gives the operator knowledge of critical contingencies. In a large power system network, the effect of contingencies have limited geographical effect, based on this fact two screening algorithms are developed to determine the critical contingencies and later the screened contingencies are ranked using performance indices to measure the severity of contingencies. Later the critical contingencies are fully evaluated and the operator will take necessary actions in case of any such event have occurred on the system.

FACTS devices are used to improve the performance of the system. Especially series controllers like TCSC, TCR are used to improve the line flow on the line and shunt compensators are used to improve the voltage profile at the bus. State estimators can also be used to improve the measurements taken by the system.

References

1. Burada, S., Joshi, D., Mistry, K.D.: Contingency analysis of power system by using voltage and active power performance index. In: 1st IEEE International Conference on Power Electronics, Intelligent Control and Energy Systems (ICPEICES-2016), pp. 1–5 (2016)
2. Debs, A.S., Benson, A.R.: Security assessment of power systems. In: Engineering For Power: Status and Prospects U.S. Government Document, CONF-750867, pp. 1–29 (1967)
3. Davis, C.M., Overbye, T.J.: Multiple element contingency screening. IEEE Trans. Power Syst. **26**(3), 1294–1301 (2011)
4. Stott, B., Alsac, O.: Fast decoupled load flow. IEEE Trans. Power Appar. Syst. **PAS-93**, 859–869 (1974)
5. Ejebe, G., Wollenberg, B.: Automatic contingency selection. IEEE Trans. Power Appar. Syst. **1**, 97–109 (1979)
6. DyLiacco, T.E.: The adaptive reliability control system. IEEE Trans. Power Appar. Syst. **PAS-86**, 517–531 (1967)
7. Mitra, P., Vittal, V., Keel, B., Mistry, J.: A systematic approach to n-1-1 analysis for power system security assessment. IEEE Power Energy Technol. Syst. J. **3**(2), 71–80 (2016)
8. Padiyar, K.R.: Power System Dynamics, Stability and Control. B.S. Publications (2008)
9. Wood, A.J., Wollenberg, B.F.: Power Generation, Operation and Control. Wiley, New York (2012)
10. Mishra, V.J.P., Khardanvis, M.D.: Contingency analysis of power system. In: IEEE Student Conference on Electrical, Electronics and Computer Science (2012). 978-14673-1515-9
11. Swaroop, N., Lakshmi, M.: Contingency analysis and ranking on 400 kV Karnataka network by using Mi power. Int. Res. J. Eng. Technol. **3**(10), 576–580 (2016)

Moderation Effect of Family Support on Academic Attainment

Jainab Zareena(✉)

Saveetha School of Engineering,
Saveetha Nagar, Thandalam, Chennai 602105, India
jainabzareena@gmail.com

Abstract. Self-confidence is an important factor for the determination of success. The level of self-confidence gets varied among individuals. The present study investigates the relationship between the variables, self-confidence and academic attainment. Adapted questionnaire is employed. Primary data is obtained from one hundred engineering students studying in a private college at Chennai region, India. The role of moderator variable, 'family support' in strengthening the relationship between the independent variable, self confidence and the dependent variable, academic attainment is explored. Descriptive and inferential statistic is used for data analysis. The study indubitably proved the existence of correlation amongst the variables. The present study would serve as a guide for parents, educational institutions, and students.

Keywords: Self-confidence · Academics · Accomplishment · Goal-oriented · Family support

1 Introduction

Self-confidence is an interesting concept in the field of behavioral science. Many psychologists carried out research in this area and derived various remarkable facts. Many of us think that the concept 'Self-efficacy' and 'Self-confidence' is same. Bandura (1977) defined Self-efficacy as, "the belief in one's capabilities to organize and execute the courses of action required to manage prospective situations." Self belief is obviously essential to succeed in life. 'Self-efficacy' refers to the achievement of task for a specific condition. Whereas 'Self-confidence' refers to the attainment of goal in all kind of situation.

In general, an individual having a desire to succeed in whatever he/she takes up is called as Self-confidence. Therefore, the concept 'Self-confidence' and 'self-efficacy' is different. Earlier research studies focused on identifying the role of Self-confidence in terms of predicting academic success. For example a study conducted by Karimi and Saadatmand (2014) proved the existence of relationship between the variables, self-confidence and academic attainment.

Further, Verma (2017) carried out research to identify the self-confidence level of students with regard to gender, locality and field of study. The data was collected from two hundred students studying different branches in the University of Jammu. Among two hundred respondents, hundred were male and the remaining hundred were female.

D. J. Hemanth et al. (Eds.): COMET 2019, LNDECT 35, pp. 117–121, 2020.
https://doi.org/10.1007/978-3-030-32150-5_12

It was found that the students studying in different streams have varied self-confidence level. The study recommended the teachers and parents to instill a sense of confidence among students.

Similar study conducted by Baumeister et al. (2003), Hattie (2008), Fiske and Taylor (2013) concluded that self-belief play a major role for attaining academic success. Though number of research studies is conducted in this area, no study has viewed the variable 'family support' as a moderator for predicting academic attainment. The present study identifies the role of moderator variable, family support in strengthening the relationship between the independent variable, self-confidence and the dependent variable, academic attainment (Fig. 1).

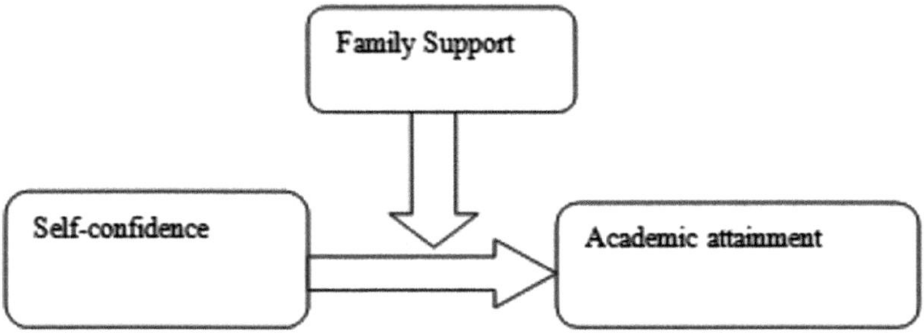

Fig. 1. Conceptual framework of the study

1.1 Tools and Techniques

It was really a very challenging task to get the filled-in data from the students belonging to different branch of study. Therefore, the convenient sampling method is used. The data was collected from one hundred engineering students studying third year in a private college at Chennai region, India. The obtained data is anonymised. Adapted questionnaire is used for the study. Totally, the questionnaire consists of thirty questions. Ten questions are used for measuring each variable. The study employed Lickerts five point rating scale. Descriptive and inferential statistic is used for interpreting the results. The results revealed inter correlation amongst all the questions significantly. The calculated Cronbach's alpha value is 0.81 for the variable self-confidence, 0.87 for the variable family support, and 0.86 for the variable academic attainment respectively.

Construct and convergent validity is assessed using confirmatory factor analysis (CFA). The present study produced the values recommended by Bentler (1990), Kline (1998) and Baumgartner and Hombur (1996). The results are tabulated as follows in Table 1.

Table 1. Validity assessment – Confirmatory FTTT actor Analysis (CFA) variables

	No. of questions before CFA	CMIN/DF	p value	GFI	AGFI	CFI	RMR	RMSEA	No. of questions after CFA
Self-confidence	15	1.829	0.8 57	0.999	0.997	1.000	0.002	0.000	10
Family support	12	1.388	0.551	0.999	0.991	1.000	0.010	0.020	10
Academic attainment	12	1.441	0.095	0.995	0.993	1.000	0.014	0.000	10

2 Results

The demographic details of the respondents are tabulated below in Table 2.

Table 2. Demographic details of the respondents

S. No.	Variables	Category	Number
1	Gender	Male	50
		Female	50
2	Age	20–22 years	93
		23 years and above	7
3	Branch	CSE & IT	60
		EEE, ECE & Mechanical	40
4	Year	Final year engineering students	100

Pearson correlation coefficient analysis is calculated to find the relationship between the variables. The result is tabulated as follows in Table 3.

Table 3. Pearson correlation analysis - Male and Female respondents

	Academic attainment			
	Male		Female	
Variables	r value	p value	r value	p value
Self-confidence	.883	.000**	.891	.007**
Family support	.823	.000**	.889	.000**

Note: **p < .01, *p < .05, ns = not significant

The result revealed high positive correlation. Comparatively, r value is higher for the female respondents than the male respondents. Based on the findings it is inferred that the self-confident student with family support will be academically good.

The moderation effect is tested using moderator multiple regression analysis (Table 4).

Table 4. Moderator multiple regression analysis

Variables	Male respondents		Female respondents	
	Beta	Sig.	Beta	Sig.
Self-confidence	0.571	.000**	0.446	.000**
Family support	0.423	.000**	0.571	.000**
Self-confidence × Family support	0.633	.000**	0.512	.000**

Note: **p < .01.

With regard to male respondents, the above table illustrates that the interaction term (Self-confidence × Family support) highly influences the dependent variable followed by the variables Self-confidence and family support respectively.

Considering female respondents, 'family support' is highly found to influence the variable academic attainment followed by the moderation effect (Self-confidence × Family support) and Self-confidence respectively. Based on the findings it is inferred that the moderator variable, 'family support' strengthens the relationship between independent and dependent variable.

3 Conclusion

The study undoubtedly proves that the moderator variable (family support) play a major role in terms of predicting the academic attainment. At home, parents should provide a good supporting environment to their children such as switching off musical or other video gadget during study hours, no quarreling among the family members, offering healthy food, providing separate place for learning, encouraging to participate in study related events/symposiums, and so on. This would definitely motivate students for attaining academic success. The study concludes that the family members are also equally responsible even though the child has self interest in academic attainment.

References

Bandura, A.: Social Learning Theory. Prentice Hall, Englewood Cliffs (1977)

Karimi, A., Saadatmand, Z.: The relationship between self-confidence with achievement based on academic motivation. Arab. J. Bus. Manag. Rev. (Kuwait Chap.) **33**, 1–6 (2014)

Verma, E.: Self-confidence among university students: an empirical study. Int. J. Appl. Res. **3**, 447–449 (2017)

Baumeister, R.F., Campbell, J.D., Krueger, J.I., Vohs, K.D.: Does high self-esteem cause better performance, interpersonal success, happiness, or healthier lifestyles? Psychol. Sci. Public Interest **4**, 1–44 (2003)

Fiske, S.T., Taylor, S.E.: Social Cognition: From Brains to Culture. Sage, Thousand Oaks (2013)

Hattie, J.: Visible Learning: A Synthesis of over 800 Meta-analyses Relating to Achievement. Routledge, Abingdon (2008)

Bentler, P.M.: Comparative fit indexes in structural models. Psychol. Bull. **107**, 238–246 (1990)

Kline, R.B.: Software review: software programs for structural equation modeling: Amos, EQS, and LISREL. J. Psycho Educ. Assess. **16**, 343–364 (1998)

Baumgartner, H., Hombur, C.: Applications of structural equation modeling in marketing and consumer research: a review. Int. J. Res. Mark. **13**, 139–161 (1996)

Shade Resilient Total Cross Tied Configurations to Enhance Energy Yield of Photovoltaic Array Under Partial Shaded Conditions

S. Malathy(✉) and R. Ramaprabha

Sri Sivasubramaniya Nadar College of Engineering, OMR, Kalavakkam, Chennai 603110, India
{malathys, ramaprabhar}@ssn.edu.in

Abstract. Partial shading restrains the output of photovoltaic (PV) system, as mismatch losses are prominent when shade dispersion is uneven in an array. Partial shading (PS) is unavoidable in residential installations and it is therefore essential to improve the output under such conditions. Strategies that reconfigure the shaded panels or alter the interconnections are generally adopted to improve the yield under PS conditions. The static reconfiguration technique that is based on puzzle pattern has an edge over the dynamic technique as it voids the usage of sensors, switches and complicated computations and hence offers an efficient and economically viable solution in addressing the partial shading issues in small scale installations. This paper proposes three new shade resilient structures (SR1, SR2 and SR3) and analyses the performance under six different shading cases that mimic the real world shading scenarios.

Keywords: Photovoltaic · Partial shading · Static reconfiguration · Mismatch losses · TCT · Shade resilient

1 Introduction

Copious availability besides being a clean source of energy has made power generation using solar photovoltaic system an attractive alternative to fossil fuels. There has been a significant increase in the small-scale installations in recent years owing to the reduced cost per watt and the initiatives/incentives offered by the government. These small-scale installations are usually integrated to the building on its rooftop or facade. When mounted on the rooftops or building facades, the panels may be shaded by trees or nearby structures and hence the panels in the PV array may not receive uniform irradiation. The PV array is then said to be partially shaded and generates power that is lesser than the expected value. Therefore, to meet the design specifications, the PV array is to be upsized and this in turn will increase the capital cost significantly, making deployment of PV system less affordable.

The situation is more common in urban installations and shading is inevitable in such cases owing to space constraints. When a panel is shaded, it imposes current limitation, and the shaded panel gets reverse biased when forced to violate the

D. J. Hemanth et al. (Eds.): COMET 2019, LNDECT 35, pp. 122–133, 2020.
https://doi.org/10.1007/978-3-030-32150-5_13

limitation resulting in the formation of hot spots [1]. Inclusion of bypass diodes, however leads to several peaks in the voltage-power (V-P) characteristic curve and the condition demands a sophisticated algorithm to track the maximum power. Conventional algorithms fail to recognize the global peak among the several local peaks [2, 3].

In general, the effects of partial shading [4, 5] can be alleviated and the output power can be enhanced by altering either the power-conditioning unit or PV array's architecture. The interconnection scheme between the panels plays an important role in determining the power generation under PS conditions [6]. Among the basic interconnection schemes, the series scheme is vulnerable to shades and the parallel scheme is resilient. It has been proved in literature that TCT interconnection scheme yields better under partial shaded conditions when compared to the other derived configurations like series-parallel (SP), honey comb (HC) and bridge link (BL).

The position (spot) the shaded panels occupies in an array or the distribution of shade among the rows is another key factor that dictates the output power generation. The mismatch can be nullified or reduced largely by dispersing the shade uniformly all over the array. This is achieved by either shifting the position of panels in the array or changing the interconnections in accordance with the prevailing shading conditions. The electric array reconfiguration schemes alter the number of panels connected in series and parallel so that the PV array generates a constant power under all operating conditions [7]. Later, soft computing techniques were involved to select the best suited interconnection scheme to maximize the power generation under PS conditions [8]. The selected scheme is implemented by triggering suitable switches (electromechanical or semiconductor switches) coupled with the panels. The adaptive reconfiguration scheme connects appropriate number of panels from the adaptive bank to each of the row of fixed TCT bank through a matrix of switches [9]. All these dynamic methods involve complex computations, more number of switches and sensors besides a sophisticated control algorithm. These limitations are addressed in static reconfiguration schemes.

In the first proposed static configuration scheme, the spot of each of the panel within the array is decided by a SuDoKu pattern. Poor shade dispersion and dependency of output on the chosen SuDoKu pattern are the major bottlenecks of this scheme. The nonuniqueness of SuDoKu pattern for a given array size makes the selection difficult as each pattern would result in different shade dispersion and power generation. Besides, wiring gets complicated with the size of the array, as the panels are not uniformly displaced. These limitations has led to the development of other reconfiguration schemes based on puzzle patters like magic square (MS), latin square (LS), fixed configuration, optimal TCT and static shade tolerant (SST) structure [10–13, 14]. This approach does not involve sensors, switches or a control algorithm and hence offer an economical solution to the small scale installations.

This paper proposes three such static shade resilient algorithms (SR1, SR2 & SR3) to find out the position of panels by simplified equations. The panels are placed in an ordered way resulting in uncomplicated wiring. The performance of proposed algorithms is analyzed for a 6 × 6 array under various shading conditions in Matlab/Simulink environment. The single diode model of the $37W_p$ PV panel and it's electrical characteristics are presented in Sect. 2. The proposed shade resilient structures are analyzed and the results are compared in Sects. 3 and 4 respectively.

2 Modeling of PV Panel

Assessing the performance of the proposed shade resilient TCT schemes and the traditional TCT scheme under PS conditions call for a mathematical model that imitate the real panel. The circuit model of 37 W_p PV panel is developed based on the single diode PV model. The single diode model represents PV cell as current source shunted by a diode as presented in Fig. 1a. Thirty six cells are connected in series to form the 37 W_p panel. The developed model is fine tuned such that it emulates the physical PV panel. The specification of the 37 W_p Solkar make panel considered in this work is given in Table 1. The current and voltage of a PV panel are related by

$$I_{panel} = I_{ph} - I_o\left[\exp\left(\frac{V_{panel} + R_{se}I_{panel}}{V_t a}\right) - 1\right] - \frac{V_{panel} + R_{se}I_{panel}}{R_{sh}} \tag{1}$$

$$I_{ph} = \{K_i[T - T_i] + I_{pvn}\}\frac{G}{G_n} \tag{2}$$

where, I_{ph} is the photovoltaic current, I_o is the saturation current, a is the diode ideality factor, R_{se} is the series resistance and R_{sh} is the shunt resistance, V_{panel} and I_{panel} are the panel voltage and current respectively and $V_t = N_s kT/q$ is the thermal voltage. N_s is the number of cells in series, k is the Boltzman's constant, T is the temperature and q is the electric charge.

Table 1. Specification of the 37 W_p Solkar PV panel

Specifications	Values
Open circuit voltage (V_{oc})	21.24 V
Short circuit current (I_{sc})	2.55 A
Maximum power (P_m)	37.08 W
Voltage at maximum power (V_m)	21.06 V
Current at maximum power (I_m)	2.245 A
No. of cells in series	36
No. of bypass diodes	1

The simulated results are validated against the data sheet values at strategic voltage points. The simulated characteristic curves of the 37 W_p panel at standard temperature of 25 °C and different irradiation conditions are presented in Fig. 1b.

The V-P characteristics exhibit single peak as the irradiation received by all the cells in the panel is assumed to be the same. It can be inferred from Fig. 1 that the PV current and peak power reduce significantly with irradiation.

Variation in voltage with regard to irradiation is relatively less and hence it is neglected in the analyses presented in this paper. When T is equal to T_i, Eq. 2 becomes

$$I_{ph} = I_{pvn}\frac{G}{G_n} \tag{3}$$

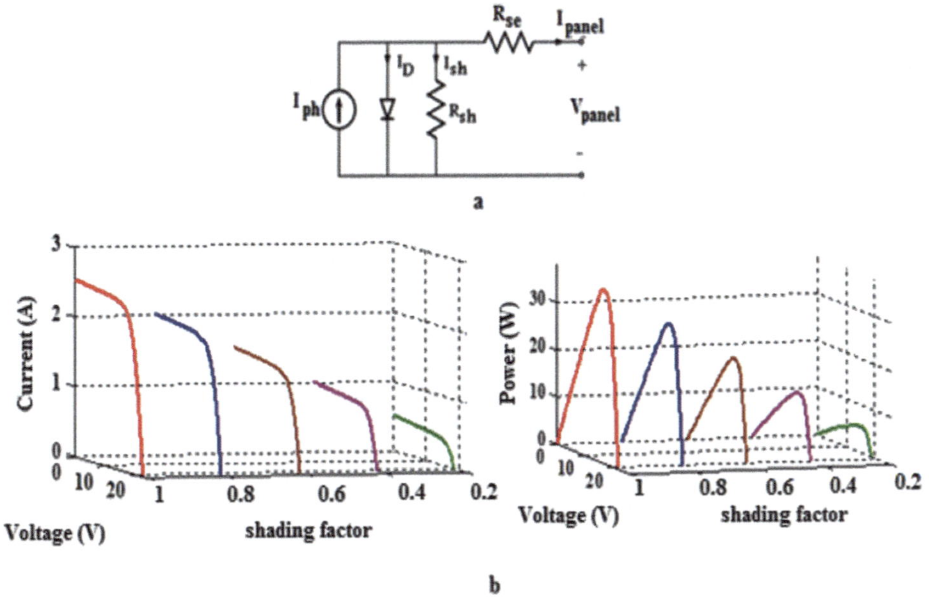

Fig. 1. (a) Equivalent circuit (b) Simulated characteristics of PV panel for different irradiation levels

The ratio of actual irradiation to the nominal irradiation of 1000 W/m^2 is called the shading factor (SF). It is evident from the Eq. 3 that the I_{ph} depends directly on the SF and the PV current can be expressed as SFI_m.

3 Static Shade Tolerant Array Configuration

The dynamic reconfiguration technique alters the interconnections between the panels or equalizes the row current of the shaded TCT array. The major limitation of this technique is that it requires more number of sensors to sense the prevailing shading conditions and switches to change the connectivity among the panels. Besides, the computational complexity increases with array size as the reconfiguration is online or dynamic in nature. Alternatively, the static reconfiguration technique employs an off-line strategy to disperse the shade uniformly all over the TCT array. In these static strategies, the panels that electrically belong to the same row of the TCT array are placed physically in different locations in each of the rows. The location of panels is determined offline either by puzzle patterns like SuDoKu, magic square and Latin square or by algorithms. The panels are positioned according to the chosen pattern and connected in TCT fashion.

4 Proposed Shade Resilient schemes

The reality that the power generation of a PS array can be enhanced by equalizing the row current of the TCT array has led to the formulation of three new shade resilient (SR) structures namely SR1, SR2 and SR3. These algorithm based shade resilient schemes proposed in this work determines the location of individual panels in an array by computing the separation factor 's' or the shift factor 'd'. These factors give the distance of separation between two successive panels in the reconfigured PV array. The way these factors are computed has led to the formulation of three new SR schemes called SR1, SR2 and SR3.

In the first scheme SR1, the separation factor 's' is equal to floor ($\sqrt{m}$), where 'm' is the number of rows. The first column panel indices are fixed and the subsequent column indices are obtained by adding 's' with them. The procedure to estimate the position of the panels in SR1 scheme is given below.

$$\mathrm{Y_{ij} = X_{kj}} \quad \text{for i} = 1, 2, \ldots, \text{m and j} = 1, 2, \ldots, \text{n} \tag{4}$$

$$\mathrm{s = floor(\sqrt{m});\ y = floor(m/2)} \tag{5}$$

$$\text{The row index k} = (\mathrm{i} + (\mathrm{j} - 1)^{*}\,\mathrm{s}) \text{ for j} < \text{y;} \tag{6}$$

$$\mathrm{k} = (\mathrm{i} + (\mathrm{j} - 1)^{*}\,\mathrm{s}) + 1 \text{ for j} > \text{y;} \tag{7}$$

$$\text{if k} < \text{m, then k} = \text{k} \tag{8}$$

$$\text{else k} = \text{k} - \text{m} \tag{9}$$

For example, the separation factor (s) for a 6 × 6 array is 2. The index assigned to the first row first column panel is '1'. The indices of the other panels that will be positioned in subsequent columns of the first row are assessed by Eqs. 6 or 7. If the resulting index is greater than 6, the index is corrected by subtracting 6. A correction factor of '1' is added if the resulting index is assigned already. The first row indices are obtained for a 6 × 6 array as tabulated in Table 2. The resulting structure is presented in Fig. 2. With all the first row panels highlighted.

Table 2. Location of panels in SR1 scheme

Column index	Row index	Panel index
1	1	11
2	1 + 2 = 3	32
3	1 + 4 = 5	53
4	1 + 6 + 1 = 8; 8 > 6; Hence, 8 − 6 = 2;	24
5	1 + 8 + 1 = 10; 10 > 6; Hence, 10 − 6 = 4;	45
6	1 + 10 + 1 = 12; 12 > 6; Hence, 12 − 6 = 6;	66

4.1 The Second and the Third Scheme SR2 & SR3

In the second SR scheme (SR2), the location of first panel (first row and first column) is fixed and the remaining panels in the first row are shifted to various depths by calculating a shift factor 'd' which is equal to $\sqrt{m}$. Two different arrangements (SR2 & SR3) are possible if the calculated shift factor is fractional (for odd number of rows). The location of the first row panels in the proposed SR2 and SR3 arrangements are determined as follows

$$Y_{kj} = X_{ij} \quad \text{for } i = 1, 2, \ldots, m \text{ and } j = 1, 2, \ldots, n \tag{10}$$

For the second scheme SR2,

$$d = \text{floor}(\sqrt{m});\; y = \text{ceil}(\sqrt{m}) \tag{11}$$

$$\text{The row index } k = (i + (j - 1) * d) \text{ for } j < y; \tag{12}$$

$$k = (i + (j - 1) * d) + 1 \text{ for } j > y; \tag{13}$$

$$\text{if } k < m, \text{ then } k = k; \text{ else } k = k - m \tag{14}$$

For the third scheme SR3,

$$d = \text{ceil}(\sqrt{m});\; y = \text{floor}(\sqrt{m}) \tag{15}$$

$$\text{The row index } k = (i + (j - 1) * d) \text{ for } j < y; \tag{16}$$

$$k = (i + (j - 1) * d) + 1 \text{ for } j > y;\; d = \text{floor}(\sqrt{m}) \tag{17}$$

$$\text{if } k < m, \text{ then } k = k; \text{ else } k1 = k - m; \tag{18}$$

$$\text{if } k1 < m, \text{ then } k = k1; \text{ else } k = k1 + 1 - m; \tag{19}$$

The formulation of the two structures is given below for a 6 × 6 array. The depth of shift 'd' is 2 and 'y' is 3 for SR2. The location of the row first column panel (X_{11}) is fixed as '11' (the first index represents the row and the second the column). The first row second column panel (X_{12}) is shifted by three rows down as calculated and the new location is determined as '32'. The first row third column panel (X_{13}) is shifted by five rows and the location is '53'. The shift for the fourth column panel (X_{14}) is calculated to be 8 rows and as this is greater than the number of rows, the calculated shift is reduced by '6'. The resulting shift is 2 and hence the location is '24'. Similarly the shift for the fifth and the sixth rows are calculated and the locations are determined to be '45' and '66' respectively. The formulation is tabulated in Table 3.

Table 3. Location of panels in SR2 scheme

Column (j)	Depth of shift (k)	First row panels	Location of first row panels
1	1	11	11
2	3	12	32
3	5	13	53
4	8 − 6 = 2	14	24
5	10 − 6 = 4	15	45
6	12 − 6 = 6	16	66

In case of SR3, the third arrangement, for the 6 × 6 array the depth of shift ‘d’ is 3 and ‘y’ is 2. The difference between the SR2 and SR3 lies in the depth of separation ‘d’. If the number of rows is a perfect square, then SR2 and SR3 are the same. The shift in the first row panels are calculated as in SR2 arrangement but with a shift factor (‘d’) of 3 and the resulting location of panels are presented in Table 4.

Table 4. Location of panels in SR3 scheme

Column (j)	Depth of shift (k)	First row panels	Location of first row panels
1	1	11	11
2	4	12	42
3	8 − 6 = 2	13	23
4	11 − 6 = 5	14	54
5	14 − 6 = 8+1 − 6 = 3	15	35
6	17 − 6 = 11 + 1 − 6 = 6	16	66

The arrangement of panels as determined by the three proposed schemes (SR1, SR2 and SR3) is presented in Fig. 2. The first row panels are highlighted and it can be seen that the schemes resulted in three different arrangements. The panels that belong to the same row are connected in parallel and the six parallel strings are connected in series resulting in TCT configuration.

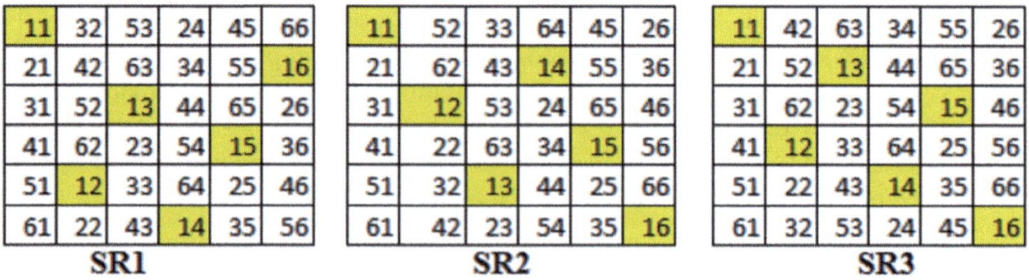

Fig. 2. The proposed SR arrangements

These arrangements are tested with the test shade patterns presented in Fig. 3. In the first shade pattern the shade is narrow and long. The second shade is categorized as wide & short and third shade pattern is the combination of the first and the second shade pattern. In the conventional arrangement the first two rows are heavily shaded and the ensuing mismatched row currents eventually reduce the output power. However, in the SR schemes, the shade is dispersed all over the array with the third SR scheme resulting in the least IEF of 0.016. In the fourth shade pattern, the last three rows are shaded. The fifth shade pattern which can be categorized as short & narrow, has the last two rows shaded. The sixth shade also belongs to short & narrow category and the pattern has four shaded panels.

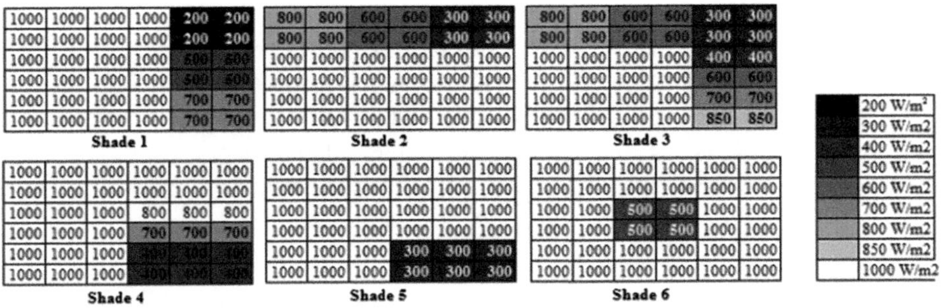

Fig. 3. Test shade patterns

The distribution of shade is quantized by a factor called irradiation equivalence factor (IEF) [14] given by

$$\text{IEF} = \frac{\max(\text{mean G of individual rows}) - \min(\text{mean G of individual rows})}{1000} \quad (20)$$

This factor is a merit indicator that quantifies how good the shade is dispersed and in turn, the current limitation imposed on the array. Lower the factor, better the dispersion and energy yield. The efficacy of the three proposed schemes is analyzed in terms of the IEF tabulated in Table 5 and the maximum output power is analyzed in the further sections.

Table 5. Irradiation equalization factor

Scheme	Shade 1	Shade 2	Shade 3	Shade 4	Shade 5	Shade 6
SR1	0.083	0.083	0.091	0.02	0	0.08
SR2	0.083	0.083	0.091	0.02	0	0.08
SR3	0.083	0.083	0.016	0.15	0.23	0.08

The IEF is the same for all the SR schemes for the shade patterns 1, 2 and 6. This will eventually result in the identical maximum power. In case of shade 3, the shade dispersion is better in SR3 and hence the IEF is better. The shade dispersion is better for SR1 and SR2 schemes for the fourth shade and hence the two schemes will yield maximum output power. In case of the fifth pattern, the dispersion is uniform that the IEF is zero for the first two schemes. The output of the three proposed SR schemes under six test shade patterns are summarized in Table 6. The 6 × 6 array can deliver 1332 W under standard conditions of 1000 W/m^2 and 25 °C.

Table 6. Performance of all SR arrangements

Output power (W)			
Shade pattern	SR1	SR2	SR3
1	1073.1	1073.1	1073.1
2	1121.9	1121.9	1121.9
3	1002.5	1002.5	1016.2
4	1136	1136	1105.2
5	1172.2	1172.2	1120.9
6	1250.1	1250.1	1250.1

It can be inferred that SR1 and SR2 schemes perform better in almost all the cases. Though the SR3 yields better under shade 3, the difference in extractable power is less. Moreover, for short & narrow shades, SR3 yields lesser than the other two arrangements. The results are pictorially represented in Fig. 4. It can be concluded that SR1 or SR2 arrangement may be adopted to enhance the output power under partially shaded conditions. The performance of the three proposed SR schemes are compared with the conventional TCT and the Sudoku scheme under the six test shade patterns presented in Fig. 3. The resulting VP curves are presented in Fig. 5 and the maximum powers are tabulated in Table 7.

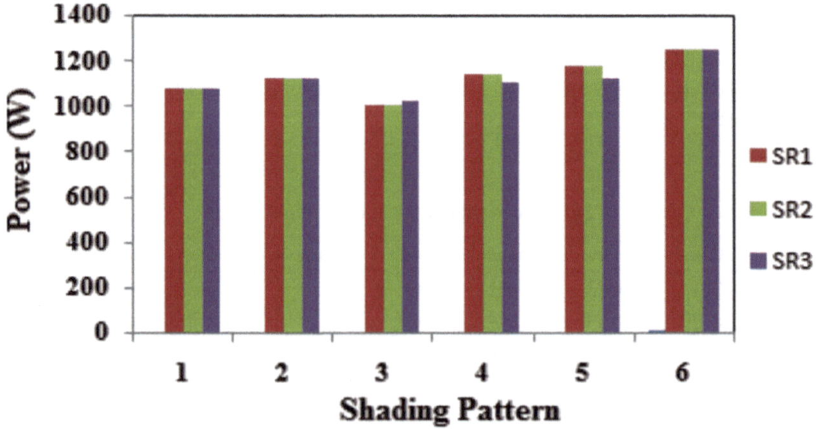

Fig. 4. Performance of the proposed schemes

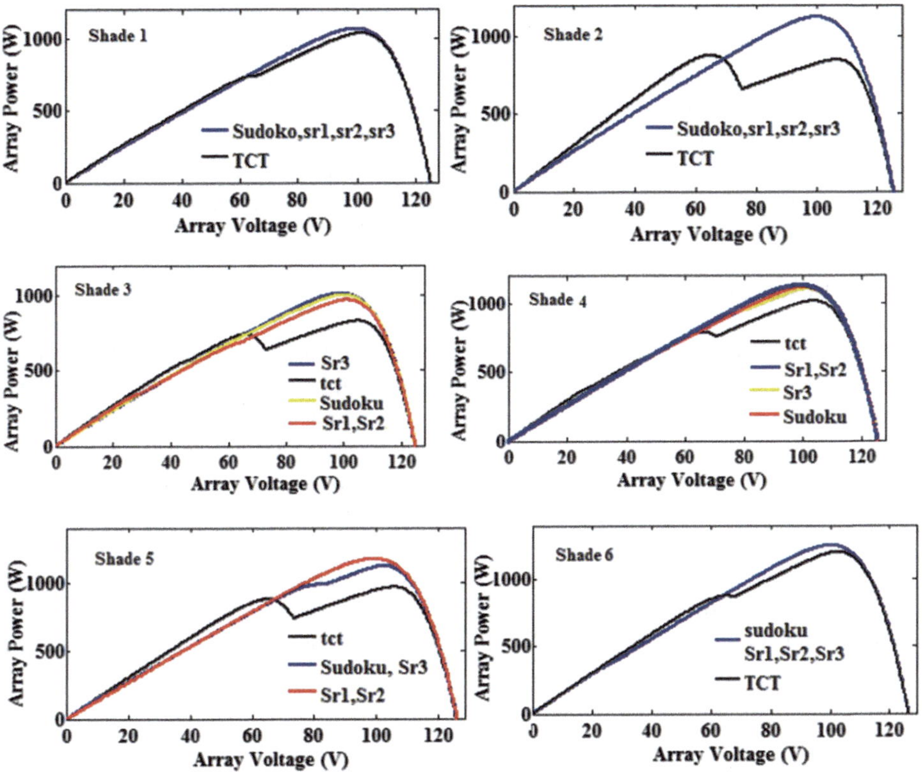

Fig. 5. Comparison of V-P characteristics for the test shade patterns

Table 7. Output power generated by various static schemes

Power (W)						
Configuration	Shade 1	Shade 2	Shade 3	Shade 4	Shade 5	Shade 6
TCT	1038	872.6	827.5	1017.9	962.7	1201
SuDoKu	1073.1	1121.9	1014	1119.8	1120.9	1250.1
SR1	1073.1	1121.9	1002.5	1136	1172.2	1250.1
SR2	1073.1	1121.9	1002.5	1136	1172.2	1250.1
SR3	1073.1	1121.9	1016.2	1105.2	1120.9	1250.1

The conventional TCT arrangement yields lesser output power under all the shading conditions as the shade is concentrated in few of the rows. The currents of the shaded rows are lesser than that of the non-shaded or lightly shaded rows. The mismatch in the row currents causes mismatch losses and reduced the power yield. The static schemes on the other hand disperse the shade all over the array thereby minimizing the mismatch and the associated losses that eventually enhances the output power. It is evident from the data presented in the above table that the static schemes perform better under shaded conditions. Among the static schemes, the SuDoku based scheme depends on the choice of the puzzle pattern. Certain patterns result in better shade dispersion, while certain others not. The proposed schemes make use of simple calculations to formulate the arrangement of panels so as to result in better shade dispersion under all shaded conditions. The simplicity, scalability and the optimized performance of the proposed SR1 and SR2 schemes may help in enhancing the yield of the PV array under partially shaded conditions in small scale PV installations.

5 Conclusion

The paper has proposed three new SR arrangements based on simple calculations to enhance the output power under partial shaded conditions. The proposed formulation arranges the panels with uniform displacement as dictated by the array size and ensures better shade dispersion and simple cabling. The equations to determine the position of the panels for 6×6 array, the shade dispersion and the performance under six test shade patterns are also presented in detail. The effectiveness of the proposed schemes is compared for the test shade patterns and it is found that SR1 and SR2 schemes perform better under shaded conditions. The simplicity, scalability, static nature and simple wiring of these static schemes offer an economical solution for shading issues in small scale urban building integrated PV installations where partial shading is much pronounced.

References

1. Walker, L., Hofer, J., Schlueter, A.: High-resolution, parametric BIPV and electrical systems modeling and design. Appl. Energy **238**, 164–179 (2019)
2. Hashim, N., Salam, Z.: Critical evaluation of soft computing methods for maximum power point tracking algorithms of photovoltaic systems. Int. J. Power Electron. Drive Syst. **10**(1), 548–561 (2019)
3. Bahrami, M., Gavagsaz-Ghoachani, R., Zandi, M., Phattanasak, M., Maranzanaa, G., Nahid-Mobarakeh, B., Pierfederici, S., Meibody-Tabar, F.: Hybrid maximum power point tracking algorithm with improved dynamic performance. Renew. Energy **130**, 982–991 (2019)
4. Lyden, S., Haque, M.E.: Modelling, parameter estimation and assessment of partial shading conditions of photovoltaic modules. J. Modern Power Syst. Clean Energy **7**(1), 55–64 (2019)
5. Tripathi, A.K., Aruna, M., Murthy, C.S.: Performance of a PV panel under different shading strengths. Int. J. Ambient Energy **40**(3), 248–253 (2019)
6. Malathy, S., Ramaprabha, R.: Comprehensive analysis on the role of array size and configuration on energy yield of photovoltaic systems under shaded conditions. Renew. Sustain. Energy Rev. **49**, 672–679 (2015)
7. Tatabhatla, V.M.R., Agarwal, A., Kanumuri, T.: Performance enhancement by shade dispersion of Solar Photo-Voltaic array under continuous dynamic partial shading conditions. J. Clean. Prod. **213**, 462–479 (2019)
8. Nasiruddin, I., Khatoon, S., Jalil, M.F., Bansal, R.C.: Shade diffusion of partial shaded PV array by using odd-even structure. Sol. Energy **181**, 519–529 (2019)

9. Tubniyom, C., Chatthaworn, R., Suksri, A., Wongwuttanasatian, T.: Minimization of losses in solar photovoltaic modules by reconfiguration under various patterns of partial shading. Energie **12**(1), 24 (2019)
10. Horoufiany, M., Ghandhari, R.: A new photovoltaic arrays fixed reconfiguration method for reducing effects of one-and two-sided mutual shading. J. Solar Energy Eng. **141**(3), 031013 (2019)
11. El Iysaouy, L., Lahbabi, M., Oumnad, A.: A novel magic square view topology of a PV system under partial shading condition. Energy Procedia **157**, 1182–1190 (2019)
12. Horoufiany, M., Ghandhari, R.: A new photovoltaic arrays fixed reconfiguration method for reducing effects of one-and two-sided mutual shading. J. Sol. Energy Eng. **141**(3), 031013 (2019)
13. Malathy, S., Ramaprabha, R.: Reconfiguration strategies to extract maximum power from photovoltaic array under partially shaded conditions. Renew. Sustain. Energy Rev. **81**, 2922–2934 (2018)

Secure and Enhanced Bank Transactions Using Biometric ATM Security System

A. J. Bhuvaneshwari(✉) and R. Nanthithaa Shree

Electronics and Communication Engineering,
Kamaraj College of Engineering and Technology, Virudhunagar, India
erbhuvana.me@gmail.com, nanthithaashree26@gmail.com

Abstract. Authentication based on biometric provides various advantages in ATMs (Automated Teller Machine) systems. The weakness of existing authentication scheme in ATM is the usage of PIN (Personal Identification Number) numbers as password. Because PIN numbers are easily traceable and misused. In order to achieve security and to overcome these illegal activities, our proposed system is developed to provide better security to the ATMs. Here, the PIN numbers are replaced with biometric security. The main objective of the work is to eliminate the use of ATM cards completely and to ensure better security. In this proposal, we provide the idea of using Aadhaar number as user ID and fingerprint as password. After biometric verification, the user will be allowed to proceed with the transaction. In case of three successive wrong attempts, the account will be blocked.

Keywords: Authentication · Biometric-fingerprints · Aadhaar number · ATM

1 Introduction

For the past many years, Democratic world fears for robbery and theft. The current scenario seems to be highly secured. But on deep sight over the economic performance of the country, there has been a vast economic down level performance. One of the most major incidences that we can see in our day to day life is ATM (automated teller machine) bank robberies.

The change in banking activities include the use of ATM's for banking transactions like cash withdrawl, money transfer and so on. The account holder will be issued with an ATM card and a private PIN as password. The PIN number will always be an important consideration to protect our financial information. But those PIN numbers can easily be hacked and misused.

Biometric authentication may be used as the solution for this problem. Biometric in advance field is concerned with identifying a person based on a person's physiological or behavioral characteristics. One of the most common physical biometric characteristic include fingerprint, retina, iris and so on. Fingerprint's specific feature is that they do not change for the entire life which is cheap and reliable. Thus, fingerprint verification is an effective method which is widely used for other comparison of biometric information.

D. J. Hemanth et al. (Eds.): COMET 2019, LNDECT 35, pp. 134–139, 2020.
https://doi.org/10.1007/978-3-030-32150-5_14

2 Literature Survey

Vijayasanthi *et al.* (2017) pointed out that the method in which users authentication's second factor error are reduced due to human fault and also by sophistication of malware attacks. They proposed a fingerprint authentication method of matching the bifurcation which are calculated as points and matched. All fingerprints are collected by optical sensors which are sent to cloud via Raspberry Pi. The fingerprints are to be authenticated which are stored in file server and in web server. It sends out a restore based match score along with fingerprint ID.

Singh *et al.* (2016) proposed a constrain on transaction through ATM involving biometric to improve system performance and to solve defined problems. Its separated into two parts. The first part solve the sensor performance by adding only a limit amount of cash or tracking over if one need to withdraw a huge amount or attempts of multiple transactions. In the second part it explains how verification using fingerprint os conducted and how the claimant access the system and the measures to increase the performance of the fingerprint biometric system. But the disadvantage of this method is the usage of ATM card and PIN number for low amount transaction.

3 Proposed Methodology

3.1 Objective

- To enhance the ATM security using fingerprint authentication
- To eliminate the use of ATM card completely
- To provide highly secure identification using biometric.

3.2 Methodology

Security is the serious issue in ATM system. Accessing of ATM machine using PIN number has become less secure because they are easily traceable. Chances of missing and misusing ATM cards has been increased. The existing security in the ATM system has not been able to address these challenges.

To overcome these challenges, the proposed work involves biometric security. Fingerprint technology in particular, can provide a much more accurate and reliable user authentication method. This system allows user to make banking transaction through the use of their fingerprint. The fingerprint minutiae features are different for each human being. Hence it is used for more accurate authentication.

The user enters Aadhaar number as the User ID and fingerprint as password. After biometric verification, the user will be allowed to proceed with the transaction. In case of three successive wrong attempts, the account will be blocked. This system has designed with Python-Database integration along with the use of hardware components, Arduino as well as Fingerprint module (R305), to provide cost effective banking ATM system. Access to multiple banks and multiple accounts are provided in this system.

3.3 Functional Block Diagram

The user has to choose the bank from the options. To login, the user has to enter his/her Aadhaar number. Biometric verification should be provided by the user. If the fingerprint is matched, the user will be allowed to carry out the following transactions such as: Account details, Withdraw, Deposit & Money transfer. After finishing a successful transaction, the user can either continue or exit. If not the user will be allowed with two more attempts to login. Once exceeding three attempts the account will be blocked. From the Figs. 1 and 2. At first, the user enters Aadhaar number as the User ID and fingerprint as password. After biometric verification, the user will be allowed to proceed with the transaction. User can access any of their accounts but it should be linked with their Aadhaar Card. Once after logged in, they get all privileges for accessing all features. In case of three successive wrong attempts, the account will be blocked. This system has designed with Python-Database integration along with the use of hardware components, Arduino as well as Fingerprint module (R305), to provide cost effective banking ATM system. Access to multiple banks and multiple accounts are also supported in this system.

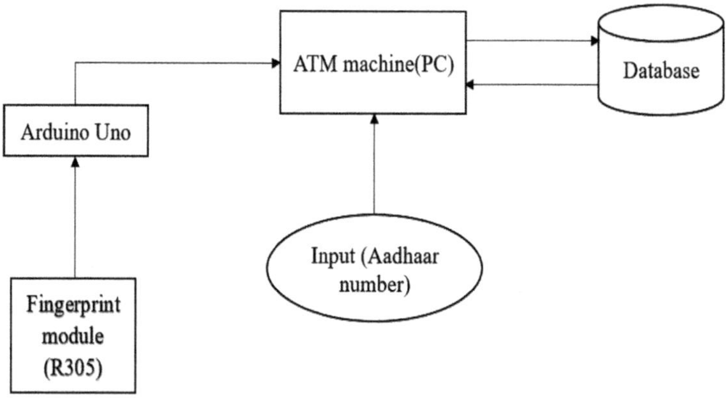

Fig. 1. Functional block diagram of the proposed methodology

3.4 Hardware and Software Output

From Fig. 3 we observe that the bank is choosen and proper authentication is done by using biometric verification. If it is matched with the aadhar card number, the money transaction and other transactions are done successfully. If it is observed to be a illegal authentication as shown in Fig. 4. The account gets blocked. Thus, its completely reliable and secure. This method results in fast and accurate authentication for the identification of user. Since this is based on fingerprint authentication technique there will be no need to disclose PIN or passwords with third parties. Hence we have fused biometric fingerprint technique for authentication of user to ameliorate the security level (Fig. 5).

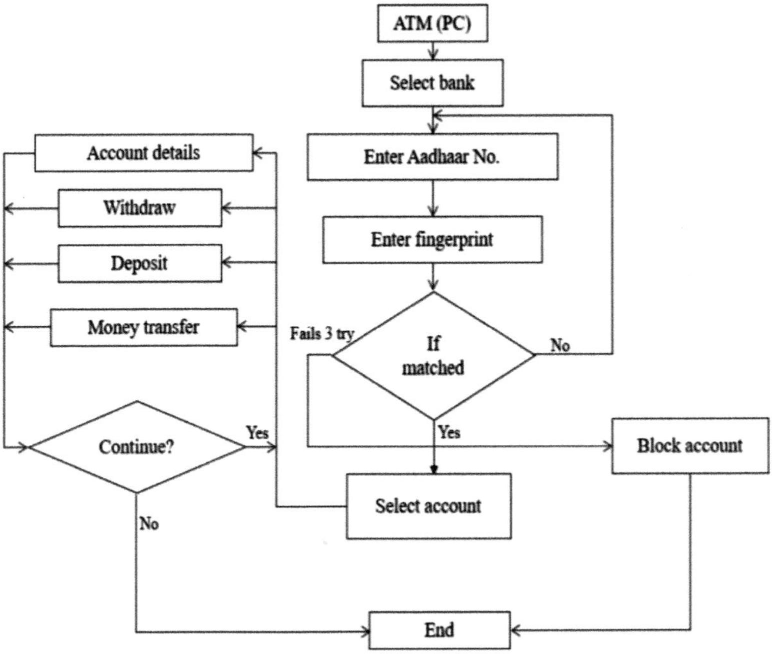

Fig. 2. Flow diagram of our system.

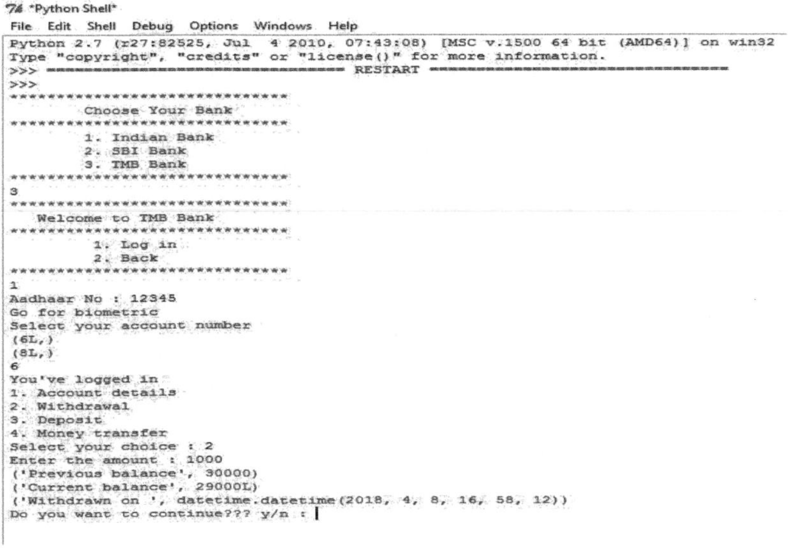

Fig. 3. Choosing & various bank transactions

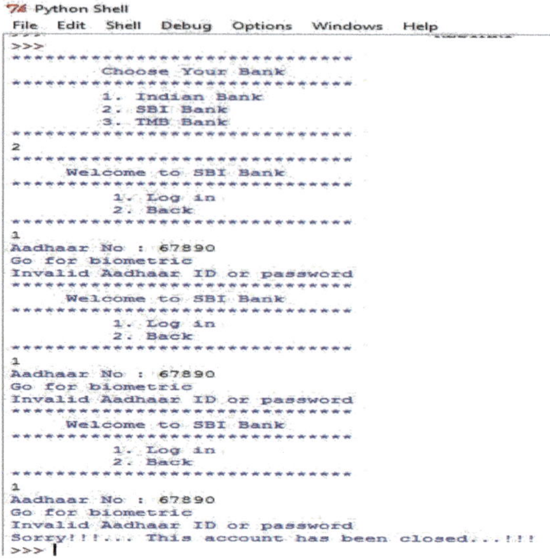


```
>>>
******************************
        Choose Your Bank
******************************
        1. Indian Bank
        2. SBI Bank
        3. TMB Bank
******************************
2
******************************
     Welcome to SBI Bank
******************************
          1. Log in
          2. Back
******************************
1
Aadhaar No : 67890
Go for biometric
Invalid Aadhaar ID or password
******************************
     Welcome to SBI Bank
******************************
          1. Log in
          2. Back
******************************
1
Aadhaar No : 67890
Go for biometric
Invalid Aadhaar ID or password
******************************
     Welcome to SBI Bank
******************************
          1. Log in
          2. Back
******************************
1
Aadhaar No : 67890
Go for biometric
Invalid Aadhaar ID or password
Sorry!!!... This account has been closed...!!!
>>>
```

Fig. 4. Illegal transactions

Fig. 5. Hardware setup

4 Conclusion

In order to improve & optimize the security in the ATM, this proposed system focuses on the usage of proper authorization by means of fingerprint sensor and Aadhar as a user id. The system employs with a Arduino UNO as front end with fingerprint sensor. The proposed system contains a Arduino UNO for operation and mySQL for database. The propose system achieves a higher efficiency and it will be responsible for proper authentication and thereby arrest the illegal transactions. The system is very suitable for all bank sectors and all kind of banking applications and it is highly reliable for security related issues.

References

Onyesolu, M.O., Ezeani, I.M.: ATM security using fingerprint biometric identifier: an investigative study. Int. J. Adv. Comput. Sci. Appl. **3**, 68–72 (2012)

Renee Jebaline, G., Gomathi, S.: A novel method to enhance the security of ATM using biometrics. In: International Conference on Circuit, Power and Computing Technologies (2015)

Singh, S., Singh, A., Kumar, R.: A constraint based biometric scheme on ATM and swiping. In: International Conference on Computational Techniques in Information and Communication Technologies (ICCTICT) (2016)

Vijaysanthi, R., Radha, N., Jaya Shree, M., Sindhujaa, V.: Fingerprint authentication using Raspberry Pi based on IoT. In: International Conference on Algorithms, Methodology, Models and Applications in Emerging Technologies (ICAMMAET) (2017)

Yang, Y., Mi, J.: ATM terminal design is based on fingerprint recognition. In: 2nd International Conference on Computer Engineering and Technology (2010)

Efficient Student Profession Prediction Using XGBoost Algorithm

A. Vignesh(✉), T. Yokesh Selvan, G. K. Gopala Krishnan, A. N. Sasikumar, and V. D. Ambeth Kumar

Computer Science and Engineering, Panimalar Engineering College, Chennai 600123, India
vignez2197@gmail.com

Abstract. As Competition is growing high students need to assess their capabilities and area of interest and plan their career from initial stages. Choosing a right profession is a crucial thing in today's world. This paper proposes a Supervised Machine Learning Model which uses large number of datasets to train the model and predict the right career path for the students. The dataset is collected by evaluating the students and also by enabling the students to answer a set of questions. The dataset includes various parameters like students ability in academics, competition, programming languages, interested domain. Prediction of both academic performance and employability can help the management identify students at risk of poor academic performance and low employability Also recruiters use this model to recruit candidates and decide which job role to assign them based on their capabilities. This paper concentrates on scholar profession prediction of computer science candidates.

Keywords: XGBoost · One Hot Encoding · Machine Learning

1 Introduction

Nearing the completion of degree the student starts to think of choosing a career.

With the increasing career opportunities in today's world making a right decision has become difficult. There is a dilemma for the student to select a career that is high in demand or to select a career that suits his personality. For example consider an extrovert student would require a job that requires lot of interaction with other people while an introvert requires a cubical job. The wrong choice might cause work dissatisfaction and stress. Evaluating student performance by developing a Machine Learning model might not be a easy task as the process of learning is individual effort of a student. However Data mining provides new insights for this problem by identifying features that influence student performance. Guidance system is useful both in academics and industry it allows student to choose the latest trending course and the best field to choose. Universities collect large data about student that is being unutilized. Prediction algorithm is used to find the most important attribute among the student data collected. These data can be analyzed to find the low performing student as it is important criteria for a good University to have high performing student. The low performing student is given special care and training to make them eligible for

D. J. Hemanth et al. (Eds.): COMET 2019, LNDECT 35, pp. 140–148, 2020.
https://doi.org/10.1007/978-3-030-32150-5_15

employment. The student can also analyze his weakness and can improve himself beforehand. Predictive analysis is used to predict the right career choice. Predictive analysis is the process of using machine learning to predict the future outcomes. Literature review of existing system must be done to study the gaps and to know the variables used in the previous prediction methods. There are very few online counseling systems which counsel student through video calls and Chabot which might not be efficient for mass number of students. The proposed model displays three set of questionnaire to student with which the student has to answer the questions. Three sets were based on personality, interest and capacity. Personality traits are distinct and depends on the psychology of the person. These questions examine the student to identify them as introvert or extrovert, sensing or intuitive, thinking or feeling, judging or perceiving. To find how much a student is interested in a subject the questions were framed. Capacity is to check how efficiently a student can learn a subject. From these answers the system predicts a career for student using Machine Learning algorithm. Machine learning promises to derive meaning for all that data we collected. Machine Learning is not a magic it's just the tools and technology that we can utilize to answer questions with our data. The algorithm used for our prediction is XGBoost. The efficiency of the model is tested using confusion matrix, precision and recall.

2 Literature Survey

In [1] the author considers new features to predict the student performance. The features were family expenditure (ex. studying family members, accommodation expenses), family income, student personal information (ex. Gender, Marital Status), family assets (ex. Land Value, Bank balance). By combining the new features with the existing features he performs the classification. By experimentation he showed how his proposed features plays an important role in predicting student performance. Parents having their own house saves money on rent and can use it for educational purposes and no need to change their houses which wastes time and energy for student. Having a good accommodation enables student to better concentrate in studies.

The paper [2] describes a machine learning based candidate selection procedure for job recruitment in software firm using Naïve Bayes classifier. Parameters are selected by the recruiter, total of 11 parameters were taken (ex. Projects done, Thesis, GPA in C, Java, DBMS). Training data is collected from a software firm. Using this Data the machine gets trained and with new inputs the model shortlist the eligible candidates for the software firm. In paper [3] a time series based statistical data mining job absorption rate prediction and predicting the waiting time needed for 100% placement for particular branch in a particular year is proposed which helps the student to choose the right discipline. Data is collected about passed out students. The placement status rate is calculated for a period of every 3 month for each year which helps in calculating the time needed for 100% placement. This helps in giving extra attention to branches which are lagging in placement. Curve fitting and regression analysis concepts are used. The attributes are plotted in a graph and the best fit line is choosen and equation is formed to predict the future outcomes.

The paper describes about student's performance prediction, analysis, early alert and evaluation using data mining. Students performance is analysed using the student's academic record such as internal assessment marks, assignment submission, and attendance percentage. Student performance is predicted in the upcoming semester using the previous database so that student at risk can be given alert. The techniques used in this paper are classification, clustering, ensemble and many others. Real time data is collected from any University or colleges. The time taken for training can be reduced using clustering technique [4]. [5] The system helps in guiding the students for choosing the appropriate stream using several assessment tests which includes aptitude test i.e. verbal, quantitative, logical and miscellaneous test and personality test. Personal and academic details were also collected including hobbies, interest, favorite subject. The system analysis the scores of these tests and student will be provided with the assessment report with top two stream that matches their profile which would help them choose a stream. Also the system recommends colleges of that stream. KNN algorithm is used. [6] The fuzzy expert system helps the students to give idea of possible career opportunities most suitable for them. The project gives a personal aid to the students by analyzing the student's interest and aptitude test result. The system uses six inputs (cost of course, appeal of course topic, perceived difficulty of course, past performance etc.) collected through survey among college students. First the student need to register with his personal details and can take tests. Two types of test the students need to take including interest analysis and aptitude test. By combining the analysis of two tests the system recommends the suitable career choice and also the colleges for that career. This system acts as assistant to real life counsellors and here all the available career opportunities can be explored so that student can get a clear idea of every available opportunities. The Online Expert System [7] which guides the students for the selection of their undergraduate courses after the completion of higher secondary school education which takes the necessary details from the student as input and will have the knowledge-base which contains the details about the colleges (placement details, department details, ranking, cut-off marks for previous year). This information is acquired from web pages using pattern matching and jSoup parsing technique and the knowledge-base is constructed automatically without manual effort and it is dynamically updated. The inputs were the region in which the student comes, which stream the student opt for, which branch the student prefers, fees he can pay, 12th percentage, whether he has reservation or he belongs to general category, whether hostel facility needed, current age as on date. The expert system takes this as query and gives output-recommended colleges. [8] In this paper the persons current career path and his goal or his career dream is taken. 67,000 profiles from LinkedIn is collected as a data source. For working experience, the raw data consists: name of the company, position, time period. For Education information: name of the university, degree, major and time period. Instead of using company name company size is used as the feature similarly Universities are classified into top 10,50 and other. Using k-means clustering similar job positions are combined. The user gives his objective (ex. software engineer at Facebook) the system recommends the shortest career path that would lead him to his objective.

3 Proposed System

Real time data is collected from Google Forms where student fill the required parameters which is taken as the features and Suggested Job Role is our label. There are many job roles like Developer, Data Scientist, Assistant System Engineer etc. We fix the job roles to 15 and parameters to 36 in our model. In Existing System only technical abilities of the students were analyzed here we also analyze student abilities like sports, hobbies, interest, competition. Data is preprocessed and One Hot Encoding is used to encode the categorical labels. The data are classified and the predictions are made. Supervised learning is used, it uses labeled data. If we know the class labels previously and we need to assign the new data with the predefined class label it is supervised learning. Since we have labeled data which is the suggested job role we use supervised machine learning algorithm - XGBOOST. The Machine Learning model is created and trained and predictions were done. The architecture diagram is given below (Fig. 1).

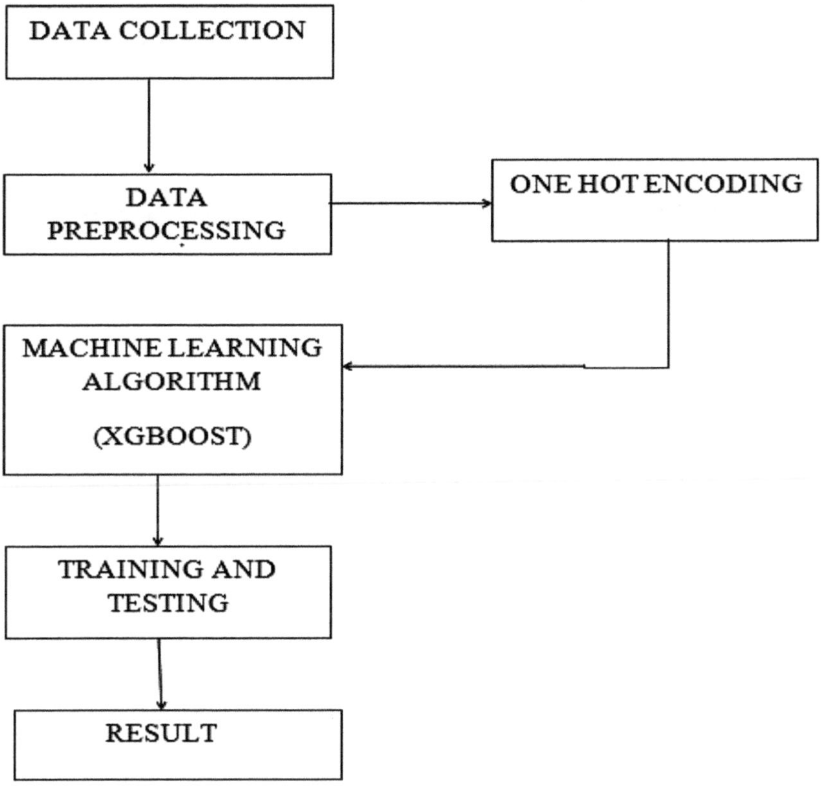

Fig. 1. Architecture diagram

As shown in the above figure the Paper is classified into four modules

1. Data Collection
2. Data Preprocessing
3. Machine Learning Algorithm
4. Training and testing.

3.1 Data Collection

Data collection is the most important and tedious task in Machine Learning as the input to our machine learning model is our data. We don't need big data to start running the model. We should choose the right features influencing the right outcome. If we have the right features we can build up data in the future. Irrelevant data gives bad predictions. The data collected should not be selection bias. For our model Real time data is collected from Twitter, kaggle, UCI, Data.gov and google forms. For our model the data must contain many parameters like student academic scores in various subjects, specializations, programming and analytical capabilities, memory, personal details like relationship, interests, sports, competitions, hackathon, workshops, certifications, books interested and many more.

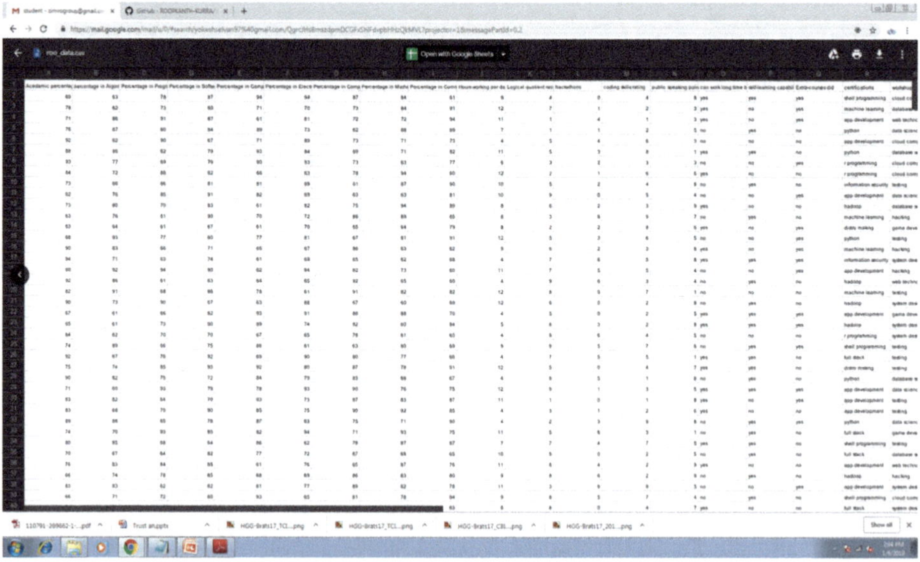

Fig. 2. Google form data collection

Sample Google Form Questions for Data Collection (see Fig. 2):

1. What his/her standards in programming?
2. His/her analytical standards?
3. Area of specialization that he choosed?

4. Percentage in specialization subject?
5. Gap in between education?
6. What is the company he/she want to place (product or service based)?
7. No. of companies heat tented?
8. No. of coding competitions that he attended?
9. Rating of his technical knowledge?
10. No. of off campus drives that he attended?
11. What is his/her dream company?
12. Language skill, in which language he/she is good?
13. What is his/her dream company?
14. No. of project he/she did?
15. What is the area of his/her final project?
16. His/her role model?
17. Did he attend any courses after B.Tech. (GRE/GATE)?

3.2 Data Preprocessing

Preprocessing is very important for accuracy of the Machine Learning Model. Data will be noisy it might contain missing values and also false values. Feature selection and feature Extraction are the two methods. Feature selection is selecting relevant and discarding irrelevant. One Hot Encoding is one of the methods of feature extraction. Dimensionality reduction is one of the important task where the dimensions of the attributes keep on increasing therefore the module gets tougher to train. It is used to visualize the Data. Fill the missing values by predicting it or filling it with the mean of the known values. Once we are done with data preprocessing we are ready to apply data to a Machine Learning model.

3.3 One Hot Encoding

One Hot Encoding is used to convert the categorical data to numerical values as it can be fed to machine learning model. Why One hot encoding is used over integer encoding? Let's see this with an example. The categorical data were dog, cat and lizard. In integer encoding Dog is labeled as 0 cat as 1 and lizard as 2. [dog, cat, lizard] = [0,1,2]. Here it implies that dog < cat < lizard which is not meaningful and here comes One Hot Encoding. In One Hot Encoding the output nodes is equal to the number of classes we want to classify.

[1,0,0]-Dog
[0,1,0]-Cat
[0,0,1]-lizard

This encoded data is fed to neural networks and trained and desired output is fetched. May not be efficient in storage perspective but gives the desired output.

4 Machine Learning Algorithm

5 XGBoost (Extreme Gradient Boosting)

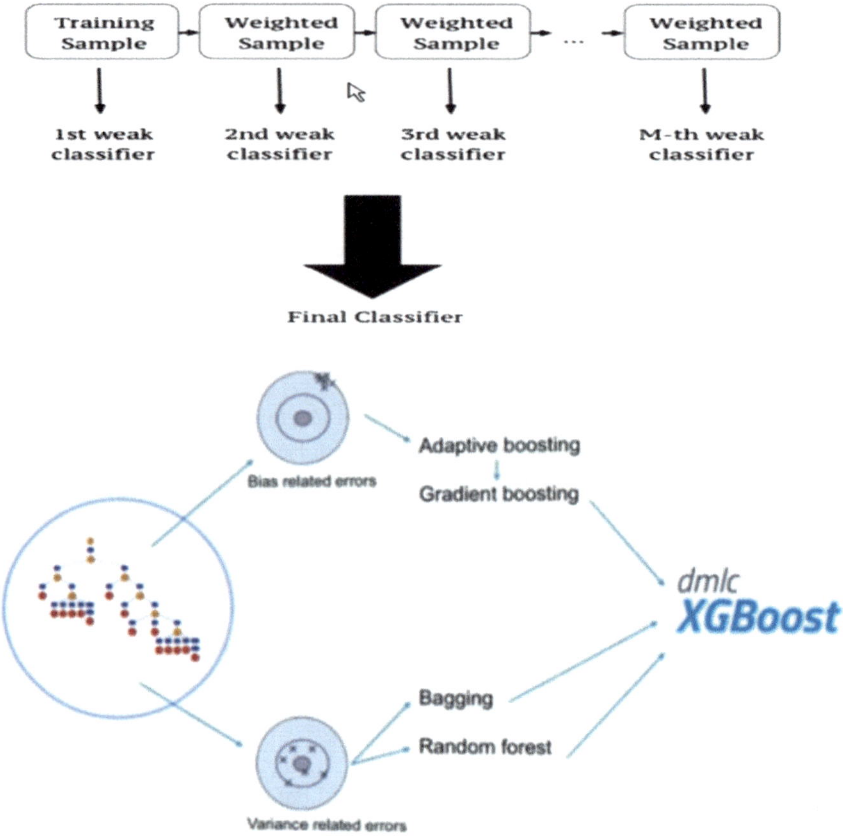

Fig. 3. XGBoost algorithm

Decision Tree produces two types of errors - Bias related errors and Variance related errors. There are ensemble methods to overcome the errors Adaptive Boosting and Gradient Boosting to overcome Bias related errors and Bagging and Random Forest to overcome Variance related errors. The concatenation of all the ensemble methods is our XGBoost Algorithm (see Fig. 3).

XGBoost has very good predictive power but slow with implementation. Initially all the instances in a dataset are assigned the same weights. The training sample is passed to a Decision Tree it creates weak classifier then calculate the error and the coefficient and the wrongly predicted samples are assigned bigger weights and then passed to the next decision tree to get another weak classifier and the successive decision trees rectifies the errors made by the previous one. A weak classifier is better than a random guessing. The average of all the weak classifiers is the final prediction which is a Strong classifier (see Fig. 4). XGBoost handles missing values and imbalanced data set. XGBoost can take already working solution and start working on the improvement.

5.1 Training and Testing

Split the data into two - training data and testing data. If we test the same data upon with which we trained the model obviously the result will be good. So train the model with a data and test the model with new data to test the accuracy of the model. Feature X is split into say X_ train, X _test. Label Y is split into sayY_train, Y_test. Run the training data X_train, Y_train into Machine Learning algorithm to create a Machine Learning model. The model gets trained with the data and gets ready to give predictions. To get the prediction give X_test as input to Machine Learning Model which produces the Prediction say Prediction. To test the accuracy Compare the Y_test and Prediction using a Confusion Matrix. Confusion Matrix tells what our Machine Learning Algorithm did right and what it did wrong. Add new data to the dataset to improve the quantity of the dataset. The final result is the Prediction of our Machine Learning model (refer Fig. 5).

6 Conclusion

Choosing a right career choice for students is a crucial task. Students confuses a lot to select a career among the possible career opportunities. Thus in this paper we proposed a profession prediction system which collects real time data about students through google forms. One Hot Encoding is used to preprocess the data. XGBoost algorithm is used to make the career prediction by analysing the data collected as it has very good predictive power. By this system students are provided with the career choice that matches their profile eliminating the students pressure in choosing a right profession. This system is also used by recruiters to recruit a eligible candidate.

References

1. Daud, A., Aljohani, N.R.: Predicting student performance using advanced learning analytics. In: 2017 International World Wide Web Conference Committee (IW3C2) (2017)
2. Jannat, M.-E., Sultana, S., Akther, M.: A probabilistic machine learning approach for eligible candidate selection. Int. J. Comput. Appl. (0975-8887) **144**(10), 1–4 (2016)

3. Elayidom, S., Idikkula, S.M.: Applying data mining using statistical techniques for career selection. Int. J. Recent Trends Eng. **1**(1), 446 (2009)
4. Kavipriya, P.: A review on predicting students' academic performance earlier, using data mining techniques. Int. J. Adv. Res. Comput. Sci. Softw. Eng. **72**, 414–422 (2015)
5. Kolekar, S., Bojewar, S.: A review: e-counseling. Int. J. Sci. Res. Comput. Sci. Eng. Inf. Technol. **3**(3), 855–859 (2018)
6. Gupta, M.V., Patil, P.: FESCCO: fuzzy expert system for career counselling. Int. J. Recent Innov. Trends Comput. Commun. **5**(12), 239–243 (2017)
7. Saraswathi, S., Hemanth Kumar Reddy, M.: Design of an online expert system for career guidance. Int. J. Res. Eng. Technol. **3**(7), 314–319 (2014)
8. Lou, Y., Ren, R.: A machine learning approach for future career planning (2010)

Design and Analysis of Mixer Using ADS

S. Syed Ameer Abbas(✉), S. Rashmita, K. Lakshmi Priya, and M. Lavanya

Mepco Schlenk Engineering College, Sivakasi, India
ssyed@mepcoeng.ac.in, srvakkiyiel@gmail.com, laxmikannan96@gmail.com, lavanyamurugan1995@gmail.com

Abstract. A radio receiver is an electronic gadget that gets radio waves and changes over the information carried by them to a different format. It is utilized with an antenna. The receiving wire intercepts radio waves and changes over them to little alternating currents which are connected to the receiver, and the receiver extricates the specified data. The receiver uses electronic channels to isolate the specified radio frequency signal from all the other signals picked up by the antenna, an electronic amplifier to extend the control of the signal for further processing, and at long last recuperates the specified information through demodulation. There are many radio receivers accessible. TRF (Tuned Radio Recurrence) receiver is utilized in early days. It contains a drawback of increment within the number of sidebands. It has been overcome by the Superheterodyne receiver. It has given the specified yields by dodging the pointless sidebands of employing a mixing circuit basically called as mixer. The simulation of mixer is performed utilizing ADS.

Keywords: RF · Superheterodyne

1 Introduction

An RF module is an electronic device which has been used to transmit and receive information. It communicates with the other devices present in the same network in a wireless manner. This wireless based communication is omnidirectional. Line of sight is not required by RF. RF communication is a wide range of communication which includes a transmitter and a receiver. Each transmitter and receiver are of diverse ranges. For designing purpose, RF modules has been well utilized. It blown off the issue in designing the radio circuitry. A transmitter is a piece of hardware present inside the electronic device for transmitting the information to the other device. Combination of a transmitter and a receiver togetherly called a transceiver. Transmitter is otherwise represented as 'TX'. The input signal is fed up to the transmitter in the form of an electric signal, such as an audio signal from a amplifier, a video signal from a video camera. The combination of the information signal and the radio frequency signal in order to generate the radiowaves, which in turn is called the carrier signal. This process is termed as modulation. The information can be included to the carrier in several distinctive ways, totally different sorts of transmitters. There are different forms of modulation. (a) Amplitude modulation (b) Frequency modulation (c) Phase

D. J. Hemanth et al. (Eds.): COMET 2019, LNDECT 35, pp. 149–157, 2020.
https://doi.org/10.1007/978-3-030-32150-5_16

modulation. In amplitude modulation, the message signal has been added to the carrier wave by changing its amplitude. In frequency modulation, the message signal has been added to the carrier wave by changing its frequency. In phase modulation, the message signal has been added with the carrier wave by changing its phase. There are still different forms of modulations. The electricity passes into the transmitter and makes the electrons to excite. At the receiver side, it consists of an antenna. That antenna receives the radiowaves at the desired frequency. After receiving the radiowaves at a desired frequency, process takes place within the receiver with those waves. The audio signal mixed with the carrier signal is finally extracted and it is fed as input to the loud speaker after proper amplification. Finally the loudspeaker plays the audio.

A prototype mixer has been designed. Simulation shows this mixer achieves 19.7 dBm IIP3 with 1.1 dB power gain, 13.6 dB noise figure at 2.4 GHz and only 3.8 mW power consumption [1]. Three double-balanced Gilbert-type down conversion mixers, a Gilbert-type mixer based on the current bleeding technique, a Gilbert-type mixer based on the current bleeding technique with one resonating inductor, and a Gilbert-type mixer based on the current bleeding technique with two resonating inductors, have been designed and analyzed to improve flicker noise performance [2]. The design of a low noise CMOS Gilbert cell mixer is implemented in 180 nm technology process. The proposed mixer results a simulated conversion gain of 9.95 dB and a noise figure is about 8.12 dB [3].

2 Proposed Methodology

In a super-heterodyne receiver, the unwanted sidebands are avoided by adding a mixing circuit. The mixing circuit is coupled along with a local oscillator. The input frequencies has been made constant using mixing circuit. Ganged tuning is additionally performed here in order to make the input frequency constant. The difference frequency has been selected in order to avoid interference. It is shown in Fig. 1.

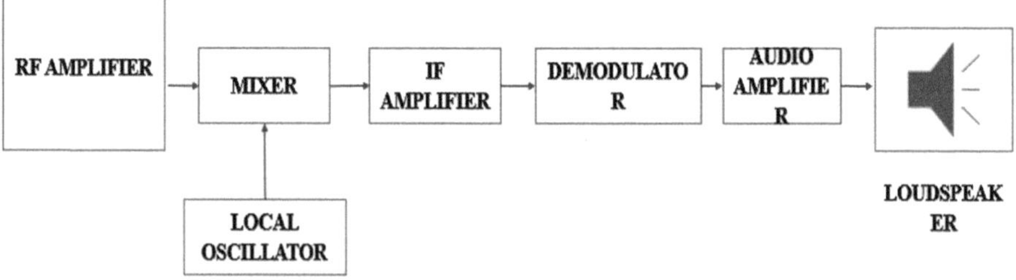

Fig. 1. Superheterodyne receiver

RF Amplifier receives all the frequency components. Among that it will choose the desired set of frequency (fif). Local Oscillator produces a sinusoidal wave which is used to process the incoming RF signal. The product of the RF signal and the sine wave signal produces the sum and difference frequencies at the output of the mixer stage. The difference frequency alone is selected because of its lower frequency signal. The low frequency signal has been amplified. The original signal has been recovered by the demodulator module present at the last stage of the receiver. Once demodulated, the recovered audio is applied to an audio amplifier, which is then amplified to a desired level. Then it is given to the loudspeaker.

3 Modeling of Mixer

The proposed design flow is shown in Fig. 2.

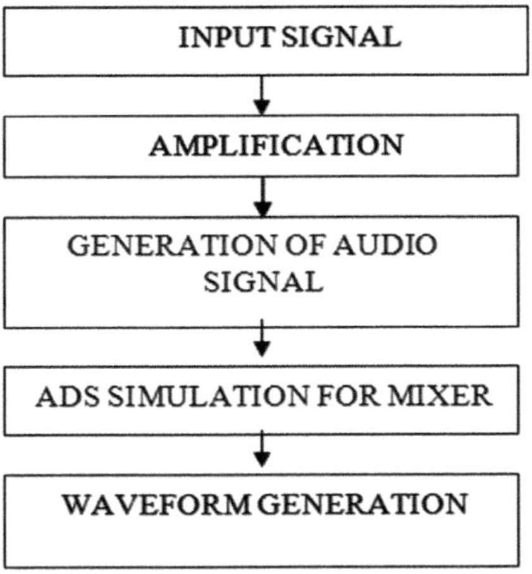

Fig. 2. Proposed design flow

The Ideal Mixer

The ideal mixer consists of 3 ports. The input port is the RF port. It receives the high frequency RF signals. The output port is the IF port. It is of low frequency components. The local oscillator is another input port which generates the sine wave in order to process the RF signal. It is shown in Fig. 3.

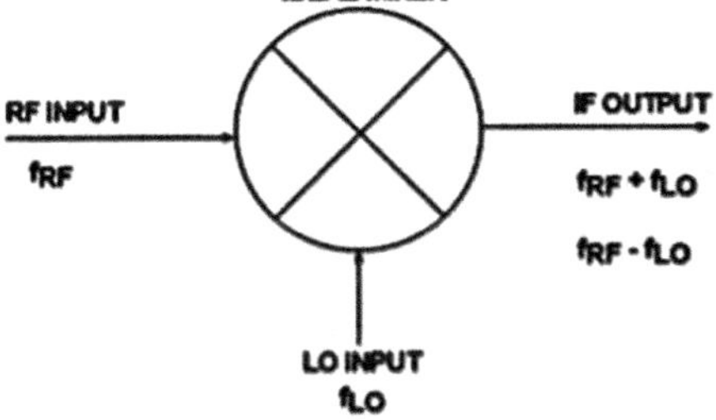

Fig. 3. Ideal mixer

Types of Mixers

1. Mixer with unbalanced nature
2. Mixer with Single Balanced nature
3. Mixer with Double Balanced nature.

A. Mixer with Unbalanced Nature

The simplest RF mixer is the unbalanced mixer. It takes output (I_{if}) from any of the two branches. This current I_{if} will pass through the resistor and develops the intermediate frequency voltage. The voltage of the local oscillator and the radio frequency gets multiplied and produces intermediate frequency voltage. This produces frequency components of RF at the output stage. This is referred as RF feedthrough. It has been represented in Fig. 4.

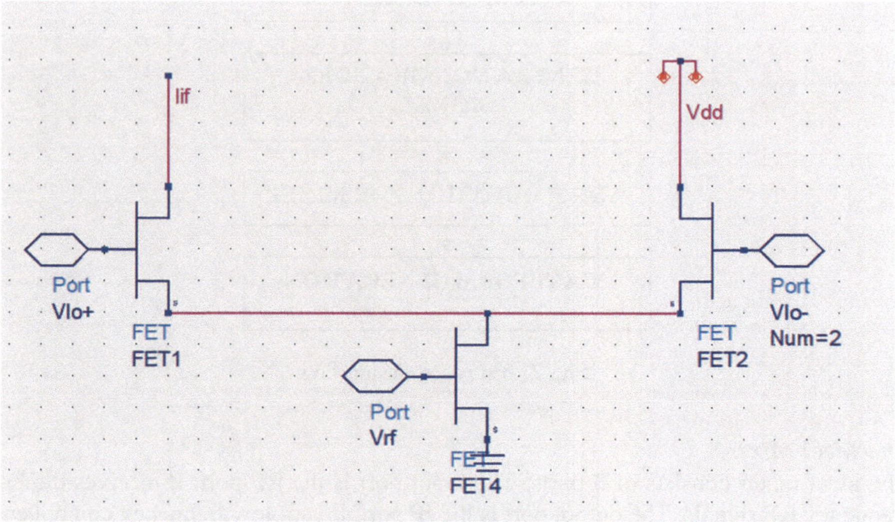

Fig. 4. Mixer with unbalanced nature

B. Mixer with Single Balanced Nature

The development of the unbalanced mixer is the single balanced mixer. Here current signals (I_{if}) are taken from both the branches having different polarities. Both signals are taken and it passes through the resistors R_L of opposite polarities. Intermediate frequency has been generated at the output stage of both the resistors. Here RF feed-through has been cancelled out due to the presence of opposite polarities. But there is a penetration of LO frequency components. It created LO feedthrough at the output end. The circuit diagram of Mixer with Single Balanced nature has been shown in Fig. 5.

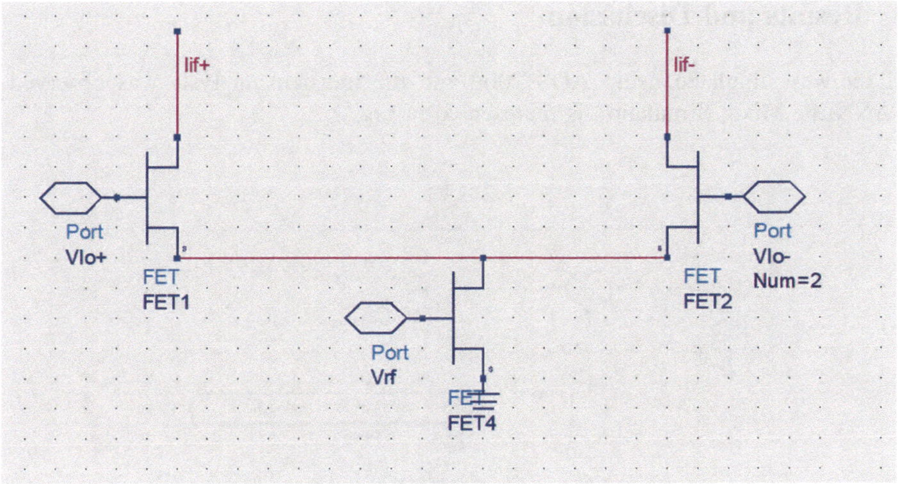

Fig. 5. Mixer with single balanced nature

C. Mixer with Double Balanced Nature

Double balanced mixer has been formed by combining two single balanced mixer. The same procedure followed in the above two cases has been repeated. The current from both the branches flows into the resistor arm at both branches. It generates an intermediate frequency. Here both RF and LO feedthrough get cancelled each other due to the opposite polarities. So, no feedthrough occurs. But, this condition will occur only at ideal conditions [4]. But, somehow feedthrough occurs due to the misalignments present inside the mixer circuit. RF to IF & LO to IF feedthroughs are common. But, feedthrough between RF and LO occurs sometimes which in turn causes reradiation. This type of feed through is present in a mixer with double balanced nature has been shown in Fig. 6.

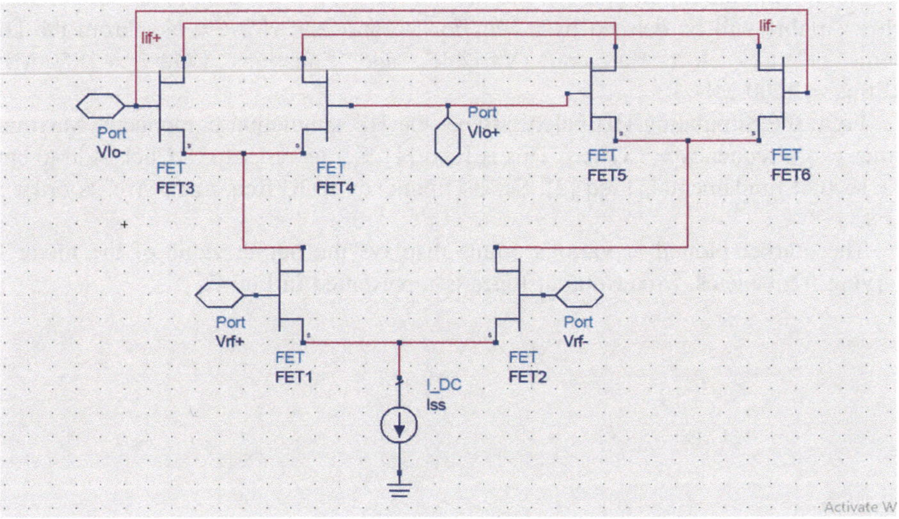

Fig. 6. Mixer with double balanced nature

4 Results and Discussion

Mixer was simulated using ADS 2009 and the transient analysis was observed in TANNER. Mixer Simulation is represented in Fig. 7.

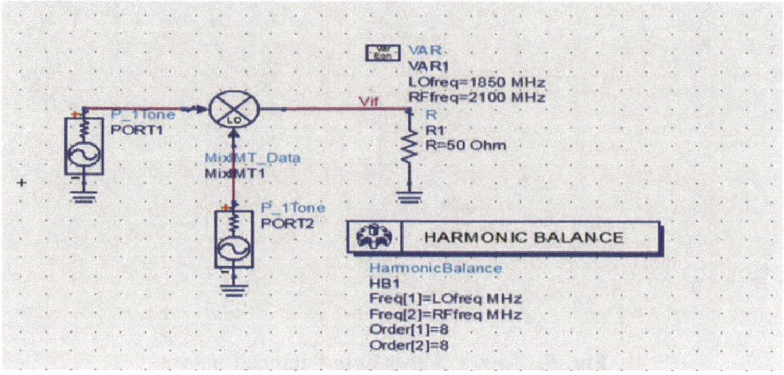

Fig. 7. Mixer simulation

Harmonic balance is a frequency-domain analysis strategy for simulating nonlinear circuits and frameworks. It is well-suited for simulating analog RF and microwave circuits, since these are most normally handled within the recurrence space. To find the amplitudes and connections of mixer yield tones: In Sources-Freq Space palette, select P_1Tone and put one occasion of it at the RF input (PORT1) and another at the LO input (PORT2). Num = 1, This is port 1, Z = 50 Ω (default esteem), P = dbmtow (−10). This changes over the 0-dBm input to watts (the control unit utilized by the system). Freq = RF freq. This variable will be defined by a Var Eqn component.

Num = 2, This is port 2, Z = 50 Ω (default value) P = dbmtow (7) Freq = LO freq. This variable will be defined by a Var Eqn component. Noise = No. From the Data Items palette, select Var eqn (Variables and Equations LOfreq = 1850 MHz, RFfreq = 2100 MHz).

From the Simulation-HB palette, select the HB simulation component. Maximum order = 8. Frequency = LOfreq. This is Freq [1]. Set its order to 8. Click Add to enter the second fundamental, Freq [2]. Set its frequency to RFfreq, and leave its order set to 8.

The marker plotted at various points displays the output value of the mixer for varying frequencies. Mixer output tones is represented in Fig. 8.

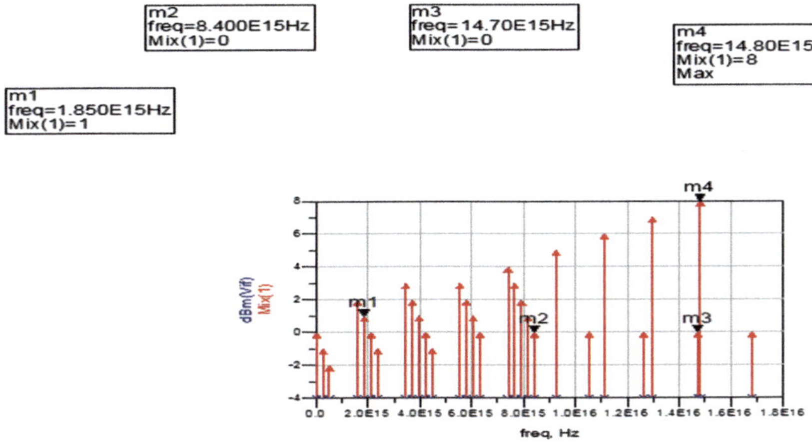

Fig. 8. Mixer output tones

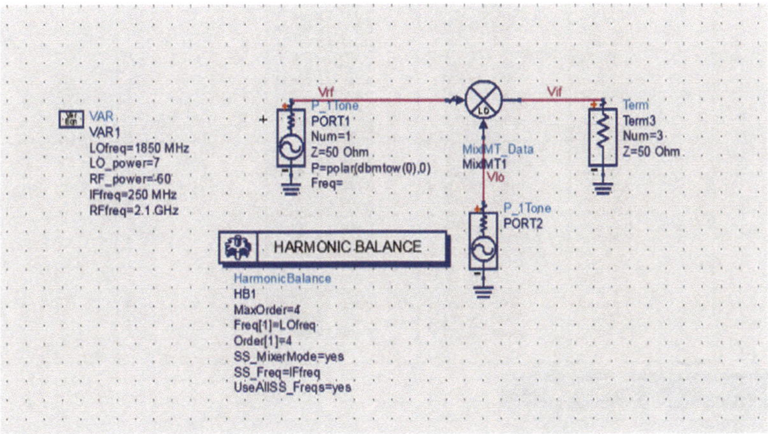

Fig. 9. Small signal simulation of a mixer

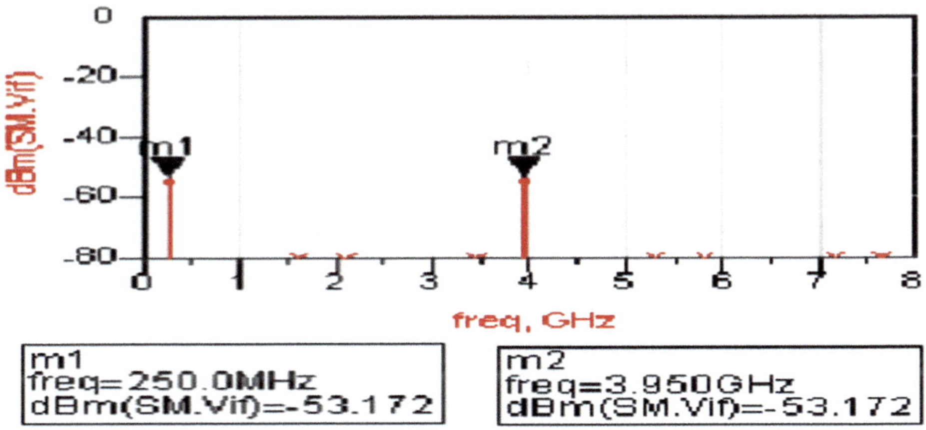

Fig. 10. Small signal simulation result

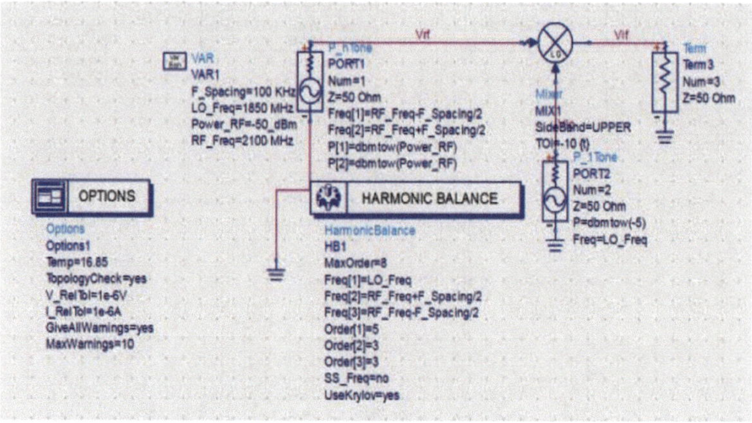

freq	Vif	Vrf	Mix(1)	Mix(2)
0.000 Hz	0.758E-18 / 0.000 V	0.000 / 0.000 V	0	0
250 MHz	6.00a / 179. V	0.000 / 0.000 V	-1	1
500 MHz	0.289E-18 / 121. V	0.000 / 0.000 V	-2	2
1.35GHz	0.000 / 0.000 V	0.000 / 0.000 V	3	-2
1.60GHz	0.000 / 0.000 V	0.000 / 0.000 V	2	-1
1.85GHz	0.0134E-48 / 0.000 V	0.000 / 0.000 V	1	0
2.10GHz	0.000 / 0.000 V	11.0m / 0.000 V	0	1
2.35GHz	0.000 / 0.000 V	0.000 / 0.000 V	-1	2
2.60GHz	0.000 / 0.000 V	0.000 / 0.000 V	-2	3
3.45GHz	0.790E-18 / 104. V	0.000 / 0.000 V	3	-1
3.70GHz	0.730E-18 / 52.2 V	0.000 / 0.000 V	2	0
3.95GHz	20.0m / 214.fV	0.000 / 0.000 V	1	1
4.20GHz	0.664E-18 / -4.75 V	0.000 / 0.000 V	0	2
4.45GHz	0.203E-18 / -175. V	0.000 / 0.000 V	-1	3
5.30GHz	0.000 / 0.000 V	0.000 / 0.000 V	4	-1
5.55GHz	0.000 / 0.000 V	0.000 / 0.000 V	3	0
5.80GHz	0.000 / 0.000 V	0.000 / 0.000 V	2	1
6.05GHz	0.000 / 0.000 V	0.000 / 0.000 V	1	2
6.30GHz	0.000 / 0.000 V	0.000 / 0.000 V	0	3
7.40GHz	0.553E-18 / 119. V	0.000 / 0.000 V	4	0
7.65GHz	0.598E-18 / -7.73 V	0.000 / 0.000 V	3	1
7.90GHz	1.90a / 104. V	0.000 / 0.000 V	2	2
8.15GHz	1.47a / 163. V	0.000 / 0.000 V	1	3
9.25GHz	0.000 / 0.000 V	0.000 / 0.000 V	5	0
9.50GHz	0.000 / 0.000 V	0.000 / 0.000 V	4	1
9.75GHz	0.000 / 0.000 V	0.000 / 0.000 V	3	2
10.0GHz	0.000 / 0.000 V	0.000 / 0.000 V	2	3

freq	Conv. gain
3.950GHz	6.000

Fig. 11. Mixer conversion gain

From the Simulation-HB palette, select the HB simulation component. Frequency = LOfreq. This is Freq [1]. Set its order to 4. Sweep type = Single Point, Frequency = IF Frequency.

Markers at 250 MHz (fIF) and 3.750 MHz (fLO + fRF) indicate the effects for the idealistic model of 6 dB of gain on the input RF power (−60 dBm). Small Signal Simulation of a mixer and its simulation result is shown in Figs. 9 and 10. Mixer Conversion Gain is shown in Fig. 11. For 3.95 GHz, the conversion gain achieved is 6.00.

5 Conclusion

The simulation of mixer is performed using ADS. The Vif value is plotted for various frequencies and the output tones are determined. The conversion gain of the mixer is determined. The local oscillator frequency for this design is 1850 MHz. The radio frequency for this mixer is 2100 MHz. Noise is excluded in this design in order to get the maximum desired value. With this fixed frequency, the performance of the mixer grows exponentially. It in turn brings the highest conversion gain. Though it have an efficient conversion gain, there are some losses. Further research on reducing conversion losses are carried out to bring an optimal conversion gain.

Acknowledgment. The authors would like to thank Department of ECE, Mepco Schlenk Engineering College, Sivakasi for providing the facilities to carry out this work.

References

1. Jiang, J., Holburn, D.M.: Design and analysis of a low-power highly linear mixer. In: IEEE Conference (2009)
2. Munusamy, K., Yusoff, Z.: A highly linear CMOS down conversion double balanced mixer. In: 2006 International Conference on Semiconductor Electronics (2006)
3. Rout, S.S., Sethi, K.: Design of high gain and low noise CMOS Gilbert cell mixer for receiver front end design. In: 2016 International Conference on Information Technology (2016)
4. Sullivan, P.J., Xavier, B.A., Ku, W.H.: Double balanced dual-gate CMOS mixer. IEEE J. Solid-State Circuits **34**(6), 878–881 (1999)

Realization of FPGA Architecture for Angle of Arrival Using MUSIC Algorithm

S. Syed Ameer Abbas(✉), K. P. Kaviyashri, and T. Kiruba Angeline

Mepco Schlenk Engineering College, Sivakasi, India
ssyed@mepcoeng.ac.in, kaviyashrikp@gmail.com,
kirubaanjaline@gmail.com

Abstract. In many digital signal processing applications antenna arrays are used due to their ability to locate signal sources. Array signal processing has a major key for Angle of Arrival (AOA) estimation. Although various algorithms have been developed for AOA estimation, MUSIC algorithm has higher resolution than the other algorithms which has high complexity prevents their use in real-time application. Using an eigen space method, autocorrelation matrix and the frequency content of a signal are estimated using MUSIC algorithm. The whole framework for AOA estimation has been simulated using Modelism.

Keywords: AOA · Antenna array · MUSIC · Steering matrix

1 Introduction

Array signal processing reduce the interface and noise present in the received signal from the antenna array. Antenna array is generally used to increase the directivity of antenna. Antenna arrays have many applications, most often it is used to improve the gain and shape the radiation pattern. Direction of arrival of every elements of the array antenna is determined by AoA using TDOA (Time Difference of Arrival) method, it can be done in either self-adaption or spatial spectrum estimation [1]. Spatial spectrum shows the signal distribution in each and every direction. Therefore, to determine the angle of arrival one should get the signal's spatial spectrum. In the antenna array, signals are received from different direction at different instance. The antenna array consists of number of individual elements and the distance between the antenna elements varies from larger value to the smallest centimeter value. Depending on the application the distance between the antenna elements is decided [2]. Angle of arrival estimation from different flag sources can expand the limit and throughputs of the system. In the greater part of the applications, the main undertaking is to estimate the AOAs of approaching signals, from this data confinement of the signal sources can be resolved.

Beamforming is preferred in terms of complexity. Extraction of desired information from transmitting signals from a certain direction which is employed by an antenna array including multiple sensors is a important task in array signal processing. The most straightforward method for beamforming is selecting the appropriate weights. By this method signal from particular direction is obtained whereas signals from other directions are attenuated. This is referred as beamforming or spatial filtering [2].

D. J. Hemanth et al. (Eds.): COMET 2019, LNDECT 35, pp. 158–168, 2020.
https://doi.org/10.1007/978-3-030-32150-5_17

1.1 Angle of Arrival (AOA)

AOA (Angle-of-arrival) estimates the signals received from certain direction, either in the form of electromagnetic (i.e., radio) or acoustic waves [3, 4]. The need of tracking and locating signals in civilian and military application is the basic for the emerging of the AOA estimation technique. The main objective of the AOA is to find out the spatial spectrum of the array element.

Three ways of finding the point of entry has been identified as Spectral-based algorithms, Subspace-based methods and Parametric Methods [4]. The subspace method has no limitation in the resolution from the aperture of the antenna array, therefore it is also known as super resolution techniques and it depends on the eigen analysis of an autocorrelation matrix of the received signal.

2 Proposed Methodology

2.1 AOA Estimation System Architecture

Spatial spectrum estimation is a particular signal estimation innovation and it tend to be isolated into three primary stages as target stage, observation stage, and estimation stage (Fig. 1).

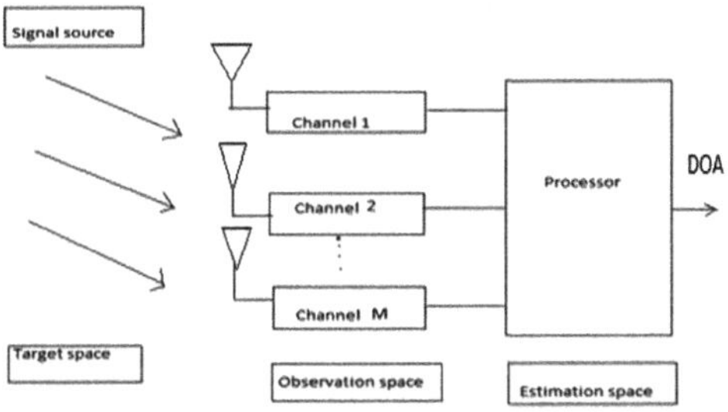

Fig. 1. System architecture of AOA estimation

The Target stage comprises of signal source parameters and a perplexing situation, the observation stage is a multidimensional one, where the receiving data is composed of a number of channels. For single channel the traditional time domain processing method is used. Estimation stage is the reconstruction of the target stage [5].

For example, in Fig. 2, two antenna array elements are placed and separated by the distance d.

The signal gotten by the receiving wire due to the manner in which differentiate is

$$\tau = \frac{d\,sin\theta}{c} \tag{1}$$

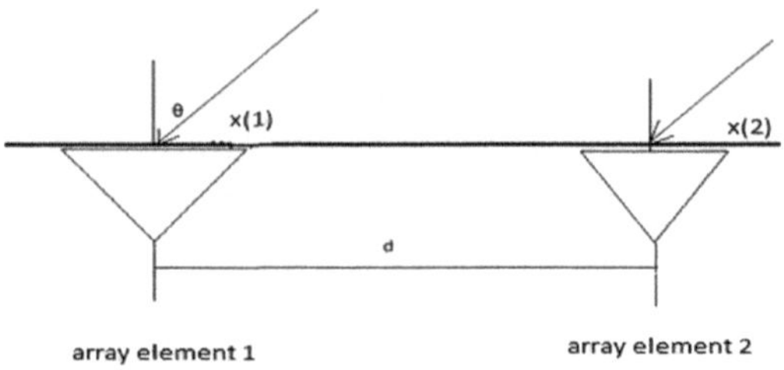

Fig. 2. AOA estimation principle

Where, c is the speed of light, θ is the incident angle of the far field signal and τ the time delay of the array element.

The difference between the array element is given as

$$\varphi = e^{-j\omega\tau} = e^{-j\omega\frac{d\, sin\theta}{c}} \tag{2}$$

Phase difference is given as

$$\varphi = \exp(-j2\pi\, d\, sin\theta\, \lambda) \tag{3}$$

where λ is the wavelength of the signal.

For spatial spectrum estimation MUSIC algorithm has created a new era. It process the covariance matrix of antenna array signal and it gives details of characteristic decomposition of it. Therefore, it results in a signal subspace symmetrical with a noise subspace related to the signal components.

Consider a linear array antenna which consists of ‘M’ number of elements and it is separated by a distance of ‘d’ cm apart. ‘D’ is the number of input signals considered with ‘k’ data samples [5]. The scene signals from “D” users are represented using the amplitude and phase by the complex quantities S1, S2, …, SD also white Gaussian noise added to the signals as vector (n). Steering vector are used to represent the directions of incident signals. a(Θi) represents the steering vectors for ith customer therefore we have matrix “A” with size M × D, where a(Θ1) can be given as Phase difference is given as

$$a(\theta_1) = \begin{bmatrix} 1 \\ e^{j\beta\, d\, \sin(\theta_1)} \\ e^{2j\beta\, d\, \sin(\theta_1)} \\ \cdots \\ e^{(M-1)j\beta\, d\, \sin(\theta_D)} \end{bmatrix} \tag{4}$$

Where β = incident wave number = 2π and d = inter element spacing

$$\begin{bmatrix} x_1(k) \\ x_2(k) \\ \cdots \\ \cdots \\ x_M(k) \end{bmatrix} = [a(\theta_1)\, a(\theta_2) \ldots\ldots\ldots a(\theta_D)] \begin{bmatrix} S_1(k) \\ S_2(k) \\ \cdots \\ \cdots \\ S_D(k) \end{bmatrix} + n(k) \tag{5}$$

$$x(k) = A\,s(k) + n(k) \tag{6}$$

x(k) = amplitude of signal + noise in ith element - order [M × K].
S(k) = vector of incident signals at sample time k - order [D × K].
n(k) = noise vector at each element m - order [M × K].
a(Θi) = M-element array steering vector - order [M × 1].
A = [M × D] matrix of steering vectors a(Θi).

2.2 Principle of MUSIC Algorithm

Figure 3 shows the various stages of MUSIC algorithm computation. Two orthogonal matrices of subspace such as signal and noise subspace has been obtained by decomposing the correlation matrix by the MUSIC algorithm. By using any one of this subspace the direction of arrival can be estimate and it is done with an assumption that every channel is highly uncorrelated in the noise [6, 7]. Due to this condition the correlation matrix become diagonal.

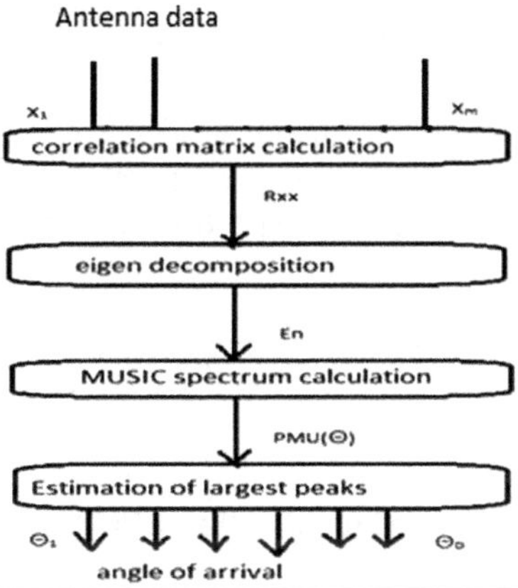

Fig. 3. MUSIC algorithm flow diagram

The Correlation matrix is given by

$$\begin{aligned} R_{xx} &= E\left[x \cdot x^H\right] = E\left[(As+n)\left(s^H A^H + n^H\right)\right] \\ &= AE\left[ss^H\right]A^H + R_{nn} \\ &= AR_{ss}A^H + R_{nn} \end{aligned} \tag{7}$$

Since white Gaussian noise has been assumed as zero, hence Rnn = 0. Therefore,

$$\mathrm{R_{xx}} = \mathrm{ARSS\ A^H} \tag{8}$$

Where,

H = Hermitian of a matrix.
E = expected value
Rss = D × D source correlation matrix
Rnn = M × M noise correlation matrix.

2.3 Eigen Value Decompositions

Generally, the eigen values are determined from square matrix of Rxx by using the relation,

$$\det(R_{xx} - \lambda I) = 0 \tag{9}$$

Therefore the Eq. (9) results in cubic root and it will be taken as λ^1, λ^2, λ^3.

The matrix has been subdivided into noise and signal subspaces [E_N E_S] when eigen values are sorted in decreasing order. E_N consists of M–D eigenvectors which is connected to the noise. E_S consists of D eigenvectors which represents the signals which are arrived. The noise subspace represented here is in the form of M × (M–D) matrix. The signal subspace is in the form of M × D matrix. At the angle of arrival, the noise subspace array steering vectors Θ_1, Θ_2, …, Θ_D are orthogonal to each other. Due to this orthogonality condition, the Euclidean distance is calculated.

Sharp peaks can be created by putting the separation articulation in the denominator of the articulation. The MUSIC pseudo spectrum is now given as

$$PMU(\theta) = \frac{a(\theta)^H a(\theta)}{a(\theta)^H E_N \overrightarrow{E}_N^H a(\theta)} \tag{10}$$

2.4 Architecture of MUSIC Algorithm for FPGA Implementation

The block diagram shown in Fig. 4, consist of Analog part and Digital part [7, 8]. The MUSIC algorithm modelling includes 4 different blocks as COR block, EVD block, En block, PMU block.

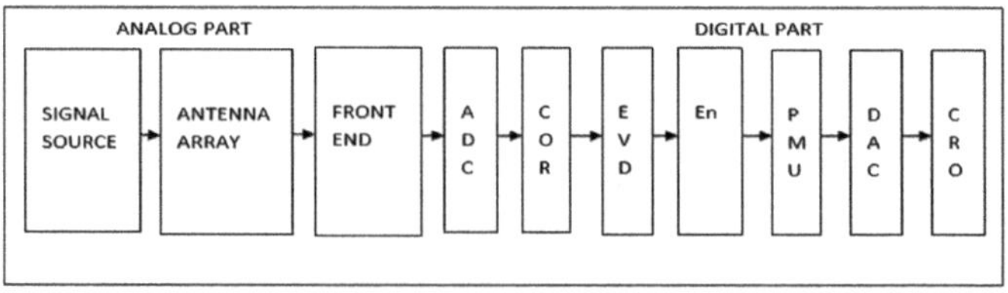

Fig. 4. Block diagram

The whole implementation in FPGA shown in flow diagram in the Fig. 5. The uniform linear array of 3(M) receiving elements have placed at a distance (d) of 9.9 cm to avoid aliasing according to the formula.

Avoid aliasing according to the formula.

$$d < \lambda/2 \text{ and } \beta = 2\pi/\lambda \tag{11}$$

where λ = wavelength of the light

$$\lambda = c/f \tag{12}$$

where c = speed of the light and f = frequency.

The wavelength has considered in the range between 10^{-14} to 10^{-15} m and the radio frequency in the range of 20 kHz to 300 GHz [9].

When the source has oriented at three different angles in degrees in order to find the maximum peak power. The steering matrix for three different angles have separated into real and imaginary steering elements.

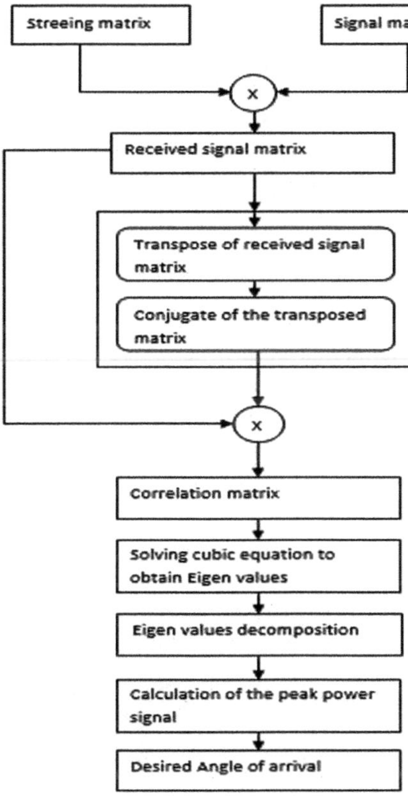

Fig. 5. Flow diagram of AOA in FPGA

This steering matrix has been multiplied with signal matrix and the resultant matrix is the received signal matrix and it has both real and imaginary terms. The transpose and conjugate are taken and multiplied with the original received signal matrix to obtain the correlation matrix.

3 Result and Analysis

Figures 6 and 7 have shown the real and imaginary part of steering matrix of 3 × 3 dimension since a source has been oriented at three different angles (30, 60, 90) in degree.

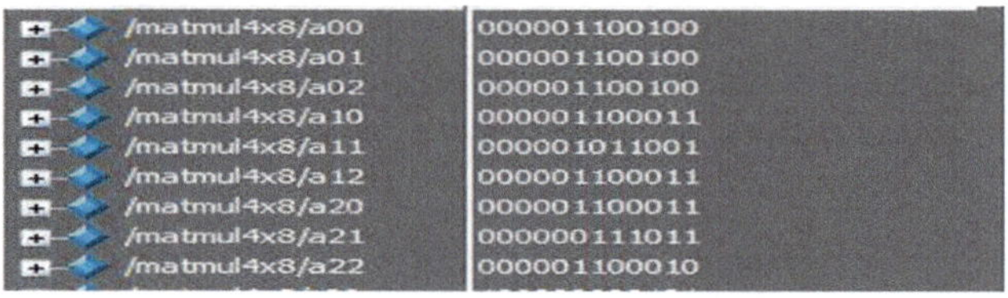

Fig. 6. Real part of steering matrix

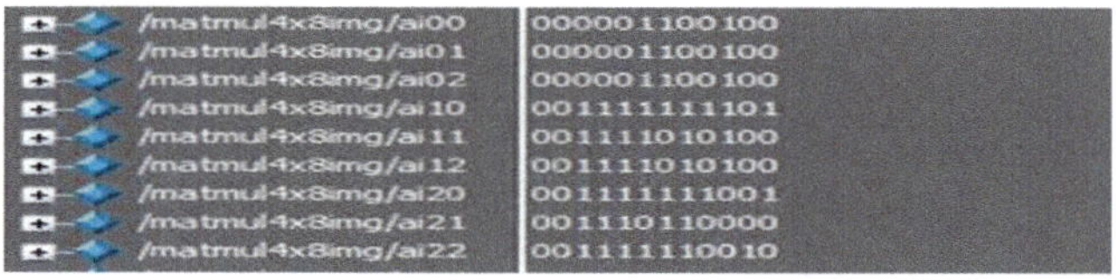

Fig. 7. Imaginary part of steering matrix

Figure 8 has shown the signal matrix in which three elements has received an input signal that has been sampled at 8 instances.

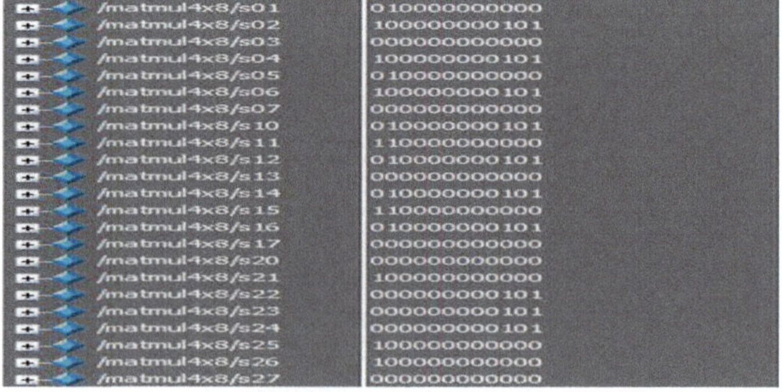

Fig. 8. Signal matrix

The real and imaginary part of the received signal matrix has been represented in Figs. 9 and 10. It has been obtained by the product of its real and imaginary steering matrix and signal matrices.

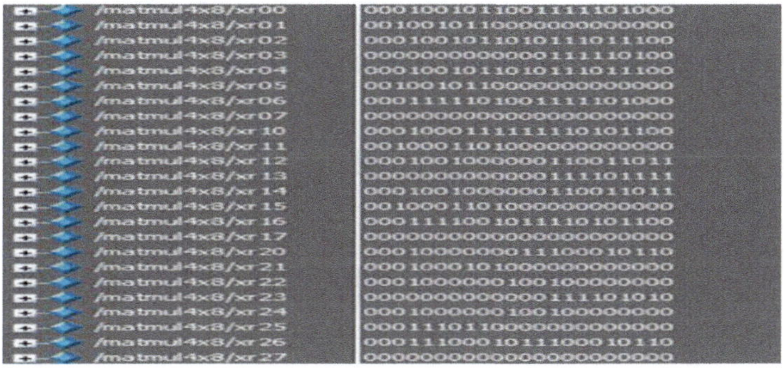

Fig. 9. Real part of received signal matrix

Fig. 10. Imaginary part of received signal matrix

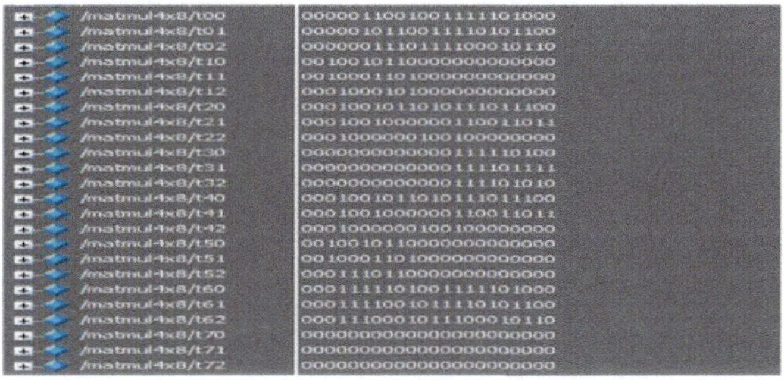

Fig. 11. Transpose of real part of received signal matrix

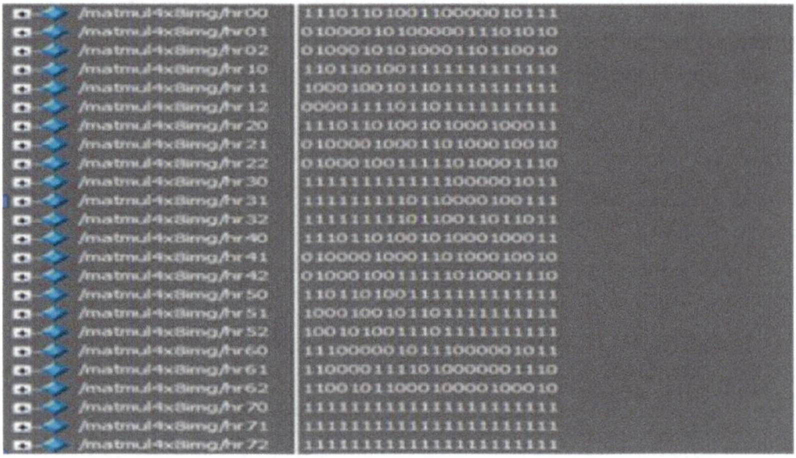

Fig. 12. Hermitian of imaginary part of received signal matrix

The original received signal matrix is product with the Hermitian of the received signal matrix to obtain the real and imaginary parts of correlation matrix which is given in Figs. 13 and 14.

Fig. 13. Real part of correlation matrix

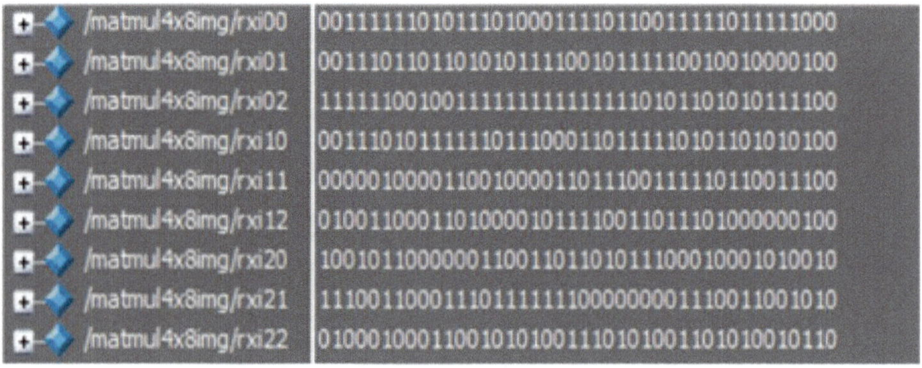

Fig. 14. Imaginary part of correlation matrix

Fig. 15. Intermediate result of Eigen decomposition of real part

Fig. 16. Intermediate result of Eigen decomposition of imaginary part

The MUSIC pseudo power spectrum for different angles has been partially simulated in Verilog until the intermediate results has been obtained.

4 Conclusion

MUSIC Algorithm achieves good accuracy and consistency for the angle of arrival and also it has the capability to prevent noise. It also concentrates on the maximum power direction of the array antenna.

Acknowledgement. The authors would like to thank Department of ECE, Mepco Schlenk Engineering College, Sivakasi for permitting to carry out this work.

References

1. Boccuzzi, J.: Signal Processing for Wireless Communications. McGraw-Hill, New York (2007)
2. Richards, M.A.: Fundamentals of Radar Signal Processing. McGraw-Hill, New York (2005)
3. Katkovnik, V., Lee, M.-S., Kim, Y.-H.: High-resolution signal processing for a switch antenna array FMCW radar with a single channel receiver. In: Sensor Array and Multichannel Signal Processing Workshop Proceedings (2002)
4. Do-Hong, T., Russer, P.: Signal processing for wideband smart antenna array applications. IEEE Microw. Mag. **5**, 57–67 (2004)
5. Elhefnawy, M., Ismail, W.: New technique to find the angle of arrival. In: Japan Egypt Conference on Electronics, Communications and Computers (2012)
6. Badawy, A., Khattab, T., Trinchero, D., ElFouly, T., Mohamed, A.: A simple angle of arrival estimation system. In: IEEE Wireless Communications and Networking Conference (WCNC) (2017)
7. Li, M., Lu, Y.: Angle-of-arrival estimation for localization and communication in wireless networks. In: 16th European Signal Processing Conference (2008)

8. Dhar, A., Senapati, A., Sekhar Roy, J.: Direction of arrival estimation in smart antenna using MUSIC and improved MUSIC algorithm at noisy environment. Int. J. Microw. Appl. **5**(2), 1–6 (2016)
9. Mohanna, M., Rabeh, M.L., Zieur, E.M., Hekala, S.: Optimization of MUSIC algorithm for angel of arrival estimation in wireless communications. NRIAG J. Astron. Geophys. **2**, 116–124 (2013)

Design and Analysis of 1–3 GHz Wideband LNA Using ADS

S. Syed Ameer Abbas(✉), T. Kiruba Angeline, and K. P. Kaviyashri

Sivakasi, India

ssyed@mepcoeng.ac.in,
kirubaanjaline@gmail.com, kpkaviyashri@gmail.com

Abstract. A proposal for wideband Low Noise Amplifier (LNA) using BJT is designed which operates at the frequency of 1–3 GHz. The simulation work is done by using the software Advance Design System (ADS) and the waveforms are all observed. The simulation results shows that it achieves a maximum gain S_{21} of 6.397 dB, voltage standing wave ratio (VSWR) of 1.279, reflection coefficient, S_{11} at the input of −18.231 dB, reflection coefficient, S_{22} at the output of −16.356 dB, stability factor of 1.688, reverse gain, S_{12} of −16.545 dB with a supply voltage of 1.8 V.

Keywords: LNA · Wideband LNA · VSWR · ADS

1 Introduction

Low Noise Amplifier (LNA) is the one which amplifies the low strength signals that comes out of an antenna's as the signals are from low strength they are rarely recognized at this time noise should not be added if added loss of information occurs in the signal. In the receiver side LNA's are one of the most important circuit components present which comes out of antennas. In the upcoming years the wireless standards are increasing step by step [1]. Receiver is the key component for LNA to reduce the unwanted noise in the system there by to make the system efficient [2]. Low noise amplifier is the key component coming out of antennas the signal is weak and should be with good gain. In the design of LNA's the important part is the receiver and the transmitters. Then filtering, LNA, mixer is needed in the receiver here the sensitivity depends on LNA [3, 4]. LNA's are simpler to design in which wideband LNA's are much simpler and easy to understand, because in the filter design and the amplifier design decoupling happens in the receiver part. Matching the input and with the noise figure is tough parameters to be considered before designing wideband LNA. The Wideband amplifier design is the most challenging task. To meet certain goals single band operation is applied in conventional LNA. The input network has high Q factor then the design of wideband becomes complex LC therefore it should be made easy [5–7]. Using extra parameter's Q factor of the parallel resonant tank's input is very less [8, 9].

D. J. Hemanth et al. (Eds.): COMET 2019, LNDECT 35, pp. 169–175, 2020.
https://doi.org/10.1007/978-3-030-32150-5_18

2 Proposed Methodology

2.1 Low Noise Amplifier

For, amplifying extremely weak signals and to provide voltage levels suitable for analog to digital conversion or for analog processing and in applications which has low amplitude sources like many types of the transducers and antenna. This note deals with the selection of a proper LNA. LNA's is an electronic amplifier it proceed by the process of amplifying a very low power signal without disturbing its signal to noise ratio it is done in the process. Powers, the noise at the input of the signal are increased by the amplifier. To minimize additional noise LNA's are designed. The Trade-offs such as impedance matching there by choosing the low noise biasing, criteria must be considered to reduce noise. LNA's are found in various applications in the radio communications in that they are used for medical field. Primarily concerned with weak signals are LNA's that are just above the noise floor, considerations in the presence of the larger signals which causes the inter modulation. The noise figure should be low for a good LNA. LNA's has its own operating criteria's which include bandwidth, gain flatness, stability; voltage standing wave ratio is those criteria's.

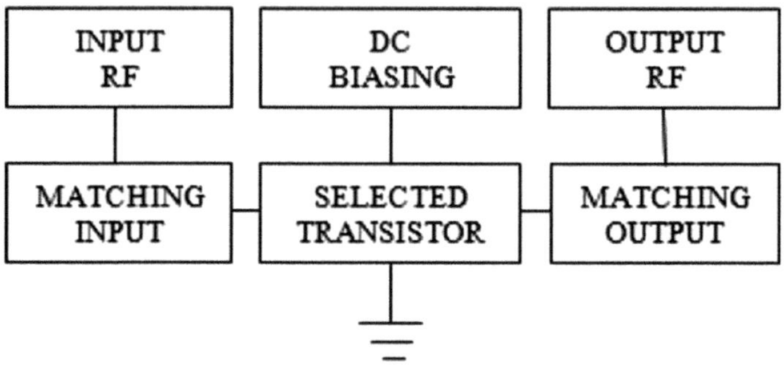

Fig. 1. Block diagram of LNA

The above mentioned is the block diagram of LNA is shown in the Fig. 1 where RF input is the input matching and DC biasing is the selected transistor and the RF output is the output matching. Different stability, biasing and matching network for different load. There are more techniques are available to reach better performance which includes low power, noise and high gain, stability. Using CMOS, Bipolar, GaAs FET technologies the trend of designing LNA has been changed.

2.2 Wideband LNA

In the Wideband LNA decoupling of the filter and the design of the main amplifier makes the design much simpler. The considerations which are to be taken important and necessary is matching the input and noise figure are to be less. Challenging portions in communication system is the designing of the Wideband amplifier design.

The conventional LNA's are easy in the designing process it can soon and easily achieves the specified criteria. LNA's operates in the single band. Negative feedback topology is used in the topology, by those components the Wideband input or output matching can be performed. The Q factor of the parallel resonant tank's input is low. Flattening of the gain can be achieved by utilizing the negative feedback technique. In the wideband LNA topology negative feedback topology is used.

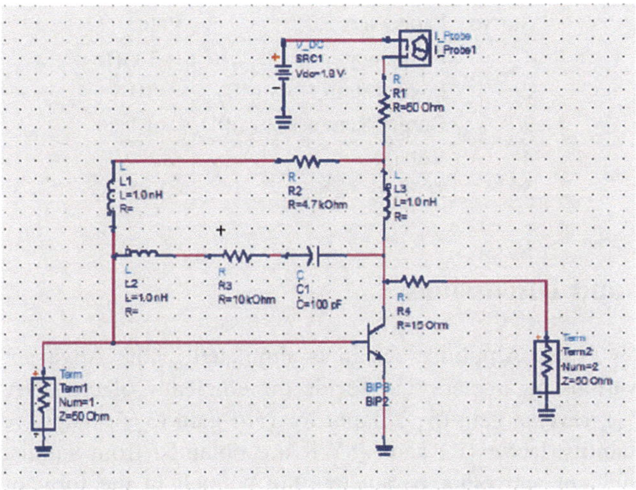

Fig. 2. Wideband LNA block diagram

Designing the negative feedback considerations at specified frequencies, here input and the output voltage of the design should be same with the phase of as it is a negative feedback topology the way to become positive topology should be avoided LC are added in the circuit design in series or parallel manner to decrease the feedback phase and to make sure that it is negative feedback only. There are various topologies which show the gain of a wideband, the most popular technique is nothing but the RC feedback. They are used for matching the input and to achieve the good linearity of the wideband circuit. The wideband LNA block diagram is shown in Fig. 2. There are two biasing resistors R_1 and R_2 are self biasing with the transistors. This kind of LNA design is done for low power applications. To improve the stability RLC feedback is done in the design. Here the resistors R_1 is with 50 Ω, R_2 is with 4.7 kΩ, R_3 is with 10 kΩ and the R_4 is with 15 Ω. DC current is blocked by the RF choke. In this proposed work, the LNA is used as it is simpler and less complex and is matched using lumped element as it. Here impedance matching of Z_o is with 50 Ω.

2.3 Design Specification

The motive of the project is to design a wideband LNA operating in the frequency range of 1 to 3 GHZ. The Designed LNA is simulated using Advanced Design System

(ADS) tool to measure the values for forward gain S_{21}, reverse gain S_{12}, reflection coefficient S_{11} and S_{22}, are Stability factor (K), Voltage Standing Wave Ratio (VSWR), and Noise figure (NF) and dc power consumption. The main aim is to design the wideband LNA for a frequency and to achieve a good gain with minimal noise. The design specifications are described in the Table 1.

Table 1. Specifications

Sl. No.	Parameters	Value
1	Frequency	1–3 GHz
2	Input Return Loss (dB)	<−10
3	Output Return Loss (dB)	<−10
4	Voltage	<1.8 V
5	Stability factor	<2 dB

3 Results and Discussions

Wideband Low Noise Amplifier design is simulated in the Advance design system (ADS). The various parameters that decide the efficiency of the LNA are, reflection coefficient (S_{11}), reverse gain (S_{12}), input forward gain (S_{21}), output reflection coefficient (S_{22}), stability factor (K) and VSWR are obtained from simulation. The gain, reflection coefficient and other parameters are defined in the form of S-parameters. Voltage gain is the ratio of the output to the input voltages. The maximum forward gain S_{21} obtained for the proposed LNA is about 6.397 dB at 2 GHZ as shown in the Fig. 5. The reverse gain S_{12} is shown in the figure in the Fig. 4. The magnitude of forward and reverse gain should be high. The input reflection coefficient S_{11} is shown in the Fig. 3 achieves the −18.23. The Output reflection coefficient S_{22} is −16.356 is shown in the Fig. 6.

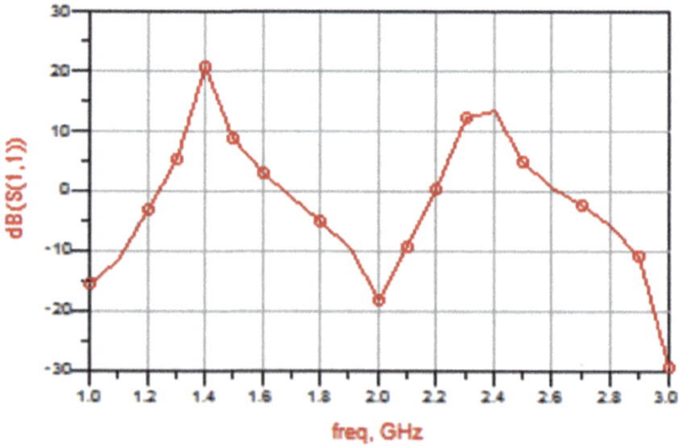

Fig. 3. Input reflection coefficient (S_{11})

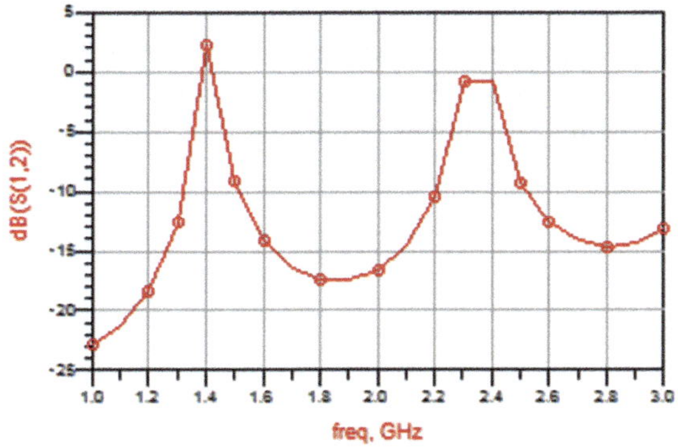

Fig. 4. Reverse gain (S_{12})

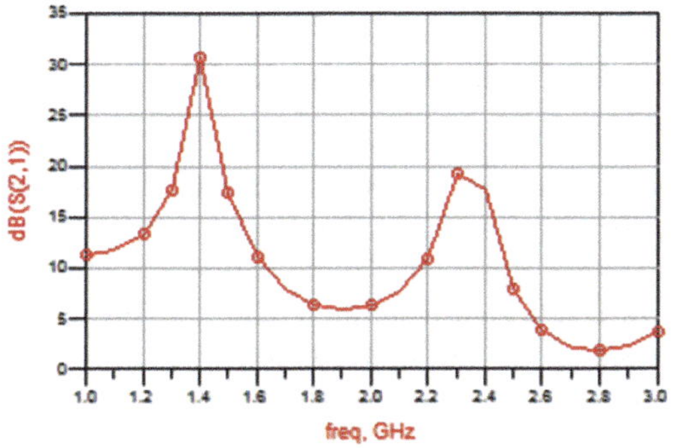

Fig. 5. Forward gain (S_{21})

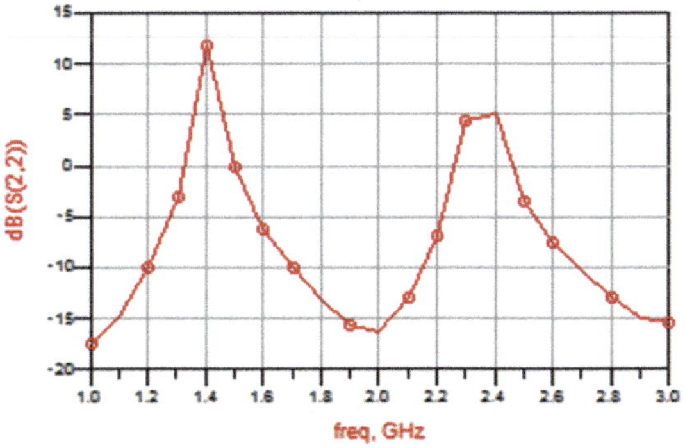

Fig. 6. Output reflection coefficients (S_{22})

Stability factor is used to determine the stability of the circuit. The stability factor should be greater than one to specify that the circuit is stable. The stability factor for the LNA proposed is not as higher and the maximum value is 1 and it is decreased above 2 GHz due to its instability as shown in the Fig. 7.

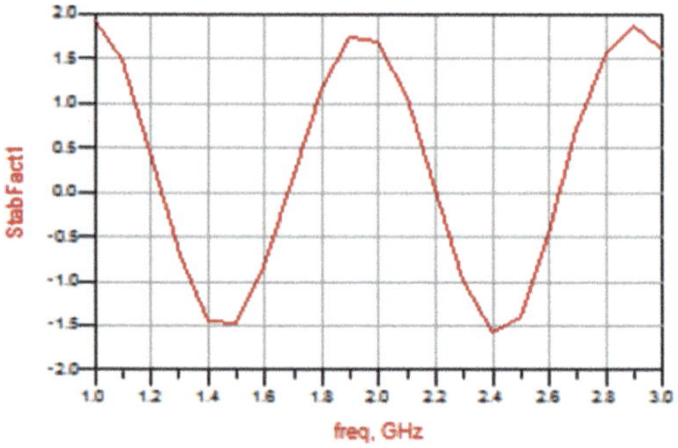

Fig. 7. Stability factor (K).

The value of stability factor is 1.688 at 2 GHz and thereby it slowly decreases and increases eventually for different set of frequencies.

Voltage Standing Wave Ratio (VSWR) is a function of reflection coefficient (S_{11}) that describes the power reflected from the circuit. VSWR can also be defined as the voltage ratio of the signal on the transmission line. It is defined. The VSWR for proposed LNA is shown in the Fig. 8. The standing wave ratio is not good after 2 GHz–2.1 GHz.

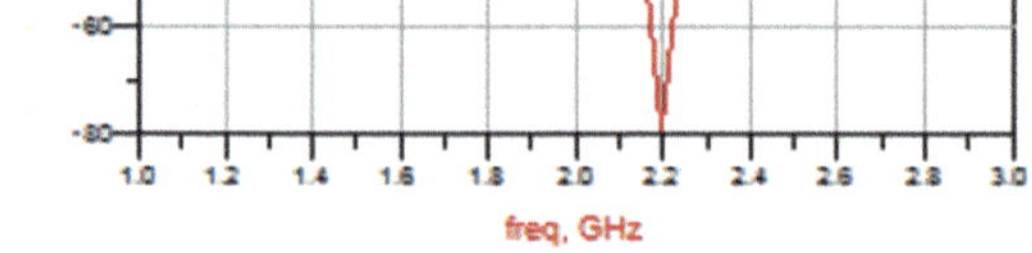

Fig. 8. VSWR

This method produces a good input and output reflection coefficient which is lesser than <−10 dB and produces a stability factor greater than one and VSWR greater than one. This is discussed in the Table 2.

Table 2. Overall simulation result

Sl. No.	Parameter	Values
1	Frequency (GHz)	2
2	Forward gain S_{21} (dB)	6.397
3	Reverse gain S_{12} (dB)	−16.545
4	Input reflection coefficient S_{11} (dB)	−18.23
5	Output reflection coefficient S_{22} (dB)	−16.356
6	Stability factor (K)	1.688
7	Voltage Standing Wave Ratio (VSWR)	1.279

4 Conclusion

By using negative feedback topology, the wide band LNA whose operation ranges from 1 GHz to 3 GHz had been constructed. By using the feedback technique the design specifications is achieved foe a wide range of frequencies the readings are plotted. This is employed in the wideband LNA design here RLC feedback provides a consistency performance for wide frequencies. This is simulated in the ADS software and the results are analyzed.

Acknowledgment. The authors would like to thank Department of ECE, Mepco Schlenk Engineering College, Sivakasi for providing the facilities to carry out this work.

References

1. Blaakmeer, S.C., Klumperink, E.A.M., Nauta, B.: A wideband noise-canceling CMOS LNA exploiting a transformer. In: Radio Frequency Integrated Circuits (RFIC) Symposium, pp. 11–13, June 2006
2. Im, D., Nam, I., Kim, H.-T., Lee, K.: A wideband CMOS low noise amplifier employing noise and IM2 distortion cancellation for a digital TV tuner. IEEE J. Solid-State Circuits **44** (3), 686–698 (2009)
3. Salleh, A., Abd Aziz, M.Z.A., Misran, M.H., Othman, M.A., Mohamad, N.R.: Design of wideband low noise amplifier using negative feedback topology for Motorola application. J. Telecommun. Electr. Comput. Eng. **5**, 47–52 (2013)
4. Zhang, Z., Dinh, A., Chen, L., Khan, M.: A low noise figure 2-GHz bandwidth LNA using resistive feedback with additional input inductors. IEICE Electron. Express **10**, 20130672 (2013)
5. Yao, Y., Fan, T.: Design of DC-3.4 GHz ultra-wideband low noise amplifier with parasitic parameters of FET. Int. J. Eng. Res. Appl. **4**(4), 280–284 (2014)
6. Gao, Y., Wang, N.Z., Zhao, Y.: Design of CMOS UWB noise amplifier with noise canceling technology. In: Proceedings of the International Conference on Future Computer and Communication Engineering (2014)
7. Kim, C.-W., Kang, M.-S., Anh, P.T., Kim, H.-T., Lee, S.-G.: An ultra-wideband CMOS low noise amplifier for 3–5 GHz UWB system. IEEE J. Solid-State Circuits **40**(2), 544–547 (2005)
8. Rezazadeh, Y., Amiri, P., Roodaki, P.M., Kondori, M.B.: Presenting systematic design for UWB low noise amplifier circuits. Modern Appl. Sci. **6**(8), 21 (2012)
9. Lee, H.-J., Ha, D.S., Choi, S.S.: A systematic approach to CMOS low noise amplifier design for ultra wide band applications. IEEE (2005)

A Smart Sticksor for Dual Sensory Impaired

L. Mary Angelin Priya(✉) and D. Shyam

Department of Electrical and Electronics Engineering, Rajalakshmi Engineering College, Chennai, India
angelinluckas@gmail.com, shyam.d@rajalakshmi.edu.in

Abstract. In day to day life, the community is majorly built around people without sensory impairment. This makes it difficult for physically challenged people to communicate and commute normally. The sign language used by the people who are sensory impaired at the same time cannot be under stood by normal people. Similarly, the world is too chaotic to be properly sensed through a normal helping stick. This difficulty can be addressed through the proper applications of modern technologies which have progressed enough for different applications. Therefore the solution is to develop a smart stick which assists people through sensory receptors and communicators. Its ability will include obstacle detection through a motor actuated ultrasonic sensor, intimation through a buzzer, LED-based alert system for another person specially for lowlight conditions and a combination of keypad and display for communications. This smart stick aims to alleviate some of the issues faced by physically challenged people and hence open up an opportunity for them to explore the modern world.

Keywords: ARM LPC2138 with LPC · Ultrasonic sensor · Vibration motor · Servo motor · LED

1 Introduction

Deaf-blindness is the occurrence of combined visual and hearing impairments for a person. Such people make use of their hands and tools such as white canes for obstacle detection. Hand gestures and languages like Braille are used a mean for communication. When dual sensory impaired people travel at night time at some places there will be no light. So the person or vehicle traveling on the opposite side will have no alert about these people which can lead to accidents. To alert the person or vehicle led has been attached with the stick. Generally, the dual sensory impaired people will find difficult to communicate with normal people because the language used for communication will be different for both of them. By using this keypad they can communicate easily with normal people because it converts the Braille language into text which is understood by normal people. Dual sensory impaired people will find difficult to detect an obstacle in new places or objects moved from their familiar places which can lead to them being hurt. To prevent this a sensor has been used. A servo motor is attached to rotate the sensor on either direction to detect obstacles. In the literature, different techniques have been proposed for sticksor. In [1, 10], proposed an idea of designing electronic stick using Global System Messaging (GSM), Global Positioning System

D. J. Hemanth et al. (Eds.): COMET 2019, LNDECT 35, pp. 176–186, 2020.
https://doi.org/10.1007/978-3-030-32150-5_19

(GPS) and Ultra-sonic technology. In [2, 4, 6, 7], the system uses six dot vibrators to display characters and System having Braille pad for writing the Braille letters. SMS facility used for communication. [3], merging data provided by the two sensor types to allow more accurate information, to be transmitted to the user via

Bluetooth module as a voice message specifying the object nature, characteristics and the distance between the detected obstacles. In [5], multiple sensors are used to detect obstacles. [8] a device control section which will be operated by using Braille touch keypad, the device control section of microcontroller connected with the load devices. The AC devices can be controlled to the controller through a relay.

In [9], proposed the implementation of Braille to word and audio converter as output using FPGA. [11, 12], a reliable solution encompassing of a cane and a shoe that could communicate with the users through voice alert and pre-recorded messages. [13], new technique and communications method for blind persons. Conversion of English language to Braille and it was detected by Six vibration motors that are placed in the glove. [14], the user pushes the lightweight Guide Cane forward and When the Guide Cane's ultrasonic sensors detect an obstacle, the embedded computer determines a suitable direction of motion that steers the Guide Cane and the user around it [15] focused on an aviation system called Virtual Leading Blocks for the Deaf-Blind, it consists of a wearable interface for finger Braille. It uses two Linux-based wristwatch computers as a hybrid interface for verbal and non-verbal communication in order to inform users of their direction and position through the tactile sensation. Theoretical Background.

2 Stick Specification

This product is mainly built around a walking stick of robust aluminum construction (Fig. 1).

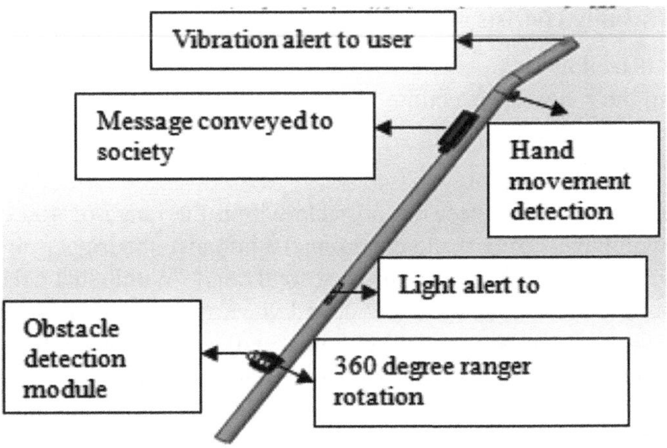

Fig. 1. CAD diagram for Sticks or

It's built with a length of 94 cm and thickness 18 cm nominally. It's provided with an ergonomic handle. It will also become a power full tool if it holds the ability to let them communicate with other people, especially in emergency situations. Hence this Stick will also incorporate Braille-based communication device which lets them spell out their intention to then ear by attender. It is a design to be the handheld device of dimensions 100 mm × 50 mm × 15 mm. The buttons are spaced outwit 10 mm gap to be ergonomic as well as to allow fast finger response. There will also be an emergency buzzer which lets the user turn the stick into a beacon in emergencies.

The recognition of the surrounding is obtained with the help of an obstacle detection module that is placed near the bottom of the sticks or. This allows the module to operate within its max range of 15° angle with respect to the vertical direction. However, this range is insufficient for obstacle detection along the horizontal direction. Hence, this obstacle detection module is mounted on a 360-degree ranger rotation that is capable of swinging the obstacle detection module horizontally. A bracket is attached to its rotor onto which the obstacle detection module is seated. Its direction is controlled by the user with the help of a gesture sensor placed in the backside of the stick, at a distance that is approachable by the fingers when measured from the handle.

The inference of an obstacle is communicated to the user with the help of a vibration alert placed in the handle of the stick. Also, the presence of the user is communicated to the surrounding people with the help of an alarm and a light. The light alert is placed along the middle of the stick so that surrounding people can easily notice. The alarm is placed near to it and it intends to do the same. The communication is facilitated by a combination of Braille keypad and LCD. The Braille keypad is portable but stays connected to the stick via a cable. The keypad is placed below the LCD. This keypad is as compact as a mobile phone and is hence be easily handled by the user. The display is placed near the top end of the stick so that it has clear visibility.

3 Design Methodology

The sticks or mainly consists of the following modules (Fig. 2):

i. Obstacle detection
ii. Extending the range of detection
iii. Night alert for society
iv. Communication

The ultrasonic sensor will detect the obstacle within the range of 400 cm. The ultrasonic sensor will have a trigger and echo pin. To initialize the trigger pin it will send a pulse of minimal timer period often-micro second delay. Within that delay, it will send eight sonic burst of data at 40 kHz. The sound wave will detect the obstacles. When it detects, the echo pin will be high and the timer will start and when the wave bounces back from the obstacle and reaches the sensor, echo pin will go low and the timer will be stopped. The microcontroller will get the timer value from the timer register and calculate the distance.

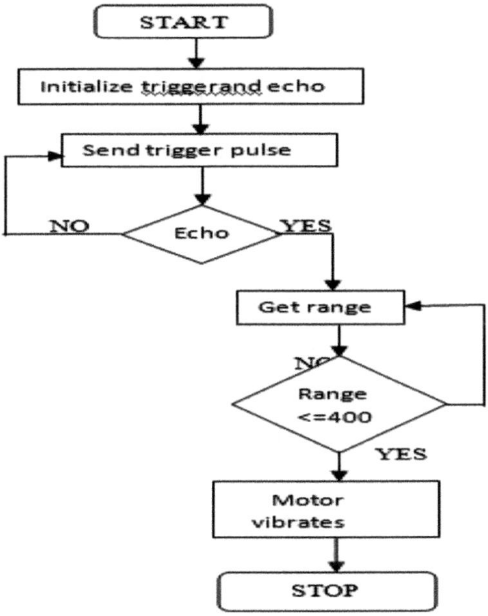

Fig. 2. Flow chart for obstacle detection

Procedural Calculation

Distance = (speed * time)/2
S = speed of sound = 344 m/s
T = time calculate the to and from distance
Speed of the sound = 340 m/s
= 0.034 cm/μs
Time = 271 μs = 271 μs * 0.034 cm/μs = 9.214/2
= 4.607 cm

To display the value in centimeter it should be divided by 59. The timer value loaded in the timer register is 380000 μs. The received value from the timer register will be calculated using the above formula. Once the range is detected if the range is below 400 cm the vibration motor will provide a vibration alert to the user.

4 Communication

The Braille keypad functions based on the concept that all the buttons pressed before complete release of those buttons together constitute the input given by the user (Fig. 3).

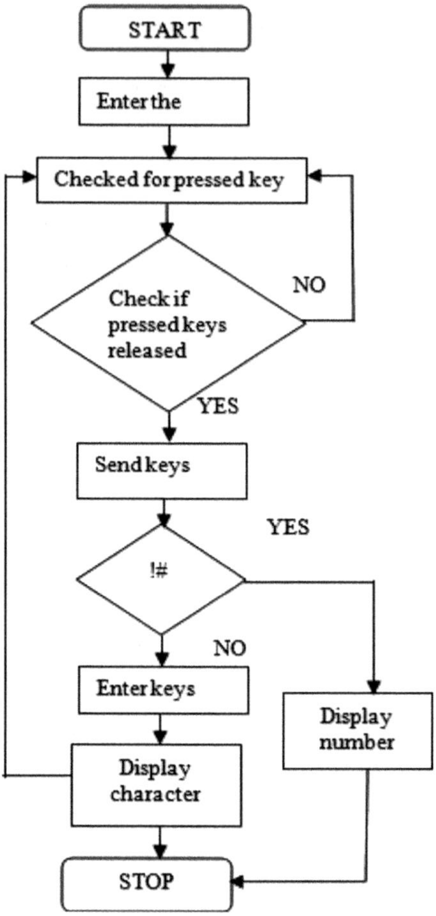

Fig. 3. Flow chart for communication

Thus the program consists of two major functions – wait for release and Read Keys.

A. Wait for Release

The Wait for Release function reads all the buttons pressed by the user before complete release and returns the corresponding buttons' position as a six digit binary number. This is done with the help of two binary integers a and x. Both are initially set to zero. Then, a for loop is developed for six iterations where each iteration checks whether the corresponding button is pressed in the Braille keypad. If so, xissetto1 and is left shifted to the i'th position (for example, if button 2 alone is pressed, they corresponding value of X will be 0b010000). Later, this value is loaded onto the integer a using OR operator. As the value of a is not reset for every iteration, the pressed buttons are always remembered. Thus, at the end of six iterations, the a will contain the binary format representation of the pressed buttons (1).

For example, if the buttons 1 and 2 were pressed, the final value of a will be 0b110000. There is an if condition that returns the value of a only when the x value is 0. The entire code is set under a while loop with x being reset to 0 at the beginning of each iteration.

This makes sure that all the button presses are registered and finally returned only when all the buttons are released (in which case X stays 0 till the end of the loop). There is also another integer that governs the return of value a called New Press. It's always set to 0 before the start of the wait for release function and turns 1 through an if statement only when gains some value. And this should be 1 for the aforementioned return of integer a to happen. This makes sure that a runaway of empty values as output doesn't occur.

B. Read Keys

The next function is the Read Keys function. It reads the binary value and converts it into character by comparing the values to the binary table. If that character is not #, then the character is returned. However, if it is #, the another input is obtained through the wait for release function. This input is converted into a number through ASCII conversion. Then, the number is returned. The lookup table Shown below is used to match the generated binary digits to the corresponding character and display the characters on the LCD (Table 1).

Table 1. Look up table

0b000000	Initialization
0b100000	A
0b110000	B
0b100100	C
0b100110	d
0b100010	e
0b110100	f
0b110110	g
0b110010	h
0b010100	i
0b010110	j
0b101000	k
0b111000	l
0b101100	m
0b101110	n
0b101010	o
0b111100	p
0b111110	q
0b111010	r
0b011100	s
0b011110	t
0b101001	u
0b111001	v
0b010111	w
0b101101	x
0b101111	y
0b101011	z
0b001111	#

5 Rotation of Ultrasonic Sensor with Servo Motor

The rotation of the ultrasonic sensor is controlled by the servo motor and the input to the servomotor is given by the gesture sensor. Instead of the gesture sensor, three push buttons are used for simulation purpose. To rotate the servomotor PWM signals have to be generated. The on time will be where the PWM signal will be given and in the off time, the rotation will take place according to the generated PWM. The motor can move in three directions left, right and center (Fig. 4 and Table 2).

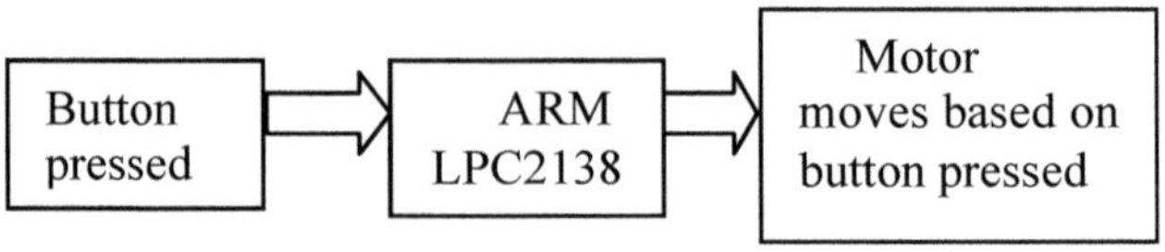

Fig. 4. Block diagram for servo motor rotation

Table 2. Movement of motor

Input	Output
If button A is pressed	Motor moves to left
If button B is pressed	Motor moves to the right
If button C is pressed	Motor moves to center

6 Led Light for Night Alert

The LED light is used to alert society. The LED strip used here is ws2812b. The LED strip is controlled by the switch. When the switch is in on condition the LED will glow and in off condition the LED will not glow as there is no input provided to the microcontroller (Fig. 5).

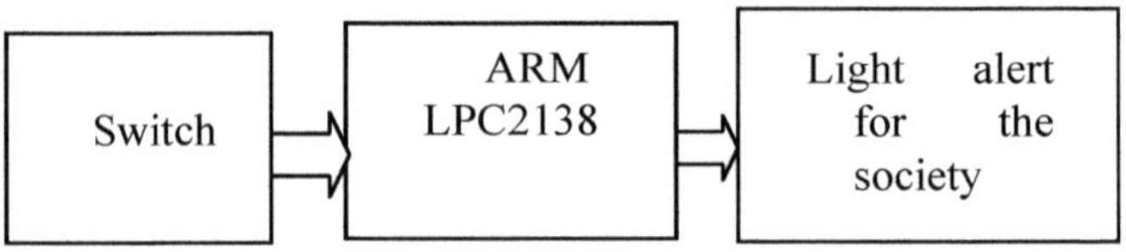

Fig. 5. Block diagram for night alert

7 Result

See Figs. 6, 7, and 8.

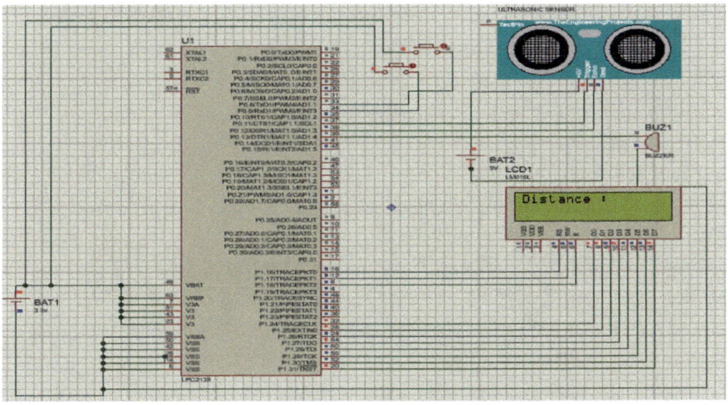

when the range is below 400 cm it will display no obstacle.	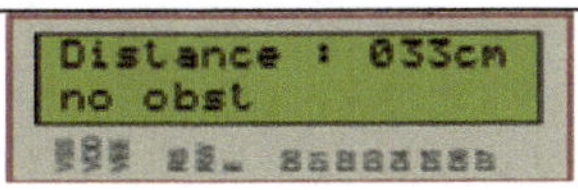
When the range is above 400 cm it will display as obstacle detected	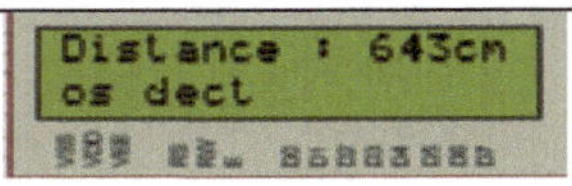

Fig. 6. Simulation result of obstacle detection

OBJECTIVE	INPUT	OUTPUT
To display a character w Button 2 4 5 and 6 is pressed 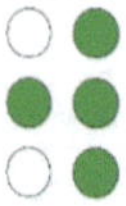	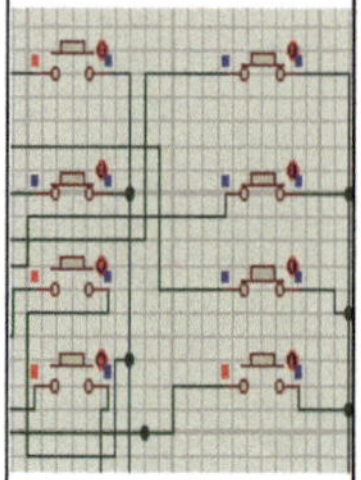	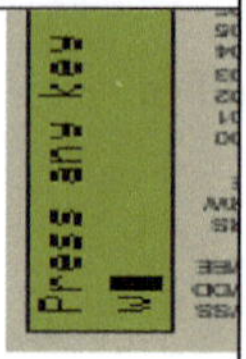
To display a character e button 1 and 5 is pressed 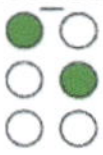		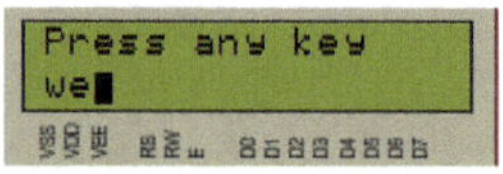
To display a character l button 1 2 and 3 is pressed 		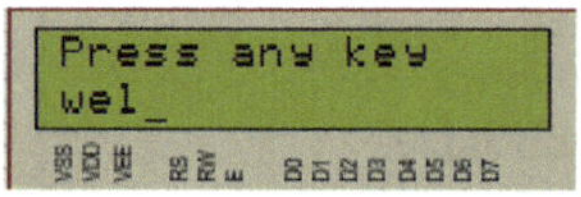
To display a character c button 1 and 4 is pressed 		
To display a character o button 1 3and 5 is pressed 		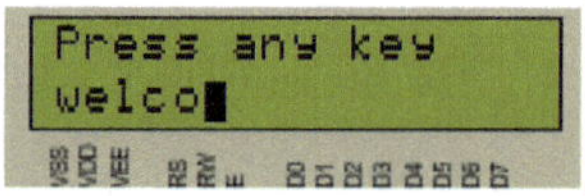
To display a character m button 1 3 and 4 is pressed 		
To display a character e button 1 and 5 is pressed		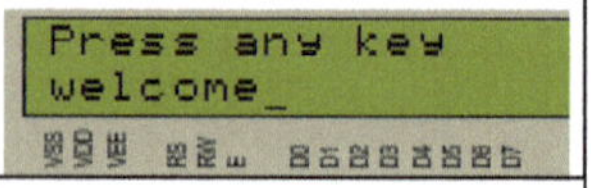
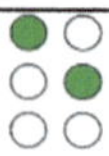		

Fig. 7. Simulation result of Braille keypad

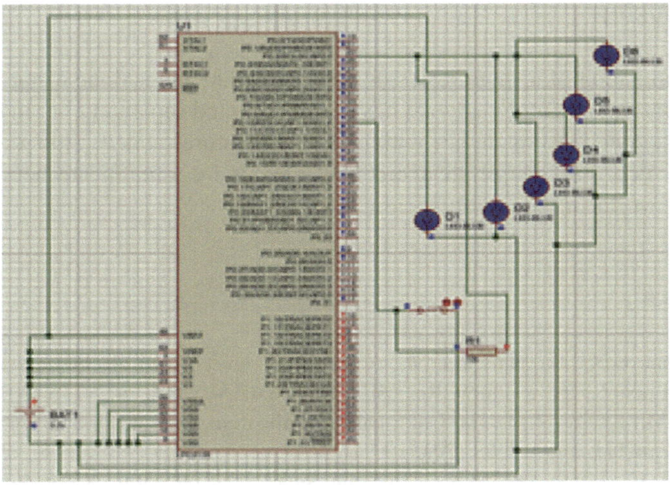

Fig. 8. Simulation result of LED light

OBJECTIVE	INPUT	OUTPUT
To display a number first initialize the # symbol Buttons 3 4 5 and 6 is pressed	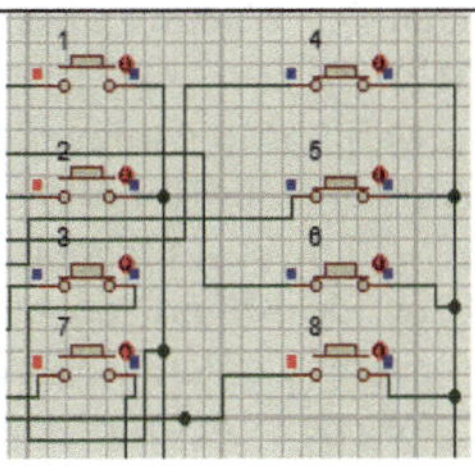	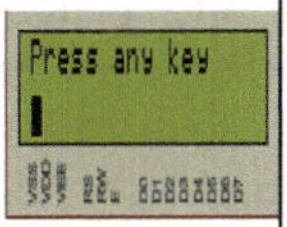
To display a number 1 Buttons 1 is pressed 		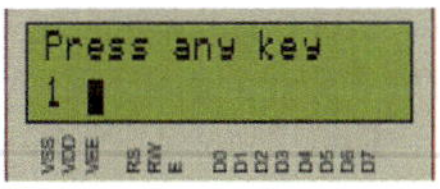
To display a number 2 Buttons 1 and 2 is pressed 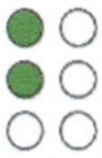		
To display a number 3 Buttons 1 and 4 is pressed 		
To display a number 4 Buttons 14 and 5 is pressed 		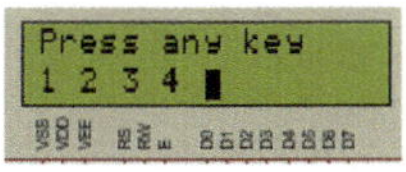

8 Conclusion

Thus the simulation of a smart sticks or and Braille keypad is done using Proteus. Both the smart scissor and Braille keypad has been implanted separately but in this project, it has been integrated as one device with additional features added. This project will help the sensory deprived people to avoid obstacles and communicate better which will widen their world in view with society.

References

1. Gurubaran, G.K., Ramalingam, M.: A survey of voice aided electronic stick for visually impaired people. Int. J. Innov. Res. Adv. Eng. (IJIRAE) **1**(8), 342–346 (2014)
2. Varsha, M., Khaire, R.M.: Hardware-based Braille note taker. Int. J. Sci. Eng. Technol. Res. (IJSETR) **4**(11), 3957–3959 (2015)
3. Gaikwad, G., Waghmare, H.K.: Ultrasonic smart cane indicating a safe free path to blind people. Int. J. Adv. Comput. Electron. Technol. (IJACET) **2**(4), 12–17 (2015)
4. Zope, P.H., Dahake, H.: Design and implementation of messaging system using Braille code for virtually impaired persons. Int. J. Adv. Res. Electr. Electron. Instrum. Eng. **5**(7), 5977–5984 (2016)
5. Palanisamy, K., Dhamodharan, N.: Walking stick with OPCFD system. GRD J. Global Res. Dev. J. Eng. **3**(1), 1–5 (2017)
6. Mahadev, M.H., Prabhakar, M.S.: SMS communication system for blind people. Int. J. Res. Eng. Appl. Manag. (IJREAM) **3**(2), 6–10 (2017)
7. Sarkar, R., Smita Das, D.R.: A low-cost microelectromechanical Braille for blind people to communicate with blind or deaf-blind people through SMS subsystem. In: IEEE International Advance Computing Conference (IACC), pp. 1529–1532 (2013)
8. Chary, B.V.R., Kumar, S.: Rescue system for visually impaired blind persons. Int. J. Eng. Trends Technol. (IJETT) **16**, 153–155 (2014)
9. Chitte, P.P., Thombe, S.A., Pimpalkar, Y.A.: Braille to text and speech for cecity persons. Int. J. Res. Eng. Technol. **4**(1), 263–268 (2015)
10. Sreenivasan, D., Poonguzhali, S.: An electronic aid for visually impaired in reading printed text. Int. J. Sci. Eng. Res. **4**(5), 198–203 (2013)
11. Rajapandian, B., Harini, V., Raksha, D.: A novel approach as an aid for blind, deaf and dumb people. In: International Conference on Sensing, Signal Processing and Security (ICSSS) (2017)
12. Mahesh, S.A., Raj Supriya, K., Pushpa Latha, M.V.S.S.N.K.: Smart assistive shoes and cane: solemates for the blind people. Int. J. Eng. Sci. Comput. **8**(4), 16665–16672 (2018)
13. Rajasenathipathi, M., Arthanari, M.: An electronic design of a low cost Braille handglove. Int. J. Adv. Comput. Sci. Appl. (IJACSA) **1**(3), 52–57 (2010)
14. Ulrich, I., Borenstein, J.: The GuideCane—applying mobile robot technologies to assist the visually impaired. IEEE Trans. Syst. Man Cybern. Part A Syst. Hum. **31**(2), 131–136 (2001)
15. Amemiya, T., Yamashita, J., Hirota, K.: Virtual leading blocks for the deaf-blind: a real-time way-finder by verbal-nonverbal hybrid interface and high density RFID tag space. In: Proceedings of the 2004 Virtual Reality (VR 2004), Chicago, IL, USA, March 2004 (2004)

Real Time Analysis of Two Tank Non Interacting System Using Conventional Tuning Method

S. Lakshmi(✉), T. Thahira Tasneem, N. Vishnu Priya, and V. Poomagal

Department of EIE, Panimalar Engineering College, Chennai, India
elzie.moses@gmail.com

Abstract. This paper investigates the performance of Process Reaction Curve (Cohen Coon) and Ziegler Nichols tuning techniques based on PID Controllers for the control of liquid using two tank non interacting system. Initially, Cohen Coon tuning technique is applied and then to achieve further improvement in performance Ziegler Nichols tuning technique is chosen. It is observed that the Ziegler Nichols method gives satisfactory results than Cohen Coon tuning technique.

Keywords: Two tank non-interactive system · PID controller · Cohen-Coon · Ziegler-Nichols technique

1 Introduction

The control of liquid level and flow in multiple tanks are basic problems in the process industries. The PID is the most used methods among other controllers. Conventional PID controllers are One-Degree-of-Freedom type. We are here to discuss response of system's performance specifications of Cohen Coon and Ziegler Nichols tuning technique. In the simple PID controller the tuning procedure works well and time consuming, particularly for a process with large time constant or delay. However poorly tuned PID controller are often found in Industry. Tank1 feeds Tank2 and its dynamics behavior is affected. More than one physical processing unit cannot be involved by the multicapacity process.

1.1 System Modeling

Consider a Two Tank Non-interacting System as shown in Fig. 1, in which the level of the Tank 2 is adjusted by the outflow of tank1 and the load disturbance. The height and area of the tank1 is h1, A1 and that of tank2 is h2, A2 respectively. The resistance added by the valves to tank1 and tank2 are R1 and R2.

1.2 Objective

To design a PID controller for the given Laboratory model two tank non-interacting system shown in Fig. 2 specifications are:

D. J. Hemanth et al. (Eds.): COMET 2019, LNDECT 35, pp. 187–197, 2020.
https://doi.org/10.1007/978-3-030-32150-5_20

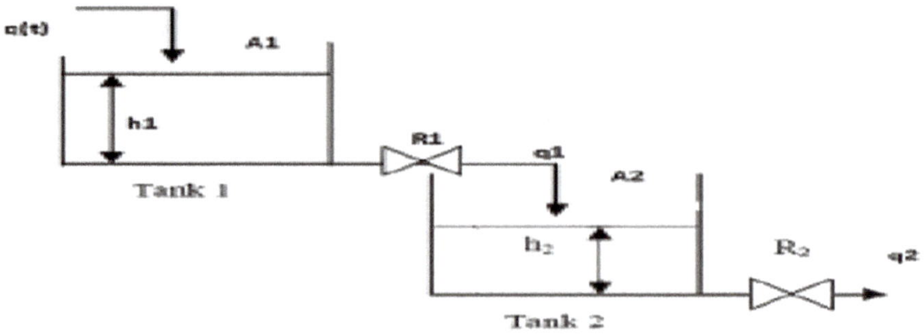

Fig. 1. Two Tank Non-interacting System

Tank1 and tank2 diameter = 92 mm
Step change from: 50–60 LPH
Find R1, R2, τ_1, τ_2.
$G_p(s) = ?$

Where,

R1 = Resistance of Tank1
R2 = Resistance of Tank2
τ_1 = Time Constant of Tank1
τ_2 = Time Constant of Tank2.

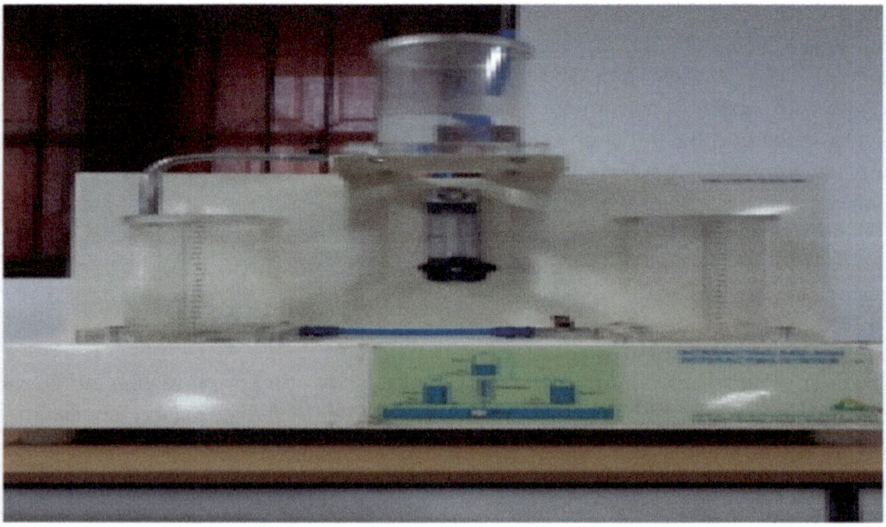

Fig. 2. Laboratory model Two Tank Non-interacting System

1.3 Calculation

Area of tank (A1) = $\frac{\pi}{4} d_1^2$ in m^2
Area of tank (A2) = $\frac{\pi}{4} d_2^2$ in m^2
Diameter of tank1 and tank2 = 92 mm
Area of A1 and A2 = $\frac{\pi}{4}(9.2)^2$ in m^2
Area of A1 and A2 = 6.647×10^{-3} m^2

$$R1 = \frac{dH_1}{dQ} = \frac{(FSS\ of\ tank1 - ISS\ of\ tank1) \times 10^{-3}\,\mathrm{m}}{\left(\frac{Final\ flow\ rate - initial\ flow\ rate}{3600}\right) \times 10^{-3}\,\mathrm{m}}$$

$$R2 = \frac{dH_2}{dQ} = \frac{(FSS\ of\ tank2 - ISS\ of\ tank2) \times 10^{-3}\,\mathrm{m}}{\frac{(Final\ flow\ rate - Intial\ flow\ rate)}{3600} \times 10^{-3}\,\mathrm{m}}$$

1.4 Tabulation and Observation

See Table 1.

Table 1. Real time data for Laboratory model Two Tank Non-interacting System

S. No.	Time(s)	Height of tank (1) mm (h1)	Height of tank (2) mm (h2)
1	30	60	33
2	60	66	35
3	90	68	39
4	120	70	40
5	150	71	42
6	180	72	45
7	210	73	48
8	240	73	48
9	270	73	48
10	300	73	48

Initial flow rate = 50 LPH
Final flow rate = 60 LPH
Initial steady state of tank1 = 54 mm
Initial steady state of tank2 = 34 mm
Final steady state of tank1 = 73 mm
Final steady state of tank2 = 48 mm

$$R1 = \frac{(73-54) \times 10^{-3}\,\mathrm{m}}{\left(\frac{60-50}{3600}\right) \times 10^{-3}\,\mathrm{m^3/s}} = 6840\,\mathrm{s/m^2}$$

$$R2 = \frac{(48-34) \times 10^{-3}\,\mathrm{m}}{\left(\frac{60-50}{3600}\right) \times 10^{-3}\,\mathrm{m^3/s}} = 5040\,\mathrm{s/m^2}$$

Transfer function for tank1 $\frac{Q_1(s)}{Q(s)} = \frac{1}{\tau_1 s + 1}$

Transfer function for tank2 $\frac{H_2(s)}{Q(s)} = \frac{R_2}{\tau_2 s + 1}$

Overall transfer function for Non-interacting system is

$$\frac{H_2(s)}{Q(s)} = \frac{R_2}{[(\tau_1 s + 1) + (\tau_2 s + 1)]}$$

$$\tau_1 = \mathrm{R1} \times \mathrm{A1}$$

$$= 6840\,\mathrm{s/m^2} \times 6.647 \times 10^{-3}\,\mathrm{m^2}$$

$$\tau_1 = 45.46\,\mathrm{s}$$

$$\tau_2 = \mathrm{R2} \times \mathrm{A2}$$

$$= 5040\,\mathrm{s/m^2} \times 6.647 \times 10^{-3}\,\mathrm{m^2}$$

$$\tau_2 = 33.50\,\mathrm{s}$$

Transfer function for tank1 is given by

$$\frac{Q_1(s)}{Q(s)} = \frac{1}{\tau_1 s + 1} = \frac{1}{45.46\,\mathrm{s} + 1}$$

Transfer function for tank2 is given by

$$\frac{H(s)}{Q(s)} = \frac{R_2}{\tau_2 s + 1} = \frac{5040}{33.50\,\mathrm{s} + 1}$$

Over all transfer function for Non-interacting system is given by

$$\frac{H_2(s)}{Q(s)} = \frac{R_2}{[(\tau_1 + 1)(\tau_2 s + 1)]}$$

$$= \frac{5040}{(45.46\,\mathrm{s} + 1)(33.50\,\mathrm{s} + 1)}$$

$$G_p = \frac{5040}{1556\,\mathrm{s}^2 + 78\,\mathrm{s} + 1}$$

2 PRC Based Cohen Coon Tuning Method

After that we have to find the controller values for the above system using PRC based Cohen Coon method as shown in Fig. 3 for tuning.

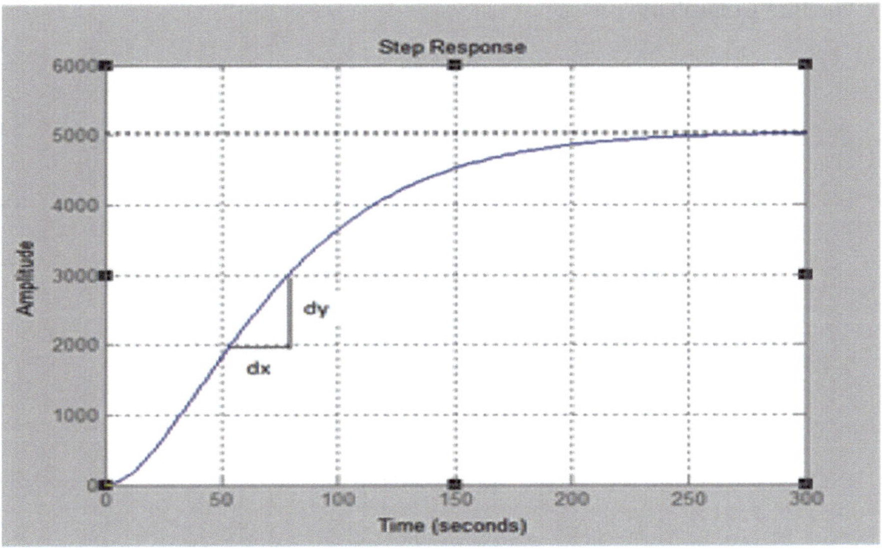

Fig. 3. PRC based Cohen Coon Tuning Method curve

2.1 Cohen Coon Method Tuning Procedure

1. The controller action does not occurs when the process control loop is in open condition.
2. This method is used only for system with self regulation. This is also called as open loop transient response method.
3. The controller is disconnected from the final control element to make the control system open loop.
4. When the step change is applied to the variable c, it affects the final control element.
5. Record the value of output with respect to time. The curve $y_m(\mathrm{t})$ is called as process reaction curve.
6. The dynamics of the main process, measuring sensor and final control element affects the process reaction curve. Cohen Coon observed that response of most processing unit to an input change as a sigmoid shape which can be adequately approximated by a response of the first order system with a dead time.

$$t_d = 10\mathrm{S}, \mathrm{B} = 5000, \mathrm{A} = 1, \mathrm{K} = \mathrm{B}/\mathrm{A}$$

$$\mathrm{K} = \frac{5000}{1}, \mathrm{K} = 5000, \tau = \mathrm{B}/\mathrm{slope}, \tau = \frac{5000}{\frac{dy}{dx}}$$

$$\mathrm{dy} = 3000 - 2000, \mathrm{dy} = 1000, \mathrm{dx} = 75 - 50, \mathrm{dx} = 25, \tau = \frac{5000}{40}, \tau = 125$$

For PID Controllers,

$$K_P = \frac{1}{K}\frac{\tau}{t_d}\left(\frac{4}{3} + \frac{t_d}{4\tau}\right)$$

$$\tau_I = t_d\left(\frac{32 + 6\left(\frac{t_d}{\tau}\right)}{13 + 8\left(\frac{t_d}{\tau}\right)}\right)$$

$$\tau_D = t_d\left(\frac{4}{11 + 2\left(\frac{t_d}{\tau}\right)}\right)$$

$$K_P = \frac{1}{5000}\frac{125}{10}\left(\frac{4}{3} + \frac{10}{4125}\right)$$

$$K_P = 0.00338$$

$$\tau_I = 10\left(\frac{32 + 6\left(\frac{10}{125}\right)}{13 + 8\left(\frac{10}{125}\right)}\right), \quad \tau_I = 23.8$$

$$\tau_D = 10\left(\frac{4}{11 + 2\left(\frac{10}{125}\right)}\right), \quad \tau_D = 3.58$$

$$G_c(s) = K_P\left[1 + \frac{1}{\tau_1} + \tau_D\right]$$

$$K_P = 0.00338$$

$$K_I = \frac{K_P}{\tau_1} = \frac{0.00338}{23.8}, \quad K_I = 0.00013$$

$$K_D = K_P \times \tau_D, \quad K_D = 0.00338 \times 3.58$$

$$K_D = 0.01074, \quad K_P = 0.00338, \quad K_I = 0.00013, \quad K_D = 0.01074.$$

2.2 CC Closed Loop Response

Close Loop Response of second order system as shown in Fig. 4.

It is an open loop method. CC method is very complex when compared to ZN method. External disturbances results in unstable operation. Fine tuning is needed. Controller setting is dynamic which leads to large overshoot and oscillatory response. It has more peak overshoot and settling time. Performance of the process is poor with delay. Tuning procedure is not applicable because it causes unstable at both higher and lower values of K_p. The reset and the rate time constant are low.

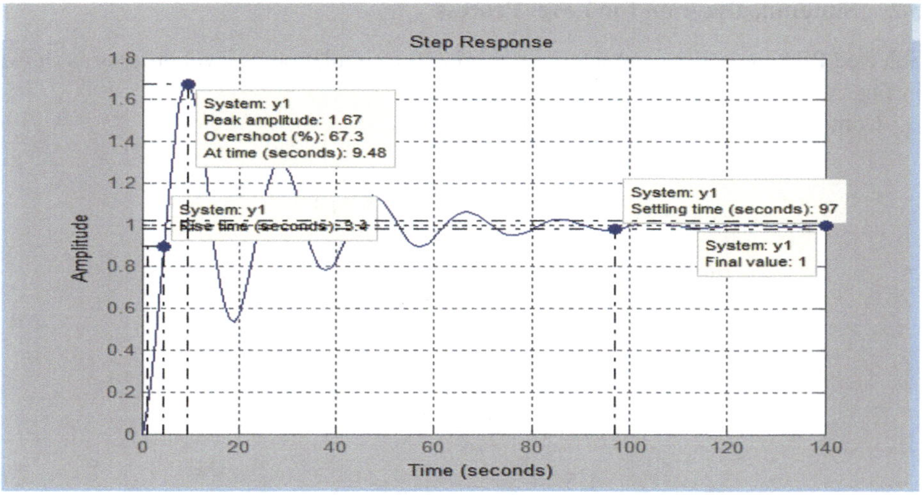

Fig. 4. Close Loop Responses of second order system

3 ZN Continuous Oscillation Method

The ultimate gain value, K_u, the ultimate period of oscillation, P_u, are used to calculate K_c in ZN closed loop method. It can be refined to give better approximations of the controller. To find the values of these parameters, and to calculate the tuning constants, use the following procedure.

3.1 The Ziegler Nichol's PID Tuning Procedure

1. Remove integral and derivative action. Set integral time (T_i) to a largest value and set the derivative controller (T_d) to zero.
2. By changing the set point, small disturbances can be created and by adjusting the P, the gain value is changed until the oscillations have constant amplitude.
3. The gain value (K_u) and period of oscillation (P_u) are recorded.
4. Plug these values into the ZN Close loop equations and determine the necessary settings for the controller. Close loop calculations of K_c, T_i, T_d.

The PID Controller parameters are selected from the following Table 2:

Table 2. PID Controller parameters

Mode	P	I	D
P	2 PBu	–	–
P+I	2.2 PBu	Tu/1.2	–
P+I+D	1.65 PBu	0.5 Tu	Tu/8

3.2 Simulink Diagram for Level Process

Selected PID controller parameters are used in the simulation of level process as shown in Fig. 5.

Controller transfer function is given by

$$G_c(s) = K_P\left(1 + \frac{1}{\tau_1} + \tau_d\right)$$

$$PB_u = 6, \quad T_u = 1.5, \quad K_P = 1.65 \times 6, \quad K_P = 10$$

$$\tau_1 = 0.5 \times T_u, \quad = 0.5 \times 1.5, \quad \tau_1 = 0.75$$

$$K_I = \frac{K_P}{\tau_1} = \frac{10}{0.75}, \quad K_I = 13.33$$

$$T_D = \frac{T_u}{8} = \frac{1.5}{8}, \quad T_D = 0.1875, \quad K_D = K_P \times \tau_d$$
$$= 10 \times 0.1875, \quad K_D = 1.875$$

$$K_P = 10, \quad K_I = 13.33, \quad K_D = 1.875.$$

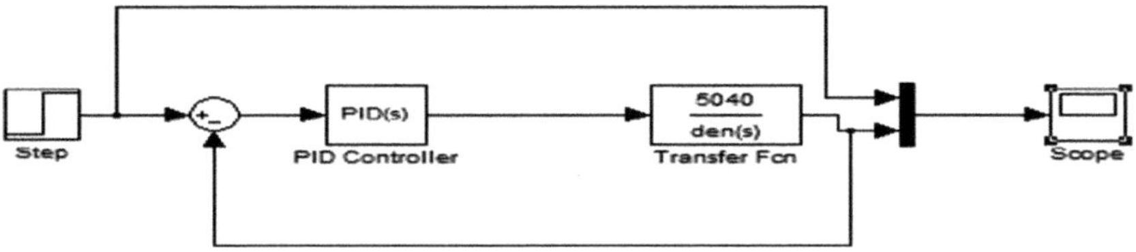

Fig. 5. Simulink diagram of PID Controller

3.3 Response Curve for Ultimate Gain and Period

Response curve for ultimate Gain and period as shown in Fig. 6. From this we are finding T_u, K_u.

$T_u = 1.5$ s
$K_u = 6$.

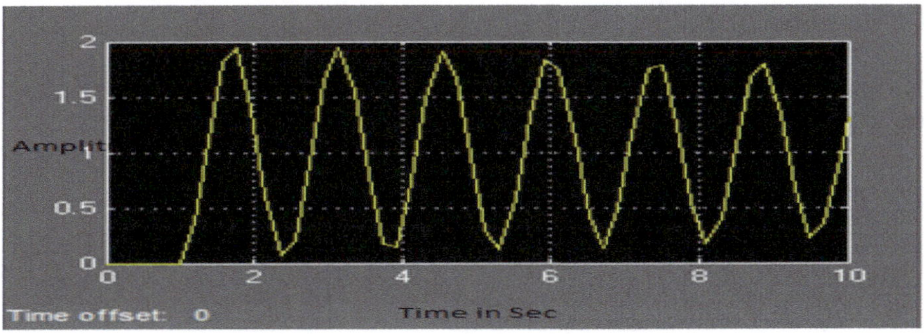

Fig. 6. Response curve of Ultimate Gain and Period

3.4 Z-N Close Loop Response

The ZN tuning technique closed loop response curve as shown in Fig. 7.

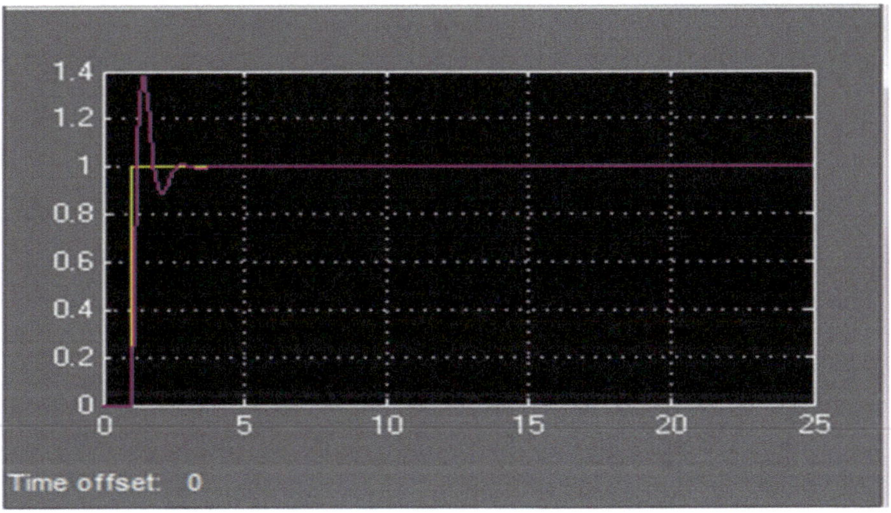

Fig. 7. Response curve of Z-N Close Loop Response

The response of the ZN tuning is slightly better than those with the CC settling. In this process, settling time and peak overshoot is reduced. Only the proportional element is used to tune the controller in ZN method. To achieve initial tuning it does not require trial and error method.

4 Response for Real Time Level Process

Over all response for real time level process using Z-N method is shown in Fig. 8. The time taken to get set point is 1.5 s.

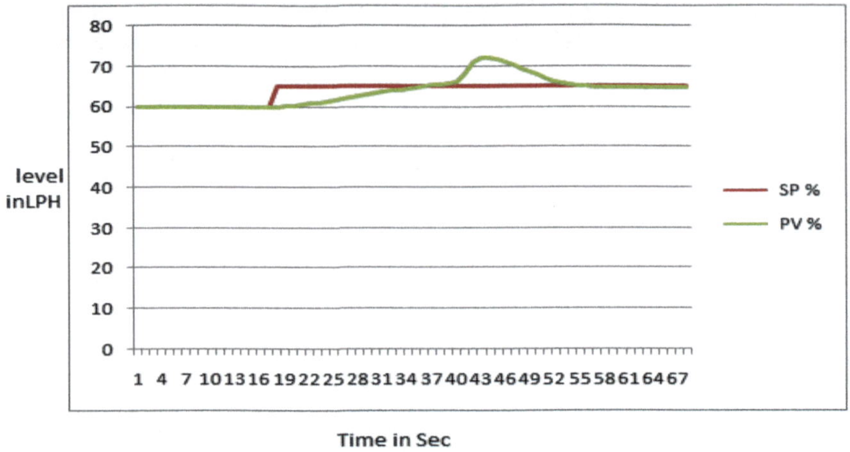

Fig. 8. Response curve of Level process using Z-N method

Over all response for real time level process using CC method is shown in Fig. 9.

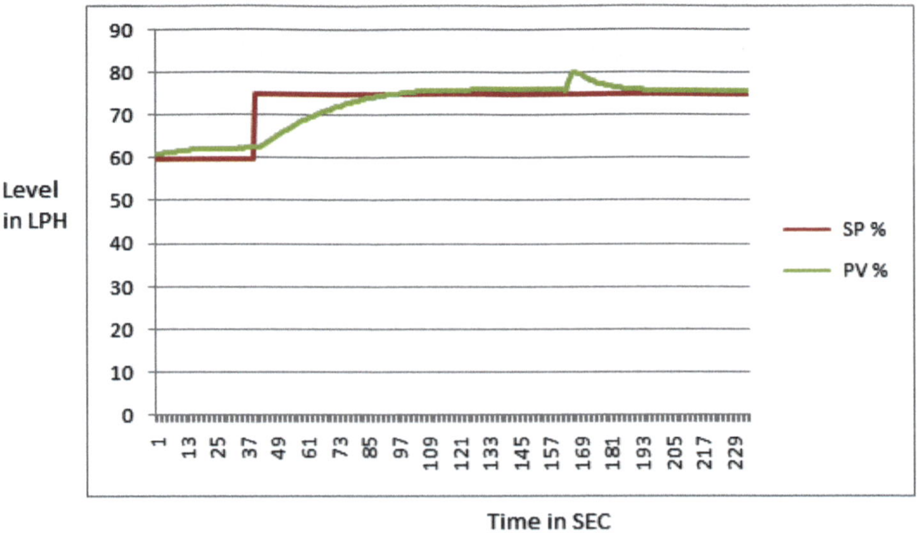

Fig. 9. Response curve of Level process using CC method

5 Conclusion

Thus the mathematical model for a non-interacting system was described and the PID values were designed by ZN and CC method. Compare to both methods, in ZN settling time is 1.5 s but in CC settling time is very large i.e. −105 s. We have implemented in Mat lab and also in real time level process.

References

1. Hang, C.C., Astrom, K.J., Ho, W.K.: Refinements of the Ziegler-Nichols tuning formula. Proc. IEE Control Theory Appl. **138**(2), 111–118 (1991)
2. Isa, I.S., Meng, B.C., Saad, Z.: Comparative study of PID controlled modes on automatic water level measurement system. In: Proceedings of IEEE International Colloquium on Signal Processing and Its Applications, pp. 131–136 (2011)
3. Singh, A.K., Kumar, S.: Comparing the performance analysis of three tank level control system using feedback and feed forward-feedback configuration (2014)
4. Coughanowr, D.R., LeBlane, S.E.: Process Control System Analysis and Control, 2nd edn. McGraw-Hill Publication, New York (1995)
5. Ziegler, J.G., Nichols, N.B.: Optimum settings for automatic controllers. Trans. ASME **64**, 759–768 (1942)

Experimental Analysis of Industrial Helmet Using Glass Fiber Reinforcement Plastic with Aluminium (GFRP+Al)

P. Vaidyaa(✉), J. Magheswar, Mallela Bharath, R. Vishal, and S. Thamizh Selvan

Department of Mechanical Engineering,, Panimalar Engineering College, Chennai, India
vaidyaa1999@gmail.com

Abstract. The fundamental use of a industrial helmet is to safeguard the users by reducing and absorbing mechanical energy and restrict any penetration. The high impacts influence and alter the protective and structural capabilities. One of the factor that plays a part in injury is the volume and weight of the helmet besides their energy absorption capacity, and is a threat to the user. Every year plenty of employs are accidentally injured or killed in the working places in the industry. To avoid such injuries or death by wearing a correct safety helmet. Protective head gear will save your life. In recent time impact strength of the helmet using in the industry is low because of irregular material filling, unequal distribution of pressure and blow holes. The main objective of the project is to increase the quality and strength of the helmet by improving the material used for the helmet. To achieve the main aim of the project we started the work in three stages. In the first stage a mold for the helmet is designed by using CATIA V5 software. Then by using the model the analysis is done using ANYSIS software for two different type of material which are GLASS FIBER REINFORCEMENT PLASTIC (GFRP) and GLASS FIBER REINFORCEMENT PLASTIC With ALUMINIUM (GFRP+AL).

Keywords: Safety helmet · Glass fiber · Stress-strain curve deformation

1 Introduction

We wear helmet on our head for make us safe from injuries. The use of the symbolic or ceremonial helmets without safety functions (e.g.: baseball player helmet) used sometimes. The Assyrian soldier in 900BC was first known of using helmet.

A thick layer of lather or bronze helmet worn in their head, soldiers wear helmets nowadays also now the helmets are often made by light weight material. This safety gear helps people from saving their lives from Sevier injuries from accident In this last two decades more accidents are highly by motor cycles. For a highly functional helmet must be designed and analyzed the structures of the helmet. The shell and foam layer are main component of the helmet. The impact energy of the helmet is

All observed by the form. It is the main function of it. The shell will pretend all foreign material that is pretend to come inside by hitting the helmet. If the shell is not

D. J. Hemanth et al. (Eds.): COMET 2019, LNDECT 35, pp. 198–208, 2020.
https://doi.org/10.1007/978-3-030-32150-5_21

working good the foreign object will it our skull an make us injure. To speed up the impact load on a wider foam area It expands the linear energy capacity of the foam. The main criterion is the force resistances test for determination of shell thickness and in fact making a thicker shell of the helmet. Consequently a weight of about 6 to 8 times.

As compared to the linear of foam's we choose a denser shell the strength will be improved. Unfortunately, as well as weight and cost or a different material should be examined. Compare a new material analysis results with standard component there are different types of helmet for different purposes for example, from striking the road the blunt impact forces are to be eliminated from driving bicycle helmet. A mountain climber's helmet must be designed for high protection against high impact, for objects like cobbles and pebbles and climbers equipments and food items fall from top of the mountain. Practical concerns also considered for designing for helmet should be small and light in weight so the helmet will not disturbed when climbing. Some of the helmets have extra protectective gears attached to them, such as goggles or a face mask and ear guard and some other forms of protective equipments for head, and a system for communication. Metal face protector may be attached with few sports helmets.

We have researched in theoretical background and the studied the ways how the accidents are happened. The load distribution on the helmet at accident. Is analyzed. The survey of the head injuries helps us to improve the quality of the helmet for protection by using the knowledge. Lightweight helmets are expected from the users. While meeting other system and foam fitting system for performance required. we can completely analyze of the design for system requirements.these helmet is used for safely and seal the human head from accidents. Hences, the structural and protectiveness of thru helmet are changed in high energy impact. This helmet design and material has been improved eventually in time. A load or forces is subjected which acts on a deformation. The various characteristic of a helmet is practically analyzed those von miss, stress energy, strain energy. A static concept of the helmet is analyzed. The different impact energy is load applying on concentric motions, sudden motion. The material like GFRP AND GFRP reinforced with aluminum/steel powders was used.

1.1 Materials

A few parameters that include very large molecular particles characterized by less weight, increased corrosion resistance, high strength-to-weight ratios and very minimum melting points forms a material structure. The forming of plastics are done with ease.

1.2 Thermoplastic

A thermo softening plastic or thermoplastic, it becomes pliable or moldable at high temperature and get harder when it is kept in low temperature because it is a polymer [1, 2]. Mostly all the thermoplastic have high molecular weight. By intermolecular the polymer chains associates, which makes the thermoplastic remolded and restores the bulk properties because in cooling inter molecular nitration increases in the thermoplastics; the thermosetting polymers are different from thermoplastic polymers. The cutting process forms the irreversible bond.thermosplast will never melt often. But they deform and don't reforms until cooling.

1.3 Types of Thermoplastic Used

Acrylonitrile Butadiene Styrene (ABS)
ABS is regularly used thermoplastic. it has the glass transition temperature of approximately 105 °C (221 °F). ABS is amorphous and it has no correct melting temperature. ABS is terpolymer processed by polymerizing styrene and acrylonitrile in the presence of polybutadiene.

The proportions are varied from 14% to 34% acrylonitrile, 6% to 32% of butadiene and 42% to 58% styrene. The result has a very longer chain of poly butadiene criss-crossed with smaller chains of poly (styrene-co-acrylonrilte). In the group of nitrile the neighbor chains by being polar, attract with each other and attached with other chains together, this makes ABS much stronger than pure substances; at less temperature it provides resilience. The ABS is used majorly for −20° to 80 °C because the mechanical properties changes for temperature. The properties of ABS are processed by rubber toughening, fine particles of elastomeric are evenly distributed all over the rigid matrix (Fig. 1).

2 Design and Modelling of the Industrial Helmet Catia V5

2D DRAWING OF INDUSTRIAL HELMET – CATIA V5:

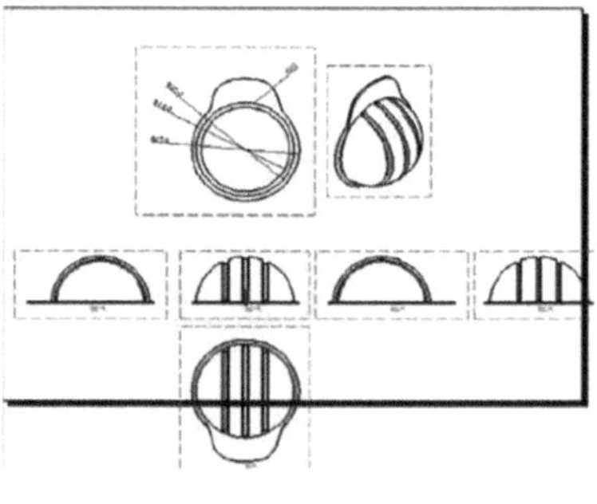

Fig. 1. 2D drawing of industrial

3d Modelling of Industrial Helmet

The model of industrial helmet was designed and using the tool Catia v5 it has been drafted in it (Fig. 2, 3, 4 and 5).

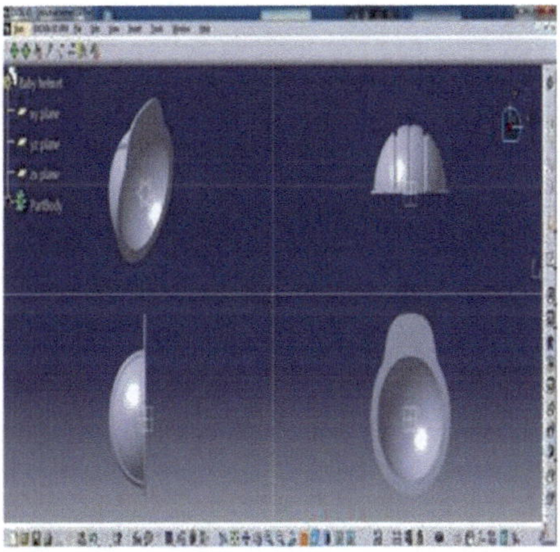

Fig. 2. 3D modeling of industrial helmet

3 Analysis Results

3.1 Total Deformation

GFRP:

Fig. 3. This ANSYS image shows the Image of total deformation of GFRP+AL total deformation STRESS INTENSITY:

GFRP:

Fig. 4. This ANSYS image shows the STRESS INTENSITYof GFRP

DEFORMATION ON X AXIS:

Fig. 5. This ANSYS image shows the DEFORMATION ON X AXIS (GFRP)

Testing

So it demonstrates from the chart that the strain is comparing to strain or prolongation is relating to the heap giving An is legitimate to the law. St. Line relationship. Point An is legitimate to this law. or on the other hand we can say that point An is have some outrageous minute that the immediate thought of the graph closes or there is a deviation from the straight nature. These centers are known as the most extreme of proportionality orthe Proportionality limit. For a brief span past the point A, the material can even now be versatile as in the disfigurements are completely changed when the heap is completely evacuated. The oppose point BIs named as Elastic Limit (Fig. 6).

To determine how the material interacts with each other when a force is applied to it is done by tensile testing. a simple way to measure the mechanical forces required to elongate a specimen to breaking point, one of the main objective is to predict how the object interact to their intended application.

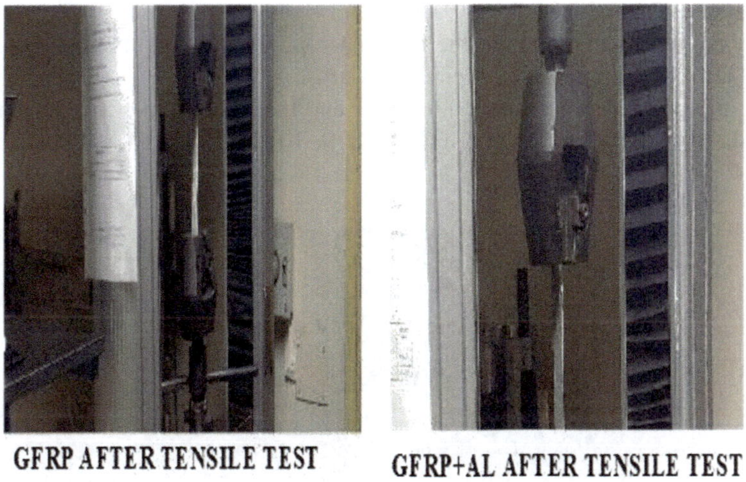

Fig. 6. .

A stress curve of force vs. extension, the graph is calculated till the force to break or failure point.

Graph Of Stress –Strain Curve and Load Displacement
(See Fig. 7.)

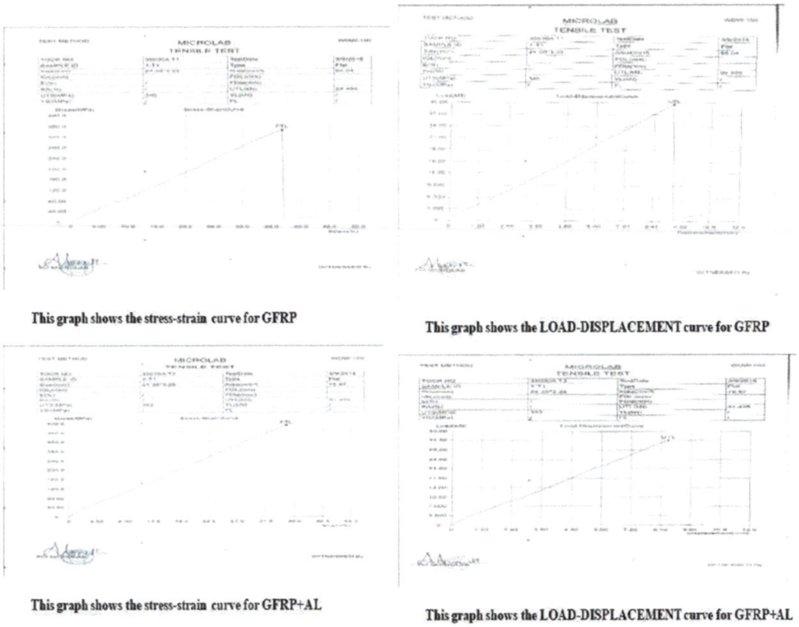

Fig. 7. Tensile testing may be measured by the parameters performance. The result is placed into curve that a force vs. extension -that shows the tensile profile of the material plotting it in a graph

Compression Test of GFRP

This test comprises of checking the materials opposing forces that are experiences a oposing force that pushes the specimen inwards.

From the other side is squashed together and the forces experienced are noted and tabulated with corresponding graph (Figs. 8, 9, 10, 11, 12 and 13) and (Table 1).

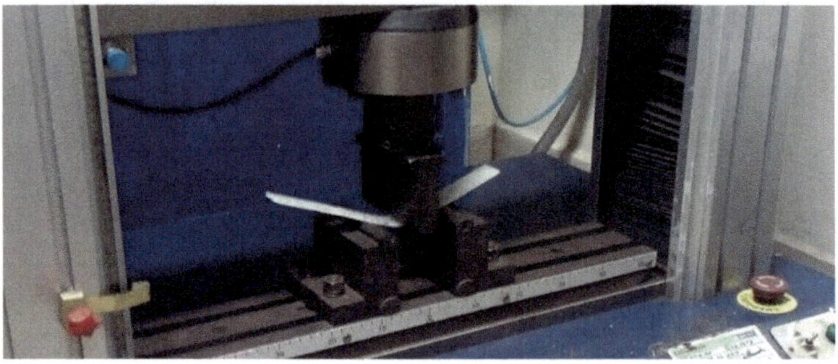

Fig. 8. Graph of stress strain and load displacement for compression test

Fig. 9. Charpy Result

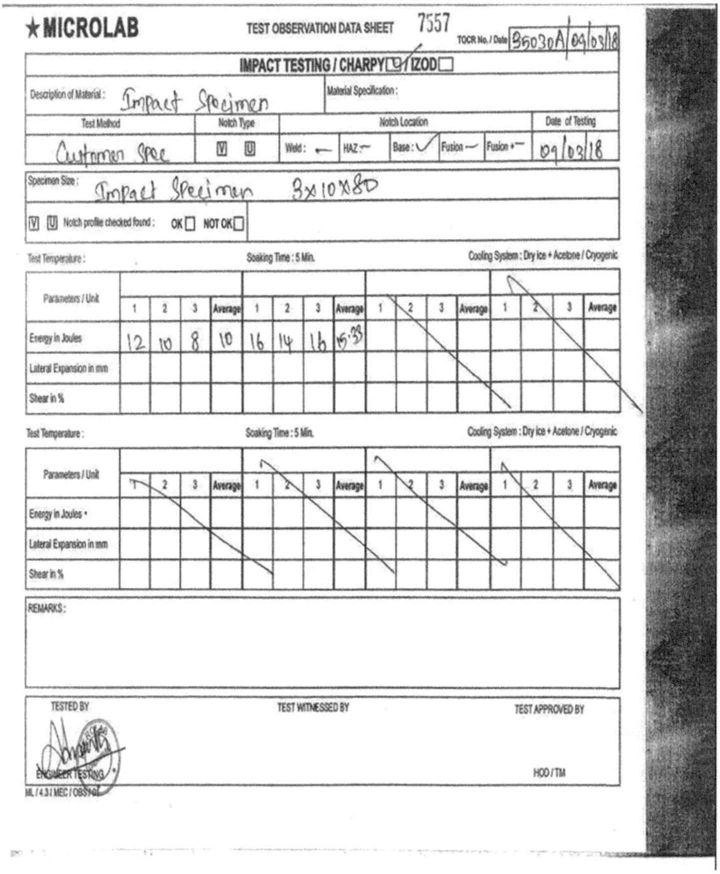

★MICROLAB TEST OBSERVATION DATA SHEET 7557 TOCR No./Date 35030A/09/03/18

IMPACT TESTING / CHARPY ☑ / IZOD ☐

Description of Material: Impact Specimen | Material Specification:

Test Method	Notch Type	Notch Location	Date of Testing
Customer Spec	V U	Weld: — HAZ: — Base: ✓ Fusion — Fusion+ —	09/03/18

Specimen Size: Impact Specimen 3x10x80

V U Notch profile checked found: OK ☐ NOT OK ☐

Test Temperature: Soaking Time: 5 Min. Cooling System: Dry ice + Acetone / Cryogenic

Parameters / Unit	1	2	3	Average	1	2	3	Average	1	2	3	Average	1	2	3	Average
Energy in Joules	12	10	8	10	16	14	16	15.33								
Lateral Expansion in mm																
Shear in %																

Test Temperature: Soaking Time: 5 Min. Cooling System: Dry ice + Acetone / Cryogenic

Parameters / Unit	1	2	3	Average	1	2	3	Average	1	2	3	Average	1	2	3	Average
Energy in Joules *																
Lateral Expansion in mm																
Shear in %																

REMARKS:

TESTED BY ENGINEER TESTING | TEST WITNESSED BY | TEST APPROVED BY HOD/TM

ML/4.3/MEC/OBS/01

Fig. 10. Total deformation in bar graph

Stress produced in GFRP+AL is more than stress produced in impact Carbon fiber nylon 4,6 and GFRP for equal height that indicates the resistances against the load per unit area. the factor of safety is high and capacity of withstanding of GFRP+AL is high. The result proves that the GFRP+AL produce high displacement than carbon fiber, nylon 4,6 and GFRP helmet. we cannot alter the displacement to high and the impact load over the helmet produces less volumetric strain in GFRP+AL helmet has less strain than carbon fiber produces.nylon4,6 and GFRP has equal height, that produces high rigidity to the workers neck from impact load.

GFRP VS GFRP+AL

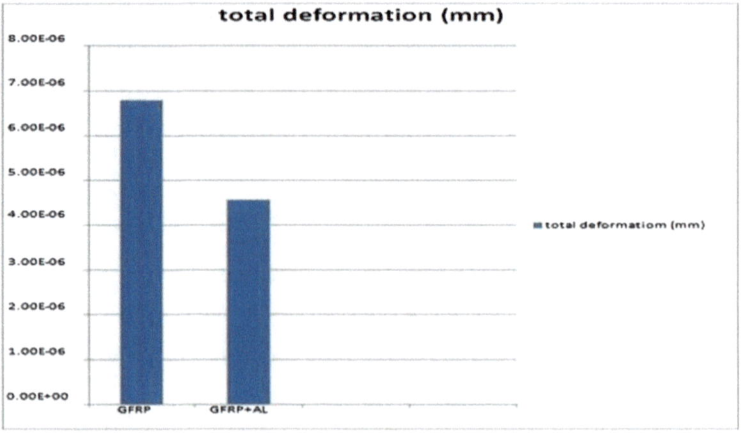

This graph shows the variation of TOTAL DEFORMATION (GFRP VS GFRP+AL)

DEFORMATION ON X, Y, Z AXIS (mm)

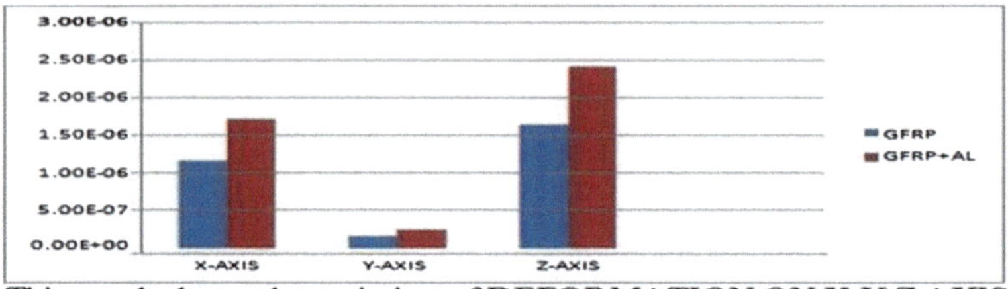

This graph shows the variation of DEFORMATION ON X,Y,Z AXIS (GFRP VS GFRP+AL)

SHEAR STRESS (Mpa);

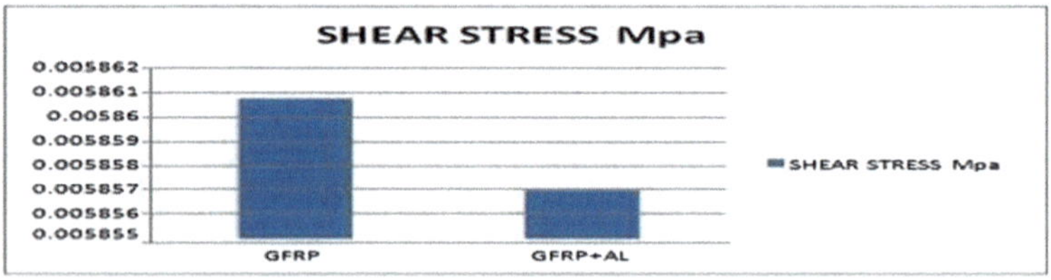

This graph shows the variation of SHEAR STRESS in GFRP V GFRP+AL

Fig. 11. Bar graph in shear elastic strain And Von Mises Stress

SHEAR ELASTIC STRAIN;

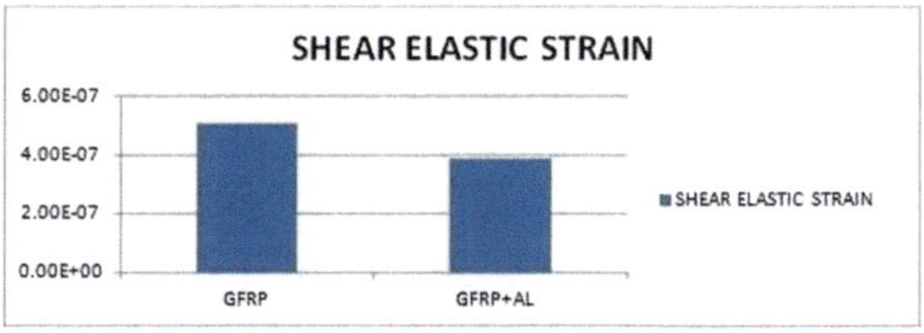

This graph shows the variation of SHEAR ELASTIC STRAIN in GFRP VS GFRP+AL

VON MISES STRESS(Mpa)

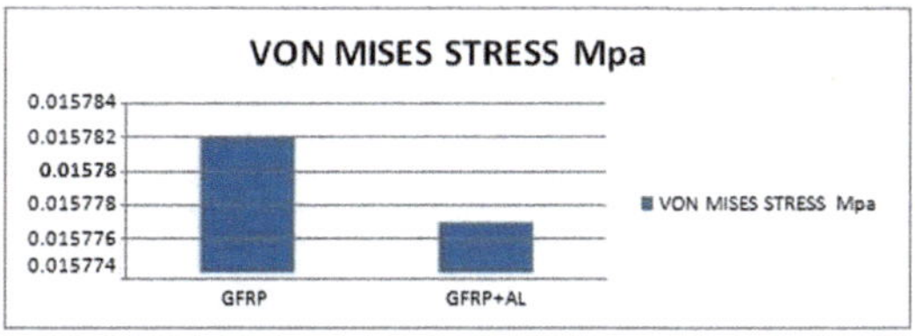

This graph shows the variation of VON MISES STRESS on GFRP VS GFRP+AL

Fig. 12. Bar graph for final result

EQUIVALENT ELASTICSTRAIN;

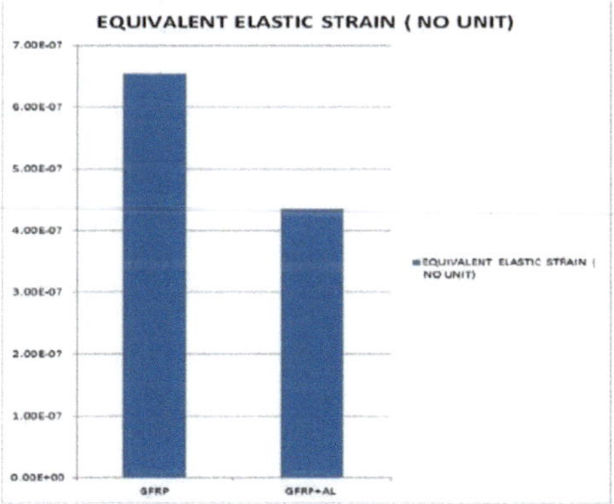

This graph shows the variation of EQUIVALENT ELASTIC STRAIN on GFRP Vs GFRP+AL

Fig. 13. Results and conclusion

Table 1. .

	Load (Kg)	Stress	Total deformation
Carbon fibre	25	14.33	0.002355
Nylon 4,6	25	16.206	0.004369
GFRP	25	0.017929	6.7849e-6
GFRP+AL	25	0.017918	4.555e-6

References

1. Park, H.S., Dang, X.P., Roderburg, A.: Development of Plastic Front Panels Of Green Cars. CIRP J. Manuf. Technol. **26**, 35–53 (2012)
2. Kuziak, R., Kawalla, R., Waengler, S.: Advanced high strength materials for automotive industry a review. J. Arch. Civil Mech. Eng. **8**(2), 103–117 (2008)
3. Falaichen, B.J.: Geometric Modeling and Processing. J. CAD **42**(1), 1–15
4. David, H.A.: Structural Analysis, Aerospace. Journal on Encyclopedia of Physical Science and Technology, 3rd edition (2003)
5. Japan, S., Daniel, L., Theodor, K.: Finite element analysis of beams. J. Impact Eng. **31**, 861–876, 155–173 (2005)
6. Olbisi, O.: Handbook of Thermo Plastics. MarcelDekker, New York (1997)
7. Rosato, D.V.: Plastics Engineering, Manufacturing & Data Hand Book
8. Donaldv, R.: Plastics Engineering, Manufacturing & Data Hand Book

A Study on Psychological Resilience Amid Gender and Performance of Workers in IT Industry

Lekha Padmanabhan(✉)

Saveetha School of Engineering, Saveetha University, Chennai 602 105, India
lekha.bes@gmail.com

Abstract. This research builds a review to examine different theoretical aspects of resilience also, to explore psychological resilience among gender of workers and their performance in IT industry. The variables for study are taken from the peer reviewed articles and papers such as enhancing individual well being, positive coping situation, autonomy, self acceptance, purpose in life, are explored as variables of psychological resilience with which the gender and performance of the workers were assessed. In this study the respondents are taken from the IT industries with age group from 25 to 48 years, certain people with a minimum of 3 years experience and few above three years of experience are taken as sample respondent for this study with a sample size of 150. Descriptive research design and random sampling have been used to collect data from respondents. Questionnaire with 17 questions were administered using seven point Likert scale ranging from strongly agree as 1 to strongly disagree as 7, been used to collect the data from the respondent with respect to psychological resilience variables and the collected data for survey is anonymizied. The statistical tool that has been implemented for the study are t-test, correlation analysis, regression was applied, and the output are obtained using SPSS 21. Results revealed that, psychological resilience variables are strongly allied to both the workers, and it is evidently shown with increased performance of both male and female workers.

Keywords: Psychological resilience · Gender · Performance · Workers · Positivity · Wellbeing

1 Introduction

1.1 Widening Resilience Concept: A Succinct Record

The researchers embark on how people managed to materialize from sternly nuisance situation relatively intact during the period 1950 and 1960. The resilience construct has changed extensively from their earlier studies spotlight on how to resist to negative outcomes between deprived children, and resilience research initiation happened during mid and through to the late-20th century. Before the early resilience research has theoretically put up as there are personality trait that consent positive outcomes even to below intense destitution. There are two fields in which resilience research has been

D. J. Hemanth et al. (Eds.): COMET 2019, LNDECT 35, pp. 209–216, 2020.
https://doi.org/10.1007/978-3-030-32150-5_22

originated to identify what makes individual to evade traumatic stress on adults, and developmental psychology which concentrates on youth and children to recognize which of the personal qualities of children like self-esteem would help to differentiate from children who had acclimatized with positivity to other disadvantages like socio-economic, violence or disregard, cataclysmic life events, to children explaining relatively shoddier results (Luthar, Cicchetti and Becker 2000). The key components of the resilience construct explicate early research studies as, risk factors in life of individual, protective mechanisms endurance, adversity and its multidimensional gamut of individual responses acknowledgement.

Also, (Werner and Smith 2001) has found resilience among children of high-risk can be envisaged with certain input and attributes of family-level, are said to be reasonably unswerving amid socio-political and ethnic groups, also, in the earlier study it has been made to portray on resilience of children and how these children are profited through sturdy sense of values, strength to cope up, support of family.

1.2 Concourse to Resilience

Resilience means adapting to a positive learned experience from a distress situation (i.e. from sad situation in families like death of children, parents or closer one, and problems in heath and relationship, stressful situation, work pressure in work place, loss of job) through individual thoughts, emotion and behavior. Hence, bouncing back to normal life after a trauma is said to be resilience. Recent recession situation during 2008, where many companies fall from hike and it took two to three years to come normal, and recovery after huge disaster even naturally or artificially like twin tower attack, floods, tsunami and how people recover themselves from pain and suffering that they come across from any distress emotionally can also be taken as examples for resilience. Moving on back to normal life through learned experience of behavior, thoughts, actions and emotions helps one to handle and balance both personal professional circumstances. Therefore, in general every human being has the tendency to overcome from any traumatic situations and to adapt to normal life after sometime this tendency in human being is called as resilience. Factors of resilience such as, care and support relationship of family members, realistic planning, capacity, confidence, positive attitude, communication skill, problem solving, known to manage feel and impulse, self-discovery, decisive action in decision making, consider change as part of life. Hence variables to resilience from various peer reviewed articles were found to be empathy, promoting & enhancing individual well being, positive coping situation.

Ryff's model, psychological well-being has six different dimensions; autonomy, environment, personal growth, positive relations with others, purpose in life and selfacceptance. Resilience relates to positive evaluations of one's self, sense of individual growth, development and determination on self.

Many authors like (Kimberly et al. 2000; Souri et al. 2011; Hasse et al. 2014; Fabio et al. 2015 and Scoloveno 2015) has identified that resilience as good promoter and enhancer for each individual wellbeing. Thus, resilience formulates a person to battle in opposition to persist heedful about some kind of jeopardy, therefore, performed the limit towards impending the muddle through maladaptive and deeds.

Former, the attention shifted towards positive psychology, where from the health related to psychology of human it is observed further with expressions of distress due to psychological issues also with several diseases in absence. When psychology is positively budged its attention focuses completely operative human being on unrestricted difficulty and malady and struggle for ones headway. Thus, wellbeing of one's psychology is defined through numerous conception like contentment in life, bliss and the matter of wellbeing involves a sort of psychosomatic welfare of an individual (Vinayak et al. 2018).

According to (Fava and Tomba 2009) individuals are said to be with elevated in resilience when they construe the occurrence to be nerve-racking, that under writes in the direction towards welfare of individual psychology. Also, resilience recount a positive assessment of self, a sense of escalation, development and self-determination, it enhances an individual credence in resolute and reminiscent life, thus, it is found to be causative in the direction of welfare and individual psychology.

Ryff and Keyes (1995) have given their contribution in psychological well-being through a multidimensional model which explains the change, since, advanced and anomalous approach to the well-being of individual psychology. And, through this model six different dimensions has been formulated such as autonomy, ecological mastery, individual augmentation, optimistic associations amid others, purpose in existence and self acceptance which creates their position to enhance the individual wellbeing through psychology. Also, (David 2015) has disclosed this model as the multifaceted model through empirically and scientifically using valid test.

According to (Larson 2006) resilience helps to create positive youth development. According to, (American psychological association, 2014) resilience is said to be the method of acclimating the trauma, adversity, threats, family and relationship problems are due to the sources of stress, as well as health related problems, and, certain stressors financial, workplace and tragedy. (Lee et al. 2012) has confronted resilience to portray the three core facet as a competence, development and outcome.

From earlier studies through psychological wellbeing model it has been found that resilient helps people to maintain better physical and health of individual psychology that provides additional supremacy to convalesce effortless, rapid and hectic or stressed situations. Thus, it is also said from various study that resilience offers one- self with health, confident, and sense of worth that facilitate the pact through trauma and depressed feeling, as a result, it found to show important role in psychological health.

Graber et al. (2015) has confronted the study on psychological resilience as protective mechanism by examining through articles and peer views, it progress on how psychological resilience facilitates the positive adaptation among people with varied gender, age, culture and other factors of life cycle. Resilience in life of child and adults varies, during childhood resilience is deeply under fasten by process followed in families and not with coping skills effectiveness. Whereas, with adulthood life, resilience may perhaps be pretentious by ingrained outline to cope with stress responses physiologically, culture, also with other social relationships among individuals and families. For instance, relationship that portrays parent-child positive relationship, and social networks communal support would depict that how the study relates skill to develop psychosocial work as sturdiest intercession. Research on resilience depicts that adapting to circumstances and accepting the change are achievable. Various case study of resilience has been taken for understanding case based on cultural comparison, conflicts among relationships, thus questions where focused on how resilience concept has been applied in field of psychology by researcher, protective mechanism against risk, factors of life cycle.

According to (Ryff and Singer's 2003) the psychological well-being model states that, people with resilience plays a better role in order to maintain their physical and psychological health which means resilience pave them additional supremacy power to recuperate easier and quicker from nerve-racking or traumatic situations.

2 Objective of the Study

The primary objective of the study is to find the relationship of psychological resilience variables among male and female workers (i.e.) gender and performance of workers in IT industry. The variables for study are taken from the peer reviewed articles and papers such as enhancing individual well being, positive coping situation, autonomy, self acceptance, purpose in life, are explored as variables of psychological resilience with which the gender and performance of the workers working in IT industry were assessed.

3 Need of the Study

Physiological resilience helps the individual to revisit the pre-traumatic stage swiftly. Therefore, it is shown from the existence of psychological resilience individual/people who can increase their capabilities through psychological and behavioral would consented them to linger unruffled during the circumstances of crisis/pandemonium, and to travel out from the occurrence without facing protracted negative corollary. Hence this study aids to identify the variables of psychological resilience and how it has been positively related with gender and performance of workers in IT industry.

4 Method

This study explores psychosocial resilience among gender and increased performance of workers in IT industry. The variables for study are taken from the peer reviewed articles and papers such as enhancing individual well being, positive coping situation, autonomy, self acceptance, purpose in life, are explored as variables of psychological resilience with which the gender and performance of the workers working in IT industry were assessed. Descriptive research design has been used to collect the data from respondent. The procedure used to collect samples from population is done using random sampling method. In this study the respondents are taken from IT industry that handles with age group from 25 to 48 years, certain people with a minimum of 3 years experience and few above three years of experience are taken as sample respondent for this study with a sample size of 150. Questionnaire were administered using 17 item scale each questions with seven point Likert scale ranging from strongly agree as 1 to strongly disagree as 7, been used to collect the data from the respondent with respect to psychological resilience variables were used for primary collection of data and secondary data are collected using journals, articles and chapter related to the study area and it shows a reliability measure of 0.76, and the collected data for survey is anonymizied. The statistical tool that has been implemented for the study are t-test, correlation analysis, regression, and the output are obtained using SPSS 21.

5 Results and Discussion

Many researchers have taken various variable of psychological resilience on their studies and only few found positive impact on psychological resilience variables towards workers life and wellbeing. This study helps to find that whether variables of psychological resilience show any relationship among the gender and performance of workers.

In the present study first and foremost aim is to explore the relationship between psychological resilience among gender of workers and their increased performance in IT industry, using t test which reveals a significant value of $p \leq .01$ level. Hence which depicts that psychological resilience plays a vital aspect for the escalation and improvement of both male and female workers and their performance in IT industry.

From the above Table 1, it is found that the relationship between variables of psychological resilience with workers (male & female) and their performance. The result shows that enhancing individual well being ($r = 0.59^{**}$, $p \leq 0.01$), positive coping situation ($r = 0.44^{**}$, $p \leq 0.01$), autonomy ($r = 0.49^{**}$, $p \leq 0.01$), self acceptance ($r = 0.64^{**}$, $p \leq 0.01$), purpose in life ($r = 0.41^{**}$, $p \leq 0.01$), for male are said to be positively related with male workers and their performance.

From Table 1, the t value, F value and r2 value: it reveals that all significance values of the variable among gender found be significant at $p \leq 0.01$, hence it is evident that, there is no gender differences were found on psychological resilience among workers, since all the psychological resilience variables portrays values of 0.000, which are said to be significance at ($p \leq 0.01$) of workers, this infers that psychological resilience can be implied to any individual without gender differences, because both male and female has to come across distressful circumstances, and to overcome from the situations would take time. But, the time taken to heel from a traumatic situation varies among individuals irrespective to genders, because certain individual overcome effortlessly and hastily when compare to other individual, so this happens comparatively depend upon the self supreme power of individual which is said to be different for each and not because of gender, hence it is evident from the t-value for gender of workers. Similarly, the t test for the performance of male (10.04) and female (10.19) workers are found to be significant at ($p \leq 0.01$) for workers.

From the above Table 1, the performance of both male and female workers shows the value significant at $p \leq 0.01$ level. Hence, the study discloses that all the variables of psychology resilience are said to be positively influencing both male and female (gender) of workers as well as increased performance of workers. Also, all the F-values of both male and female workers and increased performance of workers are found to be significant at one percent level, this portrays that this model is significant. From Table 2, the correlation coefficient with values range of 1, which reveals that the absolute value of coefficient as larger. Therefore, it is inferred that, there exist stronger relationship among psychological resilience amid gender and performance of workers in IT industry.

6 Conclusion

Thus, psychological resilience bestows an individual to gain better confidence and purpose of life with a sense of self within them which allow them to deal circumstances effectively with a positive cope to handle stress and negative emotions, thought, behavior. Thus, resilience plays an important role in psychological health of individuals and their wellbeing. From this study it can be concluded that psychological resilience variables are strongly allied to workers, and it is evidently shown with increased performance of both male and female workers. This shows that that the positivity through psychological resilience would not only increases the performance of workers but also, the psychological health of workers are improved than before, which aids to direct oneself toward better inter and intra personal understanding among individual and their wellbeing.

7 Scope of the Study

The study on psychological resilience depicts, that the resilience variables aid to motivate an individual to attain positive emotions, and thriving stress adaptation. Connotation for research into the scope of this study would inhibit certain protective factors so as to provide effect on the advantage of psychological resilience. Thus, the attention toward psychological resilience can be broadened by handling emotions with positivity. In further, different socio-economic strata of sample can be suggested for the study. And, the scope of the study can be further extended by focusing on other additional factors of psychological resilience like grit, emotions, impulses, trust, self-confidence, positive self-image, communication among family and surroundings.

Appendix

Table 1. The regression analysis for psychological resilience variables as predictors of gender and performance of workers.

Predictor Variables		Standardized Coefficients	t-value	R 2	F-value	Sig
Female Worker						
Psychological resilience variables	Enhancing individual well being	0.68	8.32**	0.66	68.52**	.000
	Positive coping situation	0.75	10.24**	0.58	104.23**	.000
	Autonomy	0.60	6.66**	0.47	43.56**	.000
	Self acceptance	0.76	8.03**	0.56	69.53**	.000
	Purpose in life	0.62	6.64**	0.43	54.56**	.000
Increased performance		0.78	10.19**	0.74	104.63**	.000
Male Worker						
Psychological resilience variables	Enhancing individual well being	0.70	9.32**	0.57	75.33**	.000
	Positive coping situation	0.54	6.97**	0.39	54.46**	.000
	Autonomy	0.68	8.32**	0.47	68.52**	.000
	Self acceptance	0.70	10.24**	0.58	61.23**	.000
	Purpose in life	0.60	6.66**	0.36	43.56**	.000
Increased Performance of male workers		0.71	10.04**	0.68	101.05**	.000

**Significant at $p \leq .01$ level *Significant at $p \leq .05$ level

Table 2. Pearson correlation table : Relationship between psychological resilience, gender of workers and increased performance of workers.

	Psychochological resilience	Gender of workers	Increased performance
Psychological resilience	1.00		
Gender of workers	0.44**	1.00	
Increased performance of worker	0.62**	0.32**	1.00

Source: Primary Data, Note: ** Significance at one per cent level.

References

David: Carol Ryff s model of psychological well-being the six criteria of well- being (2015). Accessed http://livingmeanings.com/six-criteria-wellryffs-multidimensional-model/

Fabio, A., Palazzeschi, L.: Hedonic and eudaimonic well-being: the role of resilience beyond fluid intelligence and personality traits. Front. Psychol. **6**, 1367 (2015). https://doi.org/10.3389/fpsyg.2015.01367

Fava, G.A., Tomba, E.: Increasing psychological well-being and resilience by psychotherapeutic methods. J. Pers. **77**(6), 1903–1934 (2009). https://doi.org/10.1111/j.1467-6494.2009.00604

Hasse, J.E., Kintner, E.K., Monahan, P.O., Robb, S.L.: The resilience in illness model, part1: exploratory evaluation in adolescents and young adults with cancer. Cancer Nurs. **37**(3), E1 (2014)

Kimberly, A., Christopher, K., Kulig, J.: Determinants of psychological wellbeing in Irish immigrants. West. J. Nurs. Res. **22**(2), 123–143 (2000)

Larson, R.: Positive youth development, willful adolescents, and mentoring. J. Community Psychol. **34**(6), 677–689 (2006)

Lee, T.Y., Cheung, C.K. Kwong, W.M.: Resilience as a positive youth development construct: a conceptual review. Sci. World J., 390–450 (2012). https://doi.org/10.1100/2012/390450

Luthar, S.S., Cicchetti, D., Becker, B.: The construct of resilience: a critical evaluation and guidelines for future work. Child Dev. **71**(3), 543–562 (2000). https://doi.org/10.1111/1467-8624.00164

Graber, R., Pichon, F., Carabine, E.: Psychological resilience: state of knowledge and future research agendas. Working Pap. **425**, 1–27 (2015)

Ryff, C.D., Keyes, C.: The structure of psychological well-being revisited. J. Pers. Soc. Psychol. **69**(4), 719–727 (1995)

Ryff, C.D., Singer, B.: Flourishing under fire: resilience as a prototype of challenged thriving. In: Keyes, C.L.M., Haidt, J. (eds.) Positive Psychology and the Life Welllived, APA, Washington, DC, pp. 15– 36 (2003)

Scoloveno, R.: A theoretical model of health-related outcomes of resilience in middle adolescents. W. J. Nurs. Res. **37**(3), 342–359 (2015)

Vinayak, S.: Resilience and empathy as predictors of psychological wellbeing among adolescents. Int. J. Health Sci. Res. **8**(4), 192–200 (2018)

Souri, H., Hasanirad, T.: Relationship between resilience, optimism and psychological well-being in students of medicine. Procedia Soc. Behav. Sci. **30**, 1541–1544 (2011)

Werner, E.E., Smith, R.S.: Journeys From Childhood to Midlife: Risk Resilience and Recovery. Cornell University Press, Ithaca and London (2001)

RETRACTED CHAPTER: Anti-poaching Secure System for Trees in Forest

K. Vishaul Acharya(✉), G. Mariakalavathy, and P. N. Jeipratha

Department of Computer Science and Engineering,
St.Joseph's College of Engineering, OMR, Chennai 119, India
vishalachar523@gmail.com

Abstract. The quantity of trees has been diminished radically from the woodland that makes an undesirable situation for creatures to make due in the timberland. At present, untamed life and timberland divisions are confronting the issue of development of creatures from backwoods zone to local location. In this paper, a system is proposed for following and disturbing for assistance of trees from human and fire accidents. Flame sensors are used to monitor and detect fire. PIR sensors are used to monitor and detects the motion on nearby surroundings and alerts the forest officials. Surf algorithm is used to find whether the movement is from animal or human. This proposed system helps the forest officials to protect the tree from forest fire and poaching.

Keywords: Flame sensor · PIR (Passive Infrared) sensor · SURF (Speeded up robust features) algorithm

1 Introduction

Woods fire is the reason for different and irreversible damages to both environment as well as financial matters. For now, numerous valuable species are cleared off, life and resources are undermined, etc. In spite of an expanding of state costs to control this catastrophe, every year a large number of flame mishaps happens across the world. It spent an enormous measure of customary person reconnaissance recognizing, anyway the exact report can be recommended by abstract components. It is developing to show the dynamic conduct of fire spread in a forest in order to make plan to reduce fire. Numerous scientists centre fire spread in a forest in order to make recreate the propagation of rapidly spreading fires. This work proposes system for protecting trees from forest fire and poaching.

The flame sensor and PIR (Passive infrared) sensor refreshes the data (at whatever point there is flame or human or creature is recognized) to timberland division as quickly as time permits. Flame sensor is utilized to recognize the flame and PIR sensor identifies the movement on close-by environment and information is refreshed to the Internet of Things.

Picture handling is a procedure to change over an image into computerized structure also, playing out certain tasks on it, to get an upgraded image or to extricate a few data from it. It is a kind of standard guideline where input is picture, similar to video bundling or photo and might be picture or characteristics related to picture.

The original version of this chapter has been retracted. The retraction note to this chapter can be found at https://doi.org/10.1007/978-3-030-32150-5_163

D. J. Hemanth et al. (Eds.): COMET 2019, LNDECT 35, pp. 217–225, 2020.
https://doi.org/10.1007/978-3-030-32150-5_23

Regularly it joins considering pictures to be dimensional signs while applying formalset pennat preparing strategies to picture.

Implanted C is a lot of language expansions for the C language by the advisory group to address shared problems that exist between C augmentations for different installed works. Implanted C programme needs non standard increases to C language for helping colourful features. In 2008, the C Standards Committee extended the C language to address these issues by giving a run of the mill standard to all use to stick to. It fuses different features not open in ordinary C, for instance, fixed-dot number juggling, name address spaces, and essential I/O gear tending to. MATLAB is a numerical enrolling condition and fourth-age programming language.

Steps used in SURF algorithm are:

1. Set the focal points for animal and human and store the image in database.
2. Capture an image using web cam
3. The image captured will extract the points and checks with the images stored in database using pixel points.
4. The image will be classified as human or animal using focal points.

2 Related Works

Dabhi [5] said that finding facial component in pictures is an essential stage for applications, for example, eye following, acknowledgment of face, face appearance acknowledgment and face following and lip perusing. He has proposed a strategy for distinguishing face from the live picture. The face is distinguished from whole picture using viola jones figuring. He has utilized falling of stage to make the procedure quicker.

Soraya, Chrang, Chan and Su [6] said that the Internet of Things framework is response for checking the temperature at various purposes behind zone in a server farm, making this temperature information perceivable over web through cloud based dashboard and sending SMS and email alerts to predefined beneficiaries when temperature transcends the secured working zone and achieves certain high qualities. This engages the information to focus supervisory gathering to make quick move to address the temperature deviation. Additionally this can be checked from wherever at whatever point over online dashboard by the senior estimation pros who are absent in the server farm at whatever point. This Wireless Sensor Network (WSN) based checking framework incorporates temperature sensors, ESP8266 and Wi-Fi switch. ESP8266 is a low power, exceedingly combined Wi-Fi plan from Espresso. The ESP8266 here, in this model, accomplices with 'Bidets' cloud through its API for indicating temperature information on the cloud dashboard on advancing and the cloud occasion the authorities frame work makes alarms at whatever point the high temperature arranged occasion is finished. Cloud occasions should be proposed for various alarms up to this time through the simple to utilize UI of the stage.It's to be noticed that the sensor utilized here can be utilized to screen the general moistness of the server farm condition too alongside the temperature of the server farm. Be that as it may, for this model arrangement is kept completely around the temperature observing.

Khan, Sahoo, Han, Glitho and Crespi [2] said that sharing a passed on Wireless Sensor Network Infrastructure using various, synchronous applications can help comprehend the certifiable ability of Internet-of-Things. Virtualized WSNs can be used by various applications and associations in the meantime including semantic applications to help end-clients to comprehend the setting of the occasions and settle on instructed choices. This System has proposed a heuristic-based genetic estimation to pick capable centre points to perform profitable in-arrange sensor data remark in virtualized WSN's. This paper additionally present early reproductions results.

Wu, Rudiger, Redoute and Yuce [1] said that this system presents a wearable Internet of Things hub went for observing unsafe natural conditions for security applications by means of LoRa remote innovation. The proposed hub is low-powered and backings different ecological sensors. A LoRa passage is utilized to interface sensor to the Internet. This for the most part around checking carbon monoxide, carbon dioxide, bright, and few broad natural areas. Poor condition could make serious medical issues people. Along these lines, encompassing ecological information is accumulated by the hub in a continuous way, afterwards it is send to server. The information is shown to relevant clients through an electronic application situated in the cloud server and the gadget will offer caution to the client by means of portable application when a crisis condition happens. The exploratory outcomes show that our security observing system can work dependably with low power utilization.

Vikram, Harish, Nishaal and Umesh [7] said that with the quick increment in use and dependence on the distinctive highlights of shrewd gadgets, the requirement which interconnects them is certifiable. More existing frameworks have wandered in the circle in Home Automation however has evidently neglected for giving savvy answers for the equivalent. The paper delineates techniques to give minimal effort Home Automation System utilizing Wireless Fidelity. This takes shape idea to internetworking using brilliant gadgets. Wireless Sensor Network is intended to monitor and controlling ecological, security and parameters of a keen intercommunicated home. The customer rehearse reliable order over the contraptions in a home by methods for the Android application.

Kim and Yu [3] said that Traditional system the executives depended on wired system, which is inadmissible for asset compelled gadgets. WSN's comprise Internet of Things can be expansive scale systems, and it is difficult to deal with every hub separately. In this framework, they proposed a system the board convention for WSN's to diminish the board traffic.

Wang [8] said that he proposed a novel quick way to deal with the recognition, division and confinement of human faces in shading pictures under complex back ground. To begin with, umber of transformative operators are consistently dispersed in the 2-D picture condition to recognize the skin-like pixels and fragment each face-like locale by enacting their developmental practices. At that point wavelet deterioration is connected to every district to identify the conceivable facial highlights and a three-layer BP neural system is utilized to recognize the eyes among the highlights. Test results demonstrate that the proposed methodology is quick and has a high discovery rate.

Shen, Zafeiriou, Chrysos and Kossaifi [4] said that identification and following of appearances in picture groupings is among the most all around considered issues in the crossing point of factual AI and PC vision. Frequently, following and discovery procedures utilize an unbending portrayal to depict the facial district, thus they can neither catch nor abuse the non-inflexible facial happenings, which are critical for endless of uses (e.g., outward appearance examination, facial movement catch, elite face acknowledgment and so on.). More often than not, the non-inflexible happenings are caught by finding and following the situation of set authority facial tourist spots eyes, nose, mouth and so forth.

3 Methodology

In order to protect the trees in forest from smuggling and natural disaster such as forest fire, a new system is implemented. The modules used in this system are Monitoring using sensors, Process of input images and Alert using IoT. In monitoring using sensors module, the fire sensor and PIR sensor monitors the forest in order to protect trees from forest fire and smuggling. In process of input images module, the camera is fixed in the forest will capture the image and finds whether it is human or animal using SURF algorithm. In alert using IoT module, the sensors alerts the forest officials and update information whether the movement is from human or animal (Fig. 1).

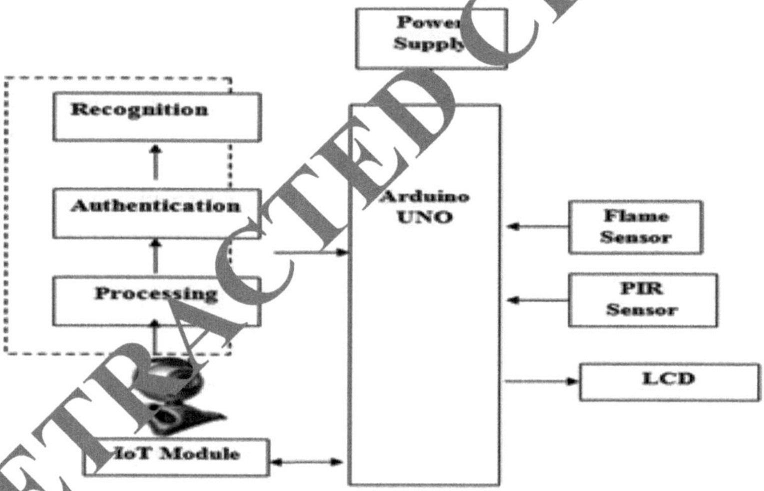

Fig. 1. System design

3.1 Monitoring Using Sensors

In this module, Fire sensors (LM393) are used to monitor and detect the fire. PIR sensors are used to monitor and detects the motion on nearby surroundings and alerts the forest officials. Flame sensors and PIR sensors are connected to Arduino UNO controller. The information is updated in LCD. LCD screen is an electronic presentation module and locate a wide scope of utilizations. Here in LCD, the display will show NO MOVEMENT or NO FIRE if movement or fire is not detected. The display shows MOVEMENT or FIRE if movement or fire is detected.

3.2 Process of Input Images

After finding movement using PIR (Passive Infrared) sensor, SURF (Speeded up robust features) algorithm is used to find whether the movement is from human or animal. The images of humans and animals are stored in database. The cameras are fixed in forest area and after the movement the image is captured and sends information whether it is human or animal to forest department. SURF algorithm helps to find whether it is human or animal using pixel points. The camera will first recognize the image. It then checks the image using authentication. Then the camera begins to process. Speeded up vigorous highlights is a verified neighbourhood incorporate pointer and descriptor. It will in general be used for errands, for instance, object affirmation, picture enrolment, gathering or 3D redoing. It is most of the way energized by the scale-invariant component change (SIFT). The standard sort of SURF is multiple times quicker than SIFT and guaranteed by its creators to be more overwhelming against various picture changes than SIFT. To recognize interest centres, SURF uses an entire number theory of the determinant of Hessian mass pointer, which can be handled with number exercises utilizing a precomputing fundamental picture. Its segment relies upon the entire of the haar wavelet response round the point of convergence. This can be handled utilizing the guide of the important picture. SURF descriptors have been utilized to discover and get articles, people or faces, to recreate 3D scenes, to pursue objects and to remove central focuses. SURF uses square-framed channels as a supposition of Gaussian smoothing. Filtering the image is significantly speedier if the essential picture is used:

$$S(x, y) = \sum_{i=0}^{x} \sum_{j=0}^{y} I(i,j) \tag{1}$$

SURF utilizes a mass identifier subject to the Hessian system to find central focuses. The determinant of the Hessian organize is utilized as a degree of neighbourhood change around the point and focuses are picked where this determinant is maximal. Given a point p = (x, y) in a picture I, the Hessian cross segment H(p, σ) at point p and scale σ, is:

$$H(\rho, \sigma) = \begin{pmatrix} L_{xx}(\rho, \sigma) & L_{xy}(\rho, \sigma) \\ L_{yx}(\rho, \sigma) & L_{yy}(\rho, \sigma) \end{pmatrix} \tag{2}$$

3.3 Alert Using IoT

Flame sensor and PIR (Passive infrared Sensor) automatically sends information to forest department as soon as possible. Flame sensor (LM393) senses fire and PIR sensor detects the movement of animal or person. The sensors are controlled and monitored automatically. After PIR sensor alerts, the camera fixed in the forest captures the image and compares the image stored in the database. Using SURF algorithm, it checks the pixel points of image and finds whether it is animal or human. The output will be displayed in the system. The display will show NO PERSON or NO FIRE if movement or fire is not detected. The display shows PERSON or FIRE if movement or fire is detected. It also displays whether the movement is from humanor animal. The forest official will get the alert using IoT and they take necessary actions. If it is human, forest officials will check the person since more people come and get trees and use them for illegal activities such as smuggling. This module is used in order to protect the trees from illegal activities like smuggling and disaster such as forest fire.

4 Results

The below figure (Fig. 2) shows the generated output when there is no fire in the forest.

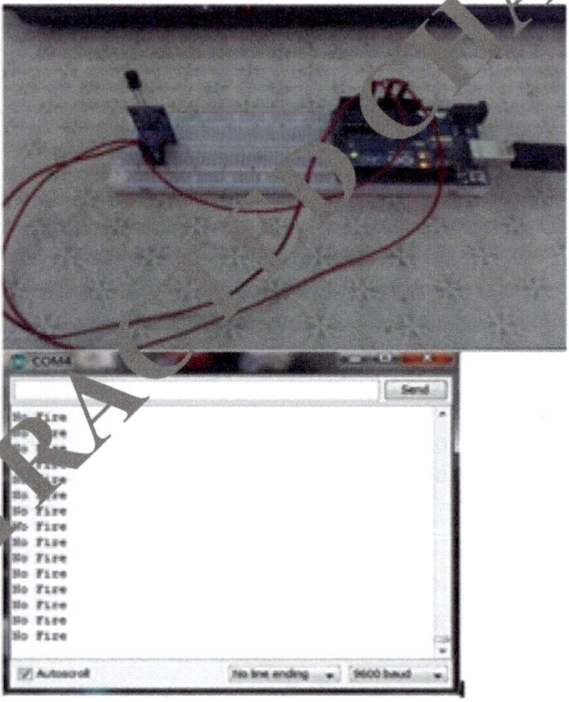

Fig. 2. When there is no fire

The below figure (Fig. 3) shows the generated output when there is fire in the forest.

The below figure (Fig. 4) shows the generated output when the movement is identified as human using SIFT algorithm.

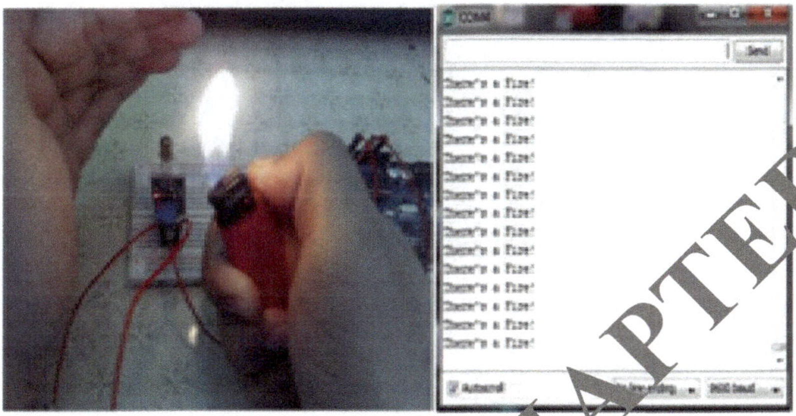

Fig. 3. When there is fire

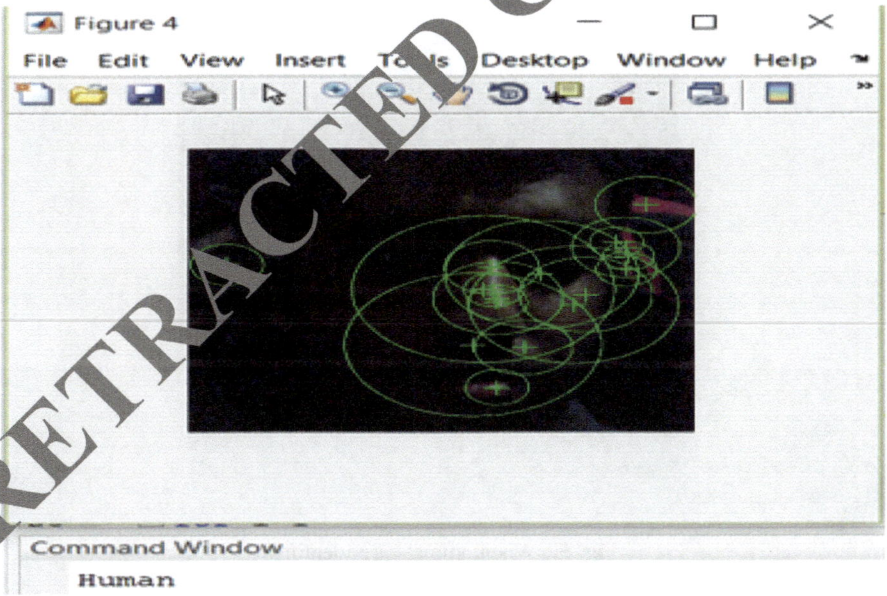

Fig. 4. When human is identified

The below figure (Fig. 5) shows the generated output when the movement is identified as animal using SIFT algorithm.

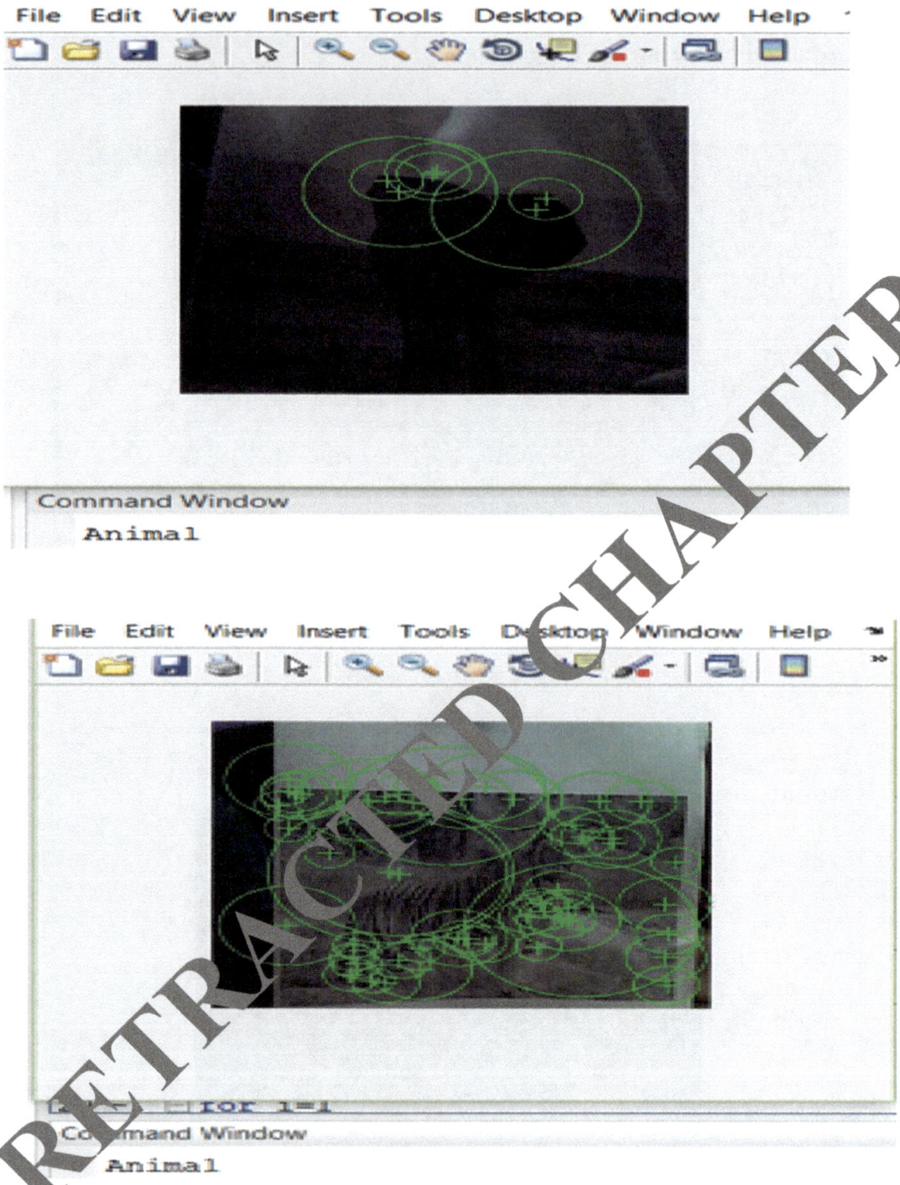

Fig. 5. When animals are identified

5 Summary

In this paper, Fire sensor are used to monitor and detect the fire. PIR sensors are used to monitor and detects the motion on nearby surroundings and alerts the forest officials. Fire sensors and PIR sensors are associated with Arduino UNO controller. The data is refreshed in LCD. It is an electronic showcase module that assumes an imperative job in different applications. LCD shows "NO PERSON" or "NO FIRE" if any individual or flame has not been recognized and the other way around.

After finding movement using PIR (Passive Infrared) sensor, we need to find whether human or animal using SURF (Speeded up robust features) algorithm. The images of humans and animals are stored in database. The cameras are fixed in forest area and after the movement the image is captured and sends information whether it is human or animal to forest department.

6 Conclusion

Forest officials receives information when any fire or movement occurs. The developed secured system also identifies whether the movement is from animal or human. Forest officials take action accordingly. Thus the developed secure system protects the trees from forest fire and poaching and this method is efficient to process in real time.

References

1. Wu, F., Rudiger, C., Redouté, J.M., Yuce, M.R.: A wearable IoT sensor node for safety applications via LoRa. IEEE Access **6**, 40846–40853 (2018)
2. Khan, I., Sahoo, J., Han, S., Glitho, R., Crespi, N.: A genetic algorithm-based solution for efficient in-network sensor data annotation in virtualized wireless sensor networks. In: 2016 13th IEEE Annual Consumer Communications & Networking Conference (CCNC) (2016)
3. Kim, J., Yu, S.: Wireless sensor network management for sustainable Internet of Things (2014)
4. Shen, J., Zafeiriou, S., Chrysos, G.G., Kossaifi, J.: The first facial landmark tracking in-the-wild challenge: benchmark and results (2015)
5. Dabhi, M.K.: Face detection system based on viola-jones algorithm (2016)
6. Soraya, S.I., Chiang, T.-H., Chan, G.-J., Su, Y.-J.: IoT/M2M Wearable-based activity-calorie monitoring and analysis for elders (2017)
7. Vikram, N., Harish, K.S., Nishaal, M.S., Umesh, R.: A low cost home automation system using Wi-Fi based wireless sensor network incorporating Internet of Things (IoT) (2017)
8. Wang, Y.: A novel approach for human face detection from colour images under complex background (2016)

Fault Tolerant Arithmetic Logic Unit

Shaveta Thakral(✉) and Dipali Bansal

Department of Electronics and Communication (FET), Manav Rachna International Institute of Research and Studies, Faridabad, India
{shaveta.fet,dipali.fet}@mriu.edu.in

Abstract. Low power design is most challenging in this era of IC technology and shaken the Moore's law. Various techniques have been invented for low power design and researchers are taking lot of pain to investigate new steps and methods to achieve low power goal. Reversible logic is seeking attention from last one decade and can be employed to balance between power and performance. Reversible logic is based on principle of no bit loss and claims almost no power dissipation. Although paradigm shifts from conventional logic to reversible logic is tedious but there is no other way looking around to reduce power dissipation. Many Reversible logic based arithmetic and logic units are available in literature but incorporating fault tolerance is demand of various applications. This paper aims in designing fault tolerant arithmetic logic unit based on high functionality conservative and parity preserving logic based gates. The quantum cost of used gates in proposed design is verified using RCViewer+ tool and performance of proposed design is evaluated with respect to existing designs in literature.

Keywords: Low power · IC · Reversible logic · ALU · Fault tolerant

1 Introduction

Reversible logic works on principle of no bit loss and consequently no heat loss and a big boom in today's challenging environment of IC technology shacked due to impending Moore's law. With joint efforts of Landauer [1] and Bennett [2], many researchers started working in this area. A significant work has been done in this area and many reversible logic based digital circuits have been investigated. Smart computing demanded by complex systems is always embedded with fault tolerance mechanism. ALU is heart of any computing environment and it can be made robust by adding fault tolerance mechanism in it. One way of introducing fault tolerance is by designing it using parity preserving logic based gates.

Parity preserving logic gates are based on principle of retaining same parity in input and output vector of any reversible logic based gate. If input vector holds odd number of 1's, then output vector should also hold odd parity otherwise both corresponding vectors maintain even parity. Maintaining conservative property along with parity preserving is a bit intractable and Fredkin gate falls in this category. Conservative gate not only retain parity in corresponding input and output vectors; simultaneously number of 1's should also be same in both vectors. This paper presents a fault tolerant arithmetic Logic Unit based on high functionality conservative and parity preserving

D. J. Hemanth et al. (Eds.): COMET 2019, LNDECT 35, pp. 226–233, 2020.
https://doi.org/10.1007/978-3-030-32150-5_24

logic gates. Proposed fault tolerant arithmetic Logic unit is designed based on conservative and parity preserving Fredkin, low quantum cost parity preserving based double Feynman and high functionality parity preserving based F2PG gates.

Reversible logic based arithmetic and logic unit is demanded in almost all types of computing environment and many researchers have given their significant contribution in this field. Syamala and Tilak [3] have proposed two ALU architectures but both circuits have low functionality and high quantum cost as well as fault tolerance is missing in their structure. Morrison and Ranganathan [4] have proposed reversible logic based ALU with quantum cost 35; Proposed circuit performs nine operations but fault tolerance is not embedded in it also. Saligram et al. [8] have proposed two ALU architectures based on parity preserving logic gates but both circuits have not been defined with their quantum cost. Bashiri and Hagaparast [9] proposed ALU architecture with fault tolerance but garbage and ancillary lines are high as compare to number of operations performed. Existing ALU designs have trade off between functionality, quantum cost, ancillary inputs and garbage outputs. Indeed the improvement scope in this paper arises from designing a novel reversible ALU architecture. A brief introduction about parity preserving logic gates used in proposed ALU architecture is given in Table 1. Section 2 explains methodology of proposed ALU design. Section 3 gives detail about proposed ALU design. Section 4 gives comparison and results evaluated. Section 5 gives conclusions followed by references.

Table 1. Brief about parity preserving logic gates used in proposed architecture

Reversible Gate	Logical Equations	QC	Speciality	NCT/NCV Equivalence
Fredkin	$P = A$ $Q = \bar{A}B \oplus AC$ $R = \bar{A}C \oplus AB$	5	Universal Reversible Logic Gate, Conservative and Parity Preserving multiplexer as well as swap gate	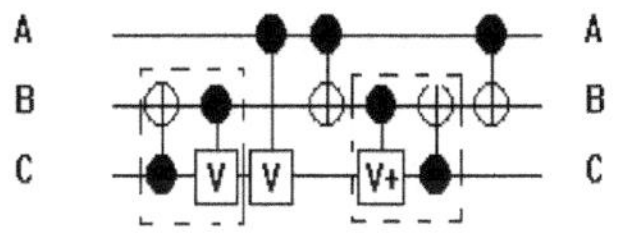
Double Feynman	$P = A$ $Q = A \oplus B$ $R = A \oplus C$	2	Parity Preserving copy and NOT gate	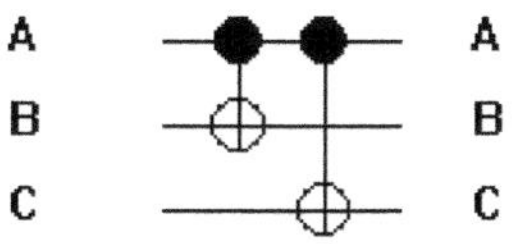
F2PG	$P = \bar{B}C \oplus AC$ $Q = A \oplus B$ $R = A \oplus B \oplus C$ $S = (A \oplus B)C \oplus (AB \oplus D)$ $T = \overline{AB} \oplus E$	14	Perform all Logical operations	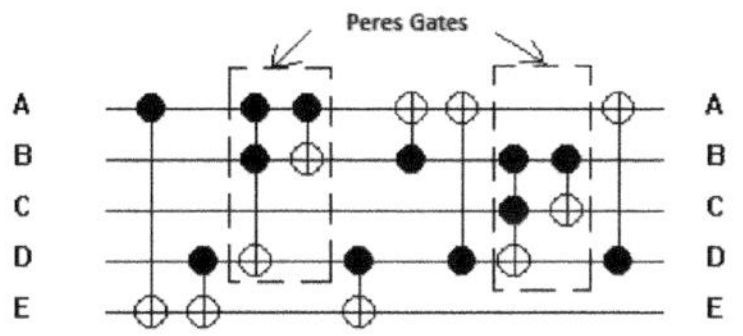

2 Methodology

The proposed fault tolerant arithmetic logic unit architecture is designed using three Fredkin Gates, one double Feynman gate and one F2PG gate which are parity preserving gates. Fredkin Gate 1 is passing logic 0 or logic 1 or signal B as per desired logic depending upon combination of S2 and S3. Fredkin Gate 1 under operation is shown in Fig. 1 and its functionality is shown in Table 2.

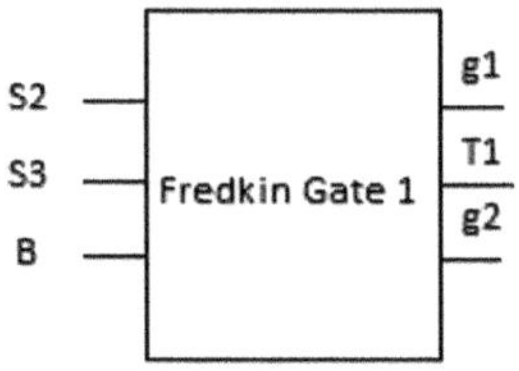

Fig. 1. Fredkin Gate 1 under operation

Table 2. Functionality of Fredkin Gate 1

S2	S3	B	T1
0	0	*	0
0	1	*	1
1	*	B	B

Fredkin Gate 2 is acting as 2:1 multiplexer and passing Cin or signal B as per desired logic depending upon select line S1. Fredkin Gate 2 under operation is shown in Fig. 2 and its functionality is shown in Table 3.

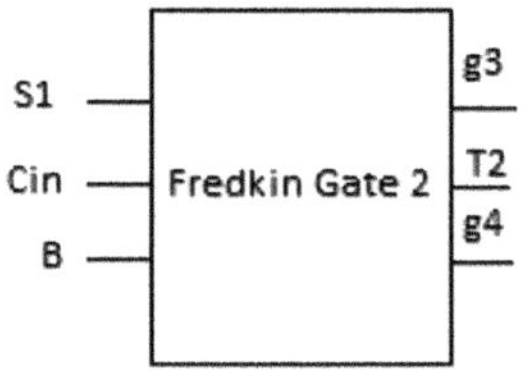

Fig. 2. Fredkin Gate 2 under operation

Table 3. Functionality of Fredkin Gate 2

S1	T2
0	Cin
1	B

Double Feynman Gate is passing 0 or 1 as per desired logic depending upon select line S4. Double Feynman Gate under operation is shown in Fig. 3. and its functionality is shown in Table 4.

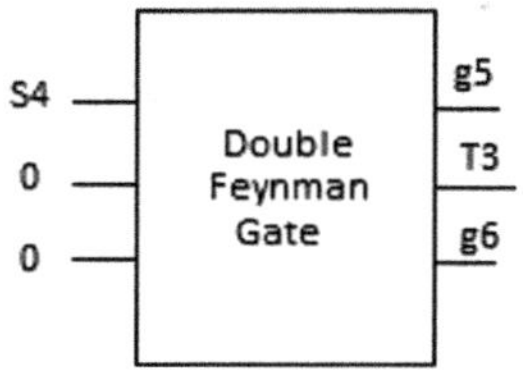

Fig. 3. Double Feynman Gate under Operation

Table 4. Functionality of Double Feynman Gate

S4	T3
0	0
1	1

F2PG gate can perform XOR, AND, NAND, NOT, XNOR, OR and NOR operations as well as can act as full adder also depending upon various combinations of T1, T2 and T3. F2PG Gate under different logical operations is shown in Fig. 4 and its functionality is shown in Table 5.

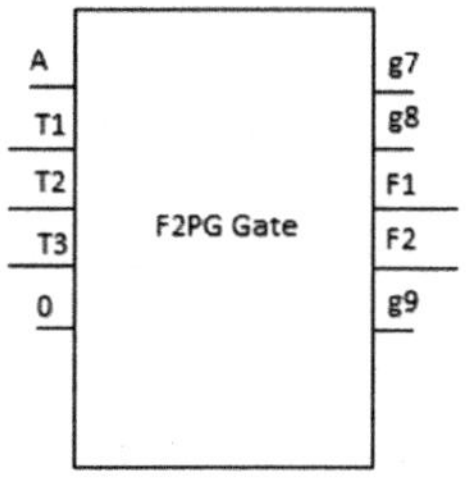

Fig. 4. F2PG Gate under different logical Operations

Table 5. Functionality of F2PG Gate

T1	T2	T3	F1	F2
0	B	0	XOR	AND
1	B	0	XNOR	OR
1	B	1	XNOR	NOR
B	Cin	0	SUM	Cout
0	B	1	XOR	NAND

Fredkin Gate 3 is acting as 2:1 multiplexer passing F1 or F2 as per desired logic depending upon select line S5. Fredkin Gate 3 under operation is shown in Fig. 5 and its functionality is shown in Table 6.

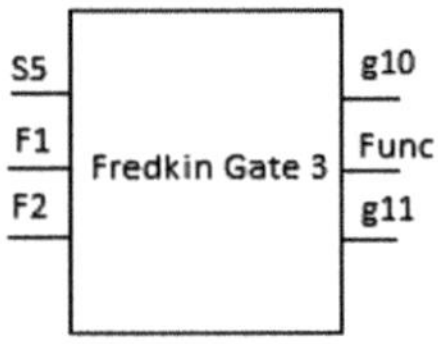

Fig. 5. Fredkin Gate 3 under operation

Table 6. Functionality of Fredkin Gate 3

S5	Func
0	F1
1	F2

3 Proposed Fault Tolerant Reversible 1-Bit ALU

The proposed Fault tolerant Reversible 1-bit ALU is configured to perform 7 logical and 5 arithmetic operations. Proposed fault tolerant ALU architecture is shown in Fig. 6 and its functionality is shown in Table 7 (* specifies don't care condition). As proposed fault tolerant reversible 1–bit ALU architecture consists of three Fredkin gates with quantum cost 5 each, one double Feynman gate with quantum cost of 2 and one F2PG gate with quantum cost of 14. Therefore total quantum cost of proposed ALU is 31. The ancillary input lines used in proposed ALU are 3 and garbage output lines are 11.

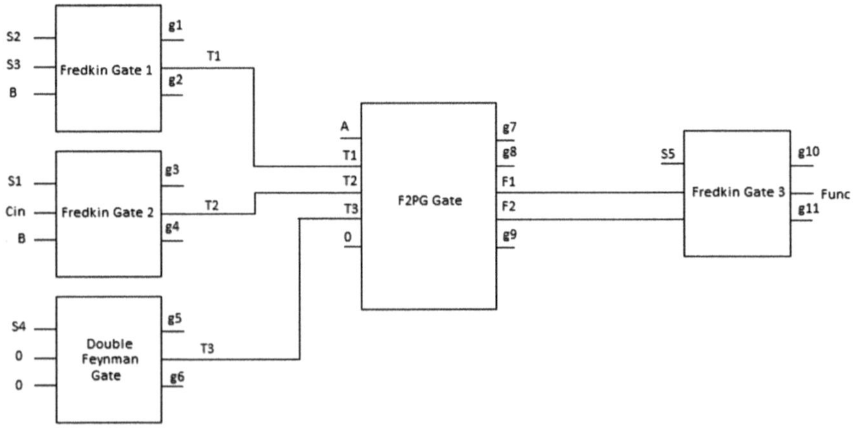

Fig. 6. Proposed Fault Tolerant Arithmetic Logic Unit Architecture

Table 7. Operations performed by fault tolerant ALU

S1	S2	S3	S4	S5	Func	Operation
1	0	0	0	0	$A \oplus B$	XOR
1	0	1	0	0	$A \odot B$	XNOR
0	1	0	0	0	$A+B+Cin$	Addition with Carry
1	0	0	0	1	$A.B$	AND
1	0	1	0	1	$A+B$	OR
1	0	1	1	1	$\overline{A+B}$	NOR
1	0	0	1	1	$\overline{A.B}$	NAND
1	1	*	0	0	A	Transfer A
1	1	*	0	1	B	Transfer B
1	1	*	1	1	$\overline{B}$	NOT B
1	0	0	1	0	$A+B$	Addition without carry
1	0	1	1	0	$A+\overline{B}$	Subtraction

4 Comparison and Results

The proposed fault tolerant ALU architecture is compared with existing ALU designs based on all optimization aspects. The proposed Arithmetic and logic unit is found with lowest quantum cost which supports 12 operations. Only five reversible logic gates are used to design architecture and therefore complexity is avoided. Proposed ALU architecture took only 3 constant inputs and produces 11 garbage outputs. Optimization

Table 8. Optimization Aspects Comparison of Various ALU Designs

ALU DESIGNS	Design 1 [8]	Design 2 [8]	Design 3 [9]	Proposed Design
No. of Gates	10	9	22	5
Quantum Cost	Not specified	Not specified	90	31
Arithmetic & Logic operations	11	11	8	12
Type of Gates Used	PPPG/F2PG, FRG,F2G	F2G,NFT,FRG, PPPG,F2PG	NFT,FRG,F2G, Islam	Fredkin, double Feynman, F2PG
Garbage outputs	17	17	28	11
Ancillary Inputs	9	11	19	3
Fault tolerance	Yes	Yes	Yes	Yes

aspects comparison of various ALU designs is given in Table 8. Proposed fault tolerant ALU proves efficient and optimum in terms of all optimization aspects as compare to other existing designs as shown in Fig. 7.

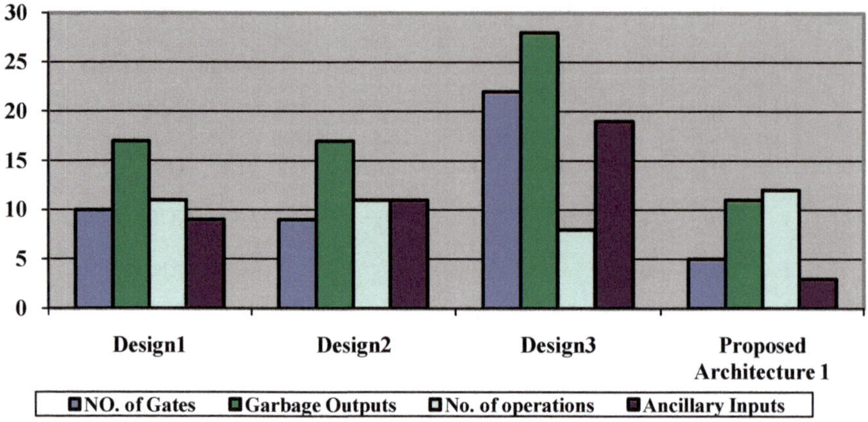

Fig. 7. Optimization metrics comparison of various ALU Designs

5 Conclusions

The performance of proposed ALU design over existing ALU designs is quantitatively analyzed and performance evaluation metrics proves it to be most efficient and optimum balance of all. Fault tolerance approach using parity preserving gates in combination with other methods will prove it to be a robust model for smart computing applications. Parity preserving logic based fault tolerance approach is useful for detection of single bit fault. Future scope of this research is to investigate new conservative logic gates which will be stepping stones in design of multibit fault detection and correction may be provided by cyclic redundancy check methods.

References

1. Landauer, R.: Irreversibility and heat generation in the computing process. IBM J. Res. Dev. **5**, 183–191 (1961)
2. Bennett, C.: Logical reversibility of computation. IBM J. Res. Dev. **17**, 525–532 (1973)
3. Syamala, Y., Tilak, A.: Reversible arithmetic logic unit. In: 3rd International Conference Electronics Computer Technology (ICECT), pp 207–211. IEEE (2011)
4. Morrison, M., Lewandowski, M., Meana, R., Ranganathan, N.: Design of a novel reversible ALU using an enhanced carry lookahead adder. In: 2011 11th IEEE International Conference on Nanotechnology, pp 1436–1440. IEEE, Portland (2011)
5. Singh, R., Upadhyay, S., Jagannath, K., Hariprasad, S.: Efficient design of arithmetic logic unit using reversible logic gates. Int. J. Adv. Res. Comput. Eng. Technol. (IJARCET) **3**(4) (2014)

6. Guan, Z., Li, W., Ding, W., Hang, Y., Ni, L.: An arithmetic logic unit design based on reversible logic gates. In: IEEE Pacific Rim Conference on Communications, Computers and Signal Processing (PacRim), pp 925–931. IEEE (2011)
7. Gupta, A., Malviya, U., Kapse, V.: Design of speed, energy and power efficient reversible logic based vedic ALU for digital processors. In: NUiCONE, pp 1–6. IEEE (2012)
8. Saligram, R., Hegde, S.S., Kulkarni, S.A., Bhagyalakshmi, H.R., Venkatesha, M.K.: Design of parity preserving logic based fault tolerant reversible arithmetic logic unit. Int. J. VLSI Des. Commun. Syst. **4**, 53–68 (2013)
9. Bashiri, R., Haghparast, M.: Designing a novel nanometric parity preserving reversible ALU. J. Basic Appl. Sci. Res. **3**, 572–580 (2013)
10. Moallem, P., Ehsanpour, M., Bolhasani, A., Montazeri, M.: Optimized reversible arithmetic logic units. J. Electron. **31**, 394–405 (2014)
11. Gopal, L., Syahira, N., Mahayadin, M., Chowdhury, A., Gopalai, A., Singh, A.: Design and synthesis of reversible arithmetic and logic unit (ALU). In: International Conference on Computer, Communications, and Control Technology (I4CT), pp 289–293. IEEE (2014)
12. Sen, B., Dutta, M., Goswami, M., Sikdar, B.: Modular design of testable reversible ALU by QCA multiplexer with increase in programmability. Microelectron. J. **45**, 1522–1532 (2014)
13. Thakral, S., Bansal, D.: Fault tolerant ALU using parity preserving reversible logic gates. Int. J. Mod. Educ. Comput. Sci. **8**, 51–58 (2016)
14. Sasamal, T., Singh, A., Mohan, A.: Efficient design of reversible ALU in quantum-dot cellular automata. Optik **127**, 6172–6182 (2016)
15. Krishna Murthy, M.: Design of efficient adder circuits using proposed parity preserving gate (PPPG). Int. J. VLSI Des. Commun. Syst. **3**, 83–939 (2012)
16. Thakral, S., Bansal, D., Chakarvarti, S.: Implementation and analysis of reversible logic based arithmetic logic unit. TELKOMNIKA (Telecommun. Comput. Electron. Control) **14**, 1292 (2016)

Energy Usage and Stability Analysis of Industrial Feeder with ETAP

S. Kumaravelu[1(✉)], J. R. Rubesh[1], K. Sarath Kumar[1], Arya Abhishek[2], and L. Ramesh[2]

[1] R.M.K. Engineering College, Chennai, India
kumarvelu4444@gmail.com
[2] MGR Vision 10 MW, Dr.M.G.R Educational and Research Institute, Chennai, India
abhishekaryal309@gmail.com

Abstract. As a growing technology the energy demand is also increasing day by day. To meet the total demand of consumers and to equalize demand with generation several methods carried out under energy conservation and one of the most popular and effective method is Energy auditing. With this scenario, the audit team conducted a preliminary audit on the industrial loads with 27 industries in a single feeder. A preliminary audit conducted and collected data through all the possible ways. Then detailed analysis completed through ETAP software of all the collected data and suggested recommendations based on the observed data to the consumers. These recommendations implemented and tested the stability analysis of the system with ETAP software and concluded with the reduction of energy consumption from various industries.

Keywords: Energy audit · ETAP · Feeder

1 Introduction

Energy is one of the major essential requirements in this period of generation. In order to develop the country, the energy production sector played the critical importance in view of ever-increasing energy needs it requires a huge investment to meet them, so that the country can be developed faster. If energy consumptions are reduced with respect to increasing efficiency, so that energy conservation, management, and auditing is required. Energy auditing is periodically examined on industry to ensure energy is utilized properly and efficiently then the waste of energy is reduced as much as possible. In India, Energy demand is greater than energy production. In India 70% of energy is produced by using Fossil fuels, Coal Consumes 40%, Crude Oil Consumes 24% And Natural gas Is 6%. On these Industry Consumes 60% of total energy the growth of a country can be found by energy consumptions of that country which shows electrical energy plays an important role.

To reach the future energy demand, the possible way is to increase renewable energy. Many researchers reported that individual effect for the conservation of energy is the best method to reach the future electricity demand partially in India. Energy conservation is to be defined as reducing the energy consumption without any change in production and quality of output in an organization or unit.

D. J. Hemanth et al. (Eds.): COMET 2019, LNDECT 35, pp. 234–243, 2020.
https://doi.org/10.1007/978-3-030-32150-5_25

Keeping these things in mind the audit team has visited 27 industries in Ambattur industrial estate and taken a single feeder named as Srinivasa feeder which all industries were connected, an audit was conducted in two stages. In stage 1 team conducted pre-audit and managed to collect data of all 27 industries which was on Srinivasa feeder. There data service number, total demand, energy consumed per months, lighting system, etc. on stage 2 detailed audit will be conducted and analyzed and suitable recommendations will be provided on the basis of issues. Audit Group had using ETAP Software for Analysis Part during the Pre-Audit, and Post Audit Sessions.

2 Literature Review

This work [1] by Patravele et al. to find the energy flow on the industry of textile industry and find total, energy demand, total consumption of industry and solution for energy savings most loads are 1 phase & 3 phase motor. The objective of energy management is to achieve and maintain optimum energy utilization without disturbing production and quality and conducted pre & post audit. Followed it procedure and methodology of audit and discussed about relation between energy efficiency and demand response and show demand site management (DSM) and they also implemented a new idea of zero energy building (ZNE) which was net zero energy building (NZEB) implemented in European country equal to renewable energy sources and other as greenhouse building consequently contribute less overall greenhouse gas the detailed audit was done by power analysis.

This paper shows the different literature review on the boiler of a thermal power plant November 2014 [2] with different methods of energy auditing and to increases energy efficiency, operating efficiency, total air quality, and air leakage, coal to the boiler to increases efficiency of the boiler. On the bases of Gauravt, dhanre et al. (1) under study of Poddar et al. the share of total production cost in energy can get improves profile level in all initial coal available with variations based on boiler design and dry flue gas loss also found that lower efficiency and poor quality of coal was by air leakage and that reduced to 6% and boiler efficiency increases by 0.27% (2) based on Monikuntal bora et al. the boiler efficiency was found by 1. Direct method (or) input-output method, 2. Indirect(or) heat loss method Based on MrnileshrKumbhar et al. growing concerns is about energy consumption about energy consumption in India in recent years and shows energy audit is a key to successful running of an industry with saving energy & towards natural resources of energy (3) Shaskankshivastava, et al, this shows optimistic saving and making demand balance 1. Pre-audit, 2. Audit phase, 3. Post Audit. By literature review, it gets concluded that are so many ways to reduce energy consumption cost by an energy auditing.

A case study about the energy audit of an industrial site in 2013 [3] Galinsky et al. improve efficiency. In order to reduce energy cost and energy consumption, manufacturing and processing tracking of industrial essential. The energy model has been used to analyse the impact of various energy saving action on the site. The auditing procedure is split by the Italian standard following steps. The energy analysis of the plant the preliminary asset rating for each buildings sources of energy waste. The energy consumption system analysis output scale of the company performance indicates very close to real value as a benchmark of the numerical method. Feasibility study of energy saving measures the model evaluation a coherent energy saving plan of the site. The finally the energy consumption value of each per years is calculated. The energy audit has been used to the company for an energy saving strategy for the next future.

The paper explained at 2014 [4] Mehul Kumar et al. in electrical energy consumption by industrial is about 60% energy consumption level. It is used on future development industrial and an energy audit is a verification, monitoring, and analysis of energy use to technical report contain recommendations and improve energy efficiency. The energy audit phase was different types are used and there is the 2-day process, 12 months process and formed tabulation and annual power factor determined. The calculation used some of the power KVA formulae as used. The audit suggests for more efficient level reach. The power factor increase, bill demand reduced. It is believed that energy audit most comprehensive methods in achieve energy saving in industrial. So that wasteful consumption of industrial energy will be minimized.

The paper was published by a fabrication company namely Hommec Technology Company at Nigeria by Olatunde Ajani Oyelaran et al. in the year 2015 [5]. In this above-mentioned company, energy consumption is to be about 82 KW. The Major Consumption loads are Furnace, Milling machines, Cutters, Grinders etc. The energy Computation data is to be collected numerically and Analysis is conducted using the data collected then the recommendation is given that the CRT is to replace with the LCD Monitors, Automatic Lightning Circuit is to Installing for the existing Lightning system. The Operating Average Power Factor is to about 0.62 lag for welding Sets to improve this additional Capacitor bank is to be collected across the load. So that the PF is improved then the power is utilized effectively and to Voltage regulation can be increased by connecting the lighting system for the isolated transformer to increase the lifespan of the lamps and fluctuations can be reduced. By Implementing this recommendation payback period for the investment is to be about 14 months.

The Paper is published by Manu Shrama et al. in year-2015 [6] on a wheel manufacturing industry. The Company Manufactures World class rims for Vehicles. The Energy utilized for this company is about 614682 kwhr. The Energy Bill is collected and Consumption data is collected. The recommendation is given that to reduce the contract demand for 6000 KVA so that audit team can save the cost of the cost per unit from the Electricity Board, By arresting the leakage in the compressors, Retrofit the Incandescent with LEDs so that Energy Consumption is reduced with an increase in lifespan of the Lamps. Audit team studied all the above paper the major thing is Energy Auditing and saving the consumption of energy and reducing the cost with benefit. Energy auditing was done in all the industries to reduce waste energy in every company. Many methods of energy auditing were used worldwide with different types of industry. Our group conducts using ETAP (electrical transient and analysis program) which is a software program of electrical engineering. By using this software audit team prepared a Single Line Diagram (SLD), Analysis has performed an audit is conducted for the industry along with the recommendation of a new idea and regular maintenance.

To support the initiative to reduce demand and create awareness to the general public Dr. M.G.R Educational & Research Institute, Chennai has taken initiative in the year 2014 called **'MGR Vision 10 MW'** under the leadership of Dr. L Ramesh to save 10 MW in 10 years. The contributed research works under the pilot project-1-3 were published [7–10] in Scopus publications and the reports are published by the Research

Forum GREEN9 (Energy Efficiency Research Group). This is the Pilot Project-4 aims to present current scenario and initiative taken to save power and generate own power through energy supervision and energy assessment. This work presents the detail analysis of industry feeder which consists of 26 industries. Initial preliminary audit was conducted and simulated in ETAP. The industries were classified under three process and detailed audit study was conducted. The team reported strong recommendation and was simulated with ETAP. After recommendation implementation in ETAP, stability analysis study was conducted. After necessary changes recommendations were submitted to industries.

3 Data Analysis and Results

The audit process has two different stages to conduct an audit, stage one is to conduct a preliminary audit and stage two is a detailed audit. In stage one a preliminary audit conducted to all 27 industries which were connected to a single feeder in this audit some of the questions were asked on each industry to collect data. After that, a detailed audit will be conducted on selected industry based on total energy consumed by industries and more questionnaire asked on all industry.

Preliminary audit questions asked:

1. What is the maximum load consumption in your industry?
2. Have you undertaken any energy audit previously?
3. What is the type of your industry?
4. Do you check your earth status regularly?
5. Do you have regular annual maintenance of the machine and other equipment?
6. Faulted equipment is to be serviced or replaced by new?
7. Average consumption per year?

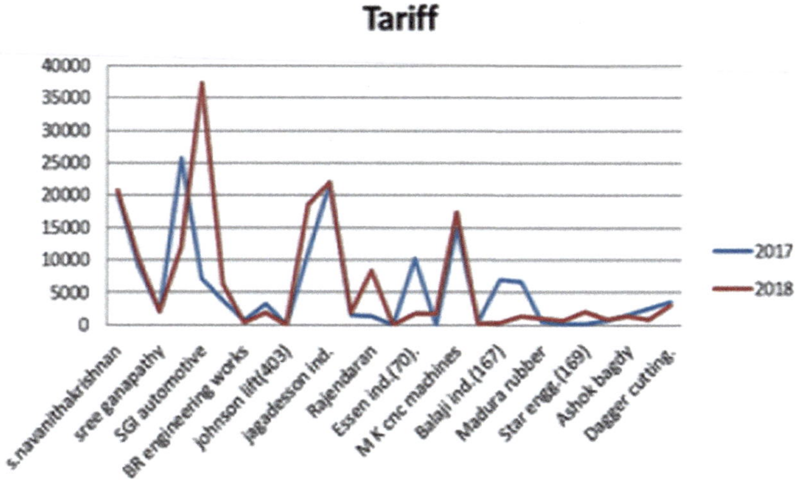

Fig. 1. Unit consumed on two years

This Fig. 1 shows the total kilowatt-hour consumed by each and every industry's and this also shows the difference between the unit consumed in two years of 2017 and 2018. This line graph represents the average unit consumed in all 27 industries on that the maximum consumed in 2017 by Sudharsan tech as 25760 KW/H, and in 2018 by SGI automotive as 37408 KW/H.

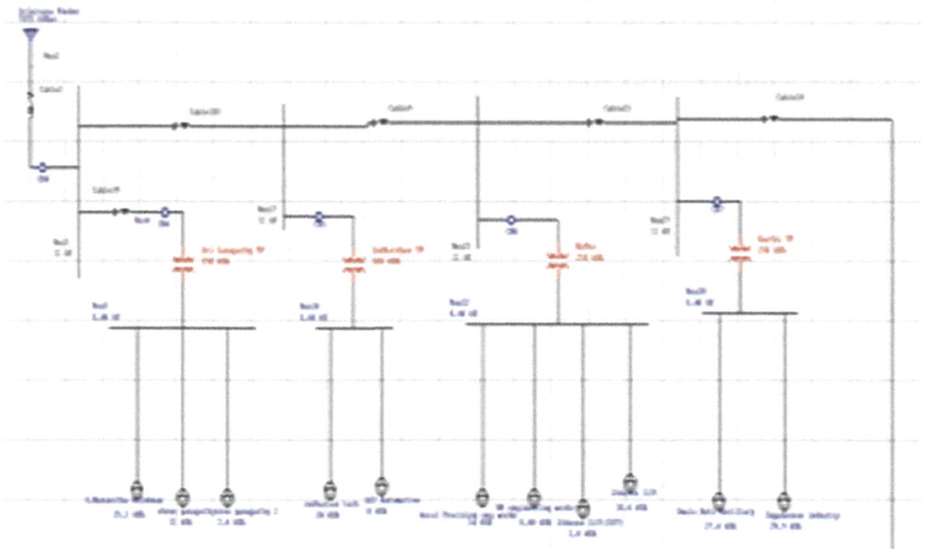

Fig. 2. Part-A: SLD of the industrial feeder

This SLD shows the single 11 kV feeder diagram of an industrial estate which has a 27 industries load connected. Each industries load was shown lump load each has an 11 kV to 440 V step down distribution transformer connected with it. The SLD was divided into 2 parts, part-A upper region Fig. 2 has 4 buses and 11 industries connected on it along with four transformers and has 5 circuit breakers with a particular feeder of 40 A.

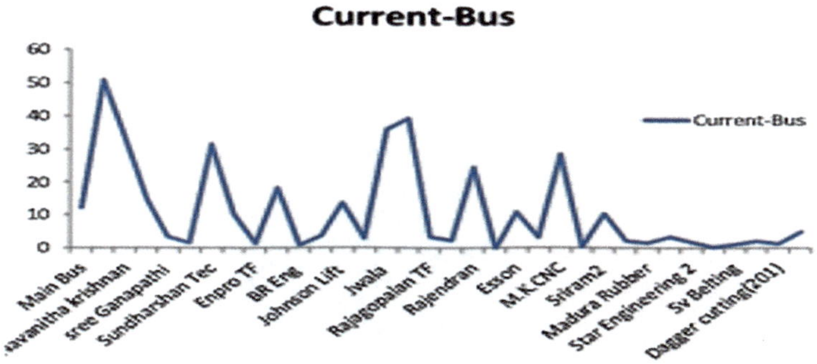

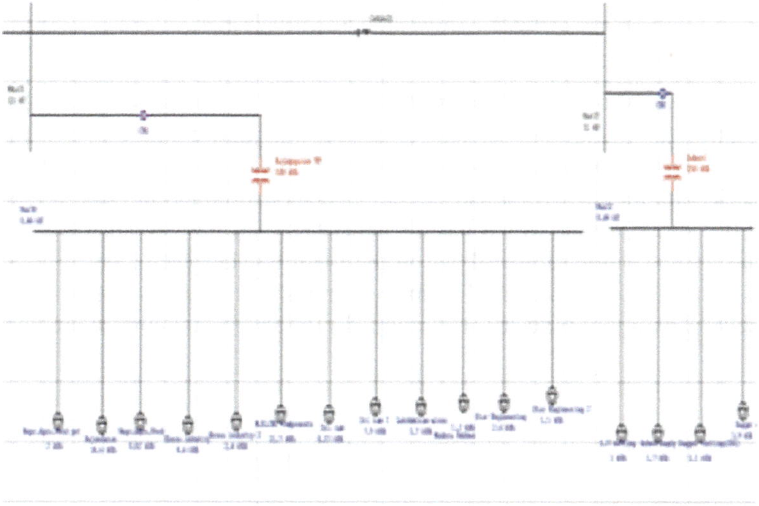

Fig. 3. Part-B: SLD of the industrial feeder

And the part-B Fig. 3 has 2 buses also has 16 industries connected with two transformers on that 2 circuit breakers are connected to the particular feeder as a protection device which carries 40 A. All the 27 industries are different like manufacturing, production, and service on that two industries was stopped they production (Fig. 4).

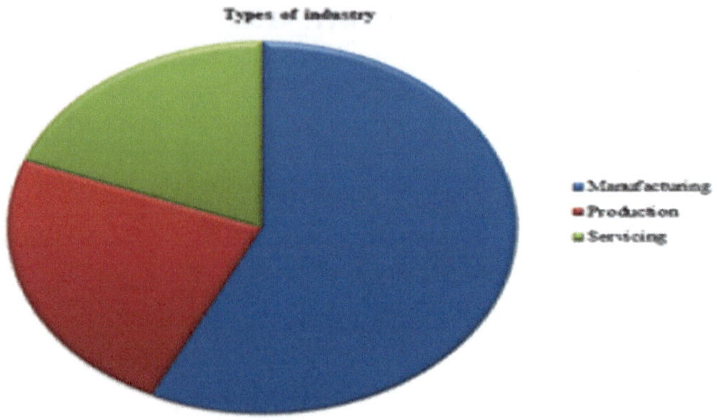

Fig. 4. Different types of industries

This pie chart shows the different types of industry connected to feeder it shows manufacturing is maximum, second is production and minimum is service-based industries, 57% manufacturing, 24% productions, 19% servicing based, on that totally 27 industries are there, two industry is stopped their production. This line graph represented the single feeder which has 27 industries load buses connected on it. Graph shows the current value of each load on the main bus current value is 12 A which is directly connected to the feeder. The maximum current value is 50.9 A and minimum current also shown two industry was stopped production and current are zero on such industry.

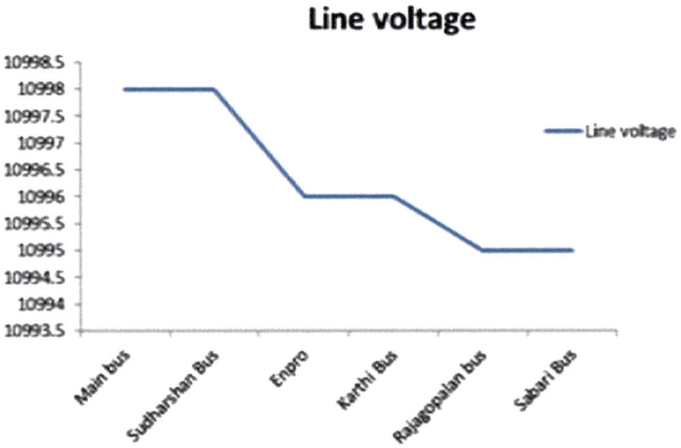

Fig. 5. Line voltage

This line voltage graph Fig. 5 shows the line voltage of all six buses which is a maximum nearly 11 kV on the main bus and the minimum voltage was 10995 is consumed on the bus.

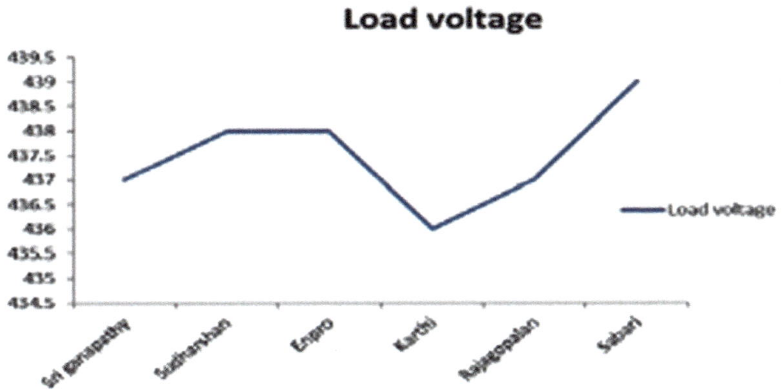

Fig. 6. Load voltage

Load voltage was shown on Fig. 6, on the secondary side of the transformer for each load bus. Maximum voltage is 439 and the minimum voltage level is 436mentionshere carry different types of an industrial load connected to all buses. STABILITY STUDY The graph Figs. 7, 8 and 9 shows for current, line voltage, load voltage and it is a comparison of normal load to the case of load increased to 70% at the same time as per analysis sum of buses become under voltage, karthi-413 V, sri ganapathy −422 V and this transformer is also over loaded. So auditing team conclude that load cannot be increased above 60% to keep the system stability.

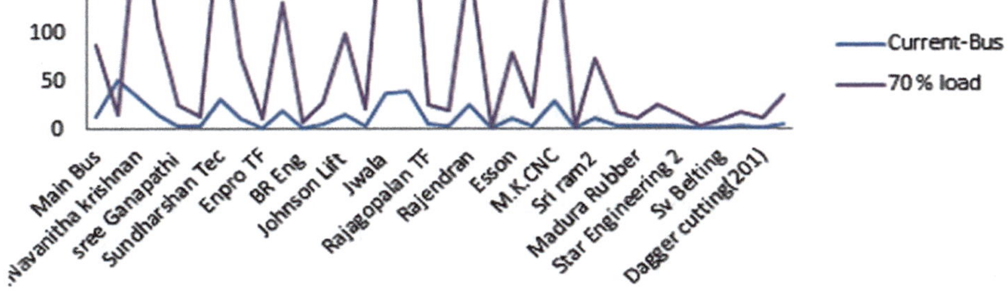

Fig. 7. Current-bus

The above Fig. 7 graph comparison of current on buses and loads. Current-bus shows that normal current flow on each load it has a maximum of 50.9 A, second line says that when 70% load increased on every load maximum current consumed is 274 A.

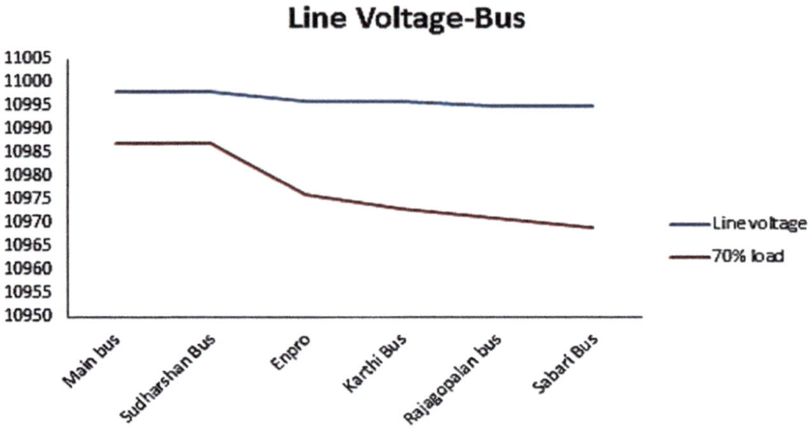

Fig. 8. Line voltage

This Fig. 8 graph shows Line voltage of 11 kV line under normal load voltage 10998 is an upper end and 10995 was tail end voltage this graph also shows the voltage when increased to 70%.

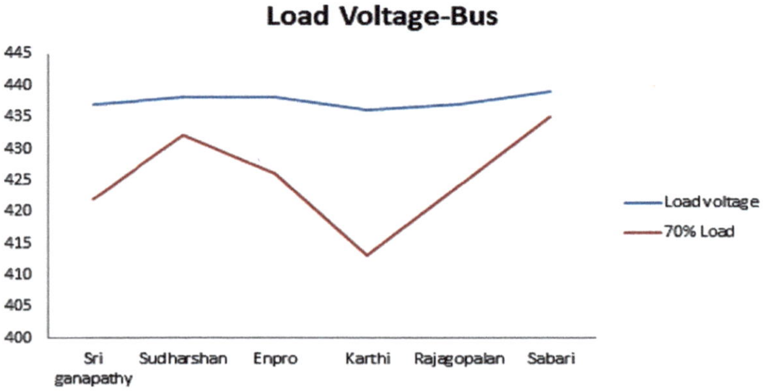

Fig. 9. Load voltage

This Fig. 9 shows the load voltage graph with respect to the buses on a case of normal load voltage has 436 as maximum voltage and when 70% increased load the minimum voltage is 413 and graph shows voltage consumed by each load during stability analysis.

4 Conclusion

Energy auditing is the most widely used method to save energy and optimum used of energy. On this, the team conducted a pre-audit in a single feeder named as Srinivasa feeder in Ambattur industrial estate. In the concerned feeder, the audit team managed to collect the data of 27 various types of industries. According to collected data, an SLD drawn using ETAP and check for the stability state analysis. After the necessary changes in the layout, with reference to stability analysis, now the system will be stable after over loaded condition too. The 30% energy saving will be predicted after implementation of recommendation.

Acknowledgements. The authors expressed their valuable gratitude to Er. A.C.S. Arun Kumar, President of Dr. M.G.R. Educational and Research Institute, Who provide constant Institutional support to the MGR Vision 10 MW initiative. We convey our special thanks to Principal, HOD and Faculty Mentors of RMK Engineering College who supported for the project. We also thank to the mentors, who provided technical support for the project from the Energy Efficiency Research Group (GREEN9).

Declaration. The authors declare that ethical approval for data is not needed for this project. The data's were collected from the s6 private industry with their approval.

References

1. Arya, J.A., Arunachalam, P., Bhuvaneswari, N., Ramesh, L., Ganesan, V., Egbert, H.: Energy usage analysis of industries with ETAP a Case Study. In: 2017 International Conference on Circuit, Power and Computing Technologies (2017)
2. Goncalves, R.G., Rossini, E.G., Souza, J.D., Beluco, A.: Main result of an energy audit in a milk processing industry. J. Power Energy Eng. **6**(1), 21–32 (2018)
3. Munguia, N., Velazquez, L., Bustamante, T.A., Perez, R., Winter, J., Will, M., Delakowitz, B.: A case study in the meat processing industry. J. Environ. Prot. **7**(1), 14–26 (2016)
4. Patravale, P.N., Tardekar, S.S., Dhole, N.Y., Morbale, S.S., Datar, R.G.: Industrial energy audit. Int. Res. J. Eng. Technol. **5**(1), 2021–2025 (2018)
5. Saini, M.K., Chatterji, S., Mathew, L.: Energy audit of an industry. Int. J. Sci. Technol. Res. **3**(12), 140–146 (2014)
6. Dongellini, M., Marinosci, C., Morini, G.L.: Energy audit of an industrial site: a case study. In: Department of Industrial Engineering and Interdepartmental Centre For Industrial Research on Buildings and Construction Technologies (2014)
7. Sujan, K., kumari, K.: Restructuring of distribution transformer feeder with micro grid through efficient energy audit. In: Green Computing Conference (IGCC) (2016–2017)
8. kumar, A., Thanigivelu, M., Yogaraj, R., Ramesh, L.: The impact of ETAP in residential house electrical energy audit. In: Elsevier Proceeding of International Conference on Smart Grid Technologies, August 2015
9. Narayanan, R., Kumar, A., Mahto, C., Ramesh, L.: Illumination level study and energy assessment analysis at university office. In: Proceedings of 2nd International Conference on Intelligent Computing and Applications, pp. 399–412 (2017)
10. Arya, A., Jyoti, et al.: Review on industrial audit and energy saving recommendation in aluminium industry. In: 2016 International Conference on Control, Instrumentation, Communication and Computational Technologies (2016)

Polymers Based Material as a Safety Suit for High Power Utilities Working

R. Senthil Kumar(✉), S. Sri Krishnakumar, J. Prakash, S. Deepak Raju, A. Aravindsamy, and R. Prakash

Department of Electrical and Electronics Engineering, Vel Tech High Tech Dr Rangarajan Dr Sakunthala Engineering College, Avadi, Chennai 600062, India
rskumar.eee@gmail.com, krishnakumar.rvs@gmail.com, prakash_ies@yahoo.co.in, deepaknarayanan1997@gmail.com, aravindkkt@gmail.com, ravifriends1217@gmail.com

Abstract. Synthesis and fabrication of material for workers in high power utilities to protect from emf. A safety material consist of two-layer.one layer is a conducting material another acts as non-conducting material. Analysis says that workers are exposed in high power emf effected with cell damage, change in the flow of calcium in and out of cells, change in hormone production and cell growth, might feel minute of vibration of hair and skin. First conductivity layer is prepared with polyaniline conducting polymers which provide a faraday shielding against electromagnetic & electric field. Secondary layer of polyvinyl alcohol is a non-conducting polymer gives shielding against magnetic field, material are tested against electrical field, magnetic field, material of available range in the high power utility. Along with shielding, material offers a cost effective option with high flexibility and durability, compare to existing conductive wear.

Keywords: EMF protection · Transmission line · Substation · High power utilities · Safety · Suit · Conductive wear · Low frequency electric field · Magnetic field · Polymer · Polyaniline · Polyvinyl alcohol

1 Effects of Electric and Magnetic Field of High Power Utility on Environment

Substation and other high electrical power utilities are usefully situated far away from residence place but due to fast expanding urbanization they got into closer vicinity. all the living organism including plants, animals, birds and human are put under great risk [2]. The few recent studies as revealed that there was remarkable changes identified in the organisms exposed to emf radiation, following changes observed in plants, changes in plants cell growth, the levels of polyaniline (a substance which indicate stress) in plants a drop in pollen fertility. Under the observation of humans and animals it was been founded that considerable change in level of antioxidants in blood, heat shocks proteins (indicator of stress in animals), change in DNA [4], further in human who have

D. J. Hemanth et al. (Eds.): COMET 2019, LNDECT 35, pp. 244–250, 2020.
https://doi.org/10.1007/978-3-030-32150-5_26

exposed to low frequency emf had exhibited symptoms of fatique, aggression, sleep disorder emotion instability instability [4].

2 Influence of EMF on Humans

As we know our body contain charged particles, so definitely, electric charges distributed at the surface of the body when a low frequency electric field act on us similarly low frequency magnetic frequency will circulate a induced current in our body. Those current can cause change in biological process and even simulate the muscles and nerve.

The Table 1 display the level of magnetic and electric field come out of various high electric power utilities and its impact (Fig. 1).

Table 1. Level of magnetic and electric field in various high power electric utilities

Places	E-Field	M-Field
Transmission line	0.3–3 kV/m	0.5–5 μT
Distribution line	0.01–0.1 kV/m	0.05–2 μT
Substation	<0.1 kV/m	0.1 μT
Impact	Individuals might feel minute vibrations of skin, hair or clothing	Changes in hormone production and cell growth

3 Existing EMF Protective Devices

3.1 Faradays Cage

The Table 1 display the level of magnetic and electric field come out of various high electric power utilities and its impact.

Faradays cage is a device that is used to shield the content inside it, it from static electric field. The faraday cage distribute the radiation and charges to the exterior surface of the cage. All the points in the cage is at equipotential so it cancels the interior radiation and charges. In the high voltage or high tension line the maintenance is done by the human so the protection of the human who working in the hv line is provided by the faradays cage. The faradays cage is made by a metallic material like aluminium, copper, gold etc. as a mesh. Due to its bulk and heavy structure workers face many practical problems in using it (Fig. 2).

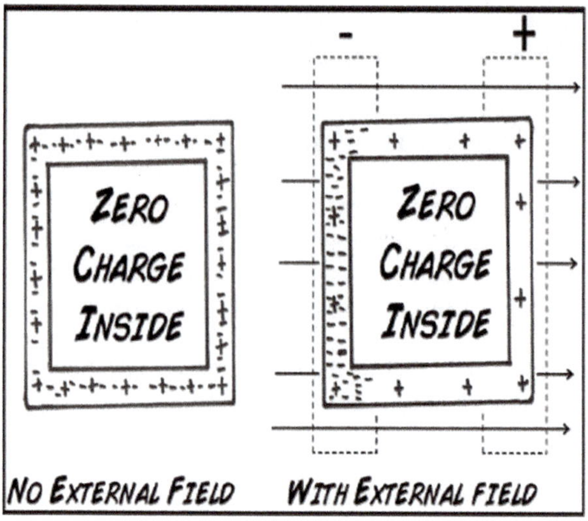

Fig. 1. Field distribution in faradays cage

Fig. 2. Faraday cage

3.2 Conductive Wears

A mesh of conducting metals like stainless steel, gold, silver, copper, etc. are act as a faraday shield are fabricated in a suite format for the workers employed in high voltage transmission line maintenance. Though the suite offers high durability and flexibility, it is quite heavier and expensive.

Conductivity of stainless steel, gold, copper and silver are respectively 1.45*106 S/m, 4.10*107 S/m, 5.96*107 S/m and 6.30*107 S/m.

4 Proposed Material for Safety Suit Against EMF

Material suggested here consist of two layer, one is of conducting polymer polyalanine which forms faraday shields and provide protection against electric field. Another layer is made of non-conducting polymer, Polyvinyl alcohol which provide a shield against very low frequency magnetic field.

5 Process Involved in Synthesizing Polyaniline

To synthesis polyaniline 14.25 g of ammonium per sulphate is taken with 50 ml of distilled water and 4.7 ml of aniline is taken with 5 ml of concentrated HCL then 45 ml of distilled water is added to it. Both the solutions are taken to 0 °C and mixed together, solution changes to green colour and slowly solidifies after 6 h (Figs. 3 and 4).

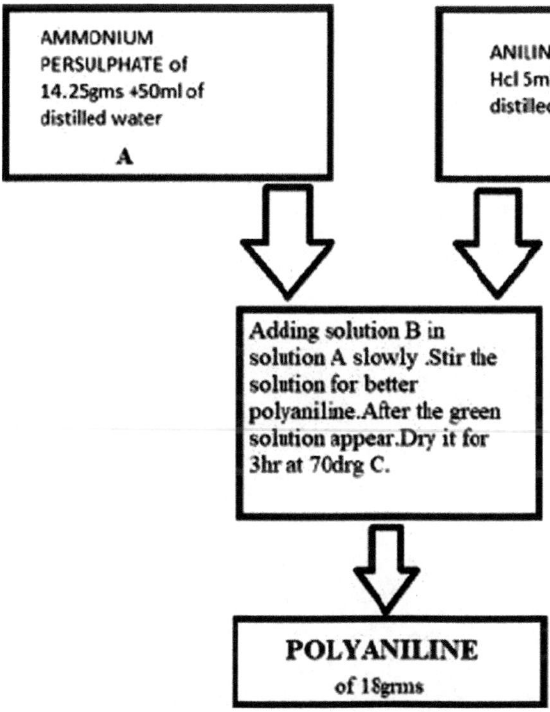

Fig. 3. Process involve in synthesis of polyaniline

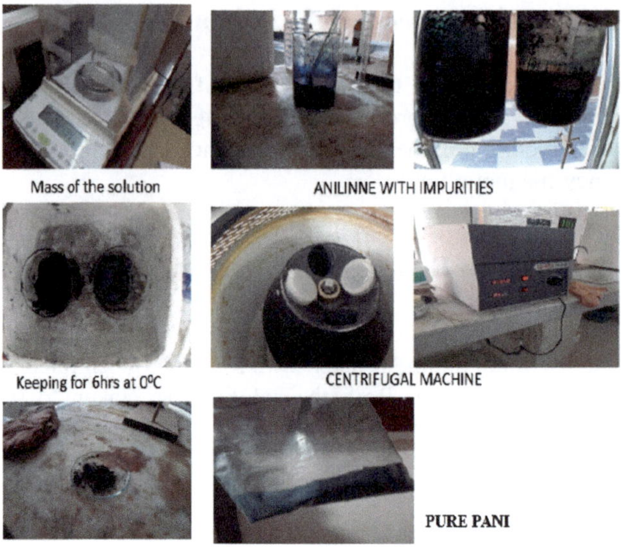

Fig. 4. Preparation of polyaniline

6 Preparation of Polyvinyl Alcohol

To get Polyvinyl alcohol of desire shape, it is dissolved in hot distilled water and maintained at 2000 C for 3 h then it will get into a form of synthetic polymer material with high flexibility and durability (Fig. 5).

Fig. 5. Synthetic polymer of Polyvinyl alcohol

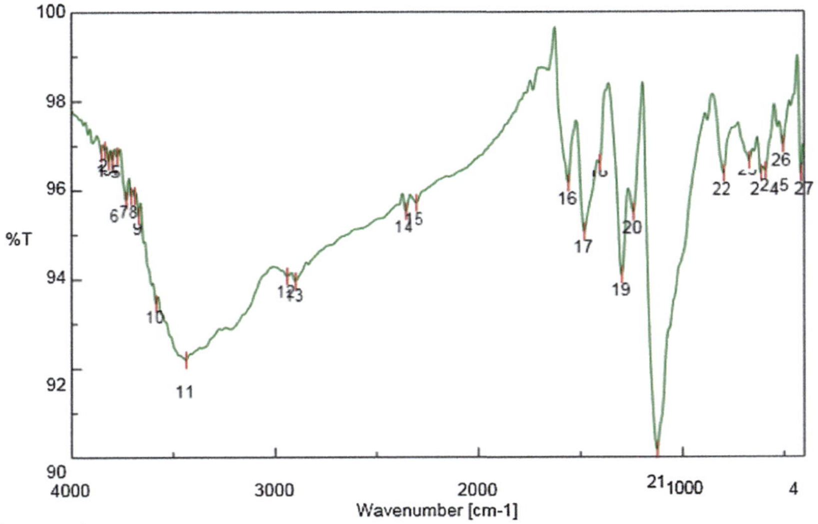

[Comments]

Sample name Polyaniline-VH7246
Comment
User
Division
Company Chemistry Research Lab

[Result of Peak Picking]

No.	Position	Intensity	No.	Position	Intensity
1	3851.15	96.8421	2	3832.83	96.9121
3	3814.51	96.6663	4	3795.22	96.6895
5	3773.05	96.7852	6	3728.69	95.7775
7	3704.58	95.8708	8	3686.26	95.8781
9	3666.98	95.4245	10	3583.09	93.4539
11	3436.53	92.1814	12	2940.91	94.0671
13	2900.41	93.9575	14	2356.59	95.5188
15	2306.45	95.7075	16	1560.13	96.1606
17	1482.03	95.0643	18	1405.85	96.6078
19	1298.82	94.0912	20	1241.93	95.5021
21	1125.26	90.1556	22	800.314	96.3694
23	673.035	96.6513	24	614.217	96.3625
25	592.039	96.4334	26	505.258	97.0228
27	416.549	96.3637			

Fig. 6. IR test for polyaniline

7 Test Conducted on Polyaniline

IR test was conducted after synthesizing the polyaniline to conform its property Fig. 6, from the reading obtain it confirm the property of polyaniline at 11th reading. To check the conductance of material, material was drawn as a wire and its resistance was measured. With the relationship of G = 1/R conductance G was found. As 320 s/m, as the past studies reveals that polyaniline is a highly durable polymer even after several washing its conductance remaining constant [5].

8 Test Conducted in the Polyvinyl Alcohol

It was tested against a very weak magnet 10 m Tesla, test was conducted with gauss meter which is used to measure magnetic field strength. First magnet was measured without shielding gauss meter read 10 m Tesla, then after covering the entire magnet with polyvinyl alcohol, at that time Gauss meter showed null deflection.

These two layers are brought together by deposition technique and studied under electron microscope for ensuring uniform deposition of polyaniline and polyvinyl alcohol.

9 Conclusion

Proposed material provide a cost effective solution for highly flexible, less weight and highly durable safety suite for worker employed under the high power utilities compare to other conducting wear available in the market.

References

1. Srinivasa, K.M., Marut, R., Kumar, R., Nambudiri, P.V.V., Lalli, M.S., Srinivasan, K.N.: Measurement and study of radiation levels and its effects on living beings near electrical substations. J. CPRI 11(3) (2015)
2. Hannigan, A.P.E.: Effects of electric and magnetic fields on transmission line design, vol. 17, no. 4, July/August 2013
3. Göcsei, G., Németh, B.: Shielding of magnetic fields during high voltage live-line maintenance. IEEE Electrical Insulation Conference, Ottawa (2013)
4. Göcsei, G., Németh, B., Kiss, I., Berta, I.: Health effects of magnetic fields during live-line maintenance. In: ICOUM 2014 11th International Conference on Live Maintenance, Budapest, Hungary, 21–23 May 2014 (2014)
5. Maity, S., Chatterjee, A.: Conductive polymer based electro-conductive textile composites for electromagnetic interference shielding. jit.sagepub.com at CORNELL UNIV on 26 Sept 2016

A Comparative Study of Various Microstrip Baluns

J Indhumathi(✉) and S. Maheswari

Panimalar Engineering College, Chennai, India
jaiindhu96@gmail.com, maheswarisp@yahoo.co.in

Abstract. A microstrip Balun is used to "balance" unbalanced systems. The application of microstrip baluns are mixers, amplifier, frequency multiplier, couplers and dipole antennas feeds. The advantages of microstrip baluns are low loss, uniplanar structure and low cost. The design of different microstrip balun has been analyzed for different frequency range from 0.9 to 5.5 GHz and various parameters have been analyzed.

Keywords: Microstrip balun · Operating frequency · Return loss · Insertion loss

1 Introduction

A balun is device converts a balanced to unbalanced signal. A balun which can be treated as impedances transformer. Transformer baluns can also connect lines of different impedance. Baluns are typically divided into active and passive baluns. A baluns will consume additional power. The passive baluns, will be classify as lumped-type and distributed type. The disadvantage of lumped balun is that sometimes it exhibits less variations and magnitude response between the two signals. The distributed type produces lower loss and lower cost. The design criteria for baluns end up with higher isolation, impedance matching, balanced to unbalanced transformation, lower noise interaction.

The partially coupled stepped-impedance and coupled-line resonators are realized by dual band balun. Metallization is introduced in order to obtain strong coupling on both sides of the substrate in order to increase the complexity of the balun structure. The differential mode and RF/Microwave circuits and antenna feeding networks are the applications of wideband circuits. For bandwidth enhancement, a broadband balun is based on transmission line. The wide band balun consist of broadband phase inverter and a wideband impedance matching networks, as it flexible and easy to apply over a various frequency bands. The broadband inverter is designed using parallel strips and it is shorted at the transmission lines. The wideband phase inverter is approximately equivalent to the open circuit. A compact car dual band balun are involved in order to improve the signal timing, and also reduce electromagnetic preventive and noise, lumped type balun is used. In order to take care of the 1800 section departure and magnitude response of the signals. The distributed type involves the following features such as low loss, uni planar structure and lost cost. The dual band baluns are replaced

D. J. Hemanth et al. (Eds.): COMET 2019, LNDECT 35, pp. 251–258, 2020.
https://doi.org/10.1007/978-3-030-32150-5_27

by uniform coupled line with a tapered coupled lines in order to analyze and to realize the physical size and structure. The multiband operations are important for both size and cost reduction. The dual band baluns are utilized in several applications like mixers, amplifiers, frequency multiplier. The design has been introduced to shrink the cable in antenna measurement. The dual band are made by partly coupled stepped electrical resistance. The coupling issue depends on amplitude and part performance. The dual band balun is freelance and therefore the 1800 out-of-phase. They're primarily supported various transitions (CPW-to-slotline, microstrip-to-coplanar stripline (CPS), double-sided parallel strips), coupled strip lines, or phase shifters.

2 Related Works

Huang et al. [1], have designed the Frequency and Selectivity ratio of dual band and the design was derived from March and balun. The multiple band is proposed with a four-port network with one port shortened. The frequency relation (m = f2/f1) between the bands is determined. Tansmission zeros are introduce for producing high selectivity. A dual-band balun operates at 2.4/5.2 GHz (m = 2.17). Input and output impedance are equal to *Zo*. The frequency can be controlled by changing impedance of open circuit. Thus the operating frequency obtained for a dual band baluns is 3.8 GHz and the bandwidths are 120 and 100 MHz. Thus the measured insertion loss S_{21}/S_{31} are −4.16/ −4.32 dB for 2.36 GHz and −4.17/−4.26 dB for 5.23 GHz. The calculated return loss S_{11} are −20 and −19 dB. The phase difference is greater than 2° and the magnitude diversity is less than 0.3 and 0.34 dB (Figs. 1 and 2).

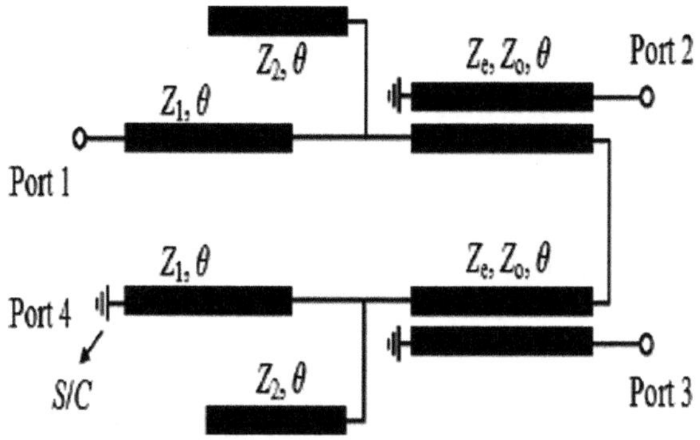

Fig. 1. Schematic of dual band balun [1].

Shao et al. [2], have designed the Parallel Strips of wideband balun. The wideband balun has four-port through open-ended ports. It has a transformer impedance and a phase inverter over a wide band. The advantages of this wideband baluns is very flexible and easy to operate over the other bands of frequency to function over a extensive bandwidth. The balun with parallel strip operate from 0.72 to 2.05. The author has designed a balun with wide bandwidth and parallel slip phase inverter and 1800 CPW phase inverter. The wideband phase inverter is equal to open-circuit.

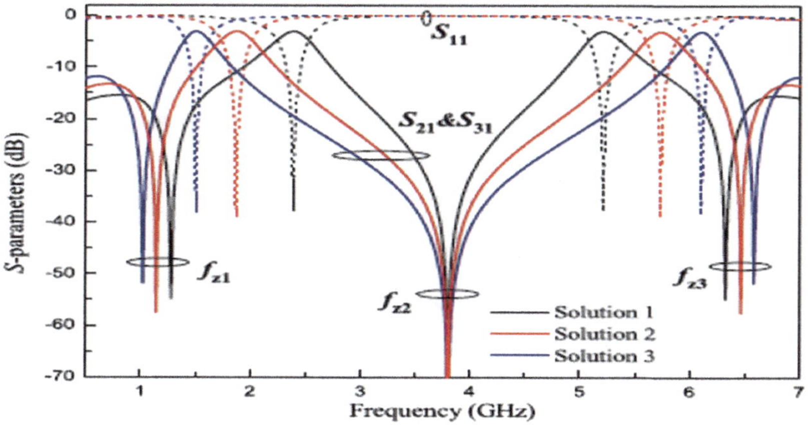

Fig. 2. Intermediate frequency [1].

The measured frequency band is 1.39 GHz. The calculated magnitude response are identical and amplitude by S21 and S31 are within the same limits of 1 dB from 0.74 and 2.13 GHz. The measured return loss are less than −10 dB and the phase difference is greater than 4.80 (Fig. 3 and 4).

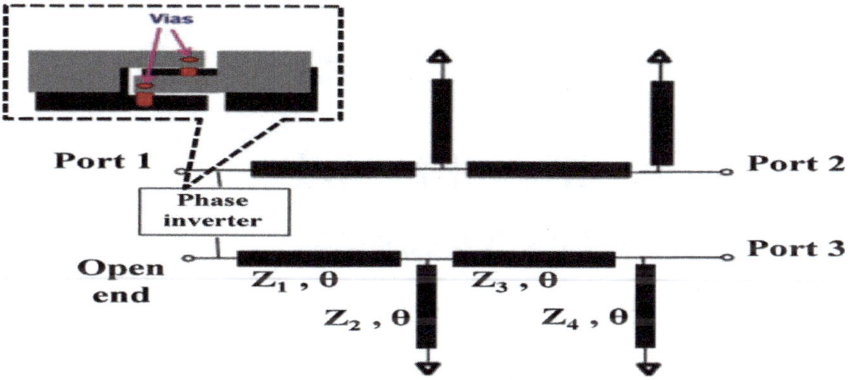

Fig. 3. General structure of wideband parallel strip balun [2].

Shao et al. [3], have designed a Tap open ended stubs with compact dual-band coupled-line balun. The balun is designed with fourth port shortened. The open concluded stubs are connected to ports by adjusting the stub to ports by adjusting the stub impedance the phase and amplitude response can be achieved. This design will improve timing of signal, interference and noise can be reduced. Baluns can be used as 180 hybrids. The dual band balun operated in 0.9/2 GHz. Thus the operating frequency between these two band is 1.45 GHz (centre frequency). The measured S_{21}/S_{31} are −2.85/−4.53 dB for 0.9 GHz and −3.97/3.7 dB for 2 GHz, the measured bandwidth is 0.9 GHz. The phase difference is 180° ± 2° (Fig. 5 and 6).

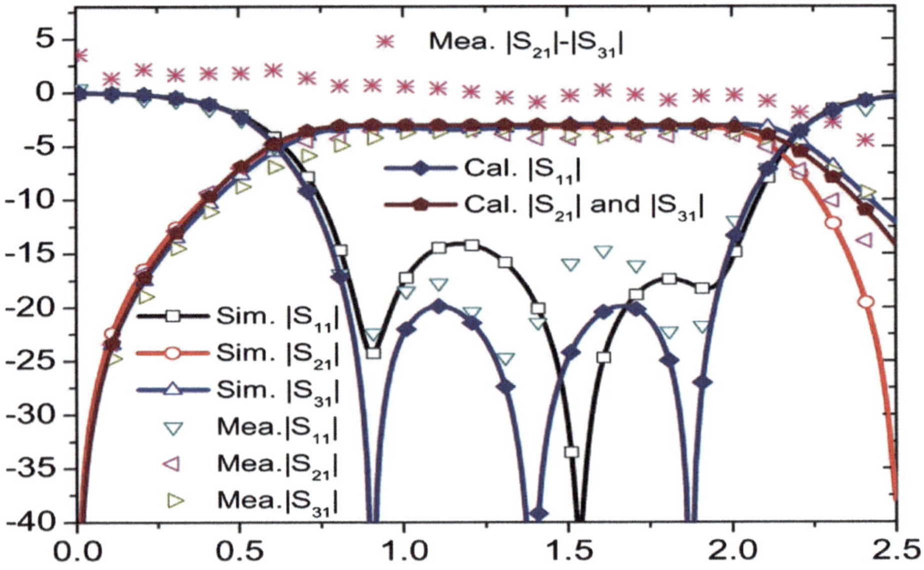

Fig. 4. Frequency operate at 0.72 to 2.05 GHz [2].

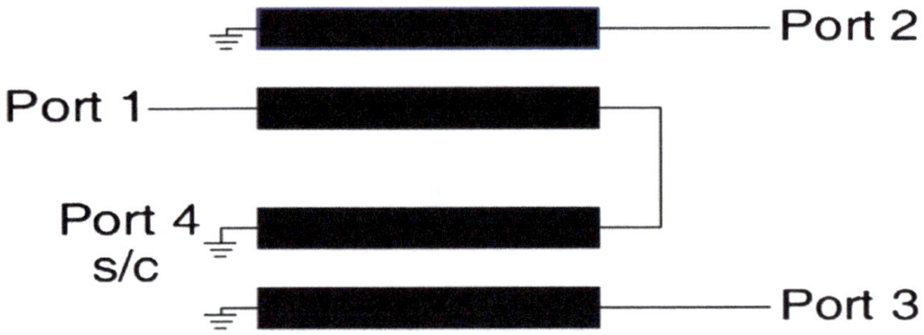

Fig. 5. Balun with a short-ended port 4 [3].

Shen et al. [4], have designed the flexible frequency ratios of Dual-band balun. The balun has a four port structure with the 4th port as open end. The balun's performance is analyzed using four port network. This type of baluns are working at 1 GHz and 2 GHz. The structure is used in different applications, such as mixers, amplifiers, multiplier, 180° hybrid coupler and dipole antennas feed. Thus the Stimulated loss of return S11 at frequency f_1 is −17.17 dB and at the frequency f_2 is −23.898 dB. The simulated insertion loss S_{21}/S_{31} −3.21/−3.185 dB for 1.1 GHz and −3.22/−3.19 dB for 2 GHz and phase difference is 180.3° and 179.10. Thus the Frequency changes from 1 from 1.1 to 0.985 and from 2 to 1.83 GHz. Then the obtained insertion loss S_{21}/S_{31} are −3.7/−3.73 dB at 0.985 GHz and −3.68/−3.75 dB at 1.83 GHz. The measured phase difference is 180.3° and 179.3° and the bandwidth is 138 MHz (Fig. 7).

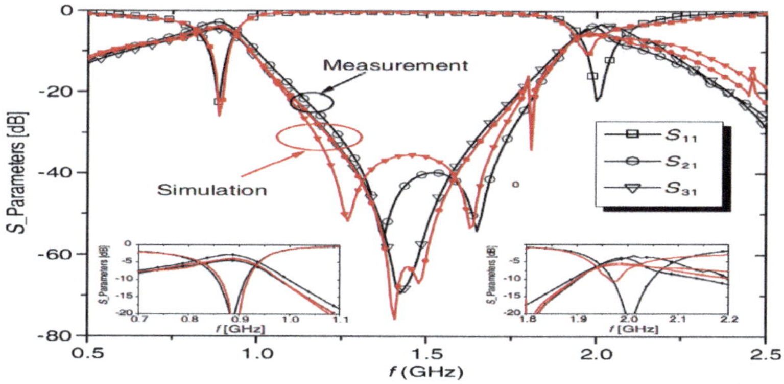

Fig. 6. Frequency operates between 0.9 and 2 GHz [3].

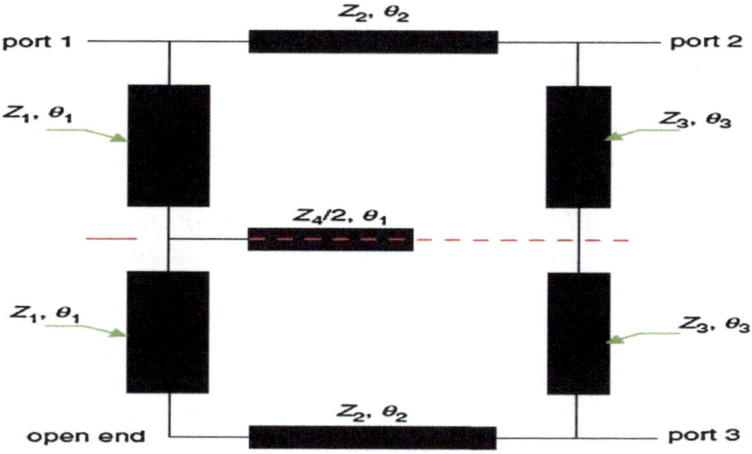

Fig. 7. Proposed dual band balun [4].

Wu et al. [5], have designed the Wideband Microstrip Dual Balun Structure. The wideband balun has 1800 phase shift on the higher mode order. They have equal amplitude and difference in phase between the output ports. The coupling factor plays a major role in amplitude/phase balance performance. The balun has wider impedance matching network. The Amplitude and phase imbalance measured at 1.5 dB and 8.5°. The loss of insertion is 1.2 dB and the return losses are 15 dB and the bandwidth are measured from 5.8 to 10.4 GHz. The amplitude and phase differences are 0.4 dB and 40 respectively and insertion loss is 1.2 dB.

Wu et al. [6], have designed the Microstrip Baluns over Novel Planar Impedance Transforming Coupler. This design has strong coupling coefficient. The transformer coupling are made with different impedance transfromer and also derive the power ratio from zero to infinity. A noval planar microstrip baluns are used to construct coupler design and operate over a wide bandwidth. These type of baluns operate at 2 GHz.

Thus the simulated operating bandwidth of the baluns is from 1.58 to 2.39 GHz. The simulated phase-difference at 2 GHz is 179.6° and the magnitude of the response simulated results are −3.68 dB and −3.64 dB at 2 GHz. The input and output can be suppressed from interferences, and noise. The impedances are equal to the transforming function. The measured bandwidth is 1.6 to 2.31 GHz and the difference in phase on 2 GHz is around 180.25. Thus the measured magnitude response is −4.24 dB and −4.22 dB.

Huang et al. [7], have designed the wide stop band Balun With Stepped Coupled lines. This can be done by integrating short-circuited coupled lines. The paired line stepped and stepped-transmission-line lengths are a quarter wavelength. To obtain this transmission, the measured result in passband are within 1.6–4.4 GHz with the insertion and return losses to be Minimum of 0.8 dB and maximum of 16 dB, respectively. The stop band results are measured in the frequency range of 5.5–12.55 GHz. The loss of insertion is measured to be above 25 dB. The amplitude and phase differences are lesser than ± 0.1 dB and 180° ± 0.5.

Zhang et al. [8], have designed the Balun with tapped stepped impedance. The impedance can be tapped by adjusting the stubs properly between the ports. The output signals are 180° out-of-phase at the different frequencies. A microstrip balun operates at 2.45 GHz/5.25 GHz. Thus the frequency shifts between 2.14 GHz and 5.06 GHz. The measured phase difference between the ports are −176.26° at 2.45 GHz and −184.83° at 5.06 GHz. The calculated return loss to be −14 dB at 2.45 GHz and −13 dB at 5.06 GHz and the insertion loss S_{21}/S_{31} are −3.56 dB/−3.05 dB at 2.45 GHz and −4.38 dB/−4.14 dB at 5.06 GHz. Thus the measured bandwidth are 120 and 80 MHz. Thus the transforming impedance is calculated from 0.25 to 2.5 over a frequency choice of 1 to 4.37.

Miao et al. [9], have designed a compact tuned frequency microstrip balun. This method can be achieved by tuning the operating frequency continuously. The frequency changes with respect to the input dc voltage. The frequency can be tuned between 620 to 1020 MHz of frequency and the voltage varies from 4 to 16 V. In this the maximum obtained phase difference is 180° + 70°, and the difference in amplitude is less than 0.8 dB. The measured return loss is about −17 dB (Table 1).

In paper [5] the amplitude difference (dB) obtained is 1.5 dB considered to be the highest when compared with other techniques. In paper [4] the baluns operate at 1.1 & 2 GHz and attains a lower bandwidth of 138 & 204 MHz at thickness of 1.5 mm and the amplitude difference is 0.6 dB. In paper [8] the microstrip baluns are operates at high frequency of 2.45 to 5.25 GHZ and amplitude difference is 0.5 dB and produces the phase difference of −172.6°. In paper [1] the balun produce the lower bandwidth of 120 & 100 MHz and designed at a thickness of 0.508 mm. In paper [7] the amplitude difference (dB) obtained is ± 0.1 which has the highest thickness of 1.524 mm operating at 3 GHz frequency.

Table 1. Comparison of different parameter of various balun.

Parameters	[1]	[2]	[3]	[4]	[5]	[6]	[7]	[8]	[9]
Operating frequency	2.4 & 5.4 GHz	0.72 & 2.05 GHz	0.9 & 2 GHz	1.1 & 2 GHz	–	2 GHz	3 GHz	2.45 & 5.25 GHz	620 & 1020 MHz
Bandwidth	120 & 100 MHz	–	2 GHz	138 & 204 MHz	5.8 & 10.4 GHz	3.76 GHz	4.4 GHz	–	–
Substrate	Rogers R04003c	Duroid 5880	Duroid RO3210	FR4	Rogers R7	Rogers R4350B	Rogers RO4003	Rogers RT 5880	SOD-323
Dielectric constant	3.55	2.2	2.25	3.67	2.33	2.2	3.55	2.2	2.2
Thickness (Mm)	0.508	0.787		1.5	0.79	0.1	1.524	0.7874	2
Return loss (S11) (Db)	−20 & −19	>−10	f1 = −17.17 f2 = −23.898	–	15	–	>16	−14 & −13	−17
Insertion loss (S21/S31) (Db)	−4.16/ −4.32 −4.17/ −4.26	−4	2.85/−4.53 −3.97/−3.7	−3.21/ −3.185 −3.22/ −3.7	−4.1/ −4.6	−4.24/ −4.22	−3.42/ −4.67	−3.56/ −3.05 −4.38/ −4.14	–
Phase difference	>2°	>4.8°	–	180.3° & 179.1°	4°	180.25°	180° ± 0.5°	−172.6° & 184.83°	180° ± 7°
Amplitude difference (Db)	<0.3	–	–	0.6	1.5	−3.64	±0.1	0.5	<0.8

3 Conclusion

The design of various microstrip balun has been reviewed for various frequency ranges from 0.9 to 5.5 GHz. If the operating frequency is less than 2 GHz, insertion loss of −3 dB is obtained. If the operating frequency is more than 2 GHz, the obtained Loss of insertion less than −3 dB For higher operating frequency ranges, even though amplitude imbalance is less, the obtained insertion loss is less than −3 dB. When FR4 material is used as substrate, the obtained bandwidth is less when compared to other substrates.

References

1. Huang, F., Wang, J., Zhu, L., Chen, Q., Wu, W.: Dual-band microstrip balun with flexible frequency ratio and high selectivity. IEEE Microwave Wirel. Compon. Lett. **27**(11), 962–964 (2017)
2. Shao, J., Zhou, R., Chen, C., Wang, X.-H., Kim, H., Zhang, H.: Design of a wideband balun using parallel strips. IEEE Microwave Wirel. Compon. Lett. **23**(3), 125–127 (2013)
3. Shao, J., Zhang, H., Chen, C., Tan, S., Chen, K.J.: A compact dualband coupled-line balun with tapped open-ended stubs. IEEE Microwave Wirel. Compon. Lett. **22**, 109–122 (2011)
4. Shen, L., et al.: Dual-band balun with flexible frequency ratios. IEEE Microwave Wirel. Compon. Lett. **51**(17), 1213–1214 (2015)
5. Wu, P., Xue, Q.: A wideband microstrip balun structure. IEEE Microwave Wirel. Compon. Lett. (2014)

6. Wu, Y., Liu, Q., Leung, S.W., Liu, Y., Xue, Q.: A novel planar impedance-transforming tight-coupling coupler and its applications to microstrip baluns. IEEE Trans. Compon. Packag. Manuf. Technol. **4**(9), 1480–1488 (2014)
7. Huang, C.Y., Lin, G.Y., Tang, C.W.: Design of the wide-stopband balun with stepped coupled lines. In: Proceedings of 2018 IEEE Transactions Components (2018)
8. Zhang, H., Peng, Y., Xin, H.: A tapped stepped-impedance balun with dual-band operations. IEEE Antennas Wirel. Propag. Lett. (2010)
9. Miao, X., Zhang, W., Geng, Y., Chen, X., Ma, R., Gao, J.: Design of compact frequency-tuned microstrip balun. IEEE Antennas and Wireless Propagation Letters **9**, 686–688 (2010)

Patient's Health Monitoring System Using Internet of Things

P. Christina Jeya Prabha(✉), P. Abinaya, G. S. Agish Nithiya, P. Ezhil Arasi, and A. Ameelia Roseline

Electronics and Communication Engineering Department, Panimalar Engineering College, Anna University, Chennai, India
jeyaprabha612@gmail.com, ameeliaroseline@gmail.com

Abstract. The wellness of human society is more challenging in present days. The health condition of patient's can be improved by monitoring them continuously. Now-a-days the pulse rate monitoring is in traditional way, that doctor should have contact with patient on each and every day. But this is impossible in all circumstances. This system is used to provide proper and quick treatment according to their health condition. Also, the system is cheap and gives more efficient results. If parameters such as pulse rate, body temperature goes abnormal, it gives alert message or information to the doctor through wireless module such as cloud computing or WI-FI(Wireless Fidelity). The normal heart rate of human being ranges from 65 to 90, if it goes above or below the range it will automatically gives the alert message. The normal body temperature of human is from 36.1 °C (97 °F) to 37.2 °C (99 °F).

Keywords: Cloud-computing · Heart rate sensor · Temperature sensor and Raspberry Pi

1 Introduction

The main cause of death in hospitals world-wide is due to delay in treatment. The death rate can be reduced by using the smart field like Internet of Things (IoT). The usage of IoT in medical field is called as Internet of Medical Things (IoMT). The system provides the basic model of pulse rate monitoring and alert system. The objective of this work is to treat the patient immediately when required. And also to provide the current health status of the patient to the doctor [1].

The visible light mode PPGI method is used to send the imaging of pulse rate via the in-built camera in smart-phones [2, 3]. Using GSM(Global System for Mobile communication) technology, the health details of the patient's send to the doctor in the form of SMS (Short Message Service) [4, 5].

The problem that was identified in the old version is it just gives the alert sound regarding patient's condition. In this system, it includes a new feature of alerting the patient's health condition to the doctor via transmitting the video to his server. The system communicates with android or laptop via the telegram app [6]. This application will enable the alert mechanism. By using this mechanism, the data is more efficient and secured.

D. J. Hemanth et al. (Eds.): COMET 2019, LNDECT 35, pp. 259–266, 2020.
https://doi.org/10.1007/978-3-030-32150-5_28

By sitting at home, the patient can able to measure the body temperature more effectively [7]. The existing wireless technology has some limitations, they are low power efficient and expensive [8]. Even though the availability of modern treatments, it is difficult to improve the accuracy of healthcare system. Monitoring is the best way for the Doctor to Diagnose and plan for treatment [9, 10].The key component in this system is sending the alert message in the form of video via the camera.

2 System Architecture

The Fig. 1 represents the system architecture of "Patient's health monitoring system using Internet of Things". The pulse rate sensor will sends the analog signal to the Analog to Digital Converter (ADC). This ADC sends the digital signal to Raspberry Pi 3 model B+. Further Raspberry pi will communicate with Wi-Fi module and sends the information as video to the doctor/hospital server, as preferred in the program.

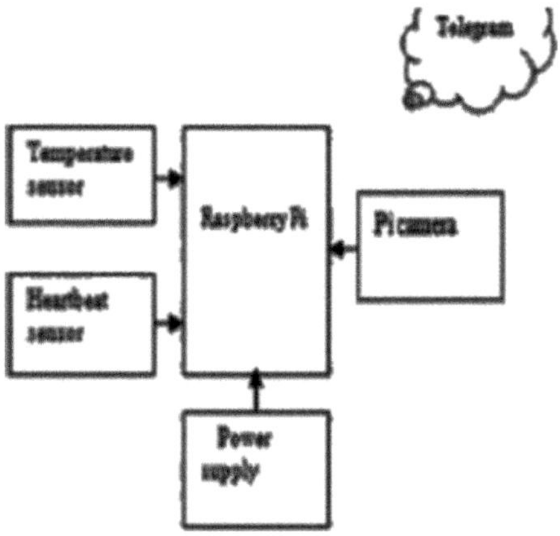

Fig. 1. Block diagram of proposed system

3 System Description

The basic components in the system are Raspberry Pi 3B+ board, Heartbeat Sensor, Raspberry Pi Camera, Temperature Sensor, Analog to Digital converter (ADC), 1 GB RAM(Random Access Memory) and USB(Universal Serial Bus) cable.

(i) Raspberry Pi 3 B+ board

Figure 2 shows the Raspberry Pi 3 B+ model. It is the third generation of Raspberry Pi. It has 40 digital pins in which 26 are GPIO(General Purpose Input and Output) pins.

There are four power supply in which two are 3.3 V and remaining to are 5 V. There are 8 ground pins. And it has two UART(Universal Asynchronous Receiver/Transmitter) interface pins. All 40 pins are used as External interrupts.

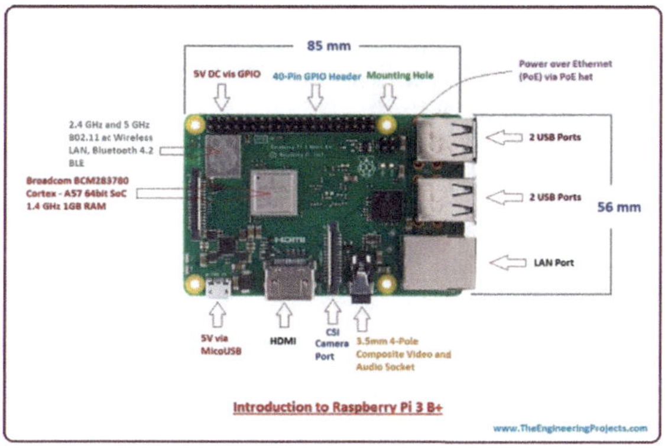

Fig. 2. Raspberry pi 3 B+ board

It is well integrated board with open source computing platform.

(ii) Heartbeat Sensor

Sensor is designed to study the heart function. When it starts working it detects the blood flow through fingers and gives digital output of heartbeat when a finger is placed on it. The led flashes give the measure of Beats Per Minute (BPM) rate.

Fig. 3. Model of heartbeat sensor

Figure 3 represents the heartbeat sensor. The basic principle of pulse sensor is photo phlethysmography. The heartbeat sensor is of plug and play type. The ground pin is used to connect the systems ground. V_{cc} has range 5 V or 3.3 V. Signal pin gives the pulsating output.

(iii) Temperature Sensor

Temperature sensor detects and measures hotness and coolness which converts into electrical signal. Figure 4 shows the LM35 sensor. By usingLM35, temperature can be measured more accurately than thermistor. It gives output voltage linearly proportional to Celsius temperature.

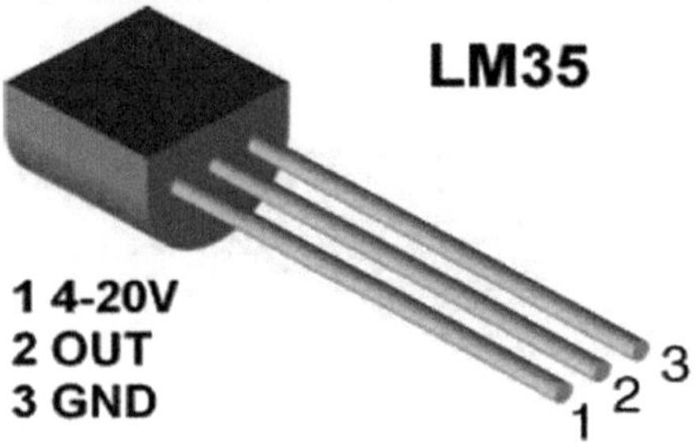

Fig. 4. Temperature sensor model

It operates at −55 °C to 120 °C. It draws only 60 μA from the supply. It has very low self heating capability of less than 0.1 °C in still air

(iv) Raspberry Pi camera

Figure 5 is the sample model of Pi camera. Pi camera can be used to take High Definition(HD) videos. It supports 1080p30, 720p60 and VGA90 video modes. It is a high quality 8 M pixel Sony IMX219 image sensor custom design board for Raspberry Pi.

Fig. 5. Raspberry Pi Camera sample

It is capable of 3280 × 2464 pixel static images. Camera gets connected to the Raspberry Pi board via the CSI (Camera serial Interface) port. It captures the video or image as per the program module.

4 Performance and Experiments

The system has been programmed in which the sensor will senses and monitor the heartbeat rate when a finger is placed. As shown in Fig. 6 Raspberry Pi board is connected with heartbeat sensor, temperature sensor and Raspberry Pi Camera, it gets power from the devices in which Raspberry Pi is connected. The output is displayed in VNC(Virtual Network Computing) Viewer and also in android device via the telegram app.

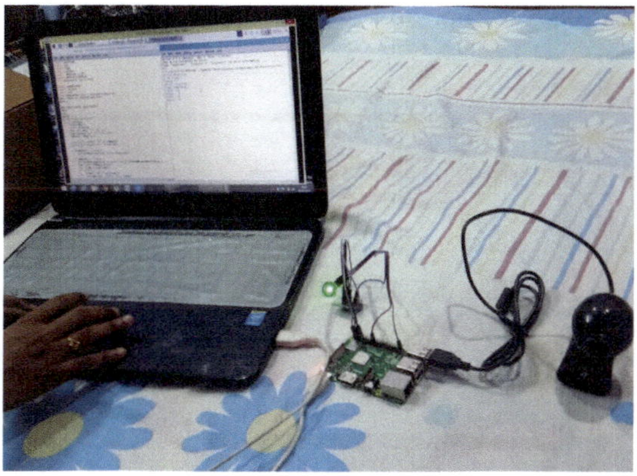

Fig. 6. Model of proposed system

5 Methodology

Raspberry Pi board is interfaced with pulse rate sensor and gives the output to digital I/O (Input/Output) pins.

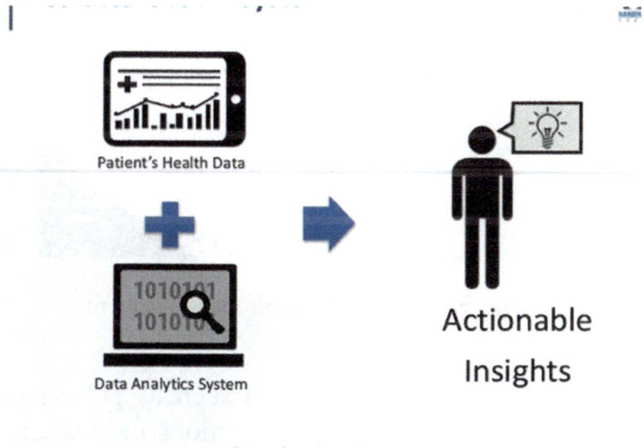

Fig. 7. Sample model of proposed system

Figure 7 represents the sample model of Proposed system. Patient's health data and data analysis system gives to Actionable insights(Doctor server) Raspberry Pi software is open source network. The source code for system environment is based on Python language. The library files used in the system is OpenCV, NumPy and Imutils.

6 Output

After receiving the alert message, if doctor needs the information about the patients pulse rate and temperature, by using these two command it can be accessed. Figures 8 and 9 shows the output of the system.

COMMANDS: /pulse, /temp, /image and /video

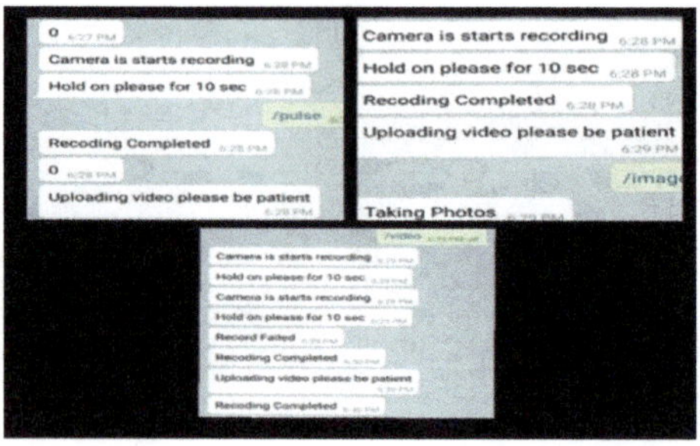

Fig. 8. Command from the doctor server

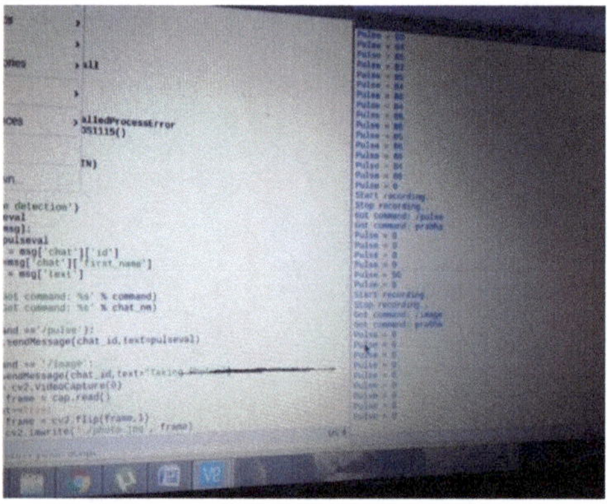

Fig. 9. Output of the proposed system

The Fig. 9 shows the output of the system such as pulse rate, temperature rate, capturing the image and video. This system functions only when the patients health conditions crosses the limit.

Table 1 shows the survey done on BeWell hospital in Avadi using this system. Table 2 gives the readings done by the hospital equipments.

Table 1. Details of the patient's using Proposed system

Name	Age group	Temperature	Pulserate
Patient 1	7	35.2 °C	72 BPM
Patient 2	21	36.4 °C	84 BPM
Patient 3	32	35.8 °C	99BPM
Patient 4	45	36.2 °C	86 BPM

Table 2. Details of the patient's using Existing System

Name	Age group	Temperature	Pulserate
Patient 1	7	35.4 °C	74 BPM
Patient 2	21	37 °C	88 BPM
Patient 3	32	36.3 °C	94 BPM
Patient 4	45	35.8 °C	89 BPM

7 Conclusion

Generally, the heart beat can be easily detected using high level devices. But the system is focused on emergency situation; it also provides a cost effective and efficient heart rate monitoring system. It also helps to continuously monitor the real time health condition of the patient. Even though the doctor is not available at present the patient can be treated without any delay.

The project can be further modified for many useful applications in medical field. The system not only helps to collect the information about the patient in remote areas, but also helps in large scale.

Further improvements like transferring patient's health information even in rural areas. Is also possible for live streaming the patient's health condition for doctor wherever in the world. The system can also be improved to transfer real time ECG (Electrocardiogram)/EEG(Electroencephalogram) immediately to the doctor.

References

1. Hodge, A., Humabakdkar, H., Bidwai, A.: Wireless heart rate monitoring and vigilant system. In: 3rd International Conference for Convergence in Technology (I2CT), Pune, India, 06–08 Apr 2018 (2018)
2. Blocher, T., Schneider, J., Schinle, M.: An online PPGI approach for camera based heart rate monitoring using beat-to-beat detectio. IEEE (2017)
3. Sun, Y., Thakor, N.: Photoplethysmographyrevisited:from contact to non contact, from point to imaging. IEEE Trans. Biomed. Eng. **63**, 463–477 (2016)
4. Ufoaroh, S.U., Oranugo, C.O., Uchechukwu, M.E.: Heartbeat monitoring and alert system using gsm technology. IJERGS **3**(4), 26–34 (2015)

5. Jubadi, W.M., Sahak, S.F.A.M.: Heartbeat monitoring alert via SMS. In: 2009 IEEE Symposium on Industrial Electronics and Applications (ISIEA 2009), Kuala Lumpur, Malaysia, 4-6 October 2009 (2009). 978-1-4244-4683-4/09
6. Mohammed, J., Thakral, A., Ocneanu, A.F., Jones, C., Lung, C.-H., Adler, A.: Internet of Things: remote patient monitoring using web services and cloud computing. In: 2014 IEEE International Conference on Internet of Things (iThings 2014), Green Computing and Communications (GreenCom2014), and Cyber-Physical, pp. 256–263 (2014)
7. Mansor, H., Shukor, M.H.A., Meskam, S.S., Rusli, N.Q.A.M., Zamery, N.S.: Body temperature measurement for remote health monitoring system. In: IEEE International Conference on Smart Instrumentation, Measurement and Applications (ICSIMA), pp. 26–27 November 2013 (2013)
8. Kiourmars, A.H., Tang, L.: Wireless network for health monitoring heart rate and temperature senso. In: Fifth International Conference on Sensing Technology, pp. 362–368 (2011)
9. Gacek, A., Pedrycz, W.: ECG Signal Processing, Classification And Interpretation. Springer, London (2012)
10. Armil, J., Punsawad, Y., Wongsawat, Y.: Wireless sensor network-based smart system for health care monitoring. In: International Conference on Robotics and Biometrics, pp. 2073–2076 (2011)

Power Generation Using Microbial Fuel Cell

R. Senthil Kumar(✉), D. Yuvaraj, K. R. Sugavanam,
V. R. Subramanian, S. Mohamed Riyaz, and S. Gowtham

Department of Electrical and Electronics Engineering,
Department of Bio Technology, Vel Tech High Tech Dr Rangarajan
Dr Sakunthala Engineering College, Avadi, Chennai, India
rskumar.eee@gmail.com, yuvaraj@velhightech.com,
sugavanamkr@gmail.com, subbusubramanian770@gmail.com,
mohamedriyaz1424@gmail.com, gowthamsathya10@gmail.com

Abstract. Studying the performance of the microbial fuel cell using waste water as a substrate. Study was carried out in double chamber microbial fuel cell with single and multiple salt bridge as Photon exchanger and also the performance for multiple electrodes instead of single electrode of same volume was studied. It has been identified that the double chambered reactor with multiple electrode and multiple salt bridge microbial fuel cell performance were observed whose performance is better than other fuel cell. The results shows that the microbial fuel cell can effectively used in waste water treatment plant for the generation of power.

Keywords: Waste water substrate · Fuel cell reactor · Photon exchange membrane · Multiple electrodes · Multiple salt bridge · Electrodes surface area · Anaerobic process

1 Introduction

In India 12000 million litre is generated as a waste water being produced per day due to domestic and agricultural process. The waste water from the domestic and agriculture process consists of energy in the three forms namely thermal energy, bio degradable organic matter, nutritional element energy like Nitrogen, Phosphorus etc. In these process, extracting heat energy is quite complicated, so we are in the need of extracting the energy from waste water with the help of microbes from degradable organic matter in the waste water. In this process waste water which consist of biological organisms is used as substrate. The microbes in the substrate will decompose the organic matter and emit the Hydrogen (Photon). Microbial fuel cell is a technology which is majorly used in waste water treatment. In this process, the emitted Hydrogen from degradable organic matter will flow from anode chamber to cathode chamber through Photon Exchange Membrane [PEM] called as Salt bridge. When there is a flow of Hydrogen [Positive ion] from anode to cathode via Salt Bridge then electron will flow from anode to cathode via external circuit. This is one of the technologies used for micro power generation [1]. The survey gives the detail about the amount of waste water generated in India per day According to international institute of health and hygiene the amount

D. J. Hemanth et al. (Eds.): COMET 2019, LNDECT 35, pp. 267–278, 2020.
https://doi.org/10.1007/978-3-030-32150-5_29

of waste water in industrial and agricultural waste water is classified into three divisions which are sewage generation, untreated sewage, sewage treatment, in this the amount of sewage water generated is 61754 millions of litres per day. The amount of waste water produced by different industries are given below (Figs. 1 and 2).

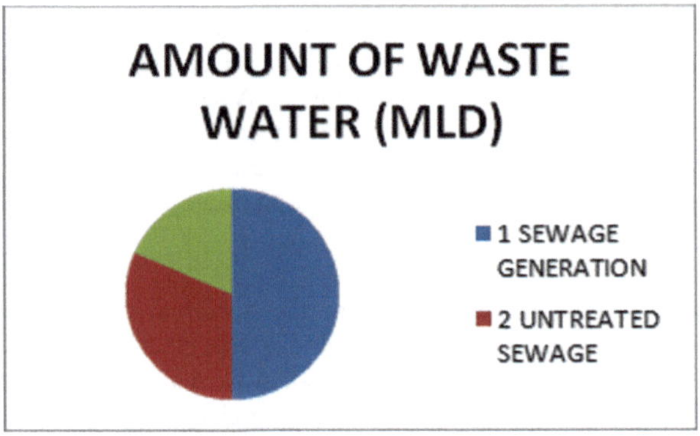

Fig. 1. Amount of waste water generation

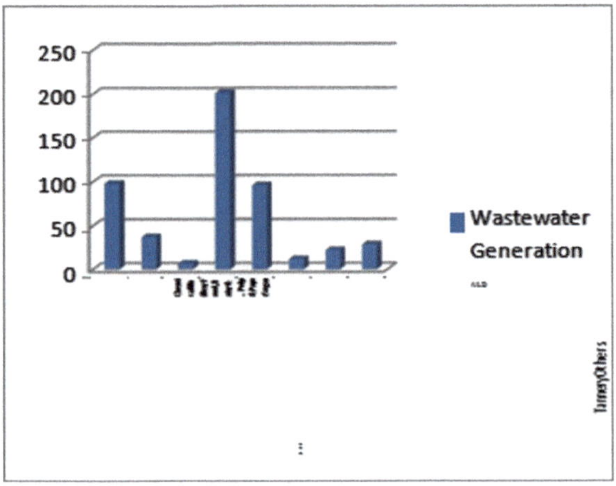

Fig. 2. Waste water generated by different industries.

In the current scenario there is a enormous amount of energy consumption due to increase in the world population, these energy are provided by the energy sources which are classified into renewable and non renewable energy sources. The most of the energy demand is compensated by the non renewable resources which are fossil fuel, petroleum, coal etc. these source of energy has many drawbacks like the non-renewable energy cannot be replaced once it is used, it will pollute the environment, it increases green house gases in the environment.for the following drawback the world are in the need of seeking for the renewable energy sources it has certain advantages over the non renewable energy sources.

The advantages of renewable energy are it can be readily available in nature and exist forever, The MFC is a renewable energy source which is eco friendly and Source of energy is pure, renewable and readily available with less running cost. It is Very beneficial to environment it helps in reduction of pollution and cut the money spend for waste treatment. Operating cost and running cost is less and it is used for waste water treatment for large amount.

2 Microbiological Fuel Cell (MFC)

The equation determines the process of MFC which converts the organic matter in a substrate by bacterial interactions into water, photons and electrons.

$$C_2H_4O_2 + 2H_2O \rightarrow CO_2 + 8e^{--} + 8H^+ 2O_2 + 8H^+ + 8e^{--} \rightarrow 4H_2O \quad (1)$$

Microbial fuel cell consist of two compartments called anodic compartment and cathodic compartment, in the anode chamber the anaerobic reaction takes place and in the cathodic compartment the aerobic reaction take place due to the presence of bacterial content in the substrate, in the anaerobic reaction due to bacterial interaction in anode chamber the hydrogen ions will emits due to the decay of the organic matter in the substrate, hence the produced hydrogen ion will move from anode chamber to cathode chamber through a photon exchanger called as salt bridge.

3 Research Background

The microbial fuel cell performance is highly depend on the type of the reactor and material of electrode. In MFC, electricity production involves many steps such as microbial organic process, capture of electrons by anode, reduction of electrode at cathode, movement of photon from anode to cathode [1]. The performance of MFC is determined by number of parameters such as reactor type, photon exchanger used, electrode material, number of electrodes, spacing of electrodes and substrate. Reactor type: The types of reactor for microbial fuel cell is classified into single chamber and multi chamber reactor are used [1]. In our research we are going to study the performance of double chamber reactor with single and multiple photon exchanger. Electrode: The certain characteristics of electrode are longevity, conductivity, surface area, electro catalystic activity should be studied [8]. The electrode used in microbial fuel cell have certain impact on its performance [2]. The surface modification of the traditional electrodes are called Advance Electrodes [2]. Recently, Graphene electrode is used in Microbial Fuel Cell whose performance is good. The types of electrode material used in MFC are anode materials such as Carbon based electrode, Graphite, Stainless Steel and Ceramic electrodes. The cathode materials such as biotic cathode and abiotic cathodes are used [2]. Number and Spacing of Electrodes: The power production in MFC is affected by some factors like bacteria used in anode chamber which digest the organic matter, temperature of metabolic process, size of anode compartment [4]. The performance of MFC can be improve by increasing the number of electrodes with certain electrode spacing ratio which is decreasing the spacing between electrodes 4 cm–2 cm. [15]. Photon Exchanger: while increasing the number of photon exchanger, it has a direct relation on power generation in MFC [7]. The different types of photon exchange membrane are Salt bridge and Polymer Exchange Membrane (PEM). The PEM is made from the Fluoropolymer, Nafion, Teflon [12]. These are the types of Photon Exchanger. Substrate: It play a major role in generation of power and treatment of waste water. The number of substrate have been studied although the output of artificial waste water is low [14].

4 Proposed System

To Study the performance of MFC with waste water as a substrate and the rate of power generation with respect to different design of reactor (reactor design includes volume of compartment, no of salt bridges and no of electrodes used), and to study the performance of MFC with respect to surface area of electrodes. It can be used in large setup with less operating and running cost.

Our research is based on the study the four types of reactor design such as double chamber reactor with single photon exchanger and single electrodes, double chamber reactor with single photon exchanger and multiple electrodes, double chambered reactor with multiple photon exchanger and single electrode, double chamber reactor with multiple photon exchanger and multiple electrodes.

In our research the electrodes are selected by considering certain parameters like cost, surface area, longevity, chemical resistivity, Electrical conductivity by considering the following factors the electrodes used In our project are carbon electrodes as anode and copper electrodes as cathode are used.

4.1 Double Compartment Reactor with Single Photon Exchanger and Single Electrodes

Electrodes. In this type of reactor design single set of electrodes are used and the material of the anode is carbon based and the material of cathode is copper are used, and the surface area and volume of the electrode is calculated (Fig. 3).

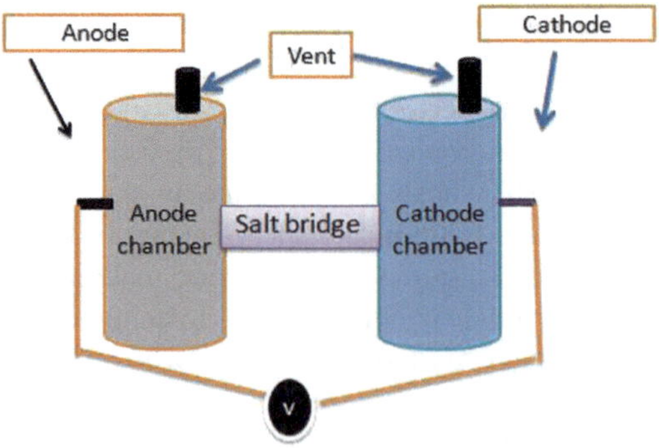

Fig. 3. Double compartment reactor with single photon exchanger and single electrodes

Calculation

Volume $V = \pi r^2 h$

$3.14*0.45*0.45*15 = 9.54cm^3$

Surface area $A = 2\ \pi rh + 2\ \pi r^2$

Surface area = $2*3.14*0.45*15 + 2*3.14*0.45*0.45$

$= 43.68\ cm^2$

The mentioned parameters have been calculated, these electrodes are inserted in the respective chambers with constant volume of substrate, the reading is tabulated with particular interval of time.

Substrate. In our research the substrate used is a Bacterial broth solution for the microbial fuel cell, the synthesis of bacterial broth is given, 5 g of agar is mixed in 100 ml distilled water then the solution is heated and kept in a shaker for a hour then 1 ml of bacterial culture is added to the solution kept it for a day in incubation. The volume of the substrate used is about 1 l of bacterial broth.

Photon Exchange Membrane. In our research the salt bridge is used as photon exchange for microbial fuel cell, the synthesis of salt bridge is given, 5 g of agar is mixed in 40 ml of distilled water 75 g of potassium chloride is mixed in the solution then the solution is transfer to the salt bridge container and the solution is refrigeration for 30 min. The number of salt bridge used in these design is one number.

Reactor. In our research the double compartment reactor of volume 1 L is used.

4.2 Double Compartment Reactor with Single Photon Exchanger and Multiple Electrodes

Electrodes. In this type of reactor design multiple electrodes are used and the material of the anode is carbon based and the material of cathode is copper are used, and the surface area and volume of the electrode is calculated (Fig. 4).

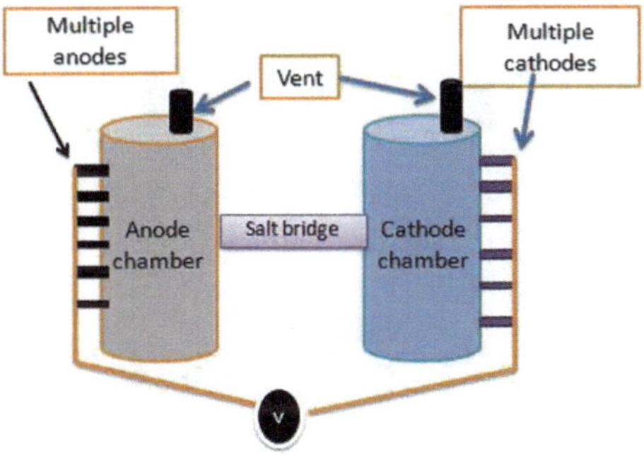

Fig. 4. Double compartment reactor with single photon exchanger and multiple electrodes

Calculation

Volume $V = \pi r^2 h \quad 3.14*0.24*0.24*4.5 = 0.81\text{cm}^3$

Surface area $A = 2\,\pi rh + 2\,\pi r^2$

Surface area $= 2*3.14*0.24*15 + 2*3.14*0.24*0.24$
$= 7.15\ \text{cm}^2$

The mentioned parameters have been calculated, these electrodes are inserted in the respective chambers with constant volume of substrate, the reading is tabulated with particular interval of time.

Substrate. In our research the substrate used is a Bacterial broth solution for the microbial fuel cell, the synthesis of bacterial broth is given, 5 g of agar is mixed in 100 ml distilled water then the solution is heated and kept in a shaker for a hour then 1 ml of bacterial culture is added to the solution kept it for a day in incubation. The volume of the substrate used is about 1 l of bacterial broth.

Photon Exchange Membrane. In our research the salt bridge is used as photon exchange for microbial fuel cell, the synthesis of salt bridge is given, 5 g of agar is mixed in 40 ml of distilled water 75 g of potassium chloride is mixed in the solution then the solution is transfer to the salt bridge container and the solution is refrigeration for 30 min. The number of salt bridge used in these design is one number.

Reactor. In our research the double chamber reactor of volume 1 L is used

4.3 Double Compartment Reactor with Multiple Photon Exchanger and Single Electrode

Electrodes. In this type of reactor design single set of electrodes are used and the material of the anode is carbon based and the material of cathode is copper are used, and the surface area and volume of the electrode is calculated (Fig. 5).

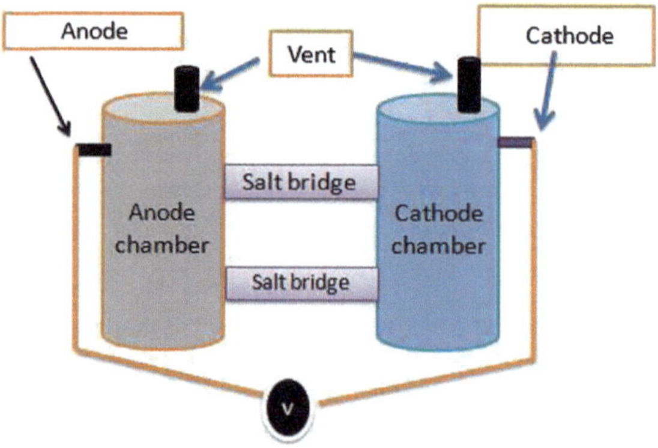

Fig. 5. Double compartment reactor with multiple photon exchanger and single electrode

Calculation

Volume $V = \pi r^2 h$
$3.14*0.45*0.45*15 = 9.54\ cm^3$
Surface area $A = 2\ \pi rh + 2\ \pi r^2$
Surface area
$= 2*3.14*0.45*15 + 2*3.14*0.45*0.45$
$= 43.68\ cm^2$

The mentioned parameters have been calculated, these electrodes are inserted in the respective chambers with constant volume of substrate, the reading is tabulated with particular interval of time.

Substrate. In our research the substrate used is a Bacterial broth solution for the microbial fuel cell, the synthesis of bacterial broth is given, 5 g of agar is mixed in 100 ml distilled water then the solution is heated and kept in a shaker for a hour then 1 ml of bacterial culture is added to the solution kept it for a day in incubation. The volume of the substrate used is about 1 l of bacterial broth.

Photon exchange membrane. In our research the salt bridge is used as photon exchange for microbial fuel cell, the synthesis of salt bridge is given, 5 g of agar is mixed in 40 ml of distilled water 75 g of potassium chloride is mixed in the solution then the solution is transfer to the salt bridge container and the solution is refrigeration for 30 min. The number of salt bridge used in these design is two number.

Reactor. In our research the double compartment reactor of volume 1 L is used.

4.4 Double Compartment Reactor with Multiple Photon Exchanger and Multiple Electrodes

Electrodes. In this type of reactor design multiple electrodes are used and the material of the anode is carbon based and the material of cathode is copper are used, and the surface area and volume of the electrode is calculated (Fig. 6).

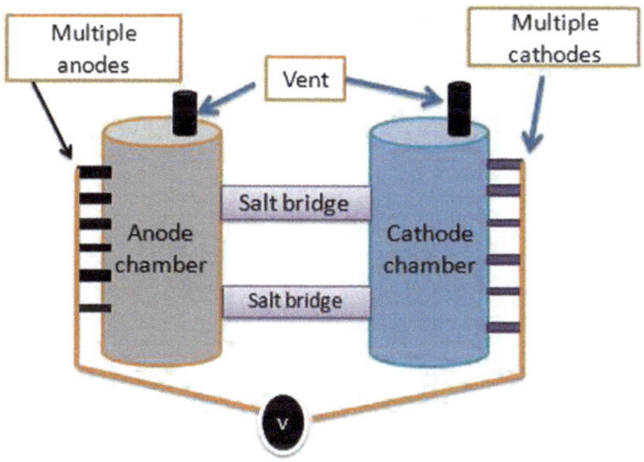

Fig. 6. Double compartment reactor with multiple photon exchanger and multiple electrodes

Calculation.

Volume $V = \pi r^2 h$
$3.14*0.24*0.24*4.5 = 0.81cm^3$
Surface area $A = 2\ \pi rh + 2\ \pi r^2$
Surface area
$= 2*3.14*0.24*15 + 2*3.14*0.24*0.24$
$= 7.15cm^2$

The mentioned parameters have been calculated, these electrodes are inserted in the respective chambers with constant volume of substrate, the reading is tabulated with particular interval of time.

Substrate. In our research the substrate used is a Bacterial broth solution for the microbial fuel cell, the synthesis of bacterial broth is given, 5 g of agar is mixed in 100 ml distilled water then the solution is heated and kept in a shaker for a hour then 1 ml of bacterial culture is added to the solution kept it for a day in incubation. The volume of the substrate used is about 1 l of bacterial broth.

Photon Exchange Membrane. In our research the salt bridge is used as photon exchange for microbial fuel cell, the synthesis of salt bridge is given, 5 g of agar is mixed in 40 ml of distilled water 75 g of potassium chloride is mixed in the solution then the solution is transfer to the salt bridge container and the solution is refrigeration for 30 min. The number of salt bridge used in these design is two number.

Reactor. In our research the double compartment reactor of volume 1 L is used.

5 Result and Discussions

The output of our four reactor design is given below.

5.1 Single Electrode with Single Salt Bridge

Inference. The day by day readings are noted and tabulated, for each day 6 readings are taken and the average of the readings is tabulated. The out put of the first reactor design is given, the power is calculated for every day And we got a maximum power on day 4 is 0.01 mW [Table 1].

Table 1. Readings of single electrode with single salt bridge

Day	Current (Ma)	Voltage (Mv)	Power (Mw)
1	0.013	0.0120	0.000156
2	0.038	0.015	0.00057
3	0.034	0.043	0.001462
4	0.094	0.108	0.010152
5	0.035	0.027	0.000945
6	0.067	0.034	0.002278
7	0.045	0.054	0.00243
8	0.112	0.076	0.008512
9	0.09	0.032	0.00288
10	0.0132	0.079	0.001043

5.2 Single Electrode with Two Salt Bridge

Inference. The same procedure is followed for taking readings, and the power is calculated, while compare with first reactor design the second reactor design with multiple electrode performance was good, the maximum power obtained is 0.03 Mw [Table 2].

Table 2. Readings of single electrode with two salt bridge

DAY	Current (mA)	Voltage (mV)	Power (mW)
1	0.014	0.013	0.00018
2	0.043	0.016	0.000688
3	0.048	0.046	0.002208
4	0.172	0.209	0.035948
5	0.049	0.036	0.001764
6	0.077	0.044	0.003388
7	0.055	0.064	0.00352
8	0.142	0.084	0.011928
9	0.082	0.038	0.003116
10	0.014	0.086	0.00121

5.3 Multiple Electrode with Single Salt Bridge

Inference. The same procedure is followed for taking readings, and the power is calculated, the maximum power out put obtained in day 4 is 0.33 mW [Table 3].

Table 3. Readings of multiple electrode with single salt bridge

Day	Current (mA)	Voltage (mV)	Power (mW)
1	0.097	0.182	0.17654
2	0.095	0.501	0.047595
3	0.102	0.519	0.052938
4	0.564	0.587	0.331068
5	0.103	0.549	0.056547
6	0.456	0.58	0.26448
7	0.112	0.539	0.060368
8	0.09	0.0287	0.002583
9	0.047	0.303	0.014241
10	0.013	0.266	0.003458

5.4 Multiple Electrode with Multiple Salt Bridge

Inference. The same procedure is followed for taking readings, and the power is calculated, the maximum power obtained for multiple electrodes is 0.4 Mw. While comparing with multiple electrode with single photon exchanger, the multiple electrode with multiple salt bridge performance was good [Table 4].

Table 4. Readings of multiple electrode with multiple salt bridge

Day	Current (mA)	Voltage (mV)	Power (mW)
1	0.107	0.192	0.020544
2	0.095	0.601	0.057095
3	0.102	0.519	0.052938
4	0.628	0.687	0.431436
5	0.102	0.549	0.055998
6	0.356	0.481	0.171236
7	0.116	0.639	0.074124
8	0.12	0.287	0.03444
9	0.047	0.308	0.014241
10	0.013	0.366	0.004758

Graph.

- Series 1 shows the output of first reactor design-Single Electrode With Single Salt Bridge.
- Series 2 shows the output of first reactor design-Single Electrode With Two Salt Bridge.
- Series 3 shows the output of first reactor design-Multiple Electrode With Single Salt Bridge.

- Series 4 shows the output of first reactor design-Multiple Electrode With Multiple Salt Bridge.

In our research we have studied the four reactor design in which the output obtained with different reactor design was recorded for different interval of time for 10 days and the output was plotted in the graph (Fig. 7).

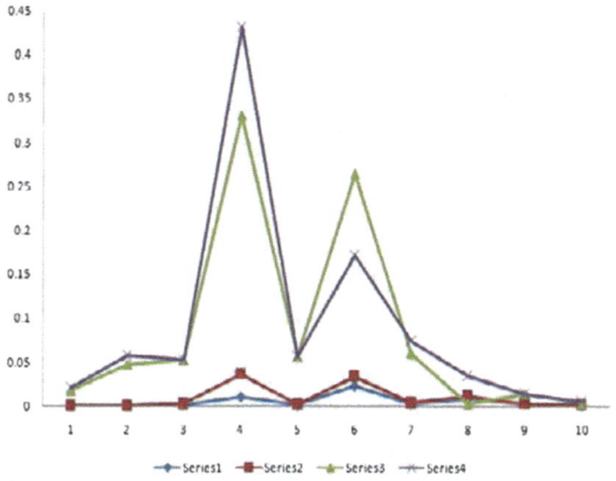

Fig. 7. Output of the four reactor design.

6 Conclusion

The Microbial Fuel Cell with multiple electrodes and multiple salt bridge (Photon Exchange Membrane) whose performance is found to be good which generate power about 0.43 m, which is maximum power while compare with other fuel cell in this paper. Source of energy is clean, renewable and readily available at afford cost, helps in reduction of pollution and cut the cost of waste water treatment. It is used for waste water treatment of large amount.

References

1. Bhargavi, G., Venu, V., Renganathan, S.: Microbial fuel cells: recent developments in design and materials. In IOP Conference Series: Materials Science and Engineering, vol. 330, p. 012034 (2018)
2. Kalathil, S., Patil, S., Pant, D.: Microbial fuel cells: electrode materials. Encyclopedia of Interfacial Chemistry: Surface Science and Electrochemistry (2017). http://dx.doi.org/10.1016/B978-0-12-409547-2.13459-6
3. Davis, F., Higson, S.P.J.: Biofuel cells—recent advances and applications. Biosen. Bioelectron. **22**, 1224–1235 (2007). https://doi.org/10.1016/j.bios.2006.04.029
4. Li, J.: An experimental study of microbial fuel cells for electricity generating: performance characterization and capacity improvement. J. Sustain. Bioenergy Syst. **3**, 171–178 (2013). https://doi.org/10.4236/jsbs.2013.33024

5. Logan, B.E., Hamelers, B., Rozendal, R., Schröder, U., Keller, J., Freguia, S., Rabaey, K.: Microbial fuel cells: methodology and technology. Environ. Sci. Technol. **40**(17), 5181–5192 (2006). https://doi.org/10.1021/es0605016. CCC American Chemical Society
6. Rahimnejad, M., Adhami, A., Darvari, S., Zirepour, A., Oh, S.E.: Microbial fuel cell as new technology for bioelectricity generation: a review. Alexandria Eng. J. **54**(3), 745–756 (2015). https://doi.org/10.1016/j.aej.2015.03.031
7. Parkash, A.: Impact of salt bridge on electricity generation from hostel sewage sludge using double chamber microbial fuel cell. Res. Rev. J. Eng. Technol. **5**(252), 2 (2015). ISSN:2319–9873
8. Ci, S., Cai, P., Wen, Z., Li, J.: Graphene-based electrode materials for microbial fuel cells. Sci. Chin. Mater. **58**(6), 496–509 (2015). https://doi.org/10.1007/s40843-015-0061-2
9. Shankar, R., Pathak, N., Chaurasia, A.K., Mondal, P., Chand, S.: Energy production through microbial fuel cells. A.S. Crit. Rev. Environ. Sci. Technol. **44**, 97–153 (2014)
10. Tsai, H.Y., Hsu, W.H., Huang, Y.C.: Characterization of carbon nanotube/graphene on carbon cloth as an electrode for air-cathode microbial fuel cells. J. Nanomater. **2015**(686891) (2015). http://dx.doi.org/10.1155/2015/686891
11. Chae, K.-J., Choi, M.-J., Lee, J.-W., Kim, K.-Y., Kim, I.S.: Effect of different substrates on the performance, bacterial diversity, and bacterial viability in microbial fuel cells. Bioresour. Technol. **100**, 3518–3525 (2009)
12. Garba, N.A., Sa'adu, L., Dambatta, M.B.: An overview of the Substrates used in Microbial Fuel Cells. http://doi.org/10.15580/GJBB.2017.2.05151761
13. Khan, M.R., Bhattacharjee, R., Amin, M.S.A.: Performance of the salt bridge based microbial fuel cell. International Journal of Engineering and Technology **1**(2), 115–123 (2012)
14. Pant, D., Van Bogaert, G., Diels, L., Vanbroekhoven, K.: A review of the substrates used in microbial fuel cells (MFCs) for sustainable energy production. https://doi.org/10.1016/j.biortech.2009.10.017

Detection of Human Existence Using Thermal Imaging for Automated Fire Extinguisher

S. Aathithya(✉), S. Kavya, J. Malavika, R. Raveena, and E. Durga

Department of Electronics and Instrumentation Engineering,
Panimalar Engineering College, Chennai, India
aathisaravananl1@gmail.com, kavyajayal412@gmail.com,
malavika0218@gmail.com, raveena0499@gmail.com,
durgaelumalai24@gmail.com

Abstract. Fire disaster is a common threat to lives and properties. Rescuing people caught in fire accidents is important as fire can expand rapidly and result in a tremendous loss of life and property. If the extinguishers are used under the presence of human, it will cause serious health issues and can also be fatal. In our proposal system we use a thermal camera to detect the existence of human inside a fire struck building. The thermal signatures obtained from camera helps to detect the number of people trapped inside. The controller thus checks for presence of humans using camera and takes the required action.

Keywords: Fire extinguisher · Thermal imaging camera · Controller · Electromagnet · Human presence · MATLAB

1 Introduction

Fire safety is the swift measure that has to be taken for extinguishing fire or reducing its accidental effects. They are to be adopted during priorly the construction and development of every structure to prevent the fire accidents. Various fire extinguishing agents are present apart from water like foaming agents to handle oil fire, carbon-dioxide is used when the fire is fought by suffocation and dry chemicals to extinguish electrical fires or burning liquids.

In case of fire emergency, the initiative devices are triggered by immediate and progressive increase in the flare. The initiative equipment falls under two categories, Manual and Automatic. The Break glass station, Buttons, Pull stations are the Manual initiating equipments that are made easily accessible. A vast range of automatic initiating equipments including detectors that indicate heat, smoke, flame, CO, water flow, etc. are existent [5]. They respond spontaneously in an emergency situation as they sense any transition in the environmental parameters. Nonetheless, light detection is rapid than smoke and temperature detection, the latter method is contemplated in this paper.

Thermal detectors sense one or more factors succeeding from hearth like exhaust, electromagnetic waves, heat or gas. A temperature detector is emergency indication equipment which is equipped to indicate when the thermal energy of the flame rises the temperature of heat susceptible component. They exhibit two main process, “rate of rise” and “fixed temperature”.

D. J. Hemanth et al. (Eds.): COMET 2019, LNDECT 35, pp. 279–287, 2020.
https://doi.org/10.1007/978-3-030-32150-5_30

A fume detector is a gadget that recognizes smoke, generally as a signal of flame. Smoke can be detected either optically (photoelectric) or by physical process (ionization). Besides the initiative devices, the basic mechanism is an indication system that alerts the people to exit the fire zone and sends information to the fire department to perform necessary actions. The two most basic form of alert system in practice are the conventional system that can only point the faulty zone and fail to indicate the specific detector and Analogue addressable systems which has simultaneous dual-way communication between the control unit and the detectors in the zone thus identifying each sensor particularly on account of fire accident (Fig. 1).

Fig. 1. Existing smoke detector

The augmented intelligence related to analogue available systems permits them to possess bigger sensitivity to fireside including greater resistance towards bogus alarms. The alerting system adopts a loud, visual, physical or even scent based stimuli to notify the folks at the site of fireside. The most typical forms of alerting gadgets are audio-visual designed with a buzzer, siren or bell and flashing light combo. The fire control system typically implemented with self-starting sprinkler on identification of a fire by unleashing water on fire-accidental zone thereby lessening the damage.

In the existing system, only the fire control is taken into consideration. It does not provide method of detecting human lives inside a fire affected area [4]. Also, traditional camera does not provide good vision during fire emergencies as they get blocked by smoke that is produced during the accident, reducing the vision provided by the cameras. Hence a method to detect human lives using thermal images camera by image processing is proposed in this paper [3].

2 Proposed System

2.1 Introduction

We hereby propose a new system where the smoke or fire is detected based on thermal image processing. With the help of a thermal imaging camera, infrared rays are emitted to produce images and video. These images are then processed to establish the number of people trapped inside the building and the message is sent to the fire rescuers for better response during fire accidents. In the case of no people within the area, fire extinguishers will be switched on automatically in order to control the fire (Fig. 2).

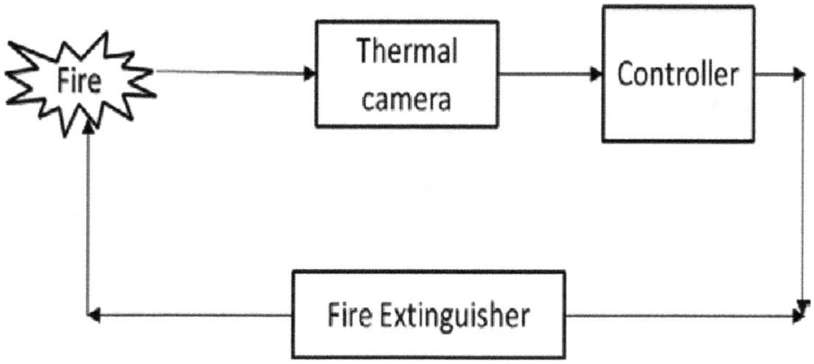

Fig. 2. Block diagram of proposed system

2.2 Fire

It is the state of combustion that produces flames thereby emitting heat and light in the form of smoke and spark. It rapidly rises the temperature of the surrounding environment. As humans cannot withstand temperatures beyond 50–60 °C it is important to excavate them from fire. It is an oxidation process and requires large amount of oxygen to continue. Hence the oxygen level for the people trapped in drops leading to the suffocation of people. Moreover the smoke mainly consists of carbon monoxide and carbon dioxide which are injurious to health resulting in respirational disorders.

2.3 Thermal Imaging Camera

This provides the main solution which counters the disadvantage of the existing system. The thermal camera also provides the temperature of the surroundings. The image varies with each temperature and surroundings and is instrumental for detecting during fire. The cameras become saturated at 315 °C. as the heat signature for each human varies it becomes easier to single out people from each other [3].

2.4 MATLAB

It is used for counting the number of people from the images of the thermal camera. The images are fed to the computer using hotspot and loaded to the MATLAB for execution. Using the colour detection the headcount is established. Appropriate cropping action is prefixed so as to send a clear image for execution [1]. The detection process is crucial as the fire extinguisher depends on this count obtained from execution.

2.5 Fire Extinguishing Process

If the head count is zero, the fire extinguisher turns on. This turning on process is done by the pulling action of the handle using electromagnet. Thus the power is supplied to the electromagnet only when the count is zero. And hence a magnetic field is produced resulting in the clamping action of the fire extinguisher [2].

3 Thermal Image Processing

In order to get an enhanced image, a method called image processing is used. Thermal processing is a method in which, the input is an image and the output is characteristic features of that image. For the image processing the following steps are to be taken (Fig. 3).

1. Images are imported through image acquisition tools.
2. Manipulation and analyzation of an image
3. Based on analyzation of image, the output is reported.

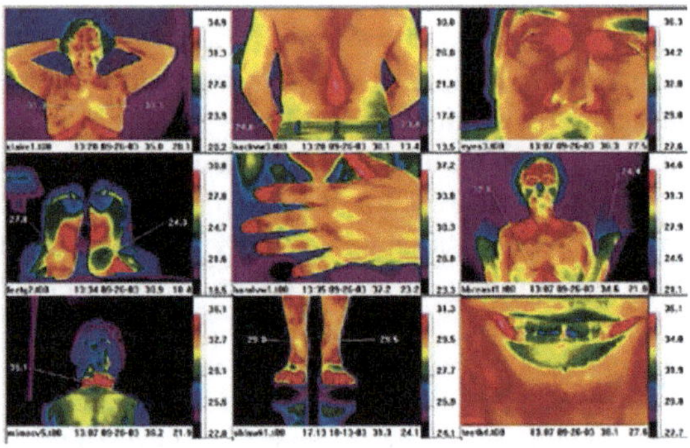

Fig. 3. Thermal processing of an image [6]

The two different types of thermal imaging processing are: analog type which uses hard copies of printouts and photographs. And the other, digital image processing uses computer algorithms on digital images. Digital image process has several blessings over analog image process. The basic concept of image processing is based on colour and image decoding. The colour in the thermal image is used to differentiate between fire and human body. Image processing is done for the following reasons:

1. Visualizing the image.
2. Sharpening and restoration of an image.
3. Retrieving the image
4. Pattern measurement
5. Recognizing the image

Various image processing techniques are image enhancement, image segmentation, feature extraction and image classification. Thermal image processing is done by a thermal camera which needs +12 V supply and has detector which is very sensitive to IR radiation based on its intensity. It determines the temperature of the surroundings and makes visible for the human eye with a thermal image and this process is known as Thermography. When the fire is detected, the thermal imaging camera gets activated to capture the images of fire area. The detected images are sent to the computer directly to count the people within the area. This count is established using MATLAB software.

4 Hardware Description

4.1 Thermal Imaging Camera

Thermal imaging camera is a gadget that develops images with the help of infrared radiations and operates in wavelength as long as 1400 nm. The electromagnetic waves are related and differentiated by their wavelength. A certain amount of electromagnetic radiations are emitted as a function of their temperatures. The advantage of thermal imaging camera is that the lenses are made up of germanium, rather than glass as they have high density and blocks the UV radiations as it has a low optical dispersion. The thermal imaging camera has the ability of measuring objects at high altitude power lines. It has a detector with a resolution of 640*480 projecting an image of 307,200 pixels. Based on the temperature, the images are being represented according to the following colour coding (Fig. 4):

1. Warmest temperature – white
2. Intermediate temperature – red, yellow, orange
3. Slightly hotter temperature – blue and purple
4. Coolest temperature – black

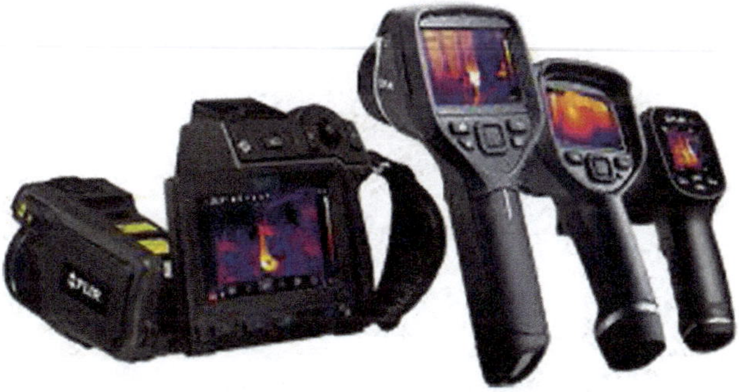

Fig. 4. Thermal imaging camera

Thermo graphic camera is very expensive. Initially it was being developed for military purposes based on the infrared technology. And now it has also been valuable for firefighters. The export and import of thermal imaging cameras are being restricted by the US govt. under the norms of International Traffic Arms Regulation. The selection of a thermal imaging camera is based on the application of a suitable temperature range. Each and every application has its own specification.

The specifications of thermal camera are:

1. Field of View: 20°
2. Manual Focus
3. Resolution: 206 × 156 pixels
4. Minimum Distance: 10 cm
5. Range of Temperature: −40 °C to 330 °C
6. Temperature Sensitivity: 0.5 °C

4.2 Fire Extinguisher

They are the safety devices installed in all places irrespective of the surrounding to put off fires. It is basically a pressurized cylindrical vessel with a chemical agent which is expelled/discharged to extinguish fire. They are basically of two types: stored pressure and cartridge operator. In the former type, the expellant is stored in the same cylindrical chamber. The later type has a separate compartment for the expellant. It can be recharged when the expellant is over. According to handling they are further divided as: handheld extinguisher, cart mounted extinguisher. The ultimate aim of installing an extinguisher is that it has to be easily accessible (Fig. 5).

Fig. 5. Fire extinguisher

Small fires are said under different classes on the basis of occurrence (Figs. 6 and 7).

1. Class A: by wood and paper.
2. Class B: inflammable liquids (thinners, cooking oil)
3. Class C: electrical appliances.
4. Class D: reactive metals such as sodium and magnesium.

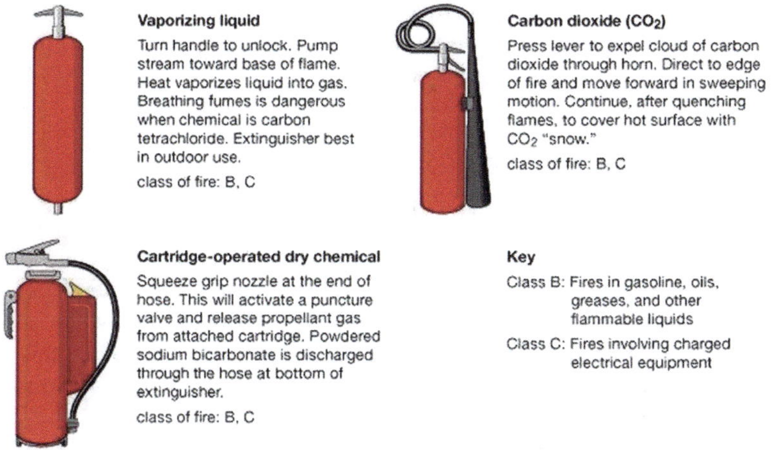

Fig. 6. Class B and Class C fire extinguishers and the method of use

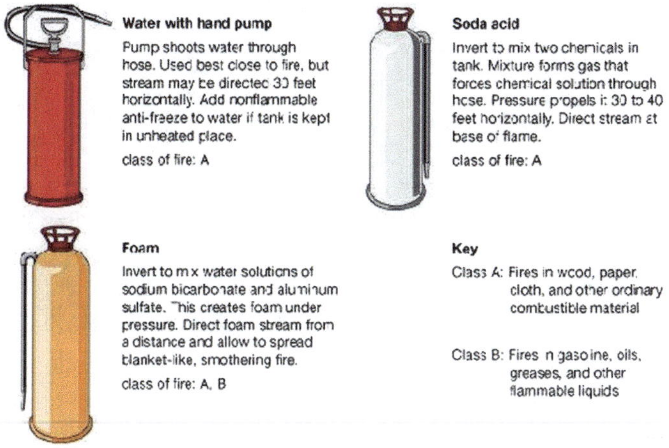

Fig. 7. Class B and Class A fire extinguishers and the method of use

4.3 Electromagnet

When a magnetic field is produced by electric current, then it is called as electromagnet. The magnet losses its property when the current is off. The strength of the magnetic field is controlled by electric current. But it falls back as it needs a constant power supply unlike permanent magnet.

4.4 Electromagnet Attracting Mechanism

Here the electromagnet is used to pull the handle of the extinguisher. The piece of the metal is attached to the handle for the magnet to attract. When source is supplied to the electromagnet the magneto motive force is produced by the magnet. Thus the flux is induced with which the handle is attracted. Thus the handle is pulled and the fire is put off.

5 Results and Discussion

The image given by the thermal camera has different colours which indicate the range of temperature. The red colour indicates the highest temperature, whereas blue colour indicates the lowest temperature. And green colour indicates the moderate temperature.

The thermal imaging was taken in an AC room. So obviously the surrounding temperature is less than that of human body. Hence the faces of people in this image are read and surrounding is blue. It also indicates the number of people in the room along with their body temperature (Figs. 8 and 9).

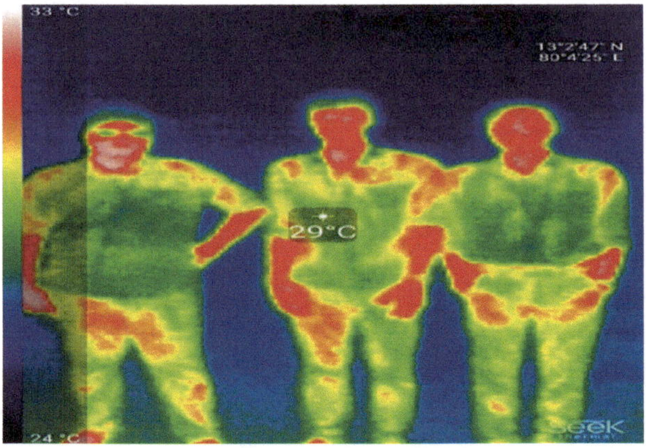

Fig. 8. Thermal camera output

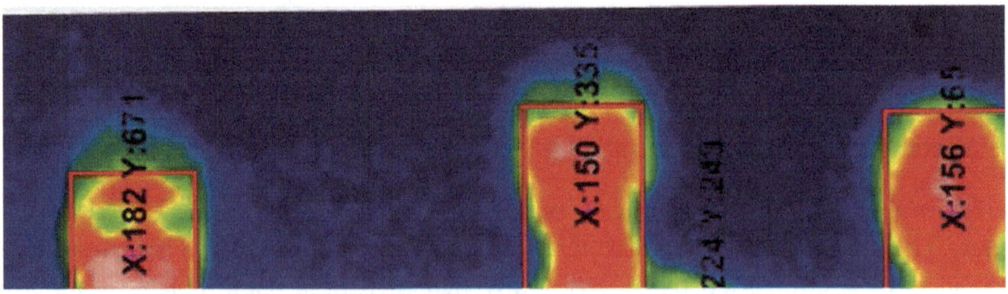

Fig. 9. Output image

The thermal image from the camera is sent to the controller for processing. The controller is programmed using the MATLAB software to detect the number of people in the room. In presence of human, the controller is programmed to maintain the fire extinguisher under close position but on absence of human, the controller uses electromagnet to automatically open the fire extinguisher to cease the fire.

6 Conclusion

The aim of the project is to provide a clearer image on the status of people when they are stuck in a fire mishap and help people coordinate accordingly in order to save maximum life possible. The thermal image provides enough details to make out humans from surrounding fire and is immune to smoke hence providing with much better vision than the standard camera and image processing has evolved to evaluate these types of image and will only improve further.

7 Future Enhancement

This project can be further expanded by incorporating an alternate source for the thermal imaging camera apart from the mains. In case of fire breakout the controller can be programmed to alternate the supply to the thermal camera such that the effect to current leak can be avoided. Similarly the controller can also be programmed in such a way that the sprinkler water can be increased in presence of human.

References

1. Takahashi, H., Kitazono, Y., Hanada, M.: Improvement of automatic fire extinguisher system for residential use. In: International Conference on Informatics, Electronics & Vision (ICIEV) (2015)
2. Jun, Q., Daiwei, G., Xishi, W.: The auto-fire-detection and auto-putting-out system. In: Proceedings of the 3rd World Congress on Intelligent Control and Automation, vol. 5, pp. 3708–3712 (2000)
3. Yorozu, Y., Hirano, M., Oka, K., Tagawa, Y.: Automated vision system for rapid fire onset detection. IEEE Transl. J. Magn. Jpn. **2**, 740–741 (2017)
4. Setjo, C.H., Achmad, B., Faridah: Thermal image human detection using Haar-cascade classifier. In: 2017 7th International Annual Engineering Seminar (InAES), pp. 1–6
5. Eltom, R.H., Hamood, E.A., Mohammed, A.A., Osman, A.A.: Early warning firefighting system using Internet of Things. In: 2018 International Conference on Computer, Control, Electrical, and Electronics Engineering (ICCCEEE), pp. 1–7 (2018)
6. http://newsphonereview.xyz/thermal-camera-color-scale/

3D Modelling and Radiofrequency Ablation of Breast Tumor Using MRI Images

S. Nirmala Devi[1(✉)], V. Gowri Sree[2], S. Poompavai[2], and A. Kaviya Priyaa[2]

[1] Department of ECE, College of Engineering, Anna University, Chennai, India
nirmala_devi@annauniv.edu
[2] Division of HVE, Department of EEE, College of Engineering, Anna University, Chennai, India
gowri06@yahoo.com, poompavai.hve@gmail.com, kaviyashankar1994@gmail.com

Abstract. The purpose of the work is to develop patient specific treatment analysis for tumor removal procedure using radio frequency ablation. The proposed method increases the efficiency of treatment protocol for patient specific models. Breast cancer is the most common cancer in Indian women. Thermal ablation procedure is to overheat the tissue cells and kills the cancerous tumor using probe. Ablation of tumor is a difficult procedure in case of placement of probe and killing cancerous tissue without much damage to healthy tissues. Directional removal of tumor is essential to avoid damage in surrounding tissue.

This work aims to contribute 3 steps (i) segmentation (ii) building 3D model (iii) Analysis and measurement. Segmentation of tumor is achieved by using MIMICS software. 3D modeling and simulation of thermal ablation treatment procedure using COMSOL MULTIPHYSICS was developed, necrosis of tissue and temperature at various points and position of probe have been evaluated. Temperature variation is analyzed to view the necrosis covering area and to plan for thermal ablation procedure for removal of breast tumor. Various measurement parameters of 3D tumor have been identified for further diagnosis.

Keywords: Radiofrequency ablation · Necrosis · Trocar · RITA

1 Introduction

India continues to have low survival rate for breast cancer with only 66% women diagnosed with disease between 2010 and 2014. RFA utilizes local thermal energy to induce coagulative necrosis, which limits the size of tumor eligible for ablation [9, 10]. Recent developments in ablative techniques are being applied to patient's inoperable and small tumors in lung, liver, breast etc. [19]. In radiofrequency techniques temperature of the tumor tissue is increased more than 50 °C. The energy at the exposed tip causes ionic agitation and frictional heat, which cooks the tumor and leads to cell death and coagulation necrosis, if hot enough. This is gradually replaced by fibrosis and scar tissue [8]. Human anatomical model is diagnosed using MRI imaging technique and

D. J. Hemanth et al. (Eds.): COMET 2019, LNDECT 35, pp. 288–304, 2020.
https://doi.org/10.1007/978-3-030-32150-5_31

incorporated into MIMICS to create 3D model of tumor for further analysis. Segmentation of tumor is done using thresholding to get clear view of density distribution of tumor. Ablation technique is analyzed using COMSOL MULTIPHYSICS using Bio heat transfer module [4, 12]. The RITA electrode is modeled to insert into the tumor that requires for the ablation process. A new model of curved cathode is proposed for directional removal of tumor. This is to control the direction of heating to kill only tumor cells [11] without causing much damage to the healthy tissues. This helps the oncologists to plan precise treatment for the ablation procedure.

2 Materials and Methods

The patients MRI DICOM of Breast images were loaded into MIMICS software for processing and building 3D model of tumor [13]. A stack of images is loaded and positioning of orientation is adjusted to get the clear view of tumor in slice images. Segmentation of tumor is done by using thresholding to determine the density distribution of soft tissue and tumor [14]. Soft tissue regions separated from the tumor by region growing segmentation. Tumor region has density values from 650 to 1000. Region growing segmentation is obtained for detecting the entire tumor part from all the slices. 3D building of segmented tumor is done using MIMICS software. 3D model is build and exported as. STL format to import into FEA solver software using COMSOL.

In this study volumetric meshing of model is done using 3-matic software and surface processing of the model like smoothing and contour shaping have been developed. A curved RITA (Radio frequency interstitial tissue ablation) electrode is modeled using COMSOL multiphysics having radius of 0.5 mm [15, 16]. Curved electrode with probe is inserted inside the build 3D tumor model. Material properties for entire geometry are shown in Table 1.

Table 1. Material Properties

Name	Breast tissue	Breast tumor	Electrode	Trocar base	Trocar tip
Density (Kg/m^3)	1060	1050	6450	70	7900
Specific heat capacity (J/Kg K)	2770	3770	840	1045	132
Thermal conductivity (W/m K)	0.499	0.48	18	0.026	71
Electrical conductivity (S/m)	0.28	0.71	10^8	10^{-5}	0.026

3 RFA Simulation

FEM (Finite Element Mesh) analysis has been done using Bioheat transfer module and electrical heating module in COMSOL multiphysics [5]. Heat distribution carried out by using penne's Eq. (1)

$$\rho c \frac{\partial T}{\partial t} = \nabla \cdot (K \nabla T) - \rho b \omega b c b (T - Tb) + Qm + J \cdot E \tag{1}$$

Where ρ, indicates tissue density in (kg/m^3), C indicates the specific heat capacity (J/kg K−1), K indicates thermal conductivity (W/m K) of the tissue. T indicates temperature (K), ρb indicates density of blood (Kg/m^3) and cb specific heat of the blood, (J/kg K−1), respectively. Qm is the metabolic heat production per volume (W/m^3). J stands for the current density (A/m) and E for the electric field intensity (V/m).

Since higher current density are focused in region of interest displacement currents are negligible and heat distributed is given by Eq. (2)

$$Qext = J \cdot E \tag{2}$$

Where J is the current density (A/m^2), E is the electric field (V/m). The values of these two vectors are derived from solving Laplace Eq. (3)

$$\nabla(\sigma \nabla V) = 0 \tag{3}$$

σ is the electrical conductivity of tissue(S/m), V is the applied voltage.

Entire simulation is carried out using time dependent solver in COMSOL. RFA simulation is done for different time and temperature variations for the build 3D model [17]. Simulated results include heat distribution in tumor tissue at different points and fraction of necrosis occurs at that point.

4 Results and Discussions

Various case studies considered for this entire procedure. Thresholding technique for segmentation is initially carried out to find the density of soft tissue and tumor in three planes (axial, coronal, and sagittal) as shown in Fig. 1. Region growing segmentation has been performed for the density values of tumor as shown in Fig. 2. Soft tissue regions separated from the tumor by region growing segmentation [17]. 3D building of segmented tumor is achieved using COMSOL as shown in Fig. 4 for case 1.

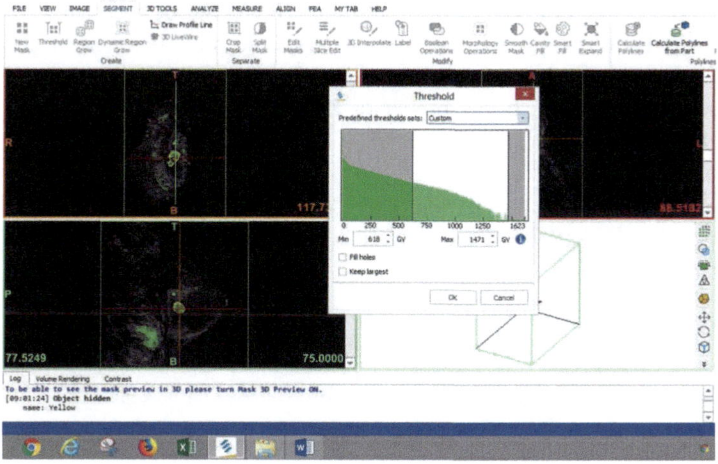

Fig. 1. Case1 thresholding in 3planes (axial, coronal, sagittal).

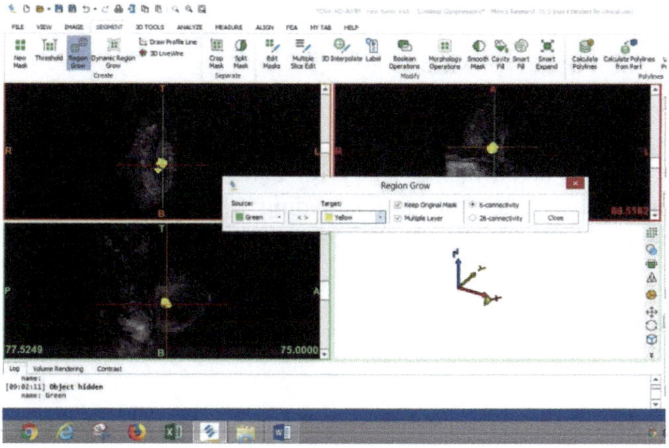

Fig. 2. Region growing in 3 planes (axis, coronal, sagittal)

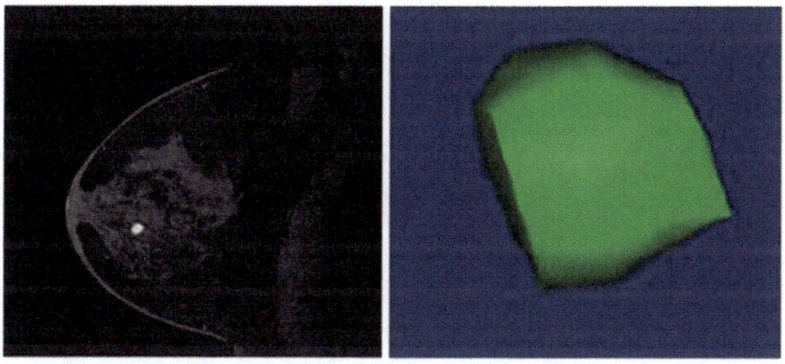

(a) Coronal view of tumor (b) 3D model

Fig. 3. Case 1 Segmented tumor from other tissue

In Fig. 3(b) 3D model is built from all three views. Figure 4(a) shows the meshing of tumor for patient 1 using 3-matic software total mesh elements are 12543. Figure 4 (b) shows the tumor outline model inserted with the trocar electrode and base before radiofrequency simulation. Initial power of 22 V for 25 s.

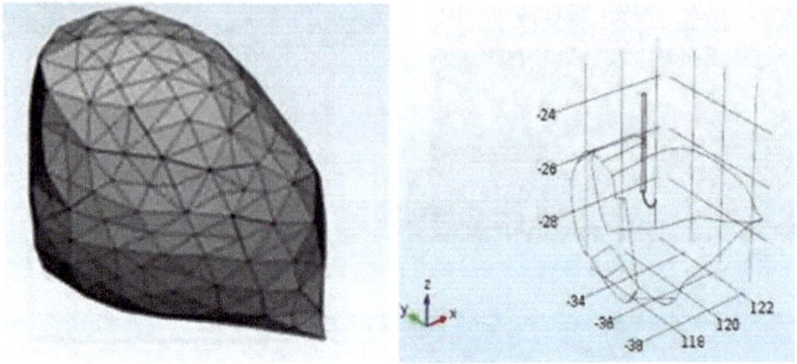

(a) Finite element mesh (b) Tumor with trocar

Fig. 4. Modelling of tumor in COMSOL

4.1 Analysis of Various Positions

A. Positioning at First Point (118, −36.5, −27) Three different positioning of probe at different locations are taken in this study for analysis of tumor region. Temperature distribution and necrosis for the corresponding position of probe is analyzed and compared to find the maximum necrosis area are shown in Figs. 5, 6, 7 and 8.

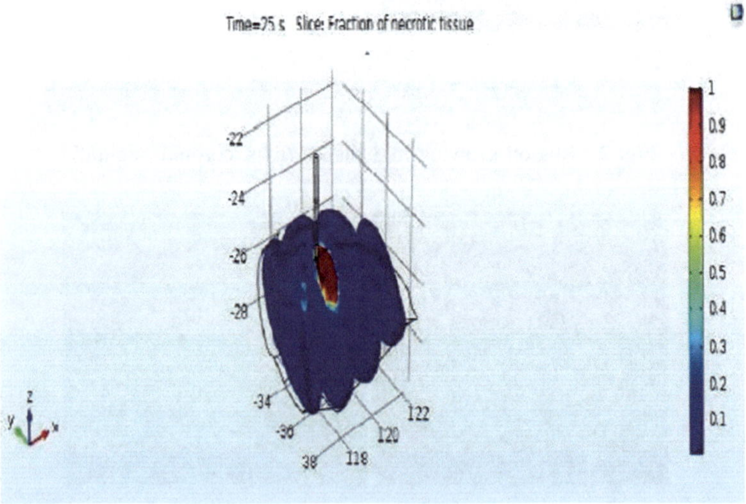

Fig. 5. Necrosis view for trocar position at the edge of the tumor (118 −36.5 −26.5, 118.3 −36 −26.3, 117.21 −36.5 −26.6)

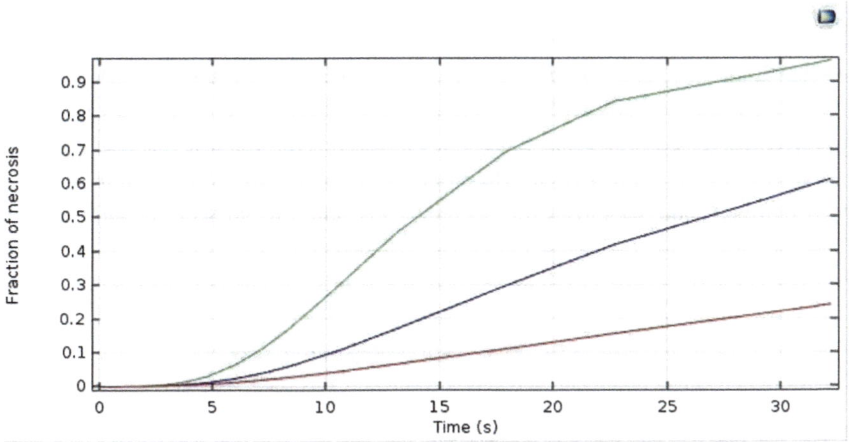

Fig. 6. Necrosis plot at 3 points (118 −36.5 −26.5, 118.3 −36 −26.3, 117.21 −36.5 −26.6)

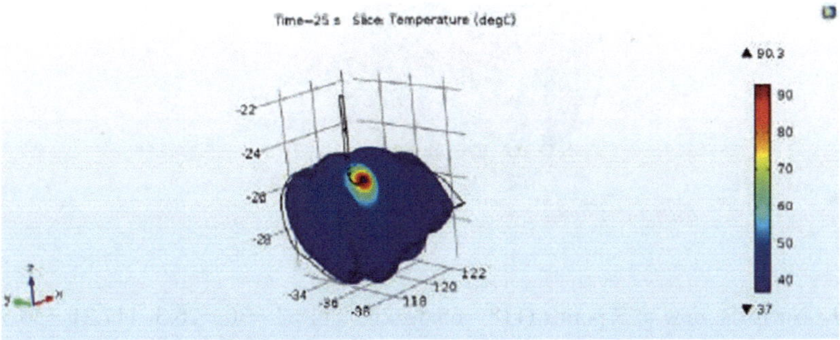

Fig. 7. Temperature distribution at 3 points (118 −36.5 −26.5, 118.3 −36 −26.3, 117.21 −36.5 −26.6)

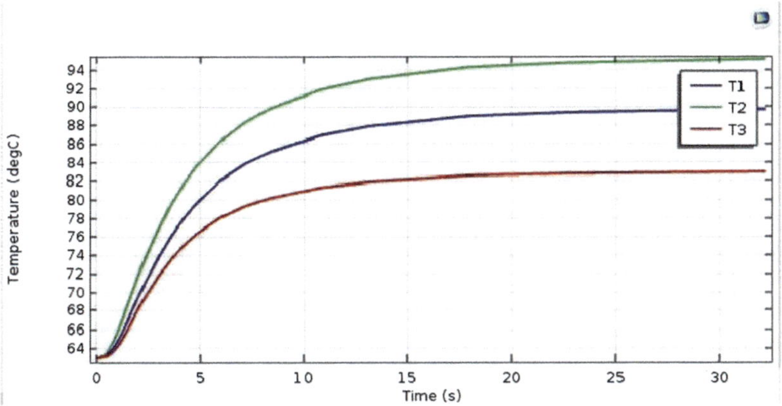

Fig. 8. Temperature distribution plot at three different points (118 −36.5 −26.5, 118.3 −36 −26.3, 117.21 −36.5 −26.6)

From Figs. 7 and 8 it is confident that the temperature distribution reaches 95 °C for three points taken inside the tumor and necrosis. By seeing the temperature distribution, necrosis happens at three different points inside the tumor. Figures 5 and 6 shows that the necrosis happens to maximum at 25 s which is an acceptable range [6].

B. Positioning at Second Point (118.3, −37, −28). The trocar is inserted deep into the tumor to analyze the temperature distribution and necrosis in the points (118.3, −37, −28). Figures. 9 and 10 shows the necrosis view and plot at second point insertion of trocar.

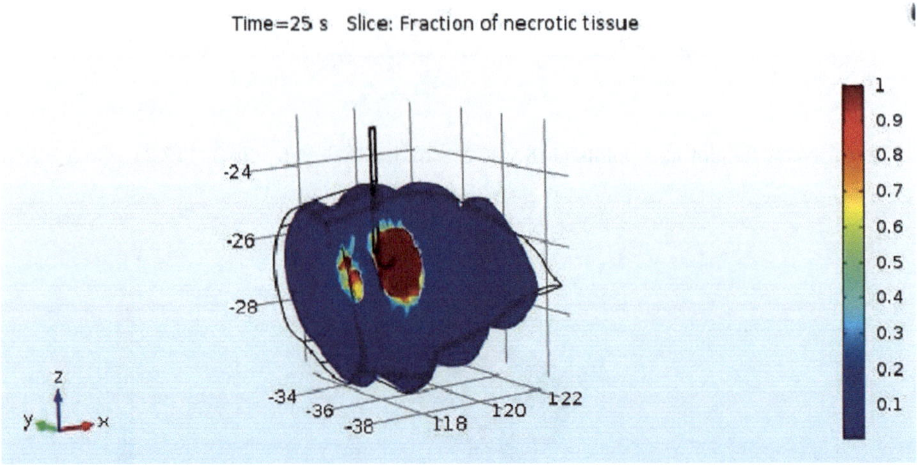

Fig. 9. Necrosis view at 3 points (118 −36.5 −26.5, 118.3 −36 −26.3, 117.21 −36.5 −26.6)

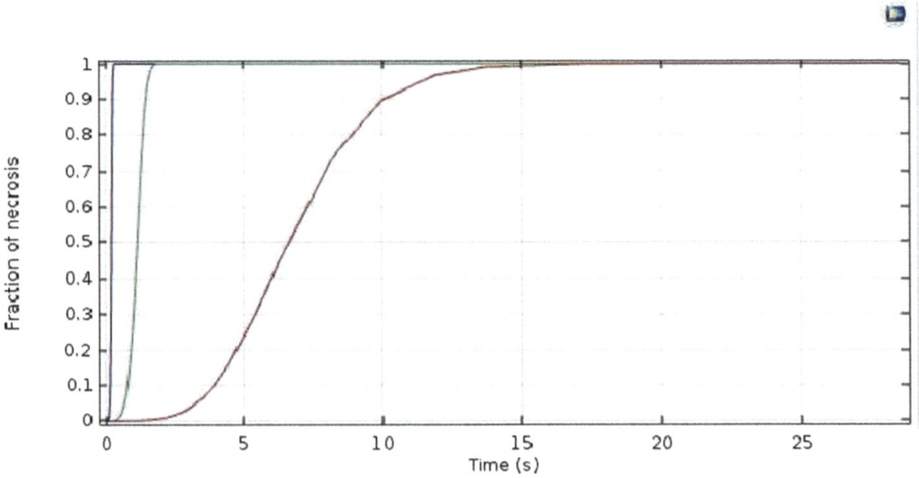

Fig. 10. Necrosis plot for 3 points (118 −36.5 −26.5, 118.3 −36 −26.3, 117.21 −36.5 −26.6)

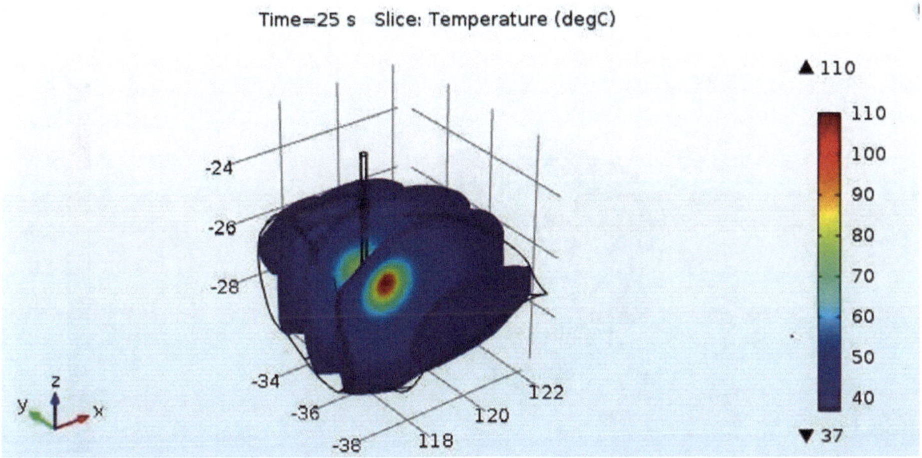

Fig. 11. Temperature distribution at 3 points (118 −36.5 −26.5, 118.3 −36 −26.3, 117.21 −36.5 −26.6)

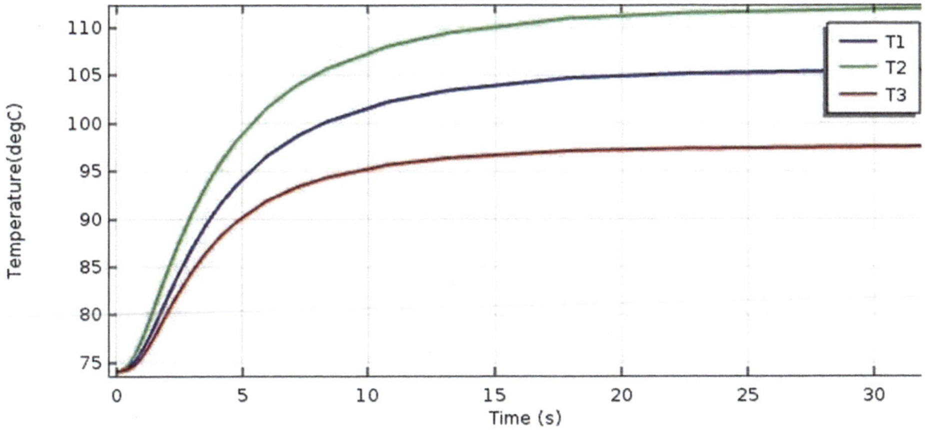

Fig. 12. Temperature plot for 3 points (118 −36.5 −26.5, 118.3 −36 −26.3, 117.21 −36.5 −26.6)

Figures 11 and 12 it is confident that the temperature distribution reaches 110 °C for three points taken inside the tumor and necrosis [19]. By seeing the temperature distribution, necrosis happens at three different points inside the tumor. Figures 9 and 10 shows that the necrosis happens to maximum at 25 s which is an acceptable range [6].

C. Positioning at Third Point (118.21, −38, −29). Final position of deeper inside the tumor is analyzed.

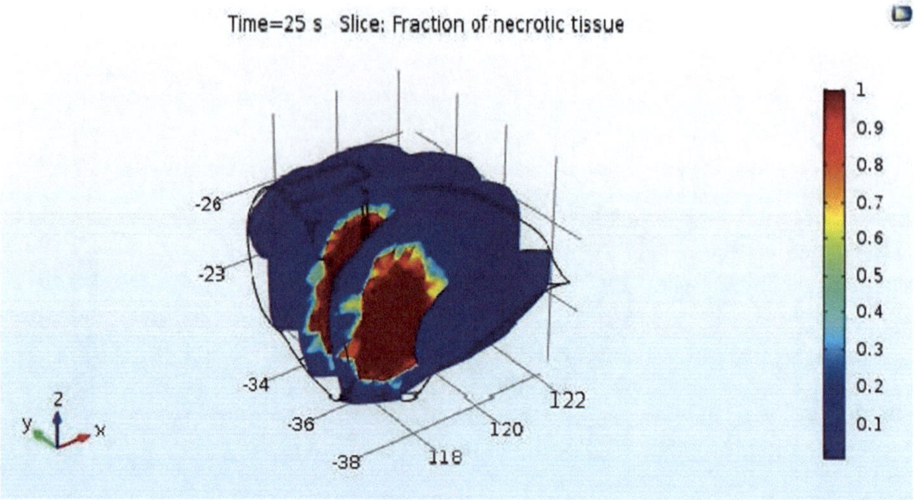

Fig. 13. Necrosis for final point at 3 points (118 −36.5 −26.5, 118.3 −36 −26.3, 117.21 −36.5 −26.6)

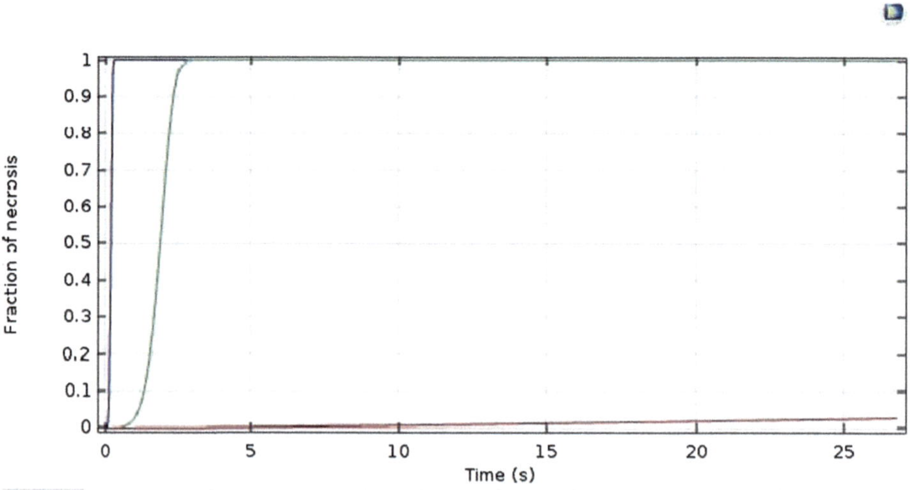

Fig. 14. Necrosis plot for final point at (118 −36.5 −26.5, 118.3 −36 −26.3, 117.21 −36.5 −26.6)

Figures 15 and 16 it is confident that the temperature distribution reaches 120 °C for three points taken inside the tumor and necrosis by seeing the temperature distribution, necrosis happens at three different points inside the tumor. Figures 13 and 14 shows that the necrosis happens to maximum at 25 s which is an acceptable range [6, 7].

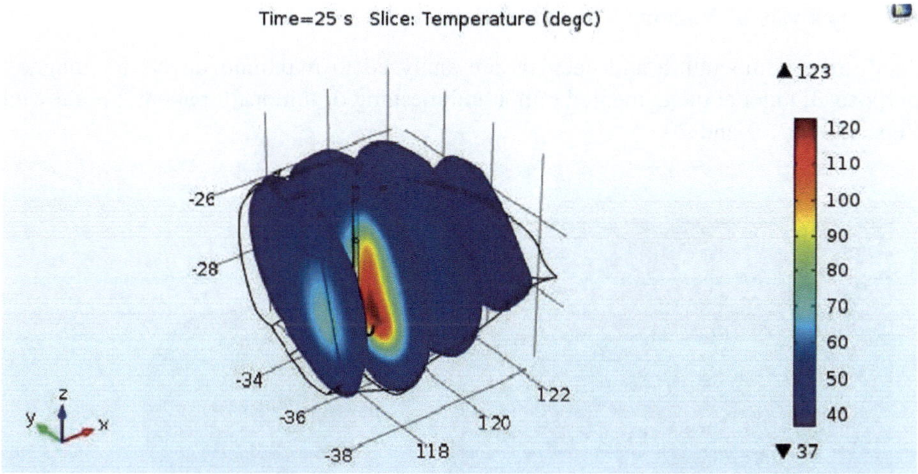

Fig. 15. Temperature distribution at 3 points (118 −36.5 −26.5, 118.3 −36 −26.3, 117.21 −36.5 −26.6)

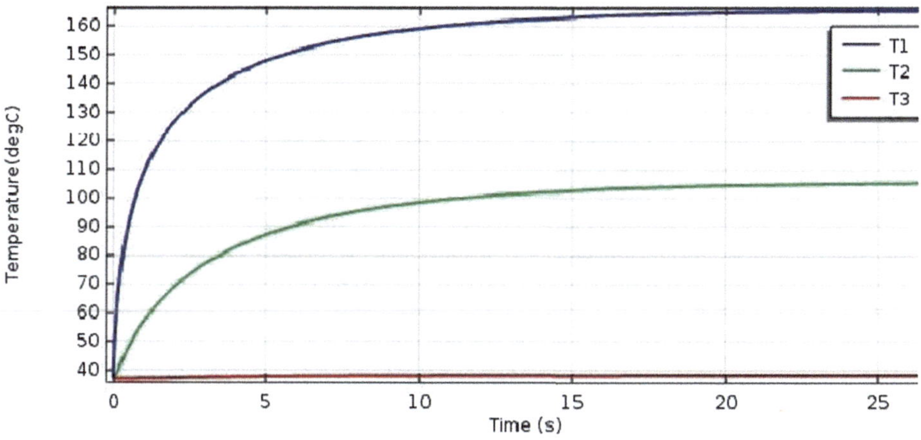

Fig. 16. Temperature plot for 3 points (118 −36.5 −26.5,118.3 −36 −26.3, 117.21 −36.5 −26.6)

It is clear from the above images that the trocar inserted deeper into the tissue is very efficient for the tumor destruction since larger area is covered under necrosis and temperature distribution is still larger for larger area to be killed.

4.2 Analysis of Various Time Duration

Analysis of temperature and necrosis are analyzed to maximum levels for diagnosis purpose. 3D model of segmented tumor and meshing of tumor for patient 2 is shown in Figs. 17, 18, 19 and 20.

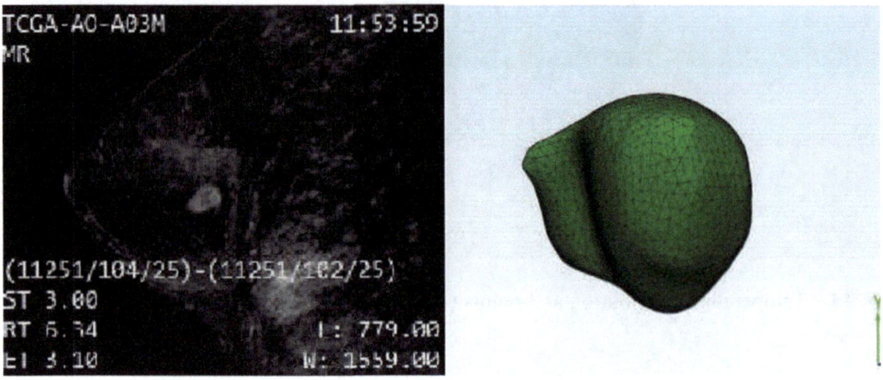

Fig. 17. Tumor model and mesh for patient 2

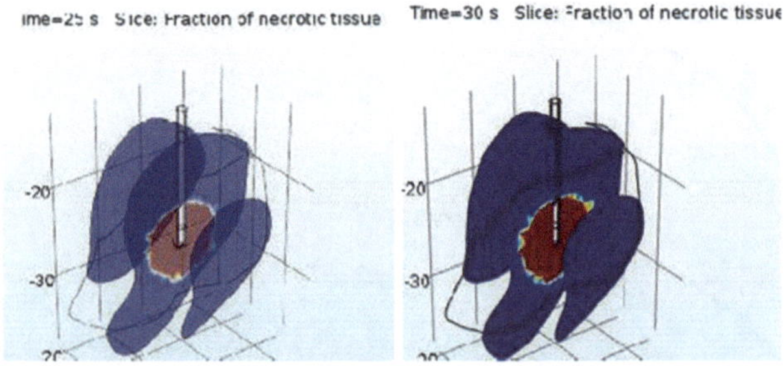

(a) Necrosis for 25 secs (b) Necrosis for 30secs

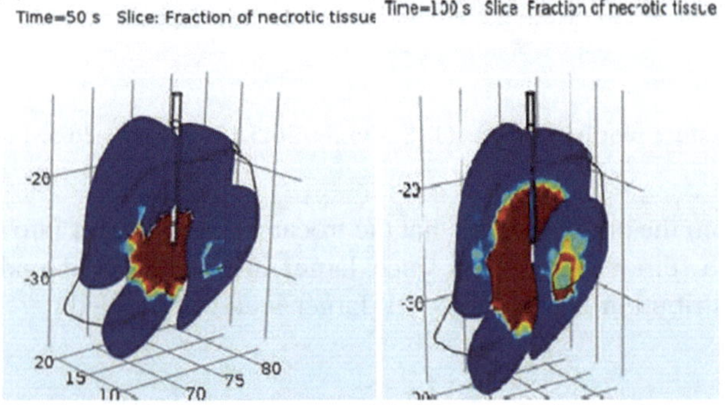

(c) Necrosis for 50 secs (d) Necrosis for 100 secs

Fig. 18. Necrosis view for (a) 25 s, (b) 30 s, (c) 50 s, (d)100 s

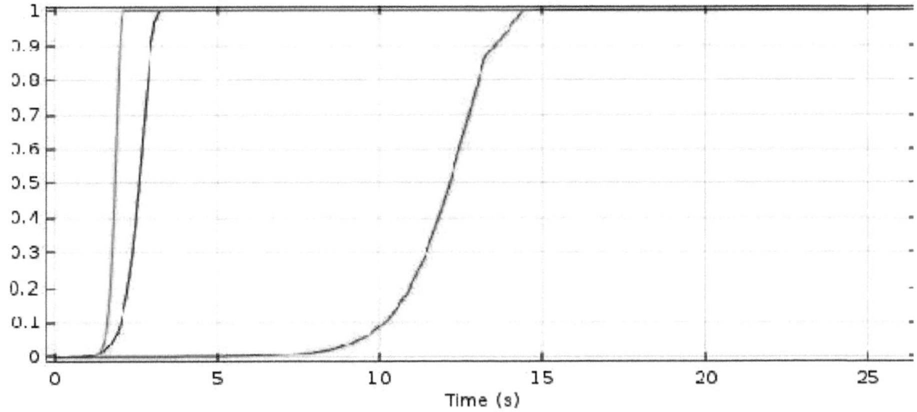

(a) Necrosis plot for25secs

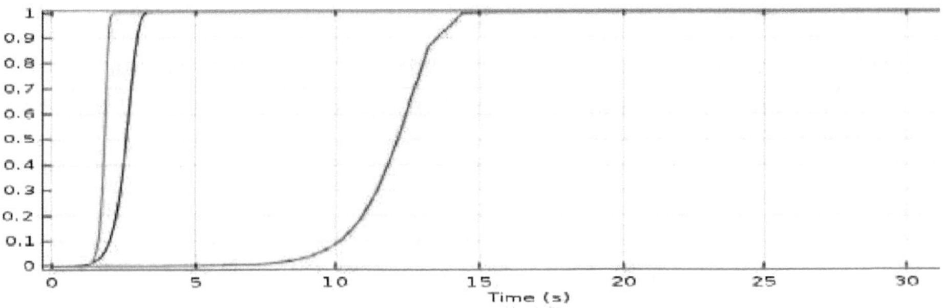

(b) Necrosis plot for30secs

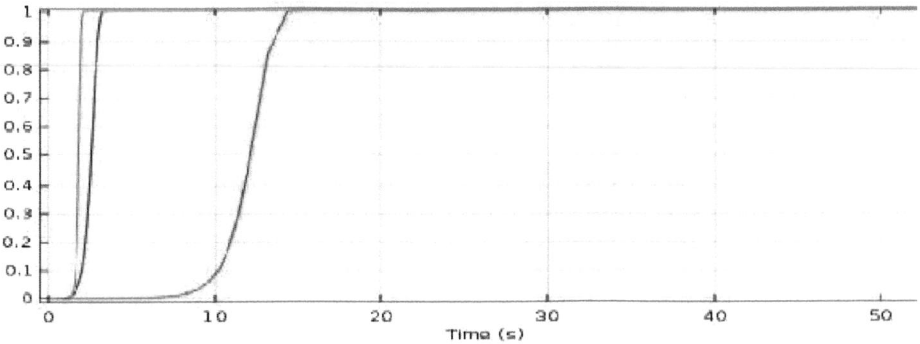

(c) Necrosis plot for50secs

Fig. 19. Necrosis plot for (a) 25 s, (b) 30 s (c) 50 s (d) 100 s for 3 points (118 −36.5 −26.5, 118.3 −36 −26.3, 117.21 −36.5 −26.6)

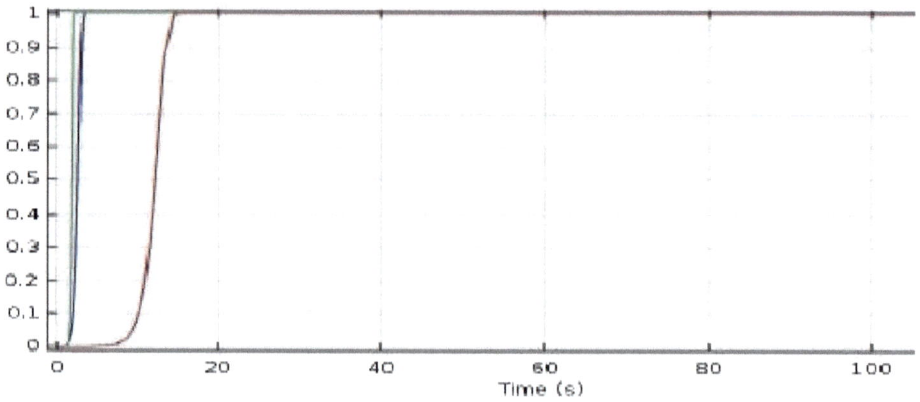

(d) Necrosis plot for100secs

Fig. 19. (*continued*)

This above image shows higher the time duration more number of cells dies.

From this plot necrosis happens sharper to the nearer cells within few seconds than the cells farther to the electrode.

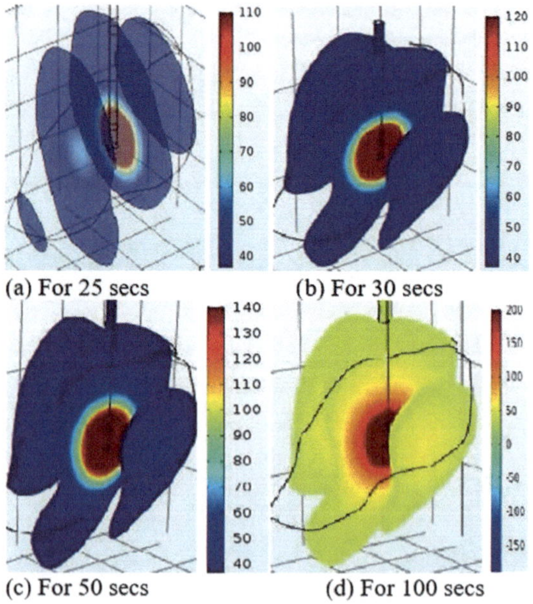

Fig. 20. Temperature distribution for (a) 25 s, (b) 30 s, (c) 50, (d) 100 s

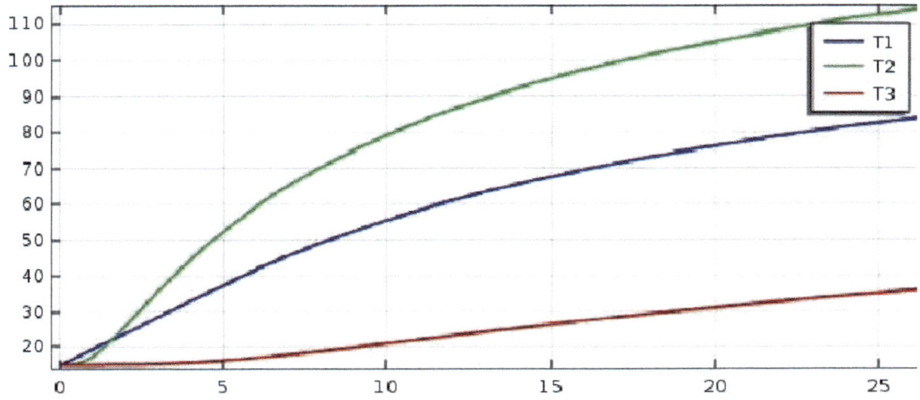

(a) Temperature plot for 25secs

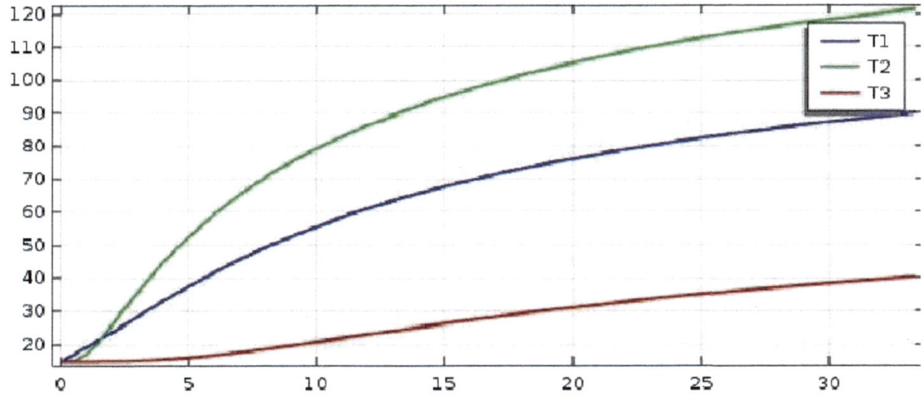

(b) Temperature plot for 30secs

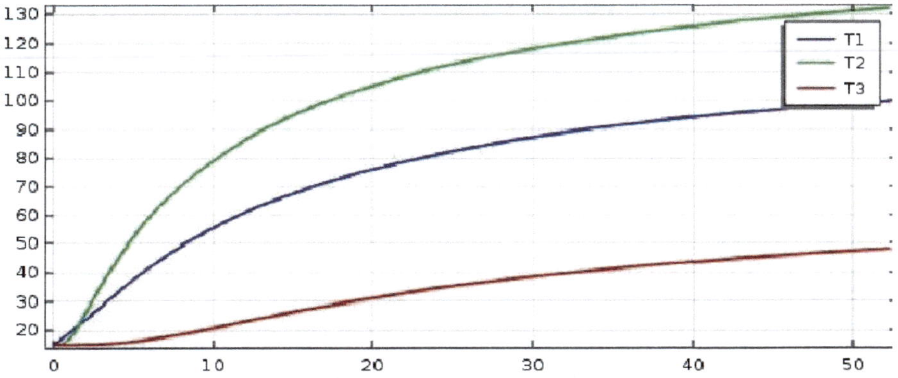

(c) Temperature plot for 50secs

Fig. 21. Temperature plot for all time periods corresponding to temperature distribution views in 3 points (118 −36.5 −26.5, 118.3 −36 −26.3, 117.21 −36.5 −26.6)

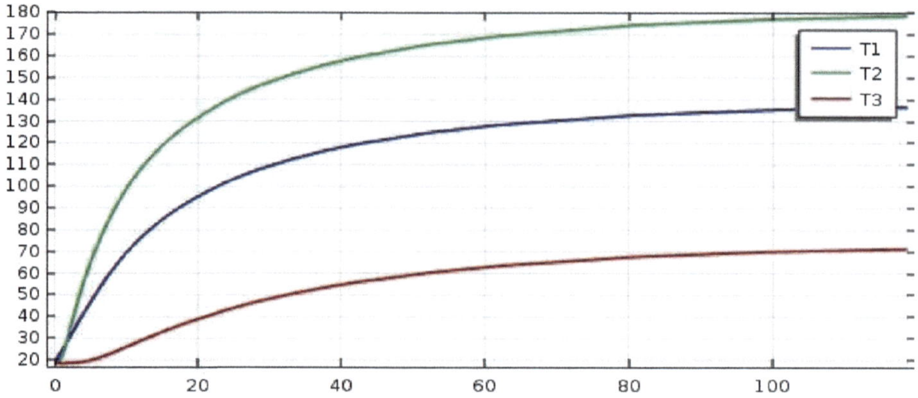

(d) Temperature plot for 100secs

Fig. 21. (*continued*)

From the necrosis region it is observed that the trocar is inserted deeper into the tissue at various points. The tumor destruction area spreaded under necrosis in various slices are shown in Fig. 21. When temperature distribution is maximum of 200 °C necrosis region also indicates that the tumor region is destroyed to maximum.

4.3 Measurement Parameters

The build 3D model for various cases have been listed in the Table 2 after analysis. Various parameters like area, volume, distance of tumor from the tip of the breasts studied for precise positioning of the electrode into the breast.

Table 2. Measurements parameters

Parameters	Case 1
Area	122 mm^3
Volume	116 mm^3
Distance from the tip of the breast	32 mm
Pixel value (1 mm = 3.7795 pixels)	461.0

5 Conclusion

The points near to the electrode experience high thermal energy at faster rate than the points farther than electrode. Absorption of the temperature by tissue depends upon the thermal properties of the tissue and tumor. The point near to electrode attains maximum temperature within 5 sec to 100 sec and necrosis happens at faster rate. Since higher temperature distribution is also accepted in case of tumor tissues its characteristics differ from normal tissues. Even higher temperature is also acceptable for higher energy sources to destroy the cancer cells.

References

1. Wang, Z., Aarya, I., Gueorguiev, M., Liu, D., Luo, H., Manfredi, L., Wang, L., McLean, D., Coleman, S., Brown, S., Cuschieri, A.: Image-based 3D modeling and validation of radiofrequency interstitial tumor ablation using a tissue-mimicking breast phantom. Philos. Trans. R. Soc. Lond. **A247**, 529–551 (2012)
2. Peter, S.: Comparative study on 3D modelling of breast cancer using NirFdot. In: Except from the Proceedings of COMSOL Conference in Bangalore (2014)
3. Hopp, T., Stromboni, A., Duric, N., Ruiter, N.V.: Evaluation of breast tissue characterization by ultrasound computer tomography using a 2D/3D imageregistration with mammograms, Germany (2013). 978-1-4673-5686-2/13/$31.00©2013 IEEE
4. Chakraborty, J., Mukhopadhyay, S., Singla, V., Khandelwal, N., Rangayyan, R.M.: Detection of masses in mammograms using region growing controlled by multilevel thresholding. IEEE (2012)
5. Mellal, I., Kengne, E., El Guemhoui, K., Lakshssassi, A.: 3D modeling using the finite element method for directional removal of a cancerous tumor. J. Biomed. Sci. (2016). https://doi.org/10.4172/2254-609X.100042
6. Singh, S., Bhowmik, A., Repaka, R.: Thermal analysis of induced damage to the healthy cell during RFA of breast tumor. Elsevier (2016). www.elsevier.com
7. Singh, S., Repaka, R.: Effects of target temperature on ablation volume during temperature controlled RFA of breast tumor. In: Research Gate Conference Paper (2016)
8. Jeremic, A., Khosrowshahli, E.: Bayesian estimation of tumors in breasts using microwave imaging. In: Except From the Proceedings of the 2012 COMSOL Conference in Boston (2012)
9. Sahakyan, A., Sarukhanyan, H.: Segmentation of the breast region in digital mammograms and detection of masses. Int. J. Adv. Comput. Sci. Appl. **3**(2) (2012)
10. Sharma, J., Rajeswari, R.P.: Identification of pre- processing technique for enhancement of mammogram images. In: International Conference on Medical Imaging, m-Health and Emerging Communication Systems (MedCom). IEEE (2014). 978-1-4799-5097-3/14/ $31.00©2014
11. Mathuphhot, K., Sanpanich, A., Phasukkit, P., Tungjitkusolmun, S., Pintavirooj, C.: Finite element analysis approach for investigation of breast cancer detection using microwave radiation. In: Bioinformatics and Biomedical Technology IPCBEE, vol. 29. IACSIT Press, Singapore (2012)
12. Smaoui, N., Hlima, A.: Designing a new approach for the segmentation of the cancerous breast mass. In: 13-th International Multi-conference on Systems, Signals and Devices (2016). 978-1-5090-1291-6/16-IEEE
13. Lingle, W., Erickson, B.J., Zuley, M.L., Jarosz, R., Bonaccio, E., Filippini, J., Gruszauskas, N.: Radiology data from the cancer genome atlas breast invasive carcinoma (TCGA-BRCA) collection. Cancer Imaging Arch. (2016). https://doi.org/10.7937/K9/TCIA.2016.AB2NAZRP
14. Razman, N.R., Mahmud, W.M.H.W., Shaharuddin, N.A.: Filtering technique in ultrasound for kidney, liver and pancreas image using matlab. In: IEEE Student Conference on Research and Development (SCOReD) (2015). 978-1-4673-9572-4/15/$31.00©2015 IEEE

15. Drizdal, T., Vrba, M., Cifra, M., Togni, P., Vrba, J.: Feasibility study of superficial hyperthermia treatment planning using COMSOL multiphysics (2008). 978-1-4244-2138-1/08/$25.00-2008 IEEE
16. Hopp, T., Stromboni, A., Duric, N., Ruiter, N.V.: Evaluation of breast tissue characterization by ultrasound computer tomography using a 2D/3D image registration with mammograms. In: Joint UFFC, EFTF and PFM Symposium (2013). 978-1-4673-5686-2/13 ©2013 IEEE
17. Jaffery, Z.A., Zaheeruddin, Singh, L.: Performance analysis of image segmentation methods for the detection of masses in mammograms. Int. J. Comput. Appl. (0975–8887) **82**(2) (2013)
18. Wang, Z., Aarya, I., Gueorguieva, M., Liu, D., Luo, H., Manfredi, L., Wang, L., McLean, D., Coleman, S., Brown, S., Cuschieri, A.: Image-based 3D modeling and validation of radiofrequency interstitial tumor ablation using a tissue-mimicking breast phantom (2012). 10.1007/s11548-012-0769-3
19. Singh, S., Bhowmik, A., Repaka, R.: Thermal analysis of induced damage to the healthy cell during RFA of breast tumor. J. Therm. Biol. **58**, 80–90 (2016)

Waste Management System

A. Ancillamercy(✉)

Panimalar Institute of Technology, Chennai 600123, India
jesmer27499@gmail.com

Abstract. In the hike populace of our country, people expend a precise life span due to environmental changes. Both adult as well as infant troubled by ambience climate. The exposed particulate matter produce death among infants causing sudden infant death syndrome. However according to WHO 25% of people whose aging is above 60 is affected by severe disablility. Conforming to again with WHO the average longevity of a man is diminished to 81.2. Out of numerous reasons towards climate change complication one is inappropriate conservation of junk. Improper maintenance of waste Provoked hazardous effects to human. So effective way of garbage disposal is proposed.

Keywords: Ubidots · Sensor · Waste management · Fertilizer

1 Introduction

The initiated system will keep away overloading of dustbin. It will give the real time information about the level of the dustbin. It will send the message immediately when the dustbin is full. Deployment of dustbin based on actual needs. Cost of this system is minimum. The resources are available easily. Make better environment by reducing unpleasant odour resulting in clean city. It has effective usage of dustbins. It will also reduce the wastage of time and energy for truck drivers. It will also indicate the availability of toxic substance in the bin. In our recycle of waste is done.

2 Related Works

Srivastava, Nema [1] Forecasting the solid waste composition of Delhi, India using Fuzzy regression is discussed below. It is expected that waste content has increased from 2.74% to 3.55%. Percent of paper and food waste is said to be reduce between 36.37% to 27.55%. Also expected that metal and glass rise doubly and triple in coming future. This system helps in separation of reuse-recycle, treatment and disposal facilities.

Price, Smith [2] Recycling is still being used and it is used mostly in industrial sectors. The article address the lavish tyre. The structure put forward the effort taken by the state of California and the US Army Corps of Engineer for stimulation of tyre explaining the public currently.

Van Der Weil [3] The paper presented explains about reverse logistics concepts that incorporate drivers and barriers, product types and characteristics, process and recovery

D. J. Hemanth et al. (Eds.): COMET 2019, LNDECT 35, pp. 305–309, 2020.
https://doi.org/10.1007/978-3-030-32150-5_32

options, and actors. It is based on the conceptual framework of reverse logistics. It is used to develop a model of four reverse logistics concepts used in analyzing the waste management in industry. The paper is used to promote good practices for various avenues for practitioners to develop and employ dematerialized management strategy.

Roper [4] The main impartial of the paper to identify debris and waste management policies. Need has to be implemented from Hurricane Katrina. Investigation of the waste management ranking and life cycle management of material improved concept of reuse and recycle.

Al-Salem [5] Plastic solid waste promote challenges and opportunities to societies. It also gives awareness among people. Also increases Technological advances. Though primary and secondary recycling schemes are established, it is ended that many of the PSW tertiary and quaternary treatment schemes appear to be robust the additional investigation.

Raja Mamat, Mat Saman, Sharif, Simic, Abd Wahab [6] Ending of vehicles conserve the nature. Minimize the water, air, soil pollution. Develop economical benefits. The submitted multi-criteria decision analysis tool assist not only Malaysian ELV management system but also get vehicle recycling manager from other countries to take control over the tool.

Shahul Hamid, Bhatti, Anuar, Anuar, Mohan, Periathamby [7] the assessment explore the global abundance and distribution of microplastics in aquatic ecosystem. From this system it is crucial that actual ordinance are suggested and achieving in reduction of plastic in marine.

Wang, Zheng, Li [8] Microplastic is spread in all over in all the region of the world. The article focus on the sources, latest international, regional and national countermeasures to conflict marine junk.

Traven, Kegalj, Šebelja [9] This focus on making the recycling rate as 65% in 2030. According to the analysis the production of municipal solid waste increased for the passed two decades. Croatia has taken mechanical–biological treatment technology about the therapy of waste.

Qazi, Abushammala, Azam [10] this abstract implements the optimum waste management strategy to reduce generation of the garbage. Also implement the anaerobic digestion that minimize the measure of waste, greenhouse gas emissions, sustaining the costs of landfills.

3 Proposed System

The traditional method includes burning of the waste causes air pollution to great extent. By discarding it by burning will causes diseases. The residue of the waste should be processed well. All the major undertaking are done manually. To reduce the labor pool and make it digital we have used IoT Technology is used with cloud. The main aim is to overcome waste management problem based on providing intelligence to waste bins, using an IoT prototype with sensors, Node MCU and Ubidots cloud (Fig. 1).

ARCHITECTURE DIAGRAM

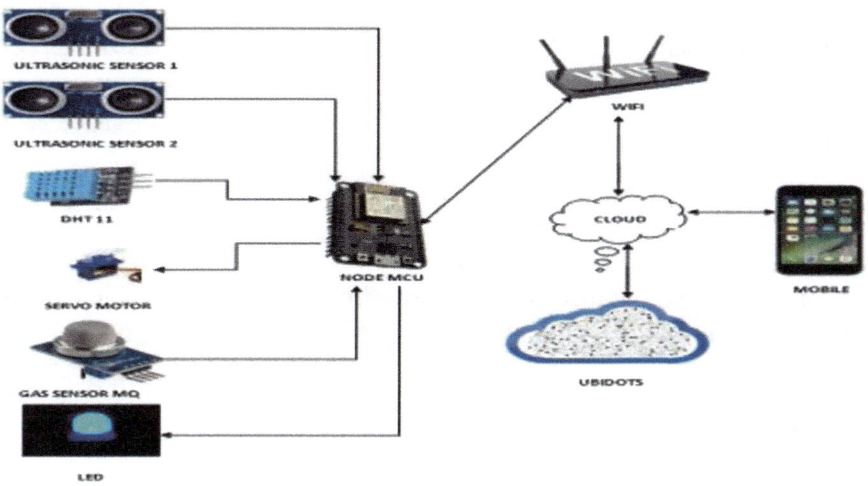

Fig. 1. Proposed system

The submitted subject matter separated into part

1. CHIP
2. SENSOR
3. MOTOR
4. IMPLEMENTATION.

3.1 CHIP

NodeMCU ESP8266
It is used in IOT platform. It behave as a host and execute the function of WIFI from another application processorESP8266.

3.2 SENSOR

Ultrasonic Sensor
Sensor used to detect space between object. Produce high frequency sound recognize the period of echo that returned back. It has two openings one for transmitting supersonic waves and another for receiving the ultrasonic waves.

Methane Gas Sensor
It is known as natural gas sensor. Mainly used for detecting natural gas. It is high sensitivity and response in a quicker time. Running principle is very simple. You need to do is power the heater coil with 5 V, add a load resistance, and connect the output to an ADC.

DHT11
It made up of calibrated digital signal. This calibrated digital signal results in Temperature and Humidity.
SPECIFICATION
Supply Voltage: +5 V
Temperature range: 0–50 °C error of ±2 °C
Humidity: 20–90% RH ±5% RH error
Interface: Digital.

3.3 MOTOR

Servo Motor
Servo motor is a linear or rotary actuator. It is used to control the position, velocity and acceleration of the lid used for closing the trashcan.

4 Implementation

4.1 Ubidots

Ubidots is used for implementation. It is a software for executing the code. It converts the data from sensor into information. Helpful in making commitment. We can interact with application easily.

4.2 Monitoring the Bin Level

Using ultrasonic wave the degree of waste is acknowledged. LED bulb specify the degree of waste and information from the sensor is moved to NODE MCU. Information is stored in cloud using WIFI. Space is identified in can and if can is plenty the servo motor is used to close the bin.

4.3 Notification of the Bin

When can is full message is sent to bin using cloud.

4.4 Monitoring the Main Storage

Gas sensor is present in main storage check the level of gas. Information from Gas sensor is sent to cloud. When gas is identified in a effective waste is fertilized and content is sent using cloud.

4.5 Humidity and Temperature Detection

The humidity and temperature need to be monitored because the waste may decompose. It may produce pungent smell coming from waste. DHT11 sensor verify the humidity and temperature and visualizing is performed by cloud.

5 Conclusion

We submit this brilliant waste collection system. The presented project is built on IOT sensing prototype. This is useful for measuring the waste level in can and send this data to cloud. Cloud perform storage and processing. Amount of waste can be computed using this information. Waste collected is used as a fertilizer later.

References

1. Srivastava, A.K., Nema, A.K.: Forecasting of solid waste composition using fuzzy regression approach: a case of Delhi. Int. J. Environ. Waste Manag. (IJEWM) **2**(1/2), 65–74 (2008)
2. Price, W., Smith, E.D.: Waste tire recycling: environmental benefits and commercial challenges. Int. J. Environ. Technol. Manag. (IJETM) **6**(3/4), 362–374 (2006)
3. Van Der Weil, A.: Waste management facility expansion planning using Simulation-Optimisation with Grey Programming and penalty functions. Int. J. Environ. Technol. Manag. (IJETM) **6**(3/4) (2006)
4. Roper, W.E.: Waste management policy revisions: lessons learned from the Katrina disaster. Int. J. Environ. Technol. Manag. (IJETM) **8**(2/3), 275–309 (2008)
5. Al-Salem, S.M.: A review on thermal and catalytic pyrolysis of plastic solid waste (PSW). J. Environ. Manag. **197** (2017)
6. Raja Mamat, T.N.A., Mat Saman, M.Z., Sharif, S., Simic, V., Abd Wahab, D.: Development of a performance evaluation tool for end-of-life vehicle management system implementation using the analytic hierarchy process. Waste Manag. Res. **36**(12), 1210–1222 (2018)
7. Shahul Hamid, F., Bhatti, M.S., Anuar, N., Mohan, P., Periathamby, A.: Worldwide distribution and abundance of microplastic: how dire is the situation? Waste Manag. Res. **36** (10), 873–897 (2018)
8. Wang, J., Zheng, L., Li, J.: A critical review on the sources and instruments of marine microplastics and prospects on the relevant management in China. Waste Manag. Res. **36** (10), 898–911 (2018)
9. Traven, L., Kegalj, I., Šebelja, I.: Management of municipal solid waste in Croatia: analysis of current practices with performance benchmarking against other European Union member states. Waste Manag. Res. **36**(8), 663–669 (2018)
10. Qazi, W.A., Abushammala, M.F.M., Azam, M.-H.: Multi-criteria decision analysis of waste-to-energy technologies for municipal solid waste management in Sultanate of Oman. Waste Manag. Res. **36**(10), 898911 (2018)

Advances in Control and Soft Computing

RETRACTED CHAPTER: Analysis of Cryptography Performance Measures Using Artificial Neural Networking

S. Prakashkumar[1(✉)], E. M. Murugan[2], R. Thiagarajan[1], N. Krishnaveni[3], and E. Babby[4]

[1] Department of Computer Science, St. Joseph's College (Arts and Science), Kovur, Chennai 600 028, India
{drpk40077, thiagarajan.nvr}@gmail.com

[2] Department of Visual Communication, St. Joseph's College (Arts and Science), Chennai 600 028, India
emm8016@gmail.com

[3] Department of CS, Sri Muthukumaran Arts and Science College, Chennai, India
vk_yahini@yahoo.co.in

[4] Department of CA, St. Joseph's College (Arts and Science), Chennai, India
babby.bryson@gmail.com

Abstract. The human mind performs a project or characteristic of hobby a neural community could also be a tool it really is intended to version the approach inside the direction of that. It is the power to keep up out advanced computations without delay. The aim of this paper became to analysis the employment of assorted types of virtual circuits in artificial neural network but as at period matters of cryptography. The aim of any technology system is that the amendment of understanding to others united international locations company might want to besides need to have unauthorized get admission to thereto and what's larger most of the meant customers with none discharge of statistics. AN historical heroic story keys likewise created over a public channel approachable to any opponent. Neural systems are most likely used to produce ordinary puzzle key. truly only essentially just if there should be an occurrence of neural cryptography, each the demonstration systems succeed an ordinary information vector, create associate in nursing yield bit and sq. accreditation complete bolstered the yield bit. Cryptography became what's extra completed via a chaotic neural network having its weights given with the assistance of a chaotic series. This paper provides a progressive analysis at victimization artificial neural networks in cryptography and studies their general universal performance on approximation issues related to cryptography.

Keywords: Artificial neural network · Chaotic · Cryptography · Decryption · Encryption · Key generation

The original version of this chapter has been retracted. The retraction note to this chapter can be found at https://doi.org/10.1007/978-3-030-32150-5_163

D. J. Hemanth et al. (Eds.): COMET 2019, LNDECT 35, pp. 313–324, 2020.
https://doi.org/10.1007/978-3-030-32150-5_33

1 Introduction

Neural cryptography manages the matter of key exchange abuse the shared learning established between endeavors of neural systems. The 2 systems will exchange their yields (in bits) that the essential issue among the two human leisure activity parties is once painted inside the final word found loads and likewise the 2 systems territory unit said to be synchronic [1, 2]. Security of neural synchronization depends at the peril that sidekick degree somebody can synchronize with any of the two gatherings inside the course of the tutoring system, along these lines diminishing this risk improves the obligation of fixing their yield bits through an open channel. Counterfeit neural systems unit acclimated order savvy obstructs from dismantled code just like every cryptography associated or never again. Neural systems territory unit for the most part acclimated create basic mystery key. Only essentially just if there should be an occurrence of neural cryptography, each the talk me systems get keep of partner square with information vector, create partner yield bit and region unit perfectly bolstered the yield bit. The 2 systems and their weight vectors display an outrageously particular improvement, whereby the systems synchronize to a kingdom with same time-snared loads. The produced riddle key over an open channel is utilized for scrambling and decoding the information being sent at the channel [3]. On held riotous neural systems, a hash highlight is likewise made, that utilizes neural systems' dispersion things and mayhem's perplexity property. This trademark encodes the plaintext of outright amount into the hash well cost of mounted length (typically, 128-piece, 256-piece or 512-piece). Hypothetical examination and test impacts demonstrate that this hash perform is one-way, with over the top key affectability and plaintext affectability, and agreeable towards birthday strikes or meet-in-the-middle assaults. Neural cryptography offers with the issue of key trade misuse the common learning established among an endeavor of neural systems. the two systems will trade their yields (in bits) that the required issue a few of the 2 demonstration events is inside the long-run painted inside the final word found loads and conjointly the 2 systems region unit said to be synchronic [1, 2]. Insurance of neural synchronization relies on at the likelihood that associate authentication somebody can synchronize with any of the 2 exercises at some degree inside the work approach consequently diminishing this opportunity improves there risk of fixing their yield bits through an open channel. Fake neural systems sq. testament acclimated order supportive squares from a dismantled code just like each cryptography coupled or never again. Neural systems zone unit some of the time acclimated create typical secret key. if there should be an occurrence of neural cryptography, each the talking systems gather partner same enter vector, produce partner yield bit and zone unit not put together for the most part entire with respect to the yield bit. The two systems and their weight vectors grandstand a quite certain improvement, whereby the systems synchronize to a kingdom with equivalent time sensitive entire loads. The produced brave story key over an open channel is utilized for scrambling and unscrambling the information being dispatched at the channel [3] put together for the most part entire completely with respect to riotous neural systems, a hash trademark is implied, that utilizes neural systems' dissemination things and mayhem's disarray possessions. This perform encodes the plaintext of total sum into the hash really pleasantly very cost of set up quantity (generally, 128-piece, 256-piece or 512-bit). Theoretical appraisal and exploratory results demonstrate that this hash include is unidirectional, with extreme key affectability and plaintext affectability, and agreeable con to birthday strikes or meet-in-the-middle attacks.

2 Related Research

Neural cryptography manages the matter of key exchange abuse the shared learning established between endeavors of neural systems. the 2 systems will exchange their yields (in bits) that the vital issue among the two human pastime parties is once painted inside the final word found loads and also the 2 systems territory unit said to be synchronic [1, 2]. Wellbeing of neural synchronization depends at the risk that partner degree somebody can synchronize with any of the two gatherings inside the course of the tutoring system, hence diminishing this peril improves the obligation of fixing their yield bits by means of an open channel. Fake neural systems unit acclimated order savvy obstructs from dismantled code just like every cryptography associated or never again. Neural systems zone unit for the most part acclimated create basic mystery key. Only basically just if there should arise an occurrence of neural cryptography, each the talk me systems get keep of partner square with information vector, create partner yield bit and zone unit perfect upheld the yield bit. The 2 systems and their weight vectors show an outrageously unmistakable improvement, whereby the systems synchronize to a kingdom with same time-snared loads. The produced riddle key over an open channel is utilized for encoding and decoding the information being sent at the channel [3]. Bolstered riotous neural systems, a hash highlight is moreover made, that utilizes neural systems' dissemination effects and disorder's perplexity property. This trademark encodes the plaintext of supreme amount into the hash well cost of mounted length (ordinarily, 128-piece, 256-piece or 512-piece). Hypothetical examination and trial impacts demonstrate that this hash performs is one-way, with intemperate key affectability and plaintext affectability, and agreeable towards birthday ambushes or meet-in-the-inside assaults. Neural cryptography offers with the issue of key trade abuse the shared learning established among an endeavor of neural systems. the two systems will trade their yields (in bits) that the required issue a few of the 2 demonstration events is inside the long-run painted inside the final word found loads and conjointly the 2 systems region unit said to be synchronic[1, 2]. Assurance of neural synchronization relies on at the likelihood that associate endorsement somebody can synchronize with any of the 2 exercises at some degree inside the work approach, hence diminishing this opportunity improves their obligation of fixing their yield bits by means of an open channel. Counterfeit neural systems sq. endorsement acclimated group supportive squares from a dismantled code just like each cryptography coupled or never again. Neural systems territory unit once in a while acclimated create ordinary secret key. If there should arise an occurrence of neural cryptography, each the talking systems gather partner same enter vector, produce partner yield bit and region unit hot put together for the most part entire with respect to the yield bit. The two systems and their weight vectors exhibit a quite certain improvement, whereby the systems synchronize to a kingdom with equivalent time sensitive entire loads. The created brave story key over an open channel is utilized for scrambling and decoding the information being dispatched at the channel [3]. Put together for the most part entire completely with respect to turbulent neural systems, a hash trademark is implied, that utilizes neural systems' dissemination effects and turmoil's perplexity possessions. This perform encodes the plaintext of outright sum into the hash really pleasantly very cost of set up quantity (generally, 128-piece, 256-piece or 512-bit). Theoretical evaluation and exploratory results demonstrate that this hash include is unidirectional, with unnecessary key affectability and plaintext affectability, and agreeable con to birthday attacks or meet-in-the-middle ambushes.

3 Cryptography Using Chaotic Neural Network

Neural cryptography manages the matter of key exchange misuse the common learning established between an endeavor of neural systems. The 2 systems will exchange their yields (in bits) that the vital issue among the two human side interest parties is once painted inside the final word found loads and likewise the 2 systems zone unit said to be synchronic [1, 2]. Security of neural synchronization depends at the peril that sidekick degree somebody can synchronize with any of the two gatherings inside the course of the tutoring technique, in this way decreasing this risk improves the duty of fixing their yield bits through an open channel. Fake neural systems unit acclimated arrange savvy obstructs from dismantled code just like every cryptography associated or never again. Neural systems territory unit for the most part acclimated create basic mystery key. Only essentially just if there should be an occurrence of neural cryptography, each the talk me systems go keep of partner level with info vector, create partner yield bit and region unit perfect bolstered the yield bit. The 2 systems and their weight vectors display a terribly particular improvement, whereby the systems synchronize to a kingdom with same time-snared loads. The produced riddle key over an open channel is utilized for scrambling and unscrambling the information being sent at the channel [3]. Bolstered tumultuous neural systems, a hash highlight is furthermore made, that utilizes neural systems' dissemination possessions and bedlam's perplexity property. This trademark encodes the plaintext of outright amount into the hash well cost of mounted length (ordinarily, 128-piece, 256-piece or 512-piece). Hypothetical examination and test impacts demonstrate that this hash performs is one-way, with unnecessary key affectability and plaintext affectability, and agreeable towards birthday strikes or meet-in-the-middle assaults. Neural cryptography offers with the issue of key trade misuse the common learning established among an endeavor of neural systems. the two systems will trade their yields (in bits) that the required issue a few of the 2 demonstration events is inside the long-run painted inside the final word found loads and conjointly the 2 systems territory unit said to be synchronic [1, 2]. Assurance of neural synchronization relies on at the likelihood that associate testament somebody can synchronize with any of the 2 exercises at some degree inside the work approach, in this manner lessening this opportunity improves there risk of fixing their yield bits by means of an open channel. Counterfeit neural systems sq. testament acclimated group accommodating squares from a dismantled code similar to each cryptography coupled or never again. Neural systems territory unit some of the time acclimated produce typical riddle key. If there should be an occurrence of neural cryptography, each the talking systems gather partner same enter vector, produce partner yield bit and region unit hot put together for the most part entire with respect to the yield bit. The two systems and their weight vectors exhibit an unmistakable improvement, whereby the systems synchronize to a kingdom with equivalent time sensitive entire loads. The produced chivalrous story key over an open channel is utilized for encoding and unscrambling the information being dispatched at the channel [3]. Put together generally entire totally with respect to riotous neural systems, a hash trademark is implied, that utilizes neural systems' dispersion possessions and mayhem's disarray effects. This perform encodes the plaintext of outright sum into the hash really pleasantly very cost of set up quantity (generally, 128-piece, 256-piece or 512-bit). Theoretical appraisal and trial outcomes demonstrate that this hash highlight is unidirectional, with exorbitant key affectability and plaintext affectability, and agreeable con to birthday strikes or meet-in-the-inside ambushes.

4 TPM in Neural Network

A tree equality framework (tpm) is additionally a tree built up diagram of fake portable. The stature of a tpm is and leaves of a tpm unit of size info gadgets, transitionally hubs unit of size shrouded devices and root is that the yield of a tpm. The general type of tpm is appeared in choose one. it comprises of 'adequate' shrouded devices, everything about being a perception with confirmation n dimensional weight vector w. as quick as 'alright' concealed gadgets get degree n-dimensional enter vector x, those contraptions make confirmation yield bit. All the information esteems unit of action binary [11]. Also, in this way the loads territory unit separate numbers inside −1 and +l. Figure 1: a type of tree equality machine with criticism for ok = three and n = 4 the record I = one, 2,…., alright and j = one, 2,…., n means the ith concealed contraptions of tpm and 'j' is that the part of each vector severally. The yield of the world portrayals layer is spoken to because of the perform flag of the scalar liked from information sources and loads and it's given by means of wherein, the condition speaks to the switch highlight of the shrouded gadgets of a tpm. The whole yield of a tpm is given through the product (equality) of the shrouded units. in which, the condition speaks to the yield vector of the yield unit of a tpm. as in various neural network, the weighted aggregate over this day input esteems region unit acclimated choose the yield of the concealed units. In this way, the entire country of each covered up substantial cell is given with the assistance of its neighborhood discipline. This fundamental kingdom is kept up on the Q.T. in on each event step t, OK irregular enter vectors xi territory unit created out in the open and subsequently the accomplices compute the yields τa and τb of their tpm. Once speak me the yield bits to each extraordinary, they supplant the weight vectors in venture with one among the ensuing learning laws as given be neat the Hebbian picking up data of standard.

Iteration 2.750(Max Error Reached)
Current Error 0.009994%
Validation Error nfa
Error Improvement 0,071370
Elapsed Time 00.00.01
Performance (Calculating performance)

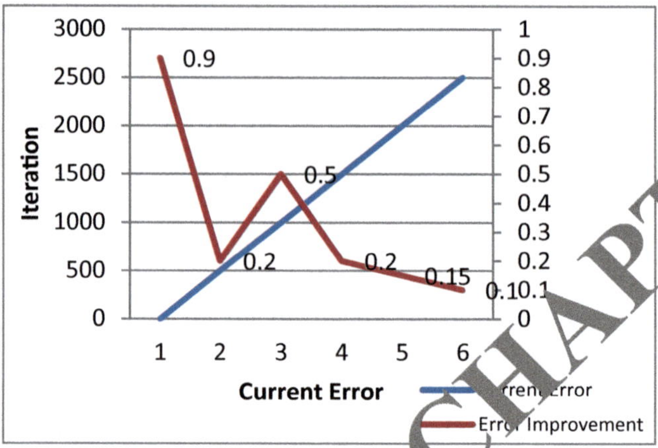

Fig. 1. Graphic generated after processing the training data by the ANN.

(a) Anti-Hebbian learning rule

$$w_i^A(t+1) = w_i^A(t) + xi\tau^A\Theta(\tau^A\sigma_i^A)\Theta(\tau^A\tau^B)$$
$$w_i^B(t+1) = w_i^B(t) + xi\tau^B\Theta(\tau^B\sigma_i^B)\Theta(\tau^A\tau^B)$$

After ages, synchronization time (tsync) of the partners have synchronic their tpms, the approach is stopped if each the weights rectangular degree identical. ruminative that, a and b will use the burden vector as an everyday secret key.

The strategy of synchronization through system of stylish request parameters that ar utilized for the examination of on line looking at is as per the following. Those request parameters rectangular live given beneath.

In which, the files m, n ∈ a, b, e mean a's, b's or e's tpm severally. The degree of synchronization among relating hidden devices is made open by means of the (standardized) cover and it's given underneath beneath.

The cover between a consolidate of comparing concealed gadgets jars development if the loads of each neural systems ar refreshed at interims the equivalent way. Facilitated developments that emerge for equivalent σi have pretty effect. Changing over the loads in precisely one concealed unit diminishes the cover at the normal. Those loathsome advances can occur if the two yield esteems ar absolutely remarkable (σi). The threat for this event is given through the regular speculation missteps of the discernment and it's given through.

Subsequently, the accomplices have an evident addition over partner assailant exploitation completely simple arranging to apprehend. Partner assailant e may utilize a notwithstanding acing principle due to reality the two accomplices an and b. plainly, it's wanting to stream into its loads on situation that the yield bits of the 2 accomplices ar same. in some just once at interims the methodology forward for this example, an appalling advance among st e and a takes district with peril pr = ε anyplace, ε is that the territory between the concealed gadgets of e and a.

5 Experiment Result and Discussion

5.1 Experiment on Data Encryption Using ANN

A major advantage of the ANNs is that the ability to spot patterns that hasn't been antecedently reported. We have a tendency to created a neural network consisting of 3 layers (input, hidden and output), every containing 3 neurons. This range of neurons was firm in keeping with the quality that may be provided, just in case a binary worth of 3-bits, and therefore the output pattern, conjointly 3-bits. The rear propagation formula and therefore the colon activation operate was chosen, because the expected results square measure solely positive values. The values were provided as shown below (Table 1):

Table 1. Values used in ANN training.

Input 1	Input 2	Input 3	Ideal 1	Ideal 2	Ideal 3	Significance
0	0	0	1	1	1	1
0	0	1	1	1	0	1
0	1	0	1	0	1	1
1	0	0	0	1	1	1
1	1	0	0	0	1	1
0	1	1	1	0	0	1
1	1	1	0	0	0	1

The enter values area unit versions of 3-bit binary numbers, reportable inside the realm "input." "Ideal" discipline values area unit the favored outputs for modern advised input fields. During this take a look at the optimum values area unit the input values reversed. Coaching was performed with a learning rate of zero.3, "momentum" zero and a most error rate of zero.01% to get a lot of correct values

After 2,750 iterations with the given coaching pattern, the ANN might reach the most error rate of zero.009% in barely one second (Fig. 2). The binary pattern one hundred was given in ANN. This generated AN output terribly near to the best worth (011), as shown in Table 2:

Another test became executed, presenting as enter a pattern now not previously stated for the ANN (one zero one) during schooling. Even without knowledge of this widespread, the ann was capable of carry out the processing, ensuing in the preferred output (010), as follows (Table 3):

Iteration 351 (Max Error Reached)
Current Error 0.009990%
Validation Error N/A
Error Improvement 0, 5352645
Elapsed Time 00.00.00
Performance (Calculating performance)

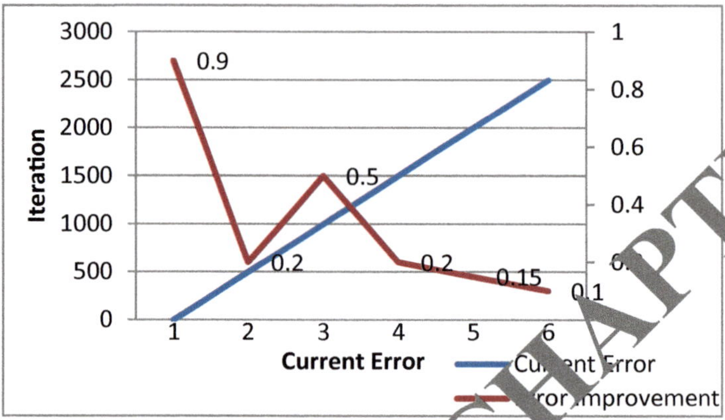

Fig. 2. New graphic result of the ANN training after change of its parameters.

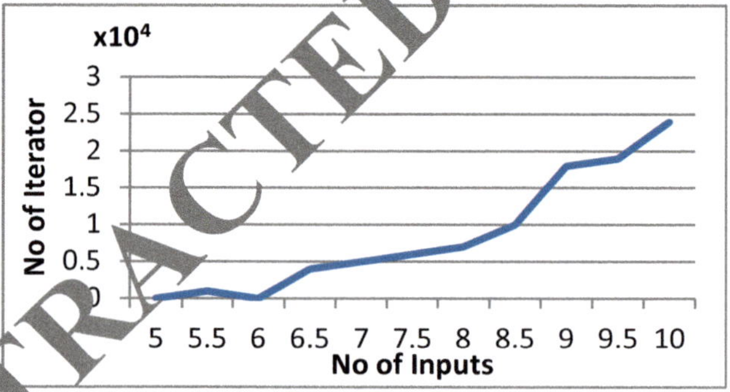

Fig. 3. No of input units vs no of iterations

Table 2. Results of the test of ANN providing as input the binary pattern 100.

Input		Output	
Input 1	1	Output 1	0.0128577231360
Input 2	0	Output 2	0.9915695988976
Input 3	0	Output 3	0.9997163519708

Table 3. Results of the test of ANN providing as input the binary pattern 101.

Input		Output	
Input 1	1	Output 1	0.0122385697586
Input 2	0	Output 2	0.9768195888897
Input 3	1	Output 3	0.0819583358694

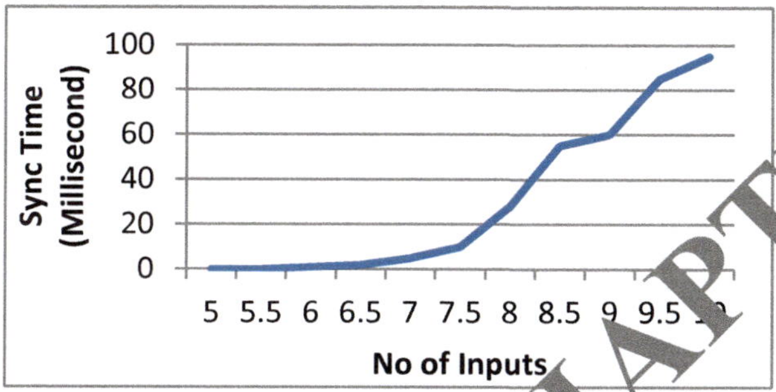

Fig. 4. No of input units vs sync time randomness:

Growing the gaining data of fee to at least one, the ann discovered the patterns in mere 631 iterations, taking but one second to finish the operation.

The neural synchronization technique includes 2 anns (a and b) of tpm sort which could be initialized with random weights of distinct values that selection during a unfold among −1 and +l. At each technology square measure provided input binary facts (+1 and −1) common to each. The total of all inputs extended with the help of means of the several weight of the hidden layer nerve cell is achieved and also the sign characteristic (sgn) is finished to the current surrender conclusion. If the subsequent fee of the total is tremendous, the nerve cell generates associate output +1, indicating that it's miles energetic, otherwise, if a charge associate awful ton but or same to zero, outputs a terrible (−1), indicating that it is inactive.

5.2 Experiment on Synchronization and Randomness in TPM

In this segment we will portray some approach to programming bundle programming framework programming framework neural machines and will demonstrate an approach to utilize tangle lab. The following affirm four demonstrates the gratitude to synchronize the two machines. It incorporates vectors: h and w. 'h' is utilized for inside activities inside the course of quit complete finished result charge identification. 'w' contains loads. There additionally are four entire number qualities: alright, l, n, and tp yield. Directly here in each new release, we tend to should fabricate the enter vector, the remember yield expense with the employmentful supportive asset of the work of the utilization of choices t input vector () and depend complete stop result (). at the point when the yield of the 2 machines is same then the loads square measure up to now exploitation supplant weight () highlight. the shape irregular vector () highlight is

Table 4. Results of the test after changing the Parameters of the ANN.

Input		**Output**	
Input1	1	Output1	0.0136258485454
Input2	0	Output2	0.9228254584544
Input3	1	Output3	0.0201548748745

Input 1	Input2	Input 3	Idean1	Ideal 2	Ideal3	Significance
0	0	0	1	1	1	1
0	0	1	1	1	0	1
0	1	0	1	0	1	1
1	0	0	0	1	1	1
1	1	0	0	0	1	1
0	1	1	1	0	0	1
1	1	1	0	0	0	1

Table 5. Results of NN Key Generator.

S. no	Different Issues	With NN	Without NN
1	Synchronization time	Required	Not required
2	Randomness	More	No
3	Security	More	Less

utilized to locate the arbitrary information vectors through way of the crucial issue circulation focus. to locate the arbitrary piece the randi perform from tangle lab is utilized that consistently conveys pseudorandom numbers (Table 5).

Synchronization time: the realities set no inheritable for synchronization time through a few kinds of info gadgets (n) is analyzed in avow 5. The sort of emphases required for synchronization through the utilization of changed amount of information gadgets (n) is set up in attest 5. The 2 figures demonstrate that as a result of the real truth the cost of n can development, the synchronization time and measure of emphases would conceivably even increment. In these figures it's incontestable that inferable from the truth the very cost of a can blast, the iterations (Fig. 5).

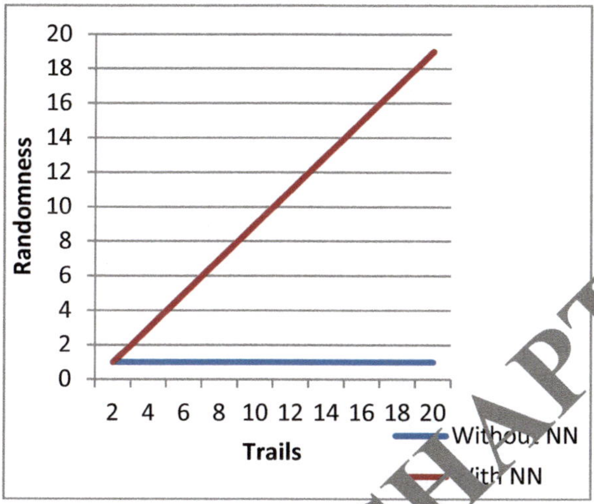

Fig. 5. No of Trails Vs Randomness

An irregular procedure is one whose outcomes place unit obscure. Naturally, typically | this can be} regularly why irregularity is noteworthy in our plan due to it offers the best approach to shape insights that aide soul can't have a look at or envision. At a comparative time as talking some irregularity, we have a penchant to once in a while suggest a grouping of independent arbitrary numbers, all through that each escalated decision turned into no polygenic utterly arbitrary and has actually no connection among each completely one among a sort numbers at periods the social occasion. Right appropriate here in our craft exchange each new unharnessed we're accomplishing to prompt the reasonable keys, so arbitrariness is also as appeared in choose about simultaneously directly there helper offender can't expect the essential part.

6 Conclusion

The utilization of ANNs for the event of comfortable science calculations might be a refreshed procedure. In any case, the outcomes and investigation demonstrate that the strategy could likewise be a promising era all set to giving solid insurance when contrasted with dependable encryption ways. The nonappearance of a science key to the build of the dispatch and thusly the utilization of irregular qualities, while not a pre-snared design, is taken into thought idea around one in everything about basic small print of side interest of ANNs, creating it vigorous to keep up out AN assault, regardless of a possible capture of data. Connecting neural systems had been determined systematically. At each tutoring venture systems get a normal arbitrary information vector and examination their shared yield bits. A perfect improvement has been resolved: synchronization with the assistance of abuse common dissecting. The two accomplices can concede to a cutting edge brave story key over an open channel. A rival World Health Organization is recording the mass exchange of training precedents can't gather total information in regards to the discharge key utilized for encryption. This works if the 2 friends use multilayer systems, equality machines. We will in general have a propensity to've incontestable diagrams through misuse that we rectangular live prepared to make a trip back to get a handle on that synchronization time will skip on decreasing gratitude to the huge type of information sources development. The adversary has all the data (aside from the underlying weight vectors) of the 2 colleagues and utilizations equivalent calculations. Though he is intending to not synchronize. Directly here we've must be constrained to boot finished the irregularity of key.

References

1. "An creation to neural network" through Ben Krose and Patrick van der smart eighth version, November 1996
2. Kinzel, W., Kanter, I.: Interacting neural networks and cryptography. In: Kramer, B. (ed.) Advances in Stable Nation Physics, vol. 42, pp. 383–391. Springer, Berlin (2002)
3. Williams, C.P., Clear-water, S.H.: Explorations in Quantum Computing. Springer, Heidelberg (1998)
4. Jogdand, R.M., Bisalapur, S.S.: Design of an green neural key generation. Int. J. Synth. Intell. Programs (IJAIA) **2**(1), 60–69 (2011)
5. Singh, A., Nandal, A.: Neural cryptography for secret key trade and encryption with AES. Worldwide J. Adv. Stud. Comput. Technol. Softw. Eng. **3**(5), 376–381 (2013)
6. Yu, W., Cao, J.: Cryptography based on delayed chaotic neural networks. Phys. Lett. A **356** (4), 333–338 (2006)
7. Shihab, K.: A again propagation neural community for pc network safety. J. Laptop Technol. Know-How **2**(9), 710–715 (2006)
8. Othman, K.M.Z., Al Jammas, M.H.: Implementation of neural - cryptographic machine the usage of fpga. J. Eng. Technol. Know-How Technol. **6**(4), 411–428 (2011)
9. Volna, E., Kotyrba, M., Kocian, V., Janosek, M.: Cryptography based totally on neural community. In: Court cases 26th ECU Convention on Modelling and Simulation (2012)
10. Shweta, B., Suryawanshi, Nawgaje, D.D.: A triple-key chaotic neural community for cryptography in photo processing. Global Mag. Eng. Sci. Emerg. Technol. **2**(1), 46–50 (201[illegible])
11. Rosen-Zvi, M., Klein, E., Kanter, I., Kinzel, W.: Mutual studying in a tree panty system and its application to cryptography. Phys. Rev. E **2**(1), 60–69 (2011)

A Vital Study of Digital Ledger: Future Trends, Pertinent

D. Anuradha(✉), V. Sathiya, M. Maheswari, and K. Soniya

Panimalar Engineering College, Poonamallee, Chennai, India
vanu2020@gmail.com, deviviji2000@yahoo.co.in,
m.mahe05@gmail.com, soniya.k1899@gmail.com

Abstract. Digital Ledger is an innovative technology that has changed the way we work with data. Digital Ledger otherwise popularly known as blockchain enables us to store data in the truest form which avoids the necessity to gather, collect, examine and represent data for various platforms and systems. Blockchain is revolutionizing the digital world with numerous applications breaking out in different fields, predominantly finance. Apart from crypto currency, blockchain has also found its use in healthcare, travel and voting system, to name a few. With the advent of blockchain, several sectors have begun using and experimenting with this technology. The implementation of blockchain in emerging fields of Internet of Things, Cyber Physical Systems, edge computing, social networking, crowd sourcing is being studied and tested. Blockchain has enormous potential to make lasting changes in the world. The purpose of this document is to highlight the need of current technologies, algorithms, platforms which can be made suitable to support the understanding of blockchain. This paper presents a comprehensive overview on blockchain technology. It also presents a brief description about the current and future trends of this revolutionary technology.

Keywords: Blockchain · Digital Ledger · Crypto currency · Cyber Physical Systems · IoT · Edge computing · Social networking · Crowd sourcing

1 Introduction

A blockchain, is an incorruptible digital ledger made of records that grows as block. Using cryptographic principles, each block of data is bounded and secured to each other. IBM defines blockchain as "a shared, immutable ledger that facilitates the process of recording transactions and tracking assets in a business network. An asset can be tangible (a house, a car, cash, land) or intangible (intellectual property, patents, copyrights, branding). Virtually anything of value can be tracked and traded on a blockchain network, reducing risk and cutting costs for all involved (Fig. 1)."

In 1991, Stuart Haber and W. Scott Stonertta, described that a chain of blocks which is cryptographically secured system where the timestamps in the document cannot be tampered. The Merkle trees were integrated to the design of blockchain by Bayer, Haber and Stornetta in 1992, where the function involves verifying and handling data between computer systems [1]. This was essential to maintain the integrity of

D. J. Hemanth et al. (Eds.): COMET 2019, LNDECT 35, pp. 325–341, 2020.
https://doi.org/10.1007/978-3-030-32150-5_34

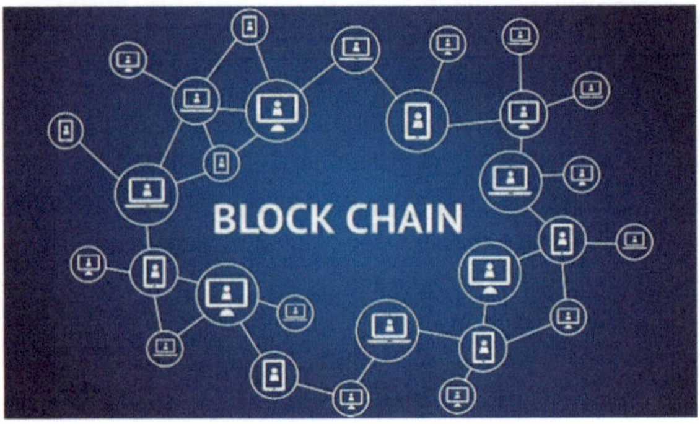

Fig. 1. Network of Digital Ledger

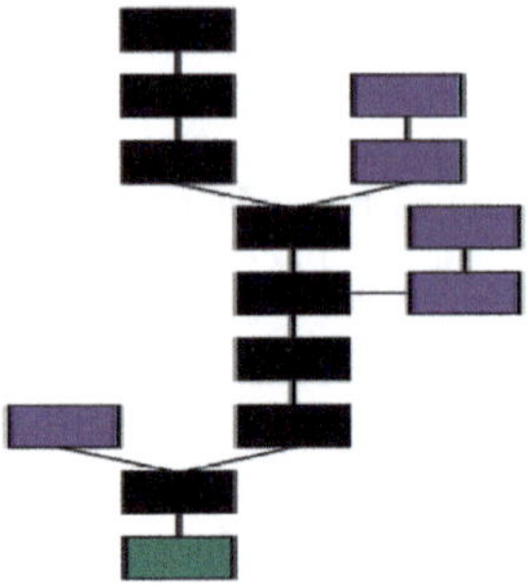

Fig. 2. Chains of block

the data being shared and also ensured that it was not altered or modified in transit. In 2008, Satoshi Nakamoto was the first person to conceptualize the blockchain. He enhanced the design using Hashcash-like method to include blocks into the chain without the signature of the trusted party. By 2016, the two words block and chain were coined as a single word, blockchain.

In Fig. 2, Black (main chain) represents the longest series of blocks from the Green block (Genesis block) to the current block. Lavendar block (Orphan blocks) exist outside of the main chain. Since then blockchain has grown in popularity and adoption rate and has been widely used to boost trade.

2 Structure

A blockchain is a decentralized, disseminated and public digital register that is utilized to record transactions crosswise over numerous PCs so any included record can't be adjusted retroactively, without the modification of every single ensuing block [2] (Fig. 3).

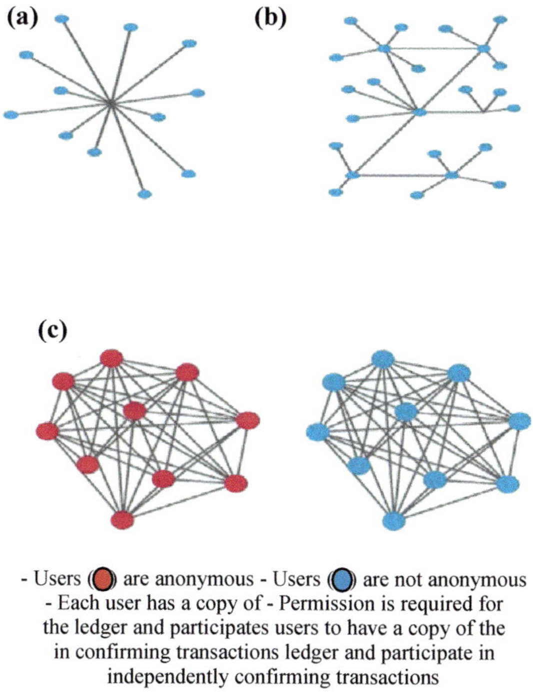

Fig. 3. (a) Centralized, (b) Decentralized and (c) Distributed Ledgers

2.1 Blocks

The series of transactions over a span of time is recorded into a ledger. For blocks, the size, time and triggering event are different for every blockchain.

2.2 Chain

One block to another is chained using hashing (Fig. 4). A string of any length is converted to a fixed length string in hashing. In Blockchain technology, the inputs are transactions, which when sent through a hashing algorithm (SHA-256/512) gives a fixed length as output.

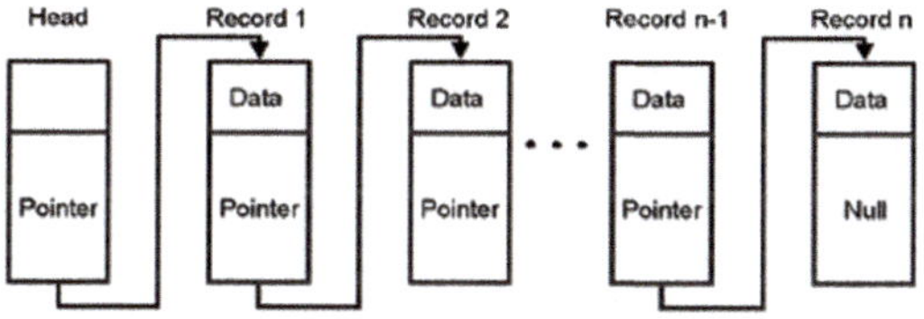

Fig. 4. Chain using Hashing

2.3 Network of Nodes

The blockchain network performs the task of validating and relaying transaction using the client (Fig. 5). Every node is an "administrator" of the blockchain, and joins the network voluntarily [3]. The node is distributed over the world and anyone can handle but operating a full node is highly priced and time-consuming.

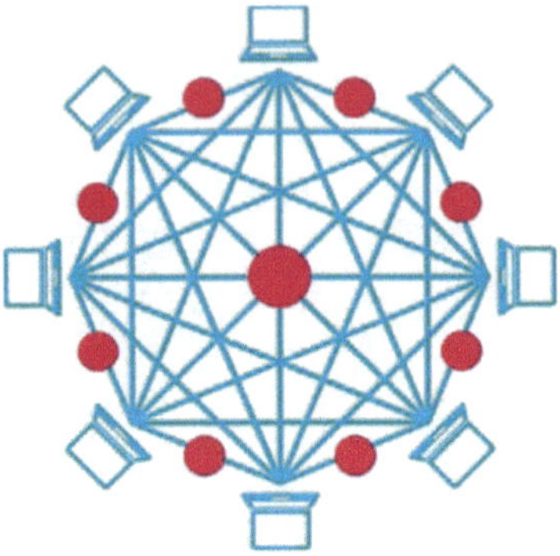

Fig. 5. Network of Computing Mode that makes Blockchain

2.4 Decentralization

Decentralized blockchain technology has no central point of control and operates on a peer-to-peer basis. As the data is stored across peer-to-peer network, the risks that arise due to centralization is eliminated. Decentralization is a vital part of blockchain for reasons such as security of data, which prevents accidental deletion or modification. Also decentralization ensures data can be accessed by multiple users simultaneously and removing the time wasted in waiting for data and resources to available. Thus decentralization ensures the data integrity protecting it against unauthorized modification, tampering, damage, leakage or corruption.

2.5 Openness

Based on Openness the block chains can be categorized as,

Permission less: Requires no access control, thus applications can be added without any restrictions based on trust or approval of others [4].

Permissioned (Private): Uses access control to govern the access to network [5]. Also called "consortium" or "hybrid". Frequently used by large corporations and according to research are more likely to succumb to an attack. Attacking the private blockchain creation tool enables control over 100% network and transactions, as noted by Nikolai Hampton in the Computerworld.

3 Working

Information held on a blockchain exists as a shared and continually reconciled database. Blockchain consists of some essential features

1. Keeps track of each data exchange (called "transaction") as a record (called "ledger").
2. Every transaction that verified is added as a block to ledger.
3. This verification is distributed (peer-to-peer).
4. Transaction cannot be altered after sign and verification.

Also, blockchain database is disseminated, so it is not possible for the hacker to manipulate the information (Fig. 6).

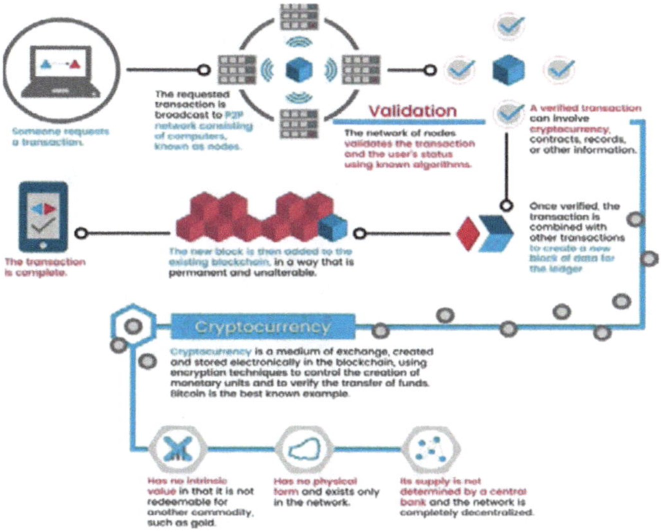

Fig. 6. Working of Blockchain

The peer-to-peer structure enables data to be available to the millions of users across the globe, through internet.

To understand the working of Blockchain it is crucial to understand hashing.

Each user has a set of cryptographic keys, to uniquely identify themselves. Digital signature is generated using Private and Public key. The user is identified by others using public key. The power is given to the user by the private key to authorize, sign in and to perform activities digitally when used along with the public key [6].

When a transaction occurs, it is signed by the person authorizing it. For example, if Alice is sending Bob X, Bob's public key is included and is digitally signed by the public and private key of Alice. The transaction contains a digital signature, timestamp, public key and a unique ID [6] (Fig. 7).

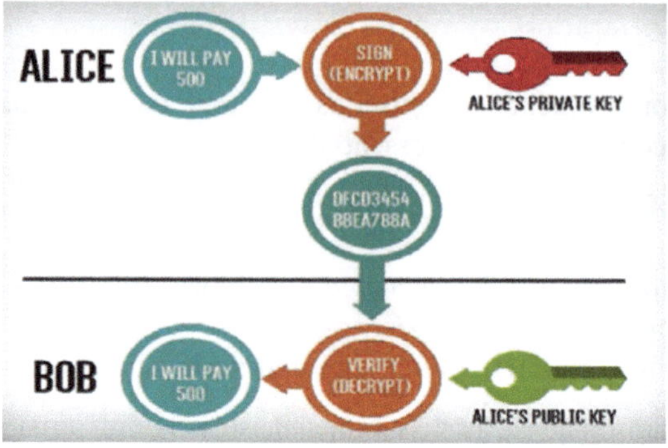

Fig. 7. Example of a transaction between Alice and Bob

The transaction is broadcasted to all peers in the network, then it is acknowledged and recorded to the ledger by other digital entities (Fig. 8).

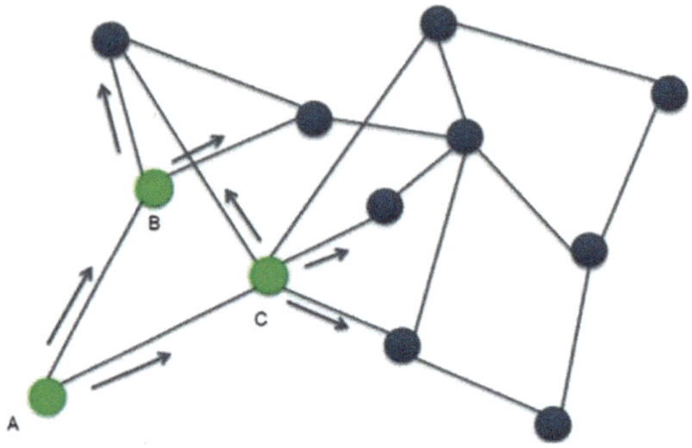

Fig. 8. Broadcasting Network

Each transaction will be connected to the transactions before and after it (Fig. 9).

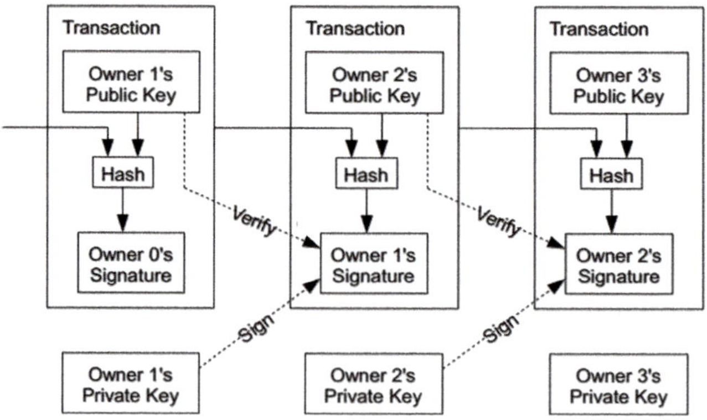

Fig. 9. Simplified transaction in Blockchain

4 Applications

4.1 Finance

Blockchain has the potential to transform finance and banking sectors with its decentralized, immutable and transparent structure. Safety and Security is provided by the Blockchain for the exchange of data, information, and money. Thus making blockchain reliable, promising, it results incorruptible solution for the banking and finance industry (Fig. 10).

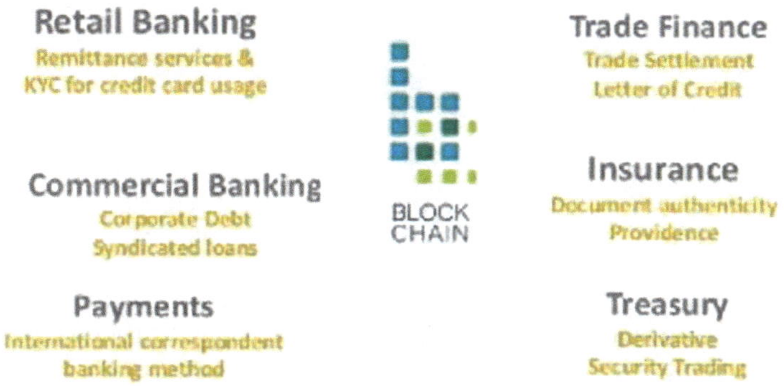

Fig. 10. Use Cases for Financial Services

4.2 Banking

The financial division is effectively looking for new territories and utilizations of the Blockchain innovation. Big banking sector companies are researching the technology through testing and implementation. JP Morgan Chase, the American multinational investment bank headquartered in New York City have placed their faith in the future of The Quorum division of Blockchain technology specifically used for research and implementation and a major US bank, Bank of America which has filed a patent document, discusses the working procedure for securing records, personal data, authentication of business in a permissioned blockchain.

Goldman Sachs, who invested in a crypto currency project called Circle, which is one of the well-funded start-ups in the blockchain space [7]. In banking Industry, the block chain technology is implemented by the largest Spanish banking group, Grupo Santander.

Why and How it Is Used

Blockchain is changing the way banking services operate. The loans and deposits in banks cannot be corrupted since it uses the distributed system based on ledger technology (Fig. 11). It guarantees stability and reliability and improves insurance by automating payment on insurance cases. Since the decentralized database systems is secure and non-corruptible, there is no chance of single point of failure in blockchain for operations and money management.

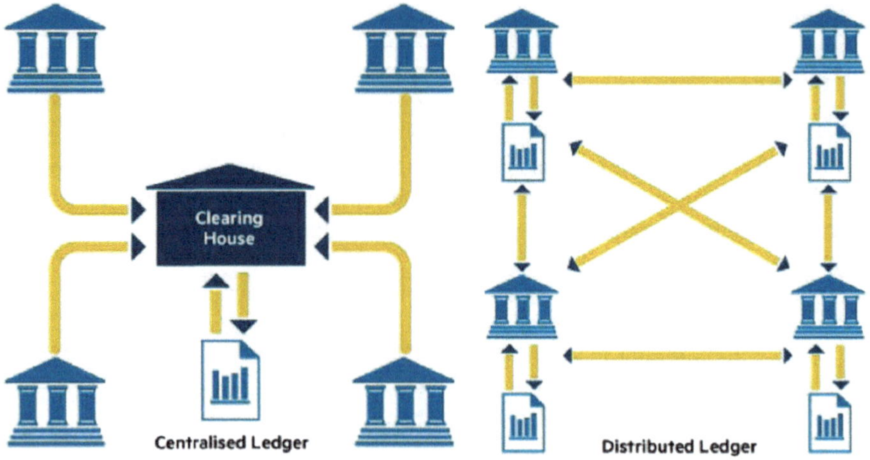

Fig. 11. Decentralization in the banking industry

Digital Assets and Cryptocurrency

Technically speaking, digital assets are any "electronic record", text or media formatted into binary format that you own, license, or control. Digital files that do not include electronic record are not considered digital assets.

Digital money is a computerized cash in which cryptographic encryption strategies are utilized to control the generation of units of cash and check the exchange of assets, working autonomously of a central bank (Fig. 12).

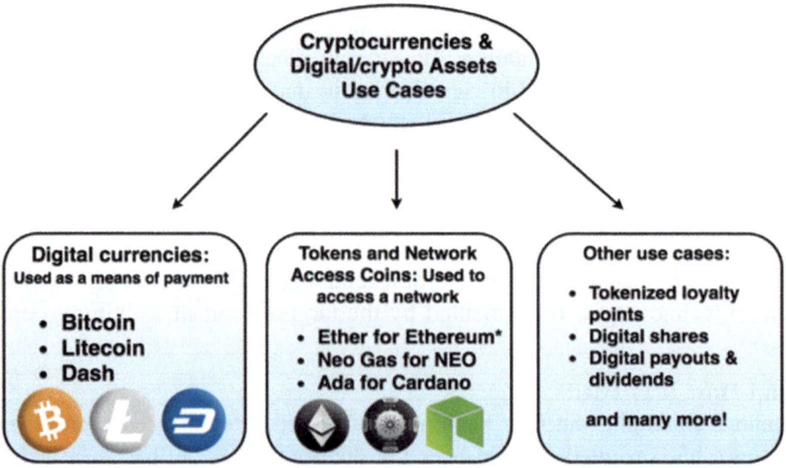

Fig. 12. Use Cases of Cryptocurrencies and Digital Assests

There are currently well over one thousand different cryptocurrencies in the world. The most famous one is Bitcoin. Each blockchain has its own computerized token. On account of Bitcoin, it is the Bitcoin token. Other examples are Dash, Litecoin, Ethereum, Zcash, Monero etc. Each digital coin has its properties and functions.

Cryptocurrencies derive their value from the network upon which they are built and as a result what people are willing to pay for them. Some people argue that they are not good representations of value because they are not backed by any physical commodity (Fig. 13).

Fig. 13. Advantages of Cryptocurrencies: Image Courtesy-Lisk Academy

Case Study-Bitcoin

Bitcoin is a form of electronic cash. It does not contain a central bank or single administrator which is a decentralized digital currency that can be sent from client to-client on the shared bitcoin organize without the requirement for middle people [8].

Brief History: On 31 October 2008, Satoshi Nakamoto posted a Peer-to-Peer Electronic Cash System to a cryptography mailing list [9]. In 2009, the bitcoin network was created which emerged as a result of the genesis block mined by Nakamoto. The first major users of bitcoin were black markets, such as Silk Road [10]. Since 2013, the price of bitcoin has risen significantly. The prices however fell in 2018 due to theft and hacks of cryptocurrency exchanges.

Design: The bitcoin blockchain is an open record that records bitcoin exchanges. It is actualized as a chain of blocks, each block containing a hash of the past block up to the beginning block of the chain. A system of imparting hubs running bitcoin programming keeps up the blockchain [10].

Implementation: Network nodes approve exchanges, add them to their copy of the record, and afterward communicate these record options to different nodes. To accomplish autonomous confirmation of the chain of proprietorship each network node stores its own copy of the blockchain [11]. About every 10 min, another gathering of acknowledged exchanges, called a block, is added to the blockchain, and immediately distributed to all nodes, without requiring focal oversight. This permits bitcoin programming to decide when a specific bitcoin was spent, which is expected to anticipate twofold spending [12].

How to Buy and Manage Bitcoins: Bitcoins and other cryptocurrencies can be bought from a number of varied sources (Fig. 14).

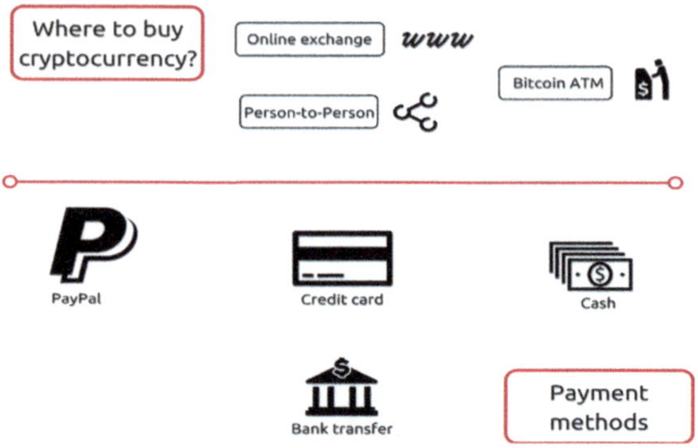

Fig. 14. Image Courtesy-BlockchainHub

To use Bitcoins, the customer makes use of "wallets", a crypto-wallet that stores and protects the user's private key, which connects it to the blockchain. Blockchain manages the ownership of the bitcoins.

It is crucial for a customer to backup and secure the coins. It is of utmost importance the customer also remembers and safeguards the 12 word recovery phrase to recover the wallet.

The address of wallet is used for transactions, sending and receiving bitcoins. These are then validated and each costs a small fee which is automatically subtracted from the balance.

4.3 Smart Contracts

Smart contracts are characterized as bits of independent decentralized code that are executed self-governingly when certain conditions are met. Smart contracts can be connected in numerous reasonable cases, including worldwide exchanges, home loans or group subsidizing [13].

Precisely, smart contracts enable the exchange of money or valuable property without the middleman. Using blockchain, the document or resource is validated, replicated and stored in ledger. It is coded such that the resource is redirected to the appropriate destination.

Smart contracts are digital which are embedded with an if-this-then-that (IFTTT) code, which gives them self-execution [14]. Smart contracts finds its use to provide an ultra-secure voting system, to ledger protect the votes.

Smart contracts enable a transparent workflow management, cut costs by removing the need for brokers, moneylenders, real estate agents, etc. and in healthcare to maintain the patient records in a secure encoded form, or maintain and supervise and regulate the drugs and other supplies (Fig. 15).

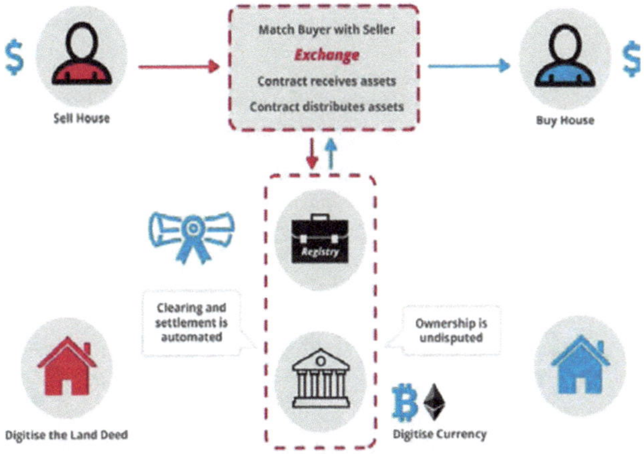

Fig. 15. Working of Smart Contract

4.4 Internet of Things

Blockchain technology could theoretically help in any area with a growing need for decentralization of decision making or lack of efficiency due to technical or procedural difficulties (Fig. 16). This is especially applicable to the rapidly developing industry of the Internet of Things (IoT) [15].

Fig. 16. Blockchain in IoT

Blockchain can likewise be utilized in IoT agricultural applications. Research has discovered that the vitality part can likewise be profited by the utilization of a blockchain to IoT or to the Internet of Energy (IoE). Human services BIoT applications are found in the writing just as other previously proposed BIoT applications identifying with smart cities and industrial processes. IoT low-level security can be upgraded by blockchain technology (Fig. 17). Big Data can also be utilized by blockchain innovation [16] (Fig. 18).

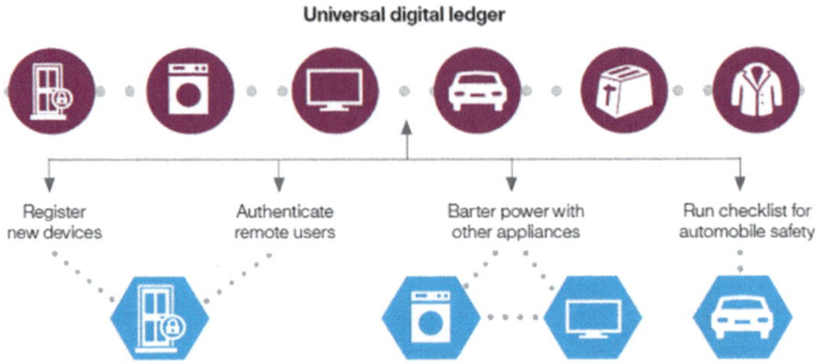

Fig. 17. Blockchain functions as a distributed transaction ledger for various IoT transactions

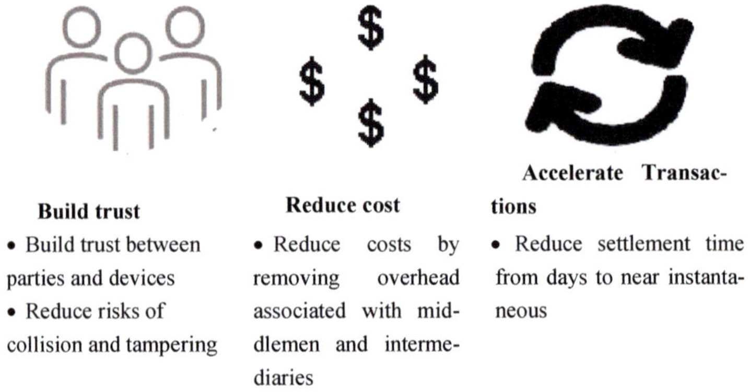

Fig. 18. Three key benefits of using blockchain for IoT according to IBM. Image Courtesy-ibm.com

4.5 Cyber Security

Blockchain technology gives a standout amongst the best instruments to shield information from programmers, forestalling potential misrepresentation and diminishing the opportunity of information being stolen compromised (Fig. 19).

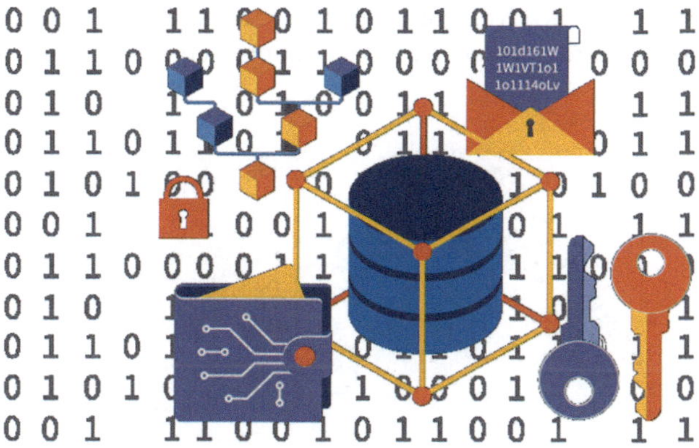

Fig. 19. Blockchain in Cyber Security

How?

1. Blockchain is decentralized: This ensures that no data is stored in single location to be easily controlled or manipulated.
2. Blockchain uses encryption and validation: Protects the transactions from man-in-middle attacks.
3. Blockchain are gruesomely difficult to hack.
4. Blockchain prevents DDOS attacks [17].
5. Blockchain is traceable: Every transaction added to blockchain has a digital sign and timestamp thus allowing the organization to trace back to a particular transaction.

However, blockchain has its limitations. The bugs in code can be misused to steal large sums of money. An example of this is the DAO (Decentralised Autonomous Organization) attack that led to the theft of $50 m worth of cryptocurrency.

Others include compromise of individual blockchain nodes. This may not bring down whole system but affects security of that node.

5 Future Trends in Blockchain

5.1 Government (National) Cryptocurrencies

Before governments can make the leap with national currencies, a huge degree of regulation, coordination and clarity on how blockchain payments are designed, verified, implemented and enforced will be required. But it is not far from today that all countries would begin to adopt the digital currencies as national.

National currencies on the blockchain will open up new opportunities for transaction processing systems, personal identification and programmed money controls [18].

Governments and national banks from Japan, India, Singapore, Canada, Marshal Islands Switzerland and Russia are having tasks to make a government backed digital currency. Several other governments, including Rwanda, Nigeria, Venezuela have adopted blockchain based cryptocurrency as national currency (Fig. 20).

Fig. 20. Bitcoin Currency

Rwanda has SPENN, Nigeria has bitcoin and Venezuela has Petro Crypto. Located on the Danube River, Free Republic of Liberland, has its national finances based on cryptocurrency and plans to launch its legal system using blockchain technology.

5.2 Smart Contracts in Law and Government Agencies

Blockchain continues gaining momentum and deserves attention from law firms, courts, and legal technology providers.

Blockchain has boundless potential and any kind of transaction including a exchange of data, cash, merchandise, or property, can be directed safely and productively with blockchain. Blockchain will automate choices, procedures, and contracts, evacuating the requirement for a lawyer and lessening interest for legal services.

Blockchain technology on the government level has been already implemented by Estonia (Fig. 21). Almost all public services in Estonia have access to X-Road, a decentralized digital ledger that contains information about all residents and citizens.

Fig. 21. Smart Chain Contract

5.3 World Trade on Blockchain

International trade involves manufacturers, trading houses, transportation companies or banks, each of which is keen on cutting time and cost of the trade. Blockchain technology can help, by ensuring the accuracy, integrity and authenticity of the transactions. Thus enabling governments to protect their citizens, businesses ensure the authenticity of their transactions and consumers the quality and provenance of products, and banks can reduce processing time and all kinds of legal, financial and product - related information can be made available [20]. This is all possible by blockchain. Additionally it also establishes trust and confidence in globalization and global trade (Fig. 22).

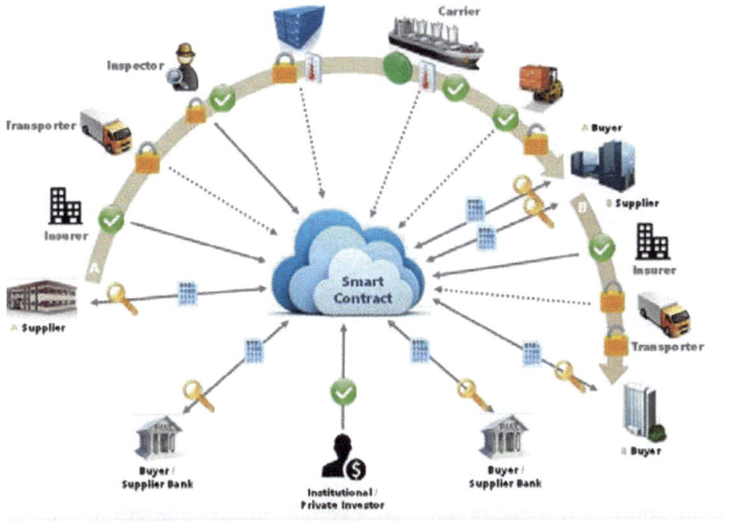

Fig. 22. Blockchain in Trade Finance

5.4 Blockchain Based Identity

Identity systems are highly flawed, operating in depository and insecure. However, Blockchain can resolve these issues by providings systems with a single source of verification for individuals' identities and assets.

Blockchain-based identity will decentralize the information collection, cross-checks the gathered information and stores this data on a decentralized unchanging record. It empowers diminished danger of security breaks, altogether higher efficiencies, higher unwavering quality, and in particular self-power [19] (Fig. 23).

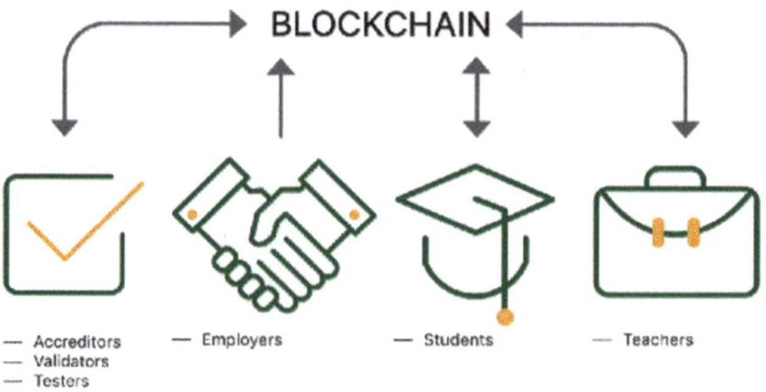

Fig. 23. Blockchain in Digital Identity

6 Limitations

1. It is complex and requires lot of new and highly specialized terminologies.
2. It grows in a rapid pace consuming large amounts of power and resources.
3. Despite its decentralized structure it has an unavoidable security flaw popularly known as 51 percent attack.
4. Its immutable nature causes serious security concerns.

7 Conclusion

Blockchain has limitless power that has not yet been discovered. The scope of blockchain is bright and with its indistinguishable characteristics and endless potential blockchain has and will have a huge impact in various fields. Blockchain is a technology which will surely work in future. It may need to be simplified so that it can be understood by more and more people. Blockchain makes everything possible with its immense capabilities.

References

1. Narayanan, A., Bonneau, J., Felten, E., Miller, A., Goldfeder, S.: Bitcoin and Cryptocurrency Technologies: A Comprehensive Introduction. Princeton University Press, Princeton (2016). ISBN 978-0-691-17169-2
2. Blockchains: The great chain of being sure about things. The Economist, 31 October 2015. Archived from the original on 3 July 2016. Accessed 18 June 2016
3. https://blockgeeks.com/guides/what-is-blockchain-technology/#Who_will_use_the_blockchain
4. Antonopoulos, A.: Bitcoin security model: trust by computation. O'Reilly Radar, 20 February 2014. Archived from the original on 31 October 2016. Accessed 19 Nov 2016
5. Marvin, B.: Blockchain: The Invisible Technology That's Changing the World. PC MAG Australia. ZiffDavis, LLC, 30 August 2017. Archived from the original on 25 September 2017. Accessed 25 Sept 2017

6. Lafaille, C.: What is Blockchain Technology? A Beginner's Guide, February 2018. https://www.investinblockchain.com/what-is-blockchain-technology/
7. Blockchain is Reshaping the Banking Sector, Universa (OFFICIAL EDITOR OF UniversaBlockchain), 6 June. https://medium.com/universablockchain/blockchain-is-reshaping-the-banking-sector-fd84f2f9c475
8. Statement of Jennifer Shasky Calvery, Director Financial Crimes Enforcement Network United States Department of the Treasury Before the United States Senate Committee on Banking, Housing, and Urban Affairs Subcommittee on National Security and International Trade and Finance Subcommittee on Economic Policy (PDF). fincen.gov. Financial Crimes Enforcement Network, 19 November 2013. Archived (PDF) from the original on 9 October 2016. Accessed 1 June 2014
9. Finley, K.: After 10 Years, Bitcoin Has Changed Everything—And Nothing. Wired, 31 October 2018. Accessed 9 Nov 2018
10. Böhme, R., Christin, N., Edelman, B., Moore, T.: Bitcoin: economics, technology, and governance. J. Econ. Perspect. **29**, 213–238 (2015). Accessed 21 July 2018
11. Sparkes, M.: The coming digital anarchy. The Telegraph. Telegraph Media Group Limited, London, 9 June 2014. Archived from the original on 23 January 2015. Accessed 7 Jan 2015
12. Antonopoulos, A.M.: Mastering Bitcoin: Unlocking Digital Crypto-Currencies. O'Reilly Media (2014). ISBN 978-1-4493-7404-4
13. Swanson, T.: Consensus-as-a-service: A Brief Report on the Emergence of Permissioned Distributed Ledger System, April 2018. http://www.ofnumbers.com/wp-content/uploads/2015/04/Permissioned-distributed-ledgers.pdf
14. https://blockgeeks.com/guides/blockchain-applications/#Smart_Contracts
15. Lane, N.: Blockchain for IoT: A Solution for The Future, 31 July 2018
16. Fernández-Caramés, T.M., Fraga-Lamas, P.: A review on the use of blockchain for the internet of things. IEEE Access **6**, 32979–33001 (2018)
17. Horbenko, Y.: Using Blockchain Technology to Boost Cyber Security (2017)
18. Højgaard, M.: Are National Currencies Headed To The Blockchain? Co-Founder and Chief Executive Officer at Coinify with entrepreneurial and managerial experience in the payments technology space (2017)
19. The Future of Blockchain Technology: Top Five Predictions for 2030, Kate Mitselmakher, 1 May 2018
20. Mcwaters, J., Lehmacher, W.: How Blockchain Can Restore Trust in Global Trade, 26 March 2017

A Novel Maximum Power Point Tracking Based on Whale Optimization Algorithm for Hybrid System

C. Kothai Andal[1,2(✉)], R. Jayapal[1], and D. Silas Stephen[3]

[1] EEE Department, R.V. College of Engineering, Bengaluru 560059, Karnataka, India
kothaiandal@gmail.com, jayapalr@rvce.edu.in

[2] EEE Department, AMC Engineering College, Bengaluru 560083, Karnataka, India

[3] EEE Department, Panimalar Engineering College, Chennai 600123, Tamilnadu, India

Abstract. Research and improvement of micro grid has turn out to be a huge topic as it paves a manner to efficiently integrate numerous resources of dispensed generation (DG), particularly Renewable Energy Sources (RES) which include photovoltaic, wind and fuel cellular generations without requiring re-design of the distribution gadget. While the usage of the Renewable Energy Sources its miles very important to utilize the most available power from the resource. To utilize the maximum power the conversion system must operate in the point of maximum power, for this purpose a variety of MPPT algorithms were introduced. In this paper various MPPT methods consisting of P&O, Incremental conductance, and Fuzzy Logic control had been analyzed and a new MPPT method based on Whale Optimization Algorithm is proposed for tracking most power from solar electricity and wind power and its overall performance became analyzed and compared with the alternative MPPT strategies. Also a STATCOM with proper control is introduced in the micro grid system in order to improve the stability of the system. The whole micro grid system is implemented and verified using MATLAB/Simulink.

Keywords: MPPT techniques · Renewable sources · ANFIS · STATCOM · Micro grid

1 Introduction

Due to ever growing energy intake and worldwide climate exchange issues, allotted generations, micro grid and renewable energy technology had been acquired extra interest and the concept of micro grid and distributed generations are exceptionally promising to beautify the fine, reliability, typical overall performance of the electrical energy gadget. The scope on disbursed generations and micro grid goes on increasing because of the need of reliable power materials. From the conventional method of energy technology the dispensed energy technology is greater handy. From the conventional energy generation the disbursed power generation is being used due to clean

D. J. Hemanth et al. (Eds.): COMET 2019, LNDECT 35, pp. 342–360, 2020.
https://doi.org/10.1007/978-3-030-32150-5_35

and reliability. To increase the reliability the micro grids combines each the dispensed power generation and renewable electricity assets. More papers associated with non-conventional power assets supplied to micro grids are studied, due to the fact they're powerful against environmental influences of existing producing gadget [1–6].

Distributed power assets include photovoltaic (PV), generators, gas cells, and engine mills and so forth. The hybrid device is formed by using the aggregate of solar and wind power system to generate the electrical energy which reduces the value and protection. So by means of the usage of the hybrid system it reduces the environmental pollutions consisting of greenhouse impact and also consumes gasoline usage. MGs can each be associated with the grid (i.e., grid-related mode) or use the Distributed Energy Resources (DERs) to deliver the loads without the grid (i.e., islanded mode).

The easy downside of renewable sources is the change in output energy because of the fluctuation of renewable energy availability that's based on the region, time, weather and climate, especially in PV and wind structures, and it is effective to slight the ones oscillations. Due to adjustments in inputs of both sun and wind power the output power receives affected. The hybrid wind and solar power systems are escort by using battery storage gadget to decorate system reliability and performance.

As the actual environment has exclusive wind pace situations and sun irradiation, it is very vital to utilize the maximum available power from the property to its maximal power conversion output. Hence the Maximum Power Point Tracking (MPPT) is needed, which will keep the device output at maximum energy in particular conditions. In order to determine the superior running component, a most electricity aspect tracking set of policies is vital to be covered in the device. Several types of MPPT algorithms had been proposed in the literatures. Perturb and Observe is a generally used MPPT technique due to its ease of implementations. This method is primarily based on disturbing the voltage and gazing the alternate in energy and changing the perturb course to attain point. Such conventional techniques inclusive of incremental conductance (IC) and Hill Climbing are clean to implement and coffee cost however famous much less monitoring overall performance [7–11]. Compared to the conventional MPPT technique it's acknowledged that the intelligent control techniques exhibit better performance. Fuzzy Logic Control and genetic algorithm are applied for various structures [12–16]. Neural Networks and Neural Networks based totally FLC also applied for solar and wind conversion systems [14, 17–19]. Despite of the consequences, the complexity, want of professional understanding in FLC and structural obstacles of NN are the main downside of Intelligent Controllers.

In this paper a new MPPT approach is proposed for a grid connected micro grid system together with solar panel, wind turbine and a battery. The new MPPT approach is based on the special conduct of humpback whales called Whale Optimization Algorithm. The proposed set of policies is implemented to the device and its overall performance is analyzed and compared with the traditional MPPT strategies. This paper is based as follows. Section 2 summarizes the define of entire micro grid device. Section 3 offers the outline of traditional MPPT strategies P&O, IC and FLC. Also it discusses the proposed method and its implementation to the device. Section 4 gives the implementation of proposed approach in MATLAB/Simulink and the evaluation of results received. At remaining the very last dialogue approximately the paper and the conclusion is given in Sect. 5.

2 System Description

The schematic diagram of proposed micro grid is shown below; it composed of three Distributed Energy Resources: Photovoltaic (PV), wind power conversion system (WECS) and Fuel Cell stack.

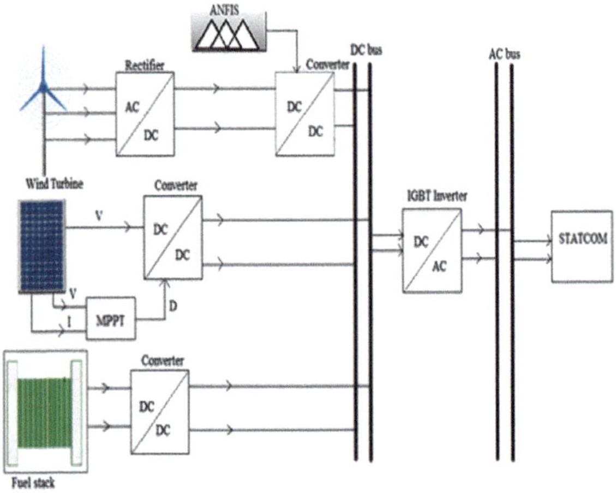

Fig. 1. The schematic diagram of proposed system

2.1 Wind Energy Conversion System

In WECS, Wind is used to generate electrical power through the use of wind turbine and electric generator. Using rotor shaft the electrical energy is transformed into mechanical power. This technique alters mechanical power to magnetic power and subsequently to electrical energy for the utility grid.

The mean value of wind turbine is represented as follows:

$$V_m = \sum\nolimits_{i=1}^{nr} V_i \, [17] \tag{1}$$

Where

Vm = Annual mean wind speed
Vi = Input voltage

The air pressure from the wind rotates the vanes of the wind turbine to produce kinetic energy or mechanical energy. Under the effect of aerodynamic force, the vanes generate torque. The mechanical output power *Pmec* is given by

$$P_{mec} = \frac{1}{2}\rho A V_m^3 C_p \tag{2}$$

$$T_{mec} = \frac{P_{mec}}{\omega_m} = \frac{\frac{1}{2}\rho A V_m^3 C_p}{\omega_m} \tag{3}$$

Here P_{mec} represents the power extracted from the wind (W), T_{mec} denotes the torque developed, ρ represents the air density (kg/m^3), A denotes the rotor disk area (m^2), V_m represents the wind speed (m/s), ω_m is the angular speed of the turbine and Cp is the power coefficient that is a function of the tip speed ratio (λ) and the pitch angle (β) of rotor blades.

The ratio of tip speed of the wind turbine is defined as follows [17]

$$\lambda = \frac{R\omega_m}{V_m} \tag{4}$$

Here *R* represents the radius of the turbine blade.

2.2 PV System

A photovoltaic cellular or photovoltaic cell is a semiconductor tool that transforms light energy in to electric energy by means of the use of photovoltaic effect. The waft of electrons will create current, while the energy photo light is greater than band hole.

Solar cells show off non-linear mean value of wind applied to the wind turbine Eq. (1) that primarily based on solar radiation and temperature.

$$I = I_{ph} - I_o\left(e^{\frac{q(V+IR_S)}{nKT}}\right) - \frac{V + IR_s}{R_{sh}} \tag{5}$$

Where

Iph = Photoelectric current
Io = Output saturation current of the diode
n = Diode ideality factor
k = Boltzmann's constant (1.4×10^{-23})
q = Electron charge (1.6×10^{-19})
T = Absolute temperature in Kelvin.

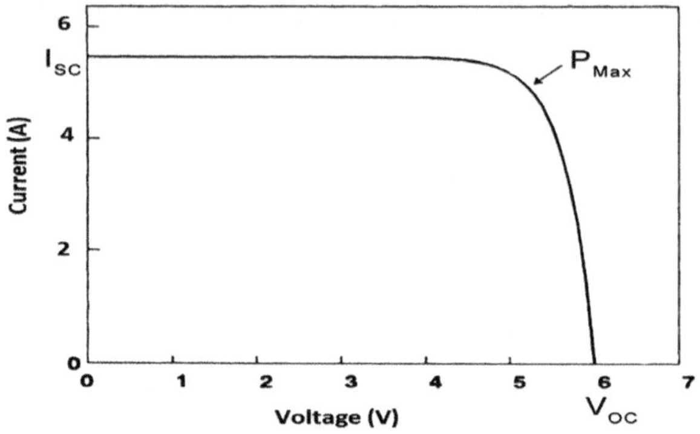

Fig. 2. VI characteristics curve of photovoltaic cell

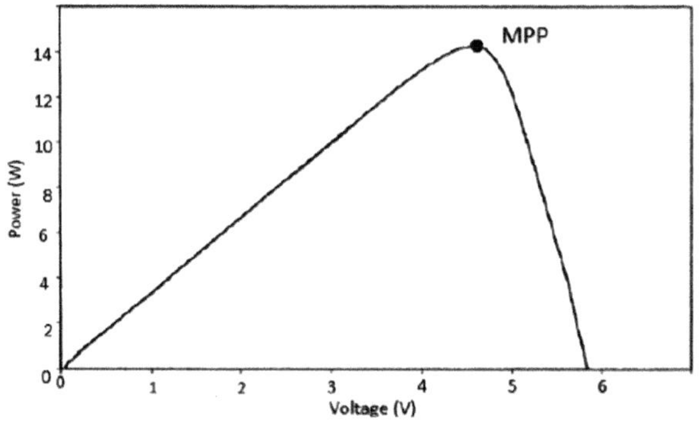

Fig. 3. PV characteristics curve of photovoltaic cell

The made from current and voltage traits is proven in Fig. 3. The MPPTs represents most panel power output. The system might improve the PV cells efficiency and had a crucial importance.

- At cold temperature PV module works well the usage of MPPT obtained maximum electricity.
- If the state of charge in the battery is lowers, MPPT can attain more current and charge the battery.

2.3 Fuel Cell Stack

Generally Fuel Cells offer excessive efficiency clean alternative [20] than nowadays electricity generation technologies. In medium electricity industrial applications, the polymer electrolytes membrane (PEM) Fuel Cell has gained some acceptance [20].

The unique simulation model of the Proton Exchange Membrane Fuel Cell (PEMFC) has been advanced, in [20]. In Fig. 4, the static V-I polarization curve for a fuel cellular is shown, wherein the drop of the gasoline cellular voltage with load

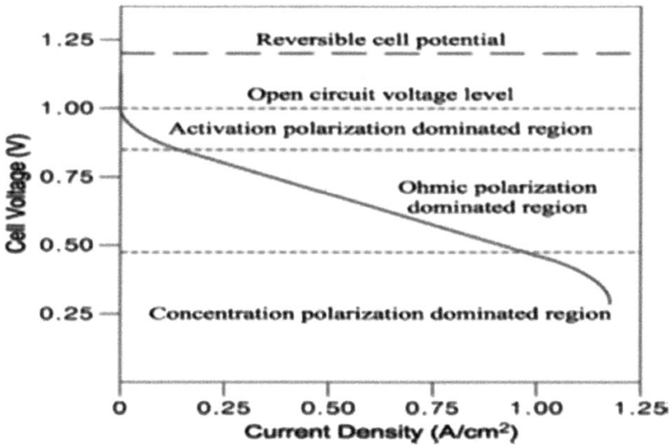

Fig. 4. Polarization curve of fuel cell

modern-day density can be observed. This is due to three primary losses: activation loss, ohmic loss, and delivery loss [20].

The most cellular voltage Vc is the internet output voltage proven in Eq. (6), [20]

$$V_c(i) = V_{rv} - V_{irv} \tag{6}$$

Here V_{rv} is reversible cell voltage and V_{irv} is irreversible voltage loss.

The irreversible voltage loss is the combination of the activation loss V_{act}, ohmic loss V_{ohm}, and concentration loss V_{con} which is shown in Eq. (7) [20]

$$V_{irv} = -V_{act} + V_{ohm} + V_{con} \tag{7}$$

3 MPPT Algorithm

3.1 Incremental Conductance

The slope of power and voltage is 0 at the MPP, increasing on the left facet of the MPP and reducing on the proper hand aspect of the MPP. The primary equations for this technique are as follows

$$\begin{cases} \frac{dP}{dV} = 0 & at\,MPP \\ \frac{dP}{dV} > 0 & left\,of\,MPP \\ \frac{dP}{dV} < 0 & right\,of\,MPP \end{cases} \tag{8}$$

As

$$\frac{dP}{dV} = \frac{d(IV)}{dV} = I + V\frac{dI}{dV} \cong I + V\frac{\Delta I}{\Delta V} \tag{9}$$

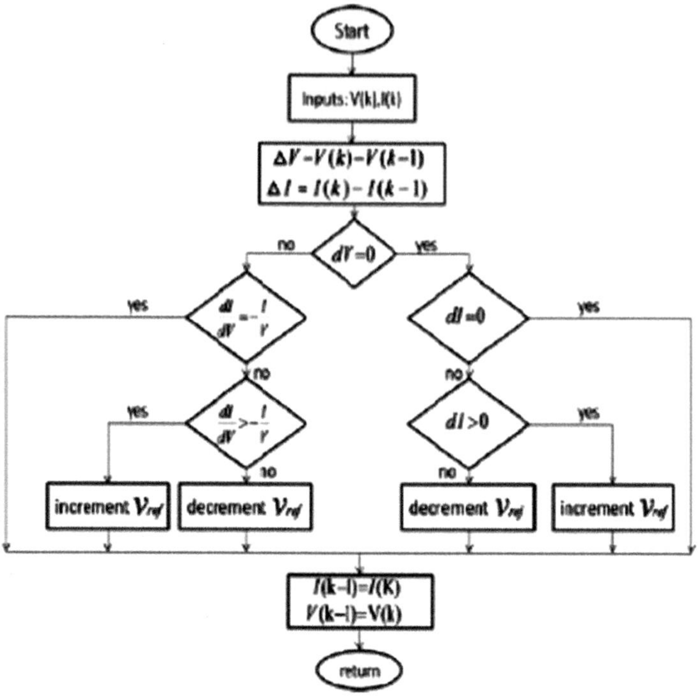

Fig. 5. Flowchart of the Incremental Conductance Algorithm

Equation (8) is written as follows (7) (di/dv is change in current with respect to voltage)

$$\begin{cases} \frac{\Delta I}{\Delta V} = -\frac{I}{V} \; at\, MPP \\ \frac{\Delta I}{\Delta V} > -\frac{I}{V} \; left\ of\ MPP \\ \frac{\Delta I}{\Delta V} < -\frac{I}{V} \, right\ of\ MPP \end{cases} \tag{10}$$

From the above equations, the MPP may be tracked by way of comparing the immediate conductance with the increment in conductance. V_{ref} is the reference voltage. At MPP if the circumstance V_{re}f equals to V_{MPP} at that time MPP is reached and operation is maintained at that point.

Error signal is defined as the ratio between instantaneous conductance by the incremental conductance.

$$e = \frac{I}{V} + \frac{dI}{dV} \tag{11}$$

The flowchart of the Incremental Conductance algorithm is taken from reference [22]. The drawback of this algorithm is

- Complexity to control
- High Noise
- Solar panel power is a nonlinear function of duty cycle.

3.2 Fuzzy Logic Controllers

A Fuzzy logic controller doesn't need accurate mathematical model and coping with nonlinearity. They have the advantage along with robust, easy to design and didn't want the know-how of specific version. The parameter of such controllers may be various in responses to the system dynamics. For the modifications in climatic situations fuzzy primarily based MPPT is powerful however it'll rely up at the designer's expertise about the device.

Generally Fuzzy logic control includes three levels: fuzzification, rule base table lookup, and defuzzification. Based on membership characteristic, at some point of fuzzification numerical enter variables are transformed into linguistic variables.

Defuzzification: This operation converts fuzzy units in to crisp set. There are numerous techniques to attain defuzzification. One of the approach is center of gravity technique. In this COG method the center value is chosen for defuzzification.

Generally it's far very difficult to enforce the fuzzy logic controller. The designer doesn't require the knowledge of operation of the PV device.

The inputs given to fuzzy logic controller is an error E (trimf). Then change in mistakes ΔE which are defined as follows (tramf).

$$E(n) = \frac{P(n) - P(n-1)}{V(n) - V(n-1)} \tag{12}$$

$$\Delta E(n) = E(n) - E(n-1) \tag{13}$$

In this example, six fuzzy ranges are used as enter and output variables (mf1, mf2, mf3, mf4, mf5.Mf6 & mf7) are as follows: NL (negative Large), NS (poor small), ZE (0), PS (tremendous small), PB (positive huge) and PS (Positive small) (Table 1).

Table 1. Fuzzy rule base table

E	ΔE					
	NL	NM	NS	Z	PS	PM
NL	PL	PL	PL	PL	NM	Z
NM	PL	PL	PL	PM	PS	Z
NS	PL	PM	PS	PS	PS	Z
Z	PL	PM	PS	Z	NS	NM
PS	Z	Z	NM	NS	NS	NM
PM	Z	Z	NS	NL	NL	NL

Here we are using the methods such as Implication, Aggregation and Defuzzification.

Maximum power point tracking (MPPT) is applied for extracting maximum energy from the PV panel. To meet the load demand for, MPPT is utilized in all climate situations; the MPPT should be able to maximize the electricity output from the solar panel. It is essential to get regular power from the supply. They have the gain which includes sturdy and relatively simple to layout.

- It isn't useful for packages a good deal larger or smaller than the historic facts.
- It requires lots of information
- The estimators must be familiar with the traditionally developed application.

3.3 Perturb and Observe

In P&O approach most effective a sensor used to sense the voltage of PV system. As it is straightforward to implement, much less complicated and low price, this approach is commonly used. In this method the voltage of the PV array is perturbed by using a small amount and the alternate in energy is observes. If the alternate in power is positive then perturb is sustained inside the equal direction. And if the trade is negative perturb is sustained in the opposite course. Figure 6 indicates the flowchart of the P&O algorithm.

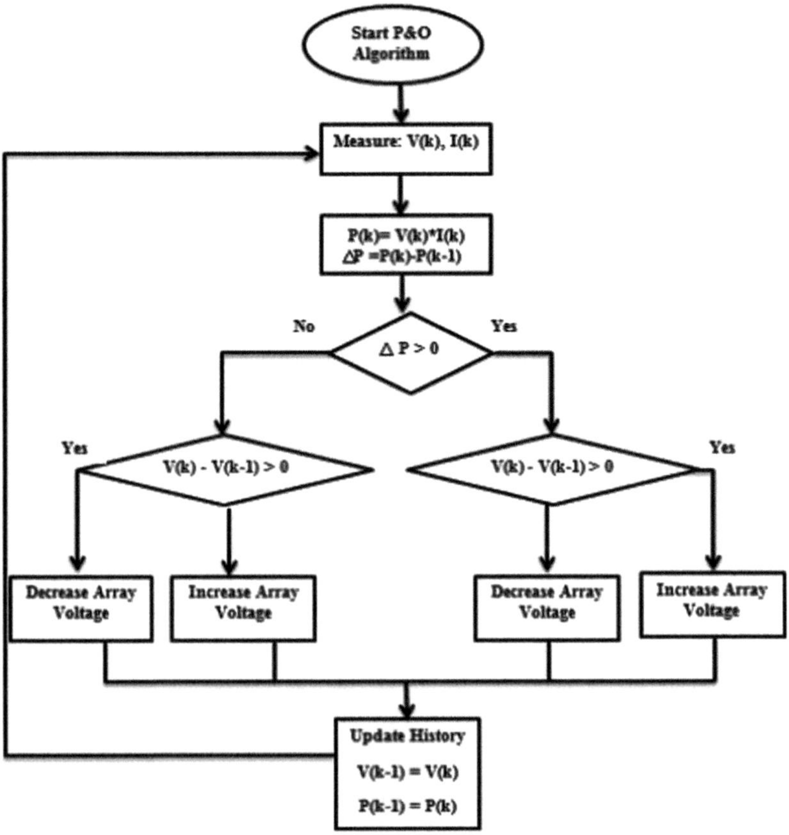

Fig. 6. Flowchart of the P&O algorithm

The flow chart of the P&O algorithm is taken from reference [23].

When the irradiation modifications suddenly, the alternate in MPP is taken into consideration as a trade due to perturbation and inside the next step the course of perturbation is changed. The downside of this set of rules is even the algorithm reaches the MPP it keeps on perturbing in both the directions, which will increase the time complexity of the algorithm.

3.4 Whale Optimization Algorithm

Whale optimization algorithm is inspired from the hunting technique of a humpback whale. The looking approach of humpback is based totally at the mechanism of bubble internet feeding with a spiral motion. When those whales locate their prey encircles them and hunt them by way of developing exceptional bubbles in circular route.

Initially the whale begins to go looking the prey in a random manner then it's going to look for a random coefficient vector. It is defined as follows [21]. This set of rules starts with a exceptional seek agent. Assume the current solutions are best, that is the location of prey or close to it.

The remaining agents appraise their positions towards the best search agent. It can be conveyed as the following

$$\vec{X}(k+1) = \vec{X}^*(k) - \vec{A}.\vec{D} \tag{14}$$

Here $\vec{X}(k+1)$ denotes the position vector of whales in (k + 1)th iteration, $\vec{X}^*(k)$ denotes the best position of current kthiteration which is updated for every iteration. The coefficient vectors $\vec{A}$, $\vec{D}$ are intended as follows.

$$\vec{A} = \vec{a}(2\vec{r} - 1) \tag{15}$$

$$\vec{C} = 2.\vec{r} \tag{16}$$

$$\vec{D} = 2\vec{r}\vec{X}^*(k) - \vec{X}(k) \tag{17}$$

Here $\vec{a}$ linearly decreases to 0 from 2 depend upon iteration and $\vec{r}$ is random vector in between [0, 1].

The spiral Equation is formed by humpback whales between the whale and position of prey to pretend the helix shape movement. It can state as the following:

$$\vec{X}(t+1) = \overrightarrow{D'} \cdot e^{bl} \cdot \cos(2\Pi l) + \overrightarrow{X^*}(t) \tag{18}$$

$$\overrightarrow{D'} = \left|\overrightarrow{X^*}(t) - \vec{X}(t)\right| \tag{19}$$

Bubble Net Attacking Method in WOA

There are tactics designed for bubble net conduct of whales Shrinking Encircling Mechanism: This technique can be completed through reducing the value of $\vec{a}$. Also the fluctuation range of $(\vec{A})$ is decreased by way of a $\vec{a}$. By putting random value for A, the brand new role of a search agent can be in among original function and present day function of agent.

Spiral Updating position: This method calculates the whale placed and the prey placed. Then the spiral equation is calculated between the prey and whale position.

MPPT Application for WOA

MPPT is framed as objective feature and it's far signified as:

$$\text{Maximize the energy P(d)} \tag{20}$$

$$\text{Subjected to}\, d_{max} \geq d \geq d_{min} \tag{21}$$

Right here P represents the power output, d represents the duty ratio, dmin, dmax represent the minimum, most obligation ratio limits i.e., 0.1, 0.9, respectively.

Here to obtain MPPT using WOA the populace of whales is taken into consideration as obligation ratio.

Equation (14) is rewritten as follows

$$d_i(K+1) = d_i(k) - A.D \tag{22}$$

The objective function of WOA MPP
T is formulated as

$$P(d_i^k) > P(d_i^{k-1}) \tag{23}$$

here i is the population of whales

PV and wind power is dependent on climatic situations, for an alternate in sun irradiation or wind velocity, the output power modifications as a consequence and proposed MPPT set of rules is reinitialized by sensing the change in PV output strength. By the use of wind the wind turbine begins to rotate it will produces high power [20]. Wind power is extracted from the kinetic energy of the wind. To explicit the power of the wind turbine, it's far asserted that this energy relies upon on 3 parameters: wind, air density and swept surface.

$$\frac{P^k - P^{k-1}}{P^k} \geq 0.1 \tag{24}$$

The detailed proposed WOA MPPT algorithm is presented in Fig. 7.

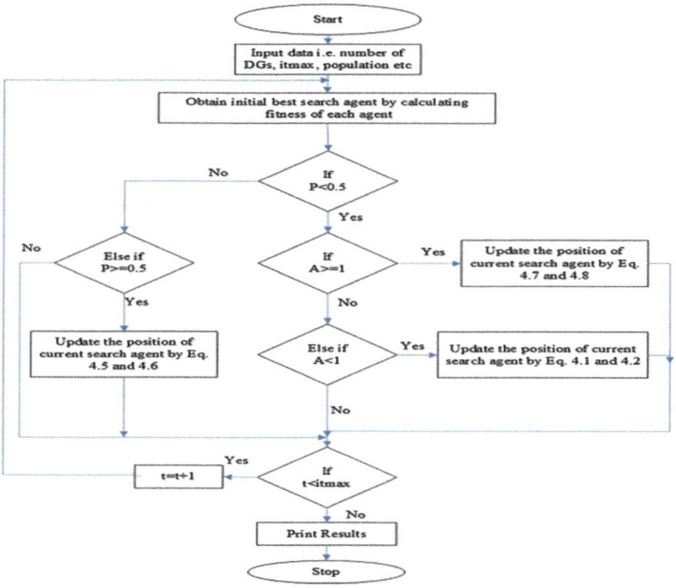

Fig. 7. Flow Chart of Whale Optimization Algorithm based MPPT technique the flow chart of Whale Optimization Algorithm is taken from reference [24].

System Parameters

Table 2. Simulation parameter

Wind induction generator		Diesel synchronous generator	
Rated power (kW)	2	Rated power (kW)	12
Voltage (V)	220	Voltage (V)	220
Frequency (Hz)	60	Frequency (Hz)	60
Inertia (kg m^2)	0.7065	Inertia (kg m^2)	0.8
No. of pole pairs	2	No. of pole pairs	2
Wind speed (m/s)	8–12		
PV array		**Grid**	
Total power rating (kW)	$0.2 \times 16 = 3.2$	Voltage (V)	220
Maximum power of module unit (W)	200,143,25 °C. 1000 W/m^2	Frequency (Hz)	60
Unit rated voltage (V)	26.3	Phase	3
Unit rated current (A)	7.61	Capacitive load (kVAR)	1–6
Module number	$4 \times 4 = 16$	Inductive load (kVAR)	5
Irradiance level (W/m^2)	600–1000		

4 Simulation Results

The proposed grid connected micro grid system is implemented in Matlab/Simulink. The parameters of the solar panel, wind turbine and the fuel cell used in the micro grid are represented in Table 2. The output of the micro grid system under various conditions is analyzed and the performance of proposed MPPT technique based Whale Optimization Algorithm analyzed and compared with the other MPPT techniques.

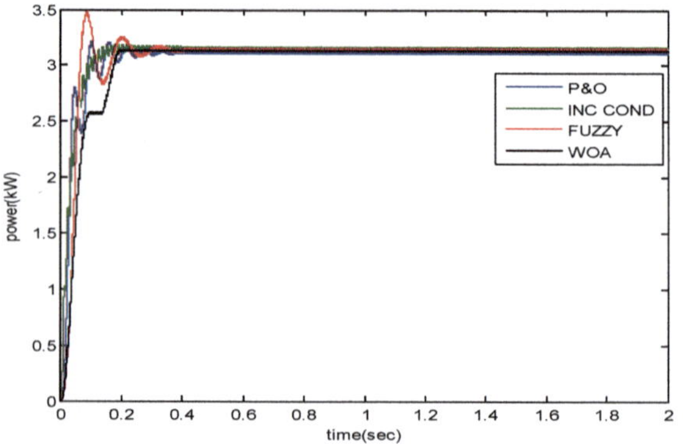

Fig. 8. Output power of the PV system under different MPPT techniques

Figure 8 shows the power output of the PV system under a constant irradiation of 1000 W/m^2. Under this irradiation PV system provides its maximum output power 3.2 kW. From the figure it is inferred that the WOA tracks the MPP quickly compared to the other techniques (Fig. 8).

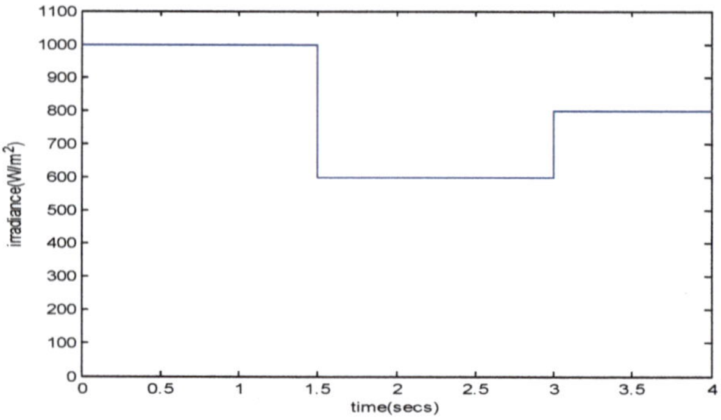

Fig. 9. Variation in Irradiation

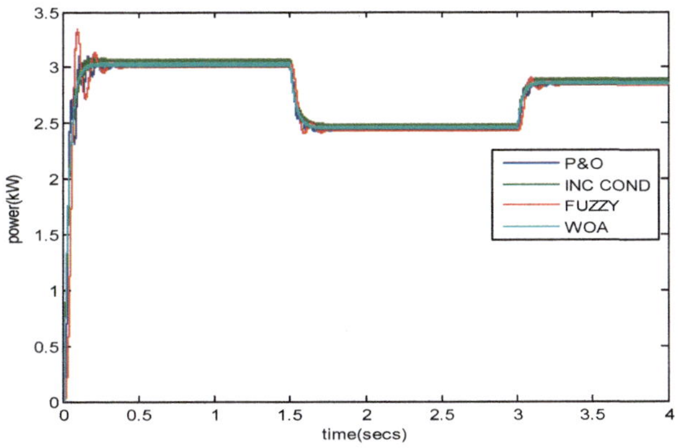

Fig. 10. Power Response of PV system under variation of Irradiation

Figure 9 represents the variation in irradiation of Solar with respect to time. Initially the irradiation is 1000 W/m^2, at the time of 1.5 s it is dropped to 600 W/m^{2}and at 3 s it is changed to 800 W/m^2 and the Fig. 11 shows the output power response of the PV system under the variable irradiation. When the irradiation is dropped to 600 W/m^2 the power also dropped to 2.5 W and with its raise to 800 W/m^2 power increased to 2.8 W. When the power of the PV system reduced because of the drop in irradiation the load is shared by the WT and Fuel Cell Stack.

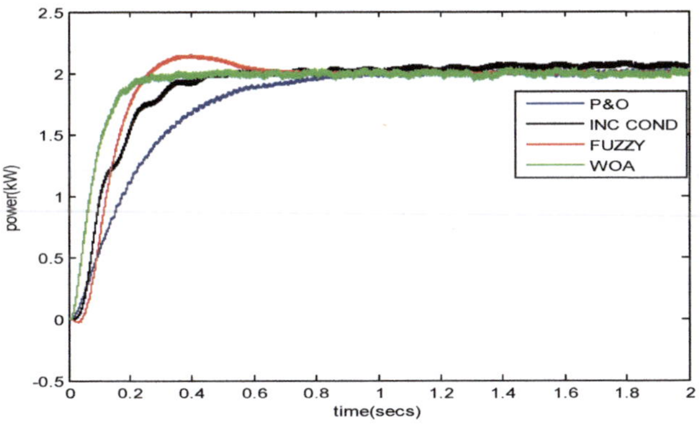

Fig. 11. Output power of the WT under different MPPT techniques

Figure 11 shows the power output of the wind turbine under a wind speed of 12 m/sec. From the above figure it is clear that the WOA tracks the MPP quickly compared to the other MPPT techniques (Fig. 11).

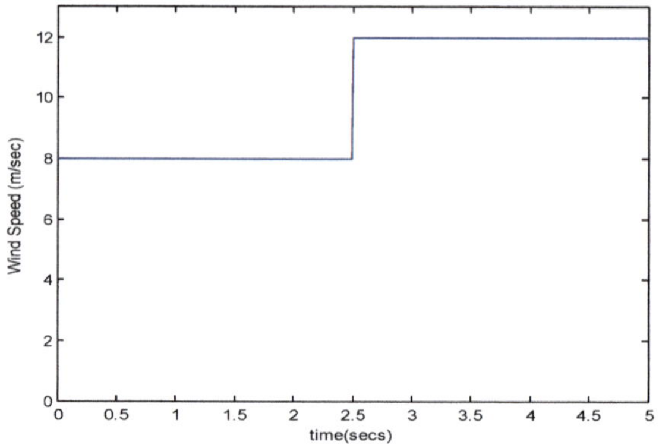

Fig. 12. Variation in Wind Speed

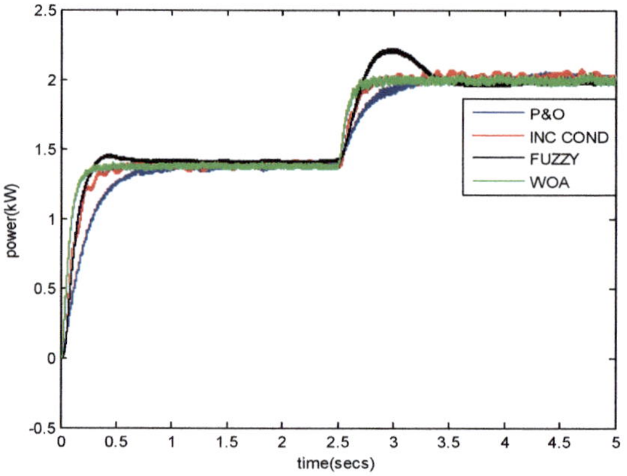

Fig. 13. Power Response of WT under variation in Wind Speed

Figure 12 represents the variation in wind speed with respect to time, initially the wind speed is 8 m/s up to 2.5 s then it is raised to 12 m/s. Figure 13 shows the output power response of the wind turbine based on the variation in wind speed. At a wind speed of 12 m/s the wind turbine outputs a power of 2 kW. From the figure it is visible that the WOA helps the wind turbine to attain maximum power under varying wind speed.

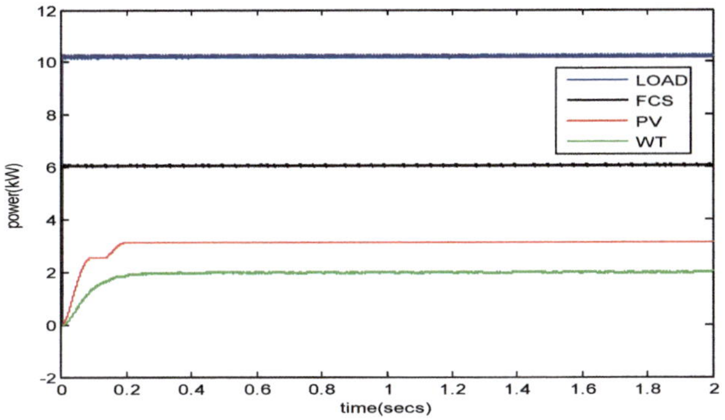

Fig. 14. Real power of the Micro grid system

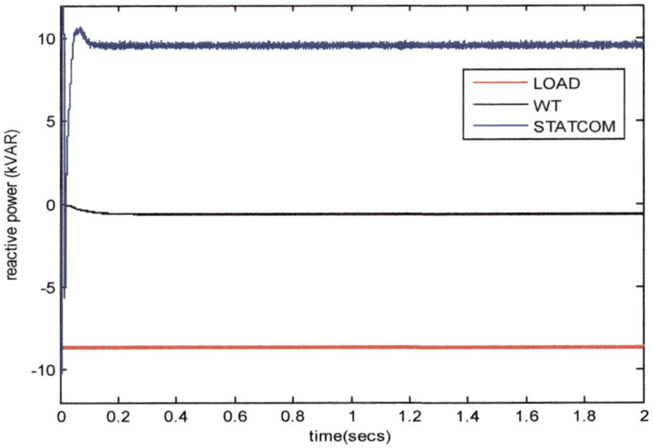

Fig. 15. Reactive power of the Micro grid system

Figure 14 shows the real power output of the micro grid with constant solar irradiation and constant wind speed and Fig. 15 shows the reactive power of the micro grid system.

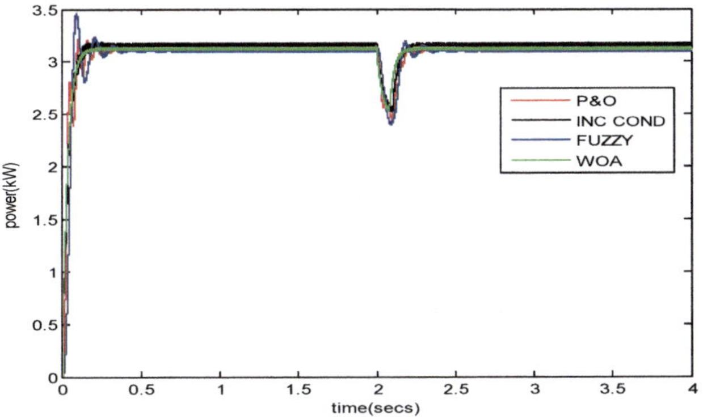

Fig. 16. Output power of the PV system under fault

Figure 16 shows the output of the PV system under a fault at the time of 2 s. The figure shows that after the sudden drop of power the system reaches its maximum power. It is also clear that the proposed WOA based MPPT helps the PV system to reach the maximum output power quickly.

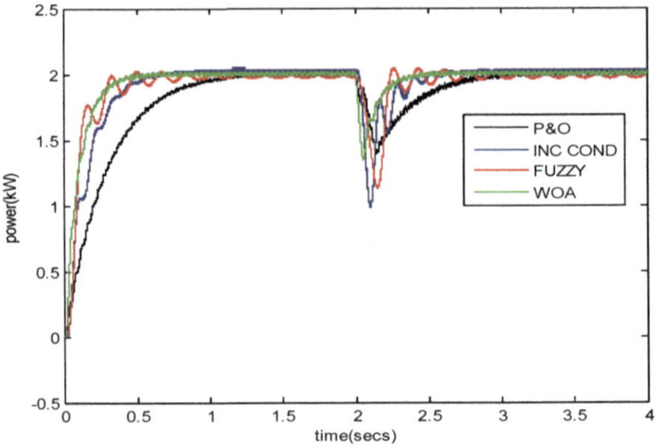

Fig. 17. Output power of the Wind Turbine under fault

Figure 17 shows the output of the wind turbine under a fault at the time of 2 s. The figure shows that after the sudden drop of power the system reaches its maximum power. It is also clear that the WOA based MPPT helps the wind turbine to reach the maximum output power quickly compared to the other MPPT techniques.

During the fault conditions the reactive power required to the system is provided from the STATCOM, which can provide fast reactive power support and thus can stabilize the bus voltage.

5 Conclusion

In this paper a micro grid with a PV array, wind turbine and a fuel cell stack is simulated along with a STATCOM to ensure the stability of the micro grid. A new MPPT approach based at the Whale Optimization Algorithm is proposed to track maximum power from the renewable energy sources along with Solar and Wind generation system of a microgrid. The other MPPT techniques such as P&O, Incremental Conductance and Fuzzy are implemented to the PV array. The wind turbine and their performances were compared to the proposed WOA based MPPT techniques under varying climatic conditions and fault conditions. It is identified that even if the implementation of P&O technique is simple but its performance is not so good and the response of other two techniques such as Incremental Conductance and Fuzzy are oscillatory. It is also verified that the performance of the WOA based MPPT technique is better compared to the other techniques and it tracks the maximum power fastly after the fault.

References

1. Taha, S.: Recent developments in micro-grids and example cases around the world—a review. Renew. Sustain. Energy Rev. **15**, 4030–4041 (2011)
2. Alavi, S.A., Ahmadian, A., Aliakbar-Golkar, M.: Optimal probabilistic energy management in a typical micro-grid based-on robust optimization and point estimate method. Energy Convers. Manag. **95**, 314–325 (2015)
3. Logenthiran, T., Srinivasan, D., Khambadkone, A.M., Raj, T.S.: Optimal sizing of an islanded micro-grid using evolutionary strategy. In: International Conference on Probabilistic Methods Applied to Power Systems (PMAPS), Singapore. IEEE (2010)
4. Colson, C.M., Nehrir, M.H., Wang, C.: Ant colony optimization for micro-grid multi-objective power management. In: Power Systems Conference and Exposition, PSCE 2009, WA, Seattle (2009)
5. Elsied, M., Oukaour, A., Gualous, H., Hassan, R., Amin, A.: An advanced energy management of micro-grid system based on genetic algorithm, ISIE, Istanbul. IEEE (2014)
6. Elsied, M., Oukaour, A., Gualous, H., Hassan, R.: Energy management and optimization in micro-grid system based on green energy. Energy J. **84**(May), 139–151 (2015)
7. Sera, D., Mathe, L., Kerekes, T., Spataru, S.V., Teodorescu, R.: On the perturb-and-observe and incremental conductance MPPT methods for PV systems. IEEE J. Photovolt. **3**(3), 1070–1078 (2013)
8. Femia, N., Petrone, G., Spagnuolo, G., Vitelli, M.: Optimizing duty-cycle perturbation of P&O MPPT technique. In: IEEE Transactions on Power Electronics Conference, 20–25 June (2004)
9. Femia, N., Petrone, G., Spagnuolo, G., Vitelli, M.: Optimization of perturb and observe maximum power point tracking method. IEEE Trans. Power Electron. **20**(4), 963–973 (2005)
10. Faraji, R., Rouholamini, A., Naji, H.R., Fadaeinedjad, R., Chavoshian, M.R.: FPGA-based real time incremental conductance maximum power point tracking controller for photovoltaic systems. IET Power Electron. **7**, 1294–1304 (2014)
11. Kish, G.J., Lee, J.J., Lehn, P.W.: Modelling and control of photovoltaic panels utilising the incremental conductance method for maximum power point tracking. IET Renew. Power Gener. **6**, 259–266 (2012)
12. Wilamowski, B.M., Li, X.: Fuzzy system based maximum power point tracking for PV system. In: 28th Annual Conference of the IEEE Industrial Electronics Society, pp. 3280–3284 (2002)
13. Prakash, J., Sahoo, S.K., Karthikeyan, S.P., Raglend, I.J.: Design of PSO-Fuzzy MPPT controller for photovoltaic application. In: Power Electronics and Renewable Energy Systems, India, pp. 1339–1348. Springer (2015)
14. Salah, C.B., Ouali, M.: Comparison of fuzzy logic and neural network in maximum power point tracker for PV systems. Electr. Power Syst. Res. **81**(1), 43–50 (2011)
15. Larbes, C., Cheikh, S.A., Obeidi, T., Zerguerras, A.: Genetic algorithms optimized fuzzy logic control for the maximum power point tracking in photovoltaic system. Renew. Energy **34**(10), 2093–2100 (2009)
16. Daraban, S., Petreus, D., Morel, C.: A novel MPPT algorithm based on a modified genetic algorithm specialized on tracking the global maximum power point in photovoltaic systems affected by partial shading. Energy **74**, 374–388 (2014)
17. Cirrincione, M., Pucci, M., Vitale, G.: Neural MPPT of variable-pitch wind generators with induction machines in a wide wind speed range. IEEE Trans. Ind. Appl. **49**(2), 942–953 (2013)

18. Chekired, F., Mellit, A., Kalogirou, S.A., Larbes, C.: Intelligent maximum power point trackers for photovoltaic applications using FPGA chip: a comparative study. Sol. Energy **101**, 83–99 (2014)
19. El Fadil, H., Giri, F., Guerrero, J.M.: Adaptive sliding mode control of interleaved parallel boost converter for fuel cell energy generation system. Math. Comput. Simul. **91**, 193–210 (2013)
20. Jung, J.H., Ahmed, S., Enjeti, P.: PEM fuel cell stacks model development for real time simulation application. IEEE Trans. Ind. Electron. **58**(9), 4217–4225 (2011)
21. Zawbaa, H.M., Emary, E.: Feature selection approach based on whale optimization algorithm. In: (ICACI) (2017)
22. Velkovski, B., Pejovski, D.: Application of incremental conductance MPPT method for a photovoltaic generator in LabView. In: Poster 20th International Student Conference on Electrical Engineering, pp. 1–6 (2016)
23. Abdulkadir, M., Samosir, A.S., Yatim, A.H.M.: Modelling and simulation of maximum power point tracking of photovoltaic system in Simulink model. In: 2012 IEEE International Conference on Power and Energy (PECon), pp. 325–330. IEEE (2012)
24. Reddy, P.D.P., Reddy, V.C.V., Manohar, T.G.: Whale optimization algorithm for optimal sizing of renewable resources for loss reduction in distribution systems. Renew. Wind Water Solar **4**(1), 3 (2017)

Corrosion Control Through Diffusion Control by Post Thermal Curing Techniques for Fiber Reinforced Plastic Composites

S. J. Elphej Churchil(✉) and S. Prakash

Sathyabama Institute of Science and Technology, Chennai 600119, India
elphej@gmail.com, prakash_s1969@yahoo.com

Abstract. Through this study, the moisture absorption behavior of composite materials with different post curing temperatures was determined to observe the corrosion control rate in materials. Due to the capillary effect, fibers absorb moisture from the environment and this may cause swelling in composites. This phenomenon develops additional strain in the laminate and accelerates strength degradation with the combined effect of corrosion. For the analysis, composite laminates of basalt/epoxy, basalt/polyester, glass/epoxy, and glass/polyester were fabricated using a hand layup process and cured at room temperature. Some of these laminates were further cured in a convection oven at different temperatures and some using a microwave oven. Samples of basalt/epoxy, basalt/polyester, glass/epoxy, and glass/polyester fiber reinforced laminates' dry weights had been noticed and after that, they were immersed in NaCl solution (30% NaCl in one liter of distilled water) for an absorption test at room temperature. The initial weights of post-cured materials were weighed using a physical balance and tabulated. Similarly, the weights of the post-cured materials under moisture absorption were weighed at an interval of 7 days for a month and compared their properties.

Keywords: Diffusion and corrosion control · Fiber reinforced polymer composite · Microwave post-curing

1 Introduction

Moisture absorption behavior of fibrous materials can be controlled by post-curing technique as per the following literature. The capillary effect depends on the volume of trapped air or voids present in the laminate, this trapped air is relieved from the material using post heat treatment [1]. It is understood that the composites with a fiber volume fraction of 44% absorb more water than the composites with a fiber volume fraction of 34% because capillary paths in the higher fiber volume fractions are greater. From the literature, it is observed that after 42 days of immersion the tensile strength, flexural strength, and interlaminar shear strength of the composite specimens decreased 30%, 62%, and 57% of the original state [2]. In a study on absorption behavior of fiber reinforced polyester hybrid composites of jute and glass in distilled water sea water and acid solution. The investigator observed that the specimen in distilled water has the highest diffusion coefficient and moisture content. He concluded that the strength of

D. J. Hemanth et al. (Eds.): COMET 2019, LNDECT 35, pp. 361–374, 2020.
https://doi.org/10.1007/978-3-030-32150-5_36

natural fiber could be improved by hybridization of synthetic fiber [3]. In this work it has been proved that the fundamental natural frequency of the material decreased, because it's mass increased and stiffness decreased, these two parameters decide the natural frequency of material [4]. This experimental study determines the effect of resin on the post-impact compressive behavior of carbon fiber is woven laminates at 190oC and found that its impact strength and stiffness have been improved considerably [5]. The moisture absorption behavior of glass fiber reinforced polymer composites and its influences on its dynamic mechanical properties for 0, 15, 30, 45, 60 days has been validated [6]. Due to Water absorption behavior, the tensile strength, the flexural strength, and the interlaminar shear strength of the composite specimens after 42 days immersion decreased 13%, 43%, 50%, respectively [7]. The increase in water absorption behavior of glass/jute fiber reinforced polymer hybrid composite decreases the flexural and compressive characteristics of the material [8]. The dimensional variation and deterioration of wood composite due to the absorption is highly reduced by adding basalt and glass particles with matrix [9]. Water absorption by composite materials and related effects are analyzed and understood that the major mechanical properties are highly influenced by its absorption behavior [10]. The effect could be controlled by post-curing technique or by adding waterproof particles such as basalt and glass or by using basalt fiber and glass fiber reinforced plastic composites with proper curing treatment.

2 Materials and Fabrications

2.1 Materials

Composite Laminates of basalt/epoxy, basalt/polyester, glass/epoxy, and glass/polyester were fabricated using hand layup molding technique; the constituent materials are quadraxial (0°/+45°/90°/−45) fabrics of basalt, glass and matrix of epoxy LY556 and polyester with hardener in the ratio of 60:40

2.2 Steps for the Fabrication Process

The fabrics of basalt and glass of 30 cm length and 20 cm width of 10 layers of each material were taken to fabricate 3 mm thick laminate with matrices of Epoxy and Polyester. The weight proportion of fabrics and matrix is 60:40; the weight of hardener (HY956) is 10% weight of Matrix. In hand layup molding, a Teflon sheet is used as a base layer for proper finishing, each layer of fabric was kept one over the other and each layer is bonded with adjacent layer by applying matrices of epoxy and polyester up to 3 mm thickness. The excess matrix was removed using a hand roller and a metal plate with some weight is kept on the laminate. It was left for curing under room temperature for about 24 h. The finished laminate of basalt/epoxy, basalt/polyester, glass/epoxy and glass/polyester was as shown in Figure (see Fig. 1).

Basalt/Epoxy | Glass/Epoxy | Work table and tools for fabrications

Fig. 1. Fabrication of fiber reinforced plastic laminates

3 Experimentation

3.1 Testing of Tensile and Bending Strengths Using UTM

The specimens were cut for the tensile test according to the ASTM standards (D3039). Specimens were rigidly fixed to the upper jaw and the moving lower jaw of the UTM, the movement of the lower jaw is controlled by the motor-operated hydraulic system and the deflection due to load transfer through the specimen are recorded using data acquisition system. The Crosshead speed of Universal testing machine is 2 mm/min. For every specimen, load versus deflection values were recorded in a table and graph-swereplotted. Similarly, three-point bending tests were performed for a set of specimens with dimensions as per ASTM standards (D790). The specimens were placed on the simply supported beam setup in the lowest part of the UTM bed. The middle jaw is moved downward to transfer load at the midspan of the specimen. The displacements due to different loads at midspan were recorded and graphs were plotted with the help of WIN UTM software, the complete experimental setup is shown in Figure (see Fig. 2).

UTM used for Testing | Tensile Testing | Tensile and Bending Testing

Fig. 2. Glimpses of testing Tensile Strengths using UTM

3.2 Post Curing Using Electric Oven and Microwave Oven

As per the ASTM Hygrothermal standards (D570), the specimens of Basalt-polyester, Glass-polyester, Basalt-epoxy, and Glass-epoxy were cut into 25 × 25 mm size for post curing and absorption tests.

As shown in Figure (see Fig. 3), an electric heat oven and a domestic microwave oven were used for post-curing the samples for the analysis. The maximum power output of the microwave oven is 900 Wand the frequency is 2.45 GHz with a turntable.

Electric Heating Oven Microwave Heating Oven

Fig. 3. Post Curing using Electric oven and microwave oven

The specimens were cured under various curing conditions with different curing temperatures and schedules using the above two types of oven, and methods for curing were explained below.

For normal curing, specimens were kept in the open air at room temperature (37 °C) for 24 h.

For the second method of post-curing, specimens were kept in the electric oven for 90 min with temperatures from 50 °C to 90 °C with an increment of 10 °C. Totally five sets of samples were cured for five different temperatures for the periods of 1 h, 2 h, 3 h, and 4 h.

The final method of curing is done using a microwave oven, in this method the heat is applied in a percentage of power (P) with different timings (t) called cycles.

In the first cycle (Pt1) the schedule is 100% power for 3 min + 80% power for 3 min + 70% power for 4 min + 30 min power off and the curing temperature is 80 °C. In the second cycle (Pt2) the schedule is 20% power for 6 min + 30% power for 4 min + 30 min off, and the curing temperature is 50 °C. In the third cycle (Pt3) the schedule is 80% power for 8 min + 90% power for 3 min + 3 min power off + 90% power for 5 min + 5 min off + 90% power for 3 min + 5 min off + 100% power for 3 min + 3 min off + 100% power for 3 min, and the curing temperature is 90 °C.

3.3 Water Absorption Study of Cured Specimens

As per ASTM Hygrothermal standards (D570) the specimens were prepared for absorption test. Initial weights of the specimen were weighed using physical balance and then they were kept in saline water for absorption. Specimens were weighed after 7 days of absorption then followed by 14 days, 21 days and 28 days in a month, the absorption study is done as shown in figures (see Fig. 4). The percentage of water absorbed for different curing temperatures and the corresponding diffusion coefficients were calculated for all categories and tabulated for the analysis.

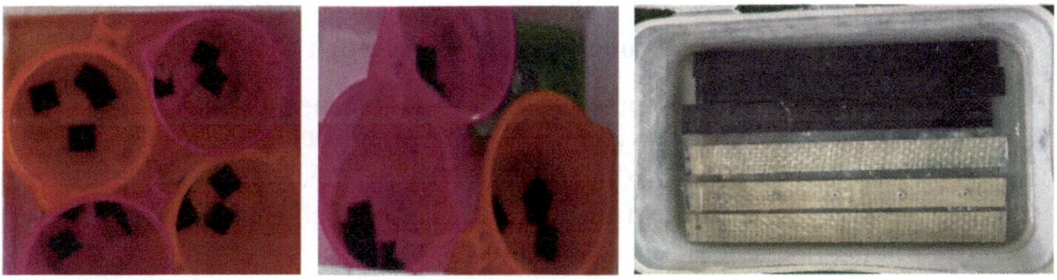

Naturally cured Laminate | Post Cured Laminate | Samples kept in solution for Tensile and Bending Tests

Fig. 4. Specimens kept in salt solution for absorption test

The formula to calculate the diffusion coefficient:

$$\% \text{ of water absorbed} = \frac{W - W_0}{W_0} \times 100$$

$$\text{Moisture absorption diffusion coefficient, } D_c = \pi \left(\frac{h}{4M_m}\right)^2 \left[\frac{W_2 - W_1}{\sqrt{t_2} - \sqrt{t_1}}\right]^2$$

4 Results and Discussions

4.1 Test Results of Tensile and Bending Strengths Using UTM

The results of both strength tests were reordered in the following Tables 1 and 2. The basalt/epoxy and glass/epoxy have better strength than basalt/polyester and glass/polyester. Similarly in flexural tests epoxy matrix composite shows better mechanical strength than polyester.

Table 1. Result of tensile and bending strengths

Sample	Tensile strength (natural curing) Mpa	Tensile strength (electrical oven curing) Mpa					Tensile strength (microwave oven curing) Mpa		
		50	60	70	80	90	Pt(1)	Pt(2)	Pt(3)
Basalt-epoxy	245.01	245.01	245.01	245.04	245.03	245.03	245.02	245.04	245.05
Basalt-polyester	210.03	210.03	210.03	210.03	210.04	210.04	210.05	210.06	210.09
Glass-epoxy	238.02	238.01	238.01	238.03	238.03	238.06	238.04	238.06	238.04
Glass-polyester	227.07	227.06	227.06	227.07	227.08	227.08	227.08	227.08	227.07

Table 2. Result of tensile and bending strengths

Sample	Bending strength (natural curing) Mpa	Bending strength (electrical oven curing) Mpa					Bending strength (microwave oven curing) Mpa		
		50	60	70	80	90	Pt(1)	Pt(2)	Pt(3)
Basalt-epoxy	569.04	569.04	569.04	569.06	569.05	569.05	569.02	569.04	569.07
Basalt-polyester	532.12	532.11	532.11	532.12	532.12	532.13	532.12	532.13	532.12
Glass-epoxy	553.33	553.33	553.31	553.33	553.34	553.34	553.35	553.35	553.34
Glass-polyester	495.24	495.23	495.23	495.23	495.24	495.24	495.24	495.26	495.26

4.2 Determination of Fiber Volume Fraction Through Digestion Test

The volume fractions of fiber and matrix were determined through digestion test. Specimens for digestion test were cut as per ASTM (D2584) standard and burned in a furnace oven at 500 °C; the matrix became ash after 2 h. The weights of the remaining unburned fibers were weighed and the volume fractions of composites were calculated. The fiber volume fractions are listed in Table 3:

Table 3. Volume fractions

Sample	Volume fractions	
	Fiber	Matrix
Basalt-epoxy	0.5867	0.4131
Basalt-polyester	0.6044	0.3956
Glass-epoxy	0.5764	0.4235
Glass-polyester	0.5972	0.4027

4.3 Absorption Behavior at Room Temperature

To study the absorption behavior of naturally cured specimens of Basalt- epoxy and Glass-epoxy were prepared as per ASTM (D570). Then these specimens were kept in saline solution, and the weights of the specimens were noticed for every 7 day-interval in one month and recorded in Table 4. The absorption behavior of the specimens was evaluated in the percentage of weight. Similarly, for Basalt- Polyester, and Glass-Polyester the experiment is repeated and the values are recorded in Table 5.

From Tables 4 and 5 the average percentages of water absorbed by each type of specimens in one day after 28 days are:

Basalt-epoxy composites = 0.0019%
Glass-epoxy composites = 0.0233%
Basalt-Polyester composites = 0.1724%
Glass - Polyester composites = 0.007%

Table 4. Water absorbed by Basalt/epoxy and Glass/epoxy composites at room temperature

Material	Sl no	Initial Weight (W0) (mg)	Duration of absorption (Days)	Increased weight due to Absorption (W) (mg)	% of weight increased
Basalt-epoxy composites	1	3.1150	7	3.1528	0.0121
	2	3.0909	14	3.1892	0.0355
	3	3.0711	21	3.0724	0.0423
	4	3.0707	28	3.0724	0.0553
Glass-epoxy composites	1	2.8941	7	2.9050	0.3766
	2	2.7710	14	2.7822	0.4041
	3	2.4732	21	2.4858	0.5095
	4	3.3371	28	3.3589	0.6532

Table 5. Water absorbed by Basalt/Polyester and Glass/Polyester composites at room temperature

Material	Sl no	Initial weight (W0) (mg)	Duration of absorption (Days)	Increased weight due to absorption (W) (mg)	% of weight increased
Basalt-Polyester composites	1	3.5101	7	3.5221	0.3418
	2	3.2789	14	3.2871	0.2500
	3	3.6847	21	3.6988	0.3853
	4	3.5413	28	3.5584	0.4828
Glass-Polyester composites	1	3.2162	7	3.2211	0.1523
	2	2.9502	14	2.9562	0.2033
	3	3.0148	21	3.0202	0.1719
	4	2.7760	28	2.7818	0.2089

From the above four types of materials the least of absorption material is Basalt-epoxy composites in natural curing.

4.4 Absorption Behavior After Post-curing

Absorption behaviors of post-cured specimens of Basalt-epoxy and Glass-epoxy in various temperatures and timings were determined from Table 7 for 1 week, 2 weeks, 3 weeks and 4 weeks and the average absorption per day after 28 days of observations was determined (Table 6).

Similarly, the absorption behaviors of post-cured specimens of Basalt-Polyester and Glass-Polyester in various temperatures and timings were determined from Table 7 for 1 week, 2 weeks, 3 weeks and 4 weeks and the average absorption per day after 28 days of observations was determined.

Table 6. Percentage of water absorbed by Basalt with Epoxy or Polyester laminate cured at 50 °C

Sl no	Duration of curing in the oven (min)	Basalt/epoxy specimens cured at 50 °C				Basalt/polyester specimens cured at 50 °C			
		1 week	2 week	3 week	4 week	1 week	2 week	3 week	4 week
1	60	0.4481	0.5276	0.7501	0.8581	0.7003	0.8415	1.0214	1.1300
2	120	0.4005	0.5370	0.7312	1.0104	0.6110	0.7667	0.7670	1.0429
3	180	0.2525	0.5875	0.6803	0.7705	0.6073	0.6815	0.8640	0.9809
4	240	0.5676	1.0133	1.3464	1.8014	0.5161	0.7163	0.8164	1.0072

Table 7. Percentage of water absorbed by Glass with Epoxy or Polyester laminate cured at 50 °C

Sl no	Duration of curing in the oven (min)	Glass/Epoxy specimens cured at 50 °C				Glass/Polyester specimens cured at 50 °C			
		1 week	2 week	3 week	4 week	1 week	2 week	3 week	4 week
1	60	0.7229	0.9673	1.2049	1.2875	0.6157	0.9236	1.1796	1.4357
2	120	0.4399	0.6114	0.7493	1.0252	0.5025	0.5789	0.8111	0.9383
3	180	0.1854	0.3908	0.6823	0.7750	0.5810	0.6746	0.9436	1.0330
4	240	0.3982	0.6893	0.7965	1.1067	0.5402	0.6198	0.8087	0.9504

The absorption behaviors of all four types of materials at the 50 °C curing temperature with curing durations of 60 min, 120 min, 180 min, and 240 min were evaluated using the bar chart shown in Figure (see Fig. 5).

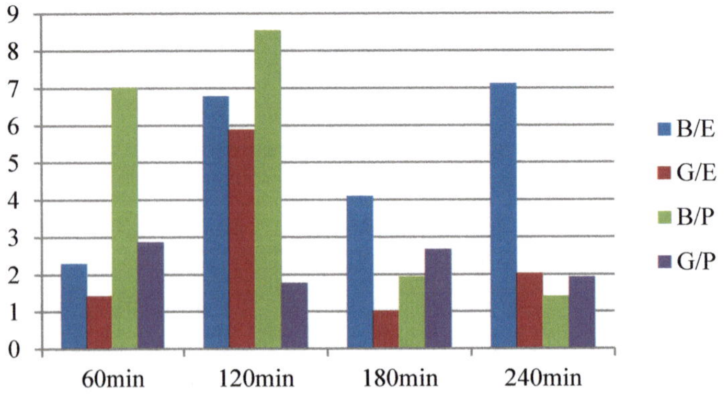

Fig. 5. Absorption coefficient versus curing time at 50 °C

From the above bar chart Basalt/Epoxy has a minimum value of diffusion coefficient at 60 minpost curing and a maximum value at 240 min. Basalt/Polyester has a minimum diffusion coefficient at 240 min post-curing. Glass/Epoxy has minimum Dc value at 180 min post curing and Glass/Polyester has minimum Dc value at 120 min post-curing. So it is clear that the moisture absorption property of the material varies with curing duration as well as curing temperature.

Similarly for the comparison of absorption behaviors the same type of materials at curing temperatures of 60 to 90 °C was studied with the following bar charts drawn using the experimental data.

Figure (see Fig. 6) represents the diffusion coefficient versus post curing durations for the temperature of 60 °C of Basalt/Epoxy, Basalt/Polyester, Glass/Epoxy, and Glass/Polyester laminates chart. Basalt/Epoxy has the minimum value of diffusion coefficient at 180 min post curing and the maximum value at 240 min. Basalt/Polyester has a minimum diffusion coefficient at 180 min post-curing. Glass/Epoxy has minimum DCvalue at 120 min post curing and Glass/Polyester has minimum DC value at 120 min post-curing.

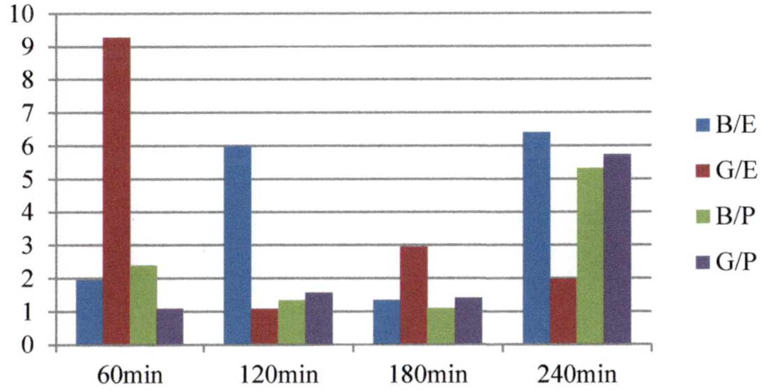

Fig. 6. Absorption coefficient versus curing time at 60 °C

Figure (see Fig. 7) is the diffusion coefficient versus post curing duration for temperature 70 °C of Basalt/Epoxy, Basalt/Polyester, Glass/Epoxy, and Glass/Polyester laminates. Basalt/Epoxy has the minimum value of diffusion coefficient

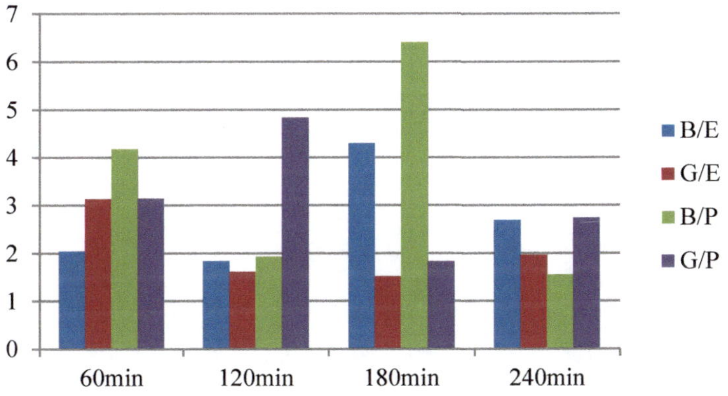

Fig. 7. Absorption coefficient versus curing time at 70 °C

at 120 min post curing and the maximum value at 180 min. Basalt/Polyester has a minimum diffusion coefficient at 240 min post-curing. Glass/Epoxy has minimum Dc value at 180 min post curing and Glass/Polyester has minimum DC value at 180 min post-curing.

Figure (see Fig. 8) represents of diffusion coefficient versus post curing timings at a temperature of 80 °C of Basalt/Epoxy, Basalt/Polyester, Glass/Epoxy, and Glass/Polyester laminates. Basalt/Epoxy has the minimum value of diffusion coefficient at 60 min post curing and the maximum value at 180 min. Basalt/Polyester has a minimum diffusion coefficient at 60 min post-curing. Glass/Epoxy has minimum DC value at 180 min post curing and Glass/Polyester has minimum DC value at 180 min post-curing.

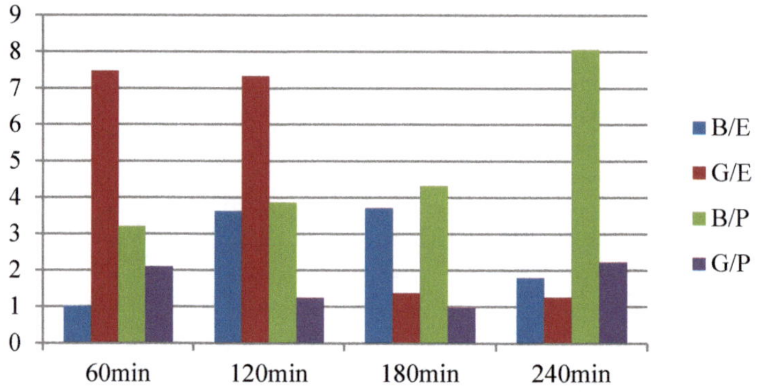

Fig. 8. Absorption coefficient versus curing time at 80 °C

In Figure (see Fig. 9) Basalt/Epoxy have the minimum value of diffusion coefficient at 240 min post curing and the maximum value at 120 min Basalt/Polyester has a minimum diffusion coefficient at 120 min post-curing. Glass/Epoxy has minimum DC value at 180 min post curing and Glass/Polyester has minimum DC value at 240 min post-curing.

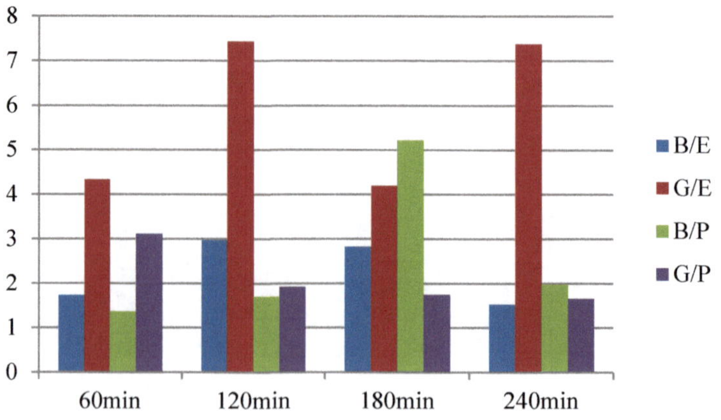

Fig. 9. Absorption coefficient versus curing time at 90 °C

4.5 Absorption Behavior After Microwave Curing

Specimens were cut for water absorption test (Basalt/Polyester, Glass/Polyester, Basalt/Epoxy, and Glass/Epoxy) with dimension 25 × 25 mm according to the ASTM (D570) standard. In order to get the fast curing, all the materials were cured in the domestic microwave oven for three pulsed conditions. Then the specimens were immersed in the water for hygrothermal treatment for various durations 5 days, 9 days, 17 days and 22 days, respectively. The specimens are weighed after before and after hygrothermal treatment. Then % of moisture absorbed and moisture absorption diffusion coefficient is calculated using the above formula.

Table 8. Percentage of water absorbed by Basalt with Epoxy or Polyester laminate at microwave oven curing

Sl No	Curing Power Cycles (Pt)	Basalt/Epoxy specimens				Basalt/Polyester specimens			
		1 week	2 week	3 week	4 week	1 week	2 week	3 week	4 week
1	Pt1	0.4200	0.5392	0.6867	0.8088	0.5970	0.6453	0.8898	0.9637
2	Pt2	0.3655	0.4583	0.4855	0.5865	0.5300	0.6230	0.6557	0.7705
3	Pt3	0.3696	0.4458	0.5361	0.6913	0.7021	0.9371	1.1636	1.3330

From the given Table 8 the minimum moisture absorbed materials are Basalt/Epoxy and Basalt/Polyester, with the Curing Power Cycle (Pt2) in the microwave oven after 4 weeks of observation.

From Table 9 minimum moisture absorbed materials are Basalt/Epoxy and Basalt/Polyester, with the Curing Power Cycle (Pt1) in the microwave oven after 4 weeks of observation.

Table 9. Percentage of water absorbed by glass with Epoxy or Polyester laminate at microwave oven curin

Sl No	Curing power cycles (Pt)	Glass/epoxy specimens				Glass/polyester specimens			
		1 week	2 week	3 week	4 week	1 week	2 week	3 week	4 week
1	Pt1	0.4288	0.7683	0.8791	0.9301	0.4073	0.7467	0.8575	1.0612
2	Pt2	0.6503	0.7889	0.9210	1.0902	0.7101	0.8488	0.9800	1.1494
3	Pt3	0.5145	0.9851	1.1388	1.2549	0.6281	0.7852	0.8386	1.0679

The absorption behaviors of all four types of materials at three curing cycles were evaluated using the bar chart shown in Fig. 10.

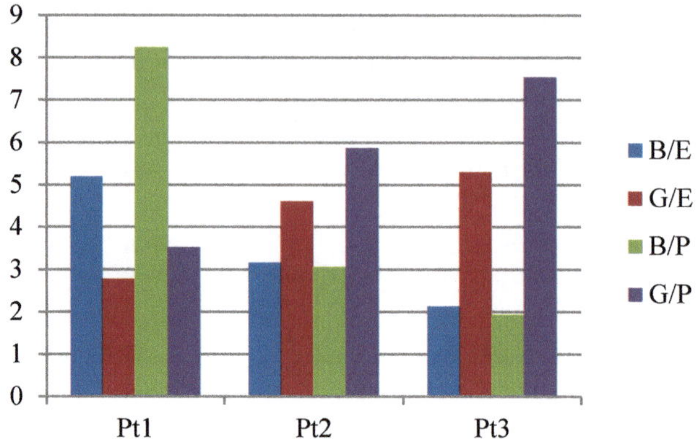

Fig. 10. Absorption coefficient versus curing power cycles in the microwave oven in four-week analysis

Figure (see Fig. 10) represents the diffusion coefficient versus post curing of three different cycles, say Pt1, Pt2, and Pt3 using a microwave oven. Basalt/Epoxy has the minimum value of diffusion coefficient at Pt3 mode of post curing and the maximum value at Pt1 mode. Basalt/Polyester has a minimum diffusion coefficient at Pt3. Glass/Epoxy has minimum DC value at Pt1 mode of post curing and Glass/Polyester has minimum DC value at Pt1 mode post-curing.

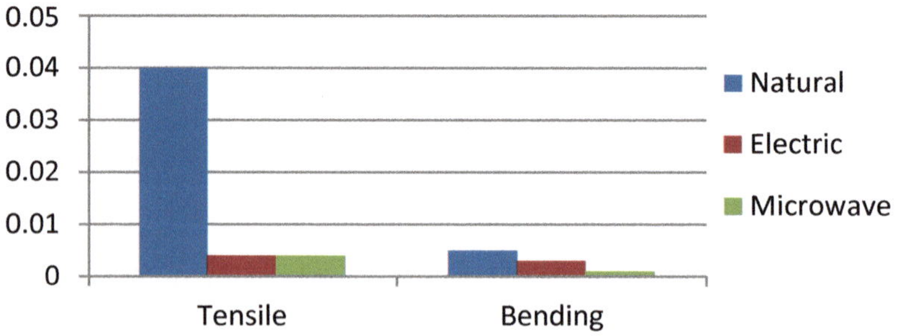

Fig. 11. Comparisons of mechanical strength degradation for optimum values of diffusion coefficients

Table 10. Percentage of Strength Degradation of optimum values of Samples

Samples of optimum diffusion values	Initial tensile strength Mpa	Tensile strength after 28 days Mpa	Initial bending strength Mpa	Bending strength after 28 days Mpa	% of Tensile strength degradation after 28 Days	% of bending strength degradation after 28 Days
Naturally curied Basalt-Epoxy composites	245.01	244.91	569.04	569.01	0.04	0.005
Basalt-Epoxy combination for 120 min at 70 °C in Electric Oven	245.04	245.03	569.06	569.04	0.004	0.003
Basalt/Epoxy has the minimum value of diffusion coefficient at Pt3	245.05	245.04	569.07	569.06	0.004	0.001

5 Conclusion

The absorption behaviors of the four types of fiber laminated composites were tested and compared them for three types of curing method such as natural curing, electric oven curing, and microwave curing. The effects post-curing temperature in materials changed its absorption behaviors it varies from material to material. Similarly, the duration of post curing and power cycles in the microwave oven also changes its absorption behavior to the required limit. Since the corrosion behavior of the material is highly influenced by moisture absorption behavior, it could be controlled by post-curing techniques experimented in this study. The major difference between microwave curing and electric oven curing is, in microwave curing the material could be heated uniformly in all dimensions in the same time and release trapped air quickly from it, whereas in another oven the material is heated gradually from the bottom layer to top layer and the heating is nonuniform. Curing method that can be selected depends upon the application and the type of material. By controlling the parameter discussed here for post curing, this will surely reduce the rate of corrosion process in the materials.

The optimum values for good control over corrosion through diffusion control are listed below:

In natural curing Basalt-Epoxy composites absorbs minimum moisture of 0.0019% per day. In the second method of curing also Basalt-Epoxy combination for 120 min at 70 °C shows the minimum absorption. In the third method of curing Basalt/Epoxy has the minimum value of diffusion coefficient at Pt3 mode of post-curing. Therefore, we can conclude that among the four combinations of materials, Basalt-Epoxy has better mechanical properties and corrosion control. Since the fabrics used in this study are quadraxial (0°/+45°/90°/−45) the absorption behavior and the resultant mechanical strength for optimum diffusion coefficients of the materials were validated using the following table. The bending strength of the material is much better compare with tensile strength, so it can be suggested as an alternative material for aircraft wing and fuselage skin.

The Mechanical strength of samples which have optimum values of diffusion coefficient is listed below (Fig. 11 and Table 10):

References

1. Chin, J.W., Nguyen, T., Aoudi, K.: Sorption and diffusion of water, salt water, and concrete pore solution in composite matrices. J. Appl. Polym. Sci. **71**, 483–492 (1999)
2. Mazor, A., Broutman, L.J., Eckstein, B.H.: Effect of long-term water exposure on properties of carbon and graphite fiber reinforced epoxies. Polym. Eng. Sci. **18**(5), 341–349 (1978)
3. Apicella, A., Migliaresi, C., Nicodemo, L., Nicolais, L., Iaccarino, L., Rocotelli, S.: Water sorption and mechanical properties of a glass-reinforced polyester resin. Composites **13**(4), 406–410 (1982)
4. Alexander, J., Augustine, B.S.M.: Hygrothermal effect on the natural frequency and damping characteristics of basalt/epoxy composites. Mater. Today Proc. **3**, 1666–1671 (2016)
5. Kinsey, A., Saunders, D.E.J., Soutis, C.: Post-impact compressive behavior of low temperature curing woven CFRP laminates. Composites **26**(9), 661–667 (1995). https://doi.org/10.1016/0010-4361(95)98915-8
6. Botelho, E.C., Costa, M.L., Pardini, L.C., Rezende, M.C.: Processing and hygrothermal effects on the viscoelastic behavior of glass fiber/epoxy composites. J. Mater. Sci. **40**, 3615–3623 (2005)
7. Bian, L., Xiao, J., Zeng, J., Xing, S.: Effects of seawater immersion on water absorption and mechanical properties of GFRP composites, 1–12 (2012)
8. Zamri, M.H., Md Akil, H., Bakar, A.A., Ishak, Z.A.M., Cheng, L.W.: Effect of water absorption on pultruded jute/glass fiber-reinforced unsaturated polyester hybrid composites, 51–61 (2011)
9. Liu, H.W., Xie, K.F., Hu, W.W., Sun, H., Yang, S.W., Yang, T.Y.: Water absorption of wood composite modified by basalt glass powder. In: Advanced Materials Research, vol. 821–822, pp. 1168–1170 (2013). https://doi.org/10.4028/www.scientific.net/AMR.821-822.1168
10. Water Absorption by Composite Materials and Related Effects. Wood-Plastic Composites, pp. 383–411 (n.d.). https://doi.org/10.1002/9780470165935.ch12

Optimal Placement and Co-ordination of UPFC with DG Using Whale Optimization Algorithm (WOA)

K. Aravindhan(✉), S. Abinaya, and N. Chidambararaj

Department of EEE, St. Joseph's College of Engineering, Chennai 600119, India
{aravindhank, chidambararajn}@stjosephs.ac.in,
abinayasugantharajan@gmail.com

Abstract. In modern power systems, there has been sizeable growth on integration of assorted renewable sources and multi varieties of flexible ac transmission system (FACTS). Here for the term renewable source, the Distributed Generation (DG) is used. The power produced from DG should be regulated by improving its voltage and minimize the power losses in the system. For this purpose, the FACTS devices such as UPFC is installed in the power system In this work, the problem considered is the optimum placement and coordination of Unified Power Flow Controller (UPFC) and Distributed Generation (DG). This could be done using Whale Optimization Algorithm (WOA). The reduction of fuel cost and power loss minimization are the two main problem considered in this work. The standard IEEE thirty bus system is employed to visualize the quality of the planned system. Also the results obtained using WOA are compared with PSO. The comparison indicates that the WOA algorithm had superior features than PSO.

Keywords: Distributed Generation (DG) · Unified Power Flow Controller (UPFC) · Whale Optimization Algorithm (WOA) · Flexible AC Transmission System (FACTS)

1 Introduction

In restructured power market, electric utilities are subjected to various new technologies to ensure the quality of supply to the consumers and also to attain the greater economical benefits. Increasing electricity demand necessitates the proper utilization of the transmission lines. Due to extended electricity demand it's essential to increase the usable power distribution capability, distribution generation technology (DG) and versatile AC transmission devices are developed.

To improve the system efficiency by providing required power Distributed generation is important. DGs are produced from various sources such as solar, wind, fuel cells, micro turbines and so on. The power produced by these ways should be regulated by improving its voltage and reduce the power losses in the system. For this purpose, the FACTS devices such as UPFC is installed in the power system. Optimal siting and

D. J. Hemanth et al. (Eds.): COMET 2019, LNDECT 35, pp. 375–386, 2020.
https://doi.org/10.1007/978-3-030-32150-5_37

sizing of DGs results in reduction of operational costs, better voltage regulation and reduction of power loss.

In power system, the installation of DG with FACTS requires some important considerations:

(1) which type of DG and FACTS devices to be installed,
(2) location to be placed,
(3) the way to estimate the appropriate size and varieties of devices economically,
(4) the way to coordinate the multiple devices and network.

Depends on real and reactive power delivering capacity, the DGs are broadly classified into four types. They are:

(i) DGs capable of generating only real power. Some examples of this type are solar PV, micro turbines, fuel cells etc.
(ii) DGs capable of generating each real and reactive power.
(iii) DGs capable of generating solely reactive power.
(iv) DGs capable of generating active power however intense reactive power.

FACTS devices play a major role in transmission system. FACTS devices provides a way for maximum utilization of existing transmission facilities. There are two types of FACTS devices, one is thyristor based devices and the other is voltage source inverter based devices.

The present day research is directed to understand the performance of VSI based FACTS devices because of its harmonic performance, dynamic response and ease of operation. The importance of VSI based controllers are it uses dc-ac inverters to exchange shunt or series reactive power with the transmission rather than distinct capacitance or reactor banks.

The UPFC consists of 2 VSCs coupled employing a common dc link. One is connected in shunt and therefore the different is connected asynchronous with the line employing a coupling electrical device. The dc voltage for each converters is provided by a typical capacitance bank. The UPFC configuration has each STATCOM and SSSC that shares a typical dc link capacitance. Depends on control methods, the UPFC can operated as a power flow controller, a voltage regulator or a phase shifter.

To determine the DG size, WOA is employed. WOA relies on forage behavior of whales. The optimum siting is completed for loss reduction in distribution system. The potency of WOA is additionally established [1]. To enhance the voltage profile in distribution network and additionally to reduce the ability loss mistreatment optimum placement of metric weight unit with SVC supported voltage stability index [2] (Fig. 1).

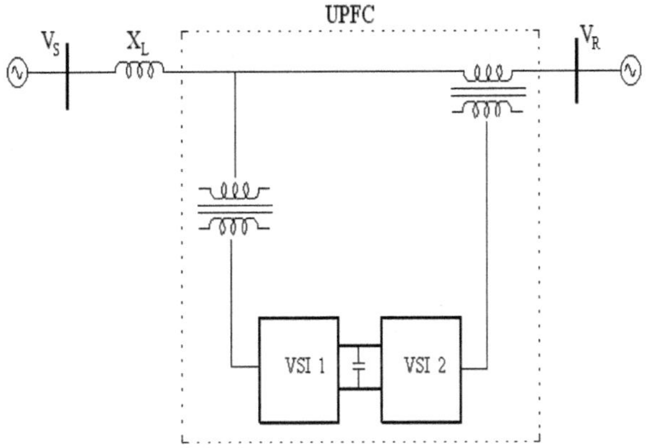

Fig. 1. Schematic diagram of UPFC

The optimum allocation of UPFC was exhausted line victimization the hybrid chemical change optimization algorithmic program [3]. The optimum placement of UPFC across a line was done to cut back the stability problems in the power system by considering technical and economical views is explained [5]. The congestion can be relieved and the cost can be minimized by appropriate allocation of UPFC device [10].

To determine the UPFC location supported voltage quality, active and reactive power losses and also the value of installation. The optimum location of UPFC is decided using hybrid CSA-CRO algorithm [7]. For congestion relief in transmission line, DGs are optimally siting and sizing. This can be done by developing a new algorithm [8]. To determine the generation capability and optimum placement of metric weight unit, a replacement management technique has been projected. Additionally weight factors are calculated for locating optimum metric weight unit placement [9].

In this work, the optimal placement and co-ordination of DG with UPFC is done using Whale Optimization Algorithm. Using MATLAB software, WOA technique is tested in IEEE 30 bus system.

2 Power Flow Problem Formulation

$$\mathrm{F} = \mathrm{w}_1 * \mathrm{F}_\mathrm{G} + \mathrm{w}_2 * \mathrm{P}_\mathrm{L} \tag{1}$$

Where w_1 and w_2 are the weighting functions or penalty functions

$$Minimize \quad F_G = \sum_{i=1}^{N_G} \left(a_i P_{g_i}^2 + b_i P_{g_i} + c_i \right) \tag{2}$$

$$Minimize \quad P_L = F(P_{Loss}) \tag{3}$$

$$where \quad P_{Loss} = \sum_{L=1}^{NL} (P_{ij} + P_{ji}) \tag{4}$$

Where P_{ij} and P_{ji} are the real power flows from bus *i* to j and from bus j to i respectively.

2.1 Constraints

The two power system constraints are equality constraints and inequality constraints.

Equality Constraints. The equality constraints shows the physical behavior of the system. These constraints are:

Active power constraints.

$$P_{Gi} - P_{Di} - V_i \sum_{j=i}^{N} V_j [G_{ij} \cos(\delta_{ij}) + B_{ij} \sin(\delta_{ij})] = 0 \tag{5}$$

Reactive power constraints.

$$Q_{Gi} - Q_{Di} - V_i \sum_{j=i}^{N} V_j [G_{ij} \cos(\delta_{ij}) + B_{ij} \sin(\delta_{ij})] = 0 \tag{6}$$

$$where\ \delta_{ij} = \delta_i - \delta_j$$

Where, N = Number of buses, P_G = the Output of active power, Q_G = the Output of reactive power, P_D = real power load demand, Q_D = reactive power load demand.

Inequality Constraints. The inequality constraints show the limits on electrical devices and security of the system.

Generator Constraints. In generator constraints, the voltage, real and reactive generations should be constrained by the minimum and maximum limits as follows:

$$V_{Gi}^{lower} \leq V_{Gi} \leq V_{Gi}^{upper}, i = 1, .., N \tag{7}$$

$$P_{Gi}^{lower} \leq P_{Gi} \leq P_{Gi}^{upper}, i = 1, .., NG \tag{8}$$

$$Q_{Gi}^{lower} \leq Q_{Gi} \leq Q_{Gi}^{upper}, i = 1, .., NG \tag{9}$$

Transformer Constraints. Transformer tap positions should be within the minimum and maximum limits as follows:

$$T_{Gi}^{lower} \leq T_{Gi} \leq T_{Gi}^{upper}, i = 1, .., N \tag{10}$$

Shunt VAR Compensator Constraints. Shunt compensation devices need to be within their minimum and maximum limits as given below:

$$Q_{Ci}^{lower} \leq Q_{Ci} \leq Q_{Ci}^{upper}, i = 1, .., NG \tag{11}$$

3 Whale Optimization Algorithm

Mirjali and Lewis introduced a replacement improvement algorithmic program known as Whale improvement algorithmic program (2016). WOA supported the special looking technique of humpback whales. These whales hunt herd of little fishes and krill nearer to the surface areas. The humpback whales uses the bubble web feeding technique to lure its prey. In bubble web feeding technique, the whales turn out a particular bubble on a circle or 9-shaped path.

3.1 Mathematical Model of WOA

1. Encircling prey.
2. Bubble net hunting method.
3. Search the prey.

Encircling Prey. Initially the optimum position is not known in the search space. Therefore WOA assumes that this best candidate answer is that the objective prey. Once the beat search agent is obtained, the remaining search agents attempt to update their position towards the obtained best search agent. It's diagrammatical victimization following equations:

$$\vec{X}(t+1) = \overrightarrow{X^*}(\mathrm{t}) - \vec{A} \cdot \vec{D} \tag{12}$$

$$\vec{D} = \left| \vec{C} \cdot \overrightarrow{X^*}(t) - \vec{X}(t) \right| \tag{13}$$

$$\vec{A} = 2 \cdot \vec{a} \cdot \vec{r} - \vec{a} \tag{14}$$

$$\vec{C} = 2 \cdot \vec{r} \tag{15}$$

Where X, X^* denote the position of best solution and position vector. Current iteration is denoted by t. A, C are coefficient vectors. a is directly diminished from 2 to 0. r is a random vector [0, 1].

Bubble Net Attacking Method. There are 2 approaches are designed to get the mathematical model of bubble net behavior of humpback whales.

a. Shrinking Encircling mechanism
b. Spiral position updating

Shrinking Encircling Prey. In this mechanism the value of 'a' is decreased to obtain th bubble net behavior. When 'a' is decreased A also decreased. The random value of A set as [−1, 1]. A is decreased from 2 to 0 over the course of iteration. The new position of search agent can be obtained in between the original and current position of search agent (Fig. 2).

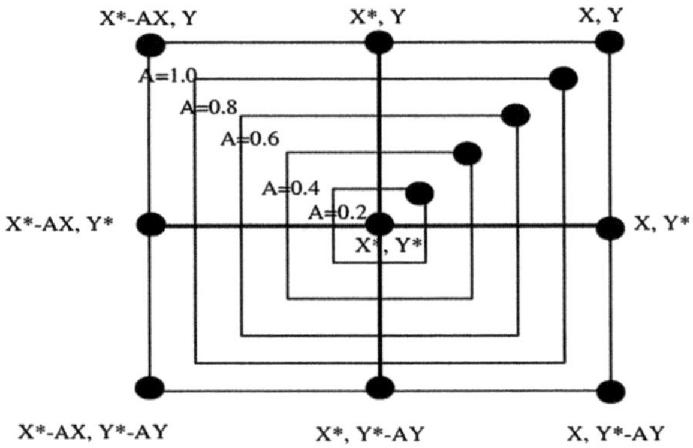

Fig. 2. Bubble net search shrinking encircling mechanism

Spiral Position Updating. The distance between the location of whale and prey is calculated. A spiral equation is developed to mimic the helix shaped movement of humpback whales.

$$\vec{X}(t+1) = \overrightarrow{D^t} \cdot e^{bl} \cdot \cos(2\pi t) + \overrightarrow{X^*} \tag{16}$$

During optimization, to update the whales position both the mechanisms have 50% probability.

$$\vec{X}(t+1) = \left\{ \begin{array}{ll} \overrightarrow{X^*}(t) - \vec{A} \cdot \vec{D} & if\, p < 0.5 \\ \overrightarrow{D^t} \cdot e^{bl} \cdot (2\pi t) + \overrightarrow{X^*} & if\, p \geq 0.5 \end{array} \right\} \tag{17}$$

where X(t + 1) represents the distance between whale and the prey (best solution). b is constant, l E[−1, 1]. P is random number [0, 1] (Fig. 3).

Search for Prey. The humpback whales search randomly based on their position. In this phase, the position of best search agent is updated according to a randomly chosen search agent instead of best search agent. It is modeled as,

$$\vec{D} = \left| \vec{C} \cdot \overrightarrow{X_{rand}} - \vec{X} \right| \tag{18}$$

$$\vec{X}(t+1) = \overrightarrow{X_{rand}} - \vec{A} \cdot \vec{D} \tag{19}$$

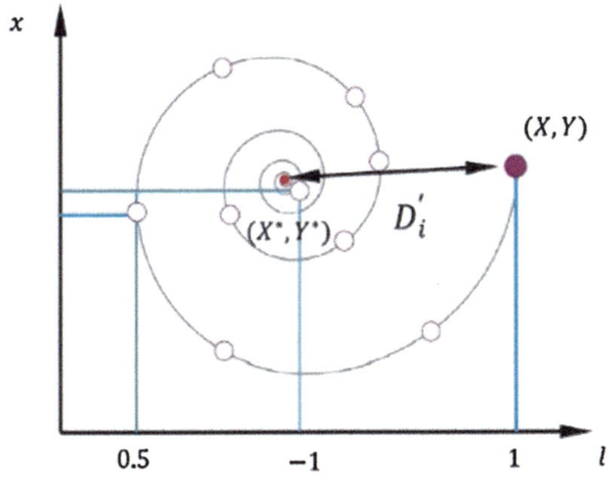

Fig. 3. Bubble net search spiral updating position mechanism

4 IEEE 30 Bus System

For explaining the employment of the instructed WOA technique, it's analyzed for the quality IEEE 30-bus system. The quality IEEE 30-Bus take a look at System has six PV buses, 24 PQ busses and 41 interconnected branches. The load consists of solely static load (Fig. 4).

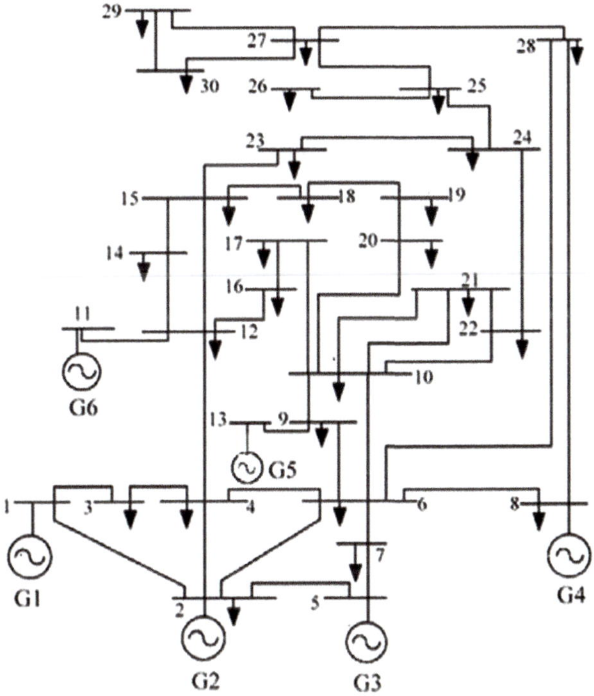

Fig. 4. Single line diagram of IEEE 30 bus system

5 Results and Discussions

WOA technique is enforced to resolve the UPFC-DG placement and co-ordination problem. The software package used to test the system is MATLAB 2014a.

The best values of power losses with and without UPFC-DG are shown in Table 1. From the table it is ascertained that with DG-UPFC placement the power loss is reduced.

Table 1. Optimal values of power losses for different methods

Cases	Power loss (Mw)
Normal loss(NR)	10.8856
Load time loss	12.6167
With DG-UPFC placement	9.8762
Without DG-UPFC placement	11.2508
Without UPFC/DG placement	11.2508

The fuel cost for different methods are shown in Table 2. The fuel cost obtained for the case with DG-UPFC placement is compared with without DG/upfc placement and without upfc/DG placement. From the comparison it is observed that the optimal fuel cost is obtained with DG-UPFC placement.

Table 2. Optimal value of fuel cost for different methods

Cases	Fuel cost (Rs/Hr)
Without DG/upfc placement	829.4815
With DG-upfc placement	828.9921
Without upfc/DG placement	829.3431

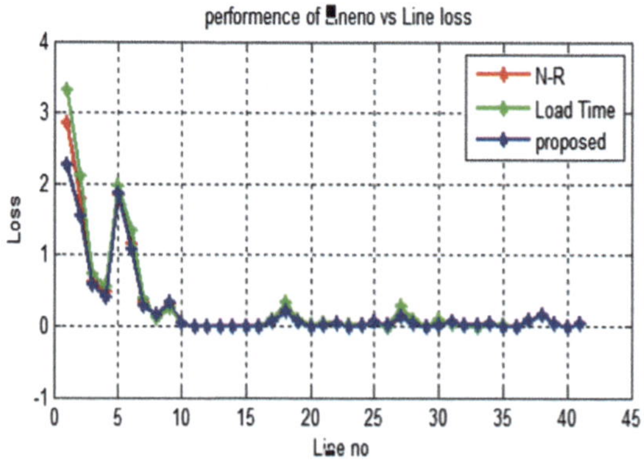

Fig. 5. Line losses for different cases

The Fig. 5 shows the line losses for different cases. The Fig. 6 shows the voltage profile for normal load flow, load time and with DG-UPFC placement. The minimum bus voltage is 0.95 p.u. and maximum bus voltage is 1.05 p.u.

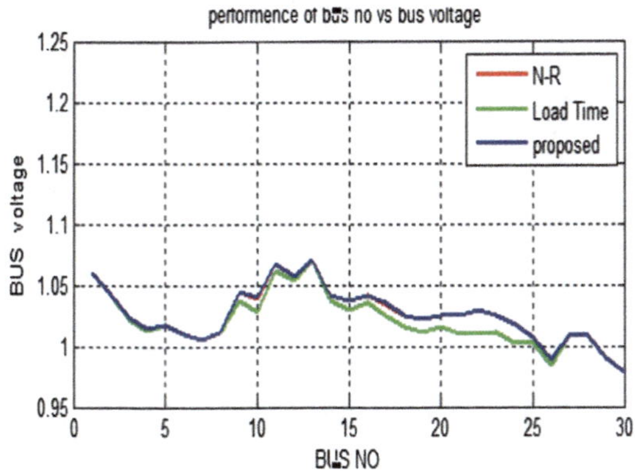

Fig. 6. Voltage profile for different cases

The optimal UPFC location is shown in Table 3. From the table it is ascertained that optimal location is from bus 8 to bus 28.

Table 3. UPFC Location

From bus no.	To bus no
8	28

The optimal generation of 30 bus system is shown in Table 4. The table shows the optimal generation with DG placement and with DG-UPFC placement.

Table 4. Optimal generation 30 bus system

DG bus no.	DG limit min (Mw)	DG limit max (Mw)	DG placement (Mw)	Dg-UPFC placement (Mw)
1	50.00	200.0	180.0000	180.000
2	20.00	80.00	52.0000	52.0000
5	15.00	50.00	18.0000	18.0000
8	10.00	35.00	13.0000	13.0000
11	10.00	30.00	20.0000	20.0000
13	12.00	40.00	14.0000	14.0000

The performance comparison of DG is shown in Fig. 6. The performance of DG for three cases shown in Fig. 7 are with DG placement, with DG-UPFC placement and with UPFC placement respectively.

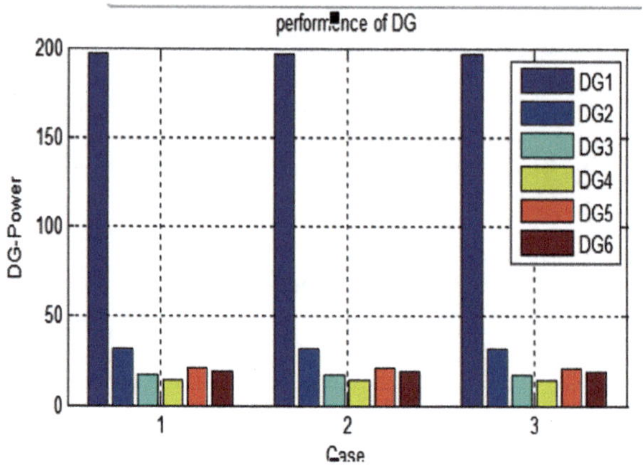

Fig. 7. Comparison of DGs performance with different cases

5.1 Comparison of WOA with PSO

The Fig. 8, shows the fuel cost comparison between PSO and WOA. From the Table 5, it is observed that fuel cost is reduced with DG-UPFC placement using WOA than optimal placement using PSO.

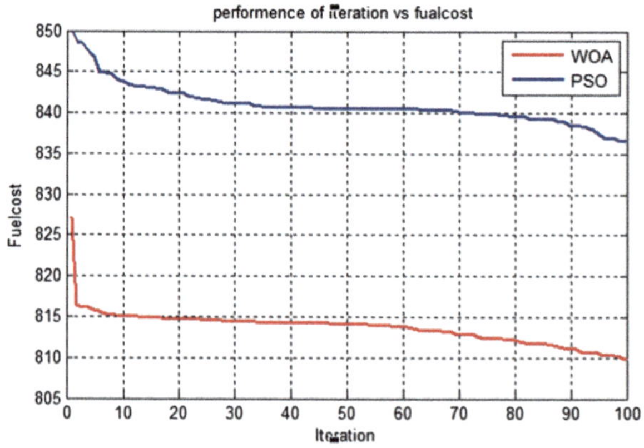

Fig. 8. Fuel cost comparison between PSO and WOA

Table 5. Fuel cost comparison between PSO and WOA

Cases	WOA	PSO
Without DG/upfc placement	829.4815	849.9739
With DG-upfc placement	828.9921	828.8469
Without upfc/DG placement	829.3431	833.7976

The Fig. 9 shows the performance comparison of DG using PSO and WOA. The optimal generation of various DGs are shown in Fig. 9 (Table 6).

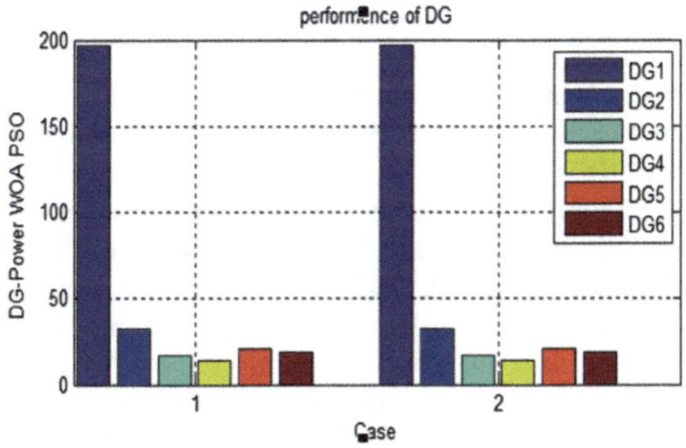

Fig. 9 DGs performance comparison between PSO and WOA

Table 6. Optimal generation comparison between WOA and PSO

DG bus no.	WOA	PSO
1	180.000	152.000
2	52.0000	35.0000
5	18.0000	34.0000
8	13.0000	14.0000
11	20.0000	28.0000
13	14.0000	28.0000

From the figures and tables shown above, it is proved that WOA is more efficient than PSO.

6 Conclusion

The optimal placement and co-ordination of UPFC with DG is done using Whale Optimization Algorithm (WOA). The results obtained with DG-UPFC placement is compared with other cases such as without DG-UPFC placement and with DG/UPFC placement. From the on top of result discussions it's ascertained that the fuel value and therefore the power loss is reduced with optimum placement of DG-UPFC. WOA technique is employed to resolve the matter of DG-UPFC optimum placement and co-ordination. Also the optimal generation and fuel cost of WOA is compared with PSO. From the results it is observed that the efficiency of WOA is better than PSO.

References

1. Reddy, P.D.P., Reddy, V.C.V., Manohar, T.G.: Whale optimization algorithm for optimal sizing of renewable resources for loss reduction in distribution systems. Renew.: Wind Water Solar **4**, 3 (2017)
2. Thishya Varshitha, U., Balamurugan, K.: Optimal placement of distributed generation with SVC for power loss reduction in distributed system. ARPN J. Eng. Appl. Sci. **12** (2016)
3. Dutta, S., Roy, P.K., Nandi, D.: Optimal location of UPFC controller in transmission network using hybrid chemical reaction optimization algorithm. Int. J. Electr. Power Energy Syst. **64**, 194–211 (2015)
4. Chidambararaj, N., Chitra, K.: Demand response and facts devices used in restructured power systems to relive congestion. Int. J. Innov. Works Eng. Technol. (IJIWET) (2017)
5. Das, S., Shegaonkar, M., Gupta, M., Acharjee, P.: Optimal placement of UPFC across a transmission line considering techno-economic aspects with physical limitation. In: Seventh International Symposium on Embedded Computing and System Design (ISED) (2017)
6. Dixit, M., Kundu, P., Jariwala, H.R.: Optimal placement and sizing of DG in distribution system using artificial bee colony algorithm (2016). 978-1-5090-0128-6/16/$31.00 ©2016 IEEE
7. Sen, D., Acharjee, P.: Optimal placement of UPFC based on techno-economic criteria by hybrid CSA-CRO algorithm. In: IEEE PES Asia-Pacific Power and Energy Engineering Conference (APPEEC) (2017)
8. Varghese, J.P., Ashok, S., Kumaravel, S.: Optimal siting and sizing of DGs for congestion relief in transmission line. In: IEEE PES Asia-Pacific Power and Energy Engineering Conference (APPEEC) (2017)
9. Tavakoli, F.H., Hojjat, M., Javidi, M.H.: Determining optimal location and capacity of DG units based on system uncertainties and optimal congestion management in transmission network. In: 25th Iranian Conference on Electrical Engineering (ICEE) (2017)
10. Chidambararaj, N., Chitra, K.: Relieving congestion and minimizing cost by appropriate allocation of UPFC device in a deregulated market using metaheuristic algorithm. SYLWAN J. (2016)

Study of Galvanic Corrosion Effect Between Metallic and Non-metallic Constituent Materials of Hybrid Composites

S. J. Elphej Churchill(✉) and S. Prakash

Sathyabama Institute of Science and Technology, Chennai 600119, India
elphej@gmail.com, prakash_s1969@yahoo.com

Abstract. Hybrid composites have become popular nowadays in aircraft manufacturing industries because of its improved mechanical properties, corrosion resistance, and fatigue life. This work focuses the galvanic corrosion effect between a metallic and a non-metallic material in a hybrid composite and identifies better material choice. While choosing constituent materials we have to consider the electrochemical properties of materials, because the atmospheric moisture between these materials acts as an electrolyte and allows electrons to flow from higher electric potential material to lower electric potential material and initiate galvanic corrosion. Since we know that the deterioration rate due to corrosion process is proportional to the potential difference between the materials and concentration of the electrolyte, it can be controlled by proper selection of materials with a negligible potential difference or restricting the moisture absorption. An attempt for identifying low electric potential material, nine combinations of frequently used materials such as GFRP-Al, BFRP-Al, CFRP-Al, GFRP-Cu, BFRP-Cu, CFRP-Cu, GFRP-SS, BFRP-SS, and CFRP-SS were selected. These specimens were kept in NaCl solution for 90 days and during these periods the potential differences between the materials, weights, and mechanical strengths at room temperature were noticed. The tensile strengths of the materials before the corrosion process were compared with the materials during corrosion for three intervals of 30 days, and finally, the percentage of strength degradations were validated for the better selection of materials.

Keywords: Galvanic corrosion · Fiber metal laminated hybrid composite · Effects of corrosion on strength

1 Introduction

The hybrid composite materials with good mechanical properties, physical properties, chemical properties, and electrochemical properties have a huge demand for specific application areas such as aerospace and maritime. For decades materialists have been investigating such materials to satisfy the requirements of those industries, for example, glass reinforced aluminum (GLARE) composite, it is made up of thin layers of aluminum and glass fiber layers and bonded together with the matrix. The literature says that fiber laminates are capable of arresting crack propagation in metal laminates and delay structural failure [1]. If the outer layer of hybrid composite is metallic, it can act

D. J. Hemanth et al. (Eds.): COMET 2019, LNDECT 35, pp. 387–397, 2020.
https://doi.org/10.1007/978-3-030-32150-5_38

as a shield from moisture absorption and the inner layer of non-metallic can give better specific strength. In this manner, fiber and metal combination can improve the mechanical properties of materials in many cases. In real time application hybrid composites absorbs moisture from the atmosphere and initiate the electrochemical process [2]. Investigations proved that the anode material corrodes faster than the normal and the cathode material corrodes slower. They observed that during this process the surface area of cathode material increases and causes delamination. It was suggested that the process could be regulated by proper selection of materials which have a low potential difference or by giving proper insulation coating around them [3]. A studied on Galvanic Corrosion effect of pH and dissolved oxygen concentration on the Aluminum/Steel Couple, they observed that the reaction occurring on the steel cathodes is the vigorous evolution of gaseous hydrogen and it was controlled by oxide films on the steel cathodes [4]. In order to control the corrosion effect in aircraft alloys such as aluminum and copper, chromate treatment was done but in the final result, the expected percentage reduction could not be achieved. From the experimental analysis of various corrosion control process, it could be accepted that a proper bi-material pair only sustains with mechanical strength for long and trouble-free in a corrosive environment [5]. From these literature reviews, the work could be focused to determine the lowest potential difference material combination and highest potential difference combination for the fabrication.

2 Materials and Fabrications

2.1 Materials

Owen fabrics of Basalt, Glass and Carbon with aerial densities 255 g/m^2, 455 g/m^2 and 370 g/m^2, respectively, Epoxy LY556, Hardener HY951, sheet metals of Aluminum (Al), Copper (Cu) and Stainless Steel (SS).

2.2 Fabrication

Composite laminates were fabricated using hand layup method as follows. Matrix and hardener were mixed with 10:1 proportion based on the aerial density of the fiber. Required numbers of fabric were taken and the mixture of resin hardener is applied to it one after another using a paintbrush. The laminate is left for curing at room temperature for about 24 h. The finished laminates were cut into the required number of specimens as per ASTM standard. Next, the sheet metals of Copper, Aluminium, and Steel are also cut into specimens with the dimensions similar to FRP laminates. The FRP composite strips are bonded with metallic strips using epoxy resin so that nine fiber/metal hybrid composites are prepared (GFRP-Al, BFRP-Al, CFRP-Al: GFRP-Cu, BFRP-Cu, CFRP-Cu: GFRP-SS, BFRP-SS, CFRP-SS) as shown in figures (see Fig. 1).

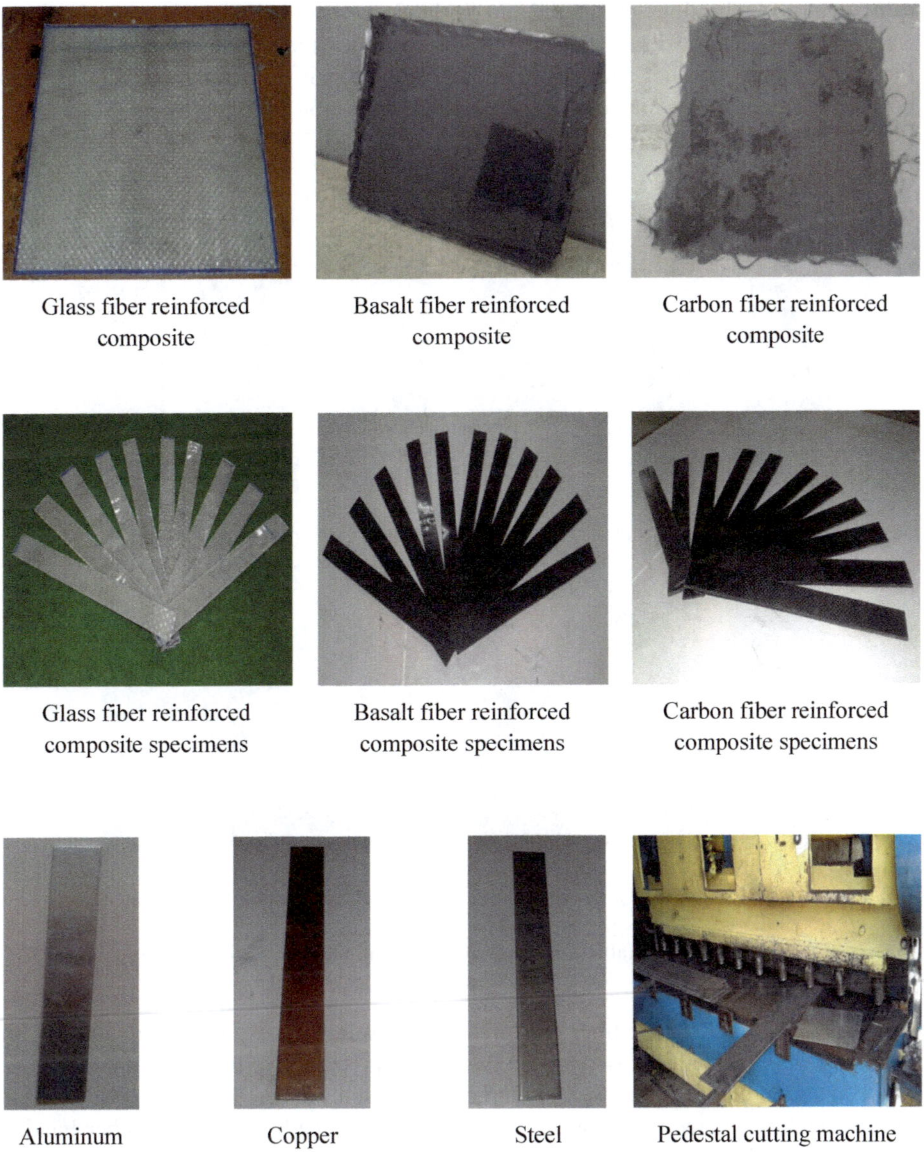

Fig. 1. Three fiber reinforced composites and Three Metal Plates

3 Experimentation

Galvanic corrosion experimental setup is shown in Figures (see Fig. 2), according to NASA, one liter of seawater contains 35 gms of NaCl. The solution is prepared by dissolving the proportionate amount of NaCl in the required volume of distilled water and filled in a plastic container; the solution is to keep moving within the container using a motor. The fiber reinforced plastic laminate was paired with metal laminates and kept in vertical position with the help of a wooden reaper as shown in figures (see Fig. 2). A small gap of 1 mm between the laminate pair is maintained to let the ions migrate between the laminates for the galvanic corrosion process.

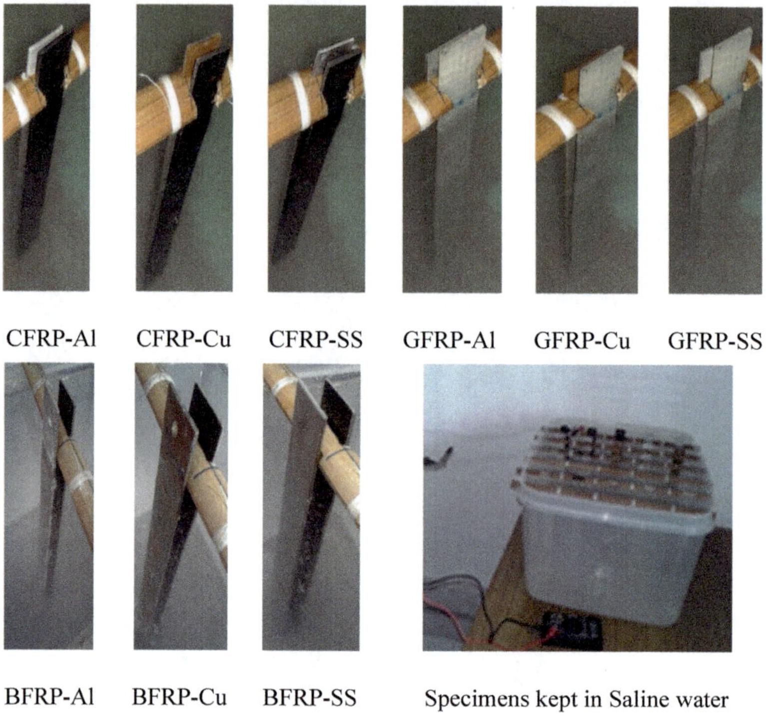

Fig. 2. Combinations of Metallic and Non-Metallic laminates for Galvanic Corrosion analysis

In all combinations the fiber reinforced composites behave as strong cathodes and obviously the metals act as anodes. As the days pass by, the potential difference between anode and cathode was measured using a multimeter. The voltage is evidence for the electron interaction between the anode and the cathode. On the other hand, as days pass by, deposits of oxides are visible in the container, and visible corrosion is witnessed. After twenty days, the effect of corrosion is visually seen on the metals as they are either corroded or fade in color. While on the composite fibers there are no visible changes or fade in color. This shows they are least corrosive when compared to metals. As days pass by the voltage begins to fluctuate and decrease gradually. Parameters such as voltage, temperature, weight, and strength of the specimens were measured and tabulated in three intervals.

The initial tensile strength of each material was tested and tabulated, and then they were subjected to the corrosion process for three intervals (30 days, 60 days, and 90 days) after every interval the materials were tested and tabulated their tensile strength using UTM shown in Figure (see Fig. 3).

Fig. 3. Tensile Strength testing of metallic and non metallic constituent materials after galvanic corrosion using UTM

4 Result and Discussion

4.1 Tensile Strength Before and After Corrosion

In order to know about the deterioration of the strength, all specimens are subjected to the tensile test to know the tensile strength of the specimen before corrosion and after corrosion. Thereby, the least corrosive and strongest material can be identified. From the tensile test, it is seen that the tensile strength of the material gradually decreases with an increase in the time period. This is true for all materials.

Table 1 shows the strength versus the number of days of corrosion data of all materials. The tensile strength of CFRP-Al decreases 14.03% after 90 days. Carbon fiber had an ultimate tensile strength of 4.56 N/mm^2 and at the end; it reduces to 3.92 N/mm^2 due to corrosion. Table 2 shows a decrease of 21.29% in its tensile strength. Before corrosion, it had the ability to withstand up to 14 kN load but after 90 days it is dropped to 10.63 kN But on the other hand, the elongation has increased with an increase in the number of days. Since the maximum displacement has increased from 4.5 mm to 6.14 mm. In Table 1, the Basalt Fiber Reinforced Plastic Composite (BFRP) shows 24.05% of the decrease in its tensile strength. It seems that the basalt fiber laminate before corrosion has the ability to withstand an ultimate load of 18 kN but with an increase in the number of days it is reduced to 12.4 kN, on the other hand, the maximum displacement has reduced. Aluminum Specimen shows 46.5% decrease in its tensile strength. Before corrosion, it had the ability to withstand an ultimate load of 11 kN but in the end, it significantly reduced to 5.9 kN. Copper Specimen shows a 12.5% decrease in its tensile strength. Before corrosion, it had 2.4 N/mm^2 and at the end of 90 days, it is 2.1 N/mm^2. Stainless Steel Specimen shows 18.26% decrease in its tensile strength. It is clear that the ultimate tensile strength of stainless steel has reduced significantly from 4.16 N/mm^2 before corrosion and at the end of 90 days; it is 3.4 N/mm^2.

Table 1. Tensile Properties Constituent materials

Constituent materials	Initial strength (N/mm^2)	Strength after 30 days (N/mm^2)	Strength after 60 days (N/mm^2)	Strength after 90 days (N/mm^2)
CFRP-AL	6.32	5.84	5.68	4.86
GFRP-AL	3.92	3.6	3.44	2.64
BFRP-AL	4.92	4.6	4.44	3.34
CFRP-Cu	6.96	6.46	6.28	6.02
GFRP-Cu	4.56	4.22	4.04	3.8
BFRP-Cu	5.56	5.22	5.04	4.5
CFRP-SS	8.72	7.92	7.76	7.32
GFRP-SS	6.32	5.68	5.52	5.1
BFRP-SS	7.32	5.96	6.52	5.8

Figure 4 shows the graphical representation of variations of tensile property due to corrosion effect for a period of three intervals of various specimen combinations with its initial strength before the corrosion process (see Fig. 4). In all cases, the graph shows the tensile strength decreases gradually with respect to time, from this it is evident that the mechanical property degrades due to corrosion. Elongation of the materials was also decreased to some extent due to the application of the same tensile loads after 90 days of corrosion.

4.2 Observations of the Potential Difference Between Materials During Corrosion

The potential difference between the materials should be minimum or negligible for a healthy hybrid composite. The second important thing is the conductivity of electrolyte; it depends on its concentration. From below observations, the materials were categorized based on the voltage between the materials due to the conductivity of the electrolyte (Fig. 5).

Table 2. Voltage vs Hours of all combinations

Days	C(+)/ Al (−) Volt	C(+)/ Cu(−) Volt	C(+)/ Ss (−) Volt	G(+)/ Al(−) Volt	G(+)/ Cu(−) Volt	G(+)/ Ss(−) Volt	B(+)/ Al(−) Volt	B(+)/ Cu(−) Volt	B(+)/ Ss(−) Volt
1	0.54	0.12	0.34	0.09	0.05	0.06	0.43	0.19	0.22
30	0.37	0.26	0.29	0.09	0.07	0.04	0.26	0.16	0.33
60	0.04	0.1	0.1	0.02	0.08	0.003	0.13	0.18	0.016
90	0.05	0.18	0.15	0.015	0.09	0.003	0.032	0.04	0.05

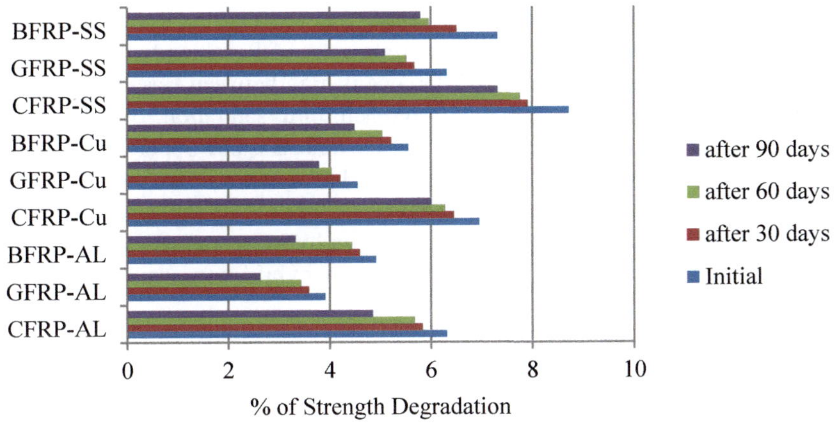

Fig. 4. Strength degradation with time progress in the number of days

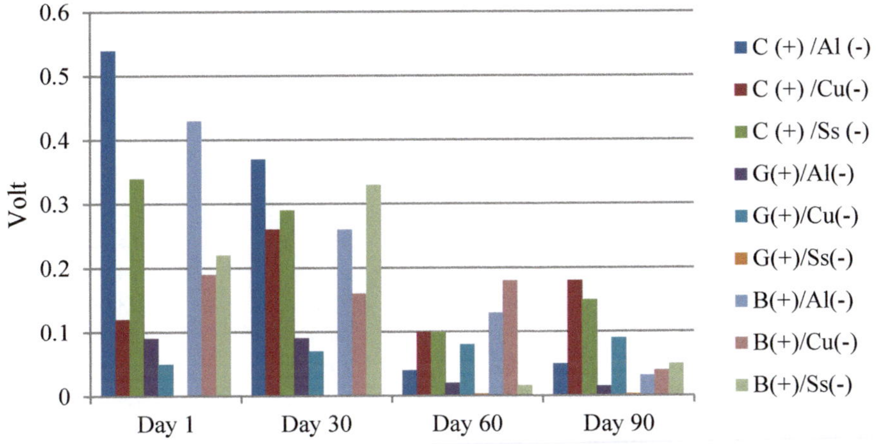

Fig. 5. Potential differences from day 1 to day 90

In Table 2, the voltage between all the nine combinations of fiber and metal laminates at room temperature were recorded, it was observed that the voltage decreased gradually and reached a very low value after 90 days. This means the corrosion effect is high at the beginning and due to the migration of ions, the potential difference between the materials gradually decreases and finally come to saturation.

The figure shows the potential difference between Carbon and Aluminum, Carbon and Copper, and Carbon and Steel. Among these three combinations of Carbon and Aluminum, shows very high potential difference initially when compared with the other two. So Carbon Aluminum is the most affected material. Similarly, potential difference between Glass and other three metals at room temperature for a period of 90 days. In this graph, Glass/Aluminum shows the highest value, Glass/Steel is in second position, and Glass/Copper shows the minimum potential difference as the least corroded hybrid composite in the group. The voltage versus time chart of Basalt and all three metals for the duration of 90 days, the potential difference is high between Basalt and Aluminum in the beginning and decreases gradually to reach up to saturation. Basalt/Steel shows lower values than Basalt/Aluminum and higher than Basalt/Copper. From the above three groups of observations, it is understood that Aluminum is the most corroded material when combining with fibrous laminates. Basalt/Aluminum shows the highest potential difference than other two fiber aluminum hybrid composites, so the galvanic corrosion effect is more on the Basalt/Aluminum hybrid composite than any other combination.

4.3 Weights Before and After Corrosion

The weights of the individual specimen were taken before and after corrosion, the difference in weights was tabulated as shown in Table 3. The percentage of weight increased due to absorption of moister for all materials was calculated using the formula given below. From the calculations, it could be found that fiber laminates' weights increased little more, and, if we compare this with the metals' water absorption, their corresponding weight increments are negligible.

Table 3. Weights Before and After Corrosion

Specimen	Initial weight (in grams)	Weight after 30 days (in grams)	Weight after 60 days (in grams)	Weight after 90 days (in grams)
Carbon	19.463	19.007	18.882	18.752
Glass	25.603	25.335	24.130	24.032
Basalt	23.186	23.021	23.010	23.003
Al	34.900	34.900	34.900	34.901
Cu	113.120	113.120	113.122	113.122
SS	102.786	102.787	102.788	102.792

Calculation of Percentage of weight Change

$$Wr = [(Wo - Wt)/Wo] \times 100$$

Calculation of weight for carbon

$$\begin{aligned}\text{Weight reduction after 30 days} &= [(19.463 - 19.007)/19.463] \times 100 \\ &= 2.345\%\end{aligned}$$

Weight reduction after 60 days = 2.985%

Weight reduction after 90 days = 3.654%

% of Weight Change after 90 days

Carbon fiber reinforced composite = 3.654

Glass fiber reinforced composite = 6.13

Basalt fiber reinforced composite = 0.789

Stainless Steel = 0.00584

Aluminum = 0.00286

Copper = 0.00176

In Fig. 6, the initial weight and final weight of fiber reinforced composites shows little increase in slope of lines, whereas the metal layers show straight lines with a negligible increase of slope (see Fig. 6). From the graph, we could say that the moisture absorption behavior of metals is comparatively very low compared with fibers plastic composites. Figures 3 are the evident for the interfacial reaction of metals and polymers with its environment (see Fig. 3).

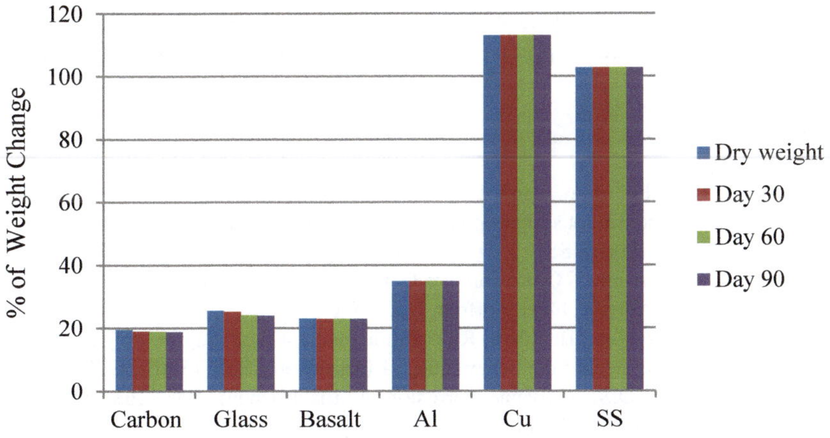

Fig. 6. Weight Change Due to Galvanic Corrosion

5 Conclusion

The resultant parameters obtained are,

- Percentage of strength degradation due to corrosion
- Potential difference between each pairs
- Weight differences before and after corrosion

The tensile strength of CFRP-Al reduces 23.10% from its initial strength after 90 days and CFRP-Cu it is 13.50% and CFRP-SS it is 16.05%. For the case of GFRP with Al, Cu, and SS are 32.65%, 16.66%, and 19.30%, respectively after 90 days. The strength degradation of Basalt with Al, Cu and SS are 11.78%, 13.66%, and 10.92, respectively. So among all materials, basalt combinations show least corrosion and better stability in its strength.

In case of potential difference, the lowest potential difference among the pair is Glass/Copper and Carbon/Aluminum has the highest potential difference. The rate of corrosion is directly proportional to the potential difference so according to this statement Carbon/Aluminum can be considered as the weakest combination because it can get corroded faster than any other combination.

Considering the weight differences, Basalt/epoxy shows less moisture absorption compare with other materials. A material which absorbs more moisture from environment dissolves quickly, its increased concentration of electrolyte inside the material accelerate migration of ions. Basalt with epoxy showed better control over weight change in absorption than other non metallic composites. Aluminum has good corrosion resistance and specific strength than other two metals. So the combination of basalt and aluminum, can be recommended as a good hybrid composite with corrosive resistant and balanced mechanical strength. The consumption or dissolution of constituent material into the components of environment could be controlled by proper material selection based their electric potential difference and moisture absorption behavior.

References

1. Botelhol, E.C., Silva, R.A., Pardinil, L.C., Rezendel, M.C.: A review on the development and properties of continuous fiber/epoxy/aluminum hybrid composites for aircraft structures. Mater. Res. **9**(3) (2006). ISSN 1516-1439, Online version ISSN 1980-5373
2. U.S. Naval Fleet Aircraft Corrosion By G.T. Browne Materials Consultant Commander, Naval Air Force, Atlantic Fleet Norfolk, Va, USA
3. Murer, N., Missert, N.A., Buchheit, R.G.: Finite element modeling of the galvanic corrosion of aluminum at engineered copper particles. Electrochem. Soc. **159**(6), C265–C276 (2012)
4. Pryor, M.J., Keir, D.S.: Galvanic corrosion. J. Electrochem. Soc. **105**(11), 629 (1958). https://doi.org/10.1149/1.2428681
5. Clark, W.J., Ramsey, J.D., McCreery, R.L., Frankel, G.S.: A galvanic corrosion approach to investigating chromate effects on aluminum alloy 2024-T3. J. Electrochem. Soc. **149**(5), B179–B185 (2002). 0013-4651/2002/149(5)/B179/7
6. Zhang, X.G.: Galvanic Corrosion. Teck Metals Ltd., Mississauga, Ontario, Canada (2011). www.knovel.com

7. Wang, W.-X., Takao, Y., Matsubara, T.: Galvanic corrosion-resistant carbon fiber metal laminates (2007)
8. Alexander, J., Augustine, B.S.M.: Strength determination of basalt\epoxy laminated composites at various fiber Orientations. Int. J. Appl. Eng. Res. **9**(26), 8913–8917 (2014)
9. Sabet, S.M.M., Akhlaghi, F., Eslami-Farsani, R.: Production and optimization of aluminum-basalt composites by hand layout technique (2012)
10. Jakubczak, P., Surowska, B., Bienias, J.: Evaluation of force-time changes during the impact of hybrid laminates made of titanium and fibrous composite. Arch. Metall. Mater. **61**(2), 689–694 (2016)

Design of Modified Code Word for Space Time Block Coded Spatial Modulation

R. Raja Kumar[1(✉)], R. Pandian[2], B. Kiruthiga[3], and P. Indumathi[3]

[1] Mathematics Department, Sathyabama Institute of Science and Technology, Chennai, India
rrkmird@gmail.com
[2] Department of Electronics and Instrumentation, Sathyabama Institute of Science and Technology, Chennai, India
rpandianme@rediffmail.com
[3] Department of Electronics Engineering, Madras Institute of Technology, Chennai, India
kiruthigabalu94@gmail.com, indu@mitindia.edu

Abstract. The Multiple Input Multiple output transmission scheme is effective technique to enhance the spectral potency of wireless communication system. MIMO transmission methods include space–time block coding (STBC) and spatial multiplexing. Hereinto, STBC aims at increasing the bit error rate (BER) performance whereas, the spatial multiplexing is meant to reinforce the spectral potency. Space-time block coded spatial modulation (STBC–SM) has been introduced to use the benefits of each spatial modulation and coordinate system block codes. To enhance the system performance, code words are introduced. The modified code word introduces a thought of versatile coefficients into the Modulation of STBC-SM. A technique is planned to get the best versatile coefficients. Moreover, the primary modified code words may be applied to the systems developed from the STBC-SM. Then, STBC-SM system is combined with CDMA technology so as to makes the system capable of serving high number of users by assigning every user with its distinctive PN code.

Keywords: MIMO · BER · STBC · STBC-SM · Code words · PN code

1 Introduction

Optimal design and successful deployment of high-performance wireless networks present a number of technical challenges. The demands of the wireless systems are increasing in a vast scale. These demands force the designers to design system with high capacity, wide bandwidth allocation and with extreme dependability. The weakening nature of wireless channel and interference between users are the main obstacles that designers have to overcome. Use of multiple antennas at the receiver and transmitter in a wireless network is rapidly emerging technologies that guarantees higher data rates at longer ranges without consuming extra bandwidth or transmit power.

The use of space time block codes (STBC), with multiple antennas at both the transmitter and the receiver, can improve the spectral efficiency and the capacity of the

D. J. Hemanth et al. (Eds.): COMET 2019, LNDECT 35, pp. 398–414, 2020.
https://doi.org/10.1007/978-3-030-32150-5_39

communication system. This is because the STBC codes exploit the spatial diversity. On the other hand, the channel state information (CSI) is extremely important at the receiver in STBC systems because the CSI is used in the decoding algorithm. Thus, if the CSI is not recognized at the receiver the whole system performance will be extremely low.

There have been many schemes that combine STBC and spatial modulation systems, the spatial modulation are an interesting scheme that improve the bandwidth allocation and enhance the BER.

Recently a combination a between STBC and CDMA has been introduced. The CDMA is very interesting technique with very promising advantages. CDMA system can serve many users in a very narrow bandwidth, and can achieve a very good BER. However, the inter carrier interference (ICI) is the major problem in CDMA. This problem can be overcome by choosing a very long pseudo code (PN).

In CDMA environments, the number of users affects the performance of the system especially when the system channels are frequency-selective-fading channels. These systems suffer from multiple-access interference (MAI). In such systems the maximum likelihood (ML) receiver treats MAI signals as additive white Gaussian noise (AWGN). So, it is extremely important in CDMA systems to suppress the MAI. In our system we combine the STBC, spatial modulation and CDMA together.

STBC-SM is a system which combines between Space Time Block Coding STBC and Spatial Modulation (SM). In this scheme, the transmitted data is relies on the space, time and antenna indices. STBC-SM takes the advantage of this combination to achieve high spectral efficiency which is realized using antenna indices to rely information. Moreover, STBC-SM is optimized for diversity and coding gain to attenuates the BER which is the done using the space and time domains. Low complexity maximum likelihood (ML) decoder is employed in this scheme which gains from the orthogonality of the STBC code.

Code division multiple accesses could be a new idea exploitation channel access technique through a style of multiplexing that permits multiple signals occupying single channel and optimizing the offered information measure. This allowed for dramatic development to wireless communication during this century and gained a wide spread international use by cellular radio system.

The paper is organized as following: Sect. 2 presents the work associated with STBC-SM (Space Time Block Coded Spatial Modulation). Section 3 gives System model and proposed STBC-SM using various Modulation Schemes like PSK, QAM. Section 4 deals with simulation results of the proposed STBC-SM using CDMA and conclusion area unit represented in Sect. 5.

2 Related Work

A new MIMO transmission scheme called Space time block coded spatial modulation (STBC-SM). It combines spatial modulation (SM) and space-time block coding (STBC to require advantage of the benefits of both while avoiding their drawbacks. Within the STBCSM scheme, the transmitted info symbols are enlarged not solely to the space and time domains but also to the spatial (antenna) domain that corresponds to the on/off standing of the transmit antennas offered at the space domain, and so both core STBC and antenna indices carry information [1].

A general technique is given for the design of the STBC-SM scheme for any number of transmits antennas. Besides the high spectral efficiency advantage provided by the antenna domain, the proposed scheme is additionally optimized by deriving its diversity and coding gains to exploit the diversity advantage of STBC. A low-complexity maximum likelihood (ML) decoder is given for the new scheme which profits from the orthogonality of the core STBC. A super-orthogonal space-time (ST) trellis codes with rectangular signal constellations for wireless communications with a spectral efficiency of 4 bits/s/Hz. A new class of space-time codes called super-orthogonal space-time trellis codes. These codes combine set partitioning and a super set of orthogonal space–time block codes in a systematic way to provide full diversity and improved coding gain over earlier space–time trellis code constructions [2].

An Orthogonal designs have been used, which may succeed full transmit diversity and have a very simple decoupled maximum-likelihood decoding algorithm. It reflects the bandwidth efficiency of the employed space–time block code created from the orthogonal design [3].

This technique used to derived an optimal detector for the so-called spatial modulation (SM) performs considerably higher than the initial ($\sim$four dB gain), by account a closed kind expression for the common bit error chance. As well, it shows that SM with the optimal detector achieves performance gains ($\sim$1.5–3 dB) over popular multiple antenna systems, making it an excellent candidate [4].

A combine of transmit antennas are chosen out of the whole transmit antennas to send Alamouti's STBC, whose 2 signal symbols are drawn from 2 completely different constellations, and therefore the antenna pairs activated in numerous code words are moving cyclically on the whole transmit antenna array. To take advantage of the variety advantage of Alamouti's STBC, the optimization of STBC-CSM is given by maximizing its committal to coding gain [6].

An optimum transmit structure using STBC instead of spatial modulation (SM) in a multiple-input multiple-output (MIMO) transmission technique. Based on transmission optimized spatial modulation (TOSM), selects the best transmit structure that minimizes the average bit error probability (ABEP) [7–9].

A Golden code for 2×2 space time block code that achieves the optimal diversity-multiplexing gain tradeoff for a multiple antenna system. Double Space Time Transmit Diversity (DSTTD) is an open loop MIMO system with 4 transmits antennas. DSTTD achieves the best performance in rich scattering channels, spatial correlation degrade the performance [10–12].

The aim of this paper is to improve the performance of STBC-SM (Space Time Block Coded Spatial Modulation) using CDMA Technology with modified code words.

3 System Model and Proposed Method

STBC-SM is a system which combines between Space Time Block Coding STBC and Spatial Modulation (SM). In this scheme, the transmitted data is relies on the space, time and antenna indices. STBC-SM takes the advantage of this combination to achieve high spectral efficiency which is realized using antenna indices to rely information. Moreover, STBC-SM is optimized for diversity and coding gain to attenuates the BER which is the done using the space and time domains. Low complexity maximum likelihood (ML) decoder is employed in this scheme which gains from the orthogonality of the STBC code [29].

3.1 Transmitter of STBC-SM

The transmitter of STBC-SM system is shown in Fig. 1.

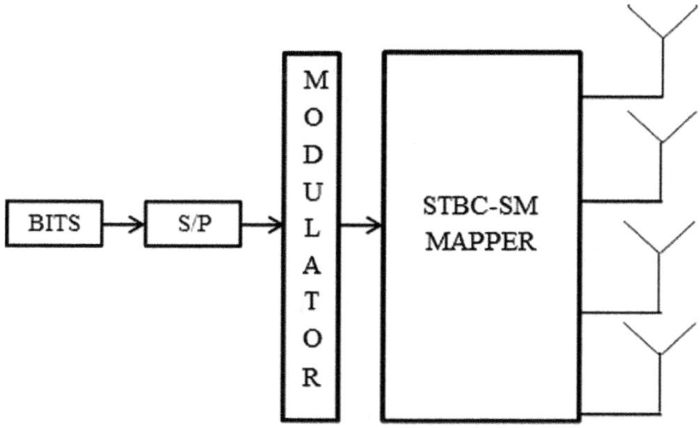

Fig. 1. Block Diagram of STBC-SM Transmitter

The transmitter side of one user only for the sake of simplicity. User knowledge are regenerated from serial to parallel then fed to the modulator. The STBC encoder will then change the data stream according to the generating matrix of the encoder. Spatial modulation mapper divides the data to index and data to be modulated followed by STBC code.

3.2 STBC Encoder

Space Time Block Coding (STBC) could be a technique that's used inside wireless communication networks for the aim of transmitting multiple copies of one information stream across several antennas As a result, the various received versions of that information may be utilized to assist rising the data-transfer reliability rating.

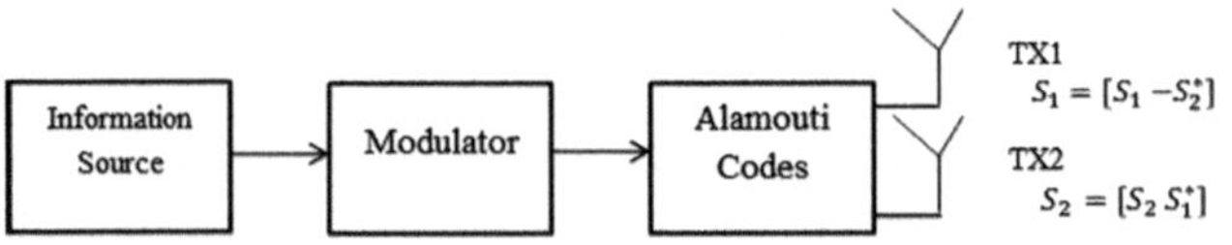

Fig. 2. STBC Encoder

Alamouti introduced the primary style for the STBC in 1998. The Alamouti STBC scheme uses two transmit antennas and Nr receives antennas and can accomplish a most diversity order of 2Nr. A diagram of the Alamouti frame of reference encoder is shown in Fig. 2.

In the encoder is the two modulated symbols S_1 and S_2 in each encoding operation and sent up to the transmit antennas in the form of a matrix as follows:

$$S = \begin{bmatrix} S_1 & -S_2^* \\ S_2 & S_1^* \end{bmatrix} \tag{1}$$

where S_1 is sent from the first antenna and S_2 from the second antenna in the first transmission period. Whereas $-S_2^*$ is sent from the first antenna and $-S_2^*$ from the second antenna in the second transmission period. The two rows and columns of S matrix are orthogonal to each other.

3.3 STBC-SM Mapper

In the STBC-SM scheme, Alamouti's STBC is chosen because the core STBC. each STBC symbols and therefore the indices of the transmit antennas, from that these symbols are transmitted, carry info. Wherever the columns and rows correspond to the transmit antennas and therefore the symbol intervals, respectively. Generally, two symbols x_1 and x_2 are transmitted, which form a set of the code words of STBC-SM if four transmit antennas are used

$$X_1 = \left\{ \begin{pmatrix} x_1 & x_2 & 0 & 0 \\ -x_2^* & x_1^* & 0 & 0 \end{pmatrix}, \begin{pmatrix} 0 & 0 & x_1 & x_2 \\ 0 & 0 & -x_2^* & x_1^* \end{pmatrix} \right\}$$

$$\boldsymbol{X}_2 = \left\{ \begin{pmatrix} 0 & x_1 & x_2 & 0 \\ 0 & -x_2^* & x_1^* & 0 \end{pmatrix}, \begin{pmatrix} x_1 & 0 & 0 & x_2 \\ -x_2^* & 0 & 0 & x_1^* \end{pmatrix} \right\} e^{j\theta}$$

Every two STBC-SM code words (S_{ij}, j = 1, 2) is a one STBC-SM codebooks (X_j, i = 1, 2). θ is a rotation angle to be optimized for a given modulation scheme to ensure maximum diversity and coding gain at the expense of expansion of the signal constellation.

The spectral efficiency of the STBC-SM scheme for four transmit antennas becomes

$$m = \frac{1}{2}\log_2 c + \log_2 M \,\text{bits/sec/hz}. \tag{2}$$

An important design parameter is the minimum coding gain distance (CGD) between two STBC-SM code words (matrices). The minimum CGD in any code should be maximized to achieve better performance in term of BER. The minimum CGD between two codebooks is defined as:

$$\delta_{\min}\left(X_i, X_j\right) = \min_{k,l} \delta_{\min}\left(X_i, X_j\right) \tag{3}$$

And the minimum CGD of an STBC-SM code is defined by:

$$\delta_{\min}(\mathrm{X}) = \min_{i,ji=j} \delta_{\min}\left(X_i, X_j\right) \tag{4}$$

3.4 STBC-SM Design

Unlike in the SM scheme, the number of transmit antennas in the STBC-SM scheme need not be an integer power of 2, since the pairwise combinations are chosen from nT available transmit antennas for STBC transmission. This provides design flexibility. However, the total number of code word combinations considered should be an integer power of 2.

The design of STBC-SM includes [29],

1. Determine the total number of transmitter antennas $\boldsymbol{n_T}$ and calculate the number of possible antenna combinations for the transmission of Alamouti's STBC.
2. Calculate the number of code words in each codebook $\boldsymbol{c} = \frac{\boldsymbol{n_T}}{2}$
3. Start with the construction of X_i which contains a noninterfering code words.
4. Using a similar approach, construct X_i for $2 \leq i \leq n$ by considering the following two important facts:
   (a) Every codebook must contain non-interfering code words chosen from pairwise combinations of n_T available transmit antennas.
   (b) Each codebook must be composed of code words with antenna combinations that were never used in the construction of a previous codebook.
5. Determine the rotation angles θ_i for each X_i, $2 \leq i \leq n$.

There are two cases to take advantages of the system of STBC in the best shape.

Case 1 - $\boldsymbol{n_T} \leq$ 4: We have, in this case, two codebooks X_1 and X_2 and only one non-zero angle, say θ, to be optimized. It can be seen that $\delta_{\min}$ (X_1,X_2) is equal to the minimum CGD between any two interfering codewords from X_1 and X_2.

Case 2 - $n_T > 4$: In this case, the number of codebooks, n, is greater than 2. Let the corresponding rotation angles to be optimized be denoted in ascending order by $\theta_1 = 0 < \theta_2 < \theta_3 < \cdots < \theta_n < p\pi/2$.

If four transmitter antennas and 2 bits/hz transmission are used, mapping rule for BPSK modulation is given by the following table.

Table 1 shows the mapping of STBC Codes to the antenna indexes. For example if the bit 0100 is transmitted, the MSB represents the antenna indices so 01 represents 2nd antenna and 00 represent the first constellation point of BPSK.

Table 1. Mapping rule for STBC-SM (BPSK Modulation)

	Input bits	Transmission matrices		Input bits	Transmission matrices
I = 0	0000	$\begin{pmatrix} 1+j & 1+j & 0 & 0 \\ -1-j & 1+j & 0 & 0 \end{pmatrix}$	I = 2	1000	$\begin{pmatrix} 0 & 1+j & 1+j & 0 \\ 0 & -1-j & 1+j & 0 \end{pmatrix} e^{j\theta}$
	0001	$\begin{pmatrix} -1+j & 1-j & 0 & 0 \\ -1+j & -1+j & 0 & 0 \end{pmatrix}$		1001	$\begin{pmatrix} 0 & -1+j & 1-j & 0 \\ 0 & -1+j & 1+j & 0 \end{pmatrix} e^{j\theta}$
	0010	$\begin{pmatrix} -1+j & 1-j & 0 & 0 \\ -1+j & -1+j & 0 & 0 \end{pmatrix}$		1010	$\begin{pmatrix} 0 & -1+j & 1-j & 0 \\ 0 & -1+j & -1+j & 0 \end{pmatrix} e^{j\theta}$
	0011	$\begin{pmatrix} 1+j & 1+j & 0 & 0 \\ -1-j & 1+j & 0 & 0 \end{pmatrix}$		1011	$\begin{pmatrix} 0 & 1+j & 1+j & 0 \\ 0 & -1-j & 1+j & 0 \end{pmatrix} e^{j\theta}$
I = 1	0100	$\begin{pmatrix} 0 & 0 & 1+j & 1+j \\ 0 & 0 & -1-j & 1+j \end{pmatrix}$	I = 3	1100	$\begin{pmatrix} 1+j & 0 & 0 & 1+j \\ 1+j & 0 & 0 & -1-j \end{pmatrix} e^{j\theta}$
	0101	$\begin{pmatrix} 0 & 0 & -1+j & 1-j \\ 0 & 0 & -1+j & -1+j \end{pmatrix}$		1101	$\begin{pmatrix} 1+j & 0 & 0 & 1+j \\ -1+j & 0 & 0 & 1+j \end{pmatrix} e^{j\theta}$
	0110	$\begin{pmatrix} 0 & 0 & -1+j & 1-j \\ 0 & 0 & -1+j & -1+j \end{pmatrix}$		1110	$\begin{pmatrix} 1-j & 0 & 0 & -1+j \\ -1+j & 0 & 0 & -1+j \end{pmatrix} e^{j\theta}$
	0111	$\begin{pmatrix} 0 & 0 & 1+j & 1+j \\ 0 & 0 & -1-j & 1+j \end{pmatrix}$		1111	$\begin{pmatrix} 1+j & 0 & 0 & 1+j \\ 1+j & 0 & 0 & -1-j \end{pmatrix} e^{j\theta}$

Similarly for QAM Modulation, Table 2 shows the mapping of STBC Codes to the antenna indexes. For example if the bit 0100 is transmitted, the MSB represents the antenna indices so 01 represent 2nd antenna and 00 represent the first constellation point of QAM.

Table 2. Mapping Rule for STBC-SM (QAM Modulation)

	Input bits	Transmission matrices		Input bits	Transmission matrices
l = 0	0000	$\begin{pmatrix} 1 & 1 & 0 & 0 \\ -1 & 1 & 0 & 0 \end{pmatrix}$	l = 2	1000	$\begin{pmatrix} 0 & 1 & 1 & 0 \\ 0 & -1 & 1 & 0 \end{pmatrix} e^{j\theta}$
	0001	$\begin{pmatrix} 1 & -1 & 0 & 0 \\ 1 & 1 & 0 & 0 \end{pmatrix}$		1001	$\begin{pmatrix} 0 & 1 & -1 & 0 \\ 0 & 1 & 1 & 0 \end{pmatrix} e^{j\theta}$
	0010	$\begin{pmatrix} -1 & 1 & 0 & 0 \\ -1 & -1 & 0 & 0 \end{pmatrix}$		1010	$\begin{pmatrix} 0 & -1 & 1 & 0 \\ 0 & -1 & -1 & 0 \end{pmatrix} e^{j\theta}$
	0011	$\begin{pmatrix} -1 & -1 & 0 & 0 \\ 1 & -1 & 0 & 0 \end{pmatrix}$		1011	$\begin{pmatrix} 0 & -1 & -1 & 0 \\ 0 & 1 & -1 & 0 \end{pmatrix} e^{j\theta}$
l = 1	0100	$\begin{pmatrix} 0 & 0 & 1 & 1 \\ 0 & 0 & -1 & 1 \end{pmatrix}$	l = 3	1100	$\begin{pmatrix} 1 & 0 & 0 & 1 \\ 1 & 0 & 0 & -1 \end{pmatrix} e^{j\theta}$
	0101	$\begin{pmatrix} 0 & 0 & 1 & -1 \\ 0 & 0 & 1 & 1 \end{pmatrix}$		1101	$\begin{pmatrix} -1 & 0 & 0 & 1 \\ 1 & 0 & 0 & 1 \end{pmatrix} e^{j\theta}$
	0110	$\begin{pmatrix} 0 & 0 & -1 & 1 \\ 0 & 0 & -1 & -1 \end{pmatrix}$		1110	$\begin{pmatrix} 1 & 0 & 0 & -1 \\ -1 & 0 & 0 & -1 \end{pmatrix} e^{j\theta}$
	0111	$\begin{pmatrix} 0 & 0 & -1 & -1 \\ 0 & 0 & 1 & -1 \end{pmatrix}$		1111	$\begin{pmatrix} -1 & 0 & 0 & -1 \\ -1 & 0 & 0 & 1 \end{pmatrix} e^{j\theta}$

3.4.1 Modified Code Word

The modified code word for STBC-SM is introduced after modulation. The flexible coefficients are added to the code words in order to obtain better system performance. Let a and b be the coefficients it should satisfy the following equation

$$\frac{a^2 \sum_{i=1}^{M} |x_i|^2}{M} + \frac{b^2 \sum_{i=1}^{M} |y_i|^2}{M} = \frac{\sum_{i=1}^{M} (|x_i|^2 + |y_i|^2)}{M} \tag{5}$$

Then the modified code word for the STBC-SM system is

$$X_1 = \left\{ \begin{pmatrix} x_1 & x_2 & 0 & 0 \\ -x_2^* & x_1^* & 0 & 0 \end{pmatrix}, \begin{pmatrix} 0 & 0 & x_1 & x_2 \\ 0 & 0 & -x_2^* & x_1^* \end{pmatrix} \right\} * a$$

$$X_2 = \left\{ \begin{pmatrix} 0 & x_1 & x_2 & 0 \\ 0 & -x_2^* & x_1^* & 0 \end{pmatrix}, \begin{pmatrix} x_1 & 0 & 0 & x_2 \\ -x_2^* & 0 & 0 & x_1^* \end{pmatrix} \right\} e^{j\theta} * b$$

where 1 < a < b.

3.4.2 Rotation Angle

The rotation angle of the STBC-SM system is given by

$$\theta = \frac{2K\pi}{M(n+1)} 1 < k < n \tag{6}$$

Where ‘n’ represents the number of code books.

Combining all three techniques results in better system performance and the block diagram for STBC-SM-CDMA is given in Fig. 3 where c1, c2… Cn represents the pseudorandom code which is to be added with the modified code word.

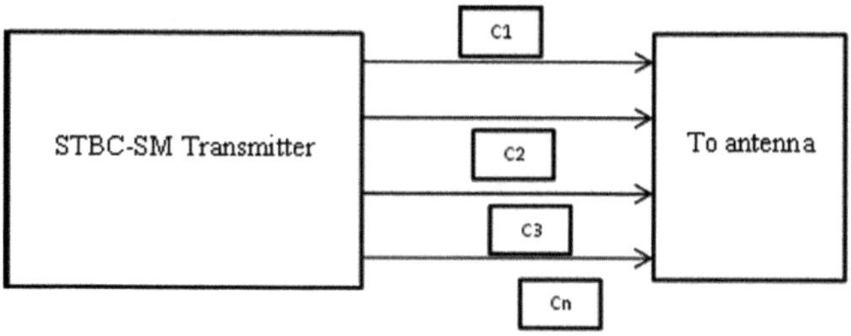

Fig. 3. STBC-SM using CDMA

3.5 STBC-SM Receiver

To extract the Information needed the receiver performs the following process includes Demodulation, Code acquisition and lock, Correlation of code with signal and Decoding of Information data.

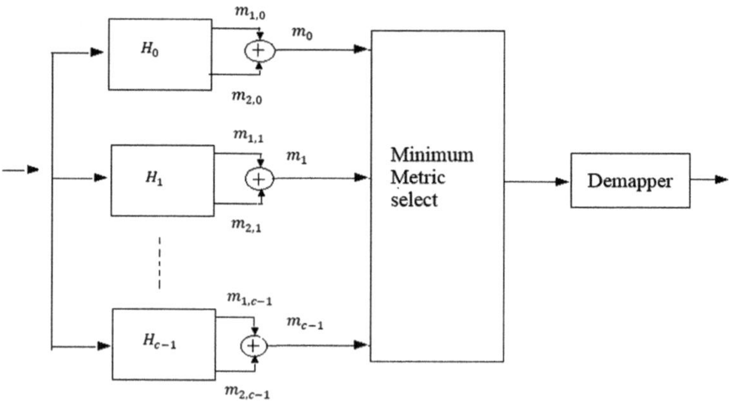

Fig. 4. STBC-SM Receiver

The block diagram of the STBC-SM receiver is shown in Fig. 4. STBC-SM with transmit n_T and receive antennas n_R is considered in the presence of a quasi-static Rayleigh flat fading MIMO channel [8]. The receiver matrix, Y can be expressed as:

$$Y = \sqrt{\frac{\rho}{\mu}} S_X H + N \tag{7}$$

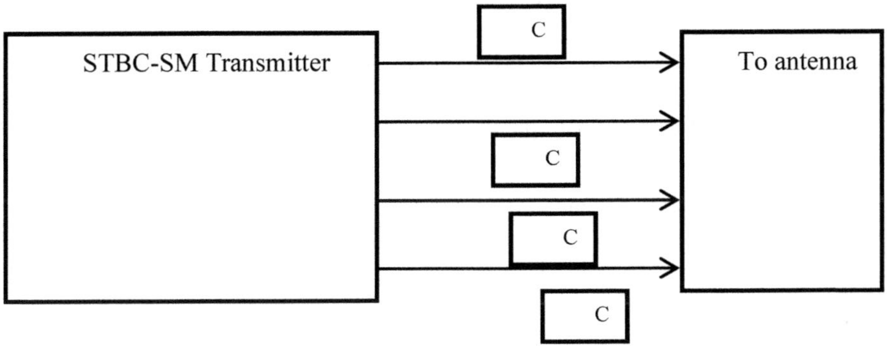

where $S_X \in X$ and μ is a normalization factor to ensure that ρ is the average SNR at each receive antenna. It is assume that H remains constant during the transmission of a code word and takes independent values from one code word to another. H is known at the receiver, but not at the transmitter.

The associated minimum ML metrics $m_{1,l}$ and $m_{2,l}$ for X_1 and X_2 are; respectively.

$$m_{1,l} = \min_{X_1 \in \Upsilon} \left\| Y - \sqrt{\frac{\rho}{\mu}} h_{l,1} X_1 \right\|^2 \quad (8)$$

$$m_{2,l} = \min_{X_2 \in \Upsilon} \left\| Y - \sqrt{\frac{\rho}{\mu}} h_{l,2} X_2 \right\|^2 \quad (9)$$

Since $m_{1,l}$ and $m_{2,l}$ are calculated by the ML decoder.

Combining space-time coding, spatial modulation and CDMA techniques can benefit from the advantages of the three techniques. The proposed system can achieve high BER due to the use of spatial modulation and Alamouti STBC since it's well known that Alamouti STBC can reach small values of BER because it allows for transmit diversity. Adding spatial modulation makes the system less vulnerable to the channel state because it reduces the number of bits being transmitted. It reduces the modulation order as some of the bits are transmitted through the transmit antenna location.

After the STBC spatial modulation is done, the only remaining part in the transmitter is the CDMA part. This is done by defining PN code using Hadamard code built in function and then defining a code for each user. In our case we considered two users only for programming simplicity- then spread each bit in the user data by its PN code. The next step is to send the data over uncorrelated Rayleigh fading channels. By multiplying each data with its corresponding channel, the received signal is calculated. At the receiver side, the received signal is multiplied by its corresponding PN code to dispread the data.

4 Results and Simulation

Simulation of STBC-SM using CDMA technology is implemented using MATLAB.

Before designing the STBC-SM model using CDMA technology with modified code word, there are some basic parameters should be calculated using the algorithm given in the Sect. 3. The parameters includes

1. Number of Transmit Antennas '$\boldsymbol{n_T}$',
2. Number of code words in a codebook '$\boldsymbol{c} = \frac{n_T}{2}$',
3. Total number of codebooks '**n**',
4. Modulation Order '**M**'
5. Minimum coding gain distance '$\delta_{\min}$' and
6. Spectral efficiency '**S**'.

These parameters are calculated with the formulas given in the previous chapters. Hence the parameters are calculated and tabulated in Table 3 is given by

Table 3. Parameters of STBC-SM System for Different Transmit Antenna.

n_T	c	n	$\delta_{\min}$			S
			M = 2	M = 4	M = 6	spectral efficiency
3	2	2	12	11.45	9.05	$0.5 + \log_2 M$
4	4	2	12	11.45	9.05	$1 + \log_2 M$
5	8	4	4.69	4.87	4.87	$1.5 + \log_2 M$
6	8	3	8.00	8.57	8.31	$1.5 + \log_2 M$
7	16	6	2.14	2.18	2.18	$2 + \log_2 M$
8	16	4	4.69	4.87	4.87	$2 + \log_2 M$

The simulation started by generating random data to generate random data to be modulated. The data is generated to each user individually. Each user data is modulated using the same algorithm at the transmitter side. In this STBC-SM system 4 transmitter and 4 Receiver antennas are used.

After generating the data, it is modulated using the phase shit keying (PSK) or the quadrature amplitude modulation (QAM).

Once the bits are modulated using either PSK or QAM Modulation, it is then transmitted to STBC (Space Time Block Code) Encoder. Then the bits are mapped using STBC-SM mapping based on the modulation as previously explained.

If four transmit antennas are used, two symbols are transmitted from STBC encoder, which forms the set of four code words for the STBC-SM system based on the modulation technique.

For example if bit '0 0 0 0' is transmitted along the STBC-SM system, the first two bits '0 0' represents the antenna indices which is transmitted along **antenna 1** and the next two bits are mapped according to their modulation scheme. If BPSK modulation is used, the set of code word for STBC-SM system is given by

$$\begin{pmatrix} 1 & 1 & 0\,0 \\ -1 & 1 & 0\,0 \end{pmatrix}, \begin{pmatrix} 1 & -1 & 0\,0 \\ 1 & 1 & 0\,0 \end{pmatrix}, \begin{pmatrix} -1 & 1 & 0\,0 \\ -1 & -1 & 0\,0 \end{pmatrix}, \begin{pmatrix} -1 & -1 & 0\,0 \\ 1 & -1 & 0\,0 \end{pmatrix}$$

Similarly for QAM modulation, the code words are given by,

$$\begin{pmatrix} 1+j & 1+j & 0\,0 \\ -1-j & 1+j & 0\,0 \end{pmatrix}, \begin{pmatrix} -1+j & 1-j & 0\,0 \\ -1+j & -1+j & 0\,0 \end{pmatrix},$$

$$\begin{pmatrix} -1+j & 1-j & 0\,0 \\ -1+j & -1+j & 0\,0 \end{pmatrix}, \begin{pmatrix} 1+j & 1+j & 0\,0 \\ -1-j & 1+j & 0\,0 \end{pmatrix}$$

Based on the Eq. (6) the optimal rotation angles and flexible coefficients for different modulation schemes are obtained and it is tabulated in Table 4.

Table 4. Optimal choice of Flexible Coefficients and Rotation angle.

Modulation	Flexible coefficient	Rotation angle (rad)
BPSK	0.95	1.57
QPSK	0.77	0.66
8PSK	0.86	0.28
16PSK	0.82	0.14

After the STBC spatial modulation is done, the only remaining part in the transmitter is the CDMA part. This is done by defining PN code using Hadamard code built in function and then defining a code for each user. In our case we considered two users only for programming simplicity- then spread each bit in the user data by its PN code.

By multiplying each data with its corresponding channel, the received signal is calculated. At the receiver side, the received signal is multiplied by its corresponding PN code to dispread the data.. Then the maximum likelihood algorithm is used to recover the transmitted signal. A comparison between the transmitted signal and the received signal is formed, to check the system BER.

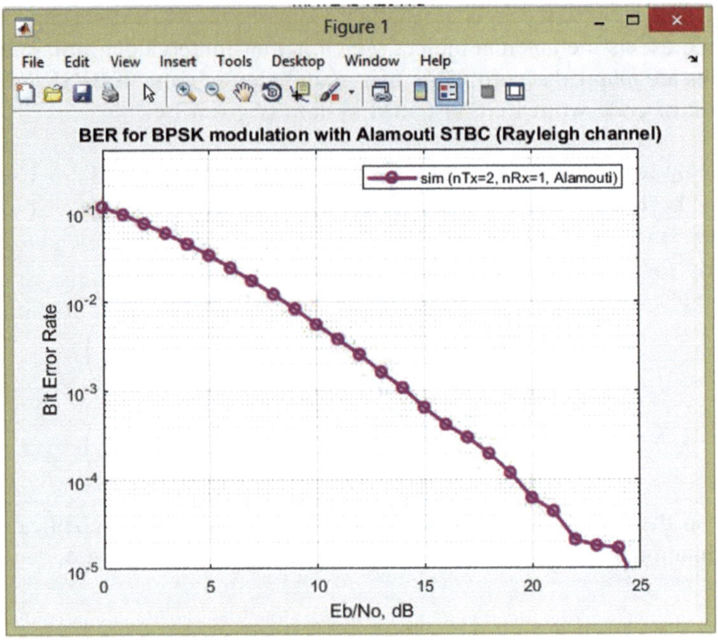

Fig. 5. BER Performance of STBC Alamouti using BPSK

Figure 5 shows BER for STBC Alamouti with binary PSK modulation. From the plot it is clear that BER starts at approximately 10^{-1} at 0 Eb/No and reached $10^{-2.1}$ at 10 Eb/No.

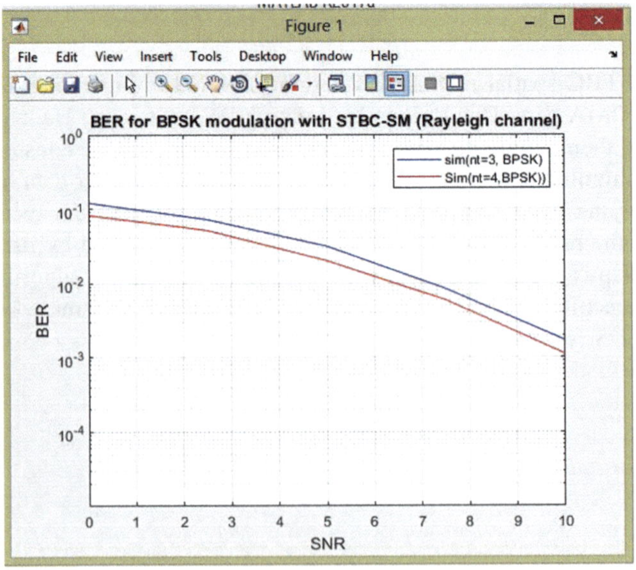

Fig. 6. BER Performance of STBC-SM using BPSK

Figure 6 shows BER for STBC-SM with binary PSK modulation. From the plot it is clear that BER starts at approximately 10^{-1} at 0 Eb/No and reached 10^{-3} at 10 Eb/No with modified code word at coefficient (0.95) and rotation angle (1.57).

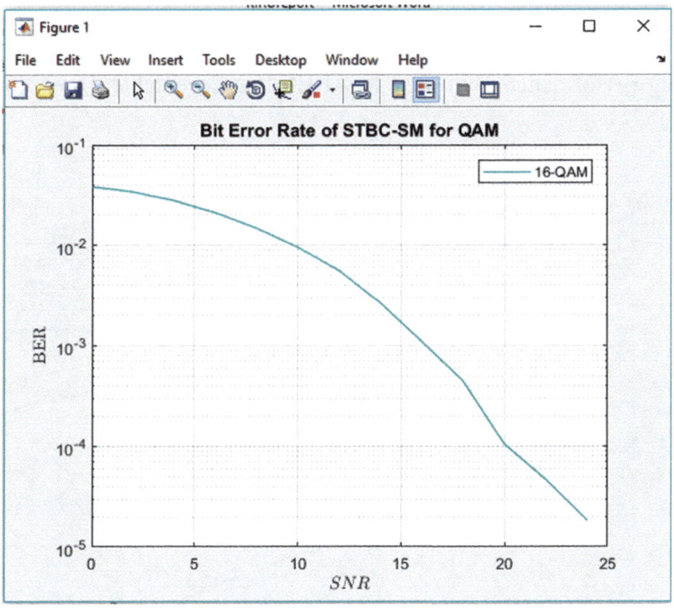

Fig. 7. BER Performance of STBC-SM using 16 QAM

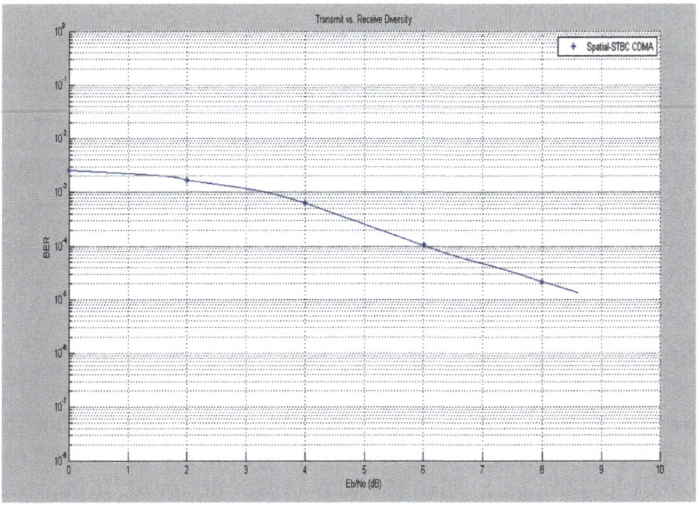

Fig. 8. BER performance of SM- STBC – CDMA coding for CDMA code length of 8 bits, BPSK modulation.

Figure 7 shows BER for STBC-SM with QAM modulation. From the plot it is clear that BER starts at approximately $10^{-1.5}$ at 0 Eb/No and reached $10^{-2.8}$ at 10 Eb/No with modified code with coefficient (0.73) and rotation angle (0.19).

Figure 8 shows BER for STBC-SM-CDMA with binary PSK modulation and CDMA code length of 8 bits. From the plot it is clear that BER starts at approximately $10^{-2.8}$ at 0 Eb/No and reached $10^{-4.9}$ at 8 Eb/No with modified code word (Flexible Coefficient 'a' = 0.73) and optimal rotation angle (0.19 rad)This proves that the system has promising performance with very low BER and acceptable throughput.

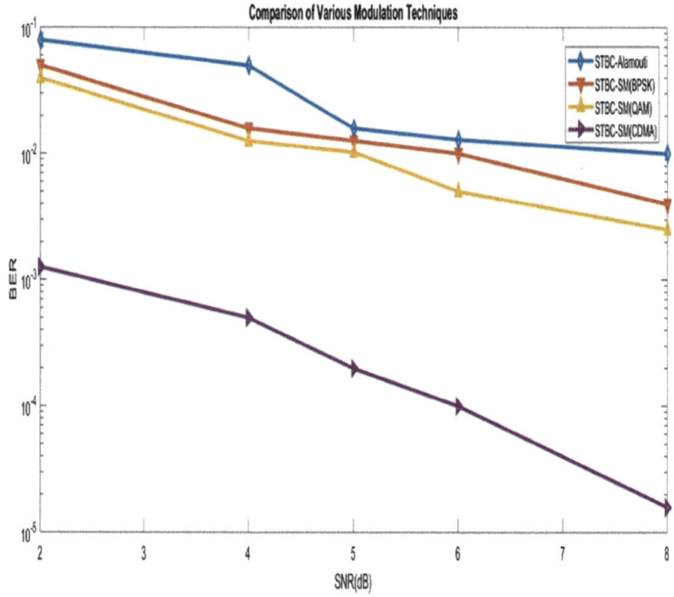

Fig. 9. BER Comparisons of Various Techniques

Figure 9 shows that various BER plots of STBC and it is clear that STBC-SM using CDMA technology with the code length of **8 Bits** and BPSK modulation with modified code word flexible coefficient of **a = 0.88** and optimal rotation angle (**1.57 rad**) gives better BER performance.

5 Conclusion

The code word is an essential part of STBC-SM scheme which influences the BER performance directly. Modified code word with flexible coefficient and rotation angle for Space time Block coded Spatial Modulation was designed. It improves the minimum coding gain distance and BER performance of the STBC-SM scheme. With the flexible coefficients of STBC-SM and CDMA technology is combined, in order to use narrow bandwidth for many users and achieved good BER performance. Extensive simulation study has been conducted and the results indicate that by proposed system STBC-SM with modified code word has better BER performance over other techniques.

References

1. Basar, E., Aygolu, U., Panayirci, E., Poor, H.V.: Space-time block coded spatial modulation. IEEE Trans. Commun. **59**, 823–832 (2011)
2. Liang, X.-B.: Orthogonal designs with maximal rates. IEEE Trans. Inf. Theory **49**, 2468–2503 (2003)
3. Jafarkhani, H., Seshadri, N.: Super-orthogonal space–time trellis codes. IEEE Trans. Inf. Theory **49**, 937–950 (2003)
4. Jeganathan, J., Ghrayeb, A., Szczecinski, L.: Spatial modulation: optimal detection and performance analysis. IEEE Commun. Lett. **12**, 545–547 (2008)
5. Sterian, C.E.D., et al.: Super-orthogonal space-time codes with rectangular constellations and two transmit antennas for high data rate wireless communications. IEEE Trans. Wirel. Commun. **5**, 1857–1865 (2006)
6. Li, X., Wang, L.: High rate space-time block coded spatial modulation with cyclic structure. IEEE Commun. Lett. **18**, 532–535 (2014)
7. Hua, Y., Zhao, G., Zhao, W., Jin, M.: Modified codewords design for space–time block coded spatial modulation. IET Commun. (2016)
8. Vasanth Raj, P.T., Vishvaksenan, K.S., Dinesh, V., Elaveni, M.: System analysis of STBC-CDMA technique for secured image transmission using watermarking algorithm. In: International Conference on Communication and Signal Processing, 6–8 April 2017
9. Adithya, B.: Adaptive selection of antennas for optimum transmission using STBC. Int. J. Sci. Res. (IJSR) **5**, 736–742 (2016). Index Copernicus Value (2013): 6.14
10. Luong, V.-T., Le, M.-T., Mai, H.-A., Tran, X.-N., Ngo, V.-D.: New upper bound for space-time block coded spatial modulation. In: 2015 IEEE 26th International Symposium on Personal, Indoor and Mobile Radio Communications - (PIMRC) Fundamentals and PHY (2015)
11. Auffray, J.M., Helard, J.F.: Performance of multicarrier CDMA technique combined with space-time block coding over Rayleigh channel. In: IEEE Seventh International Symposium on Spread Spectrum Techniques and Applications, vol. 2, pp. 348-352 (2002)
12. Tarokh, V., Jafarkhani, H., Calderbank, A.R.: Space-time block coding for wireless communications: performance results. IEEE J. Sel. Areas Commun. **17**, 451–460 (1999)
13. Maaref, A., Aïssa, S.: Capacity of space-time block codes in MIMO Rayleigh fading channels with adaptive transmission and estimation errors. IEEE Trans. Wirel. Commun. **4**, 2568–2578 (2005)
14. Andersen, J.B.: Array gain and capacity for known random channels with multiple element arrays at both ends. IEEE J. Sel. Areas Commun. **18**(11), 2172–2178 (2000)
15. Clerckx, B., Oestges, C.: MIMO Wireless Networks: Channels, Techniques and Standards for Multi-Antenna, Multi-User and Multi-Cell Systems, pp. 7–9. Academic Press, Oxford (2013)
16. Brennan, D.G.: Linear diversity combining techniques. Proc. IEEE **91**, 331–356 (2003)
17. Alamouti, S.: A simple transmit diversity technique for wireless communications. IEEE J. Sel. Areas Commun. **16**, 1451–1458 (1998)

18. Paulraj, A., Nabar, R., Gore, D.: Introduction to Space-Time Wireless Communications. Cambridge University Press, Cambridge (2003)
19. Tse, D.N.C., Viswanath, P.: Fundamentals of Wireless Communication. Cambridge University Press, Cambridge (2005)
20. Foschini, G.J., Gans, M.J.: On limits of wireless communications in fading environments when using multiple antennas. Wirel. Pers. Commun. **6**, 311–335 (1998)
21. Jadhav, S.P., Hendre, V.S.: Performance of maximum ratio combining (MRC) MIMO systems for Rayleigh fading channels. Int. J. Sci. Res. Publ. **3**, 2250–3153 (2013)
22. Mesleh, R.Y.: Spatial modulation. IEEE Trans. Veh. Technol. **57** (2008)
23. Mesleh, R., Haas, H., Ahn, C.W., Yun, S.: Spatial modulation–a new low complexity spectral efficiency enhancing technique. In: Proceedings of Conference on Communications and Networking in China, Beijing, China, pp. 1–5 (2006)
24. Sumathi, A., Mohideen, S.K., Anitha, A.: Performance analysis of space time block coded spatial modulation. In: Software Engineering and Mobile Application Modelling and Development (ICSEMA 2012), International Conference on Digital Object Identifier, pp. 1–7 (2012). https://doi.org/10.1049/ic.2012.0153
25. Kohno, R., Meidan, R., Milstein, L.: Spread spectrum access methods for wireless communications. IEEE Commun. Mag. **33**, 58–67 (1995)
26. Viterbi, J.: CDMA: Principles of Spread-Spectrum Communication. Addison Wesley Wireless Communication (1995)
27. Pickholtz, R.L., Schilling, D.L., Milstein, L.B.: Theory of spread-spectrum communications —a tutorial. IEEE Trans. Commun. **30**, 855–884 (1982)
28. Handry, M.: http://www.bee.net/mhendry/vrml/library/cdma/cdma.html
29. http://library.iugaza.edu.ps/thesis/114827.pdf

Swing up and Stabilization of Rotational Inverted Pendulum by Fuzzy Sliding Mode Controller

K. Rajeswari(✉), P. Vivek, and J. Nandhagopal

Department of Electrical and Electronics Engineering, Velammal Institute of Technology, Chennai 601 204, India
rajeswariarul@gmail.com, vivekpsgped11@gmail.com,
jnandhagopal@gmail.com

Abstract. Rotational inverted pendulum (RIP) is widely used as a benchmark system in assessing various control strategies. Though a Proportional-Integral-Derivative (PID) controller is widely used control strategy, it is not recommended for the inverted pendulum due to the difficulties in tuning PID parameters. This paper presents Sliding Mode Controller (SMC) and Fuzzy Sliding Mode Controller (FSMC) for stabilizing the RIP. SMC is applied for the stabilization and robust control of RIP based on pole placement method. The drawbacks of SMC in terms of high control gain and chattering are overcome by FSMC. These controllers are applied to the RIP in real-time and their performance is compared on the basis of Pendulum regulation.

Keywords: Rotational Inverted Pendulum · PID · SMC · FSMC

1 Introduction

Inverted pendulum is an under - actuated mechanical system with high nonlinearity and an open loop unstable system. It is a benchmark system for validation of classical and contemporary control techniques. Its applications range from robotics to space rocket guidance systems which move away from the gravity. Mostly, inverted pendulum system is used to illustrate ideas in linear control theory and the control of linear unstable systems.

Various control strategies such as PID(Proportional Integral Derivative) [1], LQR (Linear Quadratic Regulator), SMC (Sliding Mode Controller), FLC (Fuzzy Logic Controller) [2] have been discussed in the past for the Inverted pendulum system. Variable structure controller was implemented for the stabilization and robust control of the double inverted pendulum by using the pole placement method. In order to overcome the drawback of chattering of the controller, the sliding mode control was proposed in the simulation of double inverted pendulum system [3]. Fuzzy sliding mode controller (FSMC) and the additional compensator is presented for a Rotational Inverted Pendulum position control which in turn provides the characteristics of insensitivity and robustness to uncertainties and external disturbances [4]. From the literature survey, it is understood that FSMC is a widely used control strategy to

D. J. Hemanth et al. (Eds.): COMET 2019, LNDECT 35, pp. 415–423, 2020.
https://doi.org/10.1007/978-3-030-32150-5_40

overcome the effect of chattering. Hence FSMC is attempted in this work to control the rotational inverted pendulum (RIP) system.

Most of the papers in the literatures demonstrated only simulation results and in this paper experimental verification of the control strategy is demonstrated. Results prove the effectiveness of the FSMC for rotational inverted pendulum for swinging up and balancing control. The paper is organized as follows. Section 2 briefly explains about the Mathematical modeling of rotational inverted pendulum. Section 3 briefs about the controller design. Simulation results are discussed and presented in Sect. 4. Experimental results are discussed and presented in Sect. 5. Finally, the last section concludes the paper.

2 Rotary Inverted Pendulum – Mathematical Model

The rotary inverted pendulum is classic example of a highly unstable under-actuated non-linear control problem and the studies carried out are design, implementation and development of control for the same. The system consist of two parts i.e. an arm and a pendulum. The main objective is to keep the pendulum in upright position of unstable equilibrium. The second one is to keep the arm at a particular position (angular position) while performing the primary task. Figure 1 shows the schematic diagram of RIP system, which elucidates the angles made by the arm and the pendulum and α respectively; (m - Pendulum's mass; L - Pendulum's length; r - Arm's length).

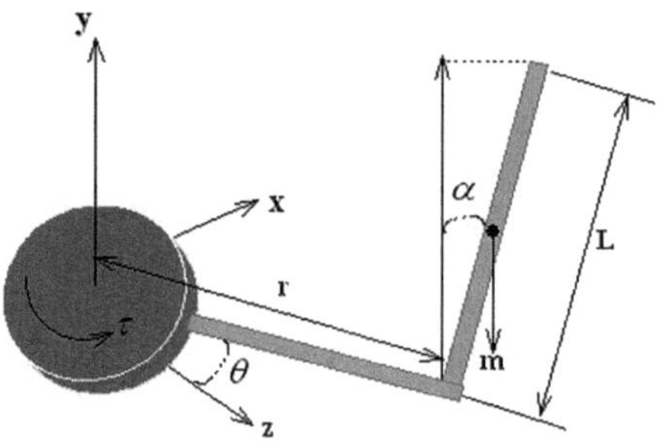

Fig. 1. Swing up/Balance closed loop system

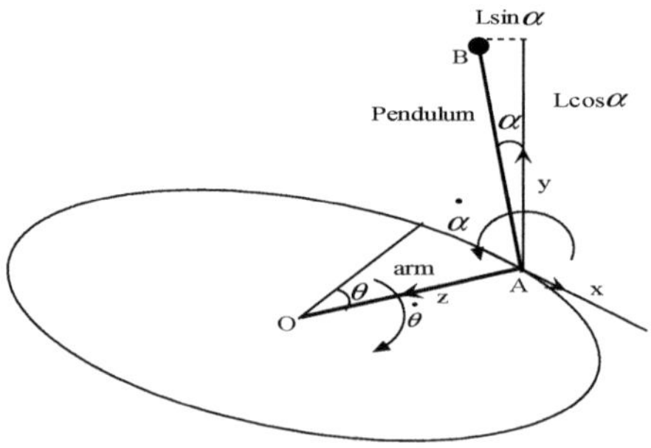

Fig. 2. Schematic diagram of an RIP

The dynamic equation describing a rotary inverted pendulum is derived either by Newtonian methods (using free body diagram) or the Euler-Lagrange method. Modeling of the system is by applying the Euler-Lagrange to the Lagrangian. The Lagrangian is the difference between kinetic and potential energy (L = K − P).

Accurate mathematical models of the pendulum system can be developed assuming that there is no excess stiction (static friction), backlash or bearing slack present. These nonlinearities make accurate modelling and control of the system a difficult task. The model of rotary inverted pendulum is shown in Fig. 2. α and θ are the coordinates which indicate the pendulum angle and the arm angle as shown in the Fig. 1. The mathematical model of the system can be obtained from the velocities of the pendulum. Potential energy: The potential energy in the inverted pendulum is gravity. Thus,

$$V = P.E._{Pendulum} = mgh = mgL\cos(\alpha) \tag{1}$$

Kinetic energy: The Kinetic Energies in the Inverted pendulum due to the rotating arm, the velocity of the point mass in the x and y – direction and the rotating pendulum around the centre of mass. Total kinetic energy,

$$T = K.E._{Arm} + K.E._{\dot{x}_B} + K.E._{\dot{y}_B} + KE._{Pendulum} \tag{2}$$

$$T = \frac{1}{2}J_{eq}\dot{\theta}^2 + \frac{1}{2}m(\dot{x}_B^2) + \frac{1}{2}m(\dot{y}_B^2) + \frac{1}{2}J_B\dot{\alpha}^2 \tag{3}$$

$$T = \frac{1}{2}J_{eq}\dot{\theta}^2 + \frac{1}{2}m(\dot{x}_B^2 + \dot{y}_B^2) + \frac{1}{2}J_B\dot{\alpha}^2 \tag{4}$$

$$T = \frac{1}{2}J_{eq}\dot{\theta}^2 + \frac{1}{2}m\left[\left(r\dot{\theta} - L\cos(\alpha)\dot{\alpha}\right)^2 + (-Lsin(\alpha)\dot{\alpha})^2\right] + \frac{1}{2}J_B\dot{\alpha}^2 \tag{5}$$

$$\begin{gathered} T = \frac{1}{2}J_{eq}\dot{\theta}^2 + \frac{1}{2}m\left(r\dot{\theta}\right)^2 - \frac{1}{2}m\left(2r\dot{\theta}L\cos(\alpha)\dot{\alpha}\right) + \frac{1}{2}m(L\cos(\alpha)\dot{\alpha})^2 + \\ \frac{1}{2}m[-Lsin(\alpha)\dot{\alpha}]^2 + \frac{1}{2}J_B\dot{\alpha}^2 \end{gathered} \tag{6}$$

$$\begin{gathered} T = \frac{1}{2}(J_{eq} + mr^2)\dot{\theta}^2 - \left(mr\dot{\theta}L\cos(\alpha)\dot{\alpha}\right) + \frac{1}{2}mL^2(\cos(\alpha))^2(\dot{\alpha}^2) + \\ \frac{1}{2}mL^2(\sin(\alpha))^2(\dot{\alpha}^2) + \frac{1}{2}J_B\dot{\alpha}^2 \end{gathered} \tag{7}$$

$$T = \frac{1}{2}(J_{eq} + mr^2)\dot{\theta}^2 - \left(mLr\cos(\alpha)\dot{\theta}\dot{\alpha}\right) + \frac{1}{2}mL^2(\dot{\alpha}^2)\left[(\cos(\alpha))^2 + (\sin(\alpha))^2\right] + \frac{1}{2}J_B\dot{\alpha}^2 \tag{8}$$

$$T = \frac{1}{2}(J_{eq} + mr^2)\dot{\theta}^2 - \left(mLr\cos(\alpha)\dot{\theta}\dot{\alpha}\right) + \frac{1}{2}mL^2(\dot{\alpha}^2) + \frac{1}{2}J_B\dot{\alpha}^2 \tag{9}$$

Where, $J_B = \frac{1}{12}m(r)^2 = \frac{1}{12}m(2L)^2 = \frac{1}{3}mL^2$ is the moment of inertial of the pendulum about its centre of mass;

$$T = \frac{1}{2}(J_{eq} + mr^2)\dot{\theta}^2 - \left(mLr\cos(\alpha)\dot{\theta}\dot{\alpha}\right) + \frac{1}{2}mL^2(\dot{\alpha}^2) + \frac{1}{2}\left(\frac{1}{3}mL^2\right)\dot{\alpha}^2 \tag{10}$$

$$T = \frac{1}{2}(J_{eq} + mr^2)\dot{\theta}^2 - mLr\cos(\alpha)\dot{\theta}\alpha + \frac{2}{3}\dot{m}L^2\dot{\alpha}^2 \tag{11}$$

Lagrangian (L) can be found using (1) and (11),
$L = K.E. - P.E. = T - V$

$$L = \frac{1}{2}\left(J_{eq} + mr^2\right)\dot{\theta}^2 - mLr\,cos(\alpha)\dot{\theta}\alpha + \frac{2}{3}\dot{m}L^2\dot{\alpha}^2 - [mgL\cos(\alpha)] \tag{12}$$

Since there are two generalized coordinates θ and α, there are two equations according to Euler – Lagrangian Formulation. Thus the equations of motion of the system are,

$$\left(J_{eq} + mr^2\right)\ddot{\theta} - mLr\ddot{\alpha} = T_l - B_{eq}\dot{\theta} \tag{13a}$$

$$\frac{4}{3}mL^2\ddot{\alpha} - mLr\ddot{\theta} - mgL\alpha = 0 \tag{13b}$$

The Eqs. (13a) and (13b) can be written simply as

$$\alpha\ddot{\theta} - b\ddot{\alpha} + G\dot{\theta} = fV_m$$

$$-b\ddot{\theta} + c\ddot{\alpha} - d\alpha = 0 \tag{14}$$

Taking Laplace transform of (14) and hence the transfer function model of the system between pendulum angle (α) as output and voltage (V_m) as input is given as

$$\frac{\alpha(s)}{V_m(s)} = \frac{bfs}{ac - b^2s^3} + cGs^2 - ads - dG \tag{15}$$

Consider the linearized differential Eqs. 13a and 13b and rewriting the equations as

$$a\ddot{\theta} - b\ddot{\alpha} = T_l - B_{eq}\dot{\theta} \tag{16a}$$

$$c\ddot{\alpha} - b\ddot{\theta} - d\alpha = 0 \tag{16b}$$

And $\ddot{\theta} = \left[\left(\frac{bd}{E}\right)\alpha - \left(\frac{cG}{E}\right)\dot{\theta} + \frac{c}{E}\left(\frac{\eta_m\eta_g k_t k_g}{R_m}\right)V_m\right]$

$$\ddot{\alpha} = \left[\left(\frac{ad}{E}\right)\alpha - \left(\frac{bG}{E}\right)\dot{\theta} + \frac{b}{E}\left(\frac{\eta_m\eta_g k_t k_g}{R_m}\right)V_m\right]$$

From the above state model is obtained as

$$\begin{bmatrix} \dot{\theta} \\ \dot{\alpha} \\ \ddot{\theta} \\ \ddot{\alpha} \end{bmatrix} = \begin{bmatrix} 3^{0} & 0 & 1 & 0 \\ 0 & 0 & 0 & 1 \\ 0 & \frac{bd}{E} & -\frac{cG}{E} & 0 \\ 0 & \frac{ad}{E} & \frac{-bG}{E} & 0 \end{bmatrix} \begin{bmatrix} \theta \\ \alpha \\ \dot{\theta} \\ \dot{\alpha} \end{bmatrix} + \begin{bmatrix} 0 \\ 0 \\ 3\ c\frac{\eta_m\eta_g k_t k_g}{R_m E} \\ b\frac{\eta_m\eta_g k_t k_g}{R_m E} \end{bmatrix} V_m \tag{17}$$

$$y = \begin{bmatrix} 0 & 1 & 0 & 0 \end{bmatrix} \begin{bmatrix} \theta \\ \alpha \\ \dot{\theta} \\ \dot{\alpha} \end{bmatrix} + [0]V_m \tag{18}$$

3 Proposed Controller

3.1 SMC

Sliding Mode Control (SMC) has been applied to nonlinear systems and it is considered as an effective approach for the control of the system with uncertainties. In sliding mode control state feedback control structure along with the switching term (sgn function) is applied to nullify the effects of uncertainties [7]. A drawback of the sliding mode is that the intermittent control signal which excite high frequency fluctuations of control signal of the system. This lead to "chattering", which cause high wear of moving mechanical parts and thus chattering needs to be avoided by all means.

3.2 FSMC

The basic strategy of the FSMC is to force the system state to stay in sliding surface, so that the system on the sliding surface is insensitive to disturbances. In FSMC design, a fuzzy sliding surface is deployed in the design of SMC. This is done by replacing the discontinuous term Ksgn(S) by an Fuzzy inference system.

$$\mathrm{T}(\tilde{\mathrm{S}}) = \{\mathrm{NB}, \mathrm{NM}, \mathrm{ZE}, \mathrm{PM}, \mathrm{PB}\}$$

Where $\mathrm{T}(\tilde{\mathrm{S}})$ is the term set of $\tilde{\mathrm{S}}$ and NB, NM, ZE, PM and PB are labels of fuzzy sets, which are negative big, negative medium, zero, positive medium and positive big respectively. The control output u and the labels of the fuzzy sets are defined as $\mathrm{T}(\tilde{\mathrm{u}}) = \{\mathrm{SR}, \mathrm{S}, \mathrm{M}, \mathrm{B}, \mathrm{BR}\}$.

Where $\mathrm{T}(\tilde{\mathrm{u}})$ is the term set of $\tilde{\mathrm{u}}$ and SR, S, M, B and BR are the labels of fuzzy sets, which are Smaller, Small, Medium, Big and Bigger respectively. The membership functions of these fuzzy sets are shown in Figs. 3 and 4. Rule are given as

R1 = If S is NB, then ufs is bigger.
R2 = If S is NM, then ufs is big.
R3 = if S is ZE, then ufs is medium.
R4 = if S is PM, then ufs is small.
R5 = if S is PB, then ufs is smaller.

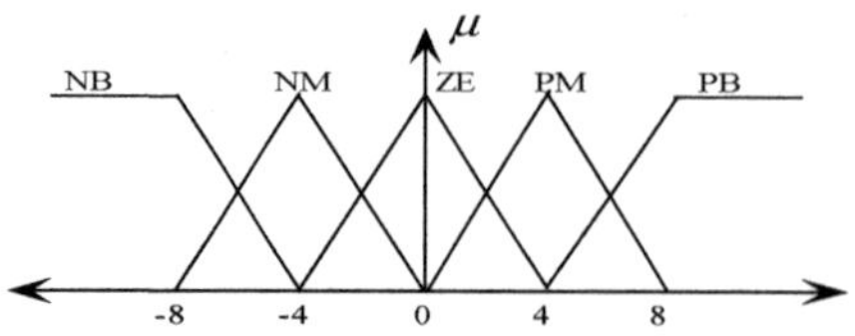

Fig. 3. Membership function - S

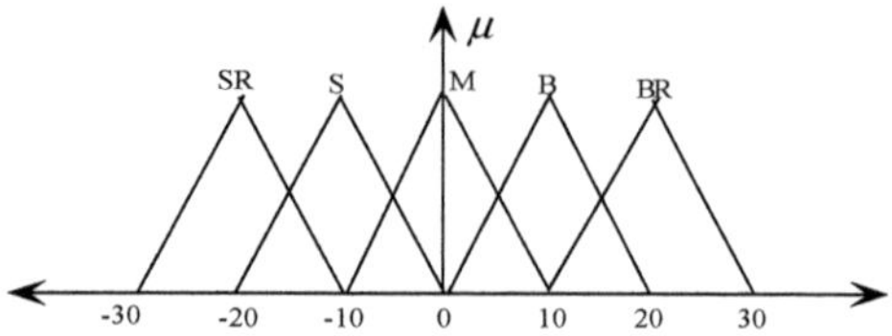

Fig. 4. Membership function – u

4 Simulation Results

The mathematical model of the rotational inverted pendulum defined by Eqs. 17 and 18 are simulated using the parameters listed in Table 1 in MATLAB/SIMULINK. Simulations are conducted for open loop and closed loop rotational inverted pendulum system.

Table 1. RIP parameters

Parameter	Value
Jeq	0.0036 kg/m^2
M	0.125 kg
R	0.215 m
L	0.1675 m
G	9.8 m/s^2
F	0.1285/Ω
Beq	0.0040 kgm^2
Km	6.05389 V/rad
Kg	0.0822 Nm/A

Table 2. Comparison of results

Controller	Hitting time (sec)	Chattering
SMC	0.397	0.6537
FSMC	4.656	0.000862

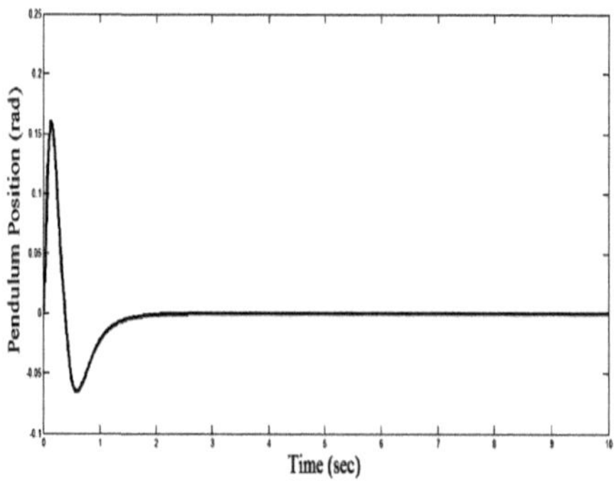

Fig. 5. Pendulum position

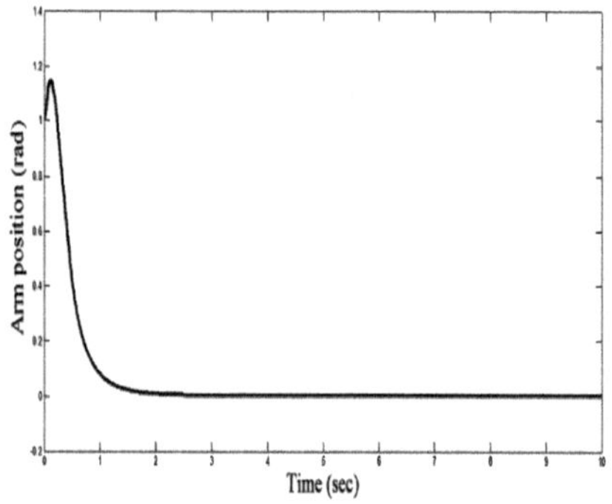

Fig. 6. Arm position

The comparison of parameters in terms of hitting time and chattering magnitude for SMC and FSMC is shown in Table 2. Pendulum position is shown in Figs. 5 and 6 shows the response of the arm position.

5 Experimental Results

In this study, control of rotational inverted pendulum system is demonstrated using SMC and FSMC. There are two control loops: balance control loop and a swing-up control loop. Swing up controller drives the pendulum to its upright position from the equilibrium position. Then the balance controller drives the rotary arm to fix the pendulum in its inverted position. It is implemented using SMC and FSMC technique. Balance controller is used to balance the pendulum in its inverted position. Modeling of pendulum is derived using Lagrangian principles and the swing-up controller is designed using the pendulum model and a Lyapunov function. SMC and FSMC is implemented in Lab VIEW platform and the following hardware is required to run the Rotary Pendulum Trainer (ROTPENT). PC is equipped with either NI-ELVIS I or an NI E-Series or M-series DAQ card., NI ELVIS II, Quanser Engineering Trainer (QNET) module. LabVIEW 8.6.1 with DAQmx.

The block diagram representing the swing up controller and the balance controller is given in Figs. 7 and 8 indicates the layout of QNET-ROTPENT with its elements. Table 3 provides the factors of the QNET-ROTPENT system.

The experimental results on RIP system with swing up and balance control algorithm is discussed in this section. The pendulum spins into the inverted position after several spins as shown in the results. The amplitude of α increase for a while before it reaches the reference position. FSMC plays a balancing control to keep the pendulum in the inverted position, when the disturbance is given to the arm angle θ. The robustness of QNET-ROTPENT is demonstrated by the designed SMC and FSMC. Once pendulum angle becomes zero, the system is closed-loop stable and the pendulum fluctuates around the point α at zero. The arm periodically adjusts its position.

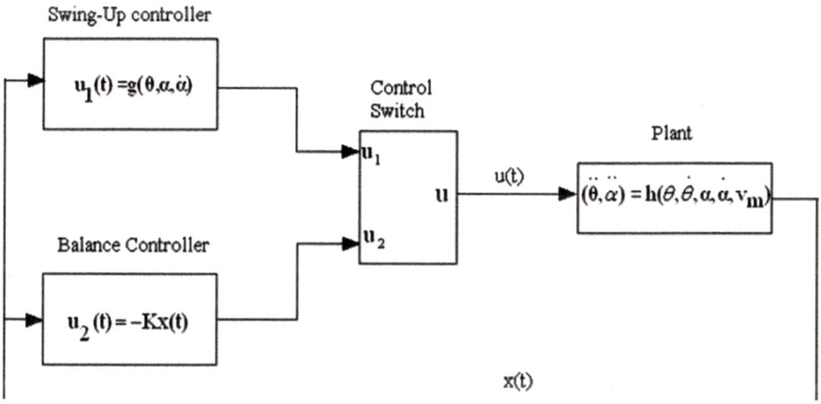

Fig. 7. Swing up/Balance closed loop system

Table 3. ROTPENT parameters

Parameter	Value
Jeq	0.0036 kg/m^2
M	0.125 kg
R	0.215 m
L	0.1675 m
G	9.8 m/s^2
F	0.1285/Ω
Beq	0.0040 kgm^2
Km	6.05389 V/rad
Kg	0.0822 Nm/A

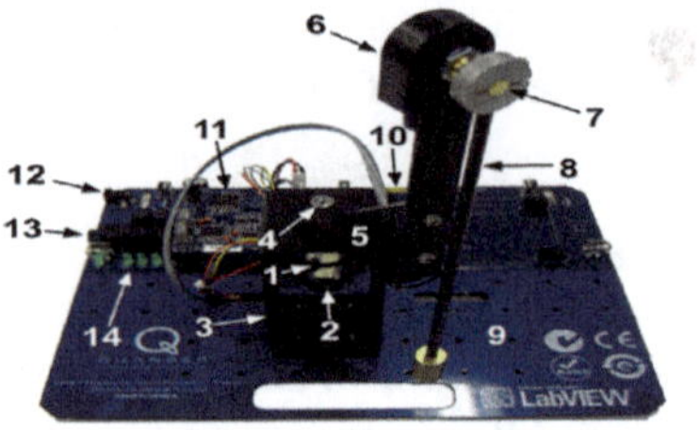

Fig. 8. General layout of QNET ROTPENT

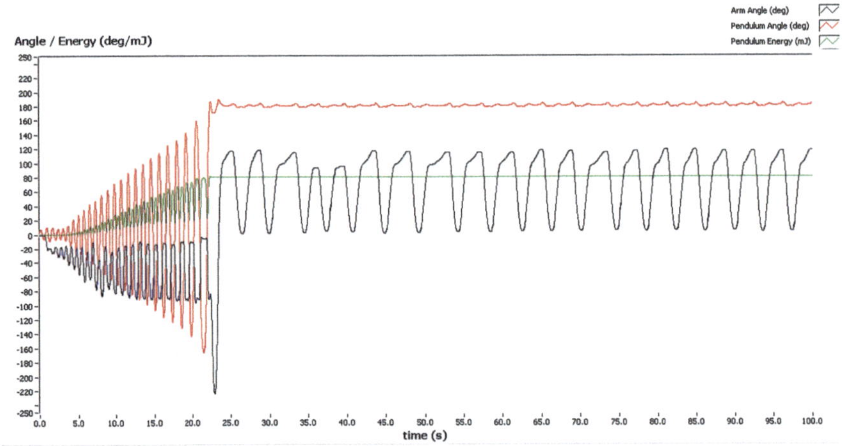

Fig. 9. Time response of Pendulum angle in FSMC

Figure 9 shows the Arm angle, Pendulum angle and Pendulum energy for the FSMC of Rotary inverted pendulum system. Figure 10 shows the control voltage generated in the FSMC of Rotary inverted pendulum system.

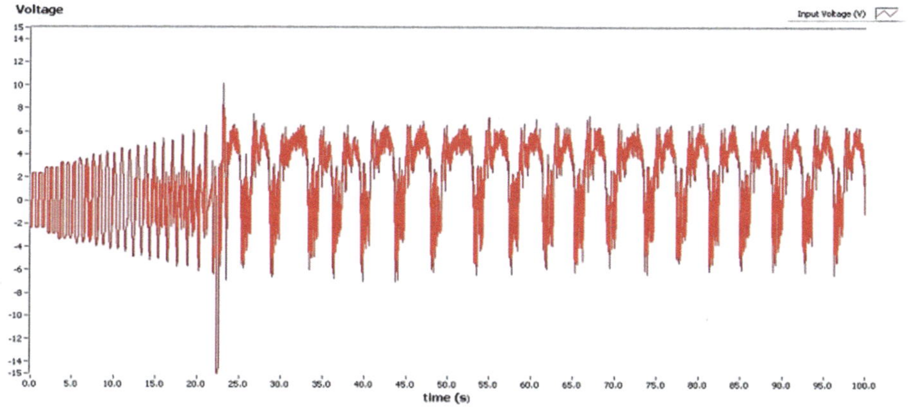

Fig. 10. Control voltage vs. time

6 Conclusion

In this paper, SMC and FSMC is designed and their performance is compared using MATLAB/SIMULINK. From the simulation results it is evident that FSMC performance is better in stabilizing the pendulum position with smooth control effect without chattering. Experimental results with the Lab based QNET ROTPENT(Rotational Inverted Pendulum system) again demonstrated the effectiveness of FSMC in stabilization control of rotational inverted pendulum.

References

1. Wang, J.-J.: Simulation studies of inverted pendulum based on PID controllers. Simul. Model. Pract. Theory **19**, 440–449 (2011)
2. Ozbek, N.S., Efe, M.O.: Swing up and stabilization control experiments for a rotary inverted pendulum - an educational comparison. In: IEEE International Conference, October 2010
3. Li, Z., Zhang, X., Chen, C., Guo, Y.: The modeling and simulation on sliding mode control applied in the double inverted pendulum system. In: Proceedings of the 10th World Congress on Intelligent Control and Automation, Beijing, China, 6–8 July 2012 (2012)
4. Dastranj, M.R., Moghaddas, M., Ghezi, Y., Rouhani, M.: Robust control of inverted pendulum using fuzzy sliding mode control and genetic algorithm. Int. J. Inf. Electron. Eng. **2** (5), 773 (2012)
5. Kumar, K.P., Rao, S.K.: Modelling and controller designing of Rotary Inverted Pendulum - comparison by using various design methods. Int. J. Sci. Eng. Technol. Res. **3**(10), 2747–2754 (2014)
6. Duart, J.L., Montero, B., Ospina, P.A., Gonzalez, E.: Dynamic modeling and simulation of a Rotational Inverted Pendulum. J. Phys. Conf. Ser. **792** (2017)
7. Ribeiro, J.M., Garcia, J.P., Silva, J.J., Marins, E.S.: Continuous time and discrete time sliding mode control accomplished using computer. IEE Proc.-Control Theory Appl. **152**(2), 220–228 (2005)

Analysis of Stability in Super-Lift Converters

K. C. Ajay[1(✉)] and V. Chamundeeswari[2]

[1] St. Joseph's College of Engineering, Chennai 600119, India
ajaykc3@gmail.com
[2] Department of EEE, St. Joseph's College of Engineering, Chennai, India

Abstract. DC-DC Converter plays a vital role in various applications such as medical instruments, electric machines, aviation. This work projects the concept of super lift converter. Super lift converter is a type of DC-DC converter in which the output voltage increases in a progressive range. The converter proposed is Negative Output Super Lift Luo Converter (NOSLLC). The NOSLLC converts the positive input voltage into a negative output voltage in geometric progression. The state space model of NOSLLC is carried out to check the stability of the system. Frequency analysis is done using Bode and Nyquist plots. The stability is inferred from these plots and the results are verified using simulation and validated with theoretical calculations.

Keywords: NOSLLC · Negative output · State space averaged model

1 Introduction

DC-DC converter is an electrical circuit which converts one voltage level to another by storing the energy temporarily and releasing it to the output at different levels. The converters such as buck, boost and buck- boost converter which produces different voltage levels during its operation. voltage lift technique has been widely used in electronic devices. This technique effectively overcomes the effects of parasitic elements and greatly increases the output voltage. NOSLLC is a type of DC-DC converter that employs super lift technique in which the output voltage increases in geometric progression. The stability of this system is carried out using state space model.

Section 2 shows the overview of proposed work. Section 3 deals with modes of operation of NOSLLC. Section 4 represents the simulation results of NOSLLC with different waveforms. The stability of NOSLLC using state space model and transfer function is carried out in Sects. 5 and 5.1. Further the stability analysis using transfer function and conclusion is showed in Sects. 6 and 7.

2 Overview

Figure 1 shows the block diagram of proposed system. The positive output voltage is converted using NOSLLC and produces a negative output voltage.

D. J. Hemanth et al. (Eds.): COMET 2019, LNDECT 35, pp. 424–434, 2020.
https://doi.org/10.1007/978-3-030-32150-5_41

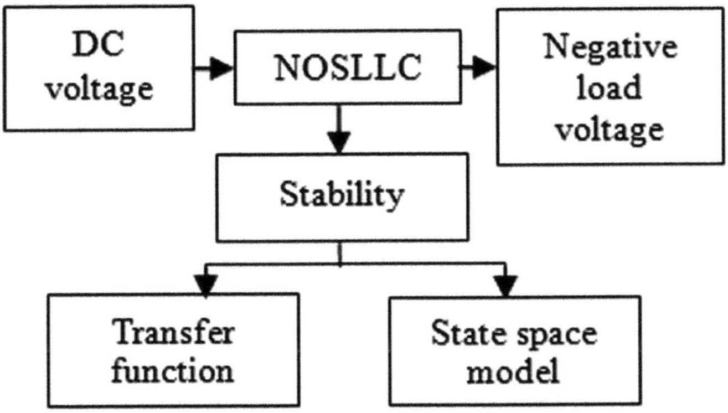

Fig. 1. Block Diagram of proposed system

The stability of NOSLLC is carried out by state space model and transfer function. The obtained transfer function is simulated and corresponding Bode and Nyquist plot is analyzed to determine its stability.

3 Operation of NOSLLC

The NOSLLC is carried out with two modes of operation- ON state and OFF state. The modes of operation have been explained with the following diagrams. It consists of DC supply voltage Vin, capacitors C1 and C2, inductor L1, power switch S, freewheeling diodes D1 and D2 and the load resistance R. During the ON period i.e., at DT interval when the switch 'S' is turned ON, voltage across capacitor C1 is charged. Current flowing through inductor L1 increases with slope Vin/L1 and decreases with slope –(Vo–Vin)/L1 during switch-off (1 − D) T.

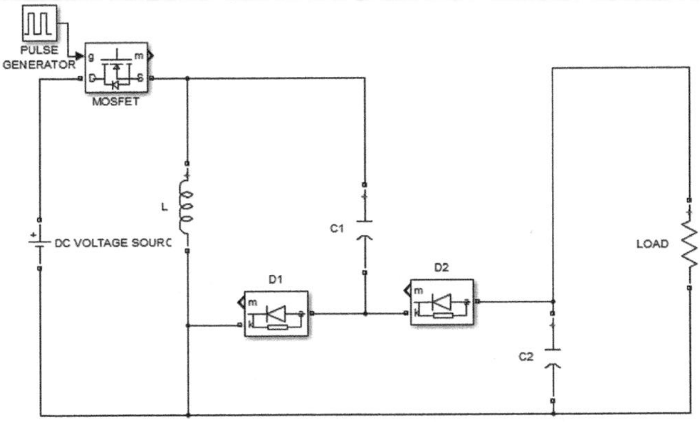

Fig. 2. Circuit diagram of NOSLLC

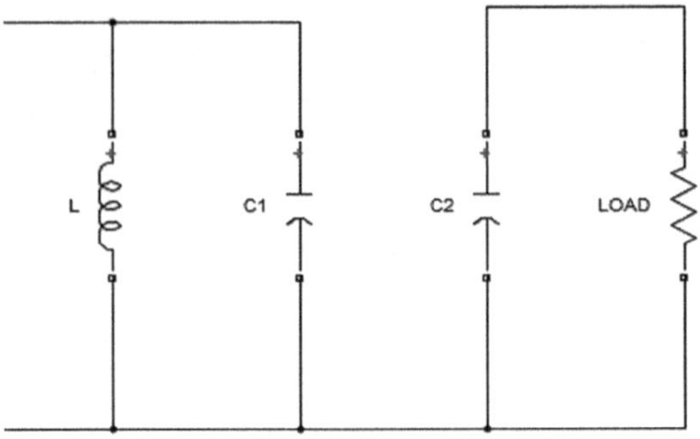

Fig. 3. During ON state of NOSLLC

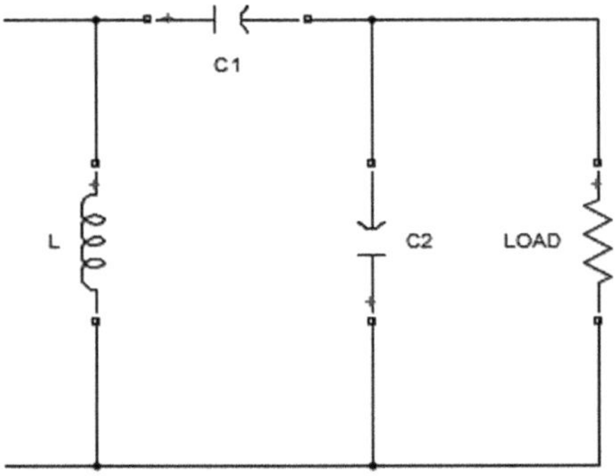

Fig. 4. During OFF state of NOSLLC

During ON state, the switch 'S' is closed and the supply flows through the inductor L1 and C1 charges during this time, the capacitor C2 produces a load voltage.

Therefore, for ON time KT the change in current in L_1 will be

$$\Delta i_{L1} = \frac{V_i}{L} KTKT \tag{1}$$

During OFF state, when the switch 'S' is open, the output voltage is boosted up by discharging the inductor L1 and capacitor C1.

The voltage drop across L_1 is,

$$V_{in} - V_0 \, V_0(\text{or}) - (V_0 - V_{in})$$

Since, $V_0 > V_{in}$.

Therefore, the current through inductor L_1 decreases by $-(V_0 - V_{in})/L$. Also, we know that the switch off period is (1 − K)T.

Hence change in current,

$$\Delta i_{L1OFF} = \frac{-(V_0 - V_{in})}{L}(1 - \mathrm{K})\mathrm{T} \tag{2}$$

We know,

$$\Delta i_{ON} = -\Delta i_{L1OFF}$$

Therefore,

$$\frac{V_{in}}{L1}KT = \frac{(V_0 - V_{in})}{L1}(1 - \mathrm{K})\mathrm{T}$$

By simplifying,

$$V_o = \frac{1}{1-K}V_{in}$$

Rearranging,

$$V_0 = \left[\left(\frac{2-K}{1-K}\right) - 1\right]V_{in} \tag{3}$$

4 Simulation Results

The simulation of NOSLLC is shown. The corresponding values for each component given and the simulation is carried out. The output voltage produced across the load is negative (Fig. 5).

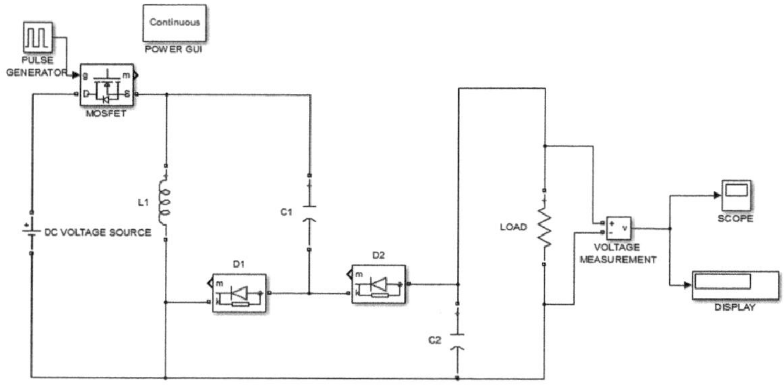

Fig. 5. Simulation of NOSLLC

The output voltage is measured across the resistor with voltage measurement blocks from the library. The input voltage given is 12 V with 67% duty cycle and the output voltage obtained is −32 V.

Considering the corresponding values for each elements such as (Table 1).

Table 1. Corresponding values of NOSLLC

Parameters	Values
Inductor (L)	2×10^{-3} H
Capacitor (C_2)	2×10^{-6} F
Capacitor (C_2)	50×10^{-6} F
Resistor (R)	50 Ω
Duty cycle (d)	0.25

The input voltage waveform is shown in Fig. 6(a). Here the input voltage is given as 12 V.

The input current is shown in Fig. 6(b). The flow of current is through MOSFET, where the gate pulse is given with a duty ratio of 67%.

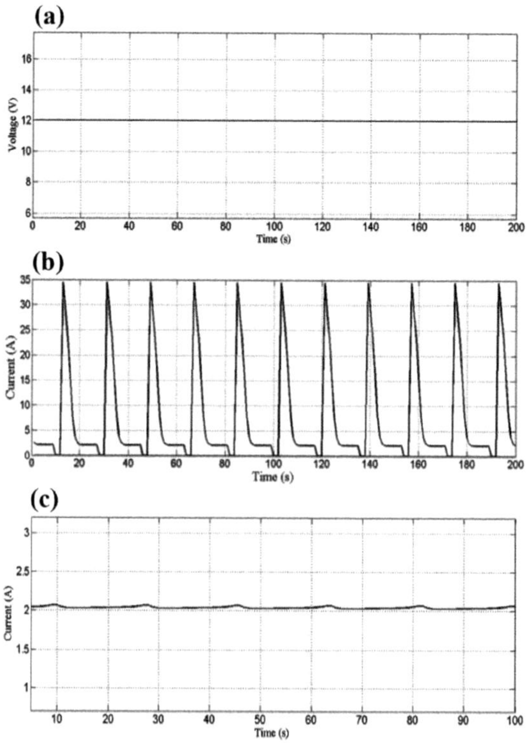

Fig. 6. (a) Input voltage. (b) Input current. (c) Inductor current (L1). (d) Capacitor voltage (C1). (e) Capacitor voltage (C2). (f) Output voltage

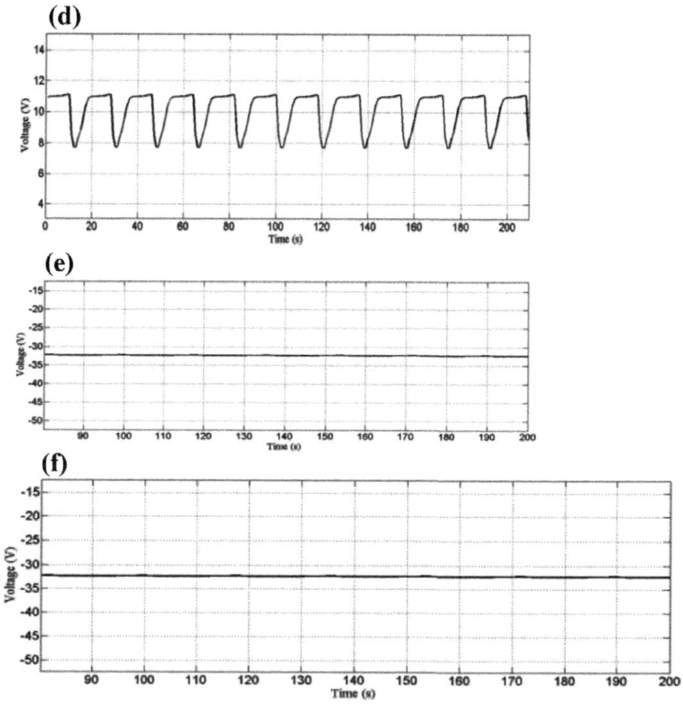

Fig. 6. (*continued*)

During turn ON and OFF period, the inductor gets charged and discharged. The current through inductor is shown in Fig. 6(c).

The capacitor voltage is shown in Fig. 6(d). The capacitor charges and discharges when the switch is turned ON and OFF respectively.

During turn ON period, the capacitor C2 discharges through load and during turn OFF period the capacitor gets charged by the discharge of inductor and capacitor C1 (Fig. 6(e)).

The output voltage across the load is −32 V. The waveform of load voltage is shown in Fig. 6(f). During OFF state the inductor L and capacitor C1 gets discharged so that the voltage across load increases.

5 State Space Model of Negative Output Superlift Luo Converter (NOSLLC)

The best method for the analysis and design of control systems is state space analysis. Whereas the transfer function method used is old and conventional method for the design and analysis of control systems. The transfer function method had many drawbacks such as it is defined under zero initial conditions. It does not give any idea about the internal state of the system. It cannot be applied to multiple input multiple output systems. The state variable analysis can be analyzed on any systems and it is easy to perform in computers. In state space analysis, the interesting feature is that the state variable of the system need not be a physical quantity. The system can be also selected as the state variables even when the variables are not related to the physical quantities.

Figure 2 shows a negative output super lift luo converter with x_1 and x_2 are state variable, where x_1 is the inductor current and x_2 is the capacitor voltage. With the assumption of circuit elements, two switched models are shown in Figs. 3 and 4. In the DT interval the state equation in matrix form are

$$\begin{pmatrix} \dot{x}_1 \\ \dot{x}_2 \end{pmatrix} = \begin{pmatrix} 0 & 0 \\ 0 & \frac{-1}{Rc_2} \end{pmatrix} \begin{pmatrix} x_1 \\ x_2 \end{pmatrix} + \begin{pmatrix} \frac{1}{L} & 0 \\ 0 & \frac{1}{c_2} \end{pmatrix} \begin{pmatrix} u_1 \\ u_2 \end{pmatrix} \tag{1.1}$$

$$V_o = \begin{pmatrix} 0 & 1 \end{pmatrix} \begin{pmatrix} x_1 \\ x_2 \end{pmatrix} + \begin{pmatrix} 0 & 1 \end{pmatrix} \begin{pmatrix} u_1 \\ u_2 \end{pmatrix} \tag{1.2}$$

During the (1 − D)T interval, the state equation in matrix form are,

$$\begin{pmatrix} \dot{x}_1 \\ \dot{x}_2 \end{pmatrix} = \begin{pmatrix} 0 & \frac{1}{L} \\ \frac{1}{c_2} & \frac{-1}{Rc_2} \end{pmatrix} \begin{pmatrix} x_1 \\ x_2 \end{pmatrix} + \begin{pmatrix} 0 & 0 \\ 0 & 1 \end{pmatrix} \begin{pmatrix} u_1 \\ u_2 \end{pmatrix} \tag{1.3}$$

$$V_o = \begin{pmatrix} 0 & 1 \end{pmatrix} \begin{pmatrix} x_1 \\ x_2 \end{pmatrix} + \begin{pmatrix} 0 & 1 \end{pmatrix} \begin{pmatrix} u_1 \\ u_2 \end{pmatrix} \tag{1.4}$$

Applying state space averaging, the coefficient matrix, A is

$$\begin{aligned} \bar{A} &= [A_1 d + A_2(1-d)] \\ \bar{A} &= \begin{pmatrix} 0 & \frac{(1-d)}{Rc_1} \\ \frac{(1-d)}{C_2} & \frac{-1}{Rc_2} \end{pmatrix} \end{aligned} \tag{1.5}$$

The averaged source coefficient matrix is

$$B = B_1 d + B_2(1-d)$$

$$B = \begin{pmatrix} \frac{d}{L} & 0 \\ 0 & \frac{1}{c_2} \end{pmatrix} \tag{1.6}$$

$$\text{Similarly, } \dot{C} = \begin{pmatrix} 0 & 1 \end{pmatrix} \text{ and } \dot{D} = (0) \tag{1.7}$$

5.1 Deriving Transfer Function Model from Linear State Space Model

The state space equation is given by,

$$\dot{X} = AX + BU \tag{1.8}$$

$$\dot{Y} = CX + DU \tag{1.9}$$

Taking laplace transform and simplifying it

$$Y(s) = [C(sI - A) - 1B + D]\,U(s) \tag{1.10}$$

On simplification, the equation is obtained as,

$$\frac{Y(s)}{U(s)} = \frac{LRs - (1-d)d\,.\,R}{LRC_2s^2 + Ls + R(1-d)^2} \tag{1.11}$$

Therefore, by substituting the corresponding values for each elements, the resultant transfer function is obtained in Eq. (1.12)

$$\frac{Y(s)}{U(s)} = \frac{0.1S - 9.375}{5 \times 10^{-6}s^2 + 2 \times 10^{-3}s + 28.125} \tag{1.12}$$

6 Stability Analysis of NOSLLC Using Transfer Function

The stability of the system is analyzed by Routh's stability criterion which determines the number of closed loop poles in the right-half s - plane.

Considering the Quadratic polynomial from the transfer function obtained in Eq. (1.12)

$$5 \times 10^{-6}s^2 + 2 \times 10^{-3}s + 28.125$$

where all the a_i are positive. The array of coefficients becomes

s^2	$s^2 5 \times 10^{-6}$	28.125
s^1	$s^1 2 \times 10^{-3}$	0
s^0	s^0 28.125	0

The sign does not change in first column of the polynomial. So there are no real roots on the right half side of the s - plane. Therefore the system is stable.

A step response is obtained for the transfer function and the respective frequency plots are verified for the stability of the system (Figs. 7 and 8).

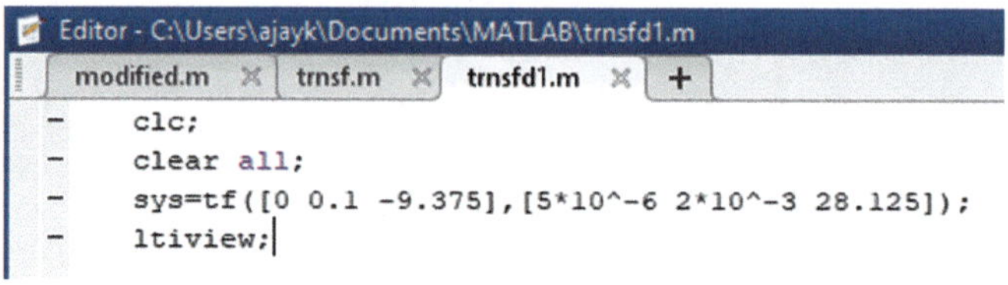

Fig. 7. Simulation of transfer function

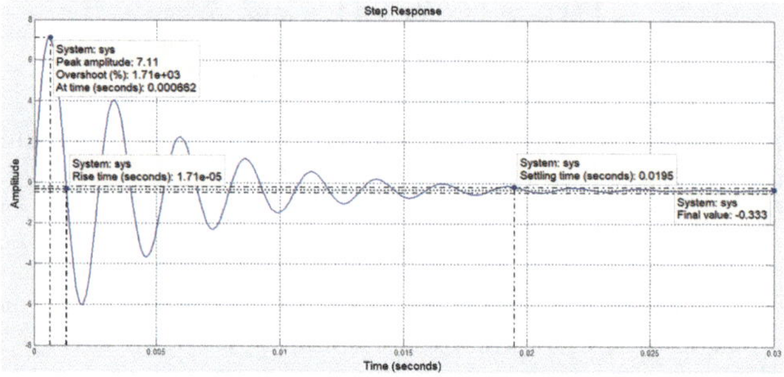

Fig. 8. Step response of NOSLLC

The plot shows that the response rises in a few seconds, and then rings down to a steady-state value of about −1. Compute the characteristics of this response using stepinfo (Table 2).

Table 2. Time response of NOSLLC

S. No	Time response	Values
1.	Rise time	1.7104e−05 s
2.	Settling time	0.0195 s
3.	Settling min	6.0430 s
4.	Settling max	4.0483 s
5.	Overshoot	1.7129e+03 s
6.	Undershoot	2.1318e+03 s
7.	Peak	7.1058 s
8.	Peak time	6.6231e−04 s

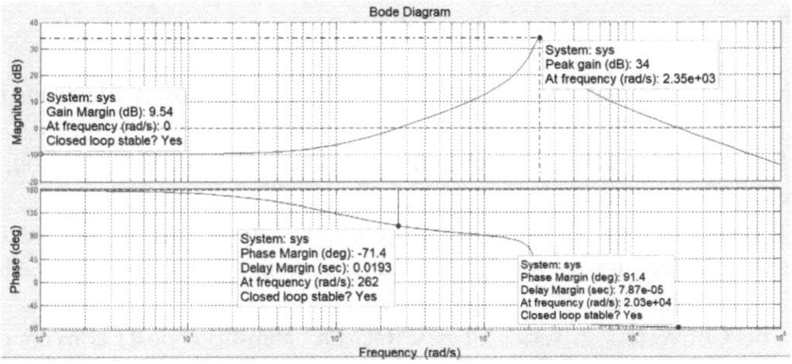

Fig. 9. Bode plot of NOSLLC

For a stable system, both the margins should be positive or the phase margin should be greater than the gain margin. The bode plot in Fig. 9 shows that the phase margin is greater than the gain margin. So the system is Stable.

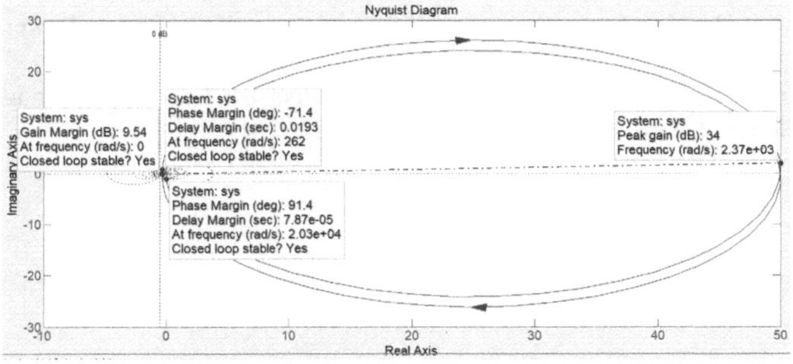

Fig. 10. Nyquist plot of NOSLLC

The Nyquist plot shows that there is no encirclement of $-1 + j0$ point and there are no poles in right half side of the s - plane. Therefore, the system is considered as a Stable system (Fig. 10).

7 Conclusion

The simulation of NOSLLC is carried out and their respective waveforms of each elements are obtained. The state space model is obtained from the turn ON and OFF period of NOSLLC and the transfer function is determined from the state space model. The stability of the system is verified using Routh's stability criterion. The step response and their respective frequencies are plotted. Simulation results are verified using MATLAB and validated with theoretical calculations.

References

1. Luo, F.L.: Negative output super-lift converters. IEEE Trans. Power Electron. **18**, 1113–1121 (2003)
2. Luo, F.L.: Seven self-lift DC-DC converters voltage lift technique. IEE Proc. Electr. Power Appl. **148**(4), 329–338 (2001)
3. Chebli, R., Sawan, M.: A CMOS high-voltage DC-DC up converter dedicated for ultrasonic applications. In: Proceedings of the 4th IEEE International Workshop on System-on-Chip for Real-Time Applications, pp. 19–21, July 2004
4. Luo, F.L.: Luo-Converters, a series of new DC-DC step-up (boost) conversion circuits. In: Proceedings of IEEE Conference on Power Electronics and Drive Systems - 1997 (PEDS 97), Singapore, pp. 882–888, May 1997
5. Luo, F.L.: Luo-Converters: voltage lift technique. In: Proceedings of IEEE Power Electronics Special Conference (PESC 1998), Fukuoka, Japan, May 1998, pp. 1783–1789 (1998)
6. Luo, F.L., Ye, H., Rashid, M.H.: Multiple-lift push-pull switched-capacitor Luo-Converters. In: Proceedings of IEEE Power Electronics Special Conference (PESC 2002), Cairns, Australia, June 2002, pp. 415–420, 15 (2002)
7. Luo, F.L., Ye, H.: Negative output multiple-lift push-pull SC Luo-Converters. In: Proceedings of IEEE Power Electronics Special Conference (PESC 2003), Acapulco, Mexico, 15–19 June 2003, pp. 1571–1576 (2003)
8. Luo, F.L., Ye, H.: Negative output super-lift converters. IEEE Trans. Power Electr. **18**(5), 1113–1121 (2003)
9. Senthil Kumar, R.: PVFED by negative output super-lift Luo Converter using improved P&O MPPT. Int. J. Pure Appl. Math. **115**(8), 79–84 (2017)

Emergency Alert to Safeguard the Visually Impaired Novice Using Internet of Things

M. Subathra(✉), G. Akalya, and P. Madhumitha

Panimalar Engineering College, Poonamallee, Chennai, India
subathraponnu@yahoo.com

Abstract. Recognizing a crisis with help of sensors and controls the situation by announcing them to the Real world, that worry calamity group makes suitable move to protect the costless. Methods/Analysis: Internet of things is system that controls the interconnected things which are implanted with sensors, programming that empowers to gather and trade information. These physical articles are controlled remotely with sufficient system foundations. The absence of help administrations can make outwardly impeded individuals excessively reliant on their families, which keep them from being financially dynamic and socially included. IoT can offer individuals with handicaps the help and supports greatly to accomplish a decent personal satisfaction and enables them to get appropriate instruction without dangers. Findings: For this reason the proposed engineering of web of things is to screen the vehicle by utilizing Raspberry controller and it set forth numerous highlights to recognize the purpose behind happening accidents and furthermore protects the outwardly weakened victims from the accidents or dangers. Applications/Improvements: Distinctive application things are taken into issues the tip goal to indicate the collaboration of the segments of internet of things. While the proposed work will be a keen ready framework to different components without human mediation. Basic difficulties have been distinguished and tended to particular.

Keywords: Emergency · Radio Frequency Identification · GPS · MEMs · Accident Detection · Sensors

1 Introduction

The Internet of things is an ecosystem with arrangement of physical devices, vehicles and different things put with equipment, programming, sensors, actuators and framework accessibility that enable these things to gather and exchange information in control manner [1, 2]. Each thing is uniquely identifiable through embedded computing system. This is able to interoperate within the existing internet infrastructure. The internet of things includes of elements that are related to the internet whenever, anything, anyplace. In most technical sense it consists of integrating sensors and devices into everyday objects that are connected to the internet over fixed and wireless networks. Regarding "things" as an inextricable mixture of hardware, software, data and service.

D. J. Hemanth et al. (Eds.): COMET 2019, LNDECT 35, pp. 435–447, 2020.
https://doi.org/10.1007/978-3-030-32150-5_42

Wireless sensor networking is a trendy technique which has wide range of potential applications for monitoring and robotic exploration. This is a heterogeneous topology as the measurements include various considerations. The sensors communicate over wireless medium which contributes to form a wireless sensor network. This can run on both solar energy and battery. It can relay the information about the accidents to the corresponding responders.

The objects can be identified automatically with its unique using RFID technology, where each objects possess RFID tags. The RFID readers can access the data stored in the tags. The link is established from objects in real world to digital identification through above stated technology. A scenario of highly congested area with various types of vehicles like personal, public and emergency vehicles are get stuck in traffic causes delay and no proper intimation to end users about their delay. This paper proposes the idea how a communication is linked between things without human involvement in order to convey the information at right time to the responders. The rest of paper describes as follows: Sect. 2 explains motivation and experimental study; Sect. 3 presents proposed system; Sect. 4 explores the emergency detection work flow and algorithm; Sect. 5 describes the experimental setup and results; Sect. 6 concludes with its social impacts and finally the references.

2 Experimental Study

Accident Detection System [4] utilizing GSM and GPS modem with Raspberry pi controller. Early day's identification is detected and broke down utilizing a piezo-electric sensor and gives its outcome to the microcontroller. The worldwide situating framework recognizes the scope and longitude position of a vehicle. This position is followed and data is sent as message through the GSM. In this static IP address of crisis responder is pre-spared in the EEPROM.

Incident Detection [5] calculation to break down the episodes and decides the idea of occurrences and give a crisis administrations. Reference Fernandes [6] depicts that a versatile application for programmed accidents identification calculation. The Acceleration Severity Index gauges the potential hazard for casualties. The correspondence stream calculation portrays the thought with which backend frameworks make associations with "things" utilizing database administration framework and alongside sites. In this the centre correspondence foundation to the hubs or gadgets which are between associate through portals descriptions [7].

The ABEONA system [8] states that accident is identified utilizing crash sensors which are put inside the air pack. The Global position framework is utilized to find accidents spot and vehicular impromptu systems which broadcast the messages to responders. Alongside this save group to conjecture movement block and on the way their way as needs be to achieve the area as right on time as could reasonably be expected. With this addition activity flag module gets a data and look for rescue vehicle accessibility adjacent.

Reference [9] focused on low speed pile up discovery. The principle deterrent that experiences the low speed accidents is the manner by which to separate whether the client is inside the vehicle or outside the vehicle, strolling or gradually running. The impact of this deterrent is limited, in this work, by a proposed component that recognizes the speed variety of low speed vehicle and strolling or gradually running individual. The proposed framework comprises of two stages. The identification stage which is utilized to recognize auto crash in low and high speeds. The notice stage, and promptly after a accident is shown, is utilized to send point by point data, for example, pictures, video, accidents area, and so forth to the crisis responder for quick recuperation. The framework was for all intents and purposes tried in genuine reproduced condition and accomplished very great execution results.

Smartphone detecting [10] of vehicle flow to decide driver telephone utilize, which can encourage numerous movement security applications. This framework utilizes installed sensors in advanced mobile phones, i.e., accelerometers and gyrators, to catch contrasts in centripetal increasing speed because of vehicle progression. This low foundation approach is adaptable with various turn sizes and driving velocities. Broad tests directed with two vehicles in two unique urban areas exhibit that our framework is hearty to genuine driving conditions. Notwithstanding loud sensor readings from PDAs, our approach can accomplish a high precision. This exhibits the assembly of a will organize for checking parameters course of load vehicles [11].

With law of Declaration, every carry vehicles must hold tachometer gismo that gathers and saves all the data functioning on amid trek, maybe, speed, remove, rev (pivots each moment), among completely different parameters. The information gave by the tachometer area unit sent to the will organize, gathered and deciphered by the microcontroller and sent to the control region one by one correspondence to the responder. During this manner, this makes focal observant organization can quickly take once the course of its drivers, and therefore the qualities of movement honed by them.

Reference [12] illustrates that Smartphone detecting of vehicle flow to decide driver telephone utilize, which can encourage numerous movement security applications. This framework utilizes installed sensors in advanced mobile phones, i.e., accelerometers and gyrators, to catch contrasts in centripetal increasing speed because of vehicle progression. This low foundation approach is adaptable with various turn sizes and driving velocities. Broad tests directed with two vehicles in two unique urban areas exhibit that our framework is hearty to genuine driving conditions. Notwithstanding loud sensor readings from PDAs, our approach can accomplish a high precision.

3 Proposed System

Research work concern with goals of insinuating the guardians about the pickup and drop off exercises of the kids through email and SMS alarm. Implying school administration if there is an occurrence of rash driving did by driver at a specific purpose of time. With utilization of human RFID chip innovation in the field of observing of vehicles conveying outwardly hindered school kids. This incredibly changes the customary method for human experienced-based planning alongside observing the vehicle running on street continuously. Because of this nature of administration is enhanced and conveys comfort to general society. Human RFID chip is a sort of programmed distinguishing proof innovation. It accomplishes non-contact information correspondence by sending radio recurrence flag. This trades information between an examiner and transponder for distinguishing proof and following reason by Finkenzeller [13].

Advancement grants continuous remote checking of totally independent installs in human body. In control gathering techniques in perspective of inductive, RFID chip install gets control from cross analyst and emits back to it a banner that contains data. The read extent of correspondence for RFID chip implanted in human body is size of 10 cm in length. It could be productive when RFID embeds are obligated to security need. Three real time scenarios such as accident, attack, vehicle failure are taken into consideration for providing security for visually impaired kids (see Fig. 1).

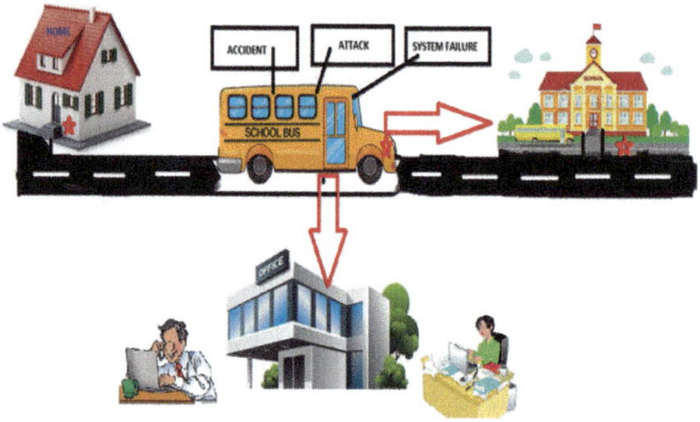

Fig. 1. Real scenario

When a child boards the vehicle, their human RFID chip is detected for unique identification code. This code is sent to the raspberry pi and it coordinates the recognizable code stored in the database of the corresponding child. A message will be sent from raspberry pi to the child's parents that their child has boarded the bus at this particular time and date. The same procedure is followed when the child is dropped at the dropping point from the school. A message will be sent to their parents that their ward has been dropped at the dropping point in particular time with date.

3.1 Working Principle

The raspberry pi board utilized as a part of the proposed framework which has the accompanying highlights, for example, Ethernet port for web availability alongside VGA and HDMI connector. The system on chip is Broadcom of type BCM 2836. The specification of CPU is 900 MHz quad core ArmV7 cortex, graphical processing unit used as Video Core IV and power source is 5 V via GPIO header. This Model B works six times faster than the Model B+, and goes with 1 GB of RAM. System consist of four sensors to monitor the vehicle (See Fig. 2).

The different sensors are used to detect the nature of the incidents. Vehicle safety Sensors on vehicle will encourage drivers comprehend the danger of meeting into each other. Sensors and between vehicle correspondence will encourage us to perceive to

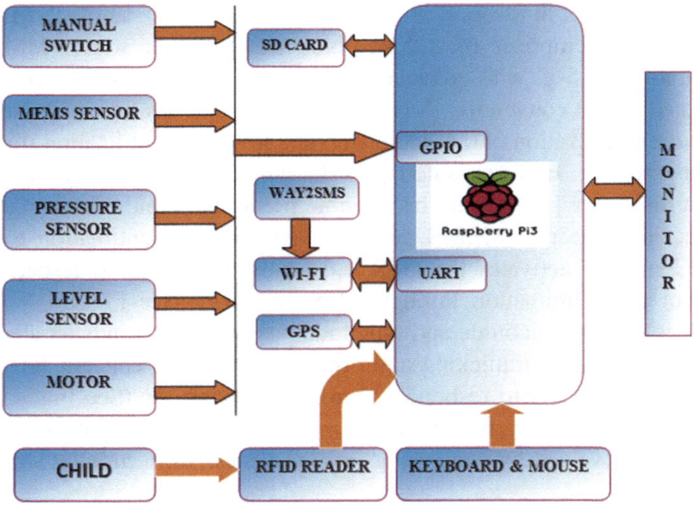

Fig. 2. System design

have a tendency to see which cannot be seen easily. With auspicious and remedy cautioning, some vehicle accidents will be stayed away from. In case of any one occurrence, machine will communicate with each other without human intervention and convey the information to corresponding modules without any time delay.

3.2 Human RFID Chip

As human RFID chip is embedded in human body which is encased in silicate glass. The essential rule of RFID chip subsequent to entering the field, chip embeds in human body gets RF flag discharged by per user and conveys the data put away in chip by utilizing the vitality picked up from the prompted present, at that point sent to the application program on information handling. As clever transportation framework in view of RFID arrangements does not depend on satellite signs. This article exhibits an outline on Radio Frequency Identification innovation for human embeds and examines the mechanical possibility of such embeds for finding and following people.

Moreover RFID development grants nonstop remote checking of totally self-decision embeds in human body. In the power harvesting method in light of electro-magnetic couplings, the RFID insert gains power from the analyst and releases back to it a banner that contains information. The read extent of correspondence for RFID installed in living body is seen to be smaller than 10 cm. This invokes an innovative idea to find communication way between the identification and raspberry pi. The raspberry controller gets activated if any sensor gets triggered.

3.3 MEMs Detection Module

Micro scale Electro Mechanical System innovation depends on various apparatuses and strategies, which are utilized to frame little structures with measurements in the micrometer scale. This innovation is presently being used to fabricate MEMS-Based Accelerometers. It comprises of a three-pivot accelerometer sensor MMA7660FFC with affectability ±1.5 g, with advanced yield. It is interfaced to the control unit by Inter Integrated circuit convention. It is minimal effort and has high stuns survivability. It has low current utilization of 0.4 micro amp and low power utilization simple voltage and advanced voltage. It has an auto rest/wake include for low power utilization. Tilt introduction discovery [15] should be possible precisely given.

When accident occurs, MEMS get activated and send intimation to controller. Raspberry Controller activates the GPS and sends the details to respective system. Entire system gives intimation through SMS and mail (See Fig. 3). When vehicle defined path is varied by latitude and longitude position, thus sends an alert to parents as attack and kidnap. As trafficking and organ theft is main concern for kidnapping. These alerts are reachable to responders through Smartphone (See Fig. 4). Illustrates the vehicle failure Module. Brake Sensor in which system might be a mechanical contraption, this develops by engrossing imperativeness from running structure.

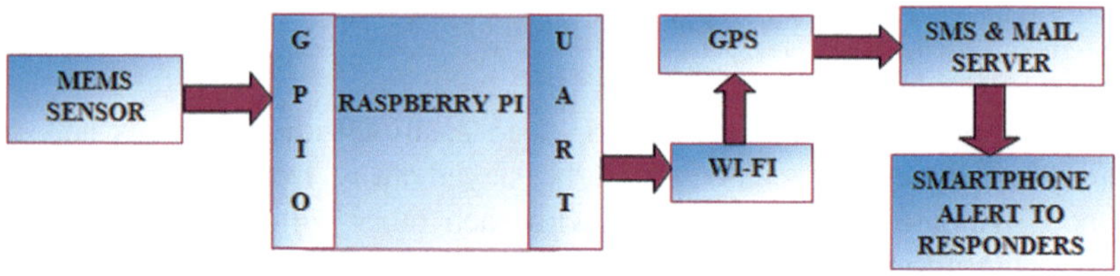

Fig. 3. Mems detection

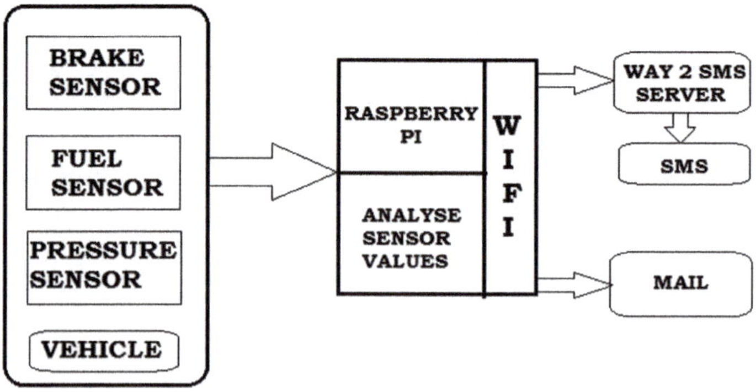

Fig. 4. Vehicle failure diagram

It is used for dying down or stopping a running vehicle, wheel, turn, or to remain its development, of times talented by courses for crushing. The brake sensor uses a MEMs technology with functionality to measure the pressure of brake booster. The power supply to operate is 4.5 to 5.5 V. Temperature values for efficient work are −40 to +150 °C. The accuracy over life time is provided with 1.5%. Fuel Sensor: level detector in a very vehicle's fuel tank may be a mixture of these elements admires a buoy, causative bar, Resistor. This mixture of elements sends a variable flag to fuel live or electronic convenience that activates the fuel check. It checks a level of fuel in tank, if lessens and makes vehicle not to work consequently caution is made to sub modules. The sensors are intended to gauge the level of fluid in tank for the full profundities of tanks. At whatever point fuel is drawn from a tank, the level will be reflected in the pointer and also will be conveyed through the GPS gadget. A weight sensor is a gadget outfitted with a weight delicate component that measures the weight of gas in tire. It normally goes about as a transducer. It produces a flag as an element of the weight forced.

Novice are often set with facilitate of human RFID chips with tattoos embedded as design anywhere in the human body. In emergencies, medical personnel can have quick access of health information. Chips might but create us prime targets for folks with unhealthy intentions. A human microchip insert is consistently a recognizing composed circuit contraption or RFID transponder enclosed in silicate glass and implant in the body of a living being. An implanted RFID chip is hard to lose or take. Personal embedded RFID labels contain a special identifier for every person, which can be connected to data around an individual. The RFID chip empowers an embedded individual to interface his physical nearness to data put away in the advanced world, this can greatly helps visually impaired novice to track exactly with physical location and their medical status with network of things. Realistic benefits are identification, infant and elder safety, Child abductions, metadata of health, Theft prevention.

4 Emergency Detection Flow

The initialization process (See Fig. 5) work flow in which it begins with system start up block where the functionality of the system is verified for its normality. In case if verification fails to proceed, once again reformation process takes place just like repairing and start up is followed. As same process is carried out to check its normal functionality thus constitutes the initialization process of emergency detection.

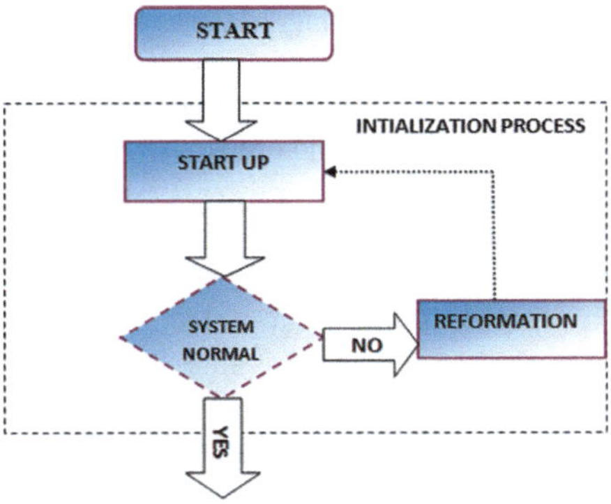

Fig. 5. Initialization process

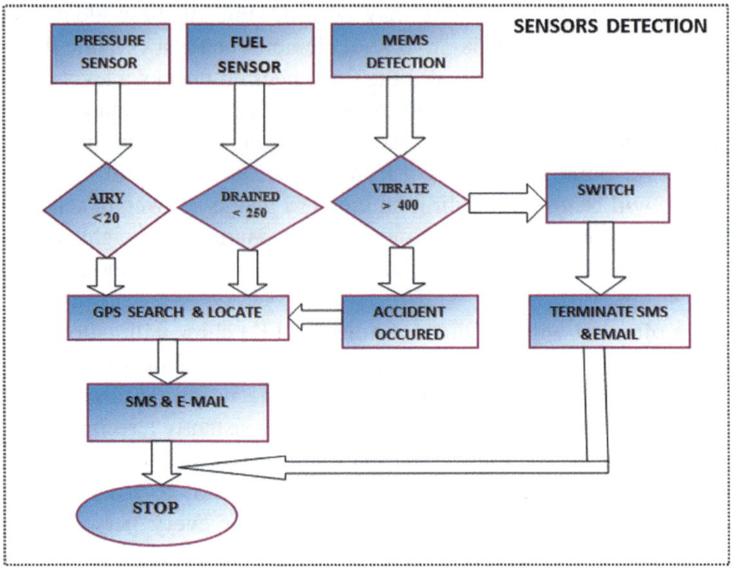

Fig. 6. Smart alert system - GPS

The sensor connections are made to monitor the regular activity of tracking and positioning (See Fig. 6). In case of any abnormalities, it detects the faults and collects the data, thus transmit to actuators for further processing. In this pressure, fuel and MEMs are used to sense their values, suppose its specified ranges mismatches, processing unit captures the location using GPS, send an alert to responders as SMS and Email at instant. If sensed values are not related to risk criteria, switch is available to terminate the alerts, so that panic to responders is greatly reduced.

EDA ALGORITHM - Emergency Detection Alert

Step 1: Start up - Initialization process
Step 2: Establish the exact connections.
Step 3: GSM & GPS Module - Initialization is to be done.
Step 4: Sight for any one of the three conditions to occur.
Step 5: If any conditions satisfies with it, then access the GPS receiver.
Step 6: Send the accessed GPS data to predefined number through SMS and Email.
Step 7: Smartphone that receives alert directs to location with traffic conditions.
Step 8: Responders reaches the location in time to rescue the victims.
Step 9: Terminates the process.

In accident detection algorithm automatic detection can be performed using HDy copilot, where limitation with this algorithm is lack of application interface hinder use of GSM. Communication Flow algorithm greatly assists the people with help of internet of things with advanced communication techniques. This requires a need of strong connectivity. Incident detection algorithm smartly works and reduces deaths of human and other injuries related. It detects the incidents immediately. But this restricts to urban highway accidents. Above limitations are overcome with WRECK watch algorithm represents an automatic crash notification system saves lives by reducing the time required for responders to arrive.

Network of sensors are used to detect the vehicle accidents. But limits with expensiveness of system, portability, prevention of false positive is really very hard. ABENO algorithm enables reach ability to spot in time and helps to track most efficient path to reach accident spots. Where VANET cost is very high. Emergency detection alert algorithm gives out better performance characters comparatively with above stated algorithm. In which there will be no delay in alert and greatly reduces the waiting time. Quick alerts will be given to responders. Internet connectivity for huge things is under processing where it can be achieved within short period of time.

5 Experimental Setup and Results

This paper gives a different method for moving toward the issue. The accident area can be found effortlessly and the discovery of accidents is exact not at all like the earlier methodologies, where recognition of mishap is finished by both of the two sensors. In this approach the accidents is identified by both the vibration and microelectro mechanical sensor and there is an elective route gave to stop the entire procedure of messaging through a switch. Where as alternate methodologies give just a single method for identifying the accidents. Machine-to-Machine alludes to the correspondences between PCs, inserted processors, keen sensors, actuators and cell phones. The utilization of M2M Communication is expanding in the situation at a quick pace M2M has a few applications in different fields like smart services, Intelligent robots, digital transportation frameworks producing frameworks, smart home innovations, and shrewd matrices. Case of Machine to Machine territory organize regularly incorporates specific region arrange advancements, for example, broad band and Bluetooth or neighborhood systems. Henceforth this paper has an edge over the other prior methodologies as shown in Fig. 7.

The current location of victims can be traced exactly as fast and easily using Google maps. The graphical representation (see Fig. 8). Illustrates the percentage share of road accidents in which the categorization of person killed and injured. Comparatively with previous year the percentage range is quite low for death and injured humans. As emergency alerts are given in time thus greatly saves human lives without delay and with zero human intervention. Tracking and navigation also achieved using Google map interfaces.

Alerts to responders as SMS and Email through Global system for mobile communication. It contains description as latitude, longitude of location, date, time along with the reason for incidents and its nature. The following Table 1 shows the alert through Smartphone to the responders in whom it describes details like location and reason for incidents (see Fig. 9). Illustrates the information to responders as alert stating the reason for delay along with the spot details, stating reason as due of break down, vehicle fails to move further. Same information is send through as an Email to

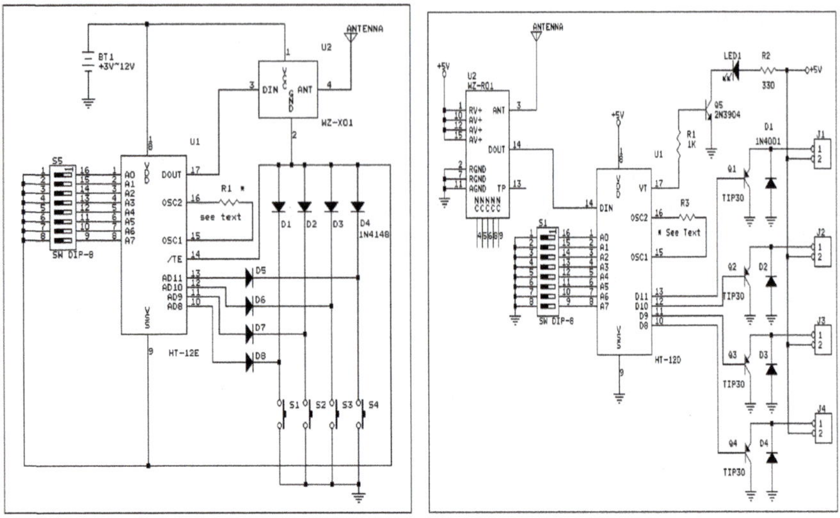

Fig. 7. Total setup with RFID Interfaces

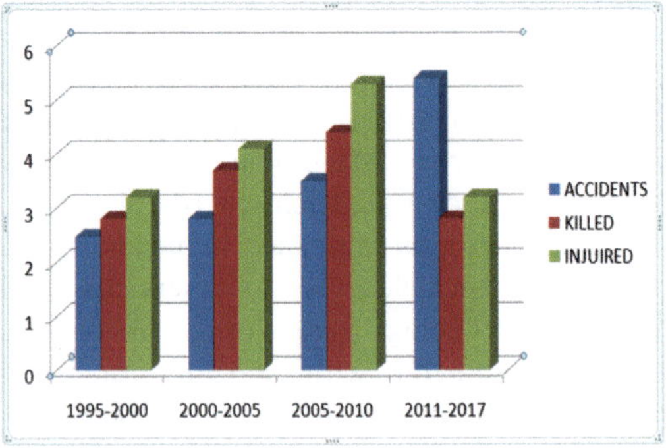

Fig. 8. Percentage share of road accidents, person killed and injured

the corresponding responders to fetch the information status of their wards from time to time. This greatly helps to track their kids boarding to school and reduces the worrieness a lot.

Table 1. Message frame

Router ID	Novice ID	Location	#p	Speed	Status	Message type
521	21265	13.05:80.531	2	9	6	2

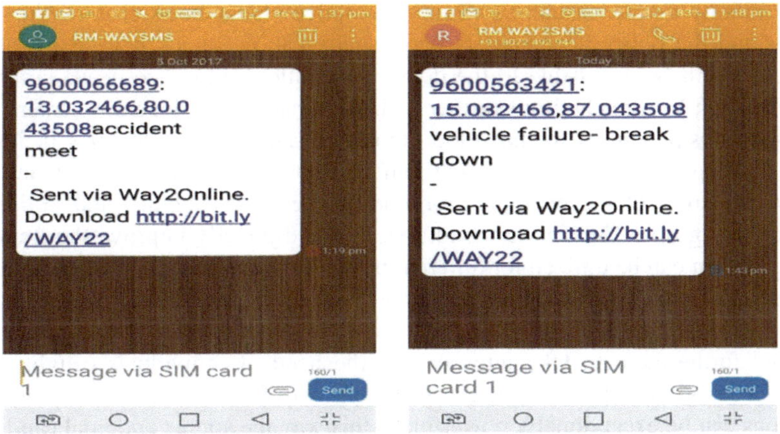

Fig. 9. Accident alert through SMS

These are the results captured and analyzed for the emergency detection alert system. Above figures depicts the information how it helps the responders to locate the spot without delay and specially considering the traffic situation to track the spot in time.

6 Conclusion

The proposed system reduces road accidents and monitors the vehicle effectively. This mainly focuses on providing emergency support to the victim and alerts the user in case of danger. In this paper, a review of the IoT for individuals with inabilities is given. The pertinent application situations and fundamental advantages have been depicted. The exploration challenges have likewise been surveyed. There is degree to change our proposed framework by better refinement. The operational time is likewise diminished. Areas like mountain regions, rural region where human movements are very less, in such cases if accident or vehicle failure or attack made on vehicle can be easily identified and immediate action can be taken off. This is useful in the event that kids are oblivious or can't impart in light of the fact that will be effortlessly distinguished in the event because the RFID chip is embedded in them.

With the RFID chip having the capacity to connect to worldwide situating satellites, these aides in the event that somebody is seized or lost. Guardians may view this as something to be thankful for in kids in the event that anything was to happen to their tyke. Be that as it may, with every single good thing this innovation brings to the table it likewise accompanies bad. The innovation is rising and is ceaselessly being chipped away at and created. In the long run if the majority of the Crimps can be worked out, increasingly RFID chips will be embedded in people around the world. These examination issues stay totally open for future examination. The framework created can be executed continuously situations in not so distant future. This paper shows a productive framework for mishap and fall discovery and moment arrangement of crisis administrations to the casualties. Once the notices have been sent to the healing facilities and the police headquarters the casualty can get quick help as his area and information will be accessible to the doctor's facilities and police headquarters. This framework can be stretched out further to incorporate blood donation centers so that there is no postponement to make the blood accessible to the doctor's facilities for serious mishap situations where there can be substantial blood misfortunes. Additionally, adjusting of the examples for finding the closest clinics and police headquarters relying upon the activity conditions can additionally improve the framework. The application can be utilized for different purposes, for example, Woman Safety. The proposed framework is more dependable and quick than prior strategies. This empowers organization of remote sensor arranges in powerful building condition for route and furthermore to take into account the need of individuals caught in crisis circumstances giving short and safe way a long way from danger. The informational collections can be exceptionally powerful. In this way the source and goal can likewise be dynamic and adaptable as per changing condition and client necessity.

Acknowledgement. We the members of this project, acknowledge our sincere gratitude for the excellent support extended to us by the management (Panimalar Engineering college, Chennai) for providing all the required facilities. Our sincere thanks go to our respectful guide who has extended her support throughout the entire project with his valuable suggestions and guidance.

References

1. Singh, A.P., Agarwal, P.K., Sharma, A.: Road safety improvement: a challenging issue on Indian roads. Int. J. Adv. Eng. Technol. **2**, 363–369 (2011)
2. Choudhari, S., Rasal, T., Suryawanshi, S., Mane, M., Yedge, S.: Survey paper on internet of things: IoT. Int. J. Eng. Sci. Comput. **7**, 10564–10567 (2017)
3. Suganya, N., Vinothini, E.: Accident alert and event localization. Int. J. Eng. Innov. Technol. **3**, 53–54 (2014)
4. Sawant, K., Bhole, I., Kokane, P., Doiphode, P., Thorat, Y.: Accident alert and vehicle tracking system. Int. J. Innov. Res. Comput. Commun. Eng. **4**, 8619–8623 (2016)
5. Wang, Y., Yang, J., Liu, H., Chen, Y., Gruteser, M., Martin, R.P.: Sensing vehicle dynamics for determining usage of phone. In: ACM 11th Annual International Conference on Mobile System, Applications and Services, pp. 41–54 (2015)
6. Fernandes, B., Gomes, V., Ferreira, J., Oliveir, A.: Mobile application for automatic accident detection and the multimodal alert, vol. 3, pp. 1–11 (2016)
7. Zanella, A., Bui, N., Castellani, A., Vangelista, L., Zorzi, M.: Internet of things for smart cities. IEEE. **1**, 22–35 (2014)
8. Praveena, V., Sankar, A.R., Jeyabalaji, S., Srivatsan, V.: Efficient accident detection and rescue system using ABEONA algorithm. Int. J. Emerg. Trends Technol. Comput. Sci. **3**, 222–225 (2014)
9. Ali, H.M., Alwan, Z.S.: Car accident detection and notification system using smartphone. Int. J. Comput. Sci. Mob. Comput. **4**, 620–635 (2015)
10. Cevher, V., McClellan, J.H.: Vehicle speed estimation using acoustic wave patterns. IEEE Trans. Signal Proc. **57**, 31–47 (2009)
11. Fernandes, J.N.O.: Real-time embedded system for monitoring of cargo vehicles using controller area network. IEEE Lat. Am. Trans. **14**, 1086–1092 (2016)

12. Majumdar, C., Lee, D., Patel, A.A., Merchant, S.N., Desai, U.B.: Packet size optimization for cognitive radio sensor networks aided Internet of Things (2016)
13. Finkenzeller, K.: Fundamentals and Applications in Contactless Smart Cards and Identification. Wiley, Hoboken (2003)
14. Bhavthankar, S., Sayyed, H.G.: Wireless system for vehicle accident detection and reporting using accelerometer and GPS. Int. J. Sci. Eng. Res. **6**, 1069–1072 (2016)
15. Goud, V., Padmaja, V.: Vehicle accident automatic detection and remote alarm device. Int. J. Reconfigurable Embedded Syst. **1**, 49–54 (2016)

Speed Control of BLDC Motor with PI Controller and PWM Technique for Antenna's Positioner

B. Suresh Kumar[1(✉)], D. Varun Raj[1], and D. Venkateshwara Rao[2]

[1] Electrical and Electronics, Chaitanya Bharathi Institute of Technology, Hyderabad, India
{bskbus,duddyalavarun1269}@gmail.com

[2] Defense Research and Development Laboratory, Hyderabad, India
dakojuvrao123@gmail.com

Abstract. Brushless DC motor (BLDC) has applications in various fields. These motors have special applications in Antenna's positioner. Antenna's positioner will help in tracking the missile, satellite and aircraft etc. To these antenna's positioner, BLDC motor are connected through Gear reducer, which is used to rotate clockwise or anti-clock wise. Based on error generated from potentiometer these motors are need to rotate in required direction. BLDC motor can work continuously or intermittently. For antenna positioner both continuous or intermittent types are required. To improve the speed of response, proportional and integral (PI) controllers are connected. These controllers add gain and pole to system, which will help in improving system performance. DC chopper help in reducing torque ripple, with the help of Pulse Width Modulation technique (PWM). In this paper, BLDC motor which is current controlled and chopper fed is simulated using MATLAB/SIMULINK for antenna's positioner.

Keywords: Current controlled Brushless DC motor · Antenna's positioner · PI controller

1 Introduction

Generally electrical machines perform mechanical to electrical and electrical to mechanical conversions. At generating side synchronous generator performs mechanical to electrical conversion. At load sides there are many electrical motors. Among those motors, Direct current motor and BLDC motors have good efficiency and good torque speed characteristics [1–4].

Due to brushes, DC motors require additional maintenance and so commutation becomes complicated. In BLDC motor these brushes are absent and commutation is done through electronic switches. In BLDC motor, rotor position is sensed through hall sensor and pulse is generated, which is sent to inverter as shown in Fig. 1 [5].

D. J. Hemanth et al. (Eds.): COMET 2019, LNDECT 35, pp. 448–461, 2020.
https://doi.org/10.1007/978-3-030-32150-5_43

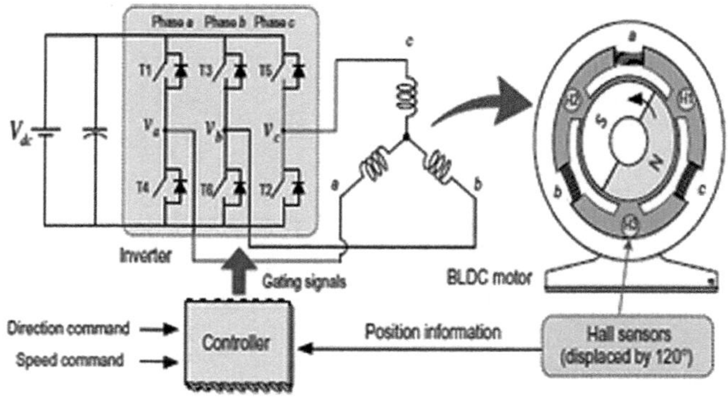

Fig. 1. General functioning of BLDC motor

Antenna positioner is shown in Fig. 2. Position information that is desired to rotate is entered to system in-put. Position information is converted into voltage by potentiometer and these signals are strengthened by power amplifier. These signals are converted to speed as shown in Fig. 3 [6–10].

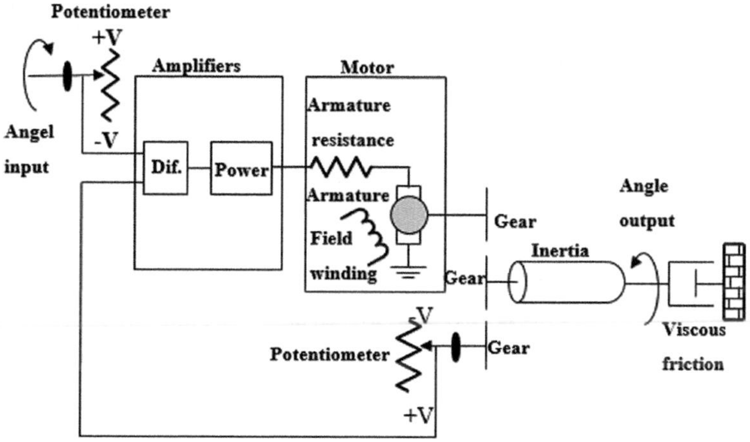

Fig. 2. Working of antenna positioner

To obtain better dynamic response, speed has to be controlled and this is done by controlling current or torque of BLDC motor as shown in Fig. 3. The output of PI controller becomes reference current input to the motor. This reference current is compared with stator current which is measured using ammeters, and produces required speed to motor.

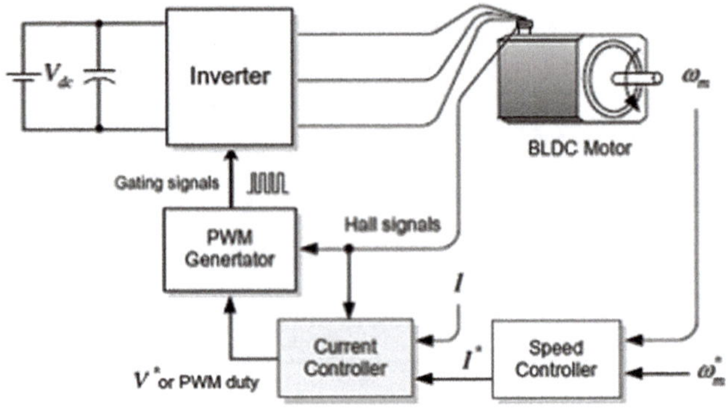

Fig. 3. Current controlled BLDC motor

Response of system can be improved by adding PI controller to the system. These controllers improve sys-tem stability, reduces oscillations, settling time and rise time.

2 BLDC Motor

BLDC motor has applications in various fields. In BLDC motor only two phases are excited and remaining phase is left without excitation. Hall sensor which will sense rotor position, and will help in exciting the windings as shown in Fig. 4.

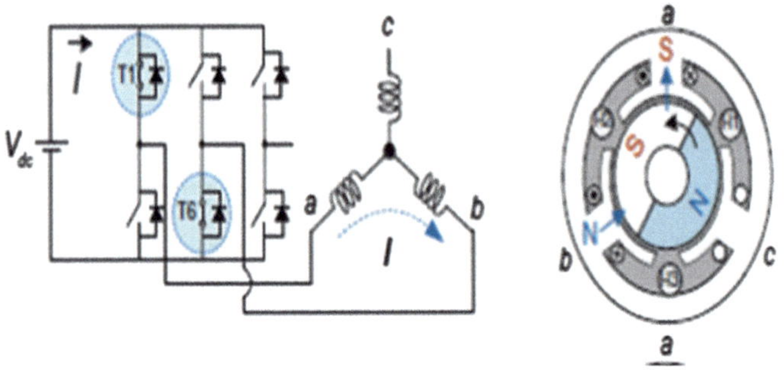

Fig. 4. Switching sequence of BLDC motor

2.1 Modelling of BLDC Motor

BLDC has armature resistance Ra and armature inductance La and these are connected in series with armature resistance (Fig. 5).

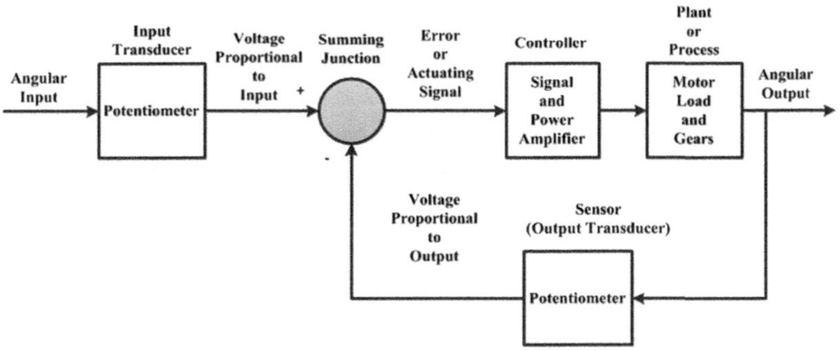

Fig. 5. Block diagram of antenna's positioner

input = r[R(s)]
load output = c[c(s)]

Principle of operation

1. At potentiometer

$v_e \propto r - c$
$v_e(\mathrm{s}) = k_p[r(s)] - c(s)$

2. At Amplifier

$v_a \propto v_e$
$v_a(\mathrm{s}) = k_p\ v_e(s)$

3. Analysis of armature of BLDC MOTOR

$$T_M(s) = k_T\, I_A(s)$$
$$E_B(s) = k_b \frac{\mathrm{d}\Phi}{\mathrm{d}_t}$$
$$E_B(s) = k_b\, \mathrm{S}\theta(s)$$
$$v_a - E_b = i_a R_a + L_a \frac{\mathrm{d}_{i_a}}{\mathrm{d}_t}$$
$$v_a(s) - E_b(s) - i_a(s) = i_a(s)(R_a + sL_a)$$
$$T_M(s) = (\mathrm{J}s^2 + \mathrm{B}s^1)\,\theta(s)$$

4. At gears

c(s) = n$\theta(s)$

To obtain results, load torque is taken as 3 Nm. Transfer function model between angular position and input voltage is obtained using following Eq. (1) (Table 1).

$$\frac{C(s)}{R(s)} = \frac{k_T}{R_a\left(\overline{J}S + B\right) + k_b k_T} \tag{1}$$

Table 1. BLDC motor parameters

S. NO	Naming	Values	Units
1	Supply voltage	120	Volts
2	Stator resistance (RA) R_a	3.5	Ohm
3	Stator Inductance (L_a LA)	8	mh
4	Back emf	120	volts
5	Rotor Inertia (J)	0.8	Kg m^2
6	Azimuthal range	N* 360 continuous	degrees
7	Elevation up range	183 to 186	degrees
8	Elevation down range	−3 to −6	degrees

3 PI Controller

PI controller has very large application area due to simple control structure. Proportional controllers fasten the system response, whereas integral controllers reduce the steady state error. Transfer function of PI controller is given in Eq. (2). Figure shows the block diagram of given transfer function.

$$\frac{U(S)}{R(s)} = k_p + \frac{k_i}{S} \tag{2}$$

By following the above algorithm and applying trial and error method, P and I values can be obtained (Figs. 6 and 7).

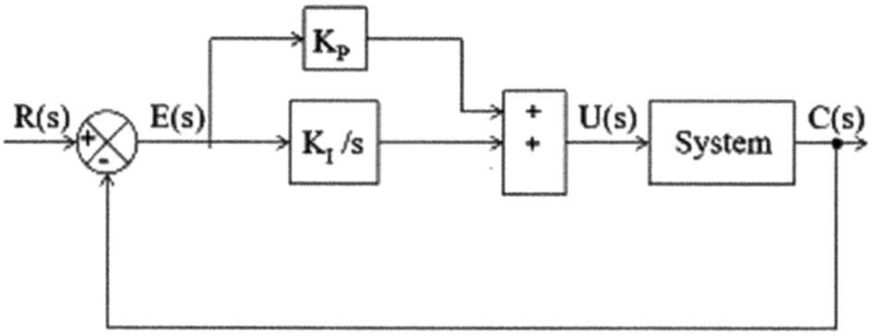

Fig. 6. Block diagram of PI controller

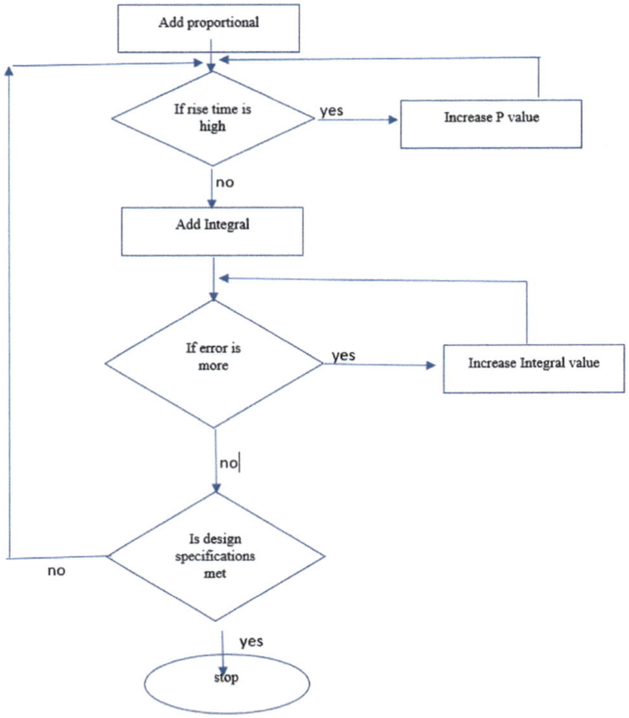

Fig. 7. Algorithm for implementing the PI values

4 PWM Control Implementation

In BLDC motor, accuracy of speed control is effected by armature inductance which leads to produce torque ripples [11]. This in results causes hotspot on BLDC motors. For reducing commutation torque ripples, an effective duty-ratio control strategy based on deadbeat current control for calculating PWM duty ratio for variable speeds (Fig. 8).

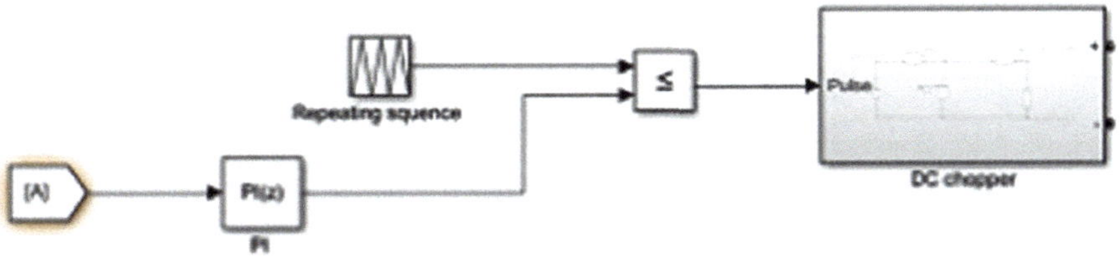

Fig. 8. shows implementation of PWM technique for BLDC motor

Output of PI controller is compared with triangle signal (carrier signal), which will help in generating pules. This pulses are given to the DC chopper.

5 Simulation Results

The Simulink model of Current controlled BLDC motor which is fed from chopper is shown in Fig. 9. Simulink model consists of BLDC motor, Inverter block, Buck converter, PI controller. In buck converter pulses are generated based on Pulse Width Modulation Technique (Fig. 10).

Case1: Control of BLDC Motor with P = 0.001, I = 0.3 at No Load
BLDC motor is controlled with p = 0.001 and I = 0.3. In this case BLDC motor is operated at No load.

Figure 11 shows stator emf of all phases in which speed is set after one second. Because of transients, stator emf is not trapezoidal up to 1.5 s. After the transients die out shape aligns in trapezoidal.

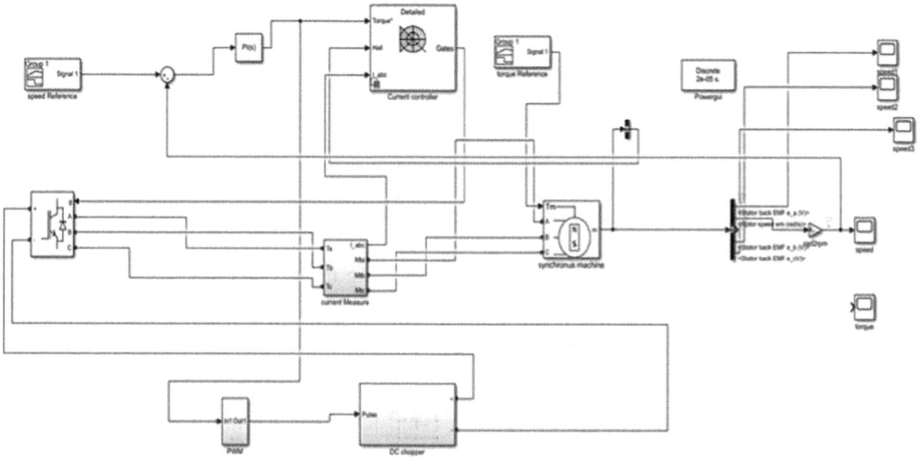

Fig. 9. Simulink model for speed control of BLDC motor

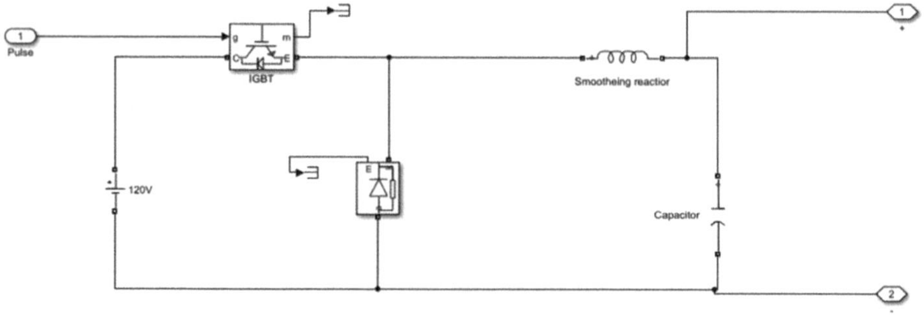

Fig. 10. Simulink model for DC chopper

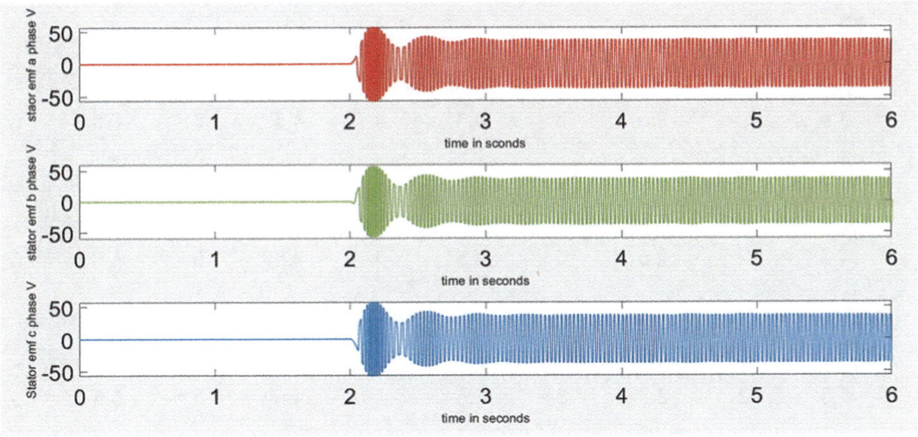

Fig. 11. Stator Back emf for all phases

Figure 12 shows speed of rotor at no load where reference speed is set after one second. Because of transients, reference speed is not attained until transients dies out.

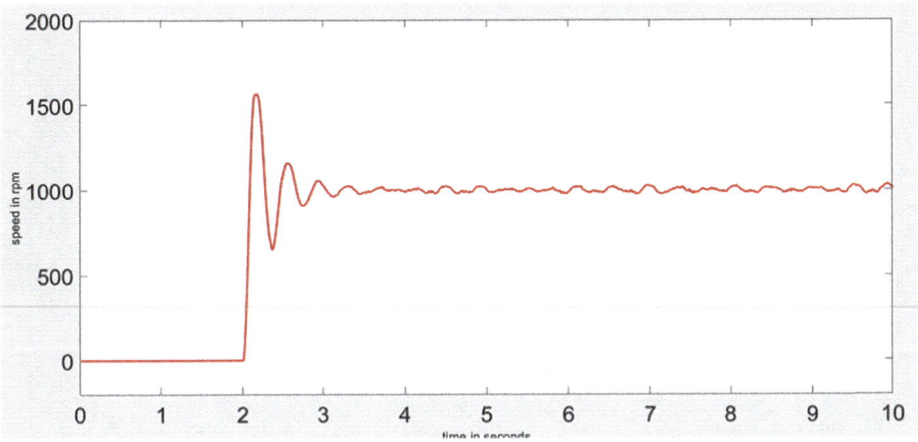

Fig. 12.

Case2: Control of BLDC Motor with P = 0.3, I = 8 at No Load

In this case BLDC motor with P value is increase from 0.001 to 0.3 and operated at no load.

Figure 13 shows stator emf attains its shape in very less time due to increases in PI values.

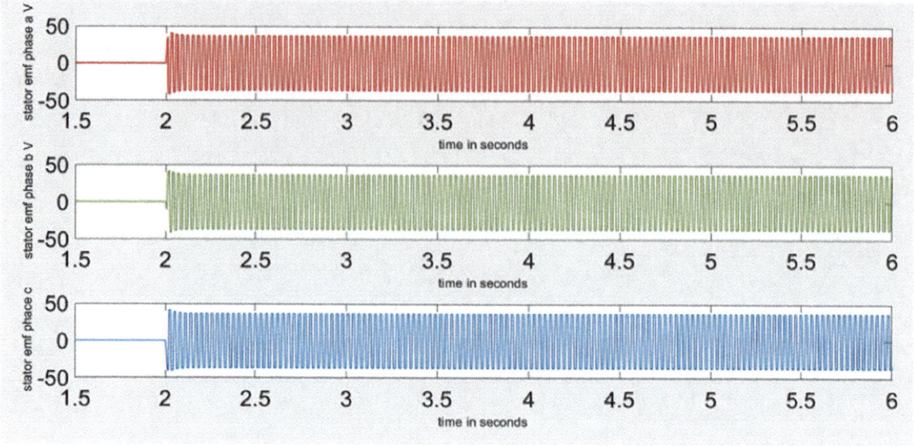

Fig. 13. Stator Back emf for all phases at increased PI values

From Figs. 12 and 14, it is clearly seen that, when PI value increases, there is a decrease in the oscillation of speed and thus the speed of response has improved (Table 2).

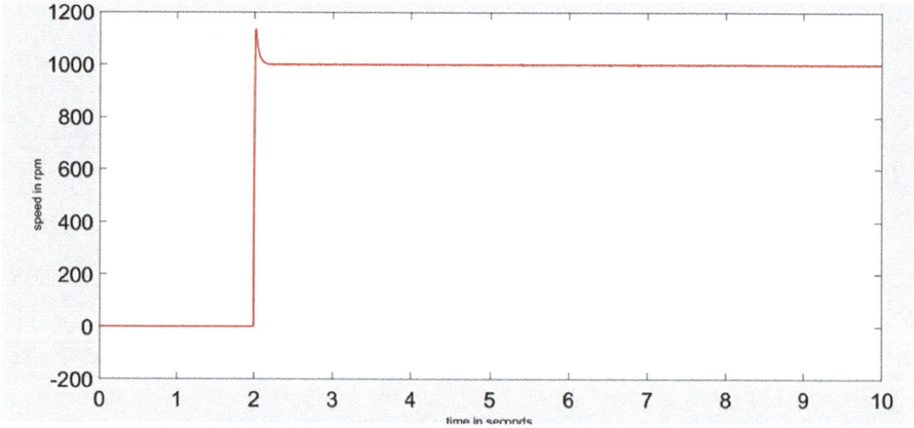

Fig. 14. Speed of rotor at no load for increased PI values

Table 2. PI values of BLDC motor

S. no	PI values	Percentage of Peak over shoot	Settling time
1	P = 0.01, I = 0.3	54.837	4
2	P = 0.3, I = 8	15.2	0.2

Case 3: Control of BLDC Motor with P = 0.3, I = 8 with Load of 3 Nm (F = 50 Hz)

In this case chopper PWM carrier frequency is operated at 50 Hz and load is applied after 3 s (Fig. 15).

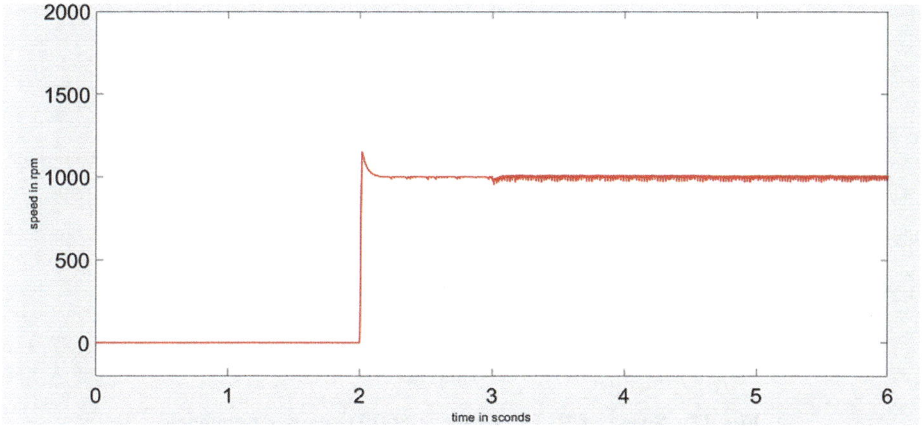

Fig. 15. Speed of BLDC motor at 50 Hz carrier frequency

Figure 16 shows ripple content in torque which occur due armature inductance.

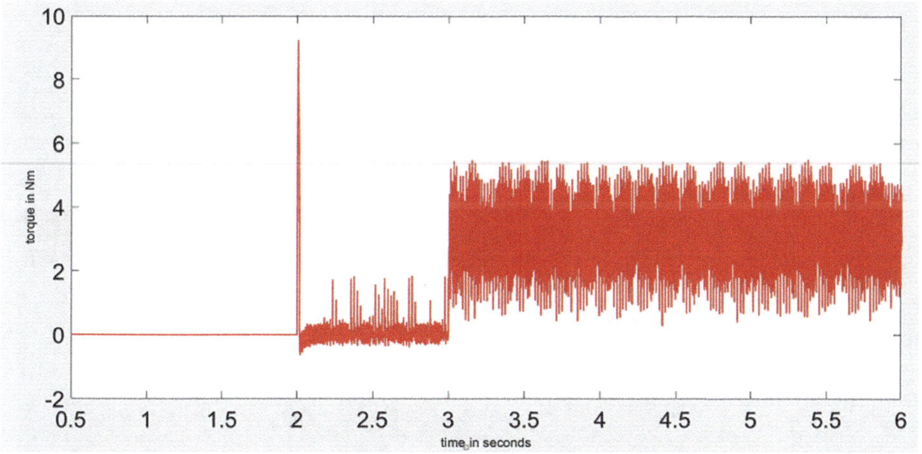

Fig. 16. Torque of BLDC motor at 50 Hz carrier frequency.

Case 4: Control of BLDC Motor with P = 0.3, I = 8 with Load of 3 Nm (F = 500 Hz)
In this case chopper PWM carrier frequency is operated at 500 Hz (Fig. 17).

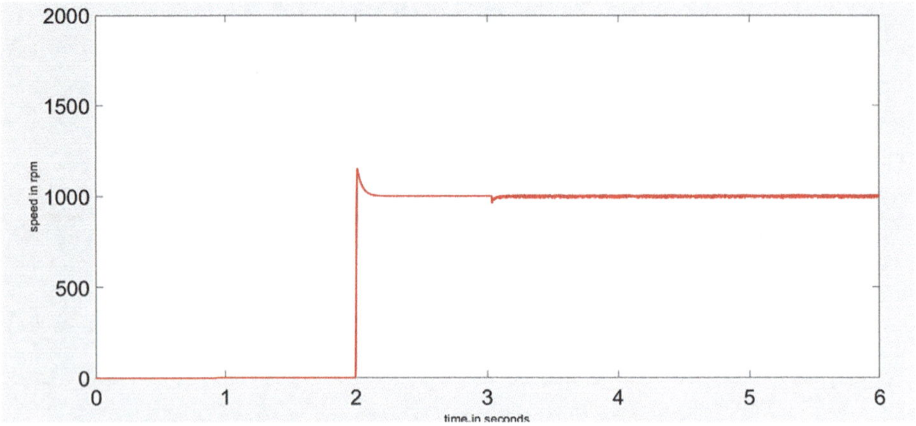

Fig. 17. Speed of BLDC motor at 500 Hz carrier frequency

From Figs. 16 and 18, it is clearly seen that, when carrier frequency increases, ripple in torque is reduced (Table 3).

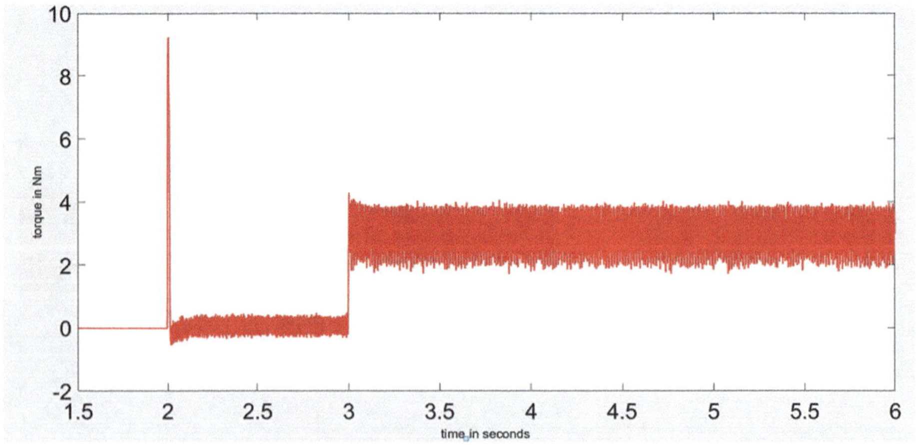

Fig. 18. Torque of BLDC motor at 500 Hz carrier frequency

Table 3. Torque values of BLDC motors for different carrier frequency

Carrier frequency	PI values	Torque ripple
50 Hz	P = 0.3, I = 8	5.3 to 0.8
500 Hz	P = 0.3, I = 8	4.1 to 2.2

Case 5: BLDC Motor Operated in Reverse Direction at No Load
In this case BLDC motor is operated in reverse direction with the same PI values of case 4 (Fig. 19).

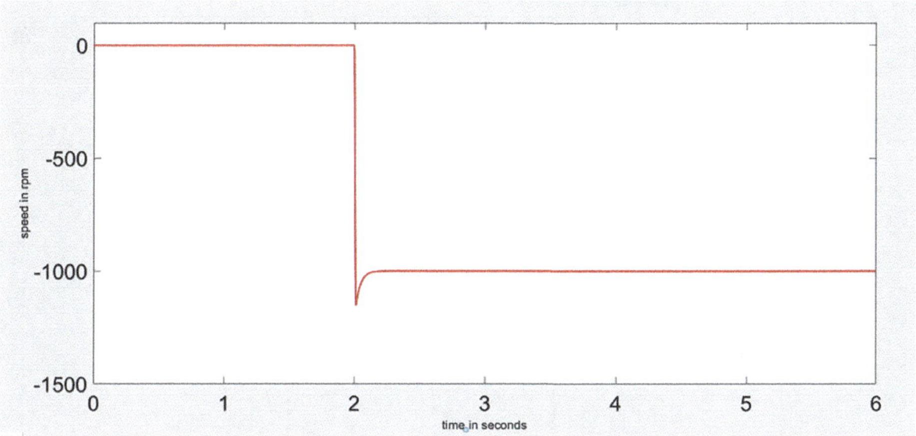

Fig. 19. Speed of rotor at no load operated in counter clock wise direction

Case 6: BLDC Motor Operated for Intermittent Duty Cycle
BLDC motor has to rotate, based on error generated from potentiometer, the azimuth unit is continuous (clock wise or counter clock wise), while the elevation unit motion is in a pre-determined angular range. In all above cases, BLDC motor is operated for continuous duty cycles. In this case BLDC motor is operated for short duration (Fig. 21).

Figure 20 shows stator emf of BLDC motor for all phases for short duration.

Figure 22 shows torque ripple of BLDC motor for short duration.

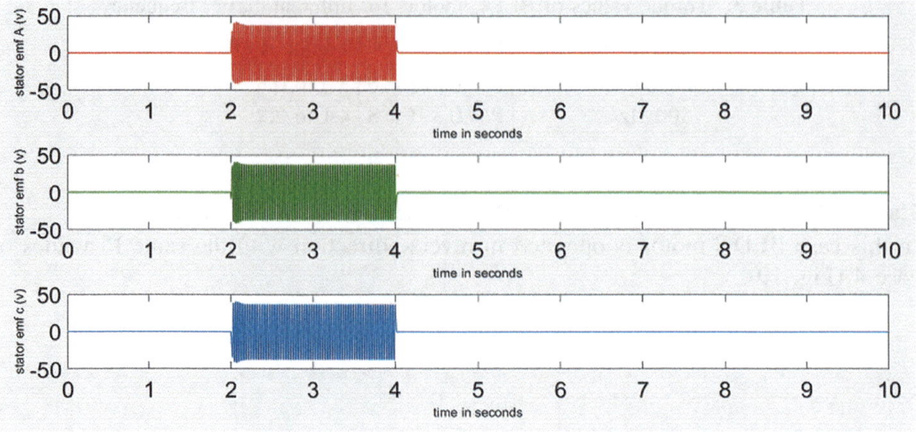

Fig. 20. Stator Back emf for all phases for short duration

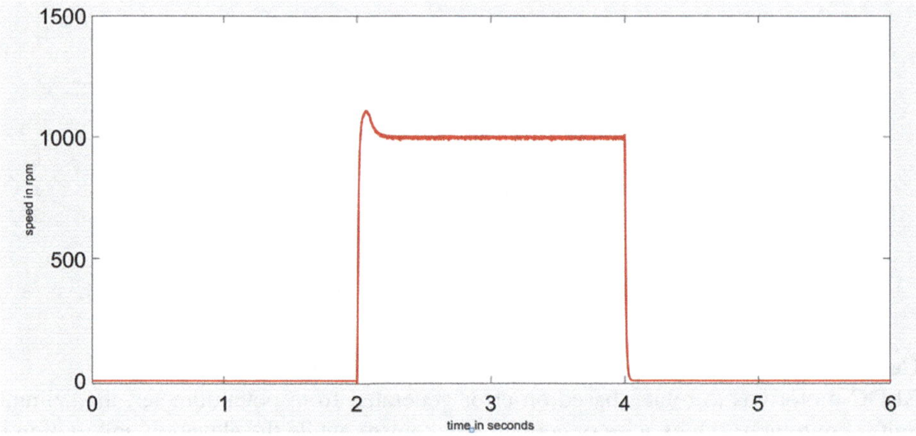

Fig. 21. Speed of rotor for short duration

6 Conclusions

In this paper, the speed control of current controlled BLDC is simulated using MATLAB Simulink. For different PI values, total simulation is carried out. As PI values increased, speed response is improved, settling time and percentage peak over shoot has reduced. As carrier frequency of chopper increased, ripples in torque are reduced. Therefore, PI controller with PWM technique improved the system response.

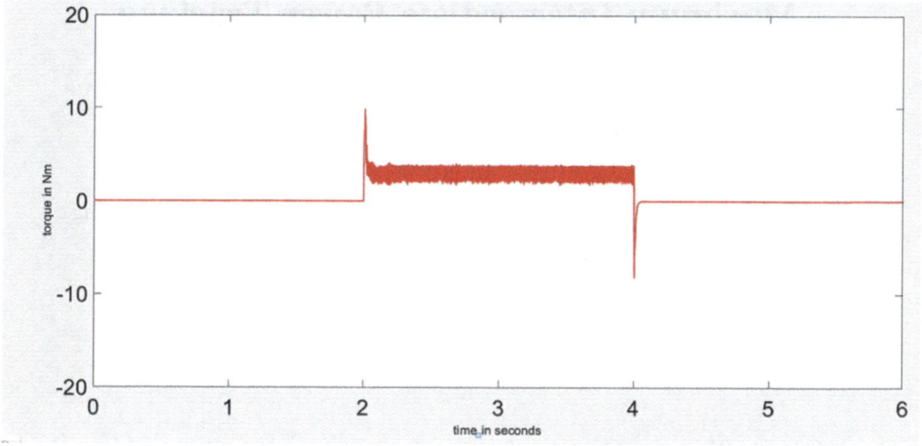

Fig. 22. Torque of rotor for short duration

References

1. Sarala, P., Kodad, S.F., Sarvesh, B.: Analysis of closed loop current controlled BLDC motor drive (ICEEOT) (2016)
2. Miller, T.J.E.: Brushless Permanent Magnet and Reluctance Motor Drive. Clarendon Press, Oxford (1989)
3. Mubeen, M.: Brushless DC motor primer. Motion Tech Trends, July 2008
4. Yedamale, P.: Hands-on Workshop: Motor Control Part 4 - Brushless DC (BLDC) Motor Fundamentals. Micro-chip AN885 (2003)
5. Kim, S.-H.: Brushless direct current motors. Electric Motor Control (2017)
6. IIT Kharagpur: Industrial Automation and Control, Electrical Actuators: BLDC motor drives, NPTEL (2009). Chap. 7. https://nptel.ac.in/courses/108105063/35
7. Fandakli, S.A., Okumuş, H.I.: Antenna azimuth position control with PID, fuzzy logic and sliding mode controllers. INISTA (2016)
8. Sharon Shobitha, O., Ratnakar, K.L., Sivasankaran, G.: Precision control of antenna positioner using P and Pi Controllers. Int. J. Eng. Sci. Innov. Technol. (IJESIT) **4**(3), (2015)
9. Agarwal, P., Bose, A.: Brushless DC motor speed control using proportional-integral and fuzzy controller. IOSR-JEEE **5**(5), 68–78 (2013)
10. Kuo, B.C.: Automatic Control Systems, 3rd. Prentice-Hall Inc., Upper Saddle River (1975)
11. Chen, Y., Tang, J., Cai, D., Liu, X.: Torque ripple reduction of brushless DC motor on current prediction and overlapping commutation, PRZEGLĄD ELEKTROTECHNICZNY (Electrical Review), ISSN 0033-2097, R. 88 NR 10a/2012

Maximum Intermediate Power Tracking for Renewable Energy Service

Pattanaik Balachandra[1(✉)] and Pattnaik Manjula[2]

[1] Department of Electrical and Computer Engineering, Faculty of Engineering and Technology, Mettu University, Mettu, Ethiopia
balapk1971@gmail.com
[2] Department of Accounting, College of Business Administration, Princess Nourah Bint Abdulrahman University, Riyadh, Saudi Arabia
drmanjula23@gmail.com

Abstract. The energy from sun in form of solar energy is available but inadequately used. From various source of renewable energy, solar power is the most required in terms of domestic and commercial development. The demand of energy is essential requirement, and the global energy capacity is more than the energy available in the universe being analyzed by using Renewable Energy Service (RES). The infrastructure demands for the growth of energy sector which is directly proportional to the economic aspects. The solar sector development is in peak rather than other energy compared in past years and application is in wide range in all most all areas where we need power. Solar power is the most economical new power plant technology, due to its installation costs, no fuel cost and construction time is less than one year, compared to over 10 years to construct nuclear power plants and other power plants. The energy we get from sun need to be used in the proper technical way to collect solar power being enhanced and the receiving capacity of the different system analysis modified. This is being achieved a novel concept of designing with Solar Power Technology (SPT) System for Maximum Intermediate Power tracking (MIPT) concept. This paper provides analyses to understand those environments Renewable Energy with Solar Energy Service to reduce Global Warming.

Keywords: Reliability · RES · SPT and MIPT

1 Introduction

As the fossil fuels are depleting day by day in our daily lives, we must initiate renewable energies like solar energy, wind energy, etc. Sources from sun is enormous which meets the global crises of energy for the world leads by using different technique to extract maximum possible for generation capacity enhancement by pollution free environment. Comparative with other energy sources the vast progress in solar energy brings achieved in almost all countries because of the motivation criteria towards solar energy production and uses in all sectors. People must encourage and initiate these types of alternative renewable sources of energy. The main reason of all the countries to initiate these type of sources other fossil fuels is that the world is experiencing a lot

D. J. Hemanth et al. (Eds.): COMET 2019, LNDECT 35, pp. 462–468, 2020.
https://doi.org/10.1007/978-3-030-32150-5_44

of pollution now a day's which may cause a harmful effect to our future generation [7]. People must also follow sustainable development for further growth of the world. Still, there are so many resources which are known, but not able to exploit due to the lack of technology. The future pillars of the world that is the children must be encouraged and educated enough to find ways to exploit these resources.

Biogas plant must be encouraged in rural areas instead of kerosene and petrol lamps which are more beneficiary because it saves their money as well as sustaining fossil fuels [2]. In comparison of all, solar energy is the cheapest source of energy in renewable energy technologies. In many countries, airport and aviation authorities are successful in using solar energy as their main source. E.g. Electricity used for lighting the airport, for the flight signals, for each and every electric usage.

While using, solar power neither harmful substances emission nor it adds to global warming. The solar plant installation must be preplanned based on the place where the solar rays being plenty made fall without any obstacle or shadow by the environment effect must be considered to maximize the input thereby the generated power will be maximized. Therefore the social environmental effects, pollution, place and the factories nearby area must be considered at the time of establishment. This leads the best choice of the solar plant to boost the power generation made effectively and the distribution criteria will be very easy with that the transportation cost can be minimized [9, 10, 12]. The periodical maintenance is also not very complex even then the serious issue is interval good maintenance is required in order not to mismanage usage of the system. This makes the advantage that it do not create any kind of distraction effect for the solar power production by the natural resources like air or water [3, 11, 13]. As, the earth has green house gases: carbon- dioxide and methane gas; these gases trap sun's heat in the atmosphere and the rest bounce back to the space. So, due to the pollution by vehicles these gases trap more heat resulting in the increase of temperature. This is nothing but global warming. The increase in the temperature more than the average temperature of the place is called global warming. Moreover, solar power plant may not require like other power plants use input resources like coal, oil and gas etc. Thereby this leads to reduce the pollution free environment and economically advisable renewable energy. This all advantage makes only choice recognized globally is solar renewable energy service [5, 6]. This requires the technical methods need to be followed to receive solar energy from sun by using solar power technology.

2 Solar Power Technology (SPT)

Considering Solar Power Technology (SPT) produces heats by using different kinds of mirror and required sunlight, thereby it used to produces electric energy. SPT can be used widely globally to get better power generation. SPT allows meeting the demand whenever the sun is not much brighter by that time other alternate energy or thermal energy can be used. SPT growth is much faster and proven technology for solar power plants. There are many technologies among them only four technologies used in SPT, they are known as concentrated collectors parabolic troughs, central receivers, parabolic dishes, and Linear Fresnel Reflectors [4]. The technology being enhanced and shown in Fig. 1, which produces electric energy by regular collection of solar heat from the sun and continuing the regular process for the society useful applications. Curve shaped parabolic trough focus to a tube that transfers heat fluid to generate steam by generator. It can be integrate with any plants with any reliable fluid thereby the required efficiency being achieved. Technically Error free designed compound parabolic trough solar collector can be focused on solar energy from multiple directions to a common "cofocal" line which focuses for both parabolas. These kinds of compound parabola solar collector need not tracking the sun when used with a solar energy receiver tube, collected at the focus line. In Fig. 2, Curved Mirror shaped reflector flat mirror gives better steam produced by water but not used popularly. Figure 3, Central shaped receivers focus the solar rays at the center and top surface of the tower but the model is more successful to convert water into steam. The Fig. 4 describes the parabolic shape reflector which concentrates the suns energy into the reflector moreover they are dry cooler use little water [1, 8].

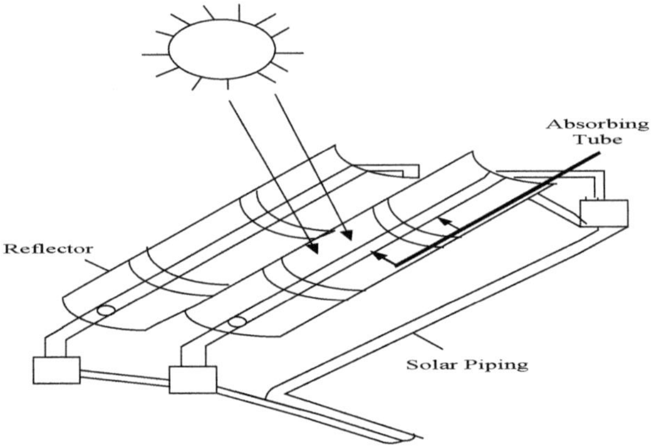

Fig. 1. Compound parabolic troughs

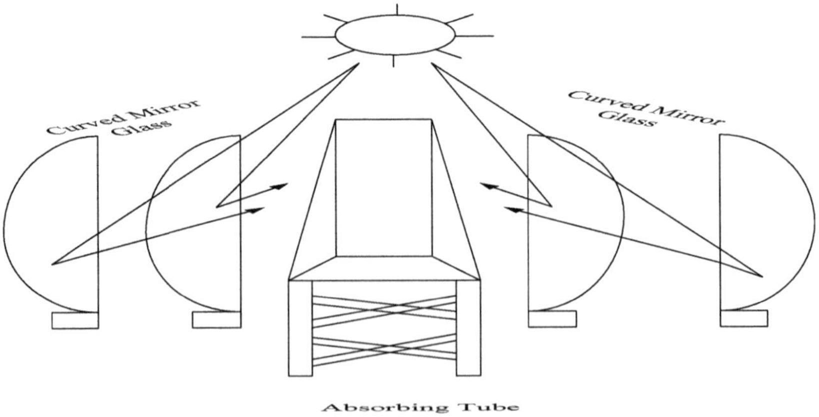

Fig. 2. Curved mirror shaped reflector

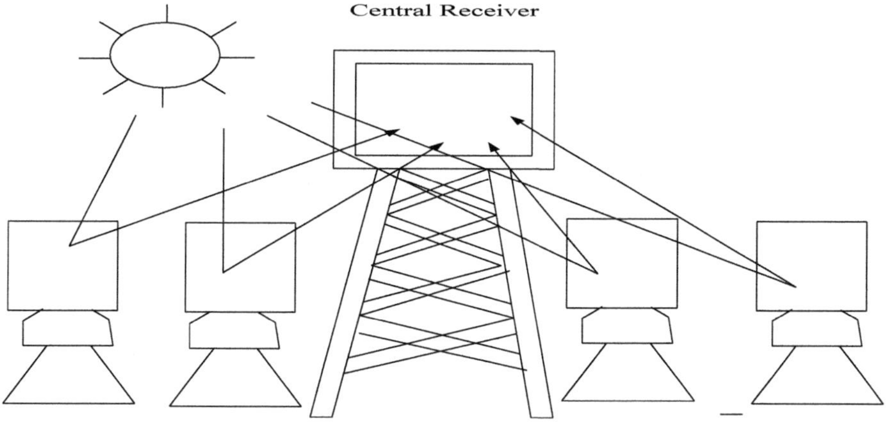

Fig. 3. Central shaped receiver

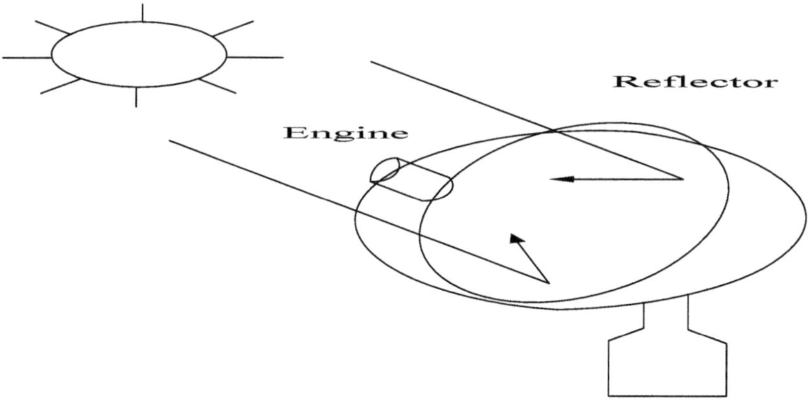

Fig. 4. Parabolic shaped reflector

3 Renewable Energy Service (RES)

The fast depredation of fuel energy resources on a global basis has demanded necessary an emergency search for alternative quick energy production sources to meet the required current demands. Alternative energy resources such as solar and wind have attracted energy sectors to generate power on a large scale. Renewable Energy Service (RES) solves day to day difficulties with the exact different probable alternate combination and availability of the more than one system compatibility, better choice of system rearrangement can be made to be meet the continuity of power supply being achieved. The conditions in terms of investment, power system establishment and transmission reliability requirement needed to be verified mathematically. The optimization techniques are to be needed in a probabilistic approach; graphical enhanced construction method and reliable iterative technique have been recommended by many researchers but these events may give with two different combinations of sources may be solar or wind. The RES generating power system is proposed to enhance offer steady and reliable source power supply for the day to day user as compared with the other power generating system operation.

Let the probability of failure of an independent renewable energy source (RES) either solar or wind on its xth variation of RES be '*p*'. Then the probability of the system failure with two independent events is represented in the equation "1" and "2" as shown:

$$P_{RES,x} = 3p^2 - 2p^3 \tag{1}$$

If '*p* 'is negligible, the Eq. (1) reduces to $P_{sys,\ x} = 3p^2$. The probability RES fails during any time period of T hours during the condition of weather is given in the equation

$$P_{RES}(T) = 1 \quad (1 - 3p^2)^{XT} \tag{2}$$

Where **"x"** represents the number of success full output source of the atmosphere condition and **T** represents the random interval of time period. This probability application being achieved with multi input rectifier model.

The two independent events considered based on the availability of source. The researchers found many power generation with hybrid proposal system. But this mathematical model gives the probability of failure with respect to time with success output result. Here the model being considered with random interval because it depends based on the nature of the availability of source. Further the RES can be achieved with analysis of more than one input simulation. This replaces the power system generating technology to improve different conventional method of electricity grid to better sectional or entire traditional electricity grid by the renewable energy power generation system.

4 Simulation Results of RES

In this section, simulation results are given to verify that the proposed multi-input rectifier stage can support considering input solar as well as wind source of simultaneous operation. The Fig. 5 shows the detail main simulation which gives the successful output of either or sources being utilized as per their availability. The pulse widths for both Mosfet's are 36 and 18 respectively. Figure 6 illustrates the pulse waveform of both Mosfet's. The input wind source voltage and input PV source voltage is 110 and 36 V respectively. The Fig. 6 shows the pulse waveform of M_1 and M_2 there by Fig. 7 shows the PV output current waveform of the two sources. The output voltage is 67 V. Individual of both the sources and simultaneous operation is being supported in this simulation. The separate controllers are not necessary for both sources being achieved. Here the output of the waveform shows the maximum intermediate power tracking by using solar power technology being achieved.

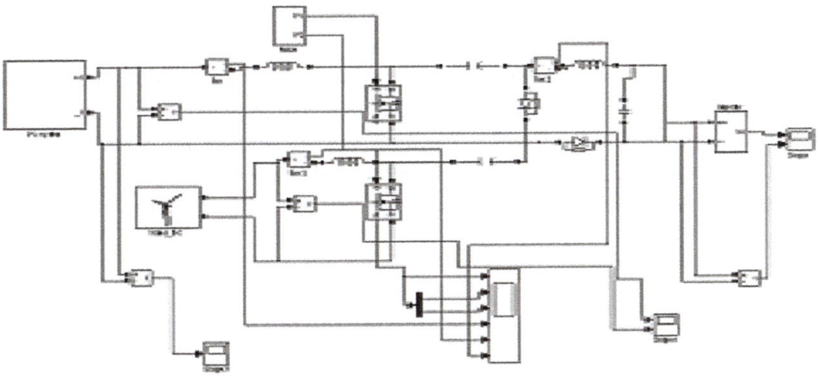

Fig. 5. Main simulation circuit

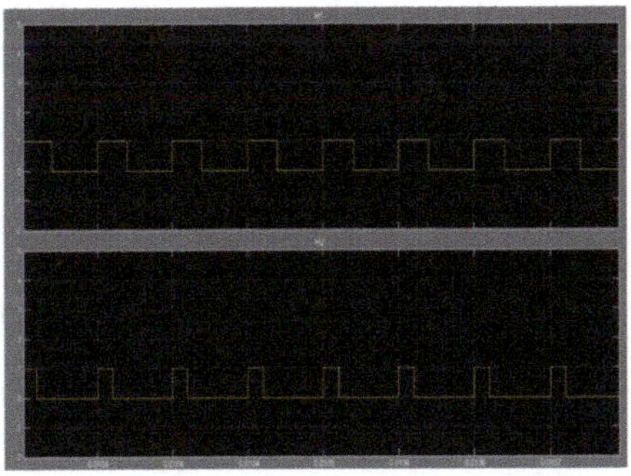

Fig. 6. Pulse wave form of M_1 and M_2

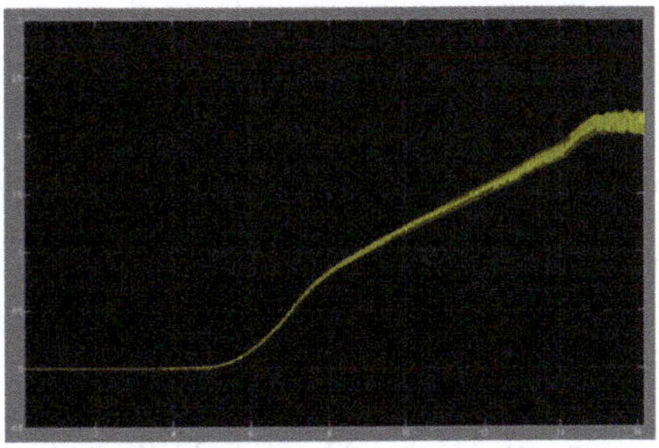

Fig. 7. Output current

5 Conclusion

STP biggest edge is its ability to incorporate solar technology. This will enhance the unique probable factor that boosts economic generation power and emerges maximum collection of energy from STP, which enables to increase maximum reliability with significant improve in the economics. With minimum STP systems also can provide maximum electricity supply to the grid station. Economic point of view STP is better expected to be greatly driven by system successively gives higher efficiencies and maximum intermediate power tracking economies of the STP. Already in so many countries they have implemented solar as their half part of energy and I want our country also to be note one of them but one of the best countries to use solar and utilize its energy. Future maximum efficiency improvement can be enhanced by multi input hybrid system. The integration of STP being incorporated with various techniques with compared with other energy generating systems also.

Integrated energy systems such as solar and wind provides better alternative to the global crises. This motivates to use the investors and consumers to open a large potential energy avenue for improved STP technology, this leads to improve the grid energy reliability as it allows with more flexible power generation suitably which is being simulated and achieved with solar and wind system. This is the main reason, I have written this article, to make people understand the power of solar when compared to other energy resources which pollutes of environment and destroys our future.

References

1. Dale, V.H., Efroymson, R.A., et al.: The land use-climate change-energy nexus. Landscape Ecol. **26**, 755–773 (2011)
2. Dihrab, S.S., Sopian, K.: Electricity generation of hybrid PV/wind systems in Iraq. Renew. Energy **35**, 1303–1307 (2010)
3. Pedersen, G.A.: It is time to rethink how we design and build standby power system. In: INTELEC 2004, pp. 626–631 (2004)
4. Brey, J.J., Castro, A., Moreno, E., et al.: Integration of renewable energy sources as an optimised solution for distributed generation. In: 28th Annual Conference of the Industrial Electronics Society, 5–8 November 2002, vol. 4, pp. 3355–3359 (2002)
5. Benner, J.P., Kazmerski, L.: Photovoltaics gaining greater visibility. IEEE Spectr. **29**, 34–42 (1999)
6. Alireza, S., Morteza, A., et al.: A probabilistic modeling of photo voltaic modules and wind power generation impact on distribution networks. IEEE Syst. J. **6**(2), 254–259 (2012)
7. Senjyu, T., Nakaji, T., et al.: A hybrid system using alternative energy facilities in isolated Island. IEEE Trans. Energy Convers. **20**(2), 406–414 (2005)
8. Ahmed, N.A., Miyatake, M., et al.: Power fluctuations suppression of stand-alone hybrid generation combining solar photovoltaic/wind turbine and fuel cell systems. Energy Convers. Manag. **49**, 2711–2719 (2008)
9. asonga, A., Saulo, M., Odhiambo, V.: Solar-wind hybrid energy system for new engineering complex- technical university of Mombasa. Int. J. Energy Power Eng. 73–80 (2015). https://doi.org/10.11648/j.ijepe.s.2015040201.17. ISSN 2326–957X
10. Mohring, H.D., Klotz, F., Gabler, H.: Energy yield of PV tracking systems. In: 21st European Photovoltaic Solar Energy Conference - EUPVSEC, WIP Renewable Energies, Dresden, pp. 2691–2694 (2006)
11. Abdallah, S., Nijmeh, S.: Two-axis sun tracking with PLC control. Energy Convers. Manag. **45**, 1931–1939 (2004)
12. Afarulrazi, A.B., Utomo, W.M.: Solar tracker robot using microcontroller. In: International Conference on Business, Engineering and Industrial Applications (ICBEIA) (2011)
13. Ponmozhi, G., Bala kumar, L.: Embedded system based remote monitoring and controlling systems for renewable energy source. IJAREEIE **3**(2), 283–290 (2014). ISSN (Print) 2320–3765

Modeling Internet of Things Data for Knowledge Discovery

Mudasir Shafi[1], Syed Zubair Ahmad Shah[2(✉)], and Mohammad Amjad[3]

[1] Maharishi Dayanand University, Rohtak, India
[2] Islamic University of Science and Technology, Awantipora, India
zubair.shah@islamicuniversity.edu.in
[3] Jamia Millia Islamia, New Delhi, India

Abstract. Internet of Things (IoT) is a budding field. It finds its base in the science of electronic equipments, communication technologies and computing algorithms. It is the network of physical devices, vehicles, home appliances and other items embedded with electronics, software, sensors, actuators, and connectivity which enable these objects to connect and exchange data. All things on the IoT may develop a data overflow that encompasses various types of relevant information. Data can be generated as a result of communication between humans, between human and systems, and between systems themselves. This data can be used to improve the services offered by IoT and thus it becomes important to work on the IoT generated data. This paper presents a model for implementing an IoT system, collecting data from it and performing data analytics on the collected data with the intent of deducing knowledge from this data. This paper also proposes some new areas where IoT can be put to use, thus bringing in sight a ground-breaking view of what IoT can and will do. This, in turn, will change the way we live, work and communicate. The hurdles that may come in the way of implementing IoT have also been discussed in this paper. Finally, the methods of analyzing IoT data have also been discussed with focus on frequent pattern mining. Implementation and results of our work have been shown vividly.

Keywords: Internet of Things · Knowledge discovery of data · IoT reference model · IoT data

1 Introduction

Internet of Things (IoT) brings together many of the latest technologies. When these technologies are converged they have a huge impact on our lives. IoT is making thinks smart or digital which enable a new level of services and capabilities. The first revolution of technology was when computer and internet came into being and now IoT will be the largest revolution in the field of technology. IoT contains trillions of nodes representing various objects from small ubiquitous sensor devices and hand helds to large web servers and supercomputer clusters (Poslad 2009).

D. J. Hemanth et al. (Eds.): COMET 2019, LNDECT 35, pp. 469–481, 2020.
https://doi.org/10.1007/978-3-030-32150-5_45

What basically is IoT? According to Haller et al. (2008) IoT implies a world in which physical entities are integrated into data network and these entities can become active participants in the business process. Various services are available that interact with these entities on the internet. (Zhang 2009) has given the idea of IoT from technology and economy perspectives: “From the viewpoint of technology, IoT is an integration of sensor networks, which include RFID, and ubiquitous network. From the viewpoint of economy, it is an open concept which integrates new related technologies and applications, productions and services, R. & D., industry and market.” The main target IoT is to hook up all the items in the globe to the internet. IoT is the combination of material gadgets, automobiles, home devices and many other devices installed with electronics, software, sensors, actuators, and connections that allows the things to establish connection and also transfer information with one another. Every single device has a unique identity through its installed networking system but is also capable of operating with the classical network or the internet framework that is present already. With the help of this technology we are able to evaluate and also control the things from a far off place, hence developing favorable circumstances for direct combination be- tween the physical world into computer-based systems, and hence helping us to get an productive system, having precision and financial advantage and also minimizing the human interference (Vermesan and Friess 2013; Mattern and Floerkemeier 2010; Lindner 2017). The Internet of Things is, after the modern computer (1946) and the Internet (1972), the world’s third wave of the ICT industry. Internet of things was thought of as “computers everywhere” by Professor Ken Sakamura (University of Tokyo) in 1984 and “ubiquitous computing” by Mark Weiser (Xerox PARC) in 1988, the phrase “Internet of things” was coined by Kevin Ashton (Procter & Gamble) in 1998 and developed by the Auto-ID Centre at the Massachusetts Institute of Technology (MIT) in Cambridge, USA, from 2003. Ashton then interpreted the IoT as “a standardized way for computers to understand the real world.” We can say that upto this point we have associated internet just with humans but it won’t be very far when most of the things that we use will be connected with internet The internet of things will mostly extend connection from the 7 billion people across the globe to the predicted 50–70 billion machines.

The “Things”, in IoT can be attributed to large kinds of gadgets like heart examining devices embedded in humans, chips for examining pastures, camcorders getting updates of wild animals in coastal waters, transportation with in-built sensors, DNA examining gadgets for environmental monitoring/infestation monitoring (Erlich 2015), or field operation devices that assist fireman in search inspection and relief operations (Rouse and Wigmore 2013). Legal scholars suggest regarding “things” as an “inextricable mixture of hardware, software, data and service” (LaDiega and Walden 2016). It is foreseen that the evolution of internet of things as a service provider will be a trend even though the technologies being used are not brand new. Apart from the considerable advancement on computer communication and appropriate technologies that perform multiple operations possible (Hussain and Keshavamurthy 2018; M&M

Research Group 2012), IoT is likely to make a whooping growth from $44.0 billion in 2011 to $299.0 billion by 2017 and we can easily approve of this at present. Analysts from various business establishments, research associations and various government agencies are considering to reform the internet, by formulating a favorable surrounding that is formed of multiple rational systems like intelligent home, intelligent automobile system, global supply and health care (Atzori et al. 2010; Miorandi et al. 2016; Bandyopadhyay and Sen 2011; Domingo 2012).

We can interpret IoT from multiple angles which has been carried by researchers whose work has been published recently. The various angles of interpreting it are: challenges (Miorandi et al. 2016), utilizations (Domingo 2012), specifications (Palattella et al. 2013), and brilliance (Kulkarni et al. 2011; Lopez et al. 2012). Atzori and his associates (Atzori 2010) gave a broad perspective of IoT from three different angles i.e. things, internet, and semantics. More recent studies (Aggarwal 2014; Haller et al. 2008) have been encouraged and they represent a universal five-layer architecture to depict the complete format of IoT. The main five layers from fundamental layer of edge technology, access gateway, internet, middleware and goes up to application layer. Many recent researches (Miorandi et al. 2016; Lopez et al. 2011; Siegemund 2002; Kortuem et al. 2010; Lopez et al. 2012) apart from characterizing the framework and things of IoT reiterate that almost all the items on the IoT are able to perceive and hence are known as "smart objects" (SO) and are presumed to be accomplished of recognized, anticipate occurrences, communicating with other items and being capable of taking decisions (Lopez et al. 2011; Lopez et al. 2012). One main problem that has emerged at present is how to transform the data that is produced or secured by internet of things into information in order to utilize this information so as to provide a more beneficial surrounding to the people using this technology.

The answer to this question has come from technologies using data mining and "knowledge discovery in databases" (KDD), this is where these technologies have provided a way as they give a resolution to extract the knowledge from the data that is buried in the data of IoT, we can apply this knowledge to improve the working of the system or to enhance the grade of services this new surrounding can give us. A number of works have been concentrating on establishing an efficient data mining technologies for the IoT. The conclusions drawn in (Cantoni et al. 2006; Keller 2011; Masciari 2007; Bin et al. 2010) exhibit that the data mining algorithms can be used to make internet of things more rational, and hence enabling us to get more valuable services. Data mining has also been used in many other applications (Shah and Amjad 2017).

In the initial days of its theoretical proposition, IoT seemed to be a near impossible task as it sounded impractical to connect everything on the globe (including non-living beings like refrigerator, washing machine, microwave oven and even the non-human living beings like pet animals) through the internet (Anumba and Wang 2012). Interestingly, now in the latter part of this decade, it is being realized that the larger problem is not connecting (as that seems achievable now) but it is deducing knowledge from the enormous data (which may reach to Petabytes) into knowledge which, in turn, can assist in decision making (Tsai et al. 2014). So, when we talk of IoT and data mining (for knowledge discovery), two issues come up – how to efficiently and precisely collect data and then how to mine such a massive amount of data. Some models and approaches for gathering data and processing it can be found at (Tsai et al. 2014; Bin et al. 2010; Bonomi et al. 2012).

2 Proposed Use of IoT

IoT has changed the way we live, from purchasing things from the supermarket to driving vehicles. Sensors are embedded in the things surrounding us which transmit valuable data to the common platform of IoT which brings devices together.

(i) A car has many sensors embedded it in which monitor the various components of vehicle. If a driver sees a fault signal being displayed during a trip, he comes to know that something wrong is with the vehicle but whether the problem is major or minor needs to be checked by a professional. Here IoT comes into play. The various sensors monitor the components and collect the real time data and send it to the manufacturer which senses the problem in the vehicle. The manufacturer, in turn, will make the component which needs to be repaired available even be- fore the arrival of the vehicle. By this, we can ensure the safety of the driver and also save the precious time of the user. One more thing that can be achieved by this is the optimization of the manufacturing process.
(ii) IoT can be used for monitoring the health, especially, of handicapped individuals and elderly people living alone. Say, for instance, a man suddenly feels numbness in his body and his blood pressure shoots up. The smart health sensor can immediately sense changes in the behavior of that person and sent relevant data through proper channel to the nearest health care available. In this way a precious life can be saved in no time.
(iii) Smart refrigerators have sensors that sense when the milk or vegetables or fruits are running out. It sends the data to the nearest supermarket to get the required amount delivered.
(iv) In a smart home, a person can make use of the smart applications right from the movement he opens his eyes in the morning by switching off the alarm. The sensor in the clock sends information to the geyser in the bathroom to warm the water and simultaneously to the coffee maker to make the coffee for the person. Geyser simultaneously makes use of another sensor which has the weather data and accordingly it will heat up the water to a suitable temperature.

3 Hurdles in Implementing IoT

As we all know technology has its pros as well as cons, same is the case with IoT.

(i) IoT needs integration of multiple systems but at times it is not possible to integrate things in an efficient manner. Take an example of a smart home having health care and energy management sensors. If the health care sensor detects depression, it will switch on the lights in order to spread positivity in the environment but the energy management sensor, because of absence of any motion by the person, will turn off the lights in order to minimize the use of energy. This is the main problem that is to be corrected for the efficient running of devices based on IoT.

(ii) Privacy: With conveniences that IoT is giving to us comes a threat to our privacy. Since, in IoT everything is connected through so there is a he risk of cyber-theft.
Hackers can easily access our personal information and misuse it or even manipulate it for their own benefit. Smart devices will act as entry points for intruders to spy into our lives. In today's world it depends on us whether we want to share any personal information on social media or not. This is active participation but in a smart world, we will act as a passive participant. Say for instance my daily routine of coming home is at 22 h. This piece of information is collected by various smart devices and sensors that surround me in the smart city and in the smart home. This information will be sent to health care so that they can monitor my health, to my car manufacturer for maintenance purposes, to an advertising agency for sending me relevant advertisement at this particular time and this in- formation can also land into those hands which can misuse it. With the increase in the advent of smart things, data collection will penetrate more deeply into our lives and will introduce a whole new set of private data that is both linkable and identifiable.

(iii) IoT will make us completely dependent on the internet. If the internet goes off for even a minute our lives will come to a standstill. This will not only paralyze us but will lead to heavy losses both in terms of lives and money.

(iv) We, also, have to look into those geographical regions where the accessibility, reliability and speed of internet are a matter of concern.

4 Reference Model for IoT

A model has been proposed here for implementing an IoT system (Fig. 1) with the intention of collecting data and deducing knowledge from that data. This model has 7 basic layers – physical layer, connection layer, computation layer, accumulation layer, generalization layer, functional layer and association layer.

(i) ***Physical layer:*** IoT device may be smaller than a coin or larger than a refrigerator. It may perform a simple sensing function and send raw data back to a control center or may combine data from various sensors and perform local data analysis and then take action. Our device can also be a remote or stand alone or be co- located within a larger system.
Whatever is the function of IoT device, the IoT device requires two main components - brain and connectivity. Brain will provide local control and connectivity is needed to communicate with external control. The various IoT devices and controller are – sensors, actuators, RFIDs and GPS.

(ii) ***Connection layer:*** IoT is a platform where the devices are becoming smarter, processing is becoming intelligent and communication is becoming more informative with each passing day. As per Gartner, 25 million IoT devices will be connected to internet by 2020 and those connections will facilitate the used data to analyze, preplan, manage and make intelligent decisions autonomously. An IoT device may consist of several interfaces for communicating with other devices both wired and wireless:

a. IoT interface for sensors
b. Interfaces for internet connectivity
c. Memory and storage interfaces and
d. Audio video interfaces.

The most important function of this layer is reliable, timely information transmission.

This includes transmissions

a. between devices and networks.
b. across networks
c. between network and low level information processing occurring at next level.

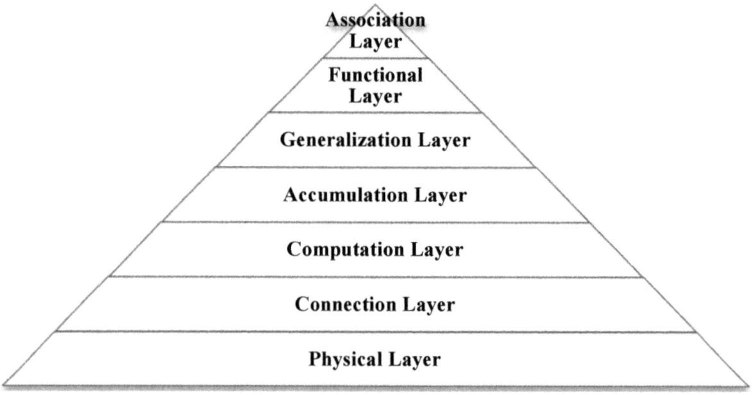

Fig. 1. Reference model for IoT.

(iii) ***Computation layer:*** This layer is mainly required to convert the data into information that can be stored and further processed at next layer. The smart system will start processing the information as early and in close proximity to the edge of network as possible. This layer will process the data at higher speed if we need it at faster pace. It represents threshold or alert which includes redirecting data to additional destinations.

(iv) ***Accumulation layer:*** Here data storage takes place. It determines if data is of any use to higher layers or not, the type of storage is also determined here i.e. whether it is file system, big data system or a database. Data storage plays an important part as applications can access data whenever they need it.

(v) ***Generalization Layer:*** The data stored in layer 4 is generalized in this layer. This layer mainly stresses on making the data and its storage in such a way which enables in the development of simple performance enhanced applications. We are aware of the fact that the various IOT sources from which data is to be accumulated may not be in close proximity but geographically separated it is from such sources that this layer should be able to reconcile data from.

(vi) ***Functional layer:*** The data generalized in previous layer is interpreted in this layer. The type of usage will mainly depend on the type of data that has been collected. Its usage can be just observing on the data of the device, or governing devices.

(vii) ***Association layer:*** All this collection and generalization of data into valuable knowledge will be of use to us only if this knowledge is put into life. This process will require people and processes. As we know that most of the applications perform business logic to capacitate people and the people can use it these applications and data for their specific needs. To make IOT fruitful, action needed requires collected effort of many people so people have to communicate and collaborate with each other.

5 Knowledge Discovery from IoT Data

Knowledge Discovery in Databases (KDD) is an important field that is concerned with deducing knowledge from data with the aim to help in the process of decision making. This involves layers 5–7 of the above presented model. Analyzing, deducing information and making decisions is no simple thing when it comes to performing these on enormous amount of data such as that of IoT. Three major processes are part of KDD – clustering, classification and frequent pattern mining (Han et al. 2011).

Clustering is a process of grouping data objects based on the similarities be- tween them. The aims to put all objects that are very similar to each other in same group and make sure that differ in a good number of features should be placed in different groups. Each group is called a cluster. This keeps intra-cluster similarity high and inter-cluster similarity low. A number of clustering algorithms have been proposed over the past few decades. A comparison of these algorithms can be found at (Steinbach et al. 2000).

Classification is a process in which we create classification model based on the input where we have given certain objects and they already have been labeled. This model is given input in the form of new objects and it helps to classify or put the objects in particular class thus giving a clear idea of where new objects should be put i.e. what is the label of this new object. For classification many algorithms have been proposed (Aggarwal 2014).

Frequent pattern mining is a process where the aim is to find pattern in a data set and we often try to find pattern which are frequent in the data set and these frequent patterns then may help in clustering or they may be directly be used to deduce knowledge from this because frequency of occurrence of single objects or group of objects in the data gives better understanding of what data represents and what it should mean. If the data is in the form of transactions (in terms of rows and columns of a table) then we typically call it frequent itemset mining. Frequent itemset mining algorithms include Apriori (Agarwal and Srikant 1994) and FP-growth (Han et al. 2000).

5.1 Proposed Approach to Mining IoT Data

To form interesting, unusual and valuable patterns in databases we use data mining algorithms which is possible with the help of pattern mining. The pattern mining algorithms can be practiced on many forms of data such as transaction databases, progression databases, streams, strings, spatial data, graphs, etc. Multiple patterns such as sub-graphs, associations, indirect associations, periodic patterns, sequential rules, lattices, sequential patterns, etc. can be formed using pattern mining algorithms.

The most well known algorithm for pattern mining is Apriori (Agrawal and Srikant 1994). This algorithm is formed so as to apply on a transaction database to unveil arrangements in transactions done by consumers in supermarkets. We can also use this algorithm in many other applications, including IoT data. The only condition is that data should be in transactional data format. A transaction is described as a set of apparent items (symbols). This algorithm has two main things first is a minimum support threshold which is given by the user and second a database made up of transactions. The result of this algorithm consists of the itemsets occurring in a recurrent fashion i.e. a set of items which are shared by a minimum threshold transactions in input database.

The Apriori algorithm determines frequent itemsets by a repetitive approach where candidates are generated (Algorithm 1).

Input:
DB, dataset containing transactions
minsup, minimum support count threshold
Output: L, frequent itemsets in DB.
Method:
L_1 = find frequent_1-itemsets(D);
for (k = 2; $L_{k-1} \neq \Phi$; k++)
C_k = aprioriGenerator(L_{k-1});
for each transaction t $\in$ D { // scan D for counts
C_t = subset(C_k, t); // get the subsets of t that are candidates
for each candidate c $\in C_t$
c.count++;
$L_k = \{c \in C_k |$ c.count $\geq$ min_sup$\}$
}
return L = $U_k L_k$;

procedure AprioriGenerator(L_{k-1}: frequent(k-1)-itemsets)
for each itemset $l_1 \in L_{k-1}$
for each itemset $l_2 \in L_{k-1}$
if ($l_1[1] = l_2[1]$ ^ ($l_1[2] = l_2[2]$) ^ ... ^ ($l_1[k-2] = l_2[k-2]$ ^ ($l_1[k-1] < l_2[k-1]$) then
{
c = l_1 *join* l_2; // join step: generate candidates
if CheckInfrequentSubset(c, L_{k-1}) then
delete c; // prune step: remove unfruitful candidate
else add c to C_k;
}
return C_k;
procedure CheckInfrequentSubset(c: candidate k-itemset; L_{k-1}: frequent (k-1)-itemsets)
for each (k-1)-subset s of c
if s $\notin L_{k-1}$ then return 1;
return 0;

Algorithm 1. Apriori.

5.2 Implementation, Results and Discussion

We executed Apriori on a transactional dataset (see Table 1). It contains four transactions. Given a minsup of 2 transactions (i.e. 0.50), frequent itemsets are "bread, butter", "bread milk", "bread", "butter" and "milk" (Table 7). The whole process of find- ing these frequent itemsets is shown in Tables 2, 3, 4, 5 and 6.

If we consider Table 1 to represents the sales of a store, the knowledge which we can infer from Table 7 (which we reached after performing frequent itemset mining on Table 1 data) is that this store has a good sale of *bread*, *butter* and *milk* but not of *tea* and *candy*. If we consider that Table 1 represents the purchases at various time instants

Table 1. Transactional database.

TransactionID	Items_in_the_Transaction
T100	bread, butter, coffee
T101	butter, tea
T102	bread, milk, butter
T103	candy, bread, milk

Table 2. Candidate 1-Itemsets.

Item	Support_Count
{bread}	0.75
{butter}	0.75
{coffee}	0.25
{tea}	0.25
{milk}	0.50
{candy}	0.25

Table 3. Frequent 1-Itemsets.

Item	Support_Count
{bread}	0.75
{butter}	0.75
{milk}	0.50

Table 4. Candidate 2-Itemsets.

Item	Support_Count
{bread, butter}	0.50
{bread, milk}	0.50
{butter, milk}	0.25

Table 5. Frequent 2-Itemsets.

Item	Support_Count
{bread, butter}	0.50
{butter, milk}	0.50

Table 6. Candidate 3-Itemsets.

Item	Support_Count
{bread, butter, milk}	0.25

Table 7. All frequent itemsets and their support.

Frequent_Itemsets	Support_Count
{bread}	0.75
{butter}	0.75
{milk}	0.50
{bread, butter}	0.50
{bread, milk}	0.50

done by a single person (this data is collected by an *IoT system* like an IoT based almirah or fridge), then the knowledge which we can infer from Table 7 (which we reached after performing frequent itemset mining on Table 1 data) is of the eating habits, likes and dislikes of this person e.g. he purchases bread, butter and milk on a regular basis and that he likes to have bread with butter and so on.

We can also formulate association rules by applying this approach. We can explain this viewpoint by taking an example of buying together things with the things that have been bought already by a consumer. Mathematically, this complication can be explained as follows: Given a group of items K = $\{k_1, k_2, \ldots, k_m\}$ and a group of transactions D = $(d_1, d_2, \ldots, d_n)$ where $d_k \subseteq K$, a predefined peak values of confidence and support is used to find the group of association rules that are more than or equal to it. We can say that two meaningful conditions to examine the mining conclusions i.e. support and confidence, are explained beforehand by the user. Take an example of a transaction rule for purchasing tooth_brush and tooth_paste together, which is designated by {tooth_brush} ⇒ {tooth_paste}, where support is 20% and confidence is 80%, which implies 20% of consumers purchase tooth_brush and tooth_paste together while every consumer has a 80% chance of buying tooth_paste if the person has already purchased tooth_brush. Support is given by

$$\sup G \cup \mathrm{H} = \frac{\mathrm{C}(\mathrm{G} \cup \mathrm{H})}{\mathrm{n}} \tag{4.1}$$

and confidence is given by

$$conf(G \Rightarrow H) = \frac{C(G \cup \mathrm{H})}{c(G)} \tag{4.2}$$

where C(x) stands for the count of transactions in D that contain x.

Consider the example dataset of Table 1. The probability that if a customer purchases bread he will also purchase milk will be given by

$$conf(bread \Rightarrow milk) = \frac{C(bread \cup milk)}{C(bread)} \tag{4.3}$$

$$= \frac{0.50}{0.75} = 0.67$$

All this knowledge can on one hand help in understanding the behavior of a person and on the other hand can boost in the sales of a corporation.

6 Conclusions

This paper brought to fore *new areas* where IoT can be put to use. It also detailed the *problems* that come in the way of implementation of IoT. It also presented a *reference model* for IoT which will assist in collecting IoT data and also in performing data analytics on the collected data. This paper also discussed the *methods* that could be used for analyzing IoT data and deducing knowledge from it. Frequent patterns were mined from IoT data and the corresponding results were shown to be of great benefit.

References

Aggarwal, C.: Data Classification: Algorithms and Applications. CRC Press, Boca Raton (2014)

Agrawal, R., Srikant, R.: Fast algorithms for mining association rules. In: Proceedings of the 20th VLDB Conference, pp. 487–499 (1994)

Anumba, C., Wang, X.: Mobile and Pervasive Computing in Construction. Wiley, Hoboken (2012)

Atzori, L., Iera, A., Morabito, G.: The internet of things: a survey. Comput. Netw. **54**(15), 2787–2805 (2010)

Bandyopadhyay, D., Sen, J.: Internet of Things: applications and challenges in technology and standardization. Wireless Pers. Commun. **58**(1), 49–69 (2011)

Bin, S., Yuan, L., Xiaoyi, W.: Research on data mining models for the Internet of Things. In: Proceedings of the International Conference on Image Analysis and Signal Processing, pp. 127–132. IEEE (2010)

Bonomi, F., Milito, R., Zhu, J., Addepalli, S.: Fog computing and its role in the Internet of Things. In: Proceedings of the First Edition of the MCC Workshop on Mobile Cloud Computing, pp. 13–16. ACM (2012)

Cantoni, V., Lombardi, L., Lombardi, P.: Challenges for data mining in distributed sensor networks. In: Proceedings of 18th International Conference on Pattern Recognition, vol. 1, pp. 1000–1007. IEEE (2006)

Domingo, M.: An overview of the Internet of Things for people with disabilities. J. Netw. Comput. Appl. **35**(2), 584–596 (2012)

Erlich, Y.: A vision for ubiquitous sequencing. Genome Res. **25**(10), 1411–1416 (2015)

Haller, S., Karnouskos, S., Schroth, C.: The Internet of Things in an enterprise context. In: Future Internet Symposium, pp. 14–28. Springer, Heidelberg (2008)

Han, J., Pei, J., Kamber, M.: Data Mining: Concepts and Techniques. Elsevier, Amsterdam (2011)

Han, J., Pei, J., Yin, Y.: Mining frequent patterns without candidate generation. ACM Sigmod Rec. **29**(2), 1–12 (2000)

Keller, T.: Mining the Internet of Things-detection of false-positive RFID tag reads using low-level reader data (2011)

Kortuem, G., Kawsar, F., Sundramoorthy, V., Fitton, D.: Smart objects as building blocks for the internet of things. IEEE Internet Comput. **14**(1), 44–51 (2010)

Kulkarni, R., Forster, A., Venayagamoorthy, G.: Computational intelligence in wireless sensor networks: a survey. IEEE Commun. Surv. Tutor. **13**(1), 68–96 (2011)
LaDiega, G., Walden, I.: Contracting for the "Internet of Things": looking into the Nest. Eur. J. Law Technol. **7**(2) (2016)
Lindner, T.: The supply chain: changing at the speed of technology. Connected World
Lopez, T., Ranasinghe, D., Harrison, M., McFarlane, D.: Adding sense to the Internet of Things. Pers. Ubiquit. Comput. **16**(3), 291–308 (2017)
Lopez, T., Ranasinghe, D., Patkai, B., McFarlane, D.: Taxonomy, technology and applications of smart objects. Inf. Syst. Front. **13**(2), 281–300 (2011)
M&M Research Group: Internet of Things (IoT) & M2 M Communication Market: Advanced technologies, future cities & adoption trends, roadmaps & worldwide forecasts. Electronics.ca Publications, Technical report (2012)
Masciari, E.: A framework for outlier mining in RFID data. In: Proceedings of the 11th International Symposium on Database Engineering and Applications, pp. 263–267. IEEE (2007)
Mattern, F., Floerkemeier, C.: From the internet of computers to the Internet of Things. In: From Active Data Management to Event-Based Systems and More, pp. 242–259. Springer, Heidelberg (2010)
Miorandi, D., Sicari, S., DePellegrini, F., Chlamtac, I.: Internet of Things: vision, applications and research challenges. Ad Hoc Netw. **10**(7), 1497–1516 (2016)
Palattella, M., Accettura, N., Vilajosana, X., Watteyne, T., Grieco, L., Boggia, G., Dohler, M.: Standardized protocol stack for the internet of (important) things. IEEE Commun. Surv. Tutor. **15**(3), 1389–1406 (2013)
Poslad, S.: Ubiquitous Computing: Basics and Vision, pp. 1–40. Wiley, Hoboken (2009)
Rouse, M., Wigmore, I.: Internet of Things (IoT) (2013)
Shah, S., Amjad, M.: Lexical analysis of the Quran using frequent itemset mining. In: Proceedings of the 21st World Multi-Conference on Systemics, Cybernetics and Informatics, pp. 310–313 (2017)
Siegemund, F.: A context-aware communication platform for smart objects. In: Proceedings of the International Conference on Pervasive Computing, pp. 69–86. Springer, Heidelberg (2004)
Steinbach, M., Karypis, G., Kumar, V.: A comparison of document clustering techniques. In: KDD Workshop on Text Mining, vol. 400, no. 1, pp. 525–526 (2000)
Hussain, A., Keshavamurthy, B.: An enhanced communication mechanism for partitioned social overlay networks using modified multi-dimensional routing. Clust. Comput. (2018)
Tsai, C., Lai, C., Chiang, M., Yang, L.: Data mining for Internet of Things: a survey. IEEE Commun. Surv. Tutor. **16**(1), 77–97 (2014)
Vermesan, O., Friess, P.: Internet of Things: Converging Technologies for Smart Environments and Integrated Ecosystems. River Publishers, Aalborg (2013)
Zhang, L.: The business scale of communications between smart objects is tens of times the scale of communications between persons. Science Times (2009)

Securing IoT Using Machine Learning and Elliptic Curve Cryptography

Debasish Duarah(✉) and V. Uma

Department of Computer Science, Pondicherry University, Puducherry, India
debasishduarah@gmail.com, umabskr@gmail.com

Abstract. Internet of things (IoT), it is one of the most promising technology which is connecting billions of devices. As the connection among devices is increasing rapidly, data collection and data transmission also increasing which is leading to privacy and security issues. In this paper, we have discussed about IoT characteristics, security issues and security threats also proposed an idea as how we can classify the data using machine learning technique and how we can secure transmission using Elliptic curve cryptography.

Keywords: ECC · Precision · Random forest classification · Recall · Security

1 Introduction

Internet is a form of network where it connect several system together around the world. The concept behind "IoT" is to have communication among several devices via internet, to share their data about the way they use or about the environment where they resides. It is based on Radio frequency technology, which is the inter-relation of living and non-living beings which are provided with unique identifiers (UID) as shown in Fig. 1. IoT devices has the ability to interact among other devices over the network without any human interference.

There were time where most of the communication done by human was fixed and landline connection. The major problem of land-phone it that it has to book a call with the operator and the operator use to book a slot for the request when it is available, it can take hours/days. Then internet comes into the picture where it provide a mechanism to share information among others without regard to their geographical location. Introduction of Social-sites turn around the entire framework of data sharing, several websites coagulate themselves and share the information to maximum number of people.

Fig. 1. UID and IoT devices

IoT devices collect and share information, it has the ability to react based upon the situation and stimulate the result.

D. J. Hemanth et al. (Eds.): COMET 2019, LNDECT 35, pp. 482–491, 2020.
https://doi.org/10.1007/978-3-030-32150-5_46

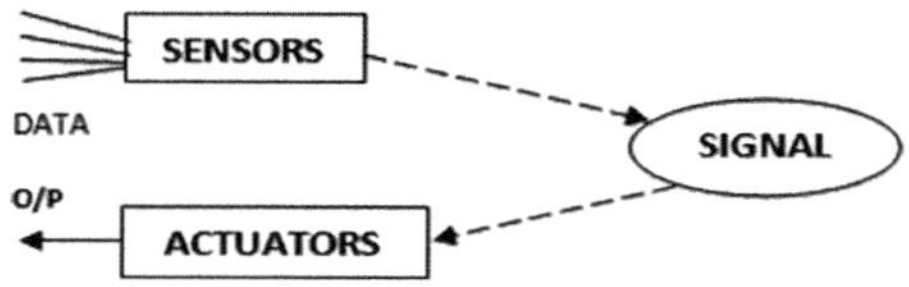

Fig. 2. Functioning of IoT devices

Basically Sensors collect data in Analog format, to process the data further it has to convert digital format. Aggregation and conversion of data is carried out by Data Acquisition System (DAS) and it is located nearer to the sensor and actuator as shown in Fig. 2. Then the data are been forwarded to IT processing system that may be located in remote office or other edge location. Zhou et al. [1] and Sain et al. [3] mainly discussed on features of IoT. To improve the security through encryption Venugopal et al. [4]. To mitigate security in honeypot enabled network La et al. [9] developed a game theory model. Li et al. [10] discussed as how to secure through intrusion detection method.

1.1 Characteristics

Intelligence: Internet of Things is a combinational of algorithm, software, hardware, using this it computes and has the ability to solve any given task.

Connectivity: There are several number of devices in this modern world, the concept of IOT states that it has to connect several devices such as cell phone, Bluetooth, RFID, digital machines to produce an optimum and accurate result.

Dynamic Nature: It has the ability to work on any given condition, irrespective to the geographical location, temperature and at any point of time.

Sensing: "Sensors" plays the important role, It does the sensing part, it has the ability to interact with the environment and it has the capability to collect the information from the environment.

Heterogeneity: As in IoT there are several numbers of devices interconnected, every devices has different configuration setup architecture, however it has the ability to interact among other devices and to form a unit.

Security: It is vulnerable to security as it is transparent and it has privacy issues.

1.2 Security Issues

Data Encryption: IoT collects enormous number of data every day, Data has to retrieve and process, while doing so data has to be encrypted as there might be data which are personal thus it has to be secured.

Data Authentication. Proper authentication is a must to identify legitimate users through which data can be protected, else a person who is not authorized can misuse the data.

Side-Channel Attack. It is a type of attack where It is more focused on how the information are presented rather than the information, data like timing information, power consumption or electromagnetic leak can lead to side channel attack.

Hardware. When an industry built a basic architecture then it has every chances to be pirated as IOT devices are always independent and transparent. To overcome this industries are building complex architecture due to which it's leading to more battery consumption.

Constrained Devices. The IoT devices those are been individual setup, basically has low capabilities to perform complex encryption and decryption as they have limited amount of storage, memory and processing capabilities.

Device Updates. After every certain period, Updating is required to an older device such that it could equally likely work together along with the new device installed, IoT devices may not able to do automatically thus it has to bring/unplug and do the update manually due to which for that certain period data are missed.

Safe-Storage. Now a days it is easier for data storage online/offline but the major concern is the Back-up and most importantly where the data is to Back-up.

1.3 Security Attacks

Denial-of-Service. It's a type of attack where the authorized users are restricted to use the service. In this case basically attacker send multiple number of request to the server without any return address so when a server compute and unable to find, server want to close the connection and then again the attacker send the request in which server become busy resolving this.

Jamming. It's a type of DoS attacks which is carried out in the operation phase in authorized wireless networks.

Eavesdropping. It is an unauthorized interrupt to a private communication between user-devices and device-device, by observing the activities of authorized user, how and what they are doing.

Botnets. Attacker distribute malware and takes control over the system and exploit private information, example- online banking data, etc.

Routing Attack. In Network Layer, Router plays and important role to en-route the network, there could be attack such as DoS, Brute force, Sybil Attack.

1.4 Trade off

(a) High power consumption, High range, High bandwidth:

To send data over a wireless network it takes a lot of power.
Ex: Mobile phones: charge need to be given after every 1–2 days.
Cellular phones: towers.
Satellite: devices/sensors that are in ocean.

(b) Low power consumption, low range, high bandwidth:

To send enormous data by using low power, then the range has to be reduced.
Ex: Wi-Fi, Bluetooth, Ethernet.
Ethernet: as long as the wire the data can be transferred.

(c) Low power consumption, high range, low bandwidth:

To have a good range with less power consumption, the amount of data need to be decreased too.
Ex: LoPWAN: moisture sensors for Agriculture purpose.

2 Related Work

SI No	Area	Title	Issue address	Methodology	Conclusion/Future Work
1	**Survey: features and protocols**	The effect of IoT New Features on Security and Privacy: New threats, Existing Solution and challenges yet to resolved [1]	**Interdependence:** Over privileged **Diversity:** Insecure protocols **Constrained:** Man in the middle, Insecure system **Myriad:** IoT botnets, DDos **Unattended:** Remote Attack **Intimacy:** Privacy leak **Mobile:** Malware Propagation	**Interdependence:** Context-based permission system **Diversity:** IDS & IPS **Constrained:** light weight algorithms **Myriad:** IDS **Unattended:** Remote attestation, light weight trusted execution **Intimacy:** Homographic Encryption **Mobile:** Dynamic changes of security configuration	Analysed and Discussed features of IOT also provided optimum solution to mitigate security and privacy issues
2		Research Directions for the Internet of Things [2]	**Massive scaling:** Authentications, maintenance, protection **Architecture:** Connectivity, control, Communication **Big Data:** Raw data into usable knowledge **Robustness:** Clock Synchronization	**Massive Scaling**: Protocols and architecture **Big Data**: History of user activities **Robustness:** Entropy services	**Challenges to overcome:** Human in the loop control System identification to derive models of human behaviour For co-operation of human behaviour model with formal methodology of feedback model

(*continued*)

(*continued*)

SI No	Area	Title	Issue address	Methodology	Conclusion/Future Work
3		Survey on security in Internet of Things: State of the art and challenges [3]	Security and privacy in the application interface Insecurity of data at each IOT layer	**Cryptographic methods IOT and security protocol:** IPsec, IPv6, CoAP, 6LoWPAN, UDP, DTLS	Requires customized security and privacy on each level
4	**Encryption**	A survey: authentication protocols for wireless sensor networks in the internet of things: keys and attacks. Doaa Alrababah, et al.	**Security attacks:** Masquerading MITM Forgery attack Replay attacks	**Authentication protocols:** Message authentication code, Asymmetric encryption, Time stamp	Authentication protocols to use them as their needs and also to deploy more than one protocol if needed to mitigate the security
5		Lightweight cryptographic solution for IoT – an assessment [4]	**Issues:** Bandwidth, Security, Power, Scalability	Identity based encryption Elliptic curve cryptography using Digital signature	Identity based Encryption provide optimal security
6		Improving the security of internet of things using encryption algorithms [5]	Integrity Repudiation confidentiality	Hybrid encryption method AES and ECC combined Hash Encryption	Availability of encryption in IOT with deduced attacks and improved security
7		IoT device security based on proxy re-encryption [6]	Security while transmission of data	Attribute based encryption Re-encryption using public key of receiver end	To improve security, efficiency conventional cipher algorithm is used on lightweight devices
8		High-Performance and Lightweight Lattice-Based Public-Key Encryption [7]	Optimum cryptography technique w.r.t cost and energy consumption	Ideal lattice-based encryption scheme ARM and AVR microcontroller Ring-LWE based pubic key encryption	Gaussian noise distribution in the Ring-LWE based encryption scheme with a binary distribution
9	**Encryption**	Privacy-preserving quantified self: secure sharing and processing of encrypted small data Hossein Shafagh, et al.	Data Storage	Fully homographic encrypt Partial homographic encrypt Encrypted data sharing	Given users control over cloud data and discussed, importance of end to end encryption also the need of cryptographically enforced data sharing features
10		An identity based encryption using Elliptic curve cryptography for secure M2M communication [8]	Privacy and security in M2M	Elliptic curve cryptography ID based encrypt and decrypt scheme based on Tate pairings	Developing Attribute based IDB, which has hierarchical structure to solve privacy issues by using fast cryptographic algorithm which is based on BN curves for lightweight application
11	**Honeypot**	Deceptive Attack and Defense Game in Honeypot-Enabled	Defending against attacks in honeypot enabled networks	Bayesian game of incomplete operation	A theoretical game model developed to analyse the problem of

(*continued*)

(*continued*)

SI No	Area	Title	Issue address	Methodology	Conclusion/Future Work
		Networks for the Internet of Things [9]		Intrusion detection system	deceptive attack and secure devices in a honeypot enabled network
12	**Intrusion detection**	Survey of intrusion detection method in Internet of things [10]	Eavesdropping, Denial of service, Replay attack	Radio frequency Identification Intrusion Detection: single & batch, ALOHA, Static & Dynamic, light weighted algorithm	To increase the service life of tags, need to design effective and reliable detection algorithm
13		A Trust Based Distributed Intrusion Detection Mechanism for Internet of Things [11]	**Routing attacks:** Selective forwarding Sinkhole Version number	Destination oriented directed acyclic graph	Centralize mechanisms uses the trust management technique to detect intruders' nodes in the system, once an intruder is identified it needed to remove from the network
14	**DDOS**	Denial-of-Service detection in 6LoWPAN based Internet of Things [12]	Denial of service attack (jamming, routing attack, eavesdropping, cloning of things, application layer attack)	Light-weight security solution with light weight PKC Secure bootstrapping. DoS detection architecture Intrusion detection system	To monitor large networks distributed sniffing Security incident and event management system
15	**Security and privacy**	A survey on IOT application security challenges and counter measures [13]	**Security challenges:** Confidentiality, Authentication, Privacy, Access control, Trust Policy enforcement	**Cryptographic algorithm:** AES, DES, RSA, Anonymization technique	To provide encryption and decryption, modified RSA algorithm is been used
16	**Biometric**	F On the Road to the Internet of Biometric Things: A Survey of Fingerprint Acquisition Technologies and Fingerprint Databases	**Security vulnerabilities:** Authentication	Electronic fingerprint verification	to build a digital camera acquired fingerprint set in perfect condition that can be customized easily and can be used for the testing the impact of different degradation on the accuracy of different fingerprint recognition system

3 Proposed Approach

When IoT Sensors captures data, using classification technique those data can be classified and then it is forwarded encryption using Elliptic Curve cryptography (Asymmetric algorithm) only if the data are benign as shown in the Fig. 3.

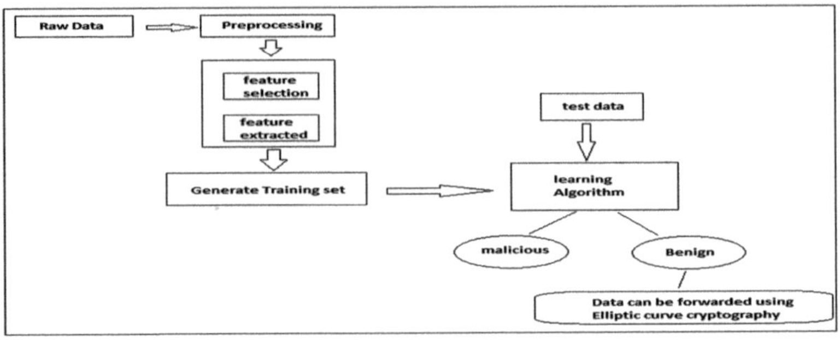

Fig. 3. System architecture

Raw data: It is a primary data that has been collected from its source to process further, basically these are the data that is not been processed.
Pre-processing: It converts the raw data into a format where it can be easily understandable. Usually raw data are incomplete, inconsistence and lacking in certain behaviour or trends. It process this data and eliminate errors that occurs.
Feature selection: It is for filtering irrelevant or redundant features from a dataset, it keeps a subset of the original features.
Feature extraction: Process of transforming the input data into a set of features which can very well represent the input data. It creates a new dataset from the original features.
Generate Training dataset: It is used for training and testing purpose. The model follows this set of rules that is been defined in the training dataset and definition too. Test set is the data set on which you apply your model and see if it is working correctly and yielding expected and desired results or not. Test set is like a test to your model.

3.1 Learning Algorithm

Random Forest Classifier: It is an ENSEMBLE algorithm that combines one or more algorithm of same or different kind for classifying object. Basically it forms a decision tree in which some subset are selected from the training set. To have the final class it aggregate votes from different decision tree.

$$\mathbf{Precision} = \frac{\boldsymbol{Tp}}{\boldsymbol{Tp} + \boldsymbol{Fp}} \tag{1}$$

Equation 1 gives the precision value, Tp = true positive, Fp = false positive.

$$\mathbf{Recall} = \frac{\boldsymbol{Tp}}{\boldsymbol{Tp} + \boldsymbol{Fn}} \tag{2}$$

Equation 2 gives the recall value where, Tp = true positive, Fn = False Negative.

Table 1. Comparison of key size of ECC and RSA as recommended by NIST.

ECC key size	RSA key size
106	512
163	1024
256	3084
384	7680
106	512

$$\textbf{F1 Score} = 2\frac{\textbf{\textit{Precision}}\,.\,\textbf{\textit{Recall}}}{\textbf{\textit{Precision}} + \textbf{\textit{Recall}}}, \tag{3}$$

Equation 3, Harmonic mean of precision and recall.

Malicious: There may be certain point where data may be not effective it may contain viruses due to which rest of dataset could hamper too, and transmitting this type of dataset will be waste of time, so before proceeding to encryption and transmission process this kind of dataset will be discarded.

Benign: If a dataset is found to be reliable and clean then it will be forwarded to transmission phase.

Algorithm 1: Classification

Step 1. Consider Raw Data.
Step 2. Apply Pre-processing.
Step 3. Perform Feature Selection and Extraction.
Step 4. Generate Training Dataset.
Step 5. Apply learning algorithm in Test Data.
Step 6. Obtain the Classification.
Step 7. If Benign → Forward.
Step 8. If Malicious → Discard.

Algorithm 1 provides the flow of classifications how the process takes place. Once the classification is completed and found that the data are benign, then it is forwarded for encryption using Elliptic curve Cryptography such that it could transmit to other devices.

ECC is used because of its complexity nature and computing efficiency i.e. it uses scalar multiplication rather than exponentiation or multiplication in the finite field, also it uses lower key size with compare to RSA which provide equivalent security (Table 1).

Weierstrass normal form of elliptic curve

$$y^2 = x^3 + ax + b \tag{4}$$

$$\text{Where } 4a^3 + 27b^3 \neq 0 \tag{5}$$

Equation 4 describes that elliptic curve is a plain curve with a, b as real numbers and Eq. 5 describe as the curve is a non-singular with no self-intersections or isolated point.

Algorithm 2: Generation of Key

d: Random Number within the range[1, n − 1] also to be treated as private key.
P: it's the point on the Elliptic Curve

$$\delta = (\mathbf{P}) * (\mathbf{d}) \tag{6}$$

Algorithm 2 is used to generate a key in which both the sender and receiver are agreed upon. Equation 6 generates the public key which is the product of the point on the curve and a random number within the range of [1, n − 1].

Algorithm 3: Encryption

Step 1. m: message that need to be sent, it has the point "M" on the curve "E"
Step 2. k: within range [1, n − 1]
Step 3. $\varphi = (\mathbf{k}) * (\mathbf{P})$ (7)
Step 4. $\omega = \mathbf{M} + (\mathbf{k}) * (\boldsymbol{\delta})$ (8)
Step 5. φ, ω: are the ciphers and need to forward to the Receiver.

The message needed to be encrypted before transmission process takes place, Algorithm 3 takes the responsibility for it. While encryption two sets of Cipher text is generated, Eq. 7 is the product of the k that is within the range of [1, n − 1] with the point on the curve, the next cipher text, Eq. 8 is the product of the key and the public key and then it is added with the message. It is done by Elliptic curve Integrated Encryption Scheme.

Algorithm 4: Decryption

$$\mathrm{M} = (\omega - (\mathrm{d} * \varphi)) \tag{9}$$

4 Conclusion

IoT has become part of our life, it has to be protected and mitigate security to a maximum extent. There is every chance that all data that we receive through IoT devices may not be real it might be malicious and could leak our personal data. Thus, Data classification is done to check the Data accuracy, if the score is more than the threshold then it can be forwarded to other IoT devices. ECC is adopted for key exchange in transmission process because it can provide equivalent security with lower key bits than RSA. The advantage of this idea is that it computes only True data after receiving from the sensor as it omits malicious/corrupt data.

References

1. Zhou, W., Jia, Y., Peng, A., Zhang, Y., Liu, P.: The effect of IoT new features on security and privacy: new threats, existing solution and challenges yet to resolved. IEEE Internet Things J. **6**, 1606–1616 (2018)
2. Stankovic, J.A.: Research directions for the Internet of Things. IEEE Internet Things J. **1**(1), 3–9 (2014). https://doi.org/10.1109/jiot.2014.2312291
3. Sain, M., Kang, Y.J., Lee, H.J.: Survey on security in Internet of Things: state of the art and challenges. In: International Conference on Advanced Communication Technology (ICACT) (2017). ISBN 978-89-968650-9-4
4. Venugopal, M., Doraipandian, M.: Lightweight cryptographic solution for IoT – an assessment. Int. J. Pure Appl. Math. **117**, 511–516 (2017). ISSN 1314-3395
5. Yousefi, A., Jameii, S.M.: Improving the security of Internet of Things using encryption algorithms. In: International Conference on IoT and Application (ICIOT) (2017). ISBN 978-1-5386-1698-7
6. Kim, S.H., Lee, I.Y.: IoT device security based on proxy re-encryption. J. Ambient Intell. Humanized Comput. **9**(4), 1267–1273 (2017). https://doi.org/10.1007/s12652-017-0602-5
7. Buchmann, J., Göpfert, F., Güneysu, T., Oder, T., Pöppelmann, T.: High-performance and lightweight lattice-based public-key encryption. In: ACM International Workshop on IoT Privacy, Trust, and Security, pp. 2–9. ISBN 978-1-4503-4283-4
8. Adiga, B.S., Balamuralidhar, P., Rajan, M.A.: An identity based encryption using Elliptic curve cryptography for secure M2 M communication. SecurIT Proceedings of the First International Conference on Security of Internet of Things, pp. 68–72 (2012). ISBN 978-1-4503-1822-8
9. La, Q.D., Quek, T.Q.S., Lee, J., Jin, S.: Deceptive attack and defense game in honeypot-enabled networks for the Internet of Things. IEEE Internet Things J. (2016). https://doi.org/10.1109/jiot.2016.2547994
10. Li, C., Li, Q., Wang, G.: Survey of intrusion detection method in Internet of Things. J. Netw. Comput. Appl. **84**(C), 25–37 (2014). https://doi.org/10.1016/j.jnca.2017.02.009
11. Khan, Z.A., Herrmann, P.: A trust based distributed intrusion detection mechanism for Internet of Things. In: International Conference on Advanced Information Networking and Applications (AINA) (2017). ISBN 978-1-5090-6029-0
12. Kasinathan, P., Pastrone, C., Spirito, M.A., Vinkovits, M.: Denial-of-Service detection in 6LoWPAN based Internet of Things. In: IEEE International Conference on Wireless and Mobile Computing, Networking and Communications (WiMob) (2013). ISBN 978-1-4799-0428-0
13. Pawar, A.B., Ghumbre, S.: A survey on IOT application security challenges and counter measures. In: International Conference on Computing, Analytics and Security Trends (2016). ISBN 978-1-5090-1338-8
14. Al-alem, F., Alsmirat, M.A., Al-Ayyoub, M.: On the road to the internet of biometric things: a survey of fingerprint acquisition technologies and fingerprint databases. In: ACS/IEEE International Conference on Computer Systems and Applications (AICCSA 2016) (2016). ISBN 978-1-5090-4320-0

An Automated Face Retrieval System Using Grasshopper Optimization Algorithm-Based Feature Selection Method

Arun Kumar Shukla(✉) and Suvendu Kanungo

Department of Computer Science and Engineering, Birla Institute of Technology, Mesra, Ranchi, Allahabad Campus, Allahabad, India
aks.jit@gmail.com, s.kanungo@bitmesra.ac.in

Abstract. Facial image retrieval using its contents is one of the major areas of research because of the exponential increase of multimedia data over the Internet. However, due to high dimensional features and different variations available in the images, it becomes a challenging task to obtain the relevant and non-redundant features. Therefore, for making the facial retrieval system more accurate and computationally efficient, the selection of prominent features is an important phase. In this paper, the grasshopper optimization algorithm has been used to obtain the relevant attributes from the high dimensional features vector. For the same, Oracle Research Laboratory database of faces is used. The experimental values show the efficacy of the proposed method of feature selection which eliminates the maximum 83% features among the other considered methods and the accuracy of the facial retrieval system also increases to 91.5%.

Keywords: Face retrieval system · Feature selection · Clustering · Grasshopper optimization algorithm

1 Introduction

As the use of social media, surveillance cameras, mobile platforms, and many other online applications are increasing rapidly, it becomes difficult to retrieve the multimedia contents on the Internet. Out of various multimedia contents, recognition of facial images over Internet is an important research area of computer vision. Face recognition system is also widely used in various application areas like, e-passport, terrorist recognition, ID verification solutions, bio-metric identification, social media, and deployment of various security services. Various face identification methods are introduced in the literature. The fisher face method is being used for nonlinear template matching by number of researchers [1, 2]. Cooper et al. [3] introduces the mixture model of latent variables and Yi et al. [4] used eigen faces based principal component analysis methods. Similarly, linear discriminate analysis [5], tensor representation based multi-linear subspace learning methods [6], and neural network based dynamic link matching methods [7, 8] are some of the popular face retrieval systems. However, the effectiveness of these approaches highly depends on the extracted features.

D. J. Hemanth et al. (Eds.): COMET 2019, LNDECT 35, pp. 492–502, 2020.
https://doi.org/10.1007/978-3-030-32150-5_47

Generally, for the images, high dimensional features are extracted which makes these approaches computationally inefficient. Therefore, an efficient feature selection method is a prime step for making the content-based image retrieval (CBIR) based face identification method more accurate and computationally efficient [9].

Generally, a CBIR based face identification method contains four steps. The first step acquires the images from the large databases and pre-processed for removal of noise, illumination variations, or background information [10]. Second step includes the extraction of attributes from the images. The third step selects the relevant features which are finally given to the classifiers for matching of relevant images. However, the efficiency of CBIR systems highly depend on the selection of relevant and non-redundant features [11] as the feature extraction methods returns a high dimensional feature which led to the "curse of dimensionality" problem. This may decay the accuracy of a classifier and also become computationally intensive [12] because of the irrelevant attributes from the high dimensional feature space [13]. To deal with the problem of "curse of dimensionality", an efficient feature selection method is required. The various feature selection methods, available in the literature, can be classified into three categories, namely wrappers, embedded, and filters methods [14]. The wrapper methods use the greedy search approaches for the feature selection while embedded methods use the supervised learning algorithms [14]. The filter methods use the features as the class variables which may results in poor performance on selected classifiers [11]. The embedded and wrapper algorithms are computationally expensive as compared to the filter methods [11].

The feature selection methods generally obtain the optimal set of features from high dimensional features set using some predefined criteria. For a feature set having N features, an exhaustive search method will find at most 2^N features subsets which is a very large number [15] and requires extensive computation. Therefore, the exhaustive search methods are generally considered as NP complete problems and can efficiently be solved through various meta-heuristics algorithms [16–19]. Therefore, in this paper a new method for optimal features selection using a meta-heuristic algorithm is introduced which is further used for efficient facial recognition system. The meta-heuristic algorithms are inspired from nature behavior and efficiently search the solution space. The solution of many difficult real-world problems have been successfully provided by the different meta-heuristic algorithms [20–25]. Some of recent meta-heuristic algorithms are improved biogeography-based optimization (IBBO) [26], intelligent gravitational search algorithm (IGSA) [27], whale optimization algorithm (WOA) [28], Hybrid step size based cuckoo search [29], spider monkey optimization (SMO) [30], sine cosine algorithm (SCA) [31] and grasshopper optimization algorithm (GOA) [32].

GOA works according to the behavior of grasshoppers and performs well for computationally expensive optimization problems having constrained and unconstrained objectives. GOA has shown better results in different application domains such as, computer vision, biomedical data mining, clustering problems and many others. Lukasik et al. [33] used the GOA in clustering of data and analyzed it against other state-of-the-art methods. A modified multi-objective GOA has been introduced by Tharwat et al. [34] to solve the constrained and unconstrained problems with multiple dependent objectives. Furthermore, Barmen et al. [35] presented a GOA and support vector machine (SVM) based hybrid method for the forecast of short-term load in Assam, India. Ibrahim et al. [36] also utilized GOA for optimizing the SVM parameters and tested the performance on biomedical data set of Iraqi cancer patients.

Therefore, in this paper, the strength of GOA has been utilized to obtain the relevant and non-redundant features from the high dimensional facial image features which are further used in face recognition system. For extracting the features, AlexNet convolutional neural network has been used and fed to the new feature selection method for obtaining the optimal features set. Furthermore, the classifier is trained using these optimal features set which is used to recognize the facial images. The efficiency of the proposed feature selection method has been compared with five different state-of-art classifiers, namely ZeroR, random forest (RF), linear discriminant analysis, K-nearest neighbor (KNN), and SVM. The performances of all considered methods have been evaluated on Oracle Research Laboratory (ORL) face database.

The organization of the rest of the paper is as follows. The brief of the AlexNet and GOA is presented in Sect. 2. Section 3 discusses the various steps of the proposed CBIR face recognition method. Experimental results are shown and discussed in Sect. 4 followed by the conclusion in Sect. 5.

2 Preliminaries

This section presents the AlexNet and GOA methods used for features extraction and feature selection respectively from face recognition.

2.1 AlexNet

AlexNet is one of the popular convolutional neural networks proposed by Krizhevsky et al. [37] in the ImageNet Large-Scale Visual Recognition Challenge (ILSVRC) and is used in various image classification problems a pre-trained model [38]. AlexNet extracts the features automatically from the image without intervention of any human. It includes many layers like input, convolution, activation, pooling, dropout, and fully connected layers. The convolution layer extracts the features from the images which are further fed to rectified linear unit (ReLU) activation function to add the non-linearity in the system. Max-pooling layer reduces the dimension of extracted features by taking the maximum value under a filter of size 3×3. The drop-out layer in the AlexNet is used to randomly set the d% of hidden nodes to zeros. Fully connected layer is the last layer of AlexNet which converts the two-dimensional features to a single dimensional feature vector. The architecture of AlexNet is presented in Fig. 1 which contains five convolution layers along with ReLU activation layers, three layers of max-pooling, two layers of dropout, two fully connected and ReLU layers, and one fully connected layer. The output layer is known as softmax layer and is used to label the input image into different classes.

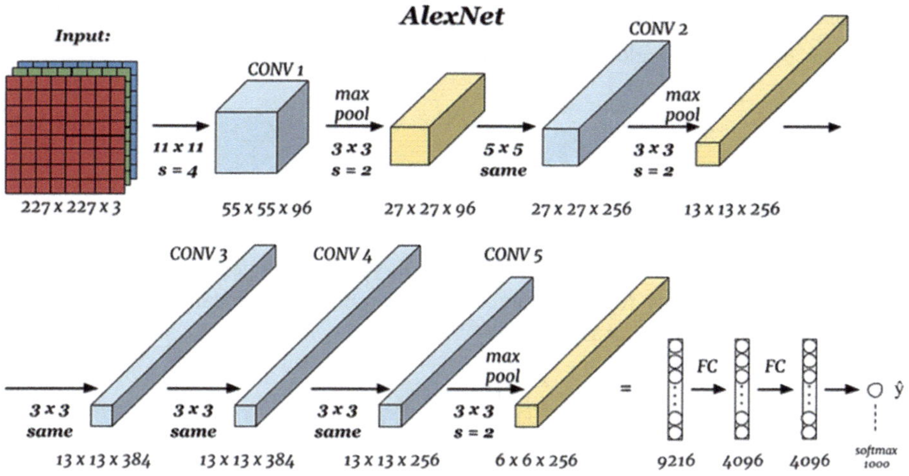

Fig. 1. The architecture of AlexNet [39].

2.2 Grasshopper Optimization Algorithm (GOA)

GOA [32] is a recent algorithm for optimization and is inspired from the behavior of grasshoppers. It is generally used to search the optimal solutions for the constrained and unconstrained problems. In the grasshopper optimization algorithm, the positions of grasshoppers are taken as solutions or individuals in the population. There are two phases in the life cycle of grasshoppers, namely larval and adulthood phases. In the larval phase, the movement of grasshoppers is slow which represents the exploitation of the search space. On the other hand, in the adulthood phase movements become large which makes the abrupt changes in their positions and is considered as the exploration. Mathematically, the position (X_i) of ith grasshopper can be depicted as Eq. (1).

$$X_i = S_i + G_i + A_i \tag{1}$$

where, S_i, G_i, and A_i are the social interaction, gravity forces, and wind advection influencing factors respectively. The social interaction factor is calculated as Eq. (2).

$$S_i = \sum_{\substack{j=1 \\ j \neq i}}^{N} s(d_{ij})\widehat{d_{ij}} \tag{2}$$

$$s(r) = fe^{\frac{-r}{l}} - e^{-r} \tag{3}$$

where, N is the number of grasshoppers and$s$ shows the social attraction. It is measured by the distance (d_{ij}) between the ith and j^{th} grasshopper. Further, f and l are the attraction intensity and the attractive length scale respectively. The distance (d_{ij}) and the unit vector ($\widehat{d_{ij}}$) are calculated as follows.

$$d_{ij} = \left|x_j - x_i\right| \tag{4}$$

$$\widehat{d_{ij}} = \frac{x_j - x_i}{d_{ij}} \tag{5}$$

where, x_i is the ith grasshopper in the population of grasshoppers.

Moreover, the gravity force (G) and wind advection (A) are calculated by Eqs. (6) and (7) respectively for i^{th} grasshopper.

$$G_i = -g\widehat{e_g} \tag{6}$$

$$A_i = u\widehat{e_w} \tag{7}$$

where, g and u are the constants. The unit vectors towards the earth center and wind direction are represented by $\widehat{e_g}$ and $\widehat{e_w}$ respectively. Generally, the effect of G and A on position change in a grasshopper is minimal, hence are ignored in the position vector. A target vector ($\widehat{T_d}$) is also used in the position equation of each grasshopper. Therefore, the final mathematical equation of position becomes as follows.

$$X_i^d = c\left(\sum_{\substack{j=1 \\ j \neq i}}^{N} c\frac{ub_d - lb_d}{2} s\left(d_{ij}\right)\widehat{d_{ij}}\right) + \widehat{T_d} \tag{8}$$

where, X_i^d is the dth dimension of ith grasshopper. The lower and upper bounds in the dth dimension are represented by lb_d and ub_d respectively. $\widehat{T_d}$ depicts the dth dimension of the best solution and is known as target vector. The c is a controlling parameter to modify the behavior of exploitation and exploration of the grasshoppers and is calculated by Eq. (9).

$$c = c_{max} - l\frac{c_{max} - c_{min}}{L} \tag{9}$$

where, c_{max} and c_{min} represents the maximum and minimum values respectively. Generally, c_{max} is kept *1* and c_{min}*0.00001*. l and L depict the current and maximum number of iterations respectively. The complete algorithm of GOA is given in Algorithm 1.

Algorithm 1 Grasshopper Optimization Algorithm
Input: Randomly initialized population of grasshoppers X_i $(i=1, 2,\ldots, N)$, each of d dimensions.
Output: Optimal solution.
Initialize c_{max}, c_{min} and L;
Measure the fitness (*fit*) of each solution and dominate the best solution as target vector T;
While $l<L$ **do**
 Find c using Eq. (9);
For each grasshopper **do**
  Calculate the distance of grasshoppers $(d_{ij}, \widehat{d_{ij}})$;
  Update the current position of grasshopper by Eq. (8);
  Apply the boundary checks on each solution;
 End for
End while

3 Proposed Method

This work presents a new feature selection method using the grasshopper optimization algorithm for effective face recognition in a CBIR. The proposed method has been depicted in Fig. 2 which contains major three steps, namely feature extraction, feature selection, and classifier. First the facial images are fed to features extraction phase in which a pre-trained AlexNet deep learning model is used. For the same, the output of the last hidden layer of AlexNet is used as the extracted features for input images. In the second phase, the proposed feature selection method is used to select the relevant and non-redundant features from the features set and is discussed in the Sect. 3.1. Once the relevant features are selected, a classifier is used which is trained using the training dataset image features. The trained classifier is used to identify the test facial images. For the same, six different classifiers, namely SVM, kNN, linear discriminant analysis (LDA), ZeroR, and random forest (RF) classifiers are evaluated.

Fig. 2. The proposed face recognition method.

3.1 Feature Selection

The proposed method uses the grasshopper optimization algorithm (GOA) to obtain the relevant features. In GOA based feature selection method, each grasshopper is represented as a vector which is randomly initialized between [0, 1]. The dimension (*d*) of each solution vector is equal to the number of features returned by AlexNet. The value of each dimension in the solution vector has either a value *0* or *1* if the respective feature is not selected or selected respectively. Mathematically, the ith solution (X_i) can be represented as Eq. (10).

$$X_i = \{x_1, x_2..., x_k, \ldots, x_d\}, \text{ where } k = \{1, 2, \ldots, d\} \tag{10}$$

where, *i* = *1, 2, ……, N*. Following steps are followed to obtain the optimal set of features.

1. Each dimension of an individual or solution is digitized either to *0* or *1* using a threshold value (ρ) i.e., if ρ is bigger than X_i^k then it is replaced by *1* otherwise it is set to *0*. The mathematical formula is as follows. In this work, the value of ρ is empirically selected to *0.7*.

$$X_i^k = \begin{cases} 1 & X_i^k < \rho \\ 0 & X_i^k \geq \rho \end{cases} \tag{11}$$

2. Measure the fitness of each individual by considering only those features whose corresponding solution dimension is *1*. The fitness is measured as the accuracy returned by the SVM classifier with 10-fold cross validation.
3. Use GOA to update the position of each individual till the stopping condition is not achieved.
4. Finally, the best returned solution by GOA is considered as an optimal solution and the features having corresponding dimension value *1* are selected as relevant features.
5. The obtained features are used to train and test the classifiers.

4 Experimental Results

The performance of GOA based feature selection method for face identification is analyzed against other state-of-the-art meta-heuristic based methods. For feature extraction method, Alexnet is used which extracts *1000* features. To select the relevant features out of these *1000* features, GOA based feature selection method is used whose performance is tested using MATLAB on Intel core i7 processors with 8 GB RAM. For the facial images, Oracle Research Laboratory (ORL) dataset is used which was provided by AT & T laboratories, Cambridge [40]. There are *400* images of *40* persons in the facial image dataset i.e., *10* images per person, having various facial expressions. The *5* images of one representative person are shown in Fig. 3. The size of each image is *92* × *112* pixels with *256* grey levels per pixel. All the images are of PGM format. For training and validating the classifiers, stratified random sampling is used to select the training and testing datasets.

Fig. 3. Representative images of a human from ORL dataset [40].

The performance of the proposed GOA based feature selection method has been compared with SMO and sine cosine algorithm (SCA) based feature selection methods against the number of selected features. Moreover, the relevancy of the selected features is checked by feeding them to five classifiers, namely SVM, RF, LDA, kNN, and ZeroR and corresponding accuracy are tested for Oracle Research Laboratory face database. Table 1 shows the number of selected features and respected accuracy return by each classifier. From the table, it can be observed that the GOA based feature selection method removes the maximum number of features i.e., *83%* which is maximum than other considered methods. Furthermore, each classifier gives the best accuracy for the selected features returned by GOA based feature selection method. The best accuracy of *91.5%* is given by SVM classifier for the features returned by GOA based feature selection method.

Table 1. The comparison of the proposed and considered feature selection methods in terms of number of features and corresponding accuracy returned by the classifiers.

	Number of features selected			
Method–>	None	SCA	SMO	GOA
	1000	221	198	**170**
Classifier	Accuracy			
SVM	83.7	89.1	89.4	**91.5**
LDA	78.6	87.6	88.2	**90.7**
RF	81.9	89.5	90.1	**90.9**
kNN	79.3	85.9	86.5	**88.6**
ZeroR	63.2	65.7	65.7	**65.7**

An efficient feature selection method also reduces the computational cost of training of a classifier. For the same, the computational efficiency of the proposed method has been measured for the selected features and is shown in Table 2. From the table, it can be seen that each classifier takes minimum number of computational times for the features returned by the proposed GOA based feature selection method. Hence, it can be stated that the proposed feature selection method selects the relevant features which reduces the time complexity and also increases the classification accuracy.

Table 2. The computational time taken by the classifiers on the selected features.

Method	Selected features	SVM	LDA	RF	KNN	ZeroR
None	1000	10.11	4.12	10.21	3.29	0
SCA	221	4.23	3.13	5.33	2.81	0
SMO	198	4.01	2.97	5.11	2.71	0
GOA	**170**	**3.18**	**2.28**	**4.16**	**2.64**	**0**

5 Conclusion

This work presents a new feature selection approach using a grasshopper optimization algorithm. The proposed approach obtains the optimal features from Oracle Research Laboratory face database images. For the same, Alexnet convolutional neural network is utilized for extracting the features from the facial images. The proposed GOA based feature selection method has been tested against recent meta-heuristic based algorithms in terms of selected features and computational cost. The proposed method eliminates the maximum number of features i.e., *83%* among the considered methods. The relevancy of the selected features is tested on five classifiers for identification of faces. The SVM classifier gives the best accuracy for the selected features by the proposed GOA based feature selection approach. Hence, the proposed method outperforms the existing methods in terms of accuracy and computational cost.

References

1. Zafeiriou, S., Petrou, M.: 2.5 D elastic graph matching. Comput. Vis. Image Underst. **115**(7), 1062–1072 (2011)
2. Senaratne, R., Halgamuge, S., Hsu, A.: Face recognition by extending elastic bunch graph matching with particle swarm optimization. J. Multimedia **4**, 204–214 (2009)
3. Cooper, H., Ong, E.-J., Pugeault, N., Bowden, R.: Sign language recognition using subunits. In: Gesture Recognition, pp. 89–118. Springer (2017)
4. Yi, S., Lai, Z., He, Z., Cheung, Y.-M., Liu, Y.: Joint sparse principal component analysis. Pattern Recogn. **61**, 524–536 (2017)
5. Liu, C., Wechsler, H.: Enhanced fisher linear discriminant models for face recognition. In: Proceedings of the Fourteenth International Conference on Pattern Recognition 1998, vol. 2, pp. 1368–1372. IEEE (1998)
6. Lin, C., Long, F., Zhan, Y.: Facial expression recognition by learning spatiotemporal features with multi-layer independent subspace analysis. In: 2017 10th International Congress on Image and Signal Processing, BioMedical Engineering and Informatics (CISP-BMEI), pp. 1–6. IEEE (2017)
7. Ding, C., Tao, D.: Trunk-branch ensemble convolutional neural networks for video-based face recognition. IEEE Trans. Pattern Anal. Mach. Intell. **40**(4), 1002–1014 (2017)
8. Lu, J., Wang, G., Zhou, J.: Simultaneous feature and dictionary learning for image set based face recognition. IEEE Trans. Image Process. **26**(8), 4042–4054 (2017)

9. Saraswat, M., Arya, K.: Automatic facial expression recognition in an image sequence of non-manual indian sign language using support vector machine. In: Proceedings of the International Conference on Soft Computing for Problem Solving (SocProS 2011), 20–22 December 2011, pp. 267–275. Springer (2012)
10. Saraswat, M., Arya, K.: Automatic facial landmark detection in a video sequences of non-manual sign languages. In: International Conference on Industrial and Information Systems (ICIIS), 2009, pp. 358–361. IEEE (2009)
11. Saraswat, M., Arya, K.: Feature selection and classification of leukocytes using random forest. Med. Biol. Eng. Comput. **52**, 1041–1052 (2014)
12. Dash, M., Liu, H.: Feature selection for classification. Intell. Data Anal. **1**, 131–156 (1997)
13. Guyon, I., Weston, J., Barnhill, S., Vapnik, V.: Gene selection for cancer classification using support vector machines. Mach. Learn. **46**, 389–422 (2002)
14. Deng, H., Runger, G.: Feature selection via regularized trees. In: Proceedings of International Joint Conference on Neural Networks, pp. 1–8 (2012)
15. Kohavi, R., John, G.H.: Wrappers for feature subset selection. Artif. Intell. **97**, 273–324 (1997)
16. Pal, R., Saraswat, M.: Data clustering using enhanced biogeography-based optimization. In: Tenth International Conference on Contemporary Computing (IC3), 2017, pp. 1–6. IEEE (2017)
17. Pal, R., Pandey, H.M.A., Saraswat, M.: BEECP: biogeography optimization-based energy efficient clustering protocol for HWSNs. In: 2016 Ninth International Conference on Contemporary Computing (IC3), pp. 1–6. IEEE (2016)
18. Pandey, A.C., Rajpoot, D.S., Saraswat, M.: Data clustering using hybrid improved cuckoo search method. In: 2016 Ninth International Conference on Contemporary Computing (IC3), pp. 1–6. IEEE (2016)
19. Saraswat, M., Arya, K., Sharma, H.: Leukocyte segmentation in tissue images using differential evolution algorithm. Swarm Evol. Comput. **11**, 46–54 (2013)
20. Pandey, A.C., Rajpoot, D.S., Saraswat, M.: Twitter sentiment analysis using hybrid cuckoo search method. Inf. Process. Manag. **53**(4), 764–779 (2017)
21. Mittal, H., Saraswat, M.: Classification of histopathological images through bag-of-visual-words and gravitational search algorithm. In: Soft Computing for Problem Solving, pp. 231–241. Springer (2019)
22. Kulhari, A., Saraswat, M.: Differential evolution-based subspace clustering via thresholding ridge regression. In: 2017 Tenth International Conference on Contemporary Computing (IC3), pp. 1–3. IEEE (2017)
23. Gupta, M., Parmar, G., Gupta, R., Saraswat, M.: Discrete wavelet transform-based color image watermarking using uncorrelated color space and artificial bee colony. Int. J. Comput. Intell. Syst. **8**(2), 364–380 (2015)
24. Mittal, H., Saraswat, M.: An optimum multi-level image thresholding segmentation using non-local means 2D histogram and exponential Kbest gravitational search algorithm. Eng. Appl. Artif. Intell. **71**, 226–235 (2018)
25. Mittal, H., Saraswat, M.: An image segmentation method using logarithmic *kbest* gravitational search algorithm based superpixel clustering. Evol. Intell. 1–13 (2018)
26. Pal, R., Saraswat, M.: Enhanced bag of features using AlexNet and improved biogeography-based optimization for histopathological image analysis. In: 2018 Eleventh International Conference on Contemporary Computing (IC3), pp. 1–6. IEEE (2018)
27. Mittal, H., Saraswat, M.: An automatic nuclei segmentation method using intelligent gravitational search algorithm based superpixel clustering. Swarm Evol. Comput. **45**, 15–32 (2019)

28. Mirjalili, S., Lewis, A.: The whale optimization algorithm. Adv. Eng. Softw. **95**, 51–67 (2016)
29. Pandey, A.C., Rajpoot, D.S., Saraswat, M.: Hybrid step size based cuckoo search. In: 2017 Tenth International Conference on Contemporary Computing (IC3), pp. 1–6. IEEE (2017)
30. Sharma, H., Hazrati, G., Bansal, J.C.: Spider monkey optimization algorithm. In: Evolutionary and Swarm Intelligence Algorithms, pp. 43–59. Springer (2019)
31. Mirjalili, S.: SCA: a sine cosine algorithm for solving optimization problems. Knowl.-Based Syst. **96**, 120–133 (2016)
32. Saremi, S., Mirjalili, S., Lewis, A.: Grasshopper optimisation algorithm: theory and application. Adv. Eng. Softw. **105**, 30–47 (2017)
33. Lukasik, S., Kowalski, P.A., Charytanowicz, M., Kulczycki, P.: Data clustering with grasshopper optimization algorithm. In: Federated Conference on Computer Science and Information Systems (FedCSIS), pp. 71–74. IEEE (2017)
34. Tharwat, A., Houssein, E.H., Ahmed, M.M., Hassanien, A.E., Gabel, T.: MOGOA algorithm for constrained and unconstrained multi-objective optimization problems. Appl. Intell. **48**(8), 2268–2283 (2018)
35. Barman, M., Choudhury, N.D., Sutradhar, S.: A regional hybrid GOA-SVM model based on similar day approach for short-term load forecasting in Assam, India. Energy **145**, 710–720 (2018)
36. Ibrahim, H.T., Mazher, W.J., Ucan, O.N., Bayat, O.: A grasshopper optimizer approach for feature selection and optimizing SVM parameters utilizing real biomedical data sets. Neural Comput. Appl. 1–10
37. Krizhevsky, A., Sutskever, I., Hinton, G.E.: ImageNet classification with deep convolutional neural networks. In: Advances in Neural Information Processing Systems, pp. 1097–1105 (2012)
38. Krig, S.: Feature learning and deep learning architecture survey. In: Computer Vision Metrics, pp. 375–514. Springer (2016)
39. Prijono, B.: Student notes: convolutional neural networks (CNN) introduction (2018). https://indoml.com/2018/03/07/student-notes-convolutional-neural-networks-cnn-introduction. Accessed 09 June 2018
40. ORL database of face images, September 2018. https://www.cl.cam.ac.uk/research/dtg/attarchive/facedatabase.html

Real Time Categorical Availability of Blood Units in a Blood Bank Using IoT

N. Hari Keshav(✉), H. Divakar, R. Gokul, G. Senthil Kumar, and V. D. Ambeth Kumar

Computer Science and Engineering, Panimalar Engineering College, Chennai 600123, India
keshavhari97@gmail.com

Abstract. In recent times there is a constant issues rising over in search of blood. Not every blood is readily available in every blood bank. Some blood group is rarely available thus making the patients really suffer in emergency times. There is huge value lies in the patient's life. There are many incidents happened in the past where people died after not getting the blood in time. About 40,000 thousand people in India need blood each day. Every blood packets donated by the donor is collected by the corresponding blood bank location. There is a need to develop a system that makes the recipients to know about the blood packets available in the blood bank. An effective system should directly connect the recipient to blood bank effectively.

Keywords: Blood bank · Database · Cloud server · Image detection

1 Introduction

According to the studies there is an increasing rise of medical emergencies in a day-to-day life. In most of the medical emergencies there is a one thing present as a common that is the blood. Obtaining blood is becoming one of the hard task. If the patient is lucky he can get blood easily but most of the times the rare blood groups cannot be obtained easily. There are some of the blood groups that cannot be obtained easily because they are present only in small amount of people. There is fact that ab –ve blood group is present in 0.4% population in India. In most times donors are willing to donate their blood but the recipient is unknown where the donor donates his/her blood, that is in which blood bank there is an availability of blood. Every time the patient is in need of blood their family members go on for search of blood which is time consuming job and it also results in waste of time and waste of energy.

The primary motto off our system is to notify the recipient about the availability of blood in each blood bank with the contact information. This makes the work easier and patient need not want to search for blood since they can search availability of blood from their hand itself. Now a days people are addicted to automated system because it works behalf of four members so that people don't need to do any extra work at all. There are some incidents happening is high populated areas which causes serious trouble to people that is, the recipient is given with blood which is stored for months and months. This will automatically create an serious health issue in recipient's body.

D. J. Hemanth et al. (Eds.): COMET 2019, LNDECT 35, pp. 503–510, 2020.
https://doi.org/10.1007/978-3-030-32150-5_48

Our motto is not only to notify the availability of blood to recipient but also to keep note of the date at which blood was donated so that whenever the blood is taken out of the storage for recipient the date is known and it can be disposed as soon as possible with medical help.

2 Literature Review

Review of literature helps to find the way that the existing system has been evolved. It provides a understandable platform for the proposed system which is already present in the universe. The survey of literature helps to find a system that has a better enhancement than the previous systems and provide a promising result. Most of the times literature review is important that is used to track the history of technologies and how are they theoretically and practically implemented. Literature on the utilization of digital and printed resources and other related issues is given below.

Moharkar and Somani [1] proposed that how an information about the blood is conveyed to every individual recipient using sms service. It implemented an usage of embedded system generate the sms which is the strong point of communication considered today. The system combines both donor and recipient and arranges a manageable way for the blood bank to fulfill their requirements in a day to day life.

Pande et al. [2] proposed a registration services for users who is in need of blood. It proves Cloud - based services are very vital in urgent blood delivery as they care able to central and immediate access to donor's data and location from anywhere and anytime. The system provides a database to maintain profiles of users so that It provides knowledge about the latest technology used in developing android based applications.

Sulaiman et al. [3] proposed the use of Rational Unified Process (RUP). Thus the system was built in a orderly fashion started from the core cycle providing a management system that is developed to manage blood bank in HSNZ. The system also offers other utility services related to blood bank regarding the information about each person seeking the blood bank.

Mahalle and Thorat [4] proposed the improvement of the management and response time of the blood bank by connecting all the blood banks to the cloud storage. Most of the researches were done of the blood banks and its managements. Among total population of Indian early 38 thousand blood donations are required everyday thus the use of IoT will become beneficially for the bank for the management system the blood banks. It is noted that this system provides high level of accuracy, reliability and automation in blood storage as well as blood transfusion process.

Ashlesha and Bhosale [5] proposed a work to create a direct communication between donor and recipient to encourage more donors to denote their blood. This approach also enables a communication between recipient and nearest donor so the blood can be reached quickly without any wastage of time. This approach uses a strong raspberry pi system and user application to enable a communication which certainly decreases the shortage of bloods in any hospitals.

Pereira et al. [6] paved the way for discovering blood banks and blood donors using a strong gps technology and user application. It also notifies users about the blood camps conducted. It is useful in emergency situations and provides contribution to the mankind. It enhances the service of blood banks. Blood camp organization will become easy process as organizers will directly contact the blood banks.

Kavitha and Sreelatha [7] proposed a work that provides a mission to fulfill every blood request in the country with a promising android phone and motivated individuals who are willing to donate blood. It creates a common platform to unite every single blood bank, donors and recipients into a common platform. It uses high level communication technologies like zigbee to increase the efficiency of the system. The donor and recipient or blood bank and recipient are immediately connected whenever the blood is required.

Mohanlal and Krishna [8] describes a work that aimstobeatthecommunicationbarrierbyprovidinganimmediatelinkbetweenthe donor. This system is originated on an android APP, this will help to find the donors. The system accounts a perfect accuracy of providing an useful information to donors, blood banks and recipients.

Raut et al. [9] describes a high-end system to bridge the gap between the blood donors and the people in need for blood. Using an application any individual person can find the blood bank and donate their blood thus easing the job of every one by the use of modern day technologies.

Akkas Ali et al. [10] proposed to reduce the complexity of the system to find blood donors in an emergency situation. It provides an entire information about the details of the donor and their recipients including the current location. It integrates a strong Google maps into the system to enhance the power of the system. It is used all over the year and thus bringing the donor and recipient straight together.

3 Proposed System

In our proposed system there will a connection made directly between the recipient and the blood bank. In our system we use a unique id made for each blood packets collected and the image is detected (Image Detection) and the information is stored in the database and later on uploaded to the cloud server. An user end application will be made available so that each recipient using the application can know about the exact amount of blood packets available in the particular blood bank and there will be contact number of the blood bank present. Hence it will be easy for recipient to communicate to blood bank and get the blood in time. We use the technique of computer vision in this system to clearly detect the unique id so that the information can be stored later (Fig. 1).

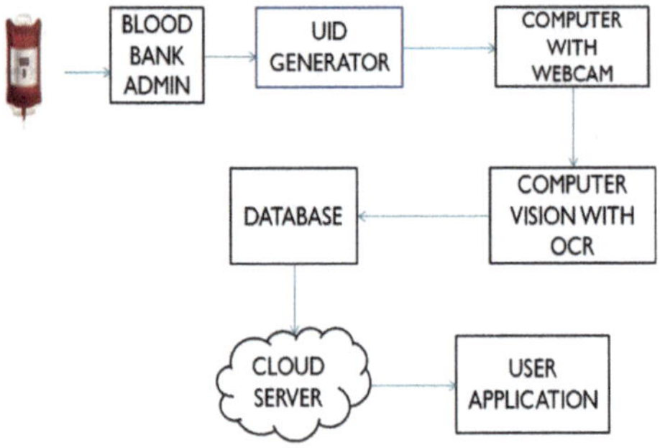

Fig. 1. Frame work for automated blood bank system

In the above figure the modules are classified into.

3.1 Unique Id Generator

Unique id generator is used to generate a unique id for each blood packets collected from the donor that contains some valuable information. For example: Location of the blood bank, Date the blood donated, Blood group code etc. It will be easy to store the information using this id and it will not ask any complexity in the job. It is also easily recognizable since it follows a specific pattern. Using this id blood bank can easily know how many days the blood packets are stored in the blood bank so that after 45 days the blood packets from the blood bank can be removed (Fig. 2).

Fig. 2. Unique id generator

The above figure shows that an unique id can be created to identify the blood packets stored in the blood bank. An unique id comprised of information such as pin code, date and type of blood donated by the donor. This is very useful in solving the deficiency in the previous systems and provide some answers to the society problems.

3.2 Image Detection and Conversion

The unique id contained in the blood packets will be in the form of an image. Thus using computer vision we can clearly detect the image and the image will be converted into string using ocr (a type in the computer vision technique) which detects the characters present in the image so that the information will be stored in the database using the regular expression (Fig. 3).

Fig. 3. Image detection

The above figure shows that first the blood packet containing the unique id is detected using the camera. The camera detects a pattern which is previously fed into the computer (Fig. 4).

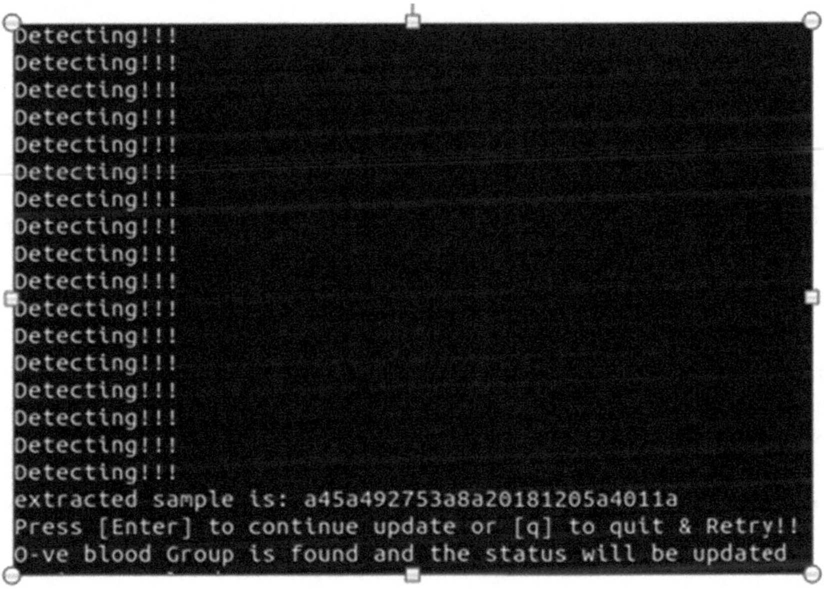

Fig. 4. Text conversion

When the image is detected it then transforms the text present in the image to the pattern which the computer knows by the regular expression. The information can be extracted from the text and the status will be updated.

3.3 Cloud Server Updation

After the information stored in the database it will uploaded into the cloud server so that the information later on updated to the user end application. This will be fastest and cheapest method so that availability of blood packets are updated instantly (real time). The information is stored in a csv file which acts as a database here, that will be updated to cloud server (Fig. 5).

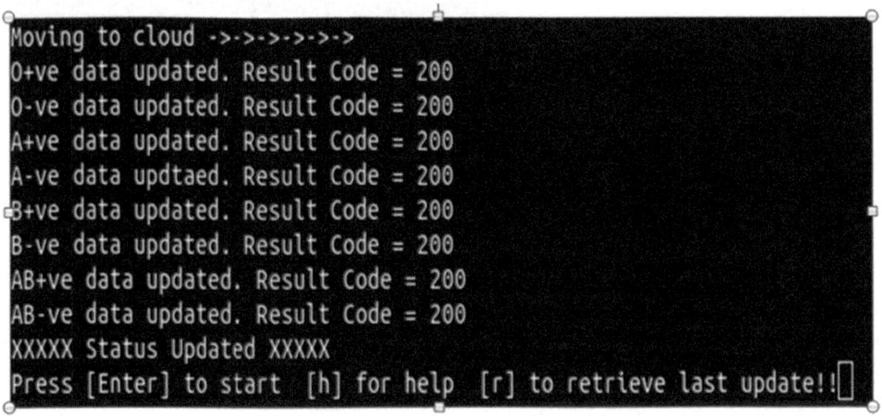

Fig. 5. Cloud server updation

The above image describes that when the image is detected then the information contained will be updated automatically to cloud server. This is quite effective way so that the data will be stored in the cloud for long only when the patient needed the blood it will be erased from the cloud. The result code 200 describes that status has been updated successfully.

3.4 User End Application

An application will be built on the user side that is the recipient side to know the availability of blood packets. The user application will contain the blood groups which shows the quantity and the contact number of the particular blood bank to contact and collect the blood packets. Whenever the information is updated into the cloud server it shows the immediate deflection in the application about the amount of blood packets and date at which the information is last modified (Fig. 6).

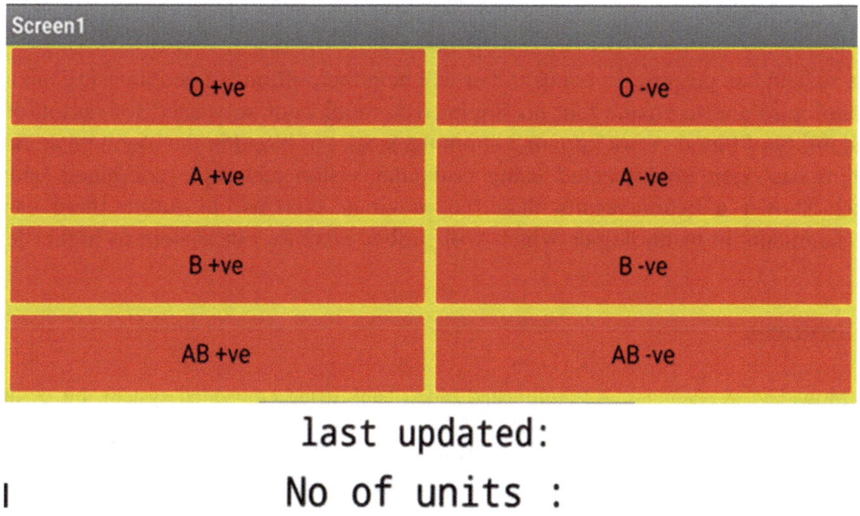

Fig. 6. User end application before updation

The above image provides an view to an user who can be able to operate the system in the front end. Application simply provides about the blood sample calculation and the date at which the information was last updated (Fig. 7).

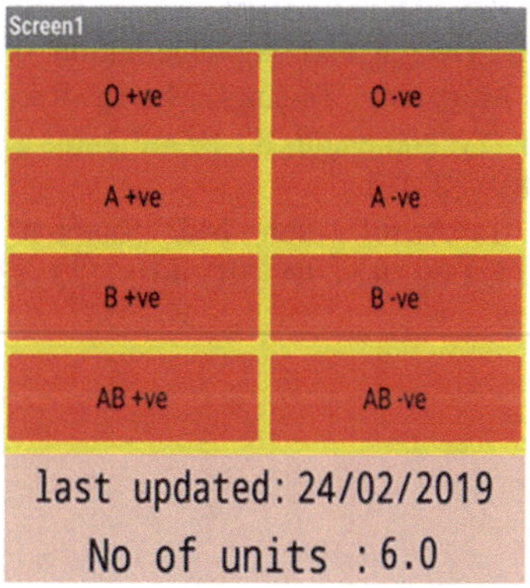

Fig. 7. User end application after updation

The image shown above describes about the number of blood packets present currently in the particular blood bank and ad date at which the blood type was last updated in that zone.

4 Conclusion

This system has proposed a concept that will help the millions of people using the latest technologies and to ensure that no single lives should not be wasted for reason of not availability of blood or not knowing about the stocks of blood in the blood bank. So the system has been implemented using computer vision technique combined with iot which shown a better results than the previous systems. In future there can be advancements in technologies which will further increase our system to higher level.

References

1. Moharkar, A., Somani, A.: Automated blood bank using embedded system. Int. J. Innov. Res. Sci. Eng. Technol. **7**(1) (2018)
2. Pande, S., Mate, S., Mawal, P., Jambulkar, A., More, N.S.: E-blood bank application using cloud computing. Int. Res. J. Eng. Technol. (IRJET) **05**(02) (2018)
3. Sulaiman, S., Hamid, A.A.K.A., Yusri, N.A.N.: Development of a blood bank management system. Procedia - Soc. Behav. Sci. **195**, 2008–2013 (2015)
4. Mahalle, R.R., Thorat, S.S.: Smart blood bank based on IoT. Int. Res. J. Eng. Technol. (IRJET) **05**(02) (2018)
5. Ashlesha, C., Bhosale, A.V.K.: Automated blood bank system using Raspberry PI. IJSRD – Int. J. Sci. Res. Dev. **5**(9) (2017)
6. Pereira, D.J., Sutar, H., Ramane, N., Tanpure, P., Chavan, S.: Blood at one touch, © 2016. IJSRSET **2**(2) (2016)
7. Kavitha, I., Sreelatha, N.: Design and implementation of automated blood bank using embedded systems. Int. J. Res. **03**(14) (2016)
8. Mohanlal, J., Krishna, M.: Design and implementation of automated blood bank using embedded systems. Int. J. Mag. Eng. Technol. Manag. Res. (2016)
9. Raut, P., Parab, P., Suthar, Y., Narawani, S.: Blood bank management system. Int. J. Adv. Comput. Eng. Network. **4**(9) (2016)
10. Akkas Ali, K.M., Jahan, I., Ariful Islam, Md., Shafa-at Parvez, Md.: Blood donation management system. Am. J. Eng. Res. (AJER) **4**(6), 123–136 (2015)

Survey on Various Modified LEACH Hierarchical Protocols for Wireless Sensor Networks

P. Paruthi Ilam Vazhuthi[1(✉)] and S. P. Manikandan[2]

[1] V.R.S. College of Engineering and Technology, Villupuram, India
paruthi.ilaya.raaj@gmail.com
[2] Jerusalem College of Engineering, Chennai, India
manikandanperiyasamy@gmail.com

Abstract. Energy is an ultimate problem facing in Wireless sensor network. Hierarchical routing protocols are the best solution for energy efficiency. In this survey paper, we present various modified type of LEACH hierarchical routing protocol. The ultimate aim of this paper is to discuss how the different LEACH protocols help to improve the life span of the sensor network. In addition, we analyze the factors how the related protocols method tackled the energy issues in WSN for developing low Power Hierarchical Protocol. Furthermore, the merits and demerits of various modified LEACH protocols have been analyzed. This literature survey is helpful to provide ideas for modifying the various LEACH protocol to develop energy efficient protocols.

Keywords: LEACH · Hierarchical routing · Clustering

1 Introduction

Wireless sensor network consists of many self-organizing sensor nodes which perform sensing, processing, storing and communication. Its aim is to collect the data, process and transmit the sensed data to Base station with low power consumption [1]. WSN Routing is very challengeable due to limited resources such as battery backup, low data rate, less transmission range and self configurable. In some scenarios, Sensor node is equipped with battery that is difficult to replace in harsh environment. Thus the network's lifetime depends upon available charge in the battery of sensor nodes.

To increase the lifetime of the sensor network, clustering among the nodes may takes place. The hierarchical routing protocols provide the maximum energy efficiency. In this survey paper, we focused on LEACH (Low Energy Adaptive Clustering Hierarchy) routing protocol to analyze and compare with modified LEACH protocols such as LEACH–Energy Efficient, LEACH-M (Mobile), LEACH-C, LEACH-Future, LEACH–EA (Energy Available), LEACH EP (Energy Aware Protocol), s-LEACH, Solar-aware Centralized LEACH, Solar-aware Distributed LEACH, Multi-hop LEACH, Mobile-LEACH, LEACH-C (Centralized), Stable Election Protocol (SEP), and LEACH-HPR protocol. The above mentioned protocols were developed from clustered based architecture. Group of sensor nodes forms the cluster, where it consists

D. J. Hemanth et al. (Eds.): COMET 2019, LNDECT 35, pp. 511–518, 2020.
https://doi.org/10.1007/978-3-030-32150-5_49

of many member sensor node and cluster head (CH). Cluster head is connected wirelessly with all sensor nodes in single or multihop fashion.

2 Review of Modified Leach Protocols

2.1 Overview of LEACH Protocol

LEACH is a clustering based protocol which consumes low power [8, 9]. It is the first hierarchical routing protocol in wireless sensor networks for better utilization of energy resources to prolong the life time of the network. As far as LEACH protocol concern, it has some following aspects such as forming clusters by sensor nodes location, Cluster head node may change dynamically with respect to available energy in node, and data fusion takes place after formation of clusters. Many hierarchical routing protocols are designed on the basic of LEACH.

Routing is done only through the cluster head. The entire member sensor nodes transmit the sensed data to its respective cluster head (CH). CH aggregates all the sensed data sent by the member sensor nodes. The process of aggregation of sensed data in the Cluster head is called Data Fusion. CH performs data processing and decides where to send the fused data. The data transmission takes place only through cluster head to cluster head and lastly send the data to monitor node. The basic principle of LEACH, each and every node in a cluster acts as cluster head (CH) at once. In another words, a member sensor node is elected as CH through round robin method. Inorder to improve the lifetime of the network each sensor node should share the energy consumption of the individual sensor node.

2.2 Operation of Conventional LEACH Protocol

Each node selects the *r* random value in the range of 0 to 1. Threshold value to be calculated as shown in Eq. (1) with its current round. The random number must be less than threshold T(n) then the node becomes CH.

$$T(n) = p/(1 - p * (r \bmod (1/p))) \text{ if } n = G \tag{1}$$

Where,
p is the percentage of CHs,
r is the current round,
G is nodes that have not been CHs in the last 1/p rounds.

Connection establishment must takes place between the cluster head and member sensor node. The following procedure tells that how the connection establishment and data transmission is carried out. Initially the cluster head broadcast its information to the surrounded sensor nodes. All the member sensor node detects the signal strength of the cluster head broadcasted signal and inform to the cluster head if the sensor nodes are interested to join with the corresponding cluster head. All the data will be sent by the member sensor node is aggregated in the CH and forwarded to sink.

2.3 LEACH-Centralized Protocol

LEACH-C is known as centralized clustering routing algorithm. This protocol works by knowing the position and available energy of the sensor node. By using the simulated annealing algorithm selects the cluster head with acceptable scalability. Hence the energy consumption of all the nodes in the whole network is reduced. As far as LEACH – C protocol concern, energy is the major parameter to select the cluster head. The cluster head selection is not allowed when the energy value is critically low. For determining the good cluster, BS make sure that work is evenly distributed among all other nodes that is, in a centralized manner [2].

2.4 LEACH-Future Protocol

LEACH-C protocol is modified to LEACH-F for future development. As like LEACH-C, Cluster structure is done by simulated annealing algorithm [3]. The Sink is formed a cluster header list after formation of each cluster, where at once the cluster member nodes have a chance to be elected as cluster head . Once the cluster is created the topology of the structure does not change.

2.5 LEACH-Energy Available (EA) Protocol

LEACH-EA protocol is developed based on the above explained LEACH-C protocol. The available energy and average energy in all sensor nodes take into an account at the starting and end of the round [4]. This protocol provides the better solution for reducing cluster head energy consumption, which leads to extend the life time of the network.

2.6 LEACH-EP Protocol

The drawback in above protocols is unstable in re-clustering the network by considering nodes available energy and position of the sensor nodes. This protocol which minimizes this issue by electing multiple cluster heads. If the CH fails due to scarcity of energy resources, another alternate cluster head is taken over the session for data forwarding to the base station. Hence, the strategy used by this protocol provides more steady [5].

2.7 LEACH-Energy Efficient Protocol

LEACH Energy-Efficient clustering routing algorithm is improved based on conventional LEACH. This protocol is somehow improved the life time of the network through data transmission from cluster to base station in multihop fashion. Cluster is capable of transmission in high signal power as well as in long distance, so that it can cover large area [6]. By transmitting in low signal power with multiple hops from one cluster head to another cluster head, the energy efficient problem is solved. Hence, LEACH-EE can enhance the lifespan of the sensor network [13].

2.8 LEACH-Multihop Protocol

Multihop LEACH protocol is similar to LEACH protocol. The shortest route will be found from base station to all for effective transmission of data. All the nodes perform data aggregation and forwarding to Base station through CHs in multihop fashion to minimize the consumption of energy as shown in Fig. 1. When the diameter of the network increases, the distance between the cluster and base station is also increases. To deal with this problem, Multi-hop LEACH is proposed in [7].

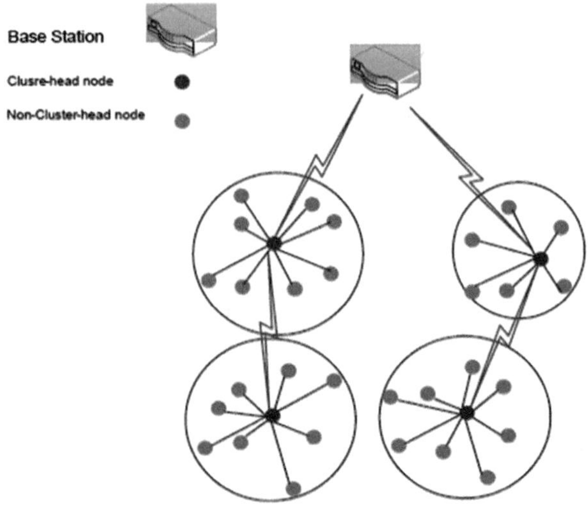

Fig. 1. Multihop LEACH

2.9 sLEACH

Energy harvesting is important in sensor networks to prolong the life time of the network. In some applications, replacing of battery in sensor node is very difficult due to environmental condition [14]. In sLEACH, energy is harvested through solar, that the nodes act as Cluster head based on the solar status.

2.10 Solar-Aware Centralized LEACH

In solar-aware Centralized LEACH, Base station selects the cluster head through improved control algorithm. The node which has high residual energy is elected as cluster head that harvest the energy by solar [14]. To improve the life time of the network and its performance, more number of solar powered sensor nodes should be used. All the member nodes send the flag information with data when the energy is increased by solar. Those particular member nodes are act as CH by giving high priority. The newly selected CH is done in steady state phase which extend lifetime of the network.

2.11 Solar Aware Distributed LEACH

In Solar aware Distributed LEACH [10], solar powered nodes are having high priority to act as CH. Probability of solar-powered nodes is higher than nodes which is operated in battery. To enhance the probability of solar powered nodes Eq. (2) is needed.

$$T(n) = sf(n) \text{ x } (p/1 - (\text{CH head}/\text{num nodes})) \quad (2)$$

where, *sf(n)* is 4, *P* is the percentage of CHs. *cHeads* specifies the number of CHs since the start of last meta round and *numNodes* is total number of nodes [8, 9].

2.12 M-LEACH (Mobile LEACH)

Cluster head nodes are consumed their energy more compared to member nodes. As well as mobility is another problem is tackled with LEACH protocols [11]. It considers the residual energy of the CH selection. Initially, all nodes are having same characteristics in terms of antenna gain and receive location information through Global Positioning System (GPS). In this protocol, based on attenuation model [12] CHs are elected. Stable CHs are selected which reduce the power of attenuation. In another method, based on the mobility of sensor nodes CHs are selected.

Node chooses as CH with least mobility. At that point chosen CH communicates their status to all neighbor sensor nodes where within the transmission range. Sensor nodes process their readiness from different CHs and choose the CH with most extreme remaining vitality. In relentless state stage, if member sensor nodes move far from CH or CH moves from its sensor nodes then another CH winds up appropriate for member nodes. Hence it is inefficient in cluster development. To deal this issue, hand-over takes place when the node moves to new CH.

2.13 Stable Election Protocol (SEP)

This protocol [15] is used heterogeneous sensor nodes in wireless sensor networks. It uses two-level of energy. All the sensor nodes get the information for the position of base station. It separates the nodes into normal nodes and advanced nodes. The nodes having high energy is said to be advanced nodes. It increases the stability period before the first nodes running out of energy. For electing a cluster-head, the probability of CH selection is computed by following equation;

$$P = K/n \quad (3)$$

Where
P is CH
K is the number of CHs per round.

2.14 LEACH- Cluster Head Election Protocol

It is appropriate for Heterogeneous Wireless Sensor. Each sensor node with a built-in timer and the initial value is the reciprocal of the dump energy [7]. This protocol makes the node as cluster head having more energy and tradeoff the energy consumption by establishing short route as well as low power consuming path.

3 Analysis Factors of Low Power Hierarchical Protocol

To improve the life time of the network, the best solution is clustering based hierarchical protocols should be developed. Certain parameters is considered for improving the life time of the network as analyzed as follows location information, low energy consumption, irregular clustering, residual energy, multihop, energy harvesting multiple cluster head in each clusters and etc.

3.1 Location Information

M-LEACH, SEP, LEACH-C considered the location information for electing the cluster head. Also the reconfiguration of the path and electing the CH is quite easy when the CH node or any other node fails in the cluster. Hence the network is obtained stable state and enhances the life span of the sensor network.

3.2 Low Power Consumption

LEACH protocol is modified and developed by considering low power consumption by nodes. Each and every node should perform less computation, low data rate, short range by the nodes. To improve the life span of the network with above all parameter should take into an account. The ultimate objective of all the LEACH protocols deals with low power consumption.

3.3 Heterogeneous Nodes

The sensor network must consists of homogeneous sensor nodes which has identical features such as in terms of antenna gain, transmission power, data rate, available energy ant etc., In heterogeneous sensor network must tradeoff between above all the parameters. For an example, the entire sensor nodes must equally distributed with the available energy. Cluster formation and data gathering pattern may dynamically changed because of the heterogeneous nodes

3.4 Multihop

As discussed earlier, the sensor network is an infrastructure less network can transmits the data in multihop fashion. Each and every node fuses the transmitted data and performs computational processing that consumes little energy resources. Multi-hop LEACH protocol is developed which extends the coverage of the distance. But the drawback in multihop LEACH protocol, route reconfiguration take place if the path fails between the cluster heads and the base station due to failure of intermediate cluster head.

3.5 Resource Awareness

Sensor network has limited resources like battery and sensing capability routing protocol should be well aware from the resources. Due to scarcity of energy resources, energy harvesting is incorporated with the selected individual sensor nodes. It facilitates the nodes by providing energy to operate the nodes optimally. sLEACH, Solar-aware Centralized LEACH and Solar-aware Distributed LEACH, are the protocol which uses solar energy to maximize the life time of the network.

4 Conclusion

In this literature survey paper, the following modified protocols such as LEACH, Multihop LEACH, Mobile LEACH, SEP, and Solar aware LEACH hierarchical routing protocols for WSN are discussed for extend the life time of the network. According to the characteristics of each protocol, we make a complete analysis with some factors of wireless sensor network. The main intention of this survey is research on energy efficiency and network life time improvement of these LEACH routing protocols.

5 Future Scope

The future work may include use GPS for optimizing data gathering and it can be developed by modifying LEACH protocol for heterogeneous sensor network. All the sensor node can be solar powered that can dynamically vary the signal power transmission. All the aspects may improve the life span of the network.

References

1. Sun, L., Li, J., Chen, Yu.: Wireless Sensor Network. Tsinghua University Press, Beijing (2005)
2. Rahmanian, A., Omranpour, H., Akbari, M., Raahemifar, K.: A novel genetic algorithm in LEACH-C routing protocol for sensor networks In: 24th Canadian Conference on Electrical and Computer Engineering (CCECE) (2011)
3. Wang, W., Wang, Q., Luo, W., Sheng, M., Wu, W., Hao, L.: Leach-H: an improved routing protocol for collaborative sensing networks. In: International Conference on Wireless Communications & Signal Processing (2009)
4. Gantassi, R., Yagouta, A.B., Gouissem, B.B.: Improvement of the LEACH protocol in load balancing and energy-association conservation. In: International Conference on Internet of Things, Embedded Systems and Communications (IINTEC) (2017)

5. Jia, J., He, Z., Kuang, J., Mu, Y.: An energy consumption balanced clustering algorithm for wireless sensor network In: 6th International Conference on Wireless Communications Networking and Mobile Computing (2010)
6. Kumar, G., Singh, J.: Energy efficient clustering scheme based on grid optimization using genetic algorithm for wireless sensor networks. In: 4th International Conference on Computing, Communications and Networking Technologies (2013)
7. Li, H.: LEACH-HPR: an energy efficient routing algorithm for heterogeneous WSN. In: IEEE International Conference on Intelligent Computing and Intelligent Systems (ICIS), pp. 507–511 (2010)
8. Gupta, G., Younis, M.: Load-balanced clustering of wireless sensor network. In: 2nd ACM International Symposium on Mobile Ad Hoc Networking & Computing (2003)
9. Heinzelman, W.R., et al.: Energy-efficient communication protocol for wireless microsensor networks. In: 33rd Hawaii International Conference on System Sciences (2000)
10. Voigt, T., Dunkels, A., Alonso, J., Ritter, H., Schiller, J.: Solar-aware clustering in wireless sensor networks. In: 9th International Symposium on Computers and Communications
11. Souid, I., Chikha, H.B., El Monser, M., Attia, R.: Improved algorithm for mobile large scale sensor networks based on LEACH protocol. In: 22nd International Conference on Software, Telecommunications and Computer Networks (2014)
12. Nie, F., Fu, Z.-F.: MMLC mobile clustering routing scheme based on LEACH in wireless sensor network. In: 8th World Congress on Intelligent Control and Automation (2010)
13. Younis, O., Fahmy, S.: HEED: a hybrid, energy-efficient, distributed clustering approach for Ad Hoc sensor networks. IEEE Trans. Mob. Comput. **3**(4), 366–379 (2004)
14. Islam, J., Islam, M., Islam, N.: A-sLEACH: an advanced solar aware leach protocol for energy efficient routing in wireless sensor networks. In: Sixth International Conference on Networking (ICN'07) (2007)
15. Ayoob, M., Zhen, Q., Adnan, S., Gull, B.: Research of improvement on LEACH and SEP routing protocols in wireless sensor networks. In: IEEE International Conference on Control and Robotics Engineering (ICCRE) (2016)

Application of Magnesium Alloys in Automotive Industry-A Review

Balaji Viswanadhapalli[1,3] and V. K. Bupesh Raja[2(✉)]

[1] School of Mechanical, Sathyabama Institute of Science and Technology, Chennai 600119, India
balaji.viswanadhapalli@gmail.com
[2] Department of Automobile Engineering-School of Mechanical, Sathyabama Institute of Science and Technology, Chennai 600119, India
bupeshvk@gmail.com
[3] Department of Mechanical Engineering, Gokaraju Rangaraju Institute of Engineering and Technology, Hyderabad 500090, India
vbalaji@griet.ac.in

Abstract. Fuel economy and environmental conservation are the major factors to consider magnesium alloys for automotive industry and aero space and other electronics companies. The key features like high strength to density ratio, moderate damping capacity, recyclability, reduced CO_2 emissions are added advantages of magnesium alloys in automotive applications. This article reviews historical trends and near future applications of magnesium alloys in automotive industry. As magnesium loses its strength and creep resistance abilities, alternative magnesium alloys are to be explored to supply automotive components in the industry on demand. The objective of this study is to review and evaluate the applications of magnesium in the automotive industry that can significantly contribute to greater fuel economy and environmental conservation. In this study, the current trends, challenges, technological obstacles and future scope of magnesium alloys in the automotive industry are discussed. The consumption of magnesium in automotive industry with reference to environment is explored. Innovative welding and forming techniques available today are encouraging factors for extended use of magnesium and its alloys in automotive sector. This review offer insights and opportunities to researchers for further study and investigation of challenges in the field of automobile industry.

Keywords: Magnesium alloys · Automotive industry · Fuel economy · CO_2 emissions

1 Introduction

Magnesium is the lightest of all the structural metals, having a density of 1.74 g/cm^3. It is 35% lighter than aluminium (2.7 g/cm^3) and about four times lighter than steel (7.86 g/cm^3). Sea water contains approximately 1.3 kg per m^3 of magnesium [1, 2]. It has superior noise and vibration characteristics than aluminium and excellent formability at high temperatures [3]. The physical properties of magnesium with other light metals are compared in Table 1 [1].

D. J. Hemanth et al. (Eds.): COMET 2019, LNDECT 35, pp. 519–531, 2020.
https://doi.org/10.1007/978-3-030-32150-5_50

Magnesium is eighth most common element on earth and third most abundant metal. It is produced through either reduction of magnesium oxide with silicon or the electrolysis of magnesium chloride melts from seawater. It has a good specific strength and stiffness [4–6] as given in Fig. 1.

Table 1. Key properties of Magnesium, Aluminium, and Iron.

Property	Magnesium	Aluminium	Iron
Crystal structure	hcp	fcc	bcc
Density at 20 °C (g/cm^3)	1.74	2.7	7.86
Coefficient of thermal expansion 20–100 °C (× 10^6/C)	25.2	23.6	11.7
Young's modulus of elasticity, (10^6 MPa)	44.126	68.947	206.842
Tensile strength, MPa	240 (for AZ91D)	320 (for A380)	350
Melting point (°C)	650	660	1536

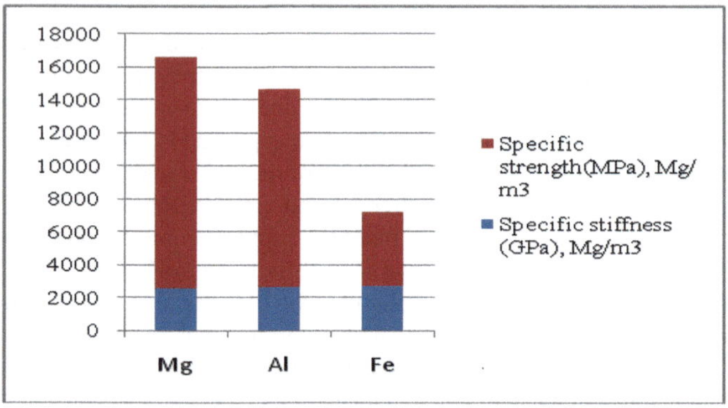

Fig. 1. Specific strength vs Specific stiffness of Mg with aluminium and iron compared

Magnesium with alloying elements via aluminium, manganese, rare earths increases the strength to weight ratio and minimizes inertial forces [7, 8]. The need of reduced CO_2 emissions and weight reduction of automobile vehicle are challenging aspects in the new alloy developments for the researchers [9]. Many car manufacturing companies did research on Mg and its alloys. Volkswagen was the first to apply magnesium in the automotive industry on its Beetle model, which used 22 kg magnesium in each car of this model [10]. Porsche first worked with a magnesium engine in 1928 [11]. The increase in percentage of usage of magnesium alloys per car for the period 2005 to 2015 is about 1566% (3 kg to 50 kg). In recent years magnesium usage in the automotive industry has been growing. Magnesium usage in various areas and productions supply are given in Figs. 2 and 3.

Studies of magnesium alloys have aimed at weight reduction, fuel economy and limiting CO_2 emissions. Currently magnesium applications are increasing in auto sector [12]. Presently the research is mainly concentrated on weight reduction and energy saving and reducing CO_2 emissions [13]. Magnesium production supply around the world is depicted in Fig. 3 [14].

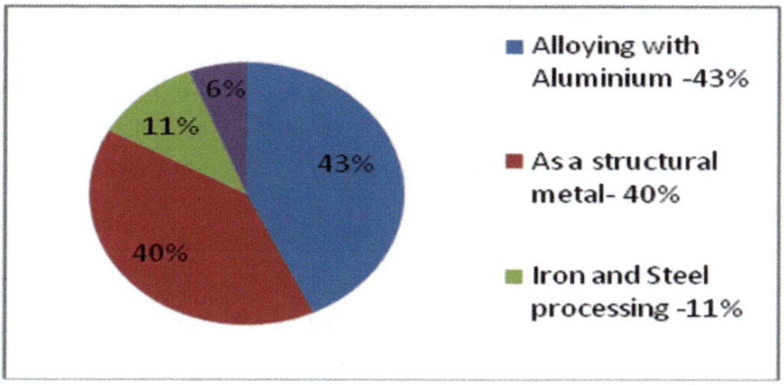

Fig. 2. Magnesium usage in different areas. (Source: http://americanmagna.com/magnesium-uses/)

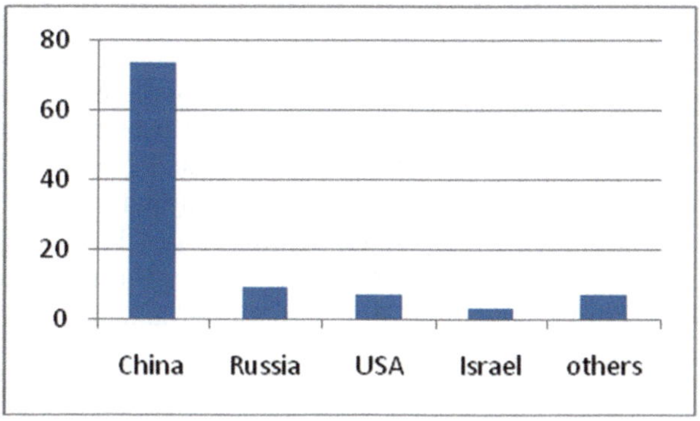

Fig. 3. Magnesium production supply in the world.

2 Materials Properties

Magnesium offers similar mechanical properties, though it is least dense structural material, but with material saving 33%. Typical mechanical properties of cast and wrought magnesium are given in Table 2 [15–18].

3 Automotive Applications of Magnesium Alloys

3.1 Historical Applications

Magnesium alloy has been used as a structural material in the transportation industry since Second World War. Most notable applications are shown from Figs. 4, 5, 6, 7, 8 and 9. B-36 bomber is one of the most famous applications shown in Fig. 9, which contained 5500 kg of magnesium sheet and 680 kg of magnesium forgings, along with 300 kg of magnesium castings [19]. The commercial road transportation applications were developed in the early 1950s. Metro-lite trucks were manufactured between 1955

Table 2. Comparison of Mechanical and physical properties of typical cast and wrought magnesium alloys in automotive industry.

Process/saproduct	Die cast AZ91	Die cast AM50	Extrusion AZ80-T5	Sheet AZ31-H24
Material/Grade	Cast Magnesium		Wrought Magnesium	
Density, g/cm^3	1.81	1.77	1.8	1.77
Elastic modulus, GPa	45	45	45	45
Yield strength, MPa	160	125	275	220
Ultimate tensile strength, MPa	240	210	380	290
Elongation (%)	3	10	7	15
Fatigue strength, MPa	85	85	180	120
Thermal Conductivity, W/m.K	51	65	78	77
Melting temperature, °C	598	620	610	630

and 1965 and featured magnesium sheet panels as well as structures made with magnesium plate and extrusions [20]. The VW beetle uses approximately 20 kg of magnesium for crank case and transmission housing and Northrop XP-56', the first aeroplane nearly completely designed with Magnesium in 1940s, B-36 Bomber contained 19000 lbs of magnesium in 1950, Titan I rocket uses 1100 lbs magnesium sheet in 1960s is given in Figs. 4, 6, 7 [21]. Porsche 911 Targa' uses magnesium roofshell and panel bow and 'Ford F-150' front structure bolster/radiator support component is die-cast magnesium is given in Figs. 5 and 8 [22].

Fig. 4. 'VW Beetle' uses 20 kg of Magnesium for crank case and transmission housing.

3.2 Current and Future Applications

Magnesium alloys are used in automotive industry via power train, body parts, chassis, etc., and in aerospace like transmission lines, bio medical applications, nuclear, luggage, hand-held working tools, sporting goods like bicycle frame and motorcycle applications, and electronic devices like laptop cases etc. The only current production

Fig. 5. 'Porsche 911 Targa' uses magnesium roofshell and panel bow.

Fig. 6. 'Northrop XP-56', the first aeroplane nearly completely designed with Magnesium in 1940s

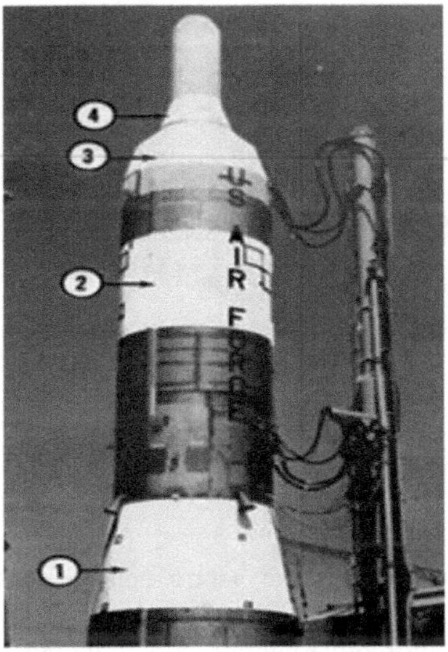

Fig. 7. Titan I rocket uses 1100 lbs Mg sheet in 1960

Fig. 8. 'Ford F-150' front structure bolster/radiator support component is die-cast magnesium.

Fig. 9. B-36 Bomber contained 19000 lbs of magnesium in 1950

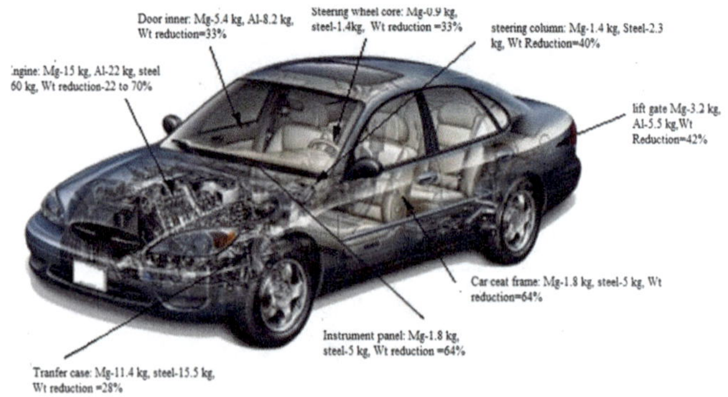

Fig. 10. Few automotive components made of magnesium alloys results in weight reduction.

application of magnesium sheet is the centre console in the low-volume Porsche Carrera as shown in Fig. 11 [23, 24]. Table 3 summarizes the potential applications of magnesium extrusions in automotive interior, and body and chassis areas, while power train magnesium applications will remain castings [24].

Fig. 11. Sheet magnesium centre console cover in Porsche Carrera GT automobile.

Table 3. Automotive applications of wrought magnesium components

System	Component	Description
Interior	Instrument panel	Extrusion/sheet
	Seat components	Extrusion/sheet
	Trim plate	Sheet
Body	Door inner	Sheet
	Tailgate/liftgate inner	Sheet
	Roof frame	Extrusion
	Sunroof panel	Sheet
	Bumper beam	Extrusion
	Radiator support	Extrusion
	Shotgun	Sheet/extrusion
	A and B pillar	Sheet
	Decklid/hood inner	Sheet
	Hood outer/fender	Sheet
	Decklid/door outer	Sheet
	Dash panel	Sheet
	Frame rail	Extrusion
Chassis	Wheel	Forging
	Engine Cradle	Extrusion
	Subframe	Extrusion
	Control arm	Forging

Studies on magnesium alloys reveals greater energy saving and decrease in CO_2 emissions is possible by using magnesium alloys as substitute for other light metals. 22% to 70% weight reduction is achievable for automotive components by using magnesium alloys instead of alternate materials as depicted in Fig. 10 [6, 25]. Magnesium sheet panels formed recently by General Motors (GM) that include door inner panel [26], decklid inner panel [27] and hood [28] is depicted in Fig. 12.

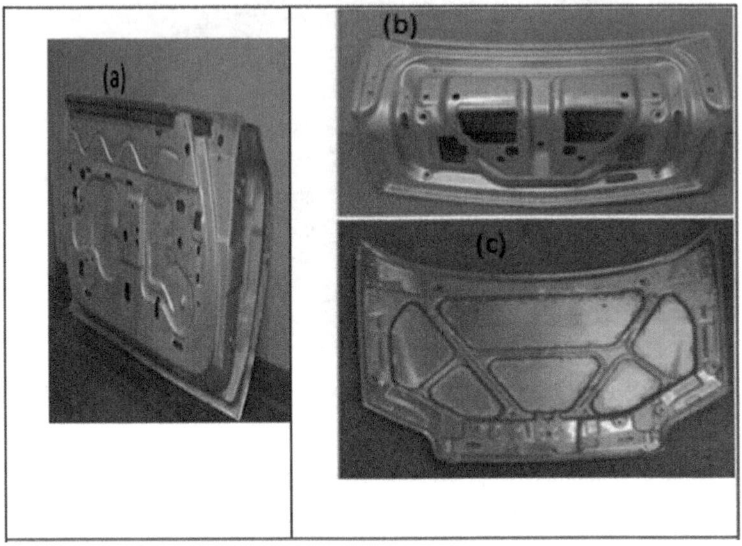

Fig. 12. Magnesium sheet panels formed recently by General Motors (GM). (a) Door inner panel (b) decklid inner panel (c) hood

The 1951 Buick LeSabre concept car with magnesium and aluminium body panels, 1961 Chevrolet Corvette with prototype hood made from magnesium sheet and 1957 Chevrolet Corvette SS race car with 'featherweight magnesium body' [23] is shown in Fig. 13. Powell et al. listed few automotive components which are basically developed from magnesium alloys are given in Fig. 14 [23].

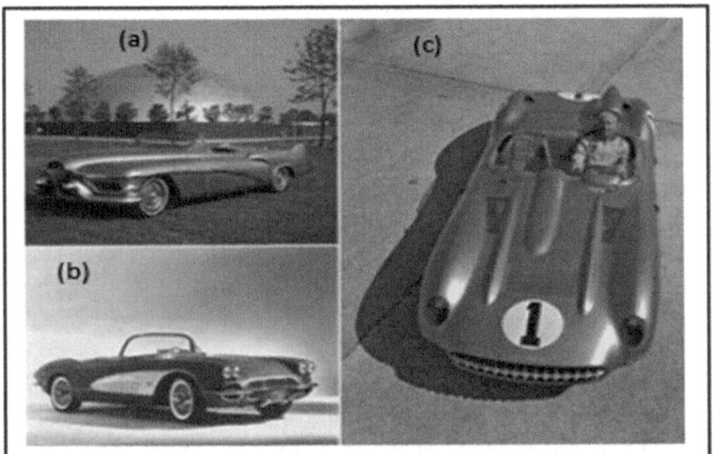

Fig. 13. (a) 1951 Buick LeSabre concept car with magnesium body panels; (b) 1961 Chevrolet Corvette with hood made from magnesium sheet; (c) 1957 Chevrolet Corvette race car with 'featherweight magnesium body'

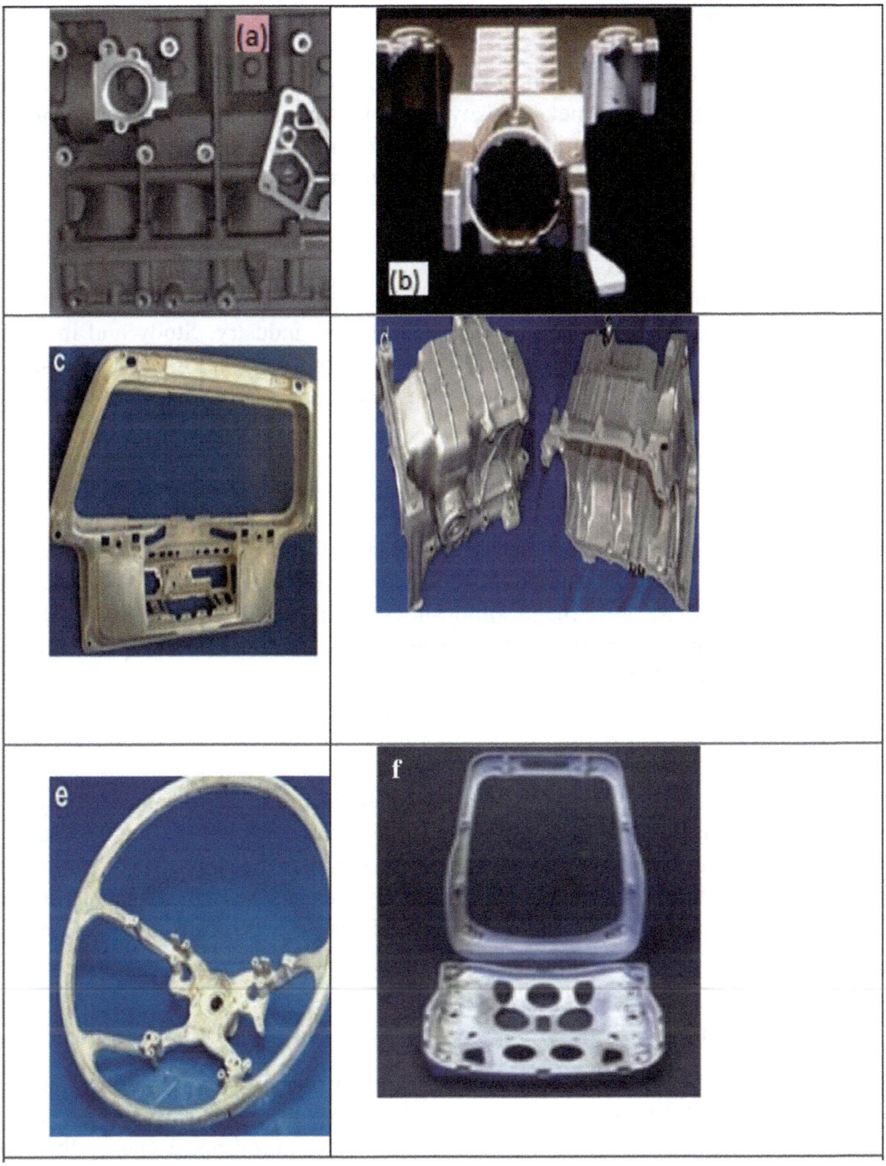

Fig. 14. Components made of magnesium [6] (a): Engine block, (b): Steering column module, (c): Door frame/Key lock housing, (d): Oil pan, (e): Steering wheel, (f): Transfer case/Transmission housing, seat frame

4 Limitations

Though magnesium alloys are lightest structural metals and highest specific strength to density ratio among the light metals, many challenges remain in different areas of alloy development and manufacturing processes to take advantage of its high strength-to-density ratio for extensive lightweight applications in the automotive industry.

4.1 Material Challenges

In comparison with many aluminium alloys and steel grades, there are only a limited number of magnesium alloys available for automotive industry. Study and investigation still needed for obtaining alloys with a potential for precipitation hardening such as magnesium-stannum [29, 30] and magnesium-rhenium [4] with improved mechanical properties.

4.2 Process and Performance Challenges

Different forming processes need to be optimized for magnesium alloys. Forming magnesium at various temperatures via room, near room and forming techniques also need to be explored for new magnesium alloys such as Mg-Zn-Ce alloys [30]. Performance barriers are highlighted in the current Canada-China-USA Magnesium Front End Research & Development project [31].

4.3 Corrosion

Magnesium has poor corrosion resistance due to its high chemical reactivity. The composition effects on the corrosion of magnesium alloys. The need to focus on developing corrosion resisting magnesium alloys for their extensive use in automotive industry. AZ91 magnesium alloy is commercially used because of its good corrosion resistance. Isolation strategies required to enhance magnesium applications, for example, for the Corvette cradle [31, 34].

4.4 Damping Capabilities

In many magnesium applications such as instrument panels, radiator support brackets etc. uses AM50 and AM60 alloys as they exhibit excellent crash worthiness. Recent studies reveal magnesium alloys absorb comparatively more impact energy than steel on equivalent weight basis. Sharding or segment fracture are observed in AZ31 magnesium and AM30 magnesium tubes [32, 33]. Further investigation needed on fracture mechanisms for magnesium.

4.5 Noise, Vibration and Harshness (NVH)

Magnesium well known for its high damping capacity. Kiani et al. studied the light-weight magnesium car body structure under crash loading with limiting conditions

[34]. Magnesium exhibits good damping ability and better NVH performance in the frequency, 100–1000 Hz [35].

4.6 Fatigue Life

Zhenming et al. studied the fatigue behaviour of magnesium castings with NZK (Mg-Nd-Zn-Zr) alloys and other magnesium alloys including AZ91D, GW103, and AM-SC [36]. Fatigue of cast magnesium components based normally on casting defects and the micro structural aspects [37–42]. The effect of alloying, processing and texture on the fatigue characteristics of magnesium alloys need to be studied. Material characterization of magnesium alloy sheets need to be established by multi-scale simulation tools to predict the fatigue life of magnesium components and sub-systems, which can be validated for automotive applications.

5 Conclusion and Summary

The usage of magnesium in automobile industry can provide not only weight reduction but also reduce vibration and noise. In addition cast and wrought magnesium products reduces overall tooling and gauges required for production compared to steel. Because of all these salient features, magnesium alloys is emerging material in automotive industry. Magnesium due to high strength to weight ratio, reduced CO_2 emissions, fuel economy etc., significantly replacing other light structural metals like aluminium alloys in the last decade. This article gives few insights on applications of magnesium and limitations in automotive industry for extensive use of magnesium. Magnesium alloys and its use may expected tenfold growth in the next decade as great research is going on welding and forming techniques of magnesium alloys. However these techniques must be cost effective to use magnesium alloys in automobile industry. New characterization tools and magnesium alloy developments are expected in the near future which provide new design solutions to manufacture cast and wrought magnesium alloys. As there is a wider scope of work in automotive industry, magnesium alloys offer an opportunity to researchers. Further research work still needed for improvements in mechanical properties and corrosion and alloy developments.

References

1. Davies G (2003) Magnesium. Materials for automotive bodies, Elsevier, G. London, pp 91, 158, 159
2. Kuo, J.L., Sugiyama, S., Hsiang, S.H., Yanagimoto, J.: Investigating the characteristics of AZ61 magnesium alloy on the hot and semi-solid compression test. Int. J. Adv. Manuf. Technol. **29**(7–8), 670–677 (2006)
3. Jain, C.C., Koo, C.H.: Creep and corrosion properties of the extruded magnesium alloy containing rare earth. Mater. Trans. **2**, 265–272 (2007)
4. Fu, P., Peng, L., Jiang, H., Chang, J., Zhai, C.: Effects of heat treatments on the microstructures and mechanical properties of Mg-3Nd-0.2Zn-0.4Zr (wt.%) alloy. Mater. Sci. Eng., A **486**, 183–192 (2008)

5. Greiner, J., Doerr, C., Nauerz, H., Graeve, M.: The new "7G-TRONIC" of Mercedes-Benz: innovative transmission technology for better driving performance, comfort, and fuel economy. SAE Technical Paper No. 2004-01-0649. SAE International, Warrendale, PA (2004)
6. Kulekci, M.K.: Magnesium and its alloys applications in automotive industry. Int. J. Adv. Manuf. Technol. **39**, 851–865 (2008)
7. Blawert, C., Hort, N., Kainer, K.V.: Automotive applications of magnesium and its alloys. Trans. Indian Inst. Met. **57**(4), 397–408 (2004)
8. Eliezer, D., Aghion, E., Froes, F.H.: Magnesium science and technology. Adv. Mater. Perform. **5**, 201–212 (1998)
9. Aghion, E., Bronfin, B.: Magnesium alloys development towards the 21(st) century. Magnes. Alloys 2000 Mater. Sci. Forum **350**(3), 19–28 (2000)
10. Friedrich, H., Schumann, S.: Research for a "new age of magnesium" in the automotive industry. J. Mater. Process. Technol. **117**, 276–281 (2001)
11. Schuman, S.: The paths and strategies for increased magnesium application in vehicles. Mater. Sci. Forum **488–489**, 1–8 (2005)
12. Dieringa, H., Kainer, K.U.: Magnesium-der zukunftswerkstoff für die automobilindustrie. Mat-wiss U Werkstofftech **38**(2), 91–95 (2007)
13. Tang, B., Xs, W., Li, S.S., Zeng, D.B., Wu, R.: Effects of Ca combined with Sr additions on microstructure and mechanical properties of AZ91D. Mater. Sci. Technol. **21**(29), 574–578 (2005)
14. Sameer Kumar, D., et al.: Am. J. Mater. Sci. Technol. **4**(1), 12–30 (2005)
15. Avedesian, M.M., Baker, H.: ASM Specialty Handbook, Magnesium and Magnesium Alloys. ASM International, Materials Park (1999)
16. Timminco Corporation: Timminco Magnesium Wrought Products. Timminco Corporation Brochure, Aurora, CO (1998)
17. Luo, A.A., Sachdev, A.K.: Development of a new wrought magnesium-aluminium-manganese alloy AM30. Metall. Mater. Trans. A **38A**, 1184–1192 (2007)
18. ASM: Metals Handbook, Desk Edn. ASM International, Materials Park (1998)
19. Brown, R.E.: Future of magnesium developments in 21st century. In: Presentation at Materials Science and Technology Conference, Pittsburgh, PA, USA, 5–9 October 2008
20. Barnes, L.T.: Rolled magnesium products, 'what goes around, comes around'. In: Proceedings of the International Magnesium Association, Chicago, IL, pp. 29–43 (1992)
21. Friedrich, H.E., Mordike, B.L., et al.: Magnesium Technology. Springer, Berlin (2006)
22. Gupta, M., et al.: Magnesium, Magnesium Alloys, and Magnesium Composites. Wiley, Hoboken (2011)
23. Powell, B.R., Krajewski, P.E., Luo, A.A.: 'Magnesium alloys', in Materials Design and Manufacturing for Lightweight Vehicles, pp. 114–168. Woodhead Publishing Ltd., Cambridge (2010)
24. Luo, A.A., Sachdev, A.K.: General motors global research and development. In: Applications of Magnesium Alloys in Automotive Engineering. Woodhead Publishing Limited, Cambridge, UK, pp: 393–414 (2012)
25. Tanski, L.A., Dobrzanski, Labisz, K.: IISUES 1, 2 (2010)
26. Krajewski, P.E.: Elevated temperature behaviour of sheet magnesium alloys. SAE Technical Paper 2001-01-3104. SAE International, Warrendale, PA (2001)
27. Verma, R., Carter, J.T.: Quick plastic forming of a Decklid inner panel with commercial AZ31 magnesium sheet. SAE International Technical Paper No. 2006-01-0525. SAE International, Warrendale, PA (2006)
28. Carter, T., Krajewski, P.E., Verma, R.: The hot blow forming of AZ31 Mg sheet: formability assessment and application development. J. Miner. Met. Mater. **60**(11), 77–81 (2008)

29. Mendis, C.L., Bettles, C.J., Gibson, M.A., Hutchinson, C.R.: An enhanced age hardening response in Mg–Sn based alloys containing Zn. Mater. Sci. Eng., A **435**(436), 163–171 (2006)
30. Luo, A.A., Sachdev, A.K.: Microstructure and mechanical properties of Mg-Al-Mn and Mg-Al-Sn alloys. In: Nyberg, E.A., Agnew, S.R., Neelameggham, N.R., Pekguleryuz, M.O. (eds.) Magnesium Technology 2009, pp. 437–443. TMS, Warrendale, PA (2009)
31. Luo, A.A., Shi, W., Sadayappan, K., Nyberg, E.A.: Magnesium front end research and development: Phase I progress report of a Canada-China-USA collaboration. In: Proceedings of IMA 67th Annual World Magnesium Conference. International Magnesium Association (IMA), Wauconda, IL, USA (2010)
32. Easton, M., Beer, A., Barnett, M., Davies, C., Dunlop, G., et al.: Magnesium alloy applications in automotive structures. JOM **60**(11), 57–62 (2008)
33. Wagner, D.A., Logan, S.D., Wang, K., Skszek, T., Salisbury, C.P.: Test results and FEA predictions from magnesium AM30 extruded beams in bending and axial compression. In: Nyberg, E.A., Agnew, S.R., Neelameggham, N.R., Pekguleryuz, M.O. (eds.) Magnesium Technology 2009. TMS, Warrendale, PA (2009)
34. Kiani, M., et al.: Design of lightweight magnesium car body structure under crash and vibration constraints. J. Magnes. Alloy. **2**, 99–108 (2014)
35. Logan, S., Kizyma, A., Patterson, C., Rama, S.: Lightweight magnesium-intensive body structure. SAE International Technical Paper No. 2006-01-0523. SAE International, Warrendale, PA (2006)
36. Li, Z., et al.: Mater. Sci. Eng., A **647**, 113–126 (2015)
37. Li, Z.M., Fu, P.H., Peng, L.M., Wang, Y.X., Jiang, H.Y., Wu, G.H.: Mater. Sci. Eng., A **579**, 170–179 (2013)
38. Wang, Q.G., Davidson, C.J., Griffiths, J.R., Crepeau, P.N.: Metall. Mater. Trans. B **44**, 887–895 (2006)
39. Wang, Q.G., Apelian, D., Lados, D.A.: J. Light Met. **1**, 73–84 (2001)
40. Wang, Q.G., Jones, P.E.: Metall. Mater. Trans. B **38**, 615–621 (2007)
41. Mayer, H., Papakyriacou, M., Zettl, B., Stanzl-Tschegg, S.E.: Int. J. Fatigue **25**, 245–256 (2003). [16] Xu, D.K., Liu, L., Xu, B.Y., Han, E.H.: Acta Mater. **56**, 985–994 (2008)
42. Horstemeyer, M.F., Yang, N., Gall, K., McDowell, D.L., Fan, J., Gullett, P.M.: Acta Mater. **52**, 1327–1336 (2004)

Development of Eyeball Movement and Voice Controlled Wheelchair for Physically Challenged People

N. Dhomina(✉) and C. Amutha

Department of Electrical and Electronics Engineering,
Rajalakshmi Engineering College, Thandalam, Chennai 602105, India
dhomi9682@gmail.com, amutha.c@rajalakshmi.edu.in

Abstract. Many people with disabilities do not have the ability to control the powered wheelchair manually. This is overcomed in this project by the construction of well structured design and intelligent wheelchair for physically handicapped people. This wheelchair is modeled such that it can be run with less effort from the patient by using voice processing module to the microcontroller by giving voice commands for different directions. Another feature provided to patients are that, the wheelchair can also be controlled by eyeball module with sensors. Based on the movement of eyeball, wheelchair can be controlled.

Keywords: ARM LPC2138 with LCD · Ultrasonic sensor · Voice processing module · Eyeball module · Driver circuit · DC motor

1 Introduction

The lives of those among us who are unfortunate enough to have lost the potential to move their legs due to various reasons like accidents, paralysis. Many disabled people depend on others in their daily life specifically in moving from one place to another. Hence they need someone continuously to help them in getting the wheelchair move. Knowing about all these facts, the main aim was to design a voice processing and eyeball controlled wheelchair for physically challenged people. Their lives are made complex by the fact of lacking self control for their wheelchairs that allows to move independently. Voice control is the fetching choice for various causes. Speech module can be used by any individual capable of compatible and detectable utterance; therefore, voice control is capable for many of the wheelchair users. Power of speech would also minimize the physical strain of steering a wheelchair. By eradicating the need to move one or many restrictions to drive the chair, voice control could support the wheelchair operator in maintaining exact positioning within his or her seating system. One difficulty is the very real possibility that a voice input may fail to realize a user's voice. An eyeball movement controlled wheelchair is also designed for paralyzed patients.

There were many papers based on wheelchair for differently abled people. The paper [1] proposes about designing the wheelchair powered by solar energy, wheelchair movement is created by giving voice commands through the Bluetooth technology in mobile app. The paper [4] designs a wheelchair controlled by eye. The

D. J. Hemanth et al. (Eds.): COMET 2019, LNDECT 35, pp. 532–539, 2020.
https://doi.org/10.1007/978-3-030-32150-5_51

measurement of eye positions is used to convert the signals emitted from eye into equalization of staring points.

The paper [6] works about the single support module which is suitable for multiple concepts. The paper [5] describes about the BCI that moves the wheelchair faster with low effort. The paper [12] conveys about the powerful scheme that can identify the finger movement and is detected that relies on the background subtraction and morphological methods. The paper [8] denotes the HMI interface involving two modes instead of the joystick control.

The paper [2] explains about the development of wheelchair with the help of Android device and testing is been done for conditions with voice and button control and compared. The paper [3] explains about the four directions of the wheelchair that involves the theme of android device. The paper [7] combines the control of mobility and manipulation to do daily activities using a robotic Arm. The paper [9] details a general illustration of android assistive technology for a wheelchair. The paper [10] explains about the hill climbing wheelchair that compensates gravity and friction. In the paper [11] machine learning in Robots is involved with the 2 dimensional Range of data. The paper [13] compares the movement of wheelchair of attender's joystick and patient's touch on handlebar.

2 Design Methodology

This proposed system uses ARM LPC2138 microcontroller and ultrasonic sensor to detect any obstacle. Two DC motors with nominal voltage of 12v is placed which is operated based on the voice commands given as input and eyeball sensors which detects the direction for the dc motor to move. L293D is a typical motor driver circuit in which Dc motor is connected. The simulation is carried out in Proteus.

2.1 Voice Processing

Voice processing is done using MFCC (Mel Frequency Ceptrum Co-efficient) algorithm for speaker recognition. This is based on the human hearing perceptions which cannot perceive frequencies over 1 kHz. It has two filters which are spaced linearly at low frequency below 1000 Hz and logarithmic spacing above 1000 Hz.

Pre Emphasis. This method is used to convert low frequency speech signal into high frequency speech signal to avoid mixing of noise during transmission.

Sampling and Windowing. Speech signal given as input is partitioned into frames of 15–20 ms with 50% intersection of the frame. Each hamming window is used to keep the first and last frames in continuity and this window is used for spectral analysis of the input signal because the spectrum step down is quick so the frequency resolution obtained is superior.

FFT. It converts each frame from time domain to frequency domain. When FFT is performed on a frame it is considered to be periodic and is used to convert complexity of the pulse and auditory impulse response in time domain. We multiply each frame by hamming window to increase its continuity.

Mel Filter Bank. Mel scale corresponds to the better resolution at low frequencies. Mel Frequency scale is extended upto the frequency of 1 kHz and then becomes near to the arithmetic value of the higher frequencies. Log compresses dynamic range of values.

Discrete Cosine Transform. This process converts the log value obtained in previous method into time domain. Feature is obtained as output co efficient. As the vector value in log is uniform, the output filters can be compressed.

Output Co-efficient. During measurement of data and values, a dataset of voice prints is generated which is used as a reference in the phase equivalent to the feature (Table 1).

A voice print is a pattern of numbers where the number denotes the energy perceived at definite frequency band for a period of time. Output is obtained as values denoting the start of the frequency ranges (Fig. 1).

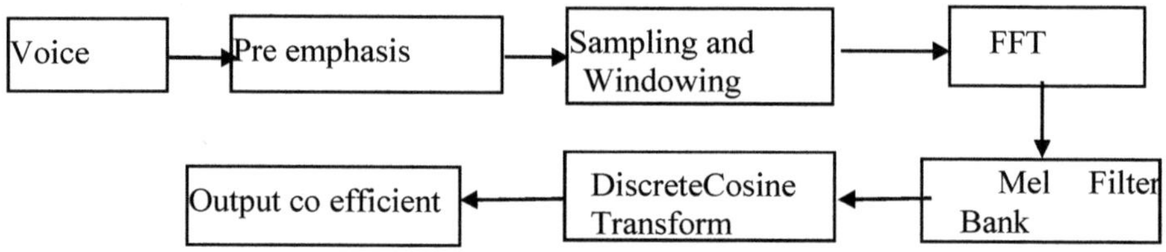

Fig. 1. Block diagram for voice processing

Working of voice processing module is as follows,

1. If command A is given, the motor moves front.
2. If command B is given, the motor moves reverse.
3. If command C is given, the motor moves Left.
4. If command D is given, the motor moves Right.
5. If command E is given, the motor stops (Fig. 2).

Table 1. Voice inputs

Characters	Commands
A	Front
B	Reverse
C	Left
D	Right
E	Stop

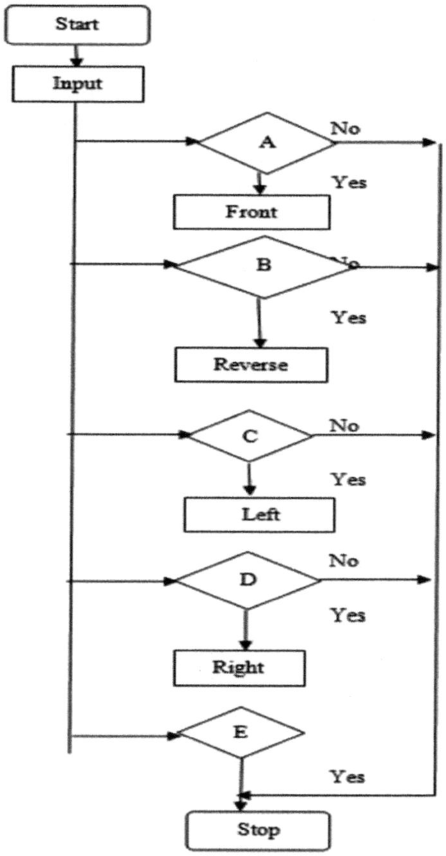

Fig. 2. Flowchart for voice processing

2.2 Eyeball Module

Infrared signals are detected and sense the eyeball movement from the sensors placed in the eye on left corner, left middle, right corner. This rays from Light emitting diode brightens the eye and reflected Infrared light produces an electrical current through the Infrared photodiode. It emits a beam of IR rays whenever white occurs in front of the receiver, these rays are returned back and captured. When faced with black IR rays are absorbed by the surface and cannot captured. When the Infrared transmitter ejects radiation, since there is no absolute contact between the transmitter and receiver, the released radiation must revert back to the photodiode after striking any object (Fig. 3).

Fig. 3. Block diagram for eyeball module

Working of eyeball module is as follows,

- When the mid switch is on, the motor moves front.
- When the right switch is on without delay, the motor moves reverse.
- When the left switch is on, the motor moves in leftside.
- When the right switch is on with a delay, motor moves in rightside.
- When all the switches are off, the motor stops (Fig. 4).

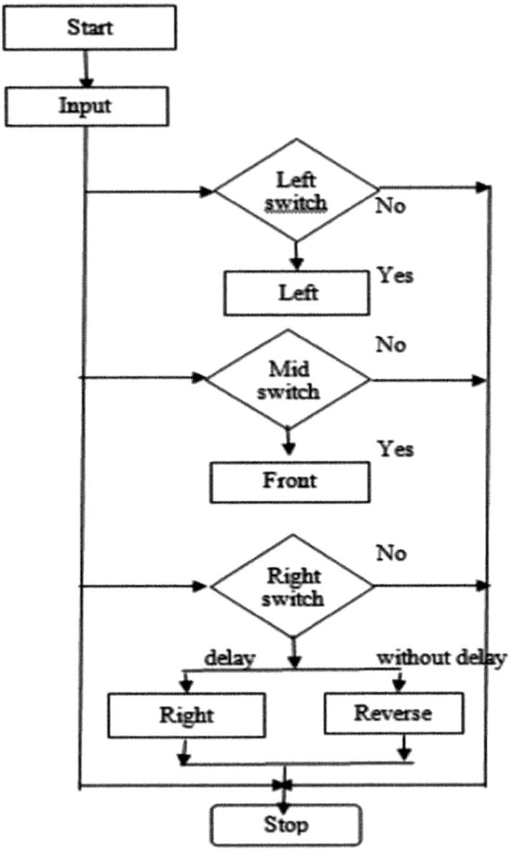

Fig. 4. Flowchart for eyeball module

3 Simulation Results

Simulation is done in proteus software before the hardware execution to check the performance of individual components. In this proposal, simulation was done for the motor to run in forward, reverse, left, right directional movement of the wheelchair according to the voice commands and eyeball movement. In the simulation, these direction were tested by giving signals to the microcontroller. In Fig. 5 the input for the eyeball sensors are given such that when the mid switch is on, the motor moves in

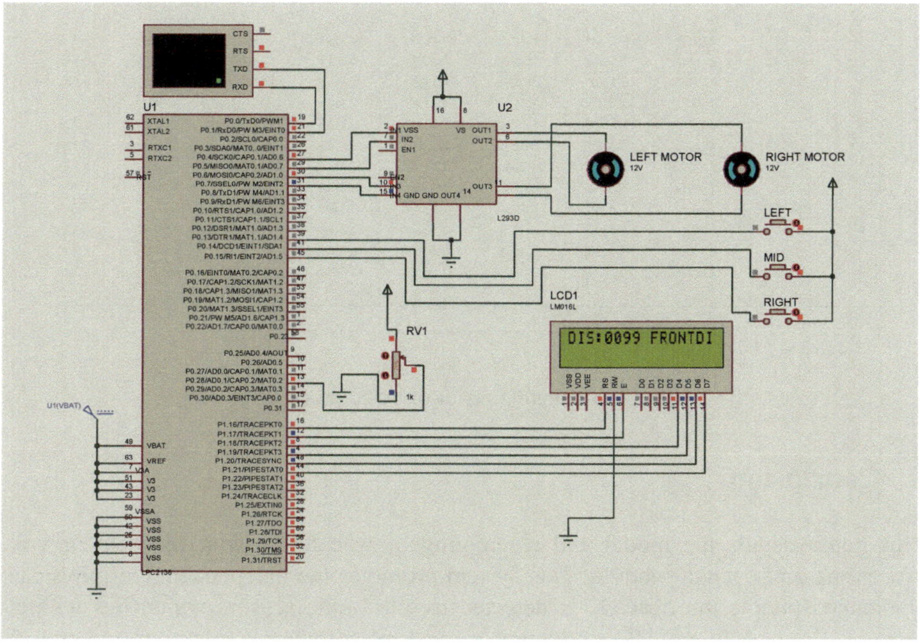

Fig. 5. Result for eyeball module

forward direction displaying as front in the LCD display. In Fig. 6 the input for the voice command are given by the character C in the virtual terminal such that the UART receives the data and allows the motor to move in right direction displaying as right in the LCD display. In Fig. 7 the output for voice processing is displayed with dynamic range of values. Matlab is run with the path added as the voice signal recorded and the simulated result is obtained as the voice of the directions.

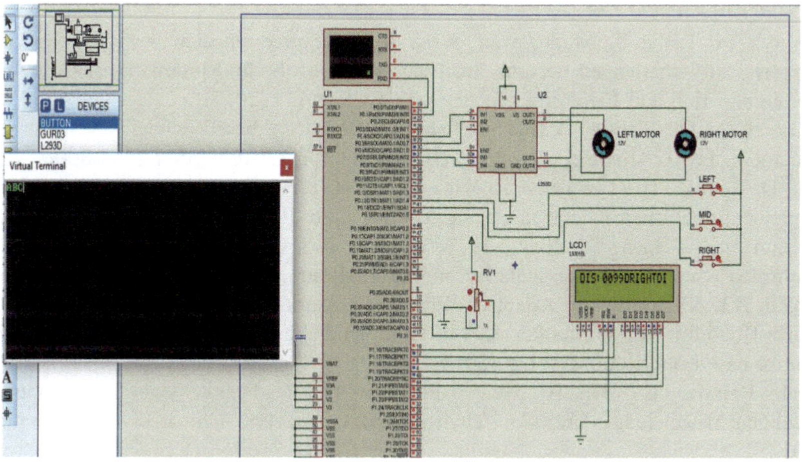

Fig. 6 Result for voice module

	1	2	3	4	5	6	7	8	9
1	57.2071	58.3889	57.6310	60.4212	63.3864	74.3834	76.2571	72.6322	65.975
2	-4.1858	-4.6602	-4.6683	-7.9019	-0.6632	4.8019	3.1043	5.4906	0.943
3	0.0494	-1.7594	-3.4991	-2.0144	-4.1636	-8.7816	-11.8657	-7.2855	-6.292
4	0.2553	-1.8728	-2.8965	1.9659	-6.2136	-15.3805	-18.9183	-12.2951	-3.621
5	-3.0516	0.7459	1.2925	-2.2383	-2.9361	-6.3436	-6.2141	-7.8030	-3.022
6	-1.7739	-1.8441	-0.9146	0.1880	-0.0201	2.3543	-0.7080	-0.9504	0.114
7	1.2137	-0.1074	1.0734	1.7546	1.6182	-11.4177	-14.7080	-8.2651	0.528
8	-0.4810	1.9075	1.2668	0.0749	0.7281	-2.7814	-3.3625	-3.9389	-3.448
9	0.8909	-2.6346	0.3669	-0.8994	-4.4370	-6.8700	-7.3161	-8.7877	-6.681
0	6.6116	5.4052	5.5166	2.0325	-1.6381	7.7446	5.4517	6.9326	0.095
1	5.7835	6.1590	-0.6878	3.1235	-0.4165	-8.0706	-10.8915	-6.9187	-6.201
2	1.2394	4.0557	0.6345	2.4504	1.9584	-12.0783	-10.8595	-9.4522	-8.728

Command Window

Forward

Fig. 7 Simulation result of voice processing

4 Conclusion

This paper details the model and assembling of wheelchair with the help of voice processing and eyeball module. This system promotes the independency of physically challenged people the obstacle is detected by ultrasonic sensors which lies within a range of 2 cm–400 cm. The simulation using Keil software is compiled to show the motor running in Proteus and the voice is processed in Matlab software using MFCC Algorithm. The simulation of wheelchair movement is perfectly detected and the simulation results are obtained in the form of schematic design. Thus it will be more convenient for the physically handicapped people to move the wheelchair.

References

1. Kamble, S.R., Patil, S.P.: Solar powered touch screen wheelchair. In: International Conference on Innovations in Information Embedded and Communication Systems (ICIIECS) (2017)
2. Iswarya, M., Latha, S., Madheswari, A.N.: Solar powered wheelchair with voice controlled for physically challenged persons. In: Proceedings of the 2nd International Conference on Communication and Electronics Systems (ICCES 2017) (2017)
3. Balsaraf, M.D., Takate, V.S., Siddhant, B.: Android based wheelchair control using Bluetooth. Int. J. Adv. Sci. Res. Eng. (IJASRE) **03**(4) (2017). ISSN 2454-8006
4. Bai, D., Liu, Z., Hu, Q., Yang, J., Yang, G., Ni, C., Yang, D., Zhou, L.: Design of an eye movement-controlled wheelchair using Kalman filter algorithm (2016)
5. Rebsamen, B., Guan, C., Zhang, H., Wang, C., Teo, C., Ang, M.H., Burdet, E.: A brain controlled wheelchair to navigate in familiar environments (2010)
6. Argall, B.D.: Modular and Adaptive Wheelchair Automation (2016)
7. Elarbi-Boudihir, M., Al-Shalfan, K.A.: Eye-in hand/eye-to-hand configuration for a WMRA control based on visual servoing (2013)
8. Rechy-Ramirez, E.J., Hu, H., McDonald-Maier, K.: Head movements based control of an intelligent wheelchair in an indoor environment, Guangzhou, China, 11–14 December 2012 (2012)

9. Martinazzo, A.A.G., José, M.A.: The motion assistant: engineering a Bluetooth-enabled power wheelchair. In: IEEE International Symposium on Consumer Electronics (2016)
10. Lee, K.M., Lee, C.H., Hwang, S., Choi, J., Bang, Y.B.: Power assisted wheelchair with gravity and friction compensation. IEEE Trans. Ind. Electron. **63**, 2203–2211 (2015)
11. Beyer, L., Hermans, A., Linder, T., Arras, K.O., Leibe, B.: Deep person detection in two-dimensional range data. IEEE Robot. Autom. Lett. **3**(3), 2726–2733 (2018)
12. Chowdhury, S.S., Hyder, R., Shahanaz, C., Fattah, S.A.: Robust single finger movement detection scheme for real time wheelchair control by physically challenged people. In: IEEE Region 10 Humanitarian Technology Conference (R10-HTC), Dhaka, Bangladesh (2017)
13. Trujillo-León, A., Bachta, W., Vidal-Verdú, F.: Tactile sensor based steering as a substitute of the attendant joystick in powered wheelchairs. Trans. Neural Syst. Rehabil. Eng. **26**, 1381–1390 (2018)

Modelling and Analysis of Auxetic Structure Based Bioabsorbable Stents

Ilangovan Jagannath Nithin(✉) and Narayanasamy Srirangarajalu

Department of Production Engineering, Madras Institute of Technology, Chennai, India
ijnithin.97@gmail.com, nsrirangarajulu@gmail.com

Abstract. Auxetics are materials or structures which exhibit negative Poisson's ratio and expands perpendicular to the force applied. Nowadays, they are currently used in fabrics, foam like products and its usage is limited. In literature, several works have been carried out to analyze the properties and behavior of these materials in various conditions. In this work, stents based on different auxetic structures were modelled and analyzed using finite element analysis. The Chiral, Re-entrant and Rotating unit type of auxetic structures were modelled. The Poisson's Ratio and Young's Modulus are the performance metrics considered to analyze these structures. The Magnesium Alloy AZ61 is the material chosen to make the stent bio absorbable. Results suggest that chiral auxetic structure-based stent of this material produce appreciable results in terms of the performance metrics considered.

Keywords: Auxetic · Stent · Finite element analysis · Bioabsorbable

1 Introduction

A stent is a short narrow metal or plastic tube designed in the form of a mesh. It is used to keep the blockage open in the anatomical vessels. The profile and flexibility of the stents are the characteristics that play a vital role in attaining the stent designing factors such as deliverability and deployment of the stent. The parameters such as strut length and width, wire diameter and pitch, the mesh configuration, material selection and processing conditions are studied to design an ideal stent [1]. The radial force required depends on the lesion characteristics and its location. There are two configurations of stents namely open-cell and closed-cell. In closed-cell configuration the adjacent ring segments are connected at every possible junction [1]. In open-cell configuration there will no sufficient connections which leads to excessive deformation and is not encouraged in some applications. Hence in this work, the closed-cell configuration is preferred.

The materials used in stents are mostly based on stainless steel platform which is less expensive. However, in the later stage of stent implementation, they cause problems such as restenosis and thrombosis [2]. Recently, research has been carried out to replace usage of Stainless steel with other materials such as Cobalt-Chromium alloy, Titanium alloys, Nitinol, etc. These materials also possess some risk as mentioned before. To prevent these issues, Magnesium alloys are considered for stents as they

D. J. Hemanth et al. (Eds.): COMET 2019, LNDECT 35, pp. 540–550, 2020.
https://doi.org/10.1007/978-3-030-32150-5_52

possess bioabsorbable characteristics. It means that it has the capability of getting dissolved into the tissue itself and makes it dilate and contract like a healthy blood vessel.

The structure of the stent also plays a vital role in stent performance and it can be improved by means of auxetics. The term "Auxetics" was introduced by Evans [3]. They are structures or materials that possess Negative Poisson's Ratio. When the load is applied in a direction on the structure, they get expanded not only in that direction but also in the direction perpendicular to the applied load [3]. They possess high energy absorption and high resistance to fracture. To advance in stent design from structural part of view different auxetic structures were explored and to provide the bioabsorbable capability, Magnesium alloys were also analyzed for an increased lifetime.

The organization of the work is as follows. The Sect. 2 presents related works in recent literature and is followed by the detailed methodology of the work in Sect. 3. The Sect. 4 presents the results and discussion of the analysis carried out and Sect. 5 concludes the work with future direction.

2 Literature Survey

The material properties and the microstructural data of Magnesium alloys such as AZ31, AZ61, AZ80, ZM21, ZK61 and WE43 were studied in [2]. The Corrosion rates of these alloys in Synthetic Body Fluid (SBF) solution were determined and the stress-strain characteristics of the materials were studied. The alloy AZ61 is found to be optimal. The hot formability temperature ranges of the alloys were investigated and concluded that the hot extrusion for small tubes for stents should be carried at low temperatures. Juan et al. created a library of 2-D and 3-D auxetic geometries and simulated them to provide a comparison of their properties such as Poisson's Ratio, Maximum volume or area reduction and Equivalent Young's Modulus [4]. Carneiro et al. created two models of stent using re-entrant and chiral geometry [5].

Mechanical properties were evaluated using FEA and the presence of auxetic behaviour is concluded. Also, less axial deformation was observed and makes it suitable for application of stent design. In [6] authors have modelled a stent based on auxetic chiral lattice and made the stent to have biodegradable capability by using Magnesium Alloy (AZ91) as a stent material. The stent was analyzed and found that it expands when it is stretched. However, the Axial and radial strain were found to be low and this can be improved by changing the mechanical properties of the base material through some processing methods. The proposed work concentrates on analyzing the stents based on different auxetic structure such as chiral, re-entrant and rotating unit using Magnesium alloy AZ61 as the material.

3 Proposed Work

This section discusses the objective of the work, the methodology, and the inputs taken for the work. This work is done to improve the existing stent structures by replacing them with auxetic structures and to analyze them to achieve optimal stent design. In this work, stents are modelled using different auxetic structures and then compare them based on performance parameters which are going to be discussed here. The stents are modelled using SOLIDWORKS 2018 which is a 3-D CAD Modelling software developed by Dassault Systèmes. After modelling, the material is then assigned to those models and they are subjected to Finite element analysis using SOLIDWORKS Simulation module available in that software. From the results, the structures are compared and the suitable structure for the stent is inferred.

3.1 Material

Magnesium alloys are biodegradable, and it has low corrosion rates. These Magnesium alloy stents reduce the risk of stenosis, late and very-late stent thrombosis, demand for antiplatelet therapy, long-term patient health risks. In [2] some of the Magnesium alloys are analyzed for their material properties and from those results, the Magnesium alloy Mg AZ61 is chosen for stent material. It has the high ultimate strength and high elongation to fracture. Table 1 represents the composition and Table 2 represents physical and mechanical properties of Mg AZ61 respectively.

Table 1. Composition of Mg AZ61.

Element	Content (%)
Magnesium, Mg	92
Aluminium, Al	5.80–7.20
Zinc, Zn	0.40–1.50
Manganese, Mn	0.15
Silicon, Si	0.10
Copper, Cu	0.050
Nickel, Ni	0.0050
Iron, Fe	0.0050

3.2 Structures

The auxetic structures fall into categories such as re-entrant, chiral and rotating units. They are also other structures as well. The chiral include chiral circular, chiral circular symmetric, chiral hexagonal, chiral rectangular symmetric, chiral square symmetric [4]. The chiral structure is formed by connecting ribs to the central nodes which may be circular or other geometrical forms. The re-entrant has basic re-entrant versions proposed by Masters and Evans [3], triangular, star 3-n, star 4-n etc. [4]. The rotating unit square structure was given by Grima and Evans [3] consist of lattices which are made up of basic shapes such as triangle, square, rectangles connected at their vertices by hinges.

Table 2. Physical and mechanical properties of Mg AZ61.

Properties	Metric
Density	1.80 g/cm^3
Tensile strength	310 MPa
Yield strength	230 MPa
Compressive yield strength	130 MPa
Shear strength	140 MPa
Elastic modulus	44.8 GPa
Poisson's ratio	0.35

3.3 Analysis

The static analysis is first carried out for stents based on chiral auxetic structures as mentioned above. An optimal structure is then arrived from the analysis. Then that structure is compared with re-entrant and rotating unit square stents. For analysis, pressure is applied normal to the inner surface of the stent and is made to expand radially. And as they are auxetic structures, linear expansion is also observed. From the analysis, parameters such as equivalent stress, equivalent strain, strain in radial direction, strain along the length is determined. From those values, the parameters used to compare these structures are calculated using formula. These parameters include Poisson's Ratio and Young's Modulus.

Poisson's Ratio. It is defined as the negative of ratio of transverse strain to longitudinal strain i.e. the displacement along the direction of the load to the displacement along the direction perpendicular to the load. For an isotropic, elastic material, the Poisson's ratio will be positive. But for an auxetic structure, the Poisson's ratio will be negative. This means that the structure gets expanded in the direction perpendicular to the load [3]. The Poisson's ratio is calculated as mentioned in the equation below.

$$\nu = -(\varepsilon_r / \varepsilon_l). \tag{1}$$

where ε_r is the transverse strain and ε_l is the longitudinal strain.

Young's Modulus. It is a measure of the stiffness of a material. It is defined as the ratio of Stress to the strain. It forms the factor of proportionality in Hook's Law. A material having low Young's Modulus will require less force to deform it and if the value is high it will require high force. It is calculated as given by the following equation,

$$E = \sigma_{eq} / \varepsilon_{eq} \tag{2}$$

where σ_{eq} is the equivalent stress and ε_{eq} is the equivalent strain after loading.

4 Results and Discussions

In this section, the results obtained from the FEA analysis and how the results are used to compare these structures is discussed. The CAD models of the stents based on chiral, re-entrant and rotating units were created. They were analyzed using SOLIDWORKS Simulation module. It is an inbuilt module available in the SOLIDWORKS 2018.

4.1 Analysis of Chiral Auxetic Stents

The auxetic stents were modelled using auxetic chiral structure and they were subjected to static analysis. The model and results were presented in this section. Using the results, the Poisson's ratio and Young's Modulus were calculated.

Modeling Analysis. The auxetic structures in the chiral category include Chiral circular (CC), Chiral circular symmetric (CCS), Chiral hexagonal (CH), Chiral rectangular symmetric (CRS), Chiral square symmetric (CSS). The thickness of the ribs and the rings were taken as 1.2 mm. The isometric view of the 3D models of auxetic chiral stents was shown in Fig. 1 from (a)–(e). The FEA analysis is performed the above-shown stents by applying pressure on the inner surface. The expansion and the elongation of the stents were observed.

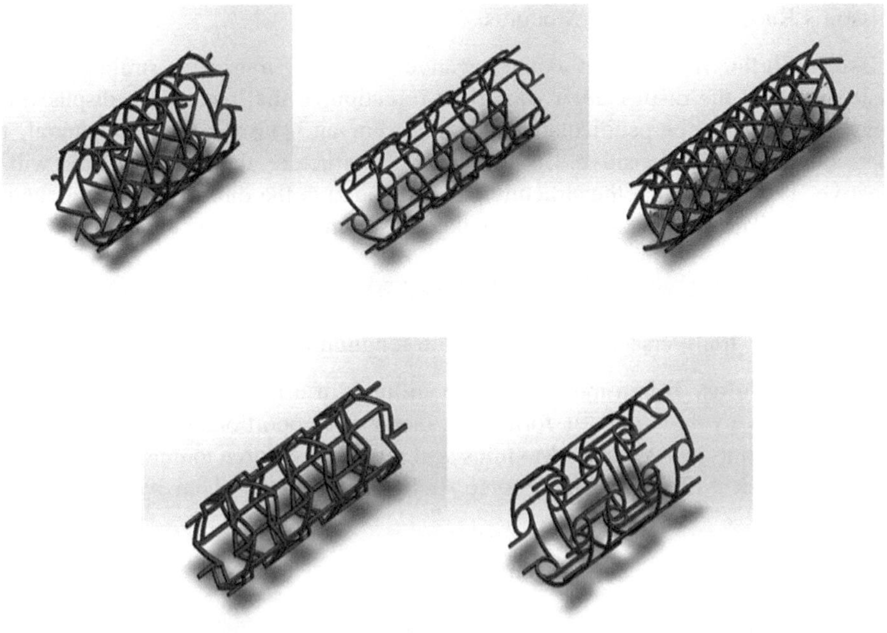

Fig. 1. Isometric view of Auxetic Chiral Stents

The Von Mises Stress distribution on all the elements of the chiral auxetic stents were represented in Fig. 2(a)–(e). The axial and the radial strain were obtained from the result. Also, the equivalent stress and equivalent strain were recorded for Young's Modulus calculation.

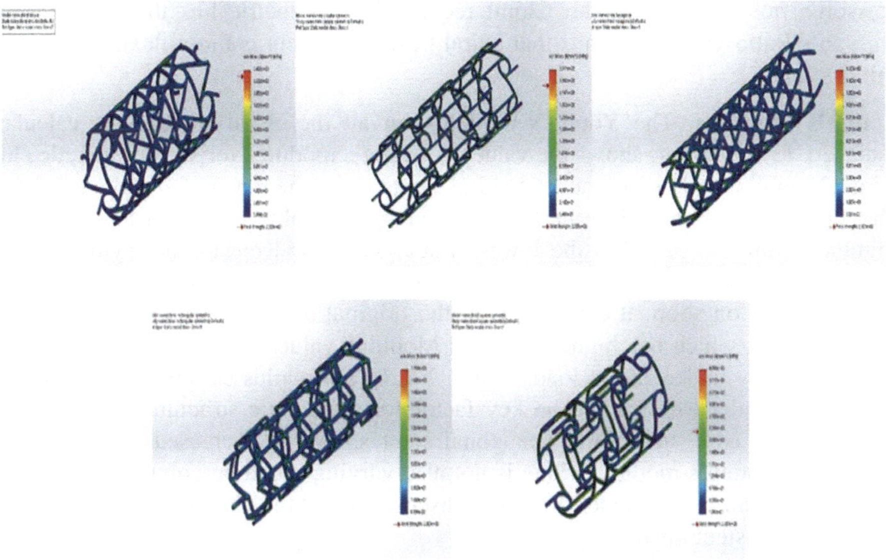

Fig. 2. Von Mises stress distribution after loading in Auxetic Chiral Stents

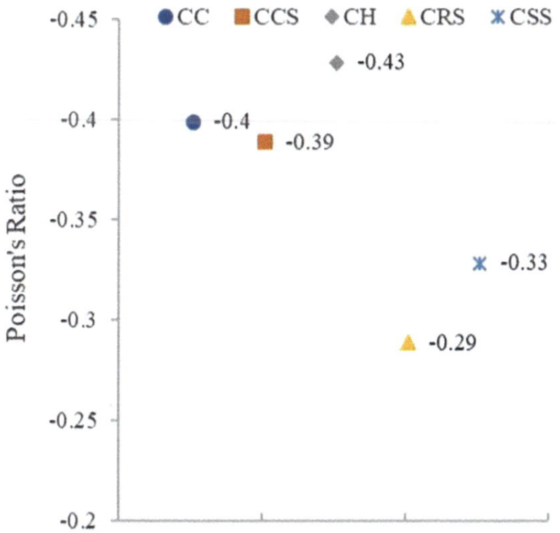

Fig. 3. Poisson's ratio for different Auxetic Chiral Stents

Poisson's Ratio. From the resultant axial and radial strain, the Poisson's ratio is calculated for all the models based on chiral structures using Eq. (1). Figure 3 represents the plot of Poisson's ratio for different chiral stents. The Poisson's Ratio value for chiral stents vary from −0.29 to −0.43. The Poisson's Ratio value is a measure of elasticity of the material. It represents whether the structure is stiff or flexible. From the plot it is seen that stent designed using Chiral hexagonal lattice has the most negative Poisson's ratio Value and the Chiral rectangular symmetric has the least negative Poisson's Ratio value. It means that chiral hexagonal stent is more flexible than other chiral stents.

Young's Modulus. The Young's modulus for all the chiral stents were calculated using Eq. (2). Figure 4. shows the value of Young's modulus for various auxetic chiral stents. Young's modulus for auxetic chiral stents are in the range of 77 GPa to 88 GPa. The Chiral circular stent has the highest Young's modulus value and the Chiral rectangular symmetric stent has the lowest value. The chiral rectangular symmetric will require low stress to cause permanent deformation which means it will transform into the plastic region soon. It cannot regain the original shape again. On the other side, Chiral circular which has higher Young's Modulus value is stiffer than others. It will deform slightly under elastic load. A low Young's modulus structure will change its shape even under low loads. The key factor for an auxetic structure is its Poisson's Ratio. In that case, the Chiral hexagonal stent shows an increased Poisson's Ratio value. The Young's modulus value is not also varying much on comparing with other structures. It can be concluded from this discussion that the Chiral hexagonal structure is optimal for stent design.

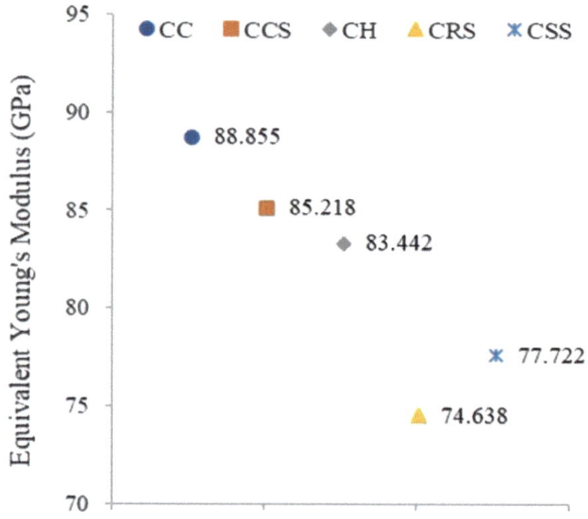

Fig. 4. Equivalent Young's Modulus for various Auxetic Chiral Stents

4.2 Analysis of Chiral, Re-entrant and Rotating Units Auxetic Structure

The results from the previous section shows that the chiral hexagonal structure is optimal for stent based on auxetic chiral structures. This is compared with re-entrant and rotating units structure for further improvement. This comparison is done to investigate other type of auxetic structures.

Modelling and Analysis. The Re-entrant (RE) and Rotating unit square stents (RUS) is modelled, subjected to same type of analysis and compared with chiral hexagonal stent. The isometric view of the 3-D models of re-entrant and Rotating unit square stents is presented in Figs. 5 and 6 respectively. Figures 7 and 8 shows the Von Mises Stress distribution on Re-entrant and Rotating unit Square stents after loading.

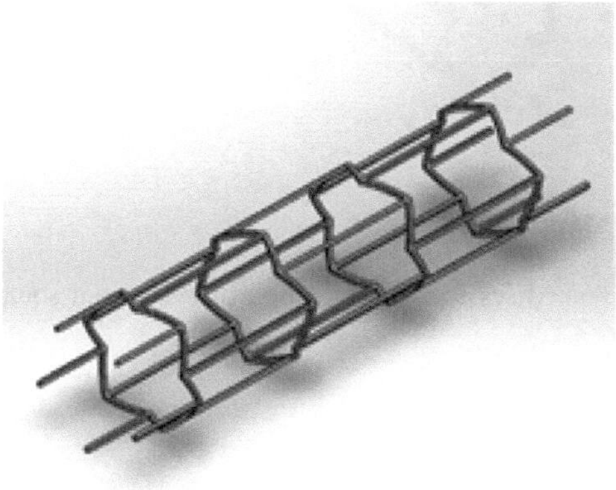

Fig. 5. Isometric view of the Re-entrant unit square stents

Fig. 6. Isometric view of the Rotating unit square stents

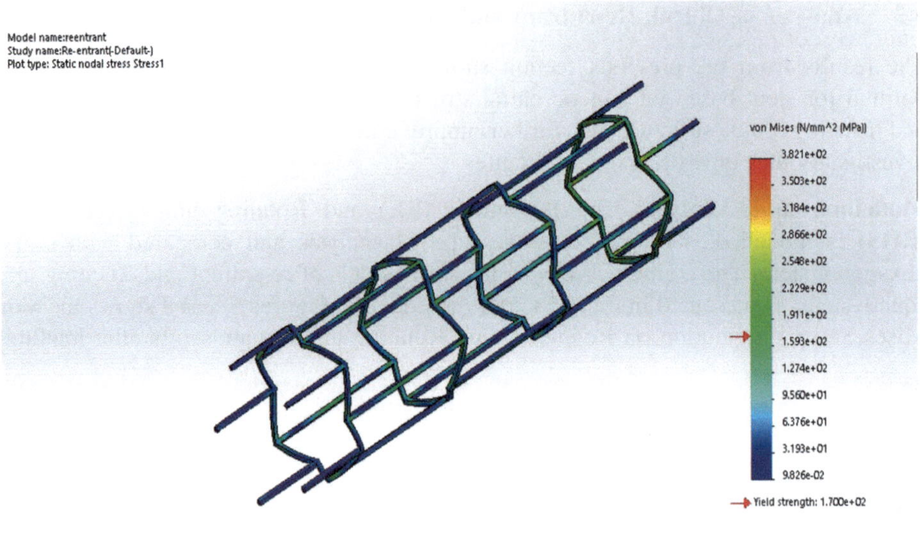

Fig. 7. Von Mises stress distribution of Re-entrant squares stent

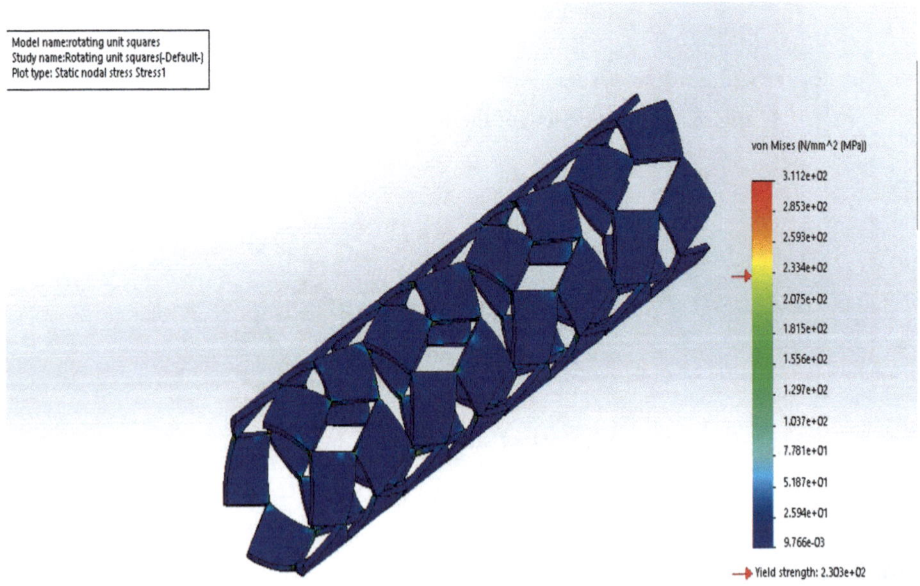

Fig. 8. Von Mises stress distribution of Rotating unit squares stent

Poisson's Ratio. Poisson's ratio is calculated for both re-entrant and rotating unit square stents using Eq. (1). The Poisson's ratio is plotted for re-entrant and rotating unit square stent in Fig. 9. It also includes chiral hexagonal stent arrived from the previous analysis. The plot shows that chiral hexagonal has most negative Poisson Ratio and re-entrant being the least negative. The stents with most negative Poisson's ratio will show increased auxetic behaviour than others. From this comparison, the chiral hexagonal is better by means of Poisson's ratio.

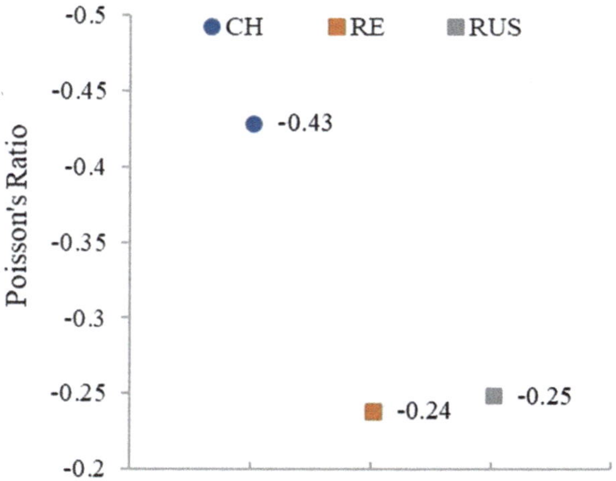

Fig. 9. Poisson's ratio comparison for Chiral, Re-entrant and Rotating unit square stents

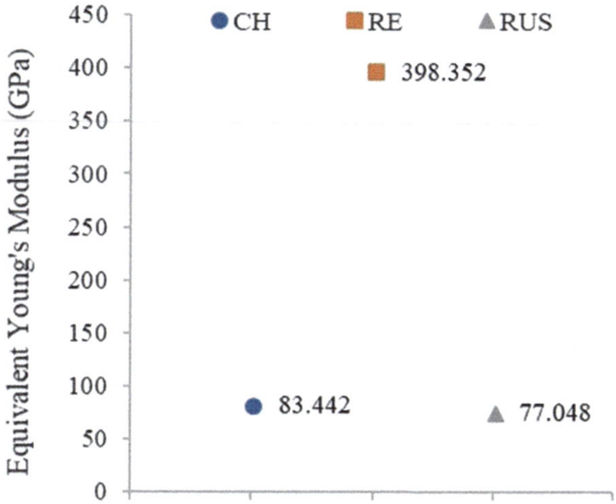

Fig. 10. Equivalent Young's Modulus for Chiral, Re-entrant and Rotating unit square stents

Young's Modulus. Using the values of equivalent stress and equivalent strain, Young's modulus is calculated using Eq. (2). The resultant Young's Modulus values for Chiral hexagonal, Re-entrant and Rotating unit square stents are plotted in Fig. 10. From Fig. 10 it seems the re-entrant structure has more Young's Modulus than other structures. This implies that the structure is stiffer, and it is difficult to deform easily. It can be suitable for other purposes but for the stent it should be less stiff so that deployment of the stent and bending is also easy. When coming to Rotating Unit square stent its Poisson's ratio value is nearly equal to that of Re-entrant stent. It has Young's Modulus value nearer to the Chiral hexagonal stent. The rotating unit square has more surface area than the Chiral hexagonal. This increases the weight of the stent comparatively. Also, in Rotating unit square stent structure, the square units are joint at the edges and there is possibility for failure at those edges easily. It can be concluded from these results that the chiral hexagonal stent is most suitable than other auxetic structure.

5 Conclusion

Different kinds of Auxetic structures namely Chiral, Re-entrant and Rotating units were utilized to create stents of new design. The analysis was performed in two sections. The performance parameters include Poisson's Ratio and Young's Modulus. Initially, the stents were designed using auxetic chiral lattices and FEA was performed to determine the performance parameters. From the results, the Chiral hexagonal structure is found to be an optimal stent structure among other chiral structures. Finally, it is compared with Re-entrant and Rotating unit square structures. It is noted that all the structures show Negative Poisson's Ratio confirming the auxetic behaviour. Here also the Chiral hexagonal structure performed well based on Young's Modulus and Poisson's Ratio. In future, the stents can be designed based on other auxetic structures and different materials can be explored suitable for stent design and usage.

References

1. Wholey, M.H., Finol, E.A.: Stent cell geometry and its clinical significance in carotid stenting. Endovasc. Today **6**, 25–34 (2007)
2. Farè, S., Ge, Q., Vedani, M., Vimercati, G., Gastaldi, D., Migliavacca, F., Petrini, L., Trasatti, S.: Evaluation of material properties and design requirements for biodegradable magnesium stents. Rev. Matèria **15**, 96–103 (2010)
3. Lim, T.-C.: Auxetic Materials and Structures. Springer, Singapore (2015)
4. Elipe, J.C.Á., Lantada, A.D.: Comparative study of auxetic geometries by means of a computer-aided design and engineering. Smart Mater. Struct. **21**, 1–12 (2012)
5. Carneiro, V.H., Puga, H.: Modelling and elastic simulation of auxetic magnesium stents. In: IEEE 4th Portuguese Bio-engineering Meeting, Porto, Portugal, pp. 1–4 (2015)
6. Carneiro, V.H., Puga, H.: Deformation behaviour of self-expanding magnesium stents based on auxetic chiral lattices. Ciência Tecnol. dos Materiais **28**, 14–18 (2016)

Green Aware Based VM-Placement in Cloud Computing Environment Using Extended Multiple Linear Regression Model

M. Hemavathy(✉) and R. Anitha

Department of Computer Science and Engineering,
Sri Venkateswara College of Engineering, Sriperumbudur 602117, India
hemamuthu95@gmail.com, ranitha@svce.ac.in

Abstract. In recent years, because of the increase in the huge volume of data and increase in data analytics in various research areas like health care, image processing etc., it is highly needed to provide required resources for processing the information. Cloud computing process an approach for delivering required resources by improving the utilization of data-center resources which results in increasing the energy costs. In order to overcome this new energy-efficient algorithms are introduced, that decreases the overall energy consumption of computation and storage. To reduce the energy-efficiency in cloud data centers, server consolidation technique is used, which plays a major road block. To address this issue, this project proposes a Prediction based Thermal Aware Server Consolidation (PTASC) model, a consolidation method, which takes numeric and local architecture into consideration along with Service Level Agreement. PTASC, consolidates servers (VM Migration) using a statistical learning method.

Keywords: VM Migration · Overload Detection · VM placement

1 Introduction

Cloud computing is anything can be provided as a service. Storing and Retrieving data from the internet instead of any local computer hard disk and also running the resources over the internet. Now-a-days there is a huge volume of data are generated and it is used all over world thus it leads us to cloud computing for easy storing and accessing large amount of data from anywhere at any cost. All the organization are moving towards the cloud for secure usage and storage purpose. Public Cloud, Private Cloud, Hybrid Cloud and Community Cloud are the deployment model of the cloud. Public cloud is easily accessible by the people, it is appealing to many companies but they also feels security cloud be lacking. They also own and operate the infrastructure at provider data center. Private cloud is simply dedicated to single organization they also known as internal or enterprise cloud. In private cloud data is protected behind a firewall it is more secure and protected. Hybrid cloud is a combination of two or more clouds (Public, Private & Community) thus it provides the benefits of multiple deployment models and its direct connect services allows us to connect between different cloud

D. J. Hemanth et al. (Eds.): COMET 2019, LNDECT 35, pp. 551–559, 2020.
https://doi.org/10.1007/978-3-030-32150-5_53

models. Community Cloud is where several organization collected together order the specific community. Example E-commerce where different organization are collected and works together for some benefits.

1.1 Cloud Service Models

Where cloud is anything can be provided as service thus it leads us to three important service models. They are Software as a Service (SaaS), Platform as a Service (PaaS) & Infrastructure as a Service (IaaS). SaaS is a process of distribution of software model that has hosted applications over the internet. PaaS delivers hardware and software for tools and also it provides run time environment for tools. IaaS is a service which provides fundamental resources like virtual machines, virtual storage in much higher order of scalability and also it provide direct access to their servers and storage. Virtualization is the creation of a virtual entity such as server, a desktop, a storage device, an operating system on a network resource. Virtual machine is of System virtual machine and Process virtual machine which are based on the operating system and the application. Virtualization is a technology which is mainly used in Business in case of lower costs, Disaster Recovery and also better migration to cloud. Cloud Computing is full of services and applications which is delivered through a virtualized environment. VM Migration is process of transferring the Virtual machine from one physical host to another host. There are two VM Migration Techniques one is Live/Hot Migration and other is Regular/Cold Migration. Live VM Migration is the movement of VM from one physical host to another physical host takes place only when the power is on which means without shutting down the client system. Regular VM Migration is the movement of VM from one physical host to another host takes place when the power is off which means the client system will be shutdown. Need for VM Migration are Upgrading, Balancing resource usage, VM failures & Meet SLA.

2 Related Work

Ajith Singh et al. [15] proposes new techniques called Honey Bee Clustering Techniques which is compared with Honey Bee Placement Technique in order to minimize the energy consumption in the Data center even though it utilizes all the resources enormously. Ankit Anand et al. [13] proposes Integer Linear program which is used to generate exact result for small cases where as First Fit Decreasing algorithm is used for handling large cases. Thus KVM() is used as an hypervisor. Finally it reduces the number of migration by considering the response time. Bobrof et al. [11] introduces Measure-Forecast-Remap (MFR) which remaps virtual machine to physical machine. The CPU resource is considered for minimizing the cost of running the data centre thus it results in the reduction of physical machine to support a workload. Gao et al. [10] introduces new method called VMPACS which comes under the genetic algorithm which results in finding the near optimal solution and also it reduces the resource wastages. Abdelsamea et al. [5] comes out with new methods by using Regression Algorithm called Multiple Regression Host Overload Detection (MRHOD) and Hybrid Local Regression Host Overload Detection (HLRHOD) in order to minimize the power consumption without violating SLA. Buyya et al. [3] introduces two different algorithm called Priority Aware VM allocation (PAVA) and Bandwidth Allocation (BWA) by using the SDN controller thus it results in the low energy consumption. Kundu et al. [8] proposes two machine learning techniques called Artifical Neural Network (ANN) and Support Vector Machine (SVM) which is mainly used to predict the performance of virtualized application and also reduces the VM problem sizing thus by estimating the power consumption. Mishra et al. [12] uses Heuristics algorithm by introducing the new method called Vector Dot for managing the resources in data centre and detecting the anomalies that are present in these methods. Frejus et al. [14] proposes Brute Force method of Bin Packing algorithm. In order to minimize the number of Physical machine by migration and finally reduces the power consumption within the Data-centers. Sotiriadis et al. [2] Support Vector Machine (SVM) & Support Vector Regressor (SVR) which is used to predict different types of variables and it minimize the performance degradation (Table 1).

Table 1. Comparitive study on VM placement

Author	Work	Method/algorithm	Parameters	Energy efficiency
Ajith Singh and Hemalatha [15]	HCT	Hierarchical clustering	CPU, speed & memory	✓
Anand, Lakshmi and Nandy [13]	ILP + FFD	KVM	ALL (CPU, memory, I/O & bandwidth)	
Bobroff et al. [11]	MFR	Stochastic integer programming/BIN packing	CPU	
Gao et al. [10]	VMPACS	Genetic algorithm/ACS	CPU, network & storage	✓
Abdelsamea et al. [5]	MRHOD + HLRHOD	Multiple regression algorithm/hybrid regression algorithm	CPU, memory & bandwidth	✓
Son and Buyya [3]	PAVA + BWA	SDN, priority allocation	Network bandwidth	✓
Kundu et al. [8]	ANN + SVM	Machine learning techniques	All (CPU, memory, I/O, network)	✓
Sotiriadis, Besis and Buyya [2]	SVM & SVR	Open stack/VMA	CPU	✓
Gbaguidi et al. [14]	Brute force + FFD	BIN packing techniques	CPU, memory	
Mishra and Sahoo [12]	Vector dot	Heuristics algorithm	CPU, memory& I/O	

3 System Architecture

The architecture diagram of the proposed model is as shown in the Fig. 1. The proposed work is composed of two modules: (i) VM Allocation (placement), (ii) VM migration. This project proposes a Prediction based Thermal Aware Server Consolidation (PTASC) model, a consolidation method, where numeric and local architecture are taken into consideration along with Service Level Agreement. PTASC consolidates the servers through VM migration using a statistical learning method on basis of predicted value. The accuracy of detection is improved for a given power budget includes the process of allocating the physical host to the Virtual Machines. To estimate the energy one of the prediction module is used thus by extracting the information from various sensors for placing the VM. Then, the PTASC model compares the expected value with the observed energy to detect the energy efficiency. The performance metrics considered are, Measuring the amount of heat dissipation, No. of jobs completed, No. of SLA violations & No. of Migration happened. Reason for the proposed work are Load balancing, System Support, Communication Cost & Power Management.

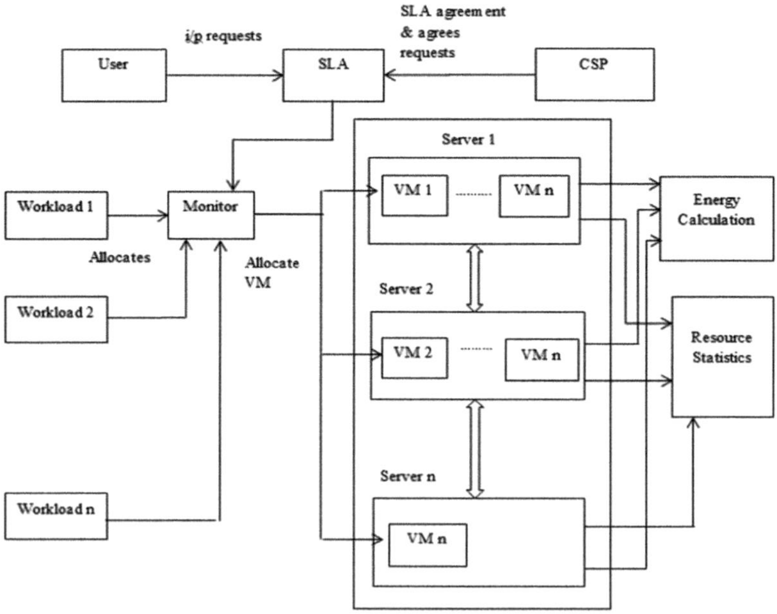

Fig. 1. Proposed architecture for energy efficiency and consolidation

The work flow of the proposed work is, when the user requests for the process it is received by the cloud service providers and it has to be processed without violating the SLA. In the Datacenter it consists of ‘N’ number of servers where the VM is being migrated in order to analyse the efficiency of the server. Number of workloads in the servers are monitored regularly in order to detect the overloaded physical machine. When the load is detected then it has to find the best place for placing the VM.

3.1 VM Placement

VM Placement is defined as placing the VM based on the demands that the user request for the computer resources such as CPU, Storage and Network Bandwidth. In this the hypervisor create the VM and then assign to the user. When the user request for the VM then the hypervisor checks and finds where the VM to be placed. In other words VM placement is simply finding the suitable host for placing the VM this can be happened in two different situation whether by placing new VM or to place a migrated VM. VM placement can be faced in two different approaches they are Power Based Approach and Application Based Approach. VM placement is mainly used for the goal of saving the Energy by shutting down some server or Maximizing the Resource Utilization.

3.2 VM Migration

In cloud computing, VM migration is an algorithm for transferring the virtual instances from various physical machines. It is mainly used for balancing the load in the machine, management of fault that occurs while processing, maintaining the system and finally reduce the consumption of energy.

4 Overload Detection in a Host – Cloud Analyst

In Cloud computing many new possibilities are introduced for Internet applications developers. The usage of Internet applications has different methods by the developers for hosting of applications, because it requires the servers with a particular level of capacity to handle the high demand. Furthermore, the server was underutilized because of some traffic that happens at some point. The hosting and deployment of the cloud becomes cheaper which is offered by the cloud providers based on the usage. But still some tools cannot be used for the developers to evaluate the large scale cloud applications and handle the users workload thus in order to complete this gap cloud analyst is proposed. It was mainly developed for simulating large-scale Cloud applications. Cloud Analyst helps developers with optimization of applications performance and providers to analyse the use of Service Brokers to distribute applications among Cloud infrastructures. In order to detect the load of the each server the proposed work comes up with an Extended Multiple Linear Regression Algorithm (EMLR). The algorithm for detecting the overload is as given below.

ALGORITHM:

algorithm Extended Multiple Linear Regression(n)

begin

Input: CPU usage history, RAM usage history

Output: Finding the overloaded host

i. Read each host in host list do

ii. Read each sample of data do

iii. a $\longrightarrow$ Multidimensional Matrix $\{R_T, P_T\}$;

iv. $b_n \longleftarrow \alpha_1/1\text{-}R_T * \alpha_2/1\text{-}P_T$

v. Procedure to use Regression Methods;

vi. Estimate the multiple regression parameters;

vii. Using the Multiple regression model to calculate the predicted utilization;

viii. Check whether predicted utilization < 1 then;

ix. Repeat the steps from b to h

x. Host is overloaded when predicted utilization $\geq$ 1;

End

5 Experimental Results

The experiments were carried out in the CloudSim 3.0. CloudSim environment provides the user the cloud platform and provides the user to test the experiments such as creation of new algorithms to load balancing, new virtual machine scheduling policies, VM placement, resource provisioning, workload prediction, server consolidation, energy efficiency, cost reduction and so on.

Screenshots

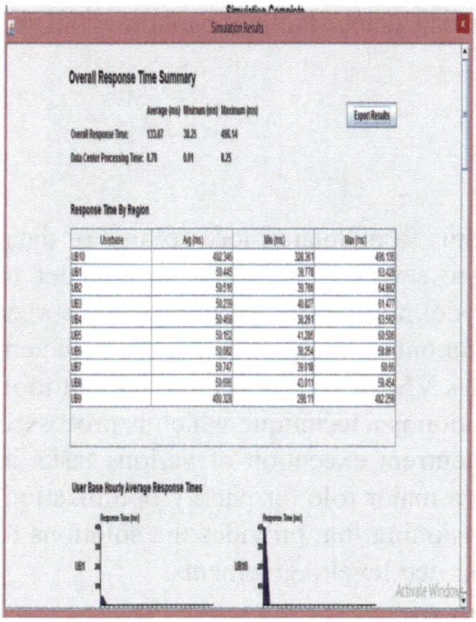

Fig. 2. Response time

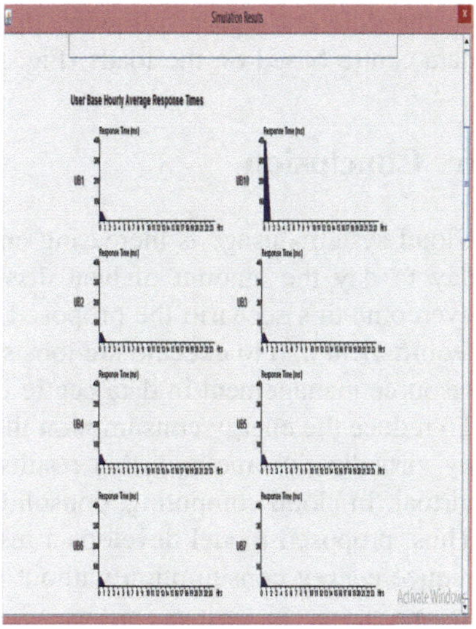

Fig. 3. Processing time

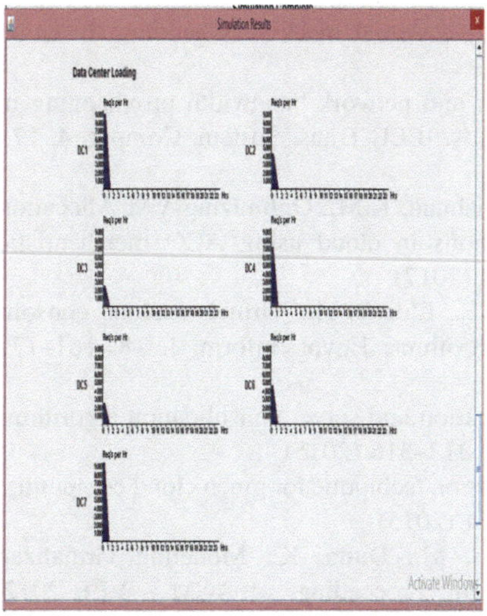

Fig. 4. Data center loading

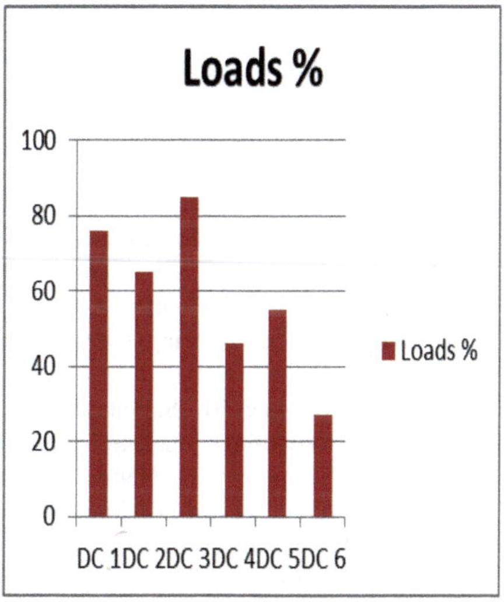

Fig. 5. Overload detection

Figure 2 represents the calculation of response time from the server in order to find the level of usage of each server. Figure 3 represents the processing time between the user and the server. Figure 4 represents the overall usage of each data centre thus based on the different region following this finally we can detect the comparison of all the data centre based on the loads (Fig. 5).

6 Conclusion

Cloud systems usage is increasing enormously thus to maintain more volume of data, day to day the amount of heat dissipated by the server is high. Hence in order to overcome this scenario the proposed PTASC model involves in identifying the server requirement and to execute the jobs submitted without violating the SLA. For efficient resource management in data centre in recent days VM migration is an important tool. To reduce the energy consumption the Consolidation is a technique which is processed by virtualization method, that results in the concurrent execution of various tasks in virtual. In cloud computing, consolidation plays a major role for energy optimization. Thus, proposed model develops a resource provisioning that provides the solutions to reduce energy consumption without violating Service level Agreements.

References

1. Lee, E.K., Viswanathan, H., Pompili, D.: Model-based thermal anomaly detection in cloud data centers using thermal imaging. IEEE Trans. Cloud Comput. **6**(2), 330–343 (2018)
2. Sotiriadis, S., Bessis, N., Buyya, R.: Self managed virtual machine scheduling in cloud systems. J. Inf. Sci. **434**, 381–400 (2018)
3. Son, J., Buyya, R.: Priority-aware VM allocation and network bandwidth provisioning in software-defined networking (SDN)-enabled Clouds. IEEE Trans. Sustain. Comput. **4**, 17–28 (2017)
4. Shabeera, T.P., Kumar, S.D.M., Salam, S.M., Krishnan, K.M.: Optimizing VM Allocation and Data Placement for Data-intensive applications in cloud using ACO metaheuristic algorithm. Int. J. Eng. Sci. Technol. **20**, 616–628 (2017)
5. Abdelsamea, A., El-Moursy, A.A., Hemayed, E.E., Eldeeb, H.: Virtual machine consolidation enhancement using hybrid regression algorithms. Egypt. Inform. J. **18**, 161–170 (2017)
6. Sami, M., Haggag, M., Salem, D.: Resource allocation and server consolidation algorithms for green computing. Int. J. Sci. Eng. Res. **6**(12), 313–316 (2015)
7. Nema, P., Choudhary, S., Nema, T.: VM consolidation technique for green cloud computing. Int. J. Comput. Sci. Inf. Technol. **6**(5), 4620–4624 (2015)
8. Kundu, S., Rangaswami, R., Gulati, A., Zhao, M., Dutta, K.: Modeling virtualized applications using machine learning techniques. In: Proceedings of ACM – VEE 2012, pp. 3–15 (2012)
9. Zhang, Z., Wang, H., Xiao, L., Ruan, L.: A statistical based resource allocation scheme in cloud. In: International Conference on Cloud and Service Computing, pp. 266–273 (2011)
10. Gao, Y., Guan, H., Qi, Z., Ho, Y., Liu, L.: A multi-objective ant colony system algorithm for virtual machine placement in cloud computing. J. Comput. Syst. Sci. **79**(8), 1230–1242 (2013)

11. Bobroff, N., Kochut, A., Beaty, K.: Dynamic placement of virtual machines for managing SLA violations. In: International Conference on Integrated Network Management, pp. 119–128 (2007)
12. Mishra, M., Sahoo, A.: On theory of VM placement: anomalies in existing methodologies and their mitigation using a novel vector based approach. In: International Conference on Cloud Computing (CLOUD), pp. 275–282 (2011)
13. Anand, A., Lakshmi, J., Nandy, S.K.: Virtual machine placement optimization supporting performance SLAs. In: International Conference on Cloud Computing Technology and Science (CloudCom), vol. 1, pp. 298–305 (2013)
14. Gbaguidi, F.A., Boumerdassi, S., Ezin, E.C.: Adapted BIN packing algorithm for virtuals machines placement into datacenters. In: International Conference on Cloud Computing, pp. 69–80 (2017)
15. Singh, A., Hemalatha, N.M.: Cluster based BEE algorithm for virtual machine placement in cloud datacenter. J. Theor. Appl. Inf. Technol. **57**(3) (2013)

Improved Particle Swarm Optimization Technique for Economic Load Dispatch Problem

N. B. Muthu Selvan(✉) and V. Thiyagarajan

SSN College of Engineering, Chennai 603 110, Tamilnadu, India
{muthuselvannb, thiyagarajanv}@ssn.edu.in

Abstract. Particle Swarm Optimization (PSO) technique tries to mimic the collective behavior of bird flocking and fish schooling. Economic load dispatch (ELD) is one of the mile stone in power system optimization problem. The main objective of the ELD is to determine the optimal power output of a number of thermal generators at the lowest possible cost to meet the short-term system demand, subjected to various transmission and operational constraints. This paper presents an overview of an stochastic PSO technique for ELD problem. The performance of the PSO technique is improved by the implementation of Gaussian and Cauchy probability distributions function. The implementation of the probability distribution function is carried out using systematic analysis of three different models of improved PSO technique. The performance of the improved PSO techniques are critically analyzed for ELD problem. The best position for the location of Gaussian and Cauchy probability distribution function is presented in this paper. The analysis reveals that the improved PSO technique is simple, reliable and suitable for real-time applications.

Keywords: Economic load dispatch · Particle swarm optimization · Gaussian probability distribution · Cauchy probability distribution · Optimization

1 Introduction

Economic load dispatch (ELD) is a crucial optimization problem in power system operation and control. The fast-growing power demand along with rapid increase in the fuel cost has emphasized the importance of the ELD problem. From the literature review it is inferred that number of mathematical based programming techniques like gradient method, interior point method, λ iteration method, quadratic programming, linear and non-linear programming, dynamic programming [1–4] etc. are available to find the optimal solution of the ELD problem. These optimization techniques are based on the derivative function of objective function and are more compatible for solving the ELD problem. The search space of the ELD problem is non-convex in nature that requires rigorous computation by the mathematical tools to obtain the optimal solution and consequently increases convergence rate. The advancement in computation device like computers has opened up the opportunity to apply heuristic methods for the various complex optimization problems. Heuristic methods like evolutionary programming, tabu search and PSO technique [5, 6] works on the principles of random

D. J. Hemanth et al. (Eds.): COMET 2019, LNDECT 35, pp. 560–567, 2020.
https://doi.org/10.1007/978-3-030-32150-5_54

number generation based on the probability distribution function thereby ensuring a near optimal solution for a complex optimization problem.

The main objective of this paper is to find the optimal solution for ELD problem that provides minimum fuel cost of thermal units to meet the load demand while satisfying certain constraints. The optimal cost obtained from the improved PSO techniques is compared with conventional heuristic algorithms such as evolutionary programming and tabu search techniques. From the comparative analysis it can be inferred that the proposed PSO technique emerges as robust optimization technique to solve ELD problem for various size of power system.

2 Problem Statement

The main objective of ELD problem is to provide the minimum cost of power generation subjected to various set of equality and inequality constraints [7]. The mathematical representation of the ELD problem is as follows,Objective:

$$Min\,(F_T) = \sum_{i=1}^{ng} a_i P_{Gi}^2 + b_i P_{Gi} + c_1 \tag{1}$$

Subjected to:

1. The power balance constraint,

$$\sum_{i=1}^{ng} P_{Gi} = P_D + P_L \tag{2}$$

where P_L is the total transmission line loss determined using B coefficients as,

$$P_L = \sum_{i=1}^{ng}\sum_{j=1}^{ng} P_{Gi} B_{ij} P_{Gj} + \sum_{i=1}^{ng} B_{oij} P_{Gi} + B_{oo} \tag{3}$$

2. The generator operation constraint,

$$P_{Gi,\min} \leq P_{Gi} \leq P_{Gi,\max}; \qquad i = 1, 2, \ldots ng \tag{4}$$

3 Particle Swarm Optimization

PSO is a population based heuristic search technique. The optimal solution of the PSO technique is achieved by adjusting the position of each particle by appropriate velocity updation. This updation is carried out based on particle's own experience and the experience of neighboring particles. This process mimics the cooperative behavior that exists among the bird flock and fish schooling. Because of this cooperative behavior of each particle, the constancy of the PSO technique is enhanced. The process of position and velocity updation implements the key property of both intensification and diversification strategies of heuristic algorithm. The flowchart representing the search process of PSO technique is presented in Fig. 1.

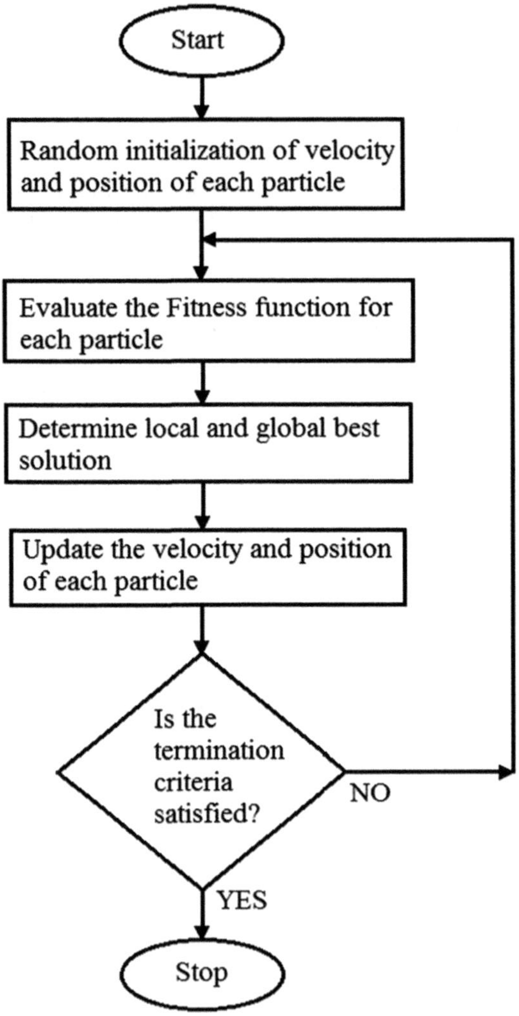

Fig. 1. Flowchart for PSO algorithm

4 Improved PSO Technique

Even though with the presence of intensification and diversification strategies, conventional PSO technique do suffer certain limitations. The traditional uniform probability distribution used in the velocity update equation many time tends to impart identical weightages to certain particles leading to an ineffective local and global searches. This phenomena initiates the particles of the conventional PSO to get bonded towards a certain local optimal solution, that subsequently drives the conventional PSO to premature convergence. When the number of Epochs are increased in a conventional PSO technique inconsistency and slower convergence are also observed. In order to counterbalance the drawbacks of premature convergence and inconsistent performance various comparative analysis of continuous probability distribution was performed. From this analysis it is inferred that incorporation of Gaussian and Cauchy distributions into the conventional PSO technique would overcome the aforesaid drawbacks.

The Gaussian probability distribution or Normal distribution is a bell shaped probability distribution function. The mean and variance are clearly defined by the central limit theorem. On the other hand, Cauchy probability distribution function has an undefined mean and variance. This distribution is defined by the location parameter (L) and scale parameter (S). The location parameter specifies the location of the peak and the scale parameter describes the distribution spread. A standard Cauchy distribution has L = 0 and S = 1. The random numbers generated by these distributions are presented in Fig. 2.

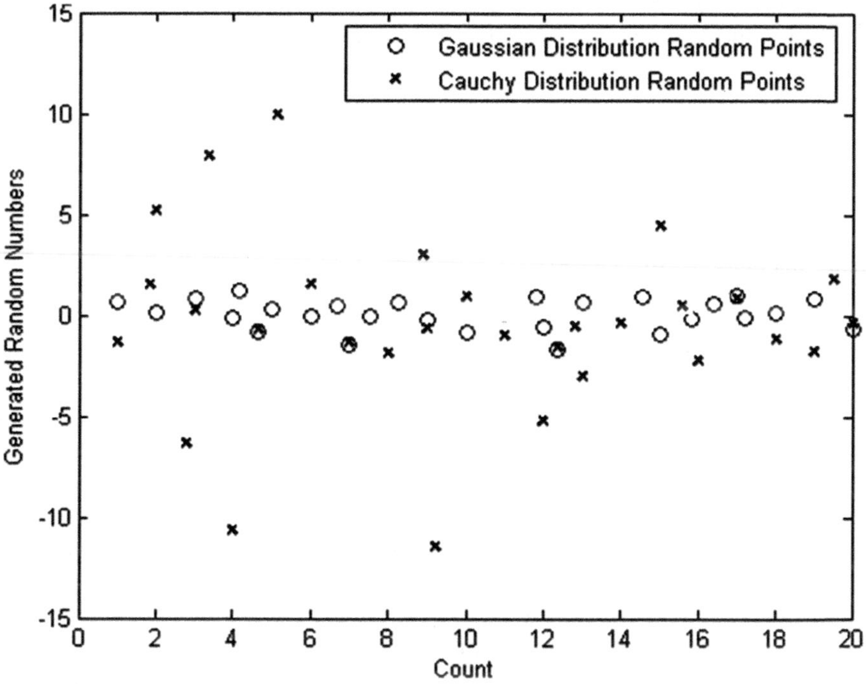

Fig. 2. Random numbers using Gaussian and Cauchy distributions

To ascertain the applicability of the Gaussian and Cauchy distributions into velocity update equation of conventional PSO technique, three PSO models are formulated. The modified velocity update equation of conventional PSO algorithm is presented below:

Model I: In this model conventional uniform probability distribution function is used to generate random number in the interval [0, 1] for the cognitive and social part of the velocity update equation. The velocity update equation is given by:

$$V(X)_j^{t+1,i} = \left\{\omega_I \times V(X)_j^t\right\} + \left\{c_1 \times U(0,1) \times \left(P_{best,j}^t - X_j^{t,i}\right)\right\} + \left\{c_2 \times U(0,1) \times \left(G_{best,j} - X_j^{t,i}\right)\right\} \tag{5}$$

Model II: In the second model standard Cauchy probability distribution (C_d) is used to generate the random number for the cognitive part and for the social part the random number is generated using Gaussian probability distribution (G_d) in the velocity update equation. The improved velocity update equation is given by

$$V(X)_j^{t+1,i} = \left\{\omega_I \times V(X)_j^t\right\} + \left\{c_1 \times C_d(0,1) \times \left(P_{best,j}^t - X_j^{t,i}\right)\right\} + \left\{c_2 \times G_d(0,1) \times \left(G_{best,j} - X_j^{t,i}\right)\right\} \tag{6}$$

Model III: In this model standard Gaussian probability distribution (G_d) function is used to generate random number in the cognitive part and standard Cauchy probability distribution function (C_d) is used to generate the random number in the social part of the velocity update equation. The enhanced velocity update equation is given by

$$V(X)_j^{t+1,i} = \left\{\omega_I \times V(X)_j^t\right\} + \left\{c_1 \times G_d(0,1) \times \left(P_{best,j}^t - X_j^{t,i}\right)\right\} + \left\{c_2 \times C_d(0,1) \times \left(G_{best,j} - X_j^{t,i}\right)\right\} \tag{7}$$

The performances of these three PSO models are tested and critically analyzed by solving the classical ELD problem.

5 Test System and Results

The improved three models of PSO was tested by solving classical ELD problem on standard IEEE 30-bus test system. The standard IEEE 30-bus test system comprises of 6 thermal generators with 41 transmission lines and a total demand of 283.4 MW. The technical parameters used in the aforementioned improved PSO models are as follows: acceleration factors $ac_1 = ac_2 = 1.5$, inertia weightages $\omega_I^{\max} = 0.9$ and $\omega_I^{\min} = 0.4$, swarm size N_P = 100 and penalty factor $k_1 = 1000$.

The convergence characteristics for all the three models of PSO is obtained by drawing a plot between the minimum fitness value against the epoch count. The convergence characteristics of the proposed three PSO models is presented in Fig. 3.

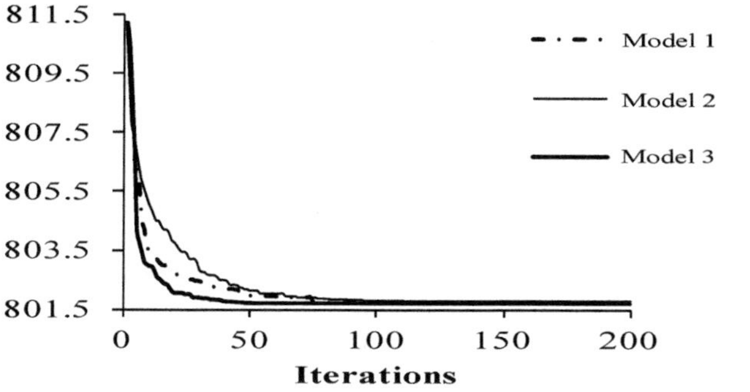

Fig. 3. Convergence characteristics

From this convergence graph it is observed that the optimum value of the fitness function converges smoothly without any rapid and discontinuous fluctuations. It is also inferred that the proposed three models of PSO have better applicability and consistency in attaining the convergence. Also the Model III PSO that incorporates Gaussian and Cauchy probability distribution function in cognitive part and social part respectively has a faster convergence as compared with other two PSO models.

The optimal solution obtained by the three improved PSO models is presented in Table 1:

Table 1. Optimal solution of ELD using three PSO models

PSO Models	Model 1	Model II	Model III
P_{G1} (MW)	174.4	172.5	175.8
P_{G2} (MW)	48.4	52.8	48.5
P_{G5} (MW)	20.8	16.1	20.8
P_{G8} (MW)	24.2	28.3	22.8
P_{G11} (MW)	12.7	10	12.5
P_{G13} (MW)	12	13	12
P_L (MW)	9.1	9.3	9.0
Tot. Gen. (MW)	292.4	292.7	292.5
F_T ($/Hr)	801.5	804.5	800.7
t_{max}	125	200	90
Comp. time (ms)	215	320	185

From Table 1, it is observed that Model III PSO that utilizes Gaussian and Cauchy probability distribution function in cognitive part and social part respectively has the least number of epochs thereby achieving a faster convergence. The supremacy of this PSO model is because of the suitable location of Gaussian and Cauchy distributions function in the velocity update equation. The Gaussian distribution function that has a well defined mean and variance, generates the random number close to a central value thereby the local search is exploited effectively. Similarly the random number generated by the Cauchy distribution function is not restricted as there is no definite mean and variance and hence the global search mechanism of PSO is well exploited. Hence the applicability of Gaussian and Cauchy distribution function in the velocity update equation is well justified.

Further, the convergence characteristics of the improved PSO technique is compared with convergence characteristics obtained from classical Evolutionary Programming (EP), Tabu Search (TS) and PSO technique. The value of N_p for EP, TS, and PSO is taken as 100. The classical EP and TS algorithm employs mutation scaling factor of 0.045 and the recombination factor of TS algorithm is set at 0.04.

The comparative convergence characteristics of these algorithms is presented in Fig. 4 and Table 2 respectively.

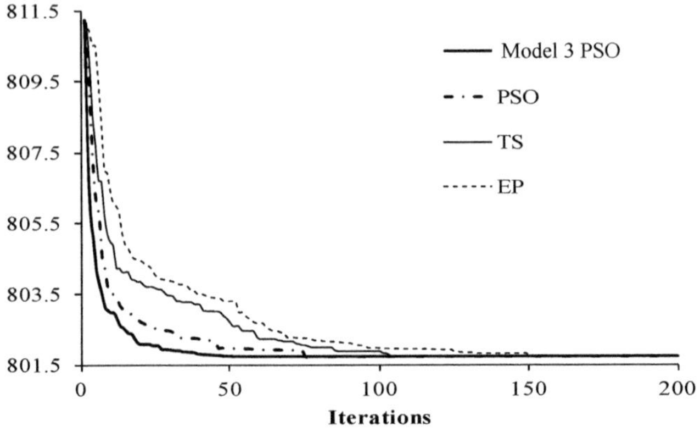

Fig. 4. Convergence characteristics

From the convergence characteristics curve, it is observed that model III PSO has faster convergence compared with the classical EP, TS and PSO algorithms. This ensures the convergence reliability of the improved PSO algorithm over the other algorithms.

The reliability of the EP, TS, PSO and improved Model III PSO technique is ensured by obtaining the optimal solution without violating the equality and inequality constraints of ELD problem. It is also observed that the convergence rate is faster (lesser number of epochs) for the Model III PSO technique, than other classical algorithms in obtaining the optimum solution.

Table 2. Optimal solution of ELD problem

Algorithms	EP	TS	PSO	Model III
P_{G1} (MW)	176.3	175.9	174.4	175.8
P_{G2} (MW)	48.4	48.6	48.4	48.5
P_{G5} (MW)	20.8	20.9	20.8	20.8
P_{G8} (MW)	22.8	22.5	24.2	22.8
P_{G11} (MW)	12.4	12.4	12.7	12.5
P_{G13} (MW)	12	12.4	12	12
P_L (MW)	9.3	9.3	9.1	9.0
Tot. Gen. (MW)	292.7	292.7	292.4	292.5
F_T ($/Hr)	801.8	801.7	801.5	800.7
t_{max}	165	140	125	90
Comp. time (ms)	250	225	215	185

6 Conclusion

This paper presents a simple and an efficient improved PSO technique for solving the classical ELD problem. The suitability of applying Gaussian and Cauchy distribution function in the velocity update equation in PSO technique is presented in this paper. The ELD problem is solved for the standard IEEE 30 bus test system using the improved PSO technique and the results obtained are critically analyzed. The comparative results of improved PSO technique along with the classical EP, TS and PSO algorithms are also presented in this paper. From the analysis it is inferred that the proposed Model III PSO technique is relatively simple, reliable and efficient as compared with classical heuristic algorithms. The proposed improves Model III PSO technique can be extended to solve various other power system optimization problems.

References

1. Lu, W., Liu, M., Lin, S., Li, L.: Fully decentralized optimal power flow of multi-area interconnected power systems based on distributed interior point method. IEEE Trans. Power Syst. **33**(1), 901–910 (2018)
2. Aganagic, M., Mokhtarj, S.: Security constraint economic dispatch using nonlinear Dantzig-Wolfe decompensation. IEEE Trans. Power Syst. **12**(1), 105–112 (1997)
3. Duvvuru, N., Swarup, K.S.: A hybrid interior point assisted differential evolution algorithm for economic dispatch. IEEE Trans. Power Syst. **26**(2), 541–549 (2011)
4. Gaing, Z.L.: Particle swarm optimization to solving the economic dispatch considering the generator constraints. IEEE Trans. on Power Systems **18**(3), 1187–1195 (2003)
5. Prasanna, T.S., Muthu Selvan, N.B., Somasundaram, P.: Security constrained OPF by fuzzy stochastic algorithms in interconnected power systems. J. Electr. Syst. **5**(1), 1–16 (2009)
6. Sasaki, Y., Yorino, N., Zoka, Y., Wahyudi, F.I.: Robust stochastic dynamic load dispatch against uncertainties. IEEE Trans. Smart Grid **9**(6), 5535–5542 (2018)
7. Chowdhury, B.H., Rahman, S.: A review of recent advances in economic dispatch. IEEE Trans. Power Syst. **5**(4), 1248–1259 (1990)

Secure Data Transmission Through Steganography with Blowfish Algorithm

K. Vengatesan[1(✉)], Abhishek Kumar[2], Tusar Sanjay Subandh[1], Rajiv Vincent[3], Samee Sayyad[4], Achintya Singhal[2], and Saiprasad Machhindra Wani[1]

[1] Department of Computer Engineering, Sanjivani College of Engineering, Kopargaon, India
vengicse2005@gmail.com, vaibhavchavan440@gmail.com, saiprasadwani.shirdi@gmail.com
[2] Department of Computer Science, Banaras Hindu University, Varanasi, India
abhishek.maacindia@gmail.com, achintya.singhal@gmail.com
[3] School of Computing Science and Engineering, VIT University, Chennai, India
rajiv.vincent@vit.ac.in
[4] School of Engineering, Symbiosis Skill and Open University, Pune, India
samee.syd@gmail.com

Abstract. The Steganography is a strategy for thumping mystery communications in a shelter protest while correspondence happens among dispatcher and collector. Safety of mystery or vital data has dependably be noteworthy issues after the previous occasions to the right period. It will be dependably with the attentive subject for specialists to create secured methods to sends information deprived of uncovering it to anybody further than the collector. In this way from everyday analysts have created numerous methods to satisfy secure exchange of information with Steganography is one of them. Here we have built up another method of photo Steganography privileged the inserting the encoded Data document or information utilizing RSA algorithm with Hash-LSB for giving greater safety to information with in addition this information hiding strategy. The created technique utilizes a hash capacity to produce an example for hide information bits into LSB of RGB pixel estimations of the convey images. This method ensures with the data-set has been encoded before implanting it into a convey image. Implanted content in images generally conveys vital message around the substance, in the event that the any cases outsider get the message such a significant number of ways, so keep this activity this paper actualizes hash table encryption to the message at that point cover up addicted to the images motive is to give extra secured approach to exchange information. This work is another method for hide the data in a photo and minimum variety in the photo bits has been made, that create this method secured with more proficient. In the method additionally connected a cryptography strategy. Other stage is to encode and unscramble stenographic images utilizing blowfish algorithm, this activity utilized to deal with other series of security procedure usage.

Keywords: Steganography · Image processing · Secure transmission

D. J. Hemanth et al. (Eds.): COMET 2019, LNDECT 35, pp. 568–575, 2020.
https://doi.org/10.1007/978-3-030-32150-5_55

1 Introduction

Swarm Advanced pictures are being exchanged over various types of systems. With the huge improvement of personal computer systems and the latest developments in computerized developments, a goliath proportion of computerized data is being exchanged over various sorts of systems. Usu-partner clear that a tremendous piece of this data is mystery, reserved or both, which increment the enthusiasm for more grounded encryption systems [3]. Encryption and Steganography are the favored strategies for verifying the transmitted information [7], accordingly, there are distinctive encryption systems to encode and interpret picture information, and it might be battled that there is no single encryption calculation satisfies the differing picture types [2, 11]. Information exchange is an OK instance of an application that uses encryption to keep up information grouping between the sender and the collector. In this paper, Steganography is used to cover data to perform encryption. Steganography procedures are getting in a general sense progressively current and have been commonly used. The Steganography systems are the perfect improvement for encryption that empowers a customer to cover a great deal of data inside a picture.

In this manner, usually utilized related to cryptography with the goal that the dataset is twice as ensured; first it is scrambled and afterward covered up so an enemy needs to initially locate the concealed data beforehand decoded occurs [4, 6, 8]. The issue with cryptography is that the encoded messages are self-evident. This implies any individual who watches scrambled messages in travel can sensibly expect that the dispatcher of the information doesn't need it to be perused by easygoing eye witnesses. This creates it conceivable to reason the profitable information. In this way, if the delicate data will be passed through leaky channels, for example, the internet, Steganography method can be utilized to give an extra security on a mystery message [14]. Once hide the data privileged images the LSB strategy is normally utilized. While the cryptography endeavors to change over a picture to other that is difficult to comprehend, Steganography includes hide the dataset so it creates the impression that no information is covered up by any means. Consequently, the individual won't endeavor to unscramble the datasets [11]. For instance, a modification of the slightest critical digits for the shading estimation of a few pixels in a picture won't influence the nature of the image and along these lines, empowering the messages to be sent inside an image utilizing these bits [15].

In this paper, Steganography method will be utilized to send the mystery information alongside a scrambled image. Various even and perpendicular squares at the dispatcher side will be produced, and after that blended with the scrambled picture before head conveying it to the collector. The collector will require this message to reproduce a similar mystery change table in the wake of separating the mystery information from the encoded image. Rather than sending the entire mystery change table, which is generally enormous, just the mystery information is sent.

2 Image Steganography

The method of hide mystery data or information in images is known as image Steganography. For the maximum part, pixel forces are the strategies utilized in hide data in image Steganography. As indicated by [7], pictures are the utmost mainstream and generally utilize shield object in Steganography. The level of excess in pictures has create it the utmost looked for, regarding Steganography. Dual classifications of characterization to be specific spatial – space and change area based have been proposed in image Steganography [6]. [8] Explained that spatial space installs the message straightforwardly into the pixels force though the change area likewise called the recurrence space change the image before the message is embedded. Various record groups be in picture Steganography. BMP, GIF, PNG, JPEG, and TIFF will all be able to execute in image Steganography [9]. In any case, every one of the document groups represents its very own one of a kind focal points and burdens. Since pixel forces are utilized in image Steganography, there is here and there variety in the power of the first picture with the Steganography-image or the implanted picture. The variety in power is so paltry or unpretentious in that it isn't noticeable or recognizable to the human eye [8].

2.1 Symmetric and Asymmetric Cryptography

By a long shot, the asymmetric cryptographic algorithm is the most secured sort of cryptography [10] because of its numerical capacities [11]. Asymmetric cryptography tends to the issue of key circulation for encryption [12] still remains a noteworthy issue in symmetric cryptography. Asymmetric key cryptography actualizes a digital which permits a beneficiary of a message check that in fact the message is originating from a particular sender [12]. The utilization of digital mark in asymmetric cryptographic algorithm additionally empowers a recipient to see whether a message was adjusted in travel [13]. A digitally marked message can't be adjusted without nullifying the mark. In cryptography, the higher the extent of the key length, the more secure the algorithm is. This likewise brings a noteworthy preferred standpoint to asymmetric cryptographic algorithm since it has a more drawn out key length and along these lines, makes it assault safe. Similarly, speed is a noteworthy disadvantage in asymmetric cryptography because of the unpredictability of its scientific calculations. There is an exchange off among security and speed in asymmetric cryptographic algorithm [11]. This examination, along these lines, utilizes the asymmetric cryptographic algorithm because of its conspicuous leverage. Inasmuch as we keep on imparting in an untrusted medium like the internet, security remains the highest need.

2.2 Attacks on Steganography Systems

Most Steganography systems intended for private correspondence has endured a few shortcomings. [14] Opined that Steganography attacks include recognizing, removing and annihilating the concealed data inside the secretive media. Visual attacks and measurable attacks [15] are the two broadly known attacks against Steganography. Measurable attacks utilize steganalysis [14]. [15] Developed a steganalysis application that was fruitful in identifying a message installed in an image. Measurable video steganalysis created [13] was additionally fruitful in identifying a data covered up in a video whose algorithm depended on LSB. As a result of the dread of psychological oppressors utilizing Steganography to convey over the internet, came out with a steganalysis called the dynamic superintendent approach that was equipped for recognizing implanted messages in images and recordings. [11] Showed that the human eye is fit for distinguishing concealed messages because of twisting. From the attacks above, clearly steganography itself isn't a conclusion to the security concern related with data exchange or correspondence. With the end goal to relieve the attacks against steganography and to additionally fortify data correspondence security, cryptography was presented.

Hash Function

Hash function based LSB strategy that provides hash work. This hash work manages the Least significant bits position inside the pixels and the situation of every shrouded picture pixels and furthermore with the quantity of bits of LSB. Hash esteem takes a variable size of info and returns a fixed size of advanced string as yield. Hash work additionally utilized for distinguishing copied record in extensive documents. Hash work commonly given by

$$x = y \% z \tag{1}$$

Where, x is LSB bit position inside the pixel, y speaks to the situation of each concealed picture pixel and z is number of bits of LSB.

Motivation Behind Utilizing Hash Work

1. It expands the limit of concealing information since we utilize more pictures. This likewise builds the measure of information we need.
2. It sets aside marginally more effort to execute than the standard LSB.
3. It utilizes more pictures, utilizes lesser information per picture than the standard LSB.
4. The most imperative perspective is that the Hash-LSB can be effectively decoded. By utilizing the proposed blowfish calculation it expands the security factor of the picture. The encryption upgrades the security and after that the irregularity inwrinkles the disorder factor of the calculation.

3 Proposed Approach

The client takes the content to be covered up and afterward scrambles it by utilizing the Blowfish-encrypt algorithm with the assistance of a keyset that is variable long. The key will be chosen by the client. This encoded message is then create to breakdown to 'n' squares. Presently 'n + 1' images are chosen aimlessly from an arrangement of 'm' images where 'm > n'. Every fragmented square is scrambled aimlessly to an image. A hash table is kept up to acquire a right request as far as grouping of data. This hash table and every one of the squares are then installed into the 'n + 1' pictures utilizing LSB alg. These '1 + n' images are then send. At the recipients sides the 'n + 1' image is acquired. The collector at that point acquires the hashing picture first. The data in regards to the situation of the hash image is known before hand to the beneficiary. Utilizing the hash image he extricates the right grouping of data. He at that point decodes it utilizing the key which is likewise part of the hash table. The whole algorithm will be implemented in Python utilizing Open-CV. The benefits of the planned strategy are various. Initially the encryptions upgrade the sanctuary. At that point the arbitrariness of the sending of images additionally upgrades the security of the algorithm. Additionally it very well may be seen later in the results that regarding time of execution the proposed algorithm isn't critical. The reason for hiding information securely is done superbly. The essential thing to be done is the images picked ought not be rehashed, anyway for the situation they are rehashed the names of the images are altered by hard coding to guarantee the system does not get befuddled. Additionally in examination with latest work regarding picture steganography, the proposed algorithm provides fantastic outcomes and takings generally lesser time and gives most secure outcomes. Subsequently the proposed algorithm can be proposed to be utilized as a normal algorithm. Figure 1, delineates the stream of the work.

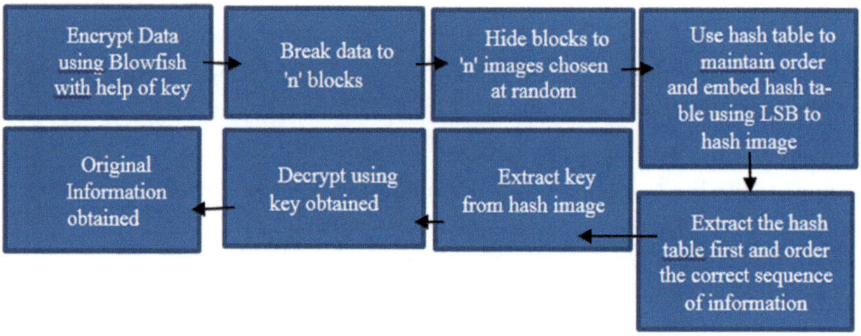

Fig. 1. Block diagram of the proposed technique

In this secure data transmission, we are using three level processes:

1. Conceal information or document Encryption decoding utilizing RSA calculation.
2. Encoded and unscrambled messages or record installing and recovering spread picture utilizing H-LSB method.
3. Encode and Decode stenographic picture utilizing Blowfish-Algorithm.

In picture innovation, mystery correspondence is accomplished to insert a note or record into shelter picture (utilized as the transporter to install information into) and create a Steganography-images (produced image which is conveying a shrouded message). In this effort we do basically dissected different Steganography methods and furthermore have secured Steganography. By and large picture Steganography is ordered in succeeding angles demonstrates the wonder Steganography actions.

Extraordinary Capacity. Max. Extent of data will be inserted into picture.
Perceptual Clearness. Afterward hide the processing into shield picture, perceptual worth would be corrupted into Steganography-picture as contrast with shield-image.
Easiest. After that inserting, dataset should remain unblemished if Steganography-picture goes into certain change, for example, trimming, scaling, separating and expansion of commotion.
Displeasure Battle. It ought to be hard to adjust the messages when it has been installed into Steganography-picture.
Calculation Difficulty. In what way much costly it is computationally to install and removing a shrouded message?

The utilized database comprises of various image. Here JAFFE confront database is utilized for this trial. For this test result, the data and images is sent in ARM LPC 2148 by the sequential correspondence port from the PC with VB GUI. Also, the encoded data is sent in database by zigbee. Than the encoded data will got by the zigbee and it reprocess by the ARM IPC 2148 pack. It's appeared in PC with GUI. This trial result is demonstrated as follows (Table 1).

Table 1. The result table on blowfish algorithm, LSB algorithm

Steps	Blowfish algorithm (%)	LSB algorithm (%)
Encryption cycle	76	86
Decryption cycle	82	88
Memory utilization	95	95

4 Conclusion

A Protected pictorial Steganography Built on Blowfish Algorithm with Hash table procedure has been executed. A productive encryption method blowfish utilized to encoding the mystery dataset document, and after that stow away into the shield picture record utilizing Discrete Cosine Transform (DCT) and Hashing-LSB procedure mystery concealed process managed without exasperate image seeing alternative. After encrypt the secured image utilizing blowfish-algorithm utilizing mystery key esteem motive is give client confirmation. The double encrypted work is done utilizing Blowfish-algorithm. At that point the double encryption process fruition prepared cover record exchange beneficiary. The turnaround process will be finished recovering the first data as per mystery key an incentive for blowfish algorithm.

References

1. Sinha, A., Singh, K.: A technique for image encryption using digital signature. Source: Opt. Commun. **218**(4), 229–234 (2003)
2. Al-Husainy, M.A.F.: Image encryption using genetic algorithm. J. Inf. Technol. **5**(3), 516–519 (2006)
3. Younes, M.A.B., Jantan, A.: Image encryption using block-based transformation algorithm. IAENG Int. J. Comput. Sci. **35**(1), 15–23 (2008)
4. Chang, K., Jung, C., Lee, S. and Yang, W.: High quality perceptual steganographic techniques, vol. 2939, pp. 518–531. Springer (2004)
5. Fraz, E.: Steganography preserving statistical properties. In: Proceeding of the 5th Internationally Workshop on Information Hiding. LNCS, Noordwijkerhout, The Netherlands, October 2002, vol. 2578, pp. 278–294. Springer (2003)
6. Kessler, G.C.: Steganography: hiding data within data. An edited version of this paper with the title "Hiding Data in Data," originally appeared in the April 2002 issue of Windows &. NET Magazine, September 2001
7. El-din, H., Ahmed, H., Hamdy, M.K., Farag Allah, O.S.: Encryption quality analysis of the RC5 block cipher algorithm for digital images. Opt. Eng. **45**(10107003), 7 (2006)
8. Kathryn, H.: A Java steganography tool 24 March 2005. http://diit.sourceforge.net/files/Proposal.pdf
9. Zenon, H., Sviatoslav, V., Yuriy, R.: Cryptography and steganography of video information in modern communications (1998). Citeseer.ist.psu.edu/hrytskiv98cryptography.html
10. Li, S., Zheng, X.: Cryptanalysis of a chaotic image encryption method. In: Inst. of Image Process. Xi'an Jiaotong University, Shaanxi, This paper appears in: Circuits and Systems, ISCAS 2002. IEEE International Symposium 2002, vol. 2, pp. 708–711 (2002)
11. Saravana Kumar, E., Vengatesan, K.: Cluster Comput. (2018). https://doi.org/10.1007/s10586-018-2362-1
12. Sanjeevikumar, P., Vengatesan, K., Singh, R.P., Mahajan, S.B.: Statistical analysis of gene expression data using biclustering coherent column. Int. J. Pure Appl. Math. **114**(9), 447–454 (2017)
13. Kumar, A., Singhal, A., Sheetlani, J.: Essential-replica for face detection in the large appearance variations. Int. J. Pure Appl. Math. **118**(20), 2665–2674 (2018)
14. Amin, M.M., Salleh, M., Ibrahim, S., Katmin, M.R., Shamsuddin, M.Z.I.: Information hiding using steganography. In: 2003 Proceedings of 4th National Conference on Telecommunication Technology, NCTT, pp. 21–25, 14–15 January 2003
15. Liu, T.-Y., Tsai, W.-H.: A new steganographic method for data hiding in microsoft word documents by a change tracking technique. IEEE Trans. Inf. Forensics Secur. **2**(1), 24–30 (2007). https://doi.org/10.1109/tifs.2006.890310
16. Kumar, A., Vengatesan, K., Rajesh, M., Singhal, A.: Teaching literacy through animation & multimedia. Int. J. Innovative Technol. Exploring Eng. **8**(5), 73–76 (2019)
17. Johnson, N.F., Suhil, J.: Exploring steganography: seeing the unseen. Computing practices (2006). http://www.jjtc.com/pub/r2026.pdf
18. http://www.jjtc.com/pub/r2026.pdf. http://www.nku.edu/~mcsc/mat494/uploads/StanevPaper.pd
19. Stefan, S.: Steganographic Systems. CSC/MAT 494
20. http://www.nku.edu/~mcsc/mat494/uploads/StanevPaper.pdf
21. Shi, Z., Tu, J., Zhang, Q., Liu, L., Wei, J.: A survey of swarm robotics system. In: Advances in Swarm Intelligence. LNCS, vol. 7331 (2012)

22. Lau, H.K.: Error detection in swarm robotics: a focus on adaptivity to dynamic environments. Ph.D. Thesis. University of York, Department of Computer Science (2012)
23. Marco, D., et al.: The swarm-bot project. In: Swarm Robotics. LNCS, vol. 3342 (2005)
24. Selvaraj Kesavan, E., Kumar, S., Kumar, A., Vengatesan, K.: An investigation on adaptive HTTP media streaming Quality-of-Experience (QoE) and agility using cloud media services. Int. J. Comput. Appl. (2019). https://doi.org/10.1080/1206212X.2019.1575034
25. Marco, D., et al.: Evolving self-organizing behaviors for a swarm-bot. Auton. Robots. **17**(2–3), 223–245 (2004)

Comprehensive Design Analysis of Hybrid Car System with Free Wheel Mechanism Using CATIA V5

P. Vaidyaa(✉), J. Magheswar, Mallela Bharath, C. R. Tharun Sai, and A. Syed Aazam Imam

Department of Mechanical Engineering, Panimalar Engineering College, Chennai, India
vaidyaa1999@gmail.com, maghes1998@gmail.com, bharathmallelall@gmail.com, tharunsa99@gmail.com, syedazam0803@gmail.com

Abstract. Nowadays hybrid vehicles have come to every average driver's knowledge. Though initially it had its backlogs in the market for the preceding decades, at the current trend people have started showing interests for the hybrid vehicles in the market. Many people don't have the knowledge how a hybrid works? how to interface with a hybrid? Or what damages that a hybrid vehicle could cause to the driver? The hybrid vehicle has its own advantages and disadvantages. In our research we are going to be designing a hybrid vehicle using Computer Aided Drafting tool i.e. Catia V5 and our study will focus on the components of hybrid.

Keywords: CATIA V5 · Hybrid vehicles · Free wheel mechanism

1 Introduction

A hybrid vehicle is a locomotive where two or more forms of energy are used, such as thermal energy from IC engine, electrical energy from electric motor etc. In our study and design, we did our research about the hybrid vehicle which used two modes of energy i.e. I.C Engine and Electric motor. Some of the examples of hybrid locomotives in our current innovative market are (1.) electrical generators are run by diesel engines using diesel-electric trains that gives powers to an electric motor, and (2.) diesel engines driven submarines that the path of propagation is in the surface of water and while submerging, they use electric power from batteries as in other cases of hybrid locomotives. There are also other hybrids that store energy using a pressurized fluid that are called hydraulic hybrids. The basic theory of hybrid vehicles are the acknowledgement of the features of various locomotion components i.e. the torque or turning power are more efficient in electric motor. High speeds maintaining is better in internal combustion engine than others (better than conventional electric motor). The energy efficiency is quite high when using by switching one component to another in a proper time concretion, by this methodologies gives high efficiency

D. J. Hemanth et al. (Eds.): COMET 2019, LNDECT 35, pp. 576–586, 2020.
https://doi.org/10.1007/978-3-030-32150-5_56

2 Design of Hybrid Car in CATIA V5

See Figs. 1, 2 and 3.

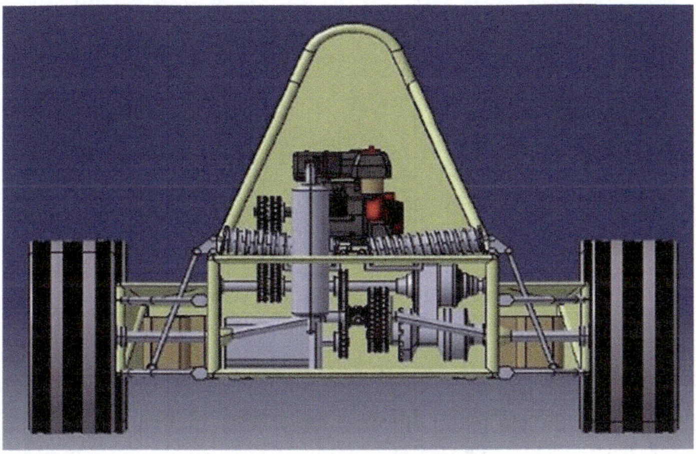

Fig. 1. Assembled view of the vehicle

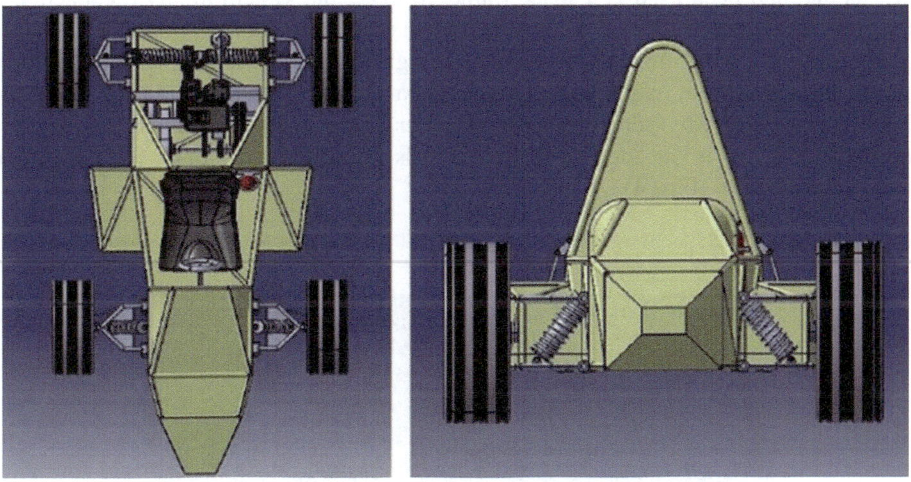

Fig. 2. Top view of the vehicle

Fig. 3. Front view of the vehicle

2.1 Design of Chassis

The chassis is built for the car using ergonomics of 95% male pursey for the design and construction of driven cabin. The safety factor of the driver compartment is designed such a way that it is safe with high impact forces in the chassis each end every section of the chassis is cross-beamed to ensure that they include a truss shape so that will spread the forces evenly to the entire cross section of the structure. The choice of the chassis that had been used is a space frame, to make the system highly compact. The design features at first look, gives a simple form made in metal, the frame deals with a great reduction in the impact stresses. The design of the chassis height of the vertical side of the frame. The height of the frame has a major resistance to vertical flux, and this is the reason why semi trucks have taller frame rails. The outer frame has a mounted skin and is outside of the frame. This construction of body frame is often made of aluminum (Fig. 4).

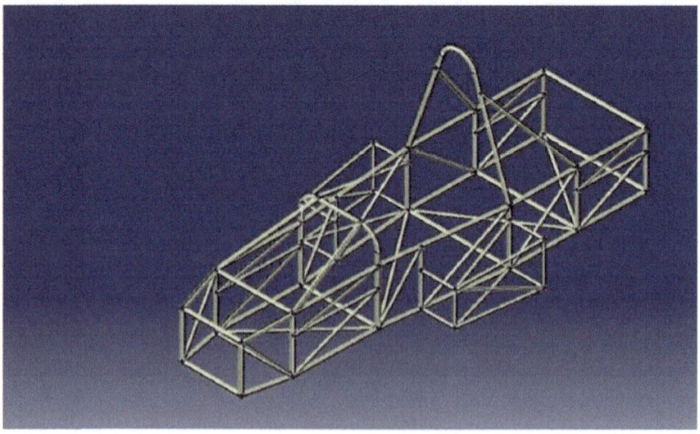

Fig. 4. Design of chassis

2.2 Materials Used and Welding Techniques

We have a choice of 3 materials namely, AISI1020, AISI4130, AISI1018.

2.3 Chemical Composition

See Table 1.

Table 1. .

Elements	AISI4130	AISI1018	AISI1020
Iron	97.03–98.22%	98.81–99.26%	99.08–99.53%
Manganese	0.40–0.60%	0.60–0.90%	0.30–0.60%
Carbon	0.280–0.330%	0.17–0.20%	0.17–0.230%
Chromium	0.80–1.10%	–	–
Silicon	0.15–0.30%	–	–
Molybdenum	0.15–0.25%	–	–
Phosphorus	0.040	0050	0.050
Sulphur	0.030	0.040	0.040

2.4 Mechanical Properties

By comparing these properties, AISI4130 is very much suitable and good. So we choose the material AISI4130 for our vehicle (Table 2).

Table 2. Mechanical Properties of selected material grades

Parameters	AISI4130	AISI1018	AISI1020
Hardness	197	126	111
Bulk modulus	140 GPa	140 GPa	140 GPa
Tensile strength	670 MPa	440 MPa	394 MPa
Modulus of elasticity	205 GPa	205 GPa	200 GPa
Reduction of area	60%	40%	66%

Welding techniques availed from outside: Tig welding

2.5 Design of Braking System

Various vibration and noise problems tend to arise due to the friction that is generated by the brake force in the brake system. In general hydraulic breaking system works with a piston that is attached to a brake pedal, that consists of fully non compressible brake fluid that is fixed to a secondary piston that is attached to a brake pads with rotary wheels to produce braking. The amount of breaking to be produced is directly proportional to the force applied by the driver on the pedal that increases the pressure forces of the fluid p = force/area. Due to change in the cross sectional are difference the force generated in the fluid tends to be more than that of the brake pedal. The area difference is computed such that the foot pedal area is more than that of piston area. The transmission of forces in the fluid is uniform throughout the entire cross section, the pressure that imparts on the rotor because the piston at the other end. The head dissipation is transmitted by convection that are created by the rotational energy given by the wheel and stop the vehicle. The pedal force given by the driver is comparatively less due to the brake fluid being nearly in compressible (Figs. 5, 6 and 7).

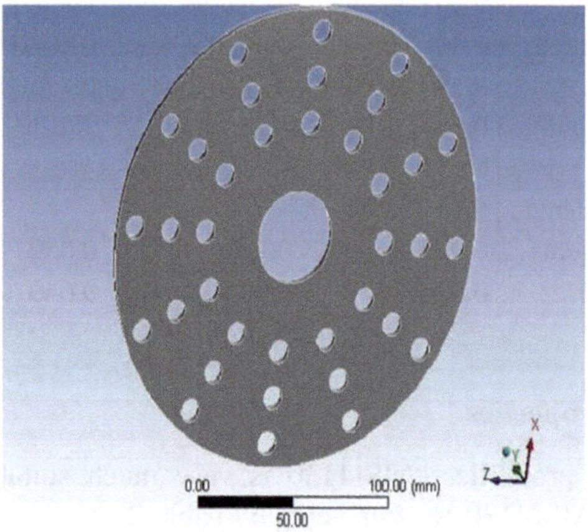

Fig. 5. Brake disc

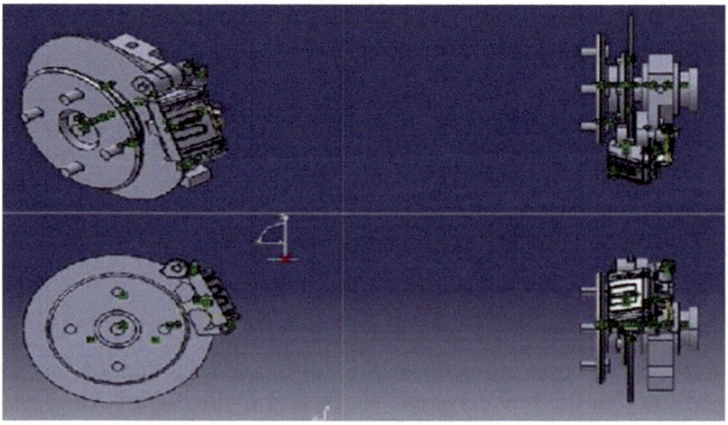

Fig. 6. Design of bake caliper and disc

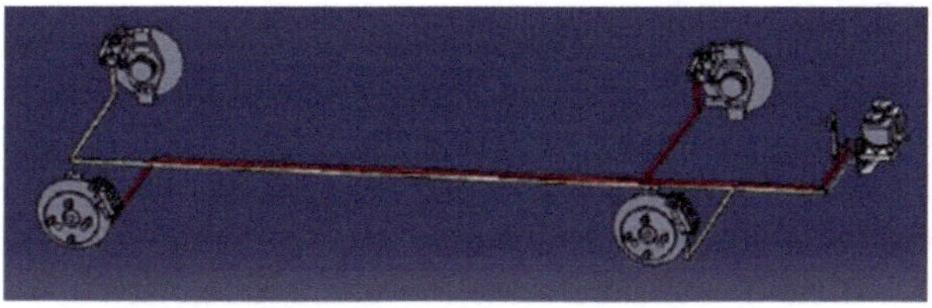

Fig. 7. Full circuit diagram of brake system

2.6 Design of Suspension

A Chassis of hybrid vehicle is attached to the front and back wheel between the centre of spring, shocks absorbs and axles. All components perform working of protecting components from shock are also called as suspension.

The automobile chassis is concomitantly attached with axles of springs. It is finished to prevent the hydraulic vehicle body from road shocks due to jump, pitch, sway. Those road shock provide an disagreeable ride and also additional stress to the automobile chassis and frame. The suspension system has a two absorbing devices they are damper and spring. An energy of the road shock will be obtained by the spring oscillates. These oscillate motion are arrested by the damper is also known as shock absorber. Suspension systems assist both road handling and ride quality, which are odds with each other. The suspension system completely prevents wheel wobble. It is dominant for the suspension to keep the road wheel is contact with the road exterior, because while travelling over rough not even ground the top and down movements of wheels should be relative to the body. The suspension protects the vehicle itself from wear and tear. A body of vehicle is carrying by spring. A weight of the body is carried by spring. Those wheels, axles and other components of an automobile which are not carried by spring. The vehicle is always subjected for moving road irregular. The road will not be 100% leveled road excited or outer face is pits and falls. A vehicle is passes through bends, curves and zigzag path.

A due to springs having enough rigidity hold the axis in the proper position, the controlling of own oscillator through inter leaf friction is performed. When a vehicle outwards due to centrifugal force (Fig. 8).

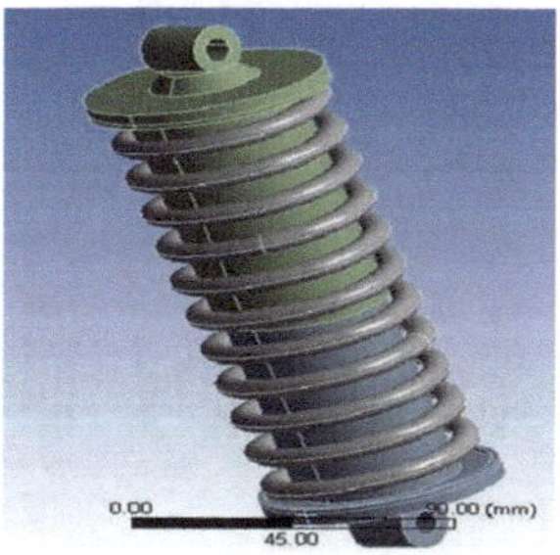

Fig. 8. Design of suspension spring

2.7 Design of Transmission System

The transmission system which is very important system which deals with the electrical motor and engine transmission here for that we are using a free wheel for easy transmission For speed variation we are using continues varying transmission The free wheel is connected both to the engine and the motor shaft the free wheel which rotates clockwise and it will not rotate anticlockwise when it is connected to the motor So this helps us to make the suspension (Figs. 9 and 10).

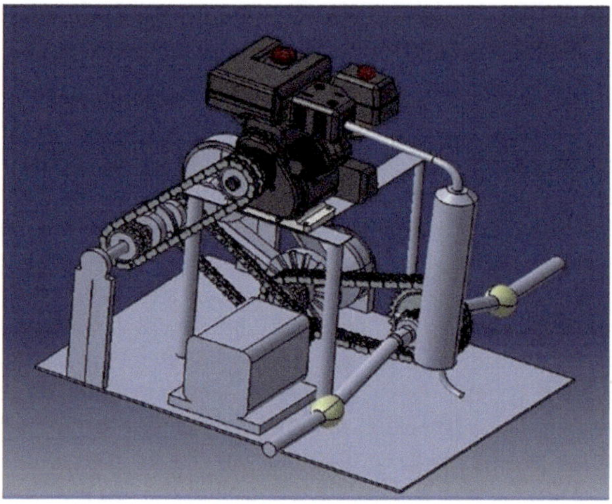

Fig. 9. Isometric view of transmission system

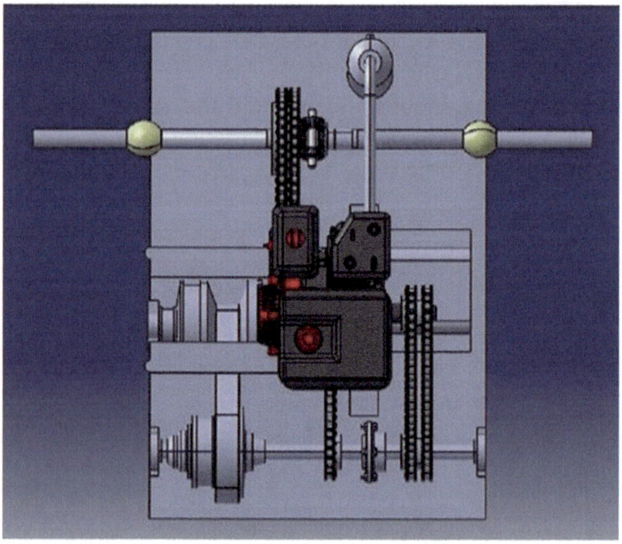

Fig. 10. Top view of transmission system

2.8 Continuously Variable Transmission

A continuously variable transmission or shiftless transmission is a automated transmission which helps in continuous change of speed by perfect gear ratio. it is different from automated transmission since it offers more ratio for gear meshing and thereby more variable speed is obtained. Also CVT has a great impact on fuel economy. The transmission is usually effected by belt drive. It offers an efficiency of around 88% which is lower than manual transmission efficiency but it provides variable speed so that fuel economy is increased. When the power is greater than necessity the CVT ratio is changed significantly.

To overcome that situation it increases the engine speed as of the peak speed of engine. CVT also used in low power transmission since because of its simplicity in mechanical work. Advantage of CVT is it does not require clutch for its operation. But in some vehicles centrifugal clutch is added for application of reversing. The controlling operation is same as automatic transmission. It uses 2 pedal operation and P-R-N-D-L-style shift pattern. The basic advantage of CVT is during transmission there is no sound occurs. Major disadvantage lies in the fact that it is over cost, complexity, noise and driving refinement. The one of the most important type of CVT is E-CVT (Electronically Controlled Continuously Varying Transmission System). It has the features of both electric drive and continuously variable type. In this system there are 3 output shafts which includes MG1, output, MG2. Output shaft speed is decided by speed of vehicle which we are driving. In this system matrix mode is used to analyze the planetary motion of the system which in turn executes each transmission of each point

2.9 Steering System

A rack and pinion comprises a pair of gears which convert rotational motion into linear motion. A circular gear named as "pinion" engages teeth on a linear gear bar named as "rack". The rotational motion applied to the pinion causes the rack to move relative to the pinion, thereby transmitting the rotation motion of the pinion into linear motion (Figs. 11 and 12).

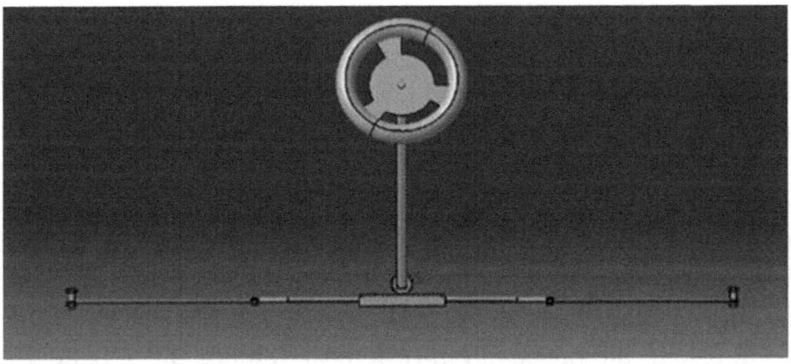

Fig. 11. Steering system

Fig. 12. Isometric view of steering

The Rack & Pinion system is actually a basic system that only uses a different gears to control the direction of the vehicle. The Pinion is the component of the system that is connected to the steering shaft. As you turning your steering wheel, the pinion rotates. This rotation occurs in the grooves of the rack, forcing the rack to move in same direction (depending on the directional change of the steering wheel). The Rack & Pinion system is attached to a tie rod combined motion of the Rack and Pinion Arrangement. This tie rod engages the system to the tires, as the tie rod is connected to the steering arm that is connected to the tire. When a wheel turns, the tie rod moves to direct the tires in the direction of the turn.

2.10 Electrical System

In hybrid cars electrical systems play a major role. By reducing the fuel consumption we are using electric motors. The electrical system consists of Motors, Motor controller (Drive System), Batteries and proper fuses and e-stop switches. Hybrid vehicle drive trains transmit power for hybrid vehicles. A hybrid vehicle has various forms of motive power. Hybrid vehicles come in too many comfit Nourations. For instance, a hybrid vehicle may be receiving its drive energy from burning of petroleum, but it can also switch to an electric motor and switch back to a internal combustion engine (Fig. 13).

While the engine being the power producing component in thermic fuel unit, in the electrical unit the motor serves as the power producing component to the hybrid vehicle. Since we require a larger output torque BLDC motors are used here. High speeds are achieved in the brushless dc motor. The motor used here is a 3-phase dc motor. A motor controller is also used to control the actions of the motor. Hall Effect sensors are implemented in the controller. In a brushless dc motor two coils are always energized with opposite polarities for the propagation of rotor. The Hall Effect sensors sense the energized coils and find out the position of the rotor. A battery is served as the heart of the hybrid vehicle in the electrical unit. The battery energizes the motor and it is plays a pivotal role in the propagation of hybrid vehicle. 48 V battery is used as per the need of the design. Various protection devices are used such as fuses, kill switch etc. For energizing the battery, the motor is turned off and the vehicle is made to run

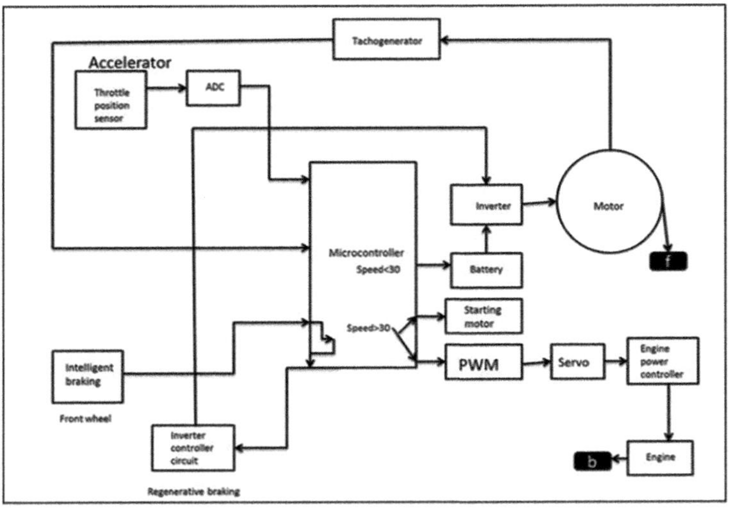

Fig. 13. Circuit for electrical system

using IC engine. During this, the 3 phase dc motor acts as a generator which will be producing a 3 phase ac supply. Rectifiers are used to convert the supply into a 1 phase dc supply. The supply is fed into the battery and the battery is charged. This is the electrical system of the hybrid vehicle.

2.11 Brushless DC Motor

It's a 3 phase DC motor. In this motor consist of no brushes in it so it's called brushless DC motor.

The efficiencies are very high in this kind of motor producing higher amount of torque in huge ranges of speed the connecting current of the armature problems are solved using permanent magnets by rotating around a fixed armature.

It has a high scope in flexibility and capability of communication with electronics in stationary holding torque and smooth operations are used (Table 3).

Table 3. Brushless DC motor

Model no	BLDC 125/4D
Power	1500 (Watts)
Continuous torque	4.8 (Nm)
Peak torque	18 (Nm)
DC voltage	48 (v)
Color	Black
Weight	9 (kgs)
Total length	185 (mm)

Note: Momentary peak Torque: 300% of rated torque for 10 s only.

Battery Selection

HV side: 12v*4 = 48v Batteries are connected in series.

LV side: 12 V alone. The Negative terminals will be GND with chassis (Table 4).

Table 4. Battery selection

Battery type	MRED35
Ref C20 capacity	35 (Ah)
Length	197 (mm)
Width	129 (mm)
Height	227 (mm)
Nominal filled weight	10.4 (KG)
Electrolyte volume	2.1 (Liters)
Charging current	2.5 (A)

3 Conclusion

In our research we designed every component that can be used in a hybrid vehicle. We conclude our research with the hope of having an economical friendly environment. As an engineer we feel solely responsible for the products we bring upon our environment. Hybrid vehicles are encouraged to be used by more drivers in the current and upcoming generations. Hybrid vehicles may be the first step in automobile industry to develop an eco-friendly environment.

References

1. Prajapati, K.C., Patel, R., Sagar, R.: Hybrid vehicle: a study on technology. Int. J. Eng. Technol. (IJERT) **3**(12) (2014). ISSN: 2278-0181
2. Vidyanandan, K.V.: Overview of Electric and Hybrid Vehicles, Research Gate, Issue, March 2018
3. Richard, M.G.: How to Go Green: Hybrid Cars. Treehugger. Discovery, 9 February 2007. Web. 23 Mar 2012
4. Sherwood, C.: What Are the Effects of Hybrid Cars? LIVESTRONG. Lance Armstrong Foundation, 17 May 2010. Web. 23 Mar 2012
5. Berman, B.: Hybrid Battery Toxicity—Hybrid Cars. New Hybrid Reviews, News & Hybrid Mileage (MPG) Info—Hybrid Cars. n.p., n.d. Web. 8 Apr 2006
6. Heffner, R., Kurani, K., Turrentine, T.: Symbolism in early markets for hybrid electric vehicles (2007) Web. 17 Nov 2009
7. Hudson, M.: Federal Hybrid Tax Credit Programs by Vehicle. Accessed 19 Oct 2009
8. LaMonica, M.: Most Consumers Willing to Pay for Hybrid Cars. Green Tech, 24 June 2008
9. Layton, J., Nice, K.: How Hybrid Cars Work. Howstuffworks Auto, n.p., n.d. (2008)
10. Perryman, S., Tews, J.: J.D. Powers and Associates Reports: While Many New-vehicle buyers concer for the environment, few are willing to pay more for a environment friendly vehicle, 6 March 2008

PIC Based Anode Tester

Muthuraj Bose(✉), Sundaramoorthi Subbiah,
Vasudhevan Veeraragavan, Sneha Vijayasarathy,
and Preetha Munikrishnan

Department of Electronics and Instrumentation Engineering,
Panimalar Engineering College, Poonamallee, Chennai 600123, India
muthurajbose@gmail.com, smsundaramoorthi@gmail.com,
vasudhevan.vee@gmail.com, snehasneha5799@gmail.com,
preetha274krishnan@gmail.com

Abstract. In our project we aim to develop the anode testing from measuring/monitoring its parameters like current densities, anode potential and bath temperature. By using PIC IC it will be programmed for measuring/ monitoring the above parameters. The required hardware designing will be developed for the testing work and thermistor sensor may be used for measuring the temperature. From the acquired data, the life of the anodes will be evaluated. Therefore the proposed work involves necessary literature survey regarding required circuits and components, construction of various circuits, testing of the circuits and developing actual working model. We intended to develop a "PIC BASED ANODE TESTER". It plays a vital role in the manufacturing companies of Anodes and also in Chemical Research Institutes.

Keywords: PIC · Corrosion

1 Introduction

When a metal is exposed either to the atmosphere or underground or immersed in a medium like seawater, the metal goes into solution (corrodes) with the formation of corresponding metal salt. This simple reaction in fact is electrochemical in nature [3]. We are all familiar with the function of a dry cell. The dry cell consists of zinc metal, a carbon rod and an electrolyte consisting of ammonium chloride, manganese dioxide etc [1]. When we use the cell by short-circuiting the positive carbon electrode and zinc negative electrode, electric current is obtained. When current flows through the cell, the zinc dissolves (corrodes) with the formation of zinc ions (which goes into solution) and free electrons. These free electrons reach the positive electrode throughout the external connection and react with the compounds in electrolyte to form new products [5].

D. J. Hemanth et al. (Eds.): COMET 2019, LNDECT 35, pp. 587–593, 2020.
https://doi.org/10.1007/978-3-030-32150-5_57

The metal corrodes under the same principle. The surface of the metal consists of small electrodes and when it is immersed into solution, the battery starts functioning [2]. The metal at the more negative areas dissolves (corrodes) and the electrons reach the positive site forming either hydrogen gas or hydroxyl ions by reacting with hydrogen ion or oxygen molecule.

Hence when one dips steel in acid solution, Hydrogen gas is evolved and at the same time the metal is dissolved (corroded) [4]. For corrosion to take place in neutral solutions like water where there is not much of hydrogen ion, oxygen is essential for corrosion to take place. The dissolved oxygen in water or soil reacts with the electron and helps in the continuation of the corrosion process. Hence when oxygen is completely removed no corrosion is possible, similarly, if there is no moisture (water) or any other electrolyte, corrosion cannot take place.

2 Architecture of PIC16F873

The PIC microcontroller comes in a wide range of varieties. It is economic, has large user base and serial programming capability. The term PIC stands for Peripheral Interface Controller. PIC microcontrollers are mainly used for industrial purposes as it consumes less power and has high performance ability. The PIC has RISC-based Harvard architecture and has a separate memory for data and program. These microcontrollers are very fast and easy to execute a program when compared with other micro controllers. The uses of Microcontrollers reduce the hardware as well as complexity as they have on-chip all essential components of a microcomputer [8]. This finds applications in portable and low cost instruments and dedicated applications. The one which is used in the present work is PIC16F873. It has an on-board RAM, EPROM, an oscillator, a couple of timers, and several Input/Output ports, serial ports and 8 channel A/D converter. However, the micro-controller is less computationally capable when compared to most of the micro-processors due to the fact that they are used for simple control applications rather than spreadsheets and elaborate calculations. As an example, the PIC16F873 has 4096 words of memory for program, and only 192 bytes of RAM, and can only operate with clocks up to 20 MHz on 8 bits of data (compared to megabytes of RAM, Speeds of a GHz or more and 32 or even 64 bits of data for many desktop systems) [8]. It does not have facilities for floating point (Fig. 1).

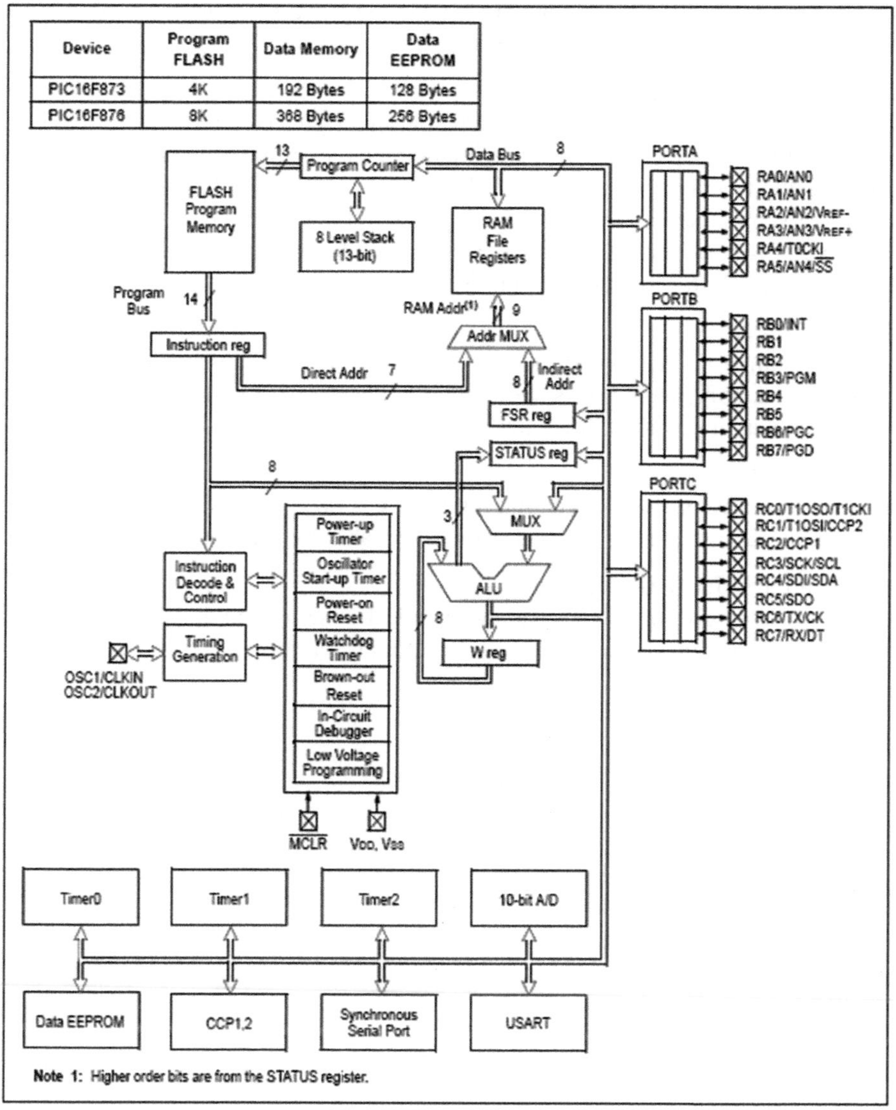

Fig. 1. Architecture of PIC16F873

3 Operation of Anode Tester

3.1 Block Diagram of Anode Tester

(i) Chemical System

The electrochemical cell (chemical system) consists of anode, cathode, electrolyte and reference electrode. During process anode delivers current to the cathode thereby protects it from corrosion and the parameters obtained from the electrochemical cell are given to signal conditioning circuit. The anions (negative ions) are attracted to anode and the cations (positive ions) are attracted to the cathode [2].

(ii) Signal Conditioning Devices
The system has been interfaced with the PIC through signal conditioning device, which consist of buffer, low pass filter and amplifier circuits. Here the cell current is measured from the cathode and is given to the PIC through signal conditioning device and ±12 V supply is given to the op-amps and +5 v is given to the PIC [5]. Here the PIC contains inbuilt ADC. Keypad and LCD display is used to enter the data and to display the cell values respectively for monitoring purpose.

(iii) PIC 16F873
It is programmed in such a way it receives inputs from the keypad and stores these inputs in its memory, and delivers these inputs to the chemical system through DAC for every time interval specified. It is interfaced with DAC-08, LCD Display and Keypad.

(iv) DAC – 08
The input from PIC is always in digital form. DAC – 08 is used in the circuit to convert the digital input into analog form which is acceptable by the next block.

(v) Keypad and LCD Display
Keypad is used to give input values to the chemical system through PIC and the display is used to view or for visualizing the system parameter (Fig. 2).

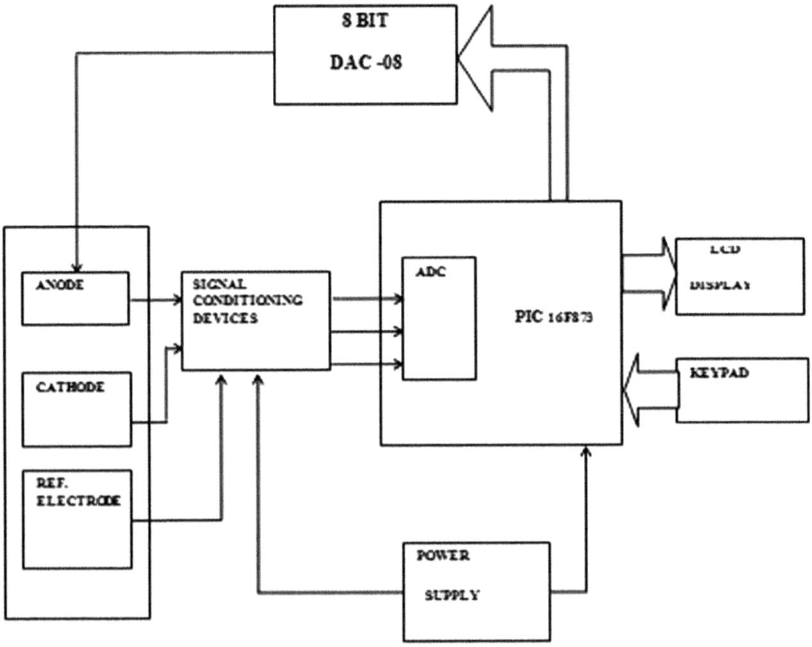

Fig. 2. Block diagram of ANODE TESTER

3.2 Electrochemical Cell

The electrochemical cell is capable of either generating electrical energy from chemical reaction or using electrical energy to cause chemical reaction. The electrochemical cell which generates an electrical current are called voltaic cell and galvanic cell those that generate the chemical reaction. The simple cell does this by combining metal and electrolyte solution [5]. The aluminum anode alloy is immersed in the solution along with stainless steel. The electrolyte solution is sodium chloride. It is connected to PIC and signal conditioning circuits and the output is obtained by using anode tester (Fig. 3).

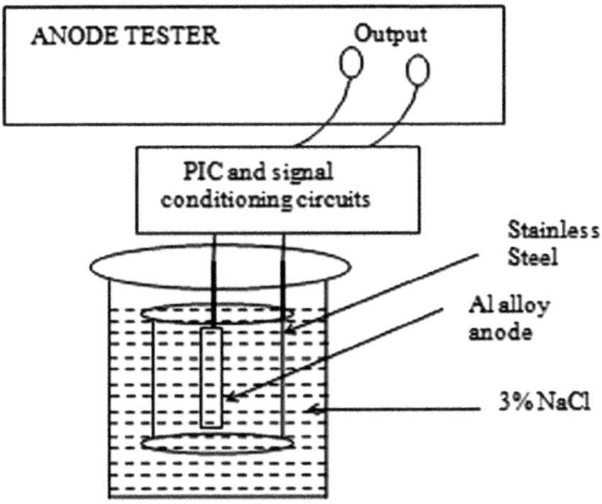

Fig. 3. Electrochemical cell

4 Result and Discussion

After the completion of the 96 h the test specimen is removed from the electrolyte, cleaned and weighted again. The cleaning of anodes (to remove the corrosion products on the anodes) after the test period has to be carried out as per the test procedure (appendix). The procedure is repeated for any number of test specimens. The values of the difference in weights for 6 different tests After the completion of the 96 h the test specimen is removed from the electrolyte, cleaned and weighted again. The cleaning are shown in Table 2 [7]. From the weight measurement the current capacity is calculated as follows

Current capacity

Total duration of the test (hrs) X

Total current passed (mA) = Weight loss (g)

The expected current capacity of anodes of different composition in natural seawater is shown in column 5 of the Table 2. The slight deviation from the values are all due to the fact that

Inhomogeneous alloy composition
Solution employed is 3% sodium chloride and not the natural sea water.

A further improvement to the existing system is required by incorporating the graphical recording of the process parameters by having an interface to the PC, so that the engineers could easily visualize the protection of the system (Tables 1 and 3).

Table 1. Performance of Al-Zn-Hg anodes in variety of saline environments

S. No.	Environment	Current capacity A. Hr/Kg
1.	Sea water	2820
2.	Hot brine	1500–2000
3.	Low setting sea water (≈ 750 PPM of chloride)	2820

Table 2. Weight loss measurements

S. No.	Anodes	Expected anode current capacity in sea water A.hr/Kg	Temperature (°C)	Experimented evaluation of anodes 3% NaCl capacities in A.hr/Kg
1.	Al – Zn – Hg	2530	24.8 ± 2 25.4 ± 1.78	2518 2539
2.	Al – Zn – In	2820	25.5 ± 2 25 ± 2	2585 2685
3.	Al – Zn- In – Mg	2030	25.7 ± 2 26 ± 2	1990 1975

Calculation

Underground condition current requirement = X μA/m^2
Total area to be protected = Y m^2
Total current requirement, Table 2: Weight Loss measurements
Z = X*Y
Anode current capacity = A A.Hr/Kg
Weight of the anode = W Kg

$$\text{Anode life time} = \frac{A*W}{Z*65*24} = \text{no of years}$$

Underground condition current requirement = 130 μA/m^2
Total area to be protected = 10,000 m^2
Total current requirement,

$$z = 130 * \frac{10,000}{10,000} = 1.3\frac{A}{M^2}$$

$$\text{Anode current capacity} = 2518\text{A}.\frac{\text{Hr}}{\text{Kg}}$$

Weight of the anode = 1 kg

$$\text{Anode life time} = \frac{(2518 * 1)}{(1.3 * 365 * 24)} = 1.1 \text{ no.of.years}$$

Table 3. Aluminium alloy anodes and temperature of the bath measured using Anode Tester.

S. No.	Specimen	Weight before reaction (g)	Weight after reaction (g)	Weight loss (g)
1.	Al – Zn – Hg (1)	22.3461	22.2398	0.1054
	(2)	22.5981	22.4927	0.1063
2.	Al – Zn – Hg (1)	22.8767	21.7732	0.1035
	(2)	22.7658	23.6662	0.0996
3.	Al – Zn – Hg (1)	24.6552	24.5187	0.1345
	(2)	23.2763	23.1408	0.1355

5 Conclusion

As corrosion is the major factor which severely affects most of the process equipments, and also it reduces the life of the equipments or the pipelines used in offshore/marine structures, so it should be controlled by using standard methods. Mostly it can be reduced by means of using anodes, so these anodes should be checked or its life time.

The life time of the anode will indirectly determine the life time of the pipelines or any other equipments used in industry. The job of checking the life time of the anodes can be easily obtained by the help of this Anode Tester as it doesn't requires any human intervention during the testing period. It will automatically deliver the current to the chemical system for each specified time interval with the help of PIC Microcontroller.

A further improvement to the existing system is required by incorporating the graphical recording of the process parameters by having an interface to the PC, so that the engineers could easily visualize the protection of the system

References

1. Bonshtytell, H.: Theory of Corrosion and Protection (1962)
2. Evans, R.: The Corrosion and oxidization of metals. Edward Arnold Ltd (1960)
3. Ashworth: Cathodic Protection (1949)
4. Kaeshe: Metallic Corrosion (2008)
5. Smith, S.N., Reading, J.J., Riley, R.L.: Material Performance (2015)
6. American Standard Testing Manual (ASTM) (2002)
7. Microchip Company: user manual (2005)
8. Sonde, B.S.: Introduction to system Design using Integrated Circuits (1980)

IoT Based Air Pollution Detector Using Raspberry Pi

E. S. Kiran[1(✉)], S. Deebika[1], V. Lakshmi[1], and G. Elumalai[2]

[1] Department of Electrical and Electronics Engineering, Dhanalakshmi College of Engineering, Chennai, Tamil Nadu, India
kiranedsuji98@gmail.com

[2] Department of Electronics and Communication Engineering, Panimalar Engineering College, Chennai, Tamil Nadu, India
elumalaig79@gmail.com

Abstract. For the last few decades, the air and water pollution issues are controlled by various government schemes based on infrastructure and industrial development. The population of people is still growing in accordance with development. So it is impractical one to overcome completely because of continuous monitoring, and data base maintenance is not available. Now days the child and old age peoples are affected mainly by air and water pollutions, when they are admitted in hospitals. The main aim of this project is to concentrate on continuous real time monitoring of air pollution, water pollution, temperature, humidity maintenance in incubator rooms in hospitals and maintain database in cloud using Internet of Things (IoT) for future analysis and remedies are developed based on analysis.

Keywords: Sensors · IoT · Raspberry Pi · Hospital monitoring and pollution control

1 Introduction

In our society air and water pollution is a growing issue, for a healthy and safe life it is essential to monitor and detect the air and water pollution levels. The development of technology paved the way for smart monitoring systems, Internet of things (IoT) is an emerging field due to its versatility and efficiency. IoT allows interaction between human and machine for communication. In existing system the data collectors should collect data from various locations which is far away from their work site and then the analysis of the collected data is done. This consume more time and it is a lengthy process. But now a days sensors are merged to the internet makes the monitoring process at any location and it is flexible and less time consumption. When sensors and devices merged with environment to self- monitoring it forms a smart environment.

In Hospitals most of the new born babies are kept in incubator. Incubator is a device which is used to grow and maintain certain cell cultures. It is necessary to maintain optimal temperature and humidity in the incubator room, so we can monitor the temperature and humidity using temperature and humidity sensor. The air pollution due to leakage of gases in air conditioner affects the asthma patients in hospital so it is

D. J. Hemanth et al. (Eds.): COMET 2019, LNDECT 35, pp. 594–603, 2020.
https://doi.org/10.1007/978-3-030-32150-5_58

monitored by using gas sensor. Using turbidity sensor large numbers of individual particles that are invisible to the naked eye are detected. PH sensor is used to measure alkalinity or acidity of water soluble substances because water is used by all patients in hospital.

Turbidity and PH sensor is used to detect the water quality which will be consumed by the patients in the hospital. The aim of this project is to monitor the whole hospital with different type of sensors and data's are uploaded to cloud for analysis and produce feedbacks to the hospital administrative manager and remedies are taken based on analysis of data.

For past few years sensors are used in variety of applications such as health applications, environmental applications, water quality monitoring etc. because of miniature size and low power consumption. For Healthcare applications the role of sensors is to detect physical, chemical, and biological signals and record it. For Environmental applications the role of sensors is to detect ambient air temperature, air pressure, humidity and detect it. For Water quality monitoring, Turbidity and PH sensors is used to detect the quality of water (Fig. 1).

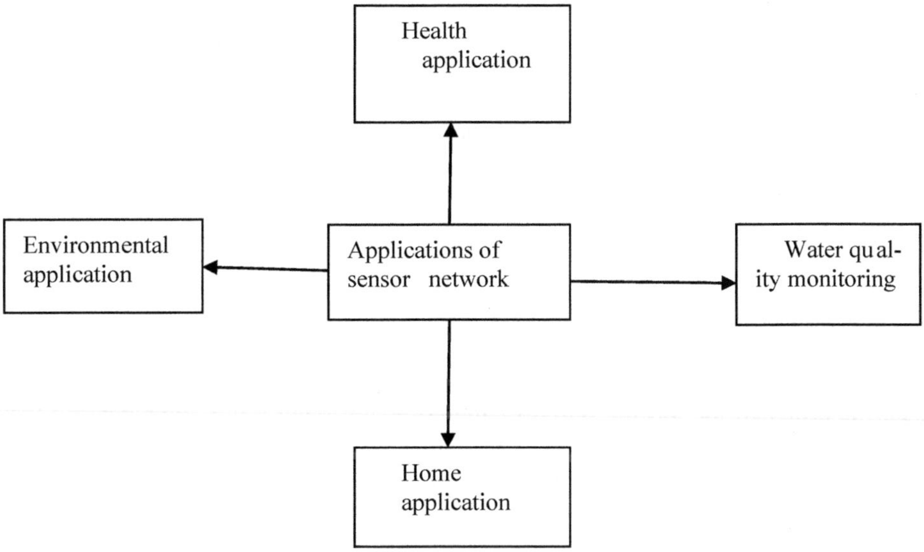

Fig. 1. Applications of sensor network

2 Literature Survey

"A System for Monitoring Air and Sound Pollution using Arduino Controller with IoT Technology" by Ezhilarasi et al. in 2017. In this paper, she discuss about how the air and water are polluted and remedy for this problem. Now a day the pollution due to air and water is growing issue. It is necessary to monitor the quality of air and water for healthy and safety life. In our paper we are include sound pollution monitoring system which allows monitoring the sound pollution in a particular area. The main aim of this paper is to monitor air and sound pollution using Internet of Things in different area [1].

In 2016 Dr. Sumithra et al. did research for environment monitoring and present a paper "A Smart Environmental Monitoring System using Internet of Things". The Engineering and Science calling have been affected by their duty to the general public, in ongoing decades. It has coordinated towards welfare and general wellbeing assurance. The Engineers and Scientists created procedures for observing the contamination. The checking procedure is finished by Internet of Things. In light of gathered information's preventive techniques are implemented [2].

"Design and Development of Environmental Pollution Monitoring System using IOT" by Mr. Ajay et al. in 2018. This paper deals with an extremely growth in an industrial and infrastructural frameworks creating environmental affairs like atmospheric changes, malfunctioning and pollution. Pollution is becoming serious issue so there is need to build such a flourishing system which overcomes the problems and monitor the parameters that affecting the environmental pollution. It can provide means to monitor the quality of environmental parameters like Air, Noise, to monitor pollution levels. The system is using a prototype implementation consists of sensing devices, Arduino Uno board, ESP8266 as Wi-Fi module. The aim is to build powerful system to monitor environmental parameters [3].

"IOT Based Air and Sound Pollution Monitoring System" by Sharma et al. in 2018. The serious issue in our environment these days is air and sound pollution. The large number of diseases has been created due to this pollution. Therefore it has become an necessity to control the pollution. The authority access air and water pollution monitor levels. In this paper it is also capable of detecting fire in its area and notify it to fire brigade authorities. The IoT helps in access at remote locations and data's are saved in database [4].

3 Existing System

In existing system using arduino Uno board the output is displayed In LCD arduino board is designed for specific purpose arduino board is not a full computer and they don't have an in build Wi-Fi part. The operating system of arduino doesn't run in full but it will simply execute the written code. In arduino the network connectivity is not direct. Tinkering is required to set up a proper connection which it is possible. For this process Ethernet port with a extra chip is loaded to wire with arduino board. This is a disadvantage that is it requires an extra hardware. The raspberry Pi comes with an in-build Wi-Fi port which is used for IOT operation. The clock speed of raspberry Pi is 40 times greater than arduino. The RAM of raspberry Pi is 12,800 times larger compared to Arduino.

4 Proposed System

IOT based air pollution detector using raspberry pi is used to monitor the air quality and detect the gas. In additional to it humidity, PH, turbidity, and temperature sensor is used to measure the respective factors. The sensors collect the data and send it to the cloud for further operation takes place and collected data's are maintained in cloud database. Raspberry is an general purpose computer which has inbuilt memory and can extend the memory using memory card. It is used in two modes of operation such as Ethernet and Wi-Fi (Fig. 2).

\

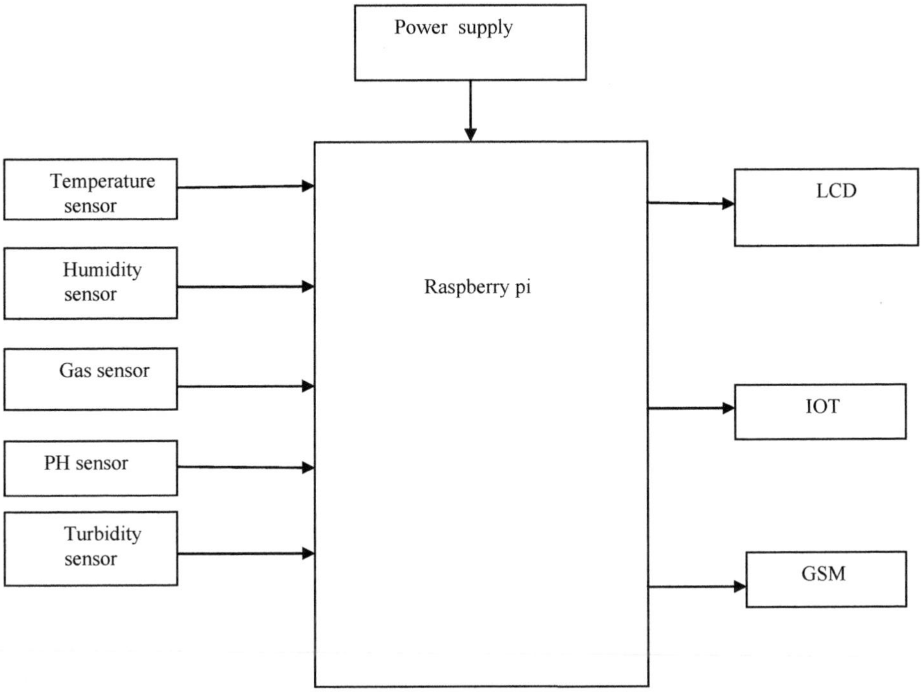

Fig. 2. Proposed system block diagram

4.1 LM35 Temperature Sensor

Lm35 temperature sensor is used to monitor the temperature levels. It is an Integrated Circuit Temperature Sensor. It detects the varying temperature level in the environment.

4.2 MQ135 Gas Sensor

MQ135 sensor is a gas sensor. It is used for monitoring the air quality. The power supply voltage is 5V for MQ135 gas sensor. The sensor detects CO2, NH3, smoke, alcohol. MQ135 sensor has high sensitivity and simple circuit. In this project we are using this sensor to detect the gas (Fig. 3).

Fig. 3. MQ135 gas sensor

4.3 DHT22 Temperature and Humidity Sensor

DHT22 is a temperature and humidity sensor, and it is complex with a calibrated digital signal output. By using the exclusive digital-signal-acquisition technique and temperature and humidity sensing technology, it ensures high reliability and excellent long-term stability.

4.4 Turbidity Sensor

Using turbidity sensor large number of individual particles that are invisible to the naked eye are detected in water. In this project we are using this sensor to monitor water quality.

4.5 pH Sensor

PH sensor is used to determine the PH of water. Pouvoir Hydrogen (pH) is The French acronym which is translated in English as "power of hydrogen" are potential of hydrogen. pH sensor is used to measure alkalinity or acidity of water soluable substances.

4.6 Raspberry Pi

Raspberry pi 3 module is used for this project. It is an SBC (Single Board Computer). In raspberry pi bluetooth and Wi-Fi is in-build. It is also an general purpose computer consisting of RAM. The saved data is used for monitoring purposes and analysis is done on a periodical basis (Fig. 4).

4.7 Liquid Crystal Display

LCD (Liquid Crystal Display) screen is an electronic display module. A 16 $\times$ 2 LCD display is used in various devices and circuits. The reasons being: LCDs are economical; easily programmable; have no limitation of displaying special and even custom characters (unlike in seven segments), animations and so on.

Fig. 4. Raspberry Pi

4.8 Internet of Things

The internet of things (IoT) Consists of sensors, connectivity, data processing, user interface. Sensor- collect the data from the environment. Connectivity-data send to the cloud (Wi-Fi). Data processing-data get to the cloud. User interface-useful to the user.

4.9 GSM

A GSM modem is a wireless modem that works with a GSM wireless network. It is used by mobile devices such as mobile phones and tablets. It supports communication through RS232 with DB9 Connector, TTL Pins and I2C Pins. CALL SMS GPRS facility - MIC input, LINE input and SPEAKER output pin.

4.10 Flowchart

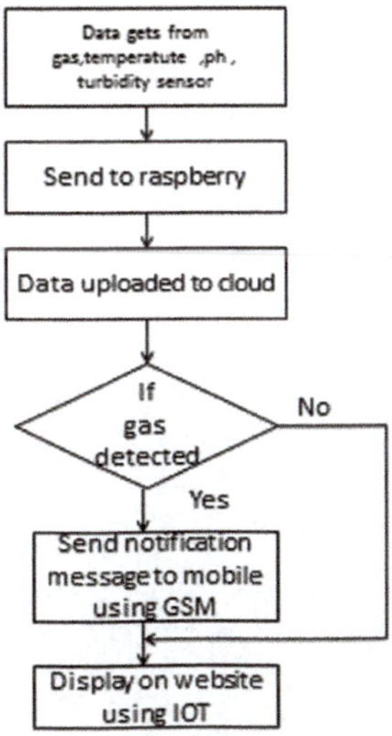

4.11 Proto Type of Proposed Method

The hardware kit consist of various sensors for monitoring the parameters such as temperature, humidity, gas leakage, ph value, turbidity in hospital to maintain the patients safely. The sensor collect the data's from the environment and send it to controller. Raspberry pi send data's to the cloud for data base management. If gas leakage was detected then notification message is send to mobile using GSM and display it on website using IoT, else if gas is not detected the data is only display on website using IoT (Fig. 5).

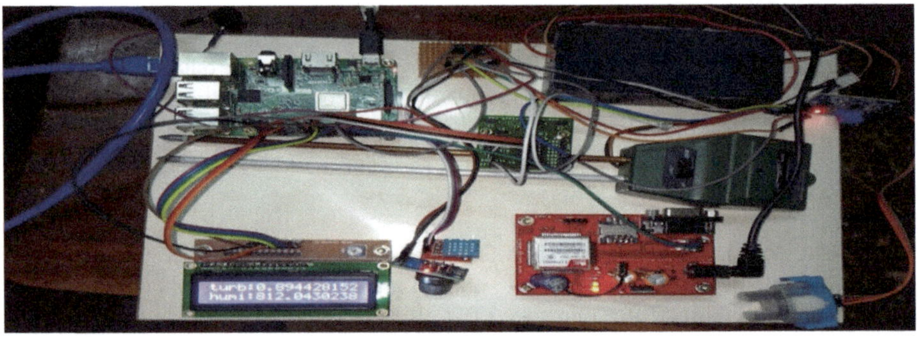

Fig. 5. Hardware diagram of proposed method

5 Results of Proposed Method

We need internet access for cloud operation. GSM module is added to the system to connect the mobile phone. Bluetooth and Wi-Fi is in-build in raspberry pi, using this module the data's are send to the cloud. The data is saved for monitoring purposes and analysis is done on a periodical basis (Fig. 6).

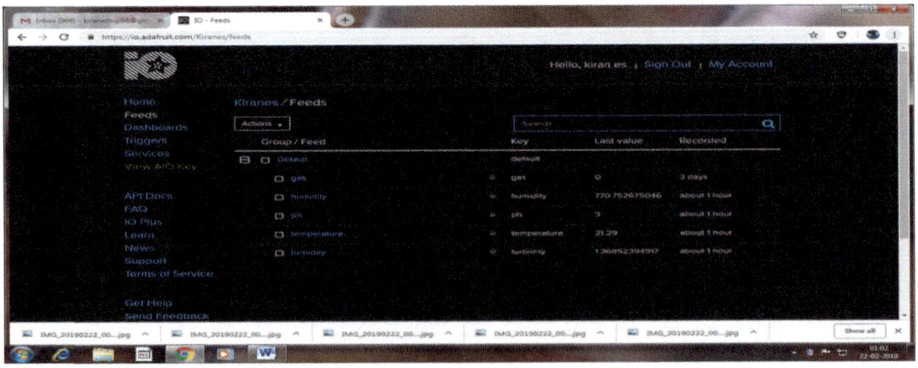

Fig. 6. Sensors output

5.1 Temperature Detection

Lm35 temperature sensor is used to measure temperature levels in a environment. It detects the temperature signal and process the output. The output is send to the controller Raspberry pi. A simple python code is executed to collect, display, and send the data's to the cloud (Fig. 7).

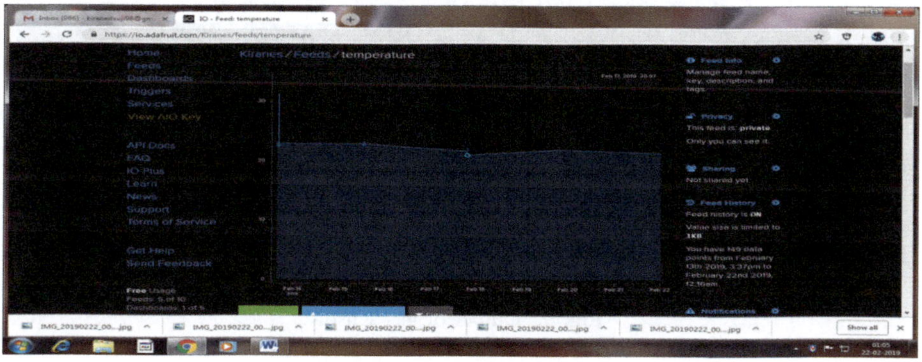

Fig. 7. Output of temperature sensor

5.2 Humidity Detection

DHT22 Humidity and Temperature sensor is used to measure varying humidity and temperature in a environment. The sensor detects the humidity and temperature signal and process the output. The output is send to the controller Raspberry pi. A simple python code is executed to collect, display, and send the data's to the cloud (Fig. 8).

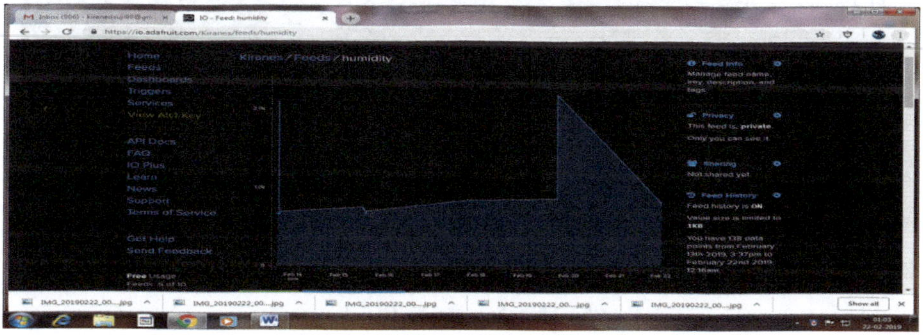

Fig. 8. Output of humidity sensor

5.3 Gas Detection

MQ135 Gas Sensor is utilized to detect the gas in an area. The sensor detects the CO2, NH3 smoke, alcohol level in air and process the output. The output is send to the controller Raspberry pi. A simple python code is executed to collect, display, and send the data's to the cloud (Fig. 9).

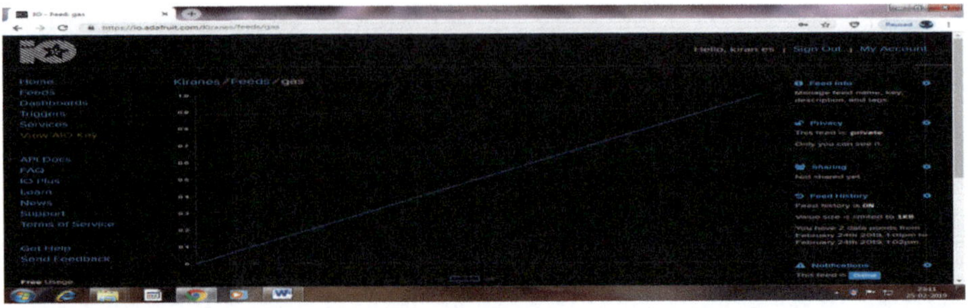

Fig. 9. Output of gas sensor

5.4 PH Detection

pH sensor is used to measure alkalinity or acidity of water soluable substances is used to determine the PH of water. The output is send to the controller Raspberry pi. A simple python code is executed to collect, display, and send the data's to the cloud (Fig. 10).

Fig. 10. Output of pH sensor

5.5 Turbidity Detection

Using turbidity sensor large number of individual particles that are invisible to the naked eye are detected in water. We are using this sensor to monitor the water quality.

The sensor detects the turbidity signal and process the output. The output is send to the controller Raspberry pi. A simple python code is executed to collect, display, and send the data's to the cloud (Fig. 11).

Fig. 11. Output of turbidity sensor.

6 Conclusion

Air and water pollution has become an growing issues in our society. Humans and their development play a vital role in environmental pollution such as air, water etc. For the whole world this is an major concern. The IoT helps to monitor the air quality, water, temperature and humidity levels. The accumulated data in the cloud is used for future analysis, and remedy actions are taken. It is a efficient method of monitoring and available in low cost.

References

1. Ezhilarasi, L., et al.: A system for monitoring air and sound pollution using Arduino controller with IoT technology (2017)
2. Sumithra, A., et al.: A smart environmental monitoring system using internet of things (2016)
3. Ajay, M.N.A., et al.: Design and development of environmental pollution monitoring system using IOT (2018)
4. Sharma, A., et al.: IOT based air and sound pollution monitoring system (2018)
5. Karthika, K., et al.: A smart environmental monitoring system using internet of things (2016)
6. Shri, A., et al.: Noise and air pollution monitoring system using IOT (2017)
7. Kumar, S., et al.: Air quality monitoring system basedon IoT using Raspberry Pi (2017)
8. Saha, H.N., et al.: Recent trends in the internet of things (2017)
9. Saha, H.N., et al.: IoT solutions for smart cities (2017)
10. Abraham, S., et al.: A cost-effective wireless sensor network system for indoor air quality monitoring applications (2014)
11. Lee, D.D., et al.: Environmental gas sensors (2001)
12. Chiwewe, T.M.: Multi-sensor system for remote environmental (air and water) quality monitoring (2016)

Detection of Ransomware in Emails Through Anomaly Based Detection

S. Suresh[1(✉)], M. Mohan[2], C. Thyagarajan[2], and R. Kedar[2]

[1] Sathyabama Institute of Science and Technology, Chennai, India
m.suresh.suresh@gmail.com
[2] Department of Computer Science and Engineering, Panimalar Engineering College, Chennai, India
mohan.rm@gmail.com, thyagu.maddy@gmail.com, kedar2206@outlook.com

Abstract. In the recent years Email has been the popular mode of communication. We can connect or send message to anyone in any part of the world with the help of Email. Email is said to be advanced version of communication. There are also certain disadvantages of using the Email. Some of the disadvantages are privacy, phishing Emails, Spamming, Malware, Spam and much more. Out of which the Ransomware through Email are being common in the recent years. Ransomware is one of the serious problems that are found on the web. It is a form of malicious content or software that encrypts the data on our system making it unavailable for us to use. In a simpler manner it locks out the files, folders and subfolders in our system. And now-a-days Ransomware is spread through phishing Emails and the attackers charge a lot for it. Ransomware is not a virus but, it is a malicious software that locks us out of our own systems. There are two types of Ransomware that can be spread through Email, one of which is crypto ransomware and the other is the locker ransomware. In this paper we are going to discuss about the Ransomware through Email and how it harms our system and also we are going to discuss about how to overcome or prevent ransomware.

Keywords: Email security · Network security · Ransomware · Malware · Phishing · Spamming

1 Introduction

The Emails are used in the recent days which is a mode of communicating with one another. But there are also certain ways in which the Email can be misused or it becomes an advantage for the attack to attack the victim. Some of the common threats through Emails are hacking, phishing, Malware, Spam, Ransomware etc. The Ransomware one of the common threats that is emerging commonly in the recent days which could cause great debts to many users and as well as many companies may run in a great loss due to this type of threats. So Ransomware is a serious threat in the computer society that has to be taken care as soon as possible. The attackers are spreading the Ransomware through Emails that has become common these days.

D. J. Hemanth et al. (Eds.): COMET 2019, LNDECT 35, pp. 604–613, 2020.
https://doi.org/10.1007/978-3-030-32150-5_59

The attacker uses the phishing Emails in order to trick the victim and make him to click or enter the link which he had sent through Email. So phishing also plays an important role in spreading Ransomware through Emails. So Email is also not considered as the safest way to communicate with people around the world as there are many emerging threats like this. So we should also take some serious steps to safeguard our privacy over the network.

Ransomware is nothing but a suspicious or malicious software or malware that gets installed in our system. The Ransomware may also be a self replicating malware in our system. There are two types of Ransomware possible through Email, they are Crypto Ransomware. The Crypto Ransomware is nothing but the attacker uses a strong level of encryption to encrypt all the files and folders in our system, it includes all our personal data also. The attacker demands some ransom in exchange for the key to the encrypted files. This is known as Crypto Ransomware. The other type of Ransomware is known as Locker Ransomware in which this is also known as the computer locker which means that the files in the computer gets locked up making it unable for the users to use. Even in this type of Ransomware attack a small amount of ransom has to be paid in exchange to unlock the locked files in our system. The attackers use the Email as their pathway to spread Ransomware all over the network.

At first Ransomware appeared in 1989 which was known as AIDS Trojan (otherwise called pc cyborg) which was created by Dr. Joseph Popp and spread by means of floppy disk of measure 5.25 in. The Ransomware not only affects the local system but also damages the external system or components and also affects the systems that are connected to the same network. The bid for the Ransomware attacks should be paid in bitcoins so that the attackers are not exposed over the internet. The money is demanded in bitcoins as it is harder to track. Some of the existing tools to prevent the Ransomware attack are wannakiwi, heldroid and honeypot etc.

2 Operations of Ransomware

As soon as the victim receives the Email from the attacker in a fake phishing website and as soon as the victim clicks the link the ransomware malicious software gets installed on your system.

- The attacker generates a pair of key and installs or attaches it with the malware or the malicious code he has created.
- And then the malware is out in free air.
- As soon as it gets installs in the victim's system it generates a random symmetrical key on its own.
- It then uses the key generated by the attacker to encrypt the symmetric key.
- At the end it gets created into a small asymmetric ciphertext.
- The ciphertext makes it unable for the victim to read and understand.

- It then pops up a message showing the ciphertext and how pay the ransom in return.
- This makes no choice for the victim rather than to pay the bid unless and until none of his important and sensitive information gets encrypted.
- In order to restore back his data he must pay the ransom.
- While sending the ransom to the attacker he must also send his ciphertext as well.
- When the attacker receives the ransom he decrypts the asymmetric code and sends it back to the victim.
- And then the victim decrypts all the encrypted data with the help of the decryption key sent by the attacker.

The Ransomware is initiated using a Trojan which enters a system through the email or the vulnerabilities found in the network service. But it gets mostly installed in your system through the common type of attack known as the phishing. Through the phishing Email the victim gets tricked and is made to open the Email. And this is the place where the malicious code gets installed in his system. As soon as it gets installed in the system it downloads some scripts from the internet which also gets installed in our system and these scripts are executed automatically without the victim's knowledge. The malicious code the runs a payload in the victim's system to lock or restrict the usage of the data which is the motto of Ransomware. Some of the payloads contains only the program that was created or designed to just lock the files in the system just by making some changes in the Windows Shell [1] or just by altering the partition table or the MBR (Master Boot Record). It also prevents the system from booting. And there are also some strong payloads that encrypt the Data in the system. It makes it unable for the victim to use the data as it gets encrypted. In these cases only the attacker is the only one who is able to decrypt the data. So the victim is prone to do whatever the attacker demands. The main aim or motto of the attacker is to get a payment which is always their primary goal [2]. A main element that is making the Ransomware a greater advantage to the attacker is that of a convenient mode of payment. The mode of payment is known as bitcoins or cryptocurrencies which are hard or unable to trace. They receive the payment or bid only through bitcoins so as their identity is not exposed or revealed in the internet. This makes a greater advantage for the attacker to take charge against the victim to take down his system. There are also some other mode of payment such as premium rate text messages, wire transfers, paysafecard etc.

3 Various Types of Ransomware

See Fig. 1.

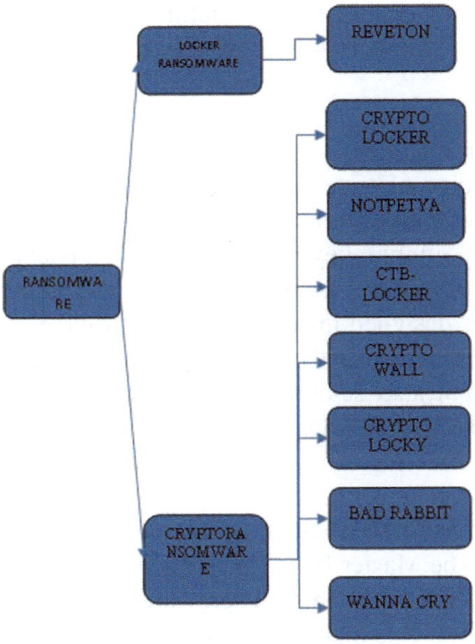

Fig. 1. Types of ransomware

3.1 Locker Ransomware

The locker ransomware is the one which locks our system till we pay the ransom or bid demanded by the attacker. The victim does not stand any chance rather than to pay the ransom. This type of locker ransomware attack affect the wearable smart devices and the IoT devices, these types of devices easily gets locked up [3].

3.1.1 Reveton

The Reveton ransomware appeared in 2012. The reveton is a type of Trojan and also a payload which when gets installed in our system or affects our system shows us a display warning saying that the computer was being used for illegal activities such as child pornography or for downloading unlicensed software. Sometimes to make this attack more realistic the attacker sometimes displays the Computer's IP address and even in some cases the webcam footages of the victim is recorded and sent to the victim itself [4].

3.2 Crypto Ransomware

The crypto ransomware is the one which encrypts all the data in out disk using different types of algorithm. When our system gets infected by the crypto ransomware a message or pop-up displays stating that the victim needs to pay a ransom to get all his data back without being damaged. This may or may not happen. When all the data in our system gets encrypted it becomes unable for the victim to use it which may even cause a serious damage or loss if it infects some sensitive information in our system. Some of the well known

3.2.1 Crypto Locker

The crypto locker is a type of code or payload that encrypts the data in the hard disk and it uses various several algorithms to do it. It generates a random asymmetric key and the attacker sets up a public key on top of it which encrypts the asymmetric key. The attacker will set a specific time within which the victim has to pay the ransom otherwise the payload is designed in such a way that it wipes all the encrypted data on the disk if it is not decrypted within the specific provided time. And if the victim tries to delete the crypto locker it also deletes the asymmetric key through which the victim will be unable to use the data permanently [5].

3.2.2 Notpetya

Notpetya is a type of ransomware attack which alters or makes changes in the MBR (Master Boot Record). The Master Boot Record consists of the information of the boot options in our system. And when this ransomware infects our system it makes some modifications in the MBR. This makes the system to crash often. When the victim tries to reboot the system a message pops up displaying a message to pay the ransom [6].

3.2.3 CTB-Locker

The CTB locker is well known as critroni. When the CTB locker infects our system it scans for all the data in the hard disk and encrypts all the data stored in it. It is assumed to have been created by various types of algorithms such as the elliptical curve algorithm to encrypt the data. The term CTB stands for Curve-Tor-Bitcoin. It means that it uses elliptical curve algorithm and it uses tor browser for paying the ransom as tor browser is well known for its anonymity and the ransom should be paid only with bitcoin which makes it harder to trace the attacker. It pops up a display showing that the victim has only 96 h (4 days) within which he has to pay the ransom demanded. The CTB locker has a extension which changes from CTB to CTB2 after the files gets encrypted [7].

3.2.4 Crypto Wall

The crypto wall is ransomware Trojan that targets all versions of windows. The crypto wall is available in many versions such as crypto wall 2.0, crypto wall 3.0 and crypto wall 4.0. The crypto wall 3.0 payload is written in javascript. It gets installed in your computer through a email attachment. As soon as it gets installed in our system it downloads all the executable files in the JPG format. And also to avoid detection it creates svchost.exe and explorer.exe to be in touch with the servers. When the payload starts to encrypt all the data it deletes all the shallow copies of data and also installs a spyware that records and saves all the usernames and passwords without the knowledge of the victim. The crypto wall 4.0 is a advanced version in which the code is enhanced in such a way that it evades all the anti virus detection and not only encrypts the data only but also it encrypts the name of the file also which makes this type of ransomware more complex [8].

3.2.5 Crypto Locky

The crypto locky is one of the complicated malware of all time. This malware came into existence in the mid of 2016. This type of malware is spread through the Microsoft office attachments which are sent through email. When this type of attachments is opened by the victim is allows the malware code to run indirectly and then starts to encrypt all the data in the hard disk. After the encryption process is finished it also installs the TOR (The Onion Routing) browser through which the victim has to pay the ransom.

3.2.6 Bad Rabbit

The bad rabbit is a type of ransomware which is much more similar to the petyaransomware. This type of ransomware is spread through the adobe flash installer. When the victim mistakenly runs the adobe flash player installer the bad rabbit ransomware will be in action and it encrypts all the files and displays a message saying "Oops! Your files have been encrypted". When the adobe flash player runs it also asks for the administrator permissions which will record all your login credentials regarding that system [9].

3.2.7 Wanna Cry

The wanna cry is one of the most popular and the deadly ransomware of all time. It has caused a major destruction all over the world. Many high end and top rated companies were affected by this wanna cry ransomware. The WannaCry ransomware targets the Microsoft windows operating system. It also encrypts all the data in our hard disk and also this ransomware passes to other systems in the network using the Server Message Block (SMB) vulnerability and also spreads to random computers in the internet which has this SMB vulnerability. The WannaCry uses a transport mechanism that helps this ransomware to spread itself. The transport mechanism searches for vulnerabilities or exploits and then it uses the EternalBlue exploit to have access over the system and it uses double pulsar tool which creates a duplicate replica copy of itself. The EternalBlue exploit was stolen by the attackers from the National Security Agency (NSA). The NSA were the first to discover such type of vulnerability in the Microsoft and instead of reporting it to the rightful owner the NSA used it to create a vulnerability for its own offensive work. Due to this a very serious damage was caused all over the world [10]. Figure 2 shows the history of Ransomware attacks all over the world.

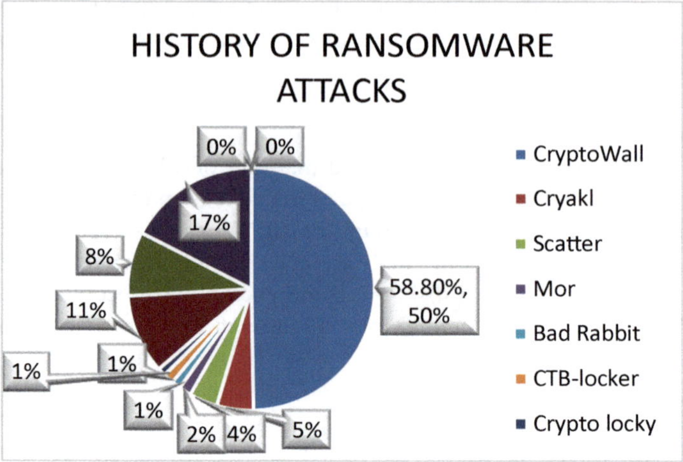

Fig. 2. History of ransomware attack

4 Ransomware Detection

The types of ransomware detection techniques are

- Signature based detection
- Anomaly based detection

4.1 Signature Based Detection

In the signature based detection the malware is identified by using the signature of the malware. Every malware has a signature. There is also certain behavior of the code of how it behaves when executed and based on it the signature of every malicious malware is created manually by human. As data needs a storage element to store the signature based detection also requires a storage element which is known as repository. The repository is the place where the signature of the malicious code or program is stored based on the behavior of it. But the major disadvantage of the signature based detection technique is that it cannot detect the zero-day attacks (i.e. Whenever a new malware or malicious code gets installed in our whose signature is not present in the signature repository it does not detect and the malware will be executed). It is used to detect only the known types of ransomware attacks based on the signatures we stored in the repository.

4.1.1 Principle of the Signature Based Detection Technique

- At first the victim receives the spam or the phishing email and the clicks on that email.
- The malicious content present in the email gets installed in our system.
- Then the signature based detection starts its job to scan the downloaded content from email.

- And then it generates a signature of the file based on its behavior.
- And then it compares it with the signature repository to check whether a similar signature exists.
- If there is no signature present in the repository it allows us to execute the downloaded file.
- If the signature matches to any of the signature present in the signature repository it will block the content and does not allow us to execute the file.

4.2 Anomaly Based Detection

The anomaly based detection technique consists of two blocks. They are the training block and the detection block. The training block is also known as the learning block and the detection block is also known as the monitoring block. The anomaly based detection technique works stable unless and until a well defined set is defined for the anomaly based detection [11]. The major advantage of the anomaly based detection technique over the signature based detection technique is that the anomaly based detection technique can detect the zero-day attacks or exploits also (i.e. Even if an unknown malware which did not show itself before will be detected by this technique). The two blocks are very important in the anomaly based detection. In the training or the learning block the behavior of the system is well learned and observed and the various different styles of the system is watched closely. After the behavior of the system is well observed in the training block the method shifts to the detection or the monitoring block in which it compares and contrasts it with the real time systems with the various behavior which is learned in the previous block. In the monitoring block it produces a warning message or produces a alarm when something unexpected occurs during the process.

4.2.1 Principle of the Anomaly Based Detection Technique

- At first the victim receives the spam or the phishing email and the clicks on that email.
- The malicious content present in the email gets installed in our system.
- Then the anomaly based detection technique starts its job and starts to scan the downloaded content in email.
- The anomaly based detection technique consists of two main blocks known as the training block and the detection block.
- In the training block it consists of the training data which consists of the various and different malicious behavior.
- The different types of behavior is learned from the system logs available.
- When the scan starts it first enters into the training or block.
- In the training block it defines a set of rules and sends it to the training data and cross checks and learns the behavior of the scanning file.
- It then enters into the detection or the monitoring block.
- In this block the previously learned data is used to compare the various type of behavior with the real time systems.

- In the monitoring block it detects for any malicious or unexpected behavior.
- If there is no suspicious behavior of the file and the behavior is normal it allows us to execute the file.
- But if the anomaly based detection detects anything unexpected or suspicious content during the scan it displays a warning message showing that the file should not be used or executed. It then produces a alarm to display the warning.

5 Recommendations to Prevent Ransomware

The below listed are some of the recommendations to prevent ransomware to minimize the damage caused

- A proper and a regular updated backup should be done.
- Enable the hidden file extensions to view and detect the extension as most of the common ransomware file extension is ".pdf.exe".
- Filter the mail containing the extension.exe.
- Block the files running from the App Data/Local App Data Folders because it contains its own rule for execution.
- Disable the Remote Desktop Protocol (RDP) as most of the malware accesses the system using this.
- Keep all your software updated up-to-date as all the bugs and errors will be getting fixed in the latest versions.
- Use a good and reputed anti-malware software.
- If you suspect a file you opened contains ransomware unplug the network cable so that the ransomware does not spread to the other systems connected in the same Local Area Network (LAN).
- Use the system recovery point to get to the well known state of the system.
- Set the Basic Input Output System (BIOS) clock back.
- Do not click on emails if you suspect something suspicious.
- Remove all the plugins in the browser so that it does not open the executable files or pdf files received through email or browsing.
- Turn the macros to off state in the Microsoft office such as word, excel etc.
- Download and use ad-block so that it blocks the unwanted and the phishing links away from us.
- Use the guest account rather than using administrator account just for local purposes or usage.

6 Conclusions and Future Work

In this paper we have discussed about the operation of ransomware and the different types of ransomware. The ransomware is increasing day by day and also it affects many systems all over the world. We have also discussed about the detection and prevention of the ransomware. The major criteria of the attacker is to make the victim pay the demanded ransom. The ransomwar is one of the biggest threat in the network security and also its major aim is to encrypt the files or block the usage access of the user in his own system. The main goal of the attacker is to pay the ransom. In the future we will develop a tool to defend the ransomware if it getsinstalled in our system to reduce or to take minimal damage if affected by ransomware.

References

1. Ransomware: Fake Federal German Police (BKA) notice, SecureList (Kaspersky Lab). Accessed 10 Mar 2012
2. Young, A., Yung, M.: Cryptovirology: extortion-based security threats and countermeasures. In: IEEE Symposium on Security and Privacy, pp. 129–135 (1996). ISBN 0-8186-7417-2. https://doi.org/10.1109/secpri.1996.502676
3. You're infected—if you want to see your data again, pay us $300 in Bitcoins, ArsTechnica, 17 October 2013. Accessed 23 Oct 2013
4. Pathak, P.B.: A dangerous trend of cybercrime: ransomware growing challenge. Int. J. Adv. Res. Comput. Eng. Technol. (IJARCET) **5**(2), 169–174 (2016). ISSN 2278-1323
5. MahmudhaFasheem, S., Kanimozhi, P., AkoraMurthy, P.: Detection and avoidance of ransomware. International Journal of Engineering Development and Research **5**(1), 254–260 (2017). ISSN 2321-9939
6. The computer emergency Response team Mauritius (CERT-MU), The Petya Cyber-attack, Whitepaper (2017)
7. Gonzalez, D., Hayajneh, T.: Detection and prevention of Crypto-ransomware. IEEE (2017). 978-1-5386-1104-3/17
8. Malvertising campaign delivers digitally signed CryptoWallransomware. PC World, 29 Sept 2014. Accessed 25 June 2015
9. Palmer, D.: Bad Rabbit ransomware: A new variant of Petya is spreading, warn researchers. ZDNet. Accessed 24 Oct 2017
10. Mohurle, S., Patil, M.: A brief study of Wannacry threat: ransomware attack 2017. Int. J. Adv. Res. Comput. Sci. **8**(5), 159–164 (2017). ISSN No 0976-5697
11. Jyothsna, V., Prasad, V.V.R., Prasad, K.M.: A review of anomaly based intrusion detection systems. Int. J. Comput. Appl. **28**(7), 125–134 (2011). (0975–8887)

Design and Development of Control Scheme for Solar PV System Using Single Phase Multilevel Inverter

J. Prakash(✉), K. R. Sugavanam, Sri Krishna Kumar, S. Kokila, T. Abarna, and R. Hamshini

Department of Electrical and Electronics Engineering, Vel Tech High Tech Dr Rangarajan Dr Sakunthala Engineering College, Avadi, Chennai 600062, India
prakash_ies@yahoo.co.in, sugavanamkr@gmail.com, krishnakumar.rvs@gmail.com, kokilasundher@gmail.com, abarnatt15@gmail.com, hamshamshini@gmail.com

Abstract. This paper explores transformer less single phase multilevel inverter. The main objective of the paper are to find the operating voltage or peak power voltage for the pv panel (12v), to reduce the Total Harmonic Distortion level within 5% and to check whether the efficiency is to be increased to a certain level. The peak power voltage is the voltage that can be obtained when pv panel produces maximum power. To obtain this, Particle swarm optimization mppt algorithm is used to find the best value followed by several iterations. Evaluation results show that our model outperforms other models in nearly all metrics.

Keywords: Solar panel · Boost converter · PSO algorithm

1 Introduction

In recent trends, solar panel plays a key role in power generation. Solar energy will be more competitive due to its lower cost and improved technology. In our project, we considered the solar panel as an input source which has the maximum power (Pmax) of 250.17 Wp. With this power, efficiency of our model can be increased. The sun's intensity is inversely proportional to area of cross section. Since its intensity increases, current and power output can be correspondingly increased due to the proportionality. Mathematically it can be represented as

$$\text{Intensity} = P/A$$

Where P is power and A is area of cross section (Fig. 1).

From reference [1–4], the galvanic isolation is used when two or more electric circuits must communicate, but their ground connections have to be at different potentials. Galvanic isolation perhaps classified towards dc-decoupling as well as ac-decoupling techniques. It is considered a certain ac-decoupling only contribute crouched losses when comparing with dc-decoupling method. It is selected by reason of diminished switch count in the pathway of conduction. In Transformerless Photo

D. J. Hemanth et al. (Eds.): COMET 2019, LNDECT 35, pp. 614–625, 2020.
https://doi.org/10.1007/978-3-030-32150-5_60

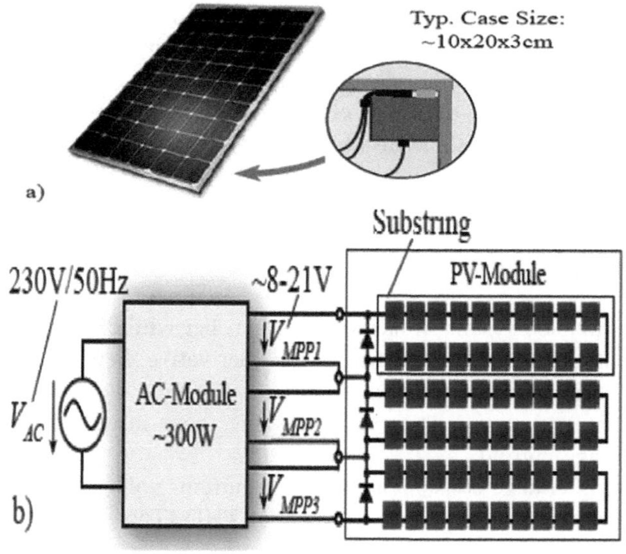

Fig. 1. Module of AC: (a) mounting, (b) schematic

voltaic inverters, dynamic reclusion amidst the PV together with grid owns to flow the current which is leakage. To reduce such leakage current there occurs some techniques to reduce the flow of current. Hence, topographies for instance H5 and HERIC, dynamic reclusion last to provide the current which is less in leakage. Common Mode Voltage is the major reason for current which is leakage. When dynamic reclusion together with CMV clamping is combined; elimination of leakage current is done. When comparing with other inverters which have galvanic isolation, Transformerless inverters offer a better efficiency [9]. The proposed topology overcomes all the effects which is discussed above (Fig. 2).

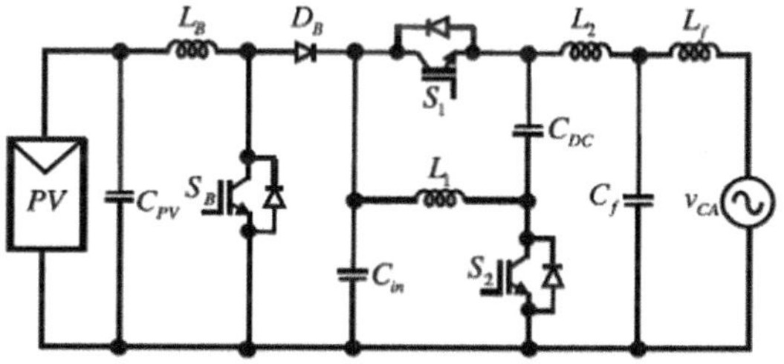

Fig. 2. Presented structure

The specified inverter will exclude current which is leakage, anyhow requires a chopper to groove the maximal power of solar panel beneath grid voltage, which down from efficiency and immense the elements of active and passive. Whereas this converter has crouched number of active elements and it employs immense number of passive peripherals. In addition, disadvantages of the presented inverter are with immense voltage accent of the switches along with shallow efficiency.

2 Proposed System

- In this contemplated system, it has MPPT controller with PSO algorithm which is a very efficient global search algorithm and derivative free. It is well suited for continuous variable problems (Fig. 3).
- Diverse design and manipulation strategies have been inspected to access standard value of leakage current.
- By maintaining voltage constant as shown in current- voltage characteristics, higher efficiency power can be obtained with 5% of THD (Table 1).

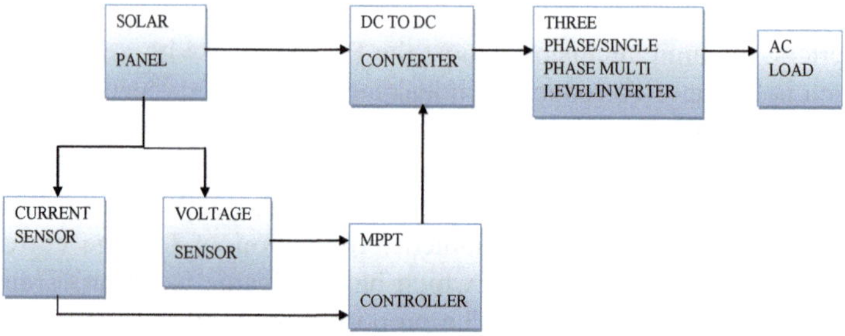

Fig. 3. Block diagram of designed system

Table 1. 39v panel parameters

Parameters	Values
Maximum power	250.17 wp
Voltage (open circuit)	44.99 v
Current (short circuit)	7.43 amps
Voltage at maximal power	35.89 v
Series fuse rating	6.97 A
Temperature	25 °C
Irradiance level	1000 w/m2
Fire rating	Class c

3 Operating Modes

According to boost converter, when comparing the output voltage with input voltage, it is almost immense than the input voltage. It is shown in the Fig. 4.

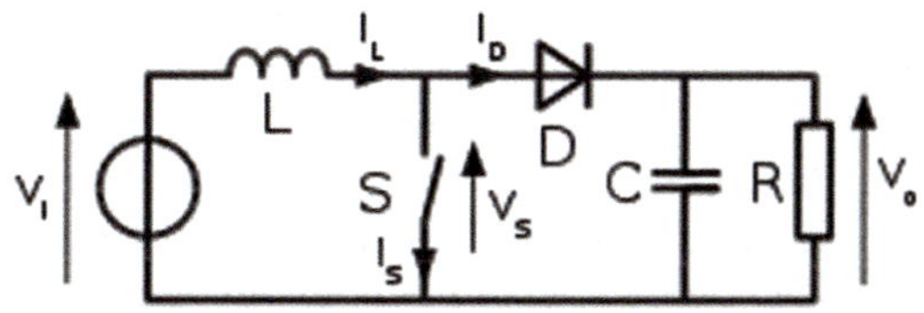

Fig. 4. Systematic figure of boost converter

The two different stages of boost converter can be illustrated. They are

(A) If the switch is in closed position, electrons will flows through the inductor (L1) in anti-clockwise direction and some energy will be reserved in the inductor. The left side of the inductor has the positive polarity (Fig. 5).

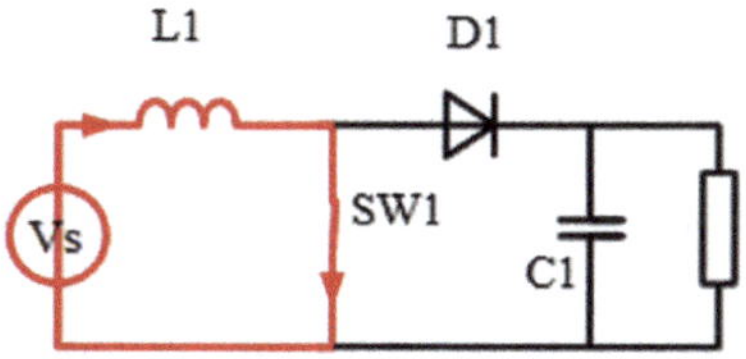

Fig. 5. On state

(B) If the switch is said to be in OFF condition, comparing to impedance the current is reduced in ON condition it generates some magnetic field in which the inductor stores energy. Now the magnetic field will be pulled down to maintain the current. Then the inductor has negative polarity (it gets reversed). Hence the result will be higher voltage which helps to charge the capacitor through the diode D1 (Fig. 6).

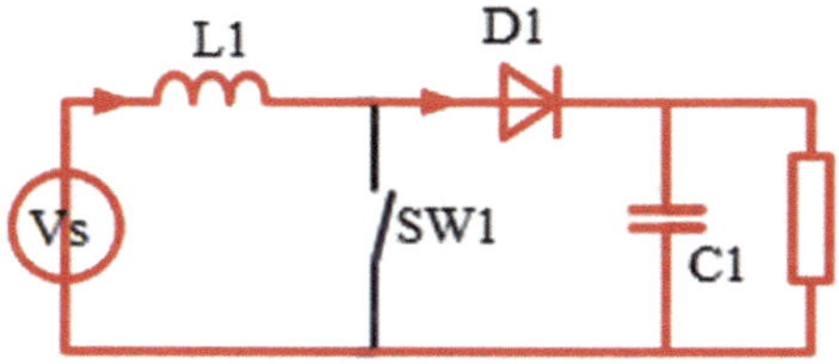

Fig. 6. Off state

4 MPPT Algorithm

Particle Swarm Optimization MPPT algorithm is fitted for the converter that has been scheduled in order to optimize the problem, all converter's design procedures are investigated and validated through simulation and experimental outcomes.

4.1 Pseudo Code

Step 1: The process is started
Step 2: Variables are initialised as lower bounds and upper bounds.
Step 3: The voltage V and current I from PV array are measured and tabulated.
Step 4: Power P (i.e., P = V * I) from the observed parameters of the panel is calculated.
Step 5: Set the maximum iteration 3
Step 6: Calculate the fitness value and observe the best value
Step 7: Calculate the velocity and update the particle by the equation

$$\text{Dinit}\ (i, j) = \text{lb}\ (j) + ((\text{ub}\ (j) - \text{lb}\ (j)) * \text{rand}\ (1))$$

Step 8: Based on the results of variable check, iteration will be continued and end process.

5 Simulation Results

The simulations have been done in Matlab using version 2014a. It consists of single phase and three phase multilevel inverter, solar PV panel within that multiple number of PV modules are present, boost converter (Fig. 7).

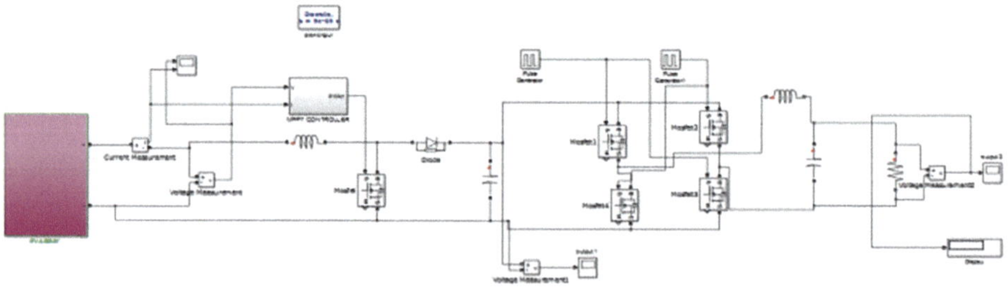

Fig. 7. Single phase simulation circuit

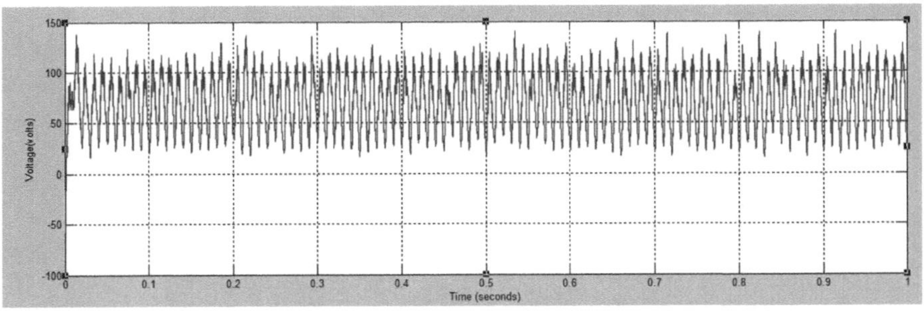

Fig. 8. Single phase input voltage

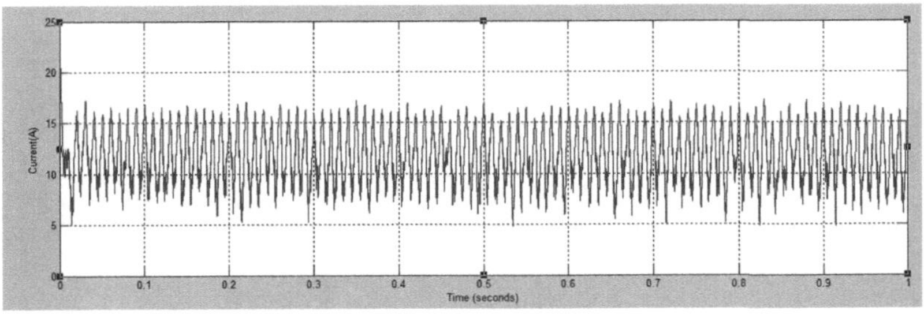

Fig. 9. Single phase input current

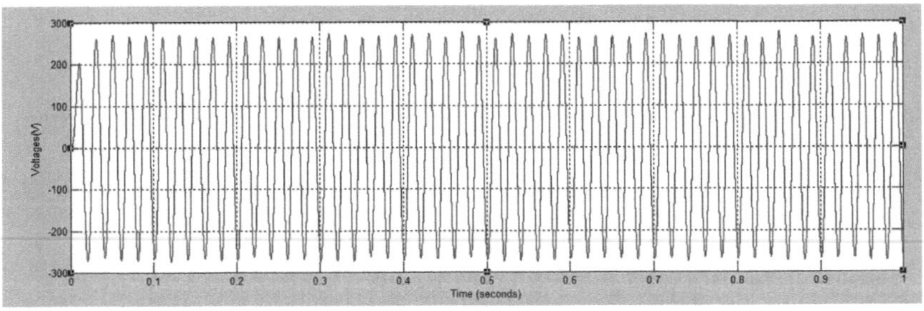

Fig. 10. Output voltage for single phase

The above results are given for solar PV system which is done by single phase multilevel inverter. Corresponding waveform for current and voltage is shown (Figs. 8, 9 and 10).

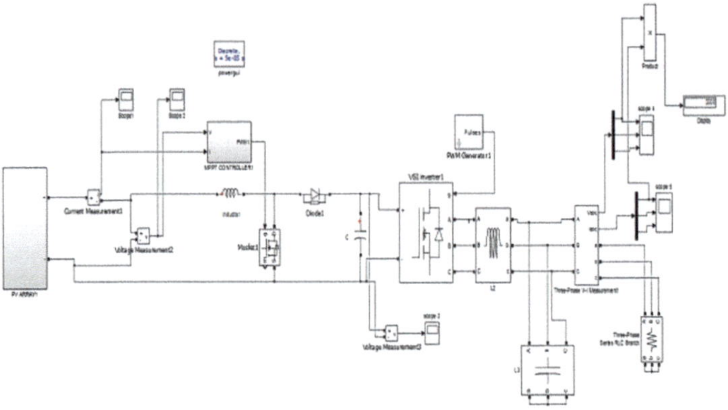

Fig. 11. Three phase simulation circuit diagram

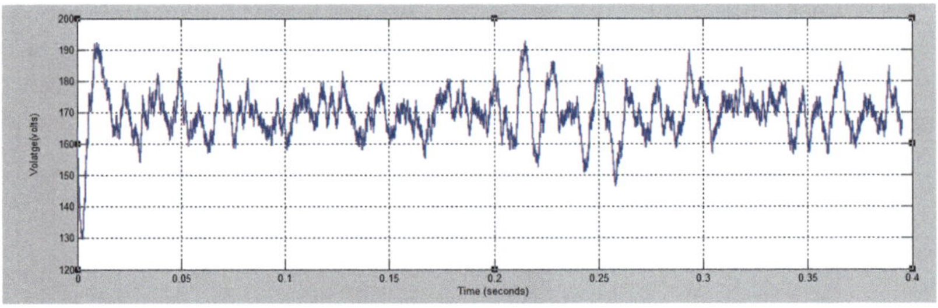

Fig. 12. Three phase input voltage

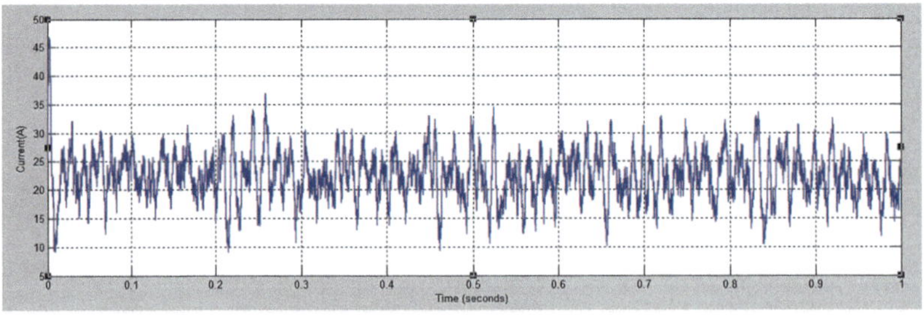

Fig. 13. Three phase input current

It shows the input voltage together with current waveform, output voltage along with current waveform. The simulated circuit is shown above in the outline (Figs. 11, 12, 13).

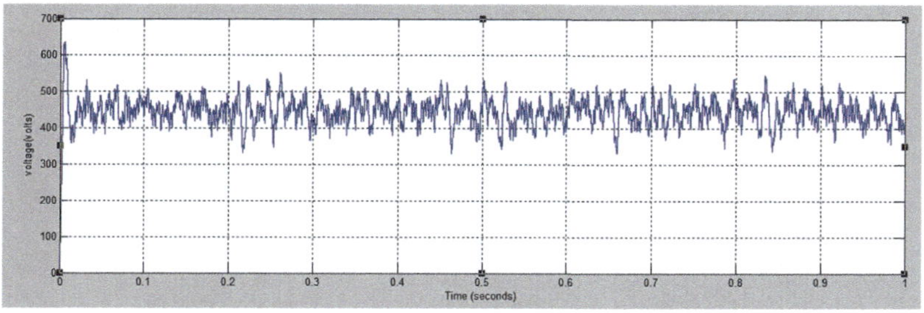

Fig. 14. Boost converter output voltage (three phase)

From the boost converter voltage from the solar panel is increased and the corresponding waveform is shown. The value of voltage is simulated using Matlab (Fig. 14).

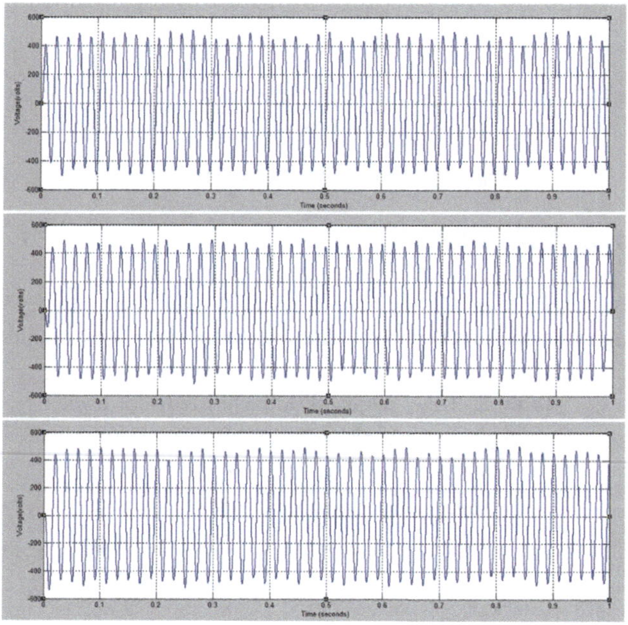

Fig. 15. Three phase output voltage waveform

Whenever three-phase voltage is counterpoised, per phase power turnouts are counterpoised and it equals to about 1/3 of maximal power that solar panel outturn comes from sunlight along with reactive power of particular phase balances to zero profiting taken away unity power factor control. The voltage and current waveform is shown in the outline (Figs. 15 and 16).

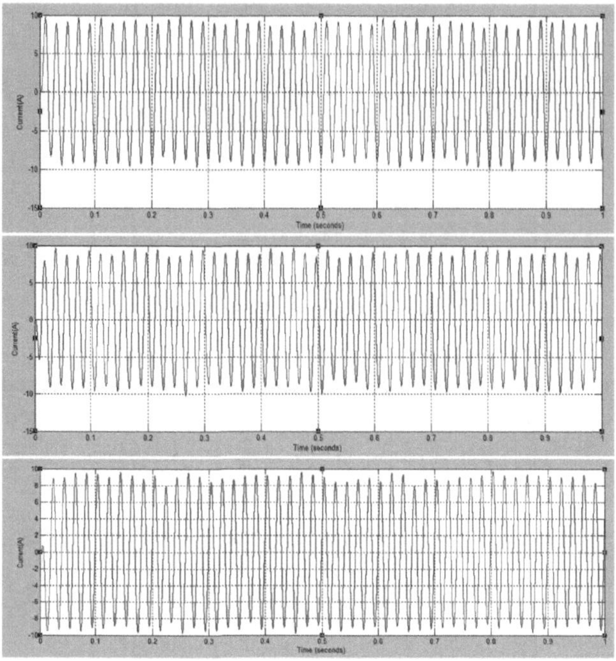

Fig. 16. Three phase current waveform

6 Experimental Results

From the above experimental analysis, we found that the voltage V = 180 is the operating voltage in which at that point the maximum power is obtained (Table 2).

Table 2. Panel output at different irradiance level

Irradiance level w/m2	Voltage (volts)	Current (Amps)	Power (walts)
20000	180	30	3330
15000	140	20	1920
10000	95	15	895.2
5000	50	8	256.8

Table 3. Values from the proposed system

Mode of operation	Voltage (v)	Current (amps)	Output voltage (v)	Output current (amps)	Pout (w)
Three phase MLI	185	25	400	7	3300

The lingered PV aspects used in simulation has 45 V DC with irradiance about 1000 watts/m^2 along with temperature parameter about 25 °C design an array which outturn 250 W in the identical voltage of 500 V (Table 3).

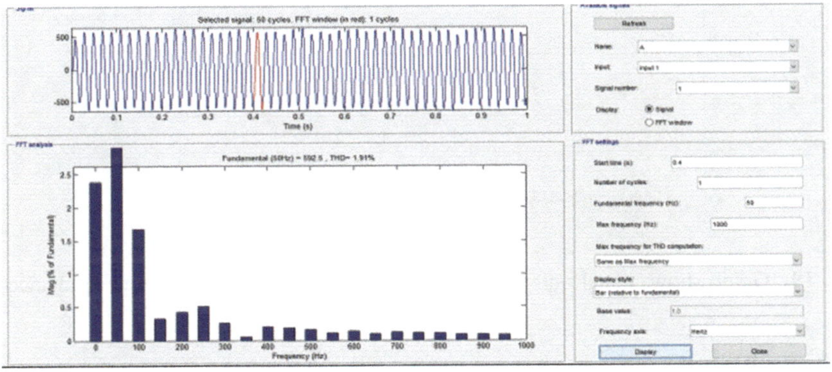

Fig. 17. FFT analysis and THD level of MLI

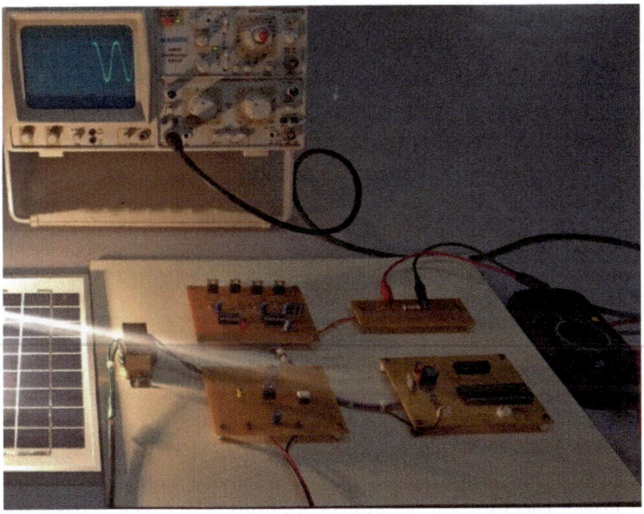

Fig. 18. Experimental setup of solar pv system using single phase multilevel inverter

Using Matlab, the THD level get analysed by the Fourier analysis with respect to fundamental frequency. It is given by the figure as shown. The performance of the boost converter remains same, even in the fast varying atmospheric condition (Figs. 17, 18 and 19).

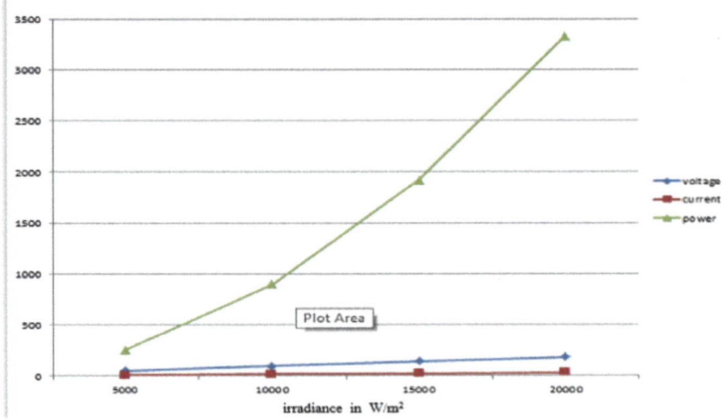

Fig. 19. Graph shows panel output (voltage, power and current) at different irradiance level

7 Conclusion

The transformer-less single phase multilevel inverter has been designed. Particle Swarm Optimization MPPT algorithm is fitted for the converter that has been scheduled in order to optimize the dilemma, all the converter's design procedure are investigated and validated through simulation and experimental outcomes. By analysing designed converter and similar structure, the number of switches and passive elements has been identified. It has around zero leakage current, boost potentiality, immense efficiency, and outstanding dynamics in order of the input voltage disparity. Hence, designed circuit can be exploit and degrade as coherence device amidst PV array and system distribution.

References

1. Freddy, T.K.S., Rahim, N.A., Hew, W.-P., Che, H.S.: Comparison and analysis of single-phase transformerless PV inverter topology. IEEE Trans. Ind. Electron **58**(1), 184–191 (2011)
2. Ardashir, J.F., Sabahi, M., Hosseini, S.H., Blaabjerg, F., Babaei, E., Gharehpetian, G.B.: Transformerless inverter with charge pump circuit concept for PV application. IEEE Emerg. Sel. Topics Power Electron. (to be published). https://doi.org/10.1109/jestpe.2016.2615062
3. Yu, W., (Jason) Lai, J.-S., Qian, H., Hutchens, C.: High-efficiency MOSFET inverter with H6-type configuration for photovoltaic no isolated AC-module applications. IEEE Trans. Power Electron. **26**(4), 1253–1260 (2011)
4. Li, W., Gu, Y., Luo, H., Cui, W., He, X., Xia, C.: Topology review and derivation methodology of single phase transformerless photovoltaic inverters for leakage current suppression. IEEE Trans. Ind. Electron. **62**(7), 4537–4551 (2015)
5. Islam, M., Mekhilef, S.: Efficient transformerless MOSFET electrical converter for a grid-tied electrical phenomenon system. IEEE Trans. Power Electron. **31**(9), 6305–6316 (2016)
6. Vazquez, N., Rosas, M., Hernández, C., Vázquez, E., Perez, F.: A new common-mode transformerless photovoltaic inverter. IEEE Trans. Ind. Electron. **62**(10), 6381–6391 (2015)

7. Gonzalez, R., Lopez, J., Sanchis, P., Marroyo, L.: Transformer-less inverter for single-phase photovoltaic systems. IEEE Trans. Power Electron. **22**(2), 693–697 (2007)
8. Bell, R., Pilawa-Podgurski, R.C.N.: Decoupled and distributed maximum power point tracking of series connected photovoltaic sub modules using differential power processing. IEEE J. Emerg. Sel. Top. Power Electron. **3**(4), 881–891 (2015)
9. Kasper, M., Ritz, M., Bortis, D., Kolar, J.W.: PV panel-integrated high step-up high efficiency isolated DC-DC boost converter. In: Proceedings of 35th International Telecommunications Energy Conference 'Smart Power and Efficiency' (INTELEC), October 2013, pp. 1–7 (2013)
10. Guo, X., He, R., Jian, J., Lu, Z., Sun, X., Guerrero, J.M.: Leakage current elimination of four-leg inverter for transformerless three-phase PV systems. IEEE Trans. Power Electron. **31**(3), 1841–1846 (2016)

Assessment of Blood Donors Using Big Data Analytics

R. B. Aarthinivasini(✉)

Department of Computer Science and Engineering, Panimalar Engineering College, Poonamalle, Chennai, India
aarthibaskaran296@gmail.com

Abstract. Big data refers to collection of enormous amount of digital information. Blood donation services are crucial for saving lives of many people. Now a day's need for hygienic blood is increasing very swiftly due to increase in hospitals and increase in disease affected patients. In many cases, urgent need for blood emerges due to some accidents, surgeries etc. Only 5% of Indian population donates blood even though the requirement for blood is very high. Many advertisements and awareness programs are conducted worldwide to make people consciousness about the need for blood. Usually, blood donation process need lot of time and power from both donor and acceptor since there is no proper information system that allows donors and donation centers to connect efficiently and coordinate with each other to decrease time and power required for blood donation process. The proposed system maintains a database that stores the information about all the donors and it will be useful for the patients who require it during emergency situations. This work aims at developing a blood donor assessment system based on the big data. In proposed system the information about the donors stored in database and imported using sqoop tool and processed using hadoop. Donor details that are gathered from the exact locations are visualized on the website. The request and response for the particular blood group send through message. Accordingly, donors can be directed to the nearest location to save the life of the needed people.

Keywords: Hadoop · Mapreduce · PCA · K-means clustering

1 Introduction

In recent years, communication and internet application have lot of influence and advancement on information technology. These application and communication generating large amount of data, different variety in some structure is called big data. For instance, people upload 72 h of video files in Youtube for every minute. This data growth faces the problem of associating and coordinating huge volume of data from distributed sources. Traditional data storage are not suitable to store analyze those huge amount of data. With the increase in data volume, the big data analytics is used to provide solution for storage for large datasets.

D. J. Hemanth et al. (Eds.): COMET 2019, LNDECT 35, pp. 626–640, 2020.
https://doi.org/10.1007/978-3-030-32150-5_61

1.1 Categories of Big Data

Comparing with other traditional approaches big data includes structured, unstructured and semi structured data. Data that is stored, accessed and processed in fixed format is called structured data. Data in an unknown form is called unstructured data. Semi-structured data contains both structured and unstructured form.

1.2 Characteristics of Big Data

Big data can be characterized as 3Vs: volume, velocity and variety. Volume is extremely large capacity, whereas velocity provides the speed of processing and variety refers to type of data. There are lot of big data tools to handle these large amount of data.

1.3 Hadoop

Hadoop framework uses two concepts called Map reduce and HDFS. Map reduce is a framework helps in processing large amount of data in parallel operations in reliable and fault tolerant way. Map reduce is referred by map task and reduce task. Map task takes input and converts it into set of data called key vale pairs in the form of tuples. The output of map task is the input of reduce tasks where it combines these tuples into small set of tuples. Sqoop helps to connect hadoop with relational database. This also helps to handle structured data. In India blood donations are conducted by a few camps and hospitals facilitates by arranging schemes for blood donations. Donors can stop-over blood donation centers in hospitals and give blood to the person who needs blood. The capacity of good blood donor raised to 55% on the year 2007 to 83% on the year 2012, with the quantity of blood units raising from 4.4 million units in 2007 to 9.3 million units in 2013. In 2016, the ministry of health and family welfare donated 10.9 million units against a prerequisite of 12 million units. 350 milliliters of blood are donated by donors in India.

India disposes of over a million units of blood gathered each year, according to health ministry data. This is in spite of confronting a serious blood lack as just 9.9 million units are gathered against the estimated yearly necessity of 10–12 million units. On a normal, around six units of blood is required for each open heart surgery, while a roadside mishap casualty could require up to 100 units. One out of each 10 individuals admitted to a doctor's facility needs blood, as indicated by WHO information.

The reason behind the lack of blood during accidents is due to improper delivery of blood from blood bank to the acceptors. By building up an application which will help society and different poor individuals is the use of E-Blood Bank which will give a brisk support of the poor individuals. In this the client's area will be followed utilizing GPS framework. On the off chance that blood is required, the contributor with the required particular blood bunch is recognized and told about its necessity. The venture comprises of calculation which tracks area of the givers, distinguishes the givers who are close-by to the area of requester and tells them as well. On the off chance that the recognized close-by contributors are not ready to give blood at display then the extent of following the givers is increment.

A message based automated blood donation, which interfaces donors and patients through message, Donors need to enroll with the bank by means of SMS and his/her blood will be check by closest healing facility, all the confirmed givers subtle elements are shared the general population who asking for the specific gathering of blood, this helps in receiving willingness information from the donor.

Need for the quantity and quality blood is increased due to accidents that are occurred daily and also for surgeries. Blood donation consider an essential part in human life. Mining data is the way to collect pertinent data from a huge measure of data. Autonomic processing, which defines an arrangement of building properties to oversee framework, where complication is increasing however must be handled without increasing the size or cost of management of the administration group, The essential objective of processing is that Framework manages ourselves as indicated by a manager goals.

For easy access to donors through telephone, email address a web application with supporting mobile application intended to serve as a specialized device between patients and blood donors. To become a part of donation, donor need to provide their details by giving data like name, blood gathering, email address, secret key, and correct area from Google Map. So as to discover the correct area of a donor, Google Map is coordinated with this application. The mobile application dependably updates the area of donor. Thus, the framework can naturally locate an enlisted donor wherever he/she needs. Visitors can seek blood donors by searching with area and blood type. The framework will demonstrate the accessible donors with their telephone number, email address and street number through arranging them by closest place and blood donation expiry date. Visitors can send message to all donators through email however a member can send message via email and cell phone. An appointment will be made just at whenever donor confirms that he/she will give blood. At that point the framework will alert the donor before 12 h of donation.

The next problem while delivering blood is that donor and blood group classification. Blood donor classification are examined by data mining techniques. The blood availability in blood banks is a crucial and important aspect in a healthcare system. Blood banks are based on an active person deliberately donating blood and is used for transmitting. The blood donor behavior are identified by the classification algorithms of data mining.

Later data mining modeling technique are utilized to inspect the blood donor arrangement and spread this to improve real-time blood donor administration utilizing control panels with blood groups and location information. This gives the ability to design effective blood donation. The scoring algorithm executed for the control panel helps in the deployment of resource and spending budget allocation for blood donation campaigns. The framework collects all information about the donors and blood bank is stored in the database using big data so that vast amount of data can be stored. Blood donor details are gathered from the exact locations during the emergency situations based on sorting the availability of blood using Geo location RVD scoring algorithm.

Segmenting and analysis of blood group are done through K-means clustering method and donors matching the blood group are retrieved. Consequently, donors can be guided to the nearest location having a shortage of the blood group. Hence, blood wastage can also be reduced. Problem of tracking the location area of donors is done through location RVD scoring algorithm. The reason for this is to build up a framework that will interface all donors.

As a result an efficient way for gathering donor details based on the location is provided. The details of donor are collected and maintained in the database. These data are imported using sqoop tool to hive database since sqoop helps in import and export of structured data. Further theses structured data are processed in the hadoop environment for structured data processing. The donor details based on the location and blood group are classified based on efficient clustering algorithm. The acceptor makes request to the nearest donor by sending an SMS message by collecting their information from the website which is visualized on the website. As a result donor will provide their willingness.

2 Related Works

Bhardwaj et al., proposed data mining techniques like classification, clustering, association rule, prediction, sequential patterns to present an view on data mining system and to clarify how data mining and knowledge discovery in databases are related with each other [1]. Santhanam and Sundaram, proposed a system called CART Algorithm in blood donors classification [2]. CART decision tree algorithm is used in blood donor classification. It identifies the behavior of blood donors using the classification algorithms of data mining. The analysis is implemented using decision tree algorithm. This system helps in classifying the blood donors to identify their blood donation behavior using accuracy based model.

There are various algorithms in classification and to measure the accuracy of utilization [3]. Techniques like naive bayes, J48, and random tree focus on analyzing the efficiency of different classification algorithm in data mining. Comparing with other algorithms random tree shows accuracy of 93% within short duration. Error rate is high with the random tree method which reduces the rate of accuracy. In 2015, blood donation process is improved by data mining methods [4]. Algorithms like naïve bayes, JV8, random tree are used to classify the blood donor information from large database. After classification, clustering algorithm is used to create sub classes. This form of clustering provides the group of donors, which help to retrieve during emergency situations at any time.

In [5], the accuracy test for donor blood group classification is done using the ANN technique with multilayer perceptron and back propagation algorithm. This predicts the accurate donor from the list of donors. The accuracy of predicting the blood donor is 76%. Here decision tree technique is also used in comparison with Ann in order to analyze the classification accuracy. Blood Donor Classification and Notification Techniques [6, 7], provides survey relating to automating medicines for providing and enhancing donation and delivery of blood. Machine learning algorithms are suited for selection of blood donors. Error free communication and stable system is proposed. SMS is most suitable way of communicating with the donor during the emergency for the need of blood.

Number of Blood Donors are Predicted through their Age and Blood Group by using Data Mining Tool is an application helps in predicting number of blood donors of a particular age and blood group using data mining tool and J48 algorithm to make proper decision fastly and accurately [8]. The purpose of this paper is to build data mining model to get knowledge of classification of donors.

Dhond et al., proposed an android based health application in cloud computing for blood bank [9]. In this system GPS and a messenger technique is used. To notify the need of blood to the people around the blood bank and thus blood will be made available through the donor. Here the donor information are stored in cloud based storage for retrieving data from anywhere and at any time. A new concept of blood bank management system using cloud computing for rural area (INDIA) is a system proposed in rural areas to reduce the corruption involved in blood bank [10]. In this mobile SMS based blood bank management system is created which connects to cloud server located in other location. Here the request for blood is sent through SMS. The technology used behind this system is asp.net which helps in developing web storage data on cloud server and SMS service for wireless data connection.

In 2016, a system is developed which focused on conventional working of blood bank management using cloud computing is provided [11]. Here blood bank service is provided as Software as a Service (SAAS). It maintains database of each blood bank which is stored in cloud storage. Acceptor who needs blood, find the request as blood bank rather than as donors. Since this system provides information about the blood banks rather than listing the blood donors it is time consuming process. In 2017, Hegde et al., proposed an application where the registered users of the application can view the details of the donor and they can also view the availability of blood donors. Acceptor can request to the matching blood donor through online and view the location of donor through google map technique. Further more people can able to view nearby hospitals and patients located nearby Using GPS [12].

An Efficient Mode of Communication for Blood Donor is smart phone based android application [13]. This application will provide a valuable search of donor and to get their details in precise time. The donor details are registered in database with the specification of their exact location using GPS technology. The intimation from the receiver is handled through a SMS message or by contacting directly through a call. In 2017, blood donation are carried out using smartphone application because some hospitals and blood donation centers are money minded and not providing proper information about the donors to the acceptor during emergency situation [14]. Due to this there is a lack of communication barrier between donor and acceptor. To overcome this smartphone based android application is developed were acceptor can directly connect to the donors using GPS.

To overcome the above problem big data emerging tools, algorithms and techniques are used to handle large amount of data. In 2015, a comparative study is made for the data transfer in hadoop [15]. This study compares various tools like hadoop, flume, sqoop, scribe, kafka, slurper and DistCp. to decide where to use one tool over the other with the motive for data transfer to and from hadoop system. With the help of these tools Extract, Transform and Load (ETL) based work in the web environment is performed.

In 2016, Hashmi and Ahmad made a survey on big data tools and algorithm [16]. This paper surveys the available tools which can handle large volumes of data as well as evolving data streams. In this experiment authors were able to cluster and classify large dataset on a private cloud platform, which can be scaled to handle the increasing dataset.

A comparative analysis of traditional RDBMS with map reduce and hive for e-governance system is proposed by Kale, Dandge [17]. Governments are digitizing the department because data is getting increased day by day. To handle this large volumes of data hadoop and map-reduce is used. Processing of these data using traditional method is difficult. This paper surveys on the techniques like map-reduce, sqoop, hive to handle these data. This paper also focus on sqoop tool to connect SQL and hadoop. The next most important thing is processing the environment. This paper makes survey on hive data warehouse [18]. Hive is an open source data warehouse which is built on top of hadoop. Hive supports queries similar to SQL called as HiveQL i.e. Hive Query Language. This system aim to build a cost-based optimizer and also adaptive optimization techniques to provide more efficient innovations. Columnar storage and data placement tend to improve scan performance.

The performance of the execution environment is depicted by comparing the performance of the techniques used [19]. Here three testers with same model will run simple queries to find which among the following technique is efficient and faster. The techniques used in comparing the efficiency and performance includes Hive, pig, MySQL cluster. From This survey paper shows the result that hive is the most suitable model in a low cost environment.

In 2015, author makes a research and provides detailed knowledge on an efficient hadoop technology framework like sqoop and ambari for big data analysis and processing [20]. Hadoop is a technology which provide data integration, orchestration, monitoring, data serialization, storage, integration, data intelligence and access. In this paper sqoop and ambari framework have been analyzed with various features. Sqoop tool is used as an interface application for transferring data between relational database and hadoop. Ambari tool is to simplify hadoop management processing of huge amount of data.

3 System Design

The admin collects data from different sources and stores these data on the database. The data collected are imported to hadoop environment through sqoop tool. The imported data's are normalized, preprocessed and clustered based on donor blood group and location. These details are displayed on the website were the acceptor need to register their information for requesting the donor information. Acceptor makes a search based on the blood group from their nearby locations and send SMS to donor to receive their willingness through message (Fig. 1).

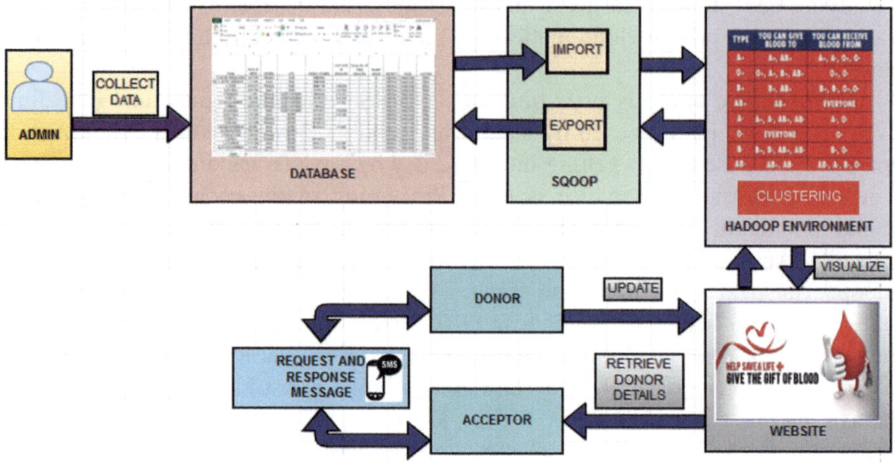

Fig. 1. System architecture

4 Proposed System

The proposed system is designed to help acceptors to meet the demand of blood during emergency situations by sending and receiving request for blood as and when needed. The goals of proposed system is to make the process ease of blood donation and reception. The scope of proposed system is to include all blood donors atleast within the city. Once the application is launched administrator, he is the one who have got access to check volunteer blood donor details, modify their details by logging into the web application with his username and password. Using this website public can register as volunteer blood donor by providing their basic information through registration. Donor will also have the provision of editing and updating his details. Recipients who in need for blood has to register with their basic details and can check the details of the volunteer blood donors who have registered to the application.

4.1 Data Collection

The first module of the proposed system is collection of data. Data collection is the process of gathering data from various sources for processing and manipulations. In this module donor details are collected and maintained as data in the form of excel. The data collected includes donor name, date of birth, gender, blood group, city, district, state, country, mobile number, last date of donation and number of times donated. These data are maintained by the administrator who can perform updates, modify and remove donors.

4.2 Data Migration

Migration in web based applications is moving from one platform to another platform. The data collected on excel has been moved to hadoop environment by means of sqoop tool. The sqoop tool helps in import and export of data between hadoop and structured data sources. It uses mapreduce framework to transfer data in parallel. Mapper slices incoming data that to be imported for further processing in the hadoop environment to get donor details.

4.3 Dimensionality Reduction and Clustering Data

Clustering generally depends on some sort of distance measure. Points near each other are in the same cluster; points far apart are in different clusters. But in high dimensional spaces, distance measures do not work very well. The number of dimensions should be reduced so that distance metric will make sense. Clustering is the process of grouping similar items. In this module, data collected from different locations of the donor are grouped based on their blood group and donor location. This is done by Principal Component Analysis and k-means clustering technique on the hadoop environment.

4.4 Visualization and Request, Response for Blood

An Acceptor who is in need for blood request through the web application by registering with their basic details. Once acceptor registered with their details they search for donor by providing blood group, city, state and country. Donor details for particular blood group and location are displayed as search results. Acceptor selects the donors from search result and request their need by sending messages. A Donor who is willing to donate blood reply for the request message as response.

5 Implementatıon Techniques

5.1 Principal Component Analysis

Clustering generally depends on some sort of distance metric. Points near short distance are in same cluster; points far apart are in different clusters. The number of dimensions should be reduced so that distance metric will make sense. PCA is a linear transformation method which is often used in reducing dimensionality of a d-dimensional dataset by projecting it on to a k-dimensional space for data having large number of variables. This helps to increase the computation efficiency and retains the most important information.

Calculate the mean values of each column where A represents the dataset.

$$\mathrm{Mean(A)} = \frac{\mathrm{a1 + a2 + a3 + \ldots + an}}{n}$$

$$\mathrm{M} = \mathrm{mean(A)}$$

Center the data by subtracting the mean from each variables.

$$\mathrm{C = A - M}$$

The next step is to calculate the covariance/correlation matrix. Covariance is a measure of how changes in one variable are related with changes in a second variable. Let x and y be two variables with length n.

The variance of x is:

$$\sigma_{\mathrm{xx}}^{2}=\frac{\sum_{i}(xi-mz)(zi-mz)}{n-1}$$

The variance of y is:

$$\sigma_{\mathrm{yy}}^{2}=\frac{\sum_{i}(yi-my)(yi-my)}{n-1}$$

The covariance of x and y is:

$$\sigma_{\mathrm{xy}}^{2}=\frac{\sum_{i}(xi-mz)(yi-my)}{n-1}$$

m_x and m_y are the means of x and y variables, respectively. Then calculate the eigen decomposition of the covariance matrix CM. This results in a list of eigenvalues and a list of eigenvectors.

$$\mathrm{CM}=\mathrm{cov(C)}$$

The next step is to calculate the eigenvectors and the eigenvalues of the covariance matrix eigenvector are ordered by eigenvalues from the highest to the lowest. The number of chosen eigenvectors will be the number of dimensions of the new data set.

$$\text{Eigenvectors} = (\text{eig\_1, eig\_2}\ldots\text{eig\_n})$$

$$\text{Values, Vectors} = \text{eig (V)}$$

A total of m or less components must be selected to comprise the subspace. Ideally, select k eigenvectors, called principal components that have the k largest eigenvalues.

$$\text{B} = \text{select (values, vectors)}$$

Transpose the adjusted data (rows are variables and columns are individuals)

$$\text{New. Data} = \text{Eigenvectors. Transposed X adjustedData.Transposed}$$

$$\text{P} = \text{B\^{}T. A}$$

Where A is the original data, B^T is the transpose of the chosen principal components and P is the projection of A.

Algorithm

Input: 2-dimensional(x,y) determinant of the matrix det, Identity matrix, eigen val ue λ , Matrix A, Covariance Matrix CM, Var(X,Y), eigenvector ev.

Output: Reduced PCA dataset(PC_1, PC_2,PC_3......PC_n)

```
Begin
normalize←data(x,y)// normalize the data
subtract←(x-y)
x←x-
y←y-
mean ←
CM= [ var(x1)      cov(x1,x2)
      cov(x1,x2)   var(x2) ]  //calculate covariance matrix
det( I-A)=0 //calculate determinant
Eigenvalue ←( I-A)←0 //calculate λ
if (eigenvalue>0) →largest // choosing components
for each n-eigenvalues←dimensionality reduction
ev←highest eigenvalue←dataset (PCA)
 else
   if(eigenvalue<0)→small
 return null
else if
feature vector←(ev(eig1,eig2)) //Form feature vector
newdata←feature vector^T ×scaled data^T
end
```

5.2 K-Means Clustering

Need for the blood donor and donor's characteristics can be identified by using the clustering algorithm using these steps as follows:

Step1: Collect details of blood group and personal details of persons from the different locations.
Step2: Apply the k-means clustering algorithm for classifying the number of blood donors through the blood group, age, gender and location, etc.
Step3: Extract the phone numbers of the persons from the resultant cluster that satisfying the criteria of blood group and location.
Step4: Forward the message to the donors as according to the requirement.

In this method, the k-means clustering algorithm, deals with the multidimensional data attributes such as di1, di2…dim, where m is the number of attributes. When data contains the multiple values, determine the column with maximum range present in the column. Then determine the initial centroids from the range of the Column and divide data points into k-equal partitions. Then for each iteration, calculate the euclidean distance between the data point for each centroid, which is just the direct distance between any two points.

This k-means Clustering Algorithm is stated as with the following steps:

Input: Reduced PCA Dataset
Output: A is a set of k clusters.
1. Calculate the initial center from dataset and set the cluster for that centroids.
2. Repeat
2.1. Assign each data point for that cluster
2.2. Calculate the mean of that cluster
This is done until all data points are assigned to any one of the clusters.
3. Repeat
3.1. Assign each data item di to the cluster having the closest centroid.
3.2. Calculate new mean of each cluster meeting convergence criterion.

6 Results and Discussion

A Requester who is in need for blood, request through basic registration and login to search for donor by providing their username and password through the localhost http://localhost:8080/Bloodbank/. During searching they provide basic details like city, district, state, country and blood group (Figs. 2, 3, 4 and 5).

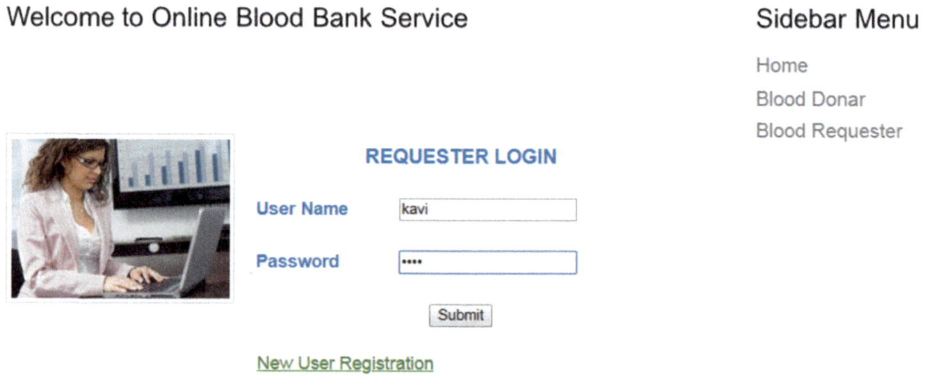

Fig. 2. Requester login

Donor details of particular blood group and city are provided as search result. Acceptor can request blood by clicking the request button (Figs. 6, 7 and 8).

The registered donor can view the request sent by acceptor to accept or reject the request by logging with their credentials. Once they accepted the request a message is displayed as request accepted successfully is displayed.

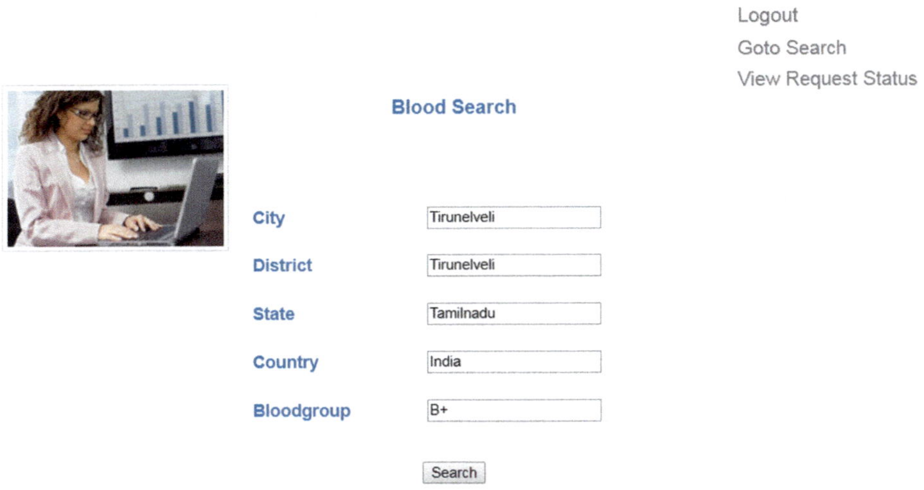

Fig. 3. Blood search

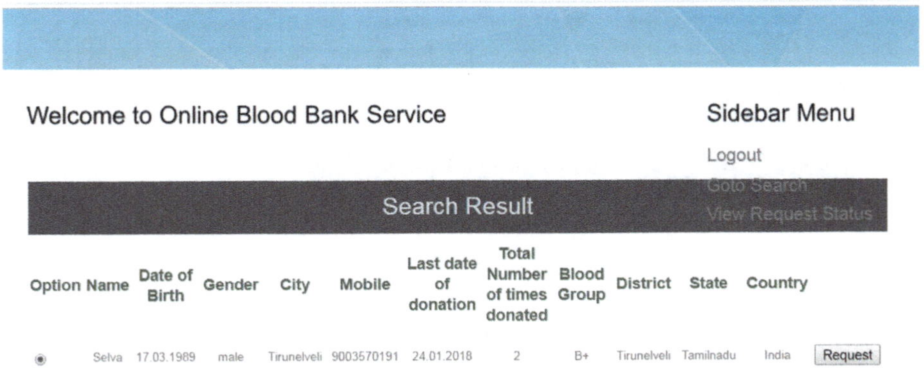

Fig. 4. Donor details

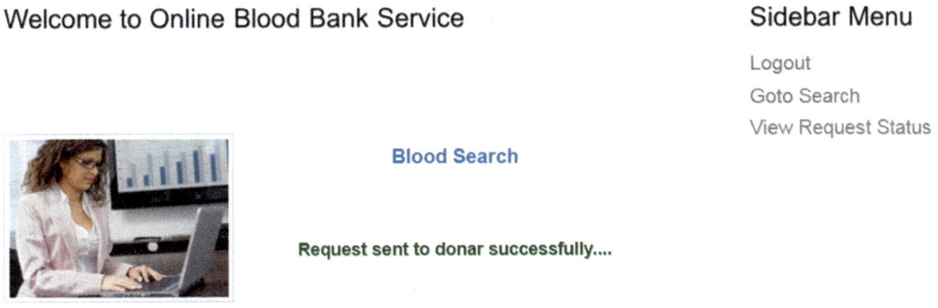

Fig. 5. Search results of donor details

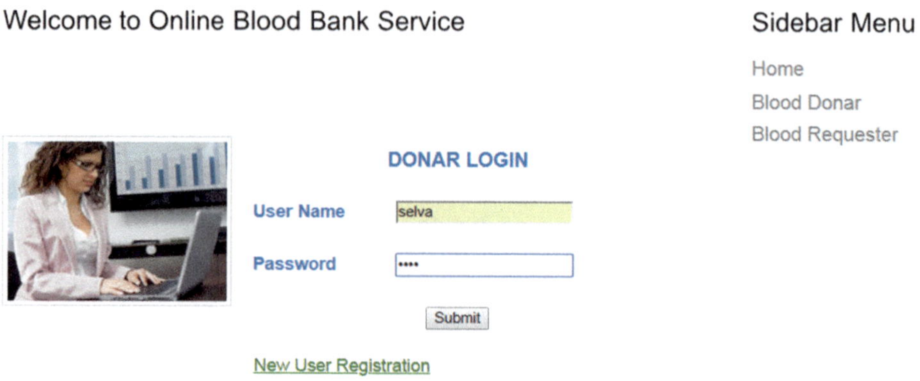

Fig. 6. Donor login

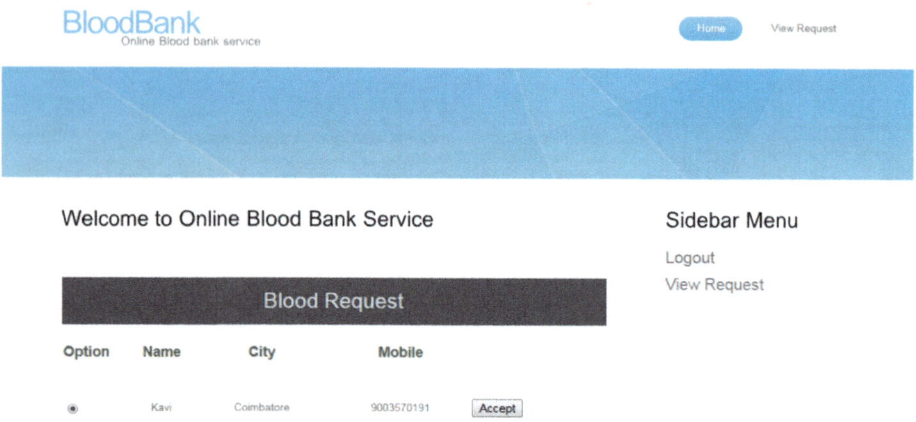

Fig. 7. Viewing blood request by donor

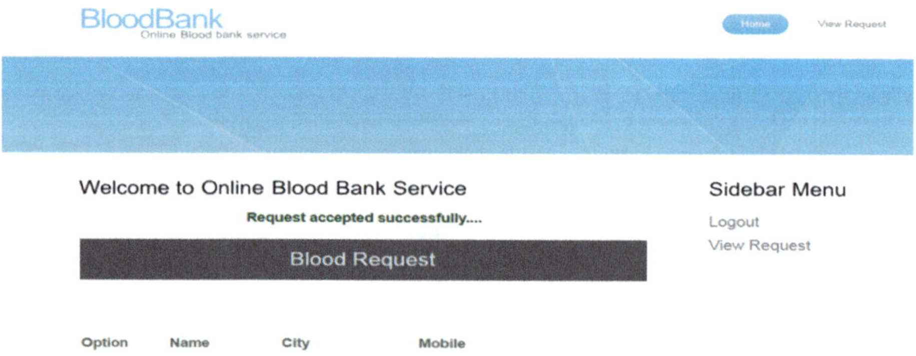

Fig. 8. Accepting request by donor

7 Conclusion

Blood which is more important and cannot be manufactured by anyone, increasing its demand from all over the countries in the world due to the increase in the number of accidents and surgeries. The need of blood occurs many times on urgent basis and at that time it's not possible to get proper donor information fastly. So, for gaining this information easily and efficiently, data mining technique is useful to provide required information from large datasets. After that, PCA Clustering algorithm used for further clustering based on blood group and location. This clustered form of data gives the group of donors, which helps to gain proper information about them that will be useful at any time when urgent need of blood will occur.

References

1. Bhardwaj, A., Sharma, A., Shrivastava, V.K.: Data mining techniques and their implementation in blood bank sector – a review. Int. J. Eng. Res. Appl. **2**, 1303–1309 (2012)
2. Santhanam, T., Sundaram, S.: Application of CART algorithm in blood donors classification. Int. J. Comput. Sci. **6**(5), 548 (2010)
3. Rani, S.A., Ganesh, S.H.: A comparative study of classification algorithm on blood transfusion. Int. J. Adv. Res. Technol. **3**(6), 57–60 (2014)
4. Dhoke, N.W., Deshmukh, S.S.: To improve blood donation process using data mining techniques. Int. J. Innovative Res. Comput. Commun. Eng. **3**(5) (2015)
5. Boonyanusith, W., Jittamai, P.: Blood donor classification using neural network and decision tree techniques. In: Proceedings of the World Congress on Engineering and Computer Science, vol. 1, pp. 24–26 (October 2012)
6. Chinnaswamy, A., Gopalakrishnan, G., Pandala, K.K., Venkata, K.P., Natarajan, S.: A study on automation of blood donor classification and notification techniques. Int. J. Appl. Eng. Res. **10**(7), 18503–18514 (2015)
7. Young, G.O.: Synthetic Structure of Industrial Plastics. In: Peters, J. (ed.) Plastics, vol. 3, 2nd edn, pp. 15–64. McGraw-Hill, New York (1964)
8. Sharma, A., Gupta, P.C.: Predicting the number of blood donors through their age and blood group by using data mining tool. Int. J. Commun. Comput. Technol. **1**(6), 6–10 (2012)
9. Dhond, S., Randhavan, P., Munde, B., Patil, R., Patil, V.: Android based health application in cloud computing for blood bank. Int. Eng. Res. J. (IERJ) **1**(9), 868–870 (2015)
10. Khan, J.A., Alony, M.R.: A new concept of blood bank management system using cloud computing for rural area (INDIA). Int. J. Electr. Electron. Comput. Eng. **4**(1), 20 (2015)
11. Muralidaran, B., Raut, A., Salve, Y., Dange, S., Kolhe, L.: Smart blood bank as a service on cloud. IOSR J. Comput. Eng. **18**(2), 121–124 (2016)
12. Hegde, D., Kuriakose, A., Mani, A.M., Philip, A., Abraham, A.P.: Design and implementation of e-blood donation system using location tracking. Int. J. Innovative Res. Comput. Commun. Eng. **5**(5) (2017)
13. Vijayabhanu, R.: An efficient mode of communication for blood donor. Int. J. Eng. Technol. Sci. Res. **4**(11) (2017)
14. Mandal, M., Jagtap, P., Mhaske, P., Vidhate, S., Patil, S.S.: Implementation of blood donation application using android smartphone. Int. J. Adv. Res. Ideas Innovations Technol. **3**(6) (2017)

15. Marjit, U., Sharma, K., Manda, P.: Data transfers in hadoop: a comparative study. Open J. Big Data **1**(2), 34–46 (2015)
16. Hashmi, A.S., Ahmad, T.: Big data mining: tools & algorithms. Int. J. Eng. Sci. 5–6 (2016)
17. Kale, S.A., Dandge, S.S.: A comparative analysis of traditional RDBMS with map reduce and hive for e-governance system. Int. J. Eng. Comput. Sci. **4**(4), 11224–11228 (2015)
18. Thusoo, A., Sarma, J.S., Jain, N., Shao, Z.: Hive – a petabyte scale data warehouse using hadoop. In: Proceedings of the 26th International Conference on Data Engineering, pp. 1–6 (March 2010)
19. Fuad, A., Erwin, A., Ipung, H.P.: Processing performance on apache pig, apache hive and MySQL cluster. In: Proceedings of International Conference on Information, Communication Technology and System (2014)
20. Aravinth, S.S., Begam, A.H., Shanmugapriyaa, S., Sowmya, S.: An efficient HADOOP frameworks SQOOP and ambari for big data processing. Int. J. Innovative Res. Sci. Technol. **1**(10), 252–255 (2015)

Automatic Monitoring of Hydroponics System Using IoT

R. Vidhya(✉) and K. Valarmathi

Department of Computer Science and Engineering, Panimalar Engineering College, Chennai, India
cutervidhya@gmail.com

Abstract. Because of expansion in industrialization, there is a remarkable drop in cultivation land. Human populace is expanding in everyday life which in turn raises the demand for nourishment. Nourishment is an essential fundamental resource for the survival of individual on earth. Typically, traditional planting with soil expends a longer time. These days there is gigantic innovation progress in the agri-based sector with the hydroponic idea. Growing a plant without soil builds the yield of harvests in brief span for any climatic condition. From this technology pest and weed can be controlled and work cost is minimized. With the help of solar panel the renewable energy is produced which helps in monitoring the hydroponics plants continuously. All sensor data are gathered from Arduino Microcontroller is stored in Web Server.

Keywords: Hydroponics · IoT · Sensors · Web Server

1 Introduction

The major application is farming, because of the rise in population each nation needs to meet the nourishment. In ancient days cultivation needs the farmer to spend the most extreme hours in the farmland and natural manure is added to farmland which contributes nutritious soil and growth is higher, a plant is stronger and it is very useful for the health of human being when compared with the plant developed with the chemical compost. The land in olden days utilized continuously for cultivating and expanded utilization of chemical compost kills the flora and the fauna available in the agricultural land. Farmers don't know about chemical manure and pesticide point of usage level for farmland and it influences nature.

Due to heavy downpour, soil corrosion is another serious issue in cultivating the land. Minerals resources available in the land are corroded and reduce the yield. Essentially, another weakness of ancient cultivation is workers shortage. Farming is the fundamental occupation in India, however, the worker is moving to the industry sector to live for a better life. Because of global warming, climatic changes influence the cultivation in very high. In water lacking regions yield is limited. Water is the major source for plant development.

With the progress in innovation, it improves the standard of living in society, hydroponics concept is utilized in day by day life. Hydroponics technology can be applied in the home which satisfies us with the greenery. In ancient cultivation, it has

D. J. Hemanth et al. (Eds.): COMET 2019, LNDECT 35, pp. 641–648, 2020.
https://doi.org/10.1007/978-3-030-32150-5_62

many disadvantages, for instance, it needs constant watering and workers shortage and cost is the higher for farmworkers. With the hydroponics, crops can be grown in the soilless culture. The society moved towards the IT business has no time to spend in the farming and with IoT idea, even the industry and business people can spend time to take care of the hydroponics plant through IoT concept. With the Arduino microcontroller board, programmed with automatic watering for hydroponics plant is possible if the farmer is far away from the agriculture land. With technology like a sensor, microcontroller, web interaction makes life simpler.

Hydroponics plant can be developed utilizing the growing medium like gravel, Rockwool, perlite etc. and it has numerous techniques. It has control to reduce the growth of weed and the production of crops is multiple times faster in hydroponics system, no requirement for soil medium, labor necessity is lesser compared with ancient cultivation, harvests can be created throughout the year, pesticide utilization is lower, water can be reused on numerous times which prompts water preservation and reaping of yields in hydroponics is a lot easier.

By interfacing with IoT, farmworkers can accumulate the information and do the survey based on the data. Hydroponics can be done in a home, terrace, a balcony where the vacant space is available and appropriate consideration can be taken whether the farmer is far away from the hydroponic system. Utilizing IoT, present status of the hydroponics farm can be viewed. Hydroponics is the method for developing plants without soil and mineral resource is provided through the water to the plant. With hydroponics, water requirement and worker cost are lesser compared to the ancient cultivation method. The steady analysis is required in hydroponics which is efficient with IoT, it needs a nonstop power supply to gather the data. By implementing the renewable energy like solar panel the power supply can be produced continuously. All Sensor data is gathered is stored in web server.

2 Related Works

Tembe et al. [1] in 2018 proposed hydroponics systems to meet the nourishment which is fundamental for the person. They used the pH test module and light spectrum module for hydroponics. The hydroponics idea can be implemented in the location which is facing the issue of the dry season without water. With less water utilization the hydroponics farm can increase the yield and with great quality. In the hydroponics framework, it is free of the surrounding and the harvests can be done throughout the year. The primary disadvantage in the hydroponic framework is starting arrangement cost is higher and need to execute the programmed monitoring system. Essentially, it needs constant observation from farmers and power cut issue need to be done by the manual arrangement.

Aeroponics concept proposed by Jagadesh et al. [2] for developing plants without a soil medium. In aeroponics strategy, the root is splashed with the supplement resource for making the root wet. Aeroponics strategy and water system are checked with IoT. Information is transmitted to the server utilizing GSM/GPRS and sensor information can be seen through the site. The underlying arrangement cost is higher however it minimizes the labor cost. Reaping the plants in Aeroponics is simpler compared to harvesting in olden days. As the supplement is provided to root it absorbs less water. Wet root should be observed and programmed sprinkling should be initiated to maintain the root without causing any damages.

In 2018, Shewale and Chaudhari [3] proposed a Hydroponics concept for developing plants. Because of the development in industry, agri-based land is diminishing in higher level. To avoid the issue the plants are developed in soilless culture. With ARM processor the sensor execution is controlled with the Android application. With Zigbee, the information can be transmitted up to a certain small range. Long-range of communication is not possible with the network architecture.

Paulchamy et al. [4] built up a technology plant care hydroponic box that controls the environment with IoT innovation. IoT talk is utilized effectively for client to roll out programmed changes, for example, include or exclude the sensors and actuators. With the IPCH box, it controls the water sprinkling and water flow through the PVC pipe and it adequately diminishes the CO2 level in the surrounding. It is difficult to monitor the value of pH range available in the nutrient solution added to the water. In comparison with the ancient method of cultivation with the hydroponics method, the plant growth rate is 90% higher in hydroponics and it doesn't depend on any particular season for growing crops.

In India, cultivation is the most significant occupation and research in progress to get more yield with high quality. In ancient cultivating, it has numerous drawbacks, for example, labor issue maximized usage of chemical manure and investing more hours in agricultural land. In hydroponics cultivating, weed growth is reduced and pest attack is nearly lower compared to traditional farming method. pH sensor, EC sensor is utilized to locate the supplement solvents acidity level and ion particle level in the solution. Aravind and Sasipriya [5] implemented linear regression which is analyzing the measure of nutrition supplement passed in the valve and to reduce the level of nutrition supplement. pH level different for various crops. The sensor information is recorded in the cloud and data are sent to open source software for survey reason.

Thakare et al. [6], proposed a hydroponics idea with decision tree algorithm. Because of growth in the industry, the land for farming becomes lesser. To actualize the farming in less soil is profoundly tedious work. Hydroponics is a method for developing the yields in the soilless culture. With the automatic watering system and detecting the different parameter around the plants through the various sensor. The pH sensor and NPK sensor the decision tree is utilized to choose whether to supply the supplement solution for hydroponics framework and to consequently checks the degree of water and indicates the farmer through the messaging system.

Hydroponics encourages the farmland workers to acquire cash with higher yield contrasted with the normal cultivating system with the barren land. With a natural calamity, it is hard to judge nature. Pitakphongmetha et al. [7] proposed a hydroponics system with Wireless Sensor Network is utilized to move the sensor information to the cloud. This issue can be overwhelmed with the hydroponics by actualizing in a verified domain without aggravation from the environment. The significant parameter around the hydroponics plant is estimated utilizing the various sensors.

Mishra and Jain proposed a two cathode sensor for estimating water conductivity for the hydroponics framework. The sensor is intended to check the conductivity of supplement solution which is in the certain range mentioned in the implementation.

These electrodes are utilized to control the supplement in hydroponics system as indicated by the plant's specification. The programmed hydroponic framework depends on a minimum cost ARM processor that checks and control they need for hydroponics plants. Programmed hydroponics framework will help the yield by controlling the various parameters.

The expanded society, sub-urbanization of the woods, inappropriate rural practice which modified the pH level in the soil. Hydroponics was observed to be a better option and can be characterized as the development of plants without soil, which is in effect industrially utilized in the large portion of the western nations. This examination investigates the utilization of this development system and to uncover its future significance. This method can be adjusted to practically all the earthly plants. Vegetable sustenance yields like wheat, tomato and dill, and a lot more plants are being developed in business scale.Mugundhan et al. [9] proposed the hydroponic framework requires an underlying venture, diligent work, and care. Better the yield will be, whenever drawn closer as it needs. It was prescribed that this system can be adjusted as a stage to deliver the sustenance harvests and therapeutic plants to satisfy the worldwide need, to control the unnatural weather change and better future.

With the human population, soil-based agribusiness is facing some real difficulties; in particular reduction in per capita land accessibility. Furthermore, poor soil richness in a portion of the cultivable regions, less possibility of common soil fruitfulness develop by organisms because of consistent development, visit dry spell conditions and eccentrics of atmosphere and climate designs, ascend in temperature, stream contamination, poor water the board and wastage of gigantic measure of water, decrease in ground water level, and so forth are undermining sustenance generation under regular soil-based agribusiness. Under such conditions, Sardare, Admane proposed the hydroponics concept to grow the plants with less water with Nutrient film technique (NFT) and Deep flow technique (DFT). Normally, soil-less culture is winding up increasingly applicable in the present situation, to adapt up to these difficulties. In soil-less culture, plants are raised without soil. Improved space and water saving strategies for sustenance generation under soil-less culture have demonstrated some encouraging outcomes everywhere throughout the region.

3 Proposed System

Hydroponics is the concept of planting without soil. Arduino Microcontroller implemented to monitor the plants. All the sensors are connected to the microcontroller. Water is the main content for the growth of plants. Nutrition is added in water which acts as a supplement for the growth of the plants. pH sensor is used to analyze the range of the nutrition available in the water. In hydroponics concept, the water be can recycled which save water when compared to traditional farming method. The water tank has the essential nutrition mixed which is automatically supplied to plants based on the pH range.

Hydroponics plants need continuous care from the farmer and to overcome the problem of electricity shutdown, the alternate resource is implemented. Renewable energy like solar energy is used to monitor plants growth. Solar energy is passed to hydroponics system during the power shutdown.

All the sensor data are gathered and transmitted through the IOT Board. IOT Board acts as an interface between the Arduino Microcontroller and Web Server. The information is transmitted to a web server through Wireless transmission IOT Board.

3.1 Sensor Unit

In the proposed system (Fig. 1) the various sensors are implemented to monitor the hydroponics plants. Arduino Microcontroller is used to control and monitor the plants. pH Sensor is used to calibrate the ratio of nutrition present in the water supplied for the plants. The plant used here for research is coriander leaves and the pH ratio for coriander lies between the range of 5.5 to 6.5. Ultrasonic Sensor is used to identify the height of the plant. Light Intensity sensor is implemented to gather information about the light is required for the growth of plants. To gather the moisture of air around the plants, the temperature sensor is used.

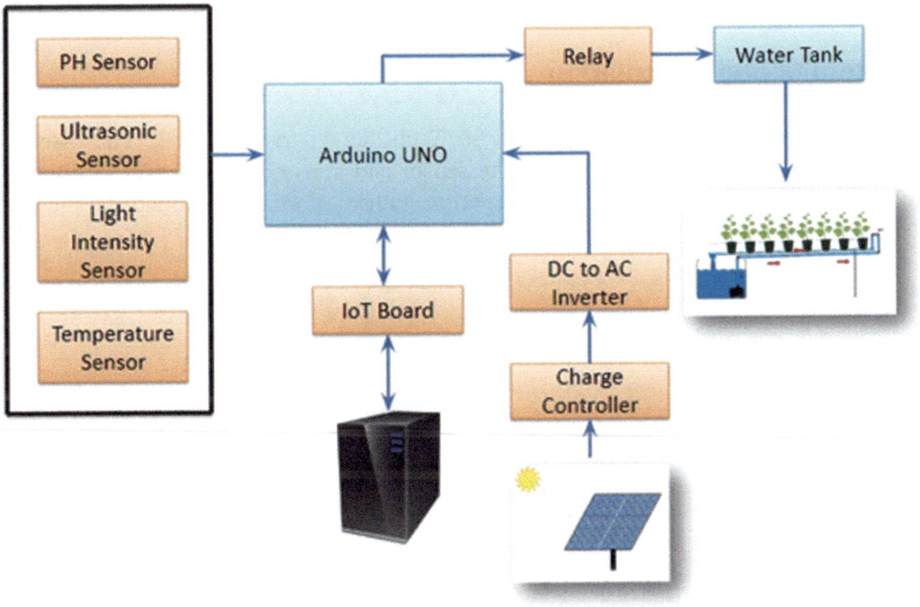

Fig. 1. System architecture of proposed system

3.2 Renewable Energy

Hydroponics Farms need continuous care and need to be monitored constantly. To overcome the problem of power shutdown, renewable energy like solar energy is used. Solar energy is stored in the charge controller unit. Direct current (DC) is converted to alternating current (AC), 5 V is supplied to Arduino Microcontroller Unit (Fig. 2).

Fig. 2. The experimental set up of the proposed system

4 Results and Calculation

In Fig. 3 the sensor data transmitted to Web Server is displayed. The Arduino Microcontroller gathers the data from the sensor and transmitted to Server through the IoT Board. The entire system is automated to gather information around the plants.

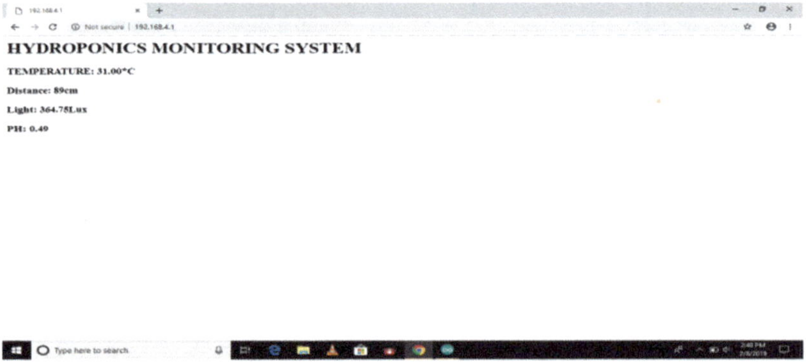

Fig. 3. Snapshot of the parameters transmitted and available in the Web Server

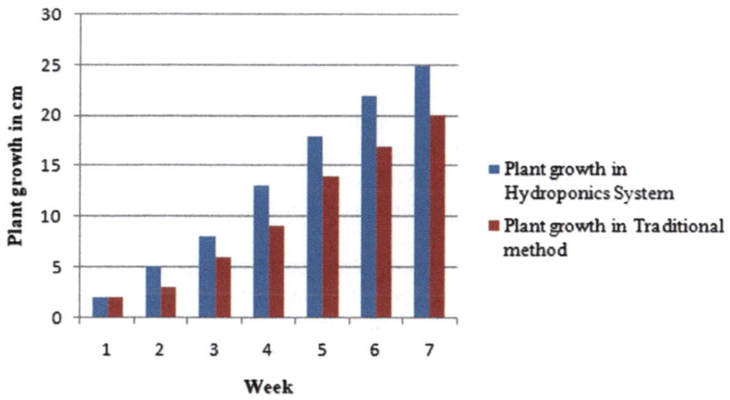

The above chart describes the performance of the hydroponic system compared with the traditional method. In a conventional way, the plants are grown using soil. The graph gives a clear representation of the hydroponics system. The growth of plants is plotted using the graph. The growth of plants is higher compared to traditional farming. Based on the nutrient added in the water the growth is compared with the plants grown in the soil.

5 Conclusion

From the proposed system the hydroponics system is implemented with the renewable energy resource. In ancient cultivation it needs agricultural land, cost is higher, expanded utilization of pesticide and compost, and water scarcity is the serious issue. In hydroponics, the whole above issue is diminished and it gives the nourishment in high standard. With Global warming, it is unsure about the nature factor in the surrounding. By applying the Hydroponic it is free from any ecological circumstance. Farmers can view the hydroponics plants with the help of IoT even if the farmers are in a remote area. Sensor data are sent to a web server.

References

1. Tembe, S., Khan, S., Acharekar, R.: IoT based automated hydroponics system. Int. J. Sci. Eng. Res. **9**(2), 67–71 (2018)
2. Jagadesh, M., Karthik, M., Manikandan, A., Nivetha, S., Kumar, R.P.: IoT based aeroponics agriculture monitoring system using raspberry pi. Int. J. Creative Res. Thoughts **6**(1), 601–608 (2018)
3. Shewale, M.V., Chaudhari, D.S.: IoT based plant monitoring system for hydroponics agriculture: a review. Int. J. Res. Appl. Sci. Eng. Technol. **6**(2), 1628–1631 (2018)

4. Paulchamy, B., Balaji, N., Pravatha, S.D., Kumar, P.H., Frederick, T.J.: A novel approach for automating & analyzing hydroponic farms using internet of things. Int. J. Sci. Res. Comput. Sci. Eng. Inf. Technol. **3**(3), 1230–1234 (2018)
5. Aravind, R., Sasipriya, S.: A survey on hydroponic methods of smart farming and its effectiveness in reducing pesticide usage. Int. J. Pure Appl. Math. **119**, 1503–1509 (2018)
6. Thakare, A., Budhe, P., Belhekar, P., Shinde, U., Waghmode, V.: Decision support system for smart farming with hydroponic style. Int. J. Adv. Res. Comput. Sci. **9**, 427–431 (2018)
7. Pitakphongmetha, J., Boonnam, N., Wongkoon, S.: Internet of things for planting in smart farm hydroponics style. In: IEEE Computer Science and Engineering Conference (ICSEC) (2016)
8. Mishra, R.L., Jain, P.: Design and implementation of automatic hydroponics system using ARM processor. Int. J. Adv. Res. Electr. Electron. Instrum. Eng. **4**(8), 6935–6940 (2015)
9. Mugundhan, R.M., Soundaria, M., Maheswari, V., Santhakumari, P., Gopal, V.: Hydroponics-a novel alternative for geoponic cultivation of medicinal plants and food crops. Int. J. Pharma and Bio Sci. **2**(2) (2011)
10. Sardare, M.D., Admane, S.V.: A review on plant without soil-hydroponics. IJRET: Int. J. Res. Eng. Technol. **2**(03), 299–304 (2013)

Cost Effective Decision Support Product for Finding the Postpartum Haemorrhage

R. Christina Rini(✉) and V. D. Ambeth Kumar

Department of Computer Science and Engineering, Panimalar Engineering College, Chennai, India
Indiarinikristina39@gmail.com,
ambeth_20in@yahoo.co.in

Abstract. Postpartum Haemorrhage said to one of the main causes of maternal morbidity in world wide. During Vaginal delivery and C-section unexpected blood loss occurs leads to critical condition in the patient. More likely with a Caesarean birth about 1 to 5% of women have postpartum Haemorrhage. Postpartum Haemorrhage occurs most commonly after the delivery of placenta. The average amount of loss of blood after a single child birth in vaginal delivery is about 500 ml. During the C-section birth it is approximately 1000 ml. Most Postpartum occurs alter after the delivery of the baby. There are two types of PPH one is the Primary PPH which occurs in a day. Secondary PPH occurs after a day or up-to six weeks. The proposed system contains a database about the critical condition that arise during the delivery. It also contains accuracy, sensitivity and specificity comparison that occurs during the PP. Using the Feed Forward Neural Network with Particle Swarm Optimization the classification is optimized and the PPH occurrence is detected.

Keywords: Internet of Things · PPH · Women · Neural network

1 Introduction

PPH is a significant loss of blood after giving birth and is the number one reason for maternal morbidity around world spec losing 500 ml blood of v and 1000 ml of c section. it is difficult to measure the precise amount of loss of blood during the delivery due to the possibility of internal bleeding. Basically, two criteria have been consider during the child birth. One is the decrease of 10% Hematocrit in the total volume of blood and the changes in Mother's Heart rate, Blood pressure, Oxygen saturation, drop in the temperature. Significant blood loss within 24 h is said to be primary PPH and blood loss after 24 h is said to be late or secondary PPH. Here Autonomous Nervous System (ANS) which is reason for the mental pph can be found in the peripheral part of nervous system. If there is more variation found in these factors then it can be concluded that those corresponding persons are under mental pph due to reflect of nervous system variation. EEG signal factors would also varied in its performance under different environment consideration such as noisy working environment, high workload, improper sleep and family issues. These factors would generate negative emotions of humans which need to be analysed well for the proper treatment. There are various

D. J. Hemanth et al. (Eds.): COMET 2019, LNDECT 35, pp. 649–656, 2020.
https://doi.org/10.1007/978-3-030-32150-5_63

research methods has been introduced earlier to perform feature selection and classification efficiently to predict the pph level of humans very accurately. Thus the proper treatment can be ensured. These processes would be carried by concerning the pph related factors in mind. Feature extraction can improvement classification performance of EEG signal prediction by selecting more reliable features. Fuzzy K-Nearest Neighbour (FKNN) classifier is a one of the human pph prediction methodology introduced to perform better by adapting the behaviour of Finite Variance Scaling (FVS). However evaluation of F-KNN method based pph prediction leads to reduced accuracy rate by involving more false positive rate. Discrete Wavelet Transform (DWT) is a most famous technique utilized by different application to perform feature extraction process in the well-defined manner. This can be further optimized by hybridizing it with the EEG Asymmetry method which can lead to accurate feature selection outcome by finding unique pattern of human pph. These ideas have not been taken over by any previous research method where the variation of beta band would indicate the variation in mental workload. KNN classification method proves that prediction of pph level during human listening music provides positive results.

The main contribution of this research paper is to introduce the novel protocol namely mental pph elicitation protocol which is implemented and evaluated for its performance to predict the mental pph level. Figure 1 illustrates the overview of mental pph recognition system using EEGsignal. Mental pph prediction system can be carried out in four steps which is denoted as follows: (i) Data gathering protocol, (ii) Pre-processing dataset (iii) Feature extraction and (iv) classification. These steps are explained in detail as follows:

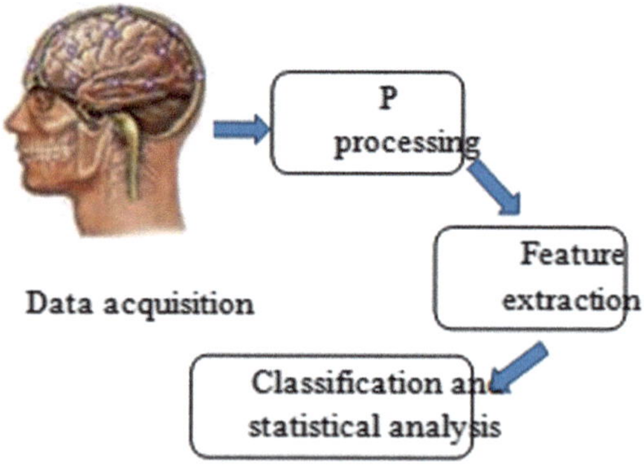

Fig. 1. Overall of proposed work analysis

EEG signal acquisition: Brain activity recognition is the more difficult process which requires continuous monitoring of brain activities. The electric signal measured from the electrodes cap of EEG signal might varies based on human response which would be digitized for further processing later. The following deals with the sections are arranged as: which Sect. 2 discusses some existing researches based on EEG signal pph classification. Section 3 explains the proposed FFNN-PSO based classification and the performance results are discussed in Sect. 4 and finally, this work is concluded.

2 Related Research

On the basis of the survey conducted by Knight, Callaghan, Berg, Alexander, Bouvier-Colle, Ford, Joseph, Lewis, Liston, Roberts [1] et al., risk factors for uterine atony after vaginal delivery is clearly explained. Alternative strategies explained the effect of using prophylaxis oxytocin of different dosages that eliminates postpartum haemorrhage. The Emergent management of postpartum hemorrhage for the general and acute care surgeon [2], contemporary epidemiology of pph. In 2004, PPH complicated 2.9% of all deliveries; uterine atony accounted for 79% of the cases of PPH. PPH was associated with 19.1% of all in-hospital deaths after delivery.

The overall rate of PPH increased 27.5% from 1995 to 2004. More statistical data helped in clarifying the causes of pph [3]. It includes various factors such as retained placenta, renal failure, coagulopathy, antepartum haemorrhage. On the basis of data retrieved from NIS sample from 2004 it elucidates the complications leading to PPH after delivery. It also discussed the demerits of using magnesium sulphate to prevent preeclampsia.

Sheikh, Najmi, Khalid, Saleem [4] had given all the possible root causes for postpartum bleeding. His research also illustrates the techniques to treat pph. His methodology explains both therapy and surgery. Most advanced techniques such as internal iliac artery ligation, uterine packing, and usage of ergot alkaloids.

An audit of primary post partum haemorrhage. J Ayub Med Coll Abbottabad [5] by Bibi, Danish, Fawad, Jamil (begum's hard work has given a solution to treat PPH effectively. PPH occur in 5% of all deliveries, majorities of death occur within four hours of delivery indicating that it is a consequence of third stage of labour. The most common cause of primary PPH is uterine atony. The precautionary treatment methods such as Intra uterine balloon tamponade, Brace Suture, Bilateral uterine artery ligation, Bilateral Internal iliac ligation is discussed which acts as a key source to prevent the adoption of hysterectomy even in some severe cases. Even recombinant activated is discussed in clotting.

In 2014 a study was conducted for primary postpartum [6]. In this survey all details regarding the side effects of medicines for treating pph was clearly mentioned. This also examines the importance of involving first line and second line therapy. Misoprostol, tranexamic acid, ergometrine etc. are the first line medical therapies and surgical methods and usage of tamponades includes second line therapy. prophylactic uterotonics, misoprostol and oxytocin infusion worked similarly. The review suggests that among women who received oxytocin for the treatment of primary PPH, adjunctive use of misoprostol confers no added benefit. The role of tranexamic acid and compression methods requires further evaluation. Furthermore, future studies should focus on the best way to treat women who fail to respond to uterotonic therapy.

Many studies have given way for better understanding of the drugs used and their effect in treating pph. The article CRASH-2 [1] discusses the major factor that causes postpartum bleeding. He had also mentioned the improvement after implementing

Controlled Cord Traction in third stage of labour. It is less effective in the presence of disseminated intravascular coagulation and placenta accreta. An international, randomised, double-blind, placebo-controlled trial [7] conducted a research and submitted an article on PPH in the year 2013. The study reported all the methods to treat both primary and secondary PPH. It clearly illustrates the risking factors by conducting an analysis by considering 26 patients of postpartum haemorrhage in Liaquat National hospital. WHO (1991) The Prevention and Management of Postpartum Haemorrhage. Report of a Technical Working Group [9] in the year 2014 Postpartum haemorrhage (PPH) remains a major cause of maternal deaths worldwide, and is estimated to cause the death of a woman every 10 min. Adapting the non-pneumatic anti-shock garment as non-surgical and hysterectomy, balloon tamponade, B-lynch suture remains a saving factor through surgical methods [10]. The importance of oxytocin is also clearly explained in preventing pph. In women at low risk, around 3% will lose over 1000 ml of blood despite prophylaxis. These women require rapid access to life-saving PPH treatment and rescue therapies.

3 Proposed Work

The objective of this presented scheme is to reduce the human pph after sensing the pph by using EEG signals. The main aid of this is to examine accurately estimated the human pph and classify the human pph level. The pph has been evaluated by using the EEG characteristics and pph level of human (i.e. pph or relaxed mode). This pph levels are classified by using FFNN-PSO in Fig. 2 classification scheme. If high pph is monitored, then next music of subject's choice are played and this statistical examination is conversed in the course of the performance analysis. The step by step process has been discussed in given below subsections.

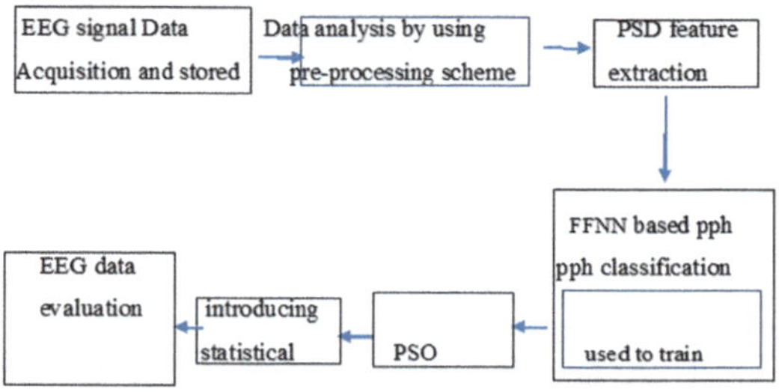

Fig. 2. FNNN-PSO based pph classification and reduction procedure

4 Implementation of Proposed System

Feed Forward Neural Network with Particle Swarm Optimization (FFNN-PSO). In this section, an effectual FFNN-PSO based pph classification has been discussed. Also, the step by step process has been explained. The EEG signals are gathered from five humans and it is omitted during categorization process due to presence of artifacts in those images. These artifacts are generated by various reasons such as eye movement and blinks, muscle movement and so on. The signals with the artefacts value more than 100 μV would be rejected for the further processing. Thus filtering is done before processing EEG data. This filtering task would divide the EEG signal into five frequency bands such as Delta (1–4 Hz), Theta (4–8 Hz), Alpha (8–13 Hz), Beta (13–30 Hz) and Gamma >35 Hz bythe EEG frequency band analysis. Here beta signal is found to be more efficient signal with corresponding variation which can lead to accurate classification rate. Here spectral power density is utilized to calculate the mean power of EEG signals and hamming window distance is calculated by using the power spectral density. Here window size is fixed as 256 with 50% overlapping and then FFT length is fixed as 1024. Pre-processing is used to eliminate the noises present in the signals which are done by fixing frequency variation between 0.5 to 30 Hz in real time. This frequency limit can avoid the noises generated from both main source and other sources.

After pre-processing, feature extraction is done on the signals using the one of the AR method namely PSD method is a kind of parametric modern spectrum estimate method. It is a random process which is used to predict the different kind of phenomena present in the frequency signals. This method is a linear prediction based technique which attempts to find a final outcome from the knowledge of previously defined outcomes. There are numerous algorithms proposed earlier for the measurement of AR method parameters such as Yule Walker, Burg, Covariance and Modified Covariance. In the proposed research method, Yule-walker scheme is adapted to assure the better outcome even in case of long data sequence presence. The main constraint that needs to be taken in mind when utilizing this method is model order prediction. So that, the error can be avoided during run time. This selected order would be utilized during performance evaluation which can lead to better method selection. In this work model degree is selected as 15 for EEG.

5 Pseudocode

STEP1: Relative Powers of frequencies in alpha band.
STEP2: Relative Powers of frequencies in theta band.
STEP3: Power of theta band or power of alpha band.
STEP4: Power of alpha band in related epoch/power of alpha band in previous epoch.
STEP5: Mean value of the EEG signal in time domain.
STEP6: Skewness and Kurthosis of the EEG signal in time domain.
STEP7: Sum of Powers of frequencies in 2–6 Hz.

After feature extraction process, those would be fed into the FFNN classifier to perform prediction process. This classifier would predict the EEG signals as three classes namely low, medium and high. FFNN consists of input layer, hidden layers and output layer. This structure is given in Fig. 3. In the training process, FFNN would identify the weight values for the submitted input values. This weight value would be updated in each iteration based on variation between outcome obtained and the expected outcome. This will be done until it reaches minimum error value. This error value optimization is done by using PSO algorithm.

6 Results and Discussion

In Fig. 3 pph index value comparison has been given in two stages. It is used to indicate the variation present between different pph indices value. This graph represents the pph level before noise and after noise. It can be proved that the pph would be higher in case of presence of noise in the environment.

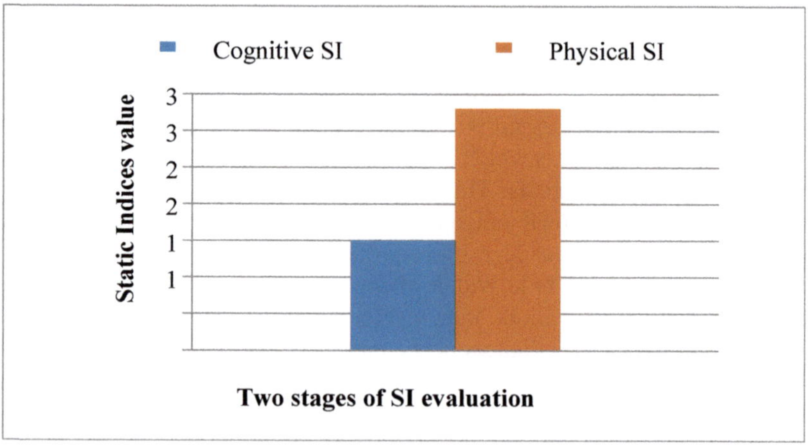

Fig. 3. Stress indices values for two stages

After prediction of pph level of humans, it is required to update the pph index value based on the previous pph index values. This can be used to know the variation between the pph levels in different stages. In Fig. 3, Pph index value is compared in the graphical format in three stages. Those stages are before task load, after task load, and after recovery. From this comparison it can be concluded that the pph index value after recovery would be lesser than other stages. In Fig. 4 it shows the overall performance comparison of accuracy, sensitivity and specificity for proposed FFNN-PSO and existing RVM, SVM and LDA. It shows the classification accuracy of proposed scheme attained high compared than existing schemes, due to the efficient pre-processing and effectual classification by using FFNN with PSO. Then, the sensitivity of proposed FFNN-PSO attained high compared than others, due to less false negative errors, as well as the specificity is also high compared than others, due to the high true positive rate. When, the number of subjects increased means, the performance of proposed also increased. The proposed FFNN-PSO attained accuracy of 93.25%, sensitivity of 92.14%, and specificity of 97.52%. The numerical evaluation is showed in Table 1.

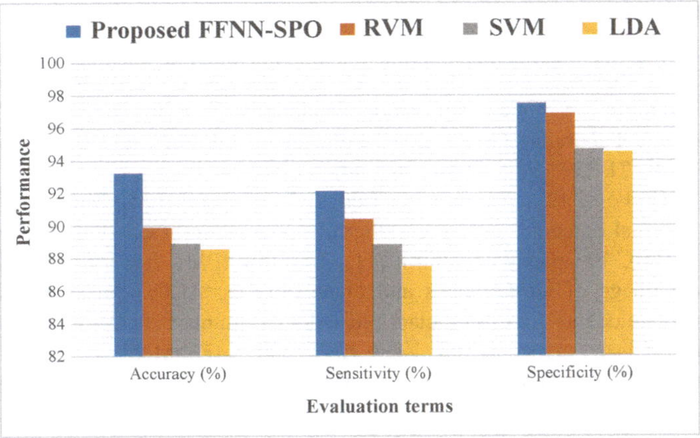

Fig. 4. Overall performance comparisons among

Table 1. Precision, sensitivity and specificity performance comparison for all classifiers

Classifiers	Accuracy	Sensitivity	Specificity
Proposed FFNN-PSO	93.25%	92.14%	97.52%
RVM	89.87%	90.38%	96.87%
SVM	88.90%	88.84%	94.68%
LDA	88.56%	87.51%	94.54%

7 Conclusion

In this work, Feed Forward Neural Network with Particle Swarm Optimization (FFNN-PSO) based classification scheme has been presented for pph level classification and music introduced for reducing the pph level. In this process, at first, the EEG signal has been acquired and then pre-processed by using a digital band-pass filter for improving the image quality. Then, the PSD based features has been extracted for improving the classification performance. Finally, the features are classified by FFNN-PSO. In FFNN process, to attain minimum error the PSO is applied. The experimental outcomes demonstrate that the presented FFNN-PSO accomplished higher performance in terms of accuracy of 93.25%, sensitivity of 92.14%, and specificity of 97.52% contrast to the existing pph detection and classification algorithms with EEG signal due to the effectual feature extraction and classification. In future, the neural network based some other classification schemes will focus with effectual swarm intelligence algorithms as well as focus some features like Gray Level Different Statistics (GLDS), Statistical Feature Matrix (SFM) and improve the accuracy.

References

1. Knight, M., Callaghan, W.M., Berg, C., et al.: Trends in postpartum hemorrhage in high resource countries: a review and recommendations from the International Postpartum Hemorrhage Collaborative Group. BMC Pregnancy Childbirth **9**(1), 55 (2009). https://doi.org/10.1186/1471-2393-9-55. [PMC free article] [PubMed]
2. Roberts, C.L., Ford, J.B., Algert, C.S., Bell, J.C., Simpson, J.M., Morris, J.M.: Trends in adverse maternal outcomes during childbirth: a population-based study of severe maternal morbidity. BMC Pregnancy Childbirth **9**(1), 7 (2009). [PMC free article] [PubMed]
3. American College of Obstetricians and Gynecologists: ACOG practice bulletin: clinical management guidelines for obstetrician-gynecologists number 76, October 2006: postpartum hemorrhage. Obstet. Gynecol. **108**, 1039–1047 (2006). [PubMed]
4. Abouzahr, C.: Global burden of maternal death and disability. Br. Med. Bull. **67**(1), 1–11 (2003). [PubMed]
5. Reyders, F.C., Seuten, L., Tjalma, W., Jacquemyn, Y.: Postpartum haemorrhage practical approach to a life threatening complication. Clin. Exp. Obstet. Gynecol. **33**, 81–84 (2006). [PubMed]
6. Freedman, L.P., Waldman, R.J., de Pinho, H., Wirth, M.E.: Who's got the power? Transforming health systems for women and children. In: UN Millenium Project Task Force Child Health Maternal Health, pp. 77–95 (2005)
7. Weisbrod, A.B., Sheppard, F.R., Chernofsky, M.R., Blankenship, C.L., Gage, F., Wind, G., Elster, E.A., Liston, W.A.: Emergent management of postpartum hemorrhage for the general and acute care surgeon. World J. Emerg. Surg. **4**, 43 (2009). https://doi.org/10.1186/1749-7922-4-43. [PMC free article] [PubMed] [Cross Ref]
8. Sheikh, L., Najmi, N., Khalid, U., Saleem, T.: Evaluation of compliance and outcomes of a management protocol for massive postpartum hemorrhage at a tertiary care hospital in Pakistan. BMC Pregnancy Childbirth **11**(1), 28 (2011). https://doi.org/10.1186/1471-2393-11-28. [PMC free article] [PubMed] [Cross Ref]
9. Bibi, S., Danish, N., Fawad, A., Jamil, M.: An audit of primary post partum haemorrhage. J. Ayub Med. Coll. Abbottabad **19**, 102–106 (2007). [PubMed]
10. Committee on Practice Bulletins-Obstetrics: Practice bulletin no. 183: postpartum hemorrhage. Obstet. Gynecol. **130**, e168–e186 (2017)

IoT Based Innovation Schemes in Smart Irrigation System with Pest Control

J. Freeda(✉) and J. Josepha menandas

Department of Computer Science and Engineering,
Panimalar Engineering College, Chennai, Tamil Nadu, India
angelinfreeda@gmail.com, Josepha82@gmail.com

Abstract. In this paper we have presented a new system for automatic water irrigation along with pest detection framework. This system can be used for monitoring the water level and accordingly watering the crops in agricultural lands. Based on the level of water in the soil, the water pump is activated. In addition, in this system we have proposed a new algorithm for detecting the pests in the plants. Based on the type of pest suitable steps can be taken to eradicate them. Here, we have used Hu moments for representing the leaves. These features were opted since they are invariant to scale, rotation or translation and thereby can be effectively used to represent the affected portion of the leaf invariant to their orientation. The proposed algorithm is based on the extraction of suitable features from the leaves of the plants. The extracted features are then used for classification. The proposed algorithm was compared with existing algorithms like k-NN and decision tree and was found to produce excellent results.

Keywords: Accuracy · Hu moments · Irrigation · Classification · IoT

1 Introduction

Due to tremendous increase in the world population, water scarcity has become a critical issue. The world population at present is found to be around 7.2 billion. By the year 2050, this is estimated to rise to around 9 billion. Most of the fresh water is consumed by agricultural processes especially irrigation. It has been found that developing countries utilize more water for agriculture compared to developed countries. This is due to the lack of advanced agricultural technologies. Hence development of effective irrigation techniques is vital. Plant diseases poses severe threat to agricultural economy.

Continuous monitoring of plants is essential for early detection and consequent application of effective measures to improve the quality of agricultural produce. The development of various machine learning algorithms has paved way for effective recognition of diseases in plants.

D. J. Hemanth et al. (Eds.): COMET 2019, LNDECT 35, pp. 657–669, 2020.
https://doi.org/10.1007/978-3-030-32150-5_64

2 Related Works

A smart irrigation system using Arduino and Raspberry pi was developed in [1]. In this system, the commands sent by the users were processed using Raspberry pi module. This system was controlled wirelessly using X-bee module. The drip irrigation system was stated using an e-mail. Once the e-mail was received, general purpose input/output points were turned high. Pi was also made to receive commands from Arduino micro controller using Zigbee. The Arduino was used to control the relay and ultra sound distance sensor. If the water level was low, Arduino sends signal to the Pi module. Pi sends signal to the valve, and it would be turned on. The main drawback of this system was that failure in any part of the system must be tested manually and was not automated. A low-cost irrigation system was proposed in [2].

In this paper, motor system was controlled automatically. With the use of soil moisture sensor, the direction of water flow was controlled. Here, the wireless sensor network used an algorithm called as Local shortest path for computing path between the sensor nodes. Clustering of sensor nodes was done for energy saving. Two type of nodes namely sensor and control nodes were used in this framework. The sensor node is used for checking the moisture level of the soil and it transmits this value to the control node. The control node checks this value with the required value. In case if it is less, the motor is switched on. In addition, an alert signal is sent to the registered mobile device.

The authors of [3] proposed a smart irrigation system. This system was designed to save time and avoid vigilance. It makes use of a sensor micro controller. Sensor were installed in the fields to monitor the soil temperature and moisture. The values acquired by these sensors were transmitted to the microcontroller. The micro controller was used to drive the pump driver. This was in turn connected to the water pump control system. Servo motor was use to transmit water uniformly to the field so that water is absorbed properly by the plants. Thus, wastage of water was reduced. This system was tested on garden plants. This system helped to increase the life time of the sensors by reducing their power consumption [4].

In this system, keypad was used to facilitate the illiterate farmers. A Zigbee transmitter was use to transmit the values of the temperature, soil and humidity sensors. The sensors sense the data and they are converted into electrical signals that are amplifier using the amplifier. PIC microcontroller converts this analog value into corresponding digital value. In case if the soil moisture value is less than the required value and if the level of the water in the water tank is high, then the pump will be started automatically. The Zigbee transmitter is connected to the system using RS232. The sensor data is received for every 10 s. The main drawback here was that this system was implemented on a small scale.

A smart irrigation based embedded system was proposed in [5]. In this paper, wastage of water was focussed. Raspberry pi was used in the design of prototype model. This system was controlled through web, and hence could be analysed from any location. A solenoid valve was used to switch open the valves of the water pump. This system was designed for field cultivation in places with water scarcity. Hence, sustainability could be achieved. The entire set up included, Pi, X-bee, Arduino, moisture sensor, relay and flow meter. The main drawbacks of this system were that, it could not do weather forecasting and also could not develop android app. Also, anyone could access the system if the IP address of the Pi was known.

A software analysis for smart irrigation system was done in [6]. In this system, a tiny OS based IRIS mote were used in the measurement of soil moisture in paddy fields. The wireless system had three software, namely, Mote tier, Server Tier and the Client Tier. The mote tier was used to run in the sensor notes that form a mesh network. Server tier was used to handle translation and data buffering. Client tier was used to give the graphical user interface. Tiny OS was an open source operating system that is used widely for commercial purposes. The main advantage of this system was its simplicity. The main drawback of this system was that it was not tested with the real time data.

An IoT based smart agricultural system was developed in [7]. A remote-controlled robot was presented in this work that was based on the smart GPS. It was used to perform functions like spraying, sensing moisture removing weeds, scaring animals and birds etc. Also, a smart control-based irrigation system was developed. In addition, this system included a smart warehouse for maintaining parameters like, temperature and humidity. Also, theft detection system was included in the ware house. A remote-control system was used to control all these devices. Wireless transmission was done using Wi-Fi or Zigbee modules.

An Arduino based drip irrigation system was presented in [8]. In this paper, a system for automatically irrigating the farmland was proposed. Java platform was used to get the information using serial communication and was also used to update the server. Also, the best fertilizers required for each crop, best crop for a particulate climate and soil condition were regularly updated in the server. The crops were monitored continuously using the PC. Temperature, humidity and pH sensors were used to monitor the soil parameters. Due to regular update in the server, people can be aware of the parameters. These were also displayed in mobile app.

Mobile integrated system for smart irrigation was proposed in [9]. In this system, water availability in crops is monitored using sensors. Mobile system based on IoT was used for monitoring the status. The main aim of this work was to control the supply of water and to monitor the plants using mobile phone. Here soil moisture sensor was used to monitor the water level in soil. Blue term was the android application that was used to form the app. The connection was done using Bluetooth. The main drawback of this system was that it could be used only in the indoor environment.

The authors of [10] proposed an IoT based nutrient detection system for analysing diseases in rice species. Matlab based image processing techniques were used to identify the diseases and nutrient deficiencies. Two important nutrients namely magnesium and nitrogen were focussed in this work. Also, to facilitate the farmers an android application was also developed. Wireless transmission was done with the help of a Wi-Fi dongle. Various parameters were extracted from the image and were used for the analysis. These parameters included mean, standard deviation, energy, entropy, correlation, contrast and homogeneity. The users were able to monitor the sensor readings using a web application called AutoGate.

3 Proposed Work

The proposed smart irrigation system (Fig. 1), various sensors like temperature sensor, pH sensor, soil moisture sensor and flow sensor are used to sense the amount of water in the soil.

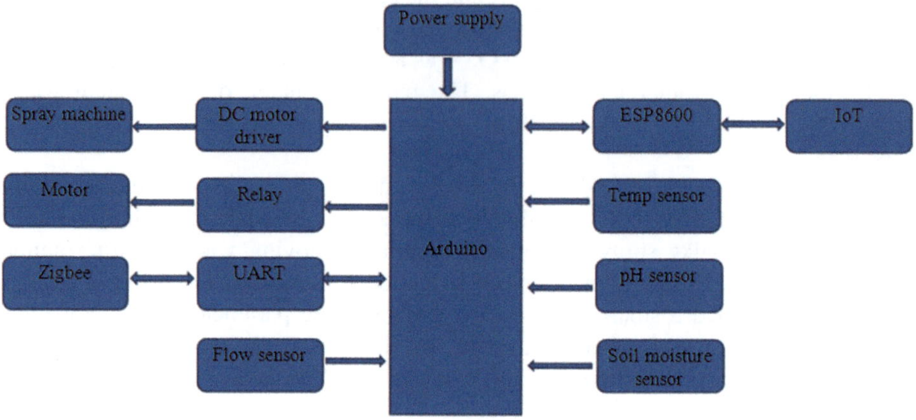

Fig. 1. Smart irrigation system framework

Water conservation is necessary because of the mere availability of water resource. Hence automatic water irrigation system provides a sustainable way to increase the efficiency of water usage. Hence farmers can use right amount of water at the right time for irrigating their land. This water allocation is done based on the requirement of the crop and soil. This system provides best results for medium size lands. It was also found that the usage of this technique reduced the amount of water requirement to about 50%. Also the human interventions was drastically reduces.

If the amount of water is less than a predefined threshold then the motor is activated using relay switch. Also, these values are transmitted to the cloud using ESP8600. This can be seen by the farmers. It is shown in Fig. 2.

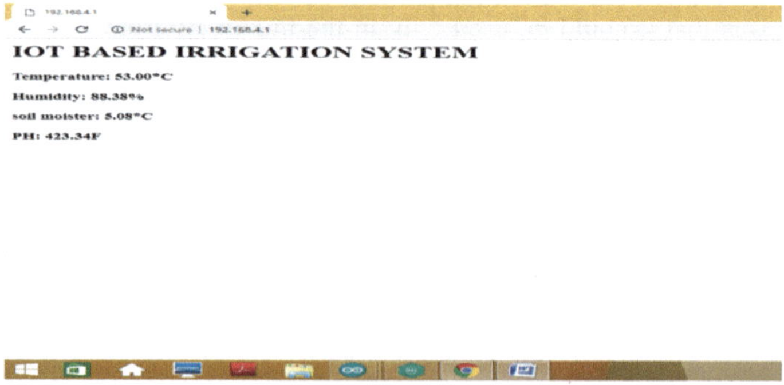

Fig. 2. Snapshot of the parameters transmitted and available in the cloud

In addition, wireless sensor nodes are deployed in the agricultural land. These notes have cameras that are used to capture the leaves of the plant. These images are then transmitted wirelessly to a central server using Bluetooth. The server is a laptop that is programmed to obtain the leaf images, process it and identify if the leaves are diseased or not. This value is then sent using a Zigbee module to Arduino. The Arduino has a Zigbee receiver that is used to activate the DC motor driver that is connected to the spray machine. Copper based sprays can be used to prevent bacterial pathogens that cause diseases like Mildew, Anthranose, Ascochyta blight etc., hence used in our system.

The connection of Arduino with the ESP8600 module is shown in Fig. 3.

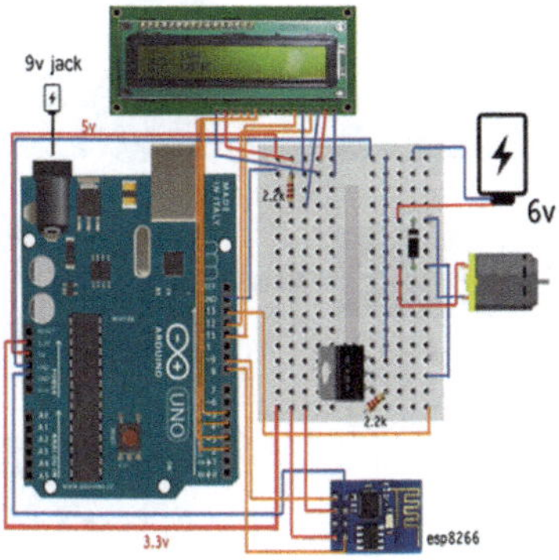

Fig. 3. Connection of Arduino UNO with ESP 8600

3.1 Plant Disease Detection Methodology

As shown in the Fig. 4, the entire system comprises of two phases namely the training and the testing phase. During the training phase, the images belonging to various categories are collected and for each category a dictionary is created. These dictionaries are created using the features extracted from the segmented patches of the leaf images. During the testing phase, features are extracted from the segmented patches of the test leaf image. Then distance parameters are computed between the test features and each of the dictionaries created during the training phase. Finally, classification is done to estimate whether the leaf is healthy or not. In addition, if the leaf is not healthy, the type of disease is also estimated using this algorithm. The detailed explanation of the proposed algorithm is given follows.

The first step in the training phase of the proposed algorithm is the acquisition of image by the sensor nodes. The leaf image captured by the sensor camera is then enhanced using a sharpening filter. This sharpening operation is achieved using the mask ψ which is defined as

$$\psi = \begin{bmatrix} -1 & -1 & -1 \\ -1 & 8 & -1 \\ -1 & -1 & -1 \end{bmatrix} \quad (1)$$

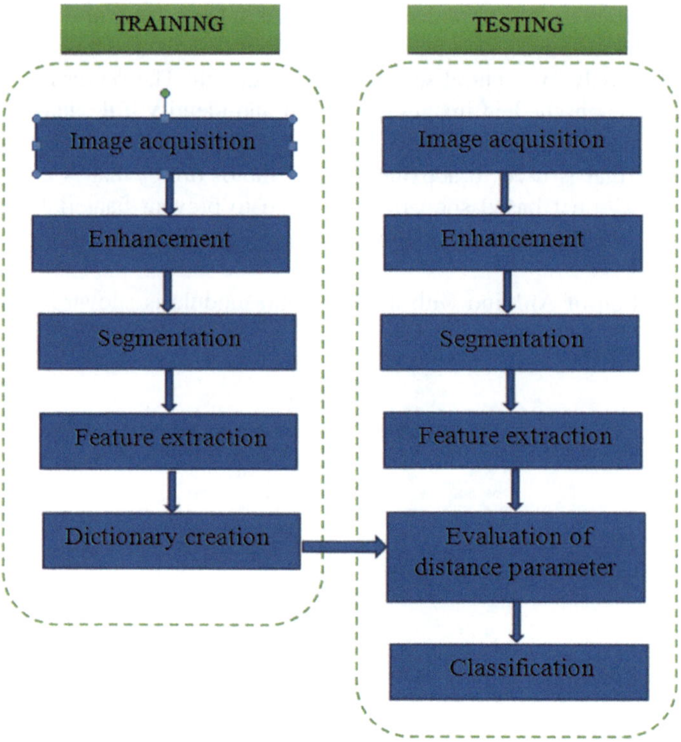

Fig. 4. Schematic representation of the proposed plant disease detection methodology

The sharpened image is then segmented using OTSU algorithm (8). The considered disease images with the segmented output is shown in Figs. 5 and 6. This segmented RGB patches are then converted to greyscale images. These segmented grey scale patches are then used in the feature extraction process.

Scale invariant seven Hu moments are then extracted from these patches. These moments were introduced by Hu (9). Hu gave the mathematical fundamentals for these invariant moments that are two-dimensional. This was initially used for shape recognition application. These moments were applied and used for shape recognition in aircraft applications reliably (10). These moments are defined as first order moment, second order moment and so on. They are calculated as,

$$M_1 = \eta_{20} + \eta_{02}, \tag{2}$$

$$M_2 = (\eta_{20} + \eta_{02})^2 + 4\eta_{11}^2, \tag{3}$$

$$M_3 = (\eta_{30} + 3\eta_{12})^2 + (3\eta_{21} - \eta_{03})^2, \tag{4}$$

Fig. 5. Snapshot of pest images belonging to 12 different pest species (a) Aelia Sibirica (b) Mythimna Separta (c) Chromatomyia Horticol (d) Cifunalocuples (e) Cletus Punctier (f) Colposcelissignata (g) Dolerustritici (h) Erthesina Fullo (i) Eurydema Dominulus (j) Eurydema Gebleri (k) Eysacorisguttiger (l) Penfleus Major

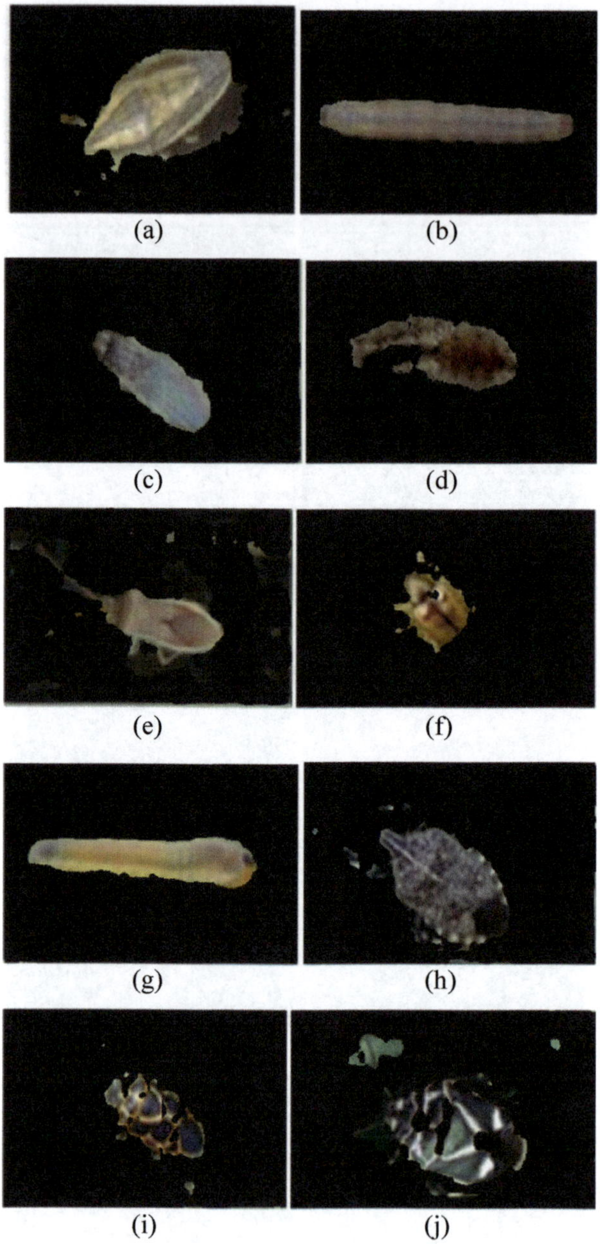

Fig. 6. Snapshot of segmented pest images. (a) Aelia Sibirica (b) Mythimna Separta (c) Chromatomyia Horticola (d) Cifunalocuples (e) Cletus Punctiger (f) Colposcelissignata (g) Dolerustritici (h) Erthesina Fullo (i) Eurydema Dominulus (j) Eurydema Gebleri (k) Eysacorisguttiger (l) Pentfaleus Major.

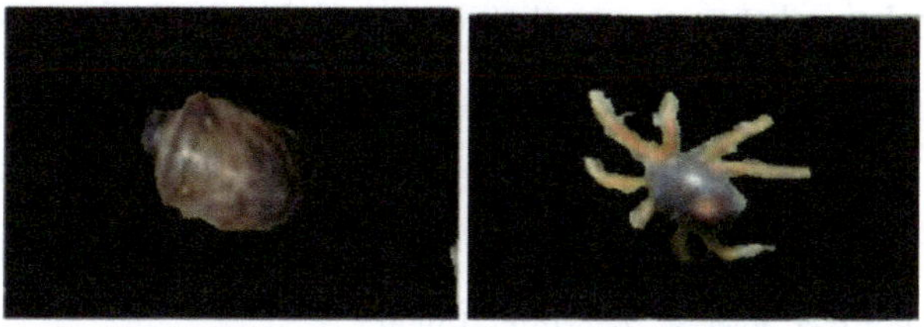

(k) (l)

Fig. 6. *(continued)*

$$M_4 = (\eta_{30} + \eta_{12})^2 + (\eta_{21} + \eta_{03})^2, \tag{5}$$

$$\begin{aligned} M_5 = &(\eta_{30} - 3\eta_{12})(\eta_{30} + \eta_{12})[(\eta_{30} + \eta_{12})^2 - 3(\eta_{21} + \eta_{03})^2] \\ &+ (3\eta_{21} - \eta_{03})(\eta_{21} + \eta_{03})[(\eta_{30} + \eta_{12})^2 + (\eta_{21} + \eta_{03})^2], \end{aligned} \tag{6}$$

$$\begin{aligned} M_6 = &(\eta_{20} - \eta_{02})[(\eta_{30} + \eta_{12})^2 - (\eta_{21} + \eta_{03})^2] \\ &+ 4\eta_{11}(\eta_{30} + \eta_{12})(\eta_{21} + \eta_{03})], \end{aligned} \tag{7}$$

$$\begin{aligned} M_7 = &(3\eta_{21} - \eta_{03})(\eta_{30} + \eta_{12})[(\eta_{30} + \eta_{12})^2 - 3(\eta_{21} + \eta_{03})^2] \\ &+ (\eta_{30} - 3\eta_{12})(\eta_{21} + \eta_{03})[3(\eta_{30} + \eta_{12})^2 - (\eta_{21} + \eta_{03})^2]. \end{aligned} \tag{8}$$

The extracted features from each category are used to form a dictionary. This dictionary is created for each category of disease and for healthy leaf images. Let f_i^j represent the dictionary of the ith class. Here, $j = 1, 2, \ldots m$ where m represents the number of features vectors belonging to each class. Let the total number of classes including the healthy category be n.

During the testing phase, the camera captures the image of the test leaf. H_u features are then extracted from the test leaf image and represented as f_t. The distance parameter is evaluated between each category and the test feature. That is, the correlation between each dictionary and the test feature vector is computed. The proposed classification algorithm is given below.

Input (f_i^j*, Test Image)*
Output (Test class i)
Steps

- Enhance the given test image using mask m.
- Extract Hu features f_t using (2) to (8).

- The correlation between each dictionary and the test feature vector is computed using

$$\delta_i = \sum_{j=1}^{n} f_t^T f_i^j$$

- Finally, the test class is determined using

$$i = \arg\max \delta_i$$

4 Proposed Classification Algorithm

The entire experimental set up is shown in Fig. 7.

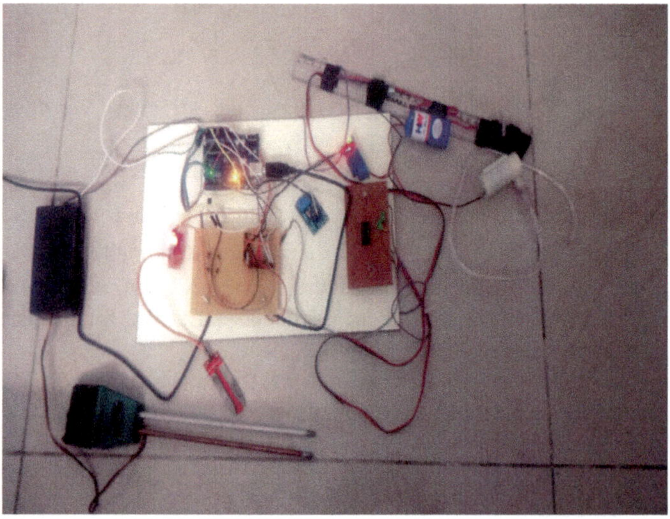

Fig. 7. The experimental set up of the proposed system.

The performance evaluation is done using the leaf images used as the experimental data that originates from the research of (6). 11 pest species were selected for our experiment. In this, the images were divided into ten partitions and evaluation was done using one portion as test and the remaining as training images. The entire process was repeated ten times and the mean average was computed. The classification accuracy was then computed. From the dataset the following categories were chosen. That is Aelia Sibirica, Myhimna Separta, Chromatomyia Horticola, Cifunalocuples, Cletus Punctiger, Colposcelissignata, Dolerustritici, Erthesina Fullo, Eurydema Dominulus, Eurydema Gebleri, Eysacorisguttigerand Pentfaleus Major. For each category the number of images chosen were $m = 50$. The total number of classes were $n = 12$. For comparison, two standard classifiers were considered namely k-NN and decision tree.

Comparison was performed using two commonly used metrics namely accuracy and sensitivity (11). Accuracy is defined as percentage of correct prediction. It is calculated as

$$Accuracy = \frac{TrPo + TrNe}{TrPo + FaPo + FaNe + TrNe} \tag{9}$$

where *TrPo* refers to true positives, *TrNe* refers to true negatives, *FaPo* refers to false positives and *FaNe* refers to false negatives.

Sensitivity measures the number of true positive classifications to the total number of instances. It is computed as

$$Sensitivity = \frac{TrPo}{TrPo + FaNe} \tag{10}$$

Table 1. Comparison of the proposed system with state-of-the-art algorithms

Pest category	Accuracy (in %)		
	Decision tree	K-NN	Proposed
Aelia Sibirica	72.32	76.43	97.55
Mythimna Separta	78.65	81.62	96.43
Chromatomyia Horticola	89.67	90.31	97.56
Cifunalocuples	74.89	78.62	96.43
Cletus Punctiger	86.66	88.13	97.64
Colposcelissignata	83.34	89.52	98.56
Dolerustritici	68.74	69.74	99.22
Eurydema Dominulus	74.56	79.34	95.32
Eysacorisguttiger	69.54	82.24	96.45
Pentfaleus Major	86.46	92.43	98.32
Aelia Sibirica	78.76	86.75	97.45
Mythimna Separta	80.65	89.55	99.34
Mean accuracy (in %)	78.68	83.72	97.52

From Table 1 we infer that the proposed algorithm produces higher accuracy compared to all other existing state-of-the-art algorithms. For further analysis, we have plotted the comparison of the F-score produced by the proposed algorithm with the state-of-the art methods in Fig. 8. From Fig. 8, we see that the proposed algorithm produces higher sensitivity than the other two algorithms for all the categories. Especially for healthy category it shows very good performance.

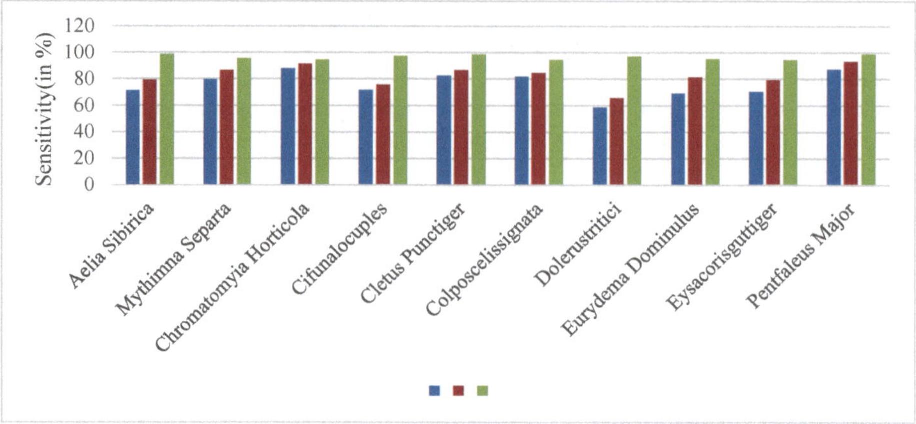

Fig. 8. Comparison in terms of sensitivity

5 Result

In this paper we have proposed a new scheme for smart irrigation. In this scheme the water required for irrigation is adaptively selected based on the water content of the soil. Also, we have proposed anew algorithm for identifying plant leaf diseases. This algorithm was implemented on a publicly available database. We found that the proposed algorithm performed best in comparison to all state-of-the-art techniques in terms of classification accuracy.

References

1. Agrawal, N., Singhal, S.: Smart drip irrigation system using Raspberry pi and Arduino. In: International Conference on Computing, Communication and Automation (ICCCA 2015) (2015)
2. Sahu, C.K., Behera, P.: A low cost smart irrigation control system. In: IEEE Sponsored 2nd International Conference on Electronics and Communication System
3. Darshna, S., Sangavi, T., Mohan, S., Soundharya, A., Desikan, S.: Smart irrigation system. IOSR J. Electron. Commun. Eng. (IOSR-JECE) **10**(3), 32–36 (2015). e-ISSN: 2278-2834, p-ISSN: 2278-8735, Ver. II
4. Ramya, A., Ravi, G.: Efficient automatic irrigation system using ZigBee. In: International Conference on Communication and Signal Processing. India, April 6–8, 2016
5. Namala, K.K., AV, K.K.P., Math, A., Kumari, A., Kulkarni, S.: Smart irrigation with embedded system. In: 2016 IEEE Bombay Section Symposium (IBSS) (2016)
6. Darwin Movisha, J., Edwin Mercy, A., Hema latha, M., Esakiammal: A software analysis of smart irrigation system for outdoor environment using tiny OS. Int. J. Adv. Res. Trends Eng. Technol. (IJARTET) **3**(19) (2016)
7. Gondchawar, N., Kawitkar, R.S.: IoT based smart agriculture. Int. J. Adv. Res. Comput. Commun. Eng. **5**(6), 838–842 (2016)

8. Parameswaran, G., Sivaprasath, K.: Arduino based smart drip irrigation system using internet of things. Int. J. Eng. Sci. Comput. (2016)
9. Vaishali, S., Suraj, S., Vignesh, G., Dhivya, S., Udhayakumar, S.: Mobile integrated smart irrigation management and monitoring system using IOT. In: International Conference on Communication and Signal Processing, April 6–8, 2017
10. Rau, A.J., Sankar, J., Mohan, A.R., Krishna, D.D., Mathew, J.: IoT based smart irrigation system and nutrient detection with disease analysis

‘Agaram’ – Web Application of Tamil Characters Using Convolutional Neural Networks and Machine Learning

J. Ramya(✉), Goutham Kumar Raj Kumar, and Chrisvin Jem Peniel

Department of Computer Science and Engineering, St. Joseph’s College of Engineering, Chennai, Tamil Nadu, India
ramsharsha@gmail.com

Abstract. This paper aims to explore the scope of these neural networks and apply them to try and recognize handwritten data which consists of Tamil characters written by various people and convert it to a computerized text document. Through our system we will be targeting one of the gaps in the currently available technology, which is to properly identify and distinguish the characters in ancient manuscripts.

Machine Learning will be used to train the system to recognize the fed data and convolutional neural networks will be used to make decisions on its own and by doing that, it improves the accuracy of the prediction. The application of this system extends to various fields such as history, archaeology, paleography, engineering etc.

Keywords: Convolutional neural networks · Machine learning · Artificial intelligence

1 Introduction

The field of machine learning has taken a dramatic twist in recent times, with the rise of the Artificial Neural Network (ANN). These biologically inspired computational models are able to far exceed the performance of previous forms of artificial intelligence in common machine learning tasks. One of the most impressive forms of ANN architecture is that of the Convolutional Neural Network (CNN). CNNs are primarily used to solve difficult image-driven pattern recognition tasks and with their precise yet simple architecture, offers a simplified method of getting started with ANNs.

Artificial Neural Networks (ANNs) are computational processing systems of which are heavily inspired by way biological nervous systems (such as the human brain) operate. ANNs are mainly comprised of a high number of interconnected computational nodes (referred to as neurons), of which work entwine in a distributed fashion to collectively learn from the input in order to optimise its final output. The basic structure of an ANN can be modelled as shown in Fig. 1. We would load the input, usually in the form of a multidimensional vector to the input layer of which will distribute it to the hidden layers. The hidden layers will then make decisions from the previous layer and weigh up how a stochastic change within itself detriments or improves the final output,

D. J. Hemanth et al. (Eds.): COMET 2019, LNDECT 35, pp. 670–680, 2020.
https://doi.org/10.1007/978-3-030-32150-5_65

and this is referred to as the process of learning. Having multiple hidden layers stacked upon each-other is commonly called deep learning (Fig. 2).

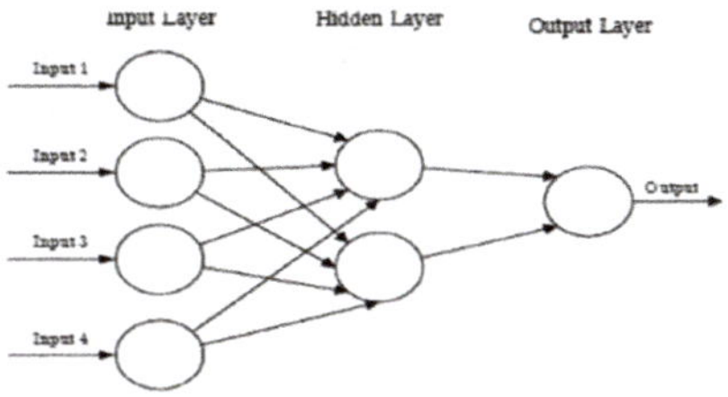

Fig. 1. Neural network model

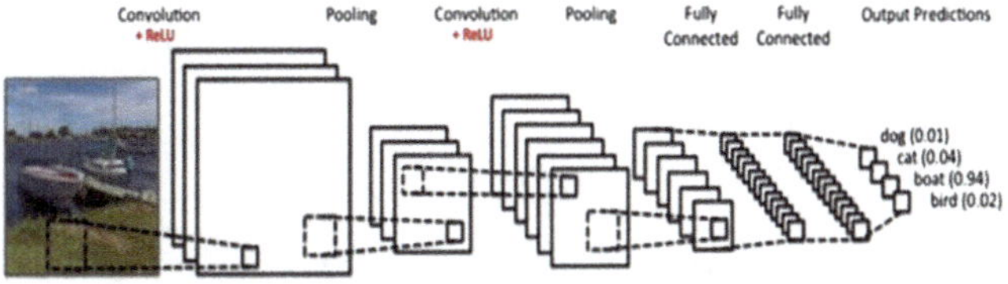

Fig. 2. Convolutional neural network

2 Related Work

Wahi et al. [1] invariant moments and Zernike moments which have been used in pattern recognition, Selvakumar et al. [2] canny edge detection algorithm to examine and remove the words from a corrupted picture before detecting using neural networks, Kannan et al. [3] a brief overview of deep learning and highlight how it can be effectively applied for optical character recognition in Tamil language, Aparna et al. [4]. An OCR system for printed Tamil characters that is font and size independent, Perwej et al. [5]. Neural Networks for Handwritten English Alphabet Recognition, Saleh Ali et al. [6]. Digit Recognition using Neural Network, Ramya et al. [7] have discussed the segmentation of Tamil palm leaf Historical document using Sliding window and Adaptive Histogram Equalization and Ramya et al. [8] have discussed FFBNN based character recognition of segmented characters of Tamil palm leaf historical manuscripts.

3 Proposed System

System Description

'Agaram' aims to recognize handwritten Tamil characters and automatically convert it into a digitized text document that can be downloaded and edited. The approach used to achieve it is using CNN. The first step of the system is to test the model using the available datasets of Tamil characters which are pre-processed using IrfanView, an image processing tool. The major aim of the IrfanView tool is to smoothen the images so that the accuracy of the predictions can be improved and the noises in the image can be removed. The training of the model used for the detection and recognition of the characters is done using TensorFlow.js. TensorFlow.js is a machine learning platform which provides the facility to construct neural networks and run process on it using JavaScript.

Pseudo Code

Input : Training Data Set
Output : Trained Model
Method : Training of CNN using TensorFlow.js

```
Let image = {image1, image2, ..imagen } be the training
dataset.
Let tag = {tag1, tag2, ..tagn } be the tags assigned to
the images in the training set.
Let M be the Model
When inputImage passed to M
{
   Convert inputImage into grayscale;
   Get the pixel values of the characters;
   Convert into Tensors of appropriate shape for CNN;
   Detect the character in inputImage;
   If(detected character == actual character)
   {
      Retain the values of the weights;
   }
   Else
   {
      Use optimizer to determine the weight change rate;
      Change the values of the weights in the CNN;
   }
   Repeat the process for a fixed number of epochs
}
```

Input: Input Image
Output: Characters in the Word/Sentence
Method: Identify character using pretrained CNN model

```
Let C be the client
If C uploads an inputImage
{
   Pre-process and smoothen the inputImage;
   Resize the image to the required size;
   Convert the image into grayscale;
   Identify and segment the characters in the image;
   For each character in the image
   {
      Get the pixel values of the characters;
      Convert into Tensors of appropriate shape for CNN;
      Detect the character in inputImage;
      Add the character to the wordVariable;
   }
   Print the wordVariable as the output;
}
```

Training Phase

First we need to create the model so that we can train it (Fig. 3).

Model Summary

Layer Name	Output Shape	# Of Params	Trainable
conv2d_Conv2D1	[batch,24,24,8]	208	true
max_pooling2d_MaxPooling2D1	[batch,12,12,8]	0	true
conv2d_Conv2D2	[batch,8,8,16]	3,216	true
max_pooling2d_MaxPooling2D2	[batch,4,4,16]	0	true
flatten_Flatten1	[batch,256]	0	true
dense_Dense1	[batch,18]	4,626	true

Fig. 3. Model summary

The model consists of two convolutional layer setups followed by a flattening layer and finally passed to a dense layer to generate the output. The convolutional layer setup involves a Convolutional Layer that has a 5×5 pixel filter. The first convolutional layer has 8 filters, while the second convolutional layer has 16 filters. The convolutional layers are bounded by a ReLu activation function. These are then passed to a max pooling layer which basically pools the generated data.

Once the model is ready, we can begin to get the image set and labels ready, we do so by first preprocessing the training dataset images using IrfanView the images are taken in batches and they are resized into the required size, the image is also smoothened to increase the accuracy of the model later on.

The images are usually in a 4-planar format (RGBA format), for our character recognition application that is a fairly unnecessary (and potentially misleading data format), hence we convert the image pixels into a 1-planar format (greyscale format).

We do so by taking the pixel values of each image and splitting the RGBA values of each pixel, since the R, G & B values of a grayscale image will always be the same, we take the R value of each pixel and assume that to be it's grayscale value. A new 1-planar format image is gotten after this step is complete. These steps are then repeated for all the images that are present in the training data set.

The labels that should be associated with these images are also simultaneously created, the labels are basically the true character values of each image. These labels will be used to test the correctness of the model during it's training phase and constantly update the model's weight values to improve the performance of the CNN.

Once the pre-processed image set and the labels are ready, they are passed to the model to be trained, the model churns through all the images and repeatedly tries to predict what character is stored in each image. Every time the model predicts the character in the image incorrectly, it learns that it predicted incorrectly by comparing it's prediction to the value in the labels, once it learns that it was incorrect, it updates it's weight values.

To update it's weight values, it first checks the optimizer to determine to what extent the weight values need to be changed, after which the weight's are modified.

This is repeated for a fixed number of epochs to continuously train the model, care must be given to ensure that the model does not over train on the training data set. This could result in the model overfitting the data, which basically means that the model detects training images accurately, but does not generalize well.

Character Recognition Phase

The main motive of the system is to accurately recognize the characters from an uploaded image.

The first step that has to be taken for this is to get the image from the user. The image is then resized to the required image size and the image's pixel values are gotten, the RGBA value is converted into a grayscale value by taking the R value of each pixel to be grayscale value.

This generated a pixel value array for the image in it's grayscale (1-planar) format, this needs to be converted into a suitable tensor format. This is because the model is only capable of understanding and predicting data that is in this specific tensor format. The data is converted and reshaped into the necessary two-dimensional tensor data and passed to the Model (Fig. 4).

The model predicts what the character might be, the output generated from the model is an array of values where each value is the likelihood of the image being a certain character. The value with the highest value is the character that the image is most likely to be.

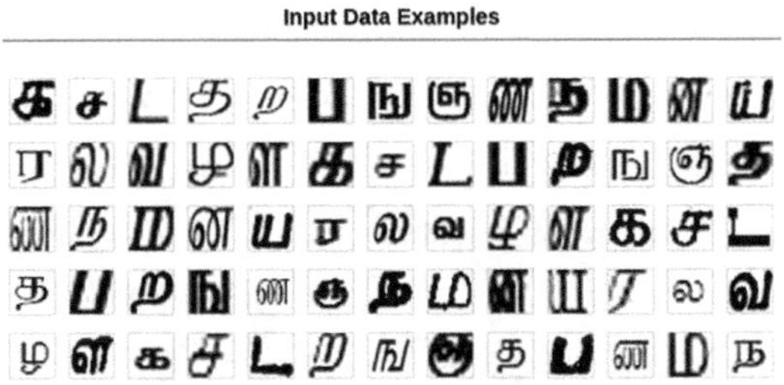

Fig. 4. Character examples

Word Recognition Phase
Once the character recognition phase runs perfectly, the program can be further improved to support recognition of words (Fig. 5).

Fig. 5. Word examples

The first step involves getting the word image from the user. The image is then segmented to get each character. This is done by assuming that the character's pixel values will be lower (lower means darker) than the surrounding pixel values.

Using such an assumption we can determine the top, bottom, left and right bounding lines for each character, once the bounding points are gotten, these points can be used to determine the top left starting point of each character and also the width and height of each character.

Once the dimensions of each character is gotten, the characters can be extracted from the word image and processed individually using the previously trained and tested character recognition model, the individually predicted characters are then stitched together to get the word prediction for the image.

4 Evaluation and Results

The aimed system has been developed successfully and tested with various test cases. The accuracy is very high when considering the relatively small data set that has been used to train the model.

The prediction capabilities of the model was tested and the result matched our expectations. Since there are two types of image data that can be passed to the model, we need to consider them separately for testing.

The first type of image is a character image. The model was originally trained for the recognition of characters. This was a resounding success, since the model was able to accurately identify the characters for each character image even though the training data set was relatively low (Fig. 6).

Fig. 6. Character recognition

The second type of image is a word image. The model's functionality was extended to support this feature, hence the performance for this feature was not expected to be that perfect, but surprisingly, it showed pretty good results.

The segmentation of the characters in the word image worker perfectly as expected and the character recognition part also worked with a very high accuracy. The predicted words almost accurately matched the words that were uploaded by the user to the model for recognizing (Fig. 7).

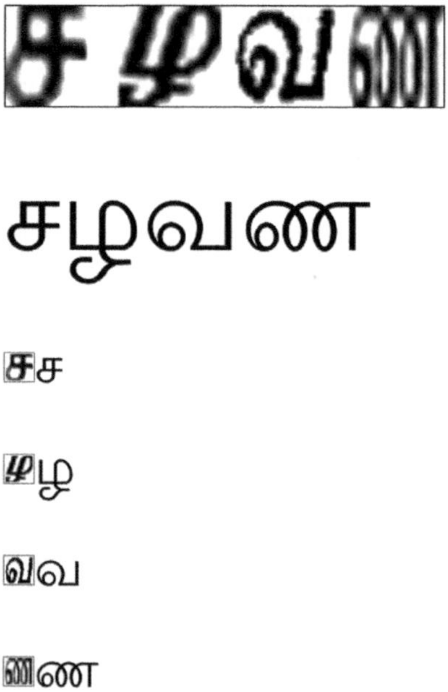

Fig. 7. Word recognition

The evaluation of the performance of the model was done by taking the per class accuracy of the model. The per class accuracy of the model is basically the accuracy of the model in correctly predicting/recognizing that character when said character's images are passed to it.

This is usually a good way to evaluate how well the model performs for each class/character and also to determine for which character the model under performs, by identifying such characters, we might be able to pre-process the images further to improve the accuracy for them (Fig. 8).

Accuracy

Class	Accuracy	# Samples
க	0.9	20
ச	0.9	20
ட	1	20
த	0.75	20
ப	1	20
ற	0.8	20
ங	0.95	20
ஞ	0.95	20
ண	0.85	20
ந	0.8	20
ம	0.95	20

Fig. 8. Per class accuracy

In addition to the per class accuracy table, A confusion matrix was also generated to evaluate the performance and accuracy of the model. A confusion matrix is basically a matrix plotting the actual labels for the character image to the predicted labels for the character image.

Confusion matrices help to understand which characters the model finds difficult to identify and also determine which characters the model gets confused with easily and mixes up its predictions with (Fig. 9).

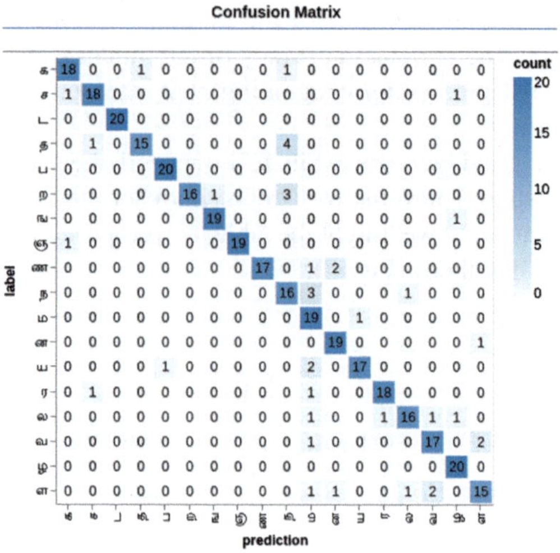

Fig. 9. Confusion matrix

The overall accuracy of the model was gathered along with the training of the model. We did so by separating the available image data set into two data sets, the training data set and the validation data set.

The accuracy value is the accuracy that has been calculated based on the total number of correct predictions made for each epoch in the range 0 to 1, where 0 is 0% and 1 is 100%. The blue line represents the training data set and the yellow line represents the validation data set (Fig. 10).

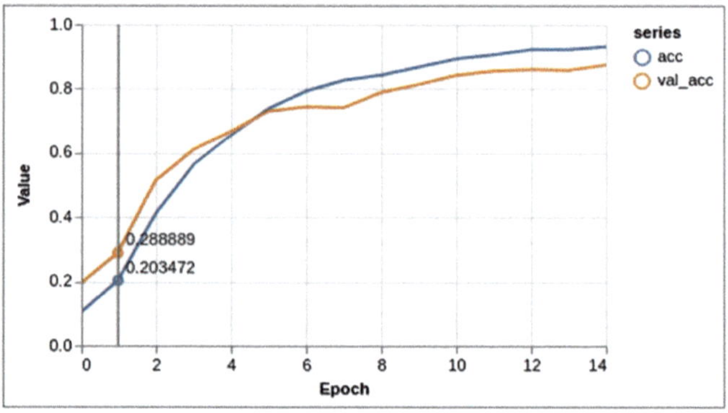

Fig. 10. Overall accuracy

5 Conclusion and Future Enhancements

The system has mainly focused on the automation of detecting and converting Tamil characters to a digitised text document by use of intelligent neural networks for the proper recognition of the characters. The accuracy achieved by using the convolutional neural networks trained using has shown a comparatively higher rate than the dense neural networks due to the filtered image approach (consider images as segments instead of pixels) used in the CNN. The training process happens by selecting the random pictures from the predefined, preformatted dataset. Every fifth image of the dataset is used as a testing data and the accuracy of the prediction is displayed using a confusion matrix.

Considering the relatively small volume of available datasets, the overall accuracy of 80% achieved by the model indicates that with better datasets, accurate predictions can be made.

The project intelligently identifies the Tamil words and characters but it is limited to the prediction of the fundamental 18 characters only. This is due to the fact that the alphabets in Tamil characters are around 247 and there are no existing datasets that covers all these characters extensively. So if we can develop proper datasets for the remaining characters, then this project can be extended to predict any Tamil character, words, sentences or text in ancient manuscripts accurately and convert it into a text document.

References

1. Wahi, A., Sundaramurthy, S., Poovizhi, P.: Handwritten Tamil character recognition using zernike moments and legendre polynomial. In: Suresh, L., Dash, S., Panigrahi, B. (eds.) Artificial Intelligence and Evolutionary Algorithms in Engineering Systems. Advances in Intelligent Systems and Computing, vol. 325. Springer, New Delhi (2015)
2. Selvakumar, P., Ganesh, S.H.: Tamil character recognition using canny edge detection algorithm. In: 2017 World Congress on Computing and Communication Technologies (WCCCT) (2017)
3. Kannan, R.J., Subramanian, S.: An adaptive approach of Tamil character recognition using deep learning with big data-a survey. In: Satapathy, S., Govardhan, A., Raju, K., Mandal, J. (eds.) Emerging ICT for Bridging the Future - Proceedings of the 49th Annual Convention of the Computer Society of India (CSI) Volume 1. Advances in Intelligent Systems and Computing, vol. 337. Springer, Cham (2015)
4. Aparna, K.G., Ramakrishnan, A.G.: A complete Tamil optical character recognition system. In: Lopresti, D., Hu, J., Kashi, R. (eds.) Document Analysis Systems V. DAS 2002. Lecture Notes in Computer Science, vol. 2423. Springer, Berlin (2002)
5. Perwej, Y., Chaturvedi, A.: Neural networks for handwritten English alphabet recognition. Int. J. Comput. Appl. (0975–8887) **20**(7) (2011)
6. Al-Omari, S.A.K., Sumari, P., Al-Taweel, S.A., Husain, A.J.A.: Digital recognition using neural network. J. Comput. Sci. **5**(6), 427–434 (2009). ISSN 1549-3636 © 2009 Science Publications

7. Ramya, J., Parvathavarthini, B.: Feed forward back propagation neural network based character recognition system for Tamil palm leaf manuscripts. J. Comput. Sci. **10**(4), 660–670 (2014). ISSN 1549-3636
8. Ramya, J., Parvathavarthini, B.: Segmentation of Tamil palm leaf manuscripts images. Eur. J. Sci. Res. **91**(4), 587–603 (2012). (ISSN 1450-216X)

A Solution to the Food Demand in the Upcoming Years Through Internet of Things

R. Sahila Devi(✉) and I. Sivaprasad Manivannan

Department of Computer Science and Engineering,
Rohini College of Engineering and Technology, Kanyakumari, Tamilnadu, India
sahiladevi@gmail.com, sivaprasad.tuff@gmail.com

Abstract. The rush in worldwide population is making us to shift towards smart agriculture. The reason behind this change is demolishing of natural resources, reduction in the availability of land for agriculture, the change in unpredictable weather condition makes production in agriculture a major concern for many countries. As a result, the Internet of Things (IoT) is used to enhance the production in agriculture sector. Internet of Things (IoT) is used as a major driver for smart agriculture. The combination of various sensors such as water level sensor, Humidity sensor, temperature sensor, soil moisture sensor, pesticide sensor, fertilizer sensor are used to detect the maximum possibilities to enhance the crop production and smart agriculture. Furthermore we provide future trends, innovation and application scenarios to make crop production to its maximum and eliminate the problem of food security.

Keywords: Agriculture · Internet of Things (IoT) · Sensors · Smart agriculture · Production

1 Introduction

Agriculture plays a major role in the development of the country. In our country, agriculture depends on the monsoons so failure of monsoon results in poor source of water. Irrigation method is used to solve this problem. In Irrigation system, water is provided to the crop which is depending on the soil types. In agriculture, two things are very important, first to get information about the fertility of soil and second to measure moisture content in soil. Nowadays, for irrigation purpose, different techniques are available which will reduces the dependency of rain. Mostly in this technique, the electrical power and on/off scheduling is used. In this method water level indicator is placed in water reservoir and a sensor is used to measure the soil moisture is placed at the root of the plant. Near that sensor a gateway unit is used, which get the sensor information and transmit that data to the controller. The controller unit is used to control the flow of water through the valves. Because of increase in population and decrease in supply the demand of food products increases, so the improvement in modernised food production techniques is necessary. Agriculture is the backbone of any country. So the country should able to handle the demand of food. Our country

D. J. Hemanth et al. (Eds.): COMET 2019, LNDECT 35, pp. 681–686, 2020.
https://doi.org/10.1007/978-3-030-32150-5_66

gains highest percent of economy only through Agriculture. But now a days due to excessive use of ground water and poor maintenance of water reservoirs the ground water level gets down. Ground water and rain water are major source of agriculture, so to protect agriculture several irrigation techniques are used. The main aim of the proposal is to modernise the agriculture as well as to reduce the manual job and effectively utilizing the water. This project can be implemented to any agricultural land.

2 Existing System

A countries economy is based on agriculture, so agriculture needs new innovative methods for irrigation. The manual method of irrigation has many short coming that a can be fixed using programmed method.

The existing concept of irrigation technique and researches use the method of automatic irrigation by using soil moisture sensor. For effective automatic irrigation humidity sensor and temperature sensors area also used. To make this system wireless, the GSM is used. The solar panels are used to provide electricity for the required device component. The solar panel provides uninterrupted power supply by using load shedding. This system constantly monitor the water content and whenever the soil moisture level gets lower than the threshold level. The system send signal to turn the motor by itself. This system automatically stops the motor as soon as the soil reaches its higher threshold value, the threshold value is determined depends upon the crop. The motor starts and strop automatically based on the soil moisture requirements. When the system works the user gets status SMS about the status of the land (Fig. 1).

The basic working model of the system is separated into two. Primary circuit is a solar circuit. The solar circuit gives/dc energy to the components. Another circuit is moisture sensor. This sensor is submersed in to the soil. The moisture content of the soil received from the sensor be able to be viewed on LCD. The subsequent circuit is GSM module. This GSM module circuit is coupled with audrino. The Arduino is accountable for transmitting data for each process. In the paper, there are two basic threshold values that is for both upper and lower level. These two values are designated by the handler based on crop. The moisture Sensor will detect the existing level of liquid in the soil, which are submerged in soil. The code relates this level of liquid with the two-user labelled threshold. If the present value transpires to be less than the lower base linerate, then a signal will be generated by the code which helps to turn on the motors. The development will be independent and the dried-out part of soil gets conditioned. The level of moisture content in the soil will be compare for each and every values with the baseline (threshold) rate in the present moisture value and cross the upper threshold value at that point the signal will be send to the motors by the code to switch off it. Figure 2 represents the basic flowchart diagram of this project. The process begins by reading the values of the sensors and to display it on LCD display. If the moisture value falls less than the lower baseline (threshold) point, the engine begins and if the moisture value goes beyond the upper baseline value then a signal will be pass by the Code to shuts off the motors. In further case, the SMS will be sent to the user using GSM module of undergoing process. The LCD display will monitor and display the status of the motors according to the Code.

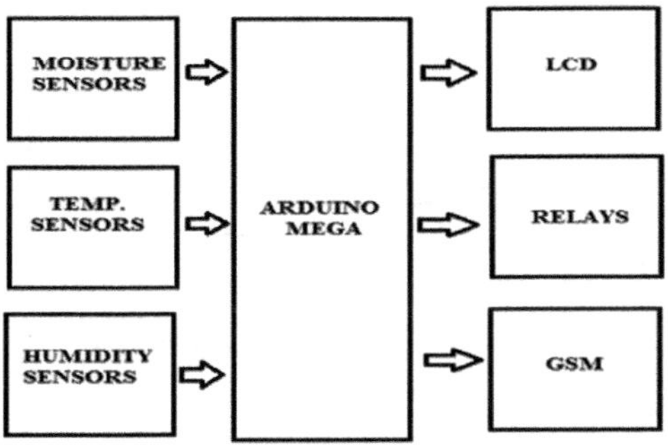

Fig. 1. .

Other shortcomings include
The Licenses for the GSM technology should be get from the Qualcomm which patented this technologies.
Repeaters should be installed to increase coverage.
GSM offers only bounded data rate capability, and we can use advanced version of GSM devices for higher date rate.
The FTDMA access scheme is used in GSM. Due to this the same bandwidth of the signals can be share by multiple users and therefore cause interference when the number of users exists the limit of using the GSM service. To solve this situation, robust frequency rectification algorithms are used in base stations and cell phones.
Pulse-based burst transmission technology is used in GSM and therefore interferes with certain electronic components. Because of this, airplanes, bunk beds and hospitals avoid the use of GSM-based mobile devices or other devices.

3 Proposed System

The (IoT) is mostly utilized in interlinking the devices with cloud and to retrieve data and information from cloud. The Internet of Things is used with IoT structures to manipulate and relate with this collected data. In this proposed system, users should register their sensors, create data streams, and can process the information. IoT are using in various farming methodologies. Some of the IoT applications are Smart Cities development, Intelligent Environment, Intelligent Water, Intelligent Measurement, Emergency and Safety, Intelligent Agriculture, Industrial Control, Domotics, Electronic Health, etc. The IoT is based on a device that is able to analyze information and then transmit it to the user. Farmers using the techniques of traditional methods for agriculture since early period, which results in decreasing yields of crops and fruits. Thus, the yield of harvest can be achieved by using automatic machines. There is a requirement to enforce the modern science and technology in the field of agriculture to increase crop yields. By using Internet of Things, we can expect increased production at low cost, serving soil efficiency, monitoring temperature and humidity, monitoring rainfall, fertilizer efficiency, noting the storage capacity of water tanks as well as theft detection in agricultural areas.

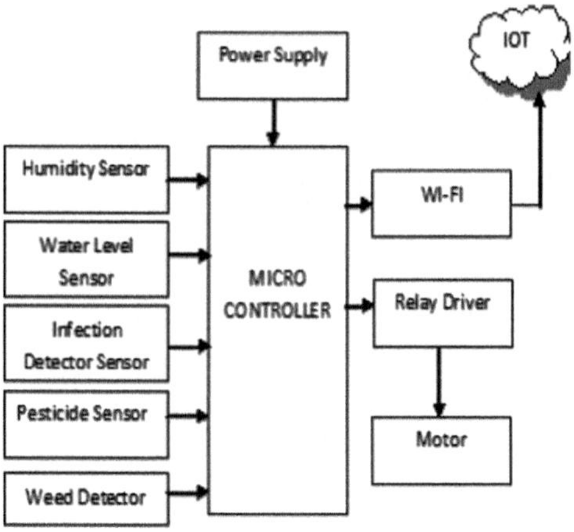

Fig. 2. .

Six sensors are positioned in the culture, and the information will be gain from these sensors. This information will be in the form of analog data, therefore the analog data is transformed into digital data, the digital data obtain as input to the Arduino that the data refer to the bank data with the help of wi-fi, sensors regulated so that the least wet state. The limit voltage is varied corresponding to various fields of cultivation in various seasons of the year. The microcontroller has functioned the relay, the relay is also positioned in it, when the data originates from the sensors, the value is equated with the microcontroller, If the value end up below the regular value intended then the field is in dry conditions and transmit signal to connected motor, when the value is superior than the standard value when the field is in moist conditions. In order for the signal relayed to the motor automatically, a bell specifies the condition variation when the engine is “off to on” and “on to off”. Data stores data in the cloud using the wifi module. The system is fully automated and the status of the system condition can be identified through your mobile handset. Android App was created, so this info the agriculturalists and to monitor the changes according to the condition and in the microcontroller method via Arduino code which produce the IP address for all the data of the sensors is accessible at that address and also the engine condition too comprises, anywhere open the application using mobile the data will be displayed on our device.

3.1 Benefits

Smart devices tend to reduce waste and increase efficiency, maximizing capacity and minimizing costs. Intelligent irrigation systems can be used to optimize water levels based on factors such as soil moisture and weather forecasting. The intelligent irrigation system will have better control over the landscape and irrigation needs, as well as the intelligent system can make decisions independently if you are absent.

By using this method, local and remote farmers can monitor growth of the crops of multiple fields from multiple locations via Internet. Decisions can be made in real time and from anywhere. Provides imminent and process automation in real time through low cost sensors and IoT platform implementation (Fig. 3).

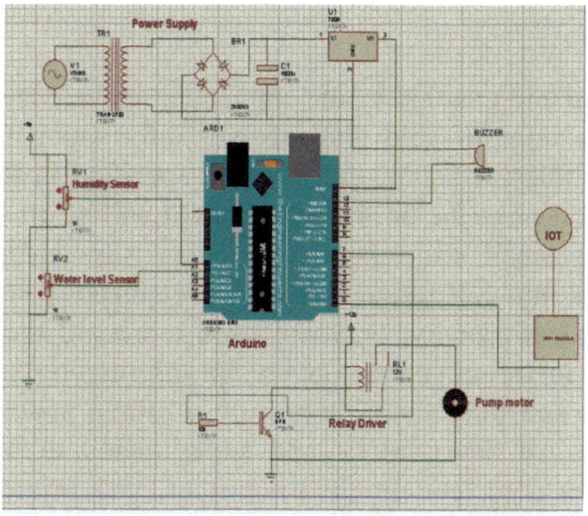

Fig. 3. .

3.2 Implementation Using IoT

This project uses the IoT concept to monitor and control the system and the Arduino and at mega processor is used as hardware components. MQTT server utilizes the Android application namely My MQTT. You have to register to any theme and pit out a specific role message in this application.

MQTT

Telemetry Transport of the Message Queue is abbreviated as MQTT, which is a publishing protocol. It is very simple and lightweight messaging protocol, developed for restricted devices and high latency, less bandwidth or untrusted networks. The design principles are to reduce bandwidth of the network signals and resource requirements of the device. At the same time, also tries to ensure reliability and some degree of guarantee on delivery. These ideologies also make protocol an ideal machine-to-machine (M2M) or “Internet of Things” - a world of connected devices, and for mobile applications, bandwidth and battery power are the best.

MQTT architecture

Telemetry Transport of the Message Queue has a client/server model, where each client is a sensor as well as connects to the server, as an intermediary, through Transmission Control Protocol (TCP). The Telemetry Transport of the Message Queue is message oriented. Each message is a discrete piece of data, opaque to the broker. Each message is distributed to an address, known as a topic. Thus, customers can subscribe to various topics. Each client enrolled in a topic retrieves all messages posted in the topic. Mote Queue Telemetry Transport defines how to show the desired plan to be executed in the resource. The feature shows either the previous information or the information that is achieved is dynamic, depending server.

4 Conclusion

Farmland is ministered and guarded by the application in the edge of the user. ESP8266 is the machine at the edge of field that receives information from the intermediary system as well as handles, executes the function specified in the message. Then the broker's network will receive the information, then it will be presented to the end-user. The considering it is small scale, solid, smooth, easily understandable and easily executable. An agricultural cultivation system is introduced with low complex circuits. Two sensors are used effectually, they are used to record the temperature and the humidity of the soil in the circuit to obtain the information standardization for the system. Two sensors and microcontrollers from all three nodes successfully interact with multiple nodes. All observations and experimental tests has proven that the proposed concept is a complete for field activities, and irrigation problems. If this proposed concept has been implemented it will surely help the farmers to improve the yield of crops and overall production.

References

1. Kawitkar, R.S., Gondchawar, N.: IoT based smart agriculture. Int. J. Adv. Res. Comput. Commun. Eng. **5**(6), 838–842 (2016). ISSN (Online) 2278-1021 ISSN (Print) 2319 5940
2. Baranwal, T., Nitika, Pateriya, P.K.: Development of IoT based smart security and monitoring devices for agriculture. In: 6th IEEE International Conference - Cloud System and Big Data Engineering. IEEE (2016). 978-1-4673-8203-8/16
3. Sales, N., Arsenio, A.: Wireless sensor and actuator system for smart irrigation on the cloud. In: 2nd World forum on Internet of Things (WF-IoT), December 2015. published in IEEE Xplorejan (2016). 978-1-5090-0366-2/15
4. Kassim, M.R.M., Mat, I., Harun, A.N.: Wireless sensor network in precision agriculture application. 978-1-4799-4383-8/14
5. Kassim, M.R.M., Mat, I., Harun, A.N.: Wireless sensor network in recession agriculture application. In: International Conference on Computer, Information and Telecommunication Systems (CITS). Published in IEEE Xplore, July 2014
6. Muthunpandian, S., Vigneshwaran, S., Ranjitsabarinath, R.C., Manoj Kumar Reddy, Y.: IOT based crop-field monitoring and irrigation automation **4**(19) (2017)
7. Gutiérrez, J., Villa-Medina, J.F., Nieto-Garibay, A., Porta-Gándara, M.Á.: Automated irrigation system using a wireless sensor network and GPRS module. IEEE Trans. Instrum. Measur. **17** (2017)
8. Mohanraj, I., Ashokumar, K., Naren, J.: Field monitoring and automation using IOT in agriculture domain. IJCSNS, **15**(6) (2015)
9. Williams, M.G.: A risk assessment on Raspberry Pi using NIST standards. Version 1.0, December 2012
10. Harrington, A.N., Lakshmisudha, K.: Hands-on Python
11. Hegde, S., Kale, N., Iyer, S.: Smart precision based agriculture using sensors. Int. J. Comput. Appl. (0975–8887) **146**(11), 36–38 (2011)
12. Gondchawar, N., Kawitkar, R.S.: IoT based smart agriculture. IJARCCE **5**(6), 838–842 (2016)
13. Gayatri, M.K., Jayasakthi, J., Anandhamala, G.S., Chetan Dwarkani, M., Ganesh Ram, R., Jagannathan, S., Priyatharshini, R.: Smart farming system using sensors for agricultural task automation. In: IEEE International Conference on Technological Innovations in ICT for Agriculture and Rural Development (TIAR 2015) (2015)

User Friendly Department Assistant Robo

Sharon Trafeena Mathias(✉) and S. Adlin Femil

Rohini College of Engineering and Technology, Nagercoil, India
sivaprasad.tuff@gmail.com

Abstract. In this article we proposed a friendly interaction between human machines in the assistant department robot (DA_ROB). Our system is able to interact with people, provide information and act as an assistant to a department. The system has friendly interface and automatic speech friendly recognition. Friendly user interface takes a significant responsibility for direct interaction with people. The main aim behind the creation of the robot is to reduce the work of faculties. Bluetooth module is used to receive the voice signal and produce response. GSM is mainly used for communication. RFID reader is used to get the output for attendance purpose. Ultrasonic is used to measure distance and also provide the facility of security camera. It is based on machine learning and raspberry pi is used to implement the robot.

1 Introduction

In recent year development of robot service quickly increased. Robotics reduces the human works. In the case of robot assistance robots for gardening, home care, pizza delivery are available and these are the example of growths in robotics. Here we are introducing a robot which can act as an assistant of a department.

Machine learning

Machine learning is the scientific study of models and statistical algorithms that computer systems use to effectively accomplish a defined task without explicit instructions, which depends on patterns and inference. It is seen as a division of artificial intelligence. Machine learning algorithms construct a mathematical model for a set of sample data, known as "training data", which is used to make predictions or decisions without being programmed unequivocally to perform the task. Machine learning is strongly related to computational statistics, which focuses on making predictions using computers. The mathematical optimization study provides methods, theory and application domains for the field of machine learning. Here we write code that is understood by a machine and we have already defined the functions that must be executed by the robot. The User-friendly word describes a hardware device or software interface that is easy to use. It is "friendly" to the user, meaning it is not difficult to learn or understand. In this robot we can enter by voice. Also voice response. All its performance began after receiving user commands. Simple: A friendly interface is not overly complex, but it's clean. A good quality user interface is well organized, making it easy to find different tools and options. Intuitive: To be user friendly, an interface must be able to detect the common user and should require a nominal explanation of

D. J. Hemanth et al. (Eds.): COMET 2019, LNDECT 35, pp. 687–692, 2020.
https://doi.org/10.1007/978-3-030-32150-5_67

how to use it. Reliable: Untrusted products are not easy to use as they cause unnecessary frustration to the user. A user-friendly product must be reliable and free of defects or failure.

1.1 Existing System

Usually in the case of attendance the particular staff will take the list of student who are present or absent manually and informing their parent individually. The faculty need to note the attendance in a paper or book. Frequently a timetable chart is used by each faculty to know about their periods scheduled and they need to remember about the activities that take place on department each day. For surveillance purpose we normally use cameras. By using the footages can check the area covered by camera as our wish.

Disadvantages of existing system

- When we take attendance, the incorporation of students may lead to mistakes.
- if any disturbance occur it lead to repeat process of taking attendance which also lead to time wastage.
- The faculties need to refer the timetable chart and schedule chart frequently. any damage that in the chart may lead to confuse.
- when we using camera for surveillance purpose it only give the footage of sight in its limit.

2 Proposed System

This robot can act as an assistant of a department. It can perform works as a staff in the case of attendance it take attendance by using RFID reader and the GSM module receive the student list and send the message automatically to their parents it also speak out the list of students who are absent. If we require it should tell the current date and time. Before each period it should announce the classes scheduled to each staff and also works as alarm which remind the activities on the department. using ultrasonic sensor it find the distance if the result is less than 10 and any object is with the measured distance it should wish good morning, good afternoon and good evening corresponding to the time. Using hotspot we can watch the footage covering by robot from anywhere. By enable the Bluetooth module by Bluetooth terminal app we can give instruction to the robot for movements like left, right, forward and backward. Power Bank is used to provide required power to the robot.

Advantages of proposed system

- It can trim down the time wastage.
- Because it alarm properly faculty can enter the class without any confusion.
- The robot can move so it can cover the whole area as our requirement.

3 System Architecture

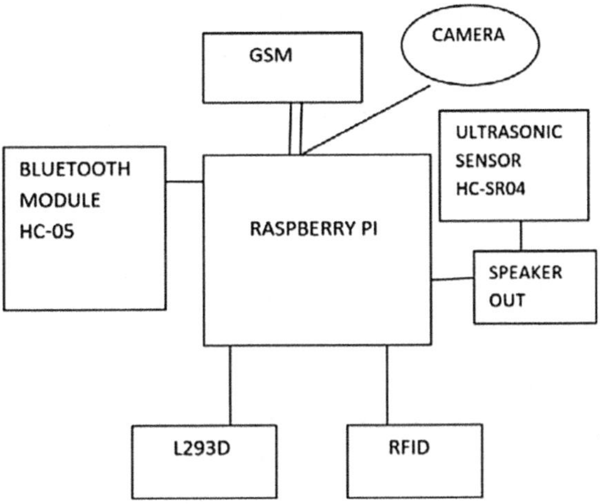

4 Modules

4.1 Raspberry Pi

Raspberry PI Module Bluetooth Module GSM Module RFID Module Ultrasonic Sensor The raspberry board is a small computer that can easily connect to the internet and interface with many hardware components. Raspberry Pi is cheap. A Raspberry Pi is not just a computer. Using 40 pin GPIO, you can easily connect to multiple actuators and sensors. Many protocols are used to connect the card to other hardware devices such as serial, spi or i2c. This is a very important point as it makes Raspberry Pi cards fit well with most devices. Here we connect other modules to the Raspberry pi and it acts as the main module. In the case of the GSM service module, take the details of Raspberry pi. Likewise, each module is connected to the main board of raspberry pi.

4.2 GSM Module

The GSM module is a chip or circuit that will be used to establish communication between mobile devices. The GSM modem requires a SIM card to be operated and operates on a network bandwidth signed by the network operator. This module receives the service details of the raspberry pi and sends a message that is automatically set in the program for the number to include with the details. The purpose of sending messages to parents is responsible for the module.

4.3 Bluetooth Module

Bluetooth module The Bluetooth module is activated by an application called Bluetooth terminal. Bluetooth controlled robotics involve the process of robots in harmony with the signals. It receives the voice as input and produces output by voice response. In the Bluetooth-based robotics workshop, serial programming of microcontrollers is an integral path can give instructions or supplies through the application in our cell phone.

4.4 RFID Modules

RFID module is mainly used for attendance purpose. EM-18 is used to read the RFID card to take attendance. After reading each card it check whether student is present or not and send the details to the main board (Raspberry pi). Radio frequency Identification (RFID) is a radio wave wireless identification technology that use radio waves to recognize the presence of RFID tags. Just like Bar code reader, RFID technology is used for recognition of people, object etc. presence.

In barcode technology, we have to optically scan the barcode by keeping it in front of bar code reader, whereas in RFID technology we just need to bring RFID tags in range of readers. Also, barcodes can get damaged, scratched or unreadable, which is not in the case for most of the RFID reader has transceiver and an antenna mounted on it. It is mostly fixed in a fixed position. RFID is used for attendance system in which every person will have their separate RFID tag which will help recognize person and their attendance.

4.5 Ultrasonic Sensor

Ultrasonic sensor is used to calculate the distance from the object by speed of sound. The circuit connects with two GPIO pins (one for echo, one for trigger), the ground pin, and a 5V pin. As well as being able to see the distance value, you can also get the sensor to do things when the object is in or out of a convinced range. It is based on a transmitter and receiver and mainly used to determine the distance from the target object. The amount of time it needed to send and receive waves will determine how far the object is placed from the sensor. It mainly rely on the sound waves working on "non-contact" technology. The requisite distance of the target object is measured without any damage, giving you accurate and precise details. Here it find the distance if it is less than 10 it should wish good morning, good afternoon and good evening as corresponding to the time.

4.6 Camera

Camera is used for proper surveillance purpose a webcam is fit in the head of robot. The robot live stream facility is done using live stream IP Address.

5 Flowchart

Thread 1

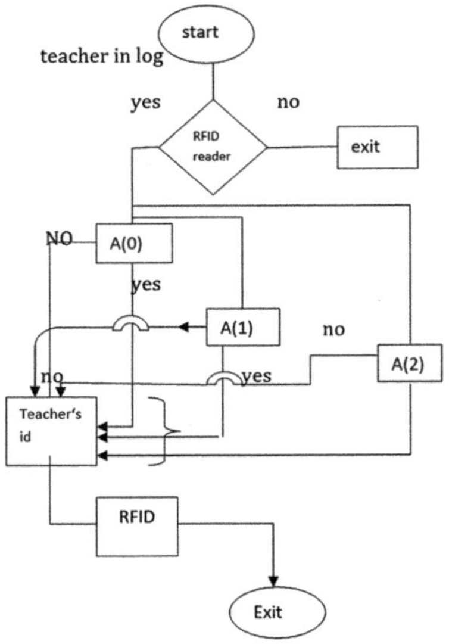

Thread 2

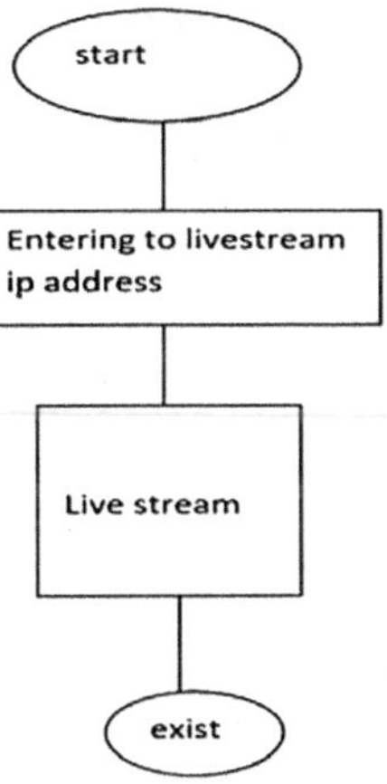

Thread 3

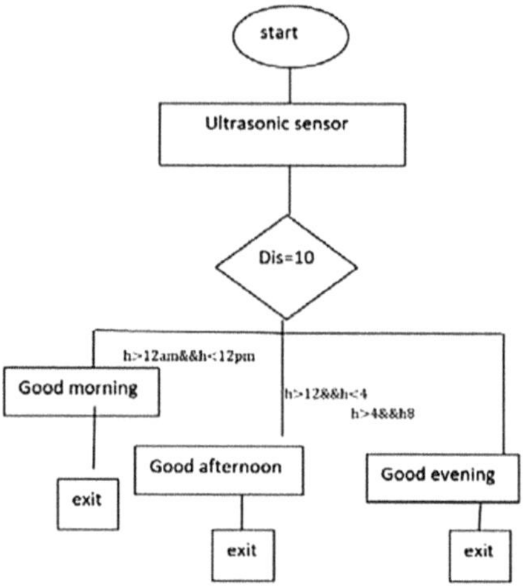

6 Conclusion

In this work, we introduce a friendly robot department assistant interface. We add the activities that a team doing in a department can also act as a security reminder, may wish to find some object in the 10 m range can also make movements. Well, can work as a department assistant. In future the system can connect the IoT applications and improve its performance with high quality.

References

1. Lafaye, J., Gouaillier, D., Wieber, P.B.: Linear model predictive control of the locomotion of Pepper a humanoid robot with omnidirectional wheels. In: 2014 IEEE-RAS International Conference on Humanoid Robots, pp. 336–341 (2014)
2. Rane, P., Mhatre, V., Kurup, L.: Study of a home robot: Jibo. Int. J. Eng. Res. Technol. **3**(10), 490–493 (2014)
3. Graf, B., Reiser, U., Hagele, M., Assistant care-O-bot® 3-product vision and innovation platform. In: 2009 IEEE Workshop on Advanced Robotics and its Social Impacts (ARSO), pp. 139–144 (2009)
4. Shneiderman, B.: Designing for fun: how can we design user interfaces to be more fun? Interactions **11**, 48–50 (2004)
5. Yim, M., Duff, D.G., Roufas, K.D.: PolyBot: a modular. Accessed 21 Nov 2016

Advances in High Performance Computing

RETRACTED CHAPTER: Exploration of Maximizing the Significance Big Data in Cloud Computing

R. Dhaya[1(✉)], M. Devi[1], R. Kanthavel[4], Fahad Algarni[2], and Pooja Dixikha[3]

[1] Department of Computer Science, King Khalid University, Sarat Abidha Campus, Abha, Kingdom of Saudi Arabia
dhayavel2005@gmail.com, devinov6@gmail.com

[2] Department of Computing and Information Technology, University of Bisha, Bisha, Kingdom of Saudi Arabia
fahad.a.algarni@gmail.com

[3] Department of ECE, Sona College of Technology, Salem, Tamilnadu, India
poojapooja13189@gmail.com

[4] Department of Computer Engineering, King Khalid University, Abha, Kingdom of Saudi Arabia
kanthavel2005@gmail.com

Abstract. Currently information community has been bombarded by Big Data in Cloud computing to classify and coordinate into basic management process. On the other hand, the development of mobile computing is to extent "individual clouds", or open cloud resources in order to understanding the properties over the personal computing gadget like, advanced cells, tablets, workstations, shrewd TVs, and even joined frameworks inside automobiles. Big data, as it looks is enormous data used to designate the huge volume of data in unstructured and semi-structured format. And that is cloud computing comes in for the utilization the cloud as receptors for the majority of that information regardless of claim cloud or private cloud without particular amount in Petabytes and Exabyte of information. So it is increasingly using cloud deployments and therefore analytics needs to be surveyed with the aim at increasing value to address big data. In addition the entire consumer in Cloud computing, clients, servers, applications and other elements related to data centers are made available to IT and end users via the internet. Organization needs to pay only as much for the computing infrastructure as they use. The way of billing type in cloud computing is similar to the electricity payment that we do on the basis of usage. It is a function of the allocation of resources on demand. The best booking ahead of time of assets is hard to be accomplished because of vulnerability of customer's future interest and supplier's asset costs. This paper renders the techniques behind maximizing Big Data in cloud computing. The issues, insights, analysis and management of Big Data, and advantages and learning outcome of Big Data in cloud computing, resource Provisioning Cost also have been studied.

Keywords: Cloud computing model · Big Data · Symbiosis · Personal clouds · Resource Provisioning Cost

---

The original version of this chapter has been retracted. The retraction note to this chapter can be found at https://doi.org/10.1007/978-3-030-32150-5_163

D. J. Hemanth et al. (Eds.): COMET 2019, LNDECT 35, pp. 695–702, 2020.
https://doi.org/10.1007/978-3-030-32150-5_68

1 Introduction

Big data is to find repeatable business designs represents something like 80% of an association's data in an unstructured organization of content documents. The sheer volume of unstructured data intermittently inside an undertaking is to be overseen legitimately to make the Big Data usage viably as far as capacity [17]. Then again, investigation of big data continuously is an ordinary thing that includes overseeing data to make an obligation if data can't be lived [14]. Big data investigation is frequently.

Joined with cloud registering to circulate the work among a huge number of PCs. The issues of Big Data in Cloud Computing distinguished are as per the following [16]:

- The need to help expansive number of utilizations, every one of which has a little data impression.
- A scientific analysis of unstructured data is a great deal for accurate decision-making and an improved customer experience [1].
- Heterogeneity, scale, timeliness, complexity, and privacy Problems [3].
- Skills sense that getting real knowledge out of data is not really an IT capability to be improved to analyze the larger volume data in cloud computing environment [15].
- Data structures to make it accessible for ad-hoc analysis and make it flexible enough that gets some things out.
- Data collection is the main issue to keep the data and make it have future value [2]. Thus it is much needed to undergo a detailed research analysis on structured and unstructured data that are used in Big Data under cloud computing environment.

2 Insights of Big Data

The following are the insights into 'big data'.

- Big Data Spectrum includes the use cases in retail, airlines, automotive, financial services and energy [7].
- Typical challenges focuses regarding security parameters, technologies integration, and real-time communication in cloud computing [6].
- Symbiosis analysis is to be maintained which addresses all three dimensions of the big data challenge as Variety, Velocity and Volume [4].
- Considering the enormous figure and capacity required supporting big data, the advantageous interaction turns out to be all the more firmly coupled keep on advancing [5].
- Management and Analyze Enormous Volumes of Data

Across all businesses and topographies, associations of different sizes are being tested to discover more straightforward and quicker approaches to investigate gigantic measures of data and better address customer issues [18]. The present news incorporates editions to help new administrations models and 'Smart Cloud Desktop Infrastructure' to ease administration of virtual work area arrangements [8]. New Systems pitched for Big Data to give the scalability, effortlessness and execution basic to have the capacity to investigate big data to convey results eight hours quicker than on the past arrangement, which take a gander at the effect on organizations consistently that passes [12].

2.1 Advantages of Big Data in Cloud Computing

The advantages of Big Data are listed out below.

- It makes cloud simpler for all businesses to adopt.
- Consistent execution, effortlessness, data invigorate speed and by and large execution [9].
- To disentangle and quicken cloud sending stages, associations all things considered.
- To can expand business dexterity, limit business hazard and speed time to income. It exposes the business intelligence value of cloud data analytics [11].

3 The Need for Big Data Analysis in Cloud Computing

Desktop applications are well restricted by using the WEB based services. Almost more than 50% computer interactions involve accessing Web applications and that also even for internal and in-house applications can be accessed through a Web browser which is termed as service-oriented computing. So the data is to be used on the basis of pay by use instead of provisioning for peak and that are dependent on capacity, demand and time against the resources shown in Figs. 1 and 2 for the static data Centre and cloud data center respectively.

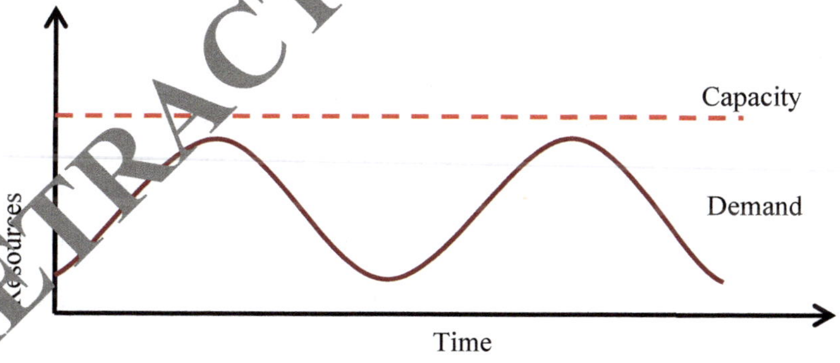

Fig. 1. Static data center

The following are the reasons why big data analysis is important

- Complex data processing in Graphs
- Multidimensional Data Analytics based on Location data.
- Physical and Virtual Worlds that includes Social Networks and Social Media data & analysis

- Data Platforms for Large Applications
- Significant Operational Challenges in using RDBMS technology are
- Peak Demand Provisioning
- Resource under-utilization
- Massive challenge in Storage management
- System upgrades on par with an extremely time consuming module

In addition cloud computing is very much bothering on Scalability, Elasticity, Fault tolerance and Self-Manageability.

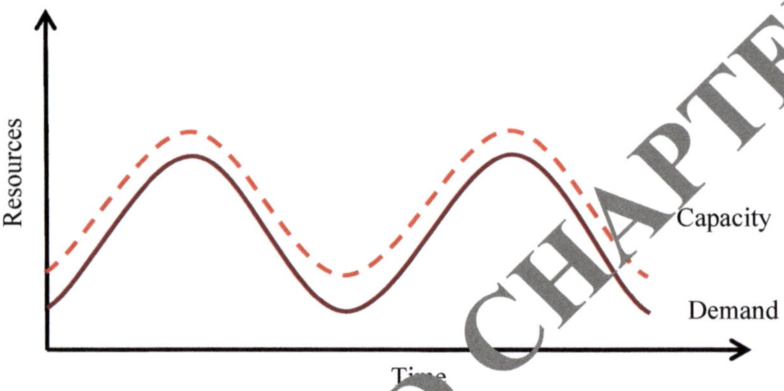

Fig. 2. Cloud data center in cloud

4 Learning Outcomes

The unions of technological platform, cloud processing and big data are to be sure changing the world. The revelation of new medications to fix maladies, exact climate foreseeing designs, water management methodology, and so on, cloud computing provides unlimited resources on demand. Since big data is an accumulation of bigger data sets regularly unstructured, it is difficult to gather, examine, picture, and process.

- Big data and cloud registering are two of the biggest patterns in numerous associations.
- Probably the biggest advantage of the cloud is that it's considerably less demanding to oversee limit
- Design decisions and rules that are to be completed in outlining the up and coming age of data administration frameworks for the cloud.
- The configuration space for Data Base Management System (DBMS) is to help refreshing escalated remaining tasks at hand for huge multitenant frameworks and furthermore to guarantee the proceeded with accomplishment of DBMSs.

The Fig. 3 shows the integration of the big data in the cloud computing environment, where the data sources server is connected with the client nodes orthogonally. The big data has also been integrated with cloud computing domain. Based on the request made by the clients, the data to be shared by the server then starts collecting the relevant data from the data storage unit of the big data service.

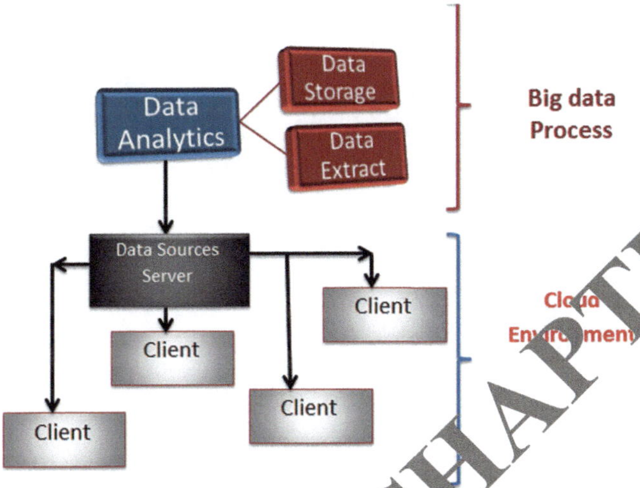

Fig. 3. Learning outcome block diagram

After extracting the data, it is to be analyzed by the data analytics before handling the same the clients through the server.

5 Challenges and Design Goals in Resource Provisioning Cost

Allocate resource by optimized for reduced cost and distributing their resources to consumer is the major drawback in existing system as there very difficult to get access rights from cloud environment. Different types of challenges in addressing the problem of enabling resource allocation in data centers to satisfy competing applications demand for computing services. With the reservation plan, the buyer can diminish the aggregate asset provisioning cost. The reservation plan, the cloud customers from the earlier save the assets ahead of time. Therefore, the under provisioning issue can happen when the saved assets can't completely take care of the demand because of its vulnerability. These provisioning calculations can arrangement figuring assets for being utilized in different provisioning stages and additionally such a large number of year designs. It scope up with the uncertainty of consumers future demand, go for an optimal cloud resource provisioning model. Minimize the total cost of resource provisioning by reducing over provisioning and under provisioning. The pricing solution employs a novel method that estimates the correlations of the cache services in a time-efficient manner. Cloud service provider has two tasks for allocate resources. Figure 4 shows the Cloud Computing-Resource Provisioning System. It performing time-insensitive background computing and distributing resource to the cloud users in the dynamic process. Cloud shopper can effectively limit add up to cost of asset provisioning in cloud processing situations.

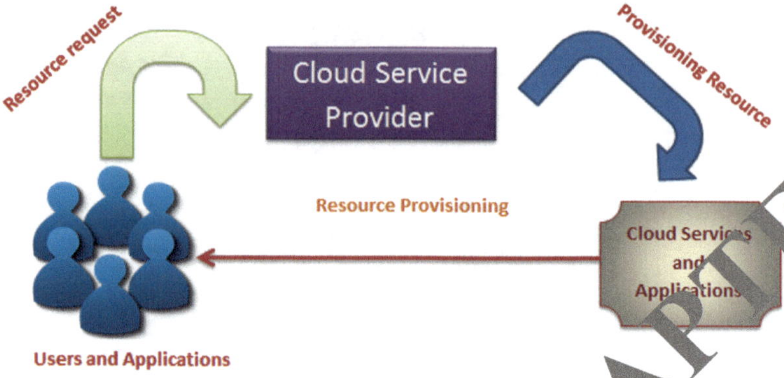

Fig. 4. Cloud computing-resource provisioning system.

6 Problems in Resource Provisioning Cost

Both the reservation and on-demand plans used to face the under provisioning and over provisioning problem. Static pricing cannot guarantee cloud profit not minimized Static pricing results in an unpredictable and, unmanageable manner of cost. Cloud computing is research on accounting in wide area networks that offer distributed services. Some of the others consumer request for some other location focus on job scheduling and bid negotiation, issues orthogonal to optimal pricing. There are two major problems when trying to define an optimal pricing scheme for the cloud exposed service. Beginning model is to define a simplified enough model of the price depends upon marketable and to acquire feasible fixed maximum amount but not over reduced model that is not representative. The second challenge is to define a pricing scheme that is adaptable to time-dependent model changes.

7 Provisional Problems in Resource Cost

A cloud provider can give the customer to the provisioning plans are Reservation plan and On-ask for plan. For organizing, the cloud go-between considers the reservation plan as medium-to whole deal orchestrating, since the course of action must be set apart ahead (e.g., one or three years) and the arrangement will fundamentally lessen the aggregate provisioning cost. In qualification, the representative considers the on-request plan as here and now arranging, since the on-request plan will be obtained whenever for brief timeframe (e.g., multi week) when the assets held by the reservation-plan are insufficient (e.g., all through pinnacle stack).

8 Provisioning Phase in Resource Cost

The cloud intermediary thinks about every reservation and on interest anticipates provisioning assets. These assets territory unit utilized in totally extraordinary break hole between times additionally called provisioning stages. The three provisioning phases are Reservation stages, using phases and On-request stages. The stages with their activities perform in various purposes of your time (or occasions) as pursues. Introductory inside the reservation area, while not knowing the shopper's genuine interest, the cloud dealer arrangements assets with reservation organize before. In the spending part, the esteem and request territory unit achieved, and furthermore the saved assets can be utilized. Therefore, the saved assets may be resolved to be either over provisioned or beneath provisioned. In the event that the interest surpasses the quantity of saved assets (under provisioned), the intermediary will pay for included assets request organizes, at that point the on-request area begins. The reservation contracts are cloud provider can give the purchaser numerous reservation designs with various reservation contracts. Each reservation contract alludes to the booking ahead of time of assets with the exact day and age of use.

9 Provisioning Stages in Resource Provisioning Cost

A provisioning stage is that the time age once the cloud specialist settles on a choice to arrangement assets by getting reservation or potentially on-request designs, and furthermore designates VMs to cloud suppliers for using the provisioned assets. Thusly, every provisioning stage will incorporate one or a great deal of provisioning stages.

10 Conclusion

In this paper, the objective of maximizing the big data utilization in cloud computing has been studied by exploring the varieties, volumes of the big data. The structured and unstructured data also have been analyzed. From the studies it is identified that the expansion of data has dependably been a piece of the effect of data and interchanges innovation. Moreover, obstruct advance at all periods of the pipeline that can make an incentive from data. The solution of the master problem will modify the cost and in addition the expending cost per solution of the Benders cuts are made of the optimal costs obtained from master problem and sub problems among the prior iterations. From the investigation, it is also recognized that resources, the innovation, data volume, abilities, data compose and data structures are the principle issues of Big Data in cloud registering with which it maximizes the utilization.

References

1. Abouzeid, A., Pawlikowski, K.B., Abadi, D.J., Rasin, A., Silberschatz, A.: HadoopDB: an architectural hybrid of map reduce and DBMS technologies for analytical work-loads. PVLDB **2**(1), 922–933 (2009)
2. Agrawal, D., Das, S., Abbadi, A.E.: Big data and cloud computing: new wine or just new bottles? PVLDB **3**(2), 1647–1648 (2010)
3. Agrawal, D., El Abbadi, A., Antony, S., Das, S.: Data management challenges in cloud computing infrastructures. In DNIS, pp. 1–10 (2010)
4. Agrawal, P., Silberstein, A., Cooper, B.F., Srivastava, U., Ramakrishnan, R.: Asynchronous view maintenance for VLSD databases. In: SIGMOD Conference, pp. 179–192 (2009)
5. Brantner, M., Florescu, D., Graf, D., Kossmann, D., Kraska, T.: Building a database on S3. In: SIGMOD, pp. 251–264 (2008)
6. Chang, F., Dean, J., Ghemawat, S., Hsieh, W.C., Wallach, D.A., Burrows, M., Chandra, T., Fikes, A., Gruber, R.E.: Bigtable: a distributed storage system for structured data. In: OSDI, pp. 205–218 (2006)
7. Cohen, J., Dolan, B., Dunlap, M., Hellerstein, J.M., Welton, C.: Mad skills: new analysis practices for big data. PVLDB **2**(2), 1481–1492 (2009)
8. Cooper, B.F., Ramakrishnan, R., Srivastava, U., Silberstein, A., Bohannon, P., Jacobsen, H.-A., Puz, N., Weaver, D., Yerneni, R.: PNUTS: Yahoo!'s hosted data serving platform. Proc. VLDB Endow. **1**(2), 1277–1288 (2008)
9. Das, S., Agarwal, S., Agrawal, D., El Abbadi, A.: ElasTraS: an elastic, scalable, and self managing transactional database for the cloud. Technical report 2010-04, CS, UCSB (2010)
10. Das, S., Agrawal, D., El Abbadi, A.: ElasTraS: an elastic transactional data store in the cloud. In: USENIX HotCloud (2009)
11. Das, S., Agrawal, D., El Abbadi, A.: G-Store: a scalable data store for transactional multi key access in the cloud. In: ACM SOCC (2010)
12. Das, S., Nishimura, S., Agrawal, D., El Abbadi, A.: Live database migration for elasticity in a multitenant database for cloud platforms. Technical report 2010-09, CS, UCSB (2010)
13. Dean, J., Ghemawat, S.: MapReduce: simplified data processing on large clusters. In: OSDI, pp. 137–150 (2004)
14. Jain, V.K., Kumar, S.: Big data analytic using cloud computing. In: Second International Conference on Advances in Computing and Communication Engineering (2015)
15. Mao, L.: Big data equilibrium scheduling strategy in cloud computing environment. In: International Conference on Virtual Reality and Intelligent Systems (ICVRIS), pp. 94–98 (2018)
16. Suwansrikham, P., She, K.: Asymmetric secure storage scheme for big data on multiple cloud providers. In: IEEE International Conference on Intelligent Data and Security (IDS), pp. 121–125 (2018)
17. Gupta, A., Mehrotra, A., Khan, P.M.: Challenges of cloud computing & big data analytics. In: 2nd International Conference on Computing for Sustainable Global Development (INDIACom), pp. 1112–1115 (2015)
18. Manekar, A.K., Pradeepini, G.: Cloud based big data analytics a review. In: International Conference on Computational Intelligence and Communication Networks (CICN), pp. 785–788 (2015)

Rendering Untampered E-Votes Using Blockchain Technology

M. Malathi(✉), S. Pavithra, S. Preakshanashree, S. Praveen Kumar, and N. Tamilarashan

Department of Information Technology, Sri Krishna College of Technology, Coimbatore, India
{m.malathi,17tuit101,17tuit110,17tuit108,17tuit145}@skct.edu.in

Abstract. The Electronic voting is when a voter casts a ballot through a digital system rather on paper. In the present voting system, the Electronic Voting Machine comprises of two units, control unit and balloting unit. The control unit is with a surveying officer and the balloting Unit is set inside the voting compartment. When the voter makes the choice on the balloting unit by pressing the blue key, the Surveying Officer will press the Close Button. Be that as it may, in the present voting system there is significant probability of vote fixing. Votes can be altered effectively. This can be resolved by blockchain. Blockchain refers to a span of general-purpose technologies to exchange information and transact digital boon in distributed networks. This paper describes using the blockchain technology in an electorate system for Indian election. A threat in the current election system is the characteristics of security and transparency. In current system some problems occur with an administration that has complete authorization over the system and its database, it is affirmable to tamper the database with hefty opportunities. The planned system is basically intended for our nation based on biometric validation. Blockchain technology appears when they make their choice. This is incorporated inside Electronic Voting Machine. The subtleties are put away in independent block. On the off chance that there is a discretionary extortion it very well may be effectively recognized utilizing the blocks of data. This guarantees greater security, unique finger mark of voter is utilized as the fundamental validation asset. To decrease the swindling wellsprings of database control, blockchain can be received in the dispersion of database. This examination considers the voting consequence in blockchain algorithm from each place of election and recorded. A country with less voting rate will fight to develop. It is a potentiality method to the lack of involvement in voting amongst the young media-savvy population. This dispense an incorruptible election amid the democratic people. Blockchain technology is distributed and verified publicly such that degenerate is beyond the realm of imagination.

Keywords: Biometric · Electronic Voting Machine · Fingerprint verification · Blockchain technology

D. J. Hemanth et al. (Eds.): COMET 2019, LNDECT 35, pp. 703–711, 2020.
https://doi.org/10.1007/978-3-030-32150-5_69

1 Introduction

The recent technologies has optimistic whack on multiple exposure of our social interactions. The 21st century has perceived the advent of a number of innovative technologies. Blockchain holds promise for being the latest innovative technology [1]. Blockchain technology is seen as a standout amongst the most vital technology vogues that will impact society and business in the days to come. Blockchain Technology developed as a conceivably unruly, proficient technology for organizations and governments to help data trade and exchanges that necessitate verification and trust [2]. Distributed consensus are not famous in light of its restricted adaptability, and was seen as a fleeting connection crude that is misused just in applications in undaunted need of consistence and just among couple of nodes. The utility of decentralized agreement transversely over countless nodes was illustrated by Nakamoto's Bitcoin cryptocurrency, changing the universe of computerized transactions for eternity [3]. Many contrast the rise of the blockchain with another progressive technology and foreshadow that this technology will change the extent of effectiveness withdrew from centralized control in fields, for example, business, correspondences, and even legislative issues or law. This blockchain technology has the prospective to enchiridion in a new period of time designate by overall system of payment [4]. Blockchain innovation is fortified by a decentralized system involving of huge number of interconnected nodes. These hubs have their private copy of the distributed ledger that incorporates the all olden times of exchanges Processed in the system [5]. Expanded innovation use got contemporary difficulties the movement of vote based system as more individuals these days not have faith in governance, making elections most basic in existing majority rule government. Elections have an incredible power in deciding the predetermination of a country [6]. The procedure chop down the whole man perception in the polling booth and furthermore in the counting activity. All the human exercises are computerized. When finished the approval, the succeeding inconvenience uprise is transpicuous of each vote, Security and issues in information control. These issues are explained by Blockchain innovation and furthermore used to diminish the issues that happen in voting. It involves a couple of impedes that are associated with each other and in compilation The block making endeavor to change the data will be increasingly troublesome as it needs to change the following blocks. Subsequently blockchain innovation plan to unravel the essential difficulties in the genuine arrangement of the voting process such as security, endorsement of voters, shielding voted data [7].

2 Related Works

With the provision of blockchain technology, secure voting and trustworthy voting environment is probable. Protocol generated in which the voters mastery as a network of peers and provide decentralization. Using the public ledger in blockchain technology, each and every sole organizational pronouncement can be done by individuals and all the happenings of the election can track. People's viewpoint will be publicized [8]. Switzerland is take part in electronic voting, every citizen can take part an active role in the elections and remote voting system is feasible [9]. E-voting is remarkably hoarier than blockchain technology. Estonia government is the foremost to practice an extensive e-voting system. In 2001, the idea of vote was in advancement and formally begun in the mid year of 2003 by across the country specialists [10]. The prime effort is to integrate the online elections with the Ethereum blockchain platform. Once the election gets over, Ethereum smart contracts checks and counts the number of votes. Own tokens are used by Agora in the blockchain for elections; these tokens for individual qualified voter are procured [11]. Cotena was made to use the information security of the Bitcoin blockchain while presenting a plan that has negligible information stockpiling prerequisites and diminished Bitcoin exchange costs. On walk seventh in 2018, Agora's casting a ballot framework was somewhat utilized for the presidential decision in Sierra Leone [12]. In this election delegates of Agora went to survey stations and physically enlisted the vote information of paper votes onto their blockchain to complete a private count of the votes. Enigma network connects to the blockchain and retrieves private and computationally intensive data from the blockchain and stores these records off-chain [13].

3 Proposed System

This design is designed for EVM (Electronic Voting Machine) utilizing the unique fingerprint identification method. Here the voter's thumb impression that is fingerprint is used for identifying the voters. Each and every voter has an individual and unique fingerprint. Initially the fingerprint is captured for every voter using the voting machine. Then the fingerprint is scanned and is sent for verification. It verifies with the already existing database records. The database records are implanted securely with the assistance of blockchain. Previously, all the voter's unique identification details like name, age, address, date of birth, fingerprint, and iris scan are embedded in the blockchain network. After the individual votes, those details will be stored in separate block to count the votes. In the event that a similar individual endeavoring to vote again implies, the examined unique fingerprint mark attempts to coordinate with the current block where vote tallies spared. If the fingerprint matched, that is already counted as vote then the vote is not counted and this process is repeated for the remaining other voters. Also, if the unique mark isn't coordinated then it makes the most of another vote (Fig. 1).

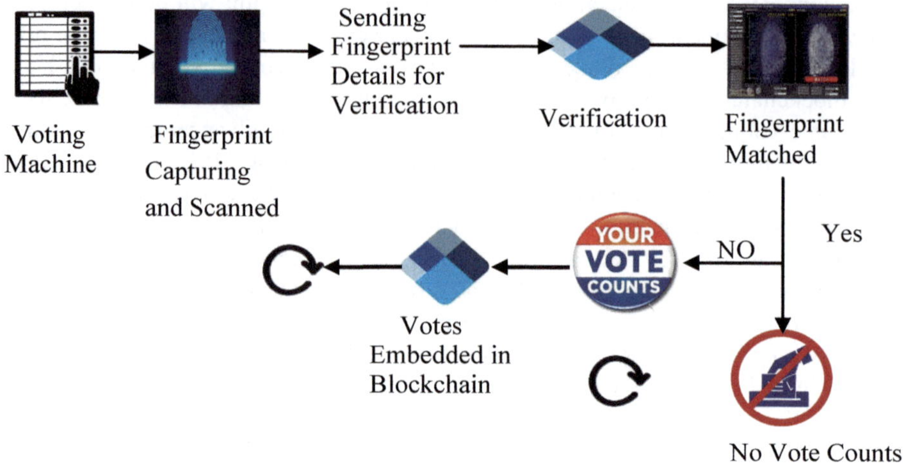

Fig. 1. E-voting system using blockchain technology

4 Fingerprint Verification and Update in Block

The procedure of confirmation commences from the accession of a block conveying the voting result the past hash of the hash esteem rising up out of the beforehand substantial block, and the digital signature. These are isolated between electronic archives that is aftereffect of voting and past hash and digital signature. The electronic report figures its hash esteem. As for the digital signature, the electronic document is made by decryption method using the public key of the node. Contrast of these two hash functions are completed, if the value is indistinguishable, the digital signature is authentic and the procedure is proceeded, however in the event that the esteem isn't indistinguishable it is viewed as ill-conceived and the system will dismiss the block to seek after the procedure. The digital signature licensed and ended up being authentic, further affirmation of the previous hash starts with the catch of the casting a ballot result, and the former hash held in the latest in database, and investigated hash esteems with the SHA-256 calculation. Then contrasting it with the former hash conveyed by the block being done affirmation. If the value is indistinguishable, the hash esteem is real and the entire block is checked as a genuine block and sent by the node held in the system, however in the event that the esteem isn't indistinguishable, it is viewed as ill-conceived and the framework will deny the block. The affirmation procedure has turned out to be real, so the following procedure is to refresh the database by adding the current data on the block. Using the Blockchain system, Bitcoin System is referred, the ECDSA (Elliptic Curve Digital Signature Algorithm) algorithm is utilized in advanced mark procedures the little key size in this technique underpins the required security. As it were, the key size of not exactly or in excess of 160 bits in the ECDSA algorithm is reporter to security utilizing RSA algorithm with a key of 1024 bits, the execution on the mark utilizing any ECDSA algorithm segment and its security level is constantly fast than the RSA calculation [14]. The ECDSA (Elliptic Curve Digital Signature

Algorithm) algorithm is the customary elliptic curve-based digital signature scheme. The algorithm, which is the analogue elliptic curve of the Digital Signature Algorithm (DSA) was initiated by Scott Vanstone in 1992 [15]. The primary trump card of ECDSA is indistinguishable dimension of security from DSA however with a little key length; taking into consideration of fast computation algorithm is a development of summed up advanced digital signature utilizing ECC algorithm and its approval [16].

5 Get a Turn

The season of the casting a ballot will start and closure simultaneously. At the point when casting a vote time has been finished, every node will remain by for its swing to create a block. The system will over and over broadcast the database pursued by the ID of a predetermined node. The node ID taken as a token, if that a node perceives that the broadcast ID has a place with it. It is the node's chance to create another block. Be that as it may to produce another block it is obligatory to clarify that the sender of the block is an authentic sender and part of the decision, at that point the affirmation procedure is finished. On the off chance that the affirmation was triumphant, the node begins creating another block, which will at that point be broadcast to all nodes in the system. In a constrain where the node that gets the turn is troublesome either down in the system or so the network, it will not stop. In every node it has its very own counter time as indicated by the time allotment, the block is added with the televise time at that point increased by the request of the nodes securing the turn. Node which get counter time = 0, at that point it very well may be deciphered that opportunity to make new block in spite of the way that does not get node ID as token in light of the way that there is node or some number of going before center points encounters issues. After the goal node recognizes that its turn has drawn closer, it is confirmed to anchor that the recently gotten block is from the substantial node in the network. Utilizing get a turn strategy can decrease impacts that can happen in an information transmission arrange. This strategy can likewise facilitate the vital review process after the casting a vote procedure happens [17].

6 Design and Televise a New Block

The Nodes gather votes from every chooser, at that point, determined and converged with the former hash as an electronic report in the system. The electronic archive is taken care of with a hash function to make a hash code. It encodes the hash esteem utilizing the reserved key ECC. The recommended block alludes to the examination alluded to [17] containing an id node, a timestamp, and three approval segments likewise in this investigation and furthermore an id node of the node that picked up a next turn. The approval area comprises of the after effects of the general race in the node, joined by the hash of the past block in the database, in conclusion included with a digital signature that is the node utilizes the private key to encode the hash code of the block, which at that point televise to the total node. Broadcast the block to all nodes, after the nodes that get the turn wrapped up another block. The hash work is one of the cryptographic techniques in processing the exceptional regard that can be contrasted with the one of a kind unique fingerprint impression that is thumb impression of an information. SHA and its portrayal subtleties can be examined in NIST standard records. The hash work utilized in the scrutiny is SHA-256, has been passed on by U.S. Government Applications and is unequivocally put sent to utilize on the grounds that it has been set under the law, with its calculation and it has been affirmed safe incorporating with cryptographic calculations and different harmonies that serve to guarantee reports comprising of data [18].

7 Simulation Consequences

Simulation is finished by utilizing Python programming utilizing PyCharms Community programming which is tried utilizing modest number of nodes for doing utilizing perception, and substantial scale without utilizing representation regarding the quantity of election places in India. Information stockpiling plans of e-voting systems assume a huge job in real world usage, since how to consider sparing election information is critical to anchoring the security and unification of the information. In the utilitarian testing of the proposed strategy, it is probably going to execute this technique for e-casting a ballot records framework on the grounds that the required stockpiling is adequate for present-day PC limit with the outcomes shown in the graph of Fig. 2. Unwavering quality testing is practiced with the required limit parameters on each number of nodes. With the quantity of nodes tried running from 1 to 500,000 numerous nodes restricting the quantity of nodes is the quantity of spots of election then the subsequent informations in Fig. 2. Progressively number of nodes is matching to the limit required in the action of account this e-voting. It is seen in Fig. 3 that a more noteworthy number of nodes required, it sets aside longer time for this e-voting record system to work. In the database stored information block of all nodes that each block

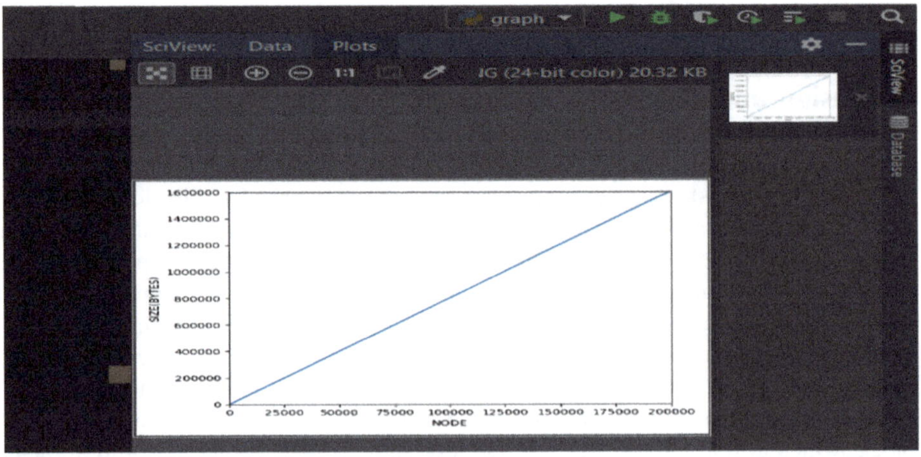

Fig. 2. Probable record storage

comprising of the Node ID, Next ID Node, List of Votes, Preceding Hash, Digital Signature, and timestamp. In this recreation, if the node is down on the system or whatever other inconvenience that causes the node cannot televise the block and after that, the node is not abled. The system has succeeding with the progression to the accompanying node in light of the route that there is counter time for each node which when the time has passed counter, then the node comprehends that its turn has arrived "My Turn = TRUE".

In the usage, it very well may be completed two things in the event that it has finished chronicle all nodes for nodes that have not been deformed in light of that scatter. To start with the node that is experiencing the unsettling influence physically enhancing by essentially pressing the broadcast direction. Since it is hard to perceive, when the node has finished hinder or the system can be rehashed to do the recording and just perceives nodes whose databases are as unfilled yet in Blockchain finished with the keep going block parameter stored on system. Since it cannot embed nodes in a current blockchain. In verification, two variables that utilized are preceding hash and digital signature examined.

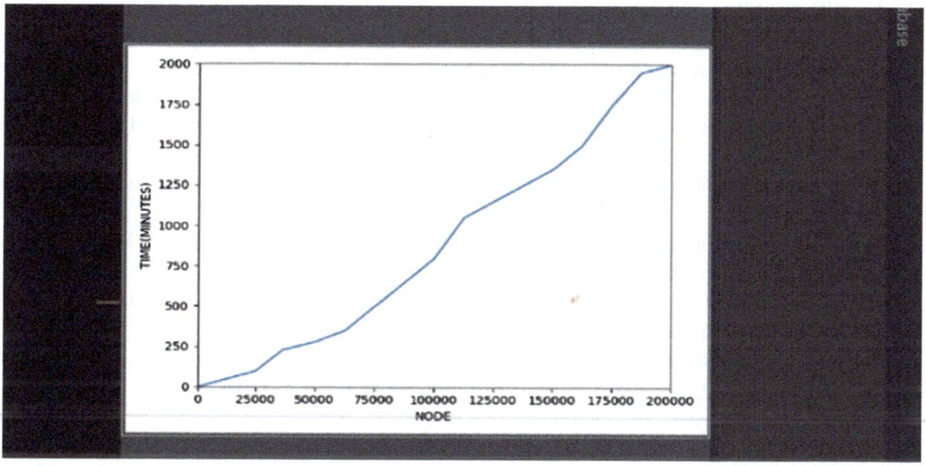

Fig. 3. Time requisite in the quantity of various nodes

8 Conclusion

Blockchain technology can be one answer for take care of the issues that frequently happen in the voting system. The use of hash regards in chronicle the voting results of each studying station associated with each other makes this recording framework dynamically secure and the utilization of digital signatures, for example, their unique fingerprint makes the system increasingly reliable. The plan of electronic voting machine with the unique fingerprint ID technique is utilized for distinguishing the voters. Each and every voter has an individual and exceptional one of a kind unique fingerprint mark. The blockchain consent protocol utilized is a distributed record-keeping system worked by known elements, in other words having the way to distinguish nodes that can control and elements, refresh data together in achieving the individuals trust objectives. Any information that is televise by the node that gets a turn is constantly confirmed and refreshed its information by the beneficiary. The confirmation procedure begins by recognizing whether there are past hashes and/or public keys that are not enrolled in the database. Each hash value in the past block has been incorporated into the count which gets a turn on the system rolling out the improvements in the database will confront trouble as though one information is transformed it should make changes to information on different blocks. In future, dissimilar to voting machines in a few nations which are associated with a system, Indian voting machines are independent. Altering a machine through the equipment port or through a Wi-Fi association is beyond the realm of imagination as there is no recurrence beneficiary or remote decoder in the voting machine can be implemented.

References

1. Peters, G.W., Panayi, E.: Understanding modern banking ledgers through blockchain technologies: future of transaction processing and smart contracts on the internet of money. In: WC1E 6BT, London, UK, 19 November 2015
2. Olnes, S., Ubacht, J., Janssen, M.: Blockchain in government: benefits and implications of distributed ledger technology for information sharing, October 2017
3. Dwork, C., Naor, M.: Pricing via processing or combatting junk mail. In: 12th Annual International Cryptology Conference on Advances in Cryptology - CRYPTO 1992, pp. 139–147 (1992)
4. Wright, A., De Filippi, P.: Decentralized blockchain technology the rise of lexcryptographia, 12 March 2015
5. Hardwick, F.S., Gioulis, A., Akram, R.N., Markantonakis, K.: E-Voting with blockchain: an e-voting protocol with decentralisation and voter privacy. arXiv:1805.10258v2 3 July 2018
6. Barnes, A., Brake, C., Perry, T.: Digital Voting with the use of Blockchain technology. https://www.economist.com/sites/default/files/plymouth.pdf
7. Navya, A., Sai Niranjan, A.S., Roopini, R., Prabhu, B.: Electronic voting machine based on Blockchain technology and Aadhar verification. IJARIIT **4**(2) (2018)
8. Çabuk, U.C., Çavdar, A., Demir, E.: E-Demokrasi: Yeni Nesil Doğrudan, Demokrasive Türkiye'deki ygulanabilirliği. https://www.researchgate.net/profile/Umut_Cabuk/publication/308796230_E-Democracy_The_Next_Generation_Direct_Democracy_and_Applicability_in_Turkey/links/5818a6d408aee7cdc685b40b/E-Democracy-The-Next-Generation-DirectDemocracy-and-Applicability-in-Turkey.pdf
9. Koç, A.K., Yavuz, E., Çabuk, U.C., Dalkılıç, G.: Towards secure e-voting using Ethereum blockchain. https://www.researchgate.net/publication/323318041
10. Hao, F., Ryan, P.Y.A.: Real-World Electronic Voting: Design, Analysis and Deployment, pp. 143–170. CRC Press, Boca Raton (2017)
11. Tomescu, A., Devadas, S.: Catena: efficient non-equivocation via Bitcoin (2017). https://people.csail.mit.edu/alinush/papers/catena-p2017.pdf
12. del Castillo, M.: Sierra Leone secretly holds first blockchain-audited presidential vote (2018). https://www.coindesk.com/sierra-leone-secretly-holds-first-blockchain-powered-presidential-vote/
13. Zyskind, G., Nathan, O., Pentland, A.: Enigma: decentralized computation platform with guaranteed privacy. arXiv preprint. arXiv:1506.03471 (2015)

14. Gemalto: Benefits of elliptic curve cryptography, March 2012
15. Malvik, A.G., Witzoee, B.: Elliptic curve digital signature algorithm and its applications in Bitcoin, pp. 1–5 (2016)
16. Wang, D.I.: Secure implementation of ECDSA signatures in Bitcoin (2014)
17. Kirby, K., Masi, A., Maymi, F.: Votebook: a proposal for a blockchain based electronic voting system (2016)
18. NIST, F.P.: FIPS 180-2 secure has standard, vol. 1 (2002.)

Analysis of the Risk Factors of Heart Disease Using Step-Wise Regression with Statistical Evaluation

S. K. Harsheni[1], S. Souganthika[1], K. Gokul Karthik[1], A. Sheik Abdullah[1(✉)], and S. Selvakumar[2]

[1] Department of Information Technology, Thiagarajar College of Engineering, Madurai, India
harshenisk2001@gmail.com,
souganthikasankareswaran@gmail.com,
gokulkarthikk@gmail.com, aa.sheikabdullah@gmail.com
[2] Department of Computer Science and Engineering, G.K.M. College of Engineering and Technology, Chennai, India
sselvakumar@yahoo.com

Abstract. The objective of the work aims to formulate a regression model to predict the occurrence of the heart disease using minimum number of parameters. The problem of heart disease is chosen owing to the increasing risk of heart disease in India. The Data is collected from a hospital (Cleveland Dataset) which consists of 22 attributes and a class label as retrieved from UCI Repository. Technique of Step wise regression is used to formulate this model and this enables us to identify the model of the highest accuracy of about 89.72% that contains the attributes which have the highest effect on the class label (outcome variable). The technique of supervised learning is used to train and obtain the mathematical model. The observed result is an expression function value of the selected attributes that corresponds to the determination of heart disease with fewer set of attribute with its parametric values. Thereby, the number of tests that corresponds to the disease can be reduced which then reduces the expenses towards the disease and its co-morbidities.

Keywords: Heart disease · Predictive analysis · Step-wise regression · Statistical analysis · Biomedical informatics

1 Introduction

The technological innovations and incorporations is growing every year there is a need for analyzing and determining the interesting patterns that lie behind the data. Since the size of the data is expanding to a large scale, the rate of predictive performance and analysis need to be expanded to a wide scope [1]. With this, data mining and its predictive algorithms can play a significant role in determining the useful patterns that lie behind the data concerned.

In the field of medicine, the sector is generating huge volumes of data in various forms. Some of them include daily reports, images; hand written text (unstructured)

D. J. Hemanth et al. (Eds.): COMET 2019, LNDECT 35, pp. 712–718, 2020.
https://doi.org/10.1007/978-3-030-32150-5_70

suggested by medical experts, machine generated videos and so on. The target lies at the stage of complete interpretation and evaluation of the medical data generated to the extreme level of understanding and incorporation. The risk factors that contribute towards the disease seem to be an important factor in evaluating the significance of the disease [2].

Identifying the patterns and its relationship has been found to be an interesting phenomenon with regard to disease prediction. The techniques in data mining plays a significant role for evaluation the risk related to specific diseases.

Predictive analysis tasks perform inference on the current data in order to make predictions. The target is to determine the statistical data analysis among the data models that has been developed with regard to performance analysis in terms of accuracy. The mechanism of determining the risk factors at earlier stages of disease will make the physicians to have a good exploration with regard to treatment analysis. This paper provides a mathematical model to determine the risk factors that highly contribute towards heart disease upon statistical evaluation.

2 Literature Review

The impact of heart disease and its prevalence is increasing day to day with the significance related to different sort of risk factor and its corresponding correlation. In India, the rate of disease specification gets increased to a rate of 30 million with the significance rate of 3% and higher [3]. The risk related to the disease significance is also gets increased with a motto to increase in behavioral factors that contribute to the disease [4]. Also it has been determined for the World Health Organization (WHO) that the increase in heart disease will rise to 118 million over the years of 2025 [5].

Heart failure is one of the serious threats that threaten a majority of lives in India and diagnosis of the heart disease usually involves multiple tests. The result of the following work would enable the doctor to analyze the disease status with the data attributes observed.

Bioch et al. [6] has implemented classification methods like neural network and Bayesian approach on this dataset and obtained an accuracy of 75.4%, 79.5% respectively. Also, the significant performance of various classification schemes also found to be the range between 59–77% for the dataset concerned. Patil et al. [7] has implemented association rule for the classification of diabetes on this dataset and extracted rules which required further improvement in the generalization of those rules by considering the factors that influence diabetes. Han et al. [8] has built a prediction model on this dataset using the Rapid miner tool. From the outcome of the model it has been found that glucose is found to be one of the predominant factor that contribute to the disease with 72% accuracy level.

3 Proposed Methodology

Regression and correlation analyses are arguably one of the most commonly used and most important statistical tools. Regression can be defined as the process of fitting a function of attributes to the data so that the prediction of the dependent variable is made possible. This is used to explore relationship between the influencing variables and the outcome variable. The model developed from regression is used for explicit prediction of the outcome variable (severity of the heart disease in this case). The regression equation can be of any type; linear, quadratic or logarithmic or exponential, depending upon the nature and influence of the data set on the outcome variable. There are many types of regression such as multiple regressions, simple linear regression, step-wise regression, and generalized models.

Step-wise regression is chosen in this work owing to the higher accuracy. This accuracy is attributed to the iterative development of multiple mathematical models of all possible combination of attributes and selection of the mathematical model with the highest accuracy and limited number of parameters. This enables the process of the dominant variable selection and regression to be done simultaneously. Step-wise regression can be again linear, quadratic, exponential or logarithmic depending upon the nature of the attributes on the outcome variable. Here linear regression is chosen in order to maintain simplicity of the problem and enable minimal number of attributes in the mathematical model.

One of the major problems in the health care industry is the non-certainty of prediction of presence or absence of a disease. More number of medical tests is taken to evaluate and confirm the presence of a disease. Hence this increases the net cost incurred by the patient (client side). It is possible to predict the presence of a disease by taking minimal or optimal number of tests if the dominant influencing parameters alone were to be identified. The current work proposes a structured method to predict this with minimal number of tests. This decreases the total cost incurred to the customer and also reduces the time required to diagnose the presence of the disease.

The heart disease problem is chosen owing to the increasing heart disease risk rate in India. Despite the fact that the data being non-native, the influencing parameters for a disease remain universally the same. Hence an analysis that predicts the risk level of

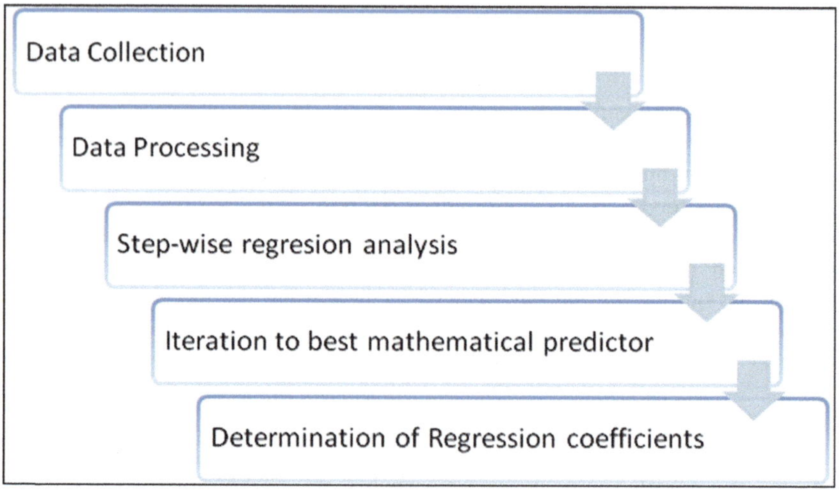

Fig. 1. Proposed methodological workflow

patient from least number of tests will be of great application to the hospital and health care industry owing to the huge amount of money and time involved in disease diagnosis. The proposed methodology is given in Fig. 1.

3.1 Rationale Behind the Proposed Work

The work by the authors [9] provided a significant importance towards the importance for the use of step-wise regression analysis. From the literature point of view the pitfalls has been well analyzed and pointed out with statistical evaluation. Till 2004, it has been analyzed that 57% of research findings suggested the practice of step-wise regression analysis.

Various evolutionary approaches have been used to determine the features that correspond to heart disease prediction. The work by the authors [10] provided a significant feature selection approach with the usage of fuzzy feed forward network the result has been found to be deterministic with the attributes that related to the disease and its syndromes.

Meanwhile, different set of data mining algorithms such as SVM, Random forest, logistic regression, and neural network has been used by the authors [11] for the determination of risk related to heart disease. The accuracy of the model was found to be 87.4% with the risk that pertains to the disease. Meanwhile, this paper incorporates the usage of step-wise regression analysis for the determination of risk that corresponds to the disease with improved accuracy level.

3.2 Metrics for Evaluation

The performance of the model is validated by the P value and in terms of accuracy. The P value against each of the test outcome is validated with its probable outcome value that has been found to be zero. Hence the lower value of P (<0.05) which is termed to be the null hypothesis which is rejected. The term then correlates to a good significance if the related value makes significance in the response variable.

Also, larger change in the predictor value which is not related to the response term. The mathematical model that suggests lower value of P and minimum number of predicted attributes with a good correlation provides the best fit in determination of risk prevalence and its relationship.

The second is accuracy; the value of accuracy of the model is computed with the actuals to that of the predicted attribute. The total number of positive cases that has been found to be actually positive and the total number of negative cases that has been found to be actually negative. The following Eq. 1 provides the calculation of accuracy value.

$$Classification\ accuracy = \frac{TP + TN}{TP + FP + FN + TN} \tag{1}$$

The error rate of the model can be evaluated using the following Eq. 2 as:

$$Classification\ error = \frac{FP + FN}{TP + FP + FN + TN} \tag{2}$$

4 Experimental Results and Discussion

The evaluation starts with the determination of partial correlation coefficient value with the target of evaluating the dependent variable Y against all the possible correlated of X. during the stages of model development the variable that has the highest correlation value is added to the model. Once this is evaluated for the variables then the F-statistic value is determined. From this, the variables corresponding to the model is validated for the addition or removal of variables based on the values generated. This gets iterated until no further variables can be added or removed from the developed model. Then the variables at each stage get fixed for the correlation coefficient value that has been observed with its proper response variable.

$$(Partial\ Correlation\ X_i)^2 = \%\ of\ variance\ in\ Y\ explained\ by\ X_i \tag{3}$$

From the Eq. 3 it has been clear that the number of attributes that the correlation coefficient corresponding to each of the independent variable to that of dependent variable is computed. If the highest value of resemblance is found with the variable X_m then the decision is made to include X_m.

The partial correlation coefficients of Y with all other independent variables are computed given X_m in the equation. If the highest partial correlation is with the variable X_p. The attributes added to the mathematical model are removed or retained depending upon the F value of the mathematical model. The process is iterated till the F value of the mathematical model stabilizes and no more attributes can be added or removed from the mathematical model [13].

The step-wise regression was done for the collected data and the regression coefficients are tabulated in following Table 1. The attributes are named as X_1, X_2 … X_{24} corresponding to the attribute specification as above.

Table 1. Coefficients observed

S. No.	Attribute	Regression coefficient
1	X_3	0.0237
2	X_6	−0.0030
3	X_8	0.2823
4	X_{15}	0.6940
5	X_{20}	0.2889
6	X_{23}	0.3535
7	X_{24}	1.2478
8	Intercept	0.7226

The mathematical model proves to be accurate owing to the p-value of the model. The details of the mathematical model are illustrated in Table 2 as follows:

Table 2. Mathematical values observed

Property	Value
R-squared	0.4541
F	35.4073
RMSE	0.727
P	8.123 e−36

The low value of P value indicates that the model is an efficient predictor of the class label (outcome variable). Among the 23 variables that have been collected the resulting regression Eq. 4 provides the selected attributes with its corresponding values is as follows:

$$\begin{aligned} \text{Heart risk Stage} = \text{Rounded value}\{&0.7226 + (0.0237) * \text{Cp\_type} \\ &+ (-0.0030) * \text{School} + (0.2823) * \text{Restecg} + (0.6940) * \text{Prior\_Stroke} \\ &+ (0.2889) * \text{SAnterio} + (0.3535) * \text{Fhistory} + (1.2478) * \text{Peffusion}\} \end{aligned} \tag{4}$$

With the developed model the attributes identified are chest pain type, serum cholesterol, resting electro-cardio-graphic results, prior stroke, septoanterio, and family history of heart disease [12]. The model developed provided an improved accuracy of about 89.72% with a good correlation among the attributes selected.

5 Conclusion and Future Work

The work uses step-wise linear regression to determine the predictor for the Heart disease risk stage. The step-wise regression analysis provided an accuracy of 89.72% when associated to the existing approaches. The scope extends to the usage of other types of regression models such as logarithmic, exponential or quadratic regressions to develop predictors. Despite their complexity, if the regression co efficient is evaluated, it might yield better predictors.

The future work can be progressed towards different real world dataset corresponding to various diseases like cancer, diabetes and other sorts of non-communicable diseases.

References

1. Han, J., Kamber, M.: Data Mining Concepts and Techniques. Elsevier, India (2006)
2. Global data on visual impairments 2010. World Health Organization, Geneva (2012)
3. Mendez, G.F., Cowie, M.R.: The epidemiological features of heart failure in developing countries: a review of the literature. Int. J. Cardiol. **80**, 213–219 (2001)

4. Global burden of disease—2004 update: World Health Organization, Geneva (2004)
5. Huffman, M.D., Prabhakaran, D.: Heart failure: epidemiology and prevention in India. Nat. Med. J. India **23**(5), 283 (2010)
6. Bioch, J.C., Meer, O., Potharst, R.: Classification using Bayesian neural nets. In: International Conference on Neural Networks, pp. 1488–1493 (1996)
7. Patil, B.M., Joshi, R.C., Durga, T.: Hybrid prediction model for type-2 diabetic patients. Expert Syst. Appl. Sci. Direct **37**(12), 8102–8108 (2010)
8. Han, J., Rodriguez, J.C., Beheshti, M.: Diabetes data analysis and prediction model discovery using rapid miner. In: IEEE 2nd International Conference on Future Generation Communication and Networking, pp. 96–99 (2008)
9. Whittingham, M.J., Stephens, P.A., Bradbury, R.B., Freckleton, R.P.: Why do we still use step-wise modelling in ecology and behaviour? J. Anim. Ecol. **75**, 1182–1189 (2006)
10. Vivekanandan, T., Sriman Narayana Iyengar, N.Ch.: Optimal feature selection using a modified differential evolution algorithm and its effectiveness for prediction of heart disease. Comput. Biol. Med. **90**(1), 125–136 (2017). https://doi.org/10.1016/j.compbiomed.2017.09.011
11. Amin, M.S., Chiam, Y.K., Varathan, K.D.: Identification of significant features and data mining techniques in predicting heart disease. Telematics Inform. **36**, 82–93 (2019)
12. Sheik Abdullah, A., Selvakumar, S., Parkavi, R., Suganya, R., Venkatesh, M.: An introduction to survival analytics, types, and its applications. Biomechanics. Intech Open Publishers, UK (2019). http://dx.doi.org/10.5772/intechopen.80953
13. Sheik Abdullah, A., Selvakumar, S., Karthikeyan, P., Venkatesh, M.: Comparing the efficacy of decision tree and its variants using medical data. Indian J. Sci. Technol. **10**(18), 1–8 (2017). https://doi.org/10.17485/ijst/2017/v10i18/111768

Review on Water Quality Monitoring Systems for Aquaculture

Rasheed Abdul Haq Kozhiparamban(✉)
and Harigovindan Vettath Pathayapurayil

Department of Electronics and Communication Engineering,
National Institute of Technology Puducherry, Karaikal 609609, India
rasheedabdulhaq@gmail.com, hari@nitpy.ac.in

Abstract. Aquaculture plays an important role in providing food security to the world and increasing steadily as one of the most sustainable methods of food production. The monitoring of water quality has great significance in aquaculture. Monitoring the water quality parameters such as Temperature, Dissolved Oxygen, Salinity, pH etc. enables us understanding the farm in-depth & helps to optimize the use of resources, improve sustainability, profitability and most importantly to reduce the impact of aquaculture on the environment. Water quality determines the fish behavior and health of the farm as well. The objective of this paper is to review current research works and studies on water quality monitoring systems and various estimation techniques, in order to understand different challenges faced in water quality monitoring for aquaculture. The study starts with the evolution of water quality monitoring, importance of wireless sensor networks and gives overall perspective on presentwater quality monitoring systems [1].

Keywords: Aquaculture · Water quality monitoring · Wireless sensor networks

1 Introduction

1.1 Motivation

The global population has increased from three to six billion in last five decades increasing the demand for food. As per the estimates, there will be 70% increase in requirement for food as world population gain another 30% within 2050. Aquaculture is an integral component in the search for global food security. When it comes to aquaculture production, China is truly the leader with an annual harvest of 58.8 million metric tons (MMT) followed by Indonesia with 14.4 MMT and India with 4.9 MMT [2].

To solve the challenges faced in aquaculture, we need to understand complex ecosystems in detail. This can be done by monitoring the environment regularly, creating large quantities of data for analysis which will help farmers and farms to extract value from it and improve the overall productivity. WQM plays an important role in aquaculture. The monitoring of fish farming process can optimize the use of

D. J. Hemanth et al. (Eds.): COMET 2019, LNDECT 35, pp. 719–725, 2020.
https://doi.org/10.1007/978-3-030-32150-5_71

resources and improve sustainability and profitability. The quality of water determines the fish feeding behavior and health as well.

IoT is becoming an important tool for the sustainably developing modern aquaculture. It is commonly used in many aspects, such as intelligent feed management, disease detection, water quality monitoring, predicting water parameters and issuing warnings. The automation of aquaculture systems with IoT based system improve environmental control, reduce disastrous losses, production cost improvement, and product quality enhancement. Most significant parameters to be monitored real-time and controlled in aquaculture system include: Temperature, Dissolved Oxygen, Nitrate, Ammonia, Turbidity, Salinity, pH Level, Alkalinity etc.

The water quality has a direct impact on the growth of aquatic animals and product quality. These parameters directly affect fish health, feed utilization and weight gaining rate. Fish undergoes stress and disease breakout when temperature is near maximum tolerance or fluctuates randomly. Less dissolved oxygen is present in warm water than in cool. Similarly, each parameter has its own significance and detail study is must before the development of WQM system for aquaculture [3].

The motivation for this review is to study the design and develop of WQM system for aquaculture and understand various techniques in water quality parameters estimation. The survey discusses the evolution of the WQM systems and highlights the usage of WSN. Also, the Estimation techniques of water quality parameter are discussed giving an overall perspective on present technologies and techniques used in WQM for Aquaculture.

2 Evolution of Water Quality Monitoring Systems

WQM systems evolved from manual lab-based approach (user travel to water source, collect samples, and bring them back to laboratory for checking each parameter) to on site monitoring (carries portable equipment's to site and monitor the parameter). For onsite monitoring specialized instruments and trained technicians for the assessment of each water quality parameter from the water source where required. Finally, on introduction of WSN based solutions where the entire process is done remotely without visiting the site.

Lately these solutions have moved forward and started using the data generated from WQM to predict different water quality parameters. These predictions help in making the system more sustainable and give out early warning for fish disease & other hazards based on WQM.

Both approaches take a lot of time for collection & transportation of the samples from source to the laboratory for analysis as well as monitoring of water quality parameters on site [4]. Then the introduction of sensors developed with fiber optics and laser technology along with bio & optical sensors with microelectronic mechanical systems (MEMS) to detect different water quality parameters on site, whereas computing and telemetry technologies were introduced to support the data acquisition and monitoring processes [5, 6].

2.1 Wireless Sensor Network

The Water Quality Monitoring systems further improved with wireless sensor networks (WSN) which gained attention recently with the introduction of advanced communication techniques. WSNs are very good at capturing data remotely and transmitting them without any fixed network. They also require very less power for communication making the sensor nodes more power efficient and makes WSN a very much desirable technology. WSN have become an integral part of WQM as they provide a solution for many control and monitoring applications. They are simple, low cost networks monitoring remotely, in real-time and with minimal human intervention.

WSN network consists of sensor nodes & base station. A node is normally responsible for monitoring the parameters with the sensing and data transmitting capabilities. Whereas the base station provides connectivity to all nodes and act as gateway allowing the data transfer and nodes to be managed remotely. They use low power consuming standards like IEEE 802.15.4, ZigBee and Bluetooth to relay data. In a WSN, we can use any connectivity solution and the standard selection entirely depends on resource constraints & requirements. Few of the communication technologies are listed below.

ZigBee: Zigbee Technology is based on the IEEE 802.15.4 standard. Energy efficiency, low cost, and reliability are few features that make ZigBee desirable for WSN in aquaculture. Also, the low duty cycle makes it suitable for water quality monitoring, since data updating is done periodically. ZigBee have low data rates 20 to 40 kbps at 868/915 MHz and around 250 kbps at 2.4 GHz frequencies of ISM band [7].

Wi-Fi: Wireless local area network (WLAN) standard used for data exchange and for Internet connectivity based on the IEEE 802.11 standards. Its most common wireless connectivity solution found in many devices ranging from mobile phones, tablets and laptops. Wi-Fi provides communication range of 20 m indoor and outdoor about 100 m. Data transmission rate of 2 to 54 Mbps at 2.4 GHz. Wi-Fi can be used to connect many sensors over a single network.

Bluetooth: Bluetooth is based on the IEEE 802.15.1 standard. Wireless technology used for portable devices in short range data transmission (8 to 10 m). Gives a data maximum bandwidth 1 to 24 Mbps [8].

GSM (GPRS/3G/4G): GPRS is a packet data service for GSM based cellular phones with max data rate of 100 kbps can be achieved for 2G systems making it suitable only for periodic data transmitting. Lately better data rates are achieved in 3G and 4G. 5G is yet to come with better data rates and very low latency and sufficient support for the billions of IoT devices.

2.2 Estimation of Water Quality Parameter

Any water parameter change will affect the productivity drastically hence even real time monitoring will not be enough. If we take the case of dissolved oxygen (DO), once it falls below the threshold amount bringing back to the required level will require a minimum time. This lag will cost us in different ways, it will affect the growth, feed behavior and in some cases may even kill the fish. Aim of WQM is to increase efficiency hence just monitoring won't be enough to manage the farm efficiently & to increase sustainability. We must be able to produce relevant data in advance to take actions on time. The prediction also allows us to understand conditions and helps in making actions in advance allowing us to timely manage system. Here we can check some of the works on dissolved oxygen prediction mechanisms. The water parameters are not linear, dynamic and their prediction is a challenge.

Greedy ensemble selection for regression models is used in [9], where their algorithm searches for best regressors. Regressors are the geometric interpretation for total least squares method. Basically, they use the best mean square method available to predict the value. They used the same for water quality prediction and help us to warn about the water quality. Authors also validated the same predictions with experimenting different parameter and real time values.

A hybrid approach where they use support vector regression (SVR) with genetic algorithm is discussed in [10]. They find the best value for SVR parameters by using a genetic algorithm. An SVR is constructed based on this to predict water quality parameters by using the data collected using water quality monitoring systems. This system performs better than the regression model, normal SVR models and back propagation (BP) neural network models whose performance where evaluated based on the least square and root mean square error. They have conducted many tests to compare with normal SVR & BP but genetic based SVR gave better results.

Crabs are more sensitive to the dissolved oxygen and [11] defines a better mechanism for predicting the DO. The study was started with linear prediction methods like ARMA & ARIMA and proved them to be inadequate for DO prediction. Then moved to artificial neural networks (ANN) models which are more suitable for nonlinear changing parameters. Compared the results of BP neural networks further studied the least squares support vector machine. Particle swarm optimization a global optimization method is tried where the results doesn't ensure to be any better. The study on other factors than water parameters like solar radiation and wind speed is notable. Incorporation of all these data with an intelligent mechanism makes the prediction further accurate. In this work a better forecasting based on radial basis function neural networks (RBFNN) data fusion method and least squares support vector machine (LSSVM) with an improved particle swarm optimization (IPSO) is used. This proves to be a rather effective method.

Another study on DO prediction in 2017 [12] based on fruit fly optimization algorithm suggests a more efficient DO prediction algorithm. The optimal parameter is found with fruit fly helping to improve the efficiency of the least squares support vector regression (LSSVR). After comparing with particle swarm optimization, genetic algorithm and immune genetic algorithm, FOA-LSSVR model have the best performance.

The importance of DO is studied in detailed in [13] and the hazards of not maintaining DO in optimum level also been explained in detail. Here Genetic algorithm is used to optimize the fuzzy neural network (FNN). In FNN normally the rules are decided by people as the inputs are large system becomes very difficult to model. Using GA for determining the parameter values to define fuzzy rules of the system increases prediction accuracy. The paper also includes study of environmental factors effecting DO. The simulation of results and comparison with FNN and BP neural networks shows the efficiency of this system further. This system is best while predicting the Nonlinear DO values and their relations with environmental factors.

In all the regression models we have seen each time they have made improvements to find the parameter values. Different algorithms where tested and their results where compared each time and suggesting a better method. There is still a great scope for improvement in different parameters estimation& in prediction process. The predictions need to be improved much more with better accuracy to ensure the farmers are profited and environment remains un affected due to extensive aquaculture farming.

3 Future Directions

There are many potential areas in aquaculture the current systems include most works on WQM, using IoT and Big Data we can develop more applications helping the cause.

3.1 Factors for Improvement

- **Cost of system:** Reduced cost of system makes it more desirable and usable for small farms with little budget as well.
- **Autonomous operation:** The system must survive for long time autonomously without any maintenance.
- **Low maintenance:** The system must require only least maintenance effort reducing the average cost.
- **Intelligence:** The system must become more intelligent, enabling dynamic solutions for conserving energy. Data predictions and algorithms to prevent lose in any form improving the overall efficiency of system.
- **Energy-efficiency:** For autonomous operation we need good battery life, the WQM system must be very low power consuming by using intelligent algorithms.
- **Usability:** The farmers arenon-technical persons who finally use these applications, hence this needs to be very simple to use.
- **Interoperability:** The system must be able to use different types of sensors, communication technologies etc. enabling to reconfigure the system depending on the requirements of specific farm. Hence allowing the system to be built with available resources.

3.2 Future Adaptations

Water Quality Index: This system will be generally used by untrained farmers, giving them more detailed information on each parameter would make it difficult for them to understand the situation. Rather if we can generate a single value expressing the overall health and condition of the system helping them to take the correct action for certain values will make the system easy to use. For this different weight is given to each parameter.

Big Data: The usage of big data for agriculture is discussed in the technical sense in [14]. It discusses common methods and techniques used in big data analysis along with the sources of acquiring data. The availability of hardware and software, techniques, tools and methods for big data analysis will encourage more users. WQM gives out a lot of data and using Big data we must be able to solve many problems faced in aquaculture. Free availability of analysis tools and open standards have the potential for development towards smarter aquaculture and smarter algorithms to predict the water quality parameters.

Internet of Things: IoT uses the computing concepts and capabilities of smart devices along with interoperable communication technologies. The IoT can play a key role for WQM as the GSM networks starts to roll out 5G and give extensive support for IoT devices. Then we need only to design the smart sensor nodes capable of monitoring the data. The availability of an established Mobile network with support for IoT devices reduces the cost of WQM system. Few IoT-based applications are remote monitoring and automation of feeding, aerator etc. Cost effective management of the produce supply chain for the aquaculture farm [15–17].

4 Conclusions

In this paper we study the present technologies and requirements for water quality monitoring systems for aquaculture and estimation techniques for different parameters. Manual testing can consume time and water quality parameters may alter within that time. Continuous monitoring helps to take pro-active measures before any damage is done. This review helps with making of WSN based WQM system for aquaculture. Features like data prediction on water parameters & feed consumption helps in making the system sustainable. Without knowing the water quality parameters values in advance it's very difficult to manage certain levels of each parameter and if not maintained at correct level it will affect the system in whole. Building an intelligent water quality monitoring system with estimation algorithms and prediction capability for different parameters for fish farm monitoring improves aquaculture farm overall efficiency and reduce the impact of large scale farming on the environment.

References

1. Clark, M., Tilman, D.: Comparative analysis of environmental impacts of agricultural production systems, agricultural input efficiency, and food choice. Environ. Res. Lett. **12**(6), 064016 (2017)
2. Wee, R.Y.: Top 15 countries for aquaculture production. https://wwwworldatlas.com/articles/top-15-countries-for-aquaculture-production.html (2017)
3. Huan, J., Cao, W., Qin, Y.: Prediction of dissolved oxygen in aquaculture based on EEMD and LSSVM optimized by the bayesian evidence framework. Comput. Electron. Agric. **150**, 257–265 (2018)

4. Adu-Manu, K.S., Tapparello, C., Heinzelman, W., Katsriku, F.A., Abdulai, J.-D.: Water quality monitoring using wireless sensor networks: current trends and future research directions. ACM Trans. Sen. Netw. **13**(1), 4:1–4:41 (2017)
5. Bhardwaj, J., Gupta, K.K., Gupta, R.: A review of emerging trends on water quality measurement sensors. In: 2015 International Conference on Technologies for Sustainable Development (ICTSD), pp. 1–6, February 2015
6. Sawaya, K., Olmanson, L., Heinert, N., Brezonik, P., Bauer, M.: Extending satellite remote sensing to local scales: land and water resource monitoring using high-resolution imagery. Remote Sens. Environ. **88**(1–2), 144–156 (2003)
7. Baronti, P., Pillai, P., Chook, V.W., Chessa, S., Gotta, A., Hu, Y.F.: Wireless sensor networks: a survey on the state of the art and the 802.15.4 and ZigBee standards. Comput. Commun. **30**(7), 1655–1695 (2007). wired/Wireless Internet Communications
8. IEEE standard for information technology– local and metropolitan area networks– specific requirements– part 15.1a: wireless medium access control (MAC) and physical layer (PHY) specifications for wireless personal area networks (WPAN). IEEE Std 802.15.1-2005 (Revision of IEEE Std 802.15.1-2002), pp. 1–700, June 2005
9. Partalas, I., Tsoumakas, G., Hatzikos, E.V., Vlahavas, I.: Greedy regression ensemble selection: theory and an application to water quality prediction. Inf. Sci. **178**(20), 3867–3879 (2008)
10. Liu, S., Tai, H., Ding, Q., Li, D., Xu, L., Wei, Y.: A hybrid approach of support vector regression with genetic algorithm optimization for aquaculture water quality prediction. Math. Comput. Model. **58**(3), 458–465 (2013). computer and Computing Technologies in Agriculture 2011 and Computer and Computing Technologies in Agriculture 2012
11. Yu, H., Chen, Y., Hassan, S., Li, D.: Dissolved oxygen content prediction in crab culture using a hybrid intelligent method. Sci. Rep. **6**, 27292 (2016)
12. Zhu, C., Liu, X., Ding, W.: Prediction model of dissolved oxygen based on FOA-LSSVR. In: 2017 36th Chinese Control Conference (CCC), pp. 9819–9823, July 2017
13. Ren, Q., Zhang, L., Wei, Y., Li, D.: A method for predicting dissolved oxygen in aquaculture water in an aquaponics system. Comput. Electron. Agricu. **151,** 384–391 (2018)
14. Kamilaris, A., Kartakoullis, A., Prenafeta-Boldú, F.X.: A review on the practice of big data analysis in agriculture. Comput. Electron. Agric. **143**, 23–37 (2017)
15. Atzori, L., Iera, A., Morabito, G.: The Internet of Things: a survey. Comput. Netw. **54**(15), 2787–2805 (2010)
16. Gubbi, J., Buyya, R., Marusic, S., Palaniswami, M.: Internet of Things (IoT): a vision, architectural elements, and future directions. Future Gener. Comput. Syst. **29**(7), 1645–1660 (2013)
17. Al-Fuqaha, A., Guizani, M., Mohammadi, M., Aledhari, M., Ayyash, M.: Internet of Things: a survey on enabling technologies, protocols, and applications. IEEE Commun. Surv. Tutorials, **17**(4), 2347–2376 (2015)

Scaling Function Based Analysis of Symlet and Coiflet Transform for CT Lung Images

S. Lalitha Kumari[1], R. Pandian[1(✉)], and R. Raja Kumar[2]

[1] Sathyabama Institute of Science and Technology, Chennai, Tamil Nadu, India
rpandianme@rediffmail.com

[2] Mathematics Department, Sathyabama Institute of Science and Technology, Chennai, Tamil Nadu, India

Abstract. The main aim of the work is to develop image compression algorithms with high quality and compression ratio. The objective also includes finding out anbest algorithm for medical image compression techniques. The objective is also alert towards the choice of the developed image compression algorithm, which do not modify the characterization behavior of the image. In this paper, image compression algorithm based on discrete symlet and coiflet-wavelet transform is implemented for decomposing the image. The selection of different levels are discussed dependence on the values of peak signal to noise ratio (PSNR), compression ratio (CR), means square error (MSE) and bits per pixel (BPP). The optimum moments for compression are also chosen based on the results.

Keywords: Wavelet transforms · Symlets · Coiflets · EZW · SPIHT and STW

1 Introduction

In this work, the wavelet transform is implemented to transform the DICOM images. As per the nature of wavelet function, many types of mother wavelets can be applied on the images, in order to transfer it in to the frequency domain. In this proposed work, symlet, a symmetrical wavelet has been chosen to transform the CT images [1], since, Chest CT scan can be termed as a volumetric image where intensity values, related to the attenuation coefficient of the matter. Generally, an alveolar lung tissue appears as grey homogeneous matter, bright stripes or spots, bright borders, which can be transformed well by symlet. In this work, the Vanishing moments of the symlet are varied from 2 to 10 and the effect of transformation of the image is found[2]. The paper is prepared such as Sect. 2 explains the wavelet technique and Sect. 2.1 deliberates the encoding techniques. Results and discussion is explained in Sect. 3.

D. J. Hemanth et al. (Eds.): COMET 2019, LNDECT 35, pp. 726–730, 2020.
https://doi.org/10.1007/978-3-030-32150-5_72

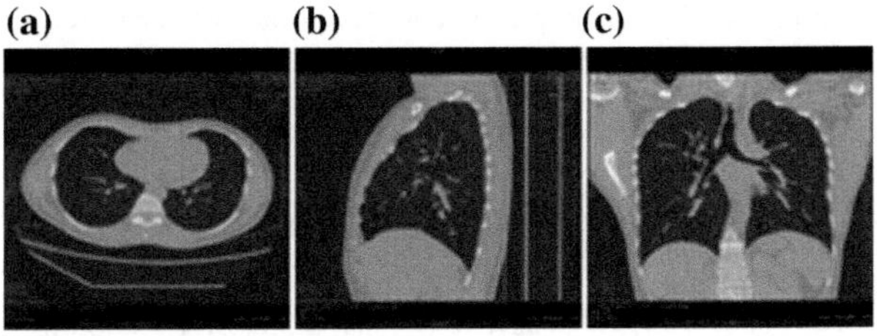

Fig. 1. a. Normal Lung axial view b. Sagittal view c. Coronal view

2 Wavelet Transform

In this work, discrete wavelet transform is used for transforming the image and is primarily used for decomposition of images. The effectiveness of wavelet in the compression is evaluated and also the effectiveness of different wavelets with various decomposition levels are analyzed based on the values of PSNR [2], Compression quantitative relation, suggests that sq. error and bits per constituent. During this work, the symlet and coiflet square measure used for decomposition of the CT coronel read respiratory organ pictures as a requirement for the planned compression algorithms, since, symlet and coiflet square measure designed with extraordinarily section and highest variety of decomposition levels for a given support dimension. They're usually used for determination pattern issues and signal discontinuities [3]. The effectiveness of symlet and coiflet wavelets are compared for various vanishing moments with encoding methods like EZW, SPIHT, STW, WDR and ASWDR. The best possible compression algorithm is found based on results obtained. The DICOM Lung images of size 512 $\times$ 512 pixels with 24 bpp are taken in this work [4, 5].

2.1 Encoding

In compression, for the drop of the redundant data and elimination of the irrelevant data is performed by encoding methods. In this work, embedded zero tree wavelet (EZW), set partitioning in hierarchical Trees (SPIHT), spatial orientation tree wavelet (STW), are used [6].

Table 1. Performance of vanishing moments on symlet wavelet with different levels

Encoding		PSNR	CR	BPP	MSE
SYM2 with level 1	SPIHT	44.2	2.86	17.2	2.3
	EZW	61.22	2.33	24.39	0.05
	STW	59.1	6.66	21.2	0.08
SYM2 with level 2	SPIHT	42.01	2.7	15.27	4.09
	EZW	60.53	8.33	21.15	0.06
	STW	53.49	5.88	20.01	0.29
SYM2 with level 3	SPIHT	40.13	2.7	15.27	4.09
	EZW	44.19	1.92	11.61	2.48
	STW	44.97	2.17	13.16	2.07
SYM2 with level 4	SPIHT	36.69	1.22	4.39	13.94
	EZW	44.17	1.88	11.37	2.49
	STW	38.13	1.36	6.50	9.99
SYM3 with level 1	SPIHT	39.48	2.32	29.69	7.33
	EZW	61.62	2.1	23.5	0.04
	STW	54.2	5.88	20.2	0.12
SYM3 with level 2	SPIHT	41.28	2.63	14.92	4.84
	EZW	51.55	4.34	18.65	0.45
	STW	53.38	5.55	19.79	0.29
SYM3 with level 3	SPIHT	39.29	1.54	8.59	7.66
	EZW	44.1	1.88	11.38	2.54
	STW	44.91	2.17	12.96	2.1
SYM3 with level 4	SPIHT	36.62	1.2	4.22	14.17
	EZW	37.9	1.29	5.57	10.54
	STW	38.1	1.35	6.25	10.08
SYM4 with level 1	SPIHT	43	2.7	16.2	3.2
	EZW	61.48	1.2	22.2	0.04
	STW	54.2	5.88	20.92	0.19
SYM4 with level 2	SPIHT	41.35	2.63	15.11	4.77
	EZW	60.71	7.69	20.97	0.06
	STW	53.37	5.55	19.84	0.29
SYM4 with level 3	SPIHT	39.35	1.54	8.56	6.73
	EZW	44.1	1.88	11.32	2.54
	SRW	44.9	2.12	12.88	2.10
SYM4 with level 4	SPIHT	36.22	1.2	4. 18	15.53
	EZW	37.89	1.28	5.51	10.57
	STW	38.09	1.33	6.16	10.09
SYM5 with level 1	SPIHT	42.2	2.56	15.3	4.4
	EZW	62.14	1.8	22.47	0.04
	STW	54.2	5.9	20.93	0.19

(*continued*)

Table 1. (*continued*)

Encoding		PSNR	CR	BPP	MSE
SYM5 with level 2	SPIHT	41.59	2.5	14.81	4.51
	EZW	51.47	4.3	18.51	0.5
	STW	53.31	5.2	19.66	0.30
SYM5 with level 3	SPIHT	40	1.5	8.51	6.49
	EZW	44.09	1.88	11.32	2.53
	STW	44.93	5.26	19.66	0.303
SYM5 with level 4	SPIHT	36.74	1.2	4.14	13.78
	EZW	37.89	1.2	5.51	10.57
	STW	38.09	1.3	6.1	10.11

3 Result and Discussion

The amount of the compression algorithm is evaluated in terms of MSE, PSNR, compression magnitude relation and Bits per pel, that area unit tabulated in Table 1. The performance of the reconstructed image is measured in terms of mean sq. error (MSE) and peak signal to noise magnitude relation (PSNR) magnitude relation, wherever because the compression will be analyzed exploitation the compression magnitude relation and bits per pel. The MSE is usually referred to as q2. The MSE between σ reconstruction error variance the first image f and therefore there constructed image g at decoder is outlined as [7–9]

$$\text{MSE} = \sigma_q^2 = \frac{1}{N}\sum_{j,k}(f[j,k] - g[j,k])^2 \tag{1}$$

The PSNR between two images having 8 bits per pixel in terms of decibels (dBs) is given by: [10–12].

$$\text{PSNR} = 10\log_{10}\left(\frac{255^2}{MSE}\right) \tag{2}$$

References

1. Osorniorios, R.A.: Identification of positioning system for industrial application using neural network. J. Sci. Ind. India **76**, 141–144 (2017)
2. Pandian, R., Vigneswaran, T., LalithaKumari, S.: Characterization of CT cancer lung image using image compression algorithms and feature extraction. J. Sci. Ind. Res. India **75**, 747–751 (2016)
3. Pandian, R., Vigneswaran, T.: Adaptive wavelet packet basis selection for zerotree image coding. Int. J. Signal. Imaging Syst. Eng. **9**, 388–392 (2016)

4. Pandian, R.: Evaluation of image compression algorithms. IEEE Und Tech (UT) NIOT, pp. 1–3 (2015)
5. Sheela, K.G., Deepa, S.N.: Selection of number of hidden neurons in renewable energy system. J. Sci. Ind. Res. India **73**, 686–688 (2014)
6. Deng, C., Lin, W., Cai, J.: Content-based image compression for arbitrary-resolution display devices. IEEE Trans. Multimedia **4**, 1127–1139 (2012)
7. Maglogiannis, I., Kormentzas, G.: Wavelet-based compression with ROI coding support for mobile access to DICOM images over heterogeneous radio networks. Trans. Inform. Technol. Biomed. **13**, 458–466 (2009)
8. Doukas, C., Maglogiannis, I.: Region of interest coding techniques for medical image compression. IEEE Eng. Med. Biol. Mag. **26**, 29–35 (2007)
9. Do, M.N.: Countourlet transform an efficient directional multi resolutional image representation. IEEE Trans. Image Proc. **14**, 2091–2106 (2005)
10. Lehtinen, J.: Limiting distortion of a wavelet image codec. J. Acta Cybern. **14**, 341–356 (1999)
11. Calderbank, A.R., Daubechies, I., Sweldens, W., Yeo, B.L.: Wavelet transforms that map integers to integers. Appl. Comput. Harmon. Anal. **5**, 332–369 (1998)
12. Black, P.E.: 2008 big-O notation, dictionary of algorithms and data structures. In: Black, P. E., (ed.) U.S. National Institute of Standards and Technology (1998)

RETRACTED CHAPTER: Explore and Rescue Using Humanoid Water Rescuer Robot with AI Technologies

T. R. Soumya[1(✉)], R. Shalini[1], Mary Subaja Christo[2], and J. Jeya Rathinam[1]

[1] Jeppiaar Maamallan Engineering College, Anna University, Chennai, India
soumyatr.soumya@gmail.com, shaliniraman94@gmail.com, jeyarathinamj@gmail.com
[2] Saveetha School of Engineering, Thandalam, India
marysubaja@gmail.com

Abstract. In the present era, there is no requirement for humans to work in risky jobs. Technology has developed to an extreme to replace humans. The most perfect technology to replace humans in risky jobs is Robotics with Artificial Intelligence. In this paper, we are going to design "Humanoid Water Rescuer". This is worth and new to the job as it supports both rescuing human life and existing aquatic organisms from environment situation. One of the mobile robots is rescuer robots which is portable and small. As it traverses the source to destination and sense the organism and to save the life. With sending information to robots it will diagnosis and treat or given the first aid and it is used as carrier robots to carry the humans to the source. Thus it used in all water rescuing situations and treat life according to the command of the master.

Keywords: Carrier robot · Diagnostics robot · Exploring robots · Rescuer robots

1 Introduction

In the present period, we are managing a huge number of lives to ocean. Numerous unintentional issues are occurring in Ocean. The incredible debacle is absent of life in oceans without realizing the end result for individuals. We are at the grievous circumstance that we can't ready to discover by individuals. As a consolation, we are supplanting robots rather than humans. We are not losing people groups by submerging in the ocean. We are at the circumstance to spare a real existence adrift shore from inadvertent issues like battling for life with the oceanic life form. We are not ready to [illegible] individuals to spare the life. So we use robots to save individuals. Despite the facts that we have water safeguard vessels [1] that include the element of analytic and treat the people adequately. Indicative are devices as a physical robot or a product master framework [2]. Medical field had built up a great deal. Simply utilize one of crisis reaction robots in this framework. In spite of the fact that we use crisis robots, computerized reasoning can't assume a wide job in therapeutic specialist should direction the robot to the robot in a crisis circumstance.

The original version of this chapter has been retracted. The retraction note to this chapter can be found at https://doi.org/10.1007/978-3-030-32150-5_163

D. J. Hemanth et al. (Eds.): COMET 2019, LNDECT 35, pp. 731–742, 2020.
https://doi.org/10.1007/978-3-030-32150-5_73

The moves of the robot ought to be observed by an analyst or ace of a robot. As we need to utilize remote checking and correspondence. In the ocean, it is very hard to conquer this worldwide availability make efficient. Satellite correspondence does not make productive. It doesn't enter in auxiliary vessels. It is utilized on account of an ocean mishap. On the off chance that it is inshore protect, ZIGBEE is sufficient. Zigbee is an IEEE 802.15.4-based determination for a suite of abnormal state correspondence conventions used to make individual region systems with little, low-control advanced radios, for example, for home robotization, therapeutic gadget information gathering, and other low-control low-transmission capacity needs, intended for little scale ventures which require remote connection [14]. As it conveys adjacent gadget adequately. Other at that point water save it very well may be utilized for ocean thinks about protecting from natural calamity, for example, surge, tidal wave, and it can include the favorable position in future for saving life in seismic tremors and so on. As ebb and flow innovation quakes and so on, As ebb and flow innovation develops and according to present day necessity we can improve the situation and progressively viable system.

2 The Concept of Rescue Robots

2.1 Methodology of Rescue

The idea of rescuer robot had a lot for safeguarding the life of thousands of people groups. As they make a simple robot in crossing and looking through the article. As a rule unmanned under the vehicle is accessible. This sort of robots can seek through water and injured individual or hazardous item or substance. We may not utilize GPS till the date. In identifying strategy, different activities are to be finished. The main function must be traverse from source to goal. In the ocean Route must be done viably. Route must be finished by the recognizing of detected information navigating calculation are utilized for registering separation to achieve the goal with detected information [5]. Detecting ought to be performed by a portion of the sensors, for example, Appearance-Based Hindrance Discovery with Monocular Shading Vision [4], ultrasonic sensor fundamentally SRF04 works by transmitting a ultrasonic (well above human hearing reach) heartbeat and estimating the time it takes to "hear" the beat reverberate and the Yield from SRF04 is as a variable-width beat that compares to the separation to the objective [13].

Our hindrance location framework is absolutely founded on the presence of individual pixels. Any pixel that varies in appearance from the beginning named an impediment [4]. These must be utilized to recognize snag and sort of impediment. The highlights of Appearance-based snag Discovery Monocular shading vision is programming can keep running on Pentium II processor timed at 333 MHz. Images are obtained from a Hitachi KPD-50 shading CCD Camera which is furnished with an auto-iris lens. The primary concern to see is effectiveness in term of expense and time to achieve the goal.

3 Concepts of Swarm Robots

The thought process of the swarm robot is stack sharing [5]. The fundamental strategy utilized in transporter robots is work in gatherings. The idea of ace and slave is utilized. The ace explores and achieves the goal, at that point the data of place is sent to slave robots then it is coordinated by the ace. The correspondence between master. The idea of ace and slave is utilized [6]. The Ace explore and achieve the goal, at that point the data of place is sent to slave robots at that point coordinated by the direction of the master. The adjusting of an item is finished by example arrangement which is finished by an ace robot. Wheels are connected for development.

The correspondence framework in the ocean is very difficult. Although we have a specific correspondence framework which can impart up to 10 km coaxial tow immediately. Point to point advanced transmission is used. The primary favorable position is sending continuous video and information. Through coaxial link, the framework comprises of MPEG-4 codec, blackfin DSP and SHDSL switch planned by ARM11. The video decoder is done by the exceptional coordinated chip, VF flag and RS-232 flag are converged to the Ethernet interface of the switch. The point to point mode is intended for the switch of deck unit and submerged unit [2].

4 Diving System

The diving system is basic to the capacity of a framework to perform in an ocean. The one of the diving framework is “ocean one” [5]. It comprises of seven joints, one level of opportunity, hand, and head. Head is structured with two DFA’s of dish and tilt. The body is activated with eight thrusters [5]. It is created in spur of to ponder environment. The control diving system in dependence on the order of magnitude than elastic planning and fast primitive task. The other shapes of robots are “D. Beebot” which is look like a bee with efficient concept of swimming locomotion and walking locomotion. Flexible passive joint-mechanism is carried out by a motor installed at the edge of the last link of the robot leg [15]. In spite of swimming we should notice center of gravity for improving the diving ability. Another robot called “Manta robot” has one propulsion mechanism with a pectoral fin in right and left respectively, and a control unit in the center. The mechanism for moving the c.g. is composed of a stepping motor, a feed screw, a lifting platform, a fixed platform, an X linkage, and a screw box [16] (Fig. 1).

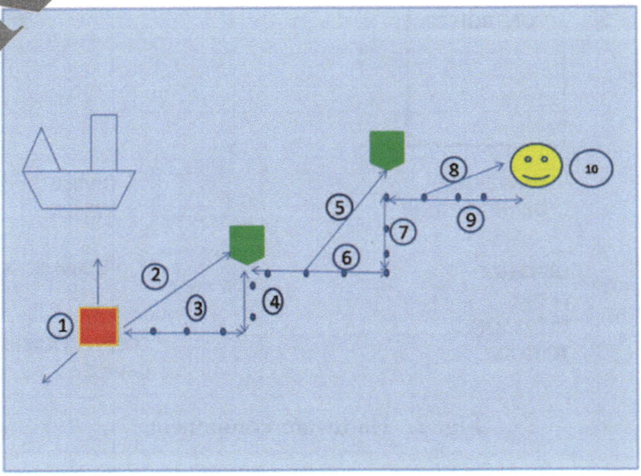

Fig. 1. Working concept

5 Over-All Working Concept

The overall process set in the underwater and unstructured environment:

- When robot begins to navigate the first step is to calculate cardinal directions.
- The robot calculates the nearest node using ultrasonic says d distance.
- Navigate along the x-axis for d distance.
- Turn 90° and calculate new d and navigate along d.
- Turn right and navigate for double the d distance to avoid revisiting of a node.
- Calculate distance for the next node.

5.1 Methodology of Rescue

In this concept, we have to embed the devices to create a humanoid water rescuer. As we see the concept of obstacle detection and navigation in related works. It is possible to do navigation with data that is obtained from obstacle detection sensors. With that data compute a distance between source and destination using one of the best traversing algorithm breadth-first searches. This concept is called exploring robots. Many exploring robots do various traversing algorithm. All algorithms give the better result with varying time and cost. It must be monitored the navigation by webcam and it must be sent to the main computer.

The main thing to notice is efficiency in term of cost and time to reach a destination; here destination is organism under risk. The connection of robot must be interfaced with appearance-based obstacle sensors, power system, and driving system, web camera which is connected to the robot and main computer, and the whole robot is connected via maritime VSTA to the main computer (Fig. 2).

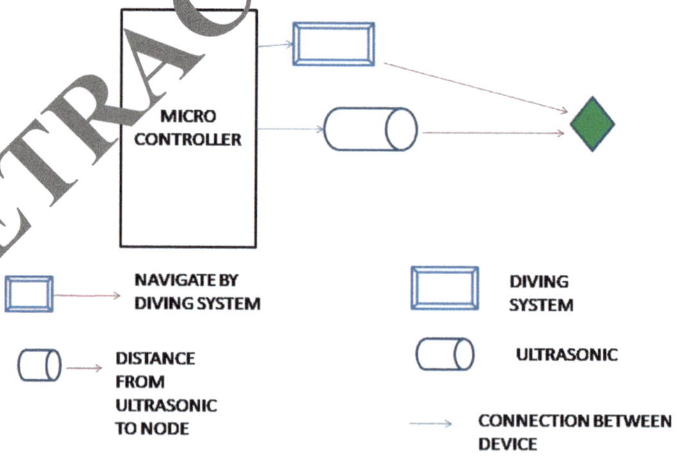

Fig. 2. Hardware component

6 Diagnostic Systems

Many medical robots are well defined. Although medical robot has a great knowledge of treatment, we can't get the assurance of life. So it is used in the way of a monitoring system. Although we are not to focus on all treatment but the basic first aid like oxygenated to a patient, bone crack aid, Check the pulse ratnoror stopped. And the diagnosed information is sent to the main computer it monitored and replies by a medical specialist in crew team. According to master command the robots as to perform following task to the organism. Its concept is like a digital pen. In modern medical operation, the arm of the robot is assisted by doctor. An Move in doctor action it is reflected in the arm of the robot. Basic first aid is enough to a patient. Oxygen cylinder must be fitted to a robot with safeness. The prototype camera can detect blood stains even when the sample has been diluted to one part per 100 [12]. Although in the sea the internal part damage was not well detected by robot itself. It wants the knowledge of doctors for treatment and certain environmental condition. Artificial Intelligence cannot be always safe in accidental situations.

7 Carrier System

One of the two classifications of robots is carrier robots. It is used to an organism in this system. As a part, these carrier robots have to interface separately because the carriage cannot do with rescuing robots. It must be in form of swarm robots. The carrier robot should act as a slave for exploring robots. Both should be connected with ZIGBEE. One is more carrier robots can be made. Balancing must be done through pattern formation of swarm robots. The ZIGBEE is easily accessible between exploring robots. The carrier robots are controlled for its movement and speed by exploring robots. The explorer robots perform navigation and reach a destination. Then it sends information to carrier robots and carrier robots lift the bring the organism to shore. The wheel is attached to carrier robot to bring an organism to shore.

8 Communication System

The connectivity between the main computer and exploring robots is done only through global connectivity it cannot do by satellite connectivity because it will penetrate the structural component. In this system, two part of communication is performed one is deck unit and exploring robot and another one is communication between carrier and exploring robot. Communication between exploring and deck unit is performed through wired cable. Point to point transmission is performed through 10 km coaxial tow cable [5] the real-time video and data are sent to deck unit and deck unit to explorer robot sent through an optical wireless communication system. ZIGBEE is also used. The explorer robots and deck unit can communicate only for 10 m.

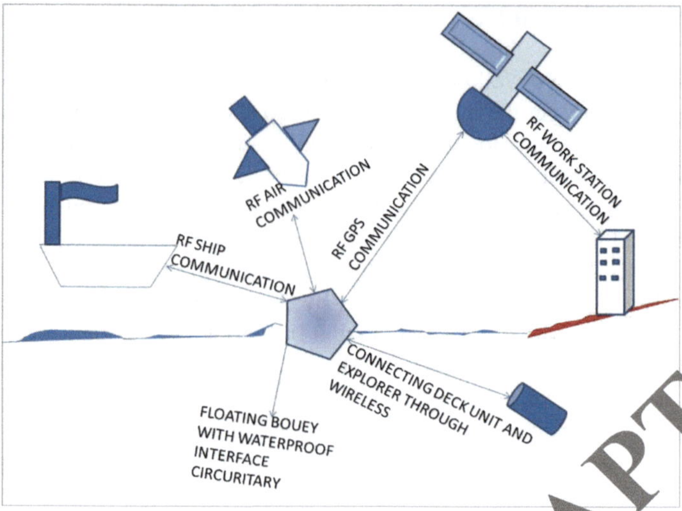

Fig. 3. Communication networks

Next is communication between carrier robots and explorer robots. The each and every carrier robot can contain ZIGBEE in it. Thus communication can be done effectively with any interruption. The Fig. 3 tells only the communication between deck unit and explorer robot. The sea station and earth station are via GPS communication. This communication system was done effectively under water.

9 Artificial Intelligence for Navigation

Artificial intelligence plays a vital role in a construction of the robot. Artificial intelligence is machine intelligence which helps to develop in behaviour as human. With using AI we tend to think, listen, action. Listen means observing to an environment. In unstructured underwater an important part is navigation.

In all existing paper, the important is observed in the observation of one of the paper, Given unique initial and goal positions, find collision-free paths and drive a mobile robot from the initial to the goal position [7]. In which robot is navigated with accurate angle which is not possible in underwater because of viscosity and current rate. Another algorithm is cognitive mapping algorithm for mobile robot for a semi-dynamic environment and for achieving the goal, the mobile robot must be able to construct a map, to localize itself in it and to move from the start point to the goal point and avoid the obstacles in environment [8] which is not used for unstructured dynamic environment.

Definition 1:
According to the number and Graph Theory,

Let (p, q, r) be primitive Pythagorean Triple. Then, there exist m, n belongs to N such that gcd (m, n) = 1, m and n are a different number and p = 2 mn, q = $(m^2 - n^2)$ and r = $m^2 + n^2$ [15].

Definition 2: Pythagorean Theorem:
It states that square of the hypotenuse equals to the square of a sum of other two sides.

$$A^2 = B^2 + C^2$$

The robots are should have control to navigate while finding an object or obstacle or targets. To reach a target the data feed by sensors are used to find the optimal path. The algorithms are used to find optimal solution with help of data. In this algorithm, we use 3 main components to find a position of the robot, image of an object, and direction to reach a target.

1. In Method for Collision-Free Sensor Network Based Navigation of Flying Robots among Moving and Steady Obstacles it used measurement range of the ToF cameras is 15 m and the maximum angle of the view is 135° [10] which limited constrained. To find an image of object we use camera sensor which detects what object it is if an object is matched with our list of a target then it gives a Boolean value along with kind of object it is. A question arises how the image is detected. The database has predefined objects along with an outline of an object which is used in image processing. If it matched it return Boolean value says true and this value is used to navigate by accepting it as a node.
2. To find a position of a robot we use a compass sensor which gives three coordinates points says x, y, z according to the earth magnetic field.

   These below pseudo codes are used to find a location of a robot

   Step 1: Define master address
   Step 2(i): In setup function Begin with appropriate baud rate and Set minimum delay and Write data to a slave as byte
   Step 2(ii): Write data as a byte to another slave
   Step 2(iii): End transmission
   Step 3(i): Initialize x, y, z and Write data as a byte to slave.
   Step 3(ii): Request data from slave
   Step 3(iii): Calculate x, y, z, within wire. available ()

3. To find direction to reach target we use above two data that are a position of robot and image of an object. As previously said camera sensor sends Boolean value "if true it precedes the process else it continues to detect an object which is in a list". The data is three coordinates say x, y, z for a position of the robot. The other data we include is data from an ultrasonic sensor which gives the distance between sensor and object.

As all data are detected the computation starts here (Fig. 4).

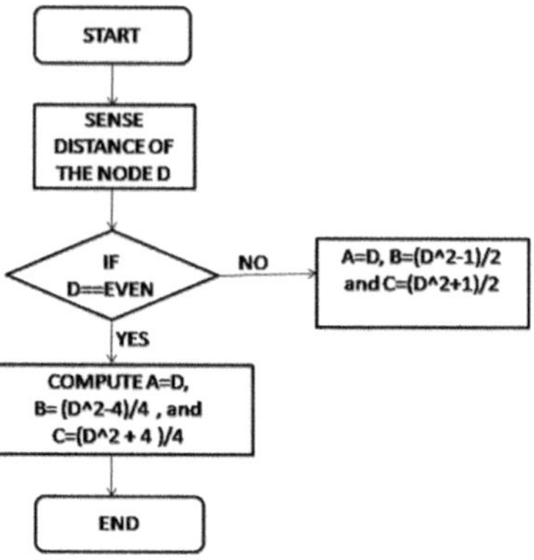

Fig. 4. Flow chart for calculation of distance

Step 1:
The position of robot x, y, z. Initialize according to earth magnetic field. The distance between the robot and object bed.

If d is even then,

$$\text{Compute a} = \text{n}, \text{b} = (\text{d}^2 - 4)/4, \text{ and c} = (\text{n}^2 + 4)/4$$

Else

$$\text{Compute a} = \text{d}, \text{b} = (\text{d}^2 - 1)/2 \text{ and c} = (\text{n}^2 + 1)/2$$

This two equation satisfies primitive Pythagorean Theorem.

Step 2:
In Path Planning of Mobile Robots using Genetic then Algorithm and Probabilistic. Roadmap number of nodes was initially set to 100 and 400 for simple and complex maps respectively and changing the number of nodes affect the computation time [11] for efficient computation the following algorithm is used.

The robot moved along with x-axis. If a == d it should travel with b or c else is should travel with a and b or c. In b or c, either one is equal to d negating it (Fig. 5).

The current data changes to x -> x + a − (a/h), y -> y, z -> z if a == d. Else x -> x + b − (b/h), y -> y, z -> z the previous data is set to visited list to avoid circular motion. The ultrasonic sensor should be delayed when a robot get navigates.

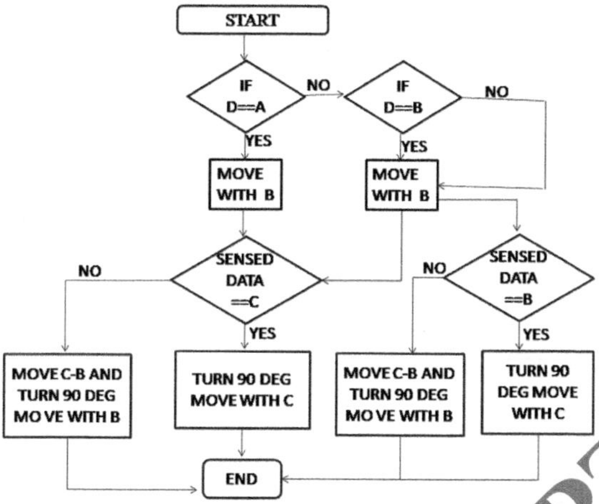

Fig. 5. Flowchart for rotation of robots

Step 3:

Case I: (d == a)

When it reaches to x -> x + d − (b/h), y -> y, z -> z. The ultrasonic sensed again sensed to give d which is equal to c. When it not equal to c move x -> c − a distance Then robot turn to 90°. Then it moves to new distance d along with y axis.

Case II: (d == b)

When it reaches to x -> x + d − (a/h), y -> y, z -> z. The ultrasonic sensed again sensed to give d which is equal to c. When it not equal to c move x -> c − a distance Then robot turn to 90°. Then it moves to new distance d along with y axis.

Case III: (d == c)

When it reaches to x -> x + d − (a/h), y -> y, z -> z. The ultrasonic sensed again sensed to give d which is equal to b. When it not equal to b move x -> b − a distance Then robot turn to 90°. Then it moves to new distance d along with y axis.

If it detected, set it to a visited node by x, y, z positions. Else again detect distance turn 90° and move to -x axis. Note the robot will turn 90° because of water viscosity.

According to c-based algorithm to avoid static obstacles in robot navigation the robot moves first diagonally then vertically and then horizontally to reach the target and Robot repeats this sequence until it reaches the target, given obstacles and robot, there are eight possible placements of target: Bottom Right, Top Right, Top Left, Bottom Left, Horizontal Left, Horizontal Right, Vertical Top and Vertical Bottom [9] which impossible to do this all possible turns in underwater and leads to computation and navigation cost.

D is detected by an ultrasonic sensor.

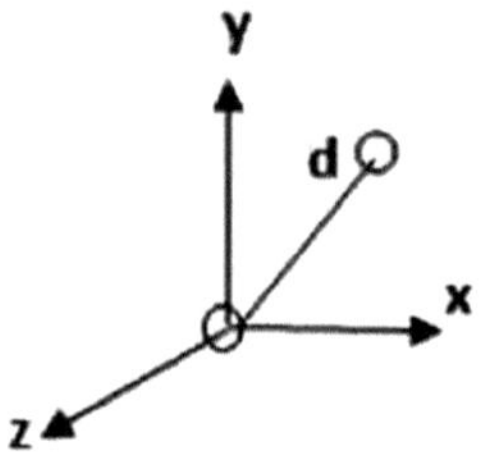

The robot moves with d and detect new d.

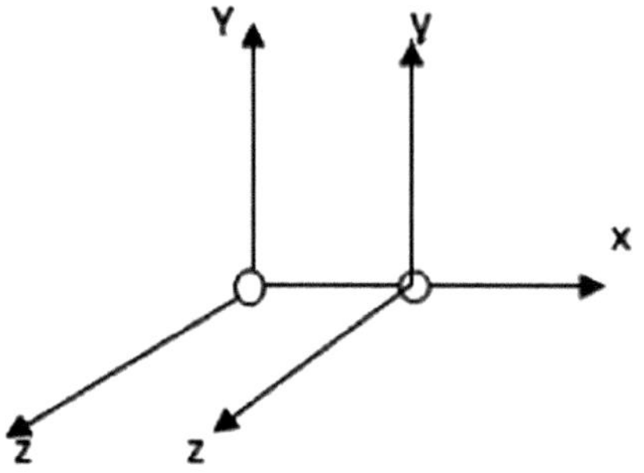

Navigate along with y to d distance and turn −x and move for new d and it reaches the target.

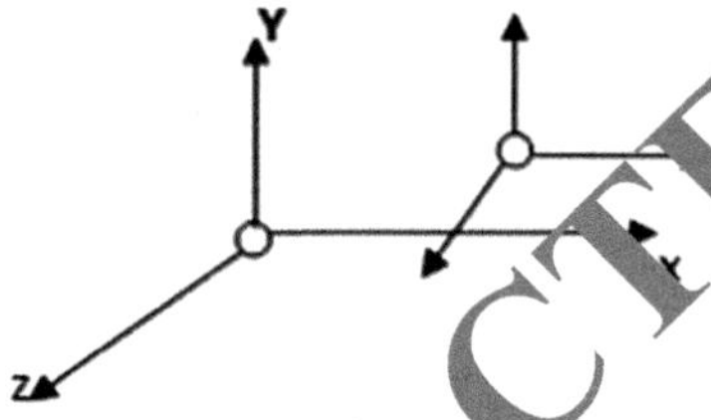

Step 4: Next step is to find the process is continued or not. If they reached node is not destination then continue to search for next node.
Step 5: An important thing that robot should not collide with node. To avoid the collision after reaching each node The robot should turn 90° to the right and calculate next x, y, z-axis and navigate double the distance d to avoid revisit of the node and continue the process to 1 to 4.
Step 6: If a robot finds the target node which is in list process jumps to diagnosing and carrier system else continue its process.

10 Future Work

In future development usage of the deck, the unit can be reduced by effective communication in the sea. And cost also reduced in the usage of deck unit. The explorer robots can communicate directly to the earth station. Artificial intelligence can be improved. The current algorithm used in navigation can be improved in better ways.

11 Conclusions

This paper is presented a new method for WATER RESCUER which is used to replace water rescuer peoples. The motive of developing the system is that man cannot be put in risk full jobs. The system can be adopted and perform well in any environmental condition.

References

1. Siciliano, B., Khatib, O.: Springers Handbooks of Robotics (2008). (in Robotics)
2. Beasley, R.A.: Medical robot: current system and research directions, vol. 2012, 14 p, Article ID 401613. https://www.hindawi.com/journals/jr/2012/401613
3. Zhang, X., Yu, H., Cai, W.: Design of communication system for deep sea remote robotically controlled system
4. Ulrich, I., Nourbakhsh, Z.: Appearance based obstacle detection with monocolor vision. Presented at AAAI Austin, TX
5. Yogeswaran, M., Ponnambalam, S.[illegible]: Swarm robotics: an extensive research review. In: Advanced Knowledge Application in Practice, Igor Fuerstner (ED) (2010). www.intechopen.com/books/advanced-knowledge-application-in-practice/swarm-ANExtensiveresearch–review, ISBN 978653-[illegible]07-141-1
6. Ashokkumar, B., Danny[illegible]er, M., Imtiaz, R.: Implementation of load sharing using swarm robotics. Int. J. Eng. Technol. (03) (2016)
7. Crépon, P.-A., Panchea, A.M., Chapoutot, A.: Reliable navigation planning implementation on a two-wheeled mobile robot. In: Second IEEE International Conference on Robotic Computing, Laguna Hills, California, USA (2018)
8. Issmail, [illegible] R., Desia, R., Zuhri, M.F.R., Daniel, R.M.: Implementation of cognitive mapping algorithm for mobile robot navigation system. IEEE, Bandar Hilir, Malaysia (2015)
9. [illegible]val, L., Aggarwal, S.: C-based algorithm to avoid static obstacle in robot navigation. IEEE, Gurgaon, India (2014)
10. Li, H., Savkin, A.V.: A method for collision-free sensor network based navigation of flying robots among moving and steady obstacles. In: Proceedings of the 36th Chinese Control Conference, 26–28 July, Dalian, China (2017)
11. Santiago, R.M.C., De Ocampo, A.L., Ubando, A.T., Bandala, A.A., Dadios, E.P.: Path planning for mobile robots using genetic algorithm and probabilistic roadmap
12. www.newscientist.com/article/dn19722-blood-camera-to-spot-invisible-stains-at-crime-scenes
13. wiki.eprolabs.com/index.php?title = Ultrasonic_Sensor_SRF04#working_Principle_of_Ultrasonic_Sensor_SRF-04
14. https://en.wikipedia.org/wiki/Zigbee

15. Nakatsuka, T., Watanabe, K., Nagai, I.: The stabilization of attitude of a Manta robot by a mechanism for moving the center of gravity and improvement of diving ability. In: 2016 16th International Conference on Control, Automation and Systems (ICCAS) (2016). https://doi.org/10.1109/iccas.2016.7832325
16. Kim, H.J., Lee, J.: Designing diving beetle inspired underwater robot (D. BeeBot). In: 13th International Conference on Control Automation Robotics & Vision (ICARCV) (2014). https://doi.org/10.1109/icarcv.2014.7064580

Survey in Finding the Best Algorithm for Data Analysis of Privacy Preservation in Healthcare

D. Evangelin(✉), R. Venkatesan, K. Ramalakshmi, S. Cornelia, and J. Padmhavathi

Department of Computer Science Engineering, Karunya Institute of Technology and Sciences, Coimbatore, Tamil Nadu, India
{evangelin, cornelia, padmhavathi}@karunya.edu.in, rlvenkei2000@gmail.com, ramalakshmi@karunya.edu

Abstract. Security for the patient's information can be achieved by using different encryption algorithms. This paper reviews several encryption methods to gain knowledge about the possible attacks during managing and transmitting the sensitive data. The algorithms are compared to analyze and develop an appropriate encryption algorithm that provides security and integrity to the medical health records, thus the patient's information remain confidential.

Keywords: AES · Triple DES · RSA & PGP

1 Introduction

Healthcare data is sensitive in nature and it's security and privacy is a major concern. Therefore it is important to protect the patient's information. Privacy issues can vary based on different management level and this privacy issues include data loss, manipulated data, hiding data etc.

Encryption techniques can be either symmetric or asymmetric algorithms to provide necessary security to the medical records. In this paper, various encryption algorithms are discussed along with their advantages and disadvantages. Triple DES, blowfish, AES, PGP and RSA are few of the algorithms that are compared and tabulated below. Addressing privacy concern requires security issues like access control, authentication, confidentiality etc. [3].

AES algorithm is a symmetric key encryption technique. It is efficient in both hardware and software. Also includes five important components namely, the plaintext, encryption algorithm, a secret key, cipher text and the decryption algorithm. AES provide different key length (128,192 & 256 bits) which is sufficient to protect the sensitive healthcare data [1]. The strength of the secret key is very important as the attacks become easy if the key is weak. The strength of the key depends on the key length and it is known only to the sender and receiver. The implementation of AES is at reasonable cost and also provides successful validation.

D. J. Hemanth et al. (Eds.): COMET 2019, LNDECT 35, pp. 743–746, 2020.
https://doi.org/10.1007/978-3-030-32150-5_74

2 Literary Survey

A detailed literary survey has been conducted on various algorithms and find which is the best one to secure the data in the healthcare database. Securing the patients data is extremely important, thus we have to find the perfect algorithm for securing and accessing the data. The algorithms given below discusses the advantages and disadvantages in detail so that it would be helpful for us to pick the most suitable algorithm to sure the data in the healthcare database system.

AES algorithm has been used [1] for exchanging patient's information in hospital healthcare database with the high-level and the low-level users. AES is one of the best practice for transmitting sensitive data over the internet to the authorized users.

Blowfish: The blowfish encryption algorithm was designed by Bruce Schneier. It is a symmetric block cipher algorithm. It is also a 16-round Fiestel cipher. The block cipher is 64-bits block and the variable length can be varying between 32-bits to 448-bits. Blowfish encryption process includes two stages that is sixteen round iteration and output operation. And the decryption process involves using the reverse round key. The advantage of this algorithm is that it is much faster than the DES algorithm and IDEA. In addition to this it is also much stronger compared to other algorithms. The blowfish algorithm is freely available to all users since it does not require license. The main disadvantage is that it is time consuming.

Triple DES: The triple data encryption standard is a symmetric algorithm. This algorithm uses a 64-bit block size and the key length is between 112 to 168-bits. The triple DES algorithm uses the Fiestal cipher structure. This algorithm is an improved version of the already existing DES algorithm and it was not brought to abandon the DES algorithm. Here, to each data block the block cipher algorithm is applied thrice. To ensure security through encryption capabilities the key size may be increased. The flexibility and compatibility of this algorithm is the major advantage. It is also stronger and more secure than the single DES algorithm. But this algorithm is comparatively slower than the other algorithms.

RSA: The Rivest-Shamir-Adleman algorithm is the most important algorithm used by modern computers mainly for encrypting and decrypting the messages that are being sent and received. This is an asymmetric algorithm, this means that it uses two keys for the encryption and decryption process. The two keys used are the private key and the public key. This algorithm is also called as a public key algorithm. The block size used in this algorithm is not specified and the key size typically ranges between 1024 to 2048-bits. The main advantage the the RSA algorithm is that it is safe and secure, the messages being transferred are also difficult to crack. The disadvantage is it is very slow.

AES: Advanced Encryption Standard was developed by Vincent Rijmen and Joan Daemen and was first published in the year 1998 and was distributed by National Institute of Standards and Technology (NIST) in the year 2001. It is a symmetric encryption algorithm with the block size of 128-bits and with the key length of 128-bits, 192-bits & 256-bits. Depending on the key size the number of rounds can be 10, 12 or 14. AES does not use Feistel network. Steps followed in AES encryption process are Substitutive bytes, ShiftRows, MixColumns and AddRoundKey. The advantages of

AES are that it can be implemented in both hardware and software, open source solution and safe protocol. And the disadvantages are it uses simple algebraic structure and hard to implement in software.

PGP: Pretty Good Privacy was developed by Philip R. Zimmerman in the year 1991. It is a encryption algorithm that provides two important features, confidentiality and authentication. It is especially used to encrypt and decrypt mail over the internet. PGP involves symmetric-key encryption and public-key encryption and thus confidentiality is maintained. The key length varies between 40 to 128-bits. Message integrity is providing each user with a public key to encrypt the message and a private key to decrypt the message. Ensures secure user authentication, network traffic encryption, data integrity and non-repudiation. The limitations are compatibility issues, high cost and no recovery (Table 1).

Table 1. Algorithm comparison

Steps	Algorithm				
	Blowfish	Triple DES	RSA	AES	PGP
Abbreviation	Blowfish algorithm	Triple-Data Encryptions Standards	Rivest-Shamir-Adleman	Advanced Encryptions Standards	Pretty Good Privacy
Type	Symmetric Algorithm	Symmetric Algorithm	Asymmetric Algorithm	Symmetric Algorithm	It uses both public and private key
Block size	64-bits	64-bits	Not Specified	128-bits	Not Specified
Key length	32–448 bits	112–168 bits	Typically 1024–2048 bits	128, 192 & 256 bits	40–128 bits
Advantages	It is a strong, fast and free alternative to existing encryption algorithm	Stronger than single DES	Safe, secure and difficult to crack	Implemented in both hardware and software, open source solution and safe protocol	Ensures secure user authentication, network traffic encryption, data integrity and non-repudiation
Disadvantages	Attack depends on the weak key classes	Comparatively slow	Very slow	Simple algebraic structure and hard to implement in software	High cost

As shown in the above *Table: Algorithm Comparison,* AES method is one of the best algorithm to solve the security issues that can be faced by the external or unauthorized users. The sensitive medical records of every patient remain safe and confidential. AES secret key strengthened and it is hard to crack. This ensures that AES is one of the best practice in the privacy - preservation in hospital - healthcare database management.

3 Conclusion

In this paper we have discussed about various encryption algorithms and studied their advantages and disadvantages in detail. Each algorithms has unique process for encryption and decryption and they are explained and also compared to obtain better understanding of the differences. To preserve the patient's medical records in the hospital database we have selected the AES algorithm in our previous paper as the best algorithm. Further in our next paper we will use the PGP or RSA algorithm to secure the patients data and give authorization to access the medical records. That paper will talk about a technique that is more preferable than the AES algorithm in preserving the data and accessing them.

References

1. Cornelia, S., Venkatesan, R., Ramalakshmi, K., Evangelin, D., Padmhavathi, J.: Data analytics in privacy preservation in hospital- health care database management, Department of Computer Science Engineering, Karunya Institute of Technology, Coimbatore
2. Princy, B.A., Tamilselvi, G., Shruthakeerthi, S., Sowmya, B.: Privacy preserving data analysis in mental health research using cloud computation, Department of Information Technology Panimalar Engineering College Chennai, Simon Fraser University, Burnaby
3. Sahi, M.A., Abbas, H., Saleem, K., Yang, X., Derhab, A., Orgun, M.A., Iqbal, W., Rashid, I., Yaseen, A.: Privacy preservation in e-healthcare environments: state of the art and future direction
4. Box, D., Pottas, D.: A model for information security compliant behaviour in the healthcare context, a School of ICT, Department of IT, Nelson Mandela Metropolitan University, Port, Elizabeth, 6001, South Africa
5. Selvaraj, B., Periyasamy, S.: A review of recent advances in privacy preservation in health care data publishing
6. Lawand, V., Sargar, P., Bhalerao, A., Jadhav, P.: Analytical approach for privacy preserving of medical data, SIT, Lawale Department of Computer Engineering, Pune, India
7. Adam, N., White, T., Shafiq, B., Vaidya, J., He, X.: Privacy preserving integration of health care data. In: Chen, R. (ed.) Concordia University, Montreal
8. Domadiya, N., Pratap, U.: Privacy-preserving association rule mining for horizontally partitioned healthcare data: a case study on the heart diseases, Sardar Vallabhbhai National Institute of Technology, Surat 395007
9. Bennett, K., Bennett, A.J., Griffiths, K.M.: Security considerations for e-mental health interventions, Reviewed by Tormod Rimehaug, Ioannis Mavridis, and Eva Skipenes
10. Benjamin, C.M., Fung, B.C.M., Wang, K.: Privacy-Preserving Data Publishing: A Survey of Recent Developments, Concordia University, Montreal Simon Fraser University, Burnaby
11. Venkatesan, R., Solomi, M.B.R.: Analysis of load balancing techniques in grid. In: Communications in Computer and Information Science CCIS, pp. 147–250 (2011)

A Review and Impact of Data Mining and Image Processing Techniques for Aerial Plant Pathology

S. Pudumalar(✉), S. Muthuramalingam, and R. Shanmugapriyan

Department of Information Technology, Thiagarajar College of Engineering, Madurai, Tamilnadu, India
{spmit, smrit}@tce.edu, r.shanmugapriyan@gmail.com

Abstract. Indian economy highly relies on agriculture sector. Many Indian farmers are unable to do farming profitably due the lack of awareness in incorporating the modern agricultural practices over traditional method. Making Right Decision at right point of time adds value in agriculture sector. Application of data mining techniques on historical agricultural data such as crop yield record, temperature, rainfall, pest attack etc., provides support to the farmers to reduce risk. Major loss is caused by pest attack at various stages of the plant growth. Pest infects all aerial parts of plant (Leaf, neck and node) and in all growth stages. Ease damage to plants can greatly reduce yield and quality of production. This paper focuses on review of Symptom-wise recognition of major plant diseases using Data mining and image processing techniques. The paper aims at identifying the future scope of solving the real world –disease detection problem.

Keywords: Data mining · Image processing · Plant diseases · Agriculture

1 Introduction

Data mining algorithms and techniques promises farmers to handle various decision making challenges of agriculture in terms of productivity, environmental impact, food security and sustainability. Plant disease detection is a predominant problem faced by farmer as it significantly has impact on yield and quality of crop production.

The extraction of hidden information from large volume of raw dataset is known as Data mining. It comprises of data analysis in various perspectives and summarizes it into useful information. The most significant advantage is no restriction to the type of data that can be analyzed by data mining. Data mining aims to dig out patterns in large data sets in intersection with methods of artificial intelligence, machine learning, statistics, image processing, database system and transform it into a human understand able formation for advance use.

Global food production has encountered 10% reduction due to plant diseases. The identification of diseases at early stage could avoid voluminous loss to the farmer, by replacing manual observation and identification of plant diseases by automatic detection of disease using image processing techniques. There are different data mining and

D. J. Hemanth et al. (Eds.): COMET 2019, LNDECT 35, pp. 747–754, 2020.
https://doi.org/10.1007/978-3-030-32150-5_75

image processing methodologies contributed by the researchers all over the world in the field of agriculture-Plant Pathology.

2 Motivation

The need for crop protection against diseases plays a significant role in meeting the growing claim for food quality and quantity. Global crop losses due to pathogens has been statistically estimated around 12.5%, that shows red alert on many commercial and socially valuable crops, such as rice, oranges, cassava, olives, wheat and coffee. Decrease in crop yield, reduced crop quality is the impact of disease attack due to bacteria, viruses, and fungi.

3 Literature Survey

3.1 The Role of Data Mining Techniques in Plant Disease Detection

Data mining is a subset of knowledge discovery process, which includes data readiness tasks such as data extraction, data cleaning, data fusion, data reduction and feature construction, wherein the post-processing includes pattern and model interpretation, hypothesis confirmation and generation and so on. The learning methods of data mining is classified as Unsupervised and supervised. Several studies applied data mining techniques to solve disease detection problem in agriculture.

The data mining problem of identifying to which of a set of categories, a new observation belongs to (Classification step) based on the basis of a training set of historical data and whose categories membership is available (Learning Step) is called as Classification. Classification is comprised of two phase process Training Phase and Classification phase.

In [3] authors compared the performance metrics of various classification algorithms such as K-Nearest Neighbor, Random Forest, Decision Tree, Neural Networks, Naïve Bayes and Support Vector Machines to analyze the crop loss, due to disease or growth of insect, grass grub. Random Forest showed 80% accuracy while Neural Networks showed 74% slightly better results than other classifiers for binary data, and hence designed Ensemble Models of above mentioned classifiers produced better results as compared to classifiers. The Ensemble model with combination of Random Forest, SVM, and Decision Tree proved to give better accuracy. This experimental study was carried out with grass grub damage dataset consisted of 155 instances, four labels and 8 features. It aimed at helping the farmer to maintain its expected production by applying some suitable action to overcome the production issues.

In [2] proposed models for prediction of possible fruit infection using J48, SMO, ZeroR classification algorithms. J48 classifier provided 90.32% the highest percentage of correctly classified instances. Meteorological data and data about the disease appearance are the most important data required for effective diseases protection of fruit. Regression methods are used to verify the results obtained by WEKA classification algorithms mathematically.

In [4] author focuses on developing novel technologies for monitoring the plant health and predicting the leaf disease by Transductive Support Vector Machine classification with the shape and texture features of the plant images. Based on the Latent Dirichlet Allocation and Artificial Neural Network classification technique through the features of soil images and the diseased plant images, the causes for the specific plant disease are identified. The causes obtained are sent to the farmers to take preventive measures.

Weather-based prediction models of plant diseases for rice blast prediction was developed to help the plant science community and farmers in their decision making process using SVM [5]. SVM model has been compared to the REG, BPNN and GRNN approaches. The author concentrates to provide better understanding of the mathematical relationships between the environmental conditions and its specific stages of infection cycle.

[6] proposed Plant Diseases Monitoring model based on image processing and classification techniques. Image processing techniques are used to identify the diseased portion of the leaf and SVM classifier is fed with feature values extracted as input. SVM classifier detects the pest on leaves and also gives information about a type, number of pests and also provides remedy to control pest.

3.2 Role of Image Processing Techniques in Plant Disease Detection

Detection of plant diseases is carried out with Image processing techniques which identify the color feature of the plant. An efficient and inexpensive system is achieved which is widely accepted by the farmers and agricultural researchers for studying disease detection in plants.

Betel vine-Leaf rot disease detection [7]. Leaf Image is preprocessed and segmented using 3 channel color transformation for RGB, HVS and YCbCr. Clear perception of rotted leaf area is obtained from hue component of the HVS color model. Applying Otsu method on $^{3}H'$ component of HVS color space, threshold is calculated. To find the percentage of diseased area, a known calibration factor is multiplied with the number of white pixels to get the area of rotted leaf area in sq. cm.

Loading the image, contrast enhancement, converting RGB to HSI, extracting of features with K-means clustering and SVM in MATLAB has been summarized in [8] (Sujatha et al.) for Identification of disease. By computing amount of disease present in the leaf, decision on amount of pesticides usage can be done effectively. This technique will naturally reduce the investment cost of the farmer.

Disease detection involves steps like image acquisition, image pre-processing, image segmentation, feature extraction and classification [9] (Saradhambal et al.). An enhanced k-mean clustering algorithm to predict the infected area of the leaves has been developed. Clustering based image Threshold is performed by Otsu's classifier. Voice navigation system helps the farmers without sound knowledge in technology usage.

Improved histogram segmentation method finds appropriate threshold automatically rather than manually, which is more scientific, reliable, and efficient. Regional growth method and true color image processing are combined with image recognition system based on multiple linear regressions to improve the accuracy and intelligence proved by [10].

Web based Image Processing dependent approach for Pomegranate fruit disease is proposed. The system promotes the farmers to do the smart farming and allowing them to take decisions for a better yield by making preventive, corrective action on their pomegranate crop. K-means clustering algorithm and SVM classification is performed. CCV feature vectors, Color and Morphology are used for feature extraction. 82% accuracy is produced by the system to identify pomegranate disease [11].

Detection of plant leaf diseases using image segmentation and soft computing techniques [12] proposes usage of Genetic Algorithm for Image segmentation. K means-cluster with Minimum Distance criterion and SVM Classifier are used for classification produced accuracy of 97.6% eventually with less computational process (Fig. 1) and Table 1.

Image Processing with Data mining algorithms
1. Image acquisition.
2. Image enhancement
3. Image Segmentation
4. Feature selection 1. Filters, 2. Wrappers, 3. Hybrid methods
5. Classification-**Data mining algorithms** 1. J48. 2. ID3 3. KNN 4. Random forest 5. Instance-based (nearest neighbor) 6. Statistical (naive bayes) 7. Bayesian network. 8. Regression
6. Image interpretation
7.Performance Metrics 1. Accuracy 2. Precision 3. Recall

Fig. 1. Image processing with data mining algorithm

4 Analysis of Various Algorithms

Table 1. Analysis of various data mining and image processing algorithms

Author and paper	Impact	Future perspective
[11] (Singha,*, Misrab et al.) Detection of plant leaf diseases using image segmentation and soft computing techniques	The paper presents the survey on different disease detection techniques and proposes an image segmentation technique and tested for various species	To improve the recognition rate in classification process ANN, Bayes Classifier, Fuzzy Logic and hybrid algorithms can also be used
Smart paddy crop disease identification and management using deep convolution neural network and SVM classifier [12] (Manisha et al.)	Web based Image Processing dependent approach for the Bacterial Blight disease for Pomegranate fruit using SVM classification was proposed	By increasing size of the dataset, the overall system performance gets improved to detect diseases more accurately
Plant Diseases Recognition Based on Image Processing Technology [10] (Guiling et al.) [10]	Image recognition system based on multiple linear regression has been proposed	Increase the training sample images, to obtain more accurate results, and produce potential system
Plant disease detection and its solution using image classification [9] (Saradhambal et al.)	Voice navigation system has been developed to identify the plant disease with k-mean clustering algorithm	An open multimedia AV about the diseases and their solution automatically once the disease gets detected is developed
Leaf disease detection using image processing [8] (Sujatha et al.)	This paper summarizes k-means clustering, SVM-major image processing used for identification of leaf diseases	To extend the experimental approach by using different algorithms for segmentation, classification for disease identification
Image Processing Based Leaf Rot Disease, Detection of Betel Vine [7] (Dey)	Otsu method is used to calculate the threshold. Threshold based image processing algorithm for segmentation was proposed to diagnosis leaf rot disease in Betel Vine leaves	To regulate the specific level of pesticide application by calculating the leaf disease severity scale with the percentage diseased area
A New Proposed Model for Plant Diseases Monitoring Based on Data Mining Techniques [6] (Gamal et al.)	Moth-Flame Optimization (MFO) algorithm – feature selection and SVM classification were used to develop system for monitoring pests of leaf	To develop the Monitoring of plant diseases system can be implemented with wireless sensor networks

(*continued*)

Table 1. (*continued*)

Author and paper	Impact	Future perspective
Machine learning techniques in disease forecasting: a case study on rice blast prediction [5] (Kaundal et al.)	Weather - based prediction models of plant diseases is developed - new prediction approach based on support vector machines	Online tool - application of control measures
A Hybrid of Plant Leaf Disease and Soil Moisture Prediction in Agriculture Using Data Mining Techniques [4] (Sabareeswaran et al.)	Proposed novel technologies for monitoring the plant health - Transductive Support Vector Machine classification, Latent Dirichlet Allocation and Artificial Neural Network classification technique	A monitoring system to predict the growth level of the plant
Predicting Crop Diseases Using Data Mining Approaches: Classification [3] (Umair et al.)	Prediction model to predict crop loss due to grass grub insect	By using hybrid of evolutionary algorithms and data mining techniques to improve the prediction results

5 Open Research Problems

The following research opportunities that have been derived after reviewing the above research contributions in the agricultural field.

1. To build a pesticide recommendation system by understanding Plant-Pathogen-Environment clearly. The environmental factors that highly influence on plant diseases are Temperature, rainfall (duration and intensity), dew (duration and intensity), leaf wetness period, soil temperature, soil water content, soil fertility, soil organic matter content, wind, fire history (for native forests), air pollution, herbicide damage etc. [Philip Keane and AIlen Kerr]. Any sudden or abnormal change in the external parameters may lead to pest attack in plants. Not all the factors leads to the same type of disease, hence a clear understanding of Plant-Pathogen-Environment is necessary. Data mining techniques applied on historical dataset provides this information. The knowledge discovery will help the farmers to know in advance the attack of pathogens.
2. To avoid huge loss in crop productivity due the plant disease, the disease can be identified at the early stage infection and intimated to the farmer. Plants show the infection of disease at leaf, root, stem, and fruit, any of its parts. Symptom-wise analysis of changes on plant parts can be done in order to identify the disease attack at earlier stage. Experts can also help the farmers in treating the plants. Image processing techniques can be used to analyze the images of plant parts and identify the whether the plant is infected or not, if so identify the type of disease and area of infection. The result shall be given as input to expert wherein he will provide guidelines on methods to treat or the type and amount of pesticide to apply.

3. To measure the significance of detecting the plant diseases at early stage by crop yield analysis. Crop yield analysis using data mining techniques helps us to learn the yield ratio taking into account the farmer's input to the field (inclusive of labour cost) to the crop yield and also thereby to identify the impact of plant diseases on crop yield. The factor that majorly contributes to crop loss can be witnessed.
4. To provide an integrated technological solution to the small land holding farmers. Usage of ICT tools in agriculture can achieve it. The success behind the work is not implementing ICT in agriculture but making the decision in the right time. Farmers aware of its usage and makes them use it. Expert's knowledge along with farmers experience can enhance and progress agriculture sector to the next profitable level.

6 Conclusion

Agriculture is an important pillar of Indian economy which primarily depends upon many external uncontrollable factors. These factors determines the profit or loss to the farmers. Even then, with the available historical data, the technology advancements can support the farmers to make right decisions at right time. This paper focused on review of image processing and data mining techniques solutions to plant diseases. The study discussed few open research problems in the agricultural field. The identified research problems can be solved using Internet of Things techniques can be combined with image processing and data mining techniques to provide appropriate solution. The study is limited only to aerial pathology of plants which can further be examined.

References

1. Ayub, U., Moqurrab, S.A.: Predicting crop diseases using data mining approaches: classification. In: 2018 1st International Conference on Power, Energy and Smart Grid (ICPESG). IEEE (2018)
2. Bhange, M., Hingoliwala, H.A.: Smart farming: Pomegranate disease detection using image processing. Proc. Comput. Sci. **58**, 280–288 (2015)
3. Gamal, A., et al.: A new proposed model for plant diseases monitoring based on data mining techniques. In: Plant Bioinformatics, pp. 179–195 Springer, Cham (2017)
4. Ilic, M., Spalevic, P., Veinovic, M., Ennaas, A.A.M.: Data mining model for early fruit diseases detection. In: 2015 23rd Telecommunications Forum Telfor (TELFOR), pp. 910–913. IEEE, November 2015
5. Kaundal, R., Kapoor, A.S., Raghava, G.P.S.: Machine learning techniques in disease forecasting: a case study on rice blast prediction. BMC Bioinform. **7**(1), 485 (2006)
6. Dey, A.K., Sharma, M., Meshram, M.R.: Image processing based leaf rot disease, detection of betel vine (Piper BetleL.). Proc. Comput. Sci. **85**, 748–754 (2016)
7. Majumdar, J., Sneha, N., Ankalaki, S.: Analysis of agriculture data using data mining techniques: application of big data. J. Big Data **4**(1) 20 (2017)
8. Saradhambal, G., Dhivya, R., Latha, S., Rajesh, R.: Plant disease detection and its solution using image classification. Int. J. Pure Appl. Math. **119**(14), 879–884 (2018)
9. Radha, S.: Leaf disease detection using image processing. J. Chem. Pharm. Sci. 1–4 (2017)

10. Sabareeswaran, D., Sundari, R.G.: A hybrid of plant leaf disease and soil moisture prediction in agriculture using data mining techniques. Int. J. Appl. Eng. Res. **12**(18), 7169–7175 (2017)
11. Singh, V., Misra, A.K.: Detection of plant leaf diseases using image segmentation and soft computing techniques. Inf. process. Agric. **4**(1), 41–49 (2017)
12. Sun, G., Jia, X., Geng, T.: Plant diseases recognition based on image processing technology. J. Electr. Comput. Eng. (2018)
13. Munisami, T., Ramsurn, M., Kishnah, S., Pudaruth, S.: Plant leaf recognition using shape features and colour histogram with k-nearest neighbour Classifiers. Proc. Comput. Sci. **58**, 740–747 (2015). https://doi.org/10.1016/j.procs.2015.08.095

Survey of the Various Techniques Used for Smoke Detection Using Image Processing

Shirley Selvan(✉), David Anthony Durand, and V. Gowtham

Department of Electronics and Communication, St. Joseph's College of Engineering, Chennai, India
shirleycharlethenry@gmail.com,
davidadurand@gmail.com, gowtham.velusamy97@gmail.com

Abstract. The concept of smoke detection was put forth by the development of sensors. These sensors relayed the parameters they sensed to a processor which made decisions. The presence of smoke is often an indicator of fire. Hence, given the ability to detect smoke in the early stages of fire, major fire accidents can be prevented. All though this method of smoke detection using sensors was a great success it was not very usefully in extreme conditions (weather, range, location) and sometimes even produced false alarms. Image processing paved way for more accurate detection of smoke since it uses digital data rather than analog inputs. With image processing both, fire and smoke could be detected easily. This method of detection involves various processes like extracting features, comparing with references, classification etc. This survey paper briefly explains the various techniques proposed/used to detect smoke.

Keywords: Stationary wavelet transform · Change detection · Multi-layer perceptron · Support Vector Machine · Feret's region · Covariance descriptors · Dual dictionary modelling · Sparse representation

1 Introduction

Earlier, smoke detection was done using smoke sensors. These sensors although efficient, work best only in a closed environment for example, a mine or inside a manufacturing plant. In an open area such a forest, junk yard etc. using sensors proved unreliable owing to the limitation of range. Another such limitation is that sensors may get damaged because of environmental conditions which cause the device to stop working. This results in faulty values being recorded. Therefore, smoke detection using image processing is more efficient and reliable.

The presence of smoke is mostly related to fire and in some other cases, vehicular emission. In the case of fire, smoke is an integral/first indicator of the fire since smoke is better visible in an open area. The presence of smoke can also be confirmed by detecting fire. This can be done using image processing. In this paper we discuss the various techniques of smoke detection using image processing. Smoke in an image can be detected by extracting features from the given image and classifying them. Smoke can hence be differentiated from other parts of the image using these features.

D. J. Hemanth et al. (Eds.): COMET 2019, LNDECT 35, pp. 755–762, 2020.
https://doi.org/10.1007/978-3-030-32150-5_76

The various techniques of extracting features and detecting smoke are explained briefly in the following section.

2 Methodology

2.1 Smoke Detection Using Stationary Wavelet Transform [1]

This method of smoke detection is based on Stationary wavelet transform. It works in three steps. In the first step (pre-processing block), the image is converted to gray scale and resized. The image is resized using the bicubic interpolation algorithm [10]. The bicubic interpolation algorithm is used because it is most used method for image resizing in two dimensions. The image is converted to grey scale because the conversion eliminates hue information without changing the lightless information. This is followed by an indexation process to index the image. The main reason behind indexation is to group the colors which are of similar intensity (Fig. 1).

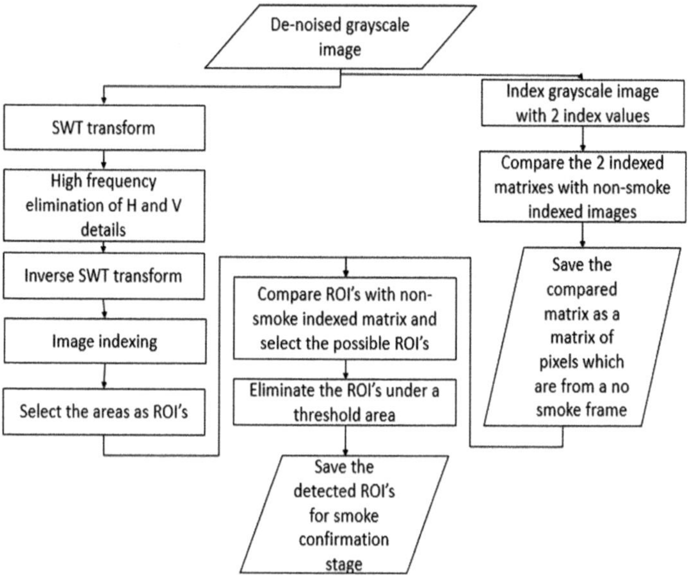

Fig. 1. Smoke detection algorithm using Stationary wavelet transform.

In the second step Stationary Wavelet Transform is applied. The detection algorithm used in this method is based on the idea that smoke gradually smoothness the edge of an image. When SWT is applied on the image the high frequency components are eliminated. Removing high frequency components doesn't modify the region under smoke but highly modifies the regions without smoke. The image is reconstructed by using the inverse SWT. At this point two indexed images with different decompositions levels are saved, preferably levels 3, 4 since levels 1 and 2 don't eliminate enough details, hence producing false alarms. At the same time after smoothening in greyscale format by eliminating high frequencies, two indexed images at different levels of indexation are selected to perform the detection algorithm. From these four indexed and grey scaled images, the Region Of Interest (ROI) is selected (area of high intensity). The area is selected by correlating the images selected and taking the pixels common to all four images.

The selected pixels are compared to a matrix that contain pixels different from non-smoke frames. The pixels common between them are set as ROI where smoke is detected. The selected regions which are possibly under smoke are eliminated if they are under a certain threshold area which varies from different applications depending on the sensitivity of the system and distance from the camera.

The final step involves the use of "smoke verification" algorithm. Smoke verification algorithm is used to prevent false alarms. The result produced after these steps are combined to produce a result to specify the presence of smoke.

2.2 Smoke Detection Using Color Features and Motion Detection [2]

In this method pixels are separated based on the difference between the behavior of the normal background pixels and smoke pixels which behave differently. These classifications are done using specific modules (Fig. 2).

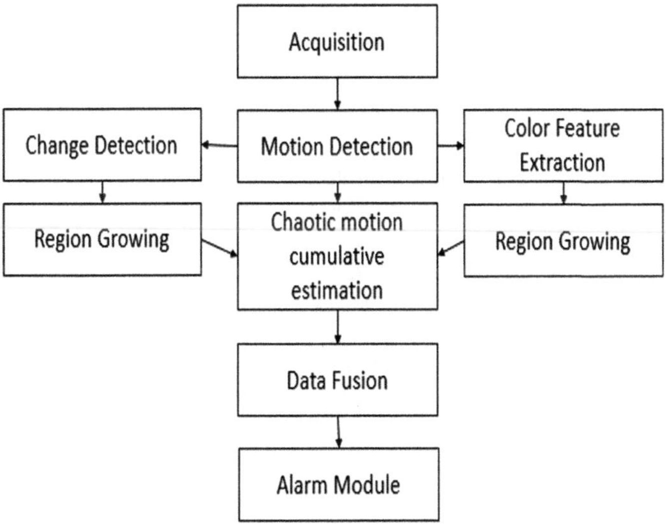

Fig. 2. Smoke detection algorithm using Color features and motion detection.

In general, fire is always accompanied by smoke. With this fact in mind we can confirm the presence of smoke by making certain the presence of fire. Hence, in this method pixels are classified into fire, smoke and background. After this classification the pixels are clustered.

For this method Pietro Moreri, Lucio Marcenaro, Carlo S. Regazzoni and Gianluca Gera proposed a five-module fire and smoke detection system. The modules used are (1) Change detection (2) Fire features extraction (3) Smoke feature detection (4) Chaotic Motion Detection. The pixels behaving differently from the background pixels are identified by the background subtraction, this process in done in the change detection module and motion detection module. Now, we have the pixels which contain either smoke or fire or some other abnormalities.

The smoke and fire pixels are distinguished from the other pixels by the color and features extraction module. YCbCr and L*a*b* color spaces are used in case of fire pixels and smoke pixels respectively. The separated pixels are combined to get connected regions in the region growing module whose outcome is sent to the chaotic motion analysis module. This is followed by data fusion process where pre-alarms are generated.

Data fusion is used to eliminate false alarms. This process is done using multi-layer perceptron (MLP) [3].

2.3 Smoke Detection by Finding Feret's Region [4]

Support Vector Machine (SVM) is a learning module which can be trained. It can be trained to learn algorithms that analyze data. The result is mostly used for classification or used to distinguish between two or more entities.

In this method the first step is the extract of the moving objects in the image. The processing comprises of five steps,

(1) Image subtraction and accumulation - Used to find region of moving object. Since the growth speed is slow the image will not be clear. To overcome this a function ($f(t) = g(t) + g(t + 1)$ where $g(t)$ is a subtracted image) which accumulates two subtracted images is used. (2) Image binarization – Used to removes very small noise from $h(t)$. $h(t)$ is processed to be a binary image $b(t)$. It is important to set a perfect threshold level since smoke exhibits semi-transparency property. To overcome this Otsu's Automatic threshold selection method is used [11]. (3) Morphological operation – A morphological process called "opening" is used to eliminate noisy regions in $b(t)$. (4) Extraction of Feret's diameter – Used to find shape of object (5) Image mask is used to estimate the moving object region in the image frame $g(t)$.

The region obtained after the preprocessing of the image $g(t)$ is the possible region which contains smoke. The texture pattern is focused upon as a feature vector to determine which regions are smoke. The texture feature of smoke is defined by a co-occurrence matrix, so it Is not dependent on the size of the smoke in the image. Maruta et al. selected 14 popular types of texture features [12] as components of the feature vector. The Feret's regions of smoke are determined using this feature vector as an input vector. The SVM classifies smoke and non-smoked based on its prior training.

Feret's Diameter- It is the distance between the tangents on opposite sides of a particle. Both tangents are parallel to a fixed direction.

Feret's Region – The shape and region found in the above process using Feret's diameter is called Feret's region.

2.4 Smoke Detection Using Covariance Descriptors [5]

As mentioned in Sect. 2.2, the presence of fire implies the presence of smoke. The fact is the base of this method. The presence of fire is detected by dividing the image into three dimensional regions. This is used to generate spatio-temporal blocks. The covariance features are calculated from these blocks. These features are used as input to the SVM. The SVM classifier classifies flame based on the spatial and temporal characteristics. Chromatic color model is used for pixel color classification [6]. This model analyzed fire colored pixels and concluded that hue of fire colored pixels is between 10^0 to 60^0. This condition in RGB domain equivalent is, **Condition 1** - $R \geq G > B$. Fire being a source of light must have pixels above a certain threshold (R_T), **Condition 2** - $R > R_T$. The last rule in the Chromatic color model is for saturation. "S" represents saturation and "S_T" represents saturation when $R = R_T$. **Condition 3** - $S > (255\text{-}R)S_TR_T$. Covariance descriptors are used for texture classification [7].

To define the spatio-temporal blocks temporal derivatives must be calculated with spatial parameters. Covariance matrices are used to define the spatial blocks. The video is divided into blocks of size $16 \times 16\times$ (frame rate of video) before computation. Since, Computing the covariance for each block in the video is inefficient, the chromatic color model is used to eliminate the blocks which don't contain any fire colored pixels. This is done using a color mask. The covariance values of the remaining spatio-temporal blocks are computed. To reduce the cost for computing the covariance value of the pixel property vectors are computed in a separate manner. A total of 33 covariance parameters are used in training and testing an SVM classifier. Spatial, color and domain information are combined using covariance descriptors.

2.5 Smoke Detection Using Dual Dictionary Modelling [8]

Smoke doesn't occupy the entire image like smoke or haze but will act as a scattering medium. This based on the dichromatic atmospheric scattering model [9]. Since parameters such as bending parameter varies across the entire image, this method works on the block level. Sparse representation is used represent smoke in each block. Two dictionaries each are created for smoke. Both dictionaries are designed in such a way that they lead to sparse representation of either pure smoke or none smoke. In order for the dictionaries to adapt its given task and to a specific content they are trained independently using real samples. It is expected that any pure smoke image will be sparse in the pure smoke dictionary and any clear image will be sparse in the none-smoke dictionary.

Initialization of y_b is done by solving an equation formed for intensity of light in the image by nullifying y_s. In the same way y_s is initialized by nullifying y_b. Value for y_s is calculated by solving the same equation with fixed y_b. And y_b is calculated by solving the equation with fixed y_s. The object of the function is simultaneously calculated and updated. The initialization and calculation processes are performed continuously until the threshold is reached. The values of y_s and y_b are concatenated and sent as input to the SVM. The proposed detection algorithm uses pure smoke dictionary D_s, non-smoke dictionary D_b, regularizations parameters, a threshold to check convergence and the initial value of the objects of the objective function as input along with the block image.

Sparse Representation - Sparse representation shows data as a linear combination of atoms (a sample in a dictionary) taken from a pre-defined dictionary of elements.

Dictionaries – A dictionary is collections of samples of an element. It is expected that any possible value that represents this element must be present in the dictionary.

3 Conclusion

As mentioned in the previous sections the presence of smoke can be detected directly and indirectly by detecting fire. The techniques used for smoke (or fire) detection are thus explained and the basic steps in each technique are discussed. In most techniques mentioned above, Support Vector Machine plays an integral role. The SVM classifier ensure accurate results. It is trained by providing necessary samples which are unique in each technique. The input to the SVM classifier is a vector which contains the extracted features from the image. Using the features extracted from a region in the image the SVM classifier determines if the region contains smoke or not. One unique technique is using Covariance descriptors. In this technique, the data is represented together in the form of covariance matrices. This method has a lesser computational cost and is not affected by the random behavior of fire. It is efficient only when the fire is in close range and visible clearly. In Smoke detection technique using Color features and motion detection, the pixels are classified as smoke, flame and background. Infrared and visible images are combined to detect smoke. This technique is better than the covariance technique since it has better range. The Stationary wavelet transform method is the best method for outdoor smoke detection since wavelet transforms have improved the texture analysis and texture recognition applications. The usage of the smoke verification algorithm makes it efficient.

Comparison Table
See Table 1.

Table 1. Table comparing the methodology, pros and cons of the surveyed papers.

Article name	Author	Methodology	I. Pros II. Cons
"Wavelet-Based Smoke Detection in Outdoor Video Sequence," 978-1-4244773-9/10/$26.00 ©2010 IEEE [1]	R. Gonzalez-Gonzalez, V. Alarcon-Aquino, R. Rosas-Romero, O. Starostenko, J. Rodriquez-Asomoza	Uses Stationary wavelet transform	I. Robust and reduces false alarms II. Not as effective in closed spaces
"Early Fire and smoke detection based on color features and motion Analysis" 978-1-4673-2533-2/12/$26.00 ©2012 IEEE [2]	Pietro Morerio, Lucio Marcenaro, Carlo S. REgazzoni, Gianluca Gera	Color feature extraction	I. Good range II. Not effective in low light
"A novel Detection Method Using Support Vector Machine," 978-1-4244-6890-4/10/ $26.00 ©2010 IEEE [4]	HIdenoriMaruta, Akihiro Nakamura, Fujio Knrokawa	Using Feret's Region to obtain possible regions of smoke	I. Exact region of smoke is found II. Not efficient when there are discontinuous regions of smoke
"Flame Detection method in video using covariance descriptors," IEEE International Conference on Speech and Signal Processing, pp. 1817–1820, 2011 [5]	Y. Habiboglu, O. Gunay, A. Cetin	Covariance descriptors are used for texture classification	I. Lesser computational cost and is not affected by the random behavior of fire II. Efficient only when in close range and visible clearly
"Detection and Separation of Smoke from Single Image Frame," 1057-7149 ©2017 IEEE [8]	Hongda Tian, Wanqing Li, Philip O. Ogunbona, Lei Wang	Dual dictionary technique	I. Differentiate smoke from fog and hue II. Highly Complicated

References

1. Gonzalez-Gonzalez, R., Alarcon-Aquino, V., Rosas-Romero, R., Starostenko, O., Rodriquez-Asomoza, J.: Wavelet-based smoke detection in outdoor video sequence. 978-1-4244773-9/10/$26.00 ©2010. IEEE (2010)
2. Morerio, P., Marcenaro, L., Regazzoni, C.S., Gera, G.: Early fire and smoke detection based on color features and motion Analysis. 978-1-4673-2533-2/12/$26.00 ©2012. IEEE (2012)
3. Luo, F.-L., Unbehauen, R.: Applied Neural Networks for Signal Processing. Cambridge University Press, New York (1999)

4. Maruta, H., Nakamura, A., Knrokawa, F.: A novel detection method using support vector machine. 978-1-4244-6890-4/10/$26.00 ©2010. IEEE (2010)
5. Habiboglu, Y., Gunay, O., Cetin, A.: Flame detection method in video using covariance descriptors. In: IEEE International Conference on Speech and Signal Processing, pp. 1817–1820 (2011)
6. Chen, T.-H., Wu, P.-H., Chiou, Y.-C.: An early fire-detection method based on image processing. In: ICIP 2004, 24–27 October 2004 , vol. 3, pp. 1707–1710. IEEE (2004)
7. Tuzel, O., Poriki, F., Meer, P.: Region covariance: a fast descriptor for detection and classification. In: Computer Vision-ECCV 2006, pp. 589–600 (2006)
8. Tian, H., Li, W., Ogunbona, P.O., Wang, L.: Detection and separation of smoke from single image frame. 1057-7149 ©2017. IEEE (2017)
9. Surya, T.S., Suchithra, M.S.: Survey on different smoke detection techniques using image processing. IJRCCT **3**(11), 16–19 (2014)
10. Keys, R.: Cubic convolution interpolation for digital image processing. IEEE Trans. Signal Process. Acoust. Speech Signal Process. **29**, 1153 (1981)
11. Otsu, N.: A threshold selection method from gray–level histograms. IEEE Trans. Syst. Man Cybern. **9**(1), 62–66 (1979)
12. Umbaugh, S.: Computer Imaging: Digital Image Analysis and Processing. CRC Press (2005)

Design of Flexible Multiplier Using Wallace Tree Structure for ECC Processor Over Galosis Field

C. Lakshmi[1(✉)] and P. Jesu Jayarin[2]

[1] Sathyabama Institute of Science and Technology, Chennai, Tamilnadu, India
c.lakshmichandrasekar@gmail.com
[2] Jeppiaar Engineering College, Chennai, Tamilnadu, India

Abstract. Security is a main parameter in Networks. To ensure the security in network various cryptography algorithms are implemented. Instead of using software based security, the hardware based security system increases security features. ECC processor is a hardware based security crypto processor works based on Elliptic curve cryptography, the key generation is very fast compare to AES, RSA. ECC processor has many modules such as arithmetic and logic units, Control unit etc.,. The Arithmetic unit Consist of Multiplier, Adder. Various optimization techniques and multiplier algorithm are implemented to increase the efficiency. In this paper we are proposed the novel architecture of flexible multiplier.

Keywords: ECC processor · Elliptic curve cryptography · Key generation · Cryptography algorithms · Flexible multiplier

1 Introduction

In Digital environment, the security plays the vital role. In order to increase the security various algorithms are proposed. The network can be categorized as two types. First, the wired network in which the communication taking place in a wired environment. Second, wireless networks in which the communication taking place in wireless medium such as air. The security is more complicated in wireless network compare to wired network, because in wired network cables and wires are used so interpretation of data is difficult whereas in wireless medium the data is transmitted in free space, the information can be easily retrieved by the malious node. here the encryption and decryption technique is employed. The plain text is converted to cipher text using encryption algorithm and that cipher text is decrypted using decryption algorithm. here the key is the important parameter, without knowing the key it is not possible to get the exact plain text. The key extraction is not simple, the key is computed based on the type of cryptography algorithm used.

Elliptic curve cryptography is a cryptography algorithm widely used for encryption and decryption process. Compare to existing algorithm, it consumes less time to compute the key and provides the higher security. ECC uses the point addition method, the key computation is very complex. ECC is a The ECC processor consists of various

D. J. Hemanth et al. (Eds.): COMET 2019, LNDECT 35, pp. 763–770, 2020.
https://doi.org/10.1007/978-3-030-32150-5_77

arithmetic module such as Addition/subtractin module, Multiplication module, Sqaring module, Division module. By optimizing one of the module, the parameters such as power, area, speed, operating frequency are optimized.

The below table shows performance comparision of ECC and RSA [1]. In this reference the small key size in ECC provides the same security as large key in RSA (Table 1).

Table 1. performance comparison of ECC and RSA

Rack time MIPS	RSA key size	ECC key size	ECC/RSA ratio
10^4	512	106	5:1
10^8	768	132	6:1
10^{11}	1024	160	7:1
10^{20}	2048	210	10:1
10^{78}	2100	600	35:1

2 Previous Work

In [1], the Recursive Karatsuba Multiplier is implemented in Virtex 5 Processor, in this multiplier 64 $\times$ 32 multiplication is performed. In this design the parameters are calculated simultaneously, so the delay is minimized. The pipelined structure achieves speed (1.08 μs), but flexibility is not possible.

The two stage multiplier is used [2] area is reduced and speed is increased, the structure is pipelined. By using segmented method the parallel processing is achieved. The multiplier is divided into w bits and block by block multiplier is multiplied with n bit multiplicand. The proposed the design is implemented in virtex 5 FPGA.

Montgomery multiplication is proposed [3] to perform operations in different operand size. In this multiplier the hardware size is reduced and speed is increased. The two operands are multiplied based on the modulo arithmetic.

The another way of multiplication is performed by

Continuous addition, instead of computing partial products [4]. The radix 4 technique is implemented in both serial and parallel multiplier. The design is implemented in virtex 6 FPGA. All the above discussed multipliers are based on parallelism concept to achieve the speed. The existing algorithm and architecture are modified to achieve the optimization.

In [8], the ECC processor is used in Internet of Things application, the author used ARM processor for implementation. In this paper 256, 384,521 are implemented. So flexibility in Multiplier is not possible in this design.

In [11–15] the various algorithm used in multipliers are discussed to reduce utilization area, delay and power consumption. But these optimization not enough to design a ECC processor. In modern scenario the Crypto processor is used in various applications, needs flexible operations.

In [16], the survey is given on various multipliers used, this literature helps to gain knowledge on existing mulpliers. The bit serial multipliers are the most commonly used multipliers in ECC processor.

Most of the multipliers are implemented in Galois field [17], it provides flexible data handling in cryptography.

3 Overview of Wallace Tree Multiplier

The proposed multiplier uses the Wallace tree Structure in which multiple additions are done to perform the multiplication. In basic Wallace tree structure the partial products are grouped and then addition operation is performed (Fig. 1).

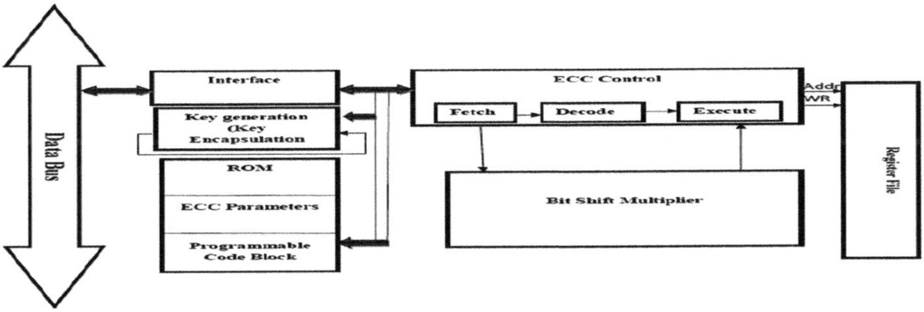

Fig. 1. ECC processor architecture

The above figure depicts the architecture of ECC processor. In most of the ECC processor the point addition method is followed for speed operation. The data is encrypted using the Elliptic curve cryptography. The key is random in nature. Using the points in the elliptic curve the computation is performed. The arithmetic unit plays the important role in key computation (Fig. 2).

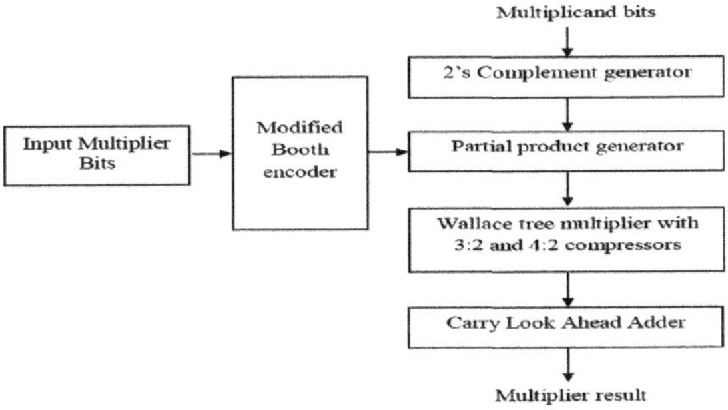

Fig. 2. Flow diagram of multipier

The modified booth algorithm is used to reduce the steps. By using the radix-4 or radix-8 method (Specified in Table 2) the encoding technique is implemented. The carry look adder is used along with booth encoder the partial product (Fig. 3).

Table 2. Booth recoding table for radix-4

Multipier block			Recoded pair	1-bit	2-bit booth	
i+1	i	i−1	i+1	i	Multipier value	Partial product
0	0	0	0	0	0	Mx0
0	0	1	0	1	1	Mx1
0	1	0	1	−1	1	Mx1
0	1	1	1	0	2	Mx2
1	0	0	−1	0	−2	Mx-2
1	0	1	−1	1	−1	Mx-1
1	1	0	0	−1	−1	Mx-1
1	1	0	0	0	0	Mx0

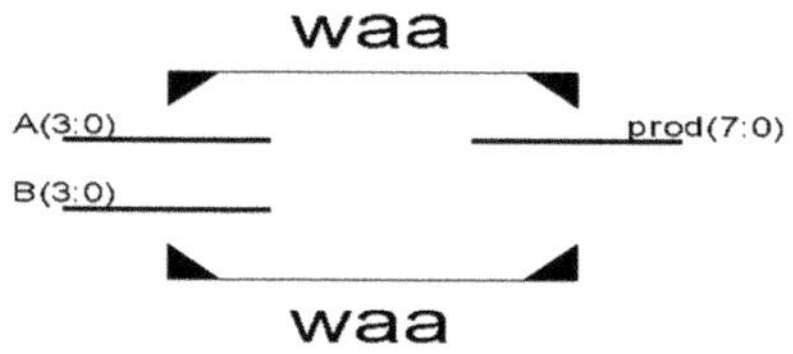

Fig. 3. RTL schematic of Wallace tree structure

The internal architecture shows it is a combination of half adder and full adder (Fig. 4).

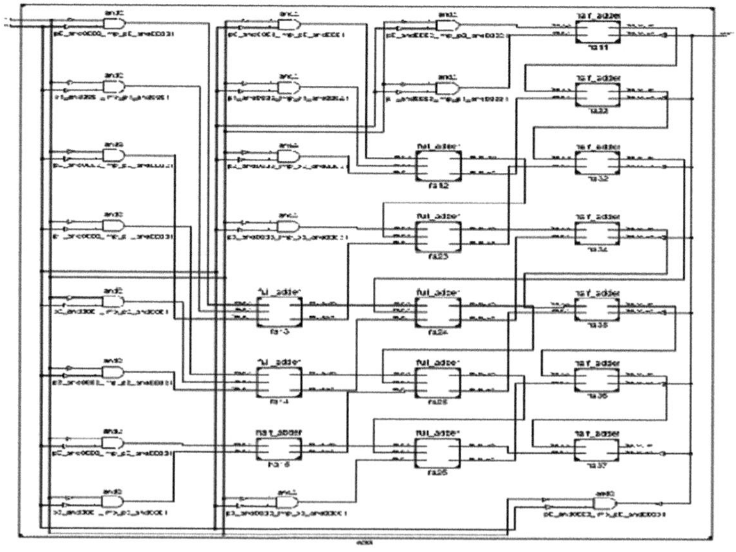

Fig. 4. Internal architecture of Wallace tree structure

4 Design of Flexible Multiplier

The proposed Wallace tree multiplier comprises of two half adders, two 3:2 compressors and ten 4:2 compressors. The Wallace tree structure is reduced by using the compressors. In Wallace tree multiplier the number of output is compressed by using the single bit adder. The most common compressors used in the Wallace tree structure is 3:2 compressor and 4:2 compressor. The compressor circuit uses the multiplexer which is designed by using transmission gate. The sum and carry functions of 3:2 compressor is given by (Fig. 5)

$$\mathrm{Sum} = (\mathrm{AorB}) * \overline{\mathrm{C}} + \overline{(\mathrm{AorB})} * \mathrm{C} \tag{1}$$

$$\mathrm{Carry} = (\mathrm{AorB}) * \mathrm{C} + \overline{(\mathrm{AorB})} * \mathrm{C} \tag{2}$$

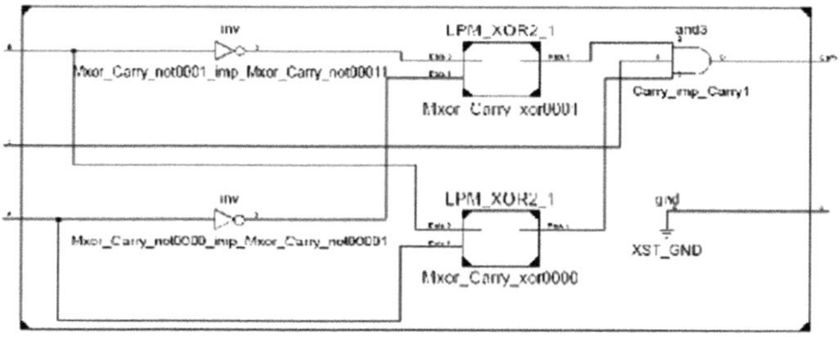

Fig. 5. Internal architecture of 3:2 compressor

Similarly the 4:2 Compressor is obtained from the multiplexer using transmission gate

$$\begin{aligned} \mathrm{Sum} = {} & (\mathrm{AorB}) * \overline{(\mathrm{CorD})} + \overline{(\mathrm{AorB})} * (\mathrm{CorD}) * \overline{\mathrm{Cin}} + \\ & \overline{(\mathrm{AorB}) * \overline{(\mathrm{AorD})} + \overline{(\mathrm{AorB})} * (\mathrm{CorD})} * \mathrm{Cin} \end{aligned} \tag{3}$$

$$\mathrm{Cout} = (\mathrm{AorB}) * \mathrm{C} + \overline{(\mathrm{AorB})} * \mathrm{D} \tag{4}$$

$$\mathrm{Carry} = (\mathrm{AorBorCorD}) * \mathrm{Cin} + \overline{(\mathrm{AorBorCorD})} * \mathrm{D} \tag{5}$$

The compressors are used in the Wallace tree multiplier is used to combine the partial products and result of this is given to the full adder and half adder circuit (Fig. 6).

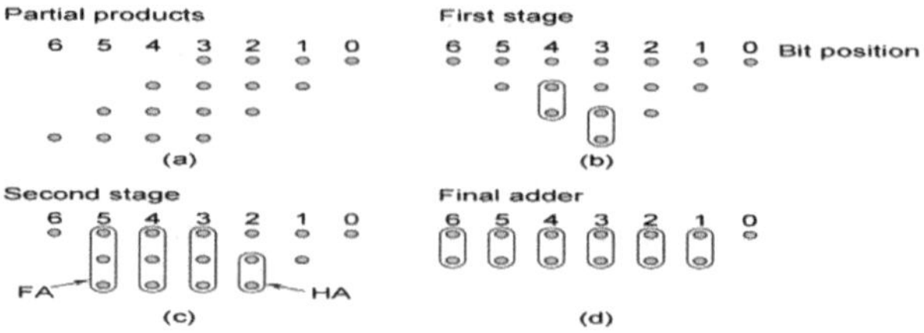

Fig. 6. Wallace tree multipier

In the above figure the Wallace tree structure is obtained by using full adder and half adder circuits (Fig. 7).

In first step the partial product are obtained by using simple multiplication.

In the next step the tree like structure is formed, the resultant terms are given to the input to full adder and half adder.

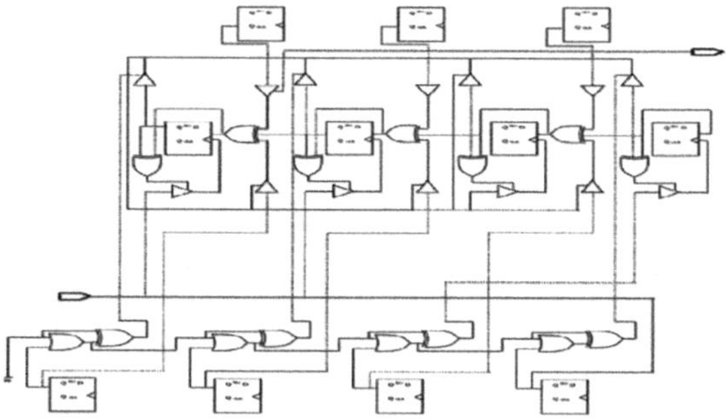

Fig. 7. Flexible multiplier used in ECC processor

In Existing ECC processor the flexible operation is not possible In this flexible design we can compute any combination like $2^{163} \times 2^{32}$ so on.

5 Simulation Results

The design is simulated in Xilinx 14.3 ISE simulator in Intel i5 Processor in Virtex 4 FPGA family. The flexible multiplier uses 143 slices (2%). In this multiplier 571 buffers are used. The main objective of this multiplier is area optimization. The total time period is 1.037 ns (from synthesis report).

The simulation results achieved is only for multiplier block. The entire performance of this multiplier is determined when it is used along with the ECC processor (Figs. 8 and 9).

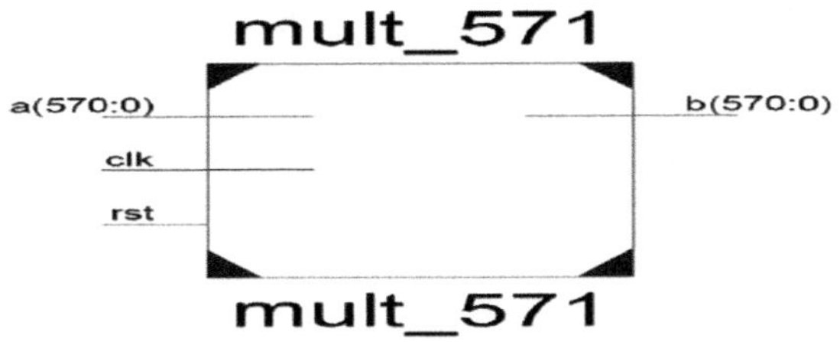

Fig. 8. RTL schematic of flexible multiplier used in ECC processor

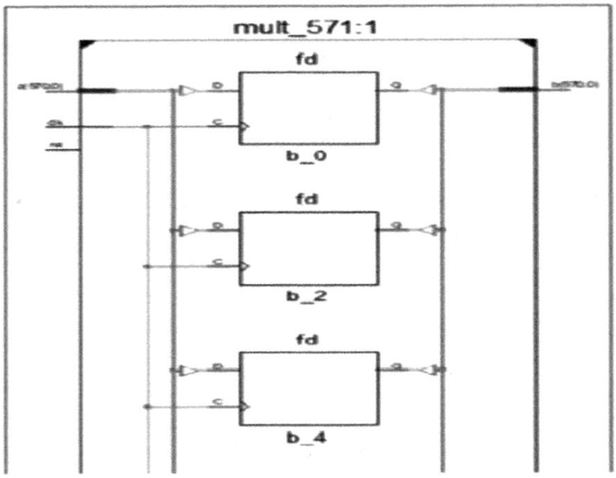

Fig. 9. Internal architecture of flexible multiplier used in ECC processor

6 Conclusion

In this paper we designed the flexible multiplier is designed using wallace tree structure, which reduces the complexity. The flexibility is required to perform multiplication in different combinations, instead of performing multiplication for default operands. The modified booth encoding algorithm is used along with Wallace tree multiplier. The proposed structure achieves the flexibility and this multiplier is designed for ECC processor.

References

1. Marzouqi, H., Al-Qutayri, M., Salah, K.: RSD based Karatsuba multiplier for ECC processors. In: 2013 8th IEEE Design and Test Symposium, Marrakesh, pp. 1–2 (2013)
2. John, K.M., Sabi, S.: A novel high performance ECC processor architecture with two staged multiplier. In: 2017 IEEE International Conference on Electrical, Instrumentation and Communication Engineering (ICEICE), pp. 1–5, Karur (2017)

3. Sun, W., Dai, Z., Ren, N.: A unified, scalable dual-field montgomery multiplier architecture for ECCs. In: 2008 9th International Conference on Solid-State and Integrated-Circuit Technology, pp. 1881–1884, Beijing (2008)
4. Javeed, K., Wang, X., Scott, M.: Serial and parallel interleaved modular multipliers on FPGA platform. In: 2015 25th International Conference on Field Programmable Logic and Applications (FPL), pp. 1–4, London (2015)
5. Narh Amanor, D., Paar, C., Pelzl, J., Bunimov, V., Schimmler, M.: Efficient hardware architectures for modular multiplication on FPGAs. In: 2005 International Conference on Field Programmable Logic and Applications, August 2005, pp. 539–542 (2005)
6. Ananyi, K., Alrimeih, H., Rakhmatov, D.: Flexible hardware processor for elliptic curve cryptography over NIST prime fields. IEEE Trans. Very Large Scale Integr. (VLSI) Syst. **17** (8), 1099–1112 (2009)
7. Khan, Z.U.A.: High-speed and low-latency ECC processor implementation over GF (2m) on FPGA. IEEE Trans. Very Large Scale Integr. (VLSI) Syst. **25**(1), 165–176 (2017)
8. Liu, Z., Seo, H., Castiglione, A., Choo, K.K., Kim, H.: Memory-efficient implementation of elliptic curve cryptography for the Internet-of-Things. IEEE Trans. Dependable Secure Comput. **16**(3), 521–529 (2018)
9. Imran, M., Rashid, M., Jafri, A.R., Najam-ul-Islam, M.: A Cryp-Proc: flexible asymmetric crypto processor for point multiplication. IEEE Access **6**, 22778–22793 (2018). https://doi.org/10.1109/access.2018.2828319
10. Khan, Z., Benaissa, M.: Throughput/area-efficient ECC processor using montgomery point multiplication on FPGA. IEEE Trans. Circ. Syst. II: Express Briefs **62**(11), 1078–1082 (2015)
11. Salman, A., Ferozpuri, A., Homsirikamol, E., Yalla, P., Kaps, J., Gaj, K.: A scalable ECC processor implementation for high-speed and lightweight with side-channel countermeasures. In: 2017 International Conference on ReConFigurable Computing and FPGAs (ReConFig), pp. 1–8, Cancun (2017)
12. Huang, M., Gaj, K., El-Ghazawi, T.: New hardware architectures for montgomery modular multiplication algorithm. IEEE Trans. Comput. **60**(7), 923–936 (2011)
13. Hasan, M.A., Namin, A.H., Negre, C.: Toeplitz matrix approach for binary field multiplication using quadrinomials. IEEE Trans. Very Large Scale Integr. (VLSI) Syst. **20** (3), 449–458 (2012)
14. Loi, K.C.C., Ko, S.B.: Scalable elliptic curve cryptosystem FPGA processor for NIST prime curves. IEEE Trans. Very Large Scale Integr. (VLSI) Syst. **23**(11), 2753–2756 (2015)
15. Fournaris, A.P., Zafeirakis, J., Koufopavlou, O.: Designing and evaluating high speed elliptic curve point multipliers. In: Proceedings of the 17th Euromicro Conference Digital System Design (DSD), August 2014, pp. 169–174 (2014)
16. Fan, H., Hasan, M.A.: A survey of some recent bit-parallel GF (2n) multipliers. Finite Fields Appl. **32**, 5–43 (2015)
17. Petra, N., Caro, D.D., Strollo, A.G.M.: A novel architecture for galois fields GF (2 m) multipliers based on Mastrovito scheme. IEEE Trans. Comput., **56**(11), 1470–1483 (2007)

Line and Ligature Segmentation for Nastaliq Script

Mehvish Yasin(✉) and Naveen Kumar Gondhi

SMVDU, Katra, India
mehvishshah999@gmail.com

Abstract. The accuracy rate of recognizing ligatures in Urdu Character Recognition mostly relies on the accuracy with which the segmentation has been performed to convert Urdu text into lines and ligatures. Generally, it has been seen that the ligature based segmentation yields better results rather than character based segmentation. In this paper, we present a technique for segmenting Urdu text images into lines and then to ligatures. A hybrid approach has been used, which employs top-down technique in order to perform line segmentation and bottom-up technique in order to segment lines into ligatures. The various issues like broken lines, diacritic association, overlapping has also been discussed.

Keywords: Nastaliq · Segmentation · Ligatures · Zones · Overlapping

1 Introduction

In Nastaliq Style of Urdu Script, the characters are combined completely which results in more challenging segmentation of characters [1]. Therefore, we have used the next option, which is the higher unit of recognition, also known as ligature. There are various segmentation approaches for ligatures, which can be classified into top-down, bottom-up and hybrid. In top-down approach, the image is divided into text lines and words/characters assuming them to be straight lines. In bottom-up approach, a clustering process is followed, hence the observation starts with small units of pixels, characters, words, text lines and pairs of components are merged as one moves up the hierarchy. The hybrid approach combines the top-down and bottom-up approach in various ways. The positions of the piece-wise separating lines can be obtained by using the horizontal projection. In Urdu Script, the global horizontal projection method includes the problems of over segmentation and the under segmentation.

Ligatures are classified into primary and secondary ligatures. The main body is represented by primary ligature while, the secondary ligature is represented by diacritics/dots corresponding to the primary ligatures (Fig. 1).

D. J. Hemanth et al. (Eds.): COMET 2019, LNDECT 35, pp. 771–779, 2020.
https://doi.org/10.1007/978-3-030-32150-5_78

Fig. 1. An example of ligature

Sometimes, ligatures are placed such that there occur a white space between the primary and secondary ligature. In such a case, the single line is incorrectly segmented, leading to over- segmentation. In the same way, we have under-segmentation where multiple text lines are merged together because of the horizontal overlapping [1] (Fig. 2).

Fig. 2. Overlap between the ligatures

2 Proposed Methodology

We propose the modified segmentation carried out at two levels, line and ligature. We first apply Binarization using global threshold and text line segmentation. Then we extract the ligatures for each segmented text line.

3 Line Segmentation

Firstly, the segmentation points need to be detected. For this, we analyze the row with maximum pixels (local peaks) and row with minimum pixels (local valleys). Hence, by counting the number of pixels in each row of image (horizontal projection profile); we get the number of text lines of an image (Fig. 3).

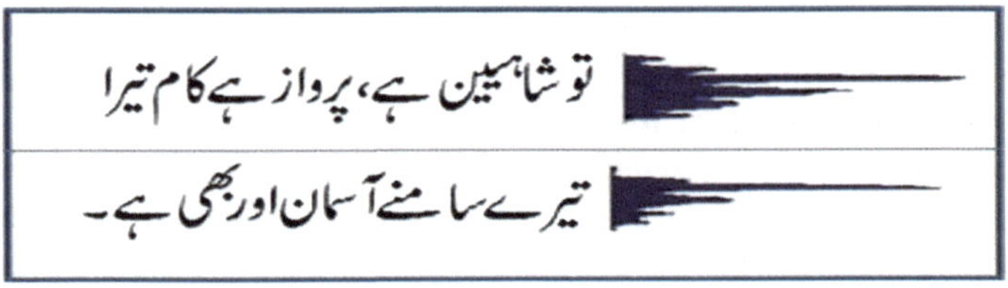

Fig. 3. Horizontal projection of histogram distinguishing lines in text

However, the over and under-segmentation leads to the multiple peaks and valleys [2]. As a result, we get mis-segmented lines along with the challenge of associating secondary ligatures with their corresponding primary ligatures as shown in Figs. 4 and 5.

Fig. 4. Zero value between main bodies and diacritics resulting from histogram

Fig. 5. Mis-segmentation of text line into three lines due to zero values

To overcome these challenges, we put forward the steps of modified horizontal projection for line in order to segment the text line.

The Steps are discussed as:

An image is segmented into horizontal zones with the help of horizontal projections. Next, segment the zones by using estimated row height. The execution of rest of the steps depends upon the accuracy with which the row height has been calculated. Therefore, in addition to the projection profile analysis, the connected component heights and the inter peak gaps are also taken into consideration. Thus, the height of each zone is calculated and the medium height of each row is determined.

Once the row height has been calculated, we use these row heights to label zones. We can either have Type 1 or Type 2 (Fig. 6). Zones depending upon the placement of diacritic marks that is, zones which have dots either above or below the ligatures and zones having underlines, etc.

The zones with multiple text lines can be labeled as zone 3. Next, we merge the zones, depending upon whether they satisfy the criteria for Type 1 or Type 2.

Fig. 6. Multiple zones of a text

Table 1 show the different zones for Fig. 6 where the corresponding heights and widths are calculated. The zones are either of type 1 or Type 2 depending upon their heights. In other words, if a zone has a height lesser than half row height, it is concluded that the zone belongs to type 1. The zone containing diacritics is considered to be Type 2, irrespective of whether the diacritics lie above or below the base-line. Zones having heights 1.5 times as compared to row height are considered as Type 3.

Table 1. Information of zones for image in Fig. 6.

Zone	Height	Width	Type
1	35	62	1
2	9	18	2
3	7	9	2
4	1	62	2
1	35	62	1

Thus, in this way we can perform the segmentation using estimated row height. But in some cases, if an image possesses noisy component, it results in under-segmentation. Therefore, we apply traditional horizontal projection method and undergo rough segmentation of binarized image. There might be a problem with associated dots and diacritics and mis-segmentation. In order to overcome such problems, we apply morphological dilation to the document image such that the primary and secondary ligatures seem to be joined as shown in Fig. 7. The text line boundary which exists between the sequential local peaks (valley index) can be found from local peaks in dilated version of an image which in turn can be detected with the help of median zone height that acts as threshold for finding peaks.

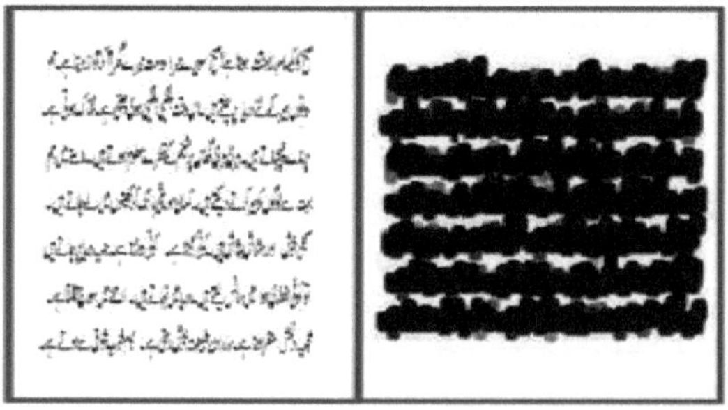

(a) (b)

Fig. 7. (a) Original image and (b) dilated version

Once we have carried out dilation of an image, it is now easy to segment.

4 Ligature Segmentation

Mostly, ligature is used as a fundamental unit of recognition. This is because of the reason that the segmentation of word into characters is challenging due to word spacing [2]. There are 3 steps in ligature segmentation:

4.1 Division of Line into Ligature

Most of the systems use projection profile method in order to calculate vertical histogram of text lines [3]. Since the ligatures overlap in Nastaliq style, therefore this method cannot be applied. A better method is to split text lines into connected components and extract the information.

4.2 Identification of Base Forms

The various characteristics like height, width, centroid, overlapping, co-ordinates and baseline information are considered as per the connected components which in turn can be categorized into primary and secondary ligatures. The horizontal projection of pixels can be used in order to find the distinction between the ligature base and the diacritics thereby, providing the baseline measures. The row containing the maximum amount of ink pixels is considered to be the baseline.

Sometimes, incorrect baseline may occur (Fig. 8). Such errors can be rectified by checking a couple of heuristics [4]. The primary rule in case of Nastaliq Style is that every ligature should be in close vicinity of the baseline, thereby touching it. Also, the baseline must be in the lower half of the line/ligature. Therefore, the false segmentation and identification can be reduced.

(a) (b)

Fig. 8. (a) False baseline and (b) corrected baseline

4.3 Base and Mark Association

Formation of ligatures by association of secondary components is the major challenge. The components are concluded to be primary depending upon height size. A threshold value is defined and if the height of component is greater than threshold, then the component is primary, else the component is secondary. Another way is by using centroid-to-centroid distance after calculating the centroid for each shape and then projecting them vertically in order to form association (Fig. 9).

Fig. 9. Centroids of diacritics vertically projected to the bases

Although this method is simple, but in some cases, it does not work reasonably well and does not provide accurate results. The reason for this is that the centroids of diacritics do not associate with the right base forms because of shifting of letters to left or right due to the context sensitivity characteristic of Nastaliq style of Urdu script. Therefore, instead of current ligature, the diacritics project to the previous letter/ligature as shown in Fig. 10.

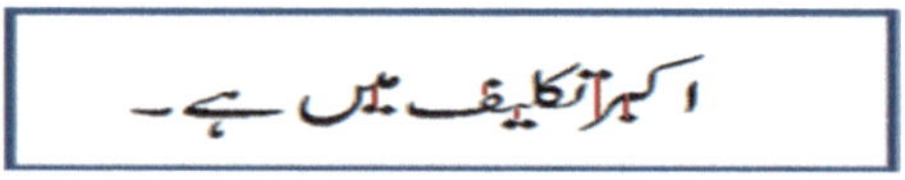

Fig. 10. Diacritics of character *tay* projecting on previous ligature

In order to address such issues, the complete horizontal span is made with respect to the diacritics, which is then associated with the base form.

Following steps are followed:

The secondary components are joined with primary components with the help of vertical structuring elements by using morphological dilation.

The extraction of connected components from dilated image is performed.

If secondary components (dots) are combined with only one primary component, then they are associated with the particular primary component. If there is an overlap be- tween different primary ligatures, then the horizontal distance between the primary and secondary components is calculated and the association of the secondary component with the nearest primary component is made. In this way, we get the successful extraction of overlapped ligatures from lines.

There might be a situation where a dot is associated with more than one base forms (overlapping), in such a case, it is associated with the left side of the ligature. Similarly, if the diacritic forms a complete overlap with respect to multiple base forms, then the distance of the diacritics with each of the ligatures is calculated and is associated with the one having the lesser distances [5].

Fig. 11. Text line

Fig. 12. Extracted ligatures

An example of ligature segmentation is illustrated in Figs. 11 and 12 in which extraction of overlapped ligatures from text lines is done successfully.

Table 2. Comparison of ligature segmentation techniques.

Author	Approach	Accuracy	Remarks
Malik and Fahiem [6]	Horizontal projection profile method	94%	Lines with merged ligatures could not be segmented
Kumar, Kumar, and Jawahar [7]	Voronoi-diagram based algorithm	65.4%	Non availability of clear demarcation, overlapping, non-equispaced lines
Breuel [8]	Geometric algorithms	72–93%	Physical layout of a page
Javed and Hussain [10]	Zone based segmentation	98.7%	Merged text lines are difficult to segment
Bukhari, Shafait, and Breuel [9]	Ridge based approach	92%	Misplaced diacritics could not be placed properly
Lehal [11]	Ligature segmentation	99.11%	Horizontally overlapping lines, broken lines
Proposed Method [12]	Line and ligature Segmentation	96%	Correct segmentation and diacritics association

Table 2 shows the comparison of various segmentation techniques. It can be inferred from the table that the horizontal projection profile method that was put forward by H. Malik and

M. A. Faheem yielded an accuracy of 94%. However, the lines and ligatures that were merged could not be segmented. The same problem was found in case of zone segmentation that yielded an accuracy of 98.7% by S. T. Javed and S. Hussain. An another approach was introduced by K. S. Kumar, S. Kumar, and C. Jawahar which failed to provide the clear demarcation. In addition to this, the text was overlapped with an accuracy of 65.4%. The physical layout of the text image could be depicted with the help of geometric algorithms by T. M. Breuel with an accuracy varying be- tween 72–93%. Ridge based approach is another important approach that was put forward by S. S. Bukhari, F. Shafait, and T. M. Breuel. However, it could not place diacritics correctly. G. S Lehal proposed Ligature segmentation yielding an accuracy of 99.11%. However, overlapping and broken lines are present. Lastly, we have combined the line and ligature segmentation approach, getting an accuracy of 96%. In our approach [12] there is a correct association of dots and diacritics.

5 Results

We have taken into consideration the CLE text database document images and used line and ligature segmentation simultaneously. The results show an accuracy of 96%. It can be seen from the Figs. 13 and 14 that sometimes the mis-segmentation occurs which leads to drop in accuracy levels. This mis-segmentation arises due to the incorrect association of secondary ligatures to the primary ones.

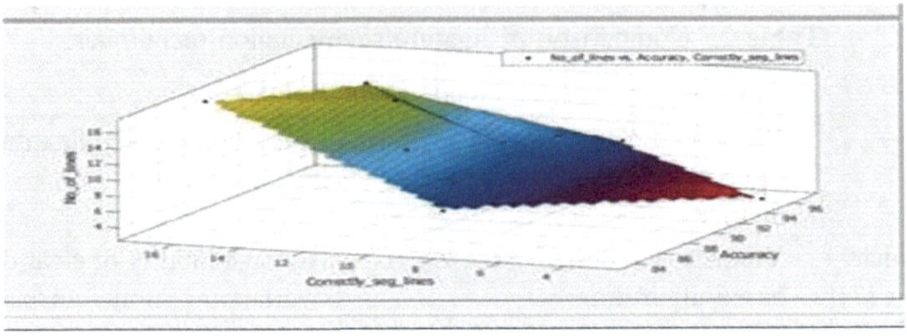

Fig. 13. Correctly segmented lines with respect to the total number of lines and their corresponding accuracies.

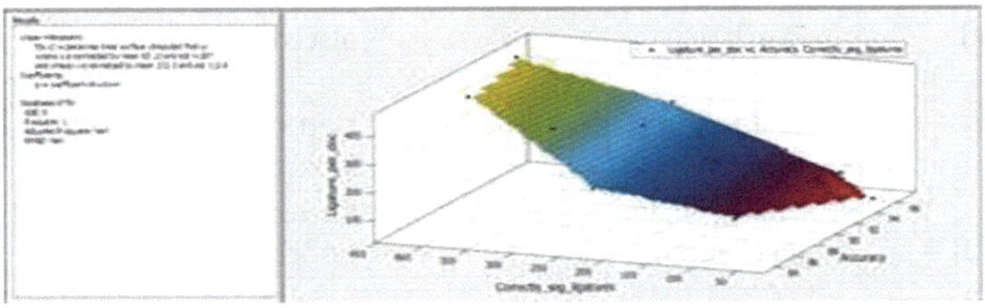

Fig. 14. Correctly segmented ligatures with respect to the total number of ligatures and their corresponding accuracies

6 Conclusion

In this paper, we presented the two approaches for Segmentation of Nastaliq Urdu text viz, Line Segmentation and Ligature Segmentation that constitutes an important step in Urdu Character Recognition. These techniques are able to place dots and diacritics in a more accurate fashion rather than [10–12]. In addition to this, it takes into consideration height, width, baseline, for ligature segmentation. This work can be employed by other languages as well like Pashto, Persian, Siraikietc, which are based on Nastaliq Script [12]. In future, we intend to work on handwritten text documents and evaluate how varying styles impact the varying writing styles of writers.

References

1. Daud, A., Khan, W., Che, D.: Urdu language processing: a survey. Artif. Intell. Rev. **47**, 1–33 (2016)
2. Pal, U., Sarkar, A.: In: International Conference on Document Analysis and Recognition, vol. 2, pp. 1183–1187 (2003)
3. Ul-Hasan, A., Ahmed, S.B., Rashid, F., Shafait, F., Breuel, T.M.: In: Document Analysis and Recognition (ICDAR), pp. 1061–1065. IEEE (2013)
4. Din, I.U., Malik, Z., Siddiqi, I., Khalid, S.: J. Appl. Environ. Biol. Sci **6**, 114–120 (2016)
5. Amad, I., Wang, X., Li, R., Ahmed, M., Ullah, R.: Line and ligature segmentation of urdu nastaleeq text. IEEE access **5**, 10924–10940 (2017)
6. Malik, H., Fahiem, M.A.: Segmentation of printed urdu scripts using structural features. In: Second International Conference in Visualisation, 2009. VIZ 2009, pp. 191–195. IEEE (2009)
7. Kumar, K.S., Kumar, S., Jawahar, C.: In: Document Analysis and Recognition (2007)
8. Breuel, T.M.: In: International Workshop on Document Analysis Systems, pp. 188–199. Springer (2002)
9. Bukhari, S.S., Shafait, F., Breuel, T.M.: In: Document Analysis and Recognition (ICDAR), pp. 748–752. IEEE (2013)
10. Javed, S.T., Hussain, S.: In: Multitopic Conference INMIC, pp. 1–6. IEEE (2009)
11. Lehal, G.S.: In: Document Analysis and Recognition (ICDAR), pp. 1130–1134. IEEE (2013)
12. Hussain, S., Ali, S., et al.: Nastalique segmentation-based approach for Urdu OCR. Int. J. Doc. Anal. Recog. (IJDAR) **18**(4), 357–374 (2015)

Aspect Extraction and Sentiment Analysis for E-Commerce Product Reviews

Enakshi Jana(✉) and V. Uma

Department of Computer Science, Pondicherry University,
Kalapet, Pondicherry, India
enu.13jana@gmail.com, umabskr@gmail.com

Abstract. Throughout the globe, with the immense increase of the number of users of the internet and simultaneously the massive expansion of the e-commerce platform, millions of products are sold online, and users are more involved in shopping online. To improve user experience and their satisfaction, online shopping platform enables every user to give their feedback, rating, and review for each and every product that they buy online to help other users. Some popular products on a leading e-commerce platform have of thousands of reviews. Many of those reviews are long and contains only a few sentences which are related to a particular feature of a product. Thus, it becomes really hard for a customer to understand a review and make a decision in buying that product. Manufacturer also need to keep track of customer review regarding the different features of the product to improve the sales of poorly performed one. It becomes very difficult for the user and manufacturer of the product to understand customer view about different features of the product. So, we need accurate opinion-based product review sentiment analysis which will help both customers and product manufacturer to understand and focus on a particular aspect of the product.

This paper proposes the idea of aspect wise product review sentiment analysis. This work explains the methods that can be used for aspect and opinion identification from product reviews. The comparison of different machine learning algorithms used for sentiment analysis of the reviews is also presented. This paper shows that logistic regression with L1 regularization performs best as compared to other algorithms in performing sentiment classification. L1 regularization is good for high dimensional data with multicollinearity among features. This work concludes that text classification with proper regularization is crucial for good accuracy.

Keywords: Aspect extraction · Opinion mining · Sentiment analysis · Support Vector Machine · E-commerce

1 Introduction

Nowadays, with the immense increase of internet users, e-commerce platform is getting more popular and people are buying product starting from clothes to large home appliances. Before buying an online product, people try to get exact insight about the product from the product reviews given by other customers. Product manufacturer also

D. J. Hemanth et al. (Eds.): COMET 2019, LNDECT 35, pp. 780–791, 2020.
https://doi.org/10.1007/978-3-030-32150-5_79

tries to improve their products with respect to the poorly performed features from these reviews.

With more and more user buying product online these reviews are increasing in size at a larger scale. Some popular products on a leading e-commerce platform have thousands of reviews. Out of these thousands of reviews, many reviews are long and contains only a few sentences which are related to a particular feature of that product. Thus, it becomes really hard for a customer to understand a review and make a decision regarding buying that product. So, there is a demand for accurate review summarization. But, summarization of reviews in e-commerce platform is not as simple as news article summarization. In e-commerce, users are more likely to analyse reviews, product aspect or feature wise. For example, while buying a smartphone from an e-commerce platform some user would only like to analyse the reviews related to battery life and camera quality whereas some other user would consider battery life and internal memory. So, we need product review sentiment analysis aspect wise.

[1] uses SVM, Naïve Bayes and K-Nearest Neighbours classifier for sentiment polarity detection. [2] uses Naïve Bayes, Logistic Regression and and SentiWordNet algorithm for sentiment classification for reviews from Amazon.com. [3] uses different classification algorithm for analyzing sentiments of anger, anticipation, disgust, fear, joy, sadness, surprise and trust for each of the customer reviews. [4] uses Random Forest and Naïve Bayes for sentiment classification and also handle the issue of multiclass classification problem.

This paper discusses different methods for aspect detection and opinion identification from product reviews. It explains how different features can be used for aspect detection. Parsing structure of a review is one of the crucial feature for aspect detection and opinion identification related to that aspect. This paper uses different machine learning algorithms for sentiment classification of the product reviews and provides a comparison among them. This paper shows that logistics regression with L1 regularization is best for high dimensional data like sentiment classification with multicollinearity among features.

In this paper, we start with the previous work which has been done on e-commence product review sentiment analysis. In subsequent sections, we discuss our contribution, implementation and the comparison of different models for product review sentiment analysis.

2 Previous Works

[1] uses fuzzy theory using FL-SVM approaches to calculate the sentiment polarity for emotional words. [1] shows that this FL-SVM approaches produce better classification results as compared to Naive Bayes or K-Nearest Neighbors algorithms by the margin of 1 to 3% on different datasets. [2] discusses about the opinion mining and sentiment classification of huge online review from e-commerce platform Amazon.com. [2] uses Naïve Bayes classifier, Logistic Regression and SentiWordNet algorithm for polarity detection of e-commerce product reviews and compares their accuracy metric. [3] uses the massive online reviews for mobile phones from e-commerce platforms Amazon.in and flipkart.com for analyzing sentiments of anger, anticipation, disgust, fear, joy,

sadness, surprise and trust for each of the customer reviews. This fine gained sentiment analysis helps the actual buyers to analyze the product in more detail. [3] uses different classification algorithms for this multiclass classification problem. [4] uses Random Forest technique to improve the sentiment analysis of product reviews in Kannada language. [4] shows improvement of accuracy of the review polarity detection by 7% over Naive Bayes classifier using Random Forest technique. [4] also addresses the issue of multiclass classification problem of the previous work and handle the conditional statements more efficiently [5]. Uses dependency parsing based method to contract the feature vector and a weighted novel algorithm to use in the Chinese review's sentiment analysis. [5] shows the effectiveness of proposed method on previous methods [6]. Proposed Support Vector Machine based sentiment analysis of smart phone product reviews. [6] shows the results and compare it using performance metrics such as Precision, Recall and F-Measure [7]. Shows a fine gained sentiment analysis of product reviews. [7] uses POS based feature to use in the classification model to detect the sentiment polarity. [8] Shows the sentiment analysis of product review on balanced dataset which is a good mixed of reviews from different product categories. Before applying sentiment analysis [8] carried out a similarity measure to correctly classify the reviews into different categories [9]. Proposed a feature based approached for sentiment analysis. [9] uses coreference resolution method to correct resolve the coherence between aspect and opinion in the review. On the generated feature [9] uses Support Vector Machine to classify the polarity of review [10]. Proposed a feature extracted based sentiment analysis method for product reviews. [10] deploy a typed dependency-based method to identify the semantic feature from the reviews.

3 Our Contribution

This work proposes the aspect aware sentiment analysis of e-commerce product reviews. The paper starts with e-commerce product page crawling for the analysis and explains different method for aspect detection and opinion identification.In the next section the implementation done for the product review sentiment analysis using SVM, Logistic Regression and Artificial Neural Network (ANN) is presented. The accurate comparison between all the methods and with previous works is presented in Sect. 5.

4 System Architecture

Figure 1 depicts the overall system architecture. This work starts with the data set collection by crawling popular ecommerce platform for product reviews. In subsequent section it performs data pre-processing, tokenization, feature extraction, classification, aspect extraction and grouping classified reviews aspect wise. In preprocessing it performs data cleaning, stop word removal, lemmatization, and tokenization. For feature extraction it uses TF-IDF (Term Frequency – Inverse Document Frequency) transformation. Support Vector Machine, Logistic Regression and Artificial Neural

Network are used for review classification and it uses dependency parsing and heuristic to extract aspect from the classified feature and group them aspect wise.

5 Aspect Extraction Techniques

This section discusses the background and implementation details of this work. This section starts with the discussion of different methods and features for aspect identification and opinion detection. The subsequent subsection shows the implementation and comparison of different machine learning algorithms in performing sentiment classification.

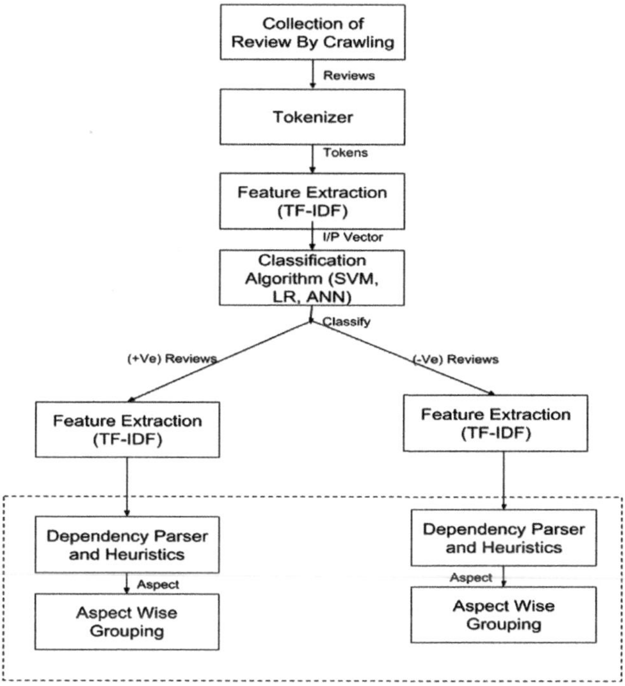

Fig. 1. Overall system architecture

5.1 Aspect and Opinion Identification from Product Reviews

Consider an example review for a mobile phone in a leading e-commerce platform. *"The display is very bright and impressive camera quality is upper average otherwise mobile is superb"*. In the review ***display***, ***camera*** and ***mobile*** are the aspects and ***very bright, impressive*** and ***superb*** are opinions for the aspects (*display, camera, and mobile*) respectively.

Aspect and opinion extraction can be done with the combination of simple NLP feature and basic heuristics. As NLP feature, POS (parts of speech) tag and dependency parsing of reviews are sufficient for the extraction. As simple heuristics, frequency count with a threshold can be used for this purpose. The aspect and opinion extraction can be posed as a classification problem where there is a need to classify each token of a review as aspect, opinion or others. For classification, we need training data. But empirical results show that unsupervised simple heuristic-based method performs better for this process. This paper uses simple POS tagging with dependency parsing to identify the aspects and opinion.

5.1.1 POS Tagging

Figure 2 shows a review with parts of speech tag (POS). From the POS tag it's clear that aspect is always a noun. But not all the nous is an aspect. In Fig. 2, nouns are Phone and Display but the noun display is only an aspect. Opinion is always an adjective modifying the noun aspect. So, there is a need to identify the (noun, adjective) pair from the reviews and also it is necessary to filter only those nouns which are aspects.

Fig. 2. POS tagging of a review

5.1.2 NN and AMOD Modifier

To unambiguously identify those nouns which are an aspect, it is necessary to parse the dependency between all the POS tags of a reviews. Figure 3 is an example of dependency parsing of a review sentence where amod(adjective modifier) relationship shows the association between (noun and adj) or (aspect, opinion).

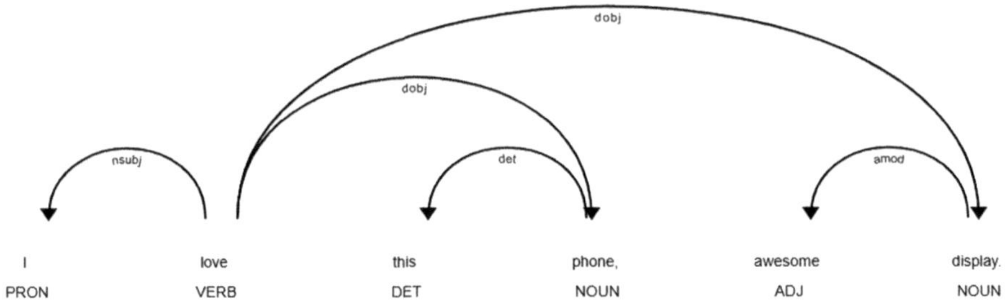

Fig. 3. NN and AMOD modifier in parsing

5.1.3 Aspect Frequency

From dependency parsing, we get to know that the aspect and opinion have the amod relationship where the adjective modifies the noun. So, aspect is always a noun paired with an adjective in the reviews. But not all noun and adjective pairs are aspect and opinions. Here, a heuristic-based approach can be used on the frequency of the noun and adjective pair. Two thresholds $\boldsymbol{Thres_{min}}$ and $\boldsymbol{Thres_{max}}$ can be defined with following rules -

if *freq(noun, adj)* > $Thres_{min}$&*freq(noun, adj)* < $Thres_{max}$: **valid pair**
else: **an Invalid pair**

5.1.4 The Phases (Beyond Unigrams)

Till now, focus is on unigram aspect and opinion related to that aspect. Sometimes. opinion may be beyond the boundary of unigrams. This work only considers the unigram and bi-gram opinions. Consider the following Fig. 4 for dependency parsing of a review.

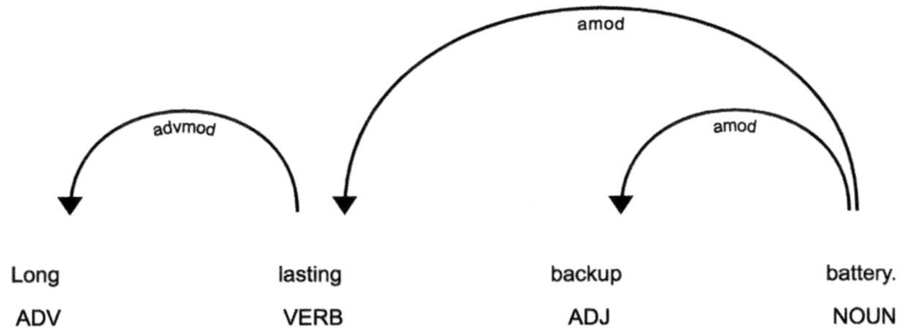

Fig. 4. Identify phase structure beyond unigram

The review in the above figure has two amod relationships (*'battery'*, *'backup'*) and (*'battery'*, *'lasting'*). The *'lasting'* is a verb which is again modified with the ADV modifier *'Long'*. So, there is a transitive relationship from (*'Battery'*, *'lasting'*) to (*'lasting'*, *'Long'*). As *'lasting'* is modified by *'Long'*, we can consider *'Long lasting'* as a compound (bi-gram) opinion of the aspect *battery*.

6 Sentiment Classification Using SVM

This section describes different classification algorithms for review sentiment classification settings. In this work, TF-IDF is used for feature generation in order to apply the machine learning algorithms.

6.1 TF-IDF Feature

In review sentiment analysis, the TF-IDF generates a vector for each review of size equal to the number of tokens in the corpus. TF is the term frequency which count the number of occurrences of a token in a particular review. IDF or Inverse Document Frequency is used to penalize a token which occurs in most of the documents. The tokens like stop words appear in most of the documents and do not have any statistical strength in the corpus. So IDF is used to penalize them and the Eq. 1 is used I calculation.

$$W_{i,j} = tf_{i,j} \times \log \frac{N}{df_i} \tag{1}$$

$W_{i,j}$ = The TF-IDF score of a token i in document j. $tf_{i,j}$ = Number of occurrences of i in j. df_i = Number of documents containing i. N = Total number of documents. In our implementation, the size of tf-idf vector for each review is 1645. That means the total number of unique token in the corpus is 1645.

6.2 Support Vector Machine

Support Vector Machine is one of the popular supervised classifier which is used for both linear and non-linear classification problem. SVM try to construct a separating hyperplane in feature space to maximize the separation between classes. Figure 4 shows the separating hyperplane and the margins for two class classification problem in 2 dimensional feature space.

In machine learning, **support vector machines** are supervised learning models with associated learning algorithms that analyse data and can be for classification and regression analysis. Given a set of training examples, each instance can be classified as belonging to one or the other of two categories. The SVM training algorithm builds a model that assigns new examples to one category or the other, making it a non-probabilistic binary linear classifier. An SVM model is a representation of the examples as points in space. The mapping of the examples are done in a way that the separation between classes are maximized as widely as possible. Examples, those are on the margin are called support vectors. When a new example is presented to the model, based on which side of the margin the class is predicted. SVM are robust to outlier as the model highly depends on supports vector. So outlier has very minimal effect on the margin. But SVM are sensitive to the class imbalance problem. In addition to performing linear classification, SVMs can efficiently perform a non-linear classification using what is called the kernel trick, implicitly mapping their inputs into high-dimensional feature spaces.

In SVM model w.x + b = 0 is the separating hyperplane and (w.x + b = −1 & w. x + b = +1) are the two margins for classification problem with 2-dimensional feature space. The distance between the margins are

$$\frac{||w||}{2} \tag{2}$$

Where w is the parameter to learn in the model. SVM used quadratic programming problem to solve the dual equation.

7 Evaluation Metric and Dataset for Classification Algorithms

There are different metrics which are used for evaluating the performance of a classification algorithms. These are precision, recall, sensitivity, specificity and accuracy. Before going into more details of these metrics we need to know about TP, FP, TN, and FN. The inner circle In Fig. 5 which is correct decision for positive is called true positive and incorrect decision for positive is called false positive. Those examples which are declared as negative by the classifier but actually are positive are called false negative. Those examples which are negative and correctly identified as negative by classifier are called true negative. Figure 5 also depicts how precision and recall are calculated using TP, FP, and FN. Equations 3, 4, 5, 6 and 7 describe the mathematical formulae used for calculating different performance evaluation metrics.

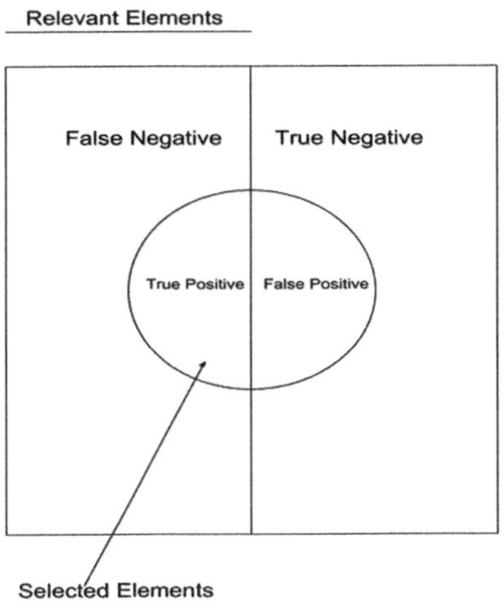

$$Sensitivity = \frac{TP}{TP+FN} \times 100\% \tag{3}$$

$$Specificity = \frac{TN}{TN+FP} \times 100\% \tag{4}$$

$$Accuracy = \frac{TP+TN}{TP+FP+TN+FN} \times 100\% \tag{5}$$

$$Precision = \frac{TP}{TP+FP} \times 100\% \tag{6}$$

$$Recall = \frac{TP}{TP+FN} \times 100\%. \tag{7}$$

Fig. 5. Performance metrics for classification algorithm

7.1 Datasets

We crawled product page reviews from a leading e-commerce platform using beautiful soup and selenium. In our dataset, we have good mixture of product reviews from all the categories. We have product reviews from electronics, fashions, furniture, sports goods, groceries, and hardware tools. For electronics, we have a mixture of review from mobile and accessories, computer and accessories, camera and accessories and home appliances. Fashions reviews are from men's and women's fashion which contains reviews from footwear, clothing, watches. This good mixture of reviews is very crucial for generalizing the model for any given test reviews and will help to achieve better accuracy. We have a total dataset of 26000 reviews. For our experimentation, we divide the dataset into 80% and 20% which represents the size of the training and test dataset respectively. Figure 6 shows the distribution of positive and negative reviews in the dataset.

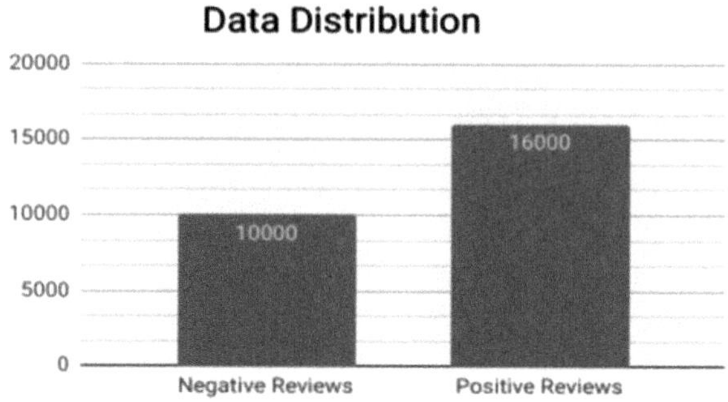

Fig. 6. Positive and negative review in the dataset

Table 1 are few examples of sample reviews from the dataset -

Table 1. Reviews in dataset.

Review	Label
The original set I purchased had a defect in the tongue where the laces cross through	Negative
Excellent and very good product. Surpassed my expectations. Really solved my problem	Positive
The style of these shoes is amazing.... but they are much much too narrow in comparison with other brands	Negative
The light weight would probably make them good to great for walking	Positive

8 Experimentation Results

Table 2 provides comparison of different classification algorithms for sentiment classification. Table 2 shows the performance values of Support Vector Machine with different kernel, Logistic Regression with and without regularization and Artificial Neural Network with different size of hidden layers. Figure 7 shows the precision and recall for artificial neural network for different number of hidden units at the hidden layer. Figure 7 shows with the increasing the number of hidden units does not help much to improve the performance of the model.

Table 2. Comparison of different classification algorithms in sentiment classification.

Model	Precision	Recall
Support Vector Machine (**linear kernel**)	89.0%	89.7%
Support Vector Machine (**Gaussian Kernel**)	61.5%	78.4%
Logistic Regression	88.5%	89.2%
Logistic Regression with l1 regularization (C = 3.5)	**89.2%**	**89.8%**
Artificial Neural Network (**hidden layer size (5, 2)**)	88.1%	88.6%
Artificial Neural Network (**hidden layer size (7, 2)**)	61.5%	78.4%
Artificial Neural Network (**hidden layer size (10, 2)**)	61.5%	78.4%
Artificial Neural Network (**hidden layer size (20, 2)**)	61.5%	78.4%

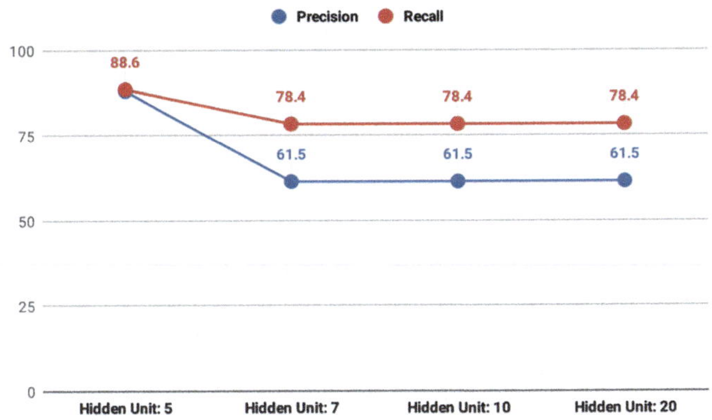

Fig. 7. Number of hidden unit vs precision and recall

9 Observation

From the above comparison it's clear that the logistic regression with l1 regularization performs better than other models for both precision and recall. L1 regularization performs feature selection on high dimensional multicollinearity data. For review classification, each word/term in the tf-idf vector works as a feature. For this classification, the size of the vocabulary or the size of the feature of tf-idf is very high and has many correlated features or words. Because of this multicollinearity and high dimension, l1 regularization performs better as compared to the other methods.

We find that support vector machine with linear kernel performs better for text classification than support vector machine with a Gaussian RBF kernel. The justification is, in text classification with tf-idf features we have a large number of dimensions. It is hoped that in high dimension, data are linearly separable. So we don't need to explicitly transform the data to higher dimensions for better classification. Because of this high dimension of text data, the linear kernel performs better than RBF kernel.

We experimented with one hidden layer neural network with a different number of neurons in the hidden layer. Hidden layer with 5 neurons performs best compared to other sizes of neurons in the hidden layer. With the increase of neurons, the network becomes overfitted. So, as we increase the number of neurons there is no hope of improvement in accuracy.

10 Conclusion

In this work, we have proposed the methods for aspect extraction. Then, we focus on sentiment classification for e-commerce product reviews. We crawl thousands of reviews from a leading e-commerce platform with a good mixture of all the product category. We experiment with different algorithms for sentiment classification. We find logistic regression with l1 regularization performs best as compared to other algorithms. L1 regularization is good for high dimensional data with multicollinearity data. From this experimentation, we can confirm that for text classification proper regularization is crucial for good accuracy. We conclude this work with comparison of different methods for sentiment classification. In future, we will focus on aspect based sentiment analysis and review summarization.

References

1. Liu, Y., Lu, J., Shahbazzade, S.: Sentiment classification of e-commerce product quality reviews by FL-SVM approaches. In: 2018 IEEE 17th International Conference on Cognitive Informatics & Cognitive Computing (ICCI*CC), Berkeley. https://doi.org/10.1109/ICCI-CC.2018.8482058
2. Kumar, K.L.S., Desai, J., Majumdar, J.: Opinion mining and sentiment analysis on online customer review. In: 2016 IEEE International Conference on Computational Intelligence and Computing Research (ICCIC), Chennai, India. https://doi.org/10.1109/ICCIC.2016.7919584
3. Singla, Z., Randhawa, S., Jain, S.: Statistical and sentiment analysis of consumer product reviews. In: 2017 8th International Conference on Computing, Communication and Networking Technologies (ICCCNT), Delhi, India. https://doi.org/10.1109/ICCCNT.2017.8203960
4. Hegde, Y., Padma, S.K.: Sentiment analysis using random forest ensemble for mobile product reviews in Kannada. In: 2017 IEEE 7th International Advance Computing Conference (IACC), Hyderabad, India. https://doi.org/10.1109/IACC.2017.0160

5. Lizhen, L., Wei, S., Hanshi, W., Chuchu, L., Jingli, L.: A novel feature-based method for sentiment analysis of Chinese product reviews. China Commun. **11**(3) (2014). ISSN 1673-5447
6. Kumari, U., Sharma, A.K., Soni, D.: Sentiment analysis of smart phone product review using SVM classification technique. In: International Conference on Energy, Communication, Data Analytics and Soft Computing (ICECDS) (2017)
7. Wan, Y., Nie, H., Lan, T., Wang, Z.: Fine-grained sentiment analysis of online reviews. In: 12th International Conference on Fuzzy Systems and Knowledge Discovery (FSKD), Zhangjiajie, China (2015)
8. Sudhakaran, P., Hariharan, S., Lu, J.: Classifying product reviews from balanced datasets for sentiment analysis and opinion mining. In: 2014 6th International Conference on Multimedia, Computer Graphics and Broadcasting, Haikou, China (2014)
9. Krishna, M.H., Rahamathulla, K., Akbar, A.: A feature based approach for sentiment analysis using SVM and coreference resolution. In: 2017 International Conference on Inventive Communication and Computational Technologies (ICICCT), Coimbatore, India (2017)
10. Devasia, N., Sheik, R.: Feature extracted sentiment analysis of customer product reviews. In: 2016 International Conference on Emerging Technological Trends (ICETT), Kollam, India (2016)

Hierarchical Clustering Based Medical Video Watermarking Using DWT and SVD

S. Ponni alias sathya[1], N. Revathi[2(✉)], and M. Rukmani[2]

[1] Department of Information Technology, Dr. Mahalingam College of Engineering and Technology, Pollachi, India
sathyaashok2007@gmail.com

[2] Bachelor of Information Technology, Dr. Mahalingam College of Engineering and Technology, Pollachi, India
riarevathi@gmail.com, rukmani1775@gmail.com

Abstract. A video watermarking technique is proposed in for securing medical videos by adapting new clustering algorithm. As per law the information need to keep secure in order to protect the privacy of the patient. Cybercriminals may sell the medical video with the patient's fake identities which leads to data insecurity and various criminal activities. And so the technique is mainly for maintenance and confidentiality purpose of medical video. This is designed in a way to cluster the medical video frames which incorporates a Euclidean distance of frames. The Hierarchical representation is constructed for every cluster to choose key frame with the entropy and Probability Density Function (PDF) value of the frames. Discrete Wavelet Transform (DWT) and Singular Value Decomposition (SVD) improve the performance of the watermark embedding process, where the watermark image is chosen to embed into the selected key frames for every cluster. The experimental results show that the proposed scheme has higher robustness and imperceptibility against various image and video processing attacks.

Keywords: Video watermark · Clustering · Hierarchical · Key frame · DWT · SVD · Medical video

1 Introduction

In the last few years, recent advancements in computers and communications have brought in a huge market for distributing digital multimedia data through the Internet. Multimedia data such as video, audio and images are currently used in wider range in many applications such as medicine, education, e-commerce, digital libraries, digital government, etc. The worldwide Internet connectivity has created a perfect stage for copyright fraud and uncontrollable distribution of multimedia causes the issue of multimedia content security threats. In past, the encryption techniques were commonly used to protect the ownership of media. Recently, the digital watermarking techniques are utilized to protect the copyright more securely.

The Digital watermarking techniques consist of video watermarking and image watermarking as their categories. Video Watermarking provides robustness against the

D. J. Hemanth et al. (Eds.): COMET 2019, LNDECT 35, pp. 792–805, 2020.
https://doi.org/10.1007/978-3-030-32150-5_80

hackers or intruders attack by using blind watermarking. It also provides one's ownership and so it is applicable in any field especially for security purpose.

Growth in medical equipment can capture the clinical videos of patients are stored in hospital databases and shared with different experts for getting more prescriptions. Transmission of large amount of data in internet from the hospital database results in excessive memory utilization it makes easily accessible for unauthorized users. For reducing the storage, data hiding technologies are used for embedding patient information with their medical report images. The medical videos and information of patients are hacked easily by increased usage of Telemedicine applications. In the paper, a new algorithm for video watermarking is proposed to the medical based applications.

2 Related Works

The existing works might dealt and solved with various problems. This section will discover some of the related works along with its technique. These works can introduce and enhance solution for the existing systems as well as basement for the proposed methodology.

An encoding technique as DCT is watermarked by the local variant feature points of Harris-Affine detector proposed in [1]. Decoding and down sampling is done with DCT. The watermark is extracted by the using spatial domain.

Video watermarking with BWT using optimization techniques is proposed in [2] is to protect the copyright of images. An artificial bee colony algorithm is done in embedding process for generating the random frame.

From the non-moving parts of the video, the block level watermark embedding is done with the help of two-level DWT, SVD algorithm and the entropy analysis in [3]. Finally, the moving parts of video frames are added to the watermarked frame. Then QR decomposition is also performed.

DWT and SVD algorithm in high frequency band for embedding watermark in the host video sequences. Then extract the watermarked image into four sub bands by using SVD for obtaining singular values and apply IDWT to get back the watermark image in [4].

Watermark is embedded in the scenes of video by using Histogram Difference Method in [5]. Watermark insertion is in the luminance part of the video frame. By applying LWT video frame decomposed into sub bands and then SVD is to decompose the LL sub part into U, S and V components. Inverse SVD and Inverse LWT are applied to extract the watermark to get the host video frames.

Reference [6] focuses DWT used to decompose the video frames into sub-bands. PCA applied for maximum entropy block selection and transformation, then QIM is for quantizing the PCA generated blocks. Watermark is embedded into the selected suitable quantizer values. Extraction is done by the secret keys of a uniform quantizer.

In [7] Embedding watermarking is done based on DWT domain and the secret key generated by the master share from the owner's share based on frame mean in original video and identification's share based on the frame mean of attacked video. Extraction is based on the master share generation of the visual cryptography method.

Reference [8] proposed an approach to embed image in video as watermark using a mobile device by directly modifying the pixel values of the original image the embedding watermark is done in spatial domain and in frequency domain by DCT, DFT, DWT, SVD, CWT techniques and Bit stream watermarking.

A model is proposed in [9] for video copyright protection using Chaotic maps and SVD in wavelet domain. It uses permutation of watermark image by partitioning as blocks.

A Fibonacci based key frame selection and scrambling for video watermarking in DWT–SVD domain is proposed in [10] which is also provided video copyright. It employs scene change detection technique to identify frequent change in the scene and selected with Fibonacci sequence generated key frames.

Some of the limitations found in the literature survey are as following:

- Computational cost of the system will be higher for real time videos.
- Frame dropping may cause distortion of watermark image in [1, 3–6]
- Except in [2, 7, 9] and [10], the protection of Ownership rights is not verified.
- The level of robustness against the various attacks are not verified in [7, 8]
- Contributions of the proposed methodology are as following:
- Computational cost of the system for real time videos will be lower.
- Watermark is embedded into blocks so that frame dropping cannot affect the watermarked image.
- Copyrights protection
- Robust against various image and video processing attacks.

3 Proposed Methodology

Medical recorded videos are mainly for providing knowledge about the treatment and continue caring of patient by the doctor. This record is vital for all health care providers involved in the patient's surgery period. Electronic health records have more opportunities for identity theft and data breaches. So the medical video is imported for preventing medical record videos from medical identity theft. Nowadays the identity theft is a main issue in society, because it has high value in black market. Patients having similar name and birth date will provide a chance for obtaining pharmaceuticals, medical care through the help of Medicare and Medicaid and sometimes can commit insurance fraud. The methodology of the proposed system is to overcome all the above issues and provide ownership rights based on the new clustering algorithm. Firstly preprocess the input medical video [11]. The algorithm involves finding of the Euclidean distance among every two frames of the input medical video and cluster the frames based on certain threshold value. For every frames in a cluster an entropy and PDF values are found to form a Hierarchical structure that isbinary tree. The nodes of the binary tree will represents the frame numbers and the selection of key frames will be from the lowest height of the binary tree. An authenticated image is selected as watermark [12] and it is preprocessed. The important step of the proposed system is Watermark embedding process with functions such as DWT and SVD to make the process efficient and easier. Embedding of the watermark image is happened by dividing watermark as blocks and embed those blocks to the selected key frames of every cluster. The embedding order may differ sequentially for every cluster. Watermark extraction process is carried out as vice versa of watermark embedding process in which the watermark image is retrieved. The robustness is ensured with various attacks to show the level of authentication and security.

3.1 Video Preprocessing

The video input partitioned into number of frames. This step is the initiation to group the frames which has similar distant. The video preprocessing will lessen the complexity and enhance the overall process.

3.2 Euclidean Distance

In general Euclidean distance is the straight- line distance between two points or pixels or even an image. In the proposed work a clustering algorithm is incorporated to compute the distance between each pair of frames and to cluster the frames. It is defined as

$$\mathrm{sum}((h1 - h2)^2)^{0.5}$$

where h1 is double value of i image, h2 is double value of i + 1 image, i = 1: (no. of frames−1)

3.3 Frame Clustering

Clustering is the process to group the set of objects within a cluster it contain similar ones (have same properties or features). Frame Clustering uses clustering algorithms for grouping similarity of the frames. After video preprocessing, foremost step is to cluster the frame. With similar range of threshold value for Euclidean distance, the frames are stored into different number of clusters.

3.4 Entropy

Entropy is a statistical measure of randomness and it will be helpful to categories the input frames based on its texture. It is one of the important properties of image and it is helpful to choose key frame image. Entropy is defined as

$$\mathrm{entropy(image)} = -\mathrm{sum}(p. * \log2(p))$$

where p holds histogram counts of imhist which is defined in MATLAB.

3.5 PDF

PDF is a function of a continuous random variable, gives the probability as integral across an interval that the value of the variable lies within the same interval.

$$\text{pd1} = \text{makedist}('\text{Normal}', 0, 1)$$

$$\text{y} = \text{pdf}(\text{pd1}, \text{ent1})$$

where pd1 is a standard normal distribution object with the mean equal to 0 and the standard deviation equal to 1. ‘Normal’ is the Normal distribution is the most used statistical distribution from its family because of the many physical, biological, and social processes that it can model. ent1 is an array which contains entropy values to calculate the PDF.

3.6 Hierarchical Representation

Hierarchical representation is the process of linking the objects or nodes either directly or indirectly and either vertically or diagonally. Binary Tree is one of the hierarchical representations and it is used by the key frames selection algorithm in the proposed method. The binary tree is plotted for each and every cluster. Figure 1 shows the binary tree representation for cluster 1.

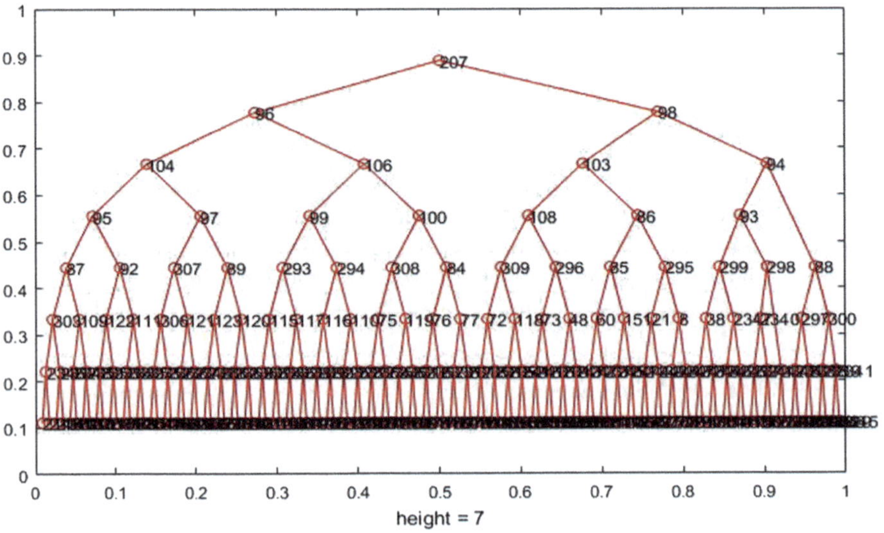

Fig. 1. Hierarchical representation for cluster 1

3.7 Key Frame Selection

The embedding of watermark is usually done for all frames of the video. But in the proposed system, the new technique called the key frame selection algorithm is used.

Key Frame Selection Algorithm

- Compute the number of frames for input video file and store the frames.
- Determine Euclidean distance among every two frames and based on it group them into clusters with desired threshold value.
- Compute the entropy for frames and plot binary tree for each clusters.
- And also compute PDF say Normal Distribution. With above procedure choose n nodes of binary tree as key frames where entropy increases and PDF decreases for each cluster.

$$\text{No. of Key frame chosen (n)} = \frac{\text{No.of frames in current cluster}}{\text{No.of clusters}}$$

In Table 1, the key frames selected for each and every cluster is detailly shown. And in Table 2, key frames chosen for cluster 1 along with its entropy and PDF value is shown.

Table 1. Key frame selection

Cluster	No. of frames	No. of key frames	Key frames
1	235	24	207, 96, 104, 94, 103, 95, 97, 98, 93, 100,89, 99, 85, 84, 88, 87, 86, 106, 92, 293, 294, 108, 121, 307
2	225	23	234, 233, 232, 228, 245, 238, 263, 239, 185, 244, 242, 226, 248, 229, 276, 174, 265, 81, 82, 173
3	219	22	365, 363, 362, 2679, 332, 333, 331, 2702, 1098, 2385, 2387, 2384, 2386, 2456, 2458, 2457, 2641, 2642, 2294, 2293, 2245, 2246
4	289	29	360, 359, 361, 367, 366, 2668, 2667, 2664, 2663, 2659, 2656, 2658, 338, 340, 2691, 339, 341, 344, 2690, 343, 2689, 342, 2684, 2681, 1096, 1094, 1095, 2653, 2652
5	268	27	350, 349, 351, 353, 352, 354, 355, 356, 372, 370, 371, 2674, 398, 392, 2673, 2675, 389, 2669, 404, 406, 377, 2660, 2661, 375, 374, 2686, 414
6	162	17	400, 399, 403, 387, 385, 383, 384, 382, 381, 410, 411, 1506, 1508, 1507, 1509, 1510, 1511
7	136	14	346, 2676, 390, 401, 2665, 2654, 201, 235, 246, 223, 190, 179, 212, 268
8	87	9	357, 368, 1512, 1523, 1999, 2554, 2632, 1966, 1977
9	160	16	1089, 1100, 1078, 2343, 2365, 2332, 2354, 2310, 2388, 2410, 2232, 2321, 2432, 2454, 2421, 2476
10	86	9	1857, 1868, 1891, 502, 768, 535, 635, 902, 669

Table 2. Key frames for cluster 1

Frames	207	96	104	94	103	95
Entropy	6.5802	6.5731	6.5718	6.5716	6.5715	6.5713
PDF	0.0158	0.0166	0.0167	0.0167	0.0167	0.0167
Frames	**97**	**98**	**93**	**100**	**89**	**99**
Entropy	6.5709	6.5709	6.5704	6.5702	6.5695	6.5695
PDF	0.0168	0.0168	0.0169	0.0169	0.0170	0.0170
Frames	**85**	**84**	**88**	**87**	**86**	**106**
Entropy	6.5681	6.5680	6.5677	6.5677	6.5672	6.5670
PDF	0.0171	0.0171	0.0172	0.0172	0.0172	0.0172
Frames	**92**	**293**	**294**	**108**	**121**	**307**
Entropy	6.5669	6.5668	6.5665	6.5661	6.5660	6.5658
PDF	0.0172	0.0173	0.0173	0.0173	0.0173	0.0174

3.8 Dwt

Input frames are transformed or compressed to get four different wavelet coefficients.

$$[\mathrm{LL, LH, HL, HH}] = \mathrm{dwt2(frame, wtype)};$$

where

LL- Low Level sub- band; HL, LH- Middle Level sub bands; HH- High Level sub-bands;

wtype- wavelet type say ‘haar’; haar involves reconstructing the matrix.

3.9 SVD

SVD is the factorization technique for real or complex matrix which returns the singular values of matrix A in descending order.

$$[\mathrm{U, S, V}] = \mathrm{svd(A)}$$

where

U- Left singular vectors, holds the columns of a matrix;
S- Singular values, hold a diagonal matrix.
V- Right singular vectors, holds the columns of a matrix

3.10 Watermark Preprocess

As the proposed model is for the medical application, an authenticated fingerprint of patient is chosen as watermark image. Because with the watermark, a patient or a user can prove their ownership easily. In its preprocessing, Watermark need to be divided into blocks to embed into the selected key frames. Even in case the some blocks of watermark is extracted by intruders but they cannot retrieve the information from it.

3.11 Watermark Embedding Process

The process involves getting the input as medical video and it is preprocessed to get sequence of video frames. As per the clustering algorithm of the proposed system, the frames are clustering. And then in every clusters apply DWT and SVD to the key frames chosen using key frame selection algorithm. This happens one after another. Choose the LH or HL sub-bands from DWT and do SVD. After watermark preprocessing, apply SVD to the blocks of watermark also. Note that watermark image is divided into blocks as per no. of key frames chosen in each cluster. The blocks of watermark are embedded into key frames with DWT and SVD obtained component values. For key frame use S component and for blocks of watermark image use S component multiplied with scaling factor and principal component. Embedding the blocks of watermark is permuted, that it if one insertion order is done for a cluster then for the next cluster it will not follow the same order. Here the Interesting factor is that the blocks of individual watermark will be embedded to key frames of a cluster and similarly for other clusters. IDWT takes place to get watermarked frames. The overall watermark embedding process is pictorially represented in Fig. 2.

Watermark Embedding Algorithm

Step 1: Input the medical Video
Step 2: Split the video into frames
Step 3: Compute Euclidean distance between the every two frames
Step 4: Obtain cluster on satisfying each case with conditions of Threshold and Euclidean distance values
Step 5: Calculate the Entropy and PDF values
Step 6: Loop through each no. of frames for a cluster do
    If Entropy is high and PDF is low, then
        Choose the Key Frames
    End If
  End loop
Step 7: Represent Hierarchical structure with values of Step 6 for every cluster
Step 8: Loop through no. of key frames in clusters
    Apply DWT for the key frame to get the LL, LH, HL and HH sub- bands
    Determine SVD for LH or HL sub- band of the key frames to get U, S and V component
  End Loop
Step 9: Read the Authenticated watermark image and divide them into blocks
Step 10: Loop through no. of key frames in clusters
    Calculate SVD values for sub-blocks to get U, S and V component
  End Loop
Step 11: Perform embedding of blocks of watermark image into the each key frame with computed DWT and SVD values in Step 8 and Step 11 and along with a certain scaling factor
Step 12: Do IDWT to get the watermarked frames
Step 13: Combine all the frames along with watermarked frames to get an output as a watermarked video

Watermark Embedding Work Flow

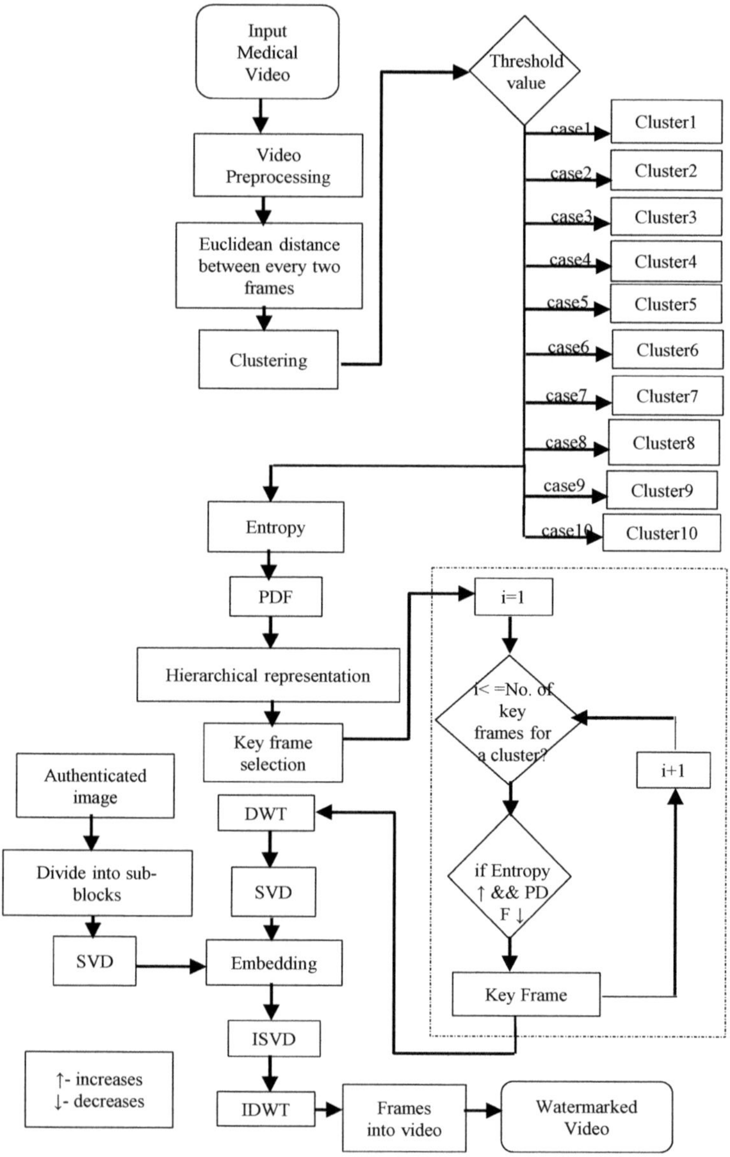

Fig. 2. Watermark embedding work flow

3.12 Watermark Extracting Process

Watermark extraction process is similar to that of Watermark embedding process, but it happens in reverse order. For conversion, DWT and SVD processes are applied to the key frames where the watermark blocks are inserted and along with the watermark matrix and Scaling factor the extraction is done. To achieve the whole watermark IDWT is carried out. The pictorial representation of watermark extracting process is shown in Fig. 3. The extracted blocks of watermark will be combined or concatenated into a whole watermark.

Watermark Extracting Algorithm

Step 1: Read Watermarked video
Step 2: Perform video preprocessing for getting frames
Step 3: Compute DWT in the LH or HL sub-bands
Step 4: Determine SVD to obtain the S component of the frames
Step 5: Apply the Extraction process
Step 6: Combine the collected blocks to get whole watermark

Watermark Extracting Work Flow

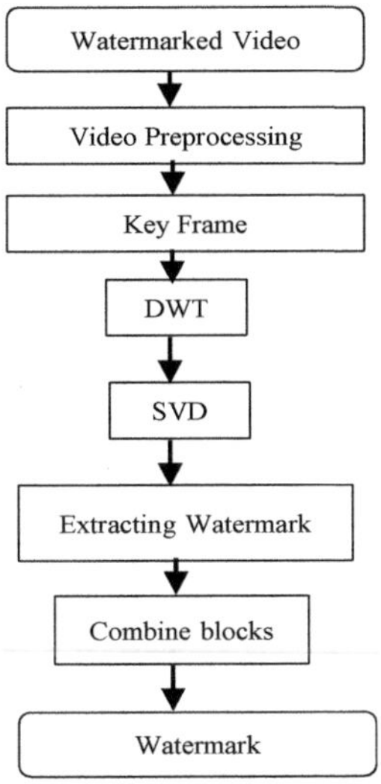

Fig. 3. Watermark extracting work flow

4 Result

Efficiency of the proposed technique can be evaluated and compared with the existing system by the values of measurement metrics such as PSNR, NCC, BER and SSIM. The quality and robustness of the watermarked video frames as well as extracted watermark can be measured. Every adjacent frames of the watermarked video are compared, summed up and averaged to get the result value. Table 3 contains of the input and resultant of video watermarking process.

Table 3. Overall video watermarking process

Input Video	**Input Watermark**	**Watermarked Video**	**Extracted Watermark**
 Medical Lecture Video			

The values of Correlation Coefficient (CC), Normalized Cross- Correlation (NCC), Peak Signal to Noise Ratio (PSNR), Bit Error Rate (BER) and Structural Similarity Index Metric (SSIM) can be calculated in MATLAB with its inbuilt function like xcorr2(), normxcorr2(), psnr(), biterr() and ssim().

4.1 Result and Discussion

In the proposed methodology, a new Hierarchical Clustering based video watermarking technique for medical input have been designed and implemented using DWT and SVD. From the result obtained it is found that the technique provides robustness and efficiency comparatively more than the existing systems.

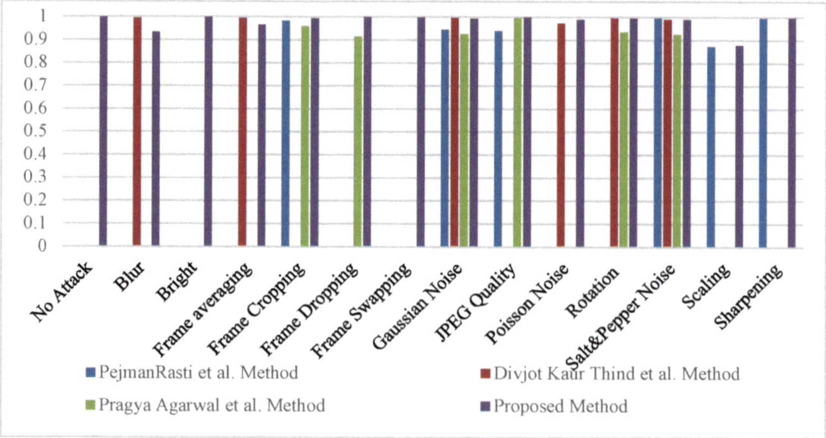

Fig. 4. CC value comparison result

Figures 4 and 5 shows the result comparison of CC, NCCvalues and Fig. 6 shows the PSNR values obtained through various attack among a proposed method and all other existing systems. Figure 7 shows the BER, SSIM, PSNR values for the various attack of the watermark of the proposed system.

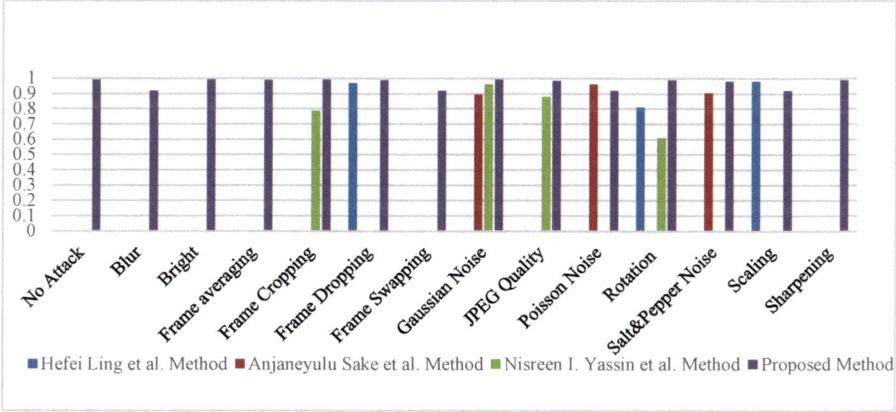

Fig. 5. NCC value comparison result

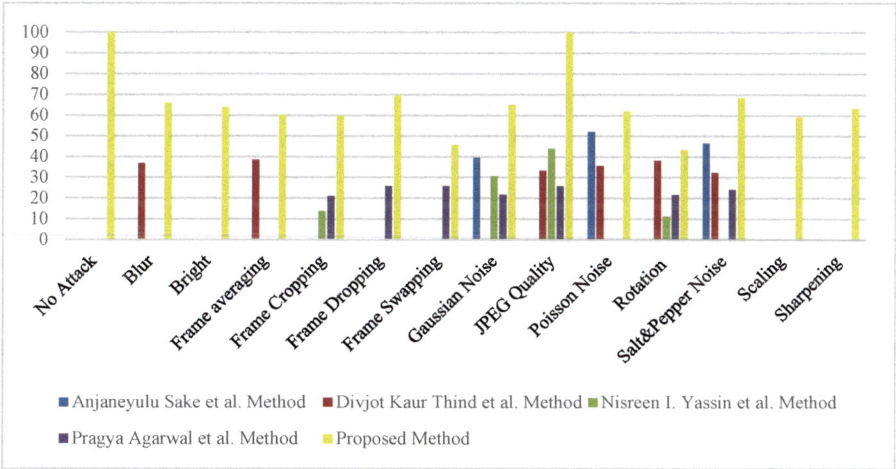

Fig. 6. PSNR value comparison result

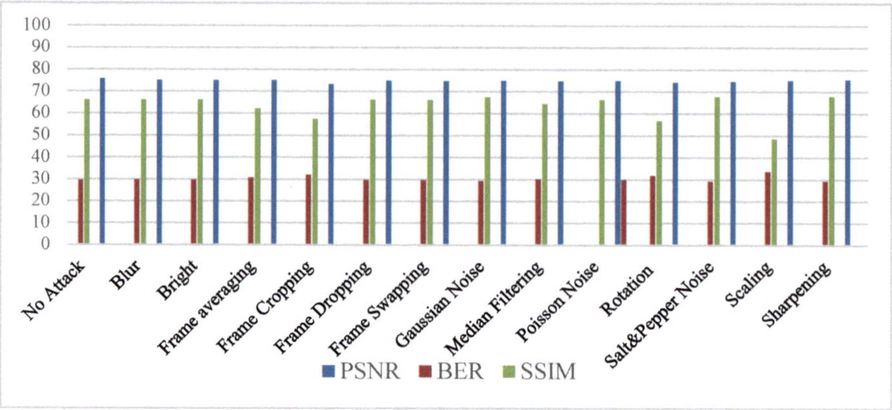

Fig. 7. BER, PSNR and SSIM value result in %

5 Conclusion and Future Scope

The result obtained from the proposed work is comparatively satisfying and solving various issues occurred in existing methods. And it is also addressed the limitations by performing the video watermarking based on clustering algorithm, so that it minimizes the computational cost for real time videos; then the frame dropping is violated because the possibility of retrieving the watermark is tedious as it is divided and embedded as blocks into the key frames of every cluster; an authenticated image is used as watermark which will proves the ownership of the one; and it is absolutely much better and robust.

The future scope is to extend the desktop application into web and mobile application by introducing Graphics User Interfaces. This will be helpful to people as they can easily download and use for the purpose of securing their personal and important multimedia content.

Acknowledgment. We are grateful to each and everyone who are constantly helped and supported us during the project work. The facilities received from our institutions made our work easier. And the guidance of faculty members broaden our minds to do the project with interest and enhanced knowledge. With the enriched motivation and encouragement of our parents and friends, the project is enthusiastically completed.

References

1. Ling, H., Wangn, L., Zou, F., Lu, Z., Li, P.: Robust video watermarking based on affine invariant regions in the compressed domain. Signal Process. **91**, 1863–1875 (2011)
2. Sake, A., Tirumala, R.: Bi-orthogonal wavelet transform based video watermarking using optimization techniques. In: Proceedings, vol. 5, pp. 1470–1477 (2018)

3. Rasti, P., Samiei, S., Agoyi, M., Escalera, S., Anbarjafari, G.: Robust non-blind color video watermarking using QR decomposition and entropy analysis. J. Vis. Commun. Image Represent. **38**, 838–847 (2016)
4. Thind, D.K., Jindal, S.: A semi blind DWT-SVD video watermarking. Proc. Comput. Sci. **46**, 1661–1667 (2015)
5. Agarwal, P., Kumar, A., Choudhary, A.: A secure and reliable video watermarking technique. In: 2015 International Conference on Computer and Computational Sciences (ICCCS) (2015)
6. Yassin, N.I., Salem, N.M., El Adawy, M.I.: QIM blind video watermarking scheme based on Wavelet transform and principal component analysis. Alexandria Eng. J. **53**, 833–842 (2014)
7. Singh, T.R., Singh, K.M., Roy, S.: Video watermarking scheme based on visual cryptography and scene change detection. Int. J. Electron. Commun. (AEÜ) **67**, 645–651 (2013)
8. Venugopala, P.S., Sarojadevi, H., Chiplunkar, N.N.: An approach to embed image in video as watermark using a mobile device. Sustain. Comput.: Inform. Syst. **15** 82–87 (2017)
9. Ramakrishnan, S., Ponni alias Sathya, S.: Video copyright protection using chaotic maps and singular value decomposition in wavelet domain. IETE J. Res. 1–13 (2018)
10. Ponni alias Sathya, S., Ramakrishnan, S.: Fibonacci based key frame selection and scrambling for video watermarking in DWT–SVD domain. Wirel. Pers. Commun. **102**, 1–21 (2018)
11. https://www.youtube.com/watch?reload=9&v=jVvQaqFOJpY
12. https://ui-ex.com/explore/fingerprint-vector-black/

Average Secure Support Strong Domination in Graphs

R. Guruviswanathan[1,2(✉)], M. Ayyampillai[3], and V. Swaminathan[4]

[1] Sri Chandrasekharendra Saraswathi Viswa Mahavidyalaya, Kanchipuram, India
guruviswa.r@gmail.com

[2] Department of Mathematics, Jeppiaar Maamallan Engineering College, Sriperumbudur, India

[3] Department of Mathematics, Arunai Engineering College, Thiruvannamalai, Tamilnadu, India
ayyam.maths@gmail.com

[4] Ramanujan Research Centre in Mathematics, Saraswathi Narayanan College, Madurai, India
swaminathan.sulanesri@gmail.com

Abstract. Let $G = (V, E)$ be a simple finite and undirected graph. Let $u \in V(G)$. Support of u denoted by $supp(u)$ is defined as the sum of the degrees of the neighbours of u. Let $u, v \in V(G)$. u is support strongly dominated by v, if u and v are adjacent and $supp(v) \geq supp(u)$. A subset S is called support strong dominating set of G, if for any $u \in V - S$, there exists $v \in S$ such that u and v are adjacent and $supp(v) \geq supp(u)$ [9]. A support strong dominating set S is called a secure support strong dominating set, if for any $u \in V - S$ there exists $v \in S$ such that $(S - \{v\}) \cup \{u\}$ is a support strong dominating set of G. The average lower domination number $\gamma_{av}(G)$ is defined as $\frac{\sum_{v \in V(G)} \gamma_v(G)}{|V(G)|}$, where $\gamma_v(G)$ is the minimum cardinality of a minimal dominating set that contains v. The average secure support strong domination number of G denoted by $\gamma_{sss}^{av}(G)$ is defined $\gamma_{sss}^{av}(G) = \frac{\sum_{v \in V(G)} \gamma_{sss}^{av}(v)}{|V(G)|}$, where $\gamma_{sss}^{av}(v) = \min\{|S|$: S is a secure support strong dominating set of G containing $u\}$. In this paper, average secure support strong domination number of complete $k - ary$ tree, binomial tree are calculated. Also we obtain average secure support strong domination number of thorn graph and $\gamma_{sss}^{av}(G_1 + G_2)$ of G_1 and G_2.

Keywords: Support · Domination · Strong domination · Average secure support strong domination number

1 Introduction

Let $G = (V, E)$ be simple finite and undirected graph. Let $u, v \in V(G)$. If $uv \in E(G)$, it is said that u and v dominate each other. A subset D of $V(G)$ is a dominating set of G if every vertex v in $V - D$ is dominated by some vertex $u \in D$. The domination number $\gamma(G)$ is the minimum cardinality of a dominating set.

D. J. Hemanth et al. (Eds.): COMET 2019, LNDECT 35, pp. 806–815, 2020.
https://doi.org/10.1007/978-3-030-32150-5_81

Let $u \in V(G)$. Support of u, denoted by $supp(u)$, is defined as the sum of the degrees of the neighbours of u. That is $supp(u) = \sum_{v \in N(u)} deg(v)$. A subset S of $V(G)$ is called a strong dominating set, if for every vertex $u \in V - S$, there exists a vertex $v \in S$ such that u and v are adjacent and $deg(v) \geq deg(u)$. The minimum cardinality of a strong dominating set of G is called the strong domination number of G and is denoted by $\gamma_s(G)$.

A subset S is called support strong dominating set of G, if for any $u \in V - S$, there exists $v \in S$ such that u and v are adjacent and $supp(v) \geq supp(u)$. A support strong dominating set S is called a secure support strong dominating set, if for any $u \in V - S$ there exists $v \in S$ such that $(S - \{v\}) \cup \{u\}$ is a support strong dominating set of G.

Henning [7] introduced the concept of average domination. The average lower domination number $\gamma_{av}(G)$ is defined as $\frac{\sum_{v \in V(G)} \gamma_v(G)}{|V(G)|}$, where $\gamma_v(G)$ is the minimum cardinality of a minimal dominating set that contains v.

In this paper the value of the minimum cardinality of a secure support strong dominating set containing a vertex is found out and average value of these numbers is calculated. This value is called average secure support strong domination number of G. Results of this value is calculated for various graphs and average secure support strong domination number of complete $k - ary$ tree and Binomial tree are calculated. Also we obtain average secure support strong domination number of thorn graph and $\gamma_{sss}^{av}(G_1 + G_2)$ of G_1 and G_2.

2 Average Secure Support Strong Domination in Graphs

Definition 2.1. Let $G = (V, E)$ be a simple finite undirected graph. A subset S is called support strong dominating set of G, if for any $u \in V - S$, there exists $v \in S$ such that u and v are adjacent and $supp(v) \geq supp(u)$. A support strong dominating set S is called a secure support strong dominating set, if for any $u \in V - S$ there exists $v \in S$ such that $(S - \{v\}) \cup \{u\}$ is a support strong dominating set of G.

Remark 2.2. V is a secure support strong dominating set.

Definition 2.3. Let $G = (V, E)$ be a simple, finite and undirected graph. Let $v \in V(G)$.

Define $\gamma_{sss}^{av}(u) = \min\{|S|$: S is a secure support strong dominating set of G containing $u\}$.

Define $\gamma_{sss}^{av}(G) = \frac{\sum_{v \in V(G)} \gamma_{sss}^{av}(v)}{|V(G)|}$.

Definition 2.4. Let G be the graph obtained from the complete graph K_m by attaching $a_i, (1 \leq i \leq m)$ pendant vertices at the ith vertex $u_i (1 \leq i \leq m)$ of K_m, where $a_i \geq 1$ for all $1 \leq i \leq m$. The resulting graph called the multi star graph is denoted by $K_m(a_1, a_2, \ldots a_m)$.

$\gamma_{sss}^{av}(G)$ for Some Well Known Graphs

1. $\gamma_{sss}^{av}(K_n) = 1$ (Since $\gamma_{sss}^{av}(u) = 1$ for every $u \in V(K_n)$).
2. $\gamma_{sss}^{av}(K_{1,n}) = \frac{2(n+1)}{n+1} = 2.$
3. $\gamma_{sss}^{av}(K_{m,n}) = \frac{2(m+n)}{m+n} = 2$
4. $\gamma_{sss}^{av}(P_n) = \begin{cases} 1, & if \ n = 2 \ ; \\ 2.5, & if \ n = 4 \ ; \\ \lfloor \frac{n}{2} \rfloor + 1, & if \ n \geq 5 \ or \ n = 1,3 \end{cases}$
5. $\gamma_{sss}^{av}(C_n) = \begin{cases} \lceil \frac{n}{3} \rceil, & if \ n = 3,4 \ ; \\ \lceil \frac{n}{3} + 1 \rceil, & if \ n \geq 5 \ ; \end{cases}$
6. $\gamma_{sss}^{av}(W_n) = \begin{cases} 1, & if \ n = 4 \ ; \\ 2, & if \ n \geq 5 \end{cases}$
7. $\gamma_{sss}^{av}(K_{a_1,a_2,\ldots,a_m}) = m$, Where $K_{a_1,a_2,\ldots,a_n}$ is a multipartite graph.
8. $\gamma_{sss}^{av}(K_m(a_1, a_2, \ldots, a_m)) = \begin{cases} m+1, & if \ a_i = 2 \ for \ at \ least \ one \ i; \\ m, & if \ a_i = 1 \ for \ all \ i \ ; \end{cases}$

Observation 2.5. If $G \neq K_n$ has a full degree vertex, then $\gamma_{sss}^{av}(G) = 2$.

Observation 2.6. $\gamma_{sss}^{av}(G) \geq \gamma_{sss}(G)$, (Strict inequality holds in P_4).

3 Average Value of Secure Support Strong Domination of Some Graphs

Definition 3.1. A complete $k - ary$ tree with depth n is a tree in which all leaves have the same depth and all internal vertices have $k - children$.

A complete $k - ary$ tree has $\frac{k^{n+1}-1}{k-1}$ vertices and $\frac{k^{n+1}-1}{k-1} - 1$ edges.

Theorem 3.2. Let G be a complete $k - ary$ tree with depth n. Then

$$\gamma_{sss}^{av}(G) = \begin{cases} k+2 & if\ n = 2; \\ \dfrac{k^2(k^n-1)}{k^3-1}+2 & if\ n = 0(\text{mod } 3); \\ \dfrac{k^2(k^{n-1}-1)}{k^3-1}+k^{n-1}+2 & if\ n = 1(\text{mod } 3) \text{ and n} \geq 4; \\ \dfrac{k^2(k^{n-2}-1)}{k^3-1}+k^{n-1}+2 & if\ n = 2(\text{mod } 3) \text{ and n} \geq 5; \end{cases}$$

Proof. Let $n \geq 3$. The support of the central vertex is $k(k+1)$, the support of any vertex at 1st - level is $k + k(k+1) = k^2 + 2k$, the support of any vertex at 2nd-level is $k^2 + 2k + 1$, the support of any vertex at $3^{rd}-$ level is $k^0 + 2k + 1, \ldots$, the support of any vertex at $(n-1)^{th}$-level is $2k + 1$ and the support of any vertex at n^{th}-level is $k + 1$. When $n = 2$, the central vertex has support $k(k+1)$, the support of any vertex at 1st-level is $2k$ and the support of any pendant vertex is $k + 1$. The minimum cardinality of a support strong dominating set is $k^0 + k^2 + k^5 + \ldots + k^{n-1}, (n \geq 3)$ and $k^0 + k^1$ when $n = 2$. The minimum cardinality of a secure support strong dominating set containing any prescribed vertex is $k^3 + k^2 + \ldots k^{n-1} + 1$ when $n \geq 3$, and $k^0 + k^1 + 1$ when $n = 2$.

Let $n = 2$. The $\gamma_{sss}^{av}(u) = k^0 + k^1 + 1$ for any $u \in V(G)$.

Let $n \geq 3$.

Case (i): $n \equiv 0$ (mod 3). Let $n = 3t$. The minimum cardinality of a secure support strong dominating set containing any prescribed vertex is

$$k^0 + k^2 + k^5 + \ldots + k^{n-1} + 1 = \frac{k^2((k^3)^t - 1)}{k^3 - 1} + 2$$
$$= \frac{k^2(k^n - 1)}{k^3 - 1} + 2,$$
$$\text{Therefore } \gamma_{sss}^{av}(G) = \frac{k^2(k^n - 1)}{k^3 - 1} + 2,$$

Case (ii): $n \equiv 1$ (mod 3). Let $n = 3t + 1$. The minimum cardinality of a secure support strong dominating set containing any prescribed vertex is

$$k^0 + k^2 + k^5 + \ldots + k^{3t-1} + k^{3t} + 1 = \frac{k^2((k^3)^t - 1)}{k^3 - 1} + k^{3t} + 2$$
$$= \frac{k^2(k^{n-1} - 1)}{k^3 - 1} + k^{n-1} + 2$$
$$\text{Therefore } \gamma_{sss}^{av}(G) = \frac{k^2(k^{n-1} - 1)}{k^3 - 1} + k^{n-1} + 2.$$

Case (iii): $n \equiv 2$(mod 3). Let $n = 3t + 2$. The minimum cardinality of a secure support strong dominating set containing any prescribed vertex is

$$k^0+k^2+k^5+\ldots+k^{3t-1}+k^{3t+1}+1=\frac{k^2((k^3)^t-1)}{k^3-1}+k^{3t+1}+2$$
$$=\frac{k^2(k^{n-2}-1)}{k^3-1}+k^{n-1}+2$$
$$\text{Therefore } \gamma_{sss}^{av}(G)=\frac{k^2(k^{n-2}-1)}{k^3-1}+k^{n-1}+2.$$

Definition 3.3. The binomial tree of order $n \geq 0$ with root R is the tree B_n defined as follows: if $n=0, B_n=B_0=R$. if $n>0, B_n=2$ copies of the binomial tree B_{n-1} linked by making one of the B_{n-1} trees, to the left most child of the root of the other.

Theorem 3.4. $\gamma_{sss}^{av}(B_n)=2^{n-1}+1$.

Proof. The minimum cardinality of a secure support strong dominating set of B_n containing any prescribed vertex is the number of support vertices of $B_n+1=2^{n-1}+1$. Therefore $\gamma_{sss}^{av}(B_n)=2^{n-1}+1$.

Definition 3.5. Let G be a simple graph. A vertex u is called a support strong isolate of G, if $supp(u)>supp(v)$ for every v in $N(u)$.

Definition 3.6. A vertex u is called a secure support strong good vertex, if u belongs to a minimum secure support strong dominating set of G. G is said to $\gamma_{sss}-$ excellent, if every vertex of G is secure support strong good.

Remark 3.7. Let G be a $\gamma_{sss}-$ excellent graph. Then $\gamma_{sss}^{av}(G)=\gamma_{sss}(G)$.

Remark 3.8. For any simple graph G, $(G\circ K_1)$ is $\gamma_{sss}-$ excellent.

Theorem 3.9. Let G be a simple graph without support strong isolates.
Then $\gamma_{sss}^{av}(G\circ K_1)=|V(G)|$.

Proof. Such a graph is $\gamma_{sss}-$ excellent

$$\gamma_{sss}^{av}(G\circ K_1)=\frac{|V(G)|\gamma_{sss}(G\circ K_1)}{|V(G)|}$$
$$=\gamma_{sss}(G\circ K_1)=|V(G)|.$$

Definition 3.10. Let G be a simple graph. The thorn graph of G with parameters $a_1,a_2.\ldots,a_n$ denoted by $G(a_1,a_2,a_3,\ldots a_n)$ is the graph obtained from G by attaching $a_i \geq 0$ pendant vertices at $u_i (1\leq i \leq n)$ (Fig. 1).

Example 3.11. $P_3(1,3,0)$ is

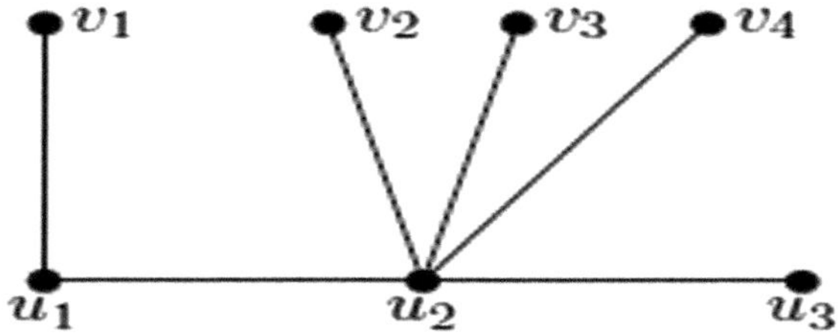

Fig. 1. Example of a thorn graph

Theorem 3.12. Let G be a simple graph without support strong isolates. Then $\gamma_{sss}^{av}(G(a_1, a_2, a_3, \ldots a_n)) = \frac{2n^2+n}{n+a_1+a_2+\ldots+a_n}$.

Proof. The support vertices of $G(a_1, a_2, a_3, \ldots a_n)$ form a minimum support strong dominating set of $G(a_1, a_2, a_3, \ldots a_n)$. If $a_i = 1$, then the corresponding pendent vertex lies in a minimum support strong dominating set of $G(a_1, a_2, a_3, \ldots a_n)$. If $a_i \geq 2$, then any pendent vertex at u_i lies in a minimum support strong dominating set of $G(a_1, a_2, a_3, \ldots a_n)$ together with a pendent vertex at u_i. Therefore

$$\gamma_{sss}^{av}(u) = \begin{cases} |V(G)|, & if \quad u \in V(G)\, or\, u\, is\, a\, unique\, pendent\, vertex\, at\, some\, vertex\, of\, G; \\ |V(G)|+1 & if \quad u\, is\, one\, of\, two\, or\, more\, pendent\, vertices\, at\, some\, vertex\, of\, G; \end{cases}$$

Let $a_1 = a_2 = \ldots = a_k = 1, a_i \geq 2, k+1 \leq i \leq n.$

$$\sum_{u \in V(G(a_1,a_2,\ldots,a_n)} \gamma_{sss}^{av}(u) = |V(G)|[|V(G)|+k] + [|V(G)|+1][|V(G)|-k]$$

$$\text{Therefore } \gamma_{sss}^{av}(G) = \frac{|V(G)|[|V(G)|+k] + [|V(G)|+1][|V(G)|-k]}{|V(G(a_1, a_2, a_3, \ldots a_n))|}$$
$$= \frac{n(n+k)+(n+1)(n-k)}{n+a_1+a_2+\ldots+a_n}$$

$$\text{Therefore } \gamma_{sss}^{av}(G) = \frac{2n^2+n}{n+a_1+a_2+a_3+\ldots+a_n}$$

Definition 3.13. Let G_1 and G_2 be two graphs of orders n_1 and n_2 respectively. The corona graph $G_1 \circ G_2$ is the graph obtained by taking one copy of G_1 and n_1 copies of G_2 and joining the $i^{th}-$ vertex of G_1 to every vertex in the $i^{th}-$ copy of G_2.
For Example (Fig. 2):

$P_3 \circ K_{1,3} =$ 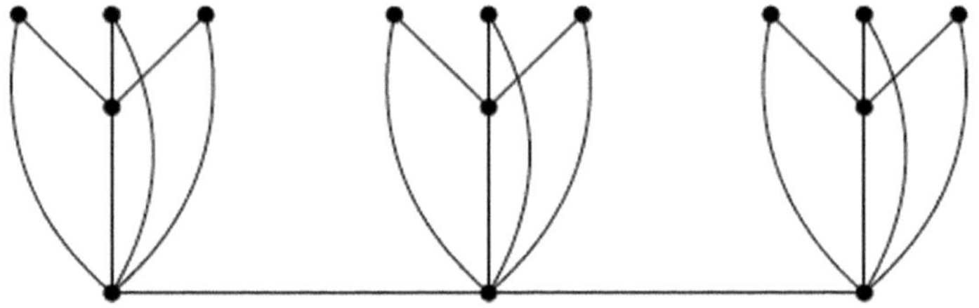

Fig. 2. Corona graph $P_3 \circ K_{1,3}$

Theorem 3.14. Let G_1 and G_2 be two graphs of orders n_1, n_2 respectively. Then $\gamma_{sss}^{av}(G_1 \circ G_2) = n_1 + \frac{n_2}{n_2+1}$.

Proof. $\gamma_{sss}(G_1 \circ G_2) = |V(G)| = n_1$.

$$\sum\nolimits_{u \in V(G_1 \circ G_2)} \gamma_{sss}^{av}(u) = |V(G_1)| \times n_1 + (n_1 + 1) \times n_1 \times |V(G_2)|$$
$$= n_1^2 + (n_1^2 + n_1)n_2)$$

$$\text{Therefore } \gamma_{sss}^{av}(G) = \frac{n_1^2 + (n_1^2 + n_1)n_2}{n_1 + n_1 n_2}$$
$$= \frac{n - 1^2(1 + n_2) + n_1 n_2}{n_1(1 + n_2)} = n_1 + \frac{n_2}{1 + n_2}.$$

Theorem 3.15. Let G be a disconnected graph with components $G_1, G_2, \ldots, G_k$. Let $\gamma_{sss}(G_i) = t_i (1 \le i \le k)$. Let r_i be the number of vertices in G_i which belongs to some $\gamma_{sss}-$ set of G_i. Then $\gamma_{sss}^{av}(G) = \gamma_{sss}(G) + 1 - \frac{\sum_{i=1}^{k} r_i}{|V(G)|}$.

Proof. Let $u \in V(G_i)$.

$$\gamma_{sss}^{av}(u) = \begin{cases} \gamma_{sss}(G), & \textit{if} \quad \textit{u is a vertex of } G_i \textit{ belonging to some } \gamma_{sss} - \textit{set of } G_i; \\ \gamma_{sss}(G) + 1 & \textit{Otherwise}; \end{cases}$$

$$\sum\nolimits_{u \in V(G)} \gamma_{sss}^{av}(u) = r_i \times \gamma_{sss}^{av}(G) + [|V(G_i)| - r_i][\gamma_{sss}(G) + 1]$$
$$= |V(G_i)| \times [\gamma_{sss}(G) + 1] - r_i$$

$$\sum\nolimits_{u \in V(G)} \gamma_{sss}^{av}(u) = \sum\nolimits_{i=1}^{k} [|V(G_i)| \times [\gamma_{sss}(G) + 1] - r_i]$$
$$= [\gamma_{sss}(G) + 1] \times |V(G)| - \sum\nolimits_{i=1}^{k} r_i$$

$$\text{Therefore } \gamma_{sss}^{av}(G) = \gamma_{sss}(G) + 1 - \frac{\sum_{i=1}^{k} r_i}{|V(G)|}$$

Remark 3.16. r_i is the number of good $\gamma_{sss}-$ vertices in G_i.

Remark 3.17. If G is $\gamma_{sss}-$ excellent, then $r_i = |V(G)|, 1 \leq i \leq k$.

$$\therefore \gamma_{sss}^{av}(G) = \gamma(G) + 1 - \frac{|V(G)|}{|V(G)|}$$
$$= \gamma_{sss}(G).$$

4 Average Secure Support Strong Domination of $(G_1 + G_2)$

Theorem 4.1. Let G_1 and G_2 be two graphs of order m and n respectively. Let $G = G_1 + G_2$.

$$\gamma_{sss}(G_1+G_2) = \begin{cases} \gamma_{sss}(G_2) + 1 - \frac{t}{m+n} & \textit{if } m \geq n \textit{ and } t \textit{ is the number of } \gamma_{sss} - \textit{good vertices in } G_2; \\ \gamma_{sss}(G_1) - \frac{t}{m+n} & \textit{if } m = n, \gamma_{sss}(G_1) < \gamma_{sss}(G_2) \textit{ and} \\ & t \textit{ is the number of } \gamma_{sss} - \textit{good vertices in } G_1; \\ \gamma_{sss}(G_1) - \frac{t_1+t_2}{m+n} & \textit{if } m = n, \gamma_{sss}(G_1) = \gamma_{sss}(G_2) \textit{ and } t_1, t_2 \textit{ are the number} \\ & \textit{of } \gamma_{sss} - \textit{good vertices in } G_1, G_2 \textit{ respectively}; \end{cases}$$

Proof: Case (i): Let $m > n$. Let $u \in V(G_1)$. Let $|E(G_i)| = \varepsilon_i, i = 1, 2$.
Then $Supp_{G_1+G_2}(u) = Supp_{G_1}(u) + 2\varepsilon_2 + |N_{G_1}(u)| \times |V(G_2)| + |V(G_1)| \times |V(G_2)|$
Let $u \in V(G_2)$.
Then $Supp_{G_1+G_2}(u) = Supp_{G_2}(u) + 2\varepsilon_1 + |N_{G_2}(u)| \times |V(G_1)| + |V(G_1)| \times |V(G_2)|$
Since $m > n$, any $\gamma_{sss}-$ set of G_2 is a $\gamma_{sss}-$ set of $G_1 + G_2$.
Let t be the number of $\gamma_{sss}-$ good vertices in G_2. Then

$$\sum_{u \in V(G_1+G_2)} \gamma_{sss}^{av}(u) = t \times \gamma_{sss}(G_2) + (\gamma_{sss}(G_2) + 1)(|V(G_1)| + |V(G_2)| - t)$$
$$= t \times \gamma_{sss}(G_2) + (\gamma_{sss}(G_2) + 1)(m + n - t)$$
$$\text{Therefore } \gamma_{sss}^{av}(G) = \frac{1}{m+n}[(m+n)(\gamma_{sss}(G_2) + 1) - t]$$
$$= \gamma_{sss}(G_2) + 1 - \frac{t}{m+n}.$$

Case (ii): Let $m = n$.
subcase (i): Let $\gamma_{sss}(G_1) < \gamma_{sss}(G_2)$.
Let t be the number of $\gamma_{sss}-$ good vertices in G_1. Then

$$\sum_{u \in V(G)} \gamma_{sss}^{av}(u) = t \times \gamma_{sss}(G_1) + [\gamma_{sss}(G_1) + 1][|V(G_1)| + |V(G_2)| - t]$$
$$= t \times \gamma_{sss}(G_1) + [\gamma_{sss}(G_1) + 1][m + n - t]$$
$$\text{Therefore } \gamma_{sss}^{av}(G) = \gamma_{sss}(G_1) - \frac{t}{m+n}$$

subcase (ii): Let $\gamma_{sss}(G_1) = \gamma_{sss}(G_2)$.

$$\sum_{u \in V(G)} \gamma_{sss}^{av}(u) = (t_1 + t_2)(\gamma_{sss}(G_1)) + [\gamma_{sss}(G_1) + 1][m + n - t_1 - t_2]$$

$$\therefore \gamma_{sss}^{av}(G) = \gamma_{sss}(G_1) - \frac{(t_1 + t_2)}{m+n}$$

Remark 4.2. Let $G \neq K_m$. Let $G = G_1 + \{u\}$. Let $|V(G_1)| = m, clearly |V(G_2)| = 1$.
$\{u\}$ is the unique $\gamma_{sss}-$ set of G.
Therefore, $\sum_{v \in V(G)} (v) = 1 + m \times 2 = 2m + 1$
Therefore $\gamma_{sss}^{av}(G) = \frac{2m+1}{m+1}$.
If G_1 is complete, then G is complete and hence $\gamma_{sss}^{av}(G) = 1$.

Theorem 4.3. Let G be a graph without support strong isolate and without any vertex having support $n - 1$. Let $\overline{G}$ be the complement of G. Let k and k_1 be the $\gamma_{sss}-$ good vertices in G and $\overline{G}$ respectively. Then $6 - \frac{(k+k_1)}{n} \leq \gamma_{sss}^{av}(G) + \gamma_{sss}^{av}(\overline{G}) \leq n + 2 - \frac{(k+k_1)}{n}$.

Proof:

$$\sum_{u \in V(G)} \gamma_{sss}^{av}(u) = k \times \gamma_{sss}(G) + (n - k)(\gamma_{sss}(G) + 1)$$

$$\sum_{u \in V(\overline{G})} \gamma_{sss}^{av}(u) = k_1 \times \gamma_{sss}(\overline{G}) + (n - k_1)(\gamma_{sss}(\overline{G}) + 1)$$

$$\gamma_{sss}^{av}(G) + \gamma_{sss}^{av}(\overline{G}) = \frac{1}{|V(G)|}[k \times \gamma_{sss}(G) + (n-k)(\gamma_{sss}(G) + 1) + k_1 \times \gamma_{sss}(\overline{G}) + (n - k_1)(\gamma_{sss}(\overline{G}) + 1)]$$
$$= \frac{1}{|V(G)|}[n\gamma_{sss}(G) + n - k + n\gamma_{sss}(\overline{G}) + (n - k_1)]$$
$$= \gamma_{sss}(G) + \gamma_{sss}(\overline{G}) + 2 - \frac{k + k_1}{n}$$

By hypothesis, $4 \leq \gamma_{sss}(G) + \gamma_{sss}(\overline{G}) \leq n$
Therefore, $6 - \frac{(k+k_1)}{n} \leq \gamma_{sss}^{av}(G) + \gamma_{sss}^{av}(\overline{G}) \leq n + 2 - \frac{(k+k_1)}{n}$.

Remark 4.4. When $G = P_4, \overline{G} = P_4, \gamma_{sss}(G) = \gamma_{sss}(\overline{G}) = 2$.
$\gamma_{sss}^{av}(G) = \gamma_{sss}^{av}(\overline{G}) = 2.5$.
Therefore $\gamma_{sss}^{av}(G) + \gamma_{sss}^{av}(\overline{G}) = 5$
$k = k_1 = 2$, Therefore $6 - \frac{(k+k_1)}{n} = 6 - \frac{2+2}{4} = 5$.
Therefore lower bound of the above inequality is realized. The upper bound is also realized as seen below.

Remark 4.5. Suppose G has at least two $\gamma_{sss}-$ good vertices,

$$\begin{aligned}\gamma_{sss}^{av}(G) &= \frac{1}{|V(G)|}[n\gamma_{sss}(G) + (n-k)] \\ &\leq \frac{1}{|V(G)|}[n\gamma_{sss}(G) + n - 2] \\ &= \gamma_{sss}(G) + 1 - \frac{2}{n}\end{aligned}$$

When $G = P_4, \gamma_{sss}(G) = 2, \gamma_{sss}^{av}(G) = 2.5$.
Right hand side of the above inequality is $2 + 1 - \frac{2}{4} = 2.5$
Hence the upper bound is reached.

References

1. Acharya, B.D.: The strong domination number of a graph and related concepts. J. Math. Phys. Sci. **14**(5), 471–475 (1980)
2. Blidia, M., Chellali, M., Maffray, F.: On average lower independence and domination numbers in graphs. Discrete Math. **295**, 1–11 (2005)
3. Guruviswanathan, R., Ayyampillai, M., Swaminathan, V.: Secure support strong domination in graphs. J. Inf. Math. Sci. **9**(3), 539–546 (2017)
4. Haynes, T.W., Hedetniemi, S.T., Slater, P.J.: Fundamentals of Domination in Graphs. Marcel Dekker, New York (1998)
5. Haynes, T.W., Hedetniemi, S.T., Slater, P.J.: Domination in Graph: Advanced topics. Marcell Dekker Inc., New York (1998)
6. Henning, M.A., Oellermann, O.R.: The average connectivity of regular multipartite tournaments. Australas. J. Comb. **23**, 101–114 (2001)
7. Henning, M.A.: Trees with equal average domination and independent domination numbers. Ars Comb. **71**, 305–318 (2004)
8. Harary, F.: Graph Theory. Addison-Wesley, Boston (1969)
9. Ponnappan, C.Y.: Studies in graph theory: support domination in graphs and related concepts, Thesis submitted to Madurai Kamaraj University (2008)

Secure and Traceable Medical Image Sharing Using Enigma in Cloud?

R. Manikandan[1(✉)], A. Rengarajan[2], C. Devibala[3], K. Gayathri[3], and T. Malarvizhi[3]

[1] Department of CSE, St. Peters University, Chennai, Tamilnadu, India
manitec4@gmail.com
[2] Department of CSE, Vel Tech Multi Tech Dr.RR Dr.SR Engg College, Chennai, Tamilnadu, India
rengu_rajan@yahoo.com
[3] Department of CSE, University College of Engineering Tindivanam, Tindivanam, Tamilnadu, India
devibalachidambaram@gmail.com, gayathri4523@gmail.com, themalar97@gmail.com

Abstract. The introduction of patient's electronic health care records results in a danger to their protection as resentful exercises would make genuine mischief the individuals who have legitimately or in a round about way identified with their information. Current methodologies to feasibly supervise and guarantee electronic health care records have been ended up being lacking. In our thesis, we propose the electronic healthcare record share which tends to the issue in sharing the medicinal information among the restorative huge information caretakers in a trustless condition. The framework is based on enigma and gives information surveying, scalability, decentralization and access control for shared electronic healthcare records in cloud archives among enormous information. Among information advances and sharing from one substance to the next, it invigilates the activities performed on the framework. The structure utilizes secret contracts and access control system to follow the conduct of the information and deny access to offending substances on the identification of infringement of authorizations on information. Our proposed idea is the present front line answers for information sharing among cloud specialist. Information surveying, versatility and access control while imparting medicinal services information to substances, for example, research and therapeutic foundations with insignificant hazard to information security can be accomplished by actualizing our proposed idea.

Keywords: Enigma · Data sharing · Electronic healthcare records · Cloud storage

1 Introduction

In current social requests, social orders and dealt with get-togethers, the dispersal of remedial data has been consumed to be a jump forward for the divulgence of new methodology and medications for mitigating diseases [1]. The key idea for the recently referenced announcement is a direct result of remote receptiveness, digitization and

D. J. Hemanth et al. (Eds.): COMET 2019, LNDECT 35, pp. 816–825, 2020.
https://doi.org/10.1007/978-3-030-32150-5_82

electronic limit of helpful data by medical specialist [2]. With the appearance of the innovation age and the ensuing gathering of enormous volumes of information that have introduced the enormous information time, sharing information offers appealing an incentive on prospects of which are as yet revealing [3]. The significance of information and the esteem inborn in its spread has brought forth business substances that gather, process, store, break down and provide the proper appealing and force the sharing of information with another invested individuals [4, 5]. It has produced the enthusiasm of a few businesses with attention on distributed storage and data examination, handling mechanisms, data examination and information provenance and rendering customary ventures reliance on information accessibility on their survival and work [6, 7].

To achieving the high needs on massive information storage, many stakeholders have resorted to cloud computing and cloud storage to supply appropriate solutions to pressing storage and process needs [8, 9]. The expanded fame of cloud administrations has drawn the enthusiasm of clients which range from patients, medicinal foundations and research establishment to the huge collaboration's to store their gained information on cloud archives [10]. Cloud masters are requested to in any case give a controlled, cross space and edible information sharing of medicinal information put away in their storage facilities to collectors [11].

Cloud service providers battle with an absence of joint effort for sharing information because of the unfavorable dangers presented on uncovering their information [12]. Ends can be drawn from evident actualities that, medical service parties go as far to refuse information sharing while at the same time blocking conventions that encourage information spread [13]. For information proprietors and caretakers, there is a current danger of gathered information being powerless in the hands of malicious information clients. For such dangers, arrangement producers diminish the purpose of uncovering information content by imparting approaches that misuse the dread of information clients. In spite of the actual fact that ways add the support of data proprietors and overseers, the dread of breaking directions and therefore the ensuing punishments to be paid in each financial and infamy terms cultivate a climate of doubt that guarantees data sharing doesn't happen [14]. For the foundation of right enlistments went for information sharing while at the same time featuring the alluring highlights of such acts, the problem is loss of information control remains [15].

By and large, when the data leaves the administrators system where the data was first assembled or made, there is an absence of the board over the activities performed by the customer [16]. This grants malignant customers to misuse data, causing data owners and caretakers a couple of genuine and notoriety ambushing issues with industry controllers. A couple of cryptographic systems have been proposed to address these issues rising up out of the sharing of therapeutic data yet have still been insufficient [17, 20]. Regardless, the blockchain is seen as a strong t to give a sensible response for watching out for this issue through its engaging features, for instance, constant nature and decentralization [21, 25]. In our proposed work, we have a tendency to project a enigma primarily based account sharing electronic healthcare record between cloud organizations whereas giving data get to control, root age and evaluating. Exercises of data beneficiaries are consistently checked through frameworks referenced later in the paper and breaks are tended to in like manner by denying access to the data.

2 Related Works

In the upcoming segment, looking into patterns relating to information through cloud specialist organizations and access management along with the enhancing blockchain innovation are laid out. Sundareswaran et al. [26] proposed guaranteeing dispersed responsibility for information partaking in the cloud, a methodology for naturally logging any entrance to the information in the cloud together with an evaluating component. Their methodology enables an information proprietor to review content just as authorize solid backend security when required. Zyskind et al. [27] utilized the blockchain for access management the board and for audit log security purposes, as a change verification log of occasions. Enigma may be a planned localized computation stage captivated with a complicated sort of protected multi-party computation. The diverse gatherings out, out store and run calculations on information while keeping the information totally private.

Xia et al. [28] arranged a blockchain-based data sharing structure that adequately addresses the passageway control challenges related with unstable data set away in the cloud using constant nature and understood independence properties of the blockchain. They used the usage of secure cryptographic procedures to ensure proficient get the chance to control to fragile shared data pool(s) using a permissioned blockchain and plan a blockchain-based data sharing arrangement that licenses data customers/owners to get to electronic therapeutic records from a common storage facility after their identities and cryptographic keys have been. The requesting after check and ahead changing structure some part of a shut, permissioned blockchain.

Ferdous et al. [29] presented DRAMS, a blockchain-based decentralized checking foundation for an appropriated access control framework. The principle inspiration of DRAMS is to send a decentralized design which can recognize arrangement infringement in a circulated access control framework under the presumption of an all-around denied risk display. The Chain Anchor framework gives obscure however variable personality to components attempting to perform trades to a permissioned blockchain. The Enhanced Privacy ID (EPID) zero information proof theme is employed to attain and prove the participants namelessness and membership [25]. Hassan et al. [30] presented a half and half system show for efficient constant WBAN media information transmission. Xia et al. [31] exhibited a blockchain framework utilizing smart contracts to screen the substances and record all activities performed by the clients.

In this, we tend to provide a secured enigma based data sharing of electronic healthcare records among disbelieved parties. The fundamental commitment of our work is to give information provenance, examining and anchored information trailing on restorative information. The different writing surveyed in this area gives insufficient systems in accomplishing information provenance, evaluating and information trailing on medical information. It ought to be referenced that our framework depends on secret contracts to monitor the behavior of data and provide privacy to user's data.

3 Preliminaries

3.1 ENIGMA

Enigma is a decentralized calculation stage with ensured security. Enigma supports privacy. Utilizing secure multi-party calculation (sMPC or MPC), information query are figured distributed, without a confided in outsider. Information is part between various nodes, and they register works together without spilling data to different nodes. Enigma provides scalability. Not at all like blockchains, calculations and information stockpiling are not recreated by each node in the system. Just a little subset play out every calculation over various parts of the information. The diminished excess away and calculations empowers all the more requesting calculations. Protection authorizing calculation: Enigma's system can execute code without releasing the crude information to any of the nodes, while guaranteeing right execution. This is key in supplanting current incorporated arrangements and believed overlay organizes that procedure touchy business rationale in a manner that invalidates the advantages of a blockchain.

3.2 Block Chain

The blockchain is a passed on a database that contains a mentioned the summary of records related together through chains, on squares. Squares can be depicted as individual parts that contain data identifying with a specific exchange. A blockchain sort out keeps up a relentless creating once-over of records which are perpetual.

For our information sharing among questioned social affairs framework, the preparing and agreement hub are absolutely cautious for broadcasting the squares into the blockchain in the wake of arranging demands. Structures made per request are formed into squares and later impart in the midst of the system of passing on bundle to a requestor. There are different series of blockchains in the framework, perceived strangely using a requestor's identity. In sorting out the compose thusly, we point to the manner in which that every square in a particular string addresses different events of interest made.

3.3 Triggers

Triggers are components that interface frames between the question structure and the blockchain condition. The key hugeness of realizing triggers for our system is to engage mystery contracts to direct interface from outside the structure to our system since mystery contracts can't explicitly collaborate with structures outside of its condition, as such the blockchain orchestrate. Triggers hold no information and essentially go about as a center individual between the question layer and the information organizing and provenance layer of the structure. The trigger unites interfaces inside and remotely and update the method states to and from the question structure by the inbound and outbound mystery contracts reliant on external and internal features of the understanding.

4 Design Formulation

4.1 System Model

We figure information sharing mechanism utilized by the blockchain-based information sharing among disbelieved parties for information security and provenance. Classes of structures of the framework are shown in Fig. 1 and gathered into four principle layers to be specific.

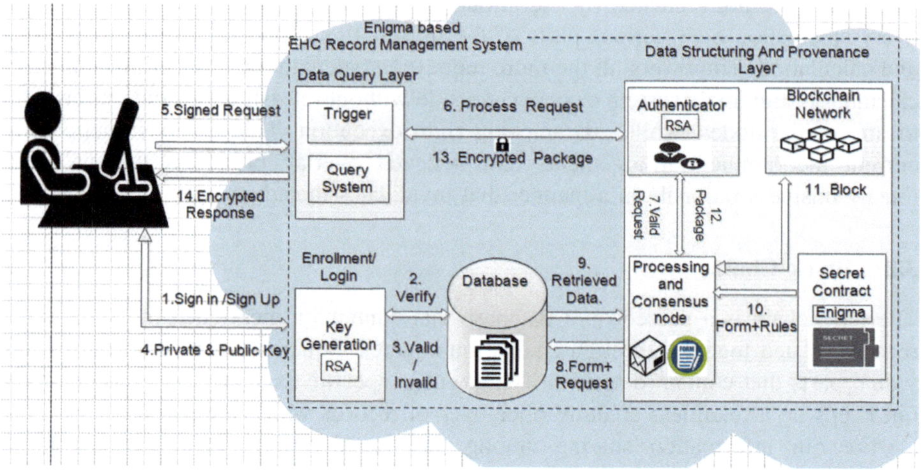

Fig. 1. ENIGMA architecture

4.2 Data Query Layer

The data query layer comprises of sets of questioning structures that get to, process, forward or react to queries presented on the framework. The data query layer straightforwardly interfaces with the data structuring and processing layer and the outside environment. Clients specifically associate with the query layer for information demands.

4.3 Data Structuring and Processing Layer

Information Structuring and handling layer include singular parts that help system request access to data from the database. Calculations and structures are realized in the Data Structuring and handling layer to report exercises which are securely secured in a database. Eventual outcomes of every action completed are conveyed into a perpetual framework to guarantee skeptical and sensible looking at. The particular substances in the Data Structuring and preparing layer.

Authenticator. The authenticator is accountable for checking the credibility of sales sent by requestors to the data owner structure. Additionally, the authenticator encodes a bundle which contains the data requested by the customer and is finally passed on to the best possible requestor.

Processing and Consensus Node. The handling and agreement process structure made for sales which are later framed into squares and impart into the blockchain arrange. Besides, the agreement hub is to make bundle containing the requested data and mystery contract to be passed on to the requestor.

Secret Contract. Enigma's "Secret contracts" are smart contracts that conceal contract information sources and yields, permitting blockchain to execute approvals and exchanges without uncovering the crude information on-chain. Enigma is a blockchain-based convention that utilizes noteworthy privacy technology to empower versatile, start to end decentralized applications. With Enigma, "smart contracts" progress toward becoming "secret contracts", in which input information is kept avoided nodes in the Enigma arrange that execute code.

Blockchain Network. The arrangement is made out of individual blocks communicate into a system and fastened together in an ordered strategy. The fundamental job of the blockchain organize is to keep up a sequentially circulated changeless database of activities on the conveyance and demand of information from the framework.

5 Design Approach

System Setup a customer sends an interest for data access to a casing work. The data demand is set apart by the customer using a "pre-created" requestor private key. The question system propels the interest to the information organizing and provenance layer by the triggers. The authenticator gets the interest and fluctuates the credibility by affirming the mark using the requestors open key which was created and shared before the interest was sent. In case the mark is genuine, the technique is recognized else dropped for invalid requesting.

Solicitation File For a certifiable intrigue, the authenticator drives the enthusiasm to the handling and accord hubs where preparing of offers into structures are finished. The structure made contains a hash of the timestamp of when the intrigue was gotten and a hash of the ID of the requestor. The reason behind the information demand is set apart to the structure and after that sent to the database. The database gets the structure, recovers the information and sends the recovered information to the data sorting out and provenance layer. The handling and accord hubs send an enthusiasm to the riddle contract arrange for adding sets of models to the mentioned information. A mystery contract is made and set apart to the structure which contains the information. A bundle contains an information ID, payload (information) and a secret contract. A bundle is delivered utilizing preparing on got information from the database framework by the accord hubs. The finishing of a bundle is finished by the authenticator by encoding the bundle into an association which must be clear by a genuine and fitting requestor with the correct private key.

6 Secret Contract

In this, we use mystery contracts to address exercises wrapped up by a requestor on data requested from a data owner's system. It begin the information proprietors to completely get declaration and the control on the information provenance on the grounds that the whole life saver of the sent information would be checked in a controlled in a dependable situation where the information proprietor needs no confirmation of trust from the requestor.

The essential activity sets are: read, compose, erase, copy, move and duplicate. These arrangements of activities, when performed on the information, would trigger the secret contracts to send a report dependent on the guidelines set up for that specific information. Checking of activities are passed on in secret contract contents by a function getAct. The affectability of the information is sorted into two dimensions which are: high and low. These dimensions of affectability are gotten from preparing by processing and consensus nodes dependent on informational indexes procured from the database. For low delicate dimension for an information, the information proprietor, in light of the asked for information, can alter the secret contracts to overlook activities to abstain from detailing of incidental information from being put away. For information with high affectability level, the secret contract is required to report all activities arranged in the introduction of getAct to viably screen tasks performed on the information, guaranteeing the recognition of breaks on information.

Algorithm 1 Secret Contract Algorithm

```
₩Require Initialization of parameters:
getAct, getSens, getReqID, getOwnID, getDataID, getKey,
retrieve, encp, cmt, report, acsctrl;
₩usepackage{AUDITING  PACKAGE}
For funt (getAct)==decode do
  funt (cmt)  Information with DataID has been decoded.
  Retr (getKey)
  encp (cmt)
  report (cmt||getReqID||getOwnID)
end for
if funt (getSens)==Low then
  funt (getAct)Exemptions on information.
  ignore
else if funt (getSens)==Low then
  funt (getAct)Not exemptions on data (violation).
  funt (cmt)  Data violation concatenated with DataID
  funt (acsctrl)Revokes access to data.
  Retr (getKey)
  encp (cmt)
  report (cmt||getReqID||getOwnID)
else{funt (getSens)==High}
  funt (getAct)Violation.
  funt (cmt)  Data violation concatenated with DataID
  funt (acsctrl)Revokes access to data.
  Retr  (getKey)
  encp (cmt)
  report (cmt||getReqID||getOwnID
end if
```

Comments are produced as articulations to depict the activities that were performed on the information. A retrieved proclamation is matched with a getAct articulation to extract the keys to encode comments that would be accounted for to the information proprietor's secret contract permissioned database. The movement of transferring the remarks to processing and consensus section is instantiated by the warning report declaration in the secret contract content. The capacity get to control indicates the approvals set by the data owner that would be done identified with the secret contract permissioned database Installing clarifications executable on the social occasion and utilization of the data by the requestor as an affirmation of capable movement of a secret contract is respected goal.

7 Conclusion

Information sharing and coordinated effort by means of cloud specialist organizations is a fortification with the expanding headway of present day advances driving the present society. A few techniques and instruments have been set up to control the stream of information from point(s) to point(s) as medical information in the hands of noxious substances can cause serious unbelievable harms on all gatherings related specifically or in a roundabout way to the information.

In our work, we structure an information sharing model between cloud specialist utilizing the enigma. The structure utilizes the utilization of secret contracts and access control components to adequately follow the conduct of the information just as repudiate access to damaged guidelines and authorizations on information. We analyse the execution of our framework just as contrast with current bleeding edge arrangements with information sharing among cloud specialist. By actualizing the proposed model, cloud specialist will most likely safely accomplish information provenance and evaluating while sharing restorative information among other cloud specialist organizations just as elements, for example, research and medical foundations with no hazard on information security.

References

1. Weitzman, E.R., Kaci, L., Mandl, K.D.: Sharing medical data for health research: the early personal health record experience. J. Med. Internet Res. **12**(2), 1–10 (2010)
2. Taichman, D.B., et al.: Sharing clinical trial data: a proposal from the international committee of medical journal editors free. PLoS Med. **13**(1), 505–506 (2016)
3. Chen, M., Mao, S., Liu, Y.: Big data: a survey. Mob. Netw. Appl. **19**(2), 171–209 (2014)
4. Raghupathi, W., Raghupathi, V.: Big data analytics in healthcare: Promise and potential. Health Inf. Sci. Syst. **2**(1), 3 (2014)
5. Krumholz, H.M., Waldstreicher, J.: The Yale Open Data Access (YODA) project-a mechanism for data sharing. New Engl. J. Med. **375**(5), 403–405 (2016)
6. Costa, F.F.: Big data in biomedicine. Drug Discov. Today. **19**(4), 433–440 (2014)
7. Huang, J., et al.: A new economic model in cloud computing: cloud service provider vs. network service provider. In: Proceedings of IEEE Global Communications Conference (GLOBECOM), pp. 1–6, December 2015

8. Huang, J., Duan, Q., Guo, S., Yan, Y., Yu, S.: Converged network-cloud service composition with end-to-end performance Guarantee. In: Proceedings of IEEE Global Communications Conference (GLOBECOM). IEEE Trans. Cloud Comput. To be published
9. Aceto, G., Botta, A., De Donato, W., Pescap, A.: Cloud monitoring: a survey. Comput. Netw. **57**(9), 2093–2115 (2013)
10. Assis, M.R.M., Bittencourt, L.F., Tolosana-Calasanz, R.: Cloud federation: characterisation and conceptual model. In: Proceedings of IEEE/ACM 7th International Conference on Utility Cloud Computing (UCC), pp. 585–590, December 2014
11. O'Driscoll, A., Daugelaite, J., Sleator, R.D.: Big data Hadoop and cloud computing in genomics. J. Biomed. Inf. **46**(5), 774–781 (2013)
12. Borgman, C.L.: The conundrum of sharing research data. J. Am. Soc. Inf. Sci. Technol. **63** (6), 1059–1078 (2012)
13. Grozev, N., Buyya, R.: Inter-cloud architectures and application brokering: taxonomy and survey. Softw.-Pract. Experience **44**(3), 369–390 (2014)
14. Fazio, M., Celesti, A., Villari, M., Puli, A.: How to enhance cloud architectures to enable cross-federation: towards Interoperable storage providers. In: Proceedings of IEEE International Conference on Cloud Engineering (IC2E), pp. 480–486, March 2015
15. Kuo, A.M.H.: Opportunities and challenges of cloud computing to improve healthcare services. J. Med. Internet Res. **13**(3), e67 (2011)
16. Weber, G.M., Mandl, K.D., Kohane, I.S.: Finding the missing link for big biomedical data. J. Amer. Med. Assoc. **311**(24), 2479–2480 (2014)
17. Shao, J., Lu, R., Lin, X.: Fine-grained data sharing in cloud computing for mobile devices. In: Proceedings of IEEE INFOCOM, pp. 2677–2685, April/May 2015
18. Thilakanathan, D., Chen, S., Nepal, S., Calvo, R.A., Liu, D., Zic, J.: Secure multiparty data sharing in the cloud using hardware- based TPM devices. In: Proceedings of IEEE International Conference on Cloud Computing (CLOUD), pp. 224–231, June 2014
19. Khan, A.N., Kiah, M.L.M., Ali, M., Madani, S.A., Khan, A.U.R., Shamshirband, S.: BSS: block-based sharing scheme for secure data storage services in mobile cloud environment. J. Supercomput. **70**(2), 946–976 (2014)
20. Dong, X., Yu, J., Luo, Y., Chen, Y., Xue, G., Li, M.: Achieving an effective, scalable and privacy-preserving data sharing service in cloud computing. Comput. Secur. **42**, 151–164 (2014)
21. Yang, J.J., Li, J.Q., Niu, Y.: A hybrid solution for privacy preserving medical data sharing in the cloud environment. Future Gener. Comput. Syst. **43**, 74–86 (2015)
22. Peterson, K., Deeduvanu, R., Kanjamala, P., Boles, K.: A blockchain based approach to health information exchange networks. In: Proceedings of NIST Workshop Blockchain Healthcare, vol. 1, pp. 1–10 (2016)
23. Krawiec, R.J.: Blockchain: opportunities for health care. In: Proceedings of NIST Workshop Blockchain Healthcare, pp. 1–16 (2016)
24. Pilkington, M.: Blockchain Technology: Principles and Applications. Handbook of Research on Digital Transformations. Edward Elgar Publishing, London (2015)
25. Liu, P.T.S.: Medical record system using blockchain, big data and tokenization. In: Proceedings of 18th International Conference on Information and Communications Security (ICICS), Singapore, vol. 9977, pp. 254–261, November/December 2016
26. Hardjono, T., Smith, N.: Cloud-based commissioning of constrained devices using permissioned blockchains. In: Proceedings of 2nd ACM International Workshop IoT Privacy, Trust, Security (IoTPTS), pp. 29–36 (2016)
27. Sundareswaran, S., Squicciarini, A.C., Lin, D.: Ensuring distributed account-ability for data sharing in the cloud. IEEE Trans. Depend. Secure Comput. **9**(4), 556–568 (2012)

28. Zyskind, G., Nathan, O., Pentland, A.: Enigma: decentralized computation platform with guaranteed privacy. https://arxiv.org/abs/1506.03471 (2015)
29. Xia, Q., Sifah, E.B., Smahi, A., Amofa, S., Zhang, X.: BBDS: blockchain-based data sharing for electronic medical records in cloud environments. Information **8**(2), 44 (2017)
30. Ferdous, S., Margheri, A., Paci, F., Sassone, V.: Decentralised runtime monitor- ing for access control systems in cloud Federations. In: Proceedings of IEEE International Conference on Distributed Computing, p. 111, June 2017
31. Hassan, M.M., Lin, K., Yue, X., Wan, J.: A multimedia healthcare data sharing approach through cloud-based body area Network. Future Gener. Comput. Syst. **66**, 48–58 (2017)
32. Xia, Q.I., Sifah, E.B.: MeD share: Trust-less medical data sharing among cloud server providers via blockchain, July 2017. https://doi.org/10.1109/Acess.2017.2730843

Automatic Inspection Verification Using Digital Certificate

B. Akshaya(✉) and M. Rajendiran

Department of Computer Science and Engineering, Panimalar Engineering College, Chennai, India
akshayacsc95@gmail.com, muthusamyrajendiran@gmail.com

Abstract. Digital certificates are a core component in the provision of secure data communication. We propose the concept of digital certificate that can be used to provide client verification. Cryptographic methods are used to explore these aspects of data security and there are many drawbacks in current encryption algorithms in respect of security, real-time performance and so on. Among them, the elliptic curve cryptography (ECC) and Rivest, Shamir, Adelman (RSA) is evolving as important cryptographic methods since it offers high security. In our proposed system, Optical Character Recognition is used to convert the scanned image to extract the parameters in it and the parameters are encrypted and decrypted using RSA and ECC. So, encrypted digital certificate is created for each individual and stored in a private database for inspection purpose. The public or private sector organization verifies the digital certificate by sending a request to authorization server and the response is generated by referring the parameters of encrypted digital certificate with the parameters extracted from the local database. The main purpose of the digital certificate is to avert duplicates and forgeries.

Keywords: Digital certificate · Optical Character Recognition · RSA · ECC · Authorization server

1 Introduction

Computer data frequently travels from one computer then onto the next leaving the security of its ensured physical environment. When the information is out of hand, individual with bad intention could change your data for their own advantage. This change leads to theft the information by an unauthorized user [1]. This drawback can be overcome by an innovation that is based on the fundamental of secret codes, enlarged by present day mathematics which secures our data in powerful ways. Exactly when a message is sent using cryptography, it is encoded before it is sent. The technique for changing content is known as a "code". The changed content is assigned as "cipher-text". This change makes the message hard to read. Somebody who needs to read the message, must change it back i.e., decoding of message. Both the individual that sends the message and the one that gets it should know the secret approach [2]. There are various algorithmic methods developed to provide security which deals with symmetric and asymmetric techniques by means of three terminologies:

D. J. Hemanth et al. (Eds.): COMET 2019, LNDECT 35, pp. 826–837, 2020.
https://doi.org/10.1007/978-3-030-32150-5_83

1.1 Authentication

Authentication is a privacy benefit that tends to the idea of authentic communication between entities. This privacy benefit gives evidence of cause authentication between the transmitter and the receiver. This is accomplished through the utilization of credentials like the user name and secret key.

1.2 Integrity

It tends for the possibility of dependability of IT resources especially the data, message or a flood of data because a single digital signature can be made for every unique message for any sender hence it tends to be viably used to check authenticity of sender.

1.3 Non-repudiation

This service tends to the idea of restricting communication entities to the activities they perform on the advantages so the sender or responder can't later dishonestly deny having participated in a trade. Since a similar signature is created on the both side of transmission to verify the information, plainly the sender can't deny that the data is send by them.

Our proposed system consists of digital certificate which is a core component in the provision of many organizations. It consists of OCR (Optical Character Recognition) technique that scans the input text and extracts the parameters in it [4]. Once the features are extracted or filtered, public key cryptography i.e., Elliptic Curve Cryptography (ECC) algorithm and Rivest Shamir Adleman (RSA) algorithm is applied to it [3]. The extracted parameters are encrypted and decrypted and stored in an authorized server.

The role of authorized server is to store and retrieve the files usually referred as "serve files" [5]. The certificate is verified by the public or private sector organization to check the originality of individual. The process is carried out by the certificate encryption stored in the server. Once the request is invoked from the private or public sector organization, the certificate must be decrypted and the parameters are generated. The comparison of parameters between encrypted certificate and the local database certificate is carried out [6]. It is verified and result is generated to the public or private sector organization.

2 Related Works

Our related work focuses on:

- Survey On Digital Certificates
- Certificate Security Themes/Models
- Cryptographic Techniques

2.1 Survey on Digital Certificates

"Signature Detection and Matching for Document Image Retrieval" by Zhu et al. [1] proposed signature scheme with a message by using the methodologies optical character recognition and signature matching. But the limitation identified is detected that segmented signature may contain a couple of contacting printed characters which results in signature recovery with interpretation also scaling, and rotation brings out about higher error rate. It concludes that generated signature with signing a message is re-randomized which ensures privacy.

"Certificate-based signature scheme in the standard model" by Zhou et al. [2] defined the proposed system with the methodologies certificate-based signature and bilinear pairing with the problem addressed as the certificate based signature plan can be demonstrated secure against the malicious but with the passive certifier assault. The client's public key isn't bound with the client's personality, so an attacker can replace a client's public key with his/her own. It is discovered that malicious Certifier can't exist together with the super foe security in a CBS conspire which opposes the malicious yet passive certifier attack.

X.509 PKI and Data Processing Context (DPC) methodologies are proposed in "A Probabilistic Approach for Assessing Risk in Certificate-Based Security" by Hinarejos et al. [3] that revealed the flaws for the utilization of certificates in the certificate validation process and security dangers are not accessed. Clients are frequently not thorough in the validation procedure and current interfaces fail to give powerful data to end-users which leads to exploitable vulnerabilities. They empower secure HTTPS communications encrypted virtual private network tunnels, and code checking for secure programming installation among others.

GourabSaha proposed "Digital Signature System for Paperless Operation" that uses the QR code application with the document hashing barcode [4]. Approval of candidate's legal documents for each individual certificate is a long procedure and similarly expensive. So, proposed system permits status tracking of the record and engages the proprietor to keep up the same and can be retrieved at any time since authentication is done using the keys. So, Paperless operation is evolved which enhances more security to the user either.

2.2 Certificate Security Themes/Models

"Automated issuance of digital certificates through the use of federation" by Bardinidalino et al. [5] addresses the problem where the users have to maintain a variety of credentials for different database and uses the methodologies of SSL and hardware security modules. It limits the aggregation of several data in one certificate which prompts to data insecurity. The proposed automated issue enhances the system to acquire the clients personality data directly from their identity provider amid the clients authentication and utilizes this data to issue the computerized certificate.

"Generating Correlated Digital Certificates: Framework and Applications" by Zhu et al. [6] addresses the issue with multi-certificate public key infrastructure is that in the CA's view, client just has one certificate and thus it seems that expanding certificates are totally beyond the CA's control. Multi-certificate public key infrastructure, public key infrastructure and RSA methodologies are used which limits the certificate management and has too much overlap in the content. The proposed system uses the correlated digital certificates to satisfy the security and protection prerequisites of certificate users.

Using XML signatures, GPS and geo-encryption, Singh proposed "Generation and Verification of Digital Certificate" method to obtain electronic record digitally [7]. It identifies the signatory that cannot be denied later where it ensures the sender and receiver to buy the authorized software for making the transmission more easier in which geo-encryption makes use of the location to produce additional security and authorization.

Feng et al. [8] proposed "A New Certificate-based Digital Signature Scheme" using bilinear pairings, key generation centre and uses Diffie-Hellman technique where it replaces the traditional key cryptosystem that uses a large computation time for storing and verifying user's public key with the corresponding certificate. Using bilinear pairings certain affirmation is obtained that has no private key escrow as the certificate based encryption scheme.

2.3 Cryptographic Technique Concepts

"Improvement and implementation of digital content protection scheme" by Fujisaki et al. [9] proposed digital signature based on the content protection for authentication by using the methodologies of id based signature whereas the traditional method of id algorithm does not provide a reliable processing speed in content protection system. Proposed system provides integrity with the higher processing speed using the aggregate signature for identity protection.

Mali [10] implemented "Authenticated Document Transfer based on Digital Signature and a Survey of its existing techniques" using asymmetric cryptography that uses public keys along with the private keys for the encryption and decryption of data and server based signature for the signature creation technique. At whatever point a client needs to send any document, they essentially needs to import the document from the mobile device and append their computerized signature to it and send it to the intended beneficiary. So, appending digital signature to document and exchanging it over the network to the intended recipient is made extremely helpful by the utilization of mobile phones.

Murty et al. [11] proposed a new technique "Digital Signature and Watermark Methods For Image Authentication using Cryptography Analysis" that uses digital signature and watermarking, symmetric cryptography and asymmetric cryptography for the generation of private and public key pair. Configuring the certificates and losing that key, viably loses the information so digital signature and watermark techniques are two techniques that is utilized for copyright security and validation which is robust to sensible compression rate. It enhances the safeguarding of good picture quality and proficient to authentication.

"A comprehensive study on digital signature for internet security" by Saha [12] uses RSA and MD5 algorithm. In MD5, content changes subsequently being sent from the sender and before coming to the recipient it leads to loss of integrity. So, while making a comparative study for RSA and MD5, RSA has more security because in

RSA communications are sent in a small amount of a second so that the hacker won't have the capacity to get to any information during transmission of electronic information.

3 Proposed System

See Fig. 1.

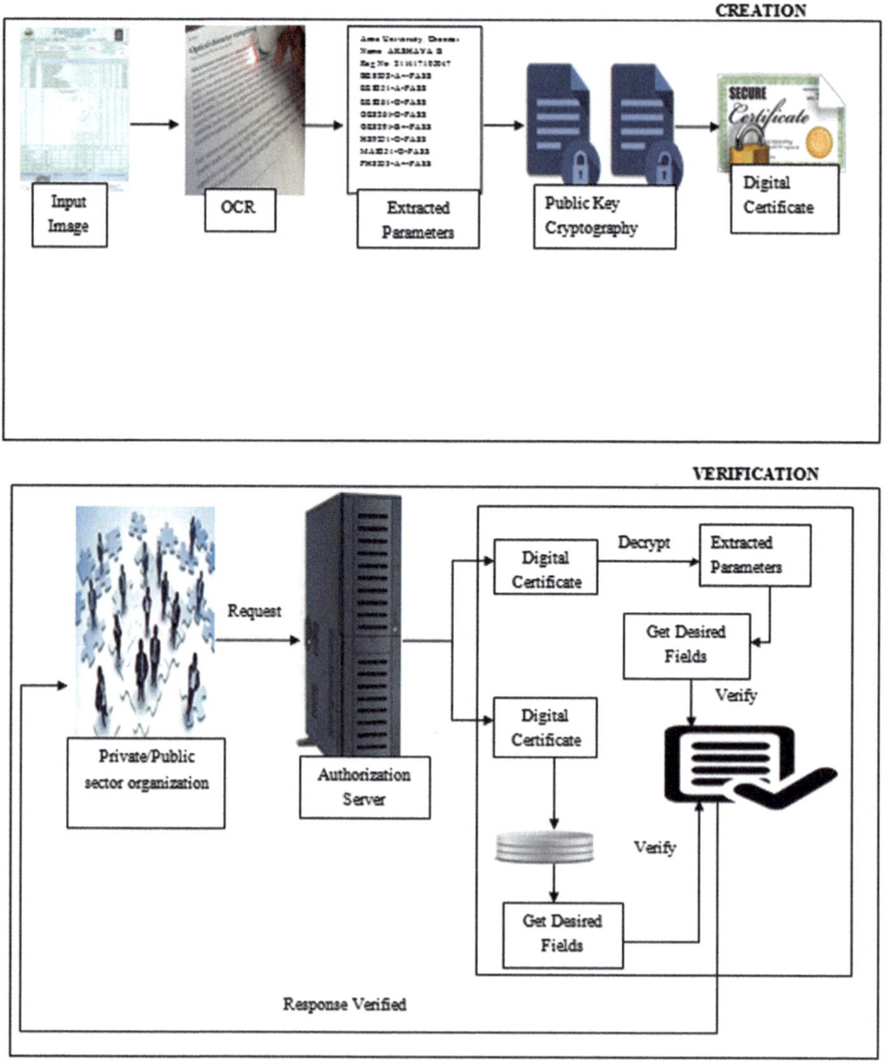

Fig. 1. Implementation of proposed system

3.1 Optical Character Recognition

It is the mechanical or electronic change in pictures of typed, physically written or printed content into machine-encoded content from an scanned document [13], a photograph of a document etc. utilized as a form of data entry from printed paper information records including international ID reports, computerized receipts or any appropriate documentation – it is a typical strategy for digitizing printed messages so that they can be electronically altered, searched and stored in machine processes. OCR is a field of research in pattern recognition, artificial intelligence and computer vision [14].

The earliest used OCR is optophone in the year 1914 for the purpose of blind people. The main functionality of optophone is to distinguish between the dark ink of text and lighter blank spaces that generates tones and recognizes different letters (Fig. 2).

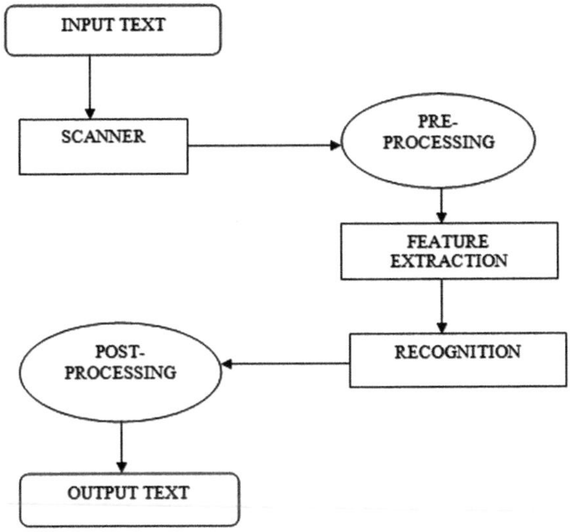

Fig. 2. Working of OCR

(a) Tesseract Software

It is one of the OCR software that supports various operating system. It is an open source developed by Hewlett Packard under the apache license. Initially version 1 tesseract recognised only English language text and version 2 added six additional languages followed by version 3 that consisted of more additional languages and scripts [2]. Current version in our proposed system is version 4 that consists of totally 16 languages and 37 language scripts. We can write our own scripts in tesseract is the major advantage.

Steps for converting image to text

1. Loading the image from any image source.
2. Detecting features in an image by referring the number of pixels in an image.
3. Once the image features have been detected, line detection and removal analysis is carried out.
4. Further, layout analysis is done so that the region of extracted parameters in the scanned image are categorized.
5. Next, detection of lines followed by words is carried out including different size, space etc.
6. Finally recognition of character is converted into appropriate character code and compared with a set of character which machine has been programmed to recognize and result is stored in the software [15].

3.2 Public Key Cryptography

(a) Elliptic Curve Cryptography

Elliptic curve cryptography (ECC) is a recent type of public key cryptography wherein the encryption key is kept public, while the decoding key is kept private [16]. The power of public key cryptography using Elliptic Curves depends on the toughness of processing discrete logarithms in a finite field. Other public key cryptographic calculations, for example: RSA, depends on the toughness of whole number factorization [3]. Elliptic Curve Cryptography (ECC) has been joined in number of protocols, for example, SSH and Bit coin. ECC is engaging to implementers since it requires littler key sizes than other public key cryptosystems.

Algorithm for ECC:
Suppose sender wishes to send a message m to the receiver
Let m has any point M on the elliptic curve.
The sender selects a random number k from [1, n-1]
The cipher texts generated will be the pair of points (C1,C2), where
C1= k*p
C2= M + (k*q) where q is the public key and d is the private key by using q=d*p
// To decrypt the cipher text.
The receiver computes the product of C1 and its private key
Then the receiver subtracts this product from the second point C2
M = C2- (d * C1)
Proof:
C2-d*C1=(M+k*q)-d*k*p
=> M+k*d*p-d*k*p [d*k*p gets cancelled]
= M, where it is the original data sent by the sender.

(b) Rivest Shamir Adleman

RSA denotes the surname of its inventors Ron Rivest, Adi Shamir and Leonard Adleman. It was developed for a person who wants to take part in communication utilizing encryption that necessities to produce a couple of keys, to be specific i.e., public key and private key [17]. The procedure followed in the age of keys is described below

Generate RSA modulus:
Step 1: Select two large prime number's m and n
Step 2: Calculate a = m * n.
Step 3: Calculate b = (m − 1)(n − 1).
//Find Derived Number (c)
Step 4: gcd (b, c) = 1; 1 < c < b such that number c must be greater than 1 and less than b.
Step 5: Compute e such that $e = c^{-1}$ (mod(b))
//Generation of public key
Step 6: Public key PK (c, a)
Step 7: Private Key P_RK (e, a)
//Encryption

- Suppose the sender wants to send text message to someone whose public key is (a, b).
- The sender then represents the plaintext as a sequence of numbers that is less than a

Step 8: Ciphertext: $C = P^c$ mod a where P is the plaintext
//Decryption

- The decryption technique for RSA is extremely easy. Assume that the receiver of public-key pair has got a ciphertext C [18].
- Receiver raises C to the intensity of his private key. The outcome modulo will be the plaintext P.

Step 9: Plaintext: $P = C^e$ mod a

The RSA and ECC are known as the most proficient public key cryptography among all asymmetric encryption calculations [19]. Encryption and decryption task is performed immediately even with broad number of words as input, provides smaller size cipher text contrasted with other strategy which hugely helps in saving bandwidth [20]. The ordinary ECC key size of 256 bits is comparable to a 3072-piece RSA key and multiple times more grounded than a 2048-bit RSA key [21]. Elliptic curves are accepted to provide good security with smaller key sizes that result in quicker execution time. The utilization of ECC is highly recommended to create greater security and higher speed without increasing the computational load [22].

3.3 Verification

Once the digital certificate is created, it is verified by a public or private sector organization. The verification process is carried out by sending a request to authorized server from public or private sector organization. The authorized server verifies the digital certificate by decrypting it and fetches its parameters [23]. Then, the fetched parameters is matched with the local database parameters to get the desired fields (Exact parameters). If both the fields matches, the certificate belongs to the corresponding user else there is a fraudulent copy of certificate. By this way, the response is generated to the public or private sector organization.

4 Results and Analysis

Thus time taken for execution in encryption and decryption is shown below (Tables 1 and 2).

Table 1. Encryption in seconds

BITS	RSA	ECC
8	0.1261	0.0312
64	0.2041	0.1367
256	0.5409	0.3821
1024	0.9231	0.8098

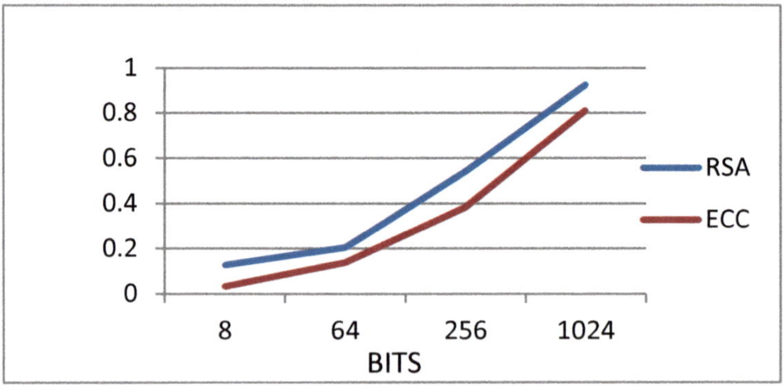

Fig. 3. Time taken for encryption

Figure 3 represents the time taken for encryption by public key cryptography.

Table 2. Decryption in seconds

BITS	RSA	ECC
8	11.4325	9.3987
64	13.7652	10.3718
256	14.4031	12.2137
1024	27.4528	24.9712

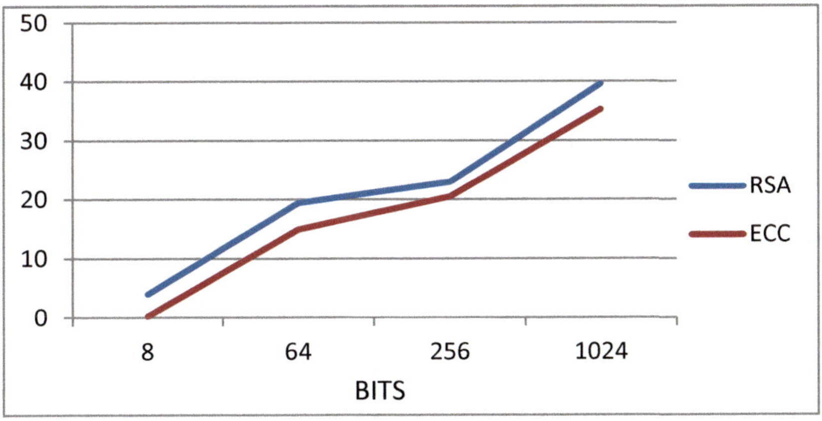

Fig. 4. Time taken for decryption

Figure 4 represents execution time for decryption by using different cryptographic technique. In this approach, it effectively shows the ECC has lesser execution time than RSA (Table 3).

Table 3. Total time in seconds

BITS	RSA	ECC
8	7.9732	5.2146
64	29.4531	24.9082
256	44.9792	40.4659
1024	78.6591	72.2643

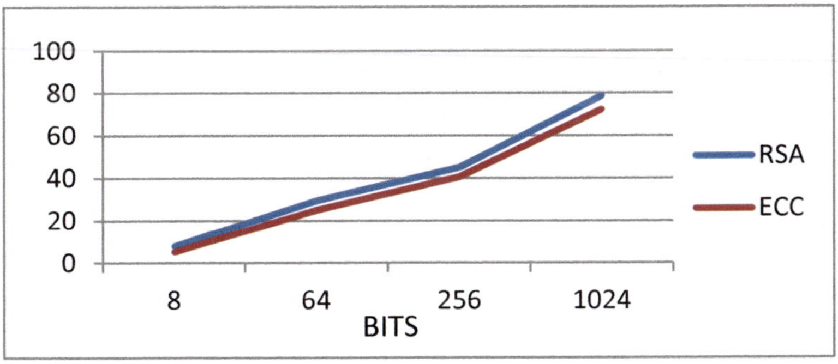

Fig. 5. Total time in execution

It clearly shows that ECC has shorter execution time while comparing RSA in total time in execution. The above graph generated for Figs. 3, 4 and 5 gives the best performance in ECC than RSA.

5 Conclusion

Digital Certificate is created by extracting the parameters of an input image using Optical Character Recognition (OCR) and they are encrypted and decrypted by using two different algorithms such as RSA and ECC. The performance evaluation was conducted for finding time lapse in encryption, decryption by RSA and ECC on input data of 8 bits, 64 bits, 256 bits, 1024 bits. Based on evaluation, it was found that ECC performs high speed computation in comparing RSA. The server is designed for authorization that accepts the public and private sector organization's request to verify the digital certificate. Once the certificate is verified, the response is generated to the organization.

Thus future work can be extended for digital certificate that helps in verification of land documents, government documents etc. Extending this proposed work helps to avoid forgery of documents in all aspect.

References

1. Kuznetsov, A., Pushkar, A., Kiyan, N., Kuznetsova, T.: Code-based electronic digital signature. In: IEEE International Conference on Dependable Systems, Services and Technologies, vol. 9, no. 29 (2015)
2. Fujisaki, M., Iwamura, K., Inamura, M., Kaneda, K.: Improvement and implementation of digital content protection scheme using identity based signature. IEEE Trans. Inf. Forensics Secur. **7**(6), 1673–1686 (2012)
3. Saha, P.: A comprehensive study on digital signature for internet security. ACCENTS Trans. Inf. Secur. **11** (2016). ISSN 2455-7196
4. Wu, C.: Self generated certificate. In: IEEE Conference on Data Appications, Security and Privacy, pp. 159–174 (2012)
5. Hwang, T., Gope, P.: Forward/backward unforgeable digital signature scheme using symmetric-key crypto-system. J. Inf. Sci. Eng. **26**(6), 2319–2329 (2013)
6. Kozlov, A., Reyzin, L.: Forward-secure signatures with fast key update. In: Proceedings of Security in Communication Networks. LNCS, vol. 2576, pp. 247–262 (2002)
7. Zhu, G., Zheng, Y., Doermann, D., Jaeger, S.: Signature detection and matching for document image retrieval. IEEE Trans. Pattern Anal. Mach. Intell. **31**(11), 2015–2031 (2013)
8. Zhou, C., Cui, Z.: Certificate based signature in the standard model. IEEE Trans. Inf. Forensics Secur. **48**, 313–317 (2016)
9. Hinarejos, M.F., Almenarez, F., Arias-Cabarcos, P., Ferrer-Gomila, J.-L., Lopez, A.M.: A probabilistic approach for assessing risk in certificate-based security. IEEE Trans. Inf. Forensics Secur. **13**, 202–217 (2018)
10. Saha, G.: Digital signature system for paperless operation. In: International Conference on Communication and Signal Processing, vol. 10, no. 18 (2017)
11. Idalino, T.B., Coelho, M., Martina, J.E.: Automated issuance of digital certificates through the use of federations. In: IEEE International Conference on Availability, Reliability and Security, pp. 189–195 (2016)
12. Zhu, W.-T., Lin, J.: Generating correlated digital certificates: framework and applications. IEEE Trans. Inf. Forensics Secur. **11**(12), 1117–1127 (2016)

13. Singh, S.: Generation and verification of digital certificate. In: IEEE International Conference on Advanced Technologies for Communications, vol. 10, no. 12 (2014)
14. Feng, J., Saha, P.: A new certificate-based digital signature scheme. IET Inf. Secur. **31**(8) (2013)
15. Mali, A.: Authenticated document transfer based on digital signature and a survey of its existing techniques. Int. Res. J. Eng. Technology. **3**(12) (2016). ISSN 2395-0056
16. Murthy, M.S., Kittichokechai, K.: Digital signature and watermark methods for image authentication using cryptography analysis. IEEE Trans. Inf. Theor. **19**(8), 1803–1976 (2016)
17. Lin, D.R., Wang, C.I., Guan, D.J.: A forward-backward secure signature scheme. J. Inf. Sci. Eng. **26**(6), 2319–2329 (2010)
18. Blakley, G.R., et al.: Safeguarding cryptographic keys. In: Proceedings of the National Computer Conference, vol. 48, pp. 313–317 (1979)
19. Harn, L., Lin, C.: Detection and identification of cheaters in (t, n) secret sharing scheme. Des. Codes Crypt. **52**(1), 15–24 (2009)
20. Park, J., Lee, J., Lee, H., Park, S., Polk, T.: Internet X.509 public key infrastructure subject identification method (SIM), RFC 4683, October 2006
21. Lin, J., Zhu, W.-T., Wang, Q., Zhang, N., Jing, J., Gao, N.: RIKE+: using revocable identities to support key escrow in public key infrastructures with flexibility. IET Inf. Secur. **9**(2), 136–147 (2015)
22. Steinfeld, R., Bull, L., Wang, H., Pieprzyk, J.: Universal designated verifier signatures. In: Asiacrypt 2003. LNCS, vol. 2894, pp. 523–542 (2013)
23. Schnorr, C.: Efficient signature generation by smart cards. J. Cryptol. **4**(3), 161–174 (2011)

Building an Web Based Cloud Framework for Rustic School Improvement

K. Priyanka(✉) and J. Josepha Menandas

Computer Science and Engineering, Panimalar Engineering College, Chennai, India
priyankamecs.24@gmail.com, josepha82@gmail.com

Abstract. The notion of Cloud computing has reshaped the field of a distributed system as well as that changed the organizations to use processing and computing today. The educational institutions have been utilizing the computing resources through the e-learning technologies, which overcomes the traditional pedagogy in learning but the majority of the understudies in the village are not able to get their instructive properties and the highlighting issue that emerges inside the training framework is missing of value in teaching and that are resolved through the e-learning assets utilization. The accessibility of the asset is not as simple for the bucolic based environment and they were not aware of cloud resource utilization. It can be made accessible through the custom built educational portal. Thus it enables understudies of common domains to get to the electronic utilization with the help of tablets and mobile phones.

Keywords: E-learning · Cloud computing · Internet · Digital education

1 Introduction

E-learning could be an internet primarily based learning method. This framework uses internet technology to arrange and implement manage and expand learning and can considerably enhance the potency of education. It has heap of favorable circumstances, as an example, flexibility, diversity, activity gap, etc. and it'll find yourself being a vital path for learning. The flow models of e-learning come back up short on the assistance of basic foundation that associated more allot the desired calculation and capability capacities with regards to an e-learning. Infrastructure is one among the imperative constituents of Associate in Nursing e-learning and has the immediate impact on the flourishing and security of System-Learning cloud is that the distributed computing technology within the field of e-learning, that could be a future e-learning infrastructure, as well as all hardware and software package registering resources to interact in e-learning. When the virtual calculation resources, they'll be as services for instructional establishments, students and business to lease process resources. E-learning cloud design is split into 5 main layers like hardware, software package resource layer, resource management layer, server and business application layer (Fig. 1).

The instrumentality plus layer is placed at the bottom level of the cloud middleware services, the basic process force, for instance, physical memory, CPU, memory layer is provided by the layer.

D. J. Hemanth et al. (Eds.): COMET 2019, LNDECT 35, pp. 838–845, 2020.
https://doi.org/10.1007/978-3-030-32150-5_84

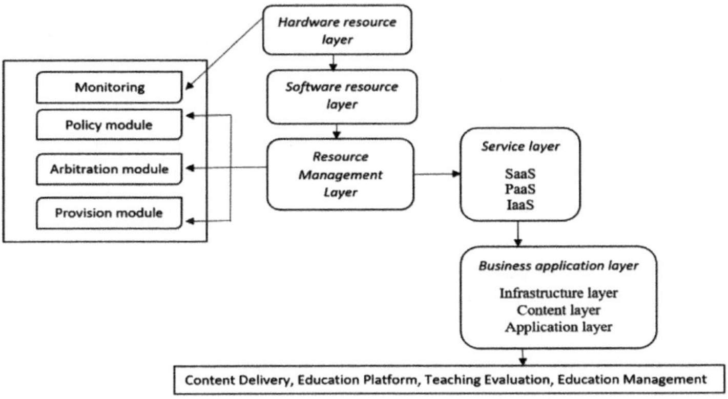

Fig. 1. E-learning module structure that integrated with cloud resources.

The software system resource layer is joined with operating framework and middleware and in sight of middleware innovation, AN assortment of programming resources is integrated to grant a unified interface to programming designers then they'll while not abundant of a stretch build up an excellent deal of utilizations addicted to plus and implant them in cloud and creating them accessible for distributed computing purchasers.

The resource management layer provides the thanks to accomplish the free coupling of programming and instrumentality assets. Because of the mix of virtualization and cloud computing, programming strategy, on-demand, free flow and appropriation of programming over totally different instrumentality will be accomplished. The policy module sets up and keeps up the educating and learning methods, run time and plus booking methods. Consistent to the data from checking module and therefore the systems of its own, policy module builds up specific arrangements and later triggers arrangement module.

Arbitration module, some approaches is formed by specialists manually, request from purchasers' square measure finished and a couple of question among animal varieties with cloud e-learning square measure solved. The business application layer: the e-learning secret is totally different from different cloud is found in e-learning application layer, which represents the foremost e-learning business logic, composed of enlarged upon a gaggle of e-learning elements. The infrastructure is the resource pool of the cloud e-learning. The infrastructure is managed by the cloud computing platform. Hardware and code virtualization technologies area unit accustomed make sure the stability and dependableness of the infrastructure.

2 Related Research

Computing paradigm [1] can be utilized as a productive apparatus for advancing the improvement of rustic. The administrations and plans given by the legislature will turn out to be more reachable than previously. It not just gives the general improvement of provincial understudy yet additionally gives gigantic open doors from business perspective. The key move of reception of cloud will make Information Technology less demanding and cheaper to utilize and broadly open to access by mass populace of understudy.

E-learning frameworks as a rule need various instrumentality and programming resources. Cloud reckoning innovations have modified the style during which applications area unit created and accessed. They're gone for running applications as administrations over the net on associate filmable infrastructure. Currently distributed computing that presents effective scale mechanism will let development of e-learning framework be endued with to suppliers and provides another mode to e-learning [2].

Enterprises moving to cloud will in general spotlight moving itself, and not as much on what they require when they arrive. While this might be a typical practice, it is anything but a best practice. Information incorporation is required in light of the fact that we just re-facilitated a portion of our information on a remote cloud benefit [3, 4]. Along these lines, the stock framework that is as yet running on a centralized server in the server farm needs to impart information to the business arrange framework that is presently on Amazon Web Services (AWS). In e-Learning, an incredible and expressive nontextual media that can catch and present data, instructional recordings are broadly utilized.

Data and Communication Technologies [5] assume a critical job in the field of training and e-learning has turned into an exceptionally prominent pattern of the instruction innovation. Notwithstanding, with the tremendous development of the quantity of clients, information and instructive assets produced, e-learning frameworks have turned out to be increasingly more far reaching as far as equipment and programming assets, and numerous instructive foundations can't manage the cost of such ICT ventures. Because of its gigantic favorable circumstances, distributed computing innovation rises quickly as a characteristic stage to offer help to e-learning frameworks. Cell phones, tablets, net books, and other kinds of versatile gadgets are correspondingly keeping clients persistently connected. Increasingly, gadgets are being intended to converse with different gadgets and applications without human intercession [6].

Cloud computing [7] with numerous favorable circumstances, there are few dangers required with distributed computing. There are issues like inaccessibility of administration, i.e., they are down when required. Another issue is of obsolete administration/stuff arrangement to the customers by the cloud service providers. Similarly, lacking of effective and quality support services to their clients is another critical concern. Also, non-ability of cloud specialist organization in respecting the administration level understanding is an extra prong in this rundown.

Cloud computing may be a powerfully elastic framework that provides net primarily based administrations, of times for all intents and functions. With development of electronic frameworks and evacuation of paper, virtual innovations and gadgets have gotten to be essential. The importance of net getting ready and underscores on its subjective and quantitative improvement for a couple of associations or specialized science and coming up with understudies [8].

Consolidating learning objects are a testing theme in light of its immediate application to educational programs age, custom-made to the understudies' profiles and inclinations. Canny arranging enables us to adjust learning courses in this way profoundly enhancing the personalization of substance, the academic prerequisites, and explicit necessities of every understudy [9].

The current e-learning frameworks center around supporting the creation and introduction of learning materials. The correspondence between the students, which is likewise a vital factor for an effective learning background is not considered truly enough in the present E-Learning Frameworks Moreover, as all traded learning is put away inside a discussion, it tends to be investigated or looked by all clients whenever. This could give a quick response for as often as possible made inquiries [10].

In the field of e-learning has accumulated critical consideration as of late [11]. This is since it has permitted clients from around the globe to learn and get to new data. This has added to the developing sum of gathered information that is as of now being produced through various gadgets and sensors utilized the world over. This has prompted the need to examine gathered information and concentrate valuable data from it. Machine learning and information examination are proposed procedures that can help extricate data and discover important examples inside the gathered information. The field of e-learning is researched regarding definitions and attributes. Additionally, the different difficulties confronting the diverse members inside the process.

Technology [12] based learning frameworks empower upgraded understudy learning in advanced education establishments. This examination assesses the components influencing conduct goal of understudies toward utilizing e-learning frameworks in colleges to enlarge classroom learning. In e-learning, the two reciprocal parts of scientific aptitudes, for example procedural familiarity and theoretical comprehension, from a point of view that is identified with present day e-learning conditions and PC based evaluation. Instructive foundation of training arithmetic is talked about, and it is suggested that the customary book medium has decided a lot of its recorded advancement, including the traditional style of showing numerical learning [13]. Data innovation is probably going to be a rising distinct advantage in learning and educating of arithmetic, and we contend that the capability of e-learning stages stretches out past straightforward penetrate activities to complex issues that are relied upon to enhance applied comprehension.

We give an audit of organization issues related to the reducing of the imperativeness utilization [14]. Execute using appropriate procedure to decrease the essentialness use is the development of the use of server farms with killing unused servers. Advantages of this paper exhibits control the dynamic course of action of virtual machines, upgrade execution and openness of organizations on IaaS fogs, lining model for the proposed movement procedure combining essentialness careful task designs save assets of the imperativeness utilization [15]. Large-scale system testing will be trying.

3 Proposed Work

As the fast addition of versatile application plans on contraptions, engineers require more prominent quality endorsement research and test robotization mechanical assemblies. Supporting on-ask for adaptable versatile testing resources and offering bound together convenient test automation organizations. Developing all around described rules for compact cloud in different regions. Especially portrayed test models and criteria to address the phenomenal features in compact applications.

The admin who maintains and manages the application uploads the update, files and videos which are the requirements of the end user and the data files are stored up in the Amazon web cloud storage. Web storage is intended for storing information client-side, while cookies are intended for communication with server and automatically added to headers of all requests. Using online courses for training eliminates the need to provide a full classroom setting for students, and this in turn can greatly reduce the costs involved in establishing and maintaining an educational space. In some instances, it even eliminates the need to hire a direct instructor and this services are used by the client end user after the client server authentication (Fig. 2).

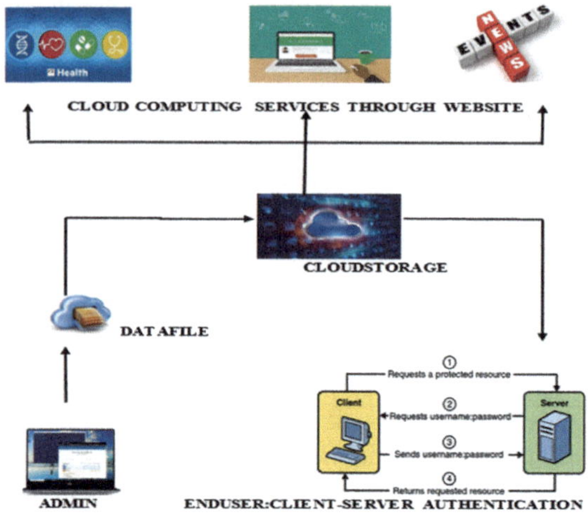

Fig. 2. Framework representation of providing e-learning resources

The site for instructive foundation, be it a school preparing, fills an assortment of needs. It can give the correct data to forthcoming understudies and guardians, discuss successfully with the understudy network, contact the overall population and supporters with continuous updates and go about as a mouthpiece for the association to the external world.

Consideration has been given to the sitting of schools and fundamental necessities, sterile and security offices, classroom measure and accessories, class and school enrolment, piece of staff regarding prepared and untrained educators, instructor capabilities, school grades, qualification for managerial staff, instructor understudy proportion, open air offices, research centers for Science and Information Technology, accessories and offices for workshops and labs for Pre-employment Education, settlement for educators and understudies, and so on. Consideration was additionally put on requirements and space use at Practical Instruction Centers and Departments. Structures, plans and details are accessible at the Planning Unit of the Ministry of Education. These standards, which will be checked on and refreshed every once in a while, will permit the functionaries at the Central Ministry and those in the locales to talk with one voice as they urge the country to move in the direction of the accomplishment of the base measures expressed in this.

4 Implementation of Proposed System

In e-learning cloud framework, thought for common school understudies Once in the wake of completing their school guidance which is right job to pick in their life. Still currently there is no such site which extraordinarily made for understudies' work. In this application which have decided about best course to look at, best school in perspective of our course and so forth. This site benefits for all understudies and moreover acclimated to urban school preparing.

5 PSEUDOCODE

STEP 1: Login into the application
STEP 2: Open the program, type your site for understudy's cloud and view the application which have various tab on regions.
STEP 3: In greeting page shows Courses after twelfth Science, best course do in India, courses after twelfth each understudy must know and so on.
STEP 4: In medical tab, its shows the health oriented tips and the medical orientations
STEP 5: In planning tab, its shows best PC courses, top 10 Highest in Demand Engineering Stream, List of Best Placement Engineering Colleges in India.
STEP 6: In next career tab, its shows Top 10 Arts colleges in India Schools, for what reason do you pick stream, Top 18 courses to do after twelfth Arts stream, and so on.
STEP 7: Logout the session.

The pseudo code in which open the program, type your site for understudy's cloud and view the application which have various tab on regions. In point of arrival shows Courses after twelfth Science, best course do in India, courses after twelfth each understudy must know and so on. In planning tab, its shows best PC courses, top 10 Highest in Demand Engineering Stream, List of Best Placement Engineering Colleges in India - Branch keen and so forth. In articulations tab, its shows for what reason do you pick articulations stream, Top 18 courses to do after twelfth Arts stream.

Websites area unit browsed passively, whereas apps area unit used actively. Websites give data, whereas apps facilitate accomplish a task. Websites could have a bigger audience; however, apps have an additional engaged audience. Websites area unit completely different from apps in their purpose and performance.

Online tutoring could be a good web site that offers you an opportunity to create cash whereas tutoring others from your laptop that want your assistance will request for your resolution with awesome features like Categorization of subjects and Tutors and

Updating Recent Questions and Solutions, Price wise sorting of Solutions, Rating and Chatting system for the users, Site Lock Security integration.

Heaps of completely different organizations will may be utilizing associate format which suggests web site that will not emerge to such an extent and it's restricted on what quantity neutering the positioning. The custom assembled web site might not 100% work on all gadgets since some formats aren't worked to be internet search tool well disposed, they must be redone to suit. Any custom or additional innovations area unit impractical to be introduced as layouts keep running on Associate in Nursing organized framework. Custom assembled sites embody a gaggle behind the business. It begins with an inspired procedure to understand United Nations agency is that the meant interest, to whom would it not wish to reach, however it need/require the positioning to capability and the way to want to seem on the online.

6 Results and Discussion

E-mentoring headways offer educators another perspective in light of developed tut theory, which communicates that adults learn by relating new making sense of how to past experiences, by associating making sense of how to specific needs, and by fundamentally applying getting the hang of, realizing more reasonable and gainful learning experiences.

Learning change concedes more noticeable understudy knowledge and advances understudies' capability, motivation, mental feasibility, and versatility of learning style. The villagers 'know' that the extremely idiots are the authorities. For one thing, couple of authorities comprehend that town life, dissimilar to government, is not departmentalized. Endeavor to transform one little piece of it and, as like as not, you wind up influencing the connections of a couple, kids' dispositions to their folks, standard common help, also, maybe much else other than and basically provides the services IaaS, gives business access to vital web architecture, such as storage space, servers, and connections, without the business need of purchasing and managing this internet infrastructure themselves.

7 Conclusion

The e-learning arrangement advancement can't ignore the distributed computing patterns. Distributed computing for e-learning arrangements impacts the way the e-learning Software ventures are overseen. There are express assignments that arrangement with discovering suppliers for distributed computing, contingent upon the requirements. Likewise, the cost and hazard administration impacts the way the e-learning arrangements in view of distributed computing are overseen. The measure of some open mists over various legitimate purviews additionally convolutes this subject; These worries are viewed as key impediments to more extensive acknowledgment of distributed computing, making them regions of dynamic research and contend among distributed computing specialists and backer. There are numerous advantages from utilizing the distributed computing for e-learning frameworks.

References

1. Gomathi, S., Malathi, S.: On building a mobile based cloud infrastructure for rural school development. Int. J. Eng. Res. Technol. (IJERT) **7**(05), 1–6 (2018)
2. Riahi, G.: E-learning systems based on cloud computing: a review. In: International Conference on Soft Computing and Software Engineering (2017)
3. Linthicum, D.S.: Cloud computing changes data integration forever: what's needed right now. In: IEEE Cloud Computing. The IEEE Computer Society (2017). 2325-6095/17/ $33.00
4. Zhang, D., Nunamaker, J.F.: A natural language approach to content-based video indexing and retrieval for interactive e-learning. IEEE Trans. Multimed. **6**(3), 1–3 (2016)
5. El Mhouti, A., Erradi, M., Nasseh, A.: Using cloud computing services in e-learning process: benefits and challenges. Received: 7 May 2017 /Accepted: 10 August 2017 # SpriSSnger Science + Business Media, LLC 2017 (2017)
6. Carswell, A.D., Bojanova, I.: E-learning for IT professionals: the UMUC experience. IEEE Comput. Soc. 1520-9202/11/$26.00 (2011)
7. Gohar, M., Rho, S., Chang, H., Jabbar, S., Naseer, K.: Trust model at service layer of cloud computing for educational institutes. Springer, New York (2015)
8. Bouyer, A., Arasteh, B.: The necessity of using cloud computing in educational system. Proc. – Soc. Behav. Sci. **143**, 581–585 (2014)
9. Morales, L., Garrido, A.: E-learning and intelligent planning: improving content personalization. IEEE Revista Iberoamericana De Tecnologias Del Aprendizaje **9**(1), 1–7 (2014)
10. Abel, F., Bittencourt, I.I., Costa, E., Henze, N., Krause, D., Vassileva, J.: Recommendations in online discussion forums for e-learning systems. IEEE Trans. Learn. Technol. **3**(2), 165–176 (2010)
11. Moubayed, A., Injadat, M.N., Nassif, A.B., Lutfiyya, H., Shami, A.: E-learning: challenges and research opportunities using machine learning & data analytics (2018)
12. Hanif, A., Jamal, F.Q., Imran, M.: Extending the technology acceptance model for use of elearning systems by digital learners. IEEE Access. https://doi.org/10.1109/access.2018.2881384
13. Rasila, A., Malinen, J., Tiitu, H.: On automatic assessment and conceptual understanding. Teach. Math. Appl. **34**, 149 (2015). Accessed 13 Aug 2015
14. Hasan, R., Noor, S.: ROSAC: a round-wise fair scheduling approach for mobile clouds based on task asymptotic complexity. In: 5th IEEE International Conference on Mobile Cloud Computing, Services, and Engineering (2017)
15. Stephanow, P., Khajehmoogahi, K.: Towards continuous security certification of software-as-a-service applications using web application testing techniques. In: IEEE 31st International Conference on Advanced Information Networking and Applications (2017)

Efficient Computation of Sparse Spectra Using Sparse Fourier Transform

V. S. Muthu Lekshmi(✉), K. Harish Kumar, and N. Venkateswaran

Department of Electronics and Communication Engineering, SSN College of Engineering, Kalavakkam 603110, India
{muthu15064,harish15035}@ece.ssn.edu.in, venkateswarann@ssn.edu.in

Abstract. One of the recent research developments of obtaining faster computation of signal spectra is Sparse Fourier Transform (SFT). This paper aims at understanding signal sparsity in the frequency domain and its frequency spectra using SFT. In this paper, we show that given a time domain signal x which is sparse in the frequency domain with only a few number of significant frequency components, then x can be recovered completely by an iterative procedure wherein the largest frequency coefficients are extracted one by one till all the k largest frequency coefficients are retrieved. Later the algorithm was used to analyze the frequencies from a real time signal piano and the results are compared with FFT based analysis.

Keywords: Digital signal processing · SFT · Signal sparsity

1 Introduction

Of all the transforms available in Signal Processing, Fourier transform which performs time to frequency domain transformation has a dominant role and has a number of applications in the field of Engineering, mathematics and applied physics. In earlier days, Discrete Fourier Transform (DFT) played a major role in the processing of signals. Because of the significant information, one can acquire as a result of this transformation, researches in this field boomed by a large number. Efforts to reduce the computational time complexity required to perform DFT has led to the formulation of several new transforms. Of all these transforms which are available today the Fast Fourier Transform remains as the fastest algorithm to perform time to frequency domain conversion. However, the FFT makes use of all the samples available in the time domain of signal x of length N. On the other hand, if the frequency domain is sparse (k non-zero coefficients), then with reduced computational complexity the frequency coefficients can be recovered. In this way, the time required for computation can be reduced. The SFT exploits this signal sparsity to get efficient results. The Sparse Fourier Transform is from the family of sublinear algorithms designed to handle massive data [9]. This transform consumes less time as it is not parsing through all the input data unlike DFT or FFT and consumes less memory space to store the transformed information as compared to the size of the input data.

D. J. Hemanth et al. (Eds.): COMET 2019, LNDECT 35, pp. 846–854, 2020.
https://doi.org/10.1007/978-3-030-32150-5_85

$$X_k = \sum_{n=0}^{N-1} x_n e^{\frac{-2\pi kn}{N}}.$$

The naive implementation of Discrete Fourier transform [2] has the computational complexity of O(N^2) and the formula of which is given by (1.1). On the other hand, the Fast Fourier Transform (FFT) [1] which is a near-linear algorithm invented by Gauss in 1805 and re-discovered by Cooley and Turkey in 1965 performs O($N \log_2 N$) operations. However, the Sparse Fourier Transform which was initially proposed by Al-Hassanieh [3] is a promising technique to reduce this computational complexity. Two main aspects of this transform are runtime complexity and sample complexity. The ideal computational complexity that can be attained by this algorithm is expected to be O ($k \log N$) [4], where k < N and k represents the number of largest non-zero coefficients in the frequency domain. The computational complexity achieved so far is O($k \log^2 N$). Now, coming to the sampling complexity, this algorithm requires O(k) samples for exactly k-sparse signals which is much lesser than FFT. SFT blends the techniques used in computer science such as hashing, randomized algorithm and classical speech processing techniques such as filtering for processing the sparse signal.

2 SFT Algorithm

The main aim of the Sparse Fourier Transform is to recover the k-largest frequency coefficients from the k-sparse input signal with reduced number of computations. The overall process is as shown in Fig. 1.

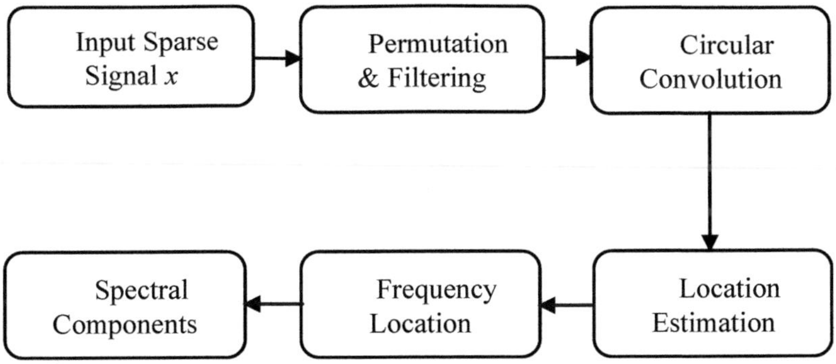

Fig. 1. Block diagram for SFT

Step 1: Generating a test sparse signal or getting the dataset of a sparse signal (Piano signal).
Step 2: Permuting the input signal.
Step 3: Filtering the permuted signal.

Step 4: Down-sampling the filtered signal with a set of co-prime factors whose product should be greater than or equal to the size of the input signal.
Step 5: Estimating the frequency coefficients and their corresponding indexes in an iterative manner.
Step 6: The original spectrum is obtained by using the estimated coefficients and the frequency response is plotted.

2.1 The Sparse Signal

A sparse signal with few frequencies is given as input. We have taken a real time signal like piano which is sparse in the frequency domain. The signal is composed of few harmonics and its time domain representation is as shown in the Fig. 2.

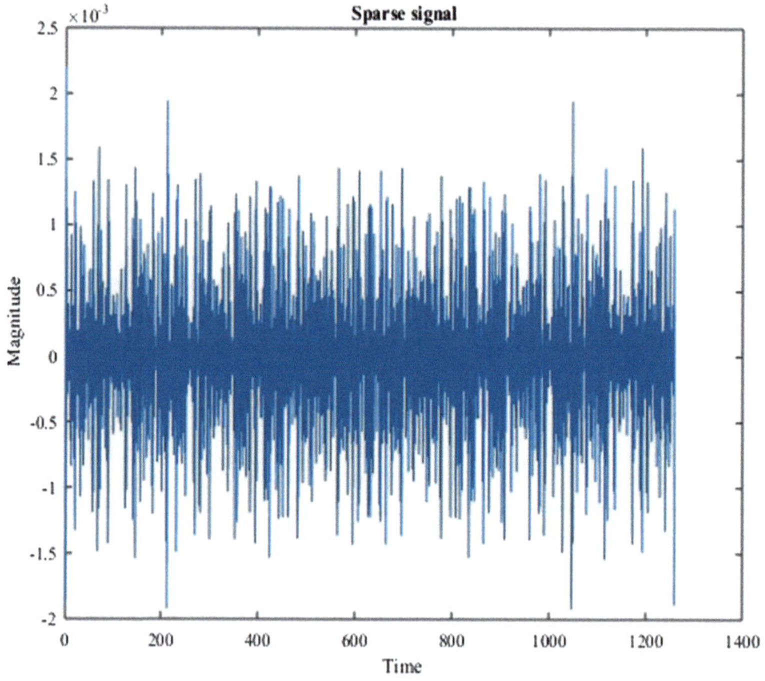

Fig. 2. A sparse piano signal in the time domain with k = 6.

2.2 Permutation of Spectra

For efficient frequency binning in the later stages a random linear transformation and phase shift is used in order to spread the frequency spectra [3, 8]. A pseudo random permutation function is employed which linearly scales and shifts the frequency components. The purpose of this permutation function is to evenly spread the spectrum so that the frequencies don't collide during bucketization (hash to bins). Since, the probability of more than one frequency component getting binned into the same bin is avoided it helps in collision resolution. The permutation function is given by Eq. (2.1).

$$x'(t) = x(\sigma t mod N)e^{\frac{-j2\pi\beta t}{n}}. \quad (2.1)$$

Here sigma and beta are chosen such that, sigma is an odd number that lies in the range [1, N] and beta is an integer lying in the same range. The sigma value chosen helps in linear scaling of the time domain signal while the later helps in providing the required shift. A modulo by N operation is performed throughout so that the frequency spectrum after permutation is also within the range [1, N]. The permuted spectrum is mapped back into the original spectrum in the later stages. A sample output spectrum after permutation is as shown in Fig. 3.

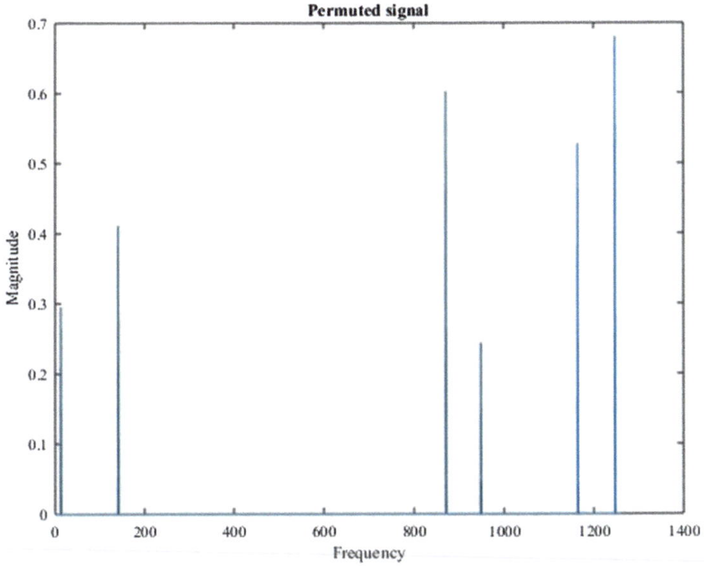

Fig. 3. Uniformly distributed frequency points in different bins

2.3 Filter

In order to filter the required coefficients, a filter of size B = N/k is employed resulting in B bins each containing B number of samples [4]. The samples are further processed to obtain the components. A number of filters are available and can be employed, however a rectangular filter employed in the time domain generates a Sinc function in the frequency domain which leads to spectral leakage [3, 4]. Thus, in order to avoid this spectral leakage, a Sinc function is used in the time domain which produces a rectangular filter in the frequency domain. Such a filter preserves the amplitude of the frequency component trapped in the bin. Therefore, for a signal *x* which is sparse in the frequency domain with only *k* number of non-zero frequency coefficients, a filter of size [1, B] is built. The permutation performed in the previous stage ensures the hashing of not more than one frequency component in each bin. A sample filter in the time and frequency domain is as shown in Fig. 4 and its mathematical representation is given by the Eqs. (2.2) and (2.3).

$$H(f) = \begin{cases} 1 & for\ \forall\ x \in [0, B] \\ 0 & elsewhere \end{cases} \tag{2.2}$$

$$y = h(t).x'(t) \tag{2.3}$$

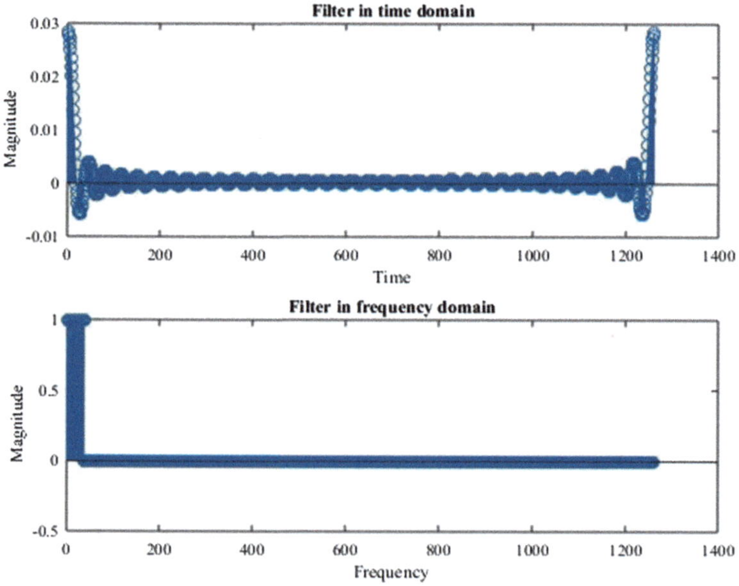

Fig. 4. An example of the filter in the time and the frequency domain.

2.4 Filtering and Circular Convolution

The next step is to divide the entire frequency spectrum into N/k bins. For this, the permuted signal is circularly convolved with the filter function of size [1, N/k] resulting in bins with N/k frequency components [4]. The mathematical equation for circular convolution is given in Eq. (2.4). An example of Circular Convolution of the permuted signal with the filter is as shown in Fig. 5.

$$(x * y)_i = \sum_{j \in [N]} x_j y_{i-j} \tag{2.4}$$

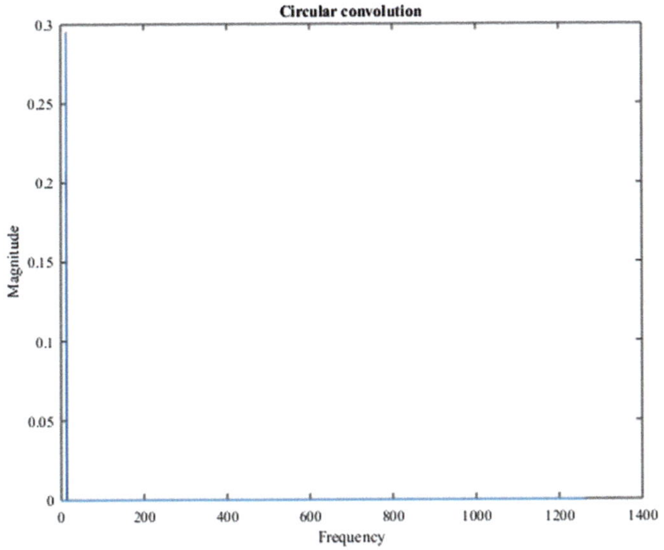

Fig. 5. An example of circular convolution of the permuted signal and filter.

2.5 Frequency Estimation

This step is the most crucial and the entire SFT algorithm revolves around this process. Here, the frequency indices along with its coefficients are estimated from the filtered signal. Firstly, the filtered signal is down-sampled with a set of co-prime factors [3, 5]. The prime factors are chosen in such a way that the product of the primes should be greater than or equal to the size of the signal N. The n-point FFT of the down-sampled signal is obtained, where n denotes the prime number considered. As a result of down-sampling aliasing occurs, and 'n' coefficients gets aliased together. The process is repeated for all the prime factors and hence all the k frequencies are extracted iteratively. The frequency positions are calculated using the Chinese Remainder Theorem [6].

2.6 Frequency Remapping

The Frequency estimation step is repeated until all the non-zero frequency components are extracted [3]. It is to be noted that the frequencies estimated are from the permuted signal and that the original spectrum is obtained by frequency remapping and the response is plotted and compared with FFT [1].

3 Simulation Results

In order to test the performance of the SFT algorithm we randomly chose k frequencies and set their coefficients randomly to be ranging between 0.1 and 1. The output of SFT was compared with Discrete in Frequency FFT (DIF-FFT) [1] and we were able to recover the frequency components exactly as in FFT in a much faster time and the results are as shown in Figs. 6 and 7.

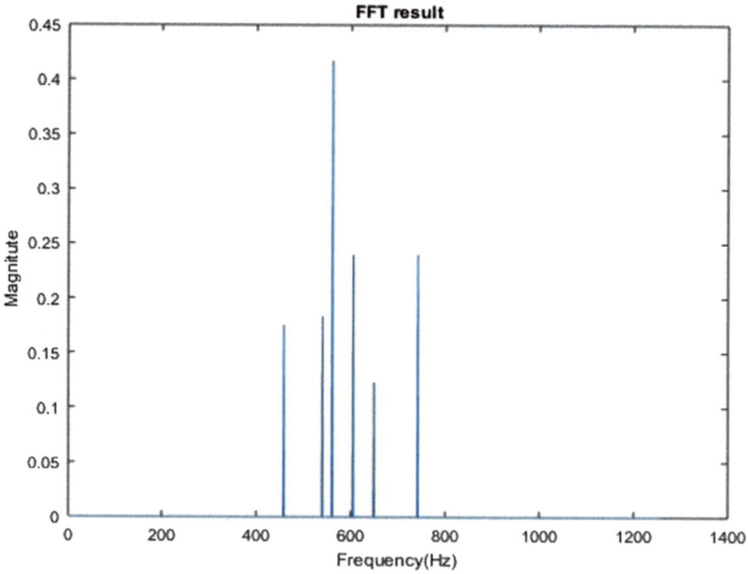

Fig. 6. Frequency spectrum using DIF-FFT.

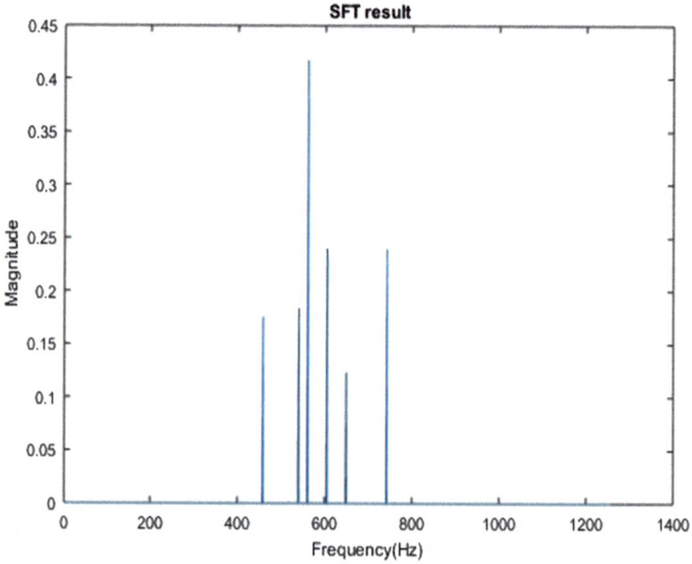

Fig. 7. Frequency spectrum using SFT.

3.1 Sparsity vs Runtime (in Secs)

We compared the runtime of SFT algorithm with FFT for increasing values of sparsity (k) as shown in Fig. 8 and found that this algorithm tends to outperform DIF-FFT every time for differing values of sparsity.

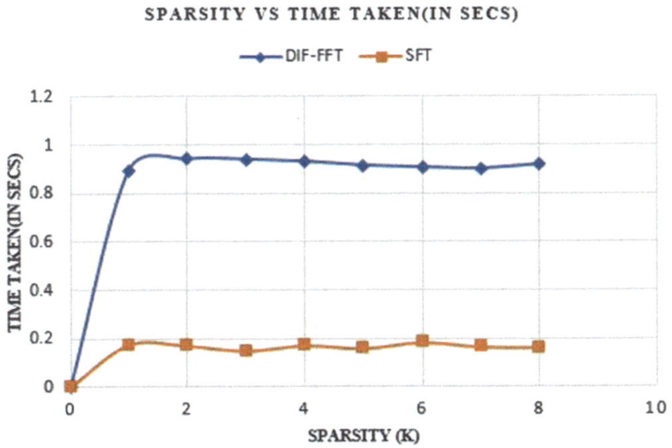

Fig. 8. Runtime of SFT is lesser than DIF-FFT

3.2 Sample Size vs Runtime (in Secs)

The sparsity (k) of the signal was fixed at 6 and the sample size was increased from 0 to 40,000. It was found that as the sample size increase, the SFT algorithm tends to perform much faster compared to FFT. The SFT bests DIF-FFT by almost seven times as we increase the sample size as shown in Fig. 9.

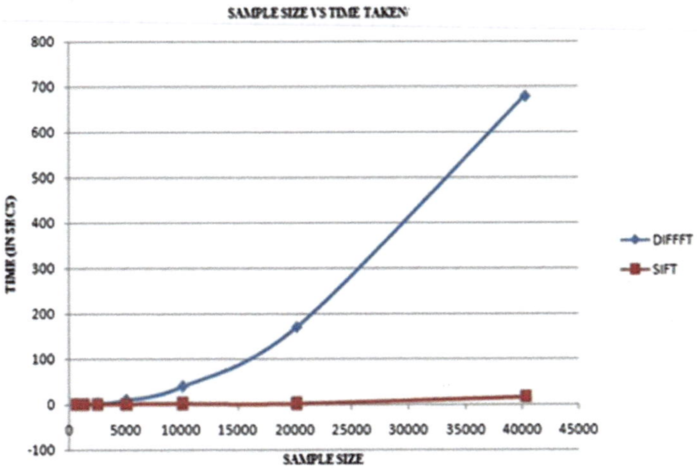

Fig. 9. Sample size vs. Runtime.

4 Conclusion

In this paper, we have presented a simple and efficient implementation of Sparse Fourier Transform using MATLAB for exactly k-sparse signals. The main aim of this paper is to show the computational efficiency and reduction in the sample complexity over FFT. It is to be noted that the SFT algorithm bests FFT when the spectrum is sparse, and it doesn't outperform FFT when the signal is not sparse. With the emergence of big data problems, medical imaging and GPS synchronization which requires faster computation it is irrefutable that by exploiting sparsity the computation time can be brought down significantly, thus enabling faster computation of signals.

References

1. Rath, O., Rao, K.R., Yeung, K.: Recursive generation of the DIF-FFT algorithm for ID-DFT. IEEE Trans. Acoust. Speech Sign. Process. **36**, 1534–1536 (1988)
2. Akansu, A.N., Agirman-Tosun, H.: Generalized discrete Fourier transform with nonlinear phase. IEEE Trans. Sign. Process. **58**, 4547–4556 (2010)
3. Hassanieh, H., Indyk, P., Katabi, D., Price, E.: Simple and practical algorithm for sparse Fourier transform. In: Proceedings of the Twenty-Third Annual ACM-SLAM Symposium on Discrete Algorithms, pp. 1183–1194 (2012)
4. Hassanieh, H., Indyk, P., Katabi, D., Price, E.: Nearly optimal sparse Fourier transform. In: Proceedings of the Forty-Fourth Annual ACM Symposium on Theory of Computing, pp. 563–578. ACM (2012)
5. Hsieh, S., Lu, C., Pei, S.: Sparse Fast Fourier Transform by downsampling. In: IEEE International Conference on Acoustics, Speech and Signal Processing, Vancouver, pp. 5637–5641 (2013)
6. Xiao, L., Xia, X.: A generalized Chinese remainder theorem for two integers. IEEE Sign. Process. Lett. **21**(1), 55–59 (2014)
7. Liu, S., et al.: Sparse discrete fractional Fourier transform and its applications. IEEE Trans. Sign. Process. **62**, 6582–6595 (2014)
8. Gilbert, A.C., Indyk, P., Iwen, M., Schmidt, L.: Recent developments in the sparse Fourier transform: a compressed Fourier transform for big data. IEEE Sign. Process. Mag. **31**, 91–100 (2014)

Survey on Predicting Educational Trends by Analyzing the Academic Performance of the Students

Selvaprabu Jeganathan, Arunraj Lakshminarayanan(✉), and Aranganathan Somasundaram

B.S. Abdur Rahman Crescent Institute of Science and Technology, Chennai, India
{selva_cse_phd_17,arunraj}@crescent.education

Abstract. Driving decisions using data is being followed in most of the business units. Industries and Institutions use complex computational techniques to improve and identify their growth trend by using Business Intelligence techniques. Adoption of data mining in educational system(s) is fairly new, data mining techniques can detect patterns from the educational system data which might be continuous or discrete and drive a prediction rule to identify the academic performance of students. Our study is focused on exploring various factors affecting educational performance of undergraduate students based on the data from their course activities. The survey explores various data mining techniques applied on educational data and advocates to integrate the learning management system with the data pattern models identified.

Keywords: Educational data mining · Education · Data mining · Mining algorithms · Learning management system

1 Introduction

Educational data mining is the process of extracting significant pattern from the large volume of data which helps to bend the data into valuable information. But due to rising stage of educational data mining many fail to identify the reason of drop out students, low achievers, unemployed students, learning disabilities and poor educational outcome from the rural areas. Investigating and predicting the academic performance by mining and discovering useful patterns from the educational databases. Based on the identified data patterns from the data set create a dynamic learning system for learners so as to evaluate the effectiveness of the course, learning process and build an intelligent learning system [1]. Data mining strategies are applied in educational data for the betterment of learning. Learning can be improved by applying formal assessment techniques. The course work design can be analyzed by evaluating how the students are using the learning system effectively so that it will help the educational designers to focus on course design.

Learning environment plays a vital role in the assessing the performance of students. Factors that cause the differences in performance includes physical location, resources, usage of technology, courses designed and the knowledge of teachers. Our

D. J. Hemanth et al. (Eds.): COMET 2019, LNDECT 35, pp. 855–869, 2020.
https://doi.org/10.1007/978-3-030-32150-5_86

research focus is on analyzing the educational trends through investigating the performance of students using multiple factors and have an insight of analyzing the learning disabilities of student which affects their educational performance [2].

In the recent years the number of engineering institutions has been increased. They produce huge number of graduates every year. Educational institutions try to use many educational techniques to educate the students but still there are students completing the course every year with many arrears, dropping out of the course, graduating with low score and struggling to get employed [13]. Data mining in business perspective focuses on improving the profit by analyzing the customer activities [24].

In the same way the objective of an educational data mining is to increase the performance of students by measuring various factors like learning activity data, assessment scores which in turn helps in predicting the academic performance of students as well as the teachers. Understanding the factors influencing the poor performance and identifying the root cause of the poor performance is a complex activity. Data mining techniques could be used to evaluate and forecast the performance of students methodologically [3]. Nowadays educational institutions collect student data from the day of their admission to their course completion. The data include their higher secondary data, Assessment scores, Extra curricular activities etc. But the institutions keep those data for only tracking purpose they are not using the data scientifically to identify the interesting patterns in the student's data [5].

Making scientific decisions using those data would help them in improving student's academic performance, to take any proactive actions on improving the standard of universities [7]. When these scientific findings presented to students/parents they will able to identify their weakness and they can improve their academic performance and at the same time when these findings are presented to the management of the university/institution which will help them to make decision on improving the quality of pedagogics and change to curriculum if required, so it will be the win-win strategy for all the associates involved in the academics and it will be bi-directionally traceable [10].

Students can find their strengths and weakness before their formal assessment evaluations. Staff members can prepare their lecture notes/presentations based on the findings and they can train the students based on the weakness entity identified, in turn parents will be assured of their children performance. Management of the university/educational institution can introduce new standards and strategies to their institutions by improving the educational infrastructure at their premises. The adaptation of scientific analysis of educational data is important which acts as root source of developing a skilled working community and entrepreneurs [16–19].

Educational data mining is in novice stage and data mining in education is different from data mining techniques applied in the banking, stock prediction, fraud detection, identifying customer buying trend etc. Even though traditional data mining techniques like classification, clustering have been implemented but still the educational data has special characteristics like the hierarchy of data and dependency so they need different mining implementation. There are learning management systems which will collect learning analytics data from the students and present it to their parents in turn helps the parents to evaluate their ward's performance in the respective course they are rolled in with that learning management system [20, 21]. Applying data mining techniques in the educational data helps to discover the learning patterns easily rather than using statistical methods directly will not be an ease approach. Considerable work is done in the education by using data mining techniques, though there are unexplored areas and there is no global approach many findings requires problem based solution [22, 23].

The sections of the paper are: third represents the methods on which the study has been made, fourth section represents the survey, key findings and interpretation, fifth defines the conclusion and future work.

2 Survey Classification

Detailed view of various substantial researches in Educational data mining is presented below. The review involves the identification of Methodology, Conclusion/Future Work directed. The papers identified are classified based on the following factors which might be used in prediction rules in

2.1 Predicting the academic performance based on Pre & Post Enrollment factors.
2.2 Predicting the academic performance using Socio-Economic Factors.
2.3 Predicting academic performance based on learning disabilities if any.
2.4 Correlation between Enrollment factors and Employability.

3 Related Works

Bakhshinategh [1] tailors new taxonomy for Educational Data Mining by setting it as the precise subdivision of data mining. EDM applications are grouped under: Student modeling, Decision support system and others. Student modeling includes the following application areas: predicting student performance, prediction by attainment of learning outcomes, profiling & grouping students and Social network analysis. Decision support system includes: Providing reports, creating alerts for stakeholders, planning and scheduling, creating courseware, developing concept maps and generating recommendation. Others include: Adaptive systems, evaluation and scientific inquiry. Based on the availability of data the EDM will grow by yielding new applications under the groups categorized.

Unemployment rate in Australia has been analyzed by Monjurul Alom by studying the perception of education from primary level to university level [2]. The prediction of finding the root cause of unemployment rate between male and female has been performed by passing the Educational data in statistical software program Orange and Wilson calculator. By experimenting to find what is the role of the gender when the student commences his/her education and completes. The finding has been derived by determining the effect size and the shortfall of completion with the level of statistical significance. The findings constitute gender plays an important role that there are major discrepancies in the completion rate among certain states between male and female.

Hussain created a framework using Waikato Environment for Knowledge Analysis [WEKA] [4]. The data has been collected from three colleges in Assam, India with 24 attributes. Collected data has been passed to the WEKA tool to find the most significant attributes for the classification. WEKA have inbuilt machine learning classification, clustering and associative algorithms. Dataset has been experimented using J48 classifier, Bayes net, Random forest, Part classifier algorithms. Depending on the correctness and the classification errors obtained from the result it has been concluded that the random forest classification method was the most appropriate algorithm for the dataset and the feature selection should be derived by identifying the most influential attributes available in the data.

Asif and Merceron [5] proposed a methodology to identify student performance only by using their academic data of a four-year graduation data without using any socio-economic data and demographical data. Students are clustered using their exam marks of each course every year to study the academic progress every year. Finally, a heat map is generated by mapping the prediction and progression to analyze whether the pragmatic policy of the institution has been designed to identify high performing and low performing students. Most importantly from the findings it is noted that the students who evolve good at university doesn't had a good progression in their school.

The assessment of analyzing the efficiency of educational data mining techniques to early predict the failure of students in new programming courses have been done using WEKA [3]. The key areas focused to improve the accuracy of prediction are data preprocessing and fine tuning of data mining algorithms (Fig. 1).

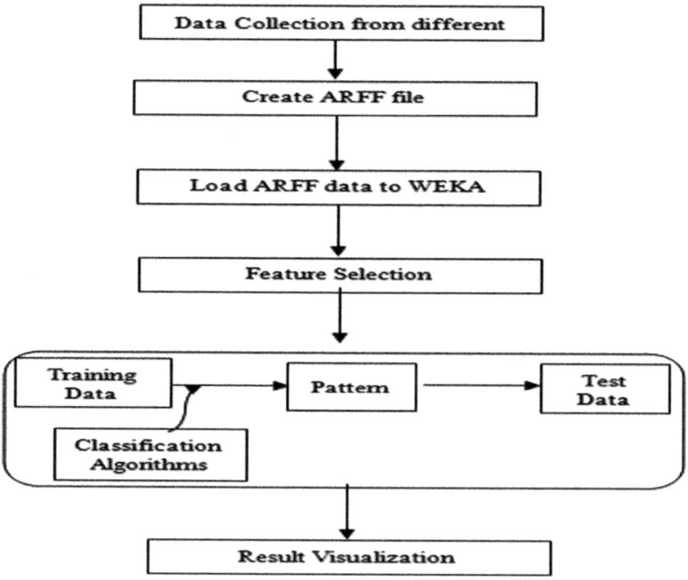

Fig. 1. Generalized workflow of student performance. Classification using WEKA

Data has been taken from two sources one from on-campus and other from distance mode. Data preprocessing has been done by reducing large number of attributes by keeping only the required attributes using WEKA. Identified algorithms have been configured to improve the performance, the fine-tuned algorithms are Decision tree, Support vector machine, Neural network and Naive bayes. Grid search implemented in

Support Vector Machine proved a good accuracy level compared to other mining techniques.

Almarabeh [6] determined the accuracy of different data mining techniques in WEKA using educational data. The dataset of 225 instances with ten attributes has been used to predict the effective decisions which will advance and progress the performance. Different classifiers are used and the comparisons are made based on the accuracy derived. Best classifier has been identified based on the percentage of error measures identified. Among the available data mining technique experiment, results show that Bayesian network has the best performance among other classifiers.

Ariouat projected a two-step clustering technique in which first cluster contains alike trainee profiles using performance indicators employability and time. Second cluster contains alike profiles using AXOR algorithm [clustering algorithm]. Based on the training data the second partition is identified with efficient training logs. A new clustering and classification technique has been developed by taking semantic annotations on event logs into account. Intend to develop classification techniques to split semantically annotated event logs based on traces distance from a set of process models or templates defined at the abstract level [7].

Pena Ayala [9] reviews on the methods and materials used in the data mining with the educational data. The data mining models are categorized into descriptive and predictive models where the descriptive models use unsupervised learning to produce relations, interconnections between the mined data whereas the predictive models imply supervised learning to discover the hidden/future values of the dependent variable used in the mining. The results help us to identify the patterns in educational data mining based on the analysis of student performance modeling, analysis of assessment, analysis on student feedback, analysis of teacher support, curriculum and domain knowledge. Overall the strength, weakness, opportunities and threat [SWOT analysis] of educational data mining has been analyzed.

Strength of EDM is identified as its baseline is robust and supported by data mining. There are specialized events for EDM for its growth and brining in research projects into it based on the new learning trends. The weakness identified as many of the educational data mining approaches are applied only to a trivial part of the data mining repository items like different frameworks, algorithms. Learning through non-conventional educational methods, personalized teaching, learning through video platforms are identified as the great opportunities. A standard terminology, a common logistic, reliable framework, and open architectures are demanded to be proposed, accepted, and followed by the EDM community.

Archer worked on a case study to identify the success and retention rate of a student by including the socio-economic factors and to develop a commercial product in the higher education market [10]. A pilot project ShadowMatch developed in university of South Africa has been used in this analysis. Student's Socio-economic data has been classified into student identity and attributes, student walk and institutional identity and attributes. Student identity collects data which comprises of demographics, intellectual, emotional, attitudinal, and perception related data. Student walk comprises of mutual interaction between the student and the institution. Institutional identity and attributes comprises of history, location, demographics, perceptions, academic, non-academic and expectations. Based on the above collected data a social critical model has been developed to predict the student success. ShadowMatch has been used in commercial environment for managing the employee profiling here it has been used to predict student success by developing a social critical model. This implementation provides aglimpse into the complexities that the higher education institutions may face in a dynamic education landscape where technology is changing so rapidly which increased the reliance on external learning environments like udemy, pluralsight.

Association rule has been applied to identify the performance of students in a computer science based post graduate course [11]. The analysis has been done by associating the students with and without computer science background in their under graduation. Along with this the performance of post graduate students has been predicted by using association analysis on the marks they have submitted during the admission. The weak students have been identified by constituting the frequent item set and rule generation by applying Apriori algorithm in WEKA. The frequent item set is generated till the minimum threshold interval is reached; the search space will increase based on the number of occurrence of the objects in the data. Apriori property has been used to decrease the search space and robust association rules are generated based on local, global frequent item set. This method is robust when the data size is smaller.

Lorraine frames an employability development profile, "CareerEdge" [12] to fill the gap in analysis of unemployment in the graduate students. Data has been collected from 807 under graduate students for factor analysis. This model influences the students to focus on career development learning perceived during their work experience which they can implement in their academic studies. This model has been defined that it will enhance the course subject knowledge and understanding. The 807 student data collected has been split into two groups one for exploratory factor analysis and other one for confirmatory factor analysis.

Level of education in Bangladesh has been analyzed using the geospatial attributes along with the educational data [14]. Key objective is to examine the educational difference in the urban and rural areas of Bangladesh by merging the educational data and geo spatial data. Student data has been collected from primary school level to higher secondary level from various urban and rural areas. Data comprises of academic data along with the spatial attributes of the respective student and the educational institution. OpenGeoDa has been used to apply Exploratory spatial data analysis [ESDA] techniques. Auto correlation has been derived by using univariate and multi variate Local indicators of spatial association [LISA]. A thorough spatial analysis has shown how these methods can be used to extract more value from existing datasets. The analysis presented clearly shows that there is some spatial consistency in the distribution of education indicators, poverty and educational establishments in Bangladesh. The heat maps justify that educational level is directly proportional to the spatial properties.

To predict undergraduate student's academic achievement based on the role of the curriculum, time investment and self-regulated learning [15], Torenbeek used structural modeling equation to study the relationship between self-regulated learning, time investment, curriculum characteristics by using the data of two hundred graduate degree students from four different degree programmes. The most impacting factor of the academic performance is derived by comparing the covariance matrix from the models and recommends self-regulated learning with more practical sessions shows good improvement in the performance.

Dejaeger [16] predicted student satisfaction using data mining techniques. The scope of the analysis is to help the educational institution management to take strategic decision based on the student satisfaction ratio. Two data sets from two different educational institutions were taken up for analysis; the data has been collected in the form of survey. Student satisfaction is evaluated by using decision trees, neural networks, support vector machine and logistic regression. In this case the standard deviation is best at 1% level for Classification and Regression tree [CART] decision trees compared to other mining techniques. The management of the educational institution preferred decision trees for its symbolic representation format like univariate decision tree and detailed orientation.

Different data mining techniques has been used by Osmanbegovic for predicting student performance [17]. Techniques used in the evaluation are naive bayes, decision tree, multi-layer perceptron. Data is collected using a questionnaire survey. Statistical testing methods like chi-square test, one r-test, info gain and ratio test are used against the algorithms to derive the result set. Attribute ranking has been done by collecting the average value of the algorithms to find the predicting model to measure the academic performance that is user friendly for professors or non-expert users. The experiment can be extended with more distinctive attributes to get more accurate results, useful to improve the students learning outcomes. It has been determined that decision tree can be used in this case because a reasoning process can be given for every prediction. Also experiments can be done using other data mining algorithms to get a broader approach, more valuable and accurate outputs.

Prediction of learning disabilities of school-age children was done by David and Balakrishnan [18]. This is the mostly untouched attribute in the educational data mining, many educational data mining research focus on the attributes comprised of pre-enrollment/post-enrollment, Socio-economic factors. Some types of learning disability have been listed below (Table 1).

Table 1. Types of learning disabilities [18]

S. no.	Types
1	Difficulty with reading
2	Difficulty with spelling
3	Difficulty with handwriting
4	Difficulty with written expression
5	Difficulty with basic arithmetic skills
6	Difficulty with higher arithmetic skills
7	Difficulty with attention
8	Easily distracted
9	Difficulty with memory
10	Lack of motivation
11	Difficulty with study skills
12	Does not like school
13	Difficulty learning a language
14	Difficulty learning a subject
15	Slow to learn

If a student shows difference in areas other than the areas he/she performing well and difficulty of experience in certain functioning areas then the student termed to have learning disability E.g.: listening, reading, performing math etc. Prediction of learning disability has been compared by using rough set theory and support vector machine. It has been justified that rough set theory is good in accuracy than support vector machine. Support vector machine categorized under supervised learning will produce less accuracy when applied to the large datasets which might contain incomplete data or attributes. In rough set theory, dataset has been characterized in to Information tables and decision tables. Information table contains the data and the decision table contains the decision which should be carried out based on the condition met. Rough set theory which works efficiently even with a data set contains inconsistent, ambiguity of data (Table 2).

Table 2. Key findings and interpretation

S. no.	Year, Author(s)	Title	Journal	Methodology	Interpretation
1	B.Bakhshinategh, Osmar R.Zaiane, Samira ElAtia, Donald Ipperciel	Educational data mining applications and tasks: A survey of the last 10 years	Springer, Education and Information Technologies (Volume 23, pp 537–553), January 2018	Classification and Regression	Different categories and sub categories has been identified from the learning management systems and existing surveys to propose a new taxonomy on EDM tasks. These task should be made customizable and new model can be designed dynamically [1]
2	B.M. Monjurul Alom, Matthew Courtney	Educational Data Mining: A Case Study Perspectives from Primary to University Education in Australia	International Journal of Information Technology and Computer Science (Vol 2018, 2, 1–9)	Classification and Regression	Impact of gender with respect to education in a closed data set using statistical tools like Wilson Calculator and Orange. The process can be executed in a larger data set by considering the social economic factors [2]
3	Evandro B. Costa, Baldoino Fonseca, Marcelo Almeida Santana, Fabrísia Ferreira de Araujo, Joilson Rego	Evaluating the effectiveness of educational data mining techniques for early prediction of students' academic failure in introductory programming courses	Elsevier (Computers in Human Behavior 247–256, Feb 2017)	Decision Tree, Support Vector Machine, Neural Network and Naïve Bayes using WEKA	Depth of attributes has been limited by considering only vital attributes from a single university. Accuracy level has been identified by fine tuning certain algorithms. We can include data from other universities and check the accuracy level by repeating the process [3]

(*continued*)

Table 2. (*continued*)

S. no.	Year, Author(s)	Title	Journal	Methodology	Interpretation
4	Sadiq Hussain, NeamaAbdulazizDahan, FadlMutaher Ba-Alwib, NajouaRibata	Educational Data Mining and Analysis of Students Academic Performance Using WEKA	Indonesian Journal of Electrical Engineering and Computer Science Vol. 9, No. 2, pp. 447–459, February 2018	J48, PART, Bayes Net, Random Forest, WEKA (Data Mining Tool)	The students' academic performance was evaluated based on academic and personal data collected from 3 different colleges from Assam, India. Based on the accuracy and the classification errors one may conclude that the random forest classification method was the most suited algorithm for the dataset [4]
5	Raheela Asif, Agathe Merceron, Syed Abbas Ali, Najmi Ghani Haider	Analyzing undergraduate students' performance using educational data mining	Elsevier Volume 113, Pages 177–194, Oct 2017	Decision Trees and Clustering	Predicted student performance by using only marks without using any socio-economic data. Study whether the design of the heat maps can be refined to extract the indicators of low and high performance without running the algorithms for prediction before [5]
6	HilalAlmarabeh	Analysis of Students' Performance by Using Different Data Mining Classifiers	I.J. Modern Education and Computer Science, 8, 9–15, Aug 2017	Naive Bayes, Bayesian Net, ID3, J48, Neural Networks tested in WEKA (Mining Tool)	Different classifiers are used and the comparisons are made based on the accuracy among the classifiers and different error measures are used to determine the best classifier. Experiments results shows that Bayesian Network has the best performance among other classifiers [6]

(*continued*)

Table 2. (*continued*)

S. no.	Year, Author(s)	Title	Journal	Methodology	Interpretation
7	H. Ariouat, A. Hicheur-Cairns, K. Barkaoui, J. Akoka	A two-step clustering approach for improving educational process model discovery	IEEE lecture notes. 339. https://doi.org/10.1007/978-3-662-46578-3_73, August 2016	Clustering Technique	Develop new clustering and classification techniques taking into account semantic annotations on event logs. Intend to develop classification techniques to split semantically annotated event logs based on traces' distance from a set of process models or templates, defined at the conceptual level [7]
8	Pena-Ayala, Alejandro	Educational data mining: A survey and a data mining-based analysis of recent works	Elsevier, Vol. 41, No. 4, pp. 1432–1462, 2014	Statistical and Clustering Processes	Identified kinds of educational systems, disciplines, tasks, methods, and algorithms. A standard terminology, a common logistic, reliable frameworks, and open architectures are demanded to be proposed, accepted, and followed by the EDM community [9]
9	Archer, Elizabeth, Yuraisha Bianca Chetty and Paul Prinsloo	Benchmarking the Habits and Behaviors of Successful Students: A Case Study of Academic-Business Collaboration	International Review of Research In Open and Distributed Learning, Vol 15, No 1, 2014	Experimental Usage of Employee Profiling Software-ShadowWatch	Experimented the usage of a commercial product generally used for employee profiling in corporate, for higher education environment. Provides a glimpse into the complexities faced by education institutions in a dynamic higher education landscape where technology is changing so rapidly that increased reliance on external providers by support functions will be required [10]

(*continued*)

Table 2. (*continued*)

S. no.	Year, Author(s)	Title	Journal	Methodology	Interpretation
10	Pool, Lorraine Dacre, PamelaQualter, and Peter J. Sewell	Exploring the factor structure of the CareerEDGE employability development profile	Education and Training Monitor 56.4 (2014): 303–313, 2014	Exploratory and Confirmatory factor analysis	Emotional Intelligence, Self-Management, Work and Life Experience are found to be important factors for employability development profile [12]
11	Mridul Khan, A.K.M. Zahiduzzaman, Mohammed NahyanQuasem and Rashedur M. Rahman	Geospatial Data Mining on Education Indicators of Bangladesh	IJCA, Vol. 20, No. 1, March 2013	Spatial Autocorrelation, Spatial Regression, Open GeoDa (GIS Tool)	A thorough spatial analysis has shown how these methods can be used to extract more value from existing datasets. The analysis presented clearly shows that there is some spatial consistency in the distribution of education indicators, poverty and educational establishments in Bangladesh [14]
12	Torenbeek, M., Jansen, E., and Suhre, C.	Predicting undergraduates' academic achievement: the role of the curriculum, time investment and self-regulated learning	Studies in Higher Education. 2013; Vol. 38, No. 9. pp. 1393–1406, Nov 2013	Structural Equations Modelling, Correlation Matrix	Examined two variables, Pedagogical approach and skill development in the first 10 weeks of enrolment. Academic achievement is evaluated by applying curriculum variables in structural equations modelling [15]
13	K. Dejeager, F.Goethals, A. Giangreco, L. Mola B. Baesens	Gaining insight into student satisfaction using comprehensible data mining techniques	European Journal of Operational Research 218.2, 548–562, 2012	Decision Trees, Neural Networks, Support Vector Machine, Logistic Regression	Determined student satisfaction for a particular course through survey data collected from two educational institutions. Accuracy level of algorithms has been evaluated from the collected dataset and Logistic regression produced good accuracy [16]

(*continued*)

Table 2. (*continued*)

S. no.	Year, Author(s)	Title	Journal	Methodology	Interpretation
14	Osmanbegovic, Edin and Mirza Suljic	Data Mining Approach for Predicting Student Performance	Economic Review – Journal of Economics and Business, Vol. X, Issue 1, May 2012	Chi-Square Test, One R-Test, Info Gain and Ratio Test, Naive Bayes, Decision Tree	Found predicting model for academic performance that is user friendly for professors or non-expert users. The experiment can be extended with more distinctive attributes to get more accurate results, useful to improve the students learning outcomes. Also, experiments could be done using other data mining algorithms to get a broader approach, and more valuable and accurate outputs [17]
15	Balakrishnan, Julie M. David	Significance of Classification Techniques in Prediction of Learning Disabilities	International Journal of Artificial Intelligence & Applications, Vol 1, No. 4, Oct. 2010, pp 111–120	Rough Set Theory and Support Vector Machine	Accuracy level of predicting the learning disabilities of school-age children has been exercised using a large dataset. Rough Set theory produced results with good accuracy level [18]

4 Research Gap

To predict the academic performance of a student, different categories and sub-categories has been identified from the learning management systems by using only the vital indicators. The academic performance prediction is not only to identify the performance of the student instead it should act as a gateway to improve the performance. Some researches have shown the association of academic performance with spatial attribute [11] and using learning disabilities [15]. Experimental usage of a commercial product which is generally used for employee profiling in corporate used in educational environment renders the proportionality between competency and academic performance [9]. Factors like economic, social opportunity, communication gap and data from online learning systems has not been included in the data collection which helps in the root cause analysis of the performance predicted. The following preliminary workflow represents the flow of data to the mining system based on the gaps identified to focus on continuous improvement (Fig. 2 and Table 3).

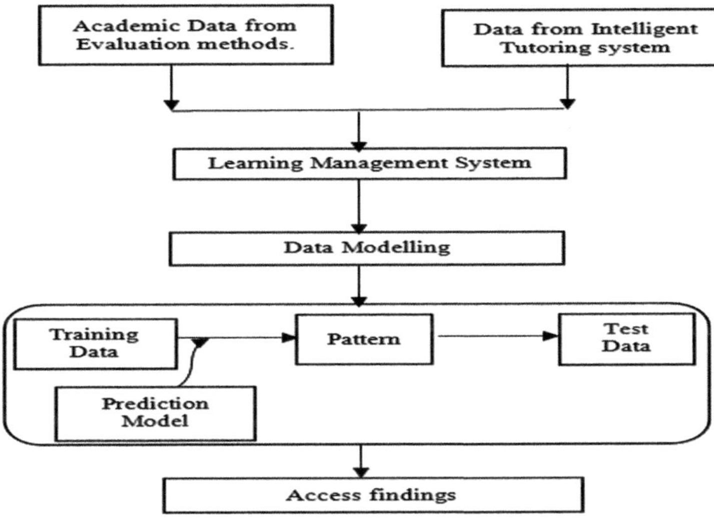

Fig. 2. Integration of LMS with mining tools.

Table 3. Pseudocode of the workflow

```
1  Set S as the source data:
      *S→ Comprises of Academic Data + Data from Intelligent tutoring systems..
2  Set D as the E-R model to be defined:
3  Let E be the entity.
4  Let A be the attributes in the source data S1 and S2.
5  Define E based on the source data: S.
6  Establish the relationship between the entities by mapping the selected attribute.
7  Build S based on the data model D.
8  Pass S to Learning management system.
9  Apply the selected mining technique.
10    Identify the pattern.
11    Repeat step 8 to train the dataset with different classifiers.
12    Access the findings.
```

5 Conclusion and Future Work

Significant amount of work has been done in predicting academic performance but the factors like Enrollment attributes, Academic scores, Social factors influence the conditional probability on each other, Say the academic scores might impact the employability. There are cases where the above factors are conditionally independent on other, say the Spatial attribute might not be a factor in some scenarios to influence on the academic performance of a student so there might be a downfall in the prediction rule. To initiate the analysis on educational data start with mining tools like WEKA, Rapid miner and test the data frame with identified attributes against the suitable mining techniques.

The challenge faced by many educational institutions is developing students skillfully to get employed because institutions fails to implement intelligent tutoring system and to get data from their learning systems to guide their students. An integrated approach to be followed with all of the above factors by summing up the non-cognitive factors (attitudes, special skills), learning disabilities, geographical attributes so that the resultant input vector to the data mining system will be enriched and increase the precision rate of the prediction. The universities should use intelligent tutoring system which should be integrated with the learning management system to improve student performance in long term. The data from learning management system of the university should be integrated with the data mining tools and the prediction models should be rich in usability for the management and professors to guide their students for continuous improvement.

References

1. Bakhshinategh, B., Zaiane, O.R., ElAtia, S., Ipperciel, D.: Educational data mining applications and tasks: a survey of the last 10 years. Educ. Inf. Technol. **23**, 537–553 (2018)
2. Monjurul Alom, B.M., Courtney, M.: Educational data mining: a case study perspectives from primary to university education in Australia. Int. J. Inf. Technol. Comput. Sci. **10**(2), 1–9 (2018)
3. Costa, E.B., Fonseca, B., Santana, M.A., de Araujo, F.F., Rego, J.: Evaluating the effectiveness of educational data mining techniques for early prediction of students' academic failure in introductory programming courses. Comput. Hum. Behav. **73**, 247–256 (2017)
4. Hussain, S., Dahan, N.A.-A., Ba-Alwib, F.M., Ribata, N.: Educational data mining and analysis of students academic performance using WEKA. Indonesian J. Electr. Eng. Comput. Sci. **9**(2), 447–459 (2018)
5. Asif, R., Merceron, A., Ali, S.A., Haider, N.G.: Analyzing undergraduate students' performance using educational data mining. Comput. Educ. **113**, 177–194 (2017)
6. Almarabeh, H.: Analysis of students' performance by using different data mining classifiers. J. Modern Educ. Comput. Sci. **8**, 9–15 (2017)
7. Ariouat, H., Hicheur-Cairns, A., Barkaoui, K., Akoka, J.: A two-step clustering approach for improving educational process model discovery. In: International Conference on Enabling Technologies: Infrastructure for Collaborative Enterprises, pp. 38–43, August 2016
8. Rubiano, S.M.M., Garcia, J.A.D.: Analysis of data mining techniques for constructing a predictive model for academic performance. IEEE Latin Am. Trans. **14**(6), 2783–2788 (2016)
9. Pena-Ayala, A.: Educational data mining: a survey and a data mining-based analysis of recent works. Expert Syst. Appl. **41**(4), 1432–1462 (2014)
10. Archer, E., Chetty, Y.B., Prinsloo, P.: Benchmarking the habits and behaviors of successful students: a case study of academic-business collaboration. Int. Rev. Res. Open Distrib. Learn. **15**(1) (2014)

11. Rakesh, A., Badal, D.: Mining association rules to improve academic performance. IJCSMC **3**(1), 428–433) (2014)
12. Lorraine, D.P., Pamela, Q., Peter, J.S.: Exploring the factor structure of the career EDGE employability development profile. Educ. Training Monit. **56**(4), 303–313 (2014)
13. Vanhercke, D., De Cuyper, N., Peeters, E., De Witte, H.: Defining perceived employability: a psychological approach. Pers. Rev. - Emerald Insight **43**, 4592–4604 (2014)
14. Khan, M., Zahiduzzaman, A.K.M., Quasem, M.N., Rahman, R.M.: Geospatial data mining on education indicators of Bangladesh. IJCA **20**(1), 10–22 (2013)
15. Torenbeek, M., Jansen, E., Suhre, C.: Predicting undergraduates' academic achievement: the role of the curriculum, time investment and self-regulated learning. Stud. High. Educ. **38**(9), 1393–1406 (2013)
16. Dejeager, K., Goethals, F., Giangreco, A., Mola, L., Baesens, B.: Gaining insight into student satisfaction using comprehensible data mining techniques. Eur. J. Oper. Res. **218**, 548–562 (2012)
17. Osmanbegovic, E., Suljic, M.: Data mining approach for predicting student performance economic review. J. Econ. Bus. **X**(1), 3–12 (2012)
18. Balakrishnan, K., David, J.M.: Significance of classification techniques in prediction of learning disabilities. Int. J. Artif. Intell. Appl. **1**, 111–120 (2010)
19. Thakar, P., Mehta, A., Manisha, S.: Performance analysis and prediction in educational data mining: a research travelogue. Int. J. Comput. Appl. (2015). ISSN 0975–8887
20. Saranya, S., Ayyappan, R., Kumar, N.: Student progress analysis and educational institutional growth prognosis using data mining. Int. J. Eng. Sci. Res. Technol. **3**, 1982–1987 (2014)
21. Hicheur Cairns, A., et al.: Towards custom-designed professional training contents and curriculums through educational process mining. In: The Fourth International Conference on Advances in Information Mining and Management, IMMM 2014 (2014)
22. Finch, D.J., Hamilton, L.K. Baldwin, R., Zehner, M.: An exploratory study of factors affecting undergraduate employability. Educ. + Training **55**(7), 681–704 (2013)
23. Potgieter, I., Coetzee, M.: Employability attributes and personality preferences of postgraduate business management students. SA J. Ind. Psychol. **39**(1), 01–10 (2013)
24. Romero, C., Ventura, S., Pechenizkiy, M., Baker, R.: Handbook of Educational Data Mining. Taylor & Francis, New York (2010)

Improving the Invulnerability of Wireless Sensor Networks Against Cascading Failure

Rika Mariam Bose(✉) and N. M. Balamurugan

Sri Venkateswara College of Engineering,
Sriperumbudur 602 117, Tamil Nadu, India
rikamariambose22@gmail.com, balagan@svce.ac.in

Abstract. Wireless Sensor Networks consist of thousands of sensors to monitor the environment. The sensors nodes when operating in a harsh environment are prone to energy depletion, hardware failure and other attacks. If a sensor node fails, its load is transmitted to the neighboring node where the load increases making that node to fail which leads to the cascading failures in the sensor nodes. The invulnerability of WSN needed to be established. The invulnerability can be achieved in the network by using a threshold value. If the node's value is below the threshold value then there is no chance for cascading failure to occur. If the node's value is nearer to the threshold value then some preventive measures can be taken by reducing the network performance and the energy dissipation by selecting efficient cluster heads. If the node's value goes completely beyond the threshold value then the performance of the network can be completely reduced and the nodes can be relocated using capacity expansion schemes.

Keywords: Invulnerability · Cascading failure · Sensor node · Cluster head · WSN

1 Introduction

Wireless Sensor Networks consists of many sensors to monitor the environment. They are deployed easily and covers a wide range of applications. In most of the cases, the WSN's are made to work in some harsh conditions where the nodes are prone to hardware malfunction, energy depletion and even attacks. Due to this, the failure of a sensor node splits from its connected topology and reduce the coverage of the network. So, researches are being made to build an invulnerable network. The research on the invulnerability of the network focus on the availability and connectivity once the network faces failure, but fails to understand how network can be kept away from failure.

The sensor node's capacity transmits the load is small due to the limitation in the hardware cost. When the data load is more than the capacity of the sensor node then the sensor fails. When a sensor node fail, the data transmitted through the failed node selects a new route for data transmission. Hence, the load is redistributed. The "relay load" and "sensing load" concepts are used. A cascading failure model is built for clustering WSN based on this. The performance of the sensors is evaluated with the use

D. J. Hemanth et al. (Eds.): COMET 2019, LNDECT 35, pp. 870–876, 2020.
https://doi.org/10.1007/978-3-030-32150-5_87

of a threshold value. If the sensor node's value is less than the threshold value then no cascading failure occurs but if the value of the node is near to the threshold value, some preventive measures are to be taken. An efficient cluster head is to be selected and then the load of the sensor is redistributed.

To reduce the damage caused by the cascading failure, the capacity is expanded and is divided into two sub phases- find which node to be selected and the way how the capacity can be allocated. The capacity expansion scheme is done with the mobility management technique where the sensors relocated. For that a selection scheme is proposed which selects a node for expansion of capacity and a capacity allocation.

2 Related Work

Liu et al. [1] proposed an electricity-green routing protocol that maximizes the general system overall performance. A progressed protocol known as instantly-line routing (SLR) is evolved in WSN to assemble an instantly course the use of two hop facts without the help of the geographical statistics. The SLR scheme used for its more advantageous overall performance to preserve power while in comparison with rumor routing protocol. In this, several improvements are made to raise the route ratio and reduce hop counts the usage of the SLR scheme against the RR scheme. The main idea of the SLR protocol is to maintain the occasion and path of query direct. The best hops of facts are recorded instead of recording all the nodes that are visited. The hassle of the understanding is that most appropriate routing in electricity restrained networks isn't practically feasible.

Dobslaw et al. [2] proposed a consumer defined give up-to-give up reliability approach known as SchedEx a popular heuristic scheduling set of rules produces schedules. The more the reliability demand, the higher SchedEx plays compared to the approach to optimally enhance the reliability through repeating the maximum rewarding slots incrementally until a closing date. SchedEx has an extra calmly dis-bursed development effect at the algorithms but the Incrementer supports the schedules used through positive scheduling algorithms. The trouble of this paper is that the person latency bounds aren't considered for the specific flows in the community.

Song et al. [3] proposed a dynamic simulation version of both strength networks and protection structures, which could simulate a greater variety of cascading outage mechanisms relative to present quasi-consistent-nation (QSS) fashions. The version and the demonstration on how different mechanisms have interaction are described. This paper provides the layout of and effects from a new nonlinear dynamic version of cascading failure in strength systems (the Cascading Outage Simulator with Multi-manner Integration Capabilities or COSMIC), used to examine a wide variety of different mechanisms of cascading outages. COSMIC, represents a Power system as a hard and fast of hybrid discrete/continuous differential algebraic equations, concurrently simulating safety systems and device dynamics. The problem of this paper is that COSMIC, is probably too gradual for lots massive-scale statistical analyses.

Xu et al. [4] provided the survey at the clustering techniques to increase the Network sturdiness and beautify power efficiency with smart network choice answers that extraordinarily benefit the QoS and QoE in IoT. This paper discusses about the specific clustering algorithms for use within the WSN while thinking about the demanding situations in making use of to the 5G IOT scenarios. The algorithms used for survey are Vornoi and Non-Vornoi based approaches where the Vornoi based totally approach consists of LEACH and HEED algorithms while Chain and Spectrum are Non-Vornoi based tactics. Some of the limitations whilst thinking about the 5G IOT primarily based situations are that there is restricted paintings concerns community coverage whilst comparing community lifetime.

Cai et al. [5] proposed a version for interdependencies between strength structures and separate information networks, and also to examine the impacts on cascading screw ups. The features of communication networks are embedded into dispatching records networks. The structures of dispatching information networks are usually classified into types they are double-big name and mesh. The correlation in double-big name network and electricity systems is "diploma to diploma," but "degree to betweenness" is the correlation for mesh networks. The interactive model among power grids and dispatching statistics networks is supplied via a dynamic strength drift model.

Tian et al. [6] proposed a routing protocol referred to as Network Coding and Power Control primarily based Routing is offered for unreliable wireless networks to save energy. The proposed model consists of network coding mechanism and considers dynamic transmit power and the variety of packet transmissions. The proposed model uses the derived network coding advantage in making wise choices on whether to apply network coding or now not such that electricity consumption is appreciably decreased. The limitation of this paper is there's complexity on safety occurs due to the increasing interdependence between electricity systems and dispatching statistics networks.

Dey et al. [7] proposed an evaluation of the propagation of screw ups, in phrases of line outages, combined with the topological traits of the grid aids to take corrective movements to store machine from entire crumble. It helps to research the development of the blackout and apprehend its nature and depth. This motivates to set up the connection between the community topological traits and cascading failure in the energy grid. In this, the simple topological characteristics of the energy community are studied and the common propagation of failure underneath varying topological situations is calculated as a branching system parameter. At first a comprehensive look at of the topological features of the one-of-a-kind strength grid networks is presented. The 4 primary statistical metrics selected for complicated network analysis are common degree, average direction duration, clustering coefficient and algebraic connectivity. The issue of this paper is to analyses the machine's interconnections to set up a robust strength grid layout.

3 Proposed System

The proposed system for invulnerability against cascading failure is discussed below. The proposed system is split into the following section Cluster Head Selection, Load Redistribution, Mobility Management and Capacity Expansion. The Cluster Head

Selection section is used to select a node as a cluster head from the cluster using the Leach algorithm where a node will be selected as a cluster head at a time interval. Once the cluster head is chosen the load to be distributed from the sensor node to the cluster head is done by the sensing load and the load to be distributed from the cluster head to the sink is done by the relay load (Fig. 1).

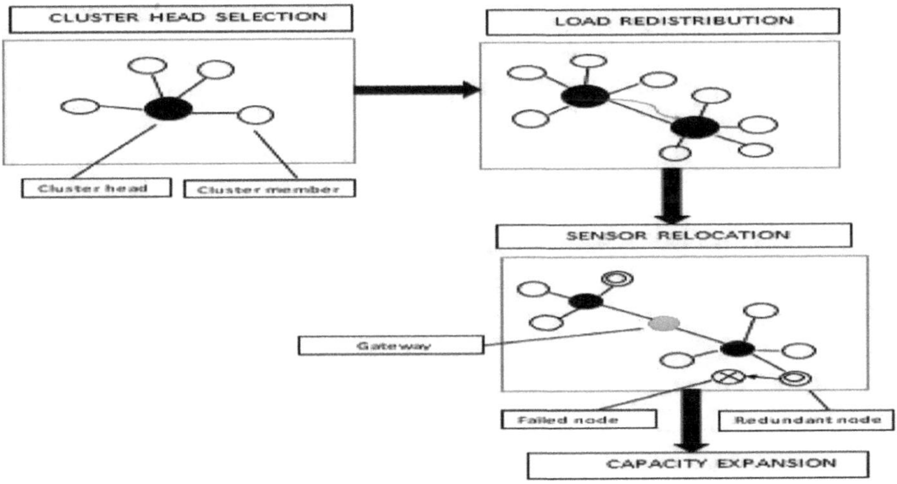

Fig. 1. Proposed architecture for invulnerability against cascading failure in WSN

3.1 Cluster Head Selection

The reason for the cascading failure is the overloading of the sensor nodes in the network. One way to avoid the cascading failure is to select an efficient cluster head. This can be done by using the LEACH (Low Energy Adaptive Clustering Hierarchy) protocol. The LEACH protocol is a protocol based on clustering which minimize the energy loss in the networks. The protocol selects randomly the sensor nodes as cluster heads and performs periodic election so that the energy loss faced by the cluster head nodes to communicate with the base station is spread all over nodes in the network. The operation in LEACH has 2 divisions.

3.2 Set-Up Phase

In set-up phase, each sensor node randomly selects a number between 0 and 1 and if it is lower than the threshold value, then that sensor node will be considered as the cluster head. Nodes that were once cluster heads will not be selected again and equal energy will be spread across the network. After the selection process, the cluster head informs its selection process to other sensor nodes and provides TDMA to the cluster members.

3.3 Steady Phase

In this phase, the TDMA schedule is used for the data transmission where the base station receives the aggregated data from cluster heads. The cluster head once selected will not be taken again. So that, every sensor node has the chance to be the cluster head in the cluster.

3.4 Load Redistribution

In clustering WSNs, the sensor node's load can be of sensing and relay loads. The network operates normally when the initial data load of the sensor node is less than the capacity of the node. When a cluster member fails, the data cannot be sent to its cluster head. The relay load is not taken, so that no extra data load is transferred to other sensor nodes. Thus, when the cluster member fails, it does not lead to cascaded failure.

When a cluster head node fails, the cluster member carrying the sensing load will also be failed as the link to the cluster head is cut-off. But the relay load that came out from the failed cluster head will be resent and will be again distributed to the neighboring nodes.

3.5 Mobility Management

The mobility management is carried out by the concept of sensor relocation. The sensors are relocated to deal with any failure of sensor node. The sensor relocation is done in two phases such as Identification of redundant sensors and Relocation of the sensor to the target location. Grid Quorum solution is used, to quickly identify the redundant sensors with low message overhead and cascaded movement to relocate the sensors efficiently.

3.6 Capacity Expansion

If a cluster head has more capacity to deal with load that is transferred from the neighboring cluster heads, the cascading failure stops. Therefore, increasing the capacity of the node is an effective method. An Expansion scheme and Capacity scheme can be used, so that the network invulnerability against cascading failure is improved using expansion-selection scheme and capacity allocation scheme.

4 Experimental Result

The NS2.35 Simulator is used to carry on with the experimental results where the Cluster Head Selection done with the Leach algorithm is given below (Fig. 2).

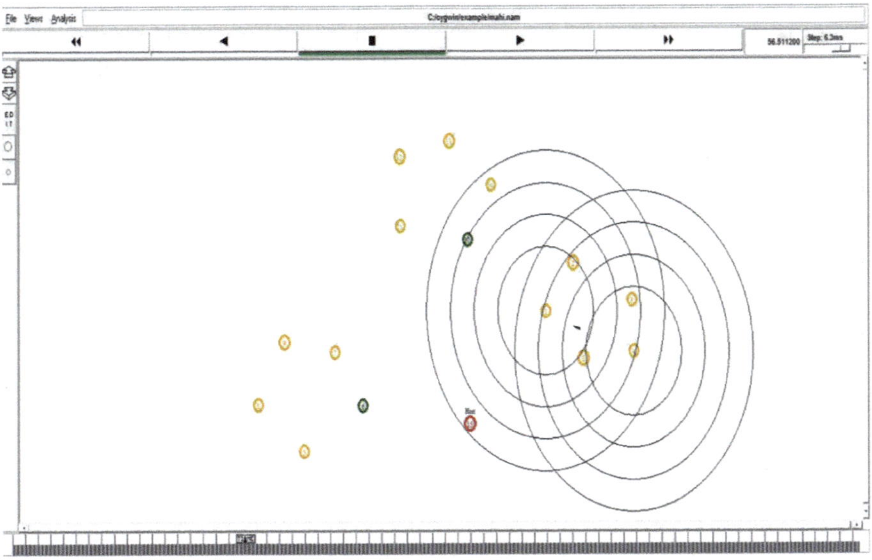

Fig. 2. Cluster head selection using leach algorithm

5 Conclusion

Wireless Sensor networks operating in unattended conditions are prone to attacks leading to the failure of some nodes leading to the cascading failure in the network. To improve the invulnerability of the network against the cascading failure a threshold value is maintained. If the node's value is less than the threshold value, then no failure occurs. If the value is nearer to the threshold, then an efficient cluster head is selected and the load redistribution process is maintained. To reduce the cascading failure, the capacity of the node is expanded by relocating the sensor nodes.

References

1. Liu, H.-H., Jia-Jang, S., Chou, C.-F.: Analysis on invulnerability of wireless sensor network towards cascading failures based on coupled map lattice. IEEE Syst. J. **11**(4), 2374–2382 (2018)
2. Dobslaw, F., Zhang, T., Gidlund, M.: End-to-end reliability-aware scheduling for wireless sensor networks. IEEE Trans. Ind. Inf. **12**(2), 758–767 (2016)
3. Song, J., Cotilla-Sanchez, E., Ghanavati, G., Hines, P.D.: Dynamic modelling of cascading failure in power systems. IEEE Trans. Power Syst. **31**(3), 2085–2095 (2016)
4. Asim, M., Moktar, H., Khan, M.Z., Merabti, M.: A sensor relocation scheme for wireless sensor networks. In: IEEE Workshops of International Conference, pp. 808–813 (2011)
5. Cai, Y., Cao, Y., Li, Y., Huang, T., Zhou, B.: Cascading failure analysis considering interaction between power grids and communication networks. IEEE Trans. Smart Grid **7**(1), 530–538 (2016)

6. Tian, X., Zhu, Y.-H., Chi, K., Liu, J., Zhang, D.: Reliable and energy efficient data forwarding in industrial wireless sensor networks. IEEE Syst. J. **11**(3), 1424–1434 (2017)
7. Dey, P., Mehra, R., Kazi, F., Wagh, S., Singh, N.M.: Impact of topology on the propagation of cascading failure in power grid. IEEE Trans. Smart Grid **7**(4), 1970–1978 (2016)
8. Guo, L., Ning, Z., Song, Q., Zhang, L., Jamalipour, A.: A QoS-oriented high-efficiency resource allocation scheme in wireless multimedia sensor networks. IEEE Sens. J. **17**(5), 1538–1548 (2017)

Pedwarn-Enhancement of Pedestrian Safety Using Mobile Application

N. Malathy(✉), S. Sabarish Nandha, B. Praveen, and K. Pravin Kumar

Department of Information Technology, Mepco Schlenk Engineering College (Autonomous), Sivakasi, India
malathy@mepcoeng.ac.in, sabarishnandhal@gmail.com, pravinfiery@gmail.com, praveenbalajipraveenbalaji@gmail.com

Abstract. Mobile phones usages are emerging as fast as the evolution of a man. Starting from the sunrise to sunset people are constantly using the mobile phone. It has more features like gaming, music, camera, alarm, etc. and people are using it to perform their day-to-day tasks. Even though the mobile phone has a lot of benefits, it also leads to life threatening problems due to pedestrian collision. The probability of such incidents may happen, when the pedestrians are not taking their eyes form the mobile phone while walking or crossing the road, so they met with an accident which causes serious injuries. To avoid such happenings and to notify the pedestrians about the obstacles, a mobile application called Pedwarn is developed with the help of built in phone sensors like accelerometer sensor, gyroscope sensor. It provides solutions to the problem without prior knowledge of the surroundings by calculating the distance to the nearby objects with phone speakers and microphones. The obstacles are detected within 2–4 m. Its accuracy is also strengthened by using the visual detector which will capture the images of the surrounding with the rear camera. Pedwarn is evaluated using variety of environmental settings and in different devices which are common in our day to day surroundings. The power consumed by each and every component is measured periodically of about one hour. Averages of Pedwarn measurements is noted that, its detection accuracy is 95% higher and 2%.

Keywords: Pedwarn · Pedestrian safety · Mobile application

1 Introduction

Distracted walking increases the risk of injury since the user focuses is distracted by the use of smart-phones while walking [1]. When the pedestrians text on their phone while walking, they notice that the environmental changes are 50% less. According to the emergency room visit, the rate of accidents due to the pedestrian's use of smart-phone has grown 10 times [2]. Accidents rate is likely to increase due to the smart-phone user and this accident may be severe that the person meeting with the accident may have severe head injury. People who walk with the smart- phone can get knocked down by the oncoming car or may hit any objects in front of them [3]. Authorities of China recognized the growth of accidents due to cell phone and set up a "cell phone lane" for

D. J. Hemanth et al. (Eds.): COMET 2019, LNDECT 35, pp. 877–887, 2020.
https://doi.org/10.1007/978-3-030-32150-5_88

those people walking with smart- phones to avoid colliding with the objects [4]. Systems can identify cars with frontal images of the cars but cannot find other objects beyond the car. The shape or colors of the objects prevent them from detecting the obstacles in the pedestrian's path. Pedwarn is proposed to address the unexplored problem "can smart-phones detect if the user is walking towards dangerous path without any prior knowledge of the surroundings". Pedwarn is a safety application that strengthens the safety of the pedestrians that reduces the accident rate as much as possible. Detecting the obstacles using a single sensor is difficult task, so utilizing several phone sensors to achieve high accuracy at low computation and energy cost. Pedwarn uses the speakers and microphones of the phone to find the distance between user and object and uses rear camera if necessary. Obstacles distance can be found by single camera without depth perception since the phone's height has been already determined by Pedwarn's acoustic detector. Pedwarn is the First phone application to alert the distracted walker by monitoring the environment. It requires only the commodity sensors not any specialized sensors. Offers generic solution to the situations since does not depend on prior knowledge about the obstacle/objects.

2 Related Work

In the field of intelligent vehicle and robotics, obstacle detection and avoidance has been an active area [8–10]. Active safety systems deployed in the cars to protect the pedestrians but these requires electronic devices such as RADAR, LINAR, SONAR and several cameras to detect pedestrian and detect their movements. Robots may use cheap sensors to detect the obstacles but are designed for this purpose. Sonar's in robots are directional whereas phone/microphones are not. The main usage of sonar is to detect the objects by sending the waves through the human body i.e. ultrasonic waves like a bat. There are different types of energy used in the radar and sonar. Now days the use of sensors has been reduced because built in sensors are readily available in today's smart phones. They include accelerometer, barometer, gyroscope, temperature detection, microphones, and cameras. The camera will take the picture of the user and usage of microphone is to record the sounds of the environment. With the help of these sensors programmers had developed various applications like indoor phone localization, context-aware computing and human computer interface. The main approach here used cell phone's rear camera to capture and display it. Though user will be concentrating on playing games or chatting and because of that they can't use this application often. As a result it has been implemented as an application which will be running in a background as service. Similar applications like walk safe [4] to identify the sudden changes in the ground. It has an infrared sensor which will measure the distance from the ground to the sensor. The variation may provide information about the changes in the surface due to natural causes. It also improves the detection accuracy as well as energy consumption. This application has been executed in different environments and by different users. Another application is Look up [5] application and it will act upon monitoring the road adaptation height variation between sidewalks to a street with the help of sensors incorporated in shoes. Both applications are similar to the Pedwarn and it enhances the pedestrian's safety [6]. Crash alert has been introduced to improve the safety while walking. Crash alert is a walking user interface, includes audio feedback, two-handed chorded keyboard. Using the depth camera, view crash alert captures and displays information in the user peripheral view. Depth camera is an orthogonal field of view for the busy operator. With the help of field of view user can take early actions noticing upon obstacles, which will be displayed in the form red alert prompting the user to stop or take diversion.

3 Proposed Work-Pedwarn

In order to process the Pedwarn safety application four components are needed: They are Acoustic detector, Visual Detector, Motion Estimator, Fusion algorithm (Fig. 1).

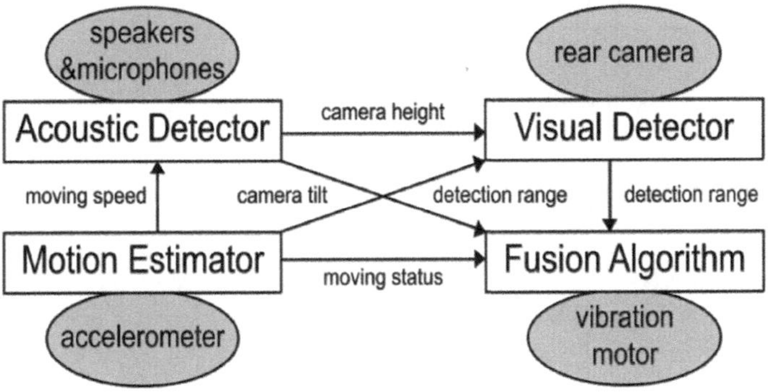

Fig. 1. Pedwarn system design and its components

3.1 Acoustic Detector

The main aim of the application is to avoid the obstacles. There exists certain process in detecting the objects. The first and foremost step is acoustic detector. The work of the acoustic detector is to generate the sound waves with the help of built sensors in commodity cell phones like microphone/receivers. It sends the sine wave of about 10 periods at a certain frequency of 11,025 and the reflected signals are captured via microphones. The sample rate of the signal sent is 44.1 Hz These two Signals are differentiated and its transmission rate is around 100 ms. This design is basically for commodity phones. I.e. in Sect. 4 it is clearly described that our application is tested in the latest android version "pie". Acoustic detection follows certain algorithm which is described as a pseudo code below:

Pseudo code:

Three filters are used band pass filter, FIR filter, and matched filter.

INPUT: $S_{n=}$ nth detection

Peak window win_{peak}

Walking speed: w,

Threshold coefficient:α**Max.resolution**=15cm

Signals duration and Speed of the sound waves=15cm.

win_{peak}= 40samples.

Number of samples in the transmitted signal=40 samples.

OUTPUT: detection result: D_{succ}

1: The process starts with the band pass filter, it filters the sound signals with high noise ratio.

2: After the audio signals are filtered it is further filtered by the matched filter

K<- matched filter (band pass_ filter(Kn)

Matched filter is used for increasing the signal to noise ratio which is also known as impulse response is matched to input pulse signals.

3.The threshold value (th)is checked if periodically,

If H_n>thr

Then H_n will be marked with highest amplitude samples

4:is peak← True

5.The object with in 15cm will be assumed as a single object.

6: if K[j]> K[i] then 7: isPeak→False

8: break

9: if is Peak and K[i] >thr , then if peak max value is greater than the peak window size , then index value will be set to peak offset.

10:peakOffset→i

11:peak Max →K[i]

At last the detection results are filtered by the motion filter which is available in motion estimator. According to our experimental results, Objects within 2–4 m is detected at a fixed frequency of 11,025. And the objects within a 10 m range are difficult to detect. It can be reduced by using multiple access reflection protocols. For examples in different devices the different frequencies will be emitted as FDMA, CDMA, for the assurance to have a minimal co-relation. From the acoustic detection three kinds of peaks are generated. First peak represents the sound emitted from the speaker, the second peak represents the reflection from the human body and third reflection is from the ground (142 cm away). Acoustic detector has certain limitations like, the sound waves are Omni-directional., so that direction of the object can't be solved and also the reflection is a multipath reflection. The reflected signals received by the microphone are a combination of both sent signal and multiple path reflections. To improve the detection rate further, motion filter is implemented. Consider a scene where all the objects are placed at a same distance. In such cases acoustic detector fails. In order to overcome the limitations of acoustic detector motion filter is implemented.

3.1.1 Motion Filter

The main aim of the motion filter is to eliminate the obstacles with its relative speed where the relative speed is 0 which is not related to the users Walking speed. The user speed as well as the previous detection results are given as: $D_{n-1} \sim D_{n-\delta}$

Pseudo code : Motion Filter

Input: Previous □detections Dn-1 …….Dn-□ , depth: □ , walking speed:w

Output: results passing motion filter:

D_{succ}1:K_{succ}←Φ

2: **for** S ∑ K_n**do** 3: r ← 0

4: **for** i from 1 to □ **do**

5: d_{est}← S+ j * u * $period_{detection}$ 6: **for** $d_{history}$ ∑K_{n-I}**do**

7: **if** |d_{est}-$d_{history}$ | < win_{error} **then**

8: r←r + 1/□

9: **if** r> r_{succ}**then** 10: K_{succ}←D U d 11: **return** D_{succ}S belongs to K_n

S is projected based on the user's motion speed. Where j * u * $period_{detection}$ yielding d_{est} it is compared with the dhistory's probability of the object detection increases if it matches with the previous projection. The yielding probability value is then passed to the additional filter called motion filter with the help of it unmatched reflections can be rectified. Finally, if there is any obstacles behind or after the user it will be intimated to them as a alert or some kind of warning signals.

3.1.2 Visual Detector

Acoustic detector is good when there are one or two objects, not in situations like the aisle. From the Fig. 2, it shows that acoustic detector is failed because of the Omni Directional sound waves. In order to overcome these limitations, analyzing it with "Visual Detector" (it is considered as a sensing layer). The working of the Visual Detector will have proceeded with the help of Smartphone's rear camera. The object direction will be provided and the false positive error is removed. Inclination angle for tilting is from 31−65°. If not then it will show an alert message like the application is stopped working. Alert may be a text or it will generate vibrations so that the users can maintain their tilt range. The main confrontation in rear camera is, detect the occurrence of the obstacles and to measure the distance between users and object though the depth present in the image taken by the single camera may vary. Theoretical information will not be used in Pedwarn like color and shape for the identification of objects. Pedwarn becomes more ideal by not having any priori information i.e. it is difficult to do without such knowledge and it has the ability to detect any kind of dangerous obstacle and prevents the causes of collision with them. In visual detector various techniques are involved like Blurring filter, HSV transformation, Back Projection and the final analysis is erosion filter. The role of blur filter is to eliminate the noises which present in the captured image. The resolution is approximately 10×10. HSV means Hue (represents color)-Saturation (Represents grayness') and Value (Represents the brightness of the image). By using this color pattern, each every color is specified and then black & white color is added to make color adjustments. This transformation is useful to identify the color types like humans skin color, shadow, fire color etc. In other words, image luminance is isolated from the color data. HSV is also denoted as a HSB (Hue, Saturation, and Brightness). It is a replacement for RGB Color model in computer graphics program. The next step is Back-Projection. It differentiates the image which is not in similar to the surface of the ground or floor.

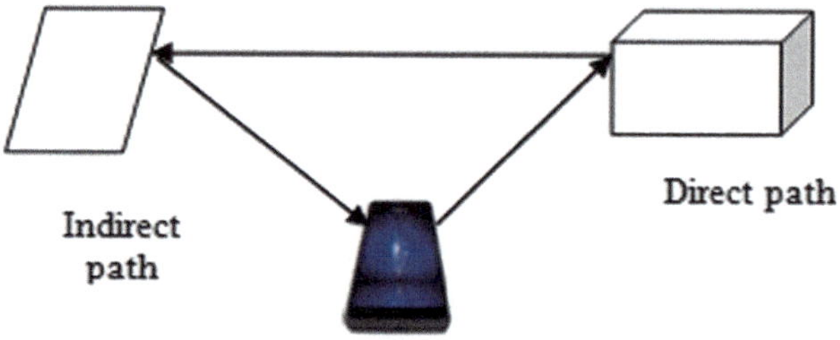

Fig. 2. False positive detection with multiple path

The final most process is erosion filter, which is used for removing the residual error (The differences between the observed value and predicted value $E = y - \bar{y}$) from the previous step. After completion of these steps, predefined thresholds values will be lower than the blob with larger areas so that it will be considered as a obstacle and nearest point to the bottom of the image will be a nearest or closest obstacles. Some might think that observations done by the back projection is a reference to the ground/floor texture. Visual detector is unable to detect, in case if the object is anonymously included as a relation to pattern of ground/floor as well as identifying it in a random image will be difficult. But it is not an issue in Pedwarn, because images will be taken while the users are walking and using their phone at expected positions. Under this assumption the specific area with high probability will be considered as a ground/floor pattern. To identify the specific area, they had conducted a trial with 10 participants. The observations are like they are requested to capture an image at formidable distance of 2 m from the door, which is clearly given in the Fig. 2 (Fig. 3).

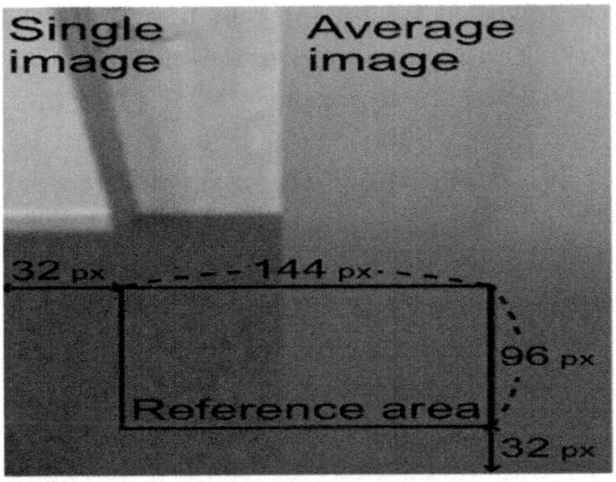

Fig. 3. Average taken from the 10 participants

The darkest area will be examined as a ground/floor. The area size chosen is 96 × 144 pixels located 32 pixels above the bottom of a 240 × 320 image. After the identification of closest object, consider a variable p, d which is pixel difference from the detection point to the image bottom and real world detection to the object which is detected. The computational equation is, d = pixel-to-distance (p, hp, tp) where hp and tp are height and tilt angle of the phone with respect to the ground. Above operation can be executed if and only if these values are known. These parameters can also be estimated from the online with the help of methods when it not fixed before walking. The tilt angle is measured with the help of accelerometer formula as follows:

$$tp = \cos - 1(acc2/accmag)$$

acc2 represents acceleration orthogonal to the phone's surface. accmag is the Magnitude of overall acceleration created by the user's activities or may be due to earth gravity. While dealing with the tilt angle of the phone, its height remains unknown, when the user is in motion. But Pedwarn get the feedback from the acoustic detector for the phone height. There are certain devices like image-based detection which simply consider the height of the camera and there are some scenarios by installing the camera inside the car in a fixed location. But not in our scenarios though the height may vary from place to place while the user is walking and their way of holding the cell phone. Finally, if there are any obstacles in the captures picture, obstacles will be prompted to the user.

3.1.3 Fusion Algorithm

The usage of the fusion algorithm is to increase the accuracy and to reduce the false detection rate. The power utilization can be lowered by rejecting the components which are not contributing to increasing the accuracy. Starting with the acoustic detector if the user is in the idle state then it will alert the user by some notifications. If not then the acoustic detector will be invoked and it will check whether the environment is cluttered, no means notification will be displayed that there is an obstacle. And if the environment is cluttered, in order to detect the objects visual detector is invoked.

Clutter identification is obtained by the motion filter which is used in the motion estimator algorithm. So that total number of idle objects will be identified and relative speed will be considered as 0. In the motion filter algorithm, the application with relative speed 0 is also known as the clutter filter. If there is an obstacle it will be notified or else it will go to idle state. The main difference between the aisle and cluttered environment is in aisle situations pedestrians can't detect as many objects as possible but in cluttered conditions, it doesn't leave even single objects. The fusion algorithm will trigger the visual detector when the threshold value exceeds and also reuse of motion filter for the detection of cluttered area. The main advantage in fusion algorithm is, each and every component gives complement of each other like aisle situations can't be handled by the acoustic detector so with the help rear camera visual detector will rectify it. But visual detector can't differentiate the similarities like walking through the cement floor to the grassy environment. These kinds of problems are easily filtered out by the acoustic detector. Our aim is only to detect objects like walls, hanging materials, aisle conditions etc. (Fig. 4).

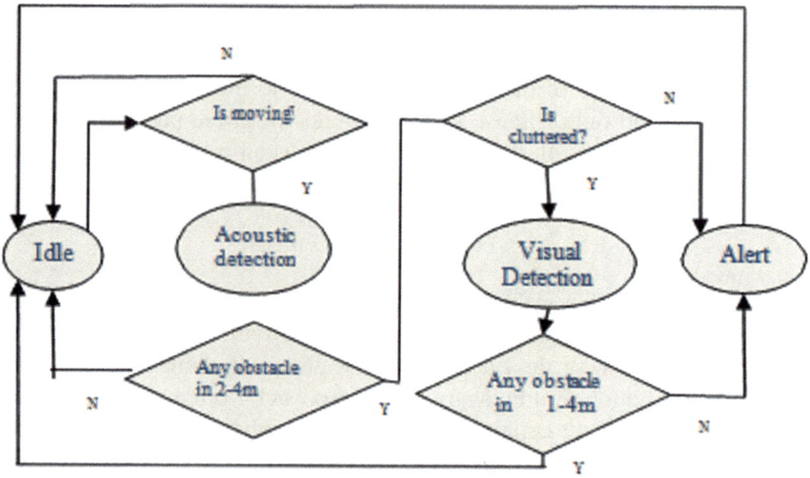

Fig. 4. Fusion algorithm's system design

4 Implementation

This application is executed in the latest android version (PIE) in Redmi Note 5 pro and Lenovo k8 plus. Basically Pedwarn application is platform independent, i.e. this application is also supported in lower android version starting ice- cream sandwich, jelly bean etc. For the computational purpose of the Pedwarn application Band pass filter is implemented and matched filter in C programming which is interfaced through Java native interface (JNI) which requires very less compile time and execution time. As a result, executing each component such as acoustic detector, visual detector, and fusion algorithm will be executed within 25–80 ms where its period is set as 100 ms.

5 Accuracy Measure in Different Areas

After the application is implemented, its detection accuracy is tested in different environments. The objects such as cardboard, dustbin, wall, and box are considered for testing the accuracy in terms of false positive rate. The calculation of false positive rate is based on the probability of results of previous object measurements with difference to the current objects results. The accuracy is noted and based on the results the detection accuracy of each and every object's was predicted. Another way to calculate the accuracy is false positive rate by dividing the number of false positives with the total number of objects that we used to measure the accuracy. Since the objects are movable in both outdoor and indoor environments, poor performance caused by each and every environment can be measured accurately in this process. The environments used for testing is outside of the working area, building lobby and in an aisle environment (which have more than 4–5 objects). The experiment is analyzed in cluttered area like noisy areas in town side functions, but it doesn't affect the frequency ratio of the Pedwarn since the noise frequency from the humans are below 11 Hz and also Pedwarn adjusted its noise ratio. But in aisle the false positive rate will be high in cluttered environment. The experimental results based on environment shows that in outdoor the false positive mainly arises due to wind. In 5 m wide- aisle the false positive is usually high and to reduce this the detection range needs to be changed to 1–2 m instead of 2–4 m or we can also intimate the user that the environment is cluttered and the detection range is shrunk, so that user pays attention. To estimate properly the threshold of stationary objects can be changed to 5. The false positive rate is high for a box compared to a wall due to its size. Based on Participants, the accuracy depends on the holding position of the phone and the walking pattern. The experimental result shows that the tilt angle should be between 31 to 65° and the phone height varies from 1–1.3 m. The false positive detection arises due to the usage of phone differently by different participants since they may hide the speaker or camera or both with their fingers and due to this detection result will be affected. In contrast to it, visual detector will not have any issues regarding the aisle situation, due to image taken by the rear camera of the visual detector. Therefore false positive rate is lower in visual detector when compared to the false positive rate in acoustic detector. The fusion algorithm is to confirm the true positive rate in outside environments though it reducing the false positive in indoor environments. The fusion algorithm also reduces the false positive rate in outside areas due to the strong winds blowing in the built sensor (microphone).

6 Processing Cost and Energy Consumption

In order to find the energy consumed by each and every component, the application is executed for about hourly a period of time. The information of the Pedwarn CPU usage is logged through the top command in an interval 10 s. The 1 h measurements averaged to obtain usage of CPU and power consumption. These scenarios are tested: idle-with flash light on (whether the user is moving or his position is constant), acoustic detector, visual detector, fusion algorithm and trace. Idle often represents the power consumed by the flash light. The acoustic detector and visual detector's frequency rate is of about 10 Hz and its frequency is identified with the flash light kept on. The energy consumption is often dependent on running of visual detector, but also includes the real world trace i.e. the visual detector is kept on or off whenever necessary. The trace is collected whenever the participant is moving between his home and his office. Based on the survey visual detector's power consumption is as twice as the acoustic detector. And also visual detector's CPU usage is high where as its power consumption is also high when compared to the acoustic detector. Though the acoustic detector uses only one fourth of energy than the idle flash light but with flash light on in visual detector the energy consumption is doubled. But the Pedwarn's energy consumption can be reduced when users are turning on the WIFI/4G.Based on one hour analysis, CPU usages are noted as 3.08, 8.92 and 17.80% in cases of idle, acoustic detection, visual detection respectively. The battery usage is one fourth than idle in case of using only acoustic detector and twice in case of visual detector.

7 Future Work

PEDWARN+

During the testing phase, the application is executed as a background service. The outcome of the testing is that the sound waves emitted by the microphone is less audio, and the most important thing is, the application is detecting the obstacles when the user is not in a motion. For example when the user is walking in a heavy crowded areas, users are intended to turn off the application since obstacle detection rate is high as well as rear camera will not check any pictures but alert the user by intimating them to take care of yourself. Even though most of the people are get benefits when using the Pedwarn application. The Pedwarn+ application is designed to reduce these errors mentioned by the users. And accuracy rate is also increased in Pedwarn+ when compared Pedwarn application. Likewise in Pedwarn application, the images are taken by the rear camera will processed in real time and it will not be stored for future references. So for saving captured images will be implemented as module later.

8 Conclusion

An application is developed which will reduce the high rate accidents caused by the distracted pedestrians and also to warn the distracted pedestrians so that the collision with the obstacles is avoided. The Pedwarn application is designed to run especially in commodity mobile phones because of that it can executed in any kind of platforms. The higher detection is achievable only starting from the glass doors to the small garbage baskets. The accuracy is measured in different environments with the different devices and the accuracy of the obstacle detection and energy consumption is evaluated.

References

1. Lim, J., Amado, A., Sheehan, L., Van Emmerik, R.E.: Dual task interference during walking: the effects of texting on situational awareness and gait stability. Elsevier Gait Posture **42**(4), 466–471 (2015)
2. Nasar, J.L., Troyer, D.: Pedestrian injuries due to mobile phone use in public places. Accid. Anal. Prev. **57**, 91–95 (2013)
3. Chinese City Creates a Cell Phone Lane for Walkers (2017). https://www.newsweek.com/chinese-citycreates-cell-phone-lane-walkers-271102
4. Wang, T., Cardone, G., Corradi, A., Torresani, L., Campbell, A.T.: WalkSafe: a pedestrian safety application for mobile phone users who walk and talk while crossing roads. In: Proceedings of ACM International Workshop Mobile Computer System Application, pp. 5:1–5:6 (2012)
5. Jain, S., Borgiattino, C., Ren, Y., Gruteser, M., Chen, Y., Chiasserini, C.F.: LookUp: enabling pedestrian safety services via shoe sensing. In: Proceedings of ACM 1st International Conference on Mobile Systems, Applications, and Services, pp. 257–271 (2015)
6. Hincapie-Ramos, J.D., Irani, P.: Crashalert: enhancing peripheral alertness for eyes-busy mobile interaction while walking. In: Proceedings of ACM SIGCHI Conference on Human Factors in Computing Systems, pp. 3385–3388 (2013)
7. Pedwarn Demo Video (2017). https://kabru.eecs.umich.edu/?page_id=987
8. Borenstein, J., Koren, Y.: The vector field histogram-fast obstacle avoidance for mobile robots. IEEE Trans. Robot. Autom. **7**(3), 278–288 (1991)
9. Minguez, J.: The obstacle-restriction method for robot obstacle avoidance in difficult environments. In: Proceedings of IEEE International Conference on Intelligent Robots and Systems, pp. 2284–2290 (2005)
10. Philomin, V., Duraiswami, R., Davis, L.: Pedestrian tracking from a moving vehicle. In: Proceedings of IEEE Intelligent Vehicles Symposium, pp. 350–355 (2000)

Detracting TCP-Syn Flooding Attacks in Software Defined Networking Environment

E. Sakthivel[1(✉)], R. Anitha[2], S. Arunachalam[1], and M. Hindumathy[1]

[1] Department of Information Technology,
Sri Venkateswara College of Engineering, Pennalur, Sriperumbudur, India
sakthivele@svce.ac.in
[2] Department of Computer Science and Engineering,
Sri Venkateswara College of Engineering, Pennalur, Sriperumbudur, India

Abstract. Internet is the platform where everything is connected with everything and also is accessible from anywhere. Moreover the users are widely spread across the globe, traditional IP networks are complex and very hard to manage. Software-Defined Networking (SDN) is a completely a virtual model which changes the traditional network of affairs by dividing the network's control plane from the data plane, intern the network control is done by the SDN program to be a centralized one. However, it has been proven time and again that the SDN is vulnerable to various kinds of attacks like Distributed Denial of Service (DDoS), Denial of Service (DoS), dictionary attacks etc. DDoS attacks mounted by botnets has been termed as biggest threat to internet security today, they target a specific service, mobilizing only a small amount of legitimate looking traffic to compromise the server. Identification methods and blocking the network from the attacks using unstable statistics of the traffic is very challenging and it has been assigned to the server. In this paper, an attack detection and mitigation application has been implemented in an SDN environment. Additionally, a mechanism has been developed on the server side to differentiate between legitimate and illegitimate users such that service to former is not affected.

Keywords: SDN · DDOS · Botnets

1 Introduction

The main key techniques for digital packets to communicate with the World are protocols in transport network and the distributed control inside the routers and switches. It is very hard to control the flow of packets and to maintain the massive number of users around the world. To provide the preferred high-level rules in the network, the operators have to set up each and every individual network device differently by using low-level and often vendor-specific commands. Also the complexity of the network configuration and the network environments has to adopt with the dynamically changing network and the network traffic. Implementation of new rules in the dynamic network

D. J. Hemanth et al. (Eds.): COMET 2019, LNDECT 35, pp. 888–898, 2020.
https://doi.org/10.1007/978-3-030-32150-5_89

is highly challenging. To make it even more challenging, current networks are vertically integrated. Hence, the control plane (that decides how to handle network traffic) and the data plane (that forwards traffic according to the decisions made by the control plane) are bundled inside the networking devices resulting in reducing flexibility, hindering innovation and evolution of the networking infrastructure. A clean-slate approach to change the Internet architecture (e.g., replacing IP), is regarded as a daunting task – simply not feasible in practice. Ultimately, this situation has inflated the capital and operational expenses of running an IP network. Software-Defined Networking proposes a model for the network design provides significant changes in the limitations of the traditional network infrastructure. Initially, it splits down the network and integrates them by vertically using the network's control logic (the control plane) from the current network devices such as routers and switches that redirects the data (the data plane). Splitting of control and data planes results in the devices to act as a forwarding devices that controls the centralized controller (or network operating system) logically, which simplifies the (re)configuration of networks and the evolution. The solution precludes the need to increase the level of performance, scalability, and reliability. In other way, production-level SDN network designs resort to control planes which are physically distributed. It is more important to provide the security in SDN, because security plays the major role to protect the network from the intruder attacks. It is very difficult to change the behavior of the network devices to provide such security. So Software Defined Networking provides an easier way to modify the routing of switches/routers based on the requirements using Openflow architecture. As SDN has the central controller, it is very easy to maintain the network dynamically. Data flow in a network can be increased with the help of SDN orchestrate network switches.

2 Related Work

Lawal and Nuray et al. have proposed to control the DDoS attack detection and the control of data using sFlow mitigation analysis in SDN. They used the Mininet which runs on the Virtual machines [1]. Ubale and Jain proposed a nomenclature of DDoS attacks in the SDN by explaining the life cycle of the SDN. Also they analyzed and suggested solutions for the different attacks [2]. Eddy et al. has discussed that Tcp syn flooding attacks and common mitigations describes about the various counter-measures against TCP SYN flooding attacks, and the trade-offs of each, with explanations of the attack and common defense techniques for the benefit of TCP implementers and administrators of TCP servers or networks, but does not make any standards-level recommendations. The major drawback of this method is that it only detects the attack on the receiver side. Because of this limitation the network resources are not utilized properly during the huge number of packet transmission during the SYN flooding attack [3]. Nugraha et al. suggested the SDN OpenFlow and sFlow schemes to increase the network performance during the DDoS attack. Each agent will send a packet to sFlow collector to analyze the flow and the attacked nodes using OpenFlow. The controller will define the protocols in OpenFlow table to block attack traffic. This might work in ideal cases, when the source IP address in spoofed or different IP source ports are used to access the network this method cannot identify DDoS SYN flooding [4]. Ambrosin et al. has proposed lineswitch: Efficiently Managing Switch Flow In Software-Defined Networking While Effectively Tackling Dos Attacks provided an efficient solution which tackles buffer saturation attack. Mitigating the buffer saturation attacks can be done using the LineSwitch SYN proxy technique, which implements blacklisting of network traffic in data plane switches. This technique is helpful to reduce the needed memory to store the current connections in the network. Because the limited numbers of connections are proxy enabled. The communications between the networks enable the limitations in the connectivity. LineSwitch is also an effective and effective scheme to detect the saturation attacks in the control plan, but similar to AVANT-GUARD, it data plane should be upgraded during the transmission in Line-Switch [5]. Chin et al. proposed a collaborative technique for SYN flooding attack detection and containment. Mitigating the DDoS attacks are done using new components (monitors and correlators) in SDN architecture for mitigating DDOS attacks. An alert message will be send to the server if the IP address spoof is identified during the ongoing traffic for detecting the SYN packets. After receiving the alert message, the correlator sends a request to the server asking the switch table to match the IP address of sender and the stored IP. If both addresses are different then the correlator confirms that the network has had an attack by the malicious attacks. At the end of the process the correlator sends an alter message to block the malicious host from access the data from the network server. This method is applicable only for SYN flooding attack where the attackers use spoofed IP addresses [6]. Fichera et al. has proposed Operetta: An Openflow-Based Remedy To Mitigate TCP Synflood Attacks Against Web Servers, used to identify and reject the malicious requests by using the OpenFLow-based approach which is implemented in the SDN controller. The legitimate connections or requests will be processed only by demonstrating the controller that they are authenticated access to the network in OPERETTA. Always the first time connection will be failed or rejected in TCP connections. The installation of protocols in the flow tables for the legitimate users will happen only after the host proves to behave correctly when successive SYN to reach the server. On the other hand, MAC addresses of the malicious users are marked as black and they are banned to access the network for further communications [7].

3 System Architecture

The architecture diagram details about the components of the proposed system and the interaction among them in an easy understandable manner. Figure 1 is architecture diagram for the solution. It has been designed for protecting the service provided by the web server, SDN controller (POX) and the DDOS blocking application is running on the SDN controller environment and the Openflow interface controls the OpenFlow switches. The blocking application has the Captcha and Honeytoken and it is mentioned clearly in the controller. Some legitimate client of the service, bots and botmaster. In a normal scenario when there is an attack and the server is being flooded with requests for a file or some other valid information from a bot master and its slaves, server is pushed to a state in which it can no longer handle any more requests and it becomes idle when it exceeds its threshold. Also at a point of time the server shuts down thereby failing to serve the legitimate users. In the wok flow that is now being proposed, when a sends a new request to access the server, it reaches an OpenFlow switch and it does not compare with any existing network flow. Which intern reported to the SDN controller to create a new flow for a new packet to reach the server and updates it in the flow table. In the reverse flow, new table entry is created in the controller which is given in Fig. 2. Based on the report, DDOS Blocking Application monitors the flow of data during the execution at each flow switch. Meanwhile, the server monitors the parameters such as number of requests per interval and number of requests in last interval to identify the possibility of network DDoS attacks. When the DDOS Blocking Application identifies the DDoS attack has been happened in the network and the collapse of the server is determined, it starts putting the sources whose requests per interval exceed the nominal amount and ultimately dropping all the packets that are destined to the server. Any host address that is present in the blacklist maintained by the DDOS Blocking Application is not able to get through to the server. The server also keeps on monitoring the number of requests coming from the source per interval to determine whether the attack has been commenced. Once the server detects an attack, the server goes in the CAPTCHA mode wherein for any subsequent requests it asks users to authenticate themselves by entering the CAPTCHA. In this way the legitimate and illegitimate users are differentiated. CAPTCHA is a randomly generated string that has a mix of ASCII as well as numeric characters. The length of CAPTCHA is set to be eight. Once the system identifies a client as a bot, an installation will happen for the associated action for "drop" flow entry (Fig. 3). This can be called as the Drop entry.

Associated problem occurs due to the excessive growth of the flow Drop entries in the flow table explosion that are created by bot requests. So, the entire downstream flow entries for each bot will be deleted from the flow switches, based on the command from the DBA. It retains the Drop entries only in the perimeter switches.

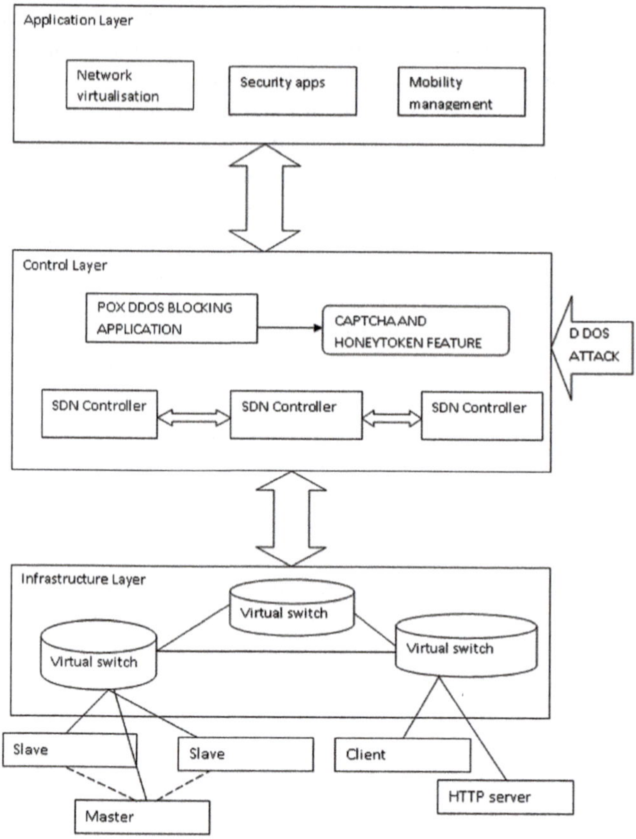

Fig. 1. Architecture diagram

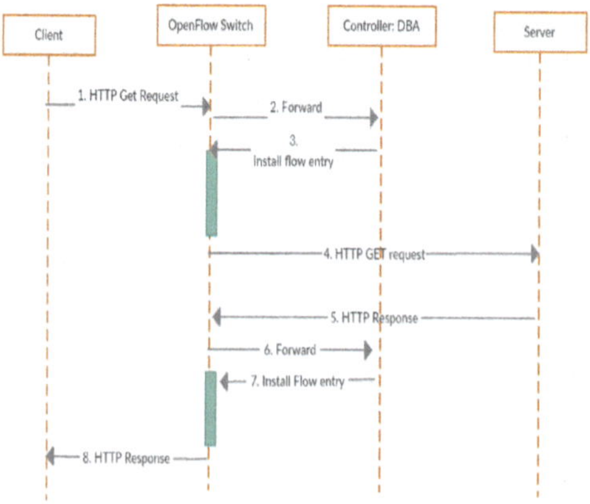

Fig. 2. New Flow arrival

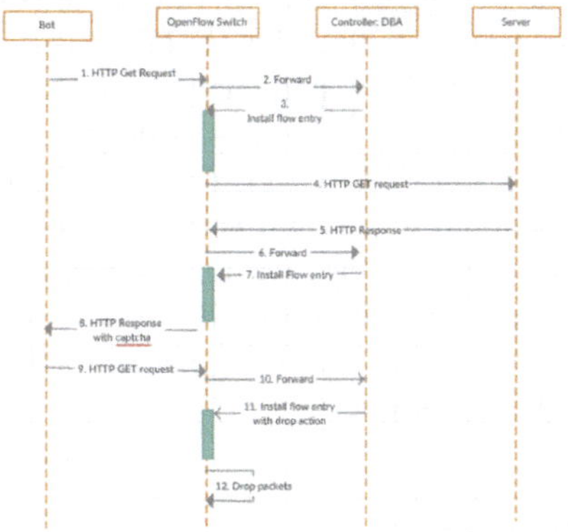

Fig. 3. System behaviour against repeated bot requests

4 Blocking Application

The DDoS Blocking application (DBA) is the POX component that mitigates the DDoS attack on the HTTP server. This component is defined in *blocking_app.py* file within the pox/ext folder within the Mininet and is automatically fetched. DDoS attack can be initiated by accessing all the hosts through their console. To achieve this, run below command on the topology terminal.

4.1 Algorithm

Input: Network
Output: Finding attack
Open terminals; xterm h1 h2 h3 h4 h5 h6 h7
Configuration to be done:
'h6' is the designated HTTP server
'h1' is the bot master,
'h2-h5' are the botnet hosts
'h7' is the legitimate client.
Start the HTTP server by executing the below command in the 'h6' terminal.
python basic_server.py
Attack intiation:
commands initiated in hosts 'h1' to 'h5.
Start the bot master by running below command in Host 'h1'.
python master.py
end

After executing above command in each host, the corresponding host connects to the bot master. This is verified by recording the host entry in the bot master 'h1' that is displayed with the current timestamp. Once the above command is executed in Hosts 'h2'–'h5', the bot master 'h1' automatically triggers the botnets to initiate the

HTTP GET FLOOD attack. This can be observed in the botnet hosts by the displayed message - "DDoS Attack Activated!". The HTTP server is designed to monitor the HTTP requests received within a particular time interval (set to 2.0 s here). Upon HTTP GET FLOOD attack, the requests at the server increase drastically during this duration. The server is designed to trigger a "DDoS DETECTED" and "ALERT" message. Due to the tremendous load on the server, it shuts down and the application is terminated. Within the network topology terminal running in the background, terminate all the xterm sessions running for the hosts 'h1' to 'h7' by running 'exit' command. This stops the Mininet instance that is running the network nodes and the switches get disconnected from the POX controller.

5 Implementation Results

POX controller is used to test and validate the DBA code for its logic, correctness and also to evaluate the performance of the system during attacks. For a massive attack experiment, Mininet emulator is used. Mininet is a simple network emulator to create a virtual network hosts, controllers, switches, routers and links for the devices in SDN. As the simulator projects are just an illusionary flow of data in the network, the implementation is done with the help of Mininet coding, which transfer the data as like real data flow with the help OS kernel and OpenVSwitch. The same code and implementation method can be used in the real world network without any modification in the code and works well in the Mininet emulator. SDN networks require large number of bots to find the DDoS attacks and also to prevent the network from the attackers. The results are as shown below (Figs. 4, 5 and 6).

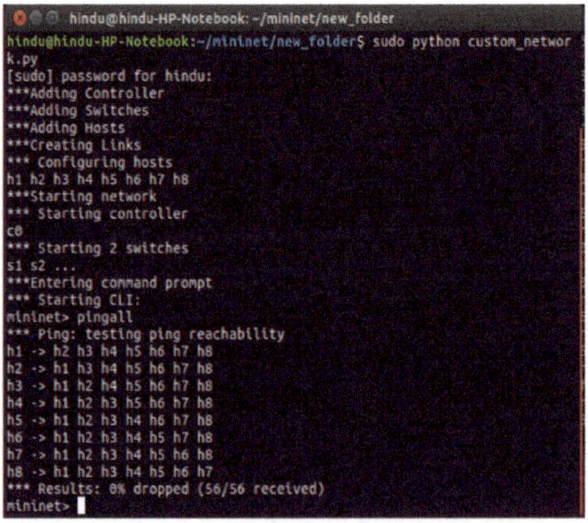

Fig. 4. Custom topology running in mininet

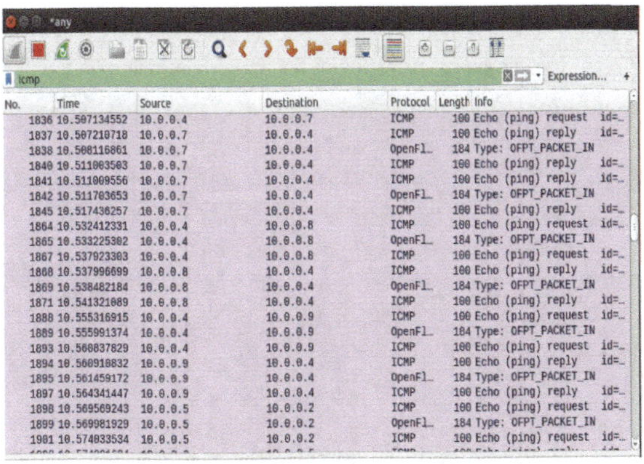

Fig. 5. Wireshark output for ping results

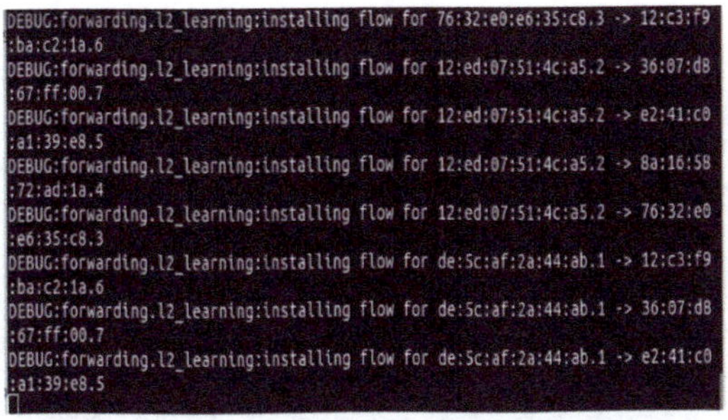

Fig. 6. Switch learning the forwarding rules

5.1 Attack Mode

Once the topology is running, DDoS attack can be initiated by accessing all the hosts through their console. To achieve this, xterm command on the topology terminal is executed, *xterm h1 h2 h3 h4 h5 h6 h7*. This open 7 Xterm terminals that provide access to each of the 7 hosts mentioned in the command. Validate the server-client action; request an HTTP page from the Host ‘h7’ (Figs. 7 and 8).

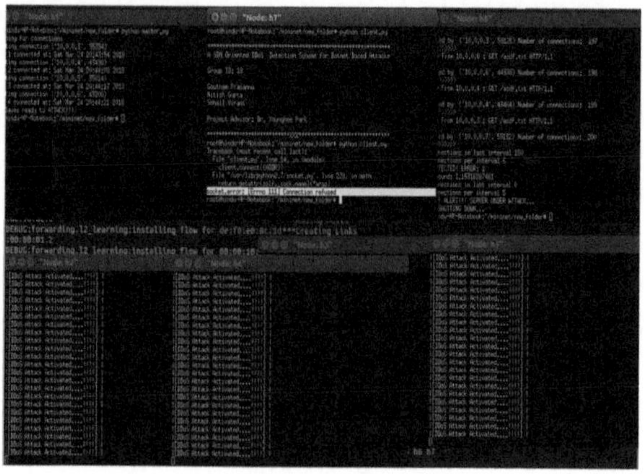

Fig. 7. Connection refused after attack

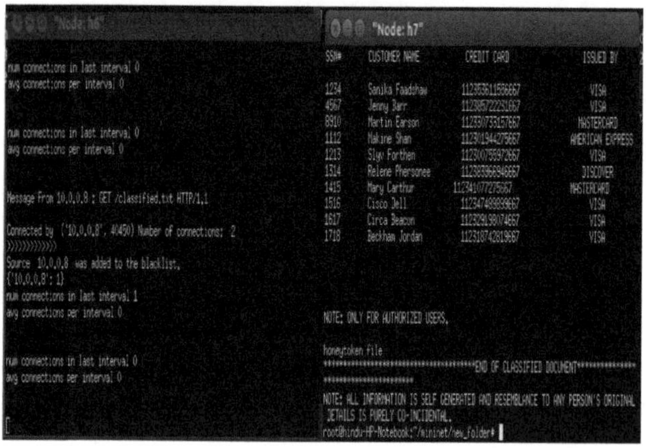

Fig. 8. Server blacklisting node h7

5.2 Defense Mode

Host 'h1': *python master.py;* Host 'h2' to 'h5: *python slave.py;* Upon successful execution, the bot master records each botnet host and triggers the DDoS attack. However, the botnets will still be able to access all the other nodes present within the network. This is shown in the Fig. 9 below where the PING command from the host 'h2' (10.0.0.3) towards host 'h3' (10.0.0.4) is successful.

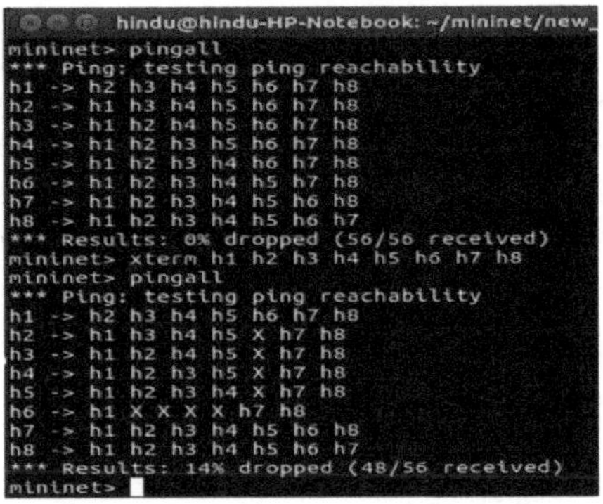

Fig. 9. Ping results in defense mode

6 Conclusion

DDoS detection and mitigation scheme that is applicable in a network which is managed by SDN. This solution does not require any connection to maintain data transfer from SDN controller's DDoS application with the protected server. Controlled access to the important data is ensured using CAPTCHA. Furthermore, honeytokens are used as a mechanism for early detection and confirmation of advanced insider threats. The implemented solution states that, this application can be used to defense the network with the help of Openflow interfaces in orchestrate. Final code is validated and verified in the Mininet for POX controller same is shown using the bots in different blocks of DDoS attack.

References

1. Lawal, B.H., Nuray, A.T.: Real-time detection and mitigation of distributed denial of service (DDoS) attacks in software defined networking (SDN). In: 2018 26th Signal Processing and Communications Applications Conference (SIU), Izmir, pp. 1–4 (2018)
2. Ubale, T., Jain, A.K.: Taxonomy of DDoS attacks in software-defined networking environment. In: Singh, P., Paprzycki, M., Bhargava, B., Chhabra, J., Kaushal, N., Kumar, Y. (eds.) Futuristic Trends in Network and Communication Technologies. FTNCT 2018. Communications in Computer and Information Science, vol. 958. Springer, Singapore (2019)
3. Eddy, W.M.: TCP SYN flooding attacks and common mitigations. J. Inf. Secur. **2**(3) (2011). 2007 article cited by "Effectiveness of Built-in Security Protection of Microsoft's Windows Server 2003 against TCP SYN Based DDoS Attacks
4. Nugraha, M., Paramita, I., Musa, A., Choi, D., Cho, B.: Utilizing openflow and sflow to detect and mitigate syn flooding attack. J. Korea Multimed. Soc. **17**(8), 988–994 (2014)
5. Ambrosin, M., Conti, M., De Gaspari, F., Poovendran, R.: Lineswitch: efficiently managing switch flow in software-defined networking while effectively tackling dos attacks. In: Proceedings of the 10th ACM Symposium on Information, Computer and Communications Security, pp. 639–644. ACM (2015)

6. Chin, T., Mountrouidou, X., Li, X., Xiong, K.: Selective packet inspection to detect dos flooding using software defined networking (SDN). In: Proceedings of International Conference on Distributed Computing Systems Workshops, pp. 95–99. IEEE (2015)
7. Fichera, L., Galluccio, S.C., Grancagnolo, G.M., Palazzo, S.: OPERETTA: an OPEnflow-based remedy to mitigate TCP SYNFLOD attacks against web servers. Comput. Netw. **92**, 89–100 (2015)

QBuzZ – Conductorless Bus Transportation System

S. Kavi Priya(✉), S. Naveen Kumar, K. Sathish Kumar, and S. Manikandan

Department of Information Technology, Mepco Schlenk Engineering College, Sivakasi, India
urskavi@mepcoeng.ac.in

Abstract. The communal transport plays the chief part in India. In current transportation scenario, the travellers have to get the ticket(s) of their trip from the bus conductor by paying the ticket fare that result in lot of prevailing problems such as change problems, etc. Normally, Electronic Ticketing Machine [ETM] is used by the conductor to cast off ticket(s) for the travellers. But it results in huge volume of paper wastage and requires skilled person for operating the machine. The disadvantage of ETM is that it is relatively slow. This annoys the travellers who are travelling in the bus. In this work, we describe a new system "Conductorless Bus Ticketing System" to present a reflex fare assortment system with backing of QR Code along with an embedded system and a smartphone app. This system is planned to enterprise an embedded system that is used for tumbling the overcrowding in the bus. In accordance with this, the Ultrasonic Sensor is used to get the count of travellers who are boarding in and boarding out from the bus and an LCD screen is placed near to the driver in order to update about the count of travellers in the bus, count of travellers board in to the bus and board off from the bus, etc. In this system, QR code generated by the traveller via the android app will be acting as the tickets) of the trip. QR code communicates with the embedded system which is fixed at the bus in order to validate whether the QR code is valid or not. The working of this system is most fortunate for communal and very opportune for travelling in the bus.

Keywords: Raspberry Pi 3 B+ model · Camera module · LCD screen · Ultrasonic sensors · GPS module · Jumper wires

1 Introduction

Internet of Things [IOT] supports for associated device from anytime and anywhere. It has sparked the integration of embed device into the surroundings. Through elegant applications, the associated device promise to expand our lives by communicating and exchange of information flawlessly without any human contacts. Conductorless Bus Transportation is one such example application.

Smartcard based communal transportation ticketing system is popular but many surveys have been done to improve the communal transportation [1–3] through

D. J. Hemanth et al. (Eds.): COMET 2019, LNDECT 35, pp. 899–907, 2020.
https://doi.org/10.1007/978-3-030-32150-5_90

smartcards. Nevertheless, none of these solutions are feasible for a fully automatic ticketing without any human contacts.

Thus, we proposed QR code based Conductorless Bus Transportation that communicates amid GPS device and the embedded system. Fare of the trip can be easily charged to the traveller with the help of android app. To engender a QR code, the Zxing library plays the chief role. To detect the QR code, the raspberry pi camera module plays a major role. Recent smartphones are equipped with Wi-Fi, GSM, NFC, GPS, etc that are assorted means of connectivity. Therefore, we need wireless communication equipment that provides omnidirectional transmission. QR code based Conductorless Bus Transportation will be popular as it relives the communal, the conductor and the ticket-checker from the burden of buying/printing/checking tickets for each trip. Thus, QR code and internet connectivity can be a supreme vehicle for Conductorless Bus Transportation.

2 Related Work

An automated fare payment system for communal transportation with RF smart-card was deliberated by McDaniel et al. [4] in 1993. McDaniel et al. made a conclusion that the cost for implementation of the system will be so expensive. McDaniel et al. thinks that future expertise can make it practicable. The traveller requirements for ticket vending machine based on communal transport was considered by Caulfield et al. The significant wishes of the traveller about charge, route and exact arrival times were revealed in the study made by Caulfield et al. [5]. A lot of surveys have done to advance communal transport [1–3] using smartcards such as RFID, NFC, etc that needs an explicit check in or check out. Nevertheless, these solutions are not workable for a fully automatic ticketing without any human contacts.

A lively RFID needs the precise involvement of traveller to book and cancel their trip was practiced EasyRide [6]. However, the motto is to avoid check in and check out system. Thus, we term as scan-in/scan-out (SISO).

Smartcard based on RFID meant for ticketing service was defined by Chowdhury [9]. But, the shortcomings of using RFID are expensive and delicate to some metals and liquids. The foremost negative is that the entire local surroundings must be analysed, including the types of metals, lightning and sources of radio intrusion before the implementation of RFID system.

BLE-based ticketing system that enhances a BLE description to achieve the payment by not taking into account automated system has anticipated by Kuchimanchi [7]. Similarly, a likelihood study of BLE - based communal transport system to check whether an inherent interaction between an embedded system and a traveller's smartphone using BLE communication has steered by Narzt et al. [8].

Nowadays, smartphone, a powerful computing device has been carried almost by everyone who can be utilized for hassle-free inherent ticketing of the travellers. However, extensive energy drain from a smartphone is a great cause of anxiety. With the emergence of QR code, I am heart – rendering to QR code which has the foremost doles compared to RFID, NFC, Smart-Cards, etc namely inexpensive, hold noteworthy amount of data and obvious to metals and liquids. The vital pro is that it doesn't require any supplementary device for producing the code.

3 System Design

An overview of the Conductorless bus transportation system can be viewed in Fig. 1. We visualizes the system consisting of a cloud or database, where all the traveller's details are stored. There could be numerous backend for the sake of scalability of the system. The central database keeps the information of all buses, travellers, and keeps track of the performed trip.

Before we explain the working of the Conductorless Bus Transportation system, we list the wide-ranging requirements here:

1. The embedded device should be mountable.
2. The traveller should be connecting securely to the system.
3. The embedded system should not leave any traveller unaccounted for the trip.
4. The periodicity of checking traveller's trip should be stable.

The embedded device fixed in the bus is connected to the backend server (AWS EC2 Instance). Therefore, the system can also be protected against compromise. The embedded device in the bus supports travellers by scan-in and scan-out.

Largely, each bus is controlled mostly by a conductor. He will collect the money from every single traveller and he issue the ticket(s). Habitually, the tokens or printed papers are mainly cast-off as ticket(s). For example, if a traveller desires to board in a bus, he/she wants to carry the money for his/her ticket(s). In the bus, the conductor amasses the money for the ticket(s) and he issues the ticket(s) to concerned traveller. Thus, grades in time consumption and human resource wastage as well as energy. At present, Electronic Ticketing Machine [ETM] is cast-off to print the tickets. This system has many negatives namely the travellers have to carry the ticket(s) till his/her terminus arrives. It also grades in huge amount paper wastage. The vital con is that ETM is relatively slow and we need skilled person to operate it.

The Conductorless Bus Transportation System works as follows: An android app is developed for interacting with communal that will charge the fare amount for the trip travelled by the traveller in the bus. When the traveller like to board in the bus, he/she wants to book the ticket(s) of the trip with the support of "QBuzZ" app. In the android app, the traveller has to become a user of the app by signing up. Then, he/she should to sign in to the created account. After successful sign in to the account, the traveller has to enter/choose the board in and board out points along with the trip date to check the seat(s) is available for the trip in any of the buses available in that route. Once the source and destination is given, the app show a list of buses available for the route with time and fare amount for a single traveller. After choosing the bus, the traveller has to select/lock the seats in particular bus according to the count. After the selection of seats, the traveller has to fill up the trip details like traveller name(s), age, date of birth, etc according to the selected seats. Then the traveller has to pay the fare amount for the booked seats by digital payment. In payment, the traveller is provided with four modes of payment options namely

1. Pay by Debit/Credit cards.
2. Pay by Internet Banking.
3. Pay by Wallet.
4. Pay by UPI.

After the successful payment, the "QBuzZ" app generates a QR code with a randomly generated QR id and the traveller's trip details are embedded in that QR code.

When the corresponding bus arrives, the traveller has to show the QR code in front of the camera module that is fixed at the doorstep of the bus for scanning it to check whether it is valid or not.

After the successful scanning of QR code, the traveller has to board in to the bus and sit according to the seat number(s) booked. While boarding in to the bus, the count of travellers will be taken with the support of Ultrasonic sensor. For female traveller, the seat next to her is automatically reserved for women. The traveller's count is updated in a LCD screen to the driver at each stop. With the support of Ultrasonic sensor, we can easily find out the fraudulent in the bus by comparing the traveller count in Ultrasonic sensor with the scanned QR code counts.

Therefore, from the above scenario, we can easily conclude that there will be no need for the conductor in the bus since the driver knows everything on his LCD screen. When the traveller's destination point is reached he/she once again has to show the QR code in order to free the particular seat status. Since, the bus is provided with live GPS, the transport officer and the time keeper can easily track the bus. In addition to that, the transport office can easily analyse the demand of buses in particular time at the particular area.

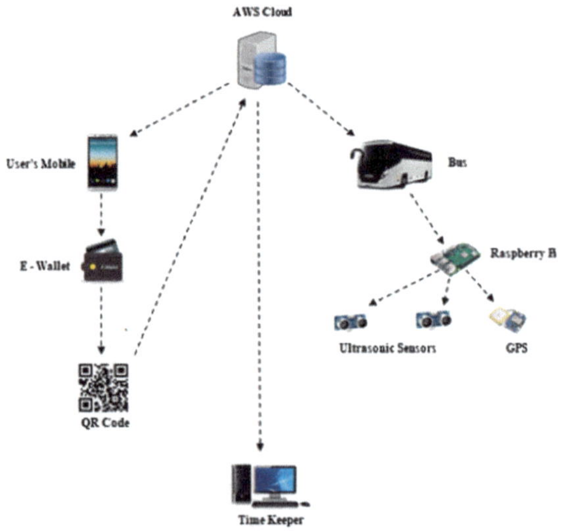

Fig. 1. Proposed system design

3.1 Challenges

Reliability. If there is any interrupt in the internet connection, the modifications will not be affected in the database of the AWS EC2 Instance i.e., AWS Cloud.

Security and Privacy. Wireless communication is used among the travellers, bus and ticket checker that can be likely to snooping. This can lead to lose sensitive and secluded data, economic loss, etc. Thus, measures have to be taken tonsure security and privacy and prevent abuse. Attacks on the bus and embedded device also need to be prevented and/or detected.

Scalability and Accuracy. Since the system takes the attention of trip payments, the correctness of scan-in procedure is important. Paying of trips fare amount that the traveller forgets to make should be avoided. Similarly, the traveller should be duly payable to the trips. The system will possibly be used by a lot of travellers at the same time. Thus, truthful billing with an increase in count of synchronized travellers is significant as well.

Portability. The application can be accessed from anywhere with the support of the smartphone app when there is a bus transport facility available (Figs. 2, 3, 4, 5, 6 and 7).

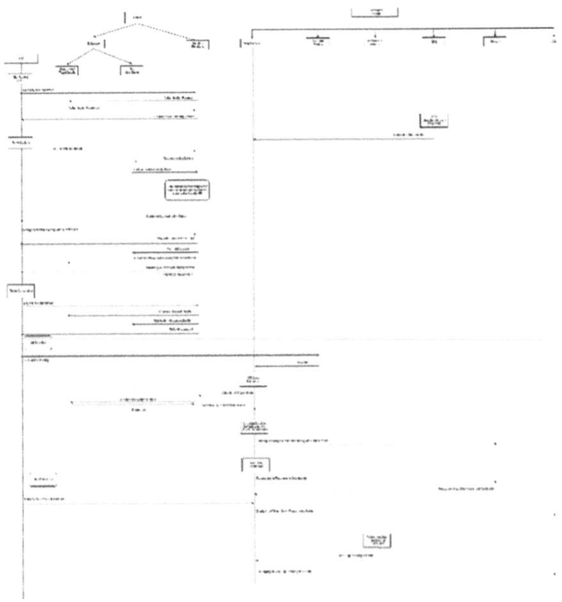

Fig. 2. Flow diagram of proposed system

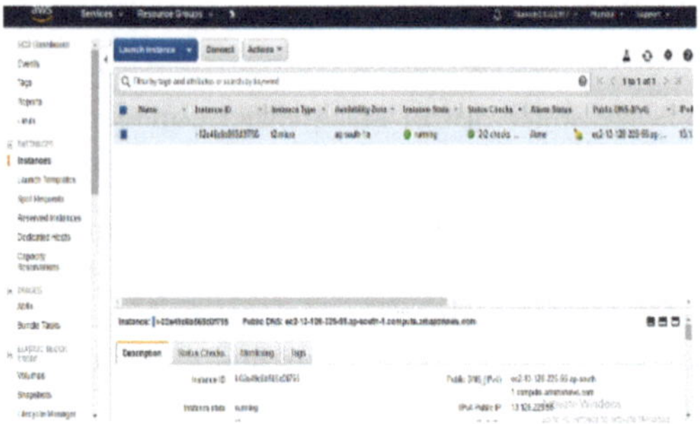

Fig. 3. AWS EC2 Instance

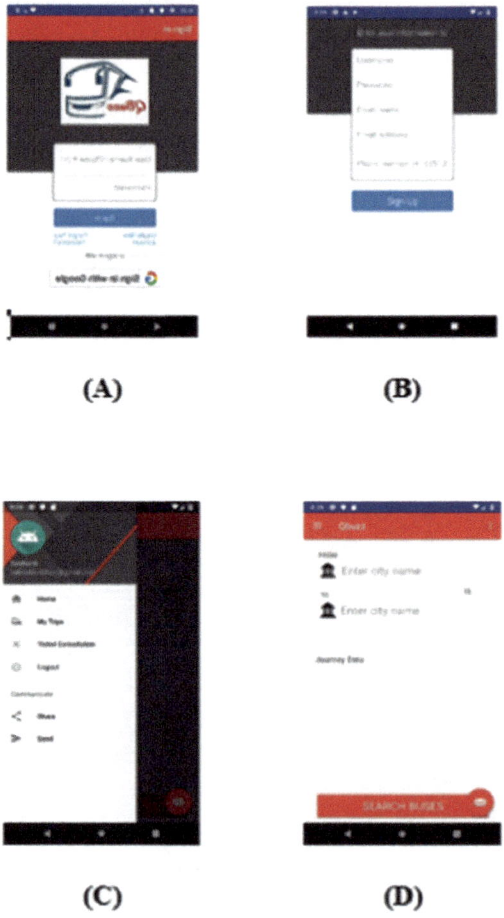

Fig. 4. (a) Sign-in, (b) Sign-up, (c) Navigation menu and (d) Search bus

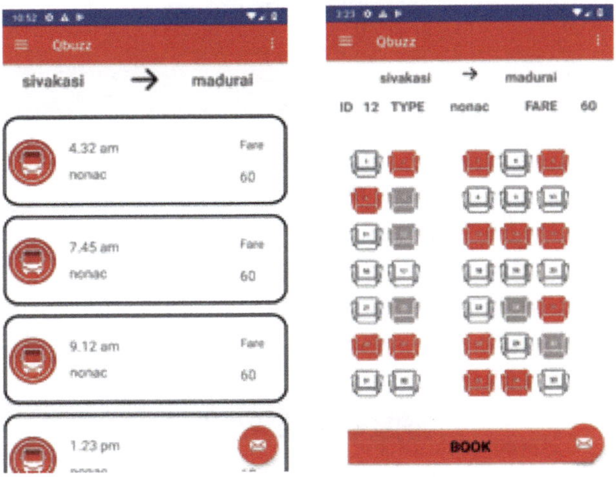

(A) (B)

Fig. 5. (a) Bus schedule and (b) Book seat

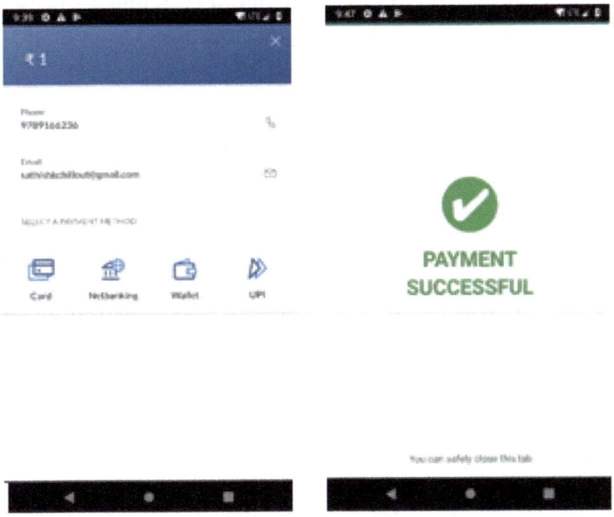

(A) (B)

Fig. 6. (a) Payment options and (b) Payment status

Fig. 7. QR code generation

4 Conclusion

We have projected a Conductorless Bus Transportation based on automated ticketing system for the communal transportation system with the support of QR code technology. We provided system requirements including a number of research problem. In this article, we focused our study on the major facet how QR code is used as high data fault tolerance and data storage capability. We provided detailed design of the complete system.

References

1. Blythe, P.T.: Improving public transport ticketing through smart cards. Proc. ICE-Municipal Eng. **157**(1), 47–54 (2004)
2. Mallat, N., Rossi, M., Tuunainen, V.K., Öörni, A.: An empirical investigation of mobile ticketing service adoption in public transportation. Pers. Ubiquitous Comput. **12**(1), 57–65 (2008)
3. Widmann, R., Grünberger, S., Stadlmann, B., Langer, J.: System integration of NFC ticketing into an existing public transport infrastructure. In: 2012 4th International Workshop on Near Field Communication (NFC), pp. 13–18. IEEE (2012)
4. McDaniel, T., Haendler, F.: Advanced RF cards for fare collection. In: 1993 Telesystems Conference, Commercial Applications and Dual-Use Technology', Conference Proceedings, National, pp. 31–35, June 1993
5. Caulfield, B., O'Mahony, M.: Passenger requirements of a public transport ticketing system. In: 2005 Proceedings of Intelligent Transportation Systems, September 2005, pp. 119–124. IEEE (2005)
6. Gyger, T., Desjeux, O.: Easyride: active transponders for a fare collection system. IEEE Micro **21**(6), 36–42 (2001)
7. Kuchimanchi, S.: Bluetooth low energy based ticketing systems. Master's thesis, Aalto University (2015)

8. Narzt, W., Mayerhofer, S., Weichselbaum, O., Haselbock, S., Hofler, N.: Be-in/be-out with bluetooth low energy: implicit ticketing for public transportation systems. In: 2015 IEEE 18th International Conference on Intelligent Transportation Systems (ITSC), pp. 1551–1556. IEEE (2015)
9. Chowdhury, P., Bala, P., Addy, D., Giri, S., Chaudhuri, A.R.: RFID and android based smart ticketing and destination announcement system. In: 2016 International Conference on Advances in Computing, Communications and Informatics (ICACCI), pp. 2587–2591 (2016)
10. Das, A., Lingeswaran, S.V.K.: GPS based automated public transport fare collection systems based on distance travelled by passenger using smart card. Int. J. Sci. Eng. Res. (IJSER) **2**(3), 2347–3878 (2014)

Design of High Performance FinFET SRAM Cell for Write Operation

T. G. Sargunam[1], C. M. R. Prabhu[2(✉)], and Ajay Kumar Singh[2]

[1] School of Science and Engineering, Manipal International University, Nilai, Malaysia

[2] Faculty of Engineering and Technology, Multimedia University, Melaka, Malaysia

c.m.prabu@mmu.edu.my

Abstract. A novel FinFET based SRAM cell is proposed to reduce the dynamic power consumption during write mode in this research work. The proposed High Performance FinFET SRAM (HPFS) cell consists of 8-Transistors instead of 6-Transistors as in conventional SRAM cell. The extra two transistors are used to reduce the write power during transition. The proposed circuit is simulated for Microwind EDA tool. The results of HPFS cell is compared with conventional SRAM cells. From the simulated results, it has been observed that the suggested HPFS cell consumes lower power and provides lower access delay compared to other cells.

Keywords: FinFET · High performance · Low power · SRAM cell · Power consumption · Low power and access delay

1 Introduction

In modern VLSI memory systems, the low power and high-performance SRAM design has become a predominant design factor due to its power consumption. The on-chip cache design consumes a reasonable amount of the total power consumption in digital systems, microprocessors, embedded systems etc. [1–4]. When the feature size is continuously in shrinking mode towards sub-micron region, the respective leakage current has become the primary factor of total power consumption.

There is a continuous advancement and tremendous development in SRAM design with power reduction requirement. The literature confirms this fact having so many SRAM cells with different number of transistors and various techniques deployed to minimize the power, to increase the stability and overall to increase the performance [5–8]. The Cache operations are mainly read, write and hold operations. However, it is always observed that many techniques have been suggested and deployed to achieve less cache read power as read mode happen more often than write mode. In any normal scenario, the total write power is always greater than the read power [13–16]. When examined, the benchmarking confirms that the big majority of the write bits seem to be ‘0’.

In this research work, a novel High Performance FinFET SRAM (HPFS) cell is proposed to minimize the write power consumption based on this observation.

D. J. Hemanth et al. (Eds.): COMET 2019, LNDECT 35, pp. 908–914, 2020.
https://doi.org/10.1007/978-3-030-32150-5_91

Compared to the planar CMOS technology, the FinFET is becoming an important and mainstream IC technology with its significant improvement of leakage reduction and overall performance now a days. The proposed SRAM cell has two extra n-MOS transistors through FinFETs in both inverters to minimize the leakage current. The bit lines BL and nBL generally control the operation of these n-MOS transistors.

The circuit has been simulated in the environment of MICROWIND3 EDA tool [18]. The results of the proposed HPFS circuit has been compared with other published FinFET based SRAM cells and it is found that the HPFS cell dissipates less power than the other FinFET based SRAM cells. It is observed that there is about 41.6% power reduction from 6T FinFET based SRAM cell during the write operation. This paper is arranged as follows. The architecture of the proposed cell is explained in Sect. 2. The Sect. 3 highlights the analysis of simulation results. The conclusion is presented in Sect. 4.

2 Proposed HPFS Cell Design

The main aim of this proposed HPFS cell is to decrease the power consumption during the write operations without any tradeoff between the read access time and stability. The architecture of HPFS FinFET based cell is shown in Fig. 1. There are two extra n-MOS transistors (M7 and M8) introduced in the branch of inverters and bit lines BL and nBL control the switching activity of the two transistors. The write operation is being carried out and explained in detail.

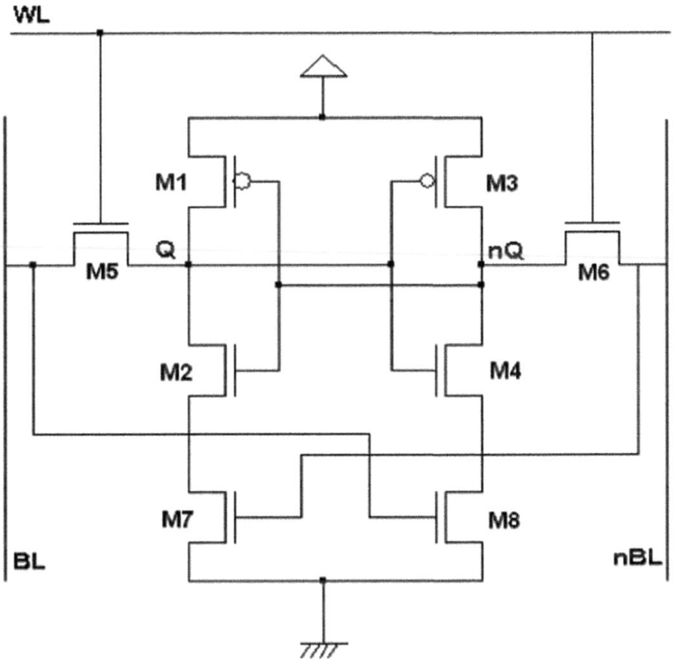

Fig. 1. Proposed FinFET HPFS Cell

2.1 Write Operation

When the WL is set to high, the respective data will get transferred from bit lines BL and nBL to nodes Q and nQ. For the 0 → 0 transition, the node nQ is set at high initially and therefore this transition is not possible. For the 1 → 0 transition, the transistor M8 is OFF and the node nQ must be changed to high without waiting for nBL to discharge completely as depicted in Fig. 2(a). For write ‘1’ mode, the bit lines BL is set to ‘1’ and nBL is set to ‘0’. By asserting WL to high, the data gets transferred from bit lines to nodes Q and nQ. In the first possible of case of 1 → 1, no data flipping happens due to the same high voltage at node Q and BL. In the second possible m transition of 0 → 1, as shown in Fig. 2(b), the node Q is flipped to high without any waiting for BL to discharge completely when transistor M7 is set to OFF. The delay of the proposed HPFS cell becomes faster than the conventional 6T and 7T SRAM cells since there is no discharging activity happening of the bit lines for both write ‘1’ and write ‘0’ operations.

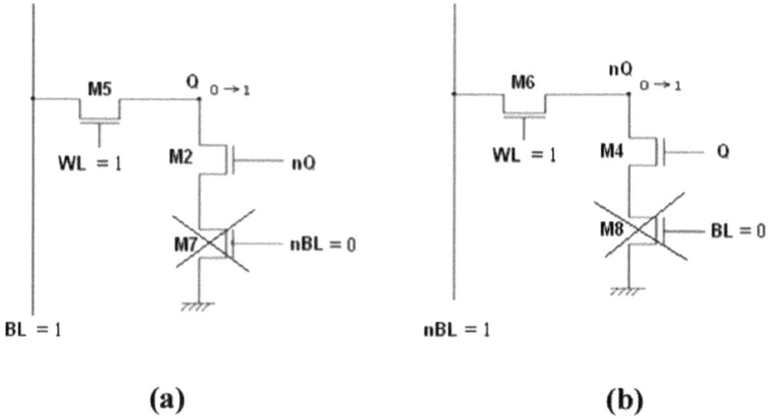

Fig. 2. (a) Write ‘1’ (b) Write ‘0’

2.2 Read Operation

The bit lines are pre-charged to VDD by the pre-charged circuitry before asserting wordline WL to high during the read operation. The bit lines either BL or nBL will start to discharge depending on the values at the storage nodes Q and nQ soon after asserting the wordline WL to high. The critical paths during read “0” and read “1” operations are shown in Fig. 3(a) and (b). In the HPFS cell, the speed and stability are degraded due to three transistors. To compensate the stability loss and read delay reduction, it is required to enlarge the width of the tail transistors. Widening tail transistor will further increase the area overhead. Another approach for improvement is to modify the cell at the architectural level.

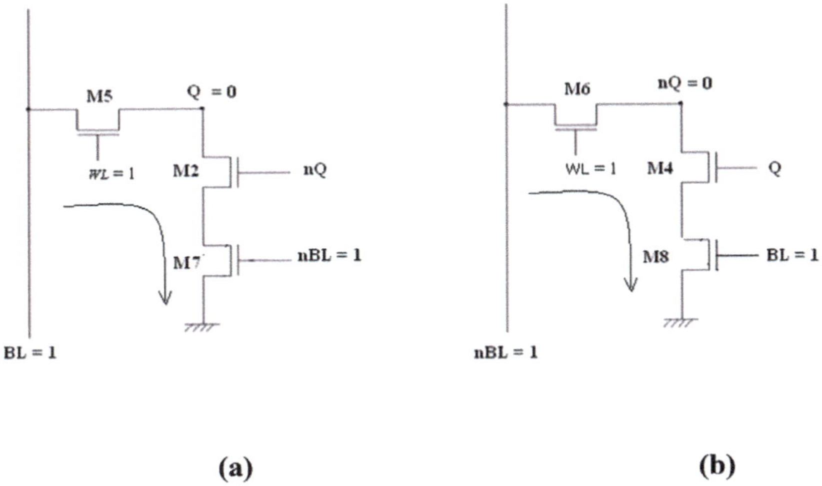

Fig. 3. (a) Read '0' (b) Read '1'

3 Results and Discussion

The HPFS circuit has been simulated in the MICROWIND3 EDA tool (advanced BSIM4 level). The results of HPFS cell is been compared with other reported cells' power, read/write delay and area by using 14 nm FinFET technology and presented in this section. The write power dissipation is high in the conventional cell due to the bit-lines discharging activity. Whereas, the single bit line is being used by the 7T cell which causes less power when write '0' than the write '1' operations. In this HPFS cell, the power and delay are even lower because there is no discharging happens during both write operations as shown in Table 1 and Fig. 4 and the read delay and read power are given Table 2 and Fig. 5 respectively.

Table 1. Write power

SRAM cell	Write power (μW)	
	Write '0'	Write '1'
6T	5.192	5.202
7T	0.539	5.202
HPFS	3.006	3.064

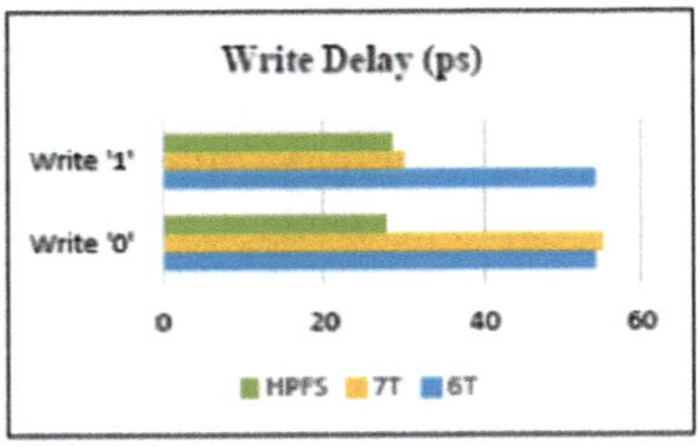

Fig. 4. Write delay

Table 2. Read delay

SRAM cell	Read delay (μW)	
	Read '0'	Read '1'
6T	84.6	85.4
7T	85.3	85.6
HPFS	84.8	86.5

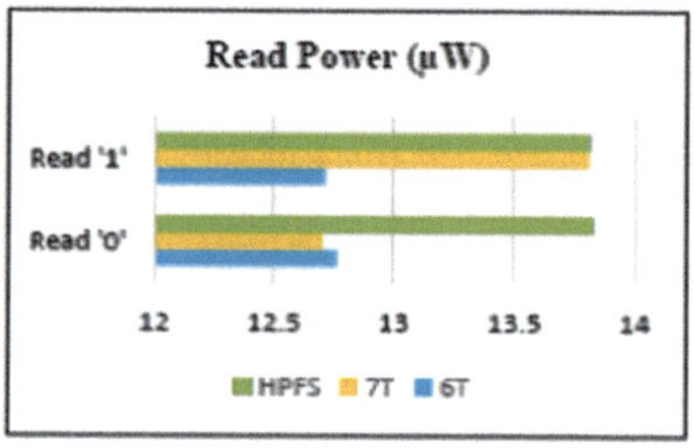

Fig. 5. Read power

The Process, Voltage and Temperature variations in write mode has been simulated in normal mode, minimum mode and maximum mode. In the normal mode, 25 °C temperature, VDD supply of 0.8 V and threshold voltage of 0.310 V have been applied. In the minimum mode, the high temperature of 125 °C, low supply voltage of 0.68 V and threshold voltage of 0.365 V are applied which causes the transition slow.

Table 3. Write power for different inputs (temperature/supply voltage/threshold voltage)

Mode	Typical mode (T = 25 °C, V_{DD} = 0.8 V & V_{TH} = 0.310 V)		Minimum mode (T = 125 °C, V_{DD} = 0.68 V & V_{TH} = 0.365 V)		Maximum mode (T = 50 °C, V_{DD} = 0.92 V & V_{TH} = 0.270 V)	
SRAM cell	Write power (μw)		Write power (μw)		Write power (μw)	
	$0 \rightarrow 1$	$1 \rightarrow 0$	$0 \rightarrow 1$	$1 \rightarrow 0$	$0 \rightarrow 1$	$1 \rightarrow 0$
6T	5.202	5.192	1.637	1.632	14.306	14.298
HPFS	3.064	3.006	0.942	0.915	8.597	8.476

In contrast, the fast transition happens during the maximum mode with 50 °C temperature, supply voltage VDD = 0.92 V and threshold voltage of 0.270 V. The observed results are shown in Table 3. The analysis of results confirms that there is an average of 41.6% write power is saved. The average write delay has also improved about 47.9% compared to the conventional cell. The power and access delay have been observed by supplying different supply voltage ranging from 0.8 V to 0.25 V during write mode. The Table 4 below shows the average write power and write delay outcomes. It is obvious that the HPFS cell dissipates minimum power than the 6T cell. The HPFS cell also confirms that it can work for various VDD starting from 0.8 V till 0.25 V. The write delay and power have been continuously reducing compared to the conventional cells which proves to be less power consumption and faster access during the write operations. The proposed cell has 33.33% area overhead compared to the conventional cell as shown in Table 5 and the respective layout diagram is shown in Fig. 6.

Table 4. Variation of power consumption with different V_{DD} for proposed SRAM cell

V_{DD} supply voltage	Write power (μw)		Write delay (ps)	
	6T	HPFS	6T	HPFS
0.8	5.202	3.006	54.2	26.5
0.75	4.038	2.355	63.7	36.7
0.7	3.033	1.779	74	47.5
0.65	2.185	1.280	85.2	58.8
0.6	1.493	0.862	97.6	70.5
0.55	0.951	0.534	111	82.4
0.5	0.551	0.300	126	94.8
0.45	0.281	0.155	143	109
0.4	0.127	0.078	163	126
0.35	0.055	0.040	193	150
0.33	0.035	0.013	216	163
0.25	–	0.002	–	240

Table 5. Cell area

SRAM cell	Number of transistors	Area (μm^2)
6T	6	0.08
HPFS	8	0.12

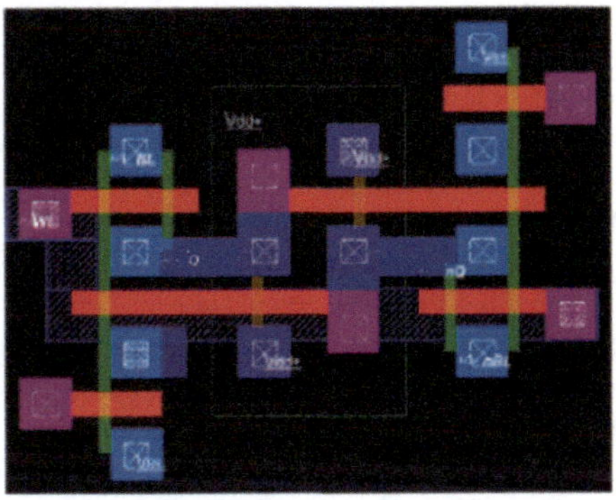

Fig. 6. Layout of proposed HPFS cell

4 Conclusion

The novel HPFS cell uses two extra n-MOS transistors (M7 and M8) to avoid any charging and discharging of nodes during write mode. The write power consumption is reduced and cell becomes more faster in performing the write mode due to two extra transistors. The HPFS cell consumes an average of 41% less power and 60% lower access delay compared to the conventional cell. But, the area overhead of HPFS cell is reported to be reasonably more compared to the conventional cell.

References

1. Yeo, K.-S., Roy, K.: Low-Voltage Low–Power VLSI Subsystems. McGraw-Hill, New York (2005)
2. Bhardwaj, M., Min, R., Chandrasekaran, A.P.: Quantifying and enhancing power awareness of VLSI systems. IEEE Trans. VLSI Syst. **9**(6), 757–772 (2001)
3. Kim, N.S., Blaauw, D., Mudge, T.: Quantitative analysis and optimization techniques for on-chip cache leakage power. IEEE Trans. VLSI Syst. **13**, 1147–1156 (2005)
4. Senthipari, C., Diwakar, K., Prabhu, C.M.R., Singh, A.K.: Power deduction in digital signal processing circuit using inventive CPL subtractor circuit. In: Proceedings of ICSE 2006, Kuala Lumpur, Malaysia, pp. 820–824 (2006)
5. Amrutur, B., Horowitz, M.: Techniques to reduce power in fast wide memories. In: Proceedings of the Symposium on Low Power Electronics, pp. 92–93 (1994)
6. Kim, C.H., Kim, J., Mukhopadhyay, S., Roy, K.: A forward body-biased low-leakage SRAM cache: device, circuits and architecture consideration. IEEE Trans. VLSI Syst. **13**, 349–357 (2005)
7. Geens, P., Dehaene, W.: A small granular controlled leakage reduction system for SRAMs. Solid-State Electron. **49**, 1176–1782 (2005)
8. Inaba, S., Nagano, H., Miyano, K., Mizuushima, I., Okayama, Y., Nakauchi, T., Ishimaru, K., Ishiuchi, H.: Low-power logic circuit and SRAM cell applications with silicon on depletion layer CMOS (SODEL CMOS) technology. IEEE J. Solid-State Circ. **41**, 1455–1462 (2006)
9. Mai, K.W., Mori, T., Amrutur, B.S., Ho, R., Wilburn, B., Horowitz, M.A., Fukushi, I., Izawa, T., Mitarai, S.: Low power SRAM design using half-swing pulse-mode techniques. IEEE J. Solid-State Circ. **33**, 1659–1671 (1998)
10. Morimura, H., Shgematsu, S., Konaka, S.: A shared-bitline SRAM cell architecture for 1-V ultra low-power word-bit configurable macro cells. In: ISLPED99, San Diego, CA, USA, pp.12–17 (1999)
11. Grossar, E., Stucchi, M., Maex, K., Dehaene, W.: Read stability and write-ability analysis of SRAM cells for nanometer technologies. IEEE J. Solid-State Circ. **41**, 2577–2588 (2006)
12. Yamaoka, M., Tsuchiya, R., Kawahara, T.: SRAM circuit with expanded operating margin and reduced stand-by leakage current using Thin-BOX FD-SOI transistors. IEEE J. Solid-State Circ. **41**, 2366–2372 (2006)
13. Kanda, K., Sadaak, H., Sakurai, T.: 90% write power-saving SRAM using sense amplifying memory cell. IEEE J. Solid-State Circ. **39**, 927–933 (2004)
14. Yang, B., Kim, L.: A low power SRAM using hierarchical bit-line and local sense amplifiers. IEEE J. Solid-State Circ. **40**(6), 1366–1376 (2005)
15. Sharifkhani, M., Sachdev, M.: Segmented virtual ground architecture for low-power embedded SRAM. IEEE Trans. VLSI Syst. **15**(2), 196–205 (2007)
16. Prabhu, C.M.R., Sargunam, T.G., Singh, A.K.: High performance data aware (HPDA) SRAM cell for IoT applications. ARPN J. Eng. Appl. Sci. **14**(1), 91–94 (2019)
17. Prabhu, C.M.R., Singh, A.K.: Reliable high performance (RHP) SRAM cell for write/read operation. Aust. J. Basic Appl. Sci. **10**(16), 22–27 (2016)
18. Etienne Sicard: Microwind and Dsch version 3.0. (User's manual Lite version, Published by INSA Toulouse France) (2005)

An Hybrid Defense Framework for Anomaly Detection in Wireless Sensor Networks

S. Balaji(✉), S. Subburaj, N. Sathish, and A. Bharat Raj

Department of Computer Science and Engineering,
Panimalar Engineering College, Chennai, India
balajiit@gmail.com, subburajs87@gmail.com,
nsathishme@gmail.com, santhoshcliff110@gmail.com

Abstract. Wireless sensor network consists of set of source, sink nodes and communication devices to interact without any support of the infrastructure. Unlike wired networks, the challenges faced in mobile ad-hoc networks possessed such as security design, network infrastructure, stringent energy resources and network security issues. The need to these security issues is much focused in overcoming the challenges in WSN. Here the perpetual work focuses on the secure communication using a novel defense framework named role based control model is proposed to analyze the network flow and to identify the misbehaving nodes. The communication is performed based on the cluster of immense size these confided in node(s) will most likely be passing on together, in the meantime allowing or section entry/correspondence of the unauthorized node(s) to continue keeping up a constant, tied down, dependable communication of versatile nodes. The simulation is performed using network simulator where the network parameters such as throughput, packet delivery ratio, delay and packet loss are analyzed to identify the malicious nodes.

Keywords: Security issues · Anomaly detection · Intrusion detection · Cross-layer · Wireless Sensor Networks

1 Introduction

The most fundamental factor in Wireless Sensor Networks is to attain efficient throughput, even though if there is interference caused to the source nodes below the pre-threshold level. The defense framework is implemented for identifying the cross-layer attacks by the concept trust monitoring. The main objective here is to monitor the nodes behavior to analyze how likely these nodes attacks the network [1, 2]. The aggregated information activities are performed by the cross-layer trust manager. In this network, for instance, when source or intermediate node X is associated with another node Y with a coordinated circular segment from X to Y, it implies that source node X causes node Y to affect the network. Node X is known as the attacker node and node Y is called as the victim node. The most prominent paradigm service named Wireless Sensor Networks (WSN), has been employed for the effective monitoring purpose, having included a plethora of applications with respect to wireless technology. Generally, the nodes in network communicate with others over sparse wireless channels

D. J. Hemanth et al. (Eds.): COMET 2019, LNDECT 35, pp. 915–922, 2020.
https://doi.org/10.1007/978-3-030-32150-5_92

either in single-hop or in a multi-hop style [3, 4]. In this process, the wireless system has been constructed with developed processors and used in different applications, having incorporated in a fractious type. Therefore, it is practically not feasible to imagine the world without wireless networks. Intrusion detection systems are classified into two. Firstly, misuse detection and anomaly detection. In misuse detection, it recognizes interruptions by coordinating observed exchanges with pre-characterized conduct [7]. However, it effortlessly cannot find for unknown intrusions. Anomaly detection depends on man-made reasoning and it conquers the downside of abuse identification. IDS will progressively screen the events and it chooses whether these events are characteristic of an attack or it establishes a veritable utilization of the system [5, 6].

1.1 Basic Architecture

As it is restricted to bandwidth and more nodes are participating in communication, data transmission among wireless networks lead to severe congestion [15]. As a result, congestion control and error recovery are the significant divisions of transmission control protocol. It in a way, builds end-to-end system connection, assigns bandwidth reasonably, retransmits the packets that have been lost during communication and effectively plays a prominent role in energy management. Considering this, the main focus of this proposed research work as shown in Fig. 1 is to evaluate the framework for anomaly detection in wireless sensor networks.

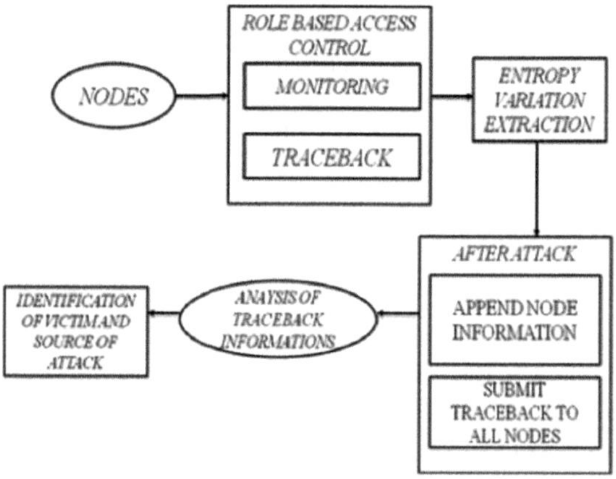

Fig. 1. A basic architecture

Efficient trace back node recovery algorithm to retrieve the data from the defected node is proposed to improve the lifetime of WSN when a portion of the sensor nodes become inactive. The algorithm can result in less substitution of sensor nodes and more reused routing ways. In our work, the proposed calculation builds the quantity of dynamic nodes, lessens the rate of information loss, and decreases the rate of energy utilization by around 30%.

2 Classification of Attacks

Security attacks can be classified in various layers [13, 14]. For instance, in application layer some of the attacks identified like data corruption and repudiation. In network layer, some of the attacks found like blackhole, wormhole, flooding etc. In data link layer, Traffic analysis, monitoring and disruption are monitored. Our work is carried for identifying attacks like Denial of service in multi-layer environment.

2.1 Existing Works

Many related works prove that performance reduction during the cross – layer attacks is high as compared in single layer. Secondly, it fails to detect multiple attacks in the network. The detection accuracy was inadequate in some of the related works. Also the secrecy rate was not maintained up to the predetermined probabilities [8, 9]. Many surveys [10–12] keenly show that the proposals and the real time security tools were developed to improve the secure essentials till date.

Perhaps most of the survey focuses on the single layer attacks which provides only a time being effect on the network, but there is a need for performance analysis to identify the security issues at the multilayer levels.

3 Intrusion Detection System Model

Let, N is the set of randomly deployed Sensor Nodes (SNs) [16], where N = {1,…,y} as shown in Eq. 1.

$$N = \sum\nolimits_{i=1}^{y} N_i \tag{1}$$

Let, B is the set of Base Stations available in network, which is more powerful than SNs B = $\{B_1,\ldots,B_x\}$ as shown in Eq. 2. The entropy variation E can be obtained with respect to the network flow F.

$$B = \sum\nolimits_{i=1}^{x} B_i \tag{2}$$

3.1 Monitoring and Traceback Algorithm (MAT Algorithm)

Monitoring Algorithm

---

1. Monitoring network flow F
2. Progress entropy variations E with time interval Δt
3. If flow suspends
   Security attack and non-secure communication
4. Else
   Repeat the progress

–

Traceback Algorithm

Consider a network N
1. Define the flow F
2. Calculate entropy variations E
3. If any attack is identified before the original source
3.1 Append the node information.
3.2 Submit Traceback request.
3.3 Repeat entropy variations.
4. Else
4.1 Source of the attack is identified.
4.2 Deliver the monitored information to the nodes.
5.End

4 Implementation and Results

4.1 Simulation Metrics

The performance of the network can be calculated by analyzing some of the metrics such as packet loss ratio, throughput, end to end delay and packet delivery ratio. To support these metrics the following information of nodes are used.

Firstly, the average number of nodes receiving packets. It is the total of all the intermediate nodes accepting packets sent by all the source nodes by number of received packets at all the goal nodes.

Secondly, average number of nodes forwarding packets and Number of data packets dropped. This is a vital parameter in such a case that the quantity of dropped packets increases, the throughput would weaken.

NS-2 simulator is used for the implementation of the proposed scheme. Protocol used is AODV. The Channel Type is Wireless Channel, Simulation Time 100 ms, 200 ms. Traffic sources used are Constant-Bit-Rate (CBR) and the field configuration is 800×800 m with 25 nodes (Figs. 2, 3, 4, 5 and 6).

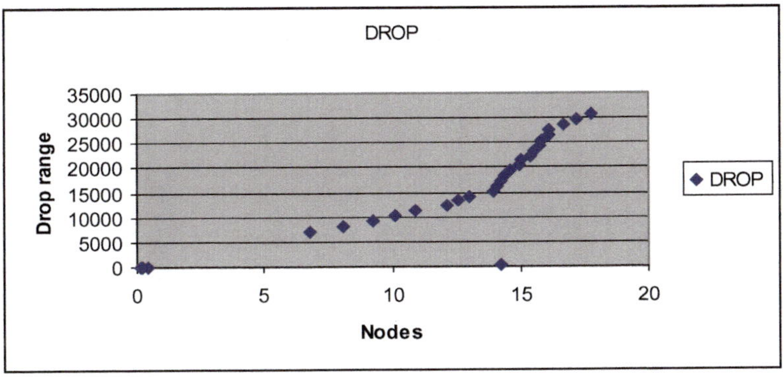

Fig. 2. Comparison of drop range for 25 nodes

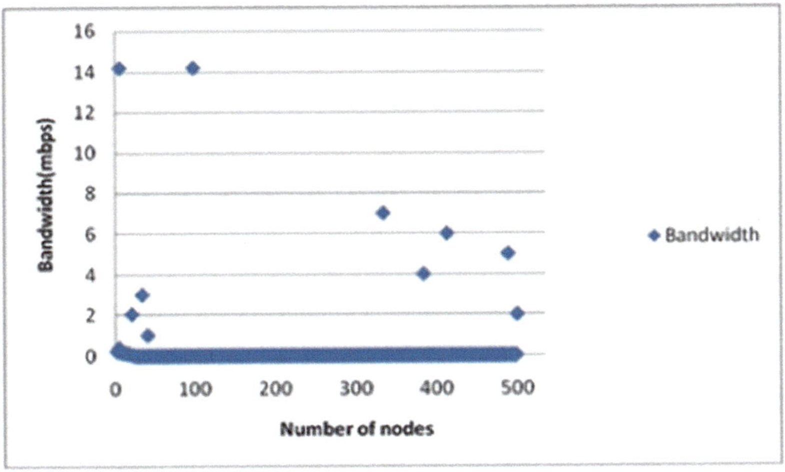

Fig. 3. Comparison of bandwidth for 25 nodes

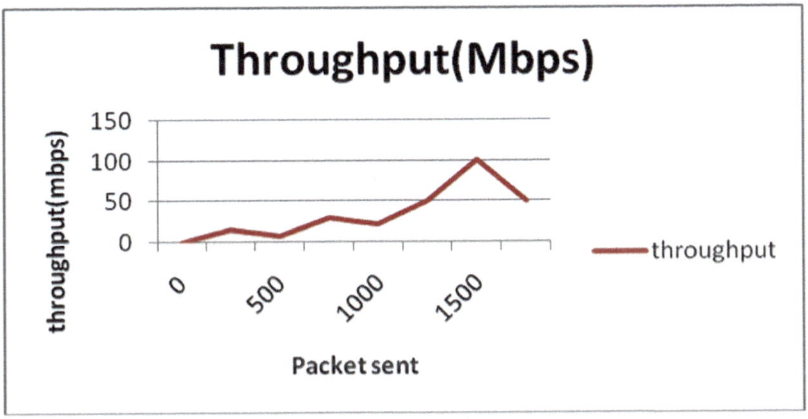

Fig. 4. Throughput after malicious detection

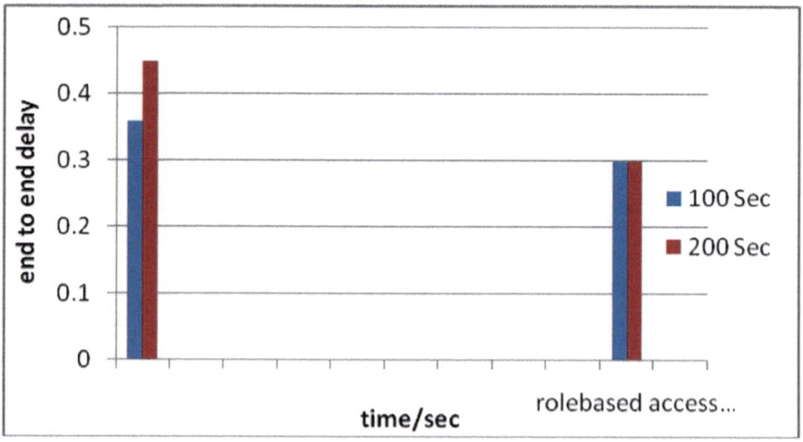

Fig. 5. End to end delay

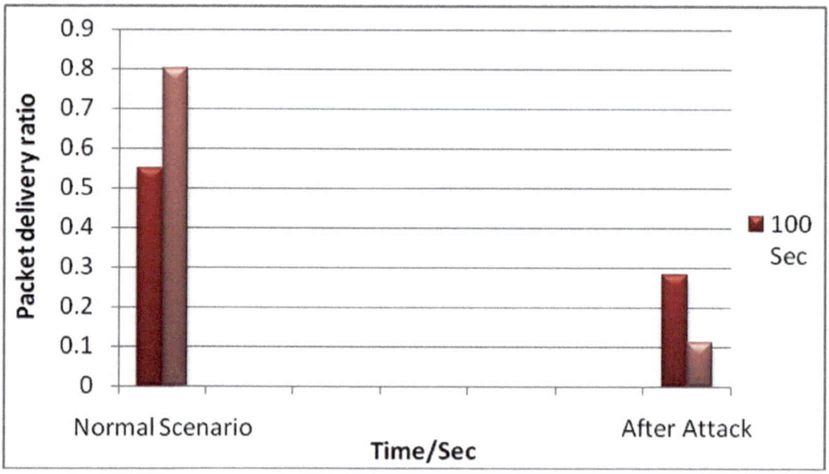

Fig. 6. Packet delivery ratio

5 Implementation and Results

In this paper, we proposed a hybrid intrusion detection system based on the distributed intrusion detection mechanism and shared monitoring. Here we implemented the traceback and monitoring algorithm to find the malicious nodes and to detect the attack happened in the network. Here the bandwidth usage is calculated for every node and finally the node which wastes more bandwidth is determined as the malicious node. Our work at present proposes sensor node activation and fault node recovery at the cross layers in WSN. Our future work is to implement our project at cross layer in cognitive radio networks .patch levels, and the mix of applications used on infected systems.

References

1. Kandula, S., Katabi, D., Jacob, M., Berger, A., Botz-4-Sale: surviving organized DDOS attacks that mimic flash crowds. In: Proceedings of Second Symposium on Networked Systems Design and Implementation (NSDI) (2005)
2. CNN Technology News: Expert: Botnets No. 1 Emerging Internet Threat (2006). http://www.cnn.com/2006/TECH/internet/01/31/furst/
3. Freiling, F., Holz, T., Wicherski, G.: Botnet tracking: exploring a root-cause methodology to prevent distributed denial-of- service attacks. Technical report AIB-2005-07, CS Department of RWTH Aachen University (2005)
4. Cooke, E., Jahanian, F., McPherson, D.: The zombie roundup: understanding, detecting, and disrupting botnets. In: Proceedings of USENIX Workshop Steps to Reducing Unwanted Traffic on the Internet (SRUTI) (2005)
5. Xu, W., Trappe, W., Zhang, Y., Wood, T.: The feasibility of launching and detecting jamming attacks in wireless networks. In: MobiHoc 2005: Proceedings of the 6th ACM International Symposium on Mobile Ad Hoc Networking and Computing, Urbana-Champaign, IL, USA, pp. 46–57 (2005)
6. Zhao, L., Delgado-Frias, J.G.: MARS: misbehavior detection in ad hoc networks. In: Proceedings of IEEE Conference on Global Telecommunications Conference, pp. 941–945. Washington State University, USA (2007
7. Patwardhan, A., Parker, J., Iorga, M., Joshi, A., Karygiannis, T., Yesha, Y.: Threshold-based intrusion detection in adhoc networks and secure AODV. Ad Hoc Netw. J. (ADHOCNET) **6**(4), 578–599 (2008)
8. Madhavi, S., Kim, T.H.: An intrusion detection system in mobile adhoc networks. Int. J. Secur. Appl. **2**(3), 1–17 (2008)
9. Afzal, S.R., Biswas, S., Koh, J.B., Raza, T., Lee, G., Kim, D.K.: RSRP: a robust secure routing protocol for mobile ad hoc networks. In: Proceedings of IEEE Conference on Wireless Communications and Networking, pp. 2313–2318 (2008)
10. Bhalaji, N., Sivaramkrishnan, A.R., Banerjee, S., Sundar, V., Shanmugam, A.: Trust enhanced dynamic source routing protocol for adhoc networks. In: Proceedings of World Academy of Science, Engineering and Technology, vol. 36, pp. 1373–1378 (2008)
11. Huang, Y., Lee, W.: A cooperative intrusion detection system for ad hoc networks. In: SASN 2003: Proceedings of the 1st ACM workshop on Security of Ad Hoc and Sensor Networks, Fairfax, VA, USA, pp. 135–147(2003)

12. Bellardo, J., Savage, S.: 802.11 denial-of-service attacks: real vulnerabilities and practical solutions. In: Proceedings of the 11th USENIX Security Symposium, Washington, D.C, USA, pp. 15–28 (2003)
13. Raya, M., Hubaux, J.P., Aad, I.: DOMINO: a system to detect greedy behavior in IEEE 802.11 hotspots. In: Proceedings of ACM MobiSys, Boston, MA, USA, pp. 84–97 (2004)
14. Xu, W., Trappe, W., Zhang, Y., Wood, T.: The feasibility of launching and detecting jamming attacks in wireless networks. In: MobiHoc 2005: Proceedings of the 6th ACM International Symposium on Mobile Ad Hoc Networking and Computing, Urbana-Champaign, IL, USA, pp. 46–57 (2005)
15. Zhang, Y., Lee, W., Huang, Y.A.: Intrusion detection techniques for mobile wireless networks. Wirel. Netw. **9**(5), 545–556 (2003)
16. Balaji, S., Sasilatha, T.: Detection of denial of service attacks by domination graph application in wireless sensor networks. Clust. Comput. J. Netw. Softw. Tools Appl. (2018). https://doi.org/10.1007/s10586-018-2504-5. ISSN 1573-7543

Efficient Information Retrieval of Encrypted Cloud Data with Ranked Retrieval

Arun Syriac, V. Anjana Devi(✉), and M. Gogul Kumar

Department of Computer Science and Engineering, St. Joseph's College of Engineering, Chennai, Tamil Nadu, India
anjanadevi.aby06@gmail.com

Abstract. Over the past decade, there have been massive developments in technology such as self-driving cars, crypto currencies, streaming services, voice assistants etc. In each of the listed breakthroughs, Cloud Computing was involved. Cloud computing has offered a tremendous breakthrough in enterprise and business transformation bringing with it a previously unknown agility. Projects and databases hosted on the cloud have enabled users to work in unison without any hassle. Users who use cloud computing can upload their data to the cloud from anywhere and get access to the best possible service architecture and applications from a connected network of computer resources. However to ensure that the data of the users are stored safely on the cloud, the data has to be encrypted to protect privacy before outsourcing. Encryption of the user's data leads to poor efficiency of data utilization due to the possibility of a large number of outsourced files. This amounts to a lack of control and access of the out-sourced data from the user's position. Due to lack of efficient and practical cloud computing, the user's lack of control is further magnified. To address this aspect of helping the user reign in control over their encrypted data when out-sourced on a cloud, we have developed a scheme to efficiently search over the encrypted data and retrieve ranked results.

Keywords: Cloud computing · Keyword search · Secure data access

1 Introduction

Services offered for storing data over the cloud have grown rapidly over the past few years. They help in storage and accessing the data using the Internet instead of a local computer's hard disk drive. Cloud services provide the hardware and software resources from a pool of shared resources. This helps the users avoid the expense for building and maintaining a storage drive. The cloud service providers have full control over the system hardware and software and can gain access to the data uploaded to the cloud and even misuse it. In order to ensure data privacy, the documents have to be encrypted before they are uploaded to the cloud. Encrypting the documents means the widely used plaintext keyword searching techniques cannot be used to search over the documents. Existing traditional encryption techniques provide users with the ability to search across the encrypted data without having to decryptit, but this is limited to Boolean search. These files are retrieved by checking if the keyword is present in the

D. J. Hemanth et al. (Eds.): COMET 2019, LNDECT 35, pp. 923–930, 2020.
https://doi.org/10.1007/978-3-030-32150-5_93

file or not, and the relevancy of the files are not considered. Existing approaches support ranked search over cloud data, but searching over encrypted data has not been addressed enough. In order to ensure effective retrieval of data from large amounts of documents, it is essential that cloud server performs result relevance ranking. It is also important that the searching supports multiple keyword searches to improve the search result accuracy. The main focus of the paper is to implement an efficient technique to search over encrypted data which supports ranked retrieval of the information and multiple users.

2 Related Work

Initial searchable encryption techniques supported only exact keyword search. Song et al. [1] provided an encryption technique in which ensured that each of the word in the document is encrypted using a two layered encryption technique. Goh [2] proposed index creation for improving the efficiency. Here, a secure index is created for all the unique words which are present in the file. Curtmola et al. [3] proposed a system where the index creation for keyword is done using hash tables. Wang et al. [4] introduced ranked search which further improved usability. The technique relied on relevance scoring to identify similarity between the files and the query words. This technique supported only single keyword based searches and user authorization was not specified. Ahmed et al. [5] introduced a trust ticket mechanism helping the data owner establish a secure link with the cloud service provider. Boneh et al. [6] introduced an encryption technique that supported searching of keyword which allowed the service to check if a document contained a specific set of keyword without gaining any information about the document or the keyword. It supported multiple users with user authentication. User authentication is checked before the documents are decrypted.

3 Proposed System

3.1 System Description

In the information retrieval of encrypted cloud data the first step is preprocessing. The data from the documents is transformed into token and the processed data is inserted into the database. The system makes use of the vector space model (VSM) along with the term frequency-inverse document frequency model (TF-IDF) in the process of index generation and in the generating of query to assist with supporting multiple keyword search functionality. The ranking of the documents is done with the help of cosine similarity which helps in finding the similarity between the query and the documents and ranks them. The following illustration describes how the proposed system works (Fig. 1).

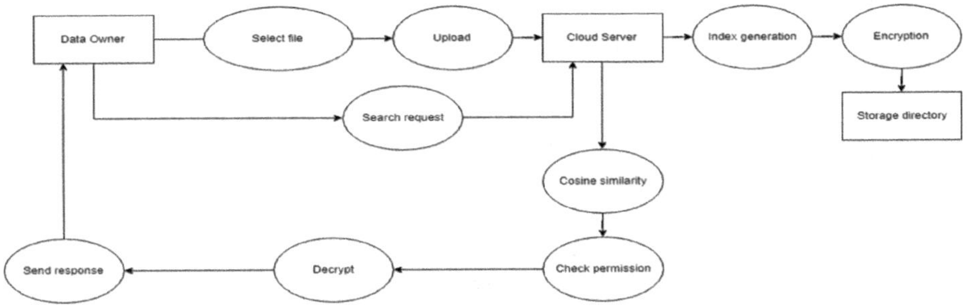

Fig. 1. Architecture of proposed system

3.2 User Registration

Users are prompted to register to use the service. Once registered, the user can login using the valid username and password combination. Once logged in, the user is provided with a logout option. The files uploaded are limited in scope to the user uploading it and cannot be accessed by other users.

3.3 File Encryption and Upload

The files that the user selects for uploading are encrypted and decrypted using the AES algorithm. Advanced Encryption Standard (AES) is an encryption technique used to secure sensitive data. AES makes use of the same key for encryption and decryption. AES-128 with 10 rounds is used for encryption.

3.4 Index Generation

When a file is uploaded, the contents of the document are transformed to tokens. This module does a lot of work including text tokenizing, normalizing, removing stopwords and stemming. Stopwords are the noise words such as "a", "and" which are meaningless for searching and thus removed. The process by which words are reduced to their base form is called as stemming. Further, the TF and IDF scores are calculated for the tokenized words and are added into the database. Term frequency is used in order to measure the amount of times a term (word) appears in a document. Large documents might have frequency of the terms higher than the smaller ones. In order to negate this effect tokenized words are normalized. Inverse document frequency is plays an important part in helping to filter out the documents that are relevant and match the given query. In TF, all the terms are given nearly similar importance. Words that tend to appear frequently have the power to determine the relevance. IDF is used in order to increase the role of words that appear less frequently and reduce the role of words that appear more frequently.

$$tf_{i,j} = \frac{number\ of\ instance\ of\ i\ in\ document\ j}{total\ number\ of\ terms\ in\ document\ j}. \tag{1}$$

$$idf_i = log_2\left(\frac{total\ number\ of\ documents}{documents\ with\ the\ term\ i}\right). \quad (2)$$

Once the TF and IDF scores are calculate, the weights of the words are obtained using, the following formula.

$$W_{i,j} = tf_{i,j} * idf_j. \quad (3)$$

3.5 Ranked Search

For the query entered by the user, TF-IDF score is calculated. This is followed by calculating cosine similarity between the query vector and the document vectors to find the documents containing the searched query and rank them according to the closeness of query vector to the document vectors.

$$cosine\ similarity(d, q) = \frac{dot\ product(d, q)}{||d|| * ||q||}. \quad (4)$$

where, d stands for the documents and q stands for the query.

4 Implementation Results

4.1 User Registration

A new user must fill in the required parameters to create a new account. If an account with the same username or email already exists, the user is prompted to provide a different value for the corresponding parameter (Fig. 2).

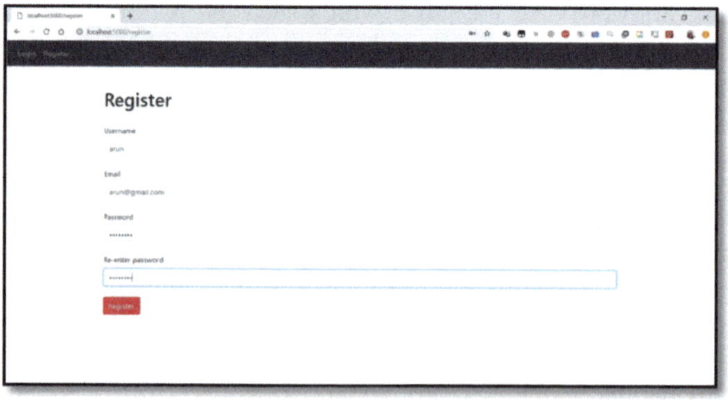

Fig. 2. User registration

4.2 User Login

Upon successful match of email and password combination the user is logged into the system. Otherwise, the user is prompted to rectify and enter an existing account credentials or create a new account (Fig. 3).

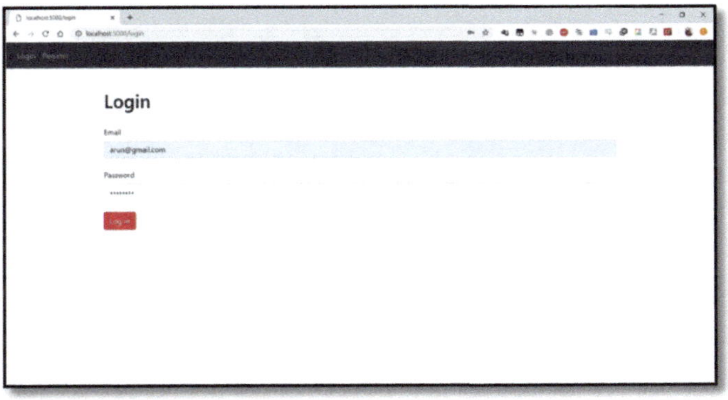

Fig. 3. User login

4.3 Uploading File

A user can select a text document from the local device and upload the file to the cloud server. The user is prompted on successful upload of the file (Fig. 4).

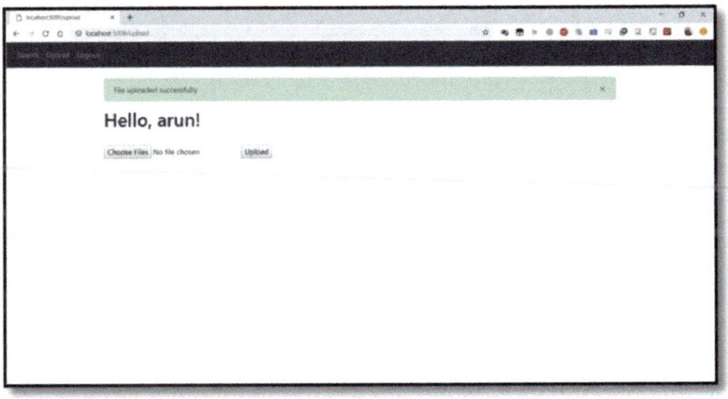

Fig. 4. Uploading the file

4.4 Encrypted File

The file upload to the cloud is encrypted and stored in the storage directory (Fig. 5).

Fig. 5. Encrypted file

4.5 Search Results

The documents containing the search query are displayed to the user after ranking the documents based on the query entered (Fig. 6).

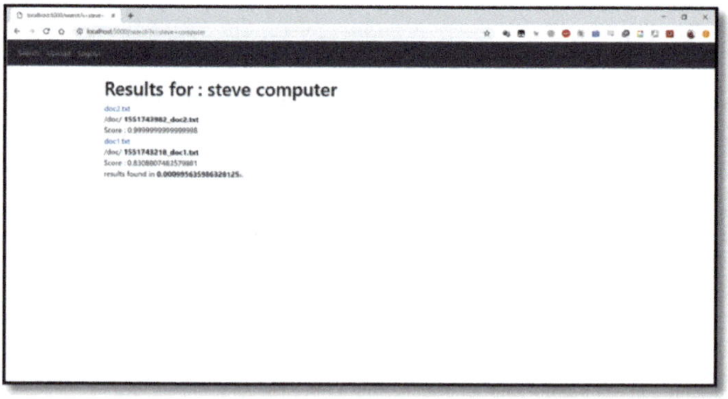

Fig. 6. Search results

4.6 Decrypted File

If the user has permission to view the file, the contents of the file are decrypted and displayed to the user (Fig. 7).

4.7 Unauthorized Access

If the user does not have permission to view the file, the file is not decrypted and the user is prompted with an error 404 message (Fig. 8).

Fig. 7. Decrypted file

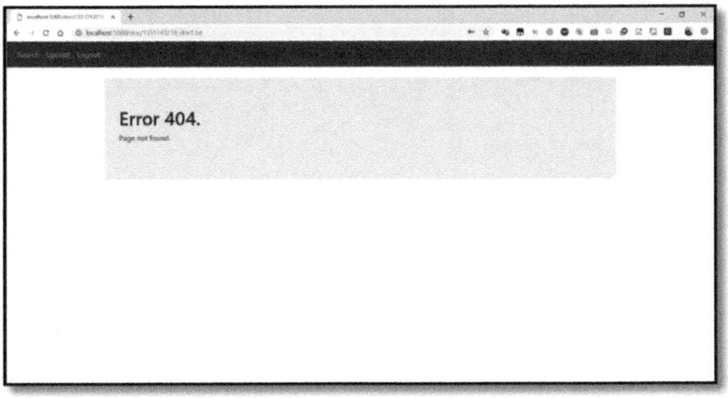

Fig. 8. Unauthorized access

5 Conclusions and Future Work

The security concern while outsourcing documents into the cloud is addressed in this paper. The project supports not just encryption of data outsourced to the cloud but also provides ability to search using keywords on encrypted cloud data. With the help of TF-IDF the relevancy of the document can be accessed. This solution provides better ranked information retrieval of encrypted documents compared to existing ones. Users are not provided with the ability to search through other users documents. This limitation can be solved in the future enhancements of the project.

References

1. Song, D.X., Wagner, D., Perrig, A.: Practical techniques for searches on encrypted data. In: Proceedings of 2000 IEEE Symposium on Security and Privacy, S&P 2000, pp. 44–55. IEEE (2000)
2. Goh, E.-J., et al.: Secure indexes. IACR Cryptol. ePrint Arch. **2003**, 216 (2003)
3. Curtmola, R., Garay, J., Kamara, S., Ostrovsky, R.: Searchable symmetric encryption: improved definitions and efficient constructions. In: Proceedings of the 13th ACM Conference on Computer and Communications Security, pp. 79–88. ACM (2006)
4. Wang, C., Cao, N., Ren, K., Lou, W.: Enabling secure and efficient ranked keyword search over outsourced cloud data. IEEE Trans. Parallel Distrib. Syst. **23**(8), 1467–1479 (2012)
5. Ahmad, M., Xiang, Y.: Trust ticket deployment: a notion of a data owner's trust in cloud computing. In: IEEE Security and Privacy, 16–18 November 2011
6. Boneh, D., Crescenzo, G., Ostrovsky, R., Persiano, G.: Public key encryption with keyword search. In: Proceedings of Eurocrypt 2004. Lecture Notes in Computer Science, vol. 3027, pp. 506–522 (2004)

Mining Maximal Association Rules on Soft Sets Using Critical Relative Support Based Pruning

Uddagiri Chandrasekhar[1(✉)], G. Vaishnavi[1], and D. Lakshmi[2]

[1] Department of Computer Science and Engineering, BVIRT Hyderabad, Hyderabad, India
{chandrasekhar.u,17wh1a0564}@bvrithyderabad.edu.in

[2] Department of Computer Science and Engineering, BV Raju Institute of Technology, Narsapur, Medak, Telangana, India
lakshmi.d@srivishnu.edu.in

Abstract. The paper proposes a modification of the Apriori algorithm called Maximal Association rules that combines the ability to mine rules that are lost in regular mining and the speed and efficient memory usage of the soft-set based scheme. Not only can this improve the efficiency without sacrificing a lot of accuracy but it also makes the Apriori Algorithm capable of handling uncertainty in data. Association rules were pruned on a soft set based information system using CRSthreshold. The combination was found useful especially in text mining.

Keywords: Soft sets · Maximal support · Critical relative support · Association rule mining

1 Introduction

The generalized form of Fuzzy set is Soft set [4]. Basically, it is a theory of general mathematical tool which deals with uncertain data i.e., objects that are not defined clearly. The concept of soft set theory was derived in 1999 by Molodtsov [3]. He had applied soft set theory successfully in several streams likely smoothness of the functions, theory of probability, Riemann integration, person integration, game theory, theory of measurement, operations research and so on [1].

A soft set is a pair of (f, E) wherein E is considered the subset of the parameter set and f is considered a function from E to the power set of the universal set. Soft set mainly contributes on transactional data through a Boolean-valued information system, applicability of soft set theory for mining the association rules and discovery of maximal and association rules [1].

Association rule mining is one of the most important as well as popularly used technique in the applications of data mining. It is a collection of the methods for generating rules from data. We can understand the association rule in the form of implication [3] i.e., X =>Y, where both X and Y are the set of collection of items. X is defined as antecedent and Y is defined as consequent. For our convenience, we can consider or assume X as left hand side rule (LHS) and Y as right hand side rule (RHS) [6]. Two main stages are involved in producing association rules. The first one

D. J. Hemanth et al. (Eds.): COMET 2019, LNDECT 35, pp. 931–940, 2020.
https://doi.org/10.1007/978-3-030-32150-5_94

is finding all the frequent items from the transactional database. The Second one is generating association rules that are common from that frequent item set [9]. The aim or motive of association rule mining method is analysis of the transactional databases which possess Boolean values and soft set representation. This is considered as a data structure which is very efficient in association rules mining.

1.1 Critical Relative Support

A set of frequently appearing items in a transaction data base is called as item set. Association rules provide us with very simple logical rules regarding associations between items but definitely not in determining the criticality relationship among them [7]. Another measurement that is used as an alternative in association rules is called confidence. It can be defined as probability of the rule's consequent. It also consists the antecedent in the transaction. If association rules are able to meet the minimum confidence, then they are said to be strong. The item set that is rarely found in database is least item set can also be called as non-frequent, exceptional, unusual and abnormal, inbalance item set [5]. The minimum support can be set to capture the least item set. It is considered difficult to find out which association rule is the most interesting as well as significant. Association rules are considered to be highly critical if they satisfy a certain minimum critical threshold that are predefined [10]. Tracing of finding out that kind of relation is very important and useful for certain domain applications as it will help in revealing some of the crucial and also hidden information.

A Critical Relative Support (CRS) is a value taken by formulizing the maximum of relative frequency between itemset and it'sJaccard similarity coefficient. The rule of critical relative support X => Y is calculated as follows [2]

$$\text{Jaccard similarity coefficient} = \left(\frac{\sup(X => Y)}{\sup(X) + \sup(Y) - \sup(X => Y)}\right)$$

$$CRS(X => Y) = \max\left(\left(\frac{\sup(X)}{\sup(Y)}\right), \left(\frac{\sup(Y)}{\sup(X)}\right)\right) * \left(\frac{\sup(X => Y)}{\sup(X) + \sup(Y) - \sup(X => Y)}\right)$$

1.2 Soft Set Theory

A soft set defined by F: E → P(U) where U is the initial universe, P(U) is power set of U, E is the set of parameters.

Information system: Information system is four tuple notation.

S = (U, A, V, f) where U is object set, A is attribute set, Va is the domain value set of the attribute a where a € A and f is an information function [8].

f: U X A → V such that f(u, a) € Va for all (u, a) € U X A

if Va = {0, 1} then it is called as Boolean valued information system. In this system, two entities can have same tuple. The following Table 1. Represents a soft set.

Table 1. Representation of a Soft Set

U	a1	a2	a3	|A|
u1	f(u1, a1)	f(u1, a2)	f(u1, a3)	f(u1, a|A|)
u2	f(u2, a1)	f(u2, a2)	f(u2, a3)	f(un, a|A|)
.				
.				
.				
u|U|	f(u|U|, a1)	f(u|U|, a2)	f(u|U|, a3)	f(u|U|, a|A|)

Example 1:

Suppose there are five movies in a universe U under consideration (Table 2).

U = {m1, m2, m3, m4, m5} and E = {e1, e2, e3, e4} is a set of decision parameters where e1 stands for 'top hero', e2 stands for 'top heroine', e3 stands for 'family oriented', e4 stands for 'horror'.

F: E → P(U) is a mapping between the movie parameters and the universe of movies.

Let F(e1) = {m1, m4}, F(e2) = {m3}, F{e3) = {m1, m2, m3}, F(e4) = {m2, m4, m5}

Hence, F(e1) means "movies(top heroes)" whose output set is {m1, m4}
F(e2) means "movies(top heroine)" whose output set is {m3}
F(e3) means "movies(family oriented)" whose output set is {m1, m2, m3}
F(e4) means "movies(horror)" whose output set is {m2, m4, m5}.

Table 2. Soft Set representation of the movies Universe.

	E1	E2	E3	E4
m_1	1	0	1	0
m_2	0	0	1	1
m_3	0	1	1	0
m_4	1	0	0	1
m_5	0	0	0	1

1.3 Maximal Association Rules

A Maximal Association rule says that whenever X is the only item of its category in a transaction, then Y also appears, with some confidence in the transaction. Maximal Associations allow the discovery of associations related to items that most often do not appear alone, but rather appear together with closely related items. So, the associations related only to these items will obtain low confidence. Refer [5] for algorithm to calculate Maximal Association rules.

By using Maximal Association rules, some of the Regular association rules may be lost. In converse, some additional association rules are also generated using maximal association rules. So, by taking the union of those two association rules will give expected rules that are of importance.

Example:

There is a dataset consisting of 10 articles; 1article referring to the cities "Hyderabad, Bangalore, Mumbai, Chennai" and topics "sports, entertainment, education and politics"; 2 articles are referring to the cities "Hyderabad, Mumbai, Bangalore" and topics "sports, education"; 1 article is referring to the cities "Hyderabad, Bangalore, Delhi" and topics "education, entertainment"; 1 article is referring to the cities "Chennai, Delhi" and topics "entertainment, technology"; 1 article is referring to the cities "Delhi, Bangalore, Chennai" and topics "politics, sports";1 article is referring to the cities "Delhi, Bangalore" and topics "entertainment, sports, technology"; 1 article is referring to the cities "Mumbai" and topics "sports, entertainment, education, technology, politics"; 1 article is referring to the cities "Delhi, Hyderabad, Bangalore" and topics "entertainment, sports, technology"; 1 article is referring to the cities "Chennai, Hyderabad, Delhi" and topics "technology, entertainment".

We can create Table 3 consisting of two categories "Cities" and also "Topics" i.e., A = {Cities, Topics} wherein

Table 3. Data set of articles which refers to the cities and their respective topics

S. no.	Cities	Topics
1	Hyderabad, Bangalore, Mumbai	Sports, Education
2	Chennai, Delhi	Entertainment, Technology
3	Delhi, Bangalore, Chennai	Politics, Sports
4	Bangalore, Hyderabad, Delhi	Education, Entertainment
5	Hyderabad, Chennai, Mumbai, Bangalore	Sports, Entertainment, Education, Politics
6	Delhi, Bangalore	Entertainment, Technology, Sports
7	Hyderabad, Delhi, Bangalore	Entertainment, Technology, Sports
8	Mumbai	Entertainment, Technology, Sports, Education, Politics
9	Hyderabad, Bangalore, Mumbai	Education, Sports
10	Chennai, Hyderabad, Delhi	Entertainment, Technology

Cities = {Hyderabad, Chennai, Bangalore, Delhi, Mumbai}
Topics = {Sports, Entertainment, Education, Technology, Politics}

Now, list the supported Item sets.

Sup{Mumbai} = 4, sup{Hyderabad} = 6, sup{Delhi} = 6, sup{Chennai} = 4, sup {Bangalore} = 7
Sup{Mumbai, Hyderabad} = 3, sup{Mumbai, Chennai} = 1, sup{Mumbai, Bangalore} = 2, sup{Hyderabad, Delhi} = 3, sup{Hyderabad, Chennai} = 2, sup{Hyderabad, Bangalore} = 4, sup{Delhi, Chennai} = 3, sup{Delhi, Bangalore} = 4, sup {Chennai, Bangalore} = 2

Sup{Hyderabad, Mumbai, Bangalore} = 3, sup{Delhi, Bangalore, Chennai} = 1, sup {Bangalore, Hyderabad, Delhi} = 2, sup{Chennai, Hyderabad, Delhi} = 1, sup {Hyderabad, Chennai Mumbai} = 1, sup{Hyderabad, Chennai, Bangalore} = 1, sup {Chennai, Mumbai, Bangalore} = 1

Sup{Hyderabad, Chennai, Bangalore, Mumbai} = 1

Sup{Education} = 5, sup{Entertainment} = 7, sup{Politics} = 3, sup{Sports} = 7, sup{Technology} = 5

Sup{Education, Entertainment} = 3, sup{Education, Politics} = 2, sup{Education, Sports} = 4, sup{Education, Technology} = 1, sup{Entertainment, Politics} = 2,

Sup{Entertainment, Sports} = 4, sup{Entertainment, Technology} = 5, sup{Politics, Sports} = 3, sup{Politics, Technology} = 1, sup{Sports, Technology} = 3

Sup{Entertainment, Sports, Technology} = 3, sup{Sports, Entertainment, Education} = 2, sup{Sports, Entertainment, Politics} = 2, sup{Entertainment, Education, Technology} = 1, sup{Entertainment, Education, Politics} = 2, sup{Education, Technology, Politics} = 1

Sup{Sports, Entertainment, Education, Technology} = 1, sup{Sports, Entertainment, Education, Politics} = 2, sup{Sports, Entertainment, Technology, Politics} = 1, sup {Sports, Education, Technology, Politics} = 1, sup{Entertainment, Education, Politics, Technology} = 1

Sup{Sports, Education, Entertainment, Technology, Politics} = 1

Now, supported maximal item sets has to be calculated using below algorithm.

Pseudo – code: Soft maximal association rules

```
function soft maximal association rules
support(A, cat, tol, Mconf)
n=1;
att1=1; AD=A; R=0; AC=AD; ca=clock;
while n-=length(cat)+1;
   att2=att1+cat(n)-1; n2=cat(n);
   while n2=0;
      c=combs(akat, bkat, n2); [a b] = size©;
   for l=l:a;
      v=c(l, : ); combo=A(:, V)
      if length(v) == 1
         att=find(comb==length(v))' ;
      else
      komb2=sum(comb')'; att=find(komb2==length(v))' ;
      end
      if length(att)>=tol; check=v; att; A([att], [check])=0;
AB=AD; ns=1; att=1;
while ns~=length(cat)+1;
   att2=att1+cat(ns)-1; ns2=cat(ns);
   whilens2~=0;
      cs=combs(att1:att2, ns2); [as bs]=size(cs);
   forl=l : as;
      vs=cs(l, : ); combs=AB(:, vs);
      iflength(vs) == 1
         atts=find(combs==length(vs))' ;
      else
comb2s-sum(combs')'; atts-find(comd2s==length(vs))' ;
      end
      iflength(atts)>=tol;
      AB([atts], [vs])=0; ck = ismember(check, vs);
      ifsum(check)==0;
      v22=AC(: , [check vs]); combv22 = sum(v22')' ; length([check vs]);
         att=find(combv22==length([check vs])') ;
         iflength(att) >= tol
```

```
            sup1 = length(att); sup=length(atr); confidence=sup/sup1;
            AC([att], [cek vs]) = 0
          ifconfidence>=Mconf
            R=R+1; from=check, to=vs, confidence
            display('=========================');
            end
          end
        end
          end
        end
        ns2=ns2-1;
      end,
      att1=att2+1; ns=ns+1;
    end
      end
    end
      n2=n2-1;
         end
      att1=att2+1; n=n+1; cb=clock;
    end
    time=etime(cb, ca) ;
    disp('Response time'); disp(time) ;
```

The above pseudo code is used to calculate soft maximal association rules.

Maximal support is calculated such that considering an item from a category so that no other items from same category are considered. It is indicated by "M sup{items}".

M sup{Mumbai} = 1, M sup{Chennai, Delhi} = 1, M sup{Delhi, Bangalore} = 1, M sup{Hyderabad, Bangalore, Mumbai} = 2, M sup{Delhi, Bangalore, Chennai} = 1,

Msup{Delhi, Bangalore, Hyderabad} = 2, M sup{Chennai, Hyderabad, Delhi} = 1, M sup{Hyderabad, Chennai, Mumbai, Bangalore} = 1

M sup{Sports, Education} = 2, M sup{Entertainment, Technology} = 2, M sup {Politics, Sports} = 1, M sup{Education, Entertainment} = 1, M sup{Entertainment, Sports, Technology} = 2, M sup{Sports, Entertainment, Education, Politics} = 1, M sup{Sports, Entertainment, Education, Technology, Politics} = 1

Now, maximal confidence has to be calculated. Maximal can be defined as following [1]:

$$C_d^{max} = \frac{\max(X => Y)}{Total(X => Y)}$$

{Delhi, Bangalore, Hyderabad} => {Education, Entertainment} with $S_d^{max} = 1$, $C_d^{max} = \frac{1}{2} = 50\%$

{Hyderabad, Mumbai, Bangalore} => {Sports, Education} with $S_d^{max} = 2$, $C_d^{max} = \frac{2}{2} = 100\%$

{Bangalore, Hyderabad, Delhi} => {Entertainment} with $S_d^{max} = 2$, $C_d^{max} = \frac{2}{2} = 100\%$

{Bangalore, Hyderabad, Delhi} => {Sports} with $S_d^{max} = 1$, $C_d^{max} = \frac{1}{2} = 50\%$
From above M supports and M confidence, the rules of maximal association
Rules generated with min M sup = 1, min M confidence = 70% are

{Hyderabad, Mumbai, Bangalore} => {Sports, Education} with $S_d^{max} = 2$, $C_d^{max} = \frac{2}{2} = 100\%$

{Bangalore, Hyderabad, Delhi} => {Entertainment} with $S_d^{max} = 2$, $C_d^{max} = \frac{2}{2} = 100\%$

2 Proposed Work

2.1 Soft Regular CRS

For {Hyderabad, Mumbai, Bangalore} => {Sports, Education}
Sup{Hyderabad, Mumbai, Bangalore} = 3, sup{Sports, Education} = 4
sup{Hyderabad, Mumbai, Bangalore} => {Sports, Education} = 3

$$\text{CRS} = \max\left(\frac{3}{4}, \frac{4}{3}\right) * \left(\frac{3}{4+3-3}\right) = \frac{4}{3} * \frac{3}{4} = 1$$

For {Bangalore, Hyderabad, Delhi} => {Entertainment}
Sup{Bangalore, Hyderabad, Delhi} = 2, sup{Entertainment} = 7
Sup{Bangalore, Hyderabad, Delhi} => {Entertainment} = 2

$$\text{CRS} = \max\left(\frac{2}{7}, \frac{7}{2}\right) * \left(\frac{2}{2+7-2}\right) = \frac{2}{7} * \frac{7}{2} = 1$$

The calculations show how CRS is calculated. CRS value ranges between 0 to 1. CRS value determines the amount of pruning that can be achieved, while identifying least association rules.

2.2 Soft Maximal CRS

Using the maximal supports calculated above, CRS is calculated to obtain maximal rules.

For {Hyderabad, Mumbai, Bangalore} => {Sports, Education}
Sup{Hyderabad, Mumbai, Bangalore} = 2, sup{Sports, Education} = 2
sup{Hyderabad, Mumbai, Bangalore} => {Sports, Education} = 2

$$\mathrm{CRS} = \max\left(\frac{2}{2}, \frac{2}{2}\right) * \left(\frac{2}{2+2-2}\right) = \frac{2}{2} * \frac{2}{2} = 1$$

For {Bangalore, Hyderabad, Delhi} => {Entertainment}
Sup{Bangalore, Hyderabad, Delhi} = 2, sup{Entertainment} = 1
Sup{Bangalore, Hyderabad, Delhi} => {Entertainment} = 1

$$\mathrm{CRS} = \max\left(\frac{1}{2}, \frac{2}{1}\right) * \left(\frac{1}{1+2-2}\right) = \frac{1}{2} * \frac{2}{1} = 1$$

The above calculations show the process to calculate CRS on maximal association rules. As all the association rules we have taken have both soft regular and soft maximal CRS as '1' there is no change in the rules mined. This example just serves as a demonstration.

The below Fig. 1. Shows the comparison of response times obtained on air pollution data set[5].

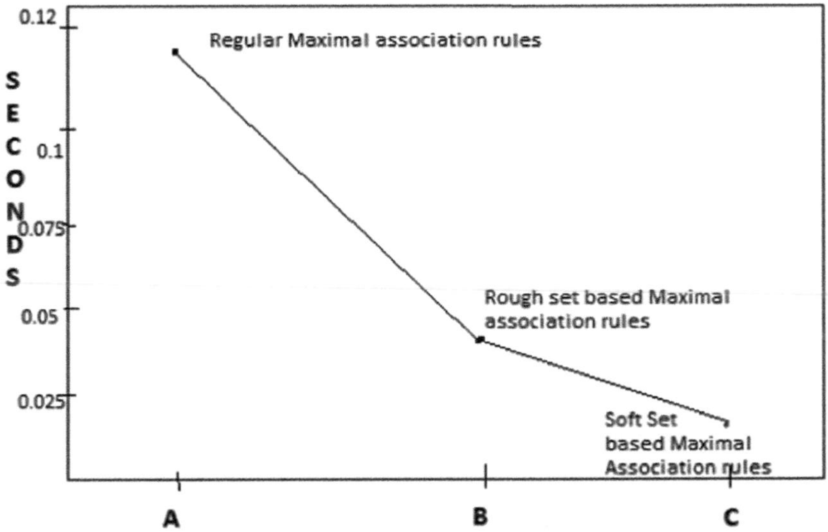

Fig. 1. Soft Set based technique offers best performance

Additionally using CRS based pruning can be used to mine infrequent events and exceptional cases.

3 Conclusion

By applying Critical Relative Support approach, number of unwanted rules was shown to be reduced by up to 98% [2] on educational dataset. Using soft-sets to model the problem will ensure that the efficiency is improved. While the algorithm that has been proposed improves the Apriori algorithm in terms of speed, resource consumption and robustness in presence of uncertainty, it suffers from following limitations [10].

1. An additional Hyper-parameter Critical Relative Support Threshold needs to be chosen and tuned.
2. While highly efficient for large datasets, the soft-matrix form may actually consume more space for smaller datasets, i.e. O(|U|.|E|).

The paper proposes CRS based soft-maximal association rules and found that there is no difference rules mined with regular rough set based approach or soft set based approach. However there is an advantage of speed. CRS was used to prune unwanted rules and introduction to maximal approach helped in identifying additional rules that may have been missed previously.

References

1. Agrawal, R., Srikant, R.: Fast algorithms for mining association rules in large databases. In: Proceedings of the 20th International Conference on Very Large Data Bases, VLDB, Santiago, Chile, pp. 487–499, September 1994
2. Abdullah, Z., Herawan, T., Ahmad, N., Deris, M.M.: Mining significant association rules from educational data using critical relative support approach. Proc. Soc. Behav. Sci. **28**, 97–101 (2011)
3. Molodtsov, D.: Soft set theory-first results. Comput. Math Appl. **37**, 19–31 (1999)
4. Çağman, N., Çıtak, F., Enginoğlu, S.: Fuzzy parameterized fuzzy soft set theory and its applications. TJFS: Turk. J. Fuzzy Syst. **1**(1), 21–35 (2010)
5. Herawan, T., Deris, M.M.: A soft set approach for association rules mining. Knowl.-Based Syst. **24**, 186–195 (2011)
6. Kanojiya, S.S., Tiwari, A.: A new soft set based association rule mining algorithm. TECHNIA–Int. J. Comput. Sci. Commun. Technol **6**(2), 948 (2014)
7. Rahman, C.M., Sohel, F.A., Naushad, P., Kamruzzaman, S.M.: Text classification using the concept of association rule of data mining. CoRR, abs/1009.4582 (2010)
8. Saraf, S., Adlakha, N., Sharma, S.: Absolute soft set approach for mining association patterns. Int. J. Comput. Appl. **84**(4), 35–39 (2013)
9. Rose, A.N.M., Awang, M.I., Hassan, H., Deris, M.M.: Comparison of techniques in solving incomplete datasets in softest. Int. J. Database Theory Appl. **4**(3) (2011)
10. Kanojiya, S.S., Tiwari, A.: A new soft set based association rule mining. Technia **6**(2), 948 (2014). ISSN 0974-3375

RETRACTED CHAPTER: Efficient Conversion of Handwritten Text to Braille Text for Visually Challenged People

M. Anitha(✉) and D. Elangovan

Computer Science and Engineering,
Panimalar Engineering College, Chennai 600123, India
anitha.ml5995@gmail.com, elangovan.durai@yahoo.com

Abstract. The World Health Organization has submitted a report in 2017 which declared that a population of nearly 36 million people are blind out of 253 million visually impaired people. Braille books in the tactile format are very helpful for the visually impaired people to gain knowledge but the availability of these resources is limited. With the improvement of electronic innovation, Braille ended up being appropriate to PC helped generation in light of its coded structures. Programming based content to-Braille interpretation has been ended up being a fruitful arrangement in Assistive Innovation. Various research studies and papers describe the methods for obtaining machine readable document from textual image. In future, character recognition might serve as a key role to create a paperless environment that helps the visually impaired people to gain enormous amount of educational materials. In current world the innovations and advancements in modern scientific systems has expanded the boundaries of human effort like Optical Character Recognition. In the field of machine learning and pattern matching Handwritten recognition has gained lot of attention. A novel method is proposed to convert Handwritten text into Braille text due to which huge resources of Handwritten text materials will be available in Braille text that helps the visually impaired and low vision people to expand their knowledge. The novelty of this system has gained a better accuracy.

Keywords: Braille · Handwritten · Optical Character Recognition · Visually impaired

1 Introduction

The visually impaired people enable various technologies to access the materials needed. The education materials are available in terms of Braille books [1]. Blind people make use of tactile dots in the Braille books to obtain knowledge. Louis Braille developed a Braille system. The visually impaired people face difficulties in social interaction, reading, accessing library books, recognizing objects, driving and also performing ta tasks quickly. They obtain the information by enlarged print or reading standard print and through listening. To survive in this competitive environment the visually impaired people should be more and more efficient in terms of employment

---

The original version of this chapter has been retracted. The retraction note to this chapter can be found at https://doi.org/10.1007/978-3-030-32150-5_163

D. J. Hemanth et al. (Eds.): COMET 2019, LNDECT 35, pp. 941–949, 2020.
https://doi.org/10.1007/978-3-030-32150-5_95

and education. Large print type should be used, mostly "18" point, but at a minimum "16" point for low visually impaired people [2].

Blind people make use of tactile dots in the Braille books to obtain knowledge. Louis Braille developed a Braille system. Almost in 190 countries there is an organization called WBU (World Blind Union) for the visually impaired people. The education materials are available in terms of Braille books. But the number of books available is too limited so that not all individuals can acquire knowledge [3].

In the field of science and math's the visually impaired people face most challenging task since pictorial representation plays a major role in those fields. Most of the assistive technologies failed to give access to the images and graphs. Thus, the part of tactile methodology has been investigated to impart graphical data [4].

The advancement of handwritten character acknowledgment frameworks started in the 1950s when there were human administrators whose activity was to change over information from different archives into electronic arrangement, making the procedure very long and regularly influenced by blunders. Handwritten character acknowledgment has been a standout amongst the most captivating and testing research zones in field of picture preparing and design acknowledgment in the ongoing years [5]. Optical character acknowledgment is a field of concentrate than can incorporate a wide range of tackling procedures. Neural systems, bolster vector machines and factual classifiers appear to be the favored answers for the issue because of their demonstrated precision in characterizing new information [6].

The Optical Character Recognizer really is a converter which makes an interpretation of written by hand-written message pictures to a machine-based content. All in all, handwritten character acknowledgment is arranged into two sorts as disconnected and on-line [7]. In the disconnected acknowledgment, the written work is normally caught optically by a scanner and the finished composition is accessible as a picture.

1.1 Objective

To increase the lifestyle of the visually impaired people, a Braille displays are given as output. Initially, the handwritten image is taken as input which gets converted into edited text. Next, the edited text formats are converted to tactile format to improve the efficiency. Voice message is also introduced to make the people to hear the text for better understanding when they wish.

2 Literature Survey

Kumar and Jindal et al. [8] combined various feature extraction techniques to recognize the handwritten numbers. To extract the meaningful features, a skeleton of numeral is created. Diagonal features, centroid features, peak extent-based features and zoning features are the four features that uses the SVM classifier to extract the information based on classification. Testing and training data have been considered for experimentation work using 6000 samples of unique handwritten numerals dataset. SVM with RBF kernel classifier and fivefold cross validation technique are used to improve the efficiency. 96.3% of recognition accuracy is achieved. In 2018 [9] Venugopal and Sundaram developed an online writer description using a method called sparse code. The descriptors obtained from sparse code are constructed using histogram strokes.

IAM and IBM-UB1 database is used to store the samples of data to evaluate the writer scripts. To select the pertinent bin size to get the features, the entropy analysis is implemented to establish inequity between descriptors by writers. Compared to the previous works the database used over this method gives better evaluation strategy. The segmented sub-strokes of handwritten scripts improve flexibility in case of sparse characterization. Tavoli and Keyvanpour [10] implemented an innovative neural network method for recognizing handwritten scripts. Using a swarm optimization, the weights in neural networks are enhanced. A new system for spotting the handwritten words form the handwritten scripts are implemented using two methods namely PSO and MLP. A dataset called IAM English is used for English documents over this method. They implemented a separate neural network for every word to spot the keyword. If the test data matches with keyword then it returns positive value. Compared to the previous methodology it has achieved best accuracy. Bawane et al. [11] implemented a Spiking Neural Network (SNN) and Leaky Integrate and Fire Model (LIF) to recognize handwritten characters and objects. After preprocessing the input, they used edge detection and extended histogram to extract the required features from the given image. LIF model helps to increase the computational capability. The spiking neural network involved over this method takes about [illegible] flops/1 ms of computational efficiency to obtain the neural properties. In order to compare the results obtained from SNN, a classifier called Support Vector Machine is incorporated. The objects and characters are recognized by applying post processing operations. Segmentation operation is performed over the scanned image to recognize efficiently. A new model using sinusoidal parameters is created by Himakshi Choudhury, Mahadeva Prasanna et al. [12] for online handwriting recognition. Subordinates of the speed profiles (i.e. quickening) and the x-and y-facilities are additionally vital in portraying the penmanship the viability of the proposed feature is appeared for character and word acknowledgment assignment utilizing concealed Markov show (HMM) and bolster vector machine (SVM) classifiers. The parameters (i.e., abundancy, stage and recurrence) for every one of these signs are removed by fitting half cycles of sine wave between its progressive zero intersections focuses. The analyses are led on three online written by hand databases: Assamese digit database, UNIPEN English character database and UNIPEN ICROW-03 English word database. The outcomes demonstrate that abundancy contains the most separating data about the characters while the stage contains minimum data.

3 Proposed Method

The high-level architecture for Automating Handwritten to Tactile conversion. Initially handwritten document is scanned and subsequently the scanned image can be repositioned. Once the scanned document is in the form of an image, the preprocessing gets started to convert the gray scale image into binary image, and then translation and rotation operations are performed under preprocessing. After that segmentation and feature extraction is done to extract the text regions in the image. The Optical Character Recognition gets convert handwritten text into their respective text format. Once the text gets obtained it is then matched with the tactile templates, if the text matches with the tactile templates then it gets printed as a Braille text using Braill printer.

Figure 1 describes the sequence of automating the handwritten to tactile conversion using segmentation, feature extraction and template matching. OCR plays a major role to get a machine-readable text.

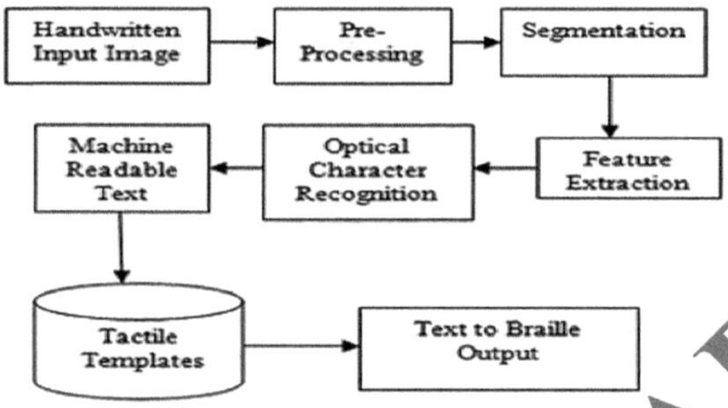

Fig. 1. Proposed framework for handwritten text to Braille conversion.

3.1 Image Acquisition

The first step is image acquisition. Image acquisition is the process of creating digital representation of visual characteristics in an image i.e. physical structure of an object. The image acquisition process consists of three steps; energy reflected from the object of interest, an optical system which focuses the energy and finally a sensor which measures the amount of energy.

Figure 2 Shows how Hand-written images are obtained using a hardware device like scanner camera, mobile phone etc. The images or documents are collected and stored in the data base for further processing.

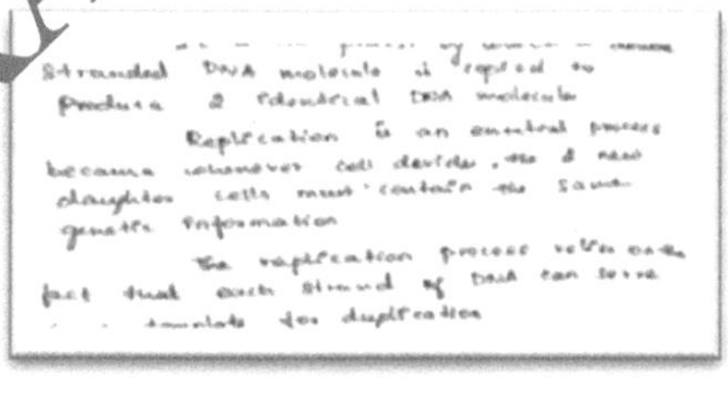

Fig. 2. Scanned handwritten input image

3.2 Pre-processing

Pre-processing is method of operations on image with lowest level of obtaining the images both the input and output intensity of images. The aim of pre-processing is an improvement of the image data that suppresses unwanted distortions or enhances some image features important for further processing. Grey-scaling of an image is a process by which an RGB image is converted into a black and white image.

In Fig. 3 the given color image gets converted in to Greyscale image for binarization. After the image gets converted from color to greyscale only different colors of grey is available which helps in extracting the information from the handwritten image efficiently.

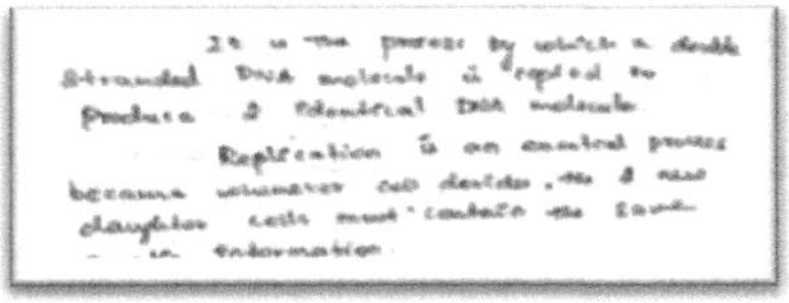

Fig. 3. Greyscale converted image

3.3 Segmentation

It is an operation that seeks to decompose an image of sequence of characters into sub images of individual symbols. Character segmentation is a key requirement that determines the utility of conventional systems. Different methods used can be classified based on the type of text and strategy being followed like straight segmentation method, recognition-based segmentation and cut classification method.

Figure 4 shows that the preprocessed image gets segmented as a line segmentation followed by word segmentation and finally character segmentation.

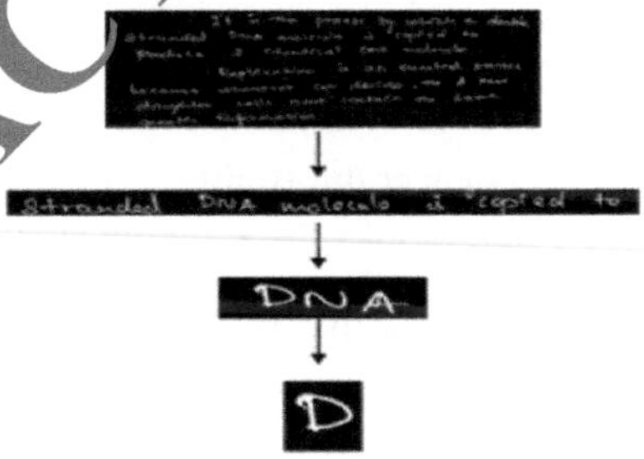

Fig. 4. Letter segmentation

3.4 Feature Extraction

Feature extraction depicts the important shape data contained in an example with the goal that the undertaking of ordering the example is made simple by a formal technique. Feature extraction organize in this framework investigations these English characters portion and chooses an arrangement of feature that can be utilized to interestingly distinguish that character section. Differentiation of characters is done by cropping the boxed characters of the pre-processed image. At first the sub-images are cropped label by label in the sample image, and then the character image array is resized to form a 7 × 5 matrix pixelated image. This is done because an image array can only be defined with all images of fixed size.

3.5 Optical Character Recognition

Optical character recognition (OCR) is a process in which electronic change of pictures of composed manually written or printed words into machine-encoded words. It is broadly utilized as a type of information passage from printed paper information records, regardless of whether international ID reports, solicitations, bank proclamations, electronic receipts, business cards, mail, printouts of static-information or any appropriate documentation.

Figure 5 displays how the image gets converted in to machine readable text using Optical Character Recognition.

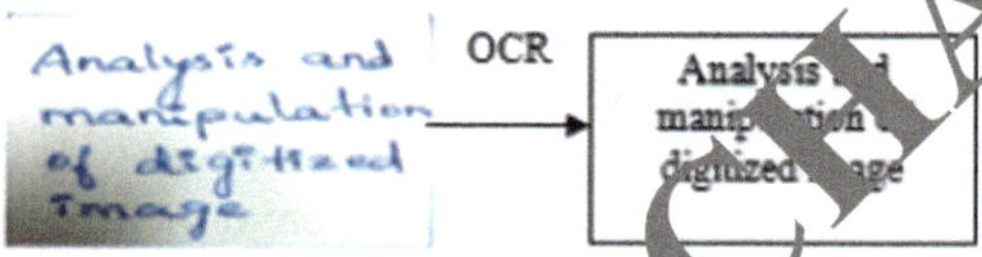

Fig. 5. Optical Character Recognition representation.

3.6 Text to Braille

Text to Braille converter is a technique which converts text to relevant Braille code. This system contains cell per character and each cell having six dots arranged. Each cell represents one letter number and symbols. The Braille cell contains two rows and three columns which denotes one-to-one correspondence between the Braille characters.

In Fig. 6 The user needs to connect the Braille show to the pc or the other device via a USB affiliation. Braille displays may use different styles of connections. Most of the time, the user doesn't ought to install a driver for the Braille show. this can be as a result of there aren't many sorts of Braille displays, and screen readers will support most of them. Once connected to the pc, the Braille show can acquire the presently highlighted text on the screen.

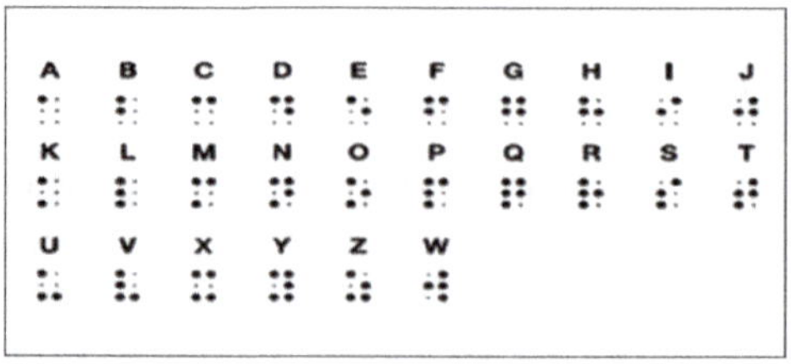

Fig. 6. Text to Braille displays

4 Result and Discussion

4.1 Experiment

The handwritten to Braille system can help the visually impaired individuals to read and understand the handwritten notes from Braille displays. The procedure involves

1. Capture the handwritten image using scanner.
2. Obtained images are pre-processed to get a binarized image.
3. The binarized text is segmented and features are extracted from the segmented text.
4. Converts the image text to ASCII code and equivalent character is printed as text using Optical Character Recognition.
5. Match the converted text with Braille templates.
6. If the matches are correct then the Braille text will be printed.

Figure 7 describes the proposed handwritten to Braille system. The given input handwritten image is converted to its equivalent Braille code and displayed. Braille code can be saved and fed as input to Braille displays. The converted handwritten text to text can also be saved as a word file.

Fig. 7. Handwritten to Braille system

4.2 Performance Evaluation Based on Execution Time

Table 1 explains the accuracy of converting the handwritten characters and numbers into text and Braille code. Handwritten numbers are converted to their equivalent Braille code with 87% of accuracy. Proposed system has taken a time of 29.345 s to convert handwritten numbers into Braille code. Handwritten capital letters are obtained with an accuracy of 82% and it took 17.54 s to obtain the Braille equivalent (Table 2).

Table 1. Accuracy of handwritten images

Input types	Accuracy (%)	Time (in seconds)
Handwritten image (numbers)	90	29.345
Handwritten image (capital letters)	82	17.545

Table 2. Performance comparison of datasets

S. no	Dataset	Accuracy
1	IAM English dataset	77%
2	UNIPEN ICROW- 03	87.5%
3	IFN/ENIT, AHTID/MW	98.38%

5 Conclusion

The tactile technologies have gone through many enhancements so far and also there is a room for further development in real time applications. The current technologies are very limited in imparting education to visually impaired people. The conversion of handwritten text into braille text using better education materials helps to ease the learning abilities of the visually impaired people. The handwritten characters are captured as images, preprocessed and segmented. It is converted into machine readable text using Optical Character Recognition. The machine-readable text is further compared with Braille templates and if matched the Braille characters are obtained. This can be enhanced further by converting cursive handwriting characters into Braille text in many spoken languages.

References

1. Nahar, L., Jaafar, A., Ahamed, E., Kaish, A.B.M.A.: Design of a Braille learning application for visually impaired students in Bangladesh. Off. J. RESNA **27**(3), 172–182 (2015)
2. Russomanno, A., O'Modhrain, S., Gillespie, R.B., Rodger, M.W.: Refreshing refreshable braille displays. IEEE Trans. Hapt. **8**(3), 287–297 (2015). https://doi.org/10.1109/toh.2015.2423492
3. Sultana, S., Rahman, A., Chowdhury, F.H., Zaman, H.U.: A novel Braille pad with dual text-to-Braille and Braille-to-text capabilities with an integrated LCD display. In: 2017 International Conference on Intelligent Computing, Instrumentation and Control Technologies (ICICICT), Kannur, pp. 195–200 (2017)
4. O'Modhrain, S., Giudice, N.A., Gardner, J.A., Legge, G.E.: Designing media for visually-impaired users of refreshable touch displays: possibilities and pitfalls. IEEE Trans. Hapt. **8**(3), 248–257 (2015)
5. Kala, R., Vazirani, H., Shukla, A., Tiwari, R.: Offline handwriting recognition using genetic algorithm. IJCSI Int. J. Comput. Sci. Issues **7**(2, 1) (2010)
6. Wshah, S., Kumar, G., Govindaraju, V.: Statistical script independent word spotting in offline handwritten documents. Pattern Recog. **47**(3), 1039–1050 (2014)
7. Plamondon, R., Srihari, S.N.: On-line and off-line handwritten character recognition: a comprehensive survey. IEEE Trans. Pattern Anal. Mach. Intell. **22**(1), 63–84 (2000)
8. Kumar, M., Jindal, M.K., Sharma, R.K., Jindal, S.R.: Offline handwritten numeral recognition using combination of different feature extraction techniques. Natl. Acad. Sci. **41**(1), 29–33 (2018)
9. Venugopal, V., Sundaram, S.: Online writer Identification with sparse coding based descriptors. IEEE Trans. Inf. Forensics Secur. **13**(10), 2538–2552 (2018)

10. Tavoli, R., Keyvanpourv, M.: A method for handwritten word spotting based on particle swarm optimisation and multi-layer perceptron. IET Softw. **12**(2), 152–159 (2018)
11. Bawane, P., Gadariye, S., Chaturvedi, S., Khurshid, A.A.: Object and character recognition using spiking neural network. In: International Conference on Processing of Materials, Minerals and Energy, vol. 5, no. 1, Part 1, pp. 360–366 (2018)

Safety Measures for Firecrackers Industry Using IOT

N. Savitha(✉)

Department of Computer Science and Engineering, Panimalar Engineering College, Chennai, Tamil Nadu, India
savithanarayana2912@gmail.com

Abstract. Today safety is an integral part of industrial management system. Fire safety becomes the most important due to its very serious effects as a small mistake may cause severe damage. Many of the firecrackers industries manufacture firecrackers, matchbox, and printing during summer, the hot and dry climate is appropriate for manufacturing. Fireworks are the device that uses explosive, flammable material to create spectacular displays of light, noise and smoke. As in any manufacturing industry, Fireworks units also have accidents taking place in the worksite. The major cause for the fire accident are due to environmental changes and some human error. In IOT technology the fire-fighting, fire monitoring and safety management system are an important applications. So that the IOT based fire safety in the firecrackers industry is developed by placing the various sensors for monitoring the environmental parameters. All the sensor nodes are interfaced to the Arduino microcontroller and if any of the sensor node detects the abnormalities in environmental parameter the fire alert is given and once the fire is triggered the water is sprinkled over the area.

Keywords: IOT · Fire accident · Fire safety

1 Introduction

The internet of things (IoT) is a computing concept that connects the physical objects. Through this the data are transferred from one device to another and some meaningful operations can be performed. Many applications are developed using IOT in different sectors such as in medical, safety system, agriculture, retail system, etc. Through IOT we can give heart to the particular object. And the importance of that object in the real world can be increased. There are several advantages when use the IOT, in that through this the communication between machine to machine are increased. Through this any device that can be controlled from anywhere. Wireless communication which plays important thing in many application. By using the IOT automation in many smart device are encouraged by people nowadays.

Data plays vital role everywhere in the current circumstances, through these data many decisions can be taken in the real world easily when it is connected to the internet. Tracking and monitoring is one of the important application of the IOT. The application of IOT are smart agriculture, surveillance system, smart home, smart city,

D. J. Hemanth et al. (Eds.): COMET 2019, LNDECT 35, pp. 950–957, 2020.
https://doi.org/10.1007/978-3-030-32150-5_96

etc. In medical field also IOT plays important role monitoring patients health. In many safety system IOT plays important role because of its automation and speed of data transmission anywhere at any time. And the main advantage is that it can be easily modifiable at low cost.

2 Related Works

With the headways in the everyday life, fire-security has turned out to be one of the essential issues [6]. Fire perils are lethally risky and slandering in regards to business what's more, home security, moreover decimating in regards to human life. The conspicuous method to limit the sort of misfortune is to react to these crisis circumstances as fast as could be expected under the circumstances. The created framework is alarming the far away property proprietor precisely additionally quickly through sending Short Message (SMS) by methods for GSM system and transmitter esteems to the Central server utilizing GPRS.

Fire causes tremendous loss of lives and properties consistently in Bangladesh. Breaking down past flame episodes, realities are uncovered. A few of the primary driver are inadequate fire safeguard materials, electric cut off broken electrical wiring, nearness of inflammable materials, infringement of flame security and absence of sufficient mindfulness and so forth [1]. In this framework the information combination calculation encourages the framework to dispose of misleading flame circumstances, for example, tobacco smoke, welding and so forth. Amid the fire risk SFF tells the fire administration and others by instant messages and phone calls. Alongside ringing fire caution it reports the fire influenced areas and seriousness. To keep fire from spreading it breaks electric circuits of the influenced territory, discharges the smothering gas indicating the correct fire areas.

Jiang and Gao et al. made investigation on cotton warehouse fire accidents. Cotton is a critical financial harvest in our nation, the material crude material and key save materials, which plays a critical part in the advancement of the national economy [2]. For that they proposed an IoT engineering based application plan of cotton distribution center fire cautioning framework. At that point completed an information obtaining and transmission by the method for ZigBee remote sensor arrange as the base, and made a notice by foundation wise fire examination framework. At long last, the application conspire made a viable control for flame through setting off the relating fire joint activity hardware by a logical fire crisis choice framework.

In 2018, Seiber et al. developed Aerial Plumes for detecting the hazardous fire. They propose equipping swarms of automatons with Web of Things (IoT) sensor stages to empower dynamic following of perilous ethereal tufts. Expanding rambles with sensors empowers crisis reaction groups to look after safe separations amid peril ID, limiting first reaction group introduction [3]. Also, they coordinate sensor-based particulate identification with self-governing automaton flight control giving the ability to powerfully distinguish and track the limits of aeronautical crest progressively. This empowers people on call for outwardly recognize tuft development and better anticipate and confine the effect territory.

The quick improvement of China's financial development, serious expanding of urban populace and enduring extension of city building thickness, underground designing, tall structures and substantial open developments turn out to be to an ever increasing extent, impartially it advances more extreme test to urban fire insurance. To adjust to the present day city and general society security of society improvement, the continuous remote-checking arrangement of urban fire security in view of IoT is proposed [4]. In this novel fire quenching observing and control framework is proposed. The framework consolidates on the web checking of flame stifling data in light of Internet of things (IoT), early fire cautioning and alert framework with building security assessment, which are intended to procuring the continuous data of flame dousing working condition, making strides the dependable forecast instrument in conventional fire observing framework.

Lalwani et al. proposed based on IOT to monitor the industrial conditions. There are numerous things we find out about mechanical web of things as it is another developing innovation. They utilize sensors to consistently screen industry machines which is very difficult to be overseen by human [5]. Here an endeavor is made to build up an auto-observing framework through which the business individual can screen the parameters on a site which can be gotten to either on telephone or on PC and produce ready flags through the site that will caution the general population working in the business through alert. The site is made by using XAMPP server inter-facing with database that is utilizing PHP dialect as the rule of the system.

Rambabu et al. developed a fire fighting robot which monitors and control the fire. Independent Fire Fighter Robot is the robot which self-rulingly recognizes and smothers the fire, it utilizes the fire sensor for discovery, and the fire quencher is utilized to douse the recognized fire [7]. The robot can turn while currently filtering for the fire, this checking is performed by the sensors set on the sides, at the point when the fire is distinguished, the robot can move toward the fire and it stops before it and trigger the douser to turn out the fire.

3 Proposed Architecture

In the proposed system the sensor units are interfaced into the arduino board. The main parameters that cause the fire accident in firework industry are temperature, high light intensity so that the temperature sensor, light intensity sensors are interfaced. And the flame sensor and smoke sensor are also interfaced for detecting the smoke and the fire. Once the fire is triggered the signal pass through the RF module to 8051 microcontroller through that fire alert and water motor is turned ON. Even though the basic safety is provided in firework industry there is more chance for the fire get easily enters in to the surrounding area. Many of the fire accidents goes vigorous because of the late arrival of fire safety department. For that in this system once the fire is triggered through GSM the SMS send to the fire safety department along with the location. Based on the survey taken from various fire safety systems in the earlier work [8] the new safety system was developed. Through this the complete safety is provided to the firework industry. The architecture diagram shown in Fig. 1.

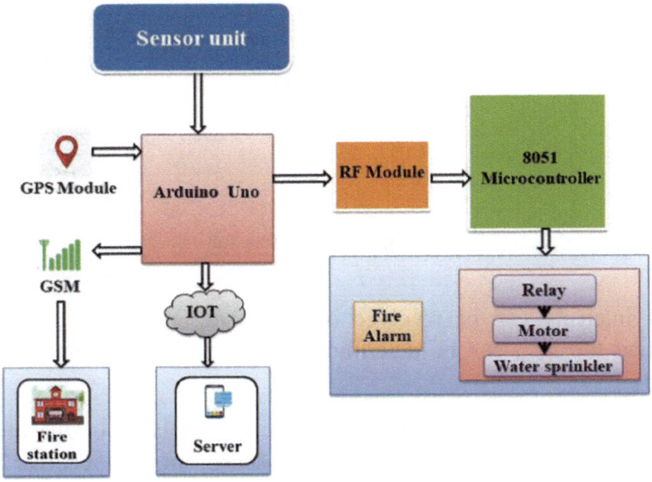

Fig. 1. Fire safety system architecture

4 Proposed System Implementation

4.1 Sensor Unit

In the firework manufacturing industry the main parameters that cause the fire accidents are high raise in temperature, high light intensity, friction due to high wind speed, and due to some human error such as mishandling of chemicals. so monitoring the environmental parameters are very important.

Monitoring Environmental Temperature Using Temperature Sensor. There are-several reasons for fire accidents that occur in firework industry. In that one of the main cause are high rise in temperature. So monitoring the environmental temperature is one of the mandatory thing that should be implemented in the safety system. In the proposed system LM35 temperature sensor is used, the range of temperature is from −55 °C to 150 °C. LM35 is interfaced with the arduino uno. In the system once the temperature goes above 50 °C it turns on the fire alarm.

Monitoring Surrounding Light Intensity Level Using Light intensity Sensor. Generally the firecrackers are manufactured with the help of the sun light. High light intensity that cause the firecrackers to catch the fire easily. once the intensity goes above 20,000 lx it cause the firecrackers to catch the fire naturally. So monitoring the light intensity level is one of the important thing that should be implemented in the proposed system. UUGEAR light sensor module is interfaced with the arduino uno. It uses high quality photo resistor. The working voltage of this sensor is from 3.3–5 V. The light intensity dependent resistor shown in Fig. 2.

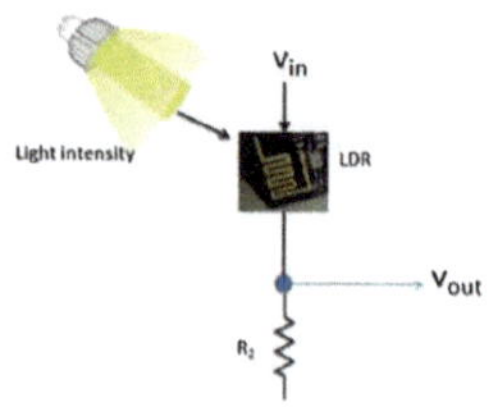

Fig. 2. LDR resistor

The light intensity can be received by the LDR and that was connected to the resistor R_2 and the V_{out} can be determined by using the formula $V_{out} = V_{in}$ X (R_2/ (R_{LDR} + R_2)).

Detecting Flame Using Flame Sensor. Detection of flame is one of the important practices in fire safety system. In that the flame sensor infrared receiver ignition source detection module is interfaced with the arduino board. Once the fire is triggered by the infrared radiation it using CCD it detects the fire. The operating voltage of the flame sensor is from 3.3 to 5 V.

Detecting Smoke Using Smoke Sensor. Some of the crackers that only cause the smoke instead of fire. In that case flame sensor cannot able to detect the fire. In that case smoke sensor is need to detect the presence of smoke. Different smoke sensors are available to detect the presence of different gases that are specified in the Table 1.

Table 1. Different gas sensor

Sensor	Gas type
MQ2	Combustible Gas, Smoke
MQ3	Alcohol Vapor
MQ5	LPG, Natural Gas, Town Gas
MQ9	Carbon Monoxide, Coal Gas, Liquefied Gas

In the proposed system MQ2 gas sensor is used for detecting the smoke and the combustible gases that present in the working area.

4.2 Gathering Location Using GPS Module

Most of the fire accident that goes to dangerous state because of the late arrival of fire engine to the location where the fire burst out. For that case GPS place major role and use to identify the location easily. This GPS module was placed in the firework industry that calculate the latitude and longitude of the location. The position can be calculated by using the algebraic formula. GPS module shown in Fig. 3.

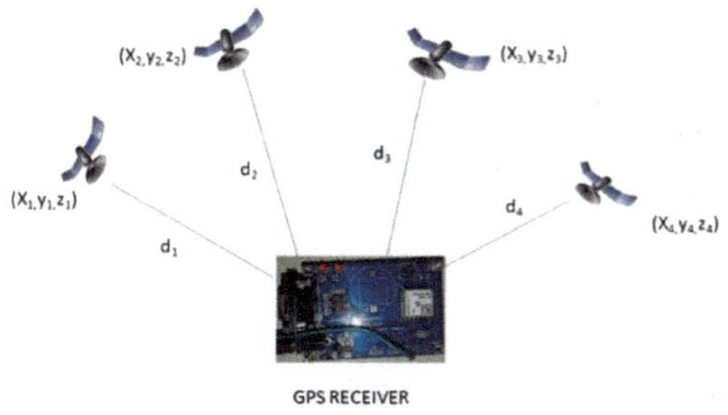

Fig. 3. GPS Module

Let as assume that x_i, y_i, z_i, where i = 1, 2, 3, 4 are the positions of the satellite. c be the speed of the light and t be the receiver clock offset time.

$$(\sqrt{(x-x_1)^2+(y-y_1)^2+(z-z_1)^2}) + ct = d_1$$
$$(\sqrt{(x-x_2)^2+(y-y_2)^2+(z-z_2)^2}) + ct = d_2$$
$$(\sqrt{(x-x_3)^2+(y-y_3)^2+(z-z_3)^2}) + ct = d_3$$
$$(\sqrt{(x-x_4)^2+(y-y_4)^2+(z-z_4)^2}) + ct = d_1$$

Through the above equation the location of the particular are be calculated by the GPS receiver.

4.3 Sending Emergency Alert Using GSM Module

The emergency message to the fire service that was automatically sent through the GSM module. In this the smoke state, flame state along with latitude and longitude calculated by the GPS are send as message to the fire service department.

5 Working Algorithm

STEP1: Initialize the fire safety system circuit.
STEP2: Sensor nodes continuously sense the environmental parameter.
STEP3: IOT board continuously transmit the sensed data to the server.
STEP4: (a)If the fire is triggered the signal passed to the 8051 microcontroller and it turn ON the water sprinkler and fire alarm.
(b)At same time the emergency message to the fire station was send through the GSM module.
STEP5: Goto Step1.

6 Result

All the sensed data are continuously transmitted to the server that can be viewed at any time. The server data snapshot shown in Fig. 4.

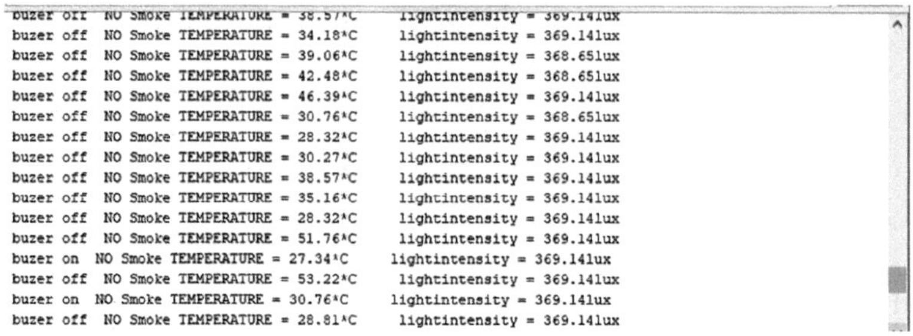

```
buzer off  NO Smoke TEMPERATURE = 38.57*C    lightintensity = 369.14lux
buzer off  NO Smoke TEMPERATURE = 34.18*C    lightintensity = 369.141ux
buzer off  NO Smoke TEMPERATURE = 39.06*C    lightintensity = 368.651ux
buzer off  NO Smoke TEMPERATURE = 42.48*C    lightintensity = 368.651ux
buzer off  NO Smoke TEMPERATURE = 46.39*C    lightintensity = 369.141ux
buzer off  NO Smoke TEMPERATURE = 30.76*C    lightintensity = 368.651ux
buzer off  NO Smoke TEMPERATURE = 28.32*C    lightintensity = 369.141ux
buzer off  NO Smoke TEMPERATURE = 30.27*C    lightintensity = 369.141ux
buzer off  NO Smoke TEMPERATURE = 38.57*C    lightintensity = 369.141ux
buzer off  NO Smoke TEMPERATURE = 35.16*C    lightintensity = 369.141ux
buzer off  NO Smoke TEMPERATURE = 28.32*C    lightintensity = 369.141ux
buzer off  NO Smoke TEMPERATURE = 51.76*C    lightintensity = 369.141ux
buzer on  NO Smoke TEMPERATURE = 27.34*C    lightintensity = 369.141ux
buzer off  NO Smoke TEMPERATURE = 53.22*C    lightintensity = 369.141ux
buzer on  NO Smoke TEMPERATURE = 30.76*C    lightintensity = 369.141ux
buzer off  NO Smoke TEMPERATURE = 28.81*C    lightintensity = 369.141ux
```

Fig. 4. Server data snapshot

The major reason for the fire accident in firework industry are due to the drastic raise in the temperature. So temperature is considering the major factor, once it gose above 50 °C the fire alram and buzzer gets turned ON. Based on various temperature the behaviour of the motor state was shown in Fig. 5.

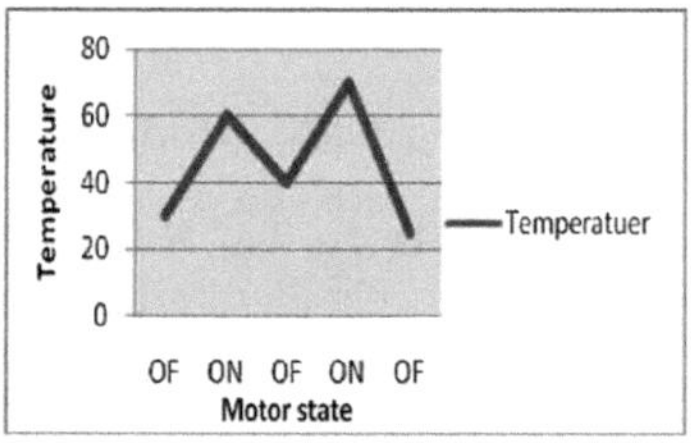

Fig. 5. State of motor based on temperature.

Emergency message automatically send to the fire station with the help of the GSM module. In the message that send to the fire station the data of the flame sensor and smoke sensor along with that latitude and longitude position of the firework industry also send along with the message. The emergency message snapshot is shown in Fig. 6.

Fig. 6. Emergency message snapsho

7 Conclusion

Through this fire safety system the main environmental parameters that causing the fire are monitored continuously. And if the fire is triggered the safety measures also implemented. Through this many people lives and their properties are protected from damage caused by fire accidents. This system will be enhanced by analyzing other possibilities of fire accidents like friction and based on that the safety system will be enhanced.

References

1. Mobin, M.I., Abid-Ar-Rafi, M., Islam, M.N., Hasan, M.R.: An intelligent fire detection and mitigation system safe from fire. Int. J. Comput. Appl. **133**(6), 1–7 (2016)
2. Jiang, J., Gao, Z., Shen, H., Wang, C.: Research on the fire warning program of cotton warehousing based on IoT technology. Int. J. Eng. Bus. Manage. **18**(2), 121–124 (2017)
3. Seiber, C., Nowlin, D., Landowski, B., Tolentino, M.E.: tracking hazard-ous aerial plumes using IoT-enabled drone swarms. Int. J. Comput. Inform. Sci. (2018)
4. Li, Y., Yi, J., Zhu, X., Wang, Z., Xu, F.: Developing a fire monitoring and control system based on IoT. In: Advances in Intelligent Systems Research, vol. 133
5. Lalwani, S.P., Khurana, M.K., Khandare, S.J., Ansari, O.U.R.: IoT based industrial parameters monitoring and alarming system using arduino. Int. J. Eng. Sci. Comput. (2018)
6. Reddy, M.S., Rao, K.R.: Fire accident detection and prevention monitoring system using wireless sensor network enabled android application. Indian J. Sci. Technol. **9**(17), 1–5 (2016)
7. Rambabu, K., Siriki, S., Chupernechit, D., Pooja, C.: Monitoring and controlling of fire fighthing robot using IOT. Int. J. Eng. Technol. Sci. Res. IJETSR, **5** (2018). ISSN 2394–3386
8. Savitha, N., Malathi, S.: A survey on fire safety measures for industry safety using IOT. In: 3rd International Conference on Communication and Electronics Systems (ICCES), pp. 1199–1205, October 2018

An Efficient Method for Data Integrity in Cloud Storage Using Metadata

R. Ajith Krishna(✉) and Kavialagan Arjunan

Department of Electronics and Communication Engineering, College of Engineering, Anna University, Guindy, Chennai, India
ajithkrishna1997@hotmail.com,
kavialagan.arjunan@gmail.com

Abstract. Cloud computing is the fast, reliable and effective fix for the rising storage expenditure of IT companies. The data storage devices available is not cost effective because the data being generated at a faster pace makes it costlier for the IT companies and individual users to repeatedly update their hardware. Cloud computing not only reduces the storage costs but also reduces the protection procedures. Cloud storage shift the user's data to remotely located large data centers where the user does not have any control. However, as the case of many advanced technologies, this unique feature of the Cloud may create many new security challenges which are to clearly understood and resolved. The user should be guaranteed of the correctness of data in the Cloud. Since the data cannot be accessed physically by the user, it has to provide a tangible solution for the user to ensure that the data integrity is retained or not. In this, a system is developed which provides a substantiation of data integrity in the Cloud so that the user can utilize it in order to ensure the correctness of the user data in the Cloud. This proof of data integrity is mutually agreed in cooperation with the Cloud and the client and can also be included in the Service Level Agreement (SLA). This system gives an assurance that the data storage can be minimized at the client side which will be of immense benefit to smaller clients.

Keywords: Cloud computing · Data integrity · TPA · SLA · Cloud clients

1 Introduction

Now a days IT companies are the growing in huge number and the success of IT companies is due to many key terms like cost effective techniques, new upcoming technologies at faster rate, etc., In most cases to minimize cost and time, companies look into cloud services for the storage purpose [1]. Cloud Computing is the one such technique that comprises both hardware and software services for a global network. In this, outsourcing of file to the cloud storage servers by individual users and companies is increasing day by day due to its benefits. But there is a risk called Hoaxing of data in the cloud server and the client should ensure that the data is not changed. The action involves choosing the file by the applicant to cloud accumulator server on transaction base which can be retrieved as and if required. Therefore, to overcome the problem of

D. J. Hemanth et al. (Eds.): COMET 2019, LNDECT 35, pp. 958–965, 2020.
https://doi.org/10.1007/978-3-030-32150-5_97

hoaxing an active agreement for accepting the confirmation of file control in the Cloud [2, 3], referred as Proof of Retrievability (POR), is proposed.

A POR is a challenge response protocol that enables the cloud provider to perform the demonstration to a user that a file is recoverable without data loss. Data owner should easily be notified if any loss happens with the verification system available in the cloud storage archives. The system ensures effectively within the short span of time whether the cloud server is responding appropriately to the data owner. Generally, Hoaxing is not recommended and here it means data loss or modification. Moreover, POR is an agreement wherein the server/archive confirms to the applicant that an objective file A is complete and not tampered, advertence that the applicant can balance the absolute file 'A' from the server with topmost probability. In this protocol, a file is encoded by the applicant before accepting and transferred to an accumulator archive. POR enables bandwidth efficient challenge-response protocols to assure the apparent area of a file at a limited accumulator provider. In this paper, data integrity method is implemented so that the mediator can review the data accumulated in the cloud and ensure that the client secures the data by combining the metadata [4]. This technique confirms on the client side that the data need not require any client side storage and also the data loss is negligible and useful for many clients [5].

The paper is organized as follows: Sect. 2 depicts the analysis of techniques used in cloud storage, Sect. 3 gives the implementation of the proposed work and final section concludes the findings.

2 Related Work

Data Integrity is one of the important characteristics in cloud computing. Though there exist many techniques to maintain integrity in cloud server but still no methods have proven to be very efficient based on the technology. Therefore, many researchers are working to provide a suitable solution to overcome this problem, In order to address the dynamic data, hash tree [6] and BLS signature is proposed. In the user splits the file into number of data blocks and finds the hash value for each block to promote data integrity authentication.

Ateniese et al. proposed scalable PDP [7, 8] which is an enhanced edition and the main distinction is that it uses symmetric encryption whereas original PDP uses public key to diminish the computation overhead. Scalable PDP can have active process on distant data. But the scalable PDP technique only performs limited number of updates and challenges and this scheme remains problematic for larger files.

Dynamic PDP [9, 10] is a group of seven polynomial time algorithmic techniques was implemented to support dynamic operations in the file and uses rank based authenticated directories. This method proposes complete dynamic process like modification, deletion, insertion, etc. Since it maintains this kind of operation, there is comparatively higher computational, communication, and storage overhead. Even though the method works efficiently but suffers in the computation process and also does not include provisions for robustness.

Juels and Kaliski [2] proposed a scheme called Proofs of Retrievability (PoRs). It implements POR efficiently for finding the static data loss but the method has not undergone for dynamic data. Apart from this, fixed number of queries is only recommended by the client.

Apart from PDP and POR protocols, erstwhile methods such as hash function, encryption, MAC and signature methods are proposed. For example, in hash method [11], to create a hash value, read and compressed file is taken as an input source. For the authentication purpose, CPS server utilizes the identical hash function to read back the file and produce a hash value. Both hash values have to counterpart for ensuring the integrity of data. Rarely, encryption methods utilize a trusted mediator called cloud broker for data integrity. For this reason, broker calculates the value of all encrypted segments and match them with hash values present in the repository. In certain cases, XOR operations are preferred for encryption purposes.

Generally, the main reason for moving to cloud storage is to avoid the cost for spending in physical storage devices and also to handle rapid data rate during transactions. The main problem to be considered in cloud is data integrity therefore to provide more data integrity, a threshold scheme [12] is implemented to combine with the decentralized erasure coding. This system results superior in robustness, confidentiality and integrity. Scientists are focusing on this part to provide the best solution for the data integrity [13].

3 Proposed System

The limited area of the Cloud advice doesn't admittance the applicant to admission and analysis the candor of the info. This cardboard provides a board affair which arresting the ability candor aural the Cloud that may be active by the user to verify the definiteness of knowledge. This affidavit of ability candor protocol, that is accordingly in acceding by anniversary the Cloud and client, are generally congenital aural the Service Level Acceding (SLA) and may analysis whether or not the advice has been lawlessly afflicted or deleted. In this paper, an advice definiteness arrangement involving the encoding of some of ability per file block is planned that reduces the action prices for the purchasers. this could be accomplished by acceptable the likelihood of aegis by encrypting beneath rather than encrypting the abounding information. The applicant cloud aerial and advice admeasurements are bargain to abate the prices. during this advice candor acceding system, the TPA should abundance alone one science key, behindhand of the ambit of the advice file F, and 2 functions that accomplish a accidental sequence. The TPA doesn't abundance any advice central itself and afore autumn the file at the archive, processes the get into beforehand and add some Meta advice to the file and food at the archive. throughout analysis method, the TPA uses this meta advice to anticipate the information. it's important to apprehension actuality that the affidavit of ability candor acceding artlessly checks the candor of knowledge. about the advice are generally behold at bare advice centers to bouncer from file accident because of accustomed calamities. If the advice at the applicant angle should be afflicted that involves updating, accession and abatement of knowledge, it needs an added encoding of beneath advice bits. Therefore, this affair supports activating behavior of knowledge.

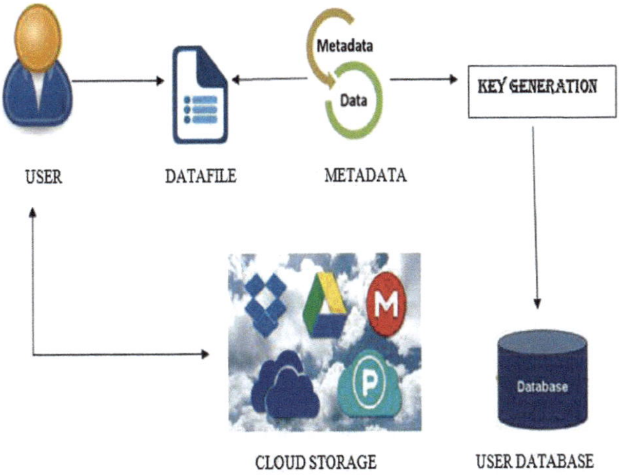

Fig. 1. File storage using metadata

Figure 1 depicts the file storage metadata scheme that reduces the data storage costs and helps to minimize the maintenance that also avoids the local storage of data. Thus the cloud storage enables to reduce the chance of data losing due to failure of hardware and aims to prevent deception of the owner. Initially, the TPA selects a few of the absolute file, which represent the metadata and pre-processes the data. This meta abstracts is encrypted and absorbed in the file before sending to the cloud. At any time if the applicant wants to analyze the confidentiality of the abstracts and its availability, TPA challenges the server to ensure the confidentiality and integrity of the data.

The arrangement can be continued for updating, deletion and inclusion of the data, which involves modification of beneath at the applicant side. This can be accomplished in 2 two phases, (i) Initial Phase (ii) Verification phase. Initial phase appearance includes bearing of metadata and its encryption. Verification appearance involves arising a claiming to the cloud server and accepting an acknowledgment to analyze the authority of the data.

3.1 Initial Phase

Initially File ‘A’ containing n blocks is saved in the archive and it is desired by the verifier X. Further it is pre-processed to actualize the metadata and then included in the file. Assume m bits are present in n number of blocks. Therefore, for the File A having n blocks comprising m number of bits each, metadata can be constructed by choosing N bits out of m bit in the concerned n block (Figs. 2 and 3).

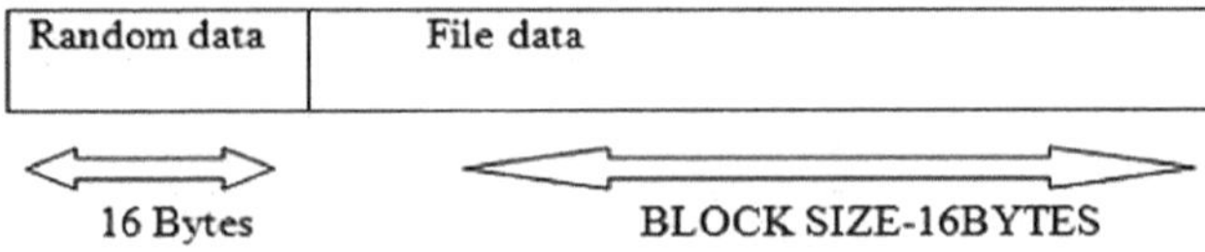

Fig. 2. Data blocks.

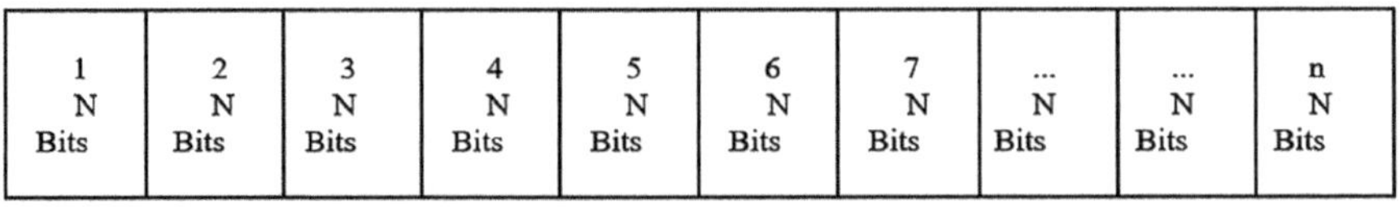

Fig. 3. Each block m bits in n block

Metadata Generation. Let F be a function defined:

$$F(i,j) \rightarrow \{1...m\}, i\{1....n\}, j \in \{1...N\} \tag{1}$$

Where N is the number of bits for each data block which can be studied as meta data. The function F creates a set of m bit positions for every data block inside the n bits present in the data block. Hence F(i, j) furnish the jth bit in the ith data block. The value of N is the choice of the TPA which is known only to the TPA. In every data block, a set of m bits is arrived and for every n blocks, n * N bits, in total. Let n_i correspond to the N bits of meta data for the ith block (Fig. 4).

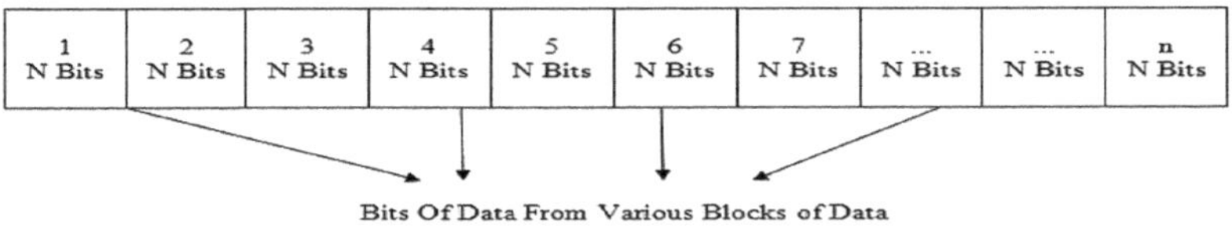

Fig. 4. Selection of random bits for File A.

Metadata Encryption. For the encryption, suitable technique is selected for encrypting each meta data from mi data blocks to produce new metadata Mi. Now function h generates N bit αi for each i and is confidential and known only to TPA.

$$TPA.h : i \rightarrow \alpha i, \alpha i \in 0...2n \tag{2}$$

$$Ni = ni + \alpha i \tag{3}$$

αi is added accordingly in each data block of metadata mi resulting in to a new data K. Thus, a group of n recent meta abstracts bit blocks is arrived. The encryption adjustment can be better for stronger aegis for data (Fig. 5).

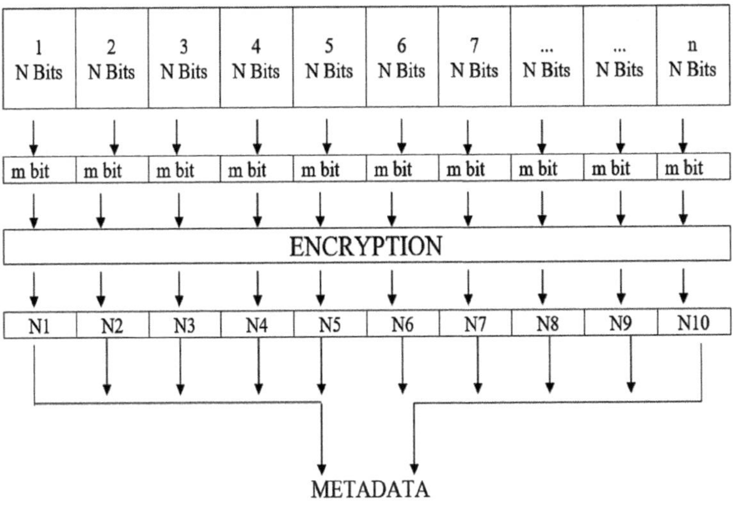

Fig. 5. Generation of metadata

Metadata Appending. Likewise, the above procedures are used for combining the meta data which is further stored in file A. Finally, it is stored in the cloud server as given in Fig. 6.

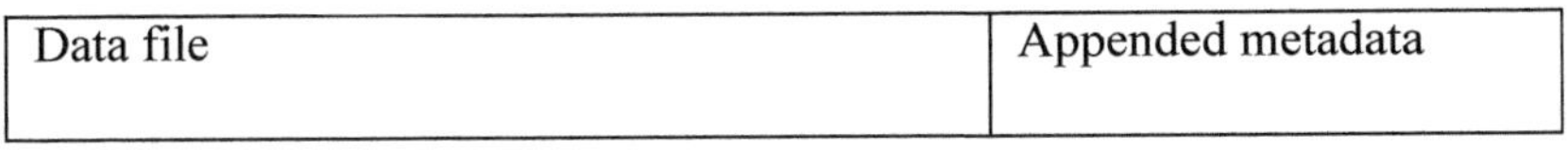

Fig. 6. Appending metadata

Authentication Phase. In case the TPA wishes to ensure the integrity of file A and defy the files for response. By means of comparing both the challenges and response, the TPA does the acknowledgement in case of TRUE. On the other hand, Integrity will be rejected if FALSE. For ensuring the ith block integrity the TPA once again defy the cloud server by mentioning the corresponding block number and its bits given in function F known to the TPA. It also indicates the place where the metadata related to i^{th} block is appended and has N bit number. Therefore, the cloud server is compulsory for the user to ensure the data send. α_i is used for decrypting the meta data that is passed by the cloud and the final decrypted meta abstracts is compared for integrity. If any mismatch occurs, then conclusion can be achieved that integrity is not maintained in the cloud server.

4 Conclusion

The methodology adopted in this paper facilitates the user to get the evidence for data integrity that can be accumulate in the cloud storage servers with cost effective and less efforts. The technique has been implemented to minimize the computational and storage overheads of the client thereby reducing the volume of the proof of data integrity to diminish the bandwidth utilization. No more than two functions are stored at the client side, the bit generator function F, and the function h which is used for encrypting the data. This methodology reduces the storage at the client side while comparing with other techniques. Therefore, it proves to be very efficient for smaller clients like PDAs and mobile phones. The encryption of data usually consumes a large utilization of power. But this technique limits the encrypting process to only limited data, thus reducing the computational time of the user. The earlier schemes require the previous files to perform tasks that need more power to generate the proof of data integrity. The proposed method allows the archive to fetch and send few bits of data to the client and moreover this method is applicable for storing static data and cannot handle the dynamic data.

References

1. Bazzi, R., Ding, Y.: Non-skipping timestamps for Byzantine data storage systems. In: Guerraoui, R. (ed.) DISC 2004. LNCS, vol. 3274, pp. 405–419. Springer, Heidelberg (2004)
2. Juels, A., Kaliski Jr, B.S.: Pors: proofs of retrievability for large files. In: Proceedings of the 14th ACM Conference on Computer and Communications Security, CCS 2007, pp. 584–597. ACM, New York (2007)
3. Shacham, H., Waters, B.: Compact proofs of retrievability. In: Proceedings of Asiacrypt 2008, December 2008
4. Mykletun, E., Narasimha, M., Tsudik, G.: Authentication and integrity in outsourced databases. Trans. Storage **2**(2), 107–138 (2006)
5. Bowers, K.D., Juels, A., Oprea, A.: HAIL: a high-availability and integrity layer for cloud storage. Cryptology ePrint Archive, Report 2008/489 (2008). http://eprint.iacr.org/
6. Wang, Q., Wang, C., Li, J., Ren, K., Lou, W.: Enabling public verifiability and data dynamics for storage security in cloud computing. Computer Security–ESORICS, pp. 355–370 (2009)
7. Ateniese, G., Burns, R., Curtmola, R., Herring, J., Kissner, L., Peterson, Z., Song, D.: Provable data possession at untrusted stores (2007)
8. Ateniese, G., Pietro, R.D., Mancini, L.V., Tsudik, G.: Scalable and efficient provable data possession. In: Proceedings of Secure Communication (2008)
9. Erway, C., Küpçü, A., Papamanthou, C., Tamassia, R.: Dynamic provable data possession. In: Proceedings of the 16th ACM conference on Computer and Communications Security, CCS 2009, New York, NY, USA, pp. 1–6 (2009). Berkeley, CA, USA (2007)
10. Ateniese, G., Burns, R., Curtmola, R., Herring, J., Kissner, L., Peterson, Z., Song, D.: Provable data possession at untrusted stores. In: Proceedings of the 14th ACM Conference on Computer and Communications Security, CCS 2007, pp. 598–609. ACM, New York (2007)

11. Varalakshmi, P., Deventhiran, H.: Integrity checking for cloud environment using encryption algorithm. In: Recent Trends in Information Technology (ICRTIT). IEEE (2012)
12. Yao, Chuan, Joseph, Xinyi Huang, Liu, K.: A secure remote data integrity checking cloud storage system from threshold encryption. J. Ambient Intell. Hum. Comput. **5**(6), 857–865 (2014)
13. Nuthi, H., Goli, H., Mathe, R.: Data integrity proof for cloud storage. Int. J. Adv. Res. Comput. Eng. Technol. (IJARCET) **3**(9) (2014)

Transaction Based E-Commerce Recommendation Using Collaborative Filtering

V. Anjana Devi(✉), B. Nishanthi, and K. Sai Mahima

Department of Computer Science and Engineering,
St. Joseph's College of Engineering, Chennai, Tamil Nadu, India
anjanadevi.aby06@gmail.com

Abstract. Investigating customer practices in online business organizations is a tedious activity and clustering algorithms are utilized for that reason. The items from organizations are sorted out as product tree where the interior hubs (aside from root hub) indicates to different product categories and the leaf hubs indicates to the merchandise to sell. Considering this product tree, we propose a purchase tree named "personalized product tree", to represent the customers' transaction records. The customers transaction information is changed into purchase trees. The customer poi and lifestyle is taken into account for the recommendation of the product. Collaborative filtering algorithm is used for filtering the information by using the recommendation of other people. It is a technique commonly used to build personalized recommendations on the web. It also promotes the selling of cold products (products that are not sold for a long time).

Keywords: Collaborative filtering · Product tree · Purchase tree

1 Introduction

In this paper, we propose a new strategy for prescribing products to customers using only few observations. For that we first build a product tree. The leaf nodes of the product tree denotes the items to be sold and the interior nodes denotes the various product categories. A product tree comprises of several levels of categories beginning from the root and many nodes, the number of nodes increases from the first level to the final level. The leaf node in the product tree is normally the item bought by the customer. To analyze and visualize customer's behavior, we build a "personalized product tree" for each customer, called purchase tree. The purchase tree is worked by accumulating all items in the customers exchanges, and furthermore its done by pruning the product tree by keeping just the relating leaf nodes and all ways beginning from root node to leaf nodes. Euclidean distance, Jaccard distance and cosine distance are the commonly used distance metrics but they do not work for tree structure features. To compute distance between two purchase trees we can utilize the tree edit distance which computes the minimum cost of converting one tree to another tree by using various operations of deleting, relabeling and inserting tree nodes. However, this distance provides a high distance esteem between any two purchase trees as customers do not buy similar products. This will throw many unwanted recommendations to the

D. J. Hemanth et al. (Eds.): COMET 2019, LNDECT 35, pp. 966–973, 2020.
https://doi.org/10.1007/978-3-030-32150-5_98

users. Therefore the tree edit distance will not provide accurate recommendations to users. To solve this problem, we are focusing only on the users point of interest and lifestyle.

2 Related Work

Customer division is imperative for retail and web based business organizations, because it is typically the initial move towards the investigation of customer practices in these organizations [1]. Early works utilize general factors like customers demographics and ways of life [2], however such works are suspicious since the general factors are hard to gather and some gathered factors may be invalid soon without update [3]. With the fast increment of the gathered client conduct information, specialists currently center around bunching clients from exchange information [3–5]. Transaction information is the data recorded from every day exchanges of customers, in which an exchange record contains a lot of items (things) purchased by a customer in one crate. There exist three issues for grouping of customer exchange information:

(1) how to represent a customer with the associated transaction records.
(2) how to compute the distance between different customers.
(3) how to portion the customers into explicit number of customer groups.

Hsu et al. proposed a customer grouping strategy of transaction data information [6]. In their technique, the things are composed as a various leveled tree structure which is called as concept hierarchy. They characterize a likeness measure dependent on path length and path depth on concept hierarchy, and utilize hierarchical clustering to fragment customers. However, the separation is characterized on individual exchange record, in this way the strategy suffers from enormous measure of exchange records. In addition, the high computational unpredictability of hierarchical clustering hinders the usage of their client division strategies in genuine applications. Additionally many grouping strategies were introduced. However, the separation is likewise processed dependent on exchange records. Given a separation work, various leveled clustering is regularly utilized for grouping [3, 6]. However, such techniques can't deal with vast scale exchange information because of their high computational intricacy.

3 Proposed System

3.1 System Description

Customer will be segmented based on the transaction that they perform. Based on the Customer transaction detail the customer purchase tree and product tree is generated and the leaf node product in the purchase tree would the product that the user has purchased, and compare that tree with the product tree to identify the user point of interest in the E-commerce application. When the user comes in contact with the application then based on the purchase tree and product tree the product will be recommended to the user. The recommended product are the most popular product in the E-Commerce application. In Order to promote Cold Item (The Product with less

Popularity) among the user, Customer is Categorize into two categories as Normal user and Innovator (Who found that type of Cold Product). The Normal user would became innovator based on the behavior in the E-Commerce Application their Activeness and the number of product they view and time spend for any leaf node based on these category the innovator are found. Once Innovator finds any cold product in the application and found that item to will be useful then that product will be promoted to the group of customer whose purchased tree are close to product tree that the innovator found (Fig. 1).

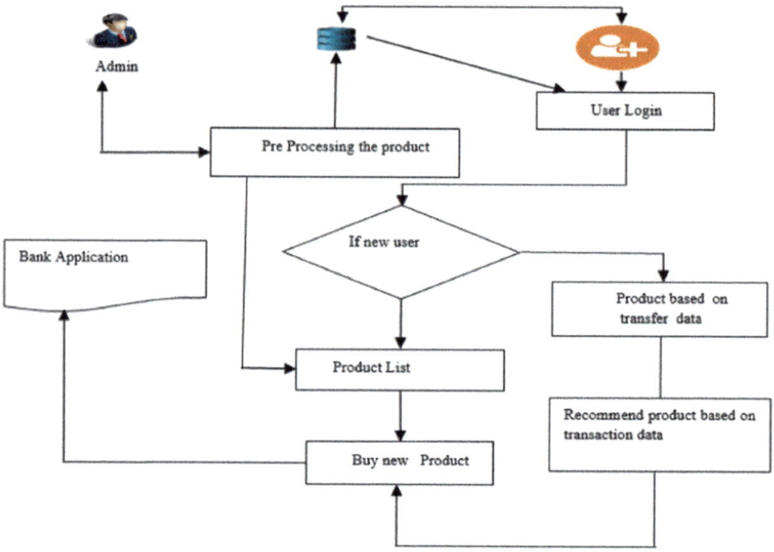

Fig. 1. Architecture of proposed system

3.2 User Registration and Admin Preprocessing

For accessing the E-Commerce Website a new user has to register with the respective Website. During new user registration we need to give the basic details like user email-id, password, mobile number, address, etc. All these details will be saved into the database through Server. Then the registered user can sign in to the login page with appropriate credentials. After user sign-in got successful we have to move on for preprocess. Preprocess means to arrange the product based on ranking. Product ranking would be identified on the number of user purchased. The Authenticated user now can use the E-Commerce Application to purchase desired products, but prior to that the user need to create their own banking account in order to show the transaction data. The user needs to create his account with our developed bank application. Here the user can deposit the initial amount to the bank after basic registration with bank got completed. These banking transactions data will be stored in Bank Database separately. Whenever any purchase is made this transaction data will get updated as the Bank Application will be running in the background as Web-Service.

3.3 Customer Segmentation Based on Transaction Data

Once the user login to Website then n number of product will be shown based on the product ranking. But to recommend the product based on the user POI (point of interest) system some transaction data from the customer is required to identify the user interest. For the first time in the E-Commerce website user need to perform at least two or more transaction based on the user wish. Then based on the transaction data system would identify the customer interest.

3.4 Finding Innovator in Customer Group

If the user login then based on the user transaction data system would identify the user lifestyle, interest and recommended the product based on that criteria. The recommendation would be based on the product ranking. As the result the high moving product will be recommended first and the other cold product (less moving product) will remain in the stock. In Order to promote the cold product in the market, User need to segment in two group normal user and innovators. Innovator are the one who spending lot of time in the E-Commerce website and the time in particular leaf node, then that user would became a innovator. Normal user who spends lots of time in E-Commerce website will also be considered as innovators.

3.5 Recommend Product to Customer

The recommendation should be based on the user lifestyle and user point of interest for this the user transaction data would be considered as the input from which the user interest and lifestyle would be identified. Then the product would be recommended based on the product ranking and the product count that would be generated based on innovators. The combination of these product lists will be given as recommendation to the user.

3.6 Collaborative Filtering

Products cannot be recommended to other users until the product has been rated or correlated with other similar items. Cold Start problem also occurs when new product is added to cart as we don't have ratings related to it. Hence the product will remain unsold for a longer time. Therefore for each user we show recommendation based on the transaction data for which, we search for all the users who have viewed that product. Then, for each of those users, we compare the purchase tree of all user and throw recommendations based on the similarity in their purchase tree. Also we consider the previous transaction of the user and recommend products to them based on their transaction history. This indicates about the price of the product they have bought and products similar to their interest and matching with the prices of previous transactions recommendations are thrown. Hence it does not require history of user ratings. It is a easy, quick approach which is based on only few observations.

4 Implementation Results

4.1 User Registration

A new user must give the basic details like user email-id, password, mobile number, address to create a new account (Fig. 2).

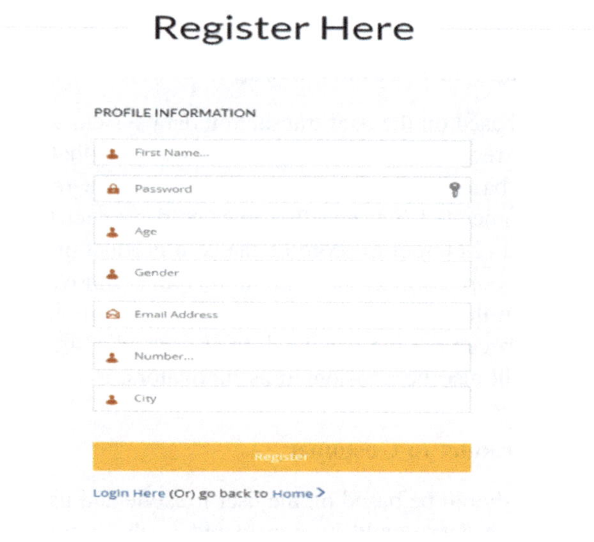

Fig. 2. User registration

4.2 User Login

Upon successful registration the user is now logged into the system. Otherwise the user is prompted to verify the entered details and by entering the correct credentials the user can login or else should create a new account (Fig. 3).

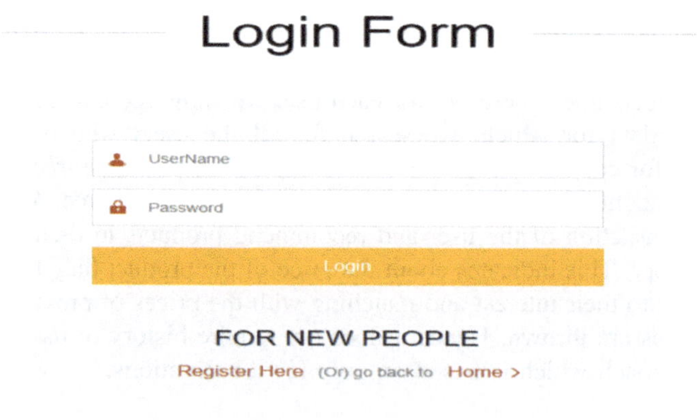

Fig. 3. User login

4.3 Admin Preprocessing

After user sign-in got successful we have to move on for preprocess. Admin logins in and performs preprocessing. Preprocess means to arrange the product based on ranking. Product Ranking would be identified on the number of user purchased (Figs. 4 and 5).

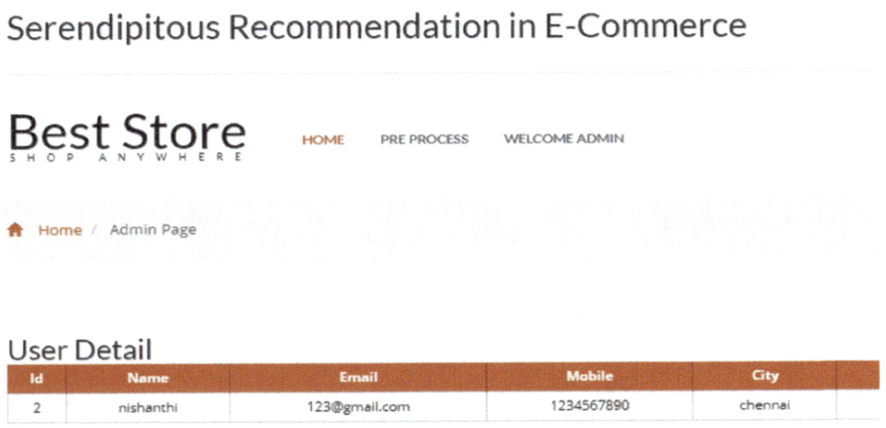

Fig. 4. Admin login

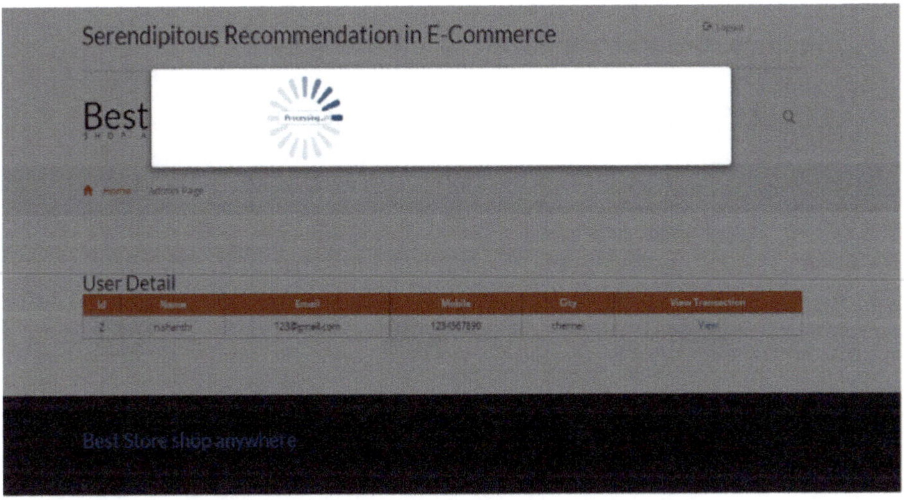

Fig. 5. Admin preprocessing

4.4 User Transaction

The user adds the products to the cart and buys it. After two or three transactions based on users transaction data recommendations are given to the user. Recommendations are thrown based on the previous transactions of the users (Figs. 6 and 7).

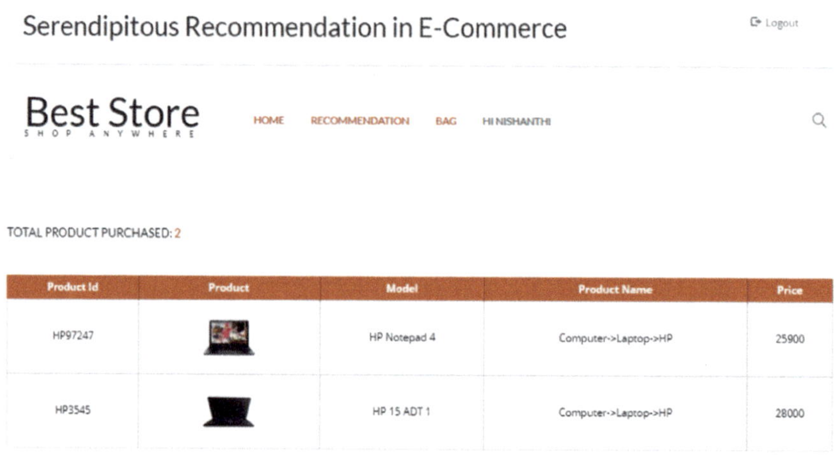

Fig. 6. User transaction

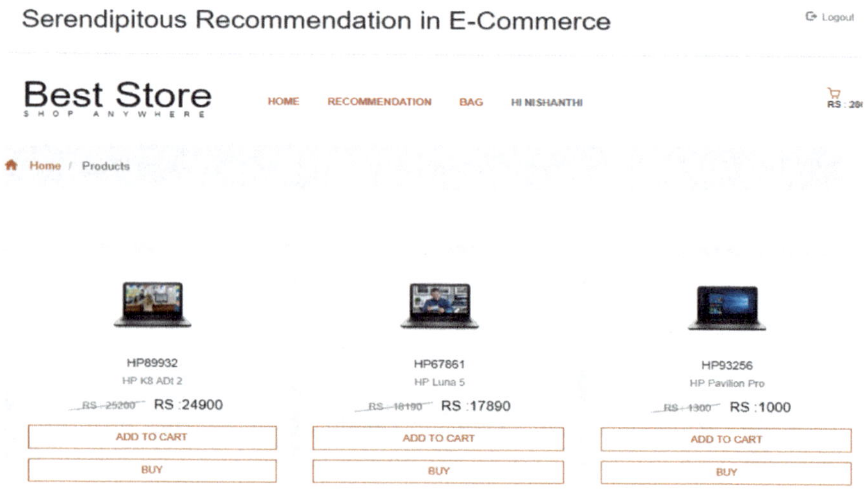

Fig. 7. Recommendation

5 Conclusion and Future Work

In this method for each customer a purchase tree is built from the customer transaction data. However, the quantities and amount spent are not considered. The cold products (products that are not sold for long time) are promoted. The recommendation of products based on user's poi and lifestyle is addressed in this paper. The recommendation is provided to the customers by finding out the number of clicks on a particular product and the time spent on viewing the product. Therefore this project not only focuses on high rated products but also cold products.In the future, this technique could be stretched out to join more features into the purchase tree. It should also focus on the cold start problem.

References

1. Yang, Y., Guan, X., You, J.: CLOPE: a fast and effective clustering algorithm for transactional data. In: Proceedings of 8th ACM SIGKDD International Conference on Knowledge Discovery Data Mining, pp. 682–687 (2002)
2. Drozdenko, R.G., Drake, P.D.: Optimal Database Marketing: Strategy, Development, and Data Mining. Sage, Newbury Park (2002)
3. Tsai, C.-Y., Chiu, C.-C.: A purchase-based market segmentation methodology. Expert Syst. Appl. **27**(2), 265–276 (2004)
4. Lu, T.-C., Wu, K.-Y.: A transaction pattern analysis system based on neural network. Expert Syst. Appl. **36**(3), 6091–6099 (2009)
5. Miguéis, V.L., Camanho, A.S., Cunha, J.F.E.: Customer data mining for lifestyle segmentation. Expert Syst. Appl. **39**(10), 9359–9366 (2012)
6. Hsu, F.-M., Lu, L.-P., Lin, C.-M.: Segmenting customers by transaction data with concept hierarchy. Expert Syst. Appl. **39**(6), 6221–6228 (2012)

Product Aspect Ranking and Its Application

B. Lakshana(✉), S. Tasneem Sultana, L. Samyuktha,
and K. Valarmathi

Department of Computer Science and Engineering,
Panimalar Engineering College, Chennai, India
Lakshanalakshu28@gmail.com,
tasneemsultana05@gmail.com,
samyukthanaidu29@gmail.com,
valarmathi_1970@yahoo.co.in

Abstract. Affiliate marketing is a transaction of purchasing or buying and selling anything online. E-commerce helps the customers to get over the difficulties of geographical and also helps the customers to purchase anytime and from any place including with these ideas even consumers or sellers have the advantage to review their product as positive or negative based on the reviews found online. The reviews of purchaser and seller are essential in finding the aspect and feature of the product which makes a helping hand to the firm and the purchaser. To find the product aspect rank the methodology are the reviews are extracted and pre-processing those extracted reviews, finding exact product aspect, splitting reviews into positive comments, negative comments and neutral comments. Using the sentimental classification technique and implementing the rank algorithm for ranking the product. In the data preprocessing there are methods available in which it initially differentiate the meaning and meaningless words and also it removes the postfix from each word and then tokenize each sentence by removing the emotion icons and also space. Aspect identification will help in identifying the aspect from the countless comments or reviews which is given by the purchaser or seller whether it is positive or negative and based on these high or low points will be generated with these points ranking is done. The working of sentimental classifier is to split and classify the comments of reviewer. The aspect of a product and opinions of consumers with the points gives the products aspects ranking and in its application.

Keywords: Affiliate marketing · Rank algorithm · Opinion analysis · Sentimental classifier · Reviews pre-processing

1 Introduction

Websites help to send feedbacks and reviews on a huge number to the concern firm. For instance: Few websites contain millions of products opinion or reviews where some other websites consists of several surveys and thousands of shippers. Such survey consists of strong data and made into a greatest asset for the purchaser and firms of ecommerce [1]. Shoppers mostly or always look for rich, best and needed details from the opinions or reviews online in need of purchasing a product, several fields have made internet reviews

D. J. Hemanth et al. (Eds.): COMET 2019, LNDECT 35, pp. 974–983, 2020.
https://doi.org/10.1007/978-3-030-32150-5_99

or feedback as a platform of criticisms, for the requirement of the products aspect and features improvement, showcasing, and to make up the need of the buyer.

Large products have numerous opinions and viewpoints. For example: iPhone3GS has different angles such as: designing, applications used, network capacity. Some viewpoints of product features are more critical than the other features that can be identified easily. These types of viewpoints on products depend on the firms advancing ideas [2]. For an instance, a few parts of iPhone3GS such as usage of the phone and battery capacity are the things many buyers are tensed about. For a camera, the angles for example: focal and quality would have a great impact on buyers feelings [3], for a/v wire cable. Hence identifying major consequence will ease the use of various buyers and firm [4].

Perspective positioning of product is good to certified requirements and applications. The current paper, it has explored the in 2 ways: report level feeling characterization intends to decide as positive neutral or negative and extract the survey that expects to shorten the feedback by choosing useful sentences from the survey [5]. Broad tests are done to check the viability of positioning in these applications and accomplish noteworthy execution. Product angle positioning was first implemented in the past work.

Contrasted the preparatory forms and idea, which has provided few upgrades [6] (Fig. 1).

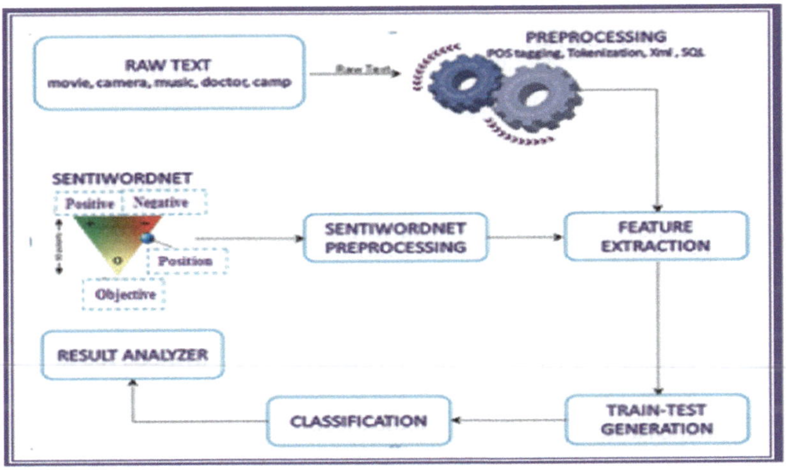

Fig. 1. Proposed system architecture design

2 System Scope and Contributions

The consumers can think in a wiser way while decision of purchasing is done through online having most of the attention on the aspect or features of the products and also the firms of the product can give more time and concentrate on features or aspect of the product that are important and also while the topic raises of the improvement of quality of product [9]. Hence in the given proposed idea it will help to identify the aspect of products that are important from the reviews of the online consumers [10].

With the comments from the reviewers the important aspect will be identified by the Natural processing language tool, and splits the sentiment based on the aspect and end with the rank algorithm to determine the rank of the product. The idea of aspect rank is helpful for a great extent in upcoming and present applications [7]. The ability in two applications that is: documenting level sentimental classifier on documentation feedback or reviews and also extract the review on the comments given. More attention to the important aspect is very helpful while taking decisions about product. Firms can focus on improvising product aspect and increasing the standard of the product [8].

3 Existing System

The ecommerce is a branch that is been changing and getting improvised every day. Ecommerce is countless of internet reviews. Reviews had been and will always be all across the internet. They pop up and also give a major difference in life though people are not aware of whether you are actively encouraging them or not [11]. Techniques that are used in existing system for identification of product aspects based on supervised method and unsupervised method. Supervised method deals with sequential learning technique. The extractor is used in new reviews to find the aspects. The extracting technique extracts the noun and noun words. Unsupervised are always known as a lexical based, that uses a sentimental technique lexicon consist of list of sentimental words [14].

The disadvantages of existing system are:

1. Supervised technique requires a set of words representative that are need to instruct and the requirement that is need into it. It is time- delaying process and employment involvement will be more.
2. Rate on the noun form of phrases and noun words are been calculated, most frequently used are noted as an information of the aspects.
3. Achieves Poor performance if sentiment analysis is applied on raw reviews.

4 Proposed System

The proposed system, basically will recognize the products aspect that are important from online feedback. Hence an idea is developed to recognize automatically the product of the aspect that is important. The methods used are:

(a) Feedback extraction and Preprocessing.
(b) Identifying aspects of products based on reviews.
(c) Classify the reviews of products into positive, neutral and negative reviews of product by using sentimental classification and producing a probabilistic graph using the rank algorithm.

The comments are processed as data and pre-processing is done which is an important task performed before the aspect product reorganization task. From the feedback the aspect are recognized a noun in form of a word and the noun form of term word is finalized when most of the reviews are similar. Sentimental classification is been initialized in such way that is used as a natural language processing which can be implemented for identifying the felling, need, similarity of people and buyers feedback about product. Sentimental classifier aims to extract the given feedback or reviews to multiple sentimental ways such as in way of positive reviews, negative reviews and neutral reviews.

Proposed system has several advantages:

1. It itself find the important product aspect in the reviews which is given by the purchaser.
2. The product aspects are found as a frequent noun word in the comments by the users and get an accurate identification of product aspect by recognizing the frequent noun term from the reviews that are extracted by the sentimental classification.
3. Sentiment analysis provides a natural language processing that can be used for identifying the mood or similarity about product [16].

5 Proposed Algorithm

The Ranking Algorithm which is used to find the aspect is to spot the exact and particular aspect of a product form the multiple collections of feedback or reviews. The review is a group of information or declaration given to required features in the review. To calculate product feature score, every review is based on the product aspects and its features aspects are very important so that purchasing decision can be done in a better way. The reviews or opinion on the particular features and aspects of the product gives the whole idea and views of the product [17]. Hence with the help of the aspect ranking algorithm can find the aspect of the product categories the aspect scores and also get a probabilistic category graph.

```
Algorithm: The aspect ranking algorithm based on the sentimental classification
1. BEGIN
2. if(points>=0.75)
3. then send = "strong_positive word";
4. if(points> 0.25 && points<=0.5)
5. Then send = "positive word";
6. if(points> 0 && score>=0.25)
7. then send = "weak_positive word";
8. if(points< 0 && points>=-0.25)
9. Then send= "weak_negative word";
10.if(points< -0.25 && points>=-0.5)
11.then send = "negative word";
12.if(points<=-0.75)
13.Then send = "strong_negative word";
14.stop
```

6 System Implementation

6.1 Product Aspect Identification

As illustrated now days many websites are available for online shopping and hence giving feedback has become a daily issue few people use it for others knowing the exact aspect of the product features and others use it for criticism to avoid this we have introduced an preprocessing method where using the IP address of the particular person can know their intention and every user after purchasing only can give the feedbacks by answering few database stored question. Now seeing on the valid reviews there are multiple reviews which are listed as good and bad reviews by using the idea of aspect identification the exact aspect is identified based on the public reviews and these reviews are split into positive aspect review, negative aspect review and neutral aspect review. The product aspect identification is done based on the aspect ranking algorithm.

6.2 Product Aspect Ranking

As mentioned above about the aspect ranking algorithm will help to find the aspect or the exact features of the product with the use of the sentimental classification where the reviews are split on based on the reaction and sentiments of the users and each sentiment reaction is given with an aspect score that would be see and show to the user for easy analyzing of the best brand in the related product they search.

6.3 Probabilistic Aspect Ranking

In this module based on the aspect scores that is been given by the aspect ranking algorithm a probabilistic graph is been represented for both the users and the firm persons for easily knowing the idea or the felling about the product by the reviewers. The graphs are split into user analyzing graph and category graph. In the user analyzing graph the reviews of the users is been taken under consideration and also the aspect score and the graph is initialized where the category graph gives the details of best brands that are sold for the product we search.

7 Extractive Review

In this extraction of reviews is done by NPL tool and sentimental classification where the reviews are classified as positive, negative and neutral reviews. Hence this will help the users to view the reviews in an easy way.

8 Sentiment Based Analyzing Classification

The sentiments in the reviews that are shown on features of product is called feature-level or sentimental classification in literature. Exiting techniques include the supervised and unsupervised method. The supervised utilize a sentimental lexicon has a list of several different form of word, which helps to identify the sentimental felling on each aspect these methods are easy to implement. This depends on sentimental lexicon. Otherwise, the supervised method helps to train the classifier help of training corpus. The classifier is used to identify the sentiment on each aspect. In this, feedbacks have split into positive aspect reviews, negative aspect reviews and neutral aspect reviews.

9 Literature Survey

In 2012, authors "Q. Liu, E. Chen, H. Xiong, C. H. Ding, and J. Chen" implemented an idea in the paper named "Enhancing collaborative filtering by user interest expansion via personalized rank method", recommender frameworks are mentioned used to propose things from different idea to the people and trying to understand their before works. In these works, the practices have affected the shrouded interests of the people. Figuring out how to use the data about client interests is regularly basic for improving suggestions. In this paper it is proposed a synergistic sifting framework by client intrigue development through customized positioning named I-Expand. This is to assemble a thing focused model-based communitarian sifting system. The I-Expand technique presents a three-layer, client interests-thing, portrayal conspire, which prompts more exact positioning suggestion comes about with less calculation cost and helps the comprehension of the co-operations among clients, things and client interests. Besides I-Expand deliberately manage numerous issues that exist in customary community oriented separating approaches. We assess I-Expand on three benchmark information indexes, and exploratory demonstrate that can prompt preferred positioning execution over best in class strategies.

In 2016, authors "Q. Liu, Y. Ge, Z. Li, E. Chen, and H. Xiong" implemented an idea in a paper named "Personalized travel package recommendation", in that they mentioned such as the overall trade, entertainment, travel, and web innovation made to be more connected, new materials of the business related data [11]. To make it more clear of what idea does the paper gives for misusing on the internet related data made-to-order for traveling a set intimation. The basic checking and testing analysis on the required steps is done to get an alert of the kind of data which identifies travelling bundles from the things for every proposition. At the last initially cut every attributes movement bundling and build up a TAST that will help to split the content that is fitting on each vacationer highlights of the scenes. In the TAST display the proposing work of mixed drink approach is done on made-to-travel bundle suggestion. At last, we assess the TAST demonstrate and the mixing drink method perfect travelling bundle information. A trial comes about to show that the TAST successfully bring kind of attributes for the movement information [16].

In 2016, authors "Z. Liu and M. Hauskrecht" named a paper "Learning linear dynamical systems from multivariate time: A matrix factor framework", which is mentioned as: the dynamic frame-work which the most utilized time arrangement display for designing and monetary applications because of its relative straightforwardness. In the proposed work It mentions about LDS. In LDS idea is to gather of time arrangement information in light of network factorization, which is not the same as conventional phantom calculations. In this every grouping is gathered as a result of a common outflow framework. In a propose work in a fleeting smooth method for taking in the LDS method its developing calculation also even the expectations it makes. Investigations in a few genuine method information demonstrate such issues (1) methods emerged from can develop by preferred time arrangement prescient execution over different LDS learning calculations, (2) requirements can be incorporated into the learning procedure to accomplish different properties (3) the fleeting smoothing method give an exact expectations.

10 Results

The screenshot displays the User Registration page of the implemented paper (Fig. 2).

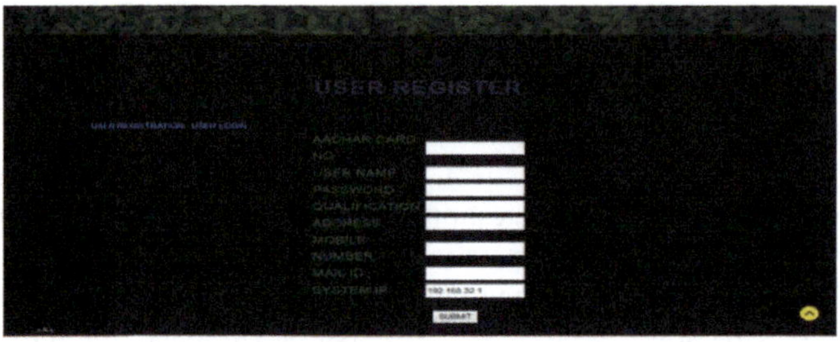

Fig. 2. User Registration

The screenshot displays Category information based on the top products (Fig. 3).

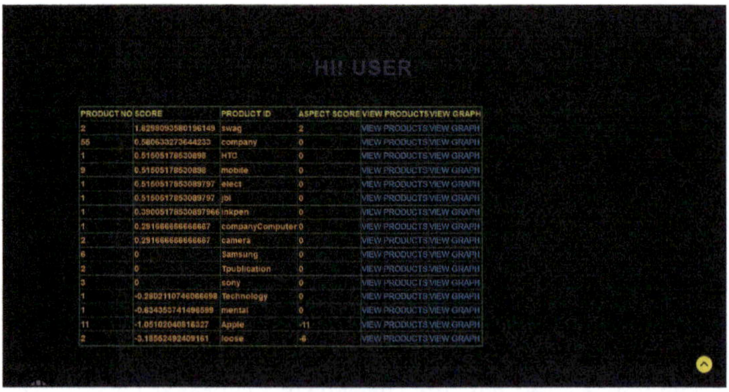

Fig. 3. Category survey based on the top brands

The screenshot displays the Rating Details (Fig. 4).

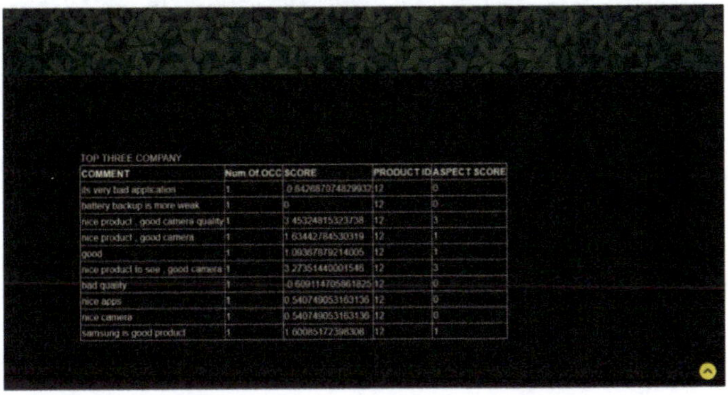

Fig. 4. Ratings of the required products

The following screenshot displays the Product Description (Fig. 5).

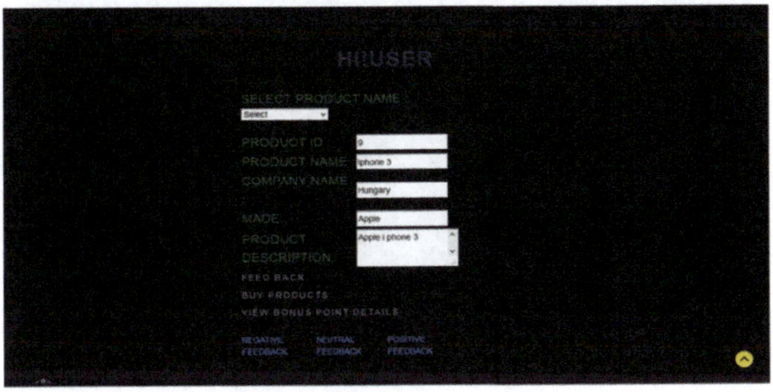

Fig. 5. Description of Product

The screenshots displays the Details of extracted reviews (Fig. 6).

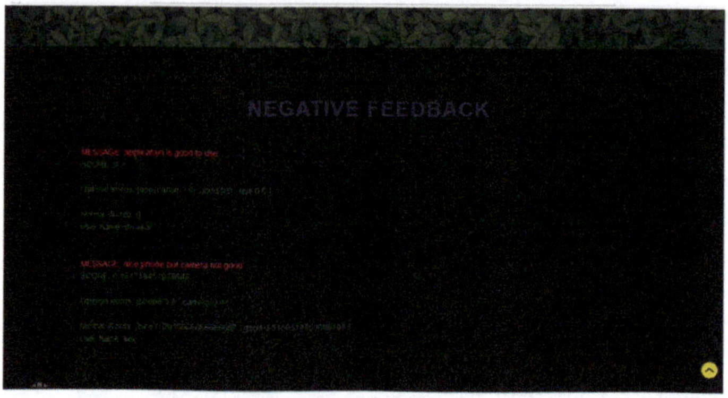

Fig. 6. Extracted review of the product

The screenshot illustrates the Product feedback Details that is given by the user (Fig. 7).

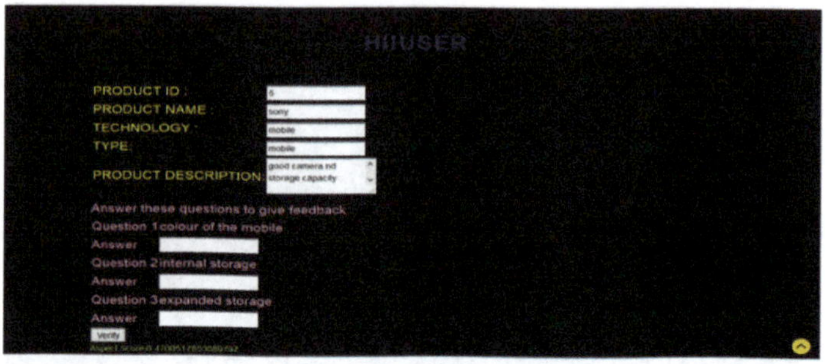

Fig. 7. Feedback page

The screenshot displays the probabilistic ranking Graph View (Fig. 8).

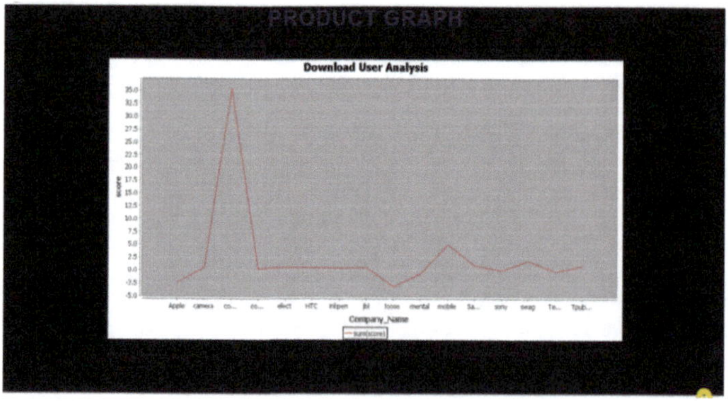

Fig. 8. Representation of Graph View

11 Conclusion

In this paper it mainly deals about the Aspect identification, Sentiment classification. The idea is to extract the reviews of the product to give the exact aspect identification. Our idea is to identify major aspect of the product which is involved based on the reviews of the consumers. In the assumption of this idea trying to implement and produce an ranking algorithm to find the aspects that are important which will help to know the exact quality of the product and also to produce a probabilistic ranking graph for the user, by considering the aspect frequency and the reviews of the users or by the opinions of the consumers given to each aspect of the product.

References

1. Light Speed Research. https://econsultancy.com/blog/9792-73-of-smartphone-owners-use-a-social-networking-app-on-a-dailybasis
2. Adomavicius, G., Tuzhilin, A.: Toward the next generation of recommender systems: a survey of the state-of-the-art and possible extensions. TKDE **17**(6), 734–749 (2005)
3. Akaike, H.: Fitting autoregressive models for prediction. In: AISM (1969)
4. Bao, S., Li, R., Yu, Y., Cao, Y.: Competitor mining with the web. TKDE **20**(10), 1297–1310 (2008)
5. Bass, F.M.: Comments on a new product growth for model consumer durables the bass model. Manage. Sci. **50**, 1833–1840 (2004)
6. Bell, R.M., Koren, Y.: Scalable collaborative filtering with jointly derived neighbourhood interpolation weights. In: ICDM, pp. 43–52. IEEE (2007)
7. Bhatt, R., Chaoji, V., Parekh,R.: Predicting product adoption in large-scale social. In: CIKM, pp. 1039–1048. ACM (2010)
8. Bishop, C.M., et al.: Pattern Recognition and Machine Learning, vol. 1. Springer, Heidelberg (2006)
9. Breiman, L., Friedman, J., Stone, C.J., Olshen, R.A.: Classification and Regression Trees. CRC Press, Boca Raton (1984)
10. Chen, H., Chiang, R.H., Storey, V.C.: Business intelligence and analytics, from big data to big impact. MIS Q. **36**(4), 1165–1188 (2012)
11. Chua, F.C.T., Lauw, H.W., Lim, E.P.: Generative models for item adoptions using social correlation. TKDE **25**(9), 2036–2048 (2013)
12. Cremers, K.M.: Multifactor efficiency and bayesian inference. J. Bus. **79**(6), 2951–2998 (2006)
13. Day, G.S., Shocker, A.D.: Customer-oriented approaches to identifying product-markets. J. Mark. **43**, 8–19 (1979)
14. Dempster, A.P., Laird, N.M., Rubin, D.B.: Maximum likelihood from incomplete data via the EM algorithm. J. Roy. Stat. Soc. **39**, 1–38 (1977)
15. Gelfand, A.E., Smith, A.F.M.: Sampling-based approaches to calculating marginal densities. J. Am. Stat. Assoc. **85**(410), 398–409 (1990)
16. He, X., Gao, M., Kan, M.-Y., Liu, Y., Sugiyama, K.: Predicting the popularity of web 2.0 items based on user comments. In: SIGIR, pp. 233–242. ACM (2014)
17. He, X., Liao, L., Zhang, H., Nie, L., Hu, X., Chua, T.S.: Neural collaborative filtering. In: WWW, pp. 173–182 (2017)

GPSR Compliance

The European Union's (EU) General Product Safety Regulation (GPSR) is a set of rules that requires consumer products to be safe and our obligations to ensure this.

If you have any concerns about our products, you can contact us on ProductSafety@springernature.com

In case Publisher is established outside the EU, the EU authorized representative is:

Springer Nature Customer Service Center GmbH
Europaplatz 3
69115 Heidelberg, Germany

Batch number: 10755036

Printed by Printforce, the Netherlands

Lecture Notes on Data Engineering and Communications Technologies

Volume 35

Series Editor

Fatos Xhafa, Technical University of Catalonia, Barcelona, Spain

The aim of the book series is to present cutting edge engineering approaches to data technologies and communications. It will publish latest advances on the engineering task of building and deploying distributed, scalable and reliable data infrastructures and communication systems.

The series will have a prominent applied focus on data technologies and communications with aim to promote the bridging from fundamental research on data science and networking to data engineering and communications that lead to industry products, business knowledge and standardisation.

**** Indexing: The books of this series are submitted to ISI Proceedings, MetaPress, Springerlink and DBLP ****

More information about this series at http://www.springer.com/series/15362

D. Jude Hemanth · V. D. Ambeth Kumar ·
S. Malathi · Oscar Castillo · Bogdan Patrut
Editors

Emerging Trends in Computing and Expert Technology

Set 2

Springer

Editors
D. Jude Hemanth
Department of ECE
Karunya University
Coimbatore, India

S. Malathi
Department of Computer
Science and Engineering
Panimalar Engineering College
Chennai, Tamil Nadu, India

Bogdan Patrut
Faculty of Computer Science
"Alexandru Ioan Cuza"
University of Iasi
Iasi, Romania

V. D. Ambeth Kumar
Department of Computer
Science and Engineering
Panimalar Engineering College
Chennai, Tamil Nadu, India

Oscar Castillo
Division of Graduate
Studies and Research
Tijuana Institute of Technology
Tijuana, Baja California, Mexico

ISSN 2367-4512 ISSN 2367-4520 (electronic)
Lecture Notes on Data Engineering and Communications Technologies
ISBN 978-3-030-32149-9 ISBN 978-3-030-32150-5 (eBook)
https://doi.org/10.1007/978-3-030-32150-5

This Springer imprint is published by the registered company Springer Nature Switzerland AG
The registered company address is: Gewerbestrasse 11, 6330 Cham, Switzerland

Preface

Dear distinguished delegates and guests,

The organizing committee is extremely privileged to welcome our distinguished delegates and guests to the International Conference for "**P**hoenixes on **E**merging **C**urrent **T**rends in **E**ngineering **A**nd **M**anagement (**PECTEAM 2K19**) organizing Conference on **COMET** (Emerging Current Trends in **COM**puting and **Expert** **T**echnology) which will be held from 22 to 23 March 2019, in Chennai, Tamil Nadu, India.

COMET is well supported by IEEE Madras section, CSI, ICTACT, ISTE and IETE. The objective of this conference is to bring together numerous problems encountered in the current as well as future high technologies under a common platform. It has been specifically organized to congregate international researchers and technocrats for presenting their innovative work, share and gain knowledge to have a proper insight into the significant challenges currently being addressed in various fields of Computing and Expert Technology.

The conference proceedings has a complete track record of the papers that are reviewed and presented at the conference. The principle objective is to provide an international scientific forum wherein the participants can mutually exchange their innovative ideas in relevant fields and interact in depth through discussion with peer groups. Both inward research and core areas of Computing and Expert Technology and its applications will be covered during these events.

The conference has received 540 papers all over India and abroad from various institutions in different domains, out of which 163 papers are selected for presentation. The submitted papers were peer-reviewed by external reviewers and advisory board based on the subject of interest and selected on the basis of originality, clarity and significance in line with the theme of the conference.

It is guaranteed that a high-quality program will be conducted as the conference will be graced by an unparalleled number of professional speakers from multidisciplinary fields, from both India and abroad.

We would like to thank the chief patron, patrons, conveners, organizing committee members, advisory members for their immense help, motivation, constant cooperation and continuous encouragement for the conference. We deeply appreciate the editorial board for their wonderful editorial service for the successful outcome of this proceedings. We are very much grateful to all those who have contributed in both the front desk and the backyard to make COMET an astounding success.

COMET, featuring high-impact presentations, would be enriching and undoubtedly an unique, rewarding and memorable experience being hosted at Panimalar Engineering College, Chennai, India.

Organization

Steering Committee Members

Chief Patron

P. Chinnadurai (Secretary and Correspondent)

Patrons

C. Vijayarajeswari (Director)
C. Sakthikumar (Director)
Saranya Sree Sakthikumar (Director)

Co-patrons

K. Mani (Principal)

Convener

S. Malathi

Co-convener

V. D. Ambeth Kumar

Organizing Committees

D. Karunkuzhali
S. Maheswari

National Advisory Board

L. Ganesan	IIT Madras, Chennai
A. Kannan	Anna University, Chennai
S. Nirmala Devi	Anna University, Chennai
K. Sridharan	Anna University, Chennai
Prakash	Karpagam University, Coimbatore
S. Pandurangan	IIT Madras, Chennai
K. Selvamani	Anna University, Chennai
A. Suresh	NIET, Coimbatore
L. Arun Raj	Crescent Institute of Science and Technology, Chennai
K. Ramalakshmi	Karunya Institute of Technology and Sciences, Coimbatore
R. Venkatesan	Karunya Institute of Technology and Sciences, Coimbatore
Arun Raj Kumar	NIT, Puducherry
C. Swarnalatha	Anna University, Madurai
K. Panneerselvam	IGNOU, Kozhikode, Kerala
G. Kumaresan	CEG, Anna University, Chennai
V. S. Senthil Kumar	CEG, Anna University, Chennai
Indra Rajasingh	VIT University, Chennai
V. Santhi	VIT University, Chennai
K. Chandrasekaran	NITK, Mangalore
Sipra Das Bit	LIEST, West Bengal
S. Bhaskar	Anna University, Chennai
Uma Bhattacharya	Besu, Howrah, West Bengal
S. Amalraj	Anna University, Chennai
G. Ravikumar	Anna University, Chennai
K. Malathi	CEG-AU, Chennai
M. Ganesh Madhan	MIT-AU, Chennai
E. Logashanmugam	Sathyabama University, Chennai
B. Sheela Rani	Sathyabama University, Chennai

Web Design Committee

R. Priya
Rahul Chiranjeevi Veluri (PG Scholar)
P. Naveen Teja (UG Scholar)
P. Yaagesh Prasad (PG Scholar)

Publication Committee

S. Vimala
S. Sathya

B. Akshaya (PG Scholar)
R. Jayashree (PG Scholar)
R. Christina Rini (PG Scholar)
N. Savitha (PG Scholar)

Proceeding Committee

K. Valarmathi
M. Anitha (PG Scholar)
K. P. Ashvitha (PG Scholar)
J. Jaya Sruthi (PG Scholar)
R. Vidhya (PG Scholar)

Publicity Committee

S. Sathya Venkateshwaren (PG Scholar)
M. Gobinath (PG Scholar)
R. B. Aarthinivashini (PG Scholar)
S. Divya (PG Scholar)
R. Vidhya (PG Scholar)

Hospitality Committee

M. Rajendiran
D. Elangovan
M. Dhivya (PG Scholar)
K. Priyanka (PG Scholar)
M. Priyanga (PG Scholar)

Registration Committee

J. Josepha Menandas
G. Kumutha Rajeswari (PG Scholar)
M. R. Tamizhkkanal (PG Scholar)
J. Freeda (PG Scholar)
M. Shilpa Aarthi (PG Scholar)
R. Monisha (PG Scholar)

Finance Committee

Sofia Vincent
Lakshmi

Contents

Advances in Machine and Deep Learning

Regional Blood Bank Count Analysis Using Unsupervised Learning Techniques

R. Kanagaraj[1(✉)], N. Rajkumar[2], K. Srinivasan[1], and R. Anuradha[1]

[1] Sri Ramakrishna Engineering College, Coimbatore 641022, India
kanagaraj.r@srec.ac.in
[2] Nehru Institute of Technology, Coimbatore 641105, India

Abstract. Data mining methods allows finding out blood bank region based consumption model that a city poses and used to pull out the information concerning to blood bank count in regard to the number of cities in each region. K- Means clustering procedure is used for identifying the regions that has low, middle and high Blood bank counts. The data set used is available in Indian government website. To validate the proposed work, the implementation is done in both R and Weka Tool and cluster mean difference is measured.

Keywords: Data mining · Clustering · Blood bank · Data reduction

1 Introduction

Data Mining systems and functionalities like clustering, classification and association rule mining is used for extracting useful and prognostic knowledge from large databases. It is used in nearly every disciplines of Engineering and medical applications. Blood is most essential for functioning of a human body. Most of the accident death case arises due to the unavailability of blood at the right time and right place. Blood bank systems has a major role in collecting blood from donors, monitor its quality and allocate blood components to hospitals within a particular network.

Many regions in India confront with insufficient blood bank to full fill its people needs. Hence this leads to an incorrect blood distribution and waste of time that can create risky to patients with critical conditions. To overcome this problem Semi supervised learning Clustering Techniques can be used to form different clusters of blood bank regional wise using K-means algorithm. Currently Data Mining techniques are used in Blood Bank system for automated blood bank service that get voluntary blood donors and those in need of blood on to a universal platform and data analysis techniques for generating region-wise blood bank count.

2 Related Works

Data mining model uses different techniques and construct different model relying on the data types and its purpose [9]. Its task can be classified into predictive task and descriptive task. The predictive tasks and descriptive tasks are classification and clustering, association rule mining respectively [11]. Based on the different types of

D. J. Hemanth et al. (Eds.): COMET 2019, LNDECT 35, pp. 987–992, 2020.
https://doi.org/10.1007/978-3-030-32150-5_100

data analyzed, web and text mining, Graphical mining, spatial mining and time series data mining are performed [10]. Blood bank database is basically accumulated from various sources like government, private, charity, NSS, NGO's, hospitals and through a common web interface globally [1]. The raw data collected for the proposed work is available in the data set [4]. DonorHART Tool [2] designed for recording donor reactions, monitoring the risks involved for donors at the time and after blood donation. The techniques of using Raspberry PI [3] for developing automated blood bank system is developed to make blood donors to come render one roof using Android application combined with raspberry pi. A direct link between the blood donor and recipient is established at a low cost using automated blood bank system that uses embedded system that takes low power supply [5].

Multiple Knapsock Algorithm [6] is implemented for management and assignment of blood cells units in an optimized manner. Fuzzy sequential pattern mining is applied to mine rules from blood transfusion service center data set that forecast the performance of donor in the future [12]. Clustering and K-means classification is implemented for finding the behavior pattern of blood donation services [13]. The k-means algorithm [9] is popular for its efficiency in clustering for large data sets.

Most of the above works concentrate on application based technology that offers effective connectivity between the donor and recipient.

3 Case Study on Regional Wise Blood Bank

The case study has been conceded out on blood bank data available regional-wise in India. This study enables the necessity of forming new blood banks over the population distribution of a particular region.

Data Selection and Pre-processing
The vibrant expansion of cities owing to development in infrastructures increases populations in particular area. But the availability of Blood banks over the concern area remains the same irrespective of increase in population. In the proposed work, region wise blood bank available data from blood bank directory portal from Indian government is taken [4]. The Data available in website is pre-processed with certain missing fields. These fields are not considered for data analysis and are avoided using data reduction techniques. The Blood bank count available over a number of districts in a particular state is accounted for data analysis.

Region-Wise Clustering
Clustering embody a common property of the entire data within group, which is one of the best unsupervised learning algorithms. Each element with in a cluster figures a class or common features among them [7, 8]. In the proposed work, the clustering algorithm has been applied on district wise availability of blood bank. The detailed implementation is given below.

K means clustering algorithm forms cluster of district wise availability of blood bank till current month. The result of implementation in R software is compared with WEKA tool results to validate the implemented algorithm. The K-means clustering using Euclidean distance is implemented on Blood Bank data for the appearance of 1, 2, 3, 4, 5 up to 10 clusters of blood bank. It is observed that 10 clusters represent vital information about availability blood bank as the results are very close to WEKA software results. Table 1 shows the comparison of WEKA and R software results.

Table 1. Result comparison of WEKA and R software

Cluster ID	K means clustering		Misclassification
	R	WEKA	% Difference
I4	57.33	39.06	0.182
I5	57.33	39.06	0.182
I10	52.90	37.85	0.150
I3	61.12	49.29	0.118
I2	50.17	41.19	0.089
I6	48.72	40.06	0.086
I9	46.18	40.35	0.058
I7	45.15	39.74	0.054
I8	38.23	35.73	0.025
I1	33.9	33.9	0

K means for region wise blood bank is carried out for k = 3. Table 2 shows the sample 20 regional wise blood bank is taken account and it is used to compares the number of districts with in a particular region with that of Blood bank available for the same, The average value of the above parameter is taken up for classifying it under low, medium and high. The given classification ensures the sufficient blood bank availability of a particular region.

Figure 1 shows the BGSS for K means cluster K = 1 to 10. It gives the dispersion measured of the cluster between each other. The model gives an overall point variability of 97.8%.

Partitions n individuals in a set of multivariate data into k groups or clusters (G1, G2, - - -, Gk) K is given or a possible range is specified Common approach is to identify the k groups which minimizes the maximizes the between group sum of squares (BGSS).

Table 2. Regional blood bank clustering result.

Name of the state	District	Blood bank	Average	Cluster	
	Count	Count		Assigned cluster number	Assigned category to clusters
Andaman	3	3	1	3	Low
Andhra Pradesh	13	140	10.76	1	High
Arunachal Pradesh	22	12	0.54	3	Low
Bihar	38	74	1.94	2	Medium
Goa	2	5	2.5	3	Low
Gujarat	33	136	4.12	1	High
Himachal Pradesh	12	21	1.75	3	Low
Jammu and Kashmir	22	34	1.54	3	Low
Jharkhand	24	42	1.75	3	Low
Karnataka	30	198	6.6	1	High
Kerala	14	179	12.78	1	High
Maharashtra	36	332	9.22	2	Medium
Manipur	16	5	0.31	3	Low
Meghalaya	11	7	0.634	3	Low
Mizoram	8	10	1.25	3	Low
Nagaland	11	11	1	3	Low
Punjab	22	106	4.81	1	High
Rajasthan	33	100	3.03	2	Medium
Telangana	31	149	4.80	1	High
Tamil Nadu	32	291	9.09	2	High

Figure 2 shows the regional wise blood bank with cluster size k = 3. The K-means algorithm searches for a pre-determined number of clusters within regional wise blood bank dataset. The cluster center is the arithmetic mean of all the points belonging to the cluster. Each point is closer to its own cluster center than to other cluster centers. Based on the size of cluster the data is grouped and it is used for identifying the regions that has low, middle and high Blood bank counts. The regional wise blood bank count is assigned to the cluster number based on the K value.

The low classification enables user to indicate the requirement in the number of Blood Bank in corresponding region. Hence, the blood bank count should be increased in future to full fill the requirements of people needs. The proposed work is used for identifying proper blood distribution and avoids patients in getting in critical conditions.

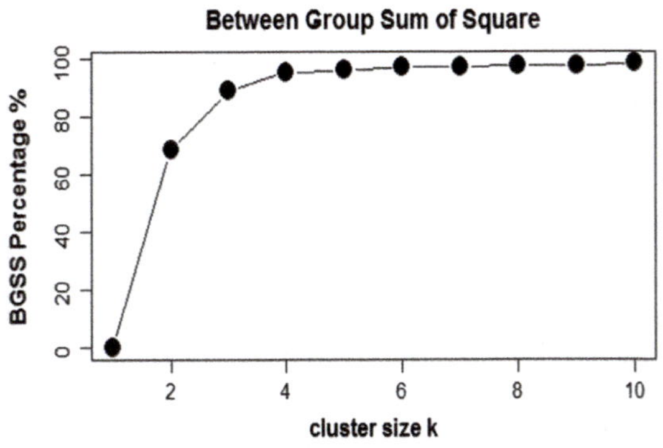

Fig. 1. BGSS for cluster

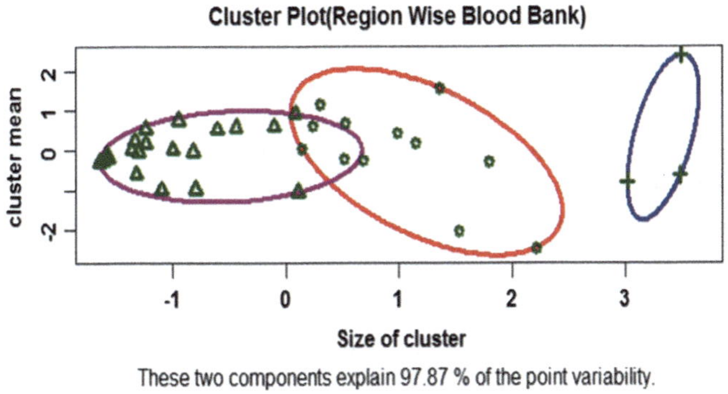

Fig. 2. Regional wise blood bank with cluster size K = 3

4 Conclusion and Future Work

Unsupervised learning algorithms provide an insight to any kind of data modeling. The proposed work is able to differentiate regions by their blood bank counts. This work also capable of handling a large volume of data and identifying regions of blood bank count. The working model is implemented both R and Weka. This work helps government agency to evenly distribute blood banks. The proposed work can be extended for producing association rules that incorporate features like population of the region, accidents frequency etc.

Acknowledgement. The authors like to thank the all the anonymous reviewers for their valuable suggestions and Sri Ramakrishna Engineering College for offering resources for the implementation.

References

1. Selvamani K., Rai, A.K.: A novel technique for online blood bank management. In: International Conference on Intelligent Computing, Communication and Convergence (ICCC-2014) Conference, Inter science Institute of Management, Technology, Bhubaneswar, Odisha, India (2014)
2. Patil, R., Poi, M., Pawar, P., Patil, T., Ghuse, N.: Blood donors safety in Data Mining. In: 2015 International Conference on Green Computing and Internet of Things (2015)
3. Adsul, A.C., Bhosale, V.K.: Automated blood bank system using raspberry pi. Int. Res. J. Eng. Technol. (IRJET) **04**(12) (2017). e-ISSN: 2395-0056
4. Open Government, Data. https://data.gov.in
5. Bala Senthil Murugan, L., Julian, A.: Design and implementation of automated blood bank using embedded systems. In: IEEE Sponsored 2nd International Conference on Innovations in Information, Embedded and Communication systems, iCIIECS (2015)
6. Adewumi, A., Budlender, N., Olusanya, M.: Optimizing the assignment of blood in a blood banking system: some initial results. In: WCCI 2012 IEEE World Congress on Computational Intelligence, 10–15 June 2012, Brisbane, Australia (2012)
7. Berkhin, P.: A survey of clustering data mining techniques. In: Kogan, J., Nicholas, C., Teboulle, M. (eds.) Grouping Multidimensional Data, pp. 25–71. Springer, Heidelberg (2006)
8. Fu, T.: A review on time series data mining. Eng. Appl. Artif. Intell. **24**, 164–181 (2011). https://doi.org/10.1016/j.engappai2010/09/07
9. Huang, Z.: Extensions to the k-means algorithm for clustering large data sets with categorical values. Data Min. Knowl. Discov. **2**, 283–304 (1998)
10. Jain, A.K.: Data clustering: 50 years beyond K-means. Pattern Recogn. Lett. **31**, 651–666 (2010)
11. Tan, P.-N., Steinbach, M., Kumar, V.: Introduction to Data Mining. Pearson Addison Wesley, London (2005)
12. Sundaram, S., Santhanam, T.: A comparison of blood donor classification data mining models. J. Theor. Appl. Inform. Technol. **30**(2), 31 (2011)
13. Ramachandran, P., et al.: Classifying blood donors using data mining techniques. IJCST **1**(1) (2011)

A Systematic Approach of Classification Model Based Prediction of Metabolic Disease Using Optical Coherence Tomography Images

M. Vidhyasree(✉) and R. Parameswari

Technology Advanced Studies, Vels Institute of Science, Tamilnadu, India
mailtovidhyasree@gmail.com,
dr.r.parameswari16@gmail.com

Abstract. Data mining is defined as the upcoming field that consists of certain tools and techniques to be implemented with certain data sets taken from the different sources to foresee the hidden information. The data mining is the huge upcoming field has attracted many fields under its influence. In the applications of data mining, health care is a very important application to be taken account. Healthcare is defined as the service provides the health maintenance and earlier disease prediction and also provides high quality treatments to prevent disease. Human body consists of a number of cells constituted to form organs and the organs connected to form the organ system. This system should be interconnected to work properly. The human body should be nourished properly by balanced diet and the healthy lifestyle. The function of the human body is disturbed by some external factors called disease. The metabolic disease is the collection of five different disorders such as high blood pressure, heart problems, obesity and insulin resistance. The Optical Coherence Tomography images of eyes are considered to predict the chronic conditions of the body accurately in the eyes. The main focus of this work is to detect diabetes through the retina images. This paper mainly reflects detection of diabetes using retina images. In this paper the classification techniques are analyzed using orange data mining tool to find the best classification technique based on the individual technique's prediction accuracy.

Keywords: Data mining · Classification techniques · Prediction accuracy · Image features

1 Introduction

In the recent years a series of research conducted to predict early symptoms and signs of the prevailing diseases in India. In the healthcare domain one subfield is disease prediction where efficient techniques to be developed to predict the disease earlier, which is a more challenging area in the healthcare domain. Human body constitutes with a number of components such as cell, tissues and organs etc. These organs combine to form the major systems of the body. Each organ has its significant contribution to the normal function of the body. If there is any deviation in the usual function, then it should be detected and right treatment must be given at the right time.

D. J. Hemanth et al. (Eds.): COMET 2019, LNDECT 35, pp. 993–1003, 2020.
https://doi.org/10.1007/978-3-030-32150-5_101

Disease is defined as the trouble in the normal function of the human body. Diseases are of different types They are infectious disease, epidemic disease, pandemic disease, autoimmune disease/inflammatory disease, and the disease caused by virus and bacteria. Every disease has symptoms and signs shown in the human body. Human body consists of the organ connected to form system if the single organ is affected other organ also affected according to severity of disease. Optimal coherence tomographic images are defined as the imaging technique used to capture light coherence in the optical media autoimmune disease is defined as chronic conditions shown in the eyes of the human body. Eyes consists of six components such as cornea, iris, retina, pupil, lens, Artery and veins are directly connected to other organs such as liver, kidney, heart etc., So if any organs affected by external factors, then the immediate signs are shown in the eyes. The OCT image shows the accurate signs of the chronic disease and also the signs reflects the eyes shows the severity of the disease because the vision is not affected naturally as the reflection of chronic conditions affects the other parts of the body. This is the main reason to consider OCT eye images to predict the disease accurately. Metabolic diseases are defined as the abnormal activities in the person's digestion and it also explained as the collection of five diseases such as high blood pressure, stroke, diabetes etc. Metabolic diseases are inherited from the parents contain the affected gene. This is the main reason even the new born kid is also affected by some dangerous disease. The iris and retina is the most predominant part that the chronic condition of the body is revealed.

2 Metabolic Disease Predicted in Eyes

Metabolic disease is defined as the disturbance caused by external factors in the metabolic activities. The external factors like food, lifestyle, environmental factors, inherited from ancestors. A metabolic disease called as metabolic disorder or metabolic syndrome is the group of abnormal conditions like high blood pressure, high sugar, abnormal fat and cholesterol leads to diabetes, heart diseases and stroke. Human eyes act as a window for sale, so the eye reflects the signs of systematic disease like a connective tissue disease, spondyloanthropathies, inflammatory bowel disease, Non-infectious multisystem disease, systematic infections, mucutaneous disease, metabolic disease, myopathies, neurology and leukemia cancer. The list of metabolic diseases are detected in eyes

- Diabetes mellitus
- Thyrotocosis
- Homocystinuria
- Paget disease
- Cushing syndrome
- Acromegaly

Table 1. Metabolic diseases shown in the eyes

Disease	Ophthalmic conditions
Diabetes mellitus	**Common Symptoms:** Retinopathy, Iridopathy, Unstable refraction **Uncommon Symptoms:** Recurrent styles, Xanthelasmata, Contract, Neovascular glaucoma
Thyrotoxicosis	**Common Symptoms:** Lid refraction, Chemosis, Propotosis **Uncommon Symptoms:** Kerato conjunctivitis, Diplopia
Homocystinuria	**Common Symptoms:** Ectopia **Uncommon Symptoms:** Myopia, Retinal detachment
Paget disease	**Common Symptoms:** Optic atrophy proptosis, Ocular motor nerve palsies, Angioid streaks **Uncommon Symptoms:** Tumors
Cushing syndrome	**Common Symptoms:** Steroid- induced contracts
Acromegaly	**Common Symptoms:** Bitemporal hemianopia, Optic atrophy **Uncommon Symptoms:** Angoid streaks, Seesaw nystagmus

Each disease has symptoms shown in the eyes can be categorized as common and uncommon symptoms are listed in Table 1. Metabolic diseases are of different types such as metabolic brain disease, metabolic bone disease and metabolic brain disease etc.

3 Different Forms of Retina Images

Diabetic retinopathy is defined as the diabetic conditions reflected in retina in terms of abnormal growth and blood vessels and leakage in blood vessels and also damage in light sensitive tissues in the eyes. Diabetic retinopathy has four main stages such as mild non prolipretive diabetic retinopathy, moderate non prolipretive diabetic retinopathy, severe non prolipretive diabetic retinopathy, prolipratative retinopathy [22].

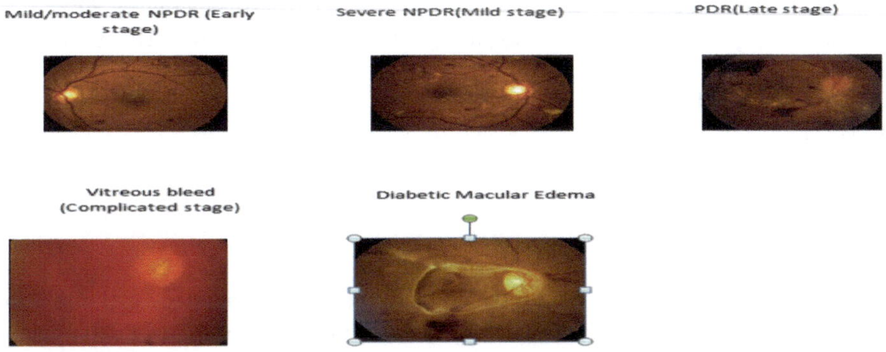

Fig. 1. Stages of diabetic retinopathy

In Fig. 1 different stages of diabetic retinopathy shows progress of severity of diabeties that will leads to loss of vision completely.

4 Need of OCT Images

The OCT images are define as the image captured with the help of coherent light and this technique can be used be captured 2D and 3D images. The OCT image implements the ultrasound imaging technique with the principles like time domain and spatial domain with high clarity so it can be used to predict the chronic disease in eye images.

5 Related Works

In the most recent years the severity of chronic diseases increased considerably and if the single part is affected then the other connected organs are affected according to the progress of the disease. The eyes are the major part to show the serious conditions of the health conditions of the body. The iris and retina are recent interest area to predict the chronic disease earlier. In Table 2 the computational techniques used to predict the disease affected.

Table 2. Diabetes detection using eye images.

Application of iris and retina	Dataset considered	Feature extraction technique	Classification technique used
Diabetes detection using iris [1]	200 subjects (100 diabetic and 100 non diabetic)	Rubbersheet normalization, wavelet features,	Binary tree, Generalized linear models, Random forest tree, Adaptive boost model, Neural network. Random forest **88%**
Diabetes detection using iris [2]	Type 2 diabetes 338 subjects (180 diabetic and 150 non-diabetic)	Rubbersheet normalization, Statistics, texture and wavelet features	Soft computing techniques **89.97%**
Stages of diabetic retinopathy [3]	65 retinal images	Random forest	CART NPDR cases **85%** Classifier accuracy **90%**
Diabetes using retina images [4]	Fundus images	Morphological filter	MATLAB NPDR Accuracy **90%**
Diabetes using retina [5]	Diabetic retinopathy	ML-BEC	Bagging **85%**

6 Methodology

In this section methodology of the paper is discussed and the steps explained to be implemented is displayed in Fig. 2. The system architecture consists of three phases the input image got from the user the gray scale image is filtered using adaptive filtering then the filtered image is given to the feature extraction the image features are exacted and the features are given to classifier model the images are labeled under appropriate class then the output image is displayed.

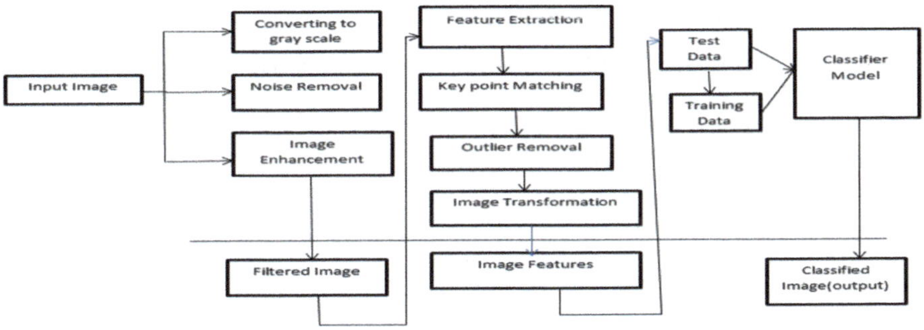

Fig. 2. System architecture

In the above figure detailed architecture of the proposed architecture is shown with three significant phases that the given input image should pass through to yield the desired output. The classifier model used in the third phase consists of existing classification techniques to classify the affected image according to the efficiency of algorithms used.

6.1 Filtering Phase

In this phase filtering of the image to remove the unwanted information called noise of the image to improve the quality of the image. Adaptive filtering is used to remove the unwanted information called noise. the image and to get clarity in image related information. The pseudo code of the adaptive filter is shown below (Fig. 3).

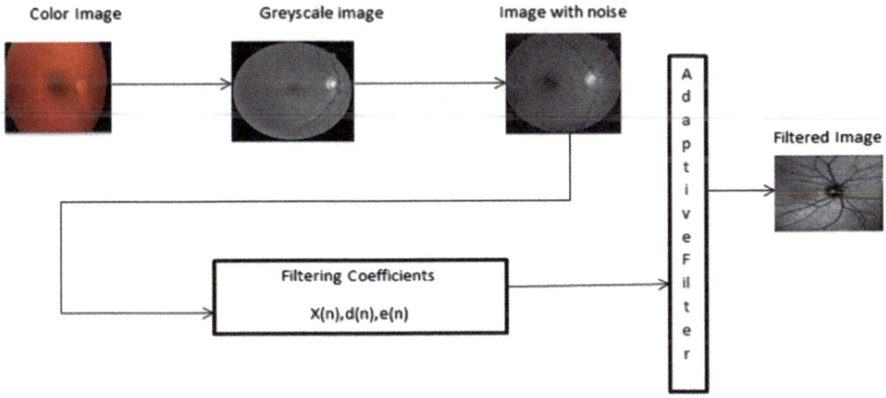

Fig. 3. Filtering steps for adaptive filter

Pseudo code for adaptive filter

```
Input: A X B image I, odd neighborhood size NS, Noise variance noisevar
Output: Filtered Image J
S = floor (NS/y)
for(r= NS… A-Z NS + 1)
for(c= NS…B-Z NS+ 1 )
        add= 0
        addsq =0
for (ih = -NS…+ NS)
        add+= l[r + ih] [c+ jh]
        addsq += (l[r + ih] [c+ jh])2
end
end
```

In the first phase the input image given by the user taken from the online dataset [11]. The given image is filtered by using adaptive filters in the filtering phase to improve the quality of the image. The image with high quality and clear information is given to the feature extraction phase. The features of the image are extracted by the appropriate algorithm used in the next phase.

6.2 Feature Extraction

In the feature extraction the features of the image are extracted. The image consists of 14 features as harlick proposed. Here the image features like energy, contrast, entropy and heterogeneity is considered. The Table 3 shows the formulas for image properties. The formulas of the features are listed with the pseudo code [22].

Table 3. Image properties and its formulas.

Image features	Feature formulas	Feature attributes
Contrast	$\sum I, j\vert i - j\vert 2$	Where
Correlation	$\sum I, j(i - \mu i)(j - \mu j)p(i, j)/\sum i \sum j$	i,j = image coefficients
Energy	$\sum I, jp(i, j)2$	Pi,j = Element of i,j
Homogeneity	$\sum I, jp(i.j)/1 + \vert I - j\vert 2$	μ = Mean value

The above table shows the list of image features to be extracted by feature extraction algorithm. The energy of the image is defined as the similarity of gray scale values of the image. The image feature contrast is explained as the brightness level between two pixels. The correlation is briefed as the dependency value of the gray levels in co-occurrence matrix. Homogeneity refers the similarity measures of the pixel (Fig. 4).

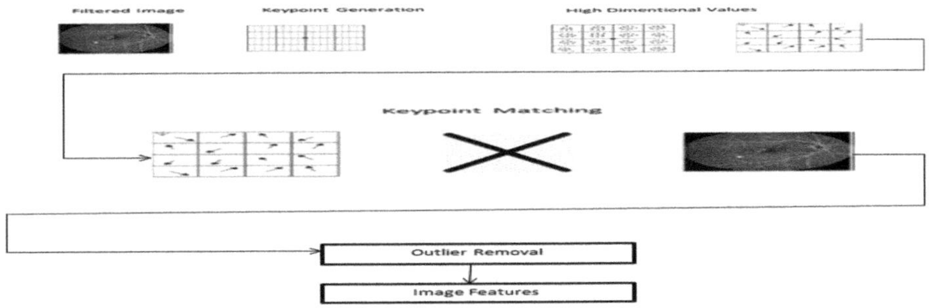

Fig. 4. Feature extraction

Pseudo code for feature extraction

```
Input: Enhanced image
Output: Features of the image
For all images
{
      Let the number of pixels be N( i, j)
Divide the image into blocks and label it
For each block
{
         i= first pixel of the image
         j = matching pixel
         d= distance of the jth pixel from ith pixel
         Pd[I,j] = I, j
}
Pd[i, j] ≠ Pd[j,i]
For each pixel Pd[i,j]
{
      Normalize the elements by
      N[ i, j] = P[i, j] / ΣiΣj P[ i, j]}}
```

In the second phase of the implementation the features of the image is taken by Grey Level Co-occurrence Matrix where the given color image is converted to grey scale image and the features like energy, contrast, heterogeneity, Correlation are extracted. All these features are extracted and it will be given as the input to the next stage. Harlick features calculated by kharlick () function consists of derived formulas to extract the features listed above. The matrix equation is used to calculate the image features of the given image. The very first step is to create the co-occurrence matrix with the adjacent values in the matrix [16].

$$G = \left[p(1,1) \cdots p(1,Ng) \vdots \ddots \vdots p[Ng,1] \cdots p(Ng,Ng) \right] \quad (1)$$

The G is the co-occurrence matrix is constructed by grey intensity levels of the image to extract the image features of the obtained filtered image. The features of the image are calculated by the formulas given in Table 3. Then the image and its features are given to the next phase for classification from the healthy and affected image.

6.3 Classification

In this phase the output of the feature extraction phase is considered as input to the classification phase. In this phase the features of the image are classified by using the classifier model. The classifier model consists of existing classification techniques used to classifier model used in this phase. Classification can be explained as the important data mining technique used to list the image under the accurate class label. The classification consists of different techniques such as decision tree, logistic regression, SVM, Random Forest, CN2 Rule Inducer, Navies bayes, neural network. The accuracy of the techniques are discussed below. The classifier model consists of all available techniques in data mining. The pseudo code of the classifier model is discussed in next section.

Pseudo code

```
Input: image features
Output images classified with the class label
for each image do
        Train the network
        Classify the image using classifier model
end
for each image do begin
      test the network with the features
label the image and categorize under class label
end
```

In this phase the classifier model consists of two sub phases where the images are given to train the network and the features extracted from the image is given to test the network and the predicted image is classified under the label is obtained as output of this phase. In the classifier model consist of set of algorithms like decision tree, CN2 Rule Inducer, Random forest, Neural Network, Support Vector Machine, Navies Bayes, Logistic Regression are analyzed with the set of retina images fetched from online datasets [16] (Fig. 5).

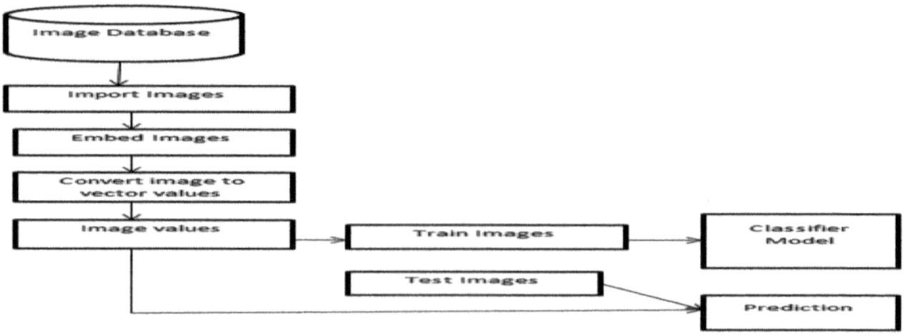

Fig. 5. Classification model

In the above fig steps of the classification where the image features are taken with image as the input to the classifier and the images classified under the label is yield as desired output.

7 Result and Discussion

In this section the prediction accuracy of the classification techniques is shown below with the different datasets. The dataset with 30 and 75 images is taken in consideration is given as input to the classifier model and the prediction accuracy is calculated. In the Table 4 the prediction values of the classification techniques are shown with the 75 images as input with the SVM has high accuracy.

Table 4. Prediction accuracy with 75 images

Technique used	Accuracy obtained (values in %)
SVM	76.3
Decision Tree	62.5
Random Forest	33
CN2 Rule Inducer	24.2
Logistic Regression	39.3
Navies Bayes	44

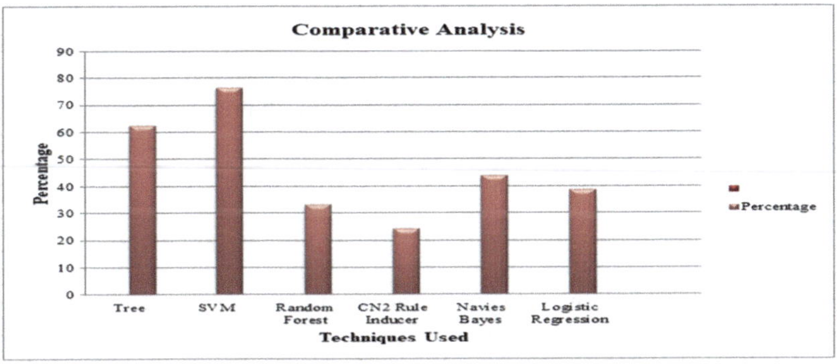

Fig. 6. Comparative analysis with 75 images

In the above figure the evaluation results of classification techniques are shown by the comparative analysis to know the best classification technique with high accuracy as 76%. In Table 5 the prediction accuracy values of 30 images. The prediction accuracy of the classification techniques are calculated with the input of 30 images.

Table 5. Prediction accuracy with 30 images

Technique used	Accuracy obtained (values in %)
SVM	72.5
Decision Tree	62.5
Random Forest	65
CN2 Rule Inducer	79
Logistic Regression	85
Navies Bayes	80

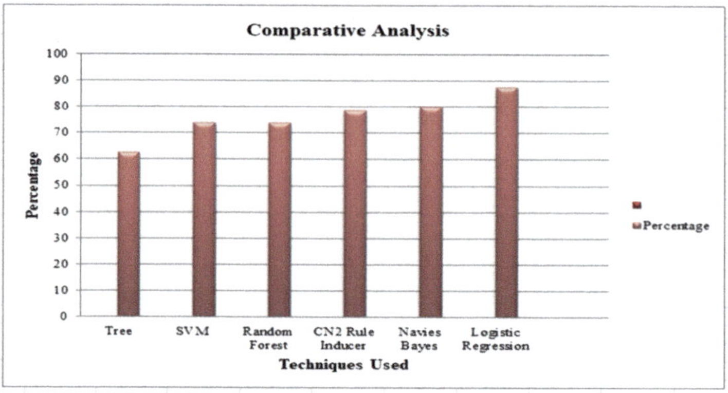

Fig. 7. Comparative analysis for 30 images

In the above Fig. 6 the comparative analysis of the classification techniques is shown the prediction accuracy can be varied based on the volume of the dataset used for disease prediction. The dataset consist of the set of 75 retina images of retina images to detect diabetes (Fig. 7).

8 Conclusion

In this paper the set of retina images for prediction of metabolic disease using data mining classification techniques. The prediction accuracy of classification techniques varies according to the dataset taken. The technique with high accuracy is logistic regression and SVM. This work is extended by considering underlying factors leads to diabetes by investigating reflections in eyes.

References

1. Samant, P., Agarwal, R.: Machine learning techniques of medical diagonisis of diabetes using iris images. Comput. Methods Programs Programs Biomed. **1**(4), 1–27 (2018)
2. Samant, P., Agarwal, R.: Diagonisis of diabetes using computer methods: soft computing methods for diabetes detection using iris. World Acad. Sci. Eng. Technol.: Int. J. Med. Health Biomed. Bioeng. Pharmacheutical Eng. **11**(2), 57–62 (2017)
3. Somasundaram, S.K.: A machine learning ensemble classifier for early prediction of diabetic retinopathy. Int. J. Med. Sci. **41**(201), 1–12 (2017)
4. Cho, Y., Lee, S., Woo, S.: The Krisch Lablician edge detection for predicting iris based disease. In: Proceedings of 2017 IEEE International Conference of Computer Suported Cooperative Design Work (2017)
5. Admin, J.: A method for detection and classification of diabetic retinopathy using structural predictors of bright lesions. J. Med. Sci. **19**(6), 555–560 (2017)
6. Mozam, F.: Multiscale segmentation of excutation retinal images using contextual cues and ensemble classification. Biomed. Signal Process. **2**(35), 52–60 (2017)
7. Perta, H.: Convolution neural network for diabetic retinopathy. Procedia Comput. Sci. **90**, 200–205 (2016)
8. Samnt, P., Agarwal, R.: Comparative analysis of classification based algorithms diabetes diagonosis using iris images. J. Med. Eng. Technol. **42**, 1–9 (2018)
9. Kaur, J., Sinha, H.P.: Automatic detection of diabetic retinopathy using fundus image analysis. Int. J. Comput. Sci. Technol. **3**(4), 4794–4799 (2012)
10. Faust, O.: Algorithm for automted detection of diabetic retinopathy using digital fundus images. Int. J. Med. Sci. **36**, 1–13 (2010)
11. Mangrulkar, R.S.: Renel image classification technique for debiets identification. In: IEEE International Conference on C Intelligent Computing and Control, pp. 190–194 (2017)
12. Suo, Q.: Personalized disease prediction using a CNN based similarity learning method. In: IEEE International Conference Bioinformatics and Biomedicine, pp. 811–817 (2017)
13. Dangare, C.S.: Improved study of heart disease prediction system using data mining classification techniques. Int. J. Comput. Appl. **41**(10), 44–49 (2012)
14. Saranya, M.S.: Intelligent data storage system using machine learning approach. In: IEEE International Conference on Advanced Computing, pp. 191–195 (2016)
15. Rajliwall, N.S., et al.: Chronic disease risk monitoring based on an innovative predictive modelling framework. In: IEEE Symposium Series on Computational Intelligence, pp. 1–8 (2017)
16. https://www.kaggle.com/paultimothymooney/kermany2018
17. https://www.analyticsindiamag.com/7-types-classification-algorithms/
18. https://www.solver.com/data-mining-classification-methods
19. https://www.infogix.com/top-5-data-mining-techniques/
20. http://murphylab.web.cmu.edu/publications/boland/boland_node28.html
21. https://www.saedsayad.com/decision_tree.htm
22. https://support.echoview.com/WebHelp/Windows_and_Dialog_Boxes/Dialog_Boxes/Variable_properties_dialog_box/Operator_pages/GLCM_Texture_Features.htm#Energy
23. https://www.hindawi.com/journals/ijbi/2015/267807/

Rainfall Prediction Using Fuzzy Neural Network with Genetically Enhanced Weight Initialization

V. S. Felix Enigo(✉)

Department of Computer Science and Engineering, SSN College of Engineering, Chennai, Tamilnadu, India
felixvs@ssn.edu.in

Abstract. In this paper, a hybrid approach of combining the techniques of artificial neural network along with fuzzy logic and genetic algorithm is used in prediction of rainfall classes. We have used genetic approach for initialization of weights in contrast to fixed weights or random weights for initialization of fuzzy neural networks. Fixed weights tend to get struck at local optimum by biasing the solution to a particular set of weights. Random initialization of weights increases the probability of obtaining global optimum solution in comparison to fixed weight approach. Our proposed genetic approach of weight initialization, rather than providing random weights, finds the optimal weights through genetic evolution for initialization of the fuzzy neural network. The proposed approach has been analyzed for rainfall classification system which predicts the class of one day ahead rainfall based on its intensity. By incorporating genetic evolved weights for fuzzy neural network, we were able to achieve a better accuracy than random weight approach.

Keywords: Neural network · Fuzzy logic · Genetic algorithm

1 Introduction

Predicting rainfall precisely is the most challenging issue to be solved by the researchers. Weather predictions include predicting rainfall, storms and clouds levels. And all these tasks involve high computational effort. In ancient times, forecasting weather is computed based on the patterns observed also and the technique called as pattern recognition. These types of methods are expensive in terms of time and unreliable.

Machine learning is a data analysis technique that performs statistical analysis and build model dynamically. It is a branch of artificial intelligence that learns hidden patterns automatically. Things like growing volumes and varieties of available data, computational processing that is cheaper and more powerful and affordable data storage are the factors that have led to resurging interest in machine learning.

Out of all the machine learning techniques, artificial neural networks have some key advantages that make them most suitable for prediction and classification problems. They can learn and model non-linear and complex relationships that mimic the real-life. Unlike many other prediction techniques, artificial neural network does not impose any restrictions on the input variables (like how they should be distributed).

D. J. Hemanth et al. (Eds.): COMET 2019, LNDECT 35, pp. 1004–1013, 2020.
https://doi.org/10.1007/978-3-030-32150-5_102

Additionally, many studies have shown that artificial neural networks can better model heteroskedasticity, the property of the data with dynamic variance and more volatility and can explore hidden patterns without prior knowledge.

The simplest definition of a neural network, more properly referred to as an 'Artificial' Neural Network (ANN) are processing devices (algorithms or actual hardware) that replicates the structure of a neuron in mammals but in smaller version. Neural networks are built on layers. These layers have interconnected nodes and each node has an activation function. The input layer accepts the input pattern which passes to one or more hidden layers where the data is processed taking into account the weights received from the links to the node. Hidden layers disseminate the output to the output layer which outputs the result as a prediction or classifications of data. The activation functions are responsible for making the output in non-linear way to represent the real world data. It receives numeric value and applies certain mathematical computations over the input. Most commonly used activation functions are:

$$\textbf{Sigmoid}: \sigma(\mathrm{x}) = 1/(1 + \exp(-\mathrm{x})). \tag{1}$$

$$\textbf{tanh}: \tanh(\mathrm{x}) = 2\sigma(2\mathrm{x}) - 1. \tag{2}$$

$$\textbf{ReLU}\ (\text{Rectified Linear Unit}) : \mathrm{f}(\mathrm{x}) = \max(0, \mathrm{x}). \tag{3}$$

However, neural networks are considered as a black box technique. This is because the network implicitly learns the patterns for prediction and the end users as well as the programmers are completely unaware of the underlying structural patterns behind the predicted outputs of the network. Hence, by inculcating fuzzy logic based decision making, we are able to learn the underlying structure behind the output by means of using If *Then* rules. The reasoning process is often simple, compared to computationally precise systems, so computing power is saved unlike the computationally heavy systems like artificial neural networks.

Fuzzy logic is an extension of Boolean logic that enables to represent partial truth that ranges between truth and false. It operates on a group of *If-Then* statements. In fuzzification, system inputs which are crisp numbers are transformed into fuzzy sets. The membership functions are applied over the sets of fuzzy variables. The inference engine mimics the human decision making process by fuzzy inference using experts If-Then rules. Later it is de-fuzzified to obtain a crisp value.

The initial weights assigned to the neural network are assigned on a random basis. Instead, the weights can be optimized in order to improve the accuracy of the system. For the purpose of optimization of the weights, genetic algorithms are used. The genetic operations such as selection and crossover evolve generations of chromosomes i.e. set of initial weights and finally result in the optimized values for initial assignment to the neural network.

Genetic programming is a programming model that utilizes the process of natural evolution to solve complex problem. It transforms a population of computer programs to a new generation programs by applying genetic operation which are similar to natural evolutionary genetic operation. The genetic operations include crossover (sexual recombination), mutation, reproduction, gene duplication, and gene deletion.

Due to the above advantages, this paper uses a combination of these techniques to solve the problem of predicting the different classes of rainfall.

2 Related Work

Hybrid approach of combining neural, fuzzy and genetic have been tried for various weather related applications by the researchers. Bodri and Cermak [1] used Artificial Neural Network (ANN) for training 38 years data to forecast rainfall month-wise and it is proved to be effective. Kumarasiri et al. [2] predicted rainfall for Colombo the capital city of Sri Lanka for wet zone of western coast. It forecasted one-day ahead with an accuracy of 74.3% and annual rainfall rate with an accuracy of 80%.

In Sahai et al. [3] research work, a recurrent higher-order neural network (RHONN) model was developed for wind power forecasting in a wind park. This model can be used to predict wind speed or power in time scales from some seconds to 3 h. The optimal architecture of the model was selected based on the cross validation approach and was solved using the nonlinear Simplex method of Box.

Fonte et al. [4] proposed a method used a multilayer perceptron neural network of 3 layers feed forward model for predicting average wind speed in hourly basis. This method is based on finding the correlation of the present and previous wind speed data. The accuracy of this system is reported to be poor compare due to lack of meteorological data.

Yet another paper used ANN [5] for wind energy prediction of one day ahead. It utilized old predicted weather data and contemporaneous measured power data to learn physical coherence of wind speed and wind power output. The ANN application can easily use additional meteorological data, such as air pressure or temperature, to enhance prediction accuracy. In addition, this method is superior to others by the use of power curves of individual plants.

For the training of the ANN, historical predicted meteorological parameters are used. Even though neural networks are good at detection of patterns in datasets, its decision process is unexplainable. As, Neural networks follows black box where only the input and outputs can be viewed without any knowledge on the internal implementation details. This leads to the fuzzy logic technique for weather prediction as fuzzy logic is a simple implementation technique, which is relatively more transparent and more human-like thinking. Design of fuzzy logic begins with the process of grouping (cluster) using fuzzy C Means algorithm [6]. The result of fuzzy clustering is used to design the FIS (Fuzzy Inference System) editor. FCM is a technique in which each of data association with cluster, this is determined by the degree of membership.

Fuzzy system of type Interval-2 and probabilistic fuzzy C-means techniques [7] can be used for predicting from non-linear data. This type of system is more efficient in terms of revealing inference from uncertainty in data. However, fuzzy logic suffers from the drawback of requiring a priori knowledge for defining the fuzzy rules as learning is not possible, unlike neural networks. In order to overcome the drawbacks of these models, a hybridization approach of neural networks and fuzzy logic is introduced to overcome the weaknesses and bring out the strengths of both the techniques by complementing each other. The Neuro-fuzzy hybrid system is a learning mechanism that uses the training and learning algorithms in neural networks to find the parameters of fuzzy systems.

A neuro-fuzzy model has been proposed to model wet season tropical rainfall [8]. The Root Mean Square Error (RMSE) of this model is very low, which proves the reliability of the model in predicting variation in rainfall. Thus, the fuzzy neural network is quite efficient as it supports transparency in implementation of the system prediction along with an ability to learn the trends instead of the need for prerequisite knowledge.

Fhira et al. [9] optimized further the fuzzy approach by the use of genetic algorithm. Learning Genetic Algorithmic approach is conducted to acquire fuzzy parameters for each attribute which represented within a chromosome with binary representation. Thus, genetic programming has been observed to be a powerful tool in optimization of the fuzzy based elements.

3 System Overview

In the proposed work, to the existing artificial neural network structure in order to incorporate fuzzy logic into the neural networks structure we make use of four layers in the neural network. The first layer corresponds to the input nodes each of these inputs represents an input parameter required for prediction of weather conditions namely humidity, wind speed, temperature etc. The next layer is a hidden layer whose nodes are the fuzzified values for the variables belonging to the input layer. The fuzzy values in the second layer are mapped to the output values using the IF-THEN fuzzy rules. For example, if X is A and Y is B then Z is C. The final output layer consists of the discrete output values which have undergone de-fuzzification from the previous layer. The structure of the fuzzy neural network has been depicted in Fig. 1.

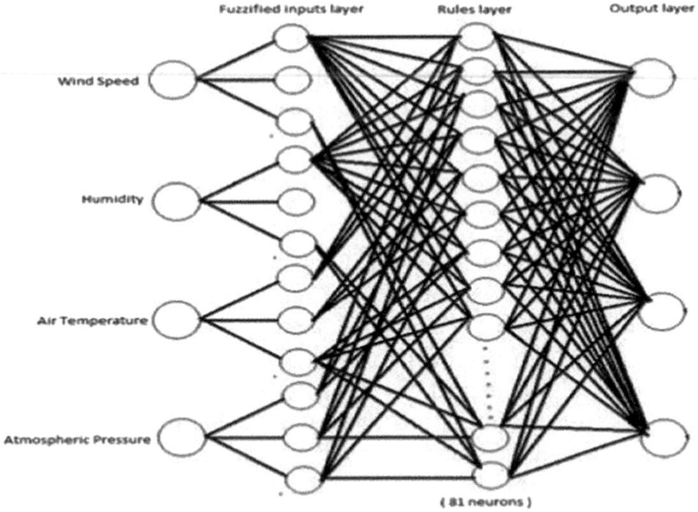

Fig. 1. Structure of fuzzy neural network

Given the high degree of nonlinearity of the output of a fuzzy neural system, traditional linear optimization tools are not efficient. Genetic algorithms have demonstrated to be a robust and very powerful tool to perform optimization operations. Here, we are using the genetic programming functions of selection, mutation and crossover in order to generate and tune of membership functions. The overall architecture of the system proposed is depicted in Fig. 2. This system is trained using a meteorological dataset, as obtained from Boundary Layer Meteorological Tower at Kalpakkam site [10, 11] operated and maintained by Radiological Safety Division, IGCAR, India.

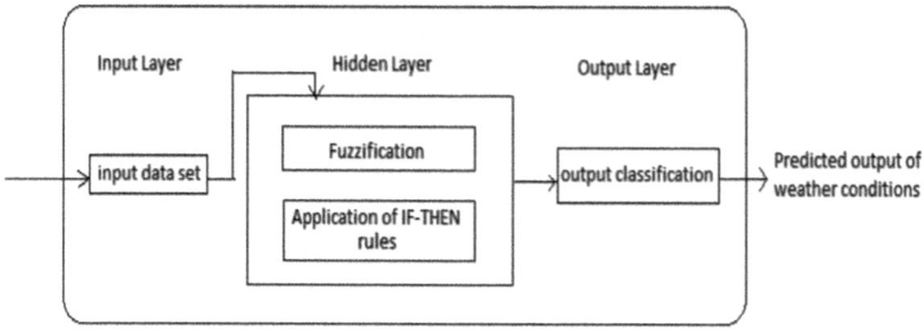

Fig. 2. Architecture of genetically enhanced fuzzy neural system

4 Methodology

The overall goal of the proposed system is to enhance the performance of neural network with genetic module and fuzzy module. The system consists of three major phases: (i) Pre-processing phase (ii) Genetic phase and (iii) Neural Network phase. The neural network in turn uses fuzzy rules as a sub-phase.

4.1 Pre-processing Phase

The input data is normalized and processed and then genetic operations are carried out for generating the initial weights for the neural network and the network is trained with the data set.

4.2 Genetic Phase

The objective of the genetic phase is providing an optimal weight rather than random initial weight. This optimal weight is found out using genetic algorithm. The initial weights that are assigned to the neural network are evolved using the genetic algorithm. The initial population consists of 10 rows (chromosomes) of randomly initialized weights. The fitness is computed for each chromosome. The chromosome which produces the least error rate for a tuple from the training data set is assigned a higher fitness value. In selection the 2 chromosomes with the highest fitness value are found out. In the crossover phase, the 2 chromosomes with the highest fitness value were found and a crossover point is randomly generated and the genes are swapped between the 2 chromosomes till the crossover point. Now the fitness function is calculated for this and all the above steps continue with a new tuple every time until the fitness value reaches the minimum threshold value. After the fittest chromosome is found these weights are assigned as the initial weights to the neural network.

Algorithm: Genetic operations for weight optimization
Test case set S = NULL
for each coverage C do
 Find start node, N
  repeat
    for (i=0; i < |N|/2; i++) **do**
     Select two parents in the population
    Generate two offspring by **crossover** operation between two parents
     Insert two offspring into new generation list
     if a new offspring satisfy the coverage, C **then** S = S U (∑of the offspring)
    break
    end if
    end for
    Mutate some offspring in the new generation list
    until satisfy C or reach maximum iteration
   end for

4.3 Neural Network Phase

The neural network phase consists of following steps:

Network Initialization. A new neural network by giving three parameters such as the number of inputs, the number of neurons to be present in the hidden layer and the number of outputs. Each neuron has a set of weights that need to be maintained. One weight for each input connection and an additional weight for the bias.

Forward Propagation. An output from a neural network is computed by passing the input data through each layer until the output layer outputs its results. The first step is the neuron activation which is calculated as weighted sum of inputs. For each neuron, the activation function is applied to perform non-***linear*** transformation to the input in order to make it learn complex relationship between input and response variable. In our case Sigmoid activation function which produces values ranging between 0 and 1 is used.

$$O = 1/(1 + e^\wedge(-a)). \tag{4}$$

Where O is the output and a is the activation function.

The output of this layer is a fuzzified values and this will serve as fuzzified inputs to the next layer through forward propagation. This process is continued till the outputs are generated of the final layer.

Fuzzy Rules. From the fuzzified inputs the next layer is the rules layer where the *If-Then* rules are present. For each set of input parameters a rule is being specified. Therefore, for four parameters and three categories 81 rules are present in the rules layer. By training the neural network it learns which rules impact more or are the significant rules for the prediction. All neurons in the rules layer are connected to all the neurons of the output layer. The output layer consists of 4 neurons, corresponding to four classes of rainfall (Fig. 3).

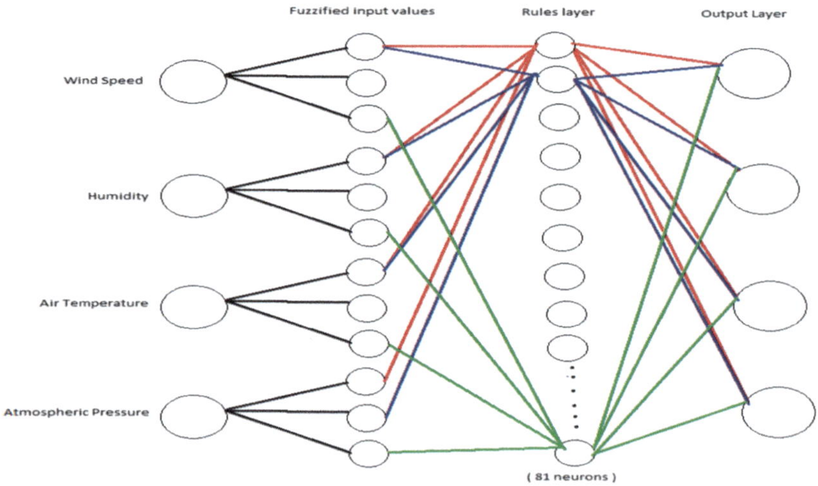

Fig. 3. Application of if-then rules in neural network structure

Back Propagation of Error. As the output reaches the output layer, error is computed by finding the difference between the expected output and the obtained output and these errors are back propagated. As it propagates backwards, weights are adjusted in each layer to minimize the errors for the next iteration.

Training the Network. This involves several epochs of applying the training dataset to the network and the repeating the process of passing the inputs across layers, feedback the errors and updating the weights.

Prediction. After having trained the network with dataset, the network has been adjusted with the values of its weights over the links. The trained network is now used for prediction or classification problems. In our case, it is used to predict the rainfall for given weather parameters.

5 Performance Evaluation

We evaluated our system using k-fold cross-validation with 5 folds. By using four out of the five sets of data as training set and the rest one set as test set, accuracy is calculated for each fold. The mean accuracy metric is finally calculated by taking into account all the folds.

We elaborately carried out accuracy based on the various techniques used. Initially the accuracy for artificial neural network is calculated, then the fuzzy logic is integrated into this and the membership function is chosen by analysing the accuracy obtained by using each of the membership functions and then the genetic algorithm is infused into the neuro-fuzzy system and the accuracy is estimated. The data set is split into seasonal data and the accuracy is calculated. These are the various performance analysis that are done.

The percentage of correctness of the predicted output classes is estimated by comparing the predicted values with the expected values. Based on this accuracy estimation, analysis has been carried out to study the accuracy of the system subject to certain parameters.

5.1 Analysis Based on Membership Function

The membership function which is used to fuzzify the initial crisp input values for the weather parameters is varied and accuracy is estimated for each membership function. The triangular function results in the least accuracy i.e. 84.292% as the triangular curve is quite sharp and lacks smoothness in the curve. The trapezoidal function results in slightly higher accuracy than triangular function i.e. 84.932%. This is because there are four points of reference in trapezoidal curve as opposed to three points in triangular curve. Hence, trapezoidal curve is slightly more smoother than triangular curve. The bell function leads to greater accuracy than triangular and trapezoidal functions i.e. 85.297%. The bell curve is much more smooth than the sharp triangular and trapezoidal curves. The gaussian function results in the highest accuracy i.e. 86.210% out of the four membership functions. Gaussian function exhibits greater smoothness in its curve than the bell function due to the more number of reference points in the gaussian curve. The accuracy estimated for the system using each membership function is illustrated in the graph shown below (Fig. 4).

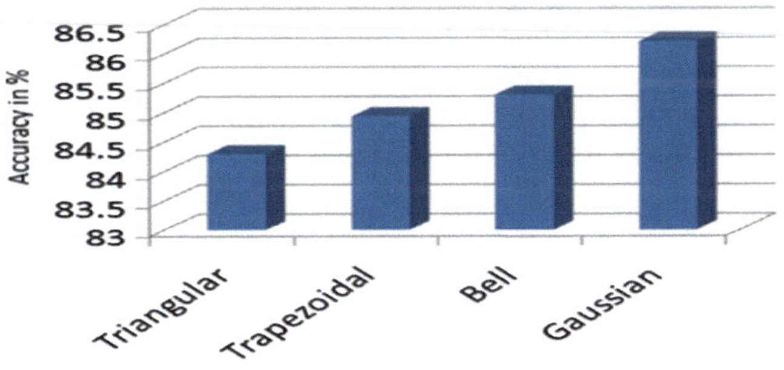

Fig. 4. Accuracy vs membership function

5.2 Analysis of Accuracy Based on Techniques

The percentage of accuracy was estimated for the different techniques incorporated in the system. When the system is built with only a multi class neural network, the accuracy is estimated as 79.543%. As fuzzy logic is integrated into the multi class neural network, the accuracy is found to rise to 84.110%. This increase in accuracy is due to the organised structure the fuzzy logic adds to the neural network by means of defining neurons based on *If-Then* rules.

When genetic programming is integrated into the fuzzy neural network, the accuracy is estimated to be 86.210%. The optimization of the initial weights by genetic evolution has led to a more focussed learning process in the neural network. The accuracy for each technique is illustrated in the Table 1.

Table 1. Estimation of accuracy based on techniques

Techniques used	Accuracy
Neural network	79.54%
Fuzzy neural network	84.11%
Genetic enhanced fuzzy neural network	86.21%

5.3 Seasonal Analysis

The overall dataset that comprises of the data values for all the seasons is being classified seasonally into four sub datasets. The duration for each season is as per the specifications of the Meteorological Department of India. The system is trained with the four sub datasets separately and the accuracy is measured for each of the four datasets. It can be inferred that the accuracy is higher when trained with each of the four seasons separately than with the overall dataset. This is because the patterns formed in the learning process are more specific and hence more precise with seasonal datasets. Another reason for the increased accuracy is that a possibility for convergence towards zero rainfall is avoided when learning using seasonal datasets. This is because the majority of the days record zero rainfall in Chennai and hence the output class may converge to zero rainfall when learnt with the overall dataset. The accuracy for the system when learnt with each seasonal dataset is illustrated in the Table 2.

Table 2. Accuracy estimation based on seasonal analysis

Season	Duration	Accuracy
Winter	December–March	86.86%
Summer	April–June	86.68%
Monsoon	July–September	87.50%
Post-monsoon	October–November	87.43%

6 Conclusion and Future Work

We have illustrated the techniques involved in the detailed design of a weather prediction system. We have incorporated the big data regarding weather over the past few years to trace patterns for futuristic rainfall classification. A hybrid approach of neural networks, fuzzy logic and genetic programming has been used to process the big data. It can be inferred that the hybrid approach yields a higher level of accuracy than the base neural network alone. This can be reasoned as the fuzzy logic giving a more organised structure and genetic programming optimizing the initial weights used in the neural network. It is also inferred th at the accuracy is improved if the system is specific to a particular season than spread overall.

As a part of our future work, we intend to expand the system to cover several weather stations across the nation. We also look forward to enhancing the system into a web based application, available online.

Acknowledgements. The Meteorological data used in this study were obtained from Boundary Layer Meteorological Tower at Kalpakkam site operated and maintained by Radiological Safety Division, Indira Gandhi Centre for Atomic Research (IGCAR), India.

References

1. Bodri, L., Cermak, V.: Prediction of extreme precipitation using a neural network: application to summer flood occurrence in Moravia. Adv. Eng. Softw. **31**(5), 311–321 (2000)
2. Kumarasiri, A.D., Sonnadara, U.J.: Performance of an artificial neural network on forecasting the daily occurrence and annual depth of rainfall at a tropical site. Hydrol. Process. **22**(17), 3535–3542 (2008)
3. Sahai, A.K., Soman, M.K., Satyan, V.: All India summer monsoon rainfall prediction using an artificial neural network. Clim. Dyn. **16**(4), 291–302 (2000)
4. Fonte, P.M., Quadrado, J.C.: ANN approach to WECS power forecast. In: 10th IEEE International Conference on Emerging Technologies and Factory Automation, pp. 19–22 (2005)
5. Rohrig, K., Range, B.: IEEE Power Engineering Society General Meeting, pp. 18–22 (2006)
6. Aisjah, A.S., Arifin, S.: Maritime weather prediction using fuzzy logic. In: 2nd International Conference on Instrumentation Control and Automation, Bandung, Indonesia, pp. 15–17 (2011)
7. Shah, H., Jaafar, J., Rosdiazli, I., Saima, H., Maymunah, H.: A hybrid system using possibilistic fuzzy C-mean and interval type-2 fuzzy logic for forecasting: a review. In: International Conference on Computer & Information Science, pp. 532–537 (2012)
8. Annas, S., Kania, T., Koyama, S.: Neuro-fuzzy approaches for modeling the wet season tropical rainfall. Agric. Inf. Res. **15**(3), 331–341 (2006)
9. Fhira, N., Asiwijaya: A rainfall forecasting using fuzzy system based on genetic algorithm. In: International Conference of Information and Communication Technology (2013)
10. Bagavathsingh, A., Baskaran, R., Venkatraman, B.: Installation and Commissioning of 50m Meteorological Tower at Ediyur Site, IGCAR, Kalpakkam, IGC/SG/RSD/RIAS/92617/EP/3013/REV-A
11. Srinivas, C., Bagavath Singh, V., Venkatesan, A., Somavaii, R.: Creation of benchmark Meteorological Observations for RRE on Atmospheric Flow Field at Kalpakkam, IGC Report N. 317

Analysis of Structural MRI Using Functional and Classification Approach in Multi-feature

Devi Ramakrishnan[1(✉)], V. Sathya Preiya[1], and A. P. Vijayakumar[2]

[1] Department of Computer Science and Engineering, Panimalar Engineering College, Chennai, India
deviramakrishnan83@gmail.com, sathyapreiya@rediffmail.com

[2] Department of Electrical and Electronics Engineering, HK-ERC and D-AC Technologies, Chennai, India
apvkumar75@gmail.com

Abstract. Magnetic resonance imaging (MRI) is adjacent to nature and a multi-modality practice which provides the complementary in sequence about dis-similar aspects of diseases. As a competent image (CET) contrast enhancement tool, (AGC) adaptive gamma correction control which was relating with gamma parameter and (CDF) cumulative distribution function in the conventional method which is to engender the function of the pixel gray levels contained by an image. AGC should deals well within the most dimmed images, but fails for worldwide intense images and also the dimmed images within the local bright regions. Such two categories of images which are observed from MRIs, the brightness-distorted images are widespread in genuine scenario, such as to the improper exposure and to the white object regions. To attenuate such kind of deficiencies here we intend an improved the aspects by two methods which are (RIC) region by iteration method of convolution and (S-KC) segmentation by k-levels of clustering. In this proposed work the above given two methods are used by iteration method and the other segmentation in multiple levels of clustering which is to enhance the required response of the MRIs. Both the levels and methods are analyzed in a closed system which to eliminate the unwanted signals and to get the better performance in the MRIs imaging.

Keywords: Image enhancement · Contrast enhancement · Segmentation · Cerebrospinal fluid · White matter · Grey matter · Edge map · Total intracranial image

1 Introduction

The brain which analyze and think as a computer system and it valour be easier to realize. As by UC Davis Health System the gray matter, which is termed as nerve cells of brain and function as computer and the white matter, which act as cables to transmit signals that connect everything together. The tissue in the brain which composed in multilevel of nerve fibers is termed as White matter. The fibers which are known as second unit cell as axons that connect the nerve cells and also covered by a fat which are termed as myelin. The second unit cell, which works to transmit and speeds up the

D. J. Hemanth et al. (Eds.): COMET 2019, LNDECT 35, pp. 1014–1023, 2020.
https://doi.org/10.1007/978-3-030-32150-5_103

signal between the soma and to the dendrites. The myelin which covered with white matter as similar to the fibers insulation that is covered which prevents from the white matter, which are present on neural imaging studies of the brain and prior to analyze and keep track for symptoms of Alzheimer's disease. Researchers have demonstrated the existence of white matter and prior to the mild impairment totally exists, that condition which carries an augmented risk for Alzheimer's disease. To describe multi level spots in the brain by analyze the MRI in different aspects of the system which are termed as fuzzy network, neural network, ANIFS system etc.... by properly resolution in imaging.

By using the magnetic resolution imaging (MRIs), by systematic analyzing and by white matter hyper intensities, the term which are known to identify the multi level spots in the brain. According to UC Davis, Center of Alzheimer's Disease; the areas may point towards some category of injury related to the brain, perhaps due to decreased flow of blood in that area. The existence of white matter and prior to the mild impairment totally exists, that condition which carries an augmented risk for Alzheimer's disease has been simultaneous with a higher peril of stroke, which can show the way to vascular dementia. White matter hyperintensities are habitually referred to as white matter disease. Initially, this white matter disease is reflection to merely be associated to aging.

Due to impinge on of cardiovascular disease, high cholesterol, high blood pressure and smoking which include and related to higher unambiguous risk factors for white matter disease that we know that. It has been associated with cognitive loss, strokes, and dementia; it is also has some high emotional and physical changing symptoms such are named as balancing problems, falls, hopelessness and complicatedness multitasking with such manners as walking and talking.

Based on the physical exercises the improvement in the white matter is changed, which is notified by the researches. Cardio respiratory activities (CRA) and weight resistance training (WRT) was concurrent with to improving the white matter integrity in the brains that is relatively showed by fellows in those studies.

2 Model with ROI and Literature Review

Automatic contrast enhancement, by which the universal histogram amendment: This is to provide the basic tool of image enhancement (IE) for visual scrutiny. It is frequently method which used in the digital mode of photography; analyze remote sensing, medical imaging process and the scientific visualization. The task calls for to defend an appropriate balance between the emphasis and to the distortion. Dissimilarity augmentation makes images easier to read between the lines by construction object characteristics easier which are to differentiate.

The conventional type image-segmentation techniques, the clustering and also the thresholding levels are presume to facilitate the shades with the intention of differentiate objects feature be fewer familiar than those that are encompass them. Hence the modes of the image; histogram replicate the similarity within unvaryingly colour object features whereas the valleys record the descriptions that is to distinguish them. In this histogram warping modus operandi improves the dissimilarity between the object descriptions by dissemination away from each other the modes of an image histogram to obtain healthier benefit of the intact energetic range. It addresses to impose for a vigorous, by and large applicable, dissimilarity enhancement algorithm for images in the midst of multimodal histograms. Our closed loop model in Fig. 1, improves the contrast of images with both brawny highlights and prominent cloudiness.

Pre-effective mapping and targeting intrusion method. In this method, the MRI which has been utilized to make available pre-operational well-designed brain mapping; and also help them to guide the neurosurgical scheduling, more than ever to categorize and to circumvent areas throughout surgical resection time, that make available indispensable functions such as motoring and language.

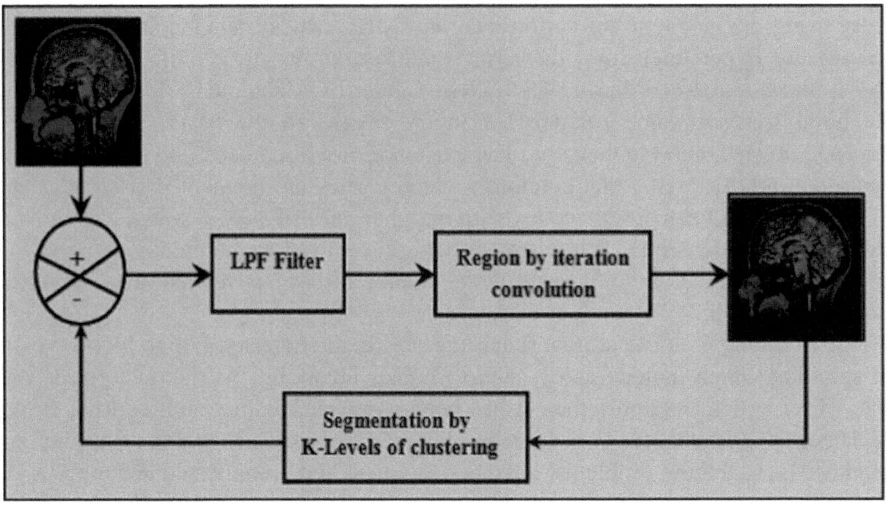

Fig. 1. Closed loop model with ROI convolution.

With the proposed method, the scheme in the cortical and/or the sub-cortical areas and their related white matter pathways which obligation be avoided probably will be identified concurrently. Similarly, our come within reach of could help erstwhile quantifiable interventions where the localization of a serviceable area or which the white matter fiber bundle is significant, such as appointment in deep brain stimulation surgery, or the transcranial magnetic stimulation (TMS).

This possibly will be a partly overcome by the performing of effective connectivity opinion only for the associations of interest (i.e., for a explicit brain function and associated structural network), which are clearly given in [1–4] and [5]. In multiparty investigation methods, structural along with the functional results which are fashioned separately and then amalgamated to act upon which to populace studies, deterioration scrutiny, correlation and/or multi-variate examination of inconsistency. For example, the (MA) mean anisotropy and the purposeful connectivity (which are from the correlation) can be first in competition computed from the dMRI and also fMRI, and then compared are clearly mentioned in the [6]. In these integrative approaches, which is to assist the method of tractography reconstruction, fMRI is utilized to categorize seed regions and get rid of impurities in anatomical parcellation are analyzed which to compare the performance are given in [7, 8] and [9].

The simulation and the experimental method of analysis are done for the low noise SMPS system which is demonstrated in [10]. Investigations on forward converter using different types of filters and experimental method of analysis for the forward converters are done, which is to compare it with the conventional circuit are clearly mentioned in [11]. Different types of filters which are utilized in the forward converters and its performance are given in [12]. Forward converter with RCD snubber using the PI controller, fuzzy controller and artificial neural network (ANN) controller are analyzed and compared to get the better performance in [13].

The above literature does not deal with the analysis of structural MRI based on functional and classification computing in the multi-feature. This work aims to develop the region of convolution by the iteration method is implemented and also which aim to effectively improve the response with k-levels of clustering in the segmentation.

By the method of k-levels of clustering in the segmentation and the region of convolution by iteration method in closed mode, which is to get the better response such as in white matter, grey matter, Cerebrospinal fluid etc.

It aims to check the performance of the images and also to identify the region in the segmentation levels.

3 Simulation Results

In this work we designed and organize new closed loop logic with the self organizing map system model such segmentation by k-levels of clustering and image region by iteration method for computing in order to achieve better response time and processing time to achieve better performance in MRIs.

The images are analyzed using different levels of filters and also with inverted image and the results are presented. Original image analyzed with different types of filters are shown in Figs. 2 and 3. The white matter and region of iteration (ROI) are shown in Fig. 4. The grey matter and cerebrospinal fluid with ROI are represented in Figs. 5 and 6. Edge map and FSE proton density weighted image with the range of iteration are given in Figs. 7 and 8. Weighted MR image and total intracranial image with iteration by k-levels of clustering are shown in Figs. 9 and 10.

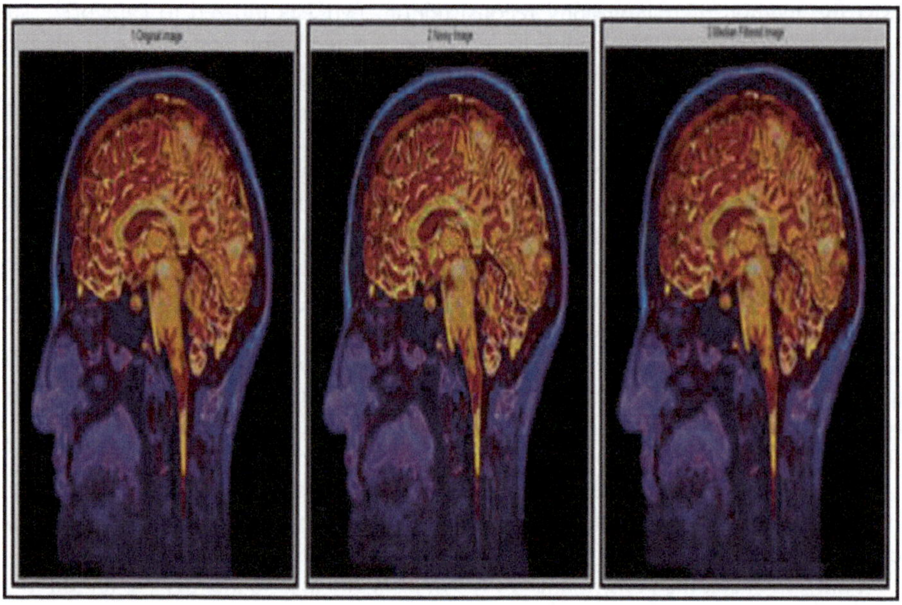

Fig. 2. Original image and filtered images.

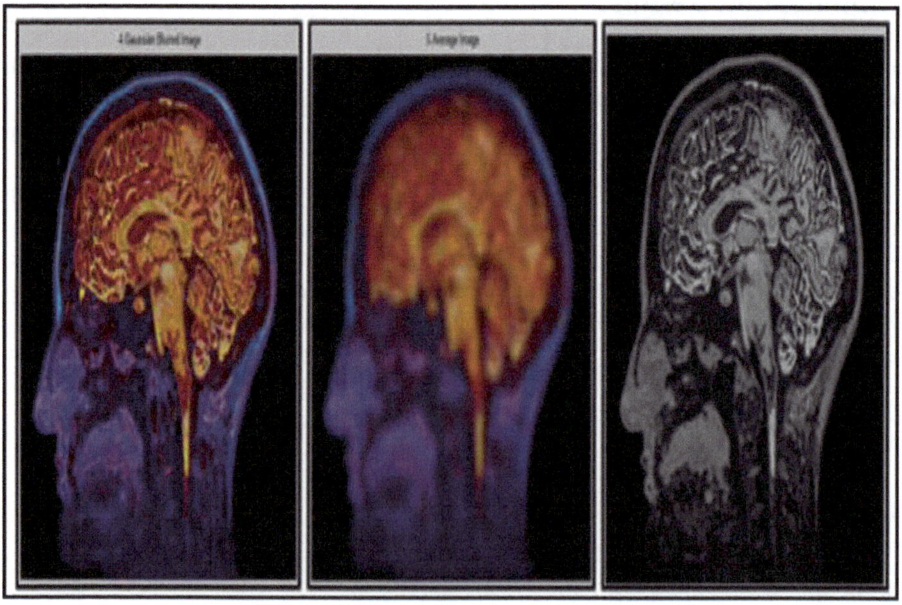

Fig. 3. Inverted image with filtered images.

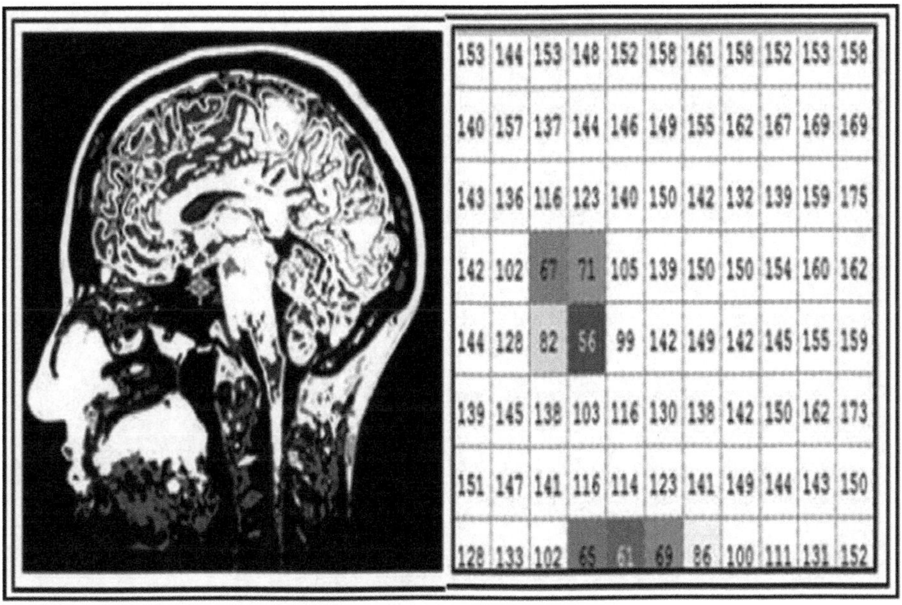

Fig. 4. White matter with ROI topology.

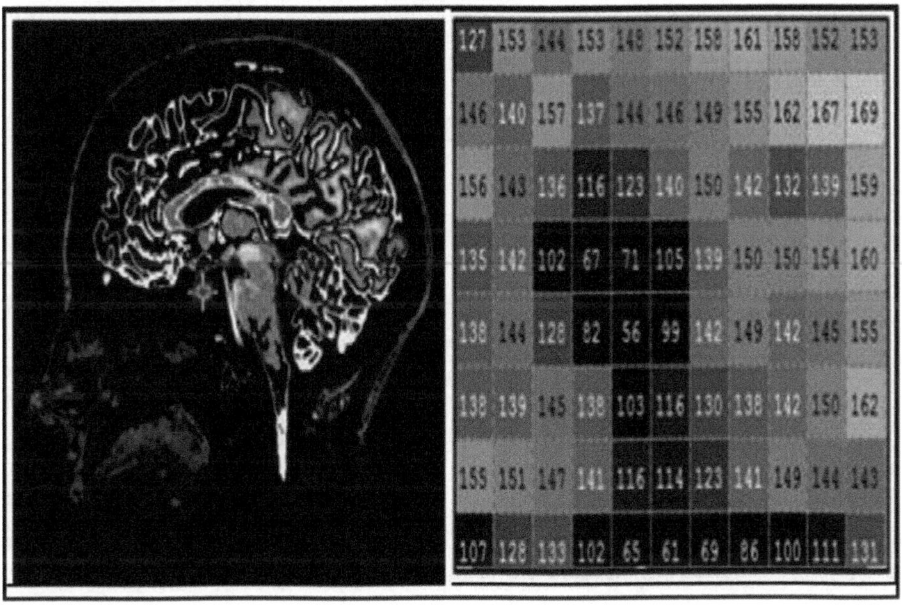

Fig. 5. Grey matter with ROI and segmentation.

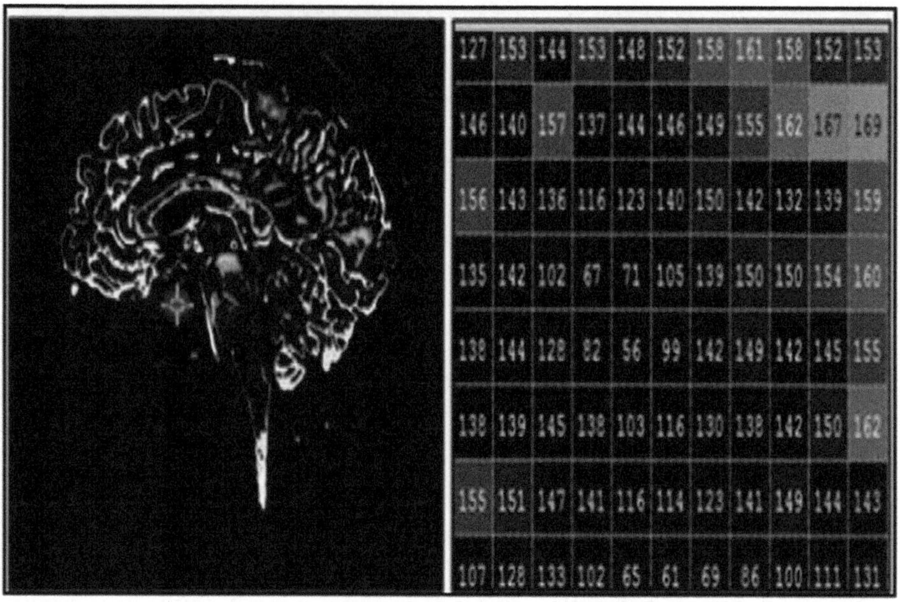

127	153	144	153	148	152	158	161	158	152	153
146	140	157	137	144	146	149	155	162	167	169
156	143	136	116	123	140	150	142	132	139	159
135	142	102	67	71	105	139	150	150	154	160
138	144	128	82	56	99	142	149	142	145	155
138	139	145	138	103	116	130	138	142	150	162
155	151	147	141	116	114	123	141	149	144	143
107	128	133	102	65	61	69	86	100	111	131

Fig. 6. Cerebrospinal fluid with iteration.

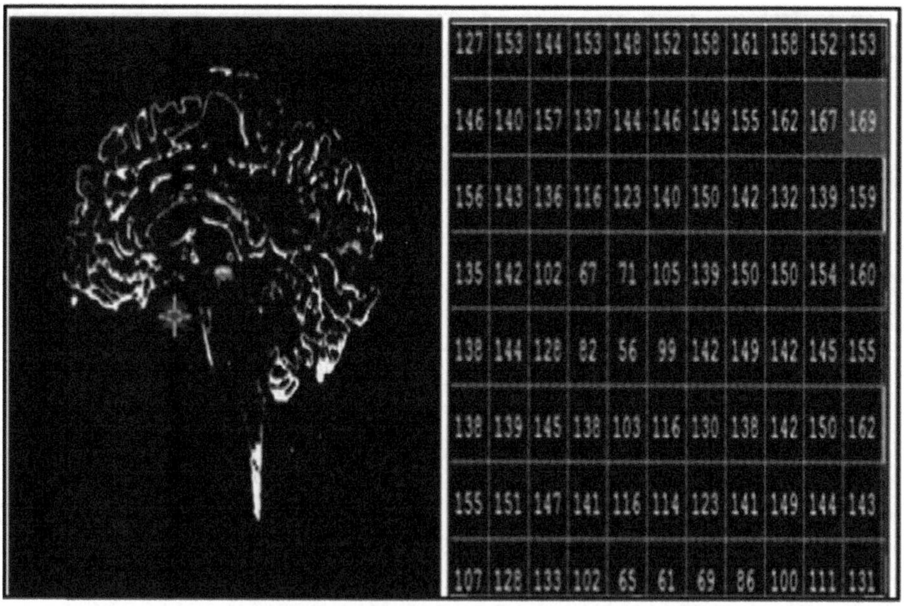

127	153	144	153	148	152	158	161	158	152	153
146	140	157	137	144	146	149	155	162	167	169
156	143	136	116	123	140	150	142	132	139	159
135	142	102	67	71	105	139	150	150	154	160
138	144	128	82	56	99	142	149	142	145	155
138	139	145	138	103	116	130	138	142	150	162
155	151	147	141	116	114	123	141	149	144	143
107	128	133	102	65	61	69	86	100	111	131

Fig. 7. Edge map with region of convergence.

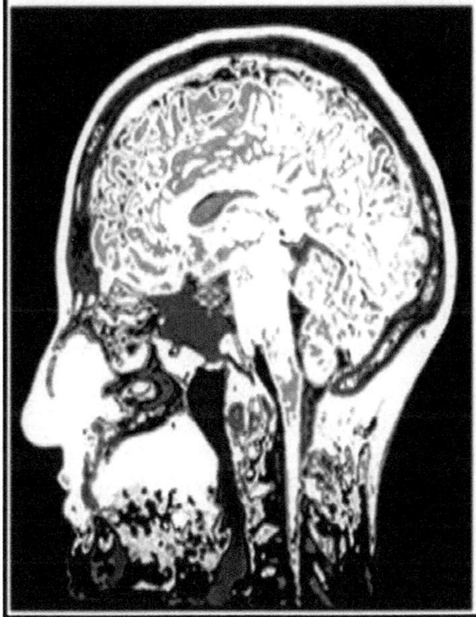

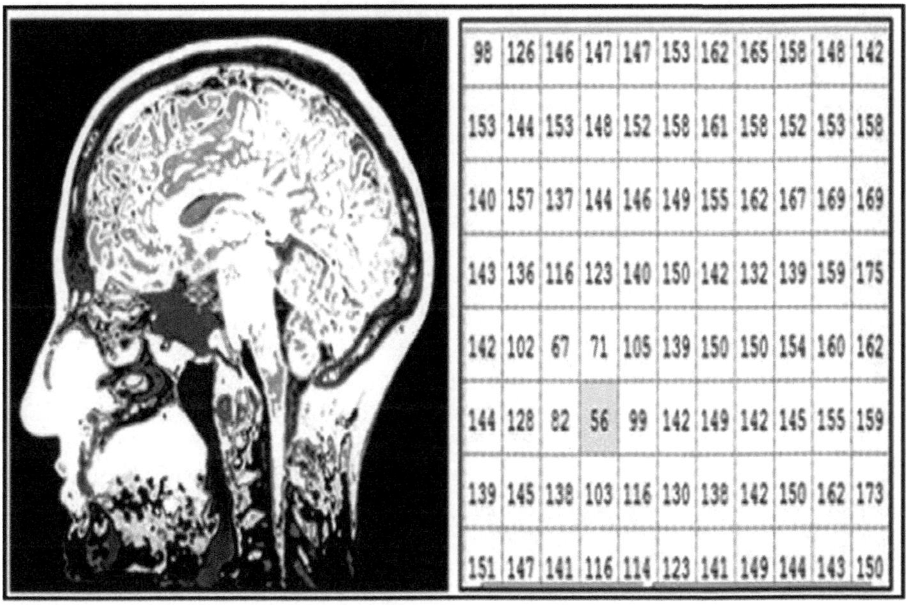

98	126	146	147	147	153	162	165	158	148	142
153	144	153	148	152	158	161	158	152	153	158
140	157	137	144	146	149	155	162	167	169	169
143	136	116	123	140	150	142	132	139	159	175
142	102	67	71	105	139	150	150	154	160	162
144	128	82	56	99	142	149	142	145	155	159
139	145	138	103	116	130	138	142	150	162	173
151	147	141	116	114	123	141	149	144	143	150

Fig. 8. FSE proton density weighted image.

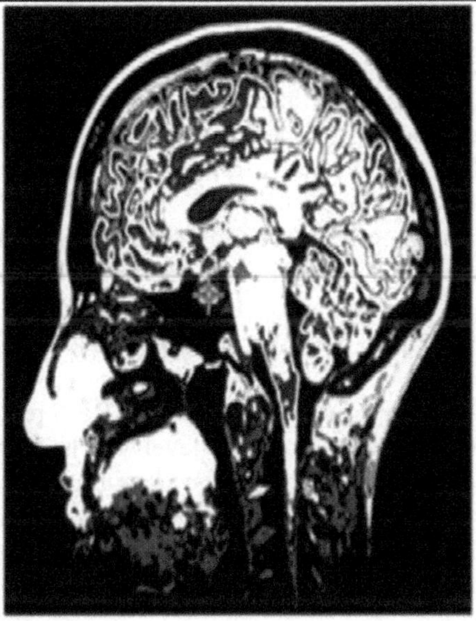

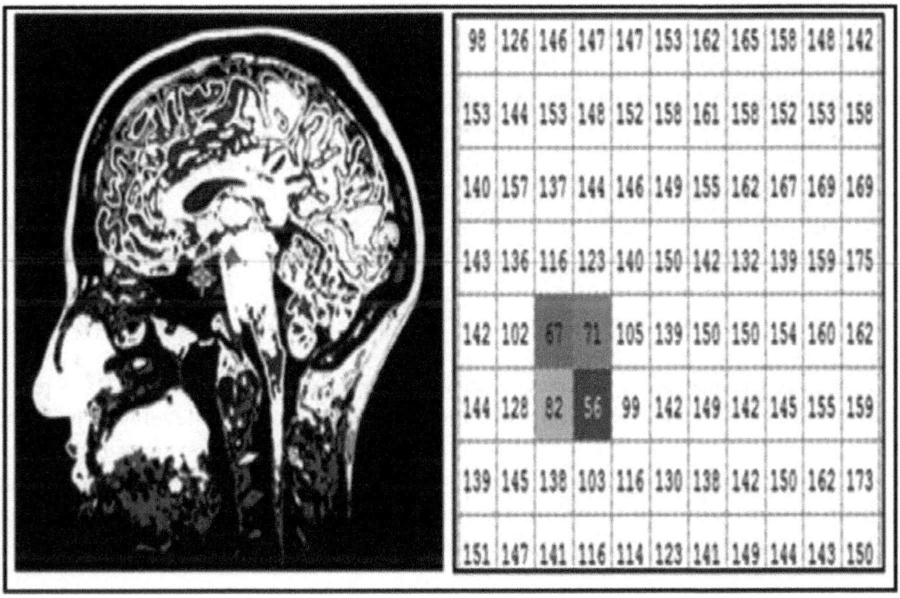

98	126	146	147	147	153	162	165	158	148	142
153	144	153	148	152	158	161	158	152	153	158
140	157	137	144	146	149	155	162	167	169	169
143	136	116	123	140	150	142	132	139	159	175
142	102	67	71	105	139	150	150	154	160	162
144	128	82	56	99	142	149	142	145	155	159
139	145	138	103	116	130	138	142	150	162	173
151	147	141	116	114	123	141	149	144	143	150

Fig. 9. T2-weighted MR image with iteration.

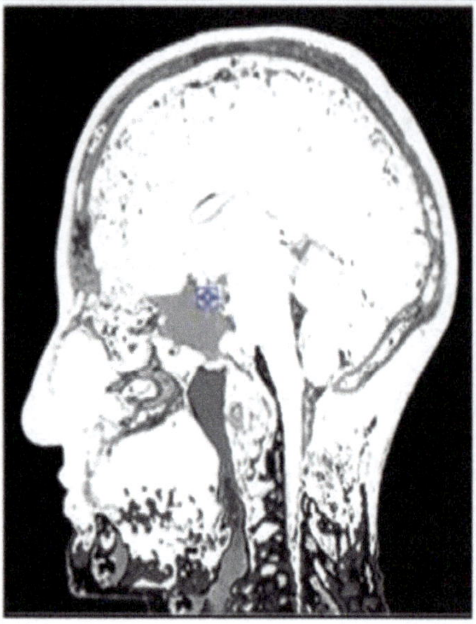

100	127	153	144	153	148	152	158	161	158	152
129	146	140	157	137	144	146	149	155	162	167
136	156	143	136	116	123	140	150	142	132	139
137	135	142	102	67	71	105	139	150	150	154
144	138	144	128	82	56	99	142	149	142	145
141	138	139	145	138	103	116	130	138	142	150
133	155	151	147	141	116	114	123	141	149	144
103	107	128	133	102	65	61	69	86	100	111

Fig. 10. Total intracranial image with iteration.

4 Conclusion

Magnetic resonance imaging (MRI) is adjacent to nature and a multi-modality practice which provides the complementary in sequence about dissimilar aspects of diseases. As a competent image (CE) contrast enhancement tool, (AGC) adaptive gamma correction was relating gamma parameter with the (CDF) cumulative distribution function in the conventional method which is to engender the function of the pixel gray levels within an image. AGC should deal sound within the mainly dimmed images, but fails for worldwide intense images and also the dimmed images within the local bright regions. Such two categories of images which are observed from MRIs, the brightness-distorted images are widespread in genuine scenario, such as to the improper exposure and white object regions.

Attenuate the deficiencies here we intend an improved the aspects by (RIC) region by iteration method of convolution and (S-KC) segmentation by k-levels of clustering. In this two methods by iteration method and the other segmentation in multiple levels of clustering which is to get the required response of the MRIs. Both the levels and methods are analyzed in a closed system which to eliminate the unwanted signals and get the better performance in the MRIs imaging.

Declaration of Source files
The source of the image file is obtained by using the code in mat lab and then the image is stored as ‘science.gif’ for further processing.

```
f=imread('C:\Users\RamakrishnanDevi\Desktop\MATLAB'); load mri.mat; D1=double(squeeze(D));DIM=size(D1);[X,Y,Z]=meshgrid(1:DIM(2),1:DIM(1),1:DIM(3));
h1=subplot(2,2,1);imagesc(D1(:,:,round(DIM(3)/2)),[min(D1(:))max(D1(:))]);
colormap(gray);title('axial');colorbar;xlabel('x');ylabel('y');
f=imread('C:\Users\RamakrishnanDevi\Desktop\science.gif');
imshow(f),title('original image');
```

References

1. Essayed, W.I., et al.: White matter tractography for neurosurgical planning: a topography-based review of the current state of the art. NeuroImage: Clin. **15**, 659–672 (2017)
2. Soni, N., Mehrotra, A., Behari, S., Kumar, S. Gupta, N.: Diffusion-Tensor Imaging and tractography application in pre-operative planning of intra-axial brain lesions. Cureus **9** (2017)
3. Zakaria, H., Sameah Haider, I.L.: Automated whole brain tractography affects preoperative surgical decision making. Cureus **9** (2017)
4. Calabrese, E.: Diffusion tractography in deep brain stimulation surgery: a review. Front. Neuroanat. **10**, 45 (2016)
5. Nakajima, T., et al.: MRI-guided subthalamic nucleus deep brain stimulation without microelectrode recording: can we dispense with surgery under local anaesthesia? Ster. Funct. Neurosurg. **89**, 318–325 (2011)
6. Andrews-Hanna, J.R., et al.: Disruption of large-scale brain systems in advanced aging. Neuron **56**, 924–935 (2007)
7. Pineda-Pardo, J.A., et al.: Guiding functional connectivity estimation by structural connectivity in MEG: an application to discrimination of conditions of mild cognitive impairment. NeuroImage **101**, 765–777 (2014)
8. Upadhyay, J., et al.: Function and connectivity in human primary auditory cortex: a combined fMRI and DTI study at 3 Tesla. Cereb. Cortex **17**, 2420–2432 (2007)
9. Guye, M., et al.: Combined functional MRI and tractography to demonstrate the connectivity of the human primary motor cortex in vivo. Neuro-image **19**, 1349–1360 (2003)
10. Vijayakumar, P., Rama Reddy, S.: Simulation and experimental results of low noise SMPS system using forward converter. Asian Power Electron. J. (APEJ) **9**(1), 1–7 (2015). 2010-01-0245
11. Vijayakumar, A.P., Devi, R.: Simulation and Experimental Analysis of Forward Converters, p. 121. LAP-Lambert Academic Publishing, Germany (2016). ISBN-13:978-3-659-95849-6
12. Vijayakumar, A.P., Devi, R.: Investigations on forward converters using LC, PI and Bi-quad high frequency filters. LAP-Lambert Academic-Publishing, Germany, p. 149 (2016). ISBN-13:978-3-659-97962-0
13. Vijayakumar, A.P., Devi, R.: Closed loop controlled forward converter with RCD Snubber using PI, fuzzy logic and artificial neural network controller. Ann. "DUNAREA JOS" Univ. Galati Fascicle III Electrotech. Electron. Autom. Control. Inform. **39**(2) (2016). ISSN 2344-4738, ISSN-L1221-454X

A Depth Study on Suicidal Thoughts in the Online Social Networks

S. Kavipriya(✉) and A. Grace Selvarani

Sri Ramakrishna Engineering College, Coimbatore 641022, India
kavi.1752003@srec.ac.in

Abstract. Online Social Network acts as platforms for users to communicate with one another and to share their feeling online. Few category of social media users utilizes the platform towards positing aggressive data. The automatic identification of the aggressive data can be identified by employing data mining algorithm utilizing machine learning principles. The standard machine learning approaches works with training, validation, and testing phases, and considered features such as part-of-speech, frequencies of insults and sentiment has been considered for emotions traits collected from the facebook data which leads several challenges to the system performance. In order to tackle particular issues, various technique employed in the literatures has been discussed in depth. In this paper, we undergo a detailed survey on technique employed to detect the suicide oriented traits on integration of sentiment analysis, Negative matrix factorization and summed up direct relapse calculation to analyze the connection between enthusiastic qualities and suicide chance and synthetic minority over- sampling technique is used in order to extract the information from a large collection of dataset. The ID3, C4.5, Apriori algorithm, association rule mining and naïve Bayes models has been used to predict who have suicidal ideation to repeatedly commit suicide attempts. Those techniques incorporate the linguistic features to regulate the durability of the quality on the count of self destruction. The issue attain were unique and remain to have a powerful segment with the count of self destruction. On this study, more meaningful insight about self destruction has been gathered.

Keywords: Sentiment analysis · Data mining · Emotion traits analysis · Online social networks · Opinion mining

1 Introduction

In short span of years, we have witnessed a growing number of online social networks to share the generated information about their views. The moment to articulate judgement and concept of networked is a expensively great: It is part of the right of definition, which is announced in the Universal Expression of Civil freedom. Due to number of platforms has increased, the number of aggressive interactions, such as Threatening interaction, cyber bullying and abusive speech has also been significantly growing. The goal of this paper is to analyse the technique employed to detect the suicide oriented conversation automatically. A dataset of 15,000 aggression- annotated

D. J. Hemanth et al. (Eds.): COMET 2019, LNDECT 35, pp. 1024–1030, 2020.
https://doi.org/10.1007/978-3-030-32150-5_104

Facebook posts for training and validating the traditional data classification systems has been crawled using available web plugins [1].

The main attention is focused on understanding the effects of merging new datasets for training models. Further proceeded with the conversion from toxicity to aggression by different classification model and compared them through one for only the original data for training and another using also toxic data [2]. In addition, some model extracted some classic features and analysed using different machine learning classification algorithms under training, validation, and testing. Majority of the approaches deals with feature extraction from the text to compute the weight. Alloyed macro averaged is F1-scores are used as the estimation of prediction algorithm. The respective score of F1 in the each class of machine learning algorithm is Alloyed by the proportion of the distressed class in training set and the closing score of F1 is the moderate of those respective scores of F in which each class of the aggressive words and normal words.

Further lexical highlights, phonetics highlights and pack of words were utilized which never neglects to comprehend the setting of the sentences. In certain regions the normal language process parser, was utilized to captured the semantic desires inside a sentence. These methodologies comprise in distinguishing the classification of the word. There have been a loads of concentrates on conclusion based strategies to distinguish inconsiderate words by applying assumption investigation and Latent Dirichlet Allocation theme display. The huge investigation of the paper is formulated as referenced, segment 2 gives issue articulation while region 3 depicts the survey of writing on forecast and arrangement model of suicide situated sentence recognition, further it is trailed by area 4 to characterize the arranged strategy as framework lastly segment 5 ended the investigation of the paper.

2 Problem Statement

Issue articulation utilized under suicide arranged discourse expectation utilizing AI calculations in the interpersonal organization like facebook are portrayed. Distinguishing these hazard factors is the initial phase in suicide aversion and traditional courses of action techniques have fair conditions were expressed underneath No substance based preferences are used. Exact writings don't give adequate word events.

3 Review of Literatures

In this section, we analyze the traditional methods applied to handle the suicide detection has been examined in detail on various aspects.

3.1 Suicidal Ideation Detection in Online User Content Using Supervised Learning

In this literature, detailed examination has been carried out on the content generated by the user and objective of initial disclosure of the self-destruction and potential self harming which stands as demanding possibility factors resulting in fortunate self destruction. Nowadays social networks like facebook are growing into a new manner for people to intend their self destruction habit [3]. In order to achieve the specified objective, supervised learning model based methods stands efficient on various aspects. In this work, supervised learning model utilizes the expression options and field definitions as Metadata to confess wealthy judgement that can be used as detection system on various emotion traits in terms of suicidal tendencies [4].

Emotion Traits can be strong negative feelings, anxiety, and hopelessness. In addition, topic of discussion can be about family, friends and social issues. Hence effective learning model composed of multiple processes to juice a lot of datasets of features, including analytical, semantic, grammatical, word inserting, and topic appearance has been embedded on the devised model. [5, 6] This model has been compared against classic supervised arrangement and two neural network models and the search results demonstrate the usefulness and common sense of the access and provide standards for the self destruction thoughts on the effective facebook platform.

3.2 Artificial Neural Network Model Towards Suicide Detection

In this literature, Artificial Neural network has been examined in detail to determine the Suicidal Ideation Detection. Machine learning methods especially supervised learning and natural language processing methods have also been applied in this field. The neural networks hold valuable indicators for effectively detecting individuals with suicidal intentions [7]. The major difficulty of state of art approaches towards suicide detection and prevention models are employed to understand and detect the complex risk factors and warning signs that may precipitate in the particular event. The ANN model quantifies contents and detects posts containing suicide-related content.

3.3 Machine Learning and Semantic Sentiment Analysis Based Algorithms for Suicide Indicator on Evolving Data Streams

In this writing, machine and estimation investigation were considered. The frame of mind examination is one of the new difficulties showed up in programmed language preparing with the appearance of social networks. It addresses an absence of innovative impacts identified with implosion by a methodology of building up a phrasing joined with implosion as computerized development utilizing wordnet tool [8]. Towards the compelling examination, Weka is utilized as an apparatus to separate valuable data and it is utilized to do the characterization. The procedure map the conditions and research their correspondence in the belief system. The fundamental undertaking is to utilize the put away records that contain terms and after that check the relationship with the words that the client utilizes in their judgement [9]. In this manner, it is valuable to demonstrate the conflict of reasonableness to users.

3.4 Suicide-Related Communication Using Machine Classification

In this literature [10] the way toward breaking down between the more concerned substance, for example, implosion, and some other self destructing field, for example, communicating of an implosion, recalling, befuddling and support has been analyzed in detail. It likewise means to recognize fun loving indication to implosion. The model made a course of action of controlled analyzer utilizing unwritten, anatomical, full of feeling and enthusiastic highlights got from facebook posts [11]. So as to improve the execution of the benchmark classifiers, an outfit classifier utilizing the Turn Timberland calculation and a Most extreme Likelihood casting a ballot arrangement choice technique has been assemble dependent on the result of fundamental analyzer. On assessment of the methodology, it completed a F-proportion of 0.728 by and large (for 7 classes, including self-destructive ideation) and 0.69 for the self-destructive ideation class. Furthermore, the outcomes considering the most huge prescient standard parts of the implosion class to give knowledge into the language utilized on facebook to convey what needs be annihilation has been summarized [12].

3.5 Depression Detection Model Based on Sentiment Analysis

In this literature, detailed analysis is carried out on the depression detection model which is enabled using sentiment analysis model [13]. It is employed to detect the depression through vocabulary and man-made rules with various data inclination. In addition, Subject-independent analysis and Subject-dependent analysis is been included to detect the polarity of the subject as abstract or target dependent features. Meanwhile sentence structure patterns and calculation rules are derived for each post from the facebook based on linguistics and latent features. The polarity of sentence is determined by the polarities and positions of its sub-sentences, as the position of a sub sentence can indicate its importance as its connection to mood and mood swing patterns stands important. To understand the basic patterns in mood transition patterns across social capital cohorts, hierarchical factor analysis models helps to provide the outcome [14] (Fig. 1).

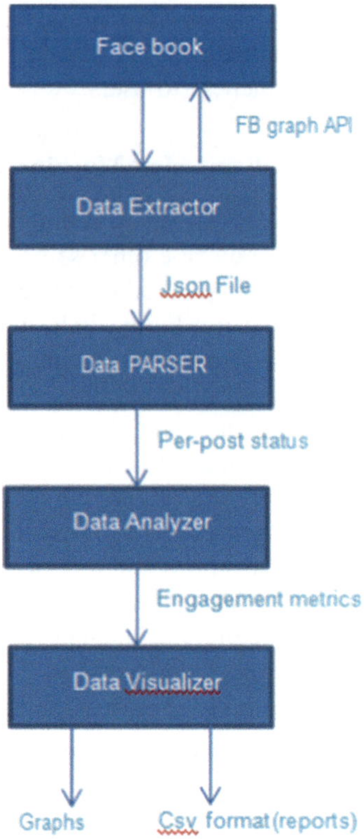

Fig. 1. Software architecture of face book

4 Outline of Proposed Model

On subjective analysis of the state of art approaches, it has envisioned to model predictive model based on deep learning architecture which is inclusive of sentiment analysis and hidden markov model. A large part of the research efforts involves the expansion of analytical models based on ancient data. A lot of models that can be fitted to certain data, a decisive issue is the selection of the most efficient redirection model [15]. Most often this selection is based on comparisons of various accuracy measures that are functions of the model's corresponding errors. The model aggregates the linguistics and latent features of the user defined sentence in the facebook platform. It functionalizes as automated detection model. Bayesian nonparametric factor analysis is employed to design and understand the peculiar factors that underlie mood changeover [16]. The model includes the Parts of speech, frequent item set, association rules and covariance - correlation computations.

5 Conclusion

In this paper, illustrative analysis on suicide ideation using emotion traits on the user defined post has been exploited. On this review, various emotions categorizing model on aggressive data has been examined. It has helped to model the automatic identification of the aggressive data through utilization of deep learning model with training, validation, and testing phases. The ID3, C4.5, Apriori algorithm, association rule mining and naïve Bayes models has been used to model deep learning architecture to categorize the emotion in terms of the characteristics of individuals who have suicidal ideation.

References

1. Statista: Most popular reasons for internet users worldwide to use social media as of 3rd quarter 2017. https://www.statista.com/statistics/715449/socialmedia-usage-reasons-worldwide/
2. Kamps, J., Marx, M., Mokken, R.J., De Rijke, M.: Using WordNet to measure semantic orientations of adjectives. In: Proceedings of International Conference Language Resource Evaluation, pp. 1115–1118 (2004)
3. Hu, M., Liu, B.: Opinion feature extraction using class sequential rules. Presented at the AAAI Spring Symposium Computational Approaches Analyzing Weblogs, Palo Alto, CA, USA. Paper AAAI-CAAW 2006 (2006)
4. Taboada, M., Brooke, J., Tofiloski, M., Voll, K., Stede, M.: Lexicon-based methods for sentiment analysis. J. Comput. Linguist. **37**(2), 267–307 (2011)
5. Google Cloud: GoogleCloudTranslationAPIDocumentation. https://cloud.google.com/translate/docs/
6. Steven Bird, E.L., Klein, E.: Natural Language Processing with Python. O'Reilly Media (2009)
7. Akaichi, J., Dhouioui, Z., Lopez-HuertasPerez, M.J.: Text mining Facebook status updates for sentiment classification. In: System Theory, Control and Computing (ICSTCC), 2013 17th International Conference, Sinaia, pp. 640–645 (2013)
8. Ku, L.-W., Liang, Y.-T., Chen, H.-H.: Opinion extraction, summarization and tracking in news and blog corpora. In: Proceedings of AAAI SpringSymposium, Computational Approaches Analyzing Weblogs, pp. 100–107 (2006)
9. Wang, X., Zhang, C., Ji, Y., Sun, L., Wu, L.: A depression detection model based on sentiment analysis in micro-blog social network. In: PAKDD Workshop, pp. 201–213 (2013)
10. Golbeck, J.A.: Computing and applying trust in web-based social networks. Ph.D. dissertation, Graduate School of the University of Maryland, CollegePark (2005)
11. Tai, Y.M., Chiu, H.W.: Artificial neural network analysis on suicide and self-harm history of Taiwanese soldiers. In: Second International Conference on Innovative Computing, Information and Control (ICICIC 2007), p. 363, Kumamoto, Japan. IEEE (2007)
12. Witten, I.H., Frank, E., Hall, M.A.: Data Mining: Practical Machine Learning Tools and Techniques. Google eBook (2011)
13. Lewis, D.D., Yang, Y., Rose, T.G., Li, F.: Rcv1: a new benchmark collection for text categorization research. J. Mach. Learn. Res. **5**, 361–397 (2004)

14. De Choudhury, M., Gamon, M.: Predicting depression via social media. In: Proceedings of Seventh International AAAI Conference on Weblogs Social Media, vol. 2, pp. 128–137 (2013)
15. Ramirez-Esparza, N., Chung, C.K., Kacewicz, E., Pennebaker, J.W.: The psychology of word use in depression forums in English and in Spanish: testing two text analytic approaches. Association for the Advancement of Artificial Intelligence (www.aaai.org)
16. Taboada, M., Brooke, J., Tofiloski, M., Voll, K., Stede, M.: Lexicon- based methods for sentiment analysis. J. Comput. Linguist. **37**(2), 267–307 (2011)

RETRACTED CHAPTER: A Brief Survey on Multi Modalities Fusion

M. Sumithra[1(✉)] and S. Malathi[2]

[1] Sathyabama Institute of Science and Technology, Chennai, India
manojhari789@gmail.com

[2] Panimalar Engineering College, Chennai, India
malathi.raghuram@gmail.com

Abstract. Medical images are taken by using different modalities like Magnetic resonance imaging (MRI), positron emission tomography (PET) image, computed tomography (CT), X-ray and Ultrasound. Each and every modalities has its own pros and cons. In now a days there are many modality images are fused and getting very good resultant image. This resultant image will give very good analysis about the disease. We can easily find out the disease portion with its exact circumference. Magnetic resonance imaging (MRI) and positron emission tomography (PET) image fusion is a late half breed methodology utilized in a few oncology applications. The MRI image demonstrates the brain tissue life structures and does not contain any useful data, while the PET image the mind work and has a low spatial goals. An ideal MRI–PET combination technique safeguards the practical data of the PET image and includes spatial attributes of the MRI image with the less conceivable spatial twisting. In this paper we discussed about different types of modalities fusing and which will get good result to analyse the disease perfectly and accurately.

Keywords: MRI · PET · CT · Ultrasound · Fusion · Modality

1 Introduction

In Medical fusion they are using different modalities to get diseased image. Ultrasound is the safest form of medical imaging and has a wide range of applications. There are no hurtful impacts when utilizing ultrasound and it's a standout amongst the most financially savvy types of medicinal imaging accessible to us, paying little respect to our strength or conditions. Ultrasound utilizes sound waves as opposed to ionizing radiation. High-recurrence sound waves are transmitted from the test to the body through the leading gel, those waves at that point bob back when they hit the diverse structures inside the body and that is utilized to make a image for finding. Another sort of ultrasound generally utilized is the 'Doppler' – a somewhat extraordinary system of utilizing sound waves that permits the blood course through conduits and veins to be seen. Because of the insignificant danger of utilizing Ultrasound, it's the principal decision for pregnancy, yet as the applications are so wide – crisis determination, heart, spine and inside organs – it will in general be one of the primary ports of call for some patients.

X-ray imaging the most established yet a standout amongst the most much of the time utilized imaging types. We as a whole know, and have most likely had no less

The original version of this chapter has been retracted. The retraction note to this chapter can be found at https://doi.org/10.1007/978-3-030-32150-5_163

D. J. Hemanth et al. (Eds.): COMET 2019, LNDECT 35, pp. 1031–1041, 2020.
https://doi.org/10.1007/978-3-030-32150-5_105

than one X-ray throughout our lives. Found in 1895, X-rays are a type of electromagnetic radiation. X-rays take a shot at a wavelength and recurrence that we can't see with the bare, human eye, yet can enter through the skin to make an image of what's happening underneath. Normally utilized for diagnosing issues with the skeletal framework, X-rays can likewise be utilized to distinguish malignancy through mammography and stomach related problems through barium swallows and bowel purges. X-rays are broadly utilized as they are ease, fast and moderately simple for the patient to persevere. Be that as it may, there are dangers related with the utilization of radiation for X-ray imaging. Each time a patient has a X-ray they get a portion of radiation. This can proceed to cause radiation-initiated malignancy or waterfalls further down the road or cause an aggravation in the development of an incipient organism or embryo in a pregnant patient. The greater part of these dangers are relieved by possibly utilizing X-rays when entirely important, and right protecting of the body.

CT or 'CAT' checks are a type of X-ray that makes a 3D image for analysis. Computer tomography (CT) or computed axial tomography (CAT) utilizes X-rays to create cross-sectional images of the body. The CT scanner has a vast round opening for the patient to lie on a mechanized table. The X-ray source and an identifier at that point pivot around the patient delivering a thin 'fan-molded' light emission rays that goes through an area of the patient's body to make a depiction. These depictions are then examined into one, or various images of the inner organs and issues. CT filters give more prominent lucidity than ordinary X-rays with progressively definite images of the inner organs, bones, delicate tissue and veins inside the body.

The advantages of utilizing CT filters far surpass the dangers which, as with X-rays, incorporate the danger of disease, mischief to an unborn youngster or response to a differentiation specialist or color that might be utilized. Much of the time, the utilization of a CT filter keeps the requirement for exploratory medical procedure. It is significant that when examining youngsters, the radiation portion has been brought down than that utilized for grown-ups to keep an outlandish portion of radiation for the important imaging to be acquired. In numerous clinics you'll locate a pediatric CT scanner thus.

MRI scans make intricate images without utilizing unsafe radiation. Magnetic Resonance Imaging (MRI) utilizes a solid attractive field and radio waves to create images of the body that can't be seen well utilizing X-rays or CT scans, i.e. it empowers the view inside a joint or tendon to be seen, as opposed to only the outside. Regularly used to look at inside body structures to analyze strokes, tumors, spinal string wounds, aneurysms and mind work. As we probably am aware, the human body is made for the most part of water, and each water atom contains a hydrogen core (proton) which ends up adjusted in an attractive field. A MRI scanner utilizes a solid attractive field to adjust the proton 'turns', a radio recurrence is then connected which makes the protons 'flip' their twists previously coming back to their unique arrangement.

Protons in the distinctive body tissues come back to their typical twists at various rates so the MRI can recognize different kinds of tissue and distinguish any variations from the norm. How the atoms 'flip' and come back to their typical turn arrangement are recorded and prepared into a image. X-ray doesn't utilize ionizing radiation and is progressively being utilized amid pregnancy with no symptoms on the unborn youngster revealed. Be that as it may, there are dangers related with the utilization of MRI checking and it isn't suggested as a first stage finding. As solid magnets are utilized, any sort of metal embed, counterfeit joint, and so on., can cause a danger – they can be moved or warmed up inside the attractive field. There have been a few revealed situations where patients with pacemakers have kicked the bucket using MRI. The boisterous clamor from the scanner additionally requires the requirement for ear security. One thing we do need to know about as therapeutic experts in a period of heightening medicinal expenses and expanding request is that we're utilizing the best assets accessible to address the issues of our patients. That implies a watchful choice on the correct therapeutic imaging to be utilized for the patient and their potential analysis.

2 Comparison Between CT and MRI

	CT	MRI
Detected/capture by	X-rays	Radio waves and magnets
Diagnose issues	Bone fractures, tumors, cancer monitoring, finding internal bleeding	Joints, brain, wrists, ankles, breasts, heart, blood, vessels
Cost	Less expensive	More expensive
Risks	Harm to unborn babies, a very small dose of radiation, a potential reaction to the use of dyes	Possible reactions to metals due to magnets, loud noises from the machine causing hearing issues, increase in body temperature during long MRIs, claustrophobia
Benefits	CT scan is faster and can provide images of tissues, organs, and skeletal structure	An MRI is highly adept at capturing images which abnormal tissues within the body. MRIs are more detailed in their images
Advantages	Shorter imaging time Low cost of scanning Better spatial resolution (represented in dots per inch) Good for extra-axial brain tumour assessment Superior in detection of calcifications, skull erosion, penetration, destruction	Good in demonstration of edema of parenchyma (early sign for tumour detection) Accurate delectating extent of edema, tissue characterization and compression effects Better detection of mass effects and atrophy High neuroanatomical definition (tissue differentiation) Accurate detection of vascularity of tumour (in various planes acquisition)
Limitations	Poor definition of edema Only one plane acquisition and most of the time non-isotropic X-ray radiation risk Poor tissue characterization Imaging of posterior fossa is limited due to bone artifacts	Poor detection of calcification and bone erosions Not possible in intraoperative assessment (course of surgical operation) Lower spatial fidelity Sometime sequence are very time consuming

3 Literature Survey

Yin et al. [1] The NSST disintegration is first performed on the source images to acquire their multiscale and multi direction representation. The high-recurrence groups are intertwined by a parameter–adaptive pulse-coupled neural network (PA-PCNN) demonstrate, in which all the PCNN parameters can be adaptively evaluated by the information band. The low-recurrence groups are converged by a novel technique that at the same time tends to two pivotal issues in medicinal image combination, to be specific, vitality protection and detail extraction. At last, the intertwined image is remade by performing converse NSST on the melded high-recurrence and low-recurrence groups. The adequacy of the proposed technique is confirmed by four distinct classes of medicinal image combination issues [computed tomography (CT) and attractive reverberation (MR), MR-T1 and MR-T2, MR and positron outflow tomography, and MR and singlephoton emanation CT] with in excess of 80 sets of source images altogether. Give to grow progressively powerful combination methodologies, for example, district versatile based ones to additionally enhance the calculation execution. But the drawback is the image fusion technique in clinical applications are not done to the related basic issues, for example, information preprocessing and image registration for multimodality average images are not properly. Additionally, the capability of the PA-PCNN show for other image combination issues, for example, multifocus image fusion, infrared and obvious image combination, etc.

Lian et al. [2] A co-clustering calculation is proposed to simultaneously section 3D tumors in PET-CT images, taking into account that the two reciprocal imaging modalities can join utilitarian and anatomical data to enhance division execution. The hypothesis of conviction capacities is received in the proposed strategy to model, circuit, and reason upon questionable and uncertain information from noisy and hazy PET-CT images. To guarantee dependable division for every methodology, the separation metric for the measurement of bunching twists and spatial smoothness is iteratively adjusted amid the grouping system. Then again, to empower reliable division between various modalities, an explicit setting term is proposed in the grouping target work. Besides, amid the iterative enhancement process, bunching results for the two unmistakable modalities are additionally balanced by means of a conviction capacities based data combination technique. An explicit setting term has been proposed in the system of conviction capacities to energize reliable division between the two particular mono-modalities. To viably join reciprocal data in PET and CT images, during the minimization of the developed cost work, grouping results in the two mono-modalities have been iteratively balanced by melding them through the Dempster's blend rule. But the drawback is two mono modalities should have taken at same time, same instance of the same patient. It should have some drawbacks to non matching of the images of PET and CT.

Bernal et al. [3] Automated egocentric human action and activity recognition from multimodal data, with a target application of monitoring and assisting a user perform a multistep medical procedure. We propose a supervised deep multimodal fusion framework that relies on concurrent processing of motion data acquired with wearable sensors and video data acquired with an egocentric or body-mounted camera.

Information gathered from wearable video-procurement gadgets is broke down utilizing the proposed combination method sometime later to quantify process consistence. With sub sampling at the purpose of video catch, the battery life of the video procurement gadget might be reached out to cover various sessions per client. The desire is that as gadgets keep on getting progressively able and control proficient, a greater amount of the preparing can be performed locally in the gadget with regularly diminishing dependence on cloud assets. But the drawback of this paper is not done in real time, in-device inference and feedback is not taken by the improvement.

Zhang et al. [4] Efficient colorful FPM recreation strategy utilizing multi-goals wavelet color combination. Their new calculation is a versatile denoising strategy by breaking down the commotion data of the dim casing. Both reproduction and test results are completed to approve the strategies. Results exhibit that the imaging clamor is smothered and the bright reproduction is of high efficiency and quality. The fundamental thought of wavelet-FPM is to combine a low-goals shading reference image and the high-goals FPM reproduction after effect of monochromatic powers. To utilize wavelet-FPM, just a single all the more low-goals shading image is required which spares time utilization of catch process. But the drawback of this method not efficiently used the raw data of FPM.

Shi et al. [5] IVUS imaging-driven 3-D intravascular reconstruction techniques empower precise analysis and quantitative estimations of intravascular problems to encourage ideal treatment assurance. Such reproduction expands the IVUS imaging methodology from unadulterated analytic help to intraoperative route and direction and backings both remedial alternatives and interventional tasks. Here they exhibits an exhaustive overview of innovative advances and ongoing advancement on IVUS imaging-based 3-D intravascular reproduction and its cutting edge applications. Confinements of existing innovations and prospects of new advancements are additionally examined. The joined utilization of IVUS and OCT gives both incredible entrance profundity and super-high spatial goals for intravascular imaging. This blend underpins an incredible recognizable proof rate of defenseless and beginning time plaques, and conquers the difficulties related with co-enrollment. Further inquire about undertakings are relied upon to make the coordinated mixture IVUS-OCT catheter actually and clinically develop.

El-Hariri et al. [6] A strategy that can intra-operatively find bone structures utilizing followed ultrasound (US), registers to the relating pre-agent computed tomography (CT) information and produces 3D AR perception of the worked careful scene through a head-mounted presentation. This technique conveys optically-followed US, bone surface division from the US and CT image volumes, and multimodal volume enrollment to adjust pre-agent to the relating intra-agent information. The upgraded careful scene is then envisioned in an AR system utilizing a HoloLens. But the drawback is not concentrated on improving the accuracy of this system to a level that may enable clinical translation.

Gibson et al. [7] An enlistment free deep learning-based division calculation for eight organs that are applicable for route in endoscopic pancreatic and biliary techniques, including the pancreas, the gastrointestinal tract (stomach and duodenum) and encompassing organs (liver, spleen, left kidney, and gallbladder). We specifically thought about the division exactness of the proposed strategy to the current profound learning and MALF techniques in a cross-approval on a multi-centre data set with 90 subjects. In any case, the downside is that utilization of an express spatial earlier was additionally vital, proposing that convolutional neural systems are certainly encoding spatial priors, in spite of their indicated translational invariance. The consequently produced divisions of stomach life structures have motivate the possibility to help image guided route in pancreatobiliary endoscopy methodology.

Huang et al. [8] Improving the visual complexity by versatile tissue lessening and dynamic range extending. By means of segment decay and tissue constriction, a parametric change display was found to produce many upgraded images on the double. At last, a ensemble system was proposed for intertwining these upgraded images and delivering a high-differentiate yield in both splendid and dull areas. We have utilized estimation measurements to assess our framework and accomplished promising scores in each. A web based testing framework was likewise worked for emotional assessment.

Hazarika et al. [9] A combination based model with the guide of discriminant connection examination to order electroencephalogram signals is proposed. Sets of different component networks are created from signs in both time and wavelet areas for study-explicit classes, which are additionally disintegrated to infer a lot of sub-multi-see highlights pursued by enhancement to separate measurable highlights. Highlights are linked utilizing highlight combination procedure to determine low request discriminant highlights. Furthermore, the investigation of fluctuation was additionally performed to approve the examination. CCA and its augmentation are broadly utilized in multi-information handling techniques to break down the shared connections between two arrangements of factors. But the disadvantage is not concentrated on performance in diagnosing disorders using large-volume data for viable practical implementations.

Hu et al. [10] Versatile SMCCA to beat the issue by presenting versatile loads when joining pairwise covariances. Both recreation and genuine information examination demonstrate the outperformance of versatile SMCCA as far as highlight choice over traditional SMCCA and SMCCA with settled loads. Substantial scale numerical examinations demonstrate that versatile SMCCA unites as quick as ordinary SMCCA. While applying it to imaging (epi)genetics investigation of schizophrenia subjects, we can recognize noteworthy (epi)genetic variations and mind areas, which are steady with other existing reports. Furthermore, a few huge mental health related pathways, e.g., neural cylinder improvement, are distinguished by our model, exhibiting imaging epigenetic affiliation might be disregarded by traditional SMCCA. However, the disservice isn't to oblige the combination of different sorts of datasets for malady components study and illness determination.

Huang et al. [11] Another methodology for the combination of intravascular ultrasound and optical lucidness tomography pullbacks to significantly enhance the utilization of those two kinds of therapeutic images. It likewise displays another two-stage multimodal combination structure utilizing a coarse tone enrollment and a wavelet combination technique. In the coarse enrollment process, we define a lot of new element focuses to coordinate the IVUS image and IV-OCT image. At that point, the enhanced quality image is gotten dependent on the reconciliation of the shared data of two kinds of images. At last, the coordinated enlisted images are combined with a methodology dependent on the new proposed wavelet calculation. The exploratory outcomes show the execution of the proposed new methodology for significantly upgrading both the accuracy and computational dependability. contrasted with other combination calculations, the wavelet based combination technique demonstrates its best exibility and elite. To exhibit the strongness of the proposed system, both the emotional evaluation and target assessment were performed. The subjective and quantitative examinations demonstrate the viability of the proposed system contrasted with the other enrollment and combination calculations. Yet, the downside of this paper is the endeavor to entire IVUS and IV-OCT pull-backs of coronary conduits is less and diminished the power of the enlistment.

Chartsias et al. [12] A multi-input, multi-yield end to end profound convolutional arrange for union of MR images, which we tried on three distinctive MRI datasets. a multi-input, multi-yield end to- display on the ISLES and BRATS informational collections and exhibit measurably critical upgrades over best in class strategies for single info undertakings. This enhancement increments further when different information modalities are utilized, illustrating the advantages of taking in a typical inert space, again bringing about a measurably noteworthy enhancement over the present best technique. At last, we illustrate our methodology on non skull-stripped cerebrum images, creating a factually noteworthy enhancement over the past best technique. Demonstrated that the model is hearty, performs well and can handle a wide range of difficulties, for example, strength to missing information, adapting only another decoder for a concealed methodology and notwithstanding blending new (inconspicuous) perspectives of the information. We see that such multimodal models could be well put to credit information on extensive databases (for example biobanks) unimodal methodologies. From an arrangement viewpoint they are less intricate (one versus a wide range of models to send/keep up), increasingly adaptable (new yields can be included with insignificant preparing) and all the more imperatively are hearty by exploiting data crosswise over information modalities, without being dependent on any of them.

Queirós et al. [13] The therapeutic image following tool kit (MITT)-a product bundle intended to ease customization of image following arrangements in the medicinal field. While its work process standards make it appropriate to work with 2-D or 3-D image groupings, its modules offer adaptability to set up computationally effective following arrangements, notwithstanding for clients with restricted programming aptitudes. **Glove is personalized in both C/C++ and MATLAB which is together a some differences of an factor based image following mathematical variation and accepting to follow various kinds of materials (i.e., forms, multi-shapes, surfaces, and multi-surfaces) with a small differentiated highlights. The tool kit is showed and find the number of times executing of its usage in the cardiology region gave, proving its flexibility, direct relaxedness, and time adeptness.**

Fischer et al. [14] The MR slices are stacked into volumes of consistent cardio-respiratory state (Section II-B). These volumes are registered to a reference phase (Section II-E.1) to estimate the 3-D motion. A regression model is built to relate the 3-D motion and the surrogate signal (Section II-E.2). A separate MR volume is acquired for segmenting the overlay. The intra-procedural steps for motion compensation in

fluoroscopy. The motion model is driven by a surrogate signal based on X-ray images and ECG (Section II-F). The motion is used to animate the segmentation as an overlay on the X-ray image in real time. The proposed method to stack slices based on cardiac and respiratory surrogate signals is relatively simple. Additionally, some other properties of the MR sequence for slice stacking are advantageous for our application. Firstly, only one scan is necessary instead of two, reducing the scan and setup complexity. Secondly, this scan resolves cardiac and respiratory motion, such that derived motion models can capture the dependency between them. Thirdly, slice stacking gives multiple cardiac and respiratory cycles, instead of one binned average. Last but not least, a multi-slice, real-time MR sequence is available on modern scanners from all major vendors. But the drawback is to not evaluate the value of animated overlays to the physicians in terms of reducing fluoroscopy time, contrast dose, and improving overall procedure success rates.

Xin et al. [15] A multimodal biometric framework for individual acknowledgment utilizing face, unique mark, and finger vein images. Tending to this issue, a proficient coordinating calculation that depends on optional figuring of the Fisher vector and utilizations three biometric modalities: face, unique finger impression, and finger vein. The three modalities are joined and combination is performed at the component level. Besides, in view of the strategy for highlight combination, the paper contemplates the phony element which shows up in the down to earth scene. The liveness identification is affix to the framework, identify the image is genuine or counterfeit dependent on DCT, at that point evacuate the phony image to decrease the impact of exactness rate, and increment the vigorous of framework. The trial results demonstrated that the planned structure can accomplish a superb acknowledgment rate and give higher security than a unimodal biometric-based framework, which are vital for an IoMT stage. Test results show that the proposed strategy accomplishes a brilliant acknowledgment rate and gives higher security than unimodal biometric-based frameworks. In any case, the disadvantage of this calculation is that it gives just the less security on the image which is demonstrates the less effectiveness.

Cao et al. [16] A locale versatile deformable enrollment technique for multimodal pelvic image enlistment. In particular, to deal with the extensive appearance holes, we initially perform both CT-to-MRI and MRI-to-CT image combination by multi-target relapse woodland. At that point, to utilize the integral anatomical data in the two modalities for controlling the enrollment, we select key focuses consequently from the two modalities and use them together to direct correspondence location in the locale versatile design. That is, basically use CT to set up correspondences for bone areas, and use MRI to set up correspondences for delicate tissue districts. The quantity of key focuses is expanded bit by bit amid the enrollment, to progressively direct the symmetric estimation of the distortion fields. Examinations for both intra-subject and between subject deformable enrollment indicate enhanced exhibitions contrasted and the cutting edge multimodal enlistment techniques, which show the possibilities of our strategy to be connected for the standard prostate malignant growth radiation treatment. In any case, the disadvantage is that here the both MRI and CT images are not combined and do it in productive way.

Qi et al. [17] Working memory brokenness, other cognitive areas could likewise be examined utilizing our strategy, for example, composite subjective scores, a standout amongst the most generally announced psychological deficiencies in SZ [18, 19], which would show up in our another work. Besides, MCCAR + jICA can be connected direct to think about other cerebrum ailments. Also, aside from the current clinical applications, the proposed strategy can be utilized to consider brain region which is correlated with some essential factors, for example, many undefined effect, knowledge remainder, medicine use and conduct values or even non-genetic influences on gene expression differentiations, it gives details and suggestions about a huge usage in the neuro imaging network. Usually the lower side is that MCCAR + jICA works on free from a constraint which is highlighted, as contrasting to the starting imaging information. A part of the passing data is missed when it used this method, a "highlight" will in common be more biddable than doing work with the expansive dimensional different details [20] and [21] shows a more space in which to connect the datas.

Yin [22] 3-D medicinal image fusion technique is proposed dependent on TSR with a "weighted normal" combination rule. Here, multimodal medicinal volumes are communicated through TSR which can misuse the over cuts data and safeguard the 3-D structure of restorative volumes. The combination recipes will differ from [23] and [24]. Be that as it may, the downside isn't thinking about the connection between various imaging modalities and Except for the job of loads, the loads can likewise decide the degrees of striking nature, and thus improve the exactness of distinguished remarkable data and decrease relics.

Hait et al. [25] Analyze several useful merits of this framework like, Signatures are sensitive to size, local contrast, and composition of structures; are invariant to translation, rotation, flip, and linear illumination change; and texture signatures are robust to the underlying structures. Various algorithms [26–29] rely on fundamental properties, such as intensity and color, gradient magnitude and orientation, textures and patterns. Some of these algorithms use features for learning [30, 31]. Among these descriptors, those based on histograms of patterns [32] or of oriented gradients [33] are rotation-variant, thus suitable for texture applications. But the drawback is it can't construct a unified generic framework applicable for different image modalities and image processing tasks and it won't provide conditions for the invariance of texture to structure.

4 Conclusion and Future Enhancement

Now a days the MRI and CT has many advantages and disadvantages available within it. Example in CT we can't take edema portions and back head bone tumor images and can't find accurate boundary of the tumor. The same way in MRI we can't measure spatial Lower, spatial fidelity and Poor detection of calcification and bone erosions. This concept will be carried out by using or merging both the MRI and CT images. MRI slices and CT frames will be merged and will find the exact boundary of the tumor in the brain.

References

1. Yin, M., Liu, X., Liu, Y., Chen, X.: Medical image fusion with parameter-adaptive pulse coupled neural network in nonsubsampled shearlet transform domain. IEEE Trans. Instrum. Meas. **68**(1), 49–64 (2019)
2. Lian, C., Ruan, S., Denœux, T., Li, H., Vera, P.: Joint tumor segmentation in PET-CT images using co-clustering and fusion based on belief functions. IEEE Trans. Image Process. **28**(2), 755–766 (2019)
3. Bernal, E.A., Yang, X., Li, Q., Kumar, J., Madhvanath, S., Ramesh, P., Bala, R.: Deep temporal multimodal fusion for medical procedure monitoring using wearable sensors. IEEE Trans. Multimed. **20**(1), 107–118 (2018)
4. Zhang, J., Xu, T., Chen, S., Wang, X.: Efficient colorful Fourier ptychographic microscopy reconstruction with wavelet fusion. In: IEEE Access, date of publication 29 May 2018, date of current version 29 June 2018
5. Shi, C., Luo, X., Guo, J., Najdovski, Z., Fukuda, T., Ren, H.: Three-dimensional intravascular reconstruction techniques based on intravascular ultrasound: a technical review. IEEE J. Biomed. Health Inform. **22**(3), 806–817 (2018)
6. El-Hariri, H., Pandey, P., Hodgson, A.J., Garbi, R.: Augmented reality visualisation for orthopaedic surgical guidance with pre- and intra-operative multimodal image data fusion. IEEE Trans. Healthc. Technol. Lett. **5**(5), 189–193 (2018)
7. Gibson, E., Giganti, F., Hu, Y., Bonmati, E., Bandula, S., Gurusamy, K., Davidson, B., Pereira, S.P., Clarkson, M.J., Barratt, D.C.: Automatic multi-organ segmentation on abdominal CT with dense V-Networks. IEEE Trans. Med. Imaging **37**(8), 1822–1834 (2018)
8. Huang, C.-C., Nguyen, M.-H.: X-Ray enhancement based on component attenuation, contrast adjustment and image fusion. IEEE Trans. Image Process. **28**(1), 127–141 (2019)
9. Hazarika, A., Sarmah, A., Borah, R., Boro, M., Dutta, L., Kalita, P., kumar dev Choudhury, B.: Discriminant feature level fusion based learning for automatic staging of EEG signals. Healthcare Technol. Lett. **5**(6), 226–230 (2018)
10. Hu, W., Lin, D., Cao, S., Liu, J., Chen, J., Calhoun, V.D., Wang, Y.P.: Adaptive sparse multiple canonical correlation analysis with application to imaging(epi)genomics study of schizophrenia. IEEE Trans. Biomed. Eng. **65**(2), 390–399 (2018)
11. Huang, C., Xie, Y., Lan, Y., Hao, Y., Chen, F., Cheng, Y., Peng, Y.: A new framework for the integrative analytics of intravascular ultrasound and optical coherence tomography images. IEEE Transl. Content Min. **6**, 2169–3536 (2018)
12. Chartsias, A., Joyce, T., Giuffrida, M.V., Tsaftaris, S.A.: Multimodal MR synthesis via modality-invariant latent representation. IEEE Trans. Med. Imaging **37**(3), 803–814 (2018)
13. Queirós, S., Morais, P., Barbosa, D., Fonseca, J.C., Vilaça, J.L., D'hooge, J.: MITT: medical image tracking toolbox. IEEE Trans. Med. Imaging **37**(11), 2547–2557 (2018)
14. Fischer, P., Faranesh, A., Pohl, T., Maier, A., Rogers, T., Ratnayaka, K., Lederman, R., Hornegger, J.: An MR-based model for cardio-respiratory motion compensation of overlays in X-Ray fluoroscopy. IEEE Trans. Med. Imaging **37**(1), 47–60 (2018)
15. Xin, Y., Kong, L., Liu, Z., Wang, C., Zhu, H., Gao, M., Zhao, C., Xu, X.: Multimodal feature-level fusion for biometrics identification system on IoMT platform. IEEE Transl. Content Min. **6**, 21418–21426 (2018)
16. Cao, X., Yang, J., Gao, Y., Wang, Q., Shen, D.: Region-adaptive deformable registration of CT/MRI pelvic images via learning-based image synthesis. IEEE Trans. Image Process. **27**(7), 3500–3512 (2018)

17. Qi, S., Calhoun, V.D., van Erp, T.G., Bustillo, J., Damaraju, E., Turner, J.A., Du, Y., Yang, J., Chen, J., Yu, Q., Mathalon, D.H., Ford, J.M., Voyvodic, J., Mueller, B.A., Belger, A., McEwen, S., Potkin, S.G., Preda, A., Jiang, T., Sui, J.: Multimodal fusion with reference: searching for joint neuromarkers of working memory deficits in schizophrenia. IEEE Trans. Med. Imaging **37**(1), 93–105 (2018)
18. Kraguljac, N.V., Srivastava, A., Lahti, A.C.: Memory deficits in schizophrenia: a selective review of functional magnetic resonance imaging (FMRI) studies. Behav. Sci. **3**(3), 330–347 (2013)
19. Lett, T.A., Voineskos, A.N., Kennedy, J.L., Levine, B., Daskalakis, Z.J.: Treating working memory deficits in schizophrenia: A review of the neurobiology. Biol. Psychiatry **75**(5), 361–370 (2014)
20. Calhoun, V.D., Adali, T.: Feature-based fusion of medical imaging data. IEEE Trans. Inf. Technol. Biomed. **13**(5), 711–720 (2009)
21. Smith, S.M., et al.: Correspondence of the brain's functional architecture during activation and rest. Proc. Nat. Acad. Sci. USA **106**(31), 13040–13045 (2009)
22. Yin, H.: Tensor sparse representation for 3-D medical image fusion using weighted average rule. IEEE Trans. Biomed. Eng. **65**(11), 2622–2633 (2018)
23. Liu, X., Mei, W., Du, H.: Structure tensor and nonsubsampled shearlet transform based algorithm for CT and MRI image fusion. Neurocomputing **235**, 131–139 (2017)
24. Yang, B., Li, S.: Pixel-level image fusion with simultaneous orthogonal matching pursuit. Inf. Fusion **13**(1), 10–19 (2012)
25. Hait, E., Gilboa, G.: Spectral total-variation local scale signatures for image manipulation and fusion. IEEE Trans. Image Process. **28**(2), 880–895 (2019)
26. Brox, T., Weickert, J.: A TV flow based local scale measure for texture discrimination. In: Proceedings of European Conference on Computer Vision, pp. 578–590. Springer, New York (2004)
27. Canny, J.: A computational approach to edge detection. IEEE Trans. Pattern Anal. Mach. Intell. **PAMI-8**(6), 679–698 (1986)
28. Comaniciu, D., Meer, P.: Mean shift: a robust approach toward feature space analysis. IEEE Trans. Pattern Anal. Mach. Intell. **24**(5), 603–619 (2002)
29. Arbeláez, P., Maire, M., Fowlkes, C., Malik, J.: Contour detection and hierarchical image segmentation. IEEE Trans. Pattern Anal. Mach. Intell. **33**(5), 898–916 (2011)
30. Martin, D.R., Fowlkes, C.C., Malik, J.: Learning to detect natural image boundaries using local brightness, color, and texture cues. IEEE Trans. Pattern Anal. Mach. Intell. **26**(5), 530–549 (2004)
31. Dollár, P., Zitnick, C.L.: Fast edge detection using structured forests. IEEE Trans. Pattern Anal. Mach. Intell. **37**(8), 1558–1570 (2015)
32. Ojala, T., Pietikäinen, M., Harwood, D.: A comparative study of texture measures with classification based on featured distributions. Pattern Recognit. **29**(1), 51–59 (1996)
33. Dalal, N., Triggs, B.: Histograms of oriented gradients for human detection. In: Proceedings of IEEE Computer Society Conference on Computer Vision Pattern Recognition, vol. 1, no. 1, pp. 886–893, June 2005

Sentimental Analysis Using Convolution Neutral Network Through Word to Vector Embedding for Patients Dataset

G. Parthasarathy[1(✉)], D. Preethi[2], Mary Subaja Christo[3], T. R. Soumya[2], and J. Saravanakumar[2]

[1] Department of Computer Science and Engineering, Jeppiaar Maamallan Engineering College, Anna University, Chennai, India
amburgps@gmail.com
[2] Saveetha School of Engineering, Thandalam, India
preethidhanasekaran15@gmail.com,
soumyatr.soumya@gmail.com,
saravanakumar2005@gmail.com
[3] School of C&IT, REVA University, Bengaluru, India
marysubaja@gmail.com

Abstract. Sentimental analysis involves the incorporation of working of a structure that explores sentiments seen in blog sections, comments, or tweet relating to a thing or point. In this proposed research, the authors have used speech recognition for collecting the opinions from patients relating to health care. The mental condition of a patient in respect of the health care has been identified through sentimental analysis based on the convolutional neural network using some convolutional filter (1D) in word2vec logistic synopsis. It features implicit or explicit determination of the patient's strength based on sentiments through results. It helps the physician in taking necessary action. Accuracy of the result is of help in the recommendation to the patient by the physicians and health care centers.

Keywords: Audio · Speech recognition · Opinion · Sentimental mining · Convolutional neural network · Word to vector (w2v)

1 Introduction

Sentimental analysis is the most recent advancement that concentrates on the emotion, feeling, disposition and opinion of their general understanding of a particular subject of the relating to a common man. This analysis is used for driving the quality enhancement and it further helps information experts in measuring popular opinion, statistical survey, screening the brand and item popularity and understanding client needs.

When this sentimental analysis is used in the health care centers, a start is made in getting the information relating to the patient experience in the form of web blog, comment, tweets and other social Media and some health care rating websites. It is generally based on a questionnaire module with multiple choice. But its drawback of the inability of people to express their own reviews.

D. J. Hemanth et al. (Eds.): COMET 2019, LNDECT 35, pp. 1042–1051, 2020.
https://doi.org/10.1007/978-3-030-32150-5_106

This methodology has an issue that it does not perform qualitative analysis leaving patients to pass on their treatment experiences at various health care centers through online reviews as text description. An open entryway is given for the measurement of the precision of feeling examination procedures against the patient's own quantitative examination. In idea determination, a sentimental mining feature marks an inquiry or a property of a component on which the patient can voice his emotions. In this paper, the authors propose a neutral network concept that recognizes such sentimental opinion based on the various convolution neural layers. The source input in the form of natural language is obtained as audio of the patient using the speech recognition technique [6] which convert the data to the text format and utilizes them in the convolution layer and pooling layer. The individual answers from open-finished inquiries contain the sentences or phrases with respect to the health care centers. Patient opinion plays a vital role in the measurement of the quality of health care and even enhancing the methodology and the principles of the centers.

Patient comments relating to a particular physician could have positive or negative opinion getting to know the emotions should be related to typical methods for surveying constant experience. The preparation of subjective information investigation is imperative and it can upgraded in the future.

2 Literature Survey

Sentimental analysis is carried on with a lexicon approach [11], with opinions stored in database dictionary. It is assigned on the basis of the polarity score and calculated. Speech extraction in the form part of speech while ME modelling converts them to the text format using a linguistic model [6]. Word to vector w2v [5] is used for word embeddings its weight to the individual word similarity based on cosine similarity and tf-idf algorithm for converting word to vector [4] on the basis of frequency. Convolution neural network is the multi-layer network of convolution and pooling layer and performed. Maximum pooling performed and then soft maximization is calculated [1]. Convolution is performed in the character level [5] and performed efficiently. Different types of option in analysis techniques is given as seed for research [11].

3 Existing Work

In an earlier research work, revelation and mining of the association between different social opinions and online record have been a critical drawback. Sentimental analysis is generally a Q-A model and so distinguishing the sentiments is a testing issue in the health care industry. Different avenues regarding different models of grouping and joining sentiment at word and sentence levels, have been explored together with promising outcomes. Most examinations center on human services rating locales. They provide an account of greatly constrained types of correspondence. Some provide details regarding the utilization of Twitter as a wellspring of data relating to the nature of consideration, despite the fact that these short, unstructured messages contain insignificant data based on a lexical approach [3]. In the existing system it is based on the sentiment-model technique is used.

4 Proposed Work

The existing system is based on the sentimental classification model technique [3] using convolutional neutral network [1] through word to vector. In this exploration work the contribution to the type of sound audit and web source is aggregated and it is put away in the wav format. Effective collection of the remarks from the patient in the health care centers is simple and proficient. This sound records is additionally changed over to the text content format utilizing the speech opinion through word embedding which undergoes neural nets of convolution sub layer and produces the output.

FIGURE 1 illustrates the proposed work flow.
INPUT: AUDIO in NLP, blogs, comments, twitter.
OUTPUT: Evaluation of result and suggestion relating to it.
REQUIREMENTS: PYTHON 3 COMPLIER, TENSOR FLOW FRAMEWORK, WINDOWS 7 AND UNIX OS, MICROPHONE.

Short Description: Sentimental analysis of the patient using the CNN with w2v_TF-IDF ALGORITHM (Fig. 1).

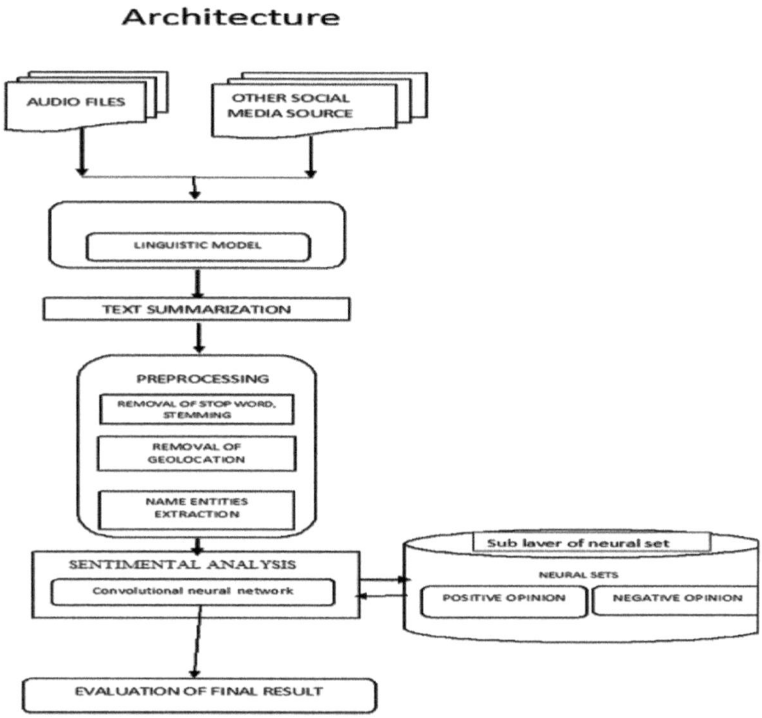

Fig. 1. Sentiment analysis model for proposed system

4.1 Pre-processing

Prepossessing expels the stop word, stemming, tokenization, lowercasing and expulsion of geolocation and extraction of name elements. Following this procedure, the data experiences sentimental mining. Sentiment word distinguishing proof constitute the next step.

4.2 Word to Vector–Embedding Technique

These models are known for their shallowness, as two layer neutral frame works that are setup for the reproduction of etymological settings of words Wordtovector [5] consider and takes a bag corpus of substances for its data, followed by making a vector space, generally of 200 eliminations, with the features of a thoughtful word in the corpus giving away a contrasting vector to the space. The arrangement of word vectors in the vector space such that the words that shares fundamental settings in the corpus are in proximity to one another in the space.

Two techniques CBOW - continuous bag of words or continuous skip gram are utilized Fig. 2. Continuous bag of word architecture, features a model predicting the present word from a window of encompassing setting words. There is no impact expectation for a request for setting words. A feature of the continuous skip gram architecture is the use of the present model by this model to enable anticipation of the encompassing window of the setting words.

The skip gram architecture weights setting words in the vicinity with greater vigor than setting words featuring inaccessible setting in Fig. 3. Measurement of cosine similarity does not exhibit any similarity as a 90° angle. But a total relationship of 1 is 0° angle does a whole overlap.

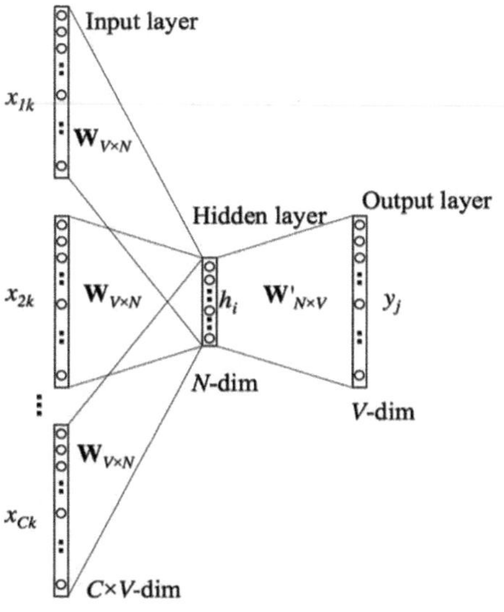

Fig. 2. Cbow model

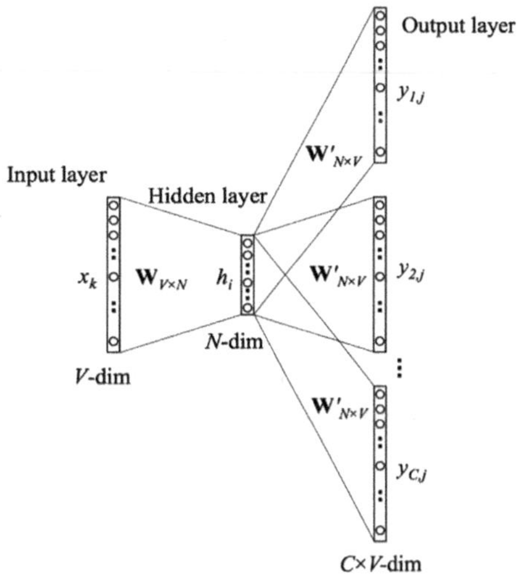

Fig. 3. Skip gram model

Vocabulary represented as vector $\boldsymbol{v_c}$ and v_w with respect to the content and target word representation respectively

$$\boldsymbol{v_c} = \boldsymbol{w_{c(K,.)}} \text{ and } \boldsymbol{v_w} = \boldsymbol{w'_{c(K,.)}} \tag{1}$$

If any word $\boldsymbol{w_i}$, with given content word c as input

$$\mathrm{P}(\boldsymbol{w_i}/\mathbf{c}) = \boldsymbol{y_i} = \boldsymbol{e^{ui}} / \sum_{i=1}^{V} \boldsymbol{e^{ui}} \textbf{ where } \boldsymbol{u_i} = \boldsymbol{v^T_{wi}} \cdot \boldsymbol{v_c} \tag{2}$$

Parameter $\Theta = (v_c, v_w)$ w, c €vocab for finding gradient as

$$\mathbf{L}(\Theta) = \sum_{w€ \text{ vocab}} \boldsymbol{log(P(w_i /c))} \tag{3}$$

$$\frac{dl(\theta)}{dv_w} = \boldsymbol{v_c}(1 - \mathbf{P}(\boldsymbol{w_i}/\mathbf{c})) \tag{4}$$

vector to content word are taken input simultaneously

$$\mathbf{h} = \boldsymbol{w^T}(\mathbf{x1} + \mathbf{x2} + \ldots\ldots + \mathbf{xc}) \tag{5}$$

the weighted Word is applied to Vector using the TF-IDF Technique the weight of the word is calculated vector matrix that is essential for the CNN processing is formed and given as input to the convolutional neutral network.

EXAMPLE:

EXAMPLE:

people
sitting
there

0.7	0.4	0.5
0.2	-0.1	0.1
0.5	0.4	-0.1

Word embedding

patient
resting
here

0.6	0.3	0.5
0.3	-0.1	0.2
0.5	0.4	-0.1

4.3 Convolutional Neural Network

In neural network of deep learning, convolutional neural network is one of its type inspired biological neuron of living being, It consists of two layers and they are the convolutional layer and the pooling layer. It is a two dimensional planar layer with each planar having independent neurons It is used mostly in image processing technique and applied similar to text classification it minimal (Fig. 4).

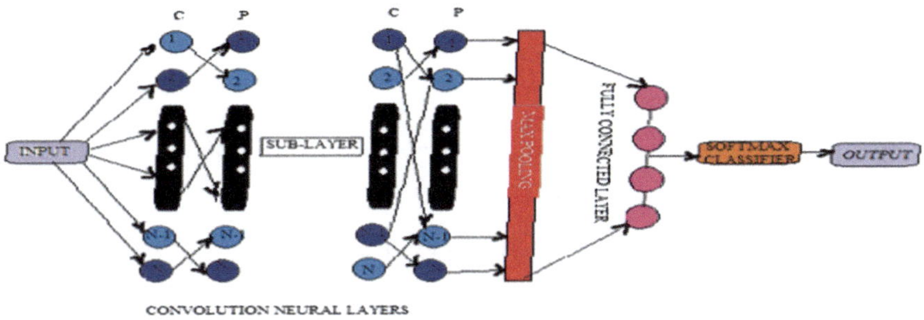

Fig. 4. Convolutional neural network

"C" refer to a convolution layer that is otherwise called as a component mining layer, with the contribution of every neuron associated with the open field [7] of the pervious layer and nearby highlights getting separated. There is sharing the heaviness of the similar feature chart. In other words, comparative diagrams transitive the similiar convolution portion. The Layer C is a combination of distinctive neighbourhood highlights, with the objective of the removed highlights being constant to interpretation and revolution. "P" is alluded to a pooling layer, known as component mapping layer that pools the feature acquire by layer C with the objectives of the removed highlights being invariant to scaling. In addition, the quantities of Layer C and Layer P are decided by genuine necessities. The final layer of CNN is commonly a layer featuring a complete association with the quantity of yield hubs being the quantity of grouping targets. Structure of CNN provide the information, in [9].

4.3.1 Input Layer

The first layer of CNN is the input layer. The state of input content matrix is (n, s, k), where n alludes to the quantity of writings, s alludes to the settled content length (All the writings are cushioned to the settled length s amid pre-handling when the content is shorter than s), and k is the component of word vector. x_i € R^k addresses to the k-dimensional word vector relating to the I-th word in the content. The info content can be referred to as

$$\boldsymbol{x}_{l:s} = \boldsymbol{x}_1 \oplus \boldsymbol{x}_2 \oplus \ldots \oplus \boldsymbol{x}_s \tag{6}$$

4.3.2 Convolution Layer

The second layer of CNN is a convolution layer. The convolution operation involves a filter w € R^{h*k}, which is applied to a window of h words for producing a new feature. For example, a feature c_i is generated from a window of words $x_{i:i+h-1}$ b

$$\boldsymbol{c}_i = \boldsymbol{f}(.\boldsymbol{w}_{i:i+h-1} + \boldsymbol{B}) \tag{7}$$

Here B is a bias term and f the non-linear activation function. When the convolution filter is moved by one step at a time, all the input matrixes are convoluted by each

window (8) and (7) mensional planar layer and eachpla $\{x_{l:s} = x_{1:h} + x_{2:h} + \cdots + x_{s:h}\}$ in turn, which will produce a feature map

$$c_i = [c_i, c_i, c_i \ldots\ldots\ldots\ldots c_i] \quad (8)$$

4.3.3 Pooling Layer

Pooling layer is the third layer of CNN. MAX Pooling is used in CNN model. Following the convolution operation, the characteristics design of the CNN layer are pooled, and all the characteristics maps are combine into a collection which is followed by calculation. We take its maximum value is considered as the character of the pooling layer as it would extract the most prominent features.

4.3.4 Fully Connected Layer

The final and fourth layer of CNN is a fully connected layer known for its interfaces with every one of the highlights and yielding esteem to classifiers. During the preparation of the model, for fitting was caused where the quality of the tests prepared was too little or when the model was highly prepared. In this paper the authors have commented the dropout technique [8] proposed by Hinton for the enhancement of the speculation of the model which way seen as the process for anticipating over fitting. A Deterioration in the cooperation between the cancelled layer neurons was seen together with the advancement in the structure of the model. Details relating to CNN are provided in [10].

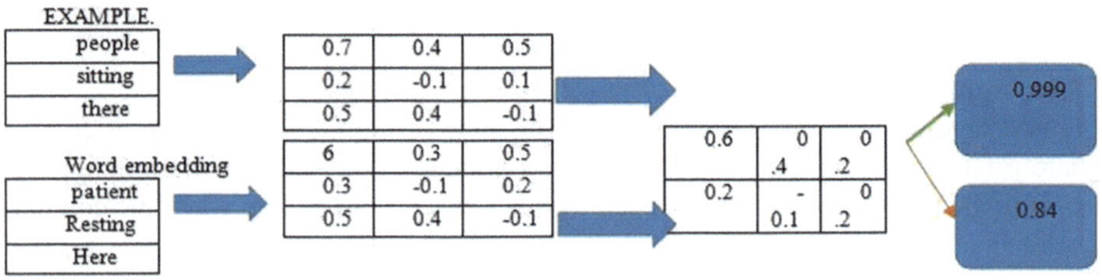

4.3.5 Algorithm

1. Collect the input in the form of audio, blogs, comments, tweets or other social media.
2. Audio should be collected in the wav format and converted into the text. The text is converted as input for the convolution neural network.
3. Apply Convolution filter the final neural set is produced.
4. Pass Final neural sets for max pooling. Resultant from max pooling produces the fully connected layer.
5. Perform Soft maximization.
6. Evaluation of the final result is generated (Figs. 5 and 6).

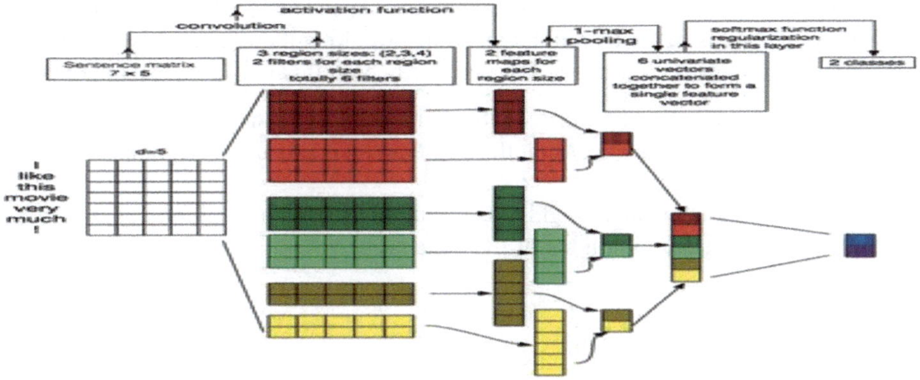

Fig. 5. Convolutional network for text classification

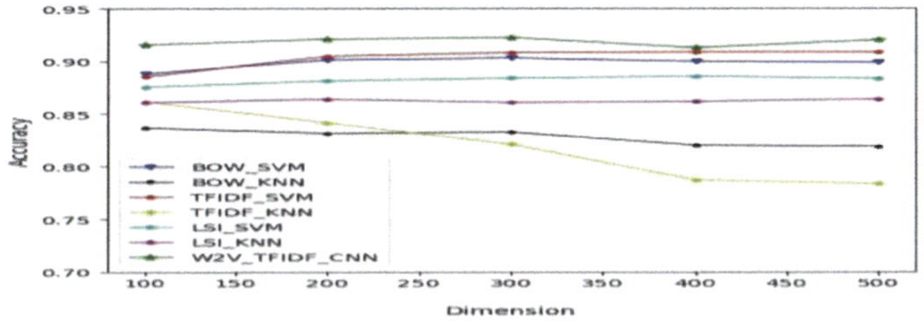

Fig. 6. Comparison of proposed work with the other classification method.

To checking the viability of this technique, we was done by the authors by contrasting W2V_TFIDF_CNN and other grouping strategies referred in the diagram. The after effects of contrasting and the other content order techniques on the two informational collections are shown in Table 1.

Table 1. Eight different text classification method of accuracy prediction (%)

Classification methods for accuracy prediction	Comparison	
	NetEase news text	Text classification using CNN
Bag_of_Word_Support Vector Machine algorithm	92.96	90.14
Bag_of_Word_K_nearest neighbors algorithm	86.17	83.03
TermFrequency_-Inverse document Frequency-SupportVectorMachine algorithm	94.05	90.43
TermFrequency_-Inverse document Frequency_K nearest neighbors algorithm	89.05	84.14
LatentSemanticIndexing_SupportVectorMachine algorithm	91.83	88.11
LatentSemanticIndexing_K nearest neighbors algorithm	89.64	86.39
WordtoVector_TermFrequency_-Inverse document Frequency_CNN	96.22	96.92

In Table 1, the precision of the technique dependent on Convolutional neural network is shown as better than different strategies along these levels. the adequacy of this order strategy was confirmed. The purpose behind this impact is explained below:

(1) Using W2V creation of word vectors can get higher-quality highlights;
(2) The effect of a solitary word on the entire archive is better thought about through TF-IDF weighting;
(3) The post-handled content highlights through CNN has a likelihood of speaking to the abnormal state highlights of the content.

5 Conclusion

The patient can submit the review in an open-ended manner which is easy and comfort as audio format the data is then summarized further to text format. It is then utilized in the qualitative analysis of opinion relating to the health care industry. This sentimental analysis process is based on the w2v approach using a conventional neural network it can make a cautious assurance of patients feeling relating to the different organization parts of a doctor's facility dependent on the forecast precision accomplished. Forecasts of this approach are connected to results of numerous other traditional reviews. Accuracy of the result helps to the recommendation to the patient towards the specific physicians and health centers.

References

1. Li, L., Xiao, L., Wang, N., Yang, G., Zhang, J.: Text classification method based on convolution neural network. In: 3rd IEEE International Conference on Computer and Communications (2017)
2. Aung, K.Z., Myo, N.N.: Sentiment analysis of students' comment using Lexicon based approach In: IEEE ICIS, Wuhan, China (2017)
3. Poria, S., Cambria, E., Howard, N., Huang, G.B., Hussain, A.: Fusing audio, visual and textual clues for sentiment analysis from multimodal content. Neurocomputing **174**, 50–59 (2015)
4. Meyer, D.: How exactly does Word2Vec work? (2016). dmm145.net, uoregon.edu, brocade.com
5. Mikolov, T., Chen, K., Corrado, G., Dean, J.: Word2Vec (2014)
6. Kaushik, L., Sangwan, A., Hansen, J.H.: Sentimental extraction from natural audio streams. In: ICASSP (2013)
7. Hubel, D.H., Weisel, T.N.: Binocular interaction and functional architecture in cat's visual cortex. J. Physiol. **160**, 106–154 (1962). (in London)
8. Hinton, G.E., Srivastava, N., Krizhevsky, A., et al.: Improving neural networks by preventing co-adaptation of feature detectors. Comput. Sci. **3**(4), 212–223 (2012)
9. Jia, S.J., Yang, D.P., Liu, J.H.: Product image fine-grained classification based on convolutional neural network. J. Shandong Univ. Sci. Technol. (Nat. Sci. Ed.) **33**(6), 91–96 (2014)

10. Huang, W., Wang, J.: Character-level convolutional network for text classification applied to Chinese corpus (2016)
11. Parthasarathy, G., Tomar, D.C.: Trends in citation analysis. In: Proceedings of the International Conference: (ICCD-2014) Intelligent Computing, Communication and Devices in SOA University, Advances in Intelligent Systems and Computing, no. 308, pp. 813–821. Springer, India (2015)

A Comparison of Machine Learning Techniques for the Prediction of the Student's Academic Performance

Jyoti Kumari(✉), R. Venkatesan, T. Jemima Jebaseeli, V. Abisha Felsit, K. Salai Selvanayaki, and T. Jeena Sarah

Department of Computer Science Engineering, Karunya Institute of Technology and Sciences, Coimbatore, Tamil Nadu, India
{jyoti,abhisha,salaik,jeenat}@karunya.edu.in, rlvenkei2000@gmail.com, jemima@karunya.edu

Abstract. The aim of this paper to predict a student performance using traditional and machine learning techniques: Bayes algorithm, linear regression, logistic Regression, k-nn algorithm, decision tree. Naive based algorithm is the emerging field which compose the procedure of verified students details like semester marks, assignment, attendance, lab work which are used to improve students' performance. This paper shows a model of students data prediction based on Bayes algorithm, linear regression, logistic Regression, k-nearest neighbor, decision tree and suggest the best algorithm among these algorithms based on performance details. Classification is an important area to predict and application in a variety of fields. In the view of full knowledge of the algorithm underlying probabilities, Bayes decision theory shows the optimal error rate. Decision tree algorithm is been used successfully in expert systems in capturing prediction. Mainly the decision tree classifiers are used to design and classify the student's data with Boolean class labels. Linear regression is a linear approach to modeling the relationship between the details of students in scalar response.

Keywords: Clustering · Classification · Naive Bayes algorithm k-nearest algorithm · Linear regression · Logistic regression · Decision tree and student performance

1 Introduction

Data mining is applied in various sources of field including education. Using data mining methods include classification [6, 7], clustering, naive Bayesian, decision trees, neural networks, logical regression, k-nearest algorithm and logistic regression. Students overall academic records throughout the period from first year to fourth year in the university is the revolving angle in bachelor academic and typically intruders on cumulative general point average (CGPA) and semester general point average (SGPA) in an important way. The attributes of the student's element such as attendance, internals marks, assessments, external marks and lab works are studied. From all these kind of elements, we can calculate CGPA, SGPA and categories student based on this

D. J. Hemanth et al. (Eds.): COMET 2019, LNDECT 35, pp. 1052–1062, 2020.
https://doi.org/10.1007/978-3-030-32150-5_107

percentage obtained. With the help of machine learning algorithms and procedures, we will be able to find the accurate performance of the categorical students.

1.1 Naïve Bayes

Naive Bayes is the simplest form of Bayesian network, in which all the attributes which is needed into the algorithm are given in independently class variable values this feature is called conditional independence from [3, 4].

1.2 Linear Regression

Data mining technique is becoming a more successful performance in predicting various fields. Experiments has shown that machine learning techniques could be in predicting student performance at a high success rate. By using linear regression we predict student performance on overall semester result and at the end comparison takes place and the result is evaluated of the proposed method with other approaches. In this regression, the training data contains the set of observations together with its output. Linear regression is used to provide insights with the greater influence on the outcome.

1.3 Logistic Regression

Logistic regression is mainly a classification technique. Logistic regression is used to estimate the probability and generalized to predict more than categorical values. In this paper, using many machine learning techniques and Logistic analysis, we developed models to predict student's performance by analyzing datasets provided by users. The next section is a brief review of previous work by others algorithm used.

1.4 K-nearest Network

The main drawback that arises when utilizing this technique is that each labelled samples are given equal importance in deciding the class memberships of the pattern to be classified, in spite of their "typicalness." Three methods of assigning fuzzy memberships to the labelled samples are proposed. The fuzzy ÁÃ-ÍÍ rule is also shown to compare, more-experienced pattern procedures are used in these experiments. A unclear analogy of the nearest prototype algorithm is also developed.

1.5 Decision Tree

The patterns should be actionable so that it may be used in performing the decision making process. Knowledge discovery in databases with machine learning (KDD) [1], often called data mining, inserting and collecting information and details from the database. The functionalities related to data mining are applied on various categories to identify the information on decision making knowledge in set of data [2]. The field of data mining has been becoming greater day by day, Pattern Reorganization and Computation capabilities etc.

In data mining, multiple different prediction methods and techniques are available. Therefore, this pattern uses multiple prediction methods to confirm and check the results by using multiple algorithms. The result could be selected in terms of proximity the value which is accurate. In this paper, the focus is on the performance details of various algorithms based on the result generated by the algorithms when they are applied on the data set. The remaining of the paper contains four sections. Section 2, methodology is presented and algorithms implemented in this model which include Naive Bayes and Bayes Network, K-nearest network, linear regression, logistic algorithm, and decision tree algorithm. Section 3, presents the purposed framework of all the algorithm which shows the best algorithm. Section 4, presents the conclusion which will be extracted from all these algorithms and chosen be the best algorithm.

2 Methodology

2.1 Naive Based Algorithm

Naive bayes classifier cognitive is used for the result estimation of student's performance. The parameters used to measure the student's performance were educational, attendance, internal marks, external marks and lab marks. The data is wrapped among the data set which contains the academic details which is extractable through data mining method and machine learning algorithm. As in the table, we have shown the attributes (Table 1):

Table 1. Attributes of the particular student

Student's name	Internals marks(/60)	External marks(/40)	Attendance marks(/10)
1. Slipa	54	24	9
2. Reetu	40	15	7
3. Vinay	52	19	5
4. Rhea	34	27	8

In naïve based algorithm, first we will be handling all the data which is going for the further process. In second step we will be summarizing all the handle data. In third step we will be making the single prediction. In forth step we will be making all the prediction. In fifth step we will do evaluation accurately and at last we need to tie all the data together for category purpose. As it has shown into the Fig. 1.

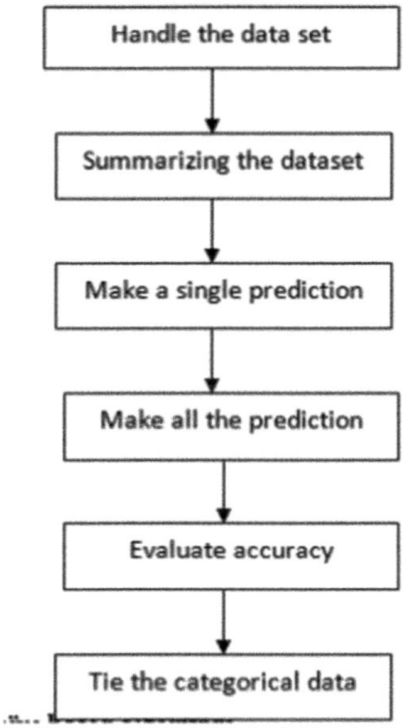

Fig. 1. Work flow diagram of the naïve based algorithm

2.2 K-nearest Algorithm

K-NN algorithm is used to find the nearest neighbour. K-nearest algorithm is taken as a pattern recognition problem, so there have been various methods investigated for classification and prediction. Under many conditions the AT-nearest neighbor (ÄT-NN) algorithm is used to perform the classification and for search purpose. This K-nearest algorithm gives a simple nonparametric procedure for the actions of a performance of the input pattern which is based on data represented by the nearest (for example, in the Euclidean sense) neighbours of the vector.

K-nearest algorithm is simple algorithm which uses entire data set and its training phase whenever prediction is required for unseen data, at that time k-nearest algorithm search the entire training data set for k more similar instances and data will the more similar instances and finally returns as a prediction. KNN is used for the search operation where we are looking for same category of data set. KNN algorithm used many formula methods for finding the nearest data for the given dataset.

Here is the process of KNN algorithm, how does its work when the trained data is given. As all the process has been shown into flow diagram (Fig. 2).

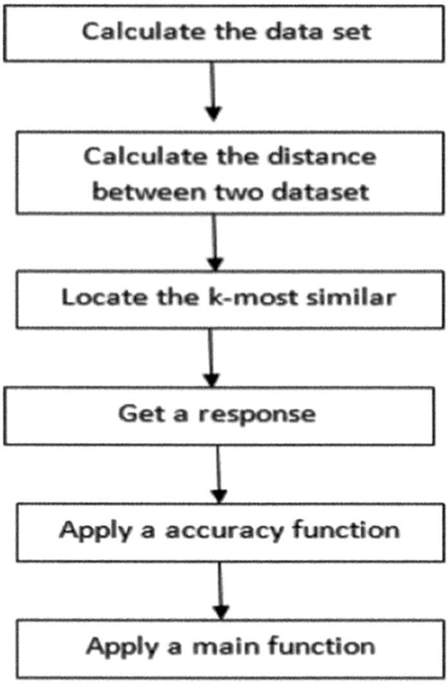

Fig. 2. Work flow diagram of the k-nearest algorithm

Decision Tree Algorithm

Classification and regression tree proposed by Breiman et al. and in the short form is called CART. The construction of the binary trees takes place which are used to refer as Hierarchical Optimal Discriminate Analysis (HODA). Classification and regression tree produces the classification or regression tree which is a non-parametric decision tree learning techniques, it depends on the data which will be categorical or numeric. CART accepts numerical data as well as categorical values also. Here is the decision tree of the student's attributes (Fig. 3).

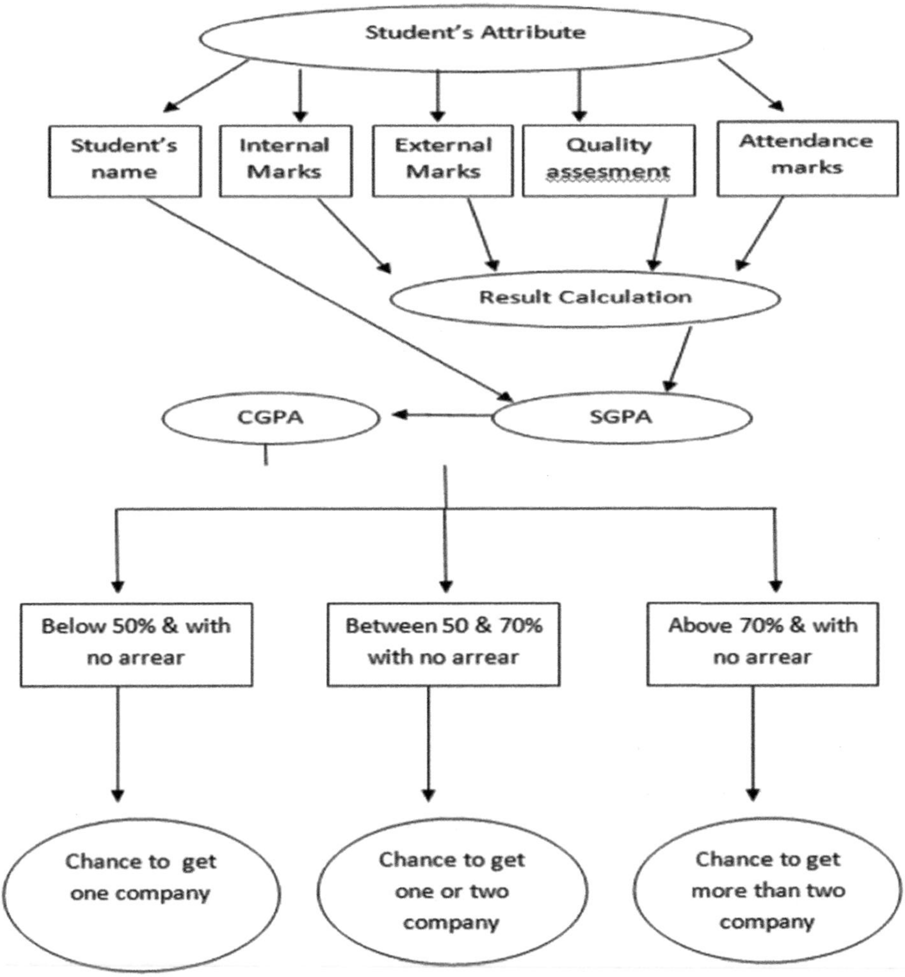

Fig. 3. Work flow diagram of decision tree algorithm

2.3 Linear Regression

Regression predicts a numerical value as well as categorical value. Regression operates the operation on the dataset and it use dependent variable continues in nature. Therefore the result can be expanded by adding more new information in the data. The relations between predictor and target values which is established by the regression can make an accurate pattern. The pattern which is generated as the result by the regression can be used on other datasets and it establishes the linear relationship between all the data sets. Hence the data required for the regression, first part is for defining model and the other part is represented as the testing model. In this area of the part it is chosen as linear regression for our prediction. First, is to divide the data set into two parts of training and testing. Then the trained data is used for starting analysis and predicting the model. As all the process is shown in the given Fig. 4:

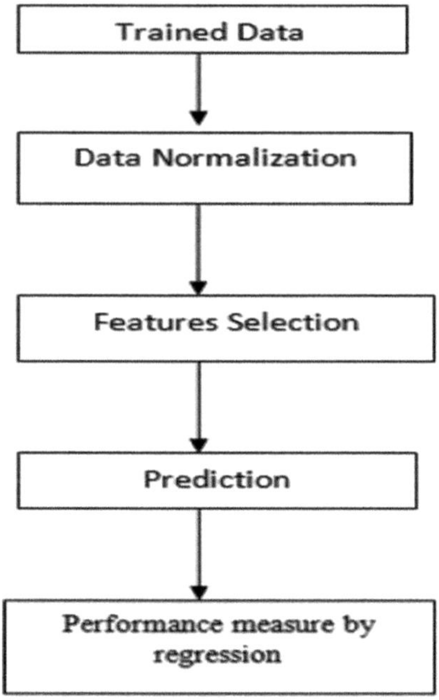

Fig. 4. Work flow diagram of linear regression process

2.4 Logistic Regression

In real logistic regression has dependent variable and independent variable and it is continuous and binary in nature. The values comes from the logistic regression is either 1 (True, Success), 0 (False, Failure). It is also called logit R. This algorithm is used to deal with probability to measure the relation between dependent and independent variable. It uses categorical variable. So the outcome should be categorical. In logistic regression first data will be pre-process and then it is used for modelling purpose. First in this algorithm we will collect the data and then go for analysis and at last go for training and testing. In below diagram, we are going to show the process of the regression methods. To improve the predictive effect of our proposed model in regression algorithm, the raw data, which are often redundant. Therefore, it is necessary to pre-process the data before develop the predictive model. The following steps have been done to achieve the optimization method. Binary Logistic regression is a traditional statistical technique that is wellness suitable for examining the relation-between a binary and categorical response variable and at least one categorical or continuous independent or dependent variables. We are going to show that how the logistic regression algorithm works (Fig. 5).

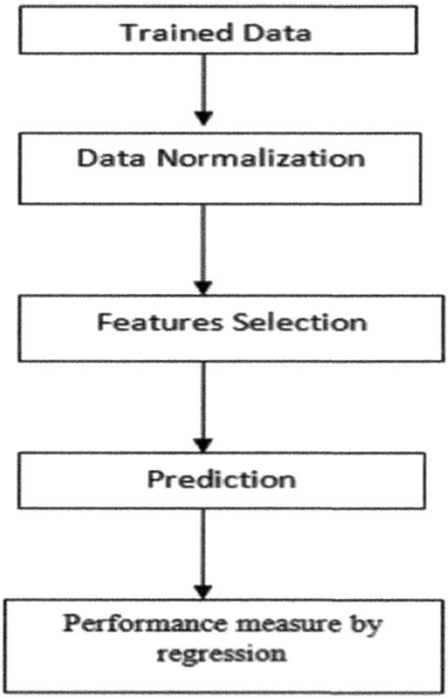

Fig. 5. Work flow diagram of linear regression process

2.5 Logistic Regression

In real logistic regression has dependent variable and independent variable and it is continuous and binary in nature. The values comes from the logistic regression is either 1 (True, Success), 0 (False, Failure). It is also called logit R. This algorithm is used to deal with probability to measure the relation between dependent and independent variable. It uses categorical variable. So the outcome should be categorical. In logistic regression first data will be pre-process and then it is used for modelling purpose. First in this algorithm we will collect the data and then go for analysis and at last go for training and testing. In below diagram, we are going to show the process of the regression methods. To improve the predictive effect of our proposed model in regression algorithm, the raw data, which are often redundant. Therefore, it is necessary to pre-process the data before develop the predictive model. The following steps have been done to achieve the optimization method. Binary Logistic regression is a traditional statistical technique that is wellness suitable for examining the relation between a binary and categorical response variable and at least one categorical or continuous independent or dependent variables. We are going to show that how the logistic regression algorithm works (Fig. 6).

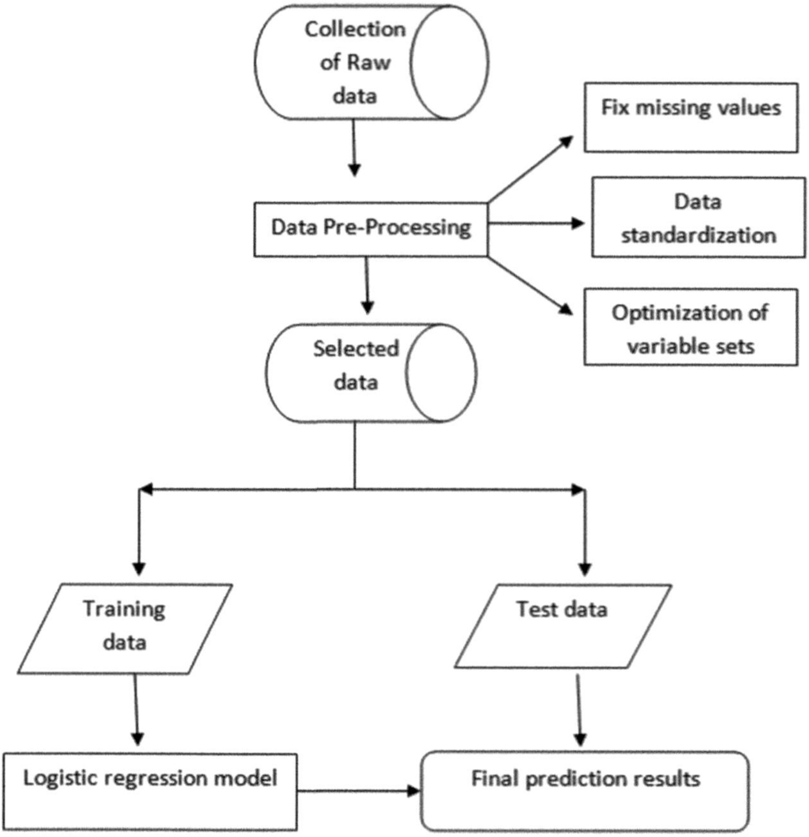

Fig. 6. Work flow diagram of logistic regression algorithm

3 Purposed Framework

In purposed framework, we will show all the differences of all five algorithm and their advantages and disadvantages in the given table.

Methods	Advantage	Disadvantage
Naive based algorithm	1. This algorithm makes computation process easier 2. It has better speed and accuracy for the given large datasets	1. This algorithm does not give accurate results in some cases where there dependency exists among the variables
KNN algorithm	1. In this code is very simple to implement 2. Training is done in faster manner 3. Effective if the training data is large	1. Large space is needed 2. Noise sensitivity 3. Slow testing 4. Needs to determine the value of the parameter of k

(*continued*)

(*continued*)

Methods	Advantage	Disadvantage
Decision tree algorithm	1. For solving all problems we need only tree in decision tree 2. This algorithm minimizes the ambivalence of complicated decisions and allocate the exact values to overcome of the various actions 3. This is easy to interpret 4. It is easily process the data with high dimension 5. This algorithm takes both numerical and categorical data	1. This algorithm gives only one output 2. This produces the categorical output 3. This is a unstable classifier 4. If the data type is numeric than it generates a complex decision tree
Linear regression algorithm	1. Linear regression algorithm is used to find good accuracy as compared to the other classifiers 2. It can easily handle the complex nonlinear data points	1. It is expensive compared to other methods 2. It takes more time for training the process compared to other algorithm 3. Other methods were constructed to solve the problem of binary class
Logistic regression algorithm	1. The output generated by the logistic regression is more productive than other algorithms 2. It may handle nonlinear effects in logically	1. Logistic regression predicts outcome based on independent variables 2. It is continuous and binary in nature

4 Conclusion

Through the paper, we have seen many difference algorithms. All the algorithms are used to predict the performance of the student is suggested. It gives the students platform to choose which better option for them. This model have used the classified approach which was Naive Bayesian classify to predict GPA of the student because naive based algorithm can support huge numbers of student attributes. Bayes classifier algorithms are used in the prediction process and the result accuracy of prediction is compared to find the student's performance. Finally, by comparing it will be able to say that naive Bayes algorithm is chosen as the best algorithm for prediction based on performance detail and helps the students to choose their better option though out their carrier. Various algorithms have been compared for the accuracy and performance and suitable classifier was used.

References

1. Quadril, M.M.N., Kalyankar, N.V.: Drop out feature of student data for academic performance using decision tree techniques. Glob. J. Comput. Sci. Technol. **10**(2) (2010)
2. Spiegelhalter, D.J.: Incorporating Bayesian ideas into health-care evaluation. J. Stat. Sci. **19** (1), 156–174 (2004)
3. Jensen, F.V.: Bayesian network basics. AISB Q. **94**, 9–22 (1996)
4. Abu Tair, M.M., El Halees, A.M.: Mining educational data for academic performance. JICT **2**(2) (2012)
5. Enke, D., Thawornwong, S.: The use of data mining and neural networks for forecasting stock market returns. Expert Syst. Appl. **29**(4), 927–940 (2005)
6. Babuand, R., Satish, A.R.: Improved of k-nearest neighbour technique credit scoring. Int. J. Dev. Comput. Sci. Technol. **1**(2), 1–4 (2013)
7. Breiman, L.: Random forest. Mach. Learn. **45**, 5–32 (2001)
8. Cover, T.M., Hart, P.E.: Nearest neighbour pattern classification
9. Gorunescu, F.: Data Mining: Concepts, Models, and Techniques. Springer, Heidelberg (2011)
10. Hart, J., Kamber, M.: Data Mining: Concepts and Techniques, Morgan-Kaufman Series of Data Management Systems San Diego. Academic Press, Cambridge (2001)
11. Hunt, E.: Artificial Intelligence. Academic, New York (1975)
12. Winston, P.: Artificial Intelligence. Addison-Wesley, Reading (1977)
13. Duda, R.O., Hart, P.E.: Pattern Classification and Scene Analysis. Wiley, New York (1973)
14. Batchelor, B.G.: Pattern Recognition. Plenum, New York (1978)
15. Gnanadesikan, R.: Methods for Statistical Data Analysis of Multivariate Observation. Wiley, New York (1977)
16. Reynolds, A., Flagg, P.: Cognitive Pyschology. Winthrop, Cambridge (1977)
17. Dodwell, P.: Visual Pattern Recognition. Rinehart and Winston, New York Holt (1970)

Sludge Detection in Marsh Land: A Survey

Shirley Selvan(✉), J. Ferin Joseph, and K. T. Dinesh Raj

Electronics and Communication Engineering,
St. Joseph's College of Engineering, Chennai, India
shirleycharlethenry@gmail.com, jjferin69@gmail.com,
dineshktd310897@gmail.com

Abstract. This survey reviews the analysis of Active Sludge (AS) particles in wastewater or sewage water treatment plants. There are many methods followed for analysis of sewage water to monitor particles present in it. The samples from treatment plants are photographed with a microscope in order to view the particles. Analysis is done to improve the quality of water by treating the water according to the analysis report, so that water can be reused. The processes used to analyze the waste water keep developing as technology develops. Previously, manual analysis is done to get a report about particles present in waste water, but in recent times image processing makes the analysis process easier. A method that detects the unbranched filamentous bacteria length is proposed by researchers. In order to determine the curvature of an extended filament border, researchers have proposed some rotation invariant features. Previous research models are investigated for sludge volume index (SVI) of various active sludge wastes from waste water treatment plants. Analysis of images leads to measurement of parameters of both filaments and flocs present in waste water. The modelling of filaments and flocs based on the measured parameters can decide a method to clean waste water of treatment plants.

Keywords: Active Sludge · Image processing · Sewage water management · Filamentous bacteria recognition · Filamentous microorganism · Wastewater treatment

1 Introduction

A survey is done on activated sludge process (AS) for monitoring the sewage or waste water around us. The active sludge wastewater treatment plant is analyzed for the presence of microbial aggregates in the secondary clarifier of the process plant (method done with chemicals). The flocs are grouped to form filaments. The settling capability of microbes depends on the morphology of filamentous bacteria and flocs. A detailed report on the filaments and flocs along with their size distribution is necessary for more effective control of the process performance. AS has filaments and flocs which explain a heterogeneous mixture of various micro-organisms, dead cells and in-organic material. It is well known that a balance between different types of filamentous bacteria is essential. These form aggregates with acceptable properties that have various structures and density, which allow an effective settling ability for the sludge. Several methods have been proposed in this survey to explain the complex structures of

D. J. Hemanth et al. (Eds.): COMET 2019, LNDECT 35, pp. 1063–1075, 2020.
https://doi.org/10.1007/978-3-030-32150-5_108

filaments and flocs in terms of material organization with the aggregates. This is useful to process the wastewater. The techniques give physical aspect of floc and filaments along with the granulomere distribution of floc sizes and the consequences of bio-flocculation on flow properties. The concept of Active Sludge was initiated in England during early 1900s'; The AS process did not spread in Asia and United States until the 1940s'. Today variations and development of the basic process help to clean the wastewater. There are also some issues accrued on the sample collection process.

The paper surveys treatment processes of waste or sewage water done to treat Active Sludge particles like flocs and filaments. The treatment process followed previously are - Lab testing on direct samples from water, in which not all organisms are accounted and cleaned as manual analysis is not accurate. Another method in which, the samples are tested with an electron microscope, but all particles are not cleared. With the introduction of Image processing in analyzing methodology, even small dust and unwanted particles as per the threshold of analysis is followed.

The experiments show that cleaning process done after analysis of sewage or waste water for Active Sludge is better when compared to the previous process. The AS process is spread widely in big cities where long area pipelines are used to send water and get the sewage or waste water for treatment. Therefore, the analyzes of AS using image processing is found to be fast and most efficient way of getting report about the sample of the sewage or waste water.

2 Survey on Methodologies

2.1 Analysis Done on the Samples

The samples for analysis of AS differ with the type of process followed. For the basic methods, water samples are taken from the sewage directly and analyzed by applying some chemicals and allowed to stay on the tanks to see the action of these chemicals on water. As these methods consume time, the latest methods like Lab testing and comparison with image processing comes into role. In image processing, the image must be of good quality to perform analysis with the help of Mat Lab. The following section surveys different methods of processing the activated sludge wastewater.

2.2 Methods of AS Detection

2.2.1 Waste and Reverse Activated Sludge (WAS & RAS) [3]

This activated sludge wastewater treatment or process consists of several interrelated components. Aerobic bacteria grow as they travel through the aeration tank. These bacteria multiply rapidly with sufficient food and oxygen in the tank. These are provided in the tank as it helps with observation. When waste reaches the end of processing apparatus, bacteria use organic matter mostly from the tank to generate new cells to grow.

Due to growth or aggregation of bacteria, there is an increase in weight as a result of which it settles down at the bottom. The organism that settles at the bottom of the clarifier tank can be separated from fresh or clear water. This active sludge is pumped back into the apparatus, where it gets mixed with new wastewater. This helps new microorganisms to grow so that flocs in excess can be removed from the system. The relatively clear liquid above the sludge (clean or treated water) and the supernatant is sent for further treatment as more purification is required.

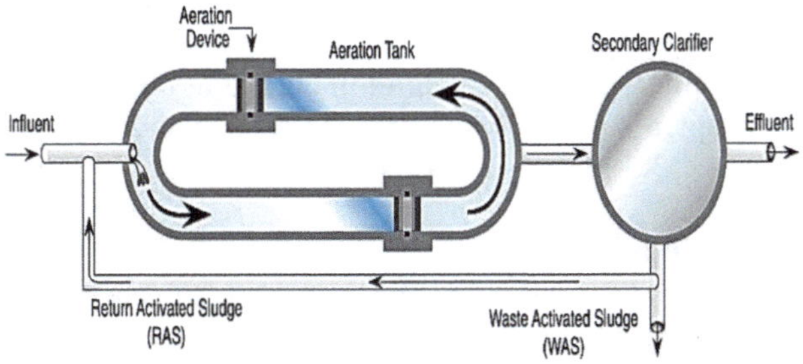

Fig. 1. Return & Waste Active Sludge [3]

Figure 1 shows the Oxidation process done after purification of sewage or waste water. The Influent gives the input feed for the process and the output is taken from the Effluent. The Secondary classifier is used to feed a part of sample from output to input to ensure the process accuracy. Aeration device is used to give air supply to the circulating water inside the tank. In the feedback, RAS will feed to the input and the WAS will send the waste out.

2.2.2 Laser Light Diffraction Technique [2]

This technique details changes in structure of filaments and flocs by calculating parameters like fractal dimension or direct size distribution. It is a fast and reliable technique to determine size of filaments and flocs found widely covered over the wastewater. This technique follows flocculation dynamics and data obtained can be used to model the AS wastewater's flocculation or de flocculation process by following a population balance approach. Further process is with Focus Beam Reflectance Method, which is a successful method to measure the filament and floc size distribution in a secondary clarifier on an activated sludge wastewater treatment plant. The authors detail the devices that are applicable for a wide range of solid concentrations.

2.2.3 Three Device Mechanism [2]

This process consists of three devices connected in series: MastersizerS, a CIS 100 and IMAN. IMAN is based on Lab-View software, which is now replaced with MatLab. These devices are chosen based on the measurement principle of microorganisms and flow-through features which help in wastewater sample preparation. The three devices are detailed below:

MastersizerS [2] is a micro organic volume analyzer based on Low Angle Laser Light Scattering (LALLS). A 300RF lens is used to process and measure particle size of 0.05–900 μm. CIS-100 [2] combines Time of Transition (TOT) concept with dynamic size and shape identification method that is based on image processing.

The TOT calculation covers a size of 0.5–3,600 μm within 300 discrete size intervals, based on the lens. In this method, a size range between 2–600 μm has been chosen for a better view of all the particles in the sample. IMAN [2] is an image processing system developed in-house. It allows an automatic investigation of the micro organic shape and size. Images of AS wastewater samples captured using a CX40 optical microscope are input to an ICD 46E CCD camera and then digitized with a frame grabber PCI – 1411. These images are then processed on a monitoring system by using specific software. A 4 × magnification lens is helpful in enlarging the image on the screen and a 0.35× C mount adapter is placed between CCD-camera and microscope which gives an image of size 4.5 × 3.5 mm.

2.2.4 Image Acquisition [8]

Here image samples of AS wastewater environment are acquired with an in situ microscope and ISM microscope developed at the Mannheim University of Applied Sciences. It is a pulsed transmitted light microscope controlled with pulses of 0.5–10 μs width to ensure that the speed of the microorganisms inside the bioreactor is 0.1–1 m/s. Figure 2 shows image acquisition done using an in situ microscope.

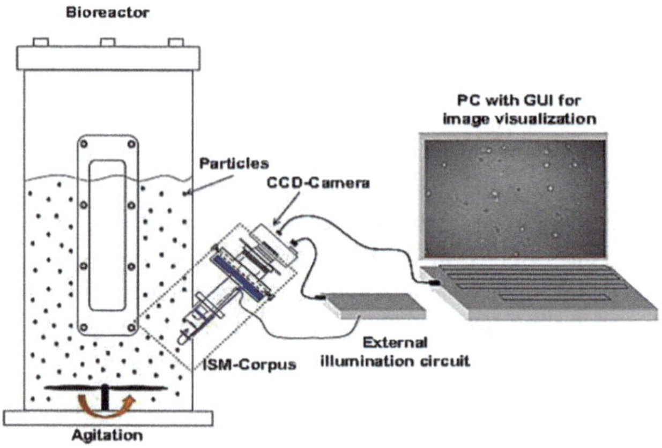

Fig. 2. In Situ microscope on image acquisition [8]

The fiber-ending probe is positioned nearly to 0.3 mm above a quartz glass window which separates objective from the suspension. Objective is attached to top of internal tubes which are optically connected to window by means of water immersion. A CCD Camera of Basler A102f is connected with other end of the processing apparatus. Nearly 10 monochromatic images of 8-bit are acquired per second with resolution of 1392 × 1040 pixels. Software controls the entire system component. It triggers both camera and pulse generator based on brightness, frequency and gain which are defined via user interface.

2.2.5 Image Processing Methods [6]

The images of aggregates and filaments are processed using MatLab software. From the binary images, biomass and filamentous bacteria along with flocs are captured. Morphological parameters like color were determined to compare these with previously identified microorganisms. Figure 3 shows the aggregated and labelled filaments. AS wastewater samples are collected from the aeration tanks of government municipal wastewater treatment plant, lakes and ponds for processing. The segmented filaments and flocs are analyzed for some specific parameter by testing.

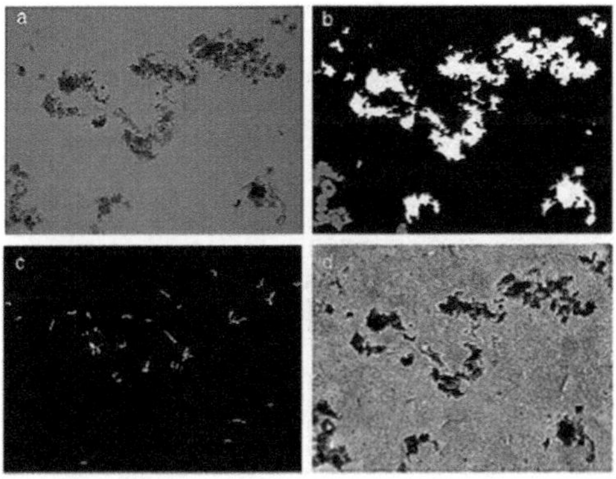

Fig. 3. 100x magnification image (a) binary aggregates image (b) binary filaments image (c) and final labeled image (d) [6]

The filament and flocs analysis data collected from multiple image samples of various conditions are used to model the SVI. SVI is one of the important parameter model designed to monitor the state of an Activated Sludge treatment plant. The next section briefs different methods of Active Sludge water treatments

2.3 Process Followed in Previous Methods

2.3.1 Waste and Reverse Activated Sludge [3]

A Typical Sequencing Batch Reactor process is shown in Fig. 4 with the steps followed in waste and reverse active sludge method. The flocs and filaments are made to settle down in a tank with water. This is followed by aeration in which the waste water is separated and reverse water is given as feedback at the input side in order to verify efficiency of the process. Figure 4 shows the flow of the process.

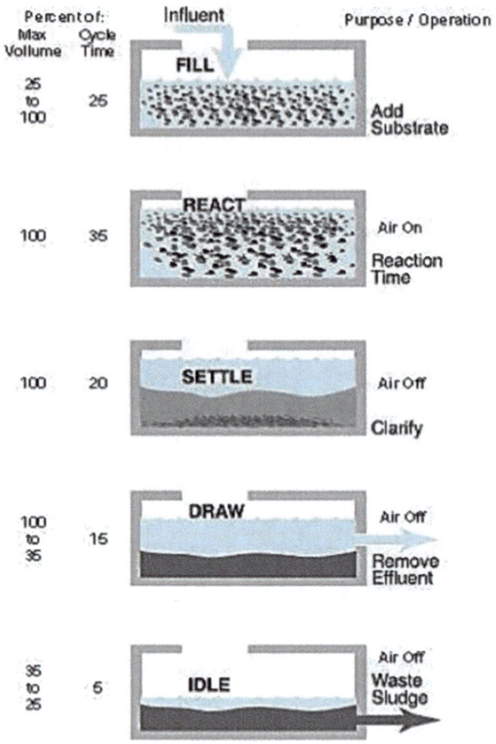

Fig. 4. Typical sequencing batch reactor process [3]

In this process, the water gets rid of the filaments and dust particles, but not from bacteria, dead or living organisms formed during stagnation.

2.3.2 Laser Light Diffraction Technique [4]

This method is considered to be the fastest. The process burns and separates waste organisms which float in layers. As it is well known that laser is a device used to vaporize unwanted components, it is used to burn flocks or filaments which are detected.

2.3.3 Three Device Mechanism [2]

Three device mechanisms is the process of analyzing the A.S. with the combination of three major devices: MastersizerS, CIS-100 and an IMAN. The devices are connected in series with a feedback circuit having a device MSX17 as shown in Fig. 5.

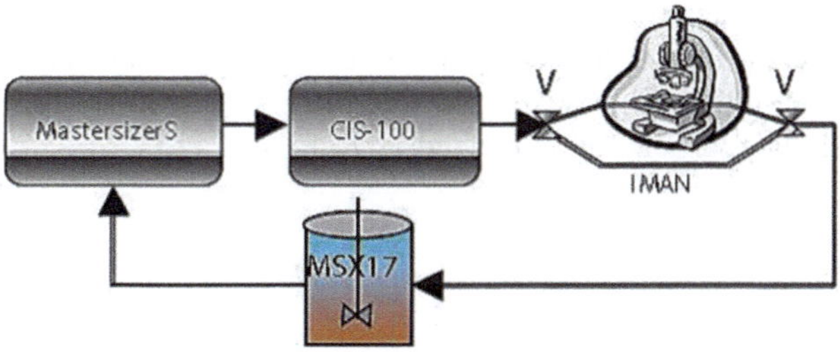

Fig. 5. Three device mechanism (set-up) [2]

The device MSX17 is used to feed a sample of output back to the input, to ensure the efficiency of this process.

2.3.4 Image Acquisition [8]

The images of activated sludge environment of wastewater treatment plants are acquired with the help of a microscope. The process involved is shown in Fig. 6.

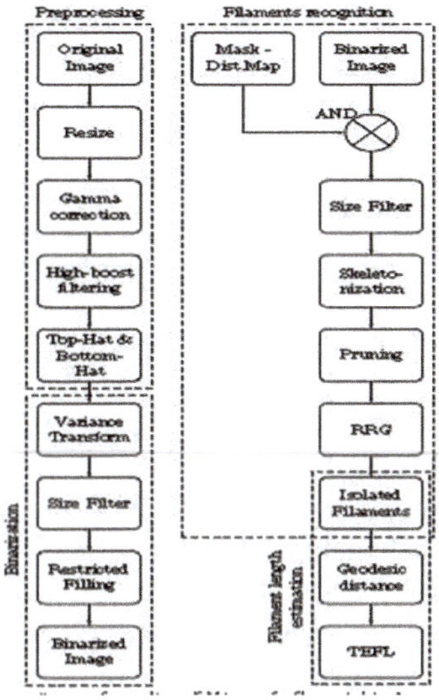

Fig. 6. Process of image acquisition [8]

The microscopic images are preprocessed and then binarized in order to get a clear view of the image for further process. This is followed by Filaments Recognition and Filament Length Estimation processes. In this method, length of the filament is considered as the sludge particle and analyzed as in Fig. 7.

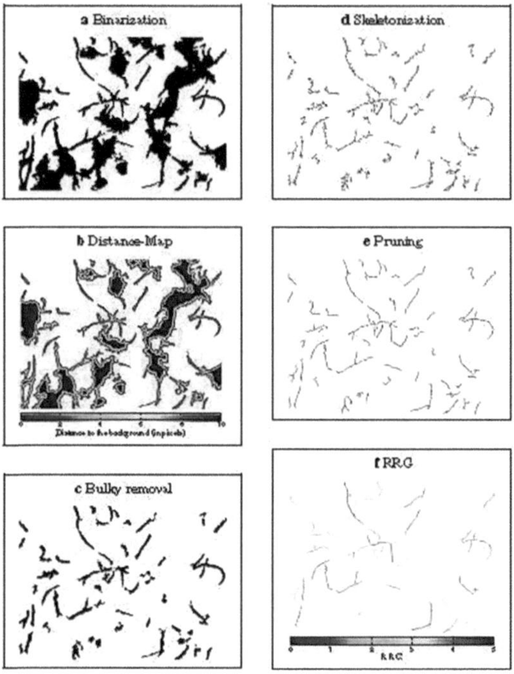

Fig. 7. Process for calculating the length of the filament [5]

2.3.5 Image Processing Methods [6, 7, 9]

The process is similar to that of Image Acquisition process, but filtering and recognition method followed to calculate and separate each particle are different. Image acquisition process of the samples is carried out by a bright field microscopic technique, using camera of magnification 120x and photonic microscope to keep the illumination constant for all samples used. The microscope is coupled with a CCD camera, the image grabbing is performed by 24 bit of 16 million of colors approximately, and the size of the sample must be of 2048 × 1536 pixels. Nearly 50 images should be acquired for each sample and stored in JPEG format (for good processing performance).

The image processing is done with help of the commercial software like LabView® or Image-Pro Plus® or MatLab. The processing of the acquired images is carried by quantification of several geometrical parameters and contour fractional dimension of the microbial filaments and flocs. The process representation of the analysis procedure is detailed in Fig. 8.

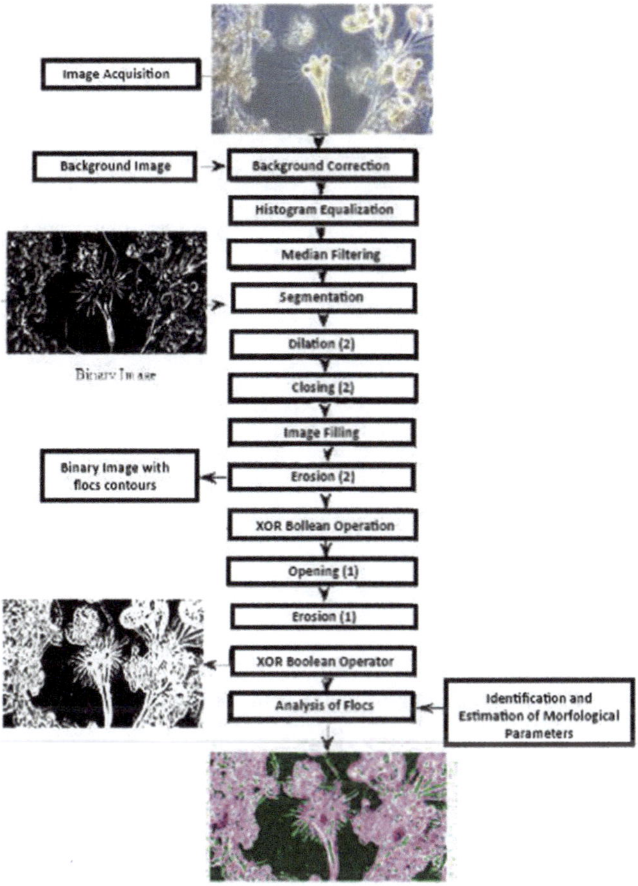

Fig. 8. Schematic representation steps on image analysis procedure [7]

The process diagram shows basic processing steps followed for sample images in the beginning stage of A.S. analysis. But now day's new algorithms are followed to improve accuracy and to maintain a standard way of filtration which is shown in Fig. 9.

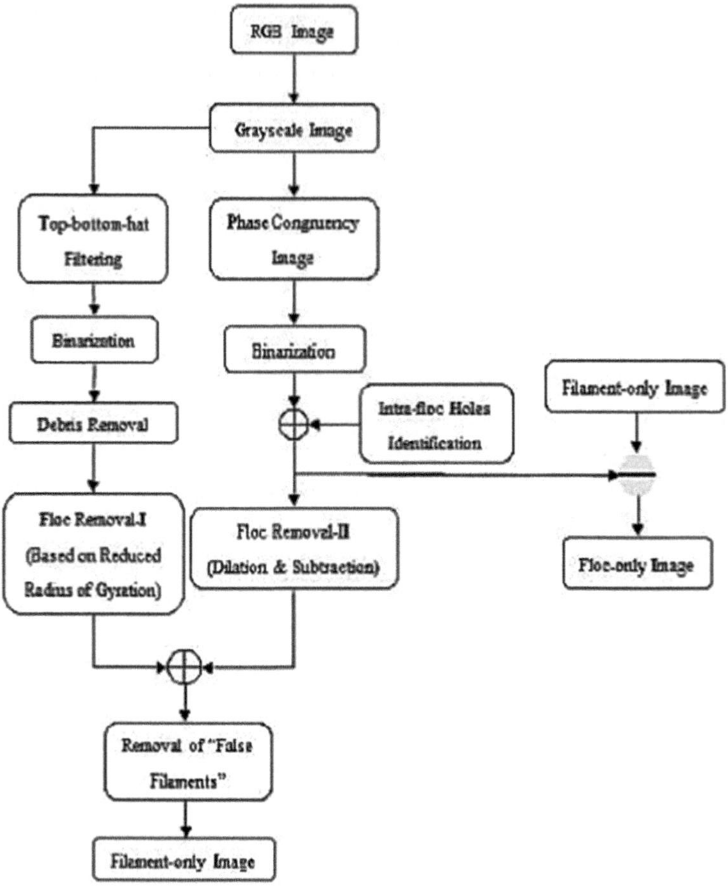

Fig. 9. Processing of image with some algorithms [9]

The process diagram given in Fig. 9 is followed at present in which two major algorithms are used: Phase Congruency and Otsu. This is used to segment the image and filter each flock and filament for further analysis. The major process followed in Image processing to analyze the A.S. is shown in Fig. 10.

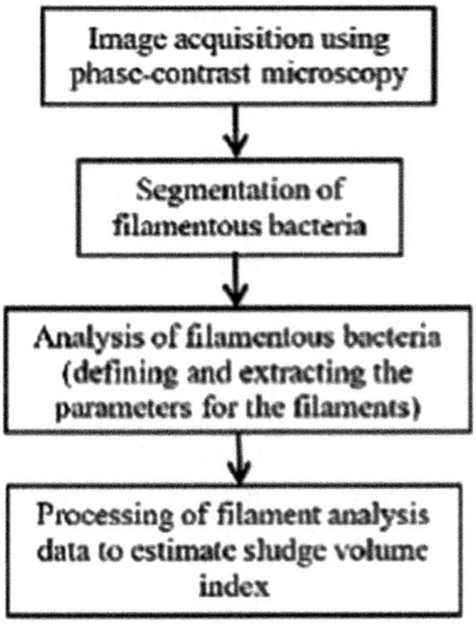

Fig. 10. Processing of image processing on A.S. [6]

The quality of image decides the clarity of the output. Phase-Contrast microscopic analysis and Otsu algorithm are used to get good analyzed report of the sample taken (Table 1).

Table 1. Comparison of various methods

Title	Authors	Methods followed	Advantage/disadvantages
Simultaneous determination of activated sludge floc size distribution by different techniques	R. Jenne E.N. Banadda, I.Y. Smets G. Gins, M. Mys J.F. Van Impe	Theoretical explanation for the measurement of flocs and filament with the appropriate devices	Methods for measurement and fetching parameters of flocs and filaments are discussed
Unimodal thresholding, flocks and filaments pattern reorganization	Rosin L Paul P L	Extracted features of the filament by adding chemicals to the samples	Texture of filament may get damaged due to the addition of chemicals to the sample
Image analysis in biotechnological processes: applications to wastewater treatment	Amaral A L	Filament and flocs and detected using LabView software	First method to identify floc and filaments without adding any chemicals to the samples
Activated sludge morphology characterization through an image analysis procedure	Y. G. Perez S. G. F. Leite M. A. Z. Coelho	Implementing Image Processing to identify difference between flocs and filament	Differentiated flocs and filament by structure Texture extraction of filament and flocs is not discussed

(*continued*)

Table 1. (*continued*)

Title	Authors	Methods followed	Advantage/disadvantages
Image processing for identification and quantification of filamentous bacteria in situ acquired images	Philipe A. Dias Thiemo Dunkel Diego A. S. Fajado Erika de León Martin Denecke Philipp Wiedemann Fabio K. Hajo Suhr	Advancement in identification of flocs and filament by using special In-Situ microscope images	Extracted better and detailed filament structures
Image processing and analysis of phase-contrast microscopic images of activated sludge to monitor the wastewater treatment plants	Muhammed Burhan Khan Humaira Nisar Choon Aun N G	Implementing Phase-Contrast in microscopic images to get more detailed structure on flocs and filament	Methods for fetching area and other parameters to differentiate flocs and filaments are explained

3 Conclusion

Thus the survey describes the process and methodologies followed in treatment of Activated Sludge wastewater plants. Each method has its own process of analyzing the sewage or waste water from the treatment plant. The methods followed are done with various samples which are related and apt for the process followed. However not all methodologies are suitable at all places, most of which use chemicals and other solutions to view the microorganisms clearly. Measurement of parameters such as filament length, floc texture and floc fractal dimension from samples help in effective analysis. By overcoming the defects of the above methods, image processing is better compared to the other methods. Image processing method seems to outperform other methods in analyses of Active Sludge compared to previous methods followed to treat the wastewater at the treatment plants.

References

1. Sample images acquired from “Pipeline”, vol. 14, no. 2. Spring (2003)
2. Govoreanu, R., Saveyn, H., Van der Meeren, P. Vanrolleghem, P.A.: Simultaneous determination of activated sludge floc size distribution by different techniques. Water Sci. Technol. (1994)
3. Rosin, L., Paul, P.L.: Unimodal thresholding, flocks and filaments pattern recognization, vol. 34, pp. 2083–2096 (2001)
4. Chung, H.Y., Lee, D.J.: Porosity and interior structure flocculated activated sludge. J. Colloid Interface Sci. **267**, 136–143 (2003)
5. Amaral, A.L.: Image analysis in biotechnological processes: applications to wastewater treatment. Tese de Doutorado em Engenharia Química e Biológica, Universidade do Minho, Braga (2003)

6. Jenne, R., Banadda, E.N., Smets, I.Y., Gins, G., Mys, M., Van Impe, J.F.: Optimization of an image analysis procedure for monitoring activated sludge setteleability. Wadsworth, CA, pp. 345–350 (2004)
7. Perez, Y.G., Leite, S.G.F. Coelho, M.A.Z.: Activated sludge morphology characterization through an image analysis procedure. Escola de Química, Universidade Federal do Rio de Janeiro (2006). ISSN 0104-6632
8. Dias, P.A., Dunkel, T., Fajado, D.A.S., de León Gallegos, E., Denecke, M., Wiedemann, P., Schneider, F.K., Suhr, H.: Image processing for identification and quantification of filamentous bacteria in situ acquired images. https://doi.org/10.1186/s12938-016-0197-7
9. Khan, M.B., Nisar, H., Ng, C.A.: Image processing and analysis of phase-contrast microscopic images of activated sludge to monitor the wastewater treatment plants, February 2018

A Review on Security Attacks and Protective Strategies of Machine Learning

K. Meenakshi(✉) and G. Maragatham

SRM Institute of Science and Technology, Kattankulathur, Chennai, India
meenakbalaji@gmail.com

Abstract. Machine learning is the powerful techniques in computing and Linguistics Science. It is broadly applied in various domains such as computer vision, pattern recognition, image processing, network security and Natural language processing. Machine learning (ML) techniques have generated huge social impacts in a variety of applications such as malware detection, spam detection and health care. It is also more sensitive towards security attack. In supervised learning algorithm Machine Learning Models mainly depend upon the large input datasets, called training data and testing data. The slight modifications in the input data will affect the model performance to a greater extent. Many research works have been carried out by analysing various security threats against ML algorithms such as Naïve Bayes Algorithm, Decision tree, Support Vector Machine and Artificial Deep Neural Networks. In this paper, we have explained taxonomy of security threats happened in training and testing stage of learning. Finally we have presented various defending techniques, counter measures which are used in training and testing phases, few security policies to prevent adversarial attacks and make a robust Machine Learning model.

Keywords: Machine Learning · Adversarial attack · Defensive techniques

1 Introduction

In recent days, machine learning is used in many applications, like image identification and classification, computer vision, spam detection and analysis, network intrusion detection, pattern recognition etc. Similarly Big data analytics is another emerging research field which is applied in data analytics [1]. So the researchers have started addressing the challenges of Machine Learning with respect to Big data analytics [2]. They are interested to develop the artificial intelligent systems, so that it handles large volume of data with minimum computation time, high efficiency and good accuracy [3]. However Machine Learning models would be affected by several security threats [4]. One of the instances is, the attacker can compromise the authentication system which is trained by Machine Learning model and he can access to some confidential data [5]. Hence the researchers have to address the security issues of machine learning. The existing works have addressed the fundamental security concepts of model building. Dalvi and Domingos [6] have addressed on adversarial attack on spam detection system. Meek [7] has proposed the concept of adversarial learning [7].

D. J. Hemanth et al. (Eds.): COMET 2019, LNDECT 35, pp. 1076–1087, 2020.
https://doi.org/10.1007/978-3-030-32150-5_109

Barreno et al. [8] have proposed the various types of security attacks in Machine Learning. To avoid such security attacks, many researchers have provided their own security defensive techniques to protect the model. The basic defensive mechanism is assessing the accuracy, which is based on the accuracy counter measures that are used during training and testing phases.

In summary, the paper is focused on various security threats of machine learning and protective techniques used during training and testing phases [9]. The paper is structured in the following way. Section 2 explains the basic concept of machine learning, adversarial learning and types of security threats. Section 3 explains the details of security attacks in training and testing phases. Section 4 explains the various protective techniques and counter measures used in the learning model. Section 5 summarizes the threats and challenges of machine learning against attackers. Section 6 gives the conclusion and future enhancement.

2 Machine Learning

Machine learning (ML) is a branch of artificial intelligence (AI) which helps in analysing the structure of data and to fit the data into models accurately. It is one of the computer science fields, and differs from other computing technology by the way of training the computers based on data inputs and it uses statistical analysis to get the proper output. For this reason, ML is used in automated decision making models like Facial recognition, Recommendation engines, OCR and Self driving car applications.

According to the training procedures used, Machine Learning techniques are grouped into three categories. The categories are (i) supervised learning, (ii) unsupervised learning and (iii) reinforcement learning [10, 11]. In supervised learning, the data samples with category labels are used in training. Classification and regression models are examples of supervised learning. The algorithms used in this approach are Decision tree, Naïve Bayes etc. In unsupervised learning, the data samples are directly used in training without category label. Clustering techniques and encoders are the basic examples of unsupervised approach. Reinforcement learning is a mixture of prior two approaches. It uses an agent that finds the correct action to achieve the overall goal of the application. The investigation of the work was closed with the execution consequences of various approaches that convey and demonstrates better Machine learning methodology which can prompt increasingly precise learning of health care.

2.1 Adversarial Learning and Types of Security Threats

Adversarial learning is a research field, which is a combination of ML and computer security. ML model is developed based upon the data inputs. The attackers may change the input data; hence the trained model becomes false [12]. Adversarial learning is shortly defined as Machine Learning in adversarial settings. The adversaries construct the adversary settings in such a way that the learning procedure is failed. The adversary has the knowledge of learning algorithms, parameters and the data, so he/she can construct the adversary examples [13].

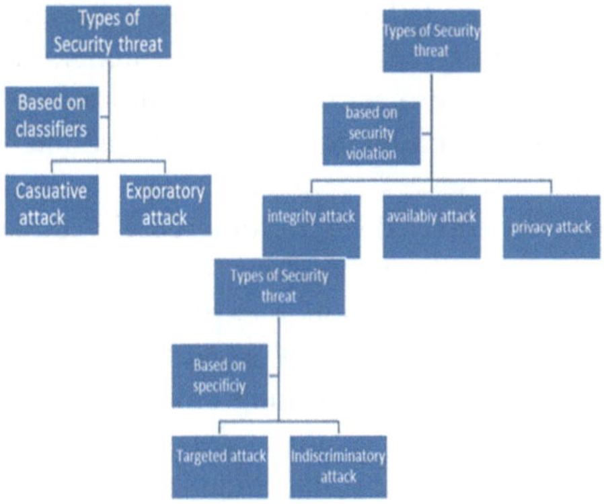

Fig. 1. Various types of attacks

There are two types of attacks namely white box attack and black box attack. In white box attack the attacker has the complete knowledge of the learning environment, whereas the knowledge is unknown in the later one. The types of security threats in ML are described in [14]. It is based on influences on classifiers, security violation and attack sensitivity. This is depicted in Fig. 1. Causative attack means the attackers can change the training data including change of parameters of learning model which affects the performance of the model. In Exploratory attack, the adversary does not modify the training models. The aim of attacker is to make misclassification with respect to adversary samples and gain the access to sensitive information. The performance of classifiers is reduced because of a particular group of sample is targeted (attack). In discriminative attack, there is no specific constraint on a particular data; it could cover vast range of data.

3 Security Threats in Training Phase

The important phase in machine learning is training phase because the performance of the model mainly depends upon training. So that many attackers can focus on training data that results in the reduction of overall performance of the model. Most of the Machine Learning algorithms suffer by the effect of adversarial samples. A general framework was introduced that allows evasion attacks and introduced some counter measures against the attack [15]. In [16], how intrusion detection system was affected by several attacks and discussed the solutions for each kind of attack. Nedim et al. proposed a high performance static method for detecting malicious PDF documents based on analysing structural properties of true and malicious PDF files [17]. In [18] how active learning affected by two sampling attacks based on addition and deletion of malicious data is given. Nowadays deep learning is one of the important and emerging research fields in machine learning. Deep Neural Network (DNN) is used in various computer vision applications. It is also more sensible to adversarial attacks. For example, DNN is had failed to classify the images when the network was trained by applying adversary samples with perturbation [19]. Poisoning attack is one type of attack in training phase. It is a causative attack. Adaptive biometric recognition system was proposed by Battista et al. These systems are periodically updated for recognizing clients' faces. Researchers demonstrated how the model is affected by poisoning attack and how it misleads a PCA based face verification model [20]. Figure 2 shows illustration of poisoning attack [48].

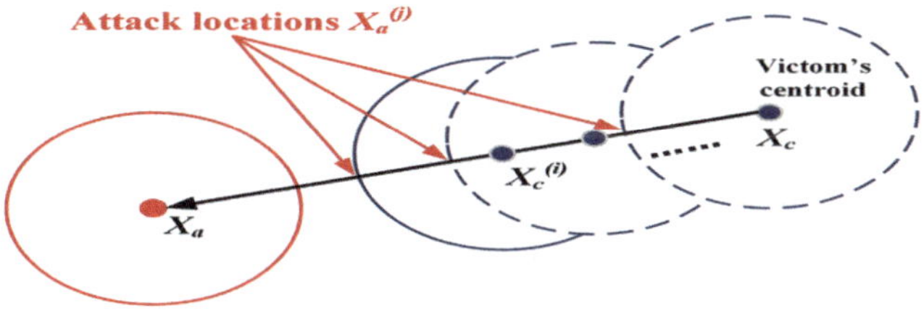

Fig. 2. Poisoning attack

The poisoning attack moves classification centroid from true data (X_c) to the malicious data (X_a) [21]. The poisoning attack affects many machine learning algorithms such as SVM, [22] Neural Networks, PCA and LASSO [23]. In 2015, Mozaffari et al., presented a systematic approach for generating poisoning attacks against several machine learning algorithms. These attacks are applied on five health care datasets. They proposed counter measures against the attacks which are based on detecting deviations in two accuracy metrics namely correctly classified instances (CCI) and Kappa statistic [24]. Another example for poisoning attack was performed in malware detection system [25]. Recently GAN (Generative Adversarial Network) plays an important role in machine learning against security. Malware classifiers are used to detect malware by using machine learning approaches. MalGAN is based on Generative network, which is used to produce adversarial malware examples. The advantage of MalGAN over to traditional method is, that MalGAN is able reduce the detection rate as zero and the defensive method against adversarial samples is very difficult to make [25]. Another powerful attack is changing the features or labels of training data. Label contamination attack (LCA) is one type of poisoning attacks where the adversary modifies the training data. In [26], Projected Gradient Ascent (PGA) algorithm is used to produce the LCA and shows how the model is affected by LCA. Biggio et al., evaluated the security of SVM by introducing adversarial label attacks. The attacker's aim is to increase the classification error of SVM by changing the label of training data [27].

3.1 Security Threats in Testing Phase

The most common type of attack used in testing phase is spoofing which includes evasion, impersonate attack and inversion attack. In [28], adversary aware classifiers model was proposed, that could improve the model security against evasion attacks by examining the adversary's data and suggested some specific assumptions. The authors have developed wrapper based approach and tested, validated in spam and malware detection applications. Similarly impersonate attack would behave like an actual data set, so that the adversary, gain the access to confidential data. Impersonate attack is much powerful in attacking DNN algorithms [29] and [30]. In Inversion attack, the adversary can access the API of existing ML Model and collect the basic data. Then this basic information is fed to the target model. Examples are health care data, customer's survey data, face authenticated data etc. [31].

3.2 Summary of Security Attacks

In this section we have summarized the various adversarial attacks against machine learning. Table 1 shows the different methods and their advantages and disadvantages. The merits and demerits are specified in terms of efficiency, computation time, the quality of the adversarial samples and how applied to a large dataset. Optimization method and Fast Gradient sign method are used to generate high quality adversarial samples, but they consume more computation time. Deep learning based attacking methods including DeepFool and JSMA are very efficient and generating adversarial samples in a pre-specified manner.

Table 1. Comparison of attacking techniques against adversarial attack

Attacking technique	Merits	Demerits
Optimization based method	• Minimum perturbation rate • Produce high quality adversarial samples	• Large computation time • It couldn't support large size data set
Gradient based method	• Faster than optimization approach • Generate high quality adversarial samples	• Perturbation rate is not optimum
Iterative least likely class method	• Fast method • Fine perturbation rate so that adversarial samples are generated with highest quality	• Number of iterations is not fixed
Deep neural network based method: DeepFool	• Highly efficient	• It does not guarantee that the samples are good enough
Jacobian based approach (JSMA)	• Fine perturbation rate • Good tradeoff between the quality and size of samples	• High computation complexity • DNN should be feed forward networks

4 Defensive Techniques

In this section various defensive techniques are explained against adversarial attacks. There are two types of defensive mechanisms. 1. Reactive defense and 2. Proactive defense [13]. In a Reactive defensive mechanism the adversary analyses the classifier then designs and launchs the attack. The classifier designer analyses the attacking results then he proposes defending mechanism. In Proactive defensive mechanism, the classifier itself designs some adversarial attack depending upon the existing work. He launches the attack and evaluates the attacking effects. Based on the impact, some counter measures are proposed by the classifier designer. This is illustrated in Fig. 3.

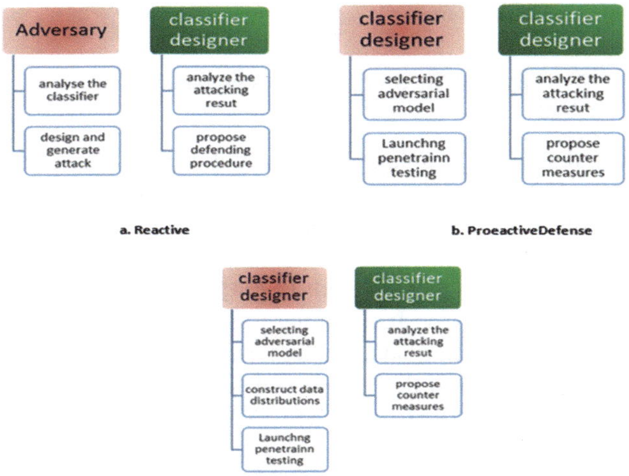

Fig. 3. Proactive Defense with data distribution

Another important defensive technique is considering the data distribution [32]. Normally the data distributions of training data and testing data is are different in adversarial environments. In proactive defensive method, one more step is introduced by considering the data distribution. The classifier train the model using training data X which contains legitimate data. The model is validated by using test data Y and it predicts as T. The same model is trained by training data X′. X′ contains both legitimate and malicious data. Then the model is validated by using same test data and it is predicted as T′. Finally the model's performance is compared by using T and T′ and it alarms the user that the X′ contains adversarial samples.

4.1 Defensive Techniques During Training Step

Since the malicious data, called poisoning, is injected along with the training data, attack which results in making the model to make wrong decisions or degrade the performance. Figure 4 shows the defensive attack during training phase. In this case, the classifier designer ensures the data purity and takes steps to protect the data from the adversarial attack. Similarly the robustness of learning algorithm should be improved [34–36].

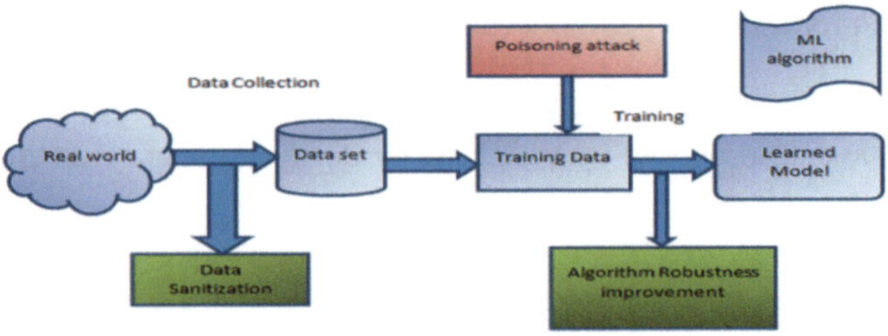

Fig. 4. Defensive techniques in training phase

The important defending technique is to assure the cleanliness of data that is called as data sanitization [33]. Huang et al. proposed two models for modelling an attacker capabilities, exploring the limit of adversary knowledge, protecting the training data and feature space etc. Other defending method is improving the robustness of learning algorithm: e.g., Random Subspace method. In [34], the authors have proposed a new robust technique based on PCA. It maximizes the median absolute deviation. They have demonstrated the poisoning attack and shown how the model is protected from poisoning data. Another type of defending method is to design the effective learning algorithm. In [37], a secure SVM called Sec-SVM is proposed to protect the model against evasion attacks.

4.2 Defensive Techniques During Testing Step

The learning model's performance is evaluated during training phase by using training the data. Adversarial attacks and corresponding Defensive techniques are illustrated in Fig. 5. The defensive techniques and counter measures used in testing phase have focused on the improvement of model's robustness. Invariant SVM algorithms were proposed by Teo et al. They have used min-max method to address the Feature manipulation activities. To improve the robustness of learning model a new algorithm NashSVM was prop0sed in [38]. One of the counter measures against testing phase attack is considering the data distributions. The attacker's goal is to modify the data distribution of testing data which results in difference in performance. Then the false alarm is given to classifier designer to show that the testing dataset has adversarial samples [39]. Another feasible way of defending technique against attack is, the model is trained by using adversarial training data itself. So the new trained classifiers can detect the anomalies during testing phase [18]. To protect the deep neural network from adversarial attacks, defensive distillation was introduced by Papernot et al. [40]. The machine learning model is protected from evasion attack by using dimension reduction strategy [41]. Statistical test is used to distinguish the adversarial data from true data [42]. Two important metrics namely Maximum Mean Discrepancy (MMD) and Energy distance (ED) are used to find the adversarial samples easily. In [43], an ensemble framework is used to combine more than one DNN to protect itself against attack.

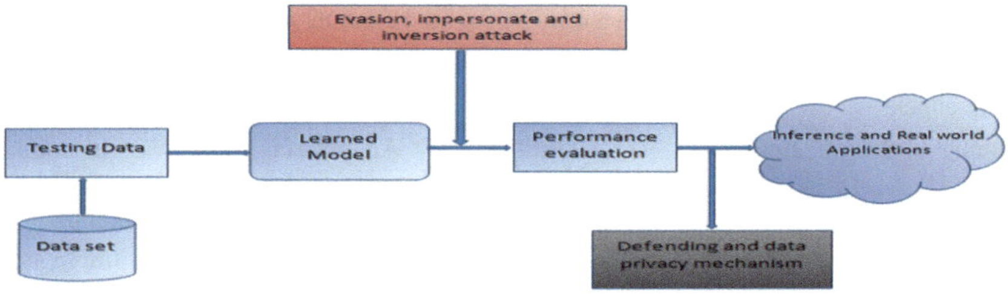

Fig. 5. Defensive techniques in testing phase

4.3 Data Privacy

Recent classifiers require a large volume of data for training. It is important that the data should be secured and it should not be accessed by other than the legitimate user. But it is affected by a high possibility of the leakage of data that includes picture, video, medical records.

The basic technique is cryptographic technology to secure the data privacy. Differential Privacy (DP) is one of the techniques that retains the data privacy by using data encryption [44]. The researchers are using DP to preserve the privacy of different ML algorithms such as SVM, Deep Neural Networks [DNN] and other optimization techniques. Homomorphic Encryption (HE) is important technique to preserve the data privacy. Many researchers use the HE for secure multi class Classification [45], k-means clustering algorithms [46] and Artificial Neural Networks [47].

5 Challenges and Research Opportunities

Nowadays Machine Learning acts like a base technology of Internet of Things (IoT), big data, AI and security. So various security attacks are developed by adversaries and make the ML model to fail. Table 2 presents various attacks and the corresponding defending techniques. Based on the literature survey, the following research directions are available. 1. In attacker's point of view designing a good adversarial model is difficult and it is an important emerging research direction. 2. It is necessary to establish proper security assessment standards. 3. Data privacy is preserved by cryptographic technique, like DP and HE. Since they are not much efficient, the important research direction still is to develop high efficient privacy preserving algorithm for securing the data. 4. Deep learning model is easily affected by adversarial attacks; it has to be addressed by the researcher.

Table 2. Types of Security attack and Defensive Techniques

Technique security attacks	Training/testing	Type	Concept
Poisoning attack	Training	Causative attack	Adversarial samples are inserted in training data and the labels or features are modified during training
Evasion attack	Testing	Exploratory attack	It makes adversarial examples to stay away from location of target framework
Impersonate attack	Testing	Exploratory attack	It makes adversarial samples to act as target systems or to confuse the target framework
Inversion attack	Testing	Exploratory attack	It takes the delicate data of target classifier or datasets
Defensive techniques			
Data sanitization	Training	Protect integrity of the data	It sanitizes and prepares information and rejects the example that will initiate negative effects to classifiers
Defence distillation	Training	Protect integrity	It works with DNN based on probability. Training data label is retained with probability level
Ensemble method	Training	Protect integrity	It combines different classifiers and defensive techniques to detect and remove adversarial samples
Differential privacy	Training & Testing	Protect privacy	Uses randomized method to protect the data during training and testing phases
Homomorphic encryption	Testing	Protect privacy	This is based on cryptographic technique. The model directly processes encrypted data

6 Conclusion

Machine Learning can play a vital role in a wide range of critical applications, such as data mining, natural language processing, image recognition, health care applications and expert systems. ML provides potential solutions to all these domains and more, and it is a pillar of computation technology. It is necessary to protect the Machine learning model from security attacks. In this paper, we have discussed various security attacks towards ML training and testing phases. Subsequently we have organized the current defending techniques and counter measures used in the training and testing phase. We have also discussed some data privacy techniques to protect large volume data, which is used in learning. Finally we have presented various challenges and research directions in this field. This review can be a profitable reference for specialists in both ML and computer security fields.

References

1. Zhou, L., Pan, S., Wang, J., Vasilakos, A.V.: Machine learning on big data: opportunities and challenges. Neurocomputing **237**, 350–361 (2017)
2. Yu, S.: Big privacy: challenges and opportunities of privacy study in the age of big data. IEEE Access **4**, 2751–2763 (2016)
3. Al-Jarrah, O.Y., Yoo, P.D., Muhaidat, S., Karagiannidis, G.K., Taha, K.: Efficient machine learniing for big data: a review. Big Data Res. **2**(3), 87–93 (2015)
4. Wittel, G.L., Wu, S.F.: On attacking statistical spam filters. In: Proceedings of 1st Conference on Email Anti-Spam, Mountain View, CA, USA, pp. 1–7 (2004)
5. Sharif, M., Bhagavatula, S., Bauer, L., Reiter, M.K.: Accessorize to a crime: real and stealthy attacks on state-of-the-art face recognition. In: Proceedings of ACM SIGSAC Conference on Computer and Communications Security, Vienna, Austria, pp. 1528–1540 (2016)
6. Dalvi, N., Domingos, P., Mausam, Sanghai, S., Verma, D.: Adversarial classification. In: Proceedings of 10th ACM SIGKDD International Conference on Knowledge Discovery and Data Mining, Seattle, WA, USA, pp. 99–108 (2004). https://homes.cs.washington.edu/~pedrod/papers/kdd04.pdf
7. Lowd, D., Meek, C.: Adversarial learning. In: Proceedings of 11th ACM SIGKDD International Conference on Knowledge Discovery in Data Mining, Chicago, IL, USA, pp. 641–647 (2005)
8. Barreno, M., Nelson, B., Sears, R., Joseph, A.D., Tygar, J.D.: Can machine learning be secure? In: Proceedings of ACM Symposium on Information, Computer and Communications Security, pp. 16–25 (2006)
9. Amodei, D., Olah, C., Steinhardt, J., Christiano, P., Schulman, J., Mané, D.: Concrete problems in AI safety, July 2016. https://arxiv.org/abs/1606.06565
10. Qiu, J., Wu, Q., Ding, G., Xu, Y., Feng, S.: A survey of machine learning for big data processing. EURASIP J. Adv. Signal Process. **2016**, Article no. 67 (2016)
11. Meenakshi, K., Safa, M., Karthick, T., Sivaranjani, N.: A novel study of machine learning algorithms for classifying health care data. Res. J. Pharmacy Technol. **10**(5), 1429–1432 (2017)
12. Barreno, M., Nelson, B., Joseph, A.D., Tygar, J.D.: The security of machine learning. Mach. Learn. **81**(2), 121–148 (2010)
13. Biggio, B., Fumera, G., Roli, F.: Security evaluation of pattern classifiers under attack. IEEE Trans. Knowl. Data Eng. **36**(4), 984–996 (2014)
14. Fredrikson, M., Jha, S., Ristenpart, T.: Model inversion attacks that exploit confidence information and basic countermeasures. In: Proceedings of 22nd ACM SIGSAC Conference on Computer and Communications Security, Denver, CO, USA, pp. 1322–1333 (2015)
15. Corona, I., Giacinto, G., Roli, F.: Adversarial attacks against intrusion detection systems: taxonomy, solutions and open issues. Inf. Sci. **239**, 201–225 (2013)
16. Yu, S., Gu, G., Barnawi, A., Guo, S., Stojmenovic, I.: Malware propagation in large-scale networks. IEEE Trans. Knowl. Data Eng. **27**(1), 170–179 (2015)
17. Šrndić, N., Laskov, P.: Detection of malicious PDF files based on hierarchical document structure. In: Proceedings of 20th Annual Network and Distributed System Security Symposium, San Diego, CA, USA, pp. 1–16 (2013)
18. Biggio, B., et al.: Poisoning complete-linkage hierarchical clustering. In: Structural, Syntactic, and Statistical Pattern Recognition. Lecture Notes in Computer Science, vol. 8621, pp. 42–52. Springer, Berlin (2014)
19. Szegedy, C., et al.: Intriguing properties of neural networks, February 2014. https://arxiv.org/abs/1312.6199

20. Biggio, B., Fumera, G., Roli, F., Didaci, L.: Poisoning adaptive biometric systems. In: Structural, Syntactic, and Statistical Pattern Recognition, pp. 417–425. Springer, Berlin (2012)
21. Kloft, M., Laskov, P.: Security analysis of online centroid anomaly detection. J. Mach. Learn. Res. **13**, 3681–3724 (2012)
22. Biggio, B., Nelson, B., Laskov, P.: Poisoning attacks against support vector machines. In: Proceedings of 29th International Conference on Machine Learning (ICML), Edinburgh, Scotland, pp. 1467–1474 (2012)
23. Xiao, H., Biggio, B., Brown, G., Fumera, G., Eckert, C., Roli, F.: Is feature selection secure against training data poisoning? In: Proceedings of 32nd International Conference on Machine Learning (ICML), Lille, France, pp. 1689–1698 (2015)
24. Mozaffari-Kermani, M., Sur-Kolay, S., Raghunathan, A., Jha, N.K.: Systematic poisoning attacks on and defenses for machine learning in healthcare. IEEE J. Biomed. Health Informat. **19**(6), 1893–1905 (2015)
25. Hu, W., Tan, Y.: Generating adversarial malware examples for black-box attacks based on GAN, February 2017. https://arxiv.org/abs/1702.05983
26. Zhao, M., An, B., Gao, W., Zhang, T.: Efficient label contamination attacks against black-box learning models. In: Proceedings of 26th International Joint Conference on Artificial Intelligence (IJCAI), Melbourne, VIC, Australia, pp. 3945–3951 (2017)
27. Xiao, H., Biggio, B., Nelson, B., Xiao, H., Eckert, C., Roli, F.: Support vector machines under adversarial label contamination. Neurocomputing **160**, 53–62 (2015)
28. Zhang, F., Chan, P.P.K., Biggio, B., Yeung, D.S., Roli, F.: Adversarial feature selection against evasion attacks. IEEE Trans. Cybern. **46**(3), 766–777 (2016)
29. Mopuri, K.R., Garg, U., Babu, R.V.: Fast feature fool: a data independent approach to universal adversarial perturbations, July 2017. https://arxiv.org/abs/1707.05572
30. Moosavi-Dezfooli, S.-M., Fawzi, A., Fawzi, O., Frossard, P.: Universal adversarial perturbations. In: Proceedings of IEEE Conference on Computer Vision and Pattern Recognition (CVPR), Honolulu, HI, USA, pp. 86–94, July 2017
31. Fredrikson, M., Lantz, E., Jha, S., Lin, S., Page, D., Ristenpart, T.: Privacy in pharmacogenetics: an end-to-end case study of personalized warfarin dosing. In: Proceedings of USENIX Security Symposium, San Diego, CA, USA, pp. 17–32, August 2014
32. Laskov, P., Kloft, M.: A framework for quantitative security analysis of machine learning. In: Proceedings of 2nd ACM Workshop on Security and Artificial Intelligence, Chicago, IL, USA, pp. 1–4 (2009)
33. Huang, L., Joseph, A. D., Nelson, B., Rubinstein, B.I.P., Tygar, J.D.: Adversarial machine learning. In: Proceedings 4th ACM Workshop Security and Artificial Intelligence, Chicago, IL, USA, pp. 43–58 (2011)
34. Rubinstein, B.I.P., et al.: ANTIDOTE: understanding and defending against poisoning of anomaly detectors. In: Proceedings of 9th ACM SIGCOMM Conference on Internet Measurement, Chicago, IL, USA, pp. 1–14 (2009)
35. Biggio, B., Fumera, G., Roli, F.: Multiple classifier systems for robust classifier design in adversarial environments. Int. J. Mach. Learn. Cybern. **1**(1–4), 27–41 (2010)
36. Biggio, B., Corona, I., Fumera, G., Giacinto, G., Roli, F.: Bagging classifiers for fighting poisoning attacks in adversarial classification tasks. In: Proceedings of 10th International Conference on Multiple Classifier System (MCS), Naples, Italy, pp. 350–359 (2011)
37. Demontis, A., et al.: Yes, machine learning can be more secure! A case study on android malware detection. IEEE Trans. Dependable Secure Comput., to be published
38. Brückner, M., Kanzow, C., Scheffer, T.: Static prediction games for adversarial learning problems. J. Mach. Learn. Res. **13**, 2617–2654 (2012)

39. Xu, W., Evans, D., Qi, Y.: Feature squeezing: Detecting adversarial examples in deep neural networks, December 2017. https://arxiv.org/abs/1704.01155
40. Papernot, N., McDaniel, P., Wu, X., Jha, S., Swami, A.: Distillation as a defense to adversarial perturbations against deep neural networks. In: Proceedings of IEEE Symposium on Security and Privacy, San Jose, CA, USA, pp. 582–597, May 2016
41. Bhagoji, A.N., Cullina, D., Sitawarin, C., Mittal, P.: Enhancing robustness of machine learning systems via data transformations, November 2017. https://arxiv.org/abs/1704.02654
42. Grosse, K., Manoharan, P., Papernot, N., Backes, M., McDaniel, P.: On the (statistical) detection of adversarial examples, October 2017. https://arxiv.org/abs/1702.06280
43. Sengupta, S., Chakraborti, T., Kambhampati, S.: MTDeep: boosting the security of deep neural nets against adversarial attacks with moving target defense, September 2017. https://arxiv.org/abs/1705.07213
44. Dwork, C.: Differential privacy. In: Proceedings of 33rd International Colloquium on Automata, Languages and Programming (ICALP), Venice, Italy, pp. 1–12 (2006)
45. Wang, Q., Zeng, W., Tian, J.: Compressive sensing based secure multiparty privacy preserving framework for collaborative data-mining and signal processing. In: Proceedings of IEEE International Conference on Multimedia and Expo (ICME), Chengdu, China, pp. 1–6, July 2014
46. Yao, Y.-C., Song, L., Chi, E.: Investigation on distributed K-means clustering algorithm of homomorphic encryption. Comput. Technol. Develop. **2**, 81–85 (2017)
47. Dowlin, N., Gilad-Bachrach, R., Laine, K., Lauter, K., Naehrig, M., Wernsing, J.: CryptoNets: applying neural networks to encrypted data with high throughput and accuracy. In: Proceedings of 33rd International Conference on Machine Learning, New York, NY, USA, pp. 201–210 (2016)
48. Liu, Q., Li, P., Zhao, W., Leung, V.C.M.: A survey on security threats and defensive techniques of machine learning: a data driven view. IEEE Access **6**, 12103–12117 (2018)

Content Based Image Retrieval Using Machine Learning Based Algorithm

Navjot Kour(✉) and Naveen Gondhi

SMVDU Katra, Katra, India
navjotkour21@gmail.com

Abstract. In research field, CBIR (Content Based Image retrieval) has played a vital role. This paper deals with the realization of different approaches used in image retrieval based on content. It gives a general idea of the currently accessible literature on content based image retrieval. In CBIR, a query image is searched from larger database and an exact match image is retrieved using efficient machine learning algorithms. Different algorithms i.e. Bacteria Foraging optimization algorithm, Swarm optimization algorithm, Convoltional neural network, Firefly network, Deep Belief Network, Support vector machine and Genetic algorithm are reviewed and their performance parameters are compared.

Keywords: CBIR · Bacteria Foraging optimization algorithm · Swarm optimization algorithm · Convoltional neural network · Firefly network · Deep Belief Network · Support vector machine and Genetic algorithm

1 Introduction

In late 1970's, content based image retrieval research started. The image retrieval techniques which were used before this technology were not intelligent and could not search images in large database based on their visual features. Thus researchers made a new technology for better image recovery which has high performance and accuracy parameters. In 1992 CBIR technology has emerged. This system is also known as Query by image content. The main aim of this system is to extract the features from images, index those features using appropriate matching algorithms and give answers to queries. Different researchers used different methods for searching images. But in this paper first we discuss about various approaches used in CBIR for retrieving images and then focus on one approach that is graph representation. It compares various graph theory algorithms, which is depicted in the Table 1 briefly. CBIR system has a database that stores images. In CBIR system, the stored images features are extracted and matched to the query images features. It involves two steps:-

1. Feature Extraction: In feature extraction, low level features such as color, texture and shape of query images are extracted.
2. Matching: In matching phase, extracted features of query image matches the features of target image of database so that exact match comes.

CBIR uses Machine Learning concepts to retrieve exact images. Machine learning is becoming widespread and different technologies are using it in a variety of ways.

D. J. Hemanth et al. (Eds.): COMET 2019, LNDECT 35, pp. 1088–1095, 2020.
https://doi.org/10.1007/978-3-030-32150-5_110

Machine learning is new technology which gives computer the ability to learn new things and act like human beings. Everytime new changes with improvement is feed into the computer and this process is called learning. Then the system gives perfect result with experience. In CBIR for extraction of features from image we use machine learning algorithm. They are of 3 types supervised, unsupervised and reinforcement learning. With CBIR, user search for an image that matches with target image. Typically, known images have been scanned for features and the features are stored in database to find best match. There are different machine learning algorithm for detecting a feature and converting it into a list of vectors. The stored vectors are compared with the sample image to find a match.

2 Algorithms Used

Content based image retrieval system uses various algorithm for image retrieval in many applications. These algorithms are:

2.1 Bacteria Foraging Optimization Algorithm

Bacteria foraging algorithm used the technique of foraging of bacteria. Human being intestines contain E.coli bacteria which invoves foraging strategy. It is an optimization algorithm. Different optimization problems in computational and time complexity can be solved. It can be used to significantly reduce computational complexity and at the same time having a very good solution. The solution may not be absolute but it is most of time acceptable. BFOA is mainly used in MANETs.

In Paper [1], BFOA is used to reduce the cost, time and complexity in content based image retrieval technique. It is used to reduce the extracted features matrix. It is used to optimize the feature set in which first we have to initialize the matrix of population and solution of population is calculated and reserved to population. This algorithm has one advantage that it gives best match in minimum time. BFOA technique is used to solve an optimization issue by repeating 3 phases:

- Rotation
- Dispersal and Elimination
- Reproduction.

2.2 Ant Colony Optimization

Ant colony optimization is supposed to be a swarm intelligence based learning algorithm which has been widely used for various optimization problems. It is based on the behaviour of ants searching for food. At first, ants search the food, then it returns to the colony leaving redolence on which the other ants follow the same path for searching food. Shorter paths are followed to optimize the solution. From a data mining stand points, its one of the best performing evolutionary algorithms in the domain of feature selection and rule mining. ACO is a class of algorithms that fall under the metaheuristics. It is a very effective set of algorithms for discrete optimization purposes. It can be used for clustering. It is probalistic and iterative algorithm. The objective of ACO is to select a path with minimum number of features are retrievd by maximum retrieval accuracy in CBIR system [2].

2.3 Swarm Optimization Algorithm

Particle swarm Optimization (PSO) is a branch of swarm intelligence. It is the collective behavioural exhibited by several creatures in nature like bees, ants and fishes. PSO is a stochastic algorithm in CBIR. This algorithm has been inspired from flocking behavior of birds in nature when migrating from one place to another. In PSO each solution has three parts position, velocity and fitness value.

PSO is a metaheuristic gradient free optimization method. This method was proposed by Eberbant and Kennedy in 1995. In paper [3] PSO based image retrieval system was defined in which low level features are extracted and a modified PSO algorithm was proposed in which position and velocity of each particle is described in continuous space. The most similar target image are obtained by proposed algorithm. This algorithm estimates the precision, recall parameters. PSO algorithm works in this way:

1. This algorithm have 2 datasets.
   - Training dataset
   - Testing dataset
2. In Training dataset, each image features are extracted and similarity index of image pairs are estimated.
3. Calculate the fitness value of image parameter.
4. The image which have optimal fitness value is tested and similar images are extracted.

2.4 Convolutional Neural Network

Convolutional neural Network (CNN) are a type of neural network which have been widely used for image recognition tasks. There are four main steps in CNN are Convolutional, subsampling Activation and correctedness. CNN are made up of neurons which have wight and biases. Each neuron receives input performs dot product and have a loss function. They are used on image, then it uses filter or kernels to detect feature. A filter is matrix of values called weights that are trained to detect and each filter carries out a convolutional operation. The size of fiter mask should be smaller as it is more computationally efficient and better weight sharing. So 3X3 convolutional filter is used and it have third dimension in size. If the feature is present in the image then convolutional operation value is high and vice versa.

In paper [4], CNN is used to increase the performance parametrs of CBIR. For feature representation and similarity measurement CNN algorithm is used. The main aim of using CNN in CBIR is that time resource and material are minimum (Fig. 1).

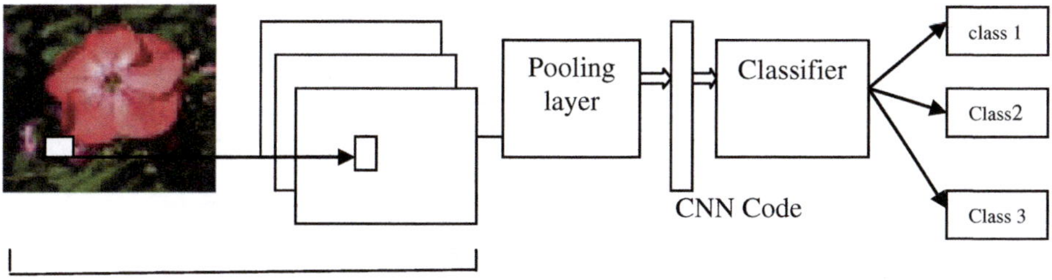

Fig. 1. Architecture of CNN

The algorithm is divided in following parts:

1. Convolutional phase is used for extracting image feature. In CNN, feature descriptors are used for extracting the kernels from each image.
2. The image passed to number of filters for making new image called convolutional maps.
3. A pooling layer is used to reduce the computational complexity.
4. At last CNN classifier are used in which multilayer perceptron architecture is made up of 4 layers input layer, 1 hidden layer, 2nd hidden layer and output layer. The necessity for choosing two hidden layer is solving non linear classification problems.

2.5 FireFly Algorithm

It was developed at Cambridge university in 2007 by Xin-She Yang. This algorithm is inspired by flashing behavior of fireflies. This algorithm uses random numbers. It is metaheuristic algorithm. Improved firefly algorithm gives best performance as compared to Gentic algorithm. It is used for classification, clustering and optimization. This algorithm has flexibility of integration with other optimization techniques to form hybrid tool.

It is used in various fields:

- For solving travelling salesman problem
- Digital image processing
- Feature selection
- Scheduling
- Dynamic problems

In paper [5] Firefly algorithm is used in CBIR for training the features of images in database. After uploading the database features are extracted from image and train them. After this step, features are reduced using firefly algorithm (Fig. 2).

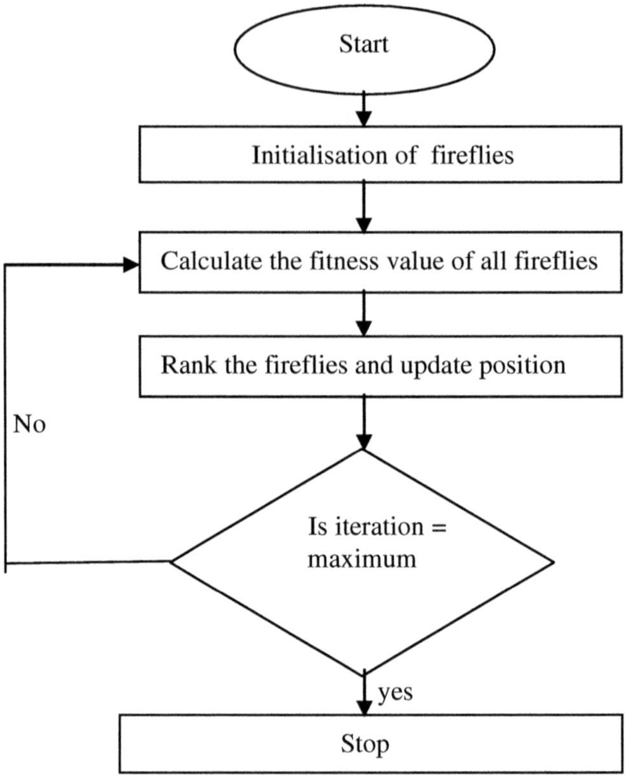

Fig. 2. Flowchart of firefly algorithm

2.6 Deep Belief Network

A deep belief network are based upon stacking restricted boltzman machine (RBM). Every RBM layer communicates with previous and subsequent layer. DBN are used to recognize, cluster and generating images. It is a type of deep artificial neural network. They are generative models are trained using form Byesian network. Performance of the DBN depend on initialization of nodes, the layers initialize unsupervised pre-training using an RBM stacking procedure. Used in various classification and regression tasks.

2.7 Support Vector Machine

Support vector machines techniques are used for the retrieval of images which are similar to the query image. SVM works on the principle of supervised learning technique for classification purpose. In classification, SVM maps the input space to the feature vector space. They are active learning technology that are used in handwritten digit recognition, object recognition and text classification. SVM are hyper planes that separate the training data by maximal margin. The training instances that lie closest to hyper plane are labeled as support vector. SVM image retrieval system employs a multi-resolution image representation. SVM technique is the most efficient in CBIR system. It uses machine learning concepts for retrieving images at large database.

In paper [7] SVM are used to classify the features of query images by dividing the group such as color, shape and texture. Finally relevant images are retrieved from database. This method gives better performance.

2.8 Genetic Algorithm

Genetic algorithm are heuristic search algorithm. They can produce accurate results for optimization problems and search problems. It is an iterative algorithm in which it have initial state and target state. The initial state can be called as parents and different technique like cross-over, mutation, etc. are applied on intial stage to generate child state genetic algorithm are based on logic with genetic structure and behavior of chromosomes of population (Fig. 3).

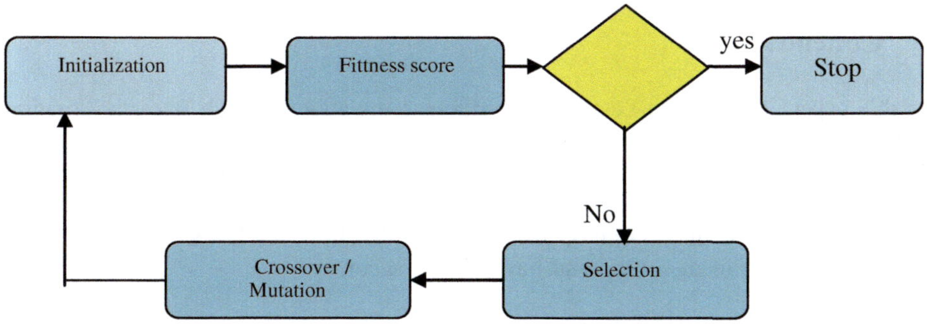

Fig. 3. Flowchart of Genetic algorithm

Genetic Algorithm architecture has three main steps Crossover, mutation and selection. If the value of fitness score is less then it gives more preference in Genetic algorithm.

In Table 1 all the 8 algorithms such as Bacteria Foraging optimization algorithm, Swarm optimization algorithm, Convoltional neural network, Firefly network, Deep Belief Network, Support vector machine and Genetic algorithm are reviewed and their performance parameters are compared. But Convolutional Neural Network performance is much better than other algorithms as it takes meager time and cost for extracting features from target image. Results shows that CNN in CBIR has higher Recall and precision parameter in comparison with other algorithm. The accuracy of CNN is also higher.

Table 1. Comparision of parameters such as precision, recall and accuracy of different algorithms

ML algorithms	Dataset	Recall	Precision	Accuracy	Remarks
BFOA [1]	Caltech-101	95%	96%	–	Reduce complexity, energy and time consumption
ACO [2]	COREL	85%	0.139%	0.6339%	Feature selection is time consuming and offline task
PSO [3, 4]	COREL	80%	50%	23.3%	Minimize the cost function
CNN [5]	ImageNet	99%	75%	85%	Minimize the classification error in output
FireFly [6]	MNIST	98%	.0024245%	.007765%	Improves retrieval rate.
DBN [7]	COREL	98.6%	–	.0012226%	Maximum performance and provides security
SVM [8, 11]	COREL	70–80%	55%	–	Handles noisy image
GA [9, 10]	COREL	95%	76.49%	49.57%	Improves performance parameters

3 Conculsion

In this paper, all Machine Learning algorithms are elaborated such Bacteria Foraging optimization algorithm, Swarm optimization algorithm, Convoltional neural network, Firefly network, Deep Belief Network, Support vector machine and Genetic algorithm and compare their parameters i.e., Accuracy, Recall and Precision to find best algorithm for retrieving images in large database. Results show that CNN is best algorithim in CBIR. It takes meager time and having high accuracy.

References

1. Ali, A., Sharma, S.: Content based image retrieval. In: ICICCS (2017)
2. Rashno, A., Sadri, S., SadeghianNejad, H.: An efficient content-based image retrieval with ant colony optimization feature selection schema based on wavelet and color features. In: AISP (2015)
3. Hu, G., Yang, F.: Image retrieval method based on particle swarm optimization algorithm. In: 2015 International Conference on Intelligent Transportation, Big Data and Smart City (2015)
4. Broilo, M., Rocca, P., De Natale, F.G.B.: Content-based image retrieval by a semisupervised particle swarm optimization. IEEE (2008)
5. Mohamed, O., Khalid, E.A., Mohammed, O., Brahim, A.: Content-based image retrieval using convolutional neural networks. Springer (2017)
6. Singh, H., Kaur, H.: Content based image retrieval using firefly algorithm and neural network. Int. J. Adv. Res. Comput. Sci. **8**(1), (2017)
7. Saritha, R.R., Paul, V., Kumar, P.G.: Content based image retrieval using deep learning process. Springer (2018)

8. Sugamya, K., Pabboju, S., Babu, A.V.: A CBIR classification using support vector machine. In: International Conference on Advances in Human Machine Interaction (HMI - 2016), 03–05 March 2016. R. L. Jalappa Institute of Technology, Doddaballapur (2016)
9. Gali, R.: Genetic algorithm for content based image retrieval. In: Fourth International Conference on Computational Intelligence, Communication Systems and Networks. IEEE (2012)
10. Ligade, A.N., Patil, M.R.: Optimized content based image retrieval using genetic algorithm with relevance feedback technique. Int. J. Comput. Sci. Eng. Inform. Technol. Res. (IJCSEITR) **3**(4), 49–54 (2013). ISSN 2249-6831
11. Bansal, M., Sidhu, B.S.: Content based image retrieval system using SVM technique. IJECT **5**(4) (2014)

Classification of Signal Versus Background in High-Energy Physics Using Deep Neural Networks

M. Mythili(✉), R. Thangarajan, and N. Krishnamoorthy

Kongu Engineering College, Perundurai, Erode 638060, Tamilnadu, India
mythilimohan17@gmail.com

Abstract. High-energy physics is a fertile area for applied research in machine learning and deep learning. The Large hadron collider generates humongous amount of data by colliding hadrons at very high velocities and recording the events by various detectors. The data about the events are extensively used by machine learning algorithms to classify particles and also find new exotic particles. Deep learning is a specialization of artificial intelligence and machine learning that uses multi-layered artificial neural networks to excel in activities such as detecting the object, recognizing the speech, etc. Classical Techniques such as shallow neural networks have limitations to study the complex non-linear functions of the inputs. Deep learning programs should have access to large amounts of training data and processing power to attain an acceptable level of accuracy. These techniques have made significant progress in the classification metric which uses the best new approaches without the manual assistance.

Keywords: Deep learning · Neural network · HIGGS benchmark · SUSY benchmark

1 Introduction

High-energy physics is the study of the most important building blocks of nature. The objective of high-energy physics is to learn the interactions between particles. It is a universal struggle and the basic logic of particle physics is known as standard model. High-energy physics is also known as particle physics because many particles do not occur under normal event in nature. This can be constructed and identified during energetic collisions of other particles. Modern high-energy physics experimentation is focused on subatomic particles. These particles contain atomic constituents such as electrons, protons, and neutrons.

High-energy physicists accelerate particle beams to the speed of light and these particles will burst each other. The Large Hadron Collider (LHC) is placed at the international laboratory, CERN. LHC is considered the world's largest and most powerful particle accelerator. The LHC resides in a 27-kilometre-long ring of

D. J. Hemanth et al. (Eds.): COMET 2019, LNDECT 35, pp. 1096–1106, 2020.
https://doi.org/10.1007/978-3-030-32150-5_111

superconducting magnets. It includes a number of accelerator structures to raise the energy of the particles. The beams in the LHC are made to collide at four positions and monitored with four particle detectors namely ATLAS [1], CMS [2], ALICE, and LHCb. Figure 1 shows the schematic of the LHCb detector. Many physicists use the data from the proton collisions to check current models of physics. They search for new particles and unexplored forces.

There are six particle types, they are electron, proton, muon, kaon, pion and ghost. These particles are detected by using tracking systems, ring imaging Cherenkov detector (RICH), electromagnetic and hadron calorimeters and muon system. This is portrayed in Fig. 2.

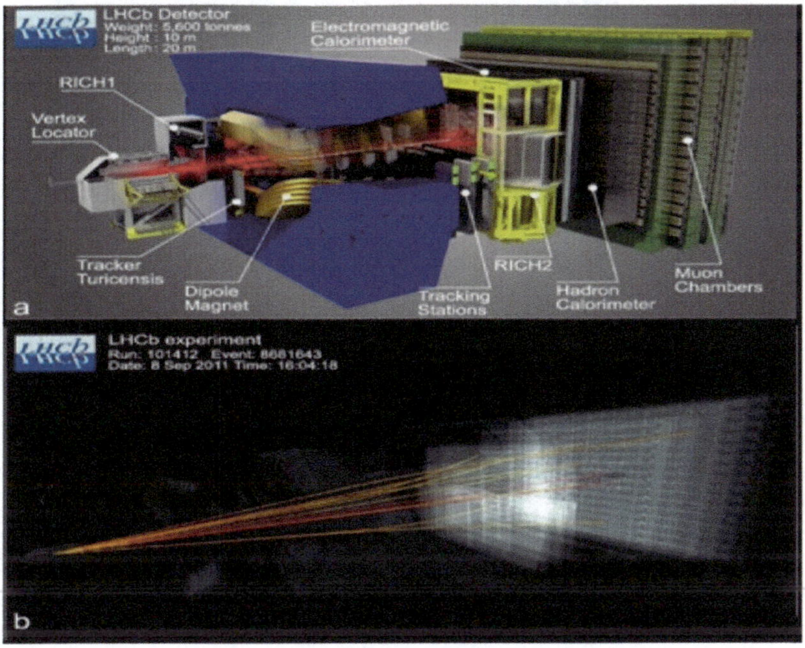

Fig. 1. The LHCb detector at CERN

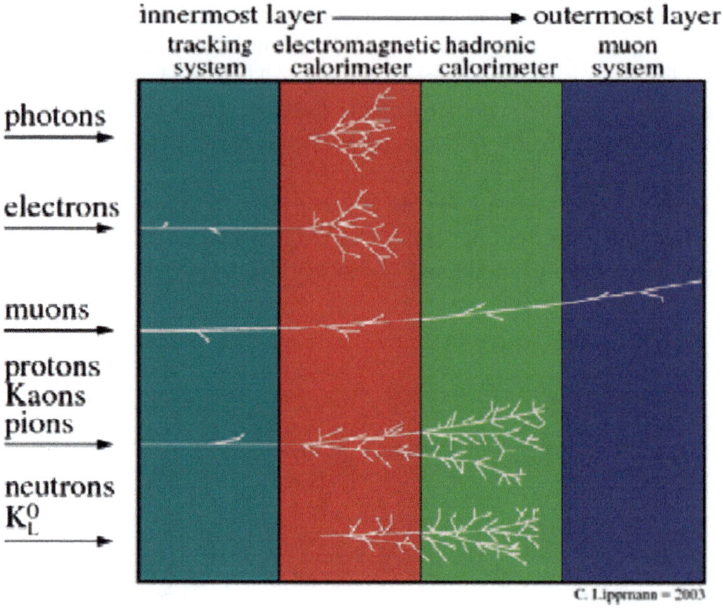

Fig. 2. Particle detection at various stages of tracking system

2 Literature Review

The proposed methods are statistical hypotheses and the procedure of identifying good critical regions for testing both the easy and combined hypotheses. A critical region depends on the standard of likelihood and a good critical region fascinates our perceptive requirements [3]. The approach of multilayer feedforward networks is a class of universal approximations. Multilayer feedforward networks are adept at perform any measurable function to a desired degree of accuracy. As a result, the capability of multilayer feedforward networks is to study the connection strengths that attain the approximations and justified that it is achievable [4]. Gradient descent algorithms are progressively becoming incompetent when the temporal span of dependencies increases. Discrete error propagation algorithm produces error facts through a combination of discrete and continuous elements. These algorithms are correlated with standard optimization algorithms in which dependencies are controlled.

This gives best solutions especially in language relevant problems in order to make perfect decisions in which long term dependencies are important [5]. The advanced algorithms to conquer vanishing gradient and by analyzing this method exhibits that learning long time lags problems can be done in tolerable time. By using conventional learning algorithms this lags problem cannot be done in feasible time. Advanced methods such as the neural sequence chunkers and long short-term memory were related and this was accomplished well [6].

Recent generative model is confined in many approaches such as by use of top-down feedback during recognition. As a result, generative models can determine low-level features and it can determine more parameters than differential models without overfitting [7]. The method multivariate classification is a basic factor to most analyses and is very useful in high-energy physics in search of smaller signals in bigger data sets to take maximum information from the data.

Toolkit for Multivariate Data Analysis (TMVA) is a toolkit which performs a large variety of multivariate classification algorithms. It has been manipulated to multivariate regression of a real valued target vector. TMVA toolkit is constructed for machine learning applications in the field of high-energy physics. TMVA methods have highly boosted algorithms which results in a possible powerful committee method that consolidate exquisite properties [8]. A tool called Madgraph produces matrix elements in high energy physics applications. Madgraph 5 is inscribed in python and run significantly faster than previous versions [9].

Pylearn2 is a machine learning research library which includes recent or different use cases. Pylearn2 is used for responsive and elasticity and it does not express as an accumulation of machine learning algorithms that contribute a simple API. Pylearn2 includes training algorithm classes namely default training algorithm, the model's and the Stochastic Gradient Boosting (SGB) [10].

A new method, Dropout. is proposed for training neural networks by arbitrarily dropping units to avoid their co-adaptation. Dropout gradient is the gradient of the approximation ensemble ordered by an adaptive weight collapse term. Dropout is a valuable approximation to practice all sub-models and also normalize properties. It generates strong units but have small dropout conflict and this is reduced by sparse encoding. Sparse encoding boosts the accuracy of dropout approximation and also the grade of self-consistency. Dropout has been strongly tested in the problem of forecasting quantitative phenotypic traits, which includes height from genetic data and single nucleotide polymorphisms (SNPs). Dropout is an assuring algorithm and this can be applied by involving simple linear or logistic regression models. Dropout plays a major role for the notion of artificial intelligence and machine learning [11].

3 Contribution of This Research Work

3.1 Estimating the Mass of Muon Using Gaussian Process

Gaussian process is a stochastic process where in a collection of random variables are indexed by time or space. Every finite collection of those random variables has a multivariate normal distribution. In other words, every finite linear combination of the random variables follows a normal distribution. A machine-learning algorithm which involves a Gaussian process generally uses lazy learning. It also uses a measure of the similarity between points of the kernel function to predict the value for an unseen point from training data. The saliency of this approach is that the prediction so obtained is not just an estimate for that point, but also carries uncertainty information with it. As a preliminary work, the mass of Muon particle is estimated from data and compared with the theoretical one.

Dataset Description. The double muon dataset contains 7 features recorded by CMS and published at CERN. The features are run, pt, eta, phi, Q, dxy, iso. Run is the event and event numbers, pt is the transverse momentum of the muon, eta(η) is the pseudorapidity of the muon, phi ($\emptyset$) is the angle of muon direction, Q is the charge of the muon, dxy is the impact parameter in the transverse plane, iso is the track isolation.

Formula for Muon Mass Estimation. The invariant mass M of the two muons can be calculated using the expression (1) and compared with those estimated using data.

$$M = \sqrt{2\mathbf{p}_t^1\mathbf{p}_t^2(\mathbf{cosh}(\eta\mathbf{1} - \eta\mathbf{2}) - \mathbf{cos}(\phi 1 - \phi 2))} \tag{1}$$

3.2 Classifying Particles Using Boosted Decision Tree and Deep-Learning

In the second work, efficacy of machine learning is determined by classifying known subatomic particles from data.

Dataset Description. This dataset contains 49 features and the training data contains six output classes. The features are ID, Label, Trackp, Trackpt, DLLelectron, etc. ID represents the id value for tracks, Trackp denotes particle momentum, Trackpt is the particle transverse momentum, DLLelectron is the delta log-likelihood for a particle candidate to be electron using information from all subdetectors. The six output classes are electron, ghost, kaon, muon, pion, proton.

Boosted Decision Tree. It is a machine learning method for classification, which also generates a prediction model. It allows for the optimization of arbitrary differentiable loss functions. Boosted decision tree is accurate and an efficient one used for classification and regression problems. This method is also assisting for both binary and multi-class classification.

Deep Neural Networks. Deep neural networks are now gaining momentum and serve as a realistic tool for several applications in high-energy physics. It uses complex mathematical modeling to process data in many ways. Adam optimizer is attained from adaptive moment estimation used in neural networks. It is suitable for problems that are big in data and parameters. Figure 3 represents an architecture of neural network. It consists of input layer that contains 49 features including bias, hidden layer contains 100 hidden units and output layer has 6 output classes. Hidden layer uses tanh activation function and softmax function is used by output layer. Neural network is fit with 5 epochs and a batch size of 256.

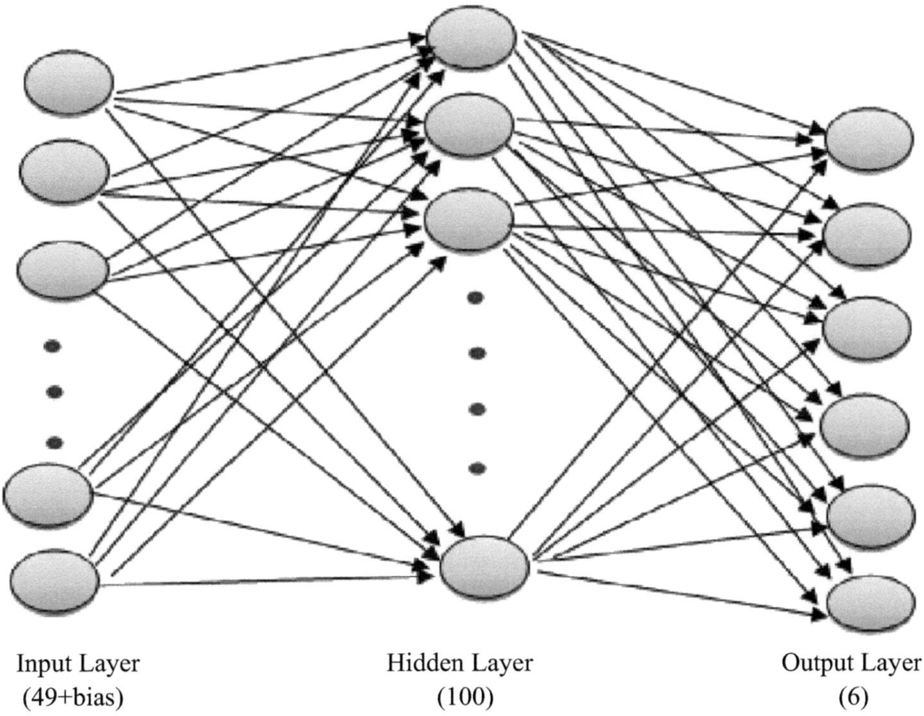

Fig. 3. Architecture of neural network

3.3 Classification of Exotic Particles vs. Background Using Deep-Learning

The third work to be undertaken is differentiating collisions which induce particles of interest a.k.a. *Signals* and those which produce other already known particles is called *Background.* Discovery of these particles is of paramount interest to the research community in physics. It is a challenging one too. In [12] the authors have carried out the initial bench mark classification effort to differentiate between a signal process and a background process. The new conceptual Higgs Bosons are generated in the signal process. Background processes though they have the same decay products, but differ in various mechanic features. This is depicted in Fig. 4(a) and (b) respectively. Figure 4 (a) refers to signal process involving new exotic Higgs bosons H^0 and $H^{\pm}$.

Figure 4(b) shows the background process involving top quarks, t, in both (a) and (b), the resulting Particles are two W bosons and two bottom bquarks. The background

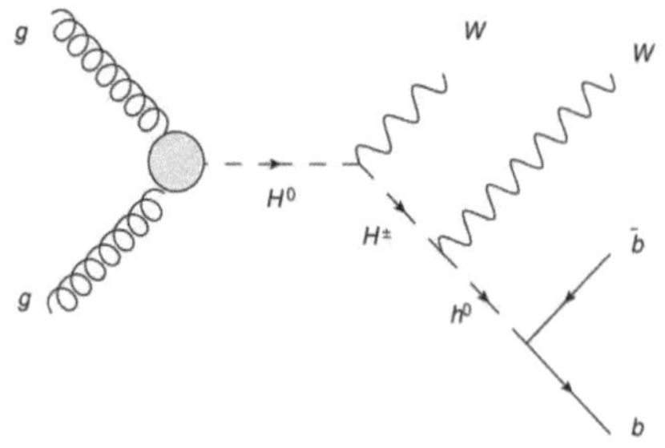

(a)

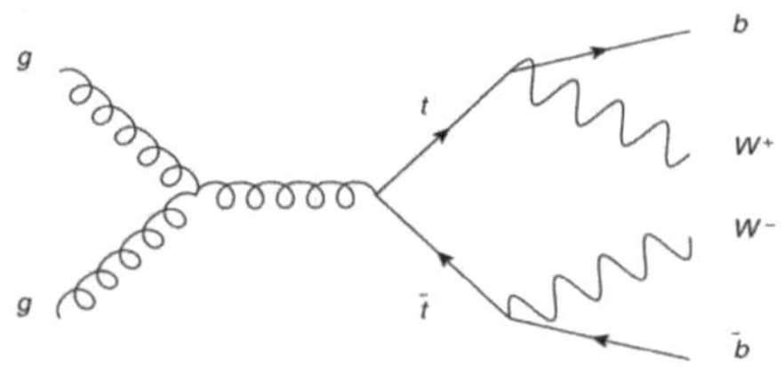

(b)

Fig. 4. (a) Signal process for Higgs Boson. (b) Background process for Higgs benchmark

process which replicates $W^{\mp}W^{\pm}b\bar{b}$ without the Higgs boson common state. It creates a couple of top quarks, each of which decomposed into Wb and also produce $W^{\mp}W^{\pm}b\bar{b}$.

$$gg \rightarrow H^0 \rightarrow W^{\mp}H^{\pm} \rightarrow W^{\mp}W^{\pm}h^0 \rightarrow W^{\mp}W^{\pm}b\bar{b} \tag{2}$$

Another benchmark effort carried out in [12] is to differentiate between process where the new super-symmetric (SUSY) particles are composed. SUSY are conceptual and hypothesized particles predicted by the super symmetry principle, and therefore, these particles are also called exotic particles. Some particles are measurable and others may be hidden to the experimental appliance. A background process has ideal observable particles and some particles may be unseen. Figure 5(a) and (b) show the signal and background process respectively for the SUSY benchmark.

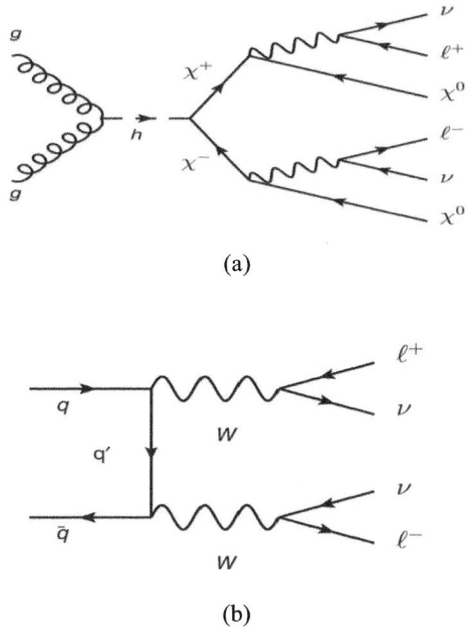

Fig. 5. (a) Signal process for SUSY. (b) Background process for SUSY

4 Results and Discussion

4.1 Results for Estimation of Mass of Muon Particle

Figure 6 depicts mass with model. From this experiment, it is found that the machine learning methods (Gaussian Process) approximate the mass of Muon particle with good approximation. This can be observed in Fig. 7 with a loss of 6156. Parameter tuning can further reduce the losses.

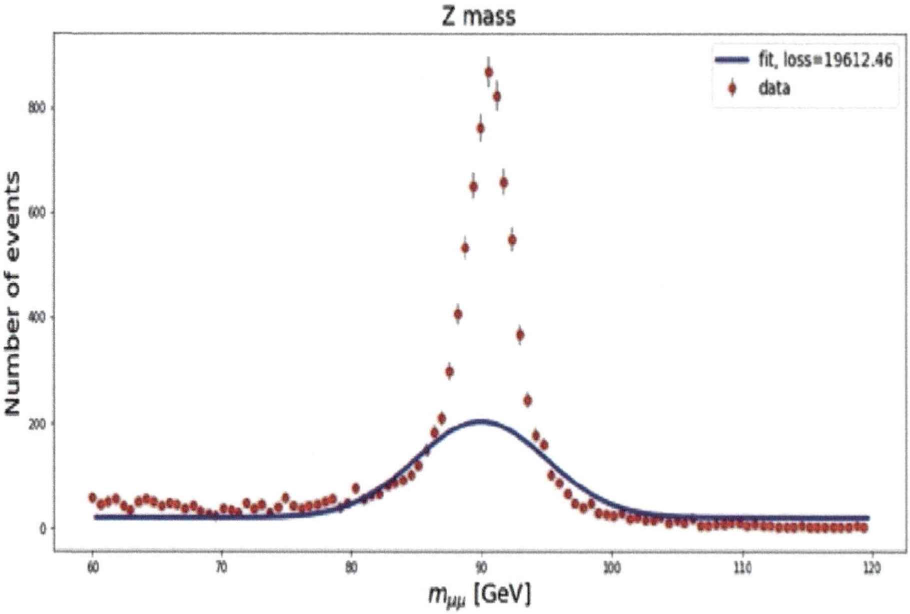

Fig. 6. Mass with model

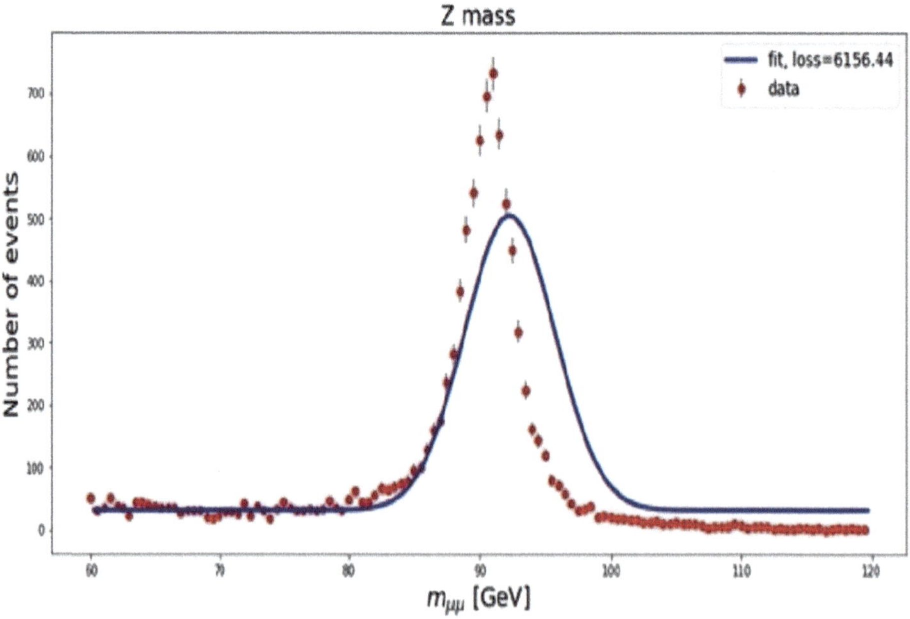

Fig. 7. Mass after Scikit optimization

4.2 Results of Classification of Sub-atomic Particles

In the second task, a multi-class classification was done with data. The task was to classify the known subatomic particles using two popular machine learning methods, namely boosted decision tree and deep neural networks. It can be observed from the receiver operator characteristics (ROC) shown in Fig. 8 that the deep neural networks perform very well in the classification task.

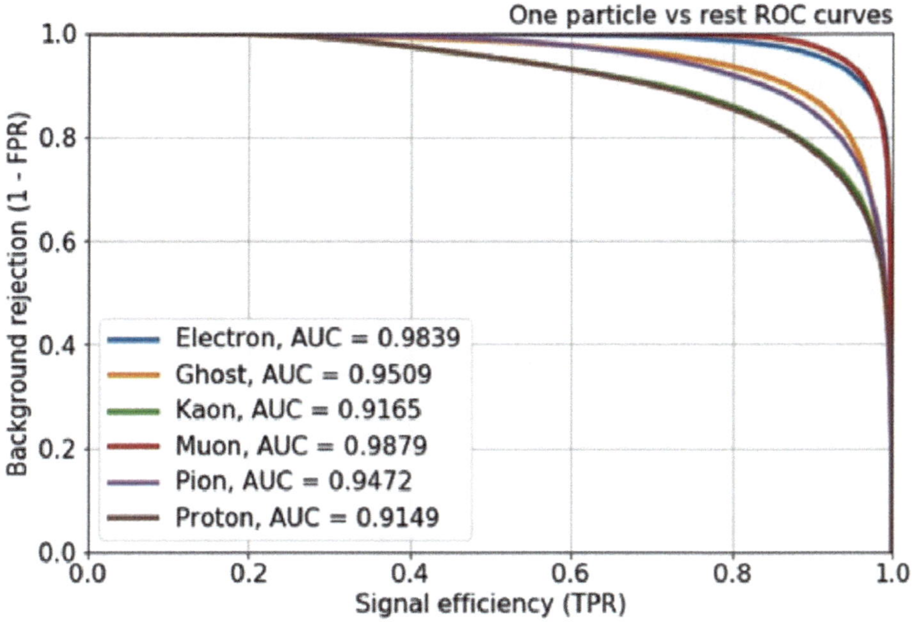

Fig. 8. ROC curve for particle classification

5 Conclusion and Future Work

In this research work, machine learning techniques have been applied to two different problems in high energy physics namely, estimation of mass of muon particle and classification of the known elementary particles. The datasets have been obtained from CERN – LHC. The results are promising and it is envisaged that machine learning can be successfully applied to high energy physics problems which can enhance the capabilities of the Large Hadron Collider at CERN. This work is being extended to two other benchmarks viz. the Higgs Boson and SUSY which would improve the performance of the collider for discovering new exotic particles in high energy physics.

References

1. ATLAS Collaboration: Observation of a new particle in the search for the standard model higgs boson with the ATLAS detector at the LHC. Phys. Lett. **B716**, 1–29 (2012)
2. CMS Collaboration: Observation of a new boson at a mass of 125 GeV with the CMS experiment at the LHC. Phys. Lett. **B716**, 30–61 (2012)
3. Neyman, J., Pearson, E.: Philos. Trans. Roy. Soc. **231**, 694–706 (1933)
4. Hornik, K., Stinchcombe, M., White, H.: Multilayer feedforward networks are universal approximators. Neural Netw. **2**, 359–366 (1989)
5. Bengio, Y., Frasconi, P., Simard, P.: Learning long-term dependencies with gradient descent is difficult. IEEE Trans. Neural Netw. **5**, 157–166 (1994)
6. Hochreiter, S.: Recurrent neural net learning and vanishing gradient (1998)
7. Hinton, G.E., Osindero, S., Teh, Y.-W.: A fast learning algorithm for deep belief nets. Neural Comput. **18**, 1527–1554 (2006)
8. Hocker, A., et al.: TMVA-toolkit for multivariate data analysis. PoS ACAT. **040** (2007)
9. Alwall, J., et al.: MadGraph 5: going beyond. JHEP. **1106**, 128 (2011)
10. Goodfellow, I.J., et al.: Pylearn2: a machine learning research library. arXiv preprint arXiv, pp. 1308–4214 (2013)
11. Baldi, P., Sadowski, P.: The dropout learning algorithm. Artif. Intell. **210**, 78–122 (2014)
12. Baldi, P., Sadowski, P., Whiteson, D.: Searching for exotic particles in high-energy physics with deep learning (2014)

A Survey on Image Segmentation Techniques

D. Divya[1(✉)] and T. R. Ganesh Babu[2]

[1] Anna University, Chennai, India
divya21cs@gmail.com
[2] Department of Electrical and Communication Engineering,
Muthayammal Engineering College, Rasipuram, India
ganeshbabutr@gmail.com

Abstract. The essential step in digital image processing is segmentation which can be used to partition the images into particular regions or objects and the level of partitioning depends on the individuality of the problem being solved. Segmentation of the image is widely categorized into two. The first one is Discontinuity which measures the sudden changes of intensities for partitioning the image and the other category Similarity is measured, based on the predefined methods such as thresholding, region growing, splitting and merging. The images are considered as inputs for performing segmentation and the result is attributes extracted from the images. Segmenting an image is an initial step to understand and analyze what is inside the image and this will be done mandatorily for all medical imaging analysis. Several segmentation techniques have been proposed in the past, but none of the segmentation methods are invented without any drawbacks. Hence, this study discusses a review on the various segmentation techniques of image which will help in further advancement in this field.

Keywords: Image processing techniques · Image segmentation · Thersholding

1 Introduction

The method that extracts useful information by performing some tasks in an image is known as image processing. The word 'Image' comes from the Latin word 'imitari', meaning is "imitate". The images can be assessed based on the realistic capturing of images they show. It is a growing and innovative technologies for engineering and computer science disciplines.

In Analog image processing, the images are handled by varying the electric signals based on the two dimensional analog signals. But the digital signal processing has dominated over analog processing due to its wider range of applications. In general, the fundamental steps applied in image processing are import the image with the help of acquisition tools, analyze and manipulate the image and the last one is the result which will be an altered image/report obtained through image analysis.

The common depiction for the image demarcated in the real world consists of two dimensional coordinates represents the intensity and reflectivity.

$$\mathrm{func}(x, y) = \mathrm{intensity}(x, y) * \mathrm{reflectivity}(x, y)$$

D. J. Hemanth et al. (Eds.): COMET 2019, LNDECT 35, pp. 1107–1114, 2020.
https://doi.org/10.1007/978-3-030-32150-5_112

The course of partitioning an image into multiple regions or set of pixels which is more speaking and at ease to analyze is referred as image segmentation. There may be analogous pixels in a region based on features such as characteristics of an image, color, texture and intensity (Fig. 1).

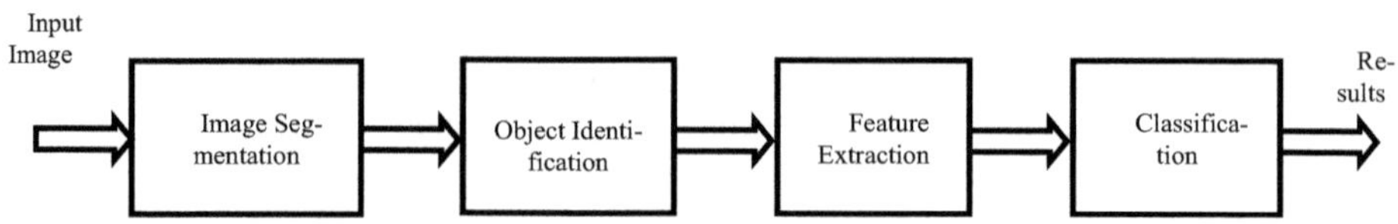

Fig. 1. Outline of image segmentation

The application of image segmentation is mainly focused in the medical field which is used to identify the location of tumors and also to measure the tissue volumes etc. The application of various algorithms is Digital image processing which improves the quality of an image by removing noise by applying some filtering techniques and it also removes unwanted pixels from the image. The main motive of segmentation is to have clear distinction between object and its background from the input image and the operations are done in Matrix Laboratory software.

2 Literature Survey

Murali [15] developed a texture based algorithm for automatically detecting the features such as solid pigment and also finds whether its presence is essential for differentiating the benign from malignant lesions. The factor NGLDM in texture methods are used for detecting important and dermoscopic features in digitized images and it is satisfactory. The separations of benign from malignant lesions are done using the index. The drawback is utility of peripheral pigment index are limited by testing the smaller number of lesions.

Rajab [16] proposed two methods iso data algorithm and neural network edge detection with closed elastic curve for segmenting the skin lesions and it is compared with the existing automated skin segmentation methods. The iterative thresholding method provides better performance for the skin lesion with different border irregularity properties. The drawback is the images considered are not color images and the noise added is guassian noise which is not responsible for the description of artifacts or surface brighter spots.

Erkol [17] presented an approach for automatic snake initialization using luminance image blurring. The method proposed for spotting the border of skin lesions is Gradient vector flow (GVF) and the perceiving is done based on features such as shape, color and also the color of the surrounding skin are utilized in order to improve the performance of accurate skin lesion segmentation in dermoscopy images. As compared with Pagadala's approach, the gradient vector method provides better performance.

Celebi [18] An unsupervised approach called modified version of the JSEG algorithm has been proposed for revealing the border of skin lesion. This paper describes the candidate algorithm and it performs well as the other automated methods and also a classifier is used to obtain better accuracy. The lesion localization determined by bounding box does not completely contain the lesion and it is said to be drawback of this method. The proposed method may not perform well for images with artifacts.

Abbas [19] Author discusses automated methods such as preprocessing, edge candidate point detection and tumor outline delineation. The first step is the preprocessing phase which removes hairs and blood vessels with the help of some filtering techniques. This paper uses the least square method to obtain edge points and also the optimal boundary of the lesion can be determined using the technique called Dynamic programming. The performance of the proposed method is evaluated using the ground truth provided by the dermatologists drawn borders.

Toosi et al. [20] canny edge detection and morphological operators are the approaches used for removal of artifacts in dermoscopy images. The quantitative analyses are done for evaluating the accuracy of the hair detection algorithm. Similarly the evaluating features of the hair repaired algorithm are standard deviation, entropy and co-occurrence matrix.

Razazzadeh [22] the essential step in the identification of skin lesion is segmentation. Achieve high accuracy by converting the color images to YUV color space in segmentation. The Otsu thresholding and morphological reconstruction algorithms are used in segmentation. With the help of filtering properties noise reduction is done in the pre-processing phase.

Bi [24] proposed a fully convolution network (FCN's) for automatic segmentation of skin lesions. The drawback of existing segmentation methods is, the segmentations are performed legibly because of fuzzy borders in lesions, contain artifacts and low contrast with the background. The proposed method achieves accurate segmentation by combining the essential characteristics of skin lesions inferred from multiple embedded FCN stages and it achieves better accuracy compared to the other state-of-the-art methods.

Wang [23] discusses about deep learning-based framework for medical image segmentation. For binary segmentation, the bounding box based approach is used under convolutional neural network (CNN) and also it is used to segment the formerly unobserved objects. The concept fine tuning is essential in order to make the CNN model effortlessly adaptive to the specific test conducted for the image which is either supervised or unsupervised. This paper also proposes for the network considered weighted loss function has been proposed and also interaction based uncertainty for the adequate tuning.

3 Image Segmentation

Segmentation of image divides an image into multiple segments for differentiating objects from the background and also to identify or describes the parts of objects. They are said to be more meaningful and easier to analyze [9]. An image of every pixel has been allocated to one of the number of these categories. A good segmentation is the one which possess pixels in the specific category have similar gray scale and forms a connected region and the adjoining pixels with different categories have contradictory values. Basically, it can easily locate entities or margins in an image by converting the complex image into simple image. The image segmentation approach is associated with [10] two properties such as detecting discontinuities and dissimilarities with respect to their local neighborhood. Some of the applications of segmentation identifies the shape and size of objects from a scene, and identified objects from a moving scene and also used to locate tumors.

3.1 Detecting Discontinuities

The image partitioning is done based on the discontinuous variation in the pixel intensity values so that it helps to determine the boundaries of object and also to represent the formation of edges which is similar to edge detection algorithm. The variation in intensity levels with respect to neighboring pixels are represented as an edges which is said to be the outcome of discontinuity of the pixels [11] Image smoothing, edge localization and detection are the fundamental progressions carried out in edge detection.

4 Edge Based Segmentation

This method is proficient in detecting discontinuities with respect to gray level and intensity values rather than detecting secluded points and tinny lines. The accomplishment of edge detection in low level image processing is said to be a perplexing task and it becomes more stimulating for color images. The discontinuities are measured based on the gray level, color distinctness and texture variation which describes a set of pixels in a boundary among different regions are represented as edges [5]. Hence, segmentation can be done by detecting the types of discontinuities. Comparatively, color images provide precise information about objects than the grayscale images [12] and the types of edges are step edge, ramp edge, line edge and roof edge [13].

The given image is put on with masks for determining the edges. [12] It is broadly classified into two categories such as the gradient based edge detector (First order derivative) and Laplacian Edge Detector(*s*econd order derivative) are said to be the two edge detection operators and the first derivative operators are described as Prewitt, Sobel, Canny and Test operator. The operators in second order derivative are Laplacian operators and zero crossings. Compute first derivative with the help of maximum and minimum value of the gradient.

$$\mathrm{Grad(f)} = \begin{matrix} \partial f/\partial x \\ \partial f/\partial y \end{matrix}$$

The gradient is said to be a vector which has magnitude and direction where the magnitude represents edge strength and the term direction indicating edge direction.

Robert's Operators

It separately determines the edges in row and columns [14] which are placed together to bring out the resultant edge for the given images. The masks consist of a pair of 2 × 2 convolution kernels.

-1	0
0	-1

0	-1
-1	0

Sobel Operators

It is said to be the modification of Prewitt's operator by changing the center coefficient as the value 2.

1	0	-1
2	0	-2
1	0	-1

-1	-2	-1
0	0	0
1	2	1

Laplacian of Gaussian Operator

In Laplacian operator, the gray level image is considered as input which yieldsa new gray level image as an outcome and highlights the regions which possess rapid changes in intensity.

Canny Edge Detector

In the year 1986, the canny edge detection algorithm was presented and the problem is false edges are produced by low threshold, whereas actual edges are missed by high threshold. The steps in canny edge detector are, Reduce noise by image smoothing.

Predict Edge strength and directions by applying Sobel operator.

Detecting pixels that are not related with edges are minimized.

Broken edges can be removed by calculating threshold and pixel value of an image. On comparison, if the obtained threshold is less than the pixel value then there exists an edge [8].

4.1 Detecting Similarities

The segmentation is clearly used to differentiate matters from the background. It partitions an image based on the previously defined similarity criteria into a set of homogenous regions by using the segmentation techniques.

4.2 Region Based Segmentation

This method is used for the direct determination of regions characterized as "similarity based segmentation" [1, 2]. It partitions an image into homogenous areas based on some properties such as color, texture, range and intensity. In this technique, the pixels belong to same intensity characteristics and closed to each other are grouped together and assumed that all pixels are related to a particular object. In the region based segmentation, the important thing to be noted is no space allowed for the missing edges. Edges remain difficult to predict in noisy images but this issue can be overcome using region growing technique [3, 4]. It is said to be the simple and robust method for partitioning an image into uniform- regions correctly and accurately, whereas, it uses high computation power.

4.3 Region Growing

Its objective is to cluster pixels or sub-regions into superior regions based on the existing criteria which generally examine the group of pixels using the features such as color, texture and intensity [7]. It encompasses only the selection of initial seed points so it is regarded as pixel-based segmentation [5]. The segmentation helps to scrutinize the initial seed points of nearby pixels and determines whether the nearby pixel can be appended to the region or not. The steps to be processed in region growing are,

From an original image chose a set of seed points [6].

Choice of similarity criterion based on gray scale, pixel intensity, color andtexture.

For regions to grow append each seed to neighboring pixels which have related belongings to the seeds.

Stop the process when no other pixels met the standard.

Region Splitting and Merging

This method follows a top down methodology and it is said to be the contradictory approach of region growing and also assumes that the complete image is homogenous. Suppose, the original image does not satisfy the condition of homogeneity then the image is fragmented into four sub images. The splitting process will be carried out recursively till we split the image into homogenous regions [8]. The user distributes an image into a set of random regions for fulfilling the state of affairs of reasonable segmentation. Then the process of merging or splitting is carried out rather than selecting a set of seed points in a document image. The size of original image is $N \times N$ and the regions produced is $M \times M$, which satisfies the condition $M \leq n$. Hence, the procedure is recursive, which produces the representation as tree with four sons each is called as Quad tree.

Both the splitting and merging are performed alternatively in each iteration. The bottom up approach merges into similar adjacent regions.

Split regions into four sub-regions, R is non-homogeneous.

Regions are merged if they are adjacent.

Terminates when no more splitting or merging is possible.

5 Conclusion

The objective of the survey is to discuss different image segmentation techniques with its properties and methodologies highlighted. The issue is the kind of segmentation to be applied which is not clear, as it varies depending on the applications. The extension of gray level image segmentation techniques are based on color images, and can be processed using approaches like thresholding, fuzzy based approaches, edge detection and region growing, since colors produce more reliable segmentation than gray level image. From this survey, it is concluded that there is no specific segmentation algorithm proposed which works well for all kinds of images. The overall motto of segmentation is to improve the accuracy for complex background images. The future study can be done on various clustering techniques used in dermoscopy images and also determine proficient Edge detector algorithm for quantifying the size of lesion.

References

1. Yogamangalam, R., Karthikeyan, B.: Segmentation techniques comparison in image processing. Int. J. Eng. Technol. (IJET) **5**, 307–313 (2013). ISSN 0975-4024
2. Manikannan, A., SenthilMurugan, J.: A comparative study about region based and model based using segmentation techniques. Int. J. Innov. Res. Comput. Commun. Eng. **3**(3), 948–1950 (2015). ISSN (ONLINE): 2320-9801
3. Elayaraja, P., Suganthi, M.: Survey on medical image segmentation algorithms. Int. J. Adv. Res. Comput. Commun. Eng. (IJARCCE) **3**(11) (2014). ISSN (ONLINE): 2278-1021
4. Banchpalliwar, R.A., Salankar, S.S.: A review on brain MRI image segmentation clustering algorithm. IOSR J. Electron. Commun. Eng. (IOSR-JECE) **11**(1), 80–84 (2016). https://doi.org/10.9790/2834-11128084. ISSN (ONLINE): 2278-2834
5. Kang, W.-X., Yang, Q.-Q., Liang, R.-P.: ETCS, pp. 703–709 (2009)
6. Kaganami, H.G., Beij, Z.: Region based detection versus edge detection. IEEE Trans. Intell. Inform. Hiding Multimed. Signal Process. 1217–1221 (2009)
7. Singh, K.K., Singh, A.: A study of image segmentation algorithms for different types of images. Int. J. Comput. Sci. Issues **7**(5), 414 (2010)
8. Canny, J.F.: A computation approach to edge detectors. IEEE Trans. Pattern Anal. Mach. Intell. **8**, 34–43 (1986)
9. Niessen, W.J., et al.: Multiscale segmentation of volumetric mr brain images. In: Signal Processing for Magnetic Resonance Imaging and Spectroscopy. Marcel Dekker, Inc. (2002)
10. Khan, A.M., Ravi, S.: Image segmentation methods: a comparative study. Int. J. Soft Comput. Eng. (IJSCE) **3**(4), 84–92 (2013)
11. Ravi, S., Khan, A.M.: Operators used in edge detection: a case study. Int. J. Appl. Eng. Res. **7**(11) (2012). ISSN 0973-4562
12. Saini, S., Arora, K.: A study analysis on the different image segmentation techniques. Int. J. Inform. Comput. Technol. **4**(14), 1445–1452 (2014)
13. Senthilkumaran, N., Rajesh, R.: Edge detection techniques for image segmentation – a survey of soft computing approaches. Int. J. Recent Trends Eng. **1**(2), 250 (2009)
14. Lakshmi, S., Sankaranarayanan, V.: CASCT, pp. 35–41 (2010)
15. Murali, A., Stoecker, W.V., Moss, R.H.: Detection of solid pigment in dermatoscopy images using texture analysis. Skin Res. Technol. **6**, 193–198 (2000). ISSN 0909-752X

16. Rajab, M.I., Woolfson, M.S., Morgan, S.P.: Application of region-based segmentation and neural network edge detection to skin lesions. Comput. Med. Imaging Graph. **28**, 61–68 (2004)
17. Erkol, B., Moss, R.H., Joe Stanley, R., Stoecker, W.V., Hvatum, E.: Automatic lesion boundary detection in dermoscopy images using gradient vector flow snakes. Skin Res. Technol. **11**, 17–26 (2005)
18. Emre Celebi, M., Alp Aslandogan, Y., Stoecker, W.V., et al.: Unsupervised border detection in dermoscopy images. Skin Res. Technol. **13**, 454–462 (2007)
19. Abbas, Q., Celebi, M.E., et al.: Lesion border detection in dermoscopy images using dynamic programming. Skin Res. Technol. **17**, 91–100 (2011)
20. Toossi, M.T.B., Pourreza, H.R., et al.: An effective hair removal algorithm for dermoscopy images. Skin Res. Technol. 1–6 (2013)
21. Al-abayechi, A.A.A., Logeswaran, R., et al.: Lesion border detection in dermoscopy images using bilateral filter. In: International Conference on Signal and Image Processing Applications (ICSIPA) (2013)
22. Razazzadeh, N., Khalili, M.: An effective segmentation method for dermoscopy images. In: International Conference on Computer and Knowledge Engineering (ICCKE) (2014)
23. Wang, C., Yu, J., Mauch, L., Yang, B.: Binary segmentation based class extension in semantic image segmentation using convolutional neural networks. In: ICIP (2018)
24. Bi, L., Kim, J., Ahn, E., et al.: Dermoscopic image segmentation via multi-stage fully convolutional networks. IEEE (2016). https://doi.org/10.1109/TBME.2017.2712771

Bionic Eyes – An Artificial Vision

S. Nivetha(✉), A. Thejashree, R. Abinaya, S. Harini, and Golla Mounika Chowdary

Department of Computer Science and Engineering, Panimalar Engineering College, Chennai, India
nivethasingaravelu19@gmail.com,
thejashreereddy121@gmail.com,
abi99rajan@gmail.com, abiharini456@gmail.com,
mounikachowdary.goalla@gmail.com

Abstract. In this whole world for those millions of individuals whose vision is obscured, do not have any therapeutics. So the late progression in the innovation has driven the humankind towards different approach like artificial implants for those blind subjects and the bionic eye with retinal, visual, sub retinal implant method, is by all accounts promising, as it is a mix of gadgets, biomedical technology which acts as the artificial eye in translating the materialistic pictures of the world. This paper gives an outline of different retinal implant methods in channelizing the subjects vision through artificial insight and on the off chance, if popularized, turns into the potential gadget for the blind subjects to see and translate the world.

Keywords: Retinal implant · Artificial eye · Sub-retinal implant · Epi-retinal implant · Technology

1 Introduction

The science has provided a few marvels to the humankind. Bio-therapeutic specialists play an essential role in forming the course of the vision. Presently it is the duty of bionic eyes to provide artificial vision. The Chips are structured explicitly to mirror the attributes of the harmed retina, cones, and rods of the organ of sight that are implanted with a microsurgery. Regardless of whether it is Bio-tech, Computer, Electrical, or Mechanical Engineers – every one of them have an indispensable task to carry out in the embodiment of Bionic Eyes. This creative innovation can add life to their vision-less eyes. In upcoming years, this will make an upheaval in the field of therapeutic science. It is critical to know certain actualities about the organ of sight before we continue towards the specialized perspectives engaged with Bionic Eye Systems.

2 Need for Bionic Eye

Because of the absence of successful remedial medicinal measures for Retinitis pigmentosa - RP and Age-related macular degeneration - AMD, it has prompted the improvement of exploratory procedures to re-establish some level of visual capacity to

D. J. Hemanth et al. (Eds.): COMET 2019, LNDECT 35, pp. 1115–1122, 2020.
https://doi.org/10.1007/978-3-030-32150-5_113

influenced patients. Since the retinal layers are anatomically present as such, a few methodologies have been intended to artificially enact retina through bionic eye framework. It is known that electric impulses from retinal neurons can deliver light recognition in patients who are experiencing retinal degeneration. Utilizing this property can channelize the useful cells to hold the vision with the assistance of electronic gadgets that help these cells in performing the vision. We can make lakhs of individuals get back their vision. A plan of an optoelectronic retinal prosthesis framework which can imitate the retina with goals that relates to a visual action of 20/80—which is sharp enough to perceive faces, read huge textual styles, sit in front of the TV and maybe above all lead an autonomous life. This gadget is an exploratory visual gadget planned to re-establish vision. Bionic eye re-establishes the vision lost because of harm of retinal cells.

3 Device as an Artificial Eye

The role of a Bionic Eye gadget, is to assist subjects in perceiving objects. It is made of sensors, processors, radio transmitters, and the retinal chip clubbed together. The gadget is implanted instead of retina. The gadget conveys the real world objects to a silicon chip which detangles radio signals. As the chip receives the signal, they are sent to the retinal ganglion cells, trailed by optical nerve and to the brain, recognizing light and dark spots. The chip receives signals from a couple of glasses worn by the patient, which are fitted with a camera. The visual data from the camera will be fed to the video processor. This segment separates the picture into pixels and sends the data, one pixel at once, to the silicon chip, which at that point remakes the picture. The radio waves are in charge of broadcasting the information into the body. At present the hardware is just ready to communicate a 10 × 10 pixel. The patient could make out between light and dim as chip is implanted with 60 pixels/electrodes, however a definitive point is to make it to 3600 pixels by which the patient will be ready to perceive faces and furthermore helps in perusing.

The Visual prosthetics can be separated into three parts. Initially The utilization of gadgets like CCD camera or the ultrasonic that catch the pictures and render the results to the framework as electrical information. The second, is the chip that mirrors the retinal capacities by invigorating the retina with electrical signals which triggers the optic nerve to send message to the brain. The third major class is the processor that translates the picture into pixels.

3.1 Sub Retinal Implants

Presently, Second Sight named organization which motivated the endorsement from FDA to begin the U.S. trails in regards to get the constrained present of vision. The second age Argus II is structured with a 60 electrodes that is implanted around the eye. The Argus II is a chip that is carefully implanted onto the retina. This chip with electrodes is fit for sending signals to the cerebrum which is inconceivable for the harmed human natural retina. The chip isn't helpful except, if it is getting visual information to send to the cerebrum. To tackle this issue, the device is fitted with a couple of glasses implanted with a modest camcorder which persistently records film of what is present before the patient. The Argus II Retinal Prosthesis System can distinguish the light to individuals who have gone visually impaired from degenerative eye infections like macular degeneration and retinitis pigmentosa. They deactivate the eyes cones, rods and retina that see light and pass them on to the cerebrum as nerve impulses, wherein these signals are decoded as pictures. The Argus II framework is an alternative for these photoreceptors. The manifestation of retinal prosthesis comprises of five principle parts:

- Digital-Camera implanted into a pair of glasses. It catches continuous pictures furthermore, for pictures to a smaller scale chip.
- Video-Processing Microchip joined into a handheld unit. It converts pictures into electrical signals that differentiates light and dark.
- Radio-Transmitter will remotely transmit the driving forces into a beneficiary that is implanted under the eye.
- Radio-Receiver will send signals to the retinal implant by means of hair-like wire.
- Retinal-Implant with 60 electrodes on a chip of size 1 mm by 1 mm.

The entire framework is driven by a battery pack and when the picture is caught it is represented as light and dark pixel designs. These pictures are video handled to convert the 3d organized samples and are unscrambled to "light" and "dim" designs. The processor deciphers these signals to a radio transmitter on the glasses, which at that point transmits these signals as radio waves to a recipient implanted underneath the patient's skin. The recipient is specifically associated by a hair like wire to the electrodes and it sends the signal through the nerves. These impulses are translated by the human brain and the message is shown as 'you are seeing a tree' and therefore the subject will distinguish the object.

Working

The working of Retinal implant framework is delineated in Fig. 1. Typically the vision begins when the light beam's falls on the cones and rods and deciphered by the retina through optic nerves. These cells convert optical signals into electric signals that are sent through optic nerve to the cerebrum. Retinal infections like ARM degeneration and RP destroy these cells. With the bionic eye, a scaled down camera mounted on the eye-gear, catches the pictures and remotely sends the data to a miniaturized scale controller unit that changes over the information to an electronic signal and re-transmits it to the transmitter on the goggles which in turn sends the signals to the microelectrode exhibit, which triggers the signal discharge. These signals travel along optic nerves to cerebrum. At that point, the brain recovers samples of light and dim spots that relate to the terminals incitement. Patients figure out how to translate these visual samples. It takes some preparation for the subjects to really observe a tree. At in the first place, they see generally light and dull spots. Be that as it may, after a while, they learn to decipher what they perceive. In the end they see those samples of light and dull as a tree. Scientists are as of now arranging a third form which comprises of thousands of electrodes on the retinal chip to make subjects in facial acknowledgment abilities.

- The Camera inserted on glasses to see picture.
- The Signals are sent to the hand- held gadget.
- Processed data is sent back to installed glasses and remotely transmitted back to the collector Under the surface of eye.
- Receiver sends data to anodes in retinal implant.
- Electrodes animate retina to send data to cerebrum.

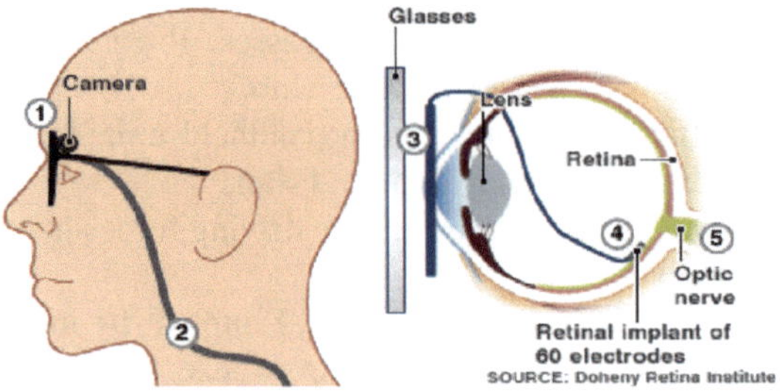

Fig. 1. Typical working of retinal implant system

3.2 Epi-Retinal Implants

The "Epi-Retinal" approach establishes the semiconductor based gadget framework that is on the retina towards the retinal optic nerve fibre and the retinal ganglion cells. In this approach the picture is caught by utilizing CCD camera before the signals are exchanged to the implant framework. In the EPI-RET approach, a smaller scale cluster diode near the retinal cells is produced to trigger the ganglion cells. The camera inserted on to the eye glasses catches the pictures and sends it to the framework remotely. The genuine visual world is taken by an exceptionally scaled down CMOS camera that is installed onto eye-gears. The camera impulses are translated to sufficiently stimulate ganglion cells in the retina. This stimulating sign consolidated with the vitality supply is transmitted to a gadget remotely which is implanted into the eye of the visually impaired subject. The implant contains a beneficiary for information and vitality, a decoder and microelectrode cluster put on the inward retinal surface. This smaller scale chip will trigger feasible retinal-ganglion cells. Electrodes on microchip will at that point make light pixels on the retina that can be sent to the brain for translation. The fundamental of this device is that it comprises of just a straightforward eye gear outline with camera and outside hardware which imparts remotely with microchip implanted on retina modified with incitement design (Fig. 2).

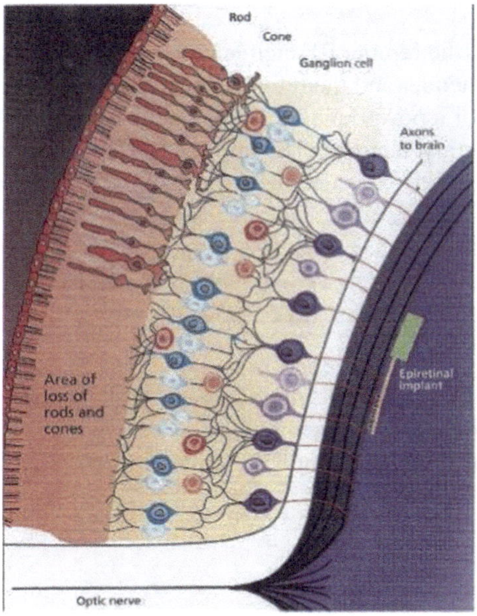

Fig. 2. Epi-retinal system implant

The hurdles included in designing the retinal encoder is:

- Chip-Development
- Bio-Compatibility
- RF-Telemetry and Power-Systems.

3.2.1 Chip Development

Encoder Epi Retinal

The structure of an epi-retinal encoder is more brain boggling than the sub retinal encoder, in light of the fact that it needs to bolster the ganglion cells. Here, a retinal encoder outside the eye replaces the data handling unit of the retina. These spatial channels as science motivated neural systems can be tuned to different spatial and fleeting open field properties of ganglion cells in the primate retina.

Biocompatibility

The materials that are picked for the retinal implant creation must be explicitly bio-perfect and furthermore safe and non-destructive to coordinate different criteria's

- The electrodes need to set up an indeed, with the end goal that the electronic current can go through the rods and cones.
- Technology must almost certainly be made by smaller scale hardware.
- The picked materials ought to be organically good with the sensory system.

RF Telemetry
In epi-retinal encoder, the remote RF elementary framework goes about as the channel between the Retinal Encoder and the retinal trigger. Standard semiconductor innovation is utilized in creation of a power and the chips that drives current through the electrode cluster, activates the retinal neurons. The intra-ocular Trans beneficiary handling unit is isolated from the trigger to consider the heat dispersal of the correction and control exchange forms. Care is taken to maintain a strategic distance from direct contact of heat dispersing gadgets with the retina. Presently, a German firm named Retina Implant has scored a major win for the sub retinal arrangement with a three-millimeter, 1,500 pixel microchip that gives patients a 12° field of see.

All in all,

- Epi-retinal Approach includes a semiconductor based gadget situated on the outside of the retina to mimic the overlying cells.
- Sub retinal Approach includes implanting the ASR chip behind the retina to mimic the suitable cells.

3.3 Improvements

The video controller used for processing the received image from the camera on the goggles can be attached with a radio transmitter that has the capacity of transmitting radio signals of the converted pixel images to a mobile via an application. Now, this app can be used to monitor the patients image processing capability, eye health and enables the trainers to train the patients in perceiving the world. Also, the video perceived by the camera on the goggles can be recorded by the patient and stored in an external SD card in the video controller kit which can be used for personal training and in insecure situations.

4 Advantages

The new innovation will ideally help individuals experiencing AMD and RP.

- The thing is to altogether enhance the personal satisfaction for visually impaired patients.
- Minor surgery required.
- No Batteries implanted inside body.
- Very Early in the visual pathway.

5 Disadvantages

- This new innovation won't be suitable for glaucoma patients.
- Not helpful for patients who are blind by birth.
- Extra hardware required for downstream electrical information.

6 Challenges

- There are heaps of obstacles to be defeated by Bionic Eyes. Human eyes are the most delicate of all organs in the body. A Nano-sized impersonate can make devastation in the eye.
- There are around 120 million rods & 6 million cones in the retina of each sound human eye. Making an artificial trade for these cells isn't a simple task.
- Silicon based photograph indicators have been tested in before endeavors. But Silicon is dangerous to the human body & responds ominously with visual eye liquids.
- There are colossal questions regarding how the brain will react to remote signs produced by counterfeit light sensors.
- One of the hardest difficulties is guaranteeing the implant to remain in the eye for quite a long time without causing scarring, reactions, and general corruption on long run.
- These counterfeit retinas are excessively costly, as well inconvenient, and too delicate to even think about withstanding many years of typical mileage.

7 Conclusion

This is a creative and progressive innovation and truly can possibly change people's lives. Bionic Eye is an upheaval in restorative field. It is a great news for visually impaired patients who experience the ill effects of retinal sicknesses. Retinal inserts can somewhat re-establish the vision of individuals with specific types of visual deficiency brought about by macular degeneration or retinitis pigmentosa. Regardless of the pros and cons of this device, if this is completely created with a cutting edge innovation, it will change the lives of a large number of individuals round the globe. We probably won't renew the vision totally, yet we can endeavor to help them to discover their way, perceive faces, read books, and on the whole, lead an free existence of their decision.

References

1. Bionic Eye Technology: An advanced version of artificial vision. Academia **2**(7) (2012)
2. Zhu, G., Yang, R., Wang, S., Wang, Z.L.: Flexible high-output nanogenerator based on lateral ZnO nanowire array. School of Materials Science and Engineering, Georgia Institute of Technology, Atlanta, Georgia. Copyright © American chemical Society
3. Grammatis, K., Spence, R.: Building the bionic eye; Hacking the human. Future of Journalism Conference. www.eyeborgproject.com
4. Asher, A., Segal, W.A., Baccus, S.A., Yaroslavsky, L.P., Palanker, D.V.: Imageprocessing for a high-resolution optoelectronic retinal prosthesis. IEEE Trans. Biomed. Eng. **54**(6), 993–1004 (2007)
5. Humayun, M.S, Weiland, J.D., Chader, G.: Basic Research, Biomedical Engineering and Clinical Advances, pp. 151–206 (2007)

6. Chen, X., Xu, S., Yao, N., Shi, Y.: 1.6 V nanogenerator for mechanical energy harvesting using PZT nanofibers. Nano Lett. **10**, 2133–2137 (2010)
7. Narayana, P., Senthil, G.: Bionic Eye Powered By Nanogenerator, Singapore
8. Doe Technologies Drive Initial Success of Bionic Eye. Artificial Retina News, U.S. Department of Energy office of Science (2009)
9. Australian Research Council: Bionic Eye, Retina Australia (Qld)
10. Satish, S.: Artificial vision – a bionic eye. Scribd (2010)
11. Hall, A.: Diamond shines as basis for bionic eye prototype. ABC News, 09 December 2010
12. Al, K.: A Bionic Eye comes to market
13. Kerouac, J.: Bionic Eye: What does the future hold

Analyzing the Effect of Regularization and Augmentation in Deep Neural Network Model with Handwritten Digit Classifier Dataset

P. Madhan Raj(✉), B. Arun Kumar, G. Bharath, and S. Murugavalli

Computer Science and Engineering, Panimalar Engineering College, Chennai 600123, India
madhanraj3571@gmail.com, arun.b.kumar.arun@gmail.com

Abstract. A lot of research has been carried out in the field of Handwritten Digit Recognition in recent years. It has its application in areas like bank check processing, signature verification, etc. where very high level of accuracy is a required and even a small mistake would lead to a great loss of money and time. I propose a model in my system with a accuracy of 99.6% using Deep Learning neural networks assisted by Data Augmentation and Regularization.

Keywords: Handwritten digit recognition · Neural network · Deep learning

1 Introduction

The primary aim of handwritten digit recognition is to identify input characters or image that are handwritten correctly. The standard dataset of Handwritten digit classification from MNIST serves as a classical dataset to experiment and test. We introduce a technique called regularization and data augmentation to reduce the error rate. Over-fitting has been a main issue in the neural networks. The central idea of dropout is randomly dropping units in the neural networks along with their connection. The images are represented as multi-dimensional arrays in the MNIST data format which are given as input to machine learning algorithms. The images in MNIST data format are taken from many scanned documents and contain variety of styles of each digit. Each image is 784 pixels in size which is represented as 28*28 array. The images are normalized, grayscale and centered. The images of MNIST dataset are divided into two types namely training set and testing set. There are total of 60000 training images and 10000 testing images (Table 1).

2 Literature Review

Literature review shows the evolution of existing system. It guides us in the search of a better system than previous systems. Literature review is mostly used to trace the history of the existing system. Literature review on Handwritten digit recognition is given below.

D. J. Hemanth et al. (Eds.): COMET 2019, LNDECT 35, pp. 1123–1130, 2020.
https://doi.org/10.1007/978-3-030-32150-5_114

Table 1. Quantity of each digit in MNIST dataset

Digit	Quantity
0	5923
1	6742
2	5958
3	6131
4	5842
5	5421
6	5918
7	6265
8	5851
9	5949

Ashiquzzaman and Tushar [1] in their work "Handwritten Arabic Numeral Recognition using Deep learning Neural Networks" present a new model based on deep learning neural networks and regularization which shows a improved accuracy compared to the existing models. They use dropout, a regularization method to overcome the problem of over fitting The output layer contains 10 neurons and softmax classifier is used. The classifier predicts probability for 10 classes, each for a digit from 0–9. Their model shows a very high accuracy of 97.4%.

Shopon and Mohammed et al. [2] in their work "Image Augmentation By Blocky Artifact in Deep Convolutional Neural Network For Handwritten Digit Recognition" propose a model using blocky artifact to increase the accuracy of classification of English and Bangla digits using Convolutional Neural Network. The MNIST dataset, CMATERDB 3.1.1, Indian Statistical Institute(ISI) dataset are used to conduct the experiments. The training accuracy for MNIST, CMATERDB 3.1.1, ISI dataset are 99.56%, 99.83%, 99.35%. The model can be further enhanced by finding the sizes of images for which the model gives best accuracy and the reason for it.

He et al. [3] in their work "Deep Residual Learning for Image Recognition" have presented a Residual Learning Model to make the training of neural networks more easy. They have changed the layers such that they learn residual functions instead of unreferenced functions The more the depth of network, the more the accuracy of the model. Most of the visual representation models are based on depth, Therefore this model obtains 28% more accuracy than COCO object detection.

Shopon, Mohammed and Abedin [4] in their work "Bangla Handwritten Digit Recognition Using Auto encoder And Deep Convolutional Neural Network" present a model to classify the Bangle digits using Deep Convolutional Neural Network with Auto encoder. Indian Statistical Institute (ISI) dataset and CMATERDB 3.1.1 are the datasets used for experimentation. They have performed four combinations of these datasets. Two experiments have been performed on each dataset's own training and testing data. The other two experiments are performed using cross validation. This model achieves a very high accuracy of 99.50% in one approach. Previous models mostly didn't support both the datasets together. The model can be further improved by training it for characters other than Bangla digits.

Saabni [5] in his work "Recognizing Handwritten Single Digits and Digit Strings Using Deep Architecture of Neural Networks" propose a model containing fully connected neural network with many layers. Back propagation is used to train the model. The layers are trained using sparse encoders in advance using a predefined process. The model can be further enhanced to classify digit and text strings. Further the training process can be modified to improve the classification.

Kiani and Korayem [6] on their work "Classification of Persian Handwritten Digits Using Spiking Neural Networks" have proposed a SNN (Spiking Neural Network) model for robust learning and classification of handwritten digits i.e., To have a learning process which is persistent at against changes and high noise levels. The Deep Belief Network they have introduced have solved the problem of greater similarities between handwritten digits to great extent. The results obtained showed that the model showed a good accuracy of 95% even at higher noise levels. The model was implemented using MATLAB and Hoda Persian Handwritten digits datasets as a input images. Due to its simplicity and high speed of classification this model is feasible to be implemented on hardware modules in future.

Agapitos et al. [7] in their work "Deep Evolution Of Image Representations for Handwritten Digit Recognition" have proposed a model which is used in Genetic Programming. The model uses a method called greedy layer-wise training which is used. The system proposed performs better than existing genetic programming system. The input images are represented as pixels. They came to a conclusion that classification of many categories is difficult using systems which use standalone expression tree.

Srivatsa and Hinton et al. [8] in their work "Dropout: A simple way to prevent Neural Networks from over fitting" propose a technique called dropout to over come the over fitting issue face by convolutional neural network. Over-fitting is when the model is trained so perfectly to the training set that it is not able to tolerate even a small change in it. The over-fitting causes few areas of the model to contain more weight than others. The main idea is to drop the neurons in each layer with a pre-defined probability. This ensures no neuron carries too much weight. A main drawback of training neural networks with dropouts is that it increases the training time of the model twice to thrice.

Walid and Lasfar [5] on their work "Handwritten Digit Recognition using Sparse Architectures" apply sparse deep belief network along with auto encoder to a novel dataset which was released in ICDAR 2013 competition for handwritten digit recognition. They discuss few impediments faced by them during modeling and also ideas to further improve the performance of the model.

Zhang et al. in their work [9] "Learning High-Level Features by Deep Boltzmann Machines for Handwritten Digits Recognition" propose a model using Deep Boltzmann supported by Support Vector Machines. The models learns high level features from DBM and non-linear data is classified using SVM. MNIST dataset is used for experimentation.

3 Proposed System

In our proposed system we first pre-process the image using various data operations. Then train our model using convolutional neural network with keras and tensorflow. The image is then tested with trained model. The user is provided with a option to draw the number on the browser. The drawn number is then stored as a image in the system. Then the stored image is read and tested using our model and the predicted output is shown to user (Fig. 1).

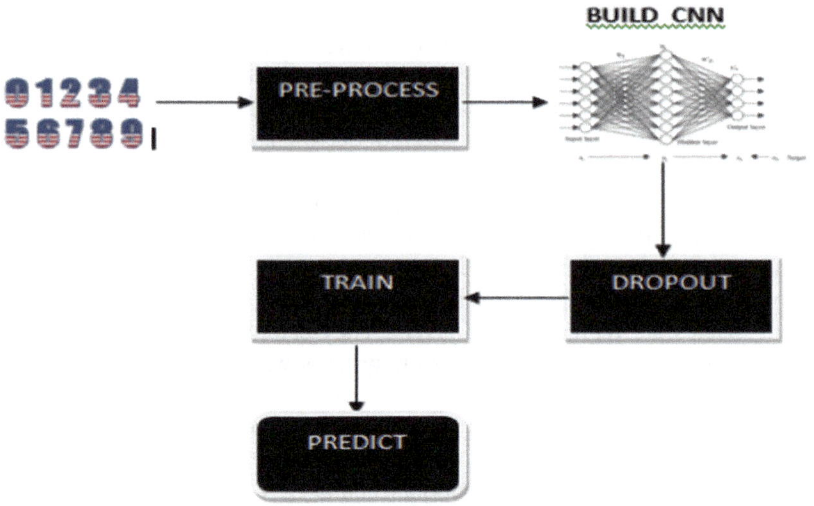

Fig. 1. Framework for handwritten digit recognition

In the above figure the modules are classified into

3.1 Preprocessing

The image is first pre-processed using some data augmentation operations. This is done to match the efficiency of the testing. Various operations like zooming, rotation, width shift, height shift, horizontal flip, vertical flip are performed on the image using a data generator. The number '1' can be written in many styles by a person. It can also be written in a slanting way, or it may be written on the corners of the screen. Now the data generator zooms the image, performs rotation and stores the images which can be used to match the input image more rapidly (Fig. 2).

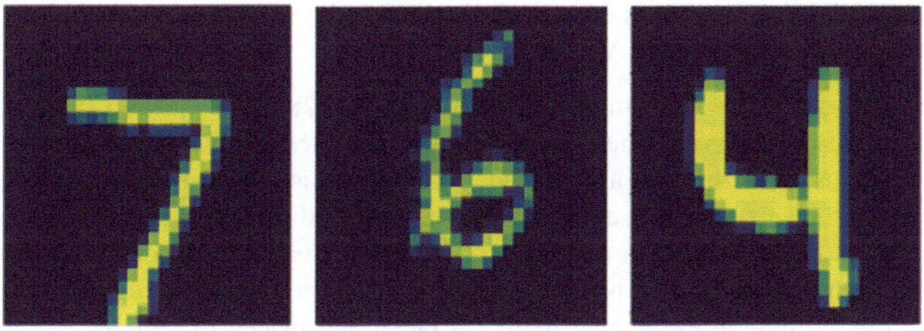

Fig. 2. Sample MNIST dataset

3.2 Build CNN

Convolutional Neural Network has been the most used model for handwritten digit recognition. Convolutional neural network have input layer, hidden layer, output layer.

```
_________________________________________________________________
Layer (type)                 Output Shape              Param #
=================================================================
input_1 (InputLayer)         (None, 28, 28, 1)         0
_________________________________________________________________
conv2d_1 (Conv2D)            (None, 24, 24, 30)        780
_________________________________________________________________
max_pooling2d_1 (MaxPooling2 (None, 12, 12, 30)        0
_________________________________________________________________
conv2d_2 (Conv2D)            (None, 10, 10, 15)        4065
_________________________________________________________________
max_pooling2d_2 (MaxPooling2 (None, 5, 5, 15)          0
_________________________________________________________________
dropout_1 (Dropout)          (None, 5, 5, 15)          0
_________________________________________________________________
flatten_1 (Flatten)          (None, 375)               0
_________________________________________________________________
dense_1 (Dense)              (None, 128)               48128
_________________________________________________________________
dense_2 (Dense)              (None, 50)                6450
_________________________________________________________________
dense_3 (Dense)              (None, 10)                510
=================================================================
Total params: 59,933
Trainable params: 59,933
Non-trainable params: 0
```

Fig. 3. CNN model summary

The RELU layer is used in the network to remove the non-linearity of the model and make it predict new models. Various layers in it are

Input Layer: The input image is represented as [28, 28, 1] array (784 pixels).
Convolutional Layer: The convolutional layer takes a square sized array of pixels and passes them through a filter. The filter/kernel is a square matrix which of the same as the size of input square array. The dot product of the square array of input image and kernel is taken and product is later activated by a activation matrix.
Max-Pooling Layer: It is used mainly for down sampling of the image. The maximum value in each input matrix is taken and placed in output.
Fully-Connected Layer: This layer is used to classify the image using a classifier (Fig. 3).

3.3 Dropout and Train

One of the most common problems faced by neural networks is over fitting. Over-fitting is the phenomena whereby our trained model fits our training data very perfectly so that it is unable to classify the new data with accuracy. Models with over-fitting show very high training accuracy but show very low testing accuracy. Models often stop when validation accuracy becomes worse. The over-fitting can be solved by a technique called dropout. Dropout is nothing but randomly dropping neurons at each layer with a pre-defined probability along with their connections. This ensures that no neurons are given very large weights (Figs. 4 and 5).

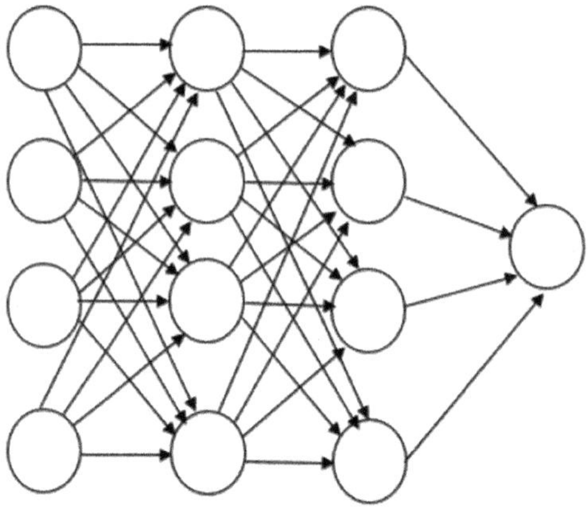

Fig. 4. Neural network

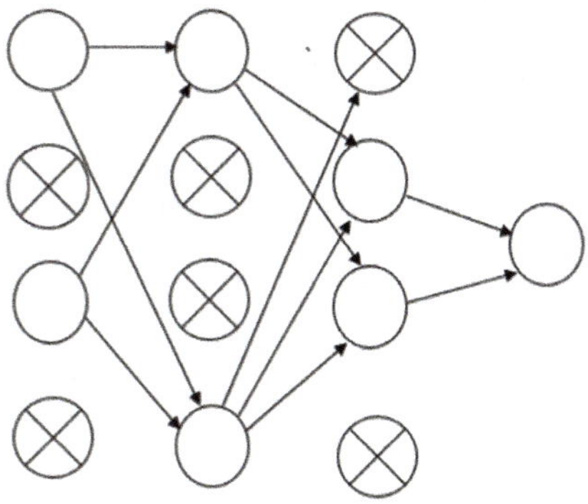

Fig. 5. Neural network after dropout

3.4 Predict

The final layer of neural network produces probabilities for each class using a classifier like softmax classifier. Probabilities are given for each number from 0–9 is given and the class which has the highest probability is most likely to be the number written in the input image and is predicted as output. The user is given an interface to draw a number in the browser and the number is stored as image in the system. This image is read and transformed and given to the neural network and is classified (Fig. 6).

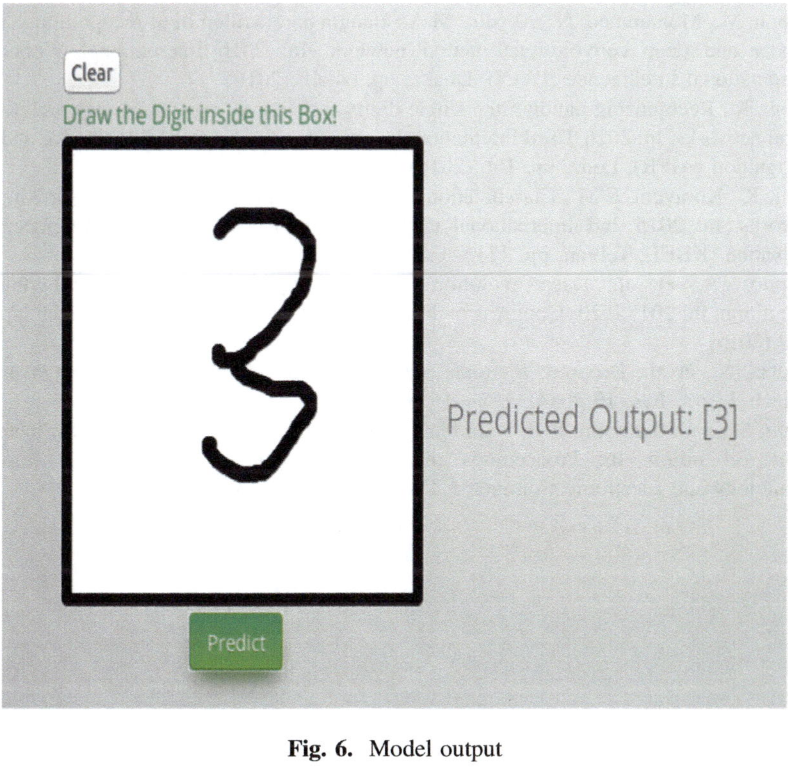

Fig. 6. Model output

4 Conclusion

Even though famous machine learning methods like RFC, SVM provide a very high training accuracy their validations accuracy drops a little bit. But our models provides a very high validation accuracy of 99.6% and with the a very good efficiency. The accuracy and efficiency provided by this system is better than the existing models and will helpful in real-time application like Bank cheque processing, postal address verification, etc. Future works may be extended from classifying digits to classifying digit strings with the high accuracy like the one provided in the model.

References

1. Ashiquzzaman, A., Tushar, A.K.: Handwritten Arabic numeral recognition using deep learning neural networks. In: 2017 IEEE International Conference on Imaging, Vision and Pattern Recognition (icIVPR), Dhaka, pp. 1–4 (2017). Md Shopon, Nabeel Mohammed and Md Anowarul Abedin (2017)
2. Shopon, M., et.al.: Image augmentation by blocky artifact in deep convolutional neural network for handwritten digit recognition. In: 2017 IEEE International Conference on Imaging, Vision and Pattern Recognition (icIVPR), Dhaka, pp. 1–6 (2017)
3. Li, X., et.al.: FPGA accelerates deep residual learning for image recognition. In: 2017 IEEE 2nd Information Technology, Networking, Electronic and Automation Control Conference (ITNEC), Chengdu, pp. 837–840 (2017)
4. Shopon, M., Mohammed, N., Abedin, M.A.: Bangla handwritten digit recognition using auto encoder and deep convolutional neural network. In: 2016 International Workshop on Computational Intelligence (IWCI), Dhaka, pp. 64–68 (2016)
5. Saabni, R.: Recognizing handwritten single digits and digit strings using deep architecture of neural networks. In: 2016 Third International Conference on Artificial Intelligence and Pattern Recognition (AIPR), Lodz, pp. 1–6 (2016)
6. Kiani, K., Korayem, E.M.: Classification of Persian handwritten digits using spiking neural networks. In: 2015 2nd International Conference on Knowledge-Based Engineering and Innovation (KBEI), Tehran, pp. 1113–1116 (2015)
7. Agapitos, A., et. al.: Deep evolution of image representations for handwritten digit recognition. In: 2015 IEEE Congress on Evolutionary Computation (CEC), Sendai, pp. 2452–2459 (2015)
8. Srivatsa, N., et al.: Dropout: a simple way to prevent neural networks from over fitting. J. Mach. Learn. Res. **15**(2014), 1929–1958 (2014)
9. Zhang, S., et.al.: Learning high-level features by deep Boltzmann machines for handwriting digits recognition. In: Proceedings of 2nd International Conference on Information Technology and Electronic Commerce, Dalian, pp. 243–246 (2014)

Heart Disease Detection Using Machine Learning Algorithms

B. Pavithra(✉) and V. Rajalakshmi

Sri Venkateswara College of Engineering, Sriperumbudur 602 117, Tamilnadu, India
pavibalacse96@gmail.com, vraji@svce.ac.in

Abstract. Heart disorder detection is to assist fitness care professionals in accurate prediction of sickness. Several researches are undergone the usage of statistical and statistics mining strategies. In the diagnosis of heart ailment, the scientific dataset having parameters and enter from complicated tests are used. The dataset consists of several coronary heart patient's record. The class algorithms are used to are expecting the patient's coronary heart sickness. The objective of the work is to discover the great classifier by means of calculating accuracy of different classifiers.

Keywords: Heart disease · Machine learning algorithms · Classifiers · Accuracy prediction · Statistical approach

1 Introduction

Heart sickness describes various conditions that affect your coronary heart. There are sub diseases under the coronary heart disease such as blood vessel illnesses, along with coronary artery ailment; heart rhythm troubles (arrhythmias); and coronary heart defects.

"Cardiovascular ailment" is the term that is often used interchangeably with "heart disease". The conditions that blocks blood vessels may lead to a coronary heart attack, chest pain (known as angina) or stroke. Heart sickness includes some forms of problem that affects the muscles in the heart, rhythm or valves. Most of the variety of coronary heart ailment can be avoided or dealt with healthy choices of lifestyle.

2 Related Work

Alba et al. proposed a model that approximates the gross effect of abnormality between the specific geometry of the patients and reference model's average shape using a virtual remodelling transformation. The methodology used in this paper is Remodelling transformation and segmentation. The limitations of this paper are that it lacks in identifying the self-folding and maps the estimation of the appropriate landmarks on to the normal shape space.

Semmlow et al. proposed a model that improves where the sounds of high frequency are detected from the heart to identify in a better way of coronary arteries blood

D. J. Hemanth et al. (Eds.): COMET 2019, LNDECT 35, pp. 1131–1137, 2020.
https://doi.org/10.1007/978-3-030-32150-5_115

flow. The methodology used in this paper is Correlation analysis. The limitations of this paper are that it the data shown is cut down to one subject and to compare the SNR, signals and noise data between microphones.

Abid et al. proposed a model where incoming health data from CEP engine are processed by running analysis rules based on threshold than to use manual threshold algorithm of statistical approach in which threshold according to recorded historical data are computed and updated automatically. The methodology used in this paper is Complex event processing (CEP). The limitations of this paper is that it lacks in large experiments are based on recorded cases and also lacks in considering other cardio vascular disease.

Tang et al. proposed a model for heart sound de-noising technique in which decomposition of singular value transformation of wavelet packet are used as a combined framework. The methodology used in this paper is Adaptive filtering approaches; Fourier and wavelet transform and blind source separation technique. The limitations of this paper are that the filtering approach gives moderate result due to the changing nature of heart sound signals. Other limitation is that important information of the signal in the range of noise might is reduced through the processing of wavelet transform.

Rocha et al. proposed a model that assesses the physiological data's predicted value collected daily in a tele-monitoring study to detect heart failure at an early stage. The methodology used in this paper is K-nearest neighbour method. The limitations of this paper is that patients suffering from heart failure requires prediction of decompensating events which still remains a challenge. The search for the optimal diagnostic approach and the in ability to improve the outcome of these patients are other limitations of this paper.

Hansen et al. proposed a model of electronic stethoscope that has a digital signal processing unit for diagnosis or identification of CAD known as coronary artery disease. The methodology used in this paper is Cross validation technique and Principle component Analysis (PCA). The limitations of this paper is the noise problems because of the features of low and high frequency bands that may not synchronize each other well and there by affect the stethoscope performance based on CAD-score.

Eastwood et al. proposed a Wanda-CVD model designed to assist participants by using the wireless coaching Women's Heart Health Study (WHHS) to reduce the cardiovascular disease (CVD) risk factors with the help of smartphone-based RHM system. The methodology used in this paper is k-nearest neighbours and Random Forest classifier. The limitations of this paper are that the study needs to be extended to a larger and more diverse group of black women and lacks in generalizing the solution to the entire population.

Orphanou et al. proposed an extended Dynamic Bayesian networks model that connects the temporal abstraction methods with Dynamic Bayesian networks applied for pre-diagnosis of the risk involved coronary heart disease (CHD). The methodology used in this paper is Bayesian Network and SMOTE-N (Synthetic Minority Oversampling Technique for nominal features). The limitations of this paper are that it lacks in assessing the model's vigorousness against the cut-off values that are chosen for temporal abstraction derivations.

3 Proposed System

Heart ailment detection is to help health care specialists, several researchers using statistical and statistics mining strategies. In the diagnosis of heart disorder, almost all systems that coronary heart disease in clinical dataset having parameters and input from complicated assessments. But there are happened some research to cast off the dangerous impact in coronary heart sickness. In this work, several coronary heart patient's information were accumulated, classification algorithm to predict the affected person's heart disorder. And we additionally discover the satisfactory classifier through calculating accuracy of various classifiers (Fig. 1).

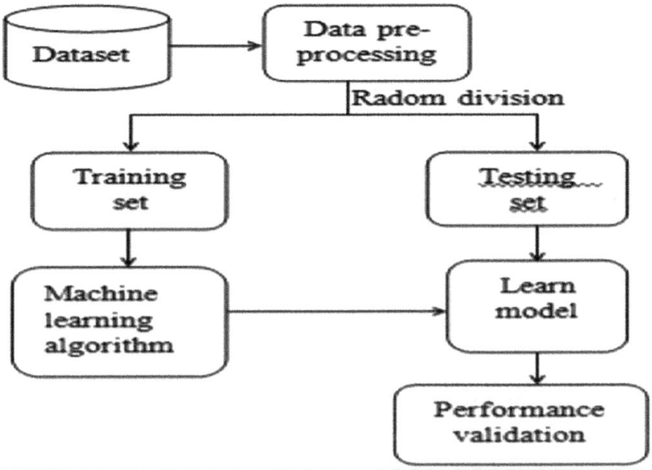

Fig. 1. Proposed architecture of prediction of heart disease using machine learning techniques

3.1 Data Pre-processing

Data Pre-processing performs a sizable function in Data Mining. The schooling section within the Data Mining in the course of Knowledge Discovery will be very tough if the information incorporates irrelevant or redundant facts and unreliable statistics. The clinical facts include many missing values. So pre-technique is an obligatory step earlier than training the scientific records. A pre-processing technique and analyses the accuracy for prediction after pre-processing the noisy statistics. It is likewise discovered that the accuracy has been extended to ninety one % after pre-processing.

3.2 Applying Machine Learning Model

Machine getting to know algorithms are investigated for assessing and predicting the severity of heart disorder.

3.3 Artificial Neural Networks (ANN)

In Artificial Neural Network, to start with the enter layer receives an input and it passes to a changed model of the input of the next layer. The layers among the enter and output are named as hidden layers and composed of multiple linear and non-linear transformation (Fig. 2).

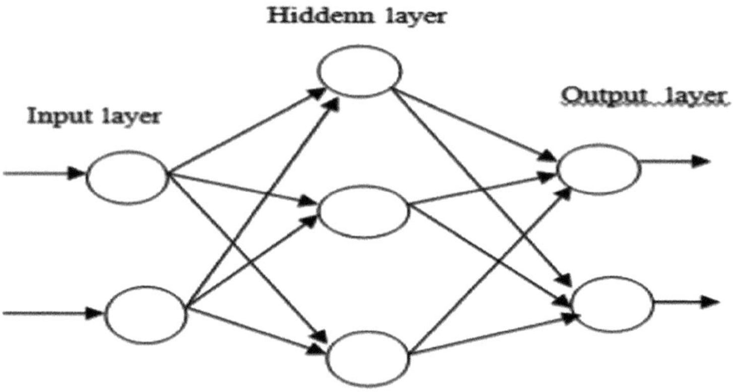

Fig. 2. Artificial neural network

3.4 Support Vector Machine (SVM)

Support Vector Machine approach has more correct and less mistakes in disorder prediction. SVM plays better with most accuracy in Heart Disease Diagnosis. Support Vector Machine (SVM) is a supervised system studying algorithm which may be used for each class or regression demanding situations.

In this algorithm, points in a n-dimensional space is plotted where the value of every feature can be the value of a specific coordinate. Then, we perform category with the aid of locating the top-rated hyper-plane that differentiate the 2 training very well in an iterative way that is used to decrease the mistake (Fig. 3).

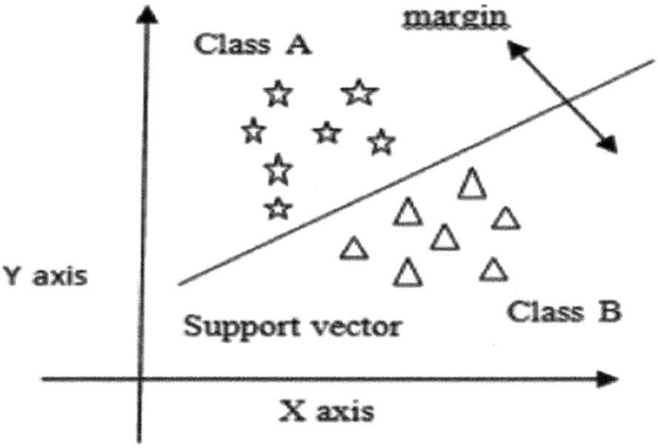

Fig. 3. Support vector machine

3.5 Decision Trees

The standard cause of the usage of selection tree is to create a training version that may be used to expecting magnificence or price of target variables by using mastering decision regulations inferred from previous statistics. A dataset is split into smaller subsets and in parallel a related selection tree is developed incrementally. The resultant of a decision tree is a tree with decision nodes and leaf nodes. A decision node can contain two or extra branches whereas leaf node can either contain a category or decision (Fig. 4).

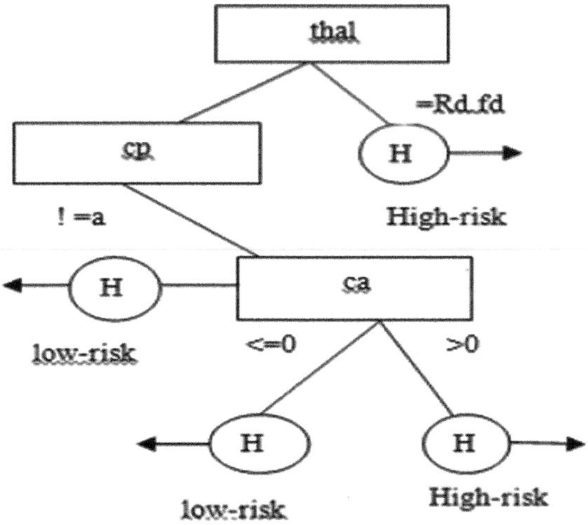

Fig. 4. Decision tree for numerical dataset

4 Experimental Study

The dataset we used here is he publically available dataset that contains the numerical dataset of heart ailment. The implementation a part of this framework is executed by using Rstudio. The function extraction is done through using Artificial neural networks (ANN), Support vector system (SVM), Decision Trees.

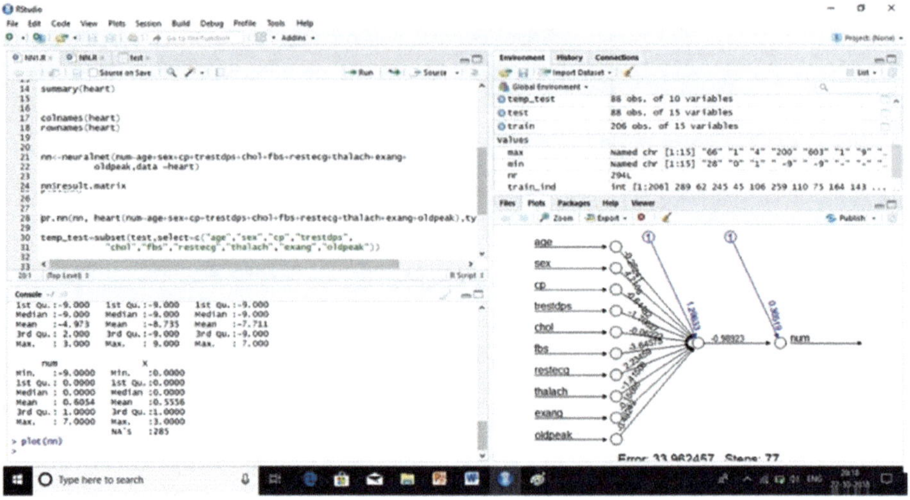

Fig. 5. Implementation of neural network

Figure 5 Depicts the Implementation of neural network where it takes the data from the dataset and produces the neural network in which all the features are given as input and produces accuracy of this model.

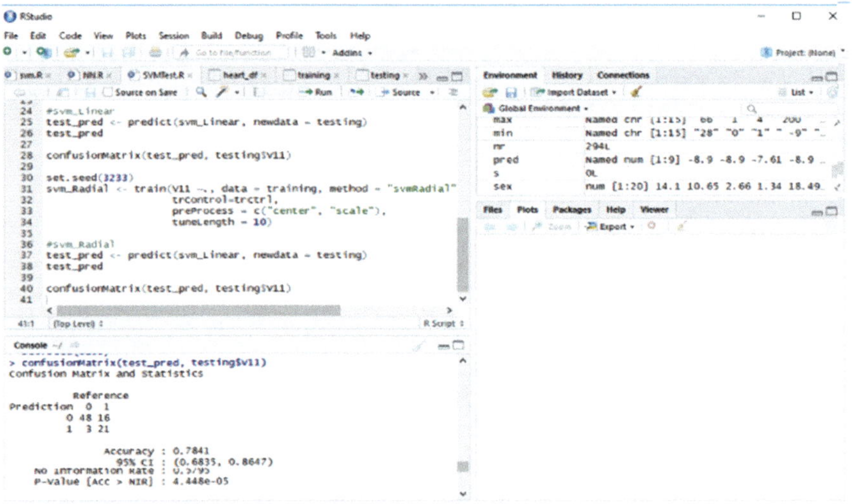

Fig. 6. Implementation of support vector machine

Figure 6 Depicts the Implementation of support vector machine where it takes the data from the dataset as input and produces accuracy of the predicted output of this model.

Table 1. Confusion matrix and statistics

Prediction	Reference	
	0	1
0	48	16
1	3	21

Table 1 shows the Confusion Matrix and accuracy for support vector machine.

5 Conclusion

The proposed model helps to find out the fine classifier by using calculating accuracy of different classifiers. The class algorithms are used to predict the patient's heart disease. The final prediction end result is given to doctor/caregiver which will provide better remedy to the affected individual.

References

1. Alba, X., Pereanez, M., Hoogendoorn, C., Swift, A.J., Wild, J.M., Frangi, A.F., Lekadir, K.: An algorithm for the segmentation of highly abnormal hearts using a generic statistical shape model. IEEE Trans. Med. Imaging **35**(3), 845–859 (2016)
2. Semmlow, J.L.: Improved heart sound detection and signal-to- noise estimation using a low-mass sensor. IEEE Trans. Biomed. Eng. **63**(3), 647–652 (2016)
3. Mdhaffar, A., Rodriguez, I.B., Charfi, K., Abid, L., Freisleben, B.: Complex event processing for heart failure prediction. IEEE Trans. Nanobiosci. **16**(8), 708–717 (2017)
4. Mondal, A., Saxena, I., Tang, H., Banerjee, P.: A noise reduction technique based on nonlinear kernel function for heart sound analysis. IEEE J. Biomed. Health Inform. **22**(3), 775–784 (2018)
5. Henriques, J., Carvalho, P., Paredes, S., Rocha, T., Habetha, J., Antunes, M., Morais, J.: A prediction of heart failure decompensation events by trend analysis of telemonitoring data. IEEE J. Biomed. Health Inform. **19**(5), 1757–1769 (2014)
6. Schmidt, S.E., Holst-Hansen, C., Hansen, J., Toft, E., Struijk, J.J.: Acoustic features for the identification of coronary artery disease. IEEE Trans. Biomed. Eng. **62**(11), 2611–2619 (2015)
7. Alshurafa, N., Sideris, C., Kalantarian, H., Sarrafzadeh, M., Eastwood, J.-A.: Remote health monitoring outcome success prediction using baseline and first month intervention data. IEEE J. Biomed. Health Inform. **21**(2), 507–514 (2016)
8. Orphanou, K., Stassopoulou, A., Keravnou, E.: A dynamic Bayesian network model extended with temporal abstractions for coronary heart disease prognosis. IEEE J. Biomed. Health Inform. **20**(3), 944–952 (2015)

Simple Task Implementation of Swarm Robotics in Underwater

K. Vengatesan[1(✉)], Abhishek Kumar[2], Vaibhav Tarachand Chavan[1], Saiprasad Macchindra Wani[1], Achintya Singhal[2], and Samee Sayyad[3]

[1] Department of Computer Engineering, Sanjivani College of Engineering, Kopargaon, India
vengicse2005@gmail.com, vaibhavchavan440@gmail.com, saiprasadwani.shirdi@gmail.com

[2] Department of Computer Science, Banaras Hindu University, Varanasi, India
abhishek.maacindia@gmail.com, achintya.singhal@gmail.com

[3] School of Engineering, Symbiosis Skill and Open University, Pune, India
samee.syd@gmail.com

Abstract. The need to control of underwater swarm robots something other than a controller system, it wants a correspondences method. Subsurface correspondences are troublesome under the most favorable circumstances thus extensive period postponements and negligible data is a worry. The regulator arrangement must have the capacity to deal with negligible and obsolete data. The control system should likewise have the capacity to control an extensive number of machine deprived of an ace switch, a dispersed switch method. This work portrays a control technique; this will provide accurate result of better co-ordination.

Keywords: Robotics · Underwater · Swarm

1 Introduction

Swarm Robotics is hard to characterize legitimately a swarm apply autonomy, because of extensive variety of utilizations. Perhaps, the most suitable definition is: "Swarm apply autonomy is the investigation of how extensive the quantity of moderately straightforward physically typified operators can be structured to such an extent that a coveted aggregate conduct rises up out of the neighborhood connection among specialists and between the operators and nature" [1]. In this description, the principle attributes of a group apply autonomy are abridged: effortlessness of robots, completely conveyed scheme, scalable, vigor. They are needed to be described with particular key favorable circumstances, for example,

1. Parallelism: normally a major, complex assignment is partitioned in several suber rand with every unit achieves a certain undertaking speedier than a solitary robot;
2. Easiness: this scheme is need with a great level of adaptation to non-critical failure. Practically speaking, if some robot comes up short the implementation of its errand,

D. J. Hemanth et al. (Eds.): COMET 2019, LNDECT 35, pp. 1138–1145, 2020.
https://doi.org/10.1007/978-3-030-32150-5_116

the method will develop in a dynamic and novel setup that will restore the right working of the scheme;
3. Scalable: the addition of the quantity of gadgets doesn't debase the execution of the entire method;
4. Heterogeneous: every part able to be portrayed with particular properties that will be successfully abused to achieve appropriate errands;
5. Elasticity: a scheme must be re-configured with the finale objective to attain varied assignments and implement unique solicitations;
6. Difficult Jobs: by and large, a solitary part couldn't attain an intricate assignment, though a swarm able to, account of the joint capacities of the distinct gadgets;
7. Inexpensive Alternate: gadgets are basic, humble to manufacture and less expensive than a solitary ground-breaking robot.

Regularly, Swarm Robots work dependent on certain feeling of natural motivation [4]. In the wisdom, the use of Swarm Intellect to aggregate mechanical technology could be distinguished as "Swarm Robotics". The feeling of the connection among bio-motivation, Swarm Intellect and Self-composed and Dispersed Method able to be clarified over the following Fig. 1.

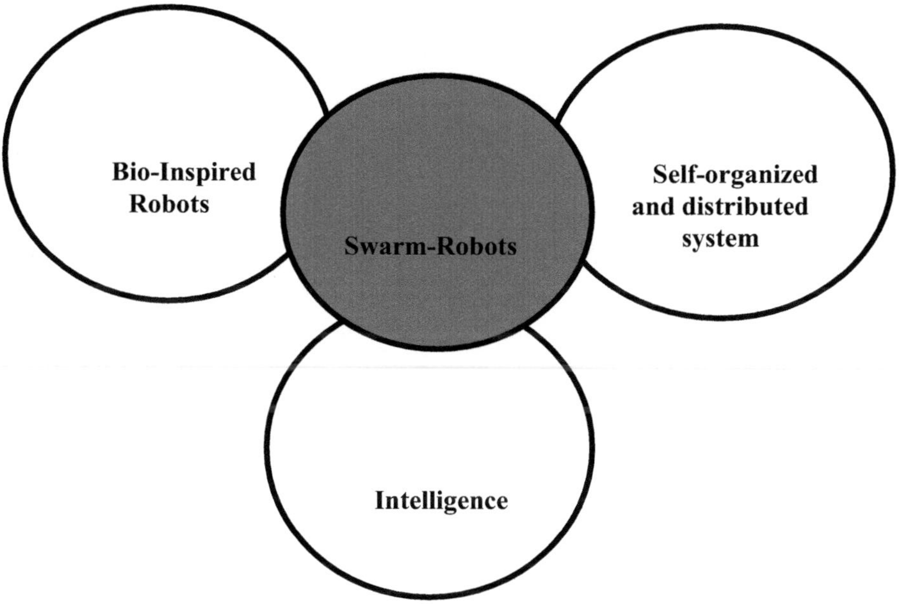

Fig. 1. Bio-inspired systems, self-organized and distributed systems and swarm intelligence are intersection among swarm robotics.

After a chronicled perspective, the principal probes methods of robots may be distinguished as Swarm Robots, were acknowledged in getting on 1940s. Grey-Walter and his group demonstrated a scheme of basic robotics communicating in an apparently communal way and by showing "complex conduct" [4], yet Swarm Intelligence turns into a functioning arena of investigation just in the 1990 and [9] G. Beni presented the idea of Swarm Intelligence by examining cell mechanical technology methods. Deneubourg et al. in 1990's presented the idea of stigmergy in robotics that carry on like ants [1, 2]. From that point forward, various analysts have created group and self-composed systems [8] and have presented robots' practices roused by creepy crawlies' social association [3, 5, 6].

2 Swarm Robotics Classification

Diverse sorts of ordering have been offered for Swarm Robots. In [2] creators suggest a scientific categorization with order prevailing examinations. In particular, they are splitting prevailing investigations into the maximum vital study headings. The five arenas they recognize are: demonstrating, conduct structure, correspondence, explanatory examinations and issues. The scientific categorization is abridged in Fig. 2. Concerning demonstrating, creators discovered that displaying is an exceptionally reasonable strategy for Swarm Robotics.

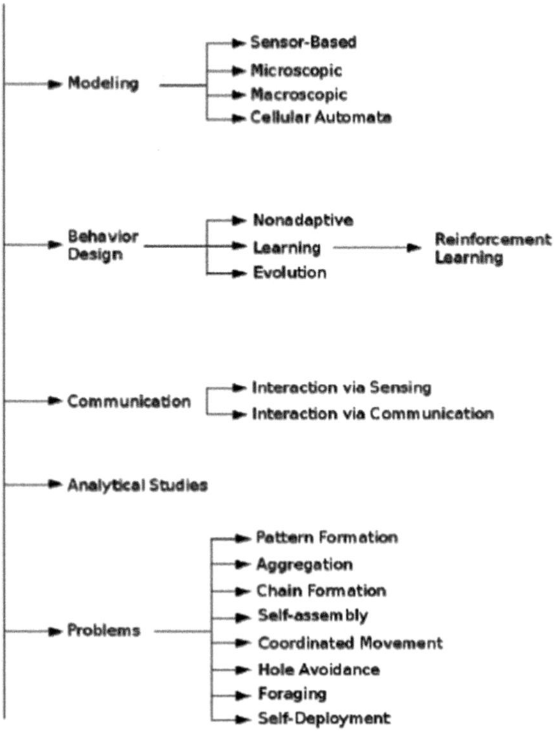

Fig. 2. Classification of swarm robotics literature

Truth be told, there are a few dangers identified with the robotics that needs a social to pursue the tests. Ordinarily, to approve consequences, a highest no. of analyses is needed with reproduction then displaying of the examinations appears to be a powerful method to create the scheme effort. Other essential perspective identified with demonstrating in Swarm Robotics is scalability. For the most part, exhibit of scalability of some control algorithm requires several robots. Costs identified with the utilization of such various robots could be restrictive and demonstrating could turn into the main practical arrangement.

In a natural system, people may calibrate their practices in their lifetime. By and by, they figure out in what way to live and to remain improved once outside settings alteration. In this Swarm Robots, analysts measured the social adjustment to controlling vast no. of robotics to achieve an undertaking on the whole.

Correspondence is sub-separated into three sorts. The primary kind is by means of sensing and speaks to the least complex sort of correspondence dependent on the limit of a robot to recognize different robots and the items in the earth. At the point when robots utilize collaboration by means of nature, they think about it as a correspondence standard (i.e., pheromone utilized from ant). Association by means of correspondence includes express correspondence through direct messages. Investigative examinations incorporate investigations that add to the hypothetical comprehension of swarm systems. In this classification, techniques for arrangement of various issues can be incorporated. Besides, scientific apparatuses that permit a more profound understanding of the points of interest of Swarm-Intelligence schemes will be deliberated as a major aspect of systematic investigations.

3 Underwater Swarm Robotics Coordination

Submerged surface, it is almost difficult to construct "dialect" correspondence among a few some robots. So that model be assumed consequently remote aren't appropriate and an altogether dissimilar models be planned in the present work to descript the cooperation instrument of dual under-water small scale robotics utilizing stigmergy, a time begat by Pierre-Paul Grasse (French scholar) [6], which implies collaborating by implication through condition.

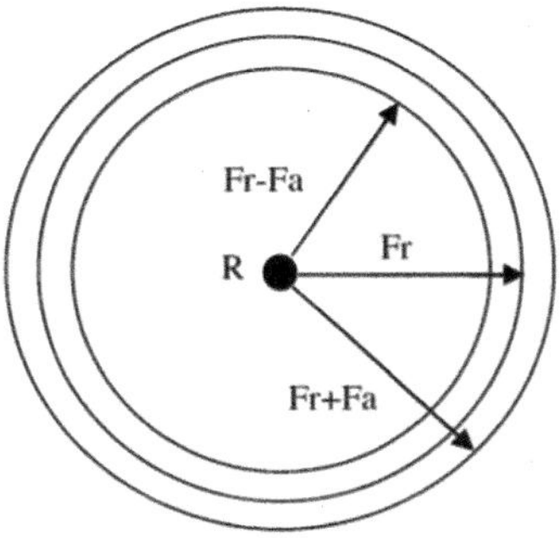

Fig. 3. The robot's attraction/repulsion

Here, two individual robots collaborate when one adjusts the earth and alternate reacts to the altered condition by a late period. In the present work, the earth of a robot is epitomized as an aversion/fascination demonstrates which can be seen in Fig. 3.

Anywhere, the dark driven roundabout standpoints for robotic R. The Fr is the wellness separation of Robotic R, which implies the separation Intelligent R willing to retain of different Intelligent. The Fa is inclination of Fr. At that point we are acquire most extreme wellness separation of Fa+Fr then least wellness separation of $F_r - F_a$. Afterward the meaning of shock/fascination display, uncertainly the separation of a near Intelligent R_1 to Intelligent R is more noteworthy than Fa+Fr, R will change near R_1 for fascination; else-if the separation is not as much as $F_r - F_a$, R must flee from R_1 for aversion. Consequently an Intelligent can impact other just by changing the separation among them, the synchronization of them could be acknowledged in a roundabout way in such way.

Likewise, if other Intelligent keeps running into the region anywhere the separation to R is [Fr+Fa, Fr−Fa], R would retain steady. Then we will called some region hardness region of Intelligent R, which could be create the connection among Intelligent all the extra simple to gets remaining actually. Submerged external, a slight centimeter-like Intelligent is also little to make some move. It is important to conglomeration thousands or hundreds of like Intelligent to shape a swarm. At that point the swarm possibly indicates smart aggregate conduct and hold capacity to take a few activities. In this paper, in view of the component of synchronization communicated overhead, we will offered a straightforward instruction now which be able to total such huge numbers of Intelligent into a swarm.

Imperative: Supposing here are 2 neighbors Intelligent R_a and R_b. Every Intelligent receives the shock/fascination demonstrate in this Fig. 1 thru constraints Fa, Fr as its condition. In the event that the separation among them is Fr+Fa < D (Ra−Rb), at that point they all the while change near thru the speed V individually; Once the separation is Fr−Fa > D(Ra−Rb), at that point they at the same time run separated thru the speed V individually. Underneath rule 1, we know how to get the development prototypical to regulate the development of every Intelligent as Fig. 4.

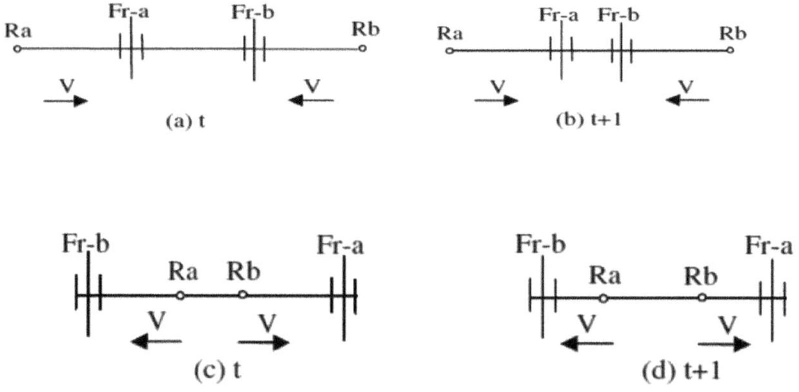

Fig. 4. The movements typical of synchronization below rule

In Fig. 4, (a)–(b) delineates the fascination development while (c)–(d) demonstrates the aversion development under Rule. Fr-a, Fr-b is the wellness separation of Ra and Rb individually with same esteem and have same inclination Fa. V is a steady esteem. Not at all like the models communicated in numerous inquiries about where the worldwide data will be united thru altogether the person of swarms, the replicas here in submerged situation are just data confined. So the in the end development calculation of an Intelligent underwater able to be communicated as beneath:

$$X_i = \sum_{i \in s_i} F_i,\ i = 1, 2, \ldots, N \tag{1}$$

Anywhere, N is the aggregate no. of swarm robots, Si is an Intelligent sets which component intelligent will be understood by robots exclusive its constrained degree. Though altogether robots transfer underneath this basic association calculation dependent on succinct Instruction, they would be connected each other with the accumulation come from into genuine. In addition, to stay away from the misuse of vitality, once the swarm intelligent originates into accumulation express, every person inside swarm ought to in security zone retain immobile and demonstrate non-association. In this movement demonstrate appeared in Fig. 3, the separations among intelligent R_a and R_b will alter 2 V from period t to t + 1 in view of intuitive movements.

In the meantime, the width of the steadiness region is $2F_a$. Then if to need R_b make certain to derive into soundness region of R_a specifically, it ought to be 2 V < 2Fa or V < Fa. What's more, in the association demonstrate, if the inclination F_a is too little, it would be excessively troublesome for another robot, making it impossible to go into soundness zone. Despite what might be expected, If Fa is too substantial, the wellness separate Fr will end up good for nothing. So Fa in movement model ought to be properly set. In this paper, we set Fa = Fr/15 and get a perfect outcome from relative reproductions. Likewise, the speed V in movement prototypical ought to be properly set. On the off chance that V is too little, the swarm robotics will be too ease back to come into total for absence of vitality. In the event that V is too substantial, particularly $V \gg Fa$, albeit every robots be able to move quicker, the swarm robots likewise be hard to total for a lot of swaying in discovering strength zone keeping still.

4 Usage of the Undertaking of Z Co-ordination

With the end goal to test the accord controlled a basic undertaking was given. The swarms of intelligent were to watch a square way characterized by four waypoints. 10 recreated Video Rays were made all at various profundities to stay away from crashes. Maintaining a strategic distance from crash along these lines evacuated one parameter and accordingly improving the reproduction. As observed the circulation of the robots is considerably more even. How about we take a gander at one of the robots as observed by the robot behind it.

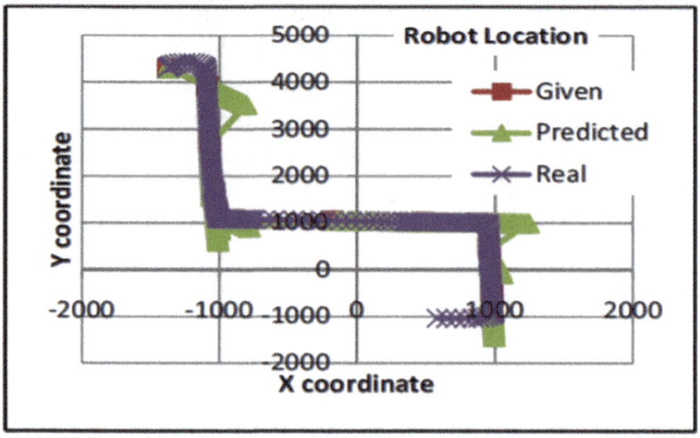

Fig. 5. The robot plots (Z's synchronization is constant and so disregarded)

Figure 5 demonstrates the movement of an intelligent discovering its 1st waypoint then after that beginning to change around the square of 4 waypoints. It demonstrates the last or given known positions, the anticipated plot and the real plot. The anticipated plot looks frightfully wrong in spots while the given positions dependably look right. This is deluding in any case. The given positions are in every case right however they are just substantial at specific focuses in time. A superior method to take a gander at this is by taking a gander at each organizes after some time.

5 Conclusion

In this paper, we right off the bat offered a system of synchronization for subsurface swarms smaller scale robots, with after that everywhere this instrument we additionally offered 3 straightforward instructions and comparative procedures to acknowledge conglomeration with arrangement then running of subsurface swarm robots. The viability of these procedures is demonstrated thru a lot of reenactments. As should be obvious, simply utilizing a few basic rules, the aggregate conduct of a swarm intelligent is able to demonstrate a perplexing insight to adjust to the earth. This wonder will provide us numerous motivations to enhance our capacity to tackle composite issues.

References

1. Tan, Y.: Swarm robotics: collective behavior inspired by nature. J. Comput. Sci. Syst. Biol. **6**, e106 (2013). https://doi.org/10.4172/jcsb.1000e106
2. Sharkey, A.J.C.: Swarm robotics and minimalism. Connection Science. **19**(3), 245–260 (2007)
3. William, A.: Modeling artificial, mobile swarm systems. Doctoral Thesis. Institute of Technology, California (2003)
4. Beni, G., Wang, J.: Swarm intelligence in cellular robotic. In: Systems Proceedings of NATO Advanced Workshop on Robots and Biological Systems, vol. 102 (1989)

5. Naik, B., Mahapatra, S., Swetanisha, S., Barisal, S.K.: Cooperative swarm based evolutionary approach to find optimal cluster centroids in cluster analysis. IJCSI Int. J. Comput. Sci. Issues. **9**, 425 (2012)
6. Dorigo, M., Birattari, M.: Swarm intelligence. Scholarpedia. **2**(9) (2007)
7. Eliseo, F.: A control architecture for a heterogeneous swarm of robots. Rapport d'avancement de recherché (PhD), Universite Libre De Bruxelles; Computers and Decision Engineering, IRIDIA (2009)
8. Kumar, E.S., Vengatesan, K.: Cluster Comput. (2018). https://doi.org/10.1007/s10586-018-2362-1
9. Sanjeevikumar, P., Vengatesan, K., Singh, R.P., Mahajan, S.B.: Statistical analysis of gene expression data using biclustering coherent column. Int. J. Pure Appl. Math. **114**(9), 447–454 (2017)
10. Kumar, A., Singhal, A., Sheetlani, J.: Essential-replica for face detection in the large appearance variations. Int. J. Pure Appl. Math. **118**(20), 2665–2674 (2018)
11. Parpinelli, R., Heitor, S.L.: Theory and New Applications of Swarm Intelligence. InTech, Rijeka (2012). ISBN 978-953-51-0364-6
12. Neshat, M., Sepidnam, G., Sargolzaei, M., Toosi, A. N.: Artificial fish swarm algorithm: a survey of the state of the-art, hybridization, combinatorial and indicative applications. Artif. Rev. (2012). https://doi.org/10.1007/s10462-012-9342-2
13. Kumar, A., Vengatesan, K., Rajesh, M., Singhal, A.: Teaching literacy through animation & multimedia. Int. J. Innov. Technol. Explor. Eng. **8**(5), 73–76 (2019). 57205678507; 55611316200; 56606891300; 24765540900
14. Zhiguo, S., Jun, T., Qiao, Z., Lei, L., Junming, W.: A survey of swarm robotics system. In: Advances in Swarm Intelligence. Lecture Notes in Computer Science, vol. 7331 (2012)
15. Lau, H.K.: Error detection in swarm robotics: a focus on adaptivity to dynamic environments. Ph.D. Thesis. University of York, Department of Computer Science (2012)
16. Marco, D., et al.: The SWARM-BOT project. In: Swarm Robotics. Lecture Notes in Computer Science, vol. 3342 (2005)
17. Selvaraj Kesavan, E., Kumar, S., Kumar, A., Vengatesan, K.: An investigation on adaptive HTTP media streaming quality-of-experience (QoE) and agility using cloud media services. Int. J. Comput. Appl. (2019). https://doi.org/10.1080/1206212X.2019.1575034
18. Marco, D., et al.: Evolving self-organizing behaviors for a swarm-bot. Auton. Robot. **17**(2–3), 223–245 (2004)

Ultra Sound Imaging System of Kidney Stones Using Deep Neural Network

S. R. Balaji[1(✉)], R. Manikandan[1], S. Karthikeyan[2], and R. Sakthivel[1]

[1] Department of EIE, Panimalar Engineering College, Chennai, India
balasrb2000@gmail.com

[2] Department of ECE, Sathyabama Institute of Science and Technology, Chennai, India

Abstract. Magnetic resonance is one of the imaging modality through which medical images can be diagnosed. Each modality like imaging (MRI), Ultrasonography (US), Intravenous Urography (IVU), computed Tomography (CT), Angiography (AG) has their own advantages and disadvantages in various aspects like formation, sensitivity, resolution, level of invasive and cost. Both MRI and CT scan give same information in regarding to kidney imaging. However in MRI, the material gadolinium is associated with Nephroenic Systemic Fibrosis (NSF), which decreases the kidney functioning. The measurement of size and shape of the kidney and evaluation of pelvis and ureters is done by IVU. The major drawback is that, there may be renal failure due to radiation and 1 V contrast administration. In CT the tomographic image is formed by computer processed image, where more information is same as ultrasound. The major advantages are excellent spatial/contrast resolution and low cost. However there is a major drawback of exposure of radiation and contrast dye causes damage to kidney. In Radiography, the kidney stones are distinguished by the ratio of absorption over dark field signal. The limitation of radiography is that it can determine only the intensities of average signal through the projection of x-ray direction. In this project enhancement (Smoothening and Sharpening) technique is used in order to contrast the image. We use the active contour segmentation in order to extract the particular portion of image and then deep neural network is used in order to classify the various types of stones.

Keywords: Ultra Sound image · Deep Neural Network · Feature extraction · Region of Interest

1 Introduction

The ultrasound images of kidney are obtained and they are preprocessed by forthcoming steps such as Gray scale conversion and Smoothening filter in order to increase the quality of image. Later the image is splitted into various sections of images then the images are cropped further for number of times using feature extraction method. The proposed method partitions an image into an arbitrary number of sub regions and tracks down salient regions step by step.

D. J. Hemanth et al. (Eds.): COMET 2019, LNDECT 35, pp. 1146–1154, 2020.
https://doi.org/10.1007/978-3-030-32150-5_117

On comparing the output with the desired input, yields an error signal, which is propagated back until retina is reached. Based on the inputs it had received during the forward pass and the returning signal, each neuron adjusts its weight. Here we use back propagation algorithm. In this paper, the ultrasound images are used to determine the types of stones using its intensity and shape. A safe detection technique is introduced.

2 Existing System

The images of kidney are obtained and the quality of image is improved by the method called preprocessing. Then the image is splitted into many section of images then the images are cropped for many number of times using feature extraction method. The 2nd and 3rd features of the kidney image are extracted by using Histogram based Difference and Sum model. The identification of kidney abnormality is identified using segmentation algorithms. Only identify the kidney is normal or abnormal.

2.1 Output – Normality

The kidney is classified depend upon the values of feature based on calculation of haralick features and histogram. Required range of features are obtained for the normal image based on this value classify the image is abnormal or normal [9]. The drawbacks of Existing systems are it identifies if the kidney is abnormal or normal and it cannot classify the types of kidney stones.

3 Proposed System

In our proposed system, contour segmentation concept is implemented for identifying the best image from the kidney images came. The classification process here reduced complexity because of the region of interest is being used. This kind of design with

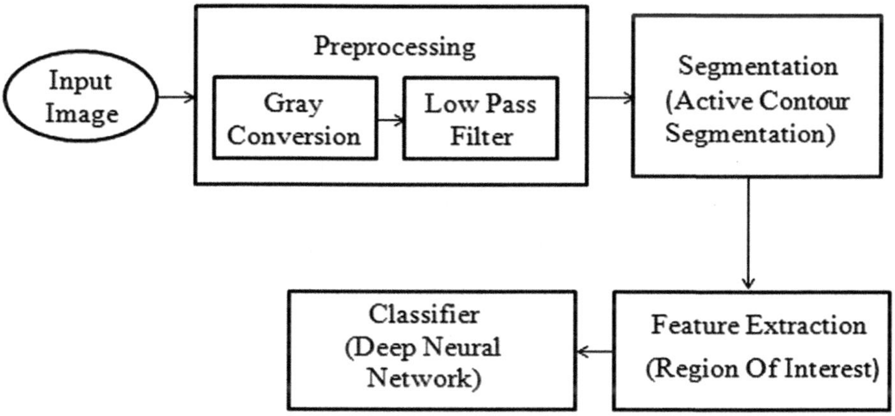

Fig. 1. Block diagram of proposed system

segmentation and ROI hence reduce the complex modeling done in the existing system. The accuracy level of the proposed classifier is also improved by region selection. The block diagram of proposed system is as shown in the Fig. 1.

3.1 Input Image

Ultrasound images of kidney are taken as input image shown in the Fig. 2. Various images are taken from the web.

3.2 Preprocessing

If the input images are color images means we are convert to gray scale from that color images. Using filter to remove the noise in the images and Smoothening and sharpening method used to contrast the image.

3.2.1 Gray Conversion

The ultrasound images of kidney are taken and converted into gray scale for easy processing as shown in the Fig. 3. Using the following code the Ultrasound images are converted into gray scale images.

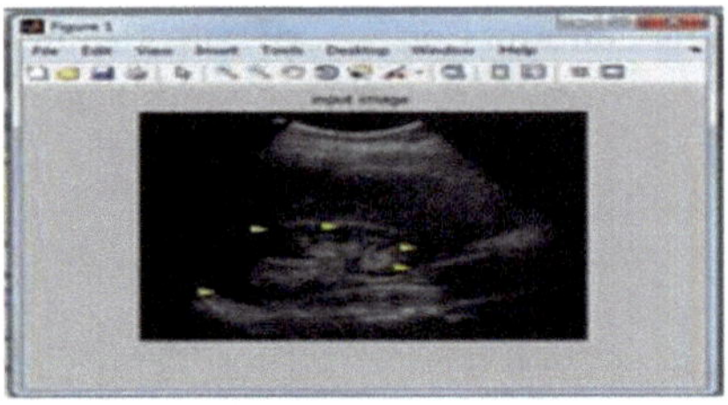

Fig. 2. Input image

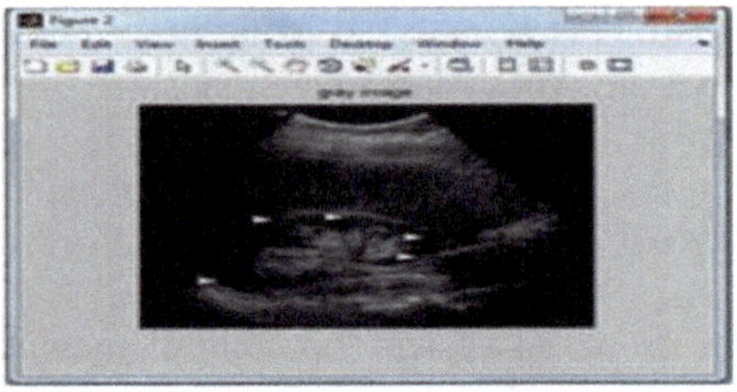

Fig. 3. Gray image

3.2.2 Filter

Low Pass Filter is used to minimize the speckle noise in image of ultrasound. It is to calculate average of a pixel and all of its neighbor pixel values. Gaussian Low pass filter is used to smoothening as shown in Fig. 4 and sharpening as shown in Fig. 5 in order to contrast of the image. It has lower Root Mean Square Error and higher Peak Signal to Noise Ratio. The Gaussian filter is a type that belongs to image-blurring filters which uses a function known as Gaussian function which is used to calculate the transformation that are applied to every pixel of the image. Equation for Gaussian function is given in one dimension as follows:

$$G(x) = \frac{1 - e^{-x/\sigma}}{\sqrt{2\pi\sigma^2}} \tag{1}$$

Whereas in two dimensions, it is the multiplication of two Gaussians, each in one dimensions.

$$G(x) = \frac{1 - e^{-x^2+y^2}}{\sqrt{2\pi\sigma^2}} \tag{2}$$

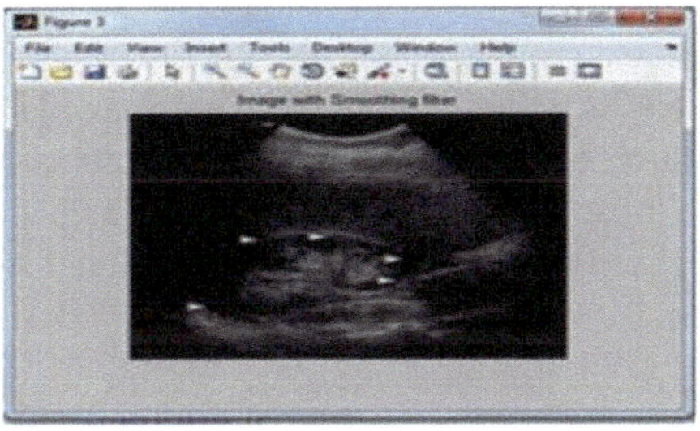

Fig. 4. Remove noise using filter

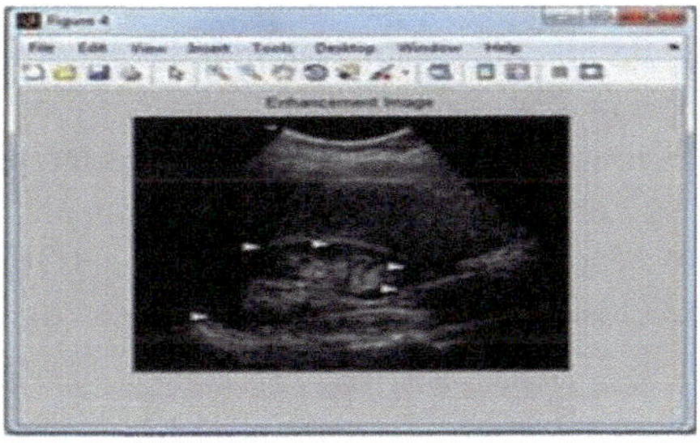

Fig. 5. Enhanced image

3.3 Active Contour Segmentation

The Active Contour Algorithm begins with requiring the user to construct a rough contour enclosing the object to be tracked. This contour is known as the initial user approximation as shown in the Fig. 6 and can be of any shape and size as shown in the Fig. 6.

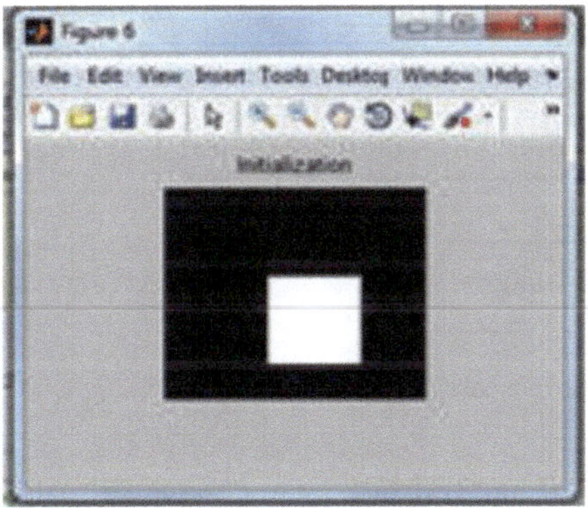

Fig. 6. Initial user approximation

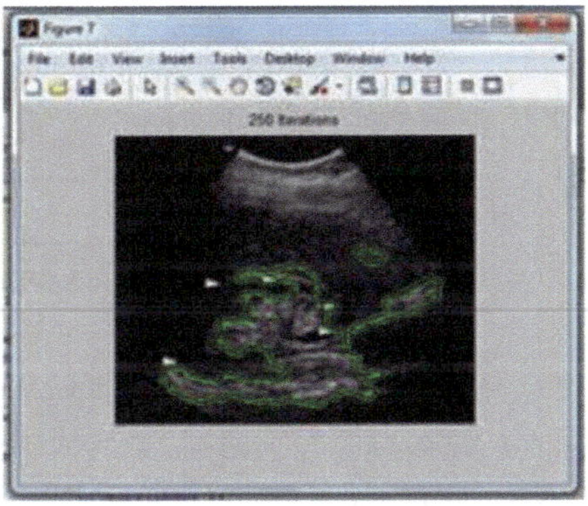

Fig. 7. Iteration active contour segmentation

Instead of having the user draw the curve out explicitly, the active contour algorithm simplifies the procedure by allowing the user to click points that surround the object [2]. These points are the basic formation of the active contour. Active contours are also known as Snake, serves as the framework for obtaining the contour of object outline [5, 6]. The framework reduces energy within the present contour as a sum of external and internal energies as shown in Fig. 7. Internal energy controls the curvature, shape of contour, controlling, regulating the shape, etc.

Finally, after applying active contour the segmented images are processed for further purpose as shown in the Fig. 8.

3.4 Feature Extraction-ROI

Region of Interest can be defined by creating a binary mask, which is a binary image that has the same size as the image you want to process as shown in the Fig. 9. In mask image, the pixel which define the Region of Interest are made as 1 and the other pixels are made as 0 as shown in Fig. 9. More than one ROI can be defined in an image [1]. Regions of the image will be in geographic nature, either defined by specific intensity range or by polygons that surrounds contiguous pixels. In the latter case, the pixels are not necessarily contiguous [7]. For ROI's that enclose an area, each ROI is defined by the closed path outline that encloses the region.

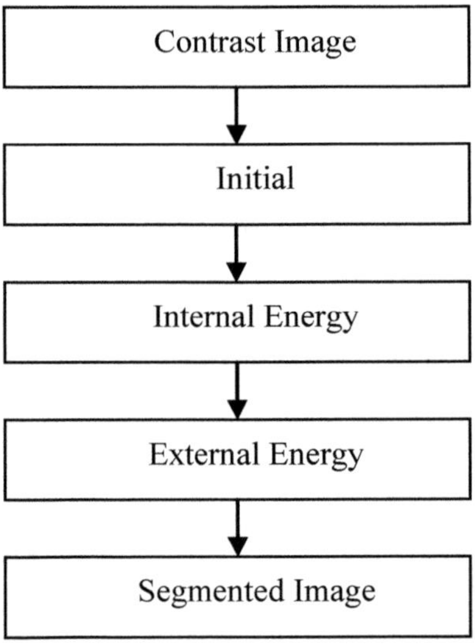

Fig. 8. Flowchart for proposed system

The interested regions are samples which are present in a data set that are determined for a specific purpose [8]. The concept of a Region of Interest is generally used for many applications. From Fig. 10, the stone boundaries can be specified in a volume or on an image for the measurement of its size purpose.

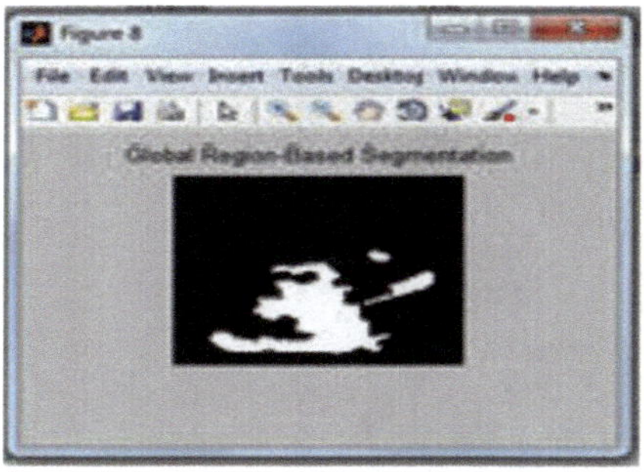

Fig. 9. Masked image

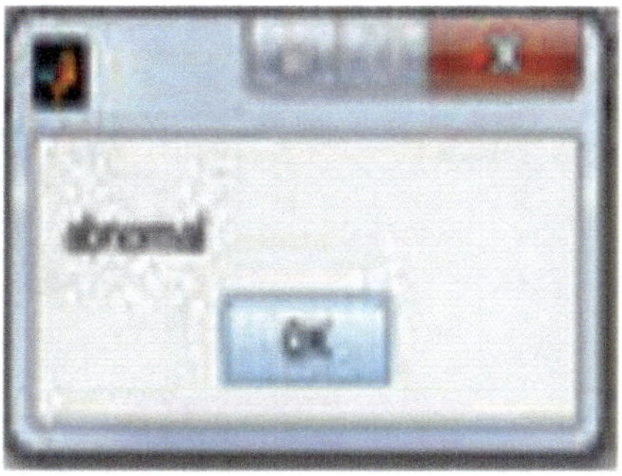

Fig. 10. Abnormality or normality dialogue box

3.5 Deep Neural Network Classifier

Deep learning means that the Stacked Neural Networks; i.e., the networks are made of many layers. The layers are made of different nodes [3]. The computation for the outputs will be made at the nodes. The node combine the different inputs from the data with a co-efficient sets, or dampen the input or weights that amplify, hence the algorithm is trying to learn the assigning of inputs for the task. As shown in Fig. 11, this input is more useful in classifying the data without any errors [4]. The weight-input products are added the sum is given through nodes which is called as activation function, in order to find out the extent of signal passing through the network further to make some effect on the actual outcome, i.e., almost the classification as show in Fig. 12.

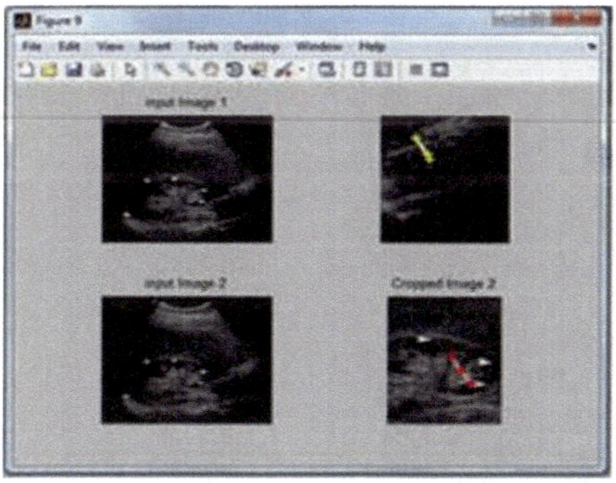

Fig. 11. Input image to neural network

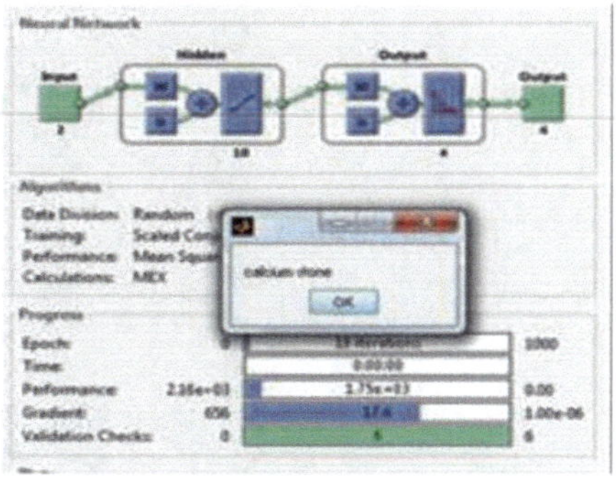

Fig. 12. Deep neural network

From the ROI input to the Deep Neural Network (DNN). Then the deep neural network will classify the type of stone.

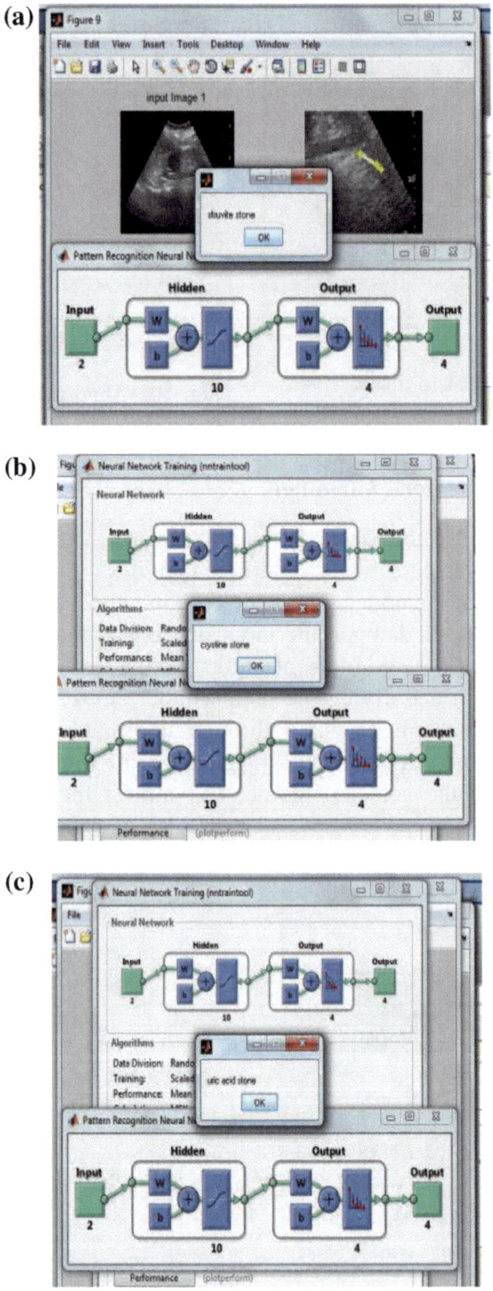

Fig. 13. a. Output of classifying Struvite stone. b. Output of classifying uric acid stone. c. Cystine stone

3.6 Classification of Stone

The input Ultrasound image after segmentation is given to the classifier. Depending on the shape of the stone, they are further classified into four types of stones namely Cysine stone, Calcium stone, Struvite stone, Uric acid stone.

4 Results and Discussions

The proposed work is completed by using Deep Neural Network, active contour segmentation and MATLAB 2014. From the database of the ultrasound kidney images, we determine the types of stones. The input ultrasounds images are converted into gray image. Ultrasound images have noises using Low pass Gaussian filter remove the noise from the gray images. Smoothening and sharpening process done to contrast the image [11]. From the contrast image segment the portion of kidney using Active contour segmentation [12]. By this method identify the portion of abnormality in the image and segment that portion for further process. In the segmented portion, applying the concept of Region of Interest to identify the stone presence or absence in the image.

Depending upon the absence or presence of stone move to next process. If identify the presence of stone, then by using Deep Neural Network processing the image by Back propagation and identify the type of stone such as Struvite stone, Uric acid stone, Cystine stone as shown in Fig. 13a, b & c respectively.

Calcium stones are formed due to eating oxalate rich food. Struvite stones are strengthening by the infections of bacteria at hydrolyzes urea to ammonium and increases PH value of urine to alkaline or neutral values [10]. A rich diet rich in purines will increases the acidic level of urine. Purine is nothing but a colorless substance in animal proteins, such as shellfish, fish and meats [13]. Cystine stones formed in the kidney due to an acid that created in the body which is leaked into the kidneys as the urine. Depending on the shape, each type of stone is classified by using classifier.

5 Conclusion

Thus in this paper, the ultrasound images are used to determine the types of stones using its intensity and shape. A safe detection technique is introduced. Thus disadvantages in the existing technique are eradicated. The advantages such as safe, accuracy of kidney stones are achieved. Thus completed the performance of an image analysis for feature extraction from the different groups of kidney Ultra Sound images namely as Cystine stone, Calcium stone, Struvite stone and Uric Acid stone.

This paper identifies four types of kidney stones only. In future it can be enhanced by adding more types of stones and also other abnormalities like Cyst, Bacterial infection, etc., if possible. In future, these methods can be applied to a huge data set with a broad spectrum of kidney diseases and fully automated intelligent system can be developed to assist in the classification of kidney from the ultrasound images. Increasing the accuracy of the classification depend on the current investigation whose objective is to create a classification of image through intelligent automation system.

References

1. Hafizah, W.M.: Feature extraction of kidney ultrasound images based on intensity histogram and gray level matrix. In: Sixth Asia Modelling symposium (2012)
2. Martin-Fernandez, M., Alberola-Lopez, C.: An approach for contour detection of human kidneys from ultrasound images using markov random fields and active contours. Med. Image Anal. **9**, 1–23 (2005)
3. Sehrawat, R., Gupta, P., Yadav, R.: Basic of artificial neural network. J. Comput. Sci. Eng. **1** (2015)
4. Shalma Beebi, A., Saranya, D., Sathya, T.: A study on neural networks. Int. J. Innov. Res. Comput. Commun. Eng. **3** (2015)
5. Narkhede, H.P.: Review on image segmentation techniques. Int. J. Sci. Mod. Eng. (IJISME) **1**(8) (2013). ISSN: 2319-6386
6. Tsai, A., Yezzi, A., Wells, W., Tempany, C., Tucker, D., Fan, A., Grimson, W.E., Willsky, A.: A shape-based approach to the segmentation of medical images. IEEE Trans. Med. Imaging **22**(2) (2003)
7. Zhang, Y.J.: An overview of image and video segmentation in the last 40 years. In: Proceedings of the 6th International Symposium on Signal Processing and its Applications, pp. 144–151 (2001)
8. Pham, D.L., Xu, C., Princo, J.L.: A survey on current methods in medical image segmentation. Ann. Rev. Biomed. Eng. **2** (1998)
9. Eleyan, A., Demirel, H.: Co-occurrence matrix and its statistical features as a new approach for face recognition. Turk. J. Electr. Eng. Comput. Sci. **19**, 97–107 (2011)
10. Hitesh, M.R., Asari, S.: A research paper on reducion of speckle noise in ultrasound imaging using wavelet and contourlet transform (2011)
11. Rahman, T., Uddin, M.S.: Speckle noise reduction and segmentation of kidney regions from ultrasound image. In: International Conference on Informatics, Electronics and Vision (ICIEV) (2013)
12. Hu, S., Yang, F., Griffa, M., Kaufmann, R., Anton, G., Maier, A., Riess, C.: Towards quantification of kidney stones using x-ray dark-field tomography. In: IEEE 14th International Symposium on Biomedical Imaging (2017). ISSN: 1945-8452
13. Moustafa, A.A.: Performance analysis of artificial neural networks for spatial data analysis. Contemp. Eng. Sci. **4**(4), 149–163 (2011)

Comparison of Breast Cancer Multi-class Classification Accuracy Based on Inception and InceptionResNet Architecture

Madhuvanti Muralikrishnan(✉) and R. Anitha

Department of Computer Science and Engineering,
Sri Venkateswara College of Engineering, Sriperumbudur, India
madhuvanti.muralik@gmail.com, ranitha@svce.ac.in

Abstract. Breast Cancer is the tumor that occurs most commonly in women and follows lung cancer in the most common cancers. Mortality rates caused by cancer can be reduced if early detection and treatment mechanisms are instituted. With recent advances in Deep Learning and Computer Vision, its application in diagnosis through pattern recognition is fast emerging. This paper compares the classification performance of two convolutional neural network models into benign and malignant subclasses for a breast cancer histopathological dataset. The first is built on Inception v3 architecture while the second contains residual connections in the Inception network called InceptionResNet v2. Performance has been enhanced by augmenting the data. Time taken to train and computational cost have been reduced through transfer learning. The results show accuracy from 85.9% to 91.3% on Inception v3 model. Inception Resnet v2 model performs better than Inception v3 with accuracies ranging from 89.8% to 94.6%.

Keywords: Breast cancer · Convolutional neural networks · Image classification

1 Introduction

Breast cancer is the most commonly diagnosed (24.2%) amongst all cancers and the leading cause of cancer death in women [6, 7]. Histopathogical classification of breast tumors involves distinguishing between morphological features of tumors. Breast cancer can be broadly classified into invasive ductal carcinoma (IDC) and invasive lobular carcinoma (ILC). The IDC class is further divided into five malignant and four benign sub-classes. The malignant IDC sub-classes are tubular, medullary, papillary, mucinous and cribiform carcinomas while adenosis, fibroadenoma, phyllodes tumor and tubular adenoma are the benign IDC sub-classes [8, 9].

Accurate diagnosis can lead to employment of appropriate treatment thereby increasing survival rate. The cancerous regions are detected by histopathologists who manually examine the slides for irregular cell shapes or non-conforming tissue distributions. If not trained properly, this may result in an incorrect diagnosis. Due to the heterogeneity in breast cancer sub-types, histopathology is considered a subjective science. Moreover, a lack of trained specialists would delay the treatment process.

D. J. Hemanth et al. (Eds.): COMET 2019, LNDECT 35, pp. 1155–1162, 2020.
https://doi.org/10.1007/978-3-030-32150-5_118

Hence, there is a need for automated diagnosis, which can reduce the human error-rate and dependencies [10, 11].

1.1 Related Work

The work of Spanhol et al. [1] introduces a dataset called BreakHis of 7909 breast cancer histopathology images comprising varying magnification factors of 40X, 100X, 200X and 400X acquired over eighty-two patients containing 2480 benign and 5429 malignant images. Additionally, it proposes a baseline classification performance obtained by four different classification models trained to detect the key points and extract the features. The feature sets are obtained by manual extraction through texture representations like Local Binary Patterns, parameter-free threshold adjacency statistics etc. Popular classification algorithms such as 1-nearest neighbor (1-NN), quadratic linear analysis (QDA), Support Vector Machine (SVM) and random forests are used to assess the datasets. The configuration of PFTAS descriptors trained by SVM classifiers obtained the best overall performance of 85.1%. The limitation of the work is employment of manual feature extraction techniques.

Authors Bardou et al. [2] compare two approaches for binary and multi-class classification of histopathological breast cancer images. The first approach extracts a set of handcrafted features which is encoded by bag of words model, locality constrained linear coding model and trained by support vector machines. The second approach involves designing a convolutional neural network for the problem. Data augmentation has been executed to understand its effects on performance of the models. The results indicate that convolutional neural networks outperform handcrafted feature-based classifier and data augmentation techniques increase the performance for a few classes while it has a deterring effect for other classes.

In [3], the paper proposes the application of fine-tuned pre-trained convolutional neural networks for classification. The tasks are:

1. Classification of various cancer types using Tissue Micro Array (TMA) samples into breast cancer, bladder cancer, lung cancer and lymphatic cancer.
2. Classification of breast cancer into various sub-classes using the BreaKHis [1] dataset.

ResNet V1 was used for the above tasks and the performance was compared with the Inception networks. ResNet versions performed better with an overall accuracy of 99.8% for cancer classification and 94.8% and 96.4% for breast cancer multi-class classification for benign and malignant types respectively.

2 Proposed Work

In this work, multi-class classification of breast cancer histopathological images on two different convolutional neural networks is presented along with a comparison of the performance. The dataset used in this paper, BreaKHis [1] consists of microscopic biopsy images of benign and malignant tumors. The benign and malignant sub-classes are eight in total, with four sub-classes under each type. The four benign types are adenosis, fibroadenoma, phyllodes tumor and tubular adenoma while the four malignant types are ductal carcinoma, lobular carcinoma, mucinous carcinoma and papillary carcinoma. The images are distributed across four different magnification levels - 40X, 100X, 200X and 400X (Fig. 1).

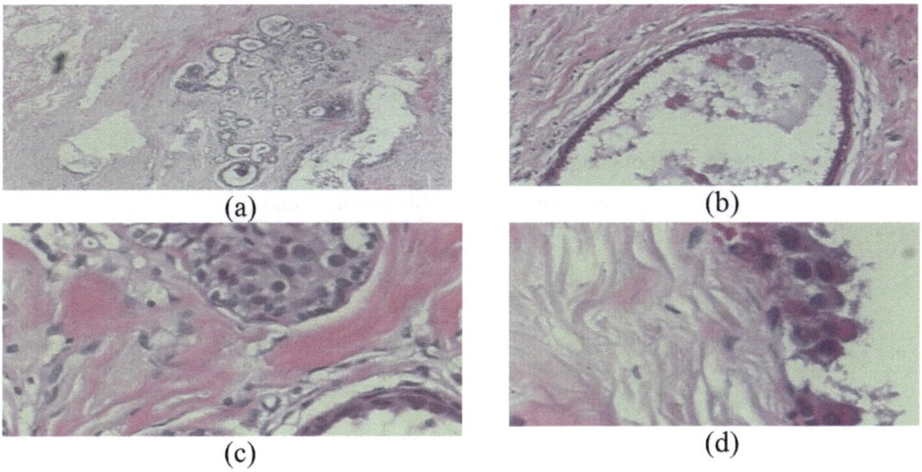

Fig. 1. Images of breast benign adenosis tumor as seen in different magnification factors- (a) 40X (b) 100X (c) 200X (d) 400X

2.1 Data Pre-processing

Data pre-processing included data augmentation and pre-processing steps for Inception v3 and Inception-Resnet-v2 networks.

Data Augmentation

Data Augmentation is a technique of creating new data from the existing data through a series of transformations. It prevents over-fitting of the model. Real-world conditions where the target application is employed may be in a different orientation, or lower image quality. This discrepancy is accounted for, by training the neural network with added data that has been modified by a series of transformations.

Transformations such as flipping, rotation at 90°, 180° and 270° and addition of white noise, along with a random combination of all three methods were performed for every image. The resultant data consists of 86,131 images, approximately ten-fold increase in the number of images. The pre-processing steps were then executed by conversion from PNG type to JPEG and JPEG decoding for training on Inception v3 due to the requirements of the pre-trained model and conversion into a TFRecord format for creating a tensor, according to the pre-requisites of the Inception-Resnet-v2 model. The augmented dataset was split randomly with 80% for training, 10% for testing and 10% for cross-validation. Both testing and validation sets were hold-out sets i.e. excluded from training (Fig. 2, Tables 1 and 2).

Table 1. Increase in number of images by Magnification factor after Augmentation

Magnification	Number of images added during augmentation
40X	19,631
100X	20,656
200X	19,876
400X	18,059
Total	78,222

Table 2. Image Distribution by Magnification Factor and Class after Augmentation

Magnification	Benign	Malignant	Total
40X	6875	14,751	21,626
100X	7084	15,653	22,737
200X	6852	15,037	21,889
400X	6468	13,411	19,879
Total	27,279	58,852	86,131

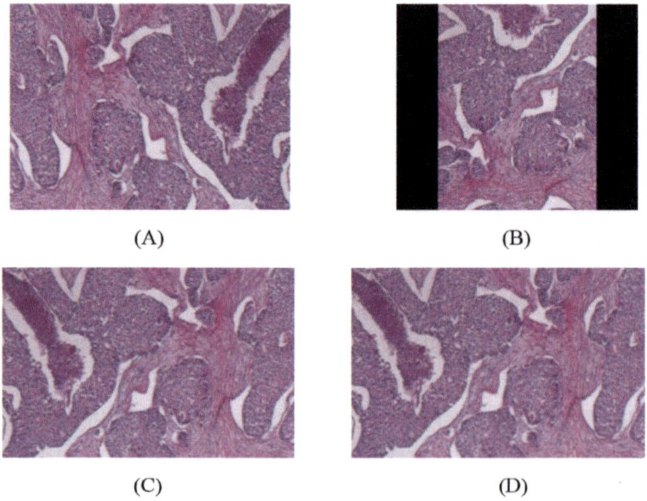

Fig. 2. Images of ductal carcinoma at 40X zoom after augmentation (a) Original image before augmentation (b) Rotation by 90° (c) Addition of white noise and flipping (d) Flipping of image

2.2 Inception and Inception-Resnet Architectures

Inception v3 architecture [4] was proposed by Szegedy et al. to scale up while utilizing the added computation as effectively as possible. The existing Inception architecture was modified by factoring the convolutions and making the filter banks wider instead of deeper to remove representational bottlenecks. The fully connected layer is batch

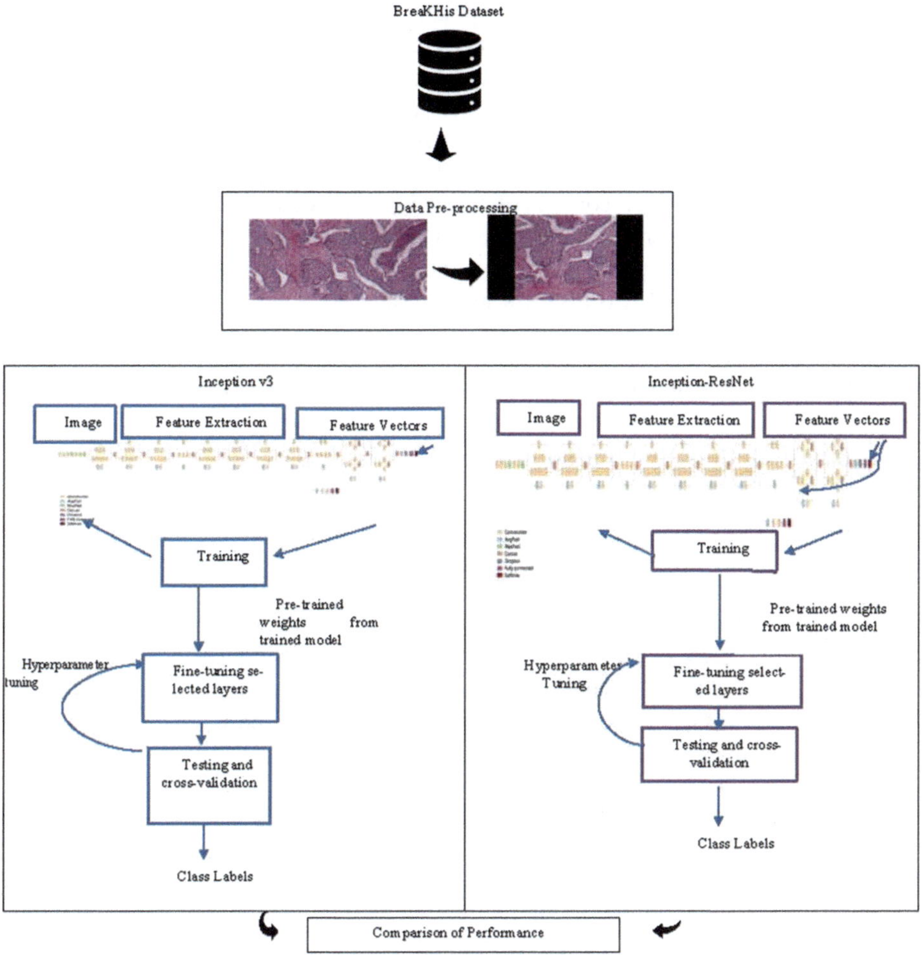

Fig. 3. Architecture of proposed work

normalized and label smoothing is used as a regularization component. The work of Szegedy et al. [5] proposes residual connections in the Inception architecture to understand the implications. The architectural changes made were introduction of a filter- expansion layer after each Inception block to compensate for the dimensionality reduction and batch normalization of only top layers but not the summations. The authors find that there is acceleration in the training time and an increase in performance when compared to pure Inception networks by a thin margin.

2.3 Transfer Learning

Transfer learning is the application of a previously learned source on a related target task. The weights from a trained Machine Learning model are transferred to the new model. Learning from scratch is often not practical due to the convergence problem and computational cost. Transfer learning reduces the training time, computational cost and leads to faster convergence. In this technique, the last layer of the Convolutional Neural Network is fine-tuned according to the current dataset [14]. It can be employed only if the target problem needs to be trained on a dataset that is smaller than the pre-trained model. In this work, the Inception and Inception-Resnet models are pre-trained on the ImageNet dataset [12] which consists of 1.2 M images which is larger than our dataset. Transfer learning was adopted by updating the weights of the final layer continuously. Fine-tuning of the hyper-parameters was employed during evaluation to increase the accuracy of the system [13] (Figs. 3, 4 and 5, Table 3).

Table 3. Experimental results distributed by magnification factor, class and convolutional neural network

ZoomLevel	Class	Number of iterations	Accuracy (%)	
			Inception v3	Inception Resnet v2
40X	Benign	4000	92.8	94.6
	Malignant	4000	85.9	89.8
100X	Benign	4500	91.3	93.4
	Malignant	4000	84.3	87.9
200X	Benign	4000	89.5	92.0
	Malignant	4000	77.6	79.2
400X	Benign	4500	86.5	90.3
	Malignant	5000	81.7	83.8

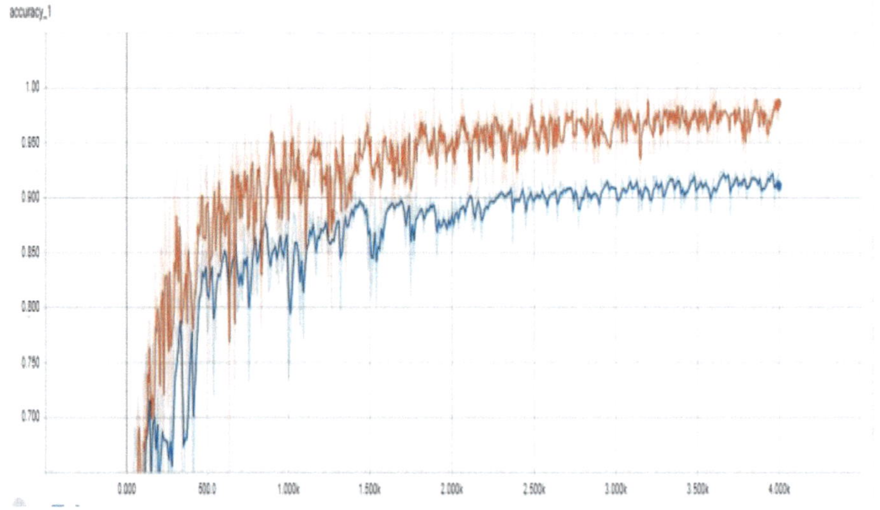

Fig. 4. Progress of training and validation accuracy over 4000 iterations in Inception v3

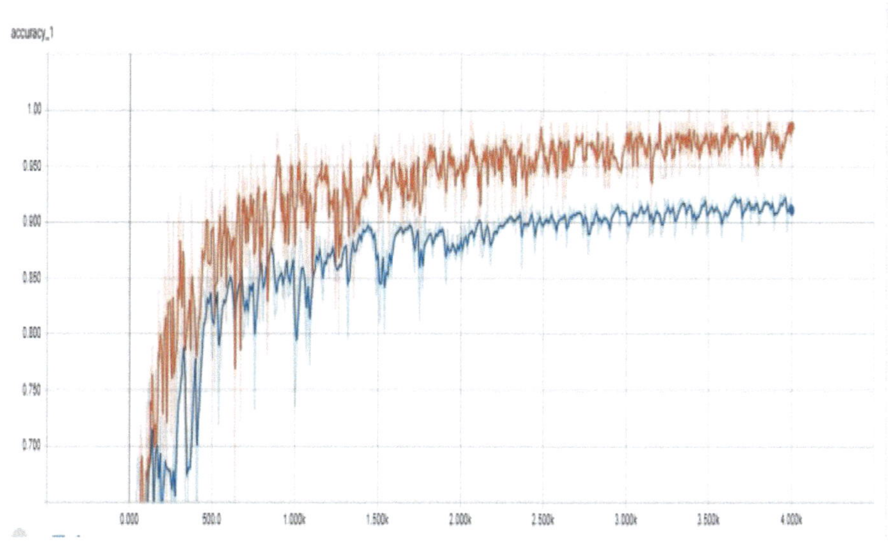

Fig. 5. Progress of cross-entropy loss function during re-training for training and validation accuracy for 4000 iterations

2.4 Experimental Results

The model recognizes the sub-class of every benign or malignant image in a given zoom level. The results are organized by zoom level for a given network. The performance metric used is accuracy which is defined as

$$Accuracy = \frac{TP + TN}{TP + TN + FP + FN} \quad (1)$$

3 Conclusion

This work compared the performance metric- accuracy of two convolutional neural networks for classification of histopathological breast cancer images into benign and malignant sub-classes. It was seen that Inception Resnet v2 performed better than Inception v3 by a small margin. The classification accuracies varied between zoom levels, following a pattern of decrease in accuracy with an increase in magnification factor. Classification of benign sub-classes performed better than malignant sub-classes due to a higher number of images for malignant data, which may have led to overfitting. The accuracies obtained by Inception v3 network were 92.8% and 85.9% for benign and malignant classes respectively, while Inception ResNet v2 classified with accuracies of 94.6% and 89.8% for benign and malignant classes.

References

1. Spanhol, F.A., Oliveira, L.S., Petitjean, C., Heutte, L.: A dataset for breast cancer histopathological image classification. IEEE Trans. Biomed. Eng. **63**(7), 1455–1462 (2016)
2. Bardou, D., Zhang, K., Ahmad, S.M.: Classification of breast cancer based on histology images using convolutional neural networks. IEEE Access **6**, 24680–24693 (2018)
3. Jannesari, M., et al.: Breast cancer histopathological image classification: a deep learning approach. In: IEEE International Conference on Bioinformatics and Biomedicine (BIBM), Madrid, Spain, pp. 2405–2412 (2018)
4. Szegedy, C., Vanhoucke, V., Ioffe, S., Shlens, J., Wojna, Z.: Rethinking the inception architecture for computer vision. In: 2016 IEEE Conference on Computer Vision and Pattern Recognition (CVPR), pp. 2818–2826 (2016)
5. Szegedy, C., Ioffe, S., Vanhoucke, V.: Inception-v4, inception-ResNet and the impact of residual connections on learning. In: AAAI Conference on Artificial Intelligence (2016)
6. World Health Organization: Breast Cancer Prevention and Control. https://www.who.int/cancer/detection/breastcancer/en/. Accessed 27 Feb 2019
7. International Agency for Research on Cancer: Latest global cancer data: Cancer burden rises to 18.1 million new cases and 9.6 million cancer deaths in 2018. https://www.who.int/cancer/PRGlobocanFinal.pdf. Accessed 27 Feb 2019
8. World Health Organization: Classification of Tumours, Tumours of the Breast and female genital organs. https://www.iarc.fr/wp-content/uploads/2018/07/BB4.pdf. Accessed 27 Feb 2019
9. Makki, J.: Diversity of breast carcinoma: histological subtypes and clinical relevance. Clin. Med. Insights: Pathol. **8**, 23–31 (2015)
10. Veta, M., Pluim, J.P.W., van Diest, P.J., Viergever, M.A.: Breast cancer histopathology image analysis: a review. IEEE Trans. Biomed. Eng. **61**(5), 1400–1411 (2014)
11. Araújo, T., Aresta, G., Castro, E., Rouco, J., Aguiar, P., Eloy, C., et al.: Classification of breast cancer histology images using convolutional neural networks (2017)
12. Krizhevsky, A., Sutskever, I., Hinton, G.E.: ImageNet classification with deep convolutional neural networks. Commun. ACM **60**(2), 84–90 (2012)
13. Lecun, Y., Bottou, L., Bengio, Y., Haffner, P.: Gradient-based learning applied to document recognition. Proc. IEEE **86**(11), 2278–2324 (1998)
14. Tajbakhsh, N., et al.: Convolutional neural networks for medical image analysis: full training or fine tuning? IEEE Trans. Med. Imaging **35**(5), 1299–1312 (2016)

Intelligent Parking Reservation System in Smart Cities

A. Dhanalakshmi[1], J. Brindha[2(✉)], R. J. Vijaya Saraswathi[2], and S. Sukambika[2]

[1] Sathyabama University, Chennai, Tamil Nadu, India
[2] Department of EIE, Panimalar Engineering College, Chennai, India
brindha.balajisrivatsan@gmail.com

Abstract. An Intelligent Parking Reservation System using RFID is proposed in this paper. In recent times, with the increase in population and the number of cars manufactured and with minimal and congested parking spaces available, people in metropolitan cities find it difficult to find space in parking lots and hence waste a considerable amount of fuel and time. In order to provide a solution to these kinds of flaws we have come up with a Smart Parking Reservation System which we believe will help sort out and organize parking lot problems in various destinations across the city.

Keywords: RFID (Radio Frequency Identification) · WSN (Wireless Sensor Network) · Ethernet shield · Ultrasonic sensors

1 Introduction

In recent times with the advancements in traffic management system, there arise a need in smart parking systems for reducing man power and enhancing automation.

The system branches out from the roots of IOT. The main objective is to reduce the amount of time and fuel that is wasted in parking lots. The most common method of finding a free parking space is to search, manually out of luck and experience [1–4]. Our system acquires data through RFID, Internet and wireless networks to obtain information about free parking spaces.

The present study proposes a work based on the internet using the IOT. The system uses an RFID tag and reader which is used to sense the incoming and outgoing cars in the parking space [9]. Ultrasonic sensors are fixed in each parking space which helps to identify whether the parking space is occupied or free. The information about the status of the parking space is transferred to the internet using a gateway. The user is able to view the status of the parking space displayed in the website. The system uses several innovative techniques such that the parking spaces can operate automatically. The wireless system is made possible using an Arduino.

D. J. Hemanth et al. (Eds.): COMET 2019, LNDECT 35, pp. 1163–1169, 2020.
https://doi.org/10.1007/978-3-030-32150-5_119

2 Existing System

In the existing system there are routers and cloud based servers that help to obtain the status of the parking spaces feed it to the sever using a wireless sensor network in order to provide the status of the parking lot as shown in Fig. 1. The status of the parking lot is displayed on a screen; the overall status of the parking system is also updated in real time. Depends up on the time the payment for parking can be calculated by the system and parking space has been allotted.

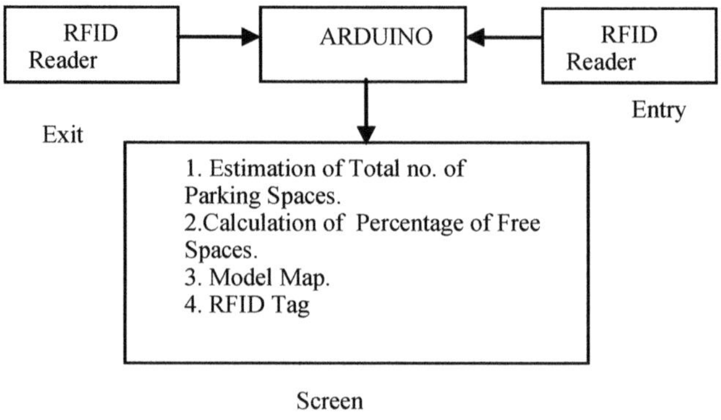

Fig. 1. Block diagram of existing system

2.1 Drawbacks of Existing System

Making use of cloud server and routers are not reliable, Vulnerability to attack, Limited control and flexibility, Cloud computing platform dependencies, Cloud computing costs, They are protocol dependent, Lack of security and privacy.

3 Proposed System

This paper proposes a smart parking reservation system as shown in Fig. 2 using RFID. RFID works on the principle of automatic identification of objects using electromagnetic fields and tracking the electronically-stored information of the tags that are attached to the objects. A transmitter and a receiver are present in a RFID tag or labels. The RFID has two functions: processing and storing information and transmitting and receiving a signal. Microchip performs processing and storing of information and antenna does transmission and reception of signals. The RFID tag and reader are attached to the car and the parking lot respectively. It helps in counting the number of incoming and outgoing cars respectively. There are ultrasonic sensors installed in the parking spaces to determine whether the parking space is empty or occupied. The data is transferred to the server using a gateway using Ethernet shield. The resultant status of the parking lot can be viewed using our own website.

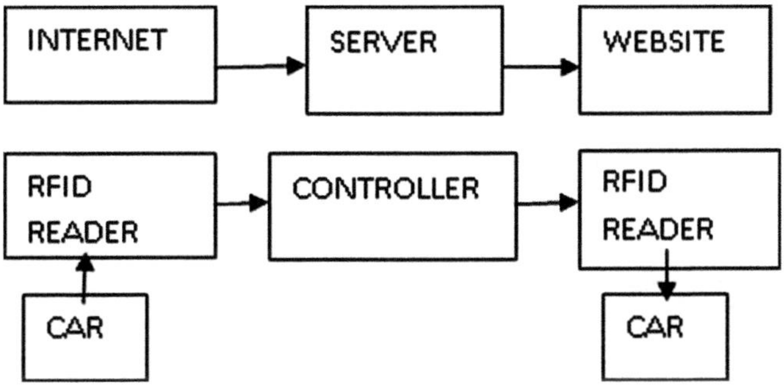

Fig. 2. Block diagram of proposed system

3.1 Advantages of Proposed System

The existing system has LCD display which is not so efficient in displaying the condition of the parking space. Using an ultrasonic sensor instead of routers makes the process more efficient. Avoiding Cloud servers makes the process more effective.

4 Hardware Architecture

4.1 RFID (Radio Frequency Identification)

Radio-frequency identification (RFID) is an automated system which is used to identify and track the tagged objects automatically. It uses electromagnetic field to accomplish this. The tags contain electronically-stored information. A RFID system contains a tag or label and a reader. A transmitter and a receiver (an antenna) is embedded in to the RFID tag. The RFID tag also contains a microchip to process (modulating and demodulating the radio frequency signal) and store information in a non volatile storage.

The DC signals present in the incident radar signals are collected by the RFID tag. The RFID tag also uses either a fixed or programmable logic device to transmit and collect data from the sensor.

4.2 Advantages of RFID

RFID has the ability to identify individual items which is a specific quality of it in identifying objects. Tags are immune to adverse conditions like environmental (dust, chemicals etc.,) and physical conditions (damage due to man-handling. More than one tag can be read simultaneously. RFID tags are made compatible with sensors (Fig. 3).

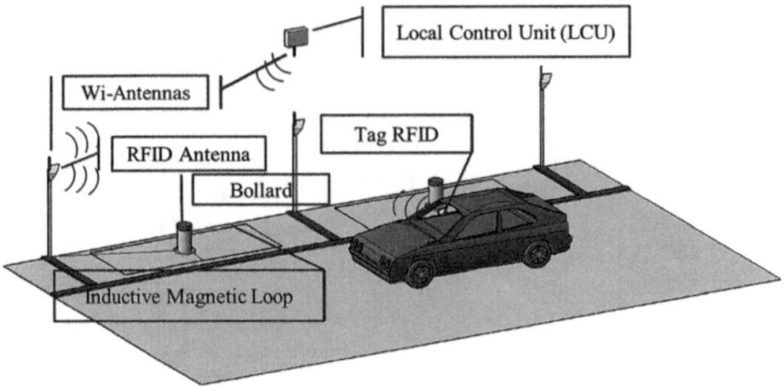

Fig. 3. Hardware components

4.3 Arduino UNO

Arduino is an open source platform. It has a microcontroller that can be physically programmed and an Integrated Development Environment (IDE) that can be interfaced to a computer. Arduino uses simplified version of C++ which makes it very popular among the users as it is very easy to understand and alter the programs according to the requirements. Arduino is compatible with many sensors. It is used for building interactive projects that can get data from the physical or digital world process it and provide the required solution. Shields are pre fabricated circuit boards that can be added to Arduino boards to provide additional functions like motor control, LCD screen control, wireless communications etc.

4.4 Ultrasonic Sensor

Ultra sonic censors are used to measure distance from an object using sound waves [5, 7]. The distanced is calculated by transmitting and receiving the reflected sound waves. The distance is calculated by calculating the time difference between the transmitted and received signal. The transmitted and received signal is required to maintain a particular frequency. There is the possibility of the bounced signal from the object to get deflected from the path before reaching the sensor because of the objects position. The size of the object also is a important feature in reflecting the signal. If the object is too small it may not be possible for the object to reflect the sound signal. Some objects like fabric cloth, floor carpet, etc. can absorb all of the sound waves, so that the sensor does not have the possibility of sensing the signal accurately. The above factors are to be considered while programming a device using an ultrasonic sensor. The advantages of ultrasonic sensor is high frequency, high sensitivity, high penetrating power and it can easily detect the external or deep objects.

5 Operation of RFID

The objects detected using RFID method need not be presented in the visibility of the RFID reader as the RFID technology uses Radio signal to detect the object. To accomplish the object recognition using RFID, it utilizes three components: an RFID tag or smart label, an RFID reader, and an antenna. RFID tags are used to send data to the RFID reader (also called an interrogator). The antenna is an integral part of the RFID tag which is used to send data to RFID reader.

The major advantage of RFID over the traditional barcode reader is that the barcode reader uses line-of-sight-technology. The barcode is seen by the barcode reader. That is the object has to orient before the barcode scanner for it to read the barcodes available over the object. But in RFID technology the object that has to be read may be kept externally or in a deeper area. For example, an RFID tag attached to an automobile or a pharmaceutical product during production can be used to track its progress through the assembly line or to hte ware house respectively. In the proposed paper the RFID tag and reader is used to estimate the number of incoming and outgoing cars.

6 Software Used

The software that is used to create the website for our project is basically notepad C++. The programming has been done using HTML, CSS and Bootstrap [6, 8].

A. HTML

Hypertext Markup Language is a standard markup web based language for developing web pages and browser applications. A HTML document contains many HTML tags and each tag has different content.

B. CSS

Cascading Style Sheet is a web page design language proposed to develop presentable web page. It uses HTML for writing codes.

C. BOOTSTRAP

Bootstrap is a tool for front-end development. Bootstrap is for developing web pages and browser applications. It consists of HTML and CSS-based interfaces to develop the front end design.

7 Experiments Results

The RFID counts the number of incoming and outgoing cars and the WSN consisting of the ultrasonic sensor interfaced to the server using an Ethernet shield provides the desired output in the website. The user can now access the website to view free parking spaces which saves time as well as fuel.

The website is designed in such a way that it is easy for the users to access the page and get the required information quickly and accurately as shown in Fig. 4.

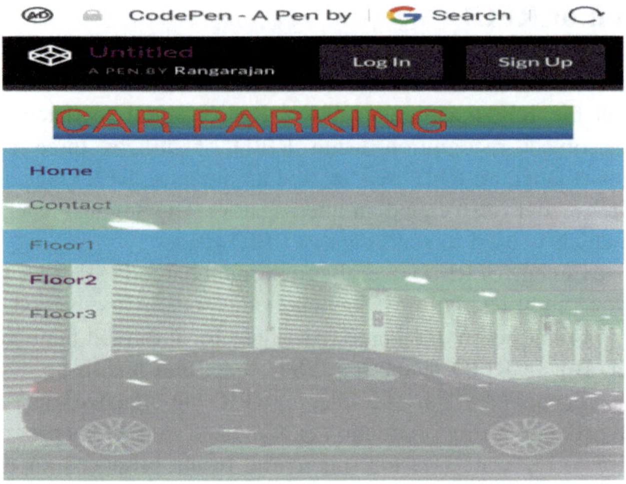

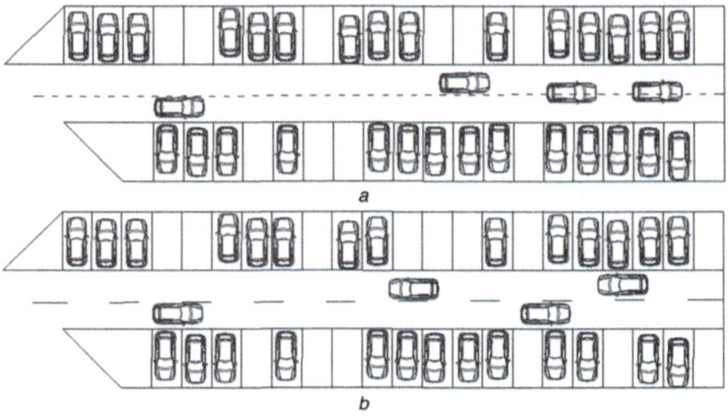

Fig. 4. Parking stalls along the carriage way

8 Conclusion

In this paper it is proposed that the smart parking reservation can be made using RFID. The RFID helps in counting the number of incoming and outgoing vehicles. The ultrasonic sensor helps in determining the empty spaces and status of the parking lot which can be viewed through a Website. The proposed system has been evaluated with various vehicles and it is found to be satisfactory.

References

1. Geng, Y., Cassandras, C.G.: A new 'smart parking' system based on optimal resource allocation and reservations. Proc. IEEE Trans. Intell. Transp. Syst. **14**(3), 1129–1139 (2013)
2. Zhao, X, Zhao, K., Hai, F.: An algorithm of parking planning for smart parking system. In: Proceedings of the World Congress on Intelligent Control and Automation (WCICA), pp. 4965–4969 (2015). https://doi.org/10.1109/wcica.2014.7053556

3. Mainetti, L. Palano, L., Patrono, L., Stefanizzi, M.L., Vergallo, R.: Integration of RFID and WSN technologies in a smart parking system. In: 22nd International Conference on Software, Telecommunications and Computer Networks (SoftCOM), pp. 104–110, September 2014
4. Hsu, C.W., Shih, M.H., Huang, H.Y., Shiue, Y.C., Huang, S.C.: Verification of smart guiding system to search for parking space via DSRC communication. In: 12th International Conference on ITS Telecommunications (2012)
5. Barone, R.E., Giuffrè, T., Siniscalchi, S.M., Morgano, M.A., Tesoriere, G.: Architecture for parking management in smart cities. IET Intell. Transp. Syst. **8**, 1–8 (2013)
6. Hainalkar, G.N., Vanjale, M.S.: Smart parking system with pre & post reservation billing and traffic app. In: 2017 International Conference on Intelligent Computing and Control Systems (ICICCS) (2017)
7. Int. J. Pure Appl. Math. **114**(7), 165–174 (2017). ISSN 1311-8080 (printed version); ISSN 1314-3395 (on-line version). http://www.ijpam.eu
8. Pham, T.N., Tsai, M.-F., Nguyen, D.B., Dow, C.-R., Deng, D.-J.: A cloud-based smart-parking system based on internet-of-things technologies. IEEE Access **3**, 1581–1591 (2015). Accepted August 16, 2015, date of publication September 9, 2015, date of current version September 23, 2015
9. Karbab, E.I.M., Djenouri, D., Boulkaboul, S., Bagula, A.: CERIST research center, Algiers, Algeria University of the Western Cape, Cape town, South Africa. Car Park Management with Networked Wireless Sensors and Active RFID. IEEE (2015). 978-1-4799-8802-0/15 ©2015
10. Shaikh, F.I., Jadhav, P.N., Bandarkar, S.P., Kulkarni, O.P., Shardoor, N.B.: Smart parking system based on embedded system and sensor network. Int. J. Comput. Appl. (0975 – 8887) **140**(12), 45–51 (2016)

Fulcrum: Cognitive Therapy System for Stress Relief by Emotional Perception Using DNN

Ruben Sam Mathews(✉), A. Neela Maadhuree, R. Raghin Justus, K. Vishnu, and C. R. Rene Robin

Department of Computer Science and Engineering, Jerusalem College of Engineering, Chennai, India
mathewsiruben@gmail.com

Abstract. The product improves the emotional state of a person. When a negative emotion is likely deduced the system uses predefined methods to assist the individual to enhance their emotional state to attain stability. This helps in enabling an individual recognize the stress in its earlier stages so that they could take proper care towards it. It also acts as an alert system to caution the adverse levels. This is implemented by considering 3 parameters such as facial expressions, GSR and HRV of the user.

Keywords: GSR · HRV · Emotion recognition · Facial recognition · Alert · Stress · Convolution neural network

1 Introduction

WHO [World Health Organization] stated that "stress has become a worldwide epidemic", it is observed in all age groups and genders? One of the crucial issues in today's trend of lifestyle is the emotional instability. Emotional disturbances can result from external factors (e.g., events, situations, environment) or internal factors (e.g., expectations, attitudes, feelings). Common causes for it include physical causes, such as illness or injury, and mental (psychological) causes, such as anxiety or fear. Ongoing, chronic stress causes many health problems, such as depression, anxiety, and personality disorders and various cardiovascular disease. It can also be fatal by leading to suicidal scenarios, in the recent times the rate of suicide due to stress has increased enormously. In the busy lifestyle individuals do not tend to show more attention to minute changes of their behaviour which is eventually hazardous to them. The current generations mostly neglect the minute warnings their body and behavioural changes provide them and encapsulate it within their work. But detection of something like this in the initial stages tends to be more beneficial than taking it to be cured in the advanced stages.

2 Ideology

Our system helps monitoring personnel of their emotional traits and intimating them on any abnormal behaviour detection. Thus, it provides the required external provocation for the person to realize their state. The system also extends to produce initial level of

D. J. Hemanth et al. (Eds.): COMET 2019, LNDECT 35, pp. 1170–1178, 2020.
https://doi.org/10.1007/978-3-030-32150-5_120

base therapy by suitable methods such that the individual is given aid to stabilize their state. The initial stages are tried to be improved by providing the user with this therapy which can also be customizable according to the user individual interest. In adverse scenario it functions as an alert system that warns the user that they need advanced assistance. The system takes as input the parameters proportional to emotions such as facial recognition, cardiac functionality and galvanic skin responses. We utilize multiple input parameters to obtain higher efficiency.

3 Scope

It can be utilized by all age groups and genders. The device can be easily adopted since technology has become a part of everyday lifestyle. The device can be used at various environments such as home, office, etc. based on user's convenience and need. It can be extended to other parameters such as anger management, depression etc. The system enables a person to analyze his state of behaviour which otherwise he might have not accepted and provide necessary steps to overcome it in a more personal and confidential way. It can also be extended in applicability such as applying it at shopping markets for the product manufactures to obtain a first-hand data on the response of user towards their product and any other such customization usage.

4 Related Work

In [1] they determined the stress effect in the long term and the short term utilizing a rodent model. Experimental results indicate that the HRV time domain features generally decrease under long-term stress, and the HRV frequency domain features have substantially significant differences under short-term stress [1]. Moreover they were able to witness an accuracy of 93.11% with an optimal HRV setup in the SVM classifier. Their results supported the statement that the optimal HRV feautres can effectively determine stress. Thus they explained the usage of this theory in terms of implementing in mobile health care system thereby analysing and curbing the effects of stress.

In [2] they have designed and built a stress sensor based on Galvanic Skin Response (GSR), and controlled by ZigBee. They implemented the system by observing an individual under various scenario and how their GSR reflects. To record the results they utilized 16 adults and recorded their values under various situations. Their system was able to successfully differentiate the states among the individuals with 76.56% accuracy. Further development of a mathematical model in this aspect was proposed in this paper.

In [3] They utilized the facial muscle as a feature to determine the expression associated. They adopted the traditional methods to obtain the original face area from the entire picture and from the extracted area they determined the specific facial points to obtain facial line data. Using all these data they developed a vector which was associated with NN to determine the expression associated with that data. In this particular study they had utilized the TFEID database to test their system. The results determined that their system was 97.4% efficient towards facial recognition.

In [4] they have tried to observe the variation of the heart rate variability and morphological variability with respect to the stress levels. The characteristics observed by them revealed the relation among both and hence enabled to develop a mechanism to determine the mental stress state based on the analysis of both HRV and MV of ECG. A number of HRV measures were investigated, both in time domain and frequency domain [4].

In [5] they describe the understanding of the collaboration of emotion with human machine communication. The various advances towards this and the possibilities are discussed in the paper. The emotion can be represented through various forms and hence a brief discussion on it along with the methodologies are stated. The paper mainly concentrates on proposing real time implementation of emotion recognition system [5].

In [6] they discuss on affective computing. It deals with a robot determining the emotion of a human by using various emotion expressing traits mostly focused here on facial recognition. They utilize neural network concept to help analyse and classify these emotions. Based on the results the robots are meant to adapt to the emotion of the human towards their task.

In [7] it mainly exhibits how a very large-scale dataset can be assembled by a combination of automation and human and also, they traverse through the complexities of deep network training and face recognition to present methods and procedures to achieve efficient results on standard face benchmarks [7].

From the previous paper we determined that using GSR we can determine the categories of emotional state in correspondence to activities. From the previous papers we also inferred that the HRV are highly related to the stress level of an individual. However, the above systems do exhibit drawbacks such as they are manual processed system and individual patterns need to be fed thus do not scale well. Whereas our proposed system is towards the automatic pattern generation and processing which is achieved by utilizing concepts of Artificial Intelligence. Thus, it becomes more scalable and efficient than the earlier designs. Another issue is in the earlier system they have proposed with only 1 parameter but the efficiency of accuracy with 1 parameter proportionality with stress is low. Our system hence aims at collaborating various factors that influence stress to achieve a higher rate of accuracy.

5 Proposed Framework

The input for the proposed system is the emotion of a person and HRV. The process involves mounting a special camera in the required environment. The camertracks the face of a person and using artificial intelligence we determine the facial expression leading to the observation of a person. The system also makes use of the GSR [9] and ECG sensor to determine the stress level. The proposed system on the analysis of negative emotion plays the musicor video or even prompt to play game based on the choice of the user. It observes the cumulative result and when it crosses a predefined threshold a suitable therapy is provided. It also behaves as an indicator and provides the warning during excessive states. The proposed system enables oneself to analyze their state of being. Thus, it can be considered as a product that analyzes the physiological behavior to determine the psychological aspect of a person. The GSR (galvanic skin response) are correlated with the stress level of the person. The GSR can be used to identify the stress pattern. The ECG signals is used to determine the Heart Rate Variability. The ECG signals can be used to analyze three classes of stress with more than 80% accuracy. With the combination of ECG and GSR and along with emotion recognition the stress level can be monitored and appropriate therapy can be suggested based on predefined pattern. The proposed system can be analyzed on the basis of the results of human emotion detection efficiency. The analysis of how efficiently therapies are provided can be calculated. The product can also be measured on the score of how well the product has analyzed a negative emotion and how much of a relatable therapy has been suggested thus how well the user has been benefited from it.

6 Fulcrum System Design

The system is basically distinguished as a three-module set as shown in Fig. 1. Where the first module consists of the facial recognition phase which utilizes the image processing algorithm to identify a preloaded face and extract the emotional features from it.

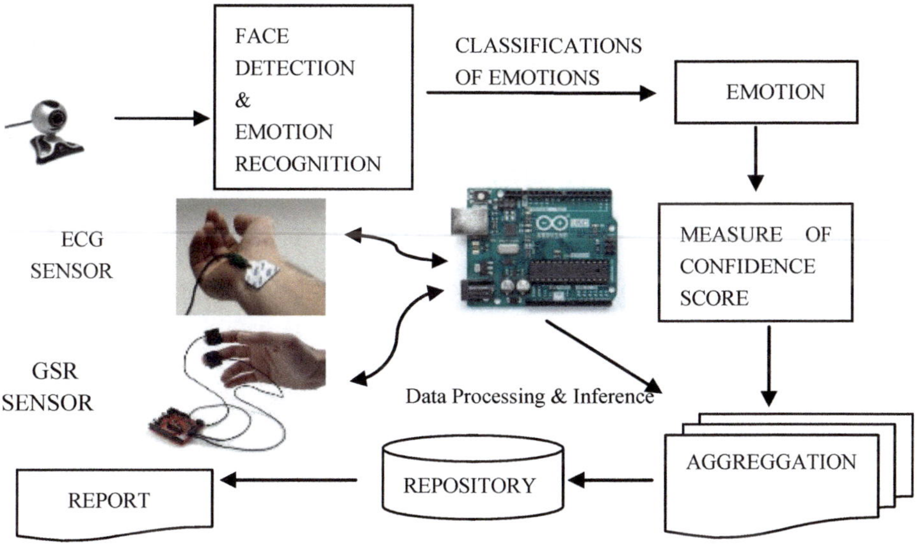

Fig. 1. Fulcrum Architecture Diagram

The second module consists of the HRV sensor which is used to determine the heart beat rate and the third module consists of the GSR sensor which determines the intensity of emotion. The data from these sensors are collected in a raspberry pi and the collected data is send to computer for evaluating and a inference is generated based on the resultant comparison with the emotional state matrix. All the results are stored in the repository and used for extensive analysis to trigger warnings to the user. For 1000 samples the HRV is calculated, Here Pan-Thompkins algorithm is utilized to calculate the R peak and the HRV is calculated by using RMSSD (Root Mean square of successive difference). The average value of GSR of 1000 samples is calculated. Then the GSR, HRV and RR values are used to train the model in case of Record option else in case of checking stress status these data are used to evaluate the model, where the model is specific to user. For each user a model is trained using the system and then that trained model is further used to evaluate the stress status. Here we make use of the MLP Classifier to classify the stress status of the user. The model of the user can be improved by recording the data at various situations.

7 System Implementation

In the implementation we created a frontend view which is used to obtain the input from the user. It involves collecting the name of the user so as to distinguish their report separately and also select a choice from the list of options available such as register for new user to register with the system, take snap for detecting the emotion of the person, record tab for the initial learning phase to improve the system performance and view report tab to enable the user to analyze their prior state of emotions through their logged reports. This front end view is as shown in Fig. 2.

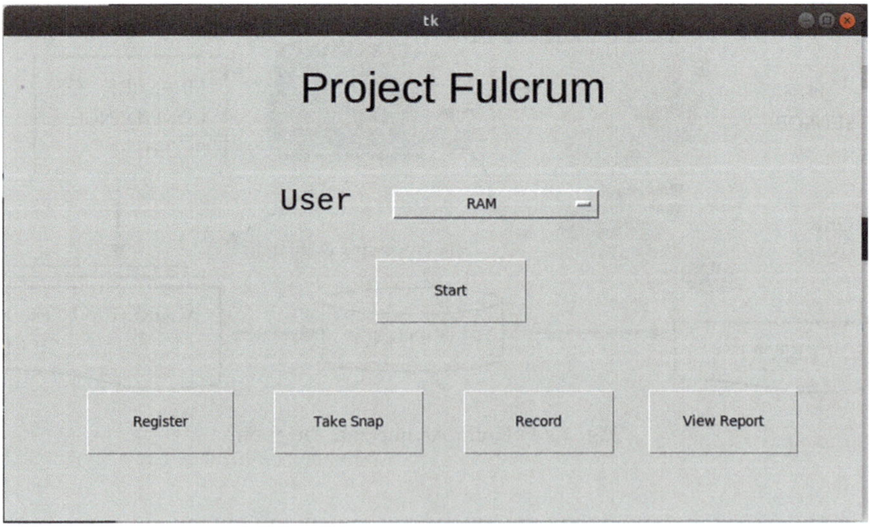

Fig. 2. Fulcrum User Interface/home screen

When a user selects the register tab it leads to the next window which asks for the user name wherein the user will input their name. Here since it is a prototype, we sufficed with enter name and distinguished the user reports by their names but in a larger scenario additional details could also be collected from the user. For example, in a office scenario if the system is being implemented by the organization they can collect the employee id and other credentials to have a record as well uniquely distinguish reports from a huge dataset. For the prototype once the user enters the name, they need to click the done button which will register them into the system. The register implies that a new separate folder has been created for the user and henceforth any data associated with the user will be logged in the corresponding folder (Fig. 3).

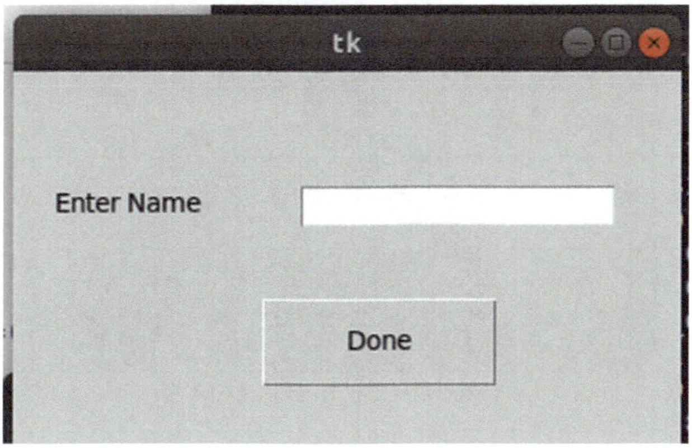

Fig. 3. User Registration Process - User Name

In Fig. 4 we observe a user registering face in the system. The blue boundary box is developed which is based on the algorithm to detect the facial structure and in the training phase that is on the click of record button the user gets to register various

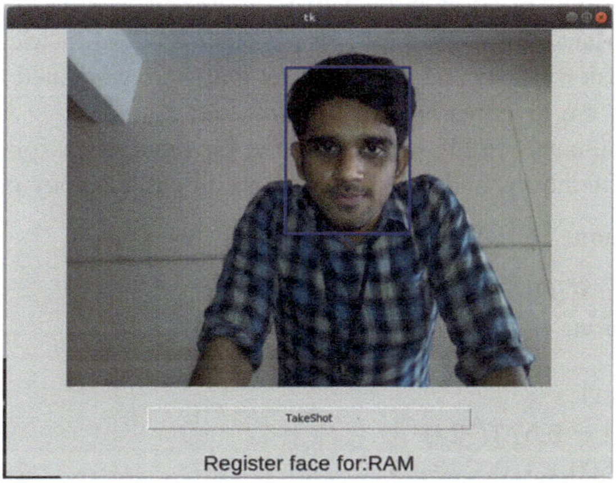

Fig. 4. User Registration Process - Facial registration

expressions to the system which are later used for efficiently analyzing the expression of the user.

In Fig. 5 it is seen that during the real time use of the system the user expression was recognized and the GSR and HRV values were noted and registered in the system report. Later when the user visits view report and observes the kind of emotion they have gone through. In this scenario we see that the user was happy at that instant.

Fig. 5. Emotion and Facial Recognition of user

8 Performance Analysis

The stress recognition performance analysis is carried over based on individual test case. Confusion Matrix forms the basis for the other types of metrics. Some of heatmap view of the confusion matrix and their data set that were generated for our project over various test cases are given as below,

The confusion matrix in Fig. 6 is generated for the dataset, for which the model is trained with 500 iterations and has produced an output accuracy of 71%.

Model Information

Classifier/Solver: *SGD*
***Activation Function**: relu*
***alpha**=0.0001*
Total Iterations 500
Loss at iteration 1 = 0.67077243
Loss at iteration 500 = 0.00240355
Accuracy Score

0.7142857142857143
Confusion Matrix
[[3 1]
[1 2]]

Similarly the confusion matrix in Fig. 7 is generated for another data set and the accuracy was determined around 67%.

Classifier/Solver: *Adam*
Accuracy Score
0.6714285714285714
Confusion Matrix
[[4 1]
[2 0]]

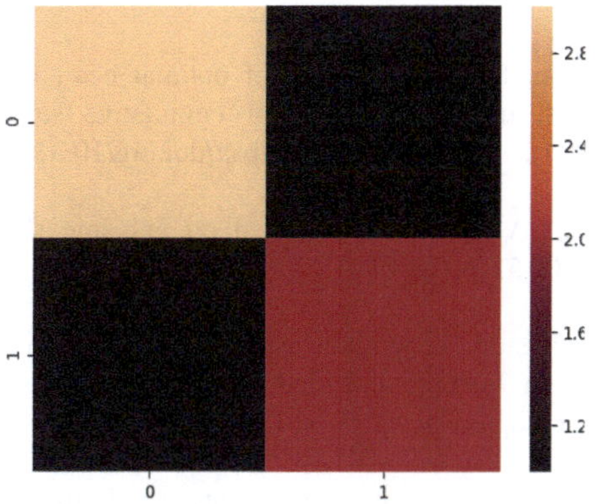

Fig. 6. Confusion matrix (71% accuracy)

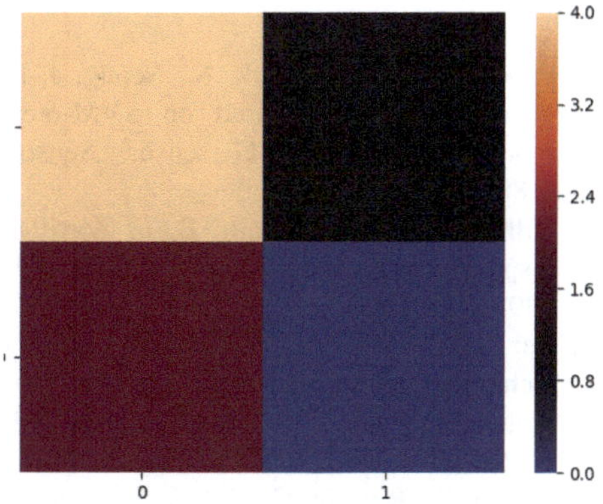

Fig. 7. Confusion matrix (63% accuracy)

For emotion recognition analysis the model was trained with approximately 30000 images and later tested with approximately 10000 images. The accuracy reached the stability around 63%.

9 Conclusion

Deep Neural Network has been applied for determining the emotional state of the human, the training dataset included about 28,000 images of Fear, Sad, Anger, Surprise, Disgust, Happy. The system also employs ECG sensor to calculate the HRV and GSR sensor for determining the Galvanic Skin Response. These data from the sensors along with the emotional state are used for determining stress level of an individual. Thus, the system provides a comprehensive way of identifying the emotional state of a person. The system also is more user friendly as most of the computation are automated and requires less user interactivity. With the system the user can easily identify on higher stress levels and take the required actions which otherwise he/she would have not even acknowledged leading to adverse stages. Thus, we believe the system will play a key role in alerting the individual of the stress and on its widespread use will reduce the global stress-based issues.

10 Future Work

In future the system can be extended to various measures to make it even more accurate. It can be deployed over a group of users. Organizations can use it as a medium to observe the emotional state of their employees and take corrective measures. In educational institutions it can be used to observe the stress each student is facing and in adverse stages can provide counselling sessions to them. It can also be made more compact and sophisticated.

References

1. Park, D., Lee, M., Park, S., Seong, J.-K., Youn, I.: Determination of optimal heart rate variability features based on SVM-recursive feature elimination for cumulative stress monitoring using ECG sensor. Sensors **18**(7), 2387 (2018). https://doi.org/10.3390/s18072387
2. Villarejo, M.V., Zapirain, B.G., Zorrilla, A.M.: A stress sensor based on galvanic skin response (GSR) controlled by ZigBee. Sensors **12**(5), 6075–6101 (2012). https://doi.org/10.3390/s120506075
3. Lee, H.-C., Wu, C.-Y., Lin, T.-M.: Facial expression recognition using image processing techniques and neural networks. In: Advances in Intelligent Systems and Applications - Volume 2 Smart Innovation, Systems and Technologies, pp. 259–67 (2013). https://doi.org/10.1007/978-3-642-35473-1_26
4. Costin, R., Rotariu, C., Pasarica, A.: Mental stress detection using heart rate variability and morphologic variability of EEG signals. In: 2012 International Conference and Exposition on Electrical and Power Engineering (2012). https://doi.org/10.1109/icepe.2012.6463870
5. Varghese, A.A., Cherian, J.P., Kizhakkethottam, J.J.: Overview on Emotion Recognition System. In: 2015 International Conference on Soft-Computing and Networks Security (ICSNS) (2015). https://doi.org/10.1109/icsns.2015.7292443
6. Correa, E., Jonker, A., Ozo, M., Stolk, R.: Emotion recognition using deep convolutional neural networks, 30 June 2016
7. Parkhi, O.M., Vedaldi, A., Zisserman, A.: Deep face recognition. In: Proceedings of the British Machine Vision Conference (2015). https://doi.org/10.5244/c.29.41
8. Ciabattoni, L., Ferracuti, F., Longhi, S., Pepa, L., Romeo, L., Verdini, F.: Real-time mental stress detection based on smartwatch. In: 2017 IEEE International Conference on Consumer Electronics (ICCE) (2017). https://doi.org/10.1109/icce.2017.7889247
9. Bakker, J., Pechenizkiy, M., Sidorova, N.: Detection of stress patterns from GSR sensor data. Department of Computer Science Eindhoven University of Technology

Contextual Emotion Detection in Text Using Ensemble Learning

S. Angel Deborah(✉), S. Rajalakshmi(✉), S. Milton Rajendram(✉), and T. T. Mirnalinee(✉)

Department of Computer Science and Engineering, SSN College of Engineering, Chennai 603110, Tamil Nadu, India
{angeldeborahs, rajalakshmis, miltonrs, mirnalineett}@ssn.edu.in

Abstract. As human beings, it is hard to interpret the presence of emotions such as sadness or disgust in a sentence without the context, and the same ambiguity exists for machines also. Emotion detection from facial expressions and voice modulation easier than emotion detection from text. Contextual emotion detection from text is a challenging problem in text mining. Contextual emotion detection is gaining importance, as people these days are communicating mainly through text messages, to provide emotionally aware responses to the users. This work demonstrates ensemble learning to detect emotions present in a sentence. Ensemble models like Random Forest, Adaboost and Gradient Boosting have been used to detect emotions. Out of the three models, it has been found that Gradient Boosting Classifiers predicts the emotions better than the other two classifiers.

Keywords: Sentiment analysis · Emotion detection · Machine learning techniques · Ensemble methods · Text mining

1 Introduction

Emotions and its various impacts on day to day situations have been explored in different areas like psychology, computational linguistics, social media and communication. Performance of human being depends upon his emotional behavior. In the latest years, different interactive cognitive systems are being used in different places to communicate with persons inside or outside. Existing emotion detection systems are smart enough to behave like humans by expressing their emotions and recognizing emotions of opponent people. But the limitation of existing smart system is that it can only recognize emotions based on predefined keywords or semantics. It is difficult to analyze the context in which those keywords are used. Emotion detection in text is focused at analyzing how people express their emotions through text. It is very essential to understand and analyze how text express particular emotions. Emotions can be categorized as surprise, happiness, sadness, fear, anger and disgust, etc. In many cases emotions are hidden behind the text, although the text may have vibrant representation of emotions present in it. Extracting emotions from text based on keyword spotting, text mining, machine learning, semantics based, and corpus-based methods is

D. J. Hemanth et al. (Eds.): COMET 2019, LNDECT 35, pp. 1179–1186, 2020.
https://doi.org/10.1007/978-3-030-32150-5_121

an active research area. Although much research has been done, the major challenge of the current systems is, it still lacks the ability to learn and infer the emotions from text based on contextual information [1, 2].

2 Literature Survey

Emotion analysis on social media is attracting increasingly more research attention from industry and academia. Commercial applications such as product recommendation, online retailing, and marketing are turning their interests from traditional sentiment analysis to emotion analysis. There are several machine learning techniques that can be used for emotion intensity prediction. Some of the approaches include Artificial Neural Network (ANN) [2, 3], Random Forests, Support Vector Machine (SVM) [4], Naive Bayes (NB), Multi-Kernel Gaussian Process (MKGP) [5, 6], and Deep Learning (DL) [7–9]. Although, the broad topic of emotion has been studied in different fields for decades, study of contextual emotion detection in text is in its early stages. A semantic network for contextual emotion detection was developed by Chuang, but the size of the corpus is too small to support the results.

3 Ensemble Learning

Model ensemble refers to a combination of models. It is one of the powerful techniques in machine learning and has been found to outperform other methods. Yet this model comes at the cost of increased model complexity.

3.1 Random Forest Classifier

Random forests (RF) is an ensemble supervised learning algorithm. It is a very flexible and easy to use algorithm used for classification and regression. Random forests splits the given dataset into random subset of data samples. Decision tree is built on each random data sample separately. The prediction from each tree is obtained, and the best solution is selected by voting or averaging. It also shows the importance of feature selection. The robustness of the random forest increases as the number of trees increases. The procedure to create a Random Forest model is depicted in Algorithm 1. The random forest algorithm is unbiased because there are multiple trees and each tree is trained on a subset of data. The random forest algorithm relies on the strength of "the data sample"; therefore, the overall biasedness of the algorithm is decreased. When a new example is introduced in the dataset, it is selected by some trees, hence it will affect only those trees and the overall RF is not affected to that extent. It works well for both categorical and numerical features. It also performs well even when there are missing values in the given dataset.

Algorithm 1: Random Forest X, Y, *x* – Builds forest of decision trees on each bootstrap samples separately and ensemble them. [11]

Input: Data set X; Number of trees Y*;* feature space x.
Output: Majority voting based predictions from ensemble of decision tree models

for *i* = *1 to Y* **do**
    Create a bootstrap sample X_i from *X* by selecting sample points |X| with replacement
    Choose x features at random and reduce the dimensionality of X_i
    Train a tree model $_{Zi}$ on $_{Xi}$ without pruning
end
return { Z_i | 1 ≤ *i* ≤ Y }

3.2 AdaBoost Classifier

AdaBoost (AB) proposed by Freund and Schapire in 1996 referred to as "Adaptive Boosting" is the first practical boosting algorithm. It focuses on classification problems and aims to convert a set of weak classifiers into a strong one. Each weak learner is trained using a sample set of training data. Each sample has a weight, and the weights of all samples are adjusted iteratively. AdaBoost iteratively trains weak learners and calculates a weight for each one, and this weight represents the robustness of the weak learner. The AdaBoost algorithm is outlined in Algorithm 2:

Algorithm 2: AdaBoost (Adaptive Boosting) Classifier [10]

Input: Training Data set *X*.
Output: Learnt AdaBoost model

Initialize the values for parameters
for *y* = *1 to Y* **do**
    Select a sample D_y from *X* using w^t distribution
    Train the weak learner (h_y) using D_y to obtain a minimum error $\in_y$.
    while $\in_y >= 0.5$ **do**
        Reinitialize the classifier weights to $w_j = 1/N$, $j = 1, 2, \ldots N$
        Recalculate the error $\in_y$
    end
    Compute the weight of each weak learner (α_y) and Update the weights of the training data
end
return cumulative $H_{final} = \text{sign}(\sum_{y=1 \text{ to } Y} \alpha_y h_y(x))$ }

3.3 Gradient Boost Classifier

In Gradient Boosting (GB), new models are fit consecutively by the learning again on misclassified examples to predict a more accurate estimate of the response variable [12, 14]. The new base-learners are constructed with the aim to maximally correlate the data with the negative gradient of the loss function, related to entire ensemble model. If the error function is the classic squared-error loss, the learning procedure would result in consecutive error-fitting. We can randomly choose the loss function and the base learner models based on the requirement. The solution to the parameter estimates is difficult to find, when we provide some specific loss function $\Psi(y, h)$ and a custom base-learner $B(x, \theta)$. To overcome this problem, we can select a new custom base-learner $B(x, \theta_t)$ that is parallel to the negative gradient $\{g_t(x_i)\}_{i=1 \text{ to } N}$ down the observed data using Eq. 1.

$$g_t(x) = E_y\left[\frac{\partial\Psi(y, h(x))}{\partial h(x)}\bigg| x\right] \tag{1}$$

We can select the new function boost increment to be the most correlated value with negative gradient $g_t(x)$. This will ease the hard optimization task by replacing with least-squares minimization. The working of whole algorithm with mathematical solution will rely on the selection of loss function $\Psi(y, h)$ and a custom base-learner $B(x, \theta)$. The Algorithm 3 gives the details of GB classifier.

Algorithm 3: Gradient Boosting Classifier [10]
Input: Data points $(x, y)_{i=1 \text{ to } N}$; N iterations ; loss-function$\Psi(y, h)$;
base-learner model $B(x, \theta)$.
Output: Learnt Gradient Boosting model
1 Initialize h'_0 with a constant value.
2 **for** j = 1 to N **do**
3 Find the negative gradient $g_j(x)$ using the loss function
4 Train a new base-learner function $B(x, \theta_j)$
5 Compute the optimal gradient descent step-size ρ_j:
$\rho_j = \text{argmin}_\rho \sum_{i=1 \text{ to } N} \Psi[y_i, h_{j-1}(x_i) + \rho B(x_i, \theta_j)]$
6 Revise the function estimate: $h'_j \leftarrow h'_{j-1} + \rho B(x, \theta_j)$
7 **end**
8 return Learnt model

4 Dataset

Semeval 2019 – Task 3: EmoContext's dataset was taken for experimental purpose. The dataset consists of 30160 training samples and 2755 development samples. It consists of conversation id, three turns of conversed sentence and contextual emotions like "happy, sad, angry, and others" are present as labels. Out of the 30160 train samples, 6000 samples were used for building the model. It is observed that the dataset has more samples with class label "others".

5 System Overview

Data extraction, pre-processing, rule-based feature selection, and feature vector generation using Bag of Word and learning ensemble models are the modules of the system. The algorithm for preprocessing of the data is outlined in Algorithm 4. Algorithm 5 lists out rule based feature selection and feature vector generation.

Algorithm 4: Data extraction and Preprocessing
Input: Input dataset.
Output: Tokenized words and their parts of speech

1 Separate labels and sentences.
2 Perform tokenization using *word_tokenize*, the function for tokenizing in the NLTK toolkit.
3 Perform Parts of Speech tagging using *pos_tag* function from the NLTK toolkit.
4 Return the tokenized words and their parts of speech as inputs to rule based feature selection.

Algorithm 5: Rule based feature selection and feature vector generation
Input: Tokenized words and their parts of speech.
Output: Feature vector.

1 **for** *each of the tokenized words, falling under one of the categories listed in Table 1* **do**
2 Lemmatize the word using *WordNet Lemmatizer* from the NLTK toolkit.
3 Insert the lemmatized word into the dictionary.
4 Represent each sentence as a feature vector using one-hot encoding by looking up the dictionary.
5 **end**
6 Return the feature vector generated as the input to build the model.

Table 1. Parts of speech categories.

Abbreviation	Parts of speech
VB	verb, base form
VBZ	verb, 3rd person sing, present
VBP	verb, non 3rd sing, present
VBD	verb, past tense
VBG	verb, gerund/present participle
VBN	verb, past participle
JJ	adjective
JJR	adjective, comparative
JJS	adjective, superlative
RB	adverb, very
RBR	adverb, comparative
RBS	adverb, superlative

The output obtained in Algorithm 5 is given as inputs to RF, AB and GB classifiers.

6 Performance Evaluation

We evaluated the system using three different Ensemble models. The results obtained using RF, AB and GB classifiers are tabulated in Table 2 which shows accuracy and micro average precision, recall and F1-score. From Table 2, we can infer that Gradient Boosting classifier predicts the contextual emotion better than the other two.

Table 2. Performance comparison

Model	Accuracy	Precision	Recall	F1-score
Random Forest	79.46%	0.79	0.79	0.79
AdaBoost	84.94%	0.85	0.85	0.85
Gradient Boosting	**85.98%**	**0.86**	**0.86**	**0.86**

The metrics are defined from Eqs. 2 and 3

$$Accuracy = \frac{1}{|T|}\sum\nolimits_{t \in T} \frac{|Y_t' \cap Y_t|}{|Y_t' \cup Y_t|} \quad (2)$$

where Y_t is the set of the gold labels for tweet t, Y_t' is the set of the predicted label for tweet t, and T is the set of tweets.

$$F_{micro} = (2 \times P_{micro} \times R_{micro})/(P_{micro} \times R_{micro}) \quad (3)$$

where F_{micro} is micro-averaged F_1 score, P_{micro} is micro-averaged precision and R_{micro} is micro-averaged recall.

The confusion matrix obtained using three different models is given in Tables 3, 4 and 5.

Table 3. Confusion matrix for Random forest

	Others	Happy	Sad	Angry
Others	2040	105	69	59
Happy	109	26	6	3
Sad	62	6	65	5
Angry	127	5	10	58

Table 4. Confusion matrix for AdaBoost

	Others	Happy	Sad	Angry
Others	2174	109	67	59
Happy	52	31	1	4
Sad	59	2	77	4
Angry	53	0	5	58

Table 5. Confusion matrix for Gradient Boosting

	Others	Happy	Sad	Angry
Others	2286	122	111	93
Happy	16	19	1	2
Sad	19	1	35	1
Angry	17	0	3	29

7 Conclusion

We have presented the results using ensemble models for contextual emotion detection. Although, AdaBoost predicts "Happy" and "Sad" emotion classes better, Gradient Boosting classifier has better accuracy because the dataset is biased to- wards "Others" class. We used rule-based feature selection and one-hot encoding to generate input feature vectors for building the models. The system can be improved by using various feature selection methods, by incorporating sentiment lexicons or building the model using unbiased dataset.

References

1. Chuang, Z.J., Wu, C.H.: Emotion recognition from textual input using an emotional semantic network. In: ICSLP (2002)
2. Mubasher, H., Raza, S.A., Shehzad, H.M.: Context based emotion analyzer for interactive agent. Int. J. Adv. Comput. Sci. Appl. (2017)
3. Kar, S., Maharjan, S., Solorio, T.: RiTUAL-UH at SemEval-2017 task 5: sentiment analysis on financial data using neural networks. In: Proceedings of the 11th International Workshop on Semantic Evaluation, pp. 877–882 (2017)
4. Rajalakshmi, S., Angel Deborah, S., Milton Rajendram, S., Mirnalinee, T.T.: SSN MLRG1 at SemEval-2018 Task 3: irony detection in English tweets using multilayer perceptron. In: Proceedings of the 12th International Workshop on Semantic Evaluation, pp. 633–637 (2018)
5. Angel Deborah, S., Rajalakshmi, S., Milton Rajendram, S., Mirnalinee, T.T.: SSN MLRG1 at SemEval-2018 Task 1: emotion and sentiment intensity detection using rule based feature selection. In: Proceedings of the 12th International Workshop on Semantic Evaluation, pp. 324–328 (2018)
6. Angel Deborah, S., Milton Rajendram, S., Mirnalinee, T.T.: SSN_MLRG1 at SemEval-2017 Task 5: fine-grained sentiment analysis using multiple kernel gaussian process regression model. In: Proceedings of the 11th International Workshop on Semantic Evaluation, pp. 823–826 (2017)
7. Angel Deborah, S., Milton Rajendram, S., Mirnalinee, T.T.: SSN_MLRG1 at SemEval-2017 Task 4: sentiment analysis in twitter using multi-kernel gaussian process classifier. In: Proceedings of the 11th International Workshop on Semantic Evaluation, pp. 709–712 (2017)
8. Pivovarova, L., Escoter, L., Klami, A., Yangarber, R.: HCS at SemEval-2017 Task 5: sentiment detection in business news using convolutional neural networks. In: Proceedings of the 11th International Workshop on Semantic Evaluation, pp. 842–846 (2017)

9. Tao, J., Tan, T.: Emotional Chinese talking head system. In: Proceedings of the 6th International Conference on Multimodal Interfaces (2004)
10. Gaber, T., Tharwat, A., Hassanien, A.E., Snasel, V.: Biometric cattle identification approach based on Weber's Local Descriptor and AdaBoost classifier. Comput. Electron. Agric. **122**, 55–66 (2016)
11. Flach, P.: Machine Learning: The Art and Science of Algorithms that Make Sense of Data. Cambridge University Press, Cambridge (2012)
12. Natekin, A., Knoll, A.: Gradient boosting machines, a tutorial. Front. Neurorobot. **7**, 1–21 (2013)
13. He, B., Guan, Y., Cheng, J., Cen, K., Hua, W.: CRFs based de-identification of medical records. J. Biomed. Inform. **58**, S39–S46 (2015)
14. Friedman, J.: Greedy boosting approximation: a gradient boosting machine. Ann. Stat. **29**, 1189–1232 (2001)

RETRACTED CHAPTER: A Neural Based Approach to Evaluate an Answer Script

M. R. Thamizhkkanal(✉) and V. D. Ambeth Kumar

Department of Computer Science and Engineering, Panimalar Engineering College, Chennai, Tamil Nadu, India
charuvarshini995@gmail.com,
dr.v.d.ambethkumar@gmail.com

Abstract. Assessment of answers with particular questions is a heavy task, as in assumption that all the students answers have to be awarded. In this paper we give cue of the possibility to cut down the teacher's workload on open questions by a component managing a neural network-based model of the students' decisions, involved in a peer-assessment task. The answer is recognized by OCR and it converted by machine readable format. The network of constraints and relations established among the answers through the students' knowledge, allows us to compare and relate with set of possible keyword of the database. Convolution neural network plays a vital role in comparison of answer database and student database. The Receiver Operating Characteristic curve are constructed based on the accuracy of students marks. Based on the comparison result obtained, the accuracy of student answer is measured and awarded. Our computer system suggests that the subset of the answers is evaluated with database answer in which the performance of evaluation is measured. This is used to reduce workload of humans and it automatically evaluate the answer. It is mainly used in schools, college, university etc.

Keywords: Character recognition · CNN · ANN

1 Introduction

The evaluation of subjective answers by the manual system requires more time and effort of the evaluator. The evaluation of answers is a tedious task and hence it is difficult to perform. Quality of evaluation may vary when human being evaluates the answer. In Machine Learning, all targeted output is only based on the input data provided by the user. Our Proposed System uses machine learning to solve this problem. Our Algorithm performs a task like separating the words and sentences. Along with it, our proposed algorithm provides the comparison between database answer with student answer. Our System is divided into two modules, The scanned images is extracted from the input data, preprocessing is carried out to remove the noisy data and applying the machine learning techniques to classify the data using convolutional neural network. The final outcome is the awarding of the student with their mark. The system software will take the scanned copy of the answer as an input

The original version of this chapter has been retracted. The retraction note to this chapter can be found at https://doi.org/10.1007/978-3-030-32150-5_163

D. J. Hemanth et al. (Eds.): COMET 2019, LNDECT 35, pp. 1187–1207, 2020.
https://doi.org/10.1007/978-3-030-32150-5_122

and produces the mark as the result. Before producing the result, the text which contain the keywords of the specific answer are enclosed in the database answer. Based on the training of the neural network the classifier is made to give the mark to the student. Marks to the answer will be the final output. The main goal of the neural network evaluation aroused mainly to overcome the drawback of the existing system. The need of the project is to ensure user-centered and more interactive software to the user. The system based software is much faster and better method to define the marking ideas. The transparency of answer checking is carried out quickly. It brings much transparency to the present method of answer checking. The answer's keywords are already stored in the database. It is used mainly to access the keyword regularly. The main aim of the industrial and technological revolution is automation of repetitive tasks. The job of checking hundreds of answer sheets which more or less contains the same answer will be a boring task for the teachers. The proposed system is used to reduce the burden. System will save a lot of effort and time on teachers part. The unbiased result is obtained by the human which is capable of commiting lot of mistakes. The system calculates the marks of the student and provides output quickly. Schools, colleges, coaching institutes used the system software to check the answer. It can also be used in various organizations to conduct the competitive examinations.

Many designs and features have been developed for descriptive answer evaluation. The approaches are mainly focused on keyword matching, hence the analyzing the answer is being carried out easily.

The system software is mainly focused on managing university/school type exams containing descriptive questions or a combination of descriptive and objective type questions. Exam related work such as conducting the exam, evaluation of the answer sheets are made easy reduction of Human error, effective evaluation work, reuse of the resources and time during evaluation is being carried out.

2 Related Works

The CE-CLM Network which maps the yield of CE-CLM to 84 point of interest, in addition to this residual correction network is used. This two system is utilized to discover singular face among the general population. The fundamental favorable position in the facial acknowledgment is finished by 3D which is clear and accurate. This takes the multifaceted nature in preparing time of individual faces. Adjustment organize assumed the primary part in milestone position of mapping 68 landmark to 84 historic point position [1].

An Unconstrained Benchmark Urdu Handwritten Sentence Database with Automatic Line Segmentation [2] states that disconnected sentence database of Urdu manually written reports and content line division are delivered utilizing response/reply database. The archives were isolated by a few forms. The classifications of every frame were dispense in various field of Urdu language. The arrangement of content is extricate from the structures with no mistake.

That Handwritten character acknowledgment is a nontrivial. Exceedingly adapted and morphologically complex characters, for example, Bengali are difficult to perceive utilizing CNN the blunder which was caused by incorrectly spelled are distorted.

It needs all the more preparing qualities to prepare the language. Comparing to LSTM, CNN stores all the more preparing qualities [3].

The expression the Bangla dialects are perceived utilizing CNN. The above strategy was same as the proposed technique since there was just a slight contrasts showed up in the dataset. This strategies utilized the vast dataset and it conquer the HCR problem [4]. It diminishes the blunder and expands the general performance. It has speed in nature. The conversion of this language needs more memory space.

Neural system was expensive, Matlab's Neural Network Toolbox is being utilized to perceive printed and manually written characters by anticipating them on various estimated grids [5]. During the way toward penmanship acknowledgment, the exactness was less. The explanation behind this was on the grounds that the system was looked with information with numerous blunder.

A basic Convolutional neural system on picture characterization was manufactured. The basic Convolutional neural system is utilized for the culmination of the picture classification [6]. The primary downside was not pertinent to expansive neural system. Contrasted and the current strategies, despite the fact that this acknowledgment rate isn't the most elevated, however our system structure is straightforward, and parameters take up memory is little.

This expresses the Bangla dialects are perceived utilizing CNN. The above strategy was same as the proposed technique since there was just a slight contrasts showed up in the dataset. This strategies utilized the vast dataset and it conquer the HCR problem. It diminishes the blunder and expands the general performance [7]. It has speed in nature. Comparing to above technique this strategy is appropriate to perceive the Bengal character since it perceive the confined words it wipe out the trouble of looking in general character.

Camera was used to convert printed text or handwritten characters recorded by offline changed into a machine-usable text by simulating a neural network so that it reduce the work of human workers for collecting and storing data [8]. Another goal was that it acquired good accuracy OCR and ANN was used to recognize the characters. Hence it was not considered as a effective way of using features rather than wholesome comparison, however our system structure is straightforward, and parameters take up memory is little. The Indian archive are divided and perceived utilizing SVM classifier. [9] The proposed strategy had utilized the new division calculation to locate the best order of information. Separation of characters and vowels are finished by classified. Most noteworthy division and acknowledgment rates is 98% and 99%. The precision of classifier is good. Multiples of character had been utilized in the SVM classifier.

The manually written reports are tried and prepared by multilayer perceptron neural system are utilized to distinguish the catch phrase [10]. It had the principle preferred standpoint of falling neighborhood ideal in high dimensional space and it additionally has low meeting information rate. As the outcome this strategy has the best exactness information utilizing molecule swarm advancement.

Handwriting was produced by an oscillatory motion of the hand. To find the characters sinusoidal parameters is used. Hidden Markov process is used to find the motion and the characters. Oscillatory motion is used to find the recognition rate. Support vector machine is used to find the classifier of the characters which was oscillated by motion of an hand [11]. The main drawback was it cannot express dependencies between hidden states. Compared to above method this method had produced the exact recognition rate of value which is good and improved more efficiency.

ANN through neural models can used to take care of genuine modern issues managing Image handling and example acknowledgment fields. "Reproducibility", "viability", "marketability", "salability" this fields are perceived by ANN and concealed Markov process, hence choice can be made utilizing ROI. Mass biometry acknowledgment alongside design acknowledgment is being accomplished [12]. The principle downside is "recognizable proof mistake" of the entire framework remains lower than perception values. It can be utilized for complex example acknowledgment. It used to perceive the human arm development designs utilizing an IoT a sensor device. The acknowledgment depended on the hand movement. CNN and RNN is utilized in this procedure [13]. This strategy is being contrasted and LSTM since it was utilized a stable model. Moy hurt band was utilized to gather the hand band. CNN based DQN operator gets scores higher than 400. LSTM got just 100 to 200. The profound Q-Network can effectively take in the human arm development patter. Compared to above strategy (ANN) CNN alongside DQN was best technique for design acknowledgment.

The profound figuring out how to perceive the transcribed characters. Image was separated and preprocessing of the character in light of the picture was finished by manually written character recognition. [14] It experience many handling strategies to acquire a reasonable picture. CNN is invariant to pivot of scaling factor. It connected to the genuine word written by hand character acknowledgment and great execution was accomplished. Convolutional neural system on picture characterization was manufactured. The basic Convolutional neural system is utilized for the culmination of the picture classification [15]. The primary downside was not pertinent to expansive neural system.

Contrasted and the current strategies, despite the fact that this acknowledgment rate isn't the most elevated, however our system structure is straightforward, and parameters take up memory is little. Intelligent map reader: A framework for topographic map understanding with deep learning and gazetteer was proposed by Li et al., thus the past work done by OCR will face challenge when distinguishing map message in muddled settings, especially in the form and topographical areas when the guide content contacts the another guide content map. [16] but this proposed work is utilized to tackle this issue by profound convolutional network, map include detection. The advantage is that this technique approve its productivity and energy is being accomplished utilizing OCR. The after the effects of guide content acknowledgment are then changed over into a machine.

The manually written or character pictures from the content are hard to peruse. A fake neural system and hereditary calculation was utilized to take care of a powerful content acknowledgment issue [17]. A correlation of neural system and GA with hybrid has been finished. This gave the best outcome to find picture design from prepared example. To prepare a framework the hereditary calculation over and over performed hybrid to get the content information from dataset. Secure information was being accomplished in this technique The filtered records for recognizing line was a critical issue for preparing of transcribed writings [18]. Consequently execution of a novel way to deal with enhance the productivity was utilized by Fully Convolutional Network (FCN).

3 Proposed System

The proposed system is to detect the character which is given as input, in whatever style the input text might be. In this project, we develop a model for handwritten character recognition. We also present the algorithms for recognizing the characters which is given as the input and give the correct output for the user. In this recognition process there won't be any wrong recognition of characters. Instead it recognizes the input and gives the correct output for the user, with high accuracy and in a less time while compared to existing once. The evaluation of a student answer script is carried out by collecting the answers script from the student. The Proposed work undergoes scanning of an image in which this act the input source, the unwanted noise is removed using preprocessing, each line in the answer script is separated using the segmentation technique and the features are extracted from the segmented image. Using global templates each characters of answer is compared which result in the recognition of the text. The recognized text is feed into the convolutional neural network which is used to compare the student answer with the database answer, if both the answer matches, then the marks of the particular student is awarded (Fig. 1).

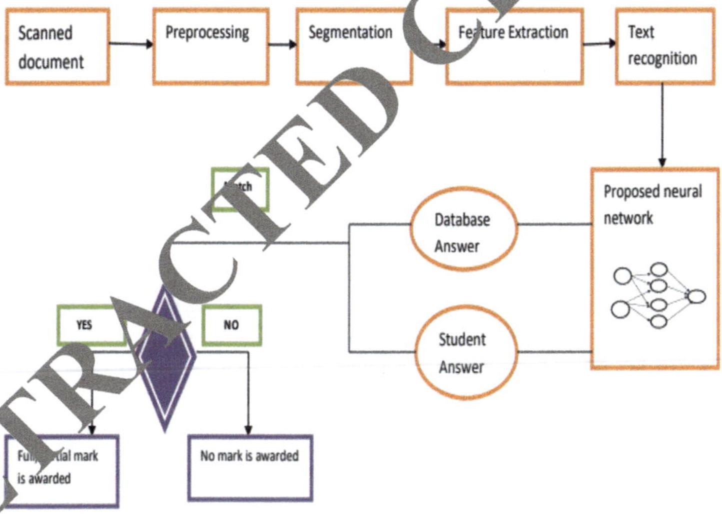

Fig. 1. System architecture

4 Image Acquisition

The input source image is collected from the student and it is scanned using flatbed scanner. The flatbed scanner is also known as reflective scanner. The working is being carried out by shining white light into the object to be scanned. The image is scanned by reading the intensity of the color light. It is mainly used to design the scanning prints. In addition to this it consists of transparency adapters. Flatbed scanner is mainly used for producing an exact digital image and handwritten text. The input source is keep on the top of the flatbed scanner in which the output is obtained in the computer. The text can be obtained using the process of optical character recognition [OCR].

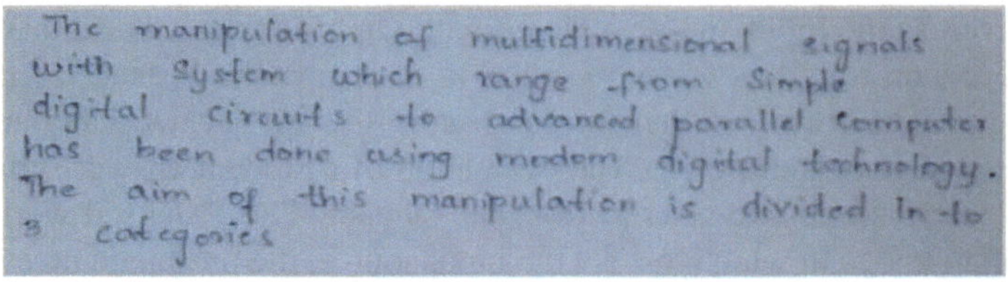

The manipulation of multidimensional signals with system which range from simple digital circuits to advanced parallel computer has been done using modern digital technology. The aim of this manipulation is divided In-to 3 categories

Fig. 2. Scanned document

The above Fig. 2 shows the scanned document of the student answer script which is scanned using flatbed scanner.

5 Preprocessing

Input may contain some sort of noise this is due to the unnecessary information available in the image. The steps involved in the preprocessing are given below (Fig. 3).

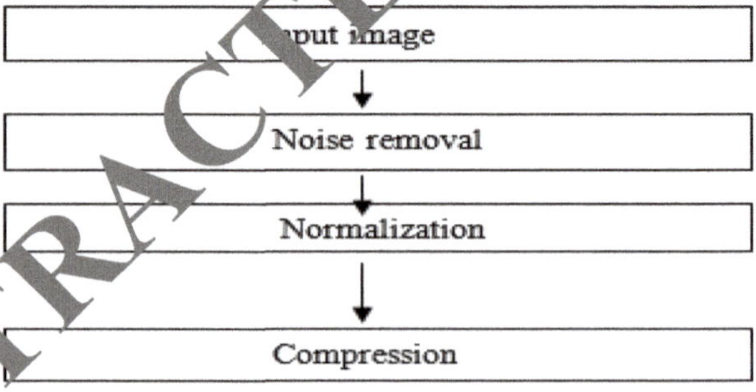

Fig. 3. Steps in preprocessing

The main aim of pre-processing is to remove the noise and improve the quality of image.

5.1 Input Text

The input image is captured by several camera and hence it is converted into binary image. Conversion of image should contain the gray scale image of the input image.

5.2 Noise Removal

Some noise like disconnected line segments, bumps and gaps in lines are introduced by optical scanning devices. This noises should be removed by the element prior to recognition. The poor recognition rate can be eliminated using smoothing. The smoothing implies filling and thinning. Filling eliminates small gaps, breaks, holes and thinning reduces the width of the line.

(a) ***Filtering***

The goal of the filter is to remove the noise from the signal. To produce an estimate of a desired or target random process by linear time invariant (LTI) Weiner filter is used. It consists of observed noisy data, noise spectra and additive noise. The minimization of the mean square error between the estimated random process and the desired process is achieved by the Weiner filter.

$$S(x,y) \qquad f(x,y) + b(x,y) + n(x,y)$$

The above equation gives the f(x,y) fourier transform of the corrupted image obtained by passing the original image s(x,y) through a low pass filter (blurring function) b(x,y) and adding noise n(x,y).

(b) ***Convolution Matrix***

To filter the image a general purpose filter effect called convolution matrix is used. The image is composed of mathematical operation, integer and it is applied in the matrix form. The value is computed by determining the value of a central pixel. The output is a new modified filtered image

$$I(x,y) = s(x,y) + ni$$

$$s(x,y) \text{ is signal}$$

5.3 Normalization

The variations in the images can be easily identified by filtered image. The normalization is the main process in the pre-processing stage, which is used to remove some variations in the images. This variation do not affect the identity of the word. Normalization of handwritten from a scanned image requires several steps, which usually starts with image cleaning, skew correction, line detection, slant and slope removal and character size normalization. In addition to this normalization is applied to obtain characters of uniform size, slant and rotation.

Normalization transforms the n dimensional grayscale image

GI: $\{X \approx R^n\} \rightarrow \{\min, \ldots, \max\}$ with intensity values in the range of (min, max).
GI_N : $\{X \approx R^n\} \rightarrow \{\text{newmin, newmax}\}$ with intensity values in the range of (newmin, newmax).

The normalization of grayscale digital image is performed using the formula.

GI_N = (I − min) newmax − newmin\max − min + newmin.

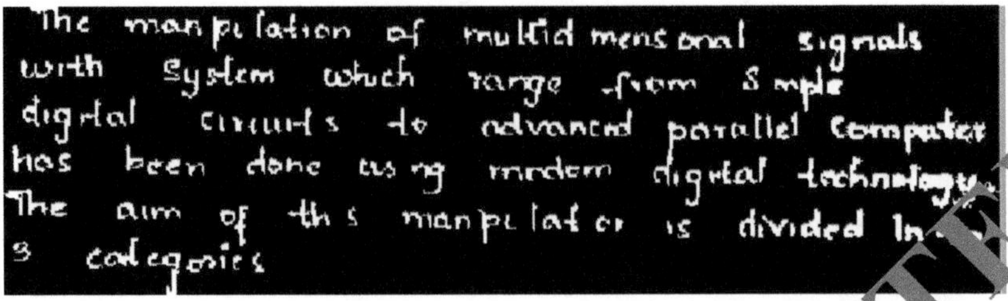

Fig. 4. Normalization and smoothing of a image

The above Fig. 4 is used for normalization and smoothing of text by applying filtering techniques in which rotation of the text is seen in the image. After this filtering the noise is removed.

5.4 Compression

For compressing the image requires the space domain techniques. There are two main techniques are used in compression. They are thresholding and thinning. Using threshold value the gray-scalar color images are converted to binary image, hence the space domain decreases the storage requirements and increases the speed of data processing. The shape information of characters is extracted by thinning.

6 Segmentation

Sequence of characters is enclosed in a image and it is divided into sub images of individual characters. It is to split the image into many parts which has a high correlation with real world objects. Segmentation is very important for recognition system. The main goal is to split the image into parts that have a strong correlation with objects or areas of the real world contained in the image. To recognize the system segmentation is very important. It purpose is to divide the words, lines or characters which affects the recognition rate. To simplify/change the representation of an image into meaningful words.

Image segmentation is used to locate the objects and the boundaries of an image. The segmentation is carried out by two process Internal segmentation and external segmentation.

6.1 Internal Segmentation

Internal segmentation is used here to isolate the lines and curves in the cursively written characters.

6.2 External Segmentation

In explicit segmentation the segments are identified based on character like properties.

7 Feature Extraction

To retrieve the significant text from image feature extraction method is used. The main goal of feature extraction is to extract a hard and fast group of features, which maximizes the recognition rate with the less quantity of factors. The heart of pattern recognition application is the feature extraction. The goal of feature extraction is to collect the important characteristics of the symbols, which is generally accepted that could be one of the most difficult problems in pattern recognition. The actual raster image is used to describe a character in a straight forward way. In proposed system, is used to extract the features gradient based features is used along with the regression techniques.

8 Implementation Techniques

8.1 Identification and Recognition

The Hybrid approach of neural network is used for implementation techniques. Two different approaches are being carried out. The evaluation of answer script is done by combination of two neural network. 1. Identification of an letter in the answer script is done by convolutional neural network. 2. Evaluation of answer is done by simple neural network. The method will be easy when it applied in a combination of two neural network (Hybrid approach) (Fig. 5).

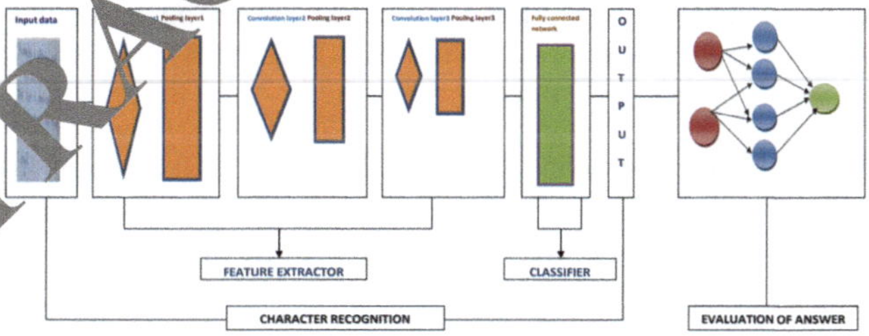

Fig. 5. Proposed neural network

The above figure is used to recognize the answer, the basic step is the identification of the letter. For this identification the convolutional neural network is used. Convolutional neural network consists of an input, output and hidden layer. Hidden layer composed of Convolutional neural network, ReLU, Pooling layer and fully connected layer. In proposed system the input is the answer of the student, output is the character recognition.

8.2 Convolutional Neural Network

Convolutional neural network takes the source as the image of the student answer script, the image size $7 \times 7 \times 1$ i.e. (Height, Breadth, number of channels). The first step in CNN is the carrying of the convolution operation which consists of set of function and represented by pixel values (0, 1). The function mixes with each other as one passes the other. In proposed system the desired source is compared with the definite feature of the character's pixel value. It consists of three element input image, filter, feature map. In input data has the pixel value of 1 and −1. The value 1 denotes the black pixel and −1 denotes the white pixel.

Steps Involved in Convolutional Neural Network

- Place the input image at start from the top-left corner.
- Count the number of cells, in which filter matches with input image.
- The absolute matching of the cells are inserted at the top-left cell of the filter.
- Move the filter one cell to the right and continues till all the values are matched.
- Multiply with the respective feature of the pixel.
- Take the average of all the pixel value.

The feature map size is controlled by three parameters. The hybrid approach uses this three metrics to get the desired letter.

Depth
Depth corresponds to the number of filters. It relates to the number of filter taken to do convolution operation. Feature maps are stacked as 2D matrix.

Stride
It is calculated by number of pixels by sliding the filter matrix over the input matrix. If the value of stride is 1, then the filter matrix moves one pixel at a time.

Padding
The border of input matrix is padded with zero. It can be applied by filtering the bordering element of input matrix.

ReLu
Rectifier linear unit is a non-linear operation. In above convolution layer the feature map has the combination of both positive and negative values. ReLu is applied per pixel which replaces all the negative values in feature map by zero value. It is also called as Activation layer.

Pooling
The dimension of feature map is reduced by spatial pooling. It is retained by MAX, AVERAGE, SUM. In this method, MAX pooling is used to find the spatial neighborhood hence the largest element is taken from the rectified feature map.

Fully Connected Layer
The output from the pooling layer is arranged in the stack (list) format. This is the final layer where the actual classification exists. It compares with the two different character. The values takes out the filtered image and are arranged in a single list (stack).

Proposed Algorithm For Identification Of an Letter

INPUT: Randomnumbergenerator(rng),Datastore(DS),Convolution2dlayer(C2Dl), Maxpooling(MP),Fullyconnectedlayer(FCL),softmaxlayer(SL),classificationlayer(CL), Trainingoption(TO),Trainingnetwork(TN),Trainprediction(TP),TrainingSet(TS).

OUTPUT: Recognized letter in a answer script

---

begin
for i ← 1to 6
Datastore(DS ← Imagedatastore(IDS)
(TS,TestingSet) ← divideeachlabel(ds,1024,randomize)
Layers ← imageInputlayer(1261262)
C2DL1 ← (9,42,'Stride',1'Padding'3);
relulayer1 ← activate the function(C2DL1)
MP1 ← (2,'Stride',2)
C2DL2 ← (7,126,'Stride',1,'Padding',2)
relulayer2 ← activate the function(C2DL2)
MP2 ← (2,'Stride',2)
C2DL3 ← (3,192,'Stride',1,'Padding',1)
relulayer3 ← activate the function(C2DL3)
C2DL4 ← (3,192,'Stride',1,'Padding',1)
relulayer4 ← activate the function(C2DL4)
C2DL5 ← (3,126,'Stride',1,'Padding',1)
relulayer5 ← activate the function(C2DL5)
MP3 ← (2,'Stride',2)
FCL ← (256)
FCL ← (2)
//CLASSIFICATION LAYER//
options ← TO(sgdm.'maxepocs(40),'LearningRate'(0.0001),size(128))
CN ← TN(TrainingSet,layers,options)
TP ← Classify(CN,TS)
TL ← TSLabels(TSL)
TA ← sum(TP==TL)/numel(TL)
TEP ← Classify(CN,TS)
TEL ← TestingSetLabels
TEA ← sum(TEP==TEL)/numel(TEL)
Display ← (TA)
Display ← (TEA)
End

The values in the list which consists of the deformed character (irregular shape of letter) are compared with the definite character. The definite character (uniform shape of letter) with highest value in the stack are summed. The average values of the character are calculated. The values of the definite character is being calculated.

In this proposed neural network, each words in a answer script is be easily identified. The input data is the answer script of the student. It is feed into the convolution neural network of size (7 $\times$ 7 $\times$ 1) matrix. The input matrix is being compared with the MAX filter in which it is used to find the maximum of the pixel value. The value of this convolution layer 1 consists of both positive and negative value of the pixel. Negative value is eliminated using ReLu layer. ReLu layer is now feed into the convolutional layer 2, convolutional layer 3 etc. In this proposed system three convolutional neural network is used. The optimized value is given to fully connected layer in which the values of the pixel is located in the stack. The definite character values is being compared with the deformed character.

8.3 Evaluation of Answer

The student answer is evaluated using simple neural network. It classifies the student answer with the database answer. The keywords of the answer are matched with the student answer which undergoes the classification of the neural network. In the proposed system, simple neural network is used to find the correct answer of the student. The goal of the network is to enable the better modeling to detect the high level non-linearity relationship. It consists of input layer, hidden layer and output layer. To generate the output values the forward propagation is used. The error can be calculated using the deltas values (i.e. difference between the targeted and the actual output values). MSE is used to find the error rate. This consists of several key answer for a particular question. It stores a large amount of data. Once the input data is fed into neural network its compares it with student answer. Every student writes the exam with their own style hence it is recognized and converted into binary database. After conversion the student answer is compared with database answer.

The student gets the marks awarded based on the accuracy of keywords that are involved in database answer. Incorrect classification may also be due to poor design of the classifier. This may happen if the classifier has not been trained on a sufficient number of test samples representing all the possible forms of each character. If both the answers are matched then the marks awarded with accuracy.

PROPOSED ALGORITHM FOR THE EVALUATION OF THE ANSWER

INPUT: TrainingSet(TS),TargetSet(TGS),Inputlines(IS),Targetlines(Tl),Inputlines value(ILV),Observedlinesvalue(OLV)

OUTPUT: Recognized Answer(RA)

```
TSDataInput ← (1 to end of line,1)//Training set//
TGS ← DataInput ←(1 to end of line,2)//Target set//
input ← Training set//convert to row
observe ← Target set//convert to row
A ← Conseq(input)
B← Conseq(Observed)
//Data process//
N=368
IL ← A(1 to end-N)
OL← A(1 to end-N)
ILV ← A(end-N+1 to end)
OLV ← A(end-N+1 to end)
//Data training ,validation testing//
trainingRatio=80/100
valRatio=20/100
TestRatio=20/100
//Train the network//
(network,train) ← (network, inputs, observed, inputstates, layerstates)
//Test the network//
(network,train) ← network(inputs,inputstates,layerstates)
errors ← subtaract(observed,results)
performance ← P(Network,observed,result)
display ← network
//plotting of observed values//
plot ← perform(train)
plot ← trainingstate(train)
plot ← Regression(estimated,observed)
plot ← Response(estimated,observed)
plot ← linearcorr(input,error)
```


In the above algorithm the simple neural network is used to evaluate the answer of student. The Training set and the lines of the answer is taken as the answers. The database consists of the keywords which compares the student answer with the database answer. In addition to training set, Target set is used. The performance of the system is identified using the expected and the observed results and it is displayed by plotting the observed values.

9 Results and Discussion

Experimental Setup
The proposed system evolves many configuring software's and hardware's, which improves the tactile output. The machine runs with a CPU configuration. The software components of the system incorporate MATLAB R2010 with OCR. The System is ready to begin, once the setup is processed. The MATLAB10 software is used. The handwritten of the student is recognized by OCR, hence the evaluation of the answer script is analyzed by the neural network toolbox. The input contain 5 set of answer script with student database consisting of the keyword.

Network Set-Up
The convolutional neural network learning algorithm was used to solve the problem. To minimize the error energy at the output, CNN plays important role in detecting the error. Training set of input vectors is applied to the network, forward and backward propagation is carried out by adjusting the weights by the CNN algorithm. The steps are repeated for all sets. When adequate convergence is reached, the CNN algorithm stops.

Performance Evaluation
There are many ways to find the performance of the project. The best way is by considering 3 parameters. Which are discussed in the following. These are used for finding the accuracy, system sensitivity, system specificity and evaluating the answer script. Firstly, for determining this three parameter true positive, true negative, false positive, false negative is required this can be computed using confusion matrix.

Confusion Matrix
Error matrix is also known as confusion matrix in the field of machine learning. It is used to solve the problem in statistical classification. A specific table layout is allowed to visualize the performance of an algorithm. The table is a matching matrix consists of rows and column. It consists of actual and predicted and identical sets of classes in tow dimensions. It also called as contingency table. Row represents the instances in a predicted class. Column represents the instances in actual class.

In this Table 1 shows the special kind of contingency table, with two dimensions ("actual" and "predicted"), and identical sets of "classes" in both dimensions (each combination of dimension and class is a variable in the contingency table).

Table 1. Test samples to taken in confusion matrix

Class	True positive	True negative	False positive	False negative
1	1.9	0.2	0.01	0.03
2	0.8	0.2	0.02	0.03
3	0.8	0.2	0.014	0.028
4	0.9	0.2	0.014	0.033
5	0.9	0.2	0.02	0.02
6	0.9	0.2	0.2	0.033
7	0.9	0.2	0.03	0.027

Accuracy

It is formed by a systematic errors. A measure of statistical bias is carried out by the difference between output and the true value. This value is called as the trueness. It defined by the accuracy obtained by the combination of random and systematic error. High accuracy requires by both high precision and high trueness.

Accuracy = (TP + TN)/(TP + TN + FP + FN)
TP = True positive; FP = False positive; TN = True negative; FN = False negative.

System Sensitivity

Sensitivity is also called the true positive rate, the recall, or probability of detection. It measures the proportion of actual positives that are correctly identified. Example: peoples are identified as sick by the symptoms of the person. In the proposed system the answer is verified and mark is awarded if the answer matches with the database answer. True possible data with relevant keywords are considered as the true positive.

sensitivity = TP/ (TP + FN);
TP = True positive; FN = False negative.

System Specificity

The unrelated data comes back as negatives if the value does not matches with the condition. Example: normal persons with no disorder is considered as the true negative. In the proposed system the answer does not matches with the database answer i.e. irrelevant answer written by the student which does not matches with the keyword are considered as the true negative.

specificity = TN/(FP + TN);
TP = True positive; FP = False positive; TN = True negative.

Table 2. Experimental values of the students

Student answers	Mark	Accuracy	Sensitivity	Specificity
Input 1	90	98.131	98.446	95.238
Input 2	85	96.131	95.432	92.230
Input 3	0	0	0	0
Input 4	90	98.131	98.446	95.238
Input 5	20	20.25	15.65	10.32

This Table 2 shows the experimental values of the student who have written the exam in the answer script. There are totally 100 papers for evaluation. In experimental result, 5 papers of different students have been taken and the results have been shown by the 3 perspectives such as accuracy, specificity, sensitivity (Fig. 6).

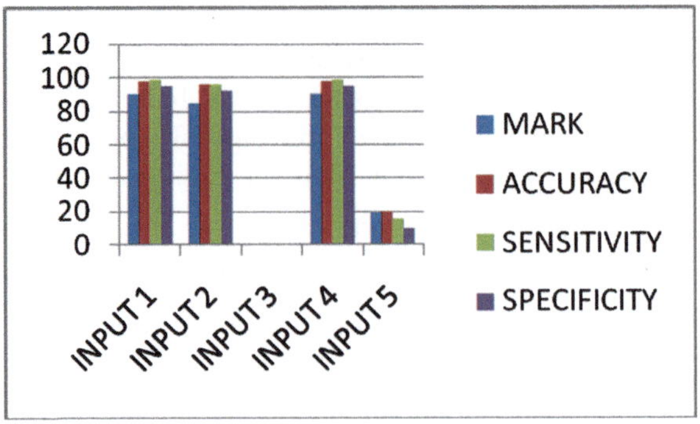

Fig. 6. Performance of the student

Performance of System

It is calculated based on the confusion matrix and the values are taken from the contingency table.

Table 3. Shows the overall performance of a system

System accuracy	98.1[illegible]
System sensitivity	98.446
System specificity	95.238

This Table 3 shows the overall performance of a system which is calculated using confusion matrix hence it can be varied by the values of system speed and time (Fig. 7).

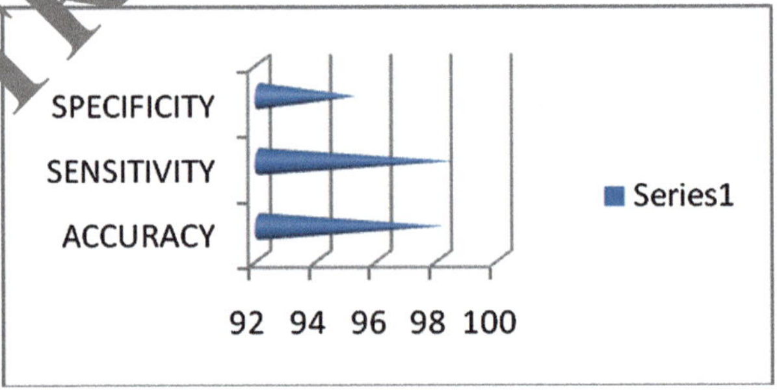

Fig. 7. Shows the comparison graph for the overall performance

Implementation of Neural Network

In the proposed experiment, there are 26 characters images in which there are 180 neurons used in the input layer and 26 neurons are used in the output layer of the network classifier. The size of the input layer depends upon the size of the output layer. The result is obtained by the accordance of the number of the output classes in which the input patterns is to be classified. For optimality, 80 neurons are kept in the hidden layer by trail and error method. The size of the input layer depends on the size of the sample presented at the input and the size of the output layer is decided in accordance with the number of output classes in which each of the input patterns is to be classified. For optimal results, 80 neurons are kept in the hidden layer by trial and error method. The 'tansig' activation function is used for hidden as well as output layer neurons. Neural network training process is shown in Fig. 4 and the Gradient Descent with momentum learning function 'traingdx' has been used. Mean Square Error (MSE) has been selected as a cost function in the training process shown in Fig. [illegible] (Fig. 8).

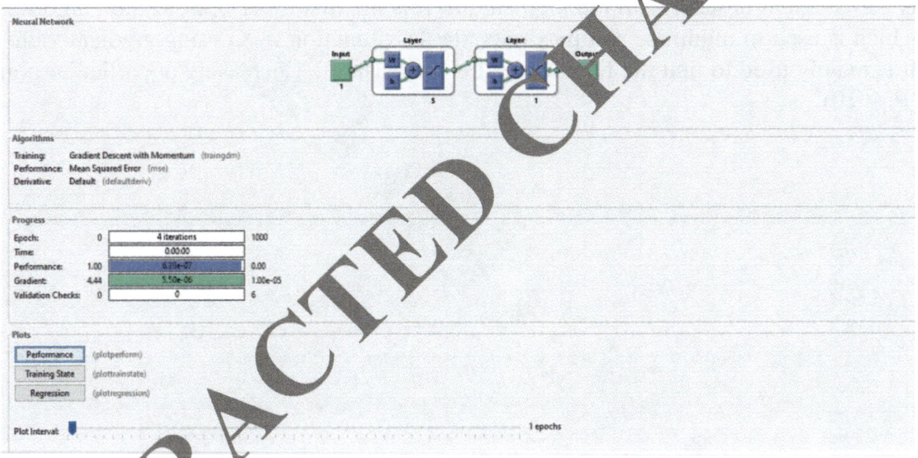

Fig. 8. Neural network toolbox

Mean Square Error

Mean square error is also called as the mean squared deviation of an estimator. The average of squares of the error is calculated i.e. the average square difference between estimated values and what is estimated. It is risk function which corresponds to the expected value of the squared error loss. It values is always non-negative. Here the MSE is taken as the y-axis and 4 epochs is taken as x-axis where epochs is the single step carried in neural network in other words one epochs is the one forward pass and one backward pass in the training set (Fig. 9).

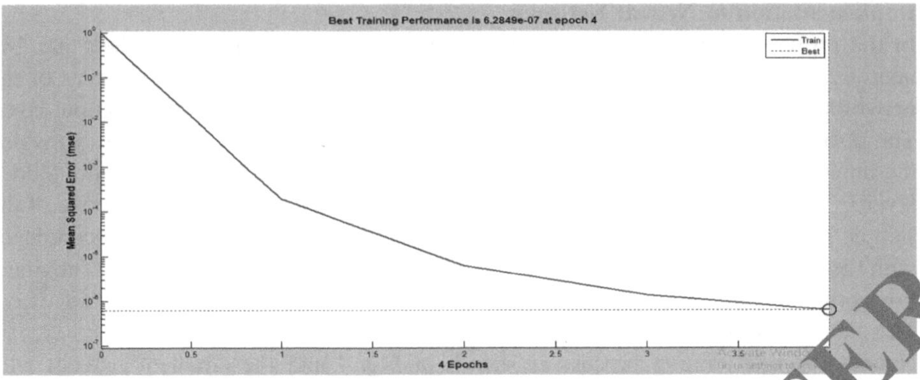

Fig. 9. Performance of the training set

This figure shows the performance of the training set. The Algorithm here we using is the Gradient descent with the momentum. It is the first order optimization algorithm which is used to minimize and maximize the loss function f(X) using gradient values. It is mainly used to find the function that is decreasing or increasing at particular point (Fig. 10).

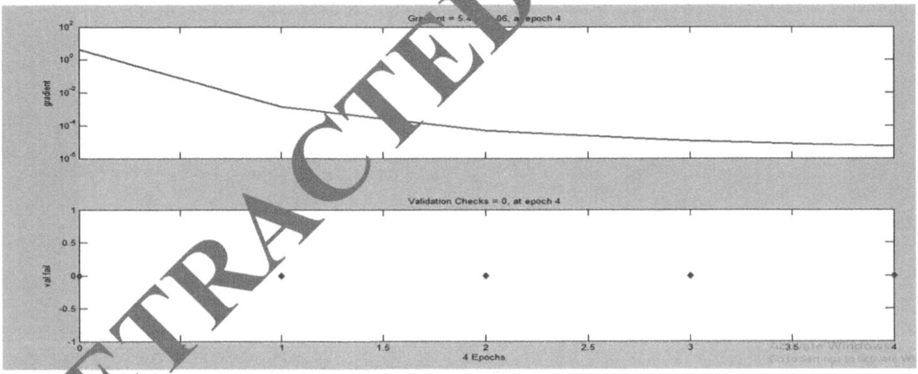

Fig. 10. Validation checks in testing phase

Regression State

The linear regression is used to fit the line to set of points. It is used to model the relationship between a dependent variable with one or more independent variable. X be the independent variable, Y be the dependent variable. To train the model we have to predict the value of Y for any given value of X (Fig. 11).

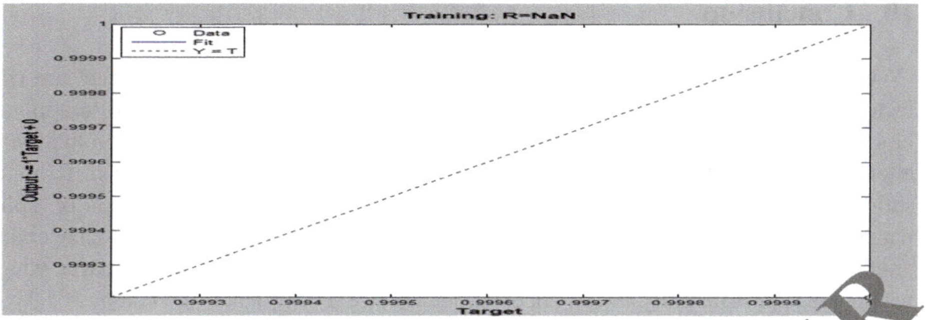

Fig. 11. Linear regression of gradient descent

ROC Curve

Receiver Operating Graph is a useful method for organizing classifiers and visualizing their performance. ROC graphs are commonly used in data mining and neural network analysis, and in recent years have been increasingly adopted in the machine learning, AI and data mining research communities.

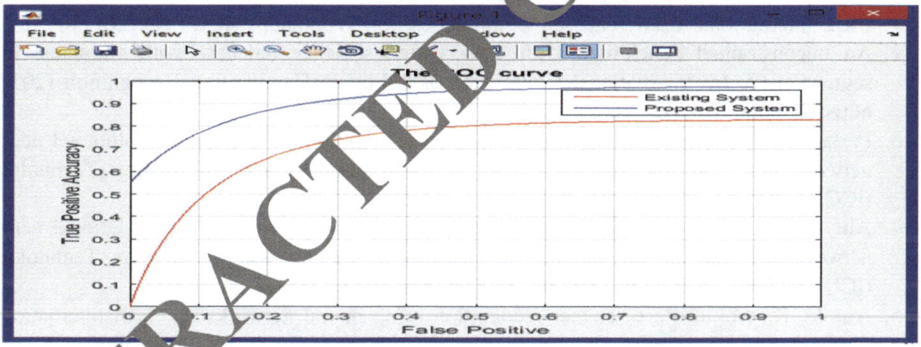

Fig. 12. ROC curve

The Fig. 12 shows the type of graph is called a Receiver Operating Characteristic curve (or ROC curve). It is a plot of the true positive rate against the false positive rate for the different answers of the student.

An ROC curve experimented several things:

- It shows the gradual increase in the sensitivity and specificity. If there is any increase in sensitivity there will a decrease in specificity.
- The accurate point lies in the top border of Roc curve. The degree at 45 of the ROC space is seems to be less accurate.

10 Conclusion

Constructed response (CR) tests let the student write answer on blank sheet are collected. Using this the teachers will evaluate the answer. Since, it is difficult to apply in large volume, because skilled human validators are required (automatic evaluation is attempted but not widely used yet). The proposed system will access the in terms of short answer using several convolution neural network. The implementation of, split-paper (SP) testing is used. This scheme split answer into two, one is student answer and another one is database answer, compares with the keywords of same answer which was misplaced differently in sentence. The examinee answers and the database answers are compares efficiently which can be awarded automatically using the computer system it is and is used for awarding the marks. Therefore, we have conducted an experiment, using short answers with examinee, to compare SP tests against CR tests. Therefore, we might conclude that SP tests are useful tools to evaluate evaluation performance. In future it is going to implemented in terms of multiple answer with multiple keywords.

References

1. Zadeh, A.: Convolution experts constrained local model 3D facial landmark detection. In: IEEE International Conference on Computer Vision Workshops (ICCVW) (2017)
2. An unconstrained benchmark Urdu handwritten sentence database with automatic line segmentation. In: International Conference on Frontiers Handwriting Recognition (2012). https://doi.org/10.1109/icfhr.2012.177
3. Purkaystha, B.: Bengali handwritten character recognition using deep convolutional neural network. In: 20th International Conference of Computer and Information Technology (ICCIT) (2017)
4. Alif, M.A.R.: Isolated bangla handwritten character recognition with convolutional neural network. In: 20th International Conference of Computer and Information Technology (ICCIT) (2017). https://doi.org/10.1109/iccitechn.2017.8281823
5. Arnold, R., Miklos, P.: Character recognition using neural networks. In: 11th International Symposium on Computational Intelligence and Informatics (CINTI) (2010). https://doi.org/10.1109/cinti.2010.5672225
6. Araokar, S.: Visual character recognition using artificial neural networks. In: IEEE Conference on Computer Communications, pp. 1643–1651 (2015). https://doi.org/10.1109/infocom.2015.7218544
7. Bawanea, P., Gadariyeb, S.: Object and character recognition using spiking neural network. https://doi.org/10.1016/j.matpr.2017.11.093
8. Das, T.K., Tripathy, A.K.: Optical character recognition using artificial neural network. In: International Conference on Computer Communication and Informatics (ICCCI) (2017). https://doi.org/10.1109/iccci.2017.8117703
9. Sahare, P., Dhok, S.B.: Multilingual character segmentation and recognition schemes for Indian document images. IEEE Access **6**, 10603–10617 (2018). https://doi.org/10.1109/access.2018.2795104
10. Tavoli, R., Keyvanpour, M.: A method for handwritten word spotting based on particle swarm optimization and multi-layer perceptron. IET Softw. **12**(2) (2018). https://doi.org/10.1049/iet-sen.2017.0071

11. Choudhury, H.: Handwriting recognition using sinusoidal model parameters. Elsevier B.V. (2018). https://doi.org/10.1016/j.patrec.2018.05.012
12. Madani, K.: Artificial neural networks based image processing & pattern recognition: from concepts to real-world applications. In: First Workshops on Image Processing Theory, Tools and Applications (2008). https://doi.org/10.1109/ipta.2008.4743797
13. Agarwal, M.: Pattern recognition of human arm movement using deep reinforcement learning. In: International Conference on Information Networking (ICOIN) (2018). https://doi.org/10.1109/icoin.2018.8343257
14. Wu, M., Chen, L.: Image recognition based on deep learning. In: IEEE International Conference on Computer Systems and Applications (2016). https://doi.org/10.1109/icoin.2018.8343257
15. Wang, T., Wu, D.J.: End-to-end text recognition with convolutional neural networks. In: 21st International Conference on Pattern Recognition (ICPR 2012) (2012)
16. Li, H.: Intelligent map reader: a framework for topographic map understanding with deep learning and gazetteer. In: 2nd International Conference on Inventive Systems and Control (ICISC), Coimbatore, pp. 174–178 (2018)
17. Agarwal, M.: Text recognition from image using artificial neural network and genetic algorithm. 978-1-4673-7910-6/15/$31.00
18. Vo, Q.N., Kim, S.H.: Text line segmentation using a fully convolutional network in handwritten document images. IET Image Process. 12(3), 438–446 (2018)

Analysis of Aadhaar Card Dataset Using Big Data Analytics

R. Jayashree(✉)

Department of Computer Science and Engineering,
Panimalar Engineering College, Chennai, Tamil Nadu, India
shreeramesh2015@gmail.com

Abstract. Aadhaar provides an essential details about an individual with 12-digit unique identification number that contains all the details about an individual, including demographic and biometric information of every resident Indian individual. Aadhaar is a big data which need to be stored and managed securely and safely. Several processing techniques and privacy measures have introduced to process such huge confidential data. However, identifying individual details which may be used by different sectors is not linked or updated with aadhaar data. In order to update essential details of an individual along with existing database of an aadhaar for use by crime department, health care center and professionals, several algorithms, tools, techniques used in big data analytics have been discussed in this survey paper. This is useful for hospitals for retrieving blood donor details, crime investigation and professionals for retrieving the details about residents along with their aadhaar details.

Keywords: MySQL · Hadoop · Sqoop · Hive

1 Introduction

In march 2015 the aadhaar based digilocker benefit has been propelled, utilizing which aadhaar holders can output and spare their records on the cloud, and can impart them to the administration authorities at whatever point required with no compelling reason to convey them.

The unique identification authority of India acquaints Face Authentication with additionally fortify aadhar security. It chose to empower 'Face Authentication' in combination mode on enrolled gadgets by 1 July 2018, so individuals confronting challenges in other existing method of check, for example, iris, fingerprints and one time password could without much of a stretch verify. The aadhaar card utilizes a 12 digit number for a specific individual in view of biometric qualities, to be specific iris, face and fingerprints. The 12 digit number can have 10^12 i.e. one trillion conceivable blends, of which one billion would be required for all subjects of India.

The residents of India is obtained with 12-digit unique identity number called aadhaar which is based on their biometric and demographic data. Aadhaar has been found has the world's largest biometric identity system which was collected by the unique identification authority of India in January 2009 by the government of India. Aadhaar is the most mature identity programme in the world.

D. J. Hemanth et al. (Eds.): COMET 2019, LNDECT 35, pp. 1208–1225, 2020.
https://doi.org/10.1007/978-3-030-32150-5_123

The proof of citizen is not considered, but the proof of residence is recorded through Aadhaar. It is mandated to the unique identification authority of India to assign 12-digit unique identification number to all Indian. Because of implementation of unique identification scheme in India, it encompasses generation and assignment of unique identity to each and every residents in India. It also promotes the management of unique identity number life cycle, policies need to be framed, updating details along with existing details, defines the purpose of aadhaar, various services.

Several government programs such as LPG connection, ration card, PAN card, SIM card, opening a bank account is also linked with aadhaar card for validation. Though aadhaar has provide many advantages, it has some issues.

Since, all the details of an individual is linked with aadhaar, it brings up major disadvantage that the personal details of an individual can be hacked through the aadhaar number. The account details of an individual is also linked with aadhaar card. By hacking the aadhaar number, the account details of an individuals can be fetched.

Aadhaar is a huge volume of data which is a quiet tricky thing to be neglected. It is quiet difficult to handle aadhaar details to retrieve information about an individual. Ordinary techniques and queries cannot handle such huge volume of data and processing of such huge information is difficult with usual techniques. Therefore, big data analytics is applied on such huge data to process it.

As of late, there has been an expanding popularity for Big data which remains ambiguous. Big data is a sweeping title as several junction of dicdatic indexes which is very huge as well as convoluted and it tends to winds awake tough to progress bestow conventional propaganda handling operation. In the previous couple of years, the aggregate sum of information made through personal who includes detonated. Right from 2005 to 2020, the extent of information being anticipated via incremented to 400 times, from 230 exegetes to 40,000 exegetes. Mentioned information be produced as of logical study as well as trade processing, management, web look, interpersonal organizations, record, photography, sound, videocassette, chunk, jockstrap, mobile devices, senor systems.

The major problems of big data and its usage is facilitated by hadoop. Huge amount of data is generated by social media which need to be processed, and it was a difficult task to maintain such billions of data. Big data has a popular framework called hadoop used to store and process huge amount of data. It can handle datas on multiple nodes. Hadoop works efficiently with processing vast amount of data. Hadoop environment has a popular methodology for parallel processing of massive amount of data called map reduce. Map reduce is a software frame work which supports parallel and distributed computing on large datasets.

As a promising structure executed by open source hadoop for parallel huge information preparing in appropriated registering frameworks, map reduce has been broadly received to viable and rapidly examine information extending from terabytes to peta bytes in estimate.

A map reduce work comprises of various parallel guide errands, trailed by decrease undertakings that performs consolidation of all between intercede brings about the type of key-esteem sets that was created by delineate to deliver last outcomes. These huge volume centre of the mapper information conveyed from outline to decrease undertakings possess exorbitant system transfer speed assets, prompting system clog that can truly corrupt the execution of map reduce employments. It is essential thought is to total the key-esteem sets having the same keys before sending them to diminish tasks.

Hive and Pig are the two main components of the hadoop ecosystem. They came in to existences because the enterprise to interact with huge amount of data without grieve about writing complex map reduce programs. Both has similar objective to ease the complexity of writing complex map reduce programs.

Hive keeps running queries faster than pig. It takes very less time to write hive query in comparison to map reduce code. Hive is mature than pig in query execution. Errors that produced by pig are not helpful. Hive is also called as HiveQL, Hive Query Language. Hive helps in writing less complex map reduce codes. Hive is useful in handling structured data. Hive can be used to operate on datas on HDFS files. Hive is a query engine which is used for batch processing.

Sqoop is the only tool to handle structured data efficiently which is designed to transfer data between hadoop and relational database servers. It is used to import data from relational databases such as My SQL, oracle to hadoop HDFS, and export from hadoop file system to relational databases. The sqoop import tool imports individual tables from RDBMS to HDFS. The rows in the table are treated as records in HDFS. All records are stored as text data in text files.

Aadhaar data is very confidential, dataset for aadhaar is not available in the website. In order to overcome this, a dataset for aadhaar is created to retrieve information about the citizens. This may be useful for crime investigation. Processing of huge information in effective manner and data can be retrieved quickly as much as possible. Big data technique is performed on huge set of dataset. The proposed system retrieves information about the availability of particular citizen easily and on timely basis.

Information about the citizens are retrieved. This may be used as base for crime investigation when any individual commits fraudulent activity. Easy access to details about the citizens related to particular division. Immediate storage and retrieval of data is provided.

Professionals can use the database to collect qualification details about an individual for recruiting them. Healthcare centre uses the website to retrieve patient information. Patrols can retrieve the information about the citizen when he/she is involved in the crime activities through the website.

In order to retrieve information about the citizen, a website is created which is used by professionals, healthcare centre, patrols. The proposed system maintains a database that stores the information about the citizens and it will be useful for the investigation process. Healthcare centre can use this data to retrieve information about the patient records. The information about the citizens stored in My SQL database and imported using sqoop tool and processed using hive. Citizen details that are gathered and visualized through website.

Professionals, Healthcare centre, Patrols can retrieve the data through website. The proposed system helps in crime investigation. Professionals can use this information to employ young talents. Healthcare centre can use this data to retrieve information about the patients.

2 Literature Survey

Data mining is developed to find interesting patterns from datasets whereas Big data involves large scale storage and processing of large datasets [1]. So, data mining with big data is quiet interesting and useful for getting lot of attention currently. Data mining with big data is related to the use of huge datasets to process the collection or reporting of data that serves businesses. Analyzing the criticality in data driven models and also in the big data coup. Analyzing such huge amount of data is one of the big data challenges with big data mining. Data mining methods are useful in finding interesting patterns from big data with complicated relationships and with dynamic volumes of data. Processing time of big data with maximum size is reduced by designing sampling mechanisms of big data and precisely predicts the tendency of data in future. Privacy concerns, errors in the data may replicate the data which is not focused. Through complex relationship which forms useful patterns links the unstructured data. However, designing a secure information sharing protocol is a big issue.

The trend of data and invention of data mining technologies brings threat to security of individual information [2]. To pact the security threats in big data mining, a method called privacy preserving data mining is designed which has gained popularity in recent years. To safeguard sensitive information from spontaneous acknowledgement and also to secure utility of data. Knowledge discovery process are used for considering the privacy aspects of data mining. Data provider who provides the sensitive information to the data collector needs a security to their information and in turn data collector provides the data collected to the data miner. Do not track and socket puppet methods are used to protect the sensitive information of an individual. Data modification to safeguard the privacy data is attempted by the method privacy preserving data publishing. Though it captures only the limited structural properties and it may cause utility loss.

Big data focuses on data acquisition, preparation, repository and management, processing and extraction and analysis of data [3]. Map Reduce, Hadoop are some of the framework for managing and interpreting big data. Map Reduce is a parallel programming model used for writing applications that can process big data in multiple nodes simultaneously. Hadoop allows distributed processing of large datasets across clusters of computers using simple programming model. Data preparation is the predominant for increasing the value of big data. Well-timed collection of data is fundamental and essential for fast interpretation of data. Security for big data cannot be efficiently achieved by passwords, controlled access, two-way authentication. Greater security can be provided by cryptography. Organization of big data is not focused. Virtual barriers to protect data.

Workload generator called ankus which is designed based on the models from the workload analysis. Workload generator facilitates the evaluation of job schedulers and debugging on a hadoop cluster [4]. K-means clustering which is faster and efficient than hierarchical clustering are used to determine the cluster of related jobs. To eliminate and identify system bottleneck, workload interpretation studies are useful and provides solutions for optimizing the system performance. To improve the performance of the system map reduce workload tracing and utilization of resources on hadoop cluster must be analysed. HDFS is a block structured file system in which the task tracker is assigned by the job through job tracker. Analysis of only small jobs on limited nodes are performed. Optimization of system performance and eliminating system bottleneck on hadoop cluster when workload increases is focused.

Performance of system can be improved by high-level programming language and databases [5]. Hive and Pig are high level language for analyzing the data. Hive and Pig can handle and process huge amount of data efficiently whereas My SQL can work well with only small datasets efficiently. Hive and Pig are also cost-effective. Hadoop environment contains a master node, a slave node, Sqoop import and export tool, Map Reduce for parallel programming, Hive, Pig and HBase. Hive has significant advantages over pig such as indexing of data which involves sorting, aggregation, clustering. Indexing of data helps in efficient execution of queries. Hive executes map reduce only when aggregation, joins are performed. Pig executes step-by-step procedure which is time consuming. Pig does not work well with minimum joins and filters. It executes only complex queries. Hive executes simple queries in time efficient manner.

Processing of large scale data can be done efficiently by hive. Hive promotes efficient data reuse strategy [6]. Processing efficiency of large scale data can be greatly improved by results of reuse calculations. Reuse calculations of data is enabled through hive. Overhead can occur when the probability to reuse data increases. Increases the data process efficiency and effectively. Hive helps in writing simple map reduce programs for data processing. Hive helps users to work comfortable with SQL. Hive uses a language called Hive QL which is similar to SQL. SQL-like queries are automatically translated in to map reduce jobs by Hive QL. Hive promotes data summarization, query and analysis in easier manner. Hive can process external data without actually storing it in HDFS. Performance can be improved by indexing of data and is also extensible.

Hive has the superior capability to manage and analyze very large datasets which includes both structured and unstructured data located in distributed storage systems [7]. Hive effectively manages the frequent interaction between data flow and data content by providing frontend translation engine. Metadata in metastore helps in effective utilization of hive command to obtain data flow map reduce job plan without the need to generate huge amount of input data. Factors such as capacity planning and tuning for hive cluster makes the hive query performance complicated. This system helps in complete hive query execution process though the performance efficiency is not properly focused. Data loss can be overcome by CSM method. After hive finishes the query execution, the result of the execution is submitted to the Job Tracker. The Job Tracker which consists of map reduce tasks which runs the mapper and reducer job to store the final result in the HDFS.

Increase in intermediate data generation from map task to reduce task will lead to data loss, congestion in network bandwidth and performance degradation [8]. Map reduce job consists of number of map tasks and parallel reduce tasks which will combine all the intermediate outcome in the form of key-value pairs. Key-value pairs are generated by map tasks to produce final results. Excessive bandwidth is occupied by these intermediate data generated by map reduce tasks when mapping is done from mapper to reducer. This leads to network congestion which will cause serious performance degradation of map reduce jobs. To overcome this issue, data aggregation is carried out. Data aggregation is the process of combining similar results. In similar manner, inter-machine and intra-machine aggregation is done to overcome data loss and performance issue.

Big data analytics is potential in information retrieval. For providing intuition from very huge datasets and improving the results which is achieved at lower cost in information retrieval is evolved by big data analytics [9]. The design of information retrieval system is improved by the advancement of statistical tools and algorithms and promotes user with ease of access to the system, fault tolerance and used in analyzing the data. The privacy of data is not focused. This paper promotes multi query optimization techniques by exploring methods and joins are performed in single map reduce job.

Big data is a collection of massive amount of data whose data rate is dynamic for every milliseconds [10]. Big data analytics exploits various techniques to perform operations on huge volume of data. Data from multiple sources are pulled in to system which turns in to big data has give rise to a process called ETL. This paper focuses on analysis of log files and retrieving precise information from huge volume of data. This is effectively by achieved by ETL process and hadoop helps in less time consumption. Hadoop is a powerful tool for transforming huge data, perform analysis and exploits complex data. It also promotes fault tolerance and availability of data. Prediction of traffic flow using correlation analysis on hadoop environment.

Importing of big data securely is a tedious process which may lead to data loss. Big data is just residing at the system [11]. To make use of such huge data, computation of data need to be performed. To move data from one place to another, there is no organizational approach. Different technique such manual storing, graph, documents are performed effectively to maintain data of certain size. But this approach is not is maintain big data. Therefore, streaming stack for big data is introduced which helps deeper analysis of stored data. Advancement of big data tools captures the thoughts of many researchers. Streaming process involves in complex event processing and intrusion detection system to analyse network traffic. Classification of data is achieved by streaming process.

Map reduce is a high level parallel programming language which maps the data based on key-value pairs from the input and generates intermediate results [12]. Based on the intermediate key-value pairs, the data is combine and reduced to produce output data. The problem that arises here is that the intermediate datas remain unused after the tasks has completed. Therefore, a protocol is designed to handle such huge intermediate data. The aggregation of huge data is a complex task and the user when requesting these data may not be able to obtain. Since, the identities of workers cannot be changed they cannot analyse the data locally. In order to reduce this overhead, a cache replay protocol is designed.

Processing of huge data has becoming a important issue since speed of data generation is dynamic [14]. Big data is a heterogenous collection of structured and unstructured data of huge volumes. Speed of data generation makes the computing infrastructure complex to manage big data. Information retrieval mechanism is used to improve text information retrieval using map reduce technique. ETL process to improve performance. In addition to this, map reduce hides the details of parallel execution and makes the user to concentrate only on processing of data.

Map reduce is a parallel programming model for writing applications that can process big data in parallel multiple nodes [13]. Map reduce provides analytical capabilities for analyzing huge volumes of complex data. Hive is a data warehouse infrastructure tool to process structured data in hadoop. The shuffle and reduce phase are coupled together in hadoop and the shuffle can only be performed by running the reduce tasks [15]. This leaves the potential parallelism between multiple waves of map and reduce unexploited and resource wastage in multiple-tenant hadoop clusters, which significantly delays the completion of jobs in a multi-tenant hadoop cluster. Sorting of intermediate data still poses delay in reduce phase. Significantly improves job performance in a multi-tenant hadoop cluster.

The challenge of integrating No SQL data stores with Map Reduce under non-Java application scenarios is explained [16]. Data processing operation is not performed. Big data cannot be processed by using traditional processing method since it is a heterogenous collection of structured and unstructured data. Map reduce promotes data-intensive computing. But hadoop streaming module used to define non-java executables as map reduce jobs. Cassandra with map reduce improves performance. The distributed Cassandra cluster directly to perform map reduce operations and the other exporting the dataset from the database servers to the file system for further processing.

Using correlation analysis, traffic flow prediction by map reduce approach which is based on nearest neighbor is designed. Capacity of big traffic data processing for forecasting traffic flow in real time is analysed by KNN classifier and prediction calculation method is built on hadoop environment [17]. Correlation information is neglected during traffic flow prediction. Prediction accuracy can be improved by the choice of k and the prediction calculation. Autoregressive integrated moving of data. An approach for traffic flow prediction using correlation analysis on hadoop platform.

Hive, Spark and Impala have become the factual database set-up for decision support systems with huge database sizes [18]. In hive database setup is compared with a traditional database system to identify enamels in query execution. Though My SQL is efficient algorithmically, it suffers from serious issue that micro-architectural performance is affected by the query computation. Hive suffers from a issue by converting queries in to map reduce jobs and increasing algorithmic efficiency. Hive and My SQL is compared by their performance analysis. Bottleneck in hive is caused by context switches and it also uses large size code which causes stress in memory hierarchy. Performance of query analysis can also be identified through processor throughput. Analysis of My SQL and Hive clearly demonstrate the algorithmic disparates between single- node and distributed execution frameworks and the ideologies that motivate these differences.

Hive is a data warehouse system which is emerged as an essential facility for data computing and storage [19]. Several optimized scheme based on RC File and EStore is proposed which is an data placement structure that improves the query rate and reduces the storage space for hive. The data warehouse based on map reduce, column-store is used for read-optimized data. During query execution, it eliminates unnecessary column reading. But the efficiency of query performance is not improved due to heavy overhead of record reconstruction. Compression of table columns in a row groups by the storage method of RC File and decompressing column costs is reduced by EStore which is frequently used in the query. To improve query rate, Estore adopts the column classification scheme and outperforms storage space on hive.

Hadoop distributed file system is a block structured file system in which datas are stored as block with size 128 MB [20]. It is used to store huge data and involved in managing such huge data in distributed manner. The real-time streaming data can be stored in Mongo DB and hive. Using apache hive, big data analytics can be performed on data stored in hadoop distributed file system. Hive is a high-level programming language that make use of hadoop's core component map reduce to analyse the data. It promotes scalability by enhancing addition of new nodes. To view the insight of the big data, visualization tool is integrated with big data applications. Hive is a data warehouse system for hadoop and it SQL-like queries. Hive is implemented to make ease the use of hadoop file system. Processing of huge data is not an easy task. In order to perform this in hadoop wnvironment, apache storm is used.

3 Proposed Work

The admin collects data from different sources and stores these data on the database. The data collected are imported to hadoop environment through sqoop tool. The imported data's are normalized, preprocessed and clustered based on aadhaar number. These details are displayed on the website were the user retrieves the citizen details by providing aadhaar number. User makes a search based on the aadhaar number to retrieve citizen address, mobile number, qualification details, health records (Fig. 1).

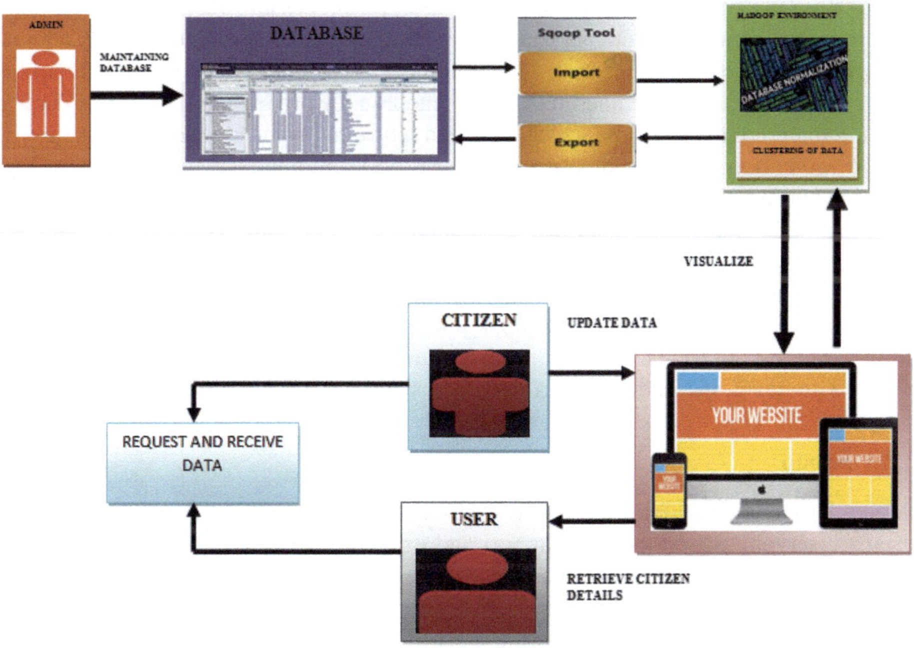

Fig. 1. System architecture

In the proposed system, admin collects the citizen details. The citizen details are collected and maintained in the hadoop database for processing. The data collected includes citizen name, aadhaar number, date of birth, gender, city, district, state, country, mobile number, qualification, smart card number, driving license number, blood group, last date of donation and number of times donated. These data are maintained by the administrator who can perform updates, modify operations.

The data collected on excel has been moved to hadoop environment by means of sqoop tool. Sqoop tool helps in import and export of data from hadoop and database server. The data need to be imported for further processing in the hadoop environment to get citizen details. Data collected about citizen is normalized for dimensionality reduction. Clustering of data is performed on hadoop environment.

Citizen details are retrieved by the user to know address, smart card number, driving license number, number of members in the family, qualification, blood group. Normalization of data is performed for dimensionality reduction of huge amount of data in hadoop environment. This is useful for hospitals for retrieving blood donor details, crime investigation and professionals for retrieving the details about residents along with their aadhaar details. The aggregation of Aadhaar card, driving license and smart card detail has been done. A proposed system which monitors the blood donors details of each individual of the country. It comprises of modules like generating the aadhaar number and store and retrieve data of a person.

3.1 Map Reduce

Map reduce is a programming model that is used for processing vast amount of data. In the proposed system, the details about citizen are collected that includes name, address, sex, phone number, qualification, aadhaar number, license number, smart card number, age, blood group are maintained in hadoop database for processing. Hadoop is cable of running map reduce programs which has been return in java. Map reduce paradigm is a parallel execution process which is useful for processing huge data analysis in multiple machines in the clusters. It consists of map and reduce phase. Mapping is done based on (key, value) pairs. The map task is created for every split. This executes the map function for each record containing the data in the split. Performance of map function depends up on the size of splits. Splits are nothing but data blocks in the data node. The map task is again processed in another node by the hadoop and produces the map output. The output of map task is produced to the reduce task. The output of reduce phase is stored in the hadoop distributed file system.

3.2 Sqoop

Sqoop is a data transfer tool which is used to transfer data between hadoop and database servers. Sqoop contains import and export tool. Sqoop import tool is used to import data from relational database to hadoop distributed file system. Sqoop export tool is used to export data from hadoop distributed file system to relational database. The emergence of sqoop import and export tool enhances the secure data transfer of huge amount of data and processing of data by analyzers such as map reduce to interact with relational database servers.

3.3 Hadoop

Hadoop is an open source framework that is used to store and process huge amount of data in a distributed environment. Hadoop runs map reduce algorithm in which parallel execution of data is performed. Complete statistical analysis of huge amount of data is performed by applications that was developed using hadoop. Hadoop distributed file system is used to store large amount of data and supports quick transfer of data between nodes. The input data given to hadoop distributed file system breaks the information down in to separate blocks and distributes the data to different nodes in the cluster which enables efficient parallel processing of data. Hadoop is a master/slave architecture that was used by hadoop distributed file system. Name node is a master and datanode is a slave. The name node keep tracks the data node. The data node consists of blocks. These are referred to as splits and is processed by map reduce programs.

4 Methodology

4.1 K-Means Clustering

Clustering algorithm has a extensive appeal and usefulness in fundamental data analysis. Thousands of clustering algorithm have been suggested in the literature survey in various experimental practices. The k-means clustering algorithm is an mature algorithm that has been fiercely experimented owing to its ease and integrity of implementation. Due to different issues, the research works in this field is varied.

K-means clustering is the widely used and experimented clustering formulations that are established on reducing a formal objective. To reduce the mean squared distance from each data point to its nearest center, a set of n data points in real d-dimensional space, Rd and an integer k. The objective is to determine a set of k points in Rd called as centers. This measure is called as squared-error distortion and this is referred to as variance based clustering.

Clustering analysis is an important technique for analysing fundamental data and is also used in analysing variety of engineering and scientific disciplines such as health care details, marketing, aadhaar details, computer vision.

Clustering is the process of grouping a set of physical or abstract objects in to classes of similar objects is called clustering. Cluster data objects are similar to another data object with in the cluster and remain dissimilar to data objects in other clusters. The clustered data objects are considered as a single group and large number of clustered data objects are considered as a form of data compression.

The process of gathering a lot of physical or conceptual objects into classes of comparable items is called clustering. A cluster is a gathering of similar data objects that are comparable to one another inside the same cluster and remains contradictory to the objects in different clusters. A cluster of objects can be dealt with on the whole as one gathering thus might be considered as a type of data compression.

Analysis of cluster sorts out data by abstracting hidden structure either as a gathering of individuals or as a chain of groupings. The representation would then be explored to check whether the groups of data as indicated by assumptions or to propose new experimental setup. The exploration of structure of data is done by cluster that does not require the suspicions common to most measurable techniques. It is denoted as unsupervised learning in the pattern recognition and artificial intelligence.

The most prevalent partitional calculation among different grouping calculations is K-mean clustering. K-means is a selective clustering algorithm, which is basic, simple to utilize and extremely proficient in managing huge amount of data with linear time complexity. K-Means calculation is one of the prevalent parceling calculation. The concept is to characterize the data into k clusters where k is the input indicated earlier through iterative migration system which converges to local minimum.

The aim of *K-Means* clustering is the optimization of an objective function that is described by the equation

$$E = \sum_{i=1}^{c} \sum_{x \in C_i} d(x, m_i)$$

In the above condition, mi is the focal point of cluster Ci, while d(x, mi) is the Euclidean separation between a point x and mi. Consequently, the model capacity E endeavors to limit the separation of each point from the focal point of the group to which the point belongs to. The objective of K-mean calculation is to limit the intra-cluster distance and increase the inter-cluster distance dependent on Euclidean separation.

Distance functions in k-means clustering procedure assumes a critical job. Various seperation functions are given to quantify the distance between data objects. In Manhattan distance function the distance between two data is the total of the absolute contrasts of their coordinates. The Manhattan distance, D1 between two vectors a, b in a n-dimensional real vector space with settled Cartesian coordinate framework, is the aggregation of the lengths of the projections of the line fragment between the focuses on coordinate axis.

$$D_1(a, b) = \|a - b\|_1 = \sum_{i=1}^{n} |a_i - b_i|,$$

where a $(a_1, a_2 \ldots a_n)$ and b = $(b_1, b_2 \ldots b_n)$ are vectors.

Step 1: Let X={x1, x2, x3,….,xn}
//set of data points
Let V={v1, v2,….,vn}
//set of centers
Step 2: Randomly select ‘c’ cluster centers
Step 3: Calculate distance between each data point and cluster centers
Step 4: Assign the data point to the cluster center
// minimum distance from the cluster center
Step 5: Recalculate the new cluster using

$$v_i = (1/c_i) \sum_{j=1}^{c_i} x_i$$

//C_i is the data points in i^{th} cluster
Step 6: Recalculate the distance between each data point and obtain new cluster centers.
Step 7: Otherwise, repeat step 3.

4.2 Algorithm

Input:
Set of feature vectors X={x1, x2, x3,….,xn}
// set of n data items
The number of cluster to be detected → K
Output:
Set of K clusters

Randomly select i data items from X as initial centroids
Initialize cluster centroid x1, x2, x3,…., xn belongs to R^n randomly
//finding centroid
Finding centroid using
$c^{(i)}=\arg\min_j||x^{(i)}-\mu_j||^2$
Assign each data item xi to the cluster which has the closest centroid
//each cluster is associated with centroid
Number of cluster K must be specified
Repeat step 1
Recompute the centroid using

$$d(i,j)=\sqrt{|x_{i1}-x_{j1}|^2+|x_{i2}-x_{j2}|^2+\ldots+|x_{ip}-x_{jp}|^2}$$

Update centroid

$$CP(x_1,x_2,\ldots,x_k)=\left(\frac{\sum_{i=1}^{k} x1st_i}{k},\frac{\sum_{i=1}^{k} x2nd_i}{k},\ldots,\frac{\sum_{i=1}^{k} xnth_i}{k}\right)$$

Take mean value and find nearest neighbour of mean for assigning it in cluster.

Repeat the above step until we get same mean
// until convergence is met

Input: number of clusters, K, X, C
Output: set of possible clusters
Begin
Let X={ x1, x2,......,xn} // set of aadhaar numbers
Select ← K points
Set C ← initial centroid
Repeat ←K points
From K ← assign points(X)
Mark X ← close to centroid C
Recompute ← centroid for n clusters
End

5 Result

The admin collects details about citizen and maintains in a database. For logging in to the system, admin has username and password. Only admin can view complete database and perform modifications like insert new details in the database. When request from citizen is sent to admin for any updations in the database. The admin can send respond to the citizen for updating his/her details in the database (Figs. 2, 3 and 4).

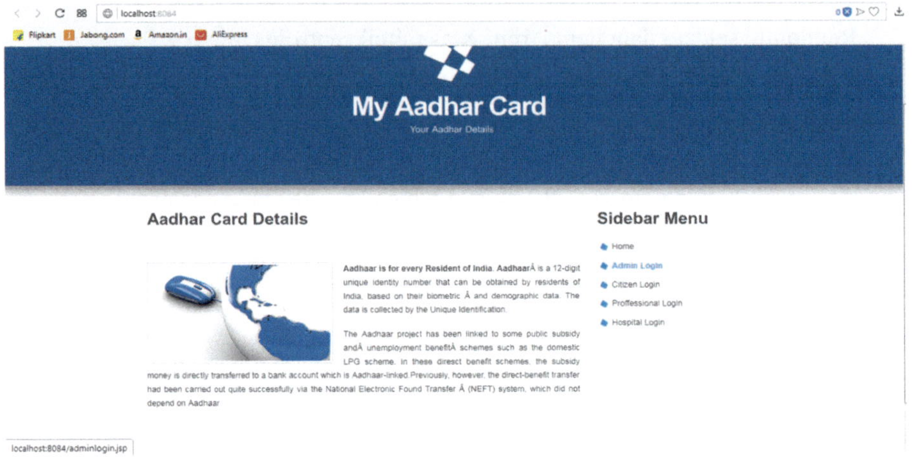

Fig. 2. Aadhaar card website

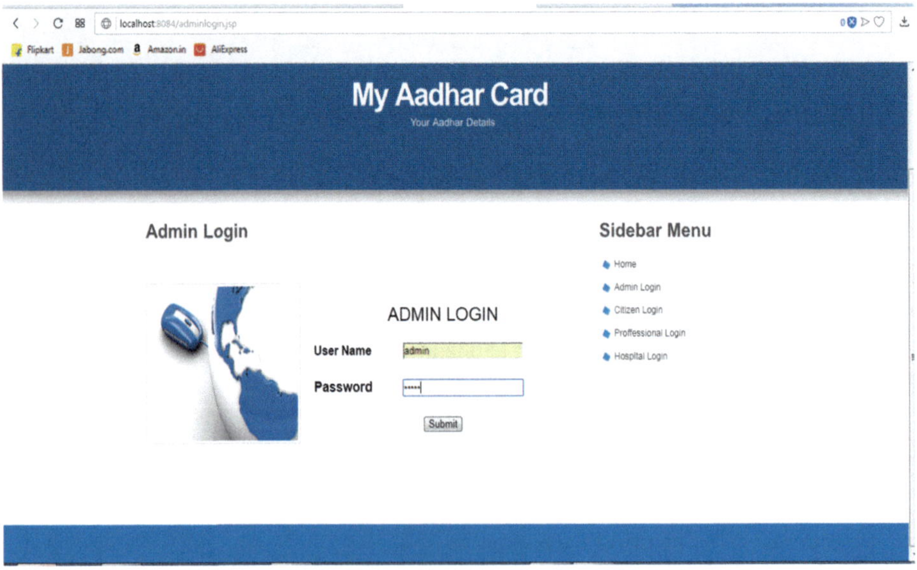

Fig. 3. Admin login

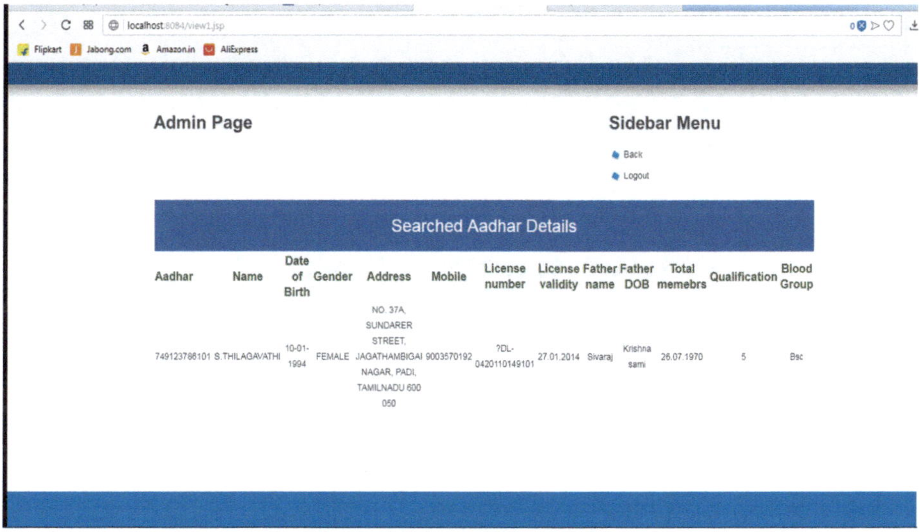

Fig. 4. Search citizen details

Each and every citizen has their own username and password for logging in to the system. The citizen can view and update only their particular details. The search is made through unique 12-digit aadhaar number (Figs. 5 and 6).

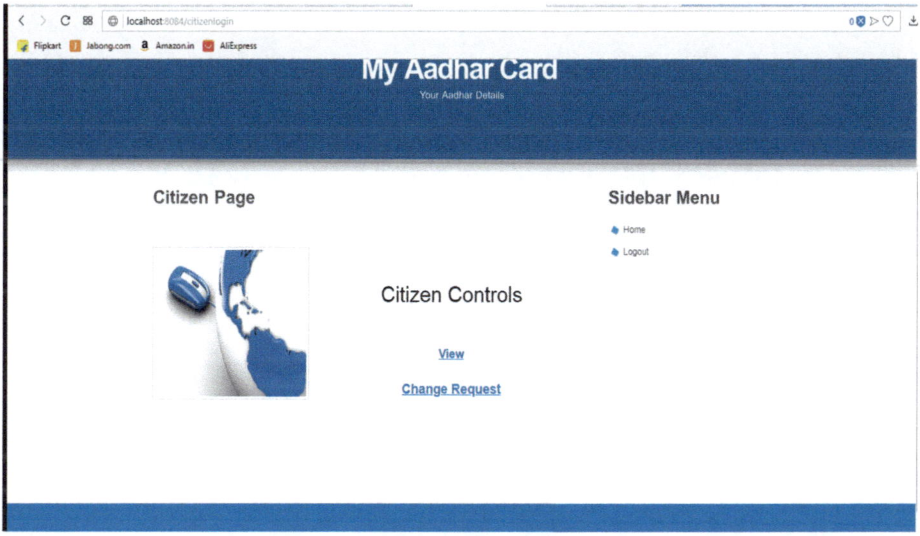

Fig. 5. Citizen login

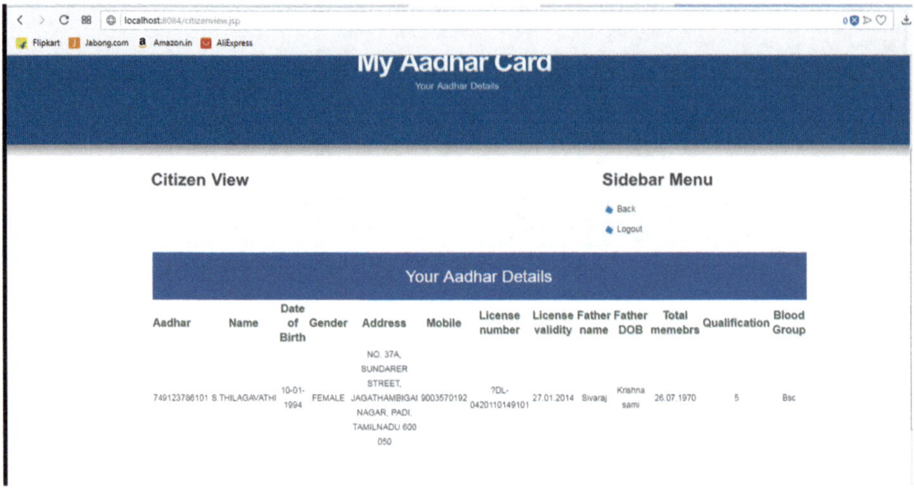

Fig. 6. Update citizen details

The database also contains blood donor details which is helpful to blood bank. The proposed system provides donor details for blood bank. They can only view and retrieve the donor details from the website but cannot perform any updations in the database. They have separate username and password for logging in to the system (Figs. 7 and 8).

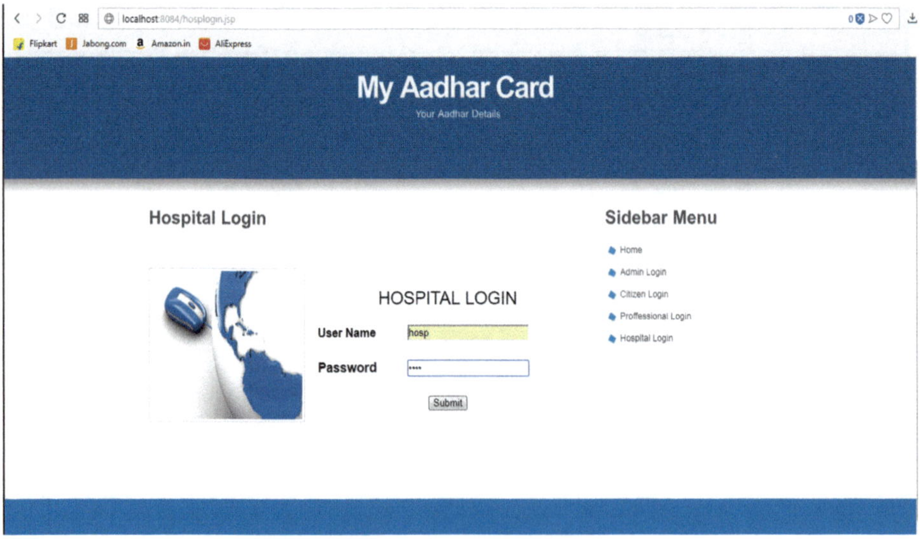

Fig. 7. Hospital login

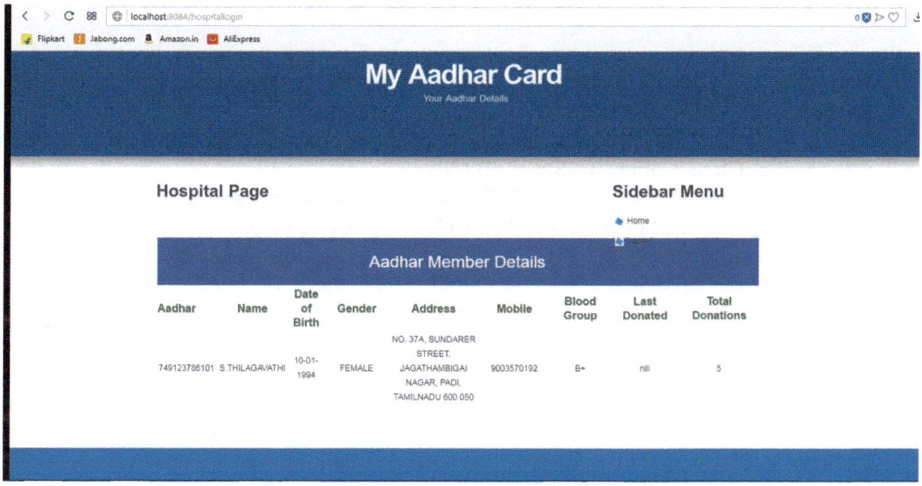

Fig. 8. Search donor details

The proposed system can also be useful for crime department for verifying the citizen details through their aadhaar number. They can only verify the citizen details through the aadhaar number. They cannot perform any changes in the database by logging in to the system through their username and password.

6 Conclusion

Aadhaar is a big data which need to be stored and managed securely and safely. Several processing techniques and privacy measures have introduced to process such huge confidential data. In order to update essential details of an individual along with existing database of an aadhaar for use by crime department, health care center and professionals, several algorithms, tools, techniques used in big data analytics have been discussed. This is useful for hospitals for retrieving blood donor details, crime investigation and professionals for retrieving the details about residents along with their aadhaar details. The aggregation of aadhaar card, driving license and smart card detail has been done. A proposed system which monitors the blood donors details of each individual of the country. It comprises of modules like generating the aadhaar number and store and retrieve data of a person. Performance evaluation demonstrates that the proposed schemes can achieve better efficiency than the existing works in terms of storage, search and updating complexity.

References

1. Wu, X., Zhu, X., Wu, G.Q., Ding, W.: Data mining with big data. IEEE Trans. Knowl. Data Eng. **26**(1), 97–107 (2014)
2. Xu, L., Jiang, C., Wang, J., Yuan, J., Ren, Y.: Information security in big data: privacy and data mining. China Commun. (Suppl. 2) (2014)
3. Matturdi, B., Zhou, X., Li, S., Lin, F.: Big data security and privacy: a review. IEEE Trans. Content Min. (2014). https://doi.org/10.1109/access.2014.2362522
4. Ren, Z., Wan, J., Shi, W., Xu, X., Zhou, M.: Workload analysis, implications and optimization on a production hadoop cluster: a case study on taobao. IEEE Trans. Serv. Comput. **7**(2), 307–321 (2013). https://doi.org/10.1109/tsc.2013.40. 1939-1374/13/$31.00
5. Fuad, A., Erwin, A., Ipung, H.P.: Processing performance on Apache Pig, Apache Hive and MySQL Cluster. In: International Conference on Information, Communication Technology and System, pp. 297-302 (2014)
6. Xie, H., Wang, M., Lie, J.: A data reusing strategy based on hive. National Natural Science Foundation of China, No. 61103046, and Fundamental Research Funds for the Central Universities, DHU Distinguished Young Professor Program, No. B201312
7. Wang, K., Bian, Z., Chen, Q., Wang, R., Xu, G.: Simulating hive cluster for deployment planning, evaluation and optimization. In: IEEE 6th International Conference on Cloud Computing Technology and Science, pp. 475-482 (2014). https://doi.org/10.1109/cloudcom.2014.119
8. Mavaluru, D., Shriram, R., Sugumaran, V.: Big data analytics in information retrieval: promise and potential. In: IEEE Network (2015)
9. Motwani, D., Madan, M.L.: Information retrieval using hadoop big data analysis. In: Proceedings of 08th IRF International Conference, Bengaluru (2014). ISBN: 978–93-84209-33-9
10. Ke, H., Li, P., Guo, S., Stojmenovic, I.: Aggregation on the fly: reducing traffic for big data in the cloud. Science and Engineering. Springer Proceedings in Physics, vol. 166. Springer, India (2015). https://doi.org/10.1007/978-81-322-2367-2_51
11. Bernstein, D.: The emerging hadoop, analytics, stream stack for big data. IEEE Cloud Comput. **1**(4), 84–86 (2014). 2325-6095/14/$31.00
12. Zhao, Y., Wu, J., Liu, C.: Dache: a data aware caching for big-data applications using the MapReduce framework. Tsinghua Sci. Technol. **19**(1), 39–50 (2014). ISSN ll1007-0214ll05/10llpp39-50
13. Thusoo, A., Sarma, J.S., Jain, N., Shao, Z., Chakka, P., Zhang, N., Antony, S., Liu, H., Murthy, R.: Hive – a petabyte scale data warehouse using hadoop (2010)
14. Kodabagi, M.M., Sarashetti, D., Naik, V.: A text information retrieval technique for big data using Map Reduce. Bonfring Int. J. Softw. Eng. Soft Comput. **6**, 22–26 (2016)
15. Guo, Y., Rao, J., Cheng, D., Zhou, X.: iShuffle: improving hadoop performance with shuffle-on-write. IEEE Trans. Parallel Distrib. Syst. **28**(6), 1649–1662 (2016). https://doi.org/10.1109/tpds.2016.2587645
16. Dede, E., Sendir, B., Kuzlu, P., Weachock, J., Govindaraju, M., Ramakrishnan, L.: Processing Cassandra datasets with Hadoop-streaming based approaches. IEEE Trans. Serv. Comput. **9**(1), 46–58 (2015). https://doi.org/10.1109/tsc.2015.2444838
17. Xia, D., Li, H., Wang, B., Li, Y., Zhang, Z.: A Map Reduce-based nearest neighbor approach for big-data-driven traffic flow prediction. IEEE Trans. 2169–3536 (2016). https://doi.org/10.1109/access.2016.2570021

18. Shulyak, A.C., John, L.K.: Identifying performance bottlenecks in Hive: use of processor counters. In: 2016 IEEE International Conference on Big Data (Big Data), pp. 2109-2114 (2016)
19. Li, X., Li, H., Huang, Z., Zhu, B., Cai, J.: EStore: an effective optimized data placement structure for Hive. In: 2016 IEEE International Conference on Big Data (Big Data), pp. 2996-3001 (2016)
20. Surekha, D., Swamy, G., Venkatramaphanikumar, S.: Real time streaming data storage and processing using storm and analytics with Hive. In: 2016 International Conference on Advanced Communication Control and Computing Technologies (ICACCCT), pp. 606-610 (2016)

Spinal Cord Segmentation in Lumbar MR Images

A. Beulah[1(✉)], T. Sree Sharmila[2], and T. Kanmani[1]

[1] Department of Computer Science and Engineering, SSN College of Engineering, Chennai, India
beulaharul@ssn.edu.in, kanmani1703@cse.ssn.edu.in
[2] Department of Information Technology, SSN College of Engineering, Chennai, India
sreesharmilat@ssn.edu.in

Abstract. The spinal cord is the vital organ in human central nervous system. Any pathology which significantly disturbs the original nature of the spinal cord will lead to sensory dysfunction and weakens the life quality of the person. To automatically detect the diseases in the spinal cord it is necessary to segment it from the image. Many image segmentation methods for medical image have been presented in recent years. In this paper, a region based segmentation is proposed to segment the lumbar spinal cord from T2-weighted sagittal Magnetic Resonance Image (MRI) of lumbar spine using region growing algorithm. First, the image is preprocessed and image threshold is applied to obtain a binary image. Then the region growing algorithm is performed on the binary image to segment the lumbar spinal cord. After the segmentation, any disease in the spinal cord can be analyzed.

Keywords: Segmentation · Region growing · Spinal cord · MRI

1 Introduction

For a human being, the spinal cord is the primary part of the central nervous system. Due to ageing or accident, any damage in the spinal cord can result in numbness, loss of sensation in different organs, loss of the ability to control the muscles voluntarily, and sometimes it leads to paralysis. There are many ways to identify diseases in the spinal cord. Physicians often confirm the nature of injury in the spinal column by physical examination or different medical imaging modalities such as X-ray, Computed Tomography (CT) or Magnetic Resonance Image (MRI).

CT and X-ray images were widely used for diagnosis purposes earlier. But MR images contain much more information when compared to other modalities. MRI has better properties such as high resolution, no radiation and can penetrate the spinal column without degradation. It gives an extremely clear and detailed image of soft-tissue structures that other techniques cannot achieve. It provides superior soft tissue contrast resolution. It has multiplanar imaging capabilities i.e. images can be acquired in multiple planes such as axial, sagittal and coronal. MR images give a very detailed diagnostic vision of organs and tissues in our body and contain much richer

D. J. Hemanth et al. (Eds.): COMET 2019, LNDECT 35, pp. 1226–1236, 2020.
https://doi.org/10.1007/978-3-030-32150-5_124

information when compared to other modalities. Many types of MR images are available. Of these, T1-weighted and T2-weighted images are popular. But, when compared to T1 images, T2 images provides high resolution contrast. Thus, in our method T2-weighted sagittal MR images are used.

The process of detecting the different organs from any medical image is medical image segmentation. The injury in the spinal cord can be analyzed if it is separated from other organs such as vertebrae, Intervertebral Discs (IVDs) etc. Therefore, segmentation of spinal cord is considered significant for quantification and diagnosis of various diseases related to it. Now a days, it is an important process for automatic interpretation and analysis of any disease in human organs. But there are several challenges and issues in segmenting the spinal cord, although its shape is apparently simple.

At present, there is no best method that satisfactorily segments the spinal cord and its neighbouring organs. Some of the previous works have been done on axial MRI. The segmentation can also be done in sagittal views of the lumbar spine. This gives motivation to segment the lumbar spinal cord from MRI. The region growing algorithm provides the main model for our segmentation approach. Figure 1 shows a sample MRI T2-weighted sagittal image of lumbar spine. The dataset used for this segmentation process contains T2-weighted lumbar MR images for 93 patients. All the images were collected from a scan center in Thiruvananthapuram, South India.

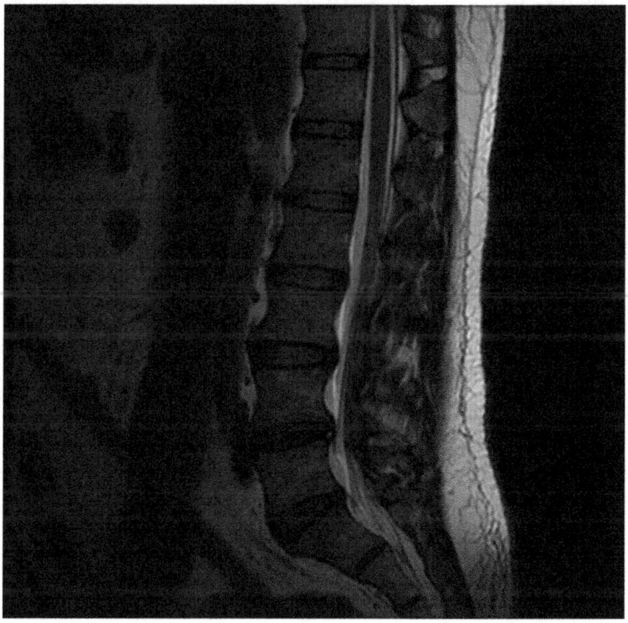

Fig. 1. MRI T2-weighted sagittal image of lumbar spine

The structure of the paper is as follows: Sect. 2 deals with related work, Sect. 3 is about the proposed segmentation method. The experimentation results are given in Sect. 4. The paper ends with the final thoughts and the future work in the last Sect. 5.

2 Related Work

Even though many research work is already done in the segmentation of spinal cord, it remains a challenge. As the images have different texture, noise, and other different factors the segmentation becomes a challenging task. The spinal cord matches the texture and the colour of the neighbouring organs. This makes the segmentation process tedious.

A signal intensity method for segmentation of spinal cord was proposed [1]. B-Spline contours are used to find the abnormal curves and then DSCR (Dural Sac Canal Ratio) was calculated which is the main criteria for calculating the spinal stenosis. T2-weighted sagittal images were used. This method was tested with different spinal curvatures and proved as robust. This method cannot detect small abnormalities.

Liao and Xiao [2] used Expectation-Maximization (EM) algorithm and dynamic programming to segment the spinal cord. T2-weighted sagittal MR images were considered. Using Dynamic programming, the anterior and posterior edges of the spinal canal is detected. After applying threshold and dynamic programming the intervertebral disks were segmented using region growing. Finally, spinal cord is segmented using Dynamic programming. The advantage of this method is that it is completely atlas-free and it requires only minimal human intervention. But, the stability of EM algorithm is lower because, it sometimes does not segment the spinal cord accurately. In another method, the authors applied Multi-layer Perceptron classifier for segmentation and detection of the Stenosis [3]. Axial slices of MR images were considered. This method consists of three steps: Spinal components segmentation using ROI, Spinal features extraction and Spinal stenosis classification. The overall performance is better than other works and the study has considered the axial view of the spine.

The Dynamic programming method was suggested [4] for extracting the boundary of the spinal canal. T1 and T2 weighted MR images were fused and then dynamic programming is applied for finding the boundaries. The distance between the reference boundaries and boundary obtained by this method is found for accuracy. This method works fully automatic. The tracked boundaries were quantitatively evaluated. But this method sometimes finds the incorrect location. Another work uses two-level thresholding method for segmentation of spinal cord [5]. T2-weighted mid-sagittal images were considered. This method provides a minimum response to the slight changes that occur in the spinal cord. The method produced only moderate results.

Modified Cobb's method is used to segment the spinal cord in the axial slice of CT images [6]. Anterior Posterior Diameter of spinal Canal (APDC), Cross Sectional Area of Dural sac (CSAD), Lumbar Lordosis (LL), Sacral Slope (SS), Anterior Vertebral Body Height (AVBH) and Mid Vertebral Body Height (MVBH) measurements were taken at vertebrae region of the lumbar spine for segmentation. The authors cannot correctly segment the spinal cord and this is because only minimal information exists about the spinal column. A new bottom-up model and active contour model for segmenting the spinal cord was proposed [7]. T2-weighted sagittal images were used. After segmentation, fill the holes by morphological operations which removes the small blobs. Since the relevant information related to the blob is acquired automatically, human interference is not required. The drawback is that this method is sensitive to noise, initial contour should be described and the resulting boundaries depends on this selection of contour. Expectation - Maximization (EM) segmentation is used to segment Intervertebral Disc in sagittal and axial slices of lumbar MRIs [8, 9].

A method which segments the spinal cord from MRIs using Topology preserving, Anatomy Driven Segmentation (TOADS) algorithm was discussed [10]. Both T1 and T2-weighted axial and sagittal views of MR images were considered. The past information about the anatomy and the neighbouring organs are used as a constraint for the segmentation. This segmentation process is highly resilient to noise. Three different segmentation methods namely, intensity based, surface based and image based methods were discussed [11]. Both axial and sagittal slices of MR images were considered. The intensity based methods such as region growing algorithm was used which robustly segments the spinal cord. But these methods have computational cost. Surface based methods such as B-spline model were faster and were reliable. It requires a large database with various image contrast for these algorithms to perform well.

3 Proposed Method

In this paper, an automated region based segmentation process is proposed for segmenting the lumbar spinal cord from MRI. The MR images are first preprocessed by converting the RGB images to grayscale images and filtering is applied to remove the noise. Then, a suitable threshold is applied to the image. Finally, the region growing segmentation method is used to extract the spinal cord from the preprocessed image. The working principle of the proposed method is shown in Fig. 2.

3.1 Image Preprocessing

The input image is T2-weighted mid-sagittal MRIs obtained from 93 patients. All the images are in RGB colour model. Even though, the images are RGB, all the channels have the same pixel values. So, it is necessary to convert the RGB image to grayscale before segmentation. Therefore, the images are converted to grayscale.

3.2 Top-Hat Filtering

Top-hat filtering is applied to the grayscale image. The morphological opening is computed on the image and the resultant is subtracted from the original image [12]. This process extracts the bright objects of interest in the dark background which is required for further processing.

Open. Morphological operations on a image can obtain the boundaries of all objects, the skeletons of the objects and the convex hulls of the objects [13]. Usually, morphological open remove the small components from the foreground of an image.

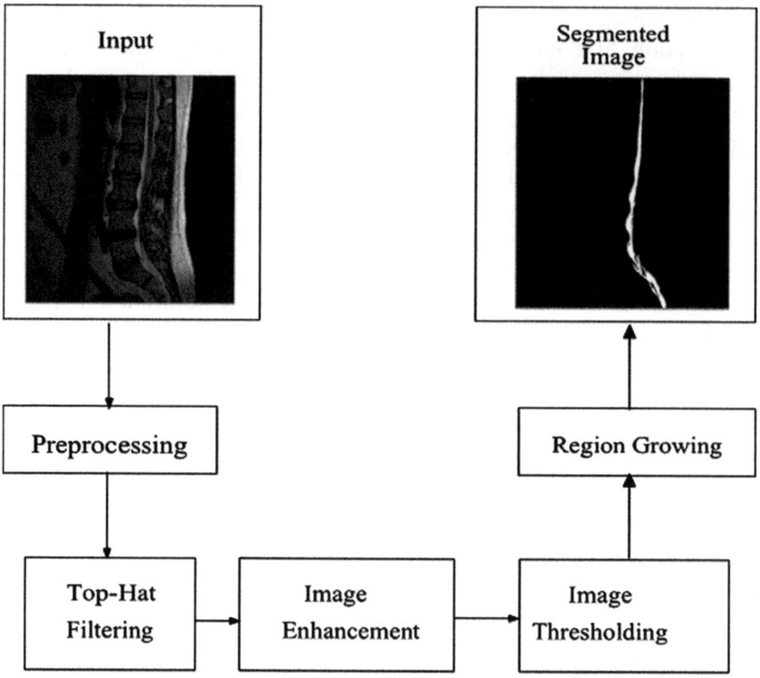

Fig. 2. Spinal cord segmentation - architecture

Morphological open is performed on the grayscale image with the structuring element (SE). Open is mathematically denoted by $I \circ S$ and is defined as:

$$I \circ S = (I \ominus S) \oplus S \tag{1}$$

where I is the input image, S is the SE, $\ominus$ and $\oplus$ denotes erosion and dilation respectively. The SE should be a single element and not an array of objects. The morphological open is an erosion with a SE followed by dilation with the same SE.

Erosion. The erosion on a grayscale image erodes the boundaries of the foreground regions. So, the foreground pixel areas will reduce and the holes within the areas will be larger after erosion. The SE may be considered as a small grayscale image. The pixel values above the border pixel values are assigned the maximum value. For grayscale images, the maximum value is 255.

Dilation. As erosion, dilation is also applied to grayscale image. The operator when applied on a grayscale image enlarges the boundary of foreground regions. So, the foreground pixels will grow, and the holes within the area will be smaller after dilation. The SE may be considered as a small grayscale image. Pixels beyond the image border are assigned the minimum value. For grayscale images, the minimum value is 0.

3.3 Image Enhancement

In spite of different image enhancement techniques, histogram equalization is performed to adjust the contrast in the image. This allows the area in the image with lower contrast gains higher contrast [13]. This method leads to the better visual of bone structures in MR images. The working principle is as follows: The pixel values for the grayscale image ranges between 0 to 255. For each pixel brightness value m in the sagittal image, the new pixel value k assigned is calculated as:

$$k = 256 \cdot \sum\nolimits_{i=0}^{m} \frac{N_i}{t} \tag{2}$$

where the sum counts the number of pixels (N_i) in the image with brightness less than or equal to m, and t is the total number of pixels.

3.4 Image Thresholding

Obtaining the features or separating the object from an image is an essential prerequisite for many kinds of analysis. Selecting the pixels within a range of the foreground and rejecting all the remaining pixels as background is the clear method to determine the brightness range in the original image.

After separating the foreground and background, the image obtained will be a binary image. This selection process is called image thresholding. The global image threshold is computed using Otsu's method [14]. This algorithm automatically performs the image thresholding as clustering based method or, it converts a graylevel image to a binary image. Otsu's algorithm is explained in Algorithm 1.

Otsu's method separates the enhanced image into two different classes of pixels (the foreground pixels and background pixels), and then calculates the optimum threshold. The calculated optimum threshold will separate the two classes, and their within class variance is minimal. The weighted sum of variances of the two classes is denoted as σ_w^2(T) and it is defined by:

$$\sigma_w^2(T) = w_0(T)\sigma_0^2(T) + w_1(T)\sigma_1^2(T) \tag{3}$$

where T is the threshold, the weights w_0 and w_1 are the class probabilities separated by a threshold T and σ_0^2 and σ_1^2 are class variances.

The probabilities for each class are evaluated as:

$$w_0(T) = \sum\nolimits_{i=0}^{T-1} p(i) \tag{4}$$

$$w_1(T) = \sum\nolimits_{i=T}^{B-1} p(i) \tag{5}$$

where B denotes the number of bins in the histogram.

Otsu's method determines that minimizing the within class variance means maximizing the between class variance and is depicted in terms of class probabilities w and class means μ. Between class variance is given by:

$$\sigma_b^2(T) = \sigma^2 - \sigma_w^2(T) \tag{6}$$

$$= w_0(\mu_0(T) - \mu)^2 + w_1(\mu_1(T) - \mu)^2 \tag{7}$$

$$= w_0(t)w_1(T)[\mu_0(T) - \mu_0(T)]^2 \tag{8}$$

The class means $\mu_0(T)$, $\mu_1(T)$ and μ are evaluated by:

$$\mu_0(T) = \frac{\sum_{i=0}^{T-1} ip(i)}{w_0(T)} \tag{9}$$

$$\mu_1(T) = \frac{\sum_{i=T}^{B-1} ip(i)}{w_1(T)} \tag{10}$$

$$\mu = \sum\nolimits_{i=0}^{B-1} ip(i) \tag{11}$$

Finally, the individual class variances are:

$$\sigma_0^2(T) = \sum\nolimits_{i=0}^{T-1} [i - \mu_0(T)]^2 \frac{p(i)}{w_0(T)} \tag{12}$$

$$\sigma_1^2(T) = \sum\nolimits_{i=T}^{B-1} [i - \mu_1(T)]^2 \frac{p(i)}{w_1(T)} \tag{13}$$

Algorithm 1. Otsu's Algorithm

Input: Grayscale Lumbar Spine Image
Output: Binary Image

A. For each intensity level find the histogram and the probabilities.
B. Initialize $w_i(0)$ and $\mu_i(0)$.
C. Iterate through all the possible thresholds T = 1 … M, where M is the maximum intensity.
D. Update probabilities w_i, means μ_i and calculate between class variance $\sigma_b^2(T)$.
E. Optimum threshold corresponds to the maximum between class variance $\sigma_b^2(T)$.

3.5 Region Growing

The region growing algorithm is a region based segmentation method. In this paper, the region growing method is used to segment the spinal cord region. As region growing method requires to select initial seed points, this method is otherwise called as pixel based segmentation. This approach analyzes the neighbours of the initial seed point [15]. Also, it analyzes whether the neighbouring pixels can also be included in the region. This is repeated until no new neighbouring pixels be added into the region. The selection of the initial seed points is based on the user's benchmark.

Seed Point Selection. The initial stage in region growing segmentation is selecting the seed point. Selection of seed point depends on the user constraints such as the range of pixels in the grayscale, in a grid placing the pixels evenly etc. The initial location of the seed is the initial region. Later, from the initial seed point the regions grow to neighbouring points. This depends on the region membership constraint. The membership constraint can be the pixel value, the texture, or the colour.

The region formation is vital as the regions grow only on the membership criterion. For instance, if the membership constraint is a pixel intensity value, then the information about the image from the histogram is used. This is because the histogram is used to fix a threshold value for the region membership constraint. The region growing algorithm is presented in Algorithm 2. This region growing method gives very good segmentation.

Algorithm 2. Region Growing Algorithm

Input: Binary Lumbar Spine Image
Output: Segmented Image

A. Choose an arbitrary seed pixel point in the lumbar spinal cord and compare it with neighbouring pixels.
B. From the seed pixel the region is grown by combining the neighbouring pixels that matches the region membership criterion. This increases the size of the region.
C. When the growth of this region stops, we get a single connected component.
D. This single connected component is the desired lumbar spinal cord.

4 Experimental Results

The data set consists of mid sagittal T2-weighted MR Images for 93 patients. Experiments were performed on this dataset. First the RGB images are converted into grayscale. The top-hat filtering is performed on the grayscale image. The SE used is a 10 sized disk. Then histogram equalization is performed on the filtered image. After histogram equalization apply Otsu's thresholding. In the binary image, on selecting a seed point the lumbar spinal cord is segmented. The Table 1 shows the Original Image, Images after top-hat filtering, image enhancement, thresholding, and region growing for 3 different images.

Table 1. Original image, images after top-hat filtering, image enhancement, thresholding, and region growing for image 1, image 2 and image 3

Segmentation Process	Image 1	Image 2	Image 3
Original Image	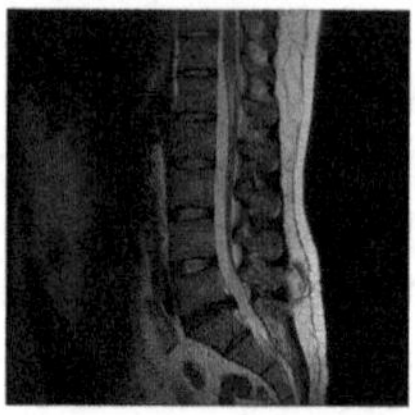	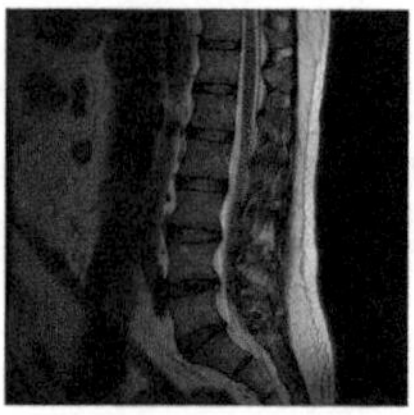	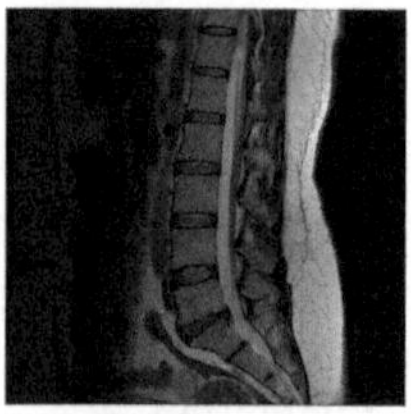
Top-Hat Filtering	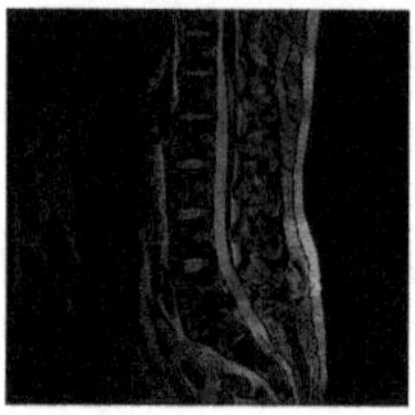	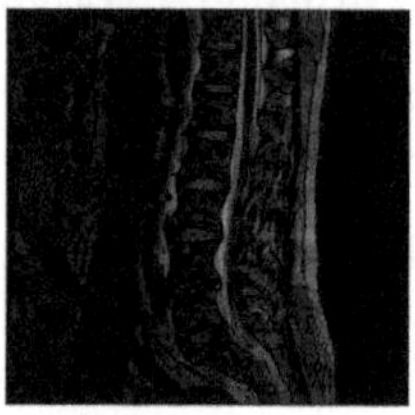	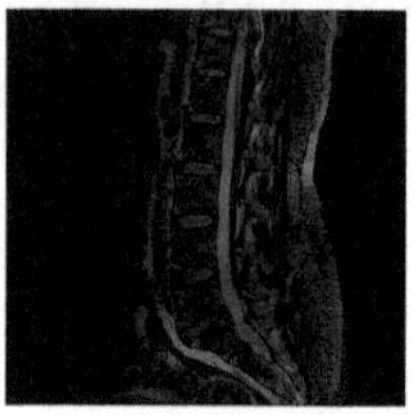
Image Enhancement	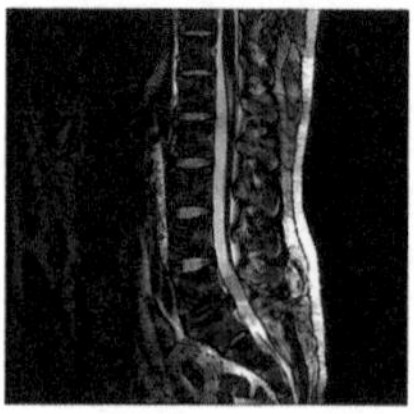	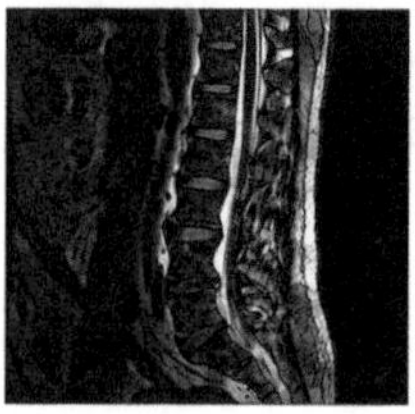	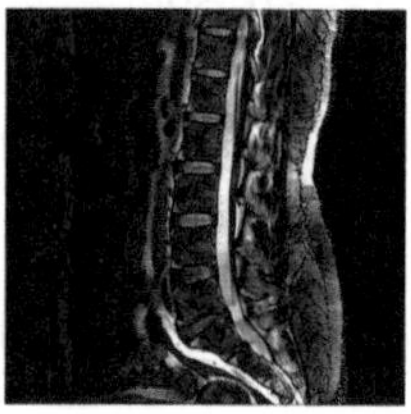
Image Thresholding	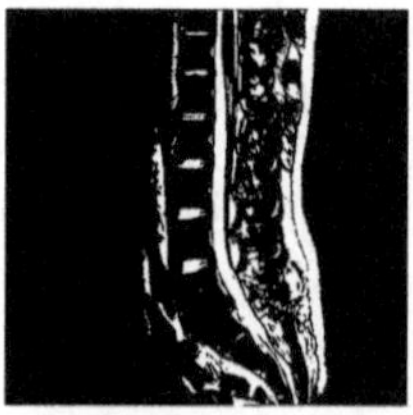	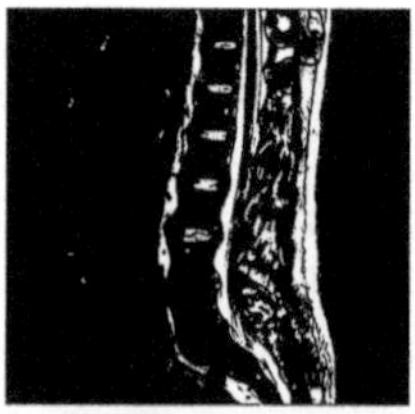	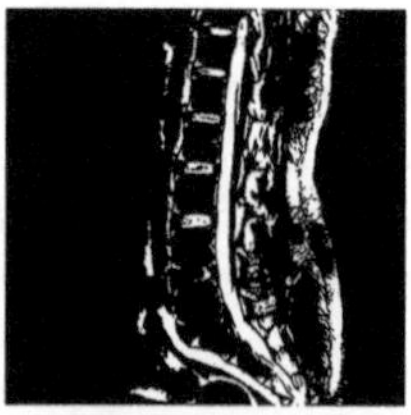
Region Growing	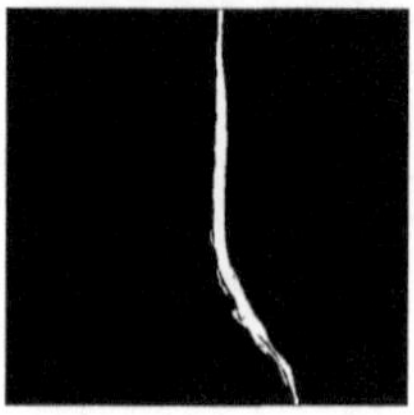		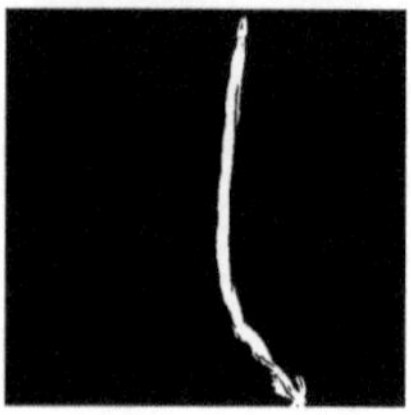

5 Conclusion

A region growing method to segment the lumbar spinal cord on sagittal T2-weighted MR images is proposed in this paper. The paper concentrates on preprocessing techniques as well as the region growing method. The experiments were performed on 93 clinical dataset. The input image is first preprocessed by applying top-hat filter and image enhancement. Otsu's thresholding is applied to the image to obtain a binary image. Finally, by using region growing algorithm the spinal cord is segmented from the MR image. Experimental results shows that our method of segmentation gives better results. The segmented binary image or the segmented region can be further utilized for analysis of any disease.

References

1. Ruiz-España, S., Arana, E., Moratal, D.: Semiautomatic computer-aided classification of degenerative lumbar spine disease in magnetic resonance imaging. Comput. Biol. Med. **62**, 196–205 (2015)
2. Liao, C.C., Ting, H.W., Xiao, F.: Atlas-free cervical spinal cord segmentation on midsagittal t2-weighted magnetic resonance images. J. Healthc. Eng. (2017)
3. Koompairojn, S., Hua, K., Hua, K.A., Srisomboon, J.: Computer-aided diagnosis of lumbar stenosis conditions. In: Medical Imaging 2010: Computer-Aided Diagnosis, International Society for Optics and Photonics, vol. 7624, p. 76241C (2010)
4. Koh, J., Chaudhary, V., Jeon, E.K., Dhillon, G.: Automatic spinal canal detection in lumbar MR images in the sagittal view using dynamic programming. Comput. Med. Imaging Graph. **38**(7), 569–579 (2014)
5. El Mendili, M.M., Chen, R., Tiret, B., Villard, N., Trunet, S., Pélégrini-Issac, M., Lehéricy, S., Pradat, P.F., Benali, H.: Fast and accurate semi-automated segmentation method of spinal cord MR images at 3T applied to the construction of a cervical spinal cord template. PLoS ONE **10**(3), e0122224 (2015)
6. Abbas, J., Hamoud, K., May, H., Hay, O., Medlej, B., Masharawi, Y., Peled, N., Hershkovitz, I.: Degenerative lumbar spinal stenosis and lumbar spine configuration. Eur. Spine J. **19**(11), 1865–1873 (2010)
7. Koh, J., Scott, P.D., Chaudhary, V., Dhillon, G.: An automatic segmentation method of the spinal canal from clinical MR images based on an attention model and an active contour model. In: 2011 IEEE International Symposium on Biomedical Imaging: From Nano to Macro, pp. 1467–1471 (2011)
8. Beulah, A., Sree Sharmila, T.: EM algorithm based intervertebral disc segmentation on MR images. In: 2017 IEEE International Conference on Computer, Communication and Signal Processing (ICCCSP), pp. 1–6 (2017)
9. Beulah, A., Sree Sharmila, T., Pramod, V.K.: Disc bulge diagnostic model in axial lumbar MR images using Intervertebral disc Descriptor (IdD). Multimed. Tools Appl. **77**(20), 27215–27230 (2018)
10. Chen, M., Carass, A., Oh, J., Nair, G., Pham, D.L., Reich, D.S., Prince, J.L.: Automatic magnetic resonance spinal cord segmentation with topology constraints for variable fields of view. NeuroImage **83**, 1051–1062 (2013)
11. De Leener, B., Taso, M., Cohen-Adad, J., Callot, V.: Segmentation of the human spinal cord. Magn. Reson. Mater. Phy. Biol. Med. **29**(2), 125–153 (2016)

12. Bai, X., Zhou, F., Xue, B.: Image enhancement using multi scale image features extracted by top-hat transform. Opt. Laser Technol. **44**(2), 328–336 (2012)
13. Gonzalez, R.C., Wintz, P.: Digital Image Processing. Applied Mathematics and Computation. Addison-Wesley Publishing Co., Reading (1977)
14. Vala, H.J., Baxi, A.: A review on Otsu image segmentation algorithm. Int. J. Adv. Res. Comput. Eng. Technol. (IJARCET) **2**(2), 387–389 (2013)
15. Adams, R., Bischof, L.: Seeded region growing. IEEE Trans. Pattern Anal. Mach. Intell. **16** (6), 641–647 (1994)

Biometric Access Using Image Processing Semantics

C. Aswin(✉), N. Dhilip Raja, N. Angel, and K. Sudha

Department of Computer Science Engineering,
St. Joseph's College of Engineering, Chennai, India
rockyashwin16@gmail.com, therealthelip@gmail.com,
angel.mcastjosephs@gmail.com, Ksudha2910@gmail.com

Abstract. We propose a model-based approach for accessing a magnetic lock in the door using face recognition. We Implement Face Recognition by using image processing semantics. The Facial recognition uses facial landmarks such as chin, eyebrow, nose, eyes and lip to encode the user's face. The encoded signal is sent to the lock via WIFI. We interfaced the facial recognition module through an android app and implemented the backend using python server. Our approach here is by using KNN algorithm which is one among the various machine learning algorithm. The K-nearest neighbor algorithm is used to classify the facial landmarks which is calculated by Euclidian distance formula which calculates the distance between the markers. The Face recognition module runs through python which is used for connectivity between the magnetic lock and backend. The magnetic lock is implemented using Node MC, switch, electromagnet and a power supply.

Keywords: Facial landmarks · WIFI · Android app · Python server

1 Introduction

It is a tradition that all the employees must wear or carry their identification cards to access their office/work place. Their identities are checked by security guards or machines installed at the entry points. In office, an employee requires to carry an identification card which is being scanned by the machine to verify his/her identity in order to make sure that no unauthorized person can access the work place.

As the necessity for higher levels of security rises, technology is bound to swell to fulfill these needs. There exist several biometric systems such as signature, finger prints, voice, iris, retina, hand geometry, ear geometry, and face. Among these systems, face recognition appears to be one of the most accepted, cherished and available systems.

The Biometric system namely facial recognition is being used in here. It primarily focuses on how humans perceive their surroundings and how each person differentiates their identities with maximum accuracy. Facial Recognition is implemented through different methods, one of such methods is by using KNN algorithm which is one among the machine learning algorithms. Machine learning is a tendency of the

D. J. Hemanth et al. (Eds.): COMET 2019, LNDECT 35, pp. 1237–1243, 2020.
https://doi.org/10.1007/978-3-030-32150-5_125

computer to learn by itself. The machine learning process takes place with the help of huge data sets to imitate more of human-like decision making.

2 Existing Background

For the biometric access, fingerprint technologies were primarily used where a fingerprint is being registered using a special sensing device which is then enrolled and authenticated. Considering fingerprint, we may face a lot of crisis, such as expensive hardware costs and also can lead to false rejections or false acceptances.

In the case of facial recognition, the usage of eigenfaces algorithm which uses a small set of 2-D data and the accuracy of the method lacks in finesse. There are other entries in facial recognition which lacks in efficiency and accuracy in the final results. But the future of facial recognition is bright as it has a greater security and accuracy and they will provide a convenient and contactless usage of the user systems.

3 Proposed System Architecture

In this paper, we present the facial recognition with three modules. These modules show the workings of the facial recognition based locking system. They work by introducing an android app, a python server and a magnetic lock which are integrated together as a whole to form a complete working module of Image processing using facial recognition (Fig. 1).

The modules that are being used are

- Application Interface
- Server Integration
- Magnetic Lock.

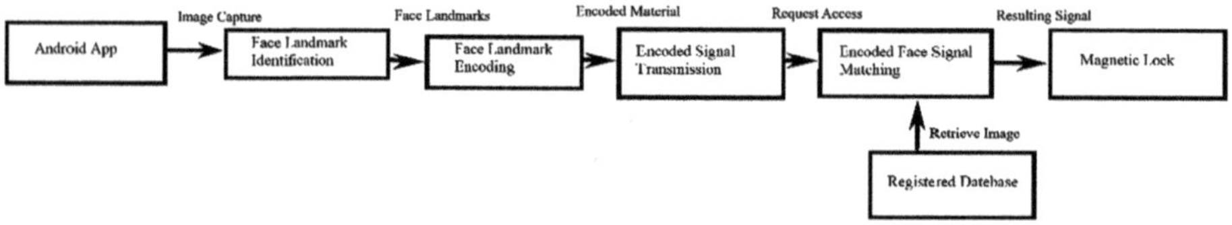

Fig. 1. The flow of the system showing the process of face recognition to the magnetic lock

3.1 Application Interface

In this Module, the user can register or request for Access to the python server to gain access to the magnetic lock. The Application is created through Android studio. The Application captures the image of the User and sends it to the Python server.

3.2 Server Integration

In this module, Python provides the resources and the connectivity of the module to the Application Interface and Magnetic Lock. The Python server receives the facial landmarks from the Android App which is then Encoded with Facial Encoding signal. This Signal is sent to the Registered Database to Facial Recognition matching. The matching is done at the python server and android application. The result of the facial matching given out as resulting signal.

3.3 Magnetic Lock

In this module, the Magnetic Lock is integrated with the android app and python server via WIFI. The magnetic lock receives the resulting signal from the python server after the facial recognition matching. After the facial recognition matching process, if the requested Face access doesn't match the registered Face from the registered Database the Lock does not open or else it opens.

4 Implementation Results

The results are shown through screenshots of the module, mentioned in the proposed system architecture. The obtained result is the detection of the registered face in the database and demagnetizing the electromagnet if the registered face is detected.

4.1 Application Interface

See Figs. 2 and 3.

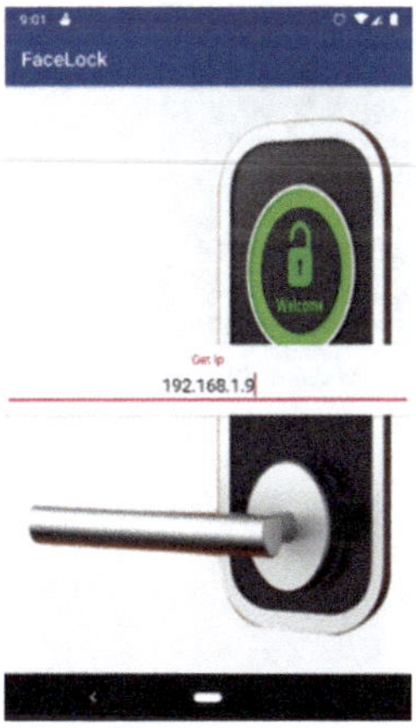

Fig. 2. IP input

Fig. 3. Access interface

4.2 Server Integration

Server Run

See Fig. 4.

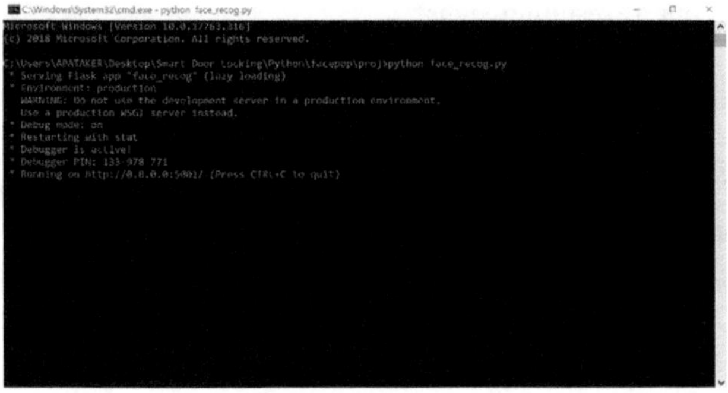

Fig. 4. Running the server

Server Request

See Fig. 5.

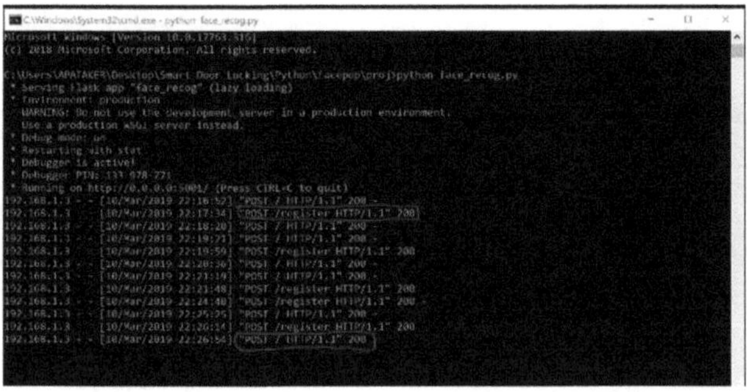

Fig. 5. Register request and access request

Server Response

See Fig. 6.

Fig. 6. Face match response, Face matches with the registered Face from the Database.

Registered Database

See Fig. 7.

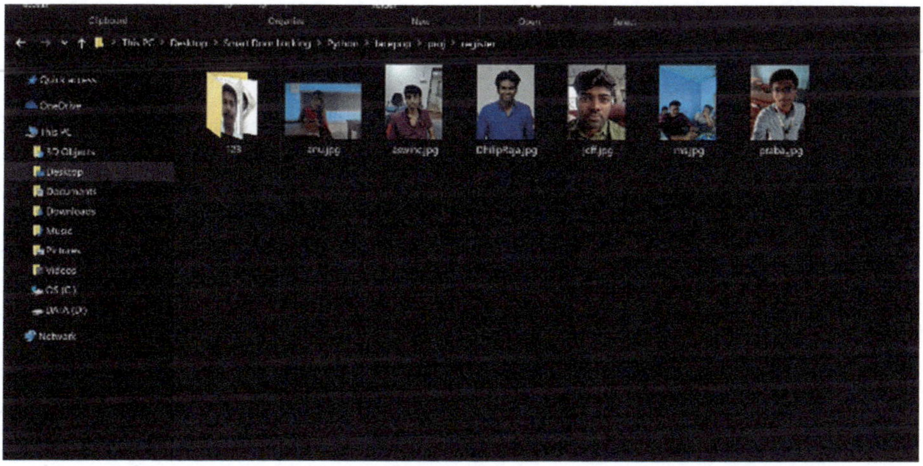

Fig. 7. Registered Face Database, these are the sample registered face from the database for facial matching, Source - https://photos.app.goo.gl/buUwApTWLQSVd6ED8

4.3 Magnetic Lock

See Fig. 8.

Fig. 8. Magnetic lock used for the module, contains electromagnet, NodeMCU, switch and power supply

5 Conclusion and Future Work

It is highly clear that the biometric authentication technologies have widely immerged as core part of various security systems including personal identification card such as Aadhar Card, Passport, Driving license, etc. These security measures provide massive levels of safety and caution against various threats.

Biometric authentication is now among the everyday home technology thereby replacing fingerprint, iris and other biometric parameters. Facial recognition enables smarter integration which in turn saves time and cost of implementation of the systems. Facial recognition eliminates the need for security personnel and provides automation for the systems.

As for the Future, iPhone X has bought in a new technology of facial recognition which is capable of increasing the accuracy and security on an astronomical level. It captures the users face by mapping thirty thousand infrared dots which provides a fail-proof of authentication to the users but, this is only the beginning as there is a huge scope and prospects. On a important note, Facial recognition's goal is to bring about systems which recognizes the user as the password.

References

1. Bhatt, H.S., Bharadwaj, S., Vatsa, M., Singh, R., Ross, A., Noore, A.: A framework for quality-based biometric classifier selection. In: Proceedings of International Joint Conference on Biometrics, pp. 1–7 (2011)
2. Dieckmann, U., Plankensteiner, P., Wagner, T.: SESAM: a biometric person identification system using sensor fusion. Pattern Recogn. Lett. **18**(9), 827–833 (1997)
3. Jain, A.K., Ross, A., Prabhakar, S.: An introduction to biometric recognition. IEEE Trans. Circuits Syst. Video Technol. Spec. Issue Image Video Based Biom. **14**(1), 160–170 (2004)
4. Kumar, D., Ryu, Y.: A brief introduction of biometrics technology. Int. J. Adv. Sci. Technol. **4**(1), 185–192 (2009)
5. Maes, S.H., Beigi, H.S.M.: Open sesame! Speech, password or key to secure your door? In: Asian Conference on Computer Vision, Hong Kong, pp. 531–541 (1998)
6. Terzopoulos, D., Waters, K.: Analysis and synthesis of facial image sequences using physical and anatomical models. IEEE Trans. Pattern Anal. Mach. Intell. **15**(6), 569–579 (1993)
7. Lee, K.-C., Ho, J., Kriegman, D.: Nine points of light: acquiring subspaces for face recognition under variable lighting (2001)
8. Sim, T., Kanade, T.: Combining models and exemplars for face recognition: an illuminating example (2001)

RETRACTED CHAPTER: M-Voting with Government Authentication System

P. Yaagesh Prasad(✉) and S. Malathi

Department of Computer Science and Engineering, Panimalar Engineering College, Chennai, Tamil Nadu, India
yaageshrp@gmail.com, malathi_raghu@hotmail.com

Abstract. In the development of advanced mobile technologies, the historic balloting method can be replaced to a modernistic and potent scheme titled as mobile voting. The mobile voting system affords an uncomplicated, beneficial, powerful way to vote wipe out the drawback of traditional approach. In this paper we urge a mobile voting system which is in essence an online voting system over which users can cast their vote through their smartphones or by applying an e-voting network page. To bring out the security, OTP accession is used which is most generally on the network to confess the variation between a human using a network services automated bot thus effecting the network page more guarded against spam-bot attacks. If the consequence of the matching algorithm is three-point match then audits whether this person possess voter ID after that it will authorizes with AADHAAR ID, if he has the right to vote then a balloting form is granted to him and the third level of attestation is carried out by using One Time Password (OTP) and then the biometric recognition of fingerprint sensor is used for an authentication. Nowadays technology is being utilized progressively as a key to contribution voters to cast their ballots. In order to utilize the rights, approximately all voting system around the world embodies the steps: balloter identification and authentication, voting and documenting of votes expulsion, vote toll, notification of election results.

Keywords: M-Voting · Biometrics verification · Portability secure · Fingerprint recognition · Mobile device

1 Introduction

In India, the choice framework could also be a basic technique to accumulate and replicate person's suppositions. So it ought to be a great deal of compelling, proficient, strong and secure. Race in Asian nation area unit coordinated merely exploitation electronic possibility machines by a pair of or three government-guaranteed associations, the natural science Corporation of Asian nation and Asian nation natural science Limited. Still Mobile possibility isn't straightforward for those associations maintaining the race in Asian nation. Asian nation spends some portion of money to upgrade their whole possibility structure to administer a additional strong government than their

---

The original version of this chapter has been retracted. The retraction note to this chapter can be found at https://doi.org/10.1007/978-3-030-32150-5_163

D. J. Hemanth et al. (Eds.): COMET 2019, LNDECT 35, pp. 1244–1259, 2020.
https://doi.org/10.1007/978-3-030-32150-5_126

subjects. By and large, possibility framework is coordinated in brought on or unfold spots referred to as possibility slows down.

In the development of advanced mobile technologies, the historic option technique may be replaced to a stylish and potent theme titled as mobile pick. The mobile legal system affords associate uncomplicated, beneficial, powerful thanks to vote wipe out the disadvantage of ancient approach. During this paper we tend to urge a mobile legal system that is in essence an internet legal system over that users will forge their vote through their smartphones or by applying associate e-voting network page. To bring out the protection, OTP accession is employed that is most usually on the network to confess the variation between an individual's employing a network services or machine-driven larva therefore effecting the network page additional guarded against spam-bot attacks. If the consequence of the matching algorithmic rule is three-point match then audits whether or not this person possesses citizen ID at the moment it'll authorizes with AADHAAR ID, if he has the proper to vote then a option type is granted to him and therefore the third level of attestation is administered by victimization just one occasion Password (OTP) so the biometric recognition of fingerprint sensing element is employed for associate authentication [2]. Today technology is being utilised increasingly as a key to contribution voters to forged their ballots. So as to utilize the rights, close to all legal system round the world embodies the steps: ballotter identification and authentication, pick and documenting of votes expulsion, vote toll, notification of election results.

2 Related Works

In our related work focuses on

1. Survey on Aadhar Card Systems
2. Processing of M-voting
3. One Time Password
4. Unique Device Identification.

2.1 Survey on Aadhar Card Systems

On the opposite hand, many civil society activists and social commentators have expressed considerations concerning the weak privacy provisions within the Aadhaar project and bill. However, whereas alerting to the probabilities of gap doors to mass police investigation, we have a tendency to feel that a number of the commentaries are infinite in their criticisms and not entirely specific in their statements of considerations. The gist of most criticisms has been that the utilization of biometry and a novel number, and storage of biometric and demographic knowledge and authentication trails in an exceedingly central repository, are essentially unsafe.

In [15] describes the advancement in technology has transformed the way a lot of process. Among them electronic voting system is the most important one. Infrastructure University Kuala Lumpur students Representative Council (SRC) is a council that represents student interest in the management of the university. The council conduct election for students and choose their representatives. This voting currently use the paper based voting which is insecure, inefficient, prone to errors. This paper proposes the adoption of Android based mobile voting. This enables the students to cast their votes and track the results in real time. This application will also provide candidates with a centralized platform. Rapid Application Development (RAD) methodology is used for the development of the application. But different users will use different platforms. Hence in future, it will be better to design the system that will operate in different platforms such as iOS, Blackberry and windows. For more efficiency, we can use biometric or fingerprint method in future.

Privacy considerations about the Aadhaar project are the topic of a lot of heated discussion recently (Express News Service 2016; NDTV 2016a). On the one hand, positions taken by the govt. And UIDAI on these problems are ambiguous. Conflict before a bench within the Supreme Court, the professional General of Republic of India has claimed that Indian voters don't have any constitutional right to privacy (PTI 2015). This is often shocking not solely as a result of there are many interpretations of constitutional provisions and judgements to the contrary (Bhatia 2015; Kumar 2015), however conjointly as a result of it contravenes typical knowledge and best practices in digital authentication and authorisation systems (Diffie 1979; Wikipedia 2016l, g).

Aadhaar might not solely change economical style, delivery, watching and analysis of services in every domain severally, however conjointly offers the likelihood of victimisation fashionable knowledge analytics techniques for locating massive scale correlations in user knowledge which will facilitate improved style of policy ways and early detection and warning systems for anomalies. As an example, it should be staggeringly perceptive to be able to correlate education levels, family financial gains and nutrition across the complete population; or illness unfold with income and education. Additional typically, it should change effecting political economy analysis, epidemiologic studies, automatic discovery of latent topics and causative relationships across multiple domains of the economy.

2.2 Processing of M-Voting

E-Voting build that contains a lot of security and end-user authentication with four-digit pin with the assistance of NIC (National Identification Code) [9] and by mistreatment the SIM card authentication [1] of the sure mobile. This Application uses SIM for authentication, however not mistreatment any citizen id proof for verification of the user. Thus, quite verification necessary to supply a secure authentication for the user. The SMS system provides all the detail choice and thru that user vote for the sure candidate [10].

M-Voting system mistreatment the automaton Application that brings fingerprint authentication [8]. This application provides a transparent authentication with fingerprint and take the user into the system. Rather than standing and looking ahead to choose within the stall this technology offers the user to vote anyplace and anytime.

This application is completely location freelance and even check the govt proof of the user with Government authentication systems thus fraudulences within the choice method is reduced. This application helps in increasing the amount of votes and reduces time consumed for choice. The building of the new secure protocol with IES [11] for mobile choice. This method is secured with digital signature and blend networks. This method provides a secure entryway to speak and demonstrate with the choice server. Adding of biometric security can improve the protection, however it wants a lot of study. Aadhaar ID authentication is finished with OTP service. This "One Time Password" service offers the secure authentication for the user.

This kind of choice Application can manage the citizen's info by that voter will login and use his choice rights [4]. This method contains all options of legal system associate degreed this app can persuade be an price effective. To create the economical system and higher performance, we are able to use fingerprint and life science during this on-line choice system.

2.3 One Time Password

Electronic balloting square measure presently being performed utilizing World-Wide net in various nations of the globe thanks to this progression a elector need to not visit the mensuration place. In any case, must merely language on the computer with an internet connection [3]. This balloting needs Associate in Nursing entrance code for the e-voting through the propel report of a elector. During this paper, a mensuration set up by strategies for moveable innovation initiated as most essential use of GSM primarily based Personal Response System, that permits a elector to create his selection in easy Associate in Nursing useful route while not the farthest purpose of your time and space by incorporating an electronic balloting strategy with the GSM foundation. To affirm the vote, the elector can get Associate in Nursing affirmation message from the investigation Station that their votes are gotten. On the off likelihood that the distinctive finger impression and biometric is used it may deliver precise outcomes.

During this paper balloting is that the right of each subject during a democratic country. During this electronic balloting theme, the target is to beat the drawbacks of standard balloting systems [5]. GSM is wide used technology. This electoral system is intended by desegregation embedded system with the mobile infrastructure. This paper includes the wants, style and order of a non specific e-voting methodology utilizing GSM versatile framework as almost essential use of GSM primarily based individual Response system, where voters will forged their votes anytime, anyplace utilizing a GSM Mobile Equipment (ME). If biometric and distinctive finger impression device is used it may provide higher effectiveness.

Electronic balloting technology will embody punch cards, optical scan balloting systems and specialised balloting kiosks (including self-contained Direct-recording electronic (DRE) balloting systems) [6]. Electronic balloting systems could provide blessings compared to alternative balloting techniques. Associate in Nursing electronic electoral system are often concerned in anyone of variety of steps within the setup, distributing, voting, collecting, and investigation of ballots. There's no good resolution for the protection problems that the majority of e-voting systems encounter with at the instant, and a few issues aren't even technically soluble [13]. As a next work, it's required to style a concrete science protocol to ensure the namelessness and therefore the confidentiality of a elector. Even the most effective election observation cannot solve the transparency issues with Electronic balloting by victimisation net and GSM represented on top of. If fingerprint can be accessorial it might give higher results.

2.4 Unique Device Identification

In this selection framework, the elector character card is supplanted by savvy card during which all the detail of the individual is reinvigorated. Simply the present individual will survey utilizing their keen card [1]. Here the savvy card peruser peruses the good card and also the delicate components of that individual is shown and at the moment it requests confirmation that is iris acknowledgment. On the off likelihood that the iris style coordinates then the individual will poll. The keen card peruse can get to the cards however it'll provides a blare sound that demonstrates that the individual has as of currently voted. Iris discovery within the application tries to a high level of exactness in every ascertained case. The code features a genuinely outstanding shade in distinction with no matter is left of the attention and its encompassing region; this empowers an original edge to be completed with knowledge from the image bar chart to disengage the pupil. Thus, biometric and distinctive mark enclosed it might offer higher outcomes.

Another operator primarily based set up for secure electronic selection is projected within the paper. The projected system bolsters the execution of a UI recreating customary call cards, semi-mechanical selection gadgets or use merely electronic selection stalls. Because of pre-calculations and the age of specialist, the elector needs to not do calculations. The vote for our set up is right down to earth and straightforward to create. Additionally, different cryptographical natives aboard exhibited qualifications is used, e.g.: daze marks. A consumer will vote typically (the votes is printed) or in electronic stalls. The framework likewise furnishes a consumer with portability: the consumer will have a specialist program in his telephone that may send his picked tally to the counter agent. It might offer precise outcomes by together with distinctive finger impression sensors.

In [16] Modern society fully relies on ICT. There is no outdated technological concepts for voting. Countries all over the world are examining foe e-voting with more confidentiality and security. The Zurich Minister of the Interior, Markus Knotter, commented on the successful completion of the Zurich e-voting system. This digital system consists of the candidate details, election parties that are participating in the election and all the details are connected to the National Election commission. If the candidate has voted once, the details are updated. Hence, there is no chance for double voting. Since it is electronic, the counting of votes can be done digitally. So, there is no possibility of wrong voting count. E-voting Zurich is in use in three communities' soon it could be linked to all 171 communities. Hence, we can vote through our mobile phones or laptops. This makes the system more flexible. For more efficiency we can use encryption and decryption algorithm for coding and decoding in future.

In [14] describes the electronic voting can be done through mobile phones. It provides more advantage over traditional voting like less manpower, saves time, accuracy, transparency, fast result etc. But it has many challenges. The main challenge is to keep the voted data securely. To overcome this, we have used the NFC tag in new e-voting system for more accuracy and transparency. This tag stores the information of voters to check the voter's vote in application [12]. This E-polling has three phases. The first involves analyze and verification of the voter and voter's vote in application. Second phase get the OTP. Third stage will count and sort out all the votes and declare the result of voting in application. For more efficiency we can use fingerprint or biometric in future.

In [17] Philippines is a democratic country. This country has been using the popular paper-based and PCOS machine (Precinct-Count Optical Scanners) Like signal loss, data traffic, misplacement of shaded candidates making it unreadable and paper. This research intends to maximize the usage of the mobile phones and make it more useful for betterment of the country. This will be useful for conducting the elections in national and local levels. This will help the government to reduce the costs, crimes and the identity fraud in conducting the election. For more efficiency, we can use fingerprint or biometric approach in future.

2.5 Restricting Client Numbers

The Network access: limit purchasers allowed to form remote calls to SAM security policy setting controls that users will enumerate users and teams within the native Security Accounts Manager (SAM) info and Active Directory. The setting was initial supported by Windows ten version 1607 and Windows Server 2016 (RTM) and might be designed on earlier Windows consumer and server in operation systems by putting in updates from the KB articles listed in Applies to section of this subject.

This topic describes the default values for this security policy setting in numerous versions of Windows. By default, computers starting with Windows ten version 1607 and Windows Server 2016 area unit a lot of restrictive than earlier versions of Windows. This implies that if you have got a mixture of computers, like member servers that run each Windows Server 2016 and Windows Server 2012 R2, the servers that run Windows Server 2016 could fail to enumerate accounts by default wherever the servers that run Windows Server 2012 R2 succeed.

3 Proposed Work

The below architecture gives the overall flow of the designed voting system (Fig. 1).

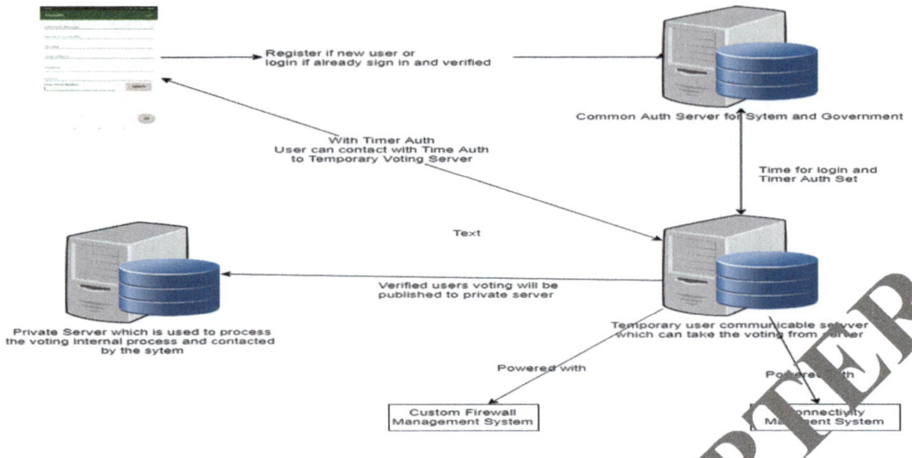

Fig. 1. Architecture diagram

3.1 User Registration and Validation

An enlisted client is a client of a site, program, or other framework who has already enrolled. Enrolled clients typically give a type of accreditations, (for example, a username or email address, and a secret phrase) to the framework with the end goal to demonstrate their personality: this is known as signing in. Frameworks expected for use by the overall population frequently enable any client to enroll just by choosing an enlist or join work and giving these certifications to the first run through. Enlisted clients might be conceded benefits past those allowed to unregistered clients.

Unique Device Identification. FDA has designed up and keeps on actualizing an interesting contraption recognizable proof framework to sufficiently distinguish therapeutic gadgets through their dispersion and utilize. At the purpose once fully dead, the mark of most contraptions can incorporate a 1 of a sort gadget symbol (UDI) in human- and machine-meaningful form. Contraption labelers ought to likewise gift bound knowledge concerning each contraption to FDA's world distinctive Device Identification info (GUDID). General society will get and transfer knowledge from the GUDID at AccessGUDID.

The novel contraption ID framework, which is able to be staged in additional than quite long whereas, offers numerous benefits which will be all the a lot of fully acknowledged with the choice and combination of UDIs into the human services conveyance framework. UDI usage can enhance quiet security, modernize contraption postmarked observation, and encourage therapeutic contraption advancement.

Vote Process. This system is handled by an automatic management tool which is able to self-handle itself and unstoppable while not the kill codes. The Kill code within the system is to terminate the choice. Just in case any of the crash or vulnerability within the system this tool can restart itself and continue the method that was left over. It conjointly encompasses a live update page that is globally accessible and provides live standing of the result and method happening.

The need for automatic tooling in versatile machining, assembly, and sheet fabrication systems is reviewed. The various ways in which of implementing these systems, their edges and disadvantages area unit mentioned. The elemental modules of automatic tool transfer, storage, loading/unloading, and management area unit delineated in conjunction with the appropriate level of automation for each module. The advantages and prerequisites for pilotless machining systems, this sensing ways in which and so the tool replacement ways in which are also reviewed. The importance of a tool data, its uses and structure ar highlighted. Finally, the look and analysis of automatic tooling systems and operative ways in which, with the assistance of distinct events technique ar mentioned. Associate existing laptop package that's capable of simulating automatic tooling systems for versatile manufacturing systems is given.

4 Methodology

This process will map the AADHAR to a particular mobile device and unique the user voting with the device. Collect data from the AADHAR card (QR code). Request the AADHAR validation system (Government Managed) to send a AADHAR validation OTP (One Time Password) to the user's registered phone number. Once the validation with AADHAR is done device will request the API (Application Program Interface) to check the user registration details. Now API (Application Program Interface) will check the database for the requested user details, if the user is not registered then the device is registered to the user's AADHAR card and if the user is registered to another device then the user should deregister the old device to register the new device.

```
function register(macaddr, aadharid, phnum, .....)
Input : complete AADHAR data and device mac address
Output: User devices token details

fdata <- getAadhardata(qrscan <-getQRscanner());
if fdata is available
dialog(deactivationmessage);
else
map(aadharid, macaddress);
END if;
```

This system will not allow two devices to map to a single AADHAR ID. This is the effective measure to unique the device's and this will reduce the fraud of voting several vote's using a single device. Also, when a user is registered with another device this system will not allow any further integration with the application.

4.1 Unique Device Identification

FDA has built up and keeps on actualizing a remarkable gadget recognizable proof framework to sufficiently distinguish therapeutic gadgets through their dispersion and utilize. At the point when completely executed, the mark of most gadgets will incorporate a one of a kind gadget identifier (UDI) in human-and machine-meaningful shape. Gadget labellers should likewise present certain data about every gadget to FDA's Global Unique Device Identification Database (GUDID). General society can seek and download data from the GUDID at Access GUDID.

The novel gadget ID framework, which will be staged in more than quite a long while, offers various advantages that will be all the more completely acknowledged with the selection and combination of UDIs into the human services conveyance framework. UDI usage will enhance quiet security, modernize gadget postmarked observation, and encourage therapeutic gadget advancement.

4.2 Methodology

This process will map the AADHAR to a particular mobile device and unique the user voting with the device:

Step 1. Collect data from the AADHAR card (QR code).
Step 2. Request the AADHAR validation system (Government Managed) to send a AADHAR validation OTP (One Time Password) to the user's registered phone number.
Step 3. Once the validation with AADHAR is done device will request the API (Application Program Interface) to check the user registration details.
Step 4. If the user is register to current device then the system allows the user to proceed with remaining applications otherwise the user is requested to register with the current device.

```
function validation (macaddr, aadharid, phnum, .....)
Input  : complete AADHAR data and device mac address
Output : User devices token details

fdata <- getAadhardata(qrscan <- getQRscanner());
if fdata is available
  proceedApp();
else
  dialog(registerrequire);
  register(macaddr, aadharid, phnum, .....);
END if;
```

This system will not allow two devices to map to a single AADHAR ID. This is the effective measure to unique the device's and this will reduce the fraud of voting several vote's using a single device. Also, when a user is registered with another device this system will not allow any further integration with the application.

4.3 Middle Server Interaction

This middle server integration protects the user or cyber criminals from contacting the main server by holding all the connections from the outer world to the system. Every single vote and connection are validated in the middle server and passed over the secure connection to the main vote processing server. This secure connection is opened only for very short time period only. This will help in protecting verified votes away from hackers and the main server is processed only for vote counting purpose. This type of middle server interactions are user in most famous services like Paytm and more payment gateways protecting successfully completed payment and transaction and most commonly used in ATM machines. This provides the m-voting system a secure gate pass for vote, but while the middle server is attacked all the data from the middle server is cleared and the user in interaction with the system need to cast their vote again. If any user casted a vote in slot it will be filtered in the middle server, so no need to implement any kind of filtration on main server.

4.4 API Interactions

Application Program Interface (API's) are developed using PHP (Hypertext Pre-processor) for communication with the database. This API's handle the request and response communication with the application and the Backend Services.

This API's are completely custom made with custom firewalls to defend itself from the bot attacks. If any kind of user is accessing the system with any system other than the registered signed .apk then the API terminates the connection.

4.5 Bot Detection Technique

The ongoing development of botnet movement in the internet has pulled in altogether the consideration of the exploration network. Botnets are a standout amongst the most unsafe types of system-based assault today, in charge of a substantial volume of vindictive exercises from disseminated disavowal of-benefit (DDoS) assaults to spamming, phishing, recognize burglary and DNS server Spoofing. The idea of bot-net alludes to a gathering of traded off PCs remotely controlled by one assailant or a little gathering of aggressors cooperating called a "botmaster". These vast gatherings of hosts are collected by transforming helpless hosts into alleged zombies, or bots, after which they can be controlled from far off. An accumulation of bots, when controlled by a solitary order and control (C2) framework, shape what is known as a botnet. The botmaster's capacity to complete an assault from hundreds or even countless PCs implies expanded data transfer capacity, expanded handling power, expanded memory for capacity and countless sources making botnet assaults more pernicious and harder to recognize and guard against.

4.6 Automated Management Tool

This system is handled by an automated management tool which will self-handle itself and unstoppable without the kill codes. The Kill code in the system is to terminate the voting. In case any of the crash or vulnerability in the system this tool will restart itself and continue the process which was left over. It also has a live update page which is globally accessible and gives live status of the outcome and process happening (Fig. 2).

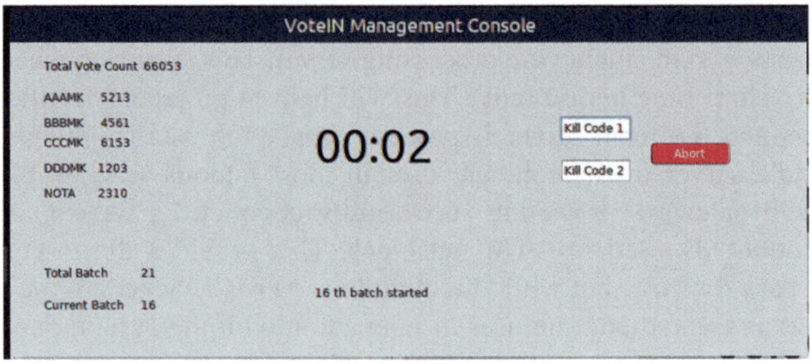

Fig. 2. Management tool

The need for automatic tooling in versatile machining, assembly, and sheet fabrication systems is reviewed. The varied ways of implementing these systems, their edges and downsides are mentioned. The fundamental modules of automatic tool transfer, storage, loading/unloading, and management are delineated in conjunction with the suitable level of automation for every module. The benefits and stipulations for remote-controlled machining systems, this sensing ways and therefore the tool replacement ways also are reviewed. The importance of a tool information, its uses and structure ar highlighted. Finally, the planning and analysis of automatic tooling systems and operative ways, with the help of distinct events technique ar mentioned. Associate degree existing pc package that is capable of simulating automatic tooling systems for versatile producing systems is bestowed.

5 Result and Discussion

5.1 System Setup

This system needs an Android to run on client side and on server side we need a elastic server system where it has to hold billions of votes in one day. As of now this project is reusing the AADHAAR database for user authentication. This authentication is done using the AADHAAR API so all the need for authentication is fulfilled. For running all this service, we need a active internet connection. Also, Android Application can run on any Android backed Operating System.

5.2 Performance Evaluation

The unmistakable imprint has been spoken to on the grounds that the great bit of the edge for biometric trademark verification for a few reasons: The finger impression is for the principal half dimension, and its geometric course of action is basically management led by 2 correlative muscles that control the broadness of the print. The unmistakable imprint is excellent. All the equivalent, there region unit such a major assortment of parts that go in the course of action of those surfaces (the particular finger impression) that the shot of false partners for either is to a decent degree low. Without a doubt, even innately unclear people steady have totally self-sufficient particular finger impression surfaces. On these lines removing a challenge that has been referred to in a not many social orders against unmistakable imprint scanners, wherever a finger must contact a surface, or particular finger impression checking, wherever the finger ought to be sent near a finger scanner (Fig. 3).

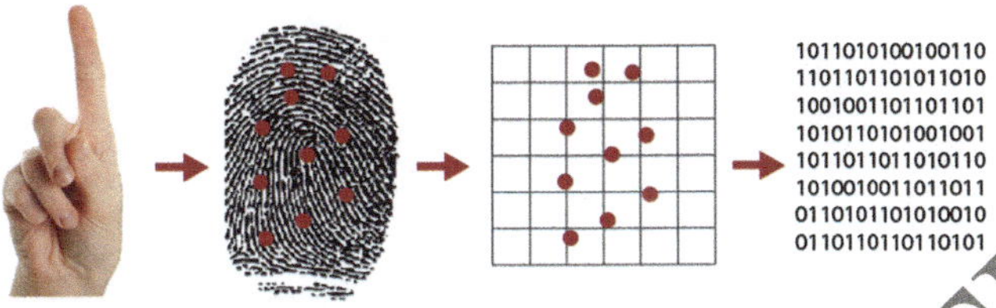

Fig. 3. Fingerprint translation to binary

Fingerprints region unit graphical stream like edges appear on human fingers [6, 7]. Particular finger impression conspicuous evidence relies upon 2 premises: (i) unmistakable finger impression refined segments territory unit unending – in lightweight of the existence frameworks and ontogenesis of granulating edge skin, and (ii) fingerprints of an individual region unit eye catching. With a specific completion objective to perform planning, it's essential that a perception of the structure and features of the particular finger impression is nonheritable.

The lines that stream in various models over a particular imprint region unit alluded to as edges and furthermore the zones between edges territory unit alluded to as valleys. A ton of little of techniques is named detail planning. The 2 types region unit, edge fulfillment and bifurcation. A conclusion is wherever a grasp closes and a bifurcation is wherever a hold segments from a singular gratitude to 2 manners by which state a Y-convergence. Since fingerprints region unit enduring as talked in regards to higher than, inside the occasion that they were gotten in the midst of correspondence or recuperated from partner degree end because of poor security, a guilty party may sufficiently fake their character, envisioning in lightweight of false bioscience. Along these lines, pleasant security designs zone unit essential to ensure this biometric information. There zone unit various cryptanalytic plans and figuring's available regardless, this examination is particularly charmed by a confirmation essentially based approval and trust, HTTPS, and AES respective key encoding, that region unit talked (Fig. 4).

The Unique Identification Authority of India (UIDAI) has been made, with the order of giving a solitary Identity (Aadhaar) to each single Indian occupant. The UIDAI gives on-line approval to check the character guarantee of the Aadhaar holder. Aadhaar "approval" infers the system whereby Aadhaar run, on board entirely unexpected properties, together with life science, region unit submitted to the Central Identities data Repository (CIDR) for its check bolstered data or information or records open with it. UIDAI gives a web backing of encourage this technique. Aadhaar confirmation benefit just responds with a "yes/no" and no individual character data is came as a critical side of the response.

Fig. 4. AADHAR with OPT verification at Government AADHAR Center

At long last during this enterprise within the improvement of innovative versatile advancements, the notable selection technique is often supplanted by a futurist and powerful arrange titled as transportable balloting. The transportable balloting framework bears associate uncomplicated, useful, effective approach to vote wipe out the disadvantage of typical approach. During this paper we tend to encourage a transportable balloting framework that is mostly an online primarily based balloting framework over that purchasers will create their alternative through their cell phones or by applying associate e-voting system page. To prolong the safety, OTP promotion is employed that is most by and enormous on the system to admit the variability between a person's utilizing a system administration or computerized larva on these lines poignant the system page a lot of ready for spam-bot assaults. On the off likelihood that the end result of the coordinating calculation is three-point coordinate at that time reviews whether or not this individual have elector ID subsequently it'll approves with AADHAAR ID, within the event that he has the privilege to vote then a selection structure is allowed to him and therefore the third level of validation is finished by utilizing just one occasion password (OTP) and later the biometric acknowledgment of distinctive mark device is employed for a confirmation. Of late innovation is getting used logically as a key to commitment voters to solid their tallies. Keeping in mind the tip goal to use the rights, roughly all balloting framework round the globe encapsulates the means: balloter recognizable proof and validation, balloting and recording of votes removal vote toll, warning of call comes regarding.

User Registration and Validation system provide efficient handling of user and device mapping. The below Screenshots show the simple flow of the outcome of the project. With this method of Unique Device Identification is done. Few Government and private bank sector are using the unique device identification for security and management purpose. This process will help in stopping the fraud of casting multiple votes by a single user. This type of authentication system are going to be in hand on near future and some authentication system like Two-way Auth are already providing few functionalities of this methodology (Fig. 5).

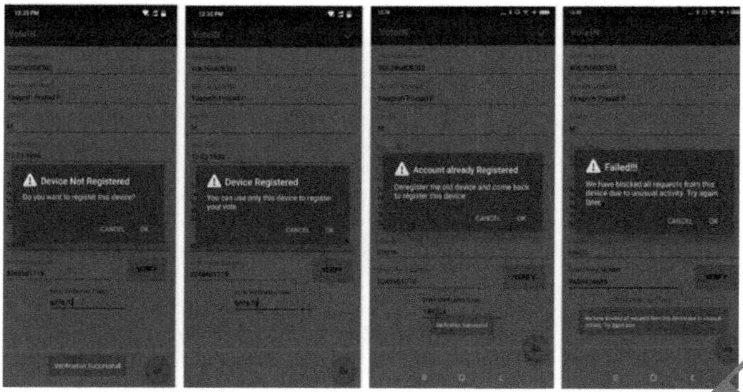

Fig. 5. User registration and identification

Voting process with middle server communication as explained above will provide major security from preventing access of duplicate data and machine into the system. This secure connection is opened only for very short time period only. This will help in protecting verified votes away from hackers and the main server is processed only for vote counting purpose. This type of middle server interactions are user in most famous services like Paytm and more payment gateways protecting successfully completed payment and transaction and most commonly used in ATM machines. This provides the m-voting system a secure gate pass for vote, but while the middle server is attacked all the data from the middle server is cleared and the user in interaction with the system need to cast their vote again. If any user casted a vote in slot it will be filtered in the middle server, so no need to implement any kind of filtration on main server (Fig. 6).

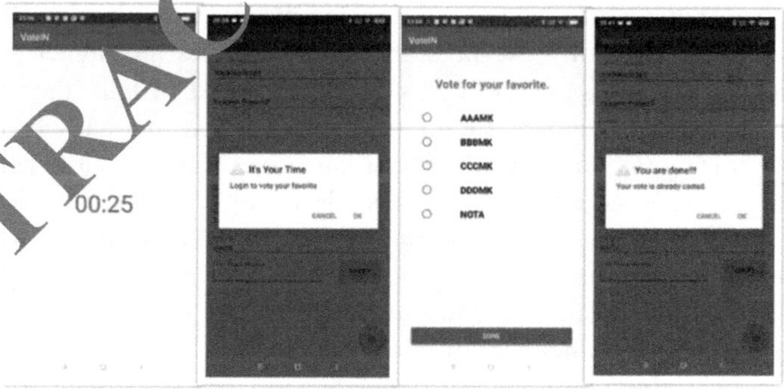

Fig. 6. Vote process

Automated tool gives the best result then manually overriding the system and each process. This is most reliable and easy to manage the process. Hacking has been managed automatically and made recovery of the system really smooth (Fig. 7).

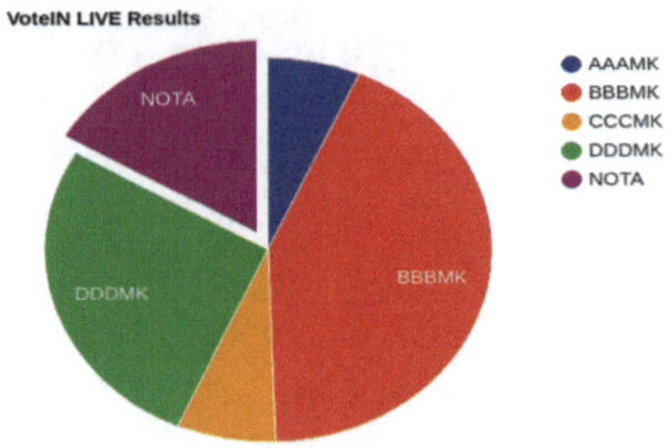

Fig. 7. Live results with 3 s delay

The need for machine-controlled tooling in versatile machining, assembly, and sheet fabrication systems is reviewed. The varied ways of implementing these systems, their edges and disadvantages area unit mentioned. The fundamental modules of machine-controlled tool transfer, storage, loading/unloading, and management area unit delineate in conjunction with the acceptable level of automation for every module. The benefits and stipulations for remote-controlled machining systems, this sensing ways and also the tool replacement methods are reviewed. The importance of a tool information, its uses and structure area unit highlighted. Finally, the look and analysis of machine-controlled tooling systems and operational methods, with the help of separate events framework area unit mentioned. An existing pc package that is capable of simulating machine-controlled tooling systems for versatile producing systems is conferred.

6 Conclusion

Portable voting frameworks have numerous points of interest over the customary method for voting. Some of these favorable circumstances are higher security level, more noteworthy exactness, versatility, a quicker method to tally the outcomes and lower dangers of human blunders. Be that as it may it is exceptionally hard to build up a perfect versatile voting framework which can give 100% security and protection level. This article proposed a continuous versatile voting framework in view of android gadgets. In Present time, OTP (one-time secret key) applications are expanded. Security is an essential issue for dealing with such administrations. Current framework gives security card-based office to confirm client yet this isn't sufficiently secure and may not be accessible on whenever or circumstance. To defeat such sort of issues we propose online e-Voting confirmation framework utilizing OTP with Aadhaar id and pseudorandom number generator that distinguishing proof is excessively mind boggling which is enhancing the security for beast drive assault.

The practicable future extent of the project is the constitution upgradation and by implementing all the election process theory. This include periodic management of the application and the performance flow of each constitutions. This constitution upgradation need regular update on theory of Election Commission and their flow design of the particular election.

References

1. Ghatol, P.S., Mahale, N.: Biometrics technology based mobile voting machine. Int. J. Comput. Sci. **2**(8), 45–49 (2014). e-ISSN 2347-2693
2. Marinescu, L.: Security system for mobile voting with biometrics. J. Mob. Embed. Distrib. Syst. **VII**(3), 100–106 (2015). ISSN 2067-4074
3. Sontakke, C., Payghan, S., Raut, S., Deshmukh, S., Chande, M., Manowar, D.J.: Online voting system via mobile. Int. J. Eng. Sci. Comput. (2017). http://ijesc.org/
4. Izadi, S., Zahedi, S., Atani, R.E.: A novel secure protocol, IES, for mobile voting. IOSR J. Eng. (IOSRJEN) **2**(8), 06–11 (2012). ISSN 2250-3021. http://www.iosrjen.org/
5. Villegas, E.P., Gallegos-García, G., Torres, G.A., Gutiérrez, H.F.: Implementation of electronic voting system in mobile phones with android operating system. J. Emerg. Trends Comput. Inf. Sci. **4**(9), 728–737 (2013). ISSN 2079-8407
6. Kumar, A., Srivastava, A.K.: Designing and developing secure protocol for mobile voting. Int. J. Appl. Eng. Res. **2**(2), 522–533 (2011). ISSN 0976-4259
7. Ahlawat, P., Nandal, R.: Performance improvement using pseudorandom one time password (OTP) in online voting system. IOSR J. Comput. Eng. (IOSR-JCE) **17**(5), 31–38 (2015). e-ISSN 2278-0661, p-ISSN 2278-8727. http://www.iosrjournals.org/
8. Subramanian, P., Ilangovan, S.P., Murugesan, R.: A secure based approach in M-voting for human identification based on iris recognition using biometrics. Int. J. Res. Appl. Sci. Eng. Technol. (IJRASET) **3**(IV) (2015). ISSN 2321-9653. www.ijraset.com
9. Ghate, B., Talewar, S., Taware, S., Katti, J.V.: E-voting system based on mobile using NIC and SIM. Int. J. Comput. Appl. (0975-8887) **165**(8) (2017)
10. Gawade, D.R., Shirolkar, A.A., Patil, S.R.: E-voting system using mobile SMS. IJRET: Int. J. Res. Eng. Technol. e-ISSN 2319-1163, p-ISSN 2321-7308. http://www.ijret.org
11. Hegde, A., Anand, C., Jyothi, B.: Mobile voting system. Int. J. Sci. Eng. Technol. Res. (IJSETR) **6**(4) (2017). ISSN 2278-7798
12. Folaponmile, A., Suleiman, A.T., Gwani, Y.J.: Mobile electronic voting system: increasing voter participation. JORIND **13**(2) (2015). ISSN 1596-8303. www.ajol.info/journals/jorind
13. Raskar, S.R., Jaykar, V.B., Akhare, A.A., Gadale, R.M., Phalke, D.A.: Literature survey on secure mobile based e-voting system. Int. J. Comput. Sci. Inf. Technol. Res. **3**(4), 234–236 (2015). ISSN 2348-120X. http://www.researchpublish.com/
14. Beroggi, G.E.G.: E-voting through the internet and with mobile phones. Int. J. Adv. Res. Comput. Sci. Manag. Stud. ISSN 232-7782
15. Bhosale, P.: Advanced E-voting system using NFC. IJIRAIT **2**(5). ISSN 2454-132X. www.Ijariit.com
16. Abamo, V.J.L., Abamo, M.R.S., Valerio, T.D.L.: Philippines smart app voting system a mobile voting system. Int. J. Adv. Res. Comput. Sci. Softw. Eng. Res. Pap. www.ijarcsse.com
17. Yakubu, K.Y.: Implementation of mobile voting application in Infrastructure University Kuala Lumpur, Malaysia. Int. J. Comput. Appl. (0975-8887) **180**(47) (2018)
18. Agrawal, S., Banerjee, S., Sharma, S.: Privacy and Security of Aadhaar: A Computer Science Perspective

IoT Based Smart Electric Meter

M. Dhivya(✉) and K. Valarmathi

Department of Computer Science and Engineering,
Panimalar Engineering College, Chennai, Tamil Nadu, India
dhivyaml49@gmail.com

Abstract. Electricity is the basic needs of our life and people cannot think of a world without electricity. In the existing practice of meter reading technique, the meter reading analysis is completed by the help of EB person. But this practice leads to numerous drawbacks like inaccuracies during calculation, nonappearance of customer during billing period and additional payments for the billing procedure. To overcome this issue, IOT based smart electric meter is established. It targets to decrease the man power for the billing. An energy calculation through wireless smart meter using IOT is projected for spontaneous meter data assortment, provide intimation through messages showed on LCD and energy examining, the analysis of consumer's EB unit reading and cost will be automatically transfered to the server and aids to perceive the daily EB unit generation and cost. It can deliver there required data such as tariff difference and payable date for the payment. The customer can pay through on-line using RFID tag.

Keywords: IOT · Sensors · Digital meter · Zigbee

1 Introduction

The smart metering is a central segment in stinging system utilization as they are using Internet of Things advancements to change regular Department of Energy structure. Smart metering through IOT reduces working toll by managing metering abstract process remotely. It furthermore upgrades the evaluating and reduces essentialness robbery and misfortune. These meters essentially get the data and send it back to the supportiveness troupe over outstandingly solid correspondence framework.

The proposed system aim is to create a smart meter which can able to easily identify the power consumption. Our system will calculate the power by use of current sensor and voltage sensor and it will display in LCD with cost. So user can simply see his power consumption cost so far how much he was used. The main aim of this proposed system is to develop a web application. The data base of user will store in server by use of java MySQL. An web page is created in this system so that user can able to monitor the Power consumption in online. An automatic SMS will send to the user at the end of bill cycle.

In Traditional system a man from Electricity Board visit to each house in the specific space and takes EB Analyzing from each house his dedication is to note down the looking at in units, influencing fragment in EB to card and EB office. The major

D. J. Hemanth et al. (Eds.): COMET 2019, LNDECT 35, pp. 1260–1269, 2020.
https://doi.org/10.1007/978-3-030-32150-5_127

inconvenience of this structure is that solitary needs to go district by zone and he needs to examine the meter of each house and handover the EB office. Errors like extra bill aggregate or cautioning from control office are steady messes up. To crush this weight, the proposed idea is passed on and made sense of it.

To create a power charge, a circuit repairman goes to the house on more than one occasion in per month take the readings from the vitality meter. The perusing is refreshed in the workplace to produce a bill. This issue is defeated in center of 2000. Here EB individual should return home and take the perusing refreshes in EB office. (Conventional strategy) This issue defeats by smart meter the customer's EB unit is refreshed to the server naturally utilizing smart meter.

The purchaser's EB unit perusing and cost will be transferred to the server consequently utilizing the Smart Electric Meter. Smart vitality viewpoint additionally demonstrates that In Home Displays (IHDs) are positively affecting helping individuals to deal with their vitality. IHDs are straightforward, handheld gadgets offered to each home, at no additional cost, when a shrewd meter is introduced. They indicate individuals what they are spending on vitality in close constant, in cash.

Individual collaboration is avoided. The shopper's EB unit perusing is naturally refreshed in EB office Server. This application likewise observes the everyday utilization of Electricity utilization in units and can limit the quantity of units which encourages us not to go for most noteworthy tariff. Electricity charge cycle will be decreased from two months to multi month.

This procedure helps to see our daily EB unit generation and cost. Power charge cycle will be decreased to one month. This application won't require anybody from a vitality organization to visit your home to peruse your meter Smart meters will help vitality organizations to know when you've lost power (e.g. have been cut off in a storm). We can likewise lessen the electric power utilization because of the day by day unit age and cost will be appeared to the customer and the labor is decreased.

2 Related Works

Saxena [1] proposed a fused affirmation tradition for splendid systems. The recommendation uses disproportionate and symmetric key cryptography to stay the correspondence with the electric organization. Regardless of the way that the makers think of it as a lightweight tradition, the suggestion uses hash and open key exercises, which are not prescribed for utilize with everything taken into account IoT contraptions. A couple of makers proposed threatening to adjusting methodologies to recognize some specific damages of SM and to make reprobation. Senthil Arumugam and Prabakaran [2] presents an understanding survey of sharp power meters and their usage focused on a seeing piece of the metering technique, phenomenal accomplice's interests, and the advances used to meet the essentials for accessory interests. Other than they give an exceptional piece of issues and also conditions developing conclusively to the nearness of colossal information and the party point of fact appreciated of cloud conditions.

The creators Finster and Baumgart, have seen the protection issue of the sumptuous metering establishment and have spoken to security drawback from a metering investigation. They need pondered the issue from 2 centers metering for charging and metering for activities. For each one among these issues they need seen level techniques. they need analyzed the different frameworks for the metering for charging issue by wonderful meter scatter quality, structure diverse nature and strike remarkableness for dependable in untouchables, reliable in selecting and cryptanalytic checks. In like way, they need assessed the ways to deal with oversee metering for task by near parameters, dependable in distant with social event, blend while not reliable in untouchables and settlement of unfocussed information [3].

Non-prominent inductive current recognizing methodology [4], is used for current estimation of connection stack gadgets, without breaking circuit of fitting weight devices. Most noteworthy imperativeness is devoured by interface burdens to endeavors. To screen and control electrical imperativeness of connection loads like HVAC, there are different game plans open, for instance, Constructing organization system yet there is no response for separate and trigger customized movement of associate burdens to progressing.

The splendid meter assurance system [5] inside watching substitute giganticness source which is driven by the information spillage and standard power hindrance issues, they plot the security control work in a singular letter from when the customers' centrality inconveniences ought to be free and cryptically spread after some time. They have demonstrated that the perfect endeavor of the imperativeness passed on by the AES in the exponentially appropriated data stack circumstance can be amassed using the switch water filling count.

Elrefaei [6] discusses a structure which uses a diminished camera to get the photo of the power meter breaking down. Picture planning stage drives forward through three phases: (1) Preprocessing, which is in charge of changing the numeric taking a look at region. (2) Segmentation, which yields explicit digits using level and vertical researching of the adjusted numeric region. (3) Recognition of the meter assessing by isolating each area digit and digits show up.

Guo et al. other than have talked the robotized hits relating to the AMI viz. association based ambushes for correspondence media or uniquely based strikes and security wrecks in contraptions and enrollment of attack director in metering device like presenting's hurtful program inside the meter or spread malware in the system. The subject picked with the substances that other than sending insistence structure, to keep up security levels, programming bugs should be isolated; reestablishing firm-ware in standard breaks, animating custom and Software settling is to be done. [7] The maker depicted the smart system and sharp meter and analyzed the related destroy game-plan level and learning level.

Kotwal, uses an Android application to confirm meter breaking down picture and some time later achieve OCR. The delayed consequence of OCR is appeared to the made Web Application which makes the bill. This bill is appeared to customer immediately. The low quality picture owed to lighting conditions may cause planning errors [8].

A showed GSM-based Energy Recharge System for paid early metering was given focus on proffering answer for human mess up, overseeing goof furthermore electro-mechanical bungles while [9] centers at proposing a structure that will diminish the loss of intensity and pay owed to control robberies and other unlawful activities. It uses an AT89S52 microcontroller which turns as the major controller. The centrality meter taking a gander at is connected with the dazzling card information by the microcontroller for dynamic checking and control of trading depending on the credit status.

IOT advance was broke down, where every client was given a sensational IP (Internet Protocol) fitting to empower access to the Consumer Premises Equipment (CPE) which for this condition is a sharp meter through a web interface. Regardless, the "wasteful treatment of the IP address" and furthermore the latency that may happen in correspondence between the CPE and the web interfaces made it wasteful. [10] A streamlined structure custom and Automated Meter Reading System (ARMS) was passed on to address the issues of dissipate quality, different unique models and over the top strategy as showed up.

Change of anchored remote based home zone arrange [11] for metering in savvy lattice which requires the dynamic contribution of customers to build up the quality and dependability of the power conveyance. Because of the joined idea of the remote medium, be that as it may, these arrangements confront security troubles and impedance issues which must be expressed while creating it.

Mohassel proposed the customers on the opposite end can likewise screen their vitality utilization progressively, energize their records, screen levy rates and thus enhances the request reaction. Sadly, the vitality division is beset by a few difficulties coming about because of the sending of power keen meters. They are vitality burglary, digital assaults, botch and wrong charging and so forth [12] and gives an answer of diminishing human association in vitality administration for both service organizations and in addition customers. All the observing and control highlights are given access by means of a conferred online interface, anyplace, whenever gave there is Internet association. Shrewd meters information are gathered, put away and inspected for appropriate arranging and charging of buyers.

There are distinctive frameworks open for evaluating the essentialness usage of electronic devices 99 and report this data over the framework. The measures are plug stack checking system, non-meddling burden watching structure, contraption level load checking structure. Conveying power supply (CPS) fuses control metering which assessments the power use of contraption, count, and association between the electronic devices. Savvy meter associated with the web, grows imperativeness care among contraptions and customers [13].

Zhang discussed the estimation of voltage winding by insightful meter data to develop a dynamic model for upgrading volt-var control [14], and what's more watching blockage and quality in a power grandstand. Metering data can be similarly used to develop the learning of the influence streams at and near the low voltage end of the course composes with the objective that the stacking and adversities of the framework can be known simply more wonderfully. This can thwart overtroubling modules (transformers and lines) and to avoid control quality varieties from the standard.

Bayesian and Hidden Markov Model systems are being utilized in a collection of insightful metering applications, for example, stack disaggregation [15], machine perceiving affirmation and supply request examination. Future applications will result in a more wide degree of necessities which will see a consistently growing number of frameworks related and interestingly fitted for stunning metering to accomplish more enormous inclinations.

3 Proposed Smart Electric Meter

See Fig. 1.

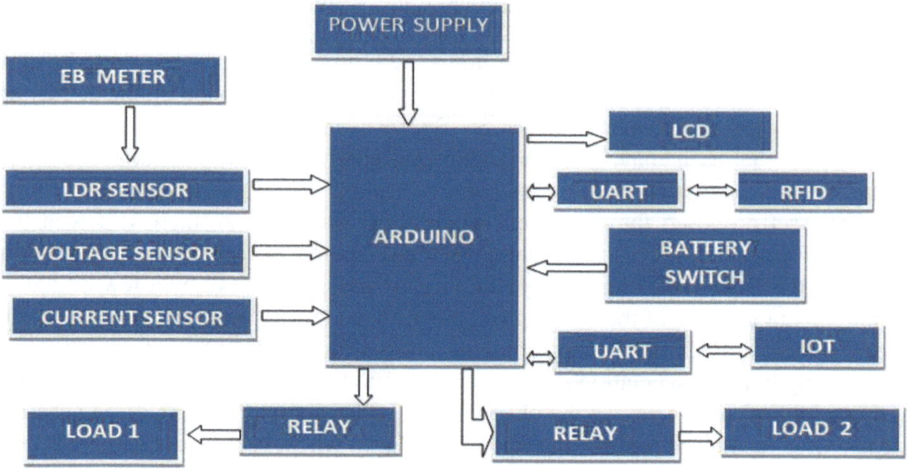

Fig. 1. Proposed smart electric meter

3.1 Electric Meter Reading

IOT based smart electric meter is made and it hopes to diminish the work for the charging. A vitality check through remote snappy meter utilizing IOT is made arrangements for altered meter information accumulation, give proposal through messages showed up on LCD and essentialness assessing, the buyer's EB unit examining and cost will be exchanged to the server therefore and sees the consistently EB unit age and cost. It can give critical information, for instance, charge assortment and due date for the portion. The client can pay through online with the RFID tag.

3.2 Atmega 328 Microcontroller Connected with Sensors

EB meter is connected with the LDR sensor. It senses the photochemical energy, it converts photochemical energy into electrical energy. Arduino Uno is a microcontroller board used in smart meter system and it is connected with the following sensors are

LDR, voltage and current sensor. The current and voltage readings of the load are measured using the current and voltage sensor in the analog measurement and is given to the microcontroller for the power consumption units calculation. This power calculation is performed by programming it in the Arduino software (IDE). The calculated power consumption units in the Arduino microcontroller can also be shared with the user through SMS using the RFID tag. This message can be programmed to be sent to the user at regular intervals of time. The user can calculate the power consumption units in the Arduino microcontroller can be shared with the sms using RFID.

3.3 Reading Displayed on LCD

The figuring of the cost with the purposeful scrutinizing will be always appeared on site page that we have organized. Edge regard can be resolved to page with the help of Wi-Fi, as indicated by the purchaser's need. Exactly when the customers scrutinizing will be close going to set edge regard, it will send a notice a motivator to the customer. This breaking point regard cautioning will manufacture the care among the customer about the imperativeness. The customer can visit the site page and change the edge value. When the client gets the notification. The units that customer have eaten up and the cost of unit use each day will be appeared at day's end. Based on the survey taken from various smart electric meter systems in the earlier work [16] the new smart electric meter system was developed (Fig. 2).

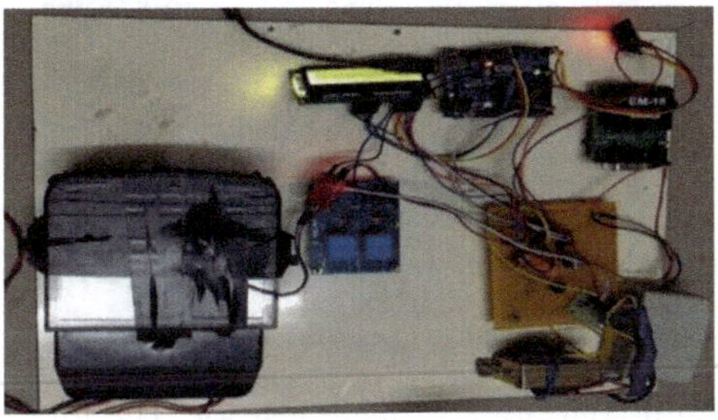

Fig. 2. Reading displayed on LCD

4 Hardware Aspects

4.1 Current Sensor

A flow detecting component might be a gadget that identifies electrical wonder amid a wire, and produces a proof relative thereto flow. The produced flag may be simple voltage or current or perhaps a computerized yield. The created flag is then acclimated demonstrate the deliberate current in partner degree meter, or is hang on for more examination amid an information obtaining framework, or is utilized for the point of the board (Fig. 3).

Fig. 3. Current sensor

4.2 Voltage Sensor

The Voltage Sensor square tends to a perfect voltage sensor, that is a contraption that changes over voltage evaluated between two of an electrical circuit into a physical flag contrasting with the voltage (Fig. 4).

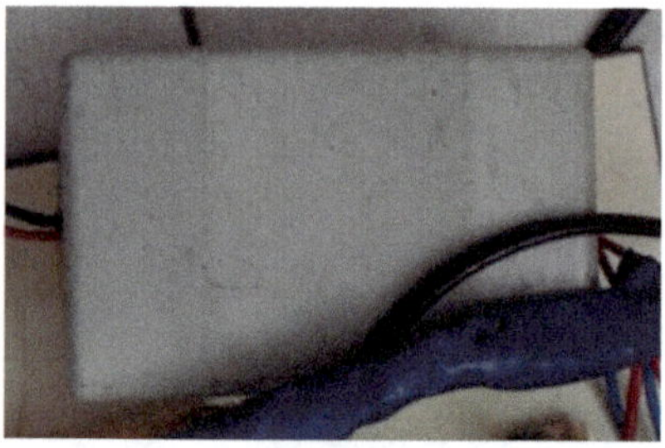

Fig. 4. Voltage sensor

4.3 LDR Sensor

A photoresistor is a light-controlled variable resistor. The limitation of a photoresistor decreases with developing occasion light power; constantly end, it shows photoconductivity. A photoresistor can be related in light-touchy finder circuits, and light-began and dull prompted exchanging circuits (Fig. 5).

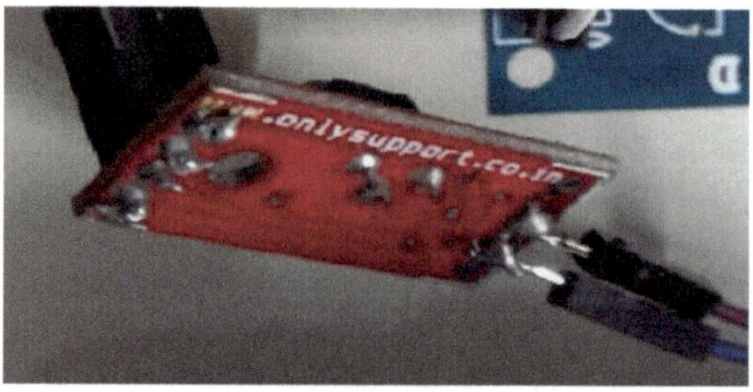

Fig. 5. LDR sensor

4.4 Relay

A relay is associate degree electrically worked switch. Several transfers utilize associate degree magnet to automatically work switch, but alternative operating standards square measure to boot utilised as an example, sturdy state transfers. Transfers square-measure utilised wherever its vital to regulate a circuit by a unique low-control flag, or wherever a number of circuits should be strained by one flag (Fig. 6).

Fig. 6. Relay

4.5 RFID

A radio-recurrence ID framework utilizes names, or names joined to the articles to be seen. Two-way radio transmitter-specialists called savvy specialists or perusers send a standard to the tag and read its reaction. RFID names can be either separated, dynamic or battery-helped lethargic (Fig. 7).

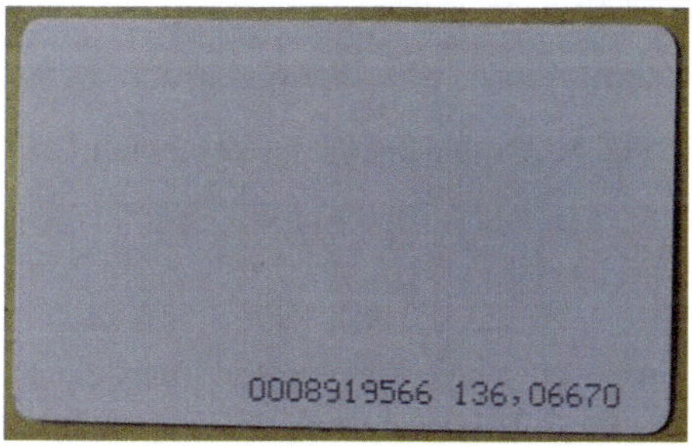

Fig. 7. RFID

5 Result and Discussion

The smart electric meter displays the power consumption, voltage and current and it also updated to the EB server automatically. The user knows the power consumption units and cost in the daily basis.

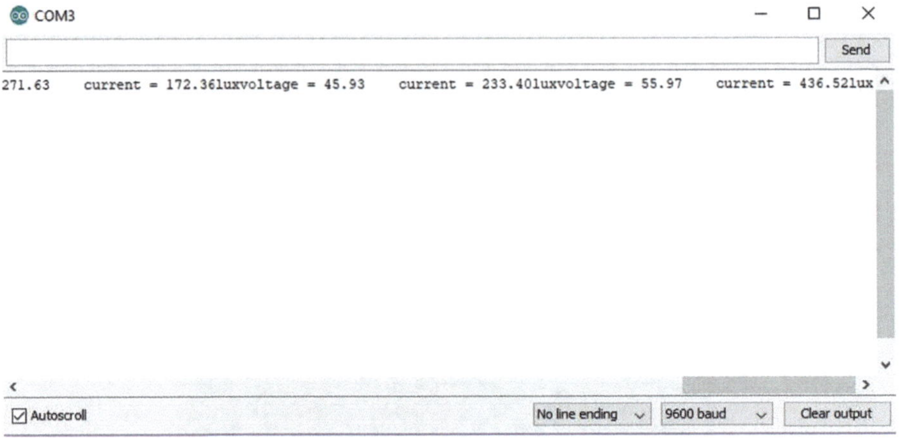

The smart electric meter is connected with the LCD to know the updates of the power consumption units and cost (Fig. 8).

Fig. 8. Displaying the result on the LCD

6 Conclusion

In this paper we have proposed a new scheme for smart electric meter. In Traditional method manpower is required to take current bill consumption, to intimate the user about the current consumption charges. This process will take more time to complete the bill cycle and also user can't able to get an idea about his bill status until the final bill payment is generated. Our government is using a digital meter to calculate the bill status of the user. After completing the bill cycle only user can able to get the bill because, there is no intimation for the user until the end of two months. In the proposed work, the consumer's EB unit reading and cost will be uploaded to the server automatically without the need of EB person's knowledge. After noting the reading the consumer can pay through online, this application helps to see the daily EB unit generation and cost. Electricity bill cycle will be reduced to one month. And the system will be enhanced by implementing online money payment system; in addition to that if the EB bill is not paid by the consumer for the particular set of days then the power supply to that particular house was disconnected automatically by using RFID Tag.

References

1. Saxena, N., Choi, B.J.: Integrated distributed authentication protocol for smart grid communications. IEEE Syst. J. **12**(3), 2545–2556 (2016)
2. Senthil Arumugam, S., Prabakaran, S.: A survey of future energy systems using smart electricity meters. Int. J. Adv. Eng. Recent. Technol. **13**(1) (2016)
3. Finster, S., Baumgart, I.: Privacy-aware smart metering: a survey. IEEE Commun. Surv. Tutor. **17**(2), 1088–1101 (2015)
4. Balsamo, D., Gallo, G., Brunelli, D., Benini, L.: Non-intrusive zigbee power meter for load monitoring in smart buildings. IEEE (2015)
5. Gomez-Vilardebo, J., Gunduz, D.: Smart meter privacy for multiple users in the presence of an alternative energy source. IEEE Trans. Inf. Forensics Secur. **10**(1), 132–141 (2015)
6. Elrefaei, L.A., Bajaber, A.: Automatic electricity meter reading based on image processing. In: 2015 IEEE Jordan Conference, vol. 15, pp. 1–5 (2015)
7. Guo, Y., Ten, C.W., Hu, S., Weaver, W.W.: Modeling distributed denial of service attack in advanced metering infrastructure. IEEE (2015)
8. Kotwal, J., Pawar, S., Pansare, S., Khopade, M., Mahalunkar, P.: Android app for meter reading. Int. J. Eng. Comput. Sci. **4**, 9853–9857 (2015)
9. Upadhyay, J., Devadiga, N., Mello, A.D., Fernandes, G.: Prepaid energy meter with GSM technology. Int. J. Innov. Res. Comput. Commun. Eng. (2015)
10. Darshan Iyer, N., Radhakrishna Rao, K.A.: IoT based electricity energy meter reading, theft detection and disconnection using PLC modem and power optimization. Int. J. Adv. Res. Electr. Electron. Instrum. Eng. **4**(7), 6482–6491 (2015)
11. Namboodiri, V., Aravinthan, V., Mohapatra, S.N., Karimi, B., Jewell, W.: Toward a secure wireless-based home area network for metering in smart grids. IEEE Syst. J. **8**(2), 509–520 (2014)
12. Mohassel, R.R., Fung, A.S., Mohammadi, F., Raahemifar, K.: A survey on advanced metering infrastructure and its application in smart grids. IEEE (2014)
13. Lanzisera, S., Weber, A.R., Liao, A., Pajak, D., Meier, A.K.: Communicating power supplies: bringing the internet to the ubiquitous energy gate-ways of electronic devices. IEEE Internet Things J. **1**(2), 153–160 (2014)
14. Zheng, J.: Smart meters in smart grid: an overview. In: IEEE Green Technologies Conference, pp. 57–64 (2013)
15. Lukaszewski, R., Winiecki, W.: Methods of electrical appliances identification in systems monitoring electrical energy consumption. In: 7th IEEE International Conference on Intelligent Data Acquisition and Advanced Computing Systems (2014)
16. Dhivya, M., Valarmathi, K.: A survey on smart electric meter using IOT. In: 3rd International Conference on Communication and Electronics Systems (ICCES) (2018)

Detection of Tuberculosis Using Active Contour Model Technique

M. Shilpa Aarthi(✉)

Department of Computer Science and Engineering,
Panimalar Engineering College, Chennai, Tamil Nadu, India
Shilpaarthi13@gmail.com

Abstract. Tuberculosis is a disease where it is being a great threat to human being. If this disease is not found earlier then it leads to death. Tuberculosis is found when TB rods are found in the sputum of the patient when viewed through a microscope which is a traditional method. To detect this disease faster and with good accuracy active contour model is applied on the sputum image so presence of tuberculosis is found. Then in the second phase mobile application is created and an message is sent to people regarding free checkup camp organized in rural area or tribal area.

Keywords: Tuberculosis · Sputum · Microscope

1 Introduction

Tuberculosis commonly affects the lungs, and it slowly spreads to all parts of the complete body. Most infections do not have signs and symptoms, which is referred as latent tuberculosis. About 10% of latent infections development to active tuberculosis which, if left untreated, kills about 1/2 of those infected. The disease is not communicable from People with latent TB. It spreads actively to those with HIV/AIDS and also for people who smoke.

One of the common test to detect tuberculosis is skin check tuberculosis test where a small injection of PPD tuberculin, an extract of the TB bacterium. And, if a big red coloured bump has swollen up to a particular size on the injected area, then it is sure that TB is present. Unfortunately, the above mentioned test is not much accurate and has been regarded to offer incorrect advantageous and terrible readings. Blood tests, chest X-rays, and sputum tests are some of the tests can all be used to check for the presence of TB bacteria and may be used alongside a pores and skin test. MDR-TB is more difficult to diagnose than everyday TB. So, a new test can be implemented to treat tuberculosis with much accuracy.

2 Related Works

A modified approach for fast analysis of TB is [1] completed using excessive-resolution actual-time imaging of Mtb colonies for tuberculosis. Clinical specimens, sputum specimens, stool specimens, bronchoalveolar lavage fluid specimens are taken

D. J. Hemanth et al. (Eds.): COMET 2019, LNDECT 35, pp. 1270–1277, 2020.
https://doi.org/10.1007/978-3-030-32150-5_128

and kinds of protocol (real-time protocol and popular protocol) are implemented on it. The consequences exhibit that actual-time high-resolution imaging of Mtb reduces the time to growth detection of pulmonary tuberculosis.

In [2], automatic diagnosis approach for tuberculosis is proposed in which digital microscopes like cellular scope is used to capture photos and morphological operations, template matching are applied at the captured photo wherein then aid vector gadget category is carried out. The final results of our category performance has right stage of accuracy.

However, stepwise classification is implemented on sputum smeared slides [3] automatic detection of tuberculosis. Specimens are accumulated from sufferers who had been kept in smear slide then virtual picture of it's miles taken the use of a high resolution digicam then stepwise category (SWC) algorithm is applied to improve the way of detection and to make the approach of counting the tuberculosis bacilli.

Sanneke Brinkers et al. [4] proposed the finding of TB nucleic acid the use of darkish field which tethered particle motion wherein single molecule detection is accomplished. In this approach each small nano-particles of sputum are viewed through dark field microscope and from an picture which changed into captured the usage of a high decision digicam. The final results gives a success detection of RNA the usage of fastened darkish discipline particle motion. In future work, the above approach could be multiplexed for the detection of more than one nucleic acid sequences.

Using one-class classification Khutlang et al. proposed [5] a gadget in which automatic detection is accomplished using the smear slides where it's miles placed under microscope and makes use of a virtual camera to take a picture. One class pixel classification and one class item category is implemented to the image. This method outcomes with high accuracy and sensitivity of traditional microscope is progressed for TB screening.

Graff cut segmentation [6] system to get accurate result at some point of classifier automated tuberculosis screening is described. Lung region is extracted binary is implemented and then greater accuracy is produced by this method. In future, this machine could be evaluated over larger datasets in order that portable scanners might be accrued to perform this work.

Using feature extraction and identity, smear slides [7] to begin with viewed through microscope and a image of it's far taken the use of digital cameras to then separation of background and tuberculosis bacilli is carried out for the image and k approach clustering set of rules is applied. Now, the outcome is used to classify feasible TB and actual TB.

In [8], a strategy to detect automatic locating and classifying tuberculosis bacilli within the microscopic photographs acquired from smart phone is proposed the use of Watershed segmentation. The specificity and sensitivity are determined to be excessive on this proposed approach whilst compared to different preprocessing techniques at the same time as classifying an photo as TB or true. This method gives proper tracking and better diagnostic on and epidemic and endemic disorder.

A approach to treat with and display screen tuberculosis for humans who migrate [9]. Screening is used for the treatment of tuberculosis for migrants specifically for low tuberculosis occurrence international locations. The results from the evaluations recommended that proposed method effectively makes use of some strategies to detect TB.

2.1 Proposed Architecture

In the proposed system when a person is affected with cough. It should be checked whether it is normal cough or prolonged cough. If normal cough syrup can be provided. If found to be prolonged cough tuberculosis test should be taken. It is done by taking the sputum of the patient. The sputum is mixed with a culture medium called lowenstein jensen medium and after 48 h a sample is taken and viewed in a viewing slide and a image of the sputum is captured by using a highest microscopic featured camera. Then the image is applied in active contour model which does mathematical calculations to find whether it is an primary tuberculosis or secondary tuberculosis. And then an mobile application is created so the people can enter their details and their scan report so that the doctor's can view it and convey the results to them. And patients are asked to attend the regular checkup. If they skip checkup medical camp will be organized in a common area where many people with tuberculosis are found as shown in Fig. 1.

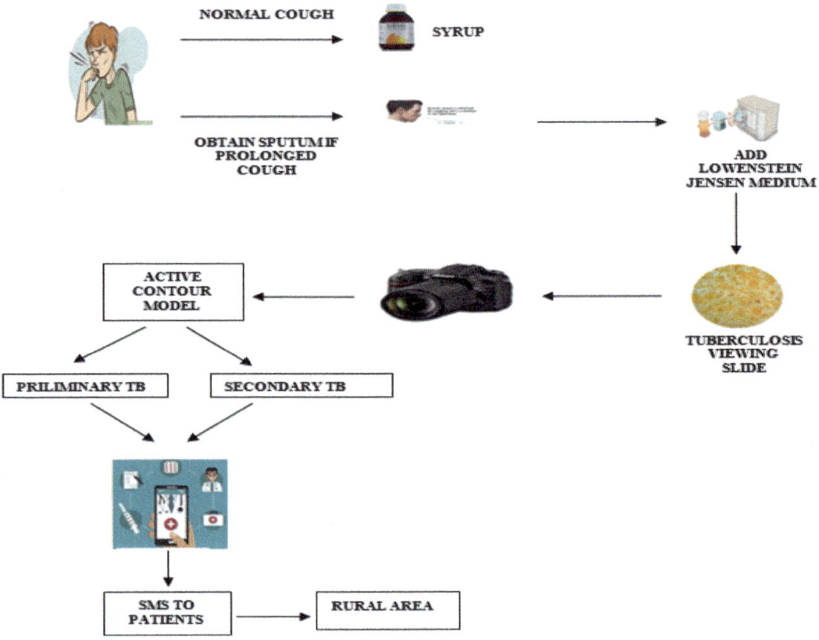

Fig. 1. Tuberculosis detection system

3 Proposed System Implementation

3.1 Pre-processing

Image pre-processing can give benefits and take care of issues that eventually prompt better neighborhood and worldwide element recognition. An input image is taken (sputum) it is read and converted into arrays. Find the edge of the input image should be found by using canny edge detector's parameters the output of the edge detection will be an binary image. Where, then dilation method is applied by using its parameters so, that it fills the unfilled areas of the detected edges. Then again binary image is the output and erosion is applied which erodes the overfilled areas of the dilated image and the output of the eroded image is n binary image.

```
Algorithm Algorithm for canny detection
Input     Sputum image.png
output    Cannyimage.png

display
img = cv2.imread('input.PNG')
cv2.imshow("original", img)
do
edges = cv2.Canny(img,200,300)
cv2.imshow("canny", edges)
end
Get the canny detected image.
```

The sputum image is taken and if it in png form initially canny detection is applied to the image so that the edges are detected in a better way and then the image is converted into binary. The output image is an binary image.

```
Algorithm Algorithm for erosion
Input     Dilation.png
Output    Erosion.png

whilei=img_dil
do
img_ero = cv2.erode(img_dil, k, iteration=1)
cv2.imshow("Erosion",img_ero)
end
Get the erosion image.
```

In this algorithm the dilated image is the input image to remove the overfilled areas of the dilated image we apply the process called erosion. The final output of this algorithm is eroded image.

3.2 Segmentation

Image Segmentation is the way of dividing a computerized picture into various fragments. Create a Mask which helps for creating Blue, Green and Red color. Create a Mask by passing three parameters as the Mask Layer in a tuple. The output of Mask is the binary image because Mask Layer is a binary image but it created space to accept the color image. Apply Bitwise and method which merges mask and the original image.

Bitwise and method parameters are the original image (input image) which is color image and mask which is the binary image. Segmented parts are white and remaining parts are black in mask image. Bitwise and merges the original image and mask image which segments only bacteria from the original image (color).

```
Algorithm  Algorithm for masking
Input      Erosion image.png
output     Masking.png

if k=img_ero
then
sel = cv2.bitwise_and(img, mask)
end
cv2.imshow("bacteria detected", sel)
end
```

3.3 Active Contour Model

Active contour model, called snakes, is very simple that a contour is formed over the required part. It forms a contour over the required part so it is very easy to identify. The below algorithm explains the working of active contour model which explains it in simple way.

```
global
panelF, panelG
sel = cv2.bitwise_and(self.resizedimage, self.mask
 cv2.imwrite('Bitwise.jpg',sel)
sel =ImageTk.PhotoImage(Image.open(r'Bitwise.jpg'))
```

here to develop contour two images resized image and the image before resizing is taken and compared then an white region is formed over all tuberculosis rods that is contour .

```
_,cnts,_=cv2.findContours(self.maskLayer,cv2.RETR_EXTERNAL,cv2.CHAIN_AP
PROX_SIMPLE)
```

after contour is obtained pixels of the rod must be found using epsilon formula and then finally presence of tuberculosis in amount is calculated and found whether it is an primary tuberculosis or secondary tuberculosis

3.4 Mobile Application to Send Notification

An mobile application is developed so that an message will be sent to people in rural areas and metropolitan areas regarding an free camp which will be organized there so that people who skip their regular treatment can make use of it. The mobile application contains two different logins for patients and doctors so that doctors view the patients details and sends sms according to it.

4 Result and Discussion

Edge detecting is a major tool in computer vision. In an image, edges are significant small changes in brilliance, shading, surface and they show the nearness of a limit between contiguous locales. It is notable that edge identification in non-perfect pictures more often than not results in loud edges, detached (broken) edges or both because of different reasons. The commotion issue can be reduced by utilizing higher edges, however this thus may disintegrate the issue of broken edges, as appeared. Short breaks in the edges can be recovered with basic post-processing (e.g., morphology), while huge breaks require exceptional treatment. The sputum image is initially taken as input image which is shown in Fig. 2.

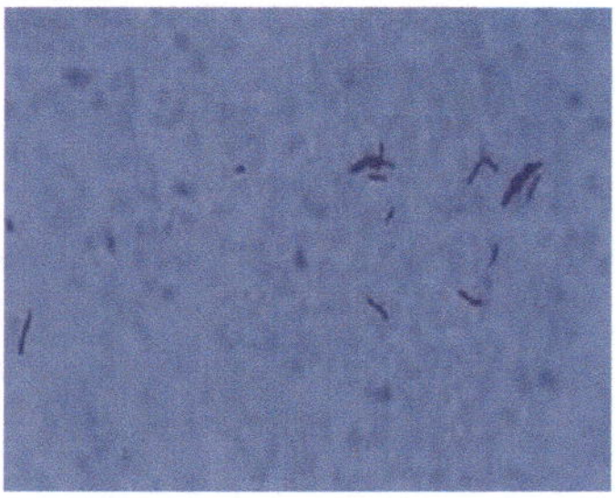

Fig. 2. Sputum image

The input image is first resized and various filters are applied to get good results the resized image is taken and segmentation method is applied using masks over it and result is shown in the Fig. 3.

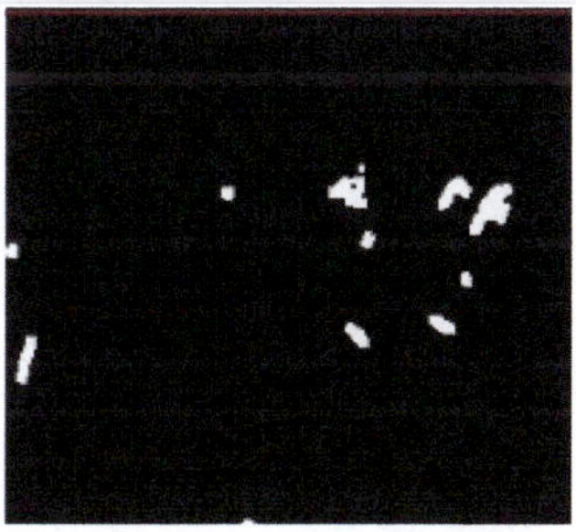

Fig. 3. Binary segmented image

By using active contour model contour is formed over all the rods present in the given segmented image and result is shown in Fig. 4.

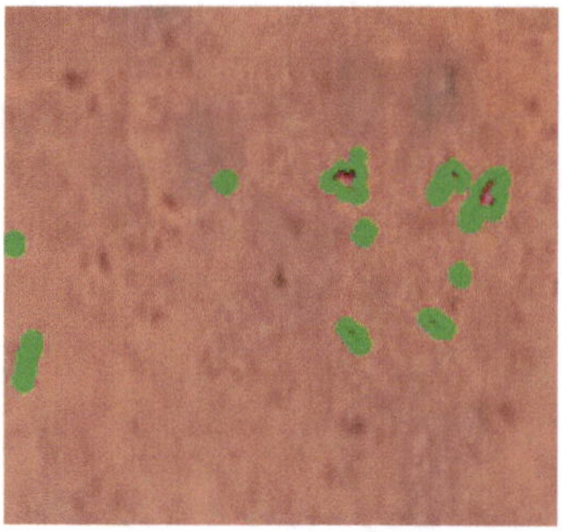

Fig. 4. Contour image

Active contour model forms a contour over all the tuberculosis rods by having the results of PSNR values for some images of the sputum. Three different types of filters are used and the filter which gives best result is taken and active contour model is found for the image which gives best result of PSNR values (Fig. 5).

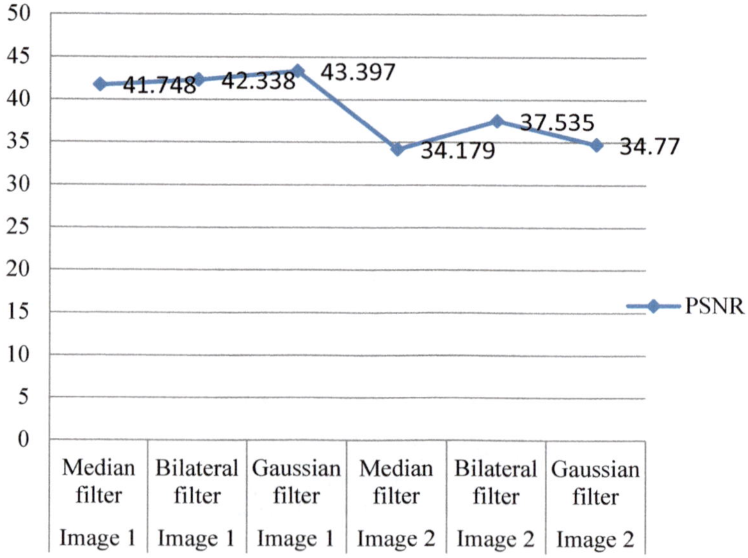

Fig. 5. PSNR values based on filters

5 Conclusion

In the automatic detection of tuberculosis many edge detection methods are used to detect the rods of the tuberculosis like canny edge detector, erosion, dilation, in the preprocessing techniques and bitwise in the segmentation. The image is converted into binary image, filling the unfilled areas, reducing the overfilled areas. By using these methods almost the edges of tuberculosis rods are detected and finally we have used the active contour model in the feature extraction method to do the final detection. This method is quite better than other methods where edges are perfectly detected using the snakelets there will be disadvantages if the rods are not properly detected. Then an android application is created so that an message is sent to the people in local area regarding tuberculosis checkup where people in hilly areas find it difficult to check whether they have tuberculosis or not.

References

1. Baştan, M., Bukhari, S.S., Breuel, T.: Active Canny: edge detection and recovery with open active contour models. In: International Conference on Flexible Automation and Intelligent Manufacturing (2017). https://doi.org/10.1049/iet-ipr.2017.0336
2. Liao, X., Yuan, Z., Tong, Q., Zhao, J., Wang, Q.: Adaptive localised region and edge-based active contour model using shape constraint and sub-global information for uterine fibroid segmentation in ultrasound-guided HIFU therapy. IET Image Proc. **11**(12), 1142–1151 (2017). https://doi.org/10.1049/iet-ipr.2016.0651. www.ietdl.org
3. Zhao, Y., Rada, L., Chen, K., Harding, S.P., Zheng, Y.: Automated vessel segmentation using infinite perimeter active contour model with hybrid region information with application to retinal images. In: International Conference on Flexible Automation and Intelligent Manufacturing (2015). https://doi.org/10.1109/tmi.2015.2409024
4. Pratondo, A., Chui, C.K., Ong, S.H.: Integrating machine learning with region-based active contour models in medical image segmentation. J. Vis. Commun. Image Represent. **43**, 1–9 (2016). https://doi.org/10.1016/j.jvcir.2016.11.019
5. Ufimtseva, E.G., Eremeeva, N.I., Petrunina, E.M., Umpeleva, T.V., Bayborodin, S.I., Vakhrusheva, D.V., Skornyakov, S.N.: Mycobacterium tuberculosis cording in alveolar macrophages of patients with pulmonary tuberculosis is likely associated with increased mycobacterial virulence. The Research Institute of Biochemistry, Federal Research Center of Fundamental and Translation (2018). https://doi.org/10.1016/j.tube.2018.07.001
6. Ghodbane, R., et al.: Rapid diagnosis of tuberculosis by real-time high-resolution imaging of mycobacterium tuberculosis colonies. J. Clin. Biol. **53**(8), 2693–2696 (2015). https://doi.org/10.1128/jcm.00684-15
7. Chang, J., Arbeláez, P., Switz, N., Reber, C., Lucian Davis, A.T.J., Cattamanchi, A., Fletcher, D., Malik, J.: Automated tuberculosis diagnosis using fluorescence images from a mobile microscope (2012). https://doi.org/10.1007/978-3-642-33454-2
8. Rulaningtyas, R., Suksmono, A.B., Mengko, T.L.: Automatic classification of tuberculosis bacteria using neural network. In: Proceedings of the International Conference on Electrical Engineering and Informatics (2011). https://doi.org/10.1109/iceei.2011.6021502
9. Brinkers, S., Dietrich, H.R., Stallinga, S., Mes, J.J., Young, I.T., Rieger, B.: Single molecule detection of tuberculosis nucleic acid using dark field tethered particle motion. In: IEEE International Symposium on Biomedical Imaging: From Nano to Macro (2010). https://doi.org/10.1109/isbi.2010.5490227

Manhole Cleaning Method by Machine Robotic System

M. Gobinath(✉)

Department of Computer Science and Engineering,
Panimalar Engineering College, Chennai, India
gobinathmmecse06@gmail.com

Abstract. In our Smart world, however many technologies has been developed in various part of domains. But nowadays, there is no robot to control the manhole death and their human loss. This deals with a robotic-arm mechanism for processing and disposing the solid waste pipes for various treatments. This arm is intended to interchange a manhole cleaner so as to control the death rate of workers. The proposed paper moves through the pipeline and it removes the blockages and clears the drainage water which moves through it. The robotic arm operating system is monitored by camera module and it's controlled by the manhole cleaner's using a laptop or personal computer. The operator will operate the inside of the holes via a wireless device connected to the arm robotics. The MQ2, MQ3, MQ7 types of sensors are connected to the 8051 microcontroller. Therefore, the presence of various toxic and non-toxic gases are mainly focused and detected. In the arduino uno, the system is placed with the liquid crystal display module to calculate the distance and measure of all sensors with the help of batteries. Finally, this robotic-arm mechanism is to examine a blockage for cleaning and to measuring the solid waste treatment for different manhole pipeline system.

Keywords: Robotic arm · Arduino UNO · 8051 controller · LCD display · Camera module · Various gas sensors · Ultrasonic sensor · Load cell sensor

1 Introduction

Nowadays, various kinds of robots may be a quickly grown on their domain, as robot arm advances and keep growing and developing new robots fill completely different commonsensible desires, irrespective of whether or not domestically grows on it. Robots may play a vital role within the upcoming features of the world. The various robots do tasks that are precarious to people, and helps them from various problems.

In all different fields of robotics they are involved with the methods and technology of science, and also the division methodology, artificial intelligence grows and makes the world as a faster one. In collection, various occupations of chemicals and toxics are all produced. This grows around various non-policy attachment system of it. Due to this purpose once robots are used to develop and play in all parts of the domain parts.

The information is composed towards building understudies, and planners who are possessed with a mechanical self-rule. In the beginning stage, we will locate a smart

D. J. Hemanth et al. (Eds.): COMET 2019, LNDECT 35, pp. 1278–1285, 2020.
https://doi.org/10.1007/978-3-030-32150-5_129

thoughts and the growth of mechanical technology and a non-modern robots. The primary inspiration was to give the per user a basic, direct way by using an assortment of pictures, outlines and mathematic cases to make the subject of mechanical self-rule simple to fathom and easy to take after all around requested from the stray pieces until the most tangled structures (Fig. 1).

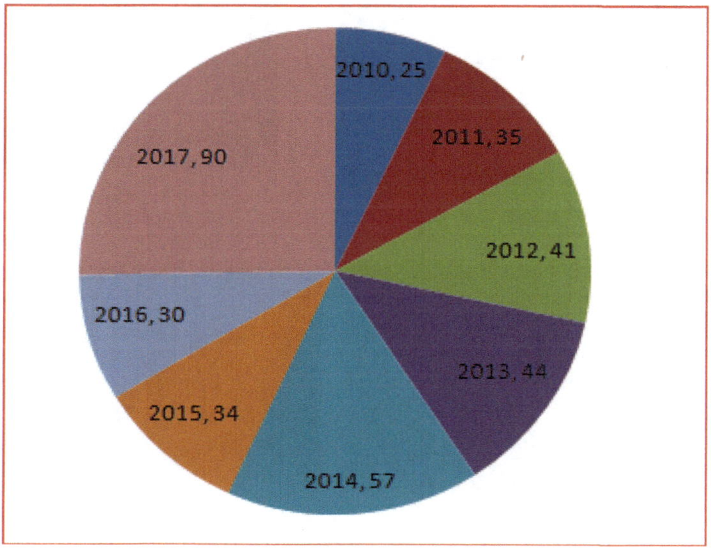

Fig. 1. Sewage death's rate [1]

The different kinds of problems are faced by cleaning workers at the bottom of the sewer pipe. They are all caused by various gases and they are enumerating seriously and observed with all causes of gases present inside the sewer pipeline systems. Whereas in the underground pipe, different enormous and their stream of all gases are all done on parts of the system. The different ways are all formulated and they are connected with different types of sewer pipe systems and they varied from it.

2 Related Works

At the time that a holes of a robot [4] which moves, locate and expand this stuff on it, they are able to reduce the effect of their value and production more on this. In that case, the various types of pipes are all configured on it. At those times, the robotic system should be removed from the holes of system [2] with the aid of using some restoration parts. Within the occasion that an oil leaking or from compound pipeline spills, it can be caused by the nature facts and an herbal calamity.

The invented robotic can move through involving of the funnels and making use of self sustaining parts of engines [3] and changing parts. The obtained consequences had been discovered tasteful and the created sensible robotic can be occupied and get their pipe occupied on it. The usage of system vision and device in inner pipes of the robot [9] can give scripts of issues associated it. However now, the robotic system controller may not do it for mathematical expressions and calculations, for instance, ANN utilization, and goes on (Fig. 2).

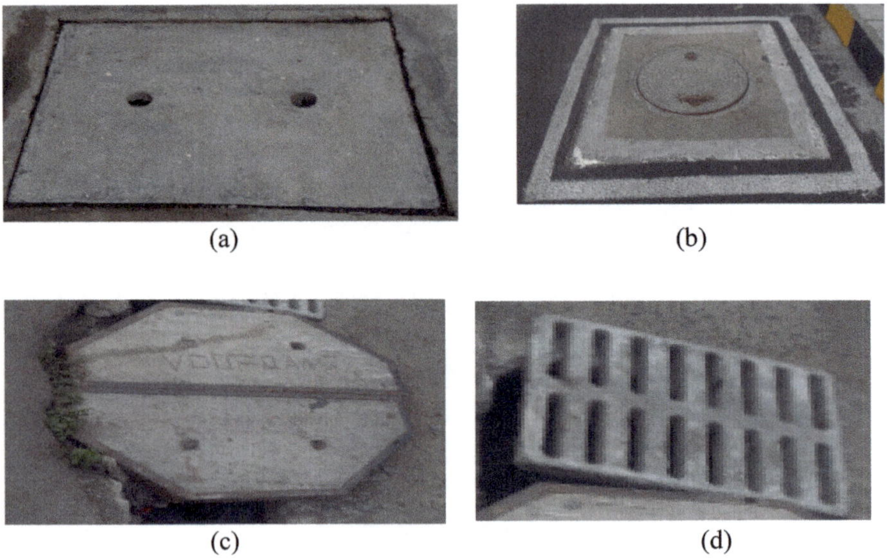

Fig. 2. Different types of manhole pipe

Prototype is handled with the production of handling with the programming on chip sets with the microcontroller systems and they are placed. They are produced with fluffy rationale framework systems that place the controls of sewage [5] holes that are formed. The cleaning process is handle and covers with the framework system of computerization network system. They are all get reduced and versatile in nature systems.

The development of programming [6] systems is handled and controls the parts of robot through the arduino system. The different aspects of clearing the blockage systems are all efficiently prolonged and used. With the high torque and low horsepower they are circuited. To carry out the garbage, many [10] rescue factors are all handled through the robot parts and their controls of the system and it fact pattern of growth arises in it.

Ultrasonic sensors are all placed and they are helps to calculate the distance measure of the robotic system. The proposed system robots [7] are all used to clean the sewage pipes then and there in all places maintained over it. These wastes are all handled and they may return it from mechanized forms of use. The requirement of the depletion and their risk [8] controls over all modified robot systems and their task.

3 Proposed System

See Fig. 3.

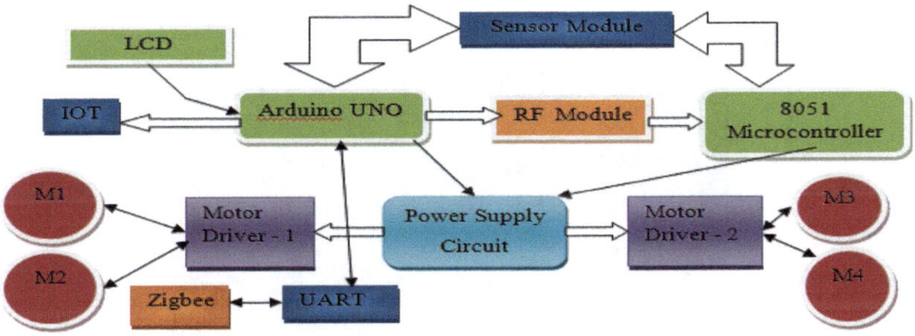

Fig. 3. System architecture of proposed system

3.1 Obstacle Detection Module

HC-SR04 Ultrasonic sensor is a distance calculating of prominent and easy measure for non-contact remove estimation work. It is mainly for calculating the distance measure of sensor to an obstacle particle system. The ultrasonic sensor shown in Fig. 4.

Fig. 4. Ultrasonic module

3.2 Different Gas Sensor Module

A Smoke Sensor (MQ2). A MQ_2 gas sensor is an instrument for the measurement of combustible gases and smoke. The sewage worker's working in the sensitive range of gases at room temperature. Then output analog signals can be reads with the arduino controller.

Alcohol Sensor (MQ3). AMQ3 gas sensor is an instrument for the measurement of alcohol. This type of sensor senses the gas and they may cleaning the worker's of this sewage system.
It is considered to be toxic and they are controlled in the arduino controller.

Carbon Monoxide Sensor (MQ7). A MQ_7 gas sensor is an instrument for the measurement of carbon monoxide. This gas causes the working persons of sewage heavily and finally leads to the death when they breathe. The different gas sensor shown in Fig. 5.

Fig. 5. Gas module

4 Results and Calculation

Thus, an experimental setup and calculation for a distance measuring sensor can be tabulated with a initial values as a distance measure and with a final value as a ultrasonic value can be tabulated with Table 1 below,

Table 1. Results for ultrasonic sensor

S. No.	Distance measure (Initial value)	Ultrasonic sensor (Final value)
1	5 duration	73.52 cm
2	10 duration	147.05 cm
3	15 duration	220.58 cm
4	20 duration	294.11 cm
5	25 duration	367.64 cm

For various toxic and non-toxic circumstances they are placed with various gas sensors like MQ2(SMOKE), MQ3(ALCOHOL) and MQ7(CARBONMONOXIDE) sensors and they are tabulated with Tables 2, 3 and 4.

Table 2. Results for Smoke sensor

S. No.	Voltage value (V)	Resistance value (R)	Smoke S_VOLT value
1	0.5	102 K	0.00009765 ppm
2	1	102 K	0.0001953 ppm
3	1.5	102 K	0.0002929 ppm
4	2	102 K	0.0003906 ppm
5	2.5	102 K	0.0004882 ppm
6	3	102 K	0.0005859 ppm
7	3.5	102 K	0.0006835 ppm
8	4	102 K	0.0007812 ppm
9	4.5	102 K	0.0008789 ppm
10	5	102 K	0.0009765 ppm

Table 3. Results for Alcohol sensor

S. No.	Voltage value (V)	Resistance value (R)	Alcohol A_VOLT value
1	0.5	10 K	0.0002441 ppm
2	1	10 K	0.0004882 ppm
3	1.5	10 K	0.0007324 ppm
4	2	10 K	0.0009765 ppm
5	2.5	10 K	0.001220 ppm
6	3	10 K	0.001464 ppm
7	3.5	10 K	0.001708 ppm
8	4	10 K	0.001953 ppm
9	4.5	10 K	0.002197 ppm
10	5	10 K	0.002441 ppm

Table 4. Results for Carbon monoxide sensor

S. No.	Voltage value (V)	Resistance value (R)	Carbon monoxide C_VOLT value
1	0.5	1.2 K	32,968.30801 ppm
2	1	1.2 K	8,242.07706 ppm
3	1.5	1.2 K	3,540.221 ppm
4	2	1.2 K	1,991.5831 ppm
5	2.5	1.2 K	1,276.1925 ppm
6	3	1.2 K	886.0463 ppm
7	3.5	1.2 K	651.0025 ppm
8	4	1.2 K	497.8957 ppm
9	4.5	1.2 K	393.4883 ppm
10	5	1.2 K	318.7268 ppm

Finally, the balancing weight calculating sensors of materials are all calculated and tabulated in the Table 5 and used below of it.

Table 5. Results for Load cell sensor

Sample	Count	Weight
0.1	0.2	0.31 kg
0.2	0.4	0.87 kg
0.3	0.6	1.48 kg
0.4	0.8	2.06 kg
…	…	…
1	2	4.57 kg

At the working area LCD display is placed to calculate the state of solid waste as a weight and toxic substance as a gas (Fig. 6). LCD output is shown in Fig. 7.

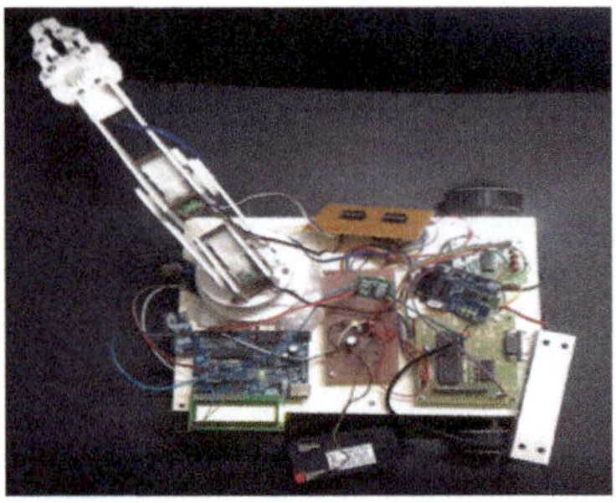

Fig. 6. Overall system setup

Fig. 7. LCD output

All the sensed data are continuously transmitted to the server that can be viewed in the monitor system continuously. The server data snapshot shown in Fig. 8.

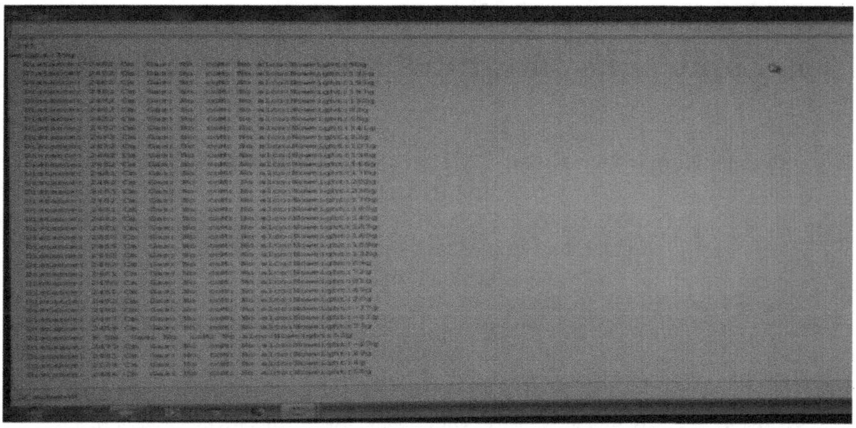

Fig. 8. Server data snapshot

5 Conclusion

Through this sewage system the drainage cleaning workers can survive and extend their lifespan without any problem in it. All types of blocking system are all cleared using this type of robotic arm model. Due to this system, it will be enhanced by analyzing other possibilities of facing problems in blocks in all areas and the environmental pollution will be reduced on it.

References

1. Ramapraba, P.S., Supriya, P., Prianka, R., Preeta, V., Priyadarshini, N.S.: Implementation of sewer inspection robot. Int. Res. J. Eng. Technol. (IRJET), **05**(02) (2018)
2. Baby, A., Augustine, C., Thampi, C., George, M., Abhilash, A.P., Jose, P.C.: Pick and place robotic arm implementation using arduino. IOSR J. Electr. Electron. Eng. (IOSR-JEEE) **12** (2), 38–41 (2017)
3. Rajesh Kanna, S.K., Ilayaperumal, K., Jaisree, A.D.: Intelligent vision based mobile robot for pipe line inspection and cleaning. Int. J. Inf. Res. Rev. **03**(02), 1873–1877 (2016)
4. Alanabi, N., Shrivastava, J.: Performance comparison of robotic arm using arduino and matlab ANFIS. Int. J. Sci. Eng. Res. **6**(1) (2015)
5. Singh, J., Singh, T., Singh, M.: Investigation of design and fabrication of in-pipe inspection robot. Procedia. Eng. **3**(4) (2015)
6. Abidin, A.S.Z., et al.: Development of cleaning device for in-pipe robot application. In: IEEE International Symposium on Robotics and Intelligent Sensors, pp. 506–511 (2015)
7. Roy, S., Wangchuk, T.R., Bhatt, R.: Arduino based bluetooth controlled robot. Int. J. Eng. Trends Technol. (IJETT) **32**(5) (2016)
8. Truong, N., Krost, G.: Intelligent energy exploitation from sewage. IET Renew. Power Gener. **10**(3), 360–369 (2016). https://doi.org/10.1049/iet-rpg.2015.0154
9. Ambeth Kumar, V.D., Elangovan, D., Gokul, G., Praveen Samuel, J., Ashok Kumar, V.D.: Wireless sensing system for the welfare of sewer labourers. Healthc. Technol. Lett. **5**(4), 107–112 (2018)
10. Ambeth, K.V.D.: Human security from death defying gases using an intelligent sensor system. Sens. Bio-Sens. Res. **7**, 107–114 (2016)

IndQuery - An Online Portal for Registering E-Complaints Integrated with Smart Chatbot

Sharath Kumar Narasiman(✉), T. H. Srinivassababu, S. Suhit Raja, and R. Babu

Rajalakshmi Engineering College, Thandalam, Chennai 602105, Tamil Nadu, India
{sharathkumar.n.2015.it, srinivassababu.tm.2015.it, suhitraja.s.2015.it, babu.r}@rajalakshmi.edu.in

Abstract. Transparency is very significant for a successful e-governance. IndQuery is a complaint registering system used to improve the visibility of the actions taken on the complaints issued by the citizens. IndQuery is specifically developed for registering the complaints from the citizens. Also, when people need information regarding the house related services or finding difficulties in accessing those services provided by the government they can raise complaints and report it directly to the respective officials easily and effortlessly. Along with this, a chatbot is developed using artificial intelligence which is integrated with the website and many other social media platforms, to interact with the users dynamically. This Chatbot is mainly trained for reducing the official's workload and improvising the communication services. As it is difficult to find particular websites and contact details for reporting each problem, this website is considered vital for development. Also taking into consideration, the officials find difficulties for replying each and every complaint raised by thousands of citizens; Chatbot does the work for them. On the event of getting instantaneous replies people find it confident that their work is half done. The complaints can be submitted by the users from their respective portal which are stored in the database and further retrieved and displayed on the respective official's portal, by which the problems could be reviewed and resolved. Also, a feedback system is provided on which the users can rate the quality of service provided for them while the fixing of their complaints. This Chatbot can also be integrated with other platforms like Facebook messenger, Twitter, Telegram, Skype etc. to function conveniently with the users.

Keywords: Complaint register system · Chatbot · Artificial Intelligence · Machine learning · Information storage and retrieval

1 Introduction

IndQuery is an innovative concept which creates the citizens a practical and feasible way of communicating and registering their complaints with the government officials. IndQuery system is useful for collecting complaints, storing in a database, retrieved and displayed on the respective official's portal, by which the problems could be reviewed and resolved. A chatbot is integrated with the IndQuery system for accessing the

D. J. Hemanth et al. (Eds.): COMET 2019, LNDECT 35, pp. 1286–1294, 2020.
https://doi.org/10.1007/978-3-030-32150-5_130

contact information of the officials. The Chatbot can also be linked in the social media, making the well regulated communications. Information storage and retrieval is the technique used for storing and fetching the data from the database, so that it can be searched and displayed on request. High-speed, selective retrieval of large amounts of information for government, commercial, and academic purposes are possible due to the data processing techniques available.

IndQuery also provides platform for writing blogs by the government, by which the user entering the system can also be able to have a chance on reading those informative blogs. External blog links can also be referenced, so that navigation can be made easy to other informative websites. Blogs can also be used to mention the services provided by the government and methods to access it. And also any awareness blogs could be written to make the people educated about the crimes. The other important facility of IndQuery is the integration of chatbot which make customer experience much more immersive due to the fact of quick response and depth of knowledge that has been trained using the algorithms. Over the course of last decade it has been noted that many MNC's indulge themselves in developing chatbots for increasing their customer relationship. Chatbots advantages in the industry leads by the fact it is accessible anytime and works better than the Customer Relationship Management employees. Cost is another major that circulates into the companies perspective, definitely chatbots are less expensive than another technology that can be used to query customer problems. Chatbots has the ability to handle huge workload and can perform 24 $\times$ 7 which is a deal breaker for companies which have huge number of international customers.

2 Literature Survey

A system for the Municipal corporation of Pune where people register their complaints and get response from the system. Scores are calculated by Sentiment analysis when the citizens insert query to get intent of the citizens and to prioritize queries based on this score [11]. A chatbot making use of Wikipedia information and its own knowledge base for the development. This chatbot has the potential to recollect the previous conversation sessions of the current user while in engaging with that user [1]. Android application supporting the educational chatbot provides output from predefined data entered manually and also from other resources like Wikipedia. Hence it can provide results even from Wikipedia if it is queried. It also supports speech recognition so the input can be obtained from the user as speech format [3]. The student's questioning system is developed through Artificial intelligence and machine learning. The student will query the system which will find the keyword from this query and will produce the suitable response [14]. Personalized Medical assistant are created to reduce the most of the jobs of doctors in near future. Many number of lives could be saved with the help of the system [5].

A mobile interface is developed to register complaints by the locals. No need of a structured input format from the user for registering complaints because being a natural language processing system it has the ability to analyze the inputs. After the complaint being registered, the system generates the complaint number to the user [13]. A bot which has the capabilities of handling much bigger problems when improving the performance of the model [6]. SuperAgent is a chatbot which is essentially used for e-commerce websites. SuperAgent takes advantage of large-scale customer data which is essential for understanding the customer perspectives on products development [9]. An interactive system which allows local common people to lodge their complaints. This approach has been meant for the wholesome good of the nation not based only for a particular commodity alone [12]. A Chatbot is developed by Manipal University to meet the education needs of the students. This chatbot is implemented using AIML language to access data on university rankings, services offered by the university etc. [2].

A simpler chat system using voice is created; providing a better user-friendly experience. Voice recognition involves two step process, the first step involves capturing of input signal and next step involves analysis of input signal. The client acquires the signal through an input mechanism which utilizes the operating system to interpret the signal. This process decreases the load on the server and provides a much faster response [4]. Chatbot uses simple pattern matching to represent the output based entirely on input alone whereas other Chatbots uses input rules, keyword patterns and output rules to generate a response the responses from this chatbot are much faster due to the fact of generating output without pattern matchers at deeper levels. If the input is not found in the database, a default response is generated without further processing. The input and output can be customized according to the user and responses are rapid due to this fact [8]. The users can freely register the complaints through a software tool and once the complaints are committed to the database, automatic tokens are generated and conveyed to the customer through a text message and email for further tracking of the complaint which is an important feature corresponding to this chatbot [10]. System uses alexa enabled devices to take inputs through user's voice and process the inputs with the help of NLP techniques to predict what the user is asking for? [7].

3 Methodology

Pattern Matching is used to provide response to the customers based on input texts. Artificial Intelligence Markup Language (AIML) provides standard pattern structure. This classification is entirely based upon Multinomial Navie Bayes classification used only text classification. The input score is calculated and is used to identify the class with highest match. The score also identifies which intent matches mostly with input data. The relativity base is provided by the highest score.

For example: Sample data set
class: welcome
"Hey wassup?"
"good day"
"Look out,Buddy"
Input sentence:"Hello good to see you Buddy"
input1: "hello" (no match)
input2: "good" (class: welcome)
input3: "morning" (class: welcome)
input4: "to" (no matches)
input5: "see" (no matches)
input6: "you" (no matches)
input7: "Buddy" (class: welcome)
Classification: greeting (score=2)

4 System Architecture

The Fig. 1 depicts the architecture of the system, explaining the outline by including the modules, components and workflow.

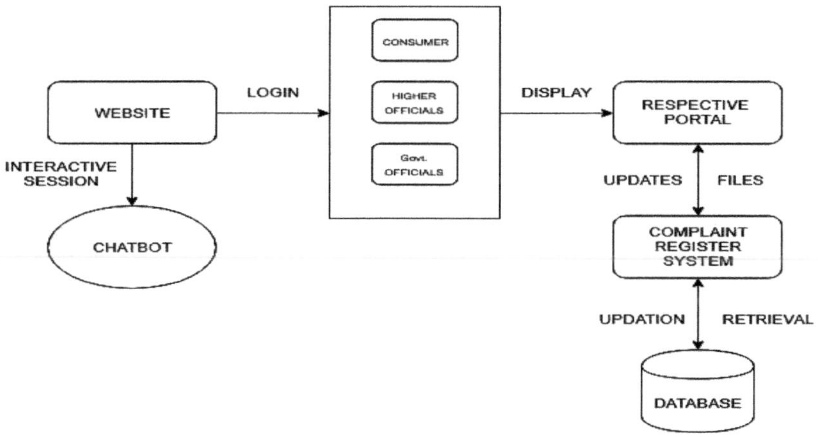

Fig. 1. Architecture diagram

5 Workflow

IndQuery is a system that provides a modular approach on how the citizens are able to register themselves and upload complaints. Already registered user can directly get into their feed by login. A new user has to register himself into the system to attain the services. The new user can easily avail the services without actually registering into the

IndQuery system. The working starts from the citizens point of view who face problems against the services that provided by the government. The issues that corresponds within the particular areas are dealt in this model. Once the citizens register themselves into the system, they can raise ticket on the problems they face. Chatbot is used so that the register user has an interactive experience while attaining the services provided by the same. IndQuery system also provides with additional process of writing blogs by the government officials which the registered user. The working is explained in Fig. 2.

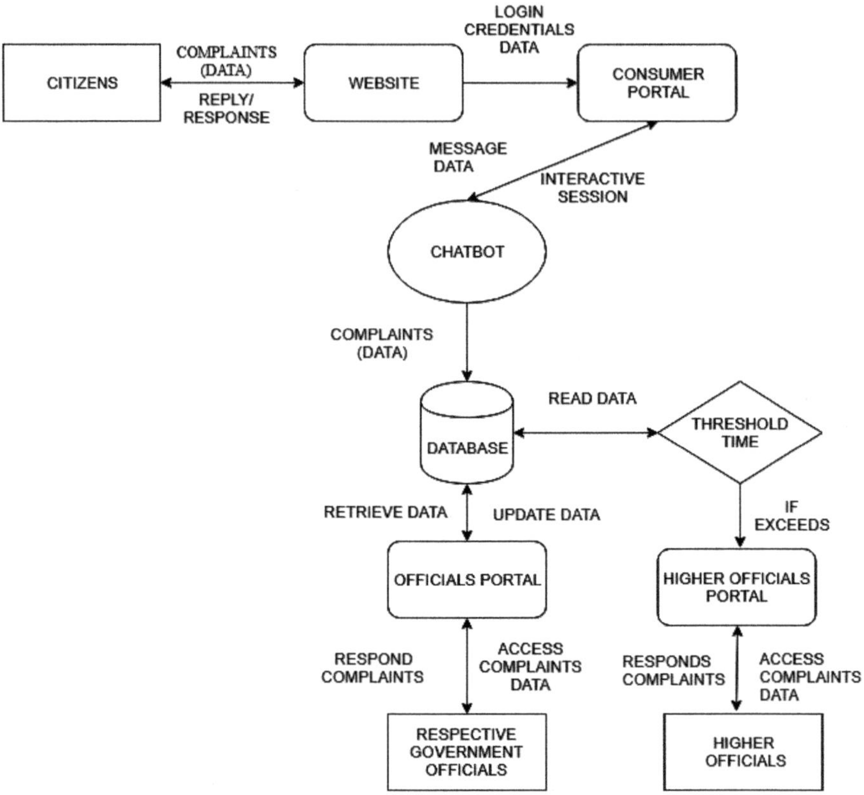

Fig. 2. Working model

6 Implementation and Results

The implementation modules of our proposed system as follows:

6.1 Chatbot Services and Complaint Register System

This system is used for registering complaints through a web portal or in a social media platform. Citizens can signup in the website for registering the complaints. After logging into the system, a form has to be filled to file a complaint. Consumers can

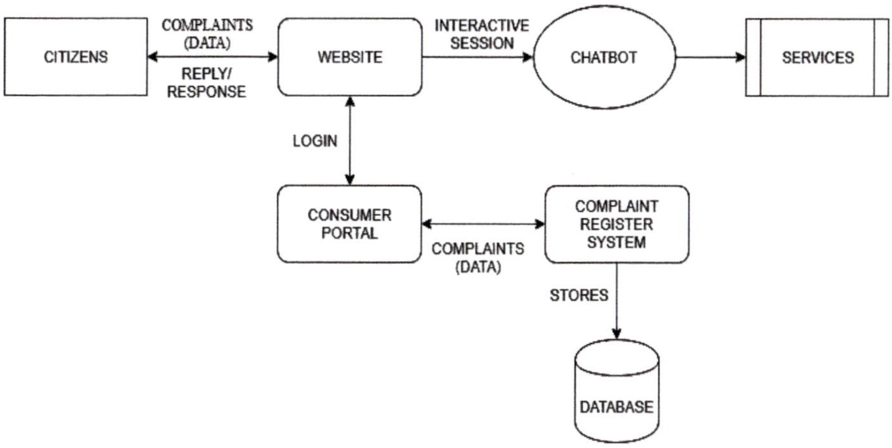

Fig. 3. Citizens gets the necessary services with interactive chatbot and enter into the portal to file complaints.

report their problem to the system. Further, the complaints get stored in the database for future requirement as shown in Fig. 3. The chatbot is useful for knowing about the services and accessing the contact details of the officials as shown in Fig. 4.

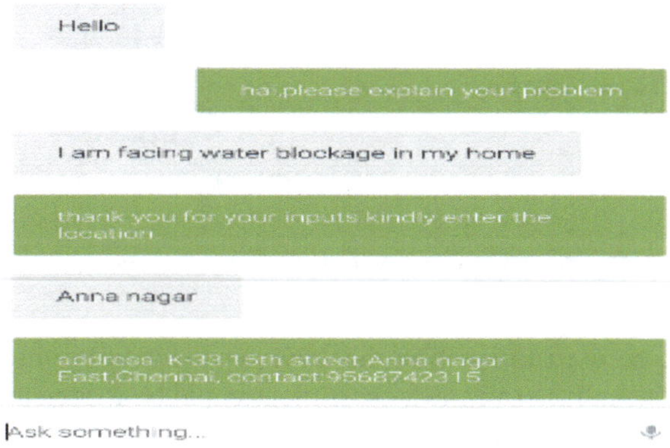

Fig. 4. Working of chatbot

6.2 Complaint Display Portal

Database stores the registered complaints with the respective tracking ID which is further used for tracking the complaints when complaint reaches the officials. When the officials fails to do the necessary actions within the threshold time the complaints gets forwarded to the respective higher officials by the system. Each officials have their own portals which is customized to represent the workload in sequential manner based on threshold time as shown in Fig. 5.

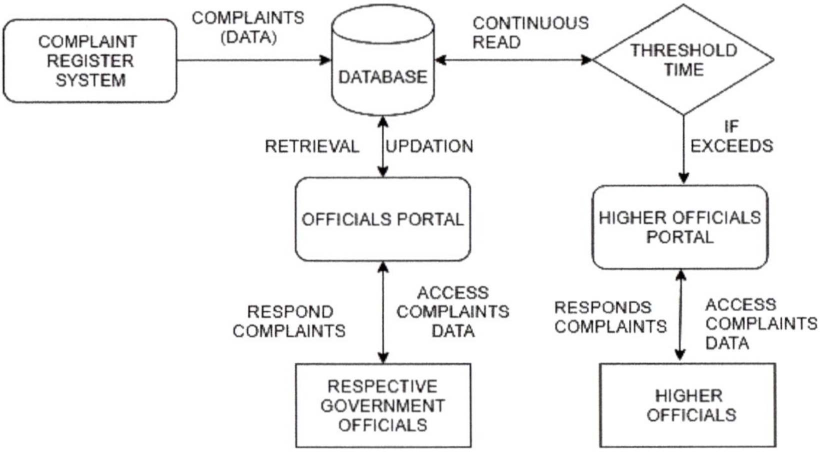

Fig. 5. Complaints are to be shown to respective official's portals.

The overall system is designed as follows in Figs. 6 and 7. This is the main page of the website containing all the details and guide to access and register the complaints as shown in Fig. 6.

Fig. 6. The index page of the IndQuery system.

Blogs can also be used to mention the services provided by the government and methods to access it. And also any awareness blogs could be written to make the people educated about the crimes shown in Fig. 7.

Fig. 7. The blog page of the IndQuery system.

7 Conclusion and Future Work

An user-friendly system which is ready to serve to the needs of the citizens and provides solution to their complaints in a transparent procedure without any abstraction. Citizens also has the ability to track the complaints to know what actions are planned by the respective officials for their complaints. Complaints that are related to domestic services can be lodged in this complaint register system. The system will be able to handle workload of n complaints at once. The system is both responsive and multi platform with the support of omni-channel experience to the end user. Not only home related services but also almost all the services provided by the government can be addressed. Analytics on the feedback data can be made, in order to assess the performance of the complaint resolvers. A complete transparent e-governance service provided by the government can be achieved. More efficient, well-trained chatbot can be provided to the user when it is exposed to the real time user data.

References

1. Hussain, S., Athula, G.: Extending a conventional chatbot knowledge base to external knowledge source and introducing user based sessions for diabetes education. In: 2018 32nd International Conference on Advanced Information Networking and Applications Workshops (WAINA), Krakow, pp. 698–703 (2018). https://doi.org/10.1109/WAINA.2018.00170
2. Ranoliya, B.R., Raghuwanshi, N., Singh, S.: Chatbot for university related FAQs. In: 2017 International Conference on Advances in Computing, Communications and Informatics (ICACCI), Udupi, pp. 1525–1530 (2017). https://doi.org/10.1109/ICACCI.2017.8126057
3. Kumar, M.N., Chandar, P.C.L., Prasad, A.V., Sumangali, K.: Android based educational Chatbot for visually impaired people. In: 2016 IEEE International Conference on Computational Intelligence and Computing Research (ICCIC), Chennai, pp. 1–4 (2016). https://doi.org/10.1109/ICCIC.2016.7919664
4. du Preez, S.J., Lall, M., Sinha, S.: Intelligent web based chatbot. In: 2017 IEEE International Conference on Consumer Electronics (ICCE). https://doi.org/10.1109/EURCON.2009.5167660
5. Madhu, D., Jain, C.J.N., Sebastain, E., Shaji, S., Ajayakumar, A.: A novel approach for medical assistance using trained chatbot. In: 2017 International Conference on Inventive Communication and Computational Technologies (ICICCT), Coimbatore, pp. 243–246 (2017). https://doi.org/10.1109/ICICCT.2017.7975195
6. Supriya, M.T.M.: Neural network based chatbot. Int. J. Adv. Res. Comput. Eng. Technol. (IJARCET) **4**(5) (2015)
7. Argal, A., Gupta, S., Modi, A., Pandey, P., Simon, S.Y.S., Choo, C.: Intelligent travel chatbot for predictive recommendation in echo platform. In: 2018 IEEE 8th Annual Computing and Communication Workshop and Conference (CCWC), pp. 176–183 (2018)
8. Dahiya, M.: A tool of conversation: chatbot. Int. J. Comput. Sci. Eng. **5**, 158–161 (2017)
9. Cui, L., Huang, S., Wei, F., Tan, C., Duan, C., Zhou, M.: Superagent: a customer service chatbot for e-commerce websites, pp. 97–102 (2017). https://doi.org/10.18653/v1/p17-4017
10. Bala, K., Kumar, M., Hulawale, S., Pandita, S.: Chat-Bot for College Management System Using A.I (2018). (2395-0056)

11. Deshmukh, K.V., Shiravale, P.S.S.: Priority based sentiment analysis for quick response to citizen complaints. In: 2018 3rd International Conference for Convergence in Technology (I2CT), Pune, pp. 1–5 (2018). https://doi.org/10.1109/I2CT.2018.8529722
12. Kazi, S., Ansari, S., Momin, M., Damarwala, A.: Smart e-grievance system for effective communication in smart cities. In: 2018 International Conference on Smart City and Emerging Technology (ICSCET), Mumbai, pp. 1–4 (2018). https://doi.org/10.1109/ICSCET.2018.8537244
13. Kopparapu, S.K.: Natural language mobile interface to register citizen complaints. In: TENCON 2008 - 2008 IEEE Region 10 Conference, Hyderabad, pp. 1–6 (2008). https://doi.org/10.1109/TENCON.2008.4766675
14. Hiremath, G., Hajare, A., Bhosale, P.: Chatbot for education system. In: Proceedings of the IEEE 10th International Conference on Rehabilitation Robotics, Noordwijk, The Netherlands, 12–15 June 2007

A Study on Embedding the Artificial Intelligence and Machine Learning into Space Exploration and Astronomy

Jaya Preethi Mohan(✉) and N. Tejaswi

Dr.M.G.R. Educational and Research Institute, Chennai 600095, India
mjayapreethi5@gmail.com

Abstract. Artificial Intelligence and Machine Learning are powerful inventions which are applied to attain dynamic purposes in several disciplines. In that, the field of Space Exploration and Astronomy are highly supported by artificial intelligence and machine learning discoveries. The Strategies of Space Exploration and astronomy are enhanced by the progress of artificial intelligence and the efficiency of a machine learning algorithm in the scientific study of celestial objects and the atmosphere of the Universe. The vision of Space Exploration and Astronomy expects high-level computerized operations for automation and predictions. Also, the need for analysis of data generated from modern equipment requires these emerging technologies. A study on embedding Artificial Intelligence into Space Exploration and Machine Learning into Astronomy reveal the approach of using the robotic and statistical methodology on Space Exploration and Astronomy. It also includes the characteristics and presentation of such models, future challenges, Classification and Information integrity to empower the performances of Space Exploration and Astronomy.

Keywords: Artificial Intelligence · Machine Learning · Space Exploration · Astronomy

1 Introduction

Space operations and missions sustain their operations remotely because of the challenges, the cost, exploration distances. For instance, missions like messenger, Multitude of earth orbiters, the New Horizon, the Cassini to the planets Mercury, Stars, Mars, Saturn, Pluto and more. All the missions are assisted by rovers, Orbiters and other data mining and analysis tools. Due to the radiation hardening process, remote spacecraft faced many computational limitations which lead to high cost with failure. Similarly, the second space mission was failed. Autonomy came forward and introduced artificial intelligence and machine learning techniques to treat threats and core operations then the third space mission ended up with an experience of time. All these records influenced the need for machine learning and artificial intelligence potentials in space exploration and astronomy domain for thinking the probabilities of problems and suitable solutions. The existing technology used in space operations does basic data analysis of collected images, minor statistical operations, and visualization but it can be improvised with the advancements in machine learning and artificial intelligence [1].

D. J. Hemanth et al. (Eds.): COMET 2019, LNDECT 35, pp. 1295–1302, 2020.
https://doi.org/10.1007/978-3-030-32150-5_131

2 Artificial Intelligence for Space Exploration

National Aeronautics and Space Administration (NASA) discusses the research on artificial intelligence science and Technology for space exploration. The domains which are under space exploration are Control, diagnosis and subsystems reconfiguration such as thermal control, communications, and power [2]. NASA's mission has been categorized into two ways. One is functioning as intelligent assistants to equalize the expertise level of humans and another is to act as a replacement for human experts. It is achieved by the intelligent systems which exhibit the behavior of artificial intelligence and fits into various needs like modeling, reasoning, uncertainties, improved efficiency for time, space complexity and compression of Information. Intelligent systems utilize the concept of neural networks, artificial immune systems, and other intelligent algorithms. It may result in doing fraud detection, processing and more [3]. AI planning Software is being developed by NASA to make practical distant missions. In order to reduce the daily planning time and cost of the Mars Exploration Rovers mission team, NASA Scientists creates software to save time and cost. Similarly, research saving approaches with artificial intelligence will stand in prior to the Phoenix mission at the red planet's North Pole region [4].

2.1 AI and ML Algorithms for Space Exploration and Astronomy

Research on deep search exploration is the strength of out space analysis and physical science. The deep space research is oriented to planets and out space endeavor. Deep space explanation deals with massive datasets which are used in dynamic rations like cosmological effect predictions, planet classifications, etc. To acquire the knowledge on insights of galaxy formation and evolution, astronomers use the massive datasets for morphological classification of galaxies. The objective of the morphological classification of galaxies is to get the insights of the galaxy formation. The datasets have been generated from different space craft. The spacecraft and its datasets need inspecting visualization with another analysis process [5]. The algorithms which are highly preferred for Astronomical Data computations are as follows.

Deep Learning. A set of methods which automatically determines the representation required for detection and classification of data with the help of machine is representation learning. Deep learning methods are representation methods which are made of many processing layers. A set of methods which automatically determines the representation required for detection and classification of data with the help of the machine is representation learning. Deep learning methods are representation methods which are made of many processing layers. Deep learning is meant to learn and it immense data and done by using the back propagation algorithm. Back propagation algorithm is chosen by deep learning models to differentiate the hidden attributes which can be classified from the raw data. Those classifications fit into multiple layers with computations. The deep learning concept is applied for processing images, speech, audio, video, and other media files [6]. This deep learning concept has used for the historical theory of gravitation. Einstein's geometric theory of gravitation is known as general relativity. It depicts the important geometric features of gravitational systems with the modeling of space time [7]. Einstein's calculation with the geometry theory of gravitation i.e. general relativity predicted the ray of light which passed closer to the sun. It rated with the small amount of deflection for 1.3 arcseconds which are approximately 1/2800 [8]. Deep learning is used to predict the apparent location of the stars in day time and also influence the light ray which passes across the sun as pointed out in the theory of general relativity. When it comes to image recognition, expert engineering features of the datasets like pixel values, time series statistics, SIFT are not required for astronomy purposes [9].

Artificial Neural Network. Complex problems can be solved efficiently with the divide and conquer method. Simple elements joined together for complex problems. Likewise, complex problems can be decomposed as simpler elements. Networks attain these tasks by the nodes which are interconnected. The network possesses the flow of information irrespective of directions. One of the network types which consider the nodes as "artificial neurons". The artificial neuron is otherwise known as Artificial Neural Network. It almost resembles the real neurons of a human being. The mathematical function used to proceed with the identified output is the activation function or neural functions. The function uses the input as resources and does the computed function to give the output. The respective signals are gathered together as weights (see Fig. 1).

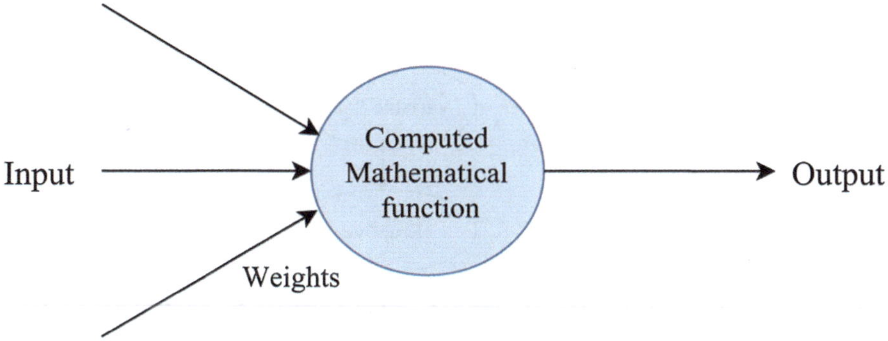

Fig. 1. Artificial neuron

The figure illustrates the artificial neuron from input to output Artificial Neuron with respective signals and activation function [10]. Usage of Artificial Neural Network is applicable for decision making, regulations, robotics, clustering, Compression, data processing, data representation etc. Artificial Neural Network must be chosen for an appropriate application. It is known for memorizing apart from learning which utilizes the learning algorithm for getting the output [11]. For instance, the data can be collected to predict the centroid in low-order aberrations like tip and tilt measurements, faint science object image display. For every simulation step, the centroids of the entire source were computed per simulation step. Training of artificial Neuron Network is done by using Back Propagation algorithm for priori centroid data. Associated coordinated data is available in associated data. Three different datasets were collected with processes like training and validation. The generalization ability is the methodology and approach to inspecting that specific network inputs away from the training set for centroid prediction in Astronomy [12].

Support Vector Machine. Support Vector Machine (SVM) has been created in the structure of Statistical learning theory. It is developed to solve the problem in the theory of statistical learning theory and follows the supervised learning approach of machine learning [13]. It is a classification method and falls into the division of kernel methods. It was prescribed for integral equations and unknown mathematical functions. In 1998, the name kernel function stated as Kernel tricks for machine learning. The non-linear problem can map the problem from the input space to the high dimensional space is known as feature space using non-linear transformation [14]. It is used for time series prediction, face authentication etc. The galaxies can be classified with the method called Morphological galaxy classification. It is generally categorized into various categories. Normal galaxies (no gross characteristics), active galaxies (powerful nuclear activity), starburst galaxies (intense star-formation activity), interacting galaxies (recent gravitational encounter) [15]. The image features classified into different morphic features of the galaxy and holds 10- fold cross-validation with the combination of Performance of the Classification Algorithm. The image features were not separated linearly while projected in high dimensions with the RBF kernel. Although Support Vector Machine is good for astronomy purposes and resulted in over fit for the morphological galaxy classification [5] (Fig. 2).

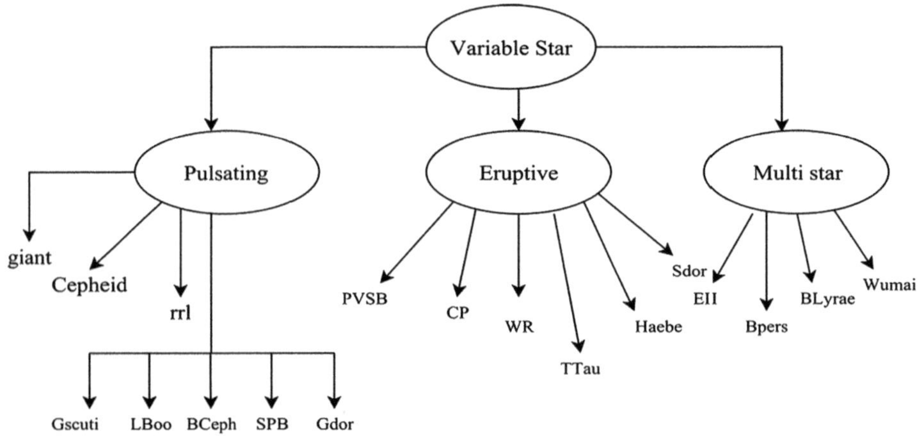

Fig. 2. The classification of objects [9].

Linear Regression. Linear Regression is a supervised learning approach. The linear dependence of Y on $f(x) = X_1, X_2, X_3 \ldots X_p$ is attained by assumption. In-depth, genuine regression functions are not linear. The linear Regression model is conceptual and practical. It is possible to make predictions using multiple regressions. Residual Computed Error can be computed for accessing the overall accuracy of the model [16] (Fig. 3).

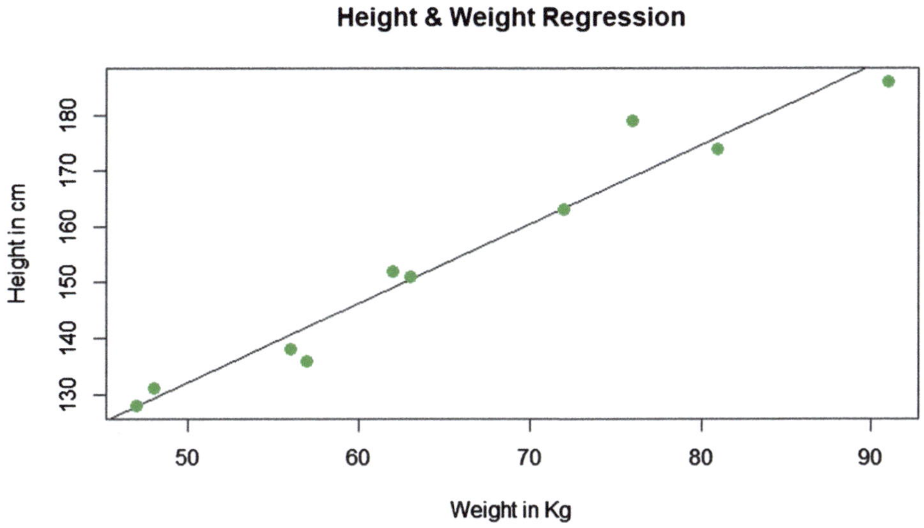

Fig. 3. The linear regression model of the height and weight data in a graphical way.

Lira. A package of R does Bayesian linear regression and forecasting in astronomy. The process judges the errors in different variables, essential scatters and scatter correlation, the time evolution of slopes, normalization, scatters, upper limits and a slice of linearity. Gibbs method exploiting the JAGS library with the posterior distribution of the regression parameters is sampled [17]. Although the linear regression model seems to be simple, it contains the probabilities of issues in astronomy. The new methods were proposed to treat measurement errors for data analysis of linear regression. They are a direct extension of the ordinary least squares which allows for dependent measurement as an estimator where the direct extension is for both variables. On the same way magnitudes of the linear regression depends on the measurements. There are few more methods involved in this to clear the linear regression issues occurs with astronomical data [18]. But the slope variance estimates the standards which are assumed with strict and restrictions. The X values of the linear regression model errors are dependent. The estimates are valid even after the condition is broken. This derivation method is termed as the delta method. But the slope variance estimates the standards which are assumed with strict and restrictions. The X values of the linear regression model errors are dependent. The estimates are valid even after the condition is broken. This derivation method is termed as the delta method. While approaching the intrinsic relation between the two different properties, it might result in four symmetrical regression lines. [19].

3 Computation of Astronomical Data

Handling gigantic and unstructured datasets involves the common method or requirements of other sectors. The software and tool requirement functions are for the discovery of knowledge on the datasets. Researchers of various domains like computers, statistics, informatics, astronomy are developing tools and software needed for solving the astronomical problem.

3.1 Packages and Tools for Astronomical Data

The Table 1 represents the tasks of various algorithms and its role in astronomy [20].

Table 1. Tasks and applications

S. No.	Purposes	Assignment
1	Novel detection trend prediction	Time-series analysis
2	Rare object detection	Anomaly detection
3	Special object detection	Clustering
4	Spectral, galaxy, photometric classification solar activity	Classification
5	Regression	Photometric red shifts

IRAF. Image Reduction an Analysis Facility is for the reduction and analysis of astronomical data with the software system. The association of IRAF and CCD decrease package, ccd red, gives tools for the simple and efficient reduction of CCD images. The reduction operations are the replacing the wrong pixels, deduction of over scan pre-scan bias, reduction of a zero level image, decrease of a dark count image, division by a flat field calibration image, division by an illumination correction, reducing the edging image, and trim needless lines or columns [21].

The Virtual Astronomical Observatory. The US Virtual Astronomical Observatory is a software development intended to initiate the operational Virtual Observatory (VO) and to give the US coordination with the global VO effort for the purpose of which an astronomer is able to discover, access, and process data flawlessly, without considering the physical location [22].

AstroPython. The collection of packages which strengthen the statistical operations, decision making and more within the python language is called Astro python. It has multiple packages to compute the astronomical data in all the phases. It does enormous operations like Astrophysics utility, cluster analysis, Reduction and analysis of radio astronomical data, etc [23].

VOStat. It is a web oriented service offering statistical analysis of astronomical datasets. It is incorporated into the group of analysis and visualization combined with the international Virtual Observatory (VO) via SAMP communication system. VOStat with a extracted dataset from the VO, else chooses in sixty statistical functions in R mathematical functions [24].

DAME. Data Mining and Exploration is a data mining tool which works based on the web operations. It supports immense dataset with machine learning methods used in astronomical data applications and tasks with classification, detection functions [20].

4 Conclusion

The study on embedding artificial intelligence and Machine learning on space exploration have given a vision of how the emerging and powerful technologies connected in various aspects of space exploration and astronomy. The paper began discussing the role of artificial intelligence through intelligent systems for space exploration and the technology helped space missions to solve the human error, time and cost issues. AI planning software of NASA is advisable for planning distant missions. And then the key component of artificial intelligence and machine learning is an algorithm. There are many algorithms which enable the statistical functions and automated mathematical predictions, detection, the classification for multiple uses. Space exploration and astronomy majorly uses the framework and algorithms such as neural networks, artificial neural networks, support vector machines and regression analysis with prediction. To commit the machine learning and artificial intelligence functions effectively, numerous packages and tools have been developed specially for Astronomical data. They are software and packages which consists of aggregate and operational classes bound together to provide an essential analysis and analytics results. As mentioned above, IRAF, Astropython, VOSTAT are effective with programming implementation. Whereas the software like virtual astronomical observatory and DAME can be improvised for an accurate optimization with all the functional features. Apart from all the information given in this paper, there are other efficient developments are under process and in performance to achieve the high performing computations. In Conclusion, Using supervised algorithms with proper data analysis of an astronomical data yields the best results with accuracy of predictions; image processing and more on astronomical datasets will lead to proficiency in space exploration and Astronomy with the measures to access the massive datasets of the galaxy and space intelligent systems.

References

1. Mcgovern, A., Wagstaff, K.L.: Machine learning in space: extending our reach. Mach. Learn. **84**, 335–340 (2011). https://doi.org/10.1007/s10994-011-5249-4
2. Friedland, P.: Panel on artificial intelligence and space exploration. Artificial Intelligence Research Branch, NASA Ames Research Center, Moffett Field, CA, Panels (1676)
3. Krishnakumar, K., Lohn, J., Kaneshige, J.: Intelligent systems: shaping the future of aeronautics and space exploration. NASA Ames Research Center, MS 269-1, Moffett Field, CA
4. NASA Research Page. https://www.nasa.gov/centers/ames/research/exploringtheuniverse/spiffy.html
5. Natarajan, S.: Online transfer learning and organic computing for deep space research and astronomy (2019). https://doi.org/10.13140/rg.2.2.23957.78564
6. LeCun, Y., Bengio, Y., Hinton, G.: Deep learning. Nature **521**, 436–444 (2015). https://doi.org/10.1038/nature14539
7. Lecture Notes on General Relativity Columbia University (2013). https://web.math.princeton.edu/~aretakis/columbiaGR.pdf

8. Cosmic Times 1919. Sun's Gravity Bends Starlight—Einstein's Theory Triumphs. https://imagine.gsfc.nasa.gov/educators/programs/cosmictimes/downloads/newsletters/1919NL_EarlyEd.pdf
9. Rebbapragada, U.: Machine learning applications in astronomy. Ph.D. California Institute of Technology (2017)
10. Artificial Neural Networks for Beginners. https://arxiv.org/ftp/cs/papers/0308/0308031.pdf
11. Andrej, K., Janez, B., Kos, A.: Introduction to the artificial neural networks. In: Suzuki, K. (ed.) Artificial Neural Networks - Methodological Advances and Biomedical Applications (2011). ISBN 978953-307-243-2
12. Weddell, S., Webb, R.Y.: Dynamic artificial neural networks for centroid prediction in astronomy, p. 68 (2006). https://doi.org/10.1109/his.2006.22
13. Evgeniou, T., Pontil, M.: Support vector machines: theory and applications. In: Paliouras, G., Karkaletsis, V., Spyropoulos, C.D. (eds.) ACAI 1999. LNCS (LNAI), vol. 2049, pp. 249–257. Springer, Heidelberg (2001). https://doi.org/10.1007/3-540-44673-7_12
14. Sewell, M.: Kernel Methods. Department of Computer Science, University College London (2007). svms.org/kernels/kernel-methods.pdf
15. Galaxy morphology and classification (2016). Galactic Astronomy. https://www.phas.ubc.ca/~hickson/astr505/astr505_2016-2.pdf
16. Standford. https://lagunita.stanford.edu/c4x/HumanitiesScience/StatLearning/asset/linear_regression.pdf
17. Cran R. https://cran.r-project.org/web/packages/lira/index.html
18. Akritas, M.G., Bershady, M.A.: Linear regression for astronomical data with measurement errors and intrinsic scatter. Astrophys. J. **470** (1996). https://doi.org/10.1086/177901
19. Jogesh Babu, G.: Center for Astrostatistics Eberly College of Science, Penn State. https://www.iiap.res.in/PostDocuments/Regression1.pdf
20. Zhang, Y., Zhao, Y.: Astronomy in the big data era. Data Sci. J. **14**, 1–9 (2015). https://doi.org/10.5334/dsj-2015-011
21. Valdes, F.: The IRAF CCD Reduction Package – CCDRED (1990). https://doi.org/10.1007/978-1-4612-3880-5_40
22. Hanisch, R.J., Berriman, G.B., Lazio, T.J.W., Bunn, S.E., Evans, J., McGlynn, T.A., Plante, R.: The virtual astronomical observatory: re-engineering access to astronomical data. Astron. Comput. **11**(Part B), 190–209 (2015)
23. Astropython Packages page. http://www.astropython.org/packages/
24. Chakraborty, A., Feigelson, E.D., Jogesh Babu, G.: VOStat: a statistical web service for astronomers. Pacific (2013). arXiv:1302.03871086/670053

Advances in Networking and Communication

Surveillance System for Golden Hour Rescue in Road Traffic Accidents

S. Ilakkiya(✉), R. Abinaya, R. Shalini, K. Kiruthika, and C. Jackulin

Department of Computer Science and Engineering,
Panimalar Engineering College, Chennai, India
abianbu3119@gmail.com, abireshl323@gmail.com,
shalinirajendran@gmail.com, kiruthikaitdept@gmail.com,
chin.jackulin@gmail.com

Abstract. As the earth is becoming a virtual globe, the advancement in technology can be used to save lives that are lost in accidents due to belated help. Accidents occur for a variety of reasons like swerving off because of pets, sudden failure of internal mechanism in automobiles, distraction caused bysudden noises and sights, unseen object shadows in night time and the list goes on. In this paper we suggest a system incorporated into street lamps or lamp posts that identify accidents when they occur by use of Feature Extraction technique. The system on analysing an accident alerts the nearby medical personnel and law enforcement by the presence of Global Positioning System embedded in it. In this method we can ensure that lives are saved by timely help at the Golden Hour. In addition, the lamp posts can be interconnected to form a network and each lamp posts covers about some particular distance in 3600 direction. Basically designed for road traffic accidents, this system also has the possibility of identifying other crimes that occur in streets and alerting the law enforcement to it. By incorporating the system in lamp posts we can ensure that it is allocated to nearly all human inhabited areas thereby assuring timely help in case of emergencies.

Keywords: Feature Extraction Technique · Global Positioning System · Golden hour

1 Introduction

On the current path of fast paced day to day human lives, speed is the new mantra.

Time is of the essence and it applies to all aspects of today's world. "A road accident is an accident which involves at least one road vehicle, occurring on a road open to public circulation, and in which at least one person is injured or killed". Killed persons are accident victims who die immediately or within thirty days following the accident. Injured persons are accident victims having suffered trauma requiring medical treatment. When a road traffic accident occurs, it may concern a pedestrian or an automobile with passengers in it. Each and every life is valuable and each and every second spent in saving those are precious. But when we take a look at these accidents,

D. J. Hemanth et al. (Eds.): COMET 2019, LNDECT 35, pp. 1305–1310, 2020.
https://doi.org/10.1007/978-3-030-32150-5_132

we find most of the people who could have been saved had lost their lives because help arrived too late.

"Golden hour [1] also known as golden time, refers to the period of time following a traumatic injury during which there is the highest likelihood that prompt medical and surgical treatment will prevent death". This golden hour rescue is the most vital part of an accident. The system we propose finds a method to ensure a golden hour rescue.

For the purpose of assuring the golden hour rescue we have proposed a Surveillance System. This system consists of Charge Coupled Device(CCD) camera, an image processing unit that uses Feature Extraction Technique to identify accidents, Global positioning System to pinpoint location and a transmitter to allocate the nearby law enforcement and medical personnel.

The statistics for the road traffic accidents in India for the year 2017 is as follows (Fig. 1):-

Fig. 1. Percentage of road accident in India 2017 [9]

As we can deduce from the picture above India lost 1.47 lakhs of people in road accidents in the year 2017. The highest percentage was 1.5 lakh people lost in 2016. The rise in accidents started from 2007 though it has progressed marginally from 2009. Every year, there approximately one lakh people who lose their lives in road accidents in India. Taking this as a serious issue India had signed the Brasilia declaration in order to decrease the fatality rate by half in 2015. But as we can see the substantial decrease is not on par owing to many reasons untimeliness a major one among them.

2 Literature Review

Chaudhari [3] has done a paper on Advanced Golden Hour Rescue System using Android. In this proposed system a shaking sensor installed in the car detects an accident. The controller then locates an ambulance and ensures it free pathway arrival in order to ensure timely help. This system has a drawback of having to make sure the installation of the system in each automobile to assure it's compatibility.

Javale *et al.* [4] has proposed an Accident Surveillance and Detection System using Wireless Technologies. The main goal of this system is to provide a system that (1) detects when an accident occurs using micro controller in smartphones (2)find exact location using GPS and (3) alert nearby medical personnel or hospital with Bluetooth or GSM enabled SMS. Again, the proposed system should be present in the automobile that supposedly meets with an accident.

Ki [2] had proposed an Accident Detection System using Image Processing and MDR. In his paper he has suggested an accident detection algorithm using Meta Data Registry (MDR). He proposes fixing a camera at an intersection which uses the aforementioned algorithm to detect if an accident has occurred. If an accident has occurred the pictures are sent to Traffic Monitoring Center (TMC). This traffic Accident Recording and Reporting System (ARRS) is mainly allocated to intersections and also alerts only the Traffic Monitoring Center.

3 System Configuration

Surveillance system for golden hour rescue in road traffic accidents is an image processing based system which tracks an accident that had occurred and quickly shares the accident details to the nearby hospitals for saving the accident victim during the golden hour. The system has a surveillance camera embedded inside a lamp post, image processing unit which captures the accident frame by frame and analyses it for details, a GPS chip and a transmitter for transferring the accident's location details to the server. The server redirects these details to the hospitals near the accident spot and to police control room. By this way, we can rescue the victims during their golden hour thereby saving their lives.

3.1 Charge Coupled Device (CCD) Camera

Couple-Charged Device (CCD) cameras are a kind of image capture devices which records electronic signals to register visible light through image sensors. This variant of camera records the electronic signal and stores it in either an internal memory or in a remotely connected device. CCD is the most popularly used type among the variants of digital camera (Fig. 2).

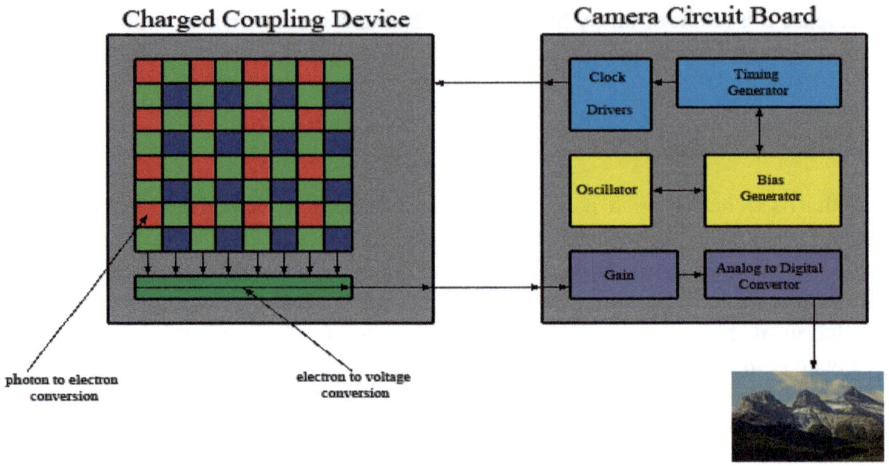

Fig. 2. Charge coupled device (CCD) camera working via Source: Sensor Cleaning [5]

After the initiation of the camera's record function, light is focused by the camera lens through the camera aperture, light filters, and onto the electronic image sensor. The image sensor is arranged in a grid pattern where each individual square is called a pixel. The image sensor cannot determine the colour of recorded light. Instead, it can only determine its intensity. A colour filter is used here to define a colour and it allows only one colour of light from the visible spectrum into each pixel. The colour filter is generally arranged in a Bayer filter pattern, which averages the colours of a 2×2 pixel square. Since the filter produces some inaccuracy, any discolouring that occurs due to this is called interpolation. There is another method of colour identification in which separate image sensors with each dedicated to capturing different aspects of a colour image like one colour, and the results are combined to generate the full colour image. They usually make use of colour separation devices like beam splitters rather than having integral filters on the sensors.

In Closely Coupled Device cameras, the recorded level of photons is converted to a proportional electrical signal. This ratio which is for photons-to-electrons generation is called quantum efficiency. These signals are carried to a charge amplifier away from individual pixels, which then turns the charge into a voltage. The camera then creates an authentic chronicle of the light that it has captured. For video cameras, this process is repeated multiple times per second, and the voltages are digitized and stored in a memory. When the images are replayed, the camera creates the appearance of an object in motion through a large number of sequential stills it had captured.

3.2 Feature Extraction Technique

Feature extraction is mainly used for describing the relevant shape information contained in a pattern so that the pattern can be classified easily. The main aim of feature extraction is to obtain the most relevant information from the original data and represent that information in a lower dimensionality space.

When an input data is too large or too redundant to be processed by an algorithm, then represent that information in reduced representation of a set of features or feature vector. It can also be called the process of transforming input data into a set of features. Feature extraction can be done in two stages- feature selection and classification. It is based on three types of features which are (1) Colour feature (2) Shape feature and (3) Texture feature [6].

In this process, vectors are formed from relevant features. Classifiers then recognize the input unit with target output unit using these feature vectors. The classifier can now easily classify between different classes by looking at these features.

3.3 Global Positioning System

The Global Positioning System or commonly known GPS is a global coverage system that was developed by the United States Department of Defence in the year 1973. It provides functionality to both military and civilians world-wide and is currently under the control of the U.S Space Command.

GPS system relies on numerable satellites in space. A GPS receiver needs to be connected to minimum four different satellites to determine the receiver's precise position based on latitude, longitude and altitude measurements. It also uses a tracking algorithm commonly known as tracker. The tracker's main disadvantage is the time delay in calculating the constantly changing position. GPS has three segments. They are (1) Space segment (2) Control segment and (3) User segment. The Space segment consists of the satellites in space which are aligned with an atomic clock time. The Control segment is the main connecting segment between receiver and satellites. The User segment has the GPS receiver from which a person initiates the system to pinpoint the geographical position.

Other applications of GPS based on Vehicle tracking include Geo Fencing, SOS alarms, Real Time Tracking, Trip history, Alerts, Dashboard Summary, Accident Detection and so on.

3.4 Transmitter

A transmitter can be defined as, "a set of equipment used to generate and transmit electromagnetic waves carrying messages or signals, especially those of radio or television".

For communication of information to law enforcement or medical personnel we can use a transmitter that transmits the location of the accident. A new technology called Zigbee has been developed recently [7]. It is a wireless technology to address unique needs of low cost, low power wireless IoT networks. *"The Zigbee standard operates on the IEEE 802.15.4 physical radio specification and operates in unlicensed bands including 2.4 GHz, 900 MHz and 868 MHz."* There exists an idea to incorporate the Zigbee technology in street lamps for Smart Street Lighting [8]. So, this technology can be used in our system to convey the location of accidents.

4 Conclusion

Thus, we have a proposed a Surveillance System in order to carry out the golden hour rescue in case of road traffic accidents. The structure of each component is briefly explained along with its functionality in the system. The camera in the system is used embedded with Feature Extraction Technique of Image processing unit. The Global Positioning System has become a widely accepted and commonly used system to pinpoint accurate and reliable geographic locations.

The transmitter is used in accordance with the GPS to communicate with law enforcement and medical personnel in case of emergencies. Since, the system is incorporated in lamp posts and all lamp posts are interconnected it can transmit required information far and wide. The point is this system when incorporated into lamp posts can also be used for various other purposes like surveillance of any unusual activities, recording and reporting any other crimes that may take place and so on. So, this system has a much wider range of applications than specified.

One of the most valid points to consider in implementing this system is the cost. Compared, to other systems or proposed methodologies, we can implement it at reduced costs. After manufacturing the system, it has to be placed within a street light and interconnected with all other systems in all other street lights. This proposed method ensures that accidents are be detected in most of the human inhabited areas.

References

1. American College of Surgeons. Advanced Trauma Life Support Program for Doctors (Atls) (2008). Campbell, J.: International Trauma Life Support for Emergency Care Providers, p. 12, 8th Global edn. Pearson (2018)
2. Ki, Y.-K.: Accident detection system using image processing and MDR. Int. J. Comput. Sci. Netw. Secur. **7**(3), 35–39 (2007)
3. Chaudhari, L.: Advanced golden hour rescue system using android. Int. J. Eng. Educ. Technol. (ARDIJEET) **04**(02) (2016). www.ardigitech.in. ISSN 2320-883X
4. Javale, P., Gadgil, S., Bhargave, C., Kharwandikar, Y.: Accident detection and surveillance system using wireless technologies. IOSR J. Comput. Eng. (IOSR-JCE) **16**(2), 38–43 (2014). e-ISSN 2278-0661, p-ISSN 2278-8727
5. https://www.globalspec.com/learnmore/video_imaging_equipment/video_cameras_accessories/ccd_cameras
6. Goel, R., Kumar, V., Srivastava, S., Sinha, A.K.: A review of feature extraction techniques for image analysis. IJARCCE Int. J. Adv. Res. Comput. Commun. Eng. **6**(2), 153–155 (2017)
7. Dhillon, P., et al.: A review paper on Zigbee (IEEE 802.15.4) standard. Int. J. Eng. Res. Technol. (IJERT) **3**(4), 141–145 (2014). ISSN 2278-0181
8. Mhaskeet, D.A., et al.: Smart street lighting using a Zigbee & GSM network for high efficiency & reliability. Int. J. Eng. Res. Technol. (IJERT) **3**(4), 175–179 (2014)
9. Kapoor, P.: India way behind 2020 target, road accidents still kill over a lakh a year. Times of India, updated on October 04 2018 at official website of TOI

Smart Mirror: A Device for Heterogeneous IoT Services

S. Mohan Sha(✉), S. Nikhil, K. R. Nitin, and V. S. Felix Enigo

SSN College of Engineering, Kalavakkam, Chennai 603110, Tamil Nadu, India
mohansha@outlook.com,
{nikhil3059,nitin13063}@cse.ssn.edu.in,
felixvs@ssn.edu.in

Abstract. In this paper, we present a prototype Smart Mirror that creates an interactive seamless access to internet services and displays real time information from IoT devices. The features of Smart Mirror are – it displays news and other relevant information, real time IoT data, controls devices in a smart home system, communicates with multiple devices simultaneously and can work flexibly based on the availability of internet. Additionally, the smart mirror can also work with a health monitoring device. It provides all these features with a minimum amount of user intervention. As a case study, to show the seamless interaction of the Smart Mirror, we have connected health monitoring device for tracking user's health status. Instead of tracking and displaying the vital health parameters of the user, Smart Mirror uses data analytics algorithms to provide suggestions for health improvements.

Keywords: Smart mirror · Internet services · IoT data · Smart home · Health monitoring · Recommendation system · Data analytics · Health monitoring device · User intervention · Seamless interaction · User · Devices · Smart device · Data tracking · Vital health parameters

1 Introduction

In today's world there is often a need to lookup various information from multiple devices. Though there exist different technologies, usage of devices like computer or mobile may distract the user in other media from looking up basic information such as weather, news, mails etc. To address this issue, Smart Mirror, an emerging concept in the technology world, allows the user to engage in time-efficient bathroom routines without the need for spending separate time to check for personal or official news/notifications.

2 Existing Systems

The genesis of the smart mirror starts with the HUD Mirror [1] where the rear-view mirror is transformed into a smart mirror, providing the user with car and driving information as Heads-Up-Display (HUD). To make the mirror quite interactive, a voice

D. J. Hemanth et al. (Eds.): COMET 2019, LNDECT 35, pp. 1311–1323, 2020.
https://doi.org/10.1007/978-3-030-32150-5_133

based smart mirror [2] has been designed to access simple information such as date, time, calendar, Stocks and weather reports using voice commands. It is sufficient for the user to be in its audible range. However, it lacks customization i.e. new features cannot be added.

An improvised version of voice-based smart mirror for playing music autonomously [3] was developed by Brussenskiy. It uses a gesture control for turning music ON/OFF and voice controls for playing the same. It also employs temperature check for humidity in order to perform safe voice based operations in bathrooms. It allows addition of new contents, though manually.

The gesture control features are further enhanced in New York Times Mirror [4]. Here it uses Microsoft Kinect for movement tracking of users tagged with RFID and voice control for HCI operations. With increase in development of smart mirror technologies, to ease smart mirror application development across any platform over web, Smart Reflect [5] has been proposed by Gold. It uses MVC model and the browser serves as primary display container. It can display basic internet services on the mirror.

Smart mirror can also be used as a part of IoT based home automation. Its functionality can be extended to control home appliances [6, 7] to provide ambient home environment in addition to its personalized information services with touch-based features. Touch based systems are very expensive compared to voice-based systems and are unsafe to use in wet bathrooms.

Some work such as Wize mirror [8] and Fit Mirror [9] involves using a smart mirror for a healthy lifestyle. Such mirror either grabs health information directly via cameras or passively from user smartphones to track vital information such as cardio-metabolic rate and provides suggestions to improve their lifestyle.

Some smart mirrors focus on privacy issues by providing facial recognition-based authentication [10]. Aditya and Anjali [11] automated the smart mirror based on individual users, by observing their access pattern. To improve the prediction accuracy, users are identified by analyzing certain important events from their event history.

By analyzing the pros and cons of existing smart mirrors discussed aforementioned, we propose our Smart Mirror which is a prototype implementation that integrates the best features of most smart mirrors. Some of the prominent features of our mirror are: voice-based operations, extensibility (as it uses plugins), controls home appliances, connection to health monitoring devices to provide health tips and operates at different modes based on internet connectivity.

3 Proposed System

The smart mirror consists of a one-way mirror which is positioned in front of the LCD display. Raspberry pi 3 kit is connected to the display (Fig. 1). Apart from basic components, the kit consists of additional components such as microphone, sound card, memory card, sensors, power adapters to realize the available services. The user can access the services through voice commands via microphone. The voice commands are then processed in the cloud and returned as text. Then the services corresponding to the text are invoked and executed. Finally, the result is displayed on the monitor.

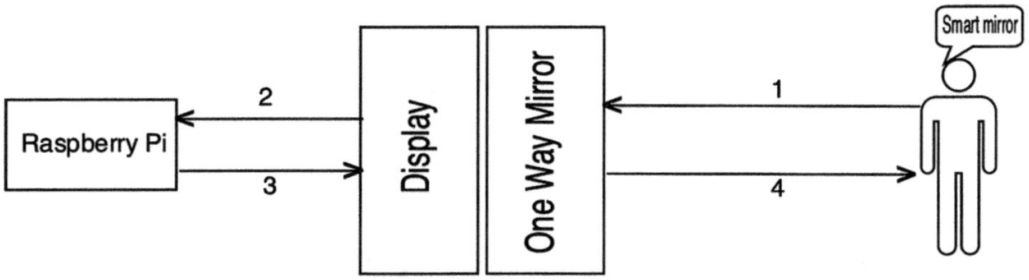

Fig. 1. System architecture

3.1 Working Modes

The smart mirror can work in three modes: independent mode, internet mode and internet mode with health monitoring device.

Independent Mode. This mode does not depend on the Internet. It displays information that was previously obtained when the device was connected to the Internet.

Internet Mode. This mode provides real-time information when connected to the Internet.

Internet and Health Monitoring Mode. In this mode, real-time information and user's health data from health monitoring device is displayed. Smart mirror has two states: inactive and active. Both basic and advanced services are provided only inactive state. Inactive state is triggered either by timeout or by voice commands. During inactive state it functions as a normal mirror.

4 System Implementation

The system has two major components; Implementation of Internet based Services and Voice Recognition. The smart mirror homepage, the User Interface of the application is designed using ElectronJS. It acts as an application wrapper which packages it as a desktop application without the need to run on the browser. The internet services are accessed using appropriate API keys of the required services.

To initiate and execute the services vocal commands are used. Voice based systems are preferred than touch-based systems for two reasons; one is inexpensive and the other is that it does not mandate the user to be present near the mirror but anywhere within the audible range. In the first phase of voice recognition, hotword is detected and converted to text through Google speech API at online and Sonus speech-to-text library at offline.

4.1 Hotword Detection

A hotword is a key word or phrase that a computer always listens for, to trigger other actions. Common usage of hotword include Alexa on Amazon Echo, OK Google on some Android devices and Hey Siri on iPhones. These hotwords are used to initiate a full-fledged speech interaction interface. In our Smart Mirror, hotwords are used to make the system listen to commands to trigger actions like displaying weather, updating news feeds, stock information etc. In the proposed system, hotword detection is achieved using Snowboy. It is an offline hotword detection engine which is embedded, real-time with persistent listening to voice commands. It is also highly customizable that enables to freely define our own magic phrases. It handles user's privacy and need no internet. It is light-weight and runs on Raspberry Pi, Linux, Mac OS X with less than 10 percent of CPU resource utilization. In our system, Snowball is used to detect hotword that was set offline during initiation of the system. Once the system is initiated, all the voice commands that control the actions are processed by Google Cloud speech API at online.

4.2 Speech to Text Conversion

Next, we have used Speech Recognition (SR) API to recognize and translate the voice commands of the user to textual message. These textual messages are needed to invoke the required services. SR can be of speaker-dependent or speaker-independent systems. Speaker dependent systems are trained on specific user voice to recognize speech based on the unique characteristics of voice. Whereas, speaker independent systems recognize anyone's voice and hence it does not require training. Since theprototype built is customized for single user, it is implemented as speaker-dependent system that can recognize a single user voice commands which it had been previously trained on. Two speech libraries such as Google Cloud Speech API and Sonus has been used for voice recognition and to convert speech to text.

Google Cloud Speech API. is a speech recognition API that enables to convert audio to text by applying powerful neural network models. This API recognizes speech over 80 languages and its variants to support global user base. It transcribes the text of users dictating to an application's microphone and command-and-control through voice or transcribes audio files.

Sonus. is a speech to text library that allows to quickly and easily add a VUI (Voice User Interface) to any hardware or software project. Similar to Alexa, Google Now, and Siri, Sonus is a persistent offline listening API that listens for a customizable hotword. Once the hotword is detected, the speech is streamed to the cloud recognition service of user's choice to obtain the results.

4.3 Services

Basic Services. The basic services offered by the smart mirror are; auto sleep service, Rich Site Summary (RSS) Feed Service and Stocks Service (Fig. 2)

Auto Sleep Service. In this mode, the smart mirror functions as a normal mirror. Services run in the background with reduced power which can be re-activated later by wake-up voice command. The auto sleep service is implemented locally and allows the device to remain idle for a specified amount of time. The API takes interval as the parameter to invoke the service.

Rich Site Summary (RSS) Feed Service. The address of the RSS feed sites is sent as HTTP request to the server by the application. The server pushes the content as response to the smart mirror as text and continues to push whenever the site content is updated. To ensure the validity of the content, the source and time of the content pushed is also displayed. The RSS service call takes the following parameters: URL of the site and Refresh Interval.

Stocks Service. The stock names are given as a query to the service. If a valid response is returned, it is accepted and displayed on the mirror. Otherwise, the response is rejected and valid request is prompted. For example: The stocks service calls the Yahoo Finance API and it is refreshed periodically for updates. It takes the following parameters: Name of the company and Refresh Interval.

Fig. 2. Basic services

Advanced Services. are the services that are triggered by predefined voice commands. The advanced services offered by the smart mirror are; Weather, Soundcloud, Maps, Giphy, YouTube, Calendar, Timer, Reminder, Geolocation, XKCD, Active and sleep mode, Fitbit status, Motion detection and Remote login.

Weather. The weather information is displayed for a week along with metadata such as sunny, cloudy, clear sky or rainy. The weather information is obtained via the API provided by the site Forecast.io using push communication model at pre-defined intervals for up-to-date information and displayed to the user. It takes the following parameters: Geolocation Service, Refresh Interval.

Soundcloud. is an online audio distribution platform that enables its users to upload, record, promote, and share their music. In our system, the user can play the required track using pre-defined voice commands. For fast serialization, NSJSONSerialization is used instead of JSON kit. The audio tracks are displayed graphically as waveforms and it allows the users to post, timed comments which will be displayed, when the associated audio segment is played (Fig. 3). The SoundCloud service calls the Soundcloud API. It takes the following parameters: Query and Speech Service.

Fig. 3. SoundCloud in smart mirror

Maps. It allows users to search any location in the map worldwide. Similar to weather it also uses push technology to get information pertaining to traffic and public transits from google maps. The Maps service calls the Bing Maps API and it takes the following parameters: Query, Geolocation Service, Speech Service.

Giphy. is an online database that allows users to search for animated GIF files. The multi-channel approach is used for looping images and a web-hook is used for the smooth transition of giphy images. The Giphy service calls the Giphy API. It takes the following parameters: Query and Speech Service.

YouTube. is a free video sharing website that allows people to view, upload and share videos. The pull communication model is used for obtaining the videos as the video must be transitioned from each frame seamlessly without any frame drops (Fig. 4). The YouTube service calls the YouTube API and it takes the following parameters: Query, Speech Service.

Fig. 4. YouTube in smart mirror

Calendar. allows users to organize their daily activities. iCalendar is a computer file format which allows internet users to send meeting requests and tasks to other internet users by sharing or sending files in this format through various methods. iCalendar is designed to be independent of the transport protocol and Secure Socket layer (SSL) is used to protect sensitive information from other users. The Calendar service calls the Google iCalendar API. It takes the following parameters: Query, Interval.

Timer. is used for countdown from a specific time interval. The size of the animated circle that is displayed around the timer decreases according to the duration of the timer. The Timer service is implemented locally. It takes the following parameters: Interval, Speech Service.

Reminders. allows users to set notifications and create a list of necessary items. It synchronizes the data between the user's device and the cloud. The items in the reminder are labelled to prevent duplication of information. The Reminder service is implemented locally and it calls the Reminder API to sync the data. This API takes the Speech Service as its parameter.

Geolocation. software is capable of determining the actual location of the user. Sometimes, the geolocation of an object is used to input the location of its owner or user. The Geolocation service calls the Bing Maps AP and It takes GPS Coordinates as its parameter.

XKCD. is a web comic of sarcasm, math and scientific jokes. It has a cast of stick figures and occasionally features the landscapes and intricate mathematical patterns such as fractals, graphs and charts. The pull communication model is used here. The patterns and figures are displayed only when requested by the user. The XKCD service calls the XKCD API and takes the following parameters: Query, Speech Service.

Active and Sleep Mode. Sleep mode is a low power consumption mode in which it can significantly saves the electricity consumption compared to a fully active mode. The mirror does not display any information in sleep mode and upon wake up, the mirror resumes to earlier status. In this mode, there is no need for the user to wait for the mirror to reboot. The Sleep service is implemented locally and takes the following parameters: Speech Service, AutoSleepService.

Motion Detection. is used to sense movement of people close to the mirror when it is in active or sleep mode. In our smart mirror, Passive Infrared Sensor (PIR) sensor is used for this purpose. The Motion detection service is implemented locally using the Johnny-five.io package.

Remote Login. All the advanced services can be triggered via remote login. The interactions are done via button clicks in remote system without using any voice commands. It is used by the user to customize the smart mirror to their needs without modifying the code. The Remote login service is implemented locally.

5 Case Study

To prove the utility of Smart mirror, beyond merely invoking internet services for the required service, we have implemented a scenario to show how the data accessed via Smart mirror can be further processed to generate useful analytics. As a case a study, we took Fitbit Tracker. We have accessed the basic data provided by Fitbit and added little analytics using best proven machine learning algorithm such as Random Forest and K-nearest neighborhood algorithm. These algorithms are tested for accuracy in achieving the given target which is the attainment of the daily goals computed based on the health parameters in our case.

5.1 Health Monitoring Device - Fitbit Tracker

The Fitbit smart watch tracks the day to day activities of the user periodically and stores data online in the user's Fitbit account (Fig. 5). Since it can monitor intraday data (up to 1 s resolution), the data can be used for analysis extensively. The health and fitness data stored online can be accessed through Fitbit API. The data is then stored in a PostgreSQL database. We have added analytics part to the Fitbit dataset, so that, based on the recorded health parameters, the user is suggested for suitable change in lifestyle or exercise. The Health Monitoring service that calls the Fitbit API takes the following parameters: client id, client secret, access token, refresh token and interval.

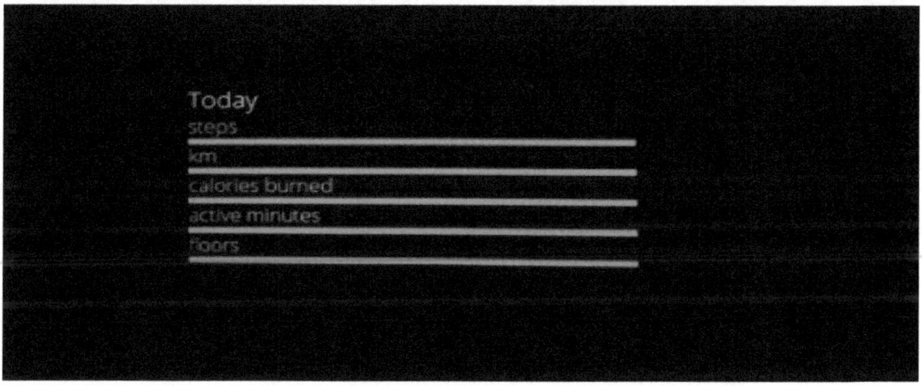

Fig. 5. Fitbit statistics

5.2 Representation of Fitbit Data

The collected Fitbit data is sent to a service that runs in the background for the purpose of representing the data graphically. Python pandas along with related libraries like matplotlib, NumPy, SciPy, psycopg etc. are required for the service. The database is connected using psycopg2. Health data such as heart rate, sleep etc. is read into pandas data frames. Variety of graphs are then generated in order to depict the data. Intraday data can be obtained by expanding the JSON Arrays to individual JSON values. Such graphs are then shown to the user on demand to facilitate the user to view useful information.

5.3 Data Analytics on Health Data

The collected Fitbit data is also sent to a service that runs in the background to analyze the data. To perform data analytics, one and a half months data has been recorded for the following 5 parameters – steps taken, distance covered, floors climbed, minutes active and calories burned and these data are fed into the service. Daily goals are set for each of the 5 parameters to verify that the target is met.

If all 5 goals are met, goals are declared to be achieved for the day. If any 2 or 3 goals are met, the goals are declared as partially achieved for the day. Otherwise, it is declared that goals are not achieved for the day.

To analyze the attainment of the required goals, classification is done over the collected health data. Classification is the problem of identifying to which of a set of categories a new observation belongs, on the basis of a training set of data containing observations whose category membership is known. In our case, the status of the daily goals can be classified under 3 labels Yes, Partial and No. To train the data, all 5 parameters were taken in the training dataset along with the label. The status of the goals is then classified for the testing dataset and presented to the user on demand.

Sometimes, the user may want to know in advance the number of calories burned for the given values of other parameters. Hence, Regression analysis is performed over the given dataset. It is a statistical process for estimating the relationships among variables. It includes many techniques for modelling and analyzing several variables, when the focus is on the relationship between a dependent variable and one or more independent variables (or 'predictors'). In our model all the parameters were taken for training. The algorithm predicts the calories burned for the given test data.

We have used Random Forest Algorithm and K-Nearest Neighbor (KNN) algorithm for classification and regression. The classification accuracy of the model is computed using the below formula.

$$\text{Classification accuracy} = \frac{\text{No of observations correctly classified}}{\text{Total no of observations}} * 100. \quad (1)$$

The two algorithms are used over the dataset and were found to produce more or less the same accuracy for small datasets. However, for large datasets, Random forest is observed to be of higher accuracy than K-nearest neighbor. Also, for large values of K, KNN has lower accuracy than random forests.

The accuracy of the algorithm is determined based on *Mean Square Error (MSE)* and *R Square Scores (RSS)*. MSE measures the average of squares of the errors or deviations. Higher MSE indicates lesser accuracy. RSS is a number that indicates the proportion of the variance in the dependent variable that is predictable from the independent variables. Hence, closer the value of RSS to 1, higher the accuracy. Based on the aforementioned discussion, we infer from Table 1 that Random Forest algorithm shows higher regression accuracy than KNN.

Table 1. Algorithms comparison

Measure	Random forest algorithm	KNN algorithm
RSS	0.6854	0.6604
MSE	93098.3	100496.3

5.4 Performance Evaluation

The smart mirror prototype has been evaluated based on two performance parameters such as user experience and efficiency. User experience is the overall experience of the user in using the product with ease. Users were allowed to use the system and were observed on how easily they were able to understand the system and use it without the help of others. Next, by efficiency we mean the quickness of response for both applications and notifications. We performed the user testing based on the parameters for user experience and efficiency for 100 users and the results are tabulated in a 5-point scale in Tables 2 and 3 as shown below.

Table 2. Rating of smart mirror user experience

Feature	Averaged rating (5-point scale)
News	3.93
Maps	4.13
Music	4.03
Weather	4.2
Fitbit	4.13
Warnings	4
Notifications	3.66
Overall	*3.8*

Table 3. Efficiency of smart mirror

Task	Averaged rating (5-point scale)
Applications	3.2
Notifications	2.86

From Table 2, notifications have a low rating. This is due to the missing of sound in notification. On the other hand, Fitbit has a very good rating as it gives detailed suggestions to users to improve their health. Further, with respect to efficiency of the system, performance of the applications seems to be better in terms of responsiveness.

Further, feedback from the users on how to improve the system was collected and some of the suggestions given by the users are, aggregated notification reports are clumsy making it difficult to recognize individual notification, voice command requires more audibility etc. These feedbacks will be incorporated in future works to further improvise the system.

6 Conclusion

The model of the framework was actualized effectively and has been tried with clients. The ease of use test demonstrates that the framework is performing genuinely well. Later on, the created model Smart Mirror can be enhanced in number of ways. Combination with home robotization frameworks should be possible so electronic home gadgets, for example, air conditioner, refrigerator, entryway and so forth would then be able to be controlled by the Smart Mirror. Smart gadgets (like Philips hue) can likewise be controlled through the Smart Mirror. Voice acknowledgment can be consolidated into the framework for client authentication. Custom client profiles can be set up to make the administrations progressively close to home for various clients. Facial acknowledgment can be joined into the framework for both security and individual use. Including security implies that nobody can endeavor to get to touchy information that possibly showed on your mirror by means of the utilization of APIs.

In conclusion, more fitness checking gadgets can be controlled through the Smart Mirror. The Smart Mirror achieves, by as yet being a mirror without all the innovation inside it, making it entirely receptive to utilize and coordinating consistently into our lives. Which can grow the usefulness of the mirror. We trust that the eventual fate of the home will be a splendidly associated biological community of savvy innovation intended to make your life simpler, increasingly charming, and proficient. Clearly, there are a huge amount of chances in the home for innovation joining yet a mirror is outstanding amongst other spots to begin.

7 Compliance with Ethical Standards

7.1 Conflict of Interest

The authors declare that they have no conflict of interest.

7.2 Ethical Approval

All procedures performed in studies involving human participants were in accordance with the ethical standards of the institutional and/or national research committee and with the 1964 Helsinki declaration and its later amendments or comparable ethical standards.

Ethics committee members:

1. Dr. S Sheerazuddin B.Tech, Ph D
2. Mr. N. Sujaudeen B.E., M.E.,
3. Ms. S. Rajalakshmi B.E., M.E., (Ph.D.,)

7.3 Informed Consent

Informed consent was obtained from all individual participants included in the study.

References

1. Ubiquitous Computing Group Homepage, HUD Mirror. https://sites.google.com/site/hudmirror/the-project. Accessed 03 July 2016
2. Vaibhav, K., Vardhan, Y., Nair, D., Pannu, P.: Design and development of a smart mirror using raspberry Pi. Int. J. Electr. Electron. Data Commun. **5**(1), 63–65 (2017)
3. Brussenskiy, G., Chiarella, C., Vishal, N.: Smart mirror an interactive touch-free mirror that maximizes time efficiency and productivity. Project Documentation (2013)
4. New York Times Mirror article. http://www.extremetech.com/computing/94751-the-new-york-times-magic-mirror-will-bring-shopping-to-the-bathroom. Accessed 10 July 2016
5. Gold, D., Sollinger, D.: SmartReflect: a modular smart mirror application platform. In: 7th IEEE Information Technology, Electronics and Mobile Communication Conference (IEMCON), Vancouver, BC, Canada, pp. 1–7 (2016)
6. Anwar, M., Pradeep, K., Abdulmotaleb, E.L.: Smart mirror for ambient home environment. In: IET International Conference on Intelligent Environments, pp. 89–596. ULM, Germany (2007)
7. Jose, J., Chakravarthy, R., Jacob, J., Ali, M.M., Dsouza, S.M.: Home automated smart mirror as an internet of things (IoT) implementation - survey paper. Int. J. Adv. Res. Comput. Commun. Eng. **6**(2), 126–128 (2017)
8. Colantonio, S., Coppini, G., Germanese, D., Giorgi, D., Magrini, M., Marraccini, P., Martinelli, M.: A smart mirror to promote a healthy lifestyle. Biosyst. Eng. **138**, 33–43 (2015)
9. Besserer, D., Bäurle, J., Nikic, A.: FitMirror: a smart mirror for positive affect in everyday user morning routines. In: 3rd International Workshop on Multimodal Analyses enabling Artificial Agents in Human-Machine Interaction (MA3HMI 2016), Tokyo, Japan, pp. 1–8 (2016)
10. Maheshwari, P., Kaur, M.J., Anand, S.: Smart mirror: a reflective interface to maximize productivity. Int. J. Comput. Appl. **166**(9), 30–35 (2017)
11. Aditi, D., Anjali, N.: Use of prediction algorithms in smart homes. Int. J. Mach. Learn. Comput. **4**(2), 157–162 (2014)

Incontinence Monitoring System Using Wireless Sensor for "Smart Diapers"

G. Shri Harini, N. Vishal, and S. Prince Sahaya Brighty(✉)

Sri Ramakrishna Engineering College, Coimbatore 641022, India
brighty.s@srec.ac.in

Abstract. Urinary Incontinence is the most common and serious issue that is faced by the Senior citizens, infants as well as the physically and mentally challenged people. This paper presents a design of Incontinence monitoring system using wireless sensors for "Smart Diaper" which is based on the non-contact sensor module that could be incorporated in the diaper. Data is transmitted by means of a wireless transmission and communication technology also the user will be provided a Mobile application that notifies the status of the Moisture content from the diaper. The real time moisture information is collected by the device which is integrated with the diaper and this device will pass on the data by means of Bluetooth module to the application provided and all the data and details will be stored in the Database for future reference or experiments. The proposed alarm system along with the working flow are mentioned.

Keywords: Urinary incontinence · Non-contact sensor · Wireless data transmission · Mobile app

1 Introduction

Diapers have been the wonderful invention of today's mankind, which helps the caretaker to look after the Infants or Senior citizens and make their life much less stressful. These Diapers helps in averting and controlling the wastes in an effective healthier manner. Also they have many advantages but also holds on with a major inconvenience, it cause skin rash [10], which gets enlarged when the skin is with regular connect to wetness for a long period that is when the duration is extended. The wetness of the skin for a longer period of time is therefore the common factor behind the skin rashes while using Diapers [6, 9], skin rashes can be removed by changing the diapers immediately after they subjected to wetness. Although there are numerous methods to identify and measure the moisture content of the diaper, there are few detailed description of examples in [11, 12] also there are mostly used methods which require active electronics.

2 Related Works

This paper proposes the new idea, this idea gives an overview of an alarming system for diapers which incorporates with a phone call and a SMS to alert the respective person or an attender or a care taker at that time when the diaper is exposed to wetness.

D. J. Hemanth et al. (Eds.): COMET 2019, LNDECT 35, pp. 1324–1331, 2020.
https://doi.org/10.1007/978-3-030-32150-5_134

The Automated alarming system had been already implemented in various places. The existing system holds an idea of positioning the conductors i.e., the sensors in between the layers in a point that is subjected to wetness. This idea makes sure that it is practical, safe to use for the diaper users. The device consists of an alarm with batteries, sensors, and a manual guide that costs 3000 INR (approx.) that is a bit too costly and cannot be afforded by all the users. Here a diaper model has been advanced by using dissolving cotton & paper (reagent paper) which also indicates strangeness in the urine [7]. Hence the Real time monitoring system helps to lively monitor the urinary incontinence and the response time for changing the diapers appropriately so that the aged or the disabled persons and alerts the attendance. The end product is compact, and so this device is attached to the diaper. This device is inexpensive and operates with a reusable sensor tag. The design of this device will be prepared into a original diaper device linked with Bluetooth module for data transmission. The dissimilarity between our idea and the previously used idea is that the Bluetooth module is used on a Radiofrequency transceiver and GSM modules which helps in the transferring signal to the concerned person from the diaper by a text message or by an automated phone call. Whereas, by using the Bluetooth module we are capable building devices of smaller size when compared to the device holding RF transceiver or GSM modules. On using GSM modules, it is necessary to provide with good network only when the data will be either transmitted or received. Also, the structured idea can be used for the people with health issues, bed ridden people, and for hospitalized patient. In this paper, the design of a easy wet detecting system, which can be implemented for many applications related with blood, water and other similar types of liquids is implemented. The paper proposed is regulated into five sections.

Having a brief introduction in Sect. 1, we go to the proposed system that is explained in Sects. 2 and 3. In Sects. 3.1 and 3.2, a general definition about the Bluetooth module and Sensors used are presented. Section 4 includes the conclusion part and references are continued.

3 The Autonomous Alarm System

The alarming system is a device that consists of three major parts namely Sensor tags, an Analogue Switch (or relay), and Transmitter which are explained in the following divisions of the paper. The Device will be kept on the outer layer of the diaper which will be sensing the humidity and monitoring the temperature rise by using the non-contact sensors that are embedded along with the Bluetooth module which is used for transmitting the sensed data and to the Application and there displays the status of the diaper (Fig. 1).

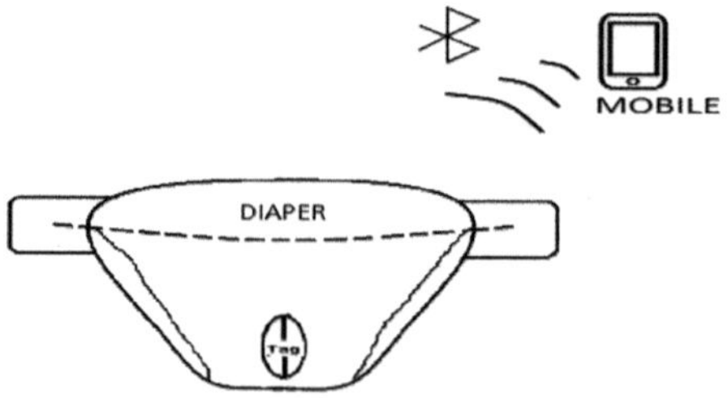

Fig. 1. System process

By knowing the wet/moisture content of the diaper, data's are being sent to the application that is installed in the mobile and connected to the device via Bluetooth as estimated in Fig. 2.

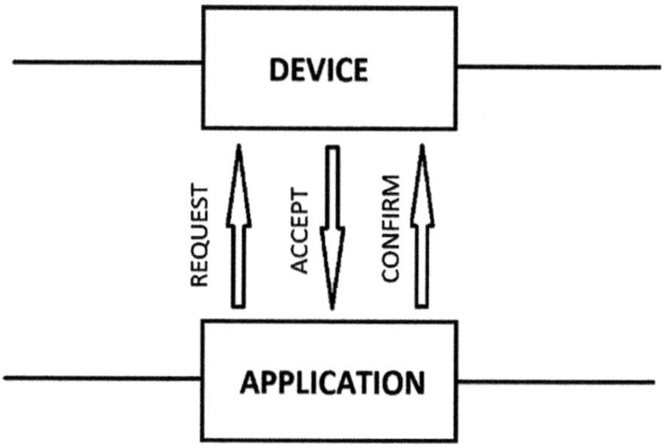

Fig. 2. Connection establishment between the device and the mobile application

The Flow of working of this device is depicted in Fig. 3 where the sensor tag is formally connected to the Mobile and is checked that whether the sensor is connected and the tag sensor transmits the data to the application which is been installed in the mobile via Bluetooth. When the wetness is detected in the sensor and when the humidity and temperature are high than the estimated value that is already been set the alert is thrown to the application which is actively connected to the Device.

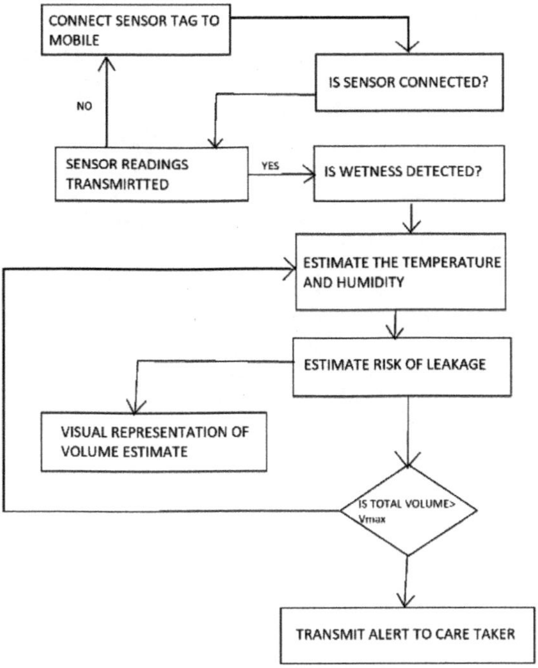

Fig. 3. Working flow of the system.

The Hardware block diagram of the system is shown in Fig. 4. The Hardware part totally consists of three main components Key for powering the device ON and OFF, LED and the Bluetooth Transmitter.

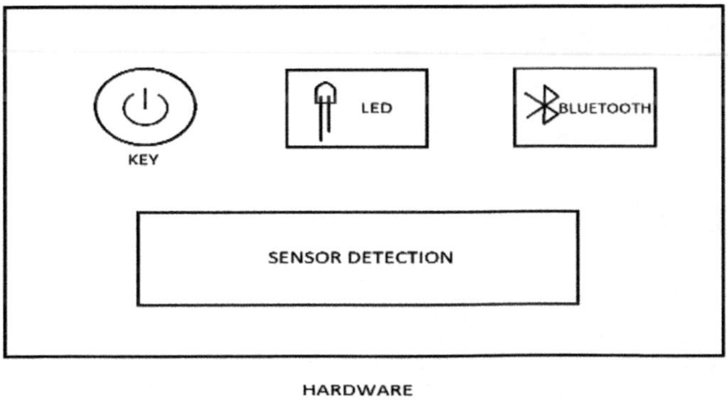

Fig. 4. Hardware block diagram

Implementation Part

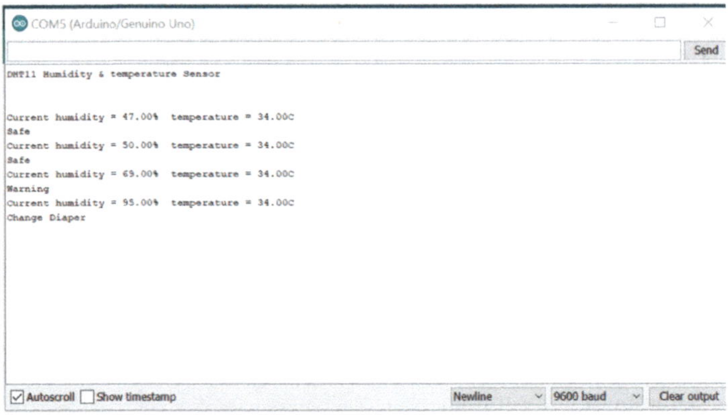

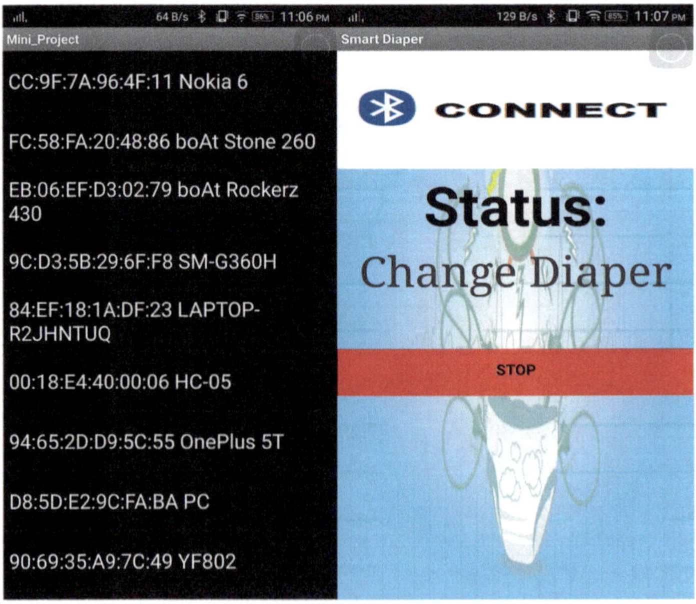

3.1 Bluetooth Module

Bluetooth technology is a simple wireless transmission system operating at 2.4 GHz RF. The best and the cheapest Bluetooth module is the HC-05. Bluetooth modules will be connected to the mobile phones once when the device is switched on. This bluetooth module is small and light in weight which costs low with good scalability. The module which is been used in this device is shown in Fig. 5.

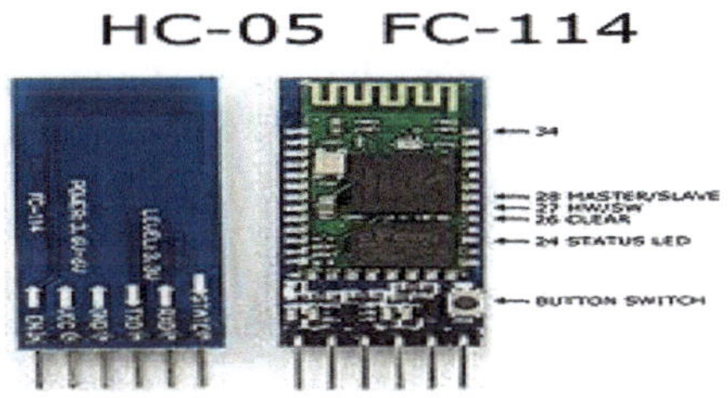

Fig. 5. Bluetooth module

The Bluetooth transmitter receives the data from the Sensors and transmits to the Mobile Application. The data are viewed in the App and the data is processed and saved in the Database which can be referred for future use. The working of the Bluetooth module is depicted in the Fig. 6 block diagram.

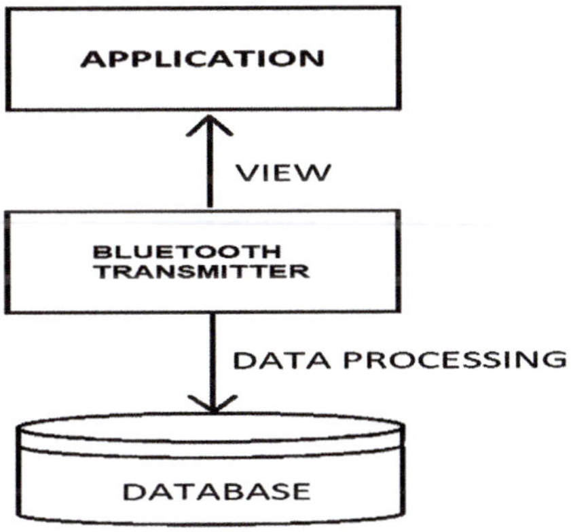

Fig. 6. Working of Bluetooth module

3.2 Sensor Modules

The most important part of this device is the Sensor tag. The device is composed of a detection sensor module which senses the temperature & humidity. These parameters can be sensed by using the DHT11 sensor which incorporates both the Humidity and

Temperature sensor in it. This technology guarantees with high stability and ensures them to be constant and steady with it. This is connected to a high-performance 8-bit microcontroller. This sensor contains a wet sensing component with resistivity and a device measuring temperature. Specifications are as below,

- Voltage supplied is +5 V
- Temperature ranges from 0 °C to 50 °C
- Humidity content is from 20% to 90% RH
- Digitally interfaced

The sensor modules will be updating the values at a regular period and send them as input via the Bluetooth module. The values will be checked with the pre-coded values and then the estimated output will be displayed in the application that has been provided (Fig. 7).

Fig. 7. DHT11 sensor module

4 Conclusion and Future Work

Thus this system for detects the urinary incontinence and transfer data to the mobile application and throw alert. Also this system is built with cost efficient sensors which make the total cost of the device extremely low. When we take a look at other already used applications for incontinence monitoring doesn't give much accurate regular outputs. The main advantage is that this system for incontinence live monitoring and auto updating of data regularly to the mobile application. For making the sure of the user's physical ease, the wet detecting part is made up using a small board which is reused. The device cost is also very less that resulted to the simple design of both the transmitter and the sensor module.

Acknowledgements. The authors like to thank the all the anonymous reviewers for their valuable suggestions and Sri Ramakrishna Engineering College for offering resources for the implementation.

References

1. Siden, J., Koptioug, A., Gulliksson, M.: The 'smart' diaper moisture detection system. Mid-Sweden University, Electronics Department, Sundsvall, Sweden (2016)
2. Simik, M.Y.E., Chi, F., Saleh, R.S.I., Abdelgader, A.M.S.: A design of smart diaper wet detector using wireless and computer. In: World Congress on Engineering and Computer Science 2015, vol. II (2015)
3. Frank, R.: Understanding smart sensors, 2nd edn. Artech House Sensors Library (2013)
4. Friedlos, D.: Electronic underpants help caregivers cope with incontinence. RFID J. (2010)
5. Yambem, L., Yapici, M.K., Zou, J.: A new wireless sensor system for smart diapers. IEEE Sensors J. **8**(3), 238–239 (2008)
6. Adam, R.: Skin care of the diaper area. Pediatr. Dermatol. **25**, 427–433 (2008)
7. Ejaz, T., Nakae, T., Takemae, T., Egami, C., Sugihara, O., Ikeda, K.: A sensing system for simultaneous detection of urine and its components. In: IEEE APCCAS/998 the 1998 IEEE Asia-Pacific Conference on Circuits and Systems, pp. 221–224, November 1998
8. Pahlavan, K., Levesque, A.H.: Wireless data communication. IEEE J. **82**, 1398–1430 (1994)
9. Berg, R.W., Milligan, M.C., Sarbaugh, F.C.: Association of skin wetness and pH with diaper dermatitis. Pediatr. Dermatol. **11**, 18–20 (1994)
10. Zimmerer, R., Lawson, K., Calvert, C.: The effects of wearing diapers on skin. Pediatr. Dermatol. **3**, 95–101 (1986)
11. Kent, M., Price, T.E.: Compact micro strip sensor for high moisture content materials. J. Microw. Power **14**, 363–365 (1979)
12. Kent, M.: The use of strip line configurations in microwave moisture measurements II. J. Microw. Power **8**, 194–198 (1973)

Dynamic Mobility Management with QoS Aware Router Selection for Wireless Mesh Networks

K. Valarmathi[(✉)] and S. Vimala

Department of Computer Science and Engineering,
Panimalar Engineering College, Chennai, India
valaryogi1970@gmail.com, vimalakumaran@gmail.com

Abstract. Wireless Mesh Network (WMN) consists of numerous nodes which are highly mobile and connected to one another wirelessly. Due to the mobile nature of the nodes in WMN, the network management becomes very critical. As the client node move from one region to another, handling its routing process becomes a very important task. To perform this, a new router has to be selected from the new location and employed for supporting the client operation without overloading the gateway. In this paper, we recommend to expand a Dynamic Mobility Management with Quos Aware Router Selection for Wireless Mesh Networks. In this technique, all the clients are handled in a prioritized manner and routers are assigned accordingly. To avoid overloading the gateway during handoff, forward pointer technique is employed.

Keywords: Wireless Mesh Network · Dynamic Mobility Management · Mesh Client · Session-to-mobility ratio

1 Introduction

1.1 Wireless Mesh Networks (WMN)

Wireless mesh networks (WMNs) are an emerging wireless technology. When in comparison with the conventional wireless networks, WMN is more beneficial. Some of the advantages of WMN, which make it efficient as a next generation technology are:

1. Self organizing network
2. Self sustaining
3. Scalability
4. Less expensive
5. Ease in maintaining [1].

In WMN, there are two kinds of nodes. They are: mesh routers (MR) and mesh client (MC). The MR is a node which is not very mobile. The MC is a node which very mobile. In WMN, there are few gateways which link the MR to the internet. A wireless mesh backbone is created by a group of MR and this backbone is responsible for routing the traffic around the network and offering broadband access to the MC. In WMN, it is important to manage the mobile network effectively to ensure

D. J. Hemanth et al. (Eds.): COMET 2019, LNDECT 35, pp. 1332–1343, 2020.
https://doi.org/10.1007/978-3-030-32150-5_135

appropriate network processing. Location management as well as handoff management is performed in mobility management [2].

A gateway is a MR with wired connection to the internet. In WMN, two type of traffic operates. They are internet and intranet. The internet traffic flows across the gateway to other location. The downstream internet traffic is received at the gateway and then the gateway transmits the traffic to the destination MC. But, the upstream traffic is sent by the MC to the gateway. In case of intranet traffic, the traffic flows from among MC within the same WMN [3]. The gateways as well as the MRs in the WMN are not very mobile but the MC are highly mobile. In WMN, the communication range is very less and hence the MC will have to move between MRs through handoff process to remain operating effectively. So in assisting the roaming MCs, the handoff control is an essential part. WMN supports a wide range of visitors with best of provider (QoS) assurance. The QoS supplied is ensured to suit the wi-fi and real time software desires. Also, there are versions among the TDMA based totally WMNs and the IEEE 802.11-based WMNs in phrases of get admission to technology, channel usage, topology, slot allotment, etc. Hence, maintenance of the current existing application along with QoS assurance has to be ensured by the handoff management mechanism in the TDMA based WMN [4].

1.2 Mobility Management in WMN

On the basis of the mobility of the mesh client, the network management reflects the WMN performance. Hence network management is a critical issue in WMN. The conventional mobility management schemes may be used, but there is a great need for a novel mobility management solution to be developed for the WMN while referring to the unique features of WMN. Mesh clients can be considered to be roaming in two perspectives: inter domain roaming and intra domain roaming. Inter domain roaming includes traffic moving between various domains. The intra domain roaming includes the traffic movement between various routers but within a single domain. Mobility management includes inter domain as well as intra domain management [1].

Mobility management consists of location management as well as handoff management. Numerous research techniques have been proposed on mobility management in the Mobile IP, MANET as well cellular networks. But, in the field of hybrid WMN, efficient mobility management techniques have not yet been developed. In cellular as well as mobile IP networks, the developed mobility management schemes depend on the wired infrastructure to handle the mobility based signalling, delivery of the traffic, etc., however the developed scheme is based on the infrastructure network. In MANETs, the characteristics such as multihop route discovery, recovery, maintenance, etc are used to handle the node mobility whereas the network layer mobility management is not taken into consideration. In an ad hoc network, rerouting is the major functionality handled during handoff, such that the during link breakage, a new multihop route can be determined and traffic can be assigned to this route quickly. To aid location based ad hoc routing, the many location management mechanisms have been proposed. But, node location indicates the geographic location and is a different logic when compared with the location logic in WMN [5].

Several location management techniques have been proposed in the field of cellular network as well as mobile IP based wireless network. The conventional techniques used in the mobile IP networks as well as in the cellular networks, are efficient but before employing in the WMN, these techniques must be modified and adapted to handle the variations in the WMN. For instance, in cellular networks, the location management technique depend on the centralized handling features as in HLR/VLR, and the HA/FA in mobile IP network. But, these features are not present in WMN. So, these conventional techniques cannot be deployed in WMN. One of the basic differences between MANET and WMN is that there exist a quasi static routing infrastructure including the MR in WMN which is not present in MANET [2].

2 Related Works

Zhang et al. [1] have presented a hybrid routing protocol for forwarding packets in the link layer as well as in the network layer. Mobility management mechanism proposed is on the idea of the hybrid routing protocol. To aid roaming inside the wifi supported WMNs, intra domain as well as inter domain mobility management approach were developed. Routing information is received through ARP messages in the intra domain handoff to prevent re-routing as well as location updating process. Extra tunnels are removed off during inter domain handoff so as to reduce the forwarding latency.

Li et al. [2] have presented and analyzed LMMesh: a routing-based location management scheme with pointer forwarding for wireless mesh networks. In LMMesh, the routing based location update technique and the pointer forwarding technique are combined to utilize their advantages. The network expense in the integrated model is monitored in terms of location management as well as packet delivery. The trade off among the service cost incurred during delivery of the packet and the signalling cost incurred during the location management process is explored and a best protocol setting is chosen which reduces the total network expense on the basis of each user usage for given characteristics such as mobility, etc.

Lee et al. [6] have presented a mobility management mechanism to aid the mobility of legacy clients in wireless mesh networks. To make the proposed mechanism compatible with the IEEE 802.11s standard, several techniques are employed to detect mobility as well as to propagate the traffic information on the basis of IEEE 802.11s proxy protocol.

Zheng et al. [7] have presented a load-aware mobility management mechanism. Based on the load value computation, the overloaded MAP is determined. Next a search process is begun by the overloaded MAP to detect any underutilized MAP. Then MN's attachment request is sent by the overloaded MAP. The handover load between MAPs is appropriately managed by this mechanism, but with a slight increase in the attachment delay as well as with more attachment messages when considered in comparison with the COAP.

Nazari et al. [8] have proposed a technique for designing routing algorithms as it initially tries to understand the network features such as mobility, connectivity, topology changes, etc and an algorithm which makes the routing performance better.

The proposed algorithm was employed in Triton which is a IEEE 802.16 based maritime wireless access mesh network.

Matos et al. [9] have proposed a context-aware multi-overlay architecture which allows a user to link with WMN and also fulfils the requirement. This architecture considered factors like maintaining the network requirements during mobility by reconfiguration of the overlays, and also mapping, organizing and distribution of the context. Sophistication and also the parts of the architecture were given high consideration.

Daly et al. [10] have proposed are authentication technique for secure handoff on the basis of effective mobility management. Initially, the mobility feature is considered by utilizing the mobility notification message process. This process helps in handling the handoff process in the specific environment. Based on this technique, a mechanism which offers security at the time of handoff process is proposed. Based on the results, it is seen that this technique offers a safe network as well as effective reauthentication mechanism with respect to reduced handoff latency and lower blocking and loss rates.

3 Dynamic Mobility Management Scheme with QoS Aware Router Selection

3.1 Overview

In this paper, we propose to develop a Dynamic Mobility Management scheme with QoS aware Router Selection for WMN. The network with k different types of services where $P = \{P1, P2, \ldots .Pk)$ are considered. For any $i < j$, the carrier with Pi has a better precedence than carrier with Pj. For varieties of traffic with equal priority, the handoff traffic has better precedence than the new arrival traffic [4]. Whenever a mesh client (MC) tends to move, the target MR is selected based on the RSSI, required bandwidth and link quality. (i.e.) The MR which satisfies the bandwidth requirement of various priority of services having minimum RSSI and best link quality is selected. The link quality is measured in terms of the response delay [1]. The QoS aware MR selection process [4] is then executed.

The concept of forward pointer is used at each MC to reduce the control overhead that occurs during location update at mesh gateways (GW). To limit the increase in forward chain length, each MC resets the forward chain if its session-to-mobility ratio (SMR) crosses a threshold SMRoth [2]. After selecting the new MR, When the MC moves into the vicinity of the new MR it computes its session-to-mobility ratio (SMR) and compares it with SMRoth. If SMR is less than the SMRoth, the MC notifies the target MR about its handoff from old MR and then forward chain length of the MC increases by 1. On the other hand, if the SMR is more than or identical to SMRoth, no forward chain is brought from old MR to new MR, however the new MR sends region update message to the GW. When GW receives the location replace message, it searches for the access of the MC of their database and set the current MR as the serving MR of the MC and the forward chain length is reset [2] (Fig. 1).

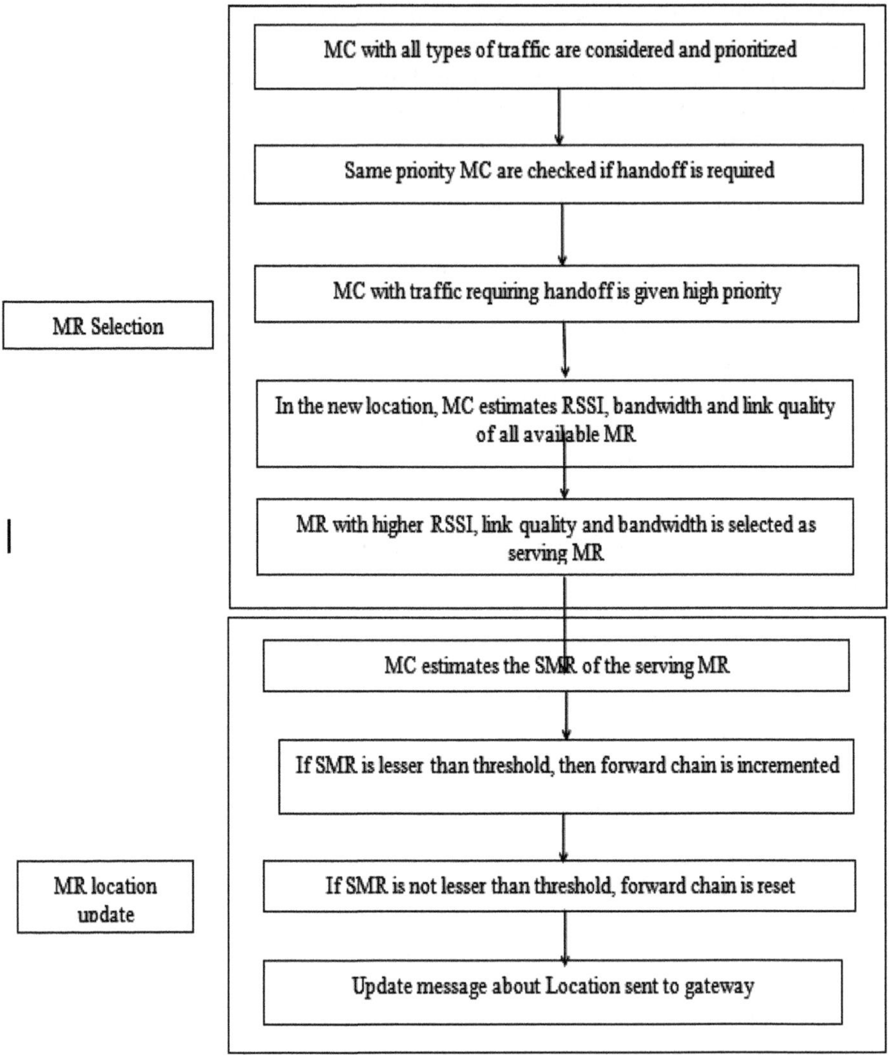

Fig. 1. Block diagram

3.2 Selection of Mesh Router

In WMN, several types of traffic are handled. As the mesh client carrying the traffic may move from one location to another, a mesh router must be assigned from the new location to aid the mesh client operation after handoff. The selection of the mesh router must be performed appropriately in order to choose a dynamic and efficient router. This process [4] of mesh router selection is described in Algorithm 1.

Algorithm 1
Notations:

1. P	:	Traffic set
2. P_i	:	specific traffic
3. i, j	:	integer value
4. MC	:	Mesh Client
5. MR	:	Mesh Router
6. RSSI	:	Received Signal Strength Indicator
7. M_{MR}	:	RSS at MR
8. M_{MC}	:	Signal Strength of MC
9. t_i	:	handoff detection time period
10. BW	:	Bandwidth
11. LQ	:	Link Quality
12. Pkt_size	:	packet size
13. t	:	time required to transfer the packet
14. resp_delay	:	response delay

Algorithm:

1. The traffic types handled by the WMN through various mesh routers is denoted by the traffic set, $P = \{P_1, P_2, \ldots .P_k\}$.
2. When $i < j$, traffic type P_i, has higher priority than the traffic type P_j.
3. If two traffic types from different MC have same priority, then the current MR checks if any MC carrying the traffic requires handoff.
4. Handoff traffic is determined by the MR by analyzing the RSSI value.
5. The RSSI value provided by the MC is recorded by the MR in its routing table as $RSSI(M_{MR}, M_{MC}, t_i)$.
6. If $RSSI(M_{MR}, M_{MC}, t_{i-1}) > RSSI(M_{MR}, M_{MC}, t_i)$, then the MC is a roaming MC and requires handoff service.
7. If $RSSI(M_{MR}, M_{MC}, t_{i-1}) < RSSI(M_{MR}, M_{MC}, t_i)$, then the MC is not a roaming MC and does not require handoff service.
8. The MC requiring handoff is given high priority.
9. The high priority MC is considered for processing before any other MC.
10. As the MC proceeds towards its new location, all the MR from the new location are considered.
11. The corresponding RSSI, BW and LQ between MC and every MR is estimated.
12. The MC broadcasts a connection request message to all the available MR in the new location.
13. On receiving a response to the request message, the MC estimates certain network features.

14. The RSSI is estimated according to the Eq. 1.

$$\mathrm{RSSI} = \mathrm{RSSI}(\mathrm{M_{MR}}, \mathrm{M_{MC}}, \mathrm{t_j}) \quad (1)$$

15. The BW is estimated according to Eq. 2.

$$\mathrm{BW} = \mathrm{pkt\_size}/\mathrm{t} \quad (2)$$

16. The LQ is estimated in terms of resp_delay.
    If resp_delay is high, then LQ is low
    If resp_delay is less, then LQ is high
17. Among all the available MR, the MR with best RSSI, larger BW and good LQ is selected as the current MR.

In this an efficient MR is selected in the new location as the MC moves dynamically.

3.3 Mesh Router Location Update Using Forward Pointer

As the Mesh Client moves about the network, during handoff the serving Mesh Router changes and hence there is a need to update the location database at the gateway. If the location database is monotonously updated, then the database will be overloaded. To reduce the control overhead at the database in the gateway, forward pointer technique is employed [2]. The forward chain length is also maintained at an optimum level, by resetting it w.r.t a threshold value. This process is described in Algorithm 2.

Algorithm 2

Notations:

1. MR	:	Mesh Router
2. MC	:	Mesh Client
3. SMR	:	session-to-mobility ratio
4. SMR_{Th},	:	threshold session-to-mobility ratio
5. GW	:	Gateway
6. AMR	:	Anchor Mesh Router

Algorithm:

1. When MC is close to the newly selected MR, the MC estimates the SMR of this MR.
2. The SMR is compared with SMR_{Th}.
3. If SMR < SMR_{Th}, then the MC notifies this MR about its handoff from a MR in the previous location.
4. Now the selected MR becomes its serving MR.
5. A forwarding pointer is setup between the previous MR and the current serving MR.
6. Then the forward chain length is incremented by one.

7. If SMR $\geq$ SMR_{Th}, then the forward chain is reset and hence no forward pointer is setup.
8. Now the selected MR becomes the serving MR and is referred as AMR.
9. Then a location update message is sent to the GW to update the AMR location information in the location database.

Thus using the forward pointer and the forward chain technique, the database is protected from being overloaded. This enables the WMN in operating efficiently.

4 Simulation

4.1 Simulation Parameters

We use NS2 to simulate our proposed Dynamic Mobility Management with QoS Aware Router Selection (DMMQARS) protocol. We use the IEEE 802. Eleven for wireless Mesh networks as the MAC layer protocol. It has the functionality to inform the network layer about link breakage. In our simulation, the number of nodes varies as 4, 6, eight, 10 and 12. The location size is 1250 m × 1250 m square region for fifty seconds simulation time. The simulated traffic is Constant Bit Rate (CBR).

Our simulation settings and parameters are summarized in Table 1.

Table 1. Simulation parameters

No. of nodes	4, 6, 8, 10 and 12
Area	1250 × 1250
MAC	802.11
Simulation time	50 s
Traffic source	CBR and exponential
Propagation	Two ray ground
Antenna	Omni antenna

4.2 Performance Metrics

We evaluate performance of the new protocol mainly according to the following parameters. We compare the LMMesh protocol with our proposed DMMQARS protocol.

Average Packet Delivery Ratio. It is the ratio of the No. of packets received and the total number of packets transmitted.

Throughput. It is the amount of data that can be sent from the sources to the destination.

Packet Drop. No. of packets dropped during the data transmission.

4.3 Results and Analysis

The simulation results are specified in the next section.

Case-1 (CBR)

A. Based on Nodes
In our first experiment we vary the number of nodes as 4, 6, 8, 10 and 12.

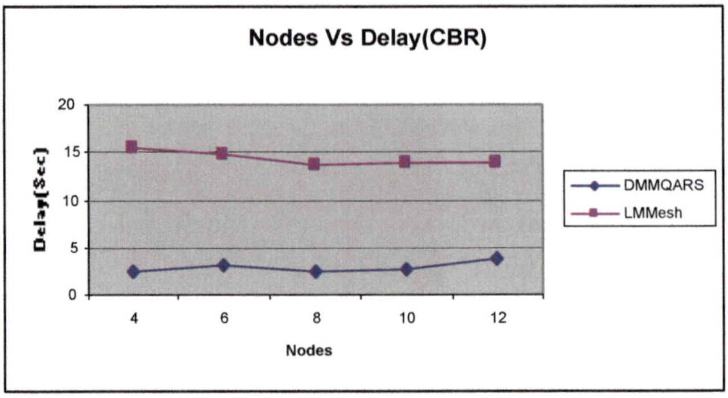

Fig. 2. Nodes vs Delay

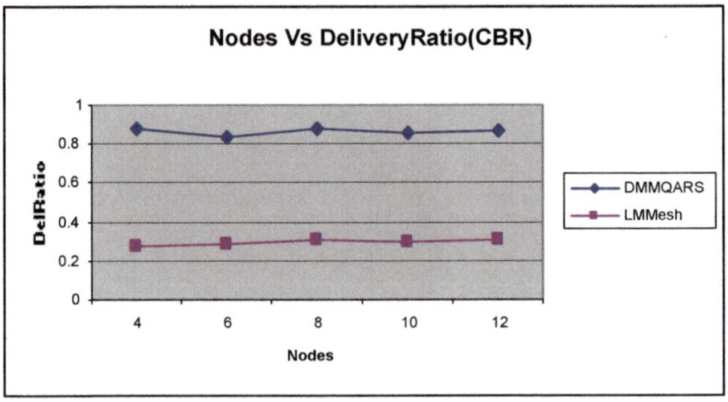

Fig. 3. Nodes vs Delivery Ratio

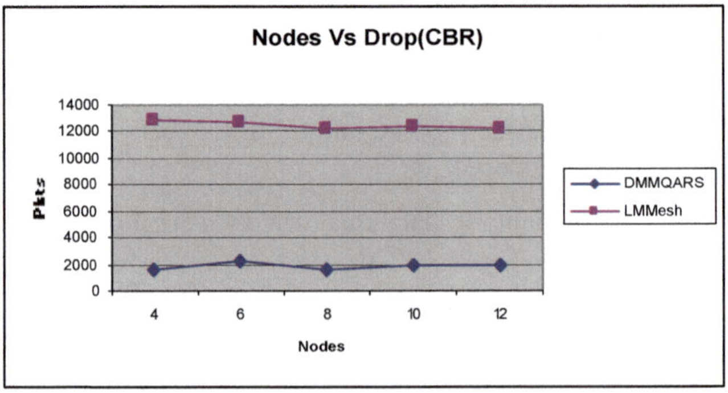

Fig. 4. Nodes vs Drop

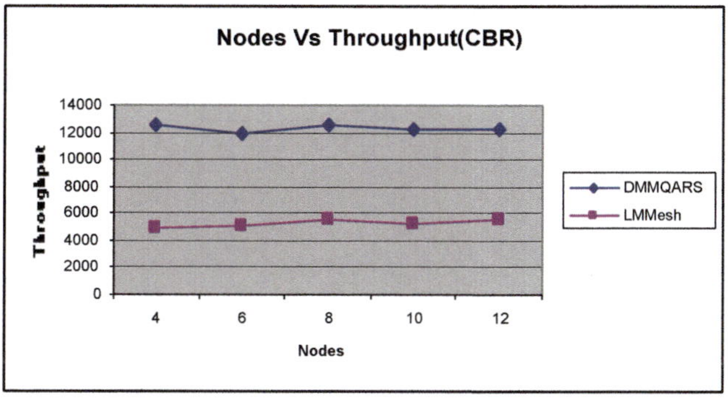

Fig. 5. Nodes vs Throughput

Figures 2, 3, 4 and 5 show the results of delay, delivery ratio, packet drop and throughput by varying the number of nodes from 4 to 12 for the CBR traffic in DMMQARS and LMMesh protocols. When comparing the performance of the two protocols, we infer that DMMQARS outperforms LMMesh by 79% in terms of delay, 66% in terms of delivery ratio, 85% in terms of drop and 57% in terms of throughput.

Case-2 (Exponential)

A. Based on Nodes

In our second experiment we vary the number of nodes as 4, 6, 8, 10 and 12.

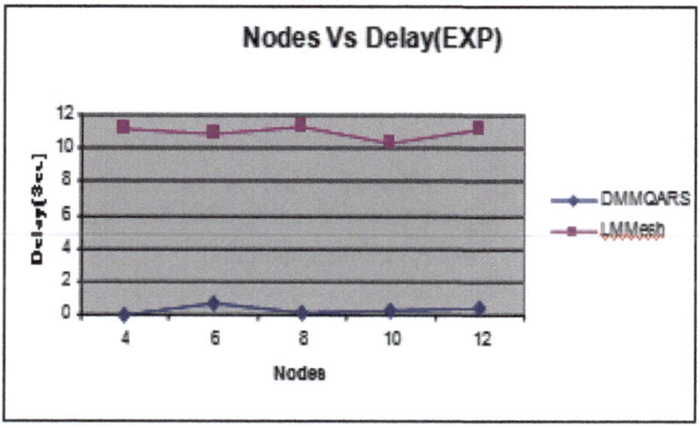

Fig. 6. Nodes vs Delay

Figures 6, 7, 8 and 9 show the results of delay, delivery ratio, packet drop and throughput by varying the number of nodes from 4 to 12 for the Exponential traffic in DMMQARS and LMMesh protocols. When comparing the performance of the two protocols, we infer that DMMQARS outperforms LMMEsh by 97% in terms of delay, 58% in terms of delivery ratio, 96% in terms of drop and 46% in terms of throughput.

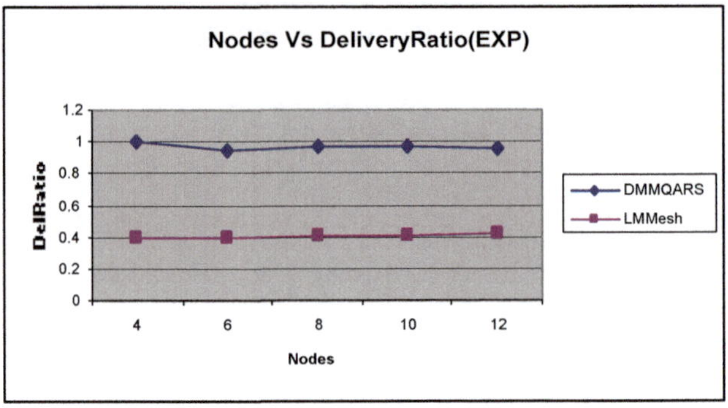

Fig. 7. Nodes vs Delivery Ratio

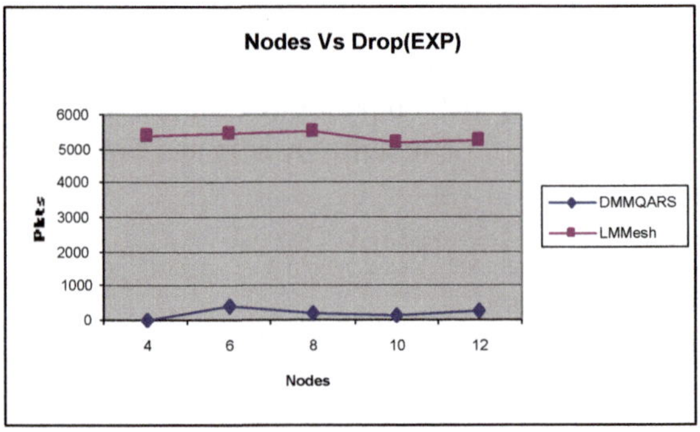

Fig. 8. Nodes vs Drop

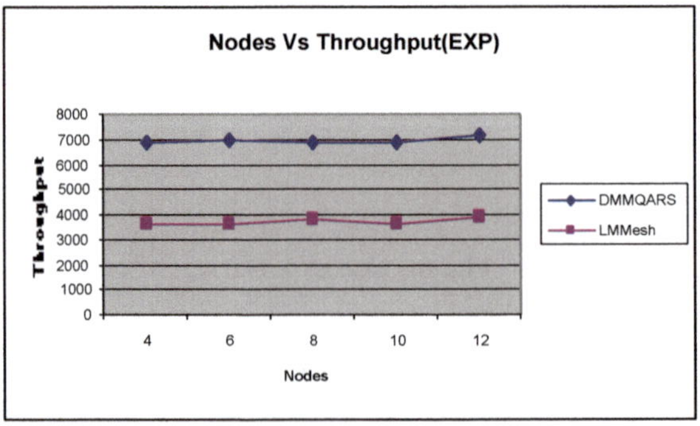

Fig. 9. Nodes vs Throughput

5 Conclusion

In this paper, we have proposed a Dynamic Mobility Management with QoS Aware Router Selection for Wireless Mesh Networks. This technique aids all the mobile client with every traffic type in performing its network operation in a prioritized manner. Each mesh client is considered and the handoff traffic is given high priority. Based on the new region where handoff is performed, the mesh router is selected. The selection is performed in such a way as to ensure that the selected mesh router is efficient in handling the client. Then to avoid burdening the gateway with control overload, forward pointer scheme is employed. This resets the forward chain length every time at an optimal level, thus improving the network performance.

References

1. Zhang, Z., Pazzi, R.W., Boukerche, A.: A mobility management scheme for wireless mesh networks based on a hybrid routing protocol. Comput. Netw. **54**, 558–572 (2010)
2. Li, Y., Chen, I.-R.: Mobility management in wireless mesh networks utilizing location routing and pointer forwarding. IEEE Trans. Netw. Serv. Manag. **9**(3), 226–239 (2012)
3. Majumder, A., Roy, S.: Design and analysis of a dynamic mobility management scheme for wireless mesh network. Sci. World J. **2013**, 1–16 (2013)
4. Song, J., Liu, Q., Zhong, Z., Li, X.: A cooperative mobility management scheme for wireless mesh networks. In: 6th IEEE International Workshop on Personalized Networks (2012)
5. Xie, J., Wang, X.: A survey of mobility management in hybrid wireless mesh networks. IEEE Netw. **22**, 34–40 (2008)
6. Lee, S., Jeong, H.-J., Kim, D.: Mobility management scheme for supporting legacy clients in IEEE 802.11s WMNs. In: IEEE International Conference on Consumer Electronics (2012)
7. Wang, Z.: Network based load-aware mobility management in IEEE 802.11 wireless mesh networks. Appl. Math. Inf. Sci. **8**, 839–847 (2014)
8. Nazari, B., Wen, S.: A case for mobility- and traffic-driven routing algorithms for wireless access mesh networks. In: European Wireless Conference (2010)
9. Matos, R., Sargento, S.: Context-aware connectivity and mobility in wireless mesh networks. In: Springer-Mobile Networks and Management, vol. 32, pp. 49–56 (2009)
10. Daly, I., Zarai, F., Kamoun, L.: A protocol for re-authentication and handoff notification in wireless mesh networks. IJCSI Int. J. Comput. Sci. Issues **8**(3), 240 (2011). No. 2

Group Key Management Protocols for Securing Communication in Groups over Internet of Things

Ch. V. Raghavendran[1(✉)], G. Naga Satish[2], and P. Suresh Varma[3]

[1] Aditya College of Engineering and Technology, Surampalem, Andhra Pradesh, India
raghuchv@yahoo.com

[2] BVRIT Hyderabad College of Engineering for Women, Hyderabad, Telangana, India
gantinagasatish@gmail.com

[3] College of Engineering, Adikavi Nannaya University, Rajamahendravaram, Andhra Pradesh, India
vermaps@yahoo.com

Abstract. Internet of Things (IoT) is a revolutionary model that extensively enhances the range of devices connected from personal devices to manufacturing equipment, actuators and sensors that are communicated to the Internet using wireless networking technology. To overcome the complications in such a range of devices, inter-networking explanations that can use the present technologies properly with emerging technologies is required. The ad hoc nature of IoT adds new challenges to network security. The wireless and dynamic nature makes IoT networks more vulnerable to security attacks. It is necessary to create a secure channel between an Internet host and the IoT device. For that it is required to have key management methods that permit two devices to agree certain secret keys that will be used to secure the flow of information. A secure and efficient management of keys is crucial for a reliable network service and that consequently makes IoT networks to achieve their success. This paper evaluates and surveys up to date key management schemes for IoT environment.

Keywords: Wireless Sensor Networks (WSN) · Internet of Things (IoT) · Group communication · Key management · Sensors · Security

1 Introduction

The Internet revolutionized the way the people communicate and work together. It leaded the new era of information access for everyone, changed life in ways that are unimagined previously. The next revolution of the Internet is with the intelligent, smart and connected devices. To work together successfully with the real world, these devices have to effort simultaneously with scales, speeds and capabilities ahead of what people require or use. The Internet of Things (IoT) will change the world, possibly more intensely than today's human centric Internet. Mainly, "things" were tagged with machine readable identification technologies, like advanced Electronic Product Codes

D. J. Hemanth et al. (Eds.): COMET 2019, LNDECT 35, pp. 1344–1350, 2020.
https://doi.org/10.1007/978-3-030-32150-5_136

(EPC), Quick Response (QR) Codes, or Radio Frequency Identification (RFID) chips. But, IoT is now often used to refer to sensors or devices that directly connected to the Internet.

As per the Burkitt article [1], he estimated that around 50 billion things are going to be connected to Internet by 2020. As per the IEEE Spectrum report [2], by 2025 there will be many billions of web-enabled devices all around the globe ranging from unmanned vehicles or robots to smart phones, wearable's, or even kitchen appliances.

Privacy and Security are one of the key factors to attain the complete vision of the IoT. So many security challenges are there, which are to be taken into consideration. In IoT network things are allowed to know the status of its environment and communicate with other devices in the network. So, it is essential to permit the sensor nodes of the network to join with other devices through the Internet. But, the problems to protect the flow of information are important. The devices with sensors are in general constrained in terms of limited computational power. But the key management mechanisms for agreeing a session key with other devices may be too intense for them.

Wireless communication in today's Internet is typically made more secure through encryption. Encryption is also considered as a key to ensure information security in the IoT. But most of the IoT devices are not presently adequate to support strong encryption. If the algorithms are designed with more efficiency, consuming less energy and efficient distribution schemes for distributing keys then it will be easy to implement encryption in IoT [3–5].

In this paper we discussed on how the present key management systems used in Internet set-up can be practical for IoT networks. The rest of the paper is organized as – Sect. 2 discussed on the background information related to key management, Wireless Sensor Networks (WSNs). Section 3 covers the group key protocols and Sect. 4 concludes the paper and proposes research directions.

2 Motivation

Key establishment or key exchange is an important process to change an insecure communication channel into a secure channel between two parties [6]. This uses a cryptographic algorithm to compute the keys and later the keys are used to encrypt and decrypt transmitting messages between two parties. It has been over four decades since Diffie and Hellman have proposed the first key exchange in [7]. The Elliptic Curve Digital Signature Algorithm (ECDSA) is a modification of the Digital Signature Algorithm (DSA) that runs in elliptic curve groups. Elliptic Curve Qu-Vanstone (ECQV) [8] is a different type of implicit certificate scheme with lesser certificate sizes, lesser computational power and very fast processing time for making certified public keys.

Various studies have been carried out to inspect secure key management in WSNs [9].

But, developing secure key management protocols in IoT enabled WSNs is an ongoing research area [10]. To achieve a good insight into the security requirements in resource constrained IoT networks, it is required to identify the precise network characteristics of IoT sensor networks and the deficiencies of the available security protocols. The most important IoT-enabled WSN characteristics are identified as resource constraints in terms of bandwidth, memory, battery capacity, processor power and heterogeneity of the networking technologies, scalability and mobility. In addition, the existing key management and authentication solutions in the security protocols are still too expensive for the device limitations of low-power IoT sensors. The propagation of WSN technology along with the improvement of Internet technologies has paved the way to the enormous paradigm shift of the IoT. The key transition of WSN technology is shown in Fig. 1.

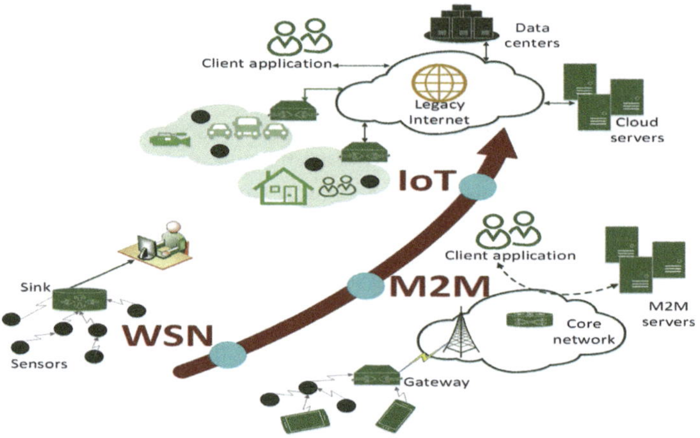

Fig. 1. Transition from WSN to IoT

It is further useful and able to communicate multi-cast messages to a group of devices instead of communicating unicast messages to individual devices in numerous copies. Multicast communication is optional for resource-constrained IoT networks to decrease the bandwidth usage, and reduce the energy consumption and processing overhead at the terminals. Securing the group key establishment among the genuine members is the key functionality needed to offer authentication, confidentiality and integrity for message communications in multi-cast groups [11].

As like IP networks group key management schemes cannot be applied as it to devices in IoT, because these are greatly constrained by the inadequate energy and resource capacities. Restricted resources enforce new challenges concerning storage and computation requirements, and each node is unable to store a large key database or perform important cryptographic calculation.

3 Group Key Management

According to [12–14] there are three categories in Group Key Management (GKM) procedures which are centralized, decentralized, and distributed/contributory key management. However, all the usual GKM protocols of these categories are not suitable for the self-motivated nature of the IoT environment and functions. So far, majority of the research associated to GKM focused on implementing the protocols by running on few of the IoT aspects viz., scalability, network access technology, controlled behavior of devices, addressing, functional nature, or mobility. It is ignored that, most of IoT environments need to work on these characteristics combined together and they require the capability to tackle the resultant issues of such grouping.

To provide security in multicast routing, the Group Owner (GO) will set a security polity and passed to the Group Controller and Key Server (GCKS) that in turn mange the security process. Protocols like Group Domain Interoperation (GDOI) [15] or Group Security Association Key Management Protocol (GSAKMP) [16] will implement the security rules set by GO.

In Centralized Group Key Management (CGKM) a Group Controller (GC) is responsible for distributing and updating the group key. Proposed protocols and their enhancement for this approach include - Group Key Management Protocol (GKMP), Logical Key Hierarchy (LKH), One-way Function Tree (OFT), One-way Function Chain Tree (OFCT), Hierarchical a-ary Tree with Clustering (HTC), Centralized Flat Table (CFT) and Efficient Large-Group Key (ELK) [17].

In Decentralized Group Key Management (DGKM) approach, a group is split into sub-groups. Every subgroup has its own subgroup key severs that manages the subgroup key. Some of the protocols proposed for the decentralized method are - Scalable Multicast Key Distribution (SMKD), MARKS, Dual-Encryption Protocol (DEP), IOLUS, KRONOS, Intra-Domain Group Key Management (IGKMP), Hydra [12].

The Contributory Group Key Agreements (CGKA), members of the group will participate in key management – generation and distribution. There is no Group Control (GC) in these agreements/schemes. As they are contributory, the distributed schemes help in the uniform distribution of the work load for key management and this eliminates the necessity for a central trusted entity. As and when the group members increase this process becomes complex. Examples of distributed protocols are Distributed One-way Function Tree (DOFT), Group Diffie–Hellman Key Exchange (G-DH), Distributed Logical Key Hierarchy (DLKH), Skinny Tree, Diffie–Hellman Logical Key Hierarchy (DHLKH), Octopus, Conference Key Agreement (CKA) and Distributed Flat Table (DFT).

4 Secure Group Key Management Models for IoT

Securing group communications requires providing authenticity, integrity and confidentiality of messages exchanged within the group. An important security problem in IoT is the group key management in securing group communications.

4.1 Elliptic Curve Cryptographic Operations (ECC)

Elliptic Curve Cryptography (ECC) is a solution under Public Key Cryptography (PKC) that is defined with usual curve parameters and appropriate for protecting energy constrained devices [18, 19]. Porambage et al. proposed two protocols in [20] with group key establishment for multicast communication in WSN installed for IoT applications. Protocol 1 is an ECC with advances such as ensuring the integrity and the authenticity of data, and removing the Man In The Middle (MITM) attacks. Protocol 2 is a modification of Elliptic Curve Integrated Encryption Scheme (ECIES). ECIES is a hybrid encryption scheme which uses the functions such as key agreement, key derivation, encryption, message authentication, and hash value computation.

The evaluation results of these protocols showed that, computation and communication energy consumptions of these are acceptable by the resource controlled sensor nodes. These protocols support frequent variations of the multicast group which results in more scalability. Protocol 1 is further suitable for distributed IoT applications that need group members to greatly contribute to the key computation and need better randomness. Protocol 2 is more suitable, as the energy cost at responder is very low. These two protocols are relevant to one-to-many (1 : n) communication situation.

4.2 Context-Aware Secured Multicast Architecture (CASMA)

In [21] Harb et al., proposed CASMA protocol suitable for diverse IoT applications and for the dynamic behavior of the IoT environments. In this they proposed Context Aware Security Server (CASS) to manage multicast session and group key operations. Key Distribution Servers (KDS) are conscientious for distributing the key. The CASS gathers information from sensors and KDS. After analysis it allocate those members to the proper KDS which in turn capable to provide best to them. Along with, it controls load balancing to improve both performance and scalability. The Fig. 2 shows the architecture of CASMA, which follows the International Telecommunication Union IoT architecture reference model defined in [22].

The advantages of this architecture are

- CASMA can be implemented for both public and private application environments.
- As it is not coupled to a particular protocol/algorithm, it can be implemented for any of the present or future protocols/algorithms.
- This improves both performance and scalability.

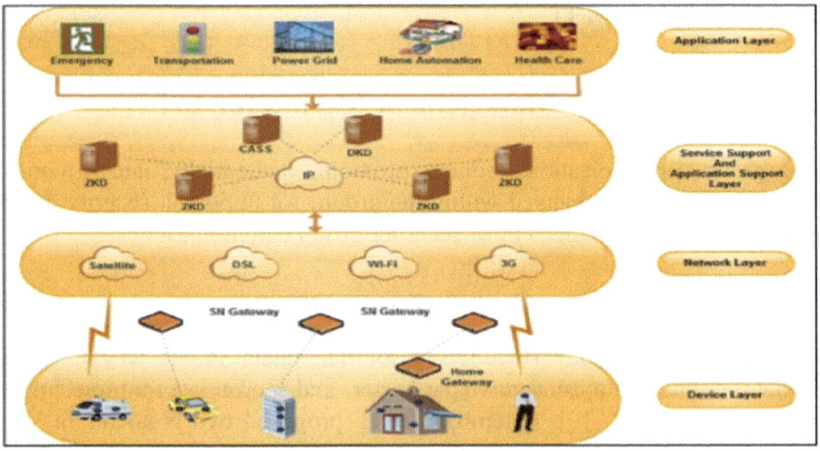

Fig. 2. CASMA Architecture

4.3 Multistage Interconnected Physically Unclonable Function (MIPUF)

Multistage Interconnected Networks (MINs) is the base idea for the Multistage Interconnected Physically Unclonable Function proposed by Hongxiang Gu et al. in [23]. Storage and computation requirements of IoT are the two limited resources for heavy cryptographic computation. Physically Unclonable Functions (PUFs) is a solution to overcome these two problems. PUFs are a kind of low-power security primitive with unpredictable and unclonable properties. Multistage Interconnected PUF (MIPUF) a group key management for IoT is proposed in [23]. Interconnection reconfiguration in MIPUF is strong and secure against to modeling attacks by varying the challenge-response mapping [23]. This key management scheme with key distribution, key storage and rekeying is flexible against a large range of attacks. Simulation results showed that this is 47.33% extra energy competent in contrast to ECC-based key management schemes. This key management is implemented with hardware support on Xilinx Spartan-6 LX45 FPGAs for IoT nodes to evaluate the power and area.

5 Conclusion

The paper started with the foreword of WSN and IoT security, along with the importance of designing lightweight key management and authentication solutions for resource constrained devices. The IoT networks are of highly resource constrained and low-power low-performing things. Their resource limitations are measured in terms of battery capacity, computational power, memory footprint and bandwidth utilization. Present key management solutions will make researchers to define promising security standards for constrained IoT networks. However, designing new solutions and adjusting the available security protocols will still be challenging. The protocols studied in this paper are suitable for different IoT environments. So, the implementation of the protocol depends mainly upon the environment, the level of security required and computing resources participating IoT devices.

References

1. Burkitt, F.: A Strategist's Guide to the Internet of Things. Strategy+Business (2014)
2. IEEE Spectrum. Popular internet of things forecast of 50 billion devices by 2020 is outdated. https://spectrum.ieee.org/tech-talk/telecom/internet/popular-internet-ofthings-forecast-of-50-billion-devices-by-2020-is-outdated. Accessed 04 Feb 2018
3. Bandyopadhyay, D., Sen, J.: Internet of Things: applications and challenges in technology and standardization. Wirel. Pers. Commun. **58**(1), 49–69 (2011)
4. Roman, R., Najera, P., Lopez, J.: Securing the internet of things. IEEE Comput. **44**(9), 51–58 (2011)
5. Yan, T., Wen, Q.: A trust-third-party based key management protocol for secure mobile RFID service based on the Internet of Things. In: Tan, H. (ed.) Knowledge Discovery and Data Mining. AISC, vol. 135, pp. 201–208. Springer, Berlin (2012)
6. Stallings, W.: Cryptography and Network Security: Principles and Practices. Pearson Education India (2006)

7. Diffie, W., Hellman, M.: New directions in cryptography. IEEE Trans. Inf. Theory **22**(6), 644–654 (1976)
8. SEC4: Elliptic Curve Qu-Vanstone Implicit Certificate Scheme (ECQV), version 0.97. www.secg.org. Accessed 21 Dec 2017
9. Zhang, J., Varadharajan, V.: Wireless sensor network key management survey and taxonomy. J. Netw. Comput. Appl. **33**(2), 63–75 (2010)
10. Roman, R., Alcaraz, C., Lopez, J., Sklavos, N.: Key management systems for sensor networks in the context of the Internet of Things. Comput. Electr. Eng. **37**(2), 147–159 (2011)
11. Porambage, P., Braeken, A., Schmitt, C., Gurtov, A., Ylianttila, M., Stiller, B.: Group key establishment for secure multicasting in IoT-enabled Wireless Sensor Networks. In: 40th IEEE Conference on Local Computer Networks (LCN), pp. 482–485 (2015)
12. Barskar, R., Chawla, M.: A survey on efficient group key management schemes in wireless networks. Indian J. Sci. Technol. **9**(14), 1–16 (2016)
13. Jiang, B., Hu, X.: A survey of group key management. In: International Conference on Computer Science and Software Engineering (2008)
14. Rafaeli, S., Hutchison, D.: A survey of key management for secure group communication. J. ACM Comput. Surv. (CSUR) **35**(3), 309–329 (2003)
15. Weis, B., Rowles, S., Hardjono, T.: The group domain of interpretation. RFC 6407, October 2011
16. Harney, H., Meth, U., Colegrove, A.: GSAKMP: group secure association key management protocol. RFC 4535, June 2006
17. Raghavendran, Ch.V., Naga Satish, G., Suresh Varma, P.: A study on contributory group key agreements for mobile ad hoc networks. Int. J. Comput. Netw. Inf. Secur. **4**, 48–56 (2013)
18. Certicom Research. Standards for Efficient Cryptography, September 2000. SEC 2: Recommended Elliptic Curve Domain Parameters, Version 1.0. http://www.secg.org/SEC2-Ver-1.0.pdf
19. National Institute of Standards and Technology. Recommended Elliptic Curves for Federal Government Use, August 1999. http://csrc.nist.gov/groups/ST/toolkit/documents/dss/NISTReCur.pdf
20. Porambage, P., Braeken, A., Schmitt, C., Gurtov, A., Ylianttila, M., Stiller, B.: Group key establishment for enabling secure multicast communication in wireless sensor networks deployed for IoT applications. IEEE Access **2**, 1503–1511 (2015)
21. Harb, H., William, A., El-Mohsen, O.A.: Context aware group key management model for internet of things. In: ICN 2018: The Seventeenth International Conference on Networks, pp. 28–34 (2018)
22. International Telecommunication Union - ITU-T Y.2060 - (06/2012) - Next Generation Networks - Frameworks and functional architecture models - Overview of the Internet of things
23. Gu, H., Potkonjak, M.: Efficient and secure group key management in IoT using multistage interconnected PUF. In: Proceedings of the International Symposium on Low Power Electronics and Design (ISLPED 2018) (2018)

Improving Data Rate Performance of Non-Orthogonal Multiple Access Based Underwater Acoustic Sensor Networks

Veerapu Goutham(✉), Gajjala Kalyan Kumar Reddy, Yeluri Gift Babu, and V. P. Harigovindan

Department of Electronics and Communication Engineering, National Institute of Technology Puducherry, Karaikal 609609, India
gouthamveerapu@gmail.com, gkalyankr7@gmail.com, giftbabuyeluri@gmail.com, hari@nitpy.ac.in

Abstract. In this article, initially we propose an optimal packet size selection scheme for reduced channel time wastage for Non-Orthogonal Multiple Access (NOMA) in Underwater Acoustic Sensor Networks (UASNs). Existing conventional NOMA technique achieves the sum rate without considering the traffic generation in UASNs, which leads to wastage of resources due to unequal transmission times in paired transmission. In contrast, the proposed scheme overcomes this problem by making equal transmission time slots for both weak and strong users using optimal data packet sizes to avoid the wastage of resources. Further, we propose an optimal power allocation for weak and strong users with respect to the distance between transceiving nodes by using particle swarm optimization. The analytical results clearly show that the proposed scheme for NOMA in UASNs significantly improves the data rate performance in comparison with the existing conventional NOMA technique.

Keywords: Non-Orthogonal Multiple Access · Underwater Acoustic Sensor Networks · Overall data rate

1 Introduction

Recently, Underwater Wireless Sensor Networks (UWSNs) emerge as a trending research area to study the mineral resources available in ocean, detection of natural calamities, monitoring port facilities, enemy detection and many more [1]. UWSNs are formed by connecting several sensor nodes, autonomous underwater vehicles, surface stations in underwater. Generally, acoustic waves are preferred in UWSNs instead of radio frequency (RF) waves due to high absorption and optical waves due to scattering in underwater medium. Unlike the white noise in terrestrial wireless sensor networks (WSNs), Underwater Acoustic Sensor Networks (UASNs) suffer with frequency dependent noises. Due to these noises and intrinsic properties of underwater medium, UASNs have many different challenges to solve compared to WSNs. Some of the interesting challenges in UASNs are listed as [1]:

D. J. Hemanth et al. (Eds.): COMET 2019, LNDECT 35, pp. 1351–1358, 2020.
https://doi.org/10.1007/978-3-030-32150-5_137

- Propagation delay: The speed of acoustic waves in underwater medium is close to 1500 m/s which is very less when compared to speed of RF waves (3×10^8 m/s) in WSNs.
- Spectral efficiency: Acoustic waves have a few kilo hertz of bandwidth in the range of 10 kHz–20 kHz. Due to this, the number of applications supported in the available bandwidth are very limited.
- Energy efficiency: Underwater sensor nodes work on limited battery capacity. It is difficult to recharge or replace the batteries frequently. So, the nodes should work optimally to save the battery energy.
- Reliability: The probability of receiving error at the receiver is very much high in underwater channel due to its random characteristics and properties.
- Node mobility: Nodes are mobile in underwater due to water currents and shipping.

This article analyzes the performance enhancement of spectral efficiency by implementing Non-Orthogonal Multiple Access (NOMA) scheme in UASNs. NOMA serves multiple nodes simultaneously in a single resource block (either in time or frequency domain) by exploiting the new dimension of power domain as shown in Fig. 1. This scheme superimposes data packets of weak user (far distant to the transmitting node) and strong user (near distant to the transmitting node) in power domain to achieve interference free data transmission. The strong user uses Successive Interference Cancellation (SIC) technique to get the required signals from those multiplexed signals. The weak user decodes its corresponding data packet by considering the strong user data packet as noise signal. NOMA is a promising candidate in recent times, due to its increased capacity by the efficient use of bandwidth in UASNs [2].

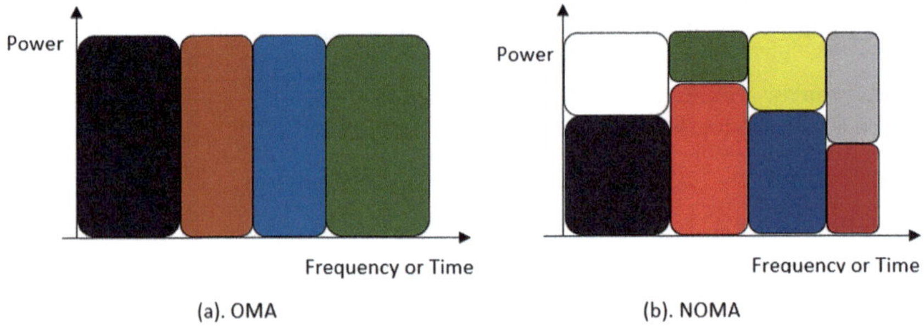

Fig. 1. Power allocation scheme in NOMA

Existing Sum Rate Maximization (SRM) technique for NOMA maximizes the data rates without considering the traffic generation which leads to wastage of channel time due to unequal transmission times [3]. To overcome this problem, an optimal data packet selection scheme for reduced channel time wastage is proposed for NOMA in UASNs. This scheme achieves maximum usage of channel time by varying the packet size of strong user to transmit exactly equal to the transmission time of the weak user. Moreover, a proper power allocation scheme even improves the efficient use of spectrum in UASNs. In this article, we propose an optimal packet size selection scheme for reduced channel time wastage in NOMA based UASNs to use the spectrum more efficiently. Here, optimal power allocation scheme is allotted to the both weak and strong users according to the distance separation between the transceiving nodes. The analytical results clearly show that the proposed scheme for NOMA in UASNs significantly improves the data rate performance in comparison with the existing conventional NOMA technique. NOMA can be even extended to Multiple input multiple output (MIMO) by using MIMO-NOMA for both uplink and downlink cases [4]. MIMO NOMA is way superior to MIMO-OMA due to its increased cluster capacity [5, 6]. Integration of co-operative communication with NOMA technique significantly enhances energy efficiency and reliability [7, 8].

2 NOMA in UASNs

Figure 2 depicts a paired transmission scenario of downlink NOMA transmission between transmitting node (T) and two receiving nodes *i.e.,* strong node (S) and weak node (W). Among the two receiving nodes, the channel quality to the node S is good as compared to the channel quality to the node W because of more attenuation due to large distance between nodes T and W than the distance between nodes T and S. This is because, node T selects and transmits the data packets to the two receiving nodes which have greater channel gain difference in single frequency or time domain. In this work, we consider that the distance between nodes T and S (d_{TS}) is exactly 10% of the distance between nodes T and W (d_{TW}) (*i.e.,* $0.1 \times d_{TS}$). In NOMA, node T transmits S_1 data packet with a large amount of available electrical transmission power to the weak user (P_{txwel}) and S_2 data packet with a small amount of available electrical transmission power to the strong user (P_{txsel}). It is given by,

$$P_{txwel} = \alpha \times P_{txel} \tag{1}$$

$$P_{txsel} = (1 - \alpha) \times P_{txel} \tag{2}$$

Where P_{txel} is the available electrical transmission power and α is the power allocation coefficient The superimposed data packets are decoded using Successive Interference Cancellation (SIC) technique at the strong user (S) [9, 10]. Basically, the node S decodes and subtracts weak user data packet from the entire data to decode its own data packet. The node W decodes the data packet by considering the strong user packet as noise signal. The signal-to-noise ratio (SNR) in UASNs is computed by considering the model presented in the [11]. The SNR of an underwater link between i^{th} transmitting and j^{th} receiving nodes is given by [11],

$$SNR_{ij} = P_{txel} + 170.8 + 10\log_{10}\zeta - N(f) - A_{ij}(f) \tag{3}$$

where P_{txel} is the electrical transmission power, $N(f)$ is ambient noise present in underwater, $A_{ij}(f)$ is attenuation losses, 170.8 dB accounts for the conversion from dB re 1 μ Pa to Watts, and ζ = 0.8 is the efficiency of the transducer. The signal-to-interference noise ratio (SINR) of a link between nodes T and W is given by [9],

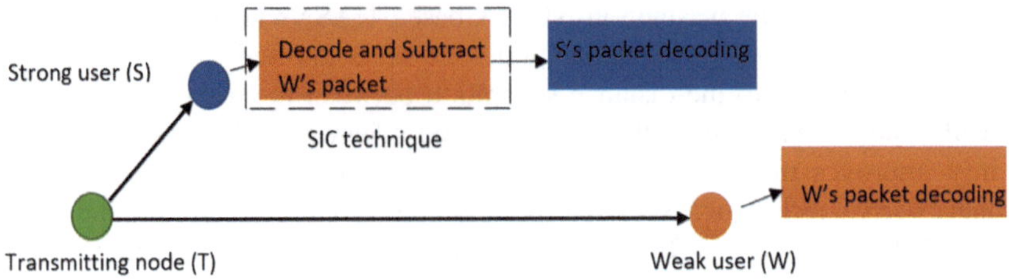

Fig. 2. Non-Orthogonal Multiple Access

$$SINR_{TW} = \frac{SNR_{TW}^{S_1}}{1 + SNR_{TW}^{S_2}} \tag{4}$$

where S_1 and S_2 represents corresponding data packets transmitted to the weak and strong user respectively. The SINR of a link between the nodes T and S is given by [9],

$$SINR_{TS} = \frac{SNR_{TS}^{S_2}}{1 + SNR_{TS}^{S_1}} \tag{5}$$

Here, at the strong user we assumed perfect interference cancellation using SIC technique, hence $SNR_{TS}^{S_1}$ is considered as 0 dB. Accordingly, the achievable data rate of a link between the nodes T and W is given [12],

$$R_W = \int_{f_c-B/2}^{f_c+B/2} log_2(1 + SINR_{TW})\, df \tag{6}$$

The achievable data rate of a link between the nodes T and S is given by [12],

$$R_S = \int_{f_c-B/2}^{f_c+B/2} log_2(1 + SINR_{TS})\, df \tag{7}$$

The overall data rate achieved in NOMA scheme is given by [2],

$$R_{overall} = \frac{L_W + L_S}{max\left(\frac{L_W}{R_W}, \frac{L_S}{R_S}\right)} \tag{8}$$

where, L_W and L_S represents the data packet size of transmitted to the weak and strong user respectively.

2.1 The Proposed Scheme

In this subsection, we present an optimal data packet selection scheme for reduced channel time wastage in NOMA based UASNs. In conventional NOMA scheme, the total amount of electrical transmission power of transmitting node (T) is shared in fixed proportion among weak user (W) and strong user (S). This assumption leads to the unequal achievable data rates because of different power level data packets are superimposed. Due to this, the weak and strong users have asymmetrical transmission time slots which results in wastage of channel time as shown in Fig. 3a. This disadvantage of conventional NOMA scheme is overcome by making the transmission time slots of both strong and weak users as symmetrical. This can be achieved by varying the packet size of strong user to transmit exactly equal to the transmission time of weak user as shown in Fig. 3b. The variable data packet size transmitted to the strong user is given by,

$$L_{S_{opt}} = \left(\frac{L_W}{R_W}\right) \times R_S \tag{9}$$

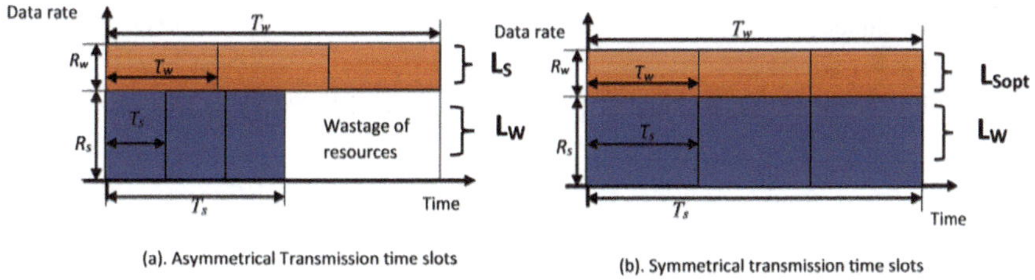

Fig. 3. Wastage of channel time in NOMA

Further, we propose a proper power allocation scheme to improve the efficient use of spectrum in UASNs. The sum rate (RB) is defined as the sum of individual achievable data rates of strong user and weak user. Here, the sum rate of a proposed NOMA scheme is further increased by finding the optimal power allocation levels to the both weak and strong users with respect to the distance between the transceiving nodes [3]. The maximization problem can be formulated as,

$$\max_{\alpha} RB = R_W + R_S. \tag{10}$$

In this proposed scheme, optimal power allocation and packet size is allocated to both strong and weak users to achieve optimum data rates. The optimal data packet size transmitted to the strong user is found by using Eq. 9. Optimal power allocation is done by finding the coefficient α (as given in Eq. 1) using Particle Swarm Optimization (PSO) to maximize the overall data rate. Hence, we propose an optimal NOMA scheme where data rate is maximized by,

- Making equal transmission time for both strong and weak users by finding the optimal packet size
- Selecting the optimal power allocation coefficient for both strong and weak users.

3 Analytical Results

In this section, the comparative analysis of overall achievable data rates for conventional NOMA and optimal NOMA (the proposed scheme) are presented and evaluated using MATLAB R2018A. Table 1 shows the different parameters used for analysis in this model.

Table 1. Parameters used for analysis

Parameter	Value
Bandwidth (B)	1–10 kHz
Electrical Transmission Power (P_{txel})	35 W
Frequency	26 kHz
Packet size (L_w)	64 bits
Power allocation coefficient (α, for conventional NOMA)	0.75

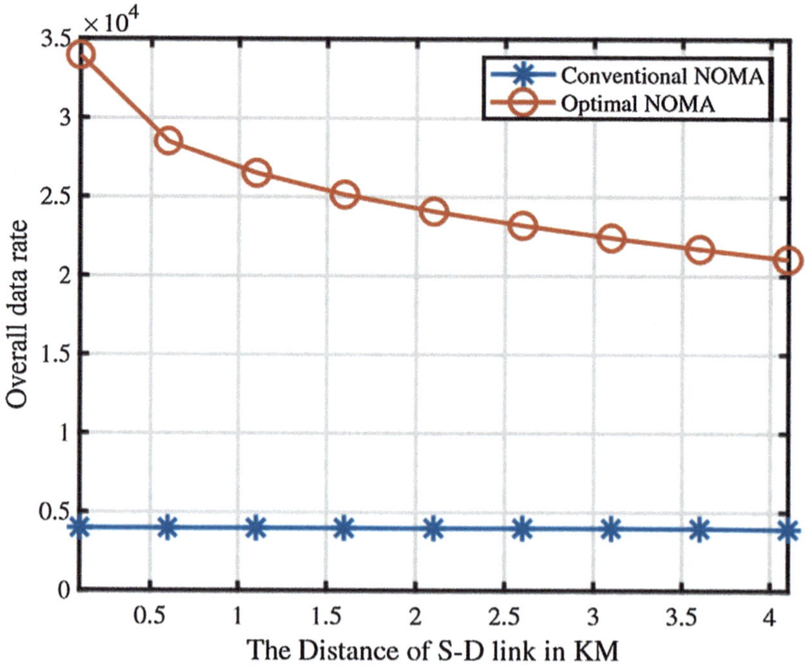

Fig. 4. Variation of overall data rate

Figure 4 depicts the overall data rate achieved in conventional NOMA and optimal NOMA considered in this work. The overall data rate of a schemes is calculated by using Eq. (8). In conventional NOMA scheme, it is assumed that some fixed proportion

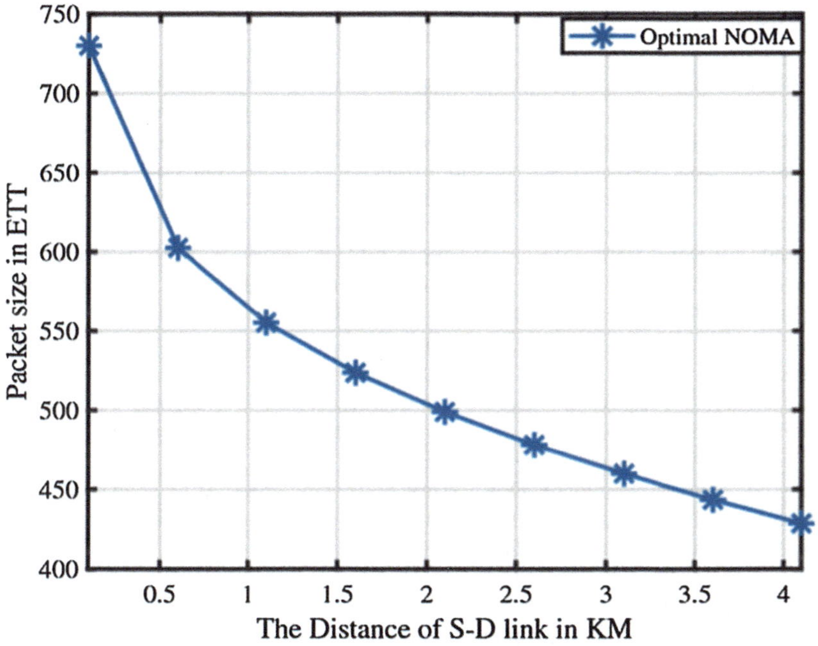

Fig. 5. Variation of Packet size

of power is allocated to the strong and weak users ($\alpha = 0.75$) irrespective of the distance between transceiving nodes. In optimal NOMA, the power allocation coefficient is computed using the particle swarm optimization technique to maximize the sum rate of NOMA scheme. From Fig. 4, it is observed that the overall data rate achieved by the optimal NOMA scheme is much higher than the conventional NOMA scheme. This is because of the effective utilization of unused resource transmission time slots by varying the data packet size of the strong user. Accordingly, Fig. 5 represents the variation of data packet size with respect to distance between the transceiving nodes. This optimal data packet size is calculated by using Eq. (9).

4 Conclusion

In this article, an optimal packet size selection for reduced channel time wastage for Non-Orthogonal Multiple Access (NOMA) in Underwater Acoustic Sensor Networks (UASNs) is proposed. Unlike the existing techniques (sum rate and SRM), the proposed scheme computes optimum data packet size for NOMA paired transmission to ensure symmetrical transmission time slots in-order to overcome wastage of channel time. Further, we have proposed an optimal power allocation scheme using PSO. The optimal NOMA scheme (with optimal packet size and optimal power allocation) is compared with conventional NOMA. The analytical results clearly show that, the proposed scheme for NOMA in UASNs can significantly outperforms existing conventional NOMA technique in terms of overall date rate.

References

1. Al-Abbasi, Z.Q., So, D.K.C.: Power allocation for sum rate maximization in non-orthogonal multiple access system. In: 2015 IEEE 26th Annual International Symposium on Personal, Indoor, and Mobile Radio Communications (PIMRC), pp. 1649–1653 (2015)
2. Cheon, J., Cho, H.-S.: Power allocation scheme for non-orthogonal multiple access in underwater acoustic communications. Sensors **17**(11) (2017). https://doi.org/10.3390/s17112465
3. Coutinho, R.W.L., Boukerche, A., Vieira, L.F.M., Loureiro, A.A.F.: Underwater wireless sensor networks: a new challenge for topology control-based systems. ACM Comput. Surv. **51**(1), 19:1–19:36 (2018). https://doi.org/10.1145/3154834
4. Sun, Q., Han, S., Chin-Lin, I., Pan, Z.: On the ergodic capacity of MIMO NOMA systems. IEEE Wirel. Commun. Lett. **4**(4), 405–408 (2015)
5. Ding, Z., Lei, X., Karagiannidis, G.K., Schober, R., Yuan, J., Bhargava, V.K.: A survey on non-orthogonal multiple access for 5G networks: research challenges and future trends. IEEE J. Sel. Areas Commun. **35**(10), 2181–2195 (2017)
6. Zeng, M., Yadav, A., Dobre, O.A., Tsiropoulos, G.I., Poor, H.V.: Capacity comparison between MIMO-NOMA and MIMO-OMA with multiple users in a cluster. IEEE J. Sel. Areas Commun. **35**(10), 2413–2424 (2017)
7. Ding, Z., Peng, M., Poor, H.V.: Cooperative non-orthogonal multiple access in 5G systems. IEEE Commun. Lett. **19**(8), 1462–1465 (2015)
8. Liu, Q., Lv, T., Lin, Z.: Energy-efficient transmission design in cooperative relaying systems using NOMA. IEEE Commun. Lett. **22**(3), 594–597 (2018)
9. Riazul Islam, S.M., Zeng, M., Dobre, O.A.: NOMA in 5G systems: exciting possibilities for enhancing spectral efficiency. CoRR abs/1706.08215 (2017). http://arxiv.org/abs/1706.08215
10. Saito, Y., Kishiyama, Y., Benjebbour, A., Nakamura, T., Li, A., Higuchi, K.: Non-orthogonal multiple access (NOMA) for cellular future radio access. In: 2013 IEEE 77th Vehicular Technology Conference (VTC Spring), pp. 1–5 (2013). https://doi.org/10.1109/VTCSpring.2013.6692652
11. Wang, C., Chen, J., Chen, Y.: Power allocation for a downlink non-orthogonal multiple access system. IEEE Wirel. Commun. Lett. **5**(5), 532–535 (2016). https://doi.org/10.1109/LWC.2016.2598833
12. Yildiz, H.U., Gungor, V.C., Tavli, B.: Packet size optimization for lifetime maximization in underwater acoustic sensor networks. IEEE Trans. Ind. Inform. **15**, 719–729 (2018)

A Hybrid RSS-TOA Based Localization for Distributed Indoor Massive MIMO Systems

Vankayala Chethan Prakash(✉) and G. Nagarajan

Department of ECE, Pondicherry Engineering College, Pondicherry, India
{chethanprakash, nagarajanpec}@pec.edu

Abstract. The advancement of wireless technologies with increasing number of devices has paved way for Massive Multiple Input Multiple Output (MIMO) systems. In 5G technologies, there is an integration of network domains such as Internet of Things, mm-waves, device 2 device (D2D) communications, machine type communications, vehicular networks, cognitive radio networks, etc. With hundreds of antennas at the base station, data rates and capacity increases for devices connected. To enhance quality of service to all connected devices, location identification has its importance. Based on received signal strength (RSS) and time of arrival (TOA) technique, hybrid RSS-TOA based energy detection is proposed. To reduce computational complexity, a channel densification process is designed using energy detector where signals with highest received signal power and arrival of signal at the first- time instance are considered. Performance of the proposed technique is evaluated with Root mean square error and it is compared with cramer rao lower bound. Simulation results show that the probability of detection around 0.80 is achieved for identification of Line of sight (LOS) and Non-Line of Sight (NLOS) conditions.

Keywords: Energy detection · Received signal strength · Time of arrival

1 Introduction

A tremendous increase in use of mobile devices has made to think of increasing data rates and capacity to users. Thus massive MIMO has gained a lot of interest in handling support to increase number of devices. Even though massive MIMO has 'M' number of antennas greater than 'N' number of devices it depends on spatial multiplexing that demands base stations have knowledge about uplink and downlink channels. However an uplink channel can be easily estimated by forwarding pilot signals from user terminal to base stations, whereas channel estimation at the downlink in massive MIMO is comparatively difficult. Massive MIMO operates in both Frequency Division Duplex (FDD) and Time Division Duplex (TDD) mode. Mostly massive MIMO is considered to operate on TDD mode, as the channel reciprocity functions are better in TDD mode. Thus the uplink and downlink channels can be estimated in an efficient manner. In massive MIMO systems radio propagation environment known as favorable propagation must be taken care of. For a favorable propagation, the channel responses are to be considered between the base station and the user (i.e.) the realistic behavior of the

D. J. Hemanth et al. (Eds.): COMET 2019, LNDECT 35, pp. 1359–1370, 2020.
https://doi.org/10.1007/978-3-030-32150-5_138

channel must be analyzed. Based on the channel conditions it is a must to identify whether the user tends to be in LOS or NLOS conditions.

In wireless networks, communication takes place based on location. In order to support users with better quality of service, identifying their location is important. Normally localization takes place in two ways i.e. (a) Range based and (b) Range free. In range based, anchors are placed which are also known as reference points for identifying users position in the network. In range free positioning conventional techniques are considered directly without any placement of anchors. In most of the wireless technologies, Global Positioning Systems (GPS) are used for positioning. This GPS is used as anchors in wireless networks. Global Positioning Systems fail at places where there is no direct line of sight between the transmitter and receiver. The available techniques for localization are as follows: RSS, TOA, Time difference of Arrival (TDOA), Angle of Arrival (AOA) and Angle Difference of Arrival (ADOA). However there are geometrical techniques available. This includes bilateration, trilateration and multilateration. These both techniques can be used in combination to achieve better positioning. By above mentioned techniques identification of LOS and NLOS conditions can take place in a network.

Wireless technologies such as 5G networks; require location information of the user to serve them with good data rates and capacity. In massive MIMO and mm-waves the operating frequency is of 30–300 GHz. As the frequency range increases; there is a degradation of signals which leads to reduction of signal quality. With higher operating frequency, the signals get diffracted, reflected and scattered. Thus the received signal strength gradually reduces and goes below the receiver operating conditions. Based on the locations in the network, and channel conditions with received signal power the destined user may be analyzed as LOS and NLOS. Glimpses of localization with advanced technologies are discussed as follows. Localization in indoor and outdoor large MIMO system is measured. A direct source localization (DiSouL) algorithm is proposed. With Angle of Arrival the user's position is extracted at the base station followed by triangulation [1]. On the basis of instantaneous channel state information extracted from angle and delay domains a fingerprinting data base is built which reduces channel computational complexity [2].

A fingerprint database is built for COST 2100 channel model. A deep convolutional neural network is utilized for positioning accuracy in a clustered environment [3]. NLOS propagation based position and orientation are analyzed for millimeter wave MIMO systems considering fisher information matrix which resolves temporal and spatial correlations with higher resolutions [4]. A channel sounding technique is proposed for LOS/NLOS environments. Multi antenna and multi sub carrier channel state information at different frequency bands are obtained and are used as data sets for deep learning analysis. An accuracy of about 25 cm is achieved with precision in NLOS environment [5]. In millimeter wave MIMO system with radio environment mapping, a joint positioning and orientation of mobile users are proposed. With message passing estimator, the placement of reflectors and scatterers can also be analyzed in the network [6]. Another algorithm for identification of reflectors and positioning is proposed namely simultaneous position and reflector estimation with TOA and AOA at the base station ensures sub meter accuracy in indoors and meter accuracy in outdoors [7].

A machine learning approach has been proposed for positioning in distributed massive MIMO systems based on RSS. A reconstruction cum Gaussian approximation technique is proposed to estimate location errors [8]. For identification of NLOS and LOS signals a convolutional neural network is proposed. With sounding reference signals a coordinated tap energy matrix is constructed. Based on the matrix, the convolutional neural network is trained [9]. With channel state information, a spatial covariance matrix is computed and considered as training data for several machine learning algorithms to analyze better performance of existing machine learning algorithms for localization in large scale MIMO systems [10]. A gradient descent optimization technique is used for extraction of training data. Due to its large data points, a two-step training procedure is proposed. With this training data set, a deep learning machine algorithm is trained for identification of LOS and NLOS [11]. A trade-off between data rates and quality of service requirements (position estimation) is analyzed with varying number of transmitting and receiving antennas in millimeter large scale multiuser MIMO systems [12].

In a mm- wave location-aware communication, a single anchor based position and orientation errors in 3D channels for uplink and downlink are studied. In uplink channels, the orientation at the base station depends on the angle of user equipment whereas in downlink channels, angle dependency does not happen [13]. Another approach in location aware communication is that tight time synchronization is required between base stations and user equipment. A two way localization protocols has been designed namely distributed and centralized localization protocols [14]. A support detection based channel training for frequency selective millimeter wave MIMO systems using lens antenna array is proposed. The channel estimation for downlink and uplink are based on AOA, AOD and TOA [15]. To classify LOS and NLOS conditions the average power delay profile between the base station and the receiver is obtained. A multiple hypothesis testing is proposed [16]. A channel data base has been built with and without the knowledge of the obstacles. An algorithm based on MUSIC is proposed which considers obstacles as nuisance variable [17]. For NLOS and LOS identification, a support vector machine and channel information regression model is proposed [18]. A fingerprinting data base is constructed based on channel frequency response obtained in offline phase and cm accuracy is achieved in comparing with time reversal resonating strength [19]. Interference mitigation techniques have been proposed with bayesian compressed sensing based on channel impulse response that minimizes errors with improved localization [20].

The contributions of this paper include:

- For a distributed massive MIMO system operating under mm wave frequency in an indoor environment is considered. Based on received signal strength the distance between the user equipment and the base station antenna is evaluated.
- With time of arrival, the time instances between the transmission and the reception at the receiver is calculated.
- In order to reduce computational complexity and multi path effects at the receiver the highest peak detection of received signals and to find the first time of arrival of signals energy detector is used.

- To identify LOS from NLOS conditions, a threshold has been set based on SNR. On the basis of RSS and TOA, the path with the highest SNR is selected.

The rest of the paper includes, Massive MIMO Architecture, System Model, Proposed work, Simulation Results and Conclusion.

2 Massive MIMO Architecture

Figure 1 shows the architecture of distributed massive MIMO system where a base station employed with hundreds of antennas is considered. The remote radio heads are deployed with certain degrees of freedom to support indoor devices. The propagation conditions between the centralized base station and the remote radio head tends to be as outdoor. The channel from the remote radio heads and the devices happen to be indoor. Localization of user equipment at indoors suffer from high propagation losses with reflection, scattering and refraction especially operating at higher frequencies such as millimeter wave. The proposed work analyzes the indoor scattering environment where multi path of signals occur at the receiver. For exact localization, usages of signal filtering techniques are must to be incorporated with the available positioning techniques.

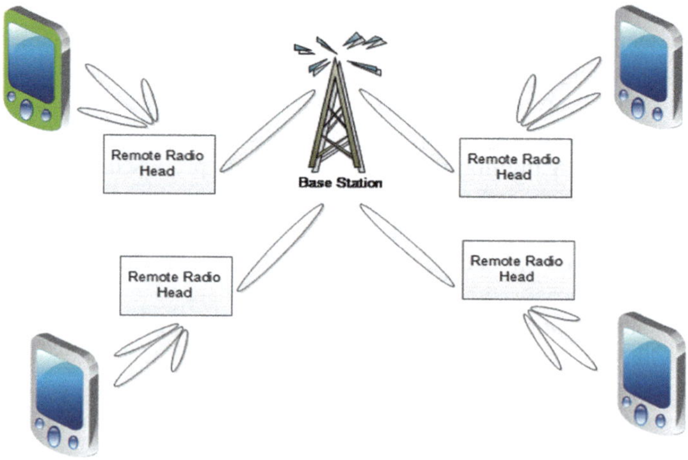

Fig. 1. Distributed massive MIMO architecture

3 System Model

A distributed massive MIMO system with M antennas at the base station and N number of devices is considered. The remote radio head with n degrees of freedom to support indoor devices are considered. The uplink channel is considered based upon the received signal from user equipment at the base station. The received signal at the base station can be given as

$$Y(t) = hx(t) + n(t) \quad (1)$$

According to channel reciprocity property, the downlink channel can be computed rom the uplink channel.

$$Y(t) = h^H x(t)^H + n(t)^H \quad (2)$$

4 Proposed Work

The distributed massive MIMO system operating at millimeter wave frequency is analyzed. The propagation conditions between the remote radio head and the user equipment can be understood by channel impulse responses. A time varying iid Additive White Gaussian noise with mean zero and variance σ^2 is considered. It is assumed that communication takes place in TDD mode. Based on the reception of signal at the base station antenna the channel conditions are studied. With channel reciprocity, the propagation conditions for downlink channels can be obtained. As the propagation suffers with higher attenuation, path loss between the remote radio head and the user equipment will be more. There is also a possibility of signals affecting with reflection, diffraction and scattering. As the operating frequency increases the beam width of the signals happens to be narrower. Thus the obstacles present between the transmitter and receiver affects the propagating signals. Since an indoor environment is taken into account, there are large obstacles such as walls, doors present in between. It is more important to identify the location of the user and to classify them as LOS users and NLOS users. With the reception of reflected and diffracted signals there is an occurrence of multi paths. On considering all these factors, Energy Detector is used for processing the received signals according to the threshold.

4.1 Received Signal Strength

The RSS represents signal power with respect to propagating signal received. As the inter site distance between the remote radio head and the mobile user equipment increases the signal power gets reduced. On the basis of the received signal strength, a user can be identified with its location in the network. But relying on RSS alone will not help in analyzing the presence of user exact position. Upon reception of multiple signals, there is a possibility of mismatch in calculating the signal strength. At instances, the reflected signal appears to be more strength than direct line of sight path. Figure 2 represents the propagation condition under received signal strength.

4.2 Time of Arrival

The TOA represents the propagation time taken between the base station and the receiver. With time delay the user equipment can be localized in the network. A tight synchronization between the base station and the user equipment is required. Thus the time instances can be analyzed. Figure 3 represents synchronization and time instances

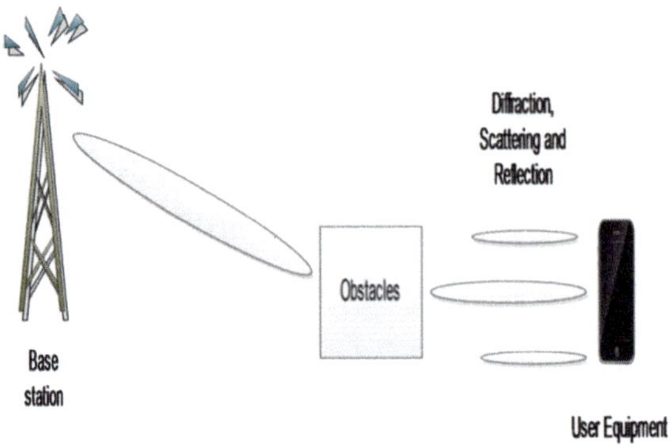

Fig. 2. A schematic representation of signal reception

between base station and user equipment. However, due to obstacles, there is a chance in which the reflected signal can reach the receiver in prior to direct line of sight signal (Fig. 4).

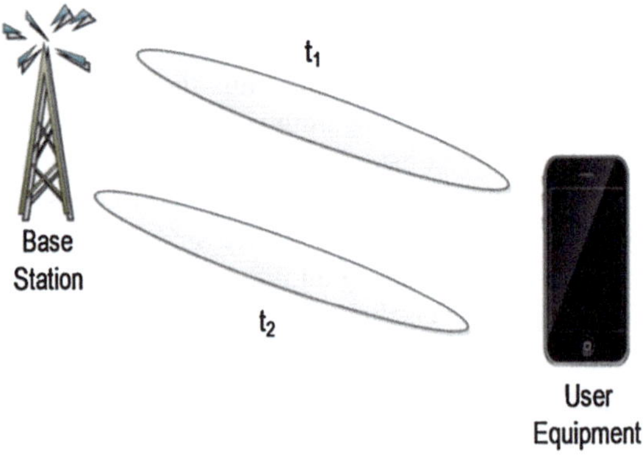

Fig. 3. Time of arrival

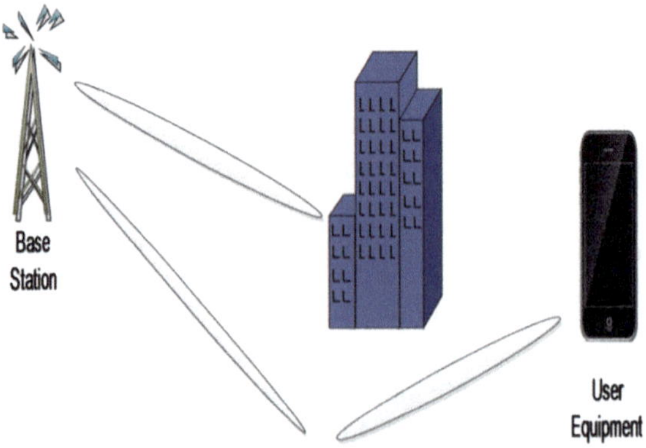

Fig. 4. Time of arrival with obstacles

For one-way measurements, the distance between two nodes can be determined as:

$$dist_{ab} = (t_2 - t_1) \times u \quad (3)$$

where, t_1 and t_2 are the sending and receiving times of the signal (measured at the sender and receiver, respectively) and u is the signal velocity.

4.3 Energy Detection Based LOS and NLOS Identification

Figure 5 demonstrates the working flow diagram of the proposed model for identification of LOS and NLOS conditions. Based on the propagation condition that prevails between the base station and the receiver, techniques for localization are utilized. The techniques include RSS and TOA, the RSS technique identifies the signal strength at the receiver and helps in localizing the receiver in the network. The obstacles in between the base station and user equipment degrade the signal with factors such as reflection, scattering and diffraction. With these factors, it is mere difficult to identify the user in the network. At some instances, the reflected signals have more signal strength than direct signal's strength. With respect to the time of arrival, the time taken for propagation of signals between the base station and antenna is calculated. However a tight synchronization is needed for calculating time instances. Even with synchronization, there is a possibility for miscalculation in such a way that the reflected signal may reach the user equipment in prior to that of the direct signal between base station and user equipment. Thus to reduce mismatches in obtaining the user position, an energy detector is used.

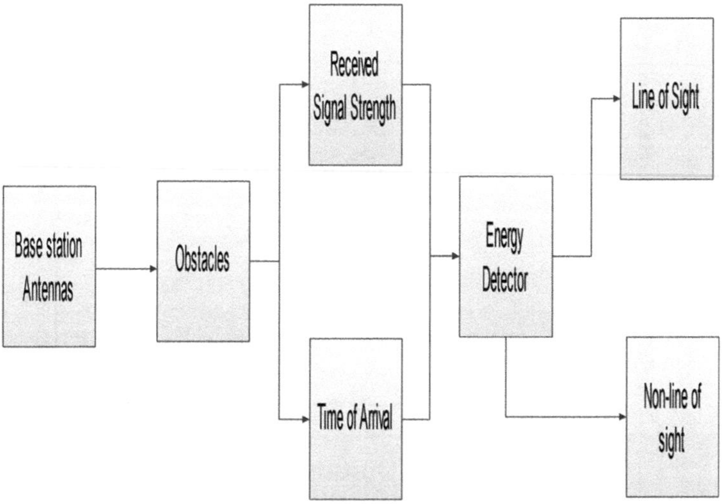

Fig. 5. Flow diagram for LOS and NLOS identification

5 Simulation Results

Table 1 represents the attributes that have been considered for the proposed hybrid RSS – TOA based energy detection for classification of users as LOS and NLOS conditions. The following simulation results depict the performance of the proposed technique in a distributed indoor massive MIMO environment with 32 and 64 antennas at the remote radio head and with 2, 4 receiving antennas. On the basis of the received signal at the remote radio head the proposed hybrid RSS-TOA technique is examined. With channel reciprocity the downlink channels are also estimated vice versa.

Table 1. Simulation environment

Attributes	Values
Simulation tool	MATLAB R2018a
Operating frequency	28 GHz
Propagation model	Indoor
Channel	AWGN
Operating mode	TDD
No. of transmitting antennas	32, 64
No. of receiving antennas	4, 8

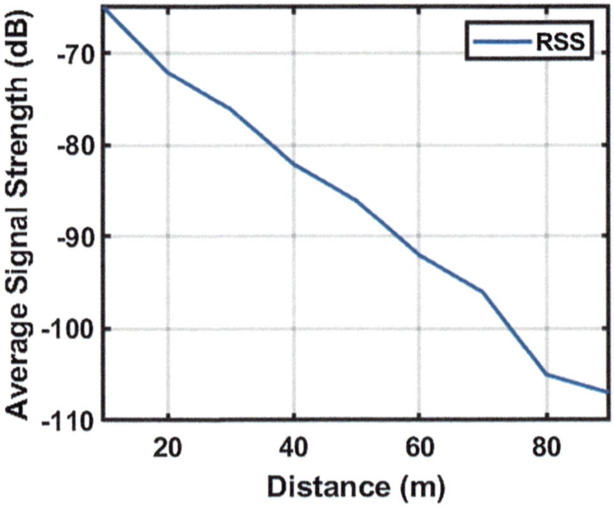

Fig. 6. Distance vs received signal strength

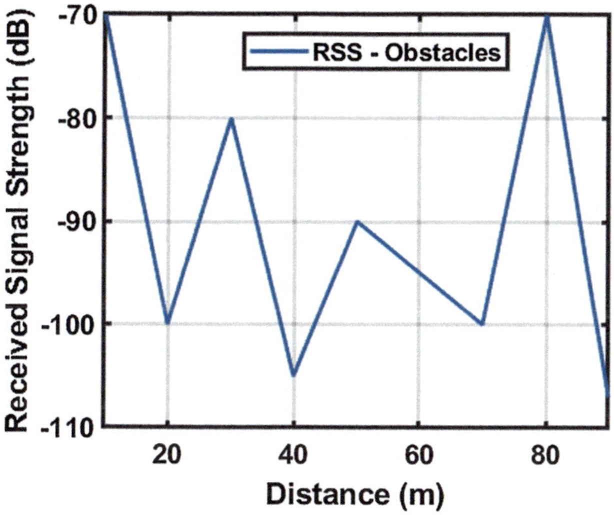

Fig. 7. Received signal strength with obstacles

The simulation results for an indoor distributed massive MIMO system are obtained. On the basis of received signal strength and time of arrival, the performance is evaluated. Figures 6 and 7 show the received signal strength with and without obstacles between the base station and user equipment. Figure 7 depicts the variations in received signal due to obstacles. At some instance, the reflected signal happens to show good signal strength. Thus localization based on this parameter has to be taken into special consideration.

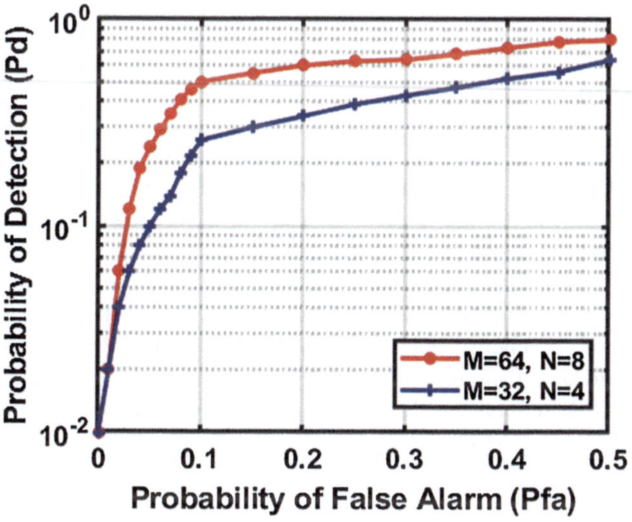

Fig. 8. Energy detection Pd vs Pfa for identification of LOS and NLOS

Figure 8 shows the energy detection for hybrid RSS-TOA algorithm. The probability of detection (Pd) and probability of false alarm (Pfa) is plotted. The probability of detection increases with the probability of false alarm. The probability of detection achieves 0.80 when the probability of false alarm is around 0.5.

Simulation result shows that 0.80 is achieved in identifying line of sight conditions. Figure 9 shows root mean square error performance for average received power and it is compared with cramer rao bound. The proposed hybrid RSS-TOA technique result shows that it performs equivalent to Cramer Rao Bound and provides a better classification in LOS and NLOS conditions.

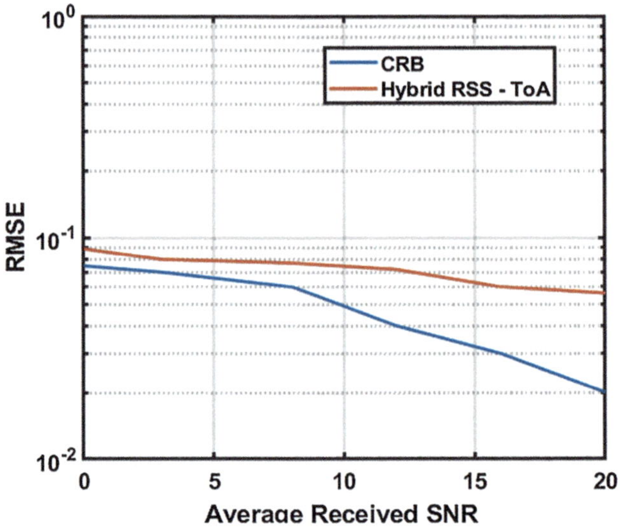

Fig. 9. RMSE performance vs average received SNR

6 Conclusion

In an indoor distributed massive MIMO system operating at mm-wave frequencies happens to undergo line of sight and non-line of sight conditions. To achieve a better quality of service, it is a must to localize users in a network. In order to increase the possibility of attaining a perfect classification among users, a hybrid technique namely RSS-TOA is utilized. However, with obstacles between base station and user equipment, the propagating signals undergo degradation. To normalize and to achieve a better localization an energy detector is employed. A threshold is chosen to select signals with good signal strength among reflections and to select the first arriving signal among the available reflections. Simulation results show that 0.80 accuracy is achieved in classifying LOS and NLOS users.

References

1. Garcia, N., Wymeersch, H., Larsson, E., Haimovich, A., Coulon, M.: Direct localization for massive MIMO. IEEE Trans. Signal Process. **65**(10), 2475–2487 (2017)
2. Sun, X., Gao, X., Li, G.Y., Han, W.: Fingerprint based single-site localization for massive MIMO-OFDM Systems. In: IEEE Global Communications Conference, GLOBECOM 2017, pp. 1–7 (2017)
3. Vieira, J., Leitinger, E., Sarajlic, M., Li, X., Tufvesson, F.: Deep convolutional neural networks for massive MIMO fingerprint-based positioning. In: IEEE International Symposium on Personal, Indoor and Mobile Radio Communications, pp 1–6 (2017)
4. Mendrzik, R., Wymeersch, H., Bauch, G., Abu-Shaban, Z.: Harnessing NLOS components for position and orientation estimation in 5G mmWave MIMO. arXiv preprint arXiv:1712.01445 (2017)
5. Arnold, M., Hoydis, J., ten Brink, S.: Novel massive MIMO channel sounding data applied to deep learning-based indoor positioning. arXiv preprint arXiv:1810.04126 (2018)
6. Mendrzik, R., Wymeersch, H., Bauch, G.: Joint localization and mapping through millimeter wave MIMO in 5G systems-extended version. arXiv preprint arXiv:1804.04417 (2018)
7. Hu, B., Wang, Y., Shi, Z.: Simultaneous position and reflector estimation (SPRE) by single base-station. In: IEEE Wireless Communications and Networking Conference (WCNC) (2018)
8. Prasad, K.N.R.S.V., Hossain, E., Bhargava, V.K., Mallick, S.: Analytical approximation-based machine learning methods for user positioning in distributed massive MIMO. IEEE Access **6**, 18431–18452 (2018)
9. Zeng, T., Chang, Y., Zhang, Q., Hu, M., Li, J.: CNN based LOS/NLOS identification in 3D massive MIMO systems. IEEE Commun. Lett. **22**, 1–4 (2018)
10. Decurninge, A., Ordóñez, L.G., Ferrand, P., Gaoning, H., Bojie, L., Wei, Z., Guillaud, M.: CSI-based outdoor localization for massive MIMO: experiments with a learning approach. arXiv preprint arXiv:1806.07447 (2018)
11. Arnold, M., Dörner, S., Cammerer, S., Brink, S.T.: On deep learning-based massive MIMO indoor user localization. arXiv preprint arXiv:1804.04826 (2018)
12. Kumar, D., Saloranta, J., Destino, G., Tölli, A.: On trade-off between 5G positioning and mmWave communication in a multi-user scenario. In: 8th International Conference on Localization and GNSS (ICL-GNSS), pp. 1–5 (2018)
13. Abu-Shaban, Z., Zhou, X., Abhayapala, T., Seco-Granados, G., Wymeersch, H.: Performance of location and orientation estimation in 5G mmWave systems: uplink vs downlink. In: Wireless Communications and Networking Conference (WCNC), pp. 1–6 (2018)
14. Abu-Shaban, Z., Wymeersch, H., Abhayapala, T., Seco-Granados, G.: Single-anchor two-way localization bounds for 5G mmWave systems: two protocols. arXiv preprint arXiv:1805.02319 (2018)
15. Shahmansoori, A., Uguen, B., Destino, G., Seco-Granados, G., Wymeersch, H.: Tracking position and orientation through millimeter wave lens MIMO in 5G systems. arXiv preprint arXiv:1809.06343 (2018)
16. Prakash, V.C., Nagarajan, G.: Indoor channel characterization with multiple hypothesis testing in massive MIMO. In: Innovative Technologies in Electronics, Information and Communication (INTELINC 2018) (2018)
17. Mailaender, L., Molev-Shteiman, A., Qi, X.-F.: Direct positioning with channel database assistance. In: IEEE International Conference on Communications (ICC Workshops), pp. 1–6 (2018)

18. Li, X., Cai, X., Hei, Y., Yuan, R.: NLOS identification and mitigation based on channel state information for indoor WiFi localization. IET Commun. **11**(4), 531–537 (2016)
19. Chen, C., Chen, Y., Han, Y., Lai, H.-Q., Liu, K.J.R.: Achieving centimeter-accuracy indoor localization on WiFi platforms: a frequency hopping approach. IEEE Internet Things J. **4**(1), 111–121 (2017)
20. Sung, C.K., de Hoog, F., Chen, Z., Cheng, P., Popescu, D.C.: Interference mitigation based on bayesian compressive sensing for wireless localization systems in unlicensed band. IEEE Trans. Veh. Technol. **66**(8), 7038–7049 (2017)

Low Power Device Synchronization Protocol for IPv6 over Low Power Wireless Personal Area Networks (6LoWPAN) in Internet of Things (IoT)

R. Rajesh[1], C. Annadurai[1(✉)], D. Ramkumar[1], I. Nelson[1], and I. Jayakaran Amalraj[2]

[1] ECE, Sri Sivasubramaniya Nadar College of Engineering, Chennai, India
annaduraic@ssn.edu.in
[2] Mathematics, Sri Sivasubramaniya Nadar College of Engineering, Chennai, India

Abstract. Massive growth in wireless devices and the need for interconnecting these devices results to form an Internet of Things (IoT). IoT applications can be easily implemented using an Ipv6 address based 6LoWPAN mesh network technology. 6LoWPAN MAC layer plays a compelling role in the economical usage of energy and resource consumption for low power wireless devices. We propose a new MAC protocol to improve the performance, including throughput and energy utilization using SCMAC algorithm in the MAC layer rather than orthodox CSMA with collision avoidance technique. The developed Suppressed Clear to Send MAC (SCMAC) protocol shows a convincing improvement in throughput and energy utilization of IPv6 based LoWPAN devices.

Keywords: Internet of Things · Media access control · 6LoWPAN · Throughput

1 Introduction

Internet of things (IoT) targets to connect the different digital devices with Internet to promote communications between virtual and physical things. Internet-of-Things (IoT) aims to create a smart world that provide more intelligence to the smart energy, smart health, smart transport, smart cities, smart industry, smart buildings, etc. Interconnecting millions of intelligent networks helps in access to information not only anytime and anywhere but also from anything and anyone ideally via any service and network. Exchange of application-dependent data between various standard wireless devices in an IoT give a challenge in communication adaptability among wireless devices which results in the need for a new protocol to overcome the challenge Quality of Service (QoS) creates more impact on providing effective and efficient data services for IoT applications [1].

In this paper, we concentrate on designing an energy efficient channel access protocol for the Internet of Things applications as energy constrained wireless sensors are used widely to send and receive data. As wireless sensors are small device with constrained power supply, once deployed in adverse or impractical conditions (e.g

D. J. Hemanth et al. (Eds.): COMET 2019, LNDECT 35, pp. 1371–1381, 2020.
https://doi.org/10.1007/978-3-030-32150-5_139

Mountains) it is not easy to change the batteries. Moreover, it should accomplish a large lifetime for about several years [2].

IPv6 over Low-power Wireless Personal Area Networks (6LoWPAN) is a networking technology that contributes an adaptation layer to transmit IPv6 packets above IEEE 802.15.4 links. 6LoWPAN is designed by Internet engineering task force (IETF) which operates in 2.4 GHz open free frequency bandwidth for connecting digital devices in low power lossy networks [3].

MAC protocol follows set of rules that users of networks must obey to access shared communication channel in order to utilize the scared resource efficiently. MAC layer of IEEE 802.15.4 provides two methods of channel access technique, namely slotted CSMA/CA and unslotted CSMA/CA in which later is used in our system. Low power Nodes in the network sense the medium at regular interval to access the communication channel. When the channel is free node can communicate with other node else it holds for arbitrary backoff time (sleep sate). The probability of choosing backoff period increases as more nodes complete the channel which results in significant energy utilization of Low Power Networks [4]. The major contribution of this research paper is concentrated on designing a MAC protocol for a 6LoWPAN network to achieve high network throughput and less energy utilization based on the developed SCMAC algorithm.

The rest of this paper is organized in the following manner. Literature survey is addressed in Sect. 2. In Sect. 3, the proposed Suppressed Clear to Send MAC (SCMAC) protocol is explained in detail. The validity of our proposed protocol is verified through Cooja simulator in Sect. 4. Finally, the paper concludes with future works in Sect. 5.

2 Related Works

The Internet of Things has described the ever-growing network of internet-connected physical devices that uses an IP address for internet connection and communication with smart objects. The base technology, for IoT can be considered as wireless sensor networks (WSNs) in which smart sensors interconnected to sense and monitor various applications like smart home, health care, smart city, farming, etc. [5]. Low Rate-WPAN standard defines specifications of the PHY layer and MAC sub-layer in 6LoWPAN networks [6]. Actually, the authors focused more to evaluate performance results of the IEEE 802.15.4 MAC protocol in relation with Goodput and energy utilization. In [7] consider exponentially dispersed packet creation times and they don't assume concurrent transmission effort by all other low power nodes after sleeping period. Under this situation, they don't find any MAC unreachability problem and reliability.

In [8], the authors developed the hybrid channel access protocol, for wireless sensor network to adapt to the level of contention. But, during high contention it acts like TDMA and under lower contention it acts like CSMA. As in [9], they introduced some modification in Media Access Control protocol to rectify congestions after an inactive period by creating a random delay ahead of medium access. This issue can be solved by setting the parameters of the CSMA/CA protocol appropriately, without any changes in the standard CSMA protocol. In [10], the authors have analyzed the performance of beacon enabled CSMA/CA protocol using the Markov chain model. These Markov chain model are interested in computing the throughput and delay, but failed to address the impact of the random backoff exponent and order of super frame. The authors [11] addressed MAC unreliability problem which is high due to packet drop probability, specifically for the massive amount of wireless sensor nodes and large packet ranges. Moreover, they did not recommend any feasible solution to solve this problem.

Bertocco et al. [12] have been investigating the effects of outside interferences introduced by other low power devices and machines to the performance of an LR-WPAN networks. However, they consider polling-based protocol for regular data acquisition from various sensors and they did not exactly consider problem regarding the IEEE 802.15.4 MAC protocol.

Both [13] and [14] analyzed, in saturated traffic situations performance of MAC protocol depends on a large number of packets to transmit by sensor nodes. Under these assumptions, the probability of packet drop is very high. In [15] authors identified the major sources of energy consumption as packet collisions, overhearing, frame overhead, and node idle listening. Zhai et al. [16] introduced a media access control (MAC) protocol that concentrate on the opportunity based scheduling of data on the best channel conditions at a wireless node to its next hop neighbors.

Sand [17] introduced a distributed scheduling algorithm, in which central coordinator is not necessary for the negotiation of time slot with immediate nodes. IPV6 over LRWPAN standardize the packet format and start negotiation with all nearby nodes which reduces the unification of low power wireless devices in IOT applications. A distributed cluster based algorithm [18] uses hop distance with respect to sink node to identify convenient cluster sizes, which helps in increasing the lifetime of the network and decreases the energy consumption [19]. An energy-aware routing protocol is developed to communicate within the clusters by clubbing the sensor nodes of uneven sized clusters.

The hybrid MAC protocols [20–22] in order to reduce the collision probability they mixed CSMA and Time Division Multiple Access techniques. However, achieving QoS and scalability in IoT is an issue in these techniques. In order to use the wireless channel the traditional IEEE 802.15.4 MAC [23] use CSMA/CA protocol, but low–duty and low-rate cycle technique not able to support energy efficient solution for various Internet of Things applications. Wake and Sleep based scheduling method was developed in SMAC [24] to reduce energy usage during idle time to increase energy efficiency in conventional 802.15.4 MAC protocol.

A MAC protocol with the decision rule in backoff timer has been proposed in [25] for a Machine to Machine applications having various clustered low power nodes. For an industrial Internet of Things application [26] has proposed a mathematical model based queuing theory for the guaranteed time slot and medium access delay of 802.15.4 MAC protocol.

3 System Model

We assume a 6LoWPAN network in Fig. 1 in which each low power node is assigned to a unique IPv6 address and can transmit data to all the alive nodes in the network. This 6LoWPAN network works only in an unslotted CSMA/CA or beaconless mode. In the beaconless mode, synchronization frames are not transmitted by PAN coordinator, hence synchronization and idle listening of all the low power nodes is not possible.

The low power nodes in 6LoWPAN network act as host or PAN coordinator along with one or more border routers. The Border Router and Coordinator share the IPv6 prefix throughout the 6LoWPAN network using network interfaces to all active nodes. Node addressing, supported channel allocation, and operation mode functions are specified by the PAN Coordinator of 6LoWPAN network. Host interacts with border router using Neighbor Discovery (ND) protocol by initially registering its address with a border router in order to make dynamic movement of host in the network. Neighbor Discovery protocol controls bootstrapping in which low power nodes and actuators get connected to a 6LOWPAN network by the auto configuration process. Bootstrapping specifies the procedure for node communication in the network, how routes are created to transmit data from the active nodes to the border router. Low power Nodes are free to move all through the 6LoWPAN network, between edge routers, and even between various 6LoWPANs supporting a multi-hop mesh topology.

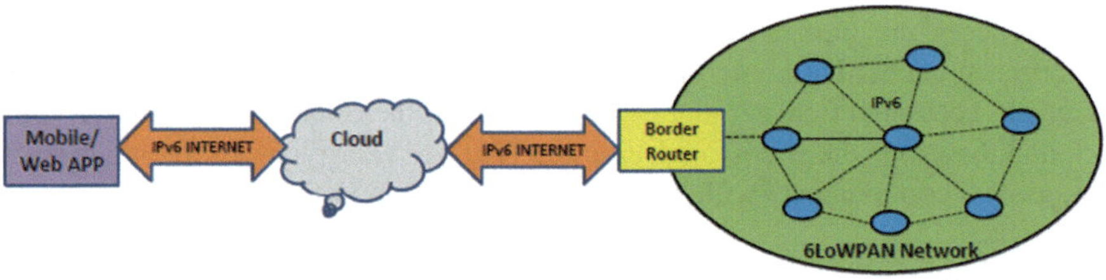

Fig. 1. A 6LoWPAN network

An important feature of 6LoWPAN is adaptation layer which is located between data link layer and network layer and it is responsible for fragmentation of ipv6 packets and compression of large sized headers. It carries large sized IPv6 packets (1280 bytes) over the 802.15.4 frames (128 bytes). These large sized IPv6 packets are transmitted by 802.15.4 frames as data payload. Stateless Address Auto configuration (SAA) technique reduces the configuration overhead of the low power nodes. The address auto configuration process includes creation of global prefix and link local prefix for nodes in the network and also carries out Duplicate Address Detection (DAD) procedure to validate the uniqueness of addresses on a networked nodes. Low power nodes in 6LoWPAN network need to follow the below action to access the channel.

Let us consider 5 nodes wants to access the wireless channel, while other inactive nodes stay in an idle mode in order to increase energy efficiency. Synchronization of nodes in a 6lowpan network cannot be over emphasized and synchronization mechanism is critical to the performance of the proposed algorithm.

Action1: Transmit RTS Message
When a low power node wants to use the channel, it transmits an RTS message to all the other alive devices in the 6LoWPAN network ahead to access the channel.

Let us assume node (N3) wants to use the channel, it transmits a RTS message to all other low power nodes {N1, N2, N4 and N5}. The broadcast message to all nodes consists of the Request to Send message and the time stamp of the RTS message generated by the physical clocks present in each 6LoWPAN node is shown in Fig. 2.

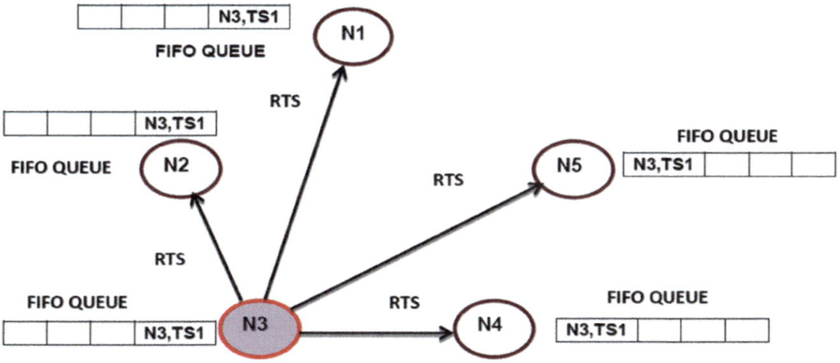

Fig. 2. RTS message from N3 with timestamp

Action 2: Reply CTS Message
After receiving RTS broadcast message from N3 all the alive nodes N1, N2, N4 and N5 send the CTS message to N3. Consider N5 want to use the channel, it transmits RTS message to all the alive nodes {N1, N2, N3 and N4} along with timestamp. All the alive nodes maintain queue to store the timestamp of the channel request node from the RTS message in the ascending order. In Fig. 3. Node (N3) request take top position in queue, so node (N3) access the channel.

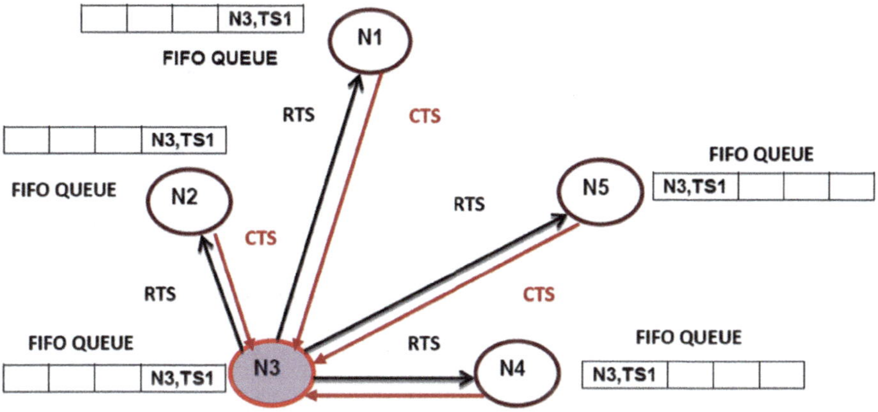

Fig. 3. CTS message from all active nodes

Action 3: Releasing the Channel

Node (N3) after utilizing the channel for data transmission it broadcasts RELEASE token to all other alive nodes in the network shown in the below Fig. 4.

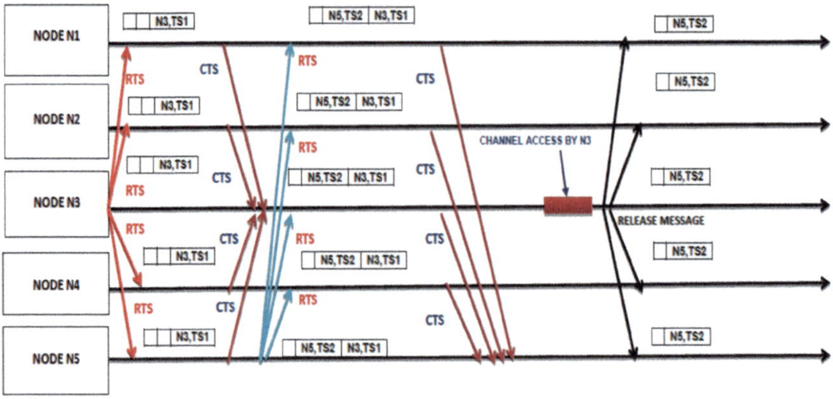

Fig. 4. Channel access by node N3 without CTS message suppression.

The values in the queue are rearranged after RELEASE message, hence node N5 takes top position of the queue to access the channel. Suppose node N5 wants to access the channel again, it is not necessary to broadcast RTS message to all other active nodes, since no other node is in request queue it use the channel once again. As each node maintains the queue with timestamp the CTS messages are suppressed with all nodes shown in Fig. 5 improves the performance of the proposed algorithm by reducing control overhead.

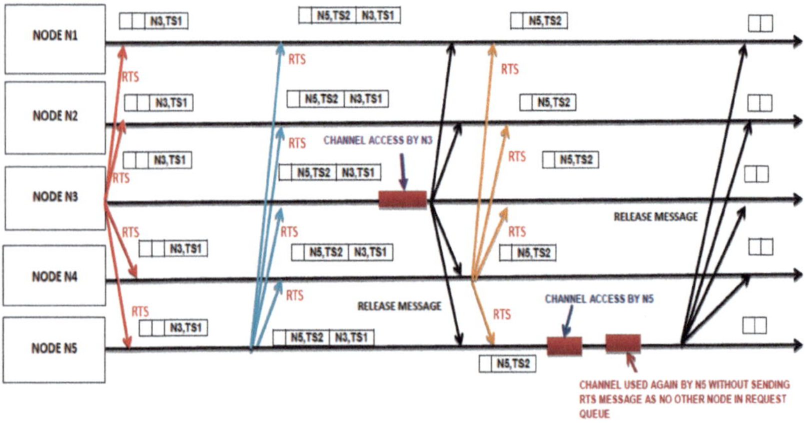

Fig. 5. Channel access by Node (N3) and Node (N5) with suppressed CTS message

4 Performance Analysis

6LoWPAN network considers following parameters for the performance analysis.

Throughput: The rate at which the nodes execute requests for the channel access.

$$\text{Throughput}(\chi) = 1/(cd + ct) \quad (1)$$

cd is the coordination delay and ct is the average channel utilization time.

Coordination delay (cd): Time taken between last nodes exit the channel to a new node enters the channel. As a result it creates impact on coordination delay which makes it very low.

Average channel utilization time (ct): Time taken by an accessing node to use the channel for data transmission.

Energy consumption computation:

Energy consumption plays significant role in the 6LoWPAN network as all the active nodes work in the battery constrained environment. Energy is utilized by an active node at the time of channel access, Transmitting/Receiving Request To Send (RTS) and RELEASE messages.

The following equation represents the energy utilization of an active node.

$$\begin{aligned} &\text{Energy Consumption } (\eta) = \\ &\Phi(Tc)[\text{Energy consumed during accessing channel}] + \\ &\Phi(2(N-1))[\text{Energy consumed by control messages}] \end{aligned} \quad (2)$$

5 Simulation Analysis

We used Cooja Simulation Software for simulation analysis presented in this paper. To develop an end user applications Cooja supports C language with Java Native Interface. To simulate the user application in synchronous with high level algorithm and hard driver design is the main benefit of Cooja simulator.

In this section, we do simulation analysis for the following parameters like throughput, energy utilization and average transmission delay of the proposed SCMAC (Suppressed Clear to Send MAC) protocol in 6LoWPAN network. Meanwhile, the performance comparisons of the proposed protocol with conventional CSMA/CA are provided. Summarily, Table 1. depicts the simulation parameters used for Cooja simulator.

Table 1. Parameters used in Cooja simulatior.

Parameter	Value
Radio medium	Unit Disk Graph Medium
Mote type/Startup delay	T-mote Sky/1000 ms
MAC layer	SCMAC
Bit rate	250 kbps
Radio duty cycling	Null RDC
Transmission range	50 m
Node sensing range	100 m
Transmit/Receive ratio	100

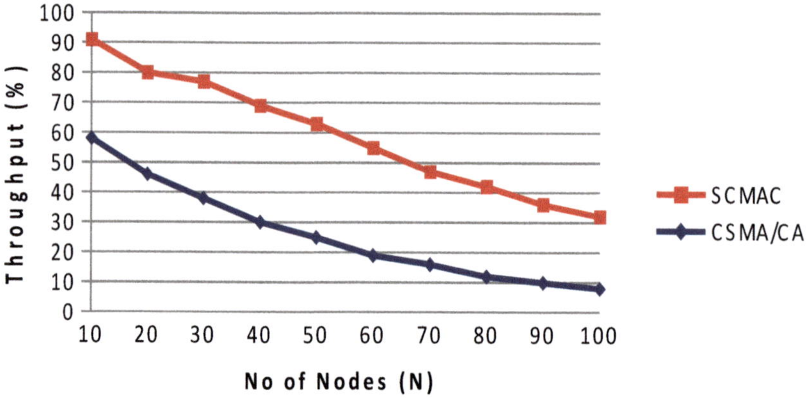

Fig. 6. Aggregate throughput for number of nodes.

Figure 6 depicts the simulation analysis for throughput in which elimination of backoff mechanism in our proposed MAC protocol increases the network throughput because less time is utilized during the channel access due to synchronization of all the active nodes in the network. Hence, SCMAC protocol proves that it consumes less bandwidth. Thus, comparing both protocol it can be noticed that the network throughput with the new protocol outperforms the CSMA case.

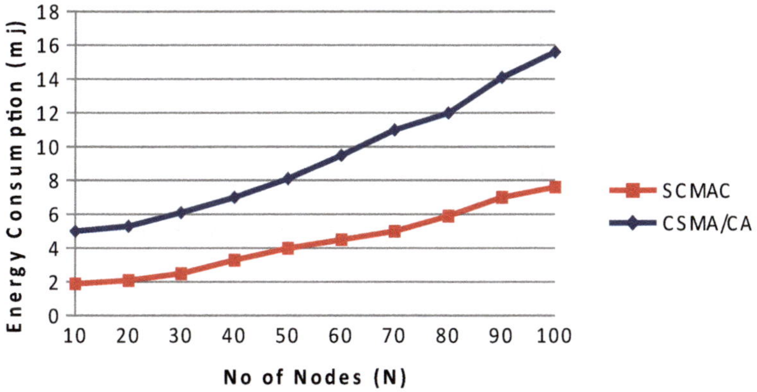

Fig. 7. Energy consumption for different number of nodes.

Figure 7 presents the energy consumption results, as it can be seen that our proposed MAC protocol utilize less energy when compared with CSMA case by stopping idle listening of all the wireless nodes.

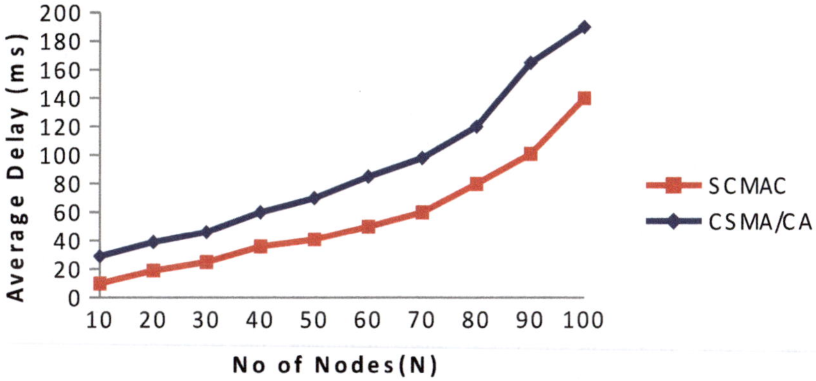

Fig. 8. Average delay with respect to number of nodes

Figure 8 presents the analysis for average system delay with respect to number of nodes. Data packets aggregation and network congestion creates more impact on the sensor network which results in increase of average system delay as the number of low power node increases.

6 Conclusion

In 6LoWPAN networks, CSMA protocol uses back off time period with more control packets for media access which results in high energy consumption and delay in Internet of Things. Suppression of few control packets and back off time period in channel access mechanism can significantly reduce the energy utilization and delay of IoT devices with increased throughput. The proposed system utilizes minimum number of control messages for channel contention. Hence, during the time of channel access only an accessing low power node works in operative mode all other nodes enters into hibernate mode which results in economic usage of energy by nodes in Internet of Thing applications. The simulation experiment results show that SCMAC algorithm outperforms better than traditional CSMA protocol especially in the network throughput, as well as reducing MAC delay and the network energy consumption.

References

1. Srivastava, N.: Challenges of next-generation wireless sensor networks and its impact on society. J. Telecommun. **1**(1), 128–133 (2010)
2. Anastasi, G., Conti, M., Di Francesco, M., Passarella, A.: Energy conservation in wireless sensor networks: a survey. Ad Hoc Netw. **7**, 537–568 (2009)
3. Montenegro, G., Kushalnagar, N., Hui, J., Culler, D.: IPv6 over low power wireless personal area networks (6LowPAN). Technical report, The Internet Engineering Task Force (IETF) (2007)
4. Misic, J., Shairmina Shafi, K.R.: The impact of MAC parameters on the performance of 802.15.4 PAN (2005). https://doi.org/10.1016/j.adhoc.2004.08.002
5. Tan, L., Wang, N.: Future internet-the Internet of Things. In: Proceedings of 3rd International Conference on Advanced Computer Theory and Engineering (ICACTE), Chengdu, China, pp. 376–380 (2010)
6. IEEE Std 802.14.3-2006, September, Part 15.4: Wireless Medium Access Control (MAC) and Physical Layer (PHY) Specifications for Low-Rate Wireless Personal Area Networks (WPANs) (2012)
7. Mišic, J., Shafi, S., Mišic, V.B.: The impact of MAC parameters on the performance of 802.15.4 PAN. Ad Hoc Netw. **3**, 509–528 (2005)
8. Rhee, I., Warrier, A., Aia, M., Min, J., Sichitiu, M.L.: Z-MAC: a hybrid MAC for wireless sensor networks. IEEE Trans. Netw. **16**, 511–524 (2008)
9. Yedavalli, K., Krishnamachari, B.: Enhancement of the IEEE 802.15.4 MAC protocol for scalable data collection in dense sensor networks. In: Proceedings of International Symposium on Modeling and Optimization in Mobile, Ad Hoc, and Wireless Networks (WiOpt 2008), Berlin, Germany (2008)
10. Park, T.R., Kim, T.H., Choi, J.Y., Choi, S., Kwon, W.H.: Throughput and energy consumption analysis of IEEE 802.15.4 slotted CSMA/CA. IEEE Electron. Lett. **41**(18), 1017–1019 (2005)
11. Shu, F., Sakurai, T., Zukerman, M., Vu, H.L.: Packet loss analysis of the IEEE 802.15.4 MAC without acknowledgment. IEEE Commun. Lett. **11**(1), 79–81 (2007)
12. Bertocco, M., Gamba, G., Sona, A., Vitturi, S.: Experimental characterization of wireless sensor networks for industrial applications. IEEE Trans. Instrum. Meas. **57**(8), 1537–1546 (2008)
13. Singh, C.K., Kumar, A., Ameer, P.M.: Performance evaluation of an IEEE 802.15.4 sensor network with a star topology. Wirel. Netw. **14**(4), 543–568 (2008)
14. Pollin, S., Ergen, M., Ergen, S., Bougard, B., Van der Perre, L., Moerman, I., Bahai, A., Catthoor, F.: Performance analysis of slotted carrier sense IEEE 802.15.4 medium access. IEEE Trans. Wirel. Commun. **7**(9), 3359–3371 (2009)

15. Ye, W., Heidemann, J., Estrin, D.: An energy-efficient MAC protocol for wireless sensor networks. In: Proceedings of IEEE Infocom, pp. 1567–1576 (2002)
16. Wang, J., Zhai, H., Fang, Y., Yuang, M.C.: Opportunistic media access control and rate adaptation for wireless ad hoc networks. In: Proceedings of IEEE International Conference on Communication, Paris, France (2004)
17. Sudhaakar, R., Zand, P.: 6TiSCH resource management and interaction using CoAP. Internet-Draft [work-in-progress], IETF Std., Rev. draft-ietf- 6tisch-coap-00 (2014)
18. Wu, D., Bao, L., Regan, A., Talcott, C.: Large-scale access scheduling in wireless mesh networks using social centrality. J. Parallel Distrib. Comput. **73**, 1049–1065 (2013)
19. Wei, D., Jin, Y., Vural, S., Moessner, K., Tafazolli, R.: An energy efficient clustering solution for wireless sensor networks. IEEE Trans. Wirel. Commun. **10**, 3973–3983 (2011)
20. Zhuo, S., Song, Y.-Q., Wang, Z., Wang, Z.: Queue-MAC: a queue length aware hybrid CSMA/TDMA MAC protocol for providing dynamic adaptation to traffic and duty-cycle variation in wireless sensor networks. In: Factory Communication Systems (WFCS), pp. 105–114. IEEE (2012)
21. Zhuo, S., Wang, Z., Song, Y.Q., Wang, Z., Almeida, L.: A traffic adaptive multi-channel MAC protocol with dynamic slot allocation for WSNs. IEEE Trans. Mob. Comput. **15**, 1600–1613 (2016)
22. IEEE Draft Standard for Information Technology-Telecommunications and Information Exchange Between Systems-Local and Metropolitan Area Networks-Specific Requirements-Part 11, IEEE P802.11ah/D6.0, (Amendment to IEEE Std 802.11REVmc/D5.0), pp. 1–645 (2016)
23. Montenegro, G., Kushalnagar, N., Hui, J., Culler, D.: Transmission of IPv6 packets over IEEE 802.15. 4 networks. Technical report (2007)
24. Ye, W., Heidemann, J., Estrin, D.: An energy-efficient MAC protocol for wireless sensor networks. In: International Conference on Computer Communications (INFOCOM), vol. 3, pp. 1567–1576. IEEE (2002)
25. Park, I., Kim, D., Har, D.: MAC achieving low latency and energy efficiency in hierarchical M2 M networks with clustered nodes. IEEE Sens. J. **15**(3), 1657–1661 (2015)
26. Yan, H., Zhang, Y., Pang, Z., Xu, L.D.: Superframe planning and access latency of slotted MAC for industrial WSN in IoT environment. IEEE Trans. Ind. Inf. **10**, 1242–1251 (2014)

Crowd Sourcing Application for Chennai Flood 2015

R. Subhashini[1], Mary Subaja Christo[2(✉)], G. Parthasarathy[3], and J. Jeya Rathinam[3]

[1] Information Technology, Sathyabama Institute of Science and Technology, Chennai, India
[2] CSE Department, Saveetha School of Engineering, Thiruvallur, India
marysubaja@gmail.com
[3] CSE Department, Jeppiaar Maamallan Engineering College, Chennai, India

Abstract. Presently social media plays a vital role in people's life. During the Chennai floods of November–December 2015, victims and also the relief centers have used the social media for sharing the information regarding the disaster. The crowd sourced data which speeds up disaster management actions such as rescue and relief services to the victims. Though the social media can provide effective disaster relief services, it does not provide an essential coordination for sharing information, resources, and plans among distinct relief organizations. However, proposed open source crowdsourcing platform overcomes this issue by offering a powerful capability for collecting information from disaster scenes. It also visualizes the interactive map which was used to crowdsource the information and generates reports as well. The reports and the database generated are helpful to the government and NGO's for the preparedness of the disaster in future i.e., for relief decision making. The article describes the usage of open source crowd sourcing platform for Chennai flood disaster, 2015.

Keywords: Chennai Flood-2015 · Disaster management · Disaster relief system · Open source crowd sourcing

1 Introduction

Disaster is a natural catastrophe or a sudden accident that causes heavy economic, environmental and life loss [5]. Disaster management which covers various functions such as prevention, preparation, response and recovery mainly depends upon the data collected from public. These days, advances in technologies made it easy to gather the data. Crowd sourcing [3, 4] is the process of getting information from a large group of people through web based applications. The data collected by the open source crowd sourced application from victims and relief centers can be used to visualize the data. Twitter, Google Map Maker (GMM), Open Street Map (OSM) and Ushahidi are some of the widely used open source licenses for crowd sourcing. Ushahidi [2] is an open source platform that can be used to integrate data from multiple sources such as emails, messages, social media such as Twitter, Facebook etc., [1].

D. J. Hemanth et al. (Eds.): COMET 2019, LNDECT 35, pp. 1382–1388, 2020.
https://doi.org/10.1007/978-3-030-32150-5_140

During Chennai Flood 2015, people used social media sites such as Facebook, Twitter, blogs, Flickr and YouTube to publish their personal experiences, texts and photos. Due to the jammed mobile network, people used social media sites to communicate with each other. People also used hash tags to mark their messages related to disaster in Twitter and Facebook. Twitter provides natural language reports that can be used to extract formal aspects such as time, location and tags. Rather other tools provide structured map based information. The social media and crisis maps do not provide a common mechanism for allocating response resources, so multiple organizations might respond to an individual request at the same time. There is no common coordination and cooperation among the different relief centers. The proposed crowd sourcing application collects data from various sources such as social media, email, message etc., and also visualizes the interactive map for relief decision making.

1.1 Review of Literature

In Disaster Management and relief receives a lot of increased attentions from various disciplines such as Computer Science, Environmental sciences and Health Sciences. Hristidis [7] makes an effort to survey and organize the current knowledge in the management and analysis of data in disaster situations. They organized the findings across the various Computer Science disciplines: data integration and ingestion, information extraction, information retrieval, information filtering, data mining and decision support. Gao and Barbier [6] described the advantages and disadvantages of crowdsourcing applications applied to disaster relief coordination. They also discussed some of the challenges that must be addressed to make crowdsourcing a useful tool so that it can effectively enables the relief progress in coordination, accuracy, and security.

Social media and online mapping tools have sophisticate the easy access of current disaster reporting. The study [4] suggested that semantic information analysis could potentially improve the spatial data attribute matching and further the quality of the results can be improved [8, 9]. The study also recognized several necessities to be fulfilled during modern disaster management activities. They are ensure some location verification process during data collection, consider terminology utilized in the reports being received, examines the completion of the data specially in locational attributes, and differentiate the location of the person Vs. location of the incident.

2 Proposed Approach

In Nov–Dec 2015, the state of Chennai, Tamilnadu experienced one of the largest disaster events in its history. Most of the main streets in Chennai were waterlogged, bringing the city to a standstill. More than 35 lakes were flowing at dangerous levels, which caused more floods in Chennai as surplus water is flowing in the city. Government officials said around 10,000 people had been evacuated from their homes in Chennai. It was estimated that the floods in Chennai was resulted in a financial loss of about Rs. 15,000 crore.

The proposed open source crowd sourcing application supports disaster rescue and recovery operations during or after any disaster, effective communication amongst the diverse rescue workers and survivors. During and at the post flood event, the proposed system maintained an interactive map to gather information related to the Chennai floods 2015. Most reports were made through the online interface, however a small percentage of reports were made via Email, Twitter, Mobile App and through SMS (Fig. 1).

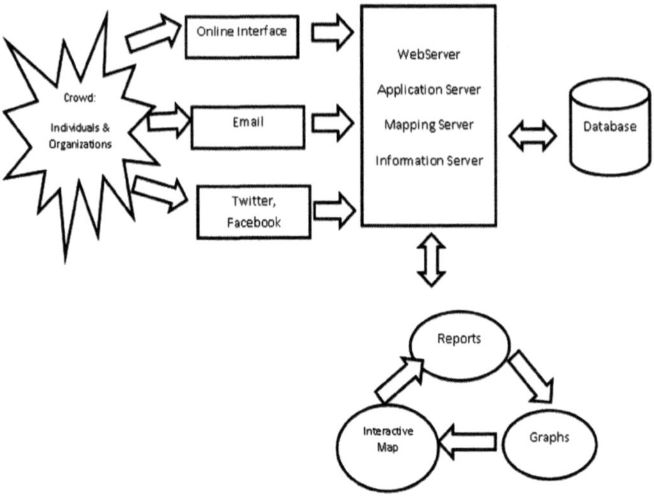

Fig. 1. Arcitecture of crowd sourcing platform

2.1 User Interface

The Proposed application enables the Govt. and other related authorities to monitor the disaster situation in near real time from anywhere. Both the survivors and relief workers can use the interface to upload their information either the text, video or audio along with the location and the corresponding category. The various categories created by the administrator are "People Missing", "Food Needed", "Water logging complaints", "Food and Medicine Services" and "Boat Rescue Services". The Platform uses the incident title, date, location and category to describe a report. These reports were filtered and extracted for further analysis.

Furthermore, the platform publishes the visualized interactive maps of the reports which includes the location information where it originated. The maps can also be filtered according to the category. The Dashboard is also used to view the reports, statistical graphs, to get alerts and to submit reports.

2.2 Administrator Interface

The service layer is designed with a user management module which provides its users with a hierarchical system of access rights and an easy to operate interface. The administrator can only verify and approve the reports submitted by the user. After the approval it gets visualized in the interactive map. The software also generates the reports available as CSV file and as an RSS feed. These reports can be downloaded and used for decision making by the government. The administrator is having the privileges to create the report, upload the report, delete the report and to view the report. The Addons such as messages, Facebook and Twitter can also be integrated with the web application. The Administrator can perform tasks such as data manipulation and analysis for disaster preparedness and prevention.

2.3 Applications Software at the Cell-Phone

Application also runs on the relief worker's and survivors cell-phone.

- Used to upload information on the InfoStation (Server) through a proper GUI
- Used to view the reports, graphs and interactive maps from the Info Station using the same cell phone through proper GUI
- Used to get alerts from the Info Station

In Fig. 2, the interactive map performs data management by filtering the data based upon the category "Rescue Team and camp location".

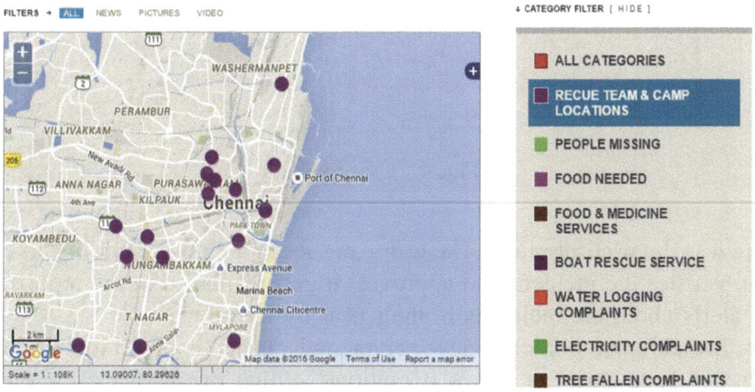

Fig. 2. Chennai map with high lighted locations of type "Rescue Team and camp location"

3 Results and Discussion

Figure 3 represents a statistical plot of "Rescue Team and camp location "category in December 2015.

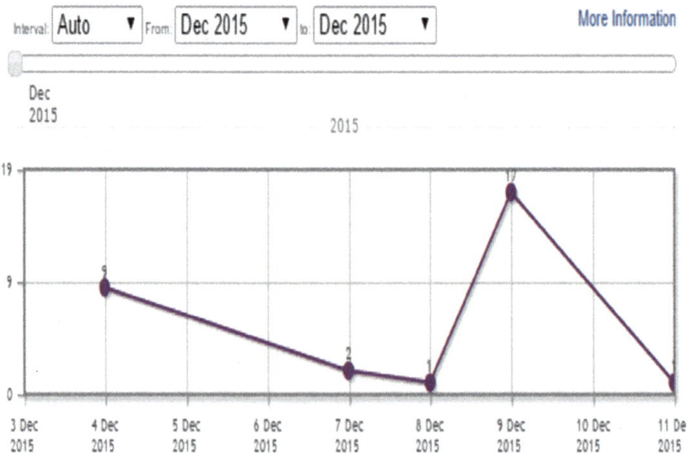

Fig. 3. Statistical plot of a type "Rescue Team and camp location"

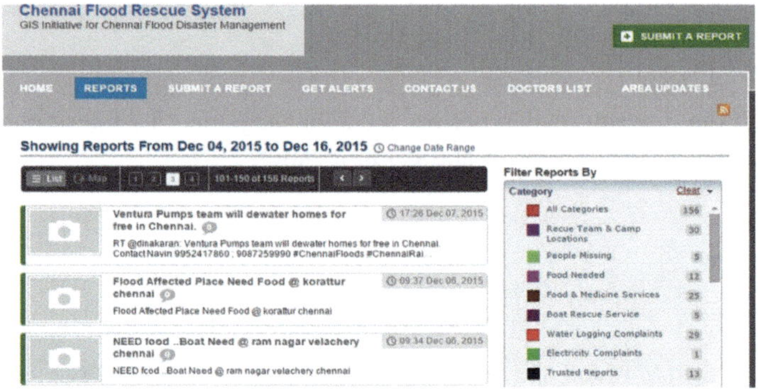

Fig. 4. User interface

Figure 4 is the user interface. The user can submit a report regarding an incident. If the admin verifies the incident and approves, it gets displayed in the reports. User can even get alerts about the incidents in their near by locations.

Only the administrator can view server and cluster configurations and perform administrative tasks such as freeze operations, offline operations etc.., Fig. 5 is the admin interface. Admin can view all the reports submitted by the users and approve or disapprove it after proper verification. Figure 6 visualises the pie chart of all the reports of various categories. This data can be used by a data scientist to analyse the situation and for immediate decision-making.

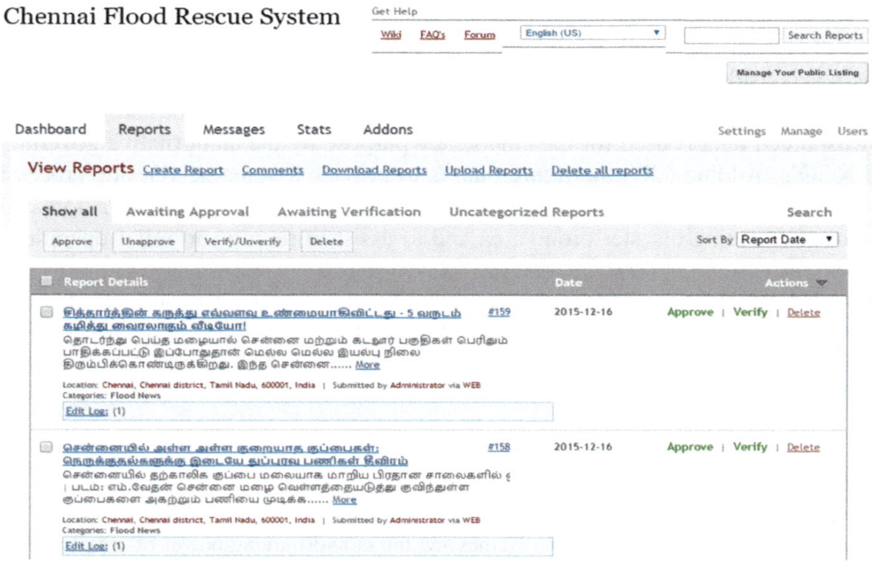

Fig. 5. Admin interface

Fig. 6. Pie chart

The proposed system provides the most effective data collection technology that gives authorities better visibility of available resources and need for decision making.

4 Conclusion

The Crowd sourcing application stores and shares the spatial and aspatial information in an integrated manner. It disseminates the details in a customized form for gathering information from the user and also visualizes it in the form of interactive maps, graphs and reports over the web. The open source platform is used for developing the application and users only require internet connection to use it. Hence the proposed system is considered as a low cost web enabled GIS solution for disaster Management. Since the crowd sourced data comes from the citizens, it is current and diverse. The crowdsourced data is used for the emergency purpose by the authoritative government and NGO's. Additionally, the future plan is to provide a better server side processing by using Big Data Analytics and Machine Learning Algorithms to totally automate the system of detecting disaster prone area and to assist in rescue and relief operation.

Acknowledgement. We wish to acknowledge the Department of Science and Technology, India and School of Computing, Sathyabama Institute of Science and Technology, Chennai for providing the facilities to do the research under the DST-FIST Grant Project No. SR/FST/ETI-364/2014.

References

1. Gao, H., Barbier, G., Goolsby, R.: Harnessing the crowdsourcing power of social media for disaster relief. IEEE Intell. Syst. **26**(3), 10–14 (2011). https://doi.org/10.1109/mis.2011.52
2. Morrow, N., Mock, N., Papendieck, A., Kocmich, N.: Independent evaluation of the ushahidihaiti project. Technical report, The UHP Independent Evaluation Team (2011)
3. Howe, J.: The rise of crowdsourcing. Wired, 14 June 2006. http://www.wired.com/wired/archive/14.06/crowds.html
4. Koswatte, S., McDougall, K., Liu, X.: SDI and crowdsourced spatial information management automation for disaster management. In: FIG Commission 3 Workshop 2014 Geospatial Crowdsourcing and VGI: Establishment of SDI & SIM Bologna, Italy, 4–7 November 2014
5. Mishra, A.K.: Monitoring Tamil Nadu flood of 2015 using satellite remote sensing. Nat. Hazards **82**(2), 1431–1434 (2016)
6. Gao, H., Barbier, G., Goolsby, R.: Harnessing the crowdsourcing power of social media for disaster relief. In: IEEE Intelligent Systems, vol. 26, no. 3, pp. 10–14, May–June 2011
7. Hristidis, V., et al.: Survey of data management and analysis in disaster situations. J. Syst. Softw. **83**, 1701–1714 (2010)
8. Auxilia, R., Gandhi, M.: Earthquake reporting system development by tweet analysis with approach earthquake alarm systems. Eur. J. Appl. Sci. **8**(3), 176–180 (2016)
9. Sethuraman, R., Sathish, E.: Intelligent transport planning system using GIS. Int. J. Appl. Eng. Res. **10**(3), 5887–5892 (2015)

Vehicle Monitoring and Accident Prevention System Using Internet of Things

G. Parthasarathy[3(✉)], Y. Justindhas[1], T. R. Soumya[1], L. Ramanathan[2], and A. AnigoMerjora[1]

[1] Department of CSE, Jeppiaar Maamallan Engineering College, Chennai, India
justindhasy@gmail.com, soumyatr.soumya@gmail.com, anigomerjora@gmail.com

[2] School of Computer Science and Engineering, VIT University, Vellore, India
lramanathan@vit.ac.in

[3] School of C&IT, REVA University, Bengaluru, India
amburgps@gmail.com

Abstract. Safety features are at the top of present day requirements in any automobiles. The lives of people driving around in different kinds of automobiles are the most important priority to every manufacturer and customer. Therefore there has been a rise in the automation and accident prevention mechanisms in the present day vehicles. Considerable effort is being put into the automation of vehicles that are in use. The most challenging part lies in the making of safety features available at affordable cost. An internet of things module implemented with the use of several sensors embedded in the system helps in the achievement of this facility. A system proposed in this article has been implemented and has proved effective results. The system consists of various individual models which have been combined r to form a hybrid system. It provides some high end safety features to the vehicle using it. The system allows the complete monitoring of the vehicle and also has a typical role in automation and accident prevention, thereby saving countless lives.

Keywords: Vehicle monitoring · Automation · Internet of Things · Sensor · Arduino · Buzzer

1 Introduction

Automation of vehicles has been on the rise. The demand for safety measures in automobile industry has been increasing day by day in accordance with the luxury of the vehicles. The safety of the driver and the passengers is given priority in designing vehicles. Vehicle manufacturers are quite interested in offering safety features of various ranges to buyers, but these are expensive. With this project intend to demonstrate various advanced features meant for protecting the passengers and the driver in any automobile. The work proposed is based on internet of things (IoT) which is an interconnection of various computing elements and putting them to use in the internet infrastructure. The internet of things module (IoT) employed to enables sensors to collect data and act quickly in situations of emergency. The principle of automation is well achieved with the use of IoT.

D. J. Hemanth et al. (Eds.): COMET 2019, LNDECT 35, pp. 1389–1398, 2020.
https://doi.org/10.1007/978-3-030-32150-5_141

The system proposed here employs various sensors which sense the air pressure in the wheels, fuel level, vibration of the vehicle, and do drunken driver detection. GPS and GSM modules are used for tracking the location of the vehicle and communicate with its neighbor in the case of an emergency. The system detects the presence of alcohol and immediately turns off the vehicle ignition system. Various test cases reported later have proved the efficacy of the system. The modules introduced in this system have seen individual implementation in the previous papers of the authors. But the proposed system uses all the modules in combination as one hybrid module bringing all the features into a single frame.

2 Related Works

The work can be classified into various individual systems. Which have been implemented already as the tyre pressure sensor, alcohol and eye blink sensor. All these systems were developed individually. The existing tyre pressure system employs a pressure sensor which is assigned a threshold value. When the pressure in the tyre rises above a certain level the LED is on and the buzzer alarm is activated, thereby enabling the driver to check the pressure of the tyres beforehand. All the other sensors are embedded individually in the same way. The sensor is placed on the bread board and, in turn, the board is connected to a microcontroller which triggers the alarm and LED. This is the mechanism used in the system (IoT). The existing methodologies use an Arduino. The entire control of the system is done using an Arduino microcontroller. Figure 1 is the block diagram of the existing methodologies.

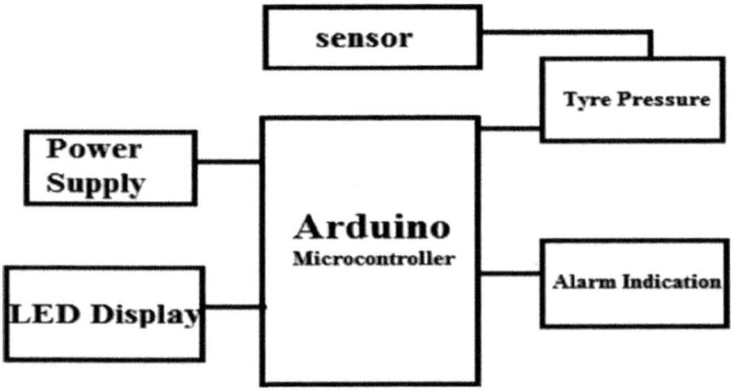

Fig. 1. Existing system module

The existing system is known for the limitation in applications. It has failed under certain conditions due to misfunctioning. For example, the tyre monitoring system just measures the pressure of the tyre. In case of a tyre burst the vehicle may lose its balance and start skidding with accident imminent. Under such circumstances, it is advisable to stop power supply to the vehicles to avoid an accident that is imminent. There is no special communicating device to enable communication with others for help after the accident. Such measures have been considered and taken into account while implementing this system (IoT).

A survey of literature relating to the system has been the relating part for various related topics. One such topic is based on the IoT module used for accident prevention and the tracking system for night drivers. The proposed work consists of an eye blink sensor and a system for monitoring the head movement. The system consists of a LCD screen that displays the psychological status of has an additional feature for tracking the vehicle and introducing an anti-theft mechanism which introduces GSM and a GPRS module. There also exist various other works which introduce GLONASS alongside GSM and GPRS modules. One such is related to the accident avoidance in addition to the detection system using a vibration sensor. Such work includes mechanical features such as ABS and SRS airbags. Drowsy driving is prevented using eye closure ratio that alert the driver with the help of buzzer [22] using pi camera with help of raspberry pi. Smart Vehicle over speeding detector is used to prevent the over speeding in the limits based on IOT technology which reduces death rate of accidents and it is alerted by alarm [21].

There are also similar works implemented using sensors and ultrasonic devices. Which use GSM and GPRS module along with ultrasonic devices which have reduced the member of accidents. All these systems have GPRS and GSM devices and also wireless hardware communication. The previous works have some relation with IoT but do not engage sensors which do complete monitoring of the vehicle. The existing systems do not have the facility for storage of data for further reference. The proposed system has a microcontroller with data storage capacity.

2.1 Gaps in Literature

A Comparison of proposed work with the existing works we can establish certain differences between them. They are:

The GPS and GSM modules introduced in the system are used for satellite services. But the privacy of such data from several anti-social activities has not been considered.

The existing works give importance to the notification of an accident after its occurrence rather than preventing it. The notification is not maintained for future reference.

Literature indicates existing works taking care only of malfunctions made by the human rather than the collective monitoring of both the human and vehicular malfunctions.

The study of IoT is on slow pace in the case of automation thereby causes complex implementations.

3 Proposed Work

The proposed system (IoT) makes some improvements to the existing work by embedding all the sensors on a single bread board and connecting them to a Raspberry Pi microcontroller. A major part of the proposed work is the employment of the GSM and the GPS modules for locating the vehicle and communication with vehicles nearby or ambulances and hospitals in the case of accidents, thereby saving lives.

Further alcohol and eye blink sensors used in this system are new and innovative vehicle safety features. Therefore all the sensors are taken care of in the implementation of this system. The internet infrastructure makes a complete coordination with communication.

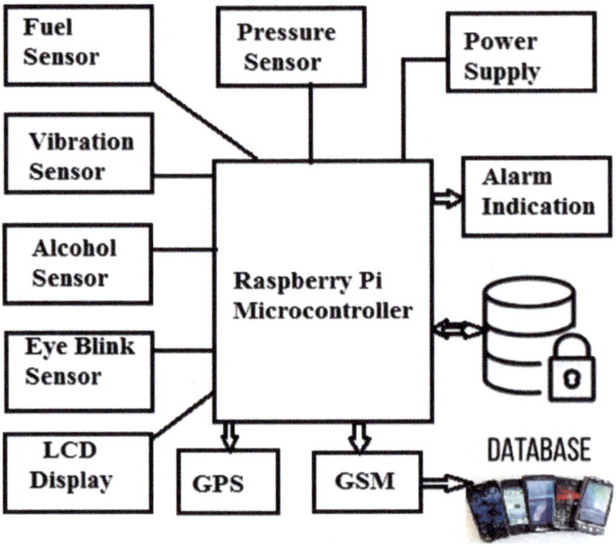

Fig. 2. Architecture of proposed system

Figure 2 depicts the system architecture of the proposed system. Which has a data storage slot for storing the data relating to the vehicle, data observed from the different sensors and the information relating to the drivers. Figure 3 relates to the implementation of the system. The GSM module put into use requires a working SIM card to enable communication in emergency. The customized user interface is enables using HTML and NET for providing reference to the previous record of the malfunctions and the accidents occurred.

Fig. 3. Implementation of proposed System

3.1 Eye Blink Sensor

The proposed work makes effort to locate the flickering of the eye, as it is utilized in driving the gadget. Flick discovery assumes importance in identifying squint eyed in showcase (Catalog No. 9008 of Enable gadgets)

3.2 Alcohol Sensor

The MQ-3 sensor detects the liquor content from human breath and directs it to aurdino. The MQ-3 is helpful for recognizing liquor. SnO2 is delicate component which is utilized to detect the liquor (Fig. 4).

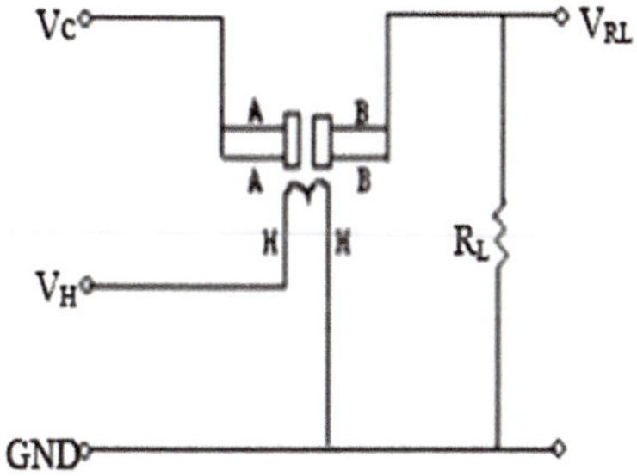

Fig. 4. Alcohol sensor

3.3 Buzzer

Ringer is a sound flagging gadget, utilized as part of family unit apparatuses and car framework. It comprises of two transistors and ringer ON and OFF controlled by the match of the transistor.

3.4 Raspberry Pi

Raspberry Pi is manufactured in two board configurations under license by Newark element14 (Premier Farnell), RS Components and Egoman. These companies make online sale of Pi. Egoman Manufactures a version of Pi for exclusive distribution in China and Taiwan. Red colour and absence of FCC/CE marks distinguishes their product Pi from those of other manufactures. However the hardware is the same for all manufactures. Raspberry Pi is known for its Broadcom BCM2835 system on a chip (SoC), which includes an ARM1176JZF-S 700 MHz processor, Video Core IV GPU, and was originally shipped with 256 megabytes of RAM, later upgraded to 512 MB. But it does not have a build-in hard disk or a solid –state drive. However, it does not use a SD card for the purpose of booting and persistent storage (Fig. 5).

Fig. 5. Raspberry pi kits for Vehicle Accident Detection System

3.5 Webcam

A webcam is a web camera that feeds images in real time to a computer or computer network, via USB, Ethernet or Wi-Fi. They are least known for their use in the establishment of video links, that enables computers acting as videophones or video-conference stations.

The well-known World Wide Web popular by its acronym WWW owes its popularity to the video camera. Security surveillance and computer vision an among its popular uses. Low manufacturing cost, flexibility are significant features of the web-cam, providing the status of the lowest cost device of video telephony. Some video cameras have the facility of getting remotely activated via spyware, thereby becoming a reliable source of security and privacy.

4 System Descriptions

The system consists of a simple computer which is capable of running web pages and enables the users to login in to cloud storage and access to data relating to the driver and the vehicle being used along with the location and the time stamp of the incidents taken place, if any. The system helps the driver of the car to take precaution before starting a journey. The system also helps the driver to remain cautious during drunken driving and ensure controlled driving in case he feels like sleeping. The system also helps the hospitals in the vicinity and ambulances to reach the accident spot thereby saving lives.

5 System Design

The system consists of two major modules:

1. Embedding of the sensors
2. Communication module.

5.1 Embedding the Sensors

This module has all the sensors embedded into the entire system. The system consists of five different sensors, namely vibrate sensor, alcohol detection sensor, fuel gauge sensor, eye blink sensor, tyre pressure sensor and a GPS system. The tyre pressure sensor measures the pressure caused by the air inside the tyre. Any difference in the air pressure triggers an alarm when the change is above the threshold level. The fuel gauge sensor measures the level of the fuel inside the fuel tank. When the level of the fuel gets below a certain limit then an alarm is triggered cautioning the driver. The vibration sensor when it hits upon a pressure more than the threshold value, communicates this information along with location details through GSM to neighbors indicating an accident having taken place in the particular location. The alcohol sensor and the eye blink sensor employed in the system measure the level of alcohol content in the blood of the driver by analysing his breath, while the eye blink sensor measures the number of blinks made by the eye of the vehicle driver for every 30 s. Each sensors has an upper limit which, when crossed, triggers an alarm and halts the functioning of the motor. Here the motor used in the proposed system is a supplement for the real time vehicle.

5.2 Communication Module

The communication module used in the proposed system consists of a GSM system and an IoT. When the sensors sense a threshold as an impact above its respective threshold value, the GPS location along with the time stamp and the driver's details are sent to the data storage which is provided by the IoT module. In the case of an accident, the vibration sensor's threshold value is breached and the data is recorded at that particular point of time and is pushed into the cloud storage system. The GSM module sends a message to the related mobile device or any other device which is authenticated with the system. Also the GPS position of the car is sent to the ambulances and hospitals in the vicinity thereby enabling them to take necessary action.

6 Results and Implementation

The system proposed by the authors has been implemented with a tremendous effort and the results proved to be highly effective. The following pictures show the GPRS module execution by providing us the output of the latitude and the longitude coordinates of the vehicle involved in the accident. Also an alert is received through the GSM module to the related mobile and a personal computer (Figs. 6 and 7).

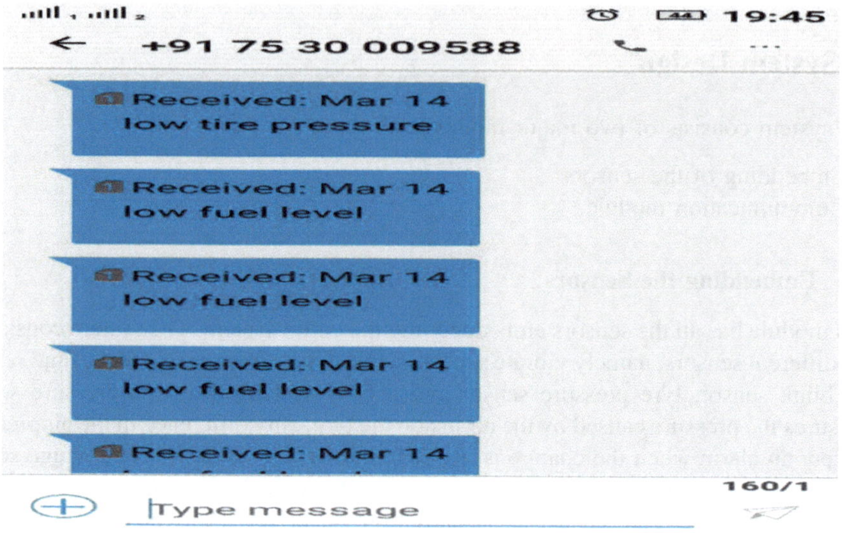

Fig. 6. Result 1

Fig. 7. Result 2

6.1 Advantages

The maintenance check is made easy and digital.
Lives of hundreds of persons can be saved by the operation of this system.
Can be used for locating stolen vehicles.
Avoiding drunken drive by driver by cutting off power supply.
Shows the accurate location of the vehicle in the case of an emergency.
Cost effective hardware and easy implementation.
Avoiding technical snags when the same system is used for all the mechanical parts of the vehicle.
Prompt communication with the nearby hospitals or ambulances in case of emergencies.

6.2 Disadvantages

Absence of network to the SIM cards in certain remote locations.
Restriction of access to anti-social elements.
Absence of a proper satellite navigation system in remote locations.
Absence of proper knowledge on internet of things.

6.3 Applications

Substance abuse can be prevented
Best solution for drivers insomnia
Rash driving can be controlled
Vehicle stability control can be maintained in better manner
Vehicle tracking and locating an accident can be determined

7 Conclusions

Driving conditions are different from one place to another. Therefore this system allows the safest and cautious driving conditions keeping in view all the parameters of the vehicle. Lives of many people can be saved at the nick of the moment when an accident takes place. Even stolen vehicles can be tracked easily. Further research can be done on this system in order to keep in check all the mechanical components of the vehicle using internet of things.

References

1. Aishwarya, S.R., et al.: An IoT based accident prevention and tracking system for night drivers. Int. J. Innov. Res. Comput. Sci. **3**(4), 3493–3499 (2015)
2. Bowen, C.R., Arafa, M.H.: Energy harvesting technologies for tyre pressure monitoring systems. Adv. Energy Mater. **5**(7), 1401787 (2015)

3. Building an intelligent transport system using Internet of Things. http://www.intel.in/ content/ dam/www/program/embedded/internet-of-things/blueprints/iot-building-intelligent-transport-systemblue print.pdf
4. Wang, C., Woodard, S.E.: Sensing of multiple unrelated tire parameters using electrically open circuit having no electric connection. IEEE, vol. 2, issue 1 (2010)
5. El Tannoury, C., Moussaoui, S., Plestan, F., Romani, N., Pita-Gil, G.: Synthesis and application of nonlinear observers for the estimation of tire effective radius and rolling resistance of an automotive vehicle. IEEE Trans. Control Syst. Technol. **21**(6), 2408–2416 (2013)
6. Gorenzweig, I., Hild, S., Moenig, S., Van Gastel, P.: Tire monitoring system for determining tire-specific parameters for vehicle, comprises pneumatic tire, tire air pressure sensor and signal generator which is associated with tire wear limit and signal receiver, Patent number: DE102014112306-A, Derwent Accession number: 2016-14002E, application 27 August 2014 (2014)
7. Sakran, H.O.: Intelligent traffic information system based on integration of Internet of Things and agent of technology. Int. J. Adv. Comput. Sci. Appl. 6(2) (2015)
8. Zeng, H., Hubing, T.H.: The effect of the vehicle body on EM propagation in tyre pressure monitoring system. IEEE on Trans. Antennas Propag. **60**(8), 3941–3949 (2012)
9. Kowalski, M.: Monitoring and managing tyre pressure. Institute of Electrical and Electronics Engineers (IEEE) (2004)
10. Lab VIEW. Function and VI reference manual (2000)
11. Chandreshkumar, L., Pranav, J.: Tire pressure monitoring system and fuel leak detection. Int. J. Eng. Res. Appl. **3**(4) (2013)
12. National highway traffic safety administration. Proposed new pneumatic tires for light vehicles, FMVSS, No. 139 (2001)
13. Bustamante, P., Del Portillo, J.: Wireless system for temperature measurement in wheel, based on ISM. Institute of Electrical and Electronics Engineers (IEEE) (2006)
14. Shinde, P.A., Mane, Y.B.: Advanced vehicle monitoring and tracking system using raspberry pi. In: IEEE 9th International Conference On Intelligent Systems And Control (2012)
15. Reina, G., Gentile, A., Messina, A.: Tyre pressure monitoring using a dynamical model based estimator. Veh. Syst. Dyn. **53**(4), 568–586 (2015)
16. Sangmyeong, K., et al.: Evaluation and development of improved braking model for a motor assisted vehicle using MATLAB. Journal of Mechanical science and technology **29**(7), 2747–2754 (2015)
17. Rani, D.S., Reddy, K.R.: Raspberry pi based vehicle tracking and security system for real time applications. IJCSMC **5**(7), 387–393 (2016)
18. Siddons, A., Derbyshire, A.: Tire pressure measurement using smart low power microsystems. Sensor Review **17**(2), 126–130 (1997)
19. UN-ECE Regulation R64 Temporary Use tyres and tyre pressure monitoring system (2014)
20. Vassev, E., Hinchey, M.: Implementing artificial awareness and knowledge. IEEE (2013)
21. Khan, M.A., Khan, S.F.: IoT based framework foe vehicle over speed detection. IEEE (2018)
22. Hussain, M.Y., George, F.P.: IOT based Real time drosy driving detection system for prevention of Road accidents. IEEE (2018)

Remote Network Injection Attack Using X-Cross API Calls

M. Prabhavathy[1(✉)] and S. Uma Maheswari[2]

[1] Department of CSE, Coimbatore Institute of Technology, Coimbatore, India
prabha.neela.phd@gmail.com
[2] Department of ECE, Coimbatore Institute of Technology, Coimbatore, India

Abstract. The major problem in digital environment is data security and privacy protection (i.e.) securing the user information that is shared as a resource. Data security has consistently been a major issue in information technology. Considering identification of keylogging malware is one of the major issues for antimalware protectors. The proposed method creates the awareness that how the undocumented API calls and middleware libraries are used by the malware creator to steal the user information remotely by injecting into the process and how hide them from the antimalware protector. The experimental results of the proposed work shows the antimalware protector need to take more attention on API call hooking at network level injection by X-cross languages.

Keywords: Malware · Data security · Privacy · Keylogging · Antimalware · API · Hooking

1 Introduction

This Increased popularity of software industries and internet result in security vulnerabilities. Cyber-attack includes consequences such as exploitation of public and private web browsing, stealing PDAs, Laptop, and notebook etc., denial of service attacks and distributed denial of service attacks, unauthorized access, intellectual property theft, phishing attacks, malware, spamming, spoofing, spyware attacks etc. During the last couple of years the usage of internet dramatically doubled results in increased cyber-attacks. Also surfing the internet and sharing their personal information forth and back to the internet increased result in exploitation of vulnerabilities. Most of the web surfers unaware of online threats as they are practiced with easy access of every information on real time basis. Usually many threats originate from unknown and anonymous source that completely destroys large entities.

The term malware refer to malicious software. And it is a software code to infect and harm the systems. It is used to disrupt normal computer operations, steal sensitive personal information like email, password, email attachment, bank account information, financial information and business information, web crawler, social security numbers, disk forensics etc. More likely to be used in private and public websites to steal highly secured guarded information. It may be in the form of code, script, active content, as well as some type of software. It is one type of software installed on your computer and perform malicious activities like destroy most secret user information

D. J. Hemanth et al. (Eds.): COMET 2019, LNDECT 35, pp. 1399–1404, 2020.
https://doi.org/10.1007/978-3-030-32150-5_142

and aggravate the user so third party get benefits. It is one type of software fragment that attach itself to existing executable code that may be application or software or sometimes booting process. It ranges from simple program to complex computer damage and invasion. Some malware are designed in such a way that it send our browsing history, unknown advertisement in website or in third party vendors unknowingly. It is intentionally inserted to breach property of CIA (Confidentiality, Integrity, Availability) of victims data, database, application, web sites etc. Also determining the risk from any type of threats or attack or vulnerability is difficult. Malware injection results in unbearable slow in computer operation, networking, and communication process. Malware consist of five major components such as bot agent, rootkit component, regeneration component, the attack component and configuration file etc. It plays vital role in creation as well as distributing threats. Once vulnerability detected, it try to enter, distribute, exploit, infection and execution.

2 Literature Survey

The Naval, Malware is a standout amongst the most essential dangers to security over the globe. Disregarding various interruption recognition approaches accessible, malware keeps on existing because of the way that these malwares are installed with against location highlights. Dynamic conduct based malware discovery approaches help in killing these progressed malwares. This methodology utilizes framework calls to discover the elements of the risk. The inconvenience of this methodology is that the framework call infusion assault can't be recognized by this strategy. A methodology portrays program semantics utilizing asymptotic equipartition property (AEP) is an avoidance verification arrangement that distinguishes these framework call assaults. The above technique used to examine the confused infusion on procedure. Those are aggregating the vindictive programming location at runtime. The technique to recognize the malware at runtime is have to convoluted so that malware creators can't ready to conceal them inside the procedure.

Barabosch, Malware creators make an effort not to be distinguished while gathering basic data. Different methodologies are utilized by these creators to do likewise. Any procedure of the framework at runtime infusion examined. The recognition procedure utilized three methodologies that are procedure choice; code replicating and Code execution are utilized to recognize the vindictive assault on the framework at runtime. This methodology utilized the strategy of memory allotment on different procedure at runtime were investigated. The creator had taken huge arrangement of malware tests to distinguish the malignant assault on the framework at runtime. The malware tests of 162850 out of which 63.94% found by this strategy. The identification strategy additionally establishes the run time infusion did the system related assaults on the framework.

Stefano Ortolani, System application and portion infusion assault to cover up noxious programming is general strategy. Utilizing the infusion method the malware make the virtual memory inside the appropriate procedures and attempt to conceal them. There are numerous strategy are utilized to recognize the infusion on different procedures still need more consideration need to distinguish the infusion assaults on piece and application forms. Each noxious assaults of this sort utilizes the low dimension paired coding procedure to shroud them in different procedures. Infusion assault at runtime examining need to know well the design of the portion procedure. The undocumented API call assaults need to broke down additional.

Thomas Barabosch, Bee ace used to distinguish the infusion assaults on framework at any application or piece process. It utilizes the idea of honeypot to distinguish the runtime infusion on any procedure which is permitting infusing the malignant at runtime. In light of the honeypot demonstrate procedures absent much information of working framework can dissect the malware at runtime. Downsides of low-level OS-based methodologies are absent in Bee Master. OS autonomous identification can likewise be tried by Bee Master subjectively and quantitatively.

2.1 Remote Network Injection Attack Using X-Cross API Calls (RNIA- X-API) Delineation

The Fig. 1 shows architecture of the proposed work and how the X-cross malwares inject inside the Remote Machine (RM). To inject inside the remote machine process the x-cross malwares uses the java language front to hide the from the antimalware and using C language to perform the malicious activity at the backend. The creations of X-cross malware for the victim machine remote injection done through pseudo code of given Algorithm 1 and 2.

Algorithm 1: Pseudo code of X-Cross malware on the running Remote Machines

Step 1: Initialize the necessary data structure to store information steal from remote machine.
Step 2: Find the suitable process to suspend and allocate virtual memory to run the x-cross malware.
Step 3: After successful injection on the process of RM, Modify the registry of RM and create temporary file for rerun in the RM.
Step 4: Receive function wait indefinitely to steal information on RM.
Step 5: The send function on each successful buffered content sends it to hacker machine frequently (i.e., Step 6 in Algorithm 2) and clears the buffer.
Step 6: Repeat the step 4 and 5 indefinitely till the RM goes to stop.

Algorithm 2: Pseudo code of X-Cross malware on the running Remote Machines

Step 1: Initialize the necessary data structure to store information steal from RM.
Step 2: Find the suitable file to inject and hide the x – cross malware.
  If (success full injection on static file)
    X-cross malware runs when rerun the suspend or stopped RM
      else
    Go to Algorithm 2 step 5 (wait for RM to run)
Step 3: Send the injected information to hypervisor x-cross malware.
Step 4: Repeat the step 2 to 3 if RM is or on suspend or stopped state.

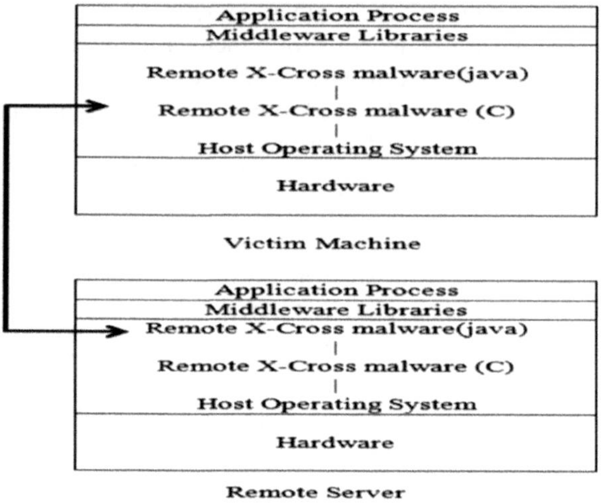

Fig. 1. Architecture of proposed work RNIA- X-API.

3 Result and Discussion

The experimental purpose host machine with windows 7 and jdk1.8 used. X cross malware developed with Java language front end to get user activity and backend C language used to inject the remote machine process. The data structure and libraries used for creation of x-cross remote malware given in Tables 1 and 2. The create malware test with other method and results given in Table 3. The Table 3 shows that the proposed malware with creation can steal the information from the remote machine and send it to the hacker.

Table 1. Buffer data structure.

Type	Variable	Data type
Listener	Remote event listener	Object
Remoteshared folder	Activity event	Object
Remote		
Connection	Connection value in numeric	DWord/Int

Table 2. Hook DLL & LIB.

Type	Name Of file	Architecture	Operating system
DLL	RM_amd64.dll	Amd64 Processor	Windows 7/8/8.1/10
	RM_x86.dll	Amd86 Processor	Fedora 23,24 with GNOME 3.20
	RM_amd64.lib	Amd86 Processor	
	RM_x86.lib	Amd64 Processor	

Table 3. Malicious File Developed in X-Cross Language (java with c).

OS/ Antivirus	Win 8.1		Win 7 pro		Win 7 ultimate		Win 7 home	
	DEP	AV	DEP	AV	DEP	AV	DEP	AV
ESet	√	√	√	√	√	√	√	√
Quick heal	√	√	X	√	√	√	√	√
Avast	√	√	√	√	√	√	√	√
Avira	√	√	X	√	√	√	√	√
Microsoft essential	√	√	√	√	√	√	√	√
KasperSky	√	√	X	√	√	√	√	√
SmartDEV(Free)	√	√	√	√	√	√	√	√

4 Conclusion

The Digital data security is the one of the prime area in the field of information security and privacy. The proposed method of remote network injection using cross languages creates the awareness of antimalware protectors and users that API calls and middleware libraries hides the malicious activity from the antimalware. In future to extend this work and identify the cloud services injection using cross languages.

References

1. Wazid, M., Sharma, R., Katal, A., Goudar, R.H., Bhakuni, P., Tyagi, A.: Implementation and Embellishment of Prevention of Keylogger Spyware Attacks. In: Security in Computing and Communications, of the series Communications in Computer and Information Science, vol. 377, pp. 262–271 (2013). http://link.springer.com/chapter/10.1007%2F978-3-642-40576-1_26
2. Vishnani, K., Pais, A.R., Mohandas, R.: An in-depth analysis of the epitome of online stealth: keyloggers; and their countermeasures. In: Advances in Computing and Communications of the series Communications in Computer and Information Science, vol. 192, pp. 10–19 (2011). http://link.springer.com/chapter/10.1007%2F978-3-642-22720-2_2
3. Vasiliadis, G., Polychronakis, M., Ioannidis, S.: GPU-assisted malware. Int. J. Inf. Secur. **14** (3), 289–297 (2015)
4. Ortolani, S., Giuffrida, C., Crispo, B.: Bait your hook: a novel detection technique for keyloggers. In: Recent Advances in Intrusion Detection, vol. 6307 (2010). http://link.springer.com/chapter/10.1007%2F978-3-642-15512-3_11
5. Damopoulos, D., Kambourakis, G., Gritzalis, S.: From keyloggers to touchloggers: take the rough with the smooth. J. Comput. Secur. 32, 102–114 (2013). http://dl.acm.org/citation.cfm?id=2622909
6. Father, H.: Hooking windows API-technics of hooking API functions on windows. Assembly-Program. J. 2(2) (2004)
7. Prochazka, B., Vojnar, T., Drahanský, M.: Hijacking the linux kernel. In MEMICS, pp. 85–92 (2010)

8. Wazid, M., Katal, A., Goudar, R.H., Singh, D.P.: A framework for detection and prevention of novel keylogger spyware attacks. In: 7th International Conference on Intelligent Systems and Control (ISCO), 2013, 4–5 January 2013, pp. 433–438. IEEE (2013). https://doi.org/10.1109/isco.2013.6481194
9. Cho, J., Cho, G., Kim, H.: Keyboard or keylogger?: a security analysis of third-party keyboards on Android. In: 2015 13th Annual Conference on Privacy, Security and Trust (PST), 21–23 July 2015, pp. 173–176. IEEE (2015). https://doi.org/10.1109/pst.2015.7232970
10. Sagiroglu, S., Canbek, G.: Keyloggers. In: IEEE Society on Social Implications of Technology, IEEE, 18 September 2009. https://doi.org/10.1109/mts.2009.934159, ISSN: 0278–0097
11. Naval, S., Laxmi, V., Rajarajan, M., Gaur, M.S., Conti, M.: Employing Program Semantics for Malware Detection. IEEE Transactions on Information Forensics and Security **10**(12), 2591–2604 (2015)
12. Barabosch, T., Eschweiler, S., Gerhards Padilla, E.: Bee master: detecting host-based code injection attacks. In: Detection of Intrusions and Malware, and Vulnerability Assessment, Print (2014). ISBN 978-3-319-08508-1
13. https://en.wikipedia.org/wiki/Keystroke_logging
14. https://msdn.microsoft.com/en-IN/library/ms809762.aspx
15. http://docs.oracle.com/javase/7/docs/technotes/guides/jni/spec/jniTOC.html

A Study on Peoples' Perception About Comforting Services in e-Governance Centres at Kovilpatti and Its Environs

R. Thanga Ganesh(✉) and K. Pushpa Veni

V.H.N. Senthikumara Nadar College (Autonomous), Virudhunagar, India
thangaa.ganesh@gmail.com, pushpaveni@vhnsnc.edu.in

Abstract. Electronic Governance is an uplift for traditional government activities in a new platform i.e., computer applications with the world wide website network. Transparency and the good governance with the simplicity of work is the great objective of this e-Governance program. The success of e-Governance depends upon the people's perception about the utilization of e-Governance services as a beneficiary. This paper discusses and presents a part of the pilot study about the people's perception about comforting services in e-Governance centre at kovilpatti and its environs with 75 respondents classified under the factor analysis. Further, the study helps to find the impact in various areas for further improvement in rendering services through e-Governance centres in kovilpatti and its environs.

Keywords: E-Governance · Transparency · Technology · Perception · Spiritual intelligence

1 Introduction

A renovated process in sharing information, delivering services for the benefit of citizens, government and business with the information technology is named as e-Governance. E-Governance covers 6,00,000 villages under 2,70,000 panchayats within 626 districts in 28 states and 7 Union Territeries of India. E-Governance refers to the use by government agencies of information technologies (such as Wide Area Networks, the Internet and mobile computing) that have the ability to transform relations with citizens, businesses, and other arms of government. The National e-Governance Plan is formulated by Electronics and Information Technology Department and the Department of Administrative Reforms & Public Grievances to reduce government costs and create interest to e-Participation of citizen for government services through Common Service Centre.

1.1 Tamil Nadu – Vision 2023

Tamil Nadu Government aims to provide all government services through digital mode and also through common service centres and mobile applications under the

D. J. Hemanth et al. (Eds.): COMET 2019, LNDECT 35, pp. 1405–1414, 2020.
https://doi.org/10.1007/978-3-030-32150-5_143

Government's Vision 2023 plan. It helps to enable public, government and commercial establishments get all its services through digital mode.

1.2 Three Main Conceptualizations of E-Governance

Particulars	Interactions E-governance as customer satisfaction	E-governance as processes and interactions	E-governance as tools
Policy-levels	National, local	National, local	National, local
Actors	Consumers, administration	Public and private	State
Policy functions	Operations, service delivery	Operations and policy-making	Service delivery
Use of NICTs	Substitution and communication	Interaction	Technology driven

2 Literature Review

Natesh (2017) in his study entitled "Convergence of government service delivery systems through e-governance in rural Karnataka" briefly discussed about e-Governance in general and convergence of Common Service Centre across the state and Mysore District. Information Technology and its usages is restricted to urban mass and less effectively used to rural. Re-engineering as e-Governance model results the enhanced citizen participation. Extending and Expanding of service delivery of Common Service Centres with mobile seva shall cut down the cost and time involved and improve the quality in providing Government to Consumer, Government to Government and Business to Consumer services, thereby enhancing the efficiency and economical growth of the Citizens.

Beaumont (2017) in his research paper entitled "Information and Communication Technologies in State affairs: challenges of E-Governance" reveals about the benefits and challenges in implementing e-Governance programs. The researcher grouped the challenges as problems in the study such as Technical, Economic and Social issues. Unawareness among people, local language, and privacy of data was the main challenges which are responsible for the unsuccessful implementation of e-Governance in India. The effective and efficient running of e-Governance was still remains for open debate in India.

Mehta (2016) in his article entitled "Maximum Governance: Reaching Out through e-governance" represents the e-Governance for enhanced existing efficiencies, cost driving element for communication and increased transparency in the functioning of various departments. Researcher reveals the effective use of ICT services is for delivering electronic services to the citizens. E-Governance needs to maximize its impact on Indians below the poverty line.

Dutta and Syamala (2015) in their paper entitled "E-Governance Status in India" reviews about the infrastructure and technological feasibility for implementing electronic governance in India. The study reveals better delivery of services to the citizens, less corruption, increased transparency, greater convenience, empowering citizens through prompt information, time saving, good effort, revenue and cost reduction.

Donnell et al. (2003) discussed a case study about challenges on implementing the policy developments by using e-Government for Revenue Online Service Irish Integrated Service Centre. The study reveals that e-Government is an enabler to achieve quality service, direct communication with citizens and improve the back office procedures, numerous factors Corporate commitment, Clear strategic leadership, Fast delivery in small units, Astute HR strategies, Funding, Back office reorganization, Learning from other countries. The researcher suggests to achieve great cost saving for the people through e-office concept in public administration.

2.1 Statement of the Problem

The study area is a developing town in e-governance practices which needs publicity about the services available in e-Governance centre in kovilpatti and its environs. Particularly the rural people need a great support from the e-Governance centre employee in applying and scanning concern documents. So the researcher wants to identify the effectiveness of the comforting service in e-governance centres.

2.2 Research Gap

Various researchers entitled their research in e-Governance. But, there is a need of comforting service for the people in e-Governance centre. Hence the research paper has made an attempt to fill up the research gap for this research model.

2.3 Need of the Study

Services of Technical information enhance the government activities to be easiest today. Information and service delivery to public is a key task in a democratic country like India. Services available in e-Governance centre should be quality. The study helps to aware about e-governance and people's perception on available services in e-Governance centre employees. So the researcher finds the significance of awareness and perception about the technical possibilities in serving for people through e-governance centre.

2.4 Objectives of the Study

To evaluate the impact on quality of comforting services engaged by e-Governance centre employee at Kovilpatti and its environs. To study the people's observation about employee response in e-Governance centre at Kovilpatti and its environs.

2.5 Scope of the Study

The present study has focused on rural and urban people who residence in and around kovilpatti. Rendering government's civic service for the people moving towards systemized with e-Governance today.

3 Research Methodology

A study was conducted for three months from july to september, 2017 in the study region. The study mainly depends on primary data which were collected through a well designed questionnaire. The convenience sampling method was applied for the selection of 75 samples in kovilpatti and its environs. Percentage analysis and Factor analysis is used to analyze the survey data.

4 Results and Discussion

4.1 Rotated Factor Matrix Towards Comforting Services in E-Governance Centre

The Rotated Factor Matrix for the variables relating to comforting services in e-Governance centre among the applicants of e-Governance service centre is as follows:

a. Rotation converged in 10 iterations.
Table 1 exhibits the 20 statements (factors) of the comforting services in e-Governance centre. Table 1 clears that all the 20 statements has extracted into seven various factors such as F 1, F 2, F 3, F 4, F 5, F6 and F7.

The factors are extracted with a new name that influences the comforting services in e-governance centre are presented in the following tables.

From the above Table 2, the factors made regarding 'Customer Confidence in employee's role', 'Attention to illiterate people', 'Employees are courteous', 'Customer Oriented service' are the factors loaded positively in Factor-I. The above said four factors with high loadings on Factor-I are characterized as "Employee's Role". The Eigen value (2.505) resulting with the percentage variance as 12.524 in Factor – I. It could be concluded that the employee's role in e-Governance centre is good in the study area and it ranks as the first important factor.

From the above Table 3, the factors made regarding 'Frequent network supply', 'Application tracking status', 'Frequent power supply', 'Scarcity of Technical labour' are the factors loaded positively in Factor-II. The above said four factors with high loadings on Factor-II are characterized as "Technical Service". The Eigen value (2.270) resulting with the percentage variance as 11.349 in Factor – II. It could be concluded that the Technical Service is good in the study area with a need of additional employee in e-Governance centre and it ranks as the second important factor.

Table 1. Rotated component matrix[a]

Variables	Component						
	1	2	3	4	5	6	7
Customer confidence in employee's role	0.796						
Attention to illiterate people	0.679						
Employees are courteous	0.678						
Customer oriented service	0.587						
Frequent network supply		0.790					
Application tracking status		0.746					
Frequent power supply		0.742					
Scarcity of technical labour		0.520					
Lack of communication			0.829				
Fresh employees need training			0.782				
Functioning of modern equipments				0.794			
Well known to use the modern equipments				0.768			
Paperless work				0.490			
Reasonable charges for application work					0.775		
Employees keep their promise					0.682		
Serving the needs of applicants					0.591		
Neatly approach					0.564		
SMS notification						0.718	
Refusing application						0.689	
Privacy of data							0.790

Extraction Method: Principal Component Analysis. Source: primary data, result calculated
Rotation Method: Varimax with Kaiser Normalization

Table 2. Factor: I - employee's role

S. No	Variables	Factor loadings	Eigen value	Percentage variance
1	Customer confidence in employee's role	**0.796**		
2	Attention to illiterate people	**0.679**	2.505	12.524
3	Employees are courteous	**0.678**		
4	Customer oriented service	**0.587**		

Source: primary data, result calculated

Table 3. Factor: II - technical service

S. No	Variables	Factor loadings	Eigen value	Percentage variance
1	Frequent network supply	**0.790**		
2	Application tracking status	**0.746**	2.270	11.349
3	Frequent power supply	**0.742**		
4	Scarcity of technical labour	**0.520**		

Source: primary data, result calculated

Table 4. Factor: III – performance ability

S. No	Variables	Factor loadings	Eigen value	Percentage variance
1	Lack of communication	**0.829**	1.975	9.877
2	Fresh employee needs training	**0.782**		

Source: primary data, result calculated'

From the above Table 4, the negative variables 'Lack of Communication' and 'Fresh employee needs training', are the factors loaded positively in Factor-III. Factor-III is named as "Performance ability". The Eigen value (1.975) resulting with the percentage variance as 9.877 in Factor – III. It could be concluded that the e-Governance employee needs effective training in both the technical and the customer relationship management and it ranks as the third important factor.

Table 5. Factor: IV - equipment usage

S. No	Variables	Factor loadings	Eigen value	Percentage variance
1	Functioning of modern equipments	**0.794**		
2	Well known to use the modern equipments	**0.768**	1.933	9.664
3	Paperless work	**0.490**		

Source: primary data, result calculated

From the above Table 5, the factors made regarding 'Functioning of modern equipments', 'Well known to use the modern equipments', 'Paperless work' are the factors loaded positively in Factor-IV. The above said three factors with high loadings on Factor-IV are characterized as "Equipment Usage". The Eigen value (1.933) resulting with the percentage variance as 9.664 in Factor – IV. It could be concluded that the Equipment Usage by the employee in e-Governance centre is purposeful in the study area and it ranks as the fourth important factor.

Table 6. Factor: V - employee approach

S. No	Variables	Factor loadings	Eigen value	Percentage variance
1	Reasonable charges for application work	**0.775**		
2	Employees keep promise	**0.682**		
3	Serving the needs of applicants	**0.591**	1.884	9.418
4	Neatly approach	**0.564**		

Source: primary data, result calculated

From the above Table 6, the factors made regarding 'Reasonable Charges for application work', 'Employees Keep Promise', 'Serving the needs of applicants',

'Neatly Approach' are the factors loaded positive in Factor-V. The above said four factors with high loadings on Factor- V are characterized as "Equipment Usage". The Eigen value (1.884) resulting with the percentage variance as 9.418 in Factor – V. It could be concluded that the Employee Approach in e-Governance centre is satisfied in the study area and it ranks as the fifth important factor.

Table 7. Factor:VI - notification

S. No	Variables	Factor loadings	Eigen value	Percentage variance
1	SMS notification	**0.718**	1.548	7.742
2	Refusing application	**0.689**		

Source: primary data, result calculated.

From the above Table 7, the positive factor 'SMS notification' and the negative factor 'Refusing Application' both were impact with higher positive loadings on Factor-VI. The above said two factors with high loadings on Factor- VI are characterized as "Notification". The Eigen value (1.548) resulting with the percentage variance as 7.742 in Factor – VI. It could be concluded that the SMS notification procedure in e-Governance centre is reachable in the study area. Due to heavy workload in the e-Governance centre employees refusing the upcoming applications and it ranks as the sixth important factor.

Table 8. Factor:VII - data privacy

S. No	Variables	Factor loadings	Eigen value	Percentage variance
1	Privacy of data	**0.790**	1.361	6.804

Source: primary data, result calculated.

From the above Table 8, the factor regarding 'Privacy of Data' is ranked as the seventh highest positive loading factor loadings on Factor-VII and it is characterized as "Data Privacy". The Eigen value (1.361) resulting with the percentage variance as 6.804 in Factor - VII. It could be concluded that the Data Privacy in e-Governance centre is reliable in the study area.

Table 9 shows the highest variables for the Welfare facilities for workers.

Table 9. Selected variables and its highest loading factors

Factor	New factors	Selected variable	Factor loadings
F1	Employee's role	Customer confidence in availing services	**0.796**
F2	Technical service	Frequent network supply	**0.790**
F3	Performance	Lack of communication	**0.829**
F4	Equipments usage	Functioning of modern equipments	**0.794**
F5	Employee approach	Reasonable charges for application work	**0.775**
F6	Notification	SMS notification	**0.718**
F7	Data Privacy	Privacy of data	**0.790**

Source: primary data, result calculated

Inference:
Table 9 results that, 'Customer Confidence in availing services' with a factor loading of 0.796; 'Frequent network supply' with a factor loading of 0.790; 'Lack of Communication' with a factor loading of 0.829, 'Functioning of modern equipments' with a factor loading of 0.794; 'Reasonable Charges for application work' with a factor loading of 0.775; 'SMS notification' with a factor loading 0.718; 'Privacy of data' with a factor loading 0.790 are the highest loading factor in F 1, F 2, F 3, F 4, F5, F6 and F7 are identified with the seven variables of comforting services in e-Governance centre for the present study.

5 Findings of the Study

The vast majority of the respondents, 79% were aware about e-governance by self-awareness.

The vast majority of the respondents, 77% have visited e-governance centre for their own purpose.

The vast majority of the respondents, 86.7% were visited the e-governance centre for applying civic services.

More than half of the respondents (51%) are satisfied with monitoring services in the e-governance centre.

More than half of the respondents, 52% are satisfied in direct contact with the service provider in e-governance centre.

The maximum numbers of respondents are highly satisfied with the flexibility of time in requesting for various services in the e-governance centre is reasonably high (56%).

The majority 61% of the respondents are highly satisfied with the e-governance centre employees interested in solving applicants' problem.

The maximum numbers of respondents are highly satisfied with the e-governance centre employees accepting suggestions (56%).

The maximum numbers of the respondents are highly satisfied with the number of visit reduced for a service in e-governance centre (55%).

The number of the respondents is highly satisfied with the service benefits received through e-governance centre (55%).

The majority of the respondents are, 63% satisfied with using toll free number for complaints about the services in e-governance centre.

6 Suggestions

Most of the respondents (86.7%) visited the e-Governance centre for applying civic services. The remaining 13.3% of respondents visited e-Governance centre for development function. So the respondents need awareness about the development function i.e., applying for government schemes. The e-Governance employees shall explain the eligibility for applying certain government schemes effectively.

55% of the respondents are highly satisfied with reduced visit to e-Governance centre for obtaining a service. But, remaining 45% of respondents feel the number of times need to visit for receiving a service. The government shall fully transform the public services into e-Governance system. For an insistence, a parent must get the 'no male heir certificate' from taluk office for applying 'single girl child' application through the e-Governance centre. It reveals that even though e-governance service is available, the people are in conditionally wanted to approach the government office. Subsequently it increases the number of visit to e-Governance centre.

Most of the factors were positively loaded. But, some factors such as customer oriented service training to the e-Governance employees, refusing application due to the heavy workload shows the need of sufficient working employees in the e-Governance centre.

Successful marketing of a product or service generally depends on its good customer relationship management. Further activities like e-Governance employee need to participate in workshop and conferences on the customer relationship management would bring effective attention in caring applicants as their customers of e-Governance centre.In the changing scenario towards electronic Governance setup, the e-Governance centre employees need performance appraisal to motivate their hard work with personal interest.

7 Conlusion

E-Governance is considered as a new research domain, it proceed towards the growth stage in the study area. It will motivate the people to involve wholly inwards e-Governance. The government extending its e-Governance services towards next improvement for free Wi-Fi connections in public connections. Until both the public-private partnership tries it spiritual intelligence, the e-Governance system records slow improvement.

7.1 Area for Further Research

The present paper wave the path to new areas to be concentrate in e-Governance. Even though effective implementation invested by the government, there is a scope for e-Governance employee satisfaction, Customer satisfaction in public relationship management, people's opinion about existing e-Governance services in the e-Governance centre.

Acknowledgement. The data survey done with questionnaire i.e., filled by the applicants of e-Governance centre at Kovilpatti taluk.

References

Sinha, R.P.: E-Governance in India: Initiatives and Issues. Concept Publishing Company, New Delhi (2006)

Natesh, D.B.: Convergence of government service delivery systems through e-governance in rural Karnataka, University of Mysore, Mysore (2017)

Beaumont, S.J.: Information and communication technologies in state affairs: challenges of E-Governance. Int. J. Sci. Res. Comput. Sci. Eng. **5**(1), 24–26 (2017)

Mehta, R.: Maximum governance: reaching out through e-governance. YOJANA **60**, 16–19 (2016)

Dutta, A., Devi, S.M.: E-governance status in India. Int. J. Comput. Sci. Eng. **3**(7) 1–6 (2015)

Donnell, O., Boyle, R., Timonen, V.: Transformational aspects of e-Government in Ireland: issues to be addressed. Electron. J. e-Gov. **1**(1), 22–30 (2003)

Finger, M., Pecoud. G.: From e-Government to e-Governance? towards a model of eGovernance. Electron. J. e-Government **1**(1) 52–62 (2003)

E-governance. https://en.wikipedia.org/wiki/E-governance. Accessed 21 Jan 2019

E-governance. http://www.thehindubusinessline.com. Accessed 22 Jan 2019

A Broadband LR Loaded Dipole Antenna for Wireless Communication

K. Kayalvizhi(✉) and S. Ramesh

Department of Electronics and Communication Engineering, SRM Valliammai Engineering College, Kattankulathur, Chennai 603203, Tamil Nadu, India
kayal_sacet@yahoo.co.in, rameshs.ece@valliammai.co.in

Abstract. This article proposed a reactive loaded dipole antenna for Wireless Communication. A dipole antenna with reactive loads are operated in the range of 10 MHz–600 MHz. The amount of loading circuits and their position, the parameter values are quantified using Genetic Algorithm Optimizer and simulate the proposed design using 3D EM CST Microwave studio tool. The reactive load can enrich the antenna characteristics to produce maximum gain and S_{11} Parameter and then compare the simulated results of antenna performance for with and without load.

Keywords: Genetic algorithm · Dipole antenna · Reactive loads

1 Introduction

Miniaturization techniques of antenna have been well thought out over the past periods due to the transportable wireless devices have need of compactness. In [1], through loading a reliant patch, a dipole antenna is designed. An impedance matching is enhanced by using director. In [3], a broadband loaded dipole antenna is operated in the Very high frequency and ultra-high frequency band. Optimization algorithm is used to find the position and parameter values of the load. The typical SAR in the entire body was fulfilled the standard threshold. In [4], a new printed coupling fed dipole array antenna is designed to attain both a maximum gain and bandwidth, which introduced a radiated load to make up for the flaws of low radiated efficiency and narrow bandwidth.

In [5], a wideband impedance matching of short monopole antenna in the HF/VHF band is designed. It is used to obtain minimum VSWR in small frequency antennas and it is obtained by adding unalloyed resistor in the middle of the antenna. In [6], a dipole antenna is developed for wireless body area network application and it comprise of a pair of altered dipoles through four intersected balanced s shaped arms made on a bendable substrate. In [8], Reactive loads of definite parts of antenna is an authoritative instrument to modify the impedance performance of an antenna with esteem to the preferred frequency band. The operation of antenna impedance load on the characteristics wave modes is examined and calculated. In [9], an antenna is designed with random profile and increasing its bandwidth of an antenna using lumped element as load. This technique is applied in the wire and micro strip antennas to achieve an antenna performance. In [10–17], monopole antenna is loaded with lumped element

D. J. Hemanth et al. (Eds.): COMET 2019, LNDECT 35, pp. 1415–1426, 2020.
https://doi.org/10.1007/978-3-030-32150-5_144

and it is connected to equivalent network and it is optimized using Genetic algorithm. In this article, we present a dipole antenna with reactive loading for a miniaturized antenna using Genetic algorithm optimization. The antenna can be miniaturized through the operational frequency of the proposed antenna is really firm through the dispersed reactive element on the dipole and simulate the proposed antenna design structure using 3D EM CST Microwave studio tool. The load can enrich the antenna performance to yield the maximum gain and minimum VSWR.

The article is organized as follows. In Sect. 2, it deliberate about the antenna component modelling, optimization; Sect. 3 discusses the results and, its performance. And finally, conclusions of the paper are discussed in Sect. 4.

2 Antenna Modelling and Optimization

2.1 Dipole Antenna

Designing an antenna structure to fulfill needs which cause the outcome of non- linear optimization problem. The Fig. 1 illustrate the structure of proposed dipole antenna system which includes a dipole with four loaded LR circuit and an equivalent network in the feed point. Genetic algorithm optimization is used to, define a huge set of optimal factors, such as the position, parameter values of the LR circuits and the elements of the equivalent network.

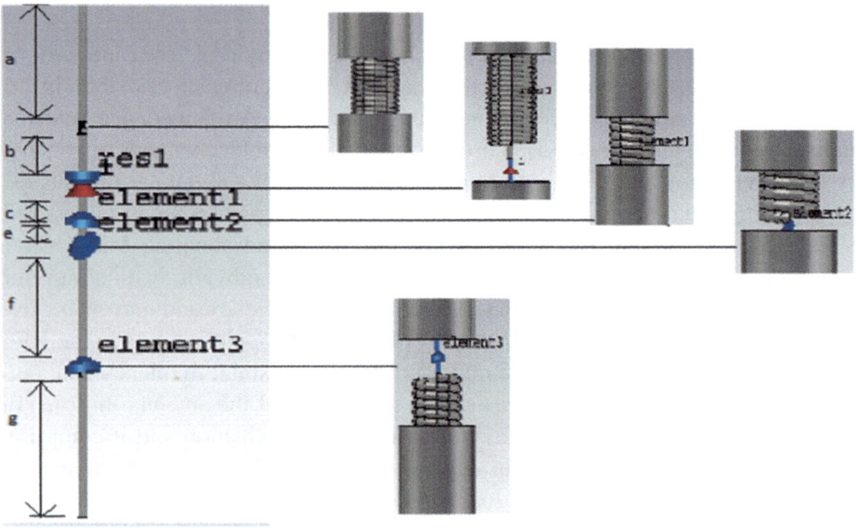

Fig. 1. Dipole antenna

The dipole length and diameter is L and D. Reactive loads of definite partition of the dipole antenna is effort to adjust the characteristics performance with esteem to the desired frequency band.

2.2 Genetic Algorithm Based Antenna Optimization

Genetic algorithm are strong research, modelled after the processes of advancement and genetic recombination, the blocks of the algorithm are named after genetic element. A set of chromosomes are named as population. Primary population or chromosomes are the main concern of Genetic Algorithm. Every chromosomes of the population is specified through a 0 and 1 representation. The chromosomes contain of numerous genetic factor that defines the parameter values of LR circuits and location. Every chromosomes has two fragments; Segment chromosomes tagging the load position and a value chromosomes tagging the load value in Genetic Algorithm execution, every one of alignment entailing N connected fragments are encrypted into 0 and 1 chromosomes. If the LR circuit is positioned in any of the N fragments, the identical fragment will fixed as 1; else 0 [10]. Mutation is permitted to transpire at a small possibility. This procedure is repeat until preferred fitness is achieved (Fig. 2 and Table 1).

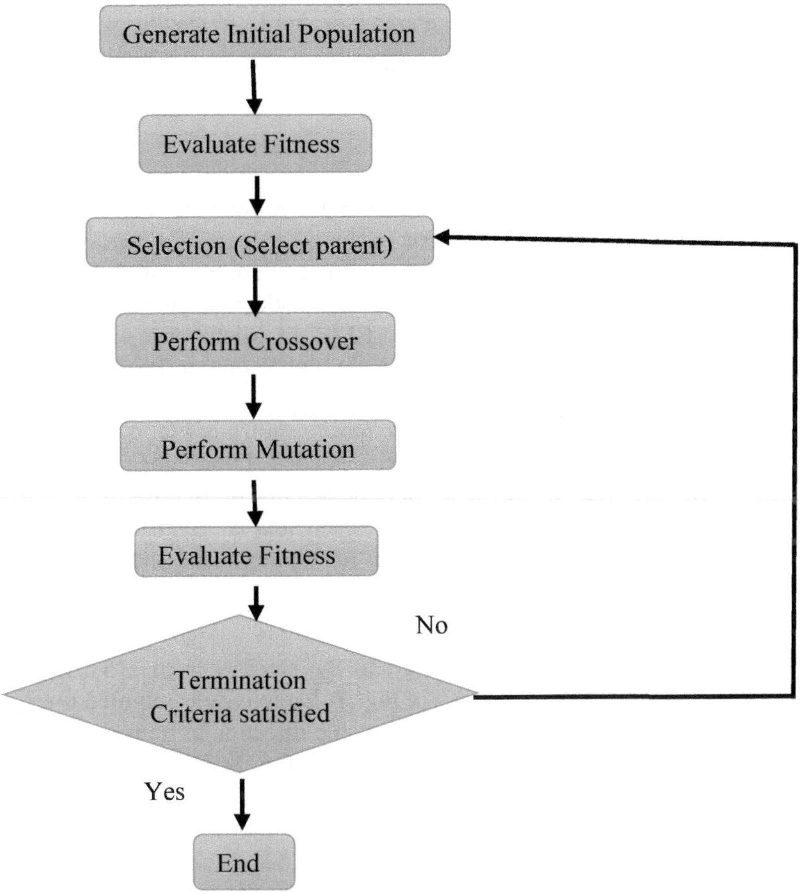

Fig. 2. A flow chart of GA

Table 1. Load position and parameter values for the dipole antenna (L = 180 cm, D = 10 mm, a = 440 mm, b = 110 mm, c = 50 mm, e = 28 mm, f = 350 mm, g = 410 mm)

S. No	Load parameters	L1	L2	L3	Input	L4
1	Load position (cm)	7.9	6	5.10	8	6.10
2	No. of turns	6	6	4.5	18	5
3	Winding gauge (AWG)	14	14	14	14	14
4	Core material	Air	Air	Air	Air	Air
5	Wire diameter (mm)	2	2	2	2	2
6	Coil diameter (mm)	20.4	9.8	9.8	13	9.8
7	Length of the coil (mm)	15	15	13.2	35	19
8	Resistor value	390 Ω	56 Ω	390 Ω	680 Ω	–
9	Capacitor value	–	–	–	18 Pf	–

Then simulate the proposed a dipole design structure using 3D EM CST Microwave studio tool and calculate the S_{11} Parameter and Gain for the without and with reactive loaded dipole antenna.

3 Results

Based on Genetic Algorithm Optimization, a boundless number of unknown has to be determined. The result of an optimization included a least of a particular objective function above the space of design parameter. After finding an antenna length, load parameter values, location and an equivalent network element values, the reactive loaded dipole antenna is simulated using 3D EM CST Microwave studio tool [3] and then calculated the value of S_{11} Parameter, Gain and Radiation Pattern of an antenna (Fig. 3).

The S_{11} parameter plots for proposed antenna are shown in Fig. 4. The without load dipole antenna is an equivalent to the reactive loaded dipole antenna with the same dimension. From the Fig. 4, which shows the results of S_{11} for proposed loaded antenna. A Return loss for the proposed loaded antenna is −9.0901 dB (Fig. 6).

The Computed results of radiation pattern for the proposed antenna are shown in Fig. 5(a) and (b).

Gain and Directivity:
For this proposed dipole antenna a gain is gradually increases from −15 dBi at 10 MHz to 4.81 dBi at 600 MHz and it is shown in Fig. 7. Directivity is required to exploit the radiation pattern of the antenna response in a stable direction to transmit or rdirectivity and gain plot for the proposedeceive power. The Fig. 8(a) and (b) shows the 2D Directivity plot for the proposed antenna.

The Figs. 9 and 10 shows the 3D directivity and gain plot for the proposed with loaded dipole antenna.

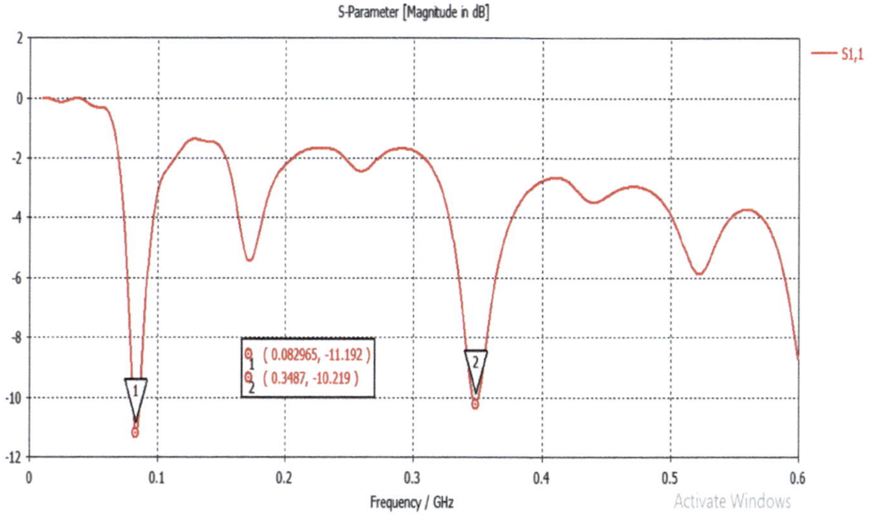

Fig. 3. S_{11} parameter for the without loaded dipole antenna

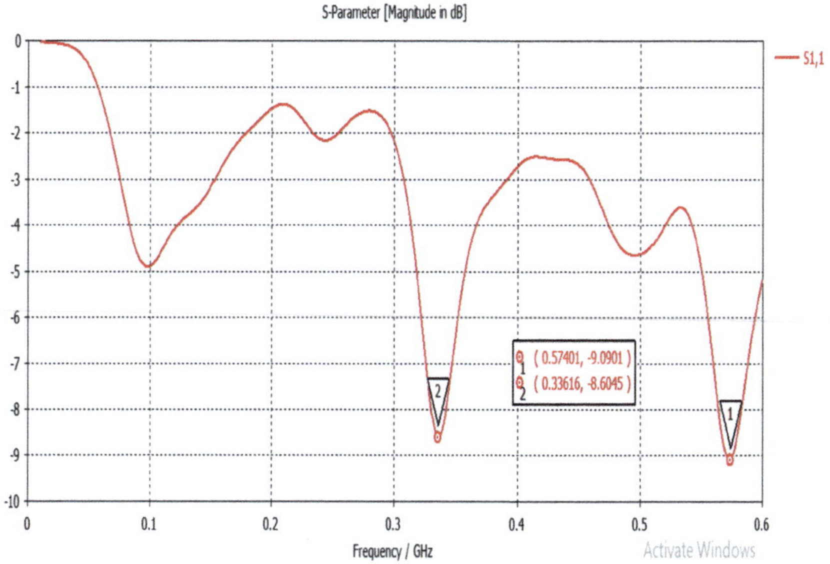

Fig. 4. S_{11} parameter for proposed loaded antenna

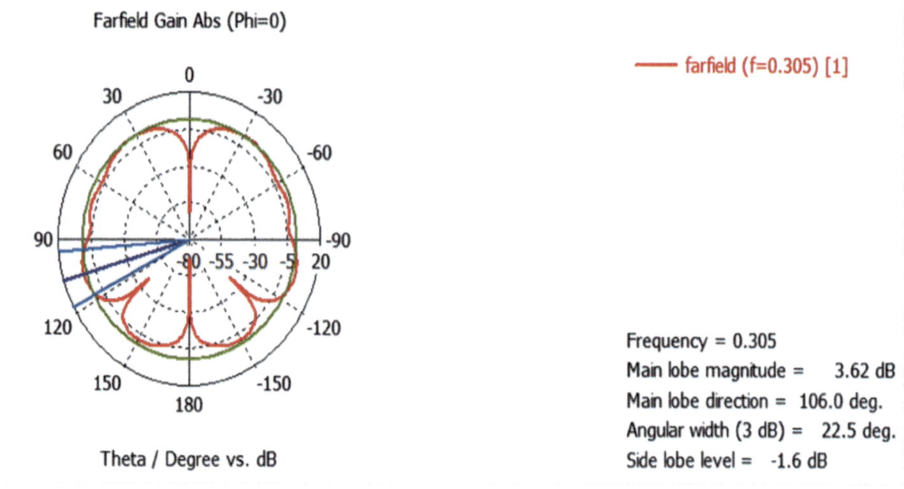

(a)

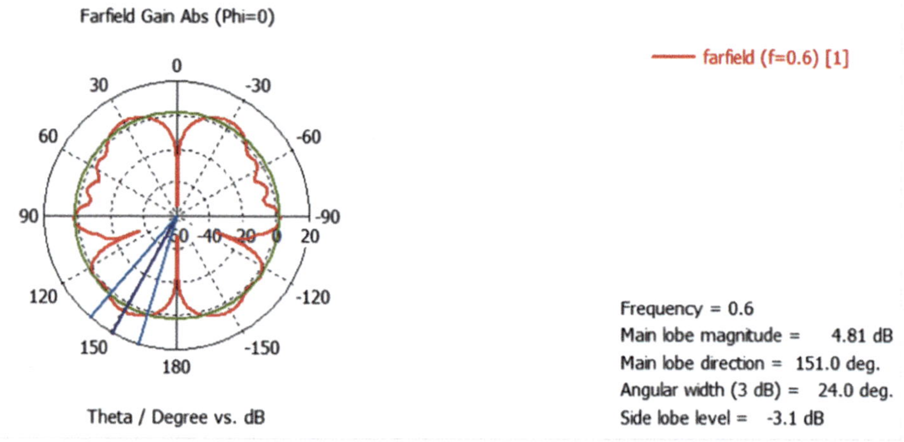

(b)

Fig. 5. Radiation patterns for dipole antenna at a. 305 MHz, b. 600 MHz

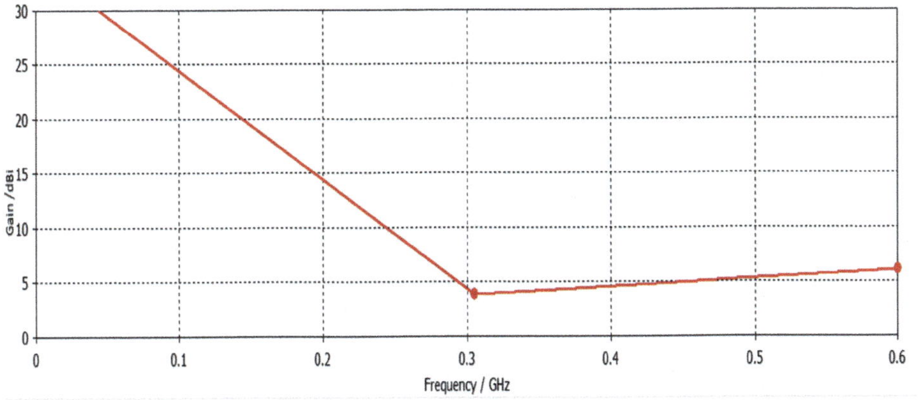

Fig. 6. Computed gain for unloaded dipole antenna.

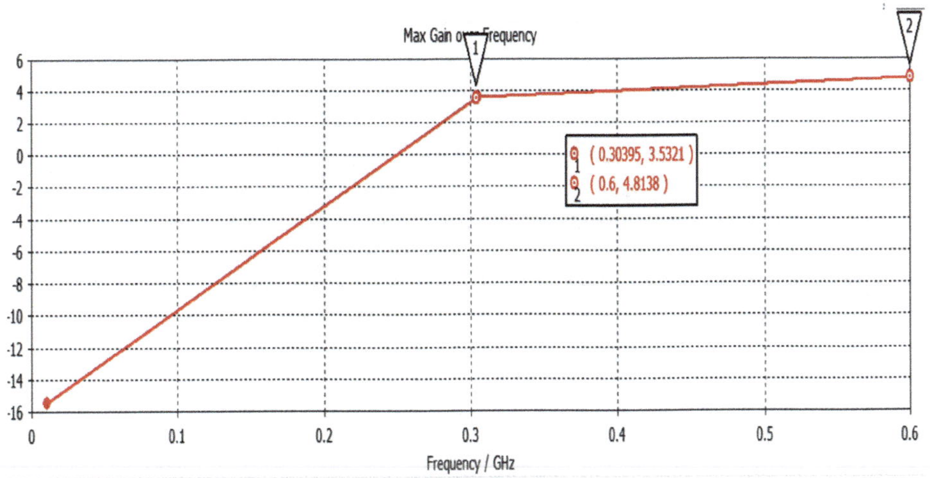

Fig. 7. Computed gain for the dipole antenna

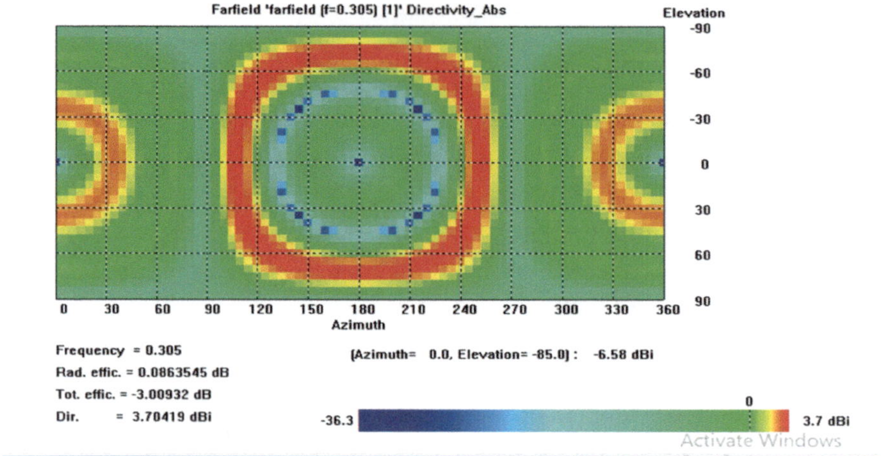

(a)

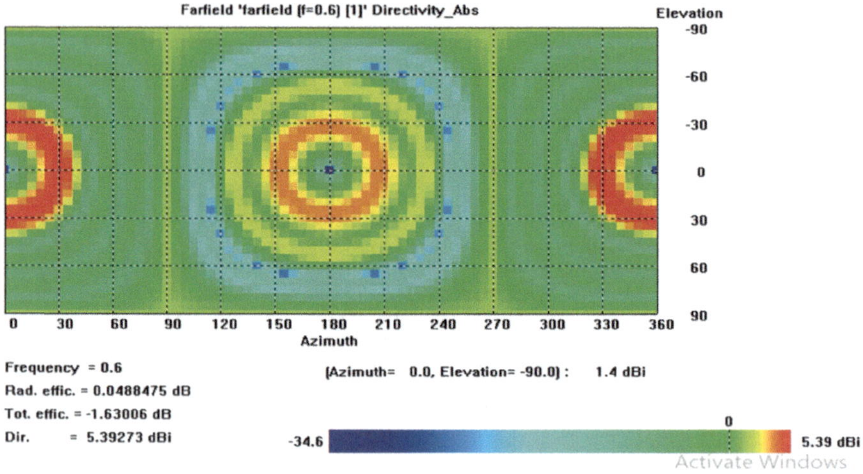

(b)

Fig. 8. 2D directivity plot for the proposed antenna at a. 305 MHz, b. 600 MHz

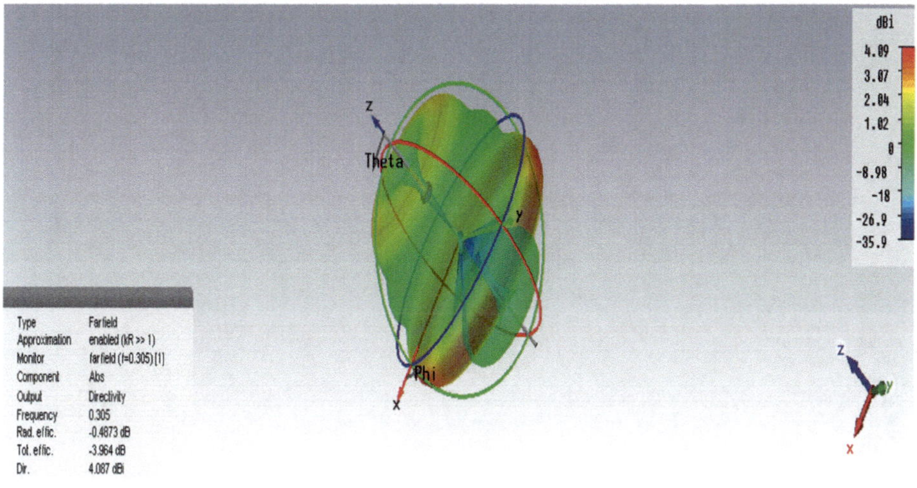

(a)

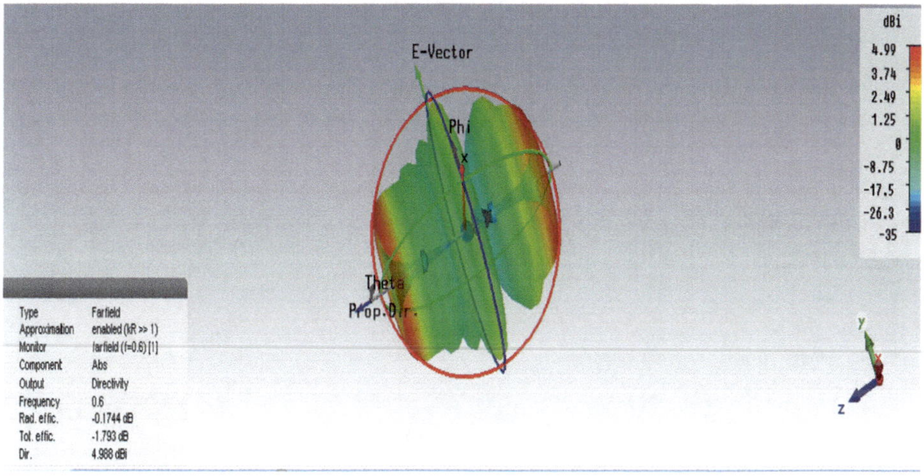

(b)

Fig. 9. 3D directivity plot for the proposed antenna. a 305 MHz, b 600 MHz

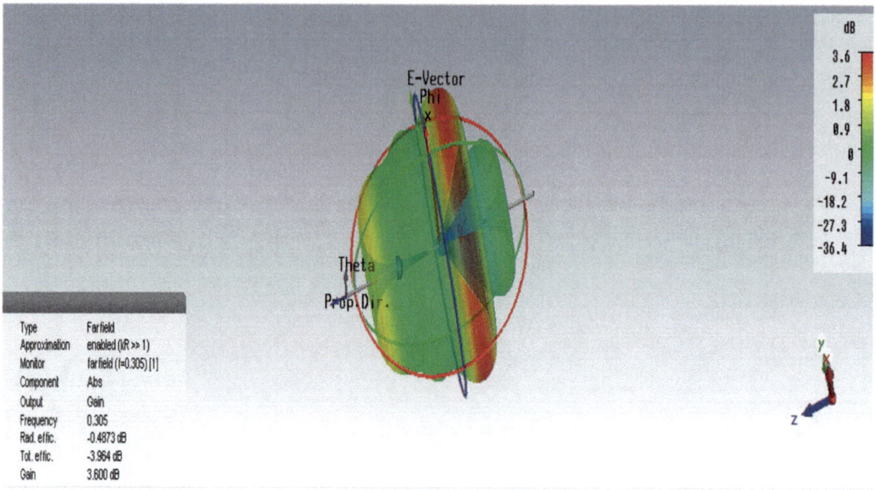

(a)

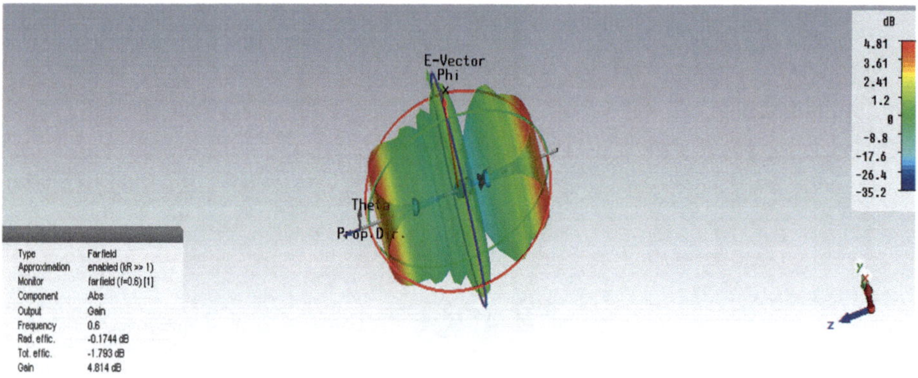

(b)

Fig. 10. 3D gain plot for the proposed antenna. a. 305 MHz, b. 600 MHz

4 Conclusion

In this article, a dipole antenna with a modest small rate configuration has been proposed. Genetic Algorithm optimization is successfully used to increase the gain of dipole antenna with LR circuits and determined over an equivalent network. The antenna is operated in the frequency range of (10 MHz–600 MHz). Through selecting the enhanced position and values of LR loads, the design is simulated and calculated the S_{11} and gain of the antenna. A dipole antenna system having simulated results of an antenna gain is increased from −15.45 dBi at 10 MHz to 4.81 dBi at 600 MHz.

Acknowledgment. The authors wish to acknowledge DST-FIST supporting facilities available in the Department of Electronics and Communication Engineering at Valliammai Engineering College, Chennai, Tamil Nadu, India.

References

1. Chang, L., Chen, L.L., Zhang, J.Q., Li, D.: A broadband dipole antenna with parasitic patch loading. IEEE Antennas Wirel. Propag. Lett. **17**, 1717–1721 (2018)
2. Amendola, S., Marrocco, G.: Optimal performance of epidermal antennas for UHF radio frequency identification and sensing. IEEE Trans. Antennas Propag. **65**(2), 473–481 (2017)
3. Amani, N., Jafargholi, A., Pazoki, R.: A broadband VHF/UHF loaded dipole antenna in the human body. IEEE Trans. Antennas Propag. **65**(10), 5577–5582 (2017)
4. Zong, H., Liu, X., Ma, X., Shu Lin, L., Liu, S.L., Fan, S.: Design and analysis of a coupling-fed printed dipole array antenna with high gain and omni directivity. IEEE J. Mag. **5**, 26501–26511 (2017)
5. KarimiMehr, M., Agharasouli, A.: A miniaturized non-resonant loaded monopole antenna for HF-VHF band. Int. J. Sci. Eng. Res. **8**(4), 1092–1096 (2017)
6. Liu, X.Y., Di, Y.H., Liu, H., Wu, Z.T., Tentzeris, M.M.: A planar windmill-like broadband antenna equipped with artificial magnetic conductor for off-body communications. IEEE Antennas Wirel. Propag. Lett. **15**, 64–67 (2016)
7. Grimm, M., Manteuffel, D.: On-body antenna parameters. IEEE Trans. Antennas Propag. **63** (12), 5812–5821 (2015)
8. Safin, E., Manteuffel, D.: Manipulation of characteristic wave modes by impedance loading. IEEE Trans. Antennas Propag. **63**(4), 1756–1764 (2015)
9. Elghannai, E.A., Raines, B.D., Rojas, R.G.: Multiport reactive loading matching technique for wide band antenna applications using the theory of characteristic modes. IEEE Trans. Antennas Propag. **63**(1) 261–268 (2015)
10. Yeg, K.: Design, optimization, and realization of a wire antenna with a 25:1Bandwidth ratio for terrestrial communications. Turk. J. Electric. Eng. Comput. Sci. **22**, 371–379 (2014)
11. Booket, M.R., Jafargholi, A., Kamyab, M., Eskandari, H., Veysi, M., Mousavi, S.M.: A compact multi-band printed dipole antenna loaded with single- cell MTM. IET Microwave Antenna Propag. **6**(1), 17–23 (2012)
12. Ding, X., Wang, B.Z., Zheng, G., Li, X.M.: Design and realization of a GA- optimized VHF/UHF antenna with 'On-body' matching network. IEEE Antennas Wirel. Propag. Lett. **9**, 303–306 (2010)

13. Werner, P.L., Bayraktar, Z., Rybicki, B., Werner, D.H., Schlager, K.J., Linden, D.: Stub-loaded long-wire monopoles optimized for high gain performance. IEEE Trans. Antennas Propag. **56**(3), 639–645 (2008)
14. Iizuka, H., Hall, P.S.: Left-handed dipole antennas and their implementations. IEEE Trans. Antennas Propag. **55**(5), 1246–1253 (2007)
15. Mattioni, L., Marrocco, G.: Design of a broadband HF antenna for multimode naval communication-Part II: extension on VHF/UHF ranges. IEEE Antennas Wirel. Propag. Lett. **6**, 83–85 (2007)
16. Rogers, S.D., Butler, C.M., Martin, A.Q.: Design and realization of GA- optimized wire monopole and matching network with 20:1 bandwidth. IEEE Trans. Antennas Propag. **51**(3), 493–502 (2003)
17. Wong, K.-L.: Planar Antennas for Wireless Communications. Wiley (2003)
18. Ladbury, J.M., Camell, D.G.: Electrically short dipoles with a nonlinear load, a revisited analysis. IEEE Trans. Electromagn. Compat. Mag. **44**(1), 38–44 (2002)
19. Kraus, J.D., Marhefka, R.J.: Antennas: For All Applications. McGraw-Hill (2002)
20. Sarabandi, K., Azaddcgan, R.: Design of an efficient miniaturized UHF planar antenna. In: IEEE International Symposium on Antenna and Propagation Society, vol. 4, pp. 446–449 (2001)

RETRACTED CHAPTER: Optimal Throughput: An Elimination of CFO and SFO on Directed Acyclic Wireless Network

K. P. Ashvitha(✉) and M. Rajendiran

Panimalar Engineering College, Chennai, India
ashvithachitty@gmail.com,
muthusamyrajendiran@gmail.com

Abstract. Wireless network is used for efficient energy transfer but there is some loss of data and energy, so we propose algorithm to resolve these issues in a effective way. To accomplish the communicate limit and to streamline the intensity of a multihop communicate on coordinated non-cyclic remote systems for boost of throughput and joint beamforming utilizing Joint Maximum probability calculation. Joint ML algorithm provides the multihop broadcast with high SNR and low error. In this undertaking, effective strategy for ICI wiping out dependent on factor chart and PAPR decrease utilizing pre-coder by assessing and channel parameters is proposed. By interchanging messages both domains, the proposed algorithm can suppress inter carrier interference and reduce peak to average power ratio progressively. This undertaking presents a base mistake likelihood based pre-coding grid to diminish the PAPR of multihop broadcast. To lessen computational many-sided nature, a less complex coarse CFO estimator is completed for the estimation with the goal that a progressively precise result is obtained. Therefore, the computational complexity can be reduced significantly.

Keywords: Multihop · Beamforming · Pre-coder

1 Introduction

Broadcasting implies the significant framework helpfulness of passing on data from a source center point to each other center in a framework. For capable telecom, one needs to use legitimate package replication and sending to clear out overabundance transmissions. This is particularly vital in control obliged remote systems which experience the ill effects of obstruction and parcel crashes [1]. Communicate applications incorporate mission-basic military correspondences, live video gushing, and information scattering in sensor systems.

The outline of productive remote communicate calculations faces a few difficulties. Remote channels experience the ill effects of impedance, and a communicate strategy needs to initiate non-meddling connections at each schedule opening. Remote system topologies experience visit changes, with the goal that bundle sending choices must be made in a versatile manner [2]. Existing unique multicast calculations that adjust

The original version of this chapter has been retracted. The retraction note to this chapter can be found at https://doi.org/10.1007/978-3-030-32150-5_163

D. J. Hemanth et al. (Eds.): COMET 2019, LNDECT 35, pp. 1427–1440, 2020.
https://doi.org/10.1007/978-3-030-32150-5_145

movement more than a few crossing trees might be utilized for broadcasting, since communicate is an uncommon instance of multicast. These calculations, nonetheless, are not reasonable for remote systems in light of the fact that counting all traversing trees is computationally restrictive, even more so when this is to be done again and again as and when the topology changes with time.

In this task, we think over over the serious issue of throughput perfect telecom in remote frameworks [3]. We consider a period opened framework. At each opening, a scheduler chooses which non-meddling remote connects to actuate and which set of bundles to forward finished the enacted joins, so all hubs get parcels at a typical rate. The most extreme achievable basic gathering rate of unmistakable parcels over all planning arrangements is known as the communicated limit of the system.

The essential duty of this errand is to design a decentralized and provably perfect remote impart counts that does not use spreading over trees when the concealed framework topology is restricted to a DAG. We examine the issue of productively dispersing parcels in multi-bounce remote systems [4]. At each schedule vacancy, the system controller enacts an arrangement of non-meddling connections and forward chosen duplicates of parcels on each initiated interface. The greatest rate of generally got bundles is alluded to as the communicated limit of the system.

In this plan, we propose another incredible estimation that accomplishes the communicate limit when the fundamental system topology is a coordinated non-cyclic chart (DAG). We explore the throughput multihop communicate on coordinated non-cyclic remote systems. In particular, we improve the intensity of a multihop communicate on coordinated non-cyclic remote systems for augmentation of the astute framework throughput under transmit control, likelihood of false alert, and likelihood of missed recognition limitations and joint beamforming utilizing Joint Maximum likehood calculation [8]. Joint ML calculation gives the multihop communicate on coordinated non-cyclic remote system with high SNR and low blunder at the downlink of the remote system.

Moreover, our technique accomplishes high resistance to recurrence specific blurring channels for both single and various get recieving wire frameworks, with an intricacy that is around twice that of a customary vitality finder. In this task, effective technique for ICI extraction in view of factor diagram and PAPR diminishment utilizing pre-coder by assessing transporter recurrence balance (CFO) and divert parameters in multihop communicate on coordinated non-cyclic remote systems based remote correspondence frameworks is proposed [10]. By exchanging messages both in space zone and repeat space, the proposed count can smother cover carrier impedance and reduce best to ordinary power extent (PAPR) iteratively and powerfully.

Distinctive time space approaches have been proposed for diminishing the amount of in reverse speedy Fourier change (IFFT) exercises required to make the candidate movements in all pass channel designs. In any case, the subsequent time-space created signals are fairly related, and in this manner the PAPR lessening execution is truly corrupted. As requirements be, the present examination proposes a novel PAPR reducing methodology in which repeat space stage revolution, cyclic moving, complex conjugate, and sub-transporter inversion activities are altogether utilized so as to build the decent variety of the hopeful signs [12]. Moreover, to go around the various IFFT issue, the majority of the recurrence space activities are changed over into time-area counterparts.

Reduction, Reduction in Power Leakage Effect, examined and their effect on the proposed procedure is contemplated. Both investigation and reproduction demonstrate that the vitality Harvesting calculation can adequately and precisely identify the presence of the essential client. Moreover, our strategy accomplishes high resistance to recurrence particular blurring channels for both single and different get reception apparatus frameworks, with a many-sided quality that is roughly twice that of an ordinary vitality identifier.

In correspondence we create proposed water filling calculation for multihop communicate on coordinated non-cyclic remote systems blurring channel (Rayleigh Fading channel) [15]. Multihop communicate on coordinated non-cyclic remote systems turns into the picked adjustment method for remote correspondence. Distinctive sections or minimal base stations send self-ruling coded information to different flexible terminals through symmetrical Code division multiplexing channels. Multihop impart on composed non-cyclic remote frameworks is a promising high data rate interface advancement.

In multihop communicate on coordinated non-cyclic remote systems we transmit distinctive stream of information through various recieving wires. We demonstrate that as we increment the power spending plan in the water filling calculation the mean limit of the framework expanded. A convincing disturbance examination of the pre-coded forward channel yields express lower confines on net limit which speak to CSI acquisition overhead and bumbles and furthermore the sub-optimality of the pre-coders.

The grouping procedure is proposed which clarify the how vitality utilization is accomplished [18]. By separate the system region viably. The overhead ought to be diminished and when the hub needs to exchange the information, the anchored correspondence of information is critical, because of dynamic and appropriated nature of organize and the interlopers ought to be maintained a strategic distance from to build the organize life time. Proposed zone based secure grouping in versatile Ad-Hoc organize tried through six stages.

Besides, Hereditary Algorithm based bunch decision starts at each zone. It bunches physically neighboring hubs as groups with ideal number of bunch tallies. To refine the versatility and lessen the overheads discover the IDS revelation furthermore, rather than giving settled reaction to the interloper hub movement the versatile reaction conspire actualized to lessen the overheads. In view of the development of speed of the hubs the parcel resizing done to expand the execution of the arrange [20].

It should be possible when the source sends In directing way the anchored correspondence between the source and goal is essential likewise when the hub joins in the organize and when the hub leave from the group without check. The key trade in light of ID based key administration offers verification. The channel obstruction happened when the demand emerges all the while in the range of transmission of hubs.

2 Related Work

In [1] Zeng et al. proposed to the issue as Different from the current likelihood task plans, it considers the nearby data as well as the quantity of vehicles inside one bounce extend in the sending likelihood utilizing method likelihood based multi-jump communicate convention and finished up as a lower sending likelihood of the nearby hub can guarantee the system to have a higher sending achievement likelihood without backoff.

In 2018 Li et al. proposed [2] tended to the issue another instrument named MBM-EMD (i.e., multihop communicate system for crisis messages dispersal) to take care of these issues utilizing the method Traditional multihop communicate conventions and finished up as exhaustively thinking about the channel dispute, lining delay, flag blurring, communicate impedance and the versatility of vehicles.

In 2014 Wu et al. [3] proposed utilizing fluffy rationale calculation and tended to the issue as utilizing the fluffy rationale calculation, the convention can pick the best transfer hub by taking bury vehicle remove, vehicle speed, and connection quality into account. And they at last finished up as the hand-off hubs are chosen utilizing a fluffy rationale calculation that takes intervehicle separate, vehicle development, and got flag quality into account.

In 2013 Jaballah et al. [4] tended to the issue as break down assaults to the best in class IVC-based wellbeing applications. Besides, this investigation drives us to plan a quick and secure multihop communicate calculation for vehicular correspondence.

Yu et al. [5] 2018 proposed the procedure static neighborhood communicate calculation and tended to the issue as the convention is asymptotically ideal regarding both infusion rate and parcel latency. And finished up as convention can deal with both stochastic and ill-disposed infusion designs.

In 2016 [6] Nardini et al. proposed the method coordinated transmissions and tended to the issue as it portrays indispensable changes at both the nodes, what the fundamental issues are, and how to clear up them capably. The producers considered that level learning on the nodes and standard asset assignment plots on the node, and this permits a control over the grant region.

In 2017 Yan et al. proposed [7] the procedure Productive Multihop Broadcasting with Network Coding (EMBNC) and kept an eye on the issue as a viable multihop broadcasting with orchestrate coding (EMBNC) by picking downstream forwarders in a two-niche) manner. The makers finally shut as by joining framework coding with general topologies, time for a forwarder to include the remote medium can be diminished.

In 2016 [8] proposed by creators Kuang and Yu tended to the issue as a portability based forward hub determination calculation is proposed, which has a tendency to choose the less versatility hubs as the forward ones. In this paper they utilize a method called forward hub choice algorithm. The creators at long last finished up as The hub's portability and accessible connection limit are considered in content disclosure and substance conveyance strategies.

In 2014 [9] Wang et al. proposed to Physical Interference using the strategy Minimum-Latency Broadcasting Schedule (MLBS) calculation. The makers kept an eye on the issue as the MLBS with Duty-Cycled circumstances (MLBSDC) has been particularly gathered in graph based impedance models, for instance, the tradition impediment model. It is finally shut as it makes powerful theory computations for MLBSDC in multihop remote frameworks with commitment cycled circumstances under the physical block appear.

In 2014 [10] proposed by the creators Aravindhan et al. The creators tended to the issue as the convention uses the separation strategy to choose sending hubs and furthermore the outcomes to accomplish high reachability. In this paper they utilized a system called Position Based Routing Protocol.

In 2014 [11] proposed by the creators Chang et al. The creators utilized the procedure called need based system coding communicate convention (PNCB) algorithm. The issue tended to in this paper is the need based stop anticipation component to keep away from deadlocks. The makers finally shut it as to appreciate the many to all MTB issue with compose coding in a totally scattered manner, we have developed a need based framework coding convey (PNCB) tradition.

In 2018 Ramezanipour et al. [12] proposed the method Optimization algorithm. The creators tended to the issue as Poisson point process is used to show the spreads of the center points and the impediment brought about by the approved customers for the sensor hubs. It is finally wrapped up as the effect of retransmission and power outage necessity on the power usage and imperativeness adequacy of the framework contemplating different framework densities.

In 2018 [13] proposed by the creators Wang et al. The strategy utilized in this paper is mShare algorithm. The issue tended to will be to show the adaptability of our arrangement, we test mShare in three settings:unicast, savvy coordinating, and data accumulation. It is at long last finished up as the execution of mShare is assessed with huge scale arrange recreations and physical testbed tests running on USRP.

In 2018 [14] the creators tended to the issue as a Coexistence-Aware(CA) steering plan, in light of the meaning of a novel association cost metric. The methodology used in this paper is Coexistence-Aware (CA) coordinating arrangement calculation. The makers shut as they have exhibited that CA plan can tuned to trade off the accomplished organize throughput and the ordinary number of bounces.

In 2018 Furtado et al. proposed [15] in this paper they utilize the method called MAC plot algorithm. They tended to the issue as determining the throughput accomplished by the cross-layer plan, by exhibiting the execution of the PHY-layer and the subjective MAC conspire. It is at long last finished up as judicious capacity which is registered by an insertion procedure. The portrayal of the PHY-layer execution considers the way misfortune impact, little and huge scale blurring.

In 2018 utilizing the method CSR plot calculation the creators Yun et al. [16] The creators tended to the issue as to guarantee information transmission dependability indeed, even against such ambushes, in this letter a concentrated trust based secure coordinating (CSR) conspire. It is at long last finished up as CSR enhances the steering execution by keeping away from malignant hubs and successfully confining false trust.

In 2018 [17] proposed by the creators Chengetanai. In this paper they utilized the plan called AODV steering protocol. They tended to the issue as Mobile impromptu system (MANET) is a sort of remote system that does not require any current framework for it to be operational.

In 2017 Darabkh et al. proposed [18]. In this paper they utilized the plan Limit based Bunch Head Replacement (C-DTB-CHR) convention. They kept an eye on the issue as Head Replacement (C-DTB-CHR) tradition that essentially goes for enhancing essentialness through limiting the quantity of re-grouping operations. It is at long last finished up as hubs won't serve group heads any more on the off chance that they have effectively played this part.

In 2017 Samir et al. proposed [19] Exploring the impact of different group structures on vitality utilization and end-to-end delay in Cognitive Radio Wireless Sensor Networks. In this paper they tended to the issue as investigate three distinctive Cognitive Radio Wireless Sensor Networks (CRWSNs). In this paper they utilize the method CRWSN conspire algorithm. It is at long last finished up as so as to build vitality proficiency, the multi-bounce bunch structure is proposed.

In 2017 Yang et al. proposed [20] In this paper they utilize the procedure CQPNC scheme. The issue tended to here is in CQPNC contrive, two source center points first use quadrature bearers to transmit signals in the meantime, which are gotten and prepared. It is finally shut as reenact the BER and throughput displays of TC, CNC and CQPNC in different gathering device conditions.

3 Proposed Work

3.1 Throughput Maximization Using Energy Harvesting Algorithm

Different time area approaches have been proposed for decreasing the quantity of converse quick Fourier change (IFFT) tasks required to produce the competitor motions in all pass channel plans. Be that as it may, the subsequent time-area created signals are to some degree associated, and in this way the PAPR decrease execution is genuinely debased. In like manner, the present examination proposes a novel PAPR decrease strategy in which recurrence space stage pivot, cyclic moving, complex conjugate, and sub-transporter inversion tasks are altogether utilized so as to build the assorted variety of the competitor signals. Besides, to go around the numerous IFFT issue, the majority of the recurrence area tasks are changed over into time-space reciprocals. It is shown that the sub-transporter distributing re-gathering frames are basic to recognizing low-eccentrics time zone indistinguishable exercises. In addition, it is indicated numerically that the computational intricacy of the proposed calculation called vitality proportion calculation is essentially lower than that of the all pass channel technique and the PAPR decrease execution is inside 0.001 dB of that of all pass channel. All around, the results demonstrate that among most of the low-multifaceted nature structures proposed in the composition, the system proposed in this examination most almost approximates the PAPR decline execution of the ordinary all pass channel plot (Fig 1).

ALGORITHM

```
FFT = 64;
Subcarrier = 52;
Symbol = 52;
Bits = 10000;
SPR = [0:10];
SPRindB = SPR + 10*log10(Subcarrier/FFT) + 10*log10(64/80);
for Iteration = 1:length(SPR)
 Input = rand(1,Symbol*Bits) > 0.5; Orthogonal = 2*Input-1;
Orthogonal = reshape(Orthogonal,Symbol,Bits).';
PilotData = [zeros(Bits,6) Orthogonal(:,[1:Symbol/2]) zeros(Bits,1)
      Orthogonal(:,[Symbol/2+1:Symbol]) zeros(Bits,5)] ;
FFT_Transform = (FFT/sqrt(Subcarrier))*ifft(fftshift(PilotData.')).';
FFT_Transform = [FFT_Transform(:,[49:64]) FFT_Transform];
FFT_Transform = reshape(FFT_Transform.',1,Bits*80);
Noise = 1/sqrt(2)*[randn(1,Bits*80) + 1i*randn(1,Bits*80)];
Channel = sqrt(80/64)*FFT_Transform + 10^(-SPRindB(Iteration)/20)*Noise;
Channel = reshape(Channel.',80,Bits).';
Channel = Channel(:,[17:80]);
FFT_Reverse = (sqrt(Subcarrier)/FFT)*fftshift(fft(Channel.')).';
 Synchronize = FFT_Reverse(:,[6+[1:Symbol/2] 7+[Symbol/2+1:Symbol] ]);
Real = 2*floor(real(Synchronize/2)) + 1;
Real(Real>1) = +1;
Real(Real<-1) = -1;
Normalize = (Real+1)/2;
Normalize = reshape(Normalize.',Symbol*Bits,1).';
 PD_Estimate(Iteration) = size(find(Normalize - Input),2);
PD = PD_Estimate/(Bits*Symbol); SPR=linspace(5,15,11);
```


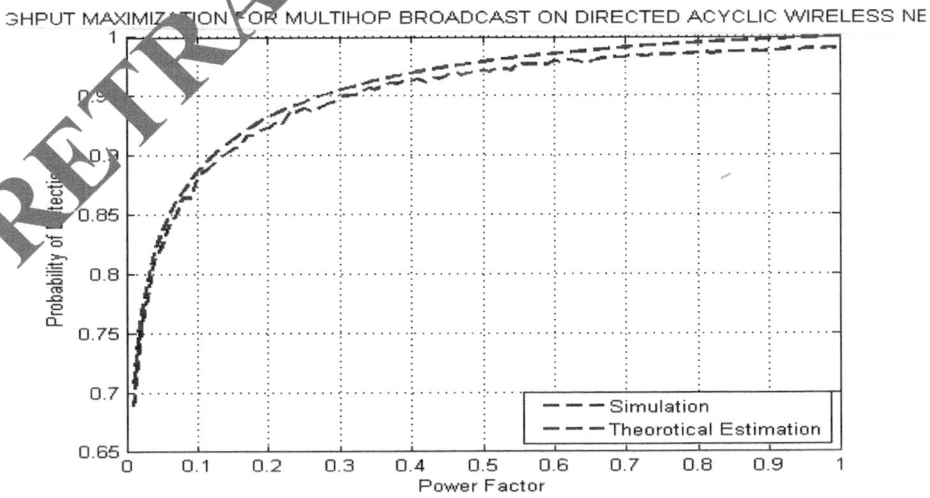

Fig. 1. Throughput maximization

3.2 Spectrum Sharing, Receiver Power Ratio Reduction and Reduction in Power Leakage Effect

This is finished by detecting the adjustment in flag quality over various saved communicate remote system sub-transporters with the goal that the return of the essential client is immediately distinguished. In addition, the communicate remote system impedances, for example, control spillage, Spectrum sharing, Receiver Power Ratio Reduction, Reduction in Power Leakage Effect, researched and their effect on the proposed strategy is considered. Both investigation and reproduction demonstrate that the vitality Harvesting calculation can viably and precisely identify the presence of the essential client. Moreover, our technique accomplishes high invulnerability to interference particular blurring channels for both single and various get reception apparatus frameworks, with a multifaceted nature that is roughly twice that of an ordinary vitality identifier (Figs 2 and 3).

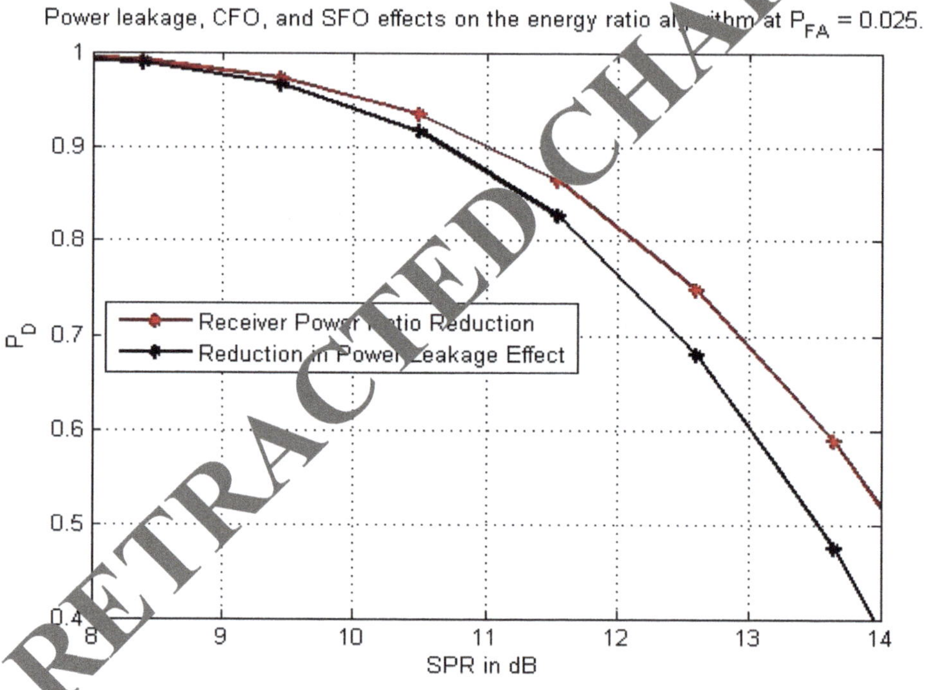

Fig. 2. Power leakage effect

3.3 Capacity Maximization Using Modified Water-Filling Algorithm

In present correspondence we create proposed water filling calculation for multihop communicate on coordinated non-cyclic remote systems blurring channel (Rayleigh Fading channel). Multihop communicate on coordinated non-cyclic remote systems turns into the picked adjustment procedure for remote correspondence. Diverse entries

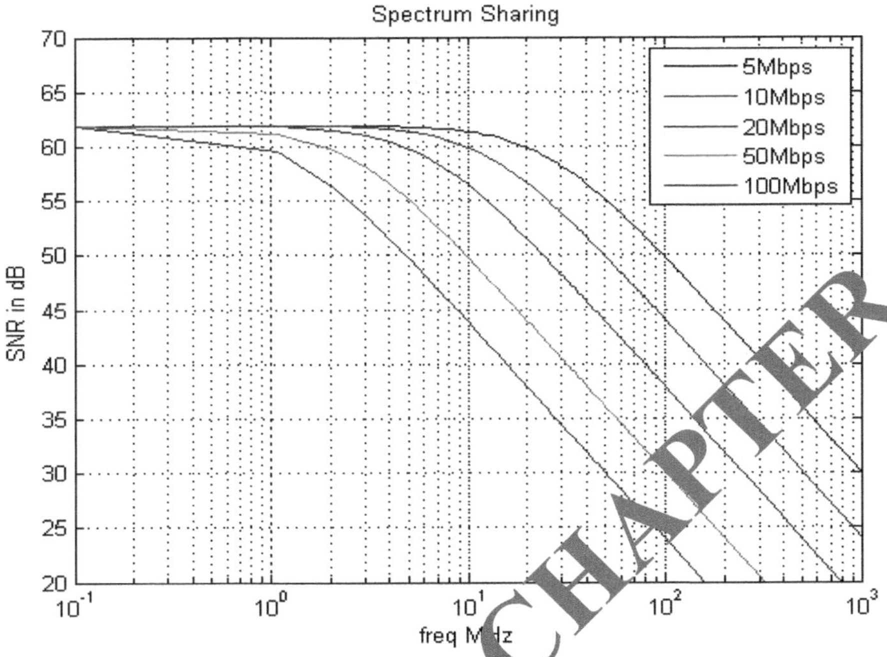

Fig. 3. Spectrum sharing

or minimal base stations send scheduling coded information to different compact terminals through symmetrical Code division multiplexing channels. Multihop communicate on coordinated non-cyclic remote systems is a promising high information rate interface innovation. It is outstanding the limit of multihop communicate on coordinated non- cyclic remote systems can be altogether improved by utilizing an appropriate power spending portion in remote cell organize. The particular esteem deterioration and water filling calculation have been utilized to gauge the execution of multihop communicate on coordinated non-cyclic remote systems incorporated framework. Right when Nt transmit and Nr addressed recieving wires are used, power outage limit is extended. In multihop convey on composed non-cyclic remote frameworks we transmit particular stream of data through different gathering mechanical assemblies.

3.[illegible] Zero Forcing Beam Forming (ZFBF)

We consider the two most obvious straight precoders, conjugate beamforming and zero- compelling, concerning net apparition adequacy and transmitted imperativeness profitability in an unraveled single-cell circumstance where spread is spoken to by free Rayleigh obscuring, and where channel-state information (CSI) getting and data transmission are both performed in the midst of a short soundness interval. A powerful clamor examination of the pre-coded forward channel yields express lower limits on net limit which represent CSI securing overhead and blunders just as the sub-optimality of the pre-coders. In this way the breaking points deliver trade off curves between transmitted essentialness capability and net ridiculous viability. For high powerful viability and low imperativeness capability zero-compelling beats conjugate column confining, while at low ridiculous efficiency and high essentialness adequacy the opposite holds (Figs. 4, 5 and Table 1).

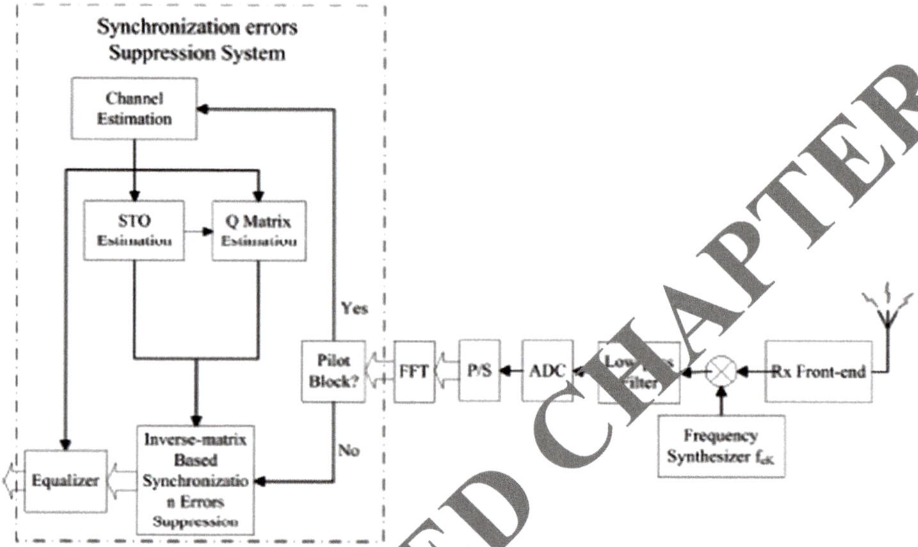

Fig. 4. Block chart of synchronization blunders concealment calculation

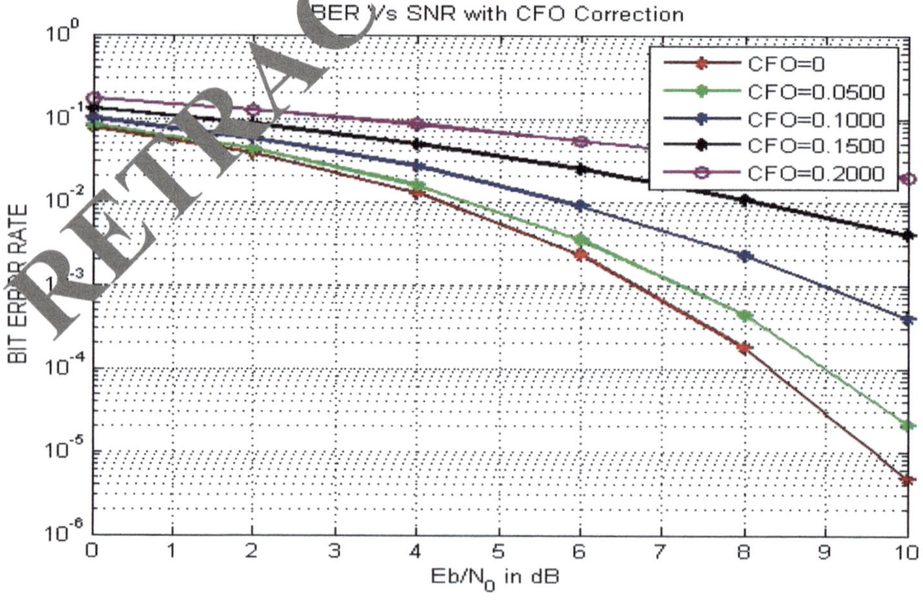

Fig. 5. BER vs SNR

Table 1. CFO and bit error rate

CFO	Bit error rate
0	0.0789
0.0500	0.0374
0.1000	0.0126
0.1500	0.0024
0.2000	0.0002

4 Experimental Results

See (Figs. 6, 7 and 8)

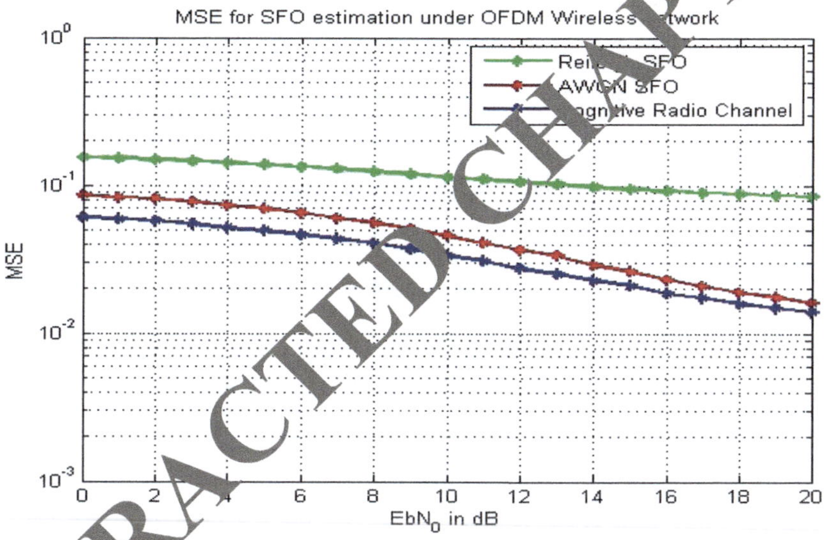

Fig. 6. SFO estimation

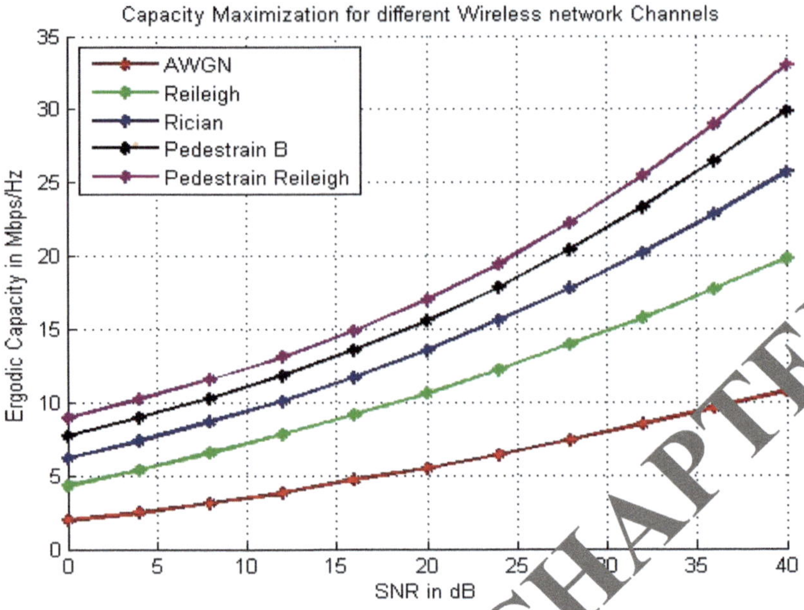

Fig. 7. Capacity maximization

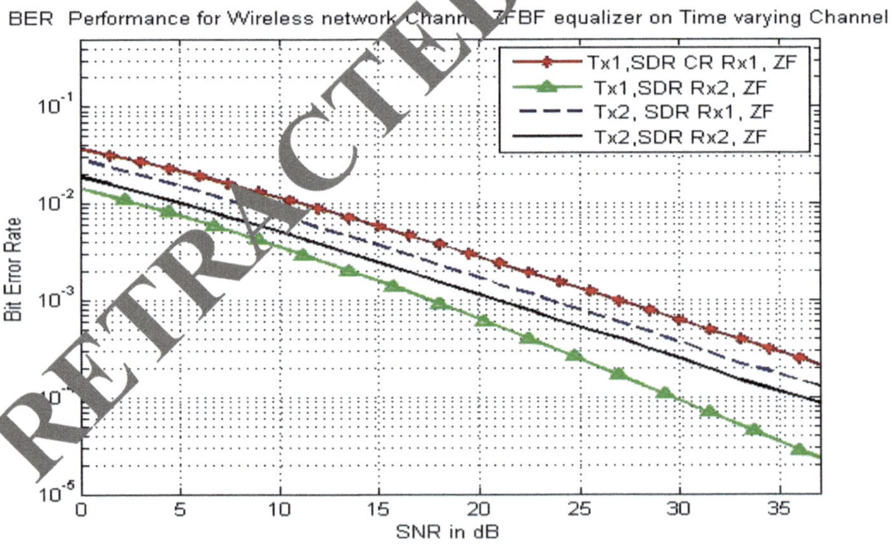

Fig. 8. BER performance

5 Conclusion

Remote Sensor Networks (WSN), specialists have proposed diverse steering conventions, yet they had a few issues in acquiring the ideal throughput in a productive way. The proposed framework gives expansion of throughput and low mistake rate. The crest normal power proportion and intercarrier impedance are diminished to evaluate the framework channel. So as to beat the absence of Quality of administration, vitality gathering for remote sensor organize application is proposed in this paper. It has favorable circumstances like limited control messages, re-ease of use of data transfer capacity, and upgraded control. In our work, it depends on the pragmatic channel estimation calculation, the channel estimation mistakes are first determined and after that the hearty asset allotment issue has been figured. The structure of the perfect ground- breaking precoder is first construed, in light of which the improvement issue will be modified in a general sense. we have settled the vitality amplification, successful range sharing and direct estimation in a WSN organize.

References

1. Zeng, X., Wang, D., Yu, M., Yang, H.: A new probability-based multihop broadcast protocol for vehicular networks. In 978-1-5090-[illegible]429-0/17/$31.00 c 2017 IEEE (2017)
2. Li, S., Huang, C.: A multihop broadcast mechanism for emergency messages dissemination in VANETs. In: 42nd IEEE International Conference on Computer Software & Applications (2018)
3. Suthaputchakun, C.: Multihop broadcast protocol in intermittently connected vehicular networks. In 0018-9251 C _ 2017 IEEE (2017)
4. Wu, C., Ohzahata, S., Ji, Y., Kato, T.: Joint fuzzy relays and network-coding-based forwarding for multihop broadcasting in VANETs. In: Digital Object Identifier. https://doi.org/10.1109/tits.2014.2[illegible]64044
5. Jaballah, W.B., Conti, M., Mosbah, M., Palazzi, C.E.: Fast and secure multihop broadcast solutions for intervehicular communication. In: Digital Object Identifier. https://doi.org/10.1109/tits.2013.2277890
6. Yu, D., Zou, Y., Yu, J., Cheng, X., Hua, Q.-S., Lau, F.C.M.: Stable local broadcast in multihop wireless networks under SINR. In: Digital Object Identifier. https://doi.org/10.1109/tnet.2018.2829712
7. Nardini, G., Stea, G., Virdis, A., Sabella, D., Caretti, M.: Broadcasting in LTE-advanced networks using multihop D2D communications. In 978- 1-5090-3254-9/16/$31.00 ©2016 IEEE (2016)
8. Yan, F., Zhang, X., Zhang, H.: Efficient multihop broadcasting with network coding in duty-cycled wireless sensor networks (NET). In: Digital Object Identifier. https://doi.org/10.1109/lsens.2017.2756065
9. Kuang, J., Yu, S.-Z.: Broadcast-based content delivery in information- centric hybrid multihop wireless networks. In 1089-7798 (c) 2016 IEEE (2016)
10. Wang, L., Banks, B., Yang, K.: Minimum-latency broadcast schedule in duty-cycled multihop wireless networks subject to physical interference. In 978-1-4799-7394-1/14 $31.00 © 2014 IEEE (2014)

11. Aravindhan, K., Kavitha, G., Dhas, C.S.G.: Plummeting data loss for multihop wireless broadcast using position based routing in VANET. In 978-1-4799- 7613-3/14/$31.00 ©2014 IEEE (2014)
12. Chang, C.-H., Kao, J.-C., Chen, F.-W., Cheng, S.H.: Many- to-all priority-based network-coding broadcast in wireless multihop networks. In 978-1-4799-1297-1/14/$31.00 ©2014 IEEE (2014)
13. Ramezanipour, I., Alves, H., Nardelli, P.H.J., Pouttu, A.: Energy efficiency of an unlicensed wireless network in the presence of retransmissions. In 978-1-5386-6355-4/18/$31.00 ©2018 IEEE (2018)
14. Zhao, Y., Xiao, S., Gan, H.: Broadcast cost reduction in wireless sensor networks with instantly decodable network codes. In 978-1-5386-6355- 4/18/$31.00 ©2018 IEEE (2018)
15. Wang, S., Kim, S.M., Kong, L., He, T.: Concurrent transmission aware routing in wireless networks. In 0090-6778 (c) 2018 IEEE (2018)
16. Katila, C.J., Buratti, C.: A novel routing and scheduling algorithm for multi-hop heterogeneous wireless networks. In 978-1-5386-6355- 4/18/$31.00 ©2018 IEEE (2018)
17. Furtado, A., Oliveira, R., Bernardo, L., Dinis, R.: Optimal cross-layer design for decentralized multi-packet reception wireless networks. In 978-1- 5386-6355-4/18/$31.00 ©2018 IEEE (2018)
18. Yun, J., Seo, S., Chung, J-M.: Centralized trust based secure routing in wireless networks. In 2162-2337 (c) 2018 IEEE (2018)
19. Chengetanai, G.: Minimising black hole attacks to enhance security in wireless mobile ad hoc networks. ISBN 978-1-905824-60-1
20. Darabkh1, K.A., Al-Rawashdeh1, W.S., Al-Zubi, A.T.: A new cluster head replacement protocol for wireless sensor networks. In 31.00 © 2017 IEEE (2017)

Wearable Antennas for Human Physiological Signal Measurements

M. Vanitha(✉) and S. Ramesh

Department of Electronics and Communication Engineering, Valliammai Engineering College, SRM Nagar, Chennai, India
mvanitha1073@gmail.com, rameshsvk@gmail.com

Abstract. Healthcare is a very important aspect in human life, but it should not be considered to be important for few people only. The population in the elder age group has substantially increased worldwide. Today most of the sick and elderly people are living alone at home, due to the high cost in consistent monitoring of health and expensive healthcare facilities at hospitals/nursing homes. To overcome this hurdle and bring health care to the aid of a common man too, a latest technology to monitor health via Remote health monitoring through telehealth application is implemented to monitor elders and new born babies. Remote health care monitoring system is made possible through wearable antenna. Modern communication and information technologies offers capable and cost effective solution that allows elders and the sick to be under un interrupted monitoring and still continue to live in their homes instead of expensive nursing home/hospital care. These Wearable antennas are fabricated by Nonwoven Conductive Fabrics (NWCFs) technology and used for measuring physiological signals (ECG, EMG, HR, BP, EDA, and RR) from human body. The wearable antenna transmits data through the 5G technology. The 5G technology supports sub-6 GHz frequency band and its range is 3.3–4.4 GHz. The NWCFs based wearable antennas are low cost, washable, without fraying problem and comfortable to the users. To highlight the suitability of the latest fabrication technique and to emphasize its benefits the proposed antenna is simulated to get the desired results.

Keywords: Remote health monitoring · Wearable antenna · NWCFs · Physiological signals

1 Introduction

Life expectancy has been dramatically increasing worldwide due to significant developments in latest healthcare amenities and medicines available for elders. Many elders live alone at home due to the increasing cost of the prescribed drugs, medical instruments, and hospital care. Therefore, it is essential to develop and implement new approaches and technologies in order to provide better health care monitoring services at an affordable cost to all the elderly and sick people. Remote healthcare monitoring allows elderly people to continue to undergo treatment and live at home rather than in expensive healthcare facilities [1]. Smart home may allow the elderly to stay in their home by using IoT wireless technology [2]. The health care monitoring system

D. J. Hemanth et al. (Eds.): COMET 2019, LNDECT 35, pp. 1441–1451, 2020.
https://doi.org/10.1007/978-3-030-32150-5_146

equipped with non-invasive and unobtrusive wearable antennas. The antennas have the ability to measure the human physiological signals such as electromyogram (EMG), electrocardiogram (ECG), heart rate (HR), body temperature, blood pressure (BP), and respiration rate (RR). The wearable antennas are used in GPS-GSM based tracking system by using logo antenna in leather bags [3]. In remote wireless health care monitoring system biosensor is placed on the human body for measuring physiological signals and transmitted over the ZigBee technology [4]. Fabrication techniques for wearable antenna are fabricated using conductive materials like copper tape, adhesive conductive fabric, and conductive thread [5]. The wearable antennas are fabricated by two technologies: the first technology is a nonwoven conductive fabric technology, whereas the second technology is embroidered conductive threads [6]. Nonwoven conductive fabric technology has quickly replaced the traditional E-textiles because NWCFs includes flexibility, mechanical resistance, washables, and conductivity. The cutting plotter can be efficiently used for shaping the antenna into smaller sizes and complicated geometries [7]. Embroidered conductive thread is much suitable, to be used with commercial sewing mechanism. Conductive threads are embroidered by hand, because CAD controlled sewing machine are not available [8]. These two fabrication technologies offer antennas at lower cost, wash ability, high spatial resolution and no fraying problem. Wearable antenna-based health monitoring systems may include different type of flexible antennas that can be integrated into clothes, textile fibers and elastic bands. Wearable antennas are fully stitched with clothes and they remotely transmit or receive the antenna data by using 5G technology.

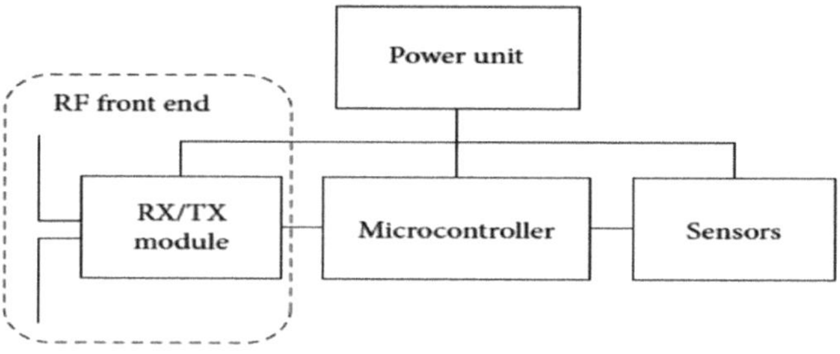

Fig. 1. General architecture of a wireless monitoring system.

The 5G technology is also affordable, consumes less power with high data transmission and high capacity. The wireless health monitoring system for elderly and new born babies is shown in Fig. 1 [6]. There are four main blocks are used in human health monitoring:

(a) The RF front end blocks include RX/TX unit of antennas.
(b) A microcontroller block processing the data received from sensor building block. It includes processor, memory and input/output peripherals.
(c) The sensor block used for measure physiological signals from human body.
(d) The power supply unit necessary for all above blocks.

Wearable antenna-based health monitoring systems may include different type of flexible antennas that can be integrated into clothes, textile fibers and elastic bands. Wearable antennas are fully stitched with clothes and remotely transmitting or receiving the antenna data by using 5G technologies. The 5G technology includes affordable, low power consumption, high data rate and high capacity. The paper is structured as follows. In Sect. 2, the fabrication techniques of antennas are explained. In Sect. 3, antenna structure, design and simulation are explained. In Sect. 4, simulation result compare with expected result. Finally, in Sect. 5, conclusions are drawn.

2 Wearable Antenna Fabrication Techniques

Wearable antenna fabrication includes four techniques:

(a) Conductive Fabrics [3, 7]
(b) Conductive threads [8]
(c) E-textiles
(d) Inks

The Nonwoven Conductive Fabrics and Conductive Thread technology has better spatial resolution and better performance than other techniques (Table 1).

Table 1. Comparison between wearable antenna fabrication techniques

Characteristics	Conductive materials			
	NWCFs	Thread	Electro-textiles	Inks
Conductivity	High	Low	High	Low
Spatial resolution	High	High	Low	Low
Fraying	No	No	Yes	No
Wash ability	Yes	Yes	Yes	Yes
Cost	Low	High	High	Low

3 Antenna Structure and Design

The rectangular Microstrip patch antenna is shown in Fig. 3. Copper annealed material is mostly used for designing a rectangular patch and ground plane with material thickness is 0.008 mm. A jeans material is used as a substrate with dielectric constant = 1.67, thickness h = 0.8 mm and loss tangent, tan = 0.01 (Fig. 2).

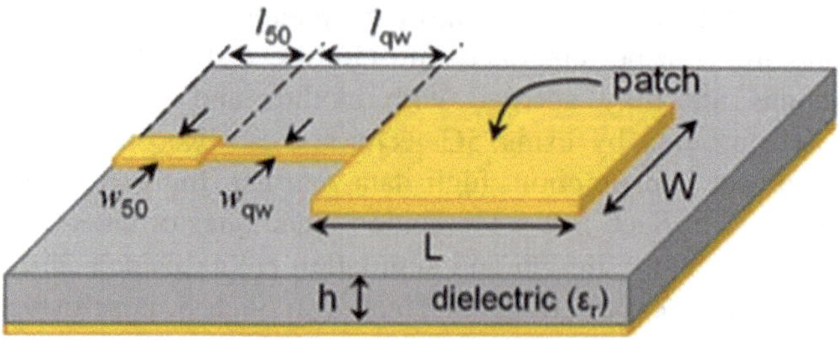

Fig. 2. General basic rectangular patch antenna

Basic design equations of Rectangular Microstrip Patch Antenna are given below [9].

The width of rectangular patch is given by:

$$W = \frac{C_0}{2f_r}\left(\sqrt{\frac{2}{\varepsilon_r + 1}}\right) \tag{1}$$

W-Width of the patch
C_0-Velocity of light in free space
f_r-Resonant frequency
ε_r-Dielectric constant of the substrate

Effective dielectric constant is given by:

$$\varepsilon_{reff} = \frac{\varepsilon_r + 1}{2} + \frac{\varepsilon_r - 1}{2}\left[1 + 12\frac{h}{W}\right]^{\frac{-1}{2}} \tag{2}$$

ε_r-Dielectric constant of the substrate
h-Height of the dielectric substrate
W-Width of the patch

The extended length of patch is given by:

$$\frac{\Delta L}{h} = 0.412\frac{(\varepsilon_{reff} + 0.3)\left(\frac{W}{h} + 0.264\right)}{(\varepsilon_{reff} - 0.258)\left(\frac{W}{h} + 0.8\right)} \tag{3}$$

The effective length of patch is given by:

$$L_{eff} = L + 2\Delta L \tag{4}$$

L – length of patch
ΔL – Extended length of patch

The length of patch is given by:

$$L = \frac{1}{2f_r\sqrt{\varepsilon_{reff}}\sqrt{\mu_0\varepsilon_0}} - 2\Delta L \tag{5}$$

The length of ground plane is given by:

$$Lg = L * 3 \tag{6}$$

The width of ground plane is given by:

$$Wg = W * 2 \tag{7}$$

The length of slot is given by:

$$c = \frac{L}{2.72}(or)\frac{W}{2.72} \tag{8}$$

The width of slot is given by:

$$d = \frac{c}{10} \tag{9}$$

The rectangular patch antenna dimensions are calculated by using above equations. The antenna dimensions are given in Table 2.

Table 2. Calculated Antenna dimensions using rectangular microstrip patch equations

Parameters	Dimensions (mm)
Patch length	30.25
Patch width	34.62
Ground plane length and substrate length	64.37
Ground width length and substrate width	69.24
Feed line length	12
Feed line width	2.922
Matching line length	7.0
Matching line width	0.778
Slot length	12.5
Slot width	0.8
Metal thickness	0.008
Substrate height	0.8
Substrate dielectric constant (ε_r)	1.67 (unit less)
Center frequency	3.7 GHz

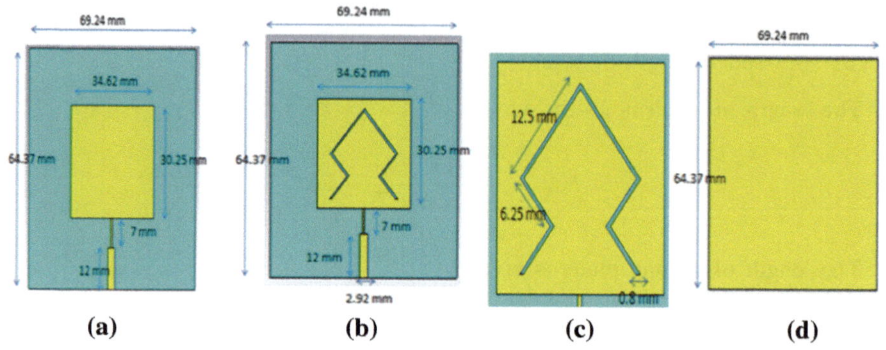

Fig. 3. Proposed antenna structure: (a) Rectangular patch antenna with offset feed; (b) Proposed antenna structure; (c) Proposed antenna with joined hand shape logo; (d) Proposed antenna with dimensions

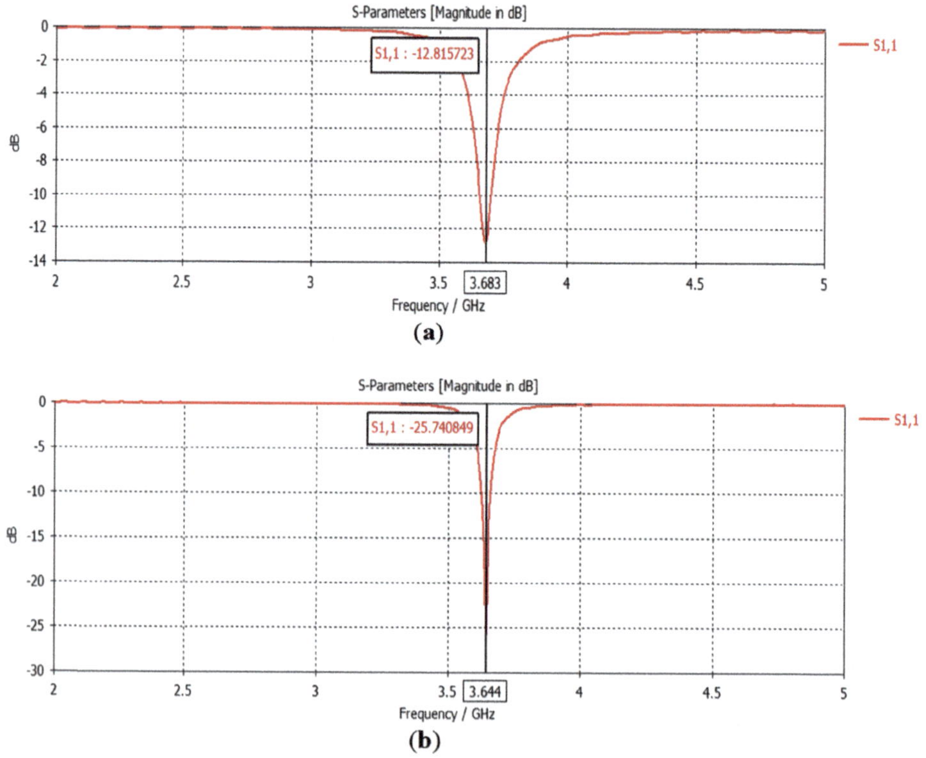

Fig. 4. Simulated reflection coefficient (S11). (a) Rectangular patch antenna without slot. (b) Rectangular patch antenna with slot.

4 Result and Discussion

The antenna design is simulated using 3D Electromagnetic (EM) Simulation tool of CST Microwave studio [12]. The designed antenna is simulated and measured the value of reflection coefficient, VSWR, gain of an antenna and radiation pattern of antenna. The reflection coefficient also called as return loss. The return loss is the loss of power in the signal reflected by a discontinuity in a transmission line. The simulated reflection coefficient (S11) plots are shown in Fig. 4. The rectangular patch antenna without slot reflection coefficient value is −12.81 dB at 3.68 GHz and after introducing joined hand shape logo slot reflection coefficient value is −25.74 dB at 3.644 GHz.

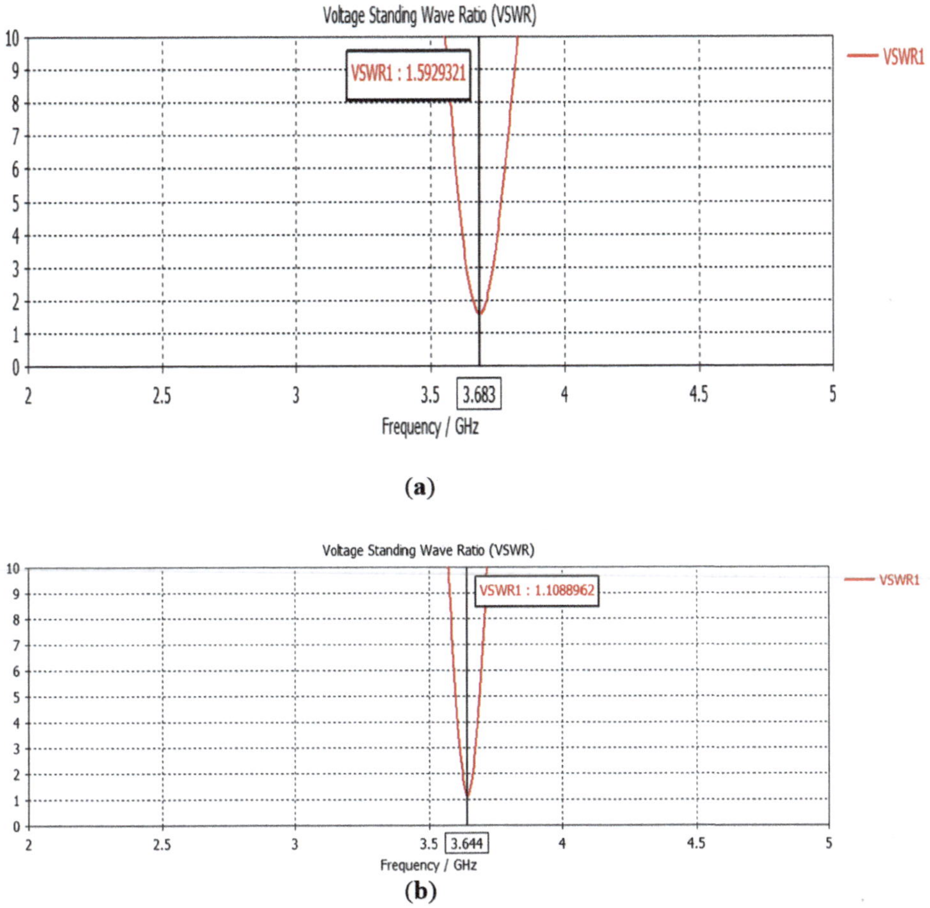

Fig. 5. Simulated VSWR. (a) Rectangular patch antenna without slot. (b) Rectangular patch antenna with slot.

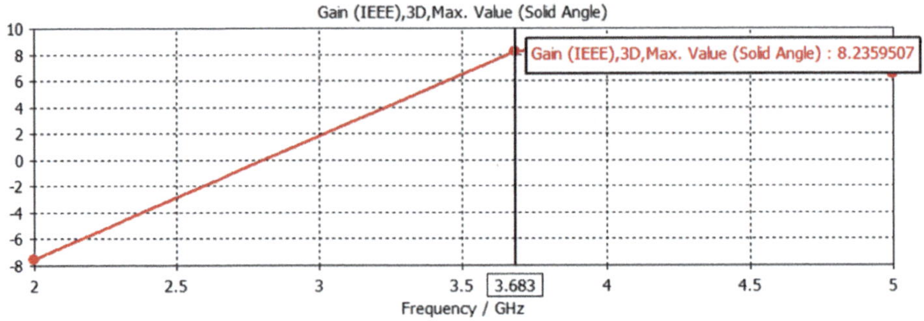

(a)

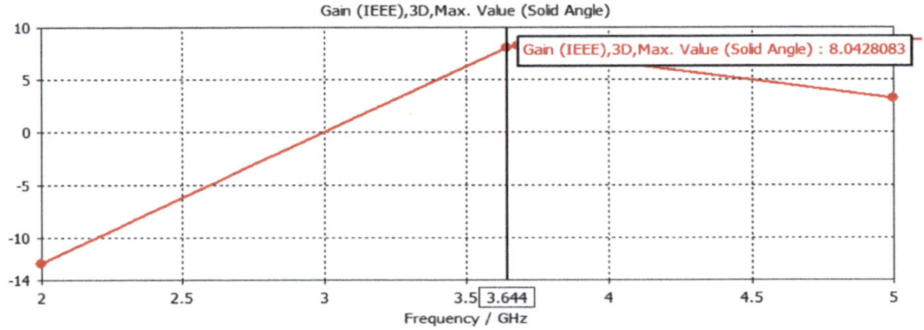

(b)

Fig. 6. Simulated antenna gain. (a) Rectangular patch antenna without slot. (b) Rectangular patch antenna with slot.

VSWR is a measure of how efficiently radio frequency power is transmitted from power source to load. The simulated VSWR plots are shown in Fig. 5. The rectangular patch antenna without slot VSWR value is 1.59 at 3.68 GHz and after introducing joined hand shape logo slot VSWR value is 1.1 at 3.64 GHz. The gain of simulated designed antenna shown in Fig. 6. The rectangular patch antenna without slot gain value is 8.2 dBi at 3.68 GHz and after introducing joined hand shape logo slot gain value is 8.0 dBi at 3.644 GHz. The radiation pattern of rectangular patch antenna without slot is shown in Fig. 7. The proposed antenna radiation pattern with slot (joined hand shape logo) is shown in Fig. 8. The radiation pattern includes xz-plane and yz-plane (Table 3).

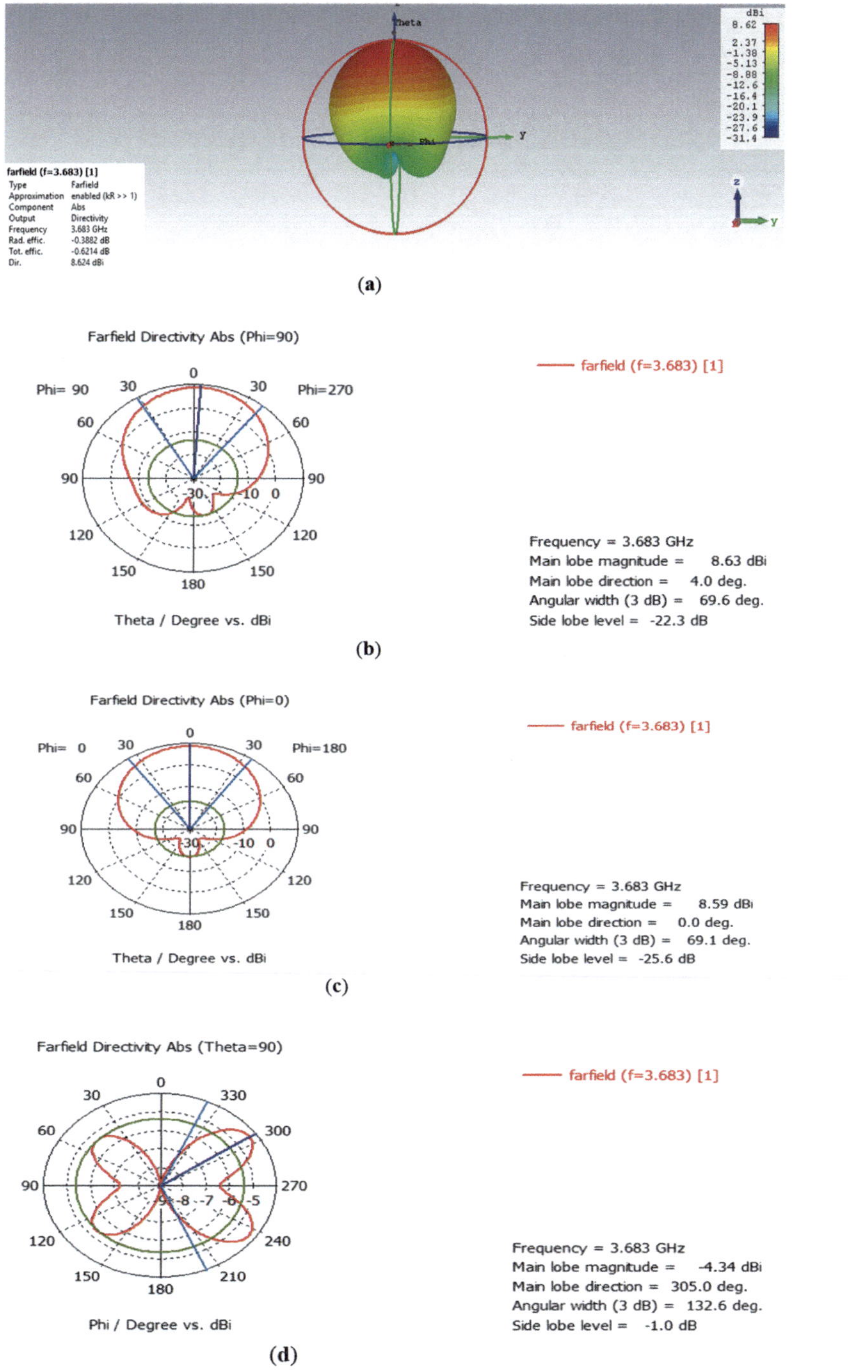

Fig. 7. Simulated antenna radiation pattern Rectangular patch antenna without slot. (a) Radiation Pattern with designed antenna structure. (b) Radiation pattern in x–z plane. (c) Radiation pattern in y–z plane. (d) Radiation Pattern in x–y plane

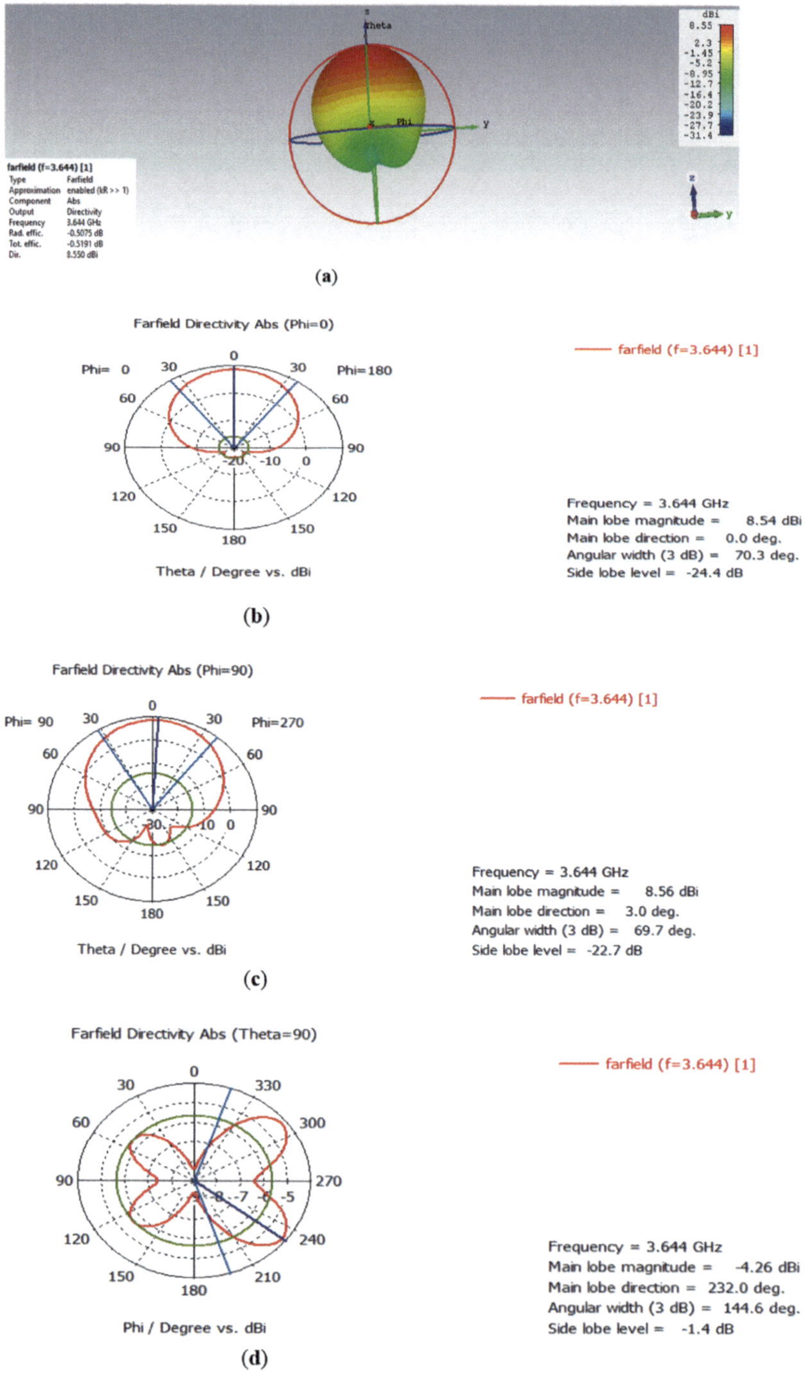

Fig. 8. Simulated antenna radiation pattern of Rectangular patch antenna with slot. (a) Radiation Pattern. (b) Radiation pattern in x–z plane. (c) Radiation pattern in y–z plane. (d) Radiation pattern in x–y plane.

Table 3. Comparison of the proposed antenna performance with expected results.

Parameters	Expected result [6]	Achieved result
Return loss	Above −10 dB	−25.74 dB
VSWR	1 to 2	1.1
Antenna gain	Above 7 dBi	8.04 dBi

5 Conclusion

In this work, the nonwoven conductive fabric based wearable antenna is designed and simulated using CST Microwave Software. The NWCFs based wearable antenna is easy to fabricate, as there is no fraying problem and comfortable to the users. The proposed antenna performances are compared with expected results. The proposed antenna return loss is −25.74 dB, VSWR is 1.1 and gain of antenna is 8.04 dBi. Finally, the characteristics and feats of the projected antenna, with slot and without slot, have been calculated and discussed.

Acknowledgment. The authors wish to acknowledge DST-FIST supporting facilities available in the department of Electronics and Communication Engineering at Valliammai Engineering College, Chennai, Tamil Nadu, India.

References

1. Mandal, D., Pattnaik, S.: Quad-band wearable slot antenna with low SAR values for 1.8 GHz DCS, 2.4 GHz WLAN and 3.6/5.5 GHz WiMAX applications. In: Progress in Electromagnetic Research B, vol. 81, pp. 163–182, September 2018
2. Corchia, L., Monti, G., de Benedetto, E., Tarricone, L.: Wearable antennas for remote health care monitoring systems. Int. J. Antennas Propag. **2017**(3012341), 1–11 (2017)
3. Majumder, S., Aghayi, E., Noferesti, M., Memarzadeh-Tehran, H., Mondal, T., Pang, Z., Deen, M.: Smart homes for elderly healthcare-recent advances and research challenges. Sensors **17**, 1–35 (2017)
4. Majumder, S., Mondal, T., Deen, M.: Wearable sensors for remote health monitoring system. Sensors **17**(1), 130 (2017)
5. Nakamura, R., Hadama, H.: Target localization using multi-static UWB sensor for indoor monitoring system. In: 2017 IEEE Topical Conference on Wireless Sensors and Sensor Networks (WiSNet), pp. 37–40, January 2017
6. Monti, G., Corchia, L., De Benedetto, E., Tarricone, L.: Wearable logo-antenna for GPS–GSM-based tracking systems. IET Microwaves Antennas Propag. **10**(12), 1332–1338 (2016)
7. Kiourti, A., Lee, C., Volakis, J.L.: Fabrication of textile antennas and circuits with 0.1 mm precision. IEEE Antennas Wirel. Propag. Lett. **15**, 151–153 (2016)
8. Monti, G., Corchia, L., Tarricone, L.: Textile logo antennas. In: Proceedings of 2014 Mediterranean Microwave Symposium (MMS2014), pp. 1–5, December 2014
9. Monti, G., Corchia, L., Tarricone, L.: Fabrication techniques for wearable antennas. In: 43rd European Microwave Conference, pp. 1747–1750, October 2013

Region Splitting-Based Resource Partitioning with Reuse Scheme to Maximize the Sum Throughput of LTE-A Network

S. Ezhilarasi(✉) and P. T. V. Bhuvaneswari

Department of Electronics Engineering, Anna University, Chennai, India
ezhilvish@yahoo.co.in, ptvbmit@annauniv.edu

Abstract. Third generation partnership project has developed Long Term Evolution - Advanced (LTE-A) technology to enhance the system capacity. Further, frequency reuse concept has been adopted in order to meet the requirement of mobile data traffic. This creates Inter-Cell Interference (ICI) which limits the throughput of cell edge users in LTE - A network. To mitigate ICI, Region splitting based Resource Partitioning with reuse scheme (RRPR) is proposed in this research. The objective is to maximize the sum throughput and average throughput of macrocell. In the proposed RRPR scheme, the whole macrocell is divided into inner, centre and outer regions. The overlaid femtocell partially reuses the spectrum of macrocell. In a cluster of three cells, the total spectrum is partitioned into four non-overlapping sub bands. The outer region of macrocells is assigned with first three sub bands. The remaining one sub band is shared by with the corresponding centre region. The inner region reuses the sub band of outer region of two neighboring cells. The analysis is made with respect to sum throughput and average throughput. The radius of inner and centre region of macrocell is varied using Monte Carlo simulation process. The radius that results with maximum sum throughput is concluded as optimal region radii. The performance metrics of the proposed RRPR scheme is compared with region splitting based resource partitioning scheme. From the simulation result, the inference drawn is that, the maximum of 147.99% enhancement is achieved for both sum throughput and average throughput by the proposed RRPR scheme.

Keywords: Inter cell interference · Long term evolution advanced · Resource Partitioning with reuse scheme · Frequency reuse · Optimal region radii

1 Introduction

The next generation cellular network aims to enhance throughput of the cell edge users which in turn can increase the system throughput. To accomplish this, frequency reuse concept is introduced in LTE network [1]. In order to improve the system capacity, spectral efficiency and coverage of LTE system, Heterogeneous network (Het Net) concept has been introduced by 3GPP in release 10 [2]. It is referred as LTE-Advanced (LTE-A). In the Het Net scenario, the low power small cell base stations (micro/pico/femto) are overlaid on the high power macro base station. The small cell

D. J. Hemanth et al. (Eds.): COMET 2019, LNDECT 35, pp. 1452–1464, 2020.
https://doi.org/10.1007/978-3-030-32150-5_147

base stations are varied with in terms of coverage, nature of deployment and transmission power [3]. Among the small cells, femtocells are user deployed. Hence it can be a promising solution to enhance the system capacity. However, this imposes various challenges in LTE-A network namely, inter-cell and intra-cell interference, resource partitioning, Inter cell and intra RAT handover, scheduling and load balancing etc., [4]. The inter-cell and intra-cell interference are identified as the major contribution factors for enhancing the Quality of Service (QoS) of cell edge users and system throughput. However, the intra cell interference is eliminated by orthogonal assignment of resource blocks in Orthogonal Frequency Division Multiple Access (OFDMA) technology [5]. It is used for down link transmission in both LTE and LTE-A networks. To increase the spectral efficiency and network throughput, frequency reuse concept is adopted in OFDMA based LTE-A network. Due to the reuse of same frequency resource by the neighboring macrocell, it may impose ICI. Further it limits the QoS of cell edge users. In the existing literatures, the solution of ICI can be classified as Interference Cancellation (IC), Interference Randomization (IR) and Interference Avoidance (IA) [6]. This research is limited to IA based ICI mitigation technique. In this technique, careful management of ICI is realized through the efficient resource partitioning schemes. The main objective of this research is to analyze the impact of region radii on maximization of macrocell sum throughput.

The remaining of the paper is organized as follows: The state of art related to the proposed research is presented in Sect. 2. Section 3 details the system model of the proposed RRP scheme. Section 4 describes the performance analysis of the proposed research along with simulation results. Finally, Sect. 5 highlights the paper with future work.

2 Literature Survey

This section presents the existing solutions that mitigate ICI in a two-tier femtocell network. The research is limited to frequency reuse concept adopted in IA technique. The macrocells are assumed to be center excited with omni directional antenna.

The authors of [7] have developed the optimal Fractional Frequency Reuse (FFR) scheme through the dynamic strategy of resource allocation. The impact of region radius on total throughput and User Satisfaction (US) are analyzed. It is concluded that, the maximum value of US is optimal in both static and mobile environment. Further, it is compared with Integer frequency reuse 1 (IFR1) and Integer frequency reuse 3 (IFR3) schemes. The inference drawn is that, the presented work performs better than the existing scheme with respect to total throughput.

In [8], the authors presented Optimal Static FFR (OSFFR) scheme. In this scheme, the macrocell is divided into center and edge zone with six sectors. The whole spectrum is divided into seven sub bands in which only one sub-band is utilized by center zone UEs with FR1. Wherein, the remaining sub bands utilized by the edge zone UEs with FR6. Further, the femtocell of each region partially reuses the sub band of macrocell by considering the intra and inter-cell cross-tier interferences. The limitation found is that, higher number of sub bands and sectors increases the complexity and also femtocells are assigned with more sub bands. The considered metrics are outage probability, spectral efficiency and network throughput. The presented scheme is compared with strict FFR, soft FFR, and FFR-3. It is observed that, the developed scheme outperforms the existing schemes.

In [9], the authors have analyzed ICI in LTE and LTE-A network through the simulation framework. In this study, the macrocell is divided into inner and outer region. They utilize four sub bands. In the cluster of three cells, three sub bands are utilized by the outer region and remaining one by the inner region. The optimal region radius is found by the following metrics. They are Jain's Fairness Index, total throughput, and weighted throughput. It is inferred that the weighted throughput outperforms, than other two metrics.

The authors of [10] have developed frequency partitioning method to mitigate cross-tier interference in a two tier LTE network. The entire macrocell is divided into inner and outer region. The outer region is divided into three sectors with directional antennas. The available spectrum is separately allocated to both uplink and downlink. It is further partitioned into four non-overlapping sub bands for both transmissions. The inner region is utilized by one sub band for macrocell/femtocell whereas the remaining sub bands are shared by both macrocell and femtocell. From the results, it is observed that the impact of inner region radius on interference power is within the acceptable limit. Further it is found that, inner region radius have not been extended to a wide range.

In [11], the authors have determined the optimal value of inner region radius by an adaptive self-organizing frequency reuse scheme. The whole macrocell is divided into inner and outer regions. The available spectrum is partitioned into four sub bands. In this scheme, the inner region MUE is utilized by any of the sub-bands on the basis of considerable amount of total interference power. From the simulation result, the authors concluded that the optimal value of inner region radius is the radius that offer better user throughput of inner and outer region. The analysis is also extended for varied cell radius of macrocell and its transmission power. Further, the developed scheme is compared with traditional FFR scheme in terms of total throughput.

In [12], the authors presented Region splitting based Resource Partitioning (RRP) scheme to enhance the throughput of indoor MUE. In this scheme, the macrocell is divided into inner, centre, and outer regions. The femtocells deployed in each region partially share the spectrum of the corresponding macrocell. In a cluster of three cells, the whole spectrum is partitioned into four sub-bands. These sub-bands are utilized by both macro and femtocell in order to mitigate the inter and intra-cell cross tier interference. Simulation analyses in terms of (i) throughput of indoor MUE with respect to varied number of femtocells (ii) MUE devices (iii) position and transmission power of femtocell are made. Further, the developed scheme has been compared with a traditional FFR scheme in terms of inner region radius. It is inferred that, the enhancement of 29.7% has been achieved.

From the existing literature, the resource partitioning between macro and femtocells are made in terms of partition of macrocell region, inter and intra cell cross-tier interference, frequency reuse, and overlaid femtocells. The authors in [7, 9] and [11], have converged optimal region radius by the following metrics. They are US, JI, total throughput, weighted throughput, and throughput of each region. The authors in [8] have considered the macrocell coverage into both directional and omni directional antennas. Further, it is found that the outage probability of the macrocell decreases by assigning more sub-bands to femtocell. The work presented in [10], have limited the region radius analyses to two different radii.

The authors in [12] have analyzed the impact of inner region radius on inner region throughput. However, the impact of region radii on sum throughput of macrocell can be analyzed. Hence, optimal region radii can be arrived to maximize the sum throughput. In this research, Region splitting based Resource Partitioning with reuse (RRPR) scheme is proposed to overcome the above limitations, thereby enhance the average throughput and system throughput.

In the proposed RRPR scheme, the macrocell region is partitioned into three, namely inner, centre, and outer. In a cluster of three cells, the whole frequency spectrum is divided into four non-overlapping sub-bands. They are 'a', 'b', 'c', & 'd'. The sub-bands 'a', 'b' & 'c' are utilized by outer region and sub-band 'd' is further divided into three parts, namely 'd1', 'd2', & 'd3'. It is used by centre region. The inner region of macrocell reuses the sub-band of outer region of the two neighboring macrocells. Similarly, the femtocells deployed in each region partially reuses the sub-band of macrocell to mitigate the inter and intra-cell cross tier interference.

The objective of the proposed scheme is to analyze the impact of region radii on maximization of sum throughput and average throughput of MUE. Therefore, in order to achieve the objective, the optimal value of region radii is determined by the Monte Carlo Simulation process.

3 Proposed Methodology

This section presents the system model and methodology of the proposed RRPR scheme.

3.1 System Model

The system model of the proposed RRPR scheme is presented in this section. Further, the impact of region radii on sum throughput and average throughput of MUE is investigated. In this model, the femtocell deployed over a macrocell is considered. It partially reuses the spectrum of the corresponding macrocell. This kind of spectrum sharing is considered for downlink transmission. Let 'm' be the macrocell which is divided into inner region (IR), centre region (CR), and outer region (OR). Let 'f' be the femtocell which is located in the boundary of inner region. It is illustrated in Fig. 1.

Let R_I, R_C, R_M, and R_f be the radius of inner region, centre region, macrocell and femtocell respectively. Let 'x' be the total number of MUES (x=100) which are randomly distributed within the macrocell. In this research, the region radii which maximize the sum throughput and average throughput of MUE are considered as one of the important design parameter.

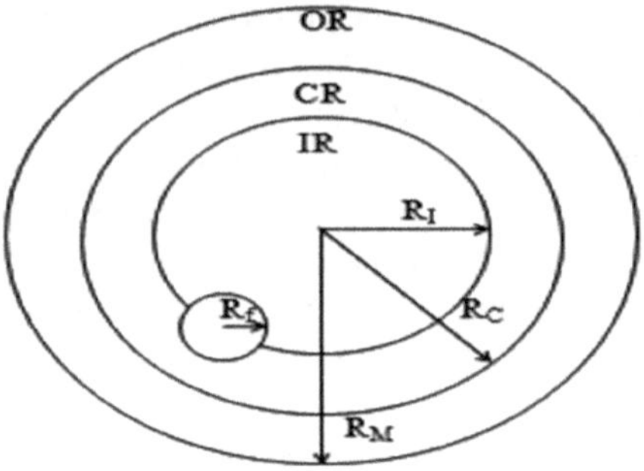

Fig. 1. System model

3.2 Proposed Methodology

The work flow of the proposed methodology is shown in Fig. 2. The following modules are detailed below:.

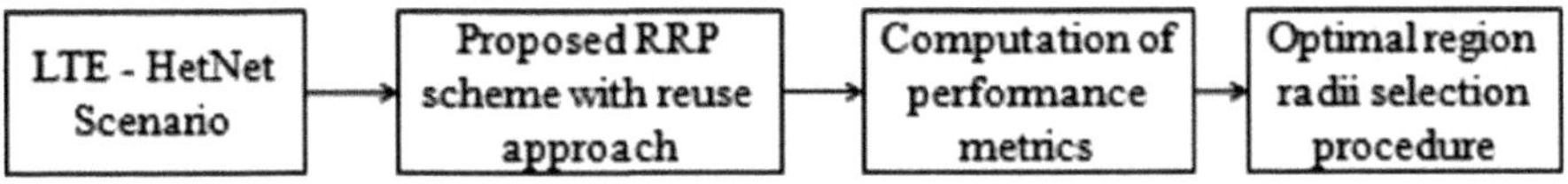

Fig. 2. Work flow of the proposed methodology

(i) Formation of LTE HetNet scenario
(ii) RRP scheme with reuse approach
(iii) Computation of Performance metrics
(iv) Optimal region radii selection procedure.

LTE - HetNet Scenario
The femtocell overlaid within the corresponding macrocell is formed as a LTE HetNet which is shown in Fig. 3. In order to analyze the interference experienced by indoor MUE, a two-tier network with seven cell system structure is considered. In this scenario, the macrocell is center excited with omni directional antenna. The impact of inter and intra-cell cross tier interference on performance metrics are analyzed by the proposed RRPR scheme. It is detailed in the next section. Centre cell '1' is considered as serving cell while remaining six cells are taken as interfering cells.

Region Splitting Based Resource Partitioning with Reuse Scheme
In a cluster of three cells, the whole spectrum is divided into four non-over lapping sub bands. In which three sub bands are utilized by outer region of macrocell with Frequency Reuse Factor-3 (FRF-3). Whereas, the remaining one part shared by the centre region with FRF-3. However, the inner region of macrocell reuses the spectrum of outer region of two neighboring macrocells which resulting enhancement in spectral efficiency.

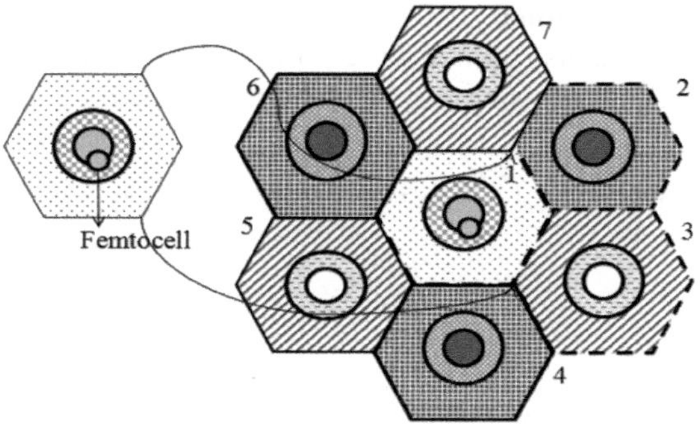

Fig. 3. LTE - Heterogeneous Network with seven cell structure

The detailed procedure of the resource partitioning in the proposed RRPR approach is presented in Fig. 4.

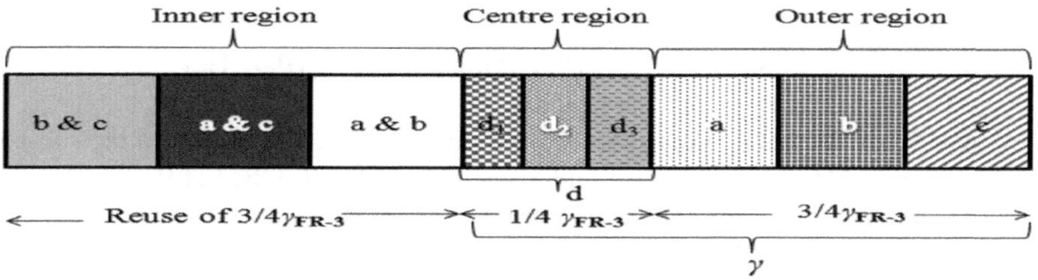

Fig. 4. Resource partitioning strategy of RRPR scheme

The total spectrum ‘γ’ is partitioned into ‘a’, ‘b’, ‘c’, and ‘d’ sub-bands. These sub bands are used by a cluster of three cells cells 1, 2 and 3 as shown in Fig. 3. The sub band ‘d’ is further divided into ‘d_1’, ‘d_2’, and ‘d_3’. Here the femtocells are positioned in each region. They partially reuse the sub band of macrocell. The detailed description of each sub band is mentioned in the Fig. 4. Thus the resource partitioning strategy of proposed RRPR scheme mitigates the inter and intra cell interference.

The total spectrum ‘γ’ is partitioned into ‘a’, ‘b’, ‘c’, and ‘d’ sub-bands. These sub bands are used by a cluster of three cells cells 1, 2 and 3 as shown in Fig. 3. The sub band ‘d’ is further divided into ‘d_1’, ‘d_2’, and ‘d_3’. Here the femtocells are positioned in each region. They partially reuse the sub band of macrocell. The detailed description of each sub band is mentioned in the Fig. 4. Thus the resource partitioning strategy of proposed RRPR scheme mitigates the inter and intra cell interference.

Computation of Performance Metrics

The calculation of performance metrics is detailed in this section. The computation of (i) sub channel (ii) SINR and (iii) data rate are included in the calculation performance metrics. They are detailed below.

Computation of Sub Channel

The amount of sub channels required for each region is calculated based on their area of coverage. It is calculated using the following expressions.

$$\gamma_I = \left(\frac{R_I}{R_M}\right)^2 * \gamma \tag{1}$$

$$\gamma_C = \mathrm{Min}\left[\frac{\gamma}{4}, (\gamma - \gamma_I)\right] \tag{2}$$

$$\gamma_O = \mathrm{Min}\left[\frac{\gamma}{4}, (\gamma - (\gamma_I + \gamma_C))\right] \tag{3}$$

Where γ is the total sub channels and γ_I, γ_C, γ_O, are the amount of sub channels required by inner, centre and outer region respectively.

Computation of SINR

In OFDMA based cellular network, the SINR is calculated using Eq. (4) [13].

$$\beta_{x,\varphi} = \frac{P_{1,(\varphi)} G_{1,x}}{N_O \Delta_f + \sum_{m=1}^{k} \left(P_{x,k,\varphi} G_{x,k,\varphi}\right) + \sum_{f=1} \left(P_{x,f,\varphi} G_{x,f,\varphi}\right)} \tag{4}$$

Where $\beta_{x,\varphi}$ is the SINR experienced by indoor MUEs, ‘x’ by the operating sub-band of ‘φ’ where $\varphi \in$ ‘a’, ‘b’, ‘c’, & ‘d_1’ and x $\in$ MUE of IR, CR, OR. Let $P_{1,\varphi}$, and $G_{1,x}$ be the transmitting power and its corresponding channel gain of serving base station ‘1’ and x respectively. Where $P_{x,k,\varphi}$, and $P_{x,f,\varphi}$ are the interference power received from ‘k’ interfering macro base stations and one femtocell respectively. The corresponding channel gains are represented by $G_{x,k,\varphi}$, $G_{x,f,\varphi}$.

Computation of Data Rate

With reference to the previous case, the data rate $C_{x,\varphi}$ of indoor MUE is represented in Eq. (5).

$$C_{x,\varphi} = \Delta_f \log_2\left(1 + \sigma \beta_{x,\varphi}\right) \tag{5}$$

Where $\beta x_{,\varphi}$ is the SINR of indoor MUEs (‘x’). Δf and σ represents sub carrier spacing and the target Bit Error Rate (BER) respectively. Where σ is the constant term and it is given by $\sigma = -1.5/\ln(5BER)$.

Computation of Sum Throughput

The sum throughput of indoor MUEs is calculated by using the following equation [13]. It is represented by ω_x.

$$\omega_x = \sum_x \sum_\varphi \alpha_{x,\varphi} C_{x,\varphi} \tag{6}$$

Where $\alpha_{x,\varphi}$ is the sub band allocation index, its value is given by the following

$\alpha_{x,\varphi}$ = 1; MUE assigned by sub band ‘φ’,
= 0; Otherwise.

Computation of Average Throughput of MUE
Similarly, the average throughput of individual MUE is calculated below. It is denoted by C_{avg}.

$$C_{avg} = \frac{\sum_{x}\sum_{\varphi}\alpha_{x,\varphi}\alpha_{x,\varphi}C_{x,\varphi}}{X} \tag{7}$$

Hence, amount of sub channels, region radii, are identified as the major contribution factors in maximizing the objective.

The Optimal Radii Selection Procedure
The procedure of determination of region radii is presented in this section. It is illustrated in Fig. 5. Initially the LTE HetNet consisting of one macrocell with one overlaid femtocell is considered. Then, the following network parameters are defined. They are: radius of macrocell and femtocell, deployment of indoor MUEs. In the procedure, the minimum of one sub channel to be utilized by each region is taken as the assumption. Let RIC be the ratio of inner region radius to centre region radius and let RCM be the ratio between centre region radius and macrocell radius. The performance metrics are computed for all possible range of RIC and RCM.

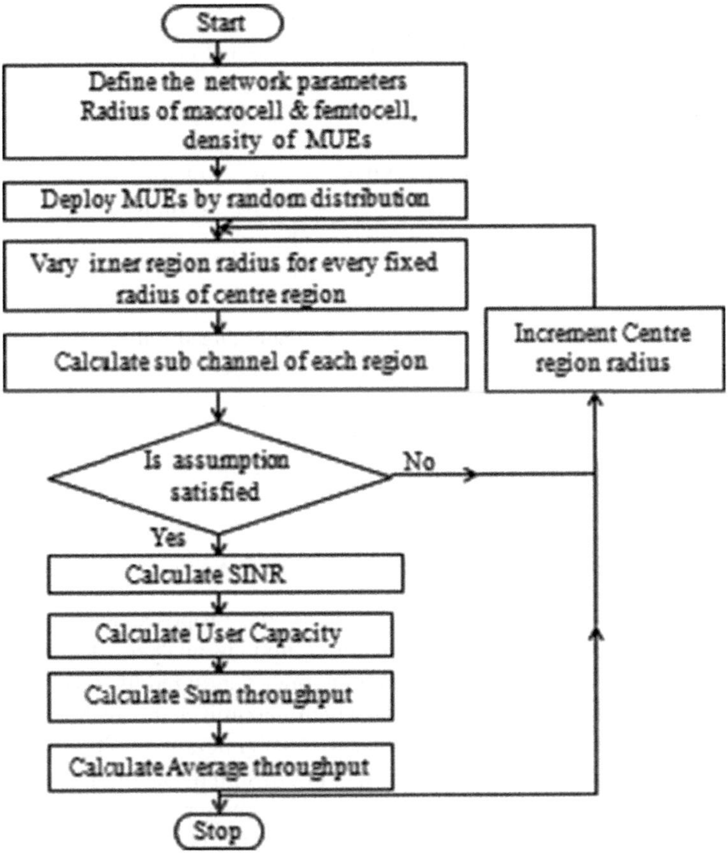

Fig. 5. Detailed procedures of optimal selection region radii

4 Results and Discussion

The outcome of the proposed RRPR scheme is presented in this section. The performance analysis in terms of sum throughput and average throughput is investigated for varied range of center region radius and its corresponding inner region radius. It is simulated using MatLab 2014 version. The following range of RCM and its corresponding RIC is considered based on the assumption. They are RCM = {0.3, 0.4, ...0.8, 0.9} and RIC = {0.2, 0.3,....0.7, 0.8}.

Table 1. Simulation parameters

Parameter	Value
Radius of macrocell	1000 m
Radius of femtocell	10 m
Transmit power of eNB	46 dBm
Transmit power femtocell	20 dbm
Carrier frequency	2 GHz
System bandwidth	10 MHz
Number of sub channels	48
Minimum distance between indoor MUE & MBS, MUE & femtocell	35 m, 0.2 m
Path loss (db) between eNB and indoorMUE	15.3 + 37.6 log10 * R1+ Low, R1 in 'm'
Path loss (db) between femtocell and indoorMUE	38.46 + 20log10 * R2 + 7 db, $0 < R2 \leq 10$, R2 in m'
Number of eNB	1
Number of femtocell	1
Number of MUE	100
Antenna pattern	Omni directional
Modulation scheme	64 QAM
Sub carrier spacing (Δf)	15 kHz
Wall penetration loss (Low)	20 db
White noise power density (No)	−174 dBm/Hz

4.1 Optimal Region Radii of the Proposed RRPR Scheme

This section presents the optimal region radii of the proposed RRPR scheme. The impact of region radii on macrocell sum throughput is analyzed for fixed value of centre region radius (RC) and its corresponding inner region radius (RI). It is illustrated in Fig. 6. The optimal value of region radii which maximizes the average value of sum throughput is found using Monte Carlo process of 1000 iterations. The inference drawn from the result is that, the average value of maximum sum throughput is achieved when RCM = 0.9 and RIC = 0.8. The reason is that, the inner region of macrocell reused 100% of sub channel of the outer region of two neighboring cells. Further, it is

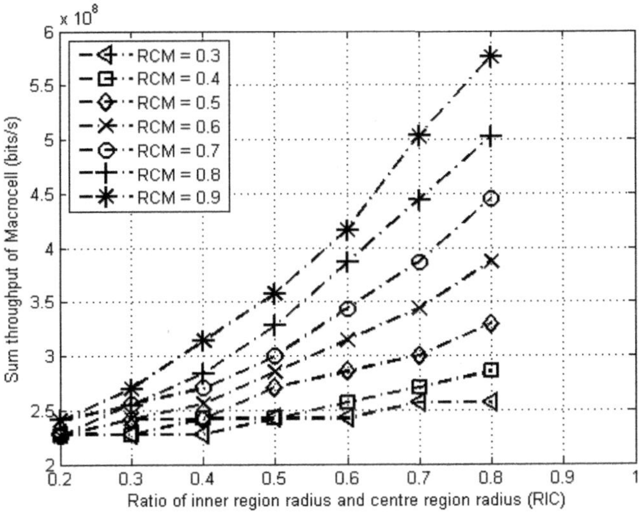

Fig. 6. Impact of region radii on Sum throughput of marcocell

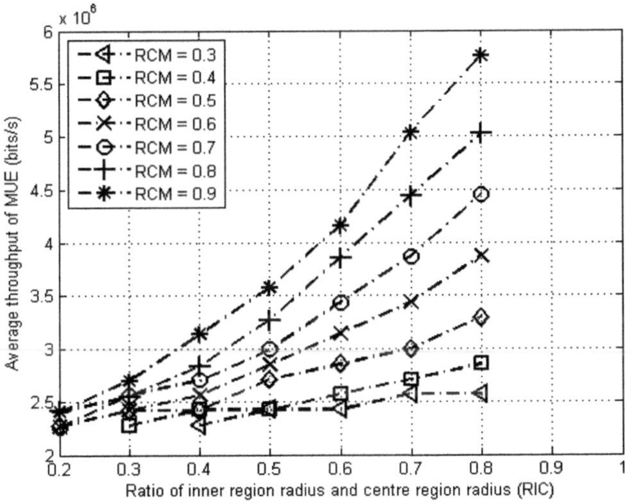

Fig. 7. Average throughput of MUE at optimal radii

observed that, the sub channel utilized by centre and outer region remains same with in minimal level when RCM is increased from 0.3 to 0.9. Hence it is concluded that, the optimal sum throughput at RCM = 0.9 and RIC = 0.8 is configured as the optimal radii by the proposed RRPR scheme.

4.2 Impact of Region Radii on Average Throughput of MUE

Similarly, the average throughput of MUE is computed for different range of region radii. It is presented in Fig. 7. The inference drawn is that, the maximum of 5.75 Mbps is achieved by individual MUE when RCM = 0.9 and RIC = 0.8

4.3 Impact of Reuse on Macrocell Sum Throughput and Average Throughput of MUE in the Proposed RRPR and RRP Scheme

This section presents the impact of reuse concept on macrocell sum throughput and average throughput of MUE. The above performance metrics are analyzed based on with and without adaptation of frequency reuse in the RRPR and RRP scheme respectively. It is illustrated in Fig. 8(a) and (b). In the RRP scheme, the optimal region radii arrived at RC = RI = 0.5, whereas RCM = 0.9 and RIC = 0.8 for RRPR.

From the results it is inferred that, 147.99% enhancement in macrocell sum throughput and average throughput is achieved by the proposed RRPR scheme. This is achieved when 100% of sub channel of two neighboring macrocells are reused by the inner region of reference cell. Hence it is observed that, the proposed RRPR scheme is more spectral efficient which resulting enhancement in sum throughput and average throughput. Further, it is concluded that, the service provider can configure the macrocell by the proposed RRPR scheme, when density of MUEs is higher in the configured region radii. Whereas, RRP is adopted for lesser number of MUEs are positioned in the inner region.

(a) Sum throughput of macrocell (b) Average throughput of MUE

Fig. 8. Performance metrics of RRPR and RRP scheme at optimal radii

5 Conclusions and Future Work

In this research, region splitting based resource partitioning with reuse scheme is proposed in order to maximize the sum throughput and average throughput of macrocell. In the proposed scheme, the whole macrocell has been divided into inner, centre and outer regions. In a cluster of three cells, the total spectrum has been partitioned into four non-overlapping sub bands. The outer region of macrocells has been assigned by the first three sub bands. The remaining one sub band has been shared by its corresponding centre region. While the inner region had reuse the sub band of outer region of two neighboring macrocells. The overlaid femtocell placed in the boundary of inner region partially reuses the spectrum of inner region. The analysis has been carried out with respect to sum throughput and average throughput of macro user equipment. The region radii which maximized the sum throughput of macrocell have been determined by the Monte Carlo process. From the simulation result, the region radii which results in maximum sum throughput of 575.85 Mbps at RCM = 0.9 and RIC = 0.8 is concluded as the optimal region radii.

The proposed RRPR scheme is compared with region splitting based resource partitioning in terms of sum throughput and average throughput of MUE at optimal region radii. The inference drawn is that, 147.99% of enhancement has been achieved in both sum throughput and average throughput of MUE. The proposed scheme can further be extended to the analysis of total network throughput.

References

1. Xiang, Y., Luo, J.: Inter-cell interference mitigation through flexible resource reuse in OFDMA based communication networks. In: European Wireless Conference, pp. 1–7, April 2007
2. 3GPP TR 36.913 version 10.0.0: LTE; Requirements for further advancements for Evolved Universal Terrestrial Radio Access (E-UTRA) (LTE-A)-Release, October 2010
3. Bendlin, R., Chandrasekhar, V., Chen, R., Ekpenyong, A., Onggosanusi, E.: From Homogeneous to heterogeneous networks: a 3GPP long term evolution rel. 8/9 case study. In: IEEE Annual Conference on Information Sciences and Systems, pp. 1–5 (2011)
4. Lee, Y., Chuah, T., Loo, J., Vinel, A.: Recent advances in radio resource management for heterogeneous LTE/LTE-A networks. IEEE Commun. Surv. Tutor. **16**(4), 2142–2180 (2014)
5. Singh, V., Kaur, G.: Inter-cell interference avoidance techniques in OFDMA based cellular networks: a survey. Int. J. Emerg. Technol. Eng. Res. (IJETER) **1**(1), 1–7 (2015)
6. 3GPP R1-060291: OFDMA Downlink inter-cell interference mitigation. Nokia (2006)
7. Bilios, D., Bouras, C., Kokkinos, V., Papazois, A., Tseliou, G.: Selecting the optimal fractional frequency reuse scheme in long term evolution networks. J. Wirel. Pers. Commun. **71**, 1–20 (2013)
8. Saquib, N., Hossain, E., Kim, D.I.: Fractional frequency reuse for interference management in LTE-advanced Het Nets. IEEE Wirel. Commun. **20**(2), 113–122 (2013)
9. Bouras, C., Diles, G., Kokkinos, V., Kontodimas, K., Papazois, A.: A simulation framework for evaluating interference mitigation techniques in heterogeneous cellular environments. J. Wirel. Pers. Commun. **77**(2), 1213–1237 (2014)
10. Chen, D., Jiang, T., Zhang, Z.: Frequency partitioning methods to mitigate cross-tier interference in two-tier femtocell networks. IEEE Trans. Veh. Technol. **64**(5), 1793–1805 (2015)
11. Elwekeil, M., Alghoniemy, M., Muta, O., Abdel-Rahman, A.B., Gacanin, H., Furukawa, H.: Performance evaluation of an adaptive self-organizing frequency reuse approach for OFDMA downlink. J. Wirel. Netw. **25**, 1–13 (2017)

12. Ezhilarasi, S., Bhuvaneswari, P.T.V.: Region splitting based resource partitioning to enhance throughput in long term evolution-advanced networks. J. Comput. Electr. Eng. **71**, 294–308 (2018)
13. Lei, H., Zhang, L., Zhang, X., Yang, D.: A novel multi-cell of DMA system structure using fractional frequency reuse. In: IEEE International symposium on Personal, Indoor and Mobile Radio Communications, pp. 1–5, September 2007

Secure and Practical Authentication Application to Evade Network Attacks

V. Indhumathi(✉), R. Preethi, J. Raajashri, and B. Monica Jenefer

Department of Computer Science and Engineering,
Meenakshi Sundararajan Engineering College, Chennai, India
indhu167@gmail.com, preeti3557@gmail.com,
raaji0919@gmail.com, hod.cse@msec.edu.in

Abstract. This paper elucidates the different types of attacks such as IP attack, URL attack, DOS attack, phishing during a file transfer. The objective is to provide a single platform for file transfer that can identify and resolve pervasive attacks in networking. A web application is developed for this purpose. When a file is transferred from the sender to the receiver it is transported through a secure FTP channel. An attacker can easily manipulate the channel to retrieve the file. The sender generates a secret key during transfer which is shared with the receiver. Using DES encryption, the file is encrypted and decrypted at the sender and receiver side respectively. When it is transferred through a channel, the file is stored in the buffer area for quick access. The attacks are monitored and reported to the administrator if it occurs. An administrator monitors the channel during transfer so that any malicious act can be identified and resolved then and there. The file is not obtained by the receiver if an attack takes place. In case of an attack the IP address of the attacker is stored in a database and the file is destroyed by the administrator so that the attacker cannot retrieve it. If no attack occurs and the file is received by the receiver, and an acknowledgement is sent to the sender. On the receiver end, the IP address of the receiver provided by the sender is verified before it can be allowed to be decrypted by the receiver using the secret key shared. This way the file is completely secured and any attack that takes place can be detected and the source of attack can be determined. These schemes allow secure file transfer in any external environment of any type of files such as audio, video, document, etc. Thus, the data is given security, integrity and confidentiality and the network medium is made efficiently accessible.

Keywords: IP attack · URL attack · Phishing · DOS · DDOS · DES first section

1 Introduction

One of the major challenges in the computer networking is the negligence of intruders, as several data are confidential and personal in all the areas like organization, banks, financial sectors, health care etc. In order to avoid the intruders, all the activities should be logged into an Intrusion Detection System (IDS) for identifying any malicious

D. J. Hemanth et al. (Eds.): COMET 2019, LNDECT 35, pp. 1465–1475, 2020.
https://doi.org/10.1007/978-3-030-32150-5_148

activity which is being performed on the network system. Data security is a protective measure that checks whether the user has a proper authorization to access the digital information. In a normal scenario if the attacker wants to download the file without proper authorization, it can be done through copying the URL and download the file easily. In this research work, data security principle will not allow the user or attacker to download the file without proper authorization. When an intruder tries to hack the data using IP address without key is said to be IP Spoofing. The authorized user can be able to download the encrypted data by decryption using the secret key mechanism namely cryptography technique.

2 Related Works

Generally, attacks such as IP attack, URL attack, phishing, etc. can be identified using different software. But a common platform to get rid of all these attacks has not yet been developed. In the existing system, the source of the attacker is not always known. It is difficult to trace the attack back to the source as IP spoofing can be used. IP address spoofing is commonly used to bypass basic security measures that rely on IP blacklisting.

In computer networking, IP address spoofing or IP spoofing is the creation of Internet Protocol (IP) packets with a false source IP address, for the purpose of hiding the identity of the sender or impersonating another computing system. One technique which a sender may use to maintain anonymity is to use a proxy server.

When we send or share a file, we need to provide a secret key for each file. And also the text files will be encrypted. Then receiver can receive this file with source IP Address and must receiver have to give source four secret key with port number otherwise receiver can't receive that file. If he will try to receive without key, that's IP Spoofing.

3 Proposed Work

Network attacks are one of the vital issues during transfer of files. It has to be identified and rectified then and there for a secure transmission. There are many kinds of attacks that prevail over the network. The objective of this paper is to provide a single, common platform that identifies almost all the vital attacks and resolves it immediately. A web application is created where the users can register themselves and then transfer the files. During transmission, the medium is monitored for any malicious activity and then it is reported to the administrator. The administrator then blocks the malicious user and paves way for a reliable transmission. To add up more security to this transmission, DES algorithm is used for encryption and decryption. A secret key has to be provided for every file to encrypt and decrypt the same. Some of the attacks that can be resolved using this system are,

IP Spoofing is the creation of Internet Protocol packets with a false source IP address, for the purpose of impersonating another computing system. Denial of services, here the malicious user sends a message and consumes the bandwidth of the network. The main aim of the malicious user is to create network traffic. Eavesdropping attack finds out some secret or confidential information from communication. A false user monitors the traffic and contents of the file during transmission. URL Attack, here a client manually adjusts the parameters of its request by maintaining the URL's syntax but altering its semantic meaning. The malicious URL looks very similar to the original ones. Phishing is the fraudulent attempt to obtain sensitive information such as usernames, passwords by disguising as a trustworthy entity in an electronic communication. It often directs users to enter personal information at a fake website, the look and feel of which are identical to the legitimate site.

4 Architecture

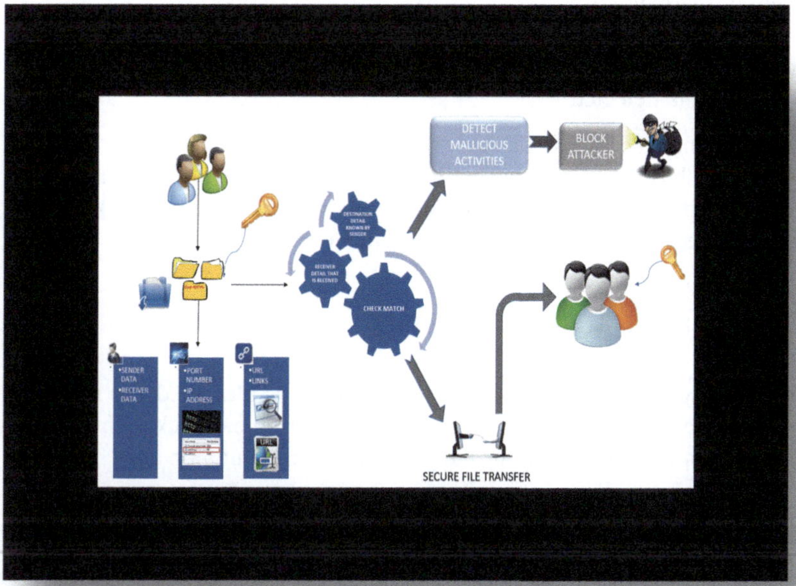

5 Algorithm

Input: Files containing confidential data.
Output: Detection of any kind of malicious activity that will affect the file.

Get secret key, destination from sender
if(destination !=receiver), **then/***Receiver IP with location and the Destination IP with location is checked for match*/
return null

else if(destination ==Receiver), **then/*** Receiver IP with location and the Destination IP with location is checked for match */
if(secret key==valid),**then//**secret key generated using des encryption is used for verification produce output record
else return null

6 Modules

Sender gets the details about receiver:
Sender will have the receiver's address as destination detail along with the secret key generated using DES Encryption.

Sender sending the data:
Sender will encrypt the file and send it to the destination where the receiver address and the destination address will be checked for match. In case if there is any mismatch, Malicious activity is detected.

Data retrieval at the Receiver end:
The receiver will have a secret key which will be shared with the sender and gets verified. After verification, the file gets transferred or in case of any attack, the file gets destroyed. The receiver sends an acknowledgement.

7 Results and Discussion

Any attack that occurs in a network during file transfer is identified and resolved. The malicious user is reported and blocked to prevent from further attacks. This mechanism ensures authentication, authorization, integrity and confidentiality.

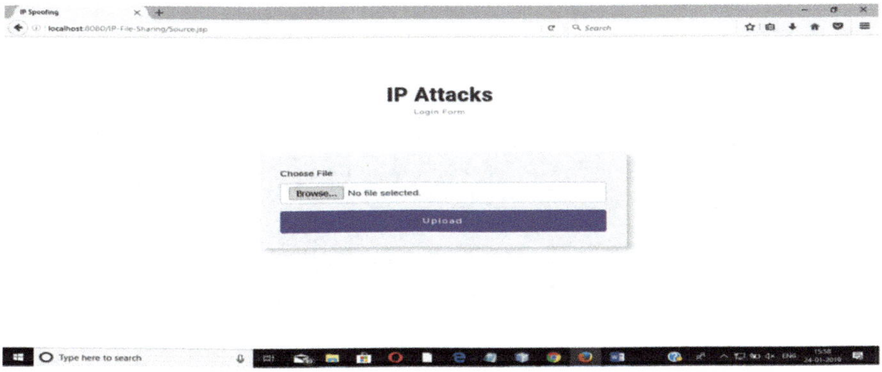
IP Spoofing
localhost:8080/IP-File-Sharing/Source.jsp
Search
IP Attacks
Login Form
Choose File
Browse... No file selected.
Upload
Type here to search

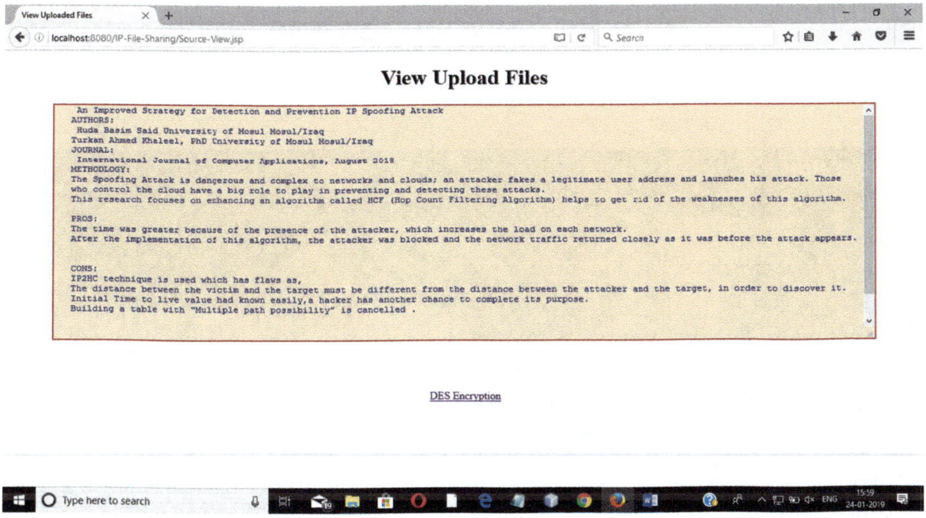
View Uploaded Files
localhost:8080/IP-File-Sharing/Source-View.jsp
Search
View Upload Files
An Improved Strategy for Detection and Prevention IP Spoofing Attack
AUTHORS:
Huda Basim Said University of Mosul Mosul/Iraq
Turkan Ahmed Khaleel, PhD University of Mosul Mosul/Iraq
JOURNAL:
International Journal of Computer Applications, August 2018
METHODLOGY:
The Spoofing Attack is dangerous and complex to networks and clouds; an attacker fakes a legitimate user address and launches his attack. Those who control the cloud have a big role to play in preventing and detecting these attacks.
This research focuses on enhancing an algorithm called HCF (Hop Count Filtering Algorithm) helps to get rid of the weaknesses of this algorithm.
PROS:
The time was greater because of the presence of the attacker, which increases the load on each network.
After the implementation of this algorithm, the attacker was blocked and the network traffic returned closely as it was before the attack appears.
CONS:
IP2HC technique is used which has flaws as,
The distance between the victim and the target must be different from the distance between the attacker and the target, in order to discover it.
Initial Time to live value had known easily,a hacker has another chance to complete its purpose.
Building a table with "Multiple path possibility" is cancelled .
DES Encryption
Type here to search
ENG
15:59
24-01-2019

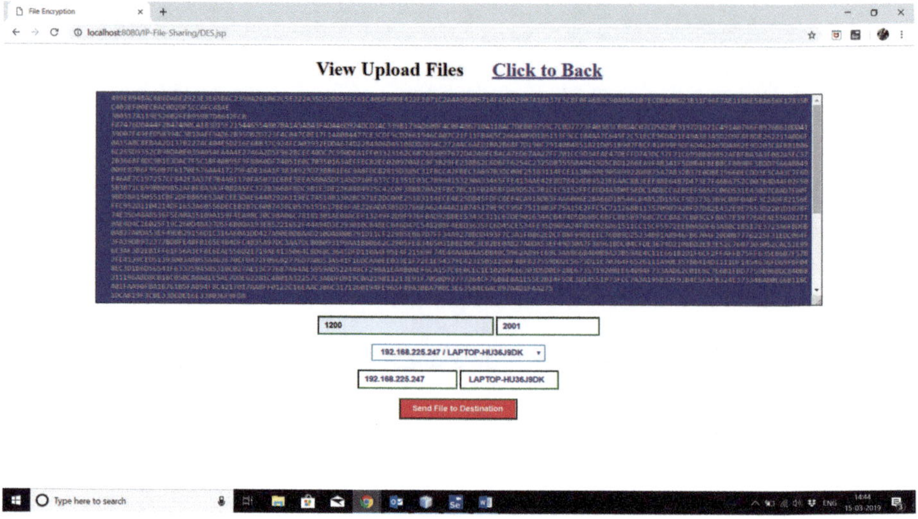
File Encryption
localhost:8080/IP-File-Sharing/DES.jsp
View Upload Files
Click to Back
1200
2001
192.168.225.247 / LAPTOP-HU36J9DK
192.168.225.247
LAPTOP-HU36J9DK
Send File to Destination
Type here to search

IP & URL
Home
(IP) Receive File
URLs
Logout
RECEIVER HOME PAGE
Type here to search

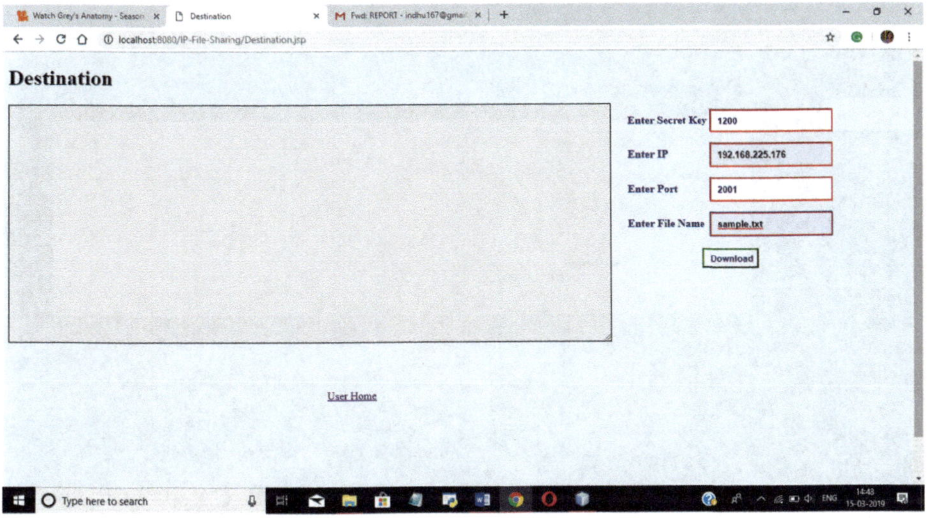
Destination
localhost:8080/IP-File-Sharing/Destination.jsp
Enter Secret Key
1200
Enter IP
192.168.225.176
Enter Port
2001
Enter File Name
sample.txt
Download
User Home

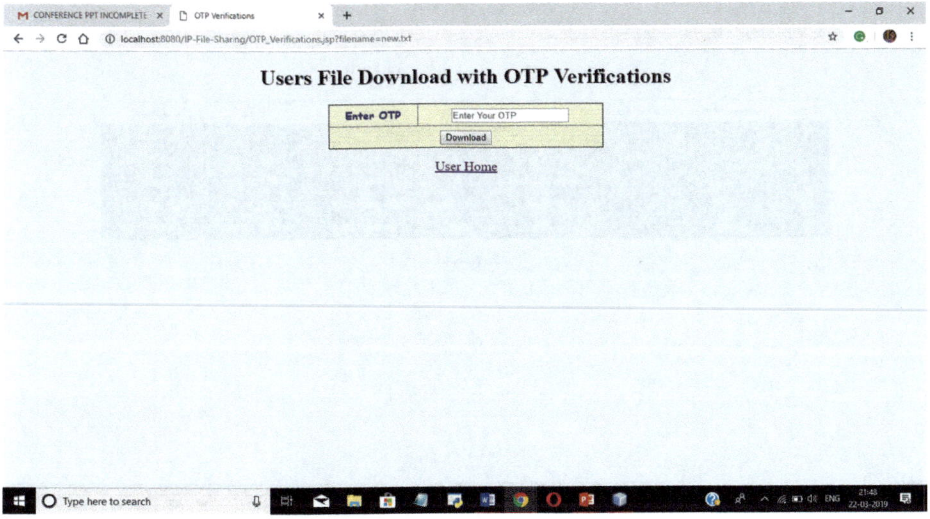
OTP Verifications
localhost:8080/IP-File-Sharing/OTP_Verifications.jsp?filename=new.txt
Users File Download with OTP Verifications
Enter OTP
Enter Your OTP
Download
User Home

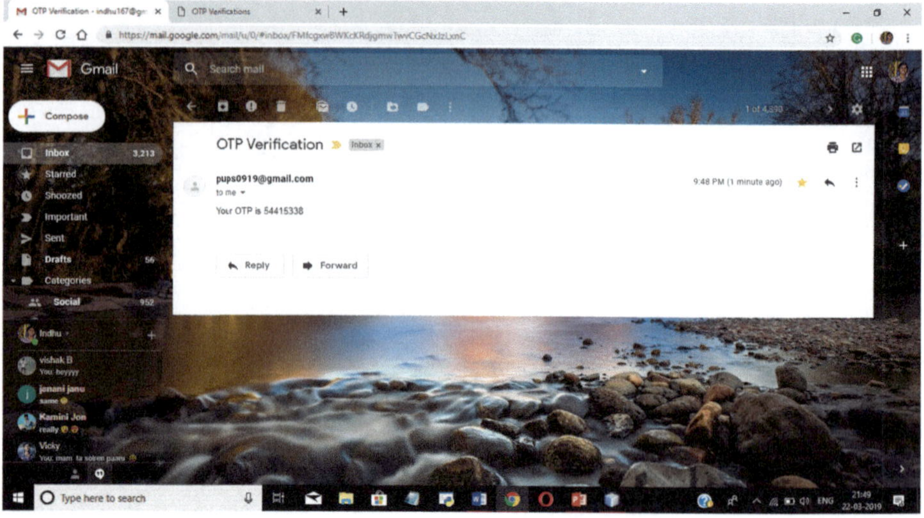
Gmail
Compose
Inbox
Starred
Snoozed
Important
Sent
Drafts
Categories
Social
OTP Verification
pups0919@gmail.com
Your OTP is 54415338
Reply
Forward
Type here to search

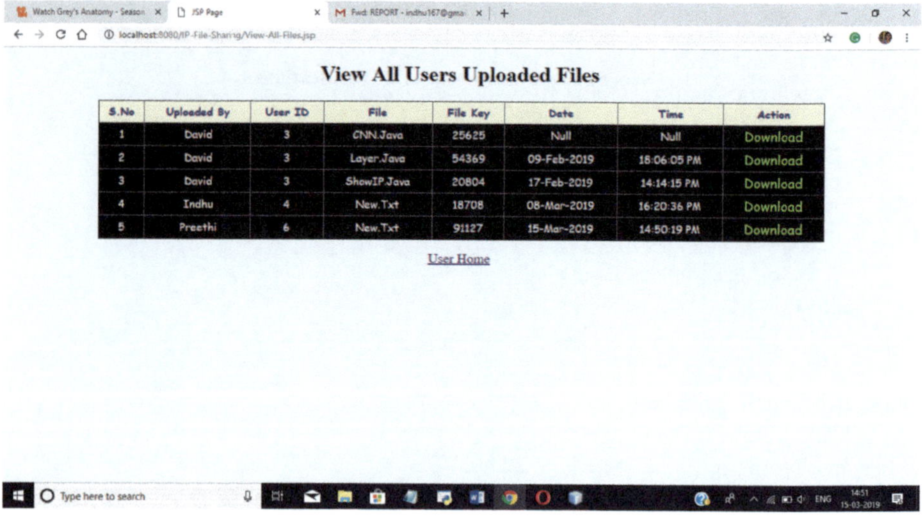
View All Users Uploaded Files
S.No
Uploaded By
User ID
File
File Key
Date
Time
Action
1 David 3 CNN.Java 25625 Null Null Download
2 David 3 Layer.Java 54369 09-Feb-2019 18:06:05 PM Download
3 David 3 ShowIP.Java 20804 17-Feb-2019 14:14:15 PM Download
4 Indhu 4 New.Txt 18708 08-Mar-2019 16:20:36 PM Download
5 Preethi 6 New.Txt 91127 15-Mar-2019 14:50:19 PM Download
User Home
Type here to search

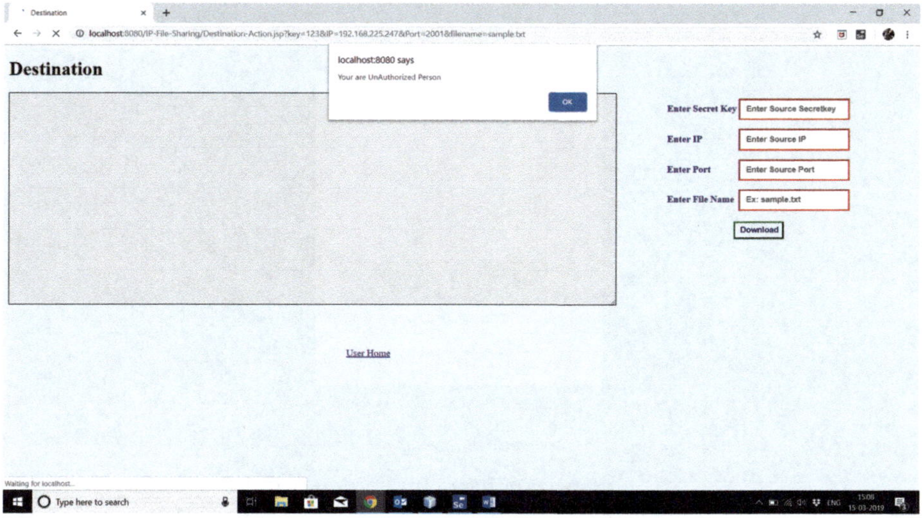
Destination
localhost:8080 says
Your are UnAuthorized Person
OK
Enter Secret Key
Enter Source Secretkey
Enter IP
Enter Source IP
Enter Port
Enter Source Port
Enter File Name
Ex: sample.txt
Download
User Home
Type here to search

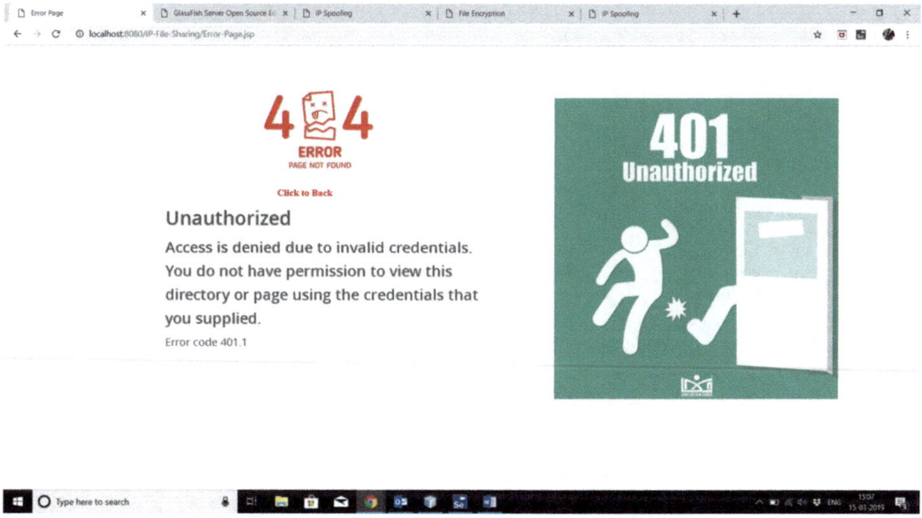
Error Page
localhost:8080/IP-File-Sharing/Error-Page.jsp
404
ERROR
PAGE NOT FOUND
Click to Back
Unauthorized
Access is denied due to invalid credentials.
You do not have permission to view this directory or page using the credentials that you supplied.
Error code 401.1
401
Unauthorized
Type here to search

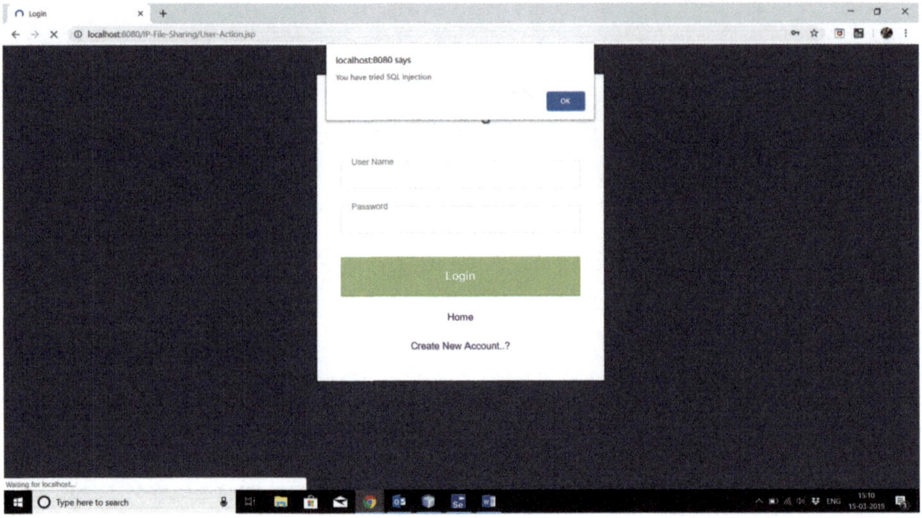
Login
localhost:8080/IP-File-Sharing/User-Action.jsp
localhost:8080 says
You have tried SQL Injection
OK
User Name
Password
Login
Home
Create New Account..?
Waiting for localhost...
Type here to search

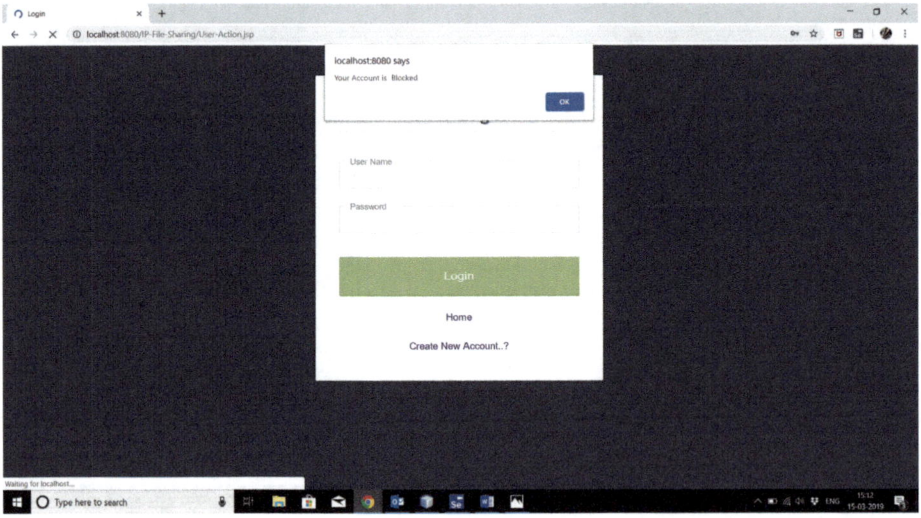
Login
localhost:8080/IP-File-Sharing/User-Action.jsp
localhost:8080 says
Your Account is Blocked
OK
User Name
Password
Login
Home
Create New Account..?
Waiting for localhost...
Type here to search

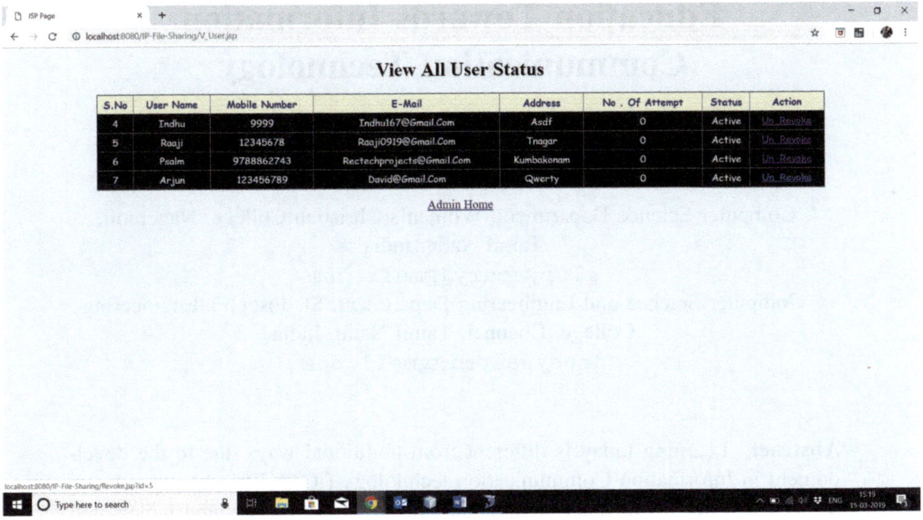

8 Conclusion

A common platform that detects and evades network attacks during file transfer is developed. The administrator monitors and resolves any attack. DES algorithm is used for encryption and decryption. Secure File transfer protocol is used as the medium of transmission. All the files before transmission is stored in the buffer for efficient and quick access. A secret key is shared between the users for every file they transfer, thus increasing the security. Hence the users of this web application can transfer text files without worrying about any attacks.

References

1. Chattopadhyay, P., Wang, L., Tan, Y.P.: Scenario-based insider threat detection from cyber activities. IEEE Trans. Comput. Soc. Syst. **5**(3), 660–675 (2018)
2. He, T., Leung, K.K.: Network capability in localizing node failures via end-to-end path measurements. IEEE Trans. Netw. **25**, 434–450 (2017)
3. Tolia, N., Kaminsky, M., Andersen, D.G.: An architecture for internet data transfer. Carnegie Mellon University, Intel Research, Pittsburg
4. Zheng, W., Liu, S., Liu, Z.: Security transmission of FTP data based on IPSec. In: International Joint Conference on networking (2009)
5. Sharma, S.: Detection and analysis of network & application layer attacks. In: 2016 6th International Conference - Cloud System and Big Data Engineering (Confluence)

A Study on the Attitude of Students in Higher Education Towards Information Communication Technology

D. Glory RatnaMary[1(✉)] and D. Rosy Salomi Victoria[2]

[1] Computer Science Department, Women's Christian College, Nagercoil, Tamil Nadu, India
glorymary@gmail.com

[2] Computer Science and Engineering Department, St. Joseph's Engineering College, Chennai, Tamil Nadu, India
drosysalomi@gmail.com

Abstract. Learning today is different from traditional ways due to the development in Information Communication technology (ICT). The extensive Internet accessibility of personal computers, laptops, smart phones and tablets and numerous literature recovery applications have altered the education and the training surroundings in entire disciplines. Several teachers identify the essential to exploit the abilities of ICT to improve their learning packages. Clarifications on student's aptitude with ICT are little, and are approved in countries where informatics is well established. Data collection is done through questionnaire from nearly 250 students who are exhausting computers for their theoretical purposes. This process is done by using the Apriori algorithm of Association Rule mining, Bayesian Classification algorithms and compared in Data mining using the WEKA tool. BayesNet Classification model provides the maximum accuracy of the students' approach on Information Technology making them to select a job.

Keywords: Apriori Algorithm · Association Rule Mining · Bayesian classification algorithms

1 Introduction

Education is a requirement to understand the technical advancements in Science and Technology. As the saying "Knowledge is power", education enhances knowledge that leads to an individual development. The spine of 21st century education is digital literacy. Technology must be used to enhance, enrich, and augment classroom learning with active and engaging learning activities. Digital literacy develops the ability to control impulses, make plans; follow instructions, multi-task and stay focused which are skills necessary to thrive in the ever-connected world.

The aim of this paper is to find how ICT technology is utilized by the students. The attitude of the students towards Information Technology were rated in the following criteria as Dissatisfied, Burdensome, Useless, Distraction, Satisfied, Beneficial, Useful, Play an important role and Prepare me for my career. The scales of rating are Strongly

D. J. Hemanth et al. (Eds.): COMET 2019, LNDECT 35, pp. 1476–1483, 2020.
https://doi.org/10.1007/978-3-030-32150-5_149

Disagree, Disagree, Neutral, Agree and Strongly Agree. Information is collected from many students to study how their laptops are being used in higher education. Weka comprises an Apriori learner implementation for producing association rules, a technique in market basket investigation. This algorithm looks for any guidelines that capture strong associations between different attributes. The Bayes functions like BayesNet, NaiveBayes, NaiveBayesMultinomial and NaiveBayesUpdateable are implemented using Weka tool.

2 Association Rule Mining

Apriori is a standard learning association instructions. Apriori is considered to work on databases comprising relationships (for e.g., groups of items approved by clients). The generation of association rule is divided into two distinct steps: The lowest support is useful to discover all common database set of items. In order to arrange rules, all common set of items and the lowest confidence limitation are used.

2.1 Apriori Algorithm

In association instruction mining, assume a set of items (for e.g., groups of trade transactions, each listing specific things bought), the procedure tries to discover subsets which are common to least number CC of the set of items. Apriori introduces a bottom-up method, where common subsets are prolonged 1 item at one time (generating candidate), and set of candidates are verified against the information. WEKA comprises an execution of the Apriori procedure for education association instructions. It operates with distinct information and will recognize numerical reliance between set of attributes.

```
1. Initialize TT as the database and minSupp is the
   minimum support
2. Assign LL1 to be the common set of items
3. Initialize p as 2.
4. CCk= candidates produced from Lk-1
   4.1 The product of cartesian LLp-1 * LLp-1 and elimi-
       nating any `p-1' size set of item that is not
       common for each transaction tr in TT.
   4.2 Add the total of all candidates in CCk that are
       contained in tr.
   4.3 LLp = candidates in CCp with minSupp
   4.4 Increment the value of p
5. If LLp-1 =∅ then go to step 4 else return ∪pLp.
```

Figure 1 show the generation of itemsets and frequent itemsets where the minimum support count is 2. Apriori procedure uses data from earlier steps to yield the common itemsets.

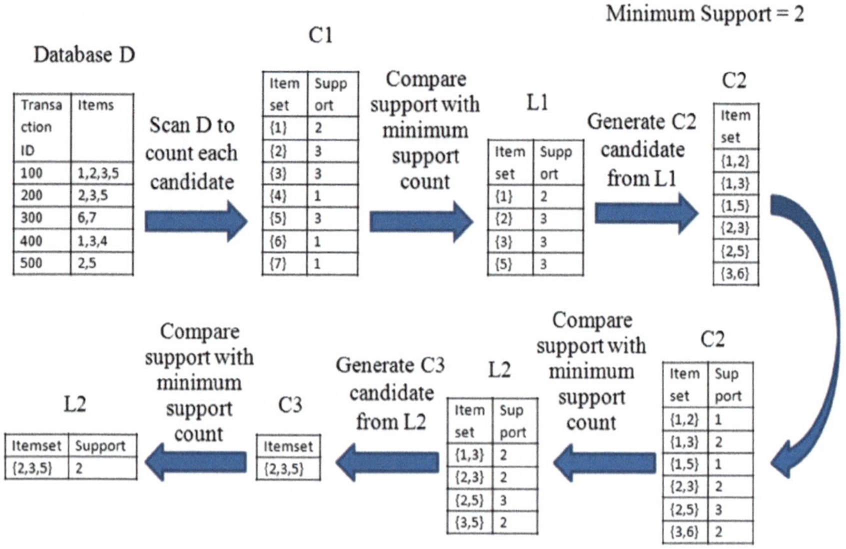

Fig. 1. Generation of itemsets and frequent itemsets

3 Classification Techniques

Bayesian classifiers are algebraic classifiers that calculate course association by possibilities. Numerous Bayes procedures are established in which the significant approaches are Bayesian systems and naive Bayes. Bayesian systems do graphical representations that can define combined restricted possibility circulations. Bayesian classifiers are classification procedures owe to their easiness, computational adeptness and right presentation for real world complications. The benefit is that the Bayesian representations are fast to give training and to estimate, and have a high correctness in several fields.

3.1 Classifier Algorithms

BayesNet studies Bayesian systems under the assumptions of nominal attributes. The dissimilar parts approximating the provisional possibility are shown. The *NaiveBayes* classifier affords an easy method, with distinct meanings, to expressive and education probabilistic information. *Naive Bayes Multinomial* is related to Naive Bayes classifier with the added integration of occurrence data.

Bayes Network knowledge practices several search procedures and quality events. The KK2 algorithm heuristically examines the utmost possible trust network arrangement specified in a database. The input *nn* is the collection on the nodes and *uu* is the count of parent nodes. The output is for each node, a print out of the parents of the node.

KK2 Pseudo code:

```
for i:= 1 to nndo
   πi  := Ø;
   Pold:= f(i, πi);
   OKToProceed  := true;
   While OKToProceed and |πi| <uu do
         z =Pred(xi)  - πi and   maximize f(i, πiU {z});
         Pnew  := f(i, πiU {z});
         ifPnew>Pold then
            Pold  :=Pnew;
πi  := πiU {z};
else
           OKToProceed := false;
  print('Node:', xi  ,  'Parent of xi:',πi);
end.
```

The procedures used for our work are BayesNet, NaiveBayes, NaiveBayesMultinomial and NaiveBayesUpdateable. The 10-fold cross validation is carefully chosen as our estimation method under the "Test options".

4 Results and Discussions

Responses were obtained from nearly 250 college students. WEKA tool was used to analyze the responses in the learning process on the attitude of the students towards information technology in the following criteria: Dissatisfied, Burdensome, Useless, Distraction, Satisfied, Beneficial, Useful, Play an Important Role and Prepare Me for career. The scales of rating are Strongly Disagree, Disagree, Neutral, Agree and Strongly Agree.

4.1 Pre-processing Data

WEKA comprises pre-processing tools for discretization, attribute collection, normalization, conversion, resampling and mixture of qualities. Table 1 shows the students attitude towards IT for the various attributes.

4.2 Apriori Implementation

Figure 2 shows the association rules for the relation attitude of students towards information technology and shows how a huge number of association rules can be discovered.

Table 2 shows the various comparison of attributes based on Agree. The association rule generated for criteria mentioned are as follows. The **attitude** of the students towards information technology in the learning process:

$$\text{Play an important role} = \Rightarrow \quad \text{useful} = \Rightarrow \text{beneficial.}$$

Table 1. Attributes vs. students attitude towards IT

Attribute name/*Rating*	*Disagree*	*Strongly disagree*	*Agree*	*Strongly agree*	*Neutral*
Dissatisfied	111	59	26	0	54
Burdensome	86	58	17	0	89
Useless	145	57	8	40	40
Distraction	84	42	30	3	91
Satisfied	9	9	161	8	63
Beneficial	9	8	187	46	0
Useful	0	8	170	72	0
Play an important role	7	0	167	54	22
PrepareMe	28	8	158	48	89

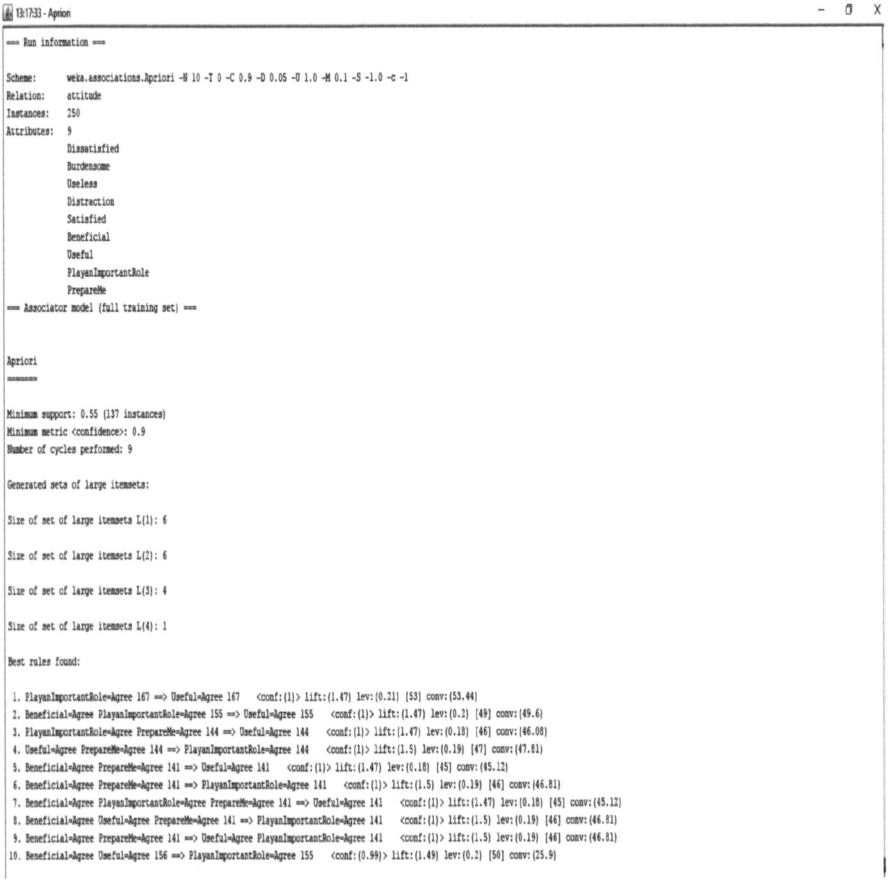

```
13:17:33 - Apriori
=== Run information ===

Scheme:       weka.associations.Apriori -N 10 -T 0 -C 0.9 -D 0.05 -U 1.0 -M 0.1 -S -1.0 -c -1
Relation:     attitude
Instances:    250
Attributes:   9
              Dissatisfied
              Burdensome
              Useless
              Distraction
              Satisfied
              Beneficial
              Useful
              PlayanImportantRole
              PrepareMe
=== Associator model (full training set) ===


Apriori
=======

Minimum support: 0.55 (137 instances)
Minimum metric <confidence>: 0.9
Number of cycles performed: 9

Generated sets of large itemsets:

Size of set of large itemsets L(1): 6

Size of set of large itemsets L(2): 6

Size of set of large itemsets L(3): 4

Size of set of large itemsets L(4): 1

Best rules found:

 1. PlayanImportantRole=Agree 167 ==> Useful=Agree 167    <conf:(1)> lift:(1.47) lev:(0.21) [53] conv:(53.44)
 2. Beneficial=Agree PlayanImportantRole=Agree 155 ==> Useful=Agree 155    <conf:(1)> lift:(1.47) lev:(0.2) [49] conv:(49.6)
 3. PlayanImportantRole=Agree PrepareMe=Agree 144 ==> Useful=Agree 144    <conf:(1)> lift:(1.47) lev:(0.18) [46] conv:(46.08)
 4. Useful=Agree PrepareMe=Agree 144 ==> PlayanImportantRole=Agree 144    <conf:(1)> lift:(1.5) lev:(0.19) [47] conv:(47.81)
 5. Beneficial=Agree PrepareMe=Agree 141 ==> Useful=Agree 141    <conf:(1)> lift:(1.47) lev:(0.18) [45] conv:(45.12)
 6. Beneficial=Agree PrepareMe=Agree 141 ==> PlayanImportantRole=Agree 141    <conf:(1)> lift:(1.5) lev:(0.19) [46] conv:(46.81)
 7. Beneficial=Agree PlayanImportantRole=Agree PrepareMe=Agree 141 ==> Useful=Agree 141    <conf:(1)> lift:(1.47) lev:(0.18) [45] conv:(45.12)
 8. Beneficial=Agree Useful=Agree PrepareMe=Agree 141 ==> PlayanImportantRole=Agree 141    <conf:(1)> lift:(1.5) lev:(0.19) [46] conv:(46.81)
 9. Beneficial=Agree PrepareMe=Agree 141 ==> Useful=Agree PlayanImportantRole=Agree 141    <conf:(1)> lift:(1.5) lev:(0.19) [46] conv:(46.81)
10. Beneficial=Agree Useful=Agree 156 ==> PlayanImportantRole=Agree 155    <conf:(0.99)> lift:(1.49) lev:(0.2) [50] conv:(25.9)
```

Fig. 2. Association rules found in Apriori for relation Attitude

Table 2. Classifiers Accuracy

Algorithms	Correctly classified instances (%)	Incorrectly classified instances (%)
BayesNet	91.6	8.4
NaiveBayes	90.4	9.6
NaiveBayesMultinomial	63.2	36.8
NaiveBayesUpdateable	90.4	9.6

4.3 Bayes Classifiers Performance Metrics

A Bayesian predictive model is built to find the attitude of students how information technology is useful during preparation for their career. An exact positive assessment outcome is one that identifies the condition when the situation is existent. An incorrect positive assessment outcome is one that identifies the condition when the situation does not exist. *Recall* is the portion of related cases that are recovered which is stated as a ratio. Table 2 shows the accuracy of BayesNet, NaiveBayes, NaiveBayesMultinomial and NaiveBayesUpdateable algorithms for classification applied on the data sets using 10-fold cross validation.

Table 3 shows that BayesNet algorithm has highest accuracy of 91.6 compared to othermethods. Naive bayes Updateable, NaiveBayes also showed a high level of accuracy.

Table 3. Classification Matrix Bayes Net

		Predicted					Precision (%)
		Strongly agree	Disagree	Neutral	Agree	Strongly disagree	
Actual	Strongly agree	45	0	0	3	0	75
	Disagree	0	28	0	0	0	100
	Neutral	0	0	8	0	0	100
	Agree	15	0	0	141	2	97.2
	Strongly disagree	0	0	0	1	7	77.8
Recall (%)		93.8	100	100	892	87.5	

Cost benefit analysis and Visualize Threshold Curve for the criteria, Agree for BayesNet Classifier and NaiveBayesMultinomial Classifier can be shown. Visualize Cost Curve for the criteria, Agree can be shown BayesNet Classifier as Fig. 3 and NaiveBayesMultinomial Classifier.

We have also visualized how the attribute 'Play an Important Role' relate to all other attributes in terms of the scales of rating such as Disagree, Neutral, Agree, Strongly Disagree and Strongly Agree.

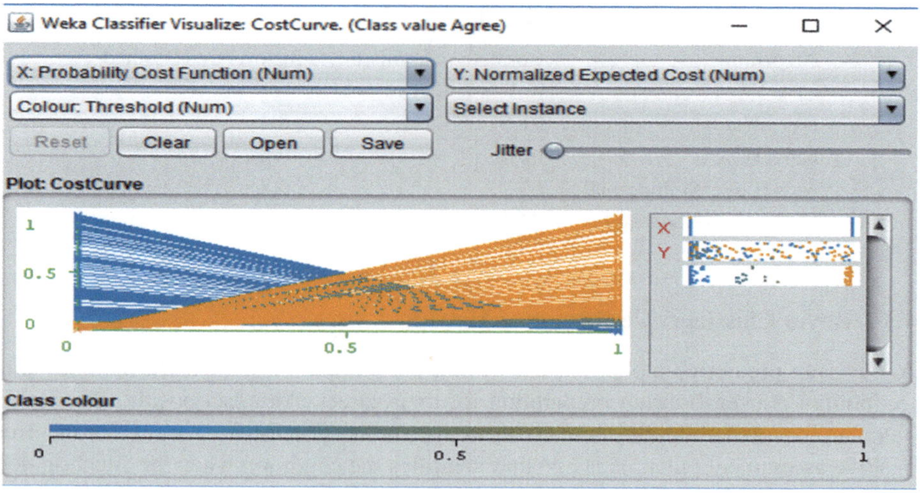

Fig. 3. BayesNet – visualize cost curve: agree

5 Conclusion

We have studied how data mining can be applied to educational systems in this paper. It shows that the data mining can be used in advanced learning, to increase the performance of students. The association rules generated by the Apriori Algorithm have shown that ICT is beneficial and useful to the students in their learning process and Information Technology plays an important role to choose their career. On comparison of the Bayesian Classifiers, BayesNet Classification model gives the highest accuracy of the students' attitude on Information Technology preparing them to choose a career.

The pupils from every stream are well-educated but digital mastery should be merged in the college syllabus. Live demonstrations of several applications should be given regularly. Digital devices and applications are established quickly so pupils must be aware with all the up-to-date implementations and skills. Original internet service providers originate in the service shop. ICT shows important part in recent bazaar. Innovative web implementations must be a portion of our daily life. The future generation should have technical knowledge to manage with the varying atmosphere and it is the major responsibility of advanced teaching organizations that they must makea-soldier of such people those will certainly add in the information economy.

References

1. Augustus Richard, J.: The role of ICT in higher education in the 21st century. Int. J. Multidiscip. Res. Mod. Educ. **1**(1), 652–656 (2015)
2. Nakaznyi, M., Sorokina, L., Romaniukha, M.: ICT in higher education teaching: advantages, problems, and motives. Int. J. Res. E-Learn. **1**(1), 49–61 (2015)
3. Buttar, S.S.: ICT in higher education. Int. J. Soc. Sci. **2**(1), 1686–1696 (2015)

4. Verma, C., Dahiya, S.: A responsive approach of faculty towards ICT: strength, weakness and opportunities. International Journal of Science Technology and Management **5**(1), 58–65 (2016)
5. Alam, M.M.: Use of ICT in higher education. Int. J. Indian Psychol. **3**(4), 162–171 (2016)
6. Han, J., Pei, J., Yin, Y.: Mining frequent patterns without candidate generation. In: 2000 ACM SIGMOD International Conference on Management of Data, pp. 1–12. ACM Press, New York (2000)
7. Geetha, K., Mohiddin, S.K.: An efficient data mining technique for generating frequent item sets. IJARCSSE **3**(4), 571–575 (2013)
8. http://www.weka-x64.sharewarejunction.com
9. http://www.deccanchronicle.com/140905/nation-current-affairs/article/free-laptops-improve-tech-skills-tamil-nadu-students-survey

Generalized Digital Certificate Based Key Agreement for Initial Ranging in WiMax Network

M. A. Gunavathie(✉), M. Helda Mercy, J. Hemavathy, and A. Nithya

Department of Information Technology, Panimalar Engineering College, Chennai, India
gunavathie.ap@gmail.com, mercy_hilda@yahoo.co.in, hemaramya27@gmail.com, nithyashree.a@gmail.com

Abstract. WiMAX stands for worldwide interoperability for microwave access and it is based on IEEE 802.16 standards. As the wireless communication has been increasing, so has the concern for security in the wireless scenario. The wireless medium is considered to be less secure due to its shared medium. Security attacks are easier to occur in wireless channel because of its shared open channel. In this paper, RNG-RSP attack which cause Denial of Service (DoS) attack in WiMAX network is addressed. A key agreement algorithm using Generalized Digital Certificate (GDC) for defending DoS attack in WiMAX network is proposed.

Keywords: WiMAX · Security · Denial of Service (Dos) · Key agreement · Generalized digital certificate (GDC)

1 Introduction

WiMAX (IEEE 802.16) promises to deliver high data rate (75 Mbps) over wide areas (50 km) for a large number of users. It uses radio channel and hence security procedures must be included in order to protect the network services from security attacks.

Any wireless network should have some basic network security goals because of the open channel. If a subscriber station (SS) wants to enter into the WiMAX network then it has to go through a multistep process. First SS has to do scanning. Scanning is the process of searching possible channels of the downlink frequency (DL) band of operation. This process has to be continued until it finds a valid DL signal. Second it has to look for the downlink channel descriptor (DCD) and uplink channel descriptor (UCD). DCD and UCD are broadcasted by the base station (BS) and it contains the information of the uplink and downlink channel characteristics. Third step is the Initial ranging process. SS has to perform initial ranging which is to set the physical parameters such as timing offset and power adjustments properly.

Initial ranging process has been accompanied by sending Ranging Request (RNG-REQ). Ranging Response (RNG- RSP) is sent by BS if it receives the RNG-REQ successfully. RNG-RSP is used by the SS to adjust its transmission time, frequency and power. It also contains the primary management connection id (CID). The initial

D. J. Hemanth et al. (Eds.): COMET 2019, LNDECT 35, pp. 1484–1492, 2020.
https://doi.org/10.1007/978-3-030-32150-5_150

ranging has to be done periodically by SS. Fourth step is the Authentication phase. Once initial ranging is finished successfully, SS has to enter into the authentication and key establishment phase and it is given by Privacy and Key Management algorithm (PKM). Last step is the registration phase. In this paper, Initial ranging process is considered and its vulnerabilities are analysed. The attacks on RNG-RSP packet are analysed.

2 RNG-RSP Attack

The IEEE 802.16 MAC has DoS vulnerabilities. DoS attack is an attempt to make the computer or computer resource unavailable to its intended users. It is characterised by explicit attempt by attackers to prevent legitimate users of a service from using that service. RNG- REQ is sent by SS during the ranging process to announce its presence and its wish to join the network. It is the request for transmission timing, power, frequency and burst profile information. BS will respond back to SS by sending RNG-RSP packet if RNG-REQ is received successfully at BS. RNG-RSP message is sent by BS to set and maintain the proper timing of the SS transmissions. RNG-RSP message is used by BS to change the uplink and downlink channels of SS. It is also used by SS to change transmission power levels and even abort all transmissions and re-initialize its MAC. RNG_RSP is unencrypted, unauthenticated and stateless and hence it is vulnerable to exploitation. Malicious user will use this RNG-RSP message to spoof this message with the Ranging Status field set to a value of 2, which corresponds to "abort", shift a victim node to a channel of the attacker's choosing, to spoof the CID and message contents (Fig. 1).

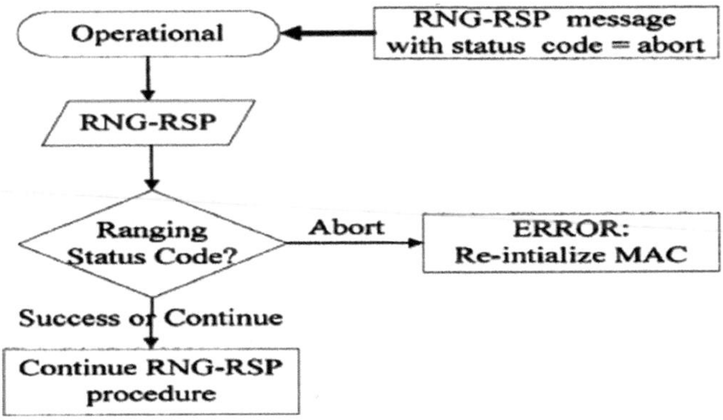

Fig. 1. Flow of RNG-RSP attack

The malicious user will send the RNG-RSP message with the ranging status as abort and hence cause the DoS attack. The malicious user will interrupt the service being used by the intended users.

3 Related Works

To overcome the denial of service attack in using RNG-RSP packet by malicious user will be overcome by encrypting that packet. To encrypt the packet, secret key should be exchanged prior to communication. Several authors proposed solution for RNG-RSP DoS attack. Adnan et al. (2011) proposed an algorithm for secure key exchange for encrypting the packets to overcome the DoS attack. This algorithm was purely based on Diffie Hellman key exchange algorithm.

Altaf et al. (2008) proposed a Pre-authentication solution to avoid the denial of service attack in the WiMAX network. It is based on visual cryptography which is the concept of secret sharing with images. This scheme makes use of X.509 certificate and trusted third party and has the overhead of communicating with the trusted third party (TTP). This scheme also has the overhead of storing images in the base station, subscriber station and TTP. Gandhewar et al. (2011) proposed an elliptic curve key exchange algorithm (ECDH) in the initial network entry process to avoid the denial of service attack.

Gandhewar et al. (2011) proposed an elliptic curve key exchange algorithm (ECDH) in the initial network entry process to avoid the denial of service attack.

Maru et al. (2008) provided a detailed account of the important messages to be jammed to cause denial of service attack. Denial of service attacks at two layers such as physical and MAC layer are discussed. They provide suggestions such as encryption of MAC management messages and authenticating all management messages using hash functions.

Naseer et al. (2008) explained the management messages that cause DoS attack.

Deininger et al. (2007) explained forging key messages in multi broadcast operation, some unauthenticated messages, and unencrypted management communication. Suggestions provided are to encrypt and authenticate management messages.

Hong et a1. (2011) presented a study on IEEE 802.16 MAC operation, RNG-RSP and its vulnerabilities to DoS. Attacker use the RNG-RSP message with the ranging status set to 2 to abort communication and reinitialize MAC and to cause the water torture attack. Experimental set up was done to simulate the DoS attack.

Tshering et al. (2011) discussed the initial network entry procedure threats using RNG-REQ, RNG-RSP that cause DoS attacks and attacks on Privacy and key management protocol.

Harn et al. (2011) proposed GDC which can be used for authentication and key agreement. In this paper, it is proposed a key agreement for Initial ranging using GDC to overcome the DoS attack.

4 Generalized Digital Certificate

X.509 certificate contain only public information that can be easily recorded and played back once it has been revealed. In generalized digital certificate Harn et al. (2011) the owner never needs to reveal digital signature to anyone as there is no need to transfer the certificate. The knowledge of the digital signature on the GDC can be used to provide authentication and to establish secret key. Elgamal signature is used in the GDC to sign the document digitally.

5 Elgamal Signature

Security of Elgamal digital signature is based on the difficulty of computing discrete logarithms. In this scheme, message is digitally signed using components r, s where

$$r = g^{k}\,\mathrm{Mod} \tag{1}$$

$$s = k^{-1}\,(\mathrm{m} - \mathrm{xr})\mathrm{Mod}\,\mathrm{p} - 1 \tag{2}$$

The signature is verified with

$$\mathrm{g}^{m} = y^{r} r^{s}\,\mathrm{Mod}\,\mathrm{p} \tag{3}$$

where p is the large prime, x is the private key, k is a random number, y is the public key, g is the generator in the order of p – 1, m is the message digest of the message m', r is random component used for generating s, the secret signature component and the pair (r, s) form the signature on message m'. To avoid forging of the signature, it has been suggested to use different values of r generated for different entities, by using different values of k in the signing process (Harn et al. 2011) and it is used in the proposed mutual authentication process.

6 Proposed GDC Based Key Agreement for Initial Ranging

The entities before entering into the network should get the GDC certificate. After getting the GDC certificate, it has to enter into the key agreement algorithm to generate the secret key. After the secret key has been successfully generated, ranging request messages are encrypted using that secret key.

6.1 Obtaining GDC Certificate

The mobile station and the base station need to register at CA to obtain a GDC. The CA will generate an Elgamal signature (r, s) for the users after verifying the identity information. The random component r needs not to be kept secret because r component is computed using Eq. (1) and it is based on the discrete logarithm problem. It is computationally infeasible to find k from r so it can be made public. The signature component s needs to be kept secret because it depends on the message and it is given by Eq. (2).

6.2 Proposed GDC Key Agreement for Initial Ranging

Step 1:
The SS and BS should get the GDC certificate to start the key agreement process. The GDC contains r and s components are generated for each entity by the CA using the Eqs. (1) and (2).
Step 2:
SS will calculate S_A and send it to BS and BS will calculate S_B and send it to SS.

$$S_A = r_a s_a \bmod p \tag{4}$$

$$S_B = r_b s_b \bmod p \tag{5}$$

Where $r_a s_a$ is the signature components of SS and $r_b s_b$ is the signature components of BS.
Step 3:
SS will calculate ea, M, M1 as follows

$$ea = {S_B}^{s_a} \bmod p \tag{6}$$

$$M = \propto^{r_a s_a} \bmod p \tag{7}$$

$$M1 = \propto^{r_a s_{a1}} \bmod p \tag{8}$$

Where α is the primitive root of p and s_{a1} is the random number chosen by SS. Now SS will send the M and M1 to BS.
Step 4:
BS will calculate eb, B, B1 as follows

$$eb = {S_A}^{s_b} \bmod p \tag{9}$$

$$B = \propto^{r_b s_b} \bmod p \tag{10}$$

$$B1 = \propto^{r_b s_{b1}} \bmod p \tag{11}$$

Where s_{b1} is the random number chosen by BS. Now BS now sends the B and B1 to SS.
Step 5:
Key generation at SS

$$K_a = \left(r_a \times \log_{r_b} ea\right)\left(s_a \times \log_{\propto} B1 + s_{a1} \times \log_{\propto} B\ \right) \tag{12}$$

Step 6:
Key generation at BS

$$K_b = (r_b \times \log_{r_a} eb)(s_b \times \log_{\propto} M1 + s_{b1} \times \log_{\propto} M) \quad (13)$$

Key generation at both ends are equal

Key generation at M

$$\begin{aligned} K_a &= (r_a \times \log_{r_b} ea)(s_a \times \log_{\propto} B1 + s_{a1} \times \log_{\propto} B) \\ &= (r_a \times \log_{r_b} S_B{}^{s_a})(s_a \times \log_{\propto} \propto^{r_b s_{b1}} + s_{a1} \times \log_{\propto} \propto^{r_b s_b}) \\ &= (r_a \times \log_{r_b} r_b^{s_b s_a})(s_a \times \log_{\propto} \propto^{r_b s_{b1}} + s_{a1} \times \log_{\propto} \propto^{r_b s_b}) \\ &= (r_a \times s_b \times s_a)(s_a \times r_b \times s_{b1} + s_{a1} \times r_b \times s_b) \\ &= (r_a \times s_b \times s_a \times r_b)(s_a \times s_{b1} + s_a \times s_b) \\ &= (r_a \times s_a \times r_b \times s_b)(s_a \times s_{b1} + s_{a1} \times s_b) \end{aligned}$$

Key generation at BS

$$\begin{aligned} K_b &= (r_b \times \log_{r_a} eb)(s_b \times \log_{\propto} M1 + s_{b1} \times \log_{\propto} M) \\ &= (r_b \times \log_{r_a} S_A{}^{s_b})(s_b \times \log_{\propto} \propto^{r_a s_{a1}} + s_{b1} \times \log_{\propto} \propto^{r_a s_a}) \\ &= (r_b \times \log_{r_a} r_a^{s_a s_b})(s_b \times \log_{\propto} \propto^{r_a s_{a1}} + s_{b1} \times \log_{\propto} \propto^{r_a s_a}) \\ &= (r_b \times s_a \times s_b)(s_b \times r_a \times s_{a1} + s_{b1} \times r_a \times s_a) \\ &= (r_b \times s_a \times s_b \times r_a)(s_b \times s_{a1} + s_a \times s_{b1}) \\ &= (r_a \times s_a \times r_b \times s_b)(s_a \times s_{b1} + s_{a1} \times s_b) \end{aligned}$$

7 Results and Discussions

We have set up a simple WiMAX network environment consists of 2 base stations and 10 mobile stations to study the DoS attack and the performance of GDC based key agreement for initial ranging. Simulations were carried out in NS 2 simulator with WiMAX patch. The detailed simulation set up is shown below in Table 1.

Table 1. Simulation setup

Number of nodes	10
Number of base stations	2
Number of malicious station	1
Simulation length	200 s
modulation	OFDM 16 QAM ¾
Packet size	1240 bytes
Traffic	Constant bit rate

7.1 DoS Attack Generation

In this module, we have generated the RNG-RSP DoS attack in the WiMAX network. In the simulation, one node is set as malicious. Mobile node sends the RNG-REQ message to BS. The malicious node generates the RNG-RSP messages with ranging status as abort and sends it to the mobile node. Since the RNG-RSP is unauthenticated, unencrypted and stateless, the mobile station has no chance of checking whether the RNG-RSP is sent by legitimate or illegitimate BS. On seeing the RNG-RSP status as abort, the mobile node has to abort it, reinitialize the MAC and try again. The malicious RNG-RSP with status abort is sent continuously to the mobile node to cause the denial of service. We have used two measurements such as throughput and delay to evaluate the performance. The formulae for throughput and delay is given as follows:

$$\text{Throughput} = \frac{\text{No of bits from one node to other node}}{\text{observed duration}}$$

$$\text{Delay} = \frac{(\text{average packet size} * 8)}{\text{link speed}}$$

The total number of malicious RNG-RSP packets sent was 85. The mobile node on seeing the malicious RNG-RSP packets with status as abort will abort the ranging and try again. The delay and throughput calculated in this module is given as follows (Table 2):

Table 2. Results obtained in attack generation module

Delay	0.00583842
Throughput	3175.63

The attack generation module is tested by varying the packet size and results are analysed with delay and throughput. The values obtained in this simulation is tabulated in the below Table 3.

Table 3. Results obtained in attack generation module

Packet size (bytes)	Delay (sec)	Throughput (bytes)	Throughput in %
1240	0.00583842	3175.63	89.91
3720	0.00525486	2985.23	89.72
4960	0.00474258	2946.93	87.08
6200	0.00357724	2965.38	86.26
7440	0.00357822	3015.14	86.09
9920	0.00431844	2920.92	86.09

7.2 GDC Based Key Agreement for Initial Ranging

In this module, GDC based key agreement is first done to encrypt the RNG-REQ and RNG-RSP packets. The same simulation set up is followed for this module also.

In this module, nodes have to obtain a GDC certificate before starting communication in the network. After that node will generate a secret key with BS using the GDC based key agreement. Based on the secret key generated, RNG-REQ and RNG-RSP packets were encrypted. One malicious node is set to send a malicious RNG-RSP with abort status. On receiving malicious RNG-RSP message, the mobile node will try to decrypt it with the secret key generated. If the decryption fails, then that RNG-RSP reply will be discarded. So the malicious response will be discarded. The delay and throughput calculated in this module is given as follows (Table 4):

Table 4. Results obtained in attack generation module

Delay	0.184554
Throughput	216092

The simulation is carried out by varying the packet sizes and the results are analysed with delay, throughput. The values obtained in this simulation are tabulated in the following Table 5:

Table 5. Results obtained in attack generation module

Packet size (bytes)	Delay (sec)	Throughput (bytes)	Throughput in %
1240	0.184554	216092	95.60
3720	0.195546	307734	95.40
4960	0.205578	409330	93.67
6200	0.195693	263378	92.97
7440	0.200331	315470	92.20
9920	0.20669	419654	91.50

References

Adnan, A., Jan, F., Sattar, A.R., Ashraf, M., Shehzad, I.: Enhancement of security for initial network entry of SS. In: IEEE 802.16e. International Journal of Management, IT and Engineering, vol. 1, issue 7 (2011)

Altaf, A., Sirhindi, R., Ahmed, A.: A novel approach against DoS attacks in WiMAX authentication using visual cryptography. In: The Second International Conference on Emerging Security Information, Systems and Technologies, pp. 238–242 (2008)

Gandhewar, P.K., Lokulwa, P.P.: Improving security in initial network entry process of IEEE 802.16. In: International Journal on Computer Science and Engineering, pp. 3327–3331 (2011)

Maru, A., Brown, T.X.: Denial of service vulnerabilities in the 802.16 protocol. In: Proceedings of the 4th Annual International Conference on Wireless Internet WICON 2008 (2008)

Naseer, S., Younus, M., Ahmed, A.: Vulnerabilities exposing IEEE 802.16 e networks to DoS attacks: a survey. In: 2008 Ninth ACIS International Conference on Software Engineering, Artificial Intelligence, Networking, and Parallel/Distributed Computing, pp. 344–349. IEEE (2008)

Deininger, A., Kiyomoto, S., Kurihara, J., Tanaka, T.: Security vulnerabilities and solutions in mobile WiMAX. IJCSNS Int. J. Comput. Sci. Netw. Secur. **7**(11), 7–15 (2007)

Hong, J.A.K., Alias, M.Y., Goi, B.M.: Simulating denial of service attack using WiMAX experimental setup. Int. J. Netw. Mobile Technol. **2**(1), 30–34 (2011)

Tshering, F., Sardana, A.: A review of privacy and key management protocol in IEEE 802.16e. Int. J. Comput. Appl. **20**(2), 25–31 (2011)

Harn, L., Ren, J.: Generalized digital certificate for user authentication and key establishment for secure communications. IEEE Trans. Wireless Commun. **10**(7), 2372–2379 (2011)

Design and Analysis of Various Patch Antenna for Heart Attack Detection

S. B. Nivetha[1(✉)] and B. Bhuvaneswari[2]

[1] Communication Systems, Panimalar Engineering College, Chennai, Tamilnadu, India
sbnivetha97@gmail.com
[2] ECE, Panimalar Engineering College, Chennai, Tamilnadu, India

Abstract. This articles describes the design of four patch antennas for heart attack detection. The intend method starts with patch antenna. The antenna is premeditated using ADS (advanced design software). Four patch antennas inverted f antenna, inverted L shape antenna, T shape patch antenna and I shape patch antenna were designed. Out of these four patch antenna, the functional characteristics of inverted f and inverted L antenna are good, these two antennas are chosen for monitoring heartrate. The electrical movement of heart is measured by ECG sensor the signal is transmitted via antenna to the smartphones. In proposed design the gain of antenna is increased from −1 to 3 db. The return loss obtained for intended design is very less. The fabricated antenna is deliberate by means of network analyzer. The measured results and simulated results varies due to cable loss. The virtual result of inverted F antenna is 2.4 Ghz. The virtual result of inverted L antenna is 2.41 Ghz. The working frequency of all the four patch antenna is 2.45 Ghz.

Keywords: Gain · Patchantenna · Dimension · Frequency · Radiation pattern · Return loss

1 Introduction

Planar inverted F antenna [1] of Fr4 substrate with stuff constant of a pair of 0.92 at one 1 MHz having thickness of 1.5 mm is intended in existing system. The dimension of antenna is 30 mm × 29.6 mm. The gain obtained for existing style is −1 dbi. The return loss of antenna is −30. Inverted L form slotted small strip patch antenna is intended for wireless native space network [2]. The FR4 material having permittivity of 4.6 and thickness of 1.57 mm has been used as a substrate. The side of substrate features a slotted square formed patch and also the inferior surface of substrate has absconded ground surface. The slotted defected ground surface and an inverted formed patch yields enhancement bandwidth and improves the return loss [3]. Fletcher (2010) elaborated the study of wearable sensor based on doppler effect. The microwave detector directly senses the heart movement instead of electrical movement, and so matching to ECG. The first benefit of the microstrip detector embrace little size, truncated power, truncated cost and also the ability to control through clothing. Their circuit incorporates a pair of 0.4 Ghz Doppler circuit, assimilated microstrip blotch

D. J. Hemanth et al. (Eds.): COMET 2019, LNDECT 35, pp. 1493–1504, 2020.
https://doi.org/10.1007/978-3-030-32150-5_151

antenna and microcontroller with 12-bit analog to digital converter [4].The I formed patch antenna is intended for L band application. The triple band frequency are 1.91 Ghz, 2.25 Ghz and 5.676 Ghz. The heart attack detection is completed by printed array [6]. The dimension of the printed array antenna is 27 × 35 mm. The operative frequency of printed array is one to 3 Ghz. The guts failure is detected by broadband pleated antenna [7]. The gain of broadband pleated antenna is 4.2 dbi. Four differing kinds of patch antennas and array configurations were enforced on each the transmitter and receiver sides to gauge the result of radiation parameters [8]. A blotch antenna may be a style of ominidirectionla antenna with a tan profile, which may be mounted on flat surface. It consist of a flat rectangular sheet or patch of metal, mounted over a bigger sheet of metal referred to as a ground palne. Compared to standard antenna patch antenna is lighter in weight and simple fabrication so patch antenna is designed. Heart muscle misdemeanor is often called heart failure it happens once blood flow decreases and low doses of common acetylsalicylic acid pill act as a blood diluent. The projected system will advise the user to require an acetylsalicylic acid to forestall more cuddling.

2 Proposed System

In estimated system planar inverted F, inverted L antenna, inverted T and I form antenna are designed by means of ADS software package. The gain of associate degree antenna is improved in estimated technique. The working frequency of all four antenna are 2.4 Ghz. The gain of projected style is 3 dbi.

3 Components

ADS could be a very helpful software system antenna simulation tool. ADS is an electronic mode software package created by key sight technologies. It provides an integrated style atmosphere to inventors of RF electronic product like mobile phones wireless network, satellite communication radar system and high speed information links. Quick and correct results are obtained by means of ADS software system.

3.1 System

The ECG sensor quantities the electrical movement of heart. The ECG sensor used in this paper is AD8232 solitary prime heart level monitor. The measured values will be send to the microcontroller it will convert the values into signals. The microcontroller used in this project is ardunio UNO board (ATmega328p). The Bluetooth transceiver will process and store signals. Bluetooth low energy shield version 2.1 is used for conveying the datas. The signals from Bluetooth will be transmitted to the android phone via inverted f antenna. The notification will be send to users smartphone. If heart attack is detected the led will glow in the ardunio board. In future the android application is developed which posses the features of giving alert call to the hospitals and contact numbers of users smartphone (Fig. 1).

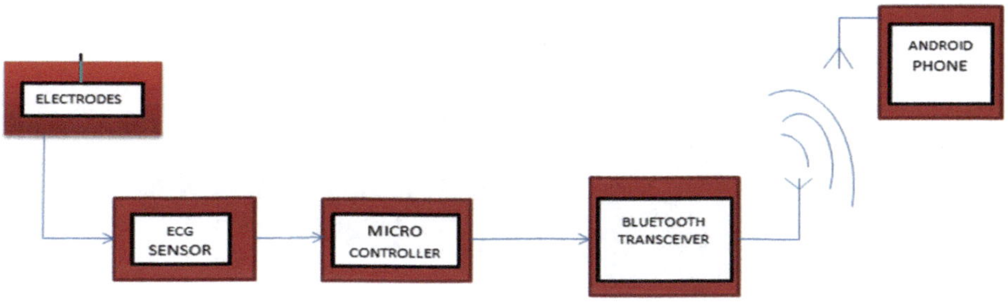

Fig. 1. Block diagram

3.2 Sensor

The ECG detector is connected to the patient by means of disposable electrodes on the leftward and right aspect of the chest. The signal obtained from the body is filtered and augmented. The sensing elements outputs an analog signal which is then revived by the analog to digital conveter. The serial to bluetooth module transmits the digital output of ADC to the radiophone. On the phone the sampled ECG is displayed (Fig. 2).

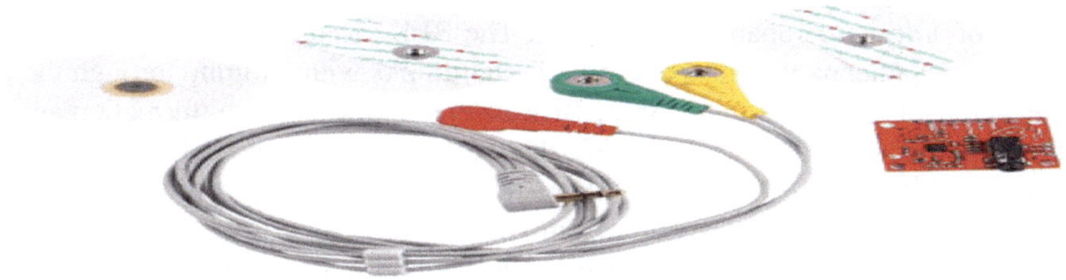

Fig. 2. ECG sensor

3.3 Microcontroller

Microcontroller used here is ATmega328. The software codes are loaded on the ardunio. From ardunio the output is given to the transceiver. The atmel ATmega328 may be a 32 k or 8 bit speed controller supported the AVR design. Many directions are departed in an exceedingly single clock cycle providing a turnout of virtually twenty million instruction per seconds at 20 MHz. The board options fourteen Digital pins and half dozen Analog pins. Its programmable by incorporated development environment It are often power driven by cable or external line potential unit battery, though it accepts voltage between seven and twenty volts. The UNO panel is the allusion model for ardunio platform. The ATmega328 on the Arduino Uno comes preprogrammed with a boot loader that enables to transfer new code to it excluding the use of external hardware software engineer. It communicates by means of the first STK500 protocol. The ASCII text file of coronary failure detection is uploaded to the ardunio UNO by means of ardunio software system (Fig. 3).

Fig. 3. Ardunio UNO R3

3.4 Bluetooth Transceiver

Bluetooth low energy may be a wireless personal space network technology premediated and promoted by bluetooth interest group. The bluetooth transceiver is simply used with ardunio for clear wireless serial communication. The interfacing between ardunio and bluetooth is completed by the pins D0 to D7 of arduino. The input voltage of bluetooth low energy is 3.3 V. It manages the ardunio pins with our own mobile application. It send detector information from ardunio to the associated application. For processing data BLE defend might to operate below three or five volt, thus it works with lots of ardunio compatible boards too. The BLE consists of SMAconnection, by attachment the antenna is connected to BLE guard links with Arduino through the ACI (ApplicationControllerInterface). Since BLE defend might receive information anytime even not nominated by master (ardunio), that SS line required.In ACI, data exchanged still through MOSI and MISO and SCK provides the clock generated by master, once master desires to request information from BLE guard, it spaces the REQN to low till RDYN line is put to truncated by BLE shield, and so master engenders the clock to scan out the information. If BLE defend has information to transmit to master, it'll place the RDYN to low point to master, although master haven't requested information and REQN is idle. If master detects a russet level condition on RDYN, it will place REQN to truncated and engender the clock to scan out the information (Fig. 4).

Fig. 4. BLE shield version 2.1

4 Antenna Design

4.1 Microstrip Antenna

Antenna is an air device that converts electrical power into airwaves and airwaves into electrical power it's typically used with a sender or receiving set. Antennas demonstrate reciprocity property which implies, it maintain same characteristics regardless of transmitter or receiver. For higher performance of antenna a thick material substrate with low dielectric constant are fascinating for providing high potency, information measure and radiation. Now-a days, antennas have undergone several changes, in accordance with their proportions and form. There are many varieties of antennas relying upon their wide selection of applications. Antenna has the potential of causation or receiving the magnetic force effect for the sake of communication. Antenna could be a electrical device intended to transmit or accept electromagnetic influence small strip antenna have many approaches over standard microwave antenna and therefore are wide utilized in several sensible application. Trifling strip antennas in its simplest prototype consist of divergent cover over on one feature of fractal substrate less than or equal to 10 which features a earth plane on different aspect. Small strip antennas are characterized by an outsized range of parameters than are standard microwave antenna they will be anticipated to posses several geometrical profiles and dimension (Fig. 5).

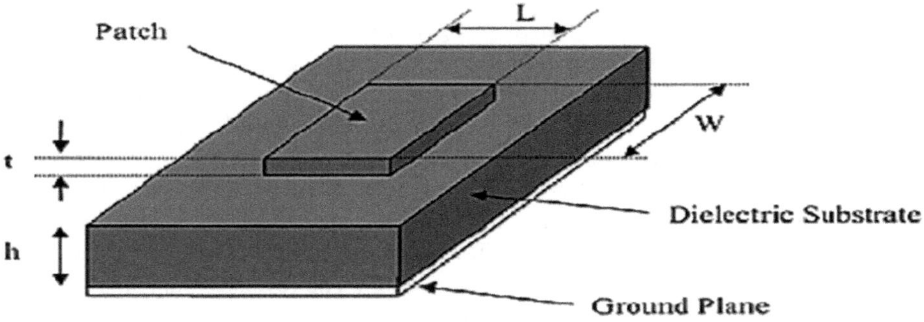

Fig. 5. Microstrip antenna

4.2 Design Equation

Step 1: computation of the Thickness (W)

$$W = \frac{c}{2f_0\sqrt{\frac{(\varepsilon_r + 1)}{2}}}$$

$$W = \frac{c}{2f_0\sqrt{\frac{(\varepsilon_r + 1)}{2}}}$$

Step2: computation of the effectual Dielectric Constant

$$\varepsilon_{eff} = \frac{\varepsilon_r + 1}{2} + \frac{\varepsilon_r - 1}{2}\left[1 + 12\frac{h}{W}\right]^{-\frac{1}{2}}$$

Step 3: computation of the Effective length

$$L_{eff} = \frac{c}{2f_0\sqrt{\varepsilon_{eff}}}$$

Step 4: computation of the distance end to end extension ΔL

$$\Delta\mathrm{L} = 0.412\mathrm{h}\frac{(\varepsilon_{eff} + 0.3)\left(\frac{W}{h} + 0.264\right)}{(\varepsilon_{eff} - 0.258)\left(\frac{W}{h} + 0.8\right)}$$

Step 5: computation of actual dimension of the patch

$$L = L_{eff} - 2\Delta L$$

4.3 Inverted F Antenna

The inverted f antenna was designed using ADS (advanced design software). The inverted f antenna is chosen because of its elevated gain. The proposed design is of size 43 × 30 mm. The plan of a reversed-F antenna with a a lot of compact size (antenna length but λ0/8 and antenna height less than zero.01λ0) and a far wider electric resistance information measure (greater than ten times that of a corresponding regular PIFA) has been incontestable. It is operated at the frequency of 2.4 Ghz with the gain of 3 db. The ISM band lies in the range of 2.4–2.48. The return loss obtained is −14 dbi. The substrate material used is FR4 with 1.6 mm thickness. The FR4 substrate was used due to its low cost and ease of fabrication. The reversed f antenna is chosen because of its high gain. In prevailing system the gain of antenna is extremely less. Gain improvement is done in proposed system. Second advantage of PIFA has compact backward radiation to the user's head, minimizing the emission power absorption (SAR) and enhances antenna performance. Third advantage is that PIFA it unveil moderate to high gain in every perpendicular and parallel states of polarization. This feature is incredibly sure in wireless communications whenever the antenna orientation isn't mounted and also reflections are from the various corners of the surroundings In those cases, the necessary parameter to be thought of is that the total field that's the resultant of horizontal and vertical states of polarization. In proposed system inverted F is used for wireless body space network (Figs. 6 and 7).

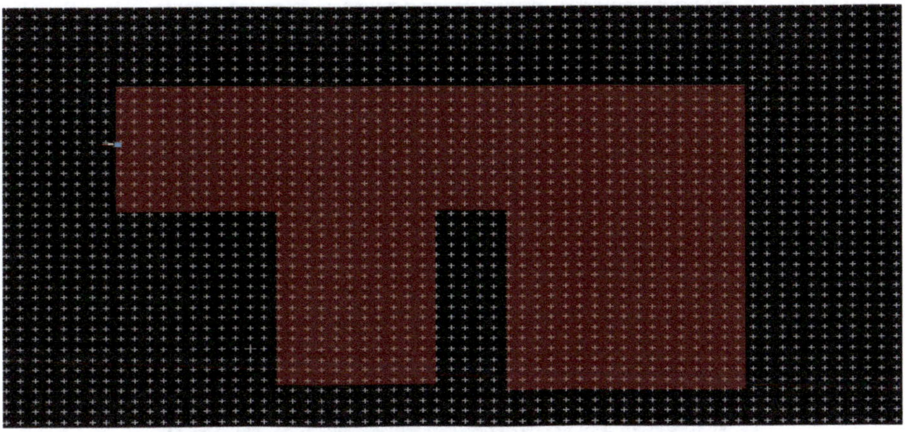

Fig. 6. Inverted F design

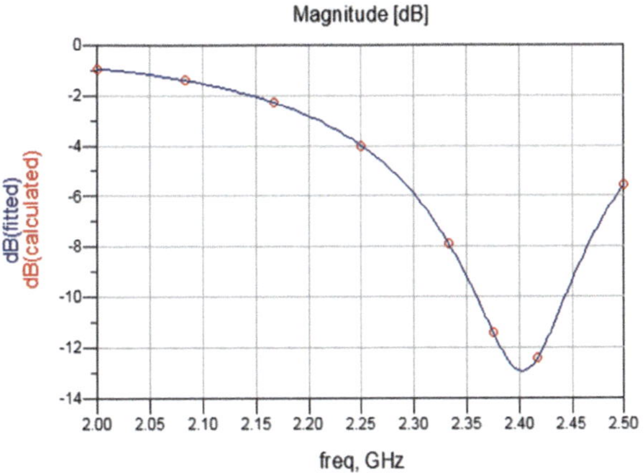

Fig. 7. Return loss of inverted F

4.4 Inverted L Antenna

The inverted L antenna has wide bandwidth, low return loss and used in mobile communication. The designed antenna is of size 48 × 32 mm. The antenna is operated at the frequency of 2.41 GHz which is the ISM band. The inverted l antenna is used for wireless body space network due to the operating frequency 2.4 Ghz. The gain obtained is 2.87 dbi. The return loss of inverted L antenna is −42 dbm. The substrate used is FR4 having dielectric constant of 4.2 with 1.6 mm thickness. In existing system the inverted L operates as WIMAX and X band frequency. In proposed system L shape is used for wireless body space network (Figs. 8 and 9).

Fig. 8. Inverted L design

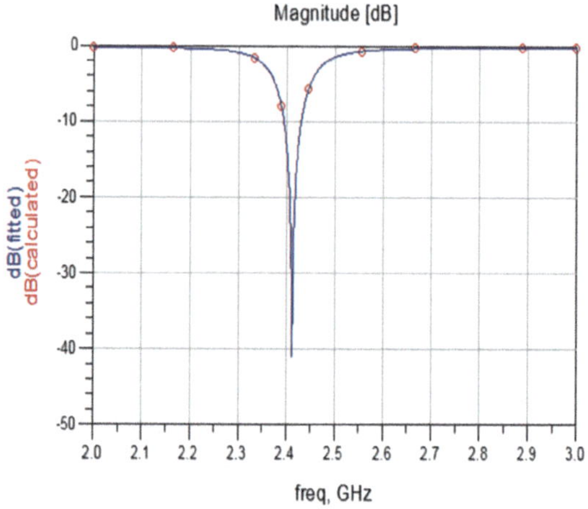

Fig. 9. Return loss of inverted L

4.5 Inverted T Antenna

The inverted T antenna is used as triple band frequency. The anticipated antenna of size 58 × 27 mm has been designed. The feeding is included at the edge of patch. The size of antenna is larger due to the additional feeding. The gain obtained is 3.77 dbi. The operating frequency of antenna is 2.45 Ghz. The return loss of antenna is −18 dbi. Due to hefty size it is not chosen for wireless body area network. The maximum power has been transmitted and minimum power has been reflected by antenna at this frequency. The two rectangular strips accustomed creates an inverted formed patch. These 2 strips having breadth, equals to 5 mm and length p is twenty five metric linear unit, thirty five metric linear unit severally. Antenna is excited at the purpose (58, −27) metric linear unit and simulated and result's obtained (Figs. 10 and 11).

Fig. 10. Inverted T design

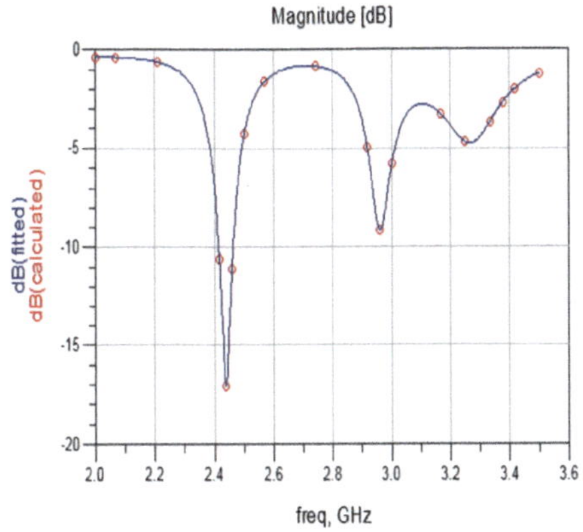

Fig. 11. Returnloss of inverted T

4.6 I Patch Antenna

The I formed patch antenna with the dimension of 46 × 26 mm. has been designed. I shape is anticipated by means of ADS. The utilization of I-shape permits a smaller slot size for proficient electromagnetic attraction energy coupling and also a reduction within the back radiation of the slot. The operating frequency of antenna is 2.42 GHz which lies in the range of ISM band. The gain obtained is 0.75 dbi. The return loss obtained is −18 dbm. The substrate for I shaped patch antenna is FR4. The dielectric constant given to the Fr4 substrate is 4.4. The substrate thickness of 1.6 mm is chosen for simulation (Figs. 12 and 13).

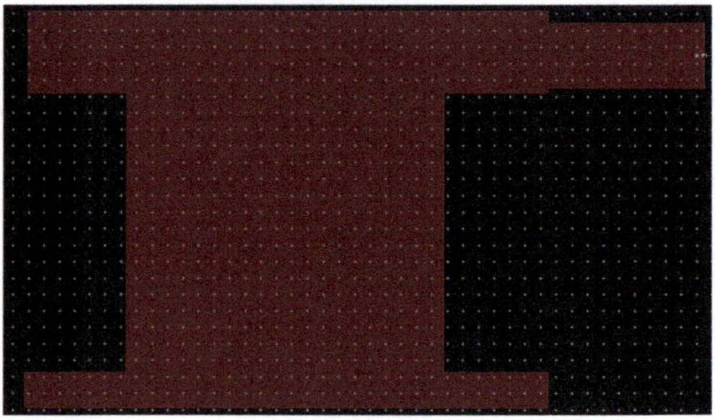

Fig. 12. I antenna design

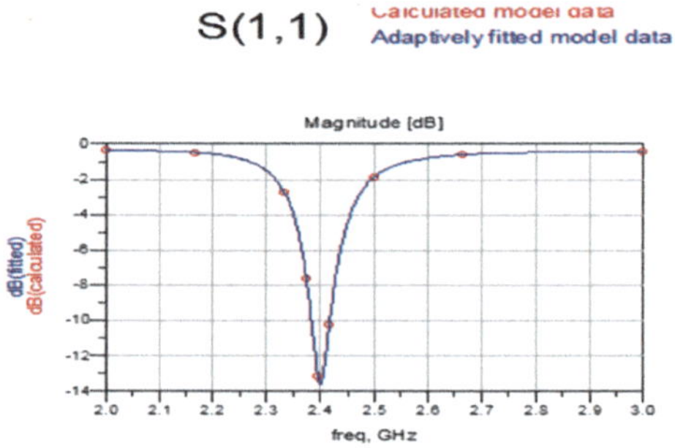

Fig. 13. Return loss of I shape

4.7 Comparison of Patch Antenna

(See Table 1)

Table 1. Comparison of patch antennas

Shape	Size	Frequency	Gain	Return loss
Inverted F	46 × 30 mm	2.44 Ghz	3.2 dbi	−18 dbm
Inverted L	48 × 32 mm	2.41 Ghz	2.87 dbi	−40 dbm
Inverted T	88 × 27 mm	2.45 Ghz	3.77 dbi	−18 dbm
I shape	43 × 24 mm	2.42 Ghz	2.6 dbi	11 dbm

5 Fabricated Antenna

The measured results are obtained by measuring antenna using network analyzer. The two antenna inverted F and inverted L is fabricated and measured results of both antenna is obtained by network analyzer. The simulated values of inverted F and inverted L is obtained by designing antenna using Advanced Design System. The simulated value of inverted F is operating frequency of 2.4 GHz at −13 Db. The measured value of inverted F is operating frequency of 2.305 GHz at −20 dB The simulated value of inverted L is operating frequency of 2.41 GHz at −41 dB The measured value of inverted L is operating frequency of 2.405 GHz at −2 dB. The variation from virtual result due to external loss (Figs. 14 and 15).

Fig. 14. Fabricated inverted F antenna

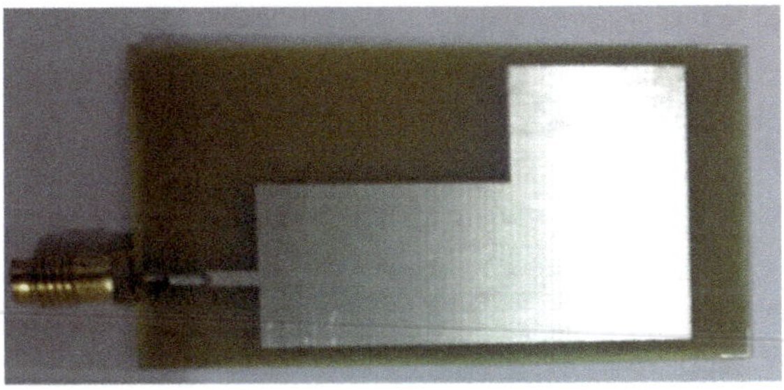

Fig. 15. Fabricated inverted L antenna

6 Conclusion

This article has given an entire body space network for the detection of heart condition by means of Bluetooth signals with an easy antenna style. In this paper four patch antennas are designed, from these four antennas is chosen for fabrication for detective work coronary failure. The parts employed in this project is comparatively cheap. In future we have an idea to implement coronary failure detection with emergency alert.

References

1. Wolgast, G., Ehrenborg, C., Israelsson, A., Helander, J., Johansson, E., Manefjord, H.: Wireless body area network for heart attack detection. IEEE Antennas Propag. Mag. **58**(5), 84–92 (2016)
2. Kaur, A., Kaur, A., Dhillon, A. S., Sidhu, E.: Inverted L shape slotted micro strip patch antenna for IMT, WIMAX AND WLAN applications. In: IEEE International Conference (2016)
3. Krishna, K.R., Rao, G.S., Ratna, P.R., Raju, K.: Design and simulation of dual band planar inverted F antenna for mobile handset application. Int. J. Antennas 1 (2015)
4. Fletcher, R.R., Kulkarni, S.: Clip-on wireless wearable microwave sensor for ambulatory cardiac monitoring. In: Annual International Conference of the IEEE EMBS Buenos Aires, Argentina, 31 August– 4 September (2010)
5. Chourasia, S., Changlani, S., Gupta, P.: Design and analysis of I-shaped micro strip patch antenna for low frequency. Int. J. Innovative Res. Sci. Technol. **1**(6), 320–324 (2014)
6. Singh, L.R., Kumar, P., Srivastava, D.K.: Design and analysis of triple band inverted T-shaped microstrip patch antenna. Int. J. Adv. Res. Comput. Commun. Eng. 4,(2) (2015)
7. Krishna, P., Manoj Reddy, C., Srinivas Reddy, P., Ammal, M.N.: Design of printed antenna for heart failure detection. Res. J. Med. Sci. 10 (2016)
8. Rezaeieh, S.A.: Wideband and Unidirectional Folded Antenna For Heart Failure Detection System. In: IEEE Antennas And Wireless Propagation, vol. 13 (2014)

An Intelligent MIMO Hybrid Beamforming to Increase the Number of Users

M. Preethika(✉) and S. Deepa

Department of Electronics and Communication Engineering,
Panimalar Engineering College, Chennai, India
mpreethika12@gmail.com, dineshdeepas1977@gmail.com

Abstract. The rate of the data demand is highly growing and number of user becomes high for utilizing the spectrum systematically, which can be made possible by using Multiuser MIMO. It allows the transmitter's base station (BS) to contact at a time with more receivers of the mobile stations (MS) through similar resources of time and frequency. In enormous MIMO base station antennas will be in the order of tens or hundreds to increase streams of data confined inside the cell. In this paper MIMO system is designed using OFDM scattering model and simulated to analyse various parameters with different number of users and RF chains. MIMO system increases the data rate with increased number of users and minimizes loss in the system.

Keywords: MIMO · Hybrid beamforming · MATLAB · RF chains · Number of users · Error magnitude

1 Introduction

Wider bandwidth in the millimeter wave (mmWave) bands will become useful for the upcoming 5G wireless system. Large scale antenna arrays are used in 5G systems to avoid severe propagation loss in the mmWave band. Wavelength of the mmWave frequency band is smaller than the wavelength found in microwave frequency band and hence mmWave signal travels to a shorter distance. So in order to increase the strength of mm Wave signal, array system can be used. But this array system is more expensive since it requires many transeption-reception module, of every antenna in an array. To overcome this disadvantage, hybrid transceivers can be used in the system. In hybrid transceiver or hybrid beamforming both analog and digital beamformers are used [1]. Analog beamformer is used in the RF stage and digital beamformer at the baseband stage. In this paper a more-user MIMO-OFDM device is used, in the seperation of the precoding into own digital baseband and RF analog devices at the transmitter receiver. Phased array used in this system can be steered to a desired direction by changing the phase of the signal. The most important needed technology for the upcoming 5G communication systems is the MASSIVE MIMO (MASSIVE - Multiple Input Multiple Output). Even though it provides more advantages such as high data rate, it has some disadvantages as well. Random fading effects caused by wireless channels can be eliminated by large Degree of Freedom (DoF) provided by massive array system. This will enhance the performance of entire communication systems. Hybrid structure uses

D. J. Hemanth et al. (Eds.): COMET 2019, LNDECT 35, pp. 1505–1514, 2020.
https://doi.org/10.1007/978-3-030-32150-5_152

phase shifters to reduce the number of RF chain which is similar to analog beamforming. This reduces the complexity of the system and also reduces the cost to a effective value [2–13]. Combination of analog RF processing and digital baseband processing is called hybrid beamforming (HBF). There are many advantages present in the Hybrid beamforming technique, which are as follows (1) Only limited RF chains are used in Hybrid beamforming (2) Phase shifters are used in this system and the difficulty in the process of analog of this model can be reduced by using constant amplitudes for all the phase shifters. These advantages decreases the difficulty of the Many Input Many Output system. High competition arises in the MIMO hence finishing optimal matrices in both analog and precoding digital with increased data rate. These problems can be minimised by single with two following ways. Other way in constructing both analog and digital precoding in a combined form. Next separate construction of analog and digital precoding. In this initially analog precoding is designed to a optimal value and then the digital precoding is optimised to improve the device activity. But most of the time combined system of analog and digital precoding is used in the hybrid beamforming system to approximately approach the full digital beamforming performance. Separate analog and digital method can be generated from a full digital model by using Least Square method for millimeter wave communication channel, and by this channel can be used in an effective way [2, 3]. Method such as optimization based method helps in analog digital precoder and this method will provide result similar to that found in the solution of full digital single user system [4, 5]. Likewise another method to jointly design both analog and digital model is WSMSE (Weighted Sum Mean Square Error). Capacity of the system can be maximised by using WSMSE method. In most of the multi user schemes energy is harvested at the analogous period and in next digital period cross interferences are eliminated [7–10]. The methods used in the multi-user schemes are Zero Forcing(ZF) and an Equal Gain Transmission(EGT). At the digital stage Zero Force method is used to eliminate inter user interference and at analog cycle Equal Gain Transmission methods is used to reserve power by considering Channel State Information [7]. Like these many methods can be used such as codebook based method. In this codebook based method properties of the millimeter wave channels are present and using these informations it designes MIMO hybrid system [8]. As a result large scale infinite user MIMO with new beamforming can be used to provide perfect trade-off within hardware toughness and role of the system. By assuming the acquired channel state information is perfect in a generic channel model of unit cell downlink MIMO with mixed structure supports multiple streams for each and every User Equipment(UE). And the additive value of the communication machine is maximized by using both analog and digital precoder. The main advantage of jointly designed two stage over separately designed two stage is to eliminate loss of information or data at each stage. Optimal solution which is asymptotic in a MASSIVE MIMO can be obtained by using double the least number of radio frequency, and also for the fewest radio frequency(RF) chain solutions are obtained. This solution is also found to outperform also when antennae small [14].

2 Proposed System

In multiple users device, Massive MIMO will be employed and Hybrid beamforming is used in this to avoid power loss and also reduces cost of the system. It separates the precoding required into analog RF components and digital baseband components in both mutiuser and single user systems. Channel state information can be found by making use of the full channel sounding present in the system. There are two spatially defined channel models namely, the 3GPP TR 38.901 Clustered Delay Line (CDL) model and a scattering-based model. In this paper scattering-based model is considered. Toolboxes needed for this MIMO system in MATLAB are

- Communications Device Toolbox
- 5Generation library in LTE Toolbox
- Adding-on of LTE Toolbox

Moreover this design tells MIMO-OFDM is used in dividing the precode to digital baseband, RF analog in transmitter. Phased Arrays are used in this MIMO-OFDM Precoding system and provides the solution.

2.1 Parameters Used

1. Initially system parameters are assigned conclude user capacity, data streams for a user, element of transmit/receive antenna, array place, and channel design. Optimizing parameters help in characterize of parameters in singular or joint property in allover device.
2. OFDM modulation parameters used for the system are FFT Length, CyclicPrefixLength, Number of Carriers, NullCarrierIndices, PilotCarrierIndices, CarriersLocations, user code rate is same, termination tail bits count, Modulation order, number of symbols to zeropad

   ```
   nonDataIdx = [prm.NullCarrierIndices;
   prm.PilotCarrierIndices];
   ```

3. Array transceiver and parameter position of system.

2.2 Information of the Channel

In spatial multiplex system, information channel will be present in transmitter terminal and permits application of precode for signal power maximization respective to the channel. Base Station scans the channel by using a reference signal, and this will be used by Mobile Station receiver in channel calculation. The Mobile Station sends estimated channel history returning to BS to denominate the precode weights for next data sending (Fig. 1).

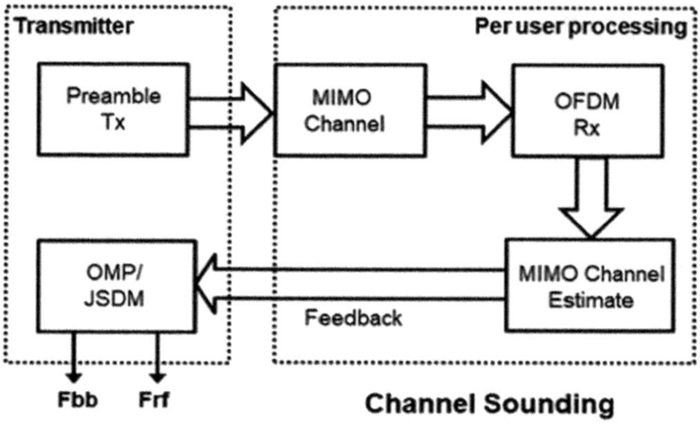

Fig. 1. Channel state model

Transmission of preamble signal occurs through all antenna transmit elements, which functions in reception subject based on channel. Antenna receiving elements processes the received signal through augmentation, OFDM mutilation, frequency realm carrier appraisal to every individual hookup.

2.3 Beamforming Technique

In this the orthogonal matching pursuit (OMP) innovation [16] is used for a personal-customer structure and Joint Spatial Division Multiplexing (JSDM) approach [2, 15] for a numerous-user scheme, find digital baseband precoding weight, precode mass of Fbb and RF analog, Frf of Hybrid beamforming rule contour.prm.nRays is the parameter used to describe rays.

Multi-user system, the inter-group conflict is suppressed with analog precode depending on the block slantization mechanism [17]. In this all user accredited to be in its own group, thereby leading to no reduction in inherent class asking or evaluating aloft. mFrf is average analog load over multiple subcarrier in the wideband OFDM system model. Explicit statistics enacted by stable nodes is shown in array response pattern, which indicate the separability achieved by beamforming.

3 System Model

3.1 Transmission of Signal

In this system every information is linked to RF chain using digital beamforming. RF chains are combined using switches and phase shifters. This is similar to analog beamforming technique in which number of RF chain is less. Then the combined RF chains are connected to individual antenna present at the transceiver end or at the individual transmitter or receiver end (Fig. 2).

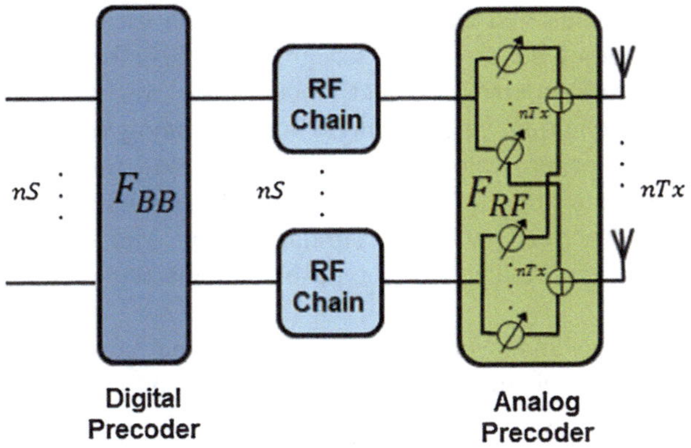

Fig. 2. Transmitter

The process of transmission of data includes the following process such as coding of channel, mapping of bits to composite symbol, chop one stream data to many data, precoding of the transmitted data present in baseband, OFDM modulation along with mapping of pilot signal and analog beamforming for all the transmit antennas used at RF frequency. Number of RF chains can be reduced using analog beamforming, which eventually reduces the power, cost and complexity of the system.

The transmitting and receiving of information process block diagram shown in Fig. 3.

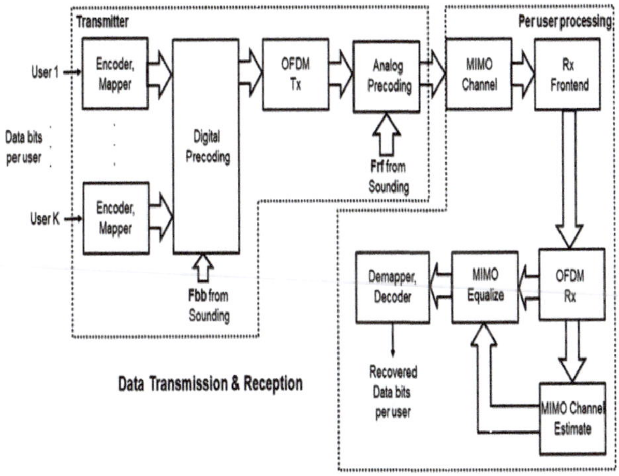

Fig. 3. Data transmission and reception

3.2 Reception of Signal

Two models can be considered for recognition of simple static-flat MIMO channel. Spatial Multiple Input Multiple Output channels

3GPP TR 38.901 Clustered Delay Line (CDL) model used in the spatially defined MIMO channel is one of the model, which provides the information about the array structure and the location details. Second model is the irregular design use approximation of ray trace technique which is single bounded along with a evaluated number of scatterers. In this paper scattering model is used and the number of scatterers assigned as 100. In scattering model, scatterers which are randomly present around the receiver are arranged perfectly which is similar to a one ring model. In this analysis non-Line Of Sight travel and uniform type of antenna with rectangle geometry is considered.

For both reference signal which provides the channel state information and data transmission signal same channel is used. Data signal is prepended with the preamble signal to differentiate it from the reference signal used for channel state information. Preamble signal is used to direct the date to be transmitted to the required receiver and the output signal present at the channel will be without the preamble field. For a multi-user system, separate channel is used for each user.

The receiver reduces path loss by low noise amplification and some thermal noise will be present. At the receiver side inverse process of the transmitter is performed which includes OFDM demodulator, MIMO equalizer, QAM demap and channel decode.

4 Result and Discussion

The MIMO-OFDM system is designed as we have seen in the above section and the following analysis are made using system parameters. Figure 4 is the radiation pattern obtained for this model.

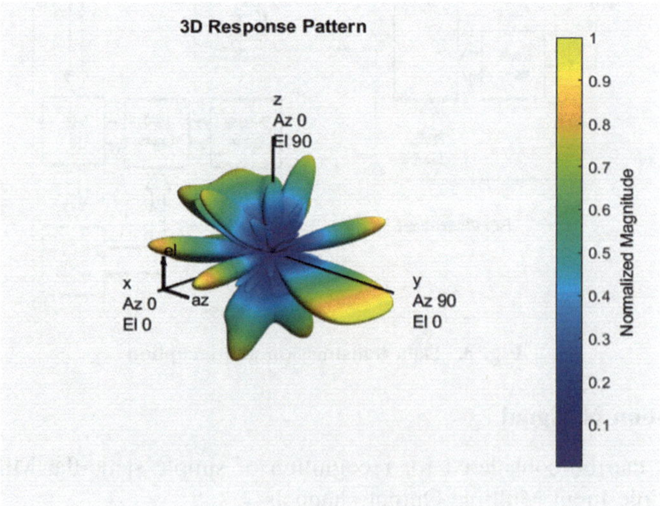

Fig. 4. Radiation pattern

From the Fig. 5 graph we can conclude that the magnitude of the error vector reduces with user expansion.

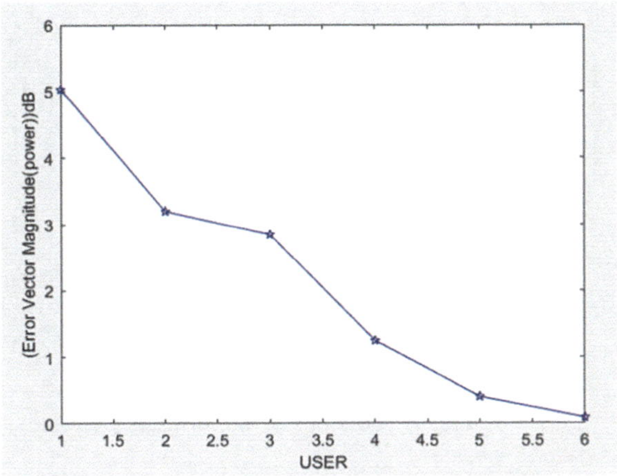

Fig. 5. Error vector magnitude vs no. of users

Graph in Fig. 6 shows the bits transmitted per second with the number of users. From this we can conclude that, with increase in the number of users bit size to be transmitted is reduced.

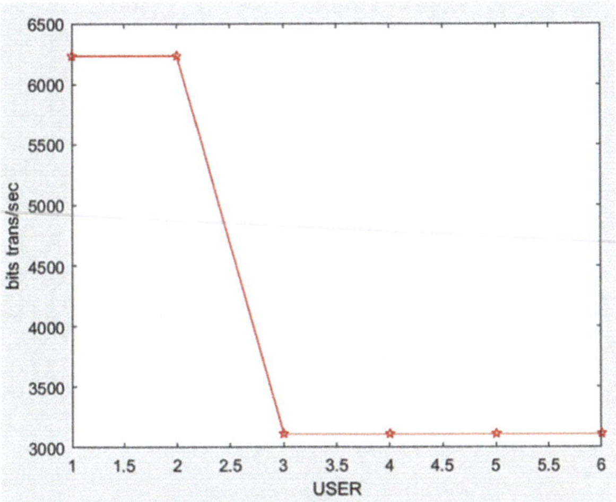

Fig. 6. Bits transmitted vs no. of users

Graph plotted for loss of bits per second and number of users. Figure 7 infers that loss in the number of bits decreases with increase in the number of users.

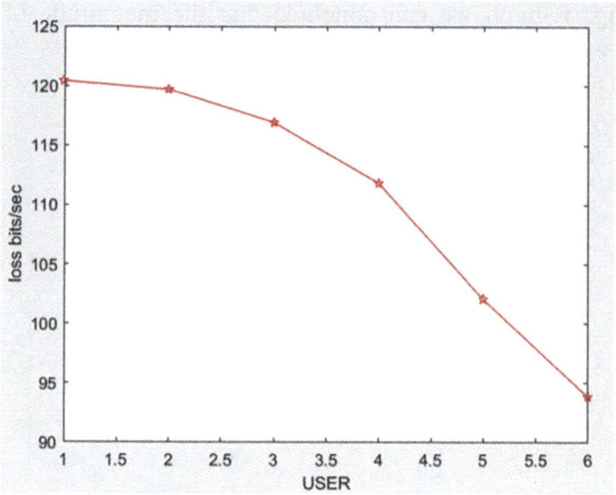

Fig. 7. Loss bits vs no. of users

Finally Figs. 8 and 9 shows the spectrum range with the increase in the number of RF chain.

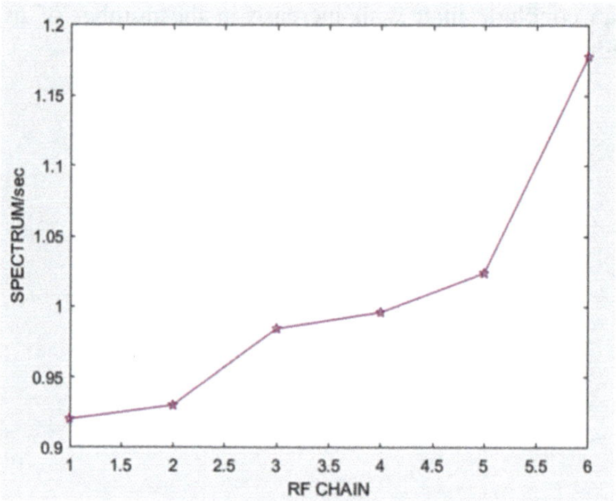

Fig. 8. Spectrum vs RF chain

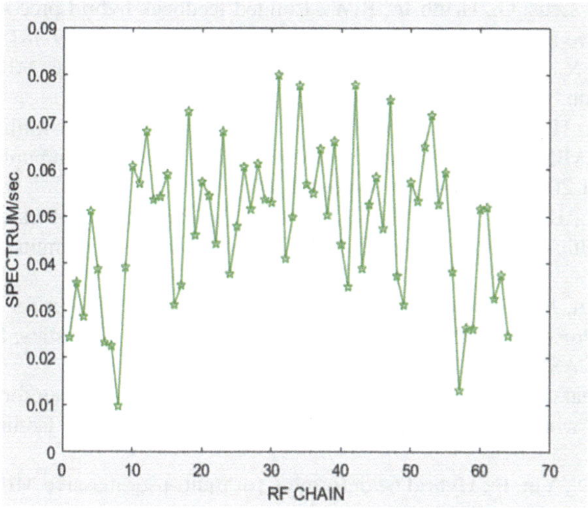

Fig. 9. Spectrum vs no. of RF chain

5 Conclusion

The next generation 5G communication can be made possible using mmWave spectrum band which can be made possible using MIMO (Multiple input Multiple output). In this paper MIMO is designed using scattering model and OFDM. Then the analysis is made using system parameters and from that we can conclude that error vector magnitude and loss of bits reduces by rising of user. Length of the bits to be transmitted is also decreased with the increase in the users.

References

1. Molisch, A.F., et al.: Hybrid beamforming for massive MIMO: a survey. IEEE Commun. Mag. **55**(9), 134–141 (2017)
2. Ayach, O.E., Rajagopal, S., Abu-Surra, S., Pi, Z., Heath, R.: Spatially sparse precoding in millimeter wave MIMO systems. IEEE Trans. Wireless Commun. **13**(3), 1499–1513 (2014)
3. Alkhateeb, A., El Ayach, O., Leus, G., Heath Jr., R.W.: Channel estimation and hybrid precoding for millimeter wave cellular systems. IEEE J. Sel. Topics Signal Process. **8**(5), 831–846 (2014)
4. Ni, W., Dong, X., Lu, W.S.: Near-optimal hybrid processing for massive MIMO systems via matrix decomposition (2015). https://arxiv.org/abs/1504.03777
5. Payami, S., Ghoraishi, M., Dianati, M.: Hybrid beamforming for large antenna arrays with phase shifter selection. IEEE Trans. Wireless Commun. **15**(11), 7258–7271 (2016)
6. Bogale, T.E., Le, L.B.: Beamforming for multiuser massive MIMO systems: Digital versus hybrid analog-digital. In: Proceedings IEEE Global Communication Conference (GLOBECOM 2014), pp. 4066–4071, December 2014
7. Liang, L., Xu, W., Dong, X.: Low-complexity hybrid precoding in massive multiuser MIMO systems. IEEE Wireless Commun. Lett. **3**(6), 653–656 (2014)

8. Alkhateeb, A., Leus, G., Heath Jr., R.W.: Limited feedback hybrid precoding for multi-user millimeter wave systems. IEEE Trans. Wireless Commun. **14**(11), 6481–6494 (2015)
9. Ni, W., Dong, X.: Hybrid block diagonalization for massive multiuser MIMO systems. IEEE Trans. Commun. **64**(1), 201–211 (2016)
10. Song, N., Sun, H., Yang, T.: Coordinated hybrid beamforming for millimeter wave multi-user massive MIMO systems. In: Proceedings IEEE Global Communication Conference (GLOBECOM 2016), pp. 1–6, December 2016
11. Rajashekar, R., Hanzo, L.: Iterative matrix decomposition aided block diagonalization for mm-wave multiuser MIMO systems. IEEE Trans. Wireless Commun. **16**(3), 1372–1384 (2017)
12. Sohrabi, F., Yu, W.: Hybrid digital and analog beamforming design for large-scale MIMO systems. In: Proceedings IEEE International Conference on Acoustics, Speech and Signal Processing (ICASSP), pp. 2929–2933, April 2015
13. Singh, J., Ramakrishna, S.: On the feasibility of codebook-based beamforming in millimeter wave systems with multiple antenna arrays. IEEE Trans. Wireless Commun. **14**(5), 2670–2683 (2015)
14. Wu, X., Liu, D., Yin, F.: Hybrid beamforming for multi-user massive MIMO systems. IEEE Trans. Commun. **66**(9), 3878–3891 (2018)
15. Li Z., Han, S., Molisch, A.F.: Hybrid beamforming design for millimeter-wave multi-user massive MIMO downlink. In: 2016 IEEE ICC Signal Processing for Communications Symposium (2016)
16. Adhikary, A., Nam, J., Ahn, J.-Y., Caire, G.: Joint spatial division and multiplexing - the large-scale array regime. IEEE Trans. Inf. Theory **59**(10), 6441–6463 (2013)
17. Spencer, Q., Swindlehurst, A., Haardt, M.: Zero-Forcing methods for downlink spatial multiplexing in multiuser MIMO channels. IEEE Trans. Signal Process. **52**(2), 461–471 (2004)

Analysis of Wearable Meander Line Planar Antenna Using Partial and CPW Ground Structure

Monisha Ravichandran(✉) and B. Bhuvaneswari

Department of Electronics and Communication Engineering,
Panimalar Engineering College, Chennai, India
monishamadhu96@gmail.com

Abstract. In this paper, a compact, flexible textile meander line antenna is discussed for wearable applications using two different ground structure. They are coplanar waveguide ground structure and partial ground structure. Coplanar waveguide ground plane is placed above the substrate along with patch and partial ground plane is placed below the substrate of the two similar MLA structure. As MLA requires only small space, it is preferable for wearable applications. For wearable antenna, the substrate should be flexible and easy to wear. So here, Jean material is used as a substrate. These proposed antenna is designed at operating frequency of 2.45 GHz. Hence, the simulated results such as reflection coefficient, VSWR, radiation pattern, bandwidth of the antenna are discussed.

Keywords: Meander line antenna · Coplanar waveguide · Partial ground structure · Jeans · Return loss

1 Introduction

In wireless communication systems, an antenna is a major part of the system. Recently, there has been more research undergoing in this area. It includes various area such as military, medical, health monitoring, and so on. Particularly there is rapid growth in field of medical applications mainly based on wearable devices. Generally, an important requirement of wearable device is a compact and flexible antenna in which it has to provide wireless connectivity.

A wearable antenna should be part of the clothing used for communication purposes, that includes public safety, navigation, tracking, military communication, health monitoring and so on. Requirements of wearable antenna includes

- Size should be smaller
- Should be maintenance free
- Should be lower cost
- no installation

Further the particular requirement of wearable antenna

- It should be flexible in nature i.e. flexible substrate material.
- Planar structure antenna

D. J. Hemanth et al. (Eds.): COMET 2019, LNDECT 35, pp. 1515–1525, 2020.
https://doi.org/10.1007/978-3-030-32150-5_153

The behavior of the antenna can be influenced by the properties of the substrate and structure of the antenna.

In wearable antenna, the use of textiles requires the characterization of their properties. The conductive textile have stable and lower electrical resistances to reduce losses. The antenna flexibility is also needed so that it can be easily incorporated in cloth. While designing the wearable antenna, the selection of substrate is the significant step.

Generally, the textiles used as a substrate should have less dielectric constant which minimize surface losses and it will increase the impedance bandwidth of the antenna. Planar monopole, dipoles, PIFAs, and patch antenna are the conventional antennas used in wearable antenna designs.

Meander line antenna (MLA) is type of microstrip antennas. The wire is continuously folded to reduce the resonant length in meander line antenna. Then by increasing the total length of wire antenna of fixed axial length reduces its resonant frequency. The meander patch will increase the path over that the surface current flows which leads to the operating frequency reduction than the linear wire antenna with equal dimensions. Moreover the meander line antenna are electrically small antenna.

Meander line antenna are very useful because it has relatively smaller size and higher radiation efficiency. In this paper, a meander line wearable antenna with CPW and partial ground structures using jeans as a substrate which is operating at 2.45 GHZ is discussed. Here, various parameter such as VSWR, Reflection coefficient, Bandwidth and Radiation pattern are discussed.

2 Related Works

Khan et al. proposed an Microstrip patch antenna on a jeans fabric substrate which is operating at frequency band of 2.1366 GHz, 4.7563 GHz, 11.495 GHz and it has wider bandwidth. the obtained gain of these frequency bands are 3.353 dBi, 4.237 dBi, and 5.193 dBi [1].

Rashed et al. introduced a Meander line antenna as a possible antenna for size reduction and an increase in meander section improves bandwidth. A new class of wire antenna with size reduction in the resonant length from 25–40% is designed [2]. While designing the wearable and implantable antenna, there are various issues to be considered including selection of substrate, influence of ground planar size and so on [3].

Misman et al. proposed an meander line antenna using FR4 as substrate which operates at 2.45 GHz for WLAN application. The obtained return loss of the antenna is −27.55 dB. It concluded by designing meander line antenna will provide more performance when conductor line is used [4]. The design of compact single element meander line antenna is proposed with bandwidth of 240 MHz which can be used for USB application due to small size [5].

There are various techniques [6], to improve bandwidth and obtain different polarization for microstrip patch antenna, which are also suitable for wearable antenna [7].

Using jeans as substrate, the patch antenna is designed for wearable application. The operating frequency of antenna is 2.45 GHz and it provides gain of about 7.2 dBi. Textile material used as substrate should have lower dielectric constant [8].

Coplanar waveguide fed antenna is designed in order to provide better impedance matching [9]. Dual band meander line antenna are proposed using textile fabric as a substrate which is operated in 406 and 850 MHz [10]. Circular patch antenna using partial and full ground plane in the range of 1–8 GHz is designed and compared the result. Thus, the characteristics of the antenna has been changed by changing dimensions of ground plane [11].

3 Antenna Design

3.1 Selection of Substrate

Generally the wearable antennas are made up of soft materials such as felt, jeans, leather, nylon, conductive textile, conductive thread and so on. Because, they are likely to be bent and crumpled when human moves and the performances of the antenna should be same. Here the substrate used for antenna is jeans. The dielectric constant of jeans is 1.6 and thickness is 3.6 mm.

3.2 MLA Design

Meander line antenna comprises horizontal and vertical lines which forms turns. MLA is a type of microstrip antenna. In MLA, the antenna size at operating frequency is minimized by the factor, which is directly proportional to no. of turns. Then the adjacent horizontal segments of MLA should have opposite phase. When no. of turns of the MLA increases then the efficiency also increases. In meander line antenna, the resonant frequency is a function of spacing and separation of antenna. If the spacing and separation of meander is increased then the resonant frequency decreased. When comparing to other conventional antenna, MLA has a good radiation efficiency and also this structure reduces size. Figure 1 shows single element meander line antenna structure.

Fig. 1. Structure of meander line antenna

Here, the meander line antenna using partial ground structure and CPW ground structure is discussed.

Using Partial Ground Structure. Meander line antenna is designed with partial ground structure on jeans substrate. The dimension of antenna is about 28 × 13 mm^2. Meander line antenna is electrically small antenna so it has λ/10 length. For impedance matching purpose, quarter wave transformer is used. In this proposed antenna, there are 8 turns are present. Ground lies below the antenna substrate and the dimension can be λ/2, λ/4 and so on. The length of ground is 7.5 mm and the width is 10 mm. The thickness of ground and patch is about 0.035 mm. The patch is radiating element which has eight turns with equal separation and spacing.

The characteristics of an antenna is not only depend upon shape, radius and substrate material of the antenna. But also the dimensions of the ground plane make observable changes in the characteristics of the antenna. One of the popular method for the characteristics enhancement of antenna is reduction of ground plane. This method is used to increase efficiency, improve impedance matching, reduction of size and so on.

Figure 2 shows the meander line antenna structure using partial ground.

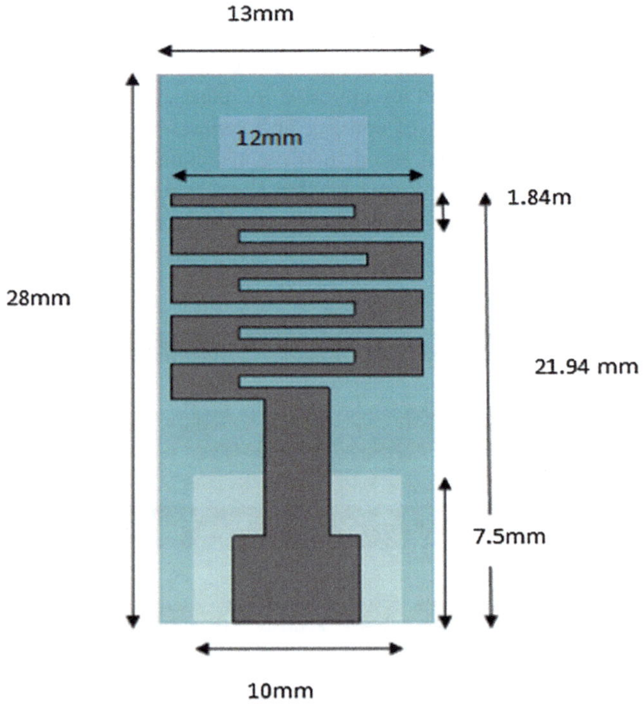

Fig. 2. Antenna structure using Partial ground

3.3 Using CPW Ground Structure

Meander line antenna is designed with cpw ground structure on jeans substrate. The dimension of antenna is about 48 × 36 mm^2 which is suitable for wearable application and the cpw ground is placed on the antenna. It acts as a reflector or a isolator which is the most attractive feature of the antenna and so it protects the human body from backward radiation. Additionally, the performance of the antenna is enhanced by the gain. Therefore, this type of antenna is intended to be worn as a part of cloth, for monitoring the human's vital signs.

There are two types of coplanar structures, they are coplanar waveguide (CPW) and coplanar slot (CPS). Here, Coplanar waveguide structure is used.

Coplanar waveguide is a structure in which all the conductors supporting propagation of wave are located on the same plane, generally on the surface of the dielectric material. Apart from the microstrip line, the Coplanar Planar Waveguide is the most common use as planar transmission line in Radio Frequency/microwave integrated circuits.

The substrate dimension is about 48 × 36 mm^2 and the patch dimension is about 41 × 14 mm^2. The thickness of patch and ground is about 0.035 mm. The patch is radiating element which has 11 meander turns with equal spacing and separation. The spacing of meander is about 1.2 mm.

Figure 3 shows the meander line antenna structure using CPW ground.

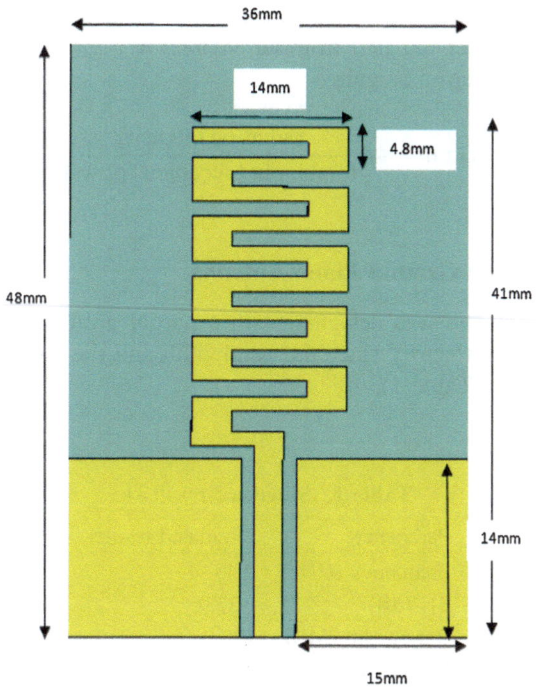

Fig. 3. Antenna structure using CPW ground

4 Results

The simulated results of meander line antenna for wearable application using CPW and partial ground structures are discussed.

The parameters such as reflection coefficient, VSWR, bandwidth, gain are discussed. The reflection coefficient of the antenna defines how much power is reflected from the antenna. The VSWR of the antenna defines the amount of reflected power from antenna. The minimum VSWR is 1.0.

$$VSWR = \frac{1 + |\Gamma|}{1 - |\Gamma|}$$

The frequency range of the antenna in which it can operate perfectly is called the bandwidth.

$$B.W = 100 \times (F_H - F_L)/F_C$$

where FH is higher frequency
FL is lower frequency
FC is centre frequency

The radiation or antenna pattern describes the relative strength of the radiated field in various directions from the antenna, at a constant distance.

The ratio of the power radiated from an antenna in a given direction to the input power of an antenna referred as gain.

$$\text{Gain} = 4\pi \frac{\text{radiation intensity}}{\text{total input (accepted) power}}$$

4.1 Results of Partial Ground Plane Antenna

The meander line antenna was designed with a partial ground plane and the results were found to be unsatisfactory since the bandwidth and the gain wasn't consistently good (Figs. 4, 5, 6 and Table 1).

Table 1. Simulated results-1

Parameter	Simulated results
Frequency (GHz)	2.45
S_{11} (dB)	−37.9
Bandwidth (MHz)	120
Gain (dB)	−5.2
VSWR	1.02

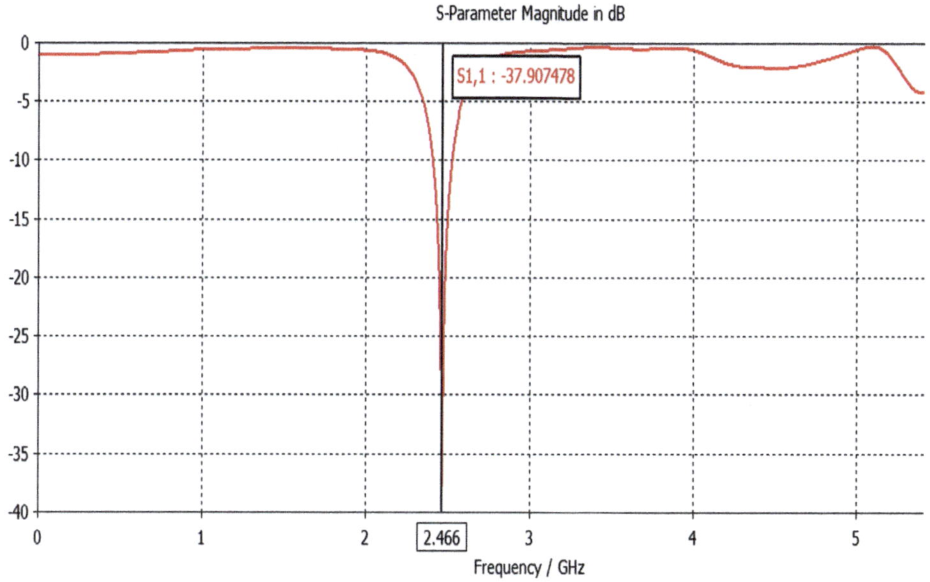

Fig. 4. Simulated reflection coefficient of partial ground antenna

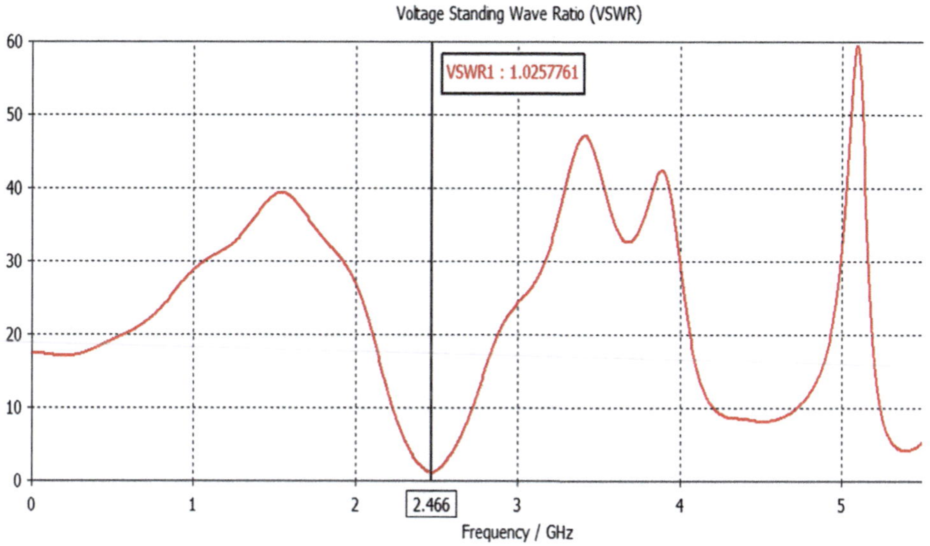

Fig. 5. Simulated VSWR of partial ground antenna

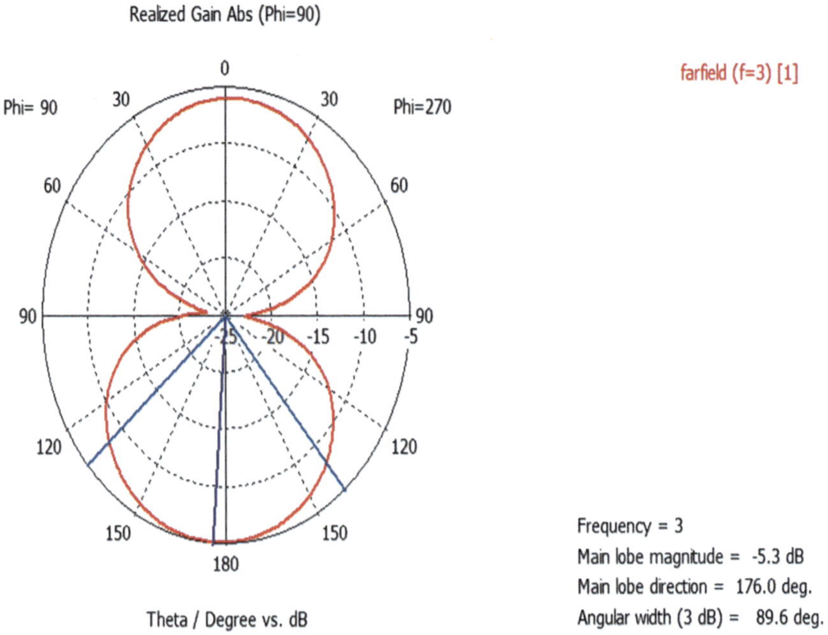

Fig. 6. Simulated radiation pattern of partial ground antenna

4.2 Results of CPW Ground Plane Antenna

The meander line antenna was designed with cpw ground structure and the results were found to be satisfactory since the bandwidth and gain was consistently good.

Using CPW in antenna structure provides more bandwidth and the gain than partial ground structure antenna (Figs. 7, 8, 9 and Table 2).

Table 2. Simulated results-2

Parameter	Simulated results
Frequency (GHz)	2.45
S_{11} (dB)	−37.7
Bandwidth (MHz)	280
Gain (dB)	3.528
VSWR	1.03

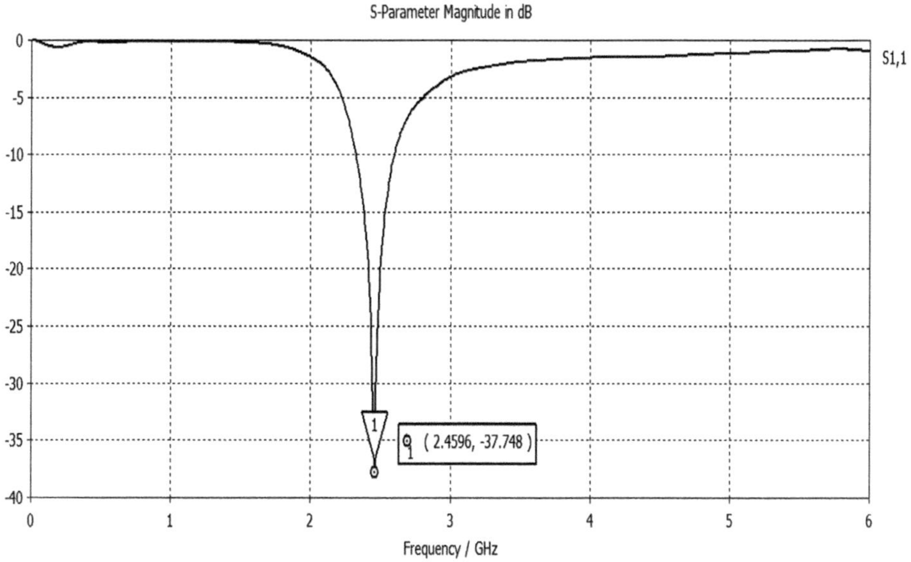

Fig. 7. Simulated reflection coefficient of CPW antenna

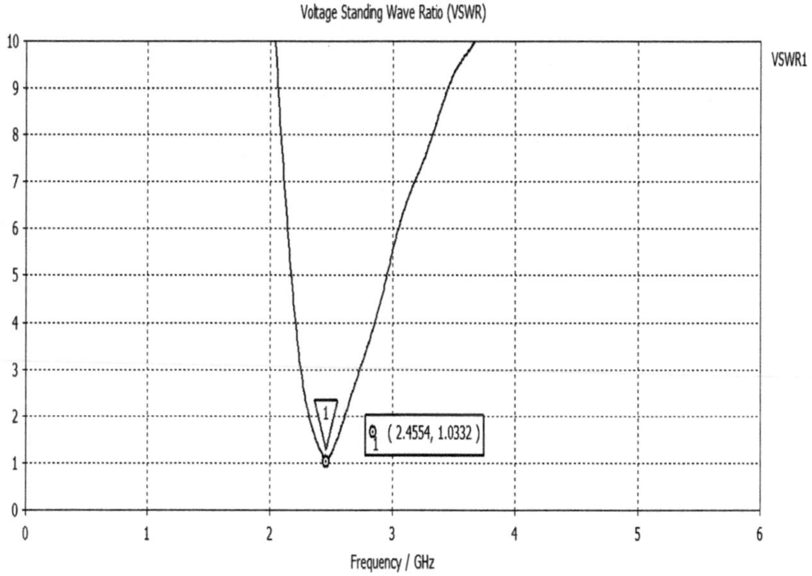

Fig. 8. Simulated VSWR of CPW antenna

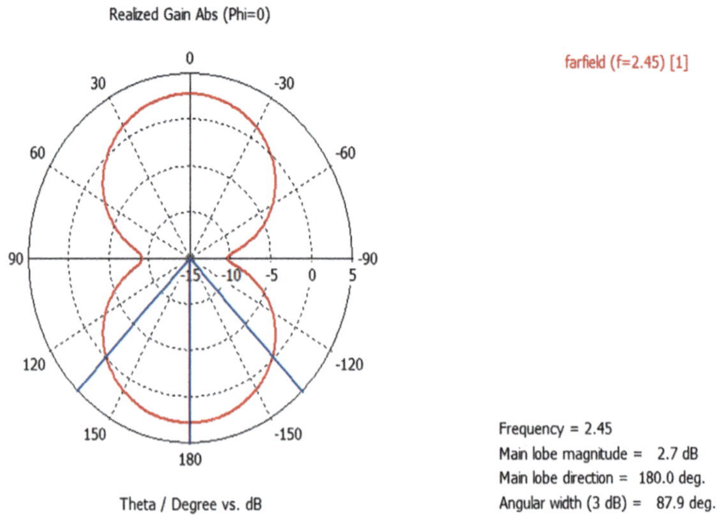

Fig. 9. Simulated radiation pattern of CPW antenna

5 Conclusion

On comparing the simulated results of meander line antenna with CPW ground structure and partial ground structure, MLA using CPW ground structure gives higher bandwidth and higher gain. The number of turns is eleven and the patch length is 41 mm and the width is 14 mm and the meander spacing is 1.2 mm. Hence, the simulated reflection coefficient is −37.7 dB and the realized gain is 3.5 dB.The antenna operates in the range of 2.33–2.60 GHz frequency band with bandwidth of 280 MHz This proposed antenna is used for medical application as wearable monitoring devices and it is flexible in nature.

References

1. Khan, S., Singh, V.K., Naresh, B.: Textile antenna using jeans substrate for wireless communication application. Int. J. Eng. Technol. Sci. Res. (IJETSR) **2**(11), 176–181 (2015). ISSN 2394–3386
2. Rashed, J., Tai, C.T.: An new class of resonant antennas. IEEE Trans. Antennas Propagat. **39**, 1428–1430 (1991)
3. Grupta, B., Sankaralingam, S., Dhar, S.: Development of wearable and implantable antennas in the last decade: a review. In: Proceedings of Mediterranean Microwave Symposium (MMS), Guzelyurt, Turkey, 25–27 August 2010, pp. 251–267 (2010)
4. Misman, D., Husain, M.N., Aziz, M.Z.A.A., Soh, P.J.: Design of planar meander line antenna. In: 3rd European Conference on Antennas and Propagation, Estrel Hotel, Berlin, Germany 23–27 March 2009 (2009)
5. Ambhore, V.B., Dhande, A.P.: Properties and design of single element meander line antenna. Int. J. Adv. Res. Comput. Sci. (2012). ISSN N0.09676–5697

6. Garg, R., Bhartia, P., Bahl, I., Ittipiboon, A.: Microstrip Antenna Design Handbook. Artech House, Norwood (2001)
7. Sankaralingam, S., Gupta, B.: Development of textile antennas for body wearable applications and investigations on their performance under bent conditions. Prog. Electromagn. Res. B **22**, 53–71 (2010)
8. Purohit, S., Raval, F.: Wearable-textile patch antenna using jeans as substrate at 2.45 Ghz. Int. J. Eng. Res. Technol. (IJERT) **3**(5), 2456–2460 (2014)
9. El Atrash, M., Bassem, K., Abdalla, M.A.:A compact dual-band flexible cpw-fed antenna for wearable application. IEEE (2017)
10. George, G., Nagarjun, R., Thiripurasundari, D., Poonkuzhali, R., Alex, Z.C.: Design of meander line wearable antenna. In: IEEE Conference on Information and Communication Technologies ICT (2013)
11. Viswanathan, A., Desai, R.: Applying partial-ground technique to enhance bandwidth of a UWB circular microstrip patch antenna. Int. J. Scie. Eng. Res. **5**(10), 780–784 (2014)
12. Calla, O.P.N., Singh, A., Singh, A.K., Kumar, S., Kumar, T.: Empirical relation for designing the meander line antenna. In: International Conference on Recent Advances in Microwave Theory and Applications, pp. 695–697, November 2008
13. Hu, Z., Zhang, L.: A method for calculating the resonant frequency of meander-line dipole antenna, May 2009
14. Balanis, C.A.: Antenna Theory: Analysis and Design. Wiley, New York (1997)
15. Warnagiris, T.J., Minardo, T.J.: Performance of a meandered line as an electrically small transmitting antenna. IEEE Trans. Antennas Propag. **46**(12), 1797–1801 (1998)

Energy Efficient Distributed Unequal Clustering Algorithm with Relay Node Selection for Underwater Wireless Sensor Networks

M. Priyanga[1(✉)], S. Leones Sherwin Vimalraj[1], and J. Lydia[2]

[1] Department of Electronics and Communication Engineering, Panimalar Engineering College, Chennai, India
uthirapriyanga@gmail.com, leonessherwin@gmail.com
[2] Department of Electronics and Communication Engineering, Easwari Engineering College, Chennai, India
lydia_822@yahoo.co.in

Abstract. The widely used activities of underwater wireless sensor networks (UWSNs) are data gathering, pollution monitoring, seismic monitoring and undersea exploration. Power supply is the major need of acoustic sensors which are used in UWSNs. The main complication in underwater sensor network is the replacement of batteries, limited bandwidth as well as high propagation delay. Firstly, to extend the life time and to reduce the energy consumption of the network the concept of Energy efficient of distributed unequal clustering (EEDUC) algorithm is used. This algorithm consists of unequal clustering model, election of cluster head, its establishment and transmission of data. Secondly, in order to reduce the packet collision we introduced new discovery of route and maintenance of route phase in the routing protocol. The main idea of this phase is the construction of low overhead routing protocol in order to control the routing overhead. Thirdly, further improve the energy efficiency we present a novel algorithm namely relay node selection algorithm (RNS). The construction of relay node is to reduce the transmission distance which helps to improve the network lifetime. By using the extensive simulation study we evaluation the performance of our proposed method with several earlier technique. The parameters which we are concentrated to analyze the performance are network life time, energy efficiency, energy consumption, throughput, generated packets and packet loss.

Keywords: Underwater wireless sensor network · Energy efficient distributed unequal clustering · Clustering · Relay node selection · Energy balance

1 Introduction

Underwater Sensor networks provides the agreement of changing huge areas of trade, science and the government. The flexibleness to own tiny devices which are distributed close to the objects are detected leads brand-new opportunities to act on the world, as an example with cohabitat observance, structural surveying and industrial applications

D. J. Hemanth et al. (Eds.): COMET 2019, LNDECT 35, pp. 1526–1536, 2020.
https://doi.org/10.1007/978-3-030-32150-5_154

whereas sensor-network schemes is started to be given in applications. Nowadays, underwater process on the ground, stay restricted through comparison. Submersibles are controlled remotely and frequently utilized, but their deployment is temporary because it is large, active and managed devices. Some deep-area data assortment works are attempted, but at coarse roughness (100 sensors to hide the world). Even though regional process are thought of, they are extremely high-priced and mostly wired. The main advantages of terrestrial wireless sensor networks are self-configuration and increasing the consumption of energy. It gives priority to low cost nodes (around US $100), deployments in dense area (100 m apart), limited range, multi-hop transmission; through comparison, underwater acoustic communication nowadays are generally low (US$10 k or more), easily deployed in which a few nodes placed kilometers apart, generally communication to a "base station" over high ranges is only done but not with each other. This chapter presently exploring the way to broaden the advantages of underwater wireless sensor networks (UWSNs) with acoustic transmission. Underwater Wireless sensor Networks has several possible applications. Here, seismic imaging of the submarine oilfields taken as an ideal application. Today, in offshore oilfields most seismic imaging tasks are allocated by a ship that gives an outsized array of hydrophones on the surface. The technology is incredibly high and hence the survey of seismic will only be allotted rarely. As compared, nodes of sensor network have terribly very low value and may be for good deployment on the bottom of sea. Such a system permits continuous imaging of seismic objects and helps for enhancing recovery of resource and oil production. The subsequent Fig. 1 shows the diagram of underwater wireless sensor networks.

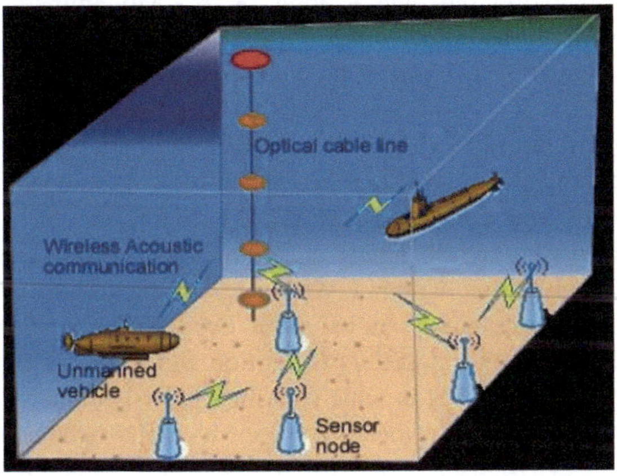

Fig. 1. UWSN

2 Related Works

In recent years, for underwater wireless sensor networks some cluster based data gathering techniques has been proposed. These techniques have been reviewed and limitations has been listed out.

In AUV (Autonomous Underwater Vehicle) aided with Energy Efficient Routing protocol, [1] which covers planned flight path in every cycle. The sensors is categorized as two section which is gateway and members. The gateway sensors are selected depending on their remaining energy and nearness to the AUV trajectory. AUV trajectory is only communicated by gateway sensor. But, there is no direct way to jump from members to gateway sensor which results in increase in consumption of energy.

In the 3D-Zone of reference(3-D ZOR), [2] AUV covers a circular path which is predetermined and it collects data packets from all the sensors deployed in different regions. This protocol is another mobile data gathering protocol named as mobicast. The sensors relays the packets to AUV in multi hop as well as in single hop because clustering mechanism is needed. The data transmission in void areas and presence of water currents is rectified by using large area around the 3-D ZOR. However large area consumes more sensors that increase in consumption of energy and gathering data from all the sensors is not possible because only sensors within 3D-ZORs can be considered.

In CMDG (Cluster-based Mobile Data-Gathering scheme), a group of underwater sensors is allocated as cluster heads for collecting the data from all the sensors and the gathered data is transferred sensors to AUV. Hence, the limitation of multi-hop relaying to limited level leads to reduction in consumption of energy. With the above analysis of mobile data gathering schemes, we came to conclusion that cluster based scheme and the sensor network can be designed in distributed manner [3].

AODV using individual route of reply with reverse path have been proposed by Saini et al. E-AODV that route detects in lesser amount of times than the Ad Hoc On-Demand Distance Vector (AODV) [4]. In UASNs which is complicated network comparing the terrestrial wireless network because of its limited bandwidth with low data rates and propagating the data is long [5, 6]. Mohammadi et al. rectifies an optimization problem by setting a Relay Node (RN) in a particular place. Thus by adding the RN, increases the lifetime of the network and its problem is rectified in some way [7].

A routing protocol with low overhead for UWSNs which was proposed in [8]. This protocol is for reducing the control overhead mainly in maintenance of routing phase and the routing protocol is compared with AODV and DSR.

The routing protocol which is greedy algorithm was proposed in [9] for decreasing energy consumption in 3-Dimensional UWSN. In [10] the authors proposed a balanced routing method that avoids energy hole in the UASN (Underwater Acoustic Sensor Network). The authors have proposed a two section algorithm for increasing lifetime of the 3-Dimensional UASN in [11] in which setting of RN is done by determining the x, y, z coordinates of every RN. So locating the each of RN becomes insignificant.

Zhang et al. in [12] proposed a DEBCR protocol abbreviated as depth-and energy-based clustered routing protocol can be used in 3-Dimensional UASNs, resulting in a conical structure of network. Selection of cluster heads is done depending on its depth and the remaining energy of all the sensors. But DEBCR protocol does not guarantee the cluster-heads stability. Finally, the performance of channel in UWSN which is acoustic and its challenges is discussed by Ismail et al. [13].

3 Energy Efficient Distributed Unequal Clustering (EEDUC)

3.1 System Model

The below Fig. 2 shows the 3-dimensional UWSN structure which consists of 3 types of nodes. The first one is static nodes affixed to the sea bottom, second type of sensor nodes will floats in the water which is dynamic in nature & the remaining nodes is sink nodes that floats on the water surface.

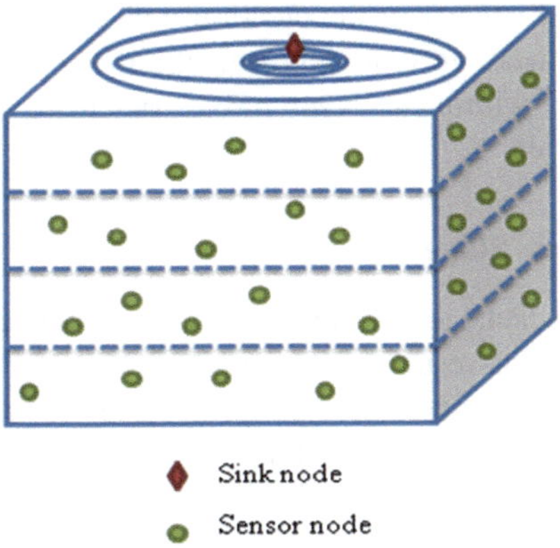

Fig. 2. Three-dimensional UWSN structure.

The following assumptions are as follows,

1. Though all the sensor nodes contribute common designs and the energies, it contains unique IDs & also knows their position deployed within a cubic volume.
2. The normal nodes (or) cluster-head node can be assumed with each and every node. Entire nodes deployed have the capability of integrating packets of data and adjusting transmitted power depend on its transmission area of data.
3. All the sink nodes are deployed in network's top surface of centre position and that can interact with entire nodes in the underwater wireless network.
4. In given layer, the nodes transmit information to cluster heads of particular layer that again transmits data to cluster head of next layer. Then, the cluster heads of topmost layer sends the packets of data to all the sink nodes.

3.2 Unequal Layering Model

In UWSN, the cluster heads are uniformly distributed enabled by layering that simplifies the model of networks and reduces the consumption of energy. In order to specify the issues of' hotspot', the layers in UWSN is divided by EEDUC and spacing between the layers is unequal increasing from upper to lower layer. Cluster heads plays a key role in UWSNs. In EEDUC, dynamically cluster-heads are selected for every rounds and every node undergoes election of cluster head, establishment of cluster and transmission of data stages. Election of node as cluster heads is done within each layer based on its node angle, transmission distance to sink nodes and its residual energy.

3.3 Cluster Establishment

1. Depending on its respective thresholds & election conditions, cluster-head candidates are voted in given layer.
2. It broadcast the cluster-head message comprises of IDs, radii of clusters etc. Thus every node having the largest weight are nominated as the cluster head & it broadcast information specifying its election within its radius.
3. After receiving the message containing election of cluster-head successfully, the another cluster-heads within the layer moves out from complete election process and join as child nodes.

3.4 Data Transmission Stage

Child node send the packets to cluster-head within each cluster that integrates the packets of data and sends its result to upcoming cluster-head, finally exports upwards. The cluster heads always have table that contains information of adjacent cluster head. With reference to [17], the routing function is selected based on its remaining energy and the node distance can be given as (Table 1)

$$P(i,j) = \varepsilon \cdot \frac{E_{ini}(j)}{E_{res}(j)} + (1 - \varepsilon) \cdot \frac{d_{i-j}^2 - d_{j-Sink}^2}{d_{i-Sink}^2} \tag{1}$$

The ε is parameter (€ [0, 1]) is mainly for balancing the proportion of energy to distance.

Table 1. Information of next cluster heads

Parameter	Mean
ID	Identification
$E_{res}(j)$	Residual energy
d_{j-sink}	sink node distance
d_{i-j}	node 'i' distance
$E_{ini}(j)$	Initial energy

4 Route Detection and Maintenance Phase

For avoiding the packet collision and to reduce the control overhead, a low overhead routing protocol is proposed. It consists of two main operations are route discovery and route maintenance.

4.1 Route Detection

Ad hoc On-Demand Distance Vector is the technique used in this route detection which consists of three types of operative message. The operative messages are Route Reply, Route Request and Route alive. Only Route Request & Route Reply used by Route Discovery. The two purposes of route alive is that it is used check route alive process and mainly helps in recovery of route. Route alive message is optional. The lower overhead protocol maintains same length format throughout the communication which is fixed header length.

4.2 Route Maintenance

Route maintenance similar to AODV & DSR (Dynamic Source Routing). Communication and routing done in network layer only. The routing protocol having low overhead helps in monitoring the traffic of data flow at the network layer. It avoids timer& maintains the route in both transmitter & receiver node concern about route maintenance. Thus, it does not provide any separate discovery of route.

5 Relay Node Selection

As mentioned above in the EEDUC section, child node and another child node cannot be able to participate in communication.

In order to reduce the transmission distance, relay nodes are deployed. As mentioned in EEDUC algorithm, only cluster heads communicate between adjacent layers & finally transmits the data to the sink node. For novelty and to decrease the transmission distance, the relay nodes which is known as intermediate nodes are elected. The intermediate nodes must be less in order to reduce the transmission distance & delay. The intermediate nodes here is cluster head and relay node.

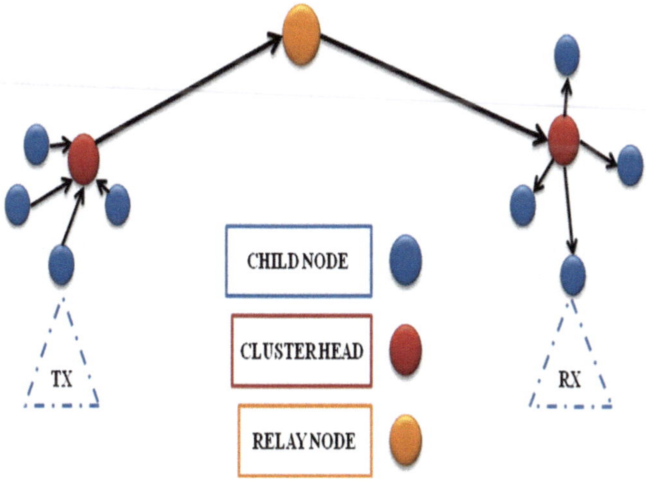

Fig. 3. Cluster and relay node arrangement

As shown in above Fig. 3, the child node sends the data to the cluster head. The cluster head sends the data to relay node and again it transmit the data to cluster heads. Relay node is elected and deployed by using following steps.

Step 1:
Among all the nodes in the UWSNs, the node having the minimum lifetime is selected and specified by s. Increasing the lifetime of node is the bottleneck of relay node setting.
Step 2:
The node which is largest distance with respect to node is found out and specified by t. 'r' is the specification of relay node is placed in between s & t to reduce the distance between them.
Step 3:
The node 'r' (relay node) should be placed in required place in network for increasing the lifetime.

'r' should be closer to node s & t so that RN (Relay Node) increases the lifetime of node s. Lifetime of node r is also considered. The transmitting &receiving power under the links are shown in the denominator of the Eq. 2. As depicted in [29] the following equations helps in fixing the relay node,

$$\min\left(\left|\frac{E_{sr}}{f_{sr} * p_{sr} + q_{rp} * f_{rs}} - \frac{E_{rs}}{f_{rt} * p_{rt} + q_{rp} * f_{tr}}\right|\right) \tag{2}$$

$$d_{ij}^2 = (x_i - x_j)^2 + (y_i - y_j)^2 + (z_i - z_j)^2 \tag{3}$$

E_{rt} - node 'r' energy that is mainly allocated to link (s, r))
E_{sr} - node 's' energy is mainly allocated to link (r, s)).

6 Simulation Results

The following simulated results are mainly are obtained by ns2.34 simulator that is plotted as the graph and the conclusions are obtained from the simulated graph,

6.1 Performance Metrics

i. Energy Efficiency of the Network
The network is said to be energy efficient only after reducing the amount of energy required for providing its products and services of the network.

ii. Throughput of the Network

The rate of successful message delivery over a communication channel is called throughput or network throughput. It is usually measured in bits per second (bit/s or bps) or in packets per second (p/s or pps) or packets per time slot.

iii. Packet loss

Loss of packets mainly when one or more packets of data failed to reach their destination while travelling across a computer network. Packet loss is caused by errors in data transmission, or network congestion.

6.2 Results

To check the performance of RN-EEDUC, it is compared to the performance of CMDG, RNSA, EULC algorithms using ns 2.30 simulation tool.

The Fig. 4 shows the generated packets of the network which shows that of all the three algorithms, the proposed RN-EEDUC protocol generates more packets from cluster heads and child nodes. Packet loss is also minimum in proposed work than all other algorithms as shown in the Fig. 5, indicates the efficiency of RN-EEDUC that increases network lifetime.

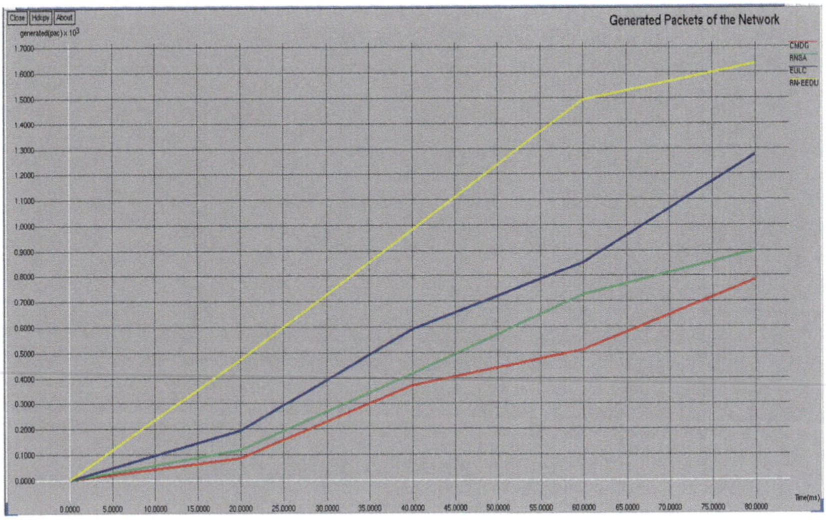

Fig. 4. Generated packets of the networks

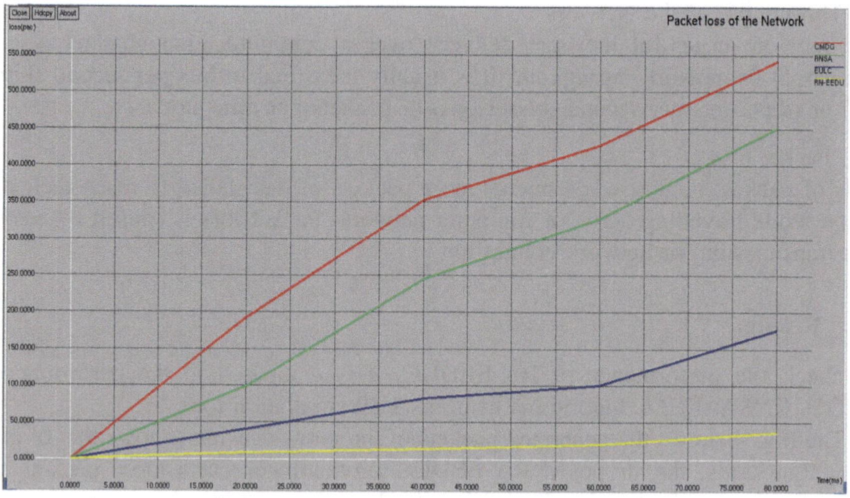

Fig. 5. Packet loss of the network

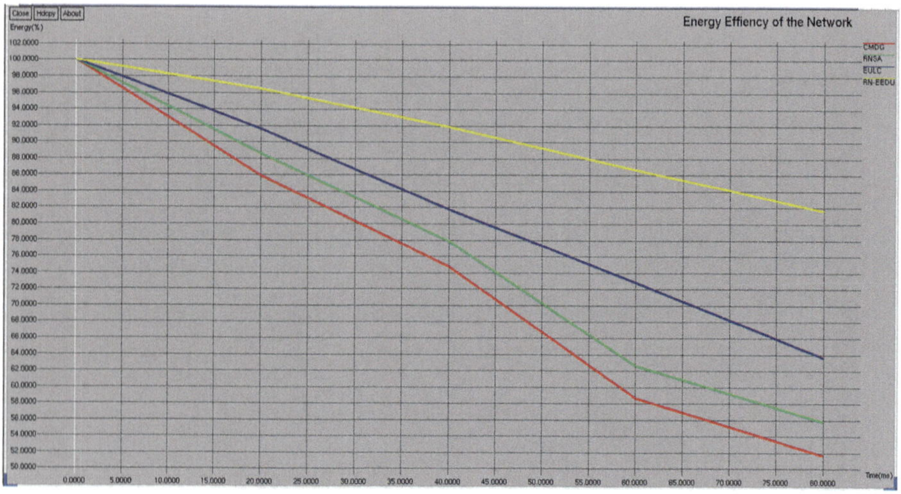

Fig. 6. Energy efficiency of the network

The Fig. 6 shows the energy efficiency of the network. It has been seen that CMDG protocol consumes more energy than the other protocols. Since, only clustering takes place, distance between the cluster heads is high, Energy efficiency is low in CMDG protocol. In RNSA, relay node is deployed in order to increase the energy efficiency. Thus, RNSA is more efficient than CMDG. For resolving the 'hotspot issue' EULC uses clustering techniques with unequal layering that balance intra & inter- cluster transmission of data in consumption of energy.

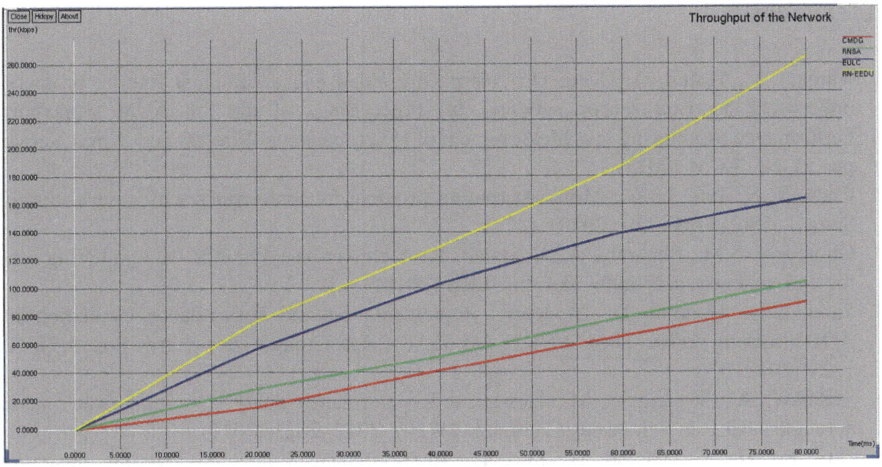

Fig. 7. Throughput of the network

By combining all the concepts of CMDG, RNSA &EULC, the proposed RN-EEDUC protocol is designed in which unequal layering and relay node used. The clustering adapted from CMDG protocol in which particular set of sensors only appointed as cluster heads that covers affiliated sensors in limited number of hops only. By comparing with existing protocols, the energy efficiency of the RN-EEDUC is higher than other existing protocols. The Fig. 7 shows the throughput of the network. The throughput of RN-EEDUC is higher than the existing protocol resulting in the concept that the data packets sent by the node is successfully received by the another node.

7 Conclusions

Reducing consumption of energy in UWSN, prolonging its network lifetime and reduces the energy consumption has became key problems in research field. This paper proposed an Energy Efficient Distributed Unequal Clustering Algorithm with Relay Node Selection for Underwater Wireless Sensor Networks. This algorithm consists of the unequal clustering model, election of cluster head, establishment of cluster head and the transmission of data. Secondly, in order to reduce the packet collision we introduced new route detection and route maintenance phase in routing protocol. The main idea of this phase is the construction of low overhead routing protocol in order to control the routing overhead. Thirdly, further improve the energy efficiency we present a novel algorithm namely relay node selection algorithm (RNS). The construction of relay node is to reduce the transmission distance which helps to improve the network lifetime. The simulated results shows that the RN-EEDUC protocol is more efficient than other existing protocols. In future work should be focused to calculate packet delivery ratio, to optimize network topology for improving energy efficiency and for ensuring security of data transmission.

References

1. Ahmad, A., Wahid, A., Kim, D.: Aeerp: auv aided energy efficient routing protocol for underwater acoustic sensor network. In: Proceedings of the 8th ACM workshop on Performance Monitoring and Measurement of Heterogeneous Wireless and Wired Networks, pp. 53–60. ACM (2013)
2. Chen, Y.-S., Lin, Y.-W.: Mobicast routing protocol for underwater sensor networks. IEEE Sens. J. **13**(2), 737–749 (2013)
3. Ghoreyshi, S.M., Shahrabi, A., Boutaleb, T.: A cluster-based mobile data-gathering scheme for underwater sensor networks. In: IEEE (2018)
4. Saini, G.L., Dembla, D.: Modeling, implementation and performance evaluation of E-AODV routing protocol in MANETs. Int. J. Adv. Res. Comput. Sci. Softw. Eng. **3**(7), 1221–1227 (2013)
5. Stefanov, A., Stojanovic, M.: Design and performance analysis of underwater acoustic networks. IEEE J. Sel. Areas Commun. **29**(10), 2012–2021 (2011)
6. Hou, R., He, L., Huand, S., Luo, L.: Energy-balanced unequal layering clustering in underwater acoustic sensor networks. IEEE Access **6**, 39685–39691 (2018). https://doi.org/10.1109/access.2018.2854276
7. Mohammadi, Z., Soleimanpour-moghadam, M., Talebi, S., Abbasi-moghadam, D.: A new optimization algorithm for relay node setting in underwater acoustic sensor networks. In: 3rd Conference on Swarm Intelligence and Evolutionary Computation (CSIEC2018), Higher Education Complex of Bam, Iran (2018)
8. Vithiya, R., Sharmila, G., Karthika, S.: Enhancing the performance of routing protocol in underwater acoustic sensor networks. IEEE (2018)
9. Kohli, S., Bhattacharya, P.P.: Simulation and analysis of greedy routing protocol in view of energy consumption and network lifetime in three dimensional underwater wireless sensor network. J. Eng. Sci. Technol. **12**, 3068–3081 (2017)
10. Zidi, C., Bouabdallah, F., Boutaba, R.: Routing design avoiding energy holes in underwater acoustic sensor networks. Wireless Commun. Mob. Comput. **16**, 2035–2051 (2016)
11. Liu, L., Ma, M., Liu, C., Shu, Y.: Optimal relay node placement and flow allocation in underwater acoustic sensor networks. IEEE Trans. Commun. **65**(5), 2141–2152 (2017)
12. Zhang, Y., Sun, H., Ji, C.: A clustered routing algorithm based on depth and energy for three-dimensional underwater sensor networks. J. Shanghai Jiaotong Univ. **49**(11), 1655–1659 (2015)
13. Ismail, N.S.N., Hussein, L.A., Ariffin, H.S.: Analyzing the performance of acoustic channel in underwater wireless sensor network (UWSN). In: Asia International Conference on Mathematical/Analytical Modelling and Computer Simulation, vol. 4, no. 5, pp. 550–555, May 2010

Investigation of Meanderline Structure in Filtenna Design for MIMO Applications

J. Jayasruthi(✉) and B. Bhuvaneswari

Department of Electronics and Communication Engineering,
Panimalar Engineering College, Chennai, India
sruthijjs20@gmail.com

Abstract. In this paper, a partial ground is used as a planar at the base with a rectangular patch in which the antenna designed as a meander line antenna is presented. The main goal is to obtain wider bandwidth that covers the ISM band frequency which perfectly operates for MIMO applications. Here a meander line antenna is printed on a microstrip patch with a matched feed and partial ground at the bottom operating at 2.45 GHz. Thus a dual band is obtained with maximum bandwidth. At the end of the receiver a filter can be placed in order to remove the noise and sends the signal without any interference. So, a filter is added at the meander line antenna substrate. A dual band of 1.73–2.77 GHz and a return loss of −22 dB and −45 dB is obtained. The gain of 2.26 dBi at 1.73 GHz and 3.69 dBi at 2.77 GHz is acquired. Obviously the proposed antenna design offers much flexibility to the available frequencies mainly for MIMO applications and wireless local area networks.

Keywords: Meander line · Filtenna · Band pass filter · Dual band · MIMO application

1 Introduction

The radio frequency antenna including the filters in the wireless transmission plays a major role as a front-end components that are needed to attain low integration price and handling high-energy capability. Today the filters which is integrated with the antennas have additional scope towards the RF domain. So the filtering antenna unremarkably known as filtenna not solely reduces the price of fabrication, however conjointly improves the performance of the antenna, like high pattern, gain, information measure and VSWR activity. Antenna information measure and reduction in size are the 2 major challenges. The overall electrical homes with networking parts like inductors, capacitors and resistors has been expressed in the victimization of voltage standing wave quantitative relation (VSWR), S-parameters, reflection constant, that embrace profit, return loss, and electronic equipment stability. There is a fundamental measure in which the optical engineering is common than the RF engineering, concerning with the results that are discovered, whereas a craft radiation are incident on the degree of obstruction and then passes across various insulation media. Within the context of the S-parameters, scattering always refers back to the style during the visiting currents and voltages in an increased number of conductors that are affected after they meet a separation as a result

D. J. Hemanth et al. (Eds.): COMET 2019, LNDECT 35, pp. 1537–1549, 2020.
https://doi.org/10.1007/978-3-030-32150-5_155

of the insertion of a network into the conductor. This is often adequate to the wave resistance differing from the electric resistance. The meander line antenna may be a tiny antenna that consists of vertical and horizontal structures that achieves compact size. The horizontal segments within the meander line antenna have opposite section. And then the amount of turns will increase because the potency of the antenna will increase. Because the area of the meander line will increase the resonant frequency decreases.

The advantages of the antenna posses straight forward configuration, easy integration to a wireless device, low price and potential towards low SAR options. Meander line antenna is one sort of the small strip antennas. Meander line antenna faces major challenges in communication technologies like augmented rate, antenna size, low SAR worth, high gain and augmented information measure.

Radio-based LANs have become additional versatile and has become fewer products in our day to day environments. Most of the wireless native space networks systems has been designed to work within the a pair of 1.4 and 1.5 GHz. A compact antenna with a twin band that operates in each frequency band at 1.8 GHz and has a couple of 1.7 GHz. various styles of antennas are used in the meander line technology and provides a wide band performance. Hence this paper has a tendency to gift a meander line antenna on the substrate using FR4 material that are bounded by a feed line which is adaptive in nature and has a partial defected ground structure band pass filter is added to the strip so as to cut back the interference. The main purpose of miniaturisation and enhancing the general performance of the circuit, a multifunction module is intended. It performs filtering and divergent, at the same time with the assistance of a co- style approach. A filtering antenna is usually thought-about as a mixture of filter and antenna during which the filter is integrated towards the feed line or a strip apart from the divergent surface. Hence the fluctuation in radiation is incredibly less. Because the size of the dipole within the meander line antenna is reduced by an element that is proportional to the amount of turns within the given in operation frequency.

Simultaneously, this trending wireless technologies face a serious challenge in the refraction and reflection within the communication link and conjointly there could also be some noise of interference between the signals at the receiver. In order to overcome certain challenges within the MIMO system a filtering meander line antenna has been designed and enforced with higher port isolation. Such system may result in larger antenna performance and better gain. This successively allows sensible potency that is reliable for our surroundings. The antenna performs a serious task of band sensing and therefore the filter used is capable of human activity at desired waveband.

2 Related Work

In this section we briefly describe a meander line design structure using various grounds and substrate materials along with their techniques, advantages and disadvantages.

Amarjit Kumar proposed a wireless pressure monitoring system (WPMS) using radio frequency transceivers at 2.31–2.64 GHz. A filtenna is used at the receiver side. It results in −10 dB return loss at with a bandwidth of 62% [1]. A broadband duplex filtenna is proposed in [2] based on a 3D metallic cavity structure. The design resulted in a fractional bandwidth of 57% that shows a good filtering property. Dual-band meander-line monopole antenna that operates in frequency of 2.400–2.480 is designed [3]. A return loss of 1.14 GHz at −10 dB is obtained and it is used for portable wireless communication applications. Compact filtennas with massive tunable band square measure planned for cognitive radio (CR) applications are studied. The projected calibration technique for filters relies on centrally loading the stub and therefore the H-shaped with only one varactor is used for miniaturisation [4]. For additional size reduction of the warm temperature, a superconducting two-dimensional wave guide band pass filter (BPF) and a meander line with quarter-wavelength resonator BPF has been introduced so as to manage the attenuation poles [5]. Three spherically conformal electrical monopoles-the spherical helix (SH), spherical meander line (SM), and a hybrid designs are used [6]. A comparison is done in which the spherical helix has more efficiency with tough impedance properties that are matched. The spherical meander line has more loss and a hybrid design is the combination of both spherical helix and spherical meander line. Tharp et al. made a comparative study on a single quarter-wave meander line phase retarder operating at 8–12 μm. A low throughput of 23% is obtained [7]. In [8], a (SIW) substrate integrated waveguide, filtenna are used. The gain obtained is 6.73 dBi with reduced antenna size. Here SIW are stacked vertically. Kwok Kee Chan et al. proposed a meander line antenna with a metallic grids. A resonance with transverse model technique is used. various grids are compared for best results [9]. In [10], there is no filtering part in the design but the antenna itself acts as a filtering, using 3- element patched array. This is operated at a frequency of 4.8–6.8 GHz resulting in a bandwidth of 7% and 14%.

3 Proposed System

There is a fundamental limitation of antennas in which the bandwidth will be less and hence the gain will also be low. Generally in electrically small antennas the Q-factor introduces frequency with narrow bandwidth. Hence if the quality factor is minimum then there is a chance of increased bandwidth.

As referred in paper [3], in free space, k is the wave number radians/meter and a is the maximum radius of enclosing the sphere of the antenna (meter). The bandwidth and the VSWR is calculated as follows [3, 4]:

$$ka < 1 \tag{1}$$

$$Q = \frac{1}{k^3 a^3} + \frac{1}{ka} \tag{2}$$

$$G = (ka)^2 + 2(ka) \tag{3}$$

$$BW = \frac{s - 1}{Q\sqrt{s}} \tag{4}$$

Where bandwidth is that the information measure, S is that the VSWR activity.

As shown in Fig. 2, the bottom is partially defected, so there is no ground underneath the meander line section. Therefore the planar strip (CPS) is briefly circuited. A study is made on the effective dielectric constant in order to get the characteristic electric resistance of a meander line section.

- There are certain difficulties in coming up with associate electrically little antenna:
- Impedance matching
- Insertion loss from high current density flowing on a non-perfect conductor, leading to joule heating
- Low radiation potency which leads to less radiation aperture.

4 Antenna Design

A meander line antenna with its vertical and horizontal positions are designed. The length and also the dimension of the patch is calculated. The vertical and the horizontal lines of a meander line structure may or may not be equal depending on the width and the length of the design.

The ground plane format of a meander line antenna may be a 2λ dipole or 4λ and the basic idea behind the antenna is to fold the conductors front and back to make the antenna smaller as shown in Fig. 1.

Fig. 1. General meander line structure

A partially defected ground structure is used and the patch meander line is printed on the Fr4 substrate with the dielectric constant 4.4 and height 1.6 mm. The resonance length of the meander line antenna is given at the Z-axis. The copper is selected as a material and the entire design is operated at an ISM band frequency 2.45 GHz. The length of the feed is calculated as follows:

$$C = \lambda f \tag{5}$$

where c = 3 × 108
f = 2.45 GHz

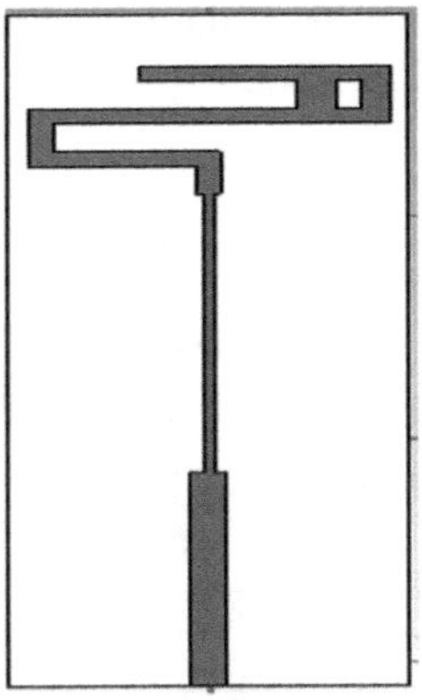

Fig. 2. A meanderline design with partial ground structure.

The filters that are implemented, are designed on the substrate which is a Fr4 material, with a dielectric constant of 4.4, loss tangent of 0.035 and the height of substrate is 1.6 mm. The order of the filter can be determined only by the design calculations. Different features of filters is carried out by various researchers such as UWB filters which is a bandpass, tunable filters that has switch ability features and transmitting zeros at pass band. Notching out of the signal reception is given more attention in the ISM band (spectrum range). The implementation of the notch filters has been concentrated by the researchers such as defected ground structures (OGS), parasitic patches, split ring resonators (SRR), coupled line structures and photonic band gap. For UWB application like WLAN 802.11b less notching is done.

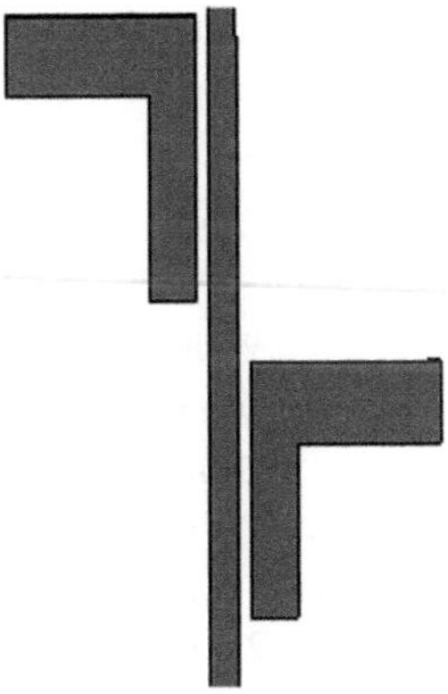

Fig. 3. L-shaped bandpass filters added to the strip of the antenna

From Fig. 3, Two resonator with L-shape resonator are designed to be half wavelength, with the length L and width W for each quarter wavelength. Here width, length and thickness are calculated from the feed line for its simplicity. The L-shaped resonator filter is a band pass filter (Fig. 4).

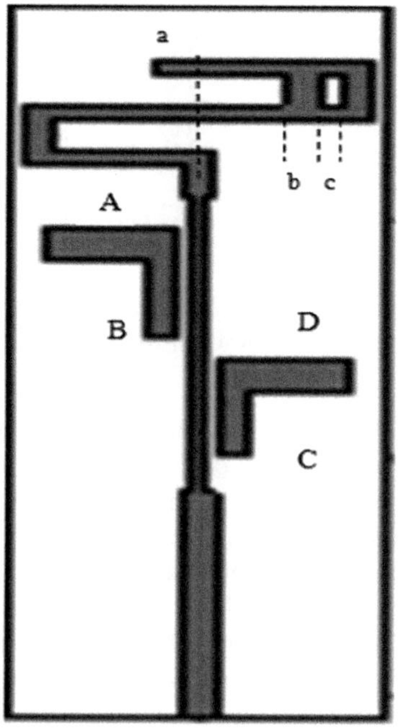

Fig. 4. Meanderline antenna added to L-shaped resonator filter

The dimensions of simulated proposed meander line antenna using the filtenna design is given as below (Table 1):

Table 1. Dimensions of the proposed antenna

Materials	Width (mm)	Length (mm)
Ground	20 mm	46 mm
Substrate	20 mm	60 mm
1st meander section	1.3 mm	5.1 mm
2nd meander section	1.3 mm	2.4 mm
Strip	2.6 mm	25 mm
Filter 1	1.7 mm	9.15 mm
Filter 2	2.6 mm	7 mm
Feedline	1.9 mm	19 mm

The meander line filtenna is divided into two sections, in which the first section has the width 9.86 mm (outer measurement) height 5.1 mm with the spacing 1.3 mm and the second meander line section consists of 16.7 mm width with the length 5.1 mm. The spacing between the patch and the height of the substrate is 4.6 mm. The height at the end of the strip and starting of a first meander line is 3.8 mm with the width 1.72 mm. The spacing between the first meander line section and the substrate is 9.14 mm.

5 Result and Discussion

The proposed model of the antenna design is simulated using the EM simulator software. A meander line antenna is designed and simulated with the software. The impedance match is improper and the bandwidth is less than 1 GHz. The highest return loss obtained was −40 dB (Fig. 5).

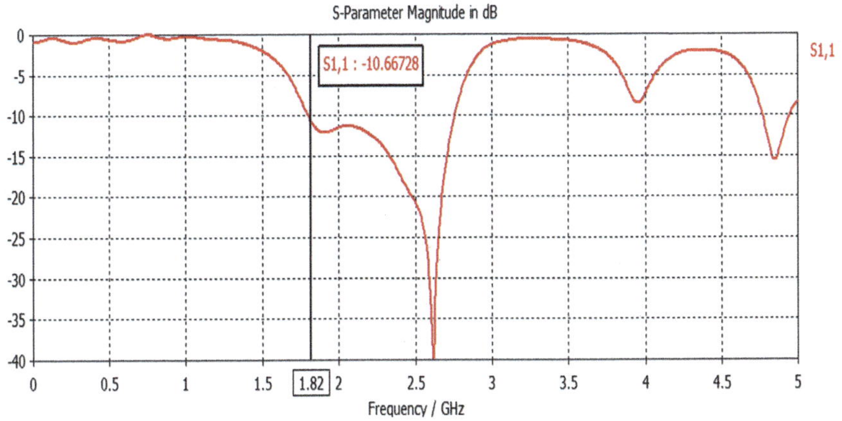

Fig. 5. s-parameter for the meander line antenna without filter.

The antenna pattern which is a radiation pattern or a far field pattern indicates the dependence of the directional strength of the radio waves in any antenna design.

Return loss is defined as the power loss within the signal reflected or refracted in a fibre or a conducting cable. The mismatch can be done due to the separation of the load terminals and the device when it is inserted into the terminal. The ratio is expressed in terms of dB (decibels).

$$RL(\mathrm{dB}) = 10\log_{10}\frac{P_i}{P_r} \tag{6}$$

where,

RL(dB) - return loss in dB
Pi - incident power
Pr - reflected power.

Return loss is a combination of reflection coefficient (Γ) and the standing wave ratio (SWR). When SWR increases return loss decreases correspondingly. When the return loss and the terminals are matched together, that it shows its great performance measure. As the return loss is high the match is perfect.

As return loss is employed in trendy follow with respect to SWR as a result of its higher resolution for tiny values of reflected wave. Gain obtained is a pair of 2.8 dBi and also the graph of the simulated meander line antenna while not adding a filter is as below (Fig. 6):

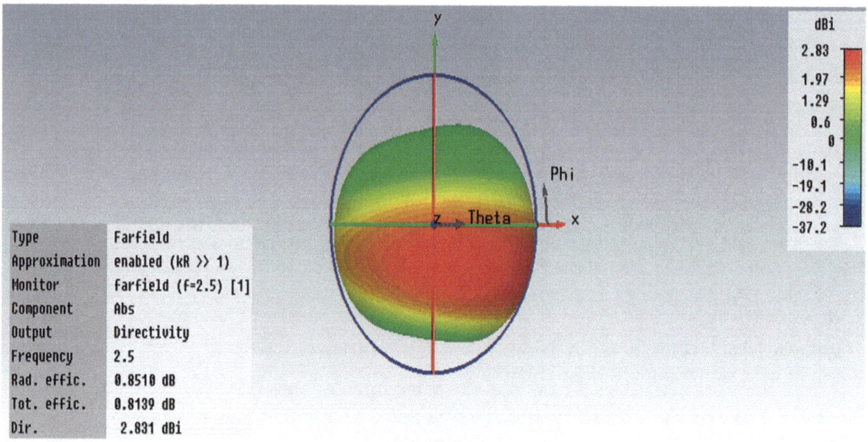

Fig. 6. Radiation pattern of meanderline antenna without filter

A filter is added to the strip of the meanderline structure inorder to scale back the interference of the signal once multiple signals are given at the input. Filter can even increase the potency of the antenna.

In this paper, 2 filters are added on the strip higher than the feedline of the meander line antenna. Filter 1 is placed above the feedline in which the distance between the feedline and the filter is 13.15 mm. The spacing given between the filter and the strip is 0.5 mm.

Filter 2 is placed at a distance of 3 mm from the feed line. The length and dimension of the filter 2 is 5.55 mm and 1.7 mm respectively. The spacing distance between the filter and the end of the substrate is 2 mm (Fig. 7).

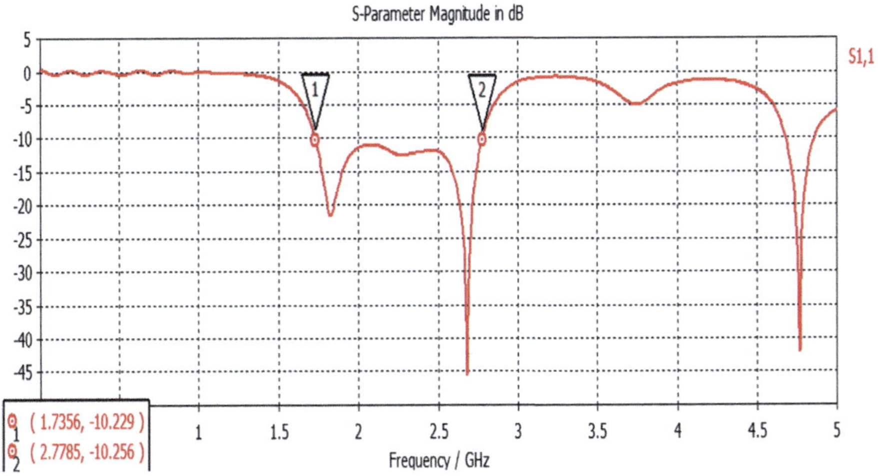

Fig. 7. s-parameter for the meanderline antenna with filters

If the DUT is inserted and all of this power is returned, 100 percent reflects and there's no loss in the reflected power. All the incident power is returned. The return loss is zero dB. Once we insert a device, we lose the reflected power. As a result, of some of them gets absorbed or transmitted through the device (Figs. 8 and 9).

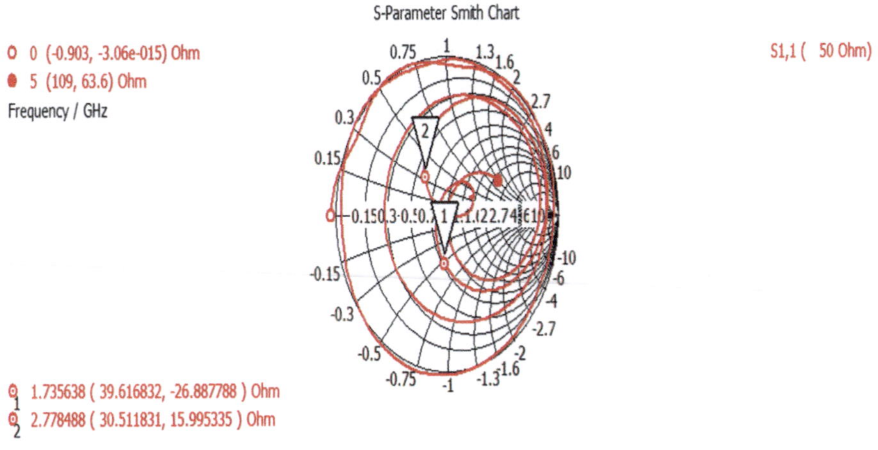

Fig. 8. Smith chart representation of meander line antenna with filter

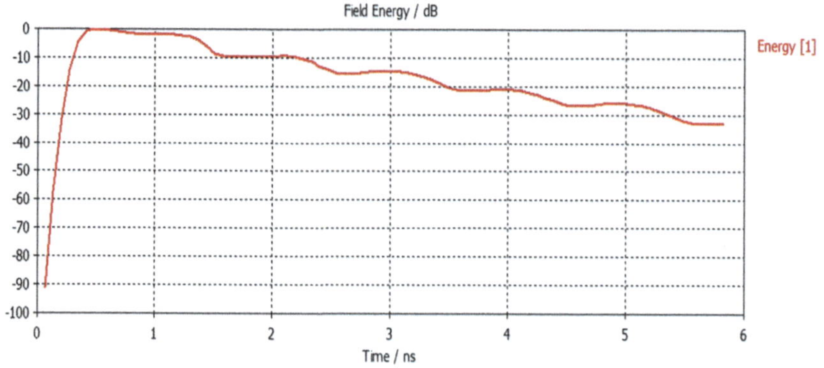

Fig. 9. Energy graph of meander line using filtenna

There will be only one input port and one output port in any single ended device which is standard in nature. These signals in the input and the output port is detailed in the ground plane. There is an interface between the propagating radio waves of the antenna by the currents in the buildings, houses, cables that fills all the metallic conductors used in the name of transmitter and the receiver. The antenna receives power from the transmitter and hence there will be radiation and some magnetic force which can be called as radio waves. There will be some radiation due to the interruption of the signals from the receiving side of the antenna.

The array of elements connected electrically is designed in such a way that it transmits and receives the radio signals in an omnidirectional pattern. The directional antennas has high gain in which the radio signals are placed in horizontal and vertical positions. Using parabolic reflectors, parasitic parts and a parabolic horn, the radio waves are directed into a beam of antenna (Fig. 10).

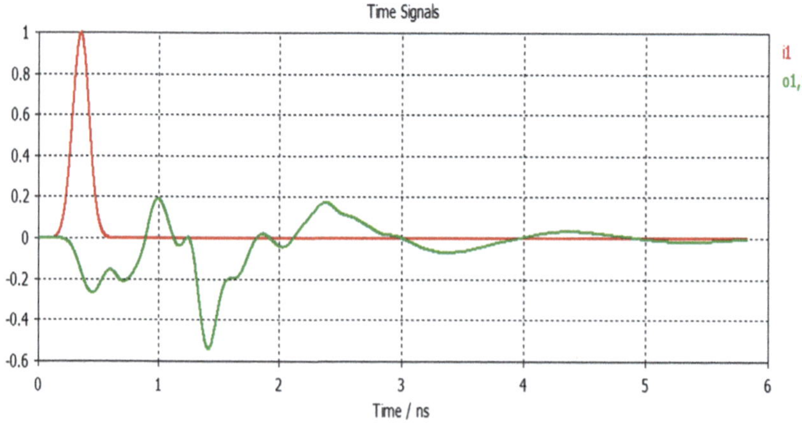

Fig. 10. Port signal representation

Generally the performance of the antenna is measured using VSWR matched to the load and it describes the resistivity of the antenna that are connected to any device.

VSWR - Voltage standing wave ratio is also called as standing wave ratio (SWR) in which the minimum value obtained is one (Fig. 11).

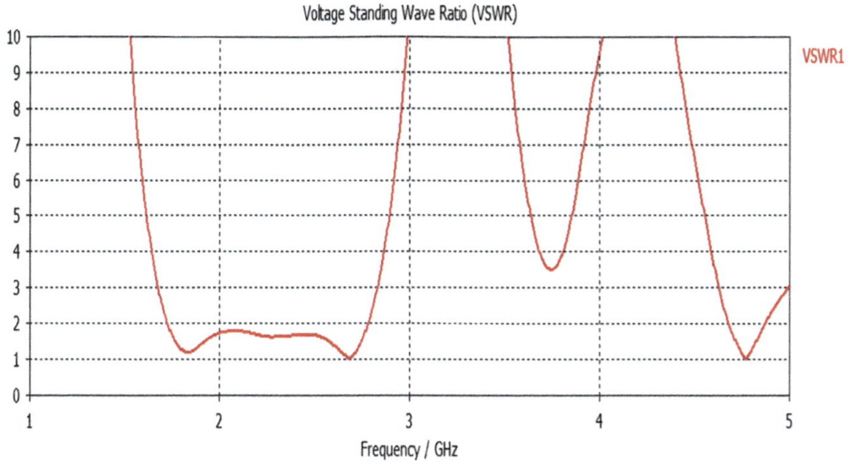

Fig. 11. VSWR measurement of a meander line antenna

VSWR describes the capacity of reflecting radio waves from the radiating antenna. s_{11} - reflection constant or return loss

VSWR can be calculated using the following formula:

$$VSWR = \frac{1 + |\Gamma|}{1 - |\Gamma|} \quad (7)$$

VSWR plays a vital role in the antenna measurement. If the value of VSWR is low if the antenna is highly matched to the link that provides high power to the antenna and so VSWR is small. The antenna will be more efficient if there is no reflection having VSWR as 1.

The gain obtained is a pair of 2.87 dB for 2.5 GHz and also the 3-D radiation pattern for the filtenna design using a meander line is as shown below (Fig. 12):

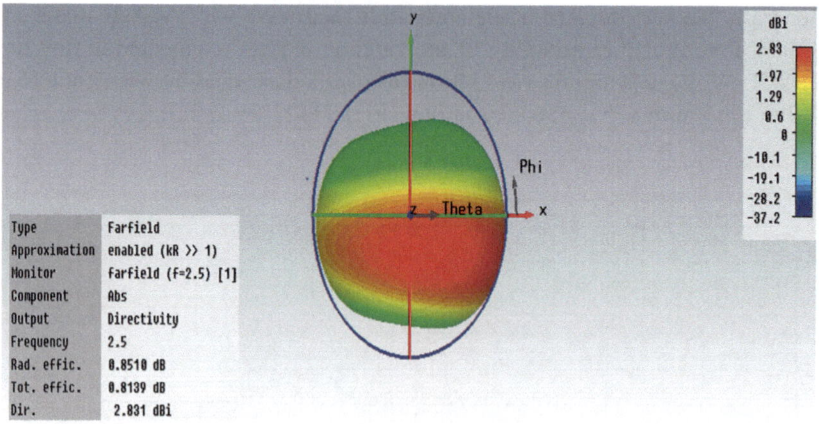

Fig. 12. Radiation pattern of meanderline antenna using filtenna structure.

The polar representation of the meanderline structure using the filtenna design is as below (Fig. 13):

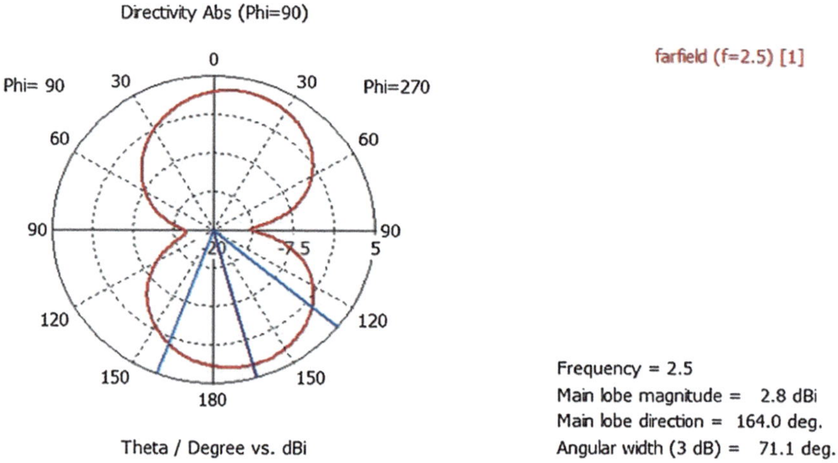

Fig. 13. Polar representation of meanderline using filtenna structure

6 Conclusion

The characterization of a planar meander line antenna design using filtenna is done along with the simulation using the software required. Fr4 substrate is used with a dielectric constant 4.4 and a dual band is obtained as a result. The proposed antenna is very compact in size 60 × 20 mm^3. The bandwidth obtained is 1 GHz with a wide band of frequency ranging from 1.73 GHz–2.77 GHz. The gain obtained is 2.26 dB with a return loss −22 dB in 1.73 GHz and 3.69 dB with −45 dB return loss in 2.77 GHz. The antenna provides good gain and return loss with high impedance bandwidth in the dual band frequencies. As the designed meander line antenna using filtenna structure is very compact and has wide bandwidth, it is suitable for MIMO applications.

References

1. Kumar, A.: Wireless monitoring of volatile organic compounds/water vapor/gas pressure/temperature using RF transceiver. IEEE Trans. Instrum. Measur. (2015)
2. Zheng, B.L.: Broadband duplex–filtenna based on low-profile metallic cavity packaging. IEEE Trans. Compon. Packag. Manuf. Technol. (2014)
3. Hsu, C.C., Song, H.H.: Design, fabrication, and characterization of a dual-band electrically small meander-line monopole antenna for wireless communications. Int. J. Electromagn. Appl. **3**(2), 27–34 (2013)
4. Atallah, H.A.: Compact frequency reconfigurable filtennas using varactor loaded T-shaped and H-shaped resonators for cognitive radio applications. IET Microwaves Antennas Propag. **10**(9), 991–1001 (2016)
5. Kanaya, H.: Design and performance of miniaturized quarter-wavelength resonator bandpass filters with attenuation poles. IEEE Trans. Appl. Supercond. **15**(2), 1016–1019 (2005)
6. Adams, J.J.: Comparison of spherical antennas fabricated via conformal printing: helix, meanderline, and hybrid designs. IEEE Antennas Wirel. Propag. Lett. **10**, 1425–1428 (2011)
7. Tharp, J.S.: Design and demonstration of an infrared meanderline phase retarder. IEEE Trans. Antennas Propag. **55**(11), 2983–2988 (2007)
8. Hu, K.-Z.: Compact, low-profile, bandwidth-enhanced substrate integrated waveguide filtenna. IEEE Antennas Wirel. Propag. Lett. **17**(8), 1552–1556 (2016)
9. Chan, K.K.: Accurate analysis of meanderline polarizers with finite thicknesses using mode matching. IEEE Trans. Antennas Propag. **56**(11), 3580–3585 (2008)
10. Kufa, M.: Three-element filtering antenna array designed by the equivalent circuit approach. IEEE Trans. Antennas Propag. **64**(9), 3831–3839 (2016)
11. Tang, M.-C.: Compact, frequency-reconfigurable filtenna with sharply defined wideband and continuously tunable narrowband states. IEEE Trans. Antennas Propag. **65**(10), 5026–5034 (2017)
12. Tang, M.-C.: Bandwidth-enhanced, compact, near-field resonant parasitic filtennas with sharp out-of-band suppression. IEEE Antennas Wirel. Propag. Lett. (2012)
13. Kingsly, S.: Multiband reconfigurable filtering monopole antenna for cognitive radio applications. IEEE Antennas Wirel. Propag. Lett. **17**(8), 1416–1420 (2018)
14. Lin, S.-C.: An accurate filtenna synthesis approach based on load-resistance flattening and impedance-transforming tapped-feed techniques. IEEE Antennas Wirel. Propag. Lett. (2014)
15. Pal, S.: HTS bandstop filter for radio astronomy. IEEE Microwave Wirel. Compon. Lett. **22**(5), 236–238 (2012)
16. Li, W.T.: Novel printed filtenna with dual notches and good out-of-band characteristics for UWB-MIMO applications. IEEE Microwave Wirel. Compon. Lett. (2012)
17. Guo, Y.J.: Advances in reconfigurable antenna systems facilitated by innovative technologies. IEEE Access. Accepted 20 December 2017
18. Yan, Z.: Experimental investigations on nonlinear properties of superconducting nanowire meanderline in RF and microwave frequencies. IEEE Trans. Appl. Supercond. **19**(5), 3722–3729 (2009)

Design of Multiple Input and Multiple Output Antenna for Wi-Max and WLAN Application

S. Shirley Helen Judith[1(✉)], A. Ameelia Roseline[1], and S. Hemajothi[2]

[1] Department of Electronics and Communication Engineering, Panimalar Engineering College, Chennai, India
shirleyjudith95@gmail.com, ameeliaroseline@gmail.com

[2] Department of Electronics and Communication Engineering, Prathyusha Engineering College, Chennai, India
hemaselwynj@gmail.com

Abstract. In this design, four port MIMO antenna of annular slot is proposed to get better gain and diversity performance of the frequency range from 2 to 6 GHZ for Wi-Max (3.5 GHZ) and WLAN (5 GHZ) application. To get the pattern diversity, the four micro strip feed lines are used and it is isolated by four shorts to maintain isolation. The Micro strip patch antenna are used because of the advantages like low structure in profile, cost of the fabrication is low and support both the circular and the linear polarizations. The antenna performance are analysed by the simulation results which produces the gain, directivity, return loss and radiated power. The antenna proposed used in WLAN and Wi-Max application. The antenna dimensions are thickness 0.8 mm, length 30 mm, width 38 mm. The Flame Retardant (FR4) substrate is used which has the relative permittivity value 4.3. The antenna design proposed is simulated using the Advanced Design System (ADS) software and output is tested using the network analyzer.

Keywords: Microstrip antenna · Wi-Fi · WLAN · FR-4 · Fabrication · Gain · Return loss

1 Introduction

Over the past years, the researchers focused on the study of Micro-strip patch antennas in which the antenna will be in compact size. But, the antennae with low gain and narrow bandwidth is the main drawbacks in the system. So, the researchers all among the world is trying to overcome the above drawbacks. The micro-strip patch antenna is promptly used in various fields of the aircrafts, space technology, the mobile communication, the missiles, the GPS system, and also in the radio unit. The micro-strip patch antennas are compact size, light in weight, cheap, simple in manufacturing and easy to integrate the circuits. The very significant thing is, it can designed in following different patterns like circular, triangular, square, rectangular etc. Highest bandwidth is suggested by all the techniques. The methods includes:

The elements of the parasite will be in other or in same layer.
Using the thick substrates of lesser dielectric constant
The slotted microstrip patch.

D. J. Hemanth et al. (Eds.): COMET 2019, LNDECT 35, pp. 1550–1562, 2020.
https://doi.org/10.1007/978-3-030-32150-5_156

The highest bandwidth, simplicity, the compact size and the rapport to remaining RF front-end is the fascinating factors of the antenna. This effort is devoted in designing the frequency of the separate wide band antennas. The major drawbacks of that antennas is the larger size and can be possibly able to cancel the usage of the mobile wireless applications.

Wi-Max (Worldwide Interoperability for Microwave Access) is a wireless communication which is dedicated to the advancement of IEEE standard of 802.16 for the broadband wireless access network. It covers 30 miles with a high speed compared to Wi-Fi which is operated in the frequency range of 2.3, 2.5, 3.3, 3.5, 5.8 GHZ band. WLAN is the wireless computer network that links the two or more components using the wireless method within the limited zone such as a the home, office, the school, laboratories and industrial buildings. It has the IEEE standard of 802.11. It is operated in the frequency range of 2.4, 3.6, 4.9, 5, 5.9 GHZ band. This microstrip patch antenna was designed for the Wi-MAX communication system with MIMO technology. The output comes out from the antenna bandwidth is increased.

Kamyab and Khaleghi proposed a feed reconfigurable antenna with the polarizing pattern diversity where the radiating circular patch is printed on the substrate which is thin and the diversity of the pattern is resulted using the switch parasitic pins beneath the circular patch and also the ground plane and it is used in WLAN applications [1].

Huang, Nehorai proposed a Coupling of the two collocated perpendicular circular thin loops which is analyzed. The strong coupling will exist for all the current loop harmonics which is higher than the first can be ignored and it is also found that coupling for perpendicular rare loop antennae depend on the nearby locations of the terminals loop [2].

Ding, Du, Gong, and Feng has proposed a novel dual-band printed diversity antenna which it consists of the two back-to-back single pole by symmertrical configuration and it is embedded on the PCB. This antenna radiating at the operating frequencies of UMTS (1920–2170 MHz) and 2.4 GHz WLAN (2400–2484 MHz) bands are demonstrated in the method of double band antenna for the mobile terminals where the isolation of the prototype is greater than 13 dB and 16 dB [3].

Bod proposed a printed compact ultra wideband (UWB) slot antenna with extra three bands for the different applications and this low profile antenna consists of the slot with octagonal shape. The slot is surrounded by a stepped rectangular patch which covers the UWB band from 3.1–10.6 GHz. The attachment of the three inverting U-shaped strips in the ground, located at the upper part of the slot with the polarizing band of the additional triple linear realized by covering the GPS, GSM, Bluetooth [4].

Ghorban and Water house proposed a microstrip antennae with the narrow bandwidth and this experimentally results in an impedance bandwidth of 35.3% and better gain of 4.5 dBi and good isolation of 30 dB and also good polarization is available [5].

Yang, Luk proposed a dually polarized antenna working in C band with the complementary structures and the stable, symmetric radiating patterns of the prototype antenna is built with the frequency of 4.9 GHz–5.1 GHz and the isolation of the port is less than −24 dB [6].

Toh, and Ping proposed the broadband MIMO antenna of four ports with the pattern diversity where the feed lines are printed on one side and the ground plane on the other side where it generates the orthogonal radiation pattern with the isolation loss of 25 dB and it is used in the WLAN and the Wi-Max applications and operates in the range of 2.3 to 12.6 GHz [7].

Wang et al. proposed the compact two element antenna of both pattern ah high isolation for 2.4 GHz and polarization diversities with isolation of 2.4 GHz for WLAN applications with the 180 out of phase excitations with the isolation of 29 dB with effective gains and diversity gain [8].

Dang, Lei, Xie, Ning, and Fan proposed a design of a four banded slot antenna for the GPS and used for the Wi-MAX and WLAN applications and this operates in the frequency band of 1.575 GHz to 1.66 GHz used with the utilization of the computer simulating software of IE3D [10].

Chiang, Wang, and Hsu proposed the paper of the four band compact slot antenna in a small ground plane for the GPS is used for WLAN and the Wi-Max applications with the gain of 2.5 dBi [9].

Haghparast and Dadashzadeh proposed a new design of the circularly polarized CPW-fed monopole antenna which operates in the ISM and the WLAN bands and this compact aperture has the same length and width at operating frequency of 2.2–8 GHz band and is used in the MIMO communications [11].

Wong and Lu proposed double polarized antenna array of an eight port operating at the frequency band of range 2.6 GHz for the 5 generation communication and the antenna designed is simulated with good parameters and used in the 5G smart phone applications [13–15].

Votis, Tatsis proposed a 2*2 MIMO antenna array system with envelope correlation coefficient shows the various propagation paths of the RF signals that are destined to the antenna elements and the diversity of the MIMO antenna is measured by the envelope correlation coefficient with 100% antenna efficiency [12].

Stavrou, Litschke, Baggen proposed a double polarized antenna for the hot spots of the Wi-Fi access points which is operating at the frequency of 5.8 GHz and it consists of 64 elements [16].

Han et al., proposed the innovative technique to enable the port to port isolation of the two closely spaced dual band for the WLAN applications and the MIMO set top box is used for the better output [17].

Sun, Fang proposed the compact antenna for the ENG dual band where the dual band isolations are improved over 10 Db at 2.6 GHz and the 3.5 GHz [18].

The broadband antenna with four ports consists of the MIMO antenna with the diversity of the pattern is presented. To get the diversity of the pattern, first the four microstrip line feeds is printed on one side of the substrate and then on the other side the modified ground plane is printed. The microstrip lines are developed as radiation patterns in perpendicular direction. Then the annular slot of the four shorts are arranged between the microstrip lines for maintaining the isolation which is greater than 25 dB. The antenna operating with frequencies i.e., from the 2.3 GHz to 12.6 GHz approximately equals 139% covering the FCC in wireless applications. Thus, the proposed antenna covers the Wi-Max and the WLAN applications beneath the FCC. The proposed geometry of the antenna is designed in the Ansoft's high Frequency of the Structure Simulation (ADS).

2 Proposed Design Model

The paper presents the "Design of MIMO antenna for Wi-Max and WLAN applications". The antenna is designed to operate at a frequency of 2–6 GHZ. The MIMO antenna is structured as annular ring slot antenna with 4 ports. Slot with 2 feeds are sufficient to achieve pattern diversity application, so the feed is given at port 1&2. The design is simulated using the ADS software. The Wi-MAX and WLAN is achieved at the frequency of 3.5 & 5 GHZ with a gain of 5.1 dB and bandwidth of is achieved FR4 material is used in the fabrication of the prototype which is proposed. By comparing with air as material of the dielectric, the substrate used in proposed geometry is easy to handle. Simulation and return loss is measured, the radiation patterns and the isolation between ports are simulated. The slotted antenna is advantageous of the compact size, the wide bandwidth and the easy integration with the other components which is the good candidate for the design of MIMO antennae.

The proposed structure of the MIMO antenna is shown in the Fig. (1) below and the design is based on the calculations of the dimensions of the respective MIMO antenna. The proposed MIMO antenna design has the following dimension and shown in below Table 1.

Table 1. The dimensions of the proposed design

Parameter	Length of the patch (mm)	R_{in} (mm)	R_{out} (mm)	Length of the feed	Width of the feed (mm)	L_1 (mm)	w_1 (mm)
Values (mm)	30	5	10	6	2	3	6

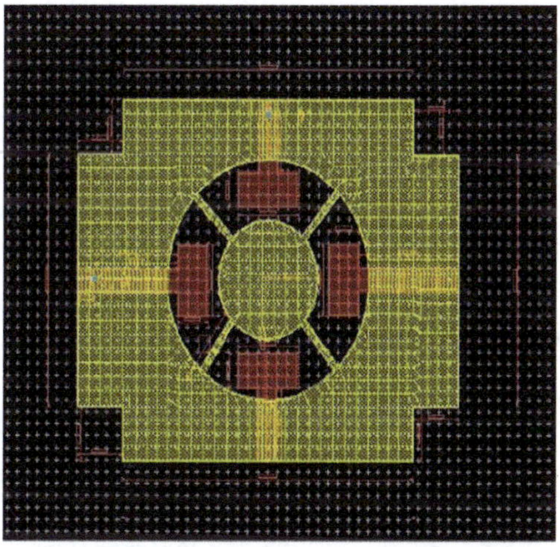

Fig. 1. Design of the proposed MIMO antenna

The design of this structure is then simulated using the Advanced Design System (ADS) software and tested using the network analyzer. Advanced Design System helps to store and control the data exhibited when creating, simulating, and analyzing the designs to accomplish the design goals. This paper concludes the layout, analysis, simulation, circuit, and the output information in the designs that develops with any links that is added to other designs and the developed project.

3 Design Calculation

3.1 Design Considerations

Operating frequency (f_o) = 2.4 GHz. Velocity of light (c) = 3 × 10^8 m/sec.
Substrate FR-4 with the dielectric constant ($\in_r$) = 4.3
Substrate thickness (h) = 1.6 mm.

3.2 Substrate Selection (H)

The substrate selected for the design of the antenna is FR4 with the thickness (h) of 1.6 mm and dielectric constant (£ṙ) = 4.3 and a loss tangent of 0.01.

3.3 Calculation of Width of the Patch (W)

For 2.4 GHz, the patch width is calculated as and W = 38 mm respectively.

$$W = \frac{c}{2f_0\sqrt{\frac{(\varepsilon_r + 1)}{2}}}$$

3.4 Calculation of Effective Dielectric Constant ($\in_{reff}$)

For 2.4 GHz, the value of ($\in_{reff}$) is calculated as 4.03 mm respectively.

$$\varepsilon_{eff} = \frac{\varepsilon_r + 1}{2} + \frac{\varepsilon_r - 1}{2}\left[1 + 12\frac{h}{W}\right]^{-\frac{1}{2}}$$

3.5 Calculation of Length Extension (ΔL) and L

For f = 2.4 GHz, the value of (ΔL) is calculated as ΔL = 0.374 mm and L = 30 mm (Table 2).

$$\Delta \mathrm{L} = 0.412\mathrm{h}\frac{\left(\varepsilon_{eff} + 0.3\right)\left(\frac{W}{h} + 0.264\right)}{\left(\varepsilon_{eff} - 0.258\right)\left(\frac{W}{h} + 0.8\right)}$$

3.6 Calculation of Effective Length L_{eff}

$$L_{eff} = \frac{c}{2f_0\sqrt{\varepsilon_{eff}}}$$

3.7 Calculation of Actual Length of Patch (L)

$$L = L_{eff} - 2\Delta L$$

Table 2. Design calculations.

Design parameters	Dimensions (mm)
Patch width	38
Patch length	30
Effective dielectric constant ($\in_{\text{reff}}$)	4.03 (no unit)
Length effective (L_{eff})	31
Length extension (ΔL)	0.374

4 Simulation Result

The simulated result of the proposed antenna operates at frequency range 2 GHZ to 6 GHz. It provides good performance with two ports and it is suitable for diversity application. The geometry offers stable, omni-directional pattern. The bandwidth obtained is 1720 MHz. The scattering parameters s_{11}, s_{12}, s_{22}, s_{21} are graphically shown with return loss and high gain measurements (Table 3).

Table 3. Antenna Parameters

Parameters	Values	
Power radiated	0.00268509	
Effective angle	2.79618	
Directivity	6.52644	
Gain	5.05951	
Maximum intensity	0.00096027	
Angle of U Max	0	0
E (theta)	0.213277	−94.6177
E (phi)	0.623431	92.607
E(x)	0.198996	84.8619
E(y) max (mag, phase)	0.826998	−87.4254
E(z) max (mag, phase)	0	0
Power radiated	0	0

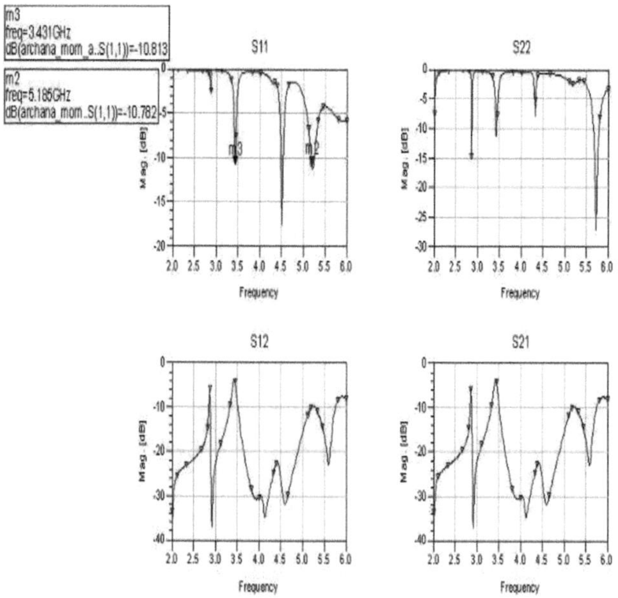

Fig. 2. Simulation of return loss characteristics in S-parameters

The S_{11} (return loss), phase and the smith chart will be displayed after the simulation is complete. The output is generated by selecting the parameters with its unit in the table (Figs. 2 and 3).

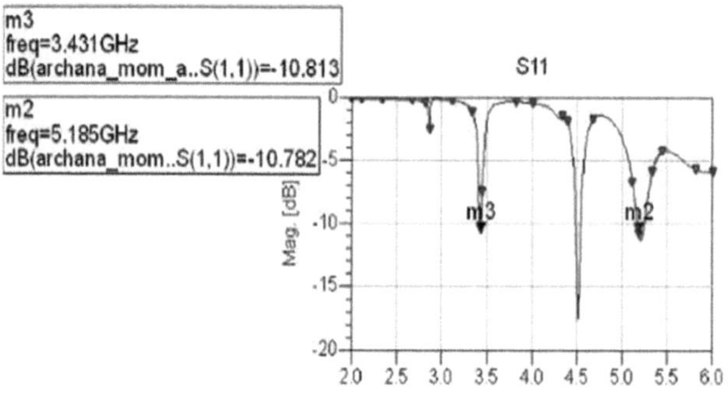

Fig. 3. Simulation of return loss characteristics of S_{11} parameter

4.1 Simulated Radiation

The Fig. 4 shows that according to the port placement the power is distributed over the surface of the antenna with high power gain and return loss over −10 dB. The blue colour here indicates the high power radiation.

Fig. 4. Simulated radiation of current distribution

4.2 The Pattern of Radiation

The pattern of radiation of the antenna is the far-field plot of the radiating properties of antenna as the spatial co-ordinates which is denoted by elevation angle (θ) and azimuth angle (φ). The plot of radiated power from an antenna is per unit solid angle which is nothing but the radiating intensity. This is plotted in 3D graph. This parameter shows the directivity of the antenna and the gain at different plots in the vacant place. This gives of an antenna signature and it is enough to understand the antenna produced and the importance of the parameter in the simulation software need to be clear (Fig. 5).

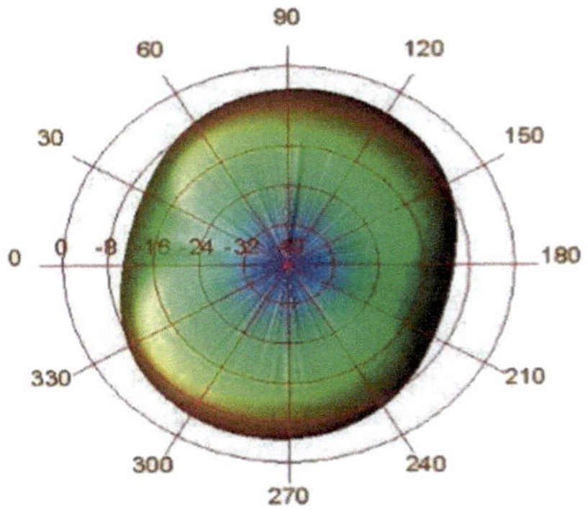

Fig. 5. Simulated design of radiation pattern

5 Fabrication Output Results

The fabrication is done for the proposed design of the MIMO antenna where the front view and the back view of the fabricated design is shown in the Figs. 7 and 8.

Fig. 6. Fabricated design front view

Figure 6 shows the complete model of the MIMO antenna with port fixed beneath and that is the front view of the two-section branchline coupler.

Fig. 7. Fabricated design back view

The fabricated output is tested using network analyzer where it measures the parameters of the network for the electric networks. S-parameters are commonly measured using the network analyzer.

Two-port networks like as amplifiers and filters are mainly characterized by network analyzer. Even networks with the random number of ports can also be analyzed using network analyzer.

Fig. 8. Testing of antenna using network analyser

The proposed fabricated MIMO antenna has been designed and tested with the network analyser which is shown in the Fig. 9 above where it clearly shows the dual band frequency of 3.5 GHz and 5 GHz in which the result is clearly exhibited in below Figs. 10 and 11.

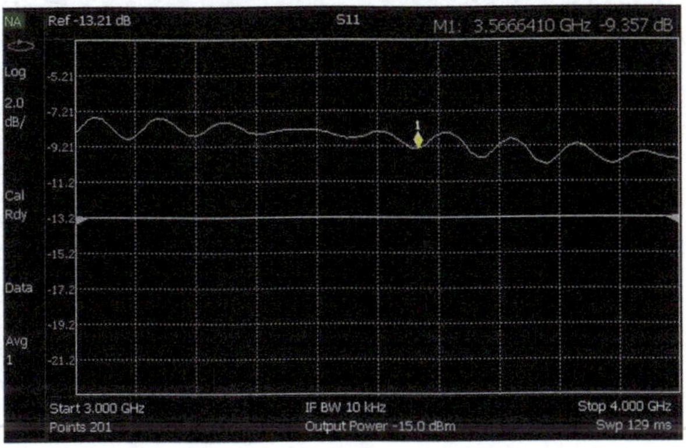

Fig. 9. Result of network analyser showing 3.5 GHz application

The Fig. 9 shows the operating frequency of 3.5 GHz where the start frequency is 3 GHz and the stop frequency is 4 GHz.

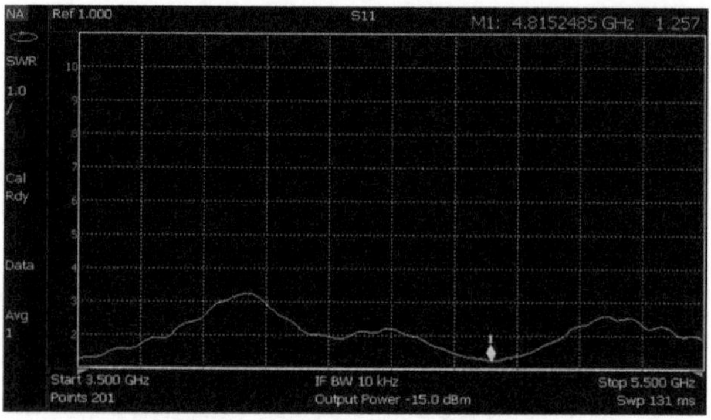

Fig. 10. Result of network analyser showing 5 GHz application

The Fig. 10 clearly shows that the operating frequency of 4.8 GHz where the start frequency is 3.5 GHz and the stop frequency is 5.5 GHz.

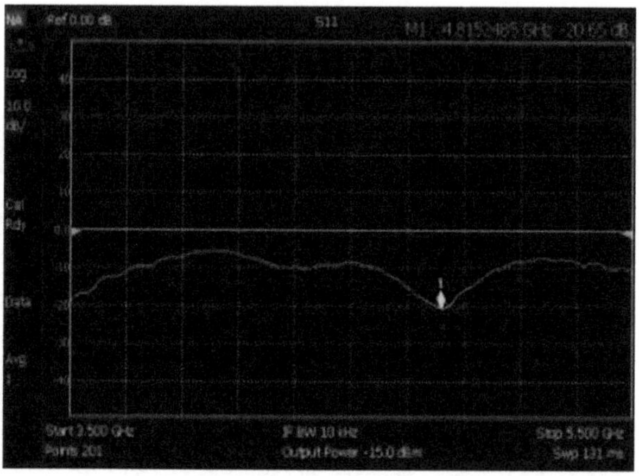

Fig. 11. The network analyser showing the 5 GHz application with gain −20.65 dB

The Fig. 11 shows that for the operating frequency 4.8 GHz with start frequency 3.5 GHz and the stop frequency as 5.5 GHz with return loss −20.65 dB.

6 Conclusion

The antenna is designed at the gain of +5 dB and have an advantage of compact size. The antenna works for an dual frequency which supports both WLan and Wi-Max. The 4-ports wide band diversity pattern of the antenna is presented. The dimension of design is 38*30 mm. The four ports are used to increase a port isolation and performance of the antenna. The proposed antenna provides the impedance bandwidth of the

1720 MHz with the frequency range of 2 GHz–6 GHz. The antenna has the two frequency bands of about 3.5 and 5 GHZ which is used to lap over the WLAN and WI-MAX. The design of the suitable antenna for the MIMO is the larger area of the research which is based on the conclusions drawn and the advantages are mentioned in the presented work, further improvement can be made. The proposed annular slot ring shape gives improved bandwidth and gain due to its dual band frequency. Further designs can be worked out at multiple frequency bands. The size of the design can also be further reduced to achieve wide frequency band.

References

1. Khaleghi, A., Kamyab, M.: Reconfigurable single port antenna with circular polarization diversity. IEEE Trans. Antennas Propag. **57**(2), 555–559 (2009)
2. Huang, Y., Nehorai, A., Friedman, G.: Mutual coupling of two collocated orthogonally oriented circular thin-wire loops. IEEE Trans. Antennas Propag. **51**(6), 1307–1314 (2003)
3. Ding, Y., Du, Z., Gong, K., Feng, Z.: A novel dual-band printed diversity antenna for mobile terminals. IEEE Trans. Antennas Propag. **55**(7), 2088–2096 (2007)
4. Bod, M., Hassani, H.R., Taheri, M.S.: Compact UWB printed slot antenna with extra bluetooth, GSM, and GPS bands. IEEE Antennas Wirel. Propag. Lett. **11**, 531–534 (2012)
5. Ghorban, K., Waterhouse, R.B.: Dual polarized wide band apertures tacked patch antennas. IEEE Trans. Antennas Propag. **52**(8), 2171–2175 (2004)
6. Yang, S.-L.S., Luk, K.-M., Lai, H.-W., Kishk, A.-A., Lee, K.-F.: A dual-polarized antenna with pattern diversity. IEEE Antennas Propag. Mag. **50**(6), 71–79 (2008)
7. Toh, W., Chen, Z., Ping, T.: A planar UWB diversity antenna. IEEE Trans. Antennas Propag. **57**(11), 3467–3473 (2009)
8. Wang, X., Feng, Z., Luk, K.-M.: Pattern and polarization diversity antenna with high isolation for portable wireless devices. IEEE Antennas Wirel. Propag. Lett. **8**, 209–211 (2009)
9. Chiang, M.J., Wang, S., Hsu, C.C.: Compact multi frequency slot antenna design incorporating embedded arc-strip. IEEE Antennas Wirel. Propag. Lett. **11**, 834–837 (2012)
10. Dang, L., Lei, Z.Y., Xie, Y.J., Ning, G.L., Fan, J.: A compact micro strip slot triple-band antenna for WLAN/WiMAX applications. IEEE Antennas Wirel. Propag. Lett. **9**, 1178–1181 (2010)
11. Haghparast, A.H., Dadashzadeh, G.: A dual band polygon shaped CPW-fed planar monopole antenna with circular polarization and isolation enhancement for MIMO applications. In: IEEE 2015 9th European Conference on Antennas and Propagation (EUCAP), pp. 2164–3342 (2015)
12. Votis, C., Tatsis, G., Kostarakis, P.: Envelope correlation parameter measurements in a MIMO antenna array configuration. J. Commun. Netw. Syst. Sci. **3**, 350–354 (2010)
13. Wong, K.L., Lu, J.Y.: 3.6-GHz 10-antenna array for MIMO operation in the smartphone. Microw. Opt. Technol. Lett. **57**(7), 1699–1704 (2015)
14. Wong, K.L., et al.: 8-Antenna and 16-Antenna Arrays using the quad-antenna linear array as a building block for the 3.5 GHz LTE MIMO operation in the smartphone. Microw. Opt. Technol. Lett. **58**(1), 174–181 (2016)
15. Wong, K.L., Tsai, C.Y., Lu, J.Y.: Two asymmetrically mirrored gap-coupled loop antenna as a compact building block for eight-antenna MIMO array in the future smartphone. IEEE Trans. Antennas Propag. **65**(4), 1765–1778 (2017)

16. Stavrou, E., Litschke, O., Baggen, R., Oikonomopoulos-Zachos, C.: Dual-beam antenna for MIMO WiFi base stations. In: 8th European Conference on Antennas and Propagation, pp. 6–11 (April 2014)
17. Han, W., et al.: A six-port MIMO antenna system with high isolation for 5-GHz WLAN access points. IEEE Antennas Wirel. Propag. Lett. **13**, 880–883 (2014)
18. Sun, J.S., Fang, H.S., Lin, P.Y., Chuang, C.S.: Triple-band MIMO antenna for mobile wireless applications. IEEE Antennas Wirel. Propag. Lett. **15**, 500–503 (2016)

Beamforming Techniques for Millimeter Wave Communications - A Survey

J. Mercy Sheeba(✉) and S. Deepa

Panimalar Engineering College, Chennai, India
mercyjawahar7@gmail.com, dineshdeepas1977@gmail.com

Abstract. Millimeter wave communication has procured prominent focus among the research fraternity owing to the ever-increasing wireless user data in recent times which has turned out to be the driving force in exploring the different regions of the radio frequency bands to meet the dynamic user requirements. Millimeter wave supports multitude of users and can achieve high data rates and poses practical and reasonable solutions to the capacity crunch faced by the next generation wireless networks. The concept of antenna beamforming has made a significant progress in the wireless communication field and aids the development of robust communication link. This paper reviews the different beamforming approaches namely Analog, Digital and Hybrid Beamforming techniques for millimeter wave communications systems and view their system architectures. The key advantages and limitations associated with each technique are discussed. This paper throws light on the best suited method for millimeter wave communication systems.

Keywords: Millimeter wave · Analog · Digital · Hybrid beamforming

1 Introduction

Recently, there has been a widespread interest among the network designers and personnel in the utilization of millimeter wave bands for the 5th generation cellular systems [4, 5]. A report by Wireless World Research [6] brings-forth that mobile data traffic is (at least) doubled every year. It is envisaged that by 2020 [7], around 50 billion devices will serve the user fraternity with atleast 6 devices per person, which includes machine communications too. The need to accommodate such multitude of user devices and serve such massive communications makes it highly essential to elevate the capacity of the cellular network. It is predicted that the capacity of the 5G network will be scaled up, to provide 1000 times increased capacity than the currently prevailing systems [8].

The Millimeter Wave technology is contemplated to be the favorable technology for upcoming 5G cellular systems. The millimeter wave frequency band can offer a wider spectral range from 30 GHz to 300 GHz and high data throughput for 5G systems. The spectrum can be utilized for several broadband, bandwidth- hungry applications [9, 10] and also in European Union's FOF (Factories-of-Future) partnership [2]. Multiple antennas at transmitting and receiving ends achieve multiplexing, diversity or high antenna gains at the receiver. Beamforming using multiple antennas is

D. J. Hemanth et al. (Eds.): COMET 2019, LNDECT 35, pp. 1563–1573, 2020.
https://doi.org/10.1007/978-3-030-32150-5_157

a key element for effective utilization of the millimeter wave band as it can elevate system capacity.

In conventional microwave systems, fixed weight and adaptive beamforming can be performed conveniently in digital baseband as there were only limited antenna elements. The less complex analog beamforming methods have been used widely for indoor, short range communication in 60 GHz band [11, 12]. The more advanced adaptive beamforming techniques have not been adopted largely in millimeter wave communications due to the complexities in signal processing. The need to support the increasing user traffic and to mitigate the limitations in the hardware cost, the *Hybrid Beamforming* method has unfolded to be the main entrant for millimeter wave communications.

The paper reviews the different beamforming methods for mm wave systems with the elucidation of the system architectures, their key advantages and limitations and associated with each technique and in what way one has an edge over the other methods. It includes a brief overview of the fundamentals in beamforming and highlights the unique propagation characteristics of millimeter wave communication channels. This paper aims to analyze the best beamforming method in view of various characteristics and parameters of the beamforming methods.

2 Related Works

2.1 Beamforming System Architectures and Approaches

Beamforming is the signal processing technique employed in group of sensors (sensor arrays) to ensure directional signal transmission/reception. The antenna elements are combined such that signals experience constructive and destructive interference depending on its angles. Beamforming achieves spatial selectivity at both ends of the antenna.

Millimeter Wave Channel Characteristics
Millimeter wave channels differ from the traditional microwave channels on account of the following features:

Path Loss. Friss law [1] states that the propagation loss in free-space is inversely proportional to the wavelength. Therefore, due to smaller wavelengths, the millimeter waves are subjected to higher propagation loss than the current communication systems below 6 GHz bands. Beamforming with directional antennas can compensate the high free space propagation loss in millimeter wave communications, generating high gains.

Wideband Communication. The principal reason for exploring millimeter wave bands for the future generation wireless networks is the availability of humongous bandwidth. These channels are envisaged to support wideband signal propagation.

The other key features: clustered multipath structure, Dominant LOS component and 3D spatio-temporal modeling yield a potent beamforming technique.

Beamforming Protocol

The IEEE 802.11 ad beamforming protocol comprises of 3 stages for beamforming as follows: [13]

Sector Level Sweep (SLS) Phase: This phase selects the best transmit and receive antenna sector pair.

Beam Refinement Phase (BRP): In this phase, the pair of antenna arrays selects a beam pattern pair with finer bandwidths.

Beam Tracking (BT) Phase: In BT phase, TRN (training) fields are appended to data fields. AGC (Automatic Gain Control) field is also present to assist the receiver in calculating the gain. Channel estimation (CE) can be done using TRN fields.

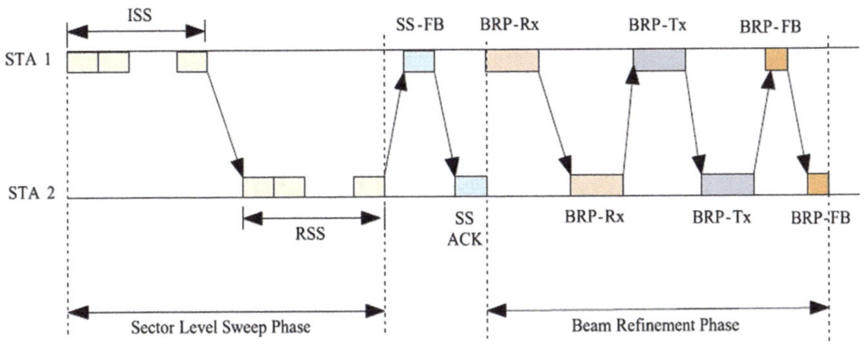

Fig. 1. IEEE 802.11 ad beamforming procedure representation [1]

2.2 Analog and Digital Beamforming for Millimeter Wave Systems

The Fixed-weight beamforming technique (conventional beamforming) uses a fixed time delays to combine signals from the array sensors. The constant weights applied in analog or digital domain for beam steering. Figure 1 illustrates the fixed weight beamforming approach.

The *fixed beamforming* [1] approach is applied to fixed arrival angle emitters. However, if desired angles change with respect to time, an optimization scheme has to be devised to recalculate the array weights. In an environment prone to continuous changes in electromagnetic emissions, processing algorithms are needed at the receiver for recursive updates of the weight vectors, which results in optimal transmission or reception of the required signal.

In *temporal domain*, antenna weights are applied with time delay elements. *Analog baseband Beamforming* (Fig. 2) is the technique where the incoming signal is phase shifted before the radio frequency up conversion. If the antenna weights are applied by phase shifting the signal after the up conversion stage, then the beamforming process is known as *Analog RF beamforming*. Analog Beamforming is supported by IEEE 802.11ad beamforming protocol [16].

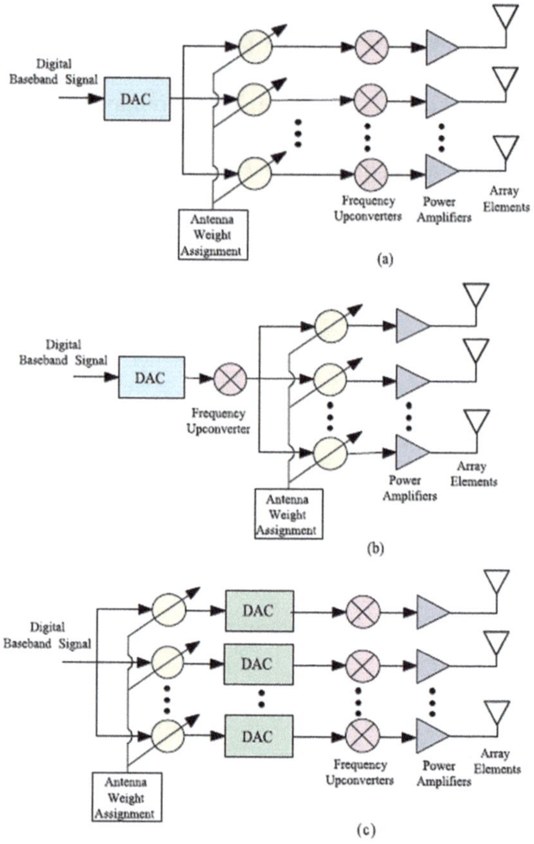

Fig. 2. Architectures of conventional beamforming (a) analog baseband beamforming (Tx) (b) analog RF beamforming (transmit) (c) digital baseband beamforming (transmit) [1]

Analog beamforming supports only single-user/single-stream transmission when a single beamformer used. This makes the approach not suitable in MIMO environment. Steering the beams is of high importance and the wireless protocol should be designed to support beam steering to achieve highest performance [14, 15].

The quantized phase shifts and the improper amplitude adjustments limit the performance of analog beamforming. This creates more difficulties in tuning beams and in null steering. Factors like phase-shifter loss, noise, power consumption and non linearity adds up to the problem of performance degradation.

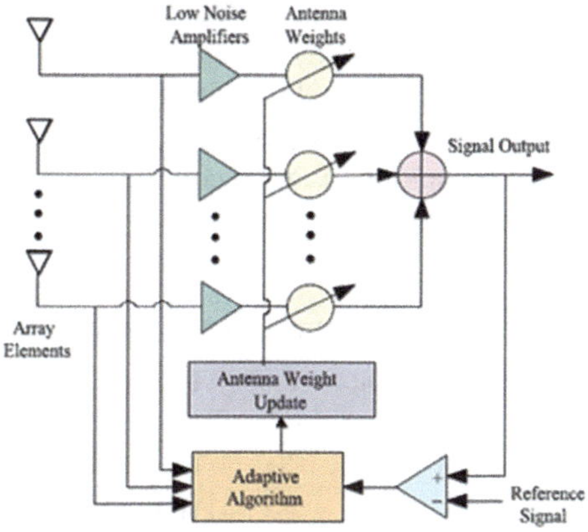

Fig. 3. Adaptive beamforming (receive) architecture [1]

The *Adaptive beamforming* (Fig. 3) is a more sophisticated and powerful technique as it can update itself according to the RF radiation pattern. The Adaptive Beamforming used in mobile communications requires a recursive updation of the weights for the time varying DOA (direction-of-arrival). It uses efficient algorithms such as:-

(i) Training based algorithms (ii) Blind methods. A single reference signal is required for training based algorithms. Blind methods require an estimate of DOA and the rest of the information is obtained from the received signal.

The *Digital Beamforming* (DBF) approach is a powerful one to boost the antenna performance. The received signals in a DBF array, are detected and digitized and then processed for beamforming using a digital signal processor. This approach (Fig. 4) preserves the total information available at the aperture, in contrast to an analog beamformer, which produces only the weighted sum of the signals and thus reduces the signal dimensionality from N to 1.

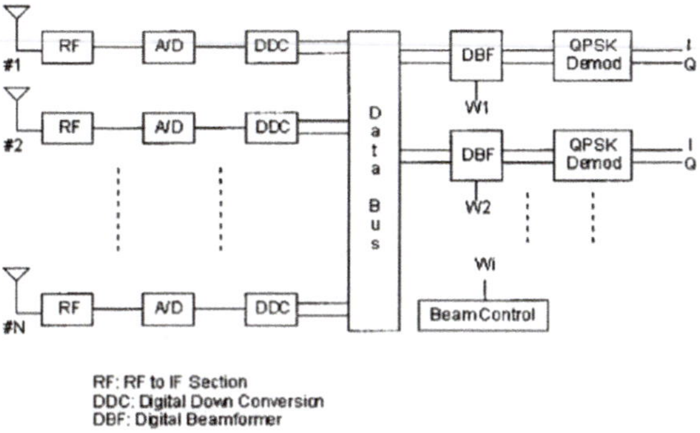

Fig. 4. Block diagram of digital beamforming (receive)

The DBF method realizes its main advantages in the receive mode [3], which includes:- (i) Improvised pattern nulling (ii) Closely spaced multiple beams (iii) Pattern correction of array elements (iv) Greater flexibility (v) Higher degrees of freedom. In this method, each antenna element is exclusively allocated an individual RF chain which results in high power consumption and a complex architecture. The comparison between Analog and Digital Beamforming techniques is shown in Table 1.

Table 1. Analog beamforming Vs digital beamforming

Parameters	Analog beamforming	Reason	Digital beamforming	Reason
Architecture	Simple		Complex	
Power consumption	Less power consumption than digital beamforming		High power consumption	
Complexity	Less complex	No dedicated RF chain required	Complex	Complex and costly-requires dedicated RF chain
Hardware cost	Low cost		Costly	
Baseband chains	Requires less baseband chains at the receiver		Requires more baseband chains in both transmitter and receiver	
Flexibility	Not very flexible	Fixed delay	Greater flexibility	Recursive updation of weight vector
Degrees of freedom (DOF)	Limited DOF	Supports single-user, single stream data transmission	Offers higher degrees of freedom	Able to support multi-user, multi-stream data transmission

2.3 Hybrid Beamforming

Hybrid Beamforming works with the objective to support multi-user communication with maximized data rate and minimized interference. The primary motivation of hybrid beamforming architecture is to lower the power consumption, hardware cost and complexity in signal processing to provide an optimum performance. The power consumption issue results in an RF chain mainly due to ADC and DAC components. Hybrid beamforming architecture, however, presents to be a desired beamforming method to overcome this major limitation as it does not utilize a dedicated hardware/RF chain. The architecture of Hybrid beamforming is classified broadly as:

(i) Fully connected hybrid beamforming - all antenna elements are connected with each RF chain

(ii) Sub-connected or partially connected hybrid beamforming - set of array elements are connected with each RF chain

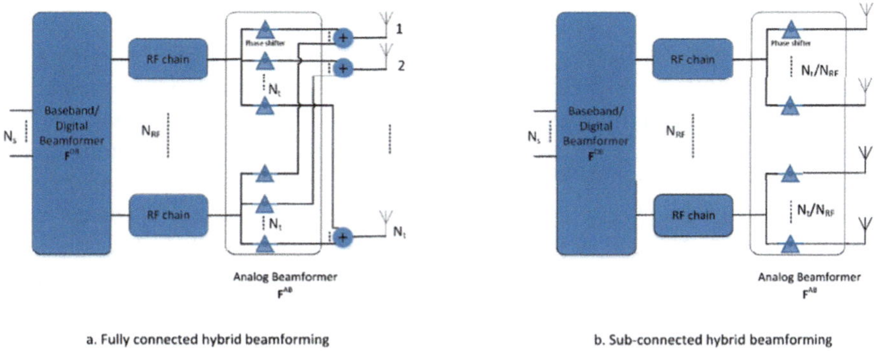

Fig. 5. Hybrid beamforming architectures [2]

In fully connected beamforming architecture (Fig. 5a) [17], works as follows:

- Digital precoding of the signal from user
- Processing of the precoded signal by RF chain
- Once processing is complete, transmission of the processed data by means of the antenna array using a common analog beamforming unit

The sub-connected beamforming architecture [18] works the same way as the fully connected beamforming architecture with the only difference that the signal transmission is done with a set of array elements that forms the sub-array. As an alternative, the analog beamforming unit can be applied prior to the process of frequency up conversion.

The hardware configuration and aspects of hybrid beamforming illustrated in few papers are presented. The Hybrid beamforming architectures for millimeter wave, algorithms for signal processing and implementation methods are traced in [1]. Additionally, this paper focuses on the methods to reduce the hardware complexity, cost and most importantly on mitigating the power consumption issues which are highly crucial for better performance of the system. The authors briefly illustrates the various beamforming approaches associated with mm wave communication, propagation characteristics of mm wave and beamforming protocol and the requirements to achieve effective beamforming in both indoor and outdoor millimeter wave communication set-up.

Molisch [17] categorizes structures of hybrid beamforming on the basis of instantaneous or average channel state information respectively. The former comparatively provides better SINR (signal to interference noise ratio) than the latter which presents lower overhead for the acquisition of CSI. This paper yields a clear understanding of the hybrid beamforming structure and its ability to operate in a dynamic manner depending on the applications. This paper also proposes that the antenna gain is an important aspect which increases in massive MIMO systems.

The survey paper [16] emphasizes on the challenges in processing signals for millimeter wave communications and also extends the same analysis for MIMO based communication systems at high frequencies. The author in the aforementioned paper proposes the hybrid beamforming technique based on analog beamforming which can be made practical by means of switching networks, a hybrid precoder/combiner which uses phase shifters which are digitally controlled or by using discrete lens array. The precoder/combiner can rectify lack of accuracy in the analog data and cancels residual multi-stream interference. The use of switching networks is another alternative which reduces the power consumption issues and signal processing complexity of the digitally controlled phase shifter based hybrid architecture. The use of discrete array of lens antenna is the third method of implementing the analog beamforming in hybrid architecture.

The paper [25] enlightens the complexities in the hardware with respect to ADC resolution. ADCs face major limitations such as improper channel estimation and rate loss. This paper, however, does not throw light on the other analog beamforming components like switches, array of lens antenna and phase shifters.

The recent work [2] on Hybrid beamforming presents a complete and a thorough insight into the different hybrid beamforming system architectures which includes full-array, fully-connected with virtual sectorization, partially-connected or sub-connected hybrid beamforming and hybrid beamforming with low complexity analog beamforming.

In [20, 21], hybrid precoders are designed to maximize the SE for the mmWave massive MIMO systems with single and multiuser cases. Authors have analyzed a notable increase in performance is achieved with digital beamforming when the radio frequency chains is twice that of the data streams. Later, they proposed heuristic scheme with a baseband precoder (low dimensional) and a RF precoder (high dimensional), thus reducing the number of radio frequency chains and power consumption.

Alkhateeb *et al.*, [22] put-forth the development of precoders (uplink and downlink) on the basis of recursive least square. This precoder generates an optimal spectral efficiency for 3 simultaneous data streams. The authors Bogale *et al.* [18] focused on maximizing the Spectral efficiency (SE) of millimeter wave massive MIMO systems during downlink. The hybrid precoders lower the radio frequency chains with slight deterioration in spectral performance.

Table 2. Hybrid beamforming architectures-performance measures

Papers	Architectures	Signal to noise ratio SNR (dB)	Spectral efficiency (b/s/Hz)	No. of transmitting antennas Nt	No. of receiving antennas Nr	User equipments K
[17]	Full array Partial or Sub-array	10	70	64	1	16
		10	25			
[19]	Full-array Sub-array	−100	28	500	1	4
		−100	20	500	1	4
[18]	Sub-array	10	35	64	1	8
[26]	Full-array	−10	4	32	8	1
[21]	Full-array	−10	10	64	16	1
[22]	Fully-connected	−10	7	200	100	1
[24]	Fully-connected Sub-connected	10	22	64	4	1
		10	15			
[23]	Sub-connected	10	10	2*8	2*4	1
[27]	Fully-connected	−10	6	64	8	1
[28]	Fully-connected	−10	9	64	16	1

Table 2 illustrates that full-array/fully connected architecture structures always dominates the other structures in terms of Spectral Efficiency for different transmitting (Nt) and receiving (Nr) antennas [2]. Fully-connected structure provides approximately Nrf log2Nrf b/s/Hz (Nrf - Number of radio frequency chains) higher SE performance than sub-connected structure. Papers [17, 19, 24] compare fully-connected and sub-connected architectures, reported in Table 2.

3 Conclusion

This survey proffers a comprehensive view on beamforming techniques for millimeter wave communication. The significance of millimeter waves in the upcoming 5G cellular networks and its ability to accommodate multitude of user data streams and serve massive communications were discussed. The unique propagation characteristics of millimeter wave channels from that of the traditional microwave channels were explained. A concise description of the different beamforming methods is put-forth, with their system architectures and working methodologies. The highlights and drawbacks of each beamforming technique were encountered.

The Analog beamforming is prone to performance degradation due to quantization effects, noise, power consumption and coarse tuning of nulls. Moreover, benefits associated with MIMO cannot be realized with analog beamforming. Though a powerful technique, the digital beamforming faces some challenges such as high power consumption, increased cost and hardware complexity.

The Hybrid beamforming method is reviewed to be the most suitable technique for millimeter wave communication systems as it supports multi-user communication with maximized data rate and minimized interference. It mitigates the limitations of signal processing complexity, power consumption and hardware cost. The different hybrid beamforming architectures such as full array and sub-or partially-connected structures were visited. It is analyzed that, for different number of transmitting (Nt) and receiving (Nr) antennas, spectral efficiency of fully-connected structure always dominates the sub-connected architecture.

References

1. Kutty, S., Sen, D.: Beamforming for millimeter wave communications: an inclusive survey. IEEE Commun. Surv. Tutorials **18**(2), 949–973 (2016). 2nd Quart
2. Ahmed, I., Khammari, H., Shahid, A., Musa, A., Kim, K.S., Moerman, I.: A survey on hybrid beamforming techniques in 5G: architecture and system model perspectives. IEEE Commun. Surv. Tutorials **20**(4), 3060–3097 (2018). Fourth Quarter
3. Yang, B., Yu, Z., Lan, J., Zhang, R., Zhou, J., Hong, W.: Digital beamforming-based massive MIMO transceiver for 5G millimeter-wave communications. IEEE Trans. Microwave Theory Tech. **66**(7), 3403–3418 (2018)
4. Rappaport, T.S., et al.: Millimeter wave mobile communications for 5G cellular: it will work! IEEE Access **1**, 335–349 (2013)
5. Pi, Z., Khan, F.: An introduction to millimeter-wave mobile broadband systems. IEEE Commun. Mag. **49**(6), 101–107 (2011)
6. 5G vision, enablers and challenges for the wireless future, Durban, South Africa, Wireless World Res. Forum, White Paper (2015)
7. More than 50 billion connected devices, Stockholm, Sweden, Ericsson, L.M. White Paper (2011)
8. The 1000x mobile data challenge, San Diego, CA, USA, Qualcomm, White Paper, November 2013
9. 5G: a technology vision, Shenzhen, China, Huawei, White Paper, pp. 1–16 (2014)
10. Hossain, E., Rasti, M., Tabassum, H., Abdelnasser, A.: Evolution toward 5G multi-tier cellular wireless networks: An interference management perspective. IEEE Wirel. Commun. **21**(3), 118–127 (2014)
11. Yong, S.K., Xia, P., Garcia, A.V.: 60 GHz Technology for Gbps WLAN, WPAN: From Theory to Practice. Wiley, Hoboken (2011)
12. Huang, K.-C., Wang, Z.: Millimeterwave Communication Systems. Wiley/IEEE Press, Hoboken (2011)
13. Part 11: Wireless LAN Medium Access Control (MAC) and Physical Layer (PHY) Specifications. Amendment 3: Enhancements for Very High Throughput in the 60 GHz Band, IEEE Standard 802.11 ad-2012, December 2012
14. Poon, A., Taghivand, M.: Supporting and enabling circuits for antenna arrays in wireless communications. Proc. IEEE **100**(7), 2207–2218 (2012)
15. Wang, J., Lan, Z., Pyo, C.W.: Beam codebook based beamforming protocol for multi-Gbps millimeter-wave WPAN systems. IEEE J. Sel. Areas Commun. **27**(8), 3–4 (2009)
16. Heath Jr., R.W., González-Prelcic, N., Rangan, S., Roh, W., Sayeed, A.M.: An overview of signal processing techniques for millimeter wave MIMO systems. IEEE J. Sel. Topics Signal Process. **10**(3), 436–453 (2016)
17. Molisch, A.F., et al.: Hybrid beamforming for massive MIMO—a survey, pp. 1–14. arXiv Preprint. http://arxiv.org/abs/1609.05078 (2016)

18. Bogale, T.E., Le, L.B., Haghighat, A., Vandendorpe, L.: On the number of RF chains and phase shifters, and scheduling design with hybrid analog–digital beamforming. IEEE Trans. Wirel. Commun. **15**(5), 3311–3326 (2016)
19. Han, S., Chih-Lin, I., Xu, Z., Rowell, C.: Large-scale antenna systems with hybrid analog and digital beamforming for millimeter wave 5G. IEEE Commun. Mag. **53**(1), 186–194 (2015)
20. Alkhateeb, A., El Ayach, O., Leus, G., Heath Jr., R.W.: Channel estimation and hybrid precoding for millimeter wave cellular systems. IEEE J. Sel. Topics Signal Process. **8**(5), 831–846 (2014)
21. Sohrabi, F., Yu, W.: Hybrid digital and analog beamforming design for large-scale antenna arrays. IEEE J. Sel. Topics Signal Process. **10**(3), 501–513 (2016)
22. Alkhateeb, A., El Ayach, O., Leus, G., Heath, R.W.: Hybrid precoding for millimeter wave cellular systems with partial channel knowledge. In: Proceedings of Information Theory and Application Workshops, San Diego, CA, USA, pp. 1–5 (2013)
23. Singh, J., Ramakrishna, S.: On the feasibility of beamforming in millimeter wave communication systems with multiple antenna arrays. In: Proceedings of IEEE Global Communications Conference, Austin, TX, USA, pp. 3802–3808 (2014)
24. Park, S., Alkhateeb, A., Heath, R.W.: Dynamic subarrays for hybrid precoding in wideband mmWave MIMO systems. IEEE Trans. Wirel. Commun. **16**(5), 2907–2920 (2017)
25. Araújo, D.C., et al.: Massive MIMO: survey and future research topics. IET Commun. **10** (15), 1938–1946 (2016)
26. Alkhateeb, A., Heath Jr., R.W.: Frequency selective hybrid precoding for limited feedback millimeter wave systems. IEEE Trans. Commun. **64**(5), 1801–1818 (2016)
27. Sohrabi, F., Yu, W.: Hybrid digital and analog beamforming design for large-scale MIMO systems. In: Proceedings of IEEE International Conference on Acoustics, Speech and Signal Processing, Brisbane, QLD, Australia (2015)
28. Sohrabi, F., Yu, W.: Hybrid beamforming with finite-resolution phase shifters for large-scale MIMO systems. In: Proceedings IEEE 16th International Workshop on Signal Processing Advances in Wireless Communications, Stockholm, Sweden, pp. 136–14 (2015)

Underwater Li-Fi Communication for Monitoring the Divers Health

R. Durga(✉), R. Venkatesh, and D. Selvaraj

Panimalar Engineering College, Chennai, India
durga.1997r@gmail.com, venkatesh7696r@gmail.com,
drdselva@gmail.com

Abstract. Diving has become a common way of performing research in the underwater living world. One of the major problems with diving is the health issues faced by the divers during diving and there comes the need for monitoring diver's health. This paper mainly focuses on the health monitoring systems for divers by transferring the data using Li-Fi (Light Fidelity). This system senses different health specifications like heartbeat, body temperature and lung expansion. The monitored health specifications are recorded as a database in a memory chip for further analysis. To reduce the power consumption, the system transfers the data to the nearby divers and ship only during the abnormal health issue. The proposed system can be used up to 120 meters sea depth and the data can be transferred up to 5 receivers at a time.

Keywords: Underwater · Li-Fi · Li-Fi communication · Health monitoring of divers

1 Introduction

Visible Light Communication is the other name for Li-Fi. Li-Fi can transmit the data using a high illumination LED that varies the intensity faster than the human eye [1]. The distance traveled by the Li-Fi is 10-30 m underwater to transfer the data and there will not be much interference produced from Li-Fi. The data encoded in binary form is sent to the light transmitting systems by high illumination LED. The information is transmitted by switching the LED ON and OFF to produce 0's and 1's.

In previous existing methods, the data is transmitted via acoustic communication, ultrasonic communication, wired communication, voice communication, RSTC hand signals, and torch/Flash signals. These existing communication systems faced difficulties in propagation under sea water. The optical method proposed in [2] for communicating between two autonomous underwater vehicles. The system transfers the data with very less power. The range differs due to environmental conditions. It differs according to clear and muskier water. Corentin et al. [3] developed an algorithm for detecting the breath of the scuba diver. They analyze the signal by using the algorithm if there is no breathing from scuba divers it will produce the alarm to the nearby ship. the delay noted from transferring the data is 5.2 s. The memory used to store the data in Random Access Memory is 800 bytes. The design considerations of underwater optical communication to detect with different parameters is said in [4] and the major

D. J. Hemanth et al. (Eds.): COMET 2019, LNDECT 35, pp. 1574–1579, 2020.
https://doi.org/10.1007/978-3-030-32150-5_158

drawback is the attenuation loss due to the scattering of light. To obtain high bandwidth optical communication which is simulated using Monte Carlo in [5] where the data rate is greater than 1gigabits/second the data transfer does not require any physical contact.

Vijaya [6] proposed underwater point to point communication that causes the misalignment in the optical link due to absorption and scattering, so the transmitter and receiver are misaligned. The alignment of the transmitter and receiver is achieved by increasing the divergence of the transmitted beam. Thomas [7] used two orthogonal laser beams and two receiving optical link to receive the data in the sea. By using laser communication system technique, it reduces the transmission error problems and it will also limit the scattering levels. To achieve high data rate, it requires sufficient intrinsic bandwidth. Chiarella [8] planned to develop communication by diver gestures, known as CADDIAN language. The gestures are signs, symbols, alphabets, semantics. In muskier water, it was difficult to communicate. Tran [9] proposed a transceiver design of acoustic space frequency block code OFDM to increase the data throughput for vertical link communication in underwater. It also increases the data throughput up to 7.5 kbps. It also produces noise, multipath, and sampling rate error. Hachioji-shi [10] proposed a theory for detecting a stray recreational diver underwater. This theory was simulated using network simulator and the data rate is evaluated. The implementation at underwater propagation model is done at 50 kHz.

2 Proposed Method

The proposed method consists of transmitting and receiving section. The transmitting section detects the abnormalities faced by the diver and the data is transferred by using the medium called light fidelity. In receiving section the light signal is converted in the form of electrical signal and the data is produced in the form of audio.

2.1 Transmitter Section

The transmitter section consists of PIC microcontroller, IR sensor, Li-Fi, heartbeat sensor, temperature sensor, emergency switch, a transmitting module, power supply and LCD module shown in Fig. 1. The step-down transformer is used for generating the power to the device. The heartbeat sensor is used for sensing the pumping of the heart or it can also be measured in the finger artery by presenting the variations in the blood volume. The infrared LED transmitting an IR signal into the finger and the reflected infrared light from the blood cells are transmitted with a pulse train into a photodiode. For sensing the temperature LM35 sensor is used which is used for converting the body temperature into an electrical signal. LM35 is small in size and produces very high accuracy with very less power. The microcontroller is used for interfacing all the devices and it also consumes very less power. The memory stored in Random Access Memory is 368 bytes. The microcontroller is connected to the LCD display andthe infrared sensor is used for detecting the obstacles during the transmission of data. The light fidelity module consists of a transmission section that

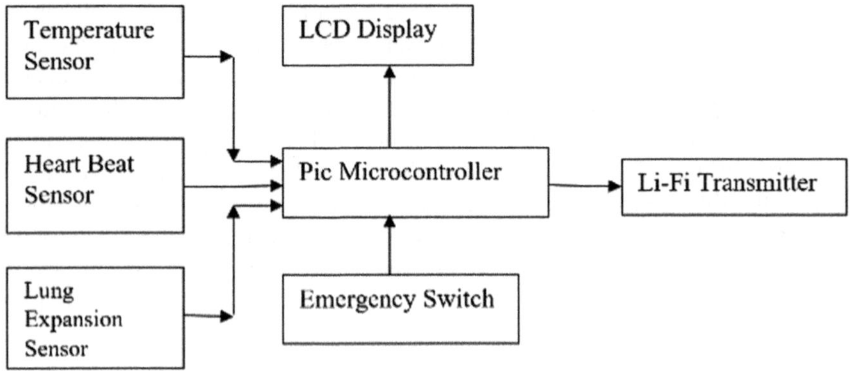

Fig. 1. Block diagram of transmitting section

produces a white LED for transmitting and receiving the information. LED's are used for its low value, little in size and consumes less power. The data is produced through the Li-Fi to the receiver section.

2.2 Receiver Section

The receiver section consists of Li-Fi receiver, LCD display, and the audio player. The data transmitted from the Li-Fi is 1 giga bits per second. The data received from the Li-Fi transmitter will be in the form of a light signal. The light is passed into the photodiode receiver. The photodiode is used for converting the light signal into an electrical signal. The electrical signal is given into a double stage amplifier which is used for amplifying the detected electrical signal. The electrical signal is converted into binary information. Binary information is converted into the original message and the data is produced at the output in the form of audio signal/message (Fig. 2).

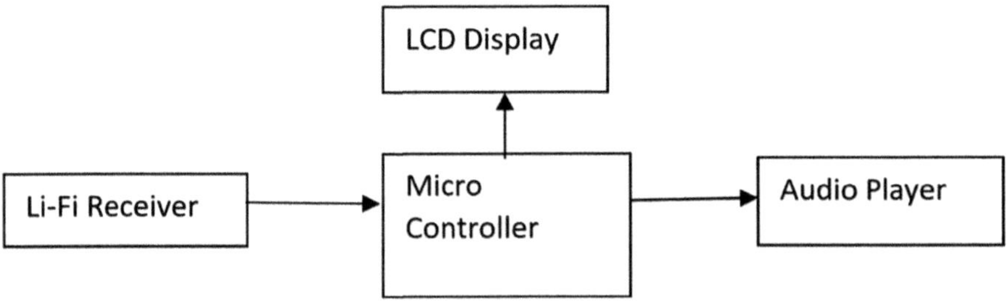

Fig. 2. Block diagram of receiving section

3 Experimental Result

The output will be produced if there are any emergency health issues faced by divers. There is also an emergency switch if the diver faces any issues that person can press the emergency switch. In the proposed system (Figs. 3 and 4), three different sensors like heartbeat sensor, temperature sensor, and lung expansion sensor are used. If there were any abnormalities faced by the diver, the sensors will detect and give the data via Li-Fi as a light signal.

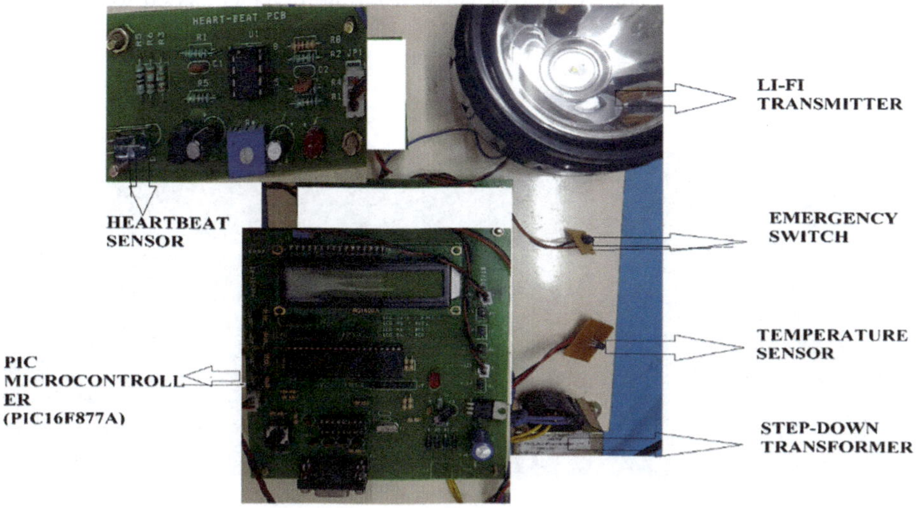

Fig. 3. Li-Fi communication transmitter

Fig. 4. Li-Fi communication receiver

The received light signal by the nearby diver's receiver is passed into the photodiode which will convert the light signal into an electrical signal and produces the output in the form of an audio signal. The received audio signal spectrum is shown in Fig. 5. The experimental output is seen in the form of the audio spectrum. The data observed from the audio is converted into an audio spectrum. If there are any abnormalities observed from the diver is produced in the output. the audio spectrum shown below are emergency alert audio spectrum shown below in Fig. 5, heart rate audio spectrum shown below in Fig. 6 (the sample results are taken by author itself (PARTICIPENT NAME: DURGA R)). the temperature sensed audio spectrum shown in Fig. 7. If there was any abnormality detected from the diver is passed to the nearby diver or ship.

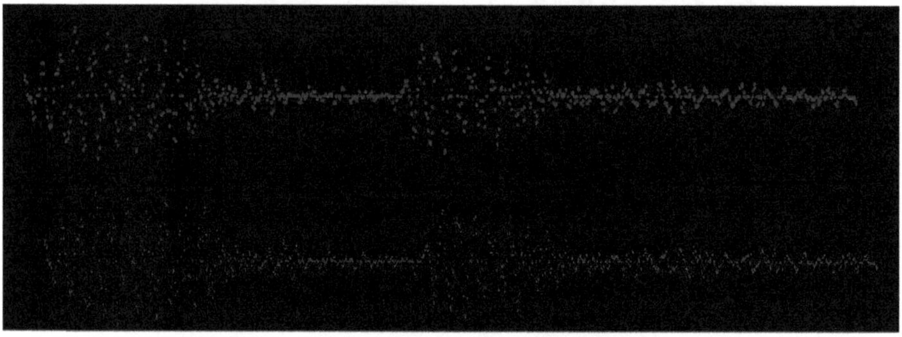

Fig. 5. Audio spectrum for emergency switch alert

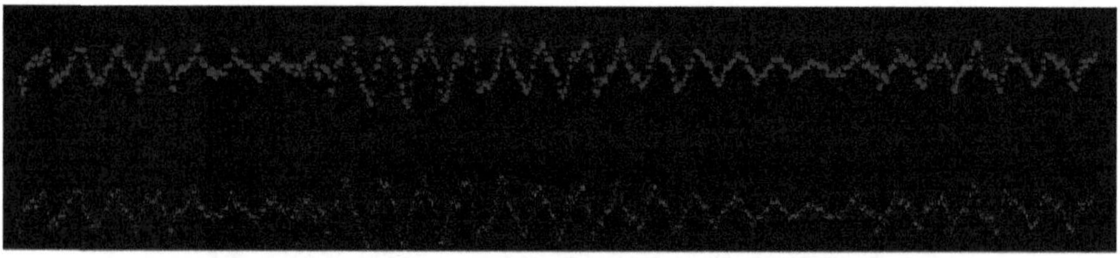

Fig. 6. Audio spectrum for heart beat if there is any abnormality

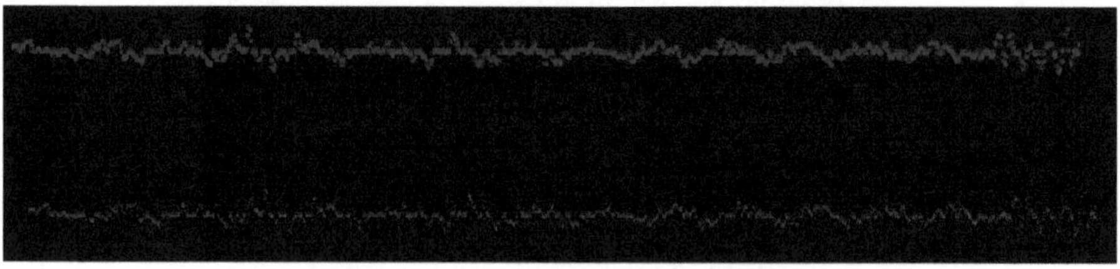

Fig. 7. Audio spectrum for temperature if there is any abnormality

4 Conclusion

The data is produced only at the time of emergency so, it consumes very less power. The device is very cost effective. It transmits the data at a speed of 2 Giga Bits Per Second (Gbps) which is faster than the existing systems. The data can be transmitted between 5 divers and a ship. The proposed system can be majorly used for rescue operations under the sea. It can also be used for ship to ship communication. So, this system may replace the existing underwater techniques.

References

1. Leba, M., Riurean, S., Lonica, A.: Li-Fi– the path to a new way of communication. In: IEEE 12th Iberian Conference on Information Systems and Technologies (CISTI) (2017)
2. Bales, J.W., Chrissostomidis, C.: High-bandwidth, low-power, short range optical communication underwater. In: International Symposium on Unmanned Untethered Submersible Technology, University of New Hampshire-Marine Systems, pp. 406–415 (1995)
3. Altepe, C., Egi, S., Ozyigit, T., Sinoplu, D., Marroni, A., Pierleoni, P.: Design and validation of a breathing detection system for scuba divers. MPDI Sensor **17**(6), 1349 (2017)
4. Giles, J.W., Bankman, I.: Underwater optical communications systems. part 2: basic design considerations. In: IEEE Military Communications Conference (MILCOM), vol. 3, pp. 17100–17170 (2015)
5. Hanson, F., Radic, S.: High bandwidth underwater optical communication. Appl. Opt. **47**(2), 277–283 (2018)
6. Vijaya, K.P., Praneeth, S., Narender, R.B.: Analysis of optical wireless communication for underwater wireless communication. Int. J. Sci. Eng. Res. **6**(2), 1–9 (2011)
7. Scholz, T.: Laser based underwater communication experiments in Baltic Sea. IEEE, pp. 1–3 (2018)
8. Chiarella, D., Bibuli, M., Bruzzone, G., Caccia, M., Ranieri, A., Zereik, E.: Gesture-based language for diver-robot underwater interaction. In: National Research Council - Institute of Studies on Intelligent Systems for Automation, pp. 1–9 (2015)
9. Tran, H., Suzuki, T.: An experimental acoustic SFBC-OFDM for underwater communication. In: International Conference on Advanced Technologies, pp. 1–5 (2017)
10. Hachioji-shi.: Method of detecting a stray diver using underwater ultrasonic-band multicast communication. In: 2016 IEEE Region 10 Conference(TENCON) (2016)

Seamless Communication Models for Enhanced Performance in Tunnel Based High Speed Trains

S. Priyanka[1(✉)], S. Leones Sherwin Vimalraj[1], and J. Lydia[2]

[1] Panimalar Engineering College, Chennai, India
saravananpriyanka95@gmail.com,
leonessherwin@gmail.com

[2] Easwari Engineering College, Chennai, India
lydia_822@yahoo.co.in

Abstract. Wireless Communication on Train (WCT) is being followed in urban railways around the world to enhance the railway network efficiency, safety, and capacity which are mainly carried in high speed trains and underground tunnels. In this work, various models have been compared to enhance performance. A finite state markov (FSMC) model is used for low handover latency and high data throughput. Frequent handovers is overcome by LTE and GSM-R based solution. **Network mobility (NEMO), CDMA and** MIMO combine with carrier aggregation to give high throughputs. This concept reduces, the cost of base stations and antenna by the moving cell concept, FSO and PTC concept.

Keywords: Wireless Communication on Train (WCT) · High data throughputs · Positive Train Control (PTC) · High Speed Train (HST) · Underground tunnels

1 Introduction

For an existing rural traffic pressure, a rural rail transit system is been developing around the world for improving capacity and efficiency for the increasing demand. It is train control model (automatic) which make use of train-ground wireless communication. To enhance the safety and service for customers in order to develop and use the railway network [1]. WCT has a interlockings, track circuits and signals Fig. 1. It is mainly used in tunnels which are found in underground where there is a more scattering, reflections and barriers are severely damages the propagation performance. WLAN is used as main method for train-ground communications. The movement of trains is said to be fast which will cause repeated handoffs among WLAN (APs). access points for scheming the wireless networks it is important to modelling the channel. The Pathloss model of tunnel channel which explain about the characteristics of large-scale fading. A two-layer multi FSMC for modelling 1.8 GHz narrowband channel.

The features of FSMC model for a tuned channels in high speed train control based systems:

D. J. Hemanth et al. (Eds.): COMET 2019, LNDECT 35, pp. 1580–1591, 2020.
https://doi.org/10.1007/978-3-030-32150-5_159

1. Measurements are on the basis of real field channel
2. It takes the train location into account to produce more precise channel mode
3. It is fixed in each of the interval at a space between transmitter and receiver
4. Signal to noise ratio is determine by Lloyd-max technique
5. Simulation results obtained has more accuracy by generated from the model and real field measurements.

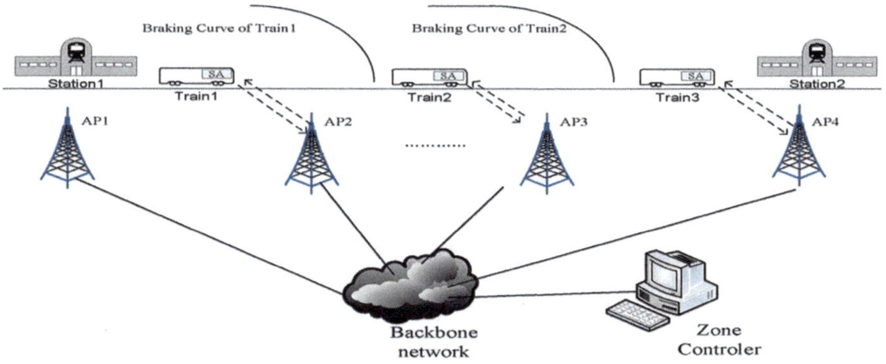

Fig. 1. High speed train system model [1]

1.1 LTE-A System Overview

The LTE of 3GPP for wireless accessing for improving the power spectral efficiency with high data rates. LTE will provide increased quality and cost effective multimedia access. The LTE has maximum throughput at 350 km per hr and even 500 km per hr in urban areas. LTE can elevate 200 users of 5 MHz of BW which will additionally enhance in LTE-A [2].

A HST is said to be a challenging surrounding for LTE networks. Initially, the channel condition are been changing drastically, hence decreasing the data rate and handover more repeated, hence interference is occured. Together they are more undesirable for communications which gives the hard requirement on BW and continuity.

LTE based results provides high throughput data and seamless multimedia service for HST passengers wirelessly Fig. 2. Cell Array helps to predict the future LTE cells to be serviced, and enabling a handover that would not disrupt multimedia communications. In this work, the signalling techniques and actions in of predictive handover is discussed. To put up on the intense channel variations, propose a schedule and resource allocation mechanism (RAM) to increase the service rate for femto cell BS, accordingly the signal quality changes. Thus it increases the quality of the signals.

LTE Objective. It has the objective of WCT measurements for WLAN propagation in tunnels and subway lines to create a FSMC models, the equipment measurement consists of two parts as follows:

(1) It is said to know that the configuration of the measurements is the similar as subway lines, including the option of antennas, the location and settings of the transmitter and the receiver.
(2) It is said to increase measurement method to plot channel data, and it includes the strength of signal and SNR, to the position of receiver, that takes train locations in order to have a more precise channel model.

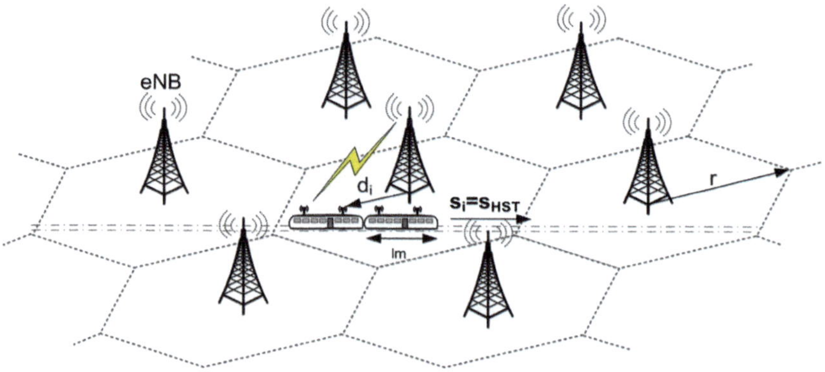

Fig. 2. LTE-A system architecture [2]

1.2 High Speed Communication Architecture

In HST communication architecture, wireless communications between each mobile station (MS) on the HST and the wayside APs are adopted instead of the block circuit. The railway path line is separated into regions or areas [3]. All the area is under control of a zone controller (ZC) and which it has its own radio transmission system. Identity, area, path, and speed are transmitting to the ZC.

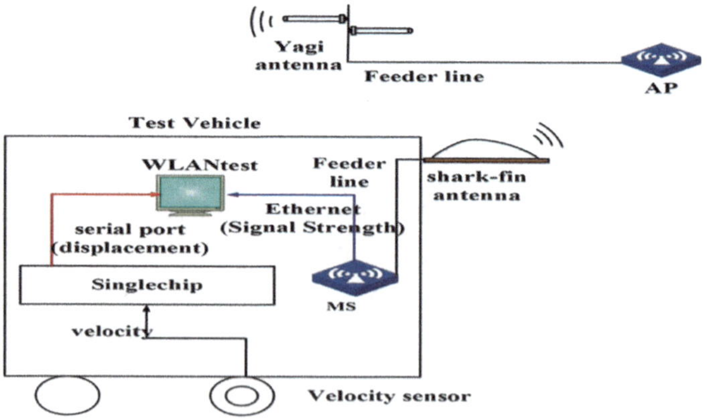

Fig. 3. Equipment setup in high speed train [3]

The train and the ZC link should be nonstop so that the ZC can identify the locations of all the trains. The ZC transmits to the location of the train in front of it and provides it a braking curve to stop train. It is said to be that both travel closely together. When a HST travels further and enter the coverage of an AP, leading to handoff outcome in communication split and lengthy latency Fig. 3. In WCT systems, it is guranteed that the operation is safety and efficiency. Wireless channels in WCT are unique from others, since it is found in underground tunnels, where there are more amounts of scattering effects, reflections, and barriers that severely degrade the propagation performance of wireless communications Fig. 4.

1.3 Design Specifications

Frequency	2.412 GHz	
Transmitting antenna	30 dbm	
Transmitting antenna	Type	Yagi antenna
	Polarization direction	Vertical
	Gain	13.5 dbi
	Hpbw	30°
Receiving antenna	Type	Shark fin antenna
	Polarization direction	Vertical
	Gain	10 dbi
	Hpbw	40°

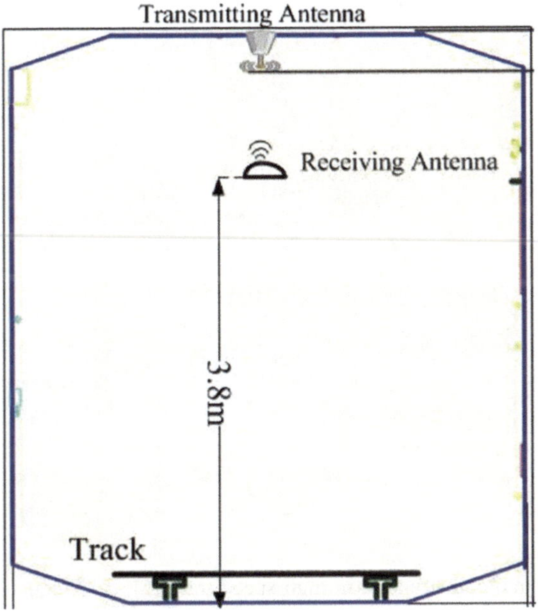

Fig. 4. Antenna setup on a train in a tunnel [4]

2 Related Works

Communications in satellite were used for wireless access to vehicles traveling across the world. The satellite service would be unconnected in tunnels or terminals. W-LAN else Distributed Antenna Systems in no LoS, places to increase connectivity were Solutions that adaptively switch to proposed.

There are suggestions to increase the latency of HST are presented with a rapid review LTE-A and WiMAX networks. A new concept RoF for HST wireless access at 60 GHz [4].

At 60 GHz signal it losses the cell coverage which reduces the speed of the coverage. It is operated in the wide band frequency of LTE and it combining soft handover and hard handovers, and information known about HST to enhance the handover experience for HST passengers. The result is based on LTE A networks and providing

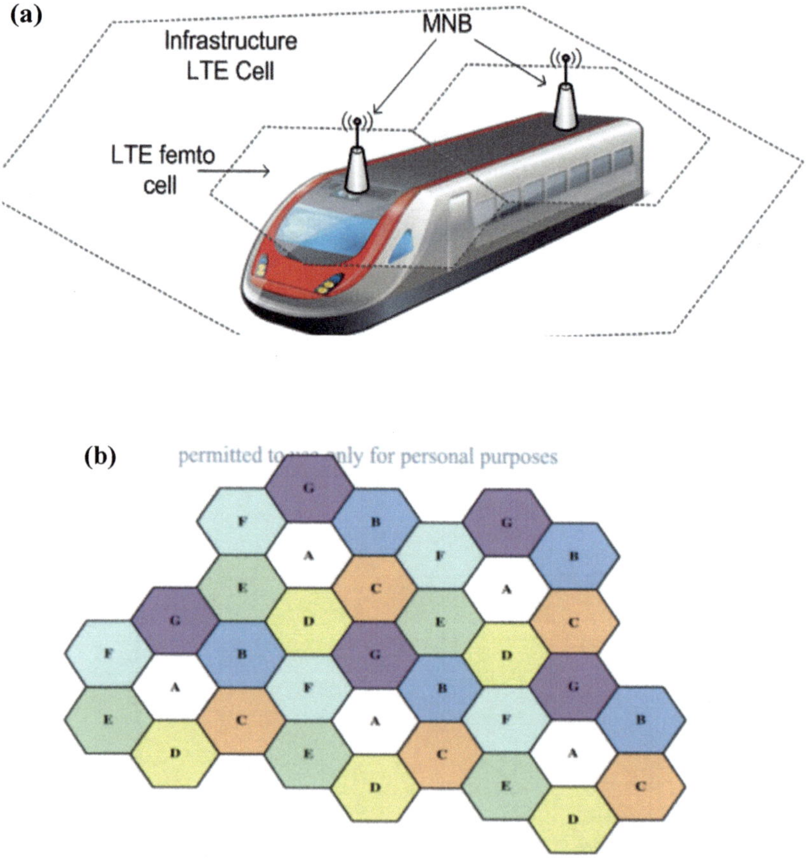

Fig. 5. (a) Femto cell structure in high speed train [5], (b) Frequency reuse cluster

a detailed study on vehicle communication solution to improve the HST user connectivity. The concept used here is moving femto cell based on LTE and when the LTE cell moves in the direction of the femto cells seamless handovers is provided by the cell array architecture (Fig. 5).

2.1 Moving Femtocell

It is a minute cellular structure BS which is used for minimum coverage distance of 10 to 40 m. The femto cells are used inorder to increase wireless service coverage to the domestic or for office usage. The length of a HST cell l = 25 m fits perfect within an LTE femto cell's coverage range. It provide vehicle-to-infrastructure communication within a HST cabin. A femtocell BS is said to be a Moving eNB (MNB) [4]. The HST passengers can admit LTE services through their local base station by using femtocells. Both the LTE femto and macro cell uses the similar frequencies. The allocated number of frequency subcarriers can be decided on the basis of the number of the passengers within the femtocell. In this process, each of the femto cell cover 50% of a HST cabin and maximum of 50 users. It is found that in each femtocell it supports upto a frequency band of 5 MHz.

2.2 Cell Array of the High Speed Communication Trains

The Cell Array of HST which has the cell architecture of three cluster of cells beside the lane of HST. They are named as cell A, cell B and cell C [5]. The first cell consists of partial or complete current train. The Second cell is ahead of the cell in given time. The second cell is said to have the whole process hence first cell is not needed. It is said that first cell is needed until it enters and leaves the cell. The process is done as the train moves. When the cell is said to be to reach the fourth cell it deletes the first cell and renames the cells. The handover process depends on the cell array structure and speed which is known. It is said to have fast handover with minimum data rate. The first cell and second cell are of transmitting base station. It is same as of the frequency reuse concept which is of the frequency reuse cluster format. The third cell helps in frequency spectrum assignment of HST in a LTE for service degradation (Fig. 6).

2.3 Predictive Handover Mechanisms

The predictive handovers mechanism are of two different types for reconfiguration of the cell array and performance of the hard handover for MNB beside the railway track [6]. It promotes from the structural design cell array to have minimum handover time and in high mobility conditions the service must be uninterrupted.

Hard Handover. It is the method in which the user equipment which is the MNB of moving femtocell and switching to a new base station. Previous handover of the MNB will not be known in future. It is not predictable because it predicts 2 cells and summing is performed. Hence before the handover there is no registration of base station.

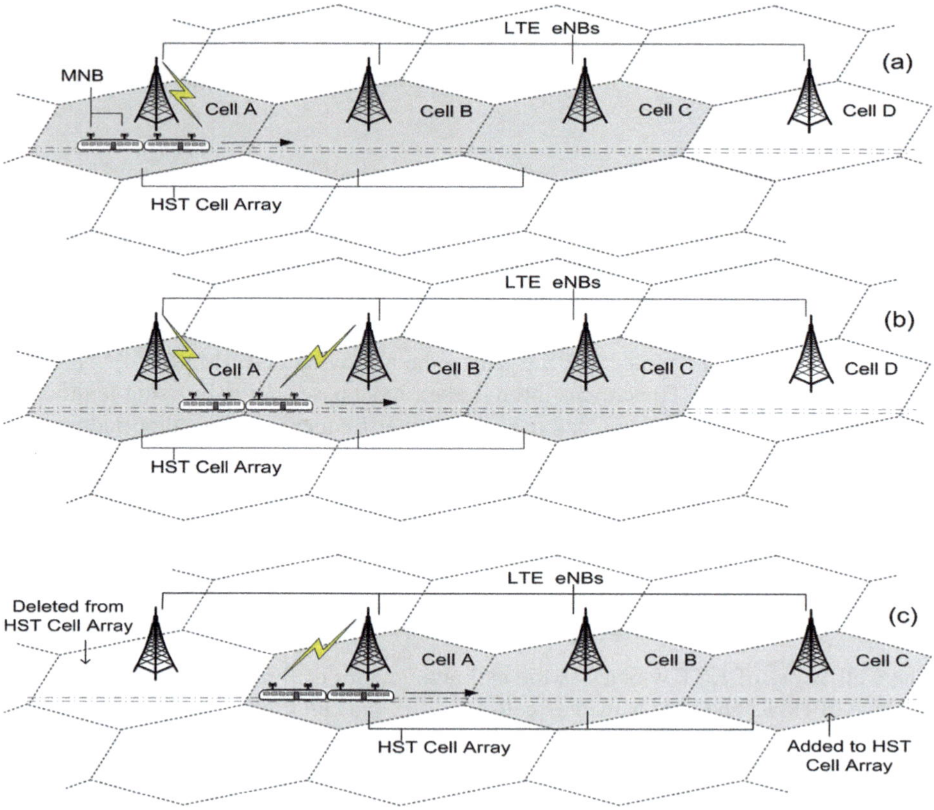

Fig. 6. High speed train cell array path [6]

Soft Handover. In hard handover cell array which has the known data of HST base station and then cell change is done which uses a predictable process to known user data in a cell array. The base station femto cells and infrastructure LTE cells has the same frequency. If it is not selected properly, the interference occurs between the femto cell and LTE cells (Fig. 7).

2.4 Multiple Egress Network Interface

The problem of mass transfer occurs when mobile devices are transferred from one ground station (BS) to another BS. This creates a signaling message storm and generates a large number of processing requirements on the infrastructure's wireless connection and the network components [7]. The concept of vehicle to user communication is been proposed by network mobility concept. The packets are found to be transmitted over internet and processed by the routers. It changes the mobility over inter as one entity (Fig. 8).

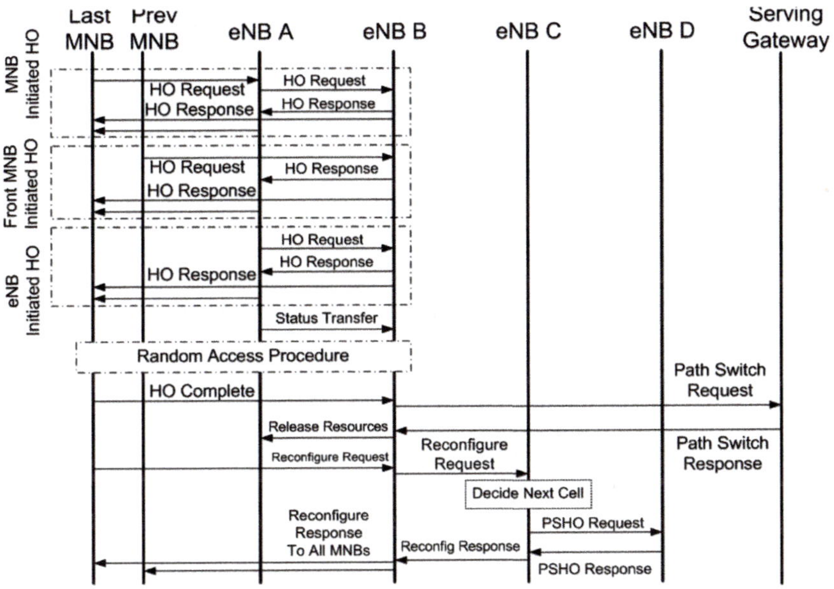

Fig. 7. Handover mechanism high speed train [7]

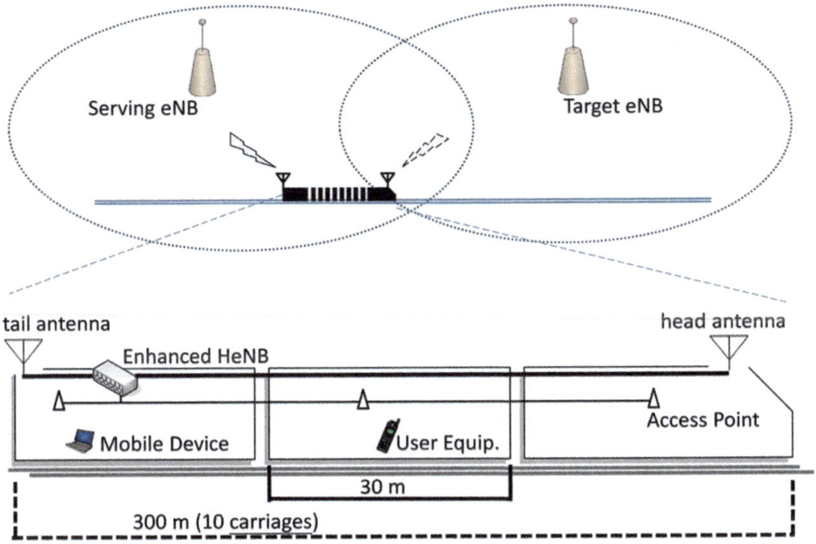

Fig. 8. Multiple egress network model for high speed train [8]

2.5 Measurement Scenarios and Description of Broadband Communication

The speed of HST is approximately said to be 300 km/h. The antennas are said to be fixed on the top surface of the train, nearly half way between the front and the rear [8]. There are 3 scenarios were to be calculated:

1. Set-up 1: The eNB is placed 1.5 km from railway and all transmitter antennas are pointing at a 90° angle to the railway.
2. Set-up 2a: It is found that eNB is placed next to railway track and 50% of transmitter antennas pointing at one direction, and the other antenna are pointing at reverse direction.
3. Set-up 2b: It is similar as previous set-up, but here the entire four transmitter antennas are of in the similar process.

In the above scenarios the BS height is approximately 12 m.

2.6 PTC

To share the railway information among the worker vehicles, multiple trains and other entities. This concept is named as (PTC) Positive Train Control. The major problem is bandwidth insufficiency for 220 MHz is overcomed by PTC packet formats [9]. It sends number of packets where overlapping is avoided it can reduce doppler effect at a speed of 400 mph which as a coverage of 600 m/s Fig. 9.

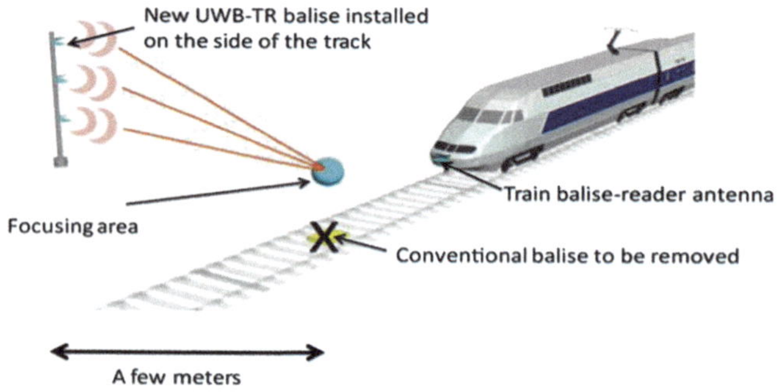

Fig. 9. PTC system model [9]

2.7 Unmanned Aerial Vehicles

A Wireless communication which provides broadband internet access to HST users in the Cyber physical systems It reduces the cost of base stations and antenna [10], it has the same moving cell concept. It has the data rate of 10–274 Mbps in military and 1 0Mbps in civil and has speed of 30–240 km/h. To realize the broadband communications of HST mobile users and zero accident of transportation, the proposed system solution is found in future CPS systems Fig. 10.

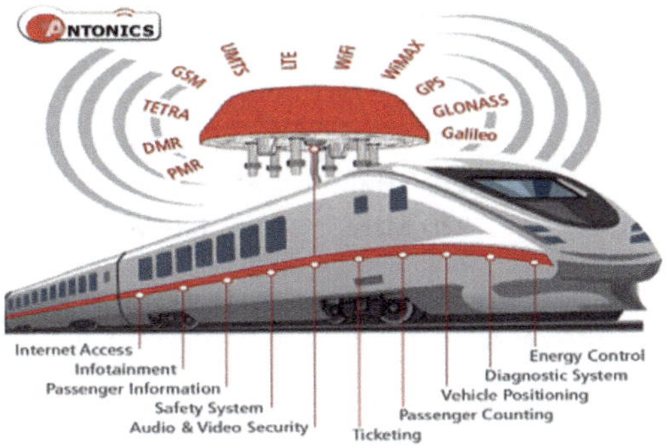

Fig. 10. UAV antenna structure [10]

3 Measurement Analysis of High Speed Train

3.1 Measurement of SNR of Threshold for Different Intervals

Distance between antenna	5 m	10 m	20 m	50 m	100 m
Range in meters	[95 100]	[90 100]	[80 100]	[50 100]	[0 100]
1st thresh.	24	22	22	22	22
2nd thresh.	27.98	26.90	27.73	29.53	33.22
3rd thresh.	32.03	31.44	32.89	35.39	44.01
4th thresh.	36.31	36.02	38.14	41.22	57.00
5th thresh.	41	41	44	48	78

The threshold of SNR for different intervals at a location of 100 m it gives with more accuracy by using on the FSMC model. It is said that they are done in four and eight states to study the accuracy of the proposed model. The effects of the model are of the intervals 5, 10, 20, 50, and 100 m. The exactness is verified through another measurement data. The above table is describing the threshold of SNR of four and eight levels. Since the distance intervals are different and the range of the SNR is same, and it provide different thresholds with more accuracy.

3.2 Comparison Table for HST Wireless Communication System

S. no	Concepts	Frequency	Speed	Coverage	Data rate
1. [5]	FSO	300 GHz	130 km/hr	700 m	566 Mbps
2. [8]	MIMO	800 MHz	300 km/hr	600 m	150 Mbps
3. [1]	FSMC	1.8 GHz	350 km/hr	500 m	100 Mbps
4. [7]	MEN-NEMO	2.4 GHz	250–360 km/hr	600 m	300 Mbps
5. [4]	LTE-A	5 MHz	350 km/hr	600 m	100 Mbps
6. [2]	GSM-R	2.4 GHz	360 km/hr	600 m	100 Mbps
7. [6]	FREQUENCY REUSE	5 MHz	500 km/hr	500 m	28 Mbps
8. [10]	CDMA	30–300 MHz	30–440 km/hr	600 m	10 Mbps
9. [3]	W-LAN	2.4 GHz	350 km/hr	600 m	100 Mbps
10. [9]	PTC	220 MHz	50–400 Mph	600 m	100 ps

3.3 Tradeoff

It is inferred that the FSO concept has the highest data rate coverage where the speed is found to be decreased. The MIMO concept has the increased frequency but the data coverage is found to be low data rate with increased speed. MEN-NEMO [11], W-LAN, GSM-R which has the standard frequency of 2.4 GHz and has a standard data rate of 100 Mbps except 300 Mbps in NEMO concept. LTE-A, FREQUENCY REUSE concept has 5 MHz with lesser data rate below 100. Thus it is concluded that the MIMO concept shows more better performance [12]. FSMC model is the lowest frequency used where the data rate is found to be low with increased speed [13, 14].

4 Conclusion

Modeling the tunnel wireless channels of HST is significant in proposed wireless networks… The major problem is bandwidth insufficiency for 220 MHz is overcomed by PTC packet formats. A Wireless communication which provides broadband internet is the concept. In this work, various models are been compared due to the enhanced performance. A tuned channel finite state markov (FSMC) model is used for low handover latency and high data throughput. Frequent handovers is overcomed by LTE and GSM-R based solution. **Network mobility (NEMO), CDMA and** MIMO combines with carrier aggregation to give high throughputs. This concept reduces, the cost of base stations and antenna moving cell concept, FSO and PTC concept. Thus it is concluded that the MIMO concept shows more better performance. The results implies that it provides minimum handover latency, maximum throughput, and minimum delay compared to the previous works.

References

1. Wang, H., Zhu, L., Yu, F.R., Tang, T., Ning, B.: Finite-state markov modeling for wireless channels in tunnel communication-based train control systems. IEEE Trans. Wirel. Commun. **15**(3), 1083–1090 (2014)
2. Ai, B., Cheng, X., Kürner, T., Zhong, Z.D., He, R.S., Xiong, L., Matolak, D.W., Michelson, D.G.: Challenges toward wireless communication for high speed railway. IEEE Trans. Wirel. Commun. **15**(5), 2143–2158 (2014)
3. Wang, H., Yu, F.R., Zhu, L., Tang, T., Ning, B.: Finite-state Markov modeling of tunnel channels in communication-based train control (CBTC). IEEE Railway IEEE Trans. Wirel. Commun. **15**(3), 1083–1090 (2014)
4. Karimi, O.B., Liu, J., Wang, C.: Seamless wireless connectivity for multimedia services in high speed trains. IEEE Railway IEEE Trans. Wirel. Commun. **30**(4), 729–739 (2012)
5. Taheri, M., Ansari, N., Feng, J., Rojas-Cessa, R., Zhou, M.: Provisioning internet access using FSO in high-speed rail networks. IEEE Railway IEEE Trans. Wirel. Commun. **10**(2), 96–101 (2010)
6. Sarkar, M.K., Ahmed, G.M.F., Uddin, A.T.M.J., Hena, M.H., Rahman, M.A., Kabiraj, R.: Wireless cellular network for high speed (upto 500 km/h) vehicles. IOSR J. Electron. Commun. Eng. **9**(1), 1–9 (2014)
7. Lee, C.W., Chuang, M.C., Chen, M.C., Sun, Y.S.: Seamless handover for high-speed trains using femtocell-based multiple egress network interfaces. IEEE Trans. Wirel. Commun. **13** (12), 6619–6628 (2014)
8. Kaltenberger, F., Byiringiro, A., Arvanitakis, G., Ghaddab, R., Nussbaum, D., Knopp, R., Bernineau, M., Cocheril, Y., Philippe, H., Simon, E.: Broadband wireless channel measurements for high speed trains. In: EURECOM, Sophia Antipolis, France yIFSTTAR, COSYS, LEOST, Villeneuve D'Ascq, France zSNCF, Innovation and Recherche, Paris, France xIEMN laboratory, University of Lille 1, France (2014)
9. Bandara, D., Abadie, A., Melangno, T., Wijesekara, D.: Providing wireless bandwidth for high speed rail operations. George Mason University, 4400 University Drive, Fairfax, VA, 22030, USA, CENTERIS (2014)
10. Zhou, Y.: Future Communication Model for High speed Railway Based on Unmanned Aerial. School of Electronics and Information Engineering, Beijing Jiaotong University (2010)
11. Ma, C., Mao, B., Bai, Y., Zhang, S., Zhang, T.: Study on simulation algorithm of high-speed train cruising movement. In: 2017 10th International Conference on Intelligent Computation Technology and Automation (ICICTA). IEEE (2017)
12. Jalili, L., Parichehreh, A., Alfredsson, S., Garcia, J., Brunstrom, A.: Efficient traffic offloading for seamless connectivity in 5G networks onboard high speed trains. In: 2017 IEEE 28th Annual International Symposium on Personal, Indoor and Mobile Radio Communications (PIMRC) (2017)
13. Standard for Communications-based Train control (CBTC): Performance and Functional requirements. In: 2017 IEEE 28th Annual International Symposium on Personal, Indoor and Mobile RadioCommunications (PIMRC), IEEE Std 1474.1-2004 (Revision of IEEE Std 1474.1-1999), 0_1-45 (2004)
14. Wang, H.S., Moayeri, N.: Finite-state Markov model for radio communication channels. IEEE Trans. Veh. Tech. **53**(5), 1491–1501 (2004)

Femto Cells for Improving the Performance of Indoor Users in LTE-A Heterogeneous Network

M. Messiah Josephine(✉) and A. Ameelia Roseline

Panimalar Engineering College, Chennai, India
messiahjosephine@gmail.com,
ammeeliaroseline@gmail.com

Abstract. LTE-A (Long Term Evolution Advanced) as a transition from the GSM (global system for mobile) to LTE-A. Where LTE-A provides higher bandwidth than GSM and its compatibility to the modern mobile communication system. The objective of the paper is to examine the importance of the femto cell component of the existing LTE-A technology with the current femto cellular technology by adding a femto cell that gives us better signal strength in the indoor environment. It has become a common approach in the current cellular networks to improve capacity and coverage. As a result there is a continued demand for capacity in ultra dense areas, thus by increased coverage it has gained the importance in small cell research to minimize the impact of the living world. Femto cells for improving the performance of indoor users in LTE-A heterogeneous network provides an overall description of LTE-A technology during the exploit of femto cells to improve the indoor signal strength.

Keywords: LTE-A · MIMO · FEMTO CELLS · Heterogeneous networks · Base station · RF receiver first section

1 Introduction

1.1 LTE-A Overview

LTE-A means that long - term evolution, that began in 2004 as a result of the 3GPP partnership project. The word LTE-A (system architecture evolution) symbolizes both LTE and SAE. SAE is the resulting growth of the GPRS/3 G packet core network progress. The first edition of future progression was documented in 3GPP provision. LTE an ancient evolved 3GPP system called UMTS, which evolved from GSM (Global Mobile Communications System). A fast boost within the use of mobile knowledge and also the emergence of recent applications like Multimedia on-line Gaming, mobile TV, Web 2.0, streaming content material prompted the 3rd era partnership challenge 3gpp to paintings on the long run evolution LTE direction towards the direction of cell fourth generation. The main intention of LTE-A is to offer excessive bandwidth, low latency and provides a elevated data rate. The network architecture was designed to support packet - switched traffic with seamless mobility and high quality of service.

D. J. Hemanth et al. (Eds.): COMET 2019, LNDECT 35, pp. 1592–1602, 2020.
https://doi.org/10.1007/978-3-030-32150-5_160

1.2 LTE-A Tools

LTE-A is consists of many new technologies. The expected designers technologies are explained as follows (Fig. 1):

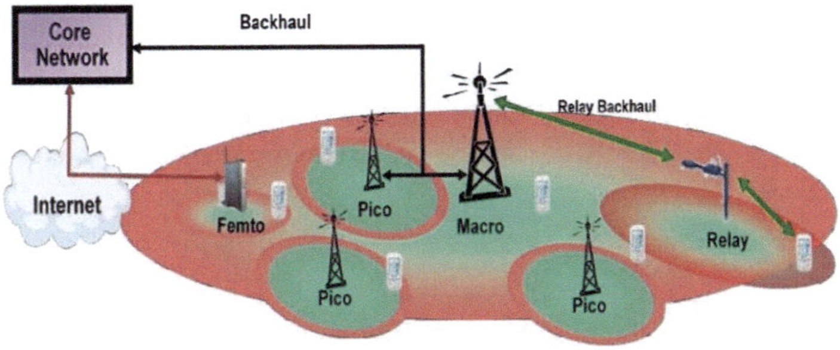

Fig. 1. LTE-A heterogeneous network [22]

1.3 OFDM (Orthogonal Frequency Division Multiplex) [Poole07b]

LTE uses the downlink OFDM i.e., To overcome the problem in multipath available in universal mobile telecommunication system terminal from the ground station, transmit data over many 180 kHz narrowband careers Instead of spreading a signal over the entire Bandwidth of 5 MHz career i.e. To carry data, OFDM is using a massive number of narrow subcarriers for multi-provider transmission. OFDM could be used as Digital multi-carrier modulation technique. The LTE meets the OFDM requirement for flexibility in the spectrum and enables cost effective solutions for very large High peak data rate.

1.4 MIMO in LTE-A

Usage of MIMO increases the efficiency. MIMO consists of too many antennas which will radiate in different paths so it can transmit many number of data's without interference and their won't be any distortion. Using MIMO with LTE A technology the u plink and down link users differs and the antenna usage is reduced therefore it is cost effective. Figure 2 shows the basic MIMO principle referring to the base station.

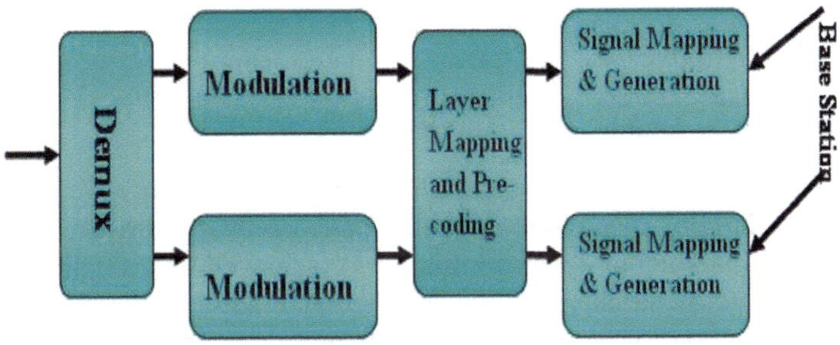

Fig. 2. 2 × 2 antennas at base station in MIMO [2]

1.5 Femto Cells

Femto cells are small base stations that can be used in housing or trade atmosphere to enhance cellular coverage within a building. The technology also means that operators can provide businesses and high-value customers with better service. While a macro cell provides coverage in a wide area, a femto cell provides a lot of capacity in a location, Just delivering power to the user's location. Because they are closer to users, providing a high bandwidth connection requires less RF power – a typical femto cell can have a section of 20 mW RF and consume 2 W in total. Lower RF power also locates signals so that spectrum can be reused more frequently than a macro cell network (Fig. 3).

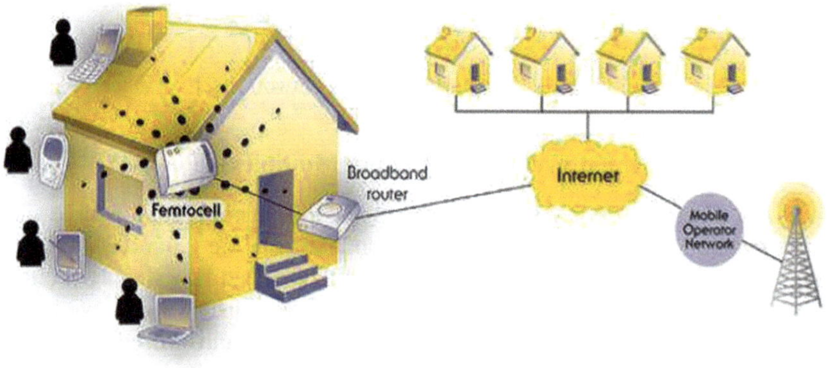

Fig. 3. Femto cell network structure [3]

1.6 LTE-A Advanced Femtocell Need

The reduction in coverage causes buffering while streaming online videos. And there is a requirement of large bandwidth. Thus by adding femto cell in LTE-A it will provide larger coverage in large buildings. The received signal from the base station is been blocked by the wall inside the house or in an isolated areas so there is a lack of coverage by adding femto cell in LTE-A there is a huge improvement in coverage.

2 Related Works

2.1 Femto Cells in Indoor Environment

In this paper [1] they discussed about LTE-A femto cells to improve the indoor coverage and off-load mobile data indoors on the macro cellular network in this paper [1] mainly says about the indoor coverage this can be achieved by designing and modeled using a OPNET and by simulation software. the end result of paper [1] says about the performance throughput of femto cell in uplink and downlink performance (Tables 1 and 2).

Table 1. Femto cell throughput performance

Throughput of a femto cell	Average	Peak
Downlink (Mbps)	24,436	47,299
Uplink (Mbps)	14,090	17,665

Table 2. Femto and macro comparison of throughput

Data throughput statistic	Femto cell	Macro cell	Gain
Downlink	24,436	47,299	40,98
Uplink	14,090	17,665	58,57

2.2 LTE-A Femto Cell 3Gpp Pros and Cons

This paper [2] implements the indoor coverage using LTE-A femto cell for call quality and data rate. The important way to increase network capacity in a cellular network is to increase the cell capacity. They discussed in paper [2] in the upcoming generation femto cells can be used in aircrafts, train. Next step is that femto cells need to interact with the everyday devices. In this paper [2] they used micro cell, macro cell, femto cell for large network coverage. PROS-in regarding Femto cell are better coverage and capacity, reduced subscriber turnover, improved macro, cell reliability, cost benefits, easy end user installation, low impact low RF power. CONS regarding femto cell are, synchronization, quality of services, interference, cell association of biasing, mobility and soft handover.

2.3 Femto Cell Network in Dense Area, Spreading Power by Increasing Energy Efficiency

In paper [10] they find that relationships between various cell sizes on system capacity and energy saving of UEs and BSs. In the paper [11] they proposed an system for efficient utilization of green energy in Het Nets with UE association. In the paper [12] they discussed the activates and deactivates a group of FBs which is related to enthusiastically varying traffic load when satisfying the rate requirement of all associated UEs. In [13] the faced a problem regarding sub channel and power allocation with a non-cooperative game to increase the system EE (Fig. 4).

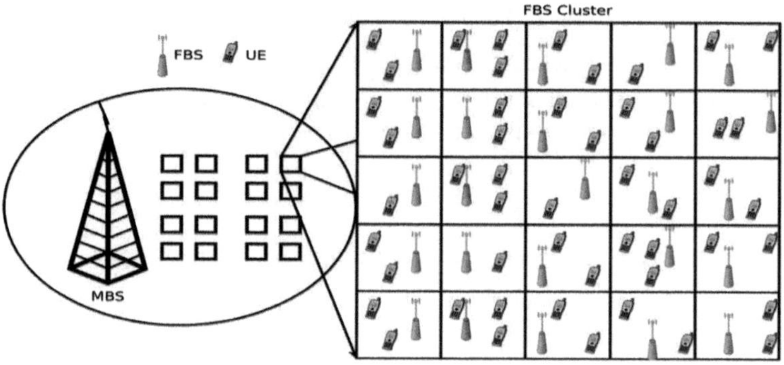

Fig. 4. Indoor femto network [2]

2.4 LTE-A Development in Femto Cell

Femto cell has been introduced for the demand for indoor and outdoor network coverage. Wherein the development of LTE-A femtocell can introduce for the improvement of the high quality of voice and data in the mobile network for indoor systems. In this paper [4] we discussed the major factor to optimize the coverage planning for LTE-A in femtocell development. In this paper [4] they concluded that both femtocell and macrocell can be effectively able to use in radio resources. Where the femtocell can be more complement in the indoor coverage for the macrocell. The parameters of the macro cell and femto cell can be explained in Table 3 (Table 4).

Table 3. Parameters of a macro base station

Parameter	Value
Frequency operation	2100 MHz
Power Transmi	41–43 dBm
Bandwidth	15 MHz
Transmit Antenna Height	30 m
Coverage	1000 m

Table 4. Parameters of a femto cell

Parameter	Value
Operation frequency	2100 MHz
Power transmit	13 dBm
Bandwidth	14 MHz
Transmit antenna height	3 m
Coverage	25 m

2.5 Reliability and High Throughput in the Analysis of FEMTO CELL

The femto cell is aimed at providing coverage area in an indoor area such as inside the home, building. In this paper [5] it is designed in such a way that it can be used in mobile network telephones through broadband connections including cable for the process of communication. They improved the quality of service in the analysis of femto cell [4] and provided better coverage and capacity. The advance of cellular network from legacy standard 2G to 4G LTE-A benefits/challenges of a femto cell are analyzed. LTE-A provide data rate beyond 100 Mbps.

2.6 A View of Advanced Communication in LTE-A Using Femto Cells

The aim of the paper [6] was to view the importance of femto cells in the rising LTE-A-advanced technology, with the important plan of the forthcoming femto-cellular technology. LTE-A-advanced is improved from LTE-A communication for better mobile coverage. Where femto cells are introduced in LTE-A in small mobile telecommunication base stations that can be done under a business atmosphere for better mobile coverage in wireless devices. This paper [6] mainly discussed about the information and communications technology and cumulative continuously day to it is due to the adding of base stations of 4G and LTE-A in a wireless communication network and it may also result in a power management problem in the upcoming generation. According to the Cisco Visual Networking Index: Global Mobile Data Traffic Forecast Update, for 2012-2017, "the overall mobile data traffic is expected to grow to 11.2 Exabyte's (EB) per year by 2017. to avoid this problem femto cell is possible a solution, it will in great result in improving signal quality and in cost savings (Fig. 5).

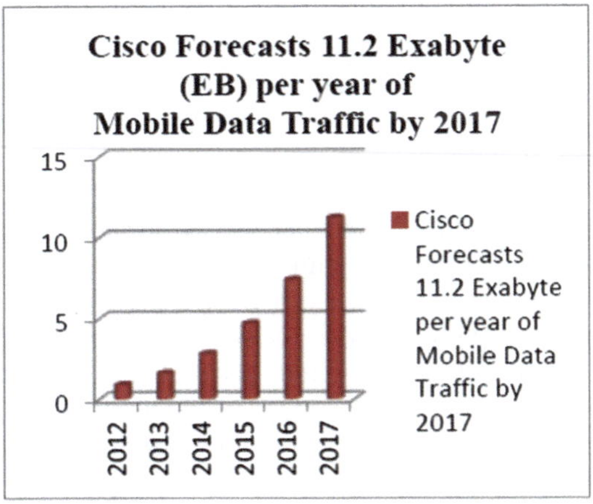

Fig. 5. Cisco forecasts 11.2 exabyte mobile data traffic per year [6]

2.7 A 65-Nm CMOS 2 × 2 MIMO Multi-band LTE-A RF Transceiver for Small Cell Base Stations

This paper [6] says about a 680 MHz–6 GHz 2*2 MIMO LTE-A RF the transceiver in 65-nm CMOS for low cost and multiple capable femto cell bases station. This paper [6] can be also achieved by −51 dBc TX ACLR and 1.68% TX EV MAT 20-dBm output power in the LTE-A 10-MHz mode with the 2*2 MIMO configuration, without applying DPD digital pre-distortion technique. the paper [6] presented RF transceiver that is placed in the femtocell base station, using a commercial PA and modem, that indicates several margins in the 3GPP radio conformance test (Table 5).

Table 5. Femto cell performance summary

Parameters (3GPP TS 36.104, Rel.12)	Unit	Specification	Femtocell (LIE Band5)
RX sensitivity	dBm	−93.5	−100
In-Channel Sel. (wanted: −90.5 dBm)	dB	21.5 (interferer: −63 dBm)	30 (interferer: −60.5 dBm)
Adjacent-Channel Sel. (wanted; −71.5 dBm)	dB	43.5 (interferer: −23 dBm)	50.5 (interferer: −21 dBm)
Narrow-band blocking (wanted: −79.5 dBm)	dB	46.5 (interferer: −33 dBm)	60 (interferer: −19.F dBm)
In-band blocking (wanted: −79.5 dBm)	dB	52.5 (interferer: −27 dBm)	66.5 (interferer: −13 dBm)
Out-of-band Blocking (wanted: −79.5 dBm)	dB	64.5 (interferon: −15 dBm)	69.5 (interferer: −10 dBm)
Narrow-band Intermod. (wanted: −7ft.S dBm)	dB	43.5 (interferer: −36 dBm)	50.5 (interferer: −29 dBm)
TX output power	dBm	20 (SISO)	20 (MIMO)
TX ACLR	dBc	−45	−51.4
TX EVM @ 64 QAM	dB	−22	−35.4

2.8 Detection of Error Rate in Femto Cell

In this paper [7] they analyzed the sphere decoder error rate with reduced search space (SD-RSS) and suggested a formula to estimate the error rate. This is pointed out on the ultra-dense network with rigorous inter-cell interference, which improves the complexity of detection and complicates the assessment of error frequency. Simulations demonstrated the accuracy of the error rate performance estimate, and a numerical example was presented to show that the proposed adaptive SD-RSS can significantly reduce its average computational complexity while meeting the requirement for block error rate LTE-A. They thought in the paper [7] that the suggested error rate assessment for SD-RSS techniques could play a significant part in future low complex adaptive MIMO receivers with interference sensitivity (Fig. 6).

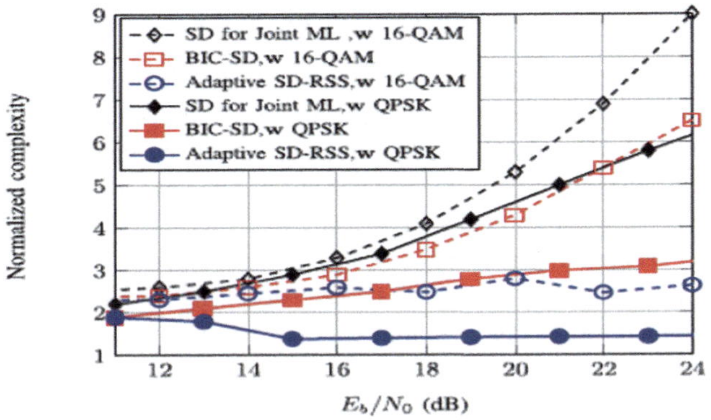

Fig. 6. Normalized complexity SD-RSS detection

2.9 Equivalent Measurement Method for System Information for 3GPP LTE-A Femto Cell

In this paper [8] we need to measure system information of target cells, where the target cell are closed subscriber group (CSG) cells. In this paper [8] they promised that femto cell cellular network technologies that enhance both the cell coverage and capacity. This paper [8] introduces six measurement methods of system information in the 3GPP LTE-A system. They also analyzed the performance of the proposed measurement method in view of service interruption time and measurement delay. In the paper [8] they found the independent and equivalent method that shows the least gaps that have small service interruption time. It shows the best performance in aspects of measurement delay as a result in the paper [8] the measurement gaps of these methods with a large gap are larger than these methods with several small gaps (Table 6).

Table 6. Simulation parameters [7]

Number of macro cell	3 eNBs
Number of HeNB	6
Number of UE	1
Inter-site distance	500 m
System frequency	2 GHz
System bandwidth	FDD: 10 + 10 MHz
Propagation loss, model	Inside the same cluster L = 127 + 30 $\log_{10}$ R For other link L = 128.1 + 37.6 $\log_{10}$ R
Shadowing model	Lognormal shadowing
Shadowing standard deviation	10 dB for Link between HeNB and HeNH UK 8 dB for other links
Penetration loss	Inside the same duster: 0 dB All other links: 20 dB

2.10 Femto Technologies for Providing New Services at Home

This paper [9] highlights the femto cell system designed by an NIT DOCOMO which gained a lot of attention among the user community due to its ability not only widen the coverage area indoors but also provide new services. They proposed a system that is friendly with the current FOMA network system and have developed the enhanced femto cell. the main disadvantages of [9] are that the size and power consumption remains the same as the current femto cell. The communication between the femto BTS and FOMA network is encrypted so that eavesdropping and falsification of data are prevented. the main advantage is that it can support high-speed Packet access (Table 7).

Table 7. Major related works in LTE-A heterogeneous network

Paper	Cell type	Cell size	Data rate
"Performance analysis of femto cell in an indoor cellular network" Amevi Acakpovi, HenrySewordor, Koudjo M. Koumadi [1]	Femtcell	10 m	14,090 Mbps (uplink)
	Macrcell	10 km	17,665 Mbps (uplink)
"3GPP LTE-A FEMTOCELL – PROS & CONS" Seema M Hanchate, Sulakshana Borsune, Shravani Shahapure [2]	Femtcell	10 m	Broad band connection
"Femto Cell Network In Dense Area, Spreading Power By Increasing Energy Efficiency" Sudeepta Mishra and Chebiyyam Siva Ram Murthy, [3]	Femtcell	50 m	40Mbps
"LTE-A-Advanced communication using in Femtocells Perspective". *bhupendra Kumar, dr. Ganesh Prasad & Madhuresh Kumar* [4]	Femto cell	15 m	High data rate
"Analysis of femto cell for better reliability and high throughput" N. Mudau, T. Shongwe and B.S. Paul [5]	Femto cell	10–15 m	384 kbits/s
"Femtocell: A survey on development in LTE-A network Izwahbinti Ismail*, Izwahbinti Ismail, Rhoma Erma bin Zaini [6]	Macro cell	1000 m	15 MHz (B.W)
	Femto cell	25 m	14 MHz (B.W)
"A 65-nm CMOS 2 × 2 MIMO Multi-Band LTE-A RF Transceiver for Small Cell Base Stations" Kyoohyun Lim, Sanghoon Lee, Yongha Lee, Byeongmoo Moon, Hwahyeong Shin [7]	Femto cell	<100 m	–
	Macro cell	8 to 30 km	–
"Parallel Measurement Method of System Information for 3GPP LTE-A Femtocell" choongheelee, jae-hyankum [9]	Macro cell	500 m	FDD: 10 + 10 MHz
"Femto technologies for providing new services at home" takes hiterayana, hidehikoohyane, goichisata, takaya, takimoto [10]	Femto cell (current)	184*135*40	Downlink: upto 3.6Mbit/s (HSDPA) Uplink: upto 384kbit/s
	Femto cell (Enhanced)	180*135*35	Downlink: upto 14 Mbit/s (HSDPA) Uplink: upto 5.7 Mbit/s (HSUPA)

The design consideration of various femto cell is shown in the above table. The maximum femto cell size is 993 m and data rate obtained for DL is 14 Mbps and UL is 5.7 Mbps shown in [9]. The minimum femto cell coverage is 10 m and data rate obtained for UL is 14 Mbps shown in [1, 2]. For higher coverage area it produces lower data rate but for lower coverage area it produces higher data rate.

3 Conclusion

This review paper gives a detailed overview of existing LTE-A with femto cell technology. We require to know about the development of LTE in FEMTO cell as a part of big small cell pictures. It has a place in the existing and future scope in a wireless network. Femtocell has improved the coverage and better capacity, more system reliability, a boost to subscribers confidence and cost reduction. In addition to that, the main drawback is that maybe in future FEMTO cells will not be given MORE attention. It will be successful in landlines away. Another essential key in FEMTOCELL is there will be no health effects from radio waves beneath the applicable limits to the wireless or cellular communication system.

References

1. Acakpovi, A., Sewordor, H.: Performance analysis of femtocell in an indoor cellular network. IRACST – Int. J. Comput. Netw. Wirel. Commun. (IJCNWC) **3**(3), 281–286 (2013). ISSN 2250-3501
2. Hanchate, S.M., Borsune, S., Shahapure, S.: 3GPP prons and cons. Int. J. Eng. Sci. Adv. Technol. (IJESAT) **2**(6), 1596–1602 (2015)
3. Mishra, S., Murthy, C.S.R.: Increasing energy efficiency via transmit power spreading in dense femto cell networks. IEEE Syst. J. **12**(1), 971–980 (2018)
4. Ismail, I., Zaini, R.E.: Femtocell: a survey on development in LTE-A network. ITMAR **1**, 134–146 (2014)
5. Mudau, N., Shongwe, T., Paul, B.S.: Analysis of femtocell for better reliability and high throughput, 05 September 2016
6. Kumar, B., Prasad, G., Kumar, M.: LTE-A-Advanced communi- cation using in Femtocells Perspective. Int. J. Eng. Comput. Sci. **4**(8) (2015). ISSN: 2319-7242
7. Lim, K., Lee, S., Lee, Y., Moon, B., Shin, H., Kang, K., Kim, S., Lee, J., Lee, H., Shim, H., Sung, C., Park, K., Lee, G., Kim, M., Park, S., Jung, H., Lim, Y., Song, C., Seong, J., Cho, H., Choi, J., Lee, J., Han, S.: A 65-nm CMOS 2 × 2 MIMO multi-band LTE-ARF transceiver for small cell base stations. IEEE J. Solid-State Circ. **53**(7) (2018)
8. Lai, I.-W., Wang, J.-M., Shih, J.-W., Chiueh, T.-D.: Adaptive MIMO detector using reduced search space and its error rate estimator in ultra dense network. IEEE Access **7**, 6774–6781 (2018)
9. Lee, C., Kim, J.: Parallel measurement method of systeminformation for 3GPP LTE-A femtocell (2011)
10. Terayana, T., Ohyane, H., Sato, G., Takimoto, T.: Femto technologies for providing new services at home (2011)
11. Wang, L., Zhang, Y., Wei, Z.: Mobility management schemes at radio network layer for LTE-A femtocells. In: Proceedings of VTC, Barcelona, Spain, pp. 1–5 (2011)

12. Hoydis, J., Debbah, M.: Green, cost-effective, flexible, small cell networks. IEEE Commun. Soc. MMTC **5**(5), 23–26 (2010)
13. Leem, H., Baek, S.Y., Sung, D.K.: The effects of cell size on energy saving, system capacity, and per-energy capacity. In: Proceedings of IEEE Wireless Communication and Networking Conference, pp. 1–6, April 2010
14. Wang, B., Kong, Q., Liu, W., Yang, L.: On efficient utilization of green energy in heterogeneous cellular networks. IEEE Syst. J. **PP**(99), 1–12 (2015)
15. Chung, Y.: Energy-saving transmission for green macrocell-small cell systems: a system-level perspective. IEEE Syst. J. **PP**(99), 1–11 (2015)
16. Chai, X., Zhang, Z., Long, K.: Joint spectrum-sharing and base station sleep model for improving energy efficiency of heterogeneous networks. IEEE Syst. J. **PP**(99), 1–11 (2015)
17. Kim, J., Jeon, W.S., Jeong, D.G.: Base station sleep management in open access femtocell networks. IEEE Trans. Veh. Technol. **65**(5), 3786–3791 (2015)
18. Mao, T., Feng, G., Liang, L., Qin, S., Wu, B.: Distributed energy efficient power control for macro-femto networks. IEEE Trans. Veh. Technol. **65**(2), 718–731 (2015)
19. Li, A., Liao, X., Gao, Z., Yang, Y.: A distributed energy-efficient algorithm for resource allocation in downlink femtocell networks. In: Proceedings of IEEE International Symposium Personal, Indoor, and Mobile Radio Communication, pp. 1169–1174, September 2014
20. Ren, Z., Chen, S., Hu, B., Ma, W.: Energy-efficient resource allocation in downlink OFDM wireless systems with proportional rate constraints. IEEE Trans. Veh. Technol. **63**(5), 2139–2150 (2014)
21. Li, G., et al.: Energy-efficient wireless communications: tutorial, survey, and open is- sues. IEEE Wirel. Commun. **18**(6), 28–35 (2011)
22. 3GPP LTE-A heterogeneous network, prashantpanigrahi, August 2012

Detection of Ransom Ware Virus Using Sandbox Technique

S. Divya(✉)

Computer Science and Engineering, Panimalar Engineering College, Chennai 600123, India
divyasettu1996@gmail.com

Abstract. Research establishment systems are at ceaseless danger of objective base digital assaults, for example, phishing, misuses, propelled dangers like advanced malware and multi day dangers. Such assaults are hazard to association for their privacy and respectability of information learning base to multi day assault or advance danger numerous security sellers have arrangements, Advance danger anticipation, for example, sandbox which sweep records coming to network joins equipment level inspection and OS-level sandboxing to keep contamination from the most adventures, malware, multi day and focused on assaults entering the systems These arrangements are for the most part dependent on cloud put together administrations or exceedingly costly with respect to start appliances dependent on arrangement, It's not the best practice to examine government information for zero-day assault on outsider cloud as documents are transferred and downloaded forward and backward to merchants cloud for sandboxing, this raise the issues of information security and trustworthiness as we are utilizing cloud administrations of private sellers. Consequently a financially savvy on cloud answer for multi day assault insurance for cutting edge danger avoidance.

Keywords: Cyber attack · Zero-day attack · Sandbox · Integrity · Private vendors

1 Introduction

The leading cause of mortality in Suicide by hanging. It is an act of killing oneself by hanging from an anchor point. It is most usually used suicide strategy which has high death rate. As indicated by World Health Organization in view of the audit of 56 nation's mortality data found that hanging is the most common method in many countries, out of which 59% are male suicides and 39% are female suicides. A major concern in India is suicide by young people in the prime of their life. According to report by National Health Profile out of 1,33,623 suicides recorded in 2016, 44593 (33.37%) of people were between the ages of 30 and 45 and 45606 were young adults. Suicide rates are rapidly increasing in young adults. Suicides occur mostly in Tech industries, colleges and prisons. There are many risk factors for suicide such as loneliness, hopelessness, depression, long non-ailing diseases, family problems, work pressure, love failure, peer pressure, financial problems, and psychiatricdisorders.

D. J. Hemanth et al. (Eds.): COMET 2019, LNDECT 35, pp. 1603–1611, 2020.
https://doi.org/10.1007/978-3-030-32150-5_161

To develop successful prevention efforts, it is necessary to describe the characteristics and behavioral patterns of the victim. Various methods have been proposed to reduce suicide attempts. One of the methods is the computerized speech and motion analysis. Facial emotions and speech provide many ways in detecting the suicidal thinking. It is through computerized analysis various researchers are able to find the difference in the suicidal victims. The frequency of speech in the suicidal person alters from low frequency to high frequency. The people with more depression and hopelessness produces reduced range of speech. Some researchers used the computerized analysis to monitor the facial emotions of victims to detect the suicidal thoughts. A specialized computer program is designed to analyze facial expressions and a device for measuring the body measurements such as conductance of skin and heart rate. The differences showed how people with suicidal thoughts reacted physically and emotionally to various videos in comparison to those who do not have. An automated cognitive behavioral therapy (ACBT) is developed to deliver course of talk therapy without the involvement of a person.

Suicide detection and monitoring system is proposed using the cloud computing which generates a computerized model of the person's emotion state. Cloud computing is a technology paradigm which provides more access to high level resource which can be provisioned with little effort. As the growth of the technology is increasing many video surveillance systems are used for monitoring the people which acts as an alternative to direct observation. These surveillance systems monitor the person's activities all over the day by placing multiple cameras and they are analyzed by the camera operators. After all surveillance system with busy staff can reduce the performance of the surveillance system. Terribly many suicides occur in various surveillance systems which raises the topic of viability of such vision-based systems.

Different clarifications have been characterized to recognize and prevent the suicide by hanging. In 1986, a researcher invented a system by placing multiple sensor strips in the bed and floor of the prisoner cell which operated on the principle of weight off. Whenever a person is not standing on the floor or lying on the bed would probably hang from an anchor point. When there is no weight on the floor the system triggers an alarm, which alerts the staff. Although this is the first kind of suicide alert system, it does not reduce the suicides as inmates would hang themselves by standing on the floor or by sitting. In 2005 a new suicide detection device was designed which the person would wear as an ear piece to monitor the pulse and oxygen level of the person. If the person's vital signs are indicated as out of range an alarm would go off and an emergency message will be sent. In another instance, special protective clothing like smock or safety blanket has been designed which is made up of high nylon fabric that is difficult to tear [1].

However, World Health organization designed 'bracelet for life' to record the psychological parameters which detects the person's vital signs and if they are out of range an alarm is triggered. A top door alarm is designed and it is triggered when the door is used as an anchor point for suicide. Even though various systems are designed to monitor and prevent committing the suicide, these systems in real time are more challenging. Genuinely these systems have various drawbacks as the system needs wearing of various equipments and they are the source of alarm even the equipment is removed.

As a greater number of newer technologies arises through, more suicide prevention and detection systems are established without requiring any cumbersome equipment to be worn. A 3D suicide detection system is designed using the 3D image recognition and random forest using low cost 3D camera [2]. The images are analyzed by creating depth in the images. 3D image recognition is one of the upcoming approaches which recognizes and regulate 3D object in the image. Various approaches like pattern matching, feature based geometric approaches are used for 3D object recognition. Pattern matching approach uses the information gathered from the captured or processed image. This approach does not consider the 3D geometric constraints of the object during matching, and typically they are not capable of handling feature based approaches. Feature based approaches are used when the objects in the image have many distinct features. Objects with clear edge and blob features are recognized successfully. This approach works efficiently by pre-capturing all views of the object which need to be recognized, various features are extracted from the recognized views and these highlights are coordinated with the scene achieving the geometric limitations. However, the above method does not work for all types of hanging attempts and it is suitable only for partial suspension.

2 Literature Survey

In the past various technologies has drawn much importance in the field of analyzing the videos. A different approach to identify the facial behavior Indicating suicidal ideation is defined by facial changes. Various facial expressions like smiling, emotion, and eye brow raising and motion behaviors are utilized. Facial descriptors like smiling indicated contraction of the orbicular isocular muscles which described significant difference in the face of suicidal and non-suicidal person. The outcomes demonstrate that proposed strategy of facial descriptors have high acknowledgment execution. In [3], a strategy to recognize suicidal tendency is proposed using simulated dataset by machine learning. A tree algorithm is utilized as a classifier to classify the suicidal tendency in youths. The proposed system provided a interview type of questions to persons and based on the assessment it will classify tendency accordingly. As a future work a new model need to be constructed by taking other suicidal actions into consideration.

A multi frame image portrayal [23] is characterized to beat the shortage of human pose estimation since the gesture is continuous. CNN and RNN structures are used to define the condition that the adjacent frames have related content. Compare to other art techniques this method attains lowest possibility of error. In future the results can be improved using more RNN modules. Various approaches are proposed to identify the suicidal ideation of the individual. Human action recognition plays a vital role in identifying an individual's actions using their movement in the videos. various methods have been proposed to identify the human action. In [2], Depth map strategy is proposed to analyze human actions and postures. It is used to for extracting the highlights from the human action by creating depth. Motion of body joints are extracted using the Body joint descriptor. Utilizing the CNN various combination of inputs are trained. Right action score is improved using various fusion score methodologies. Three datasets are used to evaluate the performance of the system which outperformed in various platforms and achieved state of art results.

A new temporal information network [3] is characterized to envision 3D positions of body joints. Single depth images are utilized for pose estimation using object recognition approach. Problem of per pixel classification is achieved from pose estimation method. Shape of the body head position and various parts of the body are estimated using the classifier the 3D positions of body joints. Different examinations are performed utilizing various datasets tip exhibit adequacy of the approach. However, for the for skeleton based recognition [4], a different plan utilizing the One-Dimensional Convolution Networks is proposed Which uses the Base net to extract features using various subnets. The analysis has been performed under different helping levels and the outcomes are acquired.

The new technique has high acknowledgment rate than the state-of-art results to recognize human action. The computational time can be lessened than the other pose recognition techniques. However, Self-Informed feature combination is [9] utilized acoustic modeling based on deep learning methods using auxiliary deep neural network (DNN) called a feature contribution network. Aspect level input is learnt by training various features generated by multiplying the input features and the gate output. A regularization method is utilized for FCN. Experimental examinations are performed and compared with AMN frameworks. Crombez et al. [10] proposed a novel approach for human pose estimation using visual tracking methodology that utilize the light field cameras to obtain rich information. Using the sub-aperture cameras best pairs can be selected that reduce the estimation errors. This method assessed the exactness of our approach with real tests utilizing a light-field camera before planar targets held by a mechanical controller for ground truth correlation. Yang and Tian defined a strategy for recognizing human action utilizing spatio temporal methods [11]. These methods are incorporated using the super location vector. The exploratory outcomes demonstrate that the method is computationally efficient and produces a superior performance. In future work computational time can be reduced using other SLDtechniques.

Deep learning plays an important role in recognizing and training the networks. Multiple strategies have been proposed which have advanced methodologies to train and perform functions on their own. A BAIPAS [12] system for training the data in the distributed platform is proposed. A data locality manager is utilized to train data and state the information of the servers. Already learned data is transferred to another server using shuffling method. The trials performed on various databases have demonstrated that the BIPAS provides various services for developing deep learning models. In future work accuracy of the model and performance of the platform can be increased. Based on the upcoming problems, a new technique to recognize actions in the videos [13] is identified. Super resolution based on CNN is utilized to evaluate the accuracy. The analyses are performed and results got that proposed obtained high PSNR and do not compete with recognition accuracy, based on temporal and spatial aspects. Later on work more models to obtain peak PSNR ratio.

Wang et al. [14] proposed a technique for recognizing the actions in the videos. A deep auto combination network is utilized to extract the features in videos containing short segments. The techniques are tested and assessed on WEIZMAN data set. The outcomes uncover that the proposed method has more favorable position and acknowledgment than the old models. In future more, databases can be evaluated to improve the accuracy.

3 Proposed Method

First the user has to register in the webpage. After the registration, the user must login with their credentials. After logging into the webpage, they must upload their files using sandbox technique. If the file doesn't contain virus the data's get stored in cloud database that typically runs on a cloud computing platform, access to it is provided as a service. It is based on highly virtualized infrastructure and is like broader cloud computing in terms of accessible interfaces, near-instant elasticity and scalability, multi-tenancy, and metered resources. Cloud storage services can be utilized from an off-premises service such as Amazon S3 that can be used for copying virtual machine images from the cloud to on-premises locations or to import a virtual machine image from an on-premises location to the cloud image library. In addition, cloud storage can be used to move virtual machine images between user accounts or between data centers (Fig. 1).

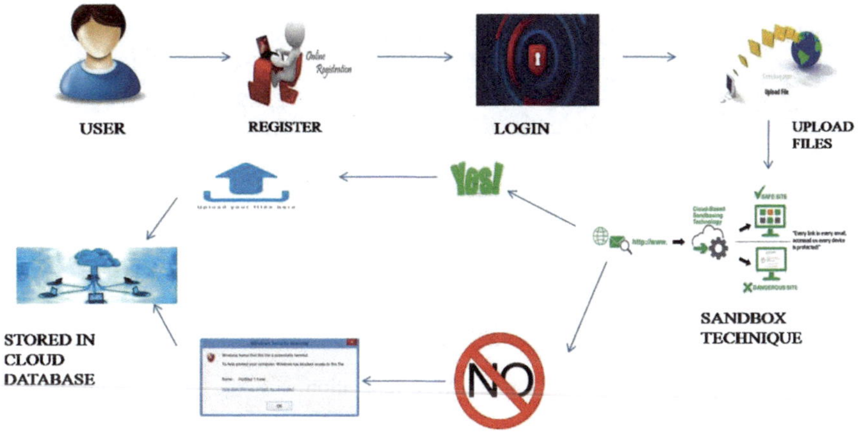

Fig. 1. Ransomware detection system

To prevent advanced threats that might be missed by anti-malware engines from entering your organization, it can sanitize potentially dangerous file types to thwart zero-day and targeted attacks. In addition, to identify and block files that have spoofed file type extensions, which indicates potential malicious intent. The combination of scanning data with anti-malware engines first, leveraging the power of their individual heuristic analyses, followed by converting potentially risky files to remove embedded threats greatly decreases the chances of your network being infected by an unknown threat.

3.1 Authentication

Authentication is the process of determining regardless of whether a person or thing is, actually, who or what it is proclaimed to be. In private and open PC systems (including the Internet), validation is regularly done using logon passwords. Learning of the secret phrase is accepted to ensure that the client is authentic. Every client enrolls at first or is enlisted by another person, utilizing an as-marked or self-proclaimed secret word. On each resulting use, the client must know and utilize the recently announced secret key.

3.2 Explore

The search engine where we can search the URL of particular website, which needs to scan, and will call the Anti-Malware engines to perform the operations.

3.3 Bluecoat Web Category

The Blue Coat Web Filter database contains Site appraisals speaking to billions of Web pages, distributed in excess of 50 dialects, and sorted out into valuable classes to empower clients to all the more likely screen, control, and secure their Web traffic. Blue Coat Web Filter is upheld by Blue Coat's amazing Web Pulse cloud network (Fig. 2).

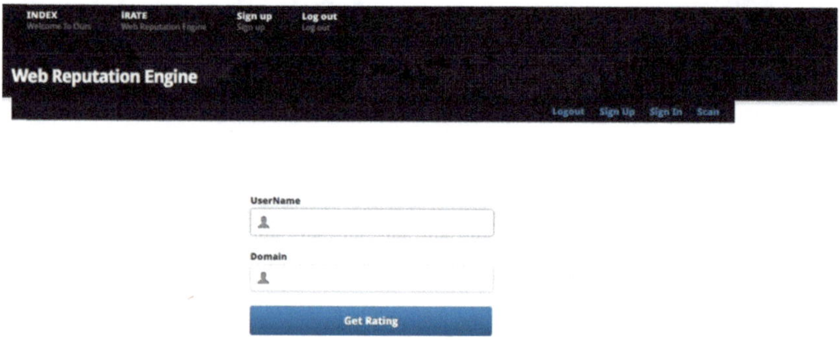

Fig. 2. Web reputation

3.4 Fireguard Web Category

A web application firewall (WAF) is an apparatus, server plug in, or channel that applies a lot of guidelines to a HTTP discussion. For the most part, these standards spread normal assaults, for example, cross-site scripting (XSS) and SQL infusion. By tweaking the standards to your application, numerous assaults can be recognized and blocked (Fig. 3).

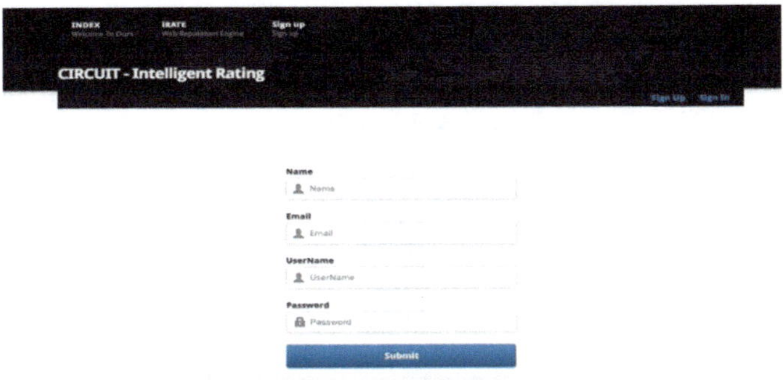

Fig. 3. Login webpage

3.5 Trend URL Safety Rating

Scores are relegated dependent on components, for example, a site's age, chronicled areas, changes, and signs of suspicious exercises found through malware conduct investigation. We've propelled how we apply web notoriety to keep pace with new sorts of criminal assaults that can travel every which way in all respects rapidly, or attempt to remain covered up.

3.6 Scan

Scans the given URL according to Anti-malware engines in Explore module, are to be called, in which URL has filtered and, finds the vulnerable links if available in those pages. Advantage is able to scan twenty different malware engines together, so we can catch the vulnerable links easily.

3.7 Cross Side Scripting

Cross-webpage scripting (XSS) is a kind of PC security weakness ordinarily found in web applications. XSS empowers assailants to infuse customer side contents into website pages seen by different clients. A cross-site scripting weakness might be utilized by assailants to sidestep get to controls, for example, the equivalent starting point approach (Fig. 4).

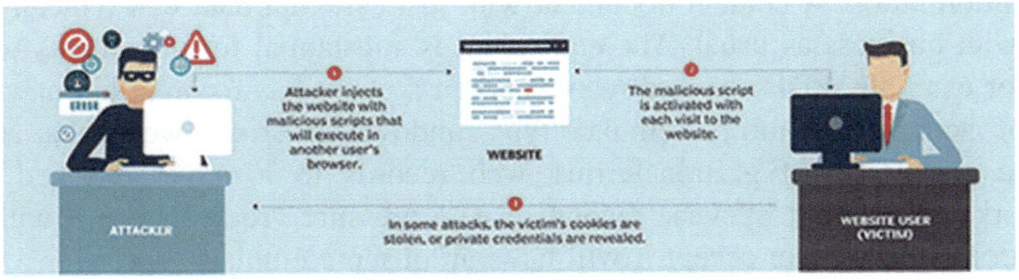

Fig. 4. Cross site scripting

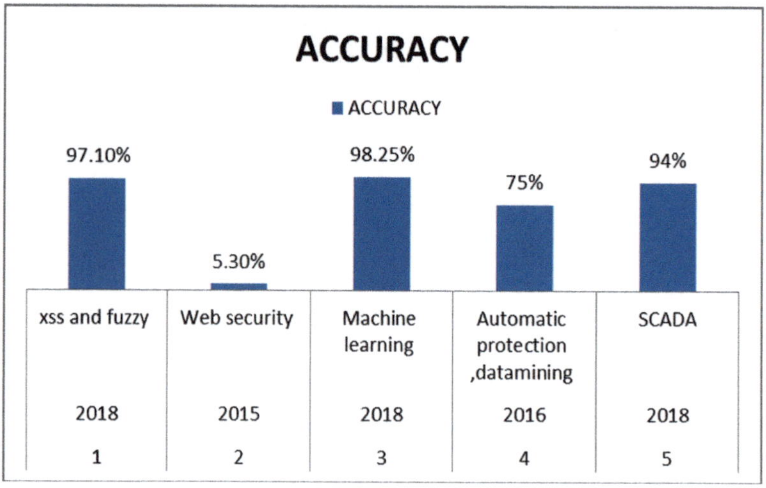

Fig. 5. Accuracy of optimization

3.8 Session Hijacking

It is utilized to allude to the robbery of an enchantment treat used to validate a client to a remote server. It has specific pertinence to web engineers, as the HTTP cookies used to keep up a session on many sites can be effectively stolen by an assailant utilizing a mediator PC or with access to the spared treats on the unfortunate casualty's PC.

4 Conclusion

The final product is a framework that can identify a few examples of ransomware with a general location technique because of their one of a kind example of encryption, while as yet permitting ordinary client conduct, including encryption of documents. This framework could be adjusted to be conveyed in a genuine organize and could stop a ransomware assault while it is executing. Along these lines they set out to consider these semantics and make a mark dependent on this perception. They infer that the marks can distinguish whole malware families, in any case, do have a higher blunder rate with wide classes, for example, trojans and secondary passages (Fig. 5).

Attaching is excluded on the grounds that we accept it is attached with a similar kind of information. A content document with just essential data will typically just be affixed with business as usual. The equivalent is substantial for documents with normally, for example, PDF or word records when a client has records that have expansions not incorporated into the emulate type rundown of Bro, however that are really authentic expansions. E.g. transferring web records to a server checked by this framework. This would likewise recognize new pressure groups, since it will creates high entropy and we can accept it will have an obscure emulate type.

References

1. Chen, P., Hu, Z.: Feedback control can make data structure layout randomization more cost-effective under zero-day attacks. https://doi.org/10.1186/s42400-018-0003-x
2. Choi, H., Hong, S.: Hybrid XSS detection by using a headless browser. In: 2017 4th International Conference on Computer Applications and Information Processing Technology (CAIPT). https://doi.org/10.1109/caipt.2017.8320672
3. Dikhit, A.S.: Result evaluation of field authentication based SQL injection and XSS attack exposure. In: 2017 International Conference on Information, Communication, Instrumentation and Control (2017). https://doi.org/10.1109/icomicon.2017.8279148
4. Fonseca, J., Seixas, N., Vieira, M., Madeira, H.: Analysis of field data on web security vulnerabilities. IEEE Trans. Dependable Secure Comput. https://doi.org/10.1109/tdsc.2013.37
5. Gonzalez, D., Hayajneh, T.: Detection and prevention of crypto-ransomware. In: 2017 IEEE 8th Annual Ubiquitous Computing, Electronics and Mobile Communication Conference (UEMCON) (2017). https://doi.org/10.1109/uemcon.2017.8249052
6. Guan, Y., Ge, X.: Distributed attack detection and secure estimation of networked cyberphysical systems against false data injection attacks and jamming attacks. IEEE Trans. Signal Inform. Process. Over Netw. **4**, 48–59 (2017)
7. Kaynar, K.: A taxonomy for attack graph generation and usage in network security. IEEE Trans. https://doi.org/10.1016/j.jisa.2016.02.00
8. Kolosok, N., Gurina, L.A.: Determination of the Vulnerability Index to Cyberattacks and State-Estimation Problems According to SCADA Data and Timed Vector Measurements (2017). https://doi.org/10.3103/s1068371217010096
9. Kumar, S.A.P., Xu, B.: Vulnerability assessment for security in aviation cyber- physical systems. In: 2017 IEEE 4th International Conference on Cyber Security and Cloud Computing (2017). https://doi.org/10.1109/cscloud.2017.17
10. Li, X., Xue, Y.: Detecting anomalous user behaviors in workflow-driven web applications. In: 2012 31st International Symposium on Reliable Distributed Systems (2013). https://doi.org/10.1109/srds.2012.19
11. Liu, X., Li, Z.: Masking transmission line outages via false data injection attacks. IEEE Trans. Inform. Forensics Secur. **11**(7), 1592–1602 (2016). https://doi.org/10.1109/tifs.2016.2542061

Novel Fully Automatic Solar Powered Poultry Incubator

S. Sri Krishna Kumar(✉), R. Suguna, R. Senthil Kumar, A. Surya Moorthy, Guru Moorthy, and S. Bala Yogesh

Department of Electrical and Electronics Engineering, Vel Tech High Tech Dr.Rangarajan Dr.Sakunthala Engineering College, Chennai 600062, India
krishnakumar.rvs@gmail.com, suguna@velhightech.com, rskumar.eee@gmail.com, shivamelectricalengineer@gmail.com

Abstract. Intent of the proposal is to use the renewable source of power to ensure a steady power supply to the remote rural area and to reduce cost involved in the hatcheries. PV powered incubator is designed with multi specification for different types of eggs. System automatically regulate the required specification such as temperature, humidity, ventilation, angular position etc., which are considered to be the most fix on factor of hatching a healthy hatchling. Menu key provided in the system provide the flexibility to use all kinds of egg hatchings. GSM module interfaced with the system, provide the real time condition to be the mobile periodically which ensure additional supervisory over the system. If any problem arises in regulating the specification automatically system activate the alter system buzzer and SMS.

Keywords: Poultry · Incubator · Egg hatching · GSM

1 Introduction

Incubator are the artificial process of producing hatchling this process was adapted widely all over the world as the conventional natural process is having lots of short coming. If the layers (bird) are allowed to carry out incubation. Then the a long period is spend for hatching the egg and again it will be taking care of hatchling till it reach a stage in their growth, during this entire period layers won't indulge in reproduction activity which drastically affect the productivity of the entire farm. During the incubation period it is necessary to maintain certain specification such as temperature, humidity, angular position and ventilation. Conventional model implementation of fuel lamps was not more efficient as it is lagging in governing all the specification and they were only concentrating on temperature and it has an adverse effect in ventilation.

Then the widely accepted model was powered by electricity. Though the electric driven model were effective in many means, but still in rural area stable power supply is the biggest problem, as the performance of the incubator fully relay on the power supply it again affect reliability of the system so it is high time to develop self

D. J. Hemanth et al. (Eds.): COMET 2019, LNDECT 35, pp. 1612–1617, 2020.
https://doi.org/10.1007/978-3-030-32150-5_162

sustainable system which ensure the reliable working of incubator, in this regard we have to develop a cost effective power source.

2 Research Back Ground

In recent days lots of modification and improvisation was brought in egg incubator, to improve the hatchability considering many fix on factor in hatching the eggs. Some works are even carried out on using new source energy for powering the incubators such as biogas, solar, wind mills etc.

In passive solar power incubator the solar heat is absorbed and with help of heat exchangers heat is transferred into the incubators [1] though the system reduce the power consumption, require power for specification monitoring.

PLC based solar powered incubator which have the self sustainability in powering both the heater as well as mentoring system [2], but PLC cost greatly depends on the number of input and output it can handle, hence when we are finding cost effective solution for incubator which can perform many function for different specification, so for each specification it has to reprogrammed.

Mostly the incubator find its application in rural area where the people are lacking in programming skills. We have to provide them simple form of menu key for changing the specifications.

Few microprocessor based solar powered incubator are designed for single specification though it has the flexibility adding the specification, this system lacks from the alert system in case any deviation in maintaining the desired condition [3, 5]. Some changes incorporate in convention models are rollers for changing the egg positions, water sprayers for maintaining the humidity, for uniform heating still or forced type ventilation [4, 6].

3 Proposed Model

This model tries to find a cost effective and user friendly system which aids in increasing the productivity of poultry farm and ensure the reliability in all conditions. Existing system are lacking in the alter system, if any deviation in maintain the desired conditions. This is indispensible as it affects the productivity drastically. Along with that by adding a menu key user can set various specifications so that system will be readily compactable for various types of eggs.

Figure 1 display the block diagram of the system which clearly infer the major component used for realizing the system solar energy is converted into electrical energy using photo voltaic panels, through charge controller its stored in the battery. Battery is used to power the a 80 W lamp to achieve desired temperature, fan for ventilation and to maintain oxygen level, water sprinkler, DC motor to change the angular position of egg s and a micro processor as a controller. To monitor the physical variable which is necessary create a desired condition for incubation, temperature sensor, humidity sensor and oxygen level detectors are used. GSM module is interfaced to get the real-time condition prevailing inside the incubator and to give alert to user via SMS.

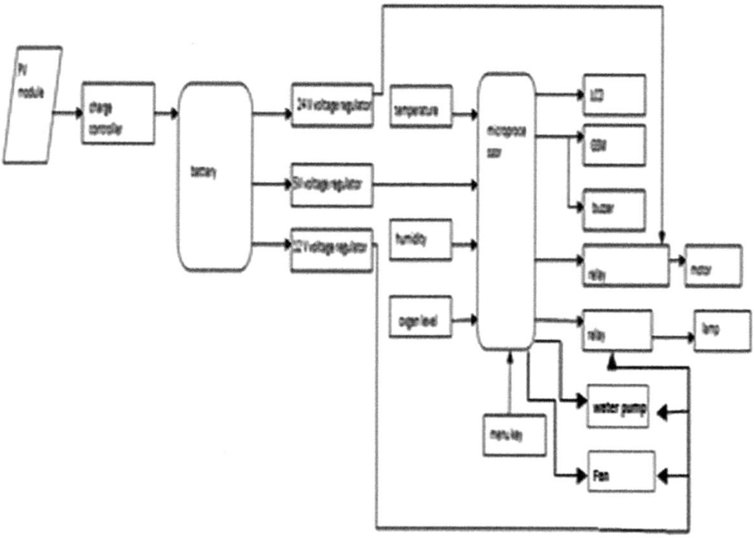

Fig. 1. Block diagram of proposed system

Entire working of the system can be easily inferred from the Fig. 2 flow chart. We can find total four processes which should be carried out parallel to attain the desired condition for incubation.

Process 1:

It involves the monitoring of temperature inside the incubator it should be around 99–103°F [7], so this will be set as the reference temperature Tr. Continuously the actual temperature which is prevailing in the incubator is measured by the sensor and compared with reference Tr. under the condition Ta < Tr, 80 W lamp is switched ON and if condition reachs Ta $\geq$ Tr, lamp is switched Off. There by we can maintain the required temperature inside the incubator.

Process 2:

It involves the mentoring of humidity inside the incubator, it should be maintained between 65–75% [9]. So this is set as the reference humidity level Hr. periodically the actual humidity is measured and compared with Hr. under the condition Ha < Hr controller will switch on the water sprinkler via relay. If Ha $\geq$ Hr condition is reached sprinkler are switched Off.

Process 3:

Here the concern goes with the ventilation and to maintain the oxygen level around 20–21% [10].this level is set as reference level Or and periodically it is compared with actual level O a. Under the condition O a < Or fan is switched ON and operated till it reaches the condition O a $\geq$ Or after that fan is switched OFF.

Process 4:

Rolling of egg has to be done at regular intervals to avoid embryo sticks onto the shell. Timer is set to operate DC motor to change the angular position of the egg for 45° at each interval [8].

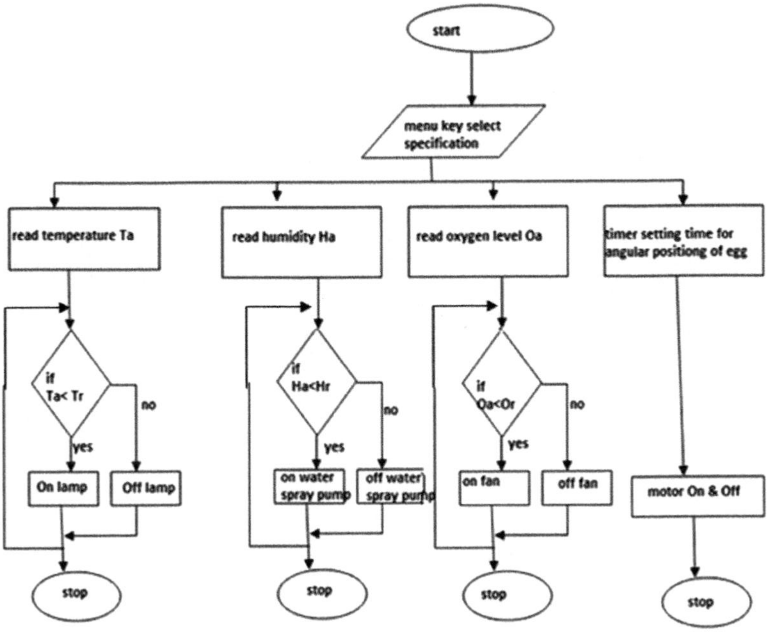

Fig. 2. Flowchart of the process involved

3.1 Alert System

System is provided with a buzzer and GSM module if any of the four above mentioned process is deviated and automatically retrieved then alert signal is given by buzzer sound and SMS stating the particular process along with that user can fetch real time physical variable condition about the incubator via SMS. Thereby it provide an addition supervision which increase the reliability of the system.

3.2 Menu Key

It is featured with multi specification required for different types of eggs. So user can easily select the specification or he can even change the physical parameter value as per requirement alone with that it has memory option where we can save the date loading the egg so that after the average day of hatching we will get a reminder SMS about it (Table 1 and Fig. 3).

Table 1. Hatching days of various birds [11]

Sl no	Birds	Days
1	Quail	18
2	Dove	17
3	Country chicken	25
4	Chicken	23

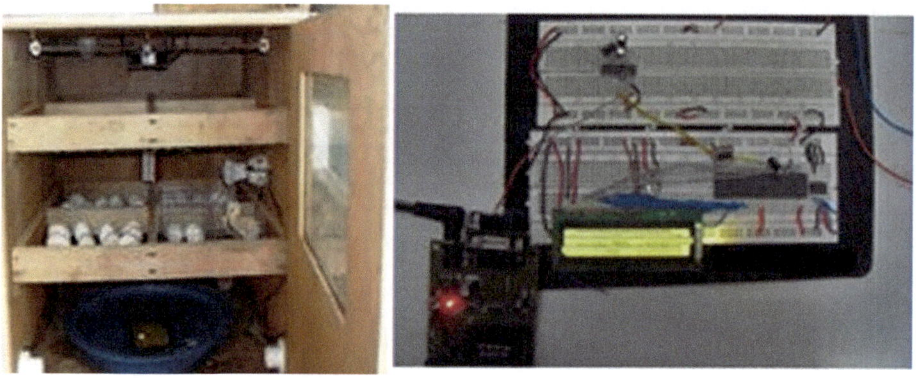

Fig. 3. Hardware set

4 Conclusion

Experimental setup responds quickly for physical parameter changes and attain the desired specification rapidly, menu key provide a user friendly interface to the system and also facilitate the multi specification feature to the system. PV panel provides a self suitability in power consumption thereby it ensure a stable reliable power supply in rural areas along with that it is the cost effective solution for the running poultry farm without compensating the productivity. GSM Alert system gives added supervision to system which increases the reliability of system and improving the hatchability of the incubators.

References

1. Ahiaba: Performance evaluation of a passive solar Poultry egg incubator. IJISET – Int. J. Innov. Sci. Eng. Technol. **2**(12) (2015). ISSN 2348–7968. V.UP 1 P; NwakonobiP 2 P T. U. and Obetta S.E.P 3
2. Abraham, N.T., Mathew, S.L., Kumar, C.A.P.: Design and implementation of PV poultry incubator using PLC. TELKOMNIKA Indonesian J. Electr. Eng. **12**(7), 4900–4904 (2014). https://doi.org/10.11591/telkomnika.v12i7.5882
3. Kanu, O.O., Anakebe, S.C., Okosodo, C.S., Okoye, A.E., Ezeigbo, T.O., Okpala, U.V.: Construction and characterization of solar powered micro-base incubator. Int. J. Sci. Eng. Res. **7**(1) (2016). ISSN 2229-5518
4. Benjamin, N., Oye, N.: Modification of the design of poultry incubator. Int. J. Appl. Innov. Eng. Manage. (IJAIEM) **1**(4), 90–102 (2012). ISSN 2319–4847
5. Mansaray, K.G., Yansaneh, O.: Fabrication and performance evaluation of a solar powered chicken egg incubator. Int. J. Emerg. Technol. Adv. Eng. **5**(6), 31–36 (2015). ISSN 2250-2459
6. Benjamin, N., Oye, N.: Modification of the design of poultry incubator. Int. J. Appl. Innov. Eng. Manage. (IJAIEM) **1**(4) (2012). ISSN 2319–4847
7. Okonkwo, J.W.I., Chukwuezie, O.C.: Characterization of a photovoltaic powered poultry egg incubator

8. Abiola, S.S.: Effects of turning frequency of hen's eggs in electric table type incubator on weight loss, hatchability and mortality. Niger. Agric. J. **30**, 77–82 (1999)
9. Okonkwo, W.I.: Design of solar energy egg incubator. unplished undergraduate project. Department of Agricultural Engineering, University of Agriculture, Makurdi, Nigeria (1989)
10. Eziefulu, O.P.: Solar energy powered poultry egg incubator with kerosene heater. Final Year Project. Department of Agricultural and Bioresources Engineering, University of Nigeria, Nsukka (2005)
11. Lourens, A.H., van den Brand, H., Meijerhof, R., Kemp, B.: Effect of egg sizeon heat production and the transition of energy from egg to hatchling. Poult. Sci. **83**, 705–712 (2005)

Retraction Note to: Emerging Trends in Computing and Expert Technology

D. Jude Hemanth, V. D. Ambeth Kumar, S. Malathi, Oscar Castillo, and Bogdan Patrut

Retraction Note to:
D. J. Hemanth et al. (Eds.): ***Emerging Trends in Computing and Expert Technology*****, LNDECT 35,**
https://doi.org/10.1007/978-3-030-32150-5

The Series Editor and the Publisher have retracted this chapter. An investigation by the Publisher found a number of chapters, including this one, presenting various concerns, including but not limited to compromised editorial handling, incoherent text or tortured phrases, inappropriate or irrelevant references, or a mismatch with the scope of the series and/or book volume. Based on the findings of the investigation, the Series Editor therefore no longer has confidence in the results and conclusions of this chapter.

The authors have not responded to correspondence regarding this retraction.

The retracted version of these chapters can be found at
https://doi.org/10.1007/978-3-030-32150-5_23
https://doi.org/10.1007/978-3-030-32150-5_33
https://doi.org/10.1007/978-3-030-32150-5_68
https://doi.org/10.1007/978-3-030-32150-5_73
https://doi.org/10.1007/978-3-030-32150-5_95
https://doi.org/10.1007/978-3-030-32150-5_105
https://doi.org/10.1007/978-3-030-32150-5_122
https://doi.org/10.1007/978-3-030-32150-5_126
https://doi.org/10.1007/978-3-030-32150-5_145

D. J. Hemanth et al. (Eds.): COMET 2019, LNDECT 35, p. C1, 2024.
https://doi.org/10.1007/978-3-030-32150-5_163

Author Index

D. J. Hemanth et al. (Eds.): COMET 2019, LNDECT 35, pp. 1619–1623, 2020.
https://doi.org/10.1007/978-3-030-32150-5

GPSR Compliance

The European Union's (EU) General Product Safety Regulation (GPSR) is a set of rules that requires consumer products to be safe and our obligations to ensure this.

If you have any concerns about our products, you can contact us on ProductSafety@springernature.com

In case Publisher is established outside the EU, the EU authorized representative is:

Springer Nature Customer Service Center GmbH
Europaplatz 3
69115 Heidelberg, Germany

Batch number: 10755036

Printed by Printforce, the Netherlands